D1188160

DICTIONARY

of Engineering Acronyms and Abbreviations

SECOND EDITION

By Harald Keller and Uwe Erb

Neal-Schuman Publishers, Inc.

New York *London*

Published by Neal-Schuman Publishers, Inc.
100 Varick St.
New York, NY 10013

Printed and bound in the United States of America.

Library of Congress Cataloging-in-Publication Data

Keller, Harald
Dictionary of engineering acronyms and abbreviations / Harald Keller, Uwe Erb. — 2nd. ed.
p. cm.
Includes bibliographical references.
ISBN 1-5570-129-9. — ISBN 1-5570-129-9.
1. Engineering—Acronyms. 2. Engineering—Abbreviations. I. Erb, Uwe. II. Title.
T11.K43 1994
620′.00148—dc20 93-11570
CIP

About the Authors

Harald Keller is an independent scientific and technical translator working on a wide range of projects for government agencies and multinational corporations. With his broad background in mechanical engineering and materials testing he is editor-in-chief of the "DC Databases of Engineering Terms."

Dr. Uwe Erb is a professor at Queen's University, Kingston, Canada, and consultant to industry. He has published widely in the field of materials science and is co-editor of the "DC Databases of Engineering Terms."

Contents

Introduction

Dictionary of Engineering Acronyms and Abbreviations, second edition, has been divided into two parts: Part A, *Engineering Acronyms and Abbreviations,* has been greatly revised and expanded to over 70,000 entries, an increase of more than 100 percent over the previous edition. Part B, *Periodical Title Abbreviations and Acronyms in Science and Engineering*, contains over 7,000 entries and is destined as a handy listing of abbreviations and acronyms for the most important engineering periodicals. In most cases it will eliminate the need for more comprehensive dictionaries, although it is not intended to be a replacement for more comprehensive publications dealing *only* with periodicals, such as Gale's *Periodical Title Abbreviations: By Abbreviation* or Ulrich's *International Periodicals Directory*.

The need for a reference book such as this is evident considering that, in spite of discontent about the use and misuse of linguistic shorthand in scientific and technical publications, new short forms continue to emerge daily. New equipment, machinery, and devices are invented and innovative techniques and processes to better produce, treat, or examine things follow suit. Most of these innovations are referred to by acronyms or abbreviations. Furthermore, technical associations and societies are founded, absorbed, renamed, or dissolved, and the short-forms used to refer to them either change or become redundant. To keep up with these developments, innovations, and changes, we have more than doubled the size of this dictionary. Beyond this goal, however, we also would like this current work and all subsequent editions to encompass acronyms and abbreviations used in the engineering sciences of both the past and the present. Therefore, we have kept some older entries for their historical value. For example, some entries refer to the USSR (Union of Soviet Socialist Republics) as country of origin, although the USSR no longer exists as a country.

This second edition of the *Dictionary of Engineering Acronyms and Abbreviations* is designed for a broad audience—for readers of literature dealing with engineering subjects, not only engineers, researchers, and scientists, but also technicians, technologists, science writers, college and university students, and even the layman will find it an extremely valuable companion when working through what can be a chaotic jungle of shorthand language. The following is an alphabetical list of some of the major fields covered:

Aeronautics and Aviation
Aerospace Engineering
Agricultural Engineering
Architecture
Artificial Intelligence
Associations and Societies
Astronomy and Astrophysics
Automation and Robotics
Automotive Engineering
Biochemistry and Biophysics
Bioengineering and Biotechnology
Biomedical Science and Engineering
Building Construction
Chemical Engineering
Chemistry
Civil Engineering
Colleges and Universities
Communication Engineering
Computer Science and Engineering
Control Engineering
Cryogenics
Degrees and Professional Titles
Design Engineering
Devices, Machinery and Equipment
Electrical Engineering
Electrochemistry
Electronics
Energy Technology
Engineering Drawing
Environmental Engineering
Experimental Techniques
Food Engineering and Technology
Forestry Sciences and Engineering
Geochemistry and Geophysics
Geophysical Engineering
Geotechnology and Geology
Government Agencies and Programs
Heating, Refrigeration and Air Conditioning
Information Technology
Institutes of Technology
Major International Currencies
Manufacturing Technology
Marine Engineering and Technology
Materials Handling
Materials Science
Mathematics

Mechanical Engineering
Metallurgical Engineering
Meteorology
Military Science and Engineering
Mining Engineering
Naval Engineering
Nuclear Science and Engineering
Oceanography
Operations Research
Pharmaceutical Engineering
Physics
Plant Engineering
Postal Engineering
Research Programs and Projects
Research Institutes and Laboratories
Safety Engineering
Technical Specifications
Textile Engineering and Technology
Transportation Engineering.

HOW TO USE THIS DICTIONARY

The entries are given in alphabetical order with uppercase letters preceding lower-case letters, e.g. "ap" [apothecaries] will be preceded by "AP" [Access Panel]. Whenever possible we have used the uppercase form of an abbreviation or acronym. Therefore, the reader will find "BAL" [Balance], although "bal" may also be acceptable, or "RADAR," although "Radar" or "radar" are also commonly found. For some entries, additional information has been added in square brackets. Such information may include: the country of origin for associations, societies, research institutes, etc.; translations from other languages for international organizations and projects; and when absolutely necessary, the particular field of engineering. Furthermore, parentheses were used to include further explanations which are not part of a particular acronym or abbreviation, e.g. "MIES" is given as "McMaster (University) Institute for Energy Studies." When necessary, parentheses are also used in cases where an acronym is used as part of the explanation, e.g. "AERA" is given as "Automated En-Route ATC (= Air Traffic Control)." Abbreviations formed by letters and numerals are listed as follows: (i) if the abbreviation does not refer to a chemical, it is listed under the corresponding letter, e.g. "3PDT" [Triple-Pole Double Throw] appears under "TPDT"; and (ii) if the abbreviation refers to a chemical substance, the numeral is disregarded, e.g. "2,4-DB" [2,4-Dichlorophenoxybutyric Acid] appears under "DB." In general, abbreviations and acronyms for chemicals, as well as their corresponding explanations, are presented with, or without square brackets, parentheses, or braces, just as they were found in the literature. Some acronyms and abbreviations are registered trademarks.

A REQUEST

In order to give the user a full overview of the many acronyms, abbreviations, initialisms, and symbols used in engineering, we have tried to make this dictionary comprehensive while, at the same time, avoiding duplications. In order to make future editions even more useful, we welcome any suggestions from our readers.

Part A

Acronyms and Abbreviations in Science and Engineering

A

A	Absorbance
	Absorbancy
	Acceptable
	Acid
	Acre
	Activity
	Adder
	Address
	Adenine
	Air
	Aircraft
	Alanine
	Alanyl
	Amacrine
	America(n)
	Ammeter
	Ampere
	Amphibian
	Amplifier
	Amplitude
	Analogous
	Analogy
	Angstrom
	Annealed
	Anode
	Answer
	Aperture
	Arch
	Arctic
	Area
	Arithmetic
	Army
	Artillery
	Array
	Asbestos
	Atlantic
	Attribute
A	Angstrom
a	absolute
	acceleration
	account
	acre
	activity
	aerial
	air
	analog
	(annus) — year
	anode
	(ante) — before
	anterior
	aqua
	are [Unit]
	asymmetric
	asynchronous
	atom
	atomic
	atto- [SI Prefix]
aΩ	abohm [Unit]
$(\mathrm{a}\Omega)^{-1}$	abmho [Unit]
AA	Absolute Alcohol
	Absolute Altitude
	Access Authorization
	Acetic Acid
	Acetic Anhydride
	Acetoacetanilide
	Acrylamide
	Acrylic Acid
	Adenylic Acid
	Adjustable Angle
	Advertising Association [UK]
	Affirmative Action
	All After
	Airborne Alert
	Airship Association [US and UK]
	Air-to-Air
	Algorithm Analysis
	All Around
	Aluminum Association [US]
	Alumni Association
	American Airlines [US]
	Amino Acid
	Aminoacetone
	Aminoacyl
	Angular Aperture
	Aniline Association [US]
	Anodized Aluminum
	Antenna Array
	Anthranilic Acid
	Anti-Aircraft
	Antiproton Accumulator
	Approximate Absolute
	Approximation Algorithm
	Archeology Abroad [UK]
	Arachidonic Acid
	Arboricultural Association [UK]
	Architectural Association [UK]
	Arithmetic Average
	Armature Accelerator
	Arsenic Analysis
	Ascorbic Acid
	Asset Amount
	Associate Administrator
	Atomic Absorption
	Atropic Acid
	Attack Announcement
	Attack Assessment
	Audio Active
	Audio Amplifier
	Author Affiliation
	Author's Alteration
	Auto Acquisition
	Automatic Answer
	Automobile Association [US]
	Autoprobe Atomization

Acronyms and Abbreviations

A/A Anti-Aircraft
A-A Air-to-Air
Aluminum-Adhesive Joint
aA abampere [Unit]
AAA Acetoacetanilide
Acetoacetic Anilide
Acrylic Acid Anhydride
Agricultural Adjustment Administration [US]
Agricultural Aircraft Association [US]
Air Force Association [US]
Alberta Association of Architects [Canada]
Amateur Astronomers Association [US]
American Academy of Allergy [US]
American Accounting Association [US]
American Airship Association [now: AA, US]
American Antarctic Association [US]
American Arbitration Association [US]
American Astronomers Association [US]
American Automobile Association [US]
Anguilla Anguilla Agglutinin
Anti-Aircraft Artillery
Appraisers Association of America [US]
Aromatic Amino Acid
Army Audit Agency [US]
Asian International Airports Association
Astronomy and Astrophysics Abstracts (Database) [Germany]
AAAA American-African Affairs Association [US]
American Association of Advertising Agencies [US]
Army Aviation Association of America [US]
AAAB American Association of Architectural Bibliographers [US]
AAAC American-Arab Affairs Council [US]
American Automatic Control Council [US]
Anti-Aircraft Artllery Command [US]
Archival Association of Atlantic Canada
Australian Army Aviation Corps
AAACC Association of American-Asian Chambers of Commerce
AAACE American Association of Agricultural College Editors [US]
AAACU Asian Association of Agricultural Colleges and Universities
AAAD Aromatic Amino Acid Decarboxylase
Association of Automotive Aftermarket Distributors [US]
AAAE American Association of Academic Editors [US]
American Association of Agricultural Engineers [US]
American Association of Airport Executives [US]
AAAF Association Aeronautique et Astronautique de France [= French Aeronautical and Astronautical Association]
AAAI Affiliated Advertising Agencies International [US]
American Academy of Allergy and Immunology [US]
American Association for Artificial Intelligence [US]
Association of Australian Aerospace Industries
AAAIA All-India Automotive and Ancillery Industries Association [India]
AAAID Arab Authority for Agricultural Investment and Development [Sudan]
AAAIP Advanced Army Aircraft Instrumentation Program [US]
AAALAC American Association for Accrediation of Laboratory Animal Care [US]
AAAM American Association of Aircraft Manufacturers
American Association for Automotive Medicine [US]
AAAOR Anti-Aircraft Artillery Operations Room
AAAP Asian-Australian Association of Animal Production Societies
AAAR American Association for Aerosol Research [US]
AAAS American Academy of Arts and Sciences [US]
American Association for the Advancement of Science [US]
AAASA Association for the Advancement of Agricultural Sciences in Africa [Ethiopia]
AAAST African Association for the Advancement of Science and Technology
AAAT Anti-Aircraft Armored Truck
AAATP Asian Alliance of Appropriate Technology Practitioners [Philippines]
AAB Advertising Advisory Board
Aircraft Accident Board
Aircraft Anti-Collision Beacon
Alliance Agricole Belge [= Belgian Agricultural Alliance]
American Association of Bioanalysts [US]
Aminoazobenzene
Army Aviation Board [US Army]
Association of Applied Biologists [Scotland]
AABC Associated Air Balance Council [US]
AABCP Advanced Airborne Command Post
AABDF Allied Association of Bleachers, Printers, Dyers and Finishers [UK]
AABE American Association of Blacks in Energy [US]
Asian Association for Biology Education
AABFS Amphibious Assault Bulk Fuel System
AABG Alpha-Acetobromoglucose
AABGA American Association of Botanical Gardens and Arboretums [US]
AABI American Association of Bicycle Importers [US]
AABM Alpha-Acetobromoglucose
Association of American Battery Manufacturers [US]
Australian Association of Britsh Manufacturers
AABNCP Advanced Airborne Command Post [US]
AABNF African Association for Biological Nitrogen-Fixation [Egypt]
AABP Acetylaminobiphenyl
Aptitude Assessment Battery Program
Australian Association of Business Publications
AABSHIL Aircraft Anti-Collision Beacon System High-Intensity Light
AAC Aeronautical Advisory Council
Aeronautical Approach Chart
African Association of Cartography [Algeria]
Afro-Asian Center [US]
Agricultural Advisory Council [UK]
Air Approach Control
Air Carbon-Arc Cutting

Airworthiness Advisory Circular [Australia]
Alaskan Air Command [US]
Alkyl Amines Council [US]
American Alpine Club [US]
Anti-Aircraft Cannon
Anti-Aircraft Command
Argon Alternating Current
Army Air Corps [US]
Arsenic Atmosphere Czochralski
Associate and Advisory Committee
Association of American Colleges [US]
Automatic Amplitude Control
Automatic Aperture Control
Automatic Approach Control
Automotive Advertisers Council [US]
Average Annual Cost
Aviation Administrative Communications
Aviation Advisory Commission

AACA Aircraft Airworthiness Certification Agency
Automotive Air Conditioning Association [now: IMACA, US]

AACB Aeronautics and Astronautics Coordinating Board [of NASA]
Australian Association of Clinical Biochemists

AACC Airport Associations Coordinating Council [Switzerland]
American Association for Clinical Chemistry [US]
American Association for Contamination Control [US]
American Association of Cereal Chemists [US]
American Automatic Control Council [US]
Area Approach Control Center
Association of Agricultural Computer Companies [US]
Automatic Approach Control Complex

AACCA Associate of the Association of Certified and Corporate Accountants

AACCLA Association of American Chambers of Commerce in Latin America

AACD Antenna Adjustable Current Distribution

AACE African Association of Correspondence Education
Airborne Alternate Command Echelon [of NATO]
American Association of Cost Engineers [US]

AACFT Army Aircraft

AACG American Association for Crystal Growth [US]

AACI Accredited Appraiser, Canadian Institute
American Association of Ceramic Industries [US]
Association of American Cancer Institutes [US]

AACIA American Association for Clinical Immunology and Allergy [US]

AACO Advanced and Applied Concept Office [US]
Arab Air Carriers Organization [Lebanon]
Assault Airlift Control Office

AACOBS Australian Advisory Council on Bibliographical Services

AACOM Army Area Communications

AACOMS Army Area Communications System [US]

AACP Advanced Airborne Command Post [US Air Force]
American Association of Colleges of Pharmacy [US]
American Association of Computer Professionals [US]
Anglo-American Council for Productivity

AACPA Asian American Certified Public Accountants [US]

AACPP Association of Asbestos Cement Pipe Producers [US]

AACR American Association for Cancer Research [US]
American Association of Clinical Research [US]
Anglo-American Cataloguing Rules

AACRAO American Association of Collegiate Registrars and Admissions Officers [US]

AACS Advanced Automatic Compilation System
Airborne Astrographic Camera System
Airways and Air Communications Service
Army Alaska Communication System [US Air Force]
Asynchronous Address Communications System

AACSCEDR Associate and Advisory Committee to the Special Committee on Electronic Data Retrieval [US]

AACSR Aluminum Alloy Constructor Steel Reinforced

AACT American Association of Candy Technologists [US]
American Association of Clinical Toxicology [US]
American Association of Commodity Traders [US]

AACV Assault Air Cushion Vehicle

AACVB Asian Association of Convention and Visitor Bureaus [Philippines]

AACW Active Adaptive Compliance Wrist

AAD Active Acoustic Device
Administrative Applications Division
Advanced Ammunitions Depot
Amino Acid Decarboxylase
Army Automation Directorate [US]
Assigned Attitude Deviation

AADA Anti-Aircraft Defended Area

AADB Army Air Defense Board [US]

AADC Advanced Avionic Digital Computer
Alberta Agriculture Development Corporation [Canada]
All-Applications Digital Computer
Anti-Aircraft Defense Commander
Army Air Defense Command [US]
Arnold Air Development Center [US Air Force]
Automatic Aeromagnetic Digital Compensator

AADCP Army Air Defense Command Post [US]

AADE American Association of Dental Publishers [US]

AADECA Asociacion Argentina de Control Automatico [= Argentinian Association for Automatic Control]

AADP Aminopyridineadeninedinucleotide Phosphate

AADS Advanced Army Defense System
Advanced Automated Directional Scanning
Airspeed and Direction Sensor
Area Air Defense System

AAE Above Airport Elevation
Aeronautical and Astronautical Engineering
Affirmative Action Employer
American Association of Engineers [US]
Asociacion Argentina de Electrotecnicos [= Argentinian Electrotechnical Association]
Army Aviation Engineer
Auto-Answer Equipment
Automatic Answering Equipment
Average Annual Earnings

AAEA African Adult Education Association
American Agricultural Editors Association [US]

AAEC Australian Atomic Energy Commission
AAED Academic American Encyclopedia Database
AAEDC American Agricultural Economics Documentation Center [US]
AAEE Aircraft and Armament Experimental Establishment [of MOD, UK]
American Academy of Environmental Engineers [US]
American Association of Economic Entomologists [US]
American Association of Electromyography and Electrodiagnosis [US]
A&AEE Aircraft and Armament Experimental Establishment [of MOD, UK]
AAeE Associate in Aeronautical Engineering
AAEER Aerophysics and Aerospace Engineering Research Report
AAEI American Association of Exporters and Importers [US]
AAEL American Association of Equipment Lessors [US]
AAELSS Active-Arm External Load Stabilization System
AAEM Acetoacetoxylethyl Methacrylate
AA/EOE Affirmative Action/Equal Opportunity Employer
AAES Advanced Aircraft Electrical System
American Association of Engineering Societies [US]
Australian Agricultural Economics Society
AAESDA Association of Architects, Engineers, Surveyors and Draftsmen of Australia
AAESWB Army Airborne Electronics and Special Warfare Board [US Army]
AA-EVP American Association — Electronic Voice Phenomena [US]
AAF Academie d'Agriculture de France [= French Academy of Agriculture]
Acetylaminofluorene
American Advertising Federation [US]
American Air Force [US]
American Architectural Foundation [US]
Army Air Field
Army — Air Force
Auxiliary Air Force
AAFA Architectural Aluminum Fabricators Association
AAFB Andrews Air Force Base [US]
Army Air Force Board [US]
Auxiliary Air Force Base
AAFC Anti-Aircraft Fire Control
Arab Aviation Finance Company
Army Air Force Center [US]
AAFCE Allied Air Forces Central Europe [of NATO]
AAFCO Association of American Feed Control Officials [US]
Association of American Fertilizer Control Officials [US]
AAFE Advanced Applications Flight Experiment [of NASA]
AAFEA Australian Airline Flight Engineers Association
AAFIF Automated Air Facility Information File [of DMAAC, US]
AAFM Association of American Feed Microscopists [US]
AAFMC Army Air Force Materiel Center
AAFNE Allied Air Forces Northern Europe [of NATO]
AAFS American Academy of Forensic Sciences [US]
AAFSE Allied Air Forces Southern Europe [NATO]
AAFSS Advanced Aerial Fire Support System
AAFTS Army Air Force Technical School
AAFU All African Farmers Union
AAG Acquisition Advisory Group
Aeromedical Airlift Group [US]
Aeronautical Assignment Group [US]
Air Adjutant General
Airports Authority Group [Canada]
Alpha-L-Acid Glycoprotein
Anthropological Association of Greece
Assistant Adjuteant-General
Association of American Geographers [US]
AAGR Air-to-Air Gunnery Range
Average Annual Growth Rate
AAgr Associate in Agriculture
AAGS American Association for Geodetic Surveying [of ACSM, US]
Association of African Geological Surveys [France]
AAGW Air-to-Air Guided Weapon
AAH Advanced Attack Helicopter
Anti-Armor Helicopter
AAHE American Association for Higher Education [US]
AAHM Association of Architectural Hardware Manufacturers
AAHO Afro-Asian Housing Organization [Egypt]
AAHQ Allied Air Headquarters
AAHS American Aviation Historical Society [US]
AAI African-American Institute
Agricultural Ammonia Institute [US]
Air Aid to Intercept
Airlines Avionics Institute [US]
American Association of Immunologists [US]
American Association of Inventors [US]
Angle of Approach Indicator
Arab-American Institute
Association of Advertisers in Ireland
Azimuth Angle Increment
AAIB Air Accidents Investigation Branch [UK]
AAIC ASEAN Aluminum Industry Club
AAIE Association for the Advancement of International Education [US]
Association of Applied Insect Ecologists [US]
AAIEE Associate of the American Institute of Electrical Engineers
AAIH American Academy of Industrial Hygiene [US]
AAII American Association for the Advancement of Invention and Innovation [US]
AAIM American Association of Industrial Management [US]
AAIMME Associate of the American Institute of Mining and Metallurgical Engineers
AAIMS An Analytical Information Management System
AAIN American Association of Industrial Nurses [US]
AAIPS American Association of Industrial Physicians and Surgeons [US]
AAIS African Association of Insect Scientists
AAITO Association of African Industrial Technology Organizations
AAJC Automatic Antijam Circuitry
AAL Above Aerodrome Level
Absolute Assembly Language [Computer Language]

	American Airlines [US]
	Ames Aeronautical Laboratory [US]
	Arctic Aeromedical Laboratory [US Air Force]
	Arctic Aerospace Laboratory [US Air Force]
	Association of Architectural Librarians [US]
	Association of Assistant Librarians [UK]
AALA	American Agricultural Law Society [US]
	American Association for Laboratory Accreditation [also: A2LA or A^2LA]
	American Automotive Leasing Association [US]
AALAS	American Association for Laboratory Animal Science [US]
AALC	Advanced Airborne Launch Center [US]
	Amphibious Assault and Landing Craft
	Amplified Automatic Level Control
AALD	Association of Agricultural Librarians and Documentalists [Netherlands]
AALDEF	Asian-American Legal Defense Education Fund
AALGCSU	American Association of Land Grant Colleges and State Universities [US]
AALL	American Association for Labor Legislation [US]
	American Association of Law Librarians [US]
AALMG	Anti-Aircraft Light Machine Gun
AALO	Association of Academic Librarians of Ontario [Canada]
AALS	Acoustic Artillery Location System
	Association for Arid Lands Studies [US]
	Association of American Library Schools [US]
AALT	Alberta Association of Library Technicians [Canada]
AALUE	Asymptotically Admissible Linear Unbiased Estimator
AAM	Airline Administrative Message
	Air-to-Air Missile
	American Academy of Mechanics [US]
	American Academy of Microbiology [US]
	American Association of Museums [US]
	Amino Acid Mixture
	Anti-Aircraft Missile
AAMA	American Agricultural Marketing Association [US]
	American Amusement Machine Association [US]
	American Architectural Manufacturers Association [US]
	Architectural Aluminum Manufacturers Association [US]
	Asian American Manufacturers Association [US]
	Automotive Accessories Manufacturers of America [US]
AAMC	Army Air Materiel Command
	Association of American Medical Colleges [US]
AAMCA	Army Advanced Materiel Concepts Agency [US]
AAME	Acetylarginine Methylester
	American Association of Microprocessor Engineers [US]
	Automated Multi-Media Exchange
AAMG	Anti-Aircraft Machine Gun
AAMGAP	Achieve as Many Goals as Possible
AAMI	American Association of Machinery Importers [US]
	Association for the Advancement of Medical Instrumentation [US]
AAM/ICOM	International Council of Museums Committee of the American Association of Museums [US]
AAML	Arctic Aeromedical Laboratory [US Air Force]
AAMRL	American Association of Medical Record Librarians [US]
AAMS	American Association of Meta-Science [US]
AAMTA	American Apparel Machinery Trade Association [US]
AAMVA	American Association of Motor Vehicle Administrators [US]
AAMX	Acetoacet-M-Xylidide
AAN	American Association of Nurserymen [US]
	Aminoacetonitrile
AANS	American Agricultural News Service [US]
AAO	Acetaldoxime
	Amino Acid Oxidase
	Anti-Air Output
	Authorized Acquisition Officer
AAOC	Anti-Aircraft Operations Center
AAODC	American Association of Oilwell Drilling Contractors [US]
AA of A	Automobile Association of America [US]
AAOHN	American Association of Occupational Health Nurses [US]
AAOM	American Association of Occupational Medicine [US]
AAOP	Antaircraft Observation Post
AAOR	Anti-Aircraft Artillery Operations Room
AAOU	Asian Association of Open Universities
AAP	Acridinylaminopropanol
	Affirmative Action Plan
	Affirmative Action Program
	Ageing Aircraft Program [of FAA, US]
	Air at Atmospheric Pressure
	Airborne Acoustic Processor
	Ambient Absolute Pressure
	Apollo Applications Program [of NASA, US]
	Association of American Publishers [US]
	Association of Aviation Psychology [US]
	Associative Array Processor
	Attached Applications Processor
	Average Annual Precipitation
9-AAP	9-Acridinylaminopropanol
AAPA	American Alfalfa Processors Association [US]
	American Association of Port Authorities [US]
	Australian Asphalt Pavement Association
AAPC	American Association of Professional Consultants [US]
	Asociacion Argentina para el Progreso de la Ciencia [= Argentinian Association for the Progress of Science]
	Australian Aluminum Production Commission
AAPCC	American Association of Poison Control Centers [US]
AAPCO	Association of American Pesticide Control Officials [US]
AAPFCO	Association of American Plant Food Control Officials [US]
AAPG	American Association of Petroleum Geologists [US]
AAPICU	American Association of Presidents of Independent Colleges and Universities [US]

AAPL Additional Programming Language
American Association of Petroleum Landsmen [US]
An Array Processing Language [Computer Language]
AAPM American Association of Physicists in Medicine [US]
AAPMA Association of Australian Port and Marine Authorities
AAPP American Association of Police Polygraphists [US]
Auxiliary Airborne Power Plant
AAPRCO American Association of Private Railroad Car Owners [US]
AAPRDTW Association for the Advancement of Policy, Research and Development in the Third World [US]
AAPS Alternative Automotive Power System
American Association for the Promotion of Science [US]
American Association of Pharmaceutical Scientists [US]
American Association of Phonetic Sciences [US]
AAPSD Alternative Automotive Power System Division [of EPA, US]
AAPSE American Association of Professors in Sanitary Engineering [US]
AAPSP Advanced Automotive Power Systems Program [US]
AAPT American Association of Physics Teachers [US]
Association of Asphalt Paving Technologists [US]
AAPTSR Australian Association for Predetermined Time Standards and Research [also: AAPTS&R]
AAPU Auxiliary Airborne Power Unit
AAPWISM American Association of Public Welfare Information Systems Management [US]
AAR Against All Risks [also: aar]
Aircraft Accident Report
Air-to-Air Refuelling
Antigen-Antibody Reaction
ASEAN Association of Radiologists
Association for Automated Reasoning [US]
Association of American Railroads [US]
Automatic Alternative Routing
Automotive Affiliated Representatives [US]
Automotive Aftermarket Retailer [US]
Average Annual Rainfall
AARA Air-to-Air Refuelling Area
AARB Air-to-Air Refuelling Boom
Australian Road Research Board
AARC Asociacion de Agricultores del Rio Caulican [= Agricultural Association of Rio Caulican, Mexico]
Australian Applied Research Center
AARDS Australian Advertising Rate and Data Service
AARINENA Association of Agricultural Research Institutions in Near East North Africa
AARL Aeronautical and Astronautical Research Laboratory [US]
AARLMP African-American Resources and Library Manpower Project [Columbia University, US]
AARM Analytical Atomic Resolution Microscope
AARN Alberta Association of Registered Nurses [Canada]
AARPS Air Augmented Rocket Propulsion System
AARR Argonne Advanced Research Reactor [also: A^2R^2 [of ANL, US]]
AARRO Afro-Asian Rural Reconstruction Organization [India]
AARS American Association of Railroad Superintendents [US]
American Association of Railway Surgeons [US]
Automatic Address Recognition System
AARU Association of Arab Universities [Jordan]
AARV Armored Aerial Reconnaissance Vehicle
AAS Academy of Applied Science [US]
Adjusted Airspeed
Advanced Antenna System
Advanced Automation System
Afghanistan Academy of Sciences
African Academy of Sciences
Air Ambulance Service
Airport Advisory Service
Alabama Academy of Science [US]
American Aeronautical Association [US]
American Astronautical Society [US]
American Astronomical Society [US]
Ancient Astronaut Society [US]
Annunciator Alarm System
Aquifer-Aquitard System
Architectural Acoustics Society [US]
Arithmetic Assignment Statement
Arkansas Academy of Science [US]
Army Air Service [US]
Arsenic Atmosphere Czochralski
Association of Architects and Surveyors [UK]
Association of Asian Studies
Atomic Absorption Spectrometer
Atomic Absorption Spectroscopy
Atomic Absorption Standard
Auckland Astronomical Society [New Zealand]
Australian Academy of Science
Australian Acoustical Society
Austrian Academy of Science
Automated Accounting System
Automatic Announcement Subsystem
AASA Academy of Arts and Sciences of the Americas [US]
AASB American Association of Small Business [US]
AASC Aerospace Application Studies Committee [of NATO]
Airworthiness Authorities Steering Committee
American Association of State Climatologists [US]
Association for the Advancement of Science in Canada
AASCO Association of American Seed Control Officials [US]
AASCS American Association of Specialized Colleges and Schools [US]
AASCU American Association of State Colleges and Universities [US]
AAScW American Association of Scientific Workers [US]
AASDMCC American Association for Small Dredging and Marine Construction Companies [US]
AASE African Association of Science Editors
American Academy of Safety Education [US]
American Academy of Sanitary Engineers [now: AAEE, US]

Australian Associated Stock Exchanges
AASF Advanced Air Striking Force [US]
AASG Association of American State Geologists [US]
AASHO American Association of State Highway Officials [US]
AASHTO American Association of State Highway and Transportation Officials [US]
AASI Association for the Advancement of Science in Israel
AASIR Advanced Atmospheric Sounder and Imaging Radiometer
AASM Association of American Steel Manufacturers [US]
AASME Associate of the American Society of Mechanical Engineers
AASMTC Association of American Steel Manufacturers Technical Committee [US]
AASO Alberta Applied Science Office [Canada]
American Association of Ship Owners [US]
AASP Advanced Automated Sample Processor
American Association of Stratigraphic Palynologists [US]
Association of African Studies Programs
AASR Airports and Airways Surveillance Radar
AASRC American Association of Small Research Companies [US]
AASRI Arctic and Antarctic Scientific Research Institute
AASSA Association of American Schools in South America
AASU All Africa Students Union [Ghana]
AASW American Association of Scientific Workers [now: USFSS]
AAT Acetoacettoluidide
Adenine/Aminopterin/Thymidine Medium
Alpha-Antitrypsin
Aminoazotoluene
Analytic Approximation Theory
Anglo-Australian Telescope [Australia]
Association of Accounting Technicians [UK]
Aat Acetobacter Aceti [Molecular Biology]
AATA All Africa Teachers Association
AATB Army Aviation Test Board [US Army]
AATBN Aromatic Amine-Terminated Butadiene/Acrylonitrile
AATC Anti-Aircraft Training Center
Army Aviation Test Command
Automatic Air Traffic Control
AATCC American Association of Textile Chemists and Colorists [US]
AATCO Army Air Traffic Coordinating Office
AATD Aviation Applied Technology Directorate [US]
AATMS Advanced Air Traffic Management System
AATO Army Air Transport Organization [US]
AATPO Association of African Trade Promotion Organizations [Morocco]
AA-tRNA Aminoacyl Transfer Ribonucleic Acid
AATT American Association for Textile Technology [US]
AATUF All-African Trade Union Federation
AAU Address Arithmetic Unit
Association of African Universities [Ghana]
Association of American Universities [US]
Association of Atlantic Universities [Canada]
Automatic Answering Unit
AAUA American Association of University Administrators [US]
AAUBO Association of Atlantic University Business Officers [Canada]
AAUG Association of Arab-American University Graduates
AAULC Association of Atlantic Universities Librarians Council [Canada]
AAUP American Association of University Professors [US]
Association of American University Presses [US]
AAUPF American Association of University Professors Foundation [US]
AAUS American Association of University Students [US]
AAUW American Association of University Women [US]
AAUWEF American Association of University Women Educational Foundation [US]
AAV Airborne Assault Vehicle
AAVCS Automatic Aircraft Vectoring Control System
AAVD Automatic Alternate Voice/Data
AAVIM American Association for Vocational Instructional Materials [US]
AAVS Aerospace Audiovisual Service
AAVSO American Association of Variable Star Observers [US]
AAVT Association of Audio-Video Technicians [US]
AAW Air-Acetylene Welding
American Association of Woodturners [US]
Anti-Air Warfare
AAWO American Automatic Weapons Association [US]
AAWH American Association for World Health [US]
AAWO Association of American Weather Observers [US]
AAWPI Association of American Wood Pulp Importers [US]
AAWS Anti-Armor Weapon System
AAWS-M Anti-Armor Weapon System — Medium
AAX Acetoacetxylidide
Automated Attendant Exchange
AAZN American Association for Zoological Nomenclature [US]
AAZPA American Association of Zoological Parks and Aquariums [US]
AB Abscisic Acid
Abstract
AB Battery
Accoustic Bridge
Adapter Booster
Address Bus
Aeronautical Board
Afterburner
Air Base
Air Blast
Air Brake
Air-Break
Alberta [Canada]
Alcian Blue
All Before
American and British (Standards)
Aminoazobenzene
Anchor Bolt
Angle Bracket
Application Block
Application Brief
Applied Biology

Arc Brazing
(Artium Baccalaureus) — Bachelor of Arts
As Brazed
Atomic Beam
Atomic Bomb
Automated Bibliography
Automatic Blowdown

A/B Acid/Base (Ratio)
Airborne

Ab Alabamine
Antibody

ABA Abscisic Acid
Active Brazing Alloy
Agaricus Bisporus Agglutinin
Air Brake Association [US]
American Bankers Association [US]
American Birding Association [US]
American Bison Association [US]
American Booksellers Association [US]
American Breed Association [US]
American Buffalo Association [Name Changed to: American Bison Association, US]
American Burn Association [US]
American Bus Association [US]
American Business Association [US]
Aminobenzoic Acid
Aminobutyric Acid
ASEAN Bankers Association [Singapore]
Association of British Archeologists
Australian Booksellers Association

ABAA Australian Business Aircraft Association

ABAC A Basic Coursewriter
Association of Balloon and Airship Constructors [US]
Association of British Aero Clubs
Association of British Aviation Consultants

ABACUS AB Atomenergi Computerized User-Oriented Services [Sweden]
Air Battle Analysis Center Utility System
Association of Bibliographic Agencies of Britain, Australia, Canada and the United States

ABAF Alcian Blue Aldehyde Fuchsin

ABAI American Boiler and Affiliated Industries [US]

abamp abampere [Unit]

ABANA Artist-Blacksmiths Association of North America [US]

ABANK Annual State Databank [of KEIS, US]

ABAR Advanced Battery Acquisition Radar
Alternate Battery Acquisition Radar

ABB Abbreviate
Abbreviation
American Board of Bioanalysts [US]

ABBA American Bee Breeder Association [US]

ABBE Advisory Board on the Built Environment [now: BRB, US]

ABBF Association of Brass and Bronze Founders

ABBIM American Brass and Bronze Ingot Manufacturers [US]

ABBM Automatic Baseband Monitor

ABBMM Association of British Brush Machinery Manufacturers

ABBR Abbreviate
Abbreviation

ABBRA American Boat Builders and Repairers Association [US]

ABC Abridged Building Classification
Academia Brasileira de Ciencias [= Brazilian Academy of Sciences]
Adaptable Board Computer
Advance Booking Charter
African Bibliographic Center
Airborne Control System
Alfred Bader Collection (of Rare Chemicals)
Already Been Converted
Alum, Blood, Charcoal and Clay Process
American Bibliographical Center [US]
American, British, and Canadian (Standards)
American Broadcasting Company [US]
American Bureau of Circulations [US]
Answer Back Code
Approach by Concept
Argentina-Brazil-Chile (Countries)
Armored Bushed Cable
Associated Builders and Contractors [US]
Association of Biotechnology Companies [US and Canada]
Association of Bituminous Contractors [US]
Association of Boards of Certification [US]
Association of British Climatologists
Atomic, Biological and/or Chemical Weapons
Audit Bureau of Circulation [US]
Australian Broadcasting Commission
Australian Broadcasting Company
Auto-Bill Calling
Automated Bleed Compensation
Automatic Bandwidth Control
Automatic Bass Compensation
Automatic Bias Compensation
Automatic Block Controller
Automatic Brightness Control

ABCA American, British, Canadian and Australian (Standards)
American Building Contractors Association [US]
American Business Communications Association [US]
Army Bureau of Current Affairs [US]
Association of Biological Collections Appraisers [US]

ABCAS Active Beacon Collision Avoidance System

ABCB Air-Blast Circuit Breaker
Air-Break Circuit Breaker
American Bottlers of Carbonated Beverages [US]

ABCC American Board of Clinical Chemistry [US]
Association of British Chambers of Commerce
Association of British Correspondence Colleges
Atomic Bomb Casualty Commission [of USAEC]

ABCCC Airborne Battlefield Command and Control Center [also: ABC^3]

ABCCTC Advanced Base Combat Communication Training Center [US]

ABCD Atomic, Biological, Chemical and Damage Control

	Association for Bridge Construction and Design [US]
	Association of Better Computer Dealers [US]
ABCI	Advanced Business Communications Incorporated
	Automotive Booster Clubs International [US]
ABCM	American Business Council of Malaysia [Malaysia]
	Association of British Chemical Manufacturers
	Association of Building Component Manufacturers [UK]
ABCMC	Automotive Battery Charger Manufacturers Council [US]
ABCO	Association of British Conference Organizers
ABCOP	Association of Boards of Certification for Operational Personnel [now: ABC, US]
ABCQ	Association of Building Contractors of Quebec [Canada]
ABCPA	Alberta Beef Cattle Performance Association [Canada]
ABCPF	Association of British Columbia Professional Foresters [Canada]
ABCS	Advisory Board for Cooperative Systems [US]
	American Business Council of Singapore [Singapore]
ABCST	American-British-Canadian Standards
ABCU	Army Bureau of Current Affairs [US]
ABD	Adhesive Bonding
	Asian Development Bank
	Association Belge de Documentation [= Belgian Association for Documentation]
	Average Business Day
ABDC	After Bottom Dead Center
ABDEA	Aminobenzyl Diethylamine
ABDL	Automatic Binary Data Link
ABDMA	Aminobenzyl Dimethylamine
ABDP	Association of British Directory Publishers
ABDR	Aircraft Battle Damage Repair
ABE	Advanced Bond Evaluator
	Aerodrome Beacon
	Arithmetic Building Element
ABEC	Aitken Bicentennial Exhibition Center [Canada]
	Annular Bearing Engineers Committee [US]
ABEI	4-Aminobutylethylisoluminol
ABEN	Association of Biologists and Ecologists of Nicaragua
ABEND	Abnormal End(ing)
ABES	Aerospace Business Environment Simulator
	Association for Broadcast Engineering Standards [US]
	Automatic Burn-in and Environmental System
ABET	Accreditation Board for Engineering and Technology [US]
ABETS	Airborne Beacon Test Set
ab ex	(ab extra) — from without
ABF	America the Beautiful Fund [US]
	Arbetarnas Bildningsforbund [= Workers Education Association, Sweden]
	Association des Bibliothecaires Francais [= Association of French Librarians]
ABFA	Azobisformamide
ABFL	Association of British Foam Laminators
ABFLO	Association of Bedding and Furniture Law Officials [US]
ABFS	Auxiliary Building Filter System
ABG	Aural Bearing Generator
ABGTS	Auxiliary Building Gas Treatment System
ABGWIU	Aluminum, Brick and Glass Workers International Union
ABH	Average Busy Hour
ABI	Abstracted Business Information
	Advance Boundary Information
	American Butter Institute [US]
	Association des Bibliotheques Internationales [= Association of International Libraries, France]
ABIA	American Boardsailing Industries Association [US]
ABI/INFORM	Abstracted Business Information/Information Needs (Database)
ABIM	Association of British Insecticide Manufacturers
ab init	(ab initio) — from the beginning
ABIPC	Abstract Bulletin of the Institute of Paper Chemistry [US]
ABIST	Autonomous Built-In Self Test
ABIX	Australian Business Index [of AUSINET]
ABK	Arrott-Belov-Kouvel Plot
ABL	Ablative
	Acetylbutyrolactone
	Architectural Block-Diagram Language
	Army Biological Laboratory [US Army]
	Atlantic Basic Language
	Atlas BASIC Language
	Automated Biological Laboratory [of NASA, US]
	Automatic Bootstrap Loader
ABLB	Alternate Binaural Loudness Balancing
ABLE	Activity Balance Line Evaluation
	Association for Biology Laboratory Education [Canada]
ABLISS	Association of British Library and Information Science Schools
ABLS	American Bryological and Lichenological Society [US]
	Association of British Library Schools
ABLUE	Asymptotically Best Linear Unbiased Estimator
ABM	Anti-Ballistic Missile
	Associate in Business Management [US]
	Asynchronous Balanced Mode
	Automated Batch Mixing
ABMA	American Boiler Manufacturers Association [US]
	American Brush Manufacturers Association [US]
	Army Ballistic Missile Agency [US]
ABMAAI	American Boiler Manufacturers Association and Affiliated Industries [now: ABMA, US]
ABMAC	Association of British Manufacturers of Agricultural Chemicals
ABMD	Air Ballistics Missile Division [of US Air Force]
ABMDA	Advanced Ballistic Missile Defense Agency [of US Army]
	Associated Building Material Distributors of America [US]
ABMEWS	Anti-Ballistic Missile Early Warning System
ABMEX	Association of British Mining Equipment Exporters
ABMIS	Anti-Ballistic Missile Intercept System
ABMS	American Bureau of Metal Statistics [US]

ABN Airborne
Australian Bibliographic Network
ABNO All But Not Only
ABNT Associacao Brasiliera de Normas Tecnicas [= Brazilian Association for Technical Standards]
ABO Astable Blocking Oscillator
ABOA Australian Bibliography on Agriculture [of AUSINET]
ABOCF Association of British Organic and Compound Fertilizers [US]
ABOI Association of British Oceanic Industries
ABOL Adviser Business-Oriented Language
ABOM American Board of Occupational Medicine [US]
ABORT Abnormal Termination
ABOS Advanced Bombardment System
ABP Actual Block Processor
Adjustable Ball Pin
Advanced Business Processor
American Business Press [Name Changed to: American Business Publishers, US]
American Business Publishers [US]
Aminobiphenyl
ABPA Aftermarket Body Parts Association [US]
American Book Producers Association [US]
Australian Book Publishers Association
ABPBC Association of Book Publishers of British Columbia [Canada]
ABPC Association of British Packing Contractors
ABPI Association of British Pharmaceutical Industry
ABPVM Association of British Plywood and Veneer Manufacturers
ABQ Associacao Brasileira de Quimica [= Brazilian Chemical Association]
ABR Abrasive
Abridged
Abridgement
Acrylic Ester-Butadiene Rubber
Automatic Band Rate
Automatic Baud Recognition
ABRC Advisory Board for the Research Councils [UK]
ABRES Advanced Ballistic Reentry System
ABRFM Association of British Roofing Felt Manufacturers
ABRO Animal Breeding Research Organization [UK]
ABRSV Abrasive
ABS Absolute (Value)
Abstract
Acrylonitrile Butadiene Styrene (Resin)
Air Bearing Surface
Air Break Switch
Alkyl Benzene Sulfonate
American Biological Society [US]
American Bladesmith Society [US]
American Bureau of Shipping [US]
Animal Behavior Society [US]
Antigen Binding Site
Anti-Lock Brake System
Association of Broadcasting Staffs [UK]
Association on Broadcasting Standards [US]
Auckland Botanical Society [New Zealand]
Automatic Bibliographic Service
Auxiliary Building Sump
Average Busy Season
abs absolute [Chemistry]
ABSA Aminobenzenesulfonic Acid
ABSAP Aminobenzenesulfamidopyridine
ABSBH Average Busy Season Busy Hour
ABSLDR Absolute Loader
ABSM Alberta Bureau of Surveying and Mapping [Canada]
Association of British Sterilizer Manufacturers
ABSMA American Bleached Shellac Manufacturers Association [US]
ABS-PA Acrylonitrile Butadiene Styrene — Polyamide (Alloy)
ABS-PC Acrylonitrile Butadiene Styrene — Polycarbonate (Alloy)
ABS-PVC Acrylonitrile Butadiene Styrene — Polyvinyl Chloride (Alloy)
ABSR Association of British Science Writers
ABS-SMA Acrylonitrile Butadiene Styrene — Styrene-Maleic Anhydride
ABST Abstract
ABSTI Advisory Board on Scientific and Technical Information [Canada]
ABSW Air-Break Switch
Association of British Science Writers
ABT Air Blast Transformer
American Board of Toxicology [US]
Association of Building Technicians [UK]
Australian Broadcasting Tribunal
abT abtesla [Unit]
ABTA Allied Brewery Traders Association
Australian British Trade Association
ABTF Airborne Task Force
ABTICS Abstract and Book Title Index Card Service [UK]
ABTS 2,2'-Azinobis(3-Ethylbenzthiazolinesulfonic Acid)
ABTSS Airborne Transponder Subsystem
ABU Asian Broadcasting Union
Asia-Pacific Broadcasting Union [Malaysia]
ABV Absolute Value
abWb abweber [Unit]
ABWE Associated Business Writers of America [US]
ABWR American Beefalo World Registry [US]
ABYC American Boat and Yacht Council [US]
ABZ Association of British Zoologists
AC Access Number
Accumulator
Acetal Polymer
Acoustic Coupler
Acrylic
Activated Complex
Active Coil
Active Current
Activity Code
Adaptive Control
Address Counter
Advanced Ceramics
Advanced Composite
Advanced Certificate
Advertising Council [US]
Advice of Charge
Advisory Circular
Advisory Committee

	Aerodynamic Center
	Agribusiness Council [US]
	Agriculture Canada
	Air Canada
	Air Carrier
	Air Commodore
	Air Conduction
	Air Conditioning
	Air Cooling
	Aircraftsman
	Airworthiness Circular
	Alaska Coalition [US]
	Algonquin College [Canada]
	Allyl Chloride
	Alpine Club
	Alternating Current
	Amacrine Cell
	American, and Canadian (Standards)
	Ammonium Citrate
	Analog Computer
	Analog Controller
	Analysis Control
	Analytical Chemistry
	Answer Complete
	Anticathode
	Anti-Coincidence Counter
	Anticyclone
	Applied Chemistry
	Applied Climatology
	Arc Cutting
	Archeological Conservancy [US]
	Arctic Circle
	Area Code
	Armored Cable
	Armor-Clad
	Armor-Clad Cable
	Army Corps
	As Cast
	Assigned Code
	Astrocyte
	Authority Code
	Author's Correction
	Autocatalysis
	Autoclave
	Autocollimator
	Autocorrelation
	Automatic Checkout
	Automatic Computer
	Automatic Control
	Auxiliary Carry
	Awaiting Connection
A-C	Alternating Current
A/C	Account
	Air Conditioning
	Aircraftsman
	Aircraftswoman
	Asbestos Cement
Ac	Acetate
	Acetyl
	Actinium
	Alto-Cumulus (Cloud)
ac	acid
	(a conto) — on account
	acres
	alternating current
	(anni currentis) — this (current) year
aC	abcoulomb [Unit]
a/c	account
a c	(a condition) — upon request
ACA	Adjacent Channel Attenuation
	Advanced Ceramics Association [US]
	Advanced Control Algorithm
	Agricultural Computer Association
	Alberta Construction Association [Canada]
	Allied Command Atlantic [of NATO]
	Amalgamated Construction Association (of British Columbia) [Canada]
	American Cartographic Association [US]
	American Chain Association [US]
	American Communications Association [US]
	American Commuters Association [US]
	American Council for Aeronautics [US]
	American Crystallographic Association [US]
	Aminocephalosporanic Acid
	Ammoniacal Copper Arsenate
	Aramament Control Agency
	Associate of the Institute of Chartered Accountants [UK]
	Association of Canadian Advertisers
	Association of Commuter Airlines [US]
	Association of Consultant Architects [UK]
	Asynchronous Communications Adapter
	Australasian Corrosion Association [Australia]
	Australian Council for Aeronautics
	Automatic Circuit Analyzer
	Automatic Communication Association
	Automatic Conference Arranger
	Average Case Analysis
7-ACA	7-Aminocephalosporanic Acid
AcA	Actinium A (= Polonium 215)
AC A	Acetic Acid
ACAA	American Coal Ash Association [US]
ACAAC	American College Admissions Advisory Center [US]
ACABQ	Advisory Committee on Administration and Budgetary Questions [of UN]
ACAC	Acetylacetonato
	Acetylacetone
	Arab Civil Aviation Council
	Asociacion Colombiana par el Avance de la Ciencia [= Colombian Association for the Advancement of Science]
	Associate Committee on Automatic Control [of NRC, Canada]
ACAD	Academic
	Academy
	Air Containment Atmosphere Dilution
	American Conference of Academic Deans [US]
ACADI	Arab Center for Agricultural Documentation Information
ACAH	Acylcholine Acyl Hydrolase
ACAI	Automatic Counting Accuracy Improvement
ACAM	Augmented Content Addressed Memory

A-CAM Anti-A-Cell Adhesion Molecule
ACAMAR Asociacion Centroamericana de Armadores [= Central American Association of Ship Owners, Guatemala]
ACAP Advanced Composite Airframe Program
Aeronautical Chart Automation Project [US]
Automatic Circuit Analysis Program
ACAR Angular Correlation of Annihilation Radiation
ACARD Advisory Council for Applied Research and Development [UK]
ACARS ARINC Communication Addressing Reporting System
ACARTSOD African Center for Applied Research Training and Social Development
ACAS Advisory Conciliation and Arbitration Service
African Commission on Agricultural Statistics [Ghana]
Airborne Collision Avoidance System
Aircraft Collision Avoidance System
ACASP Air Conditioning Analytical Simulation Package [also: A/CASP]
ACASPP American Committee to Advance the Study of Petroglyphs and Pictographs [US]
ACAST Advisory Committee on the Application of Science and Technology [of UNESCO]
ACAT Associate Committee on Air-Cushion Technology
ACAU Automatic Calling and Answering Unit
ACAV American Committee on Athropoid-Borne Viruses [US]
ACAWU Aviation, Communication and Allied Workers Union [Venezuela]
ACB Access Method Control Block
Adjusted Cost Base
Air Circuit Breaker
Application Control Block
Army Classification Battery
Association of Clinical Biochemists [UK]
Association of Custom Brokers [US]
AcB Actinium B (= Lead 211)
ACBS Accrediting Commission for Business Schools
ACC Academia de Ciencias de Cuba [= Cuban Academy of Science]
Acceptance
Account
Accompany
According
Accumulation
Accumulator
Acid Copper Chromate
Adapter Common Card
Administrative Committee on Coordination [of FAO, UN]
Advanced Carbon/Carbon
Aft Cargo Carrier
Air Control Center
Air Coordinating Committee
Allied Control Council
Amateur Computer Club [UK]
Amazonian Cooperation Council [Brazil]
American Carbon Committee [now: ACS, US]
American Chamber of Commerce [US]
American Control Conference [US]
American Copper Council [US]
American Copyright Council [US]
American Craft Council [US]
Antenna Control Console
Antigua Chamber of Commerce [of Antigua-Barbuda]
Application Control Code
Approved Capital Cost
Area Control Center
Arts and Communications Counsellor
Association of Computer Clubs [UK]
Association of Computer Consultants [now: ACCU, US]
Asynchronous Communications Controller
Atlantic Council of Canada
Australasian Corrosion Center [Australia]
Automatic Carrier Control
Automatic Chrominance Control
Automatic Color Control
Automatic Contrast Control
AcC Actinium C (= Bismuth 211)
AcC' Actinium C' (= Polonium 211)
AcC" Actinium C" (= Thallium 207)
Acc Acinetobacter Calcoaceticus [Molecular Biology]
ACCA Accelerated Capital Cost Allowance
Air Chartered Carriers Association [UK]
Air Conditioning Contractors of America [US]
Air Courier Conference of America [US]
American Chamber of Commerce in Austria
Asociacion Chilena de Control Automatico [= Chilean Association for Automatic Control]
Association of Certified and Corporate Accountants [UK]
Asynchronous Communications Control Attachment
ACCAP Autocoder-to-COBOL Conversion Aid Program
Autocoder-to-COBOL Conversion and Program
ACCAST Advisory Committee on Colonial Colleges of Arts, Science and Technology [now: COCAST, UK]
ACCB American Chamber of Commerce in Bolivia
ACCC Ad Hoc Committee for Competitive Communication [US]
Association of Canadian Community Colleges
ACCCE Association of Consulting Chemists and Chemical Engineers [also: ACC&CE, US]
ACCCI American Coke and Coal Chemicals Institute [also: AC&CCI, US]
ACCCIW Advisory Committee to the Canadian Center for Inland Waters
ACCD Accelerated Construction Completion Date
ACCE Advanced Composites Conference and Exhibition
Aircrew-Cockpit Compatibility Evaluation
American Chamber of Commerce Executives [US]
American Council for Construction Education [US]
ACCEL Acceleration
Automated Circuit Card Etching Layout
ACCES Advanced Computing and Communications Educational System [Canada]
ACCESS Accessory
Air Canada Cargo Enquiry and Service System

Aircraft Communication Control and Electronic Signaling System
Aircraft Communication Electronic Signaling System
Architects Central Constructional Engineering Surveying Service [of GLC, UK]
Architecture, Construction and Consulting Engineers Special Services [Australia]
Argonne Code Center Exchange and Storage System [of USAEC]
Automated Catalog for Computer Equipment and Software Systems [US Army]
Automated City Clerk Enhanced Search System
Automatic Computer-Controlled Electronic Scanning System

ACCET Accrediting Council for Continuing Education and Training [US]

ACCFA Agricultural Credit and Cooperative Financing Administration [Philippines]

ACCG American Committee for Crystal Growth [now: AACG, US]
American Conference on Crystal Growth [US]

ACCHAN Access Channel
Allied Command Channel [of NATO]

ACCI ASEAN Chamber of Commerce and Industry
Association of Chambers of Commerce of Ireland

ACCIS Advisory Committee for Coordination of Information Systems [of UN]

ACCJ American Chamber of Commerce in Japan

ACCL Alberta Council of College Librarians [Canada]
American College of Computer Lawyers [US]
American Council of Commercial Laboratories [US]

aCcm abcoulomb centimeter [Unit]

ACCN Academia Chilena de Ciencias Naturales [= Chilean Academy of Natural Sciences]

ACCP American Chamber of Commerce of the Philippines

ACCRA American Chamber of Commerce Researchers Association [US]

ACC-ROC American Chamber of Commerce in the Republic of China [Taiwan]

ACCS Association of Casualty and Surety Companies
Automatic Calling Card Service
Automatic Checkout and Control System

ACCSA Allied Communications Computer Security Agency [of NATO]

ACCT Account
Alliance for Coal and Competitive Transportation
American Chamber of Commerce in Thailand

ACCTG Accounting

ACCTI American Chamber of Commerce for Trade with Italy [US]

ACCTS Accounts

ACCU Association of Computer and CD-ROM Users [US]

ACCUM Accumulation

ACCUMR Accumulator

ACCW Alternating Current Continuous Wave

ACCWP Acquisition, Cataloguing and Circulation Working Party [UK]

ACD Acid Citrate Dextrose
Alarm, Control and Display
Anode-to-Cathode Distance
Antenna Control Display
Associated Construction Distributors (International) [US]
Automatic Call Distribution
Automatic Call Distributor

AcD Actinium D (= Actinium-Lead)

ACDA American Component Dealers Association [US]
Arms Control and Disarmament Agency [US]

ACDB Airport Characteristics Databank

AC-DC Alternating Current/Direct Current [also: AC/DC or ACDC]

ACD-ESS Automatic Call Distributor — Electronic Switching System

ACDI Agricultural Cooperative Development International [US]

ACDM Association of Chairmen of Departments of Mechanics [US]

ACDO Air Carrier District Office [US]

ACDPI American Cultured Dairy Products Institute [US]

ACDRI Advisory Committee on the Development of Research for Industry [of CSIR, South Africa]

ACE Academia de Ciencias Economicas [= Academy of Economic Sciences, Argentina]
Acceptance Checkout Equipment
Advanced Certificate in Education
Advanced Computing Environment
Advanced Control Experiment
Advisory Center for Education [UK]
Agricultural Communicators in Education [US]
Air-Conditioning Engineer
Air-Conditioning Engineering
Air Cushion Equipment
Airspace Coordination Element
Alberta Caravan Exhibition (Foundation) [Canada]
Allied Command Europe [of NATO]
Altimeter Control Equipment
Ambush Communication Equipment
American Council on Education [US]
American Council on the Environment [US]
Analog Contrast Enhancement
Angiotensin Converting Enzyme
Animated Computer Education
Area Control Error
Association for Computer Educators [US]
Association for Conservation Engineers [US]
Association for Cooperation in Engineering
Association of Conservation Engineers
Association of Consulting Engineers [UK]
Association for Continuing Education
Attitude Control Electronics
Audio Connecting Equipment
Automated Cable Expertise
Automated Claims Environment [USAA, US]
Automated Cost Estimate
Automatic Calling Equipment
Automatic Checkout Equipment
Automatic Clutter Eliminator
Automatic Computer Evaluation
Automatic Computing Engine
Automation Conference and Exhibition
Aviation and Computer Enthusiasts [US]
Aviation Construction Engineer

ACEA American Cotton Exporters Association [US]
Association of Consulting Engineers of Australia
Association of Cost and Executive Accountants [UK]

ACEAA Advisory Committee on Electrical Appliances and Accessories

ACEC American Consulting Engineers Council [US]
Association of Consulting Engineers of Canada

ACEC/RMF American Consulting Engineers Council/Research and Management Foundation [US]

ACEE Aircraft Energy Efficiency

ACEEE American Council for an Energy Efficient Economy [US]

ACEF Asian Cultural Exchange Foundation
Association of Commodity Exchange Firms [US]

ACEI Association of Consulting Engineers of Ireland

ACEID Asian Center for Educational Innovation Development

ACEJMC Accrediting Council on Education in Journalism and Mass Communications [US]

ACEK Association of Consulting Engineers of Kenya

ACEL Aerospace Crew Equipment Laboratory [US Navy]
Alternating Current Electroluminescent

ACELSCO Associated Civil Engineers and Land Surveyors of Santa Clara County [US]

ACEM Association of Consulting Engineers of Malaysia
Association of Consulting Engineers of Manitoba [Canada]

AcEm Actinium Emanation (= Actinon, or Radon 219) [also: Ac-Em, or An]

ACEN Association of Consulting Engineers of Norway

ACEP Advisory Committee on Export Policy [US]

ACER Australian Council for Educational Research

ACerS American Ceramic Society [US]

ACES (2-Acetamido)-2-Aminoethanesulfonic Acid
Acoustic Containerless Experiment System
Aerosol Containment Evacuation System
Air Collection and Enrichment System
Air Collection Engine System
2-[(2-Amino-2-Oxoethyl)amino]ethanesulfonic Acid
Annual Cycle Energy System
Applied Computational Electromagnetics Society [US]
Arab Center for Energy Studies
Association of Consulting Engineers of Saskatchewan [Canada]
Automated Code Evaluation System
Automatic Checkout and Evaluation System [US Air Force]

ACESA Arizona Council of Engineering and Scientific Associations

ACE-S/C Acceptance Checkout Equipment — Spacecraft

ACET Acetylene
Acetone
Advisory Committee for Electronics and Telecommunications
Association of Consulting Engineers of Tanzania

ACET A Acetic Acid

ACETM Automatic Compensation of Emissivity for Temperature Measurement
Automatic Compensation of Emissivity for Temperature Monitoring

ACF Access Control Facility
Access Cost Factor
Advanced Ceramics Facility [of MPI, Germany]
Advanced Communications Function
Agricultural Cooperative Federation [UK]
Air Combat Fighter [US Air Force]
Alternate Communications Facility
A (= Al_2O_3), C (= CaO), F (= FeO + MgO) (Diagram) [Petrology]
A (= $Al_2O_3 + Fe_2O_3 - (Na_2O + K_2O)$), C (= CaO), F (= FeO + MgO + MnO) (Diagram) [Petrology]
Area Computing Facility
ASEAN Constructors Federation
Association of Consulting Foresters [US]
Autocorrelation Function
Automobile Club of France

ACFF Air Cargo Fast Flow

ACFG Automatic Continuous Function Generation

ACFM Actual Cubic Feet per Minute [also: acfm]

ACFOA Australian Council for Overseas Aid

ACFSTA Association of Chinese Food Scientists and Technologists in America [now: CAFS, US]

ACFT Aircraft

ACG Adjacent Charging Group
Air Cargo Glider
Alternating-Current Generator
Automatic Code Generator

ACGB Aircraft Corporation of Great Britain

ACGC Asian Coordination Group Chemistry

AcGEPC Alkyl Acetylsynglycerylphosphorylcholine

ACGIH American Conference of Governmental Industrial Hygienists [US]

ACGR Associate Committee on Geotechnical Research [Canada]

ACH Acetone Cyanhydrin
Acetylcholine [also: ACh]
Adrenal Cortical Hormone
Association of Computers and Humanities
Attempts per Circuit per Hour
Automated Clearinghouse

ACHA American Council of Highway Advertisers [US]
American Cutlery Manufacturers Association [US]

ACHE Acetylcholinesterase [also: AChE]
Association for Continuing Higher Education [US]

ACHEE ASEAN Council of Higher Education Environment

ACHEMA Air-Cooled Heat Exchanger Manufacturers Association [US]
Ausstellungs-Tagung fur Chemisches Apparatewesen [= Chemical Engineering Exhibition/Congress, Germany]

ACHI Application Channel Interface

ACHP Advisory Council on Historic Preservation

AChR Acetylcholine Receptor

a Chr (ante Christum) — before Christ

a Chr n (ante Christum natum) — before Christ

ACHTR Advisory Committee for Humid Tropics Research [of UNESCO]

ACI Adjacent Channel Interference
Adjusted Calving Interval

African Culture Institute
Air Combat Intelligence
Alloy Casting Institute
American Concrete Institute [US]
Army Control Instruction
Association for Conservation Information [US]
Atlantic Canada Institute
Automatic Card Identification
Automatic Component Insertion

ACIA Asociacion Colombiana de Ingenieros Agronomos [= Colombian Association of Agricultural Engineers]
Asynchronous Communications Interface Adapter
Associated Cooperage Industries of America [US]

ACIAR Australian Center for International Agricultural Research

ACIASAO American Council of Industrial Arts and State Association Officials [now: CTEA, US]

ACIATE American Council of Industrial Arts Teacher Education [now: CTTE, US]

ACIB Air Cushion Ice-Breaking Bow

ACIC Aeronautical Chart and Information Center [US Air Force]
Alberta Council for International Cooperation [Canada]
American Committee for International Conservation [US]
Associate of the Canadian Institute of Chemistry

ACICE Authority for the Coordination of Inland Transport in Central Europe

ACID Advisory Committee on Industrial Development [Zimbabwe]
Association of Canadian Industrial Designers
Automatic Classification and Interpretation of Data

ACIFIC Aspartame Committee of the International Food Information Council [US]

ACIGY Advisory Council of the International Geophysical Year

ACIIW American Council of the International Institute of Welding [US]

ACIJ African-Caribbean Research Institute

ACIL American Council of Independent Laboratories [US]

ACIM Accident Cost Indicator Model
Axis-Crossing Interval Meter

ACIP Advisory Committee on Immunization Practice [US]
American Council of International Personnel [US]

ACIR Aviation Crash Injury Research

ACIRL Australian Coal Industry Research Laboratories

ACIS Association for Computing and Information Sciences
Automated Client Information Service

ACISUD Authority for the Coordination of Inland Transport in Southern Europe

ACITS Advisory Committee on Information Technology Standardization [Europe]

ACK Acknowledge
Acknowledgment (Character)

AcK Actinium K (= Francium 223)

ACKT Acknowledgment

ACL Academia des Ciencias de Lisboa [= Adademy of Sciences at Lisbon, Portugal]
Acetal
Advanced CMOS (= Complimentary Metal-Oxide Semiconductor) Logic
Aeronautical Computers Laboratory [US Navy]
Allis-Chalmers/Lepol Kiln
Allowable Cabin Load
Altimeter Check Location
Antigen-Carrier Lipid
Application Control Language
Applied Color Label
Association for Computational Linguistics [US]
Atlas Commercial Language
Atlas Computer Laboratories [UK]
Audit Command Language
Automated Coagulation Laboratory
Automatic Circuit Layout

ACLANT Allied Command Atlantic [of NATO]

ACLD Above Clouds

ACLG Air-Cushion Landing Gear

ACLI American Council of Life Insurance [US]

ACLM Advisory Council on Labour and Manpower (of Quebec) [Canada]

ACLO Agena Class Lunar Orbiter

ACLP Above Core Load Pad

ACLS Air-Cushion Landing System
All-Weather Carrier Landing System
American Council of Learned Societies [US]
Automated Control Landing System

ACM Active Countermeasures
Address Calculation Machine
Advanced Composite Material
Advanced Cruise Missile
Advanced Cure Monitor
Air Chief Marshall
Air Combat Maneuver
Air Composition Monitor
Alarm Control Module
Alterable Control Memory
Alternating-Current Motor
Arab Common Market
Area Contamination Monitor
Asbestos-Covered Metal
Associated Colleges of the Midwest [US]
Association for Computing Machinery [US]
Association of Crane Makers
Associative Communication Multiplexer
Atmospheric Corrosion Monitor
Automatic Clutter Mapping

aΩcm abohm centimeter [Unit]

ACMA Agricultural Cooperative Managers Association [UK]
American Cast Metals Association [US]
Associate of the Institute of Cost and Management Accountants [UK]

ACMAC Association for Computing Machinery Accreditation Committee [US]

ACMC Automotive Chemical Manufacturers Council [US]

ACME Advanced Computer for Medical Research Environment
Association of Consulting Management Engineers [US]

ACMET Advisory Council on Middle East Trade [UK]

ACM/GAMM Association for Computing Machinery/German Association for Applied Mathematics and Mechanics
ACMI Air Combat Maneuvering Instrumentation
Art and Craft Materials Institute [US]
ACMLA Association of Canadian Map Libraries and Archives
ACMMM Annual Conference on Magnetism and Magnetic Materials
ACMO Afloat Communications Management Office [of NSEC, US]
ACMOS Advanced Complimentary Metal-Oxide Semiconductor
ACMR Air Combat Maneuvering Range
ACMRI Air Combat Maneuvering Range Instrumentation
ACMRR Advisory Committee on Marine Resource Research [of FAO, UN]
ACMS Advanced Configuration Management System
Application Control/Management System
ACMSC Association for Computing Machinery Standards Committee [US]
ACMST Association for Computers in Mathematics and Science Teaching [US]
ACMU American Canoe Manufacturers Union [US]
ACN Accession Number
Acrylonitrile
Aircraft Classification Number
Artificial Cloud Nucleation
Asbestos Cloth Neck
ACNA Arctic Institute of North America [Canada]
ACNAM Associations Council of the National Association of Manufacturers Associations [US]
ACNAS Advanced Cableship Navigation Aid System
ACNBC Associate Committee on the National Building Code [of NRC, Canada]
ACNM American Cordage and Netting Manufacturers [US]
ACNOCOMM Assistant Chief of Naval Operations for Communications and Cryptology
ACNY Advertising Club of New York [US]
ACO Accounting Control Office
Architectural Conservancy of Ontario [Canada]
ASEAN Cooperative Organization
Automatic Call Origination [also: AUTOCALL or autocall]
ACOA Atlantic Canada Opportunity Agency [Canada]
ACOC Area Communications Operations Center
ACODEX Asociacion Colombiana de Exportadores [= Colombian Association of Exporters]
ACOLI Advanced Circuit Order and Layout Information
ACOM Automatic Coding System
Associate Committee on Meteorites [Canada]
Association for Convention Operations Management [US]
Awards Committee [of IEEE, US]
ACOMPLIS A Computerized London Information Service [UK]
a cond (a condition) — upon request
ACONS Automated Compressor On-Line System
ACOPP Abbreviated COBOL Preprocessor
ACOPS Advisory Committee on Pollution of the Sea [UK]
ACORC Artificial Cells and Organs Research Center [of McGill University, Canada]
ACORD Advisory Council on Research and Development [UK]
ACORN Associative Content Retrieval Network
ACOS Advisory Committee on Safety
Automatic Control System
Assistant Chief of Staff
ACOUS Acoustic(s)
ACP Accelerated Carburizing Process
Acid Phosphatase
Acyl Carrier Protein
Advanced Computational Processor
Advanced Cooperative Project [NASA]
Aerospace Computer Program [US Air Force]
African, Caribbean and/or Pacific (Countries)
Agricultural Conservation Program [US]
Airlift Command Post
Airline Control Program
Albany College of Pharmacy [US]
Allied Communications Publication
Amorphous Calcium Phosphate
Ancillary Control Processor
Anodal Closing Picture
Anticoincidence Point
Arithmetic and Control Processor
Arithmetic Control Processor
Associated Construction Publications [US]
Associate of the College of Preceptors [UK]
Association of Canadian Publishers
Association of Computer Professionals [US]
Azimuth Change Pulse
AcP Acid Phosphatase
ACPA Activated Carbon Producers Association [Belgium]
Adaptive Controlled Phased Array
American Concrete Pavement Association [US]
American Concrete Pipe Association [US]
American Concrete Pumping Association [US]
Association of Computer Programmers and Analysts [US]
AcPb Actinium-Lead (= Actinium D)
ACPD Alternating-Current Potential Drop
ACPDS Advisory Committee on Personal Dosimetry Services [of NSF, US]
ACPF Acoustic Containerless Processing Facility
Asociacion del Congreso Panamericano de Ferrocarriles [= Pan-American Railway Congress Association, Argentina]
ACPM Acoustic Chamber Position Module
Attitude Control Propulsion Motor
ACPMME Association of Concentrated and Powdered Milk Manufacturers of the EEC (= European Economic Community) [France]
ACPPA American Concrete Pressure Pipe Association [US]
ACPR Advanced Core Performance Reactor
Annular Core Pulse Reactor
ACPSM Australasian College for Physical Scientists in Medicine [Australia]
ACPTC Association of College Professors of Textile and Clothing [US]
ACP/TF Airline Control Program/Transaction Processing Facility
ACPU Auxiliary Computer Power Unit

ACR Access Control Register
Active Radar
Advanced Cracking Reactor
Aerodrome Control Radar
Aircraft Control Room
Airfield Control Radar
Alaskan Communications Region [US Air Force]
Alberta Chamber of Resources [Canada]
Alkalipolyacrylate
Alternate CPU (= Central Processing Unit) Recovery
American College of Radiology [US]
American Computer Referral (Database) [US]
Antenna Coupling Regulator
Approach Control Radar
Audio Cassette Recorder
Automatic Call Recording
Automatic Compression Regulator

ACRC Association of Commercial Records Centers [US]

ACRE Automatic Call-Recording Equipment
Automatic Checkout and Readiness

acre-ft acre-foot

acre-ft/d acre-foot per day

acre-in acre-inch

ACRES Airborne Communications Relay Station [US Air Force]

ACRI Air Conditioning and Refrigeration Institute [US]
American Cocoa Research Institute [US]

ACRIT Advisory Committee for Research on Information Transfer

ACRL Association of College and Reference Libraries [US]
Association of College and Research Libraries [of ALA, US]

ACRM American College of Radio Marketing [US]

ACRMA Air Conditioning and Refrigerating Manufacturers Association [US]

ACRS Advisory Committee on Reactor Safeguards [US]
Advisory Committee on Reactor Standards

ACRSP Association for Canadian Registered Safety Professionals

ACRT Accelerated Crucible Rotation Technique
Analysis Control Routine

ACRW American Council of Railroad Women [US]

ACS Absolute Contamination Standard
Accumulator Switch
Acrylonitrile Chlorinated Styrene
Active Control System
Administrative Control System
Adrenocortico Steroid
Advanced Ceramics Society [US]
Advanced Communications Service [US]
Advanced Computer Series
Advanced Computer Services
Advanced Computer System
Aeronautical Communications Subsystem
Agena Control System
Aircraft Control Surveillance
Alarm and Control System
Alaska Communication System [US]
Alternating Current Synchronous
Altitude Control System
American Canal Society [US]
American Cancer Society [US]
American Carbon Society [US]
American Cartographic Society [US]
American Ceramic Society [US]
American Cetacean Society [US]
American Chemical Society [US]
American Chestnut Society [US]
American Conifer Society [US]
American Copyright Society [US]
Anodal Closing Sound
Antireticular Cytotic Serum
Application Customizer Service
Applied Computer Science
Assembly Control System
Assistant Chief of Staff
Assistant Chief Statistician
Association of Commonwealth Students
Attitude Command System
Attitude Control System
Austempered Cast Steel
Australian Ceramics Society
Australian Computer Society
Auto-Calibration System
Automated Communications Set
Automatic Checkout System
Automatic Color System
Automatic Control System
Automobile Club of Switzerland
Auxiliary Code Storage
Auxiliary Core Storage
Auxiliary Cooling System

ACSA Alberta Cities Safety Association [Canada]
Allied Communications Security Agency [of NATO]
American Cotton Shippers Association [US]
Association of Collegiate Schools of Architecture [US]

ACSAD Arab Center for the Studies of Arid Zones and Dry Lands [Syria]

ACSB Amplitude Compandered Sideband

ACS/CAS American Chemical Society/Chemical Abstracts Service [US]

ACSCE Army Chief of Staff for Communications Electronics [US]

ACSF Attack Carrier Striking Force

ACSI Automotive Cooling Systems Institute [US]

ACSIL Admiralty Center for Scientific Information and Liaison [UK]

ACSL Advanced Continuous Simulation Language
Altocumulus Standing Lenticular
American Computer Science League [US]

ACSM American Congress on Surveying and Mapping [US]
Associate of the Cambourne School of Mines [US]

ACSN Appalachian Community Service Network

AcSoc Acoustical Society of America

ACSP Advanced Control Signal Processor [NASA]
Advisory Council on Scientific Policy [UK]

ACSR Aluminum Cable Steel-Reinforced
Aluminum-Clad, Steel-Reinforced
Aluminum Conductor, Steel-Reinforced

ACSR/AW	Aluminum Conductor, Steel-Reinforced using Aluminum-Clad Steel Wire
ACSR/AW	Aluminum Conductor, Steel-Reinforced using Aluminum-Clad Steel Wire
ACSR/AZ	Aluminum Conductor, Steel-Reinforced using Aluminum-Coated Steel Wire
ACSR/GA	Aluminum Conductor, Steel-Reinforced using Class A Zinc-Coated Steel Wire
ACSR/GB	Aluminum Conductor, Steel-Reinforced using Class B Zinc-Coated Steel Wire
ACSR/GC	Aluminum Conductor, Steel-Reinforced using Class C Zinc-Coated Steel Wire
ACSR/HS	Aluminum Conductor, Steel-Reinforced, High Strength
ACSR/SD	Aluminum Conductor, Steel-Reinforced, Self Damping
ACSS	Air Combat and Surveillance System
	Analog Computer Subsystem
	Association of Canadian Security Services
ACSSB	American Certified Shake and Shingle Bureau [US]
	Amplitude-Companded Single Sideband
ACSSM	Associate Committee on Soil and Snow Mechanics [Canada]
ACST	Access Time
	Acoustics
	Advisory Committee on Science and Technology [of ECOSOC, OAS]
ACSTTO	Association of Certified Survey Technicians and Technologists of Ontario [Canada]
ACSU	Association for Computer Science Undergraduates
ACSYS	Accounting Computer System
ACT	Action
	Activated Complex Theory
	Activation
	Active
	Actual
	Actuation
	Adrenocortiotropin
	Advanced Communications Technology
	Advances in Characterization and Testing
	Advisory Council on Technology
	Air Control Team
	Air Cushion Trailer
	Algebraic Compiler and Translator
	Alternative Community Telephone [Australia]
	American College of Toxicology [US]
	American College Testing Program [US]
	American Council for Turfgrass [US]
	Analog Circuit Technique
	Applied Circuit Technology
	Applied Computer Technique
	Area Correlation Tracker [US Air Force]
	Association of Commercial Television [US]
	Association of Communications Technicians [US]
	Association of Cytogenetic Technologists [US]
	Asymmetric Crystal Tomography
	Asymmetric Crystal Topography
	Australian Capital Territory
	Automatic Code Translation
ACTA	Alternative Carrier Telecommunications Association [US]
	American Cement Trade Alliance [US]
ACTE	Agkistrodon Contortrix Thrombin-Like Enzyme
	Association of Commercial and Technical Employees
ACTEC	American Coalition on Trade Expansion with Canada [US]
ACTF	Altitude Control Test Facility
ACTG	Actuating
ACTH	Adrenal Cortical Steroid Hormone
	Adrenocorticotropic Hormone
ACTION	Accepting Challenge Through Interaction with Others and Nature
ACTIS	Advanced Computed Tomography Imaging System
	Auckland Commercial and Technical Information Service [New Zealand]
ACTL	Advanced Construction Technology Laboratory [of NRC, Canada]
ACTM	American Cordage and Twine Manufacturers [now: ACNM, US]
ACTO	Association of Chief Technical Officers [UK]
	Automatic Computing Transfer Oscillator
ACTOL	Air-Cushion Takeoff and Landing
ACTP	Advanced Composite Thermoplastics
	Advanced Computer Technology Project [UK]
ACTR	Actuator
ACTRAM	Advisory Committee on the Safe Transport of Radioactive Material
ACTRAN	Analog Computer Translator
	Autocoder-to-COBOL Translating Service
	Autocoder-to-COBOL Translation
ACTRUS	Automatically-Controlled Turbine Run-up System
ACTS	Acoustic Control and Telemetry System
	Advanced Communications Technology Satellite [of NASA]
	Advanced Computed Tomography System
	African Center for Technology Studies
	All-Channel Television System
	Association of Competitive Telecommunications Suppliers
	Automatic Coin Toll Service
	Automatic Computer Telex Services
ACTSU	Association of Computer Time Sharing Users [US]
ACTT	Association of Cinematograph, Television and Allied Technicians [UK]
ACTW	Amalgamated Clothing and Textile Workers
ACTY	Actuary
ACU	Acknowledgement Unit
	Address Control Unit
	Alarm Control Unit
	Antenna Control Unit
	Arithmetic Control Unit
	Asia Currency Unit
	Association of Commonwealth Universities [UK]
	Association of Computer Users [US]
	Automatic Call Unit
AcU	Actinium U (= Uranium 235)
	Actinouranium
ACUA	Association of College and University Auditors [US]
ACUG	Atex (Computer) Commercial Users Group [US]
ACUHO-I	Association of College and University Housing Officers — International [US]

ACUI Automatic Calling Unit Interface
ACU-I Association of College Unions — International [US]
ACUIB Association of Canadian University Information Bureaus
ACUMG Association of College and University Museums and Galleries [US]
ACUP Association of Canadian University Presses
Association of College and University Printers [US]
ACUS Administrative Conference of the United States [US]
ACUTA Association of College and University Telecommunications Administrators [US]
ACUTE Accountants Computer Users Technical Exchange [US]
ACV Air-Cushion Vehicle
Alarm Check Valve
Armored Command Vehicle
ACVL Association of Cinema and Video Laboratories [US]
ACVT Air Cushion Vehicle Technician
ACW Aircraft Control Warning
Aircraftwoman
AC&W Air Control and Warning
Aircraft Control and Warning
ACWA Amalgamated Clothing Workers of America
American Clean Water Association [US]
ACWIS American Committee for the Weizmann Institute of Science [US]
ACWL Army Chemical Warfare Laboratory [US Army]
AC&WS Air Control and Warning Station
Aircraft Control and Warning Station
AcX Actinium X (= Radium 223)
ACYC Anticyclonic
ACZ Academia de Ciencias de Zaragoza [= Zaragoza Academy of Sciences, Spain]
AD Access Door
Account Directory
Active Deposit
Address Decoding
Advanced Design
Advanced Development
Advertisement
Aerodrome
Aerodynamics
Air Dose
Air Drying
Airworthiness Directive
Algorithm Design
Ampere Demand Meter
Amplitude Detection
Amplitude Distortion
Analog Divider
Analog to Digital
Anderson-Darling (Test)
Angular Distribution
Anno Domini — in the year of our lord
Anode
Anodal Deviation
Antiphase Domain
Arc Discharge
Area Drain
As Drawn
Assignment Date
ASTIA (= Armed Services Technical Information Agency) Document
Automatic Detection
Average Deviation
Awaiting Disconnection
Axial Direction
Axiomatic Development
Azeotropic Distillation
A/D Altitude/Depth
Analog-to-Digital
ad advertise
advertisement
after date
a d (a dato) — from date of issue
A$ Australian Dollar [Currency]
ADA 2-Acetamido-2-Iminodiacetic Acid
Acetonedicarboxylic Acid
Action Data Automation
Adenosine Deaminase
Adjustable Derivative Action
Advisory Area
Agricultural Development Association [UK]
Airborne Data Automation
Air Defense Area [US Army]
Air Defense Artillery
Aluminum Development Association [UK]
American Dairy Association [US]
American Dehydration Association [US]
American Dehydrators Association [now: AAPA, US]
American Dental Association [US]
Atom Development Administration
Atomic Development Authority [US]
Automatic Data Acquisition
Automatic Distillation Analyzer
ADABAS Adaptable Database System [Germany and US]
ADAC Allgemeiner Deutscher Automobil-Club [= General German Automobile Club]
Automated Direct Analog Computer
ADACC Advanced Damage Control Console
Automatic Data Acquisition and Computer Complex [US Air Force]
ADADL Ada-Based Design and Documentation Language
ADAF Advanced Design Array Radar
ADALERT Advance Alert [also: Adalert or adalert]
ADAM A Data Management System
Advanced Data Management
Advanced Direct-Landing Apollo Mission [of NASA]
Air Deflection and Modulation
Association of Distributors of Advertising Material [Scotland]
Associometrics Data Management System
Automatic Distance and Angle Measurement
Automatic Document Abstracting Method
ADAPS Anthropometric Design Assessment Program System
Automatic Display and Plotting System
ADAPSO Association of Data Processing Service Organizations [Canada/US]
ADAPT Adaption of Automatically Programmed Tools
ADAPTICOM Adaptive Communication

Adapt&Lrng Sys Adaptive and Learning Systems [Division of CS [of IEEE, US]]
ADAPTS Air-Delivered Antipollution Transfer System
ADAR Advanced Design Array Radar
Automatic Data Acquisition Routine
ADAS African Demonstration Center on Sampling Agricultural Surveys [of FAO, UN and CCTA]
Agricultural Development Advisory Service [UK]
Automatic Data Acquisition System
Automatic Dialing Alarm System
ADAT Automatic Data Accumulation and Transfer
ADATOM Adsorbed Atom [also: adatom]
ADATS Air Defense Antitank System
ADAW Arbeitsgemeinschaft Deutscher Aussteller fur Werkstoffprufung [= Study Group of German Exhibitors for Materials Testing, Germany]
ADB Activated Diffusion Bonding
Adhesive Bonding
African Development Bank [of UN]
Asian Development Bank [of UN — Philippines]
Atlantic Development Board [Canada]
Authority Database
ADBMS Available Database Management System
ADBO Authority Database Operation
ADBS Advanced Database System
Association des Documentalistes et des Bibliothecaires Specialises [= Association of Documentalists and Special Librarians, France]
ADC Advanced Digital Control
Advise Duration and Charge
Aerodrome Control
Aerospace Defense Command
Agricultural Development Council [US]
Aide-de-Camp [Military]
Airborne Digital Computer
Air Data Computer
Air Defense Command
Air Diffusion Council [US]
Allyl Diglycol Carbonate
Analog-to-Digital Conversion
Analog-to-Digital Converter
Anodal Duration Contraction
Antenna Dish Control
Applied Diamond Conference
Ardeer Double Cartridge Test
Army Dental Corps [US]
Arsenic Development Committee [France]
Association of Diving Contractors [US]
Atlantic Development Council [Canada]
Automatic Data Collection
Aviation Development Council [US]
Azodicarbonamide
A/DC Analog-to-Digital Converter
ADCA Aerospace Department Chairmen's Association [US]
Aminodecephalosporanic Acid
7-ADCA 7-Aminodecephalosporanic Acid
ADCAD Airways Data Collection and Distribution
ADCC Antibody-Dependent Cellular Cytotoxicity
Air Defense Control Center
Asynchronous Data Communications Channel
ADCCP Advanced Data Communications Control Procedure
ADCE Attitude Determination and Control Electronics
ADCI American Die Casting Institute [US]
ADCIS Association for the Development of Computer-Based Instructional Systems [US]
ADCN Aeronautical Data Communication Network
ADCOM Administrative Committee [of IEEE, US]
ADCON Address Constant
Analog-Digital Converter
ADCP Acoustic Doppler Current Profiler
ADCS Advanced Defense Communications Satellite
Attitude Determination and Control Subsystem
ADCSP Advanced Defense Communications Satellite Program [US]
ADCU Alarm Display and Control Unit
Association of Data Communications Users [US]
ADCUS Advise Customs
ADCVR Analog-to-Digital Converter
ADD Addendum
Addition
Address
Aerospace Digital Development
Automatic Diagnostic Display
ADDA Australian Database Development Association
ADDAR Automatic Digital Data Acquisition and Recording
ADDAM Air Deployable Datum Maker Buoy
ADDAS Automatic Digital Data Assembly System
ADDC Association of Desk and Derrick Clubs [US]
ADDCL Add Clause [computer programming]
ADDDS Automatic Direct Distance Dialling System
ADDER Automatic Digital Data Error Recorder
ADDIT Addition [also: ADDN]
ADD'L Additional [also: ADDL]
ADDN Automated Defense Data Network
ADDPEP Aerodynamic Deployable Decelerator Performance Evaluation Program
ADDR Adder
Address
Address Register
ADDROUT Address Out
ADDS Advanced Detector Development System
Applied Digital Data System
Automatic Data Digitizing System
ADE Address Error
Advanced Data Entry
Air Defense Emergency
Air Density Explorer
Armament Design Establishment [UK]
Association for Documentary Editing [US]
Audible Doppler Enhancer
Automated Debugging Environment
Automated Design Engineering
Automatic Drafting Equipment
ADEA Association Belge pour le Developpement Pacifique de l'Energie Atomique [= Belgian Association for the Peaceful Development of Atomic Energy]
ADECP Association of Directors of European Centers for Plastics [UK]
ADED Association of Driver Educators for the Disabled [US]
ADEE Automated Design and Engineering for Electronics
ADEKS Advanced Design Electronic Key System

ADELT Automatically Deployable Emergency Locator Transmitter
ADEM Adaptively Data Equalized Modem
Automatic Digital Electron Microscope
ADEPT Automated Direct Entry Packaging Technique
Automatic Data Extractor and Plotting Table
ADES Automatic Digital Encoding System
ADEU Automatic Data Entry Unit
ADF Acid Detergent Fiber
Aerial Direction Finding
Aerodynamnic Force
African Development Foundation
African Development Fund
Airborne Direction Finder
Annular Dark Field
Asian Development Fund
Automatic Direction Finder
Automatic Document Feed
ADFA Australian Defense Force Academy
Australian Dried Food Association
ad fin (ad finem) — to the end
ADFRF Ames-Dryden Flight Research Facility [at EAFB, US]
ADFSC Automatic Data Field Systems Command [US Army]
ADGA American Dairy Goat Association [US]
ADGB Allgemeiner Deutscher Gewerkschaftsbund [= German Federation of Trade Unions]
ADGE Air Defense Ground Environment
ADH Adhesive
Alcohol Dehydrogenase
Antidirected Hamiltonian
Antidiuretic Hormone
United Arab Emirates Dirham [Currency]
ad h l (ad hunc locum) — at the place
ad hoc (ad hoc) — for this
ADI Acceptable Daily Intake
Acoustic Data Interface
Address Incomplete
Air Distribution Institute [US]
Alternating Direction Implicit
Altitude Direction Indicator
American Documentation Institute [now: ASIS, US]
Attitude Direction Indicator
Austempered Ductile Iron
Automatic Direction Indicator
ADIC Aktueel Dokumentie en Informatie-Centrum [= Current Dokumentation and Information Center, Netherlands]
Analog-to-Digital Converter
ADICEP Association of Directors of European Centers for Plastics [UK]
ADIDAC Argonne Digital Data Acquisition Computer [of ANL, US]
ADIDS Aeronautical Digital Information Display System
ADIJ Association pour le Developpement de l'Informatique Juridique [= Association for the Development of Legal Informatics, France]
ADINA Automatic Dynamic Incremental Nonlinear Analysis
ad inf (ad infinitum) — to infinity
ad init (ad initium) — at the beginning
ad int (ad interim) — in the interim; in the meantime
ADINTELCEN Advanced Intelligence Center
ADIOS Analog-to-Digital Input/Output System
Automatic Diagnosis Input/Output System
Automatic Digital Input/Output System
ADIRS Air Data Inertial Reference System
ADIS Advanced Data Acquisition, Imaging and Storage
Air Defense Integrated System
Association for the Development of Instructional Systems [US]
Automatic Data Interchange System
ADISP Aeronautical Data Interchange System Panel
ADIT Analog-to-Digital Integration Translator
ADIZ Air Defense Identification Zone
ADJ Adjacent
Adjoining
Adjunct
Adjustment
Adjutant [also: ADJT]
ADL Applications Development Language
Automatic Data Link
ad l (ad libitum) — freely; without limit or restraint
ADLC Advanced Data Link Controller
ADLIB Adaptive Library Management System
ad lib (ad libitum) — freely; without limit or restraint
ADLIPS Automated Data Link Plotting System
ad loc (ad locum) — at the place
ADLRI Arthur D. Little Research Institute [UK]
ADLS Automatic Drag-Limiting System
ADM Activity Data Method
Adaptable Data Manager
Adaptive Delta Modulation
Address Modification
Admiral
Admiralty
Advanced Design Model
Advanced Development Model
Air Decoy Missile
Air Deployable Marker
Administration
Administrator
American Down Association [US]
Analog Data Module
Asymmetric Dimer Model
Asynchronous Disconnected Mode
Atomic Demolition Munition
Automatic Drafting Machine
ADMA Abu Dhabi Marine Areas [United Arab Emirates]
Alkyldimethyl Amine
Amusement Device Manufacturers Association [now: AAMA, US]
Aviation Distributors and Manufacturers Association [US]
ADM/ADACC Advanced Development Model/Advanced Damage Control Console
ADMD Administration Management Domain
ADMI American Dry Milk Insitute [US]
ADMIN Administration
Administrative
Administrator
ADMIRAL Automatic and Dynamic Monitor with Immediate Relocation, Allocation and Loading

ADMIRE Adaptive Decision Maker in an Information Retrieval Environment
Automated Diagnostic Maintenance Information Retrieval System
ADMIS Automated Data Management Information System
ADMR Administrator
ADMS Assistant Director of Medical Services [US Military]
Asynchronous Data Multiplexer Synchronizer
Automatic Digital Message Switch(ing)
ADMSC Automatic Digital Message Switching Center
ADMT Air-Dried Metric Ton
ADN Adiponitrile
Allgemeiner Deutscher Nachrichtendienst [= Former East German News Service]
ADNAC Air Defense of the North American Continent
ADNOC Abu Dhabi National Oil Company [United Arab Emirates]
ADO Air Defense Officer
Autodyne Oscillator
Automatic Dial-Out
ADOC Air Defense Operations Center
ADOGA American Dehydrated Onion and Garlic Association [US]
ADOIT Automatically Directed Outgoing Intertoll Trunk
ADOM Acid Deposition and Oxidants Model
ADONIS Automated Document Delivery over Networked Information Service
Automatic Digital On-Line Instrumentation System
Automatic Document On-Line Information System
ADOPE Automatic Decisions Optimizing Predicted Estimates
ADOS Advanced Disk Operating System
ADP Adapter
Adenosine Diphosphate
Administrative Data Processing
Advanced Data Processing
Advanced Display Processor
Airborne Data Processor
Air Defense Position
Airport Development Program
Amateur Data Processor
Ammonium Dihydrogen Phosphate
Association of Database Producers [UK]
Atmospheric Dew Point
Automatic Data Processing
Automatic Die Positioner
ADPA Acetone Diphenylamine
American Defense Preparedness Association [US]
Aminodiphenylamine
ADPase Adenosinediphosphatase
ADPC Agriculture Development Planning Center [of ASEAN]
Automatic Data Processing Center
ADPCM Adaptive Differential Pulse Code Modulation
Adaptive Delta Pulse Code Modulation
Association for Data Processing and Computer Management [US]
ADPE Automatic Data Processing Equipment
ADPE/S Automatic Data Processing Equipment/System
ADPESO Automatic Data Processing Equipment Selection Office
ADPG Adenosine Diphosphoglucose
ADPI American Dairy Products Institute [US]
ADPLAN Advancement Planning
ADPP Automatic Data Processing Program
ADPRP Adenosine Diphosphate Ribose Phosphate
ADPS Automatic Data Processing System
Automatic Display and Plotting System
ADPSO Association of Data Processing Service Organizations
Automatic Data Processing Selection Office
ADPT Adapter [also: ADPTR]
ADR Address
Adsorption/Desorption/Recovery
Advisory Route
Aircraft Direction Room
Applied Data Research
Audit Discrepancy Report
Aviation Design Research [US Navy]
ADRA Animal Disease Research Association [UK]
Automotive Dismantlers and Recyclers Association [US]
ADRAC Automatic Digital Recording and Control
ADRES Army Data Retrieval System [US Army]
ADRG Arc Digital Raster Graphics
ADRI Animal Disease Research Institute
ADRMP Auto-Dialed Recorded Message Player
Automated Dialer and Recorded Message Player
ADRRP Activated Dissociation Reduction Reaction Process
ADRS Address Register
Analog-to-Digital Data Recording System
Automatic Document Request Service [UK]
ADRT Analog Data Recorder Transcriber
ADS Accessory Drive System
Accurately-Defined System
Activity Data Sheet
Address
Administration of Designed Services
Administrative Data System
Advanced Data System
Advanced Debugging System
Advanced Digital System
Advanced Dressing Station
Advanced Dryer System
Aircraft Development Service [of FAA, US]
Air Defense Sector
Air Defense Ship
Air Defense System
Air Development Service [of FAA, US]
Alloy Descaling Salt
Antidiuretic Substance
Application Development System
Articulated Diving Suit
Association of Diesel Specialists [US]
Asymptotic Diffuse Scattering
Atmospheric Detritiation System
Automatic Data Set
Automatic Dependent Surveillance
Automatic Depressurization System
Automatic Duplicating System
Automatic Door Seal
ADSA American Dairy Science Association [US]

	Atomic Defense Support Agency
ADSAF	Automated Data Systems for the Army in the Field [US]
ADSAS	Air-Derived Separation Assurance System
ADSATIS	Australian Defense Science and Technology Information System
ADSC	Association of Drilled Shaft Contractors [US]
	Automatic Data Service Center
	Automatic Digital Switching Center
ADSCOM	Advanced Shipboard Communication
ADSF	Automated Directional Solidification
ads/HIgG	Antibody Adsorbed with Human IgG (= Immunoglobulin G)
ads/HS	Antibody Adsorbed with Human Serum Proteins
ADSI	Administrative Design Service Information
	Agricultural Development Service, Inc. [Philippines]
ADSIA	Allied Data System Interoperability Agency [of NATO]
ADSM	Air Defense Suppression Missile
ads/MS	Antibody Adsorbed with Mouse Serum Proteins
ADSP	Advanced Digitial Signal Processor
ADSR	Attack, Decay, Sustain and Release
ADSRI	Animal and Dairy Science Research Institute [South Africa]
ads/RS	Antibody Adsorbed with Rat Serum Proteins
ADSS	A Diesel Supply Set
	Aircraft Damage Sensing System
	Australian Defense Scientific Service
ADST	Atlantic Daylight Saving Time
ADSTAR	Automatic Document Storage and Retrieval
ADS-TP	Administrative Data System — Teleprocessing
ADSUP	Automatic Data Systems Uniform Practices
ADT	Abstract Data Type
	Adenosine Triphosphate
	Adjustable Digital Thermometer
	Admission, Discharge and Transfer
	American District Telegraph
	Application-Dedicated Terminal
	Articulated Dump Truck
	Asynchronous Data Transceiver
	Atlantic Daylight Time
	Automatic Data Translator
	Automatic Data Transmission
	Autonomous Data Transfer
	Average Daily Traffic
ADTA	American Dental Trade Association [US]
ADTC	Anglo-Dutch Trade Council [Netherlands]
ADTD	Association of Data Terminal Distributors [US]
ADTECH	Advanced Decoy Technology
ADTS	Approved Departure Times
	Automated Data and Telecommunications Service
	Automatic Data Test System
ADTSEA	American Driver and Traffic Safety Education Association [US]
ADTT	Average Daily Truck Traffic
ADTU	Auxiliary Data Translator Unit
ADU	Accumulation Distribution Unit
	Ammonium Diuranate
	Automatic Data Unit
	Automatic Dialing Unit
ADV	Acid Demand Value
	Advance
	Advertisement
ad val	(ad valorem) — according to value; to the value of
AD-VAM	Advanced Development of a Vulnerability Assessment Model [of DND, Canada]
Adv C	Advanced Certificate
ADV CHGS	Advance Charges
ADVER	Advertising
ADV FRT	Advance Freight
ADVISOR	Advanced Integrated Safety and Optimizing Computer
ADVM	Adaptive Delta Modulation Voice Modem
Adv M	Advanced Master
ADV PMT	Advance Payment
ADVT	Advertisement
ADVTG	Advertising
ADW	Aerial Distribution Wire
	Air Defense Warning
AdW	Akademie der Wissenschaften [= Academy of Sciences, Germany]
ADWA	Atlantic Deeper Waterways Association [US]
ADX	Asymmetric Data Exchange
	Automatic Data Exchange
ADXPS	Angular Dependent X-Ray Photoelectron Spectroscopy
ADXRD	Angle-Dispersive X-Ray Diffraction
AE	Accidental Error
	Acoustic Emission
	Acrylic Emulsion
	Active Aerial
	Advanced ECL (= Emitter-Coupled Logic)
	Aerial
	Aeroelasticity
	Aeroelectronics
	Aeronautical Engineer
	Aeronautical Engineering
	Aerospace Engineer
	Aerospace Engineering
	Agricultural Economist
	Agricultural Economy
	Agricultural Engineer
	Agricultural Engineering
	Airborne Electronics
	Air Escape
	Aminoethyl
	Application Engineer
	Application Engineering
	Application/Expert
	Architect Engineer
	Architectural Engineer
	Architectural Engineering
	Arithmetic Element
	Arithmetic Expression
	Army Engineer
	Attenuation Equalizer
	Atmospheric Explorer
	Atomic Energy
	Auto-Emission
	Automatic Exposure
	Automobile Engineer
	Automobile Engineering

Automotive Engineer
Automotive Engineering
Auxiliary Equation
Aviation Engineer
Aviation Engineering

A&E Accident and Emergency
Azimuth and Elevation

A/E Absorption/Emission Ratio
Absorptivity/Emissivity Ratio
Architect/Engineer

Ae Aerial

AEA Aerospace Education Association [US]
Agricultural Engineers Association [UK]
Aircraft Electronics Association [US]
Air-Entraining Agent
All-Electric Aeroplane
American Economic Association [US]
American Electronics Association [US]
American Engineering Association [US]
American Evaluation Association [US]
Association of Engineers and Architects [US]
Association of Engineers and Associates
Association of European Airlines [Belgium]
Atomic Energy Act
Atomic Energy Authority [UK]
Automotive Electric Association

AEACII Atomic Energy Advisory Committee on Industrial Information

AEAI Association Europeenne des Assureurs Industriels [= European Association of Industrial Insurers, Belgium]
Association of Engineers and Architects in Israel

AEAPS Auger Electron Appearance Potential Spectroscopy

AEB American Egg Board [US]
Atomic Energy Bureau
Auxiliary Equipment Building

AEBIG ASLIB (= Association of Special Libraries) Economics and Business Information Group [US]

AEC Agricultural Extension Center [Canada]
Airship Experimental Center [US Navy]
Aluminum Extruders Council [US]
American Engineering Council [US]
Aminoethyl Carbazole
Aminoethyl Cellulose
Aminoethyl Cysteine
Anisotropic Elastic Constant
Applied Electrochemistry
Army Education Corps [US]
Army Electronics Command [US]
Association Europeenne de Ceramique [= European Ceramics Association, France]
Association of Electronic Cottagers [US]
Atomic Energy Commission [now: ERDA, US]
Atomic Energy Council [Taiwan]
Automatic Exposure Control

AECB Association for the Export of Canadian Books
Atomic Energy Control Board [Canada]

AECC Automobile Emissions Control by Catalysts [of CEFIC, Belgium]
Asociacion Espanola de Control de Calidad [= Spanish Society for Quality Control]

AECI African Explosives and Chemical Industries

AECIP Atlantic Energy Conservation Investment Program [Canada]

AECL Advanced Emitter-Coupled Logic
Atomic Energy of Canada Limited

AECL-CANDU Atomic Energy of Canada Limited — Canada Deuterium Uranium [Canada]

AECL-CFFTP Atomic Energy of Canada Limited — Canadian Fusion Fuels Technology Project [Canada]

AECL-CRNL Atomic Energy of Canada Limited — Chalk River Nuclear Laboratory [Canada]

AECL-WNRE Atomic Energy of Canada Limited — Whiteshell Nuclear Research Establishment [Canada]

AECM Association of European Candle Manufacturers [France]
Atomic Energy Commission Manual

AECMA Association Europeenne des Constructeurs de Materiel Aerospatial [= European Association of Aerospace Equipment Manufacturers, France]

AECPR Atomic Energy Commission Procurement Regulations

AECRC Atomic Energy of Canada Research Company [of EMR, Canada]

AeCS Aero Club of Switzerland

AECT Association for Educational Communications and Technology [US]

AED Academy for Educational Development [US]
Advanced Electronic Design
ALGOL Extended for Design
Alternative Energy Division [of EMR, Canada]
Associated Equipment Distributors [US]
Association of Electronic Distributors [US]
Association of Engineering Distributors [UK]
Automated Engineering Design
United Arab Emirates Dirham [Currency]

AEDC American Economic Development Council [US]
Arnold Engineering Development Center [US]
Atlantic Engineering Design Competition

AEDF Asian Economic Development Fund

AEDP Association for Educational Data Processing [US]

AEDS Association for Educational Data Systems [now: IACE, US]
Atomic Energy Detection System

AEDU Admiralty Experimental Diving Unit [UK]

AEE Airborne Evaluation Equipment
Alliance for Environmental Education [US]
Asociacion Electrotecnica Espanola [= Spanish Electrotechnical Association]
Association of Electrical Engineers [Finland]
Association of Energy Engineers [US]
Associazione Elettrotecnica ed Elettronica [= Association for Electrotechnics and Electronics, Italy]
Atomic Energy Establishment [UK]
Autoelectronic Emission

AEEC Airlines Electronic Engineering Committee [US]
Association of European Express Carriers

AEEE Army Equipment Engineering Establishment

AEEF Association Europeenne des Exploitations Frigorifiques [= European Association of Refrigeration Enterprises, Belgium]

AEEI Associazione Elettrotecnica ed Elettronica Italiana [= Italian Association of Electrotechnics and Electronics]
AEEL Aeronautical Electronics and Electrical Laboratory [US Navy]
AEEMS Automatic Electric Energy Management System
AEEN Agence Europeenne pour l'Energie Nucleaire [= European Agency for Nuclear Energy]
AeEng Aeronautical Engineer
AEEP Association of Environmental Engineering Professors [US]
Association of European Engineering Periodicals
AEESCA Association for Engineering Education in South-Central Asia
AEESEA Association for Engineering Education in Southeast Asia
AEET Atomic Energy Establishment at Trombay [India]
AEF Aerospace Education Foundation [US]
Airfields Environment Federation [UK]
American Expeditionary Force
Aviation Engineer Force
AEFA Association of European Federations of Agro-Engineers
AEFPG Association of Engineering Firms Practicing in the Geosciences [of ASFE, US]
AEG Active Element Group
Association of Engineering Geologists [US]
Association of Exploration Geochemists [Canada]
Axial-Emission Gauge
AEGC Association Europeenne de Genie Civil [= European Association of Civil Engineers]
AEGIS Agricultural Ecological and Geographical Information System
AEGPL Association Europeenne des Gaz de Petrole Liquefies [= European Liquefied Petroleum Gas Association, France]
AEGRAFLEX Association Europeenne des Graveurs et des Flexographes [= European Association of Engravers and Flexographers, Germany]
AEGS Association of European Geological Studies
AEH Anhydroenneaheptitol
AEI Acclimatization Experience Institute [now: Institute for Earth Education, US]
Acoustic Emission Inspection
Aerial Exposure Index
Alternate Energy Institute
Alternate Enterprises Initiative Program [Canada]
Associated Electrical Industries [UK]
Associazione Elettrotecnica Italiana [= Italian Association for Electrotechnics]
Automatic Error Interrogation
Australian Educational Index [of ACER]
Average Efficiency Index
AEIA Asociacion Espanola de Informatica y Automatica [= Spanish Association of Informatics and Automation]
AEIC Association of Edison Illuminating Companies [US]
AEICP Association of Entertainment Industry Computer Professionals [US]
AEIS Aeronautical En-Route Information Service
AEJI Association of European Jute Industries
AEJMC Association for Education in Journalism and Mass Communications [US]
AEL Acceptable Exposure Limit
Aeronautical Engine Laboratory [US]
Audit Entry Language
AEM Airborne Electromagnetics
Analytical Electron Microscopy
Applications Explorer Mission
Assemblability Evaluation Method
Association of Electronic Manufacturers [US]
AEMA Asphalt Emulation Manufacturers Association [US]
AEMB Alliance for Engineering in Medicine and Biology [US]
AEMBA Advanced Executive Master of Business Administration
AEMC American Engineering Model Society [US]
Atomic Energy and Minerals Center
AEMCC Air and Expedited Motor Carriers Conference [US]
AEME Association of Export Marketing Executives [UK]
AEM-EELS Analytical Electron Microscopy — Electron Energy Loss Spectroscopy
AEM-EDS Analytical Electron Microscopy — Energy-Dispersive Spectroscopy
AEMS Airborne Electromagnetic Survey
American Engineering Model Society [US]
Association of Electronic Manufacturers
AEMSA Army Electronics Material Support Agency
AEMSAT Association of European Manufacturers of Self-Adhesive Tapes [France]
AEMSM Association of European Metal Sink Manufacturers [Scotland]
AEMT Association of Electrical Machinery Trades [UK]
AEMTM Association of European Machine Tool Merchants [UK]
AEN Advanced Electronics Network [UK]
AEO Acoustoelectric Oscillator
Air Engineer Officer
AEOC Aminoethylhomocysteine
AEOG Air Ejection Off Gas
AEP Advertising Exchange Program
Agence Europeenne de Productivite [= European Agency for Productivity [of OECD, France]]
Airports Economy Panel [of ICAO]
American Electric Power
1-(2-Aminoethyl)piperazine
Austrian Energy Forum
Averaged Evoked Potential
AEPCO Association of Economic Poison Control Officials [US]
AEPD Aminoethylpropanediol
AEPG Army Electronic Proving Ground [US]
AEPPF Albert Einstein Peace Price Foundation
AEPS Advanced Electronic Publishing System
AEPSC Atomic Energy Plant Safety Committee
AER Advanced Electric Reactor
After Engine Room
Approach End Runway
Assembly of European Regions
Association for Education by Radio [US]
Average Evoked Response
AERA American Educational Research Association [US]

Automated En-Route ATC (= Air Traffic Control)
Automotive Engine Rebuilders Association [US]

AERC Agricultural Economics Research Council [Canada]
Applied Electrostatics Research Center [Canada]

AERCB Alberta Energy Resources Conservation Board [Canada]

AERD Agricultural Engineering Research and Development
Atomic Energy Research Department [of USAEC]

AERDL Army Electronics Research and Development Laboratory [US]

AERE Association of Environmental and Resource Economists [US]
Atomic Energy Research Establishment [UK]

AERG Advanced Environmental Research Group

AERI Agricultural Economics Research Institute [UK]

AERIC Applied Economic Research and Information Center [Canada]

AERIS Airways Environmental Radar Information System

AERNO Aeronautical Equipment Reference Number

AERO Aeronautic(s)
Air Education and Recreation Organization [UK]
Alternate Energy Resources Organization [US]
Association of Electronic Reserve Officers [US]

AERODYN Aerodynamic(s)

AeroEng Aerospace Engineer [also: AeroE]

AERO ENG Aerospace Engineering

AEROF Aerological Officer

AEROMAT Advanced Aerospace Materials and Processes Conference and Exposition

AEROS Aerometric and Emissions Reporting System [of EPA, US]

AEROSAT Aeronautical Satellite System

AEROSPACECOM Aerospace Communications

AERS Atlantic Estuarine Research Society [US]

AERSG African Elephant and Rhino Specialists Group [Kenya]

AES Abrasive Engineering Society [US]
Aerodrome Emergency Service
Aeronautical Earth Station
Aerospace and Electronic Systems [IEEE Society]
Aerospace Electrical Society [US]
Agricultural Economics Society [UK]
Airways Engineering Society [US]
Alternate Energy Systems
American Electrochemical Society [US]
American Electronic Society [US]
American Electroplaters Society [now: AESF, US]
American Entomological Society [US]
American Eugenics Society [now: SSSB, US]
Apollo Experiment Support [of NASA]
Apollo Extension System [of NASA]
Applied Electrostatics
Artificial Earth Satellite [of NASA]
Asian Environmental Society
Atlantic Economic Society
Atmospheric Environment Service [US/Canada]
Atomic Emission Spectroscopy
Audio Engineering Society [US]
Auger Electron Spectroscopy
Auger Electron Spectrum
Autoelectronic Selector
Automated Environmental Station

AESA Association of Environmental Scientists and Administrators

AESAU Association of Eastern and Southern African Universities

AESC Aerospace and Electronic Systems Society [US]
American Engineering Standards Committee [US]
Automatic Electronic Switching Center

AES/DFO Atmospheric Environment Service/Department of Fisheries and Oceans [Canada]

AESE Association of Earth Science Editors [US]

AESF American Electroplaters and Surface Finishers Society [US]
Association for Electroplating and Surface Finishing

AESFS American Electroplaters and Surface Finishers Society [US]

AESG Axial-Emission Suppression Gauge

AESI Australian Earth Sciences Information System [of AUSINET]

AESMC Automotive Exhaust Systems Manufacturers Council [US]

AESMG Axial-Emission Self-Modulating Gauge

AESOP An Evolutionary System for On-Line Processing
An Experimental Structure for On-Line Planning
Automated Environmental Station for Overtemperature Probing

AESOPP An Estimator of Physical Properties (Databank) [of JUSE, Japan]

AESS Aerospace and Electronic Systems Society [of IEEE, US]

AESTE Association for Exchange of Students for Technical Experience [UK]

AES/WMO Atmospheric Environment Service/World Meteorological Organization

AET Acoustic Emission Technology
Acoustic Emission Testing
Aminoethylisothiourea Dihydrobromide
2-Aminoethylisothiouronium Bromide Hydrobromide
Automobile Engineering Technology

AETA American Embryo Transfer Association [US]

AETB Alumina-Enhanced Thermal Barrier

AETE Aerospace Engineering Test Establishment

AETR Advanced Engineering Test Reactor [US]
Advanced Epithermal Thorium Reactor [US]

AETS Association for the Education of Teachers in Science [US]

AETT Association of Educational and Training Technology [UK]

AETU All Ethiopian Trade Union

AEU Amalgamated Engineering Union [UK]
Asia Electronics Union [Japan]

AEW Airborne Early Warning

AEWB Army Electronic Warfare Board [US Army]

AEWC Airborne Early Warning and Control [also: AEW&C]

AEWES Army Engineer Waterways Experiment Station [US]

AEWIS Army Electronic Warfare Information System [US Army]

AEWTU Airborne Early Warning Training Unit

AF Accumulation Factor
Acoustic Filter
Active Filter

Address Field
Advantage Factor
Aerofoil
Africa(n)
Agriservices Foundation [US]
Air Filter
Air Force
Air France
Aldehyde Fuchsin
Allied Forces
Alternating Fields
Aminofluorene
Amplification Factor
Analog Filter
Angle Frame
Antiferroelectric(ity)
Antiferromagnetism
Antireflection Film
Antifouling
Aqua Fortis
Aramid Fiber
Arc Furnace
Arithmetic Function
Ash Fusibility
Atomic Fluorescence
Atomic Forum
Audio Frequency
Auto-Focus
Automatic Following
Auxiliary Feedwater
Availability Factor

A/F Across Flat
Air/Fuel Ratio

Af Afghani [Currency of Afghanistan]

aF abfarad [Unit]

af (anni futuri) — of next year

AFA Absorption-Filtration-Adsorption Process
Afghani [Currency of Afghanistan]
Air Force Association [US]
Air Freight Association (of America) [US]
Amateur Fencing Association [UK]
American Fan Association [US]
American Flock Association [US]
American Forestry Association [US]
American Foundrymen's Association [US]
Army Flight Activity
Asociacion Fisica Argentina [= Argentinian Association of Physics]
Association Francaise d'Astronomie [= French Association of Astronomy]

AFAA American Fire Alarm Association [US]
ASEAN Federation of Automotive Associations

AF/ACTS Air Force Advanced Computed Tomography System [US]

AFADS Advanced Forward Air Defense System

AFAL Air Force Avionics Laboratory [at WPAFB, US]

AFAMM ASEAN Federation of Agricultural Machinery Manufacturers

AFAP Atlantic Fisheries Adjustment Program [of DFO, Canada]

AFAPL Air Force Aero-Propulsion Laboratory [US]

AFAR Advanced Field Array Radar

AFARB Association of Friends of the Achievement of Richard Bellman [France]

AFAS Abteilung fur Angewandte Systemanalyse [= Division for Applied System Analysis [of KfK, Germany]]
Association Francaise pour l'Avancement des Sciences [= French Association for the Advancement of Sciences]

AFASE Association for Applied Solar Energy [US]

AFA-SEF Air Force Association — Space Education and Foundation [US]

AFaxA American Facsimile Association [also: AFAXA, US]

AFB Acoustic Feedback
Air Force Base
Antifriction Bearing
Atmospheric Fluidized Bed

AFBA Armed Forces Broadcasters Association [US]
Asociacion Farmaceutica y Bioquimica Argentina [= Argentinian Association for Pharmaceutics and Biochemistry]

AFBC Atmospheric Fluidized-Bed Combustion

AFBF American Farm Bureau Federation [US]

AFBMA Antifriction Bearing Manufacturers Association [US]

AFBMD Air Force Ballistic Missile Division [US]

AFBRF American Farm Bureau Research Foundation [US]

AFBS Acoustic Feedback System

AFC African Forestry Commission
Air Flow Controller
Air Force Cross
Air/Fuel Concentration
Alumni Fund Committee
American Forest Council [US]
(Latin-)American Forestry Commission [of FAO — UN]
Amplitude Frequency Characteristic
Aquatic Federation of Canada
Area Forecast Center
Area Frequency Coordinator
Atomic Fluid Cells
Automatic Flatness Control
Automatic Focus Control
Automatic Font Change
Automatic Frequency Control
Axial-Flow Compressor

AFCA Alberta Floor Covering Association [Canada]

AFCAC African Civil Aviation Commission [Senegal]

AFCAL Association Francaise de Calcul [= French Computing Association]

AFCALTI Association Francaise de Calcul et de Traitement de l'Information [= French Association for Computing and Information Processing]

AFCAS Automatic Flight Control and Augmentation System

AFCASI Associate Fellow of the Canadian Aeronautics and Space Institute

AFCC Available Frame Capacity Count

AFCCE Association of Federal Communications Consulting Engineers [US]

AFCE Automatic Flight Control Equipment

AFCEA Armed Forces Communications and Electronics Association [US]

AFCENT Allied Forces Central Europe [of NATO]
AFCET Association Francaise pour la Cybernetique Economique et Technique [= French Association for Cybernetics and Technology, France]
AFCFP Arab Federation of Chemical Fertilizer Producers [Kuwait]
AFCG American Fine China Guild [US]
AFCIA Armed Forces Civilian Instructors Association [US]
AFCMA Aluminum Foil Container Manufacturers Association [US]
Australian Fiberboard Container Manufacturers Association
AFCO Aluminum Foil Conference
AFCOMMSTA Air Force Communications Station [US]
AFCRL Air Force Cambridge Research Laboratory [US]
AFCS Adaptive Flight Control System
Air Force Communications System [US]
Automatic Flight Control System
Avionic Flight Control System
AFCSS Air Force Communications Support System [US]
AFCU Association of Federal Computer Users [US]
AFD Accelerated Freeze Drying
Automatic Forging Design
Axial Flux Density
Axial Flux Difference
AFDA Axial Flux Difference Alarm
AFDAC Association Francaise pour la Documentation Automatique en Chimie [= French Association for Automatic Documentation in Chemistry]
AFDASTA Air Force Data Station [US]
AFDB African Development Bank [also: AfDB]
AFDBS Association Francaise des Documentalistes et des Bibliothecaires Specialies [= French Association of Documentalists and Special Librarians [also: ADBS, France]]
AFDDA Air Force Director of Data Automation [US]
AFDIN African Development Information Network
AFDO Association of Food and Drug Officials [US]
AFDOUS Association of Food and Drug Officials of the United States
AFDREW Air Force Director of Reconnaissance and Electronic Warfare [US]
AFE Aeroassist Flight Experiment
AFEA American Farm Economic Association [US]
Aviation Facilities Energy Association [US]
AFECI Association des Fabricants Europeens de Chauffe-Bains et Chauffe-Eau Instantanes et de Chaudieres Murales au Gaz [= Association of European Manufacturers of Instantaneous Gas Water Heaters and Wall-Hung Boilers, Belgium]
AFECOGAZ Association des Fabricants Europeens d'Appareils de Controle [= European Control Equipment Manufacturers Association, Belgium]
AFEDEF Association des Fabricants Europeens d'Equipement Ferroviaires [= Association of European Railway Equipment Manufacturers, France]
AFEDES Association Francaise pour l'Etude et le Developpement des Applications de l'Energie Solaire [= French Association for Study and Development of Solar Energy Applications]
AFEE Association Francaise pour l'Etude des Eaux [= French Association for the Study of Waters]
AFEEC ASEAN Federation of Electrical Engineering Contractors
AFEI Americans for Energy Independence [US]
Arab Federation of Engineering Industries
AFEJ Asian Forum of Environmental Journalists
AFELIS Air Force Engineering and Logistics Information System [US]
AFEO ASEAN Federation of Engineering Organizations [Indonesia]
AFERA Association des Fabricants Europeens des Rubans Auto-adhesifs [= European Self-Adhesive Tape Manufacturers Association]
AFES Association Francaise pour l'Etude du Sol [= French Association for Soil Research]
AFESD Arab Fund for Economic and Social Development
AFETR Air Force Eastern Test Range [US]
AFF Above Finished Floor
Affairs
Affirmative
American Farm Foundation [US]
Army Field Force
Automatic Fast Feed
Automatic Frequency Follower
AFFDL Air Force Flight Dynamics Laboratory [US]
AFFF Aqueous Film-Forming Foam
AFFI Arab Federation of Food Industries [Iraq]
AFFIL Affiliation
AFFIRM Association for Federal Information Resources Management [US]
AFFMA ASEAN Federation of Furniture Manufacturers Associations
AFFP Air-Filed Flight Plan
AFFPI ASEAN Federation of Food Processing Industries
AFFTC Air Force Flight Test Center [US]
AFG Afghani [Currency of Afghanistan]
Analog Function Generator
Arbitrary Function Generator
AFGC American Forage and Grassland Council [US]
AFGL Air Force Geophysical Laboratory [US]
AFGM American Federation of Grain Millers [US]
AFGP Antifreeze Glycoproteins
AFGWC Air Force Global Weather Center [US]
AFHB ASEAN Food Handling Bureau
AFHF Air Force Historical Foundation [US]
AFHP Advanced Flash Hydropyrolysis
AFHRL Air Force Human Resources Laboratory [US]
AFI Address Format Identifier
American Film Institute [US]
American Forest Institute [now: AFC, US]
Assistant Flying Instructor
Association of Food Industries [US]
Automatic Fault Isolation
AFIA American Feed Industry Association [US]
American Foreign Insurance Association [US]
Association Francaise Interprofessionelle des Agrumes [= French Interprofessional Association for Tropical Fruit]
AFICCS Air Force Interim Command and Control System [US]

AFICE Association for International Cotton Emblem [Belgium]
AFICTIC Association Francaise des Ingenieurs, Chemistes et Techniciens des Industries du Cuir [= French Association of Engineers, Chemists and Technicians in the Leather Industry]
AFID American Federation of Information and Data Processing [now: AFIPS]
AFIDA Agricultural Foreign Investment Disclosure Act [US]
Asociacion de Ferias Internacionales de America [= Association of International Trade Fairs of America, Peru]
AFIL Air Filed (Flight Plan)
AFIP Armed Forces Institute of Pathology [US]
AFIPS American Federation of Information Processing Societies [US]
AFIRMS Air Force Integrated Readiness Measurement System
AFIRO Association Francaise d'Informatique et de Recherche Operationnelle [= French Association of Informatics and Operational Research]
AFIS American Forces Information Service [US]
Automatic Flight Inspection System
AFIT Air Force Institute of Technology [US]
American Fabricating Institute of Technology [US]
AFITI Association Francaise des Ingenieurs et Techniciens de l'Aeronautique [= French Association of Aeronautical Engineers and Technicians]
AFJA Association Francaise des Journalistes Agricoles [= French Association of Agricultural Journalists]
AFL Above Field Level
Abstract Family of Languages
Alberta Federation of Labour [Canada]
American Federation of Labour [US]
Association Francaise Laitiere [= French Dairy Association]
Automatic Fault Location
Afl Netherlands Antillean Florin [Currency]
AFLA Asian Federation of Library Associations
Automotive Fleet and Leasing Association [US]
AFLC Air Force Logistics Command [US]
AFL-CIO American Federation of Labour Congress of Industrial Organizations [US]
AFLCM Air Force Logistics Command Manual [US]
AFLCON Air Force Logistics Command Operations Network [US]
AFLM Accredited Farm and Land Manager
AFM Abrasive Flow Machining
Air Force Manual
Air Force Medal
Air Force Museum [at WPAFB, US]
Air Freight Motor Carriers Conference [now: AEMCC, US]
A (= Al_2O_3), F (= FeO), M (= MgO) (Diagram) [Petrology]
Analysis and Forecasting Mode
Antifriction Metal
Applied Fracture Mechanics
Atomic Force Microscopy
Automatic Flight Management
AFMA American Feed Manufacturers Association [US]
American Fiber Manufacturers Association [US]
American Furniture Manufacturers Association [US]
ASEAN Federation of Mining Associations
Association of Food Marketing Agencies (in Asia and the Pacific) [Thailand]
AFMAG Audio-Frequency Magnetics
Audio-Frequency Magnetic Fields
AFMC Aeronautical Frequency Management Committee [UK]
Armed Forces Marketing Council [US]
Asian Fluid Mechanics Committee [Japan]
Automotive Filter Manufacturers Council [US]
AFMDC Air Force Machinability Data Center [US]
Air Force Missile Development Center [US]
AFMED Allied Forces Mediterranean
AFMFIC Associated Factory Mutual Fire Insurance Companies [US]
AFMI Association of Food Marketing Institutions [now: AFMA, Thailand]
AFML Air Force Materials Laboratory [US]
AFMMP Advanced Field Monitoring Methods Program [of EPA, US]
AFMR Antiferromagnetic Resonance
AFMS Advanced Flight Management System
American Federation of Mineralogical Societies [US]
AFMTC Air Force Missile Test Center [US]
AFN All-Figure Numbering
American Forces Network
AFNE Allied Forces Northern Europe [of NATO]
AFNETF Air Force Nuclear Engineering Test Facility
AFNETSTA Air Force Networks Station [US]
AFNOR Association Francaise de Normalisation [= French Association for Standards]
AFNORTH Allied Forces Northern Europe [of NATO]
AFO Accounting and Finance Office
Advanced File Organization
Airport Fire Officer
Association of Field Ornithologists [US]
Axial Flux Offset
AFOCEL Association Foret-Cellulose [= Forest/Cellulose Association, France]
AFOMJ Association of Fat and Oil Manufacturers of Japan
AFOSR Air Force Office of Scientific Research [US]
AFP Adiabatic Fast Passage
Agence France Presse [= French News Agency]
Alternative Future Project [Norway]
Association for Finishing Processes [of SME, US]
Attached FORTRAN Processor
Automatic Flow Process
AFPA Australian Fire Protection Association
Automatic Flow Process Analysis
AFPAM Automatic Flight Planning and Monitoring
AFPAV Airfield Pavement
AFPC Armed Forces Policy Council [US]
AF/PC Automatic Frequency/Phase Controlled (Loop)
AFPL Air Force Packaging Laboratory [US]
AFPMCW Arab Federation of Petroleum, Mines and Chemical Workers
AFPRO Air Force Plant Representative Office [US]

AFP/SME Association for Finishing Processes/Society of Manufacturing Engineers [US]
AFPT Auxiliary Feed Pump Turbine
AfPU African Postal Union
AFQT Armed Forces Qualification Test [US]
AFR Advanced Fault Recognition
Africa(n)
Air Force Regulation
Air Force Reserve [US]
Air Force Route
Alternating Frequency Rejection
Amplitude Frequency Response
Application Function Routine
Automatic Field/Format Recognition
Awaiting Forward Release
Away from Reactor [Spent Fuel Storage]
AFRA American Farm Research Association [US]
AFRAA African Airlines Association [Kenya]
AFRAeS Associate Fellow of the Royal Aeronautical Society
AFRAL Association Francaise de Reglage Automatique [= French Association for Automatic Regulation]
AFRASEC Afro-Asian Organization for Economic Cooperation [Egypt]
AFRATC African Air Traffic Conference
AFRC Agricultural and Food Research Council [UK]
Armed Forces Reserve Center
AFRCE Air Force Regional Civil Engineer [US]
AFRPL Air Force Rocket Propulsion Laboratory [US]
AFRRI Armed Forces Radiobiology Research Institute [US]
AFRS Affairs
Armed Forces Radio Service
AFRSI Advanced Flexible Reusable Surface Installation
AFRTC Air Force Reserve Training Center [US]
AFRTS Armed Forces Radio and Television Service [US]
AFS Acoustic Frequency Standards
Aerodrome Fire Service
Aeronautical Fixed Service
Air Force Specialty
Air Force Station
Air Force Supply
American Fisheries Society [US]
American Foundrymen's Society [US]
Army Fire Service
Asian Fisheries Societies
Atomic Fluorescence Spectrometry
Audio-Frequency Shift
Auxiliary Fire Service
AFSA American Fire Sprinkler Association [US]
AFSAB Air Force Science Advisory Board [US]
AFSARI Automation for Storage and Retrieval of Information
AFSAT African Satellite
AFSATCOM Air Force Satellite Communications
AFSAU Association of Faculties of Science at African Universities
AFSB American Federation of Small Business [US]
AFSC Air Force Space Command [US]
Air Force Systems Command [US]
Armed Forces Staff College
Automatic Focus Control
AFSCA Amalgamated Flying Saucer Clubs of America [US]
AFSCM Air Force Systems Command Manual [US]
AFSCME American Federation of State, County and Municipal Employees [US]
AFSE Allied Forces Southern Europe [of NATO]
AFSHP Association of Federal Safety and Health Professionals [US]
AFSK Audio Frequency-Shift Keying
Auto Frequency-Shift Keying
AF/SLR Auto-Focus/Single Lens Reflex
AFSM Association of Field Service Managers
Augmented Finite-State Machine
AFSMAS Association Francophone de Spectrometrie de Masse de Solides [= French-Speaking Association of Solid Mass Spectrometry, France]
AFSMI Association of Field Service Managers, International
AFSOUTH Allied Forces Southern Europe [of NATO]
AFSR Argonne Fast Source Reactor [of ANL, US]
AFST Auxiliary Feedwater Storage Tank
AFSWC Air Force Special Weapons Center [US]
AFT Adaptive Ferroelectric Transformer
Analog Facility Terminal
Automated Funds Transfer
Automatic Fine Tuning
Axial-Flow Turbine
AFTAC Air Force Technical Applications Center [US]
American Fiber, Textile, Apparel Coalition [US]
AFTE American Federation of Technical Engineers [US]
Arab Federation for Technical Education [Iraq]
Association of Firearm and Tool Mark Examiners [US]
AFTEC Air Force Test and Evaluation Center [US]
AFTER Air Force Thermionic Engineering and Research
AFTEX ASEAN Federation of Textile Industries
AFTI Advanced Fighter Technology Integration [US Air Force]
AFTN Aeronautical Fixed Telecommunications Network [UK]
Afternoon
AFTP Association Francaise des Techniciens du Petrole [= French Association of Petroleum Technicians]
AFTRCC Aerospace Flight Test Radio Coordinating Council
AFTS Aeronautical Fixed Telecommunications Service [UK]
AFU Archives for UFO (= Unidentified Flying Object) Research [Sweden]
AFUA ARMS Inc. (Computer) Firms Users Association [US]
AFV Armored Fighting Vehicle
AFVA Armored Fighting Vehicle Association [US]
AFVMC Association of Faculties of Veterinary Medicine in Canada
AFWAL Air Force Wright Aeronautical Laboratories [US]
AFWETS Air Force's Weapon Effectiveness Testing System [US]
AFWL Air Force Weapons Laboratory [US]
AFWTR Air Force Western Test Range [US]
AG Acoustic Grating
Advanced General-Purpose
Agent General
Air Gunner
Air-to-Ground

	Algebraic Geometry
	Amplification Gas
	Antenna Gain
	Armor Grating
	Arresting Gear
	Astronomische Gesellschaft [= Astronomical Society, Germany]
	Autoleather Guild [US]
A-G	Air-to-Ground
Ag	(Agentum) — Silver
	Antigen
ag	attogram [1 ag = 10^{-18} g]
AGA	Abrasive Grain Association [US]
	American Gas Association [US]
	American Genetic Association [US]
	Asociacion General de Agricultores [= General Association of Agriculturalists, Guatemala]
	Associated Geographers of America [US]
	Australian Gas Association
AGACS	Automatic Ground-to-Air Communications Systems
AGAMP	Automatic Gain Adjusting Amplifier
AGANI	Apollo Guidance and Navigation Information
AGARD	Advisory Group for Aerospace Research and Development [of NATO]
AGB	Association of Governing Boards (of Universities and Colleges) [US]
	Asymptotic Giant Branch
	Automatic Grid Bias
AGC	Abort Guidance Computer
	Advanced Graphics Control
	Aerojet General Corporation
	Agriculture Canada
	Apollo Guidance Computer [of NASA]
	Associated General Contractors (of America) [US]
	Atlantic Geoscience Center [Canada]
	Automatic Gage Control
	Automatic Gain Control
	Automatic Gain Correction
AGCA	Association of General Contractors of America [US]
	Automatic Ground-Controlled Approach
AgCan	Agriculture Canada
AGCC	Air-Ground Communications Channel
AGCD	Association of Green Crop Driers [UK]
AGCL	Automatic Ground-Controlled Landing
AGCM	Atmospheric General Circulation Model
AGCR	Advanced Gas-Cooled Reactor
AGCRSP	Army Gas-Cooled Reactor Systems Program [US]
AGCS	Air-to-Ground Communications System
AGCY	Agency
AGD	Advanced Graduate Certificate
	Axial Gear Differential
AGDEX	Agricultural Index [UK]
AGDIC	Astro Guidance Digital Computer
AGDS	Airport Graphics Display System
	American Gage Design Standard
AGDT	Aging Date
AGE	Aerospace Ground Equipment
	Allyl Glycidyl Ether
	Asian Center for Geotechnical Engineering [Taiwan]
	Association of German Engineers [also: VDI, Germany]
	Automatic Ground Equipment
AGECON	American Bibliography of Agricultural Economics (Database) [US]
AGED	Advisory Group on Electronic Devices [US]
AGEP	Advisory Group on Electronic Parts [US]
AGET	Advisory Group on Electron Tubes [US]
AGF	Arbeitsgemeinschaft der Grossforschungsanlagen [= Working Group of the Large Research Laboratories, Germany]
	Arbeitsgemeinschaft Getreideforschung [= Working Group on Wheat Research, Germany]
	Army Ground Forces
AG-FIZ	Arbeitsgemeinschaft der Fachinformationszentren [= Working Group of Information Centers, Germany]
Ag(fod)	(Heptafluorodimethyloctanedionato)silver
AGGLUT	Agglutinating
AGGR	Aggregate
AGHS	Active Gas Handling System [of Joint European Torus]
AGI	Alliance Graphique Internationale [= International Graphics Alliance, Switzerland]
	American Geological Institute [US]
	Associazione Genetica Italiana [= Italian Genetics Association]
AGID	Association of Geoscientists for International Development [Thailand]
AG-IDL	Arbeitsgemeinschaft Information und Dokumentation — Literaturversorgung [= Working Group on Information and Documentation — Literature Supply, Germany]
AGIFORS	Airlines Group of the International Federation of Operational Research Societies
AGIL	Airborne General Illumination Light
AGILE	Autonetics General Information Learning Equipment
AGL	Above Ground Level
	Allied Geophysical Laboratory
	Association of German Librarians [Germany]
AGLINE	Agricultural On-Line (Database) [US]
AGLINET	Agricultural Libraries Information Network [US]
AGM	Air-to-Ground Missile
	Alternating Gradient Magnetometer
	Annual General Meeting
	Auxiliary General Missile
AGMA	American Gear Manufacturers Association [US]
AGMAP	Agricultural Market Assistance Program
AGN	Aerojet General Nucleonics (Reactor) [US]
	Average Group Number
AGNCS	Aluminum Gooseneck Clamp Strap
AGNES	Algorithm for Generating Structural Surrogates of English Text
AGNIS	Apollo Guidance and Navigation Industrial Support [US Army]
	Azimuth Guidance Nose-in-Stand
AGO	Alkali-Graphitic-Oxide
AGOR	Auxiliary General Oceanographic Research
AGOS	Air-Ground Operations Section
AGP	Association of Glucose Producers

AGPAT Agricultural Patents [also: AG/PAT]
AGPB Association Generale des Producteurs de Ble et Autres Cereales [= General Association of Producers of Wheat and Other Cereals, France]
AGPL Association Generale des Producteurs de Lin [= General Association of Flax Producers, France]
AGPM Associated Glass and Pottery Manufacturers [US]
Association Generale des Producteurs de Mais [= General Association of Maize Producers, France]
AGPO Association Generale des Producteurs d'Oleagineux [= General Association of Producers of Oleaginous Products, France]
AGQ Association of Geologists of Quebec [Canada]
AGR Advanced Gas-Cooled Reactor
Agriculture
Annual Growth Rate
AGRA Army Group Royal Artillery [UK]
AGRE Army Group Royal Engineer [UK]
Atlantic Gas Research Exchange
AGREE Advisory Group on Reliability of Electronic Equipment [US]
AGREP Agricultural Research Projects (Database) [of CEC, Europe]
AG-REP Agricultural Representative
AGRIC Agricultural
Agriculture [also: AGRI]
Agriculture Canada
AGRICOLA Agricultural On-Line Access (Database) [also: Agricola [of USDOA]]
AGRINET Agricultural Network [Canada]
Inter-American System for Agricultural Information
AGRINTER Inter-American Information System for the Agricultural Sciences
AGRIS Agricultural Information System [of FAO — UN]
International Information System for Agricultural Sciences and Technology [of FAO, UN]
Agritechnica International Exhibition for Agricultural Machinery and Equipment [Germany]
AGRON Agronomist
Agronomy
AGRR American Goat Record and Registry [US]
AGRU Acid Gas Removal Unit
AGS Abort Guidance System
Aircraft General Standards
Air Gunnery School [UK]
Alternating Gradient Synchrotron
American Geographical Society [US]
American Goat Society [US]
Annulus Gas System
Appalachian Geolocial Society [US]
Arizona Geological Survey [US]
Association of Graduate Schools [of Association of American Universities, US]
Automated Graphic System
Automatic Gain Stabilization
AGSAN Astronomical Guidance System for Air Navigation
AGSC Advanced Graduate Specialist Certificate
AGSP Atlas General Survey Program
AGSRO Association of Government Supervisors and Radio Officers [UK]
AGT Advanced Gas Turbine
Advanced Ground Transport
Agent
Agreement
Alberta Government Telephones [Canada]
Algorithmic Graph Theory
Association of Geology Teachers [US]
Audiographic Tele-Conference
Automotive Gas Turbine
Aviation Gas Turbine
AGTA Airport Ground Transportation Association [US]
AGTERA Advanced Gas Turbine for Engineering Research Association
AGU American Geophysical Union [US]
Arabian Gulf University
AGV Aniline Gentian Violet
Automatic Guided Vehicle
Avion a Grande Vitesse [= Hypersonic Transport Aircraft]
AGVS Automatic Guided Vehicle System
Automatic Guide Vehicle Systems [of the Materials Handling Institute, US]
AGW Actual Gross Weight
Advanced Graphic Workstation
Allowable Gross Weight
AGWSE Association of Groundwater Scientists and Engineers [US]
AGZ Actual Ground Zero
AH Acceptor Handshake
Amplifier Head
Analog Hybrid
Artificial Horizon
Available Hours
A/H Air over Hydraulic
Ah Ampere-hour
aH abhenry [Unit]
AHA American Hardboard Association [US]
American Historical Society [US]
American Hospital Association [US]
AHAM Association of Home Appliance Manufacturers [US]
American Home Appliance Manufacturers (Association) [US]
AHAMS Advanced Heavy Assault Missile System
AHB Austrian Historical Bibliography (Database) [of IMD, Austria]
AHC Alberta Housing Corporation [Canada]
American Horticultural Society [US]
Apalachian Hardwood Club [now: AHMI, US]
AHCAS Ad Hoc Committee for American Silver [US]
AHCC American-Hellenic Chamber of Commerce [Greece]
AHCM Academy of Hazard Control Management [US]
AHCS Advanced Hybrid Computer System
AHCT Advanced High-Speed Complimentary Technology
Ascending Horizon Crossing Time
AHCTL Acetylhomocysteinethiolactone
AHDGA American Hot Dip Galvanizers Association [US]
AHDRI Animal Husbandry and Dairy Research Institute [South Africa]
AHEA American Home Economics Association [US]
AHEM Association of Hydraulic Equipment Manufacturers [UK]
AHEX Association of Host Exhibitors

AHF	Architectural History Foundation [US]
	Army Historical Foundation [US]
	Asian Hospital Federation
AHFR	Argonne High Flux Reactor [of ANL, US]
AHG	Antihemophilic Globulin
AHI	Animal Health Institute [US]
	Asian Health Institute
AHIP	Advanced Helicopter Improvement Program [US]
AHITI	Animal Health and Industry Training Institute [Kenya]
AHLI	American Home Lighting Association [now: ALI, US]
AHM	Ampere-Hour Meter [also: ahm]
AHMA	American Hardware Manufacturers Association [US]
AHMI	Appalachian Hardwood Manufacturers, Inc. [US]
AHONDA	Ad Hoc Committee on New Directions of the Research and Technical Services Division [of ALA, US]
AHP	Air Horsepower
	Army Heliport
AHPA	American Honey Producers Association [US]
AHPL	A Hardware Programming Language
AHPW	Alberta Housing and Public Works [Canada]
AHQ	Allied Headquarters
AHR	Acceptable Hazard Rate
	Adsorptive Heat Recovery
	American Heritage Radio [US]
	Aqueous Homogeneous Reactor
AHRC	Australian Housing Research Center
AHRS	Altitude and Heading Reference System
AHS	Agricultural History Society [US]
	Airborne Hardware Simulator
	American Helicopter Society [US]
	American Horticultural Society [US]
AHSA	American Home Satellite Association [US]
	Art, Historical and Scientific Association [Canada]
AHSD	Authority Health and Safety Division [UK]
AHSE	Assembly, Handling and Shipping Equipment
AHSR	Air-Height Surveillance Radar
	American High Speed Rail
AHT	Aminoheptyltriazole
	Animal Health Trust [UK]
	Average Holding Time
AHTD	Association of High Tech Distributors [US]
AHTS	Anchor Handling, Tug and Supply
AHU	Air Handling Unit
	Antihalation Undercoat
AHW	Atomic Hydrogen Welding
AHWC	Ad Hoc Working Group
AI	Acquisition Institute [US]
	Address Incomplete
	Airborne Intercept
	Air-Insulated
	Air Intensifier
	Alloys Index [of METADEX]
	Amplifier Input
	Analog Input
	Annoyance Index
	Anti-Icing
	Anti-Interference
	Annoyance Index
	Applications Interface
	Area of Intersection
	Artificial Insemination
	Artificial Intelligence
	Asbestos Institute [Canada]
	Asphalt Institute [US]
	Atomics International
	Attitude Indicator
	Auto-Indexing
	Auto-Ionization
	Automatic Input
	Automation Institute [US]
A&I	Abstracting and Indexing
AIA	Aerial-Lift Industries Association [US]
	Aerospace Industries Association (of America) [US]
	Affinity Isolated Antibody
	Alberta Institute of Agrology [Canada]
	American Insurance Association [US]
	American Institute of Accountants [US]
	American Institute of Architects [US]
	American Insurance Association [US]
	American Inventors Association [US]
	Anthracite Industry Association [US]
	Apiary Inspectors of America [US]
	Archeological Institute of America [US]
	Asbestos Information Association
	Asbestos International Association [UK]
	Asociacion de Ingenieros Agronomos [= Association of Agricultural Engineers, Uruguay]
	Associate of the Institute of Actuaries
	Association of International Accountants [UK]
	Auto International Association [US]
	Automatic Image Analysis
	Automobile Importers Association [US]
	Automobile Industries Association [Canada]
	Aviation Industry Association [New Zealand]
AIAA	Aerospace Industries Association of America [US]
	American Industrial Arts Association [now: ITEA, US]
	American Institute of Aeronautics and Astronautics [US]
	Area of Intense Air Activity
	Association Interprofessionnelle de l'Aviation Agricole [= Interprofessional Association of of Agricultural Aviation, France]
	Association of International Advertising Agencies [US]
AIAC	Aerospace Industries Association of Canada
	Automotive Industries Association of Canada
AIACS	Association of International Air Courier Services [now: AICES, UK]
AIADA	American International Automobile Dealers Association [US]
AIAEE	Association for International Agricultural and Extension Education [US]
AIAF	American Institute of Architect Foundations [now: AAF, US]
	Association de l'Industrie et de l'Agriculture Francaises [= French Association of Industry and Agriculture]
AIAG	Automotive Industries Action Group [US]

AIA/NM	Asbestos Information Association/North America [US]
AIAR	American Institute for Archeological Research [US]
AIAS	American Institute of Architecture Students [US]
	Australian Institute of Agricultural Science
AIASA	American Industrial Arts Students Association [US]
AIAU	Atomic Institute of the Austrian University
AIB	American Institute of Banking [US]
	Analog Input/Output Board
	Associate of the Institute of Bankers [UK]
	Atlantic Institute of Biotechnology
	Automatic Intercept Bureau
AIBA	Agricultural Information Bank Asia
	American Industrial Bankers Association [US]
	Association of International Border Agencies [US]
AIBC	Architectural Institute of British Columbia [Canada]
AIBD	American Institute of Building Design [US]
	Asia-Pacific Institute for Broadcasting Development [Malaysia]
AIBDA	Asociacion Interamericana de Bibliothecarias y Documentalistas Agricoles [= Inter-American Association of Agricultural Librarians and Documentalists, Costa Rica]
AIBN	Azobisiobutyronitrile
AIBNRM	American Institute of Bolt, Nut and Rivet Manufacturers [now: IFI, US]
AIBS	American Institute of Biological Sciences [US]
AIBTC	Association for Information Brokerage and Technological Consultancy [Germany]
AIC	Aeronautical Information Circular
	Africa Information Center
	Agricultural Improvement Council [UK]
	Agricultural Institute of Canada
	Aircraft Industry Conference [US Army]
	American Institute of Chemists [US]
	American Institute of Conservation (of Historic and Artistic Works) [US]
	American Institute of Constructors [US]
	American Institute of Cooperation [US]
	Ammunition Identification Code
	Analytical Image Carrier
	Appraisal Institute of Canada
	Architectural Institute of Canada
	Associate of the Institute of Chemistry
	Association Internationale de la Couleur [= International Color Association, Netherlands]
	Astronomical Institute of Czechoslovakia
	Automatic Initiation Circuit
	Automatic Intercept Center
	Automatic Intersection Control
	Automobile Importers of Canada
	Automotive Information Council [US]
AICA	5-Aminoimidazole-4-Carboxamide Hydrochloride
	Association Internationale pour Calcul Analogique [= International Association for Analog Computing]
	Associazione Italiana per il Calculo Automatica [= Italian Computing Association]
AICAE	American Indian Council of Architects and Engineers [US]
AICAR	5-Aminoimidazole-4-Carboxamide-1-Ribofuranosyl
AICB	Association Internationale Contre le Bruit [= International Association against Noise, Switzerland]
AICBM	Anti-Intercontinental Ballistic Missile
AICC	Association of Independent Corrugated Converters [US]
AICCF	Association Internationale du Congres de Chemins de Fer [= International Association of Railway Congresses, Belgium]
AICCP	Association of the Institute for Certification of Computer Professionals [US]
AICE	American Institute of Chemical Engineers [US]
	American Institute of Consulting Engineers [US]
	American Institute of Crop Ecology [US]
	Associate of the Institute of Civil Engineers
AICES	Association of International Courier and Express Services [UK]
AIChE	American Institute of Chemical Engineers [also: AICHE, US]
AICMA	Association Internationale des Constructeurs de Materiel Aerospatial [= International Association of Aerospace Manufacturers, France]
AICMR	Association Internationale des Constructeurs de Materiel Roulant [= International Association of Rolling Stock Builders, France]
AICP	American Institute of Certified Planners [US]
	Association of Independent Commercial Producers [US]
AICPA	American Institute of Certified Public Accountants [US]
AICS	Associate of the Institute of Charted Ship Brokers
	Association of Independent Colleges and Schools [US]
	Association of Independent Computer Specialists [UK]
AICU	All India Cooperative Union
	Association of International Colleges and Universities [US]
AID	Agency for International Development [US]
	Aircraft Installation Delay
	Algebraic Interpretive Dialog
	American Institute of Interior Design [US]
	Architects Information Directory
	Army Intelligence Department [US]
	Artificial Insemination by Donor
	Associated Independent Distributors [now: IDA, US]
	Attention Identification
	Attention Identifier
	Augmented Index and Digest
	Auto-Interactive Design
	Automatic Identification [also: A/ID]
	Automated Inspection Device
	Automatic Imaging Device
	Automatic Information Distribution
	Automatic Interrogation Distortion
A/ID	Automatic Identification
AIDA	Analysis of Interconnected Decision Areas

Associazione Italiana per la Documentazione Avanzata [= Italian Association for Advanced Documentation]
Attention-Interest-Desire-Action (Formula)
Automatic Instrumented Diving Assembly
Automatic Intelligent Defect Analysis

AIDAB Australian International Development Assistance Bureau

AIDAS Advanced Instrumentation and Data Analysis System

AIDATS Army In-Flight Data Transmission System

AIDC Aerospace Industry Development Center [US]
American Industrial Development Council [now: AEDC, US]
Asian Industrial Development Council
Automatic Image Density Control

AIDD American Institute for Design and Drafting [US]
Auckland Industrial Development Division [New Zealand]

AIDE Airborne Insertion Display Equipment
Aircraft Installation Diagnostic Equipment
Army Integrated Decision Equipment [US]
Artificial Intelligence Development Environment
Automated Image Device Evaluation
Automated Integrated Design Engineering

AIDEA Alaska Industrial Development and Export Authority [US]

AIDEC Association for European Industrial Development and Economic Cooperation [Netherlands]

AIDF African Industrial Development Fund

AIDI Associazione Italiana di Illuminazione [= Italian Association of Illumination]
Associazione Italiana per la Documentazione e l'Informazione [= Italian Association for Documentation and Information]

AIDJEX Arctic Ice Dynamics Joint Experiment

AIDL Auckland Industrial Development Laboratory [New Zealand]

AIDO Arab Industrial Development Organization [Iraq]

AIDS Abstract Information Digest Service [US]
Acoustic Intelligence Data System
Acquired Immunodeficiency Syndrome
Advanced Interactive Debugging System
Aerospace Intelligence Data System
Agricultural Information Development Scheme [Thailand]
Airborne Integrated Data System
Aircraft Integrated Data System [of AFSC, US]
American Institute for Decision Sciences [US]
Automated Information Dissemination System
Automated Integrated Debugging System
Automated Intelligence and Data System [US Air Force]
Automatic Integrated Debugging System
Automation Instrument Data Service [UK]

AIDSCOM Army Information Data Systems Command [US]

AIE Association des Ingenieurs Electriciens [= Association of Electrical Engineers, Belgium]
Association Internationale des Entreprises d'Equipement Electrique [= International Association of Electrical Contractors, France]
Automated Insurance Environment [of USAA, US]

AIEA Agence Internationale de l'Energie Atomique [= International Agency for Atomic Energy, France]

AIECA Associated Independent Electrical Contractors of America [now: IEC, US]

AIED American Institute for Economic Development [US]

AIEE American Institute of Electrical Engineers [now: IEEE, US]
Associate of the Institute of Electrical Engineers

AIENDF Atomics International Evaluation Nuclear Data File

AIEST Association Internationale d'Experts Scientifiques du Tourisme [= International Association of Scientific Experts in Tourism, Switzerland]

AIET Average Instruction Execution Time

AIF Affiliated Inventors Foundation [US]
Arbeitsgemeinschaft Industrieller Forschungsvereinigungen [= Working Group of Industrial Research Associations, Germany]
Association Internationale Futuribles [= Futuribles International, France]
Atomic Industries Forum
Attitude Instrument Flying

AIFA Agricultural Insecticide and Fungicide Association [now: NACA, US]

AIFRB American Institute of Fishery Research Biologists [US]

AIFS African Improved Farming Scheme
American Institute for Foreign Study [US]

AIFSSF American Institute for Foreign Study Scholarship Foundation [US]

AIG Accident Investigation
Address Indicating Group
Arbeitsgemeinschaft Instandhaltung Gebaudetechnik [= Working Group on Maintenance of Technical Building Equipment [of VDMA, Germany]]
Artificial Intelligence Group [of MIT, US]
Association Internationale de Geodesie [= International Association of Geodesy, France]

AIGA American Institute of Graphic Arts [US]

AIGE Association for Individually Guided Education

AIGMF All India Glass Manufacturers

AIGS Auxiliary Inerting Gas Subsystem

AIH American Institute of Hydrology [US]

AIHA American Industrial Hygiene Association [US]

AIHC American Industrial Health Conference [US]

AIHP American Institute of the History of Pharmacy [US]

AIHSA American Insurers Highway Safety Alliance [US]

AIHTTR African Institute for Higher Technical Training and Research

AIHX Auxiliary Intermediate Heat Exchanger

AII American Interprofessional Institute [US]

AIIA Australian Information Industry Association
Australian Institute of International Affairs

AIIC Army Imagery Intelligence Corps
Association Internationale des Interpretes de Conference [= International Association of Conference Interpreters, Switzerland]

AIIE American Institute of Industrial Engineers [now: IIE, US]

AIII Association of International Industrial Irradiation [France]
AIIM Association for Information and Image Management [US]
Association of Independent Investment Managers [UK]
AIIP Association of Independent Information Professionals [US]
AIIPH All India Institute of Hygiene and Public Health
AIIS American Institute for Imported Steel [US]
AIJ Architectural Institute of Japan
AIKR Artificial Intelligence Knowledge Representation
AIL Airborne Instruments Laboratory
Aileron
Array Interconnection Logic
Artificial Intelligence Laboratory [US]
Association of International Libraries [Switzerland]
AILA American Institute of Landscape Architects [now: ASLA, US]
AILAS Automatic Instrument Landing Approach System
AILS Advanced Integrated Landing System
Automatic Instrument Landing System
AIM Abridged Index Medicus [of NLM, US]
Abstracts of Instructional Materials (Database) [now: AIM/ARM, US]
Academy for Interscience Methodology [US]
Access Isolation Mechanism
Acquisition Interface Module
Adaptive Injection Molding
Advanced Industrial Material
Advanced Information Manager
Aerial Independent Model
Age of Intelligent Machines
Air Induction Melt
Air Intercept Missile
Air-Isolated Monolithic
Airmen's Information Manual
Alarm Indication Monitor
American Innerspring Manufacturers [US]
American Institute of Management [US]
American Interactive Media [US]
Associated Information Managers [US]
Association Europeenne des Industries de Produits de Marque [= European Association of Industries of Branded Products, Belgium]
Association for Information Management [US]
Association of Information Managers [US]
Association of Innerspring Manufacturers [Name Changed to: American Innerspring Manufacturers, US]
Associazione Italiana di Metallurgica [= Italian Association for Metallurgy]
Associative Index Method
Automated Industrial Monitoring
Automatic Identification Manufacturers (International) [US]
Avalanche-Induced Migration
Awaiting Incoming Message
AIMA American Incense Manufacturers Association [US]
Association Internationale des Musees d'Agriculture [= International Association of Agricultural Museums, Denmark]
AIMACO Air Materiel Computer
AIM/ARM Abstracts of Instructional Materials and Research Materials (Database) [of OSU, US]
AIMC Academic Information Management Center [US]
AIMCAL Association of Industrial Metallizers, Coaters and Laminators [US]
AIME American Institute of Mechanical Engineers [US]
American Institute of Mining, Metallurgical and Petroleum Engineers [US]
Applied Innovative Management Engineering
AIMechE Associate of the Institution of Mechanical Engineers [UK]
AIMED Association of Independent Mailing Equipment Dealers [US]
AIMES Automated Inventory Management Evaluation System
AIMEX Australia's International Mining and Exploration Exhibition
AIMF Australasian Institute of Metal Finishing [Australia]
AIMILO Army/Industrial Material Information Liaison Office [US]
AIMIS Advanced Integrated Modular Instrument System
AIMLC Association of Island Marine Laboratories of the Caribbean [now: AMLC, Puerto Rico]
AIMLO Auto-Instructional Media for Library Orientation
AIMM Associate of the Institution of Mining and Metallurgy [UK]
Australasian Institute of Mining and Metallurgy [Australia]
Australian Institute of Mining and Metallurgy
AIMME American Institute of Mining and Metallurgical Engineers [US]
AIMMPE American Institute of Mining, Metallurgical and Petroleum Engineers [also: AIME, US]
AIMO Audibly-Instructed Manufacturing Operation
AIMP Anchored Interplanetary Monitoring Platform
AIMPE Australian Institute of Marine and Power Engineers
AIMPR Mechanical Engineers Association of Puerto Rico [US]
AIMR Association for the Improvement of the Mississippi River [US]
AIM/R Association of Manufacturers Representatives [also: AIMR, US]
AIMRA Agricultural and Industrial Manufacturers Representatives Association [US]
AIMS Advanced Integrated-Manufacturing System
Advanced Intercontinental Missile System
American Institute of Merchant Shipping [US]
Australian Institute of Marine Science
Author Index Manufacturing System
Automated Industrial Management System
Automated Information (and) Management System
Automated Inventory Management System
AIMTECH Association for Integrated Manufacturing Technology [US]
AIM-TWX Abridged Index Medicus — Teletypewriter Exchange Network [of NLM, US]
AIN Arab Institute for Navigation

AINA Arctic Institute of North America [of U of C, Canada]

AINBN Association for the Introduction of New Biological Nomenclature [Belgium]

AINSE Australian Institute of Nuclear Science and Engineering

AInstP Associate of the Institute of Physics

AIO Action Information Organization
Analog Input/Output Board

AIOA American Iron Ore Association [US]

AIOP Automatic Identification of Outward Dialing

AIOPI Association of Information Officers in the Pharmaceutical Industry [UK]

AIOU Analog Input/Output Unit

AIP Aeronautical Information Publication
Airport Improvement Program
Allied Intelligence Publication
Alphanumeric Impact Printer
Aluminum Isopropoxide
Aluminum Isopropylate
American Institute of Physics [US]
American Institute of Planners [US]
Approval in Principle
ASEAN Industrial Project
Association Internationale de Photobiologie [= International Photobiology Association, Switzerland]
Australian Institute of Petroleum
Australian Institute of Physics

AIPC Association Internationale des Palais de Congres [= International Association of Congress Centers, Yugoslavia]

AIPCEE Association des Industries du Poisson de la Communaute Economique Europeenne [= Association of the Fishery Industries of the European Economic Community, Belgium]

AIPE American Institute of Plant Engineers [US]

AIPEA Association Internationale Pour l'Etude des Argiles [= International Association for the Study of Clays, France]

AIPG American Institute of Professional Geologists [US]

AIPLA American Intellectual Property Law Association [US]

AIPO African Intellectual Property Organization

AIPPA Association Internationale pour la Protection de la Propriete Industrielle [= International Association for the Protection of Industrial Property, Switzerland]

AIPR American Institute of Pacific Relations [US]

AIPS Advanced Information Processing System
Astronomical Image Processing System

AIPT Association for International Practical Training [US]

AIQS Associate of the Institute of Quantity Surveyors [UK]

AIR Aerospace Information Report
Airborne Intercept Radar
Air Injection Reaction
Air Injection Reactor System
Air Intercept Rocket
Airline Industrial Relations (Conference) [US]
All-India Radio
American Institute of Refrigeration [US]
American Institute of Research [US]
Association for Institutional Research

AI/R Artificial Intelligence and Robotics

AIRA American Independent Refiners Association [US]

AIRAC Aeronautical Information Regulation and Control
All-Industry Research Advisory Council

AIRAD Airmen Advisory

AIRAH Australian Institute of Refrigeration, Air Conditioning and Heating

AIRCOM Air Command

AIRCON Automated Information and Reservation Computer-Oriented Network

AIR COND Air Conditioning

AIRE Australian Institution of Radio Engineers

AIREW Airborne Infrared Early Warning

AIR HP Air Horsepower

AIRCFT Aircraft

AIRI Association of Independent Research Institutes [US]

AIRIA All India Rubber Industries Association [India]

AIRL Aeronautical Icing Research Laboratory [US]

AIRLO Air Liaison Officer

AIRMIC Association of Insurance and Risk Managers in Industry and Commerce [UK]

AIRMICS Army Institute for Research in Management Information and Computer Science [US]

AIRO Association Internationale de Recherche Operationelle [= International Association of Operational Research, France]

AIRPASS Airborne Interception Radar and Pilot's Attack Sight System

AIRR Adjusted Internal Rate-of-Return

AIRS Advanced Inertial Reference Sphere
Alliance of Information and Referral Systems [US]
American Information Retrieval Service [US]
Artificial Intelligence and Robotics Society
Automatic Image Retrieval System
Automatic Information Retrieval System

AIRSIM Airways/Airport Simulation Model

AIRSS ABRES (= Advanced Ballistic Reentry System) Instrument Range Safety System

AIRXRS American Industrial Radium and X-Ray Society

AIS Academio Internacia de la Sciencoj [= International Academy of Sciences, Germany]
Accounting Information System
Advanced Information System
Advanced Instructional System
Aeronautical Information Service [of CAA, UK]
Alarm Indication Signal
Alarm Inhibit Signal
Altitude Indication System
American Interplanetary Society [Now Part of AIAA, US]
Analog Input System
Association Internationale de la Savonnerie (et de la Detergence) [= International Association of the Soap and Detergent Industry, Belgium]
Audio Integrating System
Automated Information System

Automated Injection System
Automatic Intercept System
Automatic Intercity Station
Avionics Intermediate Shop

AISA American Institute for Shipper Associations [US]
Association of International Schools in Africa

AISAG Aeronautical Information Service Automation Group [of ICAO]

AISAP Aeronautical Information Service Automation Specialist Panel [of ICAO]

AISC American Institute of Steel Construction [US]
Association of Independent Software Companies

AISE American Iron and Steel Engineers [US]
Association of Iron and Steel Engineers [US]

AISES American Indian Science and Engineering Society [US]

AISG Aeronautical Information Services Group

AISI American Iron and Steel Institute [US]

AISIF ASEAN Iron and Steel Industry Federation

AISI/SAE American Iron and Steel Institute/Society of Automotive Engineers [US]

AISP Association of Information Systems Professionals [US]

AIST Agency of Industrial Science and Technology [US]
Automatic Informational Station
Automatic Information Station

AISU Arab Iron and Steel Union

AIT Advanced Individual Training
Advanced Information Technology
Advanced Instruction Technique
Aichi Institute of Technology [Japan]
American Institute of Technology [US]
Architect-in-Training
Artificial Intelligence Technology
Asian Institute of Technology
Auto-Ignition Temperature

AITC Allylisothiocyanate
American Institute of Timber Construction [US]
Association Internationale des Traducteurs de Conference [= Association of Conference Translators, Switzerland]

AITE Aircraft Integrated Test Equipment
Automatic Intercity Telephone Exchange

AITES Association Internationale des Travaux en Souterrain [= International Tunneling Association, France]

AITI Associazione Italiano Truduttori ed Interpreti [= Italian Association of Translators and Interpreters]

AITIA American Institute of Technical Illustrators Association [US]

AITIPE Asociacion de Investigacion Tecnica de la Industria Papelera Espanola [= Association for Technical Studies of the Spanish Paper Industry, Spain]

AITIT Association Internationale de la Teinture et de l'Impression Textiles [= International Association of Textile Dyers and Printers, UK]

AITS Action Item Tracking System
Automatic Integrated Telephone System

AIU Advanced Instrumentation Unit [of NPL, US]
Alarm Interface Unit

AIV Aluminum-Intensive Vehicle

AIW Auroral Intrasonic Wave
Average Industrial Wage

AIWM American Institute of Weights and Measures

AIWRS Arctic International Wildlife Range Society

AIX Advanced Interactive Executive

AJ Analog Junction
Anti-Jamming
Area Junction
Assembly Jig

AJAI Anti-Jamming Anti-Interference

AJD Anti-Jam Display

AJM Abrasive Jet Machining
Air Jet Milling
Analog Junction Module

AK Adenylate Kinase
Alaska [US]
Aluminum Killed Steel

aka also known as

AKF A (= Al_2O_3 + Fe_2O_3 + (CaO + Na_2O), K (= K_2O), F (= FeO + MgO + MnO) (Diagram) [Petrology]

AKFM Association of Knitted Fabrics Manufacturers [US]

AKK Atomkraftkonsortiet Krangede [= Nuclear Power Study Group at Krangede, Sweden]

AKM Apogee Kick Motor

AKR Address Key Register
Auroral Kilometric Radiation

AKRO Acknowledge Receipt of

AKWIC Author and Key Word in Context

AKZ Angolan Kwanza [Currency]

AL Aclinic Line
Action Level
Additional Listing
Air League [UK]
Air Lock
Alabama [US]
Alcohol
Ames Laboratory [US]
Amplitude Limiter
Analog Link
Analysis Library
Analytical Laboratory
Analytical Limits
Approach Light
Arc Lamp
A Robot-Programming Language
Artificial Line
Assembly Language
Associated Laboratories [US]
Astronomical League [US]
Attenuation Length
Autograph Letter
Avionics Laboratory [US Air Force]

Al Aluminum

ALA Ada Language
Alighting Area
American Laminators Association [US]
American Library Association [US]
American Logistics Association [US]
Aminolevulinic Acid
Associate of the Library Association [UK]
Automobile Legal Association [US]

Ala	Alabama [US]
	Alanine
	Alanyl
ALABOL	Algorithmic and Business-Oriented Language
ALAE	Asociacion Latinoamericana de Entomologia [= Latin American Association of Entomology]
	Association of Licensed Aircraft Engineers
ALAF	Asociacion Latinoamericana de Ferrocarriles [= Latin American Association for Rolling Stock]
ALAIH	Asociacion Latinoamericana de Ictiologos y Herpetologos [= Latin American Association of Ichthyology and Herpetology]
ALA/ISAD	American Library Association — Information Science and Automation Division [US]
ALAMAR	Asociacion Latinoamericana de Armadores [= Latin American Association of Ship Owners, Uruguay]
ALAP	As Low As Practical
	Associative Linear Array Processor
ALAPCO	Association of Local Air Pollution Control Officials [US]
ALARA	As Low As Reasonably Achievable
ALARM	Air-Launched Advanced Ramjet Missile
	Automatic Light Aircraft Readiness Monitor
ALARR	Air-Launched Air Recoverable Rocket
ALARS	Association of Light Alloy Refiners and Smelters [UK]
ALART	Army Low-Speed Air Research Tasks
ALAS	Aminolevulinic Acid Synthetase
	Asynchronous Lookahead Simulator
	Automated Literature Alerting System
AlAs	Aluminum Arsenide
Alas	Alaska [US]
ALB	Anti-Lock Brakes
Alb	Albumin
ALBIS	Australian Library-Based Information System
ALBM	Air-Launched Ballistic Missile
ALBO	Automatic Line Build-Out
ALBW	Antiland Base Warfare
ALC	Adaptive Logic Circuit
	Agricultural Land Commission [UK]
	Alberta Livestock Cooperative [Canada]
	Alcohol
	Approximate Lethal Concentration
	Aromatic Liquid Crystal
	Assembly Language Coding
	Automatic Level Control
	Automatic Load Control
ALCA	American Leather Chemists Association [US]
Alcan	Alaska-Canada (Highway)
ALCAPP	Automatic List Classification and Profile Production
ALCC	Airborne Launch Control Center
	Association of London Computer Clubs [UK]
ALCE	Air Lift Control Element
ALCELL	Alcohol-Cellulose
ALCH	Approach Light Contact Height
ALCHEMI	Atom Location by Channeling Enhanced Microanalysis
ALCS	Authors Lending and Copyright Socity [UK]
ALCM	Air-Launched Cruise Missile
ALCOM	Algebraic Compiler
ALCOR	ARPA Lincoln Coherent Observable Radar [of USDOD]
ALCP	Area Local Control Panel
ALCS	Aluminum Locator-Nose Clamp Strap
ALCU	Arithmetic Logic and Control Unit
	Asynchronous Line Control Unit
ALD	Algerian Dinar [Currency]
	Analog Line Driver
	Asynchronous Line Driver
	Automated Logic Diagram
	Automatic Location Device
	Automatic Louver Damper
ALDA	Agricultural Land Development Assistance [Canada]
	Association for Librarianship and Documentation in Agriculture [Germany]
ALDEP	Automated Layout Design Program
ALDEV	African Land Development Board [Kenya]
ALDOC	Arab League Documentation Information Center
ALDP	Automatic Language Data Processing
ALDS	Apollo Launch Data System
ALE	Annual Loss Expectancy
	Atomic Layer Epitaxy
ALEAS	Asociacion Latinoamericana de Educacion Agricola [= Latin American Association for Agricultural Education]
ALECSO	Arab League Educational, Cultural and Scientific Organization
ALE/EL	Atomic Layer Epitaxy/Electroluminescent Device
ALEPH	Automatic Library Expendable Program, Hebrew University [Israel]
ALERT	Assistance for Liquid Electronic Reliability Testing
	Automated Linguistic Extraction and Retrieval Technique
	Automatic Logical Equipment Readiness Tester
ALES	American Labour Education Service [US]
ALESA	American League for Exports and Security Assistance [US]
ALESCO	Arab League Educational, Scientific and Cultural Organizations [Tunesia]
ALF	Absorption Limiting Frequency
	Alien Life Form
	Application Library File
	Automatic Letter Facer
	Automatic Line Feed
	Auxiliary Landing Field
ALFA	Air-Lubricated Free Attitude
ALFCE	Allied Land Forces Central Europe
ALFTRAN	ALGOL-to-FORTRAN Translator
ALG	Algebra(ic)
	Antilymphocytic Globulin
	Atomic Layer Growth
AlGaAs	Aluminum Gallium Arsenide
ALGM	Air-Launched Guided Missile
ALGO	Algebraic Compiler
ALGOL	Algebraic Oriented Language [Computer Language]
	Algorithmic Language [computer language]
ALGU	Association of Land Grant Universities [US]
ALH	Allan Hills (Iron Meteorite)
ALHFAM	Association for Living Historical Farms and Agricultural Museums [US]
ALHT	Apollo Lunar Hand Tool [of NASA]

ALI Alberta Laser Institute [Canada]
American Ladder Institute [US]
American Lighting Institute [US]
Asynchronous Line Interface
Automated Logic Implementation
Automotive Lift Institute [US]

ALICE Adiabatic Low-Energy Injection and Capture Experiment [also: Alice]
Airborne Line-Scan Image Processor [also: Alice]
Alaska Integrated Communications Exchange [also: Alice]

ALIGN Alignment

ALIM Air-Launched Intercept Missile

ALIN Alcoa Laboratories Information Network [US]

ALIP Annular Linear Induction Pump

ALIS Advanced Life Information System
Automated Library Information System

ALISA American League for International Security Assistance [now: ALESA, US]

ALISE Association for Library and Information Science Education [US]

ALIT Association of Laser Inspection Technologies
Automatic Line Insulation Tester

ALK Alkali(ne)

ALKY Alkalinity

ALL Accelerated Learning of Logic
Albanian Lek [Currency]
Application Language Liberator

ALLA Allied Long Lines Agency [of NATO]
Automated Laboratory Liquor Analyzer

ALLC Association for Literary and Linguistic Computing [UK]

ALL-CAPS All Capital Letters [all-caps]

Al-Li (International) Aluminum-Lithium Conference

ALLOC Allocation

ALLOW Allowance

ALLS Apollo Lunar Logistic Support [of NASA]

ALM Alarm
Applied Laboratory Method
Assembler Language for MULTICS (= Multiplexed Information and Computing Service)
Asynchronous Line Module
Asynchronous Line Multiplexer
Audiolingual Method

ALMA Alphanumeric Language for Music Analysis
American Loudspeaker Manufacturers Association [US]
Analytical Laboratory Managers Association [US]

ALMC Army Logistic Management Center [US]

ALMO African Livestock Marketing Organization [Kenya]

ALMS Aircraft Landing Measurement System
Analytic Language Manipulation System
Auxiliary Liquid Metal System

ALO Accredited Leasing Officer

ALOA Associated Locksmiths of America [US]

ALOFT Airborne Light Optical Fiber Technology

ALOG Applied Laser Optics Group

ALON
Aluminum Oxynitride (Ceramic) [also: AlON]

ALOR Advanced Lunar Orbital Rendezvous

ALOT Airborne Lightweight Optical Tracker

ALOTS Airborne Lightweight Optical Tracking System

ALP Alkaline Phosphatase
Arithmetic Logic Processor
Articulated Loading Platform
Assembly Language Preprocessor
Assembly Language Processing
Assembly Language Program
Atomic Layer Processing
Automated Language Processing
Automated Learning Process

AlP Alkaline Phosphatase

ALPA Alaskan Long Period Array

ALPAC Automated Language Processing Advisory Committee [of NAS, US]

ALPC Adaptive Linear Protective Coding

ALPHA Automated Literature Processing, Handling and Analysis

ALPHANUM Alphanumeric(s)

ALPHGR Average Linear Planar Heat Generation Rate

ALPID Analysis of Large Plastic Incremental Deformation

ALPIDP Analysis of Large Plastic Incremental Deformation, Powder Metals

ALPIDT Analysis of Large Plastic Incremental Deformation, Nonisothermal

ALPL Association of Librarians in Public Libraries [Germany]

ALPR Argonne Low Power Reactor [of ANL, US]

ALPS Advanced Linear Programming System
Advanced Liquid Propulsion System
Assembly Line Planning System
Associated Logic Parallel System
Associative Logic Parallel System
Automated Library Processing Service

ALPSP Association of Learned and Professional Society Publishers [UK]

ALQDS All Quadrants

ALR Active Line Rotation
Agricultural Land Reserve [Canada]
Allergenic Effects

ALRC Anti-Locust Research Center [UK]

ALRI Airborne Long-Range Intercept

ALRR Ames Laboratory Research Reactor [US]

ALRT Advanced Light Rapid Transit
Automated Light Rapid Transit

ALRU Automatic Line Record Update

ALS Advanced Light Source [of LBL, US]
Advanced Logistics System
Advanced Low-Power Schottky Logic
Agricultural Land Service [US]
Aircraft Landing System
Alberta Land Surveyors [Canada]
American Littoral Society [US]
American Lunar Society [US]
Antilymphocytic Serum
Approach Light System
Arithmetic and Logic Section
Automated Library System
Automatic Landing System

ALSA Area Library Service Authority

ALSB Aqueous Lithium Salt Blanket

ALSC American Lumber Standards Committee [US]

ALSCP Appalachian Land Stabilization and Conservation Program [US]
ALSEP Apollo Lunar Surface Experiment Package [of NASA]
ALSI Aluminum Silicon Alloy
ALSPEI Association of Land Surveyors of Prince Edward Island [Canada]
ALSS Advanced Location Strike System
Airborne Location and Strike System
Apollo Logistic Support System [NASA]
ALSTG Altimeter Setting
ALSTTL Advanced Low-Power Schottky Transistor-Transistor Logic [also: ALS TTL]
ALT Activated Liquid Thromboplastin
Airborne Laser Tracker
Alanine Aminotransferase
Alternate (Key)
Alteration
Alternation
Alternator
Altitude
ALTA Alberta Library Trustees Association [Canada]
Association of Local Transport Airlines
Alta Alberta [Canada]
ALTAC Algebraic Transistorized Automatic Computer
Algebraic Translator and Compiler
ALTAIR ARPA (= Advanced Research Projects Agency) Long-Range Tracking and Instrumentation Radar [of USDOD]
ALTAN Alternate Alerting Network [US Air Force]
ALTARE Automatic Logic Testing and Recording Equipment
ALTECH Alterra Laser Technologies Project [Canada]
ALTER Alternating
Alternative
ALT/GPT Alanine Aminotransferase/Glutamic-Pyruvic Transaminase
ALTM Altimeter
ALTN Alternation
Alternative
ALTRAN Algebraic Translator
Assembly Language Translator
ALTRV Altitude Reservation
ALTS Advanced Lunar Transportation System
Automated Library Technical Services [US]
Automatic Line Test Set
ALU Advanced Levitation Unit
Arithmetic-Logic Unit
Asynchronous Line Unit
Alu Arthrobacter Luteus [Molecular Biology]
ALUE Admissible Linear Unbiased Estimator
ALV Autonomous Land Vehicle
ALVIN Antenna Lobe for Variable Ionospheric Nimbus
ALWIN Algorithmic Wiswesser Notation
ALWR Advanced Light Water Reactor
ALY Alloy
AM Access Method
Address Mark
Address Mode
Address Modifier
Advanced Material
Airlock Module
Air Marshall
Air Mass (Number)
Air Melting
Air Ministry
Alternating Magnetic Field
America
Ammeter
Amorphous
Amplitude
Amplitude Modulation
Analog Monolithic
(Anno Mundi) — in the year of the world
Applied Mathematics
Applied Mechanics
Aquatic Microbiologist
Aquatic Microbiology
Arc Melting
Arithmetic Mean
(Artium Magister) — Master of Arts
Associate Member
Associative Memory
Asynchronous Modem
Asynchronous Motor
Auxiliary Memory
Am Americium
Ammonium
Amyl
am (ante meridien) — before noon [also: AM]
AM0 Air Mass 0
AM1 Air Mass 1
AMA Academy of Model Aeronautics
Acoustical Materials Association [US]
Actual Mechanical Advantage
Adhesives Manufacturers Association [US]
Aero Medical Association [US]
Agricultural Marketing Administration [US]
Air Material Area [US Air Force]
Alberta Medical Association [Canada]
Alberta Motor Association [Canada]
American Management Association [US]
American Manufacturers Association [US]
American Maritime Association [US]
American Marketing Association [US]
American Medical Association [US]
American Meteor Association [US]
American Motorcyclist Association [US]
Amorphous Magnetic Alloy
Area Minimum Altitude
Asbestos Mining Association [Canada]
Associative Memory Address
Associative Memory Array
Auckland Mathematical Association [New Zealand]
Automatic Memory Allocation
Automatic Message Accounting
Automobile Manufacturers Association [US]
AMAB Arctic Marine Advisory Board [Canada]
AMAC Automatic Material Completion
AMACUS Automated Microfilm Aperture Card Updating System [of US Army]
AMAD Activity Median Aerodynamic Diameter
Aircraft-Mounted Accessory Drive

AMADS Airframe-Mounted Auxiliary Drive System
AMA/I American Management Association International [US]
AMAIS Agricultural Materials Analysis Information Service [UK]
AMAL Amalgam
Amalgamation [also: AMALG]
AMAN Automatic Material Number
AMA/NET American Medical Association Network [US]
AMAQ 1-Amino-2-Methylanthraquinone
AMARS Air Mobile Aircraft Refueling System
AMAUS Aero Medical Association of the United States
AMARV Advanced Maneuvering Reentry Vehicle
AMAS Advanced Midcourse Active System
Automatic Material Status
AMATYC American Mathematical Association of Two-Year Colleges [US]
AMB Ambassador
Amber
Ambient
Anti-Motor Boat
Asbestos Millboard
Auto-Manual Bridge (Control)
AMBA American Malting Barley Association [US]
Association of Military Banks of America [US]
AMBCO Association of Major City Building Officials [US]
AMBCS Alaskan Meteor Burst Communication System
AMBD Automatic Multiple Blade Damper
AMBES Association of Metropolitan Borough Engineers and Surveyors [UK]
AMBIM Associate Member of the British Institute of Management
AMBIT Algebraic Manipulation by Identity Translation [Computer Language]
AMC Acceptable Means of Compliance
Account Manager Code
Acetylmethylcarbon
Aerodynamic Mean Chord
Airborne Management Computer
Aircraft Manufacturers Council
Air Materiel Command [US Air Force]
Alberta Microelectronics Center [Canada]
American Mining Congress [US]
American Movers Conference [US]
Army Materiel Command [US]
Army Medical Corps [US]
Army Missile Command [US]
Association of Management Consultants
Automated Meter Calibration
Automatic Message Counting
Automatic Mixture Control
Automatic Modulation Control
Automatic Monitoring Circuit
Auto Meter Correct
Autonomous Multiplexer Channel
Avionics Maintenance Conference [US]
AMCA Advanced Materiel Concept Agency
Air Movement and Control Association [US]
Air Moving and Conditioning Association [Name Changed to: Air Movement and Control Association, [US]
American Mosquito Control Association [US]
Asociacion de Mexico de Control Automatico [= Mexican Association of Automatic Control]
AMCB Asian Mass Communication Bulletin [of AMIC, Singapore]
AMCBMC Air Materiel Command Ballistic Missile Center [US Air Force]
AMCCW Association of Manufacturers of Chilled Car Wheels [US]
AMCEC Allied Military Communications Electronics Committee
AMCEE Association for Media-Based Continuing Education for Engineers [US]
AmCeramSoc American Ceramic Society [US]
AMCF Alkali Metal Cleaning Facility
AMCHA Aminomethylcyclohexanecarboxylic Acid
AMCHAM American Chamber of Commerce [also: AmCham]
AmChemSoc American Chemical Society [US]
AMCM Academia Mexicana de Ciencia de Materiales [= Mexican Academy of Materials Science]
AMCOM Automated Machine Control for Optimized Mining
AMCP Abnormal Milk Control Program
Allied Military Communications Panel
AMCS Airborne Missile Control System
Association for Mexican Cave Studies [US]
Association of Military Colleges and Schools [US]
AMCTB Associated Motor Carriers Tariff Bureau [US]
AMD Acid Mine Drainage
Advanced Microcomputer Device
Ambulance Manufacturers Division [of NTEA, US]
Automated Multiple Development System
AMDA Airlines Medical Directors Association
Associated Minicomputers Dealers of America [US]
Association of Medical Doctors in Asia
AMDAA Accounting Machine Dealers Association of America [later: AMMDA; now: AMDA]
AMDAC Amdahl Diagnostic Resistance Center [US]
AMDB Agricultural Machinery Development Board [UK]
AMDCB Applied Moment Double-Cantilever Beam
AMDE Association of Medical Doctors in Europe
AMDEC Agricultural Marketing Development Executive Committee [UK]
AMDEL Australian Mineral Development Laboratories
AMDM Association of Microbiological Diagnostic Manufacturers [US]
AMDMA American Metal Detector Manufacturers Association [US]
AMDS Australian MARC (= Machine-Readable Catalogue) Distribution Service
AMDSB Amplitude Modulation, Double Sideband [also: AM-DSB]
AME Advanced Modelling Extension
Amplitude Modulation Equivalent
Angle-Measuring Equipment
Association of Manufacturing Excellence [US]
Association of Membership [US]
Authorized Medical Examiner
Automatic Microfiche Editor
AMEA Association of Machinery and Equipment Appraisers [US]
AMEC Advanced Materials Engineering Center [Canada]

	Aerospace Metals Engineering Committee [of SAE, US]
AMECD	Association for Measurement and Evaluation in Counseling and Development [US]
AME/COTAR	Angle Measuring Equipment/Correlation Tracking and Ranging
AM Ed	Advanced Master of Education
AMEDD	Army Medical Department [US]
AMEDPC	Automotive Manufacturers Electronic Data Processing Council [US]
AMEDS	Army Medical Service
AMEE	Admiralty Marine Engineering Estabilishment [UK]
	Association of Managerial Electrical Executives [UK]
AMEG	Association for Measurement and Evaluation in Guidance
AMEIC	Associate Member of the Engineering Institute of Canada
AMEL	Aircraft Multiengine Land
AMEM	Association of Marine Engine Manufacturers [US]
AMEME	Association of Mining, Electrical and Mechanical Engineers [UK]
AMER	America(n)
AMER STD	American Standard
AMES	Air Ministry Experimental Station [UK]
	Applied Mechanics and Engineering Sciences [of UCSD, US]
	Association of Marine Engineering Schools [UK]
AMETA	Army Management Engineering Training Agency [US]
AMETS	Artillery Meteorological System
AMEU	Association of Municipal Electrical Utilities
AMEX	American Stock Exchange [also: Amex, US]
AMF	ACE (= Allied Command Europe) Mobile Force [of NATO]
	Advanced Manufacturing Facility
	Air Mail Facility
	Arab Monetary Fund
	Australian Mineral Foundation
	Automatic Mirror Furnace
	Automatic Mirror Facility
AMF(A)	ACE (= Allied Command Europe) Mobile Force (Air) [of NATO]
AMFEP	Association of Microbial Food Enzyme Producers [Belgium]
AMFI	Aviation Maintenance Foundation International [US]
AMFIS	American Microfilm Information Society [US]
	Automatic Microfilm Information Society
	Automatic Microfilm Information System
AMF(L)	ACE (= Allied Command Europe) Mobile Force (Land) [of NATO]
AM/FM	Amplitude Modulation/Frequency Modulation
AMFWA	Association of Midwest Fish and Wildlife Agencies [US]
AMG	Abrasives Marketing Group
	Allied Military Government
AMGRA	American Milk Goat Record Association [US]
AMHF	American Military History Foundation [now: American Military Institute, US]
AMHIC	Automatic Merchanizing Health Indystry Council
AMHS	American Material Handling Society [now: IMMS, US]
	Automated Message-Handling System
AMI	Advertising and Market Intelligence (Database) [of NYTIS, US]
	Agricultural Mechanization Institute
	Air Movement Institute [US]
	Alliance of Metalworking Industries [US]
	Alternative Mark Inversion (Signal)
	American Military Institute [US]
	American Mushroom Institute [US]
	Applied Microelectronics Institute
	Association for Multi-Image [US]
AMIA	American Metal Importers Association [US]
AMIAS	Automatic Metallurgical Image Analysis System
AMIC	Aerospace Materials Information Center [US]
	American Marine Insurance Clearinghouse [US]
	Asian Mass Communications Research and Information Center [Singapore]
AMICE	Associate Member of the Institute of Civil Engineers
AMICOM	Army Missile Command [US]
AMIDS	Airport Management and Information Display System
AMIE	Advanced Manufacturing in Electronics
AMIEE	Associate Member of the Institute of Electrical Engineers
AMIME	Asociacion Mexicana de Ingenieros Mecanicos y Electricistas [= Mexican Association of Mechanical and Electrical Engineers]
	Associate Member of the Institute of Mechanical Engineers
AMIMechE	Associate Member of the Institution of Mechanical Engineers
AMIMinE	Associate Member of the Institution of Mining Engineers
AMIN	Advertising and Marketing International Network [US]
	Arab Medical Information Network
A min	Ampere-minute [Unit]
AMINA	Association Mondiale des Inventeurs [= World Association of Inventors, Belgium]
AMInstCE	Associate Member of the Institute of Civil Engineers
AMIS	Agricultural Management Information System [of EEC, Europe]
	Aircraft Movement Information Service
	Air Movement Identification Service
	Assembly Management Information System
	Automated Management Information System
AMISC	Army Missile Command [US]
AMK	Anti-Misting Kerosene
AML	Actual Measured Loss
	Admiralty Materials Laboratory [UK]
	Advanced Manipulator Language
	Aeronautical Materials Laboratory
	A Manufacturing Language
	Amplitude-Modulated Link
	Applied Mathematics Laboratory [of DSIR, UK]
	Array Machine Language
	Automatic Modulation Limiting

	Aviation Materials Laboratory [US Army]
AMLC	Association of Marine Laboratories of the Caribbean [Puerto Rico]
	Asynchronous Multiline Controller
AMLCC	Asynchronous Multiline Communications Coupler
AMLCD	Active Matrix Liquid Crystal Display
AMM	Additional Memory Module
	Ammeter
	Ammunition
	Analog Monitor Module
	Anti-Missile Missile
	Associated Maintenance Module
AMMA	Acrylonitrile-Methylmethacrylate
AMMDA	Accounting Machine/Minicomputer Dealers Association of America [now: AMDA, US]
AMMI	American Merchant Marine Institute [US]
AMMIP	Aviation Materiel Management Improvement Program [US Army]
AMMLA	American Merchant Marine Library Association [US]
AMMM	Association of Microbiological Media Manufacturers [now: AMDM, US]
AMMO	Ammunition [also: Ammo]
AMMRC	Army Materials and Mechanics Research Center [now: AMTL, US]
AMNH	American Museum of Natural History [US]
AMNIP	Adaptive Man-Machine Nonarithmetical Information Processing
	Adaptive Man-Machine Nonnumeric Information Processing
AMNIPS	Adaptive Man-Machine Nonarithmetical Information Processing System
	Adaptive Man-Machine Nonnumeric Information Processing System
AM NIT	Ammonium Nitrate
AMNS	Artic Marine Navigation Systems [of CCG, Canada]
AMO	Aeronautics Material Officer
	Air Ministry Order [UK]
	Alternant Molecular Orbit
AMOA	Atmospheric Monitor Oxygen Analyzer
AMONO	Agricultural Machinery Operation and Management Office [PR China]
AMOR	Amorphous [also: AMORPH]
AMORT	Amortization
AMOS	Acoustic Meteorological Observing System
	Acoustic Meteorological Oceanographic Survey
	Alpha Microsystems Operating Systems [US]
	American Maritime Officers Service [US]
	Associative Memory Organizing System
	Automatic Meteorological Observing System
AMOSA	Association of Aviation Maintenance Organizations in South Africa
AMP	Adaptation Mathematical Processor
	Advanced Manned Penetrator
	Adenosine Monophosphate
	Aerospace Medical Panel [of AGARD]
	American Melting Point
	2-Amino-2-Methyl-1-Propanol
	Amperage
	Ampere
	Amplifier
	Ampoule
	Associative Memory Processor
	Assured Mail Program
amp	ampere [Unit]
3',5'-AMP	Adenosine 3',5'-Cyclic Monophosphate
A5MP	Adenosine-5-Monophosphoric Acid
AMPA	American Manganese Producers Association [US]
	American Medical Publishers Association [US]
AMPCO	Association of Major Power Consumers in Ontario [Canada]
AMPD	2-Amino-2-Methyl-1,3-Propanediol
AMPERE	Atomes et Molecules par Etudes Radio-Electriques [= Atoms and Molecules for Radioelectric Studies, Switzerland]
AMPH	Amphibian
	Amphibious
AMP HR	Ampere-hour [also: AMP-HR, amp-hr or amp hr]
AMPI	Associated Milk Producers Information
AMPIC	Atomic and Molecular Processes Information Center
AMPL	Advanced Microprocessor Programming Language
	Advanced Microprocessor Prototyping Laboratory [US]
	Amplification
	Amplifier
AMPLG	Amplidyne Generator
AMPLMG	Amplidyne Motor Generator
AMPNS	Asymmetric Multiple Position Neutron Source
AMPR	Aeronautical Manufacturers Planning Report
AMPS	Advanced Mobile Phone Service
	Automatic Message Processing System
AMPSO	Aminomethylhydroxypropanesulfonic Acid
	3-[(1,1-Dimethyl-2-Hydroxyethyl)amino]-2-Hydroxy-1-Propanesulfonic Acid
AMPSS	Advanced Manned Precision Strike System
AMPTE	Active Magnetospheric Particle Tracer Experiment
AMP-TURN	Ampere-Turn [also: amp-turn]
AMQUA	American Quaternary Association [US]
AMR	Advance Material Request
	Airport Movement Radar
	Applied Materials Research
	Atlantic Missile Range [US Air Force]
	Automated Management Report
	Automatic Message Registering
	Automatic Message Routing
	Automatic Meter Reading
	Autonomous Mobile Robot
AMRA	American Medical Record Association [US]
	Army Materials Research Agency [US]
	Automatic Meter Reading Association [US]
AMRAAM	Advanced Medium-Range Air-to-Air Missile
AMRAC	Anti-Missile Research Advisory Council [US]
AMRAD	Arda Measurements Radar
AMRC	Australian Materials Research Committee
	Automotive Market Research Council [US]
AMREF	African Medical and Research Foundation
	African Medical Research Foundation
AMRF	Automated Manufacturing Research Facility [of NIST, US]
AMRL	Aerospace Medical Research Laboratory [US]
AMRNL	Army Medical Research and Nutrition Laboratory [US]

AMRS Australian MARC (= Machine-Readable Catalogue) Record Service
AMRV Atmospheric Maneuvering Reentry Vehicle
AMS Accelerator Mass Spectrometry
Access Method Services
Acoustic Monitor System
Administrative Management Society [US]
Advanced Manufacturing System
Advanced Memory System
Administrative Management Society [US]
Aeronautical Material Specification
Aeronautical Mobile Services
Aerospace Material Specification [of SAE]
Agricultural Manpower Society [UK]
Agricultural Marketing Service [US]
Air Mail Service
Air Mass
Alma Mater Society
American Magnolia Society [now: TMS, US]
American Mathematical Society [US]
American Meteorological Society [US]
American Meteor Society [US]
American Microchemical Society [US]
American Microscopical Society [US]
American Miscellaneous Society [US]
Applied Materials Science
Army Map Service [US]
Army Medical Service
Association of Metal Sprayers [US]
Asymmetric Multiprocessing System
Atmospheric Monitor System
Attitude Measurement Sensor
Automated Manifest System
Automated Manufacturing Show (and Exhibition) [Canada]
Australian Mathematical Society
AMSA Advanced Manned Strategic Aircraft
American Meat Science Association [US]
American Metal Stamping Association [US]
Association of Medical Schools in Africa
Association of Metropolitan Sewerage Agencies [US]
AMSAT (Radio) Amateur Satellite Corporation [US]
AMSC American Marine Standards Committee [US]
Army Mathematics Steering Committee [US]
Australian Materials Science Committee
Automatic Message Switching Center
AMSE Association of Muslim Scientists and Engineers [US]
AMSL Above Mean Sea Level
AMSOC American Miscellaneous Society [US]
AMSP Allied Military Security Publications
AMSS Aeronautical Mobile Satellite Service
AMSSB Amplitude Modulation, Single-Sideband
AMSSP Aeronautical Mobile Satellite Service Panel [of ICAO]
AMST Advanced Medium STOL (= Short Takeoff and Landing) Transport
Association of Maximum Service Telecasters [US]
AMSU Advanced Microwave Sounding Unit
Auto-Manual Switching Unit
AMSU-B Advanced Microwave Sounding Unit — B
AMT Accelerated Mission Testing
Advanced Manufacturing Technology
Advanced Materials Technology
Advanced Metals Technology
Alternative Minimum Tax
Amount
Audiofrequency Magnetotellurics
Audio Magnetotellurics
AMTA Antenna Measurement Techniques Association [US]
AMTAP Advanced Manufacturing Technologies Application Program
AMTB Anti-Motor Torpedo Boat
AMTC Automated Manufacturing Training Center [Canada]
AMTCL Association for Machine Translation and Computational Linguistics [US]
AMTD Automatic Magnetic Tape Dissemination [of DDC, US]
AMTDA Agricultural Machinery and Tractor Dealers Association [UK]
American Machine Tool Distributors Association [US]
AMTE Admiralty Marine Technology Establishment
AMTEC Alkali Metal Thermoelectric Converter
Association for Media and Technology in Education in Canada
AMTEX Air Mass Transformation Experiment
AMTI Airborne Moving Target Indicator
Area Moving Target Indicator
Automatic Moving Target Indicator
AMTIDE Aircraft Multipurpose Test Inspection and Diagnostic Equipment
AMTL Army Materials Technology Laboratory [formerly: AMMRC, US]
AMTRAC Amphibian Tractor [also: amtrac]
Amphibious Tractor [also: amtrac]
AMTRAN Automatic Mathematical Translator
AMTRT Alberta Ministry of Technology, Research and Telecommunications [Canada]
AMTS Automated Materials Testing System
AMTU Advanced Materials Technology Unit [Canada]
AMU African Mathematical Union [Congo]
American Malacological Union [US]
Associated Midwestern Universities [US]
Association of Minicomputer Users [US]
Astronaut Maneuvering Unit [of NASA]
Atomic Mass Unit [also: amu]
AMUMeV Atomic Mass Unit Mega-Electronvolt
AMV Advanced Marine Vehicle
AMVER Atlantic Merchant Vessel Alert
Automated Merchant Vessel Reporting
AMWA American Medical Writers Association [US]
Association of Metropolitan Water Agencies [US]
AN Abstract Number
Accession Number
Acetonitrile
Acorn Nut
Acrylonitrile
Advanced Navigator
Air Force — Navy
Algebraic Number
Alphanumeric(s)

Aminonitrogen
Ammonium Nitrate
Army-Navy
Arrival Note
A/N Alphanumeric(s)
As Needed
An Actinon [= Actinium Emanation, or Radon 219]
ANA Antinuclear Antibody
American Nature Association [US]
3-Aminonocardicinic Acid
Army Navy Aeronautical
Article Numbering Association [UK]
Assigned Night Answer
Association of National Advertisers [US]
Association of Naval Aviation [US]
Association of Nordic Aeroclubs [Norway]
Australian Nuclear Association
Automatic Network Analyzer
Automatic Number Analysis
Automatic Number Announcer
ANACHEM Association of Analytical Chemists [US]
ANACOM Analog Computer
ANADP Association of North American Directory Publishers [US]
ANAIC Asian Network of Analytical Inorganic Chemistry
ANAL Analysis
AnalChem Analytical Chemist
ANALIT Analysis of Automatic Line Insulation Tests
ANALY CHEM Analytical Chemistry
ANALYT Analytical
ANARE Australian National Antarctic Research Expedition
ANAT Anatomical
Anatomist
Anatomy
ANATRAN Analog Translator
ANB Alpha-Naphthyl Butyrate
Australian National Bibliography [of NLA]
ANBFM Adaptive Narrow-Band FM (= Frequency Modulation) Modem
ANBLS Association of New Brunswick Land Surveyors [Canada]
ANBS Air Navigation and Bombing School
Asian Network of Biological Sciences
ANC Air Navigation Commission
Air Navigation Conference
All-Number Calling
Ancient
Army-Navy Civil Aeronautics Committee [US]
Army Nurse Corps
Automatic Nutation Control
ANCA Allied Naval Communications Agency [of NATO]
ANCAR Australian National Committee for Antarctic Research
ANCAT Abatement of Nuisances Caused by Air Transport [of ECAC, France]
ANCC Algerian National Chamber of Commerce
ANCIAWPRC Argentinian National Committee of the International Association on Water Pollution Research and Control
Australian National Committee of the International Association on Water Pollution Research and Control
Austrian National Committee of the International Association on Water Pollution Research and Control
ANCS American Numerical Control Society [US]
ANCU Airborne Navigation Computer Unit
AND Academics for Nuclear Disarmament
Air Force-Navy Design
Army-Navy Design
And Andromeda [Astronomy]
ANDA Association Nationale pour le Developpement Agricole [= National Association for Agricultural Development, France]
ANDB Air Navigation Development Board [US]
ANDIP American National Dictionary for Information Processing [US]
A-nDNA-A Anti-native DNA (= Deoxyribonucleic Acid) Antibody
Andr Andromeda [Astronomy]
ANDREE Association for Nuclear Development and Research in Electrical Engineering
ANDSA Amino Naphthol Disulfonic Acid
ANDZ Anodize
ANE Aeronautical and Navigational Electronics
ANEC American Nuclear Energy Council [US]
ANEDA Association Nationale d'Etudes pour la Documentation Automatique [= National Association for Studies in Automatic Documentation, France]
ANF Academia Nacional de Farmacia [= National Academy of Pharmacy, Brazil]
Anti-Nuclear Factor
Atlantic Nuclear Force
Atrial Natriuretic Factor
ANFCE Allied Naval Force Central Europe
ANFO Ammonium Nitrate and Fuel Oil Mixture [also: AN-FO]
ANG Air National Guard [US]
Netherlands Antillean Guilder [Currency]
ANGB Air National Guard Base
ANG-CE Air National Guard — Civil Engineering [US]
ANGOP Angolan Government News Agency
ANGUS A Navigable General-Purpose Underwater Surveyor [also: Angus]
ANH Anhydrous [also: ANHYD]
ANI Advanced Network Integration
Annular Isotropic
Automatic Number Identification
ANIA Associazione Nazionale Italiana per l'Automazione [= Italian National Association for Automation]
ANIE Associazione Nazionale Industrie Elettrotecniche [= National Association of the Electrotechnical Industry, Italy]
ANIEL Asociacion Nacional de Industrias Electronicas [= National Association of Electronics Industries, Spain]
ANILOL Aniline-Alcohol Mixture [also: anilol]
ANIM Association of Nuclear Instrument Manufacturers [US]

ANIP	Army-Navy Instrumentation Program [US]
	Army-Navy Integrated Presentation
ANIS	Annular Isotropic Source
ANL	Anneal
	Argonne National Laboratory [US]
	Automatic Noise Limiter
ANLG	Annealing
ANLOR	Angle Order
ANM	AFTN (= Aeronautical Fixed Telecommunications Network) Notification Message
ANMC	American National Metric Council [US]
ANMSA	Aminonaphthol Monosulfonic Acid
ANN	Annealing
	Annual
	Annunciator
	Answer, No-Charge
ANO	Air Navigation Order
	Alphanumeric Output
ANOCOVA	Analysis of Covariance [also: ANOCOVAR]
ANOD	Anodizing
ANOVA	Analysis of Variance [also: ANOVAR]
ANP	Aircraft Nuclear Propulsion
	Air Navigation Plan
	Allied Navigation Publication
	2-Amino-4-Nitrophenol
	Atrial Natriuretic Peptide
ANPA	American Newspaper Publishers Association [US]
ANPO	Aircraft Nuclear Propulsion Office [of USAEC]
	Alpha-Naphthylphenyloxazole
ANPOD	Antenna Positioning Device
ANPP	Army Nuclear Power Program [US]
ANPRM	Advanced Notice of Proposed Rule Making
ANPS	American Nail Producers Society [US]
ANPT	Aeronautical National (Taper) Pipe Thread
ANR	Advanced Non-Rigid (Airship)
	Alphanumeric Replacement
	Association of Neutron Radiographers [US]
	Awaiting Number Received
ANRAC	Aids Navigation Radio Control
ANRC	Animal Nutrition Research Council [US]
	Australian National Research Council
ANRIC	Annual National Information Retrieval Colloquium [US]
ANRPC	Association of Natural Rubber Producing Countries [Malaysia]
ANRT	Association Nationale de la Recherche Technique [= National Association for Technical Research, France]
ANS	Academy of Natural Sciences [US]
	Advanced Neutron Source
	Air Navigation School
	American National Standards [US]
	American Nuclear Society [US]
	Anilinenaphthalene Sulfonate
	8-Anilino-1-Naphthalenesulfonic Acid
	Answer
	Astronautical Netherlands Satellite
	Astronomical Netherlands Satellite
	Australian Nuclear Society
ANSA	Advanced Network System Architecture
	8-Anilino-1-Naphthalenesulfonic Acid
	Automatic New Structure Alert
ANSAC	American Natural Soda Ash Corporation [US]
ANSAM	Antinuclear Surface-to-Air Missile
ANSC	American National Standards Committee [now: NISO, US]
ANSCAD	Associate of Nova Scotia College of Art and Design [Canada]
ANSCR	Alphanumeric System for Classification of Recordings
ANSFMN	Accademia Nazionale dei Scienze Fisiche, Matematiche e Naturali [= National Academy of Physical, Mathematical and Natural Sciences, Italy]
ANSI	American National Standards Institute [US]
	Area of Natural and Scientific Interest
ANSI/AFBMA	American National Standards Institute/Antifriction Bearing Manufacturers Association [US]
ANSI/AGMA	American National Standards Institute/American Gear Manufacturers Association [US]
ANSI/AWS	American National Standards Institute/American Welding Society [US]
ANSI/IEEE	American National Standards Institute/Institute of Electrical and Electronic Engineers [US]
ANSI/NEMA	American National Standards Institute/National Electrical Manufacturers Association [US]
ANSI/NSF	American National Standards Institute/National Science Foundation [US]
ANSL	American National Standard Label
ANSLS	Association of Nova Scotia Land Surveyors [Canada]
ANS-NH_4	8-Anilino-1-Naphthalenesulfonic Acid, Ammonium Salt
ANSP	Academy of Natural Sciences of Philadelphia [US]
	Association of Navy Safety Professionals [US]
ANSS	American Nature Study Society [US]
ANSTI	African Network for Scientific and Technical Institutions [of UNESCO — Kenya]
ANSTO	Australian Nuclear Science and Technology Organization
ANSVIP	American National Standard Vocabulary for Information Processing [US]
ANSW	Antinuclear Submarine Warfare
ANT	Acetonitrile
	Administracion Nacional de Telecomunicacions [= National Telecommunications Administration, Paraguay]
	Antenna
Ant	Antlia [Astronomy]
ANTAC	Air Navigation Tactical Control System
ANTC	Antichaff Circuit
ANTEC	Annual Technical Conference
ANTGWDPEC	Association of National Trade Groups of Wood and Derived Products in the EEC (= European Economic Community Countries [Denmark]
ANTI	Anticoincidence
antilog	antilogarithm
anti-Rh	Anti-Rhesus
ANTIVOX	Voice-Actuated Transmitter Keyer Inhibitor
AN/TN	Aminonitrogen/Total Nitrogen Ratio
AN-TNT	Ammonium Nitrate and Trinitrotoluene Mixture
ANTS	Airborne Night Television System

	ARPA (= Advanced Research Projects Agency) Network Terminal System [of USDOD]
ANTU	Alpha-Naphthylthiourea
ANU	Airplane Nose Up
	Australian National University
ANUG	Atex (Computer System) Newspaper Users Group [US]
ANUGA	Allgemeine Nahrungs- und Genussmittel-Ausstellung [= World Food Fair and Exhibition, Germany]
ANVIS	Aviator's Night Vision Imaging System
ANVM	Association of Night Vision Manufacturers [US]
ANYST	Analyst
ANZAC	Australia and New Zealand Army Corps [also: Anzac]
ANZAAS	Australian and New Zealand Association for the Advancement of Science [also: Anzaas]
ANZSRLO	Australian and New Zealand Scientific Research Liaison Office
ANZUS	Australia-New Zealand-United States (Security Treaty)
AO	Abnormal Occurrence
	Access Opening
	Acousto-Optic(s)
	Aircraft Operator
	Amplifier Output
	Announcement of Opportunity
	Answer Only
	Army Order
	Atomic Orbital
	Atomic Oxygen
	Automated Operator
A-O	Acousto-Optic(s)
a/o	account of
AOA	Aerodrome Owners Association
	Airborne Optical Adjunct (System)
	Airport Operation Area
	American Oceanology Association [now: AOO, US]
	American Ordnance Association [US]
	At or Above
AOAA	Aminooxyacetic Acid
AOAC	Association of Official Agricultural Chemists (of North America) [US]
	Association of Official Analytical Chemists [US]
	Automobile Owners' Action Council [US]
AOAD	Arab Organization for Agricultural Development [Sudan]
AOB	Address Operation Block
	At or Below
AOC	Aerodrome Obstruction Chart
	Airline Operational Control
	Air Officer Commanding
	Air Operators Certificate [UK]
	Airport Operators Council
	American Orchid Society [US]
	Associated Overseas Countries [of EEC, Belgium]
	Association of Old Crows [US; "Old Crows" are Specialists on Electronic Warfare]
	Automatic Output Control
	Automatic Overload Circuit
	Automatic Overload Control
	Awaiting Outgoing Continuity
	Oxygen-Arc Cutting
AOCARS	African Organization for Cartography and Remote Sensing
AOCB	Any Other Competent Business
AOCI	Airport Operators Council International [US]
AOCR	Advanced Optical Character Reader
AOCS	Airline Operational Control Society [US]
	Alpha Omega Computer System
	American Oil Chemists Society [US]
	Attitude and Orbit Control System
AOCU	Arithmetic Output Control Unit
	Associative Output Control Unit
AOD	Above Ordnance Datum
	Acousto-Optic Device
	Advanced Ordnance Deport
	Argon-Oxygen Decarburization
	Arithmetic Output Data
	Army Ordnance Department
AODC	Association of Offshore Diving Contractors
AOD/VAR	Argon-Oxygen Decarburization/Vacuum Arc Remelting
AOE	Airport of Entry
	Association of Overseas Educators [US]
AOEC	Airways Operations Evaluation Center [US]
AOEW	Airplane Operating Empty Weight
AOF	Advanced Operating Facility
	Afrique Occidentale Francaise [= French West Africa]
	American Optometric Foundation [US]
A of F	Admiral of the Fleet
AOFA	Atlantic Offshore Fishermen's Association [US]
AOG	Aircraft on Ground
	Augmented Off Gas
AOI	Advance Ordering Information
	And-Or Invert
	Arab Organization for Industrialization
	Automated Operator Interface
	Automatic Optical Inspection
AOIPS	Atmospheric and Oceanic Information Processing System
AOJ	Association of Orthodox Jewish Scientists [US]
AOK	Angolan Kwanza [Currency]
AOL	Application-Oriented Language
	Atlantic Oceanographic Laboratory [Canada]
AOLO	Advanced Orbital Launch Operation
AOLS	Association of Ontario Land Surveyors [Canada]
AOMA	American Occupational Medical Association [US]
AOML	Atlantic Oceanographic and Meteorological Laboratories [of NOAA, US]
AOMR	Arab Organization for Mineral Resources
AONB	Area of Outstanding Natural Beauty
AONS	Air Observer's Navigation School
AOO	American Oceanic Organization [US]
AOP	Aerodrome Operations
	Air Observation Post
AOPA	Aircraft Owners and Pilots Association [US]
AOPA/ASF	Aircraft Owners and Pilots Association/Air Safety Foundation [US]
AOPL	Association of Oil Pipelines [US]

AOPU	Asian-Oceanic Postal Union [now: APPU, Philippines]
AOQ	Average Outgoing Quality
AOQL	Average Outgoing Quality Level
	Average Outgoing Quality Limit
AOR	Angle of Reflection
	Atlantic Ocean Region
AORB	Aviation Operational Research Branch
AORC	Association of Official Racing Chemists [US]
AORE	Army Operational Research Establishment [UK]
AORG	Army Operational Research Group [UK]
AOS	Acquisition of Signal
	Advanced Operating Service
	Advanced Operating System
	Agricultural Organization Society [UK]
	Air Observer's School
	Algebraic Operating System
	Angle of Sight
	Apparent Opening Size
	Automated Office System
	Azimuth Orientation System
AOSA	Association of Official Seed Analysts [US]
AOSC	Association of Oilwell Servicing Contractors [US]
AOSCA	Association of Official Seed Certifying Agencies [US]
AOSERP	Alberta Oil Sands Environmental Research Program [Canada]
AOSI	Alberta Oil Sands Index [of AOSIC, Canada]
AOSO	Advanced Orbiting Solar Observatory
AOSIC	Alberta Oil Sands Information Center [Canada]
AOSP	Automatic Operating and Scheduling Program
AOSTRA	Alberta Oil Sands Technology and Research Authority [Canada]
AOS/VS	Advanced Operating System/Virtual Storage
AOT	Alignment Optical Telescope
AOTT	Automatic Outgoing Trunk Test
AOTV	Aero-Assisted Orbital Transfer Vehicle
AOU	American Ornithologists Union [US]
	Apparent Oxygen Utilization
	Arithmetic Output Unit
	Associative Output Unit
	Automated Offset Unit
AP	Access Panel
	Access Permit
	Access Permittee
	Access Point
	Access Primitives
	Accounts Payable
	Acetophenone
	Adenosine Phosphoric Acid
	Administrative Processor
	Advance Payment
	After Peak
	After Perpendicular
	Air Pollution
	Airport
	Air Publication
	Air Pump
	Alkaline Permanganate
	Alkaline Phosphatase
	All-Pass Filter
	Ambient Pressure
	American Patent
	Ammonium Perchlorate
	Anionic Polymerization
	Annealed Pipe
	Annealing Point
	Anomalous Propagation
	Antarctica Project [US]
	Antennas and Propagation [IEEE Society]
	Application Program
	Applications Processor
	Applied Pharmacology
	Applied Physics
	April
	Argentinean Peso [Currency]
	Argument Programming
	Arithmetic Processor
	Arithmetic Progression
	Armor Piercing
	Array Processor
	Aseptic Processing
	Assembler Programming
	Assembly of Parties
	Associated Press [US]
	Associative Processor
	Assumed Position
	Astrophysics
	Atomic Physics
	Atomic Pile
	Atom Probe
	Attached Processor
	Author's Proof
	Automatic Polisher
	Automatic Process
	Autopilot
A/P	Account Purchase
	Attached Processor
	Authority for Purchase
ap	(anni preteriti) — of last year
	apothecaries [Unit]
4-AP	4-Aminopyridine
A-5'-P	Adenosine 5'-Monophosphate
APA	Abrus Precatorius Agglutinin
	Accidents in Private Aviation
	Accredited Public Accountant
	Acetone Producers Association [Belgium]
	Acetophenone Acetone
	Acrylamide Producers Association [US]
	Administrative Procedures Act [US]
	Advertising Photographers of America [US]
	Agricultural Precast Association [US]
	Agricultural Publishers Association [US]
	Alberta Publishers Association [Canada]
	Amalgamated Printers Association [US]
	American Parquet Association [US]
	American Pharmaceutical Association [US]
	American Pilots Association [US]
	American Planning Association [US]
	American Plywood Association [US]
	American Polygraph Association [US]
	American Poultry Association [US]

	American Press Association [US]
	American Pulpwood Association [US]
	American Pyrotechnics Association [US]
	Aminopenicillanic Acid
	Aminopropionic Acid
	Architectural Precast Association [US]
	Atlantic Publishers Association [Canada]
	Automobile Protection Association [Canada]
	Axial Pressure Angle
6-APA	6-Aminopenicillanic Acid
ApA	Adenylyladenosine
APAA	ASEAN Port Authorities Association
	Asian Patent Attorneys Association [Japan]
	Automotive Parts and Accessories Association [US]
APABC	(Institute of) Accredited Public Accountants of British Columbia [Canada]
APAC	Alkaline Permanganate Ammonium Citrate
	Asociacion Peruana para el Avance de la Ciencia [= Peruvian Association of the Advancement of Science]
APACE	Aldermaston Project for the Application of Computers to Engineering [of UKAEA]
APACHE	Accelerator for Physics and Chemistry of Heavy Elements
APACS	Airborne Position and Altitude Camera System
APACUE	Atlantic Provinces Association for Continuing University Education [Canada]
APAD	3-Acetylpyridine Adenine Dinucleotide
APADS	Automatic Programmer and Data System
APAG	American Photographic Artisans Guild [US]
	Association Europeenne des Producteurs d'Acides Gras [= European Association of Fatty Acid Producing Companies [of CEFIC, Belgium]]
	Atlantic Policy Advisory Group [of NATO]
APAIS	Australian Public Affairs Information Service [of NLA]
APAL	Array Processor Assembly Language
APALA	Asian/Pacific American Libraries Association
APAM	Array Processor Access Method
APAMM	Associacao Portuguesa dos Armadores de Marinha Mercante [= Portuguese Association of Merchant Ship Owners]
APANAC	Asociacion Panamena para el Avance de la Ciencia [= Panamanian Association of the Advancement of Science]
APAP	Acetyl-P-Aminophenol
ApApC	Adenylyladenylylcytidine
APAR	Authorized Program Analysis Report
	Automatic Processing and Recording
	Automatic Programming and Recording
APAS	Automatic Performance Analysis System
APASE	Association for the Promotion and Advancement of Science Education [Canada]
APATI	Abbreviated Precision Approach Path Indicator
APATS	Automatic Programming and Testing System
APB	Agricultural Products Board [Canada]
	Air Portable Bridge
	Antiphase Boundary
	As-Purchased Basis
	Associated Press Broadcasters [US]
APBA	Atlantic Provinces Booksellers Association [Canada]
APBF	Accredited Poultry Breeders Federation [UK]
APC	Activated Protein C
	Active Path Corrosion
	Adaptive Predictive Coding
	Advanced Personal Computer
	Advanced Pocket Computer
	Advanced Processor Card
	Advanced Professional Certificate
	Advanced Programming Course
	Aerobic Plate Counts
	Aeronautical Planning Chart
	Aeronautical Public Correspondence
	African Pacific and Caribbean Council
	Air Pollution Control
	Air Purification Control
	Alberta Press Council [Canada]
	Alternative Press Center [Canada]
	American Power Conference [US]
	Ammonium Perchlorate
	Amplitude Phase Conversion
	Approach Control
	Area Position Control
	Argon Purge Cart
	Armored Personnel Carrier
	Armor Piercing Capped
	Aromatic Polymer Composite
	Associative Processor Control
	Atlantic Press Council
	Atomic Power Constructions [UK]
	Automatic Particle Counter
	Automatic Performance Control
	Automatic Phase Control
	Automatic Position Control
	Automatic Pressure Control
	Automatic Process Control
	Automatic Program Control
	Automotive Presidents Council [US]
	Autoplot Controller
	Average Power Control
ApC	Adenylylcytidine
APCA	Air Pollution Control Association [now: AWMA, US]
	American Petroleum Credit Association [US]
	American Pollution Control Association [now: AWMA, US]
APCAC	Asia-Pacific Council of American Chambers of Commerce [Japan]
APCAS	Asia and the Pacific Commission on Agricultural Statistics [of FAO, UN — Philippines]
APCC	Advanced Physical Coal Cleaning
	Agricultural Planning and Coordinating Committee [of MOEA, Taiwan]
	Asian and Pacific Coconut Community [Indonesia]
APCChE	Asian Pacific Confederation of Chemical Engineering [Australia]
APCEF	Advanced Power Conversion Experimental Facility [of US Army]
APCHE	Automatic Programmed Checkout Equipment
APCI	Armor Piercing Capped Incendiary
	Atmospheric Pressure Chemical Ionization
APCI/MS	Atmospheric Pressure Chemical Ionization/Mass Spectrometry

APCI-T Armor Piercing Capped Incendiary with Tracer
APCL Association of Professional Color Laboratories [US]
APCM Adaptive Pulse Code Modulation
APCO Associated Public-Safety Communications Officers [US]
ApCpC Adenylylcytidylylcytidine
APCS Associative Processor Computer System
Automatic Program Control System
APC-T Armor Piercing Capped with Tracer
APCTT Asian and Pacific Center for Transfer of Technology [India]
APCVD Atmospheric Pressure Chemical Vapor Deposition
APD Aerospace Power Division [US Air Force]
Air Particulate Detector
Alloy Phase Diagram
Amplitude-Phase Diagram
Amplitude Probability Distribution
Angular Position Digitizer
Antiphase Domain
Army Pay Department
Avalanche Photodiode
Avalanche Photodiode Detector
APDA Appliance Parts Distributors Association [US]
Atomic Power Development Associates [US]
APDC Ammonium Pyrrolidine Dithiocarbamate
Apple and Pear Development Council [UK]
Asian and Pacific Development Center [Malaysia]
1-Pyrrolidinecarbodithioic Acid, Ammonium Salt
APDF Africa Project Development Facility
Asian Pacific Dental Federation
Association of Professional Design Firms [US]
APDI Asian Pacific Development Institute
APDIC Alloy Phase Diagram International Commission
APDMS Axial Power Distribution Monitoring System
APDS Armour Piercing Discarding Sabot
APDSMS Advanced Point Defense Surface Missile System [US Navy]
APDTC Ammonium Pyrrolidinedithiocarbamate
1-Pyrrolidinecarbodithioic Acid, Ammonium Salt
APDU Association of Public Data Users [US]
APE Abbreviated Plain English Language
Aminopropyl Epoxy
Antenna Positioning Electronics
Association of Professional Engineers [Canada]
APEA Association of Professional Engineers of Alberta [Canada]
Australian Petroleum Exploration Association
APEBC Association of Professional Engineers of British Columbia [Canada]
APEC All-Purpose Electronic Computer
Atlantic Provinces Economic Council [Canada]
Automated Procedures for Engineering Consultants [US]
Automotive Products Emissions Committee [US]
Automotive Products Export Council [US]
APECA American Package Express Carriers Association [US]
APEE Aminophenolethylether
APEF Association des Pays Exportateurs de Mineral de Fer [= Association of Iron Ore Exporting Countries, Switzerland]
APEG Alkaline Polyethyleneglycol
APEGGA Association of Professional Engineers, Geologists and Geophysicists of Alberta [Canada]
APEGGNWT Association of Professional Engineers, Geologists and Geophysicists of the Northwest Territories [Canada]
APEL Aeronautical Photographic Experimental Laboratory [US]
APEM Association of Professional Energy Managers [US]
Association of Professional Engineers of Manitoba [Canada]
APEN Association of Professional Engineers of Newfoundland [Canada]
APEO Association of Professional Engineers of Ontario [Canada]
APEPEI Association of the Professional Engineers of Prince Edward Island [Canada]
APER Association of Publishers Educational Representaives [UK]
APERS Antipersonnel
APES American Petroleum Equipment Suppliers [now: PESA, US]
Association of Professional Engineers of Saskatchewan [Canada]
APEX Advanced Productivity Exposition [US]
Advance Purchase Excursion (Fare)
Assembler and Process Executive
ATLAS (= Argonne Tandem Linear Accelerator System) Positron Experiment [of USDOE]
APEYT Association of Professional Engineers of the Yukon Territory [Canada]
APF Advanced Printer Function
Asian Packaging Federation [Japan]
Asphalt Plank Floor
Association of Professional Foresters [UK]
Atomic Packing Factor
Authorized Program Facility
Automatic Program Finder
Autopilot Flight Director
APFA American Pipe Fittings Association [US]
APFC American Plant Food Council [US]
American Printed Fabrics Council [US]
Asian Pacific Forestry Commission
APFCS Automatic Power Factor Control System
APFIM Atom Probe Field-Ion Microscopy [also: AP-FIM]
APFS Air Pressure Flow System
APG Advanced Plasma Gun
American Pewter Guild [US]
Aminopropyl Glass
Automatic Priority Group
Automatic Program Generation
Automatic Program Generator
Azimuth Pulse Generator
ApG Adenylylguanosine
APGA American Personnel Guidance Association [US]
American Public Gas Association [US]
Aminophenyl Glyoxalic Acid
APGC Air Proving Ground Center
APGGQ Association of Professional Geologists and Geophysicists of Quebec [Canada]
ApGpU Adenylylguanylyluridine

APGM Autonomous Precision Guided Munition
APGS Association of Professional Geological Scientists [now: AIPG, US]
APH Alkylated Phenol
APHA American Printing History Association [US]
American Public Health Association [US]
APHAZ Aircraft Proximity Hazards
APHC Aminophenol Hydrochloride
APHI Association of Public Health Inspectors [US]
APHIS Animal and Plant Health Inspection Service [US]
API Absolute Position Indicator
Addition-Reaction Polyimide
Air-Position Indicator
Alabama Polytechnic Institute [US]
Alternative Press Center [of Alternative Press Center, Canada]
American Paper Institute [US]
American Petroleum Institute [US]
American Potash Institute [US]
Arab Planning Institute
Armor Piercing Incendiary
Application Programming Interface
Applications Program Interface
Associated Photographers International [US]
Atmospheric Pressure Ionization
Automatic Priority Interrupt
APIA Asociacion Peruana de Ingenieros Agronomos [= Peruvian Association of Agricultural Engineers]
Association for the Promotion of Industry—Agriculture [France]
APIC Apollo Parts Information Center [of NASA]
APICS American Production and Inventory Control Society [US]
Atlantic Provinces Inter-University Council on the Sciences [Canada]
APIL Axial Power Imbalance Limit
APILIT American Petroleum Institute Literature (Database) [of API, US]
APIMS Atmospheric Pressure Ionization Mass Spectrometry
APIN Association for Programmed Instruction in the Netherlands
APIP Alberta Petroleum Incentive Program [Canada]
APIPAT American Petroleum Institute Patents (Database) [of API, US]
APIRG African Regional Planning and Implementation Regional Planning Group [of ICAO]
APIS Army Photographic Interpretation Section [US]
API-T Armor Piercing Incendiary with Tracer
APK Amplitude Phase Keying
APKS Amplitude Phase Keyed System
APL Aero-Propulsion Laboratory [US Air Force]
Algorithmic/Procedural Language [Computer Language]
Algorithmic Programming Language
Airplane
Anneal and Pickle Line
Applied Physics Laboratory
A Programming Language [Computer Language]
Association of Programmed Learning
Associative Programming Language
Authorized Possession Limits
Automatic Phase Lock
Average Picture Level
APLA Aminophenyl Lactic Acid
Arctic Production and Loading Atoll
Atlantic Provinces Library Association [Canada]
APLE Association of Public Lighting Engineers [US]
APLHGR Average Planning Heat Generation Rate
APLS American Plant Life Society [US]
APM Advanced Process Modeling
Air Particulate Monitor
Amplitude and Phase Modulation
Analog Panel Meter
Antenna Positioning Mechanism
Associative Principle for Multiplication
Asynchronous Packet Manager
Atom Probe Microanalysis
Autoclavable Powdered Medium
Automatic Predictive Maintenance
APMA American Pharmaceutical Manufacturers Association [US]
Automotive Parts Manufacturers Association [Canada]
APMC Alberta Petroleum Marketing Commission [Canada]
APME Aminophenol Methylether
Association of Plastic Manufacturers in Europe [Belgium]
APMHC Association of Professional Material Handling Consultants [US]
APMI American Powder Metallurgy Institute [US]
Area Precipitation Measurement Indicator
APMS Advanced Power Management System
Aquatic Plant Management Society [US]
APNB Association of Professional Engineers of New Brunswick [Canada]
APNIC Automatic Programming National Information Center [UK]
APO After Payout
Apolipoprotein
Asian Productivity Organization [Japan]
Army Post Office
Association of Professional Organizers [now: NAPO, US]
Aziridinylphosphine Oxide
APOA Arctic Petroleum Operators Association [Canada]
APOAC Asia Pacific Office Automation Control Council
APOLLO Article Procurement with On-Line Local Ordering
APOMA American Precision Optics Manufacturers Association [US]
APOS Advanced Polar Orbiting Satellite
APOTA Automatic Positioning Telemetering Antenna
APP Ammonium Polyphosphate
Apparatus
Apparently
Appendix
Application
Applied
Appointment
Apprentice
Arctic Pilot Project
Associative Parallel Processor
Atactic Polypropylene

	Auxiliary Power Plant
APPA	African Petroleum Producers Association
	Aircraft Preparation and Paint Application
	Alberta Professional Photographers Association [Canada]
	American Public Power Association [US]
	American Pulp and Paper Association [US]
	Association pour la Prevention de la Pollution Atmospherique [= Association for the Prevention of Atmospheric Pollution, France]
APPAR	Apparatus
APPC	Advanced Program-to-Program Communication
APP'D	Approved [also: APPD]
APPDA	Atlantic Provinces Power Development Act [Canada]
APPE	Association of Petrochemicals Producers in Europe [Belgium]
APPECS	Adaptive Pattern Perceiving Electronic Computer System
APPEN	Asia-Pacific People's Environment Network [Malaysia]
APPG	Adjacent Phase Pulse Generator
APPHS	Auger Peak-to-Peak Heights
APPI	Advanced Planning Procurement Information
APPITA	Australian (and New Zealand) Pulp and Paper Industry Technical Association
APPL	Aircraft Precision Position Location (Equipment)
	Appliance
	Application
	Applied
APPLE	Associative Processor Programming Language Evaluation
APPLS	Appliances
AppME	Applied Mechanics Engineer
APPMI	American Peanut Product Manufacturers, Inc. [US]
APPMSA	American Pulp and Paper Mills Superintendents Association [US]
APPN	Advanced Peer-to-Peer Networking
APPPC	Asia and Pacific Plant Protection Commission [Thailand]
APPRO	Approval
APPROX	Approximately [also: APPR]
APPS	Adenosine 3'-Phosphate-5'-Phosphosulfate
	Annual Procurement Plan and Strategy [of DSS, Canada]
APPT	Appointment
APPU	Asian-Pacific Postal Union [Philippines]
	Australian Primary Producers Union
APPX	Appendix
APQ	Available Page Queue
APR	Acoustic Paramagnetic Resonance
	Active Prominence Region [Astronomy]
	Advanced Production Release
	Airborne Profile Recorder
	Alternate Path Reentry
	Alternate Path Retry
	American Petroleum Re-Refiners [US]
	Annual Percentage Rate
	Annulus Pressure Responsive
	April
	ARF Power Reactor [US]
	Automatic Pressure Relief
APRA	Automotive Parts Rebuilders Association [US]
APRC	Automotive Public Relations Council [US]
APRD	Atmosphere Particulate Radioactivity Detector
APRES	American Peanut Research and Education Society [US]
APRF	Army Pulsed Radiation Facility [US]
APRFR	Army Pulsed Radiation Facility Reactor [US]
APRI	Absolute Rod Position Indication
	Aqua Planing Risk Indicator
APRIL	Aqua Planning Risk Indicator for Landing
	Automatically Programmed Remote Indication Logged
APRIS	Alcoa Picturephone Remote Information System [US]
APRL	Army Prosthetic Research Laboratory [US]
	Automated Primate Research Laboratory
APRM	Automatic Position Reference Monitor
	Average Power Range Monitor
APRO	Aerial Phenomena Research Organization
APRR	Association for Planning and Regional Reconstruction [UK]
APRS	American Park and Recreation Society [US]
	Association for the Preservation of Rural Scotland
	Automatic Position Reference System
APRT	Adenine Phosphoribosyltransferase
APRXLY	Approximately
APS	Abstract Preparation Sheet
	Adenosine Phosphosulfate
	Advanced Packaging System
	Advanced Photon Source [of ANL, US]
	Aircraft Prepared for Service
	Air Plasma Spraying
	Alphanumeric Photocomposer System
	American Philosophical Society [US]
	American Physical Society [US]
	American Physics Society [US]
	American Physiological Society [US]
	American Phytopathological Society [US]
	American Polar Society [US]
	American Pomological Society [US]
	American Phytopathological Society [US]
	American Purchasing Society [US]
	Aminopolystyrene
	Antennas and Propagation Society [of IEEE, US]
	Appearance Potential Spectroscopy
	Appearance Potential Spectrum
	Application Process Subsystem
	Armor-Piercing Sabot
	Array Processor Software
	Assembly Programming System
	Association de la Presse Suisse [= Swiss Press Association]
	Association of Plastics Societies [US]
	Atmospheric Plasma Spraying
	Attached Processor for Speech
	Attached Processor System
	Attended Pay Station
	Automatic Page Search
	Automatic Patching System
	Auxiliary Power System

	Auxiliary Program Storage
Aps	Apus [Astronomy]
APSA	Automatic Particle Size Analyzer
APS/AIP	(Joint) American Physical Society/American Institute of Physics (Project)
APSC	Asian Pacific Society for Cardiology
ApSci	Applied Scientist
APSDEP	Asian and Pacific Skill Development Program [of ILO, UN — Pakistan]
APSDIN	Asian and Pacific Skill Development Information Network [of APSDEP, Pakistan]
APSE	Ada Programming Support Environment [Ada is a Computer Language]
APSK	Amplitude Phase Shift Keying
APSM	Academy of Product Safety Management [US]
	Association for Physical and System Mathematics [US]
APSP	Array Processor Subroutine Package
APSR	Axial Power Shaping Rod
APSS	Automated Program Support System
APT	Advanced Passenger Train
	Advanced Passenger Transport
	Airman Proficiency Test
	Airport
	Ammonium Paratungstate
	Apartment
	Apparent Solar Time
	Armor-Piercing with Tracer [also: AP-T]
	Asia Pacific Telecommunity
	Association for Preservation Technology
	Augmented Programming Training
	Automatically Programmed Tool
	Automatic Picture Taking
	Automatic Picture Transmission
	Automation Planning and Technology
	Automatic Position Telemetering
	Automatic Process Testing
	Automatic Programmed Tool
	Automatic Programming Tool
AP-T	Armor Piercing with Tracer [also: APT]
APTA	American Public Transit Association [US]
	Atlantic Provinces Trucking Association [Canada]
APTC	Army Physical Training Corps [US]
	Atlantic Provinces Transport Commission [Canada]
APTI	Arab Petroleum Training Institute [Iraq]
	Association of Principals of Technical Institutions [UK]
APTIC	African Pyrethrum Technical Information Center
	Air Pollution Technical Information Center [US]
APTR	Advanced Pressure Tube Reactor [US]
APTS	Automatic Picture Transmission System
APTT	Activated Partial Thromboplastin Time
APTU	African Postal and Telecommunications Union [Congo]
APU	Acid Purification Unit
	Analog Processing Unit
	Arab Postal Union
	Arithmetic Processing Unit
	Asynchronous Processing Unit
	Audio Playback Unit
	Automatic Power Up
	Auxiliary Power Unit
	Auxiliary Processing Unit
ApU	Adenylyluridine
APUCOT	Automated Piezoelectric Ultrasonic Composite Oscillator Technique
APUG	AutoPrep (Computer) Users Group [US]
APUHS	Automatic Program Unit, High Speed
APULS	Automatic Program Unit, Low Speed
ApUpG	Adenylyluridylylguanosine
ApUpU	Adenylyluridylyluridine
APV	All-Purpose Vehicle
APVD	Approved
APW	Apparent Polar Wander
	Augmented Plane Wave
	Average Piece Weight
APWA	American Public Works Association [US]
APWC	Association of Professional Writing Consultants [US]
APWI	Airborne Proximity Warning Indicator
APWP	Apparent Polar Wander Path
APWR	Advanced Pressurized Water Reactor
APWSS	Asian Pacific Weed Science Society [Philippines]
APX	Appendix
AQ	Achievement Quotient
	Aminoquinoline
	Anthraquinone
	Any Quality
	Any Quantity
	(aqua) — water
	Aqueous
	Autoquote
AQAM	Air Quality Assessment Model
AQAP	Allied Quality Assurance Publications [Canada]
aq dest	(aqua destillata) — distilled water
AQE	Airman Qualification Examination
Aqil	Aquila [Astronomy]
AQL	Acceptable Quality Level
Aql	Aquila [Astronomy]
AQMG	Assistant-Quartermaster-General
AQO	Aminoquinoline Oxide
AQP	Association for Quality and Participation [US]
Aqr	Aquarius [Astronomy]
AQ REG	Aqua Regia
AQT	Acceptable Quality Test
AR	Accounts Receivable
	Accumulator Register
	Acid Resistance
	Acid-Resisting (Material)
	Acknowledgement of Receipt
	Acoustic Reflex
	Acquisition Radar
	Acrylic Resin
	Address Register
	Aerial Refueling
	Air Resistance
	Alpha Ray
	Analog Radio System
	Analytical Reagent
	Anchor Ring
	Anti-Reflection
	Anti-Resonance

Annual Return
Annual Review
Annual Reviews [US]
Applied Research
Aqua Regia
Arabia(n)
Arabic
Arc Recifier
Arithmetic Register
Arkansas [US]
Army Regulation
Arrival
As Received
As Required
Assembly and Repair
Associative Register
Attention Routine
Atomic Reactor
Atmospheric Research
Audio Range
Autoradiography
Autoregression
Aviation Radionavigation [of FCC, US]
Avionic Requirements
Awaiting Reply

Ar Argon
Aryl

ARA Aerial Rocket Artillery
Aerial Ropeway Association [UK]
Agricultural Research Administration [US]
Aircraft Replaceable Assemblies
Aircraft Research Association [UK]
Aluminum Recycling Association [US]
Amateur Rocket Association [US]
American Rental Association [US]
American Retreaders Association [US]
American Royal Association [US]
Angular Rate Assembly
Arab Roads Association [Egypt]
Asian Recycling Association [Philippines]
Associate of the Royal Academy [UK]
Associates for Radio Astronomy
Australian Robot Association
Automotive Retailers Association [Canada]

Ara-A Arabinofuranosyl Adenine

ARABC Automotive Retailers Association of British Columbia [Canada]

ARABSAT Arab Countries Regional Communications Satellite

ARAC Aerospace Research Applications Center [US]

Ara-C Arabinofuranosyl Cytosine

ARAD Airborne Radar and Doppler

ARADCOM Army Air Defense Command

ARAF Air Reserve Augmentation Flights

ARAI Automotive Research Association of India

ARAL Automatic Record Analysis Language

ARALL Aramid-Aluminum Laminate

ARAM Association of Railroad Advertising and Marketing [US]
Association of Records Administrators and Managers [US]

ARAMCO Arabian-American Oil Company

ARAN Association for the Reduction of Aircraft Noise [UK]

ARAP Aeronautical Research Associates of Princeton [US]

ARB Academie Royale de Belgique [= Royal Academy of Belgium]
Air Registration Board [UK]
Air Research Bureau
Airworthiness Requirements Board [UK]
All Routes Busy
American Research Bureau [US]
Arbitration

ARBA American Road Builders Association [now: ARTBA, US]

ARBBA American Railway Bridge and Building Association [US]

ARBOR Argonne Boiling Water Reactor [of ANL, US]

ARBS Angle Rate Bombing System

ARB U Arbitrary Units

ARC Academic Research Cooperation (Program) [Germany]
Academy of Roofing Contractors [US]
Advanced Reentry Concept
Aeronautical Research Council [UK]
Agricultural Relations Council [US]
Agricultural Research Council [UK]
Africa Resource Center
Aiken Relay Calculator
Aircraft Radio Corporation
Airworthiness Requirements Committee
Alberta Research Council [Canada]
Alternate Route Cancel
Altitude Rate Command
Ames Research Center [of NASA]
Amplitude and Rise-Time Compensation
Antireflection Coating
Architects Registration Council [UK]
Argonne Reactor Computation [of ANL, US]
Asian Research Center
Asian Resource Center
Association of Railway Communicators [US]
Atlantic Reference Center [Canada]
Atlantic Research Corporation [US]
Attached Resource Computer
Attended Resource Computer
Augmentation Research Center [US]
Automatic Relay Calculator
Automatic Remote Control
Auxiliary Roll Control
Average Response Computer

ARCA Air-Conditioning and Refrigeration Contractors of America
Asbestos Removal Contractors Association [UK]

ARCAS Automatic Radar Chain Acquisition System

ARCC Airworthiness Requirements Coordinating Committee
Animal Research and Conservation Center [now: WCI, US]

ARCCA Agricultural Research Council of Central Africa

arccos arccosine

ARCCLMA Aromatic Red Cedar Closet Lining Manufacturers Association [US]

ARCCOH Asian Regional Coordinating Committee on Hydrology
arccot arccotangent
arccsc arccosecant
ARCE Amphibious River Crossing Equipment
ARCEDEM African Regional Center for Engineering Design and Manufacturing [Nigeria]
ARCFCP Alliance for Responsible CFC (= Chlorofluorocarbon) Policy [US]
ARCG American Research Committee on Grounding [US]
ARCH Archiac
Archipelago
Architectural
Architecture
Articulated Computing Hierarchy [UK]
Arch Architect
Arch D Doctor of Architecture
ARCHE Architectural Engineering
Arch E Architectural Engineer
ARCHEDDA Architectures for Heterogeneous European Distributed Databases
ARCHEOL Archeological [also: ARCHEO]
Archeologist [also: ARCHEO]
Archeology [also: ARCHEO]
ARCHIT Architecture
ARCI American Railway Car Institute [US]
ARCIS African Regional Center for Information Science
ARCN Agricultural Research Council of Norway
ARCO Atlantic Richfield Corporation
ARCOMSAT Arab League Communications Satellite
ARCP Accredited Roofing Contractors Program [now: ARC, US]
Aerodrome Reference Code Panel [of ICAO]
ARCRL Agricultural Research Council Radiobiological Laboratory [UK]
ARCS Advanced Reconfigurable Computer System
Airline Request Communication System
Air Resupply and Communication Service
Associate of the Royal College of Science
Automated Ring Code Search
Autonomous Remotely Controlled Submersible
arcsec arcsecant
arcsin arcsine
ARCT African Regional Center for Technology [Senegal]
arctan arctangent
ARCUK Architects Registration Council of the United Kingdom
ARC/W Arc Weld
ARD Airborne Respirable Dust
Answering Recording and Dialing
Arbeitsgemeinschaft der offentlich-rechtlichen Rundfunkanstalten in Deutschland [= Working Committee of Public Broadcasting Stations in Germany]
Association of Research Directors [US]
Automated Retrieval of Data
ARDA Agriculture and Rural Development Act [Canada]
American Railway Development Association [US]
Analog Recording Dynamic Analyzer
Atomic Research and Development Authority [US]
ARDC Agricultural and Rural Development Corporation [Burma]
Air Research and Development Command [US Air Force]
Armament Research and Development Center
ARDE Armament Research and Development Establishment [now: RARDE, UK]
ARDEC Army Research, Development and Engineering Center [US]
ARDI Aviation Research Development Institute
ARDIC Association pour la Recherche et le Developpement en Informatique Chimique [= Association for Research and Development in Chemical Informatics, France]
ARDIS Army Research and Development Information System [US Army]
ARDS Advanced Remote Display Station
Artillery Regimental Data System [Canada]
ARDSA Agricultural and Rural Development Subsidiary Agreement [Canada]
ARE Activated Reactive Evaporation
Admiralty Research Establishment
Airline Revenue Enhancement
Asymptotic Relative Efficiency
Automated Responsive Environment
AREA American Railway Engineering Association [US]
American Recreational Equipment Association [US]
Army Reactor Area [US]
Association for Rational Environmental Alternatives [US]
AREC Agricultural Research and Educational Center [Lebanon]
ARED Aperture Relay Experiment Definition
AREG Apparatus Repair-Strategy Evaluation Guidelines
ARELEM Arithmetic Element Program
AREP Automated Reliability Estimation Program
ARERE Admiralty Research Establishment Rotating Arm
ARES Advanced Railroad Electronics System
ARETS Alberta Renewable Energy Test Site [Canada]
ARF Advertising Research Foundation [US]
Aquatic Research Facility
Armor Research Foundation [US]
Automatic Report Feature
ARFA Allied Radio Frequency Agency [of NATO]
ARFAA Atlantic Region Freight Assistance Act [Canada]
ARFC Automatic Radar Frequency Control
ARFF Airport Rescue Fire Fighting
ARFOR Area Forecast
ARG Air Reserve Group
American Resources Group [US]
Argentina
Argentine
Argentinian
Argument
Arg Arginine
Arginyl
Argo [Astronomy]
ARGC Australian Research Grants Committee
ARGE Arbeitsgemeinschaft der Verbande der Europaischen Schloss- und Beschlagindustrie [=

European Federation of Associations of Lock and Builders Hardware Manufacturers, Germany]

ARGONAUT Argonne Nuclear Assembly for University Training [also: Argonaut [of ANL, US]]

ARGUS Automatic Routine Generating and Updating System

ARI Agricultural Research Institute [US]
Airborne Radio Instrument
Air-Conditioning and Refrigeration Institute [US]
Aluminium Research Institute [US]
American Refractories Institute [US]
Aquatic Research Institute [US]
Australian Road Index (Database) [of ARRB]

Ari Aries [Astronomy]

ARIA Advanced Range Instrumentation Aircraft
Advanced Range Instrumented Aircraft
Apollo Range Instrumented Aircraft

ARIBA Associate of the Royal Institute of British Architects

ARIC Associate of the Royal Institute of Chemistry

ARICS Associate of the Royal Institute of Chartered Surveyors

ARIES Advanced Radar Information Evaluation System

ARINC Aeronautical Radio Incorporated [US]

ARIP Automatic Rocket Impact Predictor

ARIS Advanced Range Instrumentation Ship [of NASA]
Apollo Range Instrumentation Ship [of NASA]
Atomic Resolution Imaging System

ARIST Annual Review of Information on Science and Technology

ARISTOTLE Annual Review of Information and Symposium on the Technology of Training, Learning and Education [US]

ARIS-UHV Atomic Resolution Imaging System — Ultrahigh Vacuum

ARITH Arithmetic(s)

Ariz Arizona [US]

Ark Arkansas [US]

ARL Acceptable Reliability Level
Admiralty Research Laboratory [UK]
Aeronautical Research Laboratories [Australia]
Aerospace Research Laboratory [UK]
Applied Research Laboratories [US]
Arctic Research Laboratory [US]
Army Radiation Laboratory [US Army]
Association of Research Libraries [US]
Atlantic Research Laboratory [of NRC, Canada]
Automation Research Laboratory [Kyoto University, Japan]
Average Run Length

ARLIS Arctic Research Laboratory Ice Station [US Navy]

ARM Accumulator Read-in Module
Anhysteretic Remanent Magnetization
Anti-Radar Missile
Anti-Radiation Missile
Area Radiation Monitor
Armature
Armor
Articulated Remote Manual
Associate in Risk Management
Association of Railway Museums [US]
Association of Rotational Molders [US]
Asynchronous Response Mode
Atmosphere Radiation Monitor
Atomic Resolution Microscopy
Automated Route Management
Automatic Reel Mounting
Automatic Robotic Module
Availability, Reliability and Maintainability

ARMA American Records Managers and Administrators [US]
Asphalt Roofing Manufacturers Association [US]
Association of Records Managers and Administrators [US]
Autoregressive Moving Average

ARMAN Artificial Methods Analyst

ARMAX Autoregressive Moving Average Exogenous

ARMC Automotive Research and Management Consultants

ARMD Armored

ARMF Advanced Reactivity Measurement Facility [of INEL, US]

ARMM Association of Reproduction Materials Manufacturers [US]

ARMMS Automated Reliability and Maintainability Measurement System

ARM-PL Armor Plate

ARMS Advanced Receiver Model System
Aerial Radiological Measurements and Surveys
Aerial Radiological Measuring Survey
Amateur Radio Mobile Society [US]
Automated Record Management System

ARMT Armament

aRNA Activator Ribonucleic Acid [also: a-RNA or ARNA]

ARNAB African Research Network on Agricultural Byproducts

ARNIS Advanced Rocket Nozzle Inspection System

ARNM African Regional Network on Microbiology

ARNOT Aera Notice

ARO After Receipt of Order
Agricultural Research Organization
Air Radio Officer
All Rods Out
Army Research Office [of USDOD]
Asian Regional Office [of ICFTU]
ATS (= Air Traffice Service) Reporting Office [of FAA, US]
Automatic Recovery Option

AROD Airborne Range and Orbit Determination

ARODS Airborne Radar Orbital Determination System

AROM Alterable Read-Only Memory
Aromatic(s)
Associative Read-Only Memory

AROS Alterable Read-Only Operating System

AROU Aviation Repair and Overhaul Unit

ARP Advanced Radiator Program [of NASA]
Advanced Reentry Program
Advanced Research Project
Aeronautical Recommended Practice
Aerospace Recommended Practice
Airborne Radar Platform
Airport Reference Point
Air Raid Precautions
Aramid-Reinforced Plastics

	Argentinean Peso [Currency]
	Aviation Regulatory Proposals
	Azimuth Reset Pulse
ARPA	Advanced Research Projects Agency [of USDOD]
	Automated Radar Plotting Aid
ARPANET	Advanced Research Project Agency Network [of USDOD]
ARPE	American Registry of Professional Entomologists [US]
ARPES	Angle-Resolved Photoelectron Spectroscopy
	Angle-Resolved Photoemission Spectroscopy
	Angle-Resolved Photoemission Spectrum
ARPG	Asphalt Rubber Producers Group [US]
ARPHOS	1-Diphenylphosphino-2-Diphenylarsinoethane
ARPI	Analog Rod Position Indicator
	Automotive Refrigeration Products Institute [US]
ARPS	Aerospace Research Pilot School [US Air Force]
	Associate of the Royal Photographic Society
	Australian Radiation Protection Society
ARQ	Automatic Repeat Request
	Automatic Request for Correction
	Automatic Request for Repetition
	Automatic Response Query
ARR	Address Record Register
	Adjustable Reset Response
	Anti-Repeat Relay
	Argonne Research Reactor [of ANL, US]
	Arrangement
	Arrest(or)
	Arrival
	Association for Radiation Research [UK]
	Audio Response System
	Automatic Rerouting
	Automatic Retransmission Request
ARRA	Asphalt Recycling and Reclaiming Association [US]
ARRB	Australian Road Research Board
ARRE	Alarm Receiving and Reporting System
ARRI	Automation and Robotics Research Institute [US]
ARRL	Aeronautical Radio and Radar Laboratory
	American Radio Relay League [US]
ARRS	Aerospace Rescue and Recovery Service [US Air Force]
ARS	Active Repeater Satellite
	Advanced Reconnaissance Satellite
	Advanced Record System
	Agricultural Research Service [US]
	Air Rescue Service
	American Radium Society [US]
	American Rocket Society [Now Part of AIAZA, US]
	Amplified Response Spectrum
	Asbestos Roof Shingles
	Asian Research Service
	Automatic Recovery System
ARSA	Aeronautical Repair Station Association [US]
	Associate of the Royal Scottish Academy
	Associate of the Royal Society of Arts
ARSB	Automated Repair Service Bureau
ARSBB	Association of Railway Superintendents of Bridges and Buildings [now: ARBBA, US]
ARSC	African Remote Sensing Council
	Automatic Resolution Selection Control
ARSDP	Asian Regional Skill Development Program [now: APSDEP, Pakistan]
ARSM	Associate of the Royal School of Mines
ARSME	All-Round Shape Memory Effect
ARSO	African Regional Standardization Organization [Kenya]
ARSP	Aerospace Research Satellite Program [US Air Force]
	Aerospace Research Support Program [US Air Force]
ARSR	Air-Route Surveillance Radar
ART	Admissible Rank Test
	Advanced Reactor Technology
	Advanced Research and Technology
	Airborne Radiation Thermometer
	Alarm Reporting Telephone
	Algebraic Reconstruction Technique
	Article
	Artificial
	Artificial Resynthesis Technology
	Artillery
	Automated Request Transmission
	Automated Rotor Test System
	Automatic Range Tracker
	Automatic Reporting Telephone
	Autonomous Robotics Technology
	Average Response Time
	Average Retrieval Time
ARTA	American Reuseable Textile Association [US]
ARTBA	American Road and Transportation Builders Association [US]
ARTC	Aircraft Research and Testing Committee
	Air-Route Traffic Control
ARTCC	Air-Route Traffic Control Center [US]
ARTDO	Asian Regional Trade Development Organization
ARTE	Admiralty Reactor Test Establishment [UK]
ARTEMIS	Automatic Retrieval of Test through European Multipurpose Information Services
ARTFL	Artificial
ARTG	Azimuth Range and Timing Group
ARTI	Arab Regional Telecommunications Institute
ARTIC	A Real-Time Interface Coprocessor
	Associometrics Remote Terminal Inquiry Control System
ARTILL	Artillery
ARTOC	Army Tactical Operations Center [US]
ARTRAC	Advanced Range Testing, Reporting and Control
	Advanced Real-Time Range Control
ARTRON	Artificial Neutron
ARTS	Advanced Radar Terminal System
	Advanced Radar Traffic-Control System
	Advanced Radio Telephone Service
	American Radio Telephone System
	Automated Radar Terminal System
ARTU	Automatic Range Tracking Unit
ARTY	Artillery
ARU	Acoustical Resistance Unit
	Address Recognition Unit
	American Railway Union
	Arithmetic Unit
	Audio Response Unit

ARUBIS Angle-Resolved Ultraviolet Bremsstrahlung Isochromate Spectroscopy

ARUPS Angle-Resolved Ultraviolet Photoelectron Spectroscopy
Angle-Resolved Ultraviolet Photoelectron Spectrum

ARV Aeroballistic Reentry Vehicle
AIDS-Associated Retrovirus [Immunochemistry]
Armored Recovery Vehicle

ARVS Autonomous Rescue Vehicle System
Autonomous Research Vehicle System

ARW Air-Conditioning and Refrigeration Wholesalers [US]
Air Reserve Wing

AS Absorption Spectroscopy
Absorption Spectrum
Account Sales
Acetanisidine
Adhesion Society [US]
Advanced Schottky
Advanced System
Aeronautical Standards
Aerospace
Aerospace Standards
After Sight
Air Screw
Alloy Steel
Alongside
Aluminum Oxide — Special
Aminosalicylic Acid
Ammeter Switch
Analytical Separation
Anthranilate Synthetase
Antistatic
Antisubmarine
Aperture Stop
Application-Specific
Applied Science
Area Surveillance
Articulation Score
Ascent Stage
As Soldered
Asymmetric
Atomic Spectroscopy
Audio Signal
Australian Sceptics
Australian Standard
Austrian Schilling [Currency]
Automatic Sampling
Automatic Sprinkler
Automatic Switching
Automatic System
Automation Society [US]
Auxiliary Storage

A/S Account of Sales
Ascent Stage

As Arsenic
Alto-Stratus (Cloud)

aS absiemens [Unit]

5-AS 5-Aminosalicylic Acid

ASA Acetylsalicylic Acid
Acoustical Society of America [US]
Acrylate-Styrene-Acrylonitrile
Acrylic-Styrene-Acrylonitrile (Terpolymer)
Acrylonitrile-Styrene-Acrylate (Terpolymer)
Acrylonitrile-Styrene-Acrylonitrile
Air Security Agency
Air Services Agreement
American Snowmobile Association [US]
American Society of Agronomy [US]
American Society of Appraisers [US]
American Soybean Association [US]
American Standards Association [now: ANSI, US]
American Statistical Association [US]
American Student Association [US]
American Subcontractors Association [US]
American Supply Association [US]
Aminosalicylic Acid
Antistatic Additive
Army Security Agency [US]
Army Signal Association [US]
Asian Students Association [Hong Kong]
Association des Statisticiens Agricoles [= Association of Agricultural Statisticians, France]
Association of Southeast Asia
Astronomical Society of Australia
Atomic Scientists Association [UK]
Atomic Security Agency [US]
Atomic Sphere Approximation
Audiological Society of Australia
Automotive Service Association [US]
Average Sphere Approximation
Avicultural Society of America [US]

ASAAWE Association of South Asian Archeologists in Western Europe [UK]

ASAB Association for the Study of Animal Behaviour [UK]

ASAC Accounting Standards Authority of Canada
Asian Standards Advisory Committee

ASADA Atomic and Space Development Authority [US]

ASAE Advanced School of Automobile Engineering
American Society of Agricultural Engineers [US]
American Society of Association Executives [US]
Association Suisse pour l'Amenagement des Eaux [= Swiss Association for Water Management]

ASAIHL Association of Southeast Asian Institutions of Higher Learning [Thailand]

ASALM Advanced Strategic Air-Launched Missile

ASAM American Society for Abrasive Methods [now: AES, US]

ASAP Accelerated Surface Area and Porosimetry
Aerospace Supplier Accreditation Program
American Society of Access Professionals [US]
American Society of Aerospace Pilots [US]
American Society of Animal Production [US]
Analog System Assembly Pack
Antisubmarine Attack Plotter
Applied Systems and Personnel
Army Scientific Advisory Panel [US]
As Soon As Possible
Australian Society of Animal Production
Automated Shipboard Aerological Program [Canada]
Automated Statistical Analysis Program

ASAQS Association of South African Quantity Surveyors
ASARC Army Systems Acquisition Review Council [US]
ASAS American Society of Agricultural Sciences [US]
American Society of Animal Science [US]
ASASC Aviation Safety Authorities Steering Committee
ASAT Antisatellite
ASB Agricultural Stabilization Board [Canada]
Air Safety Board
Aluminum sec-Butoxide
Asbestos
Association of Shell Boilermakers
Association of Southeastern Biologists [US]
asb apostilb [Unit]
ASBA American Small Business Association [US]
ASBC American Safety Belt Council [US]
American Seat Belt Council [US]
American Society of Biological Chemists [now: ASBMB, US]
American Society of Biophysics and Cosmology [US]
American Society of Brewing Chemists [US]
American Standard Building Code [US]
ASBD Advanced Sea-Based Deterrent
ASBE American Society of Bakery Engineers [US]
American Society of Body Engineers [US]
ASBHCA Average Season Busy Hour Call Attempt
ASBHCC Average Season Busy Hour Call Completion
ASBI Advisory Service for the Building Industry [UK]
ASBMB American Society for Biochemistry and Molecular Biology [US]
ASBO Association of School Business Officials [US]
ASBPA American Shore and Beach Preservation Association [US]
ASBPE American Society of Business Press Editors [US]
ASBU Arab States Broadcasting Union
ASC Acetylsulfanilyl Chloride
Adhesives and Sealants Council [US]
Advanced Scientific Computer
Advertising Standards Council
Aeronautical Systems Center
Agricultural Stabilization and Cultivation
Airport Security Council [US]
Air-Powered Swing Clamp
Alberta Safety Council [Canada]
Alberta Securities Commission
American Sandblasting Contractors [US]
American Satellite Corporation [US]
American Society of Cartographers [US]
American Society for Cybernetics [US]
American Specialty Contractors [US]
American Standard Code
American Systematics Collections [US]
Applied Superconductivity Conference
Applied Surface Chemistry
Army Service Corps [US]
Assistent Sector Controller
Associated Schools of Construction [US]
Associate in Science [also: ASc]
Association of Systematic Collections [US]
Associative Structure Computer
Atlantic Systems Conference
Automatic Selectivity Control
Automatic Sensitivity Control
Automatic Sequence Control
Automatic Slip Control
Automatic Switching Center
Automatic System Controller
Automotive Sales Council [US]
Auxiliary Switch, (Normally) Closed
ASc Associate in Science [also: ASC]
ASCA American Society for Conservation Archeology [US]
American Society of Consulting Arborists [US]
Architectural Spray Coaters Association [US]
Association of Science Cooperation in Asia
Automated Spin Chemistry Analyzer
Automatic Science Citation Alerting
Automatic Subject Citation Alert
ASCAC Anti-Submarine Classification Analysis Center [US Navy]
ASCAP Aeronautical Satellite Communications Processor
American Society of Composers, Authors and Publishers [US]
ASCATS Apollo Simulation Checkout and Training System [of NASA]
ASCB American Society for Cell Biology [US]
Avionics Standard Communications Bus
ASCC Aeronautical Satellite Communications Center [US]
Air Standardization Coordinating Committee [US]
American Society for Concrete Construction [US]
Army Strategic Communications Command [US Army]
Automatic Sequence Controlled Calculator
ASCD American Society of Computer Dealers [US]
Association for Supervision and Curriculum Development [US]
ASCE American Society of Civil Engineers [US]
ASCEA American Society of Civil Engineers and Architects [US]
ASCEND Advanced System for Communications and Education in National Development
ASCENT Assembly System for Central Processor
ASCEP Australasian Society for Clinical Experimental Pharmacologists
ASCET American Society of Certified Engineering Technicians [US]
ASCF Analog Switched Capacitor Filter
ASCG Automatic Solution Crystal Growth
ASCGP Administrative Special Canadian Grains Program [Canada]
ASCII American Standard Code for Information Interchange
ASCIO Asian Oceanic Computing Industry Organization
ASCIS Australian Schools Cataloguing Information System
ASCLD American Society of Crime Laboratory Directors [US]
ASCMA American Sprocket Chain Manufacturers Association [now: ACA, US]
ASCO American Society of Composites [US]
Arab Satellite Communications Organization [Saudi Arabia]
Automatic Sustainer Cutoff

ASCOMACE Association des Constructeurs de Machines a Coudre [= Association of Sewing Machine Manufacturers [of EEC, Belgium]]
ASCOPE ASEAN Council on Petroleum [Indonesia]
ASCP American Society of Clinical Pathologists [US]
American Society of Consulting Planners [US]
Automatic System Checkout Program
ASCR Asymmetric Silicon Controlled Rectifier
ASCRT Association for the Study of Canadian Radio and Television
ASCS Agricultural Stabilization and Conservation Service [of USDA]
Aluminum Straight Clamp Strap
Area Surveillance Control System
Automatic Stabilization and Control System
ASCSA American School and Community Safety Association [US]
ASCT Address Space Control Task
ASCTA Association of Short-Circuit Testing Authorities
ASCU Association of Small Computer Users [US]
ASCUE Association of Small Computer Users in Education [US]
ASCWSA Association of Scientific Workers of South Africa
ASD Access Storage Device
Adjustable-Speed Drive
Aeronautical Systems Division [US Air Force]
Aerospace Systems Division [US Air Force]
American Society of Dowsers [US]
Associated Surplus Dealers [US]
Association of Steel Distributors [US]
Association Suisse de Documentation [= Swiss Documentation Association]
Atmospheric Sciences Department
Automatic Synchronized Discriminator
ASDA Accelerate — Stop Distance Available
ASDAR Aircraft to Satellite Data Relay
ASDC Automatic Sample Data Collection
ASDE Airport Surface Detection Equipment
American Society of Danish Engineers [US and Canada]
ASDG Aircraft Storage and Disposition Group [US Air Force]
ASDI Association for Science Documentation Information [of TIT, Japan]
Automatic Selective Dissemination of Information
ASDIC Antisubmarine Detection Investigation Committee [also: Asdic]
Armed Services Document Intelligence Center
ASDIRS Army Study Documentation and Information Retrieval System [US]
ASDL Automated Ship Data Library
ASDR Airport Surface Detection Radar
ASDSO Association of State Dam Safety Officials [US]
ASDSRS Automatic Spectrum Display and Signal Recognition System
ASDSVN Army Switched Data and Secure Voice Network
ASE Admiralty Signal Establishment [UK]
Agence Spatiale Europeenne [= European Space Agency, France]
Airborne Search Equipment
Albany Society of Engineers [US]
Alberta Stock Exchange [Canada]
Alliance to Save Energy [US]
Allowable Steering Error
Alternative Sources of Energy [US]
Altimetry System Error
Amalgamated Society of Engineers [UK]
Amplified Spontaneous Emission
Association of Science Education [UK]
Association of Scientists and Engineers [of NSSC, US]
Association of Space Explorers [US]
Association Suisse des Electriciens [= Swiss Association of Electricians]
Astronomical Society of Egypt
Automatic Spectroscopic Ellipsometry
Automatic Support Equipment
(National Institute for) Automotive Service Excellence [US]
ASEA American Solar Energy Association [US]
ASEAN Association of Southeast Asian Nations
ASEANCUPS Association of Southeast Asian Nation Countries Union in Polymer Science
ASEANIP Association of Southeast Asian Nations Institute of Physics
ASEB Aeronautics and Space Engineering Board [US]
American Society for Experimental Biology [US]
ASEC Albert Schweitzer Ecological Center [Switzerland]
American Standard Elevator Code [US]
Auburn Science and Engineering Center [US]
ASECH Acetylselenocholine [also: ASECh]
ASEE Advanced Semiconductor Equipment Exposition
American Society for Engineering Education [US]
American Society for Environmental Education [US]
Association of Supervising Electrical Engineers [UK]
ASEH American Society for Environmental History [US]
ASEL Aircraft Single Engine Land
ASEM Active Source Electromagnetic Correction
American Society for Engineering Management [US]
Analytical Scanning Electron Microscopy
ASEMC Autobody Supply and Equipment Manufacturers Council [US]
ASEMRC Allied Signal Engineered Materials Research Center [US]
ASEP American Society of Electroplated Plastics [US]
ASES American Solar Energy Society [US]
Automated Software Evaluation System
ASESA Armed Services Electronic Standards Agency [US]
ASESS Aerospace Environment Simulation System
ASET Aeronautical Services Earth Terminal
American Society for Engineering Technology [US]
American Society of EEG (= Electro-encephalographic) Technology [US]
ASE-USA Association of Space Explorers — United States of America
ASEV American Society for Enology and Viticulture [US]
ASF Air Safety Foundation [US]
American Scandinavian Foundation [US]
Ampere per Square Feet
Arab Sugar Federation [Sudan]
Army Service Forces [US]

Artificial Superlattice Film
Atlantic Salmon Federation [Canada]
Australian Science Foundation
Austrian Science Foundation
Automatic Sheet Feeder
Austrian Association of Sciences
Auxiliary Supporting Feature

ASFA Aquatic Sciences and Fisheries Abstracts [of UN]
ASFD American Society of Furniture Designers [US]
ASFDO Antisubmarine Fixed Defense Office [US]
ASFE Association of Soil and Foundation Engineers [US]
ASFE/AEFPG Association of Soil and Foundation Engineers/ Association of Engineering Firms Practicing in the Geosciences [US]
ASFG Atmospheric Sound-Focusing Gain
ASFIR Active-Swept Frequency Interferometer Radar
ASFIS Aquatic Sciences and Fisheries Information System [of UN]
ASFMRA American Society of Farm Managers and Rural Appraisers [US]
ASFPM Association of State Floodplain Managers [US]
ASFSE American Swiss Foundation for Scientific Exchange
ASFTS Airborne Systems Functional Test Stand
ASG Advanced Study Group
Aeronautical Standards Group [US]
African Seabird Group [South Africa]
Australasian Seabird Group [New Zealand]
ASGB Aeronautical Society of Great Britain
ASGCA American Society of Golf Course Architects [US]
ASGE American Society of Gas Engineers [US]
ASGLS Advanced Space-Ground Link Subsystem
ASGMT Assignment
ASGRP Atlantic Salmon Genetics Research Program [of ASF, Canada]
ASGS American Scientific Glassblowers Society [US]
ASH Assault Support Helicopter
ASHAE American Society of Heating and Air Conditioning Engineers [now: ASHRAE]
ASHC Aeronautics and Space Historical Center [France]
ASHE American Society for Hospital Engineering [US]
ASHG American Society of Human Genetics [US]
ASHI American Society of Home Inspectors [US]
ASHP American Society for Hospital Pharmacists [US]
ASHRAE American Society of Heating, Refrigeration and Air Conditioning Engineers [Formerly Subdivided into: ASHAE and ASHVE]
ASHS American Society for Horticultural Science [US]
ASHVE American Society of Heating and Ventilating Engineers [formerly also: ASH&VE; now: ASHRAE, US]
ASI Advanced Scientific Instrument
Advanced Studies Institute [of NATO]
Advanced Systems Institute [Canada]
Aeronautical Society of India
African Scientific Institute
Air Speed Indicator
Altimeter Setting Indicator
American Society of Inventors [US]
American Society of Interpreters [US]
American Standards Institute [US]
American Statistical Index (Database) [of CIS, US]
Architects and Surveyors Institute [US]
Articulated Subject Index
Astronomical Society of India
Asynchronous Series Interface
Austrian Standards Institute
Axial Shape Index

A-Si Amorphous Silicon
ASIA Automotive Service Industry Association [US]
ASIAG Association Interprofessionnelle de l'Aviation Agricole [= Interprofessional Association of Agricultural Aviation, France]
ASIB Accademia delle Scienze dell' Istituto di Bologna [= Academy of Science of the Institute of Bologna, Italy]
ASIC American Society of Irrigation Consultants [US]
Application-Specific Integrated Circuit
Asian Students Information Center
ASID Address Space Identifier
American Society for Industrial Designers [US]
American Society for Interior Designers [US]
Analytical Scanning Imaging Device
ASIDIC Association of Scientific Information and Dissemination Centers [US]
ASIDP American Society of Information and Data Processing [US]
ASIH American Society of Ichthyologists and Herpetologists [US]
ASII American Science Information Institute [US]
ASIN Agricultural Service Information Network [of NAL, US]
ASIP Aircraft Structural Integrity Program [US Air Force]
ASIRC Aquatic Sciences Information Retrieval Center [US]
ASIS Abort Sensing and Instrumentation System
American Society for Industrial Security [US]
American Society for Information Science [US]
AS/ISES American Section of the International Solar Energy Society [US]
ASIST Advanced Scientific Instruments Symbolic Translator
ASIT Adaptive Surface Interface Terminal
ASIWPCA Association of State and Interstate Water Pollution Control Administrators [US]
ASJ All-Service Jacket
Astronomical Society of Japan
ASK Access to Sources of Knowledge
Agricultural Society of Kenya
Amplitude Shift Keying
Anomalous State of Knowledge
ASKA Automatic System for Kinematic Analysis
ASKI Arbeitsgemeinschaft der Schweizerischen Kunststoffindustrie [= Working Group of the Swiss Plastics Industry]
ASKS Automatic Station Keeping System
ASKT American Society of Knitting Technologists [US]
ASL Above Sea Level
Aeronautical Structures Laboratory [US Navy]
American Sign Language
Association for Symbolic Logic [US]
Atmospheric Sciences Laboratory [US]
Available Space List

Average Signal Level
ASLA American Society of Landscape Architects [US]
ASLAP Atomic Safety and Licensing Appeal Panel [US]
ASLB Atomic Safety and Licensing Board [US]
ASLBM Air-to-Ship Launched Ballistic Missile
ASLBP Atomic Safety and Licensing Board Panel [US]
ASLE American Society of Lubrication Engineers [now: STLE, US]
ASLEEP Automated Scanning Low-Energy Electron Probe
ASLEF Amalgamated Society of Locomotive Engineers and Firemen [also: ASLE&F, UK]
ASLIB Association of Special Libraries [US]
Association of Special Libraries and Information Bureau [US]
ASLIC Association of Special Libraries and Information Centers [India]
ASLO American Society of Limnology and Oceanography [US]
Australian Scientific Liaison Office
ASLP Association of Special Libraries in the Philippines
ASLRA American Short Line Railroad Association [US]
ASLT Advanced Solid Logic Technology
ASM Advanced Surface-to-Air Missile
Airspace Management
Air-to-Surface Missile
All-Sky Monitor
American Society of Mammalogists [US]
American Society for Metals [now: ASM International, US]
American Society for Microbiology [US]
Apollo Service Module [of NASA]
Apollo Systems Manual [of NASA]
Application-Specific Module
Assembler
Association for Systems Management [US]
Asynchronous State Machine
ASMA Aerospace Medical Association [US]
ASMB Acoustical Standards Management Board [of ANSI]
ASMC Aviation Surface Material Command [US Army]
Automatic Systems Management and Control
ASMD Association of Science Museum Directors [US]
ASME American Society of Mechanical Engineers [US]
ASMEA American Society of Mechanical Engineers, Auxiliary [US]
ASME-ETCE American Society of Mechanical Engineers — Energy-Sources Technology Conference and Exhibition [US]
ASMEIGTI American Society of Mechanical Engineers International Gas Turbine Institute [US]
ASMER Association for the Study of Man-Environment Relations [US]
ASM/ESD American Society for Metals/Engineering Society of Detroit [US]
ASMFC Atlantic States Marine Fisheries Commission [US]
ASMFE Association of Soil Mechanics and Foundation Engineering
ASMFER American Society for Metals — Foundation for Education and Research [US]
ASMI Airfield Service Movement Indication
Airfield Surface Movement Indicator
ASMM American Society of Machine Manufacturers [US]
ASMMA American Supply and Machinery Manufacturers Association [US]
ASM-MSD American Society for Metals — Materials Science Division [US]
ASMO Arab Organization for Standardization in Metrology [Jordan]
ASMR Advanced Short-Takeoff-and-Landing, Medium Range
ASMS Advanced Surface Missile System
American Society for Mass Spectroscopy [US]
Automatic Search Music System
ASMT American Society for Medical Technology [US]
Assortment
ASN Abstract Syntax Notation [of ISO]
Advanced Silicon Nitride
American Society of Naturalists [US]
Atlantic Satellite Network
Average Sample Number
Asn Asparagine
Asparaginyl
ASNA American SNR (= Special Mobile Radio) Network Association [US]
ASNDT American Society for Nondestructive Testing [now: ASNT, US]
ASNE American Society of Naval Engineers [US]
ASN.1 Abstract Syntax Notation, One [of ISO]
ASNT American Society for Nondestructive Testing [US]
ASO Applied Science Office
Auxiliary Switch, (Normally) Open
ASOC Antarctica and Southern Oceans Coalition [US]
ASODDS ASWEPS (= Anti-Submarine Warfare Environmental Prediction System) Submarine Oceanographic Digital Data System [of US Navy]
ASOP Automatic Scheduling and Operating Program
Automatic Structural Optimization Program
ASOS Airport Surface Observing System
ASOVAC Asociacion Venezolana para el Avance de la Ciencia [= Venezuelan Association for the Advancement of Science]
ASOVII Asociacion Venezolana de Ingenieros Industriales [= Venezuelan Association for Industrial Engineers]
ASP Airborne SAR (= Synthetic Aperture Radar) Processor
Airborne Signal Processor
Alcoa Smelting Process
Alloy Steel Plant
Air-Speeded Post
American Selling Price
American Society for Photobiology [US]
American Society for Photogrammetry [now: ASPRS, US]
American Society of Parasitologists [US]
American Society of Perfumers [US]
American Society of Pharmacology [US]
American Society of Photogrammetry [US]
American Society of Photographers [US]
Ammunition Supply Point
Analytical Statistics Printer
Anti-Segregation Process
Asea-Stora Process

	Association of Shareware Professionals [US]
	Associative Storing Processor
	Associative Structures Package
	Astronautical Society Pacific
	Astronomical Society of the Pacific [US]
	Asymmetric Multiprocessing System
	Attached Support Processor
	Audio Signal Processing
	Automatic Schedule Procedure
	Automatic Servo Plotter
	Automatic Synthesis Program
	Average Selling Price
Asp	Asparagine Acid
	Aspartyl
ASPA	American Society for Personnel Administration [US]
	American Sod Producers Association [US]
	Association of South Pacific Airlines [Fiji]
	Australian Sugar Producers Association
ASPAC	Asian and Pacific Council
ASPA-I	American Society for Personnel Administration International [US]
ASPARTAME	Aspartylphenylalanine Methylester [also: Aspartame]
ASPB	Asian Student Press Bureau
ASpB	Arbeitsgemeinschaft der Spezialbibliotheken [= Association of Special Libraries, Germany]
ASPC	American Sheep Producers Council [US]
ASPE	American Society of Plastic Engineers [US]
	American Society of Plumbing Engineers [US]
	American Society of Professional Ecologists
	American Society of Professional Estimators [US]
ASPEC	Association of Sorbitol Producers in the European Community [Belgium]
	Automated Solid-Phase Extraction on Disposable Columns
ASPEI	Association of South Pacific Environmental Institutions
ASPEICP	Associated Schools Project in Education for International Cooperation and Peace [of UNESCO]
ASPEN	Advanced System for Process Engineering
	Arctic Ship(ping) Probability Evaluation Network [Canada]
	Asian Physics Education Network
ASPENPLAN	Asian and Pacific Energy Planning Network [Malaysia]
ASPEP	Association of Scientists and Professional Engineering Personnel [US]
ASPER	Assembly Program for Peripheral Process
	Assembly Program for Peripheral Processors
ASPH	Asphalt
ASPI	American Society for Performance Improvement [US]
	Asynchronous/Synchronous Programmable Interface
ASPIC	Author's Standard Prepress Interface Code
ASPIP	Arab Society for the Protection of Industrial Property
ASPJ	Advanced Self-Protective Jammer
	Airborne Self-Protective Jammer
ASPN	American Society of Precision Nailmakers [US]
ASPO	American Society of Planning Officials [US]
	Apollo Spacecraft Project Office [of NASA]
ASPOE	American Society of Petroleum Operations Engineers [US]
ASPP	Alloy-Steel Protective Plating
	American Society for Plant Physiologists [US]
	American Society of Picture Professionals [US]
ASPPRC	Advanced Steel Processing and Products Research Center [of CSM, US]
ASPR	Armed Services Procurement Regulations [US]
	Average Specific Polymerization Rate
ASPRS	American Society for Photogrammetry and Remote Sensing [US]
ASPS	African Succulant Plant Society
ASPT	Adaptive Signal Processing Testbed
	American Society of Plant Taxonomists [US]
ASQC	American Society for Quality Control [US]
ASQDE	American Society of Questioned Document Examiners [US]
ASR	Accumulator Shift Right
	Active Status Register
	Address Shift Register
	Airborne Surveillance Radar
	Airport Surveillance Radar
	Air Search Radar
	Air-Sea Rescue
	Alkali Silica Reaction
	Alkylene Sulfide Rubber
	Altimeter Setting Region
	Answer, Send and Receive
	Automatic Send and Receive
	Automatic Send-Receive
	Automatic Speech Recognition
	Automatic Sprinkler Riser
	Available Supply Rate
ASRA	American Society of Refrigerating Engineers [US]
	Automatic Stereophonic Recording Amplifier
ASRAAM	Advanced Short-Range Air-to-Air Missile [also: AMRAM]
ASRCT	Applied Scientific Research Corporation of Thailand
ASRD	Astronomy, Space and Radio Division [of SRC, UK]
ASRE	Admiralty Signal and Radar Establishment [UK]
	American Society of Refrigeration Engineers [now: ASHRAE]
ASRET	Affiliation of Societies Representing Engineers and Technicians [South Africa]
ASRF	American Seed Research Foundation [US]
ASRI	Academy of Sciences Research Institute [Ghana]
ASRL	Aeroelastic and Structures Research Laboratory [of MIT, US]
ASRLO	Australian Scientific Research Liaison Office
ASRM	Advanced Shuttle Rocket Motor [of NASA]
	American Association of Range Management [US]
ASRN	Assurance
ASROC	Antisubmarine Rocket
ASRPA	Army Signal Radio Propagation Agency [US]
ASRR	Academy of the Socialist Republic of Romania
ASRS	Amalgamated Society of Railway Servants
	Automated Storage and Retrieval System [also: AS/RS]
ASRWPM	Association of Semi-Rotary Wing Pump Manufacturers
ASS	Acoustical Society of Scandinavia

Aerospace Support System [of AESS]
Aerospace Surveillance System
Air Stage Service
Air Supply System
American Statistical Society [US]
Analog Switching Subsystem
Assembler
Assistant
Assurance
Automatic Selector Switch

ASSA Association for the Study of Snow and Avalanches [France]
Astronomical Society of South Africa
Austrian Solar and Space Agency

ASSASSIN Agricultural System for Storage and Subsequent Selection of Information

ASSBT American Society of Sugar Beet Technologists [US]

AS-SEED Application-Specific Self-Electrooptic-Effect Device

ASSE American Society of Safety Engineers [US]
American Society of Sanitary Engineers [US]
American Society of Swedish Engineers [US]

ASSEM Assembly

ASSESS Analytical Studies of Surface Effects of Submerged Submarines

ASSET Aerothermodynamic Structural Systems Environment Test
American Society of Scientific and Engineering Translators [US]
Association of Supervisory Staffs, Executives and Technicians [UK]

ASSILEC Association de l'Industrie Laitiere de la Communaute Europeenne [= European Community Dairy Trade Association, France]

ASSIM Assimilation

ASSL Advanced Solid-State Laser

ASSLC Advanced Solid-State Laser Conference

ASSM Association of State Supervisors of Mathematics [US]

ASSN Association [also: ASSOC]

ASSOC Associate

Assoc Prof Associate Professor

Assoc Sc Associate in Science

ASSP Acoustics, Speech and Signal Processing (Society) [of IEEE, US]

ASSR Airborne Sea and Swell Recorder
Airport Surface Surveillance Radar
Autonomous Soviet Socialist Republic

ASSS Aerospace Systems Safety Society
African Soil Science Society

ASSSDE Association of State Supervisors of Safety and Driver Education

ASST Advanced Supersonic Transport
Assessment
Assistant
American Society for Steel Treating [now: ASM International, US]

Asst Prof Assistant Professor

ASSU Air Support Signal Unit

ASSY Assembly

AST Advanced Simulation Technology
Advanced Supersonic Technology
Advanced Supersonic Transport
Advanced Surface Technology
Aerospace Technologist
Aerospace Technology
Anti-Sidetone
Army Satellite Tracking Center [US]
Aspartate Aminotransferase
Associates of Science and Technology
Astronomical Society of Tasmania
Atlantic Standard Time
Automatic Shop Tester
Auxiliary Segment Table

ASTA Aerial Surveillance and Target Acquisition
American Seed Trade Association [US]
American Surgical Trade Association [US]
Association of Short-Circuit Testing Authorities
Auckland Science Teachers Association [New Zealand]

AStA Allgemeiner Studenten-Ausschuss [= General Committee of Students, Germany]

ASTAS Antiradar Surveillance and Target Acquisition System

ASTC Advanced Semiconductor Technology Center [of IBM Inc., US]
Airport Surface Traffic Control
Association of Science and Technology Centers [US]

ASTD American Society for Training and Development [US]

ASTE American Society of Test Engineers [US]
American Society of Tool Engineers [US]
Association for Software Testing and Evaluation [US]

ASTEC Advanced Systems Technology
Antisubmarine Technical Evaluation Center [US]
Australian Science and Technology Council

ASTEM Analytical Scanning Transmission Electron Microscopy

ASTF Aero-Propulsion Systems Test Facility [US Air Force]

ASTI Applied Science and Technology Index
Association for Science, Technology and Innovation [US]

ASTIA Armed Services Technical Information Agency [now: DDC, US]

ASTIC Arctic Science and Technology Information Center

ASTIIT American Society for the Technion Israel Institute of Technology [US]

ASTINFO Asian Scientific and Technological Information Network [Thailand]
Scientific and Technical Information in Asia [of UNESCO]

ASTIS Arctic Science and Technology Information Service [of AINA, Canada]

ASTL American Society of Transportation and Logitsics [US]

ASTM American Society for Testing and Materials [US]

ASTME American Society of Tool and Manufacturing Engineers [now: Society of Manufacturing Engineers, US]

ASTMS Association of Scientific, Technical and Managerial Staffs [UK]
ASTOR Anti-Ship Torpedo
ASTOVL Advanced Short Takeoff and Vertical Landing (Aircraft)
ASTP Apollo-Soyuz Test Project [of NASA]
ASTR Astronomer
Astronomical
Astronomy
ASTRA Application of Space Techniques Relating to Aviation
Association in Scotland to Research into Astronautics
Automatic Scheduling with Time-Integrated Resource Allocation
ASTRAC Arizona Statistical Repetitive Analog Computer
ASTRAL Analog Schematic Translator to Algebraic Language
Assurance and Stabilization Trends for Reliability by Analysis of Lots
ASTREC Atomic Strike Recording System [US Air Force]
ASTRO Advanced Spacecraft Transport Reusable Orbiter
Aerodynamic Spacecraft Two-Stage Reusable Orbiter
Airspace Travel Research Organization [US]
(International) Association of State Trading Organizations (of Developing Countries) [Yugoslavia]
ASTRON Astronomer
Astronomical
Astronomy
ASTROPHYS Astrophysical
Astrophysicist
Astrophysics
ASTSSA Associated Scientific and Technical Societies of South Africa
ASTSWMO Association of State and Territorial Solid Waste Management Officials [US]
ASTT American Society of Traffic and Tranportation [now: ASTL, US]
ASU Acknowledgement Signal Unit
Appalachian State University [US]
Apparatus Slide-In Unit
Arizona State University [US]
Automatic Switching Unit
Asu Aminosuberyl
ASUG American Software Users Group [US]
ASUN Associated Students of the University of Nevada [US]
ASUPA Aluminum Sulfate Producers Association [Belgium]
ASUS Arts and Science Undergraduate Society
ASUW Antisurface Unit Warfare
ASV Aluminum-Structured Vehicle
Angle Stop Valve
Anodic Stripping Voltage
Anodic Stripping Voltammetry
Automatic Self-Verification
Avian Sarcoma Virus
ASV/CSV Anodic Stripping Voltage/Cathodic Stripping Voltage
ASVIP American Standard Vocabulary for Information Processing
ASVP Application System Verification and Transfer Program
ASVS Automatic Signature Verification System
ASW Acoustic Surface Wave
Antisubmarine Warfare
Antisubmarine Weapon
Applications Software
Automatic Socket Weld
Auxiliary Switch
ASWA American Steel Warehouse Association [now: SSCI, US]
ASWE Admiralty Surface Weapon Establishment [UK]
ASWEPS Anti-Submarine Warfare Environmental Prediction System [US Navy]
ASWG American Steel Wire Gage
AS&WG American Steel and Wire Gage
ASWM Association of State Wetland Managers [US]
ASWORG Antisubmarine Warfare Operations Research Group [US Navy]
ASWS Ammonium Sulfide Water Stripper
ASW/SCCS Antisubmarine Warfare/Ship Command and Control System
ASWSPO Antisubmarine Warfare Systems Project Office
ASXRED American Society for X-Ray and Electron Diffraction [US]
ASXRT American Society of X-Ray Technicians [US]
ASYM Association of Synthetic Yarn Manufacturers [US]
Asymmetric
Asymmetry
ASYNCH Asynchronous
ASZ American Society of Zoologists [US]
ASZD American Society for Zero Defects [now: ASPI, US]
AT Accelerating Tube
Acceptance Tag
Acceptance Test
Acoustical Tile
Address Translation
Adenine/Thymidine Medium
Advanced Technology
Aerospace Technologist
Aerospace Technology
Agricultural Technologist
Agricultural Technology
Air Temperature
Airtight
Air Transport
Alignment Telescope
Ambient Temperature
American Terms
Aminotriazole
Ampere Turn
Anti-Tank
Antithrombin
Anti-Torpedo
Applied Technology
Architectural Technician
Architectural Technologist
Architectural Technology
Arc Transmitter
Array Theory
Assay Ton

Astronomical Telescope
Atomic
Atomic Time
Audit Trail
Automated Tool
Automatic Ticketing
Automatic Translation
Automatic Transmission
Automotive Technician
Automotive Technologist
Automotive Technology
Auto-Transductor
Auto-Transformer
Auto-Throttle

At Astatine
at (technical) atmosphere
at % atomic percent
ATA Abstracts on Tropical Agriculture [of RTI, Netherlands]
Academic Travel Abroad [US]
Actual Time of Arrival
Advanced Tactical Aircraft
Advanced Tactical Attack
Advanced Transit Association [US]
Aeronautical Telecommunications Agency
African Technical Association
Air Transport Association [US]
Alberta Trucking Association [Canada]
Aloe Technology Association [US]
American Telemarketing Association [US]
American Teleport Association [US]
American Transit Association [US]
American Transport Association [US]
American Translators Association [US]
American Tree Association [US]
American Trucking Association [US]
American Tube Association [US]
Aminotriazole
Anthranilamide
Arcnet Trade Association [US]
Association of Technical Artists [US]
Asynchronous Terminal Adapter
Atlantic Treaty Association [of NATO]
Automatic Trouble Analysis
Automotive Technicians Association [US]
ata (technical) atmosphere, absolute
ATAA Air Transport Association of America [US]
Air Transport Auxiliary Association
ATAC Advisory Technical Awareness Council [US]
Air Transport Advisory Council [US]
Air Transport Association of Canada
Army Tank Automotive Center [US Army]
AT/AC Alignment Telescope/Autocollimator
ATAE Associated Telephone Answering Exchanges [now: ATSI, US]
Automotive Trade Association Executives [US]
ATAF Allied Tactical Air Force [of NATO]
ATAGR Air-to-Air Gunnery Range
ATALA Association pour le Traitement Automatique de Langue [= Association for Automatic Language Processing, France]
ATAM Association of Teaching Aids in Mathematics [UK]
ATAMS Advanced Tactical Attack/Manned System
ATAR Antitank Aircraft Rocket
ATARC Advanced Technology Alcoa Reduction Cell
ATARS Aircraft Traffic Advisory Resolution System
ATAS Academy of Television Arts and Science [US]
Advanced Technology Alert System
Association of Telephone Answering Services [US]
ATB Access Type Base
Access Type Bits
Acetylene-Terminated Bisphel
Address Translation Buffer
Advanced Technology Bomber
Air Transport Board
All Trunks Busy
Aluminum tert-Butoxide
Asphalt Tile Base
Association for Tropical Biology [US]
ATBA Automated Ticket and Boarding (Pass)
Automatic Test Break and Access
ATBCB Architectural and Transportation Barriers Compliance Board [US]
ATBM Advanced Tactical Ballistic Missile
Antitactical Ballistic Missile
ATC Acoustic Tile Ceiling
Adiabatic Toroidal Compressor
Aircraft Technical Committee
Air Temperature Control
Air Traffic Controller
Air Training Command
Air Training Corps [US]
Air Transport Command
Air Transport Committee [of ICAO]
Alcoa Technology Center [US]
Alloy-Tin Couple
American Technical Ceramics [US]
Applied Technology Council [US]
Armament Training Camp
Armament Training Center
Armored Troop Carrier
Army Training Center
Atom Count
Authorized Training Center
Automatic Temperature Compensation
Automatic Through Center
Automatic Toggle Clamp
Automatic Tool Change
Automatic Train Control
Automatic Tuning Control
Automation Training Center [US]
ATCA Advanced Tanker/Cargo Aircraft
Air Traffic Control Association [US]
Allied Tactical Communications Agency [of NATO]
Automatic Tuned Circuit Adjustment
ATCAA Automatic Tuned Circuit Adjustment Amplitude
ATCAC Air-Traffic Control Advisory Committee
ATCAS African Training Center for Agricultural Statistics [of FAO — UN]
ATCC American-Type Culture Collection [US]
ATCAP Air Traffic Control Automation Panel
ATCBI Air Transport Control Beacon Interrogator

ATCC Air Traffic Control Center
ATCCC Advanced Technical Command Control Capability
ATCDE Association of Teachers in Colleges and Departments of Education [UK]
ATCE Automatic Test and Checkout Equipment
ATCER Appropriate Technology Center — Environment [Kenya]
ATCF Axial Tube Component Feeder
ATCH Attached
ATCO Air Traffic Control Officer
ATCOS Atmospheric Composition Satellite [of NASA]
ATCP Asociacion Mexicana de Tecnicos de las Industrias de la Celulosa y del Papel [= Mexican Technical Association of the Cellulose and Paper Industries]
ATCPA Air Taxi and Commercial Pilots Association [US]
ATCRBS Air-Traffic Control Radar Beacon System
ATCS Advanced Train Control System
Air Traffic Control Service
Air Traffic Control Specialist
ATCSS Air Traffic Control Signaling System
ATCT Air Traffic Control Tower
ATD Acceptance and Takeover Date
Actual Time of Departure
Admission, Transfer and Discharge System
Advanced Technical Device
Aerospace Technology Division
Along Track Distance
American Truck Dealers [US]
Association of Tar Distillers [UK]
Asynchronous Time Dilation
Automatic Threat Detection (System)
ATDA Appropriate Technology Development Association [India]
Augmented Target Docking Adapter [of NASA]
ATDC Advanced Technology Development Center
After Top Dead Center
ATDM Asynchronous Time-Division Multiplexing
ATDMA Advanced Time-Division Multiple Access
Automotive Tool and Die Manufacturers Association [now: DTA, US]
ATDS Airborne Tactical Data System
Automatic Telemetry Decommutation System
ATE Advanced Technology Engine
Aluminum Triethyl
Air Traffic Engineer
Air-Turbo Exchanger
Artificial Traffic Equipment
Automatic Telephone Exchange
Automatic Test Equipment
Automotive Test Equipment
ATEA American Technical Education Association [US]
Army Transportation Engineering Agency [US]
ATEC Agence Transequatoriale de Communications [= Transequitorial Agency for Communications, Central Africa]
Automated Technical Control
Automatic Test Equipment Complex
Aviation Technician Education Council [US]
ATEE Acetyl Tyrosine Ethyl Ester (Monohydrate)
ATEGG Advanced Turbine Engine Gas Generator
ATEM Analytical Transmission Electron Microscopy
ATEN Association Technique pour (la Production et l'Utilisation de) l'Energie Nucleaire [= Technical Association for Nuclear Energy, France]
ATEP Asian Trade Expansion Program
ATES Airborne Trial and Evaluation Section [of CABC, Canada]
ATEWS Advanced Tactical Early Warning System
ATF Acceptance Test Facility [US]
Acid-Treated Florisil
Advanced Tactical Fighterplane
Advanced Technology Fighterplane
Advanced Toroidal Facility
Alcohol, Tobacco and Firearms Bureau [US]
American Typecasting Fellowship [US]
Australian Teachers Foundation
Automatic Test Formatter
Automatic Track Finding
Automatic Transmission Fluid
Aviation Turbine Fuel
ATFAC American Turpentine Farmers Association Cooperative [US]
ATFM Air Traffic Flow Management
ATFMU Air Traffic Flow Management Unit
ATFR Automatic Terrain-Following Radar
ATFT Additive Thin Film Technology
ATG Advanced Technology Group
Air-to-Ground
Air Transport Group [of Canadian Forces]
Air Turbine Generator
American Traders Group [US]
ATGAR Antitank Guided Air Rocket
ATGC Asian Textiles and Garments Council
ATGM Antitank Guided Missile
ATGSB Advanced Test for Graduate Studies in Business
ATH Alumina Trihydrate
ATI Access to Information
Air Technical Intelligence [US Air Force]
Appropriate Technology International
Armored Transportation Institute [US]
Army Training Instruction
Associate of the Textile Institute [UK]
Association of Technical Institutions [UK]
ATIA Access to Information Act [Canada]
ATIBT Association Technique Internationale des Bois Tropicaux [= International Technical Tropical Timber Association, France]
ATIG Alternative Technology Information Group [UK]
AT-III Antithrombin III
ATIO Association of Translators and Interpreters of Ontario [Canada]
ATIP Association Technique de l'Industrie Papetiere [= Technical Association of the Paper Industry, France]
ATIPCA Asociacion de Tecnicos de la Industria Papelera y Celulosica Argentina [= Argentinian Technical Association of the Cellulose and Paper Industries]
ATIS Air Traffic Information Service
Automatic Terminal Information Service
Automatic Traffic Information System
ATITA Air Transport Industry Training Association
ATJ Automatic Through Junction

ATJS	Advanced Tactical Jamming System
ATK	Available Tonne/Kilometer
ATL	Aeronautical Turbine Laboratory [US Navy]
	American Tariff League [US]
	Analog-Threshold Logic
	Appliance Testing Laboratory
	Anti-Thrust Law
	Artificial Transmission Line
	Atlantic
	Automated Tape Library
ATLAS	Abbreviated Test Language for All Systems
	Abbreviated Test Language for Avionics Systems
	Adaptive Test and Logic Analysis System
	Antitank Laser Assisted System
	Argonne Tandem Linear Accelerator System [of ANL, US]
	Association of Atlas Group [France]
	Automated Tape Lay-Up System
	Automatic Tabulating, Listing and Sorting (System)
ATLB	Air Transport Licensing Board [UK]
ATLIS	Army Technical Libraries and Information Systems [US]
	Automatic Tracking Laser Illumination System
ATLS	Australian Transport Literature Information System [of AUSINET]
ATLSS	Advanced Technology for Large Structural Systems
ATM	Advanced Test Module
	Airborne Thematic Mapper
	Air Turbine Motor
	Aluminum Trimethyl
	Apollo Telescope Mount [of NASA]
	Archiv fur Technisches Messen [= Archive for Metrology, Germany]
	Association of Teachers of Mathematics [UK]
	Atmosphere
	Atmospheric(s)
	Automated Teller Machine
	Auxiliary Tape Memory
	Axilrod-Teller-Moto (Theory)
atm	(standard) atmosphere
ATMA	Adhesive Tape Manufacturers Association [UK]
	American Textile Machinery Association [US]
	American Textile Manufacturers Association [US]
ATMAM	Analytical Testing Methodologies for Design with Advanced Materials
ATMC	Airspace and Traffic Management Center
	Automotive Training Managers Council [US]
ATME	Automatic Transmission Measuring Equipment
ATMG	Airspace and Traffic Management Group
ATMI	American Textile Manufacturers Institute [US]
ATMOS	Atmosphere
	Atmospheric
	Atmospheric and Ocean Research Satellite [Germany]
ATMR	Advanced Technology Medium Range
ATMRT	Advanced Technology Medium Range Transport
ATMS	Advanced Text Management System
	Association of Telephone Messaging Suppliers [US]
	Automatic Transmission Measuring System
ATN	Aeronautical Telecommunications Network [of ICAO]
	Augmented Transition Network
	Augmented Transmission Network
atn	arctangent
AT NO	Atomic Number [also: at no]
ATO	Accumulated Time Off
	African Timber Organization [Gabon]
	Agricultural Trade Office [US]
	Air Transfer Order
	Antimony-Doped Tin Oxide
	Asia Tele-Tech Organization
	Assisted Takeoff
	Automatic Trunk Office
ATOA	Air Transport Operators Association [UK]
ATOL	Assisted Takeoff and Landing
ATOLL	Acceptance, Test or Launch Language
ATOM	Apollo Telescope Orientation Mount [of NASA]
	Automatic Transmission of Mail
ATOMDEF	Atomic Defense
ATOMIC	Automatic Test of Monolithic Integrated Circuits
AtomPhys	Atomic Physicist
ATOM PHYS	Atomic Physics
ATOMS	Automated Technical Order Maintenance Sequences
ATP	Acceptance Test Procedure
	Accepted Test Procedure
	Adenosine Triphosphate
	Advanced Technology Program [of USDOC]
	Advanced Turboprop Airliner
	Airline Transport Pilot (Certificate)
	Allied Technical Publication
	Army Training Plan
	Arsenical Tough Pitch Copper
	Association of Technical Professionals [US]
	Asynchronous Transaction Processing
	Automated Test Plan
	Automatic Train Protection
A&TP	Assembly and Test Pit [of HTSF, US]
ATPase	Adenosine Triphosphatase
ATPC	Association of Tin Producing Countries [Malaysia]
ATPG	Automatic Test Pattern Generator
	Automatic Test Program Generator
ATPI	American Textbook Publishers Institute [US]
	American Transfer Printing Institute [now: ITPI, US]
ATPISO	Acetyl-Terminated Polyimidesulfone
ATPL	Airline Transport Pilots License
ATR	Advanced Terminal Reactor
	Advanced Test Reactor
	Aircraft Troble Report
	Air Traffic Regulations
	Air Transport Radio
	Air Turbo Rocket
	Answering Time Recorder
	Anti-Transmit-Receive
	Attenuated Reflection
	Attenuated Total Reference
	Attenuated Total Reflectance
	Attenuated Total Reflection
	Automatic Target Recognition
	Automatic Traffic Recorder
	Average Transfer Rate
ATRA	Advanced Transit Association

Automatic Transmission Rebuilders Association [US]

ATRAN Automatic Terrain Recognition and Navigation [also: Atran]

ATRAX Air Transportable Communications Complex

ATRC Agricultural Tools Research Center
Alberta Telecommunications Research Center [Canada]
Antitracking Control

ATRCE Advanced Test Reactor Critical Experiment

ATREM Average Time Remaining

ATREP Air Traffic Control Representative

ATREX Astrophysics Transient Explorer

ATRI Australian Turfgrass Research Institute

ATRIB Average Transfer Rate of Information Bits

ATRID Automatic Target Recognition Identification and Detection

ATRIP (International Association for the) Advancement of Teaching and Research in Intellectual Property [US]

ATRP Air Transport Regulation Panel [of ICAO]

ATRS Automated Temporary Roof Support

ATRT Anti-Transmit-Receive Tube

ATS Acetylene-Terminated Sulfone
Acquisition and Tracking System
Administrative Terminal System
Advanced Technology Satellite
Advanced Test System
Air Traffic Service [of FAA, US]
Air Traffic System
Alarm Termination Subsystem
Alliance for Traffic Safety [US]
American Technical Society [US]
American Television Society [US]
Analog Test System
Applications Technology Satellite [of NASA]
Applied Test System
Armament Training Station
Army Transport Service
Astronomical Time Switch
Audio Test Set
Austrian Schilling [Currency]
Automated Telemetry System
Automated Tracking System
Automatic Telephone System
Automatic Transport System
Automatic Trunk Synchronizer
Auto-Transformer-Starter
Auxiliary Territorial Service [US Military]
Auxiliary Tug Service

ATSA American Traffic Services Association [US]
Aminotoluene Sulfonic Acid

ATSC Advanced Television Systems Committee [US]

ATSD Airborne Traffic Situation Display

ATSDR Agency for Toxic Substances and Diseases Registry [US]

ATSI Association of Telemessaging Services International [US]

ATSIT Automatic Techniques for the Selection and Identification of Targets

ATSORA Air Traffic Services Outside Regulated Airspace [UK]

ATSP Air Transport Statistical Program [of ICAA, Canada]
Association of Technical and Supervisory Professional [US]

ATSPM Air Traffic Service Planning Manual

ATSR Along-Track Scanning Radiometer
Argonne Thermal Source Reactor [of ANL, US]

ATSS Acquisition Tracking Subsystem
Automatic Telephone Switching System

ATSSA American Traffic Safety Services Association [US]

ATSU Air Traffic Service Unit
Association of Time Sharing Users [US]

ATT Advanced Transport Technology
Attachment
Attention
Automatic Testing Technology
Automatic Toll Ticketing

AT&T American Telegraph and Telephone

ATTA American Tin Trade Association [US]

ATTAC Advanced Technologies Testing Aircraft

ATTAP Advanced Turbine Technology Application Program [US]

ATTC American Towing Tank Conference [US]
ASEAN Timber Technology Center
Automatic Transmission Test and Control

ATTCOM AT&T Communications [US]

ATTCS Automatic Takeoff Thrust Control System

ATTD Avalanche Transit-Time Device

ATTEN Attenuation
Attenuator

ATTI Arizona Transportation and Traffic Institute [US]
Association of Teachers in Technical Institutions [UK]
AT&T International

ATTIS AT&T Information System

ATTITB Air Transport and Travel Industry Training Board

ATTIX AT&T Interexchange Carrier

ATTM Attachment

ATTMCA Association of Tile, Terrazzo, Marble Contractors and Affiliates [now: NTCA, US]

ATTN Attention

ATTP Advanced Transport Technology Program

ATTRIB Attributed

ATTS Automatic Telemetry Tracking System
Automatic Tape Time Select

ATTT AT&T Technologies

ATTW Association of Teachers of Technical Writing [US]

ATU Address Translation Unit
Aerial Tuning Unit
African Telecommunications Union
Amalgamated Transit Union
Antenna Tuning Unit
Arab Telecommunications Union [Iraq]
Association Internationale des Universites [= International University Association, France]
Atlantic Telephone System
Auckland Technical University [New Zealand]
Automatic Tracking Unit
Autonomous Transfer Unit

atu	(technical) atmosphere, underpressure
ATUA	Air Transport Users Committee [UK]
ATUC	African Trade Union Confederation
	ASEAN Trade Union Council
ATURS	Automatic Traffic Usage Recording System
ATV	Abwassertechnische Vereinigung [= Association for Wastewater Technology, Germany]
	All-Terrain Vehicle
	Automatic Threshold Variation
ATW	Automatic Tube (Butt) Weld
AT/W	Atomic Hydrogen Weld
ATWA	Association of Third World Affairs
ATWE	Association of Third World Economists
ATWS	Anticipated Transient Without Scram
AT WT	Atomic Weight [also: at wt]
ATX	Automatic Telex Exchange
A2YSC	Associated Two-Year Schools in Construction [US]
ATZ	Aerodrome Traffic Zone
AU	Acadia University [Canada]
	Acousto-Ultrasonic(s)
	Agricultural Union
	Akita University [Japan]
	Alfaatih University [Libya]
	Amplifier Unit
	Angstrom Unit
	Arithmetic Unit
	Astronomical Unit
	Auburn University [US]
Au	(Aurum) — Gold
AU	Angstrom Unit
AUA	American Underground-Space Association [US]
	Association of University Architects [US]
AUBC	Association of Universities of the British Commonwealth [UK]
AUBLS	American Uniform Boiler Law Society [US]
AUBTW	Amalgamated Union of Building Trade Workers [UK]
AUC	Alberta Universities Commission [Canada]
	Algoma University College [Canada]
	Auckland University College [New Zealand]
AUCBE	Advisory Unit for Computer-Based Education [UK]
AUCBM	Arab Union for Cement and Building Materials [Syria]
AUCC	Association of Universities and Colleges of Canada
AUCCP	Arab Union for Cement and Cement Products [now: AUCBM, Syria]
AUCS	Atlantic University Computer Study
AUD	Audibility
	Audio
	Australian Dollar [Currency]
AUDACIOUS	Automatic Direct Access to Information with On-Line UDC System [of American Institute of Physics, US]
AUDAR	Autodyne Detecting and Ranging
AUDI	Arab Urban Development Institute [Saudi Arabia]
AUDIT	Automatic Unattended Detection Inspection Transmitter
AUDREY	Audio Reply
AUEW	Amalgamated Union of Engineering Workers [UK]
AUFS	American Universities Field Staff [US]
AUFSC	Association of University Forestry Schools of Canada
AUFW	Amalgamated Union of Foundry Workers [UK]
AUG	Add-On's and Upgrades
	Amdahl (Computer) Users Group [US]
	August
AUFIS	Automated Ultrasonic Flaw Inspection System
AUFWPA	Association of University Fisheries and Wildlife Program Administrators [US]
AUI	Associated Universities, Inc. [US]
	Attachment Unit Interface
AUID	Association of University Interior Design [US]
AUL	Above Upper Limit
AUM	Air-to-Underwater Missile
AUNT	Automatic Universal Translator
AUP	African Union of Physics [Nigeria]
	Arab Union of Producers
AUPS	American University Press Services [now: AAUP, US]
AUPTDE	Arab Union of Producers, Transporters and Distributors of Electricity
AUR	Association of University Radiologists [US]
Aur	Auriga [Astronomy]
AURA	Association of Universities for Research in Astronomy [US]
Auri	Auriga [Astronomy]
AURP	American Universities Research Program [US]
AURRP	Association of University Related Research Parks [US]
AUS	Austria(n)
	Austrian Schilling [Currency]
AUSA	Association of the United States Army [US]
AUSBC	ASEAN-US Business Council [Thailand]
AUSD	African Union for Scientific Development
AusIMM	Australasian Institute of Mining and Metallurgy [Australia]
	Australian Institute of Mining and Metallurgy
AUSINET	Australian Information Network
AUSMARC	Australian MARC (= Machine-Readable Catalogue)
AUSMIISL	Association of United States Members of the International Institute of Space Law
AUSS	Advanced Unmanned Search System
	Association of University Summer Sessions [US]
	Automatic Ultrasonic Scanning System
AUSSAT	Australian Satellite Network
AUST	Australia(n)
	Austria(n)
AUSTCERAM	International Ceramic Conference in Australia [also: Austceram]
AUSTL	**Australasia(n)**
	Australia(n)
AUSTPAC	Australian Packet Switching Service
AUSTRAL	Australia(n)
AUSUDIAP	Association of United States University Directors of International Agricultural Programs
AUT	Acousto-Ultrasonic Testing
	Advanced User Terminal
	Aristiole University of Thessaloniki [Greece]
	Association of University Teachers [UK]
	Author
AUTEC	Atlantic Undersea Test and Evaluation Center [US]
AUTECS	Automated Eddy Current (Inspection) System
AUTH	Author

AUTHOR Authorization [also: AUTH]
AUTO Automatic
Automation
Automaton
AUTOABSTRACT Automatic Abstract
AUTOALARM Automatic Alarm
AUTOCALL Automatic Calling [also: autocall]
Automatic Call Origination [also: autocall]
AUTOCOM Automotive Composites Conference and Exhibition
AUTOCONT Automatic Continue
AUTO CV Automatic Check Valve
AUTODIN Automatic Digital (Information) Network [of USDOD]
AUTODOC Automated Documentation
AUTOEXEC Automatic Execution
AUTOFACT Automated and Integrated Factory (Conference and Exposition) [of CASA/SME, US]
AUTO ID Automatic Identification (System)
AUTOIGN Autoignition
AUTOIGN TEMP Autoignition Temperature
AUTOMAP Automatic Machining Program
AUTOMAST Automatic Mathematical Analysis and Symbolic Translation
AUTOMEX Automatic Message Exchange Service
AUTONET Automatic Network Display
AUTOPIC Automatic Personal Identification Code
AUTOPILOT Automatic Pilot
AUTOPOLL Automatic Polling
AUTOPROG Automatic Programming
AUTOPROMPT Automatic Programming of Machine Tools
AUTOPSY Automatic Operating System
AUTO-QC Automatic Quality Control
AUTO RECL Automatic Reclosing
AUTOSATE Automated Data Systems Analysis Technique
AUTOSCRIPT Automated Systems for Composing, Revising, Illustrating and Phototypesetting
AUTOSDI Automatic Selective Dissemination of Information
AUTOSEVCOM Automatic Secure Voice Communications
AUTOSEVOCOM Automatic Secure Voice Communications
AUTOSPOT Automatic System for Positioning Tools
AUTOSTRAD Automated System for Transportation Data
AUTOSTRT Automatic Starter
AUTOSTRTG Automatic Starting
AUTOTESTCON Automobile Testing Conference [US]
AUTO TR Auto-Transformer
AUTO-TRIP Automatic Transportation Research Investigation Program
AUTOVERIFIER Automatic Verifier
AUTOVON Automatic Voice Network [of USDOD]
AUTRAN Automatic Target Recognition Analysis
Automatic Utility Translator
AUU Association of Urban Universities [US]
AUUA American UNIVAC (Computer) Users Association [US]
AUV Armored Utility Vehicle
Autonomous Underwater Vehicle
AUVS Association for Unmanned Vehicle Systems [US]
AUW All-Up-Weight
Anti-Underwater Warfare
AUWE Admiralty Underwater Weapons Establishment [UK]
AUX Auxiliary
AUXI-ATOME Societe Auxiliare pour l'Energie Atomique [= Auxiliary Society for Atomic Energy, France]
AUXIL Auxiliary
AUXS Auxiliaries
AV Active Voltage
Added Value
Air Volume
Angular Velocity
Arctic Vessel
Audiovisual
Authorized Version
Availability
Avenue
Average
Average Variability
Aviation
A/V Audio-Visual
A-V Audio-Visual
aV abvolt [Unit]
av average
avoirdupois
a/v ad valorem
a v (a vista) — at sight
AVA Absolute Virtual Address
Academy of Visual Arts [UK]
Adventitious Viral Agent
Aerodynamische Versuchsanstalt [= Research Institute for Aerodynamics, Germany]
Alberta Veterinary Association [Canada]
American Video Association [US]
American Vocational Association [US]
ASEAN Valuers Association [Malaysia]
Australian Veterinary Association
Automated Vision Association [of RIA, US]
Azimuth versus Amplitude
Ava Anabaena Variabilis [Molecular Biology]
AVAD Automatic Voice Alerting Device
AVAS Automatic VFR (= Visual Flight Rules) Advisory Service
AVASI Abbreviated Visual Approach Slope Indicator
AVBL Armored Vehicle Bridge Launcher
AVC Association of Vitamin Chemists [US]
Automatic Valve Control
Automatic Vent Control
Automatic Voltage Compression
Automatic Volume Contractor
Automatic Volume Control
AVCASA Agricultural and Veterinary Chemicals Association of South Africa
AVCG Automatic Vapor Crystal Growth
AVCOM Aviation Materiel Command [US]
AVCS Advanced Vidicon Camera System
Assistant Vice Chief of Staff
AVD Alternate Voice/Data
Aluminum Vacuum Degassing
Automatic Voice/Data
AvD Automobilverein von Deutschland [= Automobile Association of Germany]
avdp avoirdupois
AVE Aerospace Vehicle Electronics
Automatic Voltammetric Electrode

	Automatic Volume Expander
	Automatic Volume Expansion
	Avenue
AVEA	American Veterinary Exhibitors Association [US]
AVEC	Amplitude Vibration Exciter Control
AVEM	Association of Vacuum Equipment Manufacturers [US]
AVERE	Association Europeenne des Vehicules Electriques Routiers [= European Electric Road Vehicle Association, Belgium]
AVERT	Association of Volunteer Emergency Radio Teams [US]
AVF	Academie Veterinaire de France [= French Academy of Veterinary Science]
	Azimuthally Varying Field
AVG	Aminoethoxyvinylglycine
	Average
AVGAS	Aviation Gasoline [also: avgas]
AVGN	Additive Voice Gaussian Noise
AVGP	Armored Vehicle — General Purpose
AVHRR	Advanced Very-High-Resolution Radiometer
AVI	Airborne Vehicle Identification
	Automated Visual Inspection
	Aviation
AVID	Advanced Visual Information Display
AVIEM	Asociacion Venezolana de Ingenieros Electricos y Mecanicos [= Venezuelan Association of Electrical and Mechanical Engineers]
AVIONICS	Aviation Electronics [also: Avionics and avionics]
AVIOS	American Voice Input/Output Society [US]
AVIP	Association of Viewdata Information Providers [UK]
AVIRIS	Airborne Visible/Infrared Imaging Spectrometer
AVIS	Automatic Video Inspection System
AVISA	Asociacion Venezolana de Ingenieria Sanitaria y Ambental [= Venezuelan Chapter of the Inter-American Association of Sanitary Engineers]
AVL	Adel'son-Vel'skii and Landis Trees
	Automatic Vehicle Location
AVLB	Armored Vehicle Launch Bridge
AVLC	Automatic Vehicle Location and Control System
AVLINE	Audiovisual On-Line (Database) [US]
AVLS	Automatic Vehicle Locating System
AVLSI	Analog Very Large-Scale Integration
AVM	Airborne Vibration Monitoring
	Air Vice-Marshall
	Automatic Vehicle Monitoring
AVMA	Audio-Visual Management Association [US]
AVMARC	Audiovisual MARC (= Machine-Readable Catalogue)
AVMC	Association for Vertical Market Computing [US]
AVMED	Aviation Medicine
AVMR	Association of Visual Merchandise Representatives [US]
AVMRI	Arctic Vessel and Marine Research Institute
AVMS	Annulus Vacuum Maintenance System
AVNL	Automatic Video Noise Limiting
AVO	Advanced Video Option
AVOID	Airfield Vehicle Obstacle Indication Device
avoir	avoirdupois
avoir oz	avoirdupois ounce
AVOLO	Automatic Voice Link Observation
AVOSS	Added Value Operating Support System
AVP	Advanced VLSI (= Very-Large-Scale Integration) Packaging
	Attached Virtual Processor
	Avionics Panel [of AGARD]
AVPO	Axial Vapor-Phase Oxidation
AVR	Automatic Voltage Regulator
	Automatic Voice Recognition
	Automatic Volume Recognition
AVRA	Audio-Visual Research Foundation [US]
AVRAC	Agricultural and Veterinary Research Advisory Committee [East Africa]
AVRDC	Asian Vegetable Research and Development Center [Taiwan]
AVRS	Automated Vehicle Roading System
AVS	Advanced Vacuum System
	Aerospace Vehicle Simulation
	American Vacuum Society [US]
	Association of Veterinary Students [UK and Ireland]
	Automated Vision Association [US]
AVSAT	Aviation Satellite [also: AvSat]
AVSE	Armored Vehicle Survivability Enhancement
AVSF	Advanced Vertical Strike Fighter
AVSL	Association of Visual Science Librarians [US]
AVT	All Volatile Treatment
AVTA	Automatic Vocal Transaction Analysis
AVTAG	Aviation Turbine Gasoline [also: Avtag]
AVTR	Airborne Video-Cassette Tape Recorder
AVTUR	Aviation Turbine (Kerosine) [also: Avtur]
AW	Above Water
	Acid-Washed
	Acoustic Wave
	Aircraftwoman
	Air Warning
	Airworthiness
	Arc Welding
	Arming Wire
	As Welded
	Atomic Weight [also: aw]
	Automatic Weapon
A/W	Actual Weight
aW	abwatt [Unit]
AWA	Alliance of Women in Architecture [US]
	All-Weather Aircraft
	American Wilderness Alliance [US]
	American Wire Association [US]
	Association of Women in Architecture [US]
	Aviation/Space Writers Association [US]
AWAC	Airborne Warning and Control
AWACS	Airborne Warning and Control System [US]
AWADS	All-Weather Aerial Delivery System
AWAE	Automotive Wholesalers Association Executives [US]
AWAR	Area Weighted Average Resolution
AWARS	Airborne Weather and Reconnaissance System
AWASTA	African Women's Association for Scientific and Technical Advancement
AWAT	Area Weighted Average T-Number
AWB	Agricultural Wages Board [UK]
	Asian Wetland Bureau
	Australian Wheat Board
	Australian Wool Board

AWC	Active Wire Concentrator
	American Waterfowl Council [US]
	American Wood Council [US]
	American Wool Council [of ASPC, US]
	Association for Women in Computing [US]
AWCEA	American Wood Chip Export Association [US]
AWCI	American Wire Cloth Institute [US]
	Association of the Wall and Ceiling Industries — International [US]
AWCLS	All-Weather Carrier Landing System
AWCMA	American Window Covering Manufacturers Association [US]
AWCS	Air Weapons Control System
AWD	Acoustic Wave Device
	Action World Development
	All Wheel Drive
AWDA	Automotive Warehouse Distributors Association [US]
AW-DMCS	Acid-Washed — Dimethyldichlorosilane
AWDS	Ash Water Dense Suspension
AWEA	American Wind Energy Association [US]
AWES	Association of West European Shipbuilders [UK]
AWF	African Wildlife Foundation [US]
	Ausschuss fur Wirtschaftliche Fertigung [= Committee for Economic Manufacturing, Germany]
AWG	American Wire Gage
	Arbitrary Wave Generator
	Art Workers Guild [UK]
	Asian Women's Institute [Pakistan]
	Association for Women Geoscientists [US]
AWGN	Additive White Gaussian Noise
AWI	Alfred-Wegener-Institute (fur Polar- und Meeresforschung) [= Alfred Wegener Institute (for Polar and Marine Research), Germany]
	American Welding Institute [US]
	Architectural Woodwork Institute [US]
AWIC	Association for Women in Computing
AWIM	Association for Women in Mathematics
AWIS	Association of Women in Science [US]
	Aviation Weather Information Service
AWJC	American Water Jet Conference [US]
AWLS	All-Weather Landing System
AWM	Appliance Wiring Materials
	Association for Women in Mathematics [US]
AWMA	Air and Waste Management Association [US]
	American Walnut Manufacturers Association [now: FHVA/AWMA, US]
AWN	Automatic Weather Network [US Air Force]
AWO	All Weather Operation
	American Waterways Organization [US]
AWOA	American West Overseas Association [US]
AWOL	Absent Without Leave
AWOP	All-Weather Operation Panel [of ICAO]
AWOS	Airport Weather Observing System
	Automated Weather Observing System
AWP	Allied Weather Publication
	Average Weighted Pressure
AWPA	American Wire Producers Association [US]
	American Wood Preservers Association [US]
AWPB	American Wood Preservers Bureau [US]
AWPI	American Wood Preservers Institute [US]
AWRA	American Water Resources Association [US]
	Australian Welding Research Association
AWRC	Australian Wool Realization Council
AWRE	Atomic Weapons Research Establishment [UK]
AWRF	Associated Wire Rope Fabricators [US]
AWRNCO	Aircraft Warning Company [US Marines]
AWS	Agricultural Wholesale Society [UK]
	Air Warning System
	Air Weapon System
	Air Weather Service
	American War Standards [US]
	American Welding Society [US]
	Association of Women Scientists
AWSC	American Waterways Shipyard Conference [US]
AWSF	Alpha Waste Storage Facility
AWSS	Association of Women Soil Scientists [US]
AWST	Association of Women Science Teachers [UK]
AWT	Actual Work Time
	Arbeitsgemeinschaft Warmebehandlungstechnik [= Working Group on Heat Treating Technology, Germany]
	Automatic Weld Tube
AWTAC	American Welding Technology Application Center [now: AWI, US]
AWTAO	Association of Water Transportation Accounting Officers [US]
AWTE	Association for World Travel Exchange [US]
AWTP	Advanced Waste Treatment Process
AWU	Agricultural Workers Union [South Africa]
	Associated Western Universities [US]
	Association for the World University [US]
	Atomic Weight Unit [also: awu]
AWV	Ausschuss fur Wirtschaftliche Verwaltung [= Committee for Economic Administration, Germany]
AWVS	American Women's Voluntary Service [US Military]
AWW	Alert Weather Watch
AWWA	American Waterworks Association [US]
AX	Attack Experimental
	Automatic Exchange
	Automatic Transfer
AXAA	Australian X-Ray Analytical Association
AXAF	Advanced X-Ray Astrophysics Facility [of NASA]
AXD	Auxiliary Drum
AXFMR	Automatic Transformer
AXP	Axial Pitch
AYD	American Yarn Distributors [US]
AYSA	American Yarn Spinners Association [US]
AYSM	Association of Yugoslav Societies for Microbiology
AZ	Academy of Zoology [India]
	Alloyed Zone
	Arizona [US]
	Azimuth
	Azure
AZA	American Zinc Association [US]
AZAS	Adjustable Zero Adjustable Span
AZBN	Azobisisiobutyronitrile
AZDN	Azodiisobutyronitrile
AZE	Anomalous Zeeman Effect
AZEL	Azimuth Elevation
AZO	Aluminum-Doped Zinc Oxide
AZON	Azimuth Only

AZRAN Azimuth and Range
AZS Automatic Zero Set
AZT Azidodeoxythymidine
3'-Azido-3'-Deoxythymidine

B

B Bachelor
Bacillus
Bale
Balboa [Currency of Panama]
Bank
Band
Bandwidth
Bar
Base
Batch
Battery
Baume
Bay
Beam
Bel [Unit]
Belugou (Coefficient)
Bering (Standard Time)
Bit
Black [On Pencils]
Blackbody
Blank
Block
Blue
Bolivar [Currency of Venezuela]
Bond
Bone
Boron
Bottom
Braid
Brazing
Breadth
Brewster [Unit]
Brightness
British
British Thermal Unit
Broadcasting
Burning
Bus
Buffer
Byte

b bale
bar [Unit]
barn [Unit]
base
bath
battery
bay
bel [Unit]
bench
binary
blackbody
blend
boiling
bottom
brass
breadth
brick

BA Bachelor of Arts
Background Activity
Basal Area
Bayard-Alpert (Gauge)
Benzanthracene
Benzla Acetone
Benzylamine
Bend Allowance
Benzaldehyde
Binary Add(er)
Biologic Activity
Biological Abstracts [of BIOSIS, US]
Blocking Antibody
Booksellers Association [UK]
Boric Acid
Box Annealing
Braking Action
Breathing Apparatus
British Academy
British Airways
British Association (Screw Thread)
Bromo-Acetone
Buffer Amplifier
Butyl Acetate

B/A Boron/Aluminum

B-A Bayard-Alpert (Gauge) [also: B/A]

Ba Barium

BAA Bachelor of Applied Arts
Benzoyl Arginine Amide
British Acetylene Association
British Airports Authority
British Archeological Association
British Astronomical Association
Broad Agency Announcement [of USDOD]
Broadband Antenna Amplifier
Butyraniline Aldehyde

BAACMIR British Aerospace Air Combat Manoeuvering Instrumentation Range

BAAEMS British Association of Airport Equipment Manufacturers and Services

BAAR Board for Aviation Accident Research [US Army]

BAArch Bachelor of Arts in Architecture

BAAS British Association for the Advancement of Science
British Association of Accountants and Auditors

BAB Babbitt
Bundesautobahn [= German Express Highway]

BABS Blind Approach Beacon System
British Aluminum Building Service

BABT British Approval Board for Telecommunications

BAC Back Association of Canada
Background Analysis Center
Barometric Altitude Control
Binary Asymmetric Channel
Bird Association of California [US]
Bis(acryloyl)-Cystamine
Block Advisory Committee [India]
Blood Alcohol Concentration

British Airways Corporation
British Association of Chemists
British Atlantic Committee

BACAH British Association of Consultants in Agriculture and Horticulture
BACAIC Boeing Airplane Company Algebraic Interpretive Computing System
BACAN British Association for the Control of Aircraft Noise
BAcc Bachelor of Accountancy
BACE Basic Automatic Checkout Equipment
British Association of Consulting Engineers
British Association of Corrosion Engineers
Bureau of Agricultural Chemistry and Engineering [US]
BACG British Association for Crystal Growth
BACIE British Association for Commercial and Industrial Education
BACIU Bricklayers and Allied Craftsmen International Union
Back EMF Back Electromotive Force [also: back emf]
BACM British Association of Colliery Management
BACMA British Aromatic Compound Manufacturers Association
BACON British Association for Control of Aircraft Noise
BACR British Association for Cancer Research
BACS Banks Automated Clearing System
Bibliographic Access and Control System
BACTE British Association for Commercial and Technical Education
BACTERIOL Bacteriological [also: BACT]
Bacteriologist [also: BACT]
Bacteriology [also: BACT]
BADAS Binary Automatic Data Annotation System
BADC Binary Asymmetric Dependent Channel
BADEA Banque Arabique pour le Developpement Economique en Afrique [= Arab Bank for Economique Development in Africa]
BADGE Base Air Defense Ground Environment
Base Air Defense Ground Experiment
BADIC Biological Analysis Detection Instrumentation and Control
BAdmin Bachelor of Administration
BAE Beacon Antenna Equipment
Belfast Association of Engineers [Northern Ireland]
Bureau of Agricultural Economics [US]
BAe British Aerospace
BAeA British Aerobatics Association
BAEB Bituminous and Aggregate Equipment Bureau [US]
BAEC British Agricultural Export Council
British Amateur Electronics Club
BAED Bachelor of Arts in Environmental Design
BAEd Bachelor of Arts in Education
BAEE Benelux Association of Energy Economists [Europe]
Benzoyl Arginine Ethyl Ester
BAeE Bachelor of Aeronautical Engineering
BAER Brainstem Auditory Evoked Response
BAF Baffle
Bonded Acetate Fiber
BAFB Bolling Air Force Base [of US Air Force]
$Ba(FOD)_2$ Bis(1,1,1,2,2,3,3-Heptafluoro-7,7-Dimethyl-4,6-Octanedionato)barium
BAFRA British Aluminum Foil Rollers Association
BAFT Bankers Association for Foreign Trade [US]
BAG Bayard-Alpert Gauge
Bibliographic and Grouping System
Bioaccumulation Factor
British Aviation Group
BAGC Biased Automatic Gain Control
BAgE Bachelor of Agricultural Engineering
BAHS British Agricultural History Society
BAIB Beta-Aminoisobutyric Acid
BAIC Bureau of Agricultural and Industrial Chemistry [US]
BAID Bachelor of Interior Design
BAIE British Association of Industrial Editors
BAINS Basic Advanced Integrated Navigation System
BAIR Berkeley Artificial Intelligence Research
BAIT Bacterial Automated Identification Technique
BAK Baboon Kidney
Backup
BAL Balance
Balboa [Currency of Panama]
Basic Assembler Language
Bright Anneal Line
British Anti-Lewisite
Business Application Language
2,3-Dimercapto-1-Propanol
Bal Brevibacterium Albidum [Molecular Biology]
BALGOL Burroughs Algebraic Compiler
Burroughs Algorithmic Language
BALIS Bayerisches Landwirtschaftliches Informationssystem (Datenbank) [= Bavarian Agricultural Information System (Databank), Germany]
BALL Ballast
BALLAST Balanced Loading via Automatic Stability and Trim
BALLOTS Bibliographic Automation of Large Library Operations using a Time-Sharing System [now: RLIN]
BALLUTE Balloon Parachute
BALMI Ballistic Missile
BALOP Balopticon
BALPA British Airline Pilots Association
BALS Balancing Set
Blind Approach Landing System
BALUN Balanced-to-Unbalanced (Transformer)
BAM Ballistic Missile
Balloon Anisotropy Measurements
Basic Access Method
Bituminous Aggregate Mixture
Block Access Method
Broadcasting Amplitude Modulation
Bundesanstalt fur Materialprufung [= Federal Institute for Materials Testing, Germany]
BAMBI Ballistic Missile Boost Intercept
BAME Benzoyl Arginine Methyl Ester
British Aerosol Manufacturers Association
BamH Bacillus Amyloliquefaciens H [Molecular Biology]
BAMIRAC Ballistic Missile Radiation Analysis Center [US]
BAMM Balloon Altitude Mosaic Measurement
Basic Acrylic Monomer Manufacturers Association [US]

BAMO Bureau of Aeronautics Material Officer
BAMS Bachelor of Arts, Master of Science
Bulletin of the American Mathematical Society
BAMT Boric Acid Mixing Tank
BAMTM British Association of Machine Tool Merchants
BAN Best Asymptotically Normal
British Approved Name
Ban Bacillus Aneurinolyticus [Molecular Biology]
BANA Benzoyl Arginine Naphthylamide
BANCS Bell Administrative Network Communications System
BAND Bandolier
BANI Benzoyl Arginine Nitroanilide Hydrochloride
BANS Basic Air Navigation School
Bright Alphanumeric Subsystem
British Association of Numismatic Societies
BAO 2,5-Bis(4-Aminophenyl)-1,3,4-Oxadiazole
BAP Band Amplitude Product
Basic Assembler Program
Beam-Assisted Processing
Benzyl-P-Aminophenol
6-Benzylaminopurine
Biologically Active Peptides
Branch Arm Piping
British Association for Psychopharmacology
BaP Benzopyrene
BAPA British Aeromedical Practitioners Association
British Amateur Press Association
BAPC British Aircraft Preservation Council
BAPE Branch Arm Piping Enclosure
BAPL Bettis Atomic Power Laboratory [US]
BAPNA Benzoyl Arginine Nitroanilide Hydrochloride
BApplSc Bachelor of Applied Science [also: BAppS or BAppSc]
BApplStat Master of Applied Statistics [also: BAppStat]
BAPS Branch Arm Piping Shielding
BAPTA Bearing and Power Transfer Assembly
1,2-Bis(2-Aminophenoxy)ethane Tetraacetic Acid
BAQ Bundesanstalt fur Qualitatsforschung [= Federal Institute for Quality Research, Germany]
BAR Bar Address Register
Barometer
Barometric
Base Address Register
Broadcast Advertisers Report [US]
Browning Automatic Rifle
Buffer Address Register
Bureau of Aeronautics Representative [US]
bar barrel [Unit]
BARC Bangladesh Agricultural Research Council
BArch Bachelor of Architecture
BArchEng Bachelor of Architectural Engineering
BARD Bangladesh Academy for Rural Development
BARE Biased-Activated Reactive Evaporation
BARI Bangladesh Agricultural Research Institute
BARIF Banjarbaru Research Institute for Food
BARITT Barrier Injection Transit Time (Diode)
BAROM Barometer [also: BARO]
BARR Bureau of Aeronautics Resident Representative [US]
BARS Backup Attitude Reference System
Ballistic Analysis Research System
Biostatic Auroral Radar System
BARSTUR Barking Sands Tactical Underwater Test Range [Hawaii, US]
BART Baronet
Bay Area Rapid Transit [San Francisco, US]
BArt Bachelor of Arts
BARTD Bay Area Rapid Transit District [San Francisco, US]
BARV Beach Armored Recovery Vehicle
BARZREX Bartok Archives Z-Symbol Rhythm Extraction
BAS Bachelor in Agricultural Science
Bangladesh Academy of Sciences
Basic
Basic Activity Subset
Basic Auto-Oxidation Scheme
Basic Airspeed
Bibliotheks-Automatisierungssystem [= Library Automation System, Germany]
Blind Approach System
Border Agricultural Society [South Africa]
British Acoustical Society
British Antarctic Survey
Bulgarian Academy of Sciences
BASA British Architectural Students Association
BASc Bachelor of Applied Science
BASCOM BASIC Compiler
BASEC British Approval Service for Electric Cables
BASEEFA British Approval Service for Electrical Equipment in Flammable Atmospheres
BASH Booksellers Association Service House [UK]
BASI Bureau of Air Safety Investigation [Australia]
BASIC Basel Information Center for Chemistry [Switzerland]
Basic Algebraic Symbolic Interpretative Compiler
Basic Automatic Stored Instruction Computer
Battle Area Surveillance and Integrated Communications
Beginner's Algebraic Symbolic Interpretive Compiler
Beginner's All-Purpose Symbolic Instruction Code
Biological Abstracts Subjects in Context [of BIOSIS, US]
BASICPAC BASIC Processor and Computer
BASICS Battle Area Surveillance and Integrated Communication System
BASIS Bank Automated Service Information System
Battelle Automatic Research Information System [of BMI, US]
British Airways Staff Information System
Bulletin of the American Society for Information Science
Burroughs Applied Statistical Inquiry System
BASIS-E Bibliothekarisch-Analytisches System zur Informationsspeicherung und -erschliessung [= Library Analytical System for Information Storage and Retrieval, Germany]
BASJE Bolivian Air Shower Joint Experiment
BASM Bachelor of Arts, Master of Science
BAS NET Basic Network
BASRA British Amateur Scientific Research Association
BASRM British Association of Synthetic Rubber Manufacturers

BAST	Boric Acid Storage Tank
BASt	Bundesanstalt fur Strassenwesen [= Federal Institute for Road Traffic and Transportation, Germany]
BASW	Bell Alarm Switch
BASYS	Basic System
BAT	Batch
	Basic Assurance Test
	Battalion Anti-Tank
	Battery
	Biological Abstracts on Tape [of BIOSIS, US]
	Bipolar Axon Terminal
	Bond — Assembly — Test
BATA	British Air Transport Association
BAT CHG	Battery Charger
BATF	Bureau of Alcohol, Tobacco and Firearms [US]
BATMA	Bookbinding and Allied Trades Management Association [UK]
Ba(TMHD)$_2$	Bis(2,2,6,6-Tetramethyl-3,5-Heptane-dionato)barium Hydrate
BATP	Boric Acid Transfer Pump
BATRE	Battle Reconnaissance
BATS	Basic Additional Teleprocessing Support
	Bay Area Tracking System [San Francisco, US]
	Bit Access Test System
BATT	Battalion
	Batten
	Battery [also: BATTY]
BATU	Brotherhood of Asian Trade Unionists [Philippines]
BAUA	Business Aircraft Users Association [UK]
BAUD	Baudot Code
BAUFO	Bauforschungsprojekte und -berichte (Datenbank) [= Database on Civil Engineering Research Projects and Reports [of IRB, Germany]]
BAUMA	Internationale Fachmesse fur Baumaschinen und Baustoffmaschinen [= International Trade Fair for Construction and Building Materials Equipment, Germany]
BAURES	Bangladesh Agricultural University Research System
BAVIP	British Association of Viewdata Information Providers
BAW	Bulk Acoustic Wave
BAY CAND DC	Bayonette Candelabra, Double Contact
BAYS	British Association of Young Scientists
BB	Backboard
	Ball Bearing
	Baseband
	Best Best
	Blackbody
	Block Brazing
	Bottom Boundary
	Broadband
	Brush Box
	Bucket Brigade
	Building Block
	Bulk Burning
	Bulletin Board
	Bunch Block
	Bus Bar
	Double Black [On Pencils]
B&B	Bell and Bell
B-B	Butane-Butene Mixture
BBA	Bachelor of Business Administration
	Bear Biology Association [now: IABRM, Canada]
	British Bankers Association
BBAC	British Balloon and Airship Club
BBB	Better Business Bureau
	Bisbenzimidazobenzophenanthroline
BB/B	Body Bound Bolts
BBC	Bachelor of Building Construction
	British Broadcasting Corporation
	Broadband Conducted
	Bromobenzylcyanide
BBD	Barbados Dollar [Currency]
	Beam Blanking Device
	7-Benzylamino-4-Nitrobenz-2-Oxa-1,3-Diazole
	2,5-Bis(4-Biphenylyl)-1,3,4-Oxadiazole
	Bubble Bath Detector
	Bucket-Brigade Device
BBDC	Before Bottom Dead Center
BBEA	Brewery and Bottling Engineers Association
BBF	Baseband Frequency
	(International) Brotherhood of Boilermakers, Iron Shipbuilders, Blacksmiths, Forgers and Helpers [UK]
BBG	Board of Broadcasting Governors
BBGKY	Bogoliubov-Born-Green-Kirkwood-Yvon
BBIRA	British Baking Industries Research Association
BBKA	British Beekeepers Association
BBL	Basic Business Language
	Benzimidazobenzophenanthroline
bbl	barrel [Unit]
bbl/d	barrels per day [also: bbl da^{-1}]
BBM	Beam Brightness Modulation
	Break Before Make
	Bureau of Broadcast Measurement
BBMA	British Bath Manufacturers Association
BBMRA	British Brush Manufacturers Research Association
BBN	Bolt, Beranek and Newman
	Borabicyclononane
9-BBN	9-Borabicyclononane
BBO	2,5-Bis(4-Biphenylyl)oxazole
BBOT	2,5-Bis(5-tert-Butyl-2-Benzoxazolyl)thiophene
BBP	Benzyl Butyl Phthalate
	Butyl Benzyl Phthalate
BBPT	Baseband Pulse Transmission
BBQ	Benzimidazobenzisoquinolinone
BBR	Bend-Bend-Roll
	Backbody Radiation
	Broadband Radiated
BBRG	Ball Bearing
BBRR	Brookhaven Beam Research Reactor [US]
BBRU	Bituminous Binder Research Unit [South Africa]
BBS	Balanced Biaxial Stretching
	Bermuda Biological Station
	British Biophysical Association
	British Bryological Association
	Bulletin Board Service
	Bulletin Board System
BBSR	Bermuda Biological Station for Research
BBT	Bombardment
BBTA	British Bureau of Television Advertising

BBU Baseband Unit

BC Back Connected
Back Course
Basic Circle
Basic Control
Bathyconductograph
Bedford College [UK]
Before Christ
Belted Cable
Benzyl Chloride
Between Centers
Bibliographic Classification
Binary Code
Binary Comparator
Binary Counter
Biochemist
Biochemistry
Biological Chemist
Biological Chemistry
Biological Council [UK]
Bliss Classification
Blocking Capacitor
Bolt Circle
Bonded (Single) Cotton (Wire)
Bottom Chord
Bottom Contour
Breaking Current
British Columbia [Canada]
Broadcast Control
Brooklyn College [of CUNY, US]
Boundary Condition
Bus Controller
Buried Channel
Burnhampthorpe College [Canada]

BCA Battery Control Area
Beale Cypher Association [US]
Benzene Carboxylic Acid
Bliss Classification Association [UK]
British Carton Association
British College of Accountancy
Bureau of Coordination of Arabization [Morocco]

BCAA Branched-Chain Amino Acid

BCAB British Control Advisory Bureau

BCABP Bureau of Competitive Assessment and Business Policy [US]

BCAC British Columbia Aviation Council [Canada]
British Conference on Automation and Computation

BCACA British Columbia Arts and Crafts Association [Canada]

BCAE Ballarat College of Advanced Education [Australia]
Bendigo College of Advanced Education [Australia]

BCALT British Columbia Association of Library Technicians [Canada]

BCAR British Civil Airworthiness Requirements

BCAS Beacon-Based Collision Avoidance System [also: B-CAS]
Beacon Collision Avoidance System
British Columbia Agricultural Society [Canada]
British Compressed Air Society
Business Center for Academic Societies [Japan]

BCAVM British Catalogue of Audiovisual Materials

BCB Benzocyclobutene
Broadcast Band
Brown Cardboard

BCBA Beef Cattle Breeders Association [Israel]

BCC Biological Council of Canada
Block Character Check
Block Check Character
Blocked Calls Cleared
Body-Centered Cubic [also: bcc]
British Color Council
British Communications Corporation
British Copyright Council

BCCA British Columbia Construction Association [Canada]

BCCD Bulk-Channel Charge-Coupled Device

BCCF British Cast Concrete Federation

BCCL Birkbeck College Computation Laboratory [UK]

BCCRC British Columbia Cancer Research Center [Canada]

BCD Barrels per Calendar Day
Binary-Coded Decimal
Burst Cartridge Detection

BCD/B Binary-Coded Decimal/Binary

BCDC Binary-Coded Decimal Counter
British Columbia Development Corporation [Canada]

BCDIC Binary-Coded Decimal Interchange Code

BCDMOS Bipolar CMOS (= Complimentary Metal-Oxide Semiconductor) — DMOS (= Double-Diffused Metal-Oxide Semiconductor)

BCD/Q Binary-Coded Decimal/Quaternary

BCDS Bulk Chemical Delivery System

BCDSFU British Columbia Deep Sea Fishermen's Union [Canada]

BCDTA British Chemical and Dyestuff Traders Association

BCE Bachelor of Civil Engineering
Biochemical Engineer
Biochemical Engineering

BCEC Battle Creek Engineers Club

BCECC British and Central-European Chamber of Commerce

BCET Biochemical Engineering Technologist
Biochemical Engineering Technology

BCF Billion Conductor Feet
Billion Cubic Feet
Bolivar Coastal Field [Venezuela]
British Concrete Federation
Bromochlorofluoromethane
Bulked Continuous Fiber
Byte Cipher Feedback

BCFA British Columbia Federation of Agriculture [Canada]

BCFBA British Columbia Food Brokers Association [Canada]

BCFCA British Columbia Floor Covering Association [Canada]

BCF/CD Billion Cubic Feet per Calendar Day

BCFGA British Columbia Fruit Growers Association [Canada]

BCFL British Columbia Federation of Labour [Canada]

BCFMCA British Columbia Frequency Modulation Communications Association [Canada]

BCFP British Columbia Forest Products [Canada]

BCFS British Columbia Forestry Service [Canada]
BCFSK Binary-Coded Frequency-Shift Keyed
BCF/Y Billion Cubic Feet per Year
BCG Bacillus Calmette-Guerin Vaccine
Bromocresol Green
BCGA British Commercial Gas Association
British Cotton Growers Association
BCGLO British Commonwealth Geographical Liaison Office
BCH 2-Amino-2-Norbornanecarboxylic Acid
Binary-Coded Hollerith
Bits per Circuit per Hour
Block Control Header
Blocked Calls Held
Bose-Chadhuri-Hocquenghem (Code)
Bunch
BCHA British Columbia Health Association [Canada]
BCHCM Board of Certified Hazard Control Management [US]
BCHPA British Columbia Hydro and Power Authority [Canada]
BChem Bachelor of Chemistry
BCI Battery Council International [US]
Binary-Coded Information
Bit Count Integrity
Bituminous Coal Institute [US]
Broadcast Interference
BCIA British Columbia Institute of Agrologists [Canada]
BCIAWPRC Belgian Committee of the International Association on Water Pollution Research and Control
Brazilian Committee of the International Association on Water Pollution Research and Control
BCILA British Columbia Independent Loggers Association [Canada]
BCIP Belgian Center for Information Processing
Bromochloroindolyl Phosphate
BCIPPA British Cast Iron Pressure Pipe Association
BCIQ Bureau of Commodity Inspection and Quarantine [MOEA, Taiwan]
BCIRA British Cast Iron Research Association
BCISC British Chemical Industrial Safety Council
BCIT British Columbia Institute of Technology [Canada]
BCK Biochemical Kinetics
BCKA Branched Chain A-Keto Acid
BCKCD Back-Order Code
BCL Bare-Coded Label
Base Contour Line
Basic Contour Line
Battelle Columbus Laboratories [US]
Binary Cutter Location
Biological Computer Laboratory [US]
Broadcast Listener
Burroughs Common Language
Bcl Bacillus Caldolyticus [Molecular Biology]
BCLA British Columbia Library Association [Canada]
BCLMA British Columbia Lumber Manufacturers Association [Canada]
BCLN British Columbia Library Network [Canada]
BCLS British Columbia Lifeboat Society [Canada]
BCM Back Course Marker
Bank Cubic Meter [also: bcm]
Beyond Capability of Maintenance
Binary Coded Matrix
British Catalogue of Music
Bromochloromethane
BCMA British Chipboard Manufacturers Association
British Color Makers Association
British Columbia Medical Association [Canada]
BCMAF British Columbia Ministry of Agriculture and Food [Canada]
BCME Bischloromethylether
BCMEA British Columbia Maritime Employees Association [Canada]
BCML Burroughs Current Mode Logic
BC-MOSFET Buried-Channel Metal-Oxide Semiconductor Field-Effect Transistor
BCMS Boron Concentration Measuring System
BCMTA British Columbia Motor Transport Association [Canada]
BCN Beacon
Biblioteca de Congreso de la Nacion [= Library of International Congress, Argentina]
Biomedical Communications Network [US]
BCNI Business Council on National Issues
BCNU British Columbia Nurses Union [Canada]
BCO Battery Cutoff
Bill in Care of
Binary-Coded Octal
Bridge Cutoff
BCOA Bituminous Coal Operators Association [US]
BCOM Burroughs Computer Output to Microfilm
BComm Bachelor of Commerce [also: BCom]
BCompS Bachelor of Computer Science
BComSc Bachelor of Commercial Science
BCORA British Colliery Owners Research Association
BCP Best Current Practices
Bisynchronous Communications Processor
Bromocresol Purple
Bureau of Consumer Protection [US]
Burroughs Control Program
Byte Control Protocol
BCPA British Copyright Protection Association
BCPB Bromochlorophenol Blue
BCPC British Columbia Press Council [Canada]
British Crop Protection Council
BCPIT British Council for the Promotion of International Trade
BCPL Basic Combined Programming Language
BCPMA British Chemical Plant Manufacturers Association
BCPO British Commonwealth Producers Organization
BCPS Basic Call Processing System
Beam Candle Power Seconds
BCPSM Board of Certified Product Safety Management [US]
BCR Benefit-to-Cost Ratio
Bibliographic Center for Research [US]
Billing, Collecting, Remitting
Bituminous Coal Research [now: BCRNL, US]
British Columbia Railways [Canada]
Brush Contact Resistance
Buffer Control Register
Bureau of Commercial Research [UK]
BCRA British Ceramic Research Association

	British Coke Research Association
BCRC	British Columbia Research Corporation [Canada]
	British Columbia Research Council [Canada]
BCRF	British Columbia Registered Forester [Canada]
BCRG	Brine Cavity Research Group [now: SMRI, US]
BCRNL	BCR (= Bituminous Coal Research) National Laboratory [US]
BCRT	Bright Cathode-Ray Tube
BCRU	British Committee on Radiological Units
BCRUM	British Committee on Radiation Units and Measurements
BCS	Bachelor of Commercial Science
	Bachelor of Computer Science
	Bar Code Sorter
	Bardeen-Cooper-Schrieffer Theory
	Basic Combined Subset
	Basic Control System
	Best Cast Steel
	Binary Communications — Synchronous [also: BISYNC]
	Biochemical Sciences
	Biomedical Computing Society [US]
	Block Check Sequence
	Block Control Signal
	Boeing Computer Services [US]
	Boston Computer Society [US]
	Bridge Control System
	British Cartographic Society
	British Ceramic Society
	British Composites Society
	British Computer Society
	Broadcasting Station
	Business Communication System
	Business Customer Service
BCSA	British Constructional Steelwork Association
BCSC	British Columbia Safety Council [Canada]
	British Computer Services Corporation
BCSIR	Bangladesh Council of Scientific and Industrial Research
BCSO	British Commonwealth Scientific Office [US]
BCSP	Board of Certified Safety Professionals [US]
BCT	Between Commands Testing
	Body-Centered Tetragonal [also: bct]
	Bushing Current Transformer
BCTA	British Canadian Trade Association
BCTC	British Ceramic Tile Council
BCTD	Building and Construction Trades Department
BCTP	British Continental Trade Press [UK]
BCU	Binary Counting Unit
	Block Control Unit
	British Commonwealth Union
	Buffer Control Unit
BCUC	British Coal Utilities Commission
BCURA	British Coal Utilization Research Association
BCV	Barge-Carrying Vessel
BCVA	British Columbia Veterinary Association [Canada]
BCW	Buffer Control Word
	Burst Codewords
BCWA	British Clock and Watch Association
BCWF	British Columbia Wildlife Federation [Canada]
BCWL	Basic Carbonate White Lead
BCWMA	British Clock and Watch Manufacturers Association
BCWWA	British Columbia Water and Waste Association [Canada]
BD	Backward
	Bahraini Dinar [Currency]
	Band
	Barrel Distortion
	Barrels per Day
	Base Detonating
	Base Down (Light Bulb)
	Baud [also: Bd or bd]
	Benzofurane Derivative
	Binary Decoder
	Binary Divide
	Binary-to Decimal
	Blocker Deflector
	Blocking Device
	Blood Effect
	Blowing Dust
	Board
	Bond
	Bottom Down
	Bound
	Breakdown
	Brush Discharge
	Business Design
	Business Designer
	Butadiene
B/D	Bank Draft
	Binary-to-Decimal
b/d	brought down
B$	Bahamian Dollar [Currency of the Bahamas]
BDA	Bomb Damage Assessment
	Booster-Distributor Amplifier
	Bright-Dipped and Anodized
	Bund Deutscher Architekten [= Federation of German Architects]
	Bundesvereinigung der Deutschen Arbeitgeberverbande [= Federation of German Employers Associations]
BDAM	Basic Data Access Method
	Basic Direct Access Method
BDC	Binary Decimal Counter
	Bonded Double Cotton (Wire)
	Book Development Council [UK]
	Bottom Dead Center
	Bureau of Domestic Commerce
BDCB	Buffered Data/Control Bus
BDCS	Butyldimethylchlorosilane
BDD	Binary-to-Decimal Decoder
	Bureau of Dangerous Drugs [of HWC, Canada]
BD$	Barbados Dollar [Currency]
BDDT	Banque de Donnees en Toxicologie [= Databank on Toxicology [of INSERM, France]]
BDE	Baseband Distribution Equipment
	Batch Data Exchange
	Brigade
BD ELIM	Band Elimination
BDES	Batch Data Exchange Service
BDes	Bachelor of Design
BDF	Baseband Distribution Frame

	Base Detonating Fuse
BDFA	British Dairy Farmers Association
bd ft	board foot [also: bd-ft or bdft]
BDGH	Binding Head
BDGT	Budget
BDH	Bearing, Distance and Heading
BDHI	Bearing, Distance and Heading Indicator
BDHT	Blowdown Heat Transfer Program
BDI	Bearing Deviation Indicator
	Bundesverband der Deutschen Industrie [= Federation of German Industries]
	Bureau of Dairy Industry [US]
BDIA	Base Diameter
	Bund Deutscher Innenarchitekten [= Federation of German Interior Designers]
BDIAC	Battelle Defender Information Analysis Center [of BMI, US]
BDIC	Binary-Coded Decimal Interchange Code
BDIR	Bus Direction
BDL	Bundle [also: BDLE]
BDLC	Burroughs Data Link Control
BDLI	Bundesverband der Deutschen Luft-, Raumfahrt- und Ausrustungsindustrie [= Federation of German Aeronautics, Aerospace and Equipment Industries]
BDM	Bomber Defense Missile
BDMA	Benzyldimethylamine
	British Disinfectant Manufacturers Association
	Butylene Glycol Dimethacrylate
BDMS	Brewer Data Management System
BDN	Bank Draft Number
	Bell Data Network
	Blocked Deoxynucleoside
BDNA	Benzyldinonylame
BDOS	Basic Disk Operating System
	Batch Disk Operating System
BDP	Bellman's Dynamic Programming
	Bioenergy Development Program [Canada]
	Bonded, Double Paper (Wire)
	Breakdown Potential
	Business Data Processing
BDPA	Bisdiphenylenephenylallyl
	Black Data Processing Associates [US]
BDPEC	Bureau of Disease Prevention and Environmental Control [US]
BDPP	2,4-Bis(diphenylphosphino)pentane
BDPS	Beam Deflection Phase Shifter
BDPSK	Binary Differential Phase-Shift Keying
BDR	Bell Doesn't Ring
	Bi-Duplexed Redundancy
	Binary Dump Routine
	Binder
BDRY	Boundary
BDS	Benzenediazothioether
	Boards
	Bomb Disposdal Squad
	Bonded Double Silk (Wire)
	Bonds
	Bulk Data Switching
	British Display Society
	Bundesverband Deutscher Stahlhandel [= Federation of German Steel Industries]
BDSA	Benzene Disulfonic Acid
	Business and Defense Services Administration [US]
BDSC	Boride Dispersion Strengthened Copper
BDT	Bangladesh Taka [Currency]
	Binary Deck to Binary Tape
BDTA	Benzophenone Tetracarboxylic Dianhydride
BDTH	Base Down to Horizontal (Light Bulb)
BDTP	Bisdiphenyltriazinylpyridine
2,4-BDTP	2,4-Bis(5,6-Diphenyl-1,2,4-triazin-3-yl)pyridine
BDTS	Buffered Data Transmission Simulator
	Bulk Data Transfer Subsystem
BDU	Baseband Distribution Unit
	Basic Device Unit
	Basic Display Unit
	Bomb Disposal Unit
	Bundesverband der Dolmetscher und Ubersetzer [= Federation of German Interpreters and Translators]
	Bundesverband Deutscher Unternehmensberater [= Federation of German Business Consultants]
BDV	Breakdown Voltage
BDW	Buried Distribution Wire
BDWTT	Battelle Drop Weight Tear Test
BDY	Boundary
BE	Bachelor of Economics
	Bachelor of Education
	Bachelor of Engineering
	Back End
	Backscattered Electron
	Bakery Engineer
	Bakery Engineering
	Band Elimination
	Base Ejection
	Bell End
	Benchmark Experiment
	Best Estimate Model
	Bill of Exchange
	Binding Energy
	Bioengineer
	Bioengineering
	Biological Engineer
	Biological Engineering
	Biomedical Engineer
	Biomedical Engineering
	Blended Elemental Technique
	Board of Education
	Body Engineer
	Body Engineering
	Boltzmann Equation
	Bond Energy
	Boundary Element
	Bound Energy
	Breaker End
	Burst-Error
B/E	Bill of Entry
	Bill of Exchange
Be	Beryllium
Bé	Baume [Unit]
BEA	Biological Environment Affectors [US]
	British Egg Association
	British Electrical Authority [now: CEA]

	British Engineers Association
	British European Airways
	Broadcast Education Association [US]
	Bureau d'Etudes Automatismes [= Study Bureau for Control Systems, France]
	Bureau of Economic Analysis [of USDOC]
	Business Education Association [US]
BEAB	British Electrical Approvals Board
	British Electrotechnical Approvals Board
BEAC	Boeing Engineering Analog Computer
	British European Airways Corporation
	British Export Advisory Council
BEACON	British European Airways Computer Network
BEAIRA	British Electrical and Allied Industries Research Association
BEAM	Brewer Earth Atmosphere Measurements
	Building Equipment, Accessories and Materials Program [Canada]
	Burroughs Electronic Accounting Machine
BEAMA	British Electrical and Allied Manufacturers Association
BEAMOS	Beam-Accessed Metal Oxide Semiconductor
BEAMS	British Emergency Air Medical Service
BEAP	Boundary Element Application Package
BEAR	Berkeley Elites Automated Retrieval [of UCB, US]
	Biological Effects of Atomic Radiation (Committee) [of National Academy of Sciences, US]
BEARP	Beaufort Environmental Assessment Review Panel [Canada]
BEAST	Basic Experimental Automatic Syntactic Translator
BEATRIX	Breeder Experimental Matrix
	Breeder Materials Exchange Matrix
BEBC	Big European Bubble Chamber
BEBO	Bond Energy Bond Order
BEC	Beginning of the Equilibrium Cycle
	Bioenergy Council
	Book Exporters Council [of SBPA, Singapore]
	British Engineers Club
	Burst-Error Correction
BECC	Burst-Error Correction Code
BEChemEng	Bachelor of Engineering in Chemical Engineering [also: BSChemE or BEChE]
BECM	British Electrical Conduit Manufacturers
BECMA	British Electroceramic Manufacturers Association
BECO	Booster-Engine Cutoff
BECS	Basic Error Control System
	Building Energy Conservation Sector
BECTIS	Bell College Technical Information Service [UK]
BECTO	British Electric Cable Testing Organization
BED	Bachelor of Environmental Design
	Bermuda Dollar [Currency]
	Bridge Element Delay
BEd	Bachelor of Education
BEDA	Bachelor of Environmental Design in Architecture
	British Electrical Development Association
	Bureau of European Designers Associations [UK]
BEDC	Building Economic Development Committee [UK]
BEDes	Bachelor of Environmental Design
BEDS	Bachelor of Environmental Design Studies
Beds	Bedforshire [UK]
BEDT-TTF	Bis(ethylenedithio)tetrathiafulvalene
BEE	Bachelor of Electrical Engineering
BEEC	Binary Error-Erasure Channel
BEEE	Bachelor of Electrical and Electronics Engineering
BEEF	Business and Engineering Enriched FORTRAN
BEEM	Ballistic Electron Emission Microscopy
BEEN	Bureau d'Etude de l'Energie Nucleaire [= Bureau for the Study of Nuclear Energy, Belgium]
BEF	Band Elimination Factor
	Band Elimination Filter
	Blunt End Forward
	Bovine Embryonic Fluid
	British Expeditionary Force
	Buffered Emitter Follower
BEFAP	Bell Laboratories FORTRAN Assembly Program
BEHA	British Export Houses Association
BEI	Backscattered Electron Image
	British Education Index [of British Library, UK]
BEIR	Biological Effects of Ionizing Radiation
BEK	Butyl Ethyl Ketone
BEL	Belgian Franc [Currency]
	Bell (Character)
BELF	Bundesministerium fur Ernahrung, Landwirtschaft und Forsten [= Federal Ministry for Food, Agriculture and Forestry, Germany]
BELG	Belgian
	Belgium
BELINDIS	Belgian Information Dissemination Service
BELLREL	Bell Laboratories Library Real-Time Loan [US]
BEM	Bachelor of Engineering, Mining
	Bioelectromagnetics
	Boundary Element Method
BEMA	Bakery Equipment Manufacturers Association [US]
	British Essence Manufacturers Association
	Business Equipment Manufacturers Association [now: CBEMA, US]
BEMAC	British Exports Marketing Advisory Committee
BEMB	Bituminous Equipment Manufacturers Bureau [now: BAEB, US]
	British Egg Marketing Board
BEMS	Bioelectromagnetics
	Bioelectromagnetics Society [US]
BEMSA	British Eastern Merchant Shipping Association
BEM SIG	Bioelectromagnetics Special Interest Group [US]
BEN	Bureau d'Etudes Nucleaires [= Bureau for Nuclear Studies, Belgium]
BENELUX	Belgium, Netherlands and Luxembourg [also: Benelux]
BENETUG	Benelux Tandem Users Group
BENG	Bengal
BEng	Bachelor of Engineering
BEngMgt	Bachelor of Engineering and Management
BEngSc	Bachelor of Engineering and Science
BEP	Bachelor of Engineering Physics
	Back Error Propagation
	Bureau of Engraving and Printing [US]
BEPC	Beijing Electron Positron Collider [PR China]
	British Electrical Power Convention
BEPO	British Experimental Pile Operation
BEPQ	Bureau of Entomology and Plant Quarantine [US]
BER	Berliner Forschungsreaktor [= Research Reactor at Berlin, Germany]

	Bit Error Rate
BERA	Biomass Energy Research Association [US]
BER CONT	Bit Error Rate Continuous
Berks	Berkshire [UK]
BERM	Basic Encyclopedic Redundancy Media
	Bit Error Rate Monitor
BERNET	Berlin Computer Network [Germany]
BERP	British Experimental Rotor Program
BERPM	Basic Exchange Rate Planning Model
BERT	Basic Energy Reduction Technology
	Bit Error Rate Test
BERTH	Berthing
BERTS	Bit Error Rate Test Set
BES	Bachelor of Engineering Science
	Bachelor of Environmental Studies
	Basic Energy Sciences
	Basic Executive System
	Beijing Electron Synchrotron [PR China]
	Bioelectrochemical Society [Germany]
	Biological Engineering Society [US]
	Biomedical Engineering Society [US]
	N,N-Bis(2-Hydroxyethyl)aminoethanesulfonic Acid
	Black Enamel Slate
	Bose-Einstein Statistics
	Boundary Element Solver
	Brazilian Entomological Society
	British Ecological Society
BESA	British Engineering Standards Association
	Building Energy System Analysis
BESc	Bachelor of Engineering Sciences
BESRL	Behavioral Science Research Laboratory [US Army]
BESS	Bessemer
	Binary Electromagnetic Signal Signature
BESSY	Berliner Elektronen-Strahlungssynchrotron [= Berlin Electron Radiation Synchrotron, Germany]
	Bestellsystem [= German Teleordering System]
BEST	Ballastable Earthmoving Sectionized Tractor
	Ballistic Electron Schottky-Gate Transistor
	Battery Energy Storage Test
	Bioprocessing Expert System Tool
	Boehler Electroslag Topping
	Business EDP (= Electronic Data Processing) Systems Technique
	Business Electronic Systems Technique
	Business Equipment Software Technique
BET	Best Estimate of Trajectory
	Biological Engineering Technologist
	Biological Engineering Technology
	Biomedical Equipment Technician
	Blended Elemental Technique
	Brunauer, Emmett, and Teller (Method)
BETA	Battlefield Exploitation and Target Acquisition
	Bureau for Education Technology and Administration [US]
	Business Equipment Trade Association [US, and UK]
BETO	Burlington Environmental Technology Office [of WTC, Canada]
BETRA	British Export Trade Research Organization
BETT	Buildings Energy Technology Transfer
BEU	Basic Encoding Unit
	Bengal Engineering College [India]
BEV	Bevel
	Bevelled
BeV	Billion Electronvolts
BEV-BD	Bevel Board
BEX	Broadband Exchange
BF	Bachelor of Forestry
	Backface
	Back Feed
	Bad Flag
	Balance Forward
	Band Filter
	Base File
	Base Fuse
	Beat Frequency
	Blast Furnace
	Blocking Factor
	Board Foot
	Boiler Feed
	Boldface
	Both Faces
	Bottom Face
	Branching Filter
	Bright Field
	British Forces
	Brought Forward
	Butterworth Filter
B&F	Bell and Flange
	Bound and Free [also: B+F]
b/f	brought forward
BFA	Blended Friable Abrasive
	Bundesstelle fur Aussenhandelsinformation [= Federal Office for Foreign Trade Information, Germany]
BFAS	Basic File Access System
BFB	Bridge Feed Transition
BFBS	British Forces Broadcasting Service
BFCO	Band Filter Cutoff
BFD	Back Focal Distance
	Basic Floppy Disk
BFDK	Before Dark
BFE	Beam Forming Electrode
	Bromotrifluoroethylene
BFER	Base Field Effect Register
BFF	Budget Furniture Forum [of NOPA, US]
BFG	Binary Frequency Generator
BFI	Betriebsforschungsinstitut [= Institute for Plant Research [of VDEh, Germany]]
	Bismuth Formic Iodide
BFL	Back Focal Length
	Buffered FET (= Field-Effect Transistor) Logic
	Busy Flash
BFMA	British Farm Mechanization Association
	Business Forms Management Association [US]
BFMC	British Friction Materials Council
BF_3-MEA	Boron Trifluoride Monoethylamine
BFMIRA	British Food Manufacturing Industries Research Association
BFMP	British Federation of Master Printers
BFMSA	British Firework Manufacturers Safety Association
BFN	Beam Forming Network

	British Forces Network
BFO	Back Flashover
	Beat-Frequency Oscillator
BFP	Back Focal Plane
	Boiler Feed Pump
BFPA	British Fluid Power Association
BFPC	British Farm Produce Council
BFPDDA	Binary Floating Point Digital Differential Analyzer
BFPO	Bis(dimethylamino)fluorophosphine Oxide
	British Forces Post Office
BFPR	Basic Fluid Power Research Program [US]
BFR	Bridged Frequency Ringing
BFr	Belgian Franc [Currency]
BFS	Beam-Foil Spectroscopy
	Brute Force Supply
	Bundesanstalt fur Flugsicherung [= Federal Institute for Air Traffic Control, Germany]
BFSA	British Fire Services Association
bFSH	Bovine Follicle-Stimulating Hormone
BFSK	Binary Frequency Shift Keying
BFTB	Better Fabrics Testing Bureau [US]
BFUW	British Federation of University Women
BFW	Boiler Feed Water
BG	Back Gear
	Background
	Beta-Glucuronidase
	Billing Group
	Bioglass
	Birmingham Gauge
	Blue-Green (Laser)
	Board of Governors
	Bonded Goods
	Bottom Grille
	British Standard Gauge
	Bunsen-Gesellschaft [= Bunsen Society, Germany]
BGA	Barre Granite Association [US]
	Blue-Green Algae
	Brilliant Green Agar
	British Gliding Association
	Butyl Glycol Acetate
BGAV	Blue Green Algae Virus
BG/BARC	Board of Governor's Budget and Accounts Review Committee [INTELSAT]
BGC	British Gas Council
	British Glaciological Society [now: IGS, UK]
BGDE	Brigade
BGE	Butyl Glycidyl Ether
BGEN	Brigadier General [also: BGen]
BGFMA	Bridge Grid Flooring Manufacturers Association [US]
BGG	Bovine Gamma Globulin
BGHT	Bought
BGI	Bureau Gravimetrique International [= International Gravimetric Bureau [of BRGM, France]]
BGIRA	British Glass Industry Research Association
BGL	Bulgarian Lev [Currency]
Bgl	Bacillus Globigii [Molecular Biology]
BGMA	British Gear Manufacturers Association
BGMC	British Glass Manufacturers Confederation
BGN	Board on Geographic Names [US]
BGO	Bismuth Germanate
BGP	Becker, Green and Pearson Equation
BGPC	Bunsen-Gesellschaft fur Physikalische Chemie [= Bunsen Society for Physical Chemistry, Germany]
BG/PC	Board of Governor's Planning Committee [of INTELSAT]
BGRAF	Basic Graphics Software
BGRR	Brookhaven's Graphite Research Reactor [of BNL, US]
BGRV	Boost Glide Reentry Vehicle
BGS	Bachelor of General Studies
	Beta Gamma Signal
	Bombing and Gunnery School
	Bosson, Gutmann and Simmons (Equation)
	Brazilian Geochemical Society
	British Grassland Society
BGWF	British Granite and Whinstone Federation
BH	Bake Hardening
	Benzhydrol
	Binary-to-Hexadecimal
	Blasthole
	Block Handler
	Boiler House
	Brinell Hardness
	Brush Holder
	Buried Heterostructure
	Busy Hour
B/H	B (= Magnetic Induction) versus H (= Magnetizing Field)
B-H	Binary-to-Hexadecimal
BHA	Base Helix Angle
	Bottomhole Assembly
	Bus History Association [Canada]
	Butylated Hydroxyanisole
BHAB	British Helicopter Advisory Board
BHC	Benzene Hexachloride
BHCA	Busy Hour Call Attempts
BHD	Bahraini Dinar [Currency]
	Bulkhead
BHET	Bishydroxyethyl Terephthalate
BHFP	Bottomhole Flowing Pressure
BHGA	British Hang Gliding Association
BHI	British Horological Institute
	Bureau Hydrogaphique International [= International Hydrographical Bureau, Monaco]
BHJT	Bipolar Heterojunction Transistor
BHL	Busy Hour Load
BHLS	Below/Hook Lifters Section [of the Materials Handling Institute, US]
BHMA	Builders Hardware Manufacturers Association [US]
BHMT	Bishydroxymethyl Terephthalate
BHN	Brinell Hardness Number [also: Bhn]
BHP	Boiler Horsepower [also: bhp]
	Bottom Hole Pressure
	Brake Horsepower [also: bhp]
	Broken Hill Proprietary
	Bulk Handling Plant
BHPhr	Brake Horsepower-Hour
BHR	Biotechnology and Human Research
	Block Handling Routine
BHRA	British Hydromechanics Research Association

BHS	British Herpetological Society
	British Horological Society
BHSA	British Heavy Steel Association
BHSL	Basic Hytran Simulation Language
BHT	Baht [Currency of Thailand]
	Blowdon Heat Transfer
	Bottom Hole Temperature
	Butylated Hydroxytoluene
	Butylated Hydroxytoluol
Bht	Baht [Currency of Thailand]
BHU	Banaras Hindu University [India]
BHW	Boiling Heavy Water
BHyg	Bachelor of Hygiene
BI	Backward Indicator
	Base Ignition
	Basic Index
	Basicity Index
	Batch Input
	Bedford Institute [Canada]
	Billing Instructions
	Bismuth Institute [Belgium]
	Black Iron
	Blanking Input
	Break Interrupt
	Bus Interconnect
B&I	Base and Increment
Bi	Biot [Unit]
	Bismuth
BIA	Bicycle Institute of America [US]
	Boating Industry Association
	Boost, Insertion and Abort
	Brick Institute of America [US]
	British Ironfounders Association
	Buses International Association [US]
	Buyers Information Advisory
BIAA	British Industrial Advertising Association
	Bureau of Inter-American Affairs [US]
BIAC	Bio-Instrumentation Advisory Council [of AIBS, US]
	Business and Industry Advisory Committee [of OECD, France]
BIAMPA	Banque d'Information Automatisee sur les Medicaments Principes Actifs [= Automated Information Bank on Active Ingredients of Drugs [also: BIAM PA, France]]
BIAMS	Banque d'Information Automatisee sur les Medicaments Specialites [= Automated Information Bank on Special Pharmaceutical Products [also: BIAM S, France]]
BIAS	Battlefield Illumination Airborne System
	Biomedical Instrumentation Advisory Service [UK]
	Brooklyn Instutute of Arts and Sciences [US]
BIAT	British Institute of Architectural Technicians
BIATA	British Independent Air Transport Association
BIB	Balanced Incomplete Block Design
	Burn-In-Board
BIBD	Balanced Incomplete Block Design
BIBDES	Bibliographic Data Entry System [of BLAISE, UK]
BIBL	Bibliographic
	Bibliographer
	Bibliography
BIBLIOS	Book Inventory Building Library Information Oriented System [US]
BIBM	Bureau International du Beton Manufacture [= International Bureau for Precast Concrete, Belgium]
BIBO	Bounded Input, Bounded Output
BIBRA	British Industrial Biological Research Association
BIC	Beef Industry Council [of NLSMB, US]
	Biodeterioration Center [of UAB, UK]
	Bipolar Integrated Circuit
	Book Importers Council [of SBPA, Singapore]
	Bureau International des Containers [= International Container Bureau, France]
	Bureau of International Commerce
	Business Information Center
	Bus Interface Controller
BICAMS	Barnett Institute of Chemical Analysis and Materials Science [of NU, US]
BICEMA	British Internal Combustion Engine Manufacturers Association
BICEPT	Book Indexing with Context and Entry Point from Text
BICERA	British Internal Combustion Engine Research Association
BICERI	British Internal Combustion Engine Research Institute
BICINE	Bis(hydroxyethyl)glycine
BICMOS	Bipolar Complementary Metal-Oxide Semiconductor
BICROS	Binaural Contralateral Routing of Signal
BICS	British Institute of Cleaning Sciences
	Building Industry Consulting Service
BICT	Bipolar Integrated Circuit Technology
BICTA	British Investment Casters Technical Association
BID	Bachelor of Industrial Design
	Bachelor of Interior Design
	Blocker Initial Design
	Blocker Initial-Guess Design
	Blocker Initial-Guess Device
BIDAP	Bibliographic Data Processing Program
BIDCO	Built-in Digital Circuit Observer
BIDMC	Bismuth-Dimethyl Dithiocarbamate
BIDS	Burroughs Information Display System
BIE	Bachelor of Industrial Engineering
	Boundary Integral Equation
	British Institute of Engineers
	Bureau International des Expositions [= International Bureau of Exhibitions, France]
BIEE	British Institute of Electrical Engineers
BIECA	Belgium International Express Courier Association
BIEM	Boundary Integral Equation Method
BIET	British Institute of Engineering Technology
BIF	Banded Iron-Formation
	Basic in Flow
	Beef Improvement Federation [US]
	British Industries Fair
	Burundi Franc [Currency]
BIFAC	British Isles Federation of Agricultural Cooperatives
BIFET	Bipolar Compatible Field-Effect Transistor
BIFMA	Business and Institutional Furniture Manufacturers Association [US]
BIG	Bibliography and Index of Geology

BIGA Bundesamt fur Industrie, Gewerbe und Arbeit [= Federal Office for Industrie, Business and Labour, Switzerland]
BIGCHAP Bis(gluconamidopropyl)cholamide
BIGENA Bibliography and Index of Geology Exclusive of North America
BIGFET Bipolar Insulated-Gate Field-Effect Transistor
BIHFS British Institute of Hardwood Flooring Specialists
BIIA British Institute of Industrial Art
BIIL Basic Impulse Isolation Level
BIIR Bromobutyl Rubber
BIIT Bureau International d'Information pour les Telecommunications [= International Information Bureau for Telecommunications, Switzerland]
BIKAS Bibliothekskatalogsystem [= Automated Library Cataloguing System, Germany]
BIKE Biotechnologischer Informations-Knoten fur Europa [= Biotechnological Information Network for Europe [of GBF, Germany]]
BIL Basic Impulse Level
Basic Insulation Level
Block Input Length
Built-In Logic
Buried Injection Logic
BILA Battelle Institute Learning Automation [of BMI, US]
BILBO Built-in Logic Block Observability
Built-in Logic Block Observer
BILD Board for Industrial Leadership and Development [Canada]
Business Industrial Leadership Development
BILE Balanced Indicator Logical Element
BILGE Binary Load Generation
billi billion
BIM Beginning of Information Marker
Best in Match
Blade Integrity Monitor
Branch 'If' Multiplexer
British Institute of Management
Bus Interface Module
BIMAC Bistable Magnetic Core [also: BIMAG]
BIMCAM British Industrial Measuring and Control Apparatus Manufacturers Association
BIMCO Baltic and International Maritime Council [Denmark]
BIMOS Bipolar Metal-Oxide Semiconductor [also: BiMOS]
BIMRAB BuWeps-Industry Materiel Reliability Advisory Board [US Navy]
BIN Basic Identification Number
Binary
Bureau of Information on Nickel [UK]
Business Information Number
BINAC Binary Automatic Computer
BINAP Binaphthyl
2,2'-Bis(diphenylphosphino)-1,1'-Binaphthyl
BINC Building Industries National Council [UK]
BInd Bachelor of Industry
BINDT British Institute of Nondestructive Testing
BINOMEXP Binomial Expansion
BINOVC Break in Overcast
BINR Basic Intrinsic Noise Ratio
BINS Boreal Institute for Northern Studies [Canada]
BInstNDT British Institute of Nondestructive Testing
BIntArch Bachelor of Interior Architecture
BIntDes Bachelor of Interior Design
BIO Bedford Institute of Oceanography [Canada]
Bio Industry Association [UK]
Biomedical Information Processing Organization [US]
Buffered Input/Output
BIOAL Bioastronautics Laboratory
BIOALRT Bioastronautics Laboratory Research Tool
BIOCHEM Biochemical
Biochemist
Biochemistry
BIOD Bell Integrated Optical Device
BIOFOR/BIOQUAL Implementation of Biological Processes for the Forest Industry and Environmental Quality [Canada]
BIOL Biological
Biologist
Biology
Biol Gr Biological Grade [Chemistry]
BIOMAIL Bedford Institute of Oceanography Marine Advisory and Industrial Liaison Office [Canada]
BIONESS Biological (Zooplankton) Net Sampling System
BIONIC Biological and Electronic [also: Bionic, bionic]
BIONICS Biological Electronics [also: Bionics, bionics]
BIOP Buffer Input/Output Processor
BIOPHYS Biophysical
Biophysicist
Biophysics
BIOR Business Input-Output Rerun
BIORED Biological Resources Development
BIOS Baffin Island Oil Spill Project [Canada]
Basic Input-Output System
Biological Investigation of Space (Project) [of NASA]
British Intelligence Objectives Subcommittee
BIOSIS Biosciences Information Service [US]
BioSync (Structural) Biology Synchrotron Users Organization [US]
BIOT British Indian Ocean Territory
BIOTA Biological Institute of Tropical America [US]
BIOTEX Biotechnological Experiment
BIP Balanced in Plane
BASIC Interpreter Package
Binary Image Processor
Bismuth Iodoform Paraffin
Books in Print (Database)
Banco Industrial del Peru [= Industrial Bank of Peru]
BIPAC Bibliographic Procedures and Control Committee [of RLG, US]
BIPCO Built-in Place Component
BIPD Bi-Parting Doors
BIPM Bureau International des Poids et Mesures [= International Bureau of Weights and Measures, France]
BIPS Billion Instructions per Second
BIPY Bipyramidal [also: Bi-PY]
2,2'-Bipyridine

2,2'-Bipyridyl

BIR B-50 Immunoreactivity
Bioinorganic Reaction
British Institute of Radiology
Bureau International de la Recuperation [= International Reclamation Bureau, Belgium]
Bureau of Internal Revenue [US]

BIRC Bio-Integral Resource Center [US]

BIRDIE Battery Integration and Radar Display Equipment

BIRE British Institute of Radio Engineers

BIRES Broadband Isotropic Real-Time Electric-Field Sensor

BIRF Brewing Industry Research Foundation

BIRL Basic Industry Research Laboratory [of NWU, US]

BIRM Bio-Inorganic Reaction Mechanism

BIRS Basic Indexing and Retrieval System
Basic Information Retrieval System [of ISL, US]
British Institute of Recorded Sound

BIS Bank for International Settlement [Switzerland]
Basic Interchange System
Biological Information Service [US]
Bremsstrahlung Isochromat Spectroscopy
British and Irish Skeptics [Ireland]
British Ichthyological Society
British Imperial System
British Information Service
British Interplanetary Society
Brought into Service
Business Information System

BISA Bibliographic Information on Southeast Asia [University of Sydney, Australia]

BISAC Book Industry Systems Advisory Committee [of BISG, US]

BISAM Basic Indexed Sequential Access Method

Bis-AMP Bisaminomethylpropanol

BISC Berkeley Integrated Sensor Center [of UBC, US]
Business and Industry Service Center [Canada]

BISCOM Business Information System — Communications

BISCUS Business Information System — Customer Service

BISF British Iron and Steel Federation

BISFA British Industrial and Scientific Film Association
Bureau International pour la Standardisation de la Rayonne et des Fibres Synthetiques [= International Bureau for the Standardization of Man-Made Fibers, Switzerland]

BISG Book Industry Study Group [US]

BisGMA Bisphenolglycidyldimethacrylate

BISITS British Industrial and Scientific International Translation Service
British Iron and Steel Industry Translation Service

BISL British Information Service Library

BISMRA Bureau of Inter-Industrial Statistics and Multiple Regression Analysis [US]

Bis-MSB 1,4-Bis(2-Methylstyryl)benzene

Bis-NAD Bis(carbonylmethyl)nicotinamide Adenine Dinucleotide

BISNET Bank Information System Network [US]

BISNY Binary Synchronous Communication

BISPA British Independent Steel Producers Association

BISPAD Bissilylated Phenylaminodiol

BISRA British Iron and Steel Research Association

BISS Base and Installation Security System

BISSC Baking Industry Sanitation Standards Committee [US]

BIST Beijing Institute of Science and Technology [PR China]
Built-in Self Test

BIS-TRIS Bis(2-Hydroxyethyl)iminotris(hydroxymethyl)-methane [also: bis-tris]
2-Bis(2-Hydroxyethyl)-Amino-2-(hydroxymethyl)-1,3-Propanediol
2,2-Bis(hydroxymethyl)-2,2',2''-Nitrilotriethanol [also: bis-tris]
Bisiminotrismethane

BISYNC Binary Synchronous Communication
Bisynchronous [also: BISYNCH]
Bisynchronous Transmission [also: bisync]

BISYNCH Binary Synchronous Communication

BIT Backscatter Imaging Tomography
Bihar Institute of Technology [India]
Binary Digit [also: Bit or bit]
Bipolar Integrated (Circuit) Technology
Boric-Acid Injection Tank
Boron Injection Tank
Brighton Information Technology Conference
Built-In Test
Bureau International du Travail [= International Labour Bureau]

BITA British Industrial Truck Association

BITBLT Bit (= Binary Digit) Block Transfer

BITE Backward Interworking Telephony Event
Built-In Test Equipment

BITEL Bildschirm-Telefon [= Videotext Telephone, Germany]

BITL Bureau International Technique de l'ABS [= International Technical Bureau of ABS (= Acrylonitrile-Butadiene-Styrene) Producers [of CEFIC, Belgium]]

BITM Bureau International Technique du Methanol [= International Technical Bureau of Methanol Producers [of CEFIC, Belgium]]

BITN Bilateral Iterative Network

BITPI Bureau International Technique des Polyesters Insatures [= International Technical Bureau of Producers of Unsaturated Polyesters [of CEFIC, Belgium]]

BITS Birla Institute of Technology and Science [India]

bit/s bits per second

BIU Basic Income Unit
Basic Information Unit
Bus Interface Unit

BIV Best Image Voltage

BIVAR Bivariant Function Generator

BIX Binary Information Exchange

BJ Bar Joist
Bubble Jet

BJA Burlap and Jute Association [US]

BJCEB British Joint Communications and Electronics Board

BJCG British Joint Corrosion Group

BJF Batch Job Foreground

BJM Between Job Monitor

BJMA	Bangladesh Jute Mills Association
BJP	Bubble Jet Printer
BJT	Base Junction Transistor
	Bipolar Junction Transistor
BJTRA	British Jute Trades Research Association
BK	Backward
	Bank
	Bar Knob
	Black
	Block
	Book
	Brake
Bk	Barkometer
	Berkelium
BKCY	Bankruptcy
BKG	Banking
BKGRD	Background
BKKPR	Bookkeeper
BKME	Bleached Kraft Mill Effluent
BKPR	Bookkeeper
BKPT	Bankrupt [also: BKRPT]
BKR	Breaker
BKS	Books
BKSP	Backspace
BKSTS	British Kinematograph, Sound and Television Society
BKT	Basket
	Bracket
BKZ	Bernstein, Kearsley and Zapas (Model)
BL	Bale
	Barrier Layer
	Baseline
	Bell
	Bend Line
	Between Layers
	Bibliographic Level
	Bioluminescence
	Black
	Blanking
	Block
	Block Length
	Blotting
	Blue
	Bottom Layer
	Boundary Layer
	Breadth-Length
	Breech Loading
	British Library
	Building Line
	Burkitt's Lymphoma
B/L	Bill of Lading
bl	bale
	barrel [Unit]
BLA	Bachelor of Landscape Architecture
	Binary Logical Association
	Blocking Acknowledgement Signal
	British Light Aviation Center
	Bureau de Liaison des Syndicats Europeens (CEE) des Produits Aromatiques [= Liaison Bureau of the European and EEC Unions of Aromatic Products, Belgium]
BLAC	British Light Aviation Center
BLADE	Basic Level Automation of Data through Electronics
BLADES	Bell Laboratories Automatic Design System [also: BLADS, US]
BLAISE	British Library Automated Information Service
BLAM	Ballistically-Launched Aerodynamic Missile
BLAST	Blocked Asynchronous Transmission
	Boolean Logic and State Transfer
B/L att	Bill of Lading Attached
BLBS	British Library Bibliographic Services [Division of British Library]
BLBSD	British Library Bibliographic Services Division
BLC	Beef Liver Catalase
	Board Level Computer
	Boundary-Layer Concept
	Boundary-Layer Control
	British Lighting Council
BLD	Beam-Lead Device
	Binary Load Dump
BLDG	Building [also: BLD]
BLDR	Builder
BLE	Bombardment-Induced Light Emission
	Brotherhood of Locomotive Engineers
BLEND	Birmingham Loughborough Electronic Network Development [UK]
BLERT	Block Error Rate Test
BLESSED	Bell Little Electronic Symbolic System for the Electrodata
BLEU	Belgium-Luxembourg Economic Union
	Blind Landing Experimental Unit [of RAE, UK]
BLEVE	Boiling Liquid Expanding Vapor Explosion
BLF	Band Limiting Filter
	British Leather Federation
BLG	Benzyl-L-Glutamate
	Breech Loading Gun
BLI	Basic Learning Institute
BLIC	Bureau de Liaison de l'Industrie Caoutchouc [= Liaison Bureau of the Rubber Industries of the EEC, Belgium]
B-LINK	Birmingham Library and Information Network [UK]
BLIP	Background-Limited Incident Power
	Background Limited Infrared Photoconductor
BLIS	Baffle/Liner Interface Seal
	Bell Laboratories Interpretative System
BLISS	Basic Language for the Implementation of System Software
BLitt	Bachelor of Letters
BLK	Black
	Blank
	Block
	Bulk
BLK CAR	Bulk Carrier
BLKG	Blocking
BLKSH	Blackish
BLL	Below Lower Limit
BLLD	British Library Lending Division
BLLDZR	Bulldozer
BLM	Bachelor of Land Management
	Background Luminance Monitor
	Basic Language Machine
	Bilayer Lipid Membrane

Bureau of Land Management [US]
BLMRA British Leather Manufacturers Research Association
BLMSS Bayfield Laboratory for Marine Science and Surveys
BLN Balloon
BLO Below Clouds
Blocking Signal
Blower
Butyrolactone
BLOC Booth Library On-Line Circulation [of EISU, US]
BLODI Block Diagram
BLP Ball Lock Pin
Bypass Label Processor
BLPC Blocking Layer Photocell
BLR Baseline Restorer
Boiler
Breech Loading Rifle
BLRD British Library Reference Division
BLR&DD British Library Research and Development Division
BLS Bachelor of Library Science
Bureau of Labour Statistics [US]
BLSGMA British Lamp Blown Scientific Glassware Manufacturers Association
BLSH Bluish
BLSN Blowing Snow
BLSTG Blasting
BLSTG PWD Blasting Powder
BLT Bachelor of Laboratory Technology
Basic Language Translator
Borrowed Light
BL&T Blind Loaded and Traced
BLTC Bottom Loading Transfer Cask
BLTI Better Lawn and Turf Institute [now: LI, US]
BLU Basic Link Unit
Basic Logic Unit
Bipolar Line Unit
BLUE Best Linear Unbiased Estimator
BLVD Boulevard
BLWA British Laboratory Ware Association
BLX Benelux Economic Union
BLZD Blizzard
BM Bachelor of Medicine
Back Marker
Ballistic Material
Ballistic Missile
Barrels per Month
Beam
Bellows Metering Valve
Bench Mark
Bending Magnet
Bending Moment
Benelux Metallurgie [= Joint Metallurgy Organization of Belgium, Netherlands and Luxembourg]
Bill of Material
Binary Multiply
Biomaterial
Biomechanics
Biomedicine
Board Measure
Boundary Marker
Brigade Major
British Museum
Brownian Motion
Buffer Module
Bulk Modulus
Bureau of Meteorology [Australia]
Bureau of Mines [US]
Business Machine
B/M Bill of Material
BMA Bachelor of Management Arts
Bayesian Multivariate Analysis
Bicycle Manufacturers Association (of America) [US]
Biomedical Marketing Association [US]
British Manufacturers Association
British Medical Association
BMAA British Microlight Aircraft Association
BM-AGA Bureau of Mines — American Gas Association (Coal Test)
BMAR Ballistic Missile Acquisition Radar
BMath Bachelor of Mathematics
BMAW Bare Metal-Arc Welding
BMB Baltic Marine Biologists [Sweden]
British Metrication Board
BMBT Bismethoxybenzylidenebitoluidine
BMBW Bundesministerium fur Bildung und Wissenschaft [= Federal Ministry for Education and Science, Germany]
BMC Bimolecular Collision
Block Multiplexer Channel
Brittle Matrix Composite
Bubble Memory Controller
Bulk Media Conversion
Bulk Molding Compound
Burst Multiplexer Channel
Bureau of Management Consulting
BMCM Bohr-Mottelson Collective Model
BMCS Bureau of Motor Carrier Safety [US]
BMD Bahamian Dollar [Currency]
Ballistic Missile Defense
Benchmark Monitor Display System
Bermuda Dollar [Currency]
Biomechanical Device
Bubble Memory Device
BMDH Basic Minimum Descent Height
BME Bachelor of Mechanical Engineering
Bachelor of Mining Engineering
Biomechanical Engineer
Biomechanical Engineering
Biomedical Engineer
Biomedical Engineering
BMEC Ball Bearing Manufacturers Engineers Committee [US]
British Marine Equipment Council
BMEF British Mechanical Engineering Federation
BMEP Brake Mean Effective Pressure
BMES Biomedical Engineering Society [US]
BMet Bachelor of Metallurgy
BMetE Bachelor of Metallurgical Engineering
BMETO Ballistic Missiles European Task Organization
BMEWS Ballistic Missile Early Warning System
BMF Bulk Mail Facility

BMFF British Man-Made Fibers Federation
BMFT Bundesministerium fur Forschung und Technologie [= Federal Ministry for Research and Technology, Germany]
BMG Business Machine Group
BMGE Bovine Mammary Gland Epithelial
BMgtSc Bachelor of Management Science
BMI Basic Measuring Instrument
Battelle Memorial Institute [US]
Bibliography Master Index
Bismaleimide
Body Mass Index
Book Manufacturers Institute [US]
BMIC Bachelor of Microbiology
BMILS Bottom-Mounted Impact Location System
BMIS Bank Management Information System
BMIU Bricklayers and Masons Independent Union [Canada]
BMKR Boilermaker
BML Bulk Material Length
BMLA British Maritime Law Association
BMLSc Bachelor of Medical Laboratory Science
BMM Biomedical Material
BMMC Basic Monthly Maintenance Charge
BMMG British Microcomputer Manufacturers Group
BMO Ballistic Missile Office
BMOC Ballistic Missile Orientation Course
BMOM Base Maintenance and Operations Model
BMP Batch Message Processing
Biomedical Polymer
Bricklayers, Masons and Plasterers Union
Brownian Motion Process
BMR Basal Metabolic Rate
Bipolar Magnetic Region
Brookhaven's Medical Reactor [of BNL, US]
Bureau of Mineral Resources [Australia]
BMRC British Medical Research Council
BMRGG Bureau of Mineral Resources, Geology and Geophysics [Australia]
BMS Ballistic Missile Ship [US Navy]
Basic Mapping Support
Battle Management System
Biomedical System
Biometric Society [US]
Biomimetic System
Birmingham Metallurgical Society [UK]
Boeing Materials Specification
Boranemethylsulfide
British Mycological Society
Bm$ Bermuda Dollar [Currency]
BMSA Benzene Monosulfonic Acid
BMSGMA British Maize Starch and Glucose Manufacturers
BMT Base Metal Tailings
Basic Motion-Time Study
BMTA British Mining Tools Association
British Motor Trade Association
BMTD Ballistic Missile Terminal Defense
BMTFA British Malleable Tube Fittings Association
BMTI Block Mode Terminal Interface
BMTS Ballistic Missile Test System
BMTT Buffered Magnetic Tape Transfer
Buffered Magnetic Tape Transport
BMV Brome Mosaic Virus
BMVg Bundesministerium fur Verteidigung [= Federal Ministry of Defense, Germany]
BMW Beamwidth
BMWE Brotherhood of Maintenance of Way Emnployees
BMWi Bundesministerium fur Wissenschaft [= Federal Ministry for Science, Germany]
BMZ Baumusterzentrale [= Center for Industrial Design, Austria]
BN Battalion
Becklin-Neugebauer (Object) [Astronomy]
Benzonitrile
Binary Number
Boron Nitride
Branch Name
Bromonaphthalene
Bureau Number
BNA Boeing Network Architecture
British North America
Bureau of National Affairs [US]
BNAA British North America Act
BNAC British North American Committee
BNAG Bibliography of North American Geology
BNAS Bolivian National Academy of Sciences
BNB British National Bibliography [of BLBSD]
Bullet Nose Bushing
2,4,6-Tri-tert-Butylnitrosobenzene
BNBC British National Book Center
BNC Baby 'N' Connector
Bayonet Connector
Bayonet Neil-Concelman Connector
Bayonet Nut Coupling
British National Committee
BNCC British National Committee for Chemistry
BNCHBD Benchboard
BNCI Banque Nationale pour le Commerce et l'Industrie [= National Bank for Commerce and Industry, Belgium]
BNCIAWPRC Belgian National Committee of the International Association on Water Pollution Research and Control
Brazilian National Committee of the International Association on Water Pollution Research and Control
BNCM British National Committee on Materials
BNCOR British National Committee for Oceanographic Research
BNCS Board on Nuclear Codes and Standards
British Numerical Control Society
BNCSAA British National Committee on Surface-Active Agents
BNCSR British National Committee on Space Research
BNCWPC British National Committee for World Petroleum Congresses
BND Benzoylated, Naphthoylated DEAE (= Diethylaminoethyl)
Brunei Dollar [Currency of Brunei]
Bullet Nose Dowel
Bundesnachrichtendienst [= Federal Intelligence Service, Germany]

BNDO	Bureau National des Donnees Oceanique [= National Bureau for Oceanographic Data, France]
BNDS	Banque Nationale de Developpement du Senegal [= National Development Bank of Senegal]
BNEC	British National Export Council
	British Nuclear Energy Conference
BNES	British Nuclear Energy Society
BNF	Backus-Naur Form
	Backus Normal Form
	Bomb Nose Fuse
	British Nuclear Forum
	British Nutrition Foundation
BNF ABS	British Nonferrous Metals Abstracts [of BNFMTC]
BNFL	British Nuclear Fuels Limited
BNFMF	British Nonferrous Metals Federation
BNFMRA	British Nonferrous Metals Research Association
BNFMS	Bureau of Nonferrous Metals Statistics [UK]
BNFMTC	British Nonferrous Metals Technology Center
BNFSA	British Nonferrous Metals Smelters Association
BNG	Branch No Group
BNH	Burnish
BNI	Beijing Neurosurgical Institute [PR China]
BNIST	Bureau National de l'Information Scientifique et Technique [= National Bureau for Scientific and Technical Information, France]
BNL	Bell Northern Laboratories [Canada]
	Brookhaven National Laboratory [of USAEC]
BNM	Bureau National de Metrologie [= National Bureau of Metrology, France]
BNO	Backus Normal Form
BNOA	Beta-Naphthoxyacetic Acid
BNOC	British National Oil Corporation
BNPA	Beta-Nitropropionic Acid
BNPF	Beginning Negative Positive Finish
BNPP	Bis-(P-Nitrophenyl) Phosphate
BNR	Bell Northern Research [Canada]
	Bonded Negative Resistance Diode
BNRL	British Nuclear Reactors Limited
BNRS	British National Radio School
BNS	Bachelor of Natural Sciences
	Bachelor of Naval Sciences
	Bill Number Screening
	Binary Number(ing) System
BNSC	British National Space Center
BNSF	Belgian National Science Foundation
BNT	(Salt) Bath Nitriding
	Boreal Northern Titles [UA, Canada]
BNWL	Battelle Northwest Laboratories [US]
BNX	British Nuclear Export Executive
BNZS	Bipolar with N Zeros Substitution
BO	Back Order
	Beat Oscillator
	Binary-to-Octal
	Blanking Oscillator
	Blocking Oscillator
	Blow-Off
	Blow-Out
	Bond Order
	Branch Office
	Buyer's Option
B/O	Bad Order
	Branch Office
B-O	Binary-to-Octal
b/o	brought over
BOA	British Optical Association
BOAC	British Overseas Airways Corporation
BOAM	Bell Owned and Maintained
BOB	Bobbin
BOBMA	British Oil Burners Manufacturers Association
BOC	Beginning of Cycle
	Bevatron Orbit Code
	Bio-Organic Chemistry
	Block Oriented Computer
	Blowout Coil
	Bond-Order Conservation
	Bottom of Conduit
	Butoxycarbonyl
BOCA	Building Officials Conference of America
	Building Officials and Code Administrators (International) [US]
BOCES	Board of Cooperative Educational Services [US]
BOC-MP	Bond-Order Conservation — Morse Potential
BOCOL	Basic Operating Consumer-Oriented Language
BOC-ON	2-(tert-Butoxycarbonyloxyimino)-2-Phenylacetonitrile
BOCS	Bendix Optimum Configuration Satellite
BOD	Barrels of Oil per Day
	Base Ordnance Depot
	Beneficial Occupancy Date
	Biochemical Oxygen Demand
	Biological Oxygen Demand
	Bistable Optical Device
	Bottom of Duct
	Breakover Diode
BO/D	Barrels of Oil per Day [also: bo/d, BO D^{-1} or bod^{-1}]
BOD/COD	Biological Oxygen Demand/Chemical Oxygen Demand
BOE	Beginning of Extent
	Bureau of Explosives [now: HMS(BOE), US]
BOF	Basic Oxygen Furnace
BOFA	Bibliography of Agriculture [of NAL, US]
B of B	Back of Board
B of C	Bank of Canada
B of E	Board of Education
BOFGA	Biodynamic Farming and Gardening Association [US]
BOFS	Black Oil Finish Slate
BOFT	Board on Foreign Trade [of MOEA, Taiwan]
BOG	Bismuth Orthogermanate
BOH	Breakoff Height
BOHS	British Occupational Hygiene Society
BOI	Beginning of Information
	Branch Output Interrupt
BOIESA	Bureau of Oceans and International Environmental and Scientific Affairs [of USDOS]
BOL	Beginning of Life
	Bolivia(n)
BOLD	Bibliographic On-line Library Display
BOLS	Bolster
BOLT	Beam of Light Transistor

	British Columbia Organization of Library Technicians [Canada]
BOM	Basic Operating Monitor
	Bill of Materials
	Binary Order of Magnitude
BoM	Bureau of Mines [also: BOM, US]
BOMA	British Overseas Mining Association
	Building Owners and Managers Association
BOMAEC	Bureau of Mines/Atomic Energy Commission Incinerator
BOMEX	Barbados Oceanographic and Meteorological Experiment
BOMP	Base Organization and Maintenance Processor
BOMS	Bureau of Overseas Medical Service [UK]
BON	Beta-Oxynaphthoic Acid
BONAC	Broadcasting Organizations of Non-Aligned Countries [Yugoslavia]
BONUS	Boiling Nuclear Superheat Reactor
BONUS-CX	Boiling Nuclear Superheat Critical Experiment
Boo	Bootes [Astronomy]
BOOG	British Osborne Owners Group
BOOGIE	British Osborne Owners Group Information Exchange
BOOL	Boolean (Algebra)
BOOST	Booster
BOOT	Bootstrap
Boot	Bootes [Astronomy]
BOP	Balance of Payments
	Balance of Plant
	Balance of Power
	Basic Operating Program
	Basic Operator Panel
	Basic Oxygen Process
	Binary Output Program
	Bit Oriented Protocol
	Benzotriazolyl-N-oxy-tris(dimethylamino)phosphonium
	Blowout Preventer
	Bolivian Peso [Currency]
BOP-Cl	Bis(2-Oxo-3-Oxazolidinyl)phosphinic Chloride
BOPD	Barrels of Oil per Day [also: bopd]
BOPET	Biaxially-Oriented Polyethylene Terephthalate
BOPOB	Bis[2-(5-P-Biphenylyloxazoyl)]benzene
BOPP	Biaxially Oriented Polypropylene
BOPRESS	Boiler Pressure [also: bopress]
BOR	Bus Out Register
BORAD	British Oxygen Research and Development Association
BORAM	Block-Oriented Random-Access Memory
BORAX	Boiling Reactor Experiment [of INEL, US]
BORE	Beryllium Oxide Reactor Experiment [US]
BOREQ	Broadcast Request
BORIS	Book Order, Register and Invoicing System
BORSCHT	Battery, Overvoltage, Ringing, Supervision, Coding, Hybrid and Test Access
BOS	Background Operating System
	Back of Slip
	Basic Operating System
	Basic Oxygen Steel
	Basic Oxygen Steelmaking
	Batch Operating System
	Bell Operating System
	Book Order System
	Brewer Ozone Spectrophotometer
	Bundle of Sticks
	Business Office Supervisor
	Business Operating Software
BOSFET	Bidirectional Output Switch Field-Effect Transistor
BOSOR	Buckling of Shell of Revolution
BOSS	Basic Operating System Software
	Batch Operating Software System
	Behaviour of Off-Shore Structures Conference
	Benthic Organic Seston Sampler
	Boeing Operational Supervisory System
	Book of SEMI (= Semiconductor Equipment and Materials Institute) Standards
	BRE (= Building Research Establishment) On-Line Search System
	Business Opportunities Sourcing System
	Business-Oriented Software System
BOST	Basic Offshore Survival Training
BOSTI	Buffalo Organization for Social and Technological Innovation [US]
BOT	Basic Offshore Training
	Beginning of Tape
	Board of Trade [UK]
	Botanical
	Botanist
	Botany
	Bottom
BOTAC	Board of Trade Advisory Committee [UK]
BOTB	British Overseas Trade Board
BOVC	Base of Overcast
BP	Back Pressure
	Bacteriophage
	Ball Plunger
	Bandpass
	Barrel Plating
	Basal Plane
	Base Paper
	Base Plate
	Base Pointer
	Batch Processing
	Beam Processing
	Before Present
	Between Perpendiculars
	Binding Post
	Biological Processing
	Biophysics
	Bit Processor
	Black Phosphorus
	Blueprint
	Block Parity
	Boiling Point [also: bp]
	Bolivian Peso [Currency]
	Bonded, (Single) Paper (Wire)
	Boron Phosphide
	Bottom Part
	Branch Point
	Breakpoint
	British Pharmacopoeia
	Buffered Printing

	Bulk Plasmon
	Bypass
B/P	Bills Payable
	Blueprint
bp	below par
	bills payable
BPA	Bachelor of Public Administration
	Bandpass Amplifier
	Basic Probability Assignment
	Bauhinia Purpurea Agglutinin
	Biological Photographic Association [US]
	Bioprocessing Acid
	Bisphenol A
	Bonneville Power Administration [US]
	British Parachute Association
	British Ploughing Association
B/PAA	Business/Professional Advertising Association [US]
BPADCy	Bisphenol A Dicyanate
BP Add	British Pharmacopoeia Addendum
BPAM	Basic Partitioned Access Method
BPB	Bank Post Bill
	Bromophenol Blue
BPBD	Butyl Phenylbiphenylyloxadiazole
BPBF	British Paper Box Federation
BPBIRA	British Paper and Board Industry Research Association
BPC	Back Pressure Control
	Basic Pheripheral Channel
	Bilateral Private Circuit
	Binding Post Chamber
	Bonded-Phase Chromatography
	Boost Protective Cover [of NASA]
	British Pharmaceutical Codex
	British Productivity Council
	Bypass Capacitor
BPCC	2,2'-Bipyridinium Chlorochromate
	British Printing and Publishing Communications Corporation
BPCD	Barrels per Calendar Day
BPCF	British Precast Concrete Federation
BPD	Barrels per Day [also: bpd]
	Beam Positioning Drive
	Bushing Potential Device
BPDA	Bibliographic Pattern Discovery Algorithm
BPE	Bioprocess Engineering
	Boiling Point Elevation
BPEA	9,10-Bis(phenylethynyl)anthracene
BPEC	Building Products Executives Conference
BPEN	5-12-Bis(phenylethynyl)naphthacene
BPetE	Bachelor of Petroleum Engineering
BPF	Bandpass Filter
	Blueprint Files
	Bottom Pressure Fluctuation
	British Plastics Foundation
	Bromophenol Blue
BPGMA	British Pressure Gauge Manufacturers Association
BPH	Bachelor of Public Health
	Barrels per Hour
BPh	Bachelor of Philisophy
BPHA	Benzoyl Phenyl Hydroxylamine
	N-Benzoyl-N-Phenyl Hydroxylamine
BPh	Bachelor of Philosophy
BPharm	Bachelor of Pharmacy
BPhil	Bachelor of Philosophy
BPhys	Bachelor of Physics
BPI	Bits per Inch [also: bpi]
	Boost Phase Intercept
	Bytes per Inch
BPIC	Butylperoxy Isopropyl Carbonate
BPID	Book Physical Inventory Difference
BPIDP	Book Publishing Industry Development Program [Canada]
BPIF	British Printing Industries Federation
BPKT	Basic Programming Knowledge Test
BPL	Beta-Propiolactone
	Binary Program Load
	Bone Phosphate of Lime
	British Physical Laboratories
	Burroughs Program Loader
BPM	Balanced Processing Monitor
	Barrels per Month [also: bpm]
	Batch Processing Monitor
	Beam Position Monitor
	Bits per Millimeter [also: BPMM, bpm or bpmm]
BPMA	Biotechnology Purchasing Management Association [US]
	British Photographic Manufacturers Association
	British Premium Manufacturers Association
	British Printing Manufacturers Association
	British Pump Manufacturers Association
BPMC	Butylphenylmethylcarbamate
BPMD	Battelle's Project Management Division [US]
BPMF	British Pottery Manufacturers Federation
BPMM	Bits per Millimeter [also: BPM, bpm or bpmm]
BPN	Biological Polymer Network
	Breakdown Pulse Noise
BPNL	Battelle Pacific Northwest Laboratories [US]
BPO	Before Payout
	Benzoyl Peroxide
	2-(4-Biphenylyl)-5-Phenyloxazole
	British Post Office
	Bypass Orifice
BPOC	Biphenylisopropyloxycarbonyl
BPP	Bulk Paid Publication
BPPA	Book Publishers Professional Association
	British Precision Pilots Association
BPPF	Basic Program Preparation Facility
BPPM	Butyldiphenylphosphinomethylpyrrolidine Carboxylate
BPPMA	British Power Press Manufacturers Association
BPR	Bachelor of Public Relations
	Bureau of Public Roads [US]
BPRA	Book Publishers Representatives Association [UK]
	Burnable Poison Rod Assembly
BPRF	Bullet-Proof
BPRTHM	Bureau of Public Roads Transport Highway Mobilization [Canada]
BPS	Bachelor of Professional Studies
	Basic Programming Support
	Basic Programming System
	Batch Processing System
	Biophysical Society [US]

	Bits per Second [also: bps]
	Boron Phosphate Silicate
	Borophosphosilicate
	British Phycological Society
	British Printing Society
	Bulk Processing Stream
	Bulk Processing System
	Bytes per Second
BPSA	Business Products Standards Association [US]
BPSD	Barrels per Stream Day
BPSG	Boron Phosphate Silicate Glass
	Borophosphosilicate Glass
BPSI	Bits per Square Inch [also: bpsi]
BPSK	Binary Phase-Shift Keying
	Biphase Shift Keyed
BPSS	Basic Packet Switched System
BPST	2-(2'-Benzothiazolyl)-5-Styryl-3-(4'-Phthalhydrazidyl)tetrazolium
BPT	Baseband Pulse Transmission
	Bipolar Technology
	Bipolar Transistor
	Blade Passage Tones
	Borderline Pumping Temperature
B PT	Boiling Point [also: b pt]
BPTS	Bipolar Transistor Switch
B&PV	Boiler and Pressure Vessel [also: B+PV]
BPVC	Boiler and Pressure Vessel Committee [of ASME, US]
BPWR	Burnable Poison Water Reactor
BPWS	Banked Position Withdrawal Sequence
BPX	Burning Plasma Experiment
BQ	Brine Quenching
Bq	Becquerel [Unit]
BQL	Basic Query Language
	Batch Query Language
BR	Bad Register
	Ballast Resistor
	Bank Rate
	Base Register
	Basic Research
	Batch Reactor
	Beam Rocking
	Belgian Reactor [at Mol, Belgium]
	Bellows Regulating Valve
	Bend Radius
	Beta Ray
	Bills Receivable
	Biological Reactivation
	Boil-Resistant
	Bottom Register
	Brake Relay
	Branch
	Brand
	Breeder Reactor
	Britain
	British
	British Railways
	Brittle Rupture
	Bronze
	Brown
	Bruggeman-Rayleigh Formula
	Brush
	Butadiene Rubber
	Butyl Rubber
B&R	Budget and Reporting (Classification)
B/R	Bills Receivable
Br	Bromine
BRA	Base Rate Area
	Basic Rate Access
	Bee Research Association [UK]
	Beta-Resorcylic Acid
	Bioreactor Analysis
	British Refrigeration Association
	British Robot Association
BRAB	Building Research Advisory Board [now: BRB, US]
BRAC	Bangladesh Rural Advancement Committee
BRAF	Braking Action Fair
BRAG	Braking Action Good
BRAGS	Bioelectrical Repair and Growth Society [US]
BRAM	Bit Rate Agile Modem
BRAMA	British Rubber and Resin Manufacturers Association
BRAN	Braking Action Nil
BRANA	Bumper Recycling Association of North America [US]
BRANE	Bombing Radar Navigation Equipment
BRAP	Braking Action Poor
BRASS	Bottom Reflection Active Sonar System
BRASTACS	Bradford Scientific, Technical and Commercial Service [UK]
BRAZ	Brazil(ian)
BRB	Base Rate Boundary
	Benefits Review Board
	British Railway Board
	Building Research Board [US]
BRBMA	Ball and Roller Bearing Manufacturers Association
BRC	Branch Conditional
	Brazilian Cruzeiro [Currency]
	Brazilian Research Council
	Breeder Reactor Corporation [US]
BRCMA	British Radio Cabinet Manufacturers Association
BRD	Binary Rate Divider
	Bioreactor Design
	Board
	Braid
	Brunei Dollar [Currency of Brunei]
BRDA	Boxboard Research and Development Association [US]
BRDB	British Rubber Development Board
BRDC	British Research and Development Corporation
BRDG	Bituminous Research and Development Group
	Bridge
BrdU	Bromodeoxyuridine
BRE	Building Research Establishment [UK]
	Bureau of Research and Engineering
BREAM	Beaufort Region Environmental Assessment and Monitoring Program [Canada]
b rec	bills receivable
BRECOM	Broadcast Radio Emergency Communication [US Air Force]
BREG	British Rivet Export Group
BREL	Boeing Radiation Effects Laboratory [US]

BREMA	British Radio Equipment Manufacturers Association
BREN	Bare Reactor Experiment Nevada [of NASA/USDOE]
BREP	Boundary Representation
BRET	Breeder Reprocessing Engineering Test
BREV	Brevet
BRF	Band-Rejection Filter
	Bass Research Foundation [US]
	Bell Rings Faintly
	Biochemical Research Foundation [US]
	British Road Federation
BRFFI	Biochemical Research Foundation of the Franklin Institute [US]
BRG	Baud Rate Generator
	Beacon Reply Group
	Bearing
	Budget Review Group
BRGM	Bureau de Recherches Geologiques et Minieres [= Research Bureau for Geology and Mining, France]
BRGW	Brake Release Gross Weight
BRH	Bureau of Radiological Health [US]
BRI	Basic Rate Interface
	Biotechnology Research Institute [of NRC, Canada]
	Book Review Index
	Bose Research Institute [India]
	Brace Research Institute [Canada]
	Bread Research Institute [Australia]
	Building Research Institute
BRIC	Bureau de Recherche pour l'Innovation et la Convergence [= Research Bureau for Innovation and Convergence, France]
BRICS	Bureau of Research Information Control Systems [of USOE]
BRIG	Brigadier
BRIG-GEN	Brigadier-General
BRIMAFEX	British Manufacturers of Malleable Tube Fittings Export Group
BRIMARC	Brighton MARC (= Machine-Readable Catalogue) Program [UK]
BRIT	Britain
	British
	Bureau Robotique, Informatique et Telematique [= Bureau for Robotics, Informatics and Telematics, France]
BRITE	Basic Research in Industrial Technology in Europe
	Bright Radar Indicator Tower Equipment
BRITE/EURAM	Basic Research in Industrial Technology in Europe/European Research on Advanced Materials
BRK	Break
	Brick
BRKHIC	Breaks in Higher Overcast
BRKG	Brokerage
BRKR	Breaker
BRKT	Bracket
BRL	Balance Return Loss
	Ballistics Research Laboratory [US Army]
	Bureau of Research and Laboratories [Philippines]
	Butterwick Research Laboratories [UK]
BRLESC	Ballistics Research Laboratories Electronic Scientific Computer
BRLP	Burlap
BRLS	Barrier Ready Light System
BRM	Barometer
	Binary Relationship Model
BRMA	Business Records Manufacturers Association [US]
BRMCA	British Ready Mixed Concrete Association
BRN	Brown
BRNSH	Brownish
BRO	Broach
	Brother
BROM	Bipolar Read-Only Memory
BROOM	Ballistic Recovery of Orbiting Man
BROS	Brothers
BROWSER	Browsing On-Line with Selective Retrieval
BRP	Backfile Registration Project
	Beacon Ranging Pulse
	Bureau de Recherches de Petrole [= Petroleum Research Bureau, France]
BRPRA	British Rubber Producers Research Association
BRR	Battelle Research Reactor [US]
	Brookhaven Research Reactor [US]
BRRA	British Rayon Research Association
	British Refractories Research Association
BRRAMA	British Rubber and Resin Adhesive Manufacturers Association
BRRI	Bangladesh Rice Research Institute
BRRL	British Road Research Laboratory
BRS	Block Received Signal
	B-Mode Receiving Station
	Boron Recycle System
	Break Request Signal
	British Road Service Federation
	Building Research Station [UK]
	Bureau of Railroad Safety [US]
	Burma Research Society
BRS	Bibliographic Retrieval Services [US]
	Brass
BR STD	British Standard
BRT	Binary Run Tape
	Brightness
BRTA	British Road Tar Association
BR&TC	Better Roads and Transportation Council [US]
BRU	Basic Resolution Unit
	Branch Unconditional
	Brown University [US]
BRUNET	Brown University Network [US]
BRVMA	British Radio Valve Manufacturers Association
BRWM	Board on Radioactive Waste Management
BRZ	Bronze
BRZG	Brazing
BS	Bachelor of Science
	Backing Store
	Backscatter
	Backscattering Spectroscopy
	Backspace (Character)
	Backward Signalling
	Balance Sheet
	Bandeiraea Simplicifolia
	Band Spectrum
	Band Structure
	Base Stock

	Basic Sediment
	Beam Splitter
	Below Slab
	Binary Subtract(er)
	Biochemical Society [UK]
	Biological Shield
	Biological Society [US]
	Biometric Society [US]
	Biophysical Society [US]
	Biosynthesis
	Bismuth Sulphite
	Blowing Snow
	Bonded (Single) Silk (Wire)
	Both Sides
	Bottom Sediment
	Breaking Strength
	Breaking Stress
	British Standard
	Broadcasting Satellite
	Broadcast Station
	Buffer Stage
	Building Steel
	Bureau of Standards
	Bursting Strength
B&S	Beams and Stringers
	Bell and Spigot
	Brown and Sharpe (Wire) Gauge
B/S	Bill of Sale
	Bits per Second [also: b/s]
BSA	Bachelor of Science in Agriculture
	Bank Stationers Association [now: FSA, US]
	Bearing Specialists Association [US]
	Benzene Sulfonic Acid
	Bimetal Steel-Aluminum
	Biological Signal Analysis
	Biophysical Society of America
	Bis(trimethylsilyl)acetamide
	Body Surface Area
	Boston Society of Architects [US]
	Botanical Society of America [US]
	Bovine Serum Albumin
	British South Africa
	British Speleological Association
	Broadside Array
	Bund Schweizer Architekten [= Federation of Swiss Architects]
	Business Software Association [US]
BSAA	Bachelor of Science in Applied Arts
	British South American Airways
BSAC	Berkeley Sensor and Actuator Center [of UCB, US]
BSAD	Bachelor of Science in Architectural Design
BSAE	Bachelor of Science in Aeronautical Engineering
	Bachelor of Science in Architectural Engineering
BSAF	British Sulfate and Ammonia Federation
BSAg	Bachelor of Science in Agriculture
BSAgrEng	Bachelor of Science in Agricultural Engineering
BSAL	Block Structured Assembly Language
BSAM	Basic Sequential Access Method
BSAP	British Society of Animal Production
BSAR	Babcock and Wilcox Safety Analysis Report
BSArch	Bachelor of Science in Architecture
BSArchTech	Bachelor of Science in Architectural Technology
BSAS	Bachelor of Science in Architectural Studies
BSATA	Ballast, Sand and Allied Trades Association
BSA-TBS	Bovine Serum Albumin — Tris-Buffered Saline
BSB	British Standard Beam
BSBA	Bachelor of Science in Business Administration
BSBC	Bachelor of Science in Building Construction
	British Social Biology Council
BSBI	Botanical Society of the British Isles [UK]
BSBldgSci	Bachelor of Science in Building Science [also: BSBldgSc]
BSBP	British Standard Bulb Plate
BSBus	Bachelor of Science in Business
BSC	Bachelor of Science in Commerce
	Base Site Control
	Basic Message Switching Center
	Binary Symmetric Channel
	Binary Synchronous Communications
	Biological Stain Commission [US]
	Binary Synchronous Communication
	Bisynchronous Communication
	Borosilicate Crown (Glass)
	Branch or Skip on Condition
	British Shippers Council
	British Society of Commerce
	British Standard Channel
	Building Systems Council [of NAHB, US]
	Business Service Center
BSc	Bachelor of Science
BSCA	Binary Synchronous Communications Adapter
	British Sulfate of Copper Association
	Building Service Contractors Association (International) [US]
BScA	Bachelor of Science in Agriculture
BSCAC	Beaufort Sea Community Advisory Committee [Canada]
BScAgr	Bachelor of Science in Agriculture
BScAgrEng	Bachelor of Science in Agricultural Engineering
BScArch	Bachelor of Science in Architecture
BScBldgDes	Bachelor of Science in Building Design
BSCC	British Synchronous Clock Conference
BScC	Bachelor of Commercial Studies
BScCE	Bachelor of Science in Civil Engineering
	Bachelor of Science in Construction Engineering
BSCCO	Bismuth-Strontium-Calcium-Copper-Oxygen (Compound) [also: $BiSrCaCu_2O_x$]
BScCom	Bachelor of Commercial Science
BScCS	Bachelor of Science in Computer Science
BScDA	Bachelor of Science in Data Analysis
BSCE	Bachelor of Science in Civil Engineering
	Bachelor of Science in Engineering
	Bird Strike Committee Europe [Denmark]
	Boston Society of Civil Engineers [US]
BScEcon	Bachelor of Science in Economics
BScEng	Bachelor of Science in Engineering [also: BScE or BSc(Eng)]
BSCer	Bachelor of Science in Ceramics
BSCerEng	Bachelor of Science in Ceramic Engineering [also: BSCerE]
BSCerSci	Bachelor of Science in Ceramic Science
BScF	Bachelor of Science in Forestry

BScFE Bachelor of Science in Forestry Engineering
BSCFL Binary Synchronous Frame Level
BSCh Bachelor of Science in Chemistry
BSChemEng Bachelor of Science in Chemical Engineering [also: BSChemE]
BSCL Bell System Common Language
BSCM Bisynchronous Communications Macro
BScMet Bachelor of Science in Metallurgy
BScMetEng Bachelor of Science in Metallurgical Engineering
BScMnlEng Bachelor of Science in Mineral Engineering
BScMnlProcEng Bachelor of Science in Mineral Processing and Engineering
BSCN Bit Scan
BSC-NAHB Building Systems Council of the National Association of Home Builders [US]
BSComm Bachelor of Science in Commerce [also: BSCom]
BSConEng Bachelor of Science in Construction Engineering [also: BSConE]
BSConTech Bachelor of Science in Construction Technology
BSCP Biological Sciences Communication Project [US]
BScPharm Bachelor of Science in Pharmacy [also: BScPhm]
BSCRA British Steel Castings Research Association
BSCRP Bachelor of Science in City and Regional Planning
BSCS Biological Sciences Curriculum Study [US]
BSC/SS Binary Synchronous Communications/Start-Stop
BScTech Bachelor of Science and Technology
BSD Bahamian Dollar [Currency]
Ballistic Systems Division [US Army]
Barrels per Stream Day
Base Supply Depot
Beam Scan Driver
Bibliographic Services Division [of British Library, UK]
Bulk Storage Device
BSDC Binary Symmetric Dependent Channel
British Standard Data Code
BSDL Boresight Datum Line
BSDP Bibliographic Service Development Program [of CLR, US]
BSE
Bachelor of Science in Education
Bachelor of Science in Engineering
Bachelor of Science in Sanitary Engineering
Backscattered Electron
Bovine Spongiform Encephalopathy
British Shipbuilding Exports
Broadcast Satellite Experiment [Japan]
Bridge and Structural Engineer
Building Services Engineer
BSEA British Standard Equal Angle
British Steel Export Association
BSEc Bachelor of Science in Economics
BSED Bachelor of Science in Environmental Design
BSEd Bachelor of Science in Education
BSEE Bachelor of Science in Electrical Engineering
BSEET Bachelor of Science in Electrical Engineering Technology
BSELCH Buffered Selector Channel
BSEM Bachelor of Science in Engineering of Mines
BSEng Bachelor of Science in Engineering
BSERC British Science and Engineering Research Council
BSET Bachelor of Science in Engineering Technology
BSF Bachelor of Science in Forestry
Band-Stop Filter
Binational Science Foundation [Israel — US]
Bipolar Shape Formation
British Shipping Federation
British Slag Federation
British Standard Fine Thread
British Stone Federation
Bulk Shielding Facility [of ORNL, US]
BSFA British Steel Founders Association
BSFuelSc Bachelor of Science in Fuel Science
BSG Borosilicate Glass
British Standard Gauge
B&SG Brown and Sharpe Gauge
BSGeog Bachelor of Science in Geography
BSGL Branch System General License
BSH Benzenesulfonylhydrazide
bsh bushel
BSHS British Society for the History of Science
B/s/Hz Bits per second per Hertz
BSI Bandeiraea Simplicifolia Isolectin
Biochemical Society of Israel
Bit Sequence Independence
Botanical Society of Israel
British Society for Immunology
British Standards Institution
Building Stone Institute [US]
Building Systems Institute [US]
BS-I Bandeiraea Simplicifolia I
BSIC Binary Symmetric Independent Channel
BS-II Bandeiraea Simplicifolia II
BSIE Bachelor of Science in Industrial Engineering
Banking Systems Information Exchange [US]
BSIntArch Bachelor of Science in Interior Architecture
BSIR Board for Scientific and Industrial Research [Israel]
BSIRA British Scientific Instrument Research Association
BSKT Basket
BSL Bit Serial Link
Botanical Society of London [UK]
BSLA Bachelor of Science in Landscape Architecture
BSLArch Bachelor of Science in Landscape Architecture
BSM Basic Storage Module
Bit-Slice Microprocessor
Bureau of Standards and Measures [now: NIST, US]
BSMatEng Bachelor of Science in Materials Engineering [also: BSMatE]
BSMatScEng Bachelor of Science in Materials Science and Engineering [also: BSMatSciEng or BSMatSE]
BSME Bachelor of Science in Mechanical Engineering
Bachelor of Science in Mining Engineering
BSMechEng Bachelor of Science in Mechanical Engineering [also: BSMechE]
BSMed Bachelor of Science in Medicine
BSMet Bachelor of Science in Metallurgy
BSMetEng Bachelor of Science in Metallurgical Engineering [also: BSMetE]
BSMnlEng Bachelor of Science in Mineral Engineering

BSMnlProcEng Bachelor of Science in Mineral Processing and Engineering
BSMP Bit-Slice Microprocessor
BSMPT British Standard Metric Pipe Thread
BSMSP Bernoulli Society for Mathematical Statistics and Probability [Netherlands]
BSMT Basement
BSN Back-End Storage Network
Backward Sequence Number
BSNDT British Society for Nondestructive Testing
BSNH Boston Society of Natural History [US]
BSNS Buffalo Society of Natural Sciences [US]
BSO Blue Stellar Object
Bonded Silicon Oxide
BSOIW (International Association of) Bridge, Structural and Ornamental Iron Workers
BSP Backspace
Bell System Practice
Bioseparation Process
Botanical Society of Pennsylvania [US]
Boron Superphosphate
British Society of Parasitology
British Standard Pipe Thread
Bromosulphalein
Burroughs Scientific Processor
BSPA Bachelor of Science in Public Administation
BSPH Bachelor of Science in Public Health
BSPlmSc Bachelor of Science in Polymer Science
BSPMC Brake System Parts Manufacturers Council [US]
BSPS British Society for the Philosophy of Science
BSPT 2-(2'-Benzothiazolyl)-5-Styryl-3-(4'-Phthalhydrazidyl)tetrazolium
Bound-State Perturbation Theory
BSR Blip-Scan Ratio
Board of Standards Review [US]
Buffered Send/Receive
Bulk Shielding Reactor [ORNL, US]
BSRA British Ship Research Association
British Sound Recording Association
British Sugar Refiners Association
BSRAE British Society for Research in Agricultural Engineering
BSRF Borderland Sciences Research Foundation [US]
BSRFS Bell System Reference Frequency Standard
BSRIA Building Services Research and Information Association [UK]
BSRP Biological Sciences Research Paper
BSRS Bell System Repair Specification
BSS Balanced Saline Solution
Balanced Salt Solution
Basic Synchronized Subset
British Satellite Service
British Standard Specification
Broadcasting Satellite Service
Bulk Storage System
BSSC Bit Synchronizer/Signal Control
BSSG Biological Sciences Study Group
BSSM British Society for Strain Measurement
BSSMA Business Systems and Security Marketing Association [US]
BSSRS British Society for Social Responsibility in Science
Bureau of Safety and Supply Radio Services [US]
BSSS British Society of Soil Science
BSSSA British Surgical Support Suppliers Association
BST Beam-Switching Tube
Bering Standard Time
Block the SPADE Terminal Command
Bovine Somatostatin
British Standard Tee
British Standard Thread
British Standard Time
British Summer Time
Business Systems Technology
B/St Bill of Sight
BSTA British Surgical Trades Association
BSTAN Beta Solution Treated and Annealed
BSTD Bastard
BstE Bacillus Stearothermophilus ET [Molecular Biology]
BSTex Bachelor of Science in Texiles
BSTFA Bistrifluoroacetamide
Bis(trimethylsilyl)trifluoroacetamide
Bell System Technical Journal
BSTOA Beta Solution Treated and Overaged
BSTR Booster
BSU Baseband Switching Unit
Basic Sounding Unit
Beam Steering Unit
Bis(trimethylsilyl)urea
Business Service Unit
BSUA British Standard Unequal Angle
BSUG Bedford Systems Users Group [US]
BSUP Bachelor of Science in Urban Planning
BSW Botanical Society of Washington [US]
Bottom Settlings and Water
British Standard Whitworth Thread
BS&W Basic Sediment and Water
Bottom Sediment and Water
BSWM Bureau of Solid Waste Management [US]
BSXF Burst Sync Failure [SPADE]
BSY Busy
BT Bacillus Thuringiensis
Backtracker
Band Theory
Baronet
Barred Trunk
Bathythermography
Bent
Benzotriazole
Biotechnologist
Biotechnology
Block Terminal
Blue Tetrazolium
Boat
Bomb Technician
Bought
Branch Tee
Bridge Transition
British Telecom
Broadcast Technician
Broadcast Technology
Broader Term
Building Technician

	Building Technology
	Bus Tie
	Busy Tone
Bt	Bacillus Thuringiensis
bt	bought
BTA	Benzotriazole
	Benzoyltrifluoroacetone
	Book Trade Association [South Africa]
	Boring-Trepanning Association
BTAM	Basic Tape Access Method
	Basic Telecommunications Access Method
	Basic Teleprocessing Access Method
	Basic Terminal Access Method
	Batch Terminal Access Method
BTAO	Bureau of Technical Assistance Operations [of UN]
BTB	Behind-the-Bars
	Belgische Transportarbeidersbond [= Belgium Transport Workers Union]
	Bromothymol Blue
	Bus Tie Breaker
BT-BASIC	Backtracker-BASIC
BTC	Benzotrichloride
	Block Transfer Controller
	British Telecom Center
	Business and Technology Center [UK]
	Butyrylthiocholine
BTCC	Benzothiazolocarbon Cyanine
	Board of Transportation Commissioners of Canada
BTD	Binary-to-Decimal
	Bomb Testing Device
BTDA	3,3',4-4-Benzophenonetetracarboxylic Dianhydride
BTDC	Before Top Dead Center
BTDL	Back Transient Diode Logic
	Basic Transient Diode Logic
BTDU	Benzylthiodihydrouracil
BTE	Bachelor of Textile Engineering
	Baldwin, Tate and Emery Theory
	Battery Terminal Equipment
	Bench Test Equipment
	Bidirectional Transceiver Element
	Boltzmann Transport Equation
	Brake Thermal Efficiency
	Business Terminal Equipment
BTEAC	Benzyltriethylammonium Chloride
BTech	Bachelor of Technology
BTEE	Benzoyl Tyrosine Ethyl Ester
BTEMA	British Tanning Extract Manufacturers Association
BTF	Bomb Tail Fuse
	Bulk Transfer Facility
BTFLY VLV	Butterfly Valve
bt fwd	brought forward
BTG	British Technology Group
BTH	Basic Transmission Header
BThU	British Thermal Unit [also: BTU or Btu]
BTI	Bridged Tap Isolator
	British Technology Index
	British Telecommunications International
B-TIP	Battelle Technical Inputs to Planning
BTIPR	Boyce Thompson Institute for Plant Research [US]
BTL	Beginning of Tape Level
	Bell Telephone Laboratories [US]
	Between Layers
BTLE	Between-the-Lines Entry
BTLG	Bottling
BTM	Basic Time-Sharing Monitor
	Batch Time-Sharing Monitor
	Bromotrifluoromethane
BTMA	British Typewriter Manufacturers Association
	Busy-Tone Multiple Access
BTMF	Block Type Manipulation Facility
BTMSA	Bis(trimethylsilyl)acetylene
BTMSBD	1,4-Bis(trimethylsilyl)-1,3-Butadiyne
BTN	Billing Telephone Number
BTO	Bombing through Overcast
	British Trust for Ornithology
	Brussels Trade Organization [Belgium]
B to B	Back to Back
BTP	Botswana Pula [Currency]
	Branch Technical Position
	Batch Transfer Program
BTR	Behind-the-Tape Reader Board
	Better
	Bit Time Recovery
	Broadcast and Television Receivers [IEEE Group]
BTRA	Bombay Textile Research Association [India]
BTRL	British Telecom Research Laboratories
BTS	Barium Titanate/Stannate
	Batch Terminal Simulation
	Bias Temperature Stress
	Bisynchronous Terminal Support
	Brazilian Thorium Sludge
	British Telecommunications Systems
	British Tunneling Society
	Broadcast Technology Society [of IEEE, US]
	Business Telecommunications Services [Australia]
BTSC	Broadcast Television Systems Committee
BTSD	Basic Training for Skill Development
BTSP	Bootstrap
BTSS	Basic Time Sharing System
	Braille Time-Sharing System
BT ST	Billet Steel
BTT	Business Transfer Tax
BTTG	British Textile Technology Group
BTTP	British Towing Tank Panel
BTTRI	Beijing Television Technology Research Institute [PR China]
BTU	Basic Transmission Unit
	British Thermal Unit [also: Btu or BThU]
BTUC	Botswana Trade Union Center
BTU/hr	British Thermal Units per Hour [also: BTU/h]
BTU/lb	British Thermal Units per Pound
BTU/s	British Thermal Units per Second [also: BTU/sec]
BTU/sq ft	British Thermal Units per Square Foot
BTU/sq in	British Thermal Units per Square Inch
BTW	Behind-the-Wheel
BTX	Benzene, Toluene and/or Xylene
	Bildschirmtext [also: Btx; German Viewdata System]
BU	Baseband Unit
	Base Up (Light Bulb)
	Bath University [UK]
	Basic User

	Beijing University [PR China]
	Binding Unit
	Boston University [US]
	Bottom Up
	Bradley University [US]
	Brandon University [Canada]
	Bristol University [UK]
	Brock University [Canada]
	Brown University [US]
	Brunel University [UK]
	Bureau
	Buzzer
Bu	Butyl
	Butyryl
bu	bushel
BUAER	Bureau of Aeronautics [US Navy]
Bucks	Buckinghamshire [UK]
BUCLASP	Buckling of Laminated Stiffened Plates
BUD	Biological Up-Down (Probe)
BUDC	Before Upper Dead Center
BUDWSR	Brown University Display for Working Set References [US]
BUE	Built-Up Edge
BUEC	Back-Up Emergency Communications
BUF	Buffer
BUFC	British Universities Film Council
BUFORA	British Unidentified Flying Objects Research Association
BUFVA	British Universities Film and Video Council
BUGS	Brown University Graphic System [US]
BUI	Buildup Index
BUIC	Back-Up Interceptor Control
BUILD	Base for Uniform Language Definition
BUIS	Barrier Up Indicator System
BUIST	Beijing University of Iron and Steel Technology [PR China]
BUK	Burmese Kyat [Currency of Burma]
BULL	Bullet
	Bulletin
BULL PRF	Bullet-Proof
BULPAC	Bulgarian Packet Switched Network
BULTEX	Bulgarian Videotex Network
BUM	Back-Up Mode
	Break-Up Missile
BUMED	Bureau of Medicine and Surgery [US Navy]
BuMines	Bureau of Mines [also: BUMINES, BOM or BoM, US]
BUMP	Bottom Up Modular Programming
BUN	Blood Urea Nitrogen
BUNAC	British Universities North America Club [US]
BuOrd	Bureau of Ordnance [also: BUORD [US Navy]]
BUP	Bottom-Up Parsing
	British United Press
BUPCR	Bath University Program of Catalogue Research [UK]
BUPERS	Bureau of Naval Personnel [US Navy]
BUR	Back-Up Register
	Bureau
	Burmese Kyat [Currency of Burma]
BURB	Back-Up Roll Bending
BURISA	British Urban and Regional Information Systems Association
BURO	Bureau Universitaire de Recherche Operationnelle [= University Bureau for Operational Research, France]
BUR ST	Bureau of Standards
BUS	Broken-Up Structure
	Business
bus	bushel
BUSAK	Bus Acknowledgement
BUSANDA	Bureau of Supply and Accounts [US Navy]
BUSH	Bushing
BuShips	Bureau of Ships [formerly also: BUSHIPS, US; now: NSEC, US]
BusMgr	Business Manager
BUSRQ	Bus Request
BuStd	Bureau of Standards [also: BU STD, US]
BUT	Broadband Unbalanced Transformer
	Button
BUTEX	Dibutoxy Diethyl Ether
BUV	Backscatter Ultraviolet
BUWAL	Bundesamt fur Umwelt, Wald und Landschaft [= Federal Department for Environment, Forestry and Countryside, Switzerland]
BuWEPS	Bureau of Naval Weapons [also: BUWEPS, US Navy]
BUZ	Buzzer
BV	Back View
	Bochumer Verein [Vacuum Degassing Process]
	Book Value
	Breakdown Voltage
	Boundary Value
	Bureau Veritas [France]
	Busy Verification
BVA	British Veterinary Association
	British Videogram Association
BVC	Black Varnish Cambric (Insulation)
BVD	Belgische Vereinigung fur Dokumentation [= Belgian Association for Documentation]
BVE	Bivariate Exponential Distribution
	Butyl Vinyl Ether
BVMA	British Valve Manufacturers Association
BVO	Brominated Vegetable Oil
BVP	Boundary Value Problem
BVS	Bachelor of Veterinary Science
	Bibliotheksverbundsystem [= German Library Network]
BVSc	Bachelor of Veterinary Science
BVSV	Bimetal Switching Valve
BVU	Brightness Value Unit
BVW	Backward Volume Wave
BW	Bacteriological Warfare
	Bandwidth
	Biological Warfare
	Biological Wastewater
	Bloch Wall
	Both Ways
	Bow Wave
	Braided Wire
	Buried Wire
B&W	Black and White [also: B/W]
BWA	Backward Wave Amplifier

	Bandwidth Allocation
	British Waterfowl Association
	British Waterways Association
	British Waterworks Association
	British Wildlife Appeal [of RSNC, UK]
BWC	Backward Wave Converter
	Buffer Word Counter
BWCC	British Weed Control Council
BWD	Barrels of Water per Day [also: bwd, BW/D, bw/d, BW D^{-1}, bw d^{-1}]
BWDA	Bicycle Wholesale Distributors Association [US]
BWE	Bucket Wheel Excavator
BWEA	British Wind Energy Association
BWF	British Whiting Federation
BWG	Biphenyl Working Group [US]
	Birmingham Wire Gage
	British Imperial Wire Gage
BWH	Barrels of Water per Hour [also: bwh, BW/H or bw/h]
BWI	Betriebswirtschaftliches Institut [= Institute for Business Management [of VDEh, Germany]]
	Boating Writers International [US]
	British West Indies
BWIA	British West Indies Airways
BWIP	Basalt Waste Isolation Project
BWM	Backward-Wave Magnetron
	Barrels of Water per Minute [also: bwm]
BWMA	British Woodwork Manufacturers Association
BWMB	British Wool Marketing Board
BWO	Backward-Wave Oscillator
	Backwave Oscillator
BWOC	By Weight of Cement
BWP	Botswana Pula [Currency]
	Brown Wrapping Paper
BWPA	Backward-Wave Power Amplifier
	British Wastepaper Association
	British Women Pilots Association
	British Wood Preserving Association
BWPD	Barrels of Water per Day [also: bwpd]
BWPUC	British Wastepaper Utilization Council
BWR	Bandwidth Ratio
	Boiling-Water Reactor
BWRA	British Welding Research Association
BWRE	Biological Warfare Research Establishment [UK]
BWRRA	British Wire Rod Rollers Association
BWST	Borated Water Storage Tank
BWT	Backward Wave Tube
	Biological Wastewater Treatment
BWTA	British Wood Turners Association
BWTSDS	Base Wire and Telephone System Development Schedule
BWV	Back Water Valve
BX	Box
	Branch Exchange
BXB	British Crossbar
BYD	Beyond
BYMUX	Byte-Multiplexer Channel
BYP	Bypass
BYR	Buyer
BYU	Brigham Young University [US]
BZ	Brillouin Zone
Bz	Benzene
	Benzoyl
BZD	Belize Dollar [Currency of Belize]
BzOH	Benzyl Alcohol

C

C
Calm
Cancel
Candle
Capacitance
Capacitor
Capacity
Cape
Carbide
Carbon
Cast(ing)
Cathode
Celsius
Cental [Unit]
Centigrade
Chapter
Character
Chip(ping)
Chopper
Circle
Circumference
Cis (Configuration) [Chemistry]
City
Clock
Cloudy
Coefficient
Coil
Collagen
Collector
Colon [Currency of Costa Rica]
Color
Commercial
Compliance
Computer
Concentration
Concrete
Conduction
Conductor
Connection
Connector
Contact
Contrast
Control(ler)
Copyright
Corundum
Cotton
Coulomb
Coulombmeter
Counter
Couple
Coupling
Course
Crystal
Crystal Phase
Curie
Cycle
Cylinder
Cysteine
Cysteinyl
Cytosine

c
calm
calorie
cancel
candle
capacity
carat
carry
cathode
cent
centi- [SI Prefix]
centimeter
chapter
charmed quark
(circa) — approximately; about
circuit
clear
clearance
clockwise
coefficient
cognate
cold
collector
color
concentration
contact
continental air mass
copyright
core
crystalline
cubic
curie
current
cycle

ΔC
Degrees Celsius
Degrees Centigrade

CA
Cable
Cadmium Association [UK]
California [US]
Calorimetric Analysis
Canadian Army [now: CAF]
Cancel
Candle
Canonical Assembly
Capacity Assignment
Capital Asset
Carbon Arc
Cathode
Cellulose Acetate
Center of Area
Central America
Centrifugal Accelerator
Certified Accountant
Chaney Adapter
Channel Adapter
Chartered Accountant

Chemical Abstracts [of CAS, US]
Chemical Analysis
Chief Accountant
Chromic Acid
Cinnamic Acid
Circuit Analysis
Circular Aperture
Citric Acid
Civil Aviation
Clamp Assembly
Clear Aperture
Close Annealing
Cold Air
Color Analyzer
Communications Adapter
Commutative Algebra
Company Address
Compression Axis
Computer Algorithm
Computer Application
Computer Architecture
Computers and Automation
Conflict Alert
Construction Authorization
Consulting Architect
Consumers Association [UK]
Continue-Any Mode
Continuous Annealing
Contract Administrator
Contract Authorization
Contrast Analysis
Control Area
Controlled Atmosphere
Conventional Alloy
Corrective Action
Cortisone Acetate
Crab Apple [US Computer Users Group]
Cresyl Acetate
Critical Assembly
Croup-Associated Virus
Cryonics Association [US]
Current Account

C/A Close Annealing
Corrective Action
Current Account

Ca Calcium
Cathode

ca (circa) — approximately; about

C_3A Tricalcium Aluminate [Cement]

CAA Canadian Acoustical Association
Canadian Automobile Association
Caragana Arborescens Agglutinin
Center for American Archeology [US]
Central African Airways
Chinese Association of Automation [PR China]
Chromic Anodized Aluminum (Alloy)
Cicer Arietinum Agglutinin
Civil Aeronautics Administration [US]
Civil Aviation Authority [UK]
Clean Air Act [US and UK]
Commonwealth Association of Architects [UK]
Computer-Assisted Accounting
Computer-Assisted Analysis
Cost Accountants Association [UK]

CAAA Canadian Academic Accounting Association
Canadian Association of Advertising Agencies

CAAAL Classified Abstract Archive of the Alcohol Literature

CAAB Canadian Advertising Advisory Board

CAAC Civil Aviation Administration of China [PR China]

CAACE Comite des Associations d'Armateurs des Communautes Europeennes [= Organization of Shipowners of the European Communities]

CAAP Certified Advertising Agency Practitioner

CAARC Commonwealth Advisory Aeronautical Research Council [UK]

CAAS Canadian Association of Administrative Studies
Ceylon Association for the Advancement of Science [now: SLAAS]
Civil Aviation Authority of Singapore
Computer-Aided Approach Sequencing
Computer-Assisted Acquisition System
Connecticut Academy of Arts and Sciences [US]

CAASA Commercial Aviation Association of Southern Africa [also: CAA/SA, South Africa]

CAAT Canadian Academic Aptitude Test
College of Applied Arts and Technology
Computer-Assisted Audit Techniques

CAATS Canadian Automated Air Traffic System

CAB Cabin
Cabinet
Cable-Television Advertising Bureau [US]
Cabriolet
Canadian Accreditation Board [of CCPE]
Canadian Association of Broadcasters
Captive Air Bubble
Captured Air Bubble
Cellulose Acetate Butyrate
Cellulose Acetobutyrate
Civil Aeronautics Board [US]
Commonwealth Agricultural Bureau [now: CABI, UK]
Communications Adapter Board

CABATM Civil Aeronautics Board Air Transport Mobilization

CABC Canadian Airborne Center
Craftsmen's Association of British Columbia [Canada]

CABE Canadian Association of Business Economics

CABEI Central American Bank for Economic Intergration

CABET Canadian Association of Business Economics Teachers

CABI Commonwealth Agricultural Bureau International [UK]

CABLIS Current Awareness Bulletin for Librarians and Information Scientists [of BL, UK]

CABMA Canadian Association of British Manufacturers and Agencies

CABO Council of American Building Officials [US]

CaBP Calcium Binding Protein

CABRA Copper and Brass Research Association

CABS Computer-Aided Batch Searching

Computerized Annotated Bibliographic System [of University of Alberta, Canada]
Current Awareness in Biological Sciences

CABSA Copper and Brass Service Center Association [also: CBSA, US]

CAC California Avocado Commission [US]
Canadian Armored Corps
Carbon-Arc Cutting
Center for Analytical Chemistry [of NIST, US]
Center for Atmospheric Chemistry [Canada]
Cis-Anti-Cis (Geometry) [Chemistry]
Citizens Advisory Committee
Coal Association of Canada
Codex Alimentarius Commission [Joint UN Commission of FAO and WHO]
Commonwealth Aircraft Corporation [Australia]
Computer-Aided Classification
Consumers Association of Canada
Containment Atmosphere Control

CACA Canadian Agricultural Chemicals Association

CACAC Civil Aircraft Control Advisory Committee [UK]

CACAS Civil Aviation Council of the Arab States

CACB Canadian Association of Convention Bureaus
Canadian Association of Customs Brokers
Compressed-Air Circuit Breaker

CACC Civil Aviation Communications Centre [UK]

CACCI Confederation of Asian-Pacific Chambers of Commerce and Industry [Taiwan]

CACD Computer-Aided Circuit Design

CACDP California Association of County Data Processors [US]

CACDS Commonwealth Advisory Committee on Defense Science

CACHE Computer-Controlled Automated Cargo Handling Envelope

CACM Central American Common Market
Communications of the Association for Computing Machinery [US]

CACMF Central American Common Market Fund

CA CON Chemical Abstracts Condensates [of CAS, US]

CACP Cartridge-Actuated Compaction Press

CACQ Chartered Accountants Corporation of Quebec [Canada]

CACRS Canadian Advisory Council on Remote Sensing

CACS Computer-Aided Communications System
Content Addressable Computer System

CACSUSS Canadian Advisory Committee on Scientific Utilization of Space Stations

CACT Canadian Advisory Committee on Terminology

CACTS Canadian Air Cushion Technology Society

CACUL Canadian Association of College and University Libraries

CACW Central American Confederation of Workers [Costa Rica]
Core Auxiliary Cooling Water

CACWS Core Auxiliary Cooling Water System

CAD Cabling Diagram
Canadian Dollar [Currency]
Cartridge-Activated Device
Cash Aganst Disbursement
Cash Against Documents
Character Assemble Disassemble
Charged Area Development
Coincidence Axis Direction
Compensated Avalanche Diode
Computer Access Device
Computer-Aided Design
Computer-Aided Detection
Computer-Aided Dispatch
Computer-Aided Drafting
Computer Applications Digest
Containment Atmosphere Dilution

CADA Cellulose Acetate Diethylaminoacetate
Computer-Aided Design Analysis

CADAM Computer-Aided Design and Manufacturing
Computer Graphics Augmented Design and Manufacturing (System)

CADAPSO Canadian Association of Data Processing Service Organizations

CADAR Computer-Aided Design, Analysis and Reliability

CADAT Computer Aided Design and Testing

CADC Cambridge Automatic Digital Computer
Central Air Data Computer
Computer-Aided Design Center [of DTI, UK]

CAD/CAE Computer-Aided Design/Computer-Aided Engineering

CAD/CAM Computer-Aided Design/Computer-Aided Manufacturing

CAD/CAM/CAE Computer-Aided Design/Computer-Aided Manufacturing/Computer-Aided Engineering

CAD/CAT Computer-Aided Design/Computer-Aided Testing

CADD Coding and Analysis of Drillhole Data
Computer-Aided Design Drawing
Computer-Aided Drafting and Design

CADD/CAM/CAE Computer-Aided Drafting and Design/ Computer-Aided Manufacturing/Computer-Aided Engineering

CADDE Canadian Association of Deans and Directors of Education

CADDIF Computer-Aided Design Data Interchange Format

CADE Canadian Association of Drilling Engineers
Coalition Against Dangerous Exports
Computer-Aided Design Engineering
Computer-Aided Design Evaluation
Computer-Assisted Data Entry

CADEF Centro Argentino de Estudios Forestales [= Argentinian Center of Forestry Studies]

CADEP Computer-Aided Design of Electronic Products

CADES Computer-Aided Development and Evaluation System

CADET Computer-Aided Design Experimental Translator

CADETS Classroom-Aided Dynamic Educational Time-Sharing System

CADF CRT (= Cathode-Ray Tube) Automatic Direction Finding
Commutated-Antenna Direction Finder

CADFISS Computation and Data Flow Integrated Subsystem

CADI Computer-Aided Diagnostic Information

CADIA Centro Argentino de Ingenieros Agronomos [= Argentinian Center of Agricultural Engineers]

CADIC Chemical Analysis Detection Instrumentation Control

	Computer-Aided Design of Integrated Circuits
CADIF	Computer-Aided Design Instructional Facility [US]
CADIN	Canadian Integration North
CADIS	Computer-Aided Design of Information Systems
CADIZ	Canadian Air Defense Identification Zone
CADL	Cadkey Advanced Design Language
	Computer-Aided Design Language
CADM	Clustered Airfield Defeat Submunition
CAdm	Certified Administrator
CADMAT	Computer-Aided Design, Manufacture and Test
	Computer-Aided Design, Manufacturing and Testing
CADO	Central Air Documents Office [US]
CADRE	Complete ADR Environment
	Current Awareness and Document Retrieval for Engineers
CADS	Computer-Aided Design System
	Computer Analysis and Design System
	Containment Atmosphere Dilution System
	Counterfeiting, Altering, Duplicating and Simulating Currency Authenticating and Denominating System
CADSS	Combined Analog-Digital Systems Simulator
CADW	Civil Air Defense Warning (System)
CAE	Canadian Academy of Engineering
	Canadian Association of Exhibitors
	Compare Alpha Equal
	Computer-Aided Education
	Computer-Aided Engineering
	Computer-Assisted Engineering
	Computer-Assisted Enrolment
	Constant Analyzer Energy
Cae	Caelum [Astronomy]
CAEC	Canadian Architectural, Engineering and Construction Show (and Exposition)
	Central American Energy Commission [Guatemala]
CAE/CAD	Computer-Aided Engineering/Computer-Aided Design
CAE/CAM	Computer-Aided Engineering/Computer-Aided Manufacturing
CAED	Canadian Association of Equipment Distributors
CAEE	Committee on Aircraft Engine Emissions [of ICAO]
CAEF	Comite des Associations Europeennes de Fonderie [= Committee of European Foundry Associations, France]
CAEFMS	Canadian Agricultural Economics and Farm Management Society
CAEM	Canadian Association of Exposition Managers
	Controlled Atmosphere Electron Microscopy
CAeM	Commission for Aeronautical Meteorology
CAEP	Committee on Aviation Environmental Protection
CAER	Community Awareness and Emergency Response
CAES	Canadian Agricultural Economics Society
	Compressed-Air Energy Storage
CAEU	Council of Arab Economic Unity [Jordan]
CAF	Canadian Advertising Foundation
	Canadian Armed Forces
	Chemical Analysis Facility
	Cleared as Filed
	Computer-Automated Fixturing
	Content-Addressable File
	Cost and Freight
	Cost, Assurance, and Freight
	Council on Alternate Fuels [US]
	Critical Absorption Frequency
CaF	Calcium-Free
C_4AF	Tetracalcium Aluminoferrate [Cement]
CAFB	Chemically Active Fluidized Bed
CAFC	Canadian Association of Fire Chiefs
CAFCG	Constant Amplitude Fatigue Crack Growth
CAFD	Computer-Aided Filter Design
	Contact Analog Flight Display
CAFE	Computer-Aided Film Editor
	Corporate Average Fuel Economy
CAFEA	Commission on Asian and Far Eastern Affairs
CAFEE	Critical Assembly Fuel Element Exchange
CAFGE	Colombian Association of Flower Growers and Exporters
CAFM	Computer-Assisted Facilities Management
$Ca(FOD)_2$	Bis(1,1,1,2,2,3,3-Heptafluoro-7,7-Dimethyl-4,6-Octanedionato)calcium
CAFS	Canadian Association for Future Studies
	Chinese American Food Service [US]
	Content-Adressable File Store
CAFSAC	Canadian Atlantic Fisheries Scientific Advisory Committee
CAFTA	Central American Free Trade Association
CAG	Canadian Air Group [of Canadian Armed Forces]
	Canadian Association of Geographers
	Civil Aviation Group
	Computer Applications Group [of US Air Force]
	Computer Applications Group [of ASLIB, US]
	Cooperative Automation Group [of BLBS, UK]
CAGC	Clutter Automatic Gain Control
	Coded Automatic Gain Control
	Continuous Access Guided Communication
CAGE	Compiler and Assembler by General Electric
	Computerized Aerospace Ground Equipment
CAGI	Compressed Air and Gas Institute [US]
CAGS	Canadian Arctic Gas Study
	Canadian Association of Graduate Schools
	Certificate of Advanced Graduate Studies
	Chinese Academy of Geological Sciences
CAHE	Core Auxiliary Heat Exchanger
CAHS	Canadian Aviation Historical Society
CAHT	Computer Aids for Human Translation [of CMU, US]
CAHTS	Computer-Aided Heat-Treating System
CAI	Canadian Aeronautical Institute
	Centro de Automaticacion Industrial [= Center for Industrial Automation, Cuba]
	Close Approach Indicator
	Color Alteration Index
	Comite Arctique International [= International Arctic Committee, Norway]
	Compression after Impact
	Computer-Administered Instruction
	Computer-Aided Instruction
	Computer Analog Input
	Computer-Assisted Instruction
CAIBE	Chemically-Assisted Ion Beam Etching
CAIC	Caribbean Association of Industry and Commerce [Barbados]
	Computer-Assisted Indexing and Classification

CAI/CAL Computer-Aided Instruction/Computer-Aided Learning
CAIMAF Canadian Advanced Industrial Materials Forum
CAIMAW Canadian Association of Industrial, Mechanical and Allied Workers
CAIN Cataloguing and Indexing System [of NAL, US]
CAINS Carrier Aircraft Inertial Navigation System
CAINT Computer-Assisted Interrogation
CAI/O Computer Analog Input/Output
CAI/OP Computer Analog Input/Output
CAIP Canadian Artificial Intelligence Products
Civil Aircraft Inspection Procedures
Computer-Assisted Indexing Program
CAIR Conference on Artificial Intelligence and Robotics
Confidential Aviation Incident Reporting (Program) [Australia]
CAIRS Canadian Institute for Radiation Safety
Computer-Assisted Information Retrieval System
CAIS Canadian Association for Information Science
Central Abstracting and Indexing Service
Communication and Information Systems Division [of MEP, UK]
Computer-Aided Instruction System
Computer-Assisted Information Retrieval System
CAISF Chemical Abstracts Integrated Subject File [of CAS, US]
CAISIM Computer-Assisted Industrial Simulation [US Army]
CAIT Coalition for the Advancement of Industrial Technology [now: CORETECH, US]
CAJ Caulked Joint
CAK Command Access Key
Command Acknowledgment
CAL Calendar
Caliber
Calibration
Center for Applied Linguistics [US]
Certificate of Advanced Librarianship
Certificate in Applied Linguistics
Common Assembler Language
Composition Analysis Laboratory [of Colorado State University, US]
Computer-Aided Learning
Computer Animation Language
Computer-Assisted Learning
Confined Area Landing
Continuous Annealing Line
Conservation Analytical Laboratory [US]
Conversational Algebraic Language
Cornell Aeronautical Laboratory [US]
Cray Assembly Language
Cal California [US]
Kilocalorie
cal calorie
CALAS Canadian Association for Laboratory Animal Science
CaLaSOAP Calcium Lanthanum Silicate Oxyapatite
CALC Calculation
Cargo Acceptance and Load Control
Customer Access Line Charge
CaLC Calcia-Containing Lanthanum Chromite
CALCD Calculated
CALDA Canadian Airline Dispatchers Association
CALEOT Center of Applications of Lasers and Electro-Optic Technologies [US]
CALFAB Computer-Aided Layout and Fabrication
CALCG Calculating
CALCN Calculation
CALICO Computer-Assisted Language Learning and Instruction Consortium [US]
CALLS California Academic Libraries List of Serials [of CLASS, US]
CALM Catenary Anchor Leg Mooring
Collected Algorithms for Learning Machines
Computer Archive of Language Materials [of Stanford University, US]
Computer-Assisted Library Mechanization
Council of Academic Libraries of Manitoba [Canada]
CALOGSIM Computer-Assisted Logistics Simulation [US Army]
CALPA Canadian Airline Pilots Association
Cal Poly California Polytechnic State University [US]
CALR Computer-Assisted Legal Retrieval
CALRAB California Raisin Advisory Board [US]
CALS Computer-Aided Logistics Support
cals calories
Caltech California Institute of Technology [also: CIT, US]
Caltrans California Department of Transportation [also: CALTRANS]
CALURA Corporation and Labour Union Returns Act [Canada]
CALUTRON California University Cyclotron [also: Calutron]
CAM Canada Air Mail
Canadian Association of Movers
Cascade Access Method
Cellular Automata Machine
Cellulose Acetate Methacrylate
Cement Aggregate Mixture
Center for Applied Metallography [US]
Center for Aquatic Microbiology [US]
Central Address Memory
Civil Aeronautics Manual
Checkout and Maintenance
Commercial Air Movement
Communications Access Method
Computer Access Matrix
Computer-Addressed Memory
Computer-Address Matrix
Computer-Aided Manufacturing
Computer-Aided Modeling
Computer-Assisted Makeup
Computer-Assisted Microscopy
Containment Atmospheric Monitoring
Content-Addressable Memory
Continuous Air Monitor
Custom Application Module
Cybernetic Anthropomorphous Machine
Cam Camelopardalis [Astronomy]
CAMA Canadian Appliance Manufacturers Association
Canadian Automatic Merchandising Association
Centralized Automatic Message Accounting
Civil Aviation Medical Association [US]

CAMAC Center for Applied Mathematics and Advanced Computation
Computer-Automated Measurement and Control

CAMA-ONI Centralized Automatic Message Accounting — Operator Number Identification

CAMAR Common-Aperture Malfunction Array Radar
Common-Aperture Multifunction Array Radar

CAMARCO Can
adian Marine Coasting Advisory [also: Camarco]

CAMAS Confederation of African Medical Associations and Societies [Nigeria]

Cambs Cambridgeshire [UK]

CAMC Canadian Army Medical Corps
Canadian Association of Management Consultants

CAMCA Canadian-American Motor Carriers Association

CAMCOS Computer-Assisted Maintenance Planning and Control System

CAMDEC Ceramics Advanced Manufacturing Development Engineering Center [of ORNL, US]

CAMDF Canadian Agricultural Market Development Fund [of Agriculture Canada]

CAMEA Comite des Applications Militaires de l'Energie Atomique [= Committee on Military Applications of Atomic Energy, France]

CAMEL Collapsible Airborne Military Equipment Lifter
Component and Material Evaluation Loop

CAMEO Creative Audio and Music Electronics Organization [US]

CAMESA Canadian Military Electronic Standards Agency

CAMI Civil Aviation Medical Institute [US]
Coated Abrasives Manufacturers Institute [US]
Computer Applications in the Mineral/Mining Industry

CAM-I Computer-Aided Manufacturing International [US]

CAMIS Computer-Assisted Make-Up and Imaging System

CAMM Canadian Association of Medical Microbiologists
Computer-Aided Manufacturing Management
Computer-Aided Modeling and Manufacturing
Computer-Aided Modeling Machine
Council of American Maritime Museums [US]

CAMMD Canadian Association of Manufacturers of Medical Devices

CAMMS Computer-Assisted Material Management System

CAMP Cabin Air Manifold Pressure
Central Access Monitor Program
Compiler for Automatic Machine Programming
Computer-Aided Mineral Processing
Computer-Assisted Mathematics Program
Computer-Assisted Menu Planning
Computer-Assisted Movie Production
Continuous Air Monitoring Program
Controls and Monitoring Processor
Cooperative African Microform Project

cAMP Cyclic Adenosine Monophosphate

CAM-PC Cellular Automata Machine for Personal Computers

CAMRAS Computer-Assisted Mapping and Records Activities System

CAMRDC Central African Mineral Resources Development Center [Congo]

CAMROC Cambridge Radio Observatory Committee [US]

CAMRT Canadian Association of Medical Radiation Technologists

CAMS Certificate of Advanced Management Studies
Cybernetic Anthropomorphous Machine System

CAMSE (Conference and Exhibition on) Computer Applications to Materials and Molecular Science and Engineering

CAMSIM Computer-Assisted Maintenance Simulation [US Army]

CAN Canada
Canadian
Canadian Aquaculture Network
Cancel (Character)
Ceric Ammonium Nitrate
Committee on Aircraft Noise [of ICAO]
Correlation Air Navigation

CANAC Computer Assisted National ATC (= Air Traffic Control) Center [Belgium]

CANAIRDEF Canadian Air Force Defense Command

CANAIRDIV Canadian Air Force Division

CANAIRHEQ Canadian Air Force Headquarters

CANAL Command Analysis

CANAPES Canadian Acoustic Parabolic Equation System

CANARI Caribbean Natural Resources Institute

CANAS Canadian Naval Air Station

CANASA Canadian Alarm and Security Association

CANAVHED Canadian Naval Headquarters

CANBIOCON Canadian Biotechnology Conference (and Exhibition)

CANC Cancel [also: CANCL]
Cancellation

Canc Cancer [Astronomy]

CANCAM Canadian Congress of Applied Mechanics

CAN/CAM Canadian Congress of Applied Mechanics

CANCASS Canadian Command Active Sonobuoy System

CAN/CAT Canadian Cataloging Subsystem

CANCEE Canadian National Committee for Earthquake Engineering

CANCOM Canadian Satellite Communications

CAND Candelabra
Candidate
Computer-Aided Network Design

CANDE Command and Edit Language

C and F Cost and Freight [also: C&F]

C and I Cost and Insurance [also: C&I]

CANDIS Canadian Disarmament Information Service

CANDOC Canadian Documentation [also: CAN/DOC [of CISTI]]

cand phil (candidatus philosophiae) — Candidate in Philosophy

CANDU Canada Deuterium Uranium (Reactor)

CANDUBLW Canada Deuterium Uranium — Boiling Light Water

CANDUPHW Canada Deuterium Uranium — Pressurized Heavy Water

CANEL Connecticut Aircraft Nuclear Experiment [US]

CANEWS Canadian Electronic Warfare System

CANFARMS Canadian Farm Management Data System

CANFUT Canadian Future Study Group

CanLaunch Canadian Capability for Launch Site Study

CAN/LAW Canadian Computer-Assisted Legal Research System

CANMAP Canadian Marketing Assistance Program
CANMARC Canadian Machine-Readable Cataloguing [also: CAN/MARC [of NLC]]
CAN-MATE Canadian Manufacturing Advanced Technology Exchange
CANMET Canada Center for Mineral and Energy Technology [of EMR, Canada]
CAN/OLE Canadian On-Line Enquiry [of CISTI]
CANOPUS Canadian Auroral Network Open Program Unified Study
CANOZE Canadian Ozone Experiment
CANP Civil Air Notification Procedure
CANPOTEX Canadian Potash Export Association
CANQUA Canadian Quaternary Association
CANS Computer-Assisted Network Scheduling System
CANSARP Canadian Search and Rescue Planning
CAN/SDI Canadian Service for the Selective Dissemination of Information [of CISTI]
CANSG Civil Aviation Navigational Services Group
CAN/SIM Canadian Socio-Economic Information Management System
CanSIS Canadian Soil Information System [also: CAN/SIS]
CAN/SND Canadian Service for Scientific Numeric Databases
CANSPECS CANMET's (= Canada Center for Mineral and Energy Technology) Service Program for the Evaluation of Codes and Standards
CAN/TAP Canadian Technical Awareness Program
CANTASS Canadian Towed Array Sonar System
CANTAT Canadian Transatlantic Telephone [Canada/UK]
CANTO Caribbean Association of National Telecommunications Organizations
CANTRAN Cancel in Transmission
Cancel Transmission
CAN-UK Canada — United Kingdom
CANUNET Canadian University Computer Network
CANUSE Canada-United States Eastern Power Complex
CANWEC Canadian National Committee of the World Energy Conference
CAO Chief Administrative Officer
Circuit Allocation Order
Communications Authorization Order
Completed as Ordered
Computer Aided Optimization
Crimean Astrophysical Observatory [USSR]
CAOC Constant Axial Offset Control
CAOCI Commercially Available Organic Chemicals Index [of CNA, UK]
CAOD Computer-Aided Optical Design
CAODC Canadian Association of Oil-Well Drilling Contractors
COARB Civil Aviation Operational Research Branch [UK]
CAORC Council of American Overseas Research Centers [US]
CAORE Canadian Army Operation Research Establishment
CAOS Completely Automatic Operating System
CAP Canadian Assistance Plan
Canadian Association of Pathologists
Canadian Association of Physics
Capacitance
Capacitor
Capacity
Capillary
Capital (Letter)
Capitalization
Carburizing Atmosphere Process
Card Assembly Program
Cataloguing in Advanced Publication [of BLBSD, UK]
Cellulose Acetate Propionate
Chapter
Chloramphenicol
Circuit Access Point
Civil Air Patrol [US]
Civil Air Publication
Civil Aviation Publication
Cleaner Air Package
College of American Pathologists [US]
Combat Air Patrol
Common Agricultural Policy
Commonwealth Association of Planners [Canada]
Computer Access Panel
Computer-Aided Planning
Computer-Aided Production
Computer-Aided Programming
Computer-Aided Publishing
Computer-Aided Purchasing
Computer Analysts and Programmers [UK]
Computer-Assisted Processing
Computer-Assisted Production
Computerized Area Pricing
Consolidation by Atmospheric Pressure
Consolidation under Atmospheric Pressure
Consulting Assistance Program
Consumers Association of Penang [Malaysia]
Contact Approach
Continuous Airworthiness Panel [of ICAO]
Continuous Audit Program
Controlled Atmosphere Packaging
Council on Advanced Programming
Council to Advance Programming
Cryotron Associative Processor
Customer Access Panel
Cap Capricornus [Astronomy]
CAPA Canadian Aircraft Preservation Association
Canadian Association of Purchasing Agents
Commission on Asian and Pacific Affairs
Commonwealth Association of Polytechnics in Africa [Kenya]
Confederation of Asian and Pacific Accountants
CAPAC Canadian Association of Primary Air Carriers
CAPAL Computer and Photographically Assisted Learning
CAPC Canadian Army Pay Corps
Canadian Association of Professional Conservators
Civil Aviation Planning Committee
CAPCC Computer Application Process Control Committee
CAPCOM Capsule Communications
CAP COST Capital Cost
CAPD Computer-Aided Process Design
CAPDAC Computer-Aided Piping Design and Construction
CAPDM Canadian Association of Physical Distribution Management
CAPE Coalition of Aerospace Employees

Communication Automatic Processing Equipment
Computer-Assisted Patient Evaluation
Computer Assisted Planning Experiment
Computer Assisted Program Evaluation
Computer-Assisted Project Execution
Convective Available Potential Energy
Coordinated Assessment and Program Planning for Education

CAPER Canadian Association of Publishers Educational Representatives

CAPERT Computer-Assisted Program Evaluation Review Technique

CAPERTSIM Computer-Assisted Program Evaluation Review Technique Simulation

CAPHE Consortium for the Advancement of Private Higher Education [US]

CAPIC Canadian Association for Production and Inventory Control

CAPL Canadian Association of Public Libraries
Continuous Annealing and Processing Line

CAPM Computer-Aided Patient Management
Computer-Aided Production Management
Containment Atmosphere Particulate Monitor
CPU (= Central Processing Unit) Access Port Monitor

CAPMoN Canadian Air and Precipitation Monitoring Network
Canadian Atmospheric Precipitation Monitoring Network

CAPO Canadian Army Post Office

CAPOSS Capacity Planning and Operation Sequence System

CAPOSS-E Capacity Planning and Operation Sequence System — Extended Program

CAPP Commission on Air Pollution Prevention [Germany]
Computer-Aided Process Planning
Computer-Aided Production Planning
Content Addressable Parallel Processor

CAPPA Crusher and Portable Plant Association [US]

CAPPS Computer-Aided Part Programming System

CAPPS/EDGES Computer-Aided Part Programming System/ Expert DMIS (= Directory Management Information System) Graphical Editor and Simulator

CAPR Catalogue of Programs

CAPRI Card and Printer Remote Interface
Coded Address Private Radio Intercom
Computer-Aided Personal Reference Index System [of UKAEA]
Computerized Advance Personnel Requirements and Inventory
Computerized Area Pricing

CAP/RPSP College of American Pathologists/Reference Preparation for Serum Proteins [US]

CAPS Call Attempts per Second
Capital Letters
Cell Atmosphere Processing System
Center for Advanced Purchasing Studies
Certificate of Advanced Professional Studies
Civil Aviation Purchasing Service [of ICAO]
Computer-Aided Polymer Selection
Computer-Assisted Problem Solving
Control and Auxiliary Power Supply System
Courtauld's All-Purpose Simulator
Current Advances in Plant Science (Database) [UK]
3-(Cyclohexylamino)-1-Propanesulfonic Acid

CAP SCR Cap Screw

CAPSIN Civil Aviation Packet Switching Integrated Network

CAPSO 3-(Cyclohexylamino)-2-Hydroxy-1-Propane-sulfonic Acid

CAPST Capacitor Start

CAPT Captain
Center for Advanced Pyrometallurgical Technology [of University of Utah, US]

CAPTAIN Character and Pattern Telephone Access Information Network [Japan]

CAPTEX Cross-Appalachian Tracer Experiment

CAPT(N) Captain (Naval)

CAPTOR Encapsulated Torpedo

CAPT(R) Captain (Regular)

CAP'Y Capacity

CAQC Computer-Aided Quality Control [also: CAQ]

CAR Canadian Association of Radiologists
Canadian Automotive Report
Cargo
Caribbean Region
Center for Atmospheric Research [Canada]
Central African Republic
Central Apparatus Room
Channel Address Register
Channel Assignment Record
Civil Air Regulations [US]
Community Antenna Relay
Computer-Aided Repair
Computer-Aided Retrieval
Computer-Assisted Research
Computer-Assisted Retrieval
Conditioned Avoidance Response
Configuration Acceptance Review
Containment Air Recirculation
Containment Air Removal
Control Advisory Release
Corrective Action Request
Critical Angle Refractometer

CARA Cargo and Rescue Aircraft
Combat Aircrew Rescue Aircraft

CARAC Civil Aviation Radio Advisory Committee [UK]

CARAM Content-Addressable Random-Access Memory

CARAS Canadian Academy of Recording Arts and Sciences

CARAT Cargo Agents Reservation, Airwaybill Insurance and Tracking System

CARB California Air Resources Board [US]
Carbonate
Carburetor
Carburization

CARBINE Computer-Automated Real-Time Betting Information Network

CARC Canadian Agricultural Research Council
Canadian Arctic Resources Committee
Canadian Audio Research Consortium
Central America Resource Center
Chemical Agent Resistant Coating

CARCA Computer-Aided Rocking Curve Analysis

CARCAE Caribbean Regional Council for Adult Education [Barbados]

CARCS Canadian Amphibian and Reptile Conservation Society

CARD Canadian Advertising Rates and Data

Center for Agricultural and Rural Development [of Iowa State University, US]

Channel Allocation and Routing Data

Compact Automatic Retrieval Device

Compact Automatic Retrieval Display

CARDCODER Card Automatic Code System

CARDE Canadian Armament Research and Development Establishment [of DRB, Canada]

CARDI Caribbean Agricultural Research Development Institute

Cards Cardiganshire [UK]

CARE Center for Aviation Research and Education [now: NASAO, US]

Ceramic Application in Reciprocating Engines

Computer-Aided Reliability Estimation

Computer-Assisted Remediation and Education Project [Canada]

CAREIRS Conservation and Renewable Energy Inquiry and Referral Service [US]

CARES Civil Air Rescue Emergency Services

CARESS Career Retrieval Search System [of University of Pittsburgh, US]

CARETS Central Atlantic Regional Ecological Test Site

CARF Canadian Advertising Research Foundation

Canadian Amateur Radio Federation

Central Altitude Reservation Facility

Consumer Affairs and Regulatory Functions [of Office of the Assistant Secretary, US]

CARI Canadian Association of Recycling Industries

Council of Air-Conditioning and Refrigeration Industry [US]

CARIC Computerized Automation and Robotics Information Center [of MVA, US]

CARICOM Caribbean Community [Guyana]

CARIFTA Caribbean Free Trade Agreement

Caribbean Free Trade Area

CARIRI Caribbean Industrial Research Institute

CARIS Computer-Aided Resource Information System

Computerized Agricultural Research Information System [of FAO — UN]

Current Agricultural Research Information System

CARL Canadian Academic Research Libraries

Canadian Association of Research Libraries

Chemical Algorithm for Reticulation Linearization

CARM Canadian Association of Risk Managers

Carmarths Carmarthenshire [UK]

CARN Conditional Analysis for Random Networks

CARNACS Conventional Arms Reconnaissance [Canada]

CAROL Circulation and Retrieval On-Line [of JCU, Australia]

Computer-Oriented Language

CAROT Centralized Automatic Reporting on Trunks

CARP Call Accounting Reconciliation Process

Carpenter [also: CARPTR]

Carpentry [also: CARPTRY]

Committee for Acoustic Emission in Reinforced Plastics [of ASNT, US]

Computed Air Release Point

CARPAS Comision Asesora Regional de Pesca para el Atlantico Sudoccidental [= Regional Fisheries Advisory Commission for the Southwest Atlantic, Italy]

CARR Carrier

carr pd carriage paid

CARRS Close-In Automatic Route Restoral System [of NORAD]

CARS Center for Atomic Radiation Studies [US]

Coherent anti-Stokes Raman Scattering

Coherent anti-Stokes Raman Spectroscopy

Common Antenna Relay System

Community Aerodrome Radio System

Community Antenna Relay Service

Computer-Aided Routing System

Computer-Assisted Reference Service [of University of Arizona, US]

Computerized Automatic Rating System

Computerized Automotive Reporting Service

Continuous Alarm Reporting Service

CARSP Canadian Association of Road Safety Professionals

CART Cartographer

Cartography

Cartridge

Center for Advanced Resource Technology [Canada]

Central Automatic Reliability Test

Central Automated Replenishment Technique

Complete Automatic Reliability Testing

Computerized Automatic Rating Technique

CARTG Canadian Amateur Radio Teletype Group

CARTIS Computer-Aided Real-Time Inspection System

CAS Cable Activity System

Calculated Air Speed

Calibrated Air Speed

California Academy of Science [US]

California Avocado Society [US]

Call Accounting Subsystem

Center for Auto Safety [US]

Center of Atmospheric Studies [US]

Central Alarm Station

Central Amplifier Station

Certificate of Advanced Studies

Chemical Abstracts Service [of ACS, US]

Chinese Academy of Sciences [PR China]

Chinese Association for Standardization [PR China]

Chinese Astronomical Society [PR China]

Circuits and Systems (Society) [of IEEE, US]

Clean Air System

Close Air Support

Collision Avoidance System [of FAA, US]

Column Address Strobe

Commission for Atmospheric Sciences [of NAS, US]

Computer-Aided Simulation

Computer-Aided System

Computer Aid System

Computer-Assisted Steelmaking

Computerized Acquisition System

Control Augmentation System

	Control Automation System
	Controlled Airspace
	Cooperative Applications Satellite [Canada]
	Current Awareness Service
	Customer Administrative Service
	Czechoslovak Academy of Sciences
Cas	Cassiopeia [Astronomy]
CASA	Canadian Advertising and Sales Association
	Canadian Alarm and Security Association
	Canadian Automatic Sprinkler Association
	Car Audio Specialists Association [US]
	Computer-Aided Structural Analysis
	Computer and Automated Systems Association [of SME, US]
CASAA	Canadian Association of Student Awards Administrators
CAS-ACS	Chemical Abstracts Service of the American Chemical Society [US]
CASAFA	(Interunion) Commission on the Application of Science to Agriculture, Forestry and Aquaculture [Canada]
CASALS	Congress on Advances in Spectroscopy and Laboratory Sciences
CASANZ	Clean Air Society of Australia and New Zealand
CASAO	Chartered Accountant Students Association of Ontario [Canada]
CASARA	Civil Air Search and Rescue Association
CASAS	Commonwealth Association for Scientific and Agricultural Societies
CASA/SME	Computer and Automated Systems Association of the Society of Manufacturing Engineers [US]
CASAW	Canadian Association of Smelters and Allied Workers
CASB	Canadian Aviation Safety Board
	Cost Accounting Standards Board
CASC	Canadian Army Service Corps
	Canadian Automobile Sport Club
	Certified Alfalfa Seed Council [US]
CASCC	Canadian Agricultural Services Coordinating Committee
CASD	Computer-Aided System Design
CASDAC	Computer-Aided Ship Design and Construction
CAS DDS	Chemical Abstracts Service — Document Delivery Service [of ACS, US]
CASE	Citizens Association for Sound Energy
	Committee on Academic Science and Engineering [of FCST, US]
	Common Application Service Elements
	Computer-Aided Software Engineering
	Computer-Aided System Evaluation
	Computer Automated Support Equipment
	Consolidated Aerospace Supplier Evaluation
	Coordinating Agency for Suppliers Evaluation [US]
	Council for the Advancement and Support of Education [Canada]
	Counsellinmg Assistance to Small Enterprises [Canada]
	Curriculum Alignment Services for Educators [of EPIE, US]
CASEA	Canadian Association for the Study of Educational Administration
CASearch	Chemical Abstracts Search [of ACS, US]
CASES	Capabilities Assessment Expert System
CAS Files	Chemical Abstracts Service Files [of ACS, US]
CASH	Computer-Assisted Subject Headings Program [of UCSD, US]
CASI	Canadian Aeronautics and Space Institute
	Compact Airborne Spectrographic Imager
	Convenient Automotive Services Institute [US]
Ca/Si	Calcium/Silicon (Ratio)
CASIA	Chemical Abstracts Subject Index Alert [ACS, US]
CASING	Crosslinking by Activated Species of Inert Gases
CASIS	Computerized Alloy Steel Information System
CASLE	Commonwealth Association of Surveying and Land Economy [UK]
CASLIS	Canadian Association of Special Libraries and Information Services
CASMT	Central Association of Science and Mathematics Teachers [US]
CAS NUT	Castle Nut
CASP	Canadian Airspace Systems Plan
	Canadian Atlantic Storms Program
	Central American Society of Pharmacology [Panama]
CASPA	Certificate of Advanced Study in Public Administration
	Computer-Aided Sculptured Pre-Apt System
CASPAR	Cambridge Analog Simulator for Predicting Atomic Reactions
	Cushion Aerodynamics System Parametric Assessment Research
CASPPR	Canadian Arctic Shipping Pollution Prevention Regulations
CASS	Commanded Active Sonobuoy System
	Consolidated Automatic Support System
	Copper-Accelerated Salt Spray (Test)
CASSA	Coarse Analog Sun Sensor Assembly
CASSANDRA	Chromatogram Automatic Soaking Scanning and Digital Recording Apparatus
CASSI	Chemical Abstracts Service Source Index [of ACS, US]
CASSI KWOC	CASSI Keyword Out-of-Context
CASSM	Context Addressed Segment Sequential Memory
CAST	Canadian Air-Sea Transportable Brigade Group
	Canadian Air-Sea Transportable Combat Group
	Capillary Action Shaping Technique
	Chemical Abstract Searching Terminal
	Clearinghouse Announcements in Science and Technology
	Computer-Aided Solidification Technique
	Computerized Automatic Systems Tester
	Council for Agricultural Science and Technology [US]
CAST ASIA	Conference on the Application of Science and Technology to the Development of Asia
CASTI	Centers for the Analysis of Scientific and Technical Information [US]
CASTME	Commonwealth Association of Science, Technology and Mathematics Educators [UK]
CASW	Council for the Advancement of Science Writing [US]
CASWS	Close Air Support Weapon System

CAT Capacity-Activated Transducer
Carburetor Air Temperature
Catalogue
Catalysis
Catalyst
Catapult
Catechol
Category
Chemical Addition Tank
Choline Acetyltransferase
Cis-Anti-Trans (Configuration) [Chemistry]
Civil Air Transport
Clear Air Transport
Clear-Air Turbulence
College of Advanced Technology
College of Agricultural Technology
Commercial Air Transport
Compile and Test
Composition Analysis by Thickness Fringe
Computer-Aided Teaching
Computer-Aided Testing
Computer-Aided Tomography
Computer-Aided Training
Computer-Aided Translation
Computer-Aided Typesetting
Computer-Assisted Teaching
Computer-Assisted Testing
Computer-Assisted Tomography
Computer-Assisted Training
Computer Automated Test
Computer Average Transient
Computer Axial Tomography
Computerized Assisted Tomography
Computerized Axial Tomography
Conditionally Accepted Tag
Constant Analyzer Transmission
Control and Analysis Tool
Controlled Attenuator Timer
Cumulative Abbreviated Trouble (File)

CATA Canadian Advanced Technology Association
Canadian Air Transportation Administration
Commissao de Avaliacao Toxicologica dos Aditivos Alimentares [= Committee for Toxicologic Evaluation of Food Additives, Portugal]

CATAC Creative, Analytical, Technical, Alternative, Competent (Calculating)

CATAL Catalogue

CATC Canadian Air Transport Commission
Caribbean Appropriate Technology Center
Circular-Arc-Toothed Cylindrical
Commonwealth Air Transport Commission [UK]
Commonwealth Air Transport Council [UK]

CATCA Canadian Air Traffic Control Association

CATCALL Completely Automated Technique for Cataloguing and Acquisition of Literature for Libraries

CATCC Canadian Association of Textile Colorists and Chemists

CATCH Computer-Aided Testing and Checking
Computer Analysis of Thermochemical Data

CATE Center for Advanced Technology Education
Ceramic Applications in Turbine Engines
Coordinator of the Air Transport in Europe

CATED Centre d'Assistance Technique et de Documentation [= Center for Technical Assistence and Documentation, France]

CATER Collection and Analysis of Terminal Records (Service) [of ADC, US]

CATH Cathode

CATI Caribbean Aviation Training Institute
Computer-Assisted Television Instruction

CATIE Centro Agronomica Tropical Investigacion y Ensenanza [= Tropical Agriculture Research and Training Center, Costa Rica]

CATIS Computer-Aided Tactical Information System

CATLG Catalogue

CATLAS Centralized Automatic Trouble-Locating and Analysis System

CATLINE Catalogue On-Line [of National Library of Medicine, US]

CATM Canadian Achievement Test in Mathematics

Ca(TMHD)$_2$ Bis(2,2,6,6-Tetramethyl-3,5-Heptanedionato)calcium

CATNIP Computer-Assisted Technique for Numerical Index Preparation

CAT NO Catalogue Number

CATO Civil Air Traffic Operations
Compiler for Automatic Teaching Operation

CATS Center for Advanced Television Studies [US and UK]
Central Automated Transit System
Computer-Aided Teaching System
Computer-Aided Testing System
Computer-Assisted Trading System
Computer Automated Test System

CATSE Capacity of the Air Transport System [of ECEC, France]

CATSS Catalogue Support System [of U of T, Canada]

CATT Campaign Against Arms Trade [UK]
Centralized Automatic Toll Ticketing
Controlled Avalanche Transit-Time Triode
Conveyorized Automatic Tube Tester

CATTCM Canadian Achievement Test in Technical and Commercial Mathematics

CATV Cable Television
Common Antenna Television
Community Antenna Television

CAU CPU (= Central Processing Unit) Access Unit
Command/Arithmetic Unit
Compare Alpha Unequal
Controller Adapter Unit
Crypto Ancillary Unit

CAUBO Canadian Association of University Business Officers

CAUCE Canadian Association for University Continuing Education

CAUDO Canadian Association for University Development Officers

CAUFN Caution Adviced Unitil Further Notice

CAUML Computers and Automation Universal Mailing List

CAURA Canadian Association of University Research Administrators

CAUS Citicens Against UFO (= Unidentified Flying Object) Secrecy [US]

	Color Association of the United States
CAUSE	College and University Systems Exchange [US]
CAUST	Caustic
CAUT	Canadian Association of University Teachers
CAV	Cavalry
	Cavity (Resonance)
	Component Analog Video
	Composite Analog Video
	Constant Angular Velocity
	Constant Arc Voltage
CAVE	Computer Augmented Video Education
CAV-OK	Ceiling and Visibility Okay
CAVORT	Coherent Acceleration and Velocity Observations in Real Time
CAVU	Ceiling and Visibility Unlimited
CAW	Canadian Autoworkers Union
	Carbon-Arc Welding
	Channel Address Word
CAWC	Central Advisory Water Committee [UK]
CAW-G	Gas Carbon-Arc Welding
CAWS	Common Aviation Weather System
CAW-S	Shielded Carbon-Arc Welding
CAW-T	Twin Carbon-Arc Welding
CAX	Community Automatic Exchange
CAZRI	Central Arid Zone Research Institute [India]
CB	Carbon Black
	Carrier-Based
	Cataclysmic Binary
	Catch Basin
	Cathode Bias
	Cell Biologist
	Cell Biology
	Cellulose Butyrate
	Center of Buoyancy
	Central Battery
	Chlorobenzene
	Circuit Board
	Circuit Breaker
	Citizen Band (Radio)
	Clear Back
	Clinical Biochemist
	Clinical Biochemistry
	Coated Back
	Coin Box
	Coincidence (Grain) Boundary
	Common Base
	Common Battery
	Common Branch
	Communications Buffer
	Communications Bus
	Component Board
	Conduction Band
	Connecting Box
	Construction Ball
	Containment Building
	Continuous Blowdown
	Contract Bolt
	Convergent Beam
	Coupled Biquad
	Currency Bond
C/B	Cold Water/Boiling Water (Immersion Ratio)
Cb	Centibel [Unit]
	Columbium
	Cumulo-Nimbus (Cloud)
CBA	Canadian Bankers Association
	Canadian Booksellers Association
	Canadian Botanical Association
	Carboxybenzaldehyde
	Chemical Bond Approach
	Chlorobenzoic Acid
	Commonwealth Broadcasting Association [UK]
	Computer-Based Automation
	Conjugated Bile Acid
	Contact-Breaking Ammeter
	Cost-Benefit Analysis
	Council for British Archeology
CBAA	Canadian Business Aircraft Association
CBABG	Commonwealth Bureau of Animal Breading and Genetics
CBAC	Chemical-Biological Activities [of ACS, US]
CBAE	Commonwealth Board of Architectural Education [of CAA, UK]
CBAL	Counterbalance
CBARC	Conference Board of Associated Research Councils
CBAS	Chemical Bond Approach Study
C-BASIC	Commercial BASIC [Computer Language]
CBAST	Concentrated Boric Acid Storage Tank
CBAT	Central Bureau for Astronomical Telegrams [US]
CB/ATDS	Carrier-Based/Airborne Tactical Data System
CBC	Canadian Brigade Corps
	Canadian Broadcasting Corporation
	Can't Be Called
	Central Bureau of Compensation [Belgium]
	Chain Block Character
	Chemically-Bonded Ceramics
	Cipher Block Chaining
	Contact-Breaking Clock
	Continuity Bar Connector
CBCC	Chemical-Biological Coordination Center [US]
CBCS	Commonwealth Bureau of Census and Statistics [Australia]
CBCSM	Council of British Ceramic Sanitaryware Manufacturers
CBCT	Customer-Bank Communications Terminal
	Customer Banking Communications Terminal
CBD	Call Box Discrimination
	Cannabidiol
	Cash before Delivery
	Central Business District
	Commerce Business Daily
	Configuration Block Diagram
	Constant Bit Density
	Convergent Beam Diffraction
CBDB	Conference Board Database [US]
CBDC	Cape Breton Development Corporation [Canada]
CBDI	Control Red Bank Demand Indicator
CBDP	Convergent Beam Diffraction Pattern
CBDS	Circuit Board Design System
CBDST	Commonwealth Bureau of Dairy Science and Technology [UK]
CBDT	Can't Break Dial Tone
	Computer-Based Documentation and Training

CBE Centralized Branch Exchange
Chemical Beam Epitaxy
Circuit Breaker for Equipment
Citizens for a Better Environment [US]
Commander of the Order of the British Empire
Congresso Brasileira de Entomologica [= Brazilian Entomological Congress]
Companion of the Order of the British Empire
Computer-Based Education
Council of Biology Editors [US]

C&BE Computer and Business Equipment

CBEA Commonwealth Banana Exportes Association [Saint Lucia]

CBEC Canadian Book Exchange Center

CBED Convergent-Beam Electron Diffraction

CBEDP Convergent-Beam Electron-Diffraction Pattern

CBEMA Canadian Business Equipment Manufacturers Association
Computer and Business Equipment Manufacturers Association [US]

CBET Certified Biomedical Equipment Technician

CBFC Copper and Brass Fabricators Council [US]

CBG Craniofacial Biology Group [of IADR, US]

CBGA Carpathian Balkan Geological Association [Poland]

CBH Can't be Heard

CBHL Council on Botanical and Horticultural Libraries [US]

CBI Canadian Business Index
Cape Breton Island [Canada]
Charles Babbage Institute [US]
Chesapeake Bay Institute [US]
Common Batch Identification
Compound Batch Identification
Computer-Based Instruction
Computer-Based Instrumentation
Confederation of British Industry
Cooperative Business International [US]

CB&I Chicago Bridge and Iron [US]

CBIC Canadian Book Information Center
Complimentary Bipolar Integrated Circuit

CBIE Canadian Bureau of International Education

CBIN Caribbean Basin Information Network

CBIS Computer-Based Instructional System

CBIT Cape Breton Institute of Technology [Canada]

CBK Check Book

CBKP Convergent Beam Kikuchi Pattern

CBL Cement Bond Log
Chesapeake Biological Laboratory [of NRI, US]
Commercial Bill of Lading
Computer-Based Learning

CBLF Compressible Boundary Layer Flow

CBLM Cluster-Bethe Lattice Method

CBM Certified Ballast Manufacturers Association [US]
Chlorobromomethane
Commonwealth Bureau of Meteorology [Australia]
Conduction Band Minimum
Confidence Building Measure
Containerized Batch Mixer
Constant-Boiling Mixture
Conventional Buoy Mooring
Creep Brittle Material

CBMA Canadian Battery Manufacturers Association
Canadian Business Manufacturers Association
Certified Ballast Manufacturers Association

CBMC Canadian Book Manufacturing Council

CBMIS Computer-Based Management Information System

CBMM Council of Building Materials Manufacturers [US]

CBMPE Council of British Manufacturers of Petroleum Equipment

CBMS Chill Block Melt Spinning
Computer-Based Message Service
Computer-Based Message System
Conference Board of the Mathematical Sciences [US]

CBMU Canadian Board of Marine Underwriters
Current Bit Motor Unit

CBMUA Canadian Boiler and Machinery Underwriters Association

CBN Cubic Boron Nitride [also: cBN]

CBNM Central Bureau for Nuclear Measurements

CBNY Chemical Bank of New York [US]

CBO Conference of Baltic Oceanographers [Germany]
Congressional Budget Office [US]

CBOA Canadian Building Officials Association

CBOE Chicago Board Options Exchange [US]

CBOM Canadian Board of Occupational Medicine
Current Break-Off and Memory

CBOOA Cyanobenzylidene Octyloxyaniline

CBORE Counterbore

CBOSS Count Back Order and Sample Select

CBOT Chicago Board of Trade [US]

CBOTA Cape Breton Offshore Trade Association [Canada]

CBP Canadian Business Press
Construction Ball Pad
Convergent Beam Pattern

CBPAH Council for British Plastics in Agriculture and Horticulture

CBPC Canadian Book Publishers Council

CBPI Canadian Business Periodicals Index

CBPT CLIRA (= Closed Loop In-Reactor Assembly) Backup Plug Tool

CBR California Bearing Ratio
Cavity-Backed Radiator
Chemical, Biological, Radiological (Warfare)
Cloud Base Recorder
Commercial Breeder Reactor
Community Bureau of Reference [US]

CBRA Chemical, Biological and Radiological Agency [US]

CBRET Canadian Board for Registration of EEG (= Electroencephalography) Technologists

CBRI Central Building Research Institute [India]

CBRS Canadian Bond Rating Service
Chemical, Biological, and Radiological Section [of US Military]

CBRTGW Canadian Brotherhood of Railway Transport and General Workers

CBS Canadian Biochemical Society
Canadian Business Service
Call Box Station
Central Battery Signalling
Central Battery System
Chinese Biochemical Society

	Columbia Broadcasting System [US]
	Common Battery System
	Commonwealth Bureau of Soils
	N-Cyclohexyl-2-Benzothiazole Sulfenamide
CBSA	Copper and Brass Service Center Association [also: CABSA, US]
CBSE	Commonwealth Bureau of Survey Education [UK]
	Converted Backscattered Electron Secondary Electron
CBSR	Coupled Breeding Superheating Reactor [US]
CBT	Cincinnati Board of Trade [US]
	Committee for Better Transit [US]
	Computer-Based Terminal
	Computer-Based Training
	Core Block Table
CBTA	Canadian Business Telecommunications Alliance
	Cape Breton Transit Authority [Canada]
CBU	Coefficient of Beam Utilization
CBV	Contact-Breaking Voltmeter
CBW	Chemical and Biological Warfare
	Chemical and Biological Weapons
CBWA	Copper and Brass Warehouse Association [now: CBSA, US]
CBX	Cam Box
	C-Band Transponder
	Centralized Branch Exchange
	Computer Branch Exchange
	Computer-Controlled Branch Exchange
	Computerized (Private) Branch Exchange
CBZ	Carbobenzoxy
	Carbobenzyloxy
CC	Cadmium Council [US]
	Calculator
	Call Check
	Calorimetry Conference
	Canada Council
	Capsule Communicator
	Carbonaceous Chondrite
	Carbon-Carbon (Composite)
	Carriage Control
	Carrying Capacity
	Cascade Control
	Catalytic Cracking
	Catecholamine Club [US]
	Category Code
	Central Computer
	Central Control
	Centrifugal Casting
	Ceramic Coating
	Channel Command
	Channel Coordinator
	Charge Conveyor
	Chartered Cartographer
	Chemistry Consortium [US]
	Chip Carrier
	Circuit Closing
	Circumnavigators Club [US]
	Cis-Cis (Configuration) [Chemistry]
	Classification Code
	Close-Coupled
	Closed Circuit
	Closed Cup
	Closing Coil
	Cloud Chamber
	Cluster Controller
	Coarse Control
	Code Controller
	Cogeneration Council [now: CIPCA, US]
	Coiled Coils
	Coin Collect
	Cold Cathode
	Collect Call
	Colloid Chemist
	Colloid Chemistry
	Color Code
	Color Compensation
	Color Correction
	Combined Carbon
	Command Chain
	Common Collector
	Common Control
	Communication Center
	Communication Control
	Communication Controller
	Communications Canada
	Comparison Circuit
	Compression Cable
	Computer Center
	Computer Communications
	Computer Community
	Computer Conference
	Concentric Cable
	Concept Code
	Condition Code
	Conduction Current
	Constant Current
	Consular Corps
	Consulting Chemist
	Continuous Cast
	Continuous Casting
	Continuous Current
	Control Center
	Control Computer
	Control Console
	Control Counter
	Convolutional Code
	Corps Consulaire [= Consular Corps]
	Cotton Count
	Council on Competitiveness [US]
	Counter Clockwise
	Country Code
	County Council [UK]
	Coupled Channel
	Cross-Correlation
	Cross Couple
	Cursor Control
	Cyclic Code
	Cycloconverter
C&C	Command and Control
	Communication and Cognition
	Computer and Communications
C-C	Carbon-Carbon (Composite) [also: CC or C/C]

	Center to Center
Cc	Cirro-Cumulus (Cloud)
	Cumulo-Cirrus (Cloud)
cc	carbon copy
	complex conjugate
	cubic centimeter
c/c	center to center
CCA	Cam Clamp Assembly
	Canadian Cartographers Association
	Canadian Center of Architecture
	Canadian Chemical Association
	Canadian Communication Association
	Canadian Construction Association
	Canadian Consumers Association
	Capital Cost Allowance
	Carbohydrate-Containing Antibiotics
	Carrier-Controlled Approach
	Cement and Concrete Association [UK]
	Central Computer Accounting
	Central Computer Agency [of CSD, UK]
	Channel-to-Channel Adapter
	Chemical Coaters Association [US]
	Chemical Communications Association [US]
	Chemical Corps Association [US]
	Chromated Copper Arsenate
	Coagulation Control Abnormal
	Coastal Conservation Association [US]
	Cogeneration Council of America [now: CIPCA, US]
	Common Communication Adapter
	Computer and Control Abstracts
	Consumer and Corporate Affairs [Canada]
	Cooperative Communicators Association [US]
	Copper Conductors Association [UK]
	Council of Chemical Associations [now: CCAE, US]
	Crop Condition Assessment [US]
CCAAP	Central Committee for the Architectural Advisory Panels [UK]
CCAB	Canadian Circulation Audit Board
CCABC	Cancer Control Agency of British Columbia [Canada]
CCAC	Conference on Composites and Advanced Ceramics [US]
	Consumer and Corporate Affairs Canada
CCACD	Canadian Center for Arms Control and Disarmament
CCAE	Council of Chemical Association Executives [US]
CCAHC	Central Council for Agricultural and Horticultural Cooperation [UK]
CCAI	Canadian Center for Advanced Instrumentation
CC-AI	Communication and Cognition — Artificial Intelligence [also: CCAI, Belgium]
CCAM	Canadian Congress on Applied Mechanics
	Conversational Communication Access Method
CCAMLR	Commission for the Conservation of the Antarctic Marine Living Resources [Australia]
CCAP	Communications Control Application Program
CCAQ	Consultative Committee on Administrative Questions [of UN]
CCAS	Carrier-Controlled Approach System
CCASM	Canadian Council of the American Society for Metals
CCAT	Centralia College of Agricultural Technology [Canada]
	Conglutinating Complement Absorption Test
CCB	Channel Control Block
	Character Control Block
	Circuit Concentration Bay
	Coin Collecting Box
	College of Cape Breton [Canada]
	Command Control Block
	Common Carrier Bureau
	Communications Control Block
	Configuration Control Board
	Convertible Circuit Breaker
CCBA	Coupled Channel Born Approximation
CCBDA	Canadian Copper and Brass Development Association
CCBS	Clear Channel Broadcasting System
CCC	Canadian Chamber of Commerce
	Canadian Climate Center [of University of British Columbia]
	Canadian Commercial Corporation
	Canadian Committee on Cataloging
	Canadian Computer Complex
	Canadian Computer Conference
	Canola Council of Canada
	Carbon-Carbon Composite
	Car Care Council [US]
	Caribbean Conservation Corporation [US]
	Central Communications Controller
	Central Computer Complex
	Change and Configuration Control
	Chemical Coal Cleaning
	Chlorocholine Chloride
	Cis-Cis-Cis (Geometry) [Chemistry]
	Cleaved Coupled Cavity
	Command, Control and Communication
	Commodity Credit Corporation [US]
	Communications Control Console
	Comparative Capital Cost
	Computer Communication Console
	Computer Control Complex
	Convert Character Code
	Copy Control Character
	Copyright Clearance Center [US]
	Customs Charges Collectable
	Customs Cooperation Council [Belgium]
CC&C	Command, Control and Communications
C^3	Cleaved Coupled Cavity
	Command, Control and Communication
	Computer, Communications and Components
CCCC	Computerized Conferencing and Communications Center [of NJIT, US]
	Coordinating Council for Computers in Construction [US]
	Council for Car Care Centers [US]
CCCCD	Conductively-Connected Charge-Coupled Device [also: C^4D]
CCCE	California Council of Civil Engineers [US]
	Canadian Chemical Conference and Exhibition
	Commission for Certification of Consulting Engineers [US]

CCCEC Commission for Certification of Consulting Engineers in Colorado [US]
CCCE&LS California Council of Civil Engineers and Land Surveyors [US]
CCCESD Council of Chairmen of Canadian Earth Science Departments
CCCG Combined-Cycle Coal Gasification
CCCI Command, Control, Communication and Intelligence [also: C^3I]
Cyprus Chamber of Commerce and Industry
CCCL Complementary Constant-Current Logic [also: C^3L]
CCCM Center for Cement Composite Materials [US]
CCCO Committee on Climatic Changes and the Ocean [France]
CCCS Canadian Cooperative Credit Society
Command, Control and Communications System
Core Component Cleaning System
Core Component Conditioning Station
Current Controlled Current Source
CCCT Canadian Center for Creative Technology
CCD Calcite Compensation Depth
Capacitor-Charged Device
Charge-Coupled Device
Coarse Control Damper
Cold Cathode Discharge
Commonwealth Committee on Defense [now: CDC, India]
Computer-Controlled Display
Continuous-Countercurrent Decantation
Contract Completion Date
Controlled Current Distribution
Core Current Driver
Countercurrent Decantation
Countercurrent Digestion [ore leaching]
CCDA Commercial Chemical Development Association [US]
CCDC Construction Contract Development Association
CCDCG Clinical Chemistry Data Communication Group
CCDLERDSWA Commission for Controlling the Desert Locust in the Eastern Region of its Distribution in Southwest Asia [Italy]
CCDLNE Commission for Controlling the Desert Locust in the Near East [of FAO, UN — Saudi Arabia]
CCDLNWA Commission for Controlling the Desert Locust in Northwest Africa [of FAO, UN — Algeria]
CCDM Change and Configuration Control/Development and Maintenance
CCDN Corporate Consolidated Data Network [of IBM Inc.]
CCDP Canadian Continental Drilling Program
CCDS Centers for the Commercial Development of Space [of NASA]
CCDSO Command and Control Defense Systems Office
CCE Carboline Carboxylate Ethyl
Caribbean Conservation Association [Barbados]
Certified Cost Engineer
Chief Construction Engineer
Commission des Communautes Europeennes [= Commission of the European Communities, Belgium]
Communication Control Equipment
CCEA Commonwealth Council for Education Administration [UK/Australia]
CCECA Consultative Committee on Electronics for Civil Aviation [US]
CCEE Canadian Conference on Engineering Education
Committee of Concerned Electrical Engineers
CCEL Collidge Center for Environmental Leadership [US]
CCEO Caribbean Council of Engineering Organizations
CCEPC Canadian Civil Engineering Planning Committee
CCES Canadian Congress of Engineering Students
Common Control Echo Suppressor
Congress of Canadian Engineering Schools
Congress of Canadian Engineering Students
CCETT Canadian Council of Engineering Technicians and Technologists
Centre Commun d'Etudes de Television et de Telecommunications [= Center for Television and Telecommunication Studies, France]
CCEU Council on the Continuing Education Unit [US]
CCF Central Computing Facility [of NASA]
Central Communications Facility [US Air Force]
Central Control Function
Color-Correction Filter
Common Communications Format
Communications Control Field
Configuration Control Function
Cooperative Commonwealth Federation [UK]
Cross-Correlation Function
Cumulative Cost per Foot
CCFA Caribbean Cane Farmers Association [Jamaica]
CCFL Countercurrent Flow Limiting
CCFM Cryogenic Continuous-Film Memory
CCFP Closed Cup Flash Point
CCFR Constant Current Flux Reset Test Method
Coordinating Committee on Fast Reactors [of EEC, Belgium]
CCFT Controlled Current Feedback Transformer
CCG Canadian Coast Guard
Computer Communications Group
Computer-Controlled Goniometer
Constant-Current Generator
CCGC Canadian Coat Guard College
CCGCS Containment Combustion Gas Control System
CCGE Cold Cathode Gauge Experiment
CCGS Canadian Coast Guard Service
CCGT Closed-Cycle Gas Turbine
CCH Channel-Check Handler
Computerized Criminal History
Connections per Circuit per Hour
CCHS Cylinder-Cylinder-Head Sector
CCHSSA Cyprus Computer Hardware and Software Suppliers Association
CCHX Component Cooling Heat Exchanger
CCI Cadmium Council, Inc. [US]
Canadian Carpet Institute
Canadian Conservation Institute
Canadian Copyright Institute
Center for Communication [US]
Chamber of Commerce and Industries [Namibia]
Chamber of Commerce International

Chambre de Commerce Internationale [= International Chamber of Commerce, France]
Co-Channel Interface
Computer Composition International [US]
Consultative Committee International
Continuous Computation of Impact Points
Coordination Chemistry Institute [PR China]
Cotton Council International [US]

CCIA Cellular Communications Industry Association [now: CTIA, US]
Computer and Communications Industry Association [US]
Console Computer Interface Adapter

CCIBP Canadian Committee for the International Biological Program

CCIC Canadian Council for International Cooperation
Chapter Chairmen's Invitational Conference [of ASM International, US]

CCID Community Colleges for International Development [US]

CCIF Consultative Committee on International Frequencies

CCIIW Canadian Council of the International Institute of Welding

CCILMB Committee for Coordination of Investigations of the Lower Mekong Basin [Thailand]

CCIM Certified Commercial Industrial Member

CCIP Chambre de Commerce et d'Industrie du Paris [= Paris Chamber of Commerce and Industry, France]

CCIR Comite Consultatif International des Radiocommunications [= International Consultative Committee for Radio Communication [of ITU, Switzerland]]

CCIS Command and Control Information System
Command, Control and Information System
Common Channel Interoffice Signalling

CCIS/ADP Command, Control and Information Systems and Automatic Data Processing Committee [of NATO]

CCIT Central China Institute of Technology [PR China]

CCITT Comite Consultatif International Telegraphique et Telephonique [= International Telegraph and Telephone Consultative Committee [of ITU, Switzerland]]

CC-IUMRS Chinese Committee of the International Union of Materials Research Societies

CCIW Canada Center for Inland Waters

CCK Cholecystokinin

CCKF Continuity Checktone Failure [of SPADE]

CCL Capacitor-Coupled Logic
Caribbean Congress of Labour
Center for Computer/Law [US]
Coating and Chemical Laboratory
Cold-Cathode Lamp
Command Control Language
Common-Carrier Line
Common Command Language
Common Control Language
Communications Control Language
Core Current Layer
Cyber Control Language

CCLA Centenary College of Los Angeles [US]

CCLN Council Computerized Library Networks [US]

CCLOW Canadian Congress for Learning Opportunities for Women

CCLS Canadian Council of Land Surveyors
Central Committee on Lumber Standards [now: ALSC, US]

CCLW Counterclockwise

CCM Canadian Committee on MARC (= Machine-Readable Catalogue)
Call Count Meter
Center for Composite Materials [of University of Delaware, US]
Chain Crossing Model
Communications Control Module
Communications Controller Multichannel
Computer-Coupled Machine
Council of Communication Management [US]
Controlled-Carrier Modulation
Counter Countermeasures
Cross-Correlation Method

ccm cubic centimeter

CCMA Canadian Council of Management Associations
Certified Color Manufacturers Association [US]
Cyprus Clothing Manufacturers Association

CCMB Completion of Calls Meeting Busy

CCMC Commonwealth Cable Management Committee

CCMD Continuous Current Monitoring Device

CCME Canadian Council of Ministers of Education

CCMMI Congress of the Council of Mining and Metallurgical Institutions
Council of Commonwealth, Mining and Metallurgical Institutions [now: CMMI, UK]

CCMRG Commonwealth Consultative Group on Mineral Resources and Geology

CCMS Committee on Challenges of Modern Society [of NATO]

CCMSC Caribbean Common Market Standards Council [Guyana]

CCMTA Canadian Council of Motor Transport Administrators

CCMTC Cape Canaveral Missile Test Center [of NASA]

CCN Catalogue Card Number [of Library of Congress, US]
Cloud Condensation Nucleus
Computer Communication Network
Contract Change Notice

CCNDT Canadian Council for Nondestructive Testing

CC-NDT Can't Call — No Dial Tone

CCNG Computer Communications Network Group [of U of W, Canada]

CCNIM Computer-Controlled Nuclear Instrument Module

CCNR Canadian Coalition for Nuclear Responsibility
Central Commission for Navigation on the Rhine [France]
Current-Controlled Negative Resistance

CCO Conservation Council of Ontario [Canada]
Crystal-Controlled Oscillator
Current-Controlled Oscillator

CCOA Chinese Cereals and Oils Association [PR China]

CCOHS Canadian Center for Occupational Health and Safety

CCOHTA Canadian Coordinating Office for Health Technology Assessment

CCOP Constant-Control Oil Pressure
Coordinating Committee for Offshore Prospecting [of UNESCO]

CCOP/SOPAC Committee for Coordination of Joint Prospecting for Mineral Resources in South Pacific Offshore Areas [of UNESCO — Fiji]

C-CORE Center for Cold Ocean Resources Engineering [Canada]

CCOT Cycling Clutch Orifice Tube

CCP Call Control Processing
Central Canada Potash
Centrifugal Charging Pump
Channel Control Processor
China Clay Producers Trade Association [US]
Combined Cycle Plant
Committee on Commodity Problems [of UN — Italy]
Communication Control Package
Communication Control Panel
Communication Control Processor
Communication Control Program
Computer General Processing
Confederacion Cientifica Panamericana [= Pan-American Scientific Union, Argentina]
Configuration Control Panel
Console Command Processor
Console Control Package
Contract Configuration Process
Core Component Pot
Critical Compression Pressure
Cross Connection Point
Cubic Close-Packed

C&CP Corrosion and Cathodic Protection [IEEE IAS Committee]

CCPA California Canning Peach Association [US]
Canadian Chemical Producers Association
Cemented Carbide Producers Association [US]
Choline Chloride Producers Association [of CEFIC, Belgium]

CCPAB California Cling Peach Advisory Board [US]

CCPC Civil Communications Planning Committee [of NATO]

CCPD Council for Cultural Planning and Development [Taiwan]
Coupling Capacitor Potential Device

CCPE Canadian Council of Professional Engineers

CCPEOF Consulting Committee of the Professional Electro-Engineers Organization of Finland

CCPES Canadian Council of Professional Engineers and Scientists

CCPIT China Council for the Promotion of International Trade [PR China]

CCPO Conference on Charged Particle Optics

CCPS Center for Chemical Process Safety [of AIChE, US]
Consultative Committee for Postal Studies [of UPU, Switzerland]

CCR Ceiling Cavity Ratio
Center for Catalogue Research [of Bath University, UK]
Channel Command Register
Chemical Compounds Registry [of CAS-ACS, US]
Committment, Concurrency and Recovery
Computer Cassette Recorder
Condition Code Register
Control Contactor
Council for Chemical Research [US]
Crevice-Corrosion Resistance
Critical Compression Ratio
Cyclic Catalytic Reforming Process

CC&R Canadian CAD/CAM and Robotics Exhibition and Conference

CCRC Core Component Receiving Container

CCRD Canadian Council on Rural Development
Consumer and Commercial Relations Department [Canada]

CCRE Canadian Council for Research in Engineering

CCREM Canadian Council of Resource and Environment Ministers

CCRG Canadian Classification Research Group

CCRH Center for Computer Research in Humanities [University of Colorado at Boulder, US]

CCRL Combustion and Carbonization Research Laboratory [of CANMET, Canada]

CCRM Crew Commander Remote Monitor

CCRMP Canadian Certified Reference Materials Project

CCROS Card Capacitor Read-Only Storage

CCRS Canada Center for Remote Sensing

CCS Canadian Cancer Society
Canadian Ceramic Society
Canadian Computer Show
Canadian Control Status
Cartesian Coordinate System
Center Core Signal
Center for Coastal Studies [US]
Central Computer Station
C (= Hundred) Call Seconds
Change Control System
Chinese Chemical Society [PR China]
Collective Call Signal
Command, Control and Subordinate Systems
Commercial Communications Satellite [Japan]
Commitment Control System
Committee of Concerned Scientists
Common-Channel Signalling
Common Command Set
Communication Control System
Containment Cooling System
Continuous Cast Steel
Continuous Color Sequence
Continuous Commercial Service
Controlled Combustion System
Conversational Compiling System
Custom Computer System
Cyber (Computer) Credit System
Czechoslovak Chemical Society

CC&S Central Computer and Sequencer [of NASA]
Central Computer and Sequencing

CCSA Canadian Committee on Sugar Analysis
Common Carrier Special Application
Common-Channel Signalling Arrangement

	Common-Control Switching Arrangement
CCSC	Coordinating Committee for Satellite Communications [Switzerland]
CCSCC	Certificate of the Canadian Society of Clinical Chemists
CCSD	Command Communication Service Designator
CCSDS	Consultative Committee for Space Data Systems
CCSG	Computer Components and Systems Group
CCSL	Comparative Current Sinking Logic
	Compatible Current Sinking Logic
	Constrained Coincidence Site Lattice
CCSN	Community College Satellite Network [of Instructional Telecommunication Consortium, US]
CCSS	Canada Center for Space Sciences
	Common-Channel Signalling System
	Continuous Cast Stainless Steel
CCSSRRT	Continuous Cast Stainless Steel Round Robin Test
CCST	Caribbean Council for Science and Technology
	Center for Catalytic Science and Technology [of University of Delaware, US]
	Center for Computer Sciences and Technology [of NBS, US]
CCSTG	Carnegie Commission on Science, Technology and Government [US]
CCSW	Component Cooling Service Water
CCT	Canada Center for Toxicology
	Canadian Center for Toxicology
	Center-Cracked Tension Specimen
	Circuit
	Cis-Cis-Trans (Geometry) [Chemistry]
	Communications Control Team
	Complete Calls To
	Component Check Test
	Computer Compatible Tape
	Constant Current Transformer
	Continuity Check Transceiver
	Continuous Cooling Transformation
	Correlated Color Temperature
	Coupler Cut Through
	Creosote Coal Tar
	Crystal-Controlled Transmitter
CCTA	Canadian Cable Television Association
	Central Computer and Telecommunications Agency [UK]
CCTC	Canada-China Trade Council
CCTEAH	Coordinating Committee of the Trans-East African Highway [now: TEAHA, Ethiopia]
CCTest	Component Check Test
CCTF	Cylindrical Core Test Facility
CCTG	Configuration Control Task Group
CCTI	Composite Can and Tube Institute [US]
CCTL	Casing Cooling Tank Level
	Core Component Test Loop [of Argonne National Laboratory, US]
CCTMA	Closed Circuit Television Manufacturers Association [US]
CCTP	Clean Coal Technology Program
CCTR	Culham Conceptual Tokamak Reactor
CCTS	Coordinating Committee for Telecommunication by Satellite
CCTT	Canadian Council of Technicians and Technologists
CCTU	Committee of Corporate Telecommunications Users [US]
CCTV	Closed-Circuit Television
CCTV/CCD	Closed-Circuit Television/Charge-Coupled Device (Camera)
CCU	Camera Control Unit
	Canadian Commission for UNESCO
	Caribbean Consumers Union [Antigua-Barbuda]
	Central Control Unit
	Common Control Unit
	Communications Control Unit
	Computer Control Unit
	Confederation of Canadian Unions
	Contaminant Collection Unit
	Coupling Control Unit
CCUAP	Computerized Cable Upkeep Administration Program
CCUBC	Canadian Council of University Biology Chairmen
CCULP	Canada/China University Linkage Program [of AUCC, Canada]
CCUNCYP	Canadian Contingent United Nations Cyprus
CCURR	Canadian Council on Urban and Regional Research
CCV	Control Configured Vehicle
CCVS	COBOL Compiler Validation System
	Current-Controlled Voltage Source
CCW	Channel Command Word
	Channel Control Word
	Circulation Controlled Wing
	Commutatively Compound-Wound
	Component Cooling Water
	Counterclockwise
CCWBAD	Counterclockwise Bottom Angular Down
CCWBAU	Counterclockwise Bottom Angular Up
CCWBH	Counterclockwise Bottom Horizontal
CCWDB	Counterclockwise Down Blast
CCWM	Commutatively Compound-Wound Motor
CCWS	Component Cooling Water System
CCWTAD	Counterclockwise Top Angular Down
CCWTAU	Counterclockwise Top Angular Up
CCWTH	Counterclockwise Top Horizontal
CCWUB	Counterclockwise Up Blast
CCZ	Coastal Confluence Zone
CD	Cable Duct
	Cage Dipole
	Capacitor Diode
	Carbon Dating
	Card
	Carried Down
	Carrier Detect
	Cathodoluminescence
	Ceiling Diffuser
	Census Division
	Center Distance
	Charge Density
	Check Digit
	Chemical Deposition
	Circuit Description
	Circular Dichroism
	Civil Defense
	Clock Driver
	Coal Division [of EMR, Canada]

Coherent Detector
Cold Drawing
Collision Detection
Common Digitizer
Compact Disk
Composites Division [of ASM International, US]
Continental Drift
Contract Demand
Contracting Definition
Control-Rod Drive
Cord
Core Dump
Corona Discharge
Corps Diplomatique [= Diplomatic Corps]
Critical Damping
Critical Dimension
Crystal Detector
Crystal Driver
Crystal Dynamics
Current Density
Customs Declaration
Cyclodextrin

C&D Charman and Darlington Top
Collection and Delivery
Control and Display

C-D Convergence-Divergence Principle

Cd Cadmium
Cedi [Currency of Ghana]

cd candela
cord [Unit]

c/d carried down

C$ Canadian Dollar [Currency]

CDA Canada
Canadian Department of Agriculture
Capital Dividend Account
Charge Density Analysis
Coin Detection and Announcement
Command and Data Acquisition
Commercial Development Association [US]
Completely Denatured Alcohol
Computer Dealers Association [US]
Comprehensive Development Area
Conference of Defense Association
Containment Depressurization Activation
Containment Depressurization Alarm
Copier Dealers Association [US]
Copper Development Association [US]
Core Disruptive Accident
Core Dump Analysis
Cumulative Deduction Amout

CDAA Chlorodiallylacetamide

CDAP Cyanodimethylaminopyridinium

CDAPSO Canadian Data Processing Service Organization

CDAS Central Data Acquisition System
Command and Data Acquisition Station

CDB Caribbean Development Bank [Barbados]
Central Data Bank
Coal Database [UK]
Common Data Bus
Computerized Database
Current Data Bit

CDBAB California Dried Bean Advisory Board [US]

CDBN Column-Digit Binary Network

CDBS Central Database System

CDC Call Directing Character
Call Directing Code
Canadian Dairy Commission
Canadian Development Corporation
Center for Disease Control [US]
Characteristic Distortion Compensation
Civil Defense Corps [UK]
Code Directing Character
Commonwealth Defense Conference [India]
Commonwealth Development Corporation
Communications and Data Center
Composite Development Center [US]
Computer Display Channel
Configuration Data Control
Construction Design Criteria
Copper Data Center [of CDA, US]
Cryogenic Data Center [of NIST, US]

CDCC Caribbean Development Cooperation Committee

CDCCP Control Data Communications Control Procedure

CDCE Central Data Conversion Equipment

CDC-OCCE Commonwealth Defense Conference — Operation Clothing and Combat Equipment [India]

CDCR Center for Documentation and Communication Research [of WRU, US]

CDCT Centro de Documentacao Cientifica e Tecnica [= Center for Scientific and Technical Documentation, Portugal]

CDCTM Centro de Documentacion Cientifica y Tecnica de Mexico [= Mexican Center for Scientific and Technical Documentation]

CDCU Communications Digital Control Unit

CDCVR Code Converter

CDD Common Data Dictionary
Conference on Dual Distribution
Contrast-Detail-Dose (Diagram)

CDDA Conseil Departemental du Developpement Agricole [= Departmental Council for Agricultural Development, France]

CD-DA Compact Disk — Digital Audio

CDDL Conference of Directors of Danube Lines [Hungary]

CDDP Canadian Deep Drilling Project

CdDMC Cadmium Dimethyl Dithiocarbamate

Cd5MC Cadmium Pentamethylene Dithiocarbamate

CDE Canadian Depletion Expense
Canadian Development Expense
Card Data Entry
Certificate in Data Education
Control and Display Equipment

CDEC Chloroallyl Diethyldithiocarbamate

CDEE Chemical Defense Experimental Establishment [of MOD, UK]

CDES Computer Data Entry System

CDF Carrier Distribution Frame
Centered Dark Field
Class Determination and Findings
Combined Die Forging
Combined Distribution Frame
Communications Data Field

	Computer Dealers Forum [of NOPA, US]
	Confined Detonating Fuse
	Continuous Drossing Furnace
	Cooperative Development Foundation [Canada]
	Cumulative Damage Function
	Cumulative Distribution Function
CD-4	Compatible Discrete Four-Channel Sound
CDFR	Commercial Demonstration Fast Reactor
cd ft	cord foot [Unit]
CDG	Capacitor Diode Gate
CDH	Command and Data Handling
CDI	Carbodiimide
	Center for the Development of Industry [Belgium]
	Cobalt Development Institute [UK]
	Collector Diffusion Isolation
	Community Development Index
	Compact Disk — Interactive
	Comprehensive Dissertation Index
	Continuous Deionization
	Control Data Institute [US]
	Control Deviation Indicator
	Control Direction Indicator
	Course Deviation Indicator
	Cutting Die Institute [US]
CD-I	Compact Disk — Interactive
CDIB	Collector, Diffusion, Isolation, Bipolar
CDIC	Canadian Development Investment Corporation
CDIF	Component Development and Integration Facility
	Crystal Data Identification File
CDIS	Community Data Information System
CDK	Channel Data Check
CDL	Common Display Logic
	Computer Description Language
	Computer Design Language
	Computer Development Laboratory
CDLA	Computer Dealers and Lessors Association [US]
CDLC	Cellular Data Link Control
CDLRD	Confirming Design Layout Report Date
CDLS	Canadian Defense Liaison Staff
CDM	Central Data Management
	Centre de Documentation de la Mecanique [= Documentation Center for Mechanics, France]
	Code-Division Multiplexing
	Color-Division Multiplexing
	Common Data Model
	Companded Delta Modulation
	Critical Dimension Metrology
CDMA	Cartridge Direct Memory Access
	Code-Division Multiple Access
CDMS	COMRADE (= Computer-Aided Design Environment) Data Management System
CDMSCS	Committee for the Development and Management of Fisheries in the South China Sea [of IPFC, Thailand]
CDN	Canadian
	Canadian Dollar [Currency]
cDNA	Chloroplast Deoxyribonucleic Acid
	Complementary Deoxyribonucleic Acid
CDO	Community Dial Office
	P-Benzoquinone Dioxime
CDOA	Car Department Officers Association [US]
CDOIPS	Central Dispatching Organization of the Interconnected Power Systems [Czechoslovakia]
CDOM	Center for Development of Materials
CDP	Center for Design Planning [US]
	Central Data Processor
	Centralized Data Processing
	Certificate in Data Processing
	Certified Data Processor
	Checkout Data Processor
	Communications Data Processor
	Compact Disk Player
	Compressor Discharge Pressure
	Computerized Data Processing
	Correlated Data Processor
	Cresyl Diphenyl Phosphate
	Critical Decision Point
	Cytidine Diphosphate
CDPC	Canadian Database Promotion Center
	Central Data Processing Computer
	Commercial Data Processing Center
CDPIR	Crash Data Position Indicator Recorder
CD PL	Cadmium Plate
CDPM	Colored Digital Panel Meter
Cd5MC	Cadmium Pentamethylene Dithiocarbamate
CDPR	Customer Dial Pulse Receiver
CDPS	Computing and Data Processing Society
CD-PROM	Compact Disk — Programmable Read-Only Memory [also: CD PROM]
CDR	Call Data Recording
	Call Detail Recording
	Card Reader
	Command Destruct Receiver
	Commander
	Composite Damage Risk
	Conceptual Design Requirement
	Critical Design Review
	Current Directional Relay
CDRA	Canadian Drilling Research Association
	Committee of Directors of Research Associations [UK]
CDRAP	Collaborative Diesel Research Advisory Panel
CDRB	Canadian Defense Research Board
CDRE	Chemical Defense Research Establishment [UK]
CDRH	Center for Devices and Radiological Health [of FDA, US]
CDRI	Chihuahuan Desert Research Institute [US]
CDRILL	Counterdrill
CDRL	Contractual Data Requirements List
CD-ROM	Compact Disk — Read-Only Memory [also: CD ROM]
CDRR	Call Detail Recording and Reporting
CDRS	Computerized Data Retrieval System
CD-RTOS	Compact Disk — Real-Time Operating System [also: CD RTOS]
CDS	Cataloguing Distribution Service
	Centre de Documentation Siderurgique [= Documentation Center of the Iron and Steel Industry, France]
	Centre de Donnees Stellaires [= Center for Stellar Data, France]
	Charged Droplet Scrubber

Chief of Defense Staff [of Canadian Forces]
Communications and Data Subsystem
Compatible Duplex System
Cold-Drawn Steel
Compatible Duplex System
Component Disassembly Station
Comprehensive Display System
Computer Duplex System
Computerized Documentation System [of UNESCO]
Conceptual Design Study
Configuration Data Set
Container Delivery System
(Single) Cotton Double Silk (Wire)

C/DS Cache/Disk System

CDSE Computer-Driven Simulation Environment

CDSF Commercially Developed Space Facility
COMRADE (= Computer-Aided Design Environment) Data Storage Facility

cd sr candela/steradian

CDSS Computer Digital Switching System

C&DSS Communication and Data Subsystem

CDST Central Daylight Saving Time
Centre de Documentation Scientifique et Technique [= Center for Scientific and Technical Documentation [of CNRS, France]]

CDT Central Daylight Time
Conductivity, Depth and Temperature
Control Data Terminal

CDTA 1,2-Cyclohexanediaminetetraacetic Acid

CdTe Cadmium Telluride

CDTI Cockpit Display of Traffic Information

CDTL Common Data Translation Language

CDTS Centralized Digital Telecommunications System

CDTTM California Department of Transportation Test Methods [US]

CDU Cartridge Disk Unit
Central Display Unit
Color Developing Unit
Control and Display Unit
Control Display Unit
Coolant Distribution Unit
Coupling Display Unit

CDV Check Digit Verification
Compact Disk Video
Component Digital Video

CDVTPR Centre de Documentation du Verre Textile et des Plastiques Renforces [= Documentation Center for Textile Glass and Reinforced Plastics, France]

CDW Capacitor Discharge Welding
Charge Density Wave
Civil Defense Warning
Computer Data Word

CDX Control Differential Transmitter

CE Calomel Electrode
Capillary Electrophoresis
Carbon Electrode
Carbon Equivalence
Carbon Equivalent
Cathode Emission
Celestial Equator
Channel End
Chemical Engineer
Chemical Engineering
Chemical Equilibrium
Chief Engineer
Civil Engineer
Civil Engineering
Coastal Engineer
Coastal Engineering
Coefficient of Entry
Cold Emission
College of Engineering
Combustion Engineer
Combustion Engineering
Common Emitter
Communications — Electronics
Commutator End
Compulsory Expenditure
Computational Enumeration
Computer Engineer
Computer Engineering
Computer Equipment
Conducted Emission
Conservation Engineer
Conservation Engineering
Consulting Engineer
Consumer Electronics
Control Electrode
Control Engineer
Control Engineering
Controlled Environment
Corps of Engineers [US Army]
Corrosion Engineer
Corrosion Engineering
Cost Engineer
Cost Engineering
Coulomb Excitation
Council of Europe [France]
Counter Electrode
Cryogenic Engineer
Cryogenic Engineering
Crystal Engineering
Current Efficiency
Custom Engineering
Customer Engineer

C/E Calculation/Experiment

Ce Cerium

CEA Cambridge Electron Accelerator [US]
Canadian Economics Association
Canadian Education Association
Canadian Electrical Association
Canadian Export Association
Carcinoembryonic Antigen
Carnuntum Excavations [Austria]
Cement Employers Association [US]
Central Electrical Authority [UK]
Chemical Engineering Abstracts [of RSC, UK]
Circular Error Average
Coal Exporters Association (of the United States)
Combustion Engineering Association
Comite Electrotecnico Argentina [= Argentinian Electrotechnical Committee]

Commissariat a l'Energie Atomique [= Commissariat for Atomic Energy, France]
Commodity Exchange Authority [US]
Communications-Electronics Agency [US]
Competitive Equipment Analysis
Confederation Europeenne de l'Agriculture [= European Confederation of Agriculture, Switzerland]
Conservation Education Association [US]
Construction Equipment Advertisers
Consulting Engineers Association [US]
Consulting Engineers of Alberta [Canada]
Control Element Assembly
Cooperative Editorial Association [now: CCA, US]
Core Element Assembly
Council for Energy Awareness [US]
Council of Economic Advisers [US]
Cyclohexyl Ethyl Amine

CEAC Central European Analysis Commission
Committee for European Airspace Coordination [of NATO]
Consulting Engineers Association of California [US]
Control Element Assembly Computer
County Engineers Association of California [US]

CEADI Colored Electronic Altitude Director Indicator

CEAEN Centre d'Etudes pour les Applications de l'Energie Nucleaire [= Study Center for Nuclear Energy Applications, France]

CEAL Canadian Explosive Atmospheres Laboratory [of CANMET]

CEAM Concerted European Action on Magnets

CEAN Centre d'Etude pour les Applications de l'Energie Nucleaire [= Center for the Study of Nuclear Energy, Belgium]

CEANAR Commission on Education in Agriculture and Natural Resources [of NRC, US]

CEAO Communaute Economique de l'Afrique de l'Ouest [= West African Economic Community]

CEAP Canadian Energy Audit Program

CEA PRC Construction Equipment Advertisers and Public Relations Council [US]

CEARAM Commission for the Exploration of Archeological Remains of Asia Minor [Austria]

CEARC Canadian Environmental Assessment Research Council
Computer Education and Applied Research Center [US]

CEAS Center of European Agricultural Studies
College of Engineering and Applied Science

CEAU Continuing Education Achievement Unit

CEB Central Electricity Board [UK]
Council for Environmental Balance [now: NCEB, US]

CEBAF Continuous Electron Beam Accelerator Facility [US]

CEBC Consulting Engineers of British Columbia [Canada]

CEBEDEAU Centre Belge d'Etude et de Documentation des Eaux [= Belgian Center for Water Studies and Documentation]

CEBELCOR Centre Belge d'Etude de la Corrosion [= Belgian Center for Corrosion Studies]

CeBIT Welt-Centrum der Buro-, Informations- und Kommunikationstechnik [= World Center for Office, Information and Communications Technology (Exhibition and Trade Fair), Germany]

CEBM Corona, Eddy Current, Beta Ray, Microwave

CEC Canada Employment Center
Canadian Electrical Code
Canadian Engineering Competition
Caribbean Employers Confederation [Trinidad]
Cation Exchange Capacity
Center for Marine Conservation [US]
Ceramic Educational Council [US]
Charlotte Engineers Club [US]
Chiba Engineering College [Japan]
Colorado Engineering Council [US]
Combination Emission Control
Commission of the European Communities [Belgium]
Commonwealth Economic Committee [UK]
Commonwealth Engineering Conference
Commonwealth Engineers Council [UK]
Community Environmental Council [US]
Confederation Internationale des Cadres [= International Confederation of Executive and Managerial Staffs, France]
Consulting Engineers Council [UK and US]
Coordinating European Council (for the Development of Performance Tests for Lubricants and Engine Fuel) [UK]
Council for Education in the Commonwealth [UK]
Critical Electrolyte Concentration
Cryogenic Engineering Conference

CECA Consumers Energy Council of America [US]

CECAF Committee for Eastern Central Atlantic Fisheries [Senegal]

CECC California Educational Computing Consortium [US]
CENELEC Electronic Components Committee [Belgium]
Commonwealth Economic Consultative Committee [UK]
Consulting Engineers Council of Colorado [US]

CEC/CAL Consulting Engineers Council of California [US]

CECE Committee for European Construction Equipment [UK]

CECED Conseil Europeen de la Construction Electrodomestique [= European Committee of Manufacturers of Domestic Electrical Equipment, UK]

CECIF Chambre Europeenne pour le Developpement du Commerce, de l'Industrie et des Finances [= European Chamber for the Development of Trade, Industry and Finances, Belgium]

CECIMO Comite Europeen de Cooperation des Industries de la Machine-Outil [= European Committee for Cooperation of Machine Tool Industries, Belgium]

CEC/JRC Commission of the European Communities/Joint Research Center [Netherlands]

CECL Cascade-Emitter Coupled Logic

CECM Council of Eastern Caribbean Manufacturers

CEC/MINN Consulting Engineers Council of Minnesota [US]

CEC/NYS Consulting Engineers Council of New York State [US]
CECO Consulting Engineers Council of Oregon [US]
Cost Estimate Change Order
CEC-OC Canada Employment Center — On Campus
CECOM Communications-Electronics Command [US Army]
CECOPHIL Council of Engineering Consultants of the Philippines
CECRI Central Electrochemical Research Institute [India]
CECS Congress of the European Ceramic Society
CEC/TEX Consulting Engineers Council of Texas [US]
CECTK Committee for Electrochemical Thermodynamics and Kinetics [Belgium]
CECUI Consulting Engineers Council of Utah [US]
CECUA Confederation of European Computer Users Associations [Belgium]
Conference of European Computer User Association
CED Capacitance Electronic Disk
Carbon-Equivalent-Difference
Centro Elettronico di Documentazio [= Electronic Documentation Center, Italy]
Chemical Engineering Department
Cohesive Energy Density
Committee for Economic Development [US]
Constant Energy Difference
Cupriethylenediamine
Cyanoethyl Diisopropyl
CEDA Canadian Electrical Distributors Association
Central Dredging Association [Netherlands]
Committee for Economic Development of Australia
CEDAR Coupling, Energetics and Dynamics of Atmospheric Regions (Program) [US]
CEDC Canadian Engineering Data Center [of ASM International, US]
Canadian Engineering Design Competition
CEDEFOP Centre Europeen pour le Developpement de la Formation Professionelle [= European Center for the Development of Vocational Training, Germany]
CEDIC Comite Europeen des Ingenieurs-Conseils du Marche Commun [= Committee of the Consulting Engineers of the Common Market [of EEC, Belgium]]
CEDIP Canadian Exploration and Development Incentive Program
CEDM Control Element Drive Motor
CEDO Consulting and Engineering Design Organization
CEDOCAR Centre de Documentation de l'Armement [= Ordnance Documentation Center, France]
CEDOCOS Centre de Documentation sur les Combustibles Solides [= Documentation Center on Solid Fuels, Belgium]
CEDPA California Educational Data Processing Association [US]
CEDR Comite Europeen de Droit Rural [= European Committee for Agrarian Law, Belgium]
CEE Canadian Exploration Expense
Center for Environmental Education [Canada]
Competitive Entrance Examination [Kenya]
Council for Environmental Education [UK]
CEEA Communaute Europeenne de l'Energie Atomique [= European Community for Atomic Energy [of EURATOM]]
CEEAS Centre Europeen d'Etudes de l'Acide Sulfurique [= European Center for Studies of Sulfuric Acid [of CEFIC, Belgium]]
CEEC Committee of European Economic Cooperation [France]
Construction Economics European Committee [France]
CEED Community Economics and Ecological Development (Council) [US]
CEEFAX United Kingdom Teletext Service
CEELS Characteristic Electron Energy-Loss Spectroscopy
CEEMA Committee on Engineering Education in Middle Africa
CEEN Centre d'Etude de l'Energie Nucleaire [= Center for Nuclear Energy Studies, Belgium]
CEEP Centre Europeen de l'Entreprise Publique [= European Center of Public Enterprises, Belgium]
CEERI Central Electronics Engineering Research Instutute [India]
CEES Center for Energy and Environmental Studies [of Princeton University, US]
CEESA Conference of European Engineering Students Associations
CEF Cable Entrance Facility
Canadian Expeditionary Force
Carrier Elimination Filter
Critical Experiments Facility
CEFAC Civil Engineering Field Activities Center [UK]
CEFIC Conseil Europeen des Federations de l'Industrie Chimique [= European Council of Chemical Manufacturers Federations, Belgium]
CEFNO Carbethoxyformonitrile
CEFPI Council on Educational Facility Planning International
CEG Center of Exploration Geophysics [India]
Cryogenic Engineering Conference [US]
CEGA Conference of Governmental Statisticians of the Americas [of IASI, Panama]
CEGABA Carboxyethyl-Gamma-Aminobutyric Acid
CEGB Central Electricity Generating Board [UK]
CEGIS Construction Engineering Geographic Information System
CEGS Canadian Exploration Geophysical Society
CEH Chemical Economics Handbook [Stanford Research Institute, US]
Conference Europeenne des Horaires des Trains de Voyageurs [= European Passenger Train Timetable Conference, Switzerland]
CEHU College of Engineering, Hosei University [Japan]
CEI Canadian Education Index
Commission Electrotechnique Internationale [= International Electrotechnical Commission, Switzerland]
Committee for Environmental Information [US]
Communication Electronics Instructions
Computer Extended Instruction
Council of Engineering Institutions [UK]
Cycle Engineers Institute [US]

CEI-BOIS Confederation Europeenne des Industries du Bois [= European Confederation of Woodworking Industries, Belgium]
CEIC Canada Employment and Immigration Commission
CEIF Council of European Industrial Federations
Cyprus Employers and Industrialists Federation
CEIL Ceiling
CEIP Canadian Exploration Incentive Program
Carnegie Endowment for International Peace [US]
Center for Environmental Intern Programs (Fund) [US]
CEIRD Confirming Engineering Information Report Date
CEIS Caribbean Energy Information System
Center for Environmental Information Science [Japan]
CEIU Canada Employment and Immigration Union
CEL Carbon Equilibrium Loop
Carbon-Equivalent, Liquidus
Central European (Pipe-)Line
Committee on Engineering Laws [US]
Cryogenic Engineering Laboratory [of NIST, US]
CELA Canadian Environmental Law Association
CELADE Centro Latinoamericano de Demografia [= Latin American Demographic Center, Chile]
CELIMAC Comite Europeen de Liaison des Industries de la Machine a Coudre [= European Liaison Committee for the Sewing Machine Industries, Germany]
CELL Cellulose
CELP Code Excited Linear Prediction
CELRF Canadian Environmental Law Research Foundation
Cels Celsius
CELT Coherent Emitter Location Testbed
CELTE Constructeurs Europeens de Locomotives Thermiques et Electriques [= European Manufacturers of Thermal and Electric Locomotives, France]
CEM Cement
Cement Conduit
Center for Entrepreneurial Management [US]
Center for Environmental Mechanics [of CSIRO, Australia]
Cercle d'Etudes des Metaux [= Study Circle on Metals, France]
Channel Electron Multiplier
Communications-Electronics-Meteorological
Computer Education for Management
Computerized Electron Microscopy
Conventional Electron Microscope
Conversion Electron Mossbauer
Core-Excitation Model
Counter Electromotive Cell
CEMA Canadian Electrical Manufacturers Association
Cleaning Equipment Manufacturers Association [US]
Comite Europeen des Groupements de Constructeurs du Machinisme Agricole [= European Committee of Associations of Manufacturers of Agricultural Machinery, France]
Converting Equipment Manufacturers Association [US]
Conveyor Equipment Manufacturers Association [US]
CEM A Cement Asbestos
CEM AB Cement Asbestos Board
CEMAFON Comite Europeen des Materiels et Produits pour la Fonderie [= European Committee of Foundry Materials and Products, France]
CEMAP Cotton Export Market Acreage Program
CEMATEX European Committee of Textile Machinery Manufacturers [Switzerland]
CEMBS Committee for European Marine Biological Symposia [UK]
CEMBUREAU Bureau Europeen du Ciment [= European Cement Association, France]
CEMC Canadian Engineering Manpower Council
CEMF Counter Electromotive Force [also: cemf]
CEM FL Cement Floor
CEMI Canadian Engineering Manpower Inventory
Commission Europeenne de Marketing Industriel [= European Committee for Industrial Marketing, UK]
CEMIRT Civil Engineering Maintenance, Inspection, Repair and Training Team [US Air Force]
CEM MORT Cement Mortar
CEMON Customer Engineering Monitor
CEMP Center for Environmental Management and Planning [Scotland]
CEM PLAS Cement Plaster
CEM PC Cement Plaster Ceiling
CEMR Center for Energy and Mineral Resources [of TAMU, US]
CEMREL Central Midwest Regional Educational Laboratory [US]
CEMS Central Electronic Management System
Chinese Electron Microscopy Society [PR China]
Conversion Electron Mossbauer Scattering
Conversion Electron Mossbauer Spectroscopy
CEMT Conference Europeenne des Ministres des Transport [= European Conference of Ministers of Transport]
CEN Centre d'Etude de l'Energie Nucleaire [= Center for Nuclear Energy Studies, Belgium]
Comite Europeen de Normalisation [= European Standards Committee, Belgium]
Cen Centaurus [Astronomy]
CENB Consulting Engineers of New Brunswick [Canada]
CENC Centre d'Etudes Nucleaires de Cadarache [= Center for Nuclear Studies at Cadarache, France]
CENCER CEN (= Comite Europeen de Normalisation) Certification Body
CEN (= Comite Europeen de Normalisation) Certification Branch [Belgium]
CENDIT Center for Development of International Technology
CENELEC Comite Europeen de Normalisation Electrotechnique [= European Committee for Electrotechnical Standardization [also: CENEL, Belgium]]
CENFAM Centro Nazionale per la Fisica della Atmosfera e la Meteorologia [= National Center of Atmospheric Physics and Meteorology, Italy]
CENFAR Centre d'Etudes Nucleaires de Fontenay-au-Roses [= Center for Nuclear Studies at Fontenay-au-Roses, France]

CENG Centre d'Etudes Nucleaire de Grenoble [= Center for Nuclear Studies at Grenoble, France]
CEng Certified Engineer
Chemical Engineer
CENID Centro Nacional de Informcion y Documentacion [= National Center of Information and Documentation, Chile]
CENIM Centro Nacional de Investigaciones Metalurgicas [= National Center for Metallurgical Research, Spain]
CENS Centre d'Etudes Nucleaires de Saclay [= Center for Nuclear Studies at Saclay, France]
China Economic News Service [Taiwan]
CENSA Council of European and Japanese National Shipowners Associations [UK]
CENT Centigrade
Central
Centrifugal
Century
cent (centum) — hundred
CENTAG Central Army Group [of NATO]
CENT AM Central America(n)
CENTAS Central Army Group
CEN/TC CEN (= Comite Europeen de Normalisation)/ Technical Commission
CENTCON Centralized Control Facility
centihg centimeter of mercury
CENTL Central
CENTO Central European Treaty Organization [of NATO]
Cen X-3 Centaurus X-3 [Astronomy]
CEO Chief Executive Officer
Comprehensive Electronic Office
Consulting Engineers of Ontario [Canada]
CEOA Central Europe Operating Agency [of NATO]
CEOC Colloque Europeene des Organisations de Controle [= European Colloquium of Inspection Organizations]
CEOCOR Comite d'Etude de la Corrosion et de la Protection des Canalisation [= Committee for the Study of Pipe Corrosion and Protection, Belgium]
CEOT Conference on Emerging Optoelectronic Technologies
CEP Central East Pacific (Ocean Region)
Central European Pipeline
Circle of Equal Probability
Circular Error of Probability
Circular Error Probability
Citizens' Energy Project
Civil Emergency Planning [of NATO]
Civil Engineering Package
Clinical Experimental Pharmacologist
Command Executive Procedures
Computer Entry Punch
Constraint Exploiter Process
Cotton Equalization Program
Cylinder Escape Probability
Cep Cepheus [Astronomy]
CEPA Canadian Environmental Protection Act
Civil Engineering Programming Applications
(National Society for) Computer Applications in Engineering, Planning and Architecture [US]
Coupled Electron Pair Approximation
CEPAC Confederation Europeenne de l'Industrie des Pates, Papiers et Cartons [= European Confederation of Pulp, Paper and Board Industries, Belgium]
CEPACS Cargo Entry Processing and Collection System
Customs Entry Processing and Control System
CEPC Canada Civil Engineering Planning Committee
CEPCEO Comite d'Etude des Producteurs de Charbon d'Europe Occidentale [= Association of the Coal Producers of the European Community, Belgium]
CEPD Council for Economic Planning and Development [Taiwan]
CEPE Central Experimental and Proving Establishment [Canada]
Comite Europeen des Associations des Fabricants de Peinture, d'Encres, d'Imprimerie et de Couleurs [= European Committee of Paint, Printing Ink and Artists Colors Manufacturers Associations, Belgium]
CEPER Combined Engineering Plant Exchange Record
CEPEX Controlled Ecosystem Pollution Experiment
CEPO Central European Pipeline Office [of NATO]
CEPPC Central European Pipeline Policy Committee [of NATO]
CEPR Center for Energy Policy and Research [of NYIT, US]
CEPS Central Europe Pipeline System [of NATO]
Command Module Electrical Power System
Cornish Engines Preservation Society
CEPT Conference Europeene des Administrations des Postes et des Telecommunications [= European Conference of Postal and Telecommunications Administrations, Switzerland]
Council of European Post and Telegraph
CEQ Committee on Environmental Quality [of FCST, US]
Corporation of Engineers of Quebec [Canada]
Council on Environmental Quality [US]
CER Center for Estuarine Research [Canada]
Ceramic(s)
Ceramic Electrolytic Reactor
Civil Engineering Report
Combat Engineer Regiment
Complete Engineering Release
Constant Extension Rate
Coordinated Experimental Research
Cost Estimating Relationship
Critical Experiment Reactor
Cumulative Excess Returns
CERA Canadian Electronic Representatives Association
Civil Engineering Research Association [now: CIRIA, UK]
CERAP Center Radar Approach Control
CERC Civil Engineering Research Council [UK]
Coastal Engineering Research Center [US Army]
Coastal Engineering Research Council [US]
CERCA Centre de Recherche pour Combustibles Atomiques [= Research Center for Nuclear Fuels, France]
CERCLA Comprehensive Environmental Response, Compensation, and Liability Act [US]
CERDIP Ceramic Dual In-Line Package [also: Cerdip]
CerE Ceramic Engineer
CER ENG Ceramic Engineering

CERES Controlled Environment Research Laboratory [of CSIRO, Australia]
CERESIS Centro Regional de Sismologia para America del Sur [= Regional Center for Seismology for South America, Peru]
CERF Coastal Education and Research Foundation [US]
Critical Experiment Reactor Facility
CERG Chemical Ecology Research Group
Concrete-Encased Ring Ground
CERI Canadian Energy Research Institute
Center for Educational Research Innovation [of OECD, France]
Clean Energy Research Institute [US]
CERL Cambridge Electronics Research Laboratory
Canadian Explosives Research Laboratory [of CANMET]
Central Electricity Research Laboratories [US]
Ceramic Engineering Research Laboratory [Japan]
CERMET Ceramic/Metal [also: Cermet or cermet]
Ceramic-to-Metal [also: Cermet or cermet]
CERN Centre Europeen pour la Recherche Nucleaire [= European Center for Nuclear Research, Switzerland]
CERN/LEAR CERN Low-Energy Anti-Photon Ring [Switzerland]
CERO Centre d'Etudes de Recherche Operationnelle [= Study Center for Operations Research, Belgium]
CERP Confederation Europeenne des Relation Publiques [= European Confederation of Public Relations, Belgium]
CERR Center for Earth Resources Research [of MUN, Canada]
CERS Certified Energy Reduction Specialist
Commission pour l'Encouragement de la Recherche Scientifique [= Commission for the Encouragement of Scientific Research, Switzerland]
Cornell Electron Storage Ring [US]
CERT Centre d'Etudes et de Recherches de Toulouse [= Research and Study Center at Toulouse, France]
Certificate
Certified
Character Error Rate Test
Combined Environmental Reliability Test
Constant Extension Rate, Tensile
Constant Extension Rate Test
CERTICO Certification Committee [of ANSI, US]
CertPA Certificate in Public Administration
CERVED Centri Elettronica Reteconnessi Valutazione Elaborazione Dati [= Electronic Value-Added Data Network Center, Italy]
CES Capstone Engineering Society [US]
Center for Environmental Studies [of Arizona State University, US]
Centre Europeen des Silicones [= European Silicone Center [of CEFIC, Belgium]]
Cleveland Engineering Society [US]
Coast Earth Station (Satellite)
Conference of European Statisticians
Conjugated Estrogens
Constant Elasticity of Substitution
Consumer Electronics Society [of IEEE, US]
Coordinate Evaluation System
Critical Experiment Station [of WREC, US]
Cyanoethyl Sucrose
CESA Canadian Engineering Standards Association
Committee of EEC (= European Economic Community) Shipbuilders Associations [Belgium]
Concurrent Education Students Association [Canada]
CESAC Communications-Electronics Scheme Accounting and Control [US Air Force]
CESAR Canadian Expedition to Study the Alpha Ridge
CERN Electron Storage and Accumulation Ring [Switzerland]
Combustion Engineering Safety Analysis Report
CESARS Chemical Evaluation, Search and Retrieval System [US]
CESC Consulting Engineers of South Carolina [US]
CESCE Comite Europeen des Services des Conseillers [= European Committee of Consultant Services, Belgium]
CESD Composite External Symbol Directory
CESE Comparative Education Society in Europe [Belgium]
CESEMI Computer Evaluation of Scanning Electron Microscope Images
CESI Centro Elettrotecnico Sperimentale Italiano [= Italian Center for Experimental Electrotechnics]
Council for Elementary Science International [US]
CESNEF Centro di Studi Nucleari Enrico Fermi [= Enrico Fermi Center for Nuclear Studies, Italy]
CESO Canadian Executive Service Organization
Canadian Executive Service Overseas
CESR Conduction Electron Spin Resonance
CESSE Council of Engineering and Scientific Society Executives [US]
CESSIM Centre d'Etudes des Supports d'Information Medicale [= Study Center for the Support of Medical Information, France]
CESSS Council of Engineering and Scientific Society Secretaries
CET Cement Evaluation Tool
Center for Educational Technology [US]
Central European Time
Certified Electronics Technician
Certified Engineering Technician
Certified Engineering Technologist
Chemical Engineering Technologist
Chemical Engineering Technology
Civil Engineering Technician
Civil Engineering Technologist
Civil Engineering Technology
Common External Tariff
Corrected Effective Temperature
Council for Educational Technology [UK]
Critical Experiment Tank
Cumulative Elapsed Time
Cet Cetus [Astronomy]
CETATS Cetyltrimethylammonium Tosylate
CETD Calculated Estimated Time of Departure
CETDC China External Trade Development Council [Taiwan]

CETEDOC Centre de Traitement Electronique de Documents [= Center for Electronic Document Processing, Belgium]
CETIM Centre Technique des Industries Mecaniques [= Technical Center of the Machinery Industry [of CDM, France]]
CETIS Centre Europeen de Traitement de l'Information Scientifique [= European Processing Center for Scientific Information, France]
CETO Calculated Estimated Time of Overflight
CETP Canada Employment Training Program
CETR Center for Explosives Technology Research [US]
Consolidated Edison Thorium Reactor [US]
CETRA China External Trade Research Association [Taiwan]
CETS Conference on European Telecommunications Satellites
Control Element Test Stand
CETUK Council for Educational Technology for the United Kingdom
Ce-TZP Cerium-Doped Tetragonal Zirconia Polycrystal
CEU Carbonate-Evaporite Universal
Central Executive Committee
Channel Extension Unit
Communications Expansion Unit
Continuing Education Unit [US]
Construction Engineering Unit
Coupler Electronics Unit
Customs Excise Union
CEV Citrus Exocortis Viroid
Combat Engineer Vehicle
Corona Extinction Voltage
CEVAR Consumable-Electrode Vacuum-Arc Remelting
CEVM Consumable Electrode Vacuum Melting
CEW Coextrusion Welding
CEX Canadian Environmental Exposition
Central Excitation
CEY Consulting Engineers of the Yukon Territory [Canada]
CF Canadian Forces
Canadian French
Canonic Form
Can't Find
Carbofuchsin
Carbon Fiber
Carrier Frequency
Cathode Follower
Cationized Ferritin
Center Frequency
Center of Figure
Center of Flotation
Central File
Central Force
Centrifugal Force
Centripetal Force
Cerium-Free
Chebyshev Filter
Chromatic Free (Optical System)
Circuit Finder
Citrovorum Factor
Coated Front
Cold Finger
Cold Finish
Color Filter
Combined Function
Commonwealth Foundation [UK]
Compressible Flow
Concept Formulation
Conducting Furnace Black
Confinement Factor
Conservation Foundation [US]
Constant Fraction
Contemporary Force
Context Free
Continuous Flow
Control Footing
Conversion Factor
Copy Furnished
Corn Flour
Correlation Function
Corrosion Fatigue
Cosanti Foundation [US]
Cost and Freight
Council of Fellows
Counterfire
Count Forward
Creeping Flow
Cresolformaldehyde
Crest Factor
Critical Flow
Crude Fiber
Crystal Field
Cubic Feet
Curvature of Field
C&F Cost and Freight
Cf Californium
cf (confer) — compare
cubic feet
CFA California Fertilizer Association [US]
Canadian Federation of Agriculture
Canadian Field Artillery
Canadian Forces Attache
Canadian Forestry Association
Canadian Foundry Association
Caribbean Federation of Aeroclubs [Netherlands Antilles]
Carrier Frequency Alarm
Cascade-Failure Analysis
Chinese Foundrymen's Association [Taiwan]
Code of Federal Regulations
Commonwealth Forestry Association [UK]
Communaute Financiere Africaine [= African Financial Community]
Communaute Financiere Africaine Franc [Currency of Benin, Congo, Gabon, Ivory Coast, Niger, Senegal, Togo and Upper Volta]
Complete Freund's Adjuvant
Computer Family Architecture
Consumer Federation of America
Council of Iron Founders Associations
Council on Fertilizer Application [US]
Core Flood Alarm

	Cross-Field Amplifier
CFAC	Canadian Forces Air Defense Command [also: CFADC]
CFAE	Contractor-Furnished Aircraft Equipment
CFANS	Canadian Forces Air Navigation School
CFAO	Canadian Forces Administrative Order
CFAP	Copenhagen Frequency Allocation Plan
CFAR	Constant False Alarm Rate
CFAT	Carnegie Foundation for the Advancement of Teaching [US]
CFAV	Canadian Forces Arctic Vessel
CFAW	Canadian Food and Allied Workers
CFB	Canadian Forces Base
	Cipher Feedback
	Circulating Fluidized-Bed Combustion
	Coated Front and Back
	Commonwealth Forestry Bureau
	Current Feedback
CFBA	Canadian Food Brokers Association
CFBC	Circulating Fluidized-Bed Combustion
CFBR	Commercial Fast Breeder Reactor
CFBS	Canadian Federation of Biological Sciences
	Canadian Federation of Biological Societies
CFC	Canadian Forces College
	Capillary Filtration Coefficient
	Carbon-Fiber Composite
	Caribbean Food Corporation
	Central Fire Control
	Channel Flow Control
	Chlorofluorocarbon
	Coin and Fee Checking
	Colony Forming Cells
	Consolidated Freight Classification
	Continuous Flow Calorimetry
	Crossed-Film Cryotron
	Continuous Flow Calorimetry
CFCC	Canadian Forces Communications Command
CFCE	Centre Francaise du Commerce Exterieur [= French Center of Foreign Trade]
CFCF	Central Flow Control Facility
CFCM	Continuous Flow Cell Culture
CFCS	Caribbean Food Crops Society [Puerto Rico]
CFD	Compact Floppy Disk
	Computational Fluid Dynamics
	Constant Fraction Discriminator
	Cubic Feet per Day [also: cfd]
CFDA	Current File Disk Address
CFDC	Film Development Corporation
CFDD	Compact Floppy Disk Drive
CFDE	Call Failure Detection Equipment
CFDM	Companded Frequency-Division-Multiplex
CFDM/FM	Companded Frequency-Division-Multiplex/ Frequency Modulation
CFE	Canadian Forces Europe
	Certified Food Executive
	Chlorotrifluoroethylene
	Continuous Flow Electrophoresis
	Continuous Fuel Economizer
	Contractor-Furnished Equipment
	Conventional (Armed) Forces (in) Europe
	Polychlorotrifluoroethylene (Resin)
CFEE	Canadian Foundation for Economic Education
CFEM	Conference on Fossil Energy Materials [US]
CFEMC	Canadian Forces Europe Medical Center
CFER	Center for Frontier Engineering Research [Canada]
	Collector Field Effect Register
C-FER	Center for Frontier Engineering Research [Canada]
CFES	Center for Energy Systems [US]
	Continuous Flow Electrophoresis System
CFESA	Commercial Food Equipment Service Association [US]
CFF	Critical Flicker Frequency
	Critical Fusion Frequency
	Current Fault File
CFFA	Chemical Fabrics and Film Association [US]
CFFC	Combined Forced and Free Convection
CFFED	Canadaian Federation of Farm Equipment Dealers
CFFR	Cushman Foundation for Forominiferal Research [US]
CFFTP	Canadian Fusion Fuels Technology Project
CFG	Compact-Flake-Graphite Iron
	Context-Free Grammar
CFH	Cubic Feet per Hour
CFHQ	Canadian Forces Headquarters
CFHT	Canada-France-Hawaii Telescope
CFI	California Fig Institute [US]
	Call for Instruction
	Canadian Fertilizer Institute
	Ceramic Forum International [of DKG, Germany]
	Chief Flying Instructor
	Clothing and Footware Institute
	Committee on Foreign Investment
	Commonwealth Forestry Institute [UK]
	Cost, Freight and Insurance [also: cfi]
CFIA	Component Failure Impact Analysis
	Core Flood Isolation Valve Assembly
CFIAM	Canadian Forces Institute of Aviation Medicine
CFIB	Canadian Federation of Independent Business
CFIBC	Council of Forest Industries of British Columbia [Canada]
CFIC	Canadian Forestry Industries Council
	Commercial Fishing Industry Council [Canada]
CFIEI	Canadian Farm and Industrial Equipment Institute
CFIEM	Canadian Forces Institute of Environmental Medicine
CFIF	Continuous Flow Isoelectric Focusing
CFIS	California Fiscal Information System [US]
CFIUS	Committee on Foreign Investment in the United States [US]
CFL	Calibrated Focal Length
	Call Failure Signal
	Canadian Federation of Labour
	Carbon-Filament Lamp
	Clear Flight Level
	Context-Free Language
CFLA	Canadian Forces Leadership Academy [Canada]
CFLS	Canadian Forces Language School
CFM	Cathode Follower Mixer
	Cerium-Free Mischmetall
	Center Frequency Modulation
	Chlorofluoromethane
	Companded Frequency Modulation

Companding and Frequency Modulation
Confirm
Continuous Film Memory
Continuous Flow Manufacturing
Cubic Feet per Minute [also: cfm]

CFMA Construction Financial Management Association [US]
Cyprus Foodwear Manufacturers Association

CFMC Caribbean Fishery Management Council

CFMR Center for Fundamental Materials Research

CFMS Chained File Management System
Computer-Based Financial Management System

CFMU Centralized Flow Management Unit

CFN Canadian Forces Network

CFNI Caribbean Food Nutrition Institute

CFO Central Forecast Office
Chief Financial Officer
Critical Flashover

CFP Canadian Forces Publication
Canadian Forest Products
CFP Franc [Currency of French Polynesia and Tahiti]
Crystal Field Parameter

CFPA Canadian Fluid Power Association
Canadian Food Processors Association
Chlorophenyltrifluoromethyl Phenoxyacetate

CFPARU Canadian Forces Personnel Applied Research Unit

CFPD Canadian Forces Publication Depot

CFPHT Constant Fraction of Pulse Height Trigger

CFPMO Canadian Forces Project Management Office

CFPO Canadian Forces Post Office

CFPP Canadian Federation of Pulp and Paper Workers
Cold Filter Plugging Point

CFPT Canadian Forces Parachute Team

CFR Carbon Fiber Reinforced
Carbon-Film Resistor
Catastrophic Failure Rate
Center for Field Research
Center for Futures Research
Center of Field Research [US]
Code of Federal Regulations [US]
Commercial Fast-Breeder Reactor
Commercial Fast Reactor [UK]
Contact Flight Rules
Continuous Flow Reactor
Cooperative Fuel Research
Coordinating Fuel Research
Council on Foreign Relations [US]
Cumulative Failure Rate
Crash Fire Rescue
Crash, Firefighting and Rescue Services [of Transport Canada]

cfr (confer) — compare

CFRA Canadian Forces Reorganization Act

CFRC Canadian Forces Recruiting Center
Carbon-Fiber-Reinforced Composite
Continuous Fiber-Reinforced Composite

CFRE Circulating-Fuel Reactor Experiment [of ORNL, US]

CFRI Central Fuel Research Institute [India]

CFRP Carbon-Fiber-Reinforced Polymer
Carbon-Fiber-Reinforced Plastic
Comite Francais de la Recherche sur la Pollution de l'Eau [= French Committee for Research on Water Pollution, France]

CFRTI Centre Francais de Renseignements Techniques Industriels [= French Center for Technical and Industrial Information]

CFRUO Committee on the Future Role of Universities in Ontario [Canada]

CFS Call Failure Signal
Call for Service Signal
Canadian Forces Station
Canadian Forestry Service [of Environment Canada]
Carrier Frequency Shift
Center of Forensic Sciences
Central Flying School [UK]
Constant Final State
Container Freight Station
Continuous Form Stacker
Council of Fleet Specialists [US]
Cubic Feet per Second [also: cusec, or cfs]

CFSA Canadian Fire Safety Association

CFSBA Cold Finished Steel Bar Institute [US]

CF-SCAN Canadian Forces — Second Career Assistance Network

CFSEA Canadian Food Service Executives Association

CFSG Cometary Feasibility Study Group [of ESRO, France]

CFSK Coherent Frequency Shift Keying

CFSO Crystal Field Surface Orbital

CFSPL Canadian Forces Special Project Laboratory

CFSRS Canadian Forces Supplementary Radio Systems

CFSS Canadian Forces Supply System

CFSSB Central Flight Status Selection Board

CFSTI Clearinghouse for Federal Scientific and Technical Information [of USDOC]

CFSTR Continuous Flow Stirred Tank Reactor

CFT Complement-Fixation Test
Constant Fraction Trigger
Controlled Finishing Temperature
Core Flood Tank
Cray FORTRAN
Cubic Feet per Ton

CFTA Canadian Film and Television Association
Committee of Foundry Technical Associations

CFTC Commodity Futures Trading Commission [US]
Commonwealth Fund for Technical Cooperation

CFTD Constant Fraction Timing Discriminator

CFTDC Canadian Forces Training Development Center

CFTMA Caster and Floor Truck Manufacturers Association [US]

CFTRI Central Food Technological Research Institute [India]

CFTS Canadian Forces Training System

CFU Colony Forming Units
Current File User

CFUW Canadian Federation of University Women

CFWA Canadian Fruit Wholesalers Association

CFWD Canadian Foundation for World Development

CG Cascade Generator
Cathode Glow

	Center of Gravity
	Centigram
	Chain Grate
	Circuit Group
	Coarse Grain
	Coast Guard
	Code Generation
	Coincidence Gate
	Columnar Grain
	Commissary-General
	Common Ground
	Compacted Graphite Cast Iron
	Computational Geometry
	Computer Graphics
	Consul-General
	Crack Growth
	Crystal Growth
	Cyanoguanidine
cg	centigram
CGA	Canadian Gas Association
	Carrier Group Alarm
	Certified General Accountant
	Color Graphics Adapter
	Color Guild Associates [US]
	Compressed Gas Association [US]
	Contrast Gate Amplifier
	Cyprus Geographical Association
CGAA	Canadian Graphic Arts Association
	Certified General Accountants of Alberta [Canada]
CGA/EGA	Color Graphics Adapter/Enhanced Graphics Adapter
CGAMEEC	Committee of Glutamic Acid Manufacturers of the European Economic Community [France]
CGAS	Coast Guard Air Station
CGB	Ceramics and Graphite Branch [US Air Force]
	Coincidence Grain Boundary
	Commonwealth Geographical Bureau [Australia]
	Convert Gray to Binary
CGC	Calavo Growers Council [US]
	Canadian Geoscience Council
	Canadian Grain Commission
	Circuit Group Congestion
	Clock Generator Controller
	Compact Glass Column
CGCC	Chinese General Chamber of Commerce [Hong Kong]
CGD	Cosmic Gas Dynamics
CGDE	Contact Glow Discharge Electrolysis
CGE	Carriage
	Center for Global Education
	Commission Geographical Education [of IGU, Canada]
CGEC	Canadian Government Exposition Center
CGED	Caribbean Group for Cooperation in Economic Development [US]
CGEL	Cover Gas Evaluation Loop
CGF	Carrier Gas Fusion
	Continuous Glass Fiber
CGG	Continuous Galvanizing Grade (of Zinc)
CGH	Cape of Good Hope [South Africa]
	Computer-Generated Hologram
	Computer-Generated Holograph
CGHAZ	Coarse-Grained Heat-Affected Zone
CGI	Ground Control Interceptor
	Computer-Generated Image
	Computer Graphics Interface
CGIAR	Consultative Group on International Agricultural Research [US]
CGIC	Ceramics and Graphite Information Center [US Air Force]
	Compressed-Gas-Insulated Cable
CGIL	Center for Genetic Improvement of Livestock [of University of Guelph, Canada]
CGIS	Canadian Geographic Information Service [of Environment Canada]
	COMSAT (= Communications Satellite) General Integrated System
CGIT	Compressed-Gas-Insulated Transmission Line
CGL	Circling Guidance Lights
	Computer-Generated Letter
CGLBI	Canadian Geophysical Long Baseline Interferometry
CGLO	Commonwealth Geological Liaison Office
CGM	Character Generation Module
	Commission for the Geological Map of the World
	(International Conference on Joining of) Ceramics, Glass and Metal
	Computer Graphics Metafile
cgm	centigram
CGMA	Compressed Gas Manufacturers Association [US]
CGMID	Character Generation Module Identifier
CG/MOI	Center of Gravity/Moment of Inertia
CGMW	Commission for the Geological Map of the World [France]
CGP	Central Graphics Processor
	Coalition for Government Procurement [US]
	Color Graphics Printer
CGPA	Canadian Gas Processors Association
CGPC	Canadian Government Publishing Center
	Cellular General-Purpose Computer
CGPM	Conference Generale des Poids et Mesures [= General Conference on Weights and Measures, France]
CGPO	Canadian Government Printing Office
CGPS	Canadian Government Printing Service
CGPSA	Canadian Gas Processors and Suppliers Association
CGRA	Consortium General des Recherches Aeronautiques [= General Consortium for Aeronautical Research, France]
CGRAM	Clock-Generated Random-Access Memory
CGRM	Containment Gaseous Radiation Monitor
CGRP	Calcitonin Gene-Related Peptide
	Circuit Group
CGS	Canadian Geographical Society
	Canadian Geotechnical Society
	Centimeter-Gram-Second (System) [also: cgs]
	Certificate of Graduate Studies
	Chief of General Staff
	Circuit Group Congestion Signal
	Consistent Global State
	Control Guidance Subsystem
CGSA	Cellular Geographic Service Area
CGSB	Canadian General Standards Board

	Canadian Government Specifications Board
	Canadian Government Standards Board
CGSE	CGS Electrostatic System [also: cgse]
CGSET	Circuit Group Set
CGSM	CGS Electromagnetic System [also: cgsm]
CGSU	CGS Unit [also: cgsu]
CGT	Current Gate Tube
CGTO	Contracted Gaussian-Type Orbital
CGU	Canadian Geophysical Union
	Compu/Graphics Users Association [US]
CGWB	Canadian Government Wheat Board
CH	Calcium Hydroxide
	Can't Hear
	Carbohydrate
	Case Hardening
	Ceiling Height
	Central Heating
	Chain
	Chain Home (Radar)
	Channel
	Chapter
	Chief
	China
	Chinese
	Choke
	Clearinghouse
	Coal Hook
	Coastal Harbour
	Cold Work Hardening
	Conductor Head
	Confederatio Helvetica [= Swiss Confederation]
	Contention Handling
	Control Heading
	Convex Hull
	Critical Height
	Custom House
	Cyclohexane
	Cyclohexanol
C/H	Carbon/Hydrogen Ratio
Ch	Cholesteric Phase
ch	candle hour
	chain [Unit]
CHA	Canadian Historical Association
	Canadian Hydrographers Association
	Concentric Hemispherical Analyzer
	Correlation Height Analysis
	Cyclohexanol Acetate
Cha	Chameleon [Astronomy]
CHABA	Committee on Hearing and Bioacoustics [US]
ChAc	Choline Acetyltransferase
CHAD	Code for Handling Angular Data
CHAG	Chain Arrester Gear
	Compact High-Performance Aerial Gun
CHAIS	Consumer Hazards Analytical Information Service [of LGC, UK]
CHAM	Chamfer
CHAMP	Character Manipulation Procedure
	Communications Handler for Automatic Multiple Programs
	Community Health Air Monitoring Program
CHAMPS	Computerized History and Maintenance Planning System
CHAN	Channel
CHANCOM	Channel Committee [of NATO]
CHANHI	Upper Channel Corresponding to Half-Amplitude Point of a Distribution
CHANLO	Lower Channel Corresponding to Half-Amplitude Point of a Distribution
CHAP	Chapter
CHAPI	Compact Helicopter Approach Path Indicator
CHAPS	Chapters
	3-[(3-Cholamidopropyl)dimethylammonio]-1-Propanesulfonate
	Clearinghouse Automated Payments System [US]
CHAPSO	3-[(3-Cholamidopropyl)dimethylammonio]-2-Hydroxy-1-propanesulfonate
CHAR	Character
CHARA	Center for High Angular Resolution Astronomy [US]
CHARIBDIS	Chalk River Bibliographic Data Information System [of AECL, Canada]
CHART	Charter
	Computerized Hierarchy and Relationship Table
CHASM	Coventry Health and Safety Movement [UK]
CHAT	CLIRA (= Closed Loop In-Reactor Assembly) Holddown Assembly Tool
CHBA	Canadian Home Builders Association
CHC	Carbohydrate Chemistry
	Chemical Hydrogen Cracking
	Choke Coil
	Clean Harbors Cooperatives [US]
	Clearinghouse for Copyright
	Cyclohexylamine Carbonate
CHCF	Component Handling and Cleaning Facility
ChCMV	Chrysanthemum Chlorotic Mottle Viroid
CHCU	Channel Control Unit
CHD	Carbohydrate Derivative
	Chord
CHDB	Compatible High Density Bipolar
CHDC	Canadian Housing Design Council
CHDL	Computer Hardware Definition Language
CHDM	Cyclohexanedimethanol
CHE	Channel End
	Chemical Engineer
	Cylindrical Hydraulic Engineer
ChE	Chemical Engineer
	Cholinesterase [also: chE]
CHEC	Cascade Holistic Economic Consultants [US]
	Channel Evaluation and Call
	Commonwealth Human Ecology Council [UK]
CheE	Chemical Engineer
CHEF	Chemistry of High Elevation Fog
CHEM	Chemical
	Chemist
	Chemistry
CHEMDEX	Chemical Index [also: Chemdex]
ChemE	Chemical Engineer [also: ChemEng]
CHEM ENG	Chemical Engineering
CHEMFET	Chemically Selective Field-Effect Transistor
CHEMLINE	Chemical Information On-Line [US]
CHEM MET	Chemical Metallurgy
CHEMNAME	Chemical Names (Database) [of ACS, US]

CHEMRAWN Chemical Research Applied to World Needs Committee [of the USNC for IUPAC]
CHEMSDI Chemical Abstacts Selective Dissemination of Information [of ACS, US]
CHEMTREC Chemical Transportation Emergency Center [Division of CMA, US]
CHEOPS Chemical Information Systems Operators [now: EUSIDIC]
CHES 2-(N-Cyclohexylamino)ethanesulfonic Acid
Ches Cheshire [UK]
CHESNAVFAC Chesapeake Naval Division Facilities Engineering Command [US Navy]
CHESS Canadian Health Education Specialists Society
Community Health and Environmental Surveillance System
Cornell High Energy Synchrotron Source [of Cornell University, US]
CHEVMA Canadian Heat Exchanger and Vessel Manufacturers Association
CHF Chief
Coupled Hartree-Fock
Critical Heat Flux
Swiss Franc [Currency of Switzerland and Liechtenstein]
CHFIE Cordell Hull Foundation for International Education [US]
CH FWD Charges Forward
CHG Change
Charge
CHGP Charging Pump
CHI Computer/Human Interaction
CHIA Canadian Hovercraft Industries Association
CHIC Canadian Housing Information Center [of CMHC]
CHICA Community and Hospital Infection Control Association [Canada]
CHIEF Controlled Handling of Internal Executive Functions
CHIF Channel Interface
CHIL Current-Hogging Injection Logic
CHILD Cognitive Hybrid Intelligent Learning Device
Computer Having Intelligent Learning and Development
CHILL CCITT High-Level Language
ChIME Chemical Industry for Minorities in Engineering [US]
CHINT Charge Injection Transistor
CHIP Canadian Home Insulation Program
Chip Hermetically in Plastic
Cold and Hot Isostatic Pressing
C-HIP Conventional Hot Isostatic Pressing
CHIPS Clearinghouse Interbank Payment System [US]
CHIRAPHOS Bis(diphenylphosphino)butane
CHIRP Confidential Human Factors Incident Reporting System [UK]
CHK Check
CHKPT Checkpoint
CHKR Checker
CHL Current-Hogging Logic
Chl Chloroform
Chlorophyll
CHLOREP Chlorine Emergency Plan [of Chlorine Institute, US]
CHM Chairman
Chemical Machining
Chemical Milling
CHMN Chairman
CHMOS Complimentary High-Density Metal-Oxide Semiconductor
CHMR Center for Hazardous Materials Research [US]
CHMTS Components, Hybrids and Manufacturing Technology Society [of IEEE, US]
CHN Carbon, Hydrogen and Nitrogen
C/H/N Carbon/Hydrogen/Nitrogen
CHNL Channel
CHNO Carbon, Hydrogen, Nitrogen and Oxygen
C/H/N/O Carbon/Hydrogen/Nitrogen/Oxygen
CHO Carbohydrate
Carbon, Hydrogen and Oxygen
Chinese Hamster Ovary [Biochemistry]
CHOC Center for History of Chemistry [University of Philadelphia, US]
CHOICES Computerized Heuristic Occupational Information and Career Exploration System [Canada]
CHOL Common High-Order Language
CHORD Chief of Office of Research and Development
CHORI Chief of Office of Research and Inventions [US]
CHP Channel Processor
Combined Heat and Power
Continental Horsepower
Cryohydric Point
ChP Chilean Peso [Currency of Chile]
CHPAE Critical Human Performance and Evaluation
CH PPD Charges Prepaid
CHPS Characters per Second
CHQ Cheque
Chlorohydroquinone
CHR Character
c-hr candle-hour
CHRG Charge
CHRGN Character Generator
CHRIS Chemical Hazard Response Information System
CHRM Center for Holistic Resource Management [US]
CHRON Chronological
Chronology
CHRS Canadian Heritage Rivers System
Center for Hospitality Research and Service [US]
Containment Heat Removal System
CHS Calder Hall Reactor [UK]
Canadian Hydrographic Service
Characters
Charged Hard-Sphere System
Citizens for Highway Safety [US]
CHSB Construction Health and Safety Branch [of MOL, Canada]
CHSC Cambridge Higher School Certificate [UK]
CHT Call Hold and Trace
Charactron Tube
Chart
Chemical Heat Treatment
Conductive Heat Transfer
Convective Heat Transfer
Cycloheptatriene
CHU Centigrade Heat Unit [also: chu]

	Channel Unit [of SPADE]
CHUG	Canadian Honeywell Users Group
CHY	Chymotrypsinogen
CI	Call Indicator
	Card Input
	Cast Iron
	Center Island
	Center of Inertia
	Certificate of Insurance
	Chain Index
	Channel Islands
	Chemical Inspectorate [of MOD, UK]
	Chlorine Institute [US]
	Circuit Interrupter
	Colloidal Iron
	Color Index
	Combustion Institute [US]
	Comfort Index
	Communications Interface
	Composites Institute [US]
	Computer Interconnect
	Configuration Interaction
	Configuration Item
	Conservation International [US]
	Consular Invoice
	Consumers Interpol [Malaysia]
	Containment Integrity
	Control Interval
	Cordage Institute [US]
	Cost and Insurance
	Cotton Incorporated [US]
	Course Indicator
	Critical Item
	Cropping Index
	Cubic Inch [also: ci]
	Cumulative Index
C&I	Control and Indication
	Control and Instrumentation
	Cost and Insurance
C/I	Carrier-to-Interface (Ratio)
	Carrier-to-Interference (Ratio)
	Certificate of Insurance
Ci	Cirrus Cloud
	Curie
CIA	Central Intelligence Agency [US]
	Ceramics International Association [US]
	Chemical Industries Association [UK]
	Chemiluminescence Immunoassay
	Commonwealth Industries Association [UK]
	Communications Industry Association
	Communications Interrupt Analysis
	Computer Industry Association [now: CCIA, US]
	Computer Interface Adapter
	Construction Industries Association [now: CIMA, US]
	Containment Isolation A
CIAA	Canadian Industrial Arts Association
CIAC	Ceramics Information Analysis Center [of Purdue University, US]
	Construction Industry Advisory Council [US]
CIAJ	Communications Industry Association of Japan
CIAM	Computerized Integrated and Automated Manufacturing
CIAP	Climatic Impact Assessment Program
	Climatic Implications of Atmospheric Pollution
	Community-Based Industrial Adjustment Program
CIAPA	Chemical Industries Accident Prevention Association
CIAPY	Centro de Investigacion Agricolas de la Peninsula de Yucatan [= Agricultural Investigation Center of the Peninsula of Yucatan, Mexico]
CIAQR	Center for Indoor Air Quality Research [Canada]
CIAR	Canadian Institute for Advanced Research
CIAS	CCRS (= Canada Center for Remote Sensing) Image Analysis System
CIATF	Comite International des Associations Techniques de Fonderie [= International Committee of Foundry Technical Associations, Switzerland]
CIB	Centralized Intercept Bureau
	Centro de Investigaciones Basicas [= Center for Basic Research, Mexico]
	Channel Interface Bus
	COBOL Information Bulletin
	Community Inquiry Bureau
	Conseil International du Batiment pour la Recherche, l'Etude et la Documentation [= International Council for Building Research, Studies and Documentation, Netherlands]
	Containment Isolation B
	Counterfeiting Intelligence Bureau [of International Chamber of Commerce, UK]
CIBD	Centre d'Information des Banques de Donnees [= Data Bank Information Center, France]
CIBO	Council of Industrial Boiler Owners [US]
CIBS	Chartered Institution of Building Services [now: CIBSE, UK]
CIBSE	Chartered Institution of Building Services Engineers [UK]
CIC	Canada Immigration Center
	Carrier Identification Code
	Chemical Industries Council [US]
	Chemical Institute of Canada
	China Institute of Communications [PR China]
	Cogeneration Coalition
	Combat Information Center
	Command Input Buffer
	Committee on Institutional Cooperation [US]
	Common Interface Circuit
	Communication Intelligence Channel
	Community Information Center
	Computer Intelligence Corps [US Army]
	Conseil International de la Chasse et de la Conservation du Gibier [= International Council for Game and Wildlife Conservation, France]
	Construction Industry Commission [Canada]
	Consumer Information Center
	Copper/Invar/Copper Material
	Council of Independent Colleges [US]
	Counter-Intelligence Corps [US]
	Cover Integrated Cell
CICA	Canadian Industrial Communicators Assembly
	Canadian Institute of Chartered Accountants

Chicago Industrial Communications Association [US]
Competition in Contracting Act [US]
Confederation of International Contractor Associations [France]
Construction Industry Computing Association [UK]

CICAT Center for International Cooperation and Appropriate Technology

CICC Custom Integrated Circuit Conference [of IEEE, US]

CICCA Committee on International Cooperation of Cotton Associations

CICD Computer-Integrated Circuit Design

CICE Council for International Congresses of Entomology [UK]

CICHE Consortium for International Cooperation in Higher Education [US]

CICI Confederation of Information Communication Industries [UK]

CICIG Commissione Italiana del Comitato Internazionale di Geofisica [= Italian Commission of the International Committee of Geophysics]

CICIN Conference on Interlibrary Communications and Information Networks [of ALA, US]

CICIREPATO Committee for International Cooperation in Information Retrieval among Examining Patent Offices

CICP Communications Interrupt Control Program

CICRED Comite International de Cooperation dans les Recherches Nationales en Demographie [= Committee for International Cooperation in National Research in Demography, France]

CICRIS Cooperative Industrial and Commercial Reference and Information Service [UK]

CICS Canadian Industrial Computer Society
Committee for Index Cards for Standards
Customer Information Control System

CICSA Canadian Indepent Computer Services Association

CICS/VS Customer Information Control System/Virtual Storage

CICT Commission on International Commodity Trade

CID Cayman Islands Dollar [Currency]
Center for Information and Documentation [of EURATOM]
Charge-Injection Device
Collision-Induced Decomposition
Collision-Induced Dissociation
Combined Immunological Deficiency
Commercial Item Description
Communication Identifier
Component Identification
Computerized Index of Delivery Data
Criminal Investigation Department
Cubic Inch Displacement
Current-Image Diffraction Test

CIDA Canadian International Development Agency
Channel Indirect Data Addressing
Centre International de Documentation Arachnologique [= International Center for Arachnological Documentation, France]
Current Input Differential Amplifier

CIDAS Climatological Ice Data Archiving System
Conversational Interactive Digital/Analog Simulator

CIDB Canadian International Development Board
Chemie Information und Dokumentation, Berlin [= Berlin Center for Information and Documentation, Germany]

CIDC Commercial and Industrial Development Corporation

CIDE Center for International Development and Environment [US]
Commission Intersyndicale de Deshydrateurs Europeens [= European Dehydrators Association, France]

CIDEP Chemically Induced Dynamic Electron Polarization

CIDF Control Interval Definition Field

CIDIN Common ICAO (= International Civil Aviation Organization) Data Interchange Network [of UN]

CIDNP Chemically-Induced Dynamic Nuclear Polarization

CIDS Chemical Information and Data System [US Army]
Communication, Information and Documentation System
Concrete Island Drilling System

CIDST Committee for Information and Documentation on Science and Technology [UK]

CIE Center for International Education
Certificate in Electronics
Chinese Institute of Engineers [Taiwan]
Commission Internationale de l'Eclairage [= International Commission on Illumination, Austria]
Commonwealth Institute of Entomology [UK]
Computer-Integrated Engineering

cie (compagnie) — company

CIEA Committee on International Education in Agricultural Sciences [Netherlands]

CIEBAA Comite International pour l'Etude des Bauxites, d'Alumine et d'Aluminium [= International Committee for the Study of Bauxites, Alumina and Aluminum, Yugoslavia]

CIECD Council for International Economic Cooperation and Development

CIEDS Computer-Integrated Electrical Design Series

CIEE Council on International Educational Exchange [US]

CIEF Continuous Isoelectric Focusing

CIEH Charge Iterated Extended Hueckel

CIEI Center for International Environment Information [now: WEC, US]

CIELAB CIE $L^*a^*b^*$ Scale Color Difference [of CIE, Austria]

CIELUV CIE $L^*u^*v^*$ Scale Color Difference [of CIE, Austria]

CIEN Commision Interamericana de Energia Nuclear [= Inter-American Commission on Nuclear Energy [of Organization of American States]]

CIEP Canadian Institute for Economic Policy

CIER Comision de Integracion Electrica Regional [= Commission of Regional Electrical Integration, Uruguay]
Conseil International des Economies Regionales [= International Council for Local Development, France]

CIES Comparative and International Education Society [US]
Council for the International Exchange of Scholars [US]

CIETA Centre International d'Etude des Textiles Anciens [= International Center for the Study of Ancient Textiles, France]

CIE-USA Chinese Institute of Engineers — United States of America

CIF Canadian Institute of Forestry
Captive Installation Function
Central Index File
Central Information File
Central Integration Facility
Chemical Industry Federation [Finland]
Cold-Insoluble Fibrinogen
Computer Integrated Factory
Cost, Insurance and Freight [also: cif]

CIFC Cost, Insurance, Freight and Commission [also: cifc]

CIFCI Cost, Insurance, Freight, Commission and Interest [also: cifci]

CIFE Cost, Insurance, Freight and Exchange [also: cife]

CIFEG Centre International pour la Formation et les Echanges Geologiques [= International Center for Training and Exchanges in the Geosciences, France]

CIFFA Canadian International Freight Forwarders Association

CIFI Cost, Insurance, Freight and Interest [also: cifi]

CIFRI Central Inland Fisheries Research Institute [India]

CIFRR Common Instrument Flight Rules Room

CIFST Canadian Institute of Food Science and Technology

CIFT Canadian Institute of Fisheries Technology [of TUNS]

CIG Ceiling
Computer Image Generation
Cryogenic-in-Ground

CIGAS Cambridge Intercollegiate Graduate Application Scheme [UK]

CIGGT Canadian Institute for Guided Ground Transport

CIGI Canadian International Grain Institute

CIGL Centre d'Informatique General de Liege [= Center for General Informatics at Liege, Belgium]

CIGM Chemically Induced Grain Boundary Migration

CIGR Commission Internationale du Genie Rural [= International Commission of Agricultural Engineers, France]

CIGRE Conference Internationale des Grands Reseaux Electriques a Haute Tension [= International Conference on Large High-Voltage Electric Systems, France]

CIGS Chief of the Imperial General Staff

CIGTF Central Inertial Guidance Test Facility

CII Call Identity Index
Containerization and Intermodal Institute [US]
Council for Industrial Innovation

CIIA Canadian Information Industry Association
Canadian Institute of International Affairs
Commission Internationale des Industries Agricoles et Alimentaires [= International Commission for Food Industries, France]

CIIC Canadian Industrial Innovation Center
Centre d'Information de l'Industrie des Chaux et Ciments [= Information Center of the Lime and Cement Industries, France]

CIIME Committee for International Investment and Multinational Enterprises [of OECD, France]

CIIMS Canadian Information and Image Management Society

CIIR Canadian Institute of International Relations
Chlorobutyl Rubber

CIIT Canadian Institute of Industrial Technology [of NRC, Canada]
Chemical Industry Institute of Toxicology [US]

CIL Cadet Instructor List
Call Identification Line
Call Information Logging
Carbon-in-Leach
Computer-Independent Language
Current Injection Logic

C-I-L Carbon-in-Leach

CILA Centro Interamericano de Libros Academicos [= Inter-American Center of Free Academies, Mexico]

CILC Confederation Internationale du Lin et du Chanvre [= International Linen and Hemp Confederation, France]

CILE Call Information Logging Equipment

CILOP Conversion in lieu of Procurement

CILRT Containment Integrated Leak Rate Test

CILSS Comite Permanent Inter-Etats de Lutte Contre la Secheresse dans le Sahel [= Permanent Interstate Committee for Drought Control in the Sahel (Area)] [Burkina Faso]

CIM Canadian Institute of Management
Canadian Institute of Marketing
Canadian Institute of Mining and Metallurgy [also: CIMM]
Certificate in Information Management
Certificate in Management
Certified Industrial Manager
Charge Imaging Matrix
Communications Improvement Memorandum
Communications Interface Monitor
Compression Injection Molding
Computer Input Microfilm
Computer Input Multiplexer
Computer-Integrated Manufacturing
Constituent Interchange Model
Continuous Image Microfilm
Custom Injection Molding

CIMA Chartered Institute of Management Accountants [UK]
Construction Industry Manufacturers Association [US]

CIMAC Computer-Integrated Manufacturing Control
Congres International des Machines a Combustion [= International Congress on Combustion Engines, France]

Conseil International des Machines a Combustion [= International Council on Combustion Engines, France]
CIMB Construction Industry Management Board [US]
CIMCO Card Image Correction
CIM&D Canadian Integrated Manufacturing and Design Show and Conference [also: CIM/D or CIMD]
CIMD/CAEC Canadian Integrated Manufacturing and Design Show and Conference/Canadian Architectural, Engineering and Construction Exposition
CIME Chartered Institute of Marine Engineers [US]
Computer Integrated Mechanical Engineering (Show and Exhibition)
CIMechE Companion of the Institute of Mechanical Engineers
CIME-DESIGN Computer Integrated Mechanical Engineering — Design
CIMH Caribbean Institute for Meteorology and Hydrology
Comite International pour la Metrologie Historique [= International Committee for Historical Metrology, France]
CIMM Canadian Institute of Mining and Metallurgy [also: CIM]
CIMMYT Centro Internacional de Mejoramiento de Maiz y Trigo [= International Center for Corn and Wheat Improvement, Mexico]
CIMP Computer-Integrated Materials Processing
CIMPC Computer-Integrated Manufacturing Programmable Controller
CIMR Center for International Media Research
CIMS Chemical Ionization Mass Spectroscopy
Coordination and Interference Management System [of INTELSAT]
CIMT Center for Instructional Media and Technology
CIMTC Construction and Industrial Machinery Technical Committee [of SAE, US]
CIMV Cauliflower Mosaic Virus
CIMVAR CIMplicity Value Added Remarketer [CIM stands for Computer-Integrated Manufacturing]
CIN Carrier Input
Chemical Industry Notes (Database) [US]
Communication Identification Navigation
CINC Commander-in-Chief [also: C-in-C, or CinC]
CINCC Coal Industry National Consultative Council [UK]
CINCHAN Commander-in-Chief Channel [of NATO]
CINDA Computer Index for Neutron Data [of UKAEA]
CINDAS Center for Information and Numerical Data Analysis and Synthesis [US]
CINS CENTO Institute of Nuclear Science [of NATO]
CINTERFOR Centro Interamericano de Investigacion y Documentacion Sobre Formacion Profesional [= Inter-American Center for Research and Documentation on Vocational Training, Uruguay]
CIO Carrier Insertion Oscillator
Central Input/Output (Multiplexer)
Chief Information Officer
Confirming Informal Order
Congress of Industrial Organizations [US]
CIOCS Communications Input/Output Control System
CIOM Communications Input/Output Multiplexer
CIOMS Council for International Organizations of Medical Sciences [Switzerland]
CIOP Communications Input/Output Processor
CIOPORA Communaute Internationale des Obtenteurs de Plantes Ornamentales et Fruitiere a Reproduction Asexuee [= International Community of Breeders of Asexually Reproduced Fruit Trees and Ornamental Varieties, Switzerland]
CIOR Confederation Interalliee des Officiers de Reserve [= Interallied Confederation of Reserve Officers [of NATO]]
CIOS Conseil International de Organisation Scientifique [= International Council for Scientific Management, Malaysia]
CIOU Custom Input/Output Unit
CIP Canadian Institute of Planning
Carbon-in-Pulp
Cascade Improvement Program
Cast-in-Place
Cast Iron Pipe
Cataloguing in Publication
Centro Internacional de la Papa [= International Potato Center, Peru]
Clean-in-Place
Cold Isostatic Pressing
Commercial Instruction Processor
Communications Interrupt Program
Compatible Independent Peripherals
Composite Interlayer Bonding
Computer Image Processing
Current Injection Probe
C-I-P Carbon-in-Pulp
C/IP Construction/Inspection Procedure
CIPA Canadian Industrial Preparedness Association
Chartered Institute of Patent Agents [UK]
Comite International de Photogrammetrie Architecturale [= International Committee of Architectural Photogrammetry, France]
Comite International des Plastiques en Agriculture [= International Committee of Plastics in Agriculture, France]
CIPAC Collaborative International Pesticides Analytic Council [Netherlands]
Copyright, Inventions and Patent Association of Canada
CIPASH Committee on International Programs in Atmospheric Sciences and Hydrology [of NAS/NRC, US]
CIPC Canadian Institute on Pollution Control
Chloro-Isopropyl-N-Phenylcarbamate
CIPCA Cogeneration and Independent Power Coalition of America [US]
CIPE Center for International Private Enterprise [US]
CIPEC Canadian Industry Program for Energy Conservation [of EMR]
Conseil Intergouvernemental des Pays Exportateurs de Cuivre [= Intergovernmental Council of Copper Exporting Countries, France]
CIPH Canadian Institute of Plumbing and Heating
CIPHI Canadian Institute of Public Health Inspectors
CIPHONY Cipher and Telephone Equipment

CIPM Comite International des Poids et Mesures [= International Committee of Weights and Measures, France]
Council for International Progress in Management
CIPMP Commission Internationale pour la Protection de la Moselle contre la Pollution [= International Commission for the Protection of the Moselle against Pollution, France]
CIPO Bis(2-Carboisopentyloxy-3,5,6-Trichlorophenyl)oxalate
CIPOM Computers, Information Processing and Office Machines (Committee) [of Canadian Standards Association]
CIPPOCS Cast-In-Place, Push-Out Cylinders
CIPQ Centre International de Promotion de la Qualite [= International Center for Promotion of Quality, France]
CIPR Commission Internationale de Protection contre les Radiations [= International Commission of Radiological Protection, UK]
CIPRA Cast Iron Pipe Reseach Association [now: DIPRA, US]
Commission Internationale pour la Protection des Regions Alpines [= International Commission for the Protection of Alpine Regions, Liechtenstein]
CIPS Canadian Information Processing Society
CIPW Cross, Iddings, Pirsson, and Washington Classification [Petrology]
CIR Canada-India Reactor [India]
Carrier-to-Interference Ratio
Center for Industrial Research [Norway]
Characteristic Instants of Restitution
Circle
Circuit
Circular
Circulation
Circumference
Commission for Industrial Relations [UK]
Cost Information Report
Critical Isotope Reactor
Current Instruction Register
Cir Circinus [Astronomy]
CIRA Cast Iron Research Association [UK]
Committee on Industrial Research Assistance [of IRAP, Canada]
COSPAR (= Committee on Space Research) International Reference Atmosphere
CIRAC Canadian Institute for Research in Atmospheric Chemistry [of York University, Canada]
CIRB Canadian Industrial Renewal Board [of DRIE]
CIRC Center for Industrial Research and Consulting
Centralized Information Reference and Control
Circle
Circular
Circulation
Circumference
Cross-Interleaved Reed-Solomon Code
Cylindrical Internal Reflection Cell
CIRCA Computerized Information Retrieval and Current Awareness
CIRCAL Circuit Analysis
CIRCUM Circumference
CIRES Communication Instructions for Reporting Enemy Sightings
CIRF Corn Industries Research Foundation [US]
CIRGA Critical Isotope Reactor, General Atomics [US]
CIRI Canadian Industrial Risk Insurers
CIRIA Construction Industry Research and Information Association [UK]
CIRIS Complete Integrated Reference Instrumentation System
CIRK Computer Information Retrieval from Keyboards
CIRL Canadian Institute of Resources Law
CIRM Comite International de Radio Maritime [= International Committee on Maritime Radio, UK]
cir mil circular mil
CIRNAV Circumnavigate
CIRP Canadian Industrial Renewal Program
CIRQUE Conference on Industry and Resources, Queen's University Engineering [Canada]
CIRRA Canadian Industrial Relations Research Association
CIS Canadian Institute of Surveying [now: CISM]
Cargo Information System
Centre International d'Informations de Securite et d'Hygiene du Travail [= International Information Center for Industrial Safety and Health [of ILO, Switzerland]
Certificate in Information Science
Channel and Isolation Supervision
Character Instruction Set
Characteristic Isochromate Spectroscopy
Chemical Information Service [UK]
Chemical Information System
Closed Ion Source
Commercial Instruction Set
Communication and Instrumentation System
Communication Information System
Computer Imaging System
Constant Initial State
Constant Initial State Spectroscopy
Constant Injection System
Containment Isolation System
Continuous Injection System
Cooperative Independent Surveillance
Course Information System
Cranbrook Institute of Science [US]
Cue Indexing System
Current Information Selection
CISA Canadian Industrial Safety Association
Canadian Intelligence and Security Association
Casting Industry Suppliers Association [US]
CISAM Compressed Index Sequential Access Method
CISC Canadian Institute for Steel Construction
Complex Instruction Set Computer
CISCA Ceilings and Interior Systems Construction Association [US]
CISCO Compass Integrated System Compiler
CISE Consortium for International Studies Education [US]
CISEP Cellulose Industry Standards Enforcement Program [US]

CISHEC Chemical Industries Association's Safety and Health Council [UK]
CISIR Ceylon Institute of Scientific and Industrial Research [now: SLISIR]
CISM Canadian Institute of Surveying and Mapping
Centre International des Sciences Mecaniques [= International Center for Mechanical Sciences, Italy]
CISP Canadian Institute of Surveying and Photogrammetry
CISPI Cast Iron Soil Pipe Institute [US]
CISPR Comite International Special des Perturbations Radioelectriques [= International Special Committee on Radio Interference, UK]
CISR Center for Information Systems Research [of MIT, US]
Center for International Systems Research [of USDOS]
CISRC Computer and Information Science Research Center [of Ohio State University, US]
CISS Canadian Institute of Strategic Studies
Center for International Strategic Studies [of York University, Canada]
Constant Initial State Spectroscopy
CISTI Canada Institute for Scientific and Technical Information [of NRC]
CISTOD Confederation of International Scientific and Technological Organizations for Development [France]
CIT California Institute of Technology [also: Caltech, US]
Call-in-Time
Canadian Import Tribunal
Carnegie Institute of Technology [US]
Case Institute of Technology [now: CWRU, US]
Chartered Institute of Transport [UK]
Chiba Institute of Technology [Japan]
Citation
Citizen
Collection Implanted Technology
Comite International des Transports Ferroviares [= International Rail Transport Committee, Switzerland]
Compact Ignition Tokamak [US]
Compact Ignition Torus
Compressor Inlet Temperature
Computer-Integrated Testing
Computer Interface Technology
Cranfield Institute of Technology [UK]
C&IT Communications and Information Technology
CIT A Citric Acid
CITAB Computer Instruction and Training Assistance for the Blind
CITB Construction Industry Training Board [UK]
CITC Canadian Institute of Timber Construction
Canadian International Trade Classification
Committee of International Technical Cooperation [Taiwan]
Construction Industry Training Center [US]
Current Information Transfer in English
CITCA Committee of Inquiry into Technological Change in Australia
CITE Capsule-Integrated Test Equipment
Cargo-Integrated Test Equipment
Compression, Ignition and Turbine Engine
Contractor-Independent Technical Effort
Council of Institute of Telecommunications Engineers [US]
CITEP Commonwealth Industrial Training Experience Program
CITES Convention on International Trade in Endangered Species of Wild Fauna and Flora [Switzerland]
CITHA Confederation of International Trading Houses Associations [Netherlands]
CITL Canadian Industrial Transportation League
CITM Charge Injection Transistor Memory
CITS Commission Internationale Technique de Sucrerie [= International Commission of Sugar Technology, Belgium]
Computer-Integrated Testing System
Current Imaging Tunneling Spectroscopy
CITT Canadian Institute of Traffic and Transportation
CITTA Confederation Internationale des Fabricants de Tapis et de Tissus [= International Confederation of Manufacturers of Carpets and Furnishing Fabrics, France]
CITU Confederation of Independent Trade Unions [Luxembourg]
CIU Channel Interface Unit
Chlorella International Union [US]
Communications Interface Unit
Computer Interface Unit
Control Indicator Unit
CIUR Commission Internationale des Unites et des Mesures Radiologiques [= International Commission on Radiological Units and Measures, US]
CIV Center Island Vessel
City Imperial Volunteers
Civil(ian)
Containment Isolation Valve
Corona Inception Voltage
CivE Civil Engineer [also: CivEng]
CIV ENG Civil Engineering
CIW Carnegie Institution of Washington [US]
CIWS Central Instrument Warning System
Close-In Weapons System
CJ Chuck Jaw
Cold Junction
CJA (Brotherhood of) Carpenters and Joiners of America
Commonwealth Journalists Association [UK]
CJCC Commonwealth Joint Communications Committee [UK]
CJCSM Chinese Joint Committee of Societies of Materials
CJF Connecticut Joint Federation [US]
CJP Communication Jamming Processor
CJS Canadian Jobs Studies
CJTC Canada-Japan Trade Council
CK Cask
Check

	Chemical Kinetics
	Creatine Kinase
	Cytokeratin
CKAFS	Cape Kennedy Air Force Station [US]
CKAO	Coalition to Keep Alaska Oil [US]
CKD	Completely Knocked Down
	Count-Key-Data Device
CKDIG	Check Digit
CKMTA	Cape Kennedy Missile Test Annex [of NASA]
CKO	Checking Operator
CKR	Czechoslovakian Koruna [Currency]
CKS	Cell Kinetics Society [US]
cks	centistokes
CKSNI	Cape Kennedy Space Network [of NASA]
CKT	Circuit
CKW	Counterclockwise
CL	Cable Line
	Cable Link
	Carload
	Cathodoluminescence
	Ceiling Level
	Ceiling Limit
	Center Line
	Centerline Lights
	Chemiluminescence
	Claim
	Class
	Clause
	Clear
	Clearance
	Closing
	Closed Loop
	Clutch
	Coil Loading
	Command Language
	Communications Language
	Compatibility List
	Compressed Length
	Computational Linguistics
	Computer Language
	Condenser Lens
	Conference Location
	Confidence Level
	Confidence Limit
	Connecting Line
	Continuous Loading
	Contrast Liquid
	Control Language
	Control Law
	Control Leader
	Conversion Loss
	Coordinated Ligand
	Core Loss
	Corpus Luteum
	Crack Layer Model
	Current Layer
	Current Logic
	Current Loop
	Cutter Location
	Cutting Lubricant
	Cyclic Loading
C/L	Carload
Cl	Chlorine
cL	centiliter [also: cl]
cl	(citato loco) — as mentioned above
CLA	Canadian Labour Association
	Canadian Library Association
	Canadian Lumbermen's Association
	Centerline Average
	Clear and Add
	Communication Line Adapter
	Communication Link Analyzer
	Computer Law Association [US]
	Computer Lessors Association [US]
	Copyright Licensing Agency [UK]
Cla	Caryophanon Latum [Molecular Biology]
CLAC	Certificate in Library and Archives Conservation
	Christian Labour Association of Canada
	Closed Loop Approach Control
CLAF	Centro Latinoamericano de Fisica [= Latin-American Center for Physics, Brazil]
CLAIM	Center for Library and Information Management [UK]
CLAIRA	Chalk Lime and Allied Industries Research Association [now: WHRA, UK]
CLAM	Chemical Low-Altitude Missile
CLAN	Clean Air Society in the Netherlands
CLAR	Clarendon
CLARA	Cornell Learning and Recognizing Automaton [of CU, US]
CLARB	Council of Landscape Architectural Registration Boards [US]
CLAS	Clean Laminated Air-Station
	Computerized Library Acquisitions System
	Computer Library Applications Service
CLASP	Closed Line Assembly for Single Particles
	Composite Launch and Spacecraft Program
	Computer Language for Aeronautics and Space Programming [of NASA]
CLASS	California Libraries Authority for Systems and Services [US]
	Classification
	Closed Loop Accounting for Store Sales
	Composite Laminate Analysis Systems
	Computer-Based Laboratory for Automated School Systems
	Conservation Learning Activities for Science and Social Studies [of NWF, US]
	Cross-Chain Loran Atmospheric Sounding System
	Current Literature Alerting Search Service [US]
CLASSMATE	Computer Language to Aid and Simulate Scientific, Mathematical and Technical Education
CLAT	Communications Line Adapter for Teletype
CLAW	Clustered Atomic Warhead
CLAYCOP	Clay-Supported Cupric Nitrate
CLAYFEN	Clay-Supported Ferric Nitrate
CLB	Central Logic Bus
	Cone Locator Bushing
CLBA	Canadian Log Builder Association
	Canadian Long Baseline Array
CLBDA	Cyprus Land and Building Developers Association
CLC	Canadian Labour Code

Canadian Labour Congress
Caribbean Labour Congress
Central Logic Control
Circle of Least Confusion
Column Liquid Chromatography
Combinational Logic Circuit
Commonwealth Liaison Committee [UK]
Communications Line Control
Communications Link Controller
Constant Light Compensating
Containment Leakage Control
Course Line Computer
Cost of Living Council [US]
CLCA Comite de Liaison de la Construction Automobile [= Liaison Committee for the Motor Industry [of EEC, Belgium]]
CLCCR Comite de Liaison de la Construction de Carrosseries et de Remorques [= Liaison Committee of the Body and Trailer Building Industry, Germany]
CLCGM Closed Loop Cover Gas Monitor
CLCIS Closed Loop Cover and Instrumentation System
CLCM Cooperating Libraries of Central Maryland [US]
CLCOLL Clinch Collar
CLCPE California Legislative Council on Professional Engineers [US]
CLCR Controlled Letter Contract Reduction
CLCS Chinese Language Computer Society
Consequence Limiting Control System
Current Logic, Current Switching
CLCV Cold Leg Check Valve
CLD Called Line
Caprolactam Disulfide
Central Library and Documentation Branch [of ILO, Switzerland]
Chlorine Leak Detector
Cleared
Closed Loop Drive
Cloud
Constant Level Discriminator
Cooled
CLDAS Clinical Laboratory Data Acquisition System
CLDS Canada Land Data System [of Environment Canada]
Closed Loop Drive System
CLE Capillary Liquid Epitaxy
Communications Line Expander
CLEA Canadian Library Exhibitors Association
Conference on Laser Engineering and Applications [of IEEE, US]
CLEAN Commonwealth Law Enforcement Assistance Network
CLEAR Compiler, Executive Program, Assembler Routines
Components Life Evaluation and Reliability
CLECAT Comite de Liaison Europeen des Commissionnaires et Auxiliares de Transport [= European Liaison Committee of Forwarders, Belgium]
CLEFT Cleavage of Lateral Epitaxial Film for Transfer
CLEHA Conference of Local Environment Health Administrators
CLEM Cargo Lunar Excursion Module
Closed Loop Ex-Vessel Machine
Composite for the Lunar Excursion Module
CLEO Clear Language for Expressing Orders
Conference on Lasers and Electrooptics
CLEO/QELS Conference on Lasers and Electrooptics/Quantum Electronics and Laser Science (Conference)
CLEPA Comite de Liaison de la Construction d'Equipements et de Piece d'Automobiles [= Liaison Committee of Manufacturers of Motor Vehicle Parts and Equipment, UK]
CLES Certified Laboratory Equipment Specialist
CLETS California Law Enforcement Telecommunications System
CLF Capacitive Loss Factor
Clear Forward
Cutter Location File
CLFB Canadian Livestock Feed Board
CLFD Classified
CLFMI Chain Link Fence Manufacturers Institute [US]
CLG Calling Line
Catalogue
Ceiling
Controlled Lead Grade (of Zinc)
Cooperative Libraries Group [now: CAG, UK]
CLGW Cement, Lime and Gypsum Workers (International Union)
CLI Calling Line Identification
Canadian Land Inventory
Canadian Lifeboat Institution
Cathodoluminescence Imaging
Coin Level Indicator
Command Language Interpreter
Control Language Interpreter
Cutter Location Information
CLIC Canadian Law Information Council
Command Language for Interrogating Computers [of RRE, UK]
Conversational Language for Interactive Computing
CLIMAP Climate Long-Range Investigation Mapping and Prediction
CLIMATOL Climatological
Climatologist
Climatology
CLIO Conversation Language Input/Output
CLIP Cellular Logic Image Processor
Closed-Loop Incremental Positioner
Compiler Language for Information Processing
Computer Language for Information Processing
Contributions to Laboratory Investigations Program
CLIRA Closed Loop In-Reactor Assembly
CLIS Certificate of Library and Information Science
CLIST Command List
CLIV Cold Leg Isolation Valve
CLJA Closed Loop Jumper Assembly
CLK Clerk
Clock
CLKG Caulking
CLL Central Light Loss
CLM Coal-Liquid Mixture
Communications Line Multiplexer
CLMA Clinical Laboratory Management Association [US]

C/L min Carload, Minimum Weight
CLML Chicago Linear Music Language
CLMPC Canadian Labor Market and Productivity Center
CLNC Clearance
CLNG Cleaning
CLO Closet
Cornell Laboratory of Ornithology [US]
CLOAX Corrugated Laminated Coaxial Cable
CLOB Core Load Overlay Builder
CLOC Collocation
CLOS Closure
CLP Chilean Peso [Currency of Chile]
City of London Polytechnic [UK]
Clamp
Communication Line Processor
Cone Locator Pin
Contract Laboratory Program [of Environmental Protection Agency, US]
Current Line Pointer
ClPh Chlorophenyl
CLR Central Logic Rack
Clean Liquid Radwater
Clear(ance)
Clear Screen
Combined Line and Recording
Computer Language Recorder
Computer Language Research
Constant Load Rupture
Coordinating Lubricants Research
Council of Library Resources [US]
Crack-Length Ratio
Current-Limiting Reactor
Current-Limiting Resistor
CLRA Construction Labour Relations Association
CLRB Canada Labour Relations Board
CLRC Circuit Layout Record Card
Cooperative League of the Republic of China [Taiwan]
CLRI Central Leather Research Institute [UK]
CLRS Clear and Smooth
CLRU Cambridge Language Research Unit [US]
CLRV Canadian Light Rail Vehicle
CLS Canada Land Surveyor
Canadian Lumber Standard
Cask Loading Station
Cathodoluminescence Spectroscopy
Characteristic Loss Spectroscopy
Claims
Clear and Subtract
Clear Screen
Closed Loop System
Coherent-Light System
Common Language System
Communications Line Switch
Comparative Systems Laboratory [US]
Computerized Library System
Concept Learning System
Condenser Lens Supply
Constant Level Speech
Containment Leakage System
Control Launch Subsystem
CLSA Canadian Laboratory Suppliers Association
CLSD Closed
CLSG Closing
CLSI Computer Library Services, Inc. [US]
CLSM Confocal Laser Scan Microscope
CLSMDA Closed Loop System Meltdown Accident
CLSR Computer Law Service Reporter
CLSS Communication Link Subsystem
CLT Central Limit Theorem
Certificate of Laboratory Technology
Clinical Laboratory Technologist
Collect
Communications Line Terminal
Computer Language Translator
Constant Load Tensile
CLU Central Logic Unit
Circuit Line-Up
Creusot-Loire-Uddeholm (Process)
CLUE Comprehensive Loss Underwriting Exchange
CLUT Computer Logic Unit Tester
CLV Constant Linear Velocity
CLV Clevis
CLW Clockwise
College of Librarianship, Wales [UK]
CLWS Coal-Lime-Water Slurry
CM Calculating Machine
Calibrated Magnification
Calibration Magnification
Capacitor Motor
Carboxymethyl
Cards per Minute
Celestial Meridian
Celestial Mechanics
Cell Membrane
Center Matched
Center of Mass
Centimetric
Center of Mass
Central Memory
Centrifugal Moment
Ceramic Material
Certified Master
Chemical Metallurgist
Chemical Metallurgy
Chemical Milling
Chloromethane
Circular Measure
Circular Mil
City Manager
Class Master
Coal Measure
Cold Mix
Color Monitor
Command Module
Common Mode
Communications Multiplexer
Commutating Machine
Commutator Motor
Composite Material
Computational Mathematics
Computational Model

	Computer Module
	Conceptual Modelling
	Condensed Matter
	Configuration Management
	Construction and Machinery
	Consumable Electrode Remelted
	Continuity Message
	Controlled Mine Field
	Controlling Magnet
	Control Mark
	Control Mechanism
	Control Memory
	Control Monitor
	Core Memory
	Corresponding Member
	Coulombmeter
	Countermeasure
	County Municipality
	Cross-Modulation
	Cumulative
	Cylinder Mount
C/M	Communications Multiplexer
	Control Monitor
Cm	Curium
cm	centimeter
c m	circular mil
CMA	Cable Makers Association [UK]
	Calcium Magnesium Acetate
	Calcium Methanearsonate
	Canadian Manufacturers Association
	Canadian Medical Association
	Canadian Metric Association
	Canadian Mineral Analysts
	Canadian Motorcycle Association
	Canadian Museums Association
	Cast Metals Association [US]
	Census Metropolitan Areas
	Certified Management Accountant
	Chartered Mechanical Engineer
	Chemical Manufacturers Association [US]
	Circular Mil Area
	Classified Mail Address
	Closure Manufacturers Association [US]
	Colleges of Mid-America [US]
	Commonwealth Medical Association [US]
	Communications Managers Association [UK]
	Communications Market Association [US]
	Computer-Aided Microanalyzer
	Computer Management Association [US]
	Computer Monitor Adapter
	Consulting Management Engineer
	Contact-Making Ammeter
	Contract Manufacturers Association [US]
	Copper-Magnesium-Aluminum (Alloy)
	Cylindrical Mirror Analyzer
CMa	Canis Major [Astronomy]
CMAA	Construction Management Association of America [US]
	Crane Manufacturers Association of America [US]
CMACL	Composite Mode Adjective Checklist
CMAHK	Chinese Manufacturers Association of Hong Kong
CMANY	Communications Managers Association of New York [US]
CMAP	Central Memory Access Priority
CMAR	Control Memory Access Register
CMARC	Current Machine-Readable Catalogue [of Library of Congress, US]
CMB	Carbolic Methylene Blue
	Concrete Median Barrier
	Corporate Management Branch
CMBES	Canadian Medical and Biological Engineering Society
CMBG	Canadian Mechanized Brigade Group
CMC	Cable Maintenance Center
	Canada Meat Council
	Canadian Manpower Center
	Canadian Meteorological Center
	Canadian Military College
	Carboxymethyl Cellulose
	Ceramic Matrix Composite
	Certified Management Consultant
	Code for Magnetic Characters
	Collective Measures Committee [of UN]
	Command Module Computer
	Communications Channel
	Communications Mode Control
	Compliant Metal/Ceramic
	Computer-Mediated Communications
	Computer Musicians Cooperative [US]
	Contact-Making Clock
	Container Marketing Committee [of AISI, US]
	Coordinating Manual Control
	Critical Micelle Concentration [also: cmc]
	1-Cyclohexyl-3-(2-Morpholinoethyl)carbodiimide
CMCA	Canadian Masonry Contractors Association
	Character Mode Communications Adapter
	Cruise Missile Carrier Aircraft
CMCC	Canadian Museum Construction Corporation
CMCP	Canadian Museum of Contemporary Photography
	Cruise Missile Conversion Project
CMCR	Continuous Melting, Casting and Rolling
CMCS	Canadian Manufacturers of Chemical Specialties Association
	Corporate Materials Catalog System
CMCSA	Canadian Manufacturers of Chemical Specialties Association
CMCTL	Current Mode Complementary Transistor Logic
CMD	Carboxymuconolactone Decarboxylase
	Center for Management Development [of MUN, Canada]
	Cold-Mix Design
	Condensed Matter Division
	Core Memory Driver
	Command
CMDAC	Current Mode Digital-to-Analog Converter
CMDG	Commanding
CMDO	Commissioned Officer [also: CmdO]
CMDR	Commander
	Command Reject
CMDRE	Commodore
CMDS	Centralized Message Data System

	Consortium for Materials Development in Space [of UAH, US]
CME	Central Memory Extension
	Chartered Mechanical Engineers [of IME, UK]
	Chemically-Modified Electrode
	Chicago Mercantile Exchange [US]
	Chloromethyl Ether
	Common Minimum Evaluation
	Computer Measurement and Evaluation
	Concurrent Machine Environment
	Council of Ministers of Education [Canada]
	Cresyl Methylether
	Crucible Melt Extraction
CMEA	Council of Mutual Economic Assistance [also: COMECON]
CMEC	Center for Micro-Engineered Ceramics [of University of New Mexico, US]
	Council of Ministers of Education, Canada
	Crucible Melt Extraction
CMEC	Center for Micro-Engineered Ceramics [of UNM, US]
	Council of Ministers of Education, Canada
CMEQ	Corporation of Master Electricians of Quebec [Canada]
CMER	College of Mineral and Energy Resources [of WVU, US]
CMERI	Central Mechanical Engineering Research Institute [India]
CMF	Cartesian Mapping Function
	Center for Materials Fabrication
	Center for Metals Fabrication [of BMI, US]
	Central Maintenance Facility
	Coherent Memory Filter
	Common Mode Failure
	Composition Modulated Film
	Cross Modulation Factor
	Cymomotive Force [also: cmf]
cmf	circular mil-foot
CMFA	Common-Mode-Failure Analysis
CMfgE	Certified Manufacturing Engineer
CMfgT	Certified Manufacturing Technologist
CMFI	Conversion, Memory and Fault Indication
CMFLPD	Core Maximum Fraction of Limiting Power Density
CMFRI	Central Marine Fisheries Research Institute [India]
CMG	Center for Marine Geology [of Dalhousie University, Canada]
	Color Marketing Group [US]
	Commission for Marine Geology [Germany]
	Computer Management Group
	Computer Measurement Group [US]
	Computer Modeling Group
	Control Moment Gyroscope
CMHC	Canadian Mortgage and Housing Corporation
CMHEC	Carboxymethyl Hydroxyethyl Cellulose
cm Hg	centimeters of mercury [also: cmHg]
CMHI	Canadian Manufactured Housing Institute
CMHS	Canadian Military Historical Society
CMI	Canadian Motor Industries
	Can Manufacturers Institute [US]
	Caribbean Meteorological Institute [Barbados]
	Casting Metals Institute
	CEA (= Core Element Assembly) Motion Inhibit
	Cell-Mediated Immunity
	Cell Multiplication Inhibition
	Christian Michelsen Institute [Norway]
	Code Mark Inversion
	Colorado Mining Association [US]
	Comite Maritime International [= International Maritime Committee, Belgium]
	Commonwealth Mycological Institute [UK]
	Computational Mechanics Institute [UK]
	Computer-Managed Instruction
	Construction Management Institute
	Cultured Marble Institute [US]
CMi	Canis Minor [Astronomy]
CMIA	Coal Mining Institute of America [US]
	Cyprus Metal Industry Association
CMIEE	Canadian Mining and Industrial Equipment Exposition
CMIG	Canadian MAP (= Manufacturing Automation Protocol) Information Group
cmil	circular mil
C-MIRB	Committee on Megaproject Industrial and Regional Benefits
CMITP	Canada Manpower Industrial Training Program
CMJ	Canadian Mining Journal
CML	Cell-Mediated Lympholysis
	Chemical
	Commercial
	Common Mode Logic
	Computer-Managed Learning
	Critical Mass Laboratory
	Current-Mode Logic
CMLCENCOM	Chemical Corps Engineering Command [US]
CMLE	Carboxymuconate Lactonizing Enzyme
CMM	Center for Molecular Medicine [Germany]
	Commission for Marine Meteorology [of WMO, Switzerland]
	Common Memory Manager
	Communications Multiplexer Module
	Component Maintenance and Mockup
	Computerized Modular Monitoring
	Computer Main Memory
	Concentration Module Main
	Coordinate Measuring Machine
	Core Mechanical Mockup
CMMA	Concrete Mixer Manufacturers Association
CMME	Chloromethyl Methyl Ether
CMMF	Component Maintenance and Mockup Facility
CMMI	Council of Mining and Metallurgical Institutions [UK]
CMMMC	Common-Matrix Micromacrocomposite [also: CM^3C]
CMMP	Canada Manpower Mobility Program
	Carnegie Multi-Mini Processor
	Commodity Management Master Plan
	Condensed Matter and Materials Physics (Conference) [UK]
CMMS	Cation Micro Membrane Suppressor
	Computerized Maintenance Management System
CMN	Cerium Magnesium Citrate
	Cerous Magnesium Nitrate

	Commission
CMND	Command
CMO	Carboxymethyl Oxime
	Caribbean Meteorological Organization [Trinidad and Tobago]
	Common Mode Operation
CMOD	Crack Mouth Opening Displacement
CMOP	Canadian Market Opportunities Program
CMOS	Canadian Meteorological and Oceanographic Society
	Complementary Metal-Oxide Semiconductor [also: C/MOS, or COSMOS]
CMOS IC	Complementary Metal-Oxide Semiconductor Integrated Circuit
CMOSL	Complimentary Metal-Oxide Semiconductor Logic
CMOS PLD	Complimentary Metal-Oxide Semiconductor Programmable Logic Device
CMOS-RAM	Complementary Metal-Oxide Semiconductor — Random Access Memory
CMOS/SOS	Complimentary Metal-Oxide Semiconductor/Silicon on Sapphire
CMOU	Canadian Marine Officers Union
CMP	Canadian Mineral Processors [of CIM, Canada]
	Center for Materials Production [of EPRI, US]
	Center for Metal Production
	Ceramic Mold Process
	Coextrusion of Metal and Plastic
	Compounded Mobilized Planes Theory
	Condensed Matter Physics
	Configuration Management Plan
	Controlled Materials Production
	Cooperative Marketing Partner
	Cooperative Marketing Program
	Cooperative Market Participant
	Cytidine Monophosphate
CMPA	Cyprus Master-Printers Association
CMPAA	Certified Milk Producers Association of America [US]
CMP-CIM	Canadian Mineral Processors — Canadian Institute of Mining and Metallurgy
CMPE	Center for Metallurgical Process Engineering [Canada]
CMPF	Core Maximum Power Fraction
CMPLX	Complex
CMPM	Computer-Managed Parts Manufacture
CMPO	Carbomoylmethylphospine Oxide
CMPP	Chloromethylphenoxypropionic Acid
	2-(4-Chloro-O-Tolyloxy)propionic Acid
cmps	centimeter per second
CMPSS	Conference on Molecular Processes on Solid Surfaces
CMPT	Component
CMPTR	Computer [also: CMPT]
CMPX	Computerized Engineering Index [also: COMPENDEX]
CMR	Carbon Nuclear Magnetic Resonance
	CBX Management Reporter
	Centralized Mail Remittance
	Committee on Manpower Resources [US]
	Common-Mode Rejection
	Communications Moon Relay
	Communications Monitoring Report
	Competitive Market Research
	Continuous Maximum Rating
CMRA	Chemical Marketing Research Association [US]
CMRB	California Melon Research Board [US]
CMRC	Coal Mining Research Center [Canada]
CMREF	Committee on Marine Research, Education and Facilities [US]
CMRI	Carnegie Mellon Research Institute [US]
CMRR	Common-Mode Rejection Ratio
CMRS	Cellular Mobile Radio (Telecommunications) Service
	Central Mining Research Station
C-MRS	Chinese Joint Committee of the Materials Research Society
CMS	Calcium-Magnesium Silicate
	California Macadamia Society [US]
	Cambridge Monitor System
	Canadian Micrographics Society [now: CIIMS]
	Canadian Mine Services
	Caribbean Meteorological Service [now: CMO, Trinidad]
	Center for Management Systems [US]
	Center for Mass Spectrometry
	Center for Marine Studies [now: MRD of WINA, US]
	Center for Materials Science [of NIST, US]
	Centrifuge Melt Spinning
	Certificate in Museum Studies
	Certified Metrication Specialist
	Change Management System
	Chapter Management Seminar [of ASM International, US]
	Chloromethylated Polystyrene
	Chrome-Managnese-Silicon (Steel)
	Circuit Maintenance System
	Clay Minerals Society [US]
	Club Micro Son [France]
	Code Management System
	Coincidence Mossbauer Spectroscopy
	Cold Melt Steel
	Combustion and Melting System
	Common Modular Simulator
	Communications Management System
	Compiler Monitor(ing) System
	Computer Management System
	Conservation Materials and Services
	Continuous Mining System
	Conversational Monitor System
	Current-Mode Switching
	Czechoslovak Medical Society
C&MS	Consumer and Marketing Service
CMSA	Canning Machinery and Supplies Association [now: FPM&SA, US]
CMSB	Certified Metrication Specialists Board [of USMA]
CMSCI	Council of Mechanical Specialty Contracting Industries [now: ASC, US]
CMSE	Center for Materials Science and Engineering
CMSS	Controlled Mechanical Storage Systems [now: AS/RS, US]
CMSWAN	Chamber of Mines of South West Africa/Namibia [Namibia]
CMT	Cadmium Mercury Telluride

	Cassette Magnetic Tape
	Center for Management Technology [US]
	Code Matching Technique
	Computer-Managed Training
	Conversational Mode Terminal
	Corrugating Medium-Flat Test
	Council for Mineral Technology [South Africa]
CMTA	Canadian Marine Transportation Administration [of Transport Canada]
	Constant Momentum Transfer Averaging
CMTC	Composite Manufacturing Technology Center [of Alcoa Laboratories, US]
	Coupled Monostable Trigger Circuit
CMTDA	Canadian Machine Tool Distributors Association
CMTI	Celestial Moving Target Indicator
CMTM	Communications and Telemetry
CMTP	Canadian Manpower Training Program
	Continuum Mechanics of Textured Polycrystals
CMTS	Canadian Machine Tool Show (and Exhibition)
	Centralized Maintenance Test System
CMTT	Commonwealth Military Training Team
CMU	Cambridge Management Unit [UK]
	Carnegie-Mellon University [US]
	Central Michigan University [US]
	Chlorophenyldimethylurea
	Computer Memory Unit
	Control Maintenance Unit
CMV	Common-Mode Voltage
	Contact-Making Voltmeter
	Cytomegalovirus
CMVDR	Center of Marine Vessel Design and Research [of TUNS, Canada]
CMVM	Contact-Making Voltmeter
CMVPCB	California Motor Vehicle Pollution Control Board [US]
CMVSA	Canadian Motor Vehicle Safety Act
CMVSS	Canadian Motor Vehicle Safety Standard
CMVTSS	Canadian Motor Vehicle Tire Safety Standard
CMW	Centrimetric Wave
CMX	Character Multiplexer
	Concentration Module Extension
CMY	Cyan/Magenta/Yellow
CN	Canadian National Railways
	Carbon Number
	Cascade Nozzle
	Caudate Nucleus
	Cellulose Nitrate
	Center Notch
	Cetane Number
	Chain
	Chance Notice
	Citation Number
	Class Name
	Close Nipple
	Communication Network
	Company Name
	Computer Network
	Complete Name
	Consignment Note
	Contract Number
	Coordination Number
	Corporate Name
	Country Name
	Credit Note
	Cyanogen
C/N	Carrier-to-Noise (Ratio)
cn	cosine of the amplitude
CNA	Canadian Nuclear Association
	Canadian Nurses Association
	Center for Naval Analyses [US Navy]
	Centre Nucleaire des Ardennes [= Nuclear Center of the Ardennes, Belgium]
	Chemical Notation Association [UK]
	Communications Network Architecture
	Copper Nickel Alloy
	Cosmic Noise Absorption
CNAA	Committee on Nucleation and Atmospheric Aerosols [US]
	Council for National Academic Awards [UK]
CNAC	CCIS (= Common Channel Interoffice Signalling) Network Administration Center
CNAD	Conference on National Armament Directors [of NATO]
CNAS	Chemical Nomenclature Advisory Service [of LGC, UK]
	Civil Navigation Aids System
CNB	Centrale Nucleaire Belge [= Belgian Nuclear Power Station]
	Cyclical Neutral Balance
CNBMDA	Canadian National Building Materials Distribution Association
CNBr	Cyanogen Bromide
CNC	Canadian National Committee
	Carbon-Nitrogen Cycle
	Computerized Numerical Control
	Computer Numerical Control
	Condensation Nuclei Counter
CN-CA	Cellulose Nitrate — Cellulose Acetate
CNCC	Customer Network Control Center
CNC/DNC	Computer Numerical Control/Digital Numerical Control
CNCE	Council for Non-Collegiate Continuing Education
CNC/IAPS	Canadian National Committee for the International Association on the Properties of Steam [also: CNCIAPS]
CNCIAWPRC	Canadian National Committee of the International Association on Water Pollution Research and Control [also: CNC/IAWPRC]
	Chilean National Committee of the International Association on Water Pollution Research and Control
	Cyprus National Committee of the International Association on Water Pollution Research and Control
	Czechoslovakian National Committee of the International Association on Water Pollution Research and Control
CNC/IFAC	Canadian National Committee for the International Federation of Automation Control [also: CNCIFAC]
CNC/IUTAM	Canadian National Committee for the International Union of Theoretical and Applied Mechanics [also: CNCIUTAM]

CNCP Canadian National/Canadian Pacific (Telecommunications)
CN/CPT Canadian National/Canadian Pacific Telecommunications
CNCRM Canadian National Committee on Rock Mechanics
CNCS Central Navigation Control School
CNCT Connect
CNCTRC Concentric
CND Campaign for Nuclear Disarmament [UK]
Condition
Conduit
CNDA Canadian National Distribution Authority
CNDCTN Conduction
CNDCTR Conductor
CNDE Center for Nondestructive Evaluation
CNDL Conditional
CNDL EOP Conditional End of Page
CNDO Complete Neglect of Differential Overlap
CNDP Communication Network Design Program
Continuing Numerical Data Project
CNDS Condensate
CNDST Centre National de Documentation Scientifique et Technique [= National Center of Scientific and Technical Documentation, France]
CNE Canadian National Exhibition
Charge Neutrality Equation
Communications Network Emulator
Compare Numerical Equal
CNEA Comision Nacional de Energia Atomica [= National Commission of Atomic Energy, Argentina]
CNEN Comision Nacional de Energia Nuclear [= National Commission of Nuclear Energy, Mexico]
Comitato Nazionale per Energia Nucleare [= National Committee for Nuclear Energy, Italy]
CNERT Center for Natural Resources, Energy and Transport [of UN]
CNES Centre National d'Etudes Spatiales [= National Center for Space Research, France]
CNET Centre National d'Etudes des Telecommunications [= National Center for Telecommunications Studies, France]
Communications Network
CNEt Cyanoethyl
CNEUPEN Commission Nationale pour l'Etude de l'Utilisation Pacifique de l'Energie Nucleaire [= National Commission for the Study of the Peaceful Uses of Nuclear Energy, Belgium]
CNF Canadian Nature Federation
CNFRL Columbia National Fisheries Research Laboratory [US]
CNFSFACM Committee of the National Ferrous Scrap Federations and Associations of the Common Market [Belgium]
CNG Compressed Natural Gas
CNGA California Natural Gas Association [US]
CNI Center for Nuclear Information [Brazil]
Centre Nucleaire Interescaut [= Interescaut Nuclear Center, Belgium]
Changed Number Interception
Communication, Navigation and Identification
Consiglio Nazionale Ingegneri [= National Engineering Council, Italy]
Consolidated National Intervenors
CNIC Centre National d'Information Chimique [= National Center for Chemical Information, France]
Consorcio Nacional de Industriales del Caucho [= National Consortium of Rubber Industries, Spain]
CNIL Cumulative Net Investment Loss
CNIM Centro Nacional de Investigaciones Metalurgicas [= National Metallurgical Research Center, Spain]
CNIPA Committee of National Institutes of Patent Agents [UK]
CNL Circuit Net Loss
Constant Net Loss
CNLA Council on National Library Associations
CNLIA Council of National Library and Information Associations [US]
CNMA Communications Network for Manufacturing Applications [Europe]
CNMB Central Nuclear Measurements Bureau
CNMI Communications Network Management Interface
CNMP Conference on National Materials Policy [of FMS, US]
C-NMR Carbon Nuclear Magnetic Resonance
CNN Cable News Network [US]
CNOGEDC China National Oil and Gas Exploration Development Corporation
CNOOC China National Offshore Oil Corporation
CNOP Conditional Non-Operational
CNP Celestial North Pole
Communications Network Procedure
Communications Network Processor
CNPFT Centro Nacional de Pesquisa de Fruteiras de Clima Temperado [= National Research Center of Temperate Weather Fruit Trees, Brazil]
CNPI Comision Nacional de Productividas Industrial [= National Commission of Industrial Productivity, Spain]
CNPPA Commission on National Parks and Protected Areas [Canada]
CNR Canadian National Railway
Carrier-to-Noise Ratio
Composite Noise Rating
Consiglio Nazionale delle Ricerche [= National Research Council, Italy]
CNRC Canadian National Research Council
CNRM Centre National de Recherche Metallurgiques [= Center for Metallurgical Research, Belgium]
CNRN Comitato Nazionale per le Ricerche Nucleare [= National Committee for Nuclear Research [now: CNEN, Italy]]
CNRS Center National de Recherche Scientifique [= National Center for Scientific Research, France]
CNRS-ONERA CNRS Office National d'Etudes et de Recherches Aerospatiales [= National Office for Aerospace Studies and Research, France]
CNS Canadian Navigation Society
Canadian Nuclear Society
Center for Northern Studies [Canada]
Central Navigation School
Commodity News Service [US]

	Communications Navigation and Surveillance System
	Communications Network Service
	Communications Network Simulation
CNSC	Carrying Nuclear-Strike Cruiser
CNSF	Cornell National Supercomputer Facility [US]
CNSI	Canadian Newspaper Services International
CNSL	Counsel
CNSLD	Consolidated
CNSN	Canadian National Seismograph Network
CNSS	Center for National Space Study
CNT	Canadian National Telecommunications
	Celestial Navigation Trainer
	Count(er)
CNTA	Council for Nordic Teachers Associations [Denmark]
CNTL	Control
CNTOR	Contractor
CNTP	Committee for a National Trade Policy
CNTR	Container
	Contract
	Counter
CNTRL	Central
	Control(ler)
CNTRLS	Controllers
	Controls
CNTU	Confederation of National Trade Unions [Canada]
CNU	Compare Numeric Unequal
CNUCE	Centro Nazionale Universitario di Calcalo Elettronica [= National University Center for Electronic Calculation, Italy]
CNUJ	Committee for Nordic Universities of Journalism [Sweden]
CNV	Contingent Negative Variation
	Convert
CNVR	Conveyor
CNVT	Convert
CNVTR	Converter
CNY	Chinese Yuan [Currency of PR China]
CO	Carbonyl
	Center of Oscillation
	Central Office
	Chairman of Operations
	Change Order
	Change-Over
	Chief Operator
	Circuit Order
	Cleanout
	Close-Open Operation
	College of Aeronautics
	Collonial Office
	Colorado [US]
	Combined Operations
	Commanding Officer
	Communications Officer
	Company
	Conscientious Objector
	Conservation Officer
	Contracting Officer
	County
	Crystal Oscillator
	Cut Out
C/O	Carbon/Oxygen Ratio
	Cash Order
C-O	Condenser-Objective
Co	Cobalt
c/o	care of
	carried over
	cash order
COA	Certified Office Administrator
	College of Aeronautics [UK]
	Conversion of Acetyl
	Council of Agriculture [Taiwan]
CoA	Coenzyme A
COACH	Canadian Organization for Advancement of Computers in Health
COAL PRO	Coal Research Project
COAM	Coaming
	Customer Owned and Maintained (Equipment)
COAMS	Canadian Ocean Acoustic Measurement System
COAT	Coherent Optical Array Technique
	Corrected Outside Air Temperature
COAX	Coaxial (Cable) [also: coax]
COB	Close of Business
	Cut-Off Bias
CoB	Coenzyme B
COBESTCO	Computer-Based Estimating Technique for Contractors
COBI	Coded Biphase
COBIS	Computer-Based Instruction System
COBLIB	COBOL Library
COBLOC	CODAP (= Control Data Assembly Program) Language Block-Oriented Compiler
COBLOS	Computer-Based Loans System
COBOL	Common Business-Oriented Language [also: Cobol]
CO-BR	Butadiene Rubber based on Cobalt Catalyst
COBRA	Cosmic Background Radiation
COBSA	Computer Service Bureaus Association [UK]
COC	Carbon Oxide Chemisorption
	Chamber of Commerce
	Chapter Operations Committee [of ASM International, US]
	Cleveland Open Cup
	Coal Operators Conference [of CIM, Canada]
	Coded Optical Character
	Compiler Object Code
	Copper Oxychloride
COCA	Council of Ontario Contractors Associations [Canada]
COCAM	Canadian-Ontario Center for Advanced Manufacturing
COCAST	Colonial Colleges of Arts, Science and Technology [UK]
COCATRAM	Comision Centroamericana de Transporte Maritimo [= Central American Commission of Maritime Transport, El Salvador]
COCC	Character Oriented Communications Controller
COCESNA	Central American Corporation for Air Navigation Services
COCI	Consortium on Chemical Information [UK]
COCO	Committee Code

(Word) Count and Concordance Generator on Atlas [UK]
COCOT Coin-Operated Customer-Owned Telephone
COCD Canadian Ownership and Control Development (Program)
COCODE Compressed Coherency Detection
COCOM Coordinating Committee
CoCom Coordinating Committee for Multilateral Export Controls [also: COCOM]
COCORP Consortium for Continental Reflection Profiling [US]
COCRIL Council of City Research and Information Libraries [UK]
COCS Copper Oxychloride Sulfate
COD Carrier Onboard Delivery
Carrier-Operated Device
Cash on Delivery
Chemical Oxygen Demand
Clean-Out Door
Coefficient of Diffusion
Coefficient of Draft
Crack Opening Displacement
Cyclooctadiene
CODAC Coordination of Operating Data by Automatic Computer
CODAG Combined Diesel and Gas
CODAN Carrier-Operated Device Antinoise
Coded Weather Analysis [of US Navy]
CODAP Control Data Assembly Program
CODAR Coastal Ocean Dynamics Application Radar
Coherent Display Analyzing and Recording
Correlation, Detection, and Ranging [also: Codar or codar]
Correlation Display Analyzing and Recording
CODAS Control and Data Acquisition System
Customer-Oriented Data Acquisition System
CODASYL Conference on Data System Languages [US]
CODATA Committee on Data for Science and Technology [of ICSU, France]
CODAZR Committe on Desert and Arid Zones Research [US]
CODC Canadian Oceanographic Data Center
CODE Canadian Organization for Development through Education
COBOL Program Development
CODEC Coder-Decoder
CODECA Corporation for Economic Development in the Caribbean
CODED Computer Design of Electronic Devices
CODELACS Computer Developments Limited Automatic Coding System
CODES Computer Design and Education System
Computer Design and Evaluation System
CODEVER Code Verification
CODEX Codex Alimentarius Commission
CODF Crystallite Orientation Distribution Function
CODIA Colegio Dominicano de Ingenieros, Arquitectos y Agrimensores [= Dominican College of Engineers, Architects and Surveyors, Puerto Rico]
CODIC Color Difference Computer
Computer-Directed Communications
CODIL Content Dependent Information Language
Control Diagram Language
CODIPHASE Coherent Digital Phased Array System
CODIS Controlled Digital Simulator
CODIT Computer Direct to Telegraph
CODOC Cooperative Documents Project [of OULCS, Canada]
CODOG Combined Diesel or Gas
CODORAC Coded Doppler Radar Command
CODSIA Council of Defense and Space Industry Associations [US]
CODZR Committee on Desert and Arid Zone Research
COE Cab over Engine
Central Office Equipment
Coefficient of Elasticity
Coefficient of Expansion
COEC Council Operations and Exercises Committee [of NATO]
COED Char Oil Energy Development
Computer-Operated Electronic Display
COEES Central Office Equipment Engineering System
COEF Coefficient
CO_2-EOR Carbon Dioxide Enhanced Oil Recovery
COER Central Office Equipment Report
COESA Committee on Extension to the Standard Atmosphere [US]
COF Canadian Order of Foresters
Cause of Failure
Coefficient of Friction
Confusion Signal
Correct Operation Factor
Curvature of Field
Cut-Off Frequency
COFAB Committee on Finance and Banking [of ASEAN]
COFAF Committee on Food, Agriculture and Forestry [of ASEAN]
COFC Container on Flat Car
COFEB Confederation of European Bath Manufacturers [Scotland]
C of F Cost of Freight
COFFI Communications Frequency and Facility Information System [of ICAO, UN]
COFI Committee of Fisheries [of FAO — UN]
Council of (British Columbian) Forest Industries [Canada]
COFIL Core File
COFIS Canadian On-Line Financial Information Service
COFO Committee on Forestry [of UN]
CofU Council of Universities
COFRDA Canada-Ontario Forest Resource Development Agreement
COG Central Office Ground
Chip-on-Glass
Commentary Graphics
Composite Group
Computer Operations Group
COGAG Combined Gas and Gas
COGB Certified Official Government Business
COGD Circulator Outlet Gas Duct
COGECA Comite General de la Cooperation Agricole de la Communaute Europeenne [= General Committee of Agricultural Cooperation in the European Community, Belgium]

COGENE Committee on Genetic Engineering [of ICSU, France]
COGENT Compiler and Generalized Translator
Compiler, Generator and Translator
COGEO Coordinate Geometry
COGEODATA Committee on the Storage, Automatic Processing and Retrieval of Geological Data [of GSC, Canada]
COGLA Canada Oil and Gas Lands Administration
COGO Coordinate Geometry-Oriented Program
COGPE Canadian Oil and Gas Property Expense
COGR Council on Governmental Relations [US]
COGS Concordance Generation System [of U of T, Canada]
Continuous Orbital Guidance Sensor
Continuous Orbital Guidance System
COGSME Composite Group of SME (= Society of Manufacturing Engineers) [US]
COH Carboxyhemoglobin
Coefficient of Haze
COHA Canadian Occupational Health Association
COHN Certified Occupational Health Nurse
COHO Coherent Oscillator [also: Coho, coho]
COI Central Office of Information [UK]
Communication Operation Instructions
CO-I Co-Investigator
CoI Coenzyme I (= One)
COIA Canadian Ocean Industries Association
Conservative Orthopedics International Association [UK]
COIC Canadian Oceanographic Identification Center
COID Council of Industrial Design [UK]
CoII Coenzyme II (= Two)
COIME Committee on Industry, Minerals and Energy [of ASEAN]
COIN Coal Oxygen Injection Process
COBOL Indexing and Maintenance Package
Coin-Phone Operational and Information Network (System)
Committee on Information Needs [US]
Complete Operating Information
Computer and Information
Computerized Ontario Investment Network [Canada]
COINC Coincidence
COINIM Center for Information on Standardization and Metrology [of PCSMQC, Poland]
COINS Computer and Information Sciences
Counter-Insurgency
Communications Oriented Language
Computerized Information System
Computer-Oriented Language
Control in Information Systems
COL Character Outline Limit
Collect(ion)
College
Colombia(n)
Colonel
Colonial
Colony
Color
Column
Computer-Oriented Language
Cost of Living
Col Colorado [US]
Columba [Astronomy]
COLA Cooperation in Library Automation
Cost-of-Living Allowance
COL-AMCHAM Colombian-American Chamber of Commerce [Colombia]
COLASL Compiler/Los Alamos Scientific Laboratory [US]
COLC Circle of Least Confusion
COLIDAR Coherent Light Detecting and Ranging
COLINET College Libraries Information Network
COLINGO Compile On-Line and Go
COLIPA Comite de Liaison des Associations Europeennes de l'Industrie de la Parfumerie, des Produits Cosmetiques et de Toilette [= European Federation of Perfume, Cosmetics and Toiletries Industries, Belgium]
COLL Collateral
Collator
Collection
Collector
College
Colloid(al)
COLLAT Collateral
COLLOQ Colloquium
Colo Colorado [US]
colog cologarithm
COLORL Colorless
COLP Center for Oceans Law and Policy [US]
COLR Circuit Order Layout Record
COLRS CLASS On-Line Reference Service [US]
COL-SERGT Colonel-Sergeant
COLSS Core Operating Limits Supervisory System
COLT Communication Line Terminator
Computerized On-Line Testing
Computer Oriented Language Translator
Control Language Translator
Council on Library-Media Technical Assistants [US]
COM Cassette Operating Monitor
Command
Commander
Commerce
Commercial
Commission(er)
Committee
Commodore
Common
Communication
Community
Computer Output Microfiche
Computer Output Microfilm(er)
Computer Output Microform
Computer (on)to Microfiche
Computer (on)to Microfilm
Conference of Metallurgists [of CIM, Canada]
Com Coma Berenices [Astronomy]
COMA Canadian Oilfield Manufacturers Association
Computer Operations Management Association [US]
COMAC Continuous Multiple-Access Collator

COMAL Command Algorithmic Language
Common Algorithmic Language [Computer Language]
COMANSEC Computation and Analysis Section [of DRB, Canada]
COMAPS Commons Automated Publishing System [of the Canadian Parliament]
COMAR Committee on Man and Radiation [of IEEE, US]
Computer Aerial Reconnaissance
COMAS Computerized Maintenance and Administration Support
COMAT Computer-Assisted Training
COMB Combination
Combining
Combustion
Console-Oriented Model Building
COMBATEX Combat Exercises
COMBO Combination Support Chip
COMBOIS Communaute d'Interet Europeenne des Grossistes de Machine a Bois [= Community of Interest of Leading European Woodworking and Wood Processing Machinery Resellers, Germany]
COMCANLANT Command of the Canadian Atlantic
COMCM Communication Countermeasures
COMCO Computerized Operation and Maintenance Concept
COMDA Canadian Office Machinery Dealers Association
COMDEX Computer Dealer's Exposition [US]
COMDG Commanding
COMDP Cooperative Overseas Market Development Program
COMDR Commander
COMDT Commandant
COMECON Council for Mutual Economic Assistance/Aid [also: Comecon, USSR]
COMEINDORS Composite Mechanized Information and Documentation Retrieval System
COMEL Coordinating Committee for Common Market Associations of Manufacturers of Rotating Electrical Machinery [UK]
COMEPP Cornell Manufacturing Engineering and Productivity Program [US]
COMER College of Mineral and Energy Resources
COMESA Committee on Meteorological Effects of Stratospheric Aircraft [UK]
COMET Commercial Experiment Transporter [of NASA]
Computer Message Transmission
Computer-Operated Management Evaluation Technique
Council on Middle East Trade [UK]
COMETEC-GAZ Comite d'Etudes Economiques de l'Industrie du Gaz [= Economic Research Committee of the Gas Industry, Belgium]
COMETS Comprehensive On-Line Manufacturing and Engineering Tracking System
COMEX Commodity Exchange [US]
Computer and Office Machinery Exhibition
COMEXT Community External Trade Statistics [of EEC, Belgium]
COMFET Conductively-Modulated Field-Effect Transistor
COMFOR Commercial Wire Center Forecast Program
COMIT Compiler, Massachusetts Institute of Technology [US]
Computing System, Massachusetts Institute of Technology [US]
COMITEXTIL Coordinating Committee for the Textile Industries [of EEC, Belgium]
COML Commercial [also: COM'L]
Commercial Language
COMLA Commonwealth Library Association [Jameica]
COMLAB Commerce Laboratory
COMLO Compass Locator
COMLOGNET Combat Logistics Network
Communications Logistics Network
COMM Commentary
Commerce
Commercial Mission
Commission
Committee
Communications
Communications Society [of IEEE, US]
Commutator
COMMANDS Computer-Operated Marketing, Mailing and News Distribution System
COMMCEN Communications Center
COMMEN Compiler Oriented for Multiprogramming and Multiprocessing Environments
COMMEND Computer-Aided Mechanical Engineering Design System
COMM'L Commercial
COMM MUX Communications Multiplexer
COMMN Commission
COMMPUTE Computer-Oriented Music Materials Processed for User Transformation or Exchange [of SUNY, US]
COMMS Central Office Maintenance Management System
Communications Interface
COMM-STOR Communications Storage Unit
COMMUN Communication
COM/NAV Communication and Navigation Aids
COMNET Communications Network
Computer Network
International Network of Centers for Documentation and Communication Research and Policies [France]
COMP Comparative
Comparison
Compatibility
Compensation
Composition
Compound
Compression
Computer
Computer Society [of IEEE]
COMPAC Commonwealth Pacific Cable
Computer Output Microfilm Package
Computer Program for Automatic Control
COMPACT Compatible Algebraic Compiler and Translator
Computerization of World Facts [of Stanford Research Institute, US]
Computer Planning and Control Technique
Computer-Programmed Automatic Checkout and Test System
COMPANDER Compressor-Expander

COMPARE Computerized Performance and Analysis Response Evaluator
Console for Optical Measurement and Precise Analysis of Radiation from Electronics
COMPAS Committee on Physics and Society [of AIP, US]
Computer Oriented Metering, Planning and Advisory System
COMPASS Compiler-Assembler
Comprehensive Assembly System
Computer-Assisted Classification and Assignment System
COMPAY Computerized Payroll System
COMPCON Computer Convention [of IEEE, US]
COMPD Compound
COMPEL Compute Parallel
COMPENDEX Computer Engineering Index
Computerized Engineering Index [also: CMPX]
Corporate Engineering Data Exchange
COMPL Complement
Completion
COMPN Composition [also: COMPO]
COMPO Board Composition Board [also: compo board]
COMPO Image Composition Image
COMPOLE Commutating Pole [also: compole]
COMPOOL Communications Pool
COMPOSE Computerized Production Operating System Extension
COMPOW Committee on Professional Opportunities for Women
COMPR Compressor
Compreg Compregnated Wood [also: compreg]
COMPRESS Computer Research, Systems and Software
COMPROC Command Processor
COMPROG Computer Program
COMPS Commercial Materials Processing Support Facility
Computers
COMPSAC Computer Software and Applications Conference
COMP SCI Computer Science
COMPSO Computer Software and Peripherals Show
COMPSTAT European Meeting on Computational Statistics [of IASC, Netherlands]
COMPT Compartment
COMPTEL Competitive Telecommunications Association [US]
COMPUNICATIONS Computers and Communications [also: compunications]
COMR Commissioner
COMRADE Computer-Aided Design Environment
COMS Canadian Ocean Mapping System [of DFO]
COM-SAC Computer Security, Auditing and Control
COMSAT Communications Satellite Corporation [US]
COMSCAM Combat Scenario Assessment
COMSEC Communications Security
Communications Security Association [US]
COMSL Communication System Simulation Language
COMSOAL Computer Method of Sequencing Operations for Assembly Lines
COMSOC Communications Spacecraft Operation Center
COMSTAR Internal Communications Satellites [US]
COMSYL Communications System Language
COMSYS Communication System
COMT Catechol-O-Methyltransferase
COMTASS Compact Towed Sonar System
COMTEC Computer Micrographics and Technology Group [US]
COMTRAN Commercial Translator
COMZ Communications Zone
CON Conclusion
Consul
Contactor
Contrast
Control(ler)
con contra
ConA Concanavalin A
CONACYT Consejo Nacional de Ciencia y Tecnologia [= National Council for Science and Technology, Mexico]
CONAT Concrete Articulated Tower
CONBAT Converted Battalion Antitank
CONC Concentration
Concentric(ity)
Concrete
CONC B Concrete Block
CONC C Concrete Ceiling
CONCD Concentrated [also: CONC'D]
CONC F Concrete Floor
CONCH Conchological
Conchologist
Conchology
CONCN Concentration
CONCORD Concordance
COND Condenser
Condensing
Condition
Conduct
Conductivity
Conductor
CONDUIT Computers at Oregon State University, North Carolina Educational Computing Services, Dartmouth College, Iowa State University and University of Texas at Austin [US Computer Network]
CONECS Connectorized Exchange Cable Splicing
CONELRAD Control of Electromagnetic Radiation
CONF Conference
conf (confer) — compare
CONFIDAL Conjugate Filter Data Link
CONFIG Configuration
CONFLEX Conditional Reflex
CONG Congress
CONHAN Contextural Harmonic Analysis
CONICET Consejo Nacional de Investigaciones Cientificas y Tecnicas [= National Council for Scientific and Technical Research, Argentina]
CONICIT Consejo Nacional de Investigaciones Cientificas y Tecnologicas [= National Council for Scientific and Technological Research, Costa Rica]
CONIO Console Input/Output
CONIT Connector for Networked Information Transfer
Conj Conjugate
CONLIS Committee on National Library Information Systems [US]
CONN Connection

	Connector
Conn	Connecticut [US]
CONN DIA	Connection Diagram
CO/NO	Current Operator/Next Operator
CONRAD	Committee on Radiology
	Contraceptive Research and Development Program [AID, US]
CONS	Carrier-Operated Noise Suppression
	Console
	Consolidated
	Consul
CONSAL	Congress of Southeast Asian Librarians [Philippines]
CONSER	Conversion of Serials Project [Canada and US]
CONSLT	Consultant
ConSoc	Conservation Society [UK]
CONSOL	Consolidated
CONSORT	Conversation System with On-Line Remote Terminals
CONSR	Consumer
CONST	Construction
	Constant
CONSUL	Control Subroutine Language
CONT	Contact
	Content
	Continent
	Continuation
	Continue
	Contract
	Control(ler)
Contact EMF	Contact Electromotive Force [also: contact emf]
Contact PD	Contact Potential Difference [also: contact pd]
CONTAM	Contamination
CONT'D	Continued [also: CONTD]
CONTG	Containing
	Contracting
CONT HP	Continental Horsepower
CONTR	Container
	Contract
	Contraction
	Contractor
	Control(ler)
CONTRAIL	Condensation Trail [also: contrail]
CONTRAN	Control Translator
CONTRANS	Conceptual Thought, Random Net Simulation
CONTREQS	Contingency Transportation Requirements System
CONTR O	Contracting Officer
CONTROL	Controller
CONT SYS	Control Systems
CONUS	Contiguous United States
	Continental United States
CONV	Conversion
	Converter
CONVT	Convert
COO	Chairman of Operations
	Chief Operating Officer
COOKI	Coordinated Keysort Index
COOL	Checkout-Oriented Language
	Control-Oriented Language
CO-OP	Cooperation [also: COOP, Co-op or Coop]
	Cooperative [also: COOP, Co-op or Coop]
COOP M	Cooperative Measures
COORD	Coordination
	Coordinator
COOSRA	Canadian Offshore Oilspill Research Association
COP	Change Over Point
	Code of Practice
	Coefficient of Performance
	Colombian Peso [Currency]
	Common On-Line Package
	Copmmon Operational Concept
	Communication Output Printer
	Computer Operating Properly
	Computer Optimization Package
	Continuous Optimization Program
	Copper
CoP	Colombian Peso [Currency]
COPA	Canadian Office Products Association
	Copolyamide (Elastomer)
	Council on Post-Secondary Accreditation [US]
COPAC	Committee for the Promotion of Aid to Cooperatives [Italy]
	Community Patent Appeal Court [of EEC, Belgium]
	Continuous Operation Production Allocation and Control
COPAFS	Council of Professional Associations of Federal Statistics [US]
COPAG	Collision Prevention Advisory Group [of FAA, US]
COPAN	Command Post Alerting Network
COPANT	Comision Panamericana de Normas Tecnicas [= Pan-American Standards Commission, Argentina]
COPAR	Cooperative Preservation of Architectural Records [US]
COPE	Committee on Political Education [of AFL-CIO, US]
	Communications-Oriented Processing Equipment
	Console Operator Proficiency Examination
	Consumer-Optimized Product Engineering
	Copolyester
COPEP	Committee on Public Engineering Policy [of NAE, US]
COPES	Conceptually-Oriented Program in Elementary Science
	Consumer-Optimized Product Engineering System
COPESCAL	Comision de Pesca Continental para America Latina [= Commission for Inland Fisheries of Latin America, Italy]
COPI	Computer-Oriented Programmed Instruction
	Cooperative Projects with Industry
COPICS	Communications-Oriented Production Information and Control System
COPIS	Communication-Oriented Production Information System
COPOL	Copolarization [also: CO-POL]
	Council on Polytechnic Librarians [UK]
COPOLCO	Consumer Policy Committee [of ISO, Switzerland]
COP PL	Copper Plate
COPPS	Committee on Power Plant Siting [US]
COPR	Center for Overseas Pest Research [UK]
COPROD	Coproduct [also: CO-PROD]
COPS	Canadian Occupational Projections System
	Command Operations

	Component Placement System
COPSS	Committee of Presidents of Statistical Societies [US]
COPSTOC	Cornell Program for the Study of the Continents [US]
COPUL	Council of Prairie University Libraries
COPUOS	Committee on the Peaceful Uses of Outer Space [of UN]
COPUS	Canadian Organization of Part-Time University Students
COPY	Copyright
CoQ	Coenzyme Q
COR	Canadian Ownership Regulations
	Carrier-Operated Relay
	Coefficient of Reduction
	Coefficient of Restitution
	Conditioned Orientation Reflex
	Contracting Officer's Representative
	Contract Organization Review
	Cordoba [Currency of Nicaragua]
	Corner
	Correction
	Correspondence
	Corrosion
	Corrosive
CORA	Coherent Radar Array
	Conditioned Reflex Analog
	Conditioned Response Analog (Machine)
CORAD	Correlation Radar
CORAE	Chief of Operational Research and Analysis Establishment
CORAL	Class-Oriented Ring Associated Language
	Command Radio Link
	Computer On-Line Real-Time Applications Language
	Correlated Radio Link
CORAPRAN	Cobelda Radar Automatic Preflight Analysis
CORC	Cornell Computing Language [of CU, US]
CORDE	Cooperative Research, Development and Education (Project) [of FEP, Botswana]
CORDIC	Coordinate Rotation Digital Computer
CORDIS	Cold Reflex Discharge Ion Source
CORDPO	Correlated Data Printout
CORDPO-SORD	Correlated Data Printout — Separation of Radar Data
CORDS	Coherent-On Receive Doppler System
	Coordination of Record and Database System
CORE	Canadian Offshore Resources Exposition
	Coherent-On-Receive
	Cold Ocean Resources Engineering
	Committee on Research Expenditure [of CSIR, South Africa]
	Computer-Oriented Reporting Efficiency
CORELAP	Computerized Relationship Layout Planning
COREP	Combined Overload Repair Control
CORF	Committee on Radio Frequencies
CORKS	Computer-Oriented Record Keeping System
Corn	Cornwall [UK]
CORNET	Corporation Network
CORODIM	Correlation of the Recognition of Degradation with Intelligibility Measurements
CORP	Corporal
	Corporation
CORR	Correction
	Correspondence
	Correspondent
	Corrugation
CORREGATE	Correctable Gate
CORS	Canadian Operational Research Society
CORSA	Cosmic Radiation Satellite [Japan]
CORSAIR	Computer-Oriented Reference System for Automatic Information Retrieval
CORTECH	Council on Research and Technology [US]
CORTEX	Communications-Oriented Real-Time Executive
CORTS	Canada Ontario Rideau Trent Severn [Canada]
	Convert Range Telemetry Systems
COS	Cash on Shipment
	Cassette Operating System
	Centered Optical System
	Change of Subscribers
	Class of Service
	Code Operated Switch
	Communications Operating System
	Communications-Oriented Software
	Compact Operating System
	Companies
	Compatible Operating System
	Computer Operating System
	Concurrent Operating System
	Contactor Starting
	Cooper Ornithological Society [US]
	Core Operating System
	Corporation for Open Systems International [US]
	Cosmic Rays and Trapped Radiation Committee [of ESRO, France]
	Cray Operating System
cos	cosine
$\cos^{-1}$	anticosine; inverse cosine
COSA	Canadian Oil Scouts Association
	Carwash Owners and Suppliers Association [US]
COSAC	Computing Systems for Air Cargo
COSAG	Combination of Steam and Gas
$Co(SALEN)_2$	Bis(salicylidene)ethylenediamineocobalt (II)
COSAM	Cobol Shared Access Method
	Conservation of Strategic Aerospace Materials
	Co-Site Analytical Model
COSA NOSTRA	Computer-Oriented System and Newly Organized Storage-to-Retrieval Apparatus
COSAP	Cooperative On-Line Serials Acquisition Project
COSAR	Compression Scanning Array Radar
COSATI	Committee on Scientific and Technical Information [of FCST, US]
COSB	Canadian Organization of Small Business
COSBA	Computer Services and Bureau Association [UK]
COSBE	Committee for Small Business Exports [US]
COSC	Cambridge Overseas School Certificate [Lesotho]
COSCL	Common Operating System Control Language [CODASYL Committee]
COSEC	Coordinating Secretariat of National Unions of Students [Netherlands]
	Culham On-Line Single Experimental Console
cosec	cosecant

cosh	hyperbolic cosine
$cosh^{-1}$	inverse hyperbolic cosine
COSI	Committee on Scientific Information [US]
COSIE	Commission on Software Issues in the Eighties
COSINE	Committee on Computer Science in Electrical Engineering Education [of ACM, US]
COSIP	College Science Improvement Program [of NSF, US]
COSLANE	Constant Optimum Separation Lane
COSMAR	Committee on Surface Mining and Reclamation
COSMAT	Committee on the Survey of Materials Science and Engineering [US]
COSMIC	Common System for Main Interconnecting
	Computer Software Management and Information Center [of University of Georgia, US]
COSMOG	Cosmogonical
	Cosmogonist
	Cosmogony
	Cosmographer
	Cosmographical
	Cosmography
COSMOL	Cosmological
	Cosmologist
	Cosmology
COSMON	Component Open/Short Monitor
COSMOS	Coastal Multidisciplinary Oceanic System
	Complementary Metal-Oxide Semiconductor [also: CMOS or C/MOS]
	Computer-Oriented System for Management Order Synthesis
	Computer System for Mainframe Operations
	Console-Oriented Statistical Matrix Operator System
	Consortium of Selected Manufacturers Open Systems
	Cornell Simulator of Manufacturing Operations
COS/MOS	Complementary-Symmetry Metal-Oxide Semiconductor
COSOS	Conference on Self-Operating Systems [US]
COSP	Canadian Oil Substitution Program
	Central Office Signaling Panel
COSPAR	Committee on Space Research [of ICSU, France]
COSPAS	COSPAR Satellite
COSPAS/SARSAT	COSPAR Satellite/Search and Rescue Satellite
COSPEAR	Committee on Space Programs for Earth Observation [of NASA]
COSPUP	Committee on Science and Public Policy [of National Academy of Sciences, US]
COSR	Committee on Space Research [US]
COSRIMS	Committee on Research in the Mathematical Sciences [of NAS, US]
COSRO	Conical Scan On Receive Only
COSS	Computer-Optimized Storehouse System
COSSA	CSIRO Office for Space Science Applications [Australia]
COST	Cooperation in Science and Technology (Project) [Europe]
COSTAR	Conversational On-Line Storage and Retrieval
COSTDC	Committee on Science and Technology in Developing Countries [US]
COSTHA	Council on the Safe Transportation of Hazardous Articles [US]
COSTI	Center on Scientific and Technical Information [Israel]
	Committee on Scientific and Technical Information [US]
COSTPRO	Canadian Organization for the Simplification of Trade Procedures
COSWA	Conference on Science and World Affairs
COSY	Compiler System
	Compressed Symbolic Source Language
	Correction System
	Correlation Spectroscopy
COSY NMR	Correlation Spectroscopy Nuclear Magnetic Resonance
COT	Center for Office Technology [US]
	Continuity Signal
	Cotter
	Cotton
	Create Occurrence Table
	Customer-Oriented Terminal
	Cyclooctatetraene
cot	cotangent
cot^{-1}	inverse cotangent
COTAC	Committee on Transport and Communications [of ASEAN]
COTAR	Correlated Orientation Tracking and Ranging [also: Cotar or cotar]
	Correlation Tracking and Ranging [also: Cotar or cotar]
COTAR-AME	COTAR Angle Measuring Equipment
COTAR-DAS	COTAR Data Acquisition System
COTAR-DME	COTAR Distance Measuring Equipment
COTAT	Correlation Tracking and Triangulation [also: Cotat or cotat]
COTC	Canadian Officers Training Corps
	Canadian Overseas Telecommunications Corporation
coth	hyperbolic cotangent
$coth^{-1}$	inverse hyperbolic cotangent
COTI	Coordinator of Operational Technical Investigations [of Canadian Armed Forces]
COTM	Customer Owned and Telephone Company Maintained
COTR	Contracting Officer's Technical Representation
COTRAN	COBOL-to-COBOL Translator
COTT	Committee on Trade and Tourism [of ASEAN]
COT WEB	Cotton Webbing
COU	Council of Ontario Universities [Canada]
COUD-I	Collectors of Unusual Data — International [US]
coul	coulomb
COUN	Council
COUPLE	Communications-Oriented User Programming Language
COUSA	Confederation of Ontario University Staff Associations
COV	Coefficient of Variation
	Coefficient of Viscosity
	Covariance
	Cover
	Cross-Over Value
	Cut-Off Voltage

COVA Computer Code Validation
COVAR Covariance
COVER Cut-Off Velocity and Range
covers coversed sine
COVET Cooperation Venture in the Education of Teachers [Canada]
COVOA Canadian Offshore Vessel Operators Association
COW Crude Oil Washing
COWAR Coordinating Committee on Water Research [France]
COWL Cowling
COWLIS Coastal Ocean Water Level Information System [also: Cowlis]
COWPS Council on Wage and Price Stability
COZI Communications Zone Indicator
CP Calendar Process
Call Processor
Canadian Pacific
Canadian Press
Candlepower
Carbon Process
Card Punch
Carriage Paid
Cars of the Past [US]
Casing Pressure
Catalytic Process
Cathodic Protection
Cathodophosphorence
Cationic Polymerization
Cellulose Propionate
Cement Points
Center of Percussion
Center of Pressure
Center Punch
Central Processing
Central Processor
Centrifugal Pump
Certified Planner
Cesspool
Chain Procedure
Chamber Pressure
Change Proposal
Channel Program
Character Printer
Charge-Conjugation Parity
Charge Parity
Chemically Pure [also: cp]
Chemical Potential
Chemical Preparation
Chemical Processing
Chemical Properties
Chicago Pile [of ANL, US]
Chloropurine
Chromatographically-Purified
Circuit Package
Circularly Polarized
Circulating Pump
Circular Pitch
Circular Polarization
Clamping Pin
Clinical Pathologist
Clinical Pathology
Clock Pulse
Code Page
Coefficient of Performance
Coherent Potential
Cold Punching
Collision Probability
Colombian Peso [Currency]
Color Print
Color Printing
Command Post
Command Privilege
Command Processor
Command Pulse
Commercial Purity
Common Process
Communications Processor
Compacted Powder
Computer Processing
Computer Professional
Condensation Pump
Conductive Polymer
Cone Pedicel
Conference Paper
Connection Pending
Connection Point
Constant Power
Constant Pressure
Construction Permit
Consulting Planner
Continuous Path
Continuous-Path Programming
Continuous Phase
Controllable Pitch
Controllable-Pitch Propeller
Control Panel
Control Point
Control Processor
Control Program
Corrosion Protection
Counterpoise
Counterpressure
Coventry Polytechnic [UK]
Cracking Pressure
Crack Propagation
Critical Path
Cross-Ply
Cross-Polarization
Crystalline Polymer
Crystal Physics
Current Paper
Customized Processor
Cyclic Polymer
Cytidine Phosphate
Cp Casseopieum
Cyclopentadienyl
cP Continental Polar Air
centipoise
cp candlepower
centipoise
chemically pure

	commercially pure
	compare
6-CP	6-Chloropurine
C-5'-P	Cytidine 5'-Monophosphate
C£	Cyprus Pound [Currency]
CPA	California Pharmaceutical Association [US]
	California Pistachio Association [US]
	Canadian Pacific Airlines
	Canadian Petroleum Association
	Canadian Pharmaceutical Association
	Canadian Postmasters Association
	Certificate in Preservation Administration
	Certified Public Accountant
	Chemical Propulsion Abstracts [of CPIA, US]
	Chlorobenzene Producers Association [US]
	Chlorophenoxyacetic Acid
	Closest Point of Approach
	Coherent Potential Approximation
	Cold Plasma Analyzer
	Color-Phase Alternation
	Commonwealth Pharmaceutical Association [UK]
	Commutative Principle for Addition
	Compressed-Pulse Altimeter
	Computer Architecture
	Computer Performance Analysis
	Computer Press Association [US]
	Concrete Pipe Association [US]
	Concurrent Photon Amplification
	Contractors Plant Association [UK]
	Control Program Assist
	Copolar Attenuation
	Cracking Pressure Adjusted Valve
	Critical Path Analysis
	Cross Program Auditor
CpA	Cytidylyladenosine
CPAA	Cycle Parts and Accessories Association [US]
CPAC	Collaborative Pesticides Analytical Committee
CPAE	Cold Plasma Analyzer Experiment
CPAL	Containment Person Air Lock
CPAR	Construction Procurement Advancement Research (Program) [of US Army]
	Cooperative Pollution Abatement Research
CPAS	Contruction Program Administration System
CPAWS	Computer-Planning and Aircraft-Weighing Scales
CPB	Channel Program Block
	Colombo Planning Bureau [Sri Lanka]
	Contractors Pump Bureau [US]
	Corporation for Public Broadcasting [US]
CPBS	Chlorophenylbenzenesulfonate
CPBX	Computerized Private Branch Exchange
CPC	Calgary Petroleum Club [Canada]
	California Pistachio Commission [US]
	Calling Party Category
	Calling Party Control
	Canada Ports Corporation
	Canada Post Corporation
	Canadian Pallet Council
	Card-Programmed Calculator
	Center for Plant Conservation [US]
	Cerium Pyridinium Chloride
	Channel Program Command
	Chemical Protective Clothing
	Chinese Petroleum Corporation [Taiwan]
	Clock Pulse Control
	Clock Pulsed Control
	Committee for Program and Coordination
	Common Peripheral Channel
	Compound Parabolic Concentrator
	Computerized Plasma Control
	Computer Planning Consultant
	Computer Power Center
	Computer Process Control
	Computer Program Component
	Computer Programming Concepts
	Condensation Particle Counter
	Controlled-Potential Coulometry
	Core Protection Computer
	Cycle Program Control
	Cycle Program Counter
	Cyclopropylcarbinyl
CpC	Cytidylylcytidine
CPCA	Canadian Portland Cement Association
CPCCI	Conference Permanente des Chambres de Commerce et d'Industrie [= Permanent Conference of Chambers of Commerce and Industry [of EEC, Belgium]]
CPCEI	Computer Program Contract End Item
CPCEMR	Circum-Pacific Council for Energy and Mineral Resources [US]
CPCFA	Council of Pollution Control Financing Agencies
CPCGN	Canadian Permanent Committee on Geographical Names
CPCH	Calling Party Cannot Hear
CPCI	Canadian Prestressed Concrete Institute
	Computer Program Configuration Item
	CPU (= Central Processing Unit) Power Calibration Instrument
CPCRR	Cyclopropylcarbinyl Radical Rearrangement
CPCS	Check Processing Control System
CPCUG	Capital Personal Computer User Group [US]
CPD	Carboxypeptidase
	Cards per Day
	Central Pulse Distributor
	Citrate Phosphate Dextrose
	Community Planning and Development [of OAS, US]
	Compound
	Consolidated Programming Document
	Contact Potential Difference
	Critical Point Drier
	Critical Point Drying
	Cumulative Probability Distribution
CPDAMS	Computer Program Development and Management System
CPDD	Command Post Digital Display
	Conceptual Program Design Description
CPDL	Canadian Patents and Developments Limited [of DRIE]
CPDS	Compounds
	Computer Program Design Specification
CPDU	Continuous Process Development Unit [of CANMET, Canada]

CPE Canadian Painters, Etchers and Engravers
Center for Packaging Education [US]
Center for Professional Education
Central Processing Element
Central Programmer and Evaluator
Certificate of Primary Education [now: KCPE, Kenia]
Charged Particle Equilibrium
Chlorinated Polyethylene
Circular Probable Error
Computer Performance Evaluation
Computer Premises Equipment
Continuous Particle Electrophoresis
Contractor Performance Evaluation
Conventional Polyethylene
Cross-Roll Piercing Elongation
Customer Premises Equipment
Customer Provided Equipment
Custom Power Engineering
Cylindrical Photolithographic Etching

CPEA College of Physical Education Association
Concentrated Phosphate Export Association [US]
Cyprus Professional Engineers Association

CPEBS Central Processing Element Bit Slice

CPED Continuous Particle Electrophoresis Device

CPEM Conference on Precision Electromagnetic Measurements [US]

CPEQ Corporation of Professional Engineers of Quebec [Canada]

C-PET Crystallized Polyethylene Terephthalate

CPEUG Computer Performance Evaluation Users Group [of NIST, US]

CPF Calcium Phosphate Free
Canadian Patrol Frigate
Complete Power Failure
Control Program Facility

CPFF Cost plus Fixed Fee

CPFMS COMRADE (= Computer-Aided Design Environment) Permanent File Management System

CPFP Canadian Patrol Frigate Program

CPFR Calling Party Forced Release

CPFSK Continuous Phase Frequency Shift Keying

CPFT Customer Premises Facility Terminal

CPG Clock Pulse Generator
COBOL Program Generator
College Publishers Group
Controlled Pore Glass

CpG Cytidylylguanosine

CPGA Ceramic Pin Grid Array

CPH Certificate in Public Health
Characters per Hour
Close-Packed Hexagonal
2,2'-Methylene-bis-(4-Methyl-6-Cyclohexylphenol)

CPHA Canadian Ports and Harbours Association
Canadian Public Health Association

CPhA Canadian Pharmaceutical Association

C Phil Candidate in Philosophy

CPHS Containment Pressure High Signal

CPI Cable Pair Identification
Call Progress Indicator
Canadian Plastics Institute
Carbon Preference Index
Central Patents Index [UK]
Certificate in Planning Information
Characters per Inch
Chemical Processing Industries
Code Page Information
Commission Permanente Internationale Europeenne des Gaz Industriels et du Carbure de Calcium [= Permanent International European Commission on Industrial Gases and Calcium Carbide, France]
Computer-PBX (= Private-Branch Exchange) Interface
Computer Power Index
Computer Prescribed Instruction
Condensation-Reaction Polyimide
Conference Papers Index
Conference Proceedings Index
Consumer Price Index
Cotton Producers Institute [now: Cotton Inc., US]
Council for Professional Interest
Council of the Printing Industries [Canada]
Crash Position Indicator
Crop Protection Institute [US]

CPIA Canadian Photovoltaic Industries Association
Cathodic Protection Industry Association [US]
Chemical Propulsion Information Agency [of JHU, US]
Chlorinated Paraffins Industry Association [US]

CPIB Chlorophenoxyisobutyrate

CPIC Canadian Police Information Center
Chemical Process Industries of Canada
Crop Protection Institute of Canada

CPID Computer Program Integrated Document

CPIF Cost Plus Incentive Fee

CPILS Correlation Protected Instrument Landing System

CPIMA Canadian Printing Ink Manufacturers Association

CPIN Computer Program Identification Number

CPIP Computer Pneumatic Input Panel
Computer Program Implementation Process

CPIRA Copying Product and Inked Ribbon Association [US]

CPIV Comite Permanent des Industries du Verre [= Permanent Committee of the Glass Industries [of EEC, Belgium]]

CPK Creatine Phosphokinase

CPL Canadian Plastics
Capability Password Level
CAST (= Computer-Aided Solidification Technique) Programming Language
Combined Programming Language
Commercial Pilot License
Common Programming Language
Computer Program Library
Conversational Programming Language
Core Performance Log
Corporal
Critical Path Length

CPLEE Charged Particle Lunar Environment Experiment

CPLG Coupling

CPLQ Corporation of Professional Librarians of Quebec [Canada]
CPM Call Protocol Message
Carbide Powdered Metal
Cards per Minute
Central Processor Module
Certificate in Public Management
Chemical Process Modelling
Colliding-Pulse Mode-Locked
Commutative Principle of Multiplication
Computer Performance Management
Connection Point Manager
Continuous Phase Modulation
Control Path Method
Conversational Program Module
Counts per Minute [also: cpm]
Critical Path Method
Crucible Particle Metallurgy
Current Physics Microform [of AIP, US]
Current Processor Mode
Cycles per Minute [also: cpm]
CP/M Control Program for Microcomputer
CPMA Central Processor Memory Address
Computer Peripherals Manufacturers Association [US]
CPMB Concrete Plant Manufacturers Bureau [US]
CPMC Chlorophenyl-Methyl-Carbamate
CPMR Conference of Peripheral Maritime Regions [of EEC, Belgium]
CPMS Cable Pressure Monitoring System
Canadian Pest Management Society
Canadian Project Management Society
Center for Policy and Management Studies
CPN Coupon
CPO Canada Post Office
Catalytic Partial Oxidation
Chief Petty Officer
Code Practice Oscillator
Command Pulse Output
Commonwealth Producers Organization [UK]
Compulsory Purchase Order
Concurrent Peripheral Operations
CPODA Contention Priority-Oriented Demand Assignment
CPOL Communications Procedure-Oriented Language
CPP Card Punching Printer
Cast Polypropylene
Center for Plutonium Production [France]
Coal Preparation Plant
Computer Professional for Peace
Conductive Plastic Potentiometer
Containment Pressure Protection
Controllable Pitch Propeller
Current Papers in Physics (Service) [of INSPEC, UK]
Cyclopentenophrenanthrene
CPPA Canadian Pulp and Paper Association
Coal Preparation Plant Association
CPPC Caribbean Plant Protection Commission [Trinidad and Tobago]
CPPD Collaborative Program for Professional Development
CPPF Canadian Project Preparation Facility
CPPI Consultative Panel on Public Information [of UN]
CPPMA Canadian Public Personnel Management Association
CPPO Bis(2-Carbopentyloxy-3,5,6-Trichlorophenyl) Oxalate
Certified Public Purchasing Officer
CPPP Computerized Production Process Planning
CPPS Comision Permanente de Pacifico Sur [= Permanent South Pacific Commission, Colombia]
Critical Path Planning and Scheduling
CPR Cam Plate Output
Canadian Pacific Railway [now: CP]
Cardiopulmonary Resuscitation
Card Punch and Reader
Center of Polish Research
Chlorophenol Red
Coal Pile Runoff
Committee on Polar Research [of NAS/NRC, US]
Component Pilot Rework
Continuous Progress Indicator
Correspondence Pattern Recognition
Corrosion Penetration Rate
Critical Power Ratio
Cyclonene-Pyrethrin-Rotenone
CP-R Control Program — Real-Time
CPRDC Coordinated Program of Research in Distributed Computing [of SERC, UK]
CPRC Canadian Police Research Center
CPRE Council for the Protection of Rural England
CPREA Canadian Peace Research and Education Association
CPRG Computer Personnel Research Group [now: SIGCPR of ACM, US]
CPRI Canadian Peace Research Institute
Central Psi Research Institute [US]
CPRS Canadian Public Relations Society
Centralized Personnel Record System
CPS Calling Processing Subsystem
Cambridge Philosophical Society [UK]
Canadian Physiological Society
Card Programming System
Cards per Second
Cathode Potential Stabilized
Central Processing System
Central Processor System
Certified Productivity Specialist
C-Frame Profile Scanner
Characters per Second
Chinese Petroleum Society
Circuit Package Schematic
Coils per Slot
College Press Service [US]
Columns per Second
Commission on the Patent System [US]
Committee for Production Sharing [US]
Communications Processor System
Compendium of Pharmaceuticals and Specialties [of CPhA, Canada]
Conservation Protection Service [Canada]
Contour Plotting System
Controlled Path System
Control Programs Support

	Conversational Programming System
	Conversion Program System
	Counts per Second [also: cps]
	Critical Path Scheduling
	Current Population Survey [of USDOC]
	Curved Position Sensitive (Detector)
	Customer Premises System
	Cycles per Second [also: cps]
cps	centipoises
CPSA	Consumer Product Safety Act [US]
CPSC	Consumer Product Safety Commission [US]
	Consumer Protection and Safety Commission [US]
CPSCI	Central Personnel Security Clearance Index
CPSE	Counterpoise
CPSG	Common Power Supply Group
CPSK	Coherent Phase-Shift Keying
CPSM	Critical Path Scheduling Method
CPSR	Controlled Process Serum Replacement
CPSS	Certificate in Public Service Studies
	Circular Polarization Selective Surface
	Committee of Presidents of Statistical Societies
	Common Programming Support System
	Computer Power Support System
CPSSRB	Canada Public Service Staff Relations Board
CPST	Commission on Professionals in Science and Technology [US]
CPSU	California Polytechnic State University [US]
CPT	Canadian Pacific Telecommunications
	Charge, Parity, Time
	Color Picture Tube
	Control Power Transformer
	Covariant Perturbation Theory
	Critical Path Technique
CPTA	Computer Programming and Testing Activity
CPTB	Clay Products Technical Bureau [UK]
CPTE	Computer Program Test and Evaluation
CPTR	Computer
CPTS	Collidine-P-Toluenesulfonate
CPTU	Continuous Process Development Unit
	Council of Progressive Trade Unions [Trinidad]
CPU	Canadian Paperworkers Union
	Capacitive Pickup
	Central Processing Unit
	Collective Protection Unit
	Commonwealth Press Union [UK]
	Communications Processor Unit
	Computer Peripheral Unit
	Computer Processor Unit
	Control Process Unit
CpU	Cytidylyluridine
CPUID	Central Processing Unit Identification Number
CPUNCH	Counterpunch
CPUOS	Committee on the Peaceful Uses of Outer Space [of UN]
CPUS	Coalition for Peaceful Uses of Space
CPUSAC	Crafted with Pride in the USA Council [US]
CPVC	Chlorinated Polyvinyl Chloride
CPW	Commercial Projected Window
	Coplanar Waveguide
CPX	Charged Pigment Xerography
CPY	Copy
CQ	Chloroquine
	Clinical Quality
	Commercial Quality
	Controlled Quality
CQAK	Commercial Quality Aluminum-Killed Steel
CQC	Certificate in Quality Control
	Citizens for a Quieter City [US]
CQD	Come — Quick — Danger
	Customary Quick Dispatch
CQE	Cognizant Quality Engineer
CQES	Center for Quanized Electronic Structures [of UCSB, US]
CQMS	Circuit Quality Monitoring System
CQRS	Commercial Quality Rimmed Steel
CQS	Common Queue Space
	Constant-Q Spectrometer
CR	Call Reader
	Call Request
	Capacitance-Resistance
	Carbon Resistor
	Card Reader
	Carriage Return (Character)
	Carrier Recovery
	Carry Register
	Catalytic Reforming
	Cathode Ray
	Ceiling Register
	Ceylon Rupee [Currency of Sri Lanka]
	Chain Reaction
	Change Request
	Channel Request
	Chart Recorder
	Chemical Report
	Chicago Reactor [of Argonne National Laboratory, US]
	Chloroprene Rubber
	Clamp Rest
	Cold Rolling
	Collector Ring
	Command Register
	Communications Register
	Complete Round
	Composites Removal
	Conditioned Reinforcer
	Conditioned Response
	Conference Report
	Congressional Record
	Constant Rate
	Contact Resistance
	Containment Rupture
	Contract Report
	Controlled Rectifier
	Controlled Release
	Control Ratio
	Control Relay
	Control Rod
	Control Room
	Control Routine
	Cooling Rate
	Corrosion Resistance
	Cosmic Ray

Credit
Creditor
Crew
Crystalline
Crystal Rectifier
Current Rate
Current Relay

C/R Chamfer or Radius
Cr Chromium
Cruzeiro [Currency of Brazil]
cr (currentis) — this year, month, etc.
CRA California Redwood Association [US]
Canadian Residential Appraiser
Catalogue Recovery Area
Clear Rear Access
Cold-Rolled and Annealed
Colorado River Association [US]
Composite Research Aircraft
Control Rod Assembly
Corona Australis
CrA Corona Australis [Astronomy]
CRAC Careers Research and Advisory Center [UK]
CRAD Chief of Research and Development
Committee for Research into Apparatus for the Disabled
CRADA Cooperative Research and Development Agreement [US]
CRAF Civilian Reserve Air Fleet
CRAFT Changing Radio Automatic Frequency Transmission
Computerized Relative Allocation of Facilities Technique
CRAGS Chemistry Records and Grading System
CRAM Card Random Access Memory
Computerized Reliability Analysis Method
Conditional Relaxation Analysis Method
Core and Random Access Manager
CRAN Cross Scan
CRANZ Coal Research Association of New Zealand
CRAR Committee for the Recovery of Archeological Remains [US]
Control ROM (= Read-Only Memory) Address Register
CRB Community Reference Bureau (Database) [of EURATOM]
Customer Records and Billing System
CrB Corona Borealis [Astronomy]
CRBDI Control Red Bank Demand Indicator
CRBL Charles River Breeding Laboratories [US]
CRBR Clinch River Breeder Reactor [US]
Controlled Recirculation Boiling-Water Reactor
CRBR-CX Clinch River Breeder Reactor — Critical Assembly [US]
CRBRP Clinch River Breeder Reactor Plant [US]
Clinch River Breeder Reactor Project [US]
CRC Carriage Return Contact
Chemical Referral Center [US]
Civil Rights Commission [US]
Communication Relay Center
Communications Regulatory Commission [US]
Communications Research Center [Canada]
Composing Reducing Camera
Computer Robot Control
Control and Reporting Center
Coordinating Research Council [US]
Copy Research Council
Corporate Research Center
Costa Rican Colon [Currency]
Critical Reactor Component
Cumulative Results Criterion
Cyclic Redundancy Check
CRCA Canadian Research Center for Anthropology
Canadian Roofing Contractors Association
Cold-Rolled, Close-Annealed
CRCC Cyclic Redundancy Check Character
CRCGR Cyclic Redundancy Check Generator
CRCE Chicago Rice and Cotton Exchange [US]
CRCHF Crew Chief
CRCMF Circumference
CRCO Control Route Charges Office
CRCP Continuously-Reinforced Concrete Pavement
CRCPD Conference of Radiation Control Program Directors [US]
CRCTA Composite Reactor Components Test Activity
CRD Capacitor-Resistor Diode
Capital Regional District
Card Reader
Collaborative Research and Development
Comite pour la Recherche et le Developpement en Matiere d'Energie [= Committee on Energy Research and Development [of OECD, France]]
Constant Ringing Drop
Control Rod Drive
Cooperative Research and Development
CRDA Control Rod Drive Assembly
Control Rod Drop Accident
CRDEWPA Center for Research Documentation and Experimentation on Water Pollution Accidents [France]
CRDF Canadian Radio-Direction Finder
Cathode-Ray Direction Finding
CRDL Chemical Research and Development Laboratories [US Army]
CRDM Control Rod Drive Mechanism
Control Rod Drive Motor
CRDME Committee for Research into Dental Materials and Equipment
CRDS Chemical Reactions Documentation Service [UK]
CRDSD Current Research and Development in Scientific Documentation
CRE Coal Research Establishment [UK]
Commander Royal Engineers
Constant Rate-of-Extension
Controlled Residual Element
Corrosion Resistance
Current Ring End
CREA Committee on the Relation of Electricity to Agriculture [US]
CREAM Canadian Research on Exposure Assessment Modelling
CREATE Chalk River Experiment to Assess Tritium Emission [Canada]
CREB Conservation and Renewable Energy Board

CREDA Conservation and Renewable Energy Demonstration Agreement
CREDO Central Reliability Data Organization
CREO Conservation and Renewable Energy Office
CREQ Center for Research and Environmental Quality
CRES Certified Radiologic Equipment Specialist
Chinese Rare Earth Society [PR China]
Corrosion-Resistant Steel
Crescent
CRESP Coding Region Expression Selection Plasmid
CRESS Center for Research in Experimental Space Science [Canada]
Central Regulatory Electronic Stenographic System
Combined Reentry Effort for Small Systems
Computerized Reader Enquiry Service System
Computer Reader Enquiry Service System
CREST Committee on Reactor Safety and Technology [of ENEA, France]
CRESTS Courtauld's Rapid Extracting, Sorting and Tabulating System
CRETC Combined Radiation Effects Test Chamber
CREVS Control Room Emergency Ventilation System
CRF Capital Recovery Factor
Cave Research Foundation [US]
Channel Replacement Furnace Black
Clean Room Filled (Facility)
Coincidence Rangefinder
Compressor Research Facility
Computer Retailing Forum [now: CDF, US]
Conservation and Research Foundation [US]
Control Relay Forward
Correspondence Routing Form
Corticotropin-Releasing Factor
Cross-Reference Facility
Cryptographic Repair Facility [US Military]
CRFA Canadian Restaurant and Food Services Association
CRFI Custom Roll Forming Institute [US]
CRFSA Canadian Restaurant and Food Services Association
CRG Carriage
Classification Research Group [UK]
Composites Research Group
CRGR Coalition for Responsible Genetic Research [US]
CRH Channel Reconfiguration Hardware
Cold-Rolled, Hard
Corticotropin-Releasing Hormone
CRHA Canadian Railroad Historical Association
Canadian Retail Hardware Association
CRHH Cold-Rolled, Half-Hard
CR-hi Channel Request, High Priority
CRHSI Center for Research in the Hospitality Service Industry [now: CHRS, US]
CRI Canadian Research Institute
Carbohydrate Research Institute [Canada]
Caribbean Research Institute
Cement Research Institute
Central Research Institute [of EPI, Japan]
Centre de Recherches en d'Irradiations [= Center for Research in Irradiation, France]
Color Rendering Index
Committee for Reciprocity Information
Concentrated Rust Inhibitor
CRIB Computerized Resources Information Bank [of USGS]
CRID Centro di Riferimento Italiano DIANE [= Italian Reference Center for the Direct Information Access Network for Europe [of EURONET]]
CRIEPI Central Research Institute of Electric Power Institute [Japan]
CRIF Centre de Recherche Scientifique et Technique de l'Industrie des Fabrications Metalliques [= Scientific and Technical Research Center of the Metalworking Industry, Belgium]
CRIFO Civilian Research, Interplanetary Flying Objects
CRIM Clinical Research Institute of Montreal [Canada]
Composite Reaction Injection Molding
CRIME Censorship Records and Information, Middle East [US Military]
CRIMM Changsha Research Institute of Mining and Metallurgy [PR China]
CRIP Controlled Retracting Injection Point
CRIS Command Retrieval Information System
Current Research Information System [of USDOA]
CRISC (Combined) Complex and Reduced Instruction Set Computer
CRISPE Computerized Retrieval Information Service on Precision Engineering [of Cranfield Institute of Technology, UK]
CRIT Critical
CRITIC Chalk River In-Reactor Tritium Instrumented Capsule [Canada]
CRITICOMM Critical Intelligence Communications System [US Air Force]
CRJE Conversational Remote Job Entry
CRK Crank
CRKA Community Right to Know Act [US]
CRKC Crankcase
CRL Casting Research Laboratory [of Waseda University, Japan]
Center for Research in Librarianship [of U of T, Canada]
Center for Research Libraries [US]
Chalk River Laboratories [of AECL, Canada]
Chemical Research Laboratory [UK]
Coal Research Laboratories [of CANMET, Canada]
Coherent Radiation Laboratory [US]
Communications Research Laboratory [Japan and Canada]
Constant Rate-of-Loading
Construction Robotics Laboratory [US]
CRLA Canadian Railway Labour Association
College and Research Libraries Association
CR/LF Carriage Return/Line Feed
CR-lo Channel Request, Low Priority
CRLR Chemical and Radiological Laboratories [US Army]
CRM Canadian Risk Managers
Centre de Recherches Metallurgiques [= Centre for Metallurgical Research, Belgium]
Certified Record Manager
Certified Reference Material
Chemical Remanent Magnetization
Collision Risk Model
Comprehensive Resource Management

Confusion Reflector Material
Containment Radiation Monitor
Control and Reproducibility Monitor
Core Restraint Mechanisms
Counter-Radar Measures
Counter-Radar Missile
Crew Resource Management
Cross-Reacting Material
Count Rate Meter
Cyber Record Manager

CRMA Canadian Research Managers Association
Canadian Rock Mechanics Association
Commercial Refrigerator Manufacturers Association [US]

CRM/BAM Cyber Record Manager/Basic Access Method

CR-med Channel Request, Medium Priority

CRMG Calgary Rock-Mechanics Group [Canada]

CRML Coalition for Responsible Mining Law [US]

CR MOLY Chrome Molybdenum

CRMR Continuous-Reading Meter Relay

CRN Cellular Radio Network
Centre de Recherche Nucleaire [= Center for Nuclear Research, France]
Continuous Random Network

CRNL Chalk River Nuclear Laboratories [of AECL, Canada]

CRO Cathode-Ray Oscilloscope
Control Room Operator
Cosmic Ray Observatory
Copy Receipt Office

CR-OFC Corrosion-Resistant Oxygen-Free Copper

CROM Capacitive Read-Only Memory
Control Read-Only Memory

CROS Capacitor Read-Only Storage
Contra-Lateral Routing-of-Signal

CROSS Computerized Rearrangement of Subject Specialties [of BIOSIS, US]

CROSSBOW Computerized Retrieval of Organic Structures Based on Wiswesser

CROW Counter-Rotating Optical Wave

CRP Card Reader Punch
Constant Rate of Penetration
Control and Reporting Post
Corrosion-Resistant Pump
C-Reactive Protein

CRPC Canadian Research and Publication Center

CRPL Central Radio Propagation Laboratory [now: ITSA, US]

CR PL Chromium Plate

CRPM Communication Registered Publication Memorandum

CRPO Central Radio Propagation Office [US]

CRPQF Commissao Reguladora dos Productos Quimicos e Farmaceuticos [= Regulatory Committee for Chemical and Pharmaceutical Products, Portugal]

CRPT Carport

CRQ Console Reply Queuing

CRR Cathode Ray Recording
Center for Renewable Resources
Churchill Research Range [US Air Force]
Constant Relative Resolution
Constant Ringing Relay
Customer Response Representative

CRREL Cold Regions Research and Engineering Laboratory [US Army]

CRRI Central Rice Research Institute [India]

CRRL Cosmic Ray Research Laboratory [of Nagoya University, Japan]

CRS Center for Radio Science [Canada]
Center for Resource Studies [Canada]
Centralized Results System
Character Recognition System
Chemical Registry System [of ACS, US]
Chinese Rare Earth Society [PR China]
Clamp Rest Screw
Coasting Richer System
Cold-Rolled Steel
Command Retrieval System
Commercially Rapidly Solidified
Computerized Retrieval Service
Computer Recognition System
Computer Reservations System
Congressional Research Service [of Library of Congress, US]
Containment Rupture Signal
Controlled Release Society [US]
Conversion Resource (File)
Crystal Spectrometer

CRSA Control Rod Scram Accumulator

CRSC Center for Research on Scientific Communication [of JHU, US]

CRSE Course

CRSI Concrete Reinforcing Steel Institute [US]

CRSL Corporate Research Science Laboratory

CRSO Common Rolling Stock Organization

CRSP Canadian Registered Safety Professional

CRSR Center for Radiophysics and Space Research [of Cornell University, US]
Coherent Resonant Stokes Rotation

CRSS Critical Resolved Shear Stress

CRT Cathode-Ray Tube
Center for Research on Transportation [Canada]
Charactron Tube
Circuit Requirement Table
Constant Rate of Traverse
Controlled Release Technology
Cosmic Ray Telescope
Crystal Research and Technology

Crt Crater [Astronomy]

crt (currentis) — this year, month, etc.

CRTA Cost Recoverable Technical Assistance (Program) [Canada]

CR TAN LTHR Chromium Tanned Leather

CRTC Canadian Radio-Television and Telecommunications Commission
Cathode-Ray Tube Controller

CRTD Cathode-Ray Tube Drive

CRTEE Canadian Roundtable on Environment and Energy

CRTOS Cathode-Ray Tube Operating System

CRTPB Canadian Radio Technical Planning Board

CRTS Constant Return to Scale
Controllable Radar Target Simulator

CRTU Combined Receiving and Transmitting Unit
CRTV Composite Reentry Test Vehicle
CRU Card Reader Unit
Combined Rotating Unit
Commodities Research Unit [UK]
Cru Crux [Astronomy]
CRUD Chalk River Unidentified Deposit [Canada]
CRUIS Cruising
CRUS Center for Research on User Studies [of University of Sheffield, UK]
CRV Contact Resistance Variation
Curve
Crv Corvus [Astronomy]
CRVA Canadian Recreational Vehicle Association
CR VAN Chrome Vanadium
CRV DWG Curve Drawing
CRVM Cathode-Ray Voltmeter
CRW Community Radio Watch
CRWC Connecticut River Watershed Council [US]
CRWI Coalition for Responsible Waste Incineration [US]
CRWO Coding Room Watch Officer [US Navy]
CRWPC Canadian Radio Wave Propagation Committee
CRY Crystal(line)
CRYO Cryogenic(s)
CRYST Crystal(line)
Crystallize
CRYSTN Crystallization
CRYPTO Cryptographer
Cryptography
CRYPTONET Crypto-Communications Network
CS Calls per Second
Carbon Steel
Carrier System
Cascade Stack
Casein
Cast Steel
Catalysis Society [US]
Cathode Spot
Cathode Sputtering
Celestial Sphere
Center of Suspension
Center Section
Centrifugal Separator
Centro-Symmetry
Channel Status
Chartered Ship Broker
Chartered Surveyor
Check Sorter
Chemical Society
Chemical System
Chip Select
Chondroitin Sulfate
Christian Science
Circuit Switching
Circumsporozoite
Citrate Synthase
Citrate Synthetase
Civil Service
Clamp Strap
Class of Service
Coal Store
Coblentz Society (for Spectroscopy) [US]
Code Segment
Coding Specification
Cold Stage
Colepterists Society [US]
College of Science
Colloid Science
Color Specification
Color System
Combustion Science
Commercial Standard
Commercial System
Commonwealth Secretariat [UK]
Communications Satellite
Communications Services
Communications Simulator
Communications Society [of IEEE, US]
Communications System
Communication Station
Community Service
Composition Service
Computer Science
Computer Simulation
Computer Society [of IEEE, US]
Computer System
Concrete Slab
Concrete Society [UK]
Conditioned Stimulus
Conducted Susceptibility
Containment Spray
Continue-Specific Mode
Continuous Spectrum
Continuous System
Control Section
Control Set
Control Store
Control Switch
Control System
Control Systems (Society) [of IEEE, US]
Conventional System
Conveyors Section [of MHI, US]
Cooling System
Core Shift
Corporate Source
Coupled States
Crystalline Solid
Currence Sign
Current Strength
Cutting Speed
Cycle Shift
Cyclostationary
Cyclosynchrotron
(Single) Cotton (Single) Silk (Wire)
C&S Computers and Systems
C/S Call Signal
Cycles per Second
Cs Cesium
Cirro-Stratus (Cloud)
cs centistoke
c/s cycles per second [also: cps or CPS]
C_2S Dicalcium Silicate [Cement]

C_3S Tricalcium Silicate [Cement]
CSA Called Subscriber Answer
Camphorsulfonic Acid
Canadian Science Association
Canadian Shipowners Association
Canadian Space Agency
Canadian Standards Association
Caribbean Shipping Association
Cast Section Angle
Center for Sustainable Agriculture [US]
Chemical Sources Association [US]
Chemical Structure Association [UK]
Common Service Area
Common System Area
Community Service Activities [of AFL-CIO, US]
Community Service Administration
Computer Science Association
Computer Services Association [UK]
Computer System Architecture
Computer Systems Association [US]
Conference on Superconductivity and Applications
Contract Services Association (of America) [US]
Cross-Sectional Area
Cryogenic Society of America [US]
Current Science Association [India]
CSAA Canadian Society of Applied Art
CSAE Canadian Society of Agricultural Engineers
CSAM Canadian Society of Aviation Medicine
Circular Sequential Access Memory
C-Mode Scanning Acoustic Microscopy [also: C-SAM]
Conventional Scanning Acoustic Microscopy
CSAMT Controlled Source Audiofrequency Magnetotellurics
Controlled Source Audio-Magnetotelluric Tensor
CSAO Construction Safety Association of Ontario [Canada]
Council of Safety Associations of Ontario [Canada]
CSAP Canadian Society of Animal Production
CSAR Canadian Space Agency Rocket
Control Store Address Register
CSAR-1 Canadian Space Agency Rocket 1
CSAS Cove Standy Actuation Signal [US]
CSASI Canadian Society of Air Safety Investigators
CSA-SSRP Canadian Space Agency — Space Science Radioastronomy Program
CSB Called Subscriber Busy
Canada Service Bureau
Carrier and Sideband
Coherent Surface Bremsstrahlung
Companies Services Branch [of MCCR, Canada]
Concentrate Storage Building
Concrete Splash Block
Consumer Sounding-Board
CSBE Chinese Society of Biomedical Engineering [PR China]
CSBISSS Commission on Soil Biology of the International Society of Soil Science [Netherlands]
CSC Cadmium-Sulfide Cell
Cambridge School Certificate
Canada Safety Council
Canadian Society for Chemistry
Canadian Spectroscopy Conference
Canadian Symposium on Catalysis
Cartridge Storage Case
Cascade Synchronous Speed
Catalytic Surface Conversion
Central Switching Center
Centrifugal Shot Casting
Chief Sector Control
Circuit Switching Center
Cis-Syn-Cis (Configuration) [Chemistry]
Civil Service Commission [US]
Color Sub Carrier
Common Signalling Channel
Commonwealth Science Council [UK]
Commonwealth Scientific Committee
Communications Satellite Corporation
Communications Systems Center
Computer Sciences Corporation
Computer Society of Canada
Construction Specifications Canada
Core Standby Cooling
Correctional Services Canada
Course and Speed Computer
Cube-Surface Coil
Customer Support Center
csc cosecant
csc^{-1} inverse cosecant
CSCC Canadian Society of Clinical Chemists
Canadian Steel Construction Council
Certificate, Society of Clinical Chemists
Cumulative Sum Control Chart
CSCD Committee on Sugar Cane Diseases [of HSPA, US]
Common Signalling Channel Demodulator
CSCE Canadian Society for Chemical Engineering
Canadian Society for Civil Engineering
Communications Subsystem Checkout Equipment
Conference on Security and Cooperation in Europe [of NATO]
Connecticut Society of Civil Engineers [US]
CSCFN Canadian Science Committee on Food and Nutrition
csch hyperbolic cosecant
$csch^{-1}$ inverse hyperbolic cosecant
CSChE Canadian Society for Chemical Engineering
CSCM Common Signalling Channel Modem
CSCR Central Society for Clinical Research [US]
Complimentary Silicon-Controlled Rectifier
CSCRF Computer System for Crop Response to Fertilizers [of FAO — UN]
CSCS Common Signalling Channel Synchronizer
Core Standby Cooling System
CSCSI Canadian Society for Computational Studies of Intelligence
CSCT Canadian Society for Chemical Technology
CSD Census Subdivision
Charge State Distribution
Circuit Switched Data
Civil Service Department [UK]
Cold (Reactor) Shutdown
Computer Science Division
Constant Speed Drive
Controlled-Slip Differentials
CSDA Concrete Sawing and Drilling Association [US]

CSDC Circuit Switched Digital Capability
CSDD Conceptual System Design Description
CSDF Core Segment Development Facility
CSDM Continuous Slope Delta Modulation
CSDN Circuit Switched Data Network
CSDR Control Store Data Register
CSE Center for Science and Environment [India]
Certificate in Stationary Engineering
Colorado Society of Engineers [US]
Commission on Science Education
Communications Security Establishment
Computerized Shrinkage Evaluation
Computer Science and Engineering
Containment Steam Explosion
Containment Systems Experiment [US]
Control Systems Engineer
Control Systems Engineering
Core Storage Element
CSEA California State Electronics Association [US]
CSECE Canadian Society for Electrical and Computer Engineering
CSECT Control Section
CSEE Canadian Society for Electrical Engineering
Certificate, Society of Clinical Chemists
Chinese Society of Electrical and Electronics Engineers [PR China]
Committee of Stock Exchanges in the European Community [Belgium]
CSEF Current Switch Emitter Follower
CSEG Canadian Society of Exploration Geophysicists
CSEHMA Cyprus Solar and Electric Heaters Manufacturers Association
CSELT Centro Studi e Laboratori Telecommunicazione [= Study Center and Telecommunications Laboratory, Italy]
CSEM Centre Suisse d'Electronique et de Microtechnique [= Swiss Center for Electronics and Microtechnology]
Computerized Scanning Electron Microscope
Conventional Scanning Electron Microscope
CSEPA Central Station Electrical Protection Association [US]
CSERB Computer Systems and Electronics Requirements Board [UK]
CSEU Confederation of Shipbuilding and Engineering Unions [UK]
CSF Catalytic Seed Fund
Central Switching Facility
Cerebrospinal Fluid
Colony Stimulating Factor
Constant Scattering Factor Averaging
Containment Support Fixture
Continuous Streamflow
Contract Stationers Forum [of NOPA, US]
Council on Synthetic Fuels [now: Council on Alternate Fuels, US]
Cumulative Size Selection Function
Customer Support Facility
CSFA Canadian Scientific Film Association
CSFC Canadian Salt Fish Corporation
CSFE Canadian Society of Forestry Engineers
CSFGR Canadian Society for Fifth Generation Research
CSFP Commonwealth Scholarship and Fellowship Plan
CSFPCPM Chambre Syndicale des Fabricants de Papier a Cigarettes et Autres Papiers Minces [= Union of Producers of Cigarette Papers and Other Lightweight Papers, France]
CSFPN Commission on Soil Fertility and Plant Nutrition [India]
CSFS Canadian Society of Forensic Science
Computer-Simulated Fracture Surface
Continuous Streamflow Simulation
CSFTI Committee on Southern Forest Tree Improvement [US]
CSG Canada Systems Group
Casing
Constructive Solid Geometry
CSGA Canadian Seed Growers Association
Canadian Society of Graphic Arts
CSGCC Commission on Soil Genesis, Classification and Cartography [Netherlands]
CSH Calcium — Silicate — Hydrate
Called Subscriber Held
CSHCS Center for the Studies of Human Communities in Space [now: SSSP, US]
CSHP Canadian Society of Hospital Pharmacists
C-SHRP Canadian Strategic Highway Research Program
CSI Canadian Securities Institute
Chartered Surveyors Institution [UK]
Chemical Substances Index [of ACS, US]
Chlorosulfonyl Isocyanate
Coalition of Service Industries [US]
Colloquium Spectroscopicum Internationale [= International Colloquium on Spectroscopy]
Command String Interpreter
Computer Security Institute [US]
Computer Society of India
Construction Specifications Institute [US]
Construction Surveyors Institute [UK]
Consumer Satisfaction Index
Control Sequence Introducer
Current Source Inverter
CSIA Canadian Solar Industries Association
CSIC Computer System Interface Circuits
CSICC Canadian Steel Industries Construction Council
CSICOP Committee for the Scientific Investigation of Claims of the Paranormal [US]
CSIET Council on Standards for International Educational Travel [US]
CSIM Compact Short-Channel IGFET (= Insulated-Gate Field-Effect Transistor) Model
CSIMM Central South Institute of Mining and Metallurgy [PR China]
CSIO Central Scientific Instruments Organization [India]
CSIPR Comite Speciale et Internationale sur les Parasites Radiotelegraphiques [= International Special Committee on Radiotelegraphic Atmospherics [of IEC, Switzerland]]
CSIR Council for Scientific and Industrial Research [South Africa]
CSIRA Canadian Steel Industry Research Association

CSIRAC Commonwealth Scientific and Industrial Research Automatic Computer [Australia]
CSIRO Commonwealth Scientific and Industrial Research Organization [Australia]
CSIRONET Commonwealth Scientific and Industrial Research Organization Network [Australia]
CSIS Canadian Security Intelligence Service
Core Spray Injection System
CSISRS Cross-Section Information Storage and Retrieval System [of NNCSC, US]
CSIST Chung Shan Institute of Science and Technology [PR China]
CSJ Ceramic Society of Japan
Chemical Society of Japan
Crystallographic Society of Japan
CSK Countersink
Czechoslovakian Koruna [Currency]
CSK-O Countersink other Side
CSL Closest Specification Limit
Code Selection Language
Coincidence Site Lattice
Comparative Systems Laboratory [US]
Computer Sensitive Language
Computer-Simulated Language
Conceptual Schema Language
Constant Scattering Length
Continuous Sheet Lamination
Control and Simulation Language
Coordinated Science Laboratory [US]
Current Switch Logic
Cytisus Scoparius Lectin
CSLATP Canadian Society of Landscape Architects and Town Planners
CSLC Canadian Softwood Lumber Committee
CSLM Confocal Scanning Laser Microscope
CSLMA Canadian Softwood Lumber Manufacturers Association
CSLO Canadian Scientific Liaison Office
CSLP Canada Student Loans Plan
Center for Short Lived Phenomena [US]
CSLT Canadian Society of Laboratory Technologists
CSLTY Casualty
CSM Call Supervision Module
Camborne School of Mines [UK]
Canadian Society of Microbiologists
Chinese Society for Measurement
Chinese Society of Metals
Colorado School of Mines [US]
Command (and) Service Module
Commission for Synoptic Meteorology
Commission on Soil Mineralogy [France]
Communications System Monitoring
Computerized Scanning Electron Microscopy
Constant System Monitor
Continuous Sheet Memory
Continuous Slowing-Down Models
CSMA Camborne School of Mines Association [UK]
Carrier-Sense Multiple-Access
Chain Saw Manufacturing Association [now: PPEMA, US]
Chemical Specialties Manufacturers Association [US]
Communications Systems Management Association
CSMA/CD Carrier-Sense Multiple-Access with Collision Detection [also: CSMA-CD]
CSMCS Continuous Space Monte Carlo Simulation
CSME Canadian Society for Mechanical Engineering
Chinese Society of Mechanical Engineers [Taiwan]
Communications System Monitoring Equipment
CSMITH Coppersmith
CSML Continuous Self Mode Locking
CSMMFRA Cotton, Silk and Man-Made Fibers Research Association [UK]
CSMP Continuous System Modelling Program
CSN Cable Satellite Network
Circuit Switching Network
Common Services Network
Conductive Solids Nebulizer
CSNA Classification Society of North America [US]
CSNC Chemical Societies of the Nordic Countries [Sweden]
CSNDT Canadian Society for Nondestructive Testing
CSNS Canadian Science News Service
CSO Central Statistical Office [UK]
Centralized Service Observation
Chained Sequential Operation
Chief Signal Officer
Coastal States Organization [US]
Color Separation Overlay
CSOHS Canadian Society for Occupational Health and Safety
CSOV Constrained Space Orbital Variation
CSP Cellestial South Pole
Certified Safety Professional
Certified Safety Professionals [US]
Channel Substrate Planar
Chlorosulfonated Polyethylene
Coder Sequential Pulse
Commercial Subroutine Package
Communications Security Publication [US Military]
Compact Strip Production
Completely Self-Protecting (Transformer)
Continuous Sampling Plan
Control Signal Processor
Control Switching Point
Council for Scientific Policy
Crack Shape Parameter
Crystallographic Shear Plane
C/SP Communications/Symbiont Processor
CSPCA Canadian Society for the Prevention of Cruelty to Animals
CSPG Canadian Society of Petroleum Geologists
Chondroitin Sulfate Proteoglycan
CSPI Corrugated Steel Pipe Institute
CSPO Communications Satellite Project Office
CSPP Chemically-Sensitive Paper Tape (Detector)
CSR Canadian Signals Regiment
Channel Select Register
Chinese Standardized (Natural) Rubber
Committee on Space Research [France]
Common Services Rack

Continuous Speech Recognition
Continuous Stirred Reactor
Controlled Silicon Rectifier
Control Status Register
Crack-Sensitivity Ratio
Cursor

CSRA Canadian Semiotics Research Association
CSRF Canadian Synchrotron Radiation Facility
CSRG Chemical Sciences Research Group
Computer Systems Research Group [of U of T, Canada]
CSRO Compositional Short-Range Ordering
CSRP Chemical Sciences Research Paper
CSRS Canadian Symposium on Remote Sensing
Coherent Stokes-Raman Spectroscopy
Coherent Stokes-Raman Spectrum
Cooperative State Research Service [of USDA]
CSRT Canadian Society of Radiological Technicians
CSS Canada Standard Size
Canadian Space Station
Cask Support Structure
Certificate in Special Studies
Character Start Stop
Circuits and Systems Society [of IEEE, US]
Cognitive Science Society [US]
Colorado Scientific Society [US]
Command Substitution System
Communications Subsystem
Comprehensive Support Software
Computer Sharing Services [US]
Computer Software Specialist
Computer Special System
Computer Subsystem
Computer System Simulator
Containment Spray System
Continuous-Cast Stainless Steel
Control Systems Society [of IEEE, US]
Conversational Software System
Cordless Switchboard Section
Core Support Structure
Customer Support Center
Customer Switching System
CSSA Computer Society of South Africa
Concrete Society of Southern Africa
Crop Science Society of America [US]
CSSB Cedar Shake and Shingle Bureau [US]
Companded Single-Sideband
Compatible Single-Sideband
CSSB/AM Companded Single-Sideband/Amplitude Modulation
CSSBI Canadian Sheet Steel Building Institute
CSSC Canadian Steel Service Center
Construction Software Standards Council [now: NCSA, US]
CSSCI Canadian Steel Service Center Institute
CSSE Canadian Society of Safety Engineers
Conference of State Sanitary Engineers [US]
CSSG Chemical Sciences Study Group
CSSL Continuous System Simulation Language
CSSP Canadian Space Station Program
Council on Scientific Society Presidents [US]
CSSR Czechoslovak Socialist Republic
CSSRA Canadian Shipbuilding and Ship Repairing Association
CSSS Canadian Society of Soil Science
CSST Center for Solder Science and Technology [of Sandia National Laboratories, US]
Computer System Science Training
CST Canadian Scholarship Trust (Foundation)
Carrier Power Supply, Transistorized
Central Standard Time
Certified Survey Technician
Certified Survey Technologist
Channel Status Table
Cis-Syn-Trans (Configuration) [Chemistry]
Code Segment Table
Composite Science and Technology
Coordinated Storm Tracking
cSt Centistoke
C/ST Combined Station/Tower
CSTA Canadian Society of Technical Agriculturalists
Canadian Surface Transportation Administration
Coal and Slurry Technology Association [US]
CSTAR Classified Scientific and Technical Aerospace Reports [of NASA]
CSTD Center for Science and Technology for Development [of UN]
CSTG Casting
CSTHA Canadian Science and Technology Historical Association
CSTEI Center for Scientific, Technical and Economic Information [Bulgaria]
CSTEX Cost Extension
CSTF Continuous Stirred Tank Fermenter
CSTI Committee on Scientific and Technological Information [US]
CSTL Computer System Terminal Log
CSTPA Council on Soil Testing and Plant Analysts [US]
CSTPC Cassette Magnetic Tape Controller
Cost Price
CSTR Canister
Committee on Solar Terrestrial Research [US]
Continuous-Flow Stirred Tank Reactor
Continuous Stirred Tank Reactor
CSTS Combined System Test Stand
Computer Systems Technical Support
Czechoslovak Scientific and Technical Society
CSTT Contaminated Sediment Treatment Technology
CSTUN Cost per Unit
CSU Cache Store Unit
California State University [US]
Central Switching Unit
Channel Service Unit
Check Switching Unit
Circuit Switching Unit
Cleveland State University [US]
Colorado State University [US]
Common Services Unit
Continuous Speech Understanding
Combined Shaft Unit
Crude Steel Unit
Customer Service Unit

	Customer Setup
CSUA	Computer Science Undergraduate Association [US]
CSUCA	Confederacion Universitaria Centroamericana [= Confederation of Central American Universities, Puerto Rica]
CSULB	California State University, Long Beach [US]
CSUOT	Central South University of Technology [PR China]
CSUR	College of Science, University of Ryukyus [Japan]
CSUS	California State University, Sacramento [US]
CSV	Cathodic Stripping Voltage
	Cathodic Stripping Voltammetry
	Chrysanthemum Stunt Viroid
	Circuit Switched Voice
	Corona Start Voltage
CSW	Channel Status Word
	Control Switch
	Crude Steel Weight
	Custom Switch Module
CSWA	Canadian Science Writers Association
CSWPRC	Chinese Society on Water Pollution Research and Control [PR China]
CSZ	Calcia-Stabilized Zirconia
CT	Cable Test
	Caloric Theory
	Canonical Transformation
	Canton
	Carrier Telegraph (Channel)
	Carrier Transmission
	Cassette Tape
	Cell Culture Tested
	Center Tap
	Ceramic Technologist
	Ceramic Technology
	Certified Teacher
	Charge Transfer (Complex)
	Chemical Technologist
	Chemical Technology
	Ciguatoxin
	Circuit Theory
	Cis-Trans (Configuration) [Chemistry]
	Cleaning Technician
	CLEM Transporter
	Coastal Telegrapher
	Code Telegram
	Coding Theory
	Cold Trap
	Colloidal Thorium
	Color Temperature
	Commercial Translator
	Commercial Traveller
	Communications Technician
	Communications Technologist
	Communications Technology
	Communications Terminal
	Compact Tension
	Compact Toroid
	Compact Type Specimen
	Computed Tomography
	Computerized Tomography
	Computer Technologist
	Computer Technology
	Concrete Technician
	Conference Title
	Connecticut [US]
	Continuous Transformation
	Continuous Time
	Constant Teeming
	Controlled Term
	Control Tag
	Control Term
	Control Theory
	Control Transformer
	Cooling Tower
	Cordless Telephone
	Count(er)
	Counter Tube
	Counting Time
	Court
	Creep Testing
	Crosstalk [also: XT]
	Current Transformer
	Cytogenetic Technologist
	Cytogenetic Technology
C/T	Carrier-to-Noise Temperature Ratio
C-T	Concentration-Time Ratio
cT	Continental Tropical Air
ct	(centum) — one hundred
	(cum tempore) — 15 minutes after the time announced
CTA	Calculated Time of Arrival
	California Trucking Association [US]
	Canadian Transit Association
	Canadian Trucking Association
	Centre Technique de Cooperation Agricole et Rural [= Technical Center for Agricultural and Rural Cooperation, Netherlands]
	Citraconic Anhydride
	Compatibility Test Area
	Computer Traders Association [UK]
	Concrete Technicians Association [US]
	Constant-Time Anemometer
	Control Area
	Cystine Trypticase Agar
CTAB	Canadian Technology Accreditation Board
	Cetyltrimethylammonium Bromide
	Commerce Technical Advisory Board [US]
CTAF	Common Traffic Advisory Frequency [US]
CTAHE	Ceramic Technology for Advanced Heat Engines
CTAK	Cipher Text Auto Key
CTAS	Canadian Thermal Analysis Society
CTB	Code Table Buffer
	Coherent Twin Boundary
	Concentrator Terminal Buffer
CTBM	Chief Testboard Man
CTBN	Carboxyl-Terminated Butadiene Acrylonitrile
CTBS	Canadian Test of Basic Skills
CTBT	Comprehensive Test Ban Treaty [US-USSR]
CTBUH	Council on Tall Buildings and Urban Habitat [US]
CTC	Canadian Transport Commission
	Carbon Tetrachloride
	Centralized Traffic Control
	Central Training Council [UK]

Central Trust of China [Taiwan]
Channel-to-Channel
Chargeable Time Clock
Chlorotetracycline
Cis-Trans-Cis (Geometry) [Chemistry]
Communications Transistor Corporation [US]
Compact Transpiration Cooling
Conditional Transfer of Control
Counter/Timer Circuit

CTCA Canadian Telecommunications Carrier Association
Ceramic Tile Contractors Association
Channel and Traffic Control Agency
Channel-to-Channel Adapter

CTCL Community and Technical College Libraries

CTCP Clinical Toxicology of Commercial Products [Joint Database of NHL and EPA, US]

CTCS Consolidated Telemetry Checkout System

CTCU Canadian Textile and Chemical Union

CTD Charged Tape Detection
Charge-Transfer Device
Chemical Thermodynamics
Combination Thermal Drive
Conductivity, Temperature and Density
Conductivity, Temperature and Depth
Continuity Tone Detector
Cross Track Distance

CTDA Ceramic Tile Distributors Association [US]

CTDAS Canadian Trade Document Alignment System

CTDH Command and Telemetry Data Handling

CTDS Code Translation Data System

C/TDS Count/Time Data System

CTE Cable Television Engineer
Channel Translating Equipment
Charge-Transfer Efficiency
Coefficient of Thermal Expansion
Computer Telex Exchange

CTEA Canadian Telephone Employees Association
Channel Transmission and Engineering Activation
Council of Technology Education Association [US]
Cyanotriethylammonium Tetrafluoroborate

CTEF Comite Technique Europeenne pour le Fluor [= European Technical Committee for Fluorine [of CEFIC, Belgium]]

CTEM Conventional Transmission Electron Microscope

CTF Canadian Tax Foundation
Cask Tilting Fixture
Certificate
Chlorine Trifluoride
Cleanroom Technology Forum [US]
Colloquium on Thin Films
Contrast Transfer Function
Controlled Temperature Furnace
Controlled Thermonuclear Fission
Cottonseed Flour

CTFE Chlorotrifluoroethylene

CTFM Continuous-Transmission, Frequency-Modulated

CTG Cartridge
Communications Task Group [of CODASYL]

CTGC California Table Grape Commission [US]

CTGE Cartage

C13-NMR Carbon-13 Nuclear Magnetic Resonance (Database) [of INKA, Germany]

CTI Canadian Textile Institute
Canadian Trade Index
Centralized Ticket Investigation
Centre de Traitement de l'Information [= Center for Information Processing, Belgium]
Comparative Tracking Index
Computerized Tomographic Imaging
Configuration and Tuning Interface
Cooling Tower Institute [US]
Current Technology Index

CTIA Canadian Textile Importers Association
Cellular Telecommunications Industry Association [US]

CTIAC Concrete Technology Information Analysis Center [US]

CTIC Canadian Translators and Interpreters Council

CTIF Comite Technique International de Prevention et d'Extinction du Feu [= International Technical Committee for the Prevention and Extinction of Fire, Switzerland]

CTIO Cerro Tololo Inter-American Observatory [Chile]

CTIOA Ceramic Tile Institute of America [US]

CTIS Computerized Transport Information System

CTL CAGE (= Compiler and Assembler by General Electric) Test Language
Cassette Tape Loader
Checkout Test Language
Cincinnati Testing Laboratories [US]
Commercial Tape Laying Machine
Complementary Transistor Logic
Constant Time Locus
Constructive Total Loss
Core Transistor Logic

CTLM Commercial Tape Laying Machine

CTLZ Control Zone

CTM Chemithermomechanics
Communications Terminal Module
Complete Treatment Module
Continuity Transceiver Module

CTMA Canadian Tooling Manufacturers Association
Cutting Tool Manufacturers Association [now: USCTI]

CTMC Canadian Tobacco Manufacturers Council
Communication Terminal Module Controller

CTMF Ceramic Tile Marketing Federation [US]

CTMP Chemithermomechanical Pulp

CTMS Carrier Transmission Maintenance System

CTMT Combined Thermomechanical Treatment
Containment

CTN Cable Termination Network
Carton

ctn cotangent

CTNE Compania Telefonica Nacional de Espana [= National Telephone Company of Spain]

CTO Charge Transforming Operator
Chief Technical Officer
Chief Technology Officer
Commonwealth Telecommunications Organization [UK]

Corporate Telecommunications Operation
Cut-Off
CTOA Crack Tip Opening Angle
C to C Center to Center
CTOD Crack Tip Opening Displacement
C to F Center to Face
CTOL Conventional Takeoff and Landing
CTON Character to Number (Conversion)
CTOS Cassette Operating System
Cassette Tape Operating System
CTP Carboxyl-Terminated Polybutadiene
Center for Theoretical Physics [of SNU, Korea]
Central Transfer Point
Character Table Pointer
Character Translation Pointer
Charge Transforming Parameter
Chemical Treatment Pond
Columbus Telescope Project [US]
Command Translator and Programmer
Construction Test Procedure
Controlled Temperature Profile
Coordinated Test Plan
Crystal Transition Point
Cyclohexylthiophthalimide
Cytidine Triphosphate
CTPB Carboxyl-Terminated Polybutadiene Binder
CTR California Tumor Registry [US]
Carbothermal Reduction
Center
Center for Transportation Research [of ANL, US]
Centering
Central Tumor Registry
Certified Test Requirement
Certified Test Results
Civil Tilt Rotor
Collective Television Reception
Computerized Tumor Registry
Contour
Controlled Thermonuclear Reaction
Controlled Thermonuclear Reactor
Council for Tobacco Research [US]
Counter
Crack-Thickness Ratio
CTRA Coal Tar Research Association [UK]
CTRAP Customer Trouble Report Analysis System
CTRF Canadian Transport Research Forum
CTRL Control
CTRS Contrast
CTR-USA Council for Tobacco Research — USA
CTS Cable Terminal System
Cable Transmission System
Cable Turning System
Canadian Technology Satellite
Carrier Test Switch
Chemically-Treated Steel
Clear-to-Send
Committee on the Teaching of Science [of ICSU, UK]
Common Test Subroutine
Communications and Tracking Subsystem
Communications Technology Satellite [US/Canada]
Communications Terminal Synchronous
Compact Tension Specimen
Computer Telewriter System [US]
Computer Typesetting
Contract Technical Services
Contralateral Threshold Shift
Contrast Transfer Function
Controlled Thermal Severity Test
Conversational Terminal System
Conversational Time Sharing
Counter-Timer System
Courier Transfer Station [US Military]
Curved Track Simulator
CTSA Crucible and Tool Steel Association [UK]
CTSF Crack-Tip Stress Field
CTSI Central Terminal Signaling Interface
CTS/RTS Clear-to-Send/Request-to-Send
CTSS Compatible Time Sharing System
Conversational Time Sharing System
CTST Critical Trade Skills Training [Canada]
CTT Cable Trouble Ticket
Central Trunk Terminal
Character Translation Table
Cis-Trans-Trans (Geometry) [Chemistry]
College of Trade and Technology
Consejo Technica de Telecommunicacion [= Technical Council for Telecommunications, Spain]
Core-Type Transformer
CTTC Canadian Trade and Tariffs Committee
CTTE Council of Technology Teacher Education [US]
CTTF Continuous Thermal Treatment Facility
CTTL Complimentary Transistor-Transistor Logic
CTTMA Canadian Truck and Trailer Manufacturers Association
CTU Caribbean Telecommunications Union
Cartridge Tape Unit
Centigrade Thermal Unit
Central Terminal Unit
Channel Testing Unit
Communications Terminal Unit
Compatibility Test Unit
CTUC Committee on Tunneling and Underground Construction [US]
Commonwealth Trade Union Council [UK]
CTV Canadian Television
Commercial Television
Control Test Vehicle
CTWT Counterweight
CTX Continuously Variable Transaxle
CTXCO Centrex Central Office
CTXCU Centrex Customer
CTXSN Centrex System Number
CU Cambridge University [UK]
Carleton University [Canada]
Chiba University [Japan]
Chittagong University [Bangladesh]
Chongqing University [PR China]
Chungkung University [PR China]
Clemson University [US]
Close-Up
Coefficient of Utilization

	Columbia University [US]
	Common User
	Community User
	Common Update
	Concordia University [Canada]
	Construction Unit
	Consumer Union
	Control Unit
	Cornell University [US]
	Crosstalk Unit
	Crystal Unit
	Cubic
	Customer
Cu	(Cuprum) — Copper
cu	cubic
CUA	Circuit Unit Assembly
	Compugraphics Users Association
	Computer Users Association
Cuam	Cuprammonium
CUAS	Computer Utilization Accounting System
CUB	Council for UHF Broadcasting [US]
	Cubic
	Cursor Backward
CUBE	Concertation Unit for Biotechnology in Europe [also: Cube]
	Cooperating Users of Burroughs Equipment [US]
CUBOL	Computer Usage Business-Oriented Language
CUC	Coal Utilization Council [UK]
	Computer Users Committee
	Cooperative Union of Canada
CUCCA	Canadian University and College Counselling Association
cu cm	cubic centimeter
CUD	Craft Union Department [of AFL-CIO, US]
	Cursor Down
CUDAT	Common User Data
CuDBC	Copper Dibutyl Dithiocarbamate
CuDMC	Copper Dimethyl Dithiocarbamate
CuDIP	Copper Diisopropyl Dithiophosphate
CUDN	Common User Data Network
CUDOS	Continuously-Updated Dynamic Optimizing System
CUE	Computer Updating Equipment
	Control Unit End
	Cooperating Users Exchange
CUEA	Coastal Upwelling Ecosystem Analysis
CUEBS	Commission on Undergraduate Education in the Biological Sciences [of AIBS, US]
Cuen	Cupriethylenediamine
CUEP	Central Unit on Environmental Pollution [UK]
CUES	Center for Urban Environmental Studies [now: NCUES, US]
	Computer Utility Educational System
CUETS	Credit Union Electronic Transaction Service [Canada]
CUEW	Canadian Union of Educational Workers
CUF	Cumuliform
	Cursor Forward
CUFC	Consortium of University Film Centers [US]
$Cu(FOD)_2$	Bis(1,1,1,2,2,3,3-Heptafluoro-7,7-Dimethyl-4,6-Octanedionato)copper (II)
CUFOS	Center for UFO (= Inidentified Flying Object) Studies [US]
CUFT	Center for the Utilization of Federal Technology [US]
cu ft	cubic foot
cu ft/lb	cubic feet per pound
CUG	Closed User Group
	Crosfield Users Group [US]
CUH	Chinese University of Hong Kong
CUICAC	Canadian University-Industry Council on Advanced Ceramics
CUIIPS	Center Unit for International and Investment Policies Studies [Canada]
cu in	cubic inch
CUIO	Canadian Unity Information Office
CUJT	Complementary Unijunction Transistor
CULO	Cornell University Laboratory of Ornithology [now: CLO, US]
CULP	Computer Usage List Processor
CuLPCN	Copper Leucophalocyanine
CULT	Chinese University Language Translation [Hong Kong]
CUM	Cumulative
cu m	cubic meter
CUMA	Canadian Underwater Mine Apparatus
	Canadian Urethane Manufacturers Association
CUMARC	Cumulated Machine-Readable Catalogue
Cumb	Cumberland [UK]
cum div	with dividend
CUMM	Council of Underground Machinery Manufacturers
cu mm	cubic millimeter
Cum Pref	Cumulative Preference
CUMREC	College and University Machine Records Conference [US]
cu mu	cubic micron
CUNY	City University of New York [US]
CUOE	Canadian Union of Operating Engineers and General Workers
CUP	Cambridge University Press [UK]
	Canadian University Press
	Communications User Program
	Cuban Peso [Currency]
	Cursor Position
CuP	Cuban Peso [Currency]
CUPE	Canadian Union of Public Employees
CUPID	Cornell University Program for Integrated Devices
CUPM	Committee on the Undergraduate Program in Mathematics [of MAA, US]
CU PL	Copper Plate
CUPTE	Canadian Union of Professional and Technical Employees
CUPW	Canadian Union of Postal Workers
CUR	Complex Utility Routine Current
	Council on Undergraduate Research [US]
	Currency
	Current
CURE	Citizens United to Reduce Emissions [of Formaldehyde Poisining Association, US]
	Color Uniformity Recognition Equipment
	Computer User Research and Evaluation Laboratory [Canada]

Council for Unified Research and Education
CURES Computer Utilization Reporting System
CURI College-University Resource Institute [US]
CURP Certificate in Urban and Regional Planning
CURR Current
CURR ASSETS Current Assets
CURR LIABS Current Liabilities
CURS Center for Urban and Regional Studies [US]
CURTS Common User Radio Transmission System
CURV Cable-Controlled Underwater Recovery Vehicle
Cable-Controlled Underwater Research Vehicle
CUS Clean-Up System
Common User System
Coordinatively Unsaturated Site
CUSCL Customer Class
CUSEA Council for University Students of East Africa
CUSEC Canada-United States Environmental Council
cusec cubic foot per second [also: CFS or cfs]
CUSIP Committee on Uniform Security Identification Procedures [US]
CUSO Canadian Universities Service Overseas
CUSP Central Unit for Scientific Photography [of RAE, UK]
Commonly Used System Programs
CUSRPG Canada-United States Regional Planning Group [of NATO]
CUSS Computerized Uniterm Search System [US]
Cooperative Union Serials System [of OULCS, Canada]
CUSSCO College, University and School Safety Council of Ontario [Canada]
CUST Customer
CUT Chalmers University of Technology [Sweden]
Circuit Under Test
Control Unit Tester
Continuous-Use Temperature
CUTA Canadian Urban Transit Association
CUTCD Council on Uniform Traffic Control Devices [US]
CUTE Common-Use Terminal Equipment
CUTS Canadian University Travel Service
Cassette User Tape System
Computer-Utilized Turning System
CUU Cursor Up
CUW Committee on Undersea Warfare [of USDOD]
cu yd cubic yard
CV Calibrated Volume
Calorific Value
Capacitance versus Voltage
Check Valve
Chief Value
Cited Volume
Coefficient of Variation
Color Vision
Computational Vision
Computerized Vision
Computer Vision
Condensing Vacuole
Constant Arc Voltage
Constant Value
Constant Velocity (Joint)
Constant Viscosity
Constant Voltage
Consumer Video
Continuously Variable
Continuous Vulcanization
Control Valve
Counter Voltage
Crest Value
Curriculum Vitae
(Single) Cotton Varnish (Wire)
Cyclic Voltammograph
C/V Capacitance versus Voltage
C-V Current-Voltage Relationship
Capacitance — Voltage
CVA California Vehicle Act [US]
Canadian Vocational Association
Commonwealth Veterinary Association
Computerized Vibration Analysis
CVBS Composite Video Broadcast Signal
CVC Carrier Virtual Circuit
Conserved Vector Current
Constant Voltage Conditioner
Continuously Variable Crown Technology
Current Voltage Characteristics
CV/CC Constant Voltage/Constant Current
CVCM Collected Volatile Condensable Materials
CVCS Chemical and Volume Control System
CVD Cash versus Documents
Chemical Vapor Deposition
Communication Valve Development [UK]
Compact Video Disk
Countervailing Duty
Coupled Vibration-Dissociation
Current-Voltage Diagram
CVDV Coupled Vibration-Dissociation-Vibration
CVF Controlled Visual Flight
CVFC Concord Video and Film Council [UK]
CVFR Controlled Visual Flight Rules
CVGB Cable Vault Ground Bar
CVI Certified Vendor Information
Chemical Vapor Infiltration
Continuous Variation of Incidence
CVIA Computer Virus Industry Association [US]
CVIC Conditional Variable Incremental Computer
CVIM Configurable Vision Input Module
CVIS Computerized Vocational Information System
CVM Cluster Variation Method
COBOL Virtual Machine
Consumable-Electrode Vacuum Melting
CVMA Canadian Veterinary Medical Association
CVN Carrier Vehicle, Nuclear-Powered
Charpy V-Notch Test
CVn Canes Venatici [Astronomy]
CVP Containment Vacuum Pump
Corporacion Venezolana de Petroles [= Petroleum Corporation of Venezuela]
CVPO Chemical Vapor Phase Oxidation
CVR Cockpit Voice Recorder
Continuous Vertical Retort
Continuous Video Recorder
Contrast Variation Ratio
CVS Chinese Vacuum Society

Constant Volume Sampling
Cyclic Voltametric Stripping

CVSA Canadian Vehicle Safety Alliance
Commercial Vehicle Safety Alliance [US]

CVSD Continuously Variable Slope Delta-Modulation [also: CVSDM]

CVSG Channel Verification Signal Generator

CVT Chemical Vapor Transport
Committee on Vacuum Techniques [US]
Communications Vector Table
Constant Voltage Transformer
Continuously Variable Transmission
Convertible
Crystal Violet Tetrazolium

CVTR Carolians Virginia Tube Reactor [US]

CVU Constant Voltage Unit

CVW Consumable-Electrode Vacuum Melting

CW Call Waiting
Carnauba Wax
Carrier Wave
Cell Wall
Chemical Warfare
Clean Water
Clockwise
Cold Water
Cold Welding
Cold Working
Command Word
Composite Wave
Compression Wave
Compound-Wound
Constant Wattage
Continuous Wave
Control Word
Cooling Water
Coupled Wave
C-Washer

CWA Canada Water (Database) [of Environment Canada]
Canadian Waferboard Association [now: TWA, Canada]
Canadian Warehousing Association
Communication Workers of America
Constant-Wattage Autotransformer
Construction Writers Association [US]
Control Word Address

CWAA Cotton Warehouse Association of America [US]

CWAP Clean Water Action Project [US]

CWAR Continuous Wave Acquisition Radar

CWARC Canadian Workplace Automation Research Center

CWAS Contractor Weighted Average Share

CWB Canadian Welding Bureau
Canadian Wheat Board
Ceramic Wiring Board

CWBAD Clockwise Bottom Angular Down

CWBAU Clockwise Bottom Angular Up

CWC Canadian Wood Council
Communications and Electrical Workers of Canada
Communications Work Committee [Canada]
Compound Water Cyclone

CWCG Copper Wire Counterpoise Ground

CWD Creosoted Wood Duct

CWDB Clockwise Down Blast

CWDC Canadian Wood Development Council

CWDI Canadian Welding Development Institute [now: WIC, Canada]

CWED Cold Weld Evaluation Device

CWEI Canadian Wood Energy Institute

CWF Canadian Wildlife Federation
Coal-Water Fuel Technology
Construction Workers Federation [San Marino]

CWFM Continuous Wave/Frequency Modulation [also: CW/FM]

CWG Closed Waveguide
Committee for Women in Geophysics

CWI Call Waiting Indicator
China Welding Institute [PR China]
Clean World International [UK]

CWIF Continuous Wave Intermediate Frequency

CWIP Construction Work In-Progress

CWL Continuous Working Level

CWLS Canadian Well Logging Association

CWM Compound-Wound Motor

CWMC Canadian Waste Management Conference

CWME Canadian Waste Materials Exchange

CWMTU Cold-Weather Material Test Unit

CWO Capital Work Order
Cash with Order
Continuous Wave Oscillator
Custom Work Order

CWP Central West Pacific (Ocean Region)
Circulating Water Pump
Coalworker's Pneumoconiosis
Cold-Work Peak
Communicating Word Processor
Contractor Work Plan
Cooling Water System
Current Word Pointer

CWPCA Canadian Wood Pallet and Container Association

CWRA Canadian Water Resources Association

CWRC Central Water and Power Commission [India]
Chemical Warfare Review Council

CWRU Case Western Reserve University [US]

CWS Canadian Welding Society [now: WIC, Canada]
Canadian Wildlife Service
Caucus for Women in Statistics [US]
Chemical Warfare Service
Colonial Waterbird Society [Canada]
Complex Waveform Simulator
Cooperative Wholesale Society
Copper Weld Steel

CWSF Canada Wide Science Fair
Coal-Water Slurry Fuels

CWT Character Width Table
Coded Wire Tagging
Cooperative Wind Tunnel
Critical Water Temperature

cwt hundredweight

CWTAD Clockwise Top Angular Down

cwtad hundredweight, air dry

CWTAU Clockwise Top Angular Up

CWTC Chemical Waste Transportation Council [US]

CWTG Computer World Trade Group [UK]

CWTH Clockwise Top Horizontal
cwtn hundredweight net
CWU Communication Workers Union [Trinidad]
CWUB Clockwise Up Blast
CWV Continuous Wave Video
CWWA Canadian Water Well Association
CWWIA Cyprus Woodworking Industry Association
CWY Clearway
CX Central Exchange
Character Transfer
Compatible Expansion
Composite
Composite Signalling
Control Transmitter
CXA Central Exchange Area
CX-HLS Cargo Experimental, Heavy Logistic Support [US Air Force]
CXR Carrier
CXT Common External Tariff
CY Calendar Year
Copy
Country
Currency
Cycle
Cyclophoshamide
CYA Canadian Yachting Association
CYBORG Cybernetic Organism [also: Cyborg or cyborg]
CYC Cyclonic
CYCLAM 1,4,8,11-Tetraazacyclotetradecane
CYCLYD Cyclic Alkyd
Cyg Cygnus [Astronomy]
Cyg X-1 Cygnus X-1 [Astronomy]
Cyg X-3 Cygnus X-3 [Astronomy]
CYL Cylinder
CYL L Cylinder Lock
Cyn Cyanide
CYO Constant Yield Oil
CYP Cyprus Pound [Currency]
Cys Cysteine
Cysteinyl
CYTOL Cytological
Cytologist
Cytology
CZ Canal Zone [Panama]
Commutating Zone
Control Zone
Czochralski
CZC Chromated Zinc Chlorate
CZCS Coastal Zone Color Scanner
CZE Compare Zone Equal
CZR Central Zone Remelting
CZU Compare Zone Unequal

D

D	Asparagine Acid
	Aspartyl
	Dalasi [Currency of Gambia]
	Data
	Day
	December
	Decimal
	Declination
	Dee
	Delay
	Delete
	Demand
	Dendrite
	Density
	Derivative
	Descending
	Destination
	Detection
	Detector
	Deuterium
	Deviation
	Devonian [Geology]
	Dextrorotatory
	(Outside) Diameter
	Difference
	Differential
	Diffusion
	Diffusivity
	Digit(al)
	Diode
	Diopter
	Direction
	Dispersion
	Displacement
	Display
	Dissipation
	Distillation
	Distortion
	Dividend
	Division
	Doctor
	Domain
	Dong [Currency of Vietnam]
	Double
	Drain
	Drizzle
	Drop
	Drum
	Dust
d	dam
	date
	daughter [Radioactive Elements]
	day
	deci- [SI Prefix]
	decompose
	deficit
	degree
	delete
	density
	depart
	depth
	derivative
	deuteron
	dextrorotary
	(inside) diameter
	differential coefficient
	dimensional
	distance
	division
	double
	drain
	dyne [Unit]
2,4-D	2,4-Dichlorophenoxyacetic Acid
DA	Danger Area
	Data Acquisition
	Data Administrator
	Data Area
	Data Available
	Dealers Alliance [US]
	Decimal Add(er)
	Decimal-to-Analog
	Decision Altitude
	Decylamine
	Define Address
	Define Area
	Delay Amplifier
	Demand Assignment
	Department of the Army [US]
	Design Automation
	Destination Address
	Dielectric Absorption
	Differential Amplifier
	Differential Analyzer
	Digit Absorbing
	Digital-to-Analog
	Dinar [Currency of Algeria]
	Direct Access
	Direct Address
	Directional Antenna
	Directory Assistance
	Disaccommodation
	Discharge Afloat
	Discrete Address
	Display Adapter
	Dissolved Acetylene
	Doctor of Administration
	Doctor of Archeology
	Doctor of Arts
	Document Analyst
	Documents against Acceptance
	Dodecanedioic Acid
	Doesn't Answer
	Dopamine
	Double Amplitude

Double Armor
Duplex Annealing
D/A Deposit Account
Digital-to-Analog
Discharge Afloat
Documents Attached
D-A Decimal-to-Analog
Digital-to-Analog
Document Against Acceptance
Da Damkohler Number
Darcy Number
dA Deoxyadenosine
da deca- [SI Prefix]
DAA Danish Astronomical Society
Data-Access Arrangement
Diacetone Acrylamide
Diacetone Alcohol
Direct-Access Arrangement
Dominion Auto Association [Canada]
Durene Association of America [US]
DAAB Diazoaminobenzene
DAAD Deutscher Akademischer Austauschdienst [= German Academic Exchange Service]
DAAG Deputy Assistant-Adjutant-General
DAAIS Danger Area Activity Information Service
DAAM Diacetone Acrylamide
DAAS Drilling Activity Analysis System [US]
DAB Diaminobenzidine
Diaminobenzene
Dictionary of American Biography
Digital Audio Broadcasting
Display Arrangement Bits
Display Assignment Bits
Display Attention Bits
DABA Diaminobenzanilide
DABCC Diploma of American Board of Clinical Chemists
DABCO 1,4-Diazabicyclooctane
DABITC Dimethylaminoazobenzene Isothiocyanate
DABL Daisy Behavioral Language
DABPath Diploma of American Board of Pathology
DABPathAnal Diploma of American Board of Pathological Anatomy
DABPH Diploma of American Board of Public Health
DABR Diploma of American Board of Radiology
DABS Discrete Address Beacon System
DABSYL 4-(Dimethylamino)azobenzene-4'-Sulfonyl
DAC Danish CAA (= Civil Aviation Authority)
Data Acquisition
Data Acquisition and Control (System)
Data Acquisition Center
Data Analysis Computer
Data Authentication Code
Delayed Atomization Cuvette
Demand Assignment Controller
Derived Air Concentration
Design Augmented by Computer
Development Assistance Committee [of OECD, France]
Diallyl Chlorendate
Digital Amplitude Curve
Digital Analysis Converter
Digital Arithmetic Center
Digital-to-Analog Conversion
Digital-to-Analog Converter
Display Analysis Console
Distance Amplitude Compensation
Distance Amplitude Correction
DA/C Data Acquisition
DACBU Data Acquisition Control and Buffer Unit
DACC Direct-Access Communications Channel
DACE Data Acquisition and Control Executive
Data Administration Center Equipment
DACH 1,2-Diaminocyclohexane
DACM 7-Dimethylamino-4-Methyl-3-coumarinyl)-maleimide
DACOM Data Communications
Datascope Computer Output Microfilmer
DACON Data Controller
Digital-to-Analog Converter
DACOR Data Correction
Data Correlator
DACOS Data Communication Operating System
DACS Danger Area Crossing Service
Data Acquisition and Control System
Data and Analysis Center for Software
Digital Access and Cross-Connect System
Directorate of Aerospace Combat Systems
DACU Digitizing and Control Unit
DACVR Digital-to-Analog Converter
DAD Digital-Assisted Dispatching
Digital Audio Disk
Diode Array Detector
Direct-Access Device
Discharged Area Development
Drum and Display
DADB Data Analysis Database
DADC Digital Air Data Computer
DADEE Dynamic Analog Differential Equation Equalizer
DAdm Doctor of Administration
DADPS Diaminodiphenylsulfone
DADPU Diaminodiphenylurea
DADPTU Diaminodiphenylthiourea
DADS Data Acquisition Display System
Data Acquisition Display Subsystem
Diaminodiphenylsulfone
Digital Air Data System
Digitally-Assisted Dispatch System
DADSM Direct-Access Device Space Management
DADT Durability and Damage Tolerance
DADTA Durability and Damage Tolerance Assessment
DAE Diacetic Acid
DAEd Doctor of Arts in Education
DAEMON Data Adaptive Evaluator and Monitor
DAEP Directorate of Aircraft Equipment Production [UK]
Division of Atomic Energy Production [UK]
DAF Data Acquisition Facility
Denmark-America Foundation [Denmark]
Destination Address Field
Direct-Arc Furnace
Discard at Failure
Dissolved-Air Flotation
Distributed Acquisition Facility

Dry and Ash-Free
DAFA Data Accounting Flow Acceptance
DAFC Digital Automatic Frequency Control
DAFCS Digital Automatic Flight Control System
DAFM Direct-Access File Manager
Discard-at-Failure Maintenance
DAFS Deepwater Actively Frozen Seabed
Department of Agriculture and Fisheries of Scotland
Department of Agriculture and Forest Service
DAFT Digital/Analog Function Table
DAG Deputy Adjutant General
Deputy Assistant General
Development Assistance Group
Distance Amplitude Gate
DAGAS Dangerous Goods Advisory Service [of LGC, UK]
DAGC Delayed Automatic Gain Control
DAH Data Acquisition Hardware
DAHQ Di-tert-amylhydroquinone
DAI Dissertation Abstracts International
DAIM Digital Acceleration Integration Module
DAIP Diallyl Isophthalate
DAIR Direct Altitude and Identity Readout [of FAA, US]
Driver Aid, Information and Routing
Dynamic Allocation Interface Routine
DAIRS Dial Access Information Retrieval System
DAIS Defense Automatic Integrated Switch
Defense Automatic Integrated Switching System
Digital Avionics Information System
District Agricultural Improvement Station [PR China]
Doctor of Arts in Information Science
DAISY Dairy Information System [UK]
Data Acquisition and Interpretation System
Decision-Aided Information System
Digital Acoustic Imaging System
Double-Precision Automatic Interpretative System
DAIV Data Area Initializer and Verifier
DAK Decision Acknowledgement
Deny All Knowledge
Dak Dakota [US]
DAL Data-Access Line
Data Address List
Diallyl Phthalate
Direct Address Line
daL decaliter
DALK Data Link Controller
DALR Dry Adiabatic Lapse Time
DAM Damage
Data Addressed Memory
Data Association Manager
Data Association Message
Descriptor Attribute Matrix
Diallyl Maleate
Diaminomesitylene
Digital-to-Analog Multiplier
Direct-Access Memory
Direct-Access Method
Doctor of Applied Management
Dual Absorption Model
dam dekameter
DAMA Demand-Assignment Multiple Access
Dialkyl Methylamine
DAM CON Damage Control
DAMDA Dairy Appliance Manufacturers and Distributors Association [UK]
DAME Data Acquisition and Monitoring Equipment
DAMIT Data Analysis Massachusetts Institute of Technology [US]
DAMN Diaminomaleonitrile
DAMP Database of Atomic and Molecular Physics [of Queen's University, Northern Ireland]
Downrange Anti-Missile Program [US Army]
d-AMP Deoxyadenosine Monophosphate
DAMPS Data Acquisition Multiprogramming System
DAMS Defense against Missile Systems
Direct-Access Management System
DAMSU Digital Automanual Switching Unit
DAN Danish
Denmark
Distribution Analysis
daN decanewton
d and a dry and abandoned [Oil and Gas Industry]
DANDOK Danish Committee for Scientific and Technical Information and Documentation
DANRIC Department of Agriculture and Natural Resources Information Council [Philippines]
DANS Dimethylaminonaphthalenesulfonyl Chloride
DANSYL 5-Dimethylamino-1-Naphthalenesulfonyl
DANTEC Danish Center for Technical Information [Denmark]
DANTHRON 1,8-Dihydroxyanthraquinone
DAO Dioctadecyladipate
Dominion Astrophysical Observatory [Canada]
DAP Data-Access Protocol
Data Acquisition Processor
Data-Analysis Package
Deformation of Aligned Phases
Diallyl Phthalate
Diaminophenol
Diaminopimelate
Diammonium Phosphate
Diammonium Phosphate Plant
Diamylphthalate
Dibasic Ammonium Phosphate
Digital Access Point
Digital Assembly Program [of EAI, US]
Dihydroxyacetone Phosphate
Directionally Attached Piezoelectric (Element)
Distributed Array Processor
Division of Air Pollution [of USDOE]
Donor-Acceptor Pair
DAPA Diaminodipropylamine
DAPCA Development and Production Costs for Aircrafts
DAPD Directorate of Aircraft Production Development [UK]
DAPDA Dimethylaminophenylpentadienal
DAPHC Diaminophenol Hydrochloride
DAPI 4,6-Diamidino-2-Phenylindole
Diaminophenylindane
DAPIS Danish Agricultural Products Information Service
DAPN Diamylmethylpropanol

	Dimethylaminonaphthalene Sulfonylamino-ethylamino Naphthaleneoxypropanol
DAPO	Digital Advance Production Order
DAPP	Data Acquisition and Processing Program
DAPR	Digital Automatic Pattern Recognition
DAPS	Direct-Access Programming System
	Distributed Application Processing System
DApSc	Doctor of Applied Science
DAPSONE	4,4'-Diaminodiphenyl Sulfone [also: DDS]
DAPU	Data Acquisition and Processing Unit
DAQMG	Deputy Assistant Quartermaster-General
DAR	Daily Activity Report
	Damage Assessment Routine
	Data Access Register
	Defense Acquisition Radar
	Department of Applied Research
	Destination Access Register
	Diacetylresorcinol
	Directorate of Atomic Research
DARA	Deutsche Agentur fur Raumfahrt-Angelegenheiten [= German Agency for Space Related Matters]
DARC	Direct-Access Radar Channel
DArch	Doctor of Architecture
DARE	Data Retrieval System
	Document Abstract Retrieval Equipment
	Documentation Automated Retrieval Equipment
	Doppler and Range Evaluation
	Doppler Automatic Reduction Equipment
DAREOD	Damaged Airfield Reconnaissance and Explosive Ordnance Disposal
DARES	Data Analysis and Reduction System
DARLI	Digital Angular Readout by Laser Interferometry
DARME	Directorate of Armament Engineering [Canada]
DARMS	Digital Alternate Representation of Musical Symbols
DARPA	Defense Advanced Research Projects Agency [US]
DARPA/ESTO	Defense Advanced Research Projects Agency/ Electronic Systems Technology Office [of USDOD]
DARS	Digital Adaptive Recording System
	Digital Attitude and Rate System
	Digital Attitude Reference System
DARSS	Diode Array Rapid-Scan Spectrometer
DART	Daily Automatic Rescheduling Technique
	Data Accumulation and Retrieval of Time
	Data Acquisition and Recording Terminal
	Data Acquisition in Real Time
	Data Analysis Recording Tape
	Digital Automatic Readout Tracker
	Direct-Access Radio Transceiver
	Director and Response Tester
	Dual Axis Rate Transducer
	Dynamic Acoustic Response Trigger
DARTS	Digital Azimuth Range Tracking System
DAS	Danish Automation Society
	Data Access Security
	Data Acquisition System
	Data Acquisition Subsystem
	Data Analysis Software
	Data Analysis System
	Data Automation System
	Data Auxiliary Set
	Datatron Assembly System
	De-Ashing Solvent
	Delivered Alongside Ship
	Dendrite-Arm Spacing
	Department of Applied Science
	Device Access Software
	Diacetylsulfide
	Diaminostilbene
	Diaminostilbenedisulfonic Acid
	Differential Absorption and Scattering
	Digital Analog Simulator
	Digital Attenuator System
	Dimer Adatom Stacking-Fault Model
	Directory Assistance System
	Discount Access Services
	Disturbance Analysis System
	Dry, As-Molded
DASA	Defense Atomic Support Agency [of USDOD]
DASC	Defense Automotive Supply Center
	Direct Air Support Center
DASc	Doctor of Applied Science
DASCOTAR	Data Acquisition System Correlating Tracking and Ranging
DASD	Direct-Access Storage Device
DASDL	Data and Structure Definition Language
DASDR	Direct-Access Storage Dump Restore
DASF	Direct-Access Storage Facility
DASH	Direct Access Storage Handler
	Drone Anti-Submarine Helicopter
	Dual Access Storage Handling
DASHER	Dynamic Analysis of Shells of Revolution
DASL	Data-Access System Language
DASM	Direct-Access Storage Media
DASS	Demand-Assignment Signaling and Switching (Device)
	Diesel Air Start System
	Digital Access Signalling System
DAST	Diethylaminosulfur Trifluoride
	Diploma of Advanced Studies in Teaching
DASt	Deutscher Ausschuss fur Stahlbau [= German Committee for Steel Construction]
DAT	Datum
	Designation — Acquisition — Track
	Diaminotoluene
	Digital Audio Tape
	Director of Advanced Technology
	Disconnect Actuating Tools
	Disk Allocation Table
	Dynamic Address Translation
	Dynamic Address Translator
DATA	Development and Technical Assistance
	Draftsmen and Allied Technicians Associations [UK]
DATAC	Data Analog Computer
	Digital Automatic Tester and Classifier
	Digital Autonomous Terminal Access Communication
DATACOL	Data Collection System
DATACOM	Data Communication
	Data Communication Network [US Air Force]
	Data Communication Service [US]
	Data Communication System

DATAGEN Data File Generator
DATAN Data Analysis
DATANET Data Network
DATAR Digital Automatic Tracking and Ranging [also: Datar or datar]
Digital Auto-Transducer and Recorder
DATAROM Data Read-Only Memory
DATAS Data Link and Transponder Analysis System
DATCO Duty Air Traffic Control Officer
DATCOM Data Support Command [US Army]
DATD Diallyltartardiamide
DATDC Data Analysis and Technique Development Center
DATEL Data Telecommunication [also: Datel]
DATEX Data Text [also: Datex]
DAtF Deutsches Atomforum [= German Nuclear Forum]
DATICO Digital Automatic Tape Intelligence Checkout
DATR Design Approval Test Report
DATRAN Data Transmission
DATRIX Direct Access to Reference Information
DATS Danish Academy of Technical Sciences
Despun Antenna Test Satellite
Dynamic Accuracy Test Set
DATSA Depot Automatic Test System for Avionics
DATTS Data Acquision, Telecommand and Tracking Station
DAU Data Access Unit
Data Acquisition Unit
Data Adapter Unit
Disposable Adsorbent Unit
DAUG Danger Area Users Group
DAV Data above Voice
Data Valid
Deutsche Arbeitsgemeinschaft Vakuum [= German Working Group on Vacuum Technology]
DAVC Delayed Automatic Volume Control
DAVI Dynamic Antiresonant Vibration Isolator
DAVID Data Above Video System
DAVIE Digital Alphanumeric Video Insertion Equipment
DAVSS Doppler Acoustic Vortex Sensing Equipment
DAW Data Address Word
Deutsche Akademie der Wissenschaften [= German Academy of Sciences]
Dry Active Waste
DAWNS Design of Aircraft Wing Structures
DAX Data Acquisition and Control
DAZD Double Anode Zener Diode
DB Databank
Database
Data Bus
Day Book
Dead Band
Dead Beat
Decimal-to-Binary
Delayed Boiling
Delayed Broadcast
Deutsche Bundesbahn [= German Federal Railways]
Deutsche Bundesbank [= German Federal Bank]
Deutsche Bundespost [= German Federal Post]
Dichlorophenoxybutyric Acid
Diffusion Bonding
Dip Brazing
Distribution Box
Doppler-Broadening
Double-Biased (Relay)
Double Bottom
Double Braid (Wire)
Double-Break
Drilling Barge
Dry Bulb
Dumbbell
Dynamic Braking
Dynamic Breaking
D-B Decimal-to-Binary
dB decibel
2,4-DB 2,4-Dichlorophenoxybutyric Acid
$B Belize Dollar [Currency of Belize]
DBA Database Administration
Design Basis Accident
Dibenzanthracene
Dibenzylamine
Dibutyl Amine
Doctor of Business Administration
Dolichos Biflorus Agglutinin
dBa decibel, adjusted [also: dBA]
DBAAM Disk Buffer Area Access Method
DBACS Database Administrator Control System
DBAF Database Access Facility
dBa(F1A) dBa measured by set with F1A line weighting
dBa(H1A) dBa measured by set with H1A receiver weighting
DBAM Database Access Method
Database Access Module
DBAO Digital Block AND-OR Gate
dBa0 dBa at zero transmission level
DBAWG Database Administration Working Group [of CODASYL]
DBB Detector Back Bias
Detector Balanced Bias
DBC Databank COMECON [also: DB COMECON]
Database Computer
Data Bridging Capacity
Data Bus Controller
Dense Barium Crown (Glass)
Diameter Bolt Circle
Digital-to-Binary Converter
Dynamic-Breaking Contactor
dBC C-scale sound level in decibels
DBCB Database Control Block
DBCL Database Command Language
DB COMECON Databank Council for Mutual Economic Assistance
DBCP Dibromochloropropane
DBCS Database Control System
DBD Database Definition
Database Description
Database Diagnostics
Double Backscattering Diffractometer
Double-Base Diode
DBDA Design Basis Depressurization Accident
Dibenzyldodecylamine
DB/DC Database/Data Communications
DBDL Database Definition Language

DBE Dame Commander of the Order of the British Empire
Databank Eurocontrol
Design Basis Earthquake
Design Basis Event
Development Bank of Ethiopia
DBER Division of Biomedical and Environmental Research [of ERDA, US]
DBF Demodulator Band Filter
Design Basis Fault
Dynamic Beam Focus
Dynamic Beam Forming
dBf decibels referred to one femtowatt
DBFAS Digital Beam-Focusing Array Signal
DBFS Deep Bed Farming Society [US]
Dull Black Finish Slate
DBG Database Generator
DBGM Deutsches Bundesgebrauchsmuster [= German Registered Pattern or Design]
DBH Diameter at Breast Height
Dopamine Beta-Hydroxylase
DBHI Dopamine Beta-Hydroxylase Inhibitor
DBI Database Index
Database Integrity
Deutsches Bibliotheksinstitut [= German Library Institute]
Differential Bearing Indicator
Double-Byte Interleaved
DBIN Data Bus In
DBIOC Database Input/Output Control
dBJ decibel, Jerrold
DBK Deutsche Bibliotheks-Konferenz [= German Library Conference, Germany]
dBk decibels referred to one kilowatt
dB/K decibels per degree Kelvin
DBL Database Locking
Double
DBLR Doubler
DBLTG Database Language Task Group [of CODASYL]
DBM Database Management
Database Model
Data Buffer Module
Data Bus Monitor
Dead-Burnt Magnesia
Dibutyl Maleate
Direct Bombardment Mode
Direct Branch Mode
dBm decibels relative to one milliwatt
DBMC Data Bus Monitor/Controller
Di-tert-Butyl-M-Cresol
Di-tert-Butyl-Methylphenol
dB/mi decibels per mile
dBm/m² dbm per square meter
dBm/m²/Mhz dBm per square meter per megahertz
dBm(PSOPH) dBm measured by a set with psophometric weighting
DBMS Database Management Software
Database Management System
DBMSPSM Database Management System Problem Specification Model
dBm0 dBm at zero transmission level
dBm0p dBm at zero transmission level, psophometrically weighted
dBm0ps dBm at zero transmission level, psophometrically weighted for sound program
dBμV dB above one microvolt
dBμV/MHz dB above one microvolt per megahertz
DBN DEC (= Digital Equipment Corporation) Business Network
1,5-Diazabicyclonon-5-ene
DBO Drop Build-Out Capacitor
DBoA Delayed Breeder or Alternative
DBOMP Database Organization and Maintenance Processor
DBOS Disk-Based Operating System
DBP Database Processor
Deutsche Bundespost [= German Federal Post]
Deutsches Bundespatent [= German Federal Patent]
Dibutyl Phthalate
dBp decibels above 1 picowatt
DBPC Di-tert-Butyl-P-Cresol
DBPCB Database Program Communication Block [also: DB/PCB]
dBQ decibels relative to a reference voltage, measured with quasi-peak noise meter [also: dBq]
DBR Database Retrieval
Directorate of Biosciences Research [of DRB, Canada]
Distributed Bragg Reflector
Division of Building Research [of NRC, Canada]
dBr decibels relative to reference [also: dBR]
dBrap decibels above reference acoustical power [also: dBRAP]
dBrn decibels above reference noise [also: dBRN]
dBrnC decibels above reference noise, C-message weighted [also: dBRNC]
dBrnC0 decibels above reference noise, C-message weighted, at zero transmission level [also: dBRNC0]
DBS Database Supplier
Database System
Data Bridging Service
Dibutyl Sebacate
Direct Broadcast by Satellite
Direct Broadcasting System
Direct Broadcast Satellite
Division of Biological Sciences [of NRC, Canada]
Division of Biological Standards
Dodecylbenzene Sulfonate
Dominion Bureau of Statistics [now: StatCan]
Drill/Bolt/Screening Head
DBSD Double Backscattering Diffractometer
DBSH Benzene-1,3-Disulfohydrazide
DBSP Double-Based Solid Propellant
DBT Design Basis Tornado
Dry Bulb Temperature
Dry Bulb Thermometer
Ductile-to-Brittle Transition
DBTG Database Task Group [of CODASYL]
DBTT Ductile-to-Brittle Transition Temperature
Ductile-to-Brittle Transition Transformation
DBTU Dibutyl Thiourea

DBU 1,8-Diazabicycloundec-7-ene
Digital Buffer Unit
DBUT Database Update Time
DBV Deutscher Beton-Verein [= German Concrete Association]
Deutscher Bibliotheksverband [= German Library Association]
Doppler Broadening Velocity
dBV decibels relative to one Volt
DBW Design Bandwidth
dBW decibels relative to one Watt
DBX Digital Branch Exchange
dBx decibels above reference coupling
DC Damage Control
Danube Commission [Hungary]
Dartmouth College
Data Cartridge
Data Channel
Data Check
Data Classifier
Data Code
Data Collection
Data Communication
Data Control
Data Conversion
Dead Center
Decimal Classification
Decimal Code
Decorators Club [US]
Defect Control
Define Contrast
Deflector Coil
Delay Cable
Depolarization Current
Deposited Carbon
Descriptor Code
Design Change
Design Contractor
Desk Checking
Detail Condition
Device Control (Character)
Die Casting
Digital Communication
Digital Comparator
Digital Computer
Digital Controller
Digital Counter
Dimensional Coordination
Diphenylarsine Cyanide
Direct-Chill (Casting) [also: dc]
Direct Connection
Direct Coupling
Direct Current [also: dc, D-C or d-c]
Directional Coupler
Direction Cycle
Directory Clearinghouse [US]
Disk Cartridge
Disk Controller
Dispatcher Console
Display Code
Display Console
Distillation Column
District of Columbia [US]
Double Channel
Double Column
Double Contact
Double Cotton (Wire)
Double-Crucible
Driver Control
Drought Code
Duty Cycle
D&C Display and Control
D/C Double Column
Down-Converter
Drift Correction
dC Deoxycytidine
d/c double column
DCA Data Communications Administrator
Decade Counting Assembly
Defense Communications Agency [US]
Deoxycorticosterone Acetate
Device Cluster Adapter
Dichloroacetic Acid
Dichloroaniline
Digital Command Assembly
Digital Computer Association
Diploma in Computer Application
Direct Colorimetric Analysis
Directorate of Civil Aviation [UK]
Distributed Communications Architecture
Document Content Architecture
Diploma in Computer Application
Driver Control Area
Dynamic Contact Analyzer
DCAA Defense Contract Audit Agency [US]
DCAD Dichlorobenzaldehyde
DCA/DIA Document Content Architecture/Document Interchange Architecture
DCAM Data Collection Access Method
DCAOC Defense Communications Agency Operations Center [US]
DCAS Deputy Commander Aerospace System
Digital Control and Automation System
DCASA Dyers and Colorists Association of South Africa
DCAT Drug, Chemical and Allied Trade Association [US]
DCB Data Control Block
Data Control Bus
Defense Communications Board [US]
Define Control Block
Dichlorobenzene
Dichlorobutane
Dicyanobenzene
Double-Cantilever Beam
Drawout Circuit Breaker
DCBA Dichlorobenzaldehyde
Dichlorobenzoic Acid
DCBD Define Control Block Dummy
DCBRE Defense Chemical, Biological and Radiation Establishment [of DRB, Canada]
DCBRL Defense Chemical, Biological and Radiation Laboratories [now: DCBRE, Canada]
DCBS N,N-Dicyclohexyl-2-Benzothiazole Sulfenamide

DCBTF Dichlorobenzotrifluoride
DCC Data Channel Converter
Data Circuit Concentration
Data Communication Channel
Data Communications Controller
Defense Construction Canada
Department of Culture and Communications [Canada]
Destination Code Cancelled
Device Cluster Controller
Device Control Character
1,3-Dicyclohexylcarbodiimide
Digital Cross Correct
Digital Cross Current
Direct Computer Control
Direct Contact Condensation
Direct Control Channel
Display Channel Complex
District Communications Center
Double Cotton Covered (Wire)
Droplet Counter-Current Chromatograph
DCCA Design Change Cost Analysis
DCCC Data Communication Control Character
Double Current Cable Code
DCCE Department of Chemistry and Chemical Engineering
DCCEAS District of Columbia Council of Engineering and Architectural Societies [US]
DCCI Dichlorohexylcarbodiimide
DCC-MSF Direct Contact Condensation — Multistage Flash
DCCS Defense Communications Control System [US Air Force]
Digital Command Communications System
Distributed Capacity Computing System
DCCU Data Communications Control Unit
DCD Data Carrier Detect
Detector-Cooling Dewar
Digital Coherent Detector
Double Channel Duplex
Double Crystal Diffractometer
Dynamic Computer Display
DCDA Data Communication Dealers Association [US]
Dicyanodiamine
DCDAS Dicyanodiamine Sulfate
DCDC Data Communication to Disk Control
DC-DC Direct Current to Direct Current [also: DC/DC or DCDC]
DCDES Dichlorodiethyl Sulfide
DCDFM Dichlorodifluoromethane
DCDMA Diamond Core Drill Manufacturers Association [US]
DCDR Data Collection and Data Relay
DCDS Digital Control Design System
Double Cotton Double Silk (Wire)
DCE Dallas Cotton Exchange [US]
Data Circuit-Terminating Equipment
Data Communications Equipment
Data Conversion Equipment
Department of Ceramic Engineering
Department of Chemical Engineering
Department of Civil Engineering
Dichloroethane
Dichloroethylene
Die Casting Engineer
Digital Control Element
Directorate of Conservation and Environment [of DND, Canada]
Domestic Credit Expansion
DCEE Department of Civil and Environmental Engineering
DCEL Direct Current Electroluminescent
DCEN Direct Current Electrode Negative
DCEO Defense Communications Engineering Office [US]
DCEP Direct Current Electrode Positive
DCES Delaware Council of Engineering Societies [US]
DCF Data Count Field
DeCarb-Free Product
Discounted Cash Flow
Disk Control Field
Disk Controller Formatter
Distributing Center Facility
Document Composition Facility
Droplet Combustion Facility
DCFL Direct-Coupled FET (= Field-Effect Transistor) Logic
DCFN Developing Countries Farm Radio Network
DCFROR Discounted Cash Flow Rate of Return
DCG Dependent Charge Group
Design Coordination Group
Dichromated Gelatin
Diode Capacitor Gate
Direct-Current Generator
DCGD Direct-Current Glow Discharge [also: DC-GD]
DCGE Dicresyl Glycerin Ether
DCGEM Directorate of Clothing, General Engineering and Maintenance
DCH Data Channel
Dichlorohydrin
DCHA Dicyclohexylamine
DChem Doctor of Chemistry
DCHP Dicyclohexyl Phthalate
DCI Data Communication Interrogate
Defense Computer Institute [US]
Dichloroisoproterenol
1-(3',4'-Dichlorophenyl)-2-Isopropylaminoethanol
Direct Channel Interface
Directorate of Chemical Inspection [UK]
DCIB Data Communication Input Buffer
DCID Director Central Intelligence Directive [US]
DCIEM Defense and Civil Institute of Environmental Medicine [of DND, Canada]
DCIO Direct Channel Interface Option
DCIP Dichlorophenol-Indilphenol
DCIS Department of Computing and Information Sciuence
DCKP Direct Current Key Pulsing
DCL Declaration
Delayed Call Limit
Diamond Cut Lug (Refractory)
Digital Command Language
Digital Computer Laboratory
Digital Control Laboratory
Dual Current Layer
DCLC Drift Cyclotron Loss Cone (Instability)
DClChem Doctor of Clinical Chemistry

DCLU Digital Carrier Line Unit
DCM Data Communications Multiplexer
Direct Current Noise Margin
Diagnostic Control Module
Dichloromethane
Digital Capacitance Meter
Digital Command Language
Direction Cosine Matrix
Display Control Module
Direct Current Motor
Distinguished Conduct Medal
dcm dc noise margin
DCMA Dry Color Manufacturers Association [US]
DCME Dichloromethyl Ether
DCMFM Dichloromonofluoromethane
d-CMP Deoxycytidine Monophosphate
DCMS Dedicated Computer Message Switching
DCMT Decrement
DCN Data Communication Network
Deuterated Hydrocyanic Acid
Dichloronaphthalene
Distributed Computer Network
Drawing Change Notice
DCNA Data Communication Network Architecture
Dichloronitroaniline
DCO Data-Controlled Oscillator
Dehydrated Castor Oil
Digital Central Office
DCOS Data Collection Operating System
Data Communication Output Selector
Direct-Coupled Operating System
DCP Data Collection Platform
Data Communications Processor
Design Change Package
Design Criteria Plan
Dicapryl Phthalate
Dicetyl Phosphate
Dichloropentane
Dichlorophenol
Dicumyl Peroxide
Dicyclopentadiene
Diffusion Controlled Pellistor
Digital Computer Processor
Digital Computer Programming
Digital Controller Programmer
Digital Control Programmer
Direct-Current Plasma
Directly-Coupled Plasma
Distributed Communications Processor
DCPADB Defense Civil Preparedness Agency [US]
Dichloropropionaniline
Dimethyl Tetrachloroterephthalate
Distribution Common Point
DC-PBH Double-Channel Planar-Buried Heterostructure
DC-PBH-LD Double-Channel Planar-Buried Heterostructure Laser Diode
DCPC Dichlorophenyl Methyl Carbinol
Dual Channel Port Controller
DCPCM Differentially Coherent Pulse Code Modulation
DCPD Dicyclopentadiene
dCpdA Deoxycytidylyl Deoxyadenosine
dCpdG Deoxycytidylyl Deoxyguanosine
DCPG Defense Communications Planning Group [US]
DCPIP Dichlorophenolindophenol
DCPP Data Communications Preprocessor
DCPR Detailed Continuing Property Record
DCPS Digitally-Controlled Power Source
DCPSK Differentially-Coherent Phase Shift Keying
DCR Data Communication Read
Data Conversion Receiver
Data Coordinator and Retriever
Decrease
Design Change Recommendation
Design Change Request
Detail Condition Register
Dichlororesorcine
Digital Cassette Recorder
Digital Condition Register
Digital Conversion Receiver
Direct Conversion Reactor
Direct Conversion Reactor Study [US]
Direct-Current Restorer
Double Cold Reducing
Drawing Change Request
DCRA Dyers and Cleaners Research Association [UK]
DCRABS Disk Copy Restore and Backup System
DCRN Dashpot Cup Retention Nut
DCRP Direct-Current Reverse-Polarity
DCRT Data Collection Receive Terminal
DCS Data Collection System
Data Communication Subsystem
Data Communication System
Data Conditioning System
Data Control Service
Data Control System
Defense Communications System [of USDOD]
Design Control Specifications
Diagnostic Control Store
Digital Camera System
Digital Command System
Digital Communication System
Digital Computer System
Digital Cross-Connect System
Direct-Coupled System
Distributed Computing System
Distributed Control System
Double Channel Simplex
Double-Current System
DCSC Defense Construction Supply Center [of USDOD]
DCSc Doctor of Commercial Science
DCSCS Data Code and Speech Conversion Subsystem
DCSP Defense Communications Satellite Program [US]
Direct Current Straight Polarity
DC SQUID Direct Current Superconducting Quantum Interference Device
DCSS Defense Communication Terminal System
DCT Data Communications Terminal
Destination Control Table
Dichlorotoluene
Direct Current Test
Discrete Cosine Transform
Document

DCTE Data Collection and Terrain Evaluation
DCTFE Dichlorotetrafluoroethane
DCTG Decorating
DCTL Direct-Coupled Transistor Logic
DCTN Defense Commercial Television Network
DCTS Data Communication Terminal System
DCTU Directly Corrected Test Unit
DCU Data Capture Unit
Data Collection Unit
Data Command Unit
Data Communications Unit
Data Control Unit
Decade Counting Unit
Decimal Counting Unit
Device Control Unit
Dichlorourethane
Digital Control Unit
Digital Counting Unit
Disk Control Unit
Display and Control Unit
Drum Control Unit
DCUTL Direct-Coupled Unipolar Transistor Logic
DCV Direct Current, Volts
Directional Control Valve
Double Cotton Varnish (Wire)
DCW Data Communication Write
Data Control Word
Differentially Compound-Wound
Digital Chart of the World
Dynamic Channel Exchange
DCWM Differentially Compound-Wound Motor
DCWV Direct-Current Working Voltage
DCX Direct Current Experiments [of ORNL, US]
DD Data Definition (Statement)
Data Demand
Data Descriptor
Data Dictionary
Data Display
Data Division
Decimal Display
Decimal Divide
Decision Data
Deep Drawing
Deferred Delivery
Degree Day
Delay Driver
Delivered
Delivered at Docks
Density Dependent
Depth Dose
Dewey Decimal
Dichloropropane-Dichloropropane
Digital Data
Digital Display
Digital-to-Digital
Direct Dialing
Disconnecting Device
Discriminating Digit
Disk Drive
Document Delivery
Double Density
Double-Diode
Double-Distilled
Dual Damping
Due Date
Drawing Deviation
Duplex Drive
D&D Design and Development
D/D Direct Debit
D-D Demand Draft
Deuterium-Deuterium Reaction
Digital-to-Digital
dd days after date
(de dato) — from date of issue
DDA Data Differential Analyzer
Depreciation, Depletion and Amortization
Depth-Duration-Area Value
Design Data Administration
Deutscher Dampfkesselausschuss [= German Committee for Pressure Vessels]
Didecyl Adipate
Didecyldimethylammonium
Die Makers and Die Cutters Association [now: NADD, US]
Digital Dealers Association [US]
Digital Differential Analyzer
Digital Display Alarm
Direct Data Attachment
Direct Device Attachment
Direct Differential Analyzer
Direct Disk Attachment
Dodecenyl Acetate
Dynamic Dielectric Analysis
Dynamics Differential Analyzer
DD&A Depreciation, Depletion and Amortization
DDAB Didecyldimethylammonium Bromide
DDAFP Diesel Driven Auxiliary Feedwater Pump
DDAS Digital Data Acquisition System [also: D-DAS]
DDB Deutscher Dolmetscherbund [= German Confederation of Interpreters]
2,3-Dimethoxy-1,4-Bis(dimethylamino)butane
Distributed Database
Dodecylbenzene
Double Declining Balance
DDBA Dodecylbenzenesulfonate
DDBG DIMDI Database Generator [Germany]
DDBS Descriptor Database System
Digital Data Broadcast System
DDBSA Dodecylbenzenesulfonic Acid
DDC Data Distributor Center
Deck Decompression Chamber
Defense Documentation Center [of USDOD]
Degree Day, Celsius
Dewey Decimal Classification
Die Casting Development Council
Digital Data Conversion
Digital Data Converter
Digital Display Conversion
Digital Display Converter
Digital Dynamic Convergence
Direct Digital Control(ler)
Director Digital Control

	Dual Diversity Comparator
DD/C	Dual Down-Converter
DDCE	Digital Data Conversion Equipment
DDCMP	Digital Data-Communications Message Protocol
DDCS	Data Definition Control System
	Digital Data Calibration System
DDD	Defense, Description and Designation
	Defense Diamond Development [US]
	Dichlorodiphenyl-1-(Dichlorophenyl)-2,2-Dichloroethane
	Dichlorodiphenyl-1,1-Dichloroethane
	Direct Distance Dialing
	Display Decoder Driver
	6,6'-Dithiodi-2-Naphthol Disulfide
	Double-Density Disk Drive
DDDA	Dodecadienyl Acetate
	Dodecanedioic Acid
DDDM	Dihydroxydichlorodiphenylmethane
DDDOL	Dodecadienol
DDDP	Discrete Differential Dynamic Programming
DDE	Decentralized Data Entry
	Dichlorodiphenyl-1,1-Dichloroethylene
	Difference-Differential Equation
	Direct Data Entry
	Direct Design Engineering
	Director Design Engineering
	Distributed Data Entry
	Dynamic Data Exchange
DDES	Direct Data Entry Station
DDF	Database Definition File
	Degree Day, Fahrenheit
	Design Disclosure Format
	Digital Distribution Frame
DDG	Digital Data Generator
	Digital Data Group
	Digital Delay Generator
	Digital Display Generator
DDGE	Digital Display Generator Element
DDGSR	Division of the Director-General of Scientific Research [UK]
DDH	Dibromodimethylhydantoin
	Dichlorodimethylhydantoin
	Digital Data Handling
DDH&DS	Digital Data Handling and Display System
DDH_2O	Double-Distilled Water
DDI	Depth Deviation Indicator
	Dibasic Diisocyanate
	Direct Dialing-In
	Direct Digital Interface
	Dislocation-Dislocation Interaction
	Distributed Data Interface
DDIE	Direct Dialing Interface Equipment
DDIS	Development Drilling Incentive System
	Document Data Indexing Set
DDL	Data Definition Language
	Data Description Language
	Digital Data Link
	Digital Design Language
	Diode-Diode Logic
	Dispersive Delay Line
DDLC	Data Description Language Committee [of CODASYL]
DDLG	Data Definition Language Group
DDM	Data Demand Module
	Department of Data Management [US]
	Derived Delta Modulation
	Design, Drafting and Manufacturing
	Diaminodiphenylmethane
	Difference in Depth Modulation
	Digital Display Makeup
	Dihydroxydichlorodiphenylmethane
	Direct Drive Motor
	Dodecyl Mercaptan
	GDR Mark [Former Currency of German Democratic Republic]
DDMA	Disk Direct Memory Access
DDMC	Design and Drafting Management Council
DDMS	Department of Defense Manned Spaceflight [US]
DDN	Defense Data Network
	Digital Data Network
DDNAME	Data Definition Name
DDNP	Diazodinitrophenol
DDO	Direct Dialing Overseas
DDOCE	Digital Data Output Conversion Element
	Digital Data Output Conversion Equipment
DDOL	Dodecenol
7-DDOL	7-Dodecen-1-ol
D-DOPA	D-3,4-Dihydroxyphenylalanine
DDP	Department of Defense Production [now: Supplies and Services Canada]
	Dichlorodiammineplatinum
	Digital Data Processor
	Distributed Data Processing
	Dodecyl Phthalate
DDPE	Digital Data Processing Equipment
DDPL	Demand Deposit Program Library
DDPS	Digital Data-Processing System
DDPU	Digital Data-Processing Unit
DDQ	Deep Drawing Quality
	2,3-Dichloro-5,6-Dicyano-1,4-Benzoquinone
DDR	Device Dependent Routine
	Dialed Digit Receiver
	Digital Data Receiver
	Digital Data Recorder
	Double Drift Region
	Dynamic Device Reconfiguration
DDR&E	Directorate of Defense Research and Engineering [of USDOD]
DDRS	Digital Data Recording System
DDS	Data Dependent System
	Data Dictionary System
	Data Display System
	Data Dissemination System
	Dataphone Digital Service [US]
	Deep-Diving System
	Deployable Defense System
	Dewey Decimal System
	4,4'-Diaminodiphenyl Sulfone [also: DAPSONE]
	Differential and Derivative Spectrophotometry
	Digital Data Service
	Digital Data System

	Digital Display Scope
	Digital Dynamics Simulator
	Dodecenylsuccinic Anhydride
	Doppler Detection System
DDSA	Digital Data Service Adapter
	2-Dodecen-1-ylsuccinic Anhydride
	Dodecylsuccinic Anhydride
DDSS	Diffusion Dependence Shell Structure
DDT	Data Description Table
	Deduct
	Defect Detection Trial
	Deflagration to Detonation Transition
	Delayed Dailing Tone
	Design Data Transmittal
	DIBOL (= Digital's Interactive Business-Oriented Language) Debugging Technique
	Dichlorodiphenyltrichloroethane
	Digital Data Terminal
	Digital Data Transmission
	Digital Data Transmitter
	Digital Debugging Tape
	Dithiotreitol
	Doppler Data Translator
	Double-Diode-Triode
	Double Dual Tandem
	Dynamic Debugging Technique
DDTA	Derivative Differential Thermal Analysis
DDTE	Digital Data Terminal Equipment
DDTESM	Digital Data Terminal Equipment Service Module
DDTS	Digital Data Transmission System
DDTU	Digital Data Transfer Unit
DDTV	Dry Diver Transport Vehicle
DDU	Digital Distribution Unit
DDV	Deck Drain Valve
DDVP	Dimethyl Dichlorovinyl Phosphate
DDW	Dense Dislocation Wall
DDX	Digital Data Exchange
DDXF	DISOSS (= Distributed Office Support System) Document Exchange Facility
DDZ	Dimethyl-3,5-Dimethoxybenzyloxycarbonyl
DE	Data Entry
	Decision Element
	Defect Engineering
	Delaware [US]
	Deposition Efficiency
	Design Engineer
	Design Engineering
	Design Evaluation
	Device End
	Dextrose Equivalent
	Dielectric
	Diesel-Electric
	Differential Equation
	Digital Electronics
	Digital Element
	Digital Encoder
	Display Electronics
	Display Element
	Display Equipment
	District Engineer
	Division Entry
	Doctor of Engineering
	Domestic Engineer
	Domestic Engineering
	Double Entry
	Drop Electrode
	Dry Etching
	Dynamic Equation
DEA	Dark Etching Area
	Data Encryption Algorithm
	Department of Economic Affairs [UK]
	Department of External Affairs [US]
	Dielectric Absorption
	Dielectric Analysis
	Diethanolamine
	Diethylamine
	Diethylaniline
	Display Electronics Assemblies
	Drug Enforcement Administration [US]
DEAB	Diethylamine Borane
DEAC	Data Exchange Auxiliary Console
	Diethylaluminum Chloride
DEACON	Direct English Access and Control
DEAD	Diethyl Azodicarboxylate
DEAE	Diethylaminoethyl
DEAE-C	Diethylaminoethyl Cellulose
DEAE-D	Diethylaminoethyl Ether of Dextran
DEAE-RNA	Ribonucleic Acid, DEAE Salt
DEAI	Diethylaluminum Iodide
DEAL	Data Entry Application Language
	Decision Evaluation and Logic
DEANOL	Dimethylaminoethanol
	Dimethylethanolamine
DEAP	2,2-Diethoxyacetophenone
	Diffused Eutectic Aluminum Process
DEB	Data Event Block
	Data Extension Block
	Debenture
	Digital European Backbone
DEBEC	Deutsche Betriebsgesellschaft fur Drahtlose Telegraphie [= German Society for Wireless Telegraphy]
DEBS	Debentures
DEBU	DREO (= Defense Research Establishment Ottawa) Elint Browsing Utility [Canada]
DEC	December
	Decimal
	Declination
	Decoder
	Decomposition
	Decrease
	Decrement
	Development Education Center
	Dielectric Current
	2-Diethylaminoethyl Chloride Hydrochloride
	Diethylcarbinol
	Digital Electronic Circuit
	Duluth Engineers Club [US]
	Dynamic Environmental Conditioning
DECADE	Digital Equipment Corporation Automatic Design
DECAL	Decalcomania
	Desk Calculator

	Digital Equipment Corporation's Adaptation of Algorithmic Language
DECB	Data Event Control Block
DECDR	Decorder
DECEO	Defense Communication Engineering Office [US]
DECENT	Distribution of Exact Classical Energy Transfer
DECHEMA	Deutsche Gesellschaft fur Chemisches Apparatewesen [= German Society for Chemical Equipment]
DECIEM	Defense and Civil Institute of Environmental Medicine
DECIT	Decimal Digit
DECL	Declination
DECLAB	Digital Equipment Corporation Laboratory [US]
DECM	Defense Electronic Countermeasures System [US Army]
DECNET	Digital Equipment Corporation Network [US]
DECOM	Decommutator
DECOMP	Decomposition [also: DECOMPN]
DECONTN	Decontamination
DECOR	Digital Electronic Continuous Ranging
DECR	Decrease
	Decrement
DECS	Data Entry Control System
DECUS	Digital Equipment Computer Users Society [US]
DED	Data Element Dictionary
	Data Element Directory
	Dedendum
	Deduct(ion)
	Design Engineering Directorate
	Distant End Disconnect
	Doctor of Environmental Design
	Double Error Detection
DEd	Doctor of Education
DEDB	Data Entry Database
DED/D	Data Element Dictionary/Directory
DEDPU	Diethyldiphenylurea
DEDS	Data Entry and Display System
	Dual Exchangeable Disk Storage
DEDT	Department of Economic Development and Tourism [Canada]
DEDUCOM	Deductive Communicator
DEE	Data Encryption Equipment
	Department of Electrical Engineering
	Digital Evaluation Equipment
	Digital Events Evaluator
	Division of Electrical Engineering
DEEP	Data Exception Error Protection
DEER	Directional Explosive Echo Ranging
DEET	Diethyltoluamide
DEF	Data Entry Facility
	Data Extension Frame
	Defect
	Defense
	Defense Specifications [of MOD, UK]
	Defer(red)
	Definite
	Definition
DEF FN	(User-)Defined Function
DEFL	Deflection
DEF SEG	Definition Segment
DEF STD	Defense Standard [also: DEF STAN]
DEFT	Definite Time
	Dynamic Error-Free Transmission
DEFTR	Digital's Ethernet Frequency Translator
DEG	Degree
	Diethanolglycine
	Diethylene Glycol
	Directorate of Environmental Geology
deg	degree
DEGA	Diethylene Glycol Adipate
DEGaCl	Diethylgallium Chloride
DEGB	Diethylene Glycol Benzoate
DEGMBE	Diethylene Glycol Monobutyl Ether
DEGMEE	Diethylene Glycol Monoethyl Ether
DEGN	Diethylene Glycol Dinitrate
DEGS	Diethylene Glycol Succinate
DEHA	Diethylhydroxylamine
DEHPA	Di-Ethylhexylphosphoric Acid
DEI	Development Engineering Inspection
	Double Electrically Isolated
DEIS	Dielectrics and Electrical Insulation Society [of IEEE, US]
DEJF	Double End Jig Feet
DEL	Data Entry Language
	Delay
	Delete (Character)
	Delft Hydraulics Laboratory [Netherlands]
	Delineation
	Delivery
	Direct Exchange Line
Del	Delaware [US]
DEL AC	Delayed Action
DELCL	Delete Clause
DELD	Delivered
DELDT	Delivery Date
DELE	Delete
DELEX	Destroyer Life Extension Program [of Canadian Forces]
DELFIA	Dissociation-Enhanced Lanthanide Fluoroimmunoassay
DELIQ	Deliquescence
DELNI	Digital's Local Network Interconnect
DELOS	Division for Experimentation and Laboratory-Oriented Studies
DELTA	Distributed Electronic Test and Analysis
DELTIC	Delay-Line Time Compression
DELY	Delivery
DEM	Delta Modulation
	Demand
	Demodulator
	Demolition
	Demonstration
	Demurrage
	Department Engineering Materials
	Deutsche Mark [Currency of Germany]
DEMA	Data Entry Management Association [US]
	Diesel Engine Manufacturers Association [US]
	Diving Equipment Manufacturers Association [US]
DEMIST	Design Methodology Incorporating Self-Test
DEMIZ	Distance Early Warning Military Identification Zone
DEML	Demolition

DEMO Demonstration [also: demo]
DEMOD Demodulator
DEMON Decision Mapping via Optimum Network
DEMPR Digital's Ethernet Multiport Repeater
DEMPS Directorate of Engineering and Maintenance Planning Standardization
DEMS Development Engineering Management System
Digital Electronic Message Service
DEMST Demonstrate
DEMUX Demultiplexer
Demultiplexing
DEN Denmark
Density
Department of Energy [UK]
Diethylnitrosamine
Double Edge Notched
DENALT Density Altitude
DEng Doctor of Engineering
DENISE Dense Negative Ion-Beam Surface Experiment
DENR Department of Energy and Natural Resources [Canada]
DENS Density
DEnv Doctor of Environment
DEnvDes Doctor of Environmental Design
DEO Data Entry Operator
Digital End Office
DEOT Disconnect End-of-Transmission
DEP Data Entry Panel
Department
Department of Employment and Productivity [UK]
Departure
Deposit
Depot
Deputy
Design External Pressure
Diagnostic Executive Program
Dielectrophoretic
Diethyl Phthalate
Diethyl Pyrocarbonate
2,4-DEP 2,4-Dichlorophenoxyethylphosphite
DEPA Defense Electric Power Administration [US]
DEPC Diethylaminopropyl Chloride Hydrochloride
Diethylpyrocarbonate
DEPCOS Departure Coordination System [Germany]
DEPE Double Escape Peak Efficiency
DEPI Differential Equations Pseudocode Interpreter
DEPIC Dual Expanded Plastic-Insulated Conductor
DEPREC Depreciation
DEPSK Differential-Encoded Phase Shift Keying
DEPT Department
DEPU Diethylphenylurea
DEQ Dequeue
Dose Equivalent
DE/Q Design Evaluation/Qualification
DEQUE Double-Ended Queue [also: deque]
DER Declining Error Rate
Derivation
Derivative
Departure End of Runway
Diesel Engine Reduction-Drive
Directly Executable Representation
DERA Directory of Education and Research in Australia
Derb Derbyshire [UK]
DERD Display of Extracted Radar Data
DERE Dounreay Experimental Reactor Establishment [Scotland]
DEREP Digital's Ethernet Repeater
DERIV Derivation
Derivative
DERL Defense Electronics Research Laboratory [India]
DERS Data Entry Reporting System
DERV Diesel Engine Road Vehicle [also: Derv]
DES Data Encryption Standard
Data Entry System
Data Exchange System
Department of Earth Sciences
Department of Education and Science [UK]
Department of Emergency Services [US]
Department of Employment Security
Descriptor
Design
Design and Evaluation System
Desoxycholate
Diethylstilbestrol
Diethyl Sulfate
Differential Equation Solver
Digital Expansion System
Directorate of Engineering Standardization
Doctor of Engineering Science
Doctor of Environmental Studies
Dynamic Environment Simulator
DESC Data Entry System Controller
Defense Electronics Supply Center [of USDOD]
Digital Equation Solving Computer
DESCR Description
DESDMM Diethylsulfone Dimethyl Methane
DesEng Design Engineer
DES ENG Design Engineering
DESG Design [also: DESGN]
DESGR Designer
DESIG Designation
DESO Double End Shut-Off
DESRT Development and Demonstration of Site Remediation Technology (Program) [of Environment Canada]
DESS Digital Electronic Switching System
DEST Destination
DESTA Digital's Ethernet Station Adapter
DESY Deutsches Elektronen-Synchrotron [= German Electron Synchrotron]
DET Detail
Detection
Detector
Determinant
Determination
Detonator
Dielectric Testing
Diethyl Tartrate
Diethyl Telluride
Diethyltoluamide
Diethyltryptamine
Divorced Eutectic Transformation

	Double Exposure Technique
DETA	Dielectric Thermal Analyzer
	Diethylenetriamine
DETAB	Decision Table
	Design Table
DETAB-X	Design Table, Experimental
DETAP	Decision Table Processor
DETAPAC	Diethylenetriaminepentaacetic Acid
DETC	Diethylthiacarbocyanine
DET'D	Determined [also: DETD]
DETF	Data Exchange Test Facility
	Double-End Tuning Fork
DETN	Determination
DETOC	Decision Table to COBOL
DETP	Data Entry and Teleprocessing
DETRAN	Decision Translator
DETU	N,N'-Diethyl Thiourea
DETWAD	Divorced Eutectic Transformation With Associated Deformation
DEU	Data Encryption Unit
	Data Entry Unit
	Data Exchange Unit
DEUA	Diesel Engine Users Association [UK]
	Digitronics Equipment Users Association
DEUCE	Digital Electronic Universal Calculating Engine
DEUS	Data Entry, University of Saskatchewan [Canada]
DEV	Development
	Deviation
Dev	Devonshire [UK]
DEVCO	Development Committee [of ISO]
DEVEL	Development
DEVIL	Direct Evaluation of Index Languages
DEVMIS	Development Management Information System
DEVN	Deviation
DEVR	Distortion-Eliminating Voltage Regulator
DEVSIS	Development Sciences Information System [of IDRC, Canada]
DEW	Digital Electronic Watch
	Directed Energy Weapon
	Distant Early Warning
DEWIZ	Distant Early Warning Identification Zone
DEX	Data Exchange
	Deferred Execution
DEXAN	Digital Experimental Airborne Navigator
DEXPO	DEC (= Digital Equipment Corporation) Exposition [US]
DEXT	Distant End Crosstalk
DEZ	Diethyl Zinc
DF	Dark Field
	Data Field
	Dean of the Faculty
	Decimal Fraction
	Decontamination Factor
	Deep Freezing
	Deflection Factor
	Defluorination
	Degree of Freedom
	Describing Function
	Desferrioxamine
	Destination Field
	Deuterium Fluoride
	Device Flag
	Die Forming
	Differential Flotation
	Digital Filter
	Digital Fluoroscopy
	Dilution Factor
	Dirac-Fock
	Direct Flow
	Direction Finding
	Disappearing Filament
	Disk File
	Dissipation Factor
	Distribution Feeder
	Diversity Factor
	Doctor of Forestry
	Dose Factor
	Double Feeder
	Double Flag
	Double Focusing
	Double Frequency
	Douglas Fir
	Drift Flux
	Drinking Fountain
	Drive Fit
	Drop Forging
	Dual Facility
D&F	Determination and Findings
DFA	Data Flow Architecture
	Distributed Function Architecture
	Drop Forging Association [now: FIA, US]
DFAG	Double-Frequency Amplitude Grating
DFAS	Detailed Functional Application Sub-System [of CFSS, Canada]
DFAST	Dynamic File Allocation System
DFAW	Direct Fire Antitank Weapon
DFB	Data Flag Branch
	Diffusion Brazing
	Distributed Feedback
	Distribution Fuse Board
DFBM	Data Flag Branch Manager
DFBR	Data Flag Branch Register
DFBT	Dynamic Functional Board Tester
DFC	Dairy Farmers of Canada
	Data Flow Control
	Design Field Change
	Diagnostic Flowchart
	Digital-Automatic Flight Control
	Digital Flight Control
	Digital Fuel Controller
	Direct Field Costs
	Disk File Check
	Disk File Controller
	Distinguished Flying Cross
	Duty Free Confederation
DFC/ADM	Digital Fuel Controller/Advanced Development Model
DFCNV	Disk Data File Conversion Program
DFCS	Digital-Automatic Flight Control System
	Digital Flight Control System
DFCU	Disk File Control Unit
DFD	Data Flow Diagram

	(International) Data for Development (Association) [France]
	Digital Flight Display
DFDCM	Difluorodichloromethane
DFDD	Difluorodiphenyldichloroethane
DFDR	Digital Flight Data Recorder
DFDSG	Direct-Fired Downhole Steam Generator
DFE	Decision Feedback Equalizer
DFET	Depletion Field-Effect Transistor
DFEU	Disk File Electronics Unit
DFDT	Difluorodiphenyltrichloroethane
DFF	Display Format Facility
DFFR	Dynamic Forcing Function Report
DFG	Deutsche Forschungsgemeinschaft [= German Research Society]
	Diode Function Generator
DFGS	Digital Flight Guidance System
DFHP	Dislocation-Free High-Purity
DFI	Data File Interrogate
	Deep Foundations Institute [US]
	Developmental Flight Instrumentation
	Diesel Fuel Injection
	Digital Facility Interface
	Disk File Interrogate
DFL	Deutsche Forschungsanstalt fur Luft- und Raumfahrt [= German Research Institute for Aeronautics and Astronautics]
	Display Formatting Language
Dfl	Dutch Florin/Guilder [Currency of the Netherlands]
DFLD	Data Field
DFM	Data Flow Manager
	Distinguished Flying Medal
DFMSR	Directorate of Flight and Missile Safety Research [US Air Force]
DFN	Data File Number
	Deutsches Forschungsnetz [= German Research Network]
DFO	Department of Fisheries and Oceans [Canada]
	Deutsche Forschungsgesellschaft fur Oberflachenbehandlung [= German Research Society for Surface Treatment]
	Directed Format Option
	Director of Flight Operations
	Disk File Optimizer
DFOLS	Depth of Flash Optical Landing System
DFP	Diesel Fire Pump
	Diisopropyl Fluorophosphate
	Directed Fiber Preform
DFPG	Double-Frequency Phase Grating
DFPL	Data Flow Programming Language
DFPM	Disappearing Filament Pyrometer
DFR	Decreasing Failure Rate
	Disk File Read
	Dounreay Fast Reactor [Scotland]
	Dropped from Rolls
DFr	Djibouti Franc [also: DFR; Currency of Djibouti]
DFRA	Drop Forging Research Association [UK]
DFRL	Differential Relay
DFS	Department of Food Science
	Detailed Functional Specification
	Deutsche Forschungsanstalt fur Segelflug [= German Gliding Research Institute]
	Disk Filing System
	Distributed File System
	Dividends from Space [US]
	Doctor of Forest Science
	Dynamic Flight Simulator
DFSG	Direct Formed Supergroup
DFSK	Double Frequency Shift Keying
DFSR	Directorate of Flight Safety Research [US]
DFSU	Disk File Storage Unit
DFT	Deaerating Feed Tank
	Degree of Fiber Treatment
	Density Functional Theory
	Department of Fire Technology [of Southwestern Research Institute, US]
	Design for Testability
	Diagnostic Function Test
	Digital Facility Terminal
	Discrete Fourier Transform
	Draft
	Dry Film Thickness
DFTG	Drafting
DFTI	Distance from Touchdown Indicator
DFTPP	Decafluorotriphenylphosphine
DFTSMN	Draftsman
DFU	Data File Utility
	Disposable Filter Unit
DFVLR	Deutsche Forschungs- und Versuchsanstalt fur Luft- und Raumfahrt [= German Research and Test Institute for Aeronautics and Astronautics]
DFW	Diffusion Welding
	Disk File Write
DFZ	Dislocation-Free Zone
DG	Dangerous Goods
	Datagram
	Delay Generator
	Detonation Gun
	Diacylglycerol
	Diesel Generator
	Differential Gain
	Differential Generator
	Diglyme
	Diode Gate
	Directional Gyro
	Directorate-General
	Director General
	Double Girder (Crane)
	Double Glass
	Double Groove (Insulator)
	Drill Gauge
	Dry Goods
Dg	Grain Density [Crude Oil]
dG	Deoxyguanosine
dg	decigram
DGA	Dangerous Goods Advisor
	Differential Gravimetric Analysis
DGAC	Direction General Aviacion Civil [= General Civil Aviation Management, Spain]
	Direction Generale d'Aviation Civile [= General Civil Aviation Management, France]

DGB Deutsche Gesellschaft fur Betriebswirtschaft [= German Society for Industrial Engineering and Management]
Deutscher Gewerkschaftsbund [= Federation of German Trade Unions]
Disk Gap Bond
DGBAS Directorate General of Budget, Accounting and Statistics [Taiwan]
DGBC Digital Geoballistic Computer
DGCI&S Directorate General of Commercial Intelligence and Statistics [India]
DG/CS Data General (Corporation)/Communication System
DGD Deutsche Gesellschaft fur Dokumentation [= German Society for Documentation]
DG/DBMS Data General (Corporation)/Database Management System
DGDME Diethyleneglycol Dimethylether
DGDP Double-Groove, Double-Petticoat (Insulators)
DGE Diglycidyl Ether
DGEBA Diglycidyl Ether of Bisphenol A
DGES Division of Graduate Education in Science [of NSF, US]
DGF Demountable Growth Flange
Deutsche Gesellschaft fur Flugwissenschaften [= German Society for Aeronautic Sciences]
DGFH Deutsche Gesellschaft fur Holzforschung [= German Society for Wood Research]
DGfL Deutsche Gesellschaft fur Logistik [= German Society for Logistics]
DGG Deutsche Glastechnische Gesellschaft [= German Society for Glass Technology]
DGI Dental Gold Institute [US]
DGIR Department of Scientific and Industrial Research [UK]
DGIWG Digital Geographic Information Working Group [of NATO]
DG/L Data General (Corporation)/Programming Language
DGLR Deutsche Gesellschaft fur Luft- und Raumfahrt [= German Society for Aeronautics and Astronautics]
DGM Data-Grade Media
Deutsche Gesellschaft fur Metallkunde [= German Society for Metallurgy and Materials Science]
Differential Galvanometer
DGMA Deutsche Gesellschaft fur Messtechnik und Automatisierung [= German Society for Measuring Techniques and Automation]
d-GMP Deoxyguanosine Monophosphate
DGN Direccion General de Normas [= National Standards Directorate, Argentina]
DGO Deutsche Gesellschaft fur Oberflachentechnik [= German Society for Surface Technology]
DGON Deutsche Gesellschaft fur Ortung und Navigation [= German Society for Position-Finding and Navigation]
DGP Dangerous Goods Panel [of ICAO, Canada]
DGPM Direction de la Geologie et de la Prospection Miniere [= Directorate for Geology and Prospecting, Ivory Coast]
DGROUP Data Group
DGRST Delegation Generale a la Recherche Scientifique et Technique [= General Delegation on Scientific and Technological Research, France]
DGS Data Gathering System
Data Generation System
Data Ground Station
Degaussing System
Department of Geological Sciences
Display Generating System
Distance Gain Size
Doctor of Geological Sciences
DGSS Distributed Graphics Support Subroutine
DGT Digit
Directorate General de Telecommunication [= Directorate General of Telecommunication, France]
DG/TPMS Data General (Corporation)/Transaction Processing Management System
DGV Deutscher Giessereiverband [= German Foundry Association]
DGZ Desired Ground Zero
DGZfP Deutsche Gesellschaft fur Zerstorungsfreie Prufung [= German Society for Nondestructive Testing]
DH Dead Head
Decay Heat
Decimal-to-Hexadecimal
Decision Height
Delayed Hypersensitivity
Denavit-Hartenberg Process
Device Handler
Dirham [Currency of Morocco]
District Heating
Dividing Head
Document Handling
Double Heterojunction
Double Heterostructure
Dynamic Head
D-H Decimal-to-Hexadecimal
Dortmund-Horder Huttenunion [also: DH, Germany]
Duane-Hunt (Law)
Dh Dirham [Currency of Morocco]
United Arab Emirates Dirham [Currency]
dH Deutsches Hartegrad [= German Degree of Hardness]
DHA Dehydrated Humulinic Acid
Dehydroacetic Acid
Dihydroanthracene
Dihydroxyacetone
D&HAA Dock and Harbour Authorities Association [UK]
DHAEMAE Disposable Hypodermic and Allied Equipment Manufacturers Association of Europe [UK]
DHAP Dihydroxyacetone Phosphate
DHAS Dairy Herd Analysis Service (Program)
DHB Dihydroxybenzene
Dihydroxybenzoic Acid
DHBA 3,4-Dihydroxybenzylamine Hydrobromide
DHBP 4,4'-Dihydroxybenzophenone
DHC Data Handling Center
Diamond High Council [Belgium]
DHCC Decay Heat Closed Cooling
DHCF Distributed Host Command Facility

DHD Double Heat-Sink Diode
Double High Density
2HD Double High Density
DHDEE Dihydroxydiethyl Ether
DHE Data Handling Equipment
Dehydroepiandrosterone
Down-Hole Emulsification
Dump Heat Exchanger
DHEA Dehydroepiandrosterone
DHEBA Dihydroxyethylene-Bis-Acrylamide
D-HEMT Depletion-Mode High-Electron Mobility Transistor
DHEW Department of Health, Education and Welfare [US]
DHF Dirac-Hartree-Fock
DHFR Dihydrofolate Reductase
DHG Dihydroxyethylglycinate
DHHS Department of Health and Human Services [US]
DHI Deutsches Hydrographisches Institut [= German Hydrographic Institute]
Door and Hardware Institute [US]
DHIA Dairy Herd Improvement Association [US]
DHLLP Direct High-Level Language Processor
DHMS 4,4-Dihydroxy-α-Methylstilbene
DHN Decahydronaphthalene
DHP Deoxidized, High-Residual Phosphorus Copper
Dihydric Phenol
3,4-Dihydropyran
Dihydropyridine
Dihydroxypropyl
Document Handler Processor
DHP-MP Dihydroxypropyl Methylpiperazine
DHR Department of Human Resources
DHRRT Decreased Hot Rolling Reduction Treatment
DHRS Decay Heat Removal System
Direct Heat Removal System
DHS Data Handling System
Department of Health Sciences
Dihydrostreptomycin
DHSD Diesel Hybrid System Design
DHSS Department of Health and Social Security [UK]
DHT Dihydrotestosterone
Discrete Hartley Transform
Discrete Hilbert Transform
DHU Document Handler Unit
DHvA de Haas-van Alphen (Effect)
DHW Domestic Hot Water
Double Hung Windows
DHX Dump Heat Exchanger
DHXCS Dump Heat Exchanger Control System
DI Data Input
Deionization
Deionized (Water)
Demand Indicator
Department of Industry [now: DTI, UK]
Design International [US]
Device Independent
Digital Input
Direct Image
Discomfort Index
Dvorak International [US]
D/I Distinctness of Image
D-I Direct-Image (Offset System) [also: d-i]
Di Didymium
DIA Defense Intelligence Agency [of USDOD]
Design Industries Association [UK]
Diameter
Digital Input Adapter
Direct Interface Adapter
Document Interchange Architecture
Dual Interface Adapter
DIAC Defense Industry Advisory Control
Defense Industry Advisory Council [US]
Diode Alternating-Current Switch
DIACTOR Direct-Acting Regulator [also: diactor]
DIAD Digital Image Analysis and Display
Diisopropyl Azodicarboxylate
Drum Information Assembler and Dispatcher
DIADEM Dynamic International Access to Databases and Economic Models [UK]
DIAG Diagonal
Diagram
DIAGS Diagonals
Diagrams
DIAL Data for Interchange at the Application Level
Differential-Absorption Lidar
Display Interactive Assembly Language
Display Interface Assembly Language
Draper Industrial Assembly Language
Drum Interrogation, Alteration and Loading (System)
DIALGOL Dialect of ALGOL
DIAM Data Independent Architecture Model
Diameter
DIAN Digital — Analog
DIANE Digital Integrated Attack Navigation Equipment
Direct Information Access Network for Europe [of CEC, Belgium]
DIANS Digital Integrated Attack Navigation System
DIAPH Diaphragm
DIAS DIMDI Administration System [Germany]
Dublin Institute for Advanced Studies [Ireland]
Dynamic Inventory Analysis System
DIB Data Input Bus
Data Inspection Board
Diiodobutane
1,3-Diphenylisobenzofuran
DIBA Diisobutyl Adipate
Doctor of International Business Administration
DIBAC Diisobutylaluminum Chloride
DIBAH Diisobutylaluminum Hydride
DIBAL Diisobutylaluminum
DIBAL-H Diisobutylaluminum Hydride
DIBAS Dibasic
DIBITS Di-Binary Digits [also: dibits]
DIBOL Digital's Interactive Business-Oriented Language
DIBP Disobutyl Phthalate
DIBS Digital's Integrated Business System
DIC Data Insertion Converter
Detailed Interrogation Center
Differential Interference Contrast
Digital Concentrator
Digital Input Control
Digital Interchange Code

Diisopropylaminoethyl Chloride Hydrochloride
2-Dimethylaminoisopropyl Chloride Hydrochloride
Diploma of the Imperial College of Science and Technology
Disseminating Intravascular Coagulation
Doctor of the Imperial College (of Science and Technology) [UK]

DICA Daily Interest Checking Account
DICAM Data-System Interactive Communications Access Method
DICAMBA Dichloro-O-Anisic Acid (Methoxydichlorobenzoic Acid)
DICBM Defense Intercontinental Ballistic Missile
DICE Dairy and Ice Crem Equipment Association [UK]
Digital Intercontinental Conversion Equipment
DICEF Digital Communications Experimental Facility
DICIS Duane Information Center Indexing Service [US]
DICO Dissemination of Information through Cooperative Organization
DICON Digital Communication through Orbiting Needles
DICORAP Directional-Controlled Rocket-Assisted Projectile
DICORS Diver Communication Research System
DICT Dictionary
DID Data Item Description
Datamation Industry Directory
Device Identifier
Device-Independent Display
Digital Information Detection
Digital Information Display
Direct Inward Dialing
Display Interface Device
Division of Isotopes Development [of USAEC]
Drainage and Irrigation Department [Malaysia]
Drum Information Display
DIDA Diisodecyl Adipate
DIDACS Digital Data Communication System
DIDAD Digital Data Display
DIDAP Digital Data Processor
DIDM Document Identification and Description Macro
DIDO Data Input/Data Output [also: DIDO]
Digital Input/Digital Output [also: DIDO]
DIDOC Desired Image Distribution using Orthogonal Constraints
DIDOCS Device-Independent Display Operator Console Support
DIDP Diisodecyl Phthalate
DIDS Defense Integrated Data System [of DND, Canada]
Digital Information Display System
Diisothiocyanatostilbene Disulfonic Acid
Domestic Information Display System [US]
DIE Danish Institute for the International Exchange of Scientific and Literary Publications
Direct-Injection Enthalpimetry
Document of Industrial Engineering
DIEA Diisopropylethylamine
DIEGME Diethylene Glycol Monomethyl Ether
DIEN Data Input Ensemble
Diethylenetriamine
DIET Desorption Induced by Electronic Transition
DIF Data Interchange Format
Deutsches Institute zur Forderung des Industriellen Fuhrungsnachwuchses [= German Institute for the Advancement of Junior Executives]
Device Input Format
Differential Interference Microscopy
DIFET Double-Injection Field-Effect Transistor
DIFF Difference
Differential
DIFFTR Differential Time Relay
DIFKIN Diffusion Kinetics
DIFP Diisopropyl Fluorophosphate
DIFPEC Differentially Pumped Environmental Chamber
DIFU Deutsches Institut fur Urbanistik [= German Institute for Urban Planning]
DIG Design Implementation Guide
Digit
Digital Image Generated
Digital Input Gate
DIGACC Digital Guidance and Control Computer
DIGCOM Digital Computer
DIGEST Digital Geographic Exchange Standard
DIGICOM Digital Communications
Digital Communications System
DIGITAC Digital Airborne Computer
Digital Tactical Automatic Control
DIGITAR Digital Airborne Computer
DIGM Diffusion-Induced Grain Boundary Migration
DIGS Data and Information Gathering System
Disorder-Induced Gap States
Dig Sig Prog Digital Signal Processing
DIH Diploma in Industrial Health
Di-H Di-Hydrogen
Di-HETE Dihydroxyeicosatetraenoic Acid
DIHT Deutscher Industrie- und Handelstag [= Association of German Chambers of Industry and Commerce]
DII Diesel Ignition Improver
DIIC Dielectrically Isolated Integrated Circuits
DIL Dilute
Dilution
Dual-in-Line
DILEP Digital Line Engineering Program
DILIC Dual-in-Line Integrated Circuit
DILN Dilution
DILS Doppler Inertial LORAN (= Long-Range Navigation) System
DIM Data and Instruction Management
Data Interpretation Module
Device Interface Module
Dimension
DIMA Digital Image Analysis
DIMATE Depot-Installed Maintenance Automatic Test Equipment
DIMDI Deutsches Institut fur Medizinische Dokumentation und Information [= German Institute for Medical Documentation and Information]
DIMECO Dual Independent Map Encoding File of Countries [of Harvard University, US]
DIMES Defense-Integrated Management Engineering System
DIMM Data and Instruction Management Machine

DIMORPH Dimorphous
DIMOX Directed Metal Oxidation
DIMPLE Deuterium-Moderated Pile Low Energy Reactor [of AERE, UK]
DIMS Distributed Intelligence Microcomputer System
DIMUS Digital Multibeam Steering
DIN Deutsches Institut fur Normung [= German Standards Institute]
Din Dinar [Currency of Yugoslavia]
DINA Database Industry Association [of JIPDEC, Japan]
Direct Noise Amplification
Distributed Information Processing Network Architecture
DINAP Digital Network Analysis Program
DINOSEB Dinitro-sec-Butyl-Phenol
DINP Diisononyl Phthalate
DINUPS DIMDI Input and Update System [Germany]
DIO Data Input/Output
Diode
Direct Input/Output
DIOA Diisooctyl Adipate
DIOB Digital Input/Output Buffer
DIOC Digital Input/Output Control
DIOD Digital Input/Output Display (System)
DIOL Dihydric Alcohol [also: diol]
DIOP DDD (= Double-Density Disk Drive) Input/Output Processor
Diisooctyl Phthalate
Diisopropylidene Dihydroxy-bis(diphenylphosphino)butane
Dimethyldioxolane
DIOS Diisooctyl Sebacate
Distribution, Information and Optimizing System
DMA (= Direct Memory Access) Input/Output System
DIOSS Distributed Office Support System
DIOZ Diisooctyl Azelate
DIP Defense Industry Productivity (Program) [Canada]
Depth Image Processing
Design Internal Pressure
Digital Image Processing
Dimethylaminoisopropyl
Diploma
Display Information Processor
Distributed Information Processing
Dual-in-Line Package
Dual-in-Line Pin
DIPA Diisopropanolamine
DipA Diploma in Aquaculture
DIPAN Directoria da Producao Animal [= Directorate for Animal Production, Brazil]
DipBact Diploma in Bacteriology
DipBus Admin Diploma in Business Administration
DIPC 2-Dimethylaminoisopropyl Chloride
DIPCDI Diisopropylcarbodiimide
DIPD Double Inverse Pinch Device
DIPEC Defense Industrial Plant Equipment Center [of USDOD]
DipEng Diploma in Engineering
DipEngTech Diploma in Engineering Technology
DIPHOS 1,2-Bis(diphenylphosphino)ethane
Diphenylphosphinoethane
Dipl-Chem (Diplom-Chemiker) — Equivalent to: Master of Science in Chemistry
DipLibTech Diploma in Library Technology
Dipl-Ing (Diplom-Ingenieur) — Equivalent to: Master of Science in Engineering
Dipl-Ing (TH) (Diplom-Ingenieur der Technischen Hochschule) — Equivalent to: Graduate Engineer
Dipl-Kaufm (Diplom-Kaufmann) — Equivalent to: Master of Science in Business
Dipl-Kfm (Diplom-Kaufmann) — Equivalent to: Master of Science in Business
Dipl-Ldw (Diplom-Landwirt) — Equivalent to: Master of Science in Agriculture
Dipl-Phys (Diplom-Physiker) — Equivalent to: Master of Science in Physics
Dipl-Volksw (Diplom-Volkswirt) — Equivalent to: Master of Science in Economics
DipMA Diploma in Marine Affairs
DIPP Defense Industry Productivity Program [Canada]
DipPA Diploma in Public Affairs
DIPPR Design Institute for Physical Property Data [of AIChE, US]
DIPRA Ductile Iron Pipe Research Association [US]
DIP RAM Dual-in-Line Package Random-Access Memory
DIPS Development Information Processing System
N,N-Diisopropyl-2-Benzothiazole Sulfenamide
DIPSO 3-[Bis(2-Hydroxyethyl)amino]-2-Hydroxy-1-Propanesulfonic Acid
DIPT Diisopropyltartrate
DIQD Disk-Insulated Quad
DIR Data Input Register
Defense Industrial Research (Program) [of DND, Canada]
Development Inhibitor Releasing
Diffusion-Induced Recrystallization
Direction
Director
Directory
Discipline Oriented Information Retrieval
Document Information Retrieval
DIRA Danish Industrial Robot Association
DIRCOL Direction Cosine Linkage
DIR-CONN Direct-Connected
DIRECT Document Information Retrieval and Evaluation for Computer Terminals
DIRP Defense Industrial Research Program
DIRS DIMDI Information Retrieval System [Germany]
Directors
DIS Defense Investigative Service [US]
Digital Integration System
Digital Interface Subsystem
Discount
Disintegration
Dissertation Inquiry Service
Distributed Information System
Documentation Inventory System
Draft International Standard [of ISO]
Ductile Iron Society [US]
DISAC Digital Simulator and Computer
DISAM Direct and Index Sequential Access System

DISB Disburse
DISBMT Disbursement
DISBN Disband
DISC Differential Isochronous Self-Collimating Counter
Differential Scatter
Digital International Switching Center
Disability Information Services of Canada
Disconnect
Disconnection
Disconnector
Discount
Discovery
Domestic International Sales Corporations
DISCH Discharge
DISCOM Digital Selective Communications
DISCON Defense Integrated Secure Communications Network [Australia]
DISCOS Disturbance Compensation System
DISCR Discriminator
DISCT Discount
DISD Data and Information Systems Division
DISG Diagreeable
DISI Dairy Industry Society International [US]
dis/min disintegrations per minute
DISOSS Distributed Office Support System
DISP Dispatcher
Displacement
Display
Disposal
DISPLAY Digital Service Planning Analysis
DISRX Disable Receive [also: DIS RX]
DISS Digital Interface Switching System
Dissertation
Distributed Information Processing Service System
DISSD Dissolved
dis/sec disintegrations per second
DISSPLA Display Integrated Software System and Plotting Language
DIST Distance
Distillation
Distiller
Distillery
Distinguish
Distribution
District
DISTD Distilled
DISTG Distilling
DISTN Distillation
DISTNG Distinguish
DISTR Distribution
Distributor
DISTRAM Digital Space Trajectory Measurement System
DISTRG Distributing
DISTRO Distribution Rotation
DISTX Disable Transmit [also: DIS TX]
DISU Digital International Switching Unit
DIT Daido Institute of Technology [Japan]
Doctor of Industrial Technology
DITB Distributive Industry Training Board [UK]
DITC Department of Industry, Trade and Commerce [Canada]
Diisothiocyanate
DIRE Divertor in Torus Experiment
DITEC Digital Television Communication System
d-ITP Deoxyinosine Triphosphate
DITRAN Diagnostic FORTRAN
DIU Data Interchange Utility
Digital Input Unit
DIV Data in Voice
Digital Input Voltage
Divergence
Diverter
Dividend
Divider
Division
DIVA Data Input Voice Answerback
Data Inquiry Voice Answer
Digital Input Voice Answerback
Digital Inquiry Voice Answerback
DIVAD Division Air Defense
DIVD Dividend
DIVDS Dividends
DIVOT Digital-to-Voice Translator
DIVOTS Direct-Input Voice-Output Telephone System
DIY Do It Yourself
DIZ Development Impact Zone
DJ Diamond Jet
Diffusion Junction
Digital Junction
DJF Djibouti Franc [Currency of Djibouti]
DJNR Dow Jones News Retrieval [US]
DJSU Digital Junction Switching Unit
DK Dark
Deck
Direct Kinematics
dk donkey [Immunochemistry]
DKD Deutscher Kalibrierdienst [= German Calibration Service]
DKE Deutsche Elektrotechnische Kommission [= German Electrotechnical Commission]
DKFZ Deutsches Krebsforschungszentrum [= German Center for Cancer Research]
DKG Deutsche Keramische Gesellschaft [= German Ceramic Society]
dkg dekagram
DKI Data Key Idle
Deutsches Kunststoffinstitut [= German Plastics Institute]
Deutsches Kupferinstitut [= German Copper Institute]
DKK Danish Krone [Currency of Denmark, Greenland and the Faroe Islands]
dkL dekaliter [also: dkl]
dkm dekameter
DKP Dipotassium Phosphate
DKr Danish Krone [Currency of Denmark, Greenland and the Faroe Islands]
dks dekastere [Unit]
DKT Deutsche Kautschuk-Tagung [= German Rubber Conference]
DKV Deutscher Kaltetechnischer Verein [= German Refrigeration Association]

DL Data Language
Data Link
Data List
Deadload
Delay Line
Deputy Lieutenant
Developed Length
Difference Lumen
Diode Laser
Diode Logic
Discharge Lamp
Document List
Double Layer
Double Loop
Drawing List
Dual Language
Dynamic Loader

D/L Demand Loan

dL deciliter [also: dl]

DLA Data Link Adapter
Data Link Address
Defense Logistics Agency [US]
Diffusion-Limited Aggregation
Direct Lift Control
Direct-Line Attachment
Distributed-Lumped Active
Division of Library Automation [of University of California, US]
Duplex Line Control

DLC Data Link Control
Diamond-Like Carbon
Digital Logic Circuit
Digital Loop Carrier
Direct Lift Control
Duplex Line Control
Dynamic Load Characteristics

DLCC Desert Locust Control Committee [of FAO — UN]
Digital Load Cell Comparison

DLCF Data Link Control Field

DLCO Desert Locust Control Organization [of UN]

DLCO-EA Desert Locust Control Organization for Eastern Africa [of UN — Ethiopia]

DLCN Distributed Look Computer Network

DLCS Data Line Concentration System

DLD Dark-Line Defect
Delivered
Diplay List Driver

DL-DOPA DL-3,4-Dihydroxyphenylalanine

DL-DOPS DL-3,4-Dihydroxyphenylserine

DLE Data Link Escape
Direct Line Equipment

DLEA Double Leg Elbow Amplifier

DLEC Data Link Escape Character

DLES Department of Labour and Employment Services [Canada]

DLFW Department of Lands, Forests and Waters [Canada]

DLG Deutsche Landwirtschaftsgesellschaft [= German Agricultural Society]

DLGS Doppler Landing Guidance System

DLH Deutsche Lufthansa [= German Lufthansa Corporation]

DLI Data Link Interface
Defense Language Institute [US]
Delay Indefinite
Dual Link Interface

DL/I Data Language One [also: DL/1]

DLIEC Defense Language Institute, East Coast [US]

DLIR Depot Level Inspection and Repair
Downward Looking Infrared

DLIS Desert Locust Information Service [of FAO — UN]
Doctor of Library and Information Science

DLK Data Link

DLL Design Limit Load
Dial Long Lines

DLM Data Line Monitor
Dwight-Lloyd-McWane (Process)

DLMCP Distributed Look Message Communication Protocol

DLMF Drug Literature Microfilm File [of IDIS, US]

DLO Dead-Letter Office
Double Local Oscillator

DLOS Distributed Loop Operating System
Diver Lock-Out Submersible [also: DL-OS]

DLP Data Link Processor
Data Listing Program
Deoxidized, Low-Phosphorus Copper
Dynamic Low-Pass

DLPA Decorative Laminate Products Association [US]

DLPG DIMDI List Program Generator [Germany]

DLPU Data Link Processor Unit

DLR Dealer
Deutsche Forschungsanstalt fur Luft- und Raumfahrt [= German Research Institute for Aeronautics and Astronautics]

DLRG Design Layout Report Date

DLS Data Link Set
Data Link Splitter
Department of Life Sciences
Diffused Light Storage
Digital Line System
Direct Least Squares
Doctor of Library Science
Dominion Land Surveyor [now: CLS, Canada]

DLSC Defense Logistics Service Center [US]

DLSO Dial Line Service Observing

DLT Data Line Terminal
Data Line Translator
Data Link Terminal
Data Link Translator
Data-Loop Transceiver
Decision Logic Table
Decision Logic Translator
Depletion-Layer Transistor
Digital Line Termination

DLTDP Dilaurythiodipropionate

DLTM Data Link Test Message

DLTS Deep-Level Transient Spectroscopy
Deep-Level Transient Spectrum

DLTU Digital Line and Trunk Unit

DLU Data Line Unit
Digital Line Unit

DLVD Delivered

DLVL Diverted into Low Velocity Layers

DLVRY Delivery [also: DLVR]
DLVO Derjaguin, Landau, Verwey and Overbeck (Theory)
DM Dasymeter
Data Management
Data Mark
Data Memory
Data Model
Data Modelling
Decametric
Decimal Multiply
Decimetric
Delay Modulation
Delta Modulation
Demand Meter
Density Matrix
Design Manual
Design Memorandum
Deutsche Mark [Currency of Germany]
Development and Maintenance
Differential Mode
Digital Modulation
Digital Modelling
Digital Module
Digital Monolithic
Diphenylamine Chloroarsine
Direct Memory Access
Disconnected Mode
Distance Measuring
Distributing Main
Dicrete Mathematics
Discussion Memorandum
Dispersive Medium
District Municipality
Double Monochromator
Dry Matter
D&M Dressed and Matched [Lumber]
D/M Demodulation/Modulation
dm decimeter
DMA Defense Mapping Agency [US]
Dental Manufacturers of America [US]
Department of Municipal Affairs [Canada]
Dimethyl Acetal
Dimethylacetamide
Dimethylamine
Dimethylaniline
Direct Memory Access
Direct Memory Address
Direct Microassembly
Disodium Methylarsonate
Dynamic Mechanical Analysis
DMAAC Defense Mapping Agency Aerospace Center [US]
DMAB Dimethylamineborane
Dimethylaminobenzoic Acid
DMAC Defense Metals Information Center
Dimethyl Acetamide
Direct Memory Access Controller
DMACP Direct Memory Access Communications Processor
DMAD Dimethyl Acetylenedicarboxylate
DMAE Dimethylaminoethanol
Dimethylaminoethoxide
DMAEMA 2-(Dimethylamino)ethyl Methacrylate
DMAHC Defense Mapping Agency Hydrographic Center [US]
DMAI Direct Memory Access Interface
DMAL Dimethylacetal
DManSc Doctor of Management Sciences
DMAP 4-Dimethylaminopyridine
DMAPMA Dimethylaminopropylmethacrylamide
DMAPN Dimethylaminopropionitrile
DMAPS Digital Marine Acquisition and Processing System
DMAS Distribution Management Accounting System
DMB Data Management Block
4,4'-Dichloromethylbenzhydrol
Dimethoxybenzene
Disconnect and Make Busy
DMBA Dimethylbenzanthracene
DMBAS 2,5-Dimethoxy-4'-Aminostilbene
DMC Dichlorodiphenylmethyl Carbinol
4,4'-Dichloromethylbenzhydrol
Digital Microcircuit
Dimethylaminoethyl Chloride Hydrochloride
Direct Memory Channel
Direct Multiplexed Control
Dough Molding Compound
Duff Moisture Code
DMCD Dimethylcyclohexanedicarboxilate
DMCHA Dimethylcyclohexyladipate
DMCL Device/Media Control Language
DMCS Dimethyldichlorosilane
Distributed Manufacturing Control System
DMD Dimethyloxozolidinedione
DMDCS Dimethyldichlorosilane
DMDPU Dimethyldiphenylurea
DMDS Dimethyldisulfide
DMDT Dimethoxydiphenyltrichloroethane
DME Data Measuring Equipment
Department of Materials Engineering
Department of Mechanical Engineering
Department of Metallurgical Engineering
Department of Mining Engineering
Developing Market Economy
Digital Measuring Equipment
Digital Multiplex Equipment
Dimethoxyethane
Dimethyl Ether
Direct Machine Environment
Direct Measurements Explorer (Satellite)
Directorate of Mechanical Engineering [UK]
Distance-Measuring Equipment
Division of Mechanical Engineering
Dropping Mercury Electrode
Dulbecco's Modified Eagle Medium
DMEA Dimethylethylamine
DME/COTAR Distance-Measuring Equipment/Correlation Tracking and Ranging
DMED Digital Message Entrance Device
Digital Message Entry Device
DME/DP Dropping Mercury Electrode/Differential Pulse
DMedSc Doctor of Medical Science
DMEK Dimethylethylketone
DMEM Dulbecco's Modified Eagle Medium
DMEP Data Network Modified Emulator Program
Dimethoxyethylphthalate

DMERT Duplex Multiple-Environment Real-Time (Operating System)
DMET Distance-Measuring Equipment TACAN (= Tactical Air Navigation)
DMet Doctor of Metallurgy
DMEU 1,3-Dimethyl-2-Imidazolidinone
DMF Data Management Facility
Differential Matched Filter
Digital Matched Filter
Dimethylformamide
Dimethylfuran
Disk Management Facility
DMF-DMA Dimethylformamide Dimethyl Acetal
DMG Dimethylglyoxime
DMH Device Message Handler
Dimethylhydantoin
Direct Manhours
Drop Manhole
DMHF Dimethylhydantoin Formaldehyde
DMI Danish Meteorological Institute
Department of Manpower and Immigration
Department of Microbiology and Immunology
Design Management Institute [US]
Desmethylimipramine
Digital Multiplexed Interface
1,3-Dimethyl-2-Imidazolidinone
Direct Memory Interface
Director of Military Intelligence [UK]
Dnepropetrovsk Metallurgical Institute [Ukraine]
Dynamic Memory Interface
DMIRR Demand Module Integral Rocket Ramjet
DMIS Directory Management Information System
DMJTC Differential Multijunction Thermal Converter
DMK Deutsche Mark [Currency of Germany]
Dimethylketone
DML Data Manipulation Language
Demolition
Dimyristoyl Lecithin
DMLS Doppler Microwave Landing System
DMM Data Manipulation Mode
Defense Market Measures (Database) [of USDOD]
Department of Mining and Metallurgy
Digital Multimeter
Dimethylmercury
Direct Metal Mastering
Dynamical Material Modeling
DMMF Dry, Mineral-Matter Free [also: dmmf]
DMMP Dimethyl Methylphosphonate
DMN Dimension
Dimethylnitrosamine
Direction de la Meteorologie Nationale [= Directorate for National Meteorology, France]
DMNA Dimethylnitrosamine
DMNSC Digital Main Network Switching System
DMO Data Management Office
Diode Microwave Oscillator
Director of Military Operations
DMOD Delta Modulation
DMOS Data Management Operating System
Diffused Metal-Oxide Semiconductor
Diffusive Mixing of Organic Solutions
Discrete Metal-Oxide Semiconductor
Double-Diffused Metal-Oxide Semiconductor [also: D-MOS]
DMP Dimethoxypropane
Dimethylphenol
Dimethylphthalate
Direct Memory Processor
Dot Matrix Printer
Dump
DMPA 2,2-Bis(hydroxymethyl)propionic Acid
Dichlorophenylmethyl Isopropylphosphoramidothioate
Digitally-Modulated Power Amplifier
Dimethylolpropionic Acid
DMPC 3-Dimethylaminopropyl Chloride Hydrochloride
Dimyristoyl Phosphatidyl Choline
DMPD 4,5-Dimethyl-1,2-Phenylenediamine
DMPE 1,2-Bis(dimethylphosphino)ethane
Dimethoxphenylethylamine
DMPEA Dimethoxphenylethylamine
DMPO 5,5-Dimethyl-1-pyrroline-N-Oxide
DMPP 1,1-Dimethyl-4-Phenylpiperazinium Iodide
2MPPO Dimethyl-P-Phenylene Oxide
DMPR Damper
DMPRT Dual-Mode Personal Rapid Transport
DMPS 2,3-Dimercapto-1-Propanesulfonic Acid
DMPU Dimethylpropyleneurea
DMQ Direct Memory Queue
DMR Data Management Routine
Demultiplexing/Mixing/Remultiplexing
Digital Microwave Radio
Distributed Message Router
Dynamic Modular Replacement
DMRE Diploma in Medical Radiology and Electrology
DMRL Defense Metallurgical Research Laboratory [India]
DMS Database Management System
Data Management Service
Data Management System
Data Multiplex Switching
Defense Missile Systems
Dense-Media Separation
Department of Materials Science
Department of Mathematics and Statistics
Differential Maneuvering Simulator
Digital Multimeter System
Digital Multiplexing Synchronizer
Digital Multiplex Switching (System)
Diluted Magnetic Semiconductor
Dimethylsilyl
Dimethylsulfate
Dimethylsulfide
Disk Monitor System
Display Management System
Distributed Maintenance Services
Division of Materials Sciences [of USDOE]
Doctor of Medical Science
Documentation of Molecular Spectroscopy
Document Management System
Drawing Management System
Duplex Microstructure
Dynamic Mapping System

DMSc Doctor of Medical Science
DMSCC Direct Microscopic Somatic Cell Count
DMSO Dimethyl Sulfoxide
$DMSO_2$ Dimethyl Sulfone
DMSP Defense Meteorological Satellite Program [US]
DMSR Director of Missile Safety Research
DMSS Digital Multiplexing Subsystem
Digital Multiplex Switching System
DMT Digital Message Terminal
4,4'-Dimethoxytrityl Chloride
Dimethyl Terephthalate
Dimethyltryptamine
Dimethoxytrityl Chloride
Direct Memory Transfer
Dispersive Mechanism Test
DMTA Dynamic Mechanical Thermal Analyzer
DMTC Digital Message Terminal Computer
Dimethoxytrityl Chloride
DMTCNQ Dimethyl Tetracyanquinonedimethane
DMTD 2,5-Dimercapto-1,3,4-Thiadiazole
DMTR Dounreay Materials Testing Reactor [Scotland]
DMTS Department of Mines and Technical Surveys [Canada]
DMU Dictionary Management Utility
Diesel-Multiple Unit
Digital Message Unit
Dimethylurea
Distributed Microprocessor Unit
Dual Maneuvering Unit
DMUS Data Management Utility System
DMV Department of Motor Vehicles [US]
Discrete Multivibrator
Division of Motor Vehicles
DMW Decametric Wave
Decimetric Wave
Digital Milliwatt [also: DmW]
Dissimilar-Metal Weld
DMX Data Multiplexer
Direct Memory Exchange
DMZ Demilitarized Zone
DN Data Name
Decimal Number
Decimal Numbering
Decrement
Delayed Neutron
Department of the Navy [US]
Directory Number
Document Number
Down
D/N Debit Note
dn delta amplitude [Elliptic Function]
decineper
DNA Defense Nuclear Agency [US]
Deutscher Normenausschuss [= German Standards Committee]
Deoxyribonucleic Acid
Digital Network Architecture
Dinolylaniline
Does Not Apply
DNAF 10-(2',4'-Dinitrophenylazo)-9-Phenanthrol
DNAG Decade of North American Geology
DNAM Data Network Access Method
DNAP Dinitroaminophenol
4-(2',4'-Dinitrophenylazo)phenol
DNase Deoxyribonuclease
DNB Departure from Nucleate Boiling
Deutsches Nachrichtenburo [= German News Agency]
Dictionary of National Bibliography
Dinitrobenzene
Dinitrobenzoyl
DNBP Dinitrobutylphenol
DNBR Departure from Nucleate Boiling Ratio
DNBS 2,4-Dinitrobenzenesulfonic Acid, Sodium Salt
DNBSC 2,4-Dinitrobenzenesulfenyl Chloride
DNC Delayed Neutron Counting
Delayed Neutron Coupling
Dinitrocellulose
Dinitrocresol
Direct Numerical Control
Distributed Numerical Control
DNCB Dinitrochlorobenzene
DNCC Data Network Control Center
DNCCC Defense National Communications Control Center [US]
DNCIAWPRC Danish National Committee of the International Association on Water Pollution Research and Control
DNCS Distributed Network Control System
DND Department of National Defense [Canada]
dNDP Desoxyribonucleoside Diphosphate
DNDS Dinitrodiphenyl Disulfide
DNE Department of Nuclear Engineering [US]
Dinitroethane
DNFA 2,4-Dinitro-5-Fluoroaniline
DNFB 2,4-Dinitrofluorobenzene
DNHR Dynamic Non-Hierarchical Routing
DNHW Department of National Health and Welfare [Canada]
DNI Data Network Interface
Digital Noninterpolation
DNIBR Danish National Institute of Building Research
DNIC Data Net Identification Code
Digital Network Interface Circuit
DNL Differential Non-Linerarity
Dynamic Noise Limiter
DNM Delayed Neutron Monitor
dNMP Desoxyribonucleoside Monophosphate
DNMR Double Nuclear Magnetic Resonance
DNN Dinitronaphthalene
DNNSA Dinonyl Naphthalene Sulfonic Acid
DNOC Dinitro-O-Cresol
DNOCHP Dinitro-O-Cyclohexylphenol
DNODA Di(N-Octyl-N-Decyl)adipate
DNODP Di(N-Octyl-N-Decyl)phthalate
DNOSBP Dinitro-O-Sec-Butylphenol
DNP Deoxyribonucleoprotein
Dinitrophenol
Dinitrophenyl
2,4-Dinitrophenylhydrazine
Dinonyl Phthalate
Dynamic Nuclear Polarization

DNPC Dinitro-P-Cresol
2,6-DNPC 2,6-Dinitro-P-Cresol
DNPD Dinaphthylphenylenediamine
N,N'-Di-β-Naphthyl-P-Phenylene Diamine
DNPH 2,4-Dinitrophenylhydrazine
DNPM Departmento Nacional de Producao Mineral [= National Department of Mineral Products, Brazil]
DNPT N,N'-Dinitroso Pentamethylene Tetramine
DNPyr Dinitropyridyl
DNR Double Non-Return Valve
Dynamic Noise Reduction
DNRC Democritus Nuclear Research Center [Greece]
DNRE Department of Natural Resources and Energy [Canada]
DNRZ Delayed Non-Return-to-Zero
DNS 5-Dimethylamino-1-Naphthalenesulfonic Acid
Distributed Network System
Doppler Navigation System
DNSC Data Network Service Center
Digital Network Service Center
DNSE Direction Nacional del Servicio Estadistico [= National Directorate of Statistic Services, Argentina]
DNSR Director of Nuclear Safety Research
DNT Desmethylnortriplyline
Digital Network Terminator
Dinitrotoluene
DNTC 4-Dimethylamino-1-Naphthyl Isothiocyanate
dNTP Deoxyribonucleoside Triphosphate
DNTS Digital Data Exchange Network Testing System
DNZ Darned Near Zero
DO Dark Operated
Data Output
Decimal-to-Octal
Delivery Order
Design Objective
Diesel Oil
Digital Output
Dissolved Oxygen
Draw Out
D/O Delivery Order
D-O Decimal-to-Octal
do (ditto) — the same
DOA Date of Availability
Dead on Arrival
(Deliver) Documents on Acceptance (of Draft)
Department of Agriculture
Differential Operational Amplifier
Digital Output Adapter
Dioctyl Adipate
Documents on Acceptance
Dominant Obstacle Allowance
DOAE Defense Operational Analysis Establishment [UK]
DOB Data Output Bus
Date of Birth
DOBAMBC P-Decyloxybenzylidene P'-Amino-2-Methyl Butyl Cinnamate
DOBIS Dortmunder On-Line Bibliothekssystem [= Dortmund On-Line Library System [of UD, Germany]]
DOBS Disk-On-Bearing System
DOC Data Optimizing Computer
Decimal-to-Octal Conversion
Delayed-Opening Chaff
Design Office Consortium [now: CICA, UK]
Deoxycorticosterone
Department of Commerce [US]
Department of Communications [Canada]
Depth of Cut
Designated Operational Coverage
Desoxycortone
Deterministic Optimal Control
Dichromate Oxygen Consumed
Diethyloxacarbocyanine
Digital Output Control
Disk-Oriented Computer System
Direct Operating Cost
Doctor
Document
Documentation
Documentor
Dynamic Overload Control
DoC Department of Commerce [US]
DOCA Documentation Automatique [= Automatic Documentation Section [of CETIS, France]]
Deoxycorticosterone Acetate
DOCC DCA (= Defense Communications Agency) Operations Center Complex [US]
DOCDEL Document Delivery [of DIANE, Europe]
DOCFAX Document Facsimile Transmission
DOCLINE Document Ordering On-Line
DOCOCEAN Documentation Oceanique [= Documentation on Oceanography [of BNDO, France]]
DOCP Deterministic Optimal Control Problem
DOCS Disk-Oriented Computer System
Document Organization and Control System
Documents
DOCSYS Display of Chromosome Statistics System
DOCUS Display-Oriented Computer Usage System
DOD Department of Defense [also: DoD or USDOD]
Development Operations Division
Digital Optical Disk
Direct Outward Dialing
Drop on Demand
DODA Door and Operator Dealers Association [US]
DODC 3,3'-Diethyloxadicarbocyanine
DODCI Department of Defense Computer Institute [US]
DODCSC Department of Defense Computer Security Center [US]
DODD Department of Defense Directive [US]
DODGE Department of Defense Gravity Experiment [US]
DODGE-M DODGE, Multipurpose [US]
DODIS Distribution of Oceanographic Data on Isotropic Levels
DODISS Department of Defense Index of Specifications and Standards [also: DoDISS, US]
DODMAB Dioctadecyldimethylammonium Bromide
DODT Display Octal Debugging Technique
DOE Department of Education [Canada]
Department of Electronics [India]
Department of Energy [US and UK]
Department of Environment [Canada and UK]

	Design of Experiment
DoE	Department of Energy [US]
DOE/AES	Department of Environment/Atmospheric Environment Service [Canada]
DOETS	Dual-Object Electronic Tracking System
DOE(UK)	Department of Energy (UK)
DOE(US)	Department of Energy (US)
DOF	Deep Ocean Floor
	Degree of Freedom
	Department of Fisheries [US]
	Depth of Focus
	Device Operating Failure
	Device Output Format
	Dioctyl Fumarate
	Direction of Flight
DOFIC	Domain-Originated Functional Integrated Circuit
DOFP	Direct-on-Finish Process
DOH	Department of Health [Taiwan]
DOHC	Double Overhead Camshaft
DOHS	Diploma in Occupational Health and Safety
DOI	Department of Industry [now: DTI, UK]
	Department of the Interior [also: DoI, US]
	Differential Orbit Improvement
	Distinctiveness of Image
DOIO	Directly Operable Input/Output
DOIT	Database Oriented Interrogation Technique
DO/IT	Digital Output/Input Translator [also: DOIT]
DOJ	Department of Justice [also: DoJ, US]
DOL	Department of Labor [also: DoL, US]
	Director of Laboratories
	Display-Oriented Language
	Dynamic Octal Load
DOLARS	Digital Off-Line Automatic Recording System
	Doppler Location and Ranging System
DOLE	Data On-Line Editing System
	Data On-Line Editor
	Digital On-Line Editing System
	Digital On-Line Editor
DOLPHIN	Deep Ocean Logging Profiler Hydrographic Instrumentation and Navigation
DOM	Department of Mines [US]
	Design of Maintenance
	Digital Ohmmeter
	Dimethoxymethylamphetamine
	Diocytl Maleate
	Disk Operating Monitor
	Dissolved Organic Matter
	Domestic
	Dominion
	Drawn Over Mandrel
DoM	Department of Mines [US]
DOMA	Dokumentation Maschinenbau (Database) [= Documentation on Mechanical Engineering, Germany]
Dom App	Domestic Appliance
DOMSAT	Domestic Satellite [Australia]
	Domestic Satellite System
DOMSATS	Domestic Satellite Systems
DON	Delayed Order Notice
	Department of the Navy [US]
	Deuterium-Moderated Organic-Cooled Nuclear Reactor [Spain]
	Dioxynaphthalene
DONA	Decentralized Open Network Architecture
DONG	Dansk Olje Og Naturgas (Line) [= Danish Oil and Natural Gas Line]
DOP	Deliver Documents on Payment of Draft
	Designated Overhaul Point
	Detailed Operating Procedure
	Developing-Out Paper
	Dilution of Position
	Dioctyl Phthalate
	Dominican Peso [Currency of the Dominican Republic]
DOPA	3-(3,4-Dihydroxyphenyl)alanine
DOPAC	Dihydroxyphenylacetic Acid
DOPAMINE	Dihydroxyphenethylamine
DOPD	N,N'-Bis-(1-Ethyl-3-Methylpentyl)-P-Phenylene Diamine
DOPIC	Documentation of Program in Core
DOPLOC	Doppler Phase Lock
DOPS	Digital Optical Protection System
	3,4-Dihydroxyphenylserine
DOPSK	Differential Offset Phase Shift Keying
DOR	Data Output Register
	Deuterium-Moderated and Organic-Cooled Reactor
	Digital Optical Recording
	Digital Output Relay
	Division of Operating Reactors
DORA	Directorate of Operational Research and Analysis [UK]
DORAN	Doppler Range and Navigation [also: Doran or doran]
DORCMA	Door Operator and Remote Controls Manufacturers Association [US]
DORE	Defense Operational Research Establishment [of DRB, Canada]
DORF	Diamond Ordinance Radiation Facility [US]
DORIS	Direct Order Recording and Invoicing System
	Doppel-Ring-Speicher [= Double Ring Storage, Germany]
DORK	Diagnostically Optimizable Recursive Keyword
	Direct Order Recording Keyboard
	Direct On-Line Retrievable Knowledge
DORV	Deep Ocean Research Vehicle
DOS	Days on Stream
	Decision Outstanding
	Degree of Sensitization
	Density of States
	Digital Operating System
	Dioctyl Sebacate
	Disk Operating System
DOSAR	Dosimetry Applications Research Facility
DOSP	Dalhousie Ocean Studies Program [Canada]
DOSV	Deep Ocean Survey Vehicle
DOS/VS	Disk Operating System/Virtual Storage
DOS/VS-AF	Disk Operating System/Virtual Storage — Advanced Functions
DOS/VSE	Disk Operating System/Virtual Storage Extended
DOT	Deep Ocean Technology
	Deep Ocean Transponder

	Deep Ocean Trough
	Department of Trade [UK]
	Department of Transport [Canada]
	Department of Transportation [also: DoT, US]
	Designating Optical Tracker
	Digital Optical Technology
	Digital Overlay Technique
	Domain-Tip Memory
DOTC	Department of Transportation Classification [US]
DOTG	Di-O-Tolylguanidine
DOTP	Dioctyl Terephthalate
DOTS	Dynamic Ocean Track System [US]
DOTSYS	Dot System
DOTT	Di-O-Tolylthiourea
	Documentation for Translation and Terminology [of Secretary of State, Canada]
DOT/UN	Department of Transportation/United Nations
DOUG FIR	Douglas Fir
DOUG FIR-L	Douglas Fir — Lumber
DOUSER	Doppler Unbeamed Search Radar [also: Douser]
DOV	Data Over Voice
	Degree of Variance
DOVAP	Doppler Velocity and Position [also: Dovap]
DOVETT	Double Velocity Transit Time
DOWB	Deep Operating Work Board
DOXYL	Dimethyloxazolidine-N-Oxyl
	Dimethyloxazolinyloxy
DOZ	Dioctyl Azelate
doz	dozen
DP	Dash-Pot
	Data Point(er)
	Data Port
	Data Processing
	Data Processor
	Date of Publication
	Depth
	Deck Piercing
	Deep Penetration
	Deflection Plate
	Degree of Polymerization
	Deflector Plate
	Delta Pile
	Dew Point
	Dial Pulsing
	Diametral Pitch
	Diamond Pin
	Dichlorophenoxypropionate
	Differential Phase
	Differential Pressure
	Differential Pulse
	Diffraction Pattern
	Diffusion Pump
	Digital Plotter
	Digital Pulser
	Digit Present
	Disconnection Pending
	Disk Pack
	Dispersed Phase
	Display Package
	Distribution Point
	Dominican Peso [Currency of the Dominican Republic]
	Double Paper (Wire)
	Double-Pass
	Double-Pole
	Draft Proposal
	Drillpipe
	Drip-Proof
	Driving Power
	Drum Processor
	Dual-Port Bus
	Dual Property
	Dynamic Positioning
	Dynamic Pressure
	Dynamic Programming
D&P	Developing and Printing
D/P	Documents against Payment
2,4-DP	2,4-Dichlorophenoxypropionic Acid
	2,4-Dichlorprop
DPA	Data Processing Activities
	Data Protection Act [UK]
	Data Protection Agency
	Deoxidized Phosphorus Copper, Arsenical
	Department of Physics and Astronomy
	Destructive Physical Analysis
	Deutsche Presse-Agentur [= German Press Agency]
	Dial Pulse Access
	Different Premises Address
	Diphenolic Acid
	Diphenylamine
	Diploma in Pathological Anatomy
	Diploma in Public Administration
	Displacements per Atom [also: dpa]
	Doctor of Public Administration
DPAG	Dangerous Pathogens Advisory Group
DPAGE	Device Page
D-PAT	Drum-Programmed Automatic Tester
DPB	Data Processing Branch
	Defense Policy Board [US]
	1,4-Diphenyl-1,3-butadiene
	Discounted Payback
	Dodecyl Pyridinium Bromide
DPBC	Double-Pole, Back-Connected
D-PBS	Dulbecco's Phosphate Buffered Saline
DPC	Data Processing Center
	Data Processing Control
	Defense Planning Committee [of NATO]
	Departure Control
	Desert Protective Council [US]
	Destination Point Code
	Differential Phase Contrast
	Diphenyl Carbinol
	Direct Program Control
	Display Processor Code
	Division of Physical Chemistry [of ACS, US]
DPCM	Delta Pulse-Code Modulation
	Differential Pulse-Code Modulation
	Distributed Processing Communications Module
DPCSV	Differential-Pulse Cathodic Stripping Voltage
DPCT	Differential Protection Current Transformer

DPCTG Database Program Conversion Task Group [of CODASYL]
DPCX Distributed Processing Control Executive
DPD Data Processing Department
Data Processing Division
Diethyl-P-Phenylene Diamine
Diffusion Pressure Deficit
Digital Phase Difference
Digital Plane Driver
DPDA Deterministic Pushdown Automaton
Diperoxydodecanedioic Acid
DPDC Double Paper, Double Cotton (Wire)
DPDL Distributed Program Design Language
DPDT Double-Pole, Double-Throw
DPDT DB Double-Pole, Double-Throw, Double-Break
DPDT SW Double-Pole, Double-Throw Switch
DPE Data Processing Equipment
Department of Plastics Engineering
Dipiperidinoethane
Direct Plate Exposure
Distributed Processing Environment
DPed Doctor of Pedagogy
DPEK Differential Phase-Exchange Keying
DPEX Distributed Processing Executive Program
DPF Data Processing Facility
Diesel Particulate Matter
Dual Polarized Frequency
Dual Program Feature
DPFC Double-Pole, Front-Connected
DPFZ Destor-Porcupine Fault Zone [Canada]
DPG Deutsche Physikalische Gesellschaft [= German Physical Society]
Digital Pattern Generator
Diphenylguanidine
Diphosphoglycerate
Diphosphoglyceric Acid
2,3-DPG 2,3-Diphosphoglyceric Acid
DPH Department of Public Health
Diamond Pyramid Hardness
1,6-Diphenyl-1,3,5-Hexatriene
Diphenylhydantoin
Diploma in Public Health
Doctor of Public Health
DPh Doctor of Philosophy [also: DPhil]
DPI Data Processing Installation
Department of Public Information [of UN]
Device/Programmer Interface
Differential Pressure Indicator
Different Premises Information
Digital Process Instrument
Digital Pseudorandom Inspection
Diphenylimide
Dots per Inch [also: dpi]
DPIA Diethylenetriamine Producers and Importers Alliance [US]
DPI/O Data Processing Input/Output
DPIP Dichlorophenol Indophenol
Diphenyl Isophthalate
DPIS Differential Pressure Isolation Switch
DPK Diphenylketone
DPL Data Processing Language
Diploma
DPLCS Digital Propellant Level Control System
DPLL Digital Phase-Locked Loop
DPLM Domestic Public Land Mobile
DPLMRS Domestic Public Land Mobile Radio Service
DPM Data Processing Machine
Data Processing Management
Data Processing Manager
Diesel Particulate Matter
Digital Panel Meter
Diphenylmethane
Diphenylphosphinomethane
Dipivaloylmethane
Disintegrations per Minute
Distributed Presentation Management
Distributive Principle of Multiplication
Documents per Minute
DPMA Data Processing Management Association [US]
DPMC Dual Port Memory Control
DPMOAP (Society of) Data Processing Machine Operators and Programmers [US]
DPMM Dots per Millimeter [also: dpmm]
DPMS Data Project Management System
D5MTD Dipentamethylene Thiuram Disulfide
DPN Diamond Pyramid Hardness Number
Diphosphopyridine Nucleotide
Dpn Diplococcus Pneumoniae [Molecular Biology]
d(PN) Deoxypolynucleotide
DPNase Diphosphonucleosidase
DPNDA Diphenylnaphthalene Diamine
DPNH Dihydrodiphosphopyridine Nucleotide, Reduced Form
DP-NR Deproteinated Natural Rubber
DPNSS Digital Private Network Signaling System [UK]
DPO Data Processing Operations
Dial Pulse Originating
Diphenylene Oxide
Diphenyloxazole
Diphenyloxide
Double Pulse Operation
DPP Dextran Phosphate Precipitate
Digital Parallel Processor
Diphenylphosphinyl
Diphenyl Phthalate
Disposable Plotter Pen
Drip Pan Pot
DPPA Dipalmitoyl Phosphatidyl Chloride
Diphenylphosphoryl Azide
DPP-Cl Diphenylphosphinyl Chloride
DPPD N,N'-Diphenyl-P-Phenylenediamine
DPPH 2,2-Di(4-tert-Octylphenyl)-1-Picrylhydrazyl
α,α'-Diphenyl-β-Picryl Hydrazyl
DPPX Distributed Processing Programming Executive
DPR Democratic People's Republic [South Korea]
Digital Process Reporter
DPRI Disaster Prevention Research Institute
DPRR Department of Parks and Renewable Resources [Canada]
DPS Data Processing Standards
Data Processing Station
Data Processing System

	Department of Planetary Sciences
	Department of Polymer Science
	Descent Power System
	Different Premises Subscriber
	Diphenylstilbene
	Disintegrations per Second
	Disk Programming System
	Distributed Processing System
	Doctor of Professional Studies
	Document Processing System
	Double-Page Spread
	Dynamic Processing System
DPSA	Data Processing Suppliers Association [US]
	Deep Penetration Strike Aircraft
DPSK	Differential Phase-Shift Keying
DPSP	Deferred Profit Sharing Plan
DPSS	Data Processing Subsystem
	Data Processing System Simulator
	Direct Program Search System
DPST	Double-Pole, Single-Throw
DPST SW	Double-Pole, Single-Throw Switch
DP SW	Double-Pole Switch
DPT	Department
	Department of Pharmacology and Toxicology
	Dewpoint Temperature
	Dial Pulse Terminating
	Differential Polarization Telegraphy
	Different Premises Telephone
	Dinitrosopentamethylenetetramine
DPTE	Deoxidized Phosphorus Copper, Tellurium Bearing
DPTH	Diphenylthiohydantoin
DPTT	Dipentamethlene Thiuram-Tera(hexa)sulfide
	Double-Pole, Triple-Throw
DPTT SW	Double-Pole, Triple-Throw Switch
DPTU	N,N'-Diphenylthiourea
DPTX	Distributed Processing Terminal Exchange
DPTY	Deputy
DPU	Data Processing Unit
	Department of Public Utilities [US]
	Digital Patch Unit
	Diphenylurea
	Dip Pick-Up
	Disk Pack Unit
	Dust Preparation Unit
DPV	Dry Pipe Valve
DPW	Department of Public Works
DPWM	Double Pulsewidth Modulation
	Double-Sided Pulsewidth Modulation
DQ	Data Quality
	Directory Enquiry
	Direct Quenching
	Drawing Quality
DQA	Design Quality Assurance
DQAK	Drawing Quality, Aluminum-Killed
DQC	Data Quality Control
DQE	Detective Quantum Efficiency
DQM	Data Quality Monitor
DQPSK	Differential Quadrature Phase Shift Keying
DQRS	Drawing Quality Rimmed Steel
DQSK	Drawing Quality, Special-Killed
DR	Danish Reactor [Denmark]
	Darkroom
	Data Rate
	Data Receiver
	Data Recorder
	Data Recording
	Data Reduction
	Data Register
	Dead Reckoning
	Deadroom
	Debit
	Debtor
	Defense Research
	Defined Readout
	Delta Ray
	Dextrorotatory
	Differential Rate
	Diffuse Reflectance
	Digital Radiography
	Digital Radioscopy
	Digital Recording
	Digital Resolver
	Diploma in Radiology
	Direct Reduction
	Disaster Recovery
	Discrepancy Report
	Discrete Regulator
	Discrimination Radar
	Division of Research
	Divisor
	Doctor of Radiology
	Drafting Request
	Drag Reduction
	Drain
	Drawer
	Drill
	Drilling Regulation
	Drill Rod
	Drive
	Dynamic Range
D&R	Distiller and Rectifier
D/R	Deposit Receipt
	Direct/Reverse
	Downrange
Dr	Doctor
	Drachma [Currency of Greece]
dr	dram
DRA	Data Resource Administrator
	Dead Reckoning Analyzer
	Drachma [Currency of Greece]
Dra	Deinococcus Radiophilus [Molecular Biology]
	Draco [Astronomy]
DR&A	Data Reduction and Analysis
Drac	Draco [Astronomy]
Dr agr	(Doctor agronomiae) — Doctor of Agriculture
DRADS	Degradation of Radar Defense System
DRAE	Defense Research Analysis Establishment
DRAFT	Document Read and Format Translator
DRAI	Dead Reckoning Analog Indicator
DRAM	Digital Recorded Announcement Module
	Dynamic Random-Access Memory [also: dRAM]
DRAMA	Digital Radio and Multiplex Acquisition

DRAO Dominion Radio Astrophysical Observatory [Canada]
dr ap dram, apothecaries
dr av dram, avoirdupois
dr avdp dram, avoirdupois
DRAW Direct Read After Write
DRB Defense Research Board [Canada]
5,6-Dichlorobenzimidazole Riboside
DRC Damage Risk Criterion
Data Recording Control
Data Reduction Compiler
Defense Research Committee
Defense Review Committee [of NATO]
Denver Research Center
Design Rules Checking
Discontinuous Reinforced Composite
Dolphin Research Center [US]
Domestic Resource Cost
DRCL Defense Research Chemical Laboratories [Canada]
DrComSc Doctor of Commercial Science
DRCS Distress Radio Call System
Dynamically Redefinable Character Set
DRD Data Recording Device
Differential Range De-Ramp
Director of Research and Development
Direct-Reading Dosimeter
Draw-Redraw
Drill Rig Duty
DrDes Doctor of Design
D-RDF Densified Refuse-Derived Fuel [also: d-RDF]
DRD-FD Differential Range De-Ramp — Frequency Domain (Processor)
Dr disc pol (Doctor disciplinarum politicarum) — Doctor of Social Sciences
DRDTO Detection Radar Data Takeoff
DRDW Direct Read During Write
DRE Data Recording Equipment
Data Reduction Equipment
Dead Reckoning Equipment
Defense Research Establishment [of DND, Canada]
Destruction and Removal Efficiency
Digital Rebalance Electronics
Directional Reservation Equipment
Director of Research and Engineering
Dokumentationsring Elektrotechnik (Database) [= Documentation on Electrical Engineering [of ZDE, Germany]]
D/RE Disassembly/Reassembly Equipment
DREA Defense Research Establishment, Atlantic [of DND, Canada]
DREAC Drum Experimental Automatic Computer
DREAM Data Retrieval Entry and Management
DRECT Demonstration of Resource and Energy Conservation Technologies (Program) [of Environment Canada]
Development of Resource and Energy Conservation Technologies [Canada]
DRED Data Routing and Error Detecting
DREE Department of Regional Economic Expansion [later: DRIE, Canada]
DREF Data Reference System [of Environment Canada]
DREO Defense Research Establishment, Ottawa [of DND, Canada]
DREP Defense Research Establishment, Pacific [of DND, Canada]
DRES Defense Research Establishment, Suffield [of DND, Canada]
Direct Reading Emission Spectrograph
DRESS Depth-Resolved Surface Coil Spectroscopy
DRET Defense Research Establishment, Toronto [of DND, Canada]
Direct Reentry Telemetry
DRETS Direct Reentry Telemetry System
DREV Defense Research Establishment, Valcartier [of DND, Canada]
DREWS Direct Readout Equatorial Weather Satellite
dr fl dram, fluid
Dr forest (Doctor scientiae forestalium) — Doctor of Forest Science
DRFR Division of Research Facilities and Resources [US]
DR/FTIR Diffuse Reflectance/Fourier Transform Infrared Spectroscopy
DR FX Drill Fixture
DRG Division of Research Grants [of NRC, US]
DRGS Direct Readout Ground Station
Dr habil (Doctor habilitatus) — Habilitated Doctor [Europe]
Dr hc (Doctor honoris causa) — Honorary Doctor
DR HD Drill Head
DRI Data Reduction Interpreter
Dead Reckoning Indicator
Defense Research Institute [US]
Denver Research Institute [US]
Descent Rate Indicator
Desert Research Institute [US]
Direct-Reduced Iron
Document Retrieval Index
DRIC Defense Research Information Center [UK]
DRID Direct Readout Image Dissector
DRIDAC Drum Input to Digital Automatic Computer
DRIE Department of Regional and Industrial Expansion [now: ISTC, Canada]
DRIFT Diffuse Reflectance Infrared Fourier Transform
Diversity Receiving Instrumentation for Telemetry
Diversity Reliability Instantaneous Forecasting Technique
DRINC Dairy Research, Inc. [US]
DRIP (International Symposium on) Defects Recognition and Image Processing
DRIR Direct Readout Infrared
DRIVE Document Read Information Verify and Edit
Document Review into Video Entry
DRL Data Requirements Language
Data-Retrieval Language
Daytime Running Lights
Defense Research Laboratory [US]
Direct Retrieval Language
Drilling Research Laboratory
DRLG Drilling
DRM Depositional Remanent Magnetization
Detrital Remanent Magnetization
Digital Radiometer
Direction of Relative Movement

Drafting Room Manual
Dynamic Recoil Mixing
Dr med (Doctor medicinae) — Doctor of Medicine
Dr med univ (Doctor medicinae universae) — Doctor of Universal Medicine
Dr med vet (Doctor medicinae veterinariae) — Doctor of Veterinary Medicine
DRML Defense Research Medical Laboratories [of DRB, Canada]
DRMO District Records Management Officer
DRMS Data Resource Management System
DRN Drawn
Dr nat techn (Doctor rerum naturalium technicarum) — Doctor of Natural Science and Technology
DRNL Defense Research Northern Laboratories [Canada]
DRO Destructive Readout
Dielectric Resonator Oscillator
Digital Readout
Doubly Resonant Oscillator
DROD Delayed Readout Detector
Dr oec (Doctor oeconomiae) — Doctor of Economics
DROM Decoder Read-Only Memory
DROP Distribution Register of Pollutants [of EPA, US]
DROS Direct Readout Satellite
Disk Resident Operating System
DRP Data Reception Process
Data Reduction Program
Dead Reckoning Plotter
Dense Random-Packed (Structure)
Dense Random Packing
Digital Recording Process
Directional Radiated Power
Distribution Requirements Planning
Distribution Resource Planning
Dr paed (Doctor paedagogiae) — Doctor of Pedagogy
DRPC Defense Research Policy Committee [UK]
DrPH Doctor of Public Health
Dr pharm (Doctor pharmaciae) — Doctor of Pharmacy
Dr phil (Doctor philosopiae) — Doctor of Philosophy
Dr phil nat (Doctor of philosophiae naturalis) — Doctor of Natural Sciences
DRPHS Dense Random Packing of Hard Spheres
DRPI Digital Rod Position Indicator
DRPS Disk Real-Time and Programming System
Dynamic Memory Relocation and Protection System
DRQ Data Ready Queue
Data Request
DRR Department of Renewable Resources [Canada]
Digital Radar Relay
Document Release Record
Dr rer comm (Doctor rerum commercialium) — Doctor of Commercial Sciences
Dr rer hort (Doctor rerum hortensium) — Doctor of Horticultural Sciences
Dr rer mont (Doctor rerum montanarium) — Doctor of Mining Sciences
Dr rer nat (Doctor rerum naturalium) — Doctor of Natural Sciences
Dr rer techn (Doctor rerum technicarum) — Doctor of Technical Sciences
DRRL Digital Radar Relay Link
DRS Data Rate Selector
Data Recording Set
Data Retrieval System
Detection Ranging System
Device Resource (File)
Diffuse Reflectance Spectroscopy
Digital Radar Simulator
Direct Reception System
Direct Relay Satellite
Disassembly/Reassembly Station
Distributed Resource System
Document Retrieval System
Dominion Research Station [of HWC, Canada]
Double Radio Source
Dress
Dressing
Dynamic Reflectance Spectroscopy
Dr sc (Doctor scientiarum) — Doctor of Science
Dr sc agr (Doctor scientiarum agrariarum) — Doctor of Agricultural Sciences
Dr sc math (Doctor scientiarum mathematicarum) — Doctor of Mathematical Sciences
Dr sc nat (Doctor scientiarum naturalium) — Doctor of Natural Sciences
Dr sc techn (Doctor scientiarum technicarum) — Doctor of Technical Sciences
DRSN Drifting Snow
DRSS Data Relay Satellite System
DRT Data Reckoning Tracer
Data Recording Terminal
Dead Reckoning Tracer
Device Reference Table
Device Rise Time
Diode Recovery Tester
dr t dram, troy
DRTC Documentation Research and Training Center [India]
DRTE Defense Research Telecommunications Establishment [now: Communications Research Center, Canada]
DRTL Diode-Resistor-Transistor Logic
DRTM Disk Real-Time Monitor
DRTR Dead Reckoning Trainer
DRU Data Recording Unit
Data Reference Unit
DRUPA Internationale Messe fur Druck und Papier [= International Fair for Printing and Pakermaking Industries, Germany]
DRV Data Recovery Vehicle
Deep-Diving Research Vehicle
DRX Dynamic Recrystallization
DS Data Scanning
Data Security
Data Segment
Data Set
Data Sheet
Data Stream
Data Structure
Data System
Days after Sight
Debugging System

Decimal Subtract(er)
Define Storage
Degree of Substitution
Denotational Semantics
Descent Stage
Deterministic System
Deuterated Solvent
Device Selector
Dial System
Dibasic Salt
Diesel Specialist
Difference Spectrophotometry
Digital Signal
Diode Switch
Dip Soldering
Directional Solidification
Direct Sequence
Disconnect Switch
Discrete System
Disk Storage
Disk System
Dispersion Strengthening
Distributed System
Distributed Synchronization
Documents Signed
Door Switch
Doppler Sonar
Double Silk (Wire)
Double Space
Double Stranded (Wire)
Downspout
Drillship
Drive System
Drum Storage
Dynamic System

D&S Display and Storage

D/S Days after Sight
Dynamic/Static Analysis

ds decistere [Unit]

d/s days after sight

DSA Data Set Adapter
Deoxystreptamine
(Bureau Europeen d'Information pour le) Developpement de la Sante Animale [= European Information Bureau for the Development of Animal Health Products, Belgium]
Device Specific Adapter
Dial Service Assistance
Digital Serving Area
Digital Signal Analyzer
Digital Storage Architecture
Digital Subtraction Angiography
Digital Surface Analyzer
Directory System Agents
Direct Storage Access
Distributed Systems Architecture
Doppler Spectrum Analyzer
Dozenal Society of America [US]
Drillsite Supervisor Association [US]
Dynamic Screening Anomaly
Dynamic Spring Analysis
Dynamic Strain Aging

DSAA Driving School Association of America

DSAM Doppler Shift Attenuation Method

DSAN Debug Syntax Analysis

DSAP Data Systems Automation Program
Defense Systems Application Program [US]
Directory Scope Analysis Program

DSAR Data Sampling Automatic Receiver

DSARC Defense Systems Acquisitions Review Council [US]

DSAU DSI (= Digital Speech Interpolation) Signal Access Unit

DSB Data Set Block
Defense Science Board [US]
Dispersion-Strengthened Brass
Dispersion-Strengthened Bronze
Double Sideband (Modulation)
Drug Supervisory Body [Switzerland]

DSBAM Double Sideband, Amplitude Modulation [also: DSB-AM]

DSBAMRC Double Sideband, Amplitude Modulation, Reduced Carrier [also: DSB-AM-RC]

DSBEC Double Sideband, Emitted Carrier [also: DSB-EC]

DSBG Disbursing

DSBM Double Sideband Modulation

DSBRC Double Sideband, Reduced Carrier [also: DSB-RC]

DSBSC Double Sideband, Suppressed Carrier [also: DSB-SC]

DSBSCAM Double Sideband, Suppressed Carrier Amplitude Modulation [also: DSB-SC-AM]

DSBSCAM w/QM DSBSCAM with Quadrature Multiplexing [also: DSB-SC-AM w/QM]

DSBSCASK Double Sideband, Suppressed Carrier Amplitude Shift Keyed [also: DSB-SC-ASK]

DSBTC Double Sideband, Transmitted Carrier [also: DSB-TC]

DSC Data Set Controller
Data Stream Compatibility
Data Synchronizer Channel
Design Safety Criteria
Differential Scanning Calorimetry
Direct Satellite Communications
Direct Semiconductor
Direct Strip Casting
Discount
Displacement Shift Complete (Lattice)
Distinguished Service Cross
District Switching Center
Disuccinimidyl Carbonate
Doctor of Commercial Science
Double Silk Covered (Wire)

DSc Doctor of Science

DSCA Default System Control Area

DScA Doctor of Science in Agriculture

DScAdmin Doctor in Administrative Science

DSCB Data Set Control Block

DSCC Deep Space Communications Complex

DScC Doctor of Commercial Science

DSCCD Discount Code

DScCom Doctor of Commercial Science

DScEng Doctor of Science and Engineering

DSCF Digital Switched Capacitor Filter

DSCG Disodium Chromoglycate
DSCH Dual Service Channel
DSCL Displacement Shift Complete Lattice
DScMil Doctor in Military Sciences
DSCN Dispersion-Strengthened Cupro Nickel
DScNat Doctor of Natural Sciences
DSCRM Discriminator
DSCS Defense Satellite Communications System [US]
Defense System Communications Satellite
Deskside Computer System
DSCT Double Secondary Current Transformer
DScTech Doctor of Science and Technology
DSD Data-Set Definition
Defense Systems Division
Digital System Design
Digital System Diagram
Direct Service Dialing
DSDD Double-Sided, Double-Density
DSDL Data Storage Description Language
DSDP Deep Sea Drilling Project
DSDS Dataphone Switched Digital Service [US]
DSDT Data-Set Definition Table
DSE Dartmouth Society of Engineers [US]
Data Set Extension
Data Storage Equipment
Data-Switching Exchange
Data Systems Engineering
Directionally Solidified Eutectic
Direct Switching Equipment
Direct Switching Exchange
Distributed Systems Environment
DSEA Data Storage Electronic Assembly
DSECT Data Section
Dummy (Control) Section
DSEG Data Systems Engineering Group
DSEM Danish Society for Engineering Metrology
DSEP Distribution System Expansion Program
DSF Data Secured File
Directional Solidification Furnace
Disk Storage Facility
DSG Development Studies Group [Peru]
Digital Signal Generation
Double Strength Glass
DSGN Design
DSH Deactivated Shutdown Hours
DSHD Double-Sided, High-Density
DSI Dairy Society International [US]
Digital Speech Interpolation
Direct Sample Insertion
Dislocation-Solute Interaction
Dynamic System Interchange
DSID Data Set Identification
Direct Sample Insertion Device
DSIF Deep Space Instrumentation Facility [of NASA]
DSIR Department of Scientific and Industrial Research [UK]
DSIS Department of Scientific Information Services [of DND, Canada]
Digital Software Integration Station
Directorate of Scientific Information Services
DSK Disk
DSL Data Set Label
Data Simulation Language
Data Structures Language
Datura Stramonium Lectin
Deep Scattering Layer
Detroit Signal Laboratory [US Army]
Digital Simulation Language
Domestic Substances List [of American Chemical Society, US]
Drawing and Specification Listing
Dynamic Super Loudness
DSLC Data Subscriber Loop Carrier
Defense Logistics Services Center [of USDOD]
DSLO Distributed Systems Licensing Option
DSM Defense Suppression Missile
Development of Substitute Materials
Digital Simulation Model
Direct Signal Monitoring
Direct Stiffness Method
Disk Space Management
Distinguished Service Medal
Dutch State Mines
Dynamic Scattering Mode
DSMA Disodium Methanearsenate
D/SMC Dough/Sheet Molding Compound
DSMS Document Service Management System
DSMSB Die Set Manufacturers Service Bureau [US]
DSN Data Set Name
Deep Space Network [of NASA]
Distributed System Network
DSNAME Data Set Name
DSO Data Set Optimizer
Data Systems Office
Digital Storage Oscilloscope
Distinguished Service Order
Distributed System Network
DS1 Digital Signal Level One
DSORG Data Set Organization
DSOS Data Switch Operating System
DSP Defense Services Program
Defense Support Program [of US Air Force]
Desilication Product
Dibasic Sodium Phosphate
Digital Signal Processing
Digital Signal Processor
Disodium Phosphate
Distributed System Program
Downstream Processing
Dynamic Support Program
DSPB Digital Signal Processing Board
DSPI Digital Speckle-Pattern Interferometry
DSPL Display
DSPLC Display Controller
DSPM Digital Signal Processing Multiprocessor
DSPN Dispensary
DSPP Disodium Phenylphosphate
DSQD Double-Sided, Quad-Density
DSR Data Scanning and Routing
Data Set Ready
Data Storage and Retrieval
Debt Service Ratio

	Device Status Register
	Device Status Report
	Digit Storage Relay
	Digital Stepping Recorder
	Discriminating Selector Repeater
dsRNA	Double-Stranded Ribonucleic Acid
DSRT	Deletable Soft Return [also: DSRt]
DSRV	Deep-Submergence Rescue Vehicle [US Navy]
DSRW	Dry-Sand, Rubber-Wheel
DSS	Data Switching System
	Decision Support Software
	Decision Support System
	Deep-Space Station
	Deep Submergence System
	Department of Supply and Services [Canada]
	Digital Subsystem
	Digital Switching Subsystem
	2,2-Dimethyl-2-Silapentane-5-Sulfonate
	Diploma of Specialized Studies
	Director of Statistical Services
	Direct Station Selection
	Discrete Sync System
	Disk Support System
	Division of Safety Studies
	Document Storage System
	Double-Sided Scrubber
	Dynamic Support System
DSSA	Defense Security Assistance Agency
DSSB	Double Single-Sideband
DSSC	Department of the Secretary of State of Canada
	Double-Sideband Suppressed Carrier
DSSc	Diploma in Sanitary Sciences
DSSCS	Defense Special Secure Communications System
DSSD	Double-Sided, Single-Density
DSSM	Dynamic Sequencing and Segregation Model
DSSP	Deep Submergence Systems Project [US Navy]
	Defense Standardization and Specification Program [US]
DSSPTO	Deep Submergence Systems Project Technical Office [US Navy]
DSSRG	Deep Submergence Systems Research Group [US Navy]
DS-SSDA	Direct Sequence-Spread Spectrum Multiple Access
DSSV	Deep Submergence Search Vehicle [US Navy]
DST	Data Summary Tape
	Data System Test
	Daylight Saving Time
	Disk Storage Terminal
	District
	Double Summer Time
	Drill Stem Test
DS2	Digital Signal Level Two
DS3	Digital Signal Level Three
DSTCD	District Code
DSTE	Data Subcarrier Terminal Equipment
	Data Subscriber Terminal Equipment
DSTLN	Distillation
DSTN	Destination
DSTO	Defense Science and Technology Organization [Australia]
DSTR	Distributor
DStV	Deutscher Stahlbau-Verband [= German Association for Steel Construction]
DSU	Data Service Unit
	Data Storage Unit
	Data Synchronization Unit
	Data Synchronizer Unit
	Device Switching Unit
	Digital Service Unit
	Disk Storage Unit
	Drum Storage Unit
DSV	Deep Submergence Vehicle
	Digital Sum Variation
	Diving Support Vessel
	Double Silk Varnish (Wire)
DSVT	Digital Secure Voice Telephone
DSW	Data Status Word
	Device Status Word
	Direct Step On Wafer
DSX	Digital Signal Cross-Connection Equipment
	Distributed System Executive
DT	Data Tag
	Data Terminal
	Data Transfer
	Data Translator
	Data Transmission
	Date
	Dead-Time
	Decay Time
	Deflection Temperature
	Deformation Twin
	Destructive Testing
	Detection Theory
	Deuterium-Tritium Ratio
	Developed Template
	Dial Tone
	Differential Time
	Differentiating Transformer
	Diffraction Theory
	Digital Technique
	Digital Translator
	Digital Transmission
	Digit Tube
	Digroup Terminal
	Discharge Tube
	Discrete Time
	Disk-Tape
	Distributed Transaction
	Document Title
	Document Type
	Domain Theory
	Double-Throw
	Double Torsion
	Doubling Time
	Downtime
	Dravon Tube
	Drift Tube
	Drilling Technician
	Drilling Technology
	Drop Top
	Drop Tower
	Drop Tube

	Dry Ton
	Dual Tandem
	Dummy Target
	Dynamic Tear
D/T	Deuterium/Tritium (Ratio) [also: D-T]
	Disk/Tape
D-T	Deuterium-Tritium (Reaction)
D^3	Defense, Description and Designation [also: DDD]
DTA	Detailed Traffic Analysis
	Detroit Tooling Association [US]
	Development Test Article
	Differential Thermal Analysis
	Dispersion-Toughened Alumina
	Dominion Traffic Association [Canada]
DTAB	Dodecyl Trimethyl Ammonium Bromide
DTACK	Data Transfer Acknowledgement
DTAF	5-([4,6-Dichlorotriazin-2-yl]amino)fluorescein
DTAM	Descend to and Maintain
DTAN	2,2'-Dithiobis(1-aminonaphthalene)
DTARS	Digital Transmission and Routing System
DTAS	Data Transmission and Switching System
DTAU	Digital Test Access Unit
DTB	Danmarks Tekniske Bibliotek [= Technical Library of Denmark]
	Decimal-to-Binary
	Dynamic Translation Buffer
DTBP	Dedicated Total Buried Plant
	Di-tert-Butyl Peroxide
DTC	Data Transmission Center
	Data Transmission Control
	Dead-Time Compensation
	Decision Threshold Computer
	Department of Technical Cooperation [UK]
	Desert Tortoise Council [US]
	Design-to-Cost (Process)
	Detection Threshold Computer
	Differential Thermal Coating
	Diode Transistor Compound
	Discrete-Time Control
	Doppler Translation Channel
DTCP	Diode Transistor Compound Pair
DTCS	Digital Test Command System
	Discrete-Time Control System
DTCU	Data Transmission Control Unit
DTD	Data Transfer Done
	Dimethyl Tin Difluoride
DTDM	Dithio Dimorpholine
DTDMA	Distrubuted Time Division Multiple Access
DTDMAC	Ditallowdimethylammonium Chloride
DTDP	Ditridecyl Phthalate
DTDS	Digital Television Display System
DT/DT	Drop Tube/Drop Tower (Facility)
DTE	Data Terminal Emulator
	Data Terminal Equipment
	Data Transmission Equipment
	Destructive Testing Equipment
	Development Test and Evaluation
	Differential Thermal Expansion
	Digital Test Executive
	Dithioerythritol
DTech	Doctor of Technology
DTE/DCE	Data Terminal Equipment/Data Communications Equipment
DTEM	Deep Transient Electromagnetic System [also: DTEMS]
DTENT	Date of Entry
DTF	Date to Follow
	Default-the-File
	Define the File
	Definite Tape File
	Dial Tone First
	Diving Test Facility
	Dynamic Track Follower
DTFDW	Deciduous Tree Fruit Disease Workers [US]
DTFT	Discrete Time Fourier Transform
DTG	Date-Time Group
	Differential Thermogravimetry
	Display Transmission Generator
	Dynamically Tuned Gyroscope
DTI	Data Transfer Interface
	Department of Trade and Industry [UK]
	Display Terminal Interchange
	Distortion Transmission Impairment
DTIC	Defense Technical Information Center [US]
DTICA	Departmento Tecnico Interamericano de Cooperacion Agricole [= Interamerican Technical Department for Agricultural Cooperation, Chile]
DTIE	Division of Technical Information Extension [of USAEC]
DTL	Diode-Transistor Logic
DTLS	Digital Television Lightwave System
DTL/TTL	Diode-Transistor Logic/Transistor-Transistor Logic
DTM	Delay Timer Multiplier
	Device Test Module
	Digital Terrain Model
D2M	Diethylene Glycol Dimethyl Ether
DTMB	David Taylor Model Basin [US Navy]
DTMF	Dual-Tone Multifrequency
DTMS	Database and Transaction Management System
	Digital Test Monitoring System
DTN	Data Transporting Network
	Defense Telephone Network
DTNB	5,5'-Dithiobis(2-Nitrobenzoic Acid)
DTNSRDC	David Taylor Naval Ship Research and Development Center [US]
DTO	Decentralized Toll Office
	Deuterium-Tritium Oxide
	Digital Testing Oscilloscope
DTOS	N-Dimethyl Dithiocarbamyl-N'-Oxidiethylene Sulfenamide
DTP	Daily Transaction Reporting
	Data Transfer Protocol
	Desk-Top Publishing
	Diagnostic-Therapeutic Pair
	Directory Tape Processor
	Dithiophosphate
DTPA	Diethylenetriaminepentaacetate
	Diethylenetriaminepentaacetic Acid
DTPC	Desert Tortoise Preserve Committee [US]
DTPD	N,N'-Ditolyl-P-Phenylene Diamine
DTPL	Domain Tip Propagation Logic
DTPMT	Date of Payment

DTPP Demonstration Tokamak Power Plant
DTR Daily Transaction Reporting
Data Telemetering Register
Data Terminal Reader
Data Terminal Ready
Data Transfer Register
Definite Time Relay
Demand-Totalizing Relay
Diffusion Transfer
Digital Telemetering Register
Discharge-Tube Recifier
Disposable Tape Reel
Distribution Tape Reel
Document Transmittal Record
Downtime Radio
DTRC David Taylor Research Center [US Navy]
DTR/DSR Data Terminal Reader/Data Storage and Retrieval
DTRE Defense Telecommunications Research Establishment
DTRF Darlington Tritium Removal Facility [Canada]
Data Transmittal and Routing Form
DTRI Dairy Training and Research Institute [Philippines]
DTRS Development Test Requirement Specification
DT/RSS Data Transmission/Recording Subsystem
DTS Data Tape Service
Data Transmission System
Defense Telephone Service [US]
Defense Telephone System
Diffusion Total System
Digital Tandem Switch
Digital Termination System
Discrete-Time Series
Discrete-Time Signal
Discrete-Time System
Double-Throw Switch
DTSA Discrete Time Series Analysis
DTSS Dartmouth Time-Sharing System
DTS-W Defense Telephone Service, Washington [US]
DTT Data Transition Tracking
Data Transmission Terminal
Design Thermal Transient
Dithiothreitol
DTTF Deviation in Time to Failure
Digital Tape and Tape Facility
dTTP Deoxyribonucleoside Triphosphate
DTTT Dynamic Time-Temperature Transformation
DTTU Data Transmission Terminal Unit
DTU Data Terminating Unit
Data Transfer Unit
Data Transmission Unit
Digital Tape Unit
Digital Telemetry Unit
Digital Transmission Unit
Dual Toplogical Unitarization
DTUC David Thompson University Center [Canada]
DTUL Deflection Temperature under Load
DTUTL Digital Tape Unit Tape Facility
DTV Digital Television
Digital-to-Television [also: D-TV or D/TV]
DTVC Digital Transmission and Verification Converter
DTVM Differential Thermocouple Voltmeter
DTW Dual Tandem Wheel (Landing Gear)
DTWX Dial Teletypewriter Exchange
DU Dalhousie University [Canada]
Delay Unit
Depleted Uranium
Dimensioning Unit
Display Unit
Doshisha University [Japan]
Drexel University [US]
Duke University [US]
Dundee University [UK]
Dutch
Du D-Load Ultimate
DUAL Dynamic Universal Assembly Language
DUALABS Data Use Access Laboratories
DUAT Direct User Access Terminal
DUC Dense Upper Cloud
Dual-Access Utility Circuit
DUCE Denied Usage Channel Evaluator
d-UDP 2'-Deoxyuridine 5'-Diphosphate
DUE Detection of Unauthorized Equipment
DUF Diffusion under Film
DUG Datapac User Group
DUM Disk User Multi-Access Unit
DUMAND Deep Underwater Muon and Neutrino Detector
DUMC Duke University Medical Center [US]
DUMD Deep Underwater Measuring Device
d-UMP 2'-Deoxyuridine 5'-Monophosphate
DUMS Deep Unmanned Submersible
DUMU Disk User Multi-Access Unit
DUNC Deep Underwater Nuclear Counter
DUniv Doctor of the University
DUNMIRE Dundee University Numerical Method Information Retrieval System [UK]
DUNS Data Universal Numbering System
DUO Datatron Users Organization
DOS (= Disk-Operating System) under OS (= Operating System)
DUP Data User Port
Disk Utility Program
Diundecyl Phthalate
Duplicate
Dural Duraluminum
DUS Data User Station
Diagnostic Utility System
DUT Dalian University of Technology [PR China]
Delft University of Technology [Netherlands]
Device under Test
DUV Damaging Ultraviolet
Data Under Voice
Deep Ultraviolet
DV Differential Voltage
Direct Vision
Dump Valve
DVA Designed, Verified and Assigned Date
Digital Voice Announcer
Dynamic Visual Acuity
DVARS Doppler Velocity Altimeter Radar Set
DVB Divinylbenzene
DVBST Direct View Bistable Storage Tube
DVC Device

DVCCS	Differential Voltage-Controlled Current Source
DVD	Detail Velocity Display
DVDA	Dollar Volume Discount Agreement
DVES	DOD (= Department of Defense) Value Engineering Services [US]
DVESO	DOD (= Department of Defense) Value Engineering Services Office [US]
DVF	Digital Variable Frequency
DVFO	Digital Variable-Frequency Oscillator
DVFR	Defense Visual Flight Rules [US]
DVI	Digital Video Interactive [also: DV-I]
	Digital Video Interface
	Direct Voice Input
DVL	Deutsche Versuchsanstalt fur Luftfahrt [= German Research Institute for Aeronautics]
DVLPT	Development
DVM	Deutscher Verband fur Materialforschung und -prufung [= German Association for Materials Research and Testing]
	Digital Voltmeter
	Displaced Virtual Machine
	Doctor of Veterinary Medicine
DVOM	Digital Volt-Ohmmeter
DVOR	Doppler VHF (= Very High Frequency) Omnidirectional Radio Range
DVR	Design and Verification Routine
DVS	Deutscher Verband fur Schweisstechnik [= German Welding Association]
	Digital Video System
DVSc	Doctor of Veterinary Science [also: DVS]
DVST	Direct-View Storage Tube
DVTL	Dovetail
DVX	Digital Voice Exchange
DV-Xα	Discrete Variational Xα Method
DW	Daisy Wheel
	Deadweight
	Delta Wing
	Detonation Wave
	Developed Width
	Die Welding
	Distilled Water
	Double Word
	Dried Weight
	Drop Wire
	Dry Weight
	Dual Wheel (Landing Gear)
	Dumbwaiter
	Dynamic Wave
DWA	Double-Wire Armor
DWB	Designer's Workbench
	Development Workbook
DWBA	Direct-Wire Burglar Alarm
	Distorted-Wave Born Approximation
DWC	Deadweight Capacity
DWCM	Dried Weight of Cell Mass
DWED	Department of Western Economic Diversification [Canada]
	Dry Well Equipment Drain
DWFD	Dry Well Floor Drain
DWG	Drawing
DWI	Drop Weight Index
DWIA	Distorted Wave Impulse Approximation
DWICA	Deep Water Isotopic Current Analyzer
DWL	Design Water Line
	Dominant Wavelength
	Dowel
	Downwind Localizer
DWM	Destination Warning Marker
DWMI	Diamond Wheel Manufacturers Institute [US]
DWP	Daisy Wheel Printer
	Digital Waveform Pattern
DWPF	Defense Waste Processing Facility
DWR	Dry Weight Rank Method
DWS	Disaster Warning System
DWSMC	Defense Weapons Systems Management Center [of USDOD]
DWST	Demineralized Water Storage Tank
DWT	Deadweight (Tonnage)
	Drop-Weight Test
dwt	pennyweight
DWTT	Drop-Weight Tear Test
DWV	Data With Voice
	Drain, Waste and Ventilation System
	Drain, Waste and Vent Pipe
DX	Destroyer Experimental
	Direct Current Switching
	Direct Expansion Coil
	Distance
	Distance Reception
	Document Transfer
	Document Transmission
	Duplex (Repeater)
dx	differential of x
DXC	Data Exchange Control
DXE	Dixylylethane
DXF	Data Exchange Format
	Data Transfer Facility
DX/RSTS	Document Transmission/Resource Time Sharing
DXS	Data Exchange System
DXS/OS	Data Exchange System/Operating System
DXS/ST	Data Exchange System/Statement Translator
DXS/TL	Data Exchange System/Transaction Language
Dy	Dysprosium
dy	differential of y
DYANA	Dynamic Analyzer [Computer Program]
$Dy(FOD)_3$	Tris(heptafluorodimethyloctanedionato)dysprosium
DYG	Dyeing
DYN	Dynamic(s)
	Dynamo
dyn	dyne
DYNASAR	Dynamic Systems Analyzer
DYNAVIS	Dynamic Video Display System
DYNM	Dynamotor
DYP	Directory Yellow Pages
d/yr	days per year
DYSAC	Digitally Simulated Analog Computer
	Dynamic Storage Analog Computer [also: DYSTAC]
DYSTAL	Dynamic Storage Allocation Language [Computer Language]

Dy(THD)$_3$ Tris(tetramethylheptanedionato)dysprosium
DZ Depleted Zone
Dizygotic
Doctor of Zoology
Dropping Zone
dz dozen
DZ PR Dozen Pairs
DZA Doppler Zeeman Analyzer
DZD Dinar [Currency of Algeria]
DZF Dokumentationszentrale Feinwerktechnik [= Documentation Center for Precision Engineering, Germany]

E

E Earth
East(ern)
Eccentric(ity)
Edge
Edition
Effectivity
Einsteinium [also: Es]
Elasticity
Elbow
Electric(ity)
Electrode
Electrode Potential
Electromotive Force
Electron
Electronic(s)
Ellipse
Elliptical
Emergency
Emission
Emissive Power
Emitter
Enamel
End
Energy
Engine
Engineer
Engineering
English
Entrainment
Entrance
Eotvos [Unit]
Equatorial
Equipment
Equivalent
Erlang [Unit]
Error
Estimate
Evaporation
Evaporativity
Exa- [SI Prefix]
Excellence
Execution
Exponent
Exposure
Expression
Extraordinary
Extrinsic
Glutamin Acid
Glutamyl
Voltage [Symbol]

e eccentric
effective
efficiency
efficient
elastic
electric
electron
electron charge
emitter
energy
erg [Unit]
error
evaporation
excellent
exponent(ial)
exposure
base of natural logarithm

EA Easy Axis
Educational Age
Effective Address
Electric Actuator
Electrically-Alterable
Electrics Association
Electroacoustics
Electroanalysis
Elemental Analysis
Elliptical Aperture
Energy Analysis
Energy Association [US]
Engineering Analysis
Enterprise Analysis
Enthalpimetric Analysis
Enumeration Area
Ethanolamine
Event Analysis
Exhaust Air
Extended Abstract
(Department of) External Affairs [Canada]

ea each

EAA East Africa Association [UK]
Electric Auto Association [US]
Engineer in Aeronautics and Astronautics
Environmental Assessment Act
Essential Amino Acid
Ethyl Acetoacetate
Ethylene Acrylic Acid
European Accounting Association [Belgium]
European Aluminum Association [Germany]
Experimental Aircraft Association [US]

EAAA European Association of Advertising Agencies

EAAAF Experimental Aircraft Association Aviation Foundation [US]

EAAC East African Airways Corporation
European Agricultural Aviation Center
Experimental Aircraft Association of Canada

EAAE European Association of Agricultural Economists [Belgium]

EAAFRO European African Agricultural and Forestry Research Organization [Kenya]

EAAP European Association for Animal Production [Italy]

EAASH European Academy of Arts, Sciences and Humanities [France]

EAB Economic Analysis Bureau
Educational Activities Board

	Engineering Activity Board [of SAE, US]
	European American Bank
	Exclusion Area Boundary
EABRD	Electrically-Activated Bank Release Device
EAC	East African Community
	Education Affairs Committee [of ASM International, US]
	EEC (= European Economic Community) Advisory Council
	Engineering Affairs Council [of AAES, US]
	Environmental Action Coalition [US]
	Error Alert Control
	Expect Approach Clearance
	Experiment Apparatus Container
	External Affairs Canada
EACA	Eta-Amino-N-Caproic Acid
EAC-AIA	EEC (= European Economic Community) Advisory Council of the Asbestos International Association, Belgium]
EACC	Error Adaptive Control Computer
EACE	East African Certificate of Education
EACG	European Association of Exploration Geophysicists [Netherlands]
EACM	European Association for Composite Materials [France]
EACO	Engineers and Architects Council of Oregon [US]
EACPI	European Association of County Planning Institutions [Belgium]
EACRP	European-American Committee on Reactor Physics [of ENEA]
EACS	Electronic Automatic Chart System
EACSO	East Africa Common Services Organization
EAD	Electroacoustic Dewatering
	Estimated Availability Date
EADAS	Engineering and Administration Data Acquisition System
EADB	East African Development Bank [Uganda]
EADC	Ethylaluminum Dichloride
EADF	Elliptical Aperture with Dynamic Focus
EADI	Electronic Attitude Directional Indicator
	European Association of Development Research and Training Institutes [Switzerland]
EADP	European Association of Directory Publishers [Belgium]
EAE	Eastern Association of Electroencephalographers [US]
	Extended Arithmetic Element
EAEC	European Airlines Electronic Committee
	European Atomic Energy Commission
EAEG	European Association of Exploration Geophysicists [Netherlands]
EAEP	Energy Action Educational Project
EAES	European Atomic Energy Society
EAETB	Efficiency and Alternative Energy Technology Branch
EAETLFEM	European Association for the Exchange of Technical Literature in the Field of Ferrous Metallurgy [Luxembourg]
EAF	Electric Arc Furnace
	Environmental Action Foundation [US]
	Equivalent Availability Factor
EAFB	Edwards Air Force Base [US Air Force]
EAFFRO	East African Freshwater Fishery Research Organization
EAFRO	East African Fishery Research Organization
EAGE	Electrical Aerospace Ground Equipment
EAGGF	European Agricultural Guidance and Guarantee Fund
EAGLE	Elevation Angle Guidance Landing Equipment
	European Association for Grey Literature Exploitation [UK]
EAH	Engineering Association of Hawaii [US]
EAHRA	East African Railways and Harbours Administration
EAI	Engineering Advance Information
	Engineers and Architects Institute [US]
EAIMB	East African Industries Management Board [Kenya]
EAIR	Extended Area Instrumentation Radar
EAIRB	East African Industrial Research Board [Kenya]
EAIRO	East African Industrial Research Organization
EAK	Ethyl Amyl Ketone
EAL	Electromagnetic Amplifying Lens
	Equipment Air Lock
EALM	European Association of Livestock Markets [Belgium]
EAM	Electronic Accounting Machine
	Electronic Automatic Machine
	Embedded Atom Method
	Ethylene-Acrylate Copolymer
EAMCBP	European Association of Makers of Corrugated Base Papers [Germany]
EAMD	East African Meteorological Department
EAMFRO	East African Marine Fisheries Research Organization
EAMTC	European Association of Management Training Centers
EAN	Engineering Association of Nashville [US]
	European Article Number
	European Article Numbering Association [Belgium]
EANA	European Alliance of News Agencies
EANDC	European-American Nuclear Data Committee [of ENEA]
EANDRO	Electrically-Alterable Nondestructive Readout
EANHS	East African Natural History Society [Kenya]
EANPG	European Air Navigation Planning Group [of ICAO]
EAO	Egyptian Agricultural Organization
EAON	Except As Otherwise Noted
EAOS	Easy Access Ordering System
EAP	Ecological Agriculture Project [Canada]
	Emergency Action Plan
	Employee Assistance Program
	Equivalent Air Pressure
	Evoked Action Potential
	Expenditure Analysis Plan
	Experimental Activity Proposal
EAPA	European Asphalt Pavement Association [Sweden]
EAPD	Electro-Absorption Photodiode
EAPFBO	European Association of Professional Fire Brigade Officers [Belgium]
EAPH	East Africa Publishing House [Kenya]
EAPM	European Association of Personnel Management [Germany]

EAPR European Association for Potato Research [Netherlands]
EAPROM Electrically-Alterable Programmable Read-Only Memory
EAPS European Association for Population Studies [Netherlands]
EAPV Eastern Arctic Patrol Vessels [Canada]
EAR Electron Affinity Rule
Electronically Agile Radar
Employee Attitude Research
Eroded Area Rate
EARB European Airlines Research Bureau [now: AEA, Belgium]
EARC Elimination of Ambiguity in Radiotelephony Call Signs (Study Group)
Extraordinary Administrative Radio Conference
EARCCUS East African Regional Committee for the Conservation and Utilization of the Soil
EARCOM East Asia Regional Council of Overseas Schools [US]
EARI Engineering Agency for Resources Inventories [US]
EARIC East African Research Information Center
EAROM Electrically-Alterable Read-Only Memory
EAROPH East Asia Regional Organization for Planning and Housing
EARP Environmental Assessment and Review Program
EARSeL European Association of Remote Sensing Laboratories [France]
EARTS En-Route Automated Radar Tracking System
EAS Education Administation Specialist
Egyptian Academy of Sciences
Electronic Article System
Equivalent Airspeed
European Aquaculture Society [Belgium]
Extended Area Service
Experimental Army Satellite
Experiment Assurance System
Extended Area Service
Extensive Air Shower
EASA Electrical Apparatus Service Association [US]
Engineering Association of South Africa
EASB East African Settlement Board
EASC Ethylaluminum Sesquichloride
EASCOMINT Extended Air Surveillance Communications Intercept [US Air Force]
EASCON Electronics and Aerospace Systems Convention [of IEEE, US]
EASE Electrical Automatic Support Equipment
Engineering Automatic System for Solving Equations
European Association of Science Editors [UK]
EASI Electrical Accounting for the Security Industry
Estimate of Adversary Sequence Interruption
European Association of Shipping Informatics [UK]
EASIAP Engineering and Applied Sciences Industrial Affiliates Program [University of Manitoba, Canada]
EASINET European Area Sales and Information Network
EASL Engineering Analysis and Simulation Language
Experimental Assembly and Sterilization Laboratory [of NASA]
EASP Electric Arc Spraying
EAST Experimental Army Satellite, Tactical
EASTEC Eastern States Exposition Center [US]
EASTT Experimental Army Satellite Tactical Terminal
EASY Early Acquisition System
Efficient Assembly System
Engine Analyzer System
EAT Electroacoustic Testing
Estimated Arrival Time
Expected Approach Time
EATCS European Association for Theoretical Computer Science [Germany]
EATITU East African Tractor and Implement Testing Unit
EATJP European Association for the Trade in Jute Products [Netherlands]
EATMS Electro-Acoustic Transmission Measuring System
EATP European Association for Textile Polyolefins [France]
EATU Eastern African Telecommunications Union
EAU Engineer Aviation Unit
EAUG European Atex (Computer) Users Group [Netherlands]
EAUTC Engineer Aviation Unit Training Center
EAVE Experimental Autonomous Vehicle
EAVE-EAST Experimental Autonomous Vehicle — East
EAVRO East African Veterinary Research Organization [Kenya]
EAW Equivalent Average Words
EAWAG Eidgenossische Anstalt fur Wasserversorgung, Abwasserreinigung und Gewasserschutz [= Federal Institute for Water Supply, Wastewater Treatment and Water Pollution Control, Switzerland]
EAWS East African Wildlife Society [Kenya]
EAX Electronic Automated Exchange
EB Eastbound
Ebony
Edge Brightness
Electric Braking
Electron Beam
Electron Bombardment
Elementary Body
End-of-Block (Character)
Equal Brake
Evolutionary Biology
Executive Board
Exponential Born
Exposure Back
Eye Bolt
E/B Electrode per Bit
E$ Ethiopian Birr [Currency]
EBA Environmental Ballistics Associates [US]
EBAA European Business Aviation Association [Belgium]
EBAC European Bank Advisory Committee
EBAM Electron-Beam-Accessible Memory
Electron-Beam-Addressed Memory
EB ASB Ebony Asbestos
EBB Electronic Bulletin Board [Supply and Services Canada]
Extra Best Best
EBBA Eastern Bird Banding Association [US]

	Estuarine and Brackish-Water Biological Association [now: EBSA, UK]
	4-Ethoxybenzylidene-4-Butylaniline
EBC	Electron Beam Cutting
	Enamel Bonded (Single) Cotton (Wire)
	Equivalent Boron Content
EBCA	External Branch Condition Address
EBCD	Extended Binary-Coded Decimal
EBCDIC	Extended Binary-Coded Decimal Interchange Code
EBCE	Electron-Beam Control Electronics
EBCHR	Electron Beam Cold-Hearth Refining
EBCI	External Branch Condition Input
EBCSM	East Bay Council on Surveying and Mapping
EBD	Effective Billing Date
	Extrinsic Boundary Dislocation
EBDC	Enamel Bonded Double Cotton (Wire)
	Ethylenebisdithiocarbamate
EBDI	Electronic Business Data Interchange [US]
EBDP	Enamel Bonded Double Paper (Wire)
EBDS	Enamel Bonded Double Silk (Wire)
EBER	Equivalent Binary Error Rate
EBES	Electronic Banking Economics Society [US]
	Electron Beam Exposure System
EBF	Electron-Beam Focusing
	Electron Bombardment Furnace
	Externally Blown Flap
EB&F	Equipment Blockages and Failures
EBFS	Enclosure Building Filtration System
EBHC	Equated Busy-Hour Call
EBHSS	Electron Beam High Speed Scan
EBI	Equivalent Background Input
EBIB	Energy Bibliography and Index [of CEMR, US]
EBIC	Electron Beam Induced Conduction
	Electron Beam Induced Current
	European Banks International Company
EBIS	EIS (= Economic Information Systems) Business Information System [US]
	Electron-Beam Ion Source
	Ethylenebisisothiocyanate Sulfide
EBL	Electron-Beam Lithography
	Electronic Bearing Line
EBLI	Electronics Business Leading Indicator
EBM	Electromagnetic Billetmaker
	Electron-Beam Machining
	Electron-Beam Melting
	Electronic Bearing Marker
	Extended Branch Mode
EBMA	Electron Beam Microprobe Analysis
	Engine Booster Maintenance Area
EBMD	Electron Beam Mode Discharge
EBMF	Electron Beam Microfabricating System
EBNF	Extended Backus Naur Form
EBOR	Experimental Beryllium Oxide Reactor [of USAEC]
EBOR-CX	Experimental Beryllium Oxide Reactor — Critical Assembly [US]
EBP	Enamel (Single) Bonded Paper (Wire)
EBPA	Electron Beam Parametric Amplifier
EB-PVD	Electron Beam — Physical Vapor Deposition
EBR	Electron Beam Recording
	Electron Beam Remelting
	Electronic Beam Recording
	Epoxy Bridge Rectifier
	Experimental Breeder Reactor
E-BR	Emulsion Butadiene Rubber
EB&RA	Engineer Buyers and Representatives Association [UK]
EBRD	Electron Beam Rotating Disk
EBS	Electron-Beam Semiconductor
	Electron Bombarded Semiconductor
	Electronic Beam Squint-Tracking System [of SATCOM]
	Emergency Borating System
	Emergency Broadcasting System
	Enamel Bonded (Single) Silk (Wire)
EBSA	Estuarine and Brackish-Water Sciences Association [UK]
	4-Ethylbenzenesulfonic Acid
EBSP	Electron Backscatter Pattern
EBSS	Earle's Balanced Salt Solution
EBT	Eccentric Bottom Tapping
	Electroless Bath Treatment
	Elmo Bumpy Torus
EBTS	ECC Bypass Test Facility
EBU	European Broadcasting Union [Switzerland]
EBV	Epstein-Barr Virus
EBVD	Electron-Beam Vapor Deposition
EBW	Electron Beam Welding
	Exploding Bridgewire
EBW-HV	Electron Beam Welding — High Vacuum
EBW-MV	Electron Beam Welding — Medium Vacuum
EBW-NV	Electron Beam Welding — Nonvacuum
EBWR	Experimental Boiling Water Reactor [of Argonne National Laboratory, US]
EC	Earth Current
	Eastern Cedar
	Echo Controller
	Ecology Center [US]
	Economy Cartridge
	Eddy Current
	Edge Clamp
	Edge Connector
	Effective Concentration
	Electrical Conductivity
	Electrical Conductor
	Electric Circuit
	Electric Controller
	Electrification Council [US]
	Electrocapillary
	Electrochemistry
	Electrode Current
	Electrolytic Capacitor
	Electron Capture
	Electron Cloud
	Electron Cyclotron
	Electronic Calculator
	Electronic Ceramic
	Electronic Circuit
	Electronic Conductivity
	Electronic-Coupled
	Element Count
	Emission Control
	Enamel Covering

End Cell
Endothelial Cell
Engineering Ceramics
Engineering Change
Engineering Construction
Engineering Council [UK]
Engineers Club
Environment Canada
Environmental Chemistry
Enzyme Commission
Error Correction
Error Counter
Esterified Cholesterol
Estimated Concentration
Ethyl Cellulose
Ethyl Centralite
European Chapter
European Commission
European Community
Event Code
Event Count
Event Counter
Exhaust Coefficient
Experimental Chemist
Experimental Chemistry
Explorers Club [US]
Extended Control (Mode)

ec (exempli causa) — as an example
EC50 Effective Concentration 50 [also: EC_{50}]
ECA Economic Commission for Africa [Ethiopia]
Economic Cooperation Administration [US]
Electrical Contractors Association [UK]
Electronic Confusion Area
Emergency Controlling Authority
Engineering and Computer Science Association
Engineering Contractors Association [US]
Engineering Critical Assessment
Environment Conservation Authority
Environment Council of Alberta [Canada]
Erythrina Cristagalli Agglutinin
European Confederation of Agriculture [Switzerland]
Exchange Carrier Association [now: NECA, US]
ECAART European Conference on Accelerators in Applied Research and Technology [Germany]
ECAC Electromagnetic Compatibility Analysis Center [of USDOD]
European Civil Aviation Conference [France]
ECAC/US-CRS European Civil Aviation Conference/United States Working Group on Computer Reservation Systems
ECAD Electrical Computer-Aided Design
ECAFE Economic Commission for Asia and the Far East [Thailand]
ECAHTI European Committee for Agricultural and Horticultural Tools and Implements [France]
ECAM Electronic Centralized Aircraft Monitor
ECAMA European Citric Acid Manufacturers Association [of CEFIC, Belgium]
ECAP Electronic Circuit Analysis Program
Energy-Compensated Atom Probe
ECAR East Central Area Reliability Coordination Agreement
ECARS Electronic Coordinator and Readout System
Enhanced Airline Communications and Reporting System
ECASIA European Conference on Applications of Surface and Interface Analysis [Hungary]
ECASS Electronically-Controlled Automatic Switching System
ECAT European Center for Automatic Translation [Luxembourg]
ECATRA European Car and Truck Rental Association [Germany]
ECB Electrically Controlled Birefringence
Electronic Codebook
European Conference on Biomaterials
Event Control Block
ECBA European Communities Biologists Association [Germany]
ECBO European Cell Biology Organization [UK]
ECBTE European Committee for Building Technical Equipment [France]
ECC Earth Continuity Conductor
Eccentric(ity)
Economic Council of Canada
Electrocardiocorder
Electron Channeling Contrast
Electronically-Controllable Coupler
Electronic Calibration Center [of NBS, US]
Electronic Common Control
Electronic Components Conference [US]
Emergency Control Center
Emergency Core Coolant
Emergency Core Cooling
Energy Conservation Coalition
Environmental Control Council
Error Checking and Correction
Error-Correction Code
Ethyl Chlorocarbonate
European Communities Commission
Expanded Community Calling
ECCA European Coil Coaters Association [Belgium]
ECCAI European Coordinating Committee for Artificial Intelligence [France]
ECCANE East Coast Conference on Aerospace and Navigational Electronics [US]
ECCB Electronic Components Certification Board [US]
ECCC European Communities Chemistry Committee [UK]
ECCCO European Culture Collections Curators Organization [now: ECCO, UK]
ECCFPP European Conference on Controlled Fusion and Plasma Physics
ECCLS European Committee for Clinical Laboratory Standards [UK]
ECCM Eastern Caribbean Common Market
Electronic Counter-Countermeasures
European Conference on Composite Materials
ECCMF European Council of Chemical Manufacturers Federations
ECCO Engineering Command Control and Operation
Engineers Coordinating Council of Oregon [US]

European Conference of Conscripts Organization [Netherlands]
European Culture Collections Organization [UK]
ECCP Engineering Concepts Curriculum Project [US]
ECCS Electrolytic Chromium-Coated Steel
Electronic C (= Hundred) Call Seconds
Emergency Core Cooling System
ECD East Caribbean Dollar [Currency of Antigua-Barbuda, Grenada, St. Christopher Nevis, St. Kitts, St. Lucia, St. Vincent and the Grenadines]
Electrochemical Deposition
Electrochromeric Display
Electron Capture Detector
Energy Conversion Device
Engineers Club of Dayton [US]
Enhanced Color Display
Equipment Configuration Data
Error Control Device
EC$ East Caribbean Dollar [Currency of Antigua-Barbuda, Grenada, St. Christopher Nevis, St. Kitts, St. Lucia, St. Vincent and the Grenadines]
ECDC Electrochemical Diffused Collector
ECDIN Environmental Chemicals Data and Information Network [of EURATOM]
ECDM Electrochemical Discharge Machining
ECDMMRL European Committee for the Development of the Meuse and Meuse/Rhine Links [Belgium]
ECDO Electronic Community Dial Office
ECE Echo Control Equipment
Economic Commission for Europe [of UN — Switzerland]
Electrochemical Electrode
Electronic Communications Engineer
Engineering Capacity Exchange
European Commodities Exchange [France]
Export Council for Europe [UK]
ECEA Economic Community of Eastern Africa
ECEC East Carolina Engineers Club [US]
ECELR Epithermal Critical Experiment Laboratory Reactor
ECerS European Ceramic Society
ECETOC European Chemical Industry Ecology and Toxicology Center [Belgium]
ECF Echo Control Factor
Emergency Cooling Functionality
Emission Contribution Fraction
Ethyl Chloroformate
European Caravan Federation [UK]
European Commission on Forestry and Forestry Products [Italy]
European Composites Forum
European Conference on Fracture
Extracellular Fluid
ECFA European Committee for Future Accelerators
ECFCI European Center of Federations of the Chemical Industry
ECFFP European Commission on Forestry and Forest Products
ECFI East Caribbean Farm Institute
ECFM Eddy Current Flow Meter
ECG Electrocardiogram
Electrocardiography
Electrochemical Grinding
Electro-Epitaxial Crystal Growth
Engineering Consulting Group [of MOEA, Taiwan]
ECGD Export Credits Guarantee Department [of BOT, UK]
ECGF European Container Glass Federation [Belgium]
ECGS Endothelial Cell Growth Supplement
ECH Earth Coverage Horizon Measurement
Echelon
Echo Cancellation Hybrid
Eddy Current Heating
Ethylene Chlorohydrin
ECHO Electronic Computing Hospital-Oriented
European Commission Host Organization
ECI Cast Iron Electrode
Eddy Current Inspection
Electrochemical Interface
Emergency Cooling Injection
Engine Component Improvement
Equity Capital Investment
ECIM European Commission for Industrial Marketing [UK]
ECIO European Conference on Integrated Optics
EC-IOA European Committee of the International Ozone Association [France]
ECIS European Community Information Service
European Council of International Schools [UK]
ECITER Electron Cyclotron International Thermonuclear Experimental Reactor [also: EC ITER]
ECITO European Central Inland Transport Organization
ECIWA European Committee of Importers and Wholesalers Associations
ECKC Engineers Club of Kansas City [US]
ECL East Coast Laboratory [of ESSA, US]
Eddy Current Loss
Electrical Communication Laboratory [Japan]
Electrogenerated Chemiluminescence
Electronics Components Laboratory
Emitter-Coupled Logic
Equipment Component List
Execution Control Language
Executive Control Language
ECLA Economic Commission for Latin America [of UN — Chile]
ECLAC Economic Commission for Latin America and the Caribbean [of UN — Chile]
ECLAT European Computer Leasing and Trading Association
European Conference on Laser Treatment (of Materials)
ECLG European Consumer Law Group [Belgium]
ECLM Economic Community for Livestock and Meat [Burkina Faso]
ECLO Emitter-Coupled Logic Operator
ECL-TTL Emitter-Coupled Logic/Transistor-Transistor Logic
ECM Effective Calls Meter
Electric Coding Machine
Electrochemical Machining
Electrochemical Metallizing
Electrochemical Milling
Electronically-Commutated Motor
Electronic Control Module

	Electronic Countermeasures
	Engineers Club of Minnestoa [US]
	Environmentally Conscious Manufacturing
	European Common Market
	Extended Core Memory
	Extracellular Matrix
ECMA	Electronic Computer Manufacturers Association
	Embalming Chemical Manufacturers Association [US]
	Engineering College Magazines Associated [US]
	European Carton Makers Association [Netherlands]
	European Catalysts Manufacturers Association [of CEFIC, Belgium]
	European Computer Manufacturers Association [Switzerland]
ECMC	Electric Cable Makers Confederation
	European Container Manufacturers Committee [Italy]
ECME	Economic Commission for the Middle East [of UN]
	Electronic Countermeasures Equipment
ECMP	Electronic Countermeasures Program
ECMRA	European Chemical Marketing Research Association [UK]
ECMSA	Electronics Command Meteorological Support Agency [US Army]
ECMT	European Conference of Ministers of Transport [France]
ECMU	Electronic Countermeasures Upgrade
ECN	Emergency Communication Network
	Engineering Change Notice
	Equivalent Carbon Number
ECNDT	European Council for Nondestructive Testing [Italy]
ECNE	Electric Council of New England [US]
ECNG	East Central Nuclear Group [US]
ECNM	Engineers Club of Northern Minnesota [US]
ECNR	European Council for Nuclear Research
ECO	Electron-Coupled Oscillator
	Electronic Central Office [US]
	Electronic Checkout
	Electronic Contact Operator
	Emergency Control Officer
	Engineering Change Order
	Engineering Control Office
	Engineers Club of Omaha [US]
	Epichlorohydrin Copolymer
	European Coal Organization
E/CO	Ethylene/Carbon Monoxide (Copolymer)
ECOC	Engineering Club of Oklahoma City [US]
	European Conference on Optical Communication
ECODU	European Control Data Users Organization [US]
ECODUG	European Control Data Users Group
ECoG	Electrocortiogram
	Electrocortiography
ECOL	Ecological
	Ecologist
	Ecology
ECOM	Electronic Computer Originated Mail (Service) [of USPO]
	Electronics Command [US Army]
E-COM	Electronic Mail
ECOMA	European Computer Measurement Association
ECOMS	Early Capability Orbital Manned Station
ECON	Economic(s)
	Economist
	Economizer
	Economy
	Electromagnetic Emission Control
ECOO	Educational Computing Organization of Ontario [Canada]
ECOP	Extension Committee on Organization and Policy [US]
ECorA	Erythrina Corallodendron Agglutinin
EcoR	Escherichia Coli RY [Molecular Biology]
ECORQ	Economic Order Quantity
ECOS	Energy Conservation and Substitution
ECOSOC	Economic and Social Committee
	Economic and Social Council
ECOSS	European Conference on Surface Science
ECOWAS	Economic Community of West African States
ECP	Effective Cable Pair
	Electric Current Perturbation
	Electrochemical Processing
	Electromagnetic Capability Program [US Air Force]
	Electromagnetic Containerless Processing
	Electron Channeling Pattern
	Electronic Circuit Protector
	Energy Conversion Program [of CANMET, Canada]
	Engineering Change Proposal
	Engineers Club of Philadelphia [US]
	Equipment Conversion Package
	External Casing Packer
ECPA	Energy Conservation and Production Act [US]
	Energy Consumers and Producers Association [US]
ECPD	Engineers Council for Professional Development [now: ABET, US]
ECPE	European Center for Public Enterprises
ECPGB	Entrance Cable Protector Ground Bar
ECPI	Electronic Computer Programming Institute
ECPRD	European Center for Parliamentary Research and Documentation [Luxembourg]
ECPS	Electronic Compendium of Pharmaceuticals and Specialties [of CPhA, Canada]
	Expanded Control Program Store
	Extended Control Program Support
ECR	Electron Cyclotron Resonance
	Electronic Cash Register
	Electronic Control Receiver
	Electronic Control Relay
	Emergency Cooling Recirculation
	Engineering Change Request
	Engineering Change Requirement
	Error Control Receiver
	Estimate Change Request
	External Control Register
ECRC	Electricity Council Research Center
	Electronic Components Reliability Center
	Electronic Components Research Center [US]
	Engineering College Research Council [US]
ECR-CVD	Electron-Cyclotron Resonance — Chemical Vapor Deposition
ECRD	Eddy Current Resonance Digitizing
ECRH	Electron Cyclotron Resonance Heating

ECRIE European Center for Research and Information Exchange [Belgium]
ECRL Eastern Caribbean Regional Library [West Indies]
ECRO European Chemoreception Research Organization [Switzerland]
ECS Echo Control Subsystem
Ecuadorean Sucre [Currency]
Editorial Coordination Services
Electrical Control System
Electrochemical Science
Electrochemical Series
Electrochemical Society [US]
Electroconvulsive Shock
Electromagnetic Compatibility Society [of IEEE, US]
Electronic Control Switch
Embedded Computer System
Emergency Coolant System
Emission Control System
Energy Communication Services
Engine Control System
Environmental Conservation Service
Environmental Control Shroud
Environmental Control System
Etched Circuits Society [US]
European Ceramic Society
European Committee for Standardization
European Communications Satellite
Evaporation Control System
Expanded Control Store
Experimental Communication Satellite [Japan]
Extended Core Storage
ECSA European Communications Security Agency
European Computing Services Association
Exchange Carriers Standards Association [US]
Expanded Clay and Shale Association [now: LAPA, US]
ECSAMR Emergency Committee to Save America's Marine Resources [US]
ECSC European Coal and Steel and Community
European Community for Steel and Coal
ECSF Engineers Club of San Francisco [US]
ECSG Electronic Connector Study Group [US]
ECSL Engineers Club of St. Louis [US]
ECSS European Committee for the Study of Salt [France]
European Conference on Surface Science
ECST Condensate Storage Tank
ECSW Extended Channel Status Word
EC SW End Cell Switch
ECT Eddy Current Testing
Electronic Control Technology
Emission-Controlled Tomography
Engineer's Club of Toronto [Canada]
Equicohesive Temperature
Evans Clear Tunnel
ECTA Error-Correcting Tree Automation
European Cutting Tools Association [UK]
ECTC Eastern Coal Transportation Conference [US]
ECTEL European Telecommunications and Professional Electronics Industry [UK]
ECTEOLA Epichlorohydrin Triethanolamine
ECTEOLA-C Epichlorohydrin Triethanolamine Cellulose
E-CTFE Ethylene-Chlorotrifluoroethylene [also: ECFTE]
ECTL Emitter-Coupled Transistor Logic
ECTN Eastern Canada Telemetered Network
ECTR Extended Connection Table Representation
ECU East Carolina University [US]
Electronic Control Unit
Electronic Conversion Unit
Emission Control Unit
Energy Conservation and Utilization
Environmental Control Unit
Exabyte Control Unit
European Currency Unit [Currency of the European Community]
E CUBE Energy Conservation Utilizing Better Engineering [also: E Cube]
ECUCT East China University of Chemical Technology [PR China]
ECUT Energy Conservation and Utilization Technology Program [of USDOE]
ECV Enamel (Single) Cotton Varnish (Wire)
ECWA Economic Commission for Western Africa [of UN]
ECWIM European Committee of Weighing Instrument Manufacturers [France]
ECWU Energy and Chemical Workers Union
ECX Electronically-Controlled Telephone Exchange
E-CYCLE Execution Cycle
ED Earth Detector
Edition [also: EDIT]
Editor [also: EDIT]
Education
Educator
Effective Dose
Electrical Differential
Electrodecantation
Electrodeposition
Electrodialysis
Electrodynamics
Electron Density
Electron Diffraction
Electronic Design
Electronic Device
Electronic Dummy
Emergency Distance
Engine Drive
Engineering Department
Engineering Design
Engineering Designer
Engineering Draftsman
Engineering Drawing
Entry Date
Erase Display
ERIC Document [of USOE]
Error Detection
Estimated Dose
Existence Doubtful
Expanded Display
Expansion Drum
External Deflector
External Device
Extractive Distillation

	Extra-Low Dispersion Glass
ED50	Effective Dose 50 [also: ED_{50}]
EDA	Economic Development Administration [US]
	Economic Development Agreement
	Educational Development Association [UK]
	Electrical Development Association [also: BEDA, UK]
	Electronic Defense Association [also: AOC, US]
	Electronic Design Automation
	Electronic Differential Analyzer
	Electronic Digital Analyzer
	Embedded Direct Analysis
	Energy-Dispersive Analysis
	Engineering Design Activities (Agreement) [of ITER Project]
	Ethylenediamine
	European Demolition Association [Netherlands]
	European Desalination Association [Scotland]
EDAC	Error Detection and Correction
	1-Ethyl-3-(3-Dimethylaminopropyl)carbodiimide
EDAM	Educational Distributors of Manitoba [Canada]
EDANA	European Disposables and Nonwovens Association [Belgium]
EDANS	Ethylenediaminenaphthalenesulfonic Acid
1,5-EDANS	1,5-Ethylenediaminenaphthalenesulfonic Acid
EDB	Earned Depletion Base
	Economic Development Board [US and Singapore]
	Educational Data Bank
	Energy Database [of ORNL, US]
	Ethylene Dibromide
EDBD	Environmental Database Directory [of NODC, US]
EDBL	Edible
EDBS	Educational Database System
EDC	Economic Development Committee [of NEDC, UK]
	Education Development Center
	Effective Dielectric Constant
	Electronic Desk Calculator
	Electronic Digital Computer
	Emergency Decontamination Center
	Enamel Double Cotton (Wire)
	Energy Distribution Curve
	Engineering Design Consultants
	Error Detection and Correction
	Error-Detection Code
	Ethylcarbodiimide Chloride
	Ethylene Dichloride
	European Defense Community
	European Documentation Center
	Export Development Corporation [Canada]
	External Disk Channel
	External Drum Channel
EDCC	Electronic Data Council of Canada
EDCL	Electric Discharge Convection Laser
EDCl	Ethylcarbodiimide Chloride
EDCOM	Editor and Compiler
EDCPF	Environmental Data Collection and Processing Facility
EDCS	Engineering Data Control System
EDCT	Expect Departure Clearance Time
EDCV	Enamel Double Cotton Varnish (Wire)
EDCW	External Device Control Word
EDD	Electronic Data Display
	Electronic Document Delivery
	Envelope Delay Distortion
	Expected Date of Delivery
EdD	Doctor of Education
EDDA	Ethylenediaminediacetic Acid
EDDF	Error Detection and Decision Feedback
EDDHA	Ethylenediaminedi(O-Hydroxyphenylacetic Acid)
EDDQ	Extra Deep Drawing Quality
EDDUS	Electronic Data Display and Update System
EDE	Economic Development Foundation [Philippines]
	Emergency Decelerating
	Emitter Dip Effect
	External Document Exchange
EDETATE	Ethylene Diamine Tetraacetic Acid
EDEW	Enhanced Distant Early Warning
EDF	Electrophoresis Duplicating Film
	Engineering Data Form
	European Development Fund
EDFA	Erbium-Doped Fiber Amplifier
EDFP	Engine Driven Fire Pump
EDFR	Effective Date of Federal Recognition
EDG	Edge
	Electrical Discharge Grinding
	Electrodischarge Grinding
	Electrodynamic Gradient
	Emergency Diesel Generator
	Exploratory Development Goals
EDGE	Electronic Data Gathering Equipment
EDGES	Expert DMIS (= Directory Management Information System) Graphical Editor and Simulator
EDGF	Electrodynamic Gradient Freeze
EDHE	External Data Handling Equipment
EDI	Electromagnetic Discharge Imaging
	Electron Diffraction Instrument
	Electronic Data Interchange
EDIAC	Electronic Display of Indexing Association and Content
EDIB	Ethyl Diiodobrassidate
EDIC	Exploration Drilling Incentive Program [Canada]
EDICT	Engineering Departmental Interface Control Technique
	Engineering Document Information Collection Technique
EDIF	Electronic Design Interchange Format
EDIN	Electronic Data Interchange Network
EDIP	European Defense Improvement Program [of NATO]
EDIS	Elektronisches Dokumentations- und Informationssystem [= Electronic Documentation and Information Retrieval System [Germany and Switzerland]]
	Engineering Data Information System
	Environmental Data and Information Service [now: NESDIS, US]
	Exploratory Drill Incentives System
EDIT	Edition [also: ED]
	Editor [also: ED]
	Engineering Development Integration Test
EDITAR	Electronic Digital Tracking and Ranging
EDL	Electrodeless Discharge Lamp
EDLCC	Electronic Data Local Communications Central

EDLIN Line Editor (Program)
EDM Electrical Discharge Machining
Electron Discharge Machining
Engineering Data Management
Entity Data Model
EdM Master of Education
EDMA Ethylene Glycol Dimethacrylate
Extended Direct Memory Access
EDMF Euclid-IS Data Management Facilities
EDMS Electronic Device and Materials Symposium
Engineering Data Microreproduction System
Extended Data Management System
EDO Economic Development Office
Effective Diameter of Objective
Engineering Duties Only
Executive Director of Operations
EDOC Effective Date of Change
EDOS Extended Disk Operating System
EDOS-MSO Extended Disk Operating System — Multistage Operations
EDOT Effective Date of Training
EDP Educational Data Processing
Effective Depth of Penetration
Electronic Data Processing
Electrophoresis Duplicating Paper
Enterprise Development Program
Experimental Development Program
EDPA Exhibition Designers and Producers Association [US]
EDPAA Electronic Data Processing Auditors Association [US]
EDPAF Electronic Data Processing Auditors Foundation [US]
EDPC Electronic Data Processing Center
EDPE Electronic Data Processing Equipment
EDPI Electronic Data Processing Institute
EDPM Electronic Data Processing Machine
EDPS Electronic Data Processing System
EDR Electrodermal Reaction
Electrodialysis Reversal
Equivalent Direct Radiation
EDRA Environmental Design Research Association [US]
EDRCC Electronic Data Remote Communications Complex
EDRI Electronic Distributors Research Institute [US]
EDRL Effective Damage Risk Level
EDRP European Demonstration Reprocessing Plant
EDRS Engineering Data Retrieval System
ERIC Document Reproduction Service [of USOE]
EDS Editors
Electric-Discharge Sintering
Electric Drive System
Electron Devices Society [of IEEE, US]
Electronic Data System
Electronic Document Storage
Emergency Detection System
Enamel Double Silk (Wire)
Energy Data System [of Environmental Protection Agency, US]
Energy-Dispersive Spectrometry
Energy-Dispersive Spectroscopy
Engineering Data Sheet
Engineering Data System [of USDOD]
Environmental Data Service [now: NESDIS, US]
Exchangeable Disk Storage
Expert Debugging System
EdS Specialist in Education
EDSA Expert Dataflow and Static Analysis Tool
Expert Debugging System for Ada
EDSAC Electronic Delay Storage Automatic Calculator
Electronic Discrete Sequential Automatic Computer
EDSC Engineering Data Service Center [US Air Force]
EDS/EELS Energy-Dispersive Spectroscopy/Electron Energy Loss Spectroscopy
EDSG Electrooptical Data Systems Group [US]
EDSS Explosives Detection Security System
EDST Eastern Daylight Saving Time
Elastic Diaphragm Switch Technology
EDSV Enamel Double Silk Varnish (Wire)
EDT Early Decay Time
Eastern Daylight Time
Electric Discharge Tube
Electronic Data Transmission
Electronic Design Transfer
Energy Dissipation Test
Engineer Design Test
Ethylenediamine Tartrate
EDTA Ethylenediaminetetraacetate
Ethylenediaminetetraacetic Acid
EDTAN Ethylenediaminetetraacetonitrile
EDTA Na_4 Ethylenediaminetetraacetic Acid Tetrasodium Salt
EDTC Ethyldipropylthiocarbamate
EDTCC Electronic Data Transmission Communications Central
EDTN 1-Ethoxy-4-(Dichloro-1,3,5-Triazinyl)naphthalene
EDTV Extended Definition Television
EDU Electronic Display Unit
Experimental Diving Unit
EDUC Education [also: ED]
Educator [also: ED]
EDUCOM Educational Communications
EDUCTL Educational
EDUG European Data-Manager Users Group
EDUNET Education Network
EDVAC Electron Discrete Variable Automatic Compiler
Electronic Digital-Vernier Analog Computer
Electronic Discrete Variable Automatic Computer
EDW Electron Density Wave
EDWC Electrical Discharge Wire Cutting
EDWM Electrodynamic Wattmeter
EDX Energy-Dispersive X-Ray Analysis
Event Driven Executive
EDXA Energy-Dispersive X-Ray Analysis
EDXRD Energy-Dispersive X-Ray Diffraction
EDXRF Energy-Dispersive X-Ray Fluorescence
EE Earthquake Engineer
Earthquake Engineering
Electrical Engineer
Electrical Engineering
Electrically-Erasable
Electronics Engineer
Electronics Engineering
Enantiomeric Excess

End-Effector
Energy Engineer
Energy Engineering
Enolether
Environmental Engineer
Environmental Engineering
Errors Excepted
Evolutionary Ecology
Explosives Engineer
Explosives Engineering
External Environment

EEA Electronic Engineering Association [UK]
Ethylene Ethyl Acrylate
Euonymus Europaeus Acetone Powder
Euonymus Europaeus Agglutinin

EEB European Environmental Bureau [Belgium]
European Export Bank

EEBP Estacao Experimental de Biologia e Piscicultura [= Experimental Station for Biology and Fish Culture, Brazil]

EEC Electronic Engine Control
End of Equilibrium Cycle
EUROCONTROL Experimental Center
European Economic Commission
European Economic Community
Evaporative Emission Control

EECA European Electronic Component Manufacturers Association [Belgium]

EECE Emergency Ecomonic Commission for Europe

EEC/EURAM European Economic Community/European Research on Advanced Materials

EECL Emitter-Emitter Coupled Logic

EECW Emergency Exchanger Cooling Water

EED Electrical and Electronics Division
Electroexplosive Device
Electron Energy Distribution

EEDA Edmonton Economic Development Authority [Canada]

EEDB ERDA (= Energy Research and Development Adminstration) Energy Database

EEDF Electron Energy Distribution Function

EEDQ 2-Ethoxy-1-Ethoxycarbonyl-1,2-Dihydroquinoline
Ethyl 1,2-Dihydro-2-Ethoxy-1-Quinolinecarboxylate

EEE Electrical/Electronic/Electromechanical (Components)
Electrical Engineering Exposition

EEEI Energy, Economics and Environment Institute

EEET Electronic Excitation Energy Transfer

EEF Earth Ecology Foundation [US]
Engineering Employers Federation [UK]
Export Expansion Fund [Canada]

EEG Electroencephalogram
Electroencephalography
Energy and Environment Group [India]
Environmental Engineers Group [of UBC, Canada]

EEI Edison Electric Institute [US]
Environmental Equipment Institute
Essential Elements of Information

EEIB Environmental Engineering Intersociety Board [US]

EEIC Electrical/Electronics Insulation Conference [US]

EEL Electron Energy Loss
Exclusive Exchange Line

EELS Electron Energy Loss Spectroscopy
Electron Energy Loss Spectrum

EELS-EDX Electron Energy Loss Spectroscopy/Energy-Dispersive X-Ray Analysis

EEM Earth Entry Module
Electronics Engineer's Master
Electronic Engine Management
Electronic Equipment Monitoring
Emission Electron Microscopy
Excitation Emission Matrix

EEMA Electrical and Electronic Manufacturers Association [US]

EEMAC Electrical and Electronic Manufacturers Association of Canada

EEMDA Electrical/Electronics Materials Distributors Association [US]

EEMJEB Electrical and Electronic Manufacturers Joint Education Board

EEMS Enhanced Expanded Memory Specification
European Environmental Mutagen Society [Netherlands]

EEMTIC Electrical and Electronic Measurement and Test Instruments Conference

EEO Electroendosmosis
Equal Employment Opportunity
Executive Engineering Order

EEO/AA Equal Employment Opportunity/Affirmative Action (Employer)

EEOC Equal Employment Opportunity Commission [US]

EEP Electroencephalophony

EEPA Electromagnetic Energy Policy Alliance [US]

EEPC (India) Engineering Export Promotion Council [US]

EEPLD Electrically Erasable Programmable Read-Only Memory [also: E^2PLD]

EEPNL Estimated Effective Perceived Noise Level

EEPROM Electrically-Erasable Programmable Read-Only Memory [also: E^2PROM]

EER Electrolyte Electroreflectance
Explosive Echo Ranging

EERA Electrical Equipment Representatives Association [US]

EERC Earthquake Engineering Research Center [of UCR, US]

EERI Earthquake Engineering Research Institute [US]

EERJ External Expansion Ramjet

EERL Earthquake Engineering Research Laboratory [US]
Electrical Engineering Research Laboratory

EE&RM Elementary Electrical and Radio Material Training School

EEROM Electrically-Erasable Read-Only Memory [also: E^2ROM]

EES Engineering Experiment Station

EESC Erie Engineering Societies Council [US]

EESI Earth Environment Space Initiatives [of Canadian Space Agency]

EESMB Electrical and Electronics Standards Management Board [of ANSI, US]

EEUA Engineering Equipment Users Association [UK]

EET Eastern European Time

Electrical Engineering Technologist
Electrical Engineering Technology
Electronic Excitation Transfer
Electronic Exhaust Transducer
Electronics Engineering Technician
Electronics Engineering Technologist
Electronics Engineering Technology
Environmental Engineering Technologist
Environmental Engineering Technology
Equipment Engaged Tone
Estimated Elapsed Time

EETD Environmental Emergencies Technology Division [of ETC, Canada]
EETF Electronic Environmental Test Facility
EEUA Engineering Equipment Users Association
EEVT Electrophoresis Equipment Verification Test
EEW Epoxy per Equivalent Weight
EEWD Enhanced Exchange Wide Dial
EEZ Exclusive Economic Zone
EF Each Face
Edge Filter
Effective Force
Electric Furnace
Electrofining
Electroforming
Electronic Filing
Elevation Finder
Elliptic Filter
Emitter Follower
Engineering Foundation [US]
Error Factor
Error-Free
Eurodata Foundation [UK]
Evaluation Finder
Extended Facility
Extra Fine (Thread)
Extremely Fine
EFA Erbium-Doped Fiber Amplifier
Esterified Fatty Acid
EFAPI Euro-Market Federation of Animal Protein Importers
EFAPP Enrico Fermi Atomic Power Plant
EFAR Economic Feeder Administration and Relief
EFAS Electronic Flash Approach System
En-Route Flight Advisory Service
EFATCA European Federation of Air Traffic Controllers Associations
EFB Electric Flash Butt Welding
Engineering Foundation Board [US]
Error-Free Block
European Federation of Biotechnology [Germany]
EFBD Emergency Feed Baron Detector
EFBPBI European Federation of the Brush and Paint Brush Industries [Belgium]
EFBWW European Federation of Building and Woodworkers [Belgium]
EFC Electronic Frequency Control
European Federation of Corrosion [Germany]
European Forestry Commission
Expect Further Clearance
EFCC European Federation of Conference Cities
EFCE European Federation of Chemical Engineering [Germany]
EFCGU European Federation of Chemical and General Workers Union [Belgium]
EFCS European Federation of Cytology Societies [Italy]
EFCSM European Federation of Ceramic Sanitaryware Manufacturers [Italy]
EFCT European Federation of Conference Towns
EFCV Excess Flow Check Valve
EFD Engineering Flow Diagram
Equivalent Full Discharge
EFDAS Epsilon Flight Data Acquisition System
EFE Early Fuel Evaporation System
EFEM Energy Filtering Electron Microscopy
EFEO European Flight Engineers Organization
EFET Enhancement-Mode Field-Effect Transistor
EFF Effect
Effectiveness
Efficiency
European Furniture Federation
EFFBR Enrico Fermi Fast Breeder Reactor [US]
EFFCY Efficiency
EFFD Elektronik Fabricant Foreningen Danmark [= Danish Association of Electronic Products Manufacturers]
EFFE European Federation of Flight Engineers
EFFL Efflorescence [also: EFFLOR]
EFFOST European Federation of Food Science and Technology [UK]
EFG Edge-Defined, Film-Fed Growth
Electric Field Gradient
Elemental and Functional Group (Analysis)
EFI Educational Futures, Inc. [US]
Electronic Flight Instrument
Electronic Forum for Industry [UK]
Electronic Fuel Injection
Engineered, Furnished and Installed
Enrico Fermi Institute [of University of Chicago, US]
Environic Foundation International [US]
Error-Free Interval
EFIA European Fertilizer Importers Association [Belgium]
EFID Electric-Field In-Process Dressing
EFIR European Flight Information Region
EFIS Electronic Flight Instrument System
EFISP Enrico Fermi International School of Physics
EFL Effective Focal Length
Egptian Federation of Labour
Emitter-Follower Logic
Emitter-Function Logic
English as a First Language
English as a Foreign Language
Equivalent Focal Length
Error Frequency Limit
EFLA Education Film Library Association [US]
EFM Eight-to-Fourteen Modulation
Electric-Field Monitor
Extended Flygare Method
EFMA European Fertilizer Manufacturers Association [Belgium]

	European Fittings Manufacturers Association [UK]
EFMC	Elastic Fabric Manufacturers Council [of NTA, US]
	European Federation of Medical Chemistry [UK]
EFMCNTA	Elastic Fabric Manufacturers Council of the Northern Textile Association [US]
EFMD	European Foundation for Management Development [Belgium]
EFMS	Experimental Flight Management System
EFNEA	European Federation of National Engineering Associations [France]
EFNS	Educational Foundation for Nuclear Science [US]
EFO	Engineers Foundation of Ohio [US]
EFOA	European Fuel Oxygenates Association [of CEFIC, Belgium]
EFOR	Equivalent Forced Outage Rate
EFP	Electrofluid Dynamic Process
	Electronic Field Production
	Enhanced Flux Pinning
	Enrico Fermi Power Plant [Italy]
	European Federation of Parasitologists [Sweden]
	European Federation of Purchasing
	Explosively Formed Projectile
EFPD	Equivalent Full Power Days
EFPH	Equivalent Full Power Hours
EFPI	European Federation of the Plywood Industry [France]
EFPIA	European Federation of Pharmaceutical Industries Associations [Belgium]
EFPM	Effective Full Power Month
EFPS	European Federation for Productivity Services [Sweden]
EFPW	European Federation for the Protection of Waters
EFR	Emerging Flux Regions
	Eppley Foundation for Research [US]
	Error-Free Region
EFRAP	Exchange Feeder Route Analysis Program
EFS	Electronic Frequency Selection
	Error-Free Seconds
EFSIM	Explosive Foam Shock Initiation Model
EFT	Earliest Finish Time
	Electronic Funds Transfer
	Eno Foundation for Transportation [US]
EFTA	Electronic Funds Transfer Association [US]
	European Free Trade Association
EFTAPA	European Free Trade Association — Plastics Association
EFTC	European Freight Timetable Conference [Czechoslovakia]
EFTEC	European Fluorocarbon Technical Committee [Belgium]
EFTP	Error File Teaching Package
EFTPOS	Electronic Funds Transfer at Point of Sale
EFTR	Ethernet Frequency Translator
EFTS	Electronic Funds Transfer System
	Elementary Flying Training School
EFVA	Education Foundation for Visual Aids [UK]
EFW	Energy from Waste Program [Canada]
EFW/TS	Energy from Waste Transfer Station
EG	Ecological Genetics
	Economic Geography
	Edge Grain
	Electric Generator
	Electron Gas
	Electron Gun
	Engineering Geologist
	Engineering Geology
	Equipment Ground
	Ethylene Glycol
	(Single) Enamel (Single) Glass (Wire)
	Exploration Geochemist
	Exploration Geochemistry
	Exploration Geophysicist
	Exploration Geophysics
eg	(exempli gratia) — for example
EGA	Enhanced Graphics Adapter
	Ethylglycolacetate
	Evolved Gas Analysis
	Extragalactic Astronomy
EGA/CGA	Enhanced Graphics Adapter/Color Graphics Adapter
EGA/MDA	Enhanced Graphics Adapter/Monochrome Display Adapter
EGAS	European Group for Atomic Spectroscopy [Netherlands]
EGASF	European General Aviation Safety Foundation
EGATS	EUROCONTROL Guild of Air Traffic Controllers
EGBD	Extrinsic Grain Boundary Dislocation
EGBE	Ethylene Glycol Butyl Ether
EGC	Environmental Geochemistry
	Exploration Geochemistry
EGCI	Export Group for the Constructional Industries [UK]
EGCMA	European Gas Control Manufacturers Association
EGCR	Experimental Gas-Cooled Reactor [of ORNL, US]
E/GCR	Extended Group Coded Recording
EGD	Electrogasdynamics
	Environmental Graphic Designer
	Evolved Gas Detection
EGDN	Ethylene Glycol Dinitrate
EGDS	Extragalactic Distance Scale
E GER	East Germany
EGF	Electrodynamic Gradient Freeze
	Epidermal Growth Factor
	European Grassland Federation [Netherlands]
EGG	Electrogastrogram
	Electrogastrography
EGGA	European General Galvanizers Association [UK]
EGI	Electronic Gasoline Injection
	End of Group Indicator
EGIF	Equipment Group Interface
EG£	Egyptian Pound [Currency]
EGM	Electrogel Machining
	Electronic Governor Module
	European Glass Container Manufacturers [UK]
EGMBE	Ethylene Glycol Monobutylether
EGME	Ethylene Glycol Monomethylether
EGO	Eccentric Geophysical Observatory
EGOCO	European Group of Oil Companies
EGOS	European Group for Organizational Studies [Netherlands]
EGP	Egyptian Pound [Currency]
	Exploration Geophysics

EGPC Egypt General Petroleum Corporation

EGPS Extended General-Purpose Simulator

EGR Electronic Governor Regulator
Exhaust Gas Recirculation

EGRESS Emergency Global Rescue, Escape and Survival System [of NASA]
Evaluation of Glide Reentry Structural Systems

EGRS Extragalactic Radio Source

EGRSA Edible Gelatin Research Society of America [US]

EGS European Geophysical Society [Germany]

EGSA Electrical Generating Systems Association [US]

EGSC Eastern Group Supply Council [UK]

EGSCCP European Graduate Summer Course on Computational Physics

EGSE Electrical Ground Support Equipment

EGSMA Electrical Generating Systems Marketing Association [now: EGSA, US]

EGT Ethylene Glycol Bis(trichloroacetate)
Exhaust Gas Temperature

EGTA Ethylene Glycol Tetraacetic Acid
European Group of Television Advertising [UK]

EGTS Emergency Gas Treatment System

EGW Electrogas Welding

EH Effective Height
Electrohydraulics
Electron Hole
Engineering Hydrology
Extra High

EHA Electric Heating Association [US]
Ethyl Hexyl Acetate
Ethyl Hexyl Alcohol
European Helicopter Association

EHAP Employee Health Assistance Program

EHC Electrochemical Hydrogen Cracking
Electrohydraulic Control
Ethylenic Hydrocarbon

EHCAC Egyptian High Commission of Automatic Control

EHD Elastohydrodynamic(s)
Electrohydrodynamic(s)
Electron-Hole Drop

EHDA Electrohydrodynamic Atomization

EHDTV Enhanced High Definition Television

EHE External Heat Exchanger

EHEC Ethylhydroxyethyl Cellulose

E-HEMT Enhancement-Mode High-Electron-Mobility Transistor

EHEP Experimental High-Energy Physics

EHF Electrohydraulic Forming
Engineers Hall of Fame
Extra-High Frequency
Extremely High Frequency

EHFB Electrical Historical Foundation Board

EHF SATCOM Extremely High Frequency Satellite Communication

EHIS Emission History Information System [of EPA, US]

EHM Engine Health Monitoring
Extended Hueckel Method

EHO (Verein) Eisenhutte Osterreich [= Austrian Association for Iron and Steel Technology]

EHOG European Host Operators Group

EHP Effective Horsepower [also: ehp]
Electric Horsepower
Electron-Hole Pair
Electron-Hole Potential Method

EHPA Ethylhexyl Phosphoric Acid

EHPG Ethylenediamine-N,N'-Bis(2-Hydroxyphenylacetic acid)dimethyl Ester

EHPRG European High Pressure Research Group [France]

EHS Environmental Health Service
Extra High Strength

EHSI Electronic Horizontal-Situation Indicator

EHT Extended Huckel Theory
Extra-High Tension

EHTPS Extra-High-Tension Power Supply

EHV Extra-High Voltage
Extremely High Voltage

EHW Extremely High Water

EI Earth Inductor
East Indies
Electrical Instrument
Electrical Insulation
Electrogist International [now: NECA, US]
Electromagnetic Interference
Electronic Interference
Electron Impact
Electron Ionization
Employee Involvement
End Injection
Engineering Index
Engineering Information [US]
Engineering Instruction
Error Indicator
Exposure Index

EIA Electronic Industries Association [US]
Energetic Ion Analysis
Energy Information Administration (Division) [of USDOE]
Engineering Industries Association [UK]
Environmental Impact Appraisal
Enzyme Immunoassay
Equipment Interchange Association [US]

EIAA Electronic Industry Association of Alberta [Canada]

EIAC Electronic Industries Association of Canada

EIAJ Electronic Industries Association of Japan

EIAP Environmental Impact Assessment Project

EIASM European Institute for Advanced Studies in Management

EIA/TIA Electronic Industries Association/ Telecommunications Industry Association

EIB Electronics Installation Bulletin
European Investment Bank [Luxembourg]
Export-Import Bank

EIBA Ethylene Isobutyl Acrylate
European International Business Association [Belgium]

EIC Electrical Insulation Conference [now: EEIC, US]
Employment and Immigration Canada
Engineer-in-Charge
Engineering Institute of Canada
Environmental Industries Council [US]
Environment Information Center [US]
Equipment Identification Code

EICA East India Cotton Association

EICAS	Engine Indication and Crew Alerting System
EICR	Eastern Interior Coal Region
EICV	Engine Idling Control Valve
EID	Electronic Instrument Digest
	Electron-Induced Desorption
	Electron-Stimulated Ion Desorption
EIDLT	Emergency Identification Light
EIDS	Electronic Information Delivery System
EIED	Electrically-Initiated Explosive Device
EIES	Electronic Information Exchange System [US]
	Electron Impact Emission Spectroscopy
EIFAC	European Inland Fisheries Advisory Commission [of FAO — Italy]
EIGA	Engineering Industry Group Apprenticeship [UK]
EIH	Error Interrupt Handler
EIHSW	European Institute of Hunting and Sporting Weapons [Belgium]
EIII	(Association of the) European Independent Information Industry
EIIS	Energy Industry Information System
EIJC	Engineering Institutions Joint Council [UK]
EIJHE	East Indian Jute and Hessian Exchange
EIL	Electron Injection Laser
EIM	European Institute for the Media [UK]
	Excitability Inducing Material
EIMA	Electrical Insulating Materials Association
	Exterior Insulation Manufacturers Association [US]
EIM&M	Environmental Instrumentation Measurement and Monitoring
EIMO	Electronic Interface Management Office [US Navy]
EIN	Education Information Network
	European Informatics Network
E IND	East Indies
EINIS	European Integrated Network of Image and Services [France]
EIO	Extended Interaction Oscillator
EIP	Environmental Innovation Program [of Environment Canada]
	Equipment Installation Procedure
	Export Industrial Park
EIPC	European Institute of Printed Circuits [Switzerland]
EIR	Engineering Information Report
	Engineering Investigation Request
EIRB	European Investment Research Bureau
EIRD	Engineering Information Report Date
EIRMA	European Industrial Research Management Association [France]
EIRP	Effective Isotropic Radiated Power
	Equivalent Isotropically Radiated Power
EIS	Economic Information System
	Effluent Inventory System
	Electrical Induction Steel
	Electromagnetic Intelligence System [US Air Force]
	Electronic Information Service
	End-Interruption Sequence
	Engineering Index Service
	Environmental Impact Statement
	Environmental Information System
	Executive Information System
	Expanded Inband Signalling
	Extended Instruction Set
EISA	European Independent Steelworks Association [Belgium]
	Extended Industry Standard Architecture
EISA/ISA	Extended Industry Standard Architecture/Industry Standard Architecture
EISA/MCA	Extended Industry Standard Architecture/ Microchannel Architecture
EISO	Environmental Information Society of Ontario [Canada]
EISU	Eastern Illinois State University [US]
EIT	Engineer-in-Training
	European Institute of Technology
EITB	Engineering Industry Training Board [UK]
EITE	European Institute of Transuranium Elements [Germany]
EIU	Eastern Illinois University [US]
	Economic Intelligence Unit [UK]
	Equipment Inventory Update
EIVT	European Institute for Vocational Training
EIW	European Institute for Water [France]
EJ	Electronic Jamming
	Electronic Junction
EJC	Engineers Joint Council [US]
EJCC	Eastern Joint Computer Conference [US]
EJCNC	Engineers Joint Council Nuclear Congress [US]
EJCT	Engineering Joint Council Thesaurus [US]
EJECT	Ejector
EJF	Estimated Junction Frequency
E-JFET	Enhancement-Mode Junction Field-Effect Transistor
EJMA	Expansion Joint Manufacturers Association [US]
EK	Electrokinetics
EKG	Effective Kilogram
	Electrokymogram
	Electrokymography
EKN	Ecology of Knowledge Network [US]
EKS	Electrocardiogram Simulator
EKT	Electrokinetic Transducer
EKW	Electrical Kilowatts
EKY	Electrokymogram
	Electrokymography
EL	Education Level
	Elasticity
	Elastic Limit
	Electric Line
	Electroluminescence
	Electronics Laboratory
	Electron Lens
	Electrotechnical Laboratory
	Elevation
	Elongation
	Engineering Letter
	Entry Level
	Erase Line
	Exchange Line
	Executive Level
	Explosion Limit
	Extra Low
ELA	Elastomer Lubricating Agent
	European Laser Association [Netherlands]

Experimental Lake Area (Project) [of DFO, Canada]
ELAF Electron Linear Acceleration Facility
ELAH Earle's Lactalbumin Hydrolysate
ELAN Elementary Language
ELATS Expanded Litton Automated Test Set
ELBA Emergency Location Beacon Aircraft
ELC Element Count
Environmental Liaison Center [Kenya]
Exchange Line Capacity
Extra-Low Carbon Steel
ELCA European Landscape Contractors Association [Germany]
ELCB Earth-Leakage Circuit Breaker
Earth-Leakage Contact Breaker
ELCD Evaporative Loss Control Device
ELCl Extra-Low Chloride
ELCO Electrostatic Coaxial
Eliminate and Count Coding
ELCOM Electronics and Computers
ELCON Electricity Consumers Resource Council [US]
ELD Economic Load Dispatching
Edge-Lighted Display
Electroless Deposition
Electroluminescent Display
Extra-Long Distance
ELDC European Lead Development Committee [UK]
ELDO European Launcher Development Organization
ELDOR Electron-Electron Double Resonance
ELE Equivalent Logic Element
ELEC Electric(al)
Electrician
Electricity
European League for Economic Cooperation [UK]
Elec Gr Electronic Grade [Chemistry]
ELECOM Electronic Computer
ELECOMPS Electronic Components
ELECT Electrolyte
Electrolytic
ELECTN Electrician
ELECTR Electric(al)
Electrician
Electricity
Electronic(s)
ELECTROMAG Electromagnetic(s)
ELECTRON Electronic(s)
ELED Edge-Emitting Light Emitting Diode
Entry-Level Employee Development
ELEED Elastic Low-Energy Electron Diffraction
ELEM Element
ELEPLTG Electroplating
ELEV Elevation
Elevator
ELEX Electronics
ELF Electroluminescent Ferroelectricity
Electronic Logging Facility
Ellipsometry, Low-Field
Extensible Language Facility
Extremely Low Frequency
ELFA Electric Light Fittings Association [UK]
ELFE Extremely Low Frequency Effect
ELG Electrolytic Grinding
ELGMT Ejector-Launcher Guided-Missile Transporter
ELGRA European Low Gravity Research Association
ELH Enol-Lactone Hydrolase
ELI Energy Loss Image
English Language Interpreter
Environmental Law Institute [US]
Environment Liaison International [Kenya]
Equitable Life Interpreter
Extended Lubrication Interval
Extensible Language I
Extra-Low Interstitial
ELIAS Environmental Libraries Automated System [of Environment Canada]
ELINT Electromagnetic Intelligence
Electronic Intelligence
ELIP Electrostatic Latent Image Photography
ELISA Enzyme-Linked Immunosorbent Assay
ELK External Link
ELL Eccentric Leveling Lugs
ELLA Eastern Lamp and Lighting Association [US]
European Long Lines Agency [of NATO]
ELLIPT Elliptical
Ellipticity
ELM Extended Length Measure
Extra Low Mass
ELMA Electric Lamp Manufacturers Association [UK]
ELMAP Exchange Line Multiplexing Analysis Program
ELME Emitter Location Method
ELMIG Electronic Library Membership Initiative Group [of ALA, US]
ELMO European Laundry and Dry Cleaning Machinery Manufacturers Organization [Germany]
ELMS Elastic Loop Mobility System
Engineering Lunar Model Surface
Experimental Library Management System
ELNI Ethernet Local Network Interconnect
ELOISE European Large Orbiting Instrumentation for Solar Experiments
ELON East Longitude
ELONG Elongation
ELOP Extended Logic Plan
ELP Electropolishing
Element Processor
English Language Program
ELPC Electroluminescent — Photoconductive
ELPG Electric Light and Power Group
ELPH Elliptical Head
ELPHR Experimental Low-Temperature Process Heat Reactor
ELPO Electrodeposition
ELR Engineering Laboratory Report
Equal Listener Response Scale
Existing Lapse Rate
Extra Long Range
ELRAC Electronic Reconnaissance Accessory Set
ELRO Electronics Logistics Research Office [US Army]
ELS Earth Landing System
Economic Lot Size
Electroluminescence Screen
Electron Energy Loss Spectroscopy

	Element Symbol
	Emitter Location System
	Energy-Loss Spectroscopy
	Environmental Law Society [of Harvard University, US]
	Equal Load Sharing
	Error Likely Situation
ELSA	Energy Loss Spectral Analysis
ELSB	Edge-Lighted Status Board
ELSBM	Exposed Location Single Buoy Mooring
ELSEC	Electronic Security
ELSI	Extra-Large-Scale Integration
ELSIE	Electronic Letter Sorting and Indicating Equipment
	Electronic Signalling and Indicating Equipment
	Electronic Speech Information System
ELSS	Extravehicular Life Support System
EL-SSC	Electronic Switching System Control
ELSPECS	Electronic Specifications
ELT	Electrometer
	Electronic Level Transducer
	Element
	Emergency Location Transmitter
	English Language Teaching
ELTAD	Emergency Location Transmitter, Automatic Deployable
ELTAF	Emergency Location Transmitter, Automatic Fixed
ELTAP	Emergency Location Transmitter, Automatic Portable
ELTR	Emergency Location Transmitter/Receiver
ELU	Existing Carrier Line-Up
ELV	Earth Launch Vehicle
	Electrically-Operated Valve
	Enclosed-Frame Low Voltage
	Expendable Launch Vehicle
	Extra-Low Voltage
ELVAC	Electric Furnace Melting, Ladle Refining, Vacuum Degassing and Continuous Casting
ELW	Extended Linked Well
	Extreme Low Water
ELWAR	Electronic Warfare
EM	Effective Modulus
	Efficiency Modulation
	Elastic Modulus
	Electrolytic Meter
	Electromagnet(ic)
	Electromagnetism
	Electromechanics
	Electrometallurgist
	Electrometallurgy
	Electrometer
	Electrometric
	Electromigration
	Electromotor
	Electronic Mail
	Electronics Manufacturing Group [of SME, US]
	Electron Microscopy
	Electron Multiplier
	Electrophoretic Mobility
	Emission Pump
	End-of-Medium (Character)
	Engineering Management
	Engineering Manual
	Engineering Material
	Engineering Memorandum
	Engineering Model
	Engineer of Mines
	Enlisted Men
	Environmental Management
	Enzyme Mechanism
	Epitaxial Mesa
	Equimomental
	Ethoxylated Monoglyceride
	Evaluation Model
	Extensible Machine
	Extractive Metallurgist
	Extractive Metallurgy
Em	Emanation
em	(emeritus) — retired
e/m	electron charge/electron mass ratio
EMA	Effective Medium Approximation
	Electromagnetic-Acoustic Transducer
	Electronic Mail Association [US]
	Electronic Missile Acquisition
	Electronics Manufacturers Association [US]
	Electronics Materiel Agency [US Army]
	Emergency Minerals Administration [US]
	Employment Management Association [US]
	Engineered Materials Abstracts
	Engine Manufacturing Association [US]
	Environmental Management Association [US]
	Environmental Mediation Association (International) [US]
	Equipment Market Abstracts [US]
	Ethylene-Maleic Anhydride
	Ethylene Methyl Acrylate
	European Marketing Association [UK]
	European Monetary Agreement
	Excavator Makers Association
	Extended Mercury Autocoder
EMAA	Engineering Materials Achievement Award [of ASM International, US]
	Envelope Manufacturers Association of America [US]
	Ethylene Methacrylic Acid
	European Mastic Asphalt Association [France]
EMABC	Electronics Manufacturers Association of British Columbia [Canada]
EMAC	Educational Media Association of Canada
	Electronics Manufacturers Association of Canada
	European Marketing Academy [Belgium]
E-MAD	Engineer Maintenance Assembly Disassembly [US]
EMAIA	Electrical Meter and Allied Industries Association [Australia]
EMAP	Electromagnetic Array Profiling
EMAR	Experimental Memory-Address Register
EMAS	Electromagnetic Acoustic System
EMAT	Electromagnetic Acoustic (Wave) Transducer
EMATS	Emergency Mission Automatic Transmission Service
	Emergency Mission Automatic Transmission System
EMAV	Electromagnetic Relief Valve
EMB	Electrical Modernization Bureau
	Electromagnetic Braking

	Engineering in Medicine and Biology [IEEE Society]
	Eosine Methylene Blue
	Experimental Model Basin [US Navy]
EMBA	Executive Master of Business Administration
EMBC	European Molecular Biology Conference
EMBERS	Emergency Bed Request System
EMBET	Error Model Best Estimate of Trajectory
EMBH	Etched Mesa Buried Heterostructure
EMBKN	Embarkation
EMBL	European Molecular Biology Laboratory [Germany]
EMBnet	European Molecular Biology Data Network
EMBO	European Molecular Biology Organization [Germany]
	Eta-Maleimidocaproyloxysuccinimide
EMBRAPA	Empresa Brasileira de Pesquisa Agropecuaria [= Brazilian Cattle Raising Research Enterprise]
EMC	Elastomeric-Molding Compound
	Electromagnetic Capability
	Electromagnetic Casting
	Electromagnetic Compatibility
	Electromechanochemical
	Electronic Materials Conference
	Electron Microscopy Center
	Engineered Military Circuit
	Engineering Manpower Commission [of AAES, US]
	Engineering Military Circuit
	Engineering Mockup Critical Experiment
	Ensemble Monte Carlo
	Equilibrium Moisture Content
	European Mathematical Council [UK]
	European Mechanics Colloquium
	European Mechanics Committee [also: EuroMech, Germany]
	European Metals Conference [Belgium]
	European Microwave Conference
	European Military Communication
	European Muon Collaboration
	Excess Minority Carrier
	Extended Multiplier Channel
EMCA	Electronic Motion Control Association [US]
EMCAB	Electromagnetic Compatibility Advisory Board
EMCCC	European Military Communications Coordinating Committee [of NATO]
EMCF	European Monetary Cooperation Fund
EMC-FOM	Electromagnetic Compatibility Figure of Merit
EMCON	Emission Control
EMCP	Electromagnetic Compatibility Program
EMCRF	European and Mediterranean Cereal Rusts Foundation [UK]
EMCS	Electromagnetic Compatibility Standardization
	Energy Management and Controls Society
	Energy Monitoring and Control System
	Eta-Maleimidocaproyloxysuccinimide
EMCTP	Electromagnetic Compatibility Test Plan
EMCV	Encephalomyocarditis Virus
EMCWP	European Mediterranean Commission on Water Planning [Italy]
EMD	Electric-Motor Driven
	Electromagnetic Damping
	Electronic Map Display
	Extractive Metallurgy Division [of AIME, US]
EMDI	Energy Management Display Indicator
EMDP	Energy Management Development Program
EME	Earth-Moon-Earth
	Electromagnetic Energy
	Electromechanical Energy
	Environmental Mine Engineering
EMEC	Electronic Maintenance Engineering Center [US Army]
EMEFS	Eulerian Model Evaluation Field Study
EMEPS	Electronic Message Privacy System
EMER	Emerald
	Emergency
Emer	Emeritus
EMETF	Electromagnetic Environmental Test Facility [US]
EMEX	Equatorial Mesoscale Experiment
EMF	Effective Mass Filter
	Electromagnetic Field
	Electromagnetic Forming
	Electromotive Force [also: emf]
	European Management Forum
	European Monetary Fund
	Evolving Magnetic Features
EMG	Electromagnetic Generator
	Electromyogram
	Electromyography
EMH	Expedited Message Handling
EMHS	Electronic Message Handling System
EMI	Early Manufacturing Involvement
	Electromagnetic Induction
	Electromagnetic Interference [also: emi]
	End-of-Message Indicator
	Extractive Metallurgy Institute [US]
EMIC	Electronic Materials Information Center [US]
	Environmental Mutagen Information Center [of NIEHS, US]
EMICE	Electromagnetic Interference Control Engineer
EMINT	Electromagnetic Intelligence
EMIR	Electromagnetic Interference Resolution
EMI/RFI	Electromagnetic Interference/Radio Frequency Interference
EMIRS	Electrochemically-Modulated Infrared Reflectance Spectroscopy
EMIRTEL	Emirates Telecommunications Corporation [of United Arab Emirates]
EMIS	Educational Management Information System
	Electromagnetic Intelligence System
	Electronic Markets and Information System
	Electronic Materials Information Services
EMIT	Engineering Management Information Technique
EMKO	Ethyl Michler's Ketone Oxime
EML	Electromagnetic Levitator
	Electron Microscopy Laboratory
	Engineering Mechanics Laboratory [of NIST, US]
	Equipment Modification List
	Expected Measured Loss
	Exterior Metal Loss
EMLF	Eastern Mineral Law Foundation [US]
EMM	Ebers-Moll Model
	Electromagnetic Measurement
	Electromagnetic Moment

EMMA Electron Manual Metal-Arc
Electron Microscope Microanalyzer
Electron Microscopy and Microanalysis
Equitorial Mount with Mirrors for Acceleration
Extra MARC (= Machine-Readable Catalogue) Material
Eye Movement Measuring Apparatus
EMMAQUA Equitorial Mount with Mirrors for Acceleration with Water Spray
EMML Electromechanical Machine Laboratory
EMMS Electronic Mail and Message System
EMO Electromagnetic Oscillograph
Electromechanical Oscillator
Emergency Measures Organization
EMOS Earth's Mean Orbital Speed
EMP Elastomer Modified Plastomer
Electromagnetic Pulse
Electromechanical Power
Electromolecular Propulsion
Emission Pattern
End of Month Payment
EMPA Eidgenossische Materialprufungsanstalt [= Federal Laboratory for Materials Research and Testing, Switzerland]
European Maritime Pilots Association
Executive Master of Public Affairs
EMPB Emergency Mobilization Preparedness Board
EMPF Electronic Manufacturing Productivity Facility
EMPIRE Early Manned Planetary-Interplanetary Round-Trip Expedition
EMPL Employee
EMPLT Employment
EMPMD Electronic, Magnetic and Photonic Materials Division [of TMS, US]
EMPR Electromagnetic Pulse Radiation
Electromagnetic Pulse Response
Ethernet Multiport Repeater
EMPRA Emergency Multiple Person Rescue Apparatus
EMPS Electronic Message Privacy System
Ethernet Multiport Station
EMQ Economic Manufacturing Quantity
Electromagnetic Quiet
EMR Eddy-Making Resistance
Electrolytic Metal Recovery
Electromagnetic Radiation
Electromagnetic Relay
Electromagnetic Response
Electromechanical Relay
Electromechanical Research
(Department of) Energy, Mines and Resources [Canada]
Engine Mature Ratio
Executive Management Responsibility
EMRA Electrical Manufacturers Representive Association
EMRC European Medical Research Council [France]
EMRIC Educational Media Research Information Center [US]
EMRL Engineering Mechanics Research Laboratory [US]
EMRLD Excimer Moderate-Power Raman-Shifted Laser Device
EMRP Effective Monopole Radiated Power
EMRS European Materials Research Society [also: E-MRS, France]
EMS Earnings per Manshift
Earth and Mineral Sciences
Earthquake Monitoring System
Econometric Society [US]
Electromagnetic Separation
Electromagnetic Spectrum
Electromagnetic Stirring
Electromagnetic Submarine
Electromagnetic Surveillance
Electromagnetic Susceptibility
Electromechanical System
Electromotive Series
Electronic Mail System
Electronic Management System
Electronic Medical System
Electronic Message Service
Electronic Message System
Electronic Monitoring System
Electron Microscopy Society
Emergency Medical Services
Energy Management System
Engineering Management Society [of IEEE, US]
Engine Management System [US Army]
Enhanced Memory Specification
Environmental Management Service [Canada]
Environmental Monitoring System
Environmental Mutagen Society [at ORNL, US]
Equilibrium Mode Simulator
Ethylmethane Sulfonate
European Monetary System
Expanded Memory Specification
Expanded Memory Support
Export Marketing Service
Extended Memory Specification
EMSA Electron Microscopy Society of America [US]
Electronic Materiel Support Agency [US Army]
EMSC Electrical Manufacturers Standards Council [of NEMA, US]
EMSEC Emanations Security
EMSI Electron Microscopy Society of India
EMSL Environmental Monitoring Systems Laboratory [of EPA, US]
EM/SME Electronics Manufacturing Group of the Society of Manufacturing Engineers [US]
EMSS Emergency Manual Switching System
E. Mitchell Scientific Society [US]
EMT Electrical Mechanical Tubing
Electrical Metallic Tubing
Electromagnetic Theory
Electromagnetic Tubing
Electronic Maintenance Technician
Emergency Medical Technologist
Equivalent Megaton
European Mediterranean Tropo
EMTA Electromedical Trade Association [UK]
EMTECH Electromagnetic Technology
EMTP Electromagnetic Transients Program
EMTS Ethylmercury-P-Toluenesulfonanilide
EMTTF Equivalent Mean Time to Failure

EMU Electrical-Multiple Unit
Electromagnetic Unit [also: emu]
Emulator
European Monetary Unit
Expanded Memory Unit
Extravehicular Maneuvering Unit
Extravehicular Mobility Unit
EMW Electromagnetic Warfare
Electromagnetic Wave
EN Earthcare Network [US]
Electronegative
Enamel(ling)
Enforcement Notification
Equipment Number
ENA Electronic Networking Association [US]
European Neuroscience Association [Netherlands]
ENAA Epithermal Neutron Activation Analysis
ENADS Enhanced Network Administration System
ENAM Enamel
ENAR Eastern North American Region
ENB Ethylidene Norbornene
ENC Enclosure
Entanglement Network Coalition [US]
Equivalent Noise Charge
ENCA European Naval Communications Agency [of NATO]
ENCC Ente Nazionale per la Cellulosa e per la Carta [= National Agency for Cellulose and Paper, Italy]
ENCIAWPRC Egyptian National Committee of the International Association on Water Pollution Research and Control
ENCL Enclosure
ENCY Encyclopedia [also: ENCYC or ENCYCL]
END Equivalent Neutral Density
ENDA-TM Environnement et Developpement du Tiers Monde [= Environment and Development of the Third World, Senegal]
ENDC Eastern Nigeria Development Corporation
ENDOR Electron-Nuclear Double Resonance
ENDS European Nuclear Documentation System [of EURATOM]
ENE East-North-East
Enterprise Networking Event [US]
Estimated Net Energy
ENEA European Nuclear Energy Agency [of OECD]
European Nuclear Energy Agreement
European Nuclear Energy Association
ENEL Ente Nazionale per l'Energia Elettrica [= Italian National Electric Energy Agency]
ENERDEMO Conservation and Renewable Engergy Demonstration Program [Canada]
ENET Engineering Network
ENF End Notched Flexure
ENFET Enzyme-Based Field-Effect Transistor
ENFIA Exchange Network Facilities for Interstate Access [US]
ENFOR Energy from the Forest Program [Canada]
ENG Electronic News Gathering
Engine
Engineer(ing)
England
English
Engraving
Eng Engineer
ENGL England
English
ENGO Environment Non-Governmental Organizations
ENGR Engineer
Engraving
ENGRG Engineering
ENGS Engines
EngScD Doctor of Engineering Science
ENGSOC Engineering Society [also: EngSoc]
ENI Ente Nazionale Idrocarburi [= National Hydrocarbon Agency, Italy]
Ethiopian Nutrition Institute
ENIAC Electronic Numerical Integrator and Calculator
ENL Enamel(ling)
ENQ Enqueue
Enquiry (Character)
ENR Epoxidized Natural Rubber
Equivalent Noise Resistance
Excess Noise Ratio
ENRC European Nuclear Research Committee
ENRT En-Route
ENRX Enable Receive [also: EN RX]
ENS Electrostatic Nonmetallic Separator
European Nuclear Society [Switzerland]
ENSDF Evaluated Nuclear Structure Data File [of INKA, Germany]
ENSEARCH Environmental Management Association of Malaysia
ENSI Energia Nucleare Sud Italia [= Nuclear Energy for Southern Italy (Project) [of SENN, Italy]]
Equivalent Noise Sideband Input
ENSIP Engine Structural Integrity Program [US Air Force]
ENSM Ecole Nationale Superieure de Mines [= National Higher School of Mines, France]
ENSO El Nino Southern Oscillation
ENSP Engineering Specification
ENT Emergency Negative Thrust
Enter
Entrance
Equivalent Noise Temperature
ENTAC Engine-Teleguide Anti-Char
ENTC Engine Negative Torque Control
ENTD Entered
ENTELEC Energy Telecommunications and Electrical Association [US]
ENTG European NATO Training Group
ENTOMOL Entomological
Entomologist
Entomology
ENTR Entrance
ENTX Enable Transmit [also: EN TX]
ENU N-Ethyl-N-Nitrosourea
ENV Envelope
EnvE Environmental Engineer
ENVITEC Technology for Environmental Protection — International Trade Fair and Congress [Germany]
EO Editorial Operations
Electron Optics
Electro-Optic(s)
Electrosmosis

	End Office
	End-On
	Engineering Order
	Equal Opportunity
	Equivalent Orifice
	Errors and Omissions
	Ethylene Oxide
	Executive Office
	Executive Order
	Executive Organ
E-O	Electron-Optic(s)
EOA	End of Address
	Essential Oil Association [now: FMA, US]
EO/AA	Equal Opportunity/Affirmative Action (Employer)
EOAP	Earth Observation Aircraft Program [US]
EOAR	European Office of Aerospace Research
EOARDC	European Office of the Air Research and Development Command [US]
EOB	End of Block
EOBT	Estimated Off-Block Time
EOC	Emergency Operating Center
	End of Cycle
EOCI	Electric Overhead Crane Institute [now: CMAA, US]
EOCR	Experimental Organic Cooled Reactor [of INEL, US]
EOD	End of Data
	End of Dialing
	End of Document
	Every Other Day
	Explosive Ordnance Disposal
EODA	Eastern Ontario Development Association [Canada]
EODC	Eastern Ontario Development Corporation [Canada]
EODD	Electro-Optical Digital Deflector
EOE	End of Extent
	Equal Opportunity Employer
	Error and Omission Excepted
	European Options Exchange
E&OE	Error and Omission Excepted
EOEC	End of Equilibrium Cycle
EOF	End of File
	Energy Optimizing Furnace
EOG	Electrooculogram
	Electrooculography
	Electroolfactogram
EOHP	Except Otherwise Herein Provided
EOI	End of Information
	End of Inquiry
EOIC	Ethylene Oxide Industry Council [US]
EOIS	Electro-Optical Imaging System
EOJ	End of Job
EOL	Earth Observatory Laboratory
	End of Life
	End of Line
	Expression-Oriented Language
EOLM	Electron-Optical Light Modulator
EOLN	End of Line
EOLR	Electrical Objective Loudness Rating
EOLT	End of Logical Tape
EOM	End of Medium
	End of Message
	End of Month
EOP	End of Page
	End of Program
	End Output
	Even-Odd Predominance
	Exchange Offering Prospectus
	Executive Office of the President [US]
EOPC	Electro-Optical Phase Change
EOQ	Economic Order Quantity
EOQC	European Organization for Quality Control [Switzerland]
EOR	Electro-Optical Research
	End of Record
	End of Reel
	End of Run
	Enhanced Oil Recovery
	Explosive Ordnance Reconnaissance
EORA	Explosive Ordnance Reconnaissance Agent
EORS	Emergency Oil Spill Response System
EORTC	European Organization for Research on the Treatment of Cancer
EOS	Earth Observatory Satellite
	Earth Observing System
	Egyptian Organization for Standardization
	Electrical Overstress
	Electronic Office Service
	Electronic Optical System
	Electro-Optical System
	Electrophoresis Operations in Space
	End of Selection
	Equation of State
EOSAT	Earth Observation Satellite
EOT	Electrooptical Technology
	End of Tape
	End of Task
	End of Text
	End of Transmission
	Engine Order Telegraph
EOV	End of Volume
EOW	Engineering Order Wire
	Engine over the Wing
	Equal Opportunities for Women
EOX	Extractable Organic Chlorine
EOY	End of Year
EP	Echo Prospecting
	Ecole Polytechnique [= Polytechnic Institute]
	Elastic Peak
	Electrical Polarization
	Electrical Properties
	Electric Power
	Electrode Potential
	Electrolytic Polishing
	Electronic Packaging
	Electron Probe
	Electrophony
	Electrophoresis
	Electrophotography
	Electroplating
	Electropolishing
	Electropositive
	Electropulse
	Elongated Punch
	Emergency Preparedness

Emulation Program
Emulsion Polymer
End of Program
Engineering Personnel
Engineering Plastics
Environmental Physiology
Environmental Protection
Epitaxial Planar
Equipment Practise
Equipotential
Etched Plate
Ethylene Propylene
Expanded Polystyrene
Experimental Physicist
Experimental Physics
Explosion-Proof
Extended Performance
Extended Play
Extended Programmability
Extra Protection
Extra Pulse
Extreme Pressure

Ep Epoxy
E£ Egyptian Pound [Currency]
EPA Eastern Provincial Airways [Canada]
Electronic Publishing Abstracts [US]
Enhanced Performance Architecture
Environmental Protection Act [Canada]
Environmental Protection Agency [US]
Europaisches Patentamt [= European Patent Office, Germany]
European Photochemistry Association [UK]
European Productivity Agency [now: OECD, France]
Expanded Polystyrene Association
Extended Performance Architecture
EPAA Epithermal Neutron Activation Analysis
European Primary Aluminum Association [Germany]
EPA/AWMA Environmental Protection Agency/Air and Waste Management Association [US]
EPABX Electronic Private Automatic Branch Exchange
EPAC Expanded Polystyrene Association of Canada
EPAM Elementary Perceiver and Memorizer
EPASYS European Patents Administration System [of EPO, Germany]
EPA/TSCA Environmental Protection Agency/Toxic Substances Control Act [US]
EPB Earth Physics Branch [of EMR, Canada]
Environmental Periodicals Bibliography [of ESI, US]
EP-BL Ethylene Propylene Block Copolymer
EPBM Enhanced Probability-Based Matching
EPBS Earth-Pressure Balanced Shield
EPBX Electronic Private Branch Exchange
EPC Earth Potential Compensation
Easy Processing Channel (Furnace Black)
Economic Policy Committee [of OECD, France]
Editorial Processing Center
Electric Propulsion Conference
Electrolytic Photocell
Electronic Power Conditioner
Electronic Program Control
Elementary Processing Center
Emergency Planning Canada
Emergency Preparedness Canada [of DND]
Engineering, Procurement and Construction
Environmental Protection Control
Environment Protection Council [Kuwait]
European Patent Convention
Evaporative Pattern Casting
Export Publicity Council [UK]
EPCA Economic Planning and Coordination Authority [Hawaii, US]
Energy Policy and Conservation Act [US]
EPCCS Emergency Positive Control Communications System
EPCM Electropulse Chemical Machining
Engineering, Procurement and Construction Management
EPCO Engine Parts Coordinating Office
EPD Earliest Possible Date
Earliest Practicable Date
Electric Power Database
Electric Power Distribution
Electronic Proximity Detector
Excess Profits Duty
Exchange Parameter Definition
Experimental Physics Department
Extraction and Processing Division [of TMS, US]
EPDAN Elastic-Plastic Deformation Analysis
EPDCC European Pressure Die Casting Committee [UK]
EPDIC European Powder Diffraction Conference [Netherlands]
EPDM Ethylene-Propylene-Dimonomer
Ethylene-Propylene Diene Monomer
EPDT Estimated Project Duration Time
EPE Electron-Plastic Effect
Energetic Particles Explorer Satellite [of NASA]
Engineering Progress Exposition
European Conference on Power Electronics and Applications
EPEA Electrical Power Engineers Association [UK]
EPEC Emerson Programmer, Evaluator and Controller
Environmental Protection and Education Club [Kenya]
EPEX Epoxy Resin Extender
EPF Ecole Polytechnique Federale [= Federal Polytechnic Institute, Switzerland]
European Packaging Federation [Netherlands]
EPFA Established Program Financing Act
EPFB Easy Processing Furnace Black
EPFM Elastic-Plastic Fracture Mechanics
E&P Forum (Oil Industry International) Exploration and Production Forum [UK]
EPFL Ecole Polytechnique Federale de Lausanne [= Federal Polytechnic Institute of Lausanne, Switzerland]
EPFT Elastic-Plastic Fracture Toughness
EPG Electrostatic Particle Guide
Emergency Procedure Guidelines
European Press Group
Experimental Proving Ground
EPGA Emergency Petroleum and Gas Administration [US]
EPH Electric Process Heating [IEEE IAS Committee]

EPhMRA European Pharmaceutical Marketing Research Association [UK]
EPI Earth Path Indicator
Effective Pair Interaction
Electric Power Institute [of CRI, Japan]
Electronic Position Indicator
Electron Probe Instrument
Elevation Position Indicator
Emergency Position Indicator
Emulsion Polymers Institute [US]
Engineering Physics Institute
Environmental Policy Institute [US]
Epichlorohydrin
EPIA Electric Power Industry Abstracts [of EEI, US]
EPIB Emergency Position-Indicating Beacon
EPIC Earth-Pointing Instrument Carrier
Electrical Properties Information Center [US]
Electronic Photochromic Integrating CRT (= Cathode Ray Tube)
Electronic Printer Image Construction
Electronic Products Information Center
Electronic Properties Information Center [of CINDAS, US]
Evaluator Programmer Integrated Circuit
Evidence Photographers International Council [US]
Exchange Price Information Computer [UK]
Extended Performance and Increased Capability
EPIDC East Pakistan Industrial Development Council
EPIE Educational Products Information Exchange (Institute) [US]
EPIRB Emergency Position-Indicating Radio Beacon
EPIRBS Emergency Position-Indicating Radio Beacon System
EPK Ethyl Propyl Ketone
EPL Encoder Programming Language
Environmental Protection Limit
Exciter Power Logic
Exclusive Prospecting License
EPLA Electronic Precedence List Agency
EPLAF European Planning Federation [UK]
EPLAN Elastic Plastic Analysis
EPLANS Engineering, Planning and Analysis System
EPLD EISA (= Extended Industry Standard Architecture) Programmable Logic Device
Electrically Programmable Logic Device
Enhanced Programmable Logic Device
Erasable Programmable Logic Device
EPM Earth-Probe-Mars [of NASA]
Economic Performance Monitoring
Engineering Procedure Memorandum
Environmental Project Manager
Equivalent per Million [also: epm]
Ethylene-Propylene Monomer
External Polarization Modulation
EPMA Electronic Parts Manufacturers Association
Electron Probe Microanalysis
EPMAU Expected Present Multi-Attribute Utility
EPN Epoxidized Phenol Novolac
Epoxy Phenol Novolac
Ethyl-P-Nitrophenyl Phenylphosphorothioate
External Priority Number
EPNdB Effective Perceived Noise Level in Decibels
Equivalent Perceived Noise Level in Decibels
EPNL Equivalent Perceived Noise Level
EPNS Electroplated Nickel Silver
EPO Erythropoietin
European Patent Office [also: EPA, Germany]
European Patent Organization
Examination Procedure Outline
EPOA East Coast Petroleum Operators Association [US]
EPOS Electronic Point of Sale
Engineered Plastics on Screen
EPP Electromagnetic Wave Propagation Panel [of AGARD]
End Plate Potential
European Producer Price
EPPO Emergency Planning Office of Ontario [Canada]
European and Mediterranean Plant Protection Organization [France]
EPPP Employee Profit Participation Plan
EPPS Ethylpiperazinepropanesulfonic Acid
N-(2-Hydroxyethyl)-piperazine-N'-3-Propanesulfonic Acid
EPR Effective Production Rate
Einstein-Podolsky-Rosen (Experiment)
Electrical Pressure Regulator
Electrochemical Potentiokinetic Reactivation
Electronic Planning and Research
Electron Paramagnetic Resonance
Engine Pressure Ratio
Error Pattern Register
Essential Performance Requirements
Ethylene-Propylene Rubber
Evaporator Pressure Regulator
EPRI Electric Power Research Institute [US]
EPRI RDS Electric Power Research Institute — Research and Development Information System
EPROM Electrically Programmable Read-Only Memory
Erasable Programmable Read-Only Memory
EPRTCS Emergency Power Ride Through Capability System
EPS Early Production System
Electric Power Subsystem
Electric Power Supply
Electric Power System
Electromagnetic Position Sensor
Emergency Power Supply
Encapsulated Postscript
Engineering Performance Standards
Environmental Protection Service [Canada]
Equilibrium Problem Solver
Equivalent Prior Sample
European Physical Society [Switzerland]
Even Parity Select
Expandable Polystyrene
Expanded Polystyrene
EPSCG European Parliamentary and Scientific Contact Group [France]
EPSCoR Experimental Program to Stimulate Competitive Research [of NSF, US]
EPSCS Enhanced Private Switched Communications Service
EPSL Emergency Power Switching Logic

EPSP Excitatory Postsynaptic Potential
EPSS Experimental Packet Switching Service [UK]
EPT Electromagnetic Propagation Tool
Electrostatic Printing Tube
Ethylene-Poly(propylene) Terpolymer
Ethylene-Propylene Terpolymer
EPTA European Power Tool Association [Germany]
Expanded Program of Technical Assistance [of UN]
EPTC Ethyl Dipropylthiocarbamate
Ethylpropylthiocarbamate
European Petroleum Technical Corporation
EPTE Existed Prior to Entry
EPU Economic Planning Unit [Malaysia]
Electrical Power Unit
Emergency Power Unit
European Payments Union [France]
European Picture Union [Sweden]
Executive Processing Unit
EPUL Ecole Polytechnique de l'Universite de Lausanne [= Polytechnic Institute of the University of Lausanne, Switzerland]
EPUT Events per Unit Time
EPZ Export Processing Zone
EPZA Export Processing Zone Administration [Taiwan]
EQ End-Quench
Enquiry
Equal(ization)
Equalizer
Equation
Equator
Equivalency
Equivalent
eq equivalent weight [Unit]
EQCC Entry Query Control Console
EQCO European Quality Control Organization
EQL Equal
EQMT Equipment
EQN Equation
EQPT Equipment [also: EQPT]
EQS Equivalent-to-Sheated Explosive
EQSP Equally Spaced
EQUALANT Equatorial Atlantic
EQUATE Electronic Quality Assurance Test Equipment
EQUIL Equilibrium
EQUIP Equipment
Equipment Usage Information Program
EQUIV Equivalent
Equl Equuleus [Astronomy]
ER Easy to Reach
Echo Ranging
Economic Region
Effectiveness Report
Effective Resistance
Elastic Recoil
Elastoresistance
Electrical Resistance
Electrolytic Rectifier
Electrolytic Reduction
Electrolytic Refining
Electronic Ram
Electroreflectance
Electrorheology
Endoplasmic Reticulum
(Office of) Energy Research [US]
Energy Research
Energy Resources
Engineering Route
Enhanced Radiation
Environmental Report
Environmental Resource [now: EAF, US]
Error
Error Rate
Error Recovery
External Reflection
External Resistance
E&R Education and Research
Engineering and Repair
Er Erbium
ERA Economic Regulatory Administration
Electric Railroaders Association
Electric Research Association [UK]
Electronic Reading Automation
Electronics Representatives Association [US]
Electron Ring Accelerator
Energy Reduction Analysis
Energy Reorganization Act [US]
Engineering Research Associate
Equal Rights Amendment [US]
European Regional Airlines [UK]
ERAM Extende-Range Antiarmor Munition
ERAP Error Recording and Analysis Procedure
ERAS Electronic Reconnaissance Access Set
ERASER Elevated Radiation-Seeking Rocket
ERATO Exploratory Research for Advanced Technology [of JRDC, Japan]
ERB Earth Radiation Budget
Educational Records Bureau [US]
Equipment Review Board
Experiment Review Board
ERBE Earth Radiation Budget Experiment
ERBM Extended Range Ballistic Missile
ERBSS Earth Radiation Budget Satellite System
ERC Economic Research Council [UK]
Education Relations Commission [Canada]
Electronics Research Center [of NASA]
Emergency Relocation Center
Energy Research Center
Engineering Research Center
Engineering Research Council
Equatorial Ring Current
Error Retry Count
ERCB Energy Resources Conservation Board [Canada]
ERCC Engine Requirement Coordinating Committee
En-Route Control Center
Error Checking and Correction
ERCEM European Regional Conference on Electron Microscopy
ERCOT Electric Reliability Council of Texas [US]
ERCR Electronic Retina Computing Reader
ERCS Emergency Rocket Command System
Emergency Rocket Communication System

ERCSS European Regional Communications Satellite System
ERCW Emergency Raw Cooling Water
ERD Elastic Recoil Detection
Electronic Research Directorate [US Air Force]
Emergency Recovery Display
Emergency Return Device
Energy Research and Development Inventory (Database) [of ORNL, US]
Equivalent Residual Dose
ERDA Economic Regional Development Agreements [Canada]
Electrical and Radio Development Association [Australia]
Electronics Research and Development Activity [of US Army]
Energy Research and Development Adminstration [formerly: USAEC]
ERDAF Energy Research and Development in Agriculture and Food
ERDAM Energy Research and Development Adminstration Manual
ERDC Eastern Region Development Corporation [Nigeria]
ERDE Electronics and Radar Development Establishment [India]
Explosive Research and Development Establishment [UK]
ERDF European Regional Development Fund [Belgium]
ERDL Electronics Research and Development Laboratory [US Army]
Engineering Research and Development Laboratory [US]
ERDS Earth Resource Data System
Environmental Recording Data Set
ERE Edison Responsive Environment
Energy Requirement for Energy
EREP Earth Resources Experimental Package
Environmental Recording, Editing and Printing
ERES Environmental Record Editing and Statistics
ERF Epoxy Resins Formulators Division [of SPI, US]
Error Function
Estuarine Research Foundation [US]
erf error function
ERFA European Radio Frequency Administration [of NATO]
ERFC Error Function Complement
Er(FOD)$_3$ Tris(heptafluorodimethyloctanedionato)erbium
ERFPI Extended Range Floating Point Interpretative System
ERG Electroretinogram
Electroretinography
Energy Research Group [Canada]
ERGS Electronic Route Guidance System
Enroute Guidance System
Er(HFC)$_3$ Tris[(heptafluoropropylhydroxymethylene)-camphorato],Erbium
ERI Earthquake Research Institute [of University of Tokyo, Japan]
Elm Research Institute [US]
Energy Research Institute [US]
Engineering Research Institute [of Tokyo University, Japan]
Eri Eridanus [Astronomy]
ERIC Educational Research Information Center
Educational Resources Information Center [of USOE]
Electronic Remote and Independent Control
Energy Rate Input Controller
ERICA Experiment on Rapidly Intensifying Cyclones over the Atlantic [Canada]
ERICZC Educational Reptiles in Captivity Zoological Compound [US]
Erid Eridanus [Astronomy]
ERIE Environmental Resistance Inherent in Equipment
ERIS Electrostatic Reflex Ion Source
Emergency Resources Identification Equipment
Engineering Resins Information System
Exoatmospheric Reentry Vehicle Interceptor System
ERISA Employee Retirement Income Security Act [US]
ERISTAR Earth Resources Information Storage, Transformation Analysis and Retrieval [of NASA]
ERJE Extended Remote Job Entry
ERL Echo Return Loss
Energy Research Laboratories [of CANMET, Canada]
Environmental Research Laboratory
Error Location
ESSA (= Environmental Science Services Administration) Research Laboratories [US]
ERM Earth Reentry Module
Earth Return Module
Elastic-Reservoir Molding
Electronic Recording Machine
ERMA Electrical Reproduction Method of Accounting
Electronic Recording Machine Accounting
Engineering Reprographic Management Association [now: ERS, US]
Environment Resources Management Association [Canada]
Expansion Rate Measuring Apparatus
ERMCO European Ready Mixed Concrete Organization [UK]
ERN Engineering Release Notice
ERNIE Electronic Random Number Indicator Equipment
ERO Emergency Repair Overseer
Energy Research Office
Engineering Release Order
European Regional Organization [of ICFTU, Belgium]
EROC (Study Group on) En-Route Obstacle Clearance Criteria
EROM Erasable Read-Only Memory
EROP Extensions and Restrictions of Operators
EROPA Eastern Regional Organization for Public Administration
EROPS Extended Range Operations (Aircraft)
EROS Earth Resources Observation Satellite
Eliminate Range 0 (= Zero) System
Experimental Reflector Orbit Shot
EROW Executive Right of Way
ERP Effective Radiated Power
Electronics Research Paper

Elevated Release Point
Emergency Recorder Plot
Emergency Response Program [US]
Environmental Research Paper
Environmental Research Project
Equivalent Reduction Potential
Error Recovery Procedure
European Recovery Program [of UN]

ERPDB Eastern Regional Production Development Board [Nigeria]

ERPLD Extended-Range Phase-Locked Demodulator

ERPTUAC European Recovery Program Trade Unions Advisory Program

ERR Elk River Reactor [US]
Error

ERRL Eastern Regional Research Laboratory [US]

ERS Earth Regeneration Society [US]
Earth Resources Satellite [of ESA, France]
Economic Research Service [US]
Educational Research Service [US]
Electric Resistant Strain
Emergency Reporting System
Employee Relocation Service
Engineering Reprographic Society [US]
Engineers Register Study
Environmental Research Satellite [of NASA]
Environmental Resource Studies
Ergonomics Research Society [UK]
European Radio Satellite
Experimental Research Society [US]
External Regulation System

ER&S Exploratory Research and Study

ERSA Electronic Research Supply Agency

ERSI Environmental Research Systems Institute

ERSO Electronics Research and Service Organization [Taiwan]

ERS-1 Earth Resources Satellite 1 (= One) [of ESA, France]
European Radio Satellite

ERSOS Earth Resources Survey Operational Satellite

ERSP Earth Resources Survey Program
Encapsulated Ring-Shell Projector
European Remote Sensing Program

ERSR Equipment Reliability Status Report

ERSS Earth Resources Satellite System
Earth Resources Survey Satellite

ERSU Energy Research Support Unit [of SERC, UK]

ERT Electrical Resistance Thermometer
Energy Resources Technology
Estimated Repair Time
European Roundtable [Belgium]
Expected Run Time

ERTA Economic Recovery Tax Act [US]

Er(TFC)$_3$ Tris[(trifluoromethylhydroxymethylene)camphorato],-Erbium

ERTOR Effective Radiational Temperature of the Ozone-Layer Region

ERTS Earth Resources Technology Satellite [US]

ERW Electric Resistance Welding

ERWP European Railway Waggon Pool [Belgium]

ERX Electronic Remote Switching

ES Earth Station
Echo Sounding
Echo Suppressor
Econometric Society [US]
Education Society [of IEEE, US]
Elastic Scattering
Electrical Schematic
Electrochemical Society [US]
Electromagnetic Storage
Electromagnetic Switch
Electromagnetic Switching
Electronic Spectroscopy
Electronic Standard
Electronic Switch(ing)
Electron Spectroscopy
Electron Synchrotron
Electrophoresis Society [US]
Electrostatics
Electrostatic Storage
Electrostriction
Element Signal
Emission Spectroscopy
Emission Spectrum
Enamel (Single) Silk (Wire)
Energy Sampling
Engineering Support
Environmental Safety [US]
Epigraphic Society [US]
Error Satisfaction
Eutectic Solidification
Experimental Station
Expert System
External Store
Extra Segment
Extruded Shape

Es Einsteinium

ESA Ecological Society of America [US]
Electric Spark Alloying
Electric Supply Authority
Electronically Scanned Array (Antenna)
Electronic Surge Arrester
Electrostatic Analyzer
Employment Standards Administration [US]
Endangered Species Act [US]
Engineers and Scientists of America [US]
Engine Service Association [US]
Entomological Society of America [US]
Equipment Service Association [US]
Ethernet Station Adapter
EURATOM Supply Agency [of EEC, Belgium]
European Space Agency [France]
Explosive Safe Area [of NASA]
Externally Specified Address

ESAC Electronic Systems Assistance Center

ESACT European Society for Animal Cell Technology [UK]

ESAEI Electric Supply Authority Engineers Institute

ESAEINZ Electric Supply Authority Engineers Institute of New Zealand

ESAIRA Electronically-Scanned Airborne Intercept Radar

ESA-IRS European Space Agency — Information Retrieval System

ESANET European Space Agency Network
ESANZ Economic Society of Australia and Nea Zealand
ESAO European Society for Artificial Organs [Switzerland]
ESAPS Experimental Strain Analysis Processing System
ESAR Electronically-Scanned Array Radar
Electronically-Steerable Array Radar
ESARCC Endangered Species Act Reauthorization Coordinating Committee [US]
ESARS Earth Surveillance and Rendezvous Simulator
Employment Service Automated Reporting System
ESAS Engineered Safeguards Actuation System
ESATC European Space Agency Technical Center [Netherlands]
ESB Economic Stabilization Board [PR China]
Electrical Standards Board [of USASI]
Electrical Stimulation of the Brain
Engineering Societies Building
Engineering Society of Baltimore [US]
Engineering Society of Buffalo [US
European Society for Biomaterials
European Society for Biomechanics [Netherlands]
Export Services Branch [of BOT, UK]
Spanish Peseta [Currency of Spain]
ESBFCOA Eastern States Blast Furnace and Coke Oven Association [US]
ESBP European Society for Biochemical Pharmacology
E-SBR Emulsion Styrene Butadiene Rubber
ESC Echo Suppressor Control
Edison Screw Cap
Electronic Spark Control
Electronic Systems Command [US Navy]
Electroslag Casting
Electrostatic Compatibility
Engineering and Scientific Computing
Engineering Service Circuit
Engineering Society of Cincinnati [US]
Engineers and Scientists of Cincinnati [US]
Entomological Society of Canada
Environmental Stress Cracking
Environment Sensitive Cracking
Equipment Serviceability Criteria
Escape (Character)
Escudo [Currency of Portugal]
Escutcheon
European Seismological Commission [Switzerland]
European Shippers Council [Netherlands]
European Space Conference
Export Supply Center [Canada]
Esc Escudo [Currency of Portugal]
ESCA Electron Spectroscopy for Chemical Analysis
Exhibition Service Contractors Association [US]
Extended Source Calibration Area
ESCAP Economic and Social Commission for Asia and the Pacific [of UN — Thailand]
ESCAPE Expansion Symbolic Compiling Assembly Program for Engineering
ESCAR Experimental Superconducting Accelerating Ring
ESCC Engineering Standards Coordinating Committee [UK]
External Stress Corrosion Cracking
ESCD Extended System Contents Directory
ESCES Experimental Satellite Communication Earth Station [India]
ESCI European Society for Clinical Investigation [UK]
ESCMT Engineering Societies Committee for Manpower Training [US]
ESCOE Engineering Societies Commission on Energy
ESCOM Electricity Supply Commission [South Africa]
ESCPB European Society of Comparative Physiology and Biochemistry [Belgium]
ESCR Environmental Stress-Cracking Resistance
ESCS Emergency Satellite Communications System
ESCSI Expanded Shale Clay and Slate Institute [US]
ESCWA Economic and Social Commission for Asia and the Pacific
ESCWS Essential Services Cooling Water System
ESD Echo Sounding Device
Electronic Semiconductor Device
Electronic Systems Division [US Air Force]
Electron-Stimulated Desorption
Electrostatic Discharge
Electrostatic Dissipation
Electrostatic Storage Deflection
Elongated Single Domain
Emergency Shutdown
Ending Sequence, Done
Engineering Society of Detroit [US]
Environmental Services Division
Estimated Standard Deviation
European System Design
Export Supply Directorate
Extension Shaft Disconnect
External Symbol Dictionary
ESDAC European Space Data Center [of ESRO, France]
ESDIAD Electron-Stimulated Desorption Ion Angular Distribution
ESDS Entry-Sequenced Data Set
Environmental Satellite Data System
ESE East-South-East
Electrical Support Equipment
Electronic System Evaluator
Environment, Safety and Economics
ESEAFE Environmental Safety and Economic Aspects of Fusion Energy [of IEA, France]
ESEC European Symposium on Engineering Ceramics
ESEM Electron Spin Envelope Modulation
Environmental Scanning Electron Microscope
ESF Electrostatic Focusing
Electrostrictive Force
Engineered Safety Feature
Engineering Safety Feature
Ethynylphenoxysulfone
European Science Foundation [France]
Export Success Fund
Extended Super Frame
ESFA Emergency Solid Fuel Administration [US]
ESFAS Engineered Safety Features Actuation System
ESFI Epitaxial Silicon-on-Insulator
ESG Electrically-Suspended Gyroscope
Electronic Sweep Generator
Electrostatic Generator
Electrostatic Gyroscope

	Engineers and Scientists Guild [US]
	(Old) English Standard Gauge
	Ethnobotany Specialists Group [US]
	Exchange Software Generator
ESH	Electric Strip Heater
	Equivalent Standard Hours
ESHAC	Electric Space Heating and Air Conditioning Committee [of IEEE IAS]
ESHPH	European Society for the History of Photography [Belgium]
ESHT	Electroslag Hot Topping
ESHU	Emergency Ship-Handling Unit
ESI	Ecole des Sciences de l'Information [= School of Information Sciences]
	Electricity Supply Industry
	Electron Spectroscopic Imaging
	Electrostatic Induction
	End of Segment Indicator
	Engineering and Scientific Interpreter
	Environmental Science Index [of EIC, US]
	Environmental Studies Institute [of IASB, US]
	Equivalent-Step Index
	Externally-Specified Index
ESIC	Environmental Science Information Center [of NOAA, US]
ESID	Electron-Stimulated Ion Desorption
ESIG	Eugenics Special Interest Group [US]
ESIP	Employee Savings Investment Plan
ESIS	Environmentally Sensitive Investment System
	European Shielding Information Service [of EURATOM]
	European Space Information System
ESISS	Experimental Submarine Integrated Sonar System
ESIT	Egyptian Society for Information Technology
ESJCP	Engineers and Scientists Joint Committee on Pensions
ESL	Electrical Services League
	Electron Beam Switched Latch
	Engineering Societies Library
	English as a Second Language
	Essential Service Line
	Expected Significance Level
	Evans Signal Laboratory [US Army]
ESLAB	European Space Research Laboratory [of ESRO, France]
ESLO	European Satellite Launching Organization
ESM	Elastomeric Shield Material
	Electronic Support Measures
	Engineers and Scientists of Milwaukee [US]
	Equivalent Standard Minute
	Error Satisfaction Method
ESMA	Electronic Sales and Marketing Association
	Engraved Stationery Manufacturers Association [US]
ESMC	Engineering Societies Monograph Committee
E-SMC	Epoxy-Matrix Sheet Molding Compound
ESME	Excited State Mass Energy
ESMR	Electrically Scanning Microwave Radiometer
ESMS	East Sullivan Monzonitic Stock
ESMST	European Society of Membrane Science and Technology [Italy]
ESN	English Speaking Nations
	European Society for Nematologists [Scotland]
ESNE	Engineering Societies of New England [US]
ESNZ	Entomological Society of New Zealand
ESO	Economic Stabilization Office
	Emergency Support Organization
	European Southern Observatory [Germany]
ESOC	European Space Operations Center [Germany]
ESOMAR	European Society for Opinion and Marketing Research [Denmark]
ESONE	European Standards of Nuclear Electronics
ESONEC	European Standards of Nuclear Electronics Committee [Switzerland]
ESOP	Employee Stock Ownership Plan
ESP	Edge-Supported Pulling
	Edit, Save and Plot Technology
	Electrical Submersible Pump
	Electronic Short Pathfinder
	Electron Spin Polarization
	Electrosensitive Programming
	Electrosonic Profiler
	Electrostatic Precipitator
	Escudo [Currency of Portugal]
	Especially
	Evoked Synaptic Potential
	Experimental Solids Proposal
	Extended Service Plan
	Externally Supported Processor
	Extrasensory Perception
	Extravehicular Support Pack [of NASA]
E&SP	Equipment and Spare Parts
ESPA	Exhaust Systems Professional Association [US]
ESPAR	Electronically Steerable Phased Array Radar
ESPB	European Student Press Bureau [Netherlands]
ESPI	Electronic Speckle-Pattern Interferometry
ESPL	Electronic Switching Programming Language
ESPOD	Electronic Systems Precision Orbit Determination
ESPOL	Executive System Problem-Oriented Language
ESPRIT	Estimation of Signal Parameters by Rotational Invariance Techniques
	European Strategic Program for Research (and Development) in Information Technology [also: Esprit]
ESPS	Engineering Societies Personnel Service
ESR	Early Site Review
	Early Storage Reserve
	Effective Signal Radiated
	Electronic Scanning Radar
	Electronic Send Receive
	Electron Spin Resonance
	Electroslag Refining
	Electroslag Remelting
	Equipment Supervisory Rack
	Equivalent Series Resistance
	Erythocyte Sedimentation Rate
	Exchangeable Sodium Ratio
	Experimental Superheat Reactor
	External Standard Ratio
E-SR	Emulsion Synthetic Rubber
ESRANGE	European Space Range [of ESRO, France]
ESRB	European Society for Radiation Biology [Belgium]

ESRC Economic and Social Research Council [UK]
ESRF Environmental Studies Revolving Fund
European Synchrotron Radiation Facility [France]
ESRI Engineering and Statistical Research Institute
ESRIN European Space Research Institute [of ESRO, France]
ESRO European Space Research Organization [of ESA, France]
ESRP Environmental Standard Review Plan
ESRR Early Site Review Report
ESRS Early Sites Research Society [US]
ESRT Electroslag Refining Technology
ESS Earle's Salt Solution
Earth System Sciences
Echo Suppression System
Electronic Sequence Switching
Electronic Switching System
Electroslag Surfacing
Electrostatic Spraying
Emplaced Scientific Station [of NASA]
Engineered Safety Systems
Environmental Science Services
Environmental Stress Screening
Environmental Survey Satellite
European Symposium for Stereology
Event Scheduling System
ES&S Engineering Services and Safety
ESSA Economists, Sociologists and Statisticians Association
Electronic Scanning and Stabilizing Antenna
Emergency Safeguards System Activation
Endangered Species Scientific Authority
Entomological Society of Southern Africa [South Africa]
Environmental and Social Systems Analyst
Environmental Science Services Administration [US]
Environmental Survey Satellite
European Single Service Association [Switzerland]
ESSDERC European Solid-State Device Research Conference
ESSF Edmonton Space Sciences Foundation [Canada]
ESSG Engineer Strategic Studies Group [US Army]
ESSP Earliest Scram Set Point
Engineers Society of St. Paul [US]
ESSSSA Egyptian Society of Solid-State Science and Applications
ESSU Electronic Selective Switching Unit
EST Earliest Start Time
Eastern Standard Time
Electrolytic Sewage Treatment
Electronic Spark Timing
Electrostatic Storage Tube
Elementary Scattering Theory
Emerging Sciences and Technologies
Engineers Society of Tulsa [US]
Enlistment Screening Test
Establish(ed)
Establishment
Estate
Estimate
Estimator
ESTA Ethernet Station Adapter
European Security Transport Association [Belgium]
ESTAB Established
ESTB Establish
ESTBT Establishment
ESTC Electrochemical Science and Technology Center [Canada]
European Space Tribology Center [UK]
ESTEC European Space Agency Technical Center
ESRO (= European Space Research Organization) Technical Center [France]
European Space Research Technology Center [of ESRO, France]
ESTI European Space Technology Institute
ESTN Estimation
ESTO Electronic Systems Technology Office [of USDOD]
ESTRAC European Satellite Tracking, Telemetry and Telecommand Network [of ESRO, France; also: ESTRACK]
ESTS Echo Suppressor Testing System
ESTU Electronic System Test Unit
ESTV Error Statistics by Tape Volume
EST WT Estimated Weight
ESU Electrostatic Unit [also: esu]
Emergency Service Unit
Empty Signalling Unit
English Speaking Union
ESV Emergy Support Vessel
Enamel (Single) Silk Varnish (Wire)
Essential Service Value
ESVM Electrostatic Voltmeter
ESW Electroslag Welding
Error Status Word
ESWL Equivalent Single Wheel Load
ESWM Engineering Society of Western Massachusetts [US]
ESWP Engineers Society of Western Pennsylvania [US]
ESWS Earth Satellite Weapon System
Emergency Service Water System
ET Earth Tide
Eastern Time
Eddy-Current Testing
Edge-Triggered
Educational Technician
Electrical Technician
Electrical Time
Electrolytic Tank
Electronics Technician
Electronic Transformer
Electron Transfer
Electron Tube
Electron Tunneling
Embryo Transfer
EMF (= Electromotive Force) — Temperature
End of Text
Endothermic
Energy Technologist
Energy Technology
Engaged Tone
Engage Test
Engineering Technician
Engineering Technologist

	Engineering Technology
	Engineering Test
	Engineering Thermoplastics
	Environmental Technologist
	Environmental Technology
	Ephemeris Time [Astronomy]
	Equal Taper
	Erector Transporter
	Estimation Theory
	Ethylenedithiotetrathiafulvalene
	Exchange Terminal
	Exothermic
	Explosives Technology
	External Tank
	Extraterrestrial
	Extruded Tube
E/T	Everhart-Thornley (Detector)
Et	Ethyl
ETA	Educational Television Association [UK]
	Electrothermal Analyzer
	Employment and Training Administration
	Estimated Time of Arrival
	European Tube Association
	European Tugowners Association [UK]
ETAA	Eastern Townships Agricultural Association
ETA-I	Electronics Technicians Association, International [US]
et al	(et alii; et aliae) — and others
ETAC	Environmental Technical Applications Center [US Air Force]
	Equilibrium Transfer Alkylating Cross-Link
	European Trade Association for Composite Materials [Switzerland]
ETAC-l	Equilibrium Transfer Alkylating Cross-Link
ETAD	Ecological and Toxicological Association of the Dyestuffs Manufacturing Industry [Switzerland]
ETADM	Embedded Trainer Advanced Development Model
ETANN	Electrically Trainable Artificial Neural Network
ETAP	Extended Technical Assistance Program [US]
ETAPC	European Technical Association for Protective Coatings [Belgium]
ETB	End of Transmission Block
	Estimated Time of Berthing
	Ethiopian Birr [Currency]
ETBE	Ethyl Tertiary Butyl Ether
ETC	Electrical Trade Council [US]
	Electronic Time Card
	Electrothermal (Integrated) Circuit
	English Translucent China
	Energy Technology Center [US]
	Environmental Technology Center [Canada]
	Estimated Time of Completion
	Ethylene Carbonate
	European Tool Committee [Germany]
	European Translations Center [Netherlands]
	Extended Text Compositor
etc	(et cetera) — and so forth
ETCE	Energy-Sources Technology Conference and Exhibition [of ASME, US]
ETCG	Elapsed-Time Code Generator
ETD	Embedded Temperature Detector
	Equivalent Transmission Density
	Estimated Time of Departure
	Exploratory Battery Technology Development and Testing
ETDB	Energy Technology Database [of CANMET, Canada]
ETDCFRL	European Training and Development Center for Farming and Rural Life [Belgium]
ETE	Electrothermal Excitation
	Estimated Time Enroute
	Estimated Time of Ejection
ETEC	Energy Technology Engineering Center [US]
ETECG	Electronics Test Equipment Coordination Group [US]
ETER	Epichlorohydrin-Ethylene Oxide-Allylglycidyl Ether Terpolymer
ETF	Engine Test Facility
ETFE	Ethylene Tetrafluoroethylene
ETG	Electronic Target Generator
	European Training Group [of NATO]
ETGTS	Electronic Text and Graphics Transfer System
ETH	Eidgenossische Technische Hochschule [= Federal Polytechnic Institute, Switzerland]
	Ether
	Ethyl
ETH ACET	Ethyl Acetate
ETHEL	European Tritium Handling Experimental Laboratory
ETHNOL	Ethnological
	Ethnology
ETHZ	Eidgenossische Technische Hochschule Zurich [= Federal Polytechnic Institute of Zurich, Switzerland]
ETI	Education Turnkey Institute [US]
	Electrical Tool Institute [now: PTI, US]
	Equipment and Tool Institute [US]
	Extraterrestrial Intelligence
ETIC	Environmental Technology Information Center
ETIM	Elapsed Time
ETIS	European Technical Information Service
ETIS-MARFO	ETIS in Machine-Readable Form
ETL	Educational Technology Language [of UWO, Canada]
	Effective Testing Loss
	Electrical Testing Laboratories [US]
	Electrotechnical Laboratories [Japan]
	Electrotechnical Laboratory
	Ending Tape Label
	Etching by Transmitted Light
ETLS	Extruded Tunnel Lining System
ETM	Electronic Test and Measurement
	Enhanced Timing Module
	Engineering Test Model
ETMF	Elapsed Time Multiprogramming Factor
ETMOD	Environmental Tritium Model
ETMQ	6-Ethoxy-2,2,4-Trimethyl-1,2-Dihydroquinoline
ETN	Electronic Tandem Network
	Equipment Table Nomenclature
ETO	Energy Technology Office
	Estimated Time Off
	Estimated Time Over Significant Point
	European Transportation Organization

EtO Ethylene Oxide [also: ETO]
E to E End to End
E-to-E Electronic-to-Electronic
ETOG European Technical Operations Group
ETOM Electron Trapping Erasable Optical Memory
ETOPS Extended Range Twin Operations
ETOS Extended Tape Operating System
ETOX EPROM (= Erasable Programmable Read-Only Memory) Tunnel Oxide (Device)
ETP Effluent Treatment Plant
Electrolytic Tough Pitch (Copper)
Electron Transfer Process
Engineering Thermoplastics
Equivalent Top Product
Estimated Time of Penetration
ETPAE Ethyl-Terminated Polyarylene Ether
ETPB Ethyltrioxaphosphabicyclooctane
ETPO European Trade Promotion Organization
ETPS Enterprise
ETQAP Education and Training in Quality Assurance Practices [of ASQC, US]
ETR Eastern Test Range [of NASA]
Electron Transfer Reaction
Engineering Test Reactor
Estimated Time of Restore
Expected Time of Response
Experimental Test Reactor
ETRAC Educational Television and Radio Association of Canada
ETR-CX Engineering Test Reactor Critical Assembly [US]
ETRS European Textile Research Symposium
ETRTO European Tire and Rim Technical Organization [Belgium]
ETS Educational Testing Service [US]
Electrodepositors Technical Society [UK]
Electronic Tandem Switching
Electronic Telegraph System
Electronic Translator System
Electronic Test Set
Engineering Test Satellite [Japan]
Enquiry Terminal System
Environmental Technical Specification
Environmental Technology Seminar [US]
Environmental Tobacco Smoke
ETSA Ethyltrimethylsilylacetate
ETSAL Electronic Terms for Space Age Language
ETSCO Engineering and Technical Societies Council [US]
ETSD Enhanced Thermionically Supported Discharge
et seq (et sequentes; et sequentia) — and the following
ETSQ Electrical Time, Superquick
ETSS Entry Time Sharing System
ETSU East Tennessee States University [US]
Energy Technology Support Unit [of DOE, UK]
ETT Electrothermal Thrusters
ETTA Eastern Townships Textile Association
1,2-Ethanediylidenetetrakisacetic Acid
ETU Electrical Trades Union
Enhanced Telephone Unit
N,N'-Ethylene Thiourea
ETV Educational Television [US]
Electric Transfer Vehicle
Elevating Transfer Vehicle
ETVM Electrostatic Transistorized Voltmeter
ETW European Trans-Sound Wind Tunnel
ETWTC European Tire and Wheel Technical Conference [now: ETRTO, Belgium]
ETX End of Text
ETYM Etymological
Etymology
EU Edinburgh University [UK]
Ehime University [Japan]
Electronic Unit
Emory University [US]
Endotoxin Unit
End User
Entropy Unit
Execution Unit
Expected Utility
Eu Euronorm [= European Standard]
Europium
EUA Electrical Utilities Application
EUC Equivalent Uranium Content
European Union of Coachbuilders [Belgium]
Extended Unit Cell
EUCAPA European Capsules Association
EUCARPIA European Association for Research on Plant Breeding [Netherlands]
EUCATEL European Committee of Associations of Telecommunication Industries
EUCEPA European Liaison Committee for Pulp and Paper
EUCHEMAP European Committee of Chemical Plant Manufacturers
EUCLID Experimental Use Computer London Integrated Display
EUCOMED European Confederation of Medical Suppliers Associations [UK]
EUF End User Facility
Equivalent Unavailability Factor
EUFMC Electric Utilities Fleet Managers Conference
$Eu(FOD)_3$ Tris(heptafluorodimethyloctanedionato)europium
EUFODA European Foodstuffs Distributors Association
EUG European Union of Geosciences [France]
$Eu(HFC)_3$ Tris[(heptofluoropropylhydroxymethylene)camphorato],-Europium
EUJS European Union of Jewish Students [Belgium]
EUM European Mediterranean Region [of ICAO]
EUMABOIS Comite Europeen des Constructeurs de Machines a Bois [= European Committee of Woodworking Machinery Manufacturers, France]
EUM-AFTN European Mediterranean Aeronautical Fixed Telecommunications Network
EUMAPRINT European Committee of Associations of Printing and Paper Converting Machinery [France]
EUMC Enameled Utensil Manufacturers Council
EUPC Electric Utility Planning Council
EUPEPTIC Evaluation of Unitary Programs for Effecting Plural Tasks in Index Construction
EUR Europe
European
European Region [of ICAO]
EURABANK European-American Bank
EURACA European Air Carrier Assembly

EurACS European Association of Classification Societies [Belgium]
EURADH European Adhesion Congress and Exhibition
EURALARM Association of European Manufacturers of Fire and Intruder Alarm Systems [Germany]
EURAM European Research on Advanced Materials [of EEC]
EURANP European Air Navigation Plan [of ICAO; also: EUR ANP]
EURAS European Anodizers Association [Switzerland]
EURASAP European Association for the Science of Air Pollution Control [UK]
EURASIP European Association for Signal Processing [Switzerland]
EURATOM European Atomic Energy Community [also: Euratom]
EUREAU Union des Associations des Distributeurs d'Eau de Pays Membres des Communautes Europeennes [= Union of the Water Supply Associations from Countries of the European Community, Belgium]
EURECA European Research Coordination Agency [also: Eureca]
European Retrievable Carrier
EUREL Association Europeenne des Reserves Naturelles Libres [= European Association for Free Nature Reserves, Belgium]
Convention of National Societies of Electrical Engineers of Western Europe [Switzerland]
EUREM European Conference on Electron Microscopy
EUREMAIL Conference Permanente de l'Industrie Europeenne de Produits Emailles [= Permanent Conference of the European Enamelled Products Industry]
EUREX Enriched Uranium Extraction
EURFCB European Frequency Coordinating Body [of ICAO]
EURIM European Conference on Research into Information Management
EURIPA European Information Providers Association [UK]
EURIS European Information Service
EURISOTOPE EURATOM Radioisotope Information Bureau
EUROAVIA Association of European Aeronautical and Astronautical Students [Netherlands]
EUROBASE European Database
EUROBAT Association of European Battery Manufacturers [Switzerland]
EUROBIT European Association of Manufacturers of Business Machines and Data Processing Equipment [Germany]
EUROBITUME European Bitumen Association [Belgium]
EUROBUILD European Organization for the Production of New Techniques and Methods in Building [France]
EUROCAE European Organization of Civil Aviation Electronics [France]
EUROCEAN European Ocean Association [Monaco]
EUROCHIMIE Societe Europeenne pour le Traitement Chimique des Combustibles Irradies [= European Society for the Chemical Treatment of Radiant Fuels]
EUROCOMP European Computing Congress
EUROCON European Convention [of IEEE, US]
EUROCONTROL European Organization for the Safety of Air Navigation
EUROCOPI European Computer Program Information Center [of EURATOM/JRC]
EUROCORD Federation des Industries de Ficellerie et Corderie de l'Europe Occidentale [= Federation of Western European Rope and Twine Industries, France]
EUROCORR (International) European Corrosion Congress
EURODICAUTOM European Automatic Dictionary
EURODOC European Joint Documentation Service [of ESRO and EUROSPACE]
EUROFAR European Future Advanced Rotorcraft
EUROFEU European Committee of the Manufacturers of Fire Protection and Safety Equipment and Fire Fighting Vehicles [Germany]
EUROFLAG European Future Large Aircraft Group
EUROFORGE European Committee of Forging and Stamping Industries [France]
EUROGYPSUM Working Community of the European Gypsum Industry [France]
EURO-HKG Europaische Gesellschaft zur Auswertung von Erfahrungen bei Planung, Bau und Betrieb von Hochtemperaturreaktoren [= European High-Temperature Nuclear Power Stations Society, Germany]
EUROLAIT Union Europeenne du Commerce des Produits Laitiers et Derives [= European Union of Importers, Exporters and Dealers in Dairy Products, UK]
EUROMAP European Committee of Machinery Manufacturers for the Plastics and Rubber Industries
EUROMAR European Marine Research and Technology Project [of EUREKA]
EUROMECH European Mechanics Colloquium [Germany]
EuroMech European Mechanics Committee [also: EMC, Germany]
EUROMICRO European Association for Microprocessing and Microprogramming [Netherlands]
EUROMIL European Organization of Military Associations [Germany]
EUROMOT European Committee of Associations of Manufacturers of Internal Combustion Engines [Netherlands]
EURONEM European Association of Netting Manufacturers [France]
EURONET European Public Data Network
EUROPEC European Offshore Petroleum Conference and Exhibition
EUROPECHE Association des Organisations Nationales d'Entreprises de Pêche de la Communaute Economique Europeenne [= Association of National Organizations of Fishing Enterprises in the European Economic Community, Belgium]
EUROPEX European Information Center for Explosion Protection [also: EuropEx]
EUROPHOT Association Europeenne des Photographes Professionels [= European Association of Professional Photographers]
EUROPILOTE European Organization of Airline Pilots Associations
EUROPLANT European Plantmakers Committee [Belgium]

EUROPOL Intra-European Air Transport Policy [of ECAC, France]
EUROPUMP European Committee of Pump Manufacturers [UK]
EUROPUR European Association of Flexible Foam Block Manufacturers [Belgium]
EURORAD European Association of Manufacturers of Radiators [Switzerland]
EUROSAC Federation Europeenne des Fabricants de Sacs en Papier a Grande Contenance [= European Federation of Multiwall Paper Sack Manufacturers, France]
EUROSPACE European Industrial Space Research Group
EUROSTAR European Communications Satellite [UK and France]
EUROSTAT Statistical Office of the European Communities [Luxembourg]
EUROTALC Scientific Association of European Talc Industry [Belgium]
EUROTOX Comite Europeen Permanent de Recherches sur la Protection des Populations contre le Risques de Toxicite [= Permanent European Research Committee for the Protection of the Population against Toxic Substances]
EUROTRAC European Project on the Transport of Atmospheric Contaminants [of EUREKA]
EUROTRIB (International) European Congress on Tribology
EUROVISION Union Europeenne de Radiodiffusion [= European Broadcasting Union, France]
EURPISO European Union of Public Relations International Service Organization [Italy]
EUR/RAN European Regional Air Navigation Meeting [of ICAO]
EUR/TFG European Traffic Forecasting Group [of EUROCONTROL]
EURYDICE Education Information Network in the European Community [Belgium]
EUSA Electrical Utilities Safety Organization [Canada]
EUSEC Conference of Representatives from European and United States (of America) Engineering Societies
EUSIDIC European Association of Scientific Information Dissemination Centers [UK]
EUSIREF European Association of Science Information Referral Centers
European Scientific Information Retrieval Working Group [of EUSIDIC]
EUSJA European Union of Science Journalists Associations [Italy]
EUSSG European Union for the Scientific Study of Glass [Belgium]
EUT Einthoven University of Technology [Netherlands]
Equipment under Test
EUTELSAT European Telecommunications Satellite Organization [France]
$Eu(TFC)_3$ Tris[(trifluoromethylhydroxymethylene)camphorato], Europium
$Eu(THD)_3$ Tris(tetramethylheptanedionato)europium
EUTP Enhanced, Unshielded Twisted Pair (Cable)
EUTRAPLAST Committee of Plastic Converter Associations of Western Europe [Belgium]
EUV Extreme Ultraviolet
EUVE Extreme Ultraviolet Explorer
EUVEPRO European Vegetable Protein Federation [Belgium]
EUVITA Extreme Ultraviolet Telescope Array
EV Efficient Vulcanization
Efficient Vulcanizing System
Electrovalency
Electroviscosity
Equalizer Valve
Event
Expected Value
eV Electron-Volt
EVA Electronic Velocity Analyzer
Employee Volunteer Action
Error Volume Analysis
Ethylene Vinyl Acetate (Copolymer)
Extravehicular Activity
EVAP Evaporation
Evaporator
EVAPN Evaporation
EVATMI European Vinyl Asbestos Tile Manufacturers Institute
EVC Electric Vehicle Council
EVCA European Venture Capital Association [Belgium]
EVCP Event Control Block
EVCS Extravehicular Communication System [of NASA]
EVD Explosive Vapor Detector
EVDE External Visual Display Equipment [of NASA]
EVDS Electronic Visual Display Subsystem
EVE (Data) Entry and Validation Equipment
Ethyl Vinyl Ether
European Videoconferencing Experimentation
Extreme Value Engineering
EVES Environment for Verifying and Evaluating Software
EVEST Experimental Valleditos Superheat Reactor [US]
EVFM Ex-Vessel Flux Monitor
EVHM Ex-Vessel Handling Machine
EVIL Extensible Video Interactive Language
EVIST Ethic and Values in Science and Technology Program
EVK Ethyl Vinyl Ketone
Evaluation Kit
EVLN Evolution
EVM Electronic Voltmeter
Ethylene Vinylacetate Copolymer
Extended Virtual Machine
EVOH Ethylene Vinyl Alcohol (Copolymer)
EVOM Electronic Voltohmmeter
EVOP Evolutionary Operation
EVPI Expected Value of Perfect Information
EVR Electronic Video Recorder
EVS Electro-Optical Viewing System
Ethics and Values Studies
EVT Equiviscous Temperature
EVTCM Expected Value Terminal Capacity Matrix
EVTM Ex-Vessel Transfer Machine
EVX Electronic Voice Exchange
EW Early Warning
Elastic Wave
Electronic Warfare
Electroslag Welding
Electrowinning

	Entry Week
	Erftwerk (Coating Process)
	Expansion Wave
E/W	Electrowinning
EWA	European Welding Association [Netherlands]
EWASER	Electromagnetic Wave Amplification by Simulated Emission of Radiation [also: Ewaser]
EWC	Electric Water Cooler
EWCAS	Early Warning and Control Aircraft System
EWCS	European Wideband Communications System
EWEA	European Wind Energy Association [UK]
EWES	Engineering Waterways Experiment Station [US Army]
EWF	Electrical Wholesalers Federation
	Equivalent-Weight Factor
EWG	Executive Working Group
EWI	Edison Welding Institute [US]
	Executive Women International [US]
EWICST	European Workshop of Industrial Computer Systems — Technical Committee
EWL	Exchange Work List
EWMA	Exponentially Weighted Moving Average
EWO	Engineering Work Order
EWP	Exploding Wire Phenomena
EWPCA	European Water Pollution Control Association [Germany]
EWR	Early Warning Radar
	Electromagnetic Wave Resistivity
EWRC	European Weed Research Council
EWRS	European Weed Research Society [Germany]
EWRT	Electrical Women's Round Table [US]
EWS	Early Warning System
	Electronic Workstation
	Emergency Weather Station
EWSA	European Wheat Starch Manufacturers Association [Germany]
EWSM	Electronic Warfare Support Measures
EWST	Elevated Water Storage Tank
EWTAT	Early Warning Threat Analysis Display
EWTMI	European Wideband Transmission Media Improvement
EWTMIP	European Wideband Transmission Media Improvement Program
EWTR	Electronic Warfare Test Range [US Air Force]
EWWS	ESSA (= Environmental Science Services Administration) Weather Wire Service [US]
EX	Examination
	Example
	Excess
	Exchange(r)
	Execution
	Exercise
	Exponent
	Exterior
	External
EXACT	(International) Exchange of Authenticated Component Performance Test Data
EXAFS	Extended X-Ray Absorption Fine Structure (Spectroscopy)
EXAM	Examination [also: EXAMN]
EXAMR	Examiner
EXC	Excavation
	Excellent
	Exception
	Excitation
	Exciter
ExCA	Exchangeable Card Architecture**EXCELS**
	Expanded Communications Electronics System [of USDOD]
EXCESS	Extensible Expert System Shell (Technology)
EXCH	Exchange
EXCIMER	Excited Dimer [also: Excimer]
EXCIPLEX	Excited State Complex
EXCL	Exclude
	Exclusion
	Exclusive
EXCLASS	Expert Job Classification Assistant
EXCO	Executive Committee
	Exfoliation Corrosion
EXCP	Execute Channel Program
EXCTR	Exciter
EXCVTG	Excavating
EXD	Exchange Degeneracy
EXDAMS	Extended Debugging and Monitoring System
EXDC	External Data Controller
ex div	without dividend
EXEC	Execute (Statement)
	Execution
	Executive
	Executive Statement
	Executive System
	Executor
ExecMBA	Executive Master of Business Administration
ExecMPA	Executive Master of Public Administration
ExecMS	Executive Master of Science
ExecMSE	Executive Master of Science in Engineering
EXELFS	Extended Energy Loss Fine Structure
EXES	Electron-Induced X-Ray Emission Spectroscopy
EXF	Ex Factory
	External Function
EXH	Exhaust
EX-HY	Extra-Heavy
EXIM	Export-Import Bank [US]
ex int	without interest
EXIST	Existence
EXLST	Exit List
EXMETNET	Experimental Meteorological Sounding Rocket Research Network
EXNOR	Exclusive nor [also: EX-NOR]
EX O	Executive Officer [also: Ex O]
EXOR	Exclusive Or [also: EX-OR]
	Executor
EXP	Expansion
	Expense
	Experiment
	Explosion
	Exponent(ial)
	Export
	Express
	Expression
	Expulsion
exp	exponential function

EXPAC Explosion Prediction and Analysis Code
EXPEND Expenditures
EXPIO Expander Input/Output
EXPL Explosives
EXPLOR Exploration
EXPO Exposition
EXPOS Exposure
EXPS Expenses
EXPT Experiment
EXPTL Experimental
EXR Execute and Repeat
EXSA Exhibition Association of South Africa
EXSC Executive Standards Council [also: ExSC]
exsec exterior secant
EXSTA Experimental Station
EXT Extended
Extension
Exterior
External
Extinction
Extinguisher
Extract
EXTD Extended
EXTERRA Extraterrestrial Research Agency [US Army]
EXTG Exterminating
EXTN Extraction
EXTR Extrusion
EXTRADOP Extended Range DOVAP (= Doppler Velocity and Position)
EXTRN External Reference
EXW Explosion Welding
EY Entry Year
EZ Electrical Zero
EZI European Zinc Institute [Netherlands]

F

F	Facsimile
	Fahrenheit
	Fairing
	Farad
	Faraday
	February
	Feed Rate
	Fellow
	Filament
	File
	Filter
	Final
	Fine
	Fire
	Flag
	Flame
	Flat
	Fluorescence
	Fluorine
	Fluorochrome
	Focal Length
	Fog
	Force
	Formula
	Forward
	Fracture
	Frame
	Fraunhofer
	Free Energy
	French
	Frequency
	Friable
	Friction
	Friday
	Fuel
	Function
	Fuse
	Phenylalanine
	Phenylalanyl
f	failure
	fair
	farthing
	fathom
	femto- [SI Prefix]
	fetch
	filament
	final
	fine
	fire
	fixed
	flat
	florin (or guilder) [Currency of the Netherlands]
	fluid
	focal length
	focal ratio
	fog
	folio
	following
	foot
	fordable
	formed
	formula
	forward
	fragile
	fragmentation
	freezing
	frequency
	friction
	fuel
	fugacity
	full
	function
	furlong [Unit]
	fusion
ΔF	Degrees Fahrenheit
f	Florin (or Guilder) [Currency of the Netherlands]
FA	Factory Automation
	False Acceptance
	Fast Algorithm
	Fast Axis
	Fatty Acid
	Ferro-Alloy
	Fibonacci Association [US]
	Field-Accelerating (Relay)
	Field Address
	Field Availability
	Fire Alarm
	Fluorescent Antibody
	Forced-Air (Cooling)
	Formic Acid
	Fourier Analysis
	Fracture Analysis
	Free Air
	Fuel Assembly
	Full Abstraction
	Full Add(er)
	Fully-Automatic
	Furfuryl Alcohol
	Furnace Annealing
	Fuse Alarm
F/A	Fuel/Air Ratio
	Fuel Assembly
fA	femtoampere
fa	fayalite [Mineralogy]
FAA	Federacion Agraria Argentina [= Agrarian Federation of Argentina]
	Federal Aviation Administration [US]
	Free of All Average
FAAAS	Fellow of the American Academy of Arts and Sciences
	Fellow of the American Association for the Advancement of Science
FAAB	Frequency Allocation Advisory Board

FAACIA	Fellow of the American Association of Clinical Immunology and Allergy
FAAD	Forward Area Air Defense
FAAD-LOS	Forward Area Air Defense — Line-of-Sight
FAAD-LOS(H)	Forward Area Air Defense — Line-of-Sight Heavy
FAAOM	Fellow of the American Academy of Medicine
FAAP	Federal Aid Airport Program [US]
	Fundacao Armando Alvares Penteado [= Armando Alvares Penteado Foundation, Brazil]
FAAR	Forward Area Alerting Radar
FAAS	Fellow of the American Academy of Arts and Sciences
	Flame Atomic Absorption Spectroscopy
FAAS/GFAAS	Flame Atomic Absorption Spectroscopy/Graphite Furnace Atomic Absorption Spectroscopy
FAATC	**Federal Aviation Administration Technical Center [US]**
FAB	Fabrication
	Fast Atom Bombardment
Fab	Fab Fragment Specific [Immunochemistry]
F(ab)	Fragment of Antibody [Immunochemistry]
FABG	Fabricating
FABI	Federation Royale des Associations Belges d'Ingenieurs [= Royal Federation of Belgian Associations of Engineers]
FAB ISO	Fabrication Isometric (Drawing)
FABMDS	Field Army Ballistic Missile Defense System [US Army]
FABMS	Fast Atom Bombardment Mass Spectroscopy
FABS	Formulated Abstracting Service
FABU	Fuel Additive Blender Unit
FAC	Facsimile
	Fast Affinity Chromatography
	Federal Airports Corporation [Australia]
	Field Accelerator
	File Access Channel
	Floating-Point Accumulator
	Forward Air Control [US Air Force]
	Forward Air Controller
	Frequency Allocation Committee
FACA	Federation Algerienne de la Cooperation Agricole [= Algerian Federation for Agricultural Cooperation]
	Fellow of the Association of Certified Accountants
FACC	Federation of African Chambers of Commerce [Ethiopia]
FACCA	Fellow of the Association of Certified and Corporate Accountants
FACCP	Fellow of the American College of Clinical Pharmacology and Chemotherapy
FACE	(International) Federation of Associations of Computer Users in Engineering, Architecture and Related Fields [Australia]
	Federation of Associations on the Canadian Environment
	Field-Alterable Control Element
	Field Artillery Computer Equipment
FACD	Foreign Area Customer Dialling
FACES	FORTRAN Automated Code Evaluation System
FACET	Facetious
	Fluid Amplifier Control Engine Test
FACISCOM	Finance and Controller Information System Command
FACOGAZ	Union of European Manufacturers of Gas Meters [Germany]
FACP	Food and Agriculture Council of Pakistan
FACR	Fellow of the American College of Radiology
FACS	Facility Assignment Control System
	Facsimile
	Federation of American Controlled Shipping [US]
	Fine Attitude Control System
	Floating-Decimal Abstract Coding System
	Fluorescence Activated Cell Sorter
	Frequency Allocation Coordinating Subcommittee
	Fully Automatic Compiling System
FACSC	Frequency Allocation Coordinating Subcommittee, Canada
FACSFAC	Fleet Area Control Surveillance Facility
FACSI	Fast Access Coded Small Images
FACSIM	Facsimile
FACSS	Federation of Analytical Chemistry and Spectroscopy Societies [US]
FACT	Facility for the Analysis of Chemical Thermodynamics [Canada]
	Fast Access Current Text
	Federation of Automated Coding Technologies [US]
	Flexible Automatic Circuit Tester
	Flight Acceptance Composite Test
	Ford Anodized-Aluminum Corrosion Test
	Foundation for Advanced Computer Technology [US]
	Fully Automatic Cataloguing Technique
	Fully Automatic Compiler Translator
	Fully Automatic Compiling Technique
FACTS	Facsimile Transmission System
	Financing Alternative Computer Terminal System
FACTY	Factory
FAD	Flavin Adenine Dinucleotide
	Floating Add
	Floating And
	Fuel Advisory Departure
FADA	Federation of Automobile Dealers Associations
FADAC	Field Artillery Digital Automatic Computer [US Army]
FADEC	Full Authority Digital Electronic Control
FADES	Fuselage Analysis and Design Synthesis
FADIC	Field Artillery Digital Computer
FADP	Finnish Association for Data Processing
FADS	Filtered Attitude Determination System
	Force Administration Data System
	FORTRAN Automatic Debugging System
FADU	File Access Data Unit
FAE	Field Application Engineer
	Final Approach Equipment
	Fuel/Air Explosives
FAES	Federated American Engineering Societies
FAETUA	Fleet Airborne Electronics Unit, Atlantic
FAETUP	Fleet Airborne Electronics Unit, Pacific
FAF	Final Approach Fix
	Financial Accounting Federation [US]
	Fly Away Field
FAFI	Federation of the Austrian Food Industry

FAFPS Five-Axis Fiber Placement System
FAGC Fast Automatic Gain Control
FAGS Federation of Astronomical and Geophysical Data Analysis Services [France]
Federation of Astronomical and Geophysical Services [of IAPSO, US]
Fellow of the American Geographical Society
FAGT Federation of Agricultural Group Traders [UK]
FAHQMT Fully-Automatic High-Quality Machine Translation
FAHR Fahrenheit
FAI Fail As Is
Federation Aeronautique Internationale [= International Aeronautics Federation, France]
Fertilizer Association of India
Fresh Air Intake
FAIA Fellow of the American Institute of Architects
FAIC Fellow of the Agricultural Institute of Canada
Fellow of the Architectural Institute of Canada
FAICE Fellow of the Institute of Civil Engineers
FAIEE Fellow of the American Institute of Electrical Engineers [now: FIEEE]
FAIMME Fellow of the American Institute of Mining and Metallurgical Engineers
FAIR Fairing
Fast Access Information Retrieval
FAIRS Federal Aviation Information Retrieval System [US]
Food and Agricultural Information Storage and Retrieval System [of FAO, UN]
Fully Automatic Information Retrieval System
FAK File Access Key
Freight All Kinds
FAL File Access Listener
Frequency Allocation List
FALCON Fuel-Air Line Charge Ordnance Neutralizer
FALT FADAC (= Field Artillery Digital Automatic Computer) Automatic Logic Tester
FALTRAN FORTRAN-to-ALGOL Translator
FAM Familiar
Family
Family of Frequencies
Fast Access Memory
Fast Auxiliary Memory
Fermentation Analysis Module
Fibrous Aerosol Monitor
File Access Manager
Final Address Message
Fire Apparatus Manufacturers Association [US]
FAME Fatty Acid Methyl Ester
Financial Assistance for Mineral Exploration
Florida Association of Marine Explorers [US]
FAMECE Family of Military Engineer Construction Equipment
FAMEM Federation of Associations of Mining Equipment Manufacturers
FAMHM Federation of Associations of Materials Handling Manufacturers
FAMIS Factory Management Information System [UK]
FAMOS Fast Multitask Operating System
Fleet Application of Meteorological Observations for Satellites
Floating-Gate Avalanche-Injection Metal-Oxide Semiconductor
FAMOSS Fiberscopic Apparatus for Measurement of Surface Strain
FAMP Fire Alarm Monitoring Panel
FAMS Forecasting and Modelling System
FAMSO Methyl Methylsulfinylmethyl Sulfide
FAMT Fully Automatic Machine Translation
FAMU Fuel Additive Mixture Unit
FANS Future Air Navigation Systems Committee [of ICAO]
FANTAC Fighter Analysis Tactical Air Combat
FAO Finish All Over
Food and Agricultural Organization [of UN]
FAO/ECE Food and Agricultural Organization/Economic Commission for Europe [of UN]
FAOIP French Association of On-Line Information Providers
FAOMA Fellow of the American Occupational Medicine Association
FAP Facility Analysis Plan
Failure Analysis Program
Fault Analysis Process
Field Application Panel
Final Approach Point
Floating-Point Arithmetic Package
Fluoroapatite
FORTRAN Assembly Program
Frequency Allocation Panel
Furylacryloylphenylalanine
FAPA Future Aviation Professionals of America [US]
FAPC Fatty Acids Producer's Council [US]
FAPGG Furylacryloylphenylalanylglycylglycine
FAPIG First Atomic Power Industry Group [Japan]
FAPP Fractured-Area-Projection Plot
FAPRA Federation of African Public Relations Associations [Ghana]
FAPS Fellow of the American Physical Society
FAPUS Frequency Allocation Panel, United States
FAPUS-MCEB Frequency Allocation Panel, United States — Military Communications Electronics Board
FAQ Fair Average Quality
FAQS Fast Queuing System
FAR Failure Analysis Report
Federal Acquisition Regulation [US]
Federal Airworthiness Regulations [US]
Federal Aviation Regulations [US]
File Address Register
Flight Aptitude Rating
Foundation of Applied Research [US]
Fuel-Air Ratio [also: F/AR]
FARADA Failure Rate Data
FARC Field Artillery Reserve Corporation
FARE Federal Acquisition Regulation [US]
FAREGAZ Union of European Manufacturers of Gas Pressure Controllers [Germany]
FARET Fast Reactor Experiment Test
Fast Reactor Test Assembly [of Argonne National Laboratory, US]
FARR Forward Area Refuelling and Rearming
FARS Fatal Accident Reporting System
FAS Faculty of Administrative Studies
Fastener

	Federation des Architectes Suisses [= Swiss Federation of Architects]
	Federation of American Scientists
	Ferrous Aluminum Sulfate
	Field Advisory Service
	File Access Subsystem
	Final Assembly Schedule
	First and Seconds
	Floating-Point Arithmetic System
	Florida Academy of Sciences [US]
	Foreign Agricultural Service [US]
	Frame Alignment Signal
	Free Alongside Ship
	Frequency Allocations Subcommittee
	Frequency Analysis System
	Frequency Assignment Subcommittee
	Fuel Availability System [of BOM, US]
FASA	Federation of ASEAN (= Association of Southeast Asian Nations) Shipowners Associations [Malaysia]
FASAF	Filipinas Americas Science and Art Foundation [US]
FASB	Financial Accounting Standards Board [US]
FASCE	Fellow of the American Society of Civil Engineers
FASE	Federation of Acoustical Societies of Europe
	Force at Specified Elongation
	Fundamentally Analyzable Simplified English
FASEB	Federation of American Societies for Experimental Biology [US]
FASII	Federation of Associations of Small Industries in India
FASIS	Fully Automatic Syntactically-Based Indexing System
FASM	Fellow of the American Society for Metals
FASME	Fellow of the American Society of Mechanical Engineers
FASP	Fault-Tolerant Array Signal Processor
FASRA	Foundation to Assist Scientific Research in Africa [Belgium]
FASS	Federation of Associations of Specialists and Sub-Contractors [UK]
	Ford Aerospace Satellite Services
	Forward Acquisition Sensor
FASSET	Functional Advanced Satellite-Communication System for Evaluation and Test [of DND, Canada]
FAST	Facility for Automatic Sorting and Testing
	Fairchild Advanced Schottky Technology
	Fast Access Storage Technology
	Fast Automatic Shuttle Transfer
	Federazione delle Associaziones Scientifiche e Tecniche [= Federation of Scientific and Technical Associations, Italy]
	Field-Data Applications, Systems and Techniques
	Flexible Algebraic Scientific Translator
	Fluorescent Antibody Staining Technique
	Formal Auto-Indexing of Scientific Texts
	Formula and Statement Translator
	Forward Air Strike
	Fuel Assembly Stability Test
FASTAC	Furnace Aerosol Sampling Technique with Autocalibration
FASTAR	Frequency Angle Scanning, Tracking and Ranging
FASTI	Fast Access to Systems Technical Information
FASTOP	Flutter and Strength Optimization Program
FASWC	Fleet Antisubmarine Warfare Command
FAT	Factory Acceptance Test
	File Allocation Table
	Final Approach Track
	Fixed Analyzer Transmission
	Fluorescent Antibody Test
	Foreign Area Toll
	Foreign Area Translation
	Formula Assembler Translator
FATAL	FADAC Automatic Test Analysis Language
FATAR	Fast Analysis Tape and Recovery
FATCAT	Film and Television Correlation Assessment Technique
FATDL	Frequency and Time-Division Data Link
FATE	Fusing and Arming Test and Evaluation
	Fusing and Arming Test Experiment
fath	fathom
FATIMA	Fatigue Indicating Meter Attachment
FATIPEC	Federation d'Associations de Techniciens des Industries des Peintures, Vernis, Emaux et Encres d'Imprimerie de l'Europe Continentale [= Federation of the Associations of Technicians of the Paint, Varnish, Enamel and Printing Ink Industries of Continental Europe, France]
FATIS	Food and Agriculture Technical Information Service [of OECD, France]
FATR	Fixed Autotransformer
FATT	Fracture Appearance Transition Temperature
	Fracture-Area Transition Temperature
FATUREC	Federation of Air Transport User Representatives in the European Community
FAU	Florida Atlantic University [US]
	Frequency Allocation and Uses
FAUE	Friedrich-Alexander University at Erlangen [Germany]
FAUL	Five Associated University Libraries (Cooperative) [US]
FAUSTUS	Frame-Activated Unified Story Understanding System
FAV	Fast-Acting Valve
FAW	Fiber Areal Weight
	Frame Alignment Word
FAWS	Flight Advisory Weather Service
FAX	Facsimile
FAXCOM	Facsimile Communications
FB	Feedback
	F-Format, Blocked Data Set
	File Block
	Film Badge
	Fire Brigade
	Fixed Block
	Flame Black
	Flat Bottom
	Fluidized Bed
	Fluorobenzene
	Forest Biology
	Freight Bill
	Furnace Brazing
	Fuse Block

Fuse Box
F&B Fire and Bilge
FBA Farm Building Association [UK]
Fellow of the British Academy
Fiber Box Association [US]
Fixed-Block Architecture
Freshwater Biological Association [UK]
Furnace Bottom Ash
FBB Fluidized-Bed Boiler
Fusion Breeder Blanket
FBBM Federation of Building Block Manufacturers [UK]
FBC Feedback Carburetor
Feedback Control
Fixed Boundary Conditions
Fluidized-Bed Combustion
Fully-Buffered Channel
FBCAEI Federation of Buildeers, Contractors and Allied Employers of Ireland
FBCN Federation of British Columbia Naturalists [Canada]
FBCR Fluidized-Bed Combustion Residue
Fluidized-Bed Control Rod
FBD Full Business Day
Functional Block Diagram
FBDB Federal Business and Development Bank [Canada]
FBFM Feedback Frequency Modulation
FBFS Fuel Building Filter System
FBFT Flow Bias Functional Test
FBH Flat-Bottom Hole
FBHTM Federation of British Hand Tool Manufacturers
FBI Federal Bureau of Investigation [US]
Federation of British Industries [now: CBI, UK]
FBIM Fellow of the British Institute of Management
FBL Form Block Lines
FBLO Foreign Branch Liaison Office
FBM Feet Board Measure [also: fbm]
F-Format, Blocked Data Set with "M" (= Machine) Control Character
Fleet Ballistic Missile
Foreground and Background Monitor
FBMP Fleet Ballistic Missile Program
FBMWS Fleet Ballistic Missile Weapon System
FBN Food Business Network
FBO Fixed Base Operator
FBOA Fellow of the British Optical Association
FBP Final Boiling Point
Fluid-Bed Processing
FBPS Forest and Bird Protection Society [New Zealand]
FBR Fast Breeder Reactor
Fast Burst Reactor
Feedback Resistance
Fiber
Forskningsbibiloteksradet [= Council for Research Libraries, Sweden]
Full Bibliographic Record
FBRAM Federation of British Rubber and Allied Manufacturers [UK]
FBRL Final Bomb Release Line
FBS Fast Blue Salt
Fetal Bovine Serum
F-Format, Blocked Standard Data Set
FBSA F-Format, Blocked Standard Data Set with "A" (= ASCII) Control Character
FBSC Fellow of the British Society of Commerce
FBT Functional Board Tester
FBUA Franco-British Union of Architects [UK]
FBW Fly-by-Wire
FC Fail Closed
Fatigue Crack
Fault Current
Faraday Cup
Feature Control
Ferrite Core
Fiber Glass Cover
Field Coil
Field Cooling
File Code
File Conversion
Filter Circuit
Finance Committee
Find Calling Party
Fine Control
Fire Code
Fire Control
Fire Cracking
First Class
Fixed Carbon
Flexible Connection
Flight Control
Flow Control
Foam Cell
Forced Cooling
Font Cartridge
Font Change (Character)
Foot-Candle
Force Control
Forced Convection
Forest Center
Forestry Canada
Foundation Center [US]
Fractionating Column
Franklin College [US]
Free Cholesterol
Free Convection
Free Cursor
Frequency Changer
Frequency Control
Frequency Converter
Frequency Counter
Fretting Corrosion
Front Connected
Fuel Calorimetry
Fuel Cell
Fuel Cycle
Full Coordination
Function Code
Funnel Cloud
Furnace Cooling
Fuse Chamber
Fc Fc Fragment Specific [Immunochemistry]
fc footcandle
4/C Four Conductor

FCA Faraday Cup Array
Farm Credit Administration [US]
Federation of Commodity Associations [UK]
Fellow of the Institute of Chartered Accountants
Fire Control Area
Fluorescence Concentration Analyzer
Frequency Control Analysis
Frequency Control and Analysis

FCA(CAN) Fellow of the Canadian Institute of Chartered Accountants

FCAMRT Fellow of the Canadian Association of Medical Radiation Technologists

FCAPO Fellow of the College of American Pathologists

FCASI Fellow of the Canadian Aeronautics and Space Institute

FCAW Flux Cored Arc Welding

FCAW-EG Flux Cored Arc Welding — Electrogas

FCB File Control Block
Forms Control Buffer
Frequency Coordinating Body [of ICAO]
Function Control Block

FCBM Federation of Clinker Block Manufacturers [UK]

FCC Face-Centered Cubic [also: fcc]
Farm Credit Corporation
Federal Communications Commission [US]
Federal Construction Council [of National Academy of Science, US]
Feedforward/Cascade Control
Flat Conductor Cable
Flight Control Center
Flight Control Computer
Florida Citrus Commission [US]
Fluid Catalytic Cracking
Fluid-Cracking Catalyst
Food Chemical Codex
Food Contaminants Commission [US]
Frame Check Character

FCCA Forestry, Conservation Communications Association [US]

FCCC Federation of Commonwealth Chambers of Commerce [UK]

FCCMG Fellow of the Canadian College of Medical Genetics

FCCPO Federal Contract Compliance Programs Office

FCCSET Federal Coordinating Council for Science, Engineering and Technology [US]

FCCTS Federal COBOL Compiler Testing Service [US]

FCD Failure-Correction Decoding
Fine Control Damper
Frequency Compression Demodulator

FCDA Federal Civil Defense Administration [US]

FCDM Flow Control Decision Message

FCDR Failure Cause Data Report

FCE Flexible Critical Experiment
Fourier Conduction Equation

FCEA Federal Capital Equipment Authority

FCEC Federation of Civil Engineering Contractors [UK]

FCEM Flow Control Execution Message

FCF Fuel Cycle Facility [of USAEC]

FCFO Full Cycling File Organization

FCFS First-Come, First-Served

FCFT Fixed Cost, Fixed Time Estimate

FCG Facing
False Cross or Ground
Fatigue Crack Growth
Federal Coordination Group [US]

FCGA Fellow of the Canadian General Accountants Association

FCI Fatigue Crack Initiation
Fluid Controls Institute [US]
Flux Change per Inch
Fuel-Cladding Interaction
Fuel Coolant Interaction
Functional Configuration Identification

FCIA Foreign Credit Insurance Association [US]

FCIB Foreign Credit Interchange Bureau [now: FCIB/NACM, US]

FCIB/NACM Foreign Credit Interchange Bureau/National Association of Credit Management [US]

FCIC Federal Crop Insurance Corporation
Fellow of the Chemical Institute of Canada

FCICA Floor Covering Installation Contractors Association [US]

FCIM Farm, Construction and Industrial Machinery

FCIR Fatigue Crack Initiation Resistance

FCL Feedback Control Loop
Ferric Chloride Leach
Fiber Composite Laminate
Flightcrew Licensing
Format Control Language
Full Container Load

FCLA Fisheries Council of Latin America

FCLTY Facility

FCM Fault Control Module
Firmware Control Memory
First Class Mail
Floating-Carrier Modulation
Foot-Candle Meter

FCMA Fellow of the Institute of Cost and Managements

FCMD Fire Command

FCME First Class Marine Engineer

FCMI Federation of Coated Macadam Industries [UK]

FCMJ Federation of Canadian Manufacturers in Japan

FCMS Factory Control Management System

FCMV Fuel Consuming Motor Vehicle

FCN Field Change Notice

FCNP Fire Control Navigation Panel

FCO Field Change Order
Flight Clearance Office
Foreign and Commonwealth Office [UK]
Frequency-Change Oscillator
Functional Checkout

F CO Fair Copy

FCOH Flight Controllers Operations Handbook

4-C1N 4-Chloro-1-Naphthol

FCP Fatigue Crack Propagation
Fellow of the College of Preceptors
File Control Package
File Control Procedure
File Control Processor
File Control Program
Flat Concurrent Prolog

FCPC Fleet Computer Programming Center

FCPP Fuel-Cell Power Plant

FCPR Fatigue Crack Propagation Rate
Fatigue Crack Propagation Resistance

FCR Facility Change Request
Fast Ceramic Reactor
Fast Cycle Resin
Fast Cycling Resin
Final Configuration Review
Fire Control Radar
Flight Control System
Floor Cavity Ratio
Fuse Current Rating

FCRE Foundation for Cotton Research and Education [US]

FCRP Fast Ceramic Reactor Program

FCRS Flightcrew Record System

FCS Facsimile Communications System
Farm Cooperative Service [US]
Fellow of the Chemical Society
Fellow of the College of Sciences
Field Control Strain
File Control Services
Fire Control System
Fish Culture Station
Fixed Control Storage
Flexible Clamping System
Flight Control System
Frame Checking Sequence
French Chemical Society

FCSC Foreign Claims Settlement Commission

FCSCC Fellow of the Canadian Society of Clinical Chemists

FCSE Fellow of the Canadian Society of Electroencephalographers

FCSLE Forecastle

FCSPP French Committee for the Study of Paranormal Phenomena [France]

FCSRT Fellow of the Canadian Society of Radiological Technicians

FCSS Fire Control Sight System

FCST Federal Council for Science and Technology [US]
Forecast

FCT Face-Centered Tetragonal
Field-Controlled Thyristor
Filament Center Tap
File Control Table
Flow-Controller Tester
Function

FCTC Fuel Centerline Thermocouple

FCTN Function

FCTRY Factory

FCTT Fuel Cladding Transient Tester

FCU File Control Unit
Flight Control Unit
Fuel Conditioning Unit

FCVS FORTRAN Compiler Validation System

FCWA Fellow of the Institute of Cost and Works Accountants

FCWG Frequency Coordination Working Group [US]

FD Feed
Feynman Diagram
Fiber Duct
Field-Decelerating (Relay)
Field Dose
File Definition
File Description
File Directory
Finished Dialing
Finite Difference
Fire Department
Fjord
Fire Damper
Fire Department
Flange Local Distance
Flexible Disk
Flight Director
Floor Drain
Floppy Disk
Fluid Dynamics
Flux Delta
Focal Distance
Food
Forced Diffusion
Forced Draft
Forging Direction
Forward
Free Discharge
Fractional Distillation
Frequency Demodulator
Frequency Detection
Frequency Discriminator
Frequency Distance
Frequency Distribution
Frequency Diversity
Frequency Divider
Frequency Division
Frequency Domain
Frequency Doubler
Fuel Demand
Full Duplex
Functional Description
Fund
Furniture Designer

F&D Facilities and Design
Findings and Determination
Freight and Demurrage

F/D Focal Length to Diameter Ratio

Fd Ferredoxin [Biochemistry]

4-D Four-Dimensional

F$ Fiji Dollar [Currency]

FDA Final Design Approval
Food and Drug Act [Canada]
Food and Drug Administration [US]
Frequency-Domain Analysis

FDAA Federal Disaster Assistance Administration [US]
Fluorenyldiacetamide

FDAI Flight Director Attitude Indicator

FDAR Federal Department of Agricultural Research [Nigeria]

FDAS Frequency Distribution Analysis Sheet

FDAU Flight Data Acquisition Unit

FDB Fahrenheit Dry Bulb
Field Descriptor Block

Field Dynamic Braking
File Data Block
Forced Draft Blower
Forestry Data Bank [of Forestry Canada]
Functional Description Block
FDC Facsimile Data Converter
Film Development Corporation [Canada]
Fire Department Connection
Flight Director Computer
Floppy-Disk Controller
Florida Department of Citrus [US]
Fluid Die Compaction
Formation Density Content (Log)
Functional Design Criteria
Furniture Development Council [now: FIRA, UK]
FD&C Foods, Drugs, and Cosmetics Color [of FDA, US]
FDCA Food, Drug and Cosmetic Act [US]
FDCM Fluorodichloromethane
FDCS Functionally Distributed Computing System
FDD Flexible Disk Drive
Floppy Disk Drive
Formatted Data Disk
FDDI Fiber Distributed Data Interface
FDDL Field Data Description Language
Frequency Division Data Link
FDDS Flight Data Distribution System
FDE Field Decelerator
FDEM Fuel Demand Evaluation Model
FDEP Flight Data Entry Panel
Formatted Data Entry Program
FDF Fiber Distribution Frame
Flight Data File
F2F Frequency Double Frequency
FD/FF Flux Delta/Flux Flow
FDFM Flight Data and Flow Management Group [of ICAO]
Frequency Division/Frequency Modulation
FDFR Federal Department of Forestry Research [Nigeria]
FDG Fractional Doppler Gate
Funding
FDGB Freier Deutscher Gewerkschaftsbund [= Confederation of Free German Trade Unions]
FDHD Floppy Drive High Density
FDI Field Discharge
Flight Detector Indicator
Frequency Domain Instrument
FDIC Federal Deposit Insurance Corporation [US]
FDL Fast Deployment Logistics
Flight Dynamic Laboratory [US Air Force]
FDLI Food and Drug Law Institute [US]
FDLS Fast Deployment Logistics Ship [US Navy]
FDM Finite-Difference Method
Form Description Macro
Frequency-Division Multiplexing
Fused Deposition Method
Fused Deposition Modeling
FDMA Frequency-Division Multiple Access
Frequency-Domain Multiple Access
FDM/FM Frequency-Division-Multiplexed/Frequency Modulated
FDMS Factory Data-Management System
Floppy Disk Management System
FDN Foundation
FDNC Frequency-Dependent Negative Conductance
FDNR Frequency-Dependent Negative Resistance
FDOS Floppy Disk Operating System
FDP Fast Digital Processor
Fibrin Degradation Product
Fibrinogen Degradation Product
Field-Developed Program
Filter Drainage Protection
Flight Data Processing
Fluid Dynamics Panel [of AGARD]
Flying Duty Period
Form Description Program
Fructose Diphosphate
Future Data Processor
FDPC Federal Data Processing Center [US]
FDPS Flight Data Processing System
FDR Facility Development Research
Feeder
File Data Register
Final Design Review
Flight Data Recorder
Flood Damage Reduction (Program)
Frequency Dependent Rejection
Frequency Diversity Radar
Frequency Domain Reflectometry
Functional Design Requirements
F DR Fire Door
FDRAKE First Dynamic Response and Kinematics Experiment
FDRI Fluid Dynamics Research Institute [of University of Windsor, Canada]
FDRS Food Distribution Research Society [US]
FDRY Foundry
FDS Fallout Decay Simulation
Field-Discharge Switch
Filter Difference Spectrometer
Fixed-Disk Storage
Flexible Disk System
Floppy Disk System
Fluid Density Sensor
Fluid Distribution System
Foreign Agriculture Service [of USDA]
Frame Difference Signal
Frequency Division Separator
FDSR Floppy Disk Send Receive
FDSS Fine Digital Sun Sensor
FDSSA Fine Digital Sun Sensor Assembly
FDSU Flight Data Storage Unit
FDT Flowing-Gas Detonation Tube
Formatted Data Tape
Full Duplex Teletype
Functional Description Table
FDTC Fiber Drum Technical Council [US]
FDU Flexible Disk Unit
Formatter and Drive Unit
Form Description Utility
FDV Fault Detect Verification
FDVR Federal Department of Veterinary Research [Nigeria]
FDX Full Duplex

FDY Foundry

FE Ferroelectret
Ferroelectric(ity)
Field Electron
Field Emission
Field Engineer
Field Equation
Finite Element
Fire Engineer
Fire Engineering
First Edition
Fixed End
Flash Evaporation
Flight Engineer
Fluidization Engineering
Fluid Engineering
Fluoroelastomer
Fluoroethylene
Focusing Electrode
Font Error
Forest Engineer
Forest Engineering
Format Effector (Character)
Foundation Engineer
Foundation Engineering
Framing Error
Free Electron
Free End
Friends of the Everglades [US]
Front End
Functional Entity
Functional Expansion
Further Education
Fuse Element

Fe (ferrum) — Iron

FEA Failure Effect Analysis
Farmstead Equipment Association [of FIEI, US]
Federal Energy Administration [US]
Federation Europeenne des Association Aerosols [= European Federation of Aerosol Associations, Belgium]
Finite Element Analysis

FeAA Ferric Acetylacetonate

FEACO Federation Europeenne des Associations de Conseils en Organisation [= European Federation of Management Consultant Associations, France]

FEALC Federacion Espeleologica de America Latina y el Caribe [= Speleological Federation of Latin America and the Caribbean, Venezuela]

FEANI Federation Europeenne de l'Associations Nationales des Ingenieurs [= European Federation of National Associations of Engineers]

FEAO Federation of European-American Organizations

FEARO Federal Environmental Assessment Review Office

FEAT Frequency of Every Allowable Term

FEATH Feathery

FEB February
Functional Electronic Block

FEBA Federal Energy Bar Association [US]
Forward Edge of Battle Area

FEBAB Federacao Brasiliera de Associacoes de Bibliotecarios [= Brazilian Federation of Library Association]

FEBMA Federation of European Bearing Manufacturer Associations [Germany]

FEBS Federation of European Biochemical Societies [France]

FEC Faculty Exchange Center [US]
Farm Electrification Council [US]
Federal Election Commission
Floating Error Code
Food and Energy Council [now: NFEC, US]
Forward Error Control
Forward Error Correction
Front-End Computer

FECC Federation Europeene du Commerce Chimique [= European Federation of Chemical Commerce, Belgium]

FECL Feedback Emitter-Coupled Logic

FECO Fringes of Equal Chromatic Order

FECP Front End Communications Processor

FECS Federation Europeenne des Fabricants de Ceramiques Sanitaires [= European Federation of Ceramic Sanitary Equipment Manufacturers]
Federation of European Chemical Societies [UK]
Foreign Exchange Counselling System [of European American Bank]

FECT Federation of European Chemical Trade [Belgium]

FED Federal
Federation
Field-Effect Diode
Field Emission Deposition
Freeze Etching Device
Fuel Examination Facility
Fusion Engineering Device

FEDA Farm Equipment Dealers Association
Food Service Equipment Distributors Association [US]

FEDAL Failed Element Detection and Location Instrument

FEDC Federal Economic Development Coordinator
Federation of Engineering Design Consultants
Fusion Engineering Design Center [of ORNL, US]

FEDD For Early Domestic Dissemination [NASA Term]

FEDEMOA Federacion Mexicana de Organisaciones Agricoles [= Mexican Federation of Agricultural Organizations]

FEDEREL Federation des Relamineurs du Fer et de l'Acier [= Federation of Iron and Steel Relaminators [of EEC, Belgium]]

FEDES Federation Europeenne de l'Emballage Souple [= European Federation for the Flexible Packaging Manufacturers, France]

FEDEX Federal Index [US]

FEDIOL Federation de l'Industrie de l'Huilerie [= Federation of Seed Crushers and Oil Processors [of EEC, Belgium]]

FEDLINK Federal Library and Information Network [US]

FEDNET Federal Network [US]

FEDP Federal Executive Development Program

FEDREG Federal Register [US]

FEDS Fixed and Exchangeable Disk Storage

FED-STD Federal Standard [US]
FEDSTRIP Federal Standard Requisitioning and Issue Procedure
FEE Frog Embryology Experiment [of NASA]
FEED Field Electron Energy Distribution
Field Emission Energy Distribution
Floating Electrode Effect Development
FEEP Field Emission Electric Propulsion System
FEF Fast Extruding Furnace Black
Foundry Educational Foundation [US]
Friends of the Earth Foundation [US]
Fusion Energy Foundation [US]
FEFANA Federation Europeeene des Fabricants d'Adjuvants pour la Nutrition Animale [= European Federation of Manufacturers of Feed Additives, Germany]
FEFC Far Eastern Freight Conference
FEFCEB Federation Europeenne des Fabricants de Caisses et Emballages en Bois [= European Federation of Wooden Boxes and Packaging Materials]
FEFCO Federation Europeenne des Fabricants de Carton Ondule [= European Federation of Corrugated Board Manufacturers, France]
FEFO First-Ended, First-Out
FEFP Fuel Element Failure Propagation
FEFPEP Federation Europeenne des Fabricants de Palettes et Emballages en Bois [= European Association for the Flexible Packaging Industry, France]
FEFPL Fuel Element Failure Propagation Loop
FEG Field Emission Gun
FEG TEM Field Emission Gun Transmission Electron Microscope
FEHVA Federation of European Heating and Ventilating Associations [Netherlands]
FEI Field Engineering Instructions
Financial Executive Institute [US]
France-Europe International [France]
Fundacao de Ciencias Aplicadas [= Foundation of Applied Sciences, Brazil]
FEIA Flight Engineers International Association
FEIC Fellow of the Engineering Institute of Canada
FEICA Federation Europeenne des Industries de Colles Adhesifs [= Association of European Adhesives Manufacturers, Germany]
FEICRO Federation of European Industrial Cooperative Research Organizations [UK]
FEIEA Federation of European Industrial Editors Associations
FEIM Federation Europeenne des Importateurs de Machines et d'Equipements de Bureau [= European Federation of Importers of Business Equipment, Belgium]
FEL Free-Electron Laser
FELACUTI Federacion Latinoamericana de Usuarios del Transporte [= Latin American Federation of Shippers Council, Colombia]
FELP Free-Electron Laser Physics
FEM Federation Europeenne de la Manutention [= European Federation of Handling Industries, Switzerland]
Field-Effect Modified
Field (Electron) Emission Microscopy
Firmware Expansion Module
Final Effluent Monitor
Financial Evaluation Program
Finite Element Method
Finite Element Modeling
FORTRAN Enhancement Package
Fundacao de Estudos do Mar [= Sea Studies Foundation, Brazil]
FEMA Farm Equipment Manufacturers Association [US]
Federal Emergency Management Agency [US]
Fire Equipment Manufacturers Association [US]
Flavor and Extract Manufacturers Association [US]
Food Equipment Manufacturers Association [US]
Foundry Equipment and Materials Association
Foundry Equipment Manufacturers Association [now: CISA, US]
FEMAC Federation Europeeene des Fabricants d'Aliments Composes [= European Federation of Compound Animal Feeding Stuff Manufacturers, Belgium]
FEMAR Fundacao de Estudos do Mar [= Sea Studies Foundation, Brazil]
FEMF Floating Electronic Maintenance Facility [of LASL, US]
Foreign Electromotive Force
FEM/FEA Finite Element Method/Finite Element Analysis
FEMFM Federation of European Manufacturers of Friction Materials [France]
FEMIB Federation Europeenne de Synicats des Fabricants de Menuiseries de Batiment [= European Federation of Building Joinery Manufacturers, Germany]
FEMIPI Federation Europeenne des Mandataires de l'Industrie en Propriete Industrielle [= European Federation of Agents of Industrial Property, Germany]
FEMO Free-Electron Molecular Orbital
FEMP Fusion Engineering Materials Program
FEMS Federation of European Materials Societies [also: fems]
Federation of European Microbiological Societies [UK]
FEN Frequency-Emphasizing Network
FENCO Foundation of Engineering Corporations [Canada]
FEO Facility Emergency Organization
Federal Energy Office [US]
Field Engineering Order
FEOTC Federal Exporters Overseas Transport Committee [Australia]
FEP Financial Evaluation Program
Fluorinated Ethylene Propylene (Resin)
Fluorinated Perfluoroethylene-Propylene
Front-End Processor
Foundation for Education with Production [Botswana]
Fuse Enclosure Package
FEPA Federation Europeenne des Fabricants de Produits Abrasifs [= European Federation of Manufacturers of Abrasive Products, France]
FEPACE Federation Europeenne des Producteurs Autonomes et des Consommateurs Industriels d'Energie [=

European Federation of Autoproducers and Industrial Consumers of Energy, Belgium]

FEP-CORDE Foundation for Education with Production — Cooperative Research, Development and Education [Botswana]

FEPD Federation Europeenne des Parfumeurs Detaillants [= European Federation of Perfumery Retailers, France]

FEPE Federation Europeenne de la Publicite Exterieure [= European Federation of Outdoor Advertising, Belgium]
Federation Europeenne des Producteurs d'Enveloppes [= European Association of Envelope Manufacturers, Switzerland]
Federation Europeenne pour la Protection des Eaux [= European Federation for Water Pollution Control]
Full Energy Peak Efficiency

FEPEM Federation of European Petroleum Equipment Manufacturers [Netherlands]

FEPMA Federation of European Pencil Manufacturer Associations [Germany]

FER Federation of Engine Re-Manufacturers [UK]
Forward Engine Room
Foundation for Educational Research [of ASM International; also: ASMFER]
Friends of Ecological Reserves
Fusion Engineering Reactor
Fusion Experimental Reactor [Japan]

FERC Federal Energy Regulatory Commission [US]

FERD Fuel Element Rupture Detection

FERIC Forest Engineering Research Institute of Canada

FERMI Enrico Fermi Breeder Reactor Plant

Fermilab Fermi National Accelerator Laboratory [US]

FERN Forest Ecosystem Research Network

FERPIC Ferroelectric Ceramic Picture Device
Ferroelectric Photoconductor Image Camera

FERROD Ferrite-Rod Antenna [also: ferrod]

FERSI Flat Earth Research Society International [US]

FERT Fertilizer

FES Fachnormenausschuss fur Eisen und Stahl [= Standards Committee of the Iron and Steel Industry, Germany]
Federation of the European Trout and Salmon Industry [Scotland]
Fellow of the Entomological Society
Field Emission Spectroscopy
Final Environmental Statement
Finite Element Solver
Flame-Emission Spectroscopy
Florida Engineering Society [US]
Florida Entomological Society [US]
Forest Extension Services [Canada]
Forms Entry System
Fundamental Electrical Standards

FESA Federation of Engineering and Scientific Associations

FESE Field-Enhanced Secondary Emission

FESEM Field Emission Scanning Electron Microscopy [also: FE-SEM]

FESI Federation Europeenne des Syndicats d'Entreprises d'Isolation [= European Federation of Associations of Insulation Contractors, France]

FESIC Far East Seed Improvement Conference

FESPA Federation of European Screen Printers Associations

FESS Facilities Engineering and Safety Services [of CFFTP, Canada]
Flywheel Energy Storage System

FEST Federation for Education, Science and Technology [South Africa]

FESYP Federation Europeenne des Syndicats de Fabricants de Panneaux de Particules [= European Federation of Associations of Particleboard Manufacturers, Germany]

FET Federal Excise Tax
Federation of Environmental Technologists [US]
Field-Effect Transistor
Fluid Engineering Technology

FETDIP Fetlington Dual-in-Line Package

FETFE Fluorelastomer with Tetrafluoroethylene Additives

FETO Factory Equipment Transfer Order
Free Estimated Time of Overflight

FETS Far-East Trade Service
Field-Effect Transistors

FETT Field-Effect Tetrode Transistor

FEUS French Engineers in the United States

FEV Forced Expired Volume

FEWA Farm Equipment Wholesalers Association [US]

FEWIA Federation of European Writing Instruments Associations [Germany]

FEWITA Federation of European Wholesale and International Trade Associations [Belgium]

FEWMA Federation of European Window Manufacturers Associations [Germany]

FEX Foreign Exchange

FEXT Far-End Crosstalk

FF Farm Foundation [US]
Fast-Fast Wave
Fast Forward
Fiber Forming
Field Function
Filtration Fraction
Fine Furnace Black
Fixed Focus
Flip-Flop
Fluid Flow
Flux Flow
Form Feed (Character)
Fragrances Foundation [US]
Freeze Fracture
Freight Forwarder
Frequency Filter
Froth Flotation
Fuel Flow
Full Field

F&F Fire and Flushing

ff and the following pages, sections, etc.
folios

FFA Federal Fisheries Act [Canada]
Fiberglass Fabrication Association [US]

	Flammable Fabrics Acts [US]
	Flygtekniska Forsoksanstalten [= Aeronautical Research Institute, Sweden]
	Forest Farmers Association [US]
	Free Fatty Acid
	Free from Alongside
	Free from Average
FFAG	Fixed-Field Alternating Gradient
FFAR	Folding Fin Air Rocket
FFB	French Forces Broadcasting
FFC	Feed Forward Control
	Flexible Flatness Control
	Flip-Flop Complimentary
FFD	Failure Flux Density
	Fluid Flow Dynamics
	FMS (= Flight Management System) Flight Data
FFDCA	Federal Food, Drug and Cosmetics Act [US]
FFE	Ferric-Ferrous Electrode
FFEC	Field-Free Emission Current
FFF	Farm Field Foundation [US]
	Flicker Fusion Frequency
	Flight Freedoms Foundation [US]
	Foundry Educational Foundation [US]
FFFF	Fusion-Fission Fuel Factory [PR China]
FFG	Fine-Fine Grain
	Superfine Grain
FFGT	Firefighting
FFH	For Further Headings
FFHR	Fusion-Fission Hybrid Reactor
FFI	Free Fluid Index
	Fuel Flow Indicator
	Full Field Investigation
FFL	Finished Floor
	First Financial Language
	Front Focal Length
FFLA	Federal Farm Loan Association [US]
FFM	Fast File Manager
	Foundation for Microbiology [US]
	Fuel Failure Mockup Facility
FFMC	Fine Fuel Moisture Code
	Freshwater Fish Marketing Corporation
FFMS	Fast Fourier Mass Spectrometry
	Free Flight Melt Spinning
FFP	Firm Fixed Price
	First Focal Point
	Fixed Fee Procurement
FFPA	Free from Prussic Acid
FFPRI	Forest and Forest Product Research Institute [Ghana]
	Forestry and Forest Product Research Institute [Japan]
FFPS	Fauna and Flora Preservation Society [UK]
FFR	Fellow of the Faculty of Radiology
	Flat Face Rolling System
	Folded Flow Reactor
FFRI	Fruit and Food Research Institute [South Africa]
FFRM	Foundation for Fundamental Research on Matter [Netherlands]
FFRS	Federal Forest Research Station [of IUFRO, Austria]
FFSA	Field Functional System Assembly
FFSF	Fossil Fired Steam Plant
	Full Fat Soy Flour
FFT	Fast Fourier Transform
	Flash Fusion Technology
	Flicker Fusion Threshold
FFTA	Fast Fourier Transform Analyzer
	Finnish Foreign Trade Association
	Foundation of the Flexographic Technical Association [US]
FFTF	Fast Flux Test Facility [of USAEC]
FFTFPO	Fast Flux Test Facility Project Office
FFTR	Fast Flux Test Reactor
	Federal Fuel Tax Rebate [Canada]
FFTSB	Federation of Trade Fairs and Trade Shows of Benelux [France]
FFTV	Free Flight Test Vehicle
FFV	Field Failure Voltage
FG	Fiberglass
	Filament Ground
	Fine Grain
	Finite Geometry
	Fission Gas
	Flat Grain
	Flue Gas
	Foot Guards
	Foreground
	Forging
	Forward Gate
	Fourth Generation [also: 4G]
	Frame Ground
	Fraunhofer-Gesellschaft [= Fraunhofer Society, Germany]
	Friction Glaze
	Fuel Gas
	Function Generator
4G	Fourth Generation [also: FG]
FGA	Feature Group A [of ENFIA, US]
	Fellow of the Gemological Association
	Free of General Average
FGAA	Federal Government Accountants Association
FGAN	Fertilizer Grade Ammonium Nitrate
FGB	Feature Group B [of ENFIA, US]
FGC	Feature Group C [of ENFIA, US]
	Fifth Generation Computer
FGCS	Fifth Generation Computer System
	Flight Guidance and Control System
FGD	Flue Gas Desulfurization
FGETR	Federal Gasoline Excise Tax Refund [Canada]
FGFAF	Fraunhofer-Gesellschaft zur Forderung der Angewandten Forschung [= Fraunhofer Society for the Advancement of Applied Research, Germany]
FGGE	First GARP (= Global Atmospheric Research Program) Global Experiment
FGGM	Federation of Gelatine and Glue Manufacture
FGI	Federation of German Industries
	Federation of Greek Industries
FGIPC	Federation of Government Information Processing Councils [US]
FGJA	Flat Glass Jobbers Association [now: FGMA, US]
4GL	Fourth Generation Language
FGM	Fission Gas Monitor

	Functionally Gradient Materials
FGMA	Flat Glass Marketing Association [US]
FGMDSS	Future Global Maritime Distress and Safety System
FGMM	Fluxgate Magnetometer
FGN	Foreign
FGR	Fertility and Genetics Research
	Finger
	Fondazione Giorgio Ronchi [= Giorgio Ronchi Foundation, Italy]
FGS	Fellow of the Geographical Society
	Fellow of the Geological Society
FGSA	Fellow of the Geographical Society of America
	Fellow of the Geological Society of America
FGW	Floor Ground Window
FH	Fire Hose
	Flat Head
	Frequency Hopping
	Full-Hole Mining
	Fumurate Hydratase
FHA	Farmers Home Administration [US]
	Federal Highway Administration
	Federal Housing Administration [US]
FHANG	Federation of Heathrow Anti-Noise Groups [UK]
FHB	Fuel Handling Building
FHC	Federal Housing Commissioner [of OASH, US]
	Fluid Hydrostatic Cell
	Fire Hose Cabinet
FHD	Fixed-Head Disk
	Flat Head
FHDA	Fir and Hemlock Door Association [US]
FHDS	Fixed-Head Disk Storage
FHEO	Fair Housing and Equal Opportunity [of Office of the Assistant Secretary, US]
FhG	Fraunhofer Gesellschaft [= Fraunhofer Society, Germany]
FHI	Fritz-Haber Institute [Germany]
FHK	Fourier-Hermite Kernel
FHLBB	Federal Home Loan Bank Board [US]
FHNC	Fermi Hypernetted Chain
FHP	Fluid Hydrostatic Pressure
	Fractional Horsepower [also: fph]
	Friction Horsepower [also: fhp]
FHR	Fire Hose Rack
FHRF	Finney-Howell Research Foundation [US]
FH&RM	Fuel Handling and Radioactive Maintenance
FHS	Forest History Society [US]
	Forward Head Shield
FHSA	Federal Hazardous Substances Act [US]
FHSF	Fixed-Head Storage Facility
FHSR	Final Hazards Summary Report
FHT	Fully Heat-Treated
FHTA	Federated Home Timber Associations [UK]
FHVA	Fine Hardwood Veneer Association [US]
FHVA/AWMA	Fine Hardwood Veneer Association/American Walnut Manufacturers Association [US]
FHWA	Federal Highway Administration [US]
FHY	Fire Hydrant
FI	Fail in Place
	Farmland Industries [US]
	Fault Identification
	Field Intensity
	Fixed Interval
	Flow Indicator
	Foresight Institute [US]
	Formaldehyde Institute [US]
	Fourier Integral
	Franklin Institute [US]
	Fraunhofer Institute [Germany]
FIA	Factory Insurance Association
	Federal Insurance Administration
	Federated Ironworkers Association of Australia
	Federation des Industries Agricoles et Alimentaires [= Federation of the Agricultural and Food Industries, Belgium]
	Federation Internationale de l'Automobile [= International Automobile Federation, France]
	Fellow of the Institute of Actuaries
	Field Information Agency [US]
	Financial Inventory Accounting
	Flame Ionization Analysis
	Flow Injection Analysis
	Fluorescence Immunoassay
	Fluorescent Indicator Absorption
	Fluoroimmunoassay
	Forging Industry Association [US]
	Freund's Incomplete Adjuvant
	Full Interest Admitted
FIAA	Fellow of the Incorporated Association of Architects and Surveyors
FIACC	Five International Associations Coordinating Committee [Hungary]
FIAeS	Fellow of the Institute of Aeronautical Sciences
FIANATM	Federation Internationale des Associations de Negociants en Aciers [= International Federation of Associations of Steel, Tube and Metal Merchants, Switzerland]
FIAP	Federation Internationale des Architectes Paysagistes [= International Federation of Landscape Architects, France]
FIAS	Fellow of the Institute of Aeronautical Sciences
FIAT	Field Information Agency, Technical [US]
FIATA	Federation Internationale des Associations de Transitaires et Assimilies [= International Federation of Freight Forwarders Associations, Switzerland]
FIB	Federation des Industries Belges [= Federation of Belgian Industries]
	Fellow of the Institute of Biology
	File Information Block
	Focused Ion Beam
	FORTRAN Information Bulletin
	Forward Indicator Bit
	Free into Barge
	Free into Bunkers
Fib	Fibrin
FIBCA	Flexible Intermediate Bulk Container Association [US]
FIBEC	Federal Industrial Boiler Emission Control [Canada]
FIBiol	Fellow of the Institute of Biology
FIBR	Fiber
FIBS	Flight Information Billing System

FIC Federal Information Center (Program) [US]
Fellow of the Institute of Chemists
Fellow of the Institute of Commerce
First-in-Chain
Flight Information Center
Flying Instructor Course
Forest Industries Council [US]
Frequency Interference Control
FICB Federal Intermediate Credit Bank [US]
FICC Frequency Interference Control Center [US Air Force]
FICCI Federation of the Indian Chambers of Commerce and Industry
FICE Fellow of the Institution of Civil Engineers
FICON File Conversion
FICPI Federation Internationale des Conseils en Propriete Industrielle [= International Federation of Industrial Property Attorneys, Switzerland]
FICS Forecasting and Inventory Control System
FID Federation Internationale de Documentation [= International Federation of Documentation, Netherlands]
Fidelity
Field Intelligence Division [US Military]
Field Ionization Detector
Flame Ionization Detector
Forecasts-In-Depth
FIDA Federation of Industrial Development Organizations [UK]
Fonds International pour le Developpement Agricole [= International Fund for Agricultural Development [of UN]]
FIDAC Film Input to Digital Automatic Computer
FIDACSYS Film Input to Digital Automatic Computer System
FIDAS Formular-Orientiertes Interaktives Datenbanksystem [= Form-Oriented Interactive Databank System, Germany]
FIDER Foundation for Interior Design Education Research [US]
FIDIC Federation Internationale des Ingenieurs Conseils [= International Federation of Consulting Engineers, Switzerland]
FIDO Film Industry Defense Organization [UK]
Fog Investigation and Dispersal Operation
FID/OM FID Committee on Operational Machine Techniques and Systems [Netherlands]
FIDOR Fiber Building Board Development Organization [UK]
FIDS Flight Information and Display System
FID/TM FID Committee on Theory and Methods of Systems, Cybernetics and Information Networks [Netherlands]
FID/TMO FID Committee on Theory, Methods and Operation of Information Systems and Networks [Netherlands]
FIEA Federation Internationale des Experts en Automobile [= International Federation of Automobile Experts, Belgium]
FIED Field Ionization Energy Distribution
FIEE Fellow of the Institute of Electrical Engineers
FIEEE Fellow of the Institute of Electrical and Electronic Engineers
FIEI Farm and Industrial Equipment Institute [US]
FIELS Foreign Information Exchange for Life Scientists [US]
FIEP Forest Industry Energy Program [US]
FIER Foundation for Instrumentation Education and Research [of Instrument Society of America, US]
FIERA Ferrous Industry Energy Research Association [Canada]
FIERF Forging Industry Educational and Research Foundation [US]
FIFDA Fellow of the International Furnishings and Design Association
FIFE Federation Internationale des Associations de Fabricants de Produits d'Entretien [= International Federation of Associations of Manufacturers of Household Products, Belgium]
FIFO First-In, First-Out
Floating Input/Floating Output
FIFRA Federal Insecticide, Fungicide and Rodenticide Act [US]
FIG Federation Internationale des Geometres [= International Federation of Surveyors, Finland]
Figure
FORTH (Programming Language) Interest Group [US]
FIGAS Falkland Islands Government Air Service
FIGE Field Inversion Gel Electrophoresis
FIGED Federation Internationale des Grandes et Moyenne Entreprises de Distribution [= International Federation of Retail Distributors, Belgium]
FIGIEFA Federation Internationale des Grossistes, Importateurs et Exportateurs en Fournitures Automobiles [= International Federation of Automotive Aftermarket Distributors, Belgium]
FIGLU Formimino-L-Glutamic Acid Transferase
FIGS Figures
Figure Shift
FII Federation of Irish Industries
FIIA Fellow of the Institute of Industrial Administration
FIIG Federal Item Identification Guide
FIILS Full Integrity Instrument Landing System
FIIR Federal Institute of Industrial Research [Nigeria]
FIL Filament
Fillet
Fillister
FILA Fellow of the Institute of Landscape Architects
Filament
FILEX File Exchange
FIL H Fillister Head (Screw) [also: FILH]
FILL Filling
FILO First-In, Last-Out
FILS Flare-Scan Instrument Landing System
FILU Four-Bit Interface Logic Unit
FIM Fault Isolation Meter
Fellow of the Institute of Metals
Fellow of the Institute of Mining and Metallurgy
Field Inspection Manual
Field Intensity Meter
Field Ion Microscopy

	Finnish Markka [Currency]
	Full Indicator Movement
FIMA	Fellow of the Industrial Medical Association
FIMATE	Factory-Installed Maintenance Automatic Test Equipment
FIMC	Fellow of the Institute of Management Consultants
FIMCEE	Federation Internationale des Marbriers de la Communaute Economique Europenne [= International Federation of the Marble Industries in the European Economic Community, Belgium]
FIMechE	Fellow of the Institute of Mechanical Engineers [also: FIME]
FIMF	Fellow of the Institute of Metal Finishing
FIML	Full-Information Maximum Likelihood
FIMM	Fellow of the Institute of Mining and Metallurgy
FIMOS	Floating-Gate Ionization Injection Metal-Oxide Semiconductor
FIMS	Functionally-Identifiable Maintenance System
FIN	Finance
	Financial
	Finish
	Finland
	Finnish
	Futures Information Network [US]
FINAC	Fast Interline Non-Active Automatic Control
FINAL	Financial Analysis Language
FINAT	Federation Internationale des Fabricants et Transformateurs d'Adhesifs et Theromocollants sur Papier et Autres Supports [= International Federation of Manufacturers and Converters of Pressure-Sensitive and Heat-Seals on Paper and Other Base Materials, Netherlands]
FINCO	Finance Committee [of ISO, Switzerland]
FINCOM	Finance Committee [of IEEE, US]
FIND	File Interrogation of Nineteen Hundred Data
	File of Industrial Data [UK]
FINE	Fighter Inertial Navigation System
FINEFTA	Finland — European Free Trade Association
FINESSE	Fusion Integrated Nuclear Experiment Strategy Study Effort
FINFO	First-In, Not Used, First-Out
FINGLA	Fission Products in Glass (Process)
FINISTRAT	Finishing Strategies
FINS	Fishing Industry News Service [Australia]
FINSHG	Finishing
FInstP	Fellow of the Institute of Physics
FInstPet	Fellow of the Institute of Petroleum
FIO	Federacion Internacional de Oleicultura [= International Olive Oil Federation, Italy]
	Florida Institute of Oceanography [US]
	Food Investigation Organization [UK]
	Free In and Out
FIOA	File Input/Output Area
FIOP	FORTRAN Input/Output Package
FIOR	Fluid Iron Ore Reduction
FIP	Fairly Important Person
	Federal Information Processing Standards [US]
	Federation Internationale de la Precontrainte [= International Federation of Prestressed Concrete, UK]
	Field Inspection Procedure
	Finance Image Processor
	Fluorescent Indicator Panel
	Foam-in-Place
	Forestry Inceptive Program [US]
	Fully-Ionized Plasma
FIPACE	Federation Internationale des Producteurs Auto-Consommateurs Industriels d'Electricite [= International Federation of Industrial Producers of Electricity for Own Consumption]
FIPAGO	Federation Internationale des Fabricants de Papiers Gommes [= International Federation of Manufacturers of Gummed Paper, Netherlands]
FIPMT	Fraunhofer Institute of Physical Measurement Techniques [Germany]
FIPS	Federal Information Processing Standard [US]
	Federal Item Procurement Specification [US]
FIPSCAC	Federal Information Processing Standard Coordinating and Advisory Committee
FIQS	Fellow of the Institute of Quantity Surveyors
FIR	Far Infrared
	File Indirect Register
	Finite-Duration Impulse Response
	Finite-Impulse Response
	Flight Information Region
	Food Irradiation Reactor [US]
	Fuel Indicator Reading
fir	firkin [Unit]
FIRA	Foreign Investment Review Act
	Foreign Investment Review Agency
	Furniture Industry Research Association [UK]
FIRC	Forest Industries Radio Communications [now: FIT, US]
FIRD	Fast-Induced Radioactivity Decay
FIRE	Fellow of the Institute of Radio Engineers
	Flight Investigation Reentry Environment
	Forest Industry Renewable Energy Program [of EMR, Canada]
FireE	Fire Engineer [also: FireEng]
FIRETRAC	Firing Error Trajectory Recorder and Computer
FIRFD	Finite-Impulse Response Filter Design
FIRI	Fishing Industry Research Institute [South Africa]
FIRL	Faceted Information Retrieval System for Linguistics
	Franklin Institute Research Laboratories [US]
FIRM	Financial Information for Resource Management
FIRMS	Forecasting Information Retrieval of Management System
FIRPRECAN	Fire Prevention Association of Canada
FIRR	Failure and Incidents Report Review Committee [of ANSI, US]
	Federal Institute for Reactor Research [Switzerland]
FIRST	Fabrication of Inflatable Reentry Structures for Testing
	Fast Interactive Retrieval System Technology
	Federal Information Research Science and Technology Network [of COSATI, US]
	Financial Information Reporting System
	Fragment Information Retrieval of Structures
FIRT	Fertilizer Industry Round Table [US]

FIS Fachinformationssystem [= Technical Information System, Germany]
Field Information System
Flight Information Service
Floating-Point Instruction Set
Functional Interference Specification
FISA Food Industries Suppliers Association [US]
Forest Industries Safety Association [Canada]
FISC Fiscal
Foundation for International Scientific Coordination [France]
FISC YR Fiscal Year
FISH First In, Stays Here
Fully-Instrumented Submersible Housing
FISHROD Fiche Information Selectively Held and Retrieved on Demand
FIST Fault Isolation by Semi-Automatic Techniques
FIT Failure in Time
Fast Installation Technique
Fault Isolation Test
Federal Income Tax
File Information Table
File Inquiry Technique
Flexible Interface Tool
Florida Institute of Technology [US]
Forest Industries Telecommunications [US]
Functional Integration Technology
FITA Federation of International Trade Associations [US]
FITCE Federation des Ingenieurs des Telecommunications de la Communaute Europeenne [= Federation of Telecommunications Engineers of the European Community, Belgium]
FITB Fluorspar International Technical Bureau [Belgium]
FITC Fluorescein Isothiocyanate
FITE Forward Interworking Telephony Event
FITI Fabric Inspection Testing Institute [South Korea]
FITS Trifluoromethanesulfonate
FIU Facility Interface Unit
Federal Information Users [US]
Federation of Information Users [US]
Florida International University [US]
FIW Flight Input Workstation
FIX Fault Isolator and Exercizer
Fixture
FIXIT Flexible Information Exploitation Interpretative Transfer
FIXS Fixtures
FIZ Fachinformationszentrum [= Technical Information Center, Germany]
FIZ EPM Fachinformationszentrum Energie—Physik—Mathematik [= Information Center for Energy, Physics and Mathematics, Germany]
FIZ-Technik Fachinformationszentrum Technik [= Information Center for Technology, Germany]
FJ Fuel-Jet
FJCC Fall Joint Computer Conference [US]
FJD Fiji Dollar [Currency]
FJI Federal Job Information
FJSRL Frank J. Seiler Research Laboratory [US Air Force]
FK Fixture Key
FKM Fluoroelastomer
FKBG Fourdrinier Kraft Board Group [of API, US]
FKBG-API Fourdrinier Kraft Board Group of American Paper Institute [US]
FKBI Fourdrinier Kraft Board Institute [US]
FKI Fachverband Klebstoffindustrie [= Association of European Adhesives Manufacturers, Germany]
Federation of Korean Industries
FKM Forschungskuratorium Maschinenbau [= Research Board for Mechanical Engineering, Germany]
FKP Falkland Islands Pound [Currency]
FL Fault Location
Field Lens
Field Length
Field Loss
Flashing
Flight Level
Floor
Floor Line
Florida [US]
Flow Line
Fluid
Fluoroleucine
Fluorescent Lamp
Fluorescent Light
Focal Length
Focal Line
Foot-Lambert
Foreign Language
Foreign Listing
Formal Language
Formal Logic
Free Length
Frenkel and Ladd Method
Full Load
Function Language
Fusible Link
fL foot-lambert
fl fluid
FLA Fabric Laminators Association [US]
Fellow of the Library Association
Fluorescent Lighting Association
Fla Florida [US]
Fl-AAS Flame Atomic Absorption Spectroscopy
FLAC Florida Automatic Computer
FLAG Fleet Locating and Graphics
FORTRAN Load and Go
FLAGS Far North Liquids and Associated Gas System
FLAIR FORTRAN Language in Core Rapid Translator
FLAM Forward-Launched Aerodynamic Missile
FLAMR Flores Assembly Program
Forward-Looking Advanced Multilobe Radar
FLAPHO Flame Photometer
FLAR Forward-Looking Airborne Radar
FLASH P Flash Point
FLB Federal Land Bank [US]
Flow Brazing
Fondation Louis de Broglie [= Louis de Broglie Foundation, France]
FLBE Filter Band Elimination
FLBIN Floating-Point Binary
FLBP Filter Bandpass

FLBR Fusible Link Bottom Register
FLC Federal Laboratory Consortium [US]
Federal Library Committee [now: FLICC, US]
Ferroelectric Liquid Crystal
Forming Limit Curve
FLCS Fiberoptics Low-Cost System
FLD Field
Formability Limit Diagram
Forming Limit Diagram
FLDA Federal Land Development Authority
FLDEC Floating Point Decimal
FLDL Field Length
fl dr fluid dram
FLEA Flux Logic Element Array
FLEC 1-(9-Fluorenyl)ethyl Chloroformate
FLECHT Full-Length Emergency Cooling Heat Transfer
FLEEP Feeble Beep [also: fleep]
FLEX Flaw Examination
Flexible
FLF Fixed Length Field
Follow-the-Leader Feedback
FLG Flag
Flange
Flooring
FLGD Flanged
FL H Flat Head (Screw) [also: FLH]
FLHLS Flashless
FLHP Filter High Pass
FLI Field Length Indicator
Fluorescence Line Imager
Free Language Indexing
FLICC Federal Library and Information Center Committee [US]
FLIH First-Level Interrupt Handler
FLIM Fast Library Maintenance
FLIMBAL Floated Inertial Measurement Ball
FLINK Flash/Wink Signal
FLINT Floating Interpretative Language
FLIP Film Library Instantaneous Presentation
Flight-Launched Infrared Probe
Floated Lightweight Inertial Platform
Floating Index Point
Floating Instrument Platform
Floating Laboratory Instrument Platform
Floating Point Interpretative Program
FLIR Forward-Looking Infrared
FLIRT Federal Librarians Round Table [of ALA, US]
Free Language Information Retrieval Tool
FLIT Fault Location by Interpretative Testing
FLITE Federal Legal Information through Electronics [US]
FL MECH Fluid Mechanics
FLMEM Floppy Disk Memory
FLNPP Fedeal Library Network Prototype Project [US]
FLO Functional Line Organization
FL/1 Function Language 1
FLOA Frederick Law Ohnsted Association [US]
FLOC Flocculent
FLO CON Floating Container
FLODAC Fluid-Operated Digital Automatic Computer
FLOLS Fresnel Lens Optical Landing System
FLOOD Fleet Observation of Oceanographic Data [US Navy]
FLOP Floating Octal Point
Floating Point
Floating-Point Operation
FLOPS Floating-Point Operations per Second
FLOR Flourished
FLORICO Florida-Puerto Rico Submarine Cable
FloTech Florida Institute of Technology [US]
FLOTOX Floating Gate Tunnel Oxide
FLOTRAN Flowcharting FORTRAN
FLOW Flow Welding
FLOWS FAA (= Federal Aviation Administration) Lincoln Laboratory Operational Weather Studies [US]
FLOX Fluorine/Liquid Oxygen Mixture
fl oz fluid ounce
FLP Filter Low Pass
Flameproof
Foreign Language Program
FLPAU Floating Point Arithmetic Unit
FLPDC Floppy Disk Controller
FLPL FORTRAN List Processing Language
FLR Flare
Fluoroleucine Resistance
Forward-Looking Radar
FLRT Federal Librarian Round Table [US]
FLRY Flurry
FLS Fair Labor Standards
Flow Switch
Free Line Signal
FLSA Fair Labor Standards Act [US]
FLSC Flexible Linear-Shaped Charge
FLSP Flame Spraying
Fluorescein Labeled Serum Protein
FLT Fault Location Technology
Filter
Fleet
Flight
Float
Fork Lift Truck
FLTCAL Flight Calibration Procedures
FLTCK Flight Check
FltLt Flight Lieutenant
FltSgt Flight Sergeant
FLTR Filter
Fusible Link Top Register
FLTSATCOM Fleet Satellite Communication System [US]
FLU Full Line-Up
FLUC Fluctuation
FLUIDICS Fluid Dynamics
FLULC Forest Land Use Liaison Committee
FLUOR Fluorescence [also: FLUORES]
Flux Luxembourg Franc [Currency]
FLV Freund Leukemia Virus
FLVR Flavour
FLW Feetlot Waste
FLWF Feetlot Waste Fiber
Frank Lloyd Wright Foundation [US]
FM Face Measurement
Fachnormenausschuss Maschinenbau [= Standards Committee on Mechanical Engineering, Germany]
Facilities Management
Factory Mutual

	Fan Marker
	Fault Modelling
	Feedback Mechanism
	Ferrimagnet(ism)
	Ferromagnetism
	Field Magnet
	Field Manual
	Field Marshall
	File Maintenance
	File Management
	Fine Measurement
	Fire Main
	Flexural Moment
	Fluid Mechanics
	Fluorescence Microscope
	Flowmeter
	Forest Management
	Fracture Mechanics
	Freezing Mixture
	Frequency Management
	Frequency Meter
	Frequency Modulation
	Frequency Multiplex
	Frequency Multiplier
Fm	Fermium
fm	fathom
	femtometer
FMA	Fabricators and Manufacturers Association [now: FMAI, US]
	Flexicore Manufacturers Association [US]
	Flight Mode Annunciation
	Food Machinery Association [UK]
	Forest Management Area
	Fragrance Materials Association [US]
FMAC	Frequency Management Advisory Council
FMAI	Fabricators and Manufacturers Association International [US]
FM/AM	Frequency Modulation/Amplitude Modulation
FMAR	Ferromagnetic Antiresonance
FMAS	Florida Marine Aquarium Society [US]
FMB	Federation of Master Builders [UK]
	Foundation for Microbiology
FMBRA	Flour Milling and Baking Research Association [UK]
FMC	Federal Maritime Commission [US]
	Felt Manufacturers Council [US]
	Fixed Message Code
	Flatness Measuring and Control
	Flexible Manufacturing Cell
	Flexible Manufacturing Center
	Flow Microcalorimeter
	Flutter Mode Control
	Forces Mobile Command [of Canadian Forces]
	Forward Motion Compensation
	Frequency-Modulated Cyclotron [of USAEC]
	Friele-MacAdam-Chickering
FMCE	Federation of Manufacturers of Construction Equipment [UK]
FMCS	Flight Management Computer System
FMCW	Frequency-Modulated Continuous Wave
FMD	Foot-and-Mouth Disease
	Frequency Multiplexing Division
	Function Management Data
FMDS	Fleet Management Demonstration System
FMDU	Fast Multiply/Divide Unit
FMDV	Foot-and-Mouth Disease Virus
FME	Field Moisture Equivalent
	Frequency Measuring Equipment
FMEA	Failure Mode and Effect Analysis
FMECA	Failure Modes, Effects and Criticality Analysis
FMEP	Friction Mean Effective Pressure
FMES	Full Mission Engineering Simulator
F-MET-PHE	Formyl-Methionyl-Phenylalanine
FMEVA	Floating Point Means and Variance
FMFB	Frequency Modulation with Feedback
FMF	Food Manufacturers Federation [UK]
FM-FM	Frequency Modulation — Frequency Modulation
FMG	Malagasy Franc [Currency of Madagascar]
FMGS	Flight Management and Guidance System
FMHA	Farmers Home Administration [also: FmHA]
FMHS	Flexible Materials Handling System
FMI	Frequency Modulation Intercity (Broadcasting)
FMIC	Frequency Monitoring and Interference Control
FMICW	Frequency-Modulated Intermittent Continuous Wave
FMIG	Food Manufacturers Industrial Group [UK]
FMIS	Fiscal Management Information System
FMIT	Fusion Materials Irradiation Test (Facility)
Fmk	Finnish Markka [Currency]
FML	Feedback, Multiple Loop
	File Manipulation Language
FMLF	File Management Loading Facility
FMLS	Full-Matrix Least Squares
FMN	Flavin Mononucleotide
	Formation
FMO	Fleet Mail Office [US Military]
	Free-Electron Molecular Orbital
FMOC	9-Fluorenylmethoxycarbonyl
FMOC-Cl	9-Fluorenylmethoxycarbonyl Chloride
FMOC-Gly	9-Fluorenylmethoxycarbonyl Glycine
FMOC-OBT	9-Fluorenylmethylcarbonate-O-Benzotriazolyl
FMOI	First Moment of Inertia
FMP	Flight Mechanics Panel [of AGARD]
	Formation Pressure
	Fructose Monophosphate
FMPE	Federation of Master Process Engravers [UK]
FMPP	Flexible Multi-Pipeline Processor
FMPS	FORTRAN Mathematical Programming System
	Functional Mathematical Programming System
FMPT	First Material Processing Test
FMR	Ferromagnetic Resonance
	Former
	Frequency-Modulated Radar
	Frequency-Modulated Receiver
FMRL	Ford Motor Research Laboratories [US]
FMRLY	Formerly
FMRR	Financial Management Rate-of-Return
FMS	Facility Mapping System
	Facility Monitoring System
	Facsimile Mail System
	Farm Management System
	Federation of Materials Societies [US]

	File Management Supervisor
	File Management System
	Financial Management System
	First Melt Sample
	Flexible Machining System
	Flexible Manufacturing System
	Flight Management System
	Flux Monitoring System
	Foreign Military Sales
	Forms Management System
	FORTRAN Monitor System
FMSA	Fellow of the Mineralogical Society of America
FMSC	Flexible Manufacturing System Complex
FMSI	Food Machinery Service Institute [US]
	Friction Material Standards Institute [US]
FMSL	Fort Monmouth Signal Laboratory [US Army]
FMSO	Fleet Material Supply Office [US Military]
FMSR	Fast Mixed Spectrum Reactor
	Finite Mass Sum Rule
FMSWR	Flexible Mild Steel Wire Rope
FMT	Flour Milling Technology
	Flush Metal Threshold
	Format
	Frequency-Modulated Transmitter
FMTA	Farm Machinery and Tractor Trade Association [Australia]
FMTP	File Management Transaction Processor
FMU	Flow Management Unit
FMV	Fair Market Value
	Formation Volume Factor
FMVE	Perfluoro Methyl Vinyl Ether
FMVSS	Federal Motor Vehicle Safety Standard [US]
FMWA	Fixed Momentum Wheel Assembly
FMX	Frequency Modulation Transmitter
FN	False Negative
	Family Name
	Ferrite Number
	Fibronectin
	Filter Network
	Flange Nut
	Flat Nose Projectile
	Foil Normal
	Function
	Functional Network
	Futures Network [UK]
FNA	Fachnormenausschuss [= Standards Committee, Germany]
	Final Approach
	Fujitsu Network Architecture
FNAA	Fast Neutron Activation Analysis
FNAB	Federation Nationale des Artisans du Batiment [= National Federation of Construction Workers, France]
FNAL	Fermi National Accelerator Laboratory [US]
FNB	Federation Nationale du Batiment [= National Federation of Building Trades, France]
	Food and Nutrition Board [US]
FNC	Ferritic Nitrocarburizing
FNCC	Foreign Claims Commission
FNCIAWPRC	Finnish National Committee of the International Association on Water Pollution Research and Control
FNDRY	Foundry
FNF	First Normal Form
FNG	Furnishing
FNH	Flashless Nonhygroscopic (Powder)
FNM	Fachnormenausschuss Materialprufung [= Standards Committee on Materials Testing, Germany]
F-NMR	Fluorinated Nuclear Magnetic Resonance
FNP	Floating Nuclear Plant
	Front-End Network Processor
	Fusion Point
FNPA	Foreign Numbering Plan Area
FNPS	4-Fluoro-3-Nitrophenyl Sulfone
FNPT	Fusion Point
FNR	Ferromagnetic Nuclear Resonance
	File Next Register
	Ford Nuclear Reactor [University of Michigan, US]
FNRC	Food and Nutrition Research Center [Philippines]
FNRS	Fonds National de la Recherche Scientifique [= National Fund for Scientific Research, Belgium]
FNS	File Nesting Store
	Food and Nutrition Service
	Fusion Neutron Source
FNSH	Finish
FNT	Fusion Nuclear Technology
FNTU	Federation of Norwegian Transport Users
FNUG	Federation of NCR (Computer) Users Groups [US]
FNWF	Fleet Numerical Weather Facility [US Navy]
FO	Factory Order
	Fail Open
	Fanout
	Fast Operate
	Fiber-Optic(s)
	Field Office
	Field-Officer
	Flashover
	Flash Override
	Flying Officer
	Forced Oscillation
	Forced-Oil (Cooling)
	Foreign Office [UK]
	Free Overside
	Fuel Oil
F&O	(Department of) Fisheries and Oceans [Canada]
F-O	Fiber-Optic(s)
fo	folio
FOA	Forced-Oil and Forced-Air Cooling
	Foreign Operations Administration [US]
	Forsvarets Forskningsanstalt [= National Defense Research Institute, Sweden]
	Free of Average
	Free on Aircraft
FOB	Forward Operating Base
	Free on Board
	Freight on Board
	Fuel on Board
FOBFO	Federation of British Fire Organizations
FOBS	Fractional Orbital Bombardment System

FOC Chemical Flux Cutting
Fiber Optic Communications
Fiber Optic Converter
Focus
Free on Car
FOCAL Formula Calculator
FOCH Forward Channel
FOC/LAN Fiber Optic Communications/Local Area Networks (Conference and Exposition)
FOD Flight Operations Department [of CAA, UK]
Foreign-Object Damage
Free of Damage
Functional Operational Design
1,1,1,2,2,3,3-Heptafluoro-7,7-Dimethyl-4,6-Octanedionato
FOE Figure-of-Eight (Reception)
Friends of the Earth [US]
FOEB Fuel-Oil Equivalent Barrel
FOEC First-Order Elastic Constant
Fourth-Order Elastic Constant
FOEF Friends of the Earth — France
FOEHK Friends of the Earth — Hong Kong [also: FOE(HK)]
FOEI Friends of the Earth — International
FOEP Friends of the Earth — Portugal
FOF Factory of the Future
FOG First Osborne Group [US]
FOG-M Fiber-Optic Guided Missile
FOGRA (Deutsche) Forschungsgesellschaft fur Druck- und Reproduktionstechnik [= German Research Society for Printing and Reproduction Technology]
FOH Forced Outage Hours
FOHMD Fiber-Optic Helmet-Mounted Display
FOI Follow-On Intercepter
Freedom of Information
FOIA Freedom of Information Act [US]
FOIF Free Oceanographic Instrument Float
FOIL File-Oriented Interpretative Language
FOL Facility Operating License
First-Order Language
First-Order Logic
Following
fol folio
fols folios
FOLZ First-Order Laue Zone
FOM Factor of Merit
Fiber Optic Modem
Figure of Merit
FOMC Federal Open Market Commission [US]
FOMCAT Foreign Material Catalogue
FOMR First-Order Moment Reorientation
FON Federation of Ontario Naturalists [Canada]
FONA Friends of the National Arboretum [US]
FONASBA Federation of National Associations of Shipbrokers and Agents [UK]
FOO Frequency of Optimum Operation
FOOD ENG Food Engineer
Food Engineering
FOOS Force out of Service
FOPS Falling Object Protective Structure
FOPT Fiber-Optic Photon Transfer

FOQ Free on Quay
FOR Forced Outage Rate
Foreign
Forestry
Free on Rail
Friends of the River [US]
FORAC Fisheries and Oceans Research and Advisory Council [Canada]
FORACS Fleet Operational Readiness Accuracy Check Site [US Navy]
FORAST Formula Assembler Translator
FORATOM Forum Atomique Europeen [= Association of European Atomic Forums, UK]
FORBLOC FORTRAN Compiled Block-Oriented Simulation Language
FORC Formula Coder
FORCE FORTRAN Conversational Environment
FORD Floating Ocean Research and Development Station
FORDACS Fuel-Oil Route Delivery and Control System
FORDAP FORTRAN Debugging Aid Program
FORDS Floating Ocean Research and Development Station
FOREM File Organization Evaluation Model
FORESDAT Formerly Restricted Data
Forest AIDS Forest Product's Abstract Information Digest Service [of FPRS, US]
FORG Forging
FORGE File Organization Generator
FORGO FORTRAN Load and Go
FORIMS FORTRAN Oriental Information Management Program
FORMAC Formula Manipulation Compiler
FORMAT FORTRAN Matrix Abstraction Technique
FORML FORTH Modification Laboratory [FIG committee]
FORS Forschungsprojekte aus Raumordnung, Stadtebau und Wohnungswesen [= Research Projects in Urban Planning and Housing (Databank), Germany]
FORST Forestry
FORTRAN Formula Translator
Formula Translation [Computer Language]
FORTRANSIT FORTRAN and Internal Translator System
FORTRUNCIBLE FORTRAN Style Runcible
FOS Fiber-Optic Sensor
Fiber-Optic System
Final Offer Selection
First-Order Spectrum
Fisheries Organization Society [UK]
Free on Ship
Fuel-Oxygen Scrap
FOSDIC Film Optical Scanning Device for Input to Computers
FOSE Federal Office Systems Exposition
FOSS Field Operations Support System
FOT Fiber Optics Technology
Forward Transfer
Free on Truck
Frequency Optimum Traffic
FOTA Fuels Open Test Assembly
FOTP Fiber-Optic Test Procedure
FOTS Fiber-Optic Transmission System

FOUND Foundation
FOV Field of View
FOW First Open Water
Forced-Oil and Forced-Water (Cooling)
Forge Welding
Free on Wagon
FP Faceplate
False Positive
Far Point
Feedback Positive
Feedback Potentiometer
Feeding Point
Filament Power Supply
File Protect
Filter Pump
Fine Paper
Fine Particle
Finger Pin
Fire Place
Fire Point
Fire Prevention
Fireproof
Fixed Point
Fixed Price
Flame Photography
Flame Photometry
Flameproof
Flash Point
Flexible Pavements [US]
Floating Point
Fluorochrome/Protein
Fluorophosphate
Focal Point
Focal Plane
Food Processing
Food Processor
Forepump
Forward Perpendicular
Four-Pole
Free Piston
Freezing Point
French Patent
Fructose Phosphate
Full Period
Function Processor
Fundamental Parameter
Fusible Plug
Fusion Point
F/P Flat Pattern
Fluorochrome/Protein
fp foot-pound
F£ Falkland Islands Pound [Currency]
4P Four-Pole
FPA Final Power Amplifier
Fire Protection Association [UK]
Flexible Packaging Association [US]
Floating Point Accelerator
Floating Point Arithmetic
Focal Plane Array
Focal Plane Assembly
Foreign Press Association [UK]
Free of Particular Average
Freight Prepaid and Allowed
Fusion Power Associates [US]
FP&A Freight Prepaid and Allowed
FPAA Federacion Panamericana de Asociaciones de Arquitectos [= Pan-American Federation of Architects Associations, Uruguay]
FPAC Flight Path Analysis and Command
Food Services Purchasing Association of Canada
FPAH Foundation for Preservation of the Archeological Heritage [US]
FPAP Floating Point Array Processor
FPAPA Forest Products Accident Prevention Association
FPASA Fire Protection Association of Southern Africa [South Africa]
FPC Federal Petroleum Commission
Federal Power Commission [now: FERC, US]
Federal Publisher's Committee [US]
Final Processing Center
Fire Pump Control
Fish Protein Concentrate
Flexible Program Control
For Private Circulation
Free-Programmable Controller
Functional Progression Chart
Fusion Power Core
FPCE Fission Products Conversion and Encapsulation
FPCH Foreign Policy Clearinghouse
FPCI Fluid Power Consultants International [US]
FPCIL Fukui Prefectural Ceramic Industry Laboratory [Japan]
FPCS Farm Planning Computer Service [UK]
Free Polar Corticosteroid
Fuel Pool Cooling System
FPCSTL Fission Product Control Screening Test Loop
FPD Focal Plane Deviation
Full Power Days
FPDA Fluid Power Distributors Association [US]
FPDD Final Project Design Description
FPDI Food Processing Development Irradiator
4PDT Four-Pole, Double-Throw [also: FPDT]
4PDT SW Four-Pole, Double-Throw Switch [also: FPDT SW]
FPDU FTAM (= File Transfer, Access, and Management) Protocol Data Unit
FPE Fire Protection Engineer
Force-Producing Element
FPEEPM Floor Proximity Emergency Escape Path Marking
FPFA Flexible Polyurethane Foam Association [now: PFA, US]
FPG Flat-Pulse Generator
FPGA Field Programmable Gate Array
FPH Floating Point Hardware
FPhysS Fellow of the Physical Society
FPI Food Service Packaging Institute [US]
FPIA Fluorescence Polarization Immunoassay
FPIB Food Production Inspection Branch [of Agriculture Canada]
FP-IMS Fixed-Point Ion Mobility Spectrometer
FPIS Forward Propagation Ionospheric Scatter
FPL Field Processing Language
Filed Flight Plan

Final Protective Line
Forest Products Laboratory [US]
Frequency Phase Lock
Functional Problem Logging

FPLA Fair Packaging and Labeling Act [US]
Field-Programmable Logic Array

FPLC Federal-Provincial Liaison Committee [Canada]

FPLF Field Programmable Logic Family

FPM Federal Personnel Manual
Feet per Minute [also: fpm]
File Protect Mode
Flashes per Minute
Floppy-Disk Processor Mode
Frames per Minute [also: fpm]
Frequency Position Modulation
Functional Planning Matrices

FPMH Failures per Million Hours

FPMI Forest Pest Management Institute [Canada]

FPMMC Federal-Provincial Mines Ministers' Conference [Canada]

FPM&SA Food Processing Machinery and Suppliers Association [US]

FPN Fixed Pattern Noise

FPP Fine Particle Processing
First Principal Plane
Fixed Point Protocol
Floating Point Package
Floating Point Processor

FPPS Flight Plan Processing System

FPQA Fixed Portion Queue Area

FPQINA Federal Plant Quarantine Inspectors National Association [now: NAAE, US]

FPR Farm Publications Report [US]
Federal Procurement Regulations
Flat Plate Radiometer
Flight Planned Route
Floating Point Register

FPRAC Federal Prevailing Rate Advisory Committee [US]

FPRC Flying Personnel Research Committee [UK]

FPRF Fats and Proteins Research Foundation [US]
Fireproof
Forest Products Research Foundation [US]

FPRI Fellow of the Plastics and Rubber Institute
Forest Products Research Institute [Japan and Philippines]

FPRL Forest Products Research Laboratory [UK]

FPRO Federal-Provincial Relations Office [Canada]

FPROM Field Programmable Read-Only Memory

FPRS Forest Products Research Society [US]

FPS Faceplate Starter
Fauna Preservation Society [now: FFPS, UK]
Feet per Second [also: fps]
Fellow of the Pathological Society
Fellow of the Philosophical Society
Fiber Placement System
Film Performance Score
Fine Particle Society [US]
Finite Population Sampling
Fire Protection Specialist
Floating Point Systems
Flocculated Polystyrene
Fluid Power Society [US]
Focus Projection and Scanning
Foot-Pound-Second (System) [also: fps]
Frames per Second [also: fps]

FPSB Fisheries Prices Support Board

FPSE Foot-Pound-Second Electrostatic System [also: fpse]
Free Piston Stirling Engine

FPSK Frequency and Phase-Shift Keying

4PSK Quadra-Phase-Shift Keying

FPSL Fission Product Screening Loop

FPSM Foot-Pound-Second Electromagnetic System [also: fpsm]

FPSP Federal Power Support Program

4PST Four-Pole, Single-Throw [also: FPST]

FPSTR Fund to Promote Scientific and Technical Research [Burkina Faso]

4PST SW Four-Pole, Single-Throw Switch [also: FPST SW]

FPSU Foot-Pound-Second Unit [also: fpsu]

4P SW Four-Pole Switch [also: FP SW]

FPT Female Pipe Thread
Finite Perturbation Theory
Fire Protection Technician
Frame Paperfeed Transport
Full Power Trial

FPTA Forest Products Traffic Association [US]

FPTC Forest Products Trucking Council [US]

FPTF Fuel Performance Test Facility

FPTS Forward Propagation Tropospheric Scatter

FPU Floating Point Unit

FPVPC Federation of Paint and Varnish Production Clubs [US]

FPWT Fire Protection Water Tank

FPY Failures per Year
First-Pass Yield

FQCY Frequency

FQHE Fractional Quantum Hall Effect

FQL Formal Query Language

FQLM Fung's Quasi-Linear Model

FQPR Frequency Programmer

FQS Friendly Query System

FQT Frequent

FR Fabric Reinforced
Faculty of Radiology
Failure Rate
False Rejection
Fast Reactor
Fast Release
Federal Register
Federal Reserve
Ferroresonance
Felicity Ratio
Fiber(glass) Reinforced
Field Regulator
Field Reversing
Field Rheostat
File Register
Fineness Ratio
Fire Resistance
Fire-Retardant
Flame Resistance

Flame-Retardant
Flash Radiography
Flash Ranging
Flight Refuelling
Floating Regulator
Flow Regulator
Fluorescent Radiation
Forced Release
Fragment
Frame
Frame Recognition
Frame Reset
France
Free Radical
French
Frequency Recorder
Frequency Response
Friday
Full Rate
Fundamental Research
Fusion Reactor

Fr Francium
Franklin [Unit]
French Franc [Currency of France, French Guinea, Guadeloupe, Martinique, Monaco, and St. Pierre and Miquelon]

FRA Federal Radio Act [US]
Federal Railway Administration [US]
Frequency Response Analyzer

FRAC Fraction
Fractionation

FRACR Fellow of the Royal Australasian College of Radiologists

FRAeS Fellow of the Royal Aeronautical Society

FRAG Fragment
Fragmentation

FRAIC Fellow of the Royal Architectural Institute of Canada

FRAM Fellow of the Royal Academy of Medicine
Ferroelectric Random-Access Memory

FRAMATOME Societe Franco-Americaine de Constructions Atomiques [= French-American Society for Nuclear Constructions, France]

FraMCoS Fracture Mechanics of Concrete Structures

FRAME Fund for the Replacement of Animals in Medical Experiments [UK]

FRANK Frequency Regulation and Networking Keying

FRAP Fuel Rod Analysis Program

FRAP-S Fuel Rod Analysis Program — Steady-State

FRAP-T Fuel Rod Analysis Program — Transient

FRAS Fellow of the Royal Astronomical Society

FRASTA Fracture-Surface Topography Analysis

FRAT Fiber-Reinforced Advanced Titanium

FRB Burundi Franc [Currency of Burundi]
Federal Reserve Bank [US]
Fisheries Research Board [Canada]
Forschungsreaktor Berlin [= Nuclear Research Reactor at Berlin [of HMI, Germany]]

FrB Burundi Franc [Currency of Burundi]

FR BEL From Below

FRBS Fellow of the Royal Botanical Society

FRC Fabric-Reinforced Ceramics
Fairchild Research Center [US]
Fast Reaction Concept
Fast Rescue Craft
Federal Radiation Council [Now Part of EPA, US]
Federal Radio Commission [US]
Fiber-Reinforced Cement
Fiber-Reinforced Ceramics
Fiber-Reinforced Composite
Fiber-Reinforced Concrete
File Research Council [UK]
Final Routing Center
Flight Research Center [of NASA]
Free Radical Chemistry
Full Route Clearance
Functional Residue Capacity
Future Requirements Committee [US]

FRCA Fire Retardant Chemicals Association [US]

Fr CFA CFA (= Communaute Financiere Africaine) Franc [Currency of Benin, Congo, Gabon, Ivory Coast, Niger, Senegal, Togo and Upper Volta]

FRCHS Franchise

FRCI Fibrous Refractory Composite Insulation

FRCM Fellow of the Royal College of Medicine

FRCMC Fiber-Reinforced Ceramic-Matrix Composites

FRCPath Fellow of the Royal College of Pathology

FRCS Forged Radius Clamp Straps

FRCTF Fast Reactor Core Test Facility [of USAEC]

FRCVS Fellow of the Royal College of Veterinary Surgeons

FRD Fast Reaction Dynamics
Fiber Resin Development
Functional Reference Device

FRDA Forest Resource Development Agreement [Canada]

FRDC Fusion Research and Development Center [of JAERI]

FR-DLP Frame Recognition — Data Link Processor

FREC Forestry Research and Education Center [Sudan]

FREconS Fellow of the Royal Economical Society

FRED Field Recovery Epitaxial Diode
Figure Reading Electronic Device
Fractionally Rapid Electronic Device
Front End for Databases [of GTEL, US]
Fund for Rural Economic Development [Canada]

FRED-FET Field Recovery Epitaxial Diode Field-Effect Transistor

FREDI Flight Range and Endurance Data Indicator

FREE Fund for Renewable Energy and the Environment [now: Renew America, US]

FREEBD Freeboard

FREL Feltman Research and Engineering Laboratory [US Army]

FREQ Frequency
Frequent

FREQMULT Frequency Multiplier

FRES Fellow of the Royal Economic Society
Fellow of the Royal Entomological Society
File Retrieval and Editing System

FRESCAN Frequency Scanning

FRESCANNAR Frequency Scanning Radar

FRESH Foam Removal for Environmentally Safe Housing [US]

FRET Freezing Rain Endurance Test
FRF Fragrance Research Fund [US]
French Franc [Currency of France, French Guinea, Guadeloupe, Martinique, Monaco, and St. Pierre and Miquelon]
FRFA Federal Regulatory Flexibility Act [US]
FRG Federal Republic of Germany
Forschungsreaktor Geesthacht [= Research Reactor at Geesthacht, Germany]
FRGS Fellow of the Royal Geographical Society
FRGSC Fellow of the Royal Geographical Society of Canada
FRGT Freight
FRH Fire Resistant Hydraulics (Program)
FRHB Federation of Registered Housebuilders [UK]
FRHistS Fellow of the Royal Historical Society
FRHortS Fellow of the Royal Horticultural Society
FRHS Fellow of the Royal Horticultural Society
FRI Flux Reversals per Inch [also: FR/I, or FRPI]
Food Research Institute [US]
Forest Research Institute [India, New Zealand]
Friday
Fuel Research Institute [South Africa]
Fulmer Research Institute [UK]
FRIBA Fellow of the Royal Institute of British Architects
FRIC Fellow of the Royal Institute of Chemistry
Forest Research Institute of Canada
FRICS Fellow of the Royal Institute of Chartered Surveyors
FRIL Fuzzy Relational Inference Language
FRIM Forest Research Institute of Malawi
FRIMP Flexible Reconfigurable Interconnected Multiprocessor
FRINGE File and Report Information Processing Generator
FRIP Fushun Research Institute of Petrochemistry [PR China]
FRIPHH Fellow of the Royal Institute of Public Health and Hygiene
FRISB Flathead River International Study Board
FRISA Fuel Research Institute of South Africa
Frisco San Francisco [US]
FRITALUX France, Italy, Benelux Economic Union
FRJ Forschungsreaktor Julich [= Nuclear Research Reactor at Julich [of KfA, Germany]]
FRL Feltman Research Laboratory [US Army]
Filter, Regulator and Lubricator Unit
Fisheries Research Laboratory [New Zealand]
FRM Farm
Fiber-Reinforced Metal
Flow Rate Meter
Forschungsreaktor Munchen [= Nuclear Research Reactor at Munich, Germany]
Frequency Response Method
FRMetS Fellow of the Royal Meteorological Society
FRMM Flux Reversals/Millimeter [also: FR/MM, or FRPMM]
FRMN Foreman
FRMR Farmer
Field Reversed Mirror Reactor
FRMS Fellow of the Royal Microscopical Society
FRMZ Forschungsreaktor Mainz [= Research Reactor at Mainz, Germany]
FRN Floating Rate Note
Furniture
FRNP Federal Radio Navigation Plan [US]
FRNS Fellow of the Royal Numismatic Society
FRO Fire Research Organization [now: JFRO, UK]
FROG Friends of Research and Odd Gadgets [of YNF, Canada]
FROM Field Programmable Read-Only Memory
Fusible Read-Only Memory
FROPA Frontal Passage
FROSFC Frontal Surface
FROZ Frozen
FRP Fares and Rates Panel [of ICAO]
Federal Radio Navigation Plan [US]
Fiberglass-Reinforced Plastic
Fiberglass-Reinforced Polyester
Fiber-Reinforced Plastics
Fiber-Reinforced Polyester
Free-Radical Polymerization
French Patent [also: FrP]
Fuel Reprocessing Plant
FRPI Flux Reversals per Inch [also: FRI, or FR/I]
FRPMM Flux Reversals/Millimeter [also: FRMM or FR/MM]
FRPS Flux Reversals per Second [also: FRS or FR/S]
FRR Functional Recovery Routine
Rwandan Franc [Currency of Rwanda]
FRRS Full Remaining Radiation Service [US]
FRS Federal Reserve System [US]
Fellow of the Royal Society [UK and Canada]
Fiber-Reinforced Superalloy
Financial Reporting System
Fire Research Station
Flux Reversals per Second [also: FR/S or FRPS]
Font Resource (File)
Forward Ready Signal
Fragility Response Spectrum
Free-Running Speed
Fruit Research Station [New Zealand]
FRSA Fellow of the Royal Society of Arts
FRSB Frequency-Referenced Scanning Beam
FRSC Fellow of the Royal Society of Canada
FRSKGD Fauna Research Section of the Kenya Game Department
FRSM Fellow of the Royal School of Mines
Fellow of the Royal Society of Medicine
FRSNA Fellow of the Royal School of Naval Architecture
FRSS Fellow of the Royal Statistical Society
FRSTP Fire-Refined Tough Pitch Copper with Silver
FRT Finish-Rolling Temperature
Freight
FRTEF Fast Reactor Thermal Engineering Facility
FRTISO Floating Point Root Isolation
FRTP Fiber-Reinforced Thermoplastic
Fiber-Reinforced Thermosetting Plastic
Fire-Refined Tough Pitch Copper
Fraction of Rated Power
FRT PPD Freight Prepaid
FRTO Flight Radio Telephony Operator
FRU Field Replaceable Unit
FRUGAL FORTRAN Rules Used as a General Applications Language
FRUSTUM Frame-Based Unified Story-Understanding Model

FRV Feedwater Regulation Valve
FRW Friction Welding
FRWI Framingham Relative Weight Index
FRWK Framework
FRXD Fully-Automatic Reperforator-Transmitter Distributor
FRZ Frozen
FRZR Freezer
FS Factor of Safety
Faraday Society [of RSC, UK]
Far Side
Fast-Slow Wave
Fast Store
Federal Specification
Federal Standard
Feedback, Stabilized
Female Soldered
Fiber Society [US]
Field Separator (Character)
Field Service
Field Stop
Field Strength
File Segment
File Separator (Character)
Filing System
Filtration Society [UK]
Final Sector
Final Selector
Final Splice
Fine Structure
Fiscal Service
Flame Spectrum
Flash Source
Flat Slip
Floating Sign
Fluorescent Screen
Fluid Statics
Foresight
Forest Service
Forged Steel
Forskningsstiftelsen Skogsarbeten [= Forest Operations Institute, Sweden]
Fourier Series
Fraunhofer Society [also: FhG, Germany]
Free Sterols
Freeze Substitution
Frequency Shift
Frequency Standard
Frequency Synthesizer
Full Scale
Full Size
Functional Schematic
Functional Specification
Function Select
Furnace Soldering
Fusible Link
F/S Fetch and Send
Fs Fracto-Stratus (Cloud)
fs femtosecond [also: fsec]
foot-second
4S Society for Social Studies in Science [US]

FSA Farm Safety Association [Canada]
Farm Security Administration [US]
Federation Suisse des Architectes Independants [= Swiss Federation of Independent Architects]
Fellow of the Society of Actuaries
Fellow of the Society of Arts
Field Search Argument
Financial Stationers Association [US]
Fluid Sealing Association [US]
Forward Sortation Area [Canada Post]
Fracture Surface Analysis
French Society of Acoustics
Fuming Sulfuric Acid
FSAA Fellow of the Society of Accountants and Actuaries
Flat Slips All Around
Flight Simulator for Advanced Aircraft
FSAAR Fraunhofer Society for the Advancement of Applied Research [also: FGFAF, Germany]
FSAC Finnish Society of Automatic Control
FSAE Fellow of the Society of Automotive Engineers
FSAR Final Safety Analysis Report
FSB Federal Specifications Board [US]
Flat Slip on Bottom
Functional Specification Block
FSBL Fusible
FSC Farm Safety Council [Canada]
Federal Supply Class [US]
Federal Supply Classification [US]
Field Studies Council [UK]
Flame-Sprayed Coating
Flight Safety Committee [UK]
Foundation for the Study of Cycles [US]
Free Secretory Component
FSCB File System Control Block
FSCC Federal Surplus Commodities Corporation [US]
Ferrous Scrap Consumers Coalition [US]
FSCI Frequency Space Characteristic Impedance
FSCR Field Select Command Register
FSCS Forged Straight Clamp Strap
FSCT Federation of Societies for Coating Technology [US]
FSD Flying Spot Digitizer
Full Scale Deflection
Full-Size Detail
FSDC Federal Statistical Data Center
FSDE Fission Suppressed Direct Enrichment
FSDO Flight Standards District Office [US]
FS/FD Flux System/Flux Delta
FSE Fellow of the Society of Engineers
Frankfurt Stock Exchange [Germany]
FSEC Federal Software Exchange Center [US]
fsec femtosecond [also: fs]
FSEI Food Service Equipment Industry [now: FEDA, US]
FSF Fading Safety Factor
Fibrin-Stabilizing Factor
Flash Smelting Furnace
Flight Safety Foundation [US]
Forensic Sciences Foundation [US]
FSFE Flash Smelting Furnace with Electrodes
FS/FW Flow of Steam/Flow of Water
FSG First Stage Graphitization
Flight Study Group [of ICAO]

FSGB Foreign Service Grievance Board
FSGO Floating Spherical Gaussian Orbitals
FSH Follicle-Stimulating Hormone
FSHDB (World) Fishing Catch Database [of FAO — UN]
FSHRH Follicle-Stimulating Hormone-Releasing Hormone
FSI Fellow of the Surveyors Institute
FSIH Fellow of the Society of Industrial Hygiene
FSIS Fast Sample Insertion System
FSJ Fuel Society of Japan
FSK Frequency Shift Keying
FSL Fail-Safe Logic
Flexible Satellite Link
Flexible System Link
Formal Semantic Language
French as a Second Language
Full Stop Landing
FSLP First Spacelab Program
FSLIC Federal Savings and Loan Insurance Corporation [US]
FSLP First Spacelab Payload
FSM Field Strength Meter
Finite Strip Method
Finite State Machine
FSMA Frame Screen Manufacturers Association [now: SMA, US]
FSMAC Fellow of the Society of Management Accountants of Canada
FSMWO Field Service Modification Work Order
FSN Federal Stock Number
Forward Specification Number
FSOS Free-Standing Operating System
FSP Fiber Saturation Point
Flight Safety Foundation [US]
Ford Satellite Plan
Frequency-Shift Pulsing
Frequency Standard, Primary
Full-Screen Processing
Fsp Fischerella Species [Molecular Biology]
FSPE Federation of Societies of Professional Engineers
Fellow of the Society of Plastics Engineers
FSPPR Fast Supercritical Pressure Power Reactor [US]
FSPT Federation of Societies for Paint Technology [now: FSCT, US]
FSQS Food Safety and Quality Service
FSR Farming Systems Research
Fast Sodium-Cooled Reactor
Feedback Shift Register
Field Service Regulations
Field Service Representative
Field Strength Ratio
File Storage Region
First Soviet Reactor [USSR]
Force Sensing Resistor
Forward Space Record
Foundation for Scientific Relaxation
Free Spectral Range
FS&R Filling, Storage and Remelt System
FSRA Federal Sewerage Research Association [US]
FSS Fatigue Striation Spacing
Federal Supply Services
Fire Sprinkler System
Fixed Satellite Service
Flight Service Station
Flight Standards Service
Floor Service Station
Flying Spot Scanner
Forensic Science Society [UK]
Formatted System Services
FSSA Fire Suppression Systems Association [US]
FSSC Foreign Student Service Council [US]
FST Federal Sales Tax
File Status Table
Flat Slip on Top
Flat Square Tube
Flavonol 3-Sulfotransferase
Forged Steel
Frequency Shift Transmission
Functional Subassembly Tester
Functional Systems Test
FSTA Food Science and Technology Abstracts (Database) [of IFIS, UK]
FSTC Field Sound Transmission Class
Foreign Science and Technology Center [US Army]
FSTMA Firearm and Security Trainers Management Association [US]
FSTNR Fastener
FSTU Fluid Sealing Technology Unit
FSTV Flat-Square Television
FSU Facsimile Switching Unit
Field Select Unit
Field Support Unit
Final Signal Unit
Florida State University [US]
Friedrich Schiller University [Germany]
FSUC Federal Statistics User Conference [US]
FSV Fire Service Valve
Floating-Point Status Vector
FSVM Frequency Selective Voltmeter
FSW Feet of Seawater [also: fsw]
Flexible Steel Wire
Forward-Swept Wing
FSWA Federation of Sewage Works Associations [US]
FSWR Flexible Steel Wire Rope
FSZ Fully Stabilized Zirconia
FT Fatigue Testing
Fault Tolerance
Film Thickness
Fine Thermal Black
Fire Technologist
Fire Technology
Firing Tables
Flat Top
Flame Tight
Flush Threshold
Food Technologist
Food Technology
Fort
Forward Transfer
Fourier Transform
Freight
Frequency and Time Committee [of IEEE IM, US]
Frequency Tracker

	Fruit
	Full Time
	Fume Tight
	Functional Test
	Fusion Technology
F&T	Fuel and Transportation
Ft	Forint [Currency of Hungary]
ft	foot
FTA	Fault Tree Analysis
	Federation of Trade Associations [Ireland]
	Field to Advice
	Flexographic Technical Association [US]
	Foreign Trade Association [Germany]
	Free Trade Agreement [Canada/US]
FTAB	Field Tab(ulator)
FTACCC	Florida Technical Advisory Committee on Citrus Canker [US]
FTAM	File Transfer, Access, and Management
	File Transfer and Access Method
FTAS	Fast Time Analysis System
ft bm	board foot meter**FTC**
	Fast Time Constant
	Fault Tolerant Compiler
	Federal Trade Commission [US]
	Frequency Time Control
	Fusion Technology Commission [of European Economic Community]
FT&C	Functional Test and Calibration
ftc	footcandle
ft-c	foot-candle
FT-CIDEC	Fourier Transform — Chemically Induced Dynamic Electron Polarization
FTD	Flight Training Device
	Folded Triangular Dipole
	Foreign Technology Division [of USAFSC]
	Frequency Translation Distortion
FTDV	Function Table Development and Verification
FTE	FFTF (= Fast Flux Test Facility) Test Engineering
	Flight Technical Error
	Flight Test Engineer
	Fracture Transition Elastic
	Full Time Equivalent
FTESA	Foundry Trades Equipment and Supplies Association [UK]
FTET	Full-Time Equivalent Terminals
FTF	Fault Transfer Facility
	Flared Tube Fitting
FTFET	Four-Terminal Field-Effect Transistor
FTG	Fitting
	Footing
FTH	Full Tree Harvesting
fth	fathom [also: fthm]
FTI	Facing Tile Institute [US]
	Federal Tax Included
	Fellow of the Textile Institute
	Foreign Traders Index [of USDOC]
	Fourier Transform Infrared
	Flux Transitions per Inch [also: FT/I, or FTPI]
FTIAP	Footwear and Tanning Industry Adjustment Program
FTIR	Fourier Transform Infrared [also: FT-IR]
	Fourier Transform Infrared Spectroscopy [also: FT-IR]
FTIRS	Fourier Transform Infrared Reflection Spectroscopy [also: FTIRRS]
FTIU	Fault Transient Interface Unit
FTL	Fast Transient Loader
	Federal Telecommunications Laboratory [US]
	Field Transmission Loss
	Flight Time Limitations
	Full Term License
ft L	foot-Lambert [also: ft-L]
FTLB	Flight Time Limitations Board
ft lb	foot-pound [also: ft-lb]
ft lbf	foot-pound force [also: ft-lbf]
FTM	Flat Tension Mask
	Flight Test Missile
	Folded Triangular Monopole
	Frequency Time Modulation
	Functional Test Manager
FTMC	Frequency and Time Measurement Counter
FTMP	Fault Tolerant Multiprocessor System
FTMS	Federal Test Method Standards
	Fourier Transform Mass Spectrometry
FTMT	Final Thermomechanical Treatment
FTN	Fountain
FTNIR	Fourier Transform Near Infrared
FTNMR	Fourier Transform Nuclear Magnetic Resonance [also: FT-NMR]
FTO	Failed to Open
	Flying Training Oeganization
	Fruit Traffic Organization [UK]
F to F	Face to Face
FTORAFUR	5-Fluoro-1-(Tetrahydro-2-Furfuryl)uracil
FTP	Federal Test Procedure
	FFTF (= Fast Flux Test Facility) Test Procedure [of USAEC]
	Field Terminal Platform
	File Transfer Protocol
	Fixed Term Plan
	Fructose Triphosphate
	Fuel Transfer Port
	Fuel Transfer Pump
ft pdl	foot-poundal [also: ft-pdl]
FTPI	Flux Transitions per Inch [also: FTI or FT/I]
FTR	Fast Test Reactor
	Feather
	Federal Travel Regulations
	Feed per Tooth per Revolution
	Fixed Transom
	Flat Tile Roof
	Frustrated Total Reflection
	Full Text Retrieval
	Functional Test Requirement
FTRC	Federal Telecommunications Records Center
FTRIA	Flow and Temperature Removable Instrument Assembly
FTS	Facsimile Text Society [US]
	Federal Telecommunications System [US]
	Field Test Set
	Flexible Track System
	Flight Telerobotic Servicer [of NASA]

	Foot Switch
	Fourier Transform Spectrometer
	Free-Time System
	Frequency and Timing Subsystem
FTSC	Fault Tolerant Spaceborne Computer
	Federal Telecommunications Standards Committee [US]
FTT	Flight Technical Tolerance
FTTA	Federal Technology Transfer Act [US]
FTTS	Flow Through Tube Sampler
FTTU	Field Technical Training Unit
FTU	First Time Use
	Flight Test Unit
	Formazin Turbidity Unit
	Fortbildungszentrum fur Technik und Umwelt [= Educational Center for Technology and the Environment, Germany]
	Frascati Tokamak Upgrade [An Italian Nuclear Reactor]
FTZ	Foreign Trade Zone
	Free Trade Zone
FTZB	Foreign-Trade Zones Board
FU	Finsen Unit [Physics]
	Fluorouracil
	Franklin University [US]
	Fukui University [Japan]
	Fukuoka University [Japan]
	Fume Concentration
	Functional Unit
	Fuse
fu	flux unit [Astrophysics]
5-FU	5-Fluorouracil
FUB	Free University of Berlin [Germany]
FUDR	Failure and Usage Data Report
	5-Fluorodeoxyuridine
FUE	Federated Union of Employers [UK]
FUFO	Fuel Fusing Option
FUFOR	Fund for UFO (= Unidentified Flying Object) Research [US]
FUIF	Fire Unit Integration Facility
FUM	Fuming
FUNC	Function
FUNCTLINE	Functional Line Diagram
FUND	Funding
FUNLIS	Fundamentals of Library and Information Science
FUNOP	Full Normal Plot
FUNY	Free University of New York [US]
FUR	Furring
fur	furlong [Unit]
FURN	Furnish [also: FUR]
FUROL	Fuel and Road Oil (Viscosity) [also: Furol]
FUS	FORTRAN Utility System
	Fuselage
FUSE	Far Ultraviolet Spectroscopic Explorer
	Federation for Unified Science Education [US]
FUT	Future
FUTZ	Fundamental Test Zone
FUV	Far Ultraviolet
FV	Fiber Volume
	Floor Valve
	Fluid Volume
	Flux Value
	Forced Vibration
	Front View
	Full Voltage
F/V	Frequency-to-Voltage (Converter)
fV	femtovolt
FVA	Floor Valve Adapter
FVC	Forced Vital Capacity
FVD	Front Vortex Distance
FVN	Failed Vector Numbers
FVPRA	Fruit and Vegetable Preservation Research Association [US]
FvPM	Fachverband Pulvermetallurgie [= Powder Metallurgy Association, Germany]
FVR	Flexible Vocabulary Recognition
	Fuse Voltage Rating
FVRDE	Fighting Vehicles Research and Development Establishment [of MOD, US]
FVS	Flight Vehicle Systems Committee [of IEEE AESS]
FVT	Flash-Vacuum Thermolysis
FVU	File Verification Utility
FVW	Forward Volume Wave
FW	Face Width
	Fachnormenausschuss Warmebehandlungstechnik metallischer Werkstoffe [= Standards Committee on Heat Treatment of Metallic Materials, Germany]
	Feed Water
	Field Weakening
	Filament Winding
	Filament-Wound
	Firmware
	First Word
	Fixed Width
	Flash Welding
	Flat Washer
	Formula Weight
	Fresh Water
	Full Wave
FWA	File Work Area
	First Word Address
	Fluorescent Whitening Agent
	Forward Wave Amplifier
	French Water Study Association
FWB	Fahrenheit Wet Bulb
	Four Wheel Brake
FWC	Fault Warning Computer
	Fine-Cut WC (= Tungsten Carbide)
	Fourdrinier Wire Council [US]
	Full-Wave Circuit
	Fully Loaded Weight and Capacity
FWCI	Foundation of the Wall and Ceiling Institute [US]
FWD	Forward
	Free Working Distance
	Fresh Water Damage
	Front-Wheel Drive
4WD	Four-Wheel Drive [also: FWD]
FWDC	Full-Wave Direct Current
FWDG	Forwarding
FWDR	Forwarder
FWE	Finished with Engines

FWERAT Fourth World Educational and Research Association Trust [UK]
FWGE Forth Worth Grain Exchange [US]
FWHF Federation of World Health Foundations [Switzerland]
FWHM Full Width at Half Maximum
FWHP Flywheel Horsepower [also: fwhp]
FWI Fire Weather Index
FWIC Federated Women's Institute of Canada
FWL Fixed Word Length
FWM Fachnormenausschuss Werkzeugmaschinen [= Standards Committee on Machine Tools, Germany]
Fourier Wave Mixing
FWONA Free World Outside North America
FWP Fair-Witness Project [US]
Federal Water Policy [Canada]
FWPCA Federal Water Pollution Control Act [US]
Federal Water Pollution Control Administration [US]
FWQA Federal Water Quality Administration [US]
FWR Full-Wave Rectifier
FWRC Federal Water Resources Council [US]
FWRM Federation of Wire Rope Manufacturers [UK]
FWRS Fish and Wildlife Reference Service [US]
FWS Fachnormenausschuss Werkzeuge and Spannzeuge [= Standards Committee on Tools and Fixtures, Germany]
Filter Wedge Spectrometer
Fish and Wildlife Service [US]
4W SW Four-Way Switch [also: FW SW]
FWTAO Federation of Women Teachers Associations of Ontario [Canada]
FWTM Full Width at Tenth Maximum
4WTS Four Wire Terminating Set [also: FWTS]
FWTT Fixed Wing Tactical Transport
FX Fixed Aera
Foreign Exchange
F(x) Function of x [also: f(x)]
FX-CCSA Foreign Exchange — Common Circuit Switching Arrangement
FXD Fixed
FXS Fine-Focus X-Ray Series
FXT Fixed Time Call
FY Fiscal Year
FYDP Five Year Defense Program [US]
FYI For Your Information
FYM Farmyard Manure
FYP Five Year Plan
FZ Float Zone
Fluorozirconate
Freezing
Fusion Zone
FZDZ Freezing Drizzle
FZES Float Zone Experiment System
FZF Float Zone Furnace
FZFG Freezing Fog
FZG Fluorozirconate Glass
FZI Forschungszentrum Informatik [= Research Center for Informatics, Germany]
FZP Fresnel Zone Plate
FZRA Freezing Rain
FZS Fellow of the Zoological Society

G

G Gage
Gain
Galvanometer
Gamma
Ganglion
Gap
Gas
Gate
Gauge
Gauss
Generation
Generator
German(y)
Gibbs Function
Giga- [SI Prefix]
Gilbert
Girder
Glass
Globe
Globular
Globule
Glucose
Glycine
Glycolic Acid
Glycyl
Gold
Gourde [Currency of Haiti]
Graph
Gravel
Gravity
Green
Greenwich Time
Grid
Ground
Grind(ing)
Guanine
Guarani [Currency of Paraguay]
Gulf

g gage
gallon
gate
gauge
gauss
gilbert
gilt
grain [Unit]
gram
gravity
grid
group
gulf

GA Gage
Gain of Antenna
Galactic Astronomy
Gallic Acid
Gas Absorption
Gas Amplification
Gas Analysis
Gauge
General Assembly [of UN]
General Average
General Aviation
Georgia [US]
Gibberellic Acid
Glide Angle
Global Address
Glutamin Acid
Glyoxylic Acid
Go Ahead Signal
Graphic Ammeter
Grapple Adapter
Gravimetric Analysis
Gypsum Association [US]

G&A General and Administrative

G/A General Average

G-A Graphite-Adhesive (Joint)
Ground-to-Air

Ga Gallium
Georgia [US]

GAA Gravure Association of America [US]

GAAEC Graphic Arts Advertisers and Exhibitors Council [US]

GAAG Gross Actual Generation

GaAlAs Gallium Aluminum Arsenide

GAAP General Adjustment Assistance Program
Generally Accepted Accounting Principles

GAAS Generally Accepted Auditing Standards
Graphite-Furnace Atomic Absorption Spectroscopy

GaAs Gallium Arsenide

GaAsFET Gallium Arsenide Field-Effect Transistor

GaAsMMIC Gallium Arsenide Monolithic Microwave Integrated Circuit

GAATS Gander Automatic Air Traffic System [Canada]

GAB Graphic Adapter Board
Gusseted Angle Bracket

GABA Gamma-Amino-N-Butyric Acid

GABOB Gamma-Amino-Beta-Hydroxybutyric Acid

GAC Geological Association of Canada
Granular Activated Carbon
Gross Available Capacity

GAC-MAC Geological Association of Canada — Mining Association of Canada

G/A CON General Average Contribution

GACS Gun Alignment Control System

GACTAI General Arbitration Council of the Textile and Apparel Industry [US]

GAD Glutamic Acid Decarboxylase

GADDS General Area Detector Diffraction System

GaDECl Gallium Diethylchloride

G/A DEP General Average Deposit

GADL Ground-to-Air Data Link

GADO General Aviation District Office [US]

GADS Generating Availability Data System [of NERC, US]

GAEC Greek Atomic Energy Commission [Greece]

GAeF Gesellschaft fur Aerosolforschung [= Association for Aersol Research, Germany]
GAELIC Grumman Aerospace Engineering Language for Instructional Checkout
GA&ES Georgia Architectural and Engineering Society
GAESDA Graphic Arts Equipment and Supply Dealers Association [US]
GAFB Griffiss Air Force Base [US]
GAG Glycosaminoglycan
Gross Available Generation
GAGB Gemmological Association of Great Britain
GAGR Group Automatic Gain Regulator
GAHF Grapple Adapter Handling Fixture
GAI Generalization Area of Intersection
Guaranteed Annual Income
GAIA Graphic Arts Industries Association
GAIBA Gamma-Aminoisobutyric Acid
GaInAs Gallium Indium Arsenide
GAINS Gimballess Analytic Inertial Navigation System
GAIU Graphic Arts International Union
GAIT General Aviation Infrastructure Tariff [Australia]
GAJ Guild of Agricultural Journalists [UK]
GaK Galactokinase
Gal Galactose [also: GAL]
Galileo [Unit]
gal gallon
GALCIT Guggenheim Aeronautical Laboratory of California Institute of Technology [US]
gal/d gallons per day [also: gal d^{-1}]
GALL Gallery
gal/min gallons per minute [also: gal min^{-1}]
Gal-1P Galactose-1-Phosphate
GALP Good Automated Laboratory Practices
GalP Galactose Phosphate
GALPAT Galloping Pattern Memory
Gal-PUT Galactose-1-Phosphate Uridyl Transferase
gal/sec gallons per second [also: gal/s or gal sec^{-1}]
GALV Galvanizing
Galvanometer
GAM Graduate in Aerospace Mechanical Engineering
Graphic Access Method
Ground-to-Air Missile
Guided Aircraft Missile
GAMA Gas Appliance Manufacturers Association [US]
General Aviation Manufacturers Association [US]
Graphics-Assisted Management Application
GAMAL Gamma Aluminum [also: Gamal]
GAMBIT Gate Modulated Bipolar Transistor
GAME Graduate in Aerospace Mechanical Engineering
GAMI Gorham Advanced Materials Institute [US]
GAMIS General Analytical Methods Information Service [of LGC, UK]
Graphic Arts Marketing and Information Service [of PIA, US]
GAMLOGS Gamma-Ray Logs
GAMM German Association for Applied Mathematics and Mechanics
GAMPS Gander Automated Message Processing System [of ICAO, Canada]
GAMS Groupement pour l'Avancement des Methodes Spectrographiques [= Working Group for the Advancement of Spectrographic Methods, France]
GAMT German Association of Medical Technologists
GAMTA General Aviation Manufacturers and Traders Association [UK]
GAN Generalized Activity Network
Generating and Analyzing Network
Generating and Assembly Network
GANIL Grand Accelerator National A Ions Lourds
GAO General Accounting Office [US]
GAP Gas and Particle (Sampler)
General Assembly Program
Glyceraldehyde Phosphate
Graphics Application Program
GAPA Ground-to-Air Pilotless Aircraft
GAPAN Guild of Airline Pilots and Navigators [UK]
G&A PAN Gyroscope and Accelerometer Panel
GAPDH Glyceraldehyde-3-Phosphate Dehydrogenase [also: GAPD]
GAPE Guyana Association of Professional Engineers
GAPMB Ghana Agricultural Produce Marketing Board
GAPP Geometric-Arithmetic Parallel Processor
GAPSFAS Graduate and Professional School Financial Aid Service
GAPT Graphical Automatically Programmed Tools
GAR Garage
Goat Anti-Rabbit [Immunochemistry]
Growth Analysis and Review
Guided Aircraft Rocket
GARBD Garboard
GARC Graphic Arts Research Center [US]
GARD General Address Reading Device
GARDAE Gather, Alarm, Report, Display and Evaluate
GAREX Ground Aviation Radio Exchange System
GARF Graphic Arts Research Foundation [US]
GARMI General Aviation Radio Magnetic Indicator
GARP Global Atmospheric Research Program [of NAS, US]
GARTEUR Group for Aeronautical Research and Technology in Europe
GAS Gasoline
General Aviation Services
Georgia Academy of Science [US]
Getaway Special [of NASA]
Goods Acquisition System
Graphics Attachment Support
GASCan Getaway Special Canister [also: GASCAN]
GASCO General Aviation Safety Committee [UK]
GASDA Gasoline and Automotive Service Dealers Association [US]
GASGASGAS Guild of Ancient Suppliers of Gas Appliance, Skills, Gins, Accessories and Substances [US]
GASL General Applied Science Laboratory
GASMPA Group for the Advancement of Spectroscopic Methods and Physicochemical Analysis [France]
GASO Gasoline
GASP General Activity Simulation Program
Generalized Academic Simulation Program
Grand Accelerated Space Platform
Graphic Applications Subroutine Package
GASS Generalized Assembly System

GASSAR	General Atomic Standard Safety Analysis Report
GAT	Gas Aggregation Technique
	General Air Traffic
	Generalized Algebraic Translator
	Georgetown Automatic Translator [US]
	Graphic Arts Terminal
	Greenwich Apparent Time
GATAC	General Assessment Three-Dimensional Analog Computer
GATB	General Aptitude Test Battery
GATCO	Guild of Air Traffic Control Officers [UK]
GATD	Graphic Analysis of Three-Dimensional Data
GATE	GARP (= Global Atmospheric Research Program) Atlantic Tropical Experiment [of NAS, US]
	Generalized Algebraic Translator Extended
	Group to Advance Total Energy
GATF	Graphic Arts Technical Foundation [US]
GATNIP	Graphic Approach to Numerical Information Processing
GATR	Ground-to-Air Transmitter Receiver
GATS	General Acceptance Test Software
GATT	General Agreement on Tariffs and Trade [of UN]
	Ground-to-Air Transmitter Terminal
GATTIS	Georgia Institute of Technology Technical Information Service [US]
GATU	Geophysical Automatic Tracker Unit
GATV	Gemini Agena Target Vehicle [of NASA]
GAUG	Georg-August University at Gottingen [Germany]
GAW	Guaranteed Annual Wage
GAWG	General Aviation Working Group [UK]
GAWR	Gross Axle Weight Rating
GaYIG	Gallium-Doped Yttrium Iron Garnet [also: Ga-YIG]
GB	Gain Bandwidth
	General Background
	Gibbs-Bogoliubov Principle
	Gigabyte
	Grain Boundary
	Gray Body
	Great Britain
	Greenish Blue
	Grid Base
	Grid Bias
	Guard Band
	Guide Block
	Gun Branch
Gb	Gigabit
	Gilbert
GBA	Grain Boundary Allotriomorph
GBAA	German Business Aviation Association
GBARC	Great Britain Aeronautical Research Committee
GBB	Groupement Belge du Beton [= Belgian Concrete Society]
GBBA	Glass Bottle Blowers Association [US and Canada]
GBC	George Brown College [Canada]
GBCD	Grain Boundary Character Distribution
GBD	Grain Boundary Diffusion
	Grain Boundary Dislocation
GBDL	Gesellschaft fur Bibliothekswesen und Dokumentation des Landbaues [= Association for Librarianship and Documentation in Agriculture, Germany]
GBF	Geographic Base File
	Gesellschaft fur Biotechnologische Forschung [= Society for Biotechnological Research, Germany]
	Grain Boundary Flow
	Great Bear Foundation [US]
GBFR	Grain Boundary Fracture Ratio
GBH	Group Busy Hour
GBHA	Glyoxal-Bis(hydroxyanil)
GBI	Gesellschaft fur Betriebswirtschaftliche Information [= Society for Business Information, Germany]
GBL	Gibraltar Pound [Currency]
GBLIC	Gaussian Band Limited Channel
GBM	Glomerular Basement Membrane
	Grain Boundary Melting
	Grain Boundary Migration
	Green Belt Movement [Kenya]
GBO	Goods in Bad Order
	Government Business Opportunities [Canada]
GBP	Gain Bandwidth Product
	Grain Boundary Plane
	Guinea-Bissau Peso [Currency of Guinea-Bissau]
	Pound Sterling [Currency of the United Kingdom and Northern Ireland]
Gbps	Gigabits per Second
GBQ	Generating Bearing Quality
GBq	Gigabecquerel
GBRS	Groupe Belge de Recherche Sous-Marine [= Belgian Submarine Research Group]
GBS	Grain Boundary Sliding
	Grain Boundary Slip
GBSR	Graphite-Moderated Boiling and Superheating Reactor
GBT	Generalized Burst Trapping
	Ground Based Test
GC	Gas Chromatography
	Gate Circuit
	Geiger Counter
	General Circular
	General Counsel
	Geochemist
	Geochemistry
	Geochronology
	Gigacycle
	Glass Ceramics
	Glassy Carbon
	Government of Canada
	Graduated Cylinder
	Grain Cube
	Grid Current
	Group Code
	Group Connector
	Guanine-Plus-Cytosine
	Guidance Computer
	Gun Cotton
	Gyro-Compass
G&C	Guidance and Control
Gc	Gigacycle
GCA	Gain Control Driver
	General Control Approach
	Glazing Contractors Association
	Government Contract Awards [US]

	Graphic Communications Association [US]
	Ground-Controlled Approach (Radar)
g-cal	gram-calorie
GCAP	Generalized Circuit Analysis Program
GCAW	Gas Carbon-Arc Welding
GCB	General Circuit Breaker
GCBR	Gas-Cooled Breeder Reactor
GCC	Gas Chromatograph Column
	Gasification Combined Cycle
	General Channel Coordinator
	Global Community Center
	Global Competitiveness Council
	Ground Control Center
	Gulf Cooperation Council
GCCA	Graphic Communications Computer Association [now: GCA, US]
GCD	Greatest Common Denominator
	Greatest Common Divisor
GCDRA	Green Crop Driers Research Association [UK]
GCE	Galveston Cotton Exchange and Board of Trade [US]
	General Certificate of Education [UK]
	General Consumers Electronics [US]
	Glassy Carbon Electrode
	Global Change Encyclopedia
	Ground Communication Equipment
	Ground Control Equipment
GC/ECD	Gas Chromatograph/Electron Capture Detector
GC ENV	Government of Canada, Environment Canada
GCF	General Control Function
	Greatest Common Factor
	Gross Capacity Factor
GCFAP	Guidance and Control Flight Analysis Program
GCFBR	Gas-Cooled Fast Breeder Reactor
GCFI	Gulf and Caribbean Fisheries Institute [US]
GCFR	Gas-Cooled Fast (Breeder) Reactor
GCFRE	Gas-Cooled Fast Reactor Experiment
GCGR	Gas-Cooled Graphite-Moderated Reactor
GCH	Gigacharacter [= 10^9 characters]
GCHQ	Government Communications Headquarters [UK]
GCHWR	Gas-Cooled Heavy Water Moderated Reactor
GCI	Generalized Communications Interface
	Genie Climatique International [= International Union of Heat and Air Conditioning and Ventilating Contractors, France]
	Graphics Command Interpreter
	Ground-Controlled Interception
	Ground-Controlled Intercept Radar
GC-IR	Gas Chromatography Infrared
	Gas Chromatography Infrared Spectroscopy
GCIRC	Groupe Consultatif International de Recherche sur le Colza [= International Consultative Research Group on Rape Seed, France]
GCIS	Ground Control Intercept Squadron
GC-ISS	Gas Chromatography Isotope Separation System
GCIU	Graphic Communications International Union
GCL	Gas-Cooled Loop
	(International) Gas Flow and Chemical Lasers (Symposium)
	Generic Control Language
GC/LC/MS	Gas Chromatography/Liquid Chromatography/Mass Spectroscopy
GCLISP	Golden Common LISP (= List Processor)
GCM	General Circulation Model [climatology]
	Generator Coordinate Method
	Greatest Common Measure
	Ground Check Monitor
g-cm	gram-centimeter
GCMA	Government Contract Management Association of America
GCMC	Glass Ceramic Matrix Composite
GCMI	Glass Container Manufacturers Institute [now: GPI, US]
GCMS	Gas Chromatography Mass Spectroscopy [also: GC/MS]
GCMSC	George C. Marshall Spaceflight Center [of NASA]
GCN	Gauge Code Number
GCOS	General Comprehensive Operating Supervisor
	General Comprehensive Operating System
	Great Canadian Oil Sands
GCP	Geometrically Close-Packed
	Guidance and Control Panel [of AGARD]
GCPS	Gigacycles per Second [also: GCS, GC/s or Gc/s]
GCR	Galvanocutaneous Reaction
	Gas-Cooled Reactor
	General Component Reference
	General Control Relay
	Group-Coded Recording
GCRE	Gas-Cooled Reactor Experiment [of INEL, US]
GCRI	German Carpet Research Institute [also: TFI, Germany]
	Glasshouse Crops Research Institute [UK]
GCS	Game Conservation Society [US]
	Gate-Controlled Switch
	General Communications System
	General Computer Systems [US]
	Gigacycles per Second [also: GCPS, GC/s or Gc/s]
	Graphic Compatibility System
	Ground Control System
Gc/s	Gigacycles per Second [also: GCS, GCPS or GC/s]
GCSE	General Certificate of Secondary Education [UK]
GCSG	Graphic Communications Societies Group [UK]
GCT	General Classification Test
	Graphics Communications Terminal
	Greenwich Civil Time
	Greenwich Conservatory Time
	Guard Control System
GCTS	Ground Communication Tracking System
GCU	Generator Control Unit
GCV	Gross Calorific Value
GCVS	General Catalogue of Variable Stars
GCW	Global Chart of the World
GCWM	General Conference of Weights and Measures
GD	Gas Detector
	Gas Discharge
	Gas Dynamics
	Gate Driver
	Gaussian Distribution
	Glow Discharge
	Good

	Grade
	Graphic Display
	Grinding Direction
	Ground Detector
	Group Delay
	Grown Diffused
	Guard
	Gyrodynamics
Gd	Gadolinium
G$	Guyana Dollar [Currency]
GDA	General Development Agreement [now: ERDA, Canada]
	Glycol Diacetate
GDB	Global Database
GDBMS	Generalized Database Management System
GDC	Gamma-Ray, Density, and Caliper Survey
	Garage Door Council [US]
	General Design Criteria
	Generalized Dynamic Charge
	Glow Discharge Condition
	Graphic Designers of Canada
	Gross Dependable Capacity
GDCH	Glycerol Dichlorohydrin
GDCI	Gypsum Drywall Contractors Association
GDDA	General Desert Development Authority [Egypt]
GDDL	Graphical Data Definition Language
GDE	Gage Deviation
	Generalized Data Entry
	Gibbs-Duhem Equation
	Ground Data Equipment
GDF	Gas Dynamic Facility [US Air Force]
	Group Distribution Frame
$Gd(FOD)_3$	Tris(heptafluorodimethyloctandionato)gadolinium
GDG	Generation Data Group
GDH	Glutamate Dehydrogenase
	Glycerophosphate Dehydrogenase
GDH-TPI	Glycerophosphate Dehydrogenase-Triosephosphate Isomerase
GDI	Gross Domestic Investment
GdIG	Gadolinium Iron Garnet
GDIFS	Gray and Ductile Ironfounders Society [US]
GDL	Gas Dynamic Laser
	Glow Discharge Lamp
	Graphic Display Library
GDM	Global Data Manager
GDMB	Gesellschaft Deutscher Metallhutten- und Bergleute [= German Society for Metallurgy and Mining]
GDME	Glycol Dimethyl Ether
GDMS	Generalized Data Management System
	Glow Discharge Mass Spectroscopy
GDN	Garden
	Government Data Network
GDNCE	Guidance
GDNR	Gardener
GDNS	Gardens
GDO	Grid-Dip Oscillator
	Gross Domestic Output
GDOES	Glow Discharge Optical Emission Spectroscopy
GDOP	Geometric Dilution of Precision
GDOS	Glow Discharge Optical Spectroscopy
GDP	Generalized Drawing Primitive
	Glow Discharge Polymer
	Goal Directed Programming
	Graphic Data Processing
	Gross Domestic Product
	Guanosine Diphosphate
GDPG	Guanosine Diphosphoglucose
GDPS	Global Data Processing System
GDR	General Design Requirement
	German Democratic Republic
	Giant Dipole Resonance
	Ground Delay Response
GDRS	Geoscience Data Referral System
GDS	General Declassification Schedule
	Glow Discharge (Optical) Spectroscopy
	Goods
	Graphic Data System
	Graphic Design System
	Graphic Display System
	Gross Debt Service
	Guards
GDSF	Generalized Data Structure Definition Facility
GDSU	Global Digital Service Unit
GDT	Gas Decay Tank
	Generator Development Tool
	Geometric Dimensioning and Tolerancing
	Graphic Display Terminal
GD&T	Geometric Dimensioning and Tolerancing
$Gd(THD)_3$	Tris(tetramethylheptanedionato)gadolinium
GDU	Graphic Display Unit
GDX	Gated Diode Crosspoint
GE	Gas Ejection
	Gas Engineer
	Gas Engineering
	Gas Engine
	Gateway Exchange
	Gauge
	Gaussian Elimination
	Generic Element
	Genetic Engineer
	Genetic Engineering
	Geological Engineer
	Geological Engineering
	Geothermal Engineer
	Geothermal Engineering
	Glass Electrode
	Gorkov and Eliashberg Theory
	Graduate Engineer
	Greater than or Equal to
	Guanidoethyl
Ge	Germanium
GEAP	General Electric Atomic Power
	General Electric Atomic Products
GEATP	Group of Experts on Air Transport Policies [of ICAO, Canada]
GEB	General Education Board [US]
GEBCO	General Bathymetric Chart of the Oceans
GEC	Generalized Equivalent Cylinder
GECRL	General Electric Corporation Research Laboratories [UK]
GECAP	General Electric Computer Analysis Program

GECOM General(ized) Compiler
GECOS General Comprehensive Operating Supervisor
General Comprehensive Operating System
GECOT Group of Experts on Costs and Tariffs
GECRD General Electric Corporate Research and Development
GED General Educational Development
General Equivalency Diploma
GEEI General Electric Electronic Installation
GEEIA General Electronics Engineering Installation
GEEP General Electric Electronic Processor
GEERS Groupe d'Etudes Europeen des Recherches Spatiales [= European Study Group for Space Research, France]
GEESE General Electric Electronic Systems Evaluator
GEF Gravure Education Foundation [US]
GEFA Gulf-European Freight Association [US]
GEFACS Groupement des Fabricants d'Appareils Sanitaires en Ceramiques [= Group of Ceramic Sanitary Equipment Manufacturers [of EEC, Belgium]]
GEFAP Groupement Europeen des Associations Nationales de Fabricants de Pesticides [= European Federation of National Associations of Pesticide Manufacturers]
GEFOAMS General Electric Foams Optimized through Analysis and Materials Selection
GEFRC General File/Record Control
GEG General Euclidian Geometry
GEGB General Electricity Generating Board
GEI Graphics Engine Interface
GEIS General Electric Information Service
Generic Environmental Impact Statement
GEISA Gestion d'Etudes des Informations Spectrographiques Atmospheriques [= Database on Spectroscopy of Planetary Atmospheres, France]
GEISHA Geodetic Inertial Survey and Horizontal Alignment
GEJ Group of Experts on Jurisdiction [of ICAO, Canada]
GEK Geomagnetic Electrokinetograph
GEL Gelatine
Gelatinous [also: GELAT]
General Emulation Language
GELOAD General Loader
GEM General Electronics Module
General Epitaxial Monolith
Generic Experiment Module
Geostatistical Evaluation of Mines
Governmental Energy and Minerals Committee [of TMS/AIME]
(National Consortium for) Graduate Degrees for Minorities in Engineering [US]
Graphic Engine Monitor
Graphics Environment Manager
Ground-Effect Machine
Ground Electromagnetic (Survey)
Graphics Environment Manager
GEMAC General Electric Measurement and Control [also: GE/MAC]
GEMC Grain Equipment Manufacturers Council [US]
GEMCOS Generalized Message Control System
GEMM Generalized Electronics Maintenance Model
GEMS General Education Management System
General Electric Manufacturing Simulator
Global Environmental Monitoring Service [of WHO, UN]
Ground Electromagnetic Survey
GEMSIP Gemini Stability Improvement Program [of NASA]
GEMT Group of European Metallurgical Thermodynamicists [Joint Group of German, British, French and Belgian Researchers]
GEN General
Generation
Generator
GENDA General Data Analysis
GENDARME Generalized Data Reduction, Manipulation and Evaluation
GENDAS General Data Analysis and Simulation
GENERIC Generation of Integrated Circuits
GENESIS GenRad's Environment for Strategy-Independent Software
GENESYS General Engineering System
Generalized System
Graduate Engineering Education System
GENIE General Information Extractor
GENIUS Genetic Interactive Unix System
GENL General
GEN MGR General Manager
GENSLA Generic Structure Language
GENTEC Genetic Technology Databank [of GBF, Germany]
GEO Geoscience Electronics [IEEE Group]
Geostationary Orbit
GEOARCHIVE Geology Archive (Database) [UK]
GeoBase Geography Database for the United States
GEOCHEM Geochemical
Geochemist
Geochemistry
GEOCOMP Geocoding and Compositing
GEOD Geodesic
Geodesy
GEODSS Ground-Based Electro-Optical Deep-Space Surveillance System
GEOF Geological Editor of Field Notes [Computer Software]
GEOGR Geographer [also: GEOG]
Geographic(al) [also: GEOG]
Geography [also: GEOG]
GEO-IRS Geostationary Orbit — Infrared Sensor
GEOIS Geographic Information System
GEOL Geologic(al)
Geologist
Geology
GeolE Geological Engineer
GEOM Geometric(al)
Geometrician
Geometry
GeoMATE Geographic Map Attribute Enhancement [also: GEOMATE]
GEON Gyro-Erected Optical Navigation
GEOPHYS Geophysical
Geophysicist
Geophysics
GEOPS Geodetic Estimates from Orbital Perturbations of Satellites

GEOREF (World) Geographic Reference System
Geological Reference File [of AGI, US]
GEORGE General Organizational Environment
GEOS Geodetic Earth Orbiting Satellite [of NASA]
Geosynchronous Earth Observation System
GEOSAR Geosynchronous Synthetic Aperture Radar
GEOSCAN Geological Survey of Canada
Ground-Based Electronic Omnidirectional Satellite Communications Antenna
GEOSECS Geochemical Ocean Sections Study
GEOTp Geographic Township
GEP General Equivalence Point
Goddard Experiment Package [of NASA]
GEPAC General Electric Process Automation Computer [also: GE/PAC]
General Electric Programmable Automatic Comparator
GEPVP Groupement Europeen des Producteurs de Verre Plat [= European Group of Flat Glass Manufacturers, Belgium]
GER German(y)
Gross Energy Requirement
GERD Gross Expenditure on Research and Development
GEREP Generalized Equipment Reliability Evaluation Procedure
GERG Groupe Europeeene de Recherche Gazieres [= European Gas Research Group, Netherlands]
GERSIS General Electric Range Safety Instrumentation System
GERT Graphical Evaluation and Review Technique
GERTS General Electric Range Tracking System
General Electric Remote Terminal Supervisor
General Electric Remote Terminal System
General Remote Terminal System
GESAC General Electric Self-Adaptive Control System
GESAMP Group of Experts on the Scientific Aspects of Marine Pollution [Italy]
GESC Government EDP (= Electronic Data Processing) Standards Committee [Canada]
GeSi Germanium Silicide
GESMO Generic Environmental Statement on Mixed Oxides
GESO Geodetic Earth-Orbiting Satellite
GESOC General Electric Satellite Orbit Control
GESS Grinding Energy Saving System
GESSAR General Electric Standard Safety Analysis Report
GET Gas Evaporation Technique
General Equivalence-Point Titration
Genetic Engineering Technologist
Genetic Engineering Technology
Geological Engineering Technologist
Geological Engineering Technology
Ground Elapsed Time
GETAC General Electric Telemetering and Control [also: GE/TAC]
GeTe Germanium Telluride
GETEL General Electric Test Engineering Language
GETIS Ground Environment Team of the International Staff [of NATO]
GETOL General Electric Training Operational Logic
Ground Effect Takeoff and Landing
GETR General Electric Test Reactor
GETS Ground Equipment Test Set
GEV Ground Effect Vehicle
GeV Gigaelectronvolt
GEVIC General Electric Variable Increment Computer
GEWS Group on Engineering Writing and Speech [of IEEE, US]
GF Gage Factor
Gas Flow
Gas Focusing
Glass Fiber
Gold Filling
Gorilla Foundation [US]
Gradient Furnance
Gravimetric Factor
Green Function
Ground Fault
gf gram force
GFA Gasket Fabricators Association [US]
Glass Formation Ability
Graphite Furnace Atomizer
GFAAS Graphite-Furnace Atomic Absorption Spectroscopy
GFAE Government-Furnished Aeronautical Equipment
Government-Furnished Avionics Equipment
GFAP Glial Fibrillary Acidic Protein
GFAPA Glycerine and Fatty Acid Producers Association [now: G&OA, US]
GFC Gas-Filled Cable
Gel Filtration Chromatography
GFCBS Glassy-Film-Coated Boundaries
GFCI Ground Fault Circuit Interrupter
GFCM General Fisheries Council for the Mediterranean [Italy]
GFD Geophysical Fluid Dynamics
GFDE Global Force-Displacement Equation
GFDNA Grain and Feed Dealers National Association [now: NGFA, US]
GFE Government-Furnished Equipment [US]
GFE&M Government-Furnished Equipment and Material [US]
GFF Glass Fiber Filter
GFFAPA Grain, Feed and Fertilizer Accident Prevention Association
GFFAR Guided Folding Fin Aircraft Rocket
GFI Ground Fault Interrupter
Guided Fault Isolation
GFID German Foundation for International Development
GFLI General Federation of Labour in Israel
GFLOPS Giga (= Billion) Floating Point Operations per Second
GFM Gas Flowmeter
Gesellschaft fur Marktforschung [= Society for Market Research, Germany]
Government-Furnished Materials [US]
Graphics Function Monitor
GFN Global Futures Network [India]
GFO Gap-Filler Output
GFP Generalized File Processor
Glass Fiber Pulling
Government-Furnished Property [US]

GFPBBD Groupement Francaise des Producteurs de Bases et Banques de Donnees [= French Association of Database and Databank Producers]
GFQ Gradient Furnace with Quenching Device
GFR Gas-Filled Rectifier
Gas-Filled Relay
Glomerular Filtration Rate
Ground Fault Monitor
Groupe Francais de Rheologie [= French Study Group on Rheology]
GFRP Glass-Fiber-Reinforced Plastic
Graphite-Fiber-Reinforced Plastic
GFS Government Finance Statistics [of IMF, US]
GFT Gas-Filled Tube
GFTU General Federation of Trade Unions [UK]
GFW Glass Filament Wound
Ground-Fault Warning
GFY Government Fiscal Year
GG Grain Growth
G-G Ground-to-Ground
Gg Gigagram
GGG Gadolinium Gallium Garnet [also: G^3]
GGR Gas Graphite Reactor
GGM Glacial Geomorphology
GGMC Guyana Geology and Mines Commission
GGOR Gross Gas-Oil Ratio
GGPD Gross Gas Produced
GGPNA Gamma-Glutamyl-P-Nitroanilide
G Gr Great Gross [Unit]
GGSE Gravity-Gradient Stabilization Experiment [of NASA]
GGT Gamma-Glutamyltransferase [also: G-GT]
GGTP Gamma-Glutamyltranspeptidase
GGTS Gravity-Gradient Test Satellite [of NASA]
GH Growth Hormone
GHA Grassland Heritage Association [US]
Greenwich Hour Angle
GHB Gamma-Hydroxybutyrate
GHC Ghanian Cedi [Currency of Ghana]
GHCP Georgia Hospital Computer Group [US]
GHF Generalized Hartree-Fock
Gradient Heating Facility
GHF-NO-CI Generalized Hartree-Fock/Natural Orbital/ Configuration Interactions
GHIUD Global Human Information Use per Decade
GHL Growth-Hormone Loaded (Cement)
GH/LCD Guest-Host/Liquid Crystal Display
GHOST Global Horizontal Sounding Technique
GHQ General Headquarters
GHRH Growth Hormone-Releasing Hormone
GHRI Guidance System, Hybrid Radio-Inertial
GHRIH Growth Hormone-Releasing Inhibiting Hormone
GHS Gill Hematoxylin Solution
GHSR Governor's Highway Safety Representative [US]
GHz Gigahertz
GI Galvanized Iron
General Issue [US Military]
Geodesic Isotensoid
Gesellschaft fur Informatik [= Society for Informatics, Germany]
Gold Institute [US]
Governmental and Industrial
Government-Initiated
Government Initiative
Government Issue [US Military]
Graded Index
Greenpeace International [UK]
Gross Investment
gi gill [Unit]
GIA Gemmological Institute of America [US]
General Industrial Application
General Industry Applications (Committee) [of IEEE IAS, US]
Gummed Industries Association [US]
GIAM (Conference on) Global Impacts of Applied Microbiology
GIANT Genealogical Information and Name Tabulating System
GIAO Gauge-Invariant Atomic Orbitals
GIASTA Groupement International pour l'Avancement des Sciences et Techniques Alimentaires [= International Working Group for the Advancement of Food Science and Technology, France]
GIB Gibraltar
GIC General Impedance Converter
General Input Channel
General Input/Output Channel
Generalized Immittance Converter
Graphite Intercalation Compound
Guaranteed Investment Certificate
GICQ General Investment Corporation of Quebec [Canada]
GID Gesellschaft fur Information und Dokumentation [= Society for Information and Documentation, Germany]
GIDEP Government-Industry Data Exchange Program [US]
GIEWS Global Information and Early Warning System [of FAO, UN]
GIF Gravito-Inertial Force
GIFA Internationale Giessereifachmesse [= International Foundry Trade Fair, Germany]
GIFAP Groupement International des Associations Nationales de Fabricants de Produits Agrochimiques [= International Group of National Association of Manufacturers of Agrochemical Products, Belgium]
GIFAS Groupement des Industries Francaises Aeronautiques et Spatiales [= Working Group of the French Aeronautics and Space Industries]
GIFS Generalized Interrelated Flow Simulation
GIFT Gas Insulated Flow Tube
General Internal FORTRAN Translator
GIGO Garbage-In, Garbage-Out
GIIP Groupement International de l'Industrie Pharmaceutique [= International Working Group of the Pharmaceutical Industry [of EEC, Belgium]]
GILSP Good Industrial Large-Scale Practice
GIM General(ized) Information Management
Generic Interface Module
GIMC Grinding of Industrial Minerals Conference
GIMIC Guard-Ring Isolated Monolithic Integrated Circuit
GIML Generalized Information Management Language

GIMRADA Geodesy Intelligence and Mapping Research and Development Agency [US Army]
GIMU Gimballess Inertial Measuring Unit
GIN Global Information Network [US]
GINA Fachnormenausschuss Giessereiwesen [= Standards Committee for Foundry Technology, Germany]
GINETEX Groupement International d'Etiquetage pour l'Entretien des Textiles [= International Association for Textile Care Labelling, Spain]
GINO Garbage-In, Nothing-Out
Graphical Input/Output
GIO Government Information Office
GIOC Generalized Input/Output Controller
GIOP General-Purpose Input/Output Processor
GIP Gastric Inhibitory Polypeptide
Gibraltar Pound [Currency]
General Information Program
Group Interface Processor
GIPB General-Purpose Instrument Bus
GIPS Ground Information Processing System
GIPSE Gravity-Independent Photosynthetic Exchange
GIPSY General Information Processing System
Generalized Information Processing System [of USGS]
GPS (= Global Positioning System) Inferred Positioning System
GIRD General Incentive for Research and Development [Canada]
GIRI Government Industrial Research Institute [Japan]
Gray Iron Research Institute [now: ICRI, US]
GIRIN Government Industrial Research Institute at Nagoya [Japan]
GIRL Generalized Information Retrieval Language [of USDNA]
German Infrared Laboratory
Graph Information Retrieval Language
GIRLS Generalized Information Retrieval and Listing System
GIS Gas Insulated Substation
Gas Insulated Switchgear
Gas Insulated System
Generalized Information System
Geographic Information System
Geophysical Incentive System
Geoscience Information Society [of AGI, US]
Grant Information System [US]
GISD Geographic Information Systems Division [of EMR, Canada]
GISL Groupement des Industries Siderurgiques Luxembourgeoises [= Working Group of the Luxembourg Iron and Steel Industry]
GISP Greenland Ice Sheet Program
GIT General Industrial Training
Georgia Institute of Technology [US]
Gesellschaft fur Informationsvermittlung und Technologieberatung [= Association for Information Brokerage and Technological Consultancy, Germany]
Granton Institute of Technology [Canada]
Graph Isomorphism Tester
GITL Gas Insulated Transmission Line
GITSIS Georgia Institute of Technology School of Information Science [US]
GITC Glucopyranosyl Isothiocyanate
GJ Gigajoule
Grown Junction
GJE Gauss-Jordan Elimination
GJP Graphic Job Processor
GK Greek
GKS Graphical Kernel System
GL Galvanoluminescence
Gas Lamp
Gate Leads
Geodesic Line
Geographic Location
Ginsburg-Landau Theory
Glass
Glaze
Glycolipid
Gothic Letter
Grade Line
Graphics Language
Green Library (Organization) [France]
Grid Leak
Gross Leak
Ground Level
Ground Line
Growth Ledge
g/L grams per Liter
gl gill [Unit]
GLAC Glacial
GLAG Ginzburg-Landau-Abrikosov-Gorkov (Theory)
Glam Glamorganshire [UK]
GLANCE Global Lightweight Airborne Navigation Computer Equipment
GLASS Germanium-Lithium Argon Scanning System
GLB Greatest Lower Bound
GLBC Great Lakes Basin Commission
GLC Gambia Labour Congress
Gas-Liquid Chromatography
German Library Conference [also: DBK, Germany]
Greater London Council [UK]
Great Lakes Commission [US]
GLCA Great Lakes Colleges Association [US]
GLCM Ground-Launched Cruise Missile
GLDC Glutamic Decarboxylase
GLDH Glutamic Dehydrogenase
GLDP Glutamic Dephosphatase
GLEAM Graphic Layout and Engineering Aid Method
GLEEP Graphite Low-Energy Experimental Pile [UK]
GLFRC Great Lakes Forest Research Center [Canada]
GLGST Geologist
GLHS Great Lakes Historical Society [US]
GLHK Granato-Luecke Theory for High Kelvin Temperatures
GLINT Global Intelligence
GLISP Great Lakes International Surveillance Plan
GLIT Glitter(ing)
GLLD Ground Laser Locator Designator
GLMI Great Lakes Maritime Institute [US]
Gln Glutamine
Glutaminyl

GLO Goddard Launch Operations [of NASA]
GLOB Globular
GLOC Gravity-Induced Loss of Consciousness
GLOK Granato-Luecke Theory for 0 (= Zero) Kelvin
GLOMEX Global Meteorological Experiment
GLONASS Global Orbiting Navigation Satellite System [USSR]
GLOPAC Gyroscopic Lower-Power Attitude Controller
GLORIA Geological Long-Range Inclined ASDIC (= Antisubmarine Detection Investigation Committee)
Glos Gloucestershire [UK]
GLOSS Global Ocean Surveillance System
Glossary
GLOTRAC Global Tracking Network
Glouc Gloucestershire [UK]
GLP Good Laboratory Practices
GLR Gas/Liquid Ratio
GLS Glass
GLSA Great Lake Seaplane Association [US]
GLU Great Lakes United [US]
Glu Glutamic Acid
Glutamyl
GLV Gemini Launch Vehicle [of NASA]
Gloves
Glove Valve
Graphische Versuchs- und Lehranstalt [= Institute for Graphical Research and Education, Austria]
Gross Leukemia Virus
GLWR Glassware
Gly Glycine
Glycyl
Glyc Glycerine
GM Galvanometer
Gaseous Mixture
General Manager
Geomagnetic
Geometric Mean
Geomorphology
Grand Master
Gravimetry
Greenwich Meridan
Grid Modulation
Group Mark
Guided Missile
Gun Metal
Gyromagnet(ism)
G-M Geiger-Muller (Tube) [also: GM]
gm gram [also: g]
GMA Gas Metal-Arc (Welding)
Geomechanics Abstracts [of RSM, UK]
Gesellschaft fur Mess- und Automatisierungstechnik [= Society for Instrumentation and Automation [of VDI/VDE, Germany]]
Glycol Methacrylate
Gravimetric Analysis
GMAC Gas Metal-Arc Cutting
General Motors Acceptance Corporation
GMAG Genetic Manipulation Advisory Group [UK]
GMAP Generalized Macroprocessor
General Macro Assembly Program
Geometric Modeling Application Program
GMAT Graduate Management Admission Test
Greenwich Mean Astronomical Time
GMAW Gas Metal-Arc Welding
GMAW-EG Gas Metal-Arc Welding, Electrogas
GMAW-P Pulsed Gas Metal-Arc Welding
GMAW-S Short Circuiting Gas Metal-Arc Welding
GMB Good Merchantable Brand [of LME, UK]
GMBS Gamma-Maleimidobutyryloxysuccinimide
GMC Geiger-Muller Counter
Gross Maximum Capacity
Ground Movement Controller
Groundwater Management Caucus [US]
GMCC General Motors Ceramic Committee [US]
GMCM Guided Missile Countermeasures
GMD Gambian Dalasi [Currency of Gambia]
Gesellschaft fur Mathematik und Datenverarbeitung [= Society for Mathematics and Data Processing, Germany]
Guided Missiles Division [US Air Force]
GMDA Groundwater Management Districts Association [now: GMC, US]
GMDEP Guided Missile Data Exchange Program [US Navy]
GMDH Group Method of Data Handling
GMDSS Global Maritime Distress and Safety System
GME Gelatin Manufacturers of Europe [of CEFIC, Belgium]
Generalized Machine Equation
Generic Macro Expander
GMEM Glascow Modified Eagle's Medium
GMF Galactic Magnetic Field
GMFB Gang Mill Fixture Base
GMFC Gem and Mineral Federation of Canada
GMFCS Guided Missile Fire Control System
GMFP Geometrical Mean Free Path
GMG Gross Maximum Generation
GMI Gelatin Manufacturers Institute [US]
General Motors (Engineering and Management) Institute
GMIS Generalized Management Information System
Government Management Information Sciences [US]
GML Generalized Mark-Up Language
Generic Mark-Up Language
Graphic Machine Language
g/mL grams per milliliter
GMM General Matrix Manipulator
GMO Glycol Monoacetate
GMP Geometric Modelling Project [UK]
Good Manufacturing Practices
Ground Movement Planner
Guanosine Monophosphate
GMR General Modular Redundancy
Giant Monopole Resonance
Greatest Meridional Radius
Ground Mapping Radar
Ground Movement Radar
GMRC General Motors Research Center [US]
GMRL General Motors Research Laboratories [US]
Group of Mathematicians of Romance Languages [Portugal]
GMRWG Guided Missile Relay Working Group [of US Navy]

GMS Gas Mass Spectrometry
Gemini Mission Simulator [of NASA]
Generalized Main Scheduling
General Maintenance System
Geometric Modeling System
Geostationary Meteorological Satellite [Japan]
Glycerol Monostearate
Ground Maintainance Support
GMSFC George Marshall Spaceflight Center [of NASA]
GMSK Gaussian Filtered Minimum Shift Keying
GMSS Graphical Modelling and Simulation System
GMT Generalized Machine Theory
Generalized Multitasking
Glass Mat Thermoplastics
Global Money Transfer
Graphics Mouse Technology
Greenwich Mean Time
GMTC General Motors Technical Center [US]
GMV Gram-Molecular Volume [also: gmv]
Guaranteed Minimum Value
GMW Gram-Molecular Weight [also: gmw]
GM/WM Group Mark/Word Mark
GMWU General Municipal Workers Union [UK]
GN Green
Group Number
G&N Guidance and Navigation
GNAT Grade-Nine Achievement Test [Jamaica]
GNC Global Navigation Chart
Graphic Numerical Control
GND Ground
GNE Gross National Expenditure
GNF German Nuclear Forum
GNI Generation of New Ideas
GNMA Government Nation Mortgage Association
GNMS Ground Network Management System
GNP Grain Neutral Spirits
Gross National Product
GnRH Gonadotropin-Releasing Hormone
GNS Guinea Syli [Currency]
GNSI Guild of Natural Science Illustrators [US]
GNSS Global Navigation Satellite System
GO Gaussian Orbitals
General Office
General Order
Generated Output
Glycerin Oleate
Graphitic Oxide
G/O Gas/Oil Ratio
GΩ Gigohm
G&OA Glycerine and Oleochemicals Association [US]
GOAL Ground Operations Aerospace Language
GOAM Government Owned and Maintained
GOC Gas/Oil Contact
General Officer Commanding
Graphic Option Controller
Greatest Overall Coefficient
GOCI General Operator-Computer Interaction
GOCINC General Officer Commanding in Chief [also: GOC in C]
GOCO Government-Owned, Contractor-Operated
GOCR Gated-Off Controlled Rectifier
GOD Glucose Oxidase
GOE Guinea Ekwele [Currency of Equatorial Guinea]
Ground Operating Equipment
GOe Gauss Oersted
GOES Geostationary Operational Environmental Satellite
GOI Gate Oxide Integrity
GOIC (Persian) Gulf Organization for Industrial Consulting [Qatar]
GOL General Operating Language
Goal-Oriented Language
Graphic On-Line Language
GOM Grain and Oil Seeds Marketing Incentives Program [Canada]
Group Occupancy Meter
GOMAC Government Microcircuit Applications Conference
GOP General Operational Plot
GOR Gained Output Ratio
Gas/Oil Ratio
General Operational Requirement
Gross Overriding Royalty
GORI Gross Overriding Royalty Interest
GORID Ground Optical Recorder for Intercept Determination
GORX Graphite Oxidation from Reactor Excursion
GOS Gate Operating System
Global Observing System [of WMO, Switzerland]
Grade of Service
Graphics Operating System
GOSIP Government OSI (= Open Systems Interconnection) Profile [US]
GOSIPS Government OSI (= Open Systems Interconnection) Procurement Specification [US]
GOSP Gas-Oil Separation Plant
GOSS Ground Operational Support System
GOSUB Go to Subroutine
GOT Glutamic-Oxalacetic Transaminase
GOTH Gothic
GO Train Government of Ontario (Commuter) Train [Canada]
GO Transit Government of Ontario Transit System [Canada]
GOTS Gravity-Oriented Test Satellite [of NASA]
GOU Gourde [Currency of Haiti]
GOV Governor
GOV-GEN Governor-General
GOVT Government [also: GOV]
GOX Gaseous Oxygen [also: GOx or gox]
GOx Glucose Oxidase
GP Gang Punch
Generalized Programming
General Preferential Tariff
General Processor
General Purpose
Geological Prospecting
Geometric Progression
Geophysics
Glide Path
Glycerophosphate
Gold Point
Graduate in Pharmacy
Graphics Processor
Ground Protection

	Group
	Guanosine Phosphate
	Guaranteed Performance
	Guinier-Preston (Zone) [also: G-P]
	Gun Powder
gp	guinea pig [Immunochemistry]
G-5'-P	Guanosine 5'-Monophosphate
GPA	Gas Processors Association [US]
	General-Purpose Amplifier
	General-Purpose Analysis
	General-Purpose Array
	Glycerophosphoric Acid
	Grade Point Average
	Graphical PERT (= Production Evaluation and Review Technique) Analog
	Groupement de Productivite Agricole [= Group for Agricultural Productivity, France]
GPa	Gigapascal
GpA	Guanylyladenosine
GPAC	General-Purpose Analog Computer
	Government and Public Affairs Committee
	Great Plains Agricultural Council [US]
GPAD	Gallons per Acre per Day [also: gpad]
GPAS	General-Purpose Airborne Simulator
GPATS	General-Purpose Automatic Test System
GPAX	General-Purpose Automation Executive
GPB	Ground Power Breeder
GPC	Gas-Pressure Cable
	Gel-Permeation Chromatography
	General Peripheral Controller
	General-Purpose Computer
	Georgia Peanut Commission [US]
	Graphics Performance Characterization
	Gross Profit Contribution
GpC	Guanylylcytidine
GPCA	General-Purpose Communications Adapter
GPCL	General-Purpose Closed Loop
GPCP	Generalized Process Control Programming
GpCpC	Guanylylcytidylylcytidine
GPD	Gallons per Day [also: gpd]
	General Protocol Driver
	General-Purpose Discipline
	Grams per Denier [Unit]
GPDC	General-Purpose Digital Computer [also: GP/DC]
GPDH	Glucose-Phosphate Dehydrogenase
	Glyceraldehyde Phosphate Dehydrogenase
G-6-PDH	Gluclose-6-Phosphate Dehydrogenase
gpDm	Geopotential Decameter
GPDS	General-Purpose Display System
GPED	Gas Phase Electron Diffraction
GPES	Ground Proximity Extraction System
GPF	Generalized Production Function
	General Planning Forecast
	General-Purpose Furnace Black
GPG	Grains per Gallon
GpG	Guanylylguanosine
gpg	guinea pig [Immunochemistry]
GPGAP	Great Plains Gasification Associates Project [US]
GPGL	General-Purpose Graphic Language
GPH	Gallons per Hour [also: gph]
	Geophysics
	Graphite
GPI	Glass Packaging Institute [US]
	Glucose Phosphate Isomerase
	Ground Point of Intercept
	Ground Position Indicator
GPIA	General-Purpose Interface Adapter
	Generic Pharmaceutical Industry Association [US]
GPIB	General-Purpose Instrumentation Bus
	General-Purpose Interface Bus
GPIB/IEEE	General-Purpose Interface Bus/Insitute of Electrical and Electronics Engineers
GPIO	General-Purpose Input-Output
GPIS	Gemini Problem Investigation Status [of NASA]
GPL	Generalized Programming Language
	General-Purpose Language
	General-Purpose Loader
	Group Processing Logic
GPLAN	Generalized Database Planning System
GPLP	General-Purpose Linear Programming
GPM	Gallons per Minute [also: gpm]
	Generalized Perturbation Method
	General-Purpose Macrogenerator
	Geopotential Meter
	Ground Potential Model
	Groups per Message
GPMA	German Productivity and Management Association [also: RKW, Germany]
GPMS	General-Purpose Microprogram Simulator
GPNA	Glutamyl-P-Nitroanilide
GPO	General Post Office
	Government Printing Office [US]
GPO-PIA	Government Printing Office and Printing Industry of America [US]
GPOS	General-Purpose Operating System
GPP	General Plant Projects
	Geophysical Prospecting
	General Print and Punch
	Gross Provincial Product [Canada]
GPPT	Gel Precipitate
GPR	Gas Production Rate
	General-Purpose Radar
	General-Purpose Rubber
	Ground-Penetrating Radar
	Ground-Probing Radar
GPRF	General-Purpose Rocket Furnace
GPRF-G	General-Purpose Rocket Furnace — Gradient
GPRF-I	General Purpose Rocket Furnace — Isothermal
GPRMC	Groupe des Plastiques Renforces et Materiaux Composites [= Organization of Reinforced Plastics and Composite Materials, Belgium]
GPS	Gallons per Second [also: gps]
	Gas-Pressure Sintering System
	General Problem Solver
	General Processing System
	Global Positioning System
	Graphic Programming Services
GPSA	Gas Processors Suppliers Association [US]
GPSCS	General-Purpose Satellite Communications System
GPSDW	General-Purpose Scientific Document Writer
GPSP	General-Purpose Simulation Program
GPSS	General Problem Statement Simulator

	General-Purpose Simulation System
	General-Purpose Systems Simulation
GPSSN	Gas Pressure Sintered Silicon Nitride
GPT	Gas Phase Titration
	Gas Power Transfer
	Gemini Pad Test [of NASA]
	General Preferential Tariff
	General-Purpose Terminal
	Glutamic-Pyruvic Transaminase
GPT-C	Glutamic-Pyruvic Transaminase-C
GPTM	Gross Profit this Month
GPTY	Gross Profit this Year
GPU	Geopotential Unit [also: gpu]
	Ground Power Unit
GpU	Guanylyluridine
GpUpG	Guanylyluridylylguanosine
GPWS	Ground Proximity Warning System
GPX	Generalized Programming Extended
GPZ	Guinier-Preston Zone
GQE	Generalized Queue Entry
GQR	Giant Quadrupole Resonance
GR	Gamma Radiation
	Gamma Ray
	Gear
	General-Purpose Register
	General Reconnaissance
	General Register
	General Relativity
	General Reserve
	Genetic Resources
	Germanium Rectifier
	Glass-Reinforced
	Glutathione Reductase
	Government Reserve
	Grade
	Grain
	Gravitational Radiation
	Gray
	Greece
	Greek
	Grid Resistor
	Grid Return
	Group
	Guanidine Rhodanate
	Guard Ring
Gr	Graphite
gr	grain
	gram
	gross
GRA	Gamma-Ray Absorption
	Government Report Announcements (Database) [of NTIS, US]
GRACE	Graphic Arts Composing Equipment
GRAD	Generalized Remote Access Database
	General Recursive Algebra and Differentiation
	Gradient
	Graduate
	Graduate Resume Accumulation and Distribution
	Graduation
GRADB	Generalized Remote Access Database
GradDip	Graduation Diploma
GRADS	Generalized Remote Access Database System
GRAF	Graphic Addition to FORTRAN
GRAIN	Graphics-Oriented Relational Algebraic Interpreter
GRAMPA	General Analytical Model for Process Analysis
GRAN	Global Rescue Alarm Network
	Granular
GRANADA	Grammatical Non-Algorithmic Data
GRANAS	Global Radio Navigation System [Germany]
GRANIS	Graphical Natural Inference System
GRAO	Gamma-Ray Astronomy Observatory
GRAPDEN	Graphic Data Entry Unit
GRAPE	Gamma-Ray Attenuation Porosity Evaluator
GRAPH	Graphic(s)
GRAPHDEN	Graphic Data Entry
GRAPL	Graphic Application Programming Language
GRARR	Goddard Range and Range Rate [of NASA]
GRAS	Generally Recognized as Safe [of FDA, US]
GRASER	Gamma-Ray Amplification by Stimulated Emission of Radiation [also: Graser or graser]
GRASP	Generalized Read and Simulate Program
	Generalized Remote Acquisition and Sensor Processing
	Generalized Retrieval and Storage Program
	Graphical Robot Applications Stimulation Package
	Graphic Service Program
GRATIS	Generation, Reduction, and Training Input System
grav	acceleration of gravity
GRB	Gas Research Board [UK]
	Geophysical Research Board [of NRC, US]
	Government Reservation Bureau
GR BR	Great Britain [also: GR BRIT]
GRC	Geotechnical Research Center [Canada]
	Geothermal Resources Council [US]
	Graduate Research Center
	Graphite-Reinforced Composite
GRCA	Glassfiber Reinforced Cement Association [UK]
GRCDA	Governmental Refuse Collection and Disposal Association [US]
GR/CIDS	Genetic Resources/Communication, Information and Documentation System [of FAO, UN]
GRCSW	Graduate Research Center of the Southwest [US]
GRD	Gamma-Ray Detector
	Greek Drachma [Currency of Greece]
	Grind
	Ground
	Ground Detector
	Ground-Resolved Distance
GRDF	Gypsum Roof Deck Foundation [now: NRDCA, US]
GRDN	Garden
GRE	Graduate Record Examination (Board)
	Graduate Reliability Engineering
	Ground Run-Up Enclosure
GREAT	Graduate Research in Engineering and Technology
GREB	Galactic Radiation Experiment Background (Satellite)
GRED	Generalized Random Extract Device
GREMEX	Goddard Research Engineering Management Exercise [of NASA]
GREP	Graphite/Epoxy Composite [also: Gr/Ep]
GREPCO	Greenland Petroleum Consortium
GRETA	Ground Radar Emissions Training Aviator

GRF Gesneriad Research Foundation [US]
Grassland Research Foundation [US]
Gravity Research Foundation
Ground Repetition Frequency
Growth-Hormone Releasing Factor
GRG Glass-Fiber Reinforced Gypsum
(International Society for) General Relativity and Gravitation [Switzerland]
Geodesy Research Group
Graphical Rewriting Grammar
Gross Reserve Generation
GRI Gas Research Institute [US]
Grassland Research Institute [US]
Group Repetition Interval
GRID Graphic Interactive Display
GRIN Graded Refractive Index
Gradient Refractive Index
Gradient Index
Graphic Input
GRINDER Graphic Interactive Network Designer
GRINS General Retrieval Inquiry Negotiation Structure
GRIP General Retrieval of Information Program
Graphics Interactive Program
GRIPHOS General Retrieval and Information Processing for Humanities-Oriented Studies
GRIPS General Relation-Based Information Processing System
GRISA Groupe de Recherche de l'Information Scientifique et Automatique [= Research Group on Scientific and Automatic Information [of EURATOM]]
GRIST Grazing-Incidence Solar Telescope
GRIT Graduated Reduction in Tension
GRL Gamma Ray Laboratory [Germany]
GRM Generalized Reed-Muller Code
Global Range Missile
grm gram
GRN Green
GRNSH Greenish
GRO Gamma-Ray Observatory [of NASA]
Gross
GROATS Graphical Output Package for Atlas
GROM Grommet
GROUT Graphical Output
GRP Gastrin-Releasing Peptide
Glass-Reinforced Plastic
Group
Group Reference Pilot
Group Reference Point
GRR Geneva Radio Regulations
Greek Research Reactor [Greece]
Guidance Reference Release
GR&R Gage Repeatablilty and Reproducibility
Gage Reproducibility and Reliability
GRS Gamma-Ray Spectrometry
Gamma-Ray Spectroscopy
Gamma-Ray Spectrum
Gears
Generalized Retrieval System
General Radio Service [Canada]
General Register Stack
General Reporting System
German Research Society [also: DFG, Germany]
GR-S General-Purpose Synthetic Rubber
GRSS Geoscience and Remote Sensing Society [of IEEE, US]
GRT General Reactor Technology
General Relativity Theory
Gross Register Tonnage
GRTG Grating
GRTP Gamma-Ray Transition Probability
GRTS General Electric Remote Terminal Supervisor
General Electric Remote Terminal System
General Remote Terminal System
GRU Gyro Reference Unit
GRV Groove
GRW Grower
GRWT Gross Weight [also: gr wt]
GR WT Grain Weight
Gross Weight [also: gr wt]
GS Galvanized Steel
General Secretary
General Semantics
General Solution
General Staff
Generating Station
Geochemical Society [US]
Geographical Society
Geological Society [UK]
Geological Survey
Geophysical Signal
Geosynthesis
Geosynthetics
Giemsa Stain
Glaciological Society [now: IGS, UK]
Glide Slope
Gold Standard
Graduate School
Grain Size
Ground Speed
Ground Stopper
Ground Support
Group Selector
Group Separator (Character)
Guild of Surveyors [UK]
Gs Gauss
GSA General Services Administration [US]
General Services Agencies
General Syntax Analyzer
Genetics Society of America [US]
Geographical Society of America
Geological Society of America [US]
Greenhouse Supplier Association [US]
Geophysical Signal Analysis
Guaranteed Savings Account
GSAM Generalized Sequential Access Method
GSB Graphic Standards Board
GSC Gas-Solid Chromatography
Genetics Society of Canada
Geological Society of Chicago [US]
Geological Survey of Canada
Grant Selection Committee

	Group Switching Center [UK]
	Guaranteed Savings Certificate
GSCG	Ground Systems Coordination Group
GSCU	Ground Support Cooling Unit
GSD	General Supply Depot
	General Systems Division
	Generator Starter Drive
	Generic Structure Diagram
	Genetically Significant Dose
	German Society for Documentation
GSDB	Geophysics and Space Data Bulletin
GSDS	General Status Display System
	Goldstone Duplicate Standard
GSE	Graphics Screen Editor
	Ground-Support Equipment
gSE	Spike Energy Unit
GSF	Gesellschaft fur Strahlen- und Umweltforschung [= Society for Radiology and Environmental Research, Germany]
GSFC	Goddard Spaceflight Center [of NASA]
GSG	Galvanized Sheet Gauge
	Glass-Silicon-Glass
GSGG	Gadolinium-Scandium-Gallium Garnet
GSH	Glutathione, Reduced
GSHP	Ground-Source Heat Pump
GSI	Gold and Silver Institute
	Government Source Inspection
	Grand Scale Integration
	Graphic Structure Input
	Guinea Syli [Currency of Guinea]
	Gunite/Shotcrete Association [US]
GSIN	Goods and Services Identification Number [of Supply and Services Canada]
GSIS	Geographic Snow Information System [Canada]
	Group for the Standardization of Information Services [US]
GSKT	Gasket
GSL	Generalized Simulation Language
	Generation Strategy Language
	Geological Society of London [UK]
	Geotechnical Science Laboratories [Canada]
	German Society for Logistics [also: DGfL, Germany]
GSM	Generalized Sequential Machine
	General Syntactic Processor
	Geological Society of Malaysia
	Gram per Square Meter [also: gsm or g/m^2]
	Graphics System Module
	Greenough Stereomicroscope
	Group Special Mobile [Europe]
GSMB	Graphic Standards Management Board [of ANSI, US]
GSMBE	Gas-Source Molecular-Beam Epitaxy
GSMT	General Society of Mechanics and Tradesmen [US]
GSO	General Staff Officer
	Geostationary Orbit
	Geosynchronous Orbit
	Graduate Service Overseas [UK]
	Ground Safety Office
	Ground Speed Oscillator
GSOC	German Space Operations Center
GSOP	Guidance Systems Operations Plan
GSP	Geographical Society of Philadelphia [US]
	Generalized System of Preferences
	General Simulation Program
	General Syntactic Processor
	Graphic Subroutine Package
	Guidance Signal Processor
GSPA	Grain Sorghum Producers Association [US]
GSPC	Gas Scintillation Proportional Chamber
	Gas Scintillation Proportional Counter
	Graphic Standards Planning Committee
GSPR	Guidance Signal Processor Repeater
GSPO	Gemini Spacecraft Project Office [of NASA]
GSPR	Guidance Signal Processor Repeater
GSR	Galvanic Skin Response
	Glide Slope Receiver
	Global Shared Resources
	Gray Scale Recording
	Great Swamp Research [now: GSRI, US]
GSRI	Great Swamp Research Institute [US]
GSRS	General Support Rocket System
GSRTST	German Society for Rocket Technology and Space Travel
GSS	Galvanized Steel Strand
	Gamma Scintillation System
	Global Surveillance System
	Government Statistics Service [UK]
	Graduate Student Society
	Graphic Service System
	Graphic Support Software
GSSA	Grassland Society of Southern Africa [South Africa]
GSSC	Ground Support Simulation Computer
GSSDAF	Gatineau Satellite Station Data Acquisition Facility [Canada]
GSSG	Glutathione, Oxidized Form
GSSPS	Gravitationally Stabilized Solar Power System
GSSW	Gas-Shielded Stud Welding
GST	General Sales Tax
	Gesellschaft Schweizerischer Tierarzte [= Swiss Society of Veterinarians]
	Glass Science and Technology
	Glutathione-S-Transferase
	Goods and Services Tax [Canada]
	Greenwich Sidereal Time
GSTA	Ground Surveillance and Target Acquisition
GSTP	General System of Tariff Preferences
GSU	General Service Unit
	Georgia State University [US]
	Graduate Student Union
	Grain Services Union
	Guaranteed Supply Unit
GSUG	Gross Seasonal Unavailable Generation
GSV	Governor Steam Valve
	Guided Space Vehicle
GSW	Galvanized Steel Wire
	Ground Saucer Watch [US]
GSWR	Galvanized Steel Wire Rope
GSZ	Geological Society of Zimbabwe
GT	Galilean Telescope
	Game Theory
	Gas Technologist
	Gas Technology

	Gas Turbine
	Gemini-Titan [of NASA]
	Geotectonics
	Glass Transition
	Glass Tube
	Glover Tower
	Glutamyltransferase
	Gopher Tape
	Graphics Terminal
	Graph Theory
	Gravitational Theory
	Grease Trap
	Great
	Greater Than
	Gross Tons
	Ground Transmit
	Group Technology
	Group Theory
G/T	Gain-to-Noise Temperature (Ratio)
	Grams of Plutonium per (Irradiated) Ton of Uranium
gt	goat [Immunochemistry]
g/t	grams per ton
g-t	gross ton
GTA	Gallotannic Acid
	Gas Tungsten-Arc
	Gemini-Titan-Agena [of NASA]
	Genetic Toxicology Association [US]
	Glass Tempering Association [US]
	Glycerol Triacetate
	Government Telecommunications Agency [of DOC, US]
	Grading Terminal Assembly
	Graph Theoretic Algorithm
	Guam Telephone Authority
GTAC	Gas Tungsten-Arc Cutting
GTAW	Gas Tungsten-Arc Welding
GTAW-P	Pulsed Gas Tungsten-Arc Welding
GTBA	Gasoline Grade Tertiary Butylacetate
GT BR	Great Britain [also: GT BRIT]
GTC	Gain Time Constant
	Gain Time Control
	Gas Turbine Compressor
	General Trading Companies
	Genetic Thermal Cycler
	Geological Testing Consultant
	Good Till Cancelled
GTCC	Group Technology Characterization Code
GTD	Gas Turbine Division [now: IGTI of ASME, US]
	Geometrical Theory of Diffraction
	Graphic Tablet Display
	Guaranteed
GTDI	Guidelines for (International) Trade Data Interchange
GTDPL	Generalized Top Down Parsing Language
GTE	Geotechnical Engineer
	Geotechnical Engineering
	Ground Transport Equipment
	Group Translating Equipment
GTEL	General Telephone and Electronics Laboratories [US]
GTEP	General Telephone and Electronics Practice
GTEPS	General Telephone and Electronics Data Services [US]
GTF	Generalized Trace Facility
	Generalized Transformation Function
	Gravity Tube Feeder
GTG	Gas Turbine Generator
GTH	Gas-Tight High-Pressure Method
GTHTGR	Gas Turbine High-Temperature Gas-Cooled Reactor [also: GT-HTGR]
GTI	Glass Technical Institute [US]
GTIS	Gloucestershire Technical Information Service [UK]
	Ground-Based Traffic Information System
GTM	Ground Test Missile
	Gyratory Testing Machine
GTMA	Gauge and Tool Makers Association [US]
GTN	Government Telecommunications Network [UK]
GTO	Gate Turn-Off (Switch)
	Gaussian-Type Orbitals
	Government Telecommunications Organization
	Guide to Operations
GTOW	Gross Takeoff Weight
GTP	General Test Plan
	Geotechnical Project
	Graphics Transform Package
	Group Transfer Polymerization
	Guanosine Triphosphate
GTPD	Geotechnical Project Design
GTQ	Guatamala Quetzal [Currency of Guatemala]
GTR	Gas Transmission Rate
	Ground Test Reactor [US Air Force]
GTRR	Georgia Institute of Technology Research Reactor [US]
GTS	General Technical Service
	Geostationary Technology Satellite
	Global Telecommunication System [of WMO, Switzerland]
	Ground Telemetry Subsystem
GTSL	Geotechnical Science Laboratories [Canada]
GTT	Glass Transition Temperature
gtt	(guttae) — drops
GTTF	Gas-Turbine Test Facility [US]
GTU	Group Terminal Unit
GTUC	Ghana Trades Union Congress
GTV	Gas Turbine Vessel
	Gate Valve
	Gate Trigger Valve
	Ground Test Vehicle
GTW	Gross Takeoff Weight
GU	Generic Unit
	Georgetown University [US]
	Gifu University [Japan]
	Guam
	Guarantee
GUA	Guarani [Currency of Paraguay]
GUAR	Guarantee(d)
GUARDSMAN	Guidelines and Rules for Data Systems Management
GUCCIAAC	General Union of Chambers of Commerce, Industry and Agriculture for Arab Countries [Lebanon]
GUHA	General Unary Hypotheses Automation
GUI	Graphic User Interface

GUIDE Guidance for Users of Integrated Data Processing Equipment
GuK Guanylate Kinase
GULP General Utility Library Program
GUN Gunnery
GUNFO Gulf United Nuclear Fuels Corporation
GUSTO Guidance Using Stable Tuning Oscillations
GUT Grand Unified Theory
GUTS Gothenburg University Terminal System
GUUG Gross Unit Unavailable Generation
GUY (French) Guyana Space Center [of ESA, France]
GV Gas Valve
Guard Vessel
Group Velocity
G/V Conductance versus Voltage
GVA Graphic Kilovolt-Ampere Meter
GVB Gelatin Veronal Buffer
Generalized Valence Bond
GVC Glazed Vitrified Clay
GVHRR Geosynchronous Very-High-Resolution Radiometer
GVL Gravel
GVMR Gross Vehicle Mass Rating
GVT Gravity Vacuum Tube System
GVW Gross Vehicle Weight
GVWR Gross Vehicle Weight Rating
GW Gas Welding
General Warning
Gigawatt
Glass Wool
Gravity Wave
Ground Wave
Guard Wire
GWC Global Weather Center [US Air Force]
Golden West Conventions [Australia]
Ground Water Council
GWDB Groundwater Development Bureau [Taiwan]
GWE Guinea-Bissau Peso [Currency of Guinea-Bissau]
GWe Gigawatt, electrical
GWEN Ground Wave Energy Network
GWF Groundwater Flow
GWh Gigawatthour
GWI Grinding Wheel Institute [US]
Groundwater Institute [US]
GWP Guinea-Bissau Peso [Currency of Guinea-Bissau]
GWR Ground Wave Radar
GWRI Ground Water Research Institute
GWT Gross Weight
GWTD Ground Water Technology Division [now: AGWSE, US]
GWth Gigawatt, thermal
GWU Gambia Workers Union
General Workers Union
George Washington University [US]
GWUP Gesellschaft fur die Wissenschaftliche Untersuchung der Parawissenschaften [= Society for the Scientific Investigation of Parascience, Germany]
GWVSS Ground Wind Vortex Sensing System
GY Greenish Yellow
Gy Grey [Unit]
GYD Guyana Dollar [Currency]
GYFM General Yielding Fracture Mechanics
GYP Gypsum
GYR Gyration
GYRO Gyroscope
GZ Ground Zero
GZT Greenwich Zone Time

H

H	Halt
	Hard [On Pencils]
	Hardness
	Hardware
	Hatch
	Hawaii(an)
	Haze
	Head
	Heat(er)
	Heating
	Heavy
	Henry
	Hexadecimal
	Hexapole
	Hexode
	High
	Histidine
	Histidyl
	Home
	Homing
	Horizontal
	Host
	Humidity
	Hybridoma [Immunology]
	Hydrant
	Hydrodynamics
	Hydrogen
h	hail
	half
	harbor
	hard
	head
	heat
	heavy
	hecto- [SI Prefix]
	height
	high
	horizon(al)
	hot
	hour
HA	Half Add(er)
	Hard Axis
	Harmonic Analysis
	Harmonic Approximation
	Head Amplifier
	Health Advisory
	Hemadsorption Virus
	Hemagglutinin
	Heptoic Aldehyde
	High Altitude
	High Angle
	Historical Association [UK]
	Hoe Automation
	Home Address
	Horse Artillery
	Hot Air
	Hour Angle
	Hydraulic Actuator
	Hydraulic Association [UK]
	Hydroxyapatite [also: HAP]
Ha	Hahnium
ha	hectare
	(hoc anno) — this year
HAA	Height Above Airport
	Helicopter Association of America
	Helix Aspersa Agglutinin
	Historic Aircraft Association [US]
HAATC	High-Altitude Air Traffic Control
HAB	Habitat
	Home Address Block
HABA	Hydroxyazobenzenebenzoic Acid
	2-(4-Hydroxyphenylazo)benzoic Acid
HAC	Herbicide Assessment Commission [US]
	High Alumina Cement
	Horticultural Advisory Council [UK]
	Hydrogen-Assisted Cracking
	Hydroxyapatite Crystal
HACL	Harward Air Cleaning Laboratory [of Harvard University, US]
HAD	Half-Amplitude Duration
	Heat-Actuated Device
	High-Altitude Density
	High-Aluminum Defect
HADES	Hypersonic Air Data Entry System
HADR	Hughes Air Defense Radar
HADS	Hypersonic Air Data Sensor
	Hypersonic Air Data System
HAE	Hot-Air Engine
Hae	Haemophilus Aegyptius [Molecular Biology]
HAES	Hawaii Agricultural Experiment Station [US]
	High Altitude Effects Simulation
HAF	High-Abrasion Furnace Black
	High Altitude Fluorescence
HAFB	Holloman Air Force Base [US]
HAFM	Helium Accumulation Fluence Monitor
HAG	Hold for Arrival of Goods
HAI	Helicopter Association International [US]
	Hemagglutination Inhibition Assay
HAIC	Hearing Aid Industry Conference
	Hetero-Atom in Context
HAIRS	High-Altitude Infrared Source
HAIT	Hash Algorithm Information Table
HAL	Harwell Automated Library [of UKAEA]
	High Accuracy Linear (System)
	High Activity Locations
	Highly-Automated Logic
HALDIS	Halifax and District Information Service [UK]
HALL ED	Hall Effect Device
HALO	High-Altitude, Low-Opening Parachute
	High-Attitude Large Optics
HALS	Hindered Amines Liquid Stabilizer
HALSIM	Hardware Logic Simulator
HALT	Hydrate Addition at Low Temperatures
HAM	Hardware Associative Memory

	Hierarchical Access Method
	Hold-and-Modify Image
	Hydrogenic Atoms in Molecules
HAMCHAM	Haitian-American Chamber of Commerce and Industry
	Honduran-American Chamber of Commerce
HAMS	High Altitude Mapping System
	Hour Angle of the Mean Sun
HAMT	Human-Aided Machine Translation
HANDS	High-Altitude Nuclear Detection Studies [of NIST, US]
HANE	High-Altitude Nuclear Effects
HANFO	Heavy Ammonium Nitrate and Fuel Oil
HAN/LCD	Hybrid Assigned Nematic/Liquid Crystal Display
HAO	High-Altitude Observatory [of NCAR, US]
HAOSS	High-Altitude Orbital Space Station
HAP	High-Altitude Platform
	Hydroxyapatite [also: HA]
HAPDAR	Hard Point Demonstration Array Radar
HA-PE	Hydroxylapatite — Polyethylene
HAPI	Helicopter Approach Plate Indicator System
HA-PLA	Hydroxylapatite — Polylactic Acid
HAPP	House Assessment Prescription Program [Canada]
HAPPE	Honeywell Associative Parallel Processing Ensemble
HAPUB	High-Speed Arithmetic Processing Unit Board
HAQO	Hydroxyaminoquinoline Oxide
HAR	Harbor Advisory Radar
	Home Address Register
	Hydrogen Absorption Reaction
HARA	High-Altitude Radar Altimeter
	High-Altitude Resonance Absorption
HARAC	High-Altitude Resonance Absorption Calculation
HARC	Houston Area Research Center [US]
HARCO	Hyperbolic Area Coverage
HARDTS	High-Accuracy Radar Transmission System
HARDWR	Hardware
HARLID	High Angular Resolution Laser-Irradiation Detector
HARM	High-Speed Anti-Radar Missile
HARP	Halpern Antiradar Point
	High Altitude Research Probe
	High Altitude Research Project
	High Altitude Rocket Probe
	Hitachi Arithmetic Processor
HARPS	Heathrow Airport Radar Processing System [UK]
HARTRAN	Hardwell FORTRAN
HAS	Heading Altitude System
	Helicopter Avionics System [US Air Force]
	High-Angle Scattering
	Hungarian Academy of Sciences
	Hydroxylamine Acid Sulfate
HASCC	Hydrogen-Assisted Stress-Corrosion Cracking
HASCI	Human Applications Standard Computer Interface
HASG	Helicopter Airworthiness Study Group [UK]
HASL	Health and Safety Laboratory [of USAEC]
HASP	High Altitude Sampling Plane
	High Altitude Sampling Program
	High Altitude Sounding Project
	High-Level Automatic Scheduling Program
	Houston Automatic Spooling Priority System
	Houston Automatic Spooling Processor
	Houston Automatic Spooling Program
HASQ	Hardware-Assisted Software Queue
HAST	Highly-Accelerated Stress Test
HASY	Handling System
HASYLAB	Hamburger Synchrotronstrahlungslabor [= Hamburg Synchrotron Radiation Laboratory, Germany]
HAT	Height Above Touchdown
	High-Altitude Testing
	Home Area Toll
	Hypoxanthine/Aminopterin/Thymidine Medium
HATA	Hellenic Atlantic Treaty Association [Greece]
HATAPH	Hexaalkyltriamidophosphazohydride
HATREMS	Hazardous and Trace Emission System [of EPA, US]
HATS	Helicopter Advanced Tactical System
	High-Accuracy Targeting Subsystem
	High-Altitude Terrain Sensor
	Hour Angle of the True Sun
HATT	High-Speed Aeronautical Technologies Testbed
HATT-X	High-Speed Aeronautical Technologies Testbed — Experimental
HAU	Horizontal Arithmetic Unit
	Hybrid Arithmetic Unit
hav	haversine
HAW	Hawaii — Continental United States Submarine Cable
	Heavy Anti-Armor Weapon
HAWS	Heavy Anti-Armor Weapon System
HAYSTAQ	Have You Stored Answers to Questions?
HAZ	Hazard
	Heat-Affected Zone
HAZCHEM	Hazardous Material Identification System
HAZEL	Homogeneous Assembly Zero Energy Laboratory [of AERE, UK]
HAZFILE	Hazards File [of NCEC, UK]
HAZ/MZ	Heat-Affected Zone/Melted Zone
HB	Brinell Hardness
	Hard Black [On Pencils]
	Hatchback
	Hexadecimal-to-Binary
	Horizontal Baffle
	Horizontal Bridgman Growth
	Hose Bib
	Hydrogen Bomb
	Hydroxybenzene
H/B	Hexadecimal-to-Binary
Hb	Hemoglobin
HBA	Home Builders Association
	Honours Bachelor of Arts
	Hydroxybenzoic Acid
	Hydroxybutyrate
HBABA	2-(4-Hydroxybenzeneazo)benzoic Acid
HBC	Higher Binding-Energy Component
	Human Biology Council [US]
HBComm	Honours Bachelor of Commerce
HBD	Hydroxybutyrate Dehydrogenase [also: HBDH]
HBED	N,N'-Di(2-Hydroxybenzyl)ethylenediamine
HBEN	High Byte Enable
Hb F	Fetal Hemoglobin
HBFP	Hematoxylin Basic Fuchsin Pecric
HBGM	Hypersonic Boost-Glide Missile
HBM	Her/His Britannic Majesty
HBN	Hazard Beacon

	Hexagonal Boron Nitride
HBO	Heavy Batch Oven
HBR	Harbour
HBS	Hawaii Botanical Society [US]
HBScF	Honours Bachelor of Science in Forestry
HBSMAA	Hack and Band Saw Manufacturers Association of America [US]
HBSS	Hank's Balanced Salt Solution
HBT	Heterojunction Bipolar Transistor
	N-(P-Hexyloxybenzylidene)-P-Toluidene
HBV	Hepatitis B Virus
HBWR	Halden Boiling Water Reactor [Norway]
HBZ	Hochschulbibliothekszentrum [= University Library Center, Germany]
HC	Half Cell
	Halochromism
	Halochromy
	Hand Control
	Handling Capacity
	Hanging Ceiling
	Hard Copy
	Hard Cradle Balancer
	Hardware Capability
	Heat Capacity
	Heat Conduction
	Heated Coil
	Heating Coil
	Heterogeneous Catalysis
	Heuristic Concepts
	Hexadecimal Code
	High Capacity
	High Carbon
	High Compression
	Holding Coil
	Hollow Cathode
	Homogeneous Catalysis
	Horizontal Cell
	Hose Connector
	Host Computer
	Hot Cathode
	House Cable
	Humidity Control
	Hybrid Circuit
	Hybrid Coil
	Hybrid Computer
	Hydraulic Circuit
	Hydraulic Controller
	Hydraulic Coupling
	Hydrocarbon
	Hydrogen Chemisorption
H&C	Hot and Cold (Water)
H/C	Hand Carry
Hc	Hermitian conjugate
hc	(honoris causa) — honorary
HCA	Heptine Carbonic Acid
	Hexachloroacetone
	Hydrochloric Acid
	Hydroxylcarbonate Apatite
HCAA	Hellenic Civil Aviation Authority [Greece]
HCAR	Higher Committee for Agrarian Reform [Egypt]
HCB	Hexachlorobenzene
	Hollow-Cone Beam
	Hydrocarbon
HCB-CBED	Hollow-Cone Beam/Convergent-Beam Electron Diffraction
HCC	Hardware Capability Code
	Horizontal Continuous Casting
HCCH	Hexachlorocyclohexane
HCD	High Current Density
	Hollow-Cathode Discharge
	Hot-Carrier Diode
HCDA	Hypothetical Core-Disruptive Accident
HCDT	Hollow-Cathode Discharge Tube
HCE	Heater Control Electronics
	Hexachloroethane
	Hollow-Cathode Effect
	Human Caused Error
HCEX	Hypercharge Exchange
HCF	High-Cycle Fatigue
	Highest Common Factor
	Host Command Facility
HCG	Hardware Character Generator
	Hexagonal Coupling
	Horizontal Location of Center of Gravity
	Human Chorionic Gonadotropin [also: hCG]
HCH	Hexachlorocyclohexane
HCI	Host Computer Interface
HCIB	Health Computer Information Bureau
HCL	Hollow Cathode Lamp
	High, Common, Low (Relay)
	High Cost of Living
	Horizontal Centerline
HCLF	Horizontal Cask Lifting Fixture
HCM	Hydraulic Core Mockup
	Hyundai Color Monitor
HCMM	Heat Capacity Mapping Mission
	Highly Conductive Mold Media
HCMOS	High-Density Complimentary Metal-Oxide Semiconductor
	High-Speed Complimentary Metal-Oxide Semiconductor
HCMTS	High-Capacity Mobile Telecommunications System
HCOHSA	Health Care Occupational Health and Safety Association [Canada]
HCP	Hard Copy Printer
	Hepatocatalase Peroxidase
	Hexagonal Close-Packed [also: hcp]
	Host Communications Processor
HCPRU	Hot Climate Physiological Research Unit [Nigeria]
HCR	Hardware Check Routine
	Heat-Curable Rubber
HCRC	Holland College Royalty Center [Canada]
HCRS	Heritage Conservation and Recreation Service
HCS	Helium Circulator Seal
	High-Carbon Steel
	Host Composition System
	Human Chorionic Somatotropin [also: hCS]
	Hundred Call Seconds
	Hungarian Chemical Society
HCSDS	High Capacity Satellite Digital Service
HCSS	Head Compartment Support Structure
	Hospital Computer Sharing System

HCT Heater Center Tap
Hot Cathode Tube
HCTDS High Capacity Terrestrial Digital Service
HCTLS High-Speed Complimentary Transistor Low-Power Schottky
HCU Homing Comparator Unit
Hydraulic Control Unit
HCZ Hydrogen Convection Zone
HD Half Duplex
Hand
Harbor Defense
Hard Disk
Hard-Drawn
Harmonic Distortion
Head
Head Diameter
Heavy Duty
Hexadecimal-to-Decimal
Hierarchical Direct
High Density
Horizontal Distance
Hot Drawing
Hub Diameter
Hydrodynamics
H-D Hexadecimal-to-Decimal
Hunter-Driffield (Curve)
HDA Hardwood Distributors Association [US]
Head/Disk Assembly
Hexadecenylacetate
Horizontal Danger Angle
Hydrodealkylation
HDAL Hexadecenal
HDAM Hierarchical Direct Access Method
HDAOS 2-Hydroxy-3-Sulfopropyl-3,5-Dimethooxyaniline
HDAS Hybrid Data Acquisition System
HDB High Density Binary
High Density Bipolar (Code)
HDBF Heavy-Duty Business Forum [US]
HDBH High Day Busy Hour
HDBK Handbook
HDC Heavy Duty Clamp
Helium Direct-Current
Hexadecimal Code
HDD High-Density Disk
HDDA Hexadecadienylacetate
HDDD High-Density Disk Drive
HDDR High-Density Digital Recording
HDDS High-Density Data System
HDDT High-Density Digital Tape
HDEP High-Density Electronic Packaging
HDF High-Density Flexible
High-Frequency Direction Finding
Horizontal Distributing Frame
H/DF Human/Dolphin Foundation [US]
HD FRZ Hard Freeze
HDG Heading
HDGA Hot Dip Galvanizers Association [US]
HDI Head-Disk Interference
Hexamethylene Diisocyanate
High-Density Interconnect
High Dose Implantation
HDIC High-Density Interconnect Circuit
High Digital Integrated Circuit
HDL Handle
Hardware Description Language
Harry Diamond Laboratories [US Army]
High-Density Lipoprotein
High-Level Data Link
HDLC Hierarchical Data Link Control
High-Level Data Link Control
HDLG Handling
HDLS Headless
HDM Humic Degradation Matter
Hydrodensimeter
Hydrodynamic Machining
HDMA Hardwood Dimension Manufacturers Association [now: NDMA, US]
HDME Hanging Drop Mercury Electrode
HDMR High-Density Moderated Reactor
High-Density Multitrack Recording
HDMS Honeywell Distributed Manufacturing System
HDMSO Hexamethyldisiloxane
HDMSW High-Density Mach Shock Wave
HDN Harden
Hydrodenitrogenation
HDNP High-Density Nickel Powder
HDNSW High-Density Nuclear Shock Wave
HDO High Density Overlay
HDOC Handy Dandy Orbital Computer
HDODA 1,6-Hexanediol Diacrylate
HDOL Hexadecenol
HDOS Hard Disk Operating System
Heath Disk Operating System
HDP High Detonation Pressure
High Discharge Pressure
Horizontal Data Processing
HDPA Hydroxydiphenylamine
HDPE High-Density Polyethylene
HDQRS Headquarters
HDR Header
High-Density Recording
Hot Dry Rock (Method)
HDRA Heavy-Duty Representatives Association [US]
Henry Doubleday Research Association [UK]
HDRI Hannah Dairy Research Institute [Scotland]
HDRSS High Data Rate Storage System
HDS Headset
Hermes Data System
High Density Sludge
Huang Diffuse Scattering
Hundreds
Hybrid Development System
Hydrodesulfurization
HD/SCSI Hard Disk/Small Computer System Interface
HDSS Hospital Decision Support System
HDST High-Density Shock Tube
HDT Heat Deflection Temperature
Heat Deflection Test
Heat Distortion Temperature
Hydrodynamic Technology
H/D/T Hydrogen/Deuterium/Tritium (Ratio)
HDTUL Heat-Deflection Temperature under Load

HDTV High-Definition Television
High-Density Television
HDU Hard Disk Unit
HDVS High-Definition Video System
HDW Hardware
High-Pressure Demineralized Water
HDWC Hawaii Deep Water Cable [US]
HDWD Hardwood
HD WHL Hand Wheel
WDWND Head Wind
HDWRE Hardware
HDX Half Duplex
HE Hall Effect
Handling Engineer
Harmful Environment
Heat Engine
Heat Exchange(r)
Helium Embrittlement
Heating Element
Heating Engineer
Heating Engineering
Heavy Enamel (Wire)
Heptachlorine Epoxide
High Efficiency
High Explosive
Highway Engineer
Highway Engineering
Horizontal Equivalent
Housekeeping Element
Hydraulic Engineer
Hydraulic Engineering
Hydroelectric(ity)
Hydrogen Electrode
Hydrogen Embrittlement
Hyperelastic(ity)
He Helium
he (hoc est) — that is
HEA Heating Engineering Association [UK]
High-Efficiency Antireflection Coating
Horticultural Education Association [UK]
Hydroxyethyl Acrylate
Hydroxyethylamine
HEALS Honeywell Error Analysis and Logging System
HEAO High-Energy Astronomy Observatory
HEAP Helicopter Extended Area Platform
High-Explosive Armor Piercing
HEART Human Engineering Analysis and Requirements Tool
HEAT High-Explosive Antitank
Hydroxyphenyl Ethyl Aminoethyl Tetralone
HEB High-Efficiency Binding
HEBC Heavy Enamel Bonded (Single) Cotton (Wire)
HEBDC Heavy Enamel Bonded Double Cotton (Wire)
HEBDP Heavy Enamel Bonded Double Paper (Wire)
HEBDS Heavy Enamel Bonded Double Silk (Wire)
HEBIS Hessisches Bibliothekssystem [= Hessian Automated Library System, Germany]
HEBP Heavy Enamel Bonded (Single) Paper (Wire)
HEBS Heavy Enamel Bonded (Single) Silk (Wire)
HEC Heavy Enamel (Single) Cotton (Wire)
Hollerith Electronic Computer
Horowitz-Eastman-Crane Method
Hydrogen Embrittlement Cracking
Hydroxyethylcellulose
HeCd Helium-Cadmium (Laser)
HECI Human-Interface Equipment Catalog Item
HECSAGON Horowitz-Eastman-Crane Symbol Array Governed by Orthodox Notation
hectol hectoliter
HECTOR Heated Experimental Carbon Thermal Oscillator Reactor
Hot, Enriched, Carbon-Moderated, Thermal-Oscillator Reactor
HECV Heavy Enamel (Single) Cotton Varnish (Wire)
HED Horizontal Electric Dipole
HEDC Heavy Enamel Double Cotton (Wire)
Houston Economic Development Council [US]
HEDCV Heavy Enamel Double Cotton Varnish (Wire)
HEDI High Endoatmospheric Defense Interceptor
HEDL Hanford Engineering Development Laboratory [US]
HEDP High Explosive Dual Purpose
1-Hydroxyethylidene-1,1-Diphosphonic Acid
HEDS Heavy Enamel Double Silk (Wire)
High Energy Dislocation Structure
HEDSV Heavy Enamel Double Silk Varnish
HEDTA N-(2-Hydroxyethyl)ethylenediaminetriacetic Acid
HEE Hydrogen Environment Embrittlement
HEED High-Energy Electron Diffraction
HEEDTA N-Hydroxyethylethylenediaminetriacetic Acid
HEEP Health Effects of Environmental Pollution [of NLM/BIOSIS]
Highway Engineering Exchange Program [US]
HEF Heated Effluents (Database) [of Cornell University, US]
High-Elongation Furnace Black
High-Energy Forging
High-Energy Forming
High-Energy Fuel
High-Expansion Foam
Hispanic Energy Forum [US]
HEG Heavy Enamel (Single) Glass (Wire)
HEHF Hanford Environmental Health Foundation [US]
HEI Health Effects Institute [US]
Heat Exchange Institute [US]
High-Energy Ignition
Human Exploration Initiative [Canada and US]
HE-I High-Explosive Incendiary
HEIS High-Energy Ion Scattering
HEI-T High-Explosive Incendiary with Tracer
HEIX Home Economics Information Exchange [of FAO, UN]
HEL Helicopter
High-Energy Laser
HeLa Helen Lane (Cancer Cells)
HELCOM Baltic Marine Environment Protection Commission — Helsinki Commission [Finland]
HELEX Hydrogenous Exponential Liquid Experiment
HELIOPS Helicopter Operations Panel [of ICAO]
HELMS Helicopter Multifunction System
HELP Hazardous Emergency Leaks Procedure
Helicopter Electronic Landing Path
Helicopter Emergency Life-Saving Program

	Highly-Extendable Language Processor
HELPIS	Higher Education Learning Programs Information Service [of CET, UK]
HEM	Hazardous Environment Machine
	Heat Exchanger Method
	Helicopter-Borne Electromagnetics
	Helicopter Electromagnetic Survey
	Hostile Environment Machine
	Hybrid Electromagnetic (Wave)
	Hydrogen Embrittlement
HEMA	2-Hydroxyethyl Methacrylate
HEMAC	Hybrid Electromagnetic Antenna Coupler
HEMEL	Hexamethylmelamine [also: HMM]
HEM FIR	Hemlock Fir
HEMP	High Altitude Electromagnetic Pulse
HEMPA	Hexamethylphosphoramide [also: HMPA]
HEMS	Helicopter Electromagnetic Survey
	Helicopter Emergency Medical Services [UK]
HEMT	High-Electron Mobility Transistor
HeNe	Helium-Neon (Laser)
HENR	Higher Energy Nuclear Reaction
HENRE	High-Energy Neutron Reactions Experiment [of USAEC]
HEO	High Earth Orbit (Satellite)
HEOD	Hexachloro Epoxy Octahydroendohexa Dimethanonaphthalene
HEOS	Highly Eccentric Orbit Satellite
HEP	Heterogeneous Element Processor
	High-Energy Physics
	High-Explosive Penetrating
	High-Explosive Plastic Projectile
	Homogeneous Element Processor
	Human Error Probability
	Hydroelectric Power
HEPA	High-Efficiency Particulate Absolute
	High-Efficiency Particulate Air (Filter)
HEPAT	High-Explosive Plastic Antitank (Charge)
HEPC	Hydro-Electric Power Commission [Canada]
HEPCAT	Helicopter Pilot Control and Training
HEPD	Heat Engine Propulsion Division [of DOE, US]
HEPES	N-(2-Hydroxylethyl)piperazine-N'-(2-Ethanesulfonic Acid)
HEPG	High-Energy Physics Group
HEPI	High-Energy Physics Index [of DESY, Germany]
HEPL	High-Energy Physics Laboratory [of Stanford University, US]
HEPP	Hoffman Evaluation Program and Procedure
	Hydroelectric Power Plant
HEPPS	N-(2-Hydroxyethyl)piperazine-N'-(3-Propanesulfonic Acid)
HEPPSO	N-(2-Hydroxyethyl)piperazine-N'-(2-Hydroxypropanesulfonic Acid)
HEPR	Hard-Grade Ethylene Propylene Rubber
HEPS	Hydroelectric Power Station
HEPTA	N-Hydroxyethylethylenediaminetriacetic Acid
HER	Human Error Rate
	Hydrogen Evolution Reaction
Her	Hercules [Astronomy]
HERA	Hadron-Elektron-Ring-Anlage [= Hadron Electron Ring Accelerator, Germany]
	High-Explosive Rocket-Assisted
HERALD	Heterogenous Experimental Reactor, Aldermaston [of AWRE, UK]
	Highly Enriched Reactor, Aldermaston [UK]
HERAS	Hellenic Radar System [Greece]
HERF	Hazards of Electromagnetic Radiation to Fuel
	High Energy Rate Forging
	High Energy Rate Forming
HERMAN	Hierarchical Environmental Retrieval for Management Access and Networking [of BIS, US]
HERMES	Heavy Element and Radioactive Material Electromagnetic Separator [UK]
	Heuristic Mechanized Documentation Information Service [Romania]
HERO	Hazards of Electromagnetic Radiation to Ordnance
	Heath Robot
	Hot Experimental Reactor of 0 (= Zero) Power
HERP	Hazards of Electromagnetic Radiation to Personnel
HERS	Hardware Error Recovery System
Herts	Hertfordshire [UK]
HERTIS	Hertfordshire Technical Library and Information Service [UK]
Her X-1	Hercules X-1 [Astronomy]
HES	Hanford Engineering Service [US]
	Hawaiian Entomological Society [US]
	Heavy Enamel (Single) Silk (Wire)
	High Early Strength (Cement)
	High-Energy Scattering
	House Exchange System
HESS	History of Earth Sciences Society [US]
	Houston Engineering and Scientific Society [US]
HET	Heavy Equipment Transporter
	Hot Electron Transistor
HE-T	High Explosive with Tracer
HETE	Hydroxyeicosatetraenoic Acid
HETP	Height Equivalent of Theoretical Plate
	Hexaethyl Tetraphosphate
HETS	Height Equivalent to a Theoretical Stage
HEU	Highly-Enriched Uranium
	Hydroelectric Unit
HEU/Th	Highly Entiched Uranium/Thorium Fuel
HEVAC	Heating, Ventilating and Air Conditioning
HEW	(Department of) Health, Education, and Welfare [US]
HEX	Hexadecimal
	Hexagon(al)
hex	Uranium Hexafluoride
Hexa	Hexamethylenetetramine [also: HMTA]
HEXFET	Hexagonal Metal-Oxide Field-Effect Transistor
HEX HD	Hexagonal Head (Screw)
HF	Half
	Hardenability Factor
	Harmonic Function
	Hartree-Fock (Field)
	Haze Filter
	Heat Flow
	Height Finder
	Heptaline Formate
	High Frequency
	High Frontier [US]
	History File
	Homogeneous Flow

	Hydraulic Fluid
	Hyperfine
Hf	Hafnium
HFA	Hard Fiber Association [US]
	Hexafluoroacetone
	High Flow Alarm
	Homofolic Acid
	Hydrofluoric Acid
HFAC	Hexafluoroacetylacetonato
	Hexafluoroacetylacetone
	Human Factors Association of Canada
HFAR	Honduran Foundation for Agricultural Research
HFAS	Honeywell File Access System
HFB	Hartree-Fock-Bogoliubov
	Hopper-Feeder-Bolter
HFBA	Heptafluorobutyric Acid
HFBC	High-Frequency Broadcasting
hf bd	half bound
HFBR	High Flux Beam Reactor [of BNL, US]
HFBS	High-Frequency Broadcasting Schedule [of ITU, Switzerland]
HFC	High-Frequency Correction
	High-Frequency Current
HFCE	HFIR Critical Experiment [of ORNL, US]
HFDF	High-Frequency Direction Finder
	High-Frequency Distribution Frame
HFE	Human Factors in Engineering
HFEF	Hot Fuel Examination Facility
HFET	Heterojunction Field-Effect Transistor
HFF	Hartree-Fock Field
HFG	Heavy Free Gas
HFI	Helicopter Foundation International [US]
	High-Frequency Inductance
HFIAW	(International Association of) Heat and Frost Insulators and Asbestos Workers
HFIB	Hexafluoroisobutylene
HFIM	High-Frequency Instruments and Measurements [IEEE IM Committee]
HFIP	Hexafluoroisopropanol
HFIR	High-Flux Isotope Reactor [of ORNL, US]
HFITR	High Field Ignition Test Reactor
HFIW	High-Frequency Induction Welding
HFL	High Free Lift
HFM	Hazardous Fluids Module
	Heat-Flow Meter
	High Field Magnetism
	High-Frequency Microphone
HFO	Hartree-Fock-Overhauser
	High-Frequency Oscillator
HFOD	Heptafluorodimethyloctanedione
HFORL	Human Factors Operations Research Laboratory
HFP	Hexafluoropropylene
HFPE	1-Hydro-Penta-Fluoropropylene
HFPO	Hexafluoropropylene Epoxide
HFPRI	Hokkaido Forest Products Research Institute [Japan]
HFPS	High-Frequency Phase Shifter
HFR	High-Flux Reactor
	High-Frequency Recombination
HFRSc	Forged Roll Scleroscope Hardness Number, Model c
HFRSd	Forged Roll Scleroscope Hardness Number, Model d
HFRT	High-Frequency Resonance Technique
HFRW	High-Frequency Resistance Welding
HFS	Hartree-Fock-Slater
	High-Field Superconductor
	Human Factors Society [US]
	Hyperfine Structure
	Hypothetical Future Samples
HFSA	Human Factors Society of America
	Hydrofluosilicic Acid
HF-SCF	Hartree-Fock Self-Consistent Field
HFSF	High-Flux Solar Furnace
hFSH	Human Follicle Stimulating Hormone
HFSS	High-Frequency Search System
	High-Frequency Sounder System
	Hyperfine Structure Spectrum
HFST	High Flux Scram Trip
HFT	Hartree-Fock Theory
	Heat Flux Transducer
HFW	High-Frequency Wave
	Hole Full of Water
	Horizontal Full Width
HFX	High Frequency Transceiver
HG	Hand Generator
	Harmonic Generator
	Height Gauge
	High Grade
	Homopolar Generator
	Horizon Grow
	Horse Guards
	Human Gastrin
	Human Genetics
	Hydraulic Gate
	Hydrogeology
Hg	(Hydrargyrum) — Mercury
hg	hectogram
HGA	High Gain Antenna
	Hop Growers Association [US]
HGAS	High Gain Antenna System
HGC	Hartford Graduate Center [US]
	Hercules Graphics Card
HGD	Hob Generating Diameter
Hg-dUTP	5-Mercuri-2'-Deoxyuridine 5'-Triphosphate
HGE	Hydraulic Grade Elevations
HGF	Hyperglycemic-Glycogenolytic Factor
HGG	Human Gamma Globulin
HGH	Human Growth Hormone
HGI	Hardgrove Grindability Index
HGM	Human Gene Mapping
HGMS	High-Gradient Magnetic Separation
HGR	Hanger
HGPRT	Hypoxanthine-Guanine Phosphoribosyl Transferase
HGRF	Hot Gas Radiating Facility
HgSeTe	Mercury Selenium Telluride
HGT	Height
	High Group Transmit
	Hypergeometric Group Testing
Hg-UTP	5-Mercuriuridine 5'-Triphosphate
HGV	Heavy Goods Vehicle
HH	Double Hard [On Pencils]
	Half-Hard [also: 1/2 H]
	Hand-Held

	Handhole
	Hanging Handset
	Hitchhiker [of NASA]
HHA	Hexahydric Alcohol
Hha	Haemophilus Haemolyticus [Molecular Biology]
HHB	Hexahydroxybenzene
HHC	Hand-Held Computer
	Higher Harmonic Control
hhd	hogshead [Unit]
HHDN	Hexachlorohexahydro Dimethanonaphthalene
HH-G	Hitchhiker, Goddard [of NASA]
HHHMU	Hydrazine Hand-Held Maneuvering Unit
HHL	High Hazard Laboratory [of ETC, Canada]
HHLR	Horace Hardy Lestor Reactor [US]
HHLT	Hand-Held Logic Tool
HH-M	Hitchhiker, MSFC (= Marshall Spaceflight Center) [of NASA]
HHMU	Hand-Held Maneuvering Unit
HHPA	Hexahydrophthalic Anhydride
HHRR	Hand-Held Rationing Radiometer
HHS	Harris Hematoxylin Solution
	(Department of) Health and Human Services [US]
	Human Head Simulator
HHSI	High Head Safety Injection
HHT	Hydroxyheptadecatrienoic Acid
HHTK	Hand-Held Test Kit
HHTU	Hand-Held Teaching Unit
HHV	Higher Heating Value
	High Heat Value
HI	Hawaii [US]
	Heat Insulation
	Heavy Environment
	Height of Instrument
	High
	High Intensity
	Hydraulic Institute [US]
	Hydraulic Intensifier
	Hydriodic Acid
	Hydronics Institute [US]
HIA	Hawaiian Irrigation Authority [US]
	Herzberg Institute of Astrophysics [of NRC, Canada]
	Horological Institute of America [US]
	Human Interface Architecture
HIAA	Hydroxyindole Acetic Acid
HIAC	High Accuracy
HIALS	High Intensity Approach Light System
HIB	High Iron Briquetting
HIBA	Hydroxyisobutyric Acid
HIBCC	Health Industry Bar Code Council
HIBEX	High-Acceleration Boost Experiment
	High Impulse Booster Experiment
HIC	Hybrid Integrated Circuit
	Hydrogen-Induced Cracking
HI-C	High Conversion Critical Experiment
HICACOM	High-Capacity Communications
HICAT	High-Altitude Clear Air Turbulence
HICLASS	Hierarchical Classification
HICOM	High Commission
HICS	Hierarchical Information Control System
HICT	Higher Institute of Chemical Technology [Bulgaria]
HID	High Density [also: HiD]
	High-Intensity Discharge (Lamp)
	High-Interstitial Defect
	Housing Industry Dynamics (Database) [US]
HIDAM	Hierarchical Indexed Direct Access Method
HIDB	Highlands and Islands Development Board [UK]
HIDC	Hexamethylindodicarbocyanine
	Hydrogen-Induced Delayed Cracking
HIDE	Hydrogen-Induced Deformation Experiment
HIDEC	Highly-Integrated Digital Electronic Control
HIDES	High-Absorption Integrated Defense Electromagnetic System
HIDF	Horizontal Side of an Intermediate Distribution Frame
HiD/LoD	High Density/Low Density (Tariff)
HIDM	High Information Delta Modulation
HIE	Hibernation Information Exchange [now: IHS, US]
HIEME	Higher Institute of Electrical and Mechanical Engineering [Bulgaria]
HIF	Hokkaido International Foundation [of HU, Japan]
	Hot Isostatic Forging
HIFAR	High-Flux Australian Reactor [Australia]
HIFET	Heterointerface Field-Effect Transistor
HiFi	High Fidelity [also: hi-fi, HI-FI or HIFI]
HIFT	Hardware Implemented Fault Tolerance
HIG	Hermetically-Sealed Integrated Gyro
HIGFET	Heterostructure-Insulated Gate Field-Effect Transistor
HIGH	High Core Threshold
HIGHCOM	High Fidelity Compander
HIGH GASSER	High Geographic Aerospace Search Radar
HIGH TECH	High Technology
HIGHVISION	(Japanese) High-Definition Telvision System
HIHAT	High-Resolution Hemispherical Reflector Antenna Technique
HI-HICAT	High High Altitude Clear Air Turbulence
HIIS	Honeywell Institute for Information Sciences [US]
HIJ	Horological Institute of Japan
HIL	Hazardous Industrial Liquid
	High-Intensity Lighting
HILAC	Heavy-Ion Linear Accelerator [also: Hilac or hilac]
HILAN	High-Level Language [also: Hilan or hilan]
HIM	Hardware Interface Module
	Hazardous Industrial Material
HIMA	Health Industry Manufacturers Association [US]
HIMAT	Highly Maneuverable Aircraft Technology [also: HiMAT]
HIMES	Highly-Maneuverable Experimental Spacecraft [also: HiMES]
HIMAIL	Hitachi Integrated Message and Information Library
HIN	Hybrid Integrated Network
HINAS	Historic Naval Ships Association of North America [US]
Hinc	Haemophilus Influenzae Rc [Molecular Biology]
HIND	Hindustan
Hind	Haemophilus Influenzae Rd [Molecular Biology]
Hinf	Haemophilus Influenzae Serotype f [Molecular Biology]
HINIL	High-Noise-Immunity Logic [also: HiNIL]
HIO	High Input/Output
HIOMT	Hydroxyindole-O-Methyltransferase

HIOS Heath/Zenith Instrument Operating Software
High Island Offshore System
HIP Hanford Isotope Plant [US]
High-Impact Polystyrene
Hitachi Parametron Automatic Computer
Host Information Processor
Hot Isostatic Pressing [also: Hip or hip]
HIPAR High-Power Acquisition Radar
HIPD High-Intensity Powder Diffractometer
HIPERNAS High-Performance Navigation System
HIPO Hierarchy plus Input-Process-Output
HIPOT High-Potential [also: Hipot or hipot]
HIPOTT High-Potential Test [also: Hipott or hipott]
HIPRBSN Hot-Isostatic Pressed Reaction-Bonded Silicon Nitride
HIPS High-Impact Polystyrene
HIPSN Hot Isostatically Pressed Silicon Nitride
HIPS-PE High-Impact Polystyrene — Polyethylene
HIPS-PVDC High-Impact Polystyrene — Polyvinylidene Chloride
HIPS-PVDC-PE High-Impact Polystyrene — Polyvinylidene Chloride — Polyethylene
HIPS-PVDC-PP High-Impact Polystyrene — Polyvinylidene Chloride — Polypropylene
HIQSA Hydroxyiodoquinolinesulfonic Acid
HIR (Study Group on) Harmful Interference to Radio [of ICAO]
Horizontal Impulse Reaction
HIRAC High Random Access
HIRAM Highest-Position Random-Access Memory
HIRAN High-Precision Shoran [also: Hiran or hiran]
HIRD High-Intensity Radiation Device
HIRDS High-Intensity Radiation Development Laboratory [of BNL, US]
HIRES Hypersonic In-Flight Refueling System
Hi-res high-resolution [also: hi-res]
HIRF High-Intensity Reciprocity Failure
HIRIS High-Resolution Interferometer/Spectrometer
HIRL High-Intensity Runway Lights
HIRNS Helicopter Infrared Navigation System
HIRS High-Impulse Retro-Rocket System
High-Resolution Infrared Radiation Sounder
High-Resolution Infrared Sounder
HIS Helicopter Integrated System
Holographic Illumination System
Home Interactive System
Homogeneous Information Set
Honeywell Information System
Hospital Information System
His Histidine
Histidyl
HISAM Hierarchical Indexed Sequential Access Method
HISARS Hydrological Storage and Retrieval Information System [of NCSU, US]
HISDAM Hierarchical Indexed Sequential Direct Access Method
HISCC Hydrogen-Induced Stress-Corrosion Cracking
HIT Harbin Institute of Technology [PR China]
High Isolation Transformer
Himeji Institute of Technology [Japan]
Hiroshima Institute of Technology [Japan]
Hokkaido Institute of Technology [Japan]
Houston International Teleport [US]
Huazhong Institute of Technology [PR China]
Hypersonic Interference Technique
HITC Hexamethylindotricarbocyanine
HITAC Hitachi Computer Services
HITEC High-Temperature Emission Control System
HITEMP High-Temperature Engine Materials Program [of NASA]
High-Temperature Turbine Engine Program [US]
HITEX High Temperature Isotope Exchange
HITRAC High Technology Training Access
HITRESS High Test Recorder and Simulator System
HITS Hierarchical Integrated Test Simulator
Holloman Infrared Target Simulator
HIU Hydrostatic Interface Unit
HIV Human Immunodeficiency Virus
HIVOS High Vacuum Orbital Simulation
HIW Hazardous Industrial Waste
HIXE Heavy Ion Induced X-ray Emission
HJ Heterojunction
Hose Jacket
Hot Junction
Hybrid Junction
HJBT Heterojunction Bipolar Transistor
HJD Heliocentric Julian Date
HJFET Heterojunction Field-Effect Transistor
HJP Hydraulic Jet Propulsion
HK Hand Knob
Hefner-Kerze [= Hefner Candle]
Hexakinase
Knoop Hardness Number
Hong Kong
HKCS Hong Kong Computer Society
HK$ Hong Kong Dollar [also: HKD; Currency]
HKGCC Hong Kong General Chamber of Commerce
HKIE Hong Kong Institution of Engineers
HKLA Hong Kong Library Association
HKSM Henry Krumb School of Mines [of Columbia University, US]
HKTDC Hong Kong Trade Development Council
HL Half Life
Harwell Laboratory [UK]
Hazardous Liquid
Heap Leaching
Hearing Level
Height-Length
Height Loss
Herpetologists League [US]
High Level
Hinge Line
Hirth Lock
Horizontal Line
Host Language
Hot Line
hL hectoliter [also: hl]
hl (hoc loco) — here
HLA Halifax Library Association [Canada]
Helicopter Loggers Association [US]
Human Leucocyte Antigen [also: hLA]
HLAD Hearing-Lookout Assist Device

HLAF High-Level Arithmetic Function
HLAIS High-Level Analog Input System
HLAH Hank's Lactalbuminhydrolysate
HLAS Hot Line Alert System
HLB Hydrophilic-Lipophilic Balance
HLCV Hot Leg Check Valve
HLD Helium Leak Detector
Hold
HLDG Holding
HLDTL High-Level Data Transistor Logic
HLF Half
HLG Hauling
HLH Heavy Lift Helicopter
High-Level Heating
hLH Human Luteinizing Hormone
HLI Host Language Interface
HLIV Hot Leg Isolation Valve
HLL High-Level Language
High-Level Logic
HLLV Heavy Lift Launch Vehicle
HLLW High-Level Liquid Waste
HLLWT High-Level Liquid Waste Tank
HLMI High Load Melt Index
HLML High-Level Microprogramming Language
HLN Hexagonal Long Nipple
HLP High-Level Programming
HLPI High-Level Programming Interface
HLQL High-Level Query Language
HLR High-Level Representation
Holder
HLRZ Hochleistungsrechenzentrum [= High-Performance Computer Center, Germany]
HLS Heavy Logistic Support [US Air Force]
High-Level Scheduler
Hue, Lightness, Saturation
HLSC High-Level Service Circuit
HLSE High-Level, Single-Ended
HLSM Homopolar Linear Synchronous Motor
HLSRS High Level Sisal Research Station [East Africa]
HLSTO Hailstone
HLSUA Honeywell Large System Users Association [US]
HLT Halt
Heterodyne Look-Through
HLTF High Level Task Force
HLTL High-Level Transistor Logic
HLTTL High-Level Transistor-Transistor Logic
HLU House Logic Unit
HLW High-Level Waste
HLYR Haze Layer Aloft
HM Hand-Made
Hard Magnetic
Harmonic Motion
Heavy Mobile
Hectometric
Her/His Majesty
High Magnification
High-Melting
High-Modulus
Home
Hot Mix
Hydraulic Modelling
Hydromagnetic
Hydromechanical
Hydrometallurgist
Hydrometallurgy
Hypermedia
hm hectometer
HMA Hard Magnetic Alloy
Hardwood Manufacturers Association [US]
Headmasters Association
Hoist Manufacturers Association [now: HMI, US]
HMAC Hazardous Materials Advisory Council [US]
HMANA Hawk Migration Association of North America [US]
HMB 2-Hydroxy-5-Methoxybenzaldehyde
HMBA Hydroxymercurbenzoate
HMC Hamburg Messe und Congress [= Hamburg Trade Fair and Congress, Germany]
Her/His Majesty's Customs and Excise [UK]
High-Strength Molding Compound
Horizontal Motion Carriage
HMCRI Hazardous Materials Control Research Institute [US]
HMCS Her/His Majesty's Canadian Ship
HMD Hexamethylenediamine
Hot-Mix Design
Hydraulic Mean Depth
HMDA Hexamethylenediamine
HMDBA Hollow Metal Door and Buck Association [US]
HMDE Hanging Mercury Drop Electrode
HMDE/ASV Hanging Mercury Drop Electrode/Anodic Stripping Voltage
HMDE/DPCSV Hanging Mercury Drop Electrode/Differential-Pulse Continuous Stripping Voltage
HMDF Horizontal Side of Main Distribution Frame
HMDI Hydrogenated 4,4'-Diphenylmethane Diisocyanate
HMDS Hexamethyldisilazane
HMDSO Hexamethyldisiloxane
HME High Vinyl Modified Epoxy
HMF Heavy-Metal Fluoride
High-Modulus Furnace Black
Hydroxymethylfuraldehyde
HMG Hardware Message Generator
Heavy Machine Gun
Her/His Majesty's Government
Historical Metallurgy Group [now: HMS, UK]
Human Menopausal Gonadotropin
HMI Hahn-Meitner Institute [Germany]
Halogen Metallide Iodide
Hardware Monitor Interface
Hardwood Manufacturers Institute [now: HMA, US]
Her/His Majesty's Inspector [UK]
Hexamethyleneimine
Hoisting Machinery Institute
Hoist Manufacturers Institute [US]
Horizontal Motion Index
HMIS Hazardous Material Information System
HML Hard Mobile Launcher
Hardware Modelling Library
HMM Hard Magnetic Material
Hardware Multiply Module
Heavy Meromyosin
Hexamethylmelamine [also: HEMEL]

HMMS Highway Maintenance Management System
HMMWV High Mobility Multipurpose Wheeled Vehicle
HMO Hardware Microcode Optimizer
Health Maintenance Organization
Hueckel Approximation for Molecular Orbitals
Hueckel Molecular Orbitals
HMOS High-Density Metal-Oxide Semiconductor
High-Performance Metal-Oxide Semiconductor
High-Speed Metal-Oxide Semiconductor
HMOS-E High-Density Metal-Oxide Semiconductor — Erasable
High-Speed Metal-Oxide Semiconductor — Erasable
HMP Hexose Monophosphate
HMPA Hexamethylphosphoramide [also: HEMPA]
HMPG Hydroxymethoxyphenylglycol
HMPT Hexamethylphosphorous Triamide
HMR Homer
Humidity-Mixing Ratio
Hybrid Modular Redundancy
HMRB Hazardous Materials Regulation Board [US]
HMRL Harvard Materials Research Laboratory [of HU, US]
HMS Hanford Meteorology Survey [US]
Harvard Medical School [US]
Hazardous Materials System [of BOE, US]
Her/His Majesty's Ship
Historical Metallurgy Society [UK]
HMSA Hardware Manufacturers Statistical Association [now: BHMA, US]
Hawk Mountain Sanctuary Association [US]
HMS(BOE) Hazardous Materials System (Bureau of Explosives) [US]
HMSC Huntsman Marine Science Center [Canada]
HMSF Hexamethylenetetraselenafulvalene
HMSO Her/His Majesty's Stationery Office [of DTI, UK]
HMSS Hospital Management Systems Society
HMSLD Helium Mass Spectrometer Leak Detector
HMT Hexamethylenetetramine [also: HMTA or Hexa]
Hypoxanthine/Methotrexate/Thymidine Medium
HMTA Hazardous Materials Transportation Act [US]
Hexamethylenetetramine [also: HMT or Hexa]
HMTC Hexamethylenetetramine Camphorate
HMTSeF Hexamethylenetetraselenafulvalene
HMTTeF Hexamethylenetetratellurafulvalene
HMTTF Hexamethylenetetrathiofulvalene
HMW Hectometric Wave
High Molecular Weight
HMW HDPE High Molecular Weight High-Density Polyethylene
HMWK High Molecular Weight Kininogen
HMWPE High Molecular Weight Polyethylene
HMY Her/His Majesty's Yacht
HN Hemagglutinin Neuraminidase
Hexagonal Nipple
Hexagon Nut
Horn
Host-to-Network
HNA Hierachical Network Architecture
Hitachi Network Architecture
Hydroxynaphthoic Acid
HNAB Hexanitroazobenzene
H-NBR Hydrogenated Nitrile Butadiene Rubber
HNC Higher National Certificate [UK]
HNCIAWPRC Hungarian National Committee of the International Association on Water Pollution Research and Control
HNCIMU Hungarian National Committee for the International Mathematical Union
HND Higher National Degree [UK]
Hundred
HNDBK Handbook
HNED Horizontal Null External Distance
HNIL High-Noise-Immunity Logic
HNL Helium Neon Laser
Honduran Lempira [Currency of Honduras]
HNM Helicopter Noise Model
Hexanitromannite
HNPA Home Numbering Plan Area
HNPF Hallam Nuclear Power Facility
hnRNA Heterogeneous Nuclear Ribonucleic Acid
HNS Hexanitrostilbene**HNTD**
Highest Non-Toxic Dose
HNVS Helicopter Night Vision System
Hughes Night Vision System
HO Hand-Operated
Hard Overhung Balancer
Harmonic Oscillator
Head Office
Hertzian Oscillator
High Order
Home Office [UK]
House
Hybrid Orbital
Hydrographic Office [US]
Ho Holmium
ho horse [Immunochemistry]
HOAB Heptyloxyazoxybenzene
HOARS Hands-On Annotated Recorded Search
HOB Homing on Offset Beacon
Hot Ore Briquetting
HOBIS Hotel Billing Information System
HOBITS Haifa On-Line Bibilographic Text System [Israel]
HOBO Homing Official Bomb
HOBOS Homing Bomb System
HOBT 1-Hydroxybenzotriazole Hydrate [also: HOBt]
HOD Higher-Order Differentiation
HODOS Hole Drilling Operating System
HODRAL Hokushin Data Reduction Algorithm Language
HOE Height-of-Eye
Holographic Optical Element
Homing Overlay Experiment
HOF Head of Form
High-Octane Fuel
$Ho(FOD)_3$ Tris(heptafluorodimethyloctanedionato)holmium
HOFR Heat-Resistant, Oil-Resistant, Flame-Retardant
HOI Hypoiodous Acid
HOL High-Order Language
Holiday
Hollow
HOLDG Holding
HOLUG Houston On-Line Users Group [US]
HOLZ Higher Order Laue Zone
HOM

	Higher-Order Mode
HOMO	Highest Occupied Molecular Orbital
Hon	Honorable
	Honorary
	Honors
HONB	N-Hydroxy-5-Norbornene-2,3-Dicarboximide
HOOPS	Hierarchical Object-Oriented Picture System
HOP	HEDL (= Hanford Engineering Development Laboratory) Overpower
	House Operating Tape
	Hydrogen Overpotential
	Hybrid Operating Program
HOPE	Hydrogen-Oxygen Primary Extraterrestrial [of NASA]
HOPG	Highly-Oriented Pyrolytic Graphite
HOPL	History of Programming Languages
HOQ	Home Office Quote
HOR	H_2O Reactor
	Heliocentric Orbit Rendezvous
	Horizon
	Horizontal [also: HORIZ or HORZ]
HORACE	H_2O Reactor Aldermaston Critical Experiment [UK]
HOROL	Horological
	Horology
HORT	Horticultural
	Horticulture
HOS	High-Order Software
	Horizontal Obstacle Sonar
HOSA	Hydroxylamine-O-Sulfonic Acid
HOSG	Helicopter Operations Study Group [UK]
HOSP	Hospital
HOST	Hot Section Technology
HOT	Hand Over Transmitter
HOTCE	Hot Critical Experiments
$Ho(THD)_3$	Tris(tetramethylheptanedionato)holmium
HOTOL	Horizontal Takeoff and Landing
HOTRAN	Hover and Transition Simulator
HOV	High Occupancy Vehicle (Lanes)
HOW	Howitzer
HOWAQ	Hot Water Quenching
HP	Hand Pump
	Hard Pipe
	Health Physics
	Heatable Plastic
	Heat Pipe
	Heat Pump
	Heterophase Polymerization
	High-Pass (Filter)
	High Performance
	High Polymer
	High Positive
	High Power
	High Pressure
	High Purity
	Highway Patrol
	Hire Purchase
	Horizontal Polarization
	Horsepower [also: hp]
	Host Processor
	Hot Pressing
	Hydraulic Power
	Hydraulic Press
	Hydraulic Pressure
	Hydrogen Purifier
	Hydroxyproline
HPA	Helix Pomatia Agglutinin
	Heuristic Path Algorithm
	High-Performance Alloy
	High-Power Amplifier
	Holding and Positioning Aid
	Hydroxypropyl Acrylate
hPa	Hectopascal
Hpa	Haemophilus Parainfluenzae [Molecular Biology]
HPB	Health Protection Branch [of DNHW, Canada]
HPBC	Homopolar Pulse Billet Heating
HPBW	Half-Power Beamwidth
HPC	Hard-Processing Channel Black
	High Pin Count
	Hydroxypropylcellulose
HPCA	High-Performance Communications Adapter
HPCE	High-Performance Capillary Electrophoresis
HPCI	High-Pressure Coolant Injection
HPCS	High-Pressure Combustion Sintering
	High-Pressure Core Spray
	High-Pressure Core Spraying System
HPD	Hearing Protective Device
	Highest Posterior Density
	High Power Density
	Horizontal Polar Diagram
	Hough-Powell Device
HPE	High-Pressure Equipment
HPEP	High-Performance Electrophoresis
HPF	Heat Pipe Furnace
	Highest Probable Frequency
	Highest Possible Frequency
	High-Pass Filter
	Hot-Pressed Ferrite
HPFD	Hybrid Personal Floating Device
HPFL	High-Performance Fuels Laboratory [US]
HPG	Hard Page [also: HPg]
	Homopolar Generator
	Hydroxypropyl Guar
HPGA	High-Pressure Gas Atomization
HPGe	High-Purity Germanium
HP/HIP	Hot Pressing followed by Hot Isostatic Pressing
HPhr	Horsepower-Hour [also: hp hr]
HPHT	High-Pressure, High-Temperature Process
HPI	Hardwood Plywood Institute [now: HPMA, US]
	Height Position Indicator
	High-Performance Imagery
	High-Pressure Injection
HPIA	Hydroxyphenylisopropyladenosine
HP-IB	Hewlett Packard Interface Bus
HPIM	High-Performance Image Analysis
HPIR	High-Probability-of-Intercept Receiver
HPIS	High-Pressure Injection System
HPIT	High-Performance Infiltrating Technology
HPL	Human Placenta Lactogen
HPLC	High-Performance Liquid Chromatography
	High-Pressure Liquid Chromatography
HP-LEC	High-Pressure Liquid Encapsulated Czochralski

HPLPLC High-Performance Low-Pressure Liquid Chromatography
HPLR Hinge Pillar
HPM Hard-Part Machining
High-Performance Motor
Hybrid Phase Modulation
HPMA Hardwood Plywood Manufacturers Association [US]
Hydroxylpropylmethacrylate
HPMC Hydroxypropylmethylcellulose
HPMWA High-Power Microwave Amplifier
HPN High-Pass Notch
Horsepower Nominal
HPOF High-Pressure Oil-Filled (Cable)
HPOT Helipotentiometer
HPP Harward Project Physics [of HU, US]
High-Performance Polymer
High-Pressure Physics
Holding Under Promise of Payment
Hot Processing Plant
4-Hydroxypyrazolopyrimidine
HPPH 5-Hydroxylphenyl-5-Phenylhydantoin
HPPS High-Performance Paper Society [Japan]
HPRC Houston Petroleum Research Center [US]
HPRL Hoechst Pharmaceutical Research Laboratories
H-PRESS High-Pressure
HPRR Health Physics Research Reactor [of ORNL, US]
HPRT Hypoxanthine Phosphoribosyl Transferase
HPS Hanford Plant Standards [formerly: HWS]
Hazardous Polluting Substance
Health Physics Society
High-Performance Paper Society [Japan]
High Pressure Sintering
High-Pressure Sodium (Lamp)
Horizontal Pull Slipmeter
HPSC High-Pressure Self-Combustion Sintering
Hot-Pressed Silicon Carbide
HPSIS High-Pressure Safety Injection System
HPSN Hot-Pressed Silicon Nitride
HPSTC Highest Point of Single Tooth Contact
HPSW High-Pressure Service Water
HPT Head per Track [also: HP/T or HT]
High-Precision Thermostat
High-Pressure Test
High-Pressure Turbine
Horizontal Plot Table
Hydroxypyrenetrisulfonate
8-Hydroxy-1,3,6-Pyrenetrisulfonic Acid
H PT High Point
HPTE High-Performance Turbine Engine
HPTET High-Performance Turbine Engine Technology
HPTLC High-Performance Thin Layer Chromatography
HPU Hydraulic Power Unit
HP-UX Hewlett Packard Unix Operating System
HPV Human Powered Vehicle
HPW Homopolar Pulse Welding
Hot Pressure Welding
HPZ Helicopter Protected Zone
Hydridopolysilylazane
HQ Headquarters
High Quality
Hydro-Quebec [Canada]
Hydroquinone
HQDA Headquarters Department of the Army [US]
HQE Hydroquinone Electrode
HQEE Hydroquinone Di(β-Hydroxyethyl) Ether
HQI Hydro-Quebec International [Canada]
HQIR Hydro-Quebec Institute of Research [Canada]
HQO Hydroxyquinoline Oxide
HR Handling Routine
Hand Reset
Harbour
Hard to Research
Hear
Heating Resistor
Heat Regenerator
Heterogeneous Reactor
High Reduction
High Resilience
High Resistance
High Resolution
High Resonance-Damping
Hit Ratio
Hoist Ring
Holding Register
Homogeneous Reactor
Hot Rolling
Relative Humidity
Rockwell Hardness
H-R Herzsprung-Russell Diagram
hr hour
HRA Hot-Rolled and Annealed
Human Reliability Analysis
Rockwell 'A' Hardness
HRAES High-Resolution Auger Electron Spectroscopy
HRAI Heating, Refrigerating and Air Conditioning Institute
HRB Highway Research Board [US]
Rockwell 'B' Hardness
HRC Hardwood Research Council [[US]
Harmonically Related Carrier Frequency
High Rupturing Capacity
Horizontal Redundancy Check
Hypothetical Reference Circuit
Rockwell 'C' Hardness
HRCG Hexagonal Reducing Coupling
HRD Human Resources Department
Human Resources Development
Hydrometeorology Research Division [Canada]
H RD Half Round [also: 1/2 RD]
HRDP Hypothetical Reference Digital Path
HRDWD Hardwood
HRDWR Hardware
HRE Homogeneous Reactor Experiment
Hypersonic Research Engine
Rockwell 'E' Hardness
HREELS High-Resolution Electron Energy Loss Spectroscopy
High-Resolution Electron Energy Loss Spectrum
HREF Horticulture Research Experimental Farm [Canada]
HREM High-Resolution Electron Microscopy
HRES High-Resolution Electron Spectroscopy
HRF Herb Research Foundation [US]
Rockwell 'F' Hardness

HRFAX High-Resolution Facsimile
HR-FESEM High-Resolution Field Emission Scanning Electron Microscopy
HR15T Rockwell '15T' Superficial Hardness
HRG Horizontal Ribbon Growth
HRH Rockwell 'H' Hardness
HRHA Hydronic Radiant Heating Association [US]
HRI Horticultural Research Institute [US]
HRIO Height-Range Indicator Operator
Horticultural Research Institute of Ontario [Canada]
HRIR High-Resolution Infrared
High-Resolution Infrared Radiometer
HRIS Highway Research Information Service [of HRB, US]
HRL Horizontal Reference Line
Human Resources Laboratory [US Air Force]
HRLC High-Resolution Liquid Chromatography
HRLEELS High-Resolution Low-Energy Electron Loss Spectroscopy
HRM Hard-Rock Mining
Hardware Read-In Mode
HRN Hexagonal Reducing Nipple
HRNA Heterogeneous Ribonucleic Acid
HRNES Host Remote Node Entry System
HRP Hand Retractable Plunger
Heat-Resistant Phenolic
Horseradish Peroxidase
Hypergroup Reference Pilot
H&RP Holding and Reconsignment Point
HRPO Horseradish Peroxidase
HRRC Human Resources Research Center
HRRTS High-Resolution Remote Tracking Sonar
HRS Heading Reference System
High-Resolution Sensing
High-Resolution Spectrography
High-Resolution Spectrometer
Host Resident Software
Hot-Rolled Steel
Hovering Rocket System
Hydraulics Research Station [UK]
hrs hours
HRSEM High-Resolution Scanning Electron Microscopy [also: HR-SEM]
HRSI High-Temperature Reusable Surface Insulator
HRSIM High-Resolution Selected Ion Monitoring
HRSP (Association of) Human Resource Systems Professionals [US]
HRSS Host Resident Software System
HRST High-Resolution Sensing Technology
HRT Hard Return [also: HRt]
High Rate Telemetry
High-Resolution Tracker
Homogeneous Reactor Test
HRTEM High-Resolution Transmission Electron Microscopy
HRU Heading Reference Unit
Hydrological Research Unit
HRV Heat Recovery Ventilator
Hyperbaric Rescue Vehicle
Hypersonic Research Vehicle
HRX Hypothetical Reference Connection
HRXRD High-Resolution X-Ray Diffractometry
HRZ Hazard Ranking System
HRZN Horizon
HS Half-Subtract(er)
Handset
Hard Solder
Hard Sphere Model
Heating Surface
Heat Sink
Heat Switch
Hemisuccinyl
Heuristic Search
Hierarchically Structured
High Sensitivity
High Shear
High Speed
High Stability
High Strength
High Structure
Holographic Stereogram
Horizon Scanner
Horizon Sensor
Host to Satellite
Hot Stage
Hydraulic System
Hydrofoil Ship
Hydrographic Society [UK]
Hydrostatic(s)
Hydroxylamine Sulfate
Hypersensitization
Hypersonic(s)
H/S Hard/Soft Ratio
Health and Safety
HSA Hemispherical Analyzer
Herb Society of America [US]
Hidden Surface Algorithm
Holly Society of America [US]
Human Serum Albumin
Hydroponic Society of America [US]
HSAB Hard-Soft Acid-Base
HSAC Health Safety and Analysis Center
Helicopter Safety Advisory Conference [US]
High-Speed Analog Computer
HSAM Hierarchical Sequential Access Method
HSAS Houldsworth School of Applied Science [of University of Leeds, UK]
HSB High-Speed Buffer
HSBA High-Speed Bus Adapter
H-SB-BL Hydrogenated Styrene Butadiene Block Copolymer
HSBIP Hsinchu Science Based Industrial Park [Taiwan]
HSC Health and Safety Commission [UK]
Health Sciences Center
High-Speed Concentrator
Hydrocarbon Subcommittee
Hydrogen Stress Cracking
HSCC Heavy Specialized Carrier Conference [US]
Hydrogen-Assisted Stress Corrosion Cracking
Hydrogen-Induced Stress Corrosion Cracking
HSc Scleroscope Hardness Number, Model c
HSCP High-Speed Card Punch
HSCT High-Speed Civil Transport (Program) [US]

	High-Speed Commercial Transport
	High-Speed Compound Terminal
HSD	Doctor of Health and Safety
	High-Speed Data
	High-Speed Displacement
	Horizontal Situation Display
	Hot Shutdown
HSd	Scleroscope Hardness Number, Model d
HSDA	High-Speed Data Acquisition [also: HS-DA]
HSDB	High-Speed Data Buffer
HSDC	Hybrid Synchro-to-Digital Converter
HSDir	Director of Health and Safety
HSDS	Hot Spot Detection System
HSE	Health and Safety Executive [UK]
	House
	Hydrostatic Equilibrium
	Hydrostatic Extrusion
HSEL	High-Speed Selector Channel
HSF	High-Speed Flow
HSG	Horizontal Sweep Generator
	Housing
HSGPC	High-Speed Gel Permeation Chromatography
HSGT	High-Speed Ground Transport
HSGTC	High-Speed Ground Test Center
HS/HM	High-Strength/High-Modulus
HSI	Heat Stress Index
	Horizontal Situation Indicator
HSIA	Halogenated Solvents Industry Alliance [US]
HSIP	Hsinchu Science-Based Industrial Park [Taiwan]
HSIS	Human Settlements Information System
HSK	Housekeeping
HSL	Hazardous Substances List
	Highway Safety Literature [of NHTSA, US]
	Hue Saturation Luminance Monitor
HSLA	High-Speed Line Adapter
	High-Strength Low-Alloy (Steel)
HSLC	Hierarchical Storage Manager
	High-Speed Liquid Chromatography
HSM	High-Speed Mechanics
	High-Speed Memory
HSN	Highly Saturated Nitrile Rubber
	Hospital Satellite Network
HSNY	Horticultural Society of New York [US]
HSP	High-Speed Photography
	High-Speed Photometer
	High-Speed Plating
	High-Speed Printer
	Hydrostatic Pressure
HSPA	Hawaiian Sugar Planters Association [US]
HSPN	High-Speed Packet Network
HSPTR	High-Speed Paper Tape Reader
HSQ	Historical Society of Queensland [Australia]
HSR	Hardware Status Register
	High-Speed Rail
	High-Speed Reader
	High-Strength Resin
HSRA	High Speed Rail Association [US]
HSRC	Health Sciences Resource Center [of CISTI, Canada]
HSRI	Highway Safety Research Institute [US]
HSRIOP	High-Speed RAD Input/Output Processor
HSRO	High-Speed Repetitive Operation
HSRTM	High-Speed Resin-Transfer Molding
	High-Strength Resin Transfer Molding
HSS	Hierarchy Service System
	High-Speed Steel
	High-Speed Storage
	History of Science Society [US]
	Hollow Structural Section
HSSA	High-Speed Steel Association [UK]
	Hydrographic Society of South Africa
HSSCC	Hot-Salt-Induced Stress-Corrosion Cracking
HSSDS	High-Speed Switched Digital Service
HSST	Heavy Section Steel Technology
HST	Handelskammer Schweiz-Tschechoslovakei [= Chamber of Commerce Switzerland-Czechoslovakia]
	Hawaiian Standard Time
	Hawaiian Sugar Technologists [US]
	High-Speed Taxi-Way Turn-Off
	High Speed Technology
	Horizontal Seismic Trigger
	Hubble Space Telescope
	Hypersonic Transport
	Hypervelocity Shock Tunnel
HSTCO	High-Stability Temperature-Compensated Crystal Oscillator
HSTU	Hunan Science and Technology University [PR China]
HSV	Herpes Simplex Virus
	Hue Saturation Value
HSVB	High-Speed Video Bus
HSVP	High-Speed Vector Processor
HSW	Heat-Sensitive Wire
HSWA	Hazardous and Solid Waste Amendment
HSY	Hard-Sphere-Yukawa Model
HSYNC	Horizontal Synchronization
HSZD	Hermetically-Sealed Zener Diode
HT	Half Tone
	Hardness Testing
	Head per Track [also: HPT or HP/T]
	Heat
	Heat Transfer
	Heat Treat(ment)
	Heavy Truck
	Height
	Height of Target
	Heliotropin
	Hematoxylin
	High Temperature
	High Tension
	High Tide
	Holding Time
	Homing Transponder
	Horizontal Tab(ulation Character)
	Horizontal Tabulator
	Hybrid Transformer
	Hydraulic Turbine
	Hydroxytryplamine
	Hypoxanthine/Thymidine Medium
	Tritiated Hydrogen
HTA	Heavier than Air

	High-Temperature Alloy
	High-Temperature Amorphous (Resin)
	Hypophysiotropic Area
HTB	Hexadecimal-to-Binary
	Highway Tariff Bureau [now: AMCTB, US]
HTC	High-Tech Ceramics
	High-Temperature Corrosion
	High-Temperature Crystallizable
	High-Temperature Crystallization
	Horizontal Toggle Clamp
	Hybrid Technology Computer
HTCI	High-Tensile Cast Iron
HT-CVD	High Temperature Chemical Vapor Deposition
HTD	Hand Target Designator
	Heterojunction Tunneling Diode
	High-Temperature Deformation
	High Torque Drive
HTDA	High-Temperature Dilute Acid
HTDS	High-Temperature Drawing Salt
HTE	Heat Treating Exposition
	Hypergroup Translating Equipment
HTF	High-Temperature Fatigue
	Horizontal Tube Feeder
HTFFR	High-Temperature Fast-Flow Reactor
HTFMI	Heat Transfer and Fluid Mechanics Institute
HTFS	Heat Transfer and Fluid Flow Service [of UKAEA]
HTG	Haitian Gourde [Currency]
	Heating
	Hydrostatic Tank Gauging
HTGCR	High-Temperature Gas-Cooled Reactor [Australia]
HTGL	High-Temperature Gas-Dynamics Laboratory [of Stanford University, US]
HTGPF	High-Temperature General Purpose Furnace
HTGR	High-Temperature Gas-Cooled Reactor
	High-Temperature Gas Reactor
HTGR-CX	High-Temperature Gas Reactor Critical Experiment
HTGR SC/C	HTGR Steam Cycle Cogeneration
HTH	High-Test Hypochlorite
HTHP	High-Temperature/High-Pressure
HTI	Hand Tools Institute [US]
	High-Technology Intensive
	High-Temperature Impact
	High-Temperature Insulation
HTIP	Housing Technology Incentives Program [of CMHC, Canada]
HTIS	Heat Transfer Instrument System
HTL	Heat Transfer Loop
	High-Threshold Logic
	Hohere Technische Lehranstalt [= Higher Technical Institution [Germany, Switzerland and Austria]
	Hydroxyl Terminated Liquid
HTLA	High-Temperature Low-Activity
HTLDC	Hsinchu Tidal Land Development Planning Commission [Taiwan]
HTLTR	High-Temperature Lattice Test Reactor [US]
HTLV	Human T-Cell Leukemia Virus
	Human T-Cell Lymphotropic Virus
HTLV-1	Human T-Cell Lymphotropic Virus, Type 1
HTM	High Temperature Material
	Hypothesis Testing Model
HTMA	Hawaii Territorial Medical Association [US]
	Hydraulic Tool Manufacturers Association [US]
HTMP	Hydrooxytetramethylpiperidine
H-TMS	Honeywell Test Management System
HTN	Heat Treatable Nodular
HTO	High-Temperature Oxidation
	Horizontal Takeoff
	Tritiated Hydrogen Oxide
HTP	High-Temperature Performance
	High-Temperature Physics
	High-Temperature Pretreatment
	High Test Peroxide
	High Thermal Performance
	Hydrothermal Processing
	Hydroxytryptophane
HTPB	Hydroxy-Terminated Polybutadiene
HTPS	High Tension Power Supply
HTR	Hanford Test Reactor [US]
	Heater
	High-Temperature Reactor
	Hitachi Training Reactor [Japan]
	Hydrothermal Reaction
HTRB	High-Temperature Reverse Bias
HTRDA	High-Temperature Reactor Development Associates [US]
HTRE	Heat Transfer Reactor Experiment [US]
HTRI	Heat Transfer Research Institute [US]
HTS	Head Track and Selector
	Heights
	Height Telling Surveillance
	High-Temperature Shift
	High-Temperature Superconductor
	High-Tensile Steel
	High Tensile Strength
	Host to Satellite
HTSC	High-Temperature Superconductivity
	High-Temperature Superconductor
HTSEC	High-Temperature Size-Exclusion Chromatography
HTSF	High-Temperature Sodium Facility
HTSH	Human Thyroid Stimulating Hormone
HTSJ	High Temperature-Society of Japan
HTSL	Heat Transfer Simulation Loop
HT/SPC	Heat Treating/Statistical Process Control
HTSS	Hamilton Test Simulation System
	Honeywell Time-Sharing System
HTST	High-Temperature Short-Time
HTTF	High-Temperature Test Facility
HTTL	High-Power Transistor-Transistor Logic
HTTMT	High-Temperature Thermomechanical Treatment
HT TR	Heat Treat(ment)
HTTT	High-Temperature Turbine Technology
HTTVMT	High-Temperature Thermovibrational Mechanical Treatment
HTU	Harbin Technical University [PR China]
	Heat Transfer Unit
	Height of Transfer Unit
HTV	High Temperature and Velocity
HTVB	High-Temperature Vacuum Brazing
HTW	High-Temperature Water
HTX	High-Temperature Crystalline
HU	Hang-Up
	Hanyang University [Korea]

	Harvard University [US]
	Hebrew University [US]
	High Usage
	Hiroshima University [Japan]
	Hokkaido University [Japan]
	Hongshan University [PR China]
	Hosei University [Japan]
HUC	High Usage Circuit
HUD	Head-Up Display
	(Department of) Housing and Urban Development [of FHA, US]
HUDAC	Housing and Urban Development Association of Canada [now: CHBA, Canada]
HUD-FDA	(Department of) Housing and Urban Development — Federal Housing Administration [US]
HUDWAC	Head-Up Display Weapons-Aiming Computer
HUF	Hungarian Forint [Currency]
HUFSAM	Highway Users Federation for Safety and Mobility [US]
HUG	Hastech Users Group [US]
	Honeywell Users Group [US]
HUGHES-NEL	Hughes Aircraft Company — Nuclear Electronics Laboratory [US]
HUGO	Highly Unusual Geophysical Operation
HULTIS	Hull Technical Interloan Scheme [UK]
HUM	Health and Usage Monitoring
HUMRRO	Human Resources Research Office
HUND	Highly Unusual Neutron Detector
Hunts	Huntingdonshire [UK]
HUR	Heat Up Rate
HURCN	Hurricane
HUST	Huazhong University of Science and Technology [PR China]
HUT	Helsinki University of Technology [Finland]
	HEDL (= Hanford Engineering Development Laboratory) Up Transient
HV	Half Value
	Hand-Operated Valve
	Hand Valve
	Hardware Virtualizer
	Heating and Ventilation
	Heavy
	High Vacuum
	High Velocity
	High Voltage
	High Volume
	Hydraulic Valve
	Hypervelocity
	Vickers Hardness
H&V	Heating and Ventilation
H/V	Horizontal/Vertical
HVA	High Voltage Apparatus
	Homovanillic Acid
HVAB	High-Volatile A Bituminous
HVAC	Heating, Ventilation and Air Conditioning
	High Vacuum
	High-Voltage Actuator
	High-Voltage Alternating Current
HVACC	High Voltage Apparatus Coordinating Committee [of ANSI, US]
HVAP	Hypervelocity, Armor Piercing
HVAR	High-Velocity Airborne Rocket
	High-Velocity Aircraft Rocket
HVAST	Hull Vibration and Strength Analysis
HVAT	High-Velocity Antitank
HVBB	High-Volatile B Bituminous
HVBO	Heterojunction Valence-Band Offset
HVC	High-Velocity Clouds
	High-Voltage Circuit
	Horizontal-Vertical Control
HVCA	Heating and Ventilating Contractors Association [UK]
HVCB	High-Volatile C Bituminous
HVCMOS	High-Voltage Complimentary Metal-Oxide Semiconductor
HVD	High-Velocity Detonation
HVDC	High-Voltage Direct-Current
HVDCT	High-Voltage Direct-Current Transmission
HVDF	High and Very-High Frequency Direction Finding
HVE	High Voltage Engineering
	Horizontal Vertex Error
HVEM	High-Voltage Electron Microscope
HVF	High Viscosity Fuel
HVG	High-Voltage Generator
HVHMD	Holographic Visor Helmet-Mounted Display
HVI	Home Ventilating Institute [now: HVIDAMCA, US]
	Human Visual Inspection
HVIC	High-Voltage Integrated Circuit
HVIDAMCA	Home Ventilating Institute Division of the Air Movement Control Association [US]
HVIS	Hypervelocity Impact System
HVL	Half-Value Layer
HVMF	High Valency Metal Fluoride
HVMOS	High-Voltage Metal-Oxide Semiconductor
HVMS	High Voltage Mass Separator
HVOF	High-Velocity Oxy-Fuel (Process)
	High-Velocity Oxygen/Fuel System
HVOSM	Highway Vehicle Object Simulation Model
HVP	High Video Pass
	Horizontal and Vertical Position
HVPS	High-Voltage Power Supply
HVR	Hard Vertical Rotating Balancer
	Hardware Vector to Raster
	High-Vacuum Rectifier
	High-Voltage Regulator
	Home Video Recorder
HVRA	Hawaiian Volcano Research Association [US]
	Heating and Ventilating Research Association [UK]
HVS	Hard Vertical Static Balancer
	High Vacuum Seal
HVSCR	High-Voltage Selenium Cartridge Rectifier
HVSEM	High-Voltage Scanning Electron Microscopy
HVT	Half-Value Thickness
	High-Vacuum Tube
	High-Voltage Test
	Hydraulic Variable Timing
HVTP	High-Velocity Target Practice
	Hypervelocity Target Practice
HVTS	High-Volume Time Sharing
	Hypervelocity Techniques Symposium
HVY	Heavy
HW	Half Wave

	Half Width
	Hammer Welding
	Handset, Wall Model
	Hardware
	Hard Water
	Hard Wired
	Hazardous Waste
	Head Wave
	Heavy Water
	Hertzian Wave
	High Water
	Hollow
	Hot Water
	Hot Wire
H/W	Hardware
HWC	Half-Wave Circuit
	Half-Wave Current
	Hazardous Waste Containment
	Health and Welfare Canada
	Hot Water Circulation
HWCTR	Heavy-Water Component Test Reactor [US]
HWD	Half-Wave Dipole
HWE	Hot Wall Epitaxy
HWERL	Hazardous Waste Engineering Research Laboratory [of EPA, US]
HWF	Hazardous Waste Federation [US]
HWGCR	Heavy-Water-Moderated Gas-Cooled Reactor
HWI	Hardware Interpreter
HWIM	Hear what I Mean
HWL	High Water Line
HWLWR	Heavy Water Moderated Light-Water Cooled Reactor
HWM	Half-Width Method
	Hazardous Waste Management
	High Water Mark
HWOCR	Heavy Water Organic-Cooled Reactor [also: HWOR]
HWOST	High Water Ordinary Spring Tide
HWP	Half Wave Plate
	Heavy-Water Plant
HWR	Half-Wave Rectifier
	Heavy-Water Reactor
HWS	Hanford Works Specification [now: HPS]
	Hanford Works Standard [now: HPS]
	Health and Welfare Canada
	Hot-Water System
HWTC	Hazardous Waste Treatment Council [US]
HWW	Hot and Warm Working
HWWA	Hamburgisches Weltwirtschaftsarchiv [= Hamburg Archives of International Commerce, Germany]
HWY	Highway
HX	Heat Exchanger
	Trans-1,4-Hexadiene
HXSA	Hexenylsuccinic Anhydride
Hy	Henry
HYACS	Hybrid Analog-Switching Attitude Control System
HYB	Hybrid (System)
HY BALL	Hydraulic Ball
HYBLOC	Hybrid Computer Block-Oriented Compiler
HYCATS	Hydrofoil Collision Avoidance and Tracking System
HYCOL	Hybrid Computer Link
HYCOTRAN	Hybrid Computer Translator
HYD	Hydraulic(s)
	Hydrology
HYDAC	Hybrid Digital Analog Computer
HYDAPT	Hybrid Digital-Analog Pulse Timer
HYDAS	Hydroacoustic Data Acquisition System
HYDR	Hydraulic(s)
	Hydrostatic(s)
HYDRO	Hydrodynamic(s)
	Hydrostatic(s)
HYDROLANT	Hydrographic Warning — Atlantic Ocean [of NOO, US]
HYDROPAC	Hydrographic Warning — Pacific Ocean [of NOO, US]
HYDX	Hydroxide
HYFAC	Hydrogenated Fatty Acid
HYFES	Hypersonic Flight Environmental Simulator
HYG	Hygroscopic
HYL	Hojalata Y Lamina Process
HYLA	Hybrid Language Assembler
HYLO	Hybrid LORAN (= Long-Range Navigation)
HYP	Harvard, Yale, and Princeton Universities [US]
	Hyphen (Character)
	Hypotenuse
	Hypothesis
Hyp	Hydroxyproline
HYPER	Hypertapes
HYPERDOP	Hyperbolic Doppler
HYPERNET	Hyper Network
hyp log	hyperbolic logarithm
HYPO	High-Power Water-Boiler Reactor [US]
HYPSES	Hydrographic Precision Scanning Echo Sounder
HYSTAD	Hydrofoil Stabilization Device
HYSTOR	Hydrogen Storage
HYV	High-Yield Variety
HZ	Haze
Hz	Hertz
HZIMP	Horizontal Impulse
HZP	Hot Zero Power

I

I	Current [Symbol]
	Igneous
	Imaginary Number
	Immediate
	Imperial
	Impulse
	Indication
	Indicator
	Industry
	Inertia
	Information
	Initial
	Initiation
	Input
	Inside
	Inspector
	Instruction
	Instrument(ation)
	Integration
	Intensity
	Interruption
	Intrinsic
	Iodine
	Ion
	Ireland
	Iron
	Island
	Isle
	Isoleucine
	Isoleucyl
	Isotropic Phase
i	igneous
	imaginary number
	inactive
	incendiary
	inclination
	incomplete
	independent
	induced
	industrial
	initial
	inner
	insoluble
	instantaneous
	institute
	institution
	instrumental
	internal
	international
	intrinsic
	ionic
	iron
	isotopic
IA	Image Amplifier
	Image Analysis
	Image Array
	Immediate Address
	Immunoassay
	Incorporated Accountant
	Incremental Analysis
	Indexed Addressing
	Indirect Address
	Industrial Application
	Industrial Artists
	Industry Application (Society) [of IEEE, US]
	Inherent Address
	Initial Appearance
	Input Axis
	Institute of Actuaries [UK]
	Instituto de Astromnomia [= Institute of Astronomy, Mexico]
	Instruction Address
	Instrumental Analysis
	Instrumentation Amplifier
	Integrated Adaptor
	Interchange Address
	Interciencia Association [Venezuela]
	Intermediate Amplifier
	International Alphabet
	International Angstrom
	Invent America (Foundation) [US]
	Ion Accelerator
	Iowa [US]
	Irrigation Association [US]
	Itaconic Acid
I&A	Indexing and Abstracting
	(Combined) Iron and Alumina
Ia	Iowa [US]
IAA	Indoleacetic Acid
	Indol-3-ylacetic Acid
	Institut Agricole d'Algerie [= Agricultural Institute of Algeria]
	Institute for Alternative Agriculture [US]
	Institute of Administrative Accounting [UK]
	Inter-American Accountant Association [Dominican Republic]
	Interim Access Authorization
	International Academy of Architecture
	International Academy of Astronautics [France]
	International Academy of Astronautics [US]
	International Acetylene Association
	International Advertising Association [US]
	International Aerosol Association [Switzerland]
	International Aerospace Abstracts [of NASA]
	International Apple Association [US]
	International Asphalt Association [now: IWA, Belgium]
	International Association of Agriculturalists
	International Association of Astacology [US]
	Inventors Association of America [US]
	Isoamyl Acetate
	Isoamyl Amine
IAAA	International Airforwarders and Agents Association [US]

IAAB Inter-American Association of Broadcasters

IAAC International Agricultural Aviation Center [UK]
International Air Cargo Consolidators
International Antarctic Analysis Center
International Association for Analog Computation [now: IMACS]
Israel Association for Automatic Control

IAACC Ibero-American Association of Chambers of Commerce [Colombia]
Inter-Allied Aeronautical Commission of Control

IAAE Institution of Automotive and Aeronautical Engineers [now: SAE, Australasia]
International Association of Agricultural Economists [US]

IAAEES International Association for the Advancement of Earth and Environmental Sciences [US]

IAAF International Agricultural Aviation Foundation [US]

IAAGR Institute of Arctic and Alpine Geochronological Research [of University of Colorado, Boulder, US]

IAAI International Airports Authority of India
International Association of Arson Investigators [US]

IAAIP Inter-American Association of Industrial Property [Argentina]

IAAJ International Association of Agricultural Journalists

IAALD International Association of Agricultural Librarians and Documentalists [Netherlands]

IAAM International Association of Auditorium Managers [US]
International Association of Automotive Modelers

IAAO International Association of Assessing Officers [US]

IAAPEA International Association Against Painful Experiments on Animals

IAARC International Administrative Aeronautical Radio Conference [UK]

IAAS Incorporated Association of Architects and Surveyors [UK]
International Association of Agricultural Students [Sweden]

IAASE Inter-American Association of Sanitary Engineering [later: IAASEE, US]

IAASEE Inter-American Association of Sanitary and Environmental Engineering [now: IAASEES, US]

IAASEES Inter-American Association of Sanitary Engineering and Environmental Sciences [US]

IAASM International Academy of Aviation and Space Medicine [Portugal]

IAASP International Association of Airport and Seaport Police

IAB Initiation Area Discriminator
Inter-American Association of Broadcasters [Name Changed to: International Association of Broadcasting, Uruguay]
International Aquatic Board
International Association of Broadcasting [Uruguay]
Interrupt Address to Bus
Isoamyl Benzoate
Isoamyl Butyrate

IABA International Association of Aircraft Brokers and Agents [Norway]

IABC Insulation Applicators Association of British Columbia [Canada]
International Association of Business Communicators [US]

IABG International Association of Botanic Gardens [Australia]

IABLA Inter-American Bibliographical Library Association

IABM International Association of Broadcasting Manufacturers [UK]

IABO International Association of Biological Oceanography [France]

IABRM International Association for Bear Research and Management [Canada]

IABS International Association for Biological Standardization [Switzerland]

IABSE International Association for Bridge and Structural Engineering [Switzerland]

IABSOIW International Association of Bridge, Structural and Ornamental Iron Workers

IABTI International Association of Bomb Technicians and Investigators [US]

IAC Idle Air Control (Valve)
Indian Airlines Corporation
Industry Advisory Committee [US]
Information Analysis Center [US]
Instrument-on-a-Card
Integrated Avionics Computer
Intelligent Asynchronous Controller
Interactive Array Computer
Inter-American Conference
Interference Absorption Circuit
Interim Acceptance Criteria
International Academy of Ceramics
International Accounting Center
International Advisory Committee
International Aerological Commission
International Agricultural Center
International Agricultural Club [of USDA]
International Algebraic Compiler
International Analysis Code
International Apple Core
International Association for Cybernetics [Belgium]
Ion-Assisted Coating
Isoamyl Caprate

IACA Inter-American College Association [US]
International Air Carrier Association [Belgium]
International Association for Classical Archeology [Italy]

I-ACAC Inter-American Commercial Arbitration Commission [US]

IACAHP Inter-African Advisory Committee for Animal Health and Production [Kenya]

IACB Inter-Agency Consultative Board
International Advisory Committee on Bibliography [of UNESCO]
International Association of Convention Bureaus [US]

IACBC International Advisory Committee on Biological Control

IACBDT International Advisory Committee on Bibliography, Documentation and Terminology [of UNESCO]

IACC IIM (= Indian Institute of Metals) — ASM (= American Society for Metals) Cooperation Committee
International Agricultural Coordination Commission
International Association for Cell Cultures
International Association of Cereal Chemistry
International Association of Conference Centers [US]

IACChE Inter-American Confederation for Chemical Engineering

IACCI International Association of Computer Crime Investigators

IACCP Inter-American Council of Commerce and Production

IACD International Association of Clothing Designers

IACDT International Advisory Committee for Documentation and Technology [of UNESCO]

IACE International Association for Computing Education [US]

IACHE International Association of Cylindrical Hydraulic Engineers [US]

IACIT International Association of Conference Interpreters and Translators

IACITC International Advisory Committee of the International Teletraffic Congress [Denmark]

IACM International Association for Computational Mechanics

IACME Inter-American Committee for Mathematical Education

IACOMS International Advisory Committee on Marine Sciences

IACP International Association of Chiefs of Police
International Association of Computer Programmers

IACPS Inter-American Committee on Peaceful Settlement [of OAS, US]

IACR Inter-American College for Radiology
International Association for Cryptologic Research [US]

IACRDVT Inter-American Center for Research and Documentation on Vocational Training [Uruguay]

IACRS Inter-Agency Committee on Remote Sensing

IACS Indian Association for the Cultivation of Science [India]
Inertial Attitude Control System
Integrated Armament Control System
International Annealed Copper Standard
International Association of Classification Societies [UK]

IACSS International Association for Computer Systems Security [US]

IACST International Association for Commodity Science and Technology

IACT Inter-Association Commission on Tsunami [of IUGG, Belgium]

IAD Immediate Action Directive
Initial Address Designator
Initiation Area Discriminator
Integrated Automatic Documentation
Internal Aerodynamics
Internationale Arbeitsgemeinschaft fur Donauforschung [= International Working Association for Danube Research, Austria]
Inventory Available Date
Ion-Assisted (Thin-Film) Deposition

IADA Independent Automotive Damage Appraisers Association [US]

IADB Inter-American Defense Board [US]
Inter-American Development Bank

IADC Inter-American Defense College
International Association of Dredging Companies [Netherlands]
International Association of Drilling Contractors [US]
Internationale Arbeitsgemeinschaft der Archiv-, Bibliotheks- und Graphikrestauratoren [= International Association for Conservation of Books, Paper and Archival Material, Germany]

IADES Integrated Attitude Detection and Estimation System

IADIC Integration Analog-to-Digital Converter

IADIS Irish Association for Documentation and Information Services

IADIWU International Association for the Development of International World Universities [France]

IADN Integrated Atmospheric Deposition Nertwork [Canada/US]

IADP INTELSAT (= International Telecommunications Satellite Organization) Assistance and Development Program [US]
Intensive Agricultural District Program [India]

IADPPNW International Architects, Designers and Planners for the Prevention of Nuclear War

IADR International Association for Dental Research [US]

IADS International Agricultural Development Service

IAE Institute for the Advancement of Engineering [US]
Institute of Atomic Energy [PR China]
Institute of Atomic Energy [of Kyoto University, Japan]
Institute of Automobile Engineers [UK]
Integral Absolute Error
International Academy of Education
International Atomic Exposition

IAEA Inter-American Education Association
International Agricultural Exchange Association [Scotland]
International Atomic Energy Agency [of UN — Austria]
International Atomic Energy Authority

IAEAC International Association of Environmental Analytical Chemistry [Switzerland]

I-AEDANS Iodoacetamidoethyl Aminonaphthalenesulfonic Acid
N-Iodoacetyl-N'-(Sulfo-1-Naphthyl)-ethylenediamine

IAEE International Association for Earthquake Engineering [Japan]
International Association for Energy Economics [US]
International Automotive Engineering Exposition
Israel Association of Environmental Engineers

IAEG International Association for Engineering Geology [France]
IAEI International Association of Electrical Inspectors [US]
IAEKM International Association of Electronic Keyboard Manufacturers [US]
IAEL International Association of Electrical Leagues [now: ILEA, US]
IAEMS International Association of Environmental Mutagen Societies [Finland]
IAEO International Atomic Energy Organization [of UN]
IAES Institute of Aeronautical Sciences [UK]
International Academy for Environmental Safety
Ion-Induced Auger Electron Spectroscopy
IAESP Instituto Agronomico de Estado de Sao Paulo [= State Agricultural Institute of Sao Paulo, Brazil]
IAESTE International Association for the Exchange of Students for Technical Experience
IAETL International Association of Environmental Testing Laboratories [US]
IAEVG International Association for Educational and Vocational Guidance [Northern Ireland]
IAEVI International Association for Educational and Vocational Information
IAF Indirect-Arc Furnace
Initial Approach Fix
Institute for Alternative Futures [US]
Interactive Facility
Inter-American Foundation
International Aquaculture Foundation [US]
International Aeronautical Federation
International Astronautical Federation [France]
IAFC Inter-American Freight Conference [Argentina/Brazil/US]
International Association of Fire Chiefs [US]
IAFCI Inter-American Federation of the Construction Industry [Mexico]
IAFE International Association of Fairs and Exhibitions [US]
IAFES International Association for the Economics of Self-Management [Yugoslavia]
IAFF International Association of Fire Fighters
IAFG Ion-Beam-Assisted Film Growth
IAFMM International Association of Fish Meal Manufacturers [UK]
IAFP International Association for Financial Planning
IAFR Institute of Applied Forth Research [US]
IAFraMCoS International Association of Fracture Mechanics of Concrete Structures
IAFS International Association for Fire Safety
International Association of Forensic Sciences [Canada]
IAFSS International Association for Fire Safety Science
IAFWA International Association of Fish and Wildlife Agencies [US]
IAG IFIP (= International Federation for Information Processing) Administrative Group [US]
Instruction Address Register
International Applications Group [of IFIP]
International Association of Geodesy [France]
Internationale Arbeitsgemeinschaft fur den Unterrichtsfilm [= International Working Group on Educational Films, Germany]
IAGA International Association of Geomagnetism and Aeronomy [Scotland]
IAGBN International Absolute Gravity Base Network
IAGC Instantaneous Automatic Gain Control
International Association of Geochemistry and Cosmochemistry [Canada]
International Association of Geophysical Contractors [US]
IAGLP International Association of Great Lakes Ports
IAGLR International Association for Great Lakes Research [US]
IAGOD International Association on the Genesis of Ore Deposits [Czechoslovakia]
IAGP International Antarctic Glaciological Project [US]
IAgrE Institution of Agricultural Engineers [UK]
IAGS Inter-American Geodetic Survey [[US]
IAH International Association of Hydrogeologists [Netherlands]
International Association for Hydrology
IAHB International Association of Human Biologists [UK]
IAHE International Association for Hydrogen Energy [US]
IAHP International Association of Horticutural Producers
IAHR International Association for Hydraulic Research [Netherlands]
IAHS International Academy of the History of Science [France]
International Association of Housing Science
International Association of Hydrological Sciences [formerly: IASH, Netherlands]
IAI Informational Acquisition and Interpretation
Initial Address Information
International Apple Institute [US]
International Association for Identification [US]
International Automotive Institute [Monaco]
IAIA International Association for Impact Assessment [US]
IAIABC International Association of Industrial Accident Boards and Commissions [US]
IAIALAR Ibero-American Institute of Agrarian Law and Agrarian Reform [Venezuela]
IAIAS Inter-American Institute of Agricultural Sciences
IAIBS Ion-Assisted Ion-Beam Sputtering
IAICU International Association of Independent Colleges and Universities
IAIDPA International Association of Information and Documentation in Public Administration [Belgium]
IAIE International Association for Intercultural Education
IAII Inter-American Indian Institute
IAIIB International Association of Independent Information Brokers [now: AIIP, US]
IAIN International Association of Institutes of Navigation [UK]
IAIS Industrial Aerodynamics Information Service [UK]
International Association of Independent Scholars [US]
IAL Immediate Action Letter
Instrument Approach and Landing Chart

	International Algebraic Language
	International Algorithmic Language
	Investment Analysis Language
IALA	International Association of Lighthouse Authorities [France]
IALC	International Aluminum-Lithium Conference
IALD	International Association of Lighting Designers [US]
IALE	Instrumented Architectural Level Emulation
IALL	International Association for Labour Legislation [US]
	International Association of Law Libraries
IALM	International Association of Lighting Maintenance [now: NALMCO, US]
IALS	International Agency Liaison Service [of FAO, UN]
	International Association of Legal Science
IAM	Indefinite Admittance Matrix
	Initial Address Message
	Institute of Administrative Management [UK]
	Institute of Aviation Medicine [UK]
	Interactive Algebraic Manipulation
	Intermediate Access Memory
	International Academy of Management [US]
	International Association of Machinists (and Aerospace Workers)
	International Association of Meteorology
	Internationale Arbeitsgemeinschaft fur Mullforschung [= International Working Group for Waste Research, Germany]
IAMACS	International Association for Mathematics and Computers in Simulation
IAMAP	International Association of Meteorology and Atmospheric Physics [Austria]
IAMAW	International Association of Machinists and Aerospace Workers
IAMB	International Association of Microbiologists
IAMBE	International Association of Medicine and Biology of Environment [France]
IAMC	Institute for Advancement of Medical Communication [US]
	Institute of Association Management Companies [US]
IAMCA	International Association of Milk Control Agencies [US]
IAMCR	International Association for Mass Communication Research [UK]
IAMCS	International Association for Mathematics and Computers in Simulation [Belgium]
IAMDA	International Alpha Micro Dealers Association
IAME	International Association for Modular Exhibitry
IAMFE	International Association on Mechanization of Field Experiments [Norway]
IAMFES	International Association of Milk, Food and Environmental Sanitarians [US]
IAMG	International Association for Mathematical Geology [US]
IAMHIST	International Association for Audio-Visual Media in Historical Research and Education [Italy]
IAML	International Association of Music Libraries
IAMLO	International African Migratory Locust Organization [Mali]
IAMLT	International Association of Medical Laboratory Technologists [UK]
IAMP	Institute of Atomic and Molecular Physics [Netherlands]
	International Association of Mathematical Physics [US]
IAMQS	International Academy of Molecular and Quantum Sciences
IAMR	Institute of Applied Manpower Research [India]
	Institute of Arctic Mineral Resources [of University of Alaska, US]
IAMS	Institute of Advanced Manufacturing Sciences
	Integrated Academic Information Management System
	International Advanced Microlithography Society
	International Association of Microbiological Societies [now: IUMS, UK]
IAMSLIC	International Association of Marine Science Libraries and Information Centers [US]
IAMTEC	Institute of Advanced Machine Tool and Control Technology [also: IAMTCT, US]
IAMWF	Inter-American Mine Workers Federation
IAN	Instituto Agrario Nacional [= National Institute of Agriculture, Venezuela]
	Integrated Acoustic Network
	Integrated Analog Network
	Isoamyl Nitrate
IANAP	Interagency Noise Abatement Program
IANC	International Airline Navigators Council
	International Anatomical Nomenclature Committee
I and A	(Combined) Iron and Alumina
I and C	Installation and Checkout [also: I&C]
I and D	Information and Documentation [also: I&D]
IANDS	International Association for Near-Death Studies [US]
IANEC	Inter-American Nuclear Energy Commission [US]
IANI	Israel Agency for Nuclear Information
IANRP	International Association of Natural Resource Pilots [US]
IAO	In and Out
	Insurers Advisory Organization
	Internal Automation Operation
IAOD	International Academy of Optimum Dentistry
IAOPA	International Council of Aircraft Owner and Pilot Associations [US]
IAOS	International Association for Official Statistics [Netherlands]
	Irish Agricultural Organization Society
IAP	Image Array Processor
	Imaging Atom Probe
	Initial Approach Procedure
	Institute of Australian Photographers
	Institutional Assistance Program
	Internal Array Processor
	International Association of Photoplatemakers [now: IPA, US]
	International Association for Planetology [Belgium]
	Inventors Assistance Program
IAPA	Industrial Accident Prevention Association [Canada]
	Inter-American Press Association

	International Airline Passengers Association
	International Association of Parametric Analysts [US]
IAPBPPV	International Association of Plant Breeders for the Protection of Plant Varieties [Switzerland]
IAPC	International Association for Pollution Control
	International Auditing Practices Committee
IAPCO	International Association of Professional Congress Organizers [Belgium]
IAPDOI	Italian Association for the Production and Distribution of On-Line Information
IAPH	International Association of Port and Harbors [Japan]
IAPHC	International Association of Printing House Craftsmen [US]
IAPIP	International Association for the Protection of Industrial Property
IAPMO	International Association of Plumbing and Mechanical Officials [US]
IAPN	International Association of Professional Numismatists
IAPO	International Association of Physical Oceanography [now: IAPSO, US]
IAPP	Indian Association for Plant Physiology
	International Association for Plant Physiology [Australia]
IAPR	International Association for Pattern Recognition [UK]
IAPRI	International Association of Packaging Research Institutes
IAPS	International Association for the Properties of Steam [US]
IAPSO	International Association for the Physical Sciences of the Ocean [US]
IAPT	International Association for Plant Taxonomy [Germany]
IAQ	International Academy for Quality [Germany]
IAQC	International Association for Quality Circles [US]
IAQDE	Independent Association of Questioned Document Examiners [US]
IAQR	International Association on Quaternary Research
IAR	Institute of Atomic Research [US]
	Instruction Address Register
	Interrupt Address Register
	Intersection of Air Routes
	Isobaric Analog Resonance
IARA	International Aerosol Research Assembly
IARC	International Agency for Research on Cancer [of WHO, Switzerland]
	International Agricultural Research Center [of CGIAR, US]
IARD	Information Analysis and Retrieval Division [of American Institute of Physics, US]
IARI	Indian Agricultural Research Institute
IARIGAI	International Association of Research Institutes for the Graphic Arts Industry [UK]
IARM	International Association of Ropeway Manufacturers
IARMCLRS	International Agreement Regarding the Maintenance of Certain Lights in the Red Sea [UK]
IARR	International Association for Radiation Research [Netherlands]
IARU	International Amateur Radio Union
IARUS	International Association for Regional and Urban Statistics [Netherlands]
IARW	International Association of Refrigerated Warehouses [US]
IAS	Image Analysis System
	Immediate Access Storage
	Indiana Academy of Science [US]
	Indian Academy of Sciences [India]
	Indicated Air Speed
	Industry Applications Society [of IEEE, US]
	Institute for Advanced Studies
	Institute for Aerospace Studies [of U of T, Canada]
	Institute for Atmospheric Sciences [of ESSA, US]
	Institute of Advanced Studies [US Army]
	Institute of Aeronautical Sciences [US]
	Institute of Aerospace Sciences [Now Part of AIAA, US]
	Instrument Approach System
	Integrated Analytical System
	Interactive Applications Supervisor
	Interactive Applications System
	International Accountants Society
	International Accounting Standards
	International Association of Sedimentologists [Denmark]
	International Association of Siderographists
	International Audio-Visual Society [US]
	International Auditing Standards
	Iowa Academy of Sciences [US]
	Irish Archeological Society
	Islamic Academy of Sciences
	Isobaric Analog State
IASA	Insurance Accounting and Statistical Association
	International Air Safety Association
	International Alliance of Sustainable Agriculture [US]
IASB	International Academy at Santa Barbara [US]
	International Aircraft Standards Bureau
IASC	Indexing and Abstracting Society of Canada
	Inter-American Safety Council
	International Accounting Standards Committee [UK]
	International Association for Statistical Computing [Netherlands]
	International Association of Science Clubs
	International Association of Seed Crushers [UK]
IASCB	Ibero-American Society for Cell Biology [Chile]
IASD	Industrial Automation Services Division
IASF	Instrumentation in Aerospace Simulation Facilities [IEEE AESS Committee]
	International Atlantic Salmon Foundation
IASH	International Association of Scientific Hydrology [now: IAHS, Netherlands]
IASI	Inter-American Statistical Institute [Panama]
IASLIC	Indian Association for Special Libraries and Information Centers
IASM	International Association of Structural Movers [US]

IASMART International Association of Structural Mechanics and Reactor Technology
IASPEI International Association of Seismology and Physics of the Earth's Interior [UK]
IASPS International Association for Statistics in Physical Sciences
IASR Interruption Address Storage Register
IASS International Association for Shell and Spatial Structures [Spain]
International Association of Security Service [US]
International Association of Soil Science
International Association of Survey Statisticians [France]
IASSIST International Association for Social Science Information Service and Technology [Canada]
IASTED International Association of Science and Technology for Development [Canada]
IASU International Association of Satellite Users [now: IASUS, US]
IASUS International Association of Satellite Users and Suppliers [US]
IASY International Active Sun Years
IAT Information Assessment Team
Internal Average Temperature
Institute for Advanced Technology [of NIST, US]
Institute for Applied Technology
Institute of Advanced Technology
Institute of Automatics and Telemechanics [USSR]
International Association for Timekeeping
International Atomic Time
International Automatic Time
IATA International Air Transport Association [Switzerland]
International Appropriate Technology Association [US]
International Association of Trade Associations
IATAE International Accounting and Traffic Analysis Equipment
IATAL International Association of Theoretical and Applied Limnology [US]
IATC International Air Traffic Communications
International Association of Tool Craftsmen
IATM International Association for Testing Materials
International Association of Transport Museums [Switzerland]
IATME International Association of Terrestrial Magnetism and Electricity
IATN International Association of Telecomputer Networks [US]
IATP International Agricultural Training Program
IATSS International Association of Traffic Safety Sciences
IATTC Inter-American Tropical Tuna Commission [US]
IATU Inter-American Telecommunications Union [US]
IATUL International Association of Technical University Libraries [Sweden]
IAU Infrastructure Accounting Unit
Institute for American Universities [France]
International Academic Union [Belgium]
International Association of Universities [France]
International Astronomical Union [US]
IAUP International Association of University Presidents [Thailand]
IAUPD International Association for Urban Planning and Design
IAUPL International Association of University Professors and Lecturers [France]
IAUR Istituto di Automatica dell'Universita di Roma [= Institute of Automation of the University of Rome, Italy]
IAUSD Inter-American Union for Scientific Development
IAV International Association for Video — VIDION [US]
International Association of Volcanology
IAVB International Association of Visceral Biomechanics
IAVC Instantaneous Automatic Video Control
Instantaneous Automatic Volume Control
International Audiovisual Center
IAVCEI International Association of Volcanology and Chemistry of the Earth's Interior [Germany]
IAVFH International Association of Veterinary Food Hygiene
IAVG International Association for Vocational Guidance
IAVGO Industrial Accident Victims Group of Ontario [Canada]
IAVS International Association for Vegetation Science [Germany]
IAVSD International Association for Vehicle Systems Dynamics [Netherlands]
IAVTC International Audio-Visual Technical Center
IAW In Accordance With
IAWA International Association of Wood Anatomists [Netherlands]
IAWCM International Association of Wiping Cloth Manufacturers [US]
IAWE International Association for Wind Engineering [Canada]
IAWL International Association of Water Law [Italy]
IAWPR International Association on Water Pollution Research [now: IAWPRC, UK]
IAWPRC Indian Association on Water Pollution Research and Control
International Association on Water Pollution Research and Control [UK]
IAWR Internationale Arbeitsgemeinschaft der Wasserwerke im Rheineinzugsgebiet [= International Association of Waterworks in the Rhine Basin Area, Netherlands]
IAWRBA International Association of Waterworks in the Rhine Basin Area [Netherlands]
IAWS International Academy of Wood Science
Irish Agricultural Wholesale Society
IB Identifier Block
In Bond [Customs]
Inbound
Incendiary Bomb
Induction Brazing
Infinite Baffle
Information Bank
Information Bureau
Input Buffer
Input Bus
Institute of Biology

	Instruction Book
	Instruction Bus
	Interface Board
	Interface Bus
	Internal Bus
	International Baccalaureate
	Ion Beam
	Ion Bombardment
ib	(ibidem) — in the same place
IBA	Independent Broadcasting Authority [of ITV, UK]
	Indolebutyric Acid
	Industrial Biotechnology Association [US]
	Institute for Briquetting and Agglomeration [US]
	Institute of British Architects
	Institute of Business Appraisers [US]
	Interacting Boson Approximation
	International Bankers Association [US]
	International Bauxite Association [Jamaica]
	International Biographical Association
	International Biometric Association [US]
	International Broadcasting Authority
	International Bryozoology Association [France]
	Investment Bankers Association
	Isobornyl Acetate
	Isobutyl Acetate
IBAA	International Bankers Association of America [US]
	International Business Aircraft Association [US]
IBAC	International Business Aviation Council [US]
IBAD	Ion-Beam-Assisted Deposition
IBAE	Institute of British Agricultural Engineers
	Ion-Beam-Assisted Etching
IBAFG	Ion-Beam-Assisted Film Growth
IBAH	Inter-African Bureau for Animal Health
IBAHP	Inter-African Bureau for Animal Health and Production [Kenya]
IBAM	Institute of Business Administration and Management [Japan]
	Instituto Brasileiro de Administracao Municipal [= Brazilian Institute for Municipal Administration]
IBAN	Institut Belge pour l'Alimentation et la Nutrition [= Belgian Institute for Food and Nutrition]
IBB	Isobutyl Benzoate
IBBA	Inland Bird Banding Association [US]
IBBD	Instituto Brasileiro de Bibliografia e Documentacao [= Brazilian Institute of Bibliography and Documentation]
IBBH	Internationaler Bund der Bau- und Holzarbeiter [= International Federation of Building and Wood Workers, Switzerland]
IBBM	Iron Body Brass Mounted
	Iron Body Bronze Mounted
IBBS	Integrated Broadband Service
IBC	Instituto Baceteriologico de Chile [= Chilean Institute of Bacteriology]
	Integrated Block Controller
	International Banking Center
	International Biographical Center
	International Biometric Conference [US]
	International Boundary Commission [US]
	International Broadcasting Convention
	Isobutyl Carbinol
IBCA	Industry Bar Code Alliance [US]
IBCC	International Building Classification Committee [Netherlands]
	International Business Council of Canada
IBCG	Internationaler Bund Christlicher Gewerkschaften [= International Federation of Christian Trade Unions, Germany]
IBCS	International Bureau of Commercial Statistics
IBD	Institute of Business Designers [US]
	International Baccalaureate Diploma
	Intrinsic Boundary Dislocation
IBDH	Inboard Chromatography Data Handler
IBDI	International Bureau of Documentation and Information
IBE	Institute of British Engineers
	International Bureau of Education [of UNESCO — Switzerland]
	Ion Beam Epiplantation
IBEC	International Bank for Economic Cooperation
IBED	Ion Beam Enhanced Deposition
IBEDOC	International Bureau of Education Documentation and Information Systems [of UNESCO]
IBEE	International Builders Exchange Executives [US]
IBEF	International Bio-Environmental Foundation [US]
IBELCO	Institut Belge de Cooperation Technique [= Belgian Institute for Technical Cooperation]
IBEW	International Brotherhood of Electrical Workers
IBEX	International Biotechnology Exposition
IBF	Incident Bright-Field
	Institute of British Foundrymen
	Institut fur Bildsame Formgebung [= Institute for Plastic Deformation [of RWTH, Germany]]
	International Booksellers Federation [also: IBV, Austria]
IBFA	Interacting Boson-Fermion Approximation
IBFD	International Bureau of Fiscal Documentation
IBFI	International Business Forms Industries [US]
IBFMP	International Bureau of the Federation of Master Printers
IBFO	International Brotherhood of Firemen and Oilers Helpers
IBG	Institute of British Geographers
	Interblock Gap
	Internationales Buro fur Gebirgsmechanik [= International Bureau of Strata Mechanics, Poland]
IBGE	Instituto Brasileiro de Geografia e Estatistica [= Brazilian Institute of Geography and Statistics]
IBI	Intelligent Buildings Institute [US]
	Intergovernmental Bureau for Informatics [Italy]
	Intergovernmental Bureau for Information [of UN]
	International Broadcasting Institute [now: IIC, UK]
IBIB	Isobutyl Isobutyrate
IBICT	Instituto Brasileiro de Informacao em Ciencia e Tecnologia [= Brazilian Institute for Information in Science and Technology, Brazil]
ibid	(ibidem) — in the same place
IBI-ICC	Intergovernmental Bureau for Informatics — International Computation Center
IBIM	Instituto de Informacion y Documentacion en Biomedicina [= Institute for Biomedical Information and Documentation, Spain]

IBION Issue Based Indian Ocean Network
IBIS ICAO (= International Civil Aviation Organization) Bird-Strike Information System
Intelligent Business Information System
Intense-Bunched Ion Source
International Bank Information System
International Book Information Service [Netherlands]
IBJ Industrial Bank of Japan
IBLA Inter-American Bibliographical and Library Association
IBM Indian Bureau of Mines
Interacting Boson Model
Ion-Beam Modification
IBMA Independent Battery Manufacturers Association [US]
IBM CAD IBM Computer-Aided Design
IBMCUA IBM Computer Users Association [US]
IBME Institute of Biomedical Engineering [of U of T, Canada]
Instituto de Biologia y Medicina Experimental [= Institute of Experimental Biology and Medicine, Argentina]
IBMM (International Conference on) Ion Beam Modification of Materials [Germany]
Ion Beam Modification of Materials
IBM PC IBM Personal Computer
IBM PC AT IBM Personal Computer Advanced Technology
IBM PC ATE IBM Personal Computer Advanced Technology Enhanced
IBM PC ATX IBM Personal Computer Advanced Technology Expanded
IBM PC XT IBM Personal Computer Expanded Technology
IBMR International Bureau for Mechanical Reproduction
IBMS Ion Beam-Modified Surface
IBMT International Bureau of Mining and Thermophysics
IBM TSS IBM Time-Sharing System
IBMX 3-Isobutyl-1-Methylxanthine
IBN Identification Beacon
Institute Belge de Normalisation [= Belgian Standards Institute]
International Biosciences Network
IBO International Baccalaureate Organization [Switzerland]
IBOL Interactive Business-Oriented Language
IBOP International Brotherhood of Operative Pottery and Allied Workers
IBP Initial Boiling Point
Institute for Business Planning [UK]
International Biological Program [of NRC, US]
Ion-Beam Processing
IBPAT International Brotherhood of Painters and Allied Trades
IBPA International Business Press Associates
Israel Book Publishers Association
IBPBAC Isobutyl-4-(4'-Phenylbenzylideneamino)cinnamate
IBPCS International Bureau for Physicochemical Standards
IBPGR International Board for Plant Genetic Resources [Italy]
IBPH 2,2'-Isobutylidene-Bis-(4-Methyl-6-tert-Butylphenol)
IBR Institute for Basic Research [of NIST, US]
Integral Boiling Reactor
Integrated Bridge Rectifier
IBRA International Bee Research Association [UK]
IBRD International Bank for Reconstruction and Development [of UN]
IBRG International Biodeterioration Research Group [UK]
IBRL Initial Bomb Release Line
IBRO International Brain Research Organization
IBRRC International Bird Rescue Research Center [US]
IBS Institute for Basic Standards [of NIST, US]
Integrated Broadband Service
INTELSAT (= International Telecommunications Satellite Organization) Business Services
Intercollegiate Broadcasting System [US]
International Broadcasters Society
Ion Beam Splitter
Ion-Beam Synthesis
IBSA International Biotechnology Suppliers Association [US]
IBSAC Industrialized Building Systems and Components
IBSE Ion Beam Sputter Etching
IBSFC International Baltic Sea Fishery Commission [Poland]
IBSHR Integral Boiling and Superheat Reactor [US]
IBSR Interactive Bibliographic Search and Retrieval
IBT Inclined Bottom Tank
Integrated Bipolar Transistor
IBTE Institution of British Telecommunication Engineers
International Bureau of Technical Education
IBTO International Broadcasting and Television Organization
IBTT International Bureau for Technical Training
IBTTA International Bridge, Tunnel and Turnpike Association [US]
IBTU International Bureau of Transport Users
IBU International Broadcasting Union
iBu Isobutyryl
IBUPU International Bureau of the Universal Postal Union
IBV Internationale Buchhandler-Vereinigung [= International Booksellers Federation, Austria]
IBW Impulse Bandwidth
IBWC International Boundary and Water Commission [US]
IBWM International Bureau of Weights and Measures [France]
IBWS International Bureau of Whaling Statistics
IBY International Biological Year
IC Identification Code
Image Coding
Image Converter
Immediate Constituent
Immunochemistry
Imperial College [UK]
Impregnated Cable
Impregnated Carbon
Impulse Conductor
Inductance-Capacitance
Induction Coil
Industrial Chemistry
Inertial Component
Information Center
Information Circular

Information Code
Infrared Cell
Inner City
Inorganic Carbon
Inorganic Chemistry
Input Circuit
Inscribed Circle
Installed Capacity
Institute of Ceramics [UK]
Institute of Chemistry
Instruction Cell
Instruction Counter
Instrument Correction
Insulated Conductor
Integrated Circuit
Intelligent Copier
Interchange Center
Intercity (Train)
Intercommunications
Intercrystalline Corrosion
Interexchange Carrier
Interface Control
Interference Contrast
Intergranular Corrosion
Interior Communication
Internal Combustion
Internal Computer
Internal Connection
Internal Conversion
International Center
International Classification
International Colloquium
International Conference
Interrupted Current
Interspecies Communication [US]
Inverse Conduction
Investment Casting
Investment Commission
Investment Committee
Ion Chamber
Ion Chromatography
Ion Counter
Ionization Chamber
Irrigation Consultant

I&C Installation and Checkout
Instrumentation and Control

I/C In Charge of [also: i/c]
Incoming
In Command

ICA In-Circuit Analyzer
Industrial Communications Agency
Industrial Communications Association
Initial Cruise Altitude
Institute of Chartered Accountants [UK]
Integrated Circuit Amplifier
Integrated Communications Adapter
Intercompany Agreement
Intercomputer Adapter
Intermuseum Conservation Association [US]
Institute of Canadian Advertising
International Cartographic Association [Australia]
International Carwash Association [US]
International Ceramic Association [US]
International Color Authority
International Commission on Acoustics [US]
International Common Access
International Communication Agency [US]
International Communication Association [US]
International Control Agency
International Cooperation Administration [now: AID, US]
International Cooperative Alliance [Switzerland]
International Copper Association
International Council on Archives
International Council for ADP (= Automatic Data Processing) [Spain]
Inventors Club of America [US]
Islamic Cement Association
Islet Cell Autoantibodies
Isocyanic Acid
Item Control Area

ICAA Institute of Chartered Accountants in Australia
Institute of Chartered Accountants of Alberta [Canada]
Insulation Contractors Association of America [US]
International Civil Airports Association [France]
International Civil Aviation Authority [Canada]

ICAC International Civil Aviation Committee
International Conference on Analytical Chemistry
International Conference on Automated Composites
International Cotton Advisory Committee [US]
Islamic Civil Aviation Council

ICACGP International Commission on Atmospheric Chemistry and Global Pollution

ICACM International Conference on Advances in Composite Materials

ICAD Integrated Computer-Aided Design
Integrated Control and Display

ICADEM Integrated Computer-Aided Design, Engineering and Manufacturing

ICADI Inter-American Center for Agricultural Documentation and Information [of IAIAS]

ICAE International Commission of Agricultural Engineering
International Commission on Atmospheric Electricity [UK]
International Conference of Agricultural Economists

ICAEC International Confederation of Associations of Experts and Consultants [France]

ICAF Industrial College of the Armed Forces [US]
International Committee on Aeronautical Fatigue [Netherlands]

ICAI Intelligent Computer-Assisted Instructional System
International Commission for Agricultural Industries
International Conference on Artificial Intelligence

ICAIT International Conference on Advanced Information Technology

ICALEO International Congress on Applications of Lasers in Electrooptics

ICALP International Colloquium on Automata, Languages and Programming

ICALU International Confederation of Arab Labour Unions
ICAM Integrated Communications Access Method
Integrated Computer-Aided Manufacturing
International Center for the Advancement of Management Education [US]
International Confederation of Architectural Museums [France]
International Conference on Advanced Materials
ICAME International Committee on Application of Mossbauer Effect
ICAMQ International Committee on Automation of Mines and Quarries [Hungary]
ICAMPS International Conference and Exhibition on Advances in Materials and Processes [India]
ICAMRS International Civil Aviation Message Routing System
ICAMS Industrial Central Atmosphere Monitoring System
ICAN Individual Circuit Analysis
International Committee on Air Navigation
ICAND Multiplicand
ICAO Institute of Chartered Accountants of Ontario [Canada]
International Civil Aviation Organization [of UN — Canada]
ICAP Inductively-Coupled Argon Plasma
Industrial Conversion Assistance Program [of EMR, Canada]
International Conference on Atomic Physics
ICAPE International Chemical and Petroleum Engineering Exhibition [UK]
ICAQUO Inventory of Contaminants in Aquatic Organisms (Database) [of FAO, UN]
ICAR Indian Council of Agricultural Research [India]
Interface Control Action Request
International Conference on Advanced Robotics
Inventory of Canadian Agricultural Research
I-CAR Inter-Industry Conference on Auto Collision Repair [US]
ICARDA International Center for Agricultural Research in the Dry Areas [Syria]
ICARE International Center for the Advancement of Research Education
ICARVS Interplanetary Craft for Advanced Research in the Vicinity of the Sun
ICAS Independent Collision Avoidance System
Intermittent Commercial and Amateur Service
International Council of the Aeronautical Sciences [Sweden]
Institute of Chartered Accountants of Saskatchewan [Canada]
ICASE International Council of Associations for Science Education [Hong Kong]
ICASLS International Center for Arid and Semiarid Land Studies [US]
ICAST International Conference on Amorphous Semiconductor Technology
ICATO Iran Civil Aviation Training Organization
ICATU International Confederation of Arab Trade Unions
ICAV Instantaneous Crankshaft Angular Velocity
ICB Incoming Calls Barred
Institute of Canadian Bankers
Interface Control Board
Internal Common Bus
International Container Bureau
Ionized Cluster Beam
ICBA Israel Cattle Breeders Association
ICBD International Center for Brewing and Distilling [UK]
ICBM Intercontinental Ballistic Missile
ICBO International Conference of Building Officials [US]
ICBP International Council for Bird Preservation [UK]
ICBRSD International Council for Building Research, Studies and Documentation [Netherlands]
ICBT Intercontinental Ballistic Transport
ICBWR Improved Cycle Boiling Water Reactor
ICC Ignition Control Compound
Independent Channel Controller
Indian Cryogenics Council [India]
Industrial Communication Council [US]
Information Commissioner of Canada
Instrument Center Correction
Intercrystalline Corrosion
Internal Conversion Coefficient
International Association for Cereal Chemistry [Austria]
International Chamber of Commerce [France]
International Climatological Commission
International Committee for Conservation [of ICOM, France]
International Computation Center [Italy]
International Computer Center
International Computing Center [of UN]
International Conference on Communications [of IEEE, US]
International Conference on Continuous Casting
International Conference on Creep
International Congress on Catalysis
International Control Center
International Coordinating Committee
International Corrosion Council [France]
Interstate Commerce Commission [US]
Invitational Computer Conference
IC&C Invoice Cost and Charges [also: ICC]
ICCA Independent Computer Consultants Association [US]
Instituto Centroamericano de Ciencias Agricolas [= Central American Institute of Agricultural Sciences]
International Committee on Coordination for Agriculture
International Congress and Convention Association [Netherlands]
International Corrugated Case Association [France]
International Council for Commercial Arbitration [Austria]
ICCAD IEEE Conference on Computer-Aided Design [US]
ICCAIA International Coordinating Council of Aerospace Industries Associations [US]
ICCAT International Commission for the Conservation of Atlantic Tunas [Spain]
ICCATCI International Committee to Coordinate Activities of Technical Groups in the Coatings Industry [France]

ICCC Inter-Council Coordination Committee [France]
International Color Computer Club
International Conference on Coordination in Chemistry
International Council for Computer Communication [US]

ICC-CAPA International Chamber of Commerce — Commission on Asian and Pacific Affairs

ICCCS International Contamination Control Conference and Symposium [Switzerland]

ICCD Intensified Charge-Coupled Device (Camera)
International Conference on Computer Design

ICCE International Commission on Continental Erosion
International Council for Computers in Education [US]

ICCEC India Chemists and Chemical Engineers Club [now: ISICCE, US]

ICCEE International Classification Commission for Electrical Engineering

ICCF International Computing and Control Facility
International Conference on Cold Fusion

ICCG International Conference on Crystal Growth

ICCH International Commodities Clearinghouse [UK]

ICCI International Conference on Composite Interfaces

ICCICA Interim Coordinating Coomittee on International Commodity Agreements [US]

ICCICE Islamic Chamber of Commerce, Industry and Commodity Exchange [Pakistan]

ICCL International Commission on Climate

ICCM International Committee for the Conservation of Mosaics [UK]
International Conference (and Exposition) on Composite Materials
Isoconcentration Contour Migration

ICCMB International Committee for the Conservation of Mud Brick [Italy]

ICCMS International Conference on Composite Materials and Structures

ICCO International Carpet Classification Organization [Belgium]

ICCP Institute for Certification of Computer Professionals [US]
International Commission for Coal Petrology [Belgium]
International Commission on Cloud Physics
International Conference on Cataloguing Principles

ICCPPS International Conference on Ceramic Powder Processing Science

ICCR International Committee for Coal Research [Belgium]

ICCS Ice Center Communications System
Industrial Combustion Control System
Inter-Computer Communications System
International Center for Chemical Studies [Yugoslavia]
International Commission of Control and Supervision
International Conference on Composite Structures
International Council on Civil Status [France]

ICCSSSAR International Coordinating Committee on Solid State Sensors and Actuators Research [US]

ICCSTR International Coordinating Committee on Solid State Transducers Research [US]

ICC-TM Interstate Commerce Commission — Transport Mobilization [US]

ICCU Interchannel Comparison Unit

ICD Institute of Civil Defense [UK]
Integrated Circuit Design
Interface Control Drawing
International Center for Development
International Congress for Data Processing [also: ICDP]
International Cooperation Department [of MOEA, Taiwan]
Isocitrate Dehydrogenase

ICDA International Compressor Distributors Association [US]
International Cooperative Development Association

ICDB Integrated Corporate Database

ICDD International Center for Diffraction Data [of JCPDS]

ICDDB International Control Description Database

ICDES Item Class Description

ICDF Inorganic Crystallographic Data File [of INKA, Germany]
Intermediate Coupling Dirac-Fock

ICDH Isocitrate Dehydrogenase

ICDL Integrated Circuit Description Language
Internal Control Description Language

ICDM International Commission on Dynamic Meteorology

ICDO International Civil Defense Organization [Switzerland]

ICDP International Confederation for Disarmament and Peace
International Congress for Data Processing [also: ICD]

ICDR Inward Call Detail Recording

ICDS Institutional Cooperation and Development Services [Canada]

ICDT Islamic Center for Development of Trade [Morocco]

ICE Icing
Immediate Cable Equalizer
In-Circuit Emulator
Input-Checking Equipment
Insertion Communications Equipment
Institute for Chemical Education [US]
Institute of Ceramic Engineers [US]
Institute of Chartered Engineers
Institute of Control Engineering [Poland]
Institution of Civil Engineers [UK]
Insulated Cable Engineer
Integrated Circuit Engineering
Integrated Communications Architecture
Integrated Cooling for Electronics
Interactive Cost Estimating (System)
Intercity Express [Germany]
Interface Cancellation Equipment
Intermediate Cable Equalizer
Internal Combustion Engine
International Congress of Entomology [now: CICE, UK]
Ion Chromatography Exclusion

ICEA Insulated Cable Engineers Association [US]
International Commission on Environmental Assessment

ICEAM International Committee on Economic and Applied Microbiology [US]

ICEC Ice Center Environment Canada
Institute of Chartered Engineers of Canada
International Cost Engineering Council [US]
International Cryogenic Engineering Committee [Germany]
International Cryogenic Engineering Conference

ICECAN Iceland-Canada Submarine Cable

ICEC/ICMC International Cryogenic Engineering Conference/ International Cryogenic Materials Conference

ICED International Coalition for Energy Development
International Conference on Engineering Design
International Council for Educational Development
International Council on Environmental Design [US]

ICEE International Control Engineering Exposition [US]

ICEF International Conferences on Environmental Future
International Congress on Engineering and Food
International Council for Educational Films
International Federation of Chemical, Energy and General Workers Unions [Belgium]

ICEI Institution of Civil Engineers of Ireland
International Combustion Engine Institute [now: EMA, US]

ICEL International Council of Environmental Law [Germany]

ICEM Integrated Computer-Aided Engineering and Manufacturing
International Conference on Electronic Materials
International Conference on Electron Microscopy
International Congress on Electron Microscopy
International Congress on Experimental Mechanics
International Council for Education Media

ICEPAK Intelligent Classifier Engineering Package

ICEPF International Commission for the Eriksson Prize Fund [Sweden]

ICER Information Center of the European Railways
Infrared Cell, Electronically-Refrigerated
Interdepartmental Committee on External Relations

ICES Integrated Civil Engineering System
Integrated Civil Engineering System (Users Group) [US]
International Conference of Engineering Societies
International Conference of Engineering Studies
International Council for the Exploration of the Sea [Denmark]

ICESC International Committee for European Security and Cooperation [Belgium]

ICET Institute for the Certification of Engineering Technicians
International Center for Earth Tides [Belgium]
International Council on Education for Teaching [UK]

ICETK International Council of Electrochemical Thermodynamics and Kinetics

ICETT Industrial Council for Educational and Training Technology

ICF Incommunication Flip-Flop
Incremental Cost per Foot
Inertial Confinement Fusion
Integrated Control Facility
Interactive Communications Feature
Intercommunication Flip-Flop
International Casting Federation
International Congress on Fracture [Japan]
International Consultants Foundation [Sweden]
International Cotton Federation
International Crane Foundation [US]

ICFA Inland Commercial Fisheries Association [US]
Institute of Chartered Financial Analysts
International Committee on Future Accelerators

ICFC International Center for Fairs and Congresses [Belgium]

ICFM Inlet Cubic Feet per Minute

ICFMH International Committee on Food Microbiology and Hygiene [Denmark]

ICFR Interdepartmental Committee of Futures Research

ICFRM International Conference on Fusion Reactor Materials

ICFTA International Committee of Foundry Technical Associations

ICFTU International Confederation of Free Trade Unions [Belgium]

ICG Icing
Indocyanine Green
Interactive Computer Graphics
International Commission on Glass [Czechoslovakia]
International Congress on Glass
International Congress of Genetics

ICGB International Cargo Gear Bureau [US]

ICGEB International Center for Genetic Engineering and Biotechnology [Austria]

ICGI International Conference on Geoscience Information

ICGIC Icing in Clouds

ICGICIP Icing in Clouds in Precipitation

ICGIP Icing in Precipitation

ICGM International Colloquium about Gas Marketing [France]

ICGW International Commission on Groundwater

ICH Ion Channeling

ICHC International Committee for Horticultural Congresses

ICHCA International Cargo Handling and Coordination Association [UK]

IChemE Institution of Chemical Engineers [UK]

ICHENP International Conference on High Energy Nuclear Physics

ICHIB Information Center of the Hungarian Industry in Budapest [Hungary]

ICHM International Committee on Historical Metrology

ICHMT International Center for Heat and Mass Transfer [Yugoslavia]

ICHS Inter-African Committee for Hydraulic Studies [Burkina Faso]

ICHSP International Congress on High-Speed Photography [now: ICHSPP, US]

ICHSPP International Congress on High-Speed Photography and Photonics [US]
ICHT Ichthyological [also: ICHTH]
Ichthyology [also: ICHTH]
ICHTM International Congress on the Heat Treatment of Materials
ICI Imperial Chemical Industries [UK]
Intelligent Communications Interface
Inter-American Copyright Institute
Investment Casting Institute [US]
ICIA Industrial, Commercial and Institutional Accountant
International Center of Information on Antibiotics [Belgium]
International Communications Industries Association [US]
International Credit Insurance Association
International Crop Improvement Association [US]
ICIANZ Imperial Chemical Industries of Australia and New Zealand
ICIAQC International Conference on Indoor Air Quality and Climate
ICIB Indian Commercial Information Bureau
International Cargo Inspection Bureau
ICIC International Copyright Information Center [of UNESCO]
ICID International Commission on Irrigation and Drainage [India]
ICIP International Conference on Information Processing
ICIPE International Center of Insect Physiology and Ecology [Kenya]
ICIREPAT International Cooperation in Information Retrieval among Examining Patent Offices [Switzerland]
ICIS Integrated Circuit Inspection System
International Conference on Ion Sources [Germany]
ICISE International Conference on Interface Science and Engineering
ICISS Impact-Collision Ion-Scattering Spectroscopy
ICIT Instituto Cubano de Investigaciones Tecnologicas [= Cuban Institute for Technological Investigations]
ICITA International Chain of Industrial and Technical Advertising Agencies [US]
International Cooperative Investigation of the Tropical Atlantic
ICITO International Committee of the International Trade Organization [of UN]
ICITP International Conference on Impact Treatment Processes
ICJ Incoming Junction
ICL Incoming Line
Inserted Connection Loss
Intercommunication Logic
Inter-Computer Communications Logic
Interface Control Layer
Interpretative Coding Language
Irish Central Library
Isentropic Condensation Level
ICLA Indian College Library Association [India]
International Committee on Laboratory Animals
ICLAE International Council of Library Association Executives [US]
ICLARM International Center for Living Aquatic Resources Management [Philippines]
ICLCF International Conference on Low-Cycle Fatigue
ICLCP International Conference on Large Chemical Plants [Belgium]
ICLR International Committee for Lift Regulations [France]
ICM Improved Capability Missile
Incoming Message
Instruction Control Memory
Instrumentation and Communications Monitor
Integral Charge-Control Model
International Conference on Microlithography
International Congress of Mathematicians
International Congress on Mechanical Behavior of Materials [Sweden]
Ion Chromatography Module
ICMA Institute of Certified Management Accountants [US]
International Congresses for Modern Architecture
International Congress on Metalworking and Automation
ICMAS International Conference on Modem Aspects of Superconductivity [France]
ICMASA Intersociety Committee on Methods for Air Sampling and Analysis [US]
ICMC Indian-Ocean Cable Management Committee
Institute of Chemical Machine Construction [USSR]
International Cokemaking Congress
International Conference on Metallurgical Coatings
International Cryogenic Materials Conference
ICMCTF International Conference on Metallurgical Coatings and Thin Films [US]
ICME International Committee on Microbial Ecology [US]
ICMEE Institution of Certified Mechanical and Electrical Engineers [South Africa]
ICMEDC International Council of Masonry Engineers for Developing Countries [Scotland]
ICMF International Conference on (Advances in) Metal Forming Techniques
ICMG International Commission on Mibrobial Genetics
ICMI International Commission on Mathematical Instruction [UK]
ICMLT International Congress of Medical Laboratory Technicians
ICMMB International Conference on Mechanics in Medicine and Biology [US]
ICMMP International Committee on Military Medicine and Pharmacy
ICMP Interchannel Master Pulse
Internet Control Message Protocol
ICMPC International Conference on Materials and Process Characterization
ICMR Indian Council of Medical Research
International Center for Medical Research
ICMS Integrated Crop Management Services [Canada]
International Commission on Mushroom Science
ICMSF International Commission on Microbiological Specifications for Food [US]
ICMUA International Commission on Meteorology in Upper Atmosphere

ICMV Integrated Circuit Multivibrator
ICN Idle Channel Noise
Instrument Communication Network
Integrated Computer Network
ICNAF International Commission for the Northwest Atlantic Fisheries
IC/NATAS International Council — National Academy of Television Arts and Sciences [US]
ICNB International Committee on Nomenclature of Bacteria
ICNCP International Commission for the Nomenclature of Cultivated Plants [Netherlands]
ICNDT International Committee on Nondestructive Testing
International Conference on Nondestructive Testing
ICNI Integrated Communications, Navigation, and Identification
ICNND Interdepartmental Committee on Nutrition for National Defense [US]
ICNPPA International Commission on National Parks and Protected Areas [now: CNPPA, Canada]
ICNV International Committee for the Nomenclature of Viruses
ICO Inter-Agency Committee on Oceanography [US]
Intergovernmental Commission on Oceanography [UK]
International Carbohydrate Organization [Canada]
International Chemistry Office
International Commission on Oceanography
International Commission on Optics [UK]
International Computer Orphanage [US]
ICOA International Castor Oil Association [US]
ICOC Indian Central Oil Seeds Committee
ICOD International Center for Ocean Development
ICOGRADA International Council of Graphic Design Associations [UK]
ICOH International Commission on Occupational Health [Switzerland]
International Congress on Occupational Health
ICOHTEC International Committee for the History of Technology [UK]
ICOLD International Commission on Large Dams [France]
ICOM International Council of Museums [France]
ICOMAT International Conference on Martensitic Transformation [US]
ICOMC International Conferences on Organometallic Chemistry
ICOME International Committee on Microbial Ecology [US]
ICOMIA International Council of Marine Industry Associations [UK]
ICOMOS International Council on Monuments and Sites [France]
ICON Integrated Control
ICONCLASS Iconography Classification
ICONS Information Center on Nuclear Standards [of ANS, US]
ICONTEC Instituto Colombiano de Normas Technicas [= Colombian Institute for Technical Standards]
ICOP Imported Crude Oil Processing
ICOR Intergovernmental Conference on Oceanic Research [of UNESCO]
ICOS Interactive COBOL Operating System
International Committee on Onomastic Sciences
ICOSO International Committee on Outer Space Onomastics
ICOSS Inertial-Command Offset System
ICOT Institute for (New Generation) Computer Technology [Japan]
ICOTOM International Conference on Textures of Materials
ICOTT Industry Coalition on Technology Transfer [US]
ICP Incoming (Message) Process
Indicator Control Panel
Inductively-Coupled Plasma
Industrial Cooperation Program
Initial Connection Protocol
Instrument Calibration Procedure
Intelligent Communications Processor
Interconnected Processing
Interim Compliance Panel
International Center of Photography [US]
International Classification of Patents
International Commission for Palynology [now: IFPS, Netherlands]
International Communication Planning
International Computer Programs
International Control Plan
Intrinsically Conductive Plastics
Inventory Control Point
Investment Corporation of Pakistan
Ion-Carburizing Calculation Program
ICPA International Cooperative Petroleum Association [US]
International Cotton Producers Association
ICP-AES Inductively-Coupled Plasma Atomic Emission Spectroscopy
ICPAM International Center for Pure and Applied Mathematics [France]
ICPBR International Commission for Plant-Bee Relationships [Denmark]
ICPC International Cable Protection Committee [UK]
International Coal Preparation Congress
Interrange Communications Planning Committee
ICPCSH International Conference on Physics and Chemistry of Semiconductor Heterostructures
ICPDATA International Commodity Production Data [of UNSO]
ICP/DCP Inductively-Coupled Plasma/Directly-Coupled Plasma
ICPE International Commission on Physics Education [of IUPAP]
ICPEMC International Commission for Protection against Environmental Mutagens and Carcinogens [Netherlands]
ICP-ES Inductively-Coupled Plasma Emission Spectroscopy [also: ICPES]
ICPI International Conference on Polyimides
ICPIG International Conference on Phenomena in Ionized Gases
ICPLC International Commission for the Protection of Lake Constance [Switzerland]
ICPM International Commission for Plant Raw Materials
International Commission on Polar Meteorology
International Conference on Polymers in Medicine

ICPMP International Commission for the Protection of the Moselle against Pollution [France]
ICP-MS Inductively-Coupled Plasma Mass Spectroscopy [also: ICPMS]
ICPN International Council of Plant Nutrition [Netherlands]
ICPNS International Conference on the Physics of Noncrystalline Solids [UK]
ICPO International Criminal Police Organization [also: INTERPOL, UK]
ICPP Idaho Chemical Processing Plant [of USAEC]
International Commission on Plasma Physics
ICPR International Conference on Production Research
ICPRAP International Commission for the Protection of the Rhine against Pollution [Germany]
ICPS Industrial and Commercial Power Systems [IEEE IAS Committe]
Interconnected Power System [also: IPS]
International Conference on the Properties of Steam
International Congress of Photographic Sciences
ICPSI International Conference on Plasma Surface Interactions
ICPTUR International Conference for Promoting Technical Uniformity on Railways [Switzerland]
ICPVT International Council for Pressure Vessel Technology [US]
ICQSA International Conference on Quanititative Surface Analysis
ICR Indirect Control Register
Inductance-Capacitance-Resistance
Initial Concentrated Rubber
Input Control Register
Institute for Chemical Research [of Kyoto University, Japan]
Institute for Cooperative Research
International Commission on Rheology
International Congress of Radiology
International Council for Reprography
Interrupt Control Register
Ion Cyclotron Resonance
Iron-Core Reactor
ICRA Industrial Chemical Research Association [US]
Interagency Committee on Radiological Assistance
International Conference on Robotics and Automation
ICRAF International Council for Research in Agroforestry [Kenya]
ICRC Indian Cancer Research Institute
International Conference on Robotics in Construction [US]
ICRDA Independent Cash Register Dealers Association [US]
ICRF Imperial Cancer Research Fund [UK]
Ion Cyclotron Range of Frequencies
ICRH Institute for Computer Research in the Humanities [US]
Ion Cyclotron Resonance Heating
ICRI International Communications Research Institute [Japan]
International Crops Research Institute
Iron Casting Research Institute [US]
ICRICE International Center of Research and Information on Collective Economy [Switzerland]
ICRISAT International Crops Research Institute for the Semi-Arid Tropics [India]
ICRM Institute of Certified Records Managers [US]
International Committee on Radionuclide Metrology
ICRO International Cell Research Organization [France]
ICRP International Commission on Radiological Protection [UK]
ICRPMA International Committee for Recording the Productivity of Milk Animals [Italy]
ICRS Index Chemicals Registry System [of ISI, US]
International Commission on Radium Standards
International Conference on Residual Stresses
ICRSDT International Commission on Remote Sensing and Data Transmission
ICRU International Commission on Radiation Units and Measurements [US]
ICS Immunochemistry System
Indian Ceramic Society
Indian Chemical Society
Infinity Color-Corrected System
Inland Computer Service [US]
Input Control System
Institute of Chartered Shipbrokers [UK]
Institite of Chartered Surveyors
Institute of Computer Science
Integrated Communication System
Integrated Controller Software
Integrated Control System
Integration Control System
Interactive Communications System
Intercarrier Sound System
Intercommunication System
Interlinked Computerized System
Intermittent Control System
Internal Countermeasures Set
International Chamber of Shipping [UK]
International Chemical Society
International Chemometrics Society [Belgium]
International Cogeneration Society [US]
International Commission on Stratigraphy
International Coronelli Society [Austria]
International Correspondence Schools [US]
Interphone Control Station
Interpretative Computer Simulation
Iron Casting Society
Isolation Containment Spray
Israel Chemical Society
ICSA In-Core Shim Assembly
International Council for Scientific Agriculture
ICSAB International Civil Service Advisory Board
ICSAR Interdepartment Committee on Search and Rescue
ICSB International Committee on Systematic Bacteriology [Canada]
International Council for Small Business [US]
ICSBA International Committee on the Study of Bauxite, Alumina and Aluminum
ICSC Interim Communications Satellite Committee [now: INTELSAT, US]

	International Center for Safety Communication
	International Civil Service Commission [US]
	International Conferences on Solution Chemistry
	Inter-Ocean Canal Study Commission [US]
ICSCC	Intercrystalline Stress Corrosion Cracking
ICSCI	International Center for Soil Conservation Information
ICSEAF	International Commission for the Southeast Atlantic Fisheries [Spain]
ICSEM	International Commission for the Scientific Exploration of the Mediterranean Sea [Monaco]
ICSEP	International Center for the Solution of Environmental Problems [US]
ICSFS	International Conference on Solid Films and Surfaces
ICSH	Interstitial-Cell-Stimulating Hormone
ICSHB	International Committee for Standardization in Human Biology
ICSHM	International Conference on the Science of Hard Materials
ICSHT	International Center for Science and High Technology [of ICTP, Italy]
ICSI	International Commission on Snow and Ice [of IAHS, Netherlands]
	International Conference on Scientific Information
ICSIC	International Conference on Shallow Impurity Centers
ICSID	International Council of Societies of Industrial Designers [Finland]
ICSISM	Institute for Cooperative Study of International Seafood Markets [now: IIFET, US]
ICSM	International Committee of Scientific Management
	International Conference on Structural Mechanics
	International Conference on Synthetic Metals
ICSMA	International Conference on the Strength of Metals and Alloys
ICSMP	Interactive Continuous Systems Modelling Program
ICSMRT	International Conference on Structural Mechanics in Reactor Technology
ICSOS	International Conference on Structure of Surfaces
ICSP	Interagency Council on Standards Policy
	International Committee on Science Photography
	International Conference on Shot Peening
ICSPRO	Inter-Secretariat Committee for Scientific Programs Relating to Oceanography
ICSPS	International Council for Science Policy Studies
ICSPTF	International Conference on Structure and Properties of Thin Films
ICSS	Instrumentation and Control Society of Singapore
	International Committee for Shell Structures [now: IASS, Spain]
	International Conference on Solid Surfaces
ICSSVM	International Commission for Small Scale Vegetation Maps [India]
ICST	Imperial College of Science and Technology [UK]
	Institute for Chemical Science and Technology
	Institute for Computer Sciences and Technologies
	Institute of Corrosion Science and Technology [UK]
ICSTDCS	International Conference on the Science and Technology of Defect Control in Semiconductors
ICSTI	International Center for Scientific and Technical Information [USSR]
	International Council for Scientific and Technical Information [France]
ICSTM	Imperial College of Science, Technology and Medicine [UK]
	International Conference on Scanning Tunneling Microscopy
ICSU	Independent Canadian Steelworkers Union
	International Council of Scientific Unions [France]
ICSUAB	International Council of Scientific Unions Abstracting Board [now: ICSTI, France]
ICSU-CTS	Committee on the Teaching of Science of the International Council of Scientific Unions [UK]
ICSW	International Commission on Surface Water
ICSWOA	International Center for Scientific Work Organization in Agriculture
ICT	Image Creation Terminal
	In-Circuit Tester
	Incoming Trunk
	Industrial Ceramics Technology
	Industrial Computed Tomography
	Institute of Circuit Technology [UK]
	Institute of Clay Technology [UK]
	Institute of Computer Technology
	Insulating Core Transformer
	Integrated-Circuit Technology
	Integrated Computer Telemetry
	Interaction Control Table
	Interactive Command Test
	International Coal Trade
	International Commission on Trichinellosis [Poland]
	International Computers and Tabulators
	International Council of Tanners [UK]
	International Critical Tables
IC/T	Integrated Computer/Telemetry
ICTA	International Center for Typographical Arts
	International College of Tropical Agriculture [India]
	International Confederation for Thermal Analysis [Israel]
	International Confederation of Technical Agriculturalists
ICTAA	Imperial College of Tropical Agriculture Association [UK]
ICTAM	International Congress on Theoretical and Applied Mechanics
ICTB	International Conference on Tall Buildings [US]
ICTC	Interdepartmental Committee on Toxic Chemicals
	International Cooperative Training Center [US]
IC/TC	Intelligent Color/Trash Coordinator
ICTE	Inertial Component Test Equipment
ICTED	International Cooperation in Transport Economics Documentation (Information Network) [of ECMT, France]
ICTF	International Conference on Thin Films
ICtl	Industrial Control [IEEE IAS Committee]
ICTP	International Center for Theoretical Physics [Italy]
	International Conference on the Technology of Plasticity

ICTPA International Conference on Titanium Products and Applications [of TDA, US]
ICTP-IUPAP International Center for Theoretical Physics — International Union of Pure and Applied Physics
ICTR International Center for Technical Research
ICTS Intermediate Capacity Transport System
International Center for Transportation Studies [Italy]
Isothermal Capacitance Transient Spectroscopy
ICTTT International Congress on Technology and Technology Transfer
ICTU Independent Canadian Transit Union
Irish Congress of Trade Unions
ICTV International Committee on Taxonomy of Viruses [UK]
ICTVTR Islamic Center for Technical and Vocational Training and Research
ICU Indicator Control Unit
Instruction Control Unit
Integrated Control Unit
Intensive-Care Unit
Interface Connection Unit
Interface Control Unit
International Chemistry Union
International Communication Union
International Communication Unit
ICUAE International Congress on University Adult Education
ICUMSA International Commission for Uniform Methods of Sugar Analysis [Australia]
ICV Initial Chaining Value
Internal Correction Voltage
ICVAN International Committee on Veterinary Anatomical Nomenclature [Austria]
ICVGE International Conference on Vapor Growth Epitaxy
ICVM International Conference on Vacuum Metallurgy
ICW In Connection With
Initial Condition Word
Intake Cooling Water
Interface Control Word
Interrupted Carrier Wave
Interrupted Continuous Wave
ICWA Institute of Cost and Works Accountants [UK]
International Coil Winding Association [US]
ICWES International Conference of Women Engineers and Scientists
ICWG International Clubroot Working Group [Scotland]
ICWM International Committee of Weights and Measures
ICWP Interstate Conference on Water Policy [US]
ICWQ International Commission on Water Quality
ICWRS International Commission on Water Resources Systems
ICWS International Cooperative Wholesale Society
ICWU International Chemical Workers Union
ICXOM International Congress on X-Ray Optics and Microanalysis
ICY International Cooperation Year
ICYYLM International Commission on Yeast and Yeast-Like Microorganisms [France]
ICZN International Commission on Zoological Nomenclature [UK]

ID Idaho [US]
Idea
Identification
Identity
Image Digitization
Image Distector
Improvement District
Indicating Device
Induced Draft
Industrial Design
Industrial Designer
Industrial Development
Industrial Drive
Industrial Drying
Infective Dose
Information Distributor
Initial Deflection
Inner Diameter
Inside Diameter
Inside Dimension
Instruction Decoder
Insulation Displacement
Integral Dose
Intelligence Department
Intelligent Digitizer
Interactive Debugging
Interdendritic
Interdigital
Interferometer and Doppler
Interior Department
Interior Design
Interior Designer
Intermediate Description
Internal Diameter
Iraqi Dinar [Currency]
Isodynamic
Isothermal Desorption
Isotope Dating
Isotope Dilution
Item Description
Item Documentation
I&D Information and Documentation
Integrate and Dump (Detection)
I/D Instruction/Data
id (idem) — the same
IDA Iminodiacetic Acid
Independent Distributors Association [US]
Industrial Developers Association
Industrial Diamond Association [US]
Input Data Assembler
Institute for Defense Analysis [US]
Integrated Digital Access
Integrated Digital Avionics
Integrated Disk Adapter
Intelligent Data Access
Interactive Debugging Aid
Interconnect Device Arrangement
International Database Association
International Desalination Association [US]
International Development Association [of UN]

	International Diamond Association (of America) [US]
	International Distribution Association [US]
	Intrusion Detection Alarm
	Ionospheric Dispersion Analysis
Ida	Idaho [US]
IDAC	Import Duties Advisory Committee [UK]
	Industrial Developers Association of Canada
IDACA	International District Heating and Cooling Association [US]
IDAFSA	International Defense and Aid Fund for Southern Africa
IDAL	Indirect Data Access List
IDAM	Indexed Direct Access Method
IDAP	Industrial Design Assistance Program
	Interactive Data Access System
IDAPS	Image Data-Processing System
IDAS	Information Displays Automatic Drafting System
	Intelligent Data Acquisition System
IDAST	Interpolated Data and Speech Transmission
IDAW	Indirect Data Address Word
IDB	Industrial Development Board [of UN]
	Industrial Development Bureau [of MOEA, Taiwan]
	Input Data Buffer
	Integrated Database
	Inter-American Development Bank
	Islamic Development Bank [Saudi Arabia]
IDBMS	Integrated Database Management System
	International Data Base Management Association [US]
IDBP	Industrial Development Bank of Pakistan
IDC	Image Dissector Camera
	Impulse-Driven Clock
	In Due Course
	Industrial Development Corporation [South Africa]
	Insulation Displacement Connector
	Intangible Drilling and Development Costs
	Integrated Disk Control
	Interior Design Society [US]
	Internal Data Channel
	International Dairy Committee
	International Data Corporation [Canada]
	International Design Conference in Aspen [US]
	International Diamond Council [Belgium]
	International Documentation Center [Sweden]
	Internationale Dokumentationsgesellschaft fur Chemie [= National Society for Chemical Documentation, Germany]
IDCA	International Development Corporation Agency [US]
IDCAS	Industrial Development Center for Arab States [Egypt]
IDCC	Integrated Data Communications Controller
	International Data Coordinating Center
	International Data Communications Center
IDCCC	International Dredging Conference Coordinating Committee [Netherlands]
IDCDA	Independent Dealer Committee Dedicated to Action [US]
IDCHE	Intergovernmental Documentation Center on Housing and Environment [France]
IDCJ	International Development Center, Japan
IDCMA	Independent Data Communications Manufacturers Association [US]
IDCNA	Insulation Distributor Contractors National Association [now: NICA, US]
IDCNS	Interdivisional Committee on Nomenclature and Symbols [of IUPAC]
IDCP	International Data Collecting Platform
IDCR	International Decade of Cetacean Research
IDCS	Image Dissector Camera System
	Instrumentation/Data Collection System
	International Digital Channel Service
IDCSP	Initial Defense Communication Satellite Project [US]
IDD	Integrated Data Dictionary
	International Direct Dialing
IDDD	International Direct Distance Dialing [US]
IDDF	Intermediate Digital Distribution Frame
IDDP	International Dairy Development Program [of FAO — UN]
IDDRG	International Deep-Drawing Research Group [UK]
IDDS	International Digital Data Service
IDE	Institute for Developing Economics [Japan]
	Interactive Data Entry
IDEA	Ideas Deserving Exploratory Analysis [of NRC, US]
	Industrial Designers Excellence Award
	Innovation Development for Employment Advancement
	Integrated Digital Electric Aircraft
	Interactive Data Entry Access
	International Data Exchange Association
IDEALS	Ideal Design of Effective and Logical Systems
IDEAS	Integrated Design and Engineering Automated System
	Integrated Design/Engineering/Architectural System
	International Data Exchange for Aviation Safety [of ICAO]
IDEC	Interior Design Educators Council [US]
IDECO	International Development Education Committee of Ontario [Canada]
IDEEA	Information and Data Exchange Experimental Activities
IDEF	ICAM (= Integrated Computer-Aided Manufacturing) Definition Method
	Integrated System Definition Language
$IDEF_0$	ICAM (= Integrated Computer-Aided Manufacturing) Definition, Version Zero
IDEM	Interdepartmental Electronic Mail [UK]
IDEMA	International Disk Drive Equipment and Materials Association
IDEMS	Integrated Diagnostic Engine Monitoring System
IDEN	Instituto de Engerharia Nuclear [= Institute of Nuclear Engineering, Brazil]
	Interactive Data Entry Network
IDENT	Identification
	Identify
IDEP	International Data Exchange Program [US]
	Interservice Data Exchange Program [of USDOD]
IDERA	International Development Education Resources Association
IDES	Integrated Defense System

	Integrated Design and Engineering System
IDEX	Initial Defense Experiment
IDF	Image Description File
	Incident Dark-Field
	Industrial Development Fund
	Institut pour le Developpement Forestier [= Institute for the Development of Forestry, France]
	Integrated Data File
	Interior Design Institute [Canada]
	Intermediate Distribution Frame
	International Dairy Federation [Belgium]
	International Development Foundation [US]
	International Drilling Federation [US]
	Inverse Document Frequency
IDFL	Integral Diode FET (= Field-Effect Transistor) Logic
IDFT	Inverse Discrete Fourier Transform
IDFTA	International Dwarf Fruit Trees Association [US]
IDG	Individual Drop Glider
	Inspector of Degaussing
	Integrated Drive Generator
IDH	Isocitrate Dehydrogenase
IDHA	International District Heating Association [now: IDHCA, US]
IDHCA	International District Heating and Cooling Association [US]
IDI	Improved Data Interchange
	Indian Development Institute
	Intelligent Dual Interface
	International Development Institute
IDIAS	Ice Data Integration Analysis System
IDIB	Industrial Diamond Information Bureau [UK]
IDIC	Industrial Development and Investment Center [of MOEA, Taiwan]
	Islamic Documentation Information Center
IDICT	Instituto de Documentacion e Informacion Cientifica y Tecnica [= Institute for Scientific and Technical Information and Documentation, Cuba]
IDIIOM	Information Display Incorporated Input-Output Machine
IDIN	International Development Information Network
IDIOT	Instrumentation Digital On-Line Transcriber
IDIS	International Dairy Industries Society
	Iowa Drug Information Service [of University of Iowa, US]
IDL	ICAD (= Integrated Computer-Aided Design) Design Language
	Information Description Language
	Instruction Definition Language
	Interactive Data Language
	Interdisciplinary Research Laboratory
IDLH	Immediate Danger to Life and Health
IDLIS	International Desert Locust Information Service
IDLRS	International Data Library and Reference Service [also: IDL&RS]
IDM	Industrial Data Management
	Integral and Differential Monitoring
	Intelligent Data Mapper
	Interactive Data Machine
	Interdiction Mission
IDMA	International Diamond Manufacturers Association
IDMAS	Interactive Database Manipulator and Summarizer
IDMH	Input Destination Message Handler
IDMS	Integrated Database Management System
	Integrated Data Management System
ID-MS	Isotope Dilution Mass Spectrometry [also: IDMS]
IDMS/R	Integrated Data Management System/Relation
IDN	Integrated Digital Network
	Intelligent Data Network
	Industrial Development Organization [of UN]
IDO	International Development Office
	Iterative Discrete On-Axis
IDOC	Inner Diameter of Outer Conductor
	Internal Dynamic Overload Control
	International Documentation and Communication Center
IDOCS	Intrusion-Detection System
ID/OD	Inside Diameter/Outside Diameter (Ratio)
IDOE	International Decade of Ocean Exploration
IDOS	Interactive Disk-Operating System
	Interrupt Disk-Operating System
IDP	Industrial Data Processing
	Inosine Diphosphate
	Input Data Processor
	Institute of Data Processing [UK]
	Integrated Data Processing
	Interdigit Pause
	Intermodulation Distortion Percentage
IDPC	Integrated Data Processing Center
IDPI	International Data Processing Institute
IDPM	Institute of Data Processing Management [UK]
IDPS	Interactive Direct Processing System
IDPS/LF	Interactive Direct Processing System/Large File
IDQ	Industrial Development Quotient
IDR	Indonesian Rupiah [Currency]
	Industrial Data Reduction
	Information Dissemination and Retrieval
	Inspection Discrepancy Report
	Institute of Delphinid Research [now: DRC, US]
	Intermediate Design Review
IDRC	Industrial Development Research Council [US]
	International Development Research Center [Canada]
IDS	Identification Section
	Idle Signal Unit
	Image-Dissector Scanner
	Inclined Driveshaft
	Information Display System
	Instrument Development Section
	Integrated Data Store
	Intelligent Display System
	Interactive Design Software
	Interactive Display System
	Interim Decay Storage
	International Development Services
IDSA	Indian Dairy Sciences Association
	Industrial Designers Society of America [US]
	Industrial Development Subsidiary Agreement
	International Development Service of America [US]
	International Diving Schools Association [US]
IDSC	International Die Sinkers Conference

IDSCP Initial Defense Satellite Communications Project [US]
IDSCS Initial Defense Satellite Communications System
IDSS ICAM (= Integrated Computer-Aided Manufacturing) Decision Support System
IDST Information and Documentation on Science and Technology
IDT Instrumented Drop Tube
Intelligent Data Terminal
Interdigital Transducer
Isodensiotracer
IDTC International Driving Tests Committee [France]
IDTS Instrumentation Data Transmission System
IDTSC Instrumentation Data Transmission System Controller
IDU Idle Signal Unit
Industrial Development Unit
Interface Data Unit
International Development Unit
5-Iodo-2'-Deoxyuridine
IDUR Iododeoxyuridine
IDW Institut fur Dokumentationswesen [= Institute for Documentation, Germany]
IDX Intelligent Digital Exchange
IE Ice Engineer
Ice Engineering
Illuminating Engineer
Illuminating Engineering
Image Enhancement
Index Error
Industrial Engineer
Industrial Engineering
Industrial Ergonomics
Information Engineer
Information Engineering
Infrared Emission
Initial Equipment
Inspection and Enforcement
Institution of Engineers [Australia and India]
Institution of Electronics [UK]
Instrument Engineer
Instrument Engineering
Institute of Energy [UK]
Intermediate Electrode
Interrupt Enable
Ion Exchange
Isoelectric
Isothermal Expansion
I&E Information and Education
Internal and External
ie (id est) — that is
IEA Information Engineering Association
Institute for Energy Analysis [UK]
Institute of Economic Affairs [UK]
Institute of Engineers of Australia
Institute of Environmental Action [Name Changed to: Urban Initiatives, US]
Institution of Engineers of Australia
Instituto de Energia Atomica [= Institute for Atomic Energy, Brazil]
International Economic Association
International Electrical Association
International Energy Agency [of OECD, France]
International Entomological Association
International Ergonomics Association [Finland]
International Exchange Association [US]
International Exhibitors Association [US]
IEAR IEA (= Instituto de Energia Atomica) Reactor [Brazil]
IE-AUST Institution of Engineers of Australia
IEB International Education Board
International Environmental Bureau
International Executive Board
IE-B Institute of Engineers — Bangladesh
IEC Independent Electrical Contractors [US]
Industrial Electrification Council [now: TEC, US]
Information Exchange Center
Integrated Electronic Component
Integrated Equipment Components
Intergraph Education Center
Interexchange Carrier
International Education Center
International Egg Commission [UK]
International Electrotechnical Commission [Switzerland]
Ion-Exchange Chromatography
Ion-Exchange Conference
Isotropic Elastic Constant
I&EC Industrial and Engineering Chemistry [of ACS, US]
IECA International Erosion Control Association [US]
IECC International Express Carriers Conference
IECE Institute of Electronic Communications Engineers
IECEE International Electrotechnical Commission System for Conformity Testing to Standards for Safety of Electrical Equipment [Switzerland]
IECEJ Institute of Electronic Communications Engineers of Japan
IECF International European Construction Federation [France]
IECI Industrial Electronics and Control Instrumentation [IEEE Group]
IECN Instituto Ecuatoriano de Ciencias Naturales [= Ecuador Institute of Natural Sciences]
IECQA International Electrotechnical Commission Qualification Assessment
IECR Institute of Engineering Cybernetics and Robotics [Bulgaria]
IED Individual Effective Dose
Institution of Engineering Designers [UK]
Interactive Electronic Display
IEDC International Energy Development Corporation
IEDD Institution of Engineering Draftsmen and Designers [UK]
IEDM International Electron Devices Meeting [of IEEE, US]
IEE Initial Environment Evaluation [Canada]
Institute for Earth Education [US]
Institute for Environmental Education [US]
Institute of Electrical Engineers [Japan]
Institution of Electrical Engineers [UK]
International Electrology Education
IEEC Industrial Electrical Equipment Council

Inter-African Electrical Engineering College
IEEE Institute of Electrical and Electronic Engineers [US]
IEEE-CS Institute of Electrical and Electronic Engineers — Computer Society
IEEIE Institution of Electrical and Electronic Incorporated Engineers [UK]
IEEJ Institute of Electrical Engineers of Japan
IEEMA Indian Electrical and Electronics Manufacturers Association [India]
IEEP Institute of European Environmental Policy [Germany]
International Environmental Education Program [of UNESCO]
IEES Imaging Electron Energy Spectrometer
IEETE Institution of Electrical and Electronics Technicians and Engineers [UK]
IEF Instruction Execution Function
International Environmental Facility
Isoelectric Focusing
IEFC International Emergency Food Committee [of FAO, UN]
IEFR International Emergency Food Reserve
IEFUA International Electronic Facsimile Users Association
IEG Information Exchange Group [of NATO]
IEH Industrial Electric Heating
IEI Implantation-Enhanced Interdiffusion
Industrial Education Institute [US]
Institute of Electrical Inspectors [Australia]
Institution of Engineering Inspection [UK]
Institution of Engineers of India
Institution of Engineers of Ireland
International Educator's Institute [US]
International Enamellers Institute [UK]
IEIC Institution of Engineers-in-Charge [UK]
IEICE Institute of Electronics, Information and Communication Engineers [Japan]
IEICEJ Institute of Electronics, Information and Communication Engineers of Japan
IEIJ Illuminating Engineering Institute of Japan
IEIS Integrated Engine Instrument System
IEJE Institut Economique et Juridique de l'Energie [= Institute for Economic and Legals Aspects of Energy, France]
IELA International Exhibition Logistics Associates [Switzerland]
IEM Interim Examination and Maintenance
Isocyanatoethylmethacrylate
IEMA Indian Electrical Manufacturers Association
International Explosive Metalworkers Association
IEM CELL Interim Examination and Maintenance Cell
IEMT International Electronic Manufacturing Technology
IEMTF Interim Examination and Maintenance Training Facility
IEO Integrated Electronic Office
IEP Image Edge Profile
Immunoelectrophoresis
Institute for Ecological Policies
Instrumentation for the Evaluation of Pictures
International Energy Program
Irish Pound [Currency]
Isoelectric Point
IEPA Independent Electron Pair Approximation
IEPG Independent European Program Group
IEPRC International Electronic Publishing Research Center [UK]
IEPS International Electronics Packaging Society [US]
International Electronic Packaging Symposium
IER Institute for Econometric Research [US]
Institute for Environmental Research [of ESSA, US]
Institute of Engineering Research [US]
IERD Industry Energy Research and Development (Program) [of CANMET, Canada]
IERE Institution of Electronic and Radio Engineers [Australia]
Institution of Electronic and Radio Engineers [Now Part of IEE, UK]
International Electrical Research Exchange
IERF International Education and Research Foundation [US]
IERS International Earth Rotation Service
IES Illuminating Engineering Society [US, UK and Australia]
Incoming Echo Suppressor
Industrial Electronics Society [US]
Information Exchange System
Institute for Earth Sciences [of ESSA, US]
Institute of Environmental Sciences [US]
Institute of Environmental Studies [Canada]
Institution of Engineers and Shipbuilders [UK]
Institution of Engineers of Singapore
Institution of Environmental Sciences [UK]
Integral Error Squared
Internal Environmental Simulator
International Ecology Society [US]
Intrinsic Electric Strength
Inverted Echo Sounder
Iowa Engineering Society [US]
IESC International Executive Service Corps
IESL Institution of Engineers of Sri Lanka
IESNA Illuminating Engineering Society of North America [US]
IESNEC Institution of Engineers and Shipbuilders of the Northeast Coast [UK]
IESS Institution of Engineers and Shipbuilders in Scotland
IET Inelastic Electron Tunneling
Initial Engine Test
Institute of Engineers and Technicians [UK]
Instrumentation Engineering Technician
Instrumentation Engineering Technologist
Instrumentation Engineering Technology
IETA International Electrical Testing Association [US]
IETC Interagency Emergency Transportation Committee
IETE Institution of Electronics and Telecommunications Engineers [India]
IETEJ Institute of Electronics and Telecommunications Engineers of Japan
IETG International Energy Technology Group [of OECD, France]
IETS Inelastic Electron Tunneling Spectroscopy
Inelastic Electron Tunneling Spectrum

IEU International Ecosystems University — Forum International [US]
IEV International Electrotechnical Vocabulary [of IEC, Switzerland]
IEX Ion Exchange Chromatography
I/EX Instruction Execution
IEXTRU International Conference on Extrusion
IF Importance Factor
Incompressible Flow
Industrial Furnace
Information
Information Feedback
Institute for Fermentation [Japan]
Instruction Field
Interface
Interferon
Intermnediate-Approach Fix
Intermediate Frequency
Interstitial Free (Steel)
Inventrepreneur Forum [US]
Inverted File
Inviscid Flow
Ion Focusing
Ionization Front
Isothermal Flow
I/F Interface
I-F Intermediate Frequency
Interstitial-Free (Steel)
IFA Immunofluorescence Analysis
Information Flow Analysis
Institute for Atomic Energy [Norway]
Integrated File Adapter
International Federation of Accountants [US]
International Federation of Airworthiness [UK]
International Fertilizer Industry Association [France]
International Financial Accountants [UK]
International Fiscal Association
International Fructose Association
Ionization Front Accelerator
IFAA International Federation of Advertising Agencies [US]
International Flow Aids Association [US]
International Furniture Accessory Association
IFABC International Federation of Audit Bureaus of Circulation
IFAC Integrated Flexible Automation Center
International Federation of Accountants [US]
International Federation of Automatic Control [Austria]
International Food Additives Council [US]
IFACE Interface Element
IFAD International Fund for Agricultural Development [of UN — Italy]
IFAI Industrial Fabrics Association International [US]
IFALPA International Federation of Airline Pilot Associations [UK]
IFAM Initial — Final Address Message
Institute fur Angewandte Materialforschung [= Institute for Applied Materials Research [of Fraunhofer Institute, Germany]]
Inverted File Access Method
IFAN Internationale Foderation der Ausschusse Normenpraxis [= International Federation for the Application of Standards, Switzerland]
IFAP International Federation of Agricultural Producers [France]
IFAPA International Foundation of Airline Passenger Associations [Switzerland]
IFAPWE Institute of Ferroalloy Producers in Western Europe [Switzerland]
IFARD International Federation of Agricultural Research Systems for Development [Netherlands]
IFATCA International Federation of Air Traffic Controllers Associations [Ireland]
IFATCC International Federation of Associations of Textile Chemists and Colorists
IFATE International Federation of Aerospace Technology and Engineering
International Federation of Airworthiness Technology and Engineering [now: IFA, UK]
IFATSEA International Federation of Air Traffic Safety Electronic Associations [UK]
IFATU International Federation of Arab Trade Unions
IFAW International Fund for Animal Welfare
IFAWPCA International Federation of Asian and Western Pacific Contractors Associations [Philippines]
IFAX International Facsimile Service
IFAXA International Facsimile Association [also: IFaxA, US]
IFB International Film Bureau
Invitation for Bids
IFBA International Fire Buff Associates
IFBPW International Federation of Business and Professional Women
IFBSO International Federation of Boat Show Organizers [US]
IFBWW International Federation of Building and Wood Workers [Switzerland]
IFC Image Flow Computer
Industrial Finance Corporation
Information Collector
Instantaneous Frequency Correlation
Institute of Forest Conservation
Institut Francais du Caoutchouc [= French Rubber Institute]
International Facilitating Committee
International Finance Corporation [US]
International Fisheries Commission
International Forging Congress
International Formulation Committee [of ICPS]
IFCAA International Fire Chiefs Association of Asia [Japan]
IFCATI International Federation of Cotton and Allied Textile Industries
IFCB International Federation of Cell Biology [Canada]
IFCC International Federation of Clinical Chemistry [Finland]
IFCCTE International Federation of Commercial, Clerical and Technical Employees
IFCE Institut Francais des Combustibles et de l'Energie [= French Institute of Combustibles and Energy]

	International Federation of Consulting Engineers
IFCF	Integrated Fuel Cycle Facility
IFCI	Industrial Finance Corporation of India
IFCM	International Federation of Christian Metalworkers
IFCMU	International Federation of Christian Miner's Unions
IFCN	Interfacility Communication Network
IFCP	Institute for Financial Crime Prevention [US]
IFCR	International Foundation for Cancer Research
IFCS	In-Flight Checkout System
	International Federation of Computer Sciences
IFCT	Industrial Finance Corporation of Thailand
	Institute of Fine Chemical Technology [USSR]
IFCTU	International Federation of Christian Trade Unions
IFD	Incipient Fire Detector
	Indentation Force Deflection
	Instantaneous Frequency Discriminator
	Internationale Foderation des Dachdeckerhandwerks [= International Federation of Roofing Contractors, Germany]
	International Federation for Documentation [now: IFID, Netherlands]
IFDA	International Furnishings and Design Association [US]
IFDC	International Fertilizer Development Center [US]
IFDO	International Federation of Data Organizations (for the Social Sciences) [Netherlands]
IFE	Institute for Fluitronics Education [US]
	Intelligent Front End
IFEAT	International Federation of Essential Oils and Aroma Trades [UK]
IFEES	International Federation of Electroencephalographic Societies
IFEMS	International Federation of Electron Microscope Societies
IFEP	Integrated Front-End Processor
IFER	International Federation of Engine Reconditioners [France]
	International Federation of Railway Advertising Companies [UK]
IFERS	International Flat Earth Research Society [now: FERSI, US]
IFES	Image Feature Extraction System
	International Field Emission Symposium
IFF	Identification Friend or Foe
	If And Only If
	Institute for Fermentation [Japan]
	Intensity Fluctuation Factor
	Interchange File Format
	International Flying Farmers [US]
	Ionized Flow Field
	Isoelectric Focusing Facility
IF/F	Identification Friend/Foe
IFFA	Institut fur Forstliche Arbeitswissenschaft [= Institute of Forestry Sciences, Germany]
	Internationale Fleischwirtschaftliche Fachmesse [= International Meat Trade Fair, Germany]
	International Frozen Food Association [US]
IFF/ATCRBS	Identification Friend or Foe/Air-Traffic Control Radar Beacon System
IFF/SIF	Identification Friend or Foe/Selective Identification Feature
IFGI	International Federation of Graphical Industries
IFGL	Initial File Generation Language
IFH	Interferon, Human
IFHE	International Federation of Hospital Engineering
IFHP	International Federation of Housing and Planning
IFHPM	International Federation of Hydraulic Platform Manufacturers
IFHT	International Federation for Heat Treatment (and Surface Engineering)
IFI	Industrial Fasteners Institute [US]
	Information for Industry [US]
	International Fabricare Institute [US]
	International Fastener Institute
	International Federation of Interior Architects and Interior Designers [Netherlands]
IFIA	Intermountain Forest Industry Association [US]
	International Federation of Inventors Associations [Sweden]
	International Federation of Ironmongers and Iron Merchants Associations
	International Fence Industry Association [US]
	International Fertilizer Industry Association [France]
IFIAS	International Federation of Institutes of Advanced Studies
IFIC	International Ferrocement Information Center
	International Food Information Council [US]
	Investment Funds Institute of Canada
IFID	International Federation for Information and Documentation [Netherlands]
IFIEC	International Federation of Industrial Energy Consumers [Switzerland]
IFIF	International Federation of Interior Designers
IFIFR	International Federation of International Furniture Removers [Belgium]
IFILE	Interface File
IFIP	International Federation for Information Processing [Switzerland]
IFIPC	International Federation of Information Processing Congresses
IFIPS	International Federation of Information Processing Societies [now: IFIP]
IFireE	Institution of Fire Engineers [UK]
IFIS	International Financial Intelligence Service [US]
	International Food Information Service [of CBDST, UK]
IFITU	Indian Federation of Independent Trade Unions
IFJ	International Federation of Journalists
IFK	Internal Flow Kinematics
IFKT	International Federation of Knitting Technologists [Switzerland]
IFL	Institute of Fluorescent Lighting [UK]
	Interfacility Link
	Internationale Fachmesse fur Ledertechnik [= International Leather Engineering Exhibition, Germany]
	International Frequency List
IFLA	International Federation of Landscape Architects [France]

International Federation of Library Associations and Institutions [Netherlands]
IFLASC International Federation of Latin American Study Centers [Mexico]
IFLIPS Integrated Flight Prediction System
IFLOT Intermediate Focal Length Optical Tracer
IFM Induction Flowmeter
Integrating Frequency Meter
Interactive File Manager
IFMA Industrial Furnace Manufacturers Association [now: IHEA, US]
International Facility Management Association [US]
International Farm Management Association [UK]
International Federation of Margarine Associations [Belgium]
International Food Service Manufacturers Association [US]
IFMBE International Federation for Medical and Biological Engineering [Canada]
IFME International Federation for Medical Electronics
International Federation of Municipal Engineers [UK]
IFMIS Intelligent Fire Management Information System
IFMMS International Federation of Mining and Metallurgical Students
IFMR Instantaneous Frequency Measurement Receiver
IFN Information
Interferon
IFNA International FidoNet (Electronic Mail System) Association [US]
IFNAES International Federation of the National Associations of Engineering Students
IFNSA International Federation of the National Standardization Associations
IFO Identifiable Flying Object
Identified Flying Object
International FORTRAN Organization [US]
Interplanetary Flying Object
IFOAM International Federation of Organic Agriculture Movements [Germany]
IFORS International Federation of Operational Research Societies [Denmark]
IFOS Ion Formation from Organic Solids
IFOSA International Federation of Stationers Association
IFOV Instantaneous Field of View
IFP Institut Francais du Petrole [= French Institute of Petroleum]
Integrated File Processor
International Federation of Purchasing
IFPA IFPS (= Integrated Flight Plan Processing System) Area
Inter-American Federation of Personnel Administration
International Fire Photographers Association [US]
IFPCW International Federation of Petroleum and Chemical Workers
IFPE Institute for Fluid Power Education [now: IFE, US]
IFPG International Frequency Planning Group
IFPI International Federation of the Phonographic Industry
International Federation of the Photographic Industry
IFPITB Inorganic Feed Phosphates International Technical Bureau [of CEFIC, Belgium]
IFPL ICAO (= International Civil Aviation Organization) Flight Plan
IFPM In-Flight Performance Monitor
International Federation of Physical Medicine
IFPMA International Federation of Pharmaceutical Manufacturers Associations [Switzerland]
IFPMM International Federation of Purchasing and Materials Management [Switzerland]
IFPP International Federation of the Periodicals Press [UK]
IFPR IFPS (= Integrated Flight Plan Processing System) Region
IFPRA International Federation of Park and Recreation Administration [UK]
International Federation of Public Relations Association
IFPS Interactive Financial Planning System
International Federation of Palynological Societies [Netherlands]
Integrated Flight Plan Processing System
IFPTA International Forest Products Transport Association
IFPTE International Federation of Professional Technical Engineers
IFPVS International Federation of Phonogram and Videogram Societies
IFPW International Federation of Petroleum Workers
IFPWA International Federation of Public Warehousing Associations [France]
IFPZ IFPS (= Integrated Flight Plan Processing System) Zone
IFR Image-to-Frame Ratio
Increasing Failure Rate
In-Flight Refueling
Instantaneous Frequency Receiver
Instrument Flight Rules
Interface Register
Internal Function Register
International Federation of Robotics
IFRA International Fragrance Association [Switzerland]
IFRB International Frequency Registration Board [of ITU, Switzerland]
IFRF International Flame Research Foundation [UK]
IFRU Interference Frequency Rejection Unit
Interference Rejection Unit
IFS Institute of Fusion Studies [of UTA, US]
Interactive File Sharing
Interactive Flow Simulator
Interchange File Separator
Intermediate Frequency Strip
International Federation of Surveyors
International Financial Statistics [of IMF, US]
International Fluidics Services
International Foundation for Science [Sweden]
Investment Feasibility Studies
Ionospheric Forward Scatter
IFSA International Fuzzy Systems Association [US]

IFSCC International Federation of Societies of Cosmetic Chemists [UK]
IFSEA International Federation of Scientific Editors Associations [US]
IFSEM International Federation of Societies for Electron Microscopy [US]
IFSM Inter-Fuel Substitution Model
IFSMA International Federation of Shipmasters Associations [UK]
IFSR International Federation for Systems Research [Austria]
IFSS International Flight Service Station
IFST Institute of Food Science and Technology [UK]
International Federation of Shorthand and Typewriting
IFSTA International Fire Service Training Association [US]
IFSTAD Islamic Foundation for Science, Technology and Development [Saudi Arabia]
IFT Institute of Food Technologists [US]
Interfacial Tension
Intermediate Frequency Transformer
International Federation of Translators [Belgium]
International Foundation for Telemetering [US]
International Foundation for Time-Sharing
IFTA In-Flight Thrust Augmentation
Institut Francais des Transports Aeriens [= French Institute of Air Transport]
International Federation of Teachers Associations
International Federation of Television Archives [Spain]
IFTAC Inter-American Federation of Touring and Automobile Clubs [Argentina]
IFTC International Federation of Thermalism and Climatism [France]
International Film and Television Council [Italy]
IFTDO International Federation of Training and Development Organizations [UK]
IFTE Intermediate Forward Test Equipment
IFTF Institute for the Future [US]
IFTS In-Flight Test System
IFToMM International Federation for the Theory of Machines and Mechanisms [Czechoslovakia]
IFTUTW International Federation of Trade Unions of Transport Workers [Belgium]
IFTWA International Federation of Textile Workers Associations
IFU Instruction Fetch Unit
IFUC Inter-Provincial Farm Union Council [Canada]
IFUNO Indian Federation of United Nations Association
IFUW International Federation of University Women [Switzerland]
IFVME Inspectorate of Fighting Vehicles and Mechanical Equipment [of MOD, UK]
IFVTCC Internationale Foderation der Vereine der Textilchemiker und Coloristen [= International Federation of Associations of Textile Chemists and Colorists, Switzerland]
IFWEA International Federation of Workers Educational Associations
IFWRI Institute of the Furniture Warehousing and Removing Industry [UK]
IFYGL International Field Year for the Great Lakes
IG Imperial Gallon
Impulse Generator
Induction Generator
Inductor Generator
Inertial Guidance
Institution of Geologists [UK]
Instructor's Guide
Ion Gun
Ionization Gauge
Isotope Geology
Ig Immunoglobulin
IGA Inner Gimbal Axis
Intergovernmental Agreement
Intergranular Attack
International General Aviation
International Geographical Association
International Geothermal Association
IgA Immunoglobulin A
IGAAS Integrated Ground/Airborne Avionics System
IGAEA International Graphic Arts Education Association [US]
I GAL Imperial Gallon
IGAS International Graphic Arts Society [US]
International Graphoanalysis Society
IGB Intermediate Gearbox
International Gravimetric Bureau [France]
IGBD Intrinsic Grain Boundary Dislocation
IGBP International Geosphere/Biosphere Program [Germany]
IGBT Insulated Gate Bipolar Transistor
IGC Inspectorate General of Customs [of MOF, Taiwan]
Institute for Graphic Communication [US]
Intergovernmental Copyright Committee [France]
Intergranular Corrosion
International Geochemical Congress
International Geological Congress
International Geophysical Committee
International Grassland Congress
IGCAR Indira Gandhi Center for Atomic Research [India]
IGCC Integrated Gasification Combined Cycle
Intergovernmental Copyright Committee
Insulating Glass Certification Council [US]
IGCG Inertial Guidance and Calibration Group [US Air Force]
IGCI Industrial Gas Cleaning Institute [US]
IGCP International Geological Correlation Program [of UNESCO — France]
IGE Institution of Gas Engineers [UK]
Isopropyl Glycidyl Ether
IgE Immunoglobulin E
IGES Initial Graphics Exchange Specification
International Geochemical Congress
International Graphics Exchange Standard
IGES/DXF Initial Graphics Exchange Specification/Data Exchange Format
IGES/PDES Initial Graphics Exchange Specification/Product Design Exchange Specification
IGESUCO International Ground Environment Subcommittee [of NATO]
IGF Inert Gas Fusion

Insulin-Like Growth Factor
Intergranular Fracture
International Genetics Federation [US]
International Graphical Federation

IGFA International Game Fish Association
IGFET Insulated-Gate Field-Effect Transistor
IGHP Innovative Guided Hypervelocity Projectile
IgG Immunoglobulin G [Immunochemistry]
IGGA International Grooving and Grinding Association [also: IG&GA, US]
IgG frac Immunoglobulin G Fraction of Antiserum [Immunochemistry]
IGGT Institute for Guided Ground Transport
IGHIA International Garden Horticultural Industry Association [US]
IGI Industrial Guest Investigator
IGIA Interagency Group for International Aviation [of FAA, US]
IGIC Isoconductive Gradient Ion Chromatography
IGIP Internationale Gesellschaft fur Ingenieurpadagogik [= International Society for Engineering Education, Austria]
IGKB Internationale Gewasserschutzkommission fur den Bodensee [= International Commission for the Protection of Lake Constance, Switzerland]
IGL Interactive Graphics Language
IGM Intergalactic Medium
Internationale Gesellschaft fur Moorforschung [= International Society for Marsh Research, Germany]
IgM Immunoglobulin M
IGMAA International Gas Model Airplane Association [US]
IGN Ignition
Institut Geographique National [= National Geographical Institute, France]
International Geographic Institute
IGNITOR Ignition Torus
IGO Inter-Governmental Organization
IGOR Intercept Ground Optical Recorder [also: Igor]
IGOSS Integrated Global Ocean Station System
IGPC Impregnated Gas-Pressure Cable
IGPCE International Great Plains Conference of Entomologists
IGPM Institut fur Geometrie und Praktische Mathematik [= Institute for Geometry and Practical Mathematics [of RWTH, Germany]]
IGPP Institute of Geophysics and Planetary Physics [of SIO, US]
IGR Institute for Groundwater Research [of University of Waterloo, Canada]
IGRC International Gas Research Conference
IGROUP Inspection Group
IGRP International Genetic Resources Program [now: RAFI, US]
IGR&P Inert Gas Receiving and Processing
IGS Immunogold Staining
Inert Gas-Shielded Welding
Inertial Guidance System
Information Generator System
Information Group Separator
Instrument Guidane System
Integrated Geophysical System
Integrated Graphics System
Interactive Graphics System
Interchange Group Separator
Internationale Gesellschaft fur Stereologie [= International Society for Stereology, Germany]
International Glaciological Society [UK]
Irish Graphical Society
IGSC International Gold and Silver Conference
IGSCC Intergranular Stress-Corrosion Cracking
IGSHPA International Ground Source Heat Pump Association [US]
IGSPS International Gold and Silver Plate Society
IGSS Immunogold Silver Staining
IGT Institute of Gas Technology [US]
Insulated Gate Transistor
Intelligent Graphics Terminal
IGTC International Gas Turbine Center [now: IGTI of ASME, US]
IGTI International Gas Turbine Institute [of ASME, US]
IGU International Gas Union
International Geographical Union [Canada]
IGWMC International Ground Water Modeling Center [US]
IGWR Institute for Groundwater Research [of University of Waterloo, Canada]
IGY International Geophysical Year
IH Immersion Heater
Induction Heating
Industrial Hygiene
Industrial Hygienist
Industrialized Housing
Interaction Handler
Interrupt Handler
IHACE International Heating and Air-Conditioning Exposition
IHAS Integrated Helicopter Avionics System
IHB International Hydrographic Bureau [now: IHO, Monaco]
IHC Industrial Hygiene Conference [US]
Intergranular Hot Cracking
Interstate Highway Capability [US]
IHCMCB International Hazard Control Manager Certification Board [US]
IHCP Induction Hardened Chrome Plating
International Handbook on Coal Petrology
IHD International Hydrological Decade
IHE Institute for the Human Environment [US]
Institution of Highway Engineers [UK]
IHEA Industrial Heating Equipment Association [US]
IHEB International Heat Economy Bureau
IHEP Institute of High Energy Physics
IHES Illinois Horticultural Experiment Stations [US]
IHET Industrial Heat Exchanger Technology
IHF Industrial Hygiene Foundation [US]
Inhibit Halt Flip-Flop
Institute of High Fidelity
Intermediate High Frequency
IHFA Industrial Hygiene Foundation of America [US]
IHFBC International High Frequency Broadcasting Conference
IHFC International Heat Flow Commission

IHFM Institute of High-Fidelity Manufacturers

IHGA International Hop Growers Association [Yugoslavia]

IHIS Integrated Hospital Information System

IHK Industrie- und Handelskammer [= Chamber of Industry and Commerce, Germany]
Internationale Handelskammer [= International Chamber of Commerce, Germany]

IHMA Industrial Housing Manufacturers Association [US]

IHO Institute of Human Origins [US]
International Health Organization
International Hydrographic Organization [Monaco]

IHP Indicated Horsepower [also: ihp]
Inositol Hexaphosphate
Institut Henri Poincare [= Henri Poincare Institute, France]
International Hydrological Program [of UNESCO — France]

IHPA International Hardwood Products Association [US]

IHPH Indicated Horsepower-Hour [also: IHP HR, IHPhr or ihph]

IHPTEP Integrated High-Performance Turbine Engine Program [US]

IHPTET Integrated High Performance Turbine Engine Technology (Program) [of USDOD]
Intermediate High Performance Turbine Engine Technology

IHPVA International Human Powered Vehicle Association [US]

IHR Institute of Historical Research [UK]

IHS Indiana Horticultural Society [US]
Information Handling Service [US]
Institute for Hydrogen Studies
Integrated Hospital System
Intensity, Hue, Saturation
International Hibernation Society [US]
International Hydrofoil Society [UK]

IHSB Industrial Health and Safety Branch [of MOL, Canada]

IHSI Induction Heating Stress Improvement

IHSM Institute for High Speed Mechanics [of Tohoku University, Japan]

IHTS Intermediate Heat Transport System

IHTU Interservice Hovercraft Trials Unit [of MOD, UK]

IHVE Institution of Heating and Ventilating Engineers [UK]

IHVS Intelligent Highway Vehicle System

IHW Industrial and Hazardous Waste
International Halley Watch [of JPL, US]

IHX Intermediate Heat Exchanger

IHXGV Intermediate Heat Exchanger Guard Vessel

II Image Intensifier
Industrial Imaging
Information Interchange
Innovators International [US]
Interrupt Inhibit
Ion Implantation

IIA Information Interchange Architecture
Information Industry Association [US]
Institute of Internal Auditors [US]
Intelligence Industries Association [US]
International Institute of Agriculture
International Inventors Association

IIAA Institute of Interamerican Affairs

IIAES Ion-Induced Auger Electron Spectroscopy

IIAILS Interim Integrated Aircraft Instrumentation and Letdown System

IIAR International Institute of Ammonia Refrigeration [US]

IIAS Interactive Instructional Answering System
Inter-American Institute of Agricultural Sciences
International Institute of Administrative Sciences [Belgium]

IIASA International Institute for Applied Systems Analysis [Austria]

IIASES International Institute for Aerial Survey and Earth Sciences [Netherlands]

IIB Institut International des Brevets [= International Patent Institute]
Institute of International Bankers [US]
International Investment Bank [USSR]

IIBA International Institute for Bioenergetic Analysis
International Intelligent Buildings Association [US]

IIBBR International Institute for Biological and Botanical Research

IIBCA Instituto de Investigaciones en Biomedicina y Ciencias Aplicadas [= Institute Biomedical Research and Applied Science, Venezuela]

IIBEM Indian Institute of Biochemistry and Experimental Medicine

IIC Impact Insulation Class
India International Center
Information Interchange
Institute for Instrumentation and Control [Australia]
Instituto de Ingenieiros de Chile [= Chilean Institute of Engineers]
Inter-Integrated Circuits [also: I^2C]
International Institute for Conservation of Historic and Artistic Work [UK]
International Institute for Cotton [Belgium]
International Institute of Communications [UK]
International Institute of Conservation
International Investment Corporation
Integrated Information Center
Islamic Investment Company

IICA Inter-American Institute for Cooperation on Agriculture [Costa Rica]

IICAF Institute for International Collaboration in Agriculture and Forestry [Czechoslovakia]

IICB International Import Custom Brokers

IICC International Institute for Commercial Competition [Belgium]

IICCG International Institute of Conservation — Canadian Group

IICL Institute of International Container Lessors [US]

IICMFA Integrated Information Center of the Ministry of Foreign Affairs [Saudi Arabia]

IICS International Interactive Communications Society [US]

IICT Instituto de Investigaciones Cientificas y Tecnologicas [= Institute for Scientific and Technical Investigations, Argentina]

IICY International Investment Corporation for Yugoslavia
IID Independent Identically Distributed
Intermittent-Integrated Doppler
IIDA Individualized Instruction for Data Access
Instituto Interamericano de Direito de Autor [= Inter-American Copyright Institute, Brazil]
IIDC Institute for International Development and Cooperation
IIDCT Instituto de Informacion y Documentacion en Ciencia y Tecnologia [= Institute for Information and Documentation in Science and Technology, Spain]
IIDQ 2-Isobutoxy-1-Isobutoxycarbonyl-1,2-Dihydroquinoline
Isobutyl 1,2-Dihydro-2-Isobutoxyl-1-Quinoline Carboxylate
IIE Institute of Industrial Engineers [US]
Institute of International Education [US]
IIEA International Institute for Environmental Affairs [now: CIDE, US]
IIED International Institute for Environment and Development [now: CIDE, US]
IIEE Ion-Induced Electron Emission
IIEIC International Institute Examinations Inquiry Committee [UK]
IIES Industrial Innovation Extension Service [of NYSSTF, US]
International Institute of the Environment Society
IIF Institute of International Finance [US]
International Institute of Forecasters [US]
IIFET International Institute of Fisheries Economics and Trade [US]
IIFSO International Islamic Federation of Student Organizations [US]
IIFT Indian Institute of Foreign Trade
IIHR Iowa University Institute of Hydraulic Research [US]
IIHS Insurance Institute for Highway Safety [US]
III International Institute of Interpreters [US]
IIIC International Institute for Intellectual Cooperation
International Irrigation Information Center
IIICR International Institute of Interdisciplinary Cycle Research [US]
IIIF Impurity-Induced Intergranular Fracture
IIIL Isoplanar Integrated Injection Logic [also: I3L, or I^3L]
IIL Integrated Injection Logic [also: I^2L]
IILE Ion-Induced Light Emission
IILP International Institute for Lath and Plaster [US]
IILS International Institute for Labour Studies
IIM Indian Institute of Management
Indian Institute of Metals
Institute for Information Management [US]
IIMA Industrial Instruments Manufacturing Association
IIMT International Institute of Milling Technology [UK]
IINCE International Institute for Noise Control Engineering
I Inf Sci Institute of Information Scientists [UK]
IINS Interuniversity Institute of Nuclear Sciences [Belgium]
IINSE International Institute of Nuclear Science and Engineering
IIOC Intelligent Input/Output Channel
IIOE International Indian Ocean Expedition
IIOP Integrated Input/Output Processor
IIP Immigrant Investor Program [of EIC, Canada]
Implementation and Installation Plan
Index of Industrial Production
Indian Institute of Petroleum
Instantaneous Impact Point
IIPE Institution of Incorporated Plant Engineers [UK]
IIPER International Institution of Production Engineering Research
IIPP International Institute for Promotion and Prestige [Switzerland]
IIPS Interactive Instructional Presentation System
IIR Imaging Infrared
Infinite-Duration Impulse Response
Infinite-Impulse Response
Institute of Interdisciplinary Research
Institute of Intergovernmental Relations
Institute of Intermodal Repairers [US]
Institute of International Research
Integrated Instrumentation Radar
International Institute of Refrigeration [France]
International Inventors Registry [US]
Isobutylene-Isoprene Rubber
Isotactic Isoprene Rubber
Resistance Heating [also: I^2R]
IIRA International Industrial Relations Association [Switzerland]
IIRB Institut International de Recherches Betteravieres [= International Institute for Sugar Beet Research, Belgium]
IIRC Interrogation and Information Reception Circuit
IIRFD Infinite-Impulse Response Filter Design
IIRI International Industrial Relations Institute
IIRS Institute for Industrial Research and Standards [US]
IIS Indian Institute of Science
Institute of Industrial Science [of University of Tokyo, Japan]
Institute of Industrial Supervisors
Institute of Information Scientists [UK]
Interactive Instructional System
International Isotope Society [US]
Intrinsic Instruction Set
IISBR International Institute of Sugar Beet Researchers [Belgium]
IISEE International Institute of Seismology and Earthquake Engineering [Japan]
IISI International Iron and Steel Institute [Belgium]
IISL International Institute of Space Law [Netherlands]
IISLS Improved Interrogator Sidelobe Suppression
IISN Institut Interuniversitaire des Sciences Nucleaires [= Inter-University Institute of Nuclear Sciences, Belgium]
IISO Institution of Industrial Safety Officers
IISP International Institute of Site Planning
IISR Indian Institute of Sugar Cane Research

IISRP International Institute of Synthetic Rubber Producers [US]
IISS Integrated Information Support System
International Institute for Strategic Studies [UK]
International Institute for the Science of Sintering [Yugoslavia]
IISSC International Industrial Symposium (and Exhibition) on the Supercollider
IIST Institute of Information Science and Technology [of TDU, Japan]
Institution of Instrumentation Scientists and Technologists [India]
IIT Iligan Institute of Technology [Philippines]
Illinois Institute of Technology [US]
Indian Institute of Technology [India]
Industrial Information Transfer
Institute of Industrial Technicians [UK]
Instituto de Investigaciones Tecnologicas [= Institute for Technological Investigations, Colombia]
Israel Institute of Technology
IITA Inland International Trade Association [US]
International Institute of Tropical Agriculture [Nigeria]
IITB Indian Institute of Technology, Bombay [India]
IITD Indian Institute of Technology, Delhi [India]
IITK Indian Institute of Technology, Kanpur [India]
Indian Institute of Technology, Kharagpur [India]
IITM Indian Institute of Technology, Madras [India]
IITRAN Illinois Institute of Technology Translator
IITRI Illinois Institute of Technology Research Institute [US]
IITS International Institute of Theoretical Sciences
IITV Image Intensified Television
IIU Instruction Input Unit
IIW International Institute of Welding [UK]
IIWG International Industry Working Group [Canada]
IIWPA International Information/Word Processing Association [formerly: IWPA, US; now: AISP, US]
IIX Ion-Induced X-Rays
IJAJ International Jitter Antijam
IJC International Joint Commission [US]
International Joint Conference
IJCAI International Joint Conference on Artificial Intelligence [US]
IJCRAB International Joint Commission Research Advisory Board
IJJU Intentional Jitter Jamming Unit
IJMA Indian Jute Millers Association
IJMARI Indian Jute Millers Association Research Institute
IJO International Jute Organization [Bangladesh]
IJP Ink Jet Printer
Internal Job Processing
IJPA International Jelly and Preserve Association [US]
IJS Interactive Job Submission
IK Inverse Kinematics
IKBS Intelligent Knowledge-Based System
IKE Ion Kinetic Energy
IKK Internationale Fachausstellung Kalte- und Klimatechnik [= International Refrigeration and Air Conditioning Exhibition, Germany]
IKL Intersecting Kikuchi Lines
IKO Institute van Kernph Ouder [Netherlands]
Ikr Icelandic Krona [Currency]
IKRA International Kirlian Research Association [US]
IKSR Internationale Kommission zum Schutze des Rheins gegen Verunreinigung [also: ICPRAP, Germany]
IKT Internationale Kautschuk-Tagung [= International Rubber Conference, Germany]
IL Idle
Illinois [US]
Imaging Library
Including Loading
Incorrect Length
Indication Lamp
Injection Luminescence
Insertion Loss
Instruction List
Interface Loop
Interior Length
Interline
Intermediate Language
Interrupt Level
Ionoluminescence
Iron Loss
I&L Installation and Logistics
I/L Import License
Il Illinium (= Promethium)
ILA Institute of Landscape Architects [UK]
Intelligent Line Adapter
International Language for Aviation
International Longshoreman's Association
Iterative Linear Algebra
ILAAS Integrated Light Aircraft Attack System
Integrated Light Attack Avionics System
ILAAT Interlaboratory Air-to-Air Missile Technology
ILAB Bureau of International Labour Affairs [US]
ILAC International Laboratory Accreditation Conference
ILACS Integrated Library Administration and Cataloguing System [Netherlands]
ILAFA Instituto Latinoamericano del Fierra y el Acero [= Latin American Iron and Steel Institute, Chile]
ILAI Italo-Latin American Institute [Italy]
ILAP Industry and Labour Adjustment Program
ILAR Institute of Laboratory Animal Resources
ILAS Instrument Landing Approach System
Interrelated Logic Accumulating Scanner
Isotrace Laboratory for Analytical Services [Canada]
ILB Initial Load Block
Inner Lead Bond
International Labour Board [US]
International Liaison Bureau [UK]
ILBM Interleave Bit Map
ILBT Interrupt Level Branch Table
ILC Initial Launch Capability
Inrevocable Letter of Credit
Instruction Length Code
Instruction Length Counter
Instruction Location Counter
Integrated Laminating Center
International Lifeboat Conference

	Ion/Liquid Chromatography
ILCA	International Labor Communications Association [US]
	International Launching Class Association
	International Lightning Class Association
ILCCS	Integrated Launch Control and Checkout System
ILCMP	International Liaison Committee on Medical Physics
ILCOP	International Liaison Committee of Organizations for Peace
ILD	Injection Laser Diode
	International Labour Documentation [of ILO, UN]
ILDA	Industrial Lighting Distributors of America [US]
ILE	Institute of Locomotive Engineers [UK]
	Interfacing Latching Element
Ile	Isoleucine
	Isoleucyl
ILEA	International League of Electrical Association [US]
ILEED	Inelastic Low-Energy Electron Diffraction
ILEM	Inter-Library Electronic Mail
ILENP	Institute of Low-Energy Nuclear Physics [PR China]
ILF	Inductive Loss Factor
	Infra Low Frequency
	Input Loading Factor
	Integrated Lift Fan
	Interlaminar Failure
	Intralow Frequency [also: ilf]
IL/F	Imaging Library/FORTRAN
ILFI	International Labour Film Institute
ILGB	International Laboratory of Genetics and Biophysics
ILHMFLT	International Laboratory for High Magnetic Fields and Low Temperatures
ILI	Indiana Limestone Institute (of America) [US]
	Instant Lunar Ionosphere
	Inter-African Labour Institute
ILIC	International Library Information Center [of University of Pittsburgh, US]
ILID	Institut fur Landwirtschaftliche Information und Dokumentation [= Institute for Agricultural Information and Documentation, Germany]
ILIMA	International Licensing Industry and Merchandisers Association [US]
ILIP	In-Line Instrument Package
ILIR	In-House Laboratories Independent Research
ILIS	Intelligent Landscape Integrated System
ILL	Illustration [also: ILLUS or ILLUST]
	Institute Laue-Langevin [France]
	Inter-Library Loan
Ill	Illinois [US]
ILLIAC	Illinois Institute of Advanced Computation [of University of Illinois, US]
	(University of) Illinois Integrator and Automatic Computer
ILLINET	Illinois Library Network [US]
ILLUM	Illumination
ILLUS	Illustration [also: ILL or ILLUST]
ILM	Information Logic Machine
	Intermediate Language Machine
ILMA	Independent Lubricant Manufacturers Association [US]
ILMAC	International Congress and Fair for Laboratory, Measuring and Automation Techniques in Chemistry
ILMC	Indiana Labor Management Council [US]
	International Light Metal Congress
ILO	Individual Load Operation
	Injection-Locked Oscillator
	International Labor Office [of UN — Switzerland]
	International Labor Organization [of UN — Switzerland]
ILP	Intermediate Language Processor
	Intermediate Language Program
	International Lithosphere Program
ILPA	International Labor Press Association [now: ILCA, US]
ILPC	International Linen Promotion Commission [US]
ILPD	Intergranular Liquid Phase Distribution
ILPF	Ideal Low-Pass Filter
ILR	Institute of Library Research [of University of California, US]
	Instruction Location Register
ILRA	International Laboratory for Marine Radioactivity
	International Lactic Acid Research Association
ILRI	Indian Lac Research Institute
	International Institute for Land Reclamation and Improvement [Netherlands]
ILRT	Integrated Leakage Rate Test
	Intermediate Level Reactor Test
ILS	Ideal Liquidus Structures
	Incoherent-Light System
	Instrument Landing System
	Instrument Line Shape
	Integrated Library System
	Integrated Logistic Support
	Interactive Laboratory System
	Interlaminar Shear
	International Latitude Service
	International Line Selector
	International Lunar Society [US]
	Interrupt Level Subroutine
	Ionization Loss Spectroscopy
	Israeli Shekel [Currency]
ILSAP	Instrument Landing System Approach
ILSF	Intermediate Level Sample Flow
ILSI	International Life Sciences Institute
ILSMT	Integrated Logistic Support Management Team
ILSO	Incremental Life Support Operation
ILSRO	Interstare Land Sales Registration Office [US]
ILSRP	Instrument Landing System Reference Point
ILSS	Interlaminar Shear Strength
ILSTAC	Instrument Landing System and TACAN (= Tactical Air Navigation)
ILSW	Interrupt Level Status Word
ILTA	Independent Liquid Terminals Association [US]
ILTMS	International Leased Telegraph Message Switching (Service) [UK]
ILTS	Institute of Low Temperature Science [of Hokkaido University, Japan]
ILU	Illinois University [US]
ILW	Industrial Liquid Waste
	Intermediate-Level Waste

ILWU International Longshoremen's and Warehousemen's Union

ILZRO International Lead and Zinc Research Organization [US]

ILZSG International Lead and Zinc Study Group [US]

IM Imaginary Part
Imperial Measure
Induction Meter
Induction Motor
Industrial Mathematics
Industry Motion Picture
Information Management
Information Manager
Ingot Metallurgy
Injection Molding
Inner Marker
Installation and Maintenance
Institute of Metallurgy
Institute of Meteorology
Institute of Metrology [USSR]
Instruction Memory
Instrumentation
Instrumentation and Measurements [IEEE Group]
Integrated Manufacturing
Integrated Modem
Integrating Meter
Intensity Modulation
Interceptor Missile
Interface Module
Interference Microscopy
Intermediate Missile
Intermediate Modulus
Intermodulation
Interrupt Mask
Ion Microscope
Isomagnetic
Isometric
Item Mark

I&M Inspection and Maintenance

I/M Ingot Metallurgy
Inspection and Maintenance

Im Imaginary Part [Mathematics]

IMA Ideal Mechanical Advantage
Industrial Medical Association [US]
Input Message Acknowledgement
Interdisciplinary Master of Arts
International Magnesium Association [US]
International Management Association [UK]
International MIDI (= Musical Instrument Digital Interface) Association [US]
International Military Archives [US]
International Milling Association
International Mineralogical Association [Germany]
International Mohair Association [UK]
International Mycological Association [UK]
Invalid Memory Address
Inventory of Marine Activities (Database) [of SPIW, Thailand]
Irish Medical Association
Islamic Medical Association

IMAC Illinois Microfilm Automated Cataloguing [of ISL, US]

IMACA International Mobile Air Conditioning Association [US]

IMACS International Association for Mathematics and Computers in Simulation [Belgium]

IMADS Integrated Machinery Analysis and Diagnostic System

IMAG Imaginary

IMAGE Interactive Menu-Assisted Graphics Environment

IMAM International Meeting on Advanced Materials [of Materials Research Society, US]

IMANF Institute of Manufacturing [UK]

IMAP Institute for Materials and Advanced Processing [of University of Illinois, US]
Integrated Mechanical Analysis Project [US]

IMarE Institute of Marine Engineers [UK]

IMAS Industrial Management Assistance Survey [US Air Force]

IMAT International Masonry Apprenticeship Trust [now: IMIAT, US]

IMAW International Molders and Allied Workers Union

IMB Intermodule Bus
Internationale Messe fur Bekleidungsmaschinen [= International Fair for Clothing Manufacturing Machines, Germany]
Institute of Molecular Biology

IMBA International Master of Business Administration

IMC Image Motion Compensation
Institute of Management Consultants [UK]
Institute of Measurement and Control [UK and New Zealand]
Instructional Materials Center
Instrument Meteorological Conditions
Integrated Maintenance Concept
Integrated Microcircuit
Integrated Multiplexer Channel
Intelligent Matrix Control
Interactive Medical Communications
Interactive Module Controller
Interface Module Cabinet
Intermediate Metal Conduit
Intermetallic Compound
Intermetallic-Matrix Composite
Internal Model Control
International Information Management Congress [US]
International Maintenance Center
International Maritime Committee
International Materials Conference
International Micrographic Congress [Name Changed to: International Information Management Congress, US]
Israel Materials Conference

IMCA Internal Model Control Approach

IMCC Institute of Management Consultants of Canada
Interstate Mining Compact Commission [US]

IMCEA International Military Club Executives Association [US]

IMCO Institute of Management Consultants of Ontario [Canada]

Intergovernmental Maritime Consultative Organization [US]
International Maritime Consultative Organization
International Maritime Countries Organization [now: IMO, UK]

IMCOS International Meteorological Consultant Service
IMCS International Metal Container Section [of MHI, US]
IMCP Instituto de Madera, Celulosa y Papel [= Institute of Wood, Cellulose and Paper, Mexico]
IMCSMHI International Metal Container Section of the Material Handling Institute [US]
IMD Institute for Marine Dynamics [of NRC, Canada]
Institute of Metals Division [of AIME, US]
Institut fur Maschinelle Dokumentation [= Institute for Automated Documentation, Austria]
Intercept Monitoring Display
Intermodulation Distortion
International MTM (= Methods-Time Measurement) Directorate [Sweden]
IMDA International Map Dealers Association [US]
IMDC Interceptor Missile Direction Center
IMDG International Maritime Dangerous Goods (Code)
IMDM Iscove's Modified Dulbecco Medium
IMDO Installation and Materiel District Office [of FAA, US]
IMDS International Meat Development Scheme [of FAO — UN]
International Microform Distribution Service
IMDT Immediate
IME Institute of Makers of Explosives [US]
Institute of Marine Engineers [UK]
Institute of Mining Engineers
Institute of Municipal Engineering [of APWA, US]
Institution of Mechanical Engineers [UK]
Instituto Militar de Engenharia [= Military Institute of Engineering, Brazil]
International Magnetosphere Explorer
International Microcomputer Exhibition
IMEA Incorporated Municipal Electrical Associations [UK]
IME/APWA Institute of Municipal Engineering/American Public Works Association [US]
IMEC Inter-University Microelectronics Center [Belgium]
IMECC Independent Metallurgical Engineering Consultants of California [slso: IMECCA, US]
IMechE Institution of Mechanical Engineers [UK]
IMEG International Management Engineering Group
IMEKO Internationale Messtechnische Konfederation [= International Measurement Confederation, Hungary]
IMEMME Institution of Mining Electrical and Mining Mechanical Engineers [UK]
IMEP Indicated Mean Effective Pressure
IMet Institute of Metals [UK]
IMEX International Mail Exchange
IMF Institute of Metal Finishing [UK]
International Marketing Federation
International Metalworkers Federation [Switzerland]
International Monetary Fund [US]
International Motorcycle Federation
IMFI Industrial Mineral Fiber Institute [US]
IMFP Inelastic Mean Free Path
IMG Imigration
IMGC Instituto di Metrologie 'G. Coloneti' [= G. Coloneti Institute of Metrology, Italy]
IMGCN Integrated Missile Ground Control Network
IMH Inlet Manhole
Institute of Materials Handling [UK]
IMI Ignition Manufacturers Institute
Improved Manned Interceptor
Institute for Marine Information [US]
Institute of the Motor Industry [UK]
Intermediate Machine Instruction
International Maintenance Institute [US]
International Market Intelligence
International Masonry Institute [US]
International Metaphysical Institute [UK]
Invention Marketing Institute [US]
Irish Management Institute
IMIA International Machinery Insurers Association [Germany]
International Medical Informatics Association [Canada]
IMIAT International Masonry Institute Apprenticeship and Training [US]
IMIC Industrial Minerals International Congress
International Medical Information Center [Japan]
IMIF International Maritime Industries Forum [UK]
IMINCO Iran Marine International Oil Company
IMinE Institution of Mining Engineers [UK]
IMIP Industrial Modernization Incentive Program
IMIR Interceptor Missile Interrogation Radar
IMIS Integrated Management Information System
IMIT Instituto Mexicano de Investigationes Tecnologicas [= Mexican Institute for Technological Investigations]
IMITAC Image Input to Automatic Computer
IML Incoming Matching Loss
Information Manipulation Language
Initial Machine Load
Inner Mold Line
Inside Mold Line
Interactive Maintenance Language
Intermediary Music Language
Intermediate Language
International Microgravity Laboratory [of NASA]
IML-1 First International Microgravity Laboratory [of NASA]
IMM Institute of Mathematic Machines [Poland]
Institution of Mining and Metallurgy [UK]
Integrated Maintenance Management
Integrated Maintenance Manual
Intelligent Memory Manager
International Monetary Market [US]
IMMA International Motorcycle Manufacturers Association [France]
Ion Microprobe Mass Analysis
IMME Institute of Mining and Metallurgical Engineers
IMMOA International Mercantile Marine Officers Associations [UK]
IMMP Integrated Maintenance Management Plan

IMMPS Institute for the Mechanics of Metal-Polymer Systems [USSR]
IMMRC International Mass Media Research Center
IMMS International Material Management Society [US]
IMMT Integrated Maintenance Management Team
IMMUN Immunological
Immunologist
Immunology
IMO International Maritime Organization [UK]
International Meteorological Organization [now: WMO, Switzerland]
International Miners Organization
IMOS Interactive Multiprogramming Operating System
Ion-Implanted Metal-Oxide Semiconductor
IMP Ice-Motion Package
Impact
Imperfection
Imperial
Import(er)
Improvement
Impulse
Industrial Management Plan
Information Management Program
Injection into Microwave Products
Inosine Monophosphate
Inosinic Acid
Insoluble Metaphosphate
Integrated Message Processor
Integrated Microprocessor
Integrated Microwave Products
Intelligent Machine Prognosticator
Intelligent Message Processor
Interface Message Processor
Interindustry Management Program
Intrinsic Multiprocessing
Interplanetary Monitoring Platform
IMPA Information Management and Processing Association
International Maritime Pilots Association [UK]
International Master Printers Association
International Motor Press Association [US]
IMPACT Immunization Monitoring Program, Active (Program) [of HWC, Canada]
Integrated Managerial Programming Analysis Control Technique
Inventory Management Program and Control Technique
IMPATT Impact Avalanche Transit Time (Diode)
Impact Ionization Avalanche Transit Time (Diode)
IMPC International Mineral Processing Congress
IMPCM Improved Capability Missile
Instituto Mexicano del Petroleo, Catalisis y Materiales [= Mexican Institute of Petroleum, Catalysis and Materials]
IMPCON Inventory Management and Production Control
IMPE International Meeting on Petroleum Engineering
IMPERF Imperfect(ion)
IMPG Importing
Impregnation
IMP GAL Imperial Gallon [also: imp gal]
IMPHOS Institut Mondial du Phosphate [= World Phosphate Institute, Morocco]
IMPI International Microwave Power Institute [US]
IMPL Implement
Implementation Language
Initial Microprogram Load
IMPN Importation
IMPR Improve(ment)
IMPREG Impregnated
IMPRESS Interdisciplinary Machine Processing for Research and Education in the Social Sciences
IMPRV Improvement
IMPS Imports
Integrated Master Programming and Scheduling System
Integrated Modular Panel System
Intelligent Management Programming System
Interface Message Processor
International Microprogrammers Society
IMPT Implement
Important
IMPTR Importer
IMPTS Improved Programmer Test Station
IMR Institute for Materials Research [of NIST, US]
Institute for Materials Research [Japan]
Institute of Metal Repair [US]
Institute of Mineral Research
Integrated Microscopy Resource [of University of Wisconsin, US]
Intelligent Machine Research
Internal Mold Release
Interruption Mask Register
Irreducible Matrix Representation
IMRA Incentive Manufacturers Representatives Association [US]
Independent Motocycle Retailers of America [US]
Infrared Monochromatic Radiation
International Manufacturers Representatives Association [US]
International Marine Radio Association
IMRADS Information Management Retrieval and Dissemination System
IMRAMN International Meeting on Radio Aids to Marine Navigation
IMRC International Marine Radio Committee
IMRF Independent Manufacturers Representatives Forum [US]
IMRI Industrial Materials Research Institute [of NRC, Canada]
IMRP International Meeting on Radiation Processing [Canada]
IMS Indian Mathematical Society
Industrial Management Service Group [of ORTECH, Canada]
Industrial Management Society [US]
Industrial Management System
Industrial Mathematics Society [US]
Inertial Measuring Unit
Information Management System
Institute for Marine Science [US]
Institute of Management Specialists [UK]

Institute of Materials Science
Institute of Mathematical Statistics [US]
Institut fur Mikroelektronische Schaltungen und Systeme [= Institute for Microelectronic Circuits and Systems [of Fraunhofer Institute, Germany]]
Instructional Management System
Instrumentation and Measurement Society [of IEEE, US]
Integrated Mechanical System
Integrated Meteorological System [US Army]
Interactive Market System [US]
Intermagnetic Shield
International Magnetospheric Studies
International Metallographic Society [US]
International Military Staff
International Mountain Society [US]
Inventory Management and Simulator
Ion Mobility Spectrometer
Ion Mobility Spectroscopy

IMSA International Medical Sciences Academy
International Municipal Signal Association [US]

IMSC International Mobile Satellite Conference

IMSE Institute for Materials Science and Engineering [of NBS, US]

IMSI Information Management System Interface

IMSL International Mathematical and Statistical Libraries

IMSP Integrated Mass Storage Processor

IMSR Interplanatary Mission Support Requirements

IMSSCE Interceptor Missile Squadron and Supervisory Control Equipment

IMSSS Interceptor Missile Squadron Supervisory Station

IMST International Mushroom Society for the Tropics [Hong Kong]

IMSU Integrated Mass-Storage Unit

IMS/VS Information Management System/Virtual Storage

IMT Immediate
Immersion Tank (Test) System
Immersion Testing
Industrial Materials Technologist
Industrial Materials Technology
Institute of Metal and Technology [of Dalian Maritime University, PR China]
Instituto Maua de Tecnologia [= Maua Institute of Technology, Brazil]
Integrated Micro-Image Terminal
Inter-Machine Trunk
Intermediate Tape
Ion Microtomography

IMTA Intensive Military Training Area
International Mass Transit Association [US]
International Marine Transport Association [US]

IMTC Instrumentation/Measurement Technology Conference [of IEEE, US]

IMTEG (European) ILS (= Instrument Landing System)/ MLS (= Microwave Landing System) Transition Group [of ICAO]

IMTS Improved Mobile Telephone Service
Improved Mobile Telephone System
International Machine Tool Show

IMU Increment Memory Unit
Inertial Measurement Unit
Instruction Memory Unit
International Mathematical Union [Finland]
International Metal Union [Switzerland]

IMun&CyE Institution of Municipal and County Engineers [UK]

IMunE Institution of Municipal Engineers [UK]

IMVIC Indole, Methyl Red, Voges-Proskauer, and Citrate Test [Biotechnology]

IMVS Institute of Medical and Veterinary Science [Australia]

IMW International Map of the World

IMWA International Mine Water Association [Spain]

IMX Instruction Memory Exchange [US]

IN Indiana [US]
Indigo
Inflammatory Response
Inlet
Input
Institute of Navigation [US]
Interference-to-Noise (Ratio)
Interior
Internal

In Indium

in inch

INA Initial Approach
Institute of Nautical Archeology [US]
Institution of Naval Architects [UK]
Integrated Network Architecture
International Normal Atmosphere

INAA Instrumental Neutron Activation Analysis

INAG Ionospheric Network Advisory Group

INAH Isonicotinic Acid Hydrazide
Isonicotinic Acid Hydrazine

INAOE Instituto Nacional de Astrofisica, Optica y Electronica [= National Institute for Astrophysics, Optics and Electronics, Mexico]

INAS Inertial Navigation and Attack System

INB Institut National du Bois [= National Wood Institute, France]

INBD Inboard
Inbound

INC Inclination
In Cloud
Inclusion
Inclusive
Incoming
Incorporate(d)
Incorporation
Increase
Increment
Input Control System
International Numismatic Commission

INCA International Newspaper and Color Association [Switzerland]

INCAP Institute of Nutrition for Central America and Panama

INCB International Narcotics Control Board [of UN — Austria]

INCC International Network Controlling Center

INCDT Incident

INCE Institute of Noise Control Engineering [US]
International Network in Chemical Education

INCEND Incendiary
INCH Integrated Chopper
INCIAWPRC Indian National Committee of the International Association on Water Pollution and Control
INCIN Incinerator
INCINC International Copyright Information Center [US]
INCIRS International Communication Information Retrieval System [of University of Florida, US]
INCL Inclosure
Include [also: INCLD]
Inclusion
Inclusive
INCO International Chamber of Commerce
INCODA International Congress of Dealers Associations [US]
INCOFILT International Consortium of Filtration Research Group
INCOLSA Indiana Cooperative Library Services Authority [US]
INCOM International Symposium on Manufacturing Technology [Canada]
INCOMEX International Computer Exhibition
INCOMPL Incomplete
INCONCRYO International Congress on Cryogenics
INCOPAC International Consumer Policy Advisory Committee [also: InCoPAC]
INCOR Indian National Committee on Oceanic Research [India]
Intergovernmental Conference on Oceanographic Research
Israel National Committee for Oceanographic Research
INCOSPAR Indian National Committee for Space Research [India]
INCOT In-Core Test Facility
INCPT Intercept
INCR Increase
Increment
Interrupt Control Register
INCRA International Copper Research Association [US]
INCRE Increment
INCUM Indiana Computer Users Meeting [US]
IND Independent
Index
India(n)
Indication
Indicator
Indigo
Indirect
Induction
Industrial
Industry
International Number Dialing
Investigative New Drug
Ind Indiana [US]
INDA International Nonwoven and Disposables Association [US]
INDC Indicate
INDE ITER (= International Thermonuclear Experimental Reactor)/NET (= Next European Torus) Design and Engineering
IndE Industrial Engineer [also: IndEng]
INDEF Indefinite
IND ENG Industrial Engineering
INDEP Independent
INDEX Inter-NASA Data Exchange
INDIAN Interplanar Distances and Angles
INDIC Indicated
INDIS Industrial Information System
INDITECNOR Instituto Nacional de Investigaciones Tecnologicas y Normalisacion [= National Institute for Technological Investigations and Standardization, Chile]
INDN Indication
INDO Intermediate Neglect of Differential Overlap
INDOR Internuclear Double Resonance
INDREG Inductance Regulator
INDS In-Core Nuclear Detection System
INDTR Indicator-Transmitter
Induced EMF Induced Electromotive Force [also: induced emf]
INDUSTR Industrial [also: INDUSTL]
INDV Individual
INDY Industry [also: IND'Y]
INE Institution of Nuclear Engineers [UK]
Instituto Nacional de Estadistica [= National Statistical Institute, Spain]
IN&EA International Nuclear and Energy Association [also: INEA, US]
INED Institut National d'Etudes Demographiques [= National Institute for Demographic Studies, France]
INEL Idaho National Engineering Laboratory [US]
Idaho Nuclear Engineering Laboratory [US]
Inelastic(ity)
International Exhibition of Industrial Electronics
INEOA International Narcotic Enforcement Officer Association [US]
INEPT Insensitive Nuclei Enhanced by Polarization Transfer
INET Intelligent Network Simulator
INETI Instituto Nacional de Engeharia e Tecnologia Industrial [= National Institute of Engineering and Industrial Technology, Portugal]
INEWS Integrated Electronic Warfare System
INF Infantry
Inferior
Information
Instituto Nacional de Farmacologia [= National Pharmacological Institute, Brazil]
Intermediate-Range Nuclear Forces (Treaty) [US-USSR]
inf infimum [Mathematics]
INFAC Instrumented Factory [of IIT, US]
InFACT Integrated Flexible Assembly Cell Technology
INFANT Iroquois Night Fighter and Night Tracker
INFCE International Nuclear Fuel Cycle Evaluation
INFCO Information Committee [of ISO, Switzerland]
INFIC International Network of Feed Information Centers [Australia]
INFINA Informatik fur die Industrielle Automation [= Congress on Informatics for Industrial Automation, Germany]
INFIRS Inverted File Information Retrieval System [of CIS, UK]

INFL Inflammability
Influence
INFLAM Inflammable
INFN Instituto Nazionale de Fisico Nucleare [= National Institute of Nuclear Physics, Italy]
INFO Information
Information Network and File Organization
INFOCEN Information Center
INFOCOM Information Communications [US]
INFOES In-Flight Operational Evaluation of Space Systems
INFOFILM International Information Film Service
INFOHOST Information on Hosts (Database) [Germany]
INFOL Information-Oriented Language
INFOLAC Information for Latin American Countries (Project) [of UN]
INFOMAT Information on Materials and Coatings
INFONET Information Network [UK]
INFOPAC Pacific Bell Information System [US]
INFOR Information-Oriented Language
INFOTERM International Information Center for Terminology [Austria]
INFOTEX Information via Telex [US]
INFRAL Information Retrieval Automatic Language
INFUS Infusible
ING Integrated Ground
Intense Neutron Generator
International Newspaper Group [US]
INGA Interactive Graphics Analysis
INGAA Interstate Natural Gas Association of America [US]
INGO International Nongovernmental Organization
InGPS Indoleglycerolphosphate Synthetase
INGRES Interactive Graphic and Retrieval System
INH Isonicotinic Hydrazide
in Hg inches of mercury
INHIGEO International Commission on the History of the Geological Science [France]
INICHAR Institut National de l'Industrie Charbonniere [= National Institute of the Coal Industry, Belgium]
INIS International Nuclear Information Service [of IAEA, Austria]
INIT Initial(ization)
Initiation
INJ Injection
INKA Information System Karlsruhe [Germany]
INKA-CONF Information System Karlsruhe — Conference Announcements [Germany]
INKA-CORP Information System Karlsruhe — Corporations [Germany]
INKA-DATACOMP Information System Karlsruhe — Data Compilations [Germany]
INKA-MATH Information System Karlsruhe — Database on Mathematics [Germany]
INKA-NUCLEAR Information System Karlsruhe — Database on Nuclear Science and Technology [Germany]
INKA-PHYS Information System Karlsruhe — Database on Physics [Germany]
INL Internal Noise Level
INLA International Nuclear Law Association [Belgium]
in lb inch-pound [also: in-lb]
INLC Initial Launch Capability
in lim (in limine) — at the outset
in loc (in locum) — in the place of
INM International Nautical Mile
INMAP Independent Microlelectronics Applications
INMARSAT International Maritime Satellite Organization [UK]
INMC International Network Management Center
in/min inches per minute [also: in min^{-1}]
INMM Institute of Nuclear Materials Management [US]
INN Institudo Nacional de la Nutricion [= National Institute of Nutrition, Argentina]
INO Indian Ocean Region [of ICAO]
Iterative Natural Orbital
INOP Inoperative
INORG Inorganic
INORG CHEM Inorganic Chemistry
INOSHAC Indian Ocean and Southern Hemisphere Analysis Center
INP If Not Possible
Inert Nitrogen Protection
Input
Institute of National Planning [Egypt]
Integrated Network Processor
Intelligent Network Processor
InP Indium Phosphide
INPADOC International Patent Documentation Center [Austria]
INKA (= Information System Karlsruhe) — Patent Documentation
INPC Isopropyl-N-Phenylcarbamate
INPE Instituto de Pesquisas Espaciais [= Brazilian Space Agency]
INPFC International North Pacific Fisheries Commission [Canada]
INPI Institut National de la Propriete Industrielle [= National Institute of Industrial Property, France]
INPL Institute National Polytechnique de Lorraine [= National Polytechnic Institute of Lorraine, France]
INPO Institute of Nuclear Power Operations [US]
INPR In Progress
INPT Institut National Polytechnique de Toulouse [= National Polytechnic Institute of Toulouse, France]
INQ Inquire
INQUA International Union for Quaternary Research [Switzerland]
INR Impact Noise Reduction
Indian Rupee [Currency of India]
Institut National de Radiodiffusion [= National Institute of Broadcasting, Belgium]
Interference-to-Noise Ratio
INRA Institute for Natural Resources in Africa
Institut National de la Recherche Agronomique [= National Institute for Agricultural Research, France]
INRAT Institute National de la Recherche Agronomique de Tunisie [= Tunesian National Institute for Agricultural Research]
INREQ Information on Request
in/rev inches per revolution [also: IPR or ipr]
INRIA Institut National de Recherche en Information et Automatique [= National Center for Information and Automation Research, France]

INRNE Institute of Nuclear Research and Nuclear Energy [Bulgaria]
INRO International Natural Rubber Organization [Malaysia]
International Naval Research Organization [US]
INROADS Information on Roads (Database) [of ARRB, Australia]
INROWASP In Rotating Water Spinning Process
INRV Institut National des Recherche Veterinaire [= National Institute of Veterinary Research, Belgium]
INS Immigration and Naturalization Service [US]
Indian Navy Ship
Inertial Navigation System
Information Network System [Japan]
Insert(ion)
Insoluble
Inspection
Inspector
Institute for Naval Studies
Institute for Nuclear Studies [Japan]
Insulation
Insurance
Integrated Networking System
International Navigation System
International News Service
Interstation Noise Suppression
Investors News Service
Ion Neutralization Spectroscopy
Iron Soldering
ins inches
in/s inches per second [also: in/sec or in sec^{-1}]
INSA Indian National Science Academy [India]
International Shipowners Association [Poland]
INSAR Instruction Address Register
INSAT India Satellite
INSATRAC Interception with Satellite Tracking
INSCIR Institut National Superieur de Chimie Industrielle de Rouen [= National Higher Institute for Industrial Chemistry of Rouen, France]
INSDOC Indian National Scientific Documentation Center [India]
in/sec inches per second [also: in sec^{-1}]
INSEE Institut National de la Statistique et des Etudes Economiques [= National Institute for Statistics and Economical Studies, France]
INSERM Institut National de la Sante et de la Recherche Medicale [= National Institute of Health and Medical Research, France]
INSFT Insufficient
INSG Institut National Polytechnique de Grenoble [= National Polytechnic Institute of Grenoble, France]
INSIS Inter-Institutional Integrated Services Information System [of the Commission of the European Communities]
INSITE Integrated Sensor Interpretation Techniques
INSJ Institute for Nuclear Studies, Japan
InSn Indium Antimonide
INSOL Insoluble
INSP Inspection
Inspector
INSPEC Information Services: Physics, Electrical and Electronics, and Computers and Control [of IEE, UK]
INSPEX Engineering Inspection and Quality Control Conference and Exhibition
INSPN Inspection
INSPR Inspector
INST Instantaneous
Instructor
Instrument
Institute
Institution
INSTAB Information Service on Toxicology and Biodegradability [of WPRL, UK]
Instability
INSTAL Instalment
INSTAR Inertialess Scanning, Tracking and Ranging
INSTARS Information Storage and Retrieval System
INSTBY Instability
InstCE Institute of Civil Engineers
InstE Institute of Electronics
INSTL Installation
Instalment
INSTLN Installation
InstM Institute of Metals [UK]
INSTMC Institute of Measurement and Control [UK]
InstME Institute of Mechanical Engineers
Institute of Mining Engineers
InstMM Institute of Mining and Metallurgy [UK]
INSTN Institution
Instruction
Institut National des Sciences et Techniques Nucleaires [= National Institute of Nuclear Science and Technology, France]
INSTOC Institute for the Study of the Continents [US]
INSTOP Institut National Scientifique et Technique d'Oceanographie et de Peche [= National Scientific and Technical Institute of Oceanography and Fishery, France]
InstP Institute of Physics
InstPet Institute of Petroleum Engineers
INSTR Instruction
Instrument
InstR Institute of Refrigeration [UK]
InstWE Institute of Water Engineers [UK]
INSULN Insulation
INT Institute for Nuclear Theory [of University of Washington, US]
Instituto Nacional de Technologia [= National Institute of Technology, Brazil]
Integer
Integration
Interest
Interference
Interior
Intermediate
Internal
International
Interphone
Interrupt
Interrogation

Interruption
Intersection
Interval
Iodonitrotetrazolium Violet
2-(4-Iodophenyl)-3-(4-Nitrophenyl)-5-Phenyltetrazolium Chloride
Isaac Newton Telescope
Iterative Numerical Technique

INTA Instituto Nacional de Tecnica Aerospacial [= National Institute of Aerospace Technology, Spain]
INTAG International Technology Management Advisory Group
INTAMEL International Association of Metropolitan City Libraries
INTAMIC International Microcircuit Card Association [France]
INTC Industrial Nuclear Technology Conference
INTCHG Interchangeability
INTCO International Code of Signals
INTCOM International Liaison Committee [of IEEE, US]
INTCP Intercept
INTD Institut National des Techniques de la Documentation [= National Institute of Documentation Techniques, France]
INTEBRID (Monolithic) Integrated and Hybrid Circuitry
INTEC Interface Technology
INTECOL International Association for Ecology [US]
INTECOM International Council for Technical Communication [UK]
International Society of Technical Communication [Denmark]
INTEL Integrated Electronics
Intelligence
INTEL-ED International Tele-Education [US]
INTELEVENT International Televent [US]
INTELPOST International Telecommunications Post [of INTELSAT]
INTELSAT International Telecommunications Satellite Organization [also: Intelsat, US]
INTEN Intensity
INTER Intermediate
Intermittent
Interruption
INTERALIS International Advanced Life Information System
INTERATOM Internationaler Atomreaktorbau [= International Construction of Nuclear Reactors, Germany]
INTERCOLOR International Commission for Fashion and Textile Colors [France]
INTERCOM Intercommunicating System [also: Intercom or intercom]
Intercommunication [also: Intercom or intercom]
INTERCOMP Intercompany
INTERCON International Convention [of IEEE, US]
INTERCOOP International Association of Consumer Cooperatives [Denmark]
INTERFRIGO International Railway-Owned Companies for Refrigerated Transport [Switzerland]
INTERGALVA International Galvanizing Conference
INTERGU Internationale Gesellschaft fur Urheberrechte [= International Copyright Society, Germany]
INTERKAMA Internationaler Kongress mit Austellung fur Mess- und Automatisierungstechnik [= International Congress with Exhibition for Instrumentation and Automation, Germany]
INTERLAINE Comite des Industries Lainieres [= Committee of the Wool Textile Industry [of EEC, Belgium]]
INTERMAG International Magnetics Conference [of IEEE, US]
INTERMARC International Machine-Readable Catalogue
INTERMARGEO International Organization for Marine Geology [USSR]
Internal EMF Internal Electromotive Force [also: internal emf]
INTERNET International Project Management Association [Switzerland]
INTERP Interpreter
INTERPACK International Fair for Packaging Machinery, Packaging Materials and Confectionery Machinery [Germany]
INTERPLAN International Group for Studies in National Planning
INTERPOL International (Criminal) Police Organization [also: Interpol, UK]
INTERPLAS International Plastics Exhibition and Conference
INTERSPUTNIK International Organization of Space Communication [USSR]
INTERTANKO International Association of Independent Tankers Owners [Norway]
INTERTEL International Television Federation
INTFU Interface Unit
INTG Integral
Integrated
INTGEN Interpreter Generator
INTI Instituto Nacional de Tecnologia Industrial [= National Institute of Industrial Technology, Argentina]
INTIM Interrupt and Timing
INTIME Interactive Textual Information Management Experiment
INTIP Integrated Information Processing
INTIPS Integrated Information Processing System
INTIST International Institute for Science and Technology
INTL Internal
INT'L International [also: INTL]
INTLK Interlock
INTMT Intermittent
INTOP International Operations Simulation
INTOR International Tokamak Reactor
International Torus
INTP Interpret
INTPER Interpreter
INTPHTR Interphase Transformer
INTR Interior
Interrupt
INTRAFAX Closed Circuit Facsimile System
INTRAN Input Translator
INTREDIS International Tree Disease Register [of USFS]
INTREX Information Transfer Experiment [of MIT, US]
INTRG Interrogator
INTRO Introduction
INTRP Interrupt
Interruption
INTS Intense

INTSF Intensify
INTSFY Intensify
INTUG International Telecommunications Users Group [UK]
INU Inertial Navigation Unit
INucE Institute of Nuclear Engineers [UK]
INV Invention
Inventor
Inventory
Inversion
Inverter
Investment
Invoice
INVAL Invalid
INVEST Investment
INVOF In the Vicinity Of
INVRN Inversion
INVT Inventory
Investment
INWATS Inward Wide Area Telephone Service
INWG International Network Working Group [of IFIPS]
in wg inch water gauge
INX Index Character
IO Information Organization
In Order
Input/Output
Institute for Oceanography [of ESSA, US]
Integrated Optics
Interpretive Operation
I/O Input/Output [also: IO or I-O]
Io Ionium (= Thorium 230)
IOA Input/Output Adapter
Institute of Outdoor Advertising [US]
International OMEGA (Radio Navigation System) Association [US]
International Ozone Association [US]
IOAP Internally-Oxidized Alloy Powder
IOAU Input/Output Access Unit
IOB Input/Output Buffer
Institute of Building [UK]
Interorganization Board [of UN]
IOBB International Organization for Biotechnology and Bioengineering [Guatemala]
IOBC Indian Ocean Biological Center
International Organization for Biological Control of Noxious Animals and Plants [France]
IOBFR Input/Output Buffer
IOBS Input/Output Buffering System
IOC Inclusion of the Overlap Charges
Indian Ocean Commission [Mauritius]
Indirect Operating Costs
Initial Operational Capability
Initial Operating Capability
Initial Orbiting Capability [of NASA]
Input/Output Channel
Input/Output Connector
Input/Output Control
Input/Output Controller
Input/Output Converter
Institute of Chemistry [UK]
Integrated Optical Circuit
INTELSAT (= International Telecommunications Satellite Organization) Operations Center [US]
Intergovernmental Oceanographic Commission [of UNESCO]
International Organization Committee
International Ornithological Congress [New Zealand]
Inter-Office Communication
I/OC Input/Output Controller
IOCA Independent Oil Compounders Association [now: ILMA, US]
IOCC Input/Output Control Center
Input/Output Control Command
Interstate Oil Compact Commission [US]
IOCD International Organization for Chemical Sciences in Development [Mexico]
IOCE Input/Output Control Element
IOCG Industrial Oil Consumers Group [US]
International Organization on Crystal Growth
IOCM Input/Output Control Module
Interim Operational Contamination Monitor (Experiment) [Space Shuttle]
IOCOM India-Malaysia Submarine Cable
IOC-ω Inclusion of the Ovcerlap Charges in the Omega
IOCP Input/Output Connection Panel
IOCS Input/Output Control Subroutine
Input/Output Control System
IOCTR Input/Output Controller
IOCTF IOC (= Input/Output Control) TDMA (= Time-Division Multiple Access) Facility
IOCU International Organization of Consumers Unions [Netherlands]
IOCWMC International Organizing Committee of the World Mining Congress
IOCV International Organization of Citrus Virologists [US]
IOD Identified Outward Dialing
Information on Demand
Input/Output Device
Institute of Directors [UK]
IODE International Oceanographic Data Exchange
IODS International Ocean Disposal Symposium [US]
IOE Indian Ocean Expedition
Industrial and Operations Engineer
Institute of Energy [UK]
International Organization of Employers
IOEH Institute of Occupational and Environmental Health [Canada]
IOF Independent Order of Foresters
Input/Output Front-End
Institute of Fuel [UK]
Interactive Operations Facility
International Oceanographic Foundation [US]
IOFC Indian Ocean Fishery Commission [Italy]
IOFI International Organization of the Flavour Industry [Switzerland]
I of M Isle of Man [UK]
I of W Inspector of Works
Isle of Wight [UK]
IOGA Industry-Organized, Government-Approved
IOGEN Input/Output Generation
IOH Input/Output Handler

IOI International Ocean Institute [Malta]
International Ozone Institute [now: IOA, US]
IOIE International Organization of Industrial Employers
IOIH Input/Output Interrupt Handler
IOITBAG International Oil Industry TBA (= Tires, Batteries, Accessories) Group [Canada]
IOJ Institute of Journalists [UK]
IOL Instantaneous Overload
Intra-Ocular Lens
IOLA Input/Output Line Adapter
Input/Output Link Adapter
IOLC Input/Output Link Controller
IOLIM International On-Line Information Meeting [UK]
IOLS Input/Output Label System
IOM Including Other Minerals
Input/Output Multiplexer
Institute of Medicine [US]
Institute of Metals [UK]
Institute of Occupational Medicine [UK]
Institute of Office Management [UK]
International Organization for Mycoplasmology [US]
Isle of Man [UK]
I/OM Input/Output Multiplexer
IOMA International Oxygen Manufacturers Association [US]
IOMP Input/Output Microprocessor
International Organization for Medical Physics [UK]
IOMTR International Organization for Motor Trades and Repairs [Netherlands]
IOMVM International Organization of Motor Vehicle Manufacturers [France]
ION Institute of Navigation [US]
Ionosphere and Aural Phenomena Advisory Committee [of ESRO, France]
IOO Input/Output Operation
IOOC International Olive Oil Council [Spain]
IOOP Input/Output Operation
IOP Ibero-American Organization of Pilots [Mexico]
Industrial Opportunities Program
Input/Output Processor
Institute of Packaging [UK]
Institute of Petroleum
Institute of Physics [UK]
Institute of Plumbing [UK]
Institute of Printing [UK]
Integrated Optics Processor
International Organization of Paleobotany [UK]
IOPAB International Organization for Pure and Applied Biophysics
IOPB International Organization of Plant Biosystematists [Switzerland]
IOPC International Oil Pollution Compensation Fund
IOPG Input Output Processor Group
IOPH International Office of Public Health
IOPKG Input/Output Package
IOPS Input/Output Programming System
IOQ Input/Output Queue
Institute of Quarrying [UK]
IOR Index of Refraction
Indian Ocean Region
Industrially Oriented Research
Input/Output Register
Institute for Operational Research [UK]
IORB Input/Output Record Block [also: IORCB]
IOREQ Input/Output Request
IORT Incremental Oil Revenue Tax
IOS Input/Output Selector
Input/Output Subsystem
Input/Output Supervisor
Input/Output System
Institute of Oceanographic Sciences [UK]
Institute of Ocean Sciences
Interactive Operating System
International Organization for Standardization
International Organization for Succulent Plant Study [UK]
Investors Overseas Service
IOSA Integrated Optic Spectrum Analyzer
International Oil Scouts Association [US]
Irish Offshore Services Association
IOSC International Oxygen Steelmaking Congress
IOSCD International Organization for Scientific Cooperation and Development
IOSEWR International Organization for the Study of the Endurance of Wire Ropes [France]
IOSS Input/Output Subsystem
IOSTE International Organization of Science, Technology and Education
IOT In-Orbit Test Antenna
Input/Output and Transfer
Input/Output Transfer
Input/Output Trunk
Institute of Transport [UK]
IOTA Information Overload Testing Apparatus
International Occultation Timing Association [US]
IoTA Institute of Transport Administration [UK]
IOTA/ES International Occultation Timing Association — European Section [Germany]
IOTCG International Organization for Technical Cooperation in Geology [USSR]
IOT&E Initial Operation Test and Evaluation
IOTG Input/Output Task Group [of CODASYL]
IOU Immediate Operation Use
Input/Output Unit
Input/Output Utility
I Owe You
IOUBC Institute of Oceanography, University of British Columbia [Canada]
IOUS Input/Output Utility Subsystem
IOVC In the Overcast
IOVST International Organization for Vacuum Science and Technology
IOW Institute of Welding
IOZ Internal Oxidation Zone
IP Ice Point
Identification Point
Identification of Position
Igneous Petrology
Ignition Point
Image Processing
Imaginary Part
Imipramine

Impact Printer
Incisoproximal
Index of Performance
Induced Polarization
Industrial Park
Industrial Production
Information Processing
Information Processor
Information Professional
Information Provider
Initial Phase
Initial Point
Inosine Phosphate
Input Processor
Institute of Physics [Taiwan]
Institute of Petroleum [UK]
Instruction Pointer
Instruction Processor
Integer Prpgramming
Intellectual Property
Interdigital Pause
Interface Processor
Intermediate Pressure
Internal Pressure
Internet Protocol
Interplanetary
Ionic Polymer
Ion Plating
Ion Pump
Item Processing
Powder Injection

I&P Indexed and Paged
Inerting and Preheating

I/P Input
I (= Current) to P (= Pressure)
Irregular Input Process

I-P Intermediate-Pressure

I-5'-P Inosine 5'-Monophosphate

IPA Illinois Pharmaceutical Association [US]
Indiana Pharmaceutical Association [US]
Industrial Perforators Association [US]
Information Processing Architecture
Information Processing Association [Israel]
Information-Technology Promotion Agency [Japan]
Institute for Polyacrylate Absorbents [US]
Institute of Public Affairs [Australia]
Institut fur Produktionstechnik [= Institute for Production Technology [of Fraunhofer Institute, Germany]]
Integrated Peripheral Adapter
Integrated Photodetection Assembly
Integrated Printer Adapter
Intermediate Power Amplifier
International Paleontological Association [US]
International Patent Agreement
International Permafrost Association
International Pharmaceutical Abstracts [of ASHP]
International Phonetic Alphabet
International Phonetic Association [UK]
International Photographers Association
International Pipe Association [US]
International Platform Association
International Prepress Association [US]
International Press Association
International Publishers Association [Switzerland]
Iowa Pharmaceutical Association [US]
Isepentenyl Adenosine
Isophthalic Acid
Isopropyl Acetate
Isopropyl Alcohol
Isopropyl Amine

IPAA Independent Petroleum Association of America [US]
International Pesticide Applicators Association [US]

IPAC Independent Petroleum Association of Canada
Institute of Public Administration of Canada
Institut Polytechnique de l'Afrique Centrale [= Polytechnic Institute of Central Africa]
International Peace Academy Committee [US]

IPAD Integrated Program for Aerospace Vehicle Design
International Plastics Association Directors

IPADAE Integrated Passive Action Detection Acquisition Equipment

IPAE Isopropylaminoethanol

IPAI International Primary Aluminum Institute [UK]

IPAM Isopropylacrylamide

IPAR Intra-Pulse Analysis Receiver

IPARA International Publishers Advertising Representatives Association

IPB Illustrated Parts Breakdown
Integrated Processor Board
Interprocessor Buffer
Isopropyl Benzenc

IPBAM International Permanent Bureau of Automotive Manufacturers

IPBC International Power Beam Conference

IPC Idaho Potato Commission [US]
Independent Control Point
Industrial Personal Computer
Industrial Process Control
Industry Planning Council [US]
Industry Policy Council [US]
Information Processing Center
Information Processing Code
Institute for Interconnecting and Packaging Electronic Circuits [US]
Institute for Personal Computing [US]
Institute of Paper Chemistry [US]
Institute of Paper Conservation [UK]
Institute of Printed Circuits [Name Changed to: Institute for Interconnecting and Packaging Electric Circuits, US]
Instituto de Plasticos y Caucho [= Institute for Plastics and Rubber, Spain]
Integrated Peripheral Channel
Integrated Printer Adapter
Integrated Process Control
Integrated Program for Commodities [of UN]
Intermediate Processing Center
Intermittent Position Control
International Patent Classification
International Photosynthesis Committee [Sweden]

International Poplar Commission [Italy]
International Pressure Conference
International Programmable Control Conference (and Exposition)
Interprocess Communication
Interprocess Controller
Interprocess Coupler
Interprocessor Communication
Ion-Pair Chromatography
Isoproyl Carbinol
Isopropyl Cresol
Isopropyl-N-Phenylcarbamate

Ipc Isopinocampheyl

IPCA Industrial Pest Control Association [UK]
International Petroleum Cooperative Alliance

Ipc$_2$BCl Chlorodiisopinocampheylborane

(IpcBH$_2$)CTMEDA Monoisopinocampheylborane Tetramethylethylenediamine

IPCCS Information Processing in Command and Control Systems

IPCEA Insulated Power Cable Engineers Association [now: ICEA, US]

IPCF Interprocess Communication Facility

IPCR Institute of Physical and Chemical Research

IPCRI Israel Palestine Center for Research Information

IPCS Institution of Professional Civil Servants [UK]
Interactive Problem Control System
International Program on Chemical Safety [Switzerland]

IPD Insertion Phase Delay
Isophoronediamine

IPDC International Program for Development of Communications [of UNESCO]

IPDD Initial Project Design Description

IPDE Identify, Predict, Decide, Execute

IPDG Isopropyl-(β-D-)Thiogalactopyranoside

IPDH In-Service Planned Derated Hours

IPDI Isophorone Diisocyanate

IPDP Industrial Programmed Data Processor
Intervals of Pulsations of Diminishing Period

IPDU Internet Protocol Data Unit

IPE Industrial Plant Equipment
Information Processing Equipment
Institute of Power Engineers
Institution of Plant Engineers [UK]
Institution of Production Engineering [UK]
Institution of Professional Engineers [New Zealand]
International Petroleum Exchange
International Political Economy
Interpret Parity Error
Inverse Photoemission Experiment
Isopropoxyethanol

IPENZ Institution of Professional Engineers of New Zealand

IPEP International Permanent Exhibition of Publications

IPETE International Petroleum Equipment and Technology Exhibition

IPETEX Institute of Petroleum Working Group on Petroleum Exploration Training [UK]

IPF Indicative Planning Figures
Inherent Power Factor
International Pharmaceutical Federation
International Powerlifting Federation

IPFA International Professional Security Association [UK]

IPFC Indo-Pacific Fishery Commission [Thailand]

IPFM Integral Pulse Frequency Modulation

IPG Information Policy Group [of OECD, France]
Interactive Presentation Graphics

IPGS Industrial Postgraduate Scholarship [of NSERC, Canada]

IPH Inches per Hour [also: iph]
International Association of Paper Historians [Germany]

IPHC International Pacific Halibut Commission [US/Canada]

IPHE Institution of Public Health Engineers [UK]

IPI Individually Presented Instruction
Industrial Production Index
Institute of Polymer Industry [Japan]
Integrated Permits and Inspections
Intelligent Printer Interface
Interchemical Printing Ink
International Pesticide Institute [US]
International Potash Institute [Switzerland]
International Press Institute

IpI Inosylylinosine

IPIECA International Petroleum Industry Environmental Conservation Association [UK]

IPIN Instituto Panamericano de Ingeniera Naval [= Pan-American Institute of Naval Engineers, Brazil]

IPIP International Personhood of Iliterate Programmers [US]

IPIRA Indian Plywood Industries Research Association

IPIRI Indian Plywood Industries Research Institute

IPL Image Processing Laboratory [of NASA]
Information Processing Language
Initialize Program Load
Initial Program Load(er)
Inner Plexiform Layer
Integrated Payload
Interrupt Priority Level

IPLA Interstate Producers Livestock Association [US]

IPlantE Institute of Plant Engineers [UK]

IPLC International Private Leased Circuit [UK]

IPLCA International Pipeline Contractors Association [France]

IPLO Institute of Professional Librarians of Ontario [Canada]

IPL-V Information Processing Language V

IPM Illuminations per Minute
Impulses per Minute
Inches per Minute [also: ipm]
Inches Penetration per Month
Incidental Phase Modulation
Incident Power Meter
Industrial Productivity Monitoring
Input Position Map
Input Position Mapper
Institute for Practical Mathematics [Germany]
Institute of Personnel Management [UK]
Intelligent Processing of Materials
Interference Prediction Model

	Internal Polarization Modulation
	Interpersonal Message
	Interruptions per Minute
IPMA	In-Plant Printing Management Association [US]
	International Personnel Management Association [US]
IPMI	International Powder Metallurgy Institute [US]
	International Precious Metals Institute [US]
IPMS	International Polar Motion Service
IPN	Initial Priority Number
	Initial Processing Number
	Inspection Progress Notification
	Instant Private Network
	Instituto Polytecnico Nacional [= National Polytechnic Institute, Mexico]
	Internal Priority Number
	Interpenetrating Network
	Interpenetrating Polymer Network
IPNJ	Industrial Photographers of New Jersey [US]
IPNL	Integrated Perceived Noise Level
IPNS	Intense Pulsed Neutron Source [of Argonne National Laboratory, US]
IPO	Indolephenoloxidase
IPOD	International Phase of Ocean Drilling
IPOEE	Institution of Post Office Electrical Engineers [UK]
IP/OP	Input/Output Interface
IPOT	Inductive Potentiometer [also: I-POT]
IPP	Imaging Photopolarimeter
	Impact Prediction Point
	In-Plant Plus Program [of ASM International, US]
	Institute of Plasma Physics
	Integrated Plotting Package
	Interface Package Process
	International Phototelegraph Position
	Inter-Processor Process
IPPA	Independent Professional Painting Contractors Association (of America) [US]
	Indo-Pacific Prehistory Association [Australia]
	International Printing Pressmen and Assistants Union
IPPC	Industrial Promotion and Productivity Center
IPPD	N-Isopropyl-N'-Phenyl-P-Phenylene Diamine
IPPDSEU	International Plate Printers, Die Stampers and Engravers Union [Canada and US]
IPPF	Instruction Preprocessing Function
IPPJ	Institute of Plasma Physics, Japan
IPPNW	International Physicians for the Prevention of Nuclear War [UK]
IPPS	Institute of Physics and the Physical Society [UK]
IPPTA	Indian Pulp and Paper Technical Association [India]
IPQC	In-Process Quality Control
IPR	Inches per Rack [textiles]
	Inches per Revolution [also: ipr or in/rev]
	Inflow Performance Relationship
	In Pulse to Register
	Institute for Polymer Research [Canada]
	Institute of Population Registration [UK]
	Institute of Public Relations [UK]
iPr	Isopropyl
$(iPr)_2$	Diisopropyl
IPRA	International Peace Research Association [Netherlands]
	International Public Relations Association [Switzerland]
IPRC	International Personal Robot Congress (and Exposition)
IPRE	International Professional Association for Environmental Affairs [Belgium]
IPRO	International Patent Research Office
IProdE	Institution of Production Engineers [UK]
IPRT	Institute for Physical Research and Technology [of Iowa State University, US]
IPS	Image Processing System
	Impact Polystyrene
	Impact Predictor System
	Impulses per Second [also: ips]
	Inches per Second [also: ips]
	Incorporated Phonographic Society [UK]
	Information Processing Society [Japan]
	In Pulse to Sender
	Installation Performance Specification
	Institute of Purchasing and Supply [UK]
	Instructions per Second
	Instrumentation Power Supply
	Instrumentation Power System
	Integrated Photosector
	Integrated Power System
	Integrated Process System
	Intelligent Printing System
	Intelligent Programming System
	Interconnected Power System
	Interconnected Power Systems [US]
	Interim Policy Statement
	Interior Pipe Size
	International Paleontological Society [US]
	International Palm Society [US]
	International Peat Society [Finland]
	International Phycological Society [US]
	International Physical Society [Italy]
	International Pipe Standard
	International Planetarium Society [US]
	International Pressure Society [US]
	Interruptions per Second
	Invariant Plane Strain
	Ionospheric Prediction Service
	Iron Pipe Size
	Israel Physical Society [Israel]
i-PS	Isotactic Polystyrene
IPSA	International Passenger Ship Association
IPSB	Interprocessor Signal Bus
IPSE	Integrated Programming Support Environment
	Integrated Project Support Environment
IPSEP	International Project for Soft Energy Paths
IPSF	International Pharmaceutical Students Federation [Israel]
IPSFC	International Pacific Salmon Fisheries Commission
IPSJ	Information Processing Society of Japan
IPSM	Institute of Physical Sciences in Medicine [UK]
IPSMCB	International Product Safety Management Certification Board [US]
IPSO	Interface Peripheral Standard Olivetti

IPSOC Information Processing Society of Canada
IPSP Inhibitory Postsynaptic Potential
IPSS International Packet Switching Service [UK]
International Packet Switching System
International Power Sources Symposium [UK]
Inter-Processor Signaling System
IPSSB Infomation Processing Systems Standards Board [of ANSI, US]
IPST International Practical Scale of Temperature
Israel Program for Scientific Translations
IPT Inches per Tooth [also: ipt]
Indexed, Paged and Titled
Internal Pipe Thread
International Pipe Thread
Isopropyl Toluene
IPTA International Patent and Trademark Association [US]
IPTC International Press Telecommunications Council [UK]
IPTG Isopropylthiogalactopyranoside
IPTM Interval Pulse Time Modulation
IPTO Information Processing Technologies Office [of ARPA, US]
IPTS International Practical Temperature Scale [of NIST, US]
IPU Institute of Public Utilities [US]
Instruction Processing Unit
International Paleontological Union [now: IPS, US]
Interprocessor Unit
Islamic Press Union
IpU Inosylyluridine
IPV Improve
IPVC Irradiated Polyvinyl Chloride
IPWF International Public Works Federation
IPY Inches Penetration per Year
International Polar Year
IQ Institute of Quarrying [UK]
Insured Quality
Intelligence Quotient
iq (idem quod) — the same as
IQA Institute of Quality Assurance [UK]
Instituto de Qualidade Alimentar [= Institute of Food Quality, Portugal]
Instituto de Quimico Agricola [= Institute of Agricultural Chemistry, Brazil]
Irish Quality Association
IQB Instituto de Quimico Biologica [= Institute of Biological Chemistry, Brazil]
IQC International Quality Center [of EOQC, Switzerland]
IQD Iraqi Dinar [Currency]
IQEC International Quantum Electronics Conference [of IEEE, US]
iqed (id quod erat demonstrandum) — that which was to be proved
IQF Interactive Query Facility
IQI Image Quality Indicator
IQL Information Query Language
Interactive Query Language
IQM Input Queue Manager
IQMH Input Queue Message Handler
IQPF International Quick Printing Foundation
IQPP Interactive Query Preprocessor
IQR Interquartile Range
IQS Institute of Quantity Surveyors [UK]
International Quality Study [Canada]
IQSY International Quiet Sun Year
IR Image Restoration I (= Current) — R (= Resistance)
Index Register
Inductive Resistor
Industrial Relations
Industrial Research
Industrial Robot
Informal Report
Information Retrieval
Infrared
Infrared Radiation
Inland Revenue
Inside Radius
Insoluble Residue
Instantaneous Relay
Instruction Register
Instrument Rating
Instrument Reading
Insulation Resistance
Intelligence Ratio
Interference Refractometer
Intermediate Register
Internal Representation
Internal Resistance
Interrogator-Responder
Interrupt Register
Interrupt Request
Ionizing Radiation
Ireland
Irish
Isoprene Rubber
I/R Instrument Rating
Ir Iridium
IRA Information Resource Administration
Input Reference Axis
Institute of Registered Architects [UK]
Internal Reflection Attachment
International Reprographics Association
International Rubber Association [Malaysia]
IR-A Infrared, Range A [from 780 to 1400 nm]
IRABOIS Institut de Recherches Appliquees au Bois [= Institute for Applied Wood Research, France]
IRAC Information Resource and Analysis Center [US]
Infrared Array Camera
Interdepartmental Radio Advisory Committee [US]
International Records Administration Conference [US]
IRACQ Infrared Acquisition Radar
Instrumentation Radar and Acquisition Panel
IRAD Independent Research and Development
IRAH Infrared Alternate Head
IRAM Indexed Random Access Memory
Institute of Research and Application of Development Methods [France]
IRAMS Infrared Automatic Mass Screening

IRAN Inspection and Repair as Necessary
IRANDOC Iranian Documentation Center
IRAP Industrial Research Assistance Program [of NRC, Canada]
Interagency Radiological Assistance Plan [of USAEC]
Interagency Radiological Assistance Program
IRAS Information Retrieval Advisory Services [UK]
Infrared Absorption Spectroscopy
Infrared Astronomical Satellite [of NASA]
IRASA International Radio Air Safety Association
IRATE Interim Remote Air Terminal Equipment
IRB Infinitely Rigid Bear
Infrared Brazing
Institute of Radiation Breeding [Japan]
Interruption Request Block
IR-B Infrared, Range B [from 1.4 to 3 μm]
IRBEL Indexed References to Biomedical Engineering Literature [of NIMR, UK]
IRBIC Infrared Beam Induced Contrast
Infrared Beam Induced Current
IRBM Intermediate-Range Ballistic Missile
IRBO Infrared Homing Bomb
IRC Incremental Related Carrier
Industrial Relations Center
Industrial Relations Councelors [US]
Industrial Relations Council [Canada]
Industrial Research Chairs (Program) [of NSERC, Canada]
Industrial Reorganization Corporation
Information Resource Center
Information Retrieval Center
Infrared Camera
Inland Revenue Commissioner [UK]
Instant Response Chromatography
Institute for Research in Construction [of NRC, Canada]
Interdisciplinary Research Center [of ICSTM, UK]
Intermediate Routing Center
International Radiation Commission [UK]
International Record Carrier
International Reference Center (for Community Water Supply and Sanitation) [Netherlands]
International Rescue Commission
International Research Council
International Rice Commission [Italy]
International Rubber Conference [Germany]
Iron Ring Compressor
IR-C Infrared, Range C [from 3 μ to 1 mm]
IRCA Immigration Reform and Control Act [US]
International Radio Club of America
International Railway Congress Association [Belgium]
International Remodeling Contractors Association [US]
IRCC Instrument Repair and Calibration Center [Thailand]
International Radio Consultative Committee [US]
IRCHMB International Research Center on Hydraulic Machinery Beijing [PR China]
IRCIHE International Referral Center for Information Handling Equipment [of UNESCO]
IRCL International Research Center on Lindane [Belgium]
IRCM Infrared Countermeasures
IRCN Institut de Recherche de la Construction Navale [= Research Institute for Naval Construction, France]
IRCO International Rubber Conference Organization [UK]
IRCOBI International Research Council on the Biokinetics of Impacts [France]
IRCOL Institute for Information Retrieval and Computational Linguistics [Israel]
IRCS International Research Communications System [UK]
IRCT Institut de Recherche du Coton et des Textiles Exotiques [= Research Institute for Cotton and Exotic Textiles, France]
IRCWD International Reference Center for Water Disposal
IRD Infrared Detector
International Resource Development [US]
IR&D Independent Research and Development
Industrial Research and Development
IRDA Industrial Research and Development Authority [UK]
IRDB Information Retrieval Databank
IRDC International Research Development Center
International Rubber Development Committee [UK]
IRDF Indentation Residual Deflection Force
IRDIA Industrial Research and Development Incentives Act [Canada]
IRDO Intermediate Retention of Differential Overlap
IRDOME Infrared Dome [also: Irdome or irdome]
IRDP Industrial and Regional Development Program [Canada]
Industrial Research Development Program
IRDS International Road Documentation Scheme [of OECD, France]
IRE Institute of Radio Engineers [now: IEEE, US]
Institute of Refractories Engineers [UK]
Institution of Radio Engineers [UK]
Instrument Rating Examiner
Internal Reflection Element
Ireland
IREB Intense Relativistic Electron Beam
IREC Irrigation Research and Extension Advisory Committee [Australia]
IRED Infrared
Infrared Emitting Diode
IREDA International Radio Electrical Distributors Association
IREE Institute of Radio and Electric Engineers [Australia]
IREEA Institute of Radio Engineering, Electronics and Automation [USSR]
IREP Integrated Reliability Evaluation Program
Interim Reliability Evaluation Program
IREQ Institut de Recherche en Electricite du Quebec [= Institute of Electrical Research of Quebec, Canada]
IRES Infrared Emission Spectroscopy
IREX International Research and Exchanges Board [US]

Iron-Ring Experiment

IRF Industrial Research Fellowship
Input Register Full
International Road Federation [US]

IRFA Institut de Recherches sur les Fruits et Agrumes [= Research Institute for Fruits and Tropical and Subtropical Fruits, France]

IRFC Intermediate Range Function Test

IRFITS Infrared Fault Isolation Test System

IRFNA Inhibited Red Fuming Nitric Acid

IRFPA Infrared Focal-Plane Array (Producibility Program) [US Army]

IR FPA Infrared Focal-Plane Array

IRG Information Retrieval Group
International Research Group
Interrange Instrumentation Group
Interrecord Gap

IRGA Infrared Gas Analyzer

IRgA International Reprographics Association [US]

IRGRD International Research Group on Refuse Disposal

IRGWP International Research Group on Wood Preservation [Sweden]

IRH Inductive Recording Head

IRHA International Rural Housing Association

IRHD International Rubber Hardness Degrees

IRI IBEC (= International Bank for Economic Cooperation) Research Institute [US]
Industrial Research Institute [US and Japan]
Institution of the Rubber Industry [UK]
International Robomation/Intelligence
Iranian Rial [Currency]
Islamic Research Institute

IRIA Infrared Information and Analysis Center [US]

IRICON Infrared Vidicon Tube

IRID Iridescent [also: IRIDESC]

IRIG Interrange Instrumentation Group [of USDOD]

IRII Industrial Research Institute of Ishikawa [Japan]

IRIP Industrial Research Institutes Program

IRIS Industrial Relations Information Service
Infrared Information Symposia [US]
Infrared Interferometer Spectrometer
Instant Response Information System
Instructional Resources Information System [of Ohio State University, US]
Integrated Radar Imaging System
Integrated Reconnaissance Intelligence System
Interactive Recorded Information Service [UK]
International Research Information Service [US]
Interrogation and Location

IRJE Interactive Remote Job Entry

IRL Industrial Research Laboratory
Information Retrieval Language

IRLA Independent Research Libraries Association [US]

IRLCO International Red Locust Control Organization

IRLCO-CSA International Red Locust Control Organization for Central and Southern Africa [Zambia]

IRLS International Red Locust Control Service
Interrogation, Recording and Location Subsystem
Interrogation, Recording and Location System [of NASA]

IRM Image Rejection Mixer
Information Resources Management
Infrared Measurement
Inorganic Reaction Mechanism
Institute of Risk Management [UK]
Interim Research Memorandum
Intermediate Range Monitor
International Resource Management [US]
Iodine Radiation Monitor
Isothermal Remanent Magnetization

IRMA Indian Refractory Makers Association [India]
Information Revision and Manuscript Assembly
Infrared Miss-Distance Approximeter
International Rail Makers Association

IRMB Institute Royal Meteorologique de Belgique [= Royal Meteorological Institute of Belgium]

IRMC International Radio-Maritime Committee [UK]

IRMGARD Information Resources Management Group and Regulatory Directorate

IRMI Industrial Research Materials Institute
International Risk Management Institute

IRMIS Integrated Resource Management Information System [Canada]

IRMP Infrared Measurement Program
Interservice Radiation Measurement Program

IRMS Information Retrieval and Management System

IRN Internal Routing Network
International Rivers Network [US]

IRNP Identifying Research Needs Program [of ASME, US]

IRO Industrial Relations Office

IROD Instantaneous Readout Detector

IRODP International Registry of Organization Development Professionals [US]

IROR Incremental Rate of Return
Internal Rate of Return

IROS Increased Reliability Operational System
Instant Response Order System [US]

IRP Infrared Photography
Initial Receiving Point
International Research Program
International Routing Plan

IR£ Irish Pound [Currency]

IRPA Irrigation Pump Administration [Philippines]
International Radiation Protection Association [France]

IRPAS Infrared Photoacoustic Spectroscopy

IRPBDS Infrared Photothermal Beam Deflection Spectroscopy

IRPI Individual Rod Position Indicator

IRPM Infrared Physical Measurement

IRPP Institute for Research on Public Policy

IRPTC International Register of Potentially Toxic Chemicals

IRQ Interrupt Request

IRR Institute for Reactor Research [Switzerland]
Integral Rock Ramjet
Internal Rate of Return
Interrupt Return Register
Iranian Rial [Currency]
Israel Research Reactor

IRRA Industrial Relations Research Association [US]

IRRAD Infrared Range and Detection

Irradiation
IRRAS Infrared Reflection-Absorption Spectroscopy
IRRB International Rubber Research Board
IRRD International Road Research Documentation [of OECD, France]
IRRDB International Rubber Research Development Board [UK]
IRREG Irregular
IRRI International Rice Research Institute [Philippines]
IRRL Information Retrieval Research Laboratory [of University of Illinois, US]
IRRMP Infrared Radar Measurement Program
IRRS Infrared Reflection Spectroscopy
IR-RS Infrared Reflow — Solderable
IRRT Institution of Rail and Rapid Transit [UK]
IRS Independent Research Service
Inertial Reference System
Infinitely Rigid System
Information Receiving Station
Information Retrieval Service [of ESA, France]
Information Retrieval System
Infrared Soldering
Infrared Spectroscopy
Infrared Spectrum
Inquiry and Reporting System
Insulated Return System
Interchange Record Separator
Intermediate Reference System
Internal Reflection Spectroscopy
Internal Revenue Service [US]
International Referral System (Database) [of UNEP, Kenya]
International Repeater Station
Interrecord Separator
Investment Removal Salt
Irrigation Research Station [New Zealand]
Isotope Removal Service
Isotope Removal System
IRSA International Radiator Standards Association [US]
IRSC Inter-Regional Subject Coverage
IRSCS Inter-Regional Subject Coverage Scheme
IRSE Institution of Railway Signal Engineers [UK]
IRSFC International Rayon and Synthetic Fibers Committee [France]
IRSG Information Retrieval Specialists Group [of British Computer Society, UK]
International Rubber Study Group [UK]
IRSIA Institut pour l'Encouragement de la Recherche Scientifique dans l'Industrie et l'Agriculture [= Institute for the Promotion of Scientific Research in Industry and Agriculture, Belgium]
IRSID Institut de Recherches de la Siderurgie [= Research Institute of the Iron and Steel Industry, France]
IRSS Inertial Reference Stabilization System
Intelligent Remote Station Support
IRST Infrared Search and Track
IRSTD Infrared Search and Target Designation System
IRSTS Infrared Search-Track System
IRT Index Return (Character)
Infrared Telescope
Infrared Tracker
In-Reactor Thimble
Institut de Recherche des Transports [= Institute for Transportation Research, France]
Institute of Reprographic Technology [UK]
Interrogator-Responder-Transponder
Interrupted Ring Tone
IRTA International Reciprocal Trade Association [US]
IRTC Infrared Thermocouple
International Road Tar Conference
International Round Table Conference
IRTCES International Research Training Center for Erosion and Sedimentation
IRTE Institute of Road Transport Engineers [UK]
IRTO International Radio and Television Organization
IRTS Interim Recovery Technical Specification
International Radio and Television Society [US]
IRTU Intelligent Remote Terminal Unit
IRTV Information Retrieval Television [US]
IRTWG Interrange Telemetry Working Group
IRU Inertial Reference Unit
Intel Real-Time Users Group [also: iRUG]
International Radium Unit
International Raiffeisen Union [Germany]
International Reference Unit
International Road Transport Union [Switzerland]
IRV Interrupt Request Vector
IRVR Instrumented Runway Visual Range
IRW Indirect Reference Word
Institute of Rural Water
Inverted Rib Waveguide
IRWA International Right of Way Association [US]
IRX Interactive Resource Executive
IS Impact Strength
Incomplete Sequence (Relay)
Indexed Sequential
Indian Standard
Induction Soldering
Industrial Society [UK]
Information Science
Information Separation
Information Separator
Information Service
Information Storage
Information System
Infrasonic(s)
Input System
In Service
Installation Start
Institute of Science
Institute of Statisticians [UK]
Instruction Set
Insulated System
Interference Spectroscope
Interference Suppressor
Internal Shield
International Standard
International Symposium
Interval Signal
Ion-Beam Synthesis
Ion Source

Ion Spectroscopy
Ion Spectrum
Island
Isle
Isolating Switch
Isomer Shift
Israeli Shekel [Currency]

I&S Inspection and Survey

ISA Image Sequence Analysis
Independent Scholar of Asia
Independent Signcrafters of America [US]
Industry Standard Architecture
Information Science Abstracts [US]
Information System Access
Information Systems Association [US]
Inorganic Sampling and Analysis
Instrument Society of America [US]
Instrument Subassembly
Intersecting Storage and Acceleration
International Schools Association
International Silk Association [France]
International Silo Association [US]
International Society of Appraisers [US]
International Society of Arboriculture [US]
International Standard Atmosphere
International Studies Association [US]
International Sugar Agreement
Interrupt Storage Area
Ion Scattering Analysis
Islamic Shipowners Association

ISAAS Indian Society for Afro-Asian Studies

ISAB Institute for the Study of Animal Behavior [UK]
International Scholastic Advisory Bureau [UK]

ISABGR International Society for Animal Blood Group Research [Australia]

ISAC Industrial Sector Advisory Committee [US]
International Scientific Agricultural Council
International Security Affairs Committee [US]

ISAD Information Science and Automation Division [of ALA, US]

ISADS Innovative Strategic Aircraft Design Studies

ISAE Indian Society of Agricultural Economics

ISAF Intermediate Super Abrasion Furnace Black

ISAGA International Simulation and Gaming Association [US]

ISAGUG International Software AG Users Group [US]

ISAH Integrated System for Automated Hydrography

ISAL Information System Access Line

ISAM Indexed Sequential Access Method
Integrated Switching and Multiplexing
International Society for Aerosols in Medicine [Austria]

ISAMA International Scientific Association for Micronutrients in Agriculture

ISAO International Society for Artificial Organs

ISAP Information Sort and Predict
International Society for Asphalt Pavements

ISAR Information Storage and Retrieval
International Society for Astrological Research
Inverse Synthetic Aperture Radar

ISARD International School for Agriculture and Resource Development

ISAS Illinois State Academy of Science [US]
Institute for Spectrochemistry and Applied Spectroscopy
Institute of Space and Aeronautical Sciences [Japan]
Institute of Space and Atmospheric Science [Canada]
International Schools Association [Switzerland]
International Screen Advertising Association
International Society of African Scientists [US]

ISASD International Symposium on Aeroelastics and Structural Dynamics

ISASI International Society of Air Safety Investigators

ISAST International Society for the Arts, Sciences and Technology [US]

ISAT Interrupt Storage Area Table

ISATA International Symposium on Automotive Technology and Automation

ISAW International Society of Aviation Writers

ISB Independent Sideband
Infrared Security Barrier
International Society of Biomechanics
International Society of Biometeorology [Switzerland]
International Society of Biorheology [Germany]
International Symposium on Bioceramics

ISBA International Sea-Bed Authority
International Society of British Advertisers

ISBB International Society of Bioclimatology and Biometeorology

ISBC International Small Business Congress

ISBD International Standard Bibliographic Description

ISBF Interactive Search of Bibliographic Files

ISBL Information System Base Language

ISBN International Standard Book Number

ISBO Islamic States Broadcasting Organization [Saudi Arabia]

ISBP International Society for Biochemical Pharmacology

ISBS Integrated Small Business Software

ISC Idle Speed Control (Motor)
Indirect Semiconductor
Information Society of Canada
Initial Slope Circuit
Instruction Staticizing Control
Integrated Storage Control
Intercompany Services Coordination
International Seismological Center [UK]
International Sericultural Commission [France]
International Service Carrier
International Society for Chronobiology [US]
International Society of Chemotherapy [Germany]
International Society of Citriculture [US]
International Society of Cryptozoology [US]
International Standards Council
International Student Conference
International Sugar Council
International Switching Center
Intersystem Communication

ISCA Industrial Specialty Chemical Association [US]

International Standards Coordination Association
International Standards Steering Committee for Consumer Affairs
Inter-Society Committee on Methods for Ambient Air Sampling and Analysis
ISCAC International Superconductor Applications Convention
ISCAN Inertialess Steerable Communication Antenna
ISCAS International Symposium on Circuits and Systems [of IEEE, US]
ISCB International Society for Clinical Biostatistics [UK]
International Society of Cell Biology
ISCC Intercrystalline Stress Corrosion Cracking
Intergranular Stress Corrosion Cracking
International Semiconductor Conference
International Service Coordination Center
International Strata Control Conference
Inter-Society Color Council [US]
Inter-Society Committee on Corrosion
Interstate Solar Coordination Council [US]
ISCC-NBS Inter-Society Color Council — National Bureau of Standards
ISCDS International Stop Continental Drift Society [US]
ISCE International Society of Chemical Ecology [US]
ISCED International Standard Classification of Education [of UNESCO]
ISCES International Society of Complex Environmental Studies
International Symposium on Condensation and Evaporation of Solids
ISCET International Society of Certified Electronics Technicians [US]
ISCEV International Society for Clinical Electrophysiology and Vision
ISCFB International Society of Craniofacial Biology
ISCIE Institute of Systems, Control and Information Engineers [Japan]
ISCLC International Symposium of Column Liquid Chromatography
ISCLT International Society of Clinical Laboratory Technologists
ISCM International Society of Cybernetic Medicine
ISCMA International Superphosphate and Compound Manufacturers Association
ISCO International Soil Conservation Organization
ISCOM Indian Satellite for Communication Technology [India]
ISCP International Society for Clinical Pathology
ISCRE International Symposium on Chemical Reaction Engineering
ISCRP International Society of City and Regional Planners
ISCS International Scientific Cooperative Service
ISCT Inner Seal Collar Tool
IScT Institute of Science and Technology
ISC/USO Intercompany Services Coordination/Universal Service Order
ISD Induction System Deposit
Information Structure Design
Information System Development
Initial Selection Done
Intermediate Storage Device
Internal Symbol Dictionary
International Society of Differentiation [US]
International Subscriber Dialing
International Symbol Dictionary
IS&D Integrate Sample and Dump
ISDA International Systems Dealers Association [US]
ISDB International Society of Developmental Biologists [France]
ISDD Integrated Systems Development Department
ISDF Intercrystalline Structure Distribution Function
Intermediate Sodium Disposal Facility
ISDG Information Science Discussion Group [UK]
ISDN Integrated Services Digital Network
International Society for Developmental Neuroscience [US]
International Standard Data Network
ISDOS Information System Design by Optimization System
ISDS Inadvertent Separation and Destruct System
Integrated Ship Design System [US Navy]
Integrated Software Development System
Integrated Switched Data Service
International Serials Data System [of UNISIST — France]
ISDSI Insulated Steel Door Systems Institute [US]
ISDSN Integrated Services Digital Satellite Network
ISDT Institute of Shaft Drilling Technicians [US]
Integrated Services Digital Terminal
ISDX Integrated Services Digital Exchange
ISE Induced Secondary Electron
Institute for Software Engineering [now: IIM, US]
Institute of Space Engineering [Canada]
Institut fur Solare Energiesysteme [= Institute for Solar Energy Systems [of Fraunhofer Institute, Germany]]
Institution of Sanitary Engineers [UK]
Institution of Structural Engineers [UK]
In-System Evaluator
Interface Science and Engineering
International Society of Electrochemistry [Austria]
International Society of Electrostimulation
International Submarine Engineering
Interrupt System Enable
Intersystem Emulator
Ion-Selective Electrode
I&SE Installation and Service Engineering
ISEA Industrial Safety Equipment Association [US]
ISEC International Solvent Extraction Conference
International Statistics Educational Center
International Symposium on Engineering Ceramics
ISECSI International Society for Educational, Cultural and Scientific Interchanges
ISEE International Society for Environmental Education
International Sun-Earth Explorer
ISEL Institute of Shipping Economics and Logistics [also: ISL, Germany]
ISEM International Society for Ecological Modelling [Denmark]
ISEO Institute of Shortening and Edible Oils [US]
ISEP International Society for Educational Planning
International Society for Evolutionary Protistology [UK]

International Symposium on Environmental Pollution
ISEPP International Sun-Earth Physics Program
ISER Integral Systems Experimental Requirements
ISES International Ship Electric Service Association [UK]
International Society of Explosives Specialists
International Solar Energy Society [Australia]
ISESCO Islamic Educational, Scientific and Cultural Organization
ISET Institut fur Solare Energieversorgungstechnik [= Institute for Solar Energy Technology, Germany]
IS&EU International Stereotypers and Electrotypers Union
ISF Imperial Smelting Furnace
Individual Store and Forward
Industrial Space Facility
Information Systems Factory
International Science Foundation
International Shipping Federation [UK]
International Society of Fat Research
ISFA Institute of Shipping and Forwarding Agents [UK]
International Scientific Film Association
ISFC International Scholarship Fund Committee
ISFD Integrated Software Functional Design
ISFE International Societies of Flying Engineers [US]
ISFET Ion-Selective Field-Effect Transistor
ISFL International Scientific Film Library
ISFM Indexed Sequential File Manager
ISFMS Indexed-Sequential File Management System
ISFNT International Symposium on Fusion Nuclear Technology [Japan]
ISFPP International Symposium on Fine Particles Processing
ISFR International Society for Fluoride Research [US]
ISFSI Independent Spent Fuel Storage Installation
International Society of Fire Service Instructors
ISG Imperial Standard Gallon
Instrumentation Selection Guide
Inter-Subblock Gap
Isolated Ground
ISGA International Study Group for Aerograms
ISGE International Society for Geothermal Engineering
ISGI International Service of Geomagnetic Indices
ISGO International Society of Geographic Ophthalmology
ISGOTT International Safety Guide for Oil Tankers and Terminals
ISGSH International Study Group on Steroid Hormones
ISGSR International Society for General Systems Research [now: ISSS, US]
ISH International Society of Hematology
International Trade Fair: Sanitation — Heating — Air Conditioning [Germany]
ISHAM International Society of Human and Animal Mycology
ISHC International Society of Heterocyclic Chemistry
ISHM International Society for Hybrid Microelectronics [US]
ISHRA Iron and Steel Holding and Realization Agency [UK]
ISHS International Society for Horticultural Science [Netherlands]
ISI Indian Standards Institution [India]
Indian Statistical Institute [India]
Initial Spread Index
In-Service Inspection
Institute for Scientific Information [US]
Internally Specified Index
International Safety Institute
International Standards Institute
International Statistical Institute [Netherlands]
Intersymbol Interference
Ion Signal for Imaging
Iron and Steel Institute [UK]
Israel Standards Institute
ISIA International Snowmobile Industry Association [US]
ISIC International Solvay Institute of Chemistry
International Standard Industrial Classification [of UN]
ISICCE International Society of Indian Chemists and Chemical Engineers [US]
ISID International Society of Interior Designers [US]
ISIF International Symposium on Integrated Ferroelectrics [US]
ISIFM International Society of Industrial Fabric Manufacturers [US]
ISI/ISTP&B Institute for Scientific Information/Index to Scientific and Technical Proceedings and Books [US]
ISIJ Iron and Steel Institute of Japan
ISILT Information Science Index Language Text
ISIMM International Society for the Interaction of Mechanics and Mathematics [Germany]
ISIP Indexed Security Investment Plan
Instantaneous Shut-In Pressure
ISIPBMP International Symposium on Interfacial Phenomena in Biotechnology and Materials Processing
ISIR Institute of Scientific and Industrial Research [of Osaka University, Japan]
International Society for Interferon Research
International Society of Invertebrate Reproduction [UK]
International Symposium on Industrial Robots
ISIS Integrated Scientific Information System
Integrated Set of Information System [of ILO, Switzerland]
Internally Switched Interface System
International Satellite for Ionospheric Studies [US/Canada]
International Science Information Service [US]
International Shipping Information Service
International Student Information System
Ion-Beam Synthesis in Semiconductors
ISISS International Summer Institute in Surface Science
ISiT Institut fur Siliziumtechnologie [= Institute for Silicon Technology [of Fraunhofer Institute, Germany]]
ISITB Iron and Steel Industry Training Board [UK]
ISJ Infrared Society of Japan
ISJTA Intensive Student Jet Training Area [US]
ISK Insert Storage Key
Instruction Space Key
Internacia Scienca Kolegio [= International College of Scientists, Italy]

	Icelandic Krona [Currency]
ISKA	International Saw and Knife Association [US]
ISL	Illinois State Library [US]
	Information Search Language
	Information System Language
	Information Systems Laboratory [of Michigan State University, US]
	Institut fur Seeverkehrswirtschaft und Logistik [= Institute of Shipping Economics and Logistics, Germany]
	Instructional Systems Language
	Integrated Schottky Logic
	Intelligent Systems Laboratory [of PARC, US]
	Interactive Simulation Language
	Intersatellite Link
	Intersystem Link
	Island
ISLA	International Survey Library Association [US]
ISLIC	Israel Society for Special Libraries and Information Centers
ISLLSS	International Society for Labour Law and Social Security [of ILO, Switzerland]
ISLS	Improved Sidelobe Suppression
ISLTC	International Society of Leather Trades Chemists
ISM	Industrial, Scientific and Medical (Equipment)
	Information Systems for Management
	Institute of Sanitation Management [now: EMA, US]
	Insulation System Module
	Interactive Surface Modeling
	International Software Marketing [US]
	Interpretive Structural Modeling
	Inverse-Speed Motor
ISMA	International Satellite Monitoring Agency
	International Shipmasters Association
	International Superphosphate Manufacturers Association
	International Symposium on Mining in the Arctic
	International Symposium on Mining with Backfill
ISMB	International Society of Mathematical Biology [France]
ISMC	International Switching Maintenance Center
ISMDA	Independent Sewing Machine Dealers Association [US]
ISMEC	Information Service in Mechanical Engineering
ISMED	International Symposium on Molecular Electronic Devices
ISMEX	International Shoe Machinery Exhibition
ISMH	Input Source Message Handler
	International Society of Medical Hydrology and Climatology [France]
ISML	Intermediate System Mockup Loop
ISMLS	Interim Standard Microwave Landing System
ISMM	International Society for Mini- and Microcomputers [Canada]
ISMMS	Intrinsically Safe Mine Monitoring System
ISMS	Image Store Management System
	International Society for Mushroom Science [UK]
ISMX	Integrated Subrate Data Multiplexer
ISN	International Society of Neurochemistry [Norway]
ISNET	Inter-Islamic Network in Space Sciences and Technology
ISNTA	International Staple, Nail and Tool Association [also: ISN&TA, US]
ISO	Individual System Operation
	Infrared Space Observatory [of ESA, France]
	Intergalactic Sysop Alliance [US]
	International Science Organization
	International Standards Organization [Switzerland]
	International Sugar Organization [UK]
	Isotropic
ISO/ASA	International Standards Organization/American Standards Association
ISOC	Individual System/Organization Cost
ISOCaRP	International Society of City and Regional Planners [Netherlands]
ISO-CMOS	Isolated Fully Recessed Complimentary Metal-Oxide Semiconductor
ISODATA	Iterative Self-Organizing Data Analysis Technique
ISOF	International Society of Ocular Fluorophotometry
ISOL	Isolation
ISOLD	Isolated
ISOM	Isometric
ISONET	International Standards Organization Network Committee
ISORID	International Information System on Research in Documentation [of UNESCO]
ISOS	International Southern Ocean Studies
ISOSC	International Society for Soilless Culture [Netherlands]
ISOT	International School of Offshore Technology
ISOTH	Isothermal
ISO WD	Isolation Ward
ISP	Imperial Smelting Process
	Independent Study Program
	Indexed Sequential Processor
	Industrial Security Plan
	Institute of Sewage Purification [UK]
	Instruction Set Processor
	Interferometer Software Package
	Internationally Standardized Profile
	International Society for Photogrammetry [now: ISPRS, Japan]
	International Society of Postmasters [Canada]
	International Society of Psychophysics
	Italian Society of Physics
ISPA	International Small Printers Association
	International Society of Parametric Analysts [US]
ISPABX	Integrated Services Private Automatic Branch Exchange
ISPC	International Sound-Program Center
	International Symposium on Plasma Chemistry
ISPE	Illinois Society of Professional Engineers [US]
	International Society of Pharmaceutical Engineers [US]
	International Society of Planetarium Educators [now: IPS, US]
ISPEC	Insulation Specification
ISPEMA	Industrial Safety Personnel Equipment Manufacturers Association
ISPF	International Science Policy Foundation [UK]
ISPhS	International Society of Phonetic Sciences [US]
ISPM	International Society of Plant Morphologists [India]

ISPMB International Society for Plant Molecular Biology [US]
ISPO International Statistical Programs Office [of USDOC]
ISPP Information System for Policy Planning
In-Service Professional Program
ISPRS International Society of Photogrammetry and Remote Sensing [Japan]
ISPS Instruction Set Processor Specifications
ISPT Istituto Superiore Poste e Telecomunicazioni [= Italian Institute for Postal and Telecommunications]
ISPWP International Society for the Prevention of Water Pollution [UK]
ISR Image Storage Retrieval
Index to Scientific Reviews [of ISI, US]
Induction Skull Remelting
Inductive Source Resistivity
Information Storage and Retrieval
Innovative Systems Research
Institute for Standards Research [US]
Institute of Seaweed Research [UK]
Interrupt Service Routine
Intersecting Storage Ring
International Society of Radiology [US]
IS&R Information Storage and Retrieval
ISRC International Standard Recording Group
International Student Research Center
ISRCSC Inter-Service Radio Components Standardization Committee [UK]
ISRD Information Storage, Retrieval and Dissemination
ISRF International Sugar Research Foundation
ISRG International Space Research Group [US]
ISRHAI International Secretariat for Research on the History of Agricultural Implements [Denmark]
ISRI Institute of Scrap Recycling Industries [US]
ISRIC International Soil Reference Information Center
ISRM Information System Resource Manager
International Society for Rock Mechanics [Portugal]
ISRO Indian Space Research Organization [India]
ISRP International Society of Respiratory Protection
ISRRS International Symposium on Research Reactor Safety
ISRRT International Society of Radiographers and Radiological Technicians [Canada]
ISRS International Safety Rating System
International Symposium on the Reactivity of Solids
ISRT Invisible Soft Return [also: ISRt]
ISRU International Scientific Radio Union
ISS Ideal Solidus Structures
Image Sharpness Scale
Index of Specifications and Standards [of DOD, US]
Information Sending Station
Information Services Seminar [US]
Information Sharing System
Information Storage System
Information Systems Specialists (Office) [of Library of Congress, US]
Input Subsystem
Institute for Strategic Studies
Institute of Systems Science [of NUS, Singapore]
Instrument Society of Sweden
Integrated Storage System
Intelligent Support System
Interlaminar Shear Strength
International Schools Services [US]
International Seaweed Symposium [Norway]
International Seismological Summary
International Society for Stereology [Germany]
International Steamboat Society
International Student Service [US]
Interrupt Safety System
Interrupt Service Subroutines
Ionosphere Sounding Satellite [Japan]
Ion-Scattering Spectroscopy
Ion-Scattering Spectrum
Iron and Steel Society [of AIME, US]
Isotope Separation System
Issue
ISSA Information Systems Security Association [US]
International Sanitary Supply Association [US]
International Ship Suppliers Association [UK]
International Slurry Seal Association [US]
International Society of Scientists-Artists [now: ISAST, US]
International Society of Stress Analysts [US]
ISS-AIME Iron and Steel Society of the American Institute of Mining, Metallurgical and Petroleum Engineers [US]
ISSC Interdisciplinary Surface Science Conference
International Ship Structures Conference
ISSCB International Society for Sandwich Construction and Bonding
ISSCC International Solid-State Circuits Conference [US]
ISSCG International Summer School on Crystal Growth
ISSCT International Society of Sugar Cane Technologists [Brazil]
ISSE International Sun Earth Explorer
ISSLIC Israel Society of Special Libraries and Information Centers [Israel]
ISSLS International Symposium on Subscriber Loops and Services
ISSM Institute of Safety and Systems Management [of University of Southern California, US]
ISSMB Information Systems Standards Management Board [of ANSI, US]
ISSMFE International Society for Soil Mechanics and Foundation Engineering [UK]
ISSN Integrated Special Services Network
International Standard Serial Number
ISSO Information System Security Officer
ISSOL International Society for the Study of the Origin of Life [US]
ISSP Institute of Solid-State Physics
International Summer School of Physics
ISSPIC International Symposium on Small Particles and Inorganic Clusters [US]
ISSS International Society for the Study of Symbols
International Society for the Systems Sciences [US]
International Society of Soil Science [Netherlands]
International Symposium on Surface Science

ISSSB International Symposium on Separation Science and Biotechnology
ISSSC International Summer School on Solidification and Casting
International Symposium on Solid State Chemistry
ISSSE International Society of Statistical Science in Economics [US]
ISST International Society for the Study of Time [US]
ISSUE Information System Software Update Environment
IST Impact Surface Treatment
Industry, Science and Technology
Information Science and Technology
Innovative Science and Technology
Institute of Science and Technology [UK]
Integrated Services Telephone
Integrated Switching and Transmission
Integrated Switching Technique
Integrated System Test
Integrated System Transformer
International Skelton Tables
International Society on Toxicology [US]
International Standard Thread
IS&T Imaging Science and Technology
ISTA Independent Software Testing Association [now: ASTE, US]
Indian Scientific Translators Association
International Seed Testing Association [Switzerland]
International Society for Technology Assessment
International Special Tooling Association [Germany]
ISTAHC International Society for Technology Assessment in Health Care
ISTAR Image Storage, Translation and Reproduction
ISTAT International Society of Transport Aircraft Traders [US]
ISTC Industry, Science and Technology Canada [formerly: DRIE]
Institute of Scientific and Technical Communicators [UK]
Integrated System Test Complex
International Steam Tables Conference [now: ICPS]
International Student Travel Confederation [Switzerland]
International Switching and Testing Center [UK]
ISTDA Institutional and Service Textile Distributors Association [US]
ISTE International Society for Tropical Ecology [India]
ISTEA Iron and Steel Trades Employers Association [UK]
ISTEC International Superconductivity Technology Center
ISTERH International Society of Trace Element Research in Humans
ISTF Integrated Servicing and Test Facility
International Society of Tropical Foresters [US]
ISTFA International Society for Testing and Failure Analysis
ISTH Isthmus
ISTIC Institute of Scientific and Technical Information of China
ISTIM Interchange of Scientific and Technical Information in Machine Language
ISTIS International Science and Technology Information Service
ISTM International Society for Testing Materials
ISTN Integrated Switching and Transmission Network
ISTP&B Index to Scientific and Technical Proceedings and Books [of ISI, US]
ISTR Indexed Sequential Table Retrieval
ISTRA Interplanetary Space Travel Research Association [UK]
ISTRC International Society of Tropical Root Crops
ISTRO International Soil Tillage Research Organization [Netherlands]
ISTS Institute for Space and Terrestrial Science [Canada]
International Shock Tube Symposium
International Symposium on Space Technology and Science
ISTU Islamic States Telecommunications Union
ISTVS International Society for Terrain-Vehicle Systems [US]
ISU Idaho State University [US]
Independent Signal Unit
Indiana State University [US]
Inertial Sensor Unit
Initial Signal Unit
Instruction Storage Unit
Interface Sharing Unit
Interface Switching Unit
International Scientific Union
International Space University [Canada]
Iowa State University [US]
ISUP Integrated Services User Part
ISUST Iowa State University of Science and Technology [US]
ISV Independent Software Vendor
In Situ Vitrification
Instantaneous Speed Variation
International Scientific Vocabulary
International Society of Videographers [US]
ISVD Information System for Vocational Decisions
ISVS International Society for Vegetation Science [now: IAVS, Germany]
ISVTNA International Symposium on Vacuum Technology and Nuclear Applications
ISW Industrial Solid Waste
ISWA Insect Screening Weavers Institute [US]
International Science Writers Association
International Solid Wastes and Public Cleansing Association [France]
ISWG Imperial Standard Wire Gage
ISWM International Society of Weighing and Measurement [US]
ISWL Isolated Single Wheel Load
ISWS Illinois State Water Survey [US]
ISWU International Society of Wang (Computer System) Users [US]
ISX Impurity Study Experiment
IT Image Tube
Immediate Transportation
Impact Test
Impulse Turbine

Incomplete Translation
Indent Tab(ulation Character)
Index Terms
Industrial Technician
Industrial Technologist
Industrial Technology
Industry Telephone
Information Technologist
Information Technology
Information Theory
Initial Tension
Input Terminal
Input Translator
Institute of Technology
Institute of Tribology [UK]
Insulating Transformer
Instrumentation Technician
Instrumentation Technologist
Instrumentation Technology
Instrument Test
Instrument Tree
Instrument Transformer
Intelligent Terminal
Interfacial Tension
Internal Thread
Internal Translator
International Table
International Tolerance
Interrogator-Transponder
Inter-Toll (Trunk)
In Transit
Isometric Transition
Isothermal
Isothermal Transformation
Italian
Italic
Italy
Item Transfer

I&T Inspection and Test

ITA Independent Telephone Association [US]
Independent Television Authority [UK]
Industrial Technical Adviser
Industrial Technology Adviser
Industrial Transport Association
Industrial Truck Association [US]
Industry and Trade Administration
Information Technologies Association [now: ITAMA, US]
Institut de Transport Aerien [= Institute of Air Transport, France]
Institute of Traffic Administration [UK]
Institute for Telecommunication and Aeronomy [of ESSA, US]
Instrumentation Technology Associates
Interface Test Adapter
Intermodal Transportation Association [US]
International Tape/Disk Association [US]
International Tar Conference
International Teleconferencing Association
International Telegraph Alphabet
International Teletraffic Congress
International Thermographers Association [US]
International Tin Association
International Tire Association [US]
International Trade Administration [of USDOC]
International Traders Association [US]
International Tube Association [UK]
International Tunneling Association [France]
International Typographic Association
Interstate Towing Association [US]

ITABC Independent Telephone Association of British Columbia [Canada]

ITAC Information Technology Association of Canada
Inter-Agency Textile Administrative Committee [US]
International Trade Affairs Committee

ITACS Integrated Tactical Air-Control System

ITAE Integrated Time and Absolute Error

ITAL Instituut voor Toepassing van Atoomenergie in de Landbouw [= Institute for Nuclear Energy in Agriculture, Netherlands]
Italian
Italic
Italy

ITAMA Information Technology Acquisition and Marketing Association [US]

ITAP Information Technology Advisory Panel [UK]

IT&AP Inspection/Test and Analysis Plan

ITAR International Traffic in Arms Regulations

ITARS Integrated Terrain Access and Retrieval System

ITAVS Integrated Testing Analysis and Verification System

ITB Incoming Trunk Busy
Industry Training Board [UK]
Intergrowth Tungsten Bronze
Intermediate Test Block
Intermediate Text Block (Character)
Internal Transfer Bus
International Time Bureau

ITBTP Institut Technique du Batiment et des Travaux Publics [= Technical Instutute of Buildings and Public Works, France]

ITC Immense Technology Commitment
Inclusive Tour Charter
Independent Telephone Company
Independent Television Commission [UK]
Industrial Training Council [UK]
(Ministry of) Industry, Trade and Commerce [Canada]
Inland Transport Committee [of UN]
Instructional Telecommunications Consortium [US]
Integrated Temperature Control System
Integrated Terminal Controller
Intense Training Course
Interagency Testing Committee
Inter-American Telecommunications Network
Interdata Transaction Controller
Intermediate Toll Center
International Tar Conference [France]
International Technology Council [of BRB, US]
International Telemetering Conference
International Teletraffic Congress
International Television Center

International Test Conference [US]
International Tin Council [UK]
International Toki Conference (on Plasma Physics) [Japan]
International Trade Center [Switzerland]
International Trade Commission [US]
International Trade Council [US]
International Training Center
International Translators Center [Netherlands]
International Tunneling Congress
Interval Time Control
Investment Tax Credit
Ionic Thermoconductivity
Ionic Thermocurrents
Italian Tile Center

IT&C Industry, Trade and Commerce

ITCA International Technical Caramel Association [US]
International Typographical Composition Association [now: TIA, US]
Isothiocyanic Acid

IT/CA International Tele/Conferencing Association [US]

IT cal International Steam Table Calorie

ITCAS International Training Center for Aerial Survey

ITCB International Textiles and Clothing Bureau

ITCC International Technical Communications Conference [of STC, US]
Interstate Truckload Carriers Conference [US]

ITCG International Trade Communications Group [Canada]

ITCI International Tree Crops Institute [US]

ITCL Item Class

ITCLC Item Class Code

ITCPN International Technical Conference on the Protection of Nature [France]

ITC/REE (Ministry of) Industry, Trade and Commerce and Regional Economic Expansion [now: DRIE, Canada]

ITCS Industrial Trade and Consumer Show
Integrated Temperature Control System
Integrated Thermionic Circuits

ITCSA Institute of Technical Communicators of Southern Africa

ITCWRM International Training Center for Water Resource Management

ITCZ Intertropical Convergence Zone

ITD Institute of Training and Development [UK]

ITDA Independent Truck Drivers Association [US]
International Tape/Disk Association

ITDB International Trade Data Bank

ITDC International Trade Development Committee [US]

ITDE Interchannel Time Displacement Error

ITDG Intermediate Technology Development Group [UK]

ITDM Intelligent Time-Division Multiplexer

ITDN Integrated Telephone and Data Network

ITDSC Item Description

ITDR Institute for Training and Demographic Research [Cameroon]

ITE Institute of Telecommunications Engineers
Institute of Television Engineers
Institute of Traffic Engineers [Name Changed to: Institute of Transportation Engineers, US]
Institute of Transportation Engineers [US]
Integrated Test Equipment
Intercity Transportation Efficiency
International Telephone Exchange

ITEA International Technology Education Association [US]
International Test and Evaluation Association [US]

ITEA-CS International Technology Education Association — Council for Supervisors [US]

ITEC Information Technology Center

ITEJ Institute of Television Engineers of Japan

ITEM Interference Technology Engineer's Master

ITEMS Imaging Technologies and Evolving Management Systems
Incoterm Transaction Entry Management System

ITEO International Trade and Employment Organization

ITEP Institute of Theoretical and Experimental Physics

ITER International Thermonuclear Experimental Reactor

ITER EDA ITER Engineering Design Activities (Agreement)

ITES Instituto Tecnologico de Estudios Superiores [= Technical Institute of Higher Studies, Mexico]

ITEWS Integrated Tactical Electronic Warfare System

ITF Institut Textile de France [= Textile Institute of France]
Integrated Test Facility
Interactive Terminal Facility
Interface File
International Transport Workers Federation [UK]
Interstitial Transfer Facility

ITFS Instructional Television Fixed Service [US]

ITFTRIA Instrument Tree Flow and Temperature Removal Instrument Assembly

ITG Informationstechnische Gesellschaft [= Society for Information Technology [of VDE, Germany]]
Information Theory Group [of IEEE, US]

ITGLWF International Textile, Garment and Leather Workers Federation

ITH In-the-Hole Drilling

ITI Industrial Technology Institute [US]
Industrial Training Institute
Inspection/Test Instruction
Institute for Technical Interchange [Hawaii, US]
Interactive Terminal Interface
Intermittent Trouble Indication
International Technology Institute [US]

ITIA Industrial Tungsten Industry Association [UK]

ITIC International Tsunami Information Center [Hawaii, US]

ITIRC IBM (Inc.) Technical Information Retrieval Center [US]

ITIS Insect Toxicologists Information Service [Netherlands]

ITL Industrial Test Laboratory [US Navy]
Integrated Transfer Launch
Intermediate Text Language
Intermediate Transfer Language
Inverse Time Limit
Italian Lira [Currency of Italy and Vatican City]

ITM Indirect Tag Memory
Information Transfer Module

ITMA International Tanning Manufacturers Association

	Irradiation Test Management Activity
ITMC	International Transmission Maintenance Center
ITMF	International Textile Manufacturers Federation [Switzerland]
ITMJ	Incoming Trunk Message Junction
ITMT	Intermediate Thermomechanical Treatment
ITN	Independent Television News (Service) [of ITV, UK]
	Integrated Teleprocessing Network
ITNL	Internal
ITNS	Integrated Tactical Navigation System
ITNSA	Item Net Sales Amount
ITO	Indium-Tin Oxide
	International Trade Organization [of UN]
ITOA	Independent Terminal Operators Association [US]
I-TOOA	Independent Truck Owner/Operator Association [US]
ITOPF	International Tanker Owners Pollution Federation [UK]
ITOS	Improved Television and Infrared Observation Satellite
	Improved TIROS (= Television and Infrared Observation Satellite) Operational Satellite
	Interactive Terminal Operating System
	Iterative Time Optimal System
ITP	Inosine Triphosphate
	Inspection Test Procedure
	Institute for Theoretical Physics [of UCSB, US]
	Integral Thermal Process
	Integrated Test Program
	Interactive Terminal Protocol
	Interstitial Thickening Process
ITPA	Independent Telephone Pioneers Association
	International Truck Parts Association [US]
ITPC	International Television Program Center
ITPI	International Transfer Printing Institute [US]
ITPO	International TOGA (= Tropical Ocean Global Atmosphere) Project Office [of WMO, Switzerland]
ITPP	Institute of Technical Publicity and Publications
ITPR	Infrared Temperature Profile Radiometer
ITPS	Integrated Teleprocessing System
ITR	Ignition Test Reactor
	In-Core Thermionic Reactor
	Instrument Test Rig
	Integrated Telephone Recorder
	INTELSAT (= International Telecommunications Satellite Organization) Test Record
	Interactive Teleprocessing System
	Interactive Text Processing System
	Interrupt Control Unit
	Inverse Time Relay
	Ion Transfer Reaction
	Isolation Test Routine
ITRC	International Tin Research Council [UK]
ITRI	Industrial Technology Research Institute [Taiwan]
	International Tin Research Institute [UK]
ITRIA	Instrument Tree Removable Instrument Assembly
ITRM	Inverse Thermoremanent Magnetization
ITRPF	International Tire, Rubber and Plastics Federation [UK]
ITS	Idaho Test Station
	Industrial Trade Show
	Information Theory Society [now: ITG of IEEE, US]
	Information Transmission System
	Insertion Test Signal
	Institute of Telecommunication Sciences [US]
	Insulation Test Specification
	Integrated Test System
	Integrated Tracking System
	Integrated Trajectory System
	Intelligent Terminal System
	Intelligent Test System
	Interactive Terminal Support
	Interactive Training System
	Interface Test Set
	Intermarket Trading System
	International Telecommunications Society
	International Temperature Scale
	International Tesla Society [US]
	International Trade Secretariat
	International Trade Show
	International Turfgrass Society [US]
	Inter-Time Switch
	Invitation to Send
	Ion Trap System
ITSA	Institute for Telecommunication Sciences and Aeronomy [of ESSA, US]
	International Thermal Spray Association
ITSC	International Telecommunications Satellite Consortium [US]
	International Telecommunications Services Complex
	International Telephone Services Center
ITSG	International Tin Study Group
IT/SP	Instrument Tree/Spool Piece
ITSS	International Team for Studying Sintering [now: IISS, Yugoslavia]
ITSU	Information Technology Standards Unit [of DTI, UK]
ITT	Impact Transition Temperature
	Institute of Textile Technology [US]
	International Telephone and Telegraph [US]
	International Towing Tank Conference [Japan]
	Intertoll Trunk
ITTC	International Towing Tank Conference
ITTFL	International Telephone and Telegraph Federal Laboratories [US]
ITTO	International Tropical Timber Organization
ITU	Ikutoku Technical University [Japan]
	International Technical University
	International Telecommunication Union [of UN — Switzerland]
	International Typographical Union
	Istanbul Technical University [Turkey]
ITUG	International Tandem (Computer) Users Group [US]
ITUSA	Information Technology Users Association [UK]
ITV	Improved TOW (= Tube-Launched, Optically-Tracked, Wire-Guided) Vehicle
	Independently Targeted Vehicle
	Independent Television [UK]
	Industrial Television
	Instructional Television
	Interactive Television

ITVA International Television Association [US]
ITVS International Television Symposium
ITZN International Trust for Zoological Nomenclature
IU Ibaraki University [Japan]
Indiana University [US]
Information Unit
Input Unit
Instruction Unit
Instrumentation Unit
Interface Unit
Interference Unit
International Unit
Iwate University [Japan]
IUA International Union of Architects
IUAA International Union of Advertisers Associations [now: WFA, Belgium]
IUAI International Union of Aviation Insurers [UK]
IUAJ International Union of Agricultural Journalists
IUAO International Union for Applied Ornithology [Germany]
IUAPPA International Union of Air Pollution Prevention Associations [UK]
IUAS International Union in Agricultural Sciences
IUB International Union of Biochemistry [US]
International Universities Bureau
IUBS International Union of Biological Sciences [France]
IUC International Union of Chemistry
International Union of Crystallography [UK]
International University Consortium [US]
Inter-University Council
IUCADC Inter-Union Commission of Advice to Developing Countries [Canada]
IUCAF Inter-Union Commission on Allocation of Frequencies
IUCED Inter-Union Commission of European Dehydrators [France]
IUCFA Inter-Union Commission on Frequency Allocations (for Radio Astronomy and Space Science) [US]
IUCN International Union for Conservation of Nature (and Natural Resources) [Switzerland]
IUCNPSG International Union for Nature and Natural Resources — Primate Specialists Group [Switzerland]
IUCr International Union of Crystallography [UK]
IUCRM Inter-Union Commission on Radio Meteorology
IUCS Inter-Union Commission on Spectroscopy
IUCST Inter-Union Commission on Science Teaching
IUCSTP Inter-Union Commission on Solar Terrestrial Physics [now: SCOSTEP, US]
IUCSTR Inter-Union Commission on Solar Terrestrial Relationships [US]
IUD Institute for Urban Design [US]
IUDH In-Service Unplanned Derated Hours
IUDH1 In-Service Unplanned Derated Hours, Class 1
IUDZG International Union of Directors of Zoological Gardens [Canada]
IUE International Ultraviolet Explorer
International Union of Electrical, Radio and Machine Workers
IUEC International Union of Elevator Constructors [UK]
IUEF International University Exchange Fund
IUF International University Foundation [US]
IUFoST International Union of Food Science and Technology [Ireland]
IUFRO International Union of Forestry Research Organizations [Austria]
IUG ICES (= Integrated Civil Engineering System) Users Group [US]
Intercom Users Group [US]
IUGB International Union of Game Biologists [Poland]
IUGG International Union of Geodesy and Geophysics [Belgium]
IUGRI International Union of Graphic Reproduction Industries [France]
IUGS International Union of Geological Sciences [Norway]
IUHPS International Union of History and Philosophy of Science
IUHS International Union of History of Science
IUI International Union of Interpreters
IUIN International Union for Inland Navigation [France]
IUJ International University of Japan
IUL Information Utilization Laboratory [of University of Pittsburgh, US]
IULCS International Union of Leather Chemists Societies
IULD International Union of Lorry Drivers [Germany]
IULEC Inter-University Labor Education Committee [US]
IULTCS International Union of Leather Technologists and Chemists Societies
IUMI International Union of Marine Insurance [Scotland]
IUMP International Union of Master Painters [Germany]
International Upper Mantle Project
IUMRS International Union of Materials Research Societies
IUMS International Union of Microbiological Societies [UK]
IUMSBD International Union of Microbiological Societies — Bacteriological Division [UK]
IUNDH In-Service Unit Derate Hours
IUNS International Union of Nutritional Sciences
IUOE International Union of Operating Engineers [UK]
IUP Indiana University of Pennsylvania [US]
Installed User Program
Irish University Press
IUPAB International Union of Pure and Applied Biophysics [Hungary]
IUPAC International Union of Pure and Applied Chemistry [UK]
IUPAP International Union of Pure and Applied Physics [Sweden]
IUPESM International Union for Physical and Engineering Sciences in Medicine [Canada]
IUPHAR International Union of Pharmacology [UK]
IUPIP International Union for the Protection of Industrial Property
IUPIW International Union of Petroleum Industrial Workers
IUPN International Union for the Protection of Nature
IUPPS International Union of Prehistoric and Protohistoric Sciences [Belgium]
IUPS International Union of Physiological Sciences [France]
IUR International Union of Radioecologists [Belgium]

	International Union of Railways [also: UIC, France]
IURC	International Union for Research of Communication [Switzerland]
IUREP	International Uranium Resources Evaluation Project
IURP	Integrated Unit Record Processor
IURS	International Union of Radio Sciences [US]
IUS	Inertial Upper Stage
	Information Unit Separator
	Initial Upper State
	Interchange Unit Selector
	Interchange Unit Separator
	Interim Upper Stage
	International Union of Speleology [Austria]
	International Union of Students [Czechoslovakia]
	Inter-University Seminar
IUSF	International Union for Surface Finishing [UK]
	International Union of Societies of Foresters [Canada]
IUSM	International Union of Surveys and Mapping
IUSSI	International Union for the Study of Social Insects [Netherlands]
IUSSP	International Union for the Scientific Study of Population [Belgium]
IUTAM	International Union of Theoretical and Applied Mechanics [Germany]
IUTDMM	International Union of Tool, Die and Mold Makers
IUTOX	International Union of Toxicology
IUTS	Inter-University Transit System [Canada]
IUVSTA	International Union for Vacuum Science, Technique and Applications [US]
IUWA	International Union of Women Architects [France]
IUWPM	Independent University, Washington-Paris-Moscow
IV	I (= Current) — V (= Voltage)
	Intermediate Vacuum
	Intermediate Voltage
	Intravenous
	Inverter
	Iodine Value
	Isovalerianic Acid
I/V	Instrument/Visual
I-V	I (= Current) — V (= Voltage)
IVA	**Intervehicular Activity**
IVACG	International Vitamin A Consultative Group
IVAR	Insertion Velocity Adjust Routine [of NASA]
IVAS	International Veterinary Acupuncture Society
IVC	Interactive Videodisk Consortium [US]
	Intermediate Velocity Cloud
	International Vacuum Congress
IVCE	International Video and Communications Exhibition
IVC/ICSS	International Vacuum Congress/International Conference on Solid Surfaces [Netherlands]
IVD	Image Velocity Deceptor
	Innovative Vehicle Design
	Ion Vapor Deposition
IVDG	Innovative Vehicle Design Group
IVDS	Independent Variable Depth Sonar
IVDT	Integrated Voice-Data Terminal
IVDW	Integrated Voice/Data Workstation
IVE	Institute of Vitreous Enamellers [UK]
	Isobutyl Vinyl Ether
	Isocyanate Vinyl Ester
IVETA	International Vocational Education and Training Association
IVF	Institut foer Verkstadsteknisk Forskning [= Institute for Laboratory Technique Research, Sweden]
IVFRC	In VFR (= Visual Flight Rules) Condition
IVG	Interrupt Vector Generator
IVGA	Israel Vegetable Growers Association
IVHM	In-Vessel Handling Machine
IVHM-EM	In-Vessel Handling Machine — Engineering Model
IVHS	Intelligent Vehicle Highway System
IVHU	In-Vessel Handling Unit
IVI	Incremental Velocity Indicator
	Interactive Visual Interface
IVIA	Interactive Video Industry Association [US]
	International Videotext Industry Association
IVIC	Instituto Venozolano de Investigaciones Cientificas [= Venezuelan Institute for Scientific Investigations]
IVIPA	International Videotext Information Providers Association
IVM	Initial Virtual Memory
	Interface Virtual Machine
IVMS	Integrated Voice Messaging System
IVMU	Inertial Velocity Measurement Unit
IVO	Improved Virtual Orbitals
IVP	Initial Value Problem
	Initial Vapor Pressure
	Installation Verification Procedure
	Installation Verification Program
IVPA	Independent Video Programmers Association [US]
IVPO	Inside Vapor-Phase Oxidation
IVR	Induction Voltage Regulator
	Instrumental Visual Range
	Integrated Voltage Regulator
IVRG	International Verticillium Research Group [of University of Guelph, Canada]
IVRI	Indian Veterinary Research Institute
IVS	Indian Vacuum Society
	In-Vessel Storage
IVSA	International Veterinary Students Association [Netherlands]
IVSI	Instantaneous Vertical Speed Indicator
IVSM	In-Vessel Storage Module
IVSU	International Veterinary Students Union
IVT	Integrated Video Terminal
IVV	Independent Verification and Validation [also: IV&V]
IVVS	In-Vessel Vehicle System
IVW	Informationsstelle uber die Verbreitung von Werbetragern [= Information Office for the Distribution of Advertising Materials and Media, Germany]
IW	Index Word
	Induction Welding
	Industrial Waste
	Inside Wire
	Isle of Wight [UK]
IWA	International Waterproofing Association [Belgium]
	International Wheat Agreement
	International Woodworkers of America

IWAHMA	Industrial Warm Air Heater Manufacturers Association
IWAP	International Watershed Advocacy Project [now: IRN, US]
IWB	Instruction Word Buffer
IWC	International Welding Conference
	International Whaling Commission [UK]
	International Whaling Convention [US]
	International Wheat Council [UK]
	International Wildlife Coalition [US]
IWCA	Inside Wiring Cable
IWCC	International Wrought Copper Council [UK]
IWCI	Industrial Wire Cloth Institute [now: AWCI, US]
IWCS	Integrated Wideband Communications System
IWCS/SEA	Integrated Wideband Communications System/ Southeast Asia
IWD	Inland Waters Directorate [Canada]
IWDA	Independent Wire Drawers Association
IWE	Institution of Water Engineers [UK]
IWEM	Institution of Water and Environmental Management [UK]
IWES	International Waste Energy System
IWF	Industry Workers Federation [San Marino]
IWFNA	Inhibited White Fuming Nitric Acid
IWFS	Industrial Waste Filtering System
IWG	Iron Wire Gauge
IWGA	International Wheat Gluten Association [US]
IWHS	Institute of Works and Highways Superintendents [UK]
IWI	Inventors Workshop International [now: IWIEF, US]
IWIEF	Inventors Workshop International Education Foundation [US]
IWIM	Institut fur Wissenschaftsinformation in der Medizin [= Institute for Scientific Information in Medicine, Germany]
IWM	Industrial Waste Management
	Institute of Works Managers [UK]
	Institut fur Werkstoffmechanik [= Institute for Materials Mechanics [of Fraunhofer Institute, Germany]]
IWMA	Institute of Weights and Measures Administration [UK]
	International Wire and Machinery Association [UK]
IWP	Intelligent Work in Process
	Interim Working Party
IWPA	International Word Processing Association [later: IIWPA, US; now: AISP, US]
	International Work Platform Association
IWPC	Institute of Water Pollution Control [UK]
IWPO	International Word Processing Organizations
IWR	Institute for Wildlife Research [US]
	Isolated Word Recognition
IWRA	International Water Resources Association [US]
	International Wild Rice Association [US]
IWRB	International Waterfowl Research Bureau [now: IWWRB, UK]
IWRC	IBM (Inc.) Watson Research Center [US]
	Independent Wire Rope Core
	International Wildlife Rehabilitation Council [US]
IWRI	International Waterfowl Research Institute [UK]
IWRPF	International Waste Rubber and Plastics Federation
IWRS	International Wood Research Society
IWS	Institute of Water Study [UK]
	Institute of Wood Science [UK]
	Institute of Work Studies
	Integrated Work Sequence
	International Wool Secretariat [UK]
IWSA	International Water Supply Association [UK]
IWSc	Institute of Wood Science [UK]
IWSG	International Wool Study Group [UK]
IWS/IT	Integrated Work Sequence/Inspection Traveller
IWSP	Institute of Work Study Practitioners
IWSS	International Weed Science Society [US]
IWTA	Inland Water Transport Authority [Pakistan]
IWtC	International Wheat Council
IWTO	International Wool and Textile Organization [Belgium]
IWTRC	International Wool Textile Research Conference
IWU	Isolation Working Unit
IWW	Industrial Workers of the World [US]
IWWA	International Water Works Association [Israel]
	International Wild Waterfowl Association [US]
IWWK	Institut fur Weltwirtschaft an der Universitat Kiel [= Institute for World Economy of the University at Kiel, Germany]
IWWRB	International Waterfowl and Wetlands Research Bureau [UK]
IX	Index Register
	Inter-Exchange
	Intersystem Crossing
	Inverted Index
	Ion Exchange
IXA	Ion-Excited X-Ray Analysis
IXAE	International X-Ray Astrophysics Explorer
IXC	Inter-Exchange Channel
IXEE	International X-Ray and Extreme Ultraviolet Explorer
IXSD	International Telex Subscriber Dialing
IXT	Interaction Crosstalk
IYF	International Youth Federation (for Environmental Studies and Conservation) [Denmark]
IYNF	International Young Nature Friends
IYSWIM	If You See What I Mean
IZ	Isolation Zone
IZA	International Zeolite Association
IZAA	Independent Zinc Alloyers Association [US]
IZE	International Association of Zoo Educators [US]

J

J	Jack
	January
	Job
	Joule
	Journal
	Jump
	Junction
	Jurassic [Geology]
	Jute
j	jack
	join
JA	Japan Academy [in Tokyo]
	Journal Announcement
	Jump Address
J/A	Joint Account
JAA	Joint Airworthiness Authority [UK]
JAALD	Japan Association of Agricultural Librarians and Documentalists
JAC	Joint Advisory Coomittee
JACC	Joint Automatic Control Conference
JACE	Joint Alternate Command Element
JACOLA	Joint Airports Committee of Local Authorities
JAD	Jamaican Dollar [Currency]
JAEC	Japan Atomic Energy Commission
JAEIP	Japan Atomic Energy Insurance Pool
JAERI	Japan Atomic Energy Research Institute
JAF	Job Accounting Facility
JAFA	Japan Auto-Focus Association
JAFAE	Japan Auto-Focus Association in Europe
JAFNA	Joint Air Force — NASA [US]
JAGOS	Joint Air-Ground Operations System
JAI	Job Accounting Interface
JAICI	Japan Association for International Chemical Information
JAIEG	Joint Atomic Information Exchange Group [of USDOD]
JAIF	Japan Atomic Industrial Forum
JAL	Japan Airlines
JALPAS	Japan Airlines Passenger Autoprocessing System
JALPG	Joint Automatic Language Processing Group
JAM	Jamaica(n)
	Japanese Association for Microbiology
JAMA	Japan Automobile Manufacturers Association
JAMASS	Japanese Medical Abstracts Scanning System [of IMIC, Japan]
JAMS	Japan Association for Mathematical Sciences
	Job Activities Management System
JAMSAT	Japanese Satellite for Amateur Radio
JAMTS	Japan Association of Motor Trade and Service
JAN	January
	Joint Army — Navy (Program) [US]
JANAF	Joint Army — Navy — Air Force [US]
JANAIR	Joint Army — Navy Aircraft Instrument Research [of USDOD]
JANAP	Joint Army — Navy — Air Force Publication [US]
JANNAF	Joint Army — Navy — NASA — Air Force [also: JAN-NAF]
JANOT	Joint Army — Navy Ocean Terminal [US]
JANS	Joint Army — Navy Specification [of USDOD]
JAP	Japan(ese)
Japan GCR	Japan Gas-Cooled Reactor
Japan TAPPI	Japan Technical Association of the Pulp and Paper Industry [also: Japan Tappi]
JAPATIC	Japan Patent Information Center
JAPEX	Japan Petroleum Exploration
JAPIA	Japan Auto Parts Industries Association
JAPIO	Japan Patent Information Organization
JAR	Japanese Association of Refrigeration
	Joint Airworthiness Requirements
	Jump Address Register
JARC	Joint Avionics Research Committee
JARI	Japan Association of Railway Industries
JARL	Japan Amateur Radio League
JARRP	Japan Association for Radiation Research on Polymers
JAS	Japanese Amateur Radio Satellite
	Joint Airmiss Section [UK]
JASA	Joint Airworthiness Steering Committee [Europe]
JASC	Japan Academic Societies Center
	Japan — America Student Conference [US]
	Japan Sea Cable
JASDF	Japan Air Self-Defense Force
JASO	Japan Standards Organization
JAST	Japan Association of Sugar Technologists
JAT	Job Accounting Table
JATCRU	Joint Air Traffic Control Radar Unit
JATIS	Japan Technical Information Service
JATO	Jet-Assisted Takeoff (Engine)
JAWG	Joint Airmiss Working Group [UK]
JB	Jet Barrier
	Junction Box
JBA	John Burroughs Association [US]
JBD	James Brake Decelerometer
JBI	James Brake Index
JBMA	John Burroughs Memorial Association [now: JBA, US]
JBMMA	Japan Business Machine Makers Association
JBNQA	James Bay and Northern Quebec Agreement [Canada]
JC	Johnson Counter
	Joint Commission
	Joint Committee
	Joint Conference
JCA	Joint Commission on Accreditation
JCAB	Japanese Civil Aviation Bureau
JCAE	Joint Committee on Atomic Energy [US]
JCAM	Joint Commission on Atomic Masses
JCAR	Joint Commission on Applied Radioactivity [France]
JCASR	Joint Committee on Avionic Systems Research
JCAU	Joint Commission on Accreditation of Universities
JCB	Job Control Block
	Joint Communications Board
JCC	Job Control Card
	Joint Communications Center

	Joint Computer Conference [US]
	Joint Conference Committee
	Joint Consultative Committee [UK]
	Joint Control Center
JCCCOMNET	Joint Coordination Center Communications Network
JCEC	Joint Communications Electronics Committee [US]
JCEG	Joint Communications Electronics Group
JCENS	Joint Communications-Electronics Nomenclature System
JCET	Joint Committee on Educational Television [US]
JCEWG	Joint Communications Electronics Working Group [of NATO]
JCF	Joint Coordinating Forum
JCG	Joint Coordinating Group
JCGS	Joint Center for Graduate Study
JCHARS	Joint Commission on High Altitude Research Stations [Switzerland]
JCHPME	Joint Commission on Higher Professional Medical Education
JCI	Joint Communications Instruction
JCIE	Joint Center for International Exchange
JCII	Japan Camera Inspection Institute
JCL	Job Command Language
	Job Control Language
JCMB	Joint Committee on Medicine and Biology [of IEEE and ISA]
JCMR	Japan Congress on Materials Research
JCN	Junction
JCNMT	Joint Committee of Nordic Marine Technology [Finland]
JCNNSRC	Joint Committee of the of the Nordic Natural Science Research Council [Sweden]
JCNPS	Joint Committee on Nuclear Power Standards
JCP	Job Control Program
	Joint Committee on Printing [of US Congress]
	Junction Call Processing
JCPDS	Joint Committee on Powder Diffraction Standards [US]
JCPDS-ICDD	Joint Committee on Powder Diffraction Standards — International Center for Diffraction Data [US]
JCPS	Junction Call Processing Subsystem
JCPWG	Joint Certification Procedures Working Group [Europe]
JCRR	Joint Commission on Rural Reconstruction [PR China/US]
JCRT	Joint Center for Radiation Therapy [of Harvard Medical School, US]
JCS	Job Control Statement
	Joint Chiefs of Staff [of USDOD]
	Joint Coordinate System
JCSAN	Joint Chiefs of Staff Alerting Network [of USDOD]
JCSTR	Joint Commission on Solar and Terrestrial Relationships
JCT	Jerusalem College of Technology [Israel]
	Job Control Table
	Journal Control Table
	Junction
JCTFI	Joint Committee for Training in the Foundry Industry
JCTN	Junction
JCU	James Cook University [Australia]
JCULS	Joint Committee on the Union List of Serials [US]
JD	Joined
	Jordanian Dinar [Currency]
J$	Jamaican Dollar [Currency]
JDA	Japan Defense Agency
JDC	Job Description Card
JDES	Joint Density of Electronic States
JDFC	Joint Danube Fisheries Commission [Czechoslovakia]
JDL	Japan Digital Laboratory
	Job Description Library
	Job Description Language
	Job Descriptor Language
JDREMC	Joint Departmental Radio and Electronic Measurements Committee [UK]
JDS	Job Data Sheet
JE	Jet Engine
	Job Entry
	Junction Exchange
	June
JEA	Joint Endeavor Agreement
	Jordan Engineers Association
JEAN	JOSS (= Johnniac Open-Shop System) Based Expression Analyzer for Nineteen-Hundred
JEBM	Jet Engine Base Maintenance
JEBM-RR	Jet Engine Base Maintenance Return Rate
JEC	Joint Economic Committee [US]
JECA	Joint Engineers Council of Alabama [US]
JECC	Japan Electronic Computer Center
	Joint Egyptian Cotton Committee
	Joint Electronic Components Conference
JECL	Job Entry Control Language
JECS	Job Entry Central Service
JEDEC	Joint Electron Device Engineering Council [US]
JEDTC	Joint Electron Device Tube Council [now: TEPAC, US]
JEEP	Joint Establishment Experimental Pile [Norway]
JEIA	Joint Export-Import Agency [US/UK]
JEIDA	Japan Electronic Industry Development Association
JEIPAC	JICST Electronic Information Processing Automatic Computer
JEL	Johnson Elastic Limit
JEM	Japan Experiment Module
	Jet Engine Modulation
	Joint Endeavor Manager
JEMC	Joint Engineering Management Conference
JEMIC	Japan Electric Meters Inspection Corporation
JEMIMA	Japan Electric Measuring Instruments Manufacturers Association
JEMRB	Joint European Medical Research Board
JEN	Junta de Energia Nuclear [= Nuclear Energy Authority, Spain]
JENER	Joint Establishment for Nuclear Energy Research
JEOCN	Joint European Operations Communications Network
JEOL	Japan Electro-Optics Laboratories
JEPIA	Japan Electronic Parts Industry Association
JEPOSS	Javelin Experimental Protection Oil Sands System
JEPP	Joint Emergency Planning Program
JEPS	Job Entry Peripheral Service
JEQ	Jump Equal

JERC Japan Economic Research Center
Japan-Europe Economic Research Center
Joint Electronic Research Committee
JERS Japanese Earth Resource Remote-Sensing Satellite
JES Job Entry Subsystem
Job Entry System
John Ericsson Society [US]
Joint Environmental Service
Joint Environmental Simulator
JESAC Joint Engineering Student Activity Committee
JESSI Joint European Silicon Submicron Initiative [of EUREKA]
Junior Engineers and Scientists Summer Institute [US]
JET Joint Enroute Terminal
Joint European Torus (Reactor)
JETDS Joint Electronics Type Designation System
JETEC Joint Electron Tube Engineering Council [US]
JETR Japan Engineering Test Reactor [Japan]
JETRO Japan External Trade Organization
JETS Job Executive and Transport Satellite
Junior Engineering Technical Society [US]
JF Junction Frequency
Junctor Frame
JFA Japan Fisheries Agency
JFAP Joint Frequency Allocation Panel
JFCA Japan Fine Ceramics Association
JFCB Job File Control Block
JFCC Japan Federation of Culture Collections
JFEA Japan Federation of Employers Associations
JFET Junction Field-Effect Transistor
JFKSC John F. Kennedy Space(flight) Center [also: KSC, US]
JFL Joint Frequency List
JFMIP Joint Financial Management Improvement Program [US]
JFN Job File Number
JFP Joint Frequency Panel
JFRO Joint Fire Research Organization [UK]
JFS Japan Foundrymen's Society
Jumbo (Group) Frequency Supply
JFSE Japan Federation of Small Enterprises
JG Junction Grammar
J/g Joule per gram
JGA Jojoba Growers Association [US]
JGCR Japan Gas-Cooled Reactor
JGF Junctor Grouping Frame
JGN Junction Gate Number
JGOFS Joint Global Ocean Flux Study
JH Juvenile Hormone
JHD Joint Hypocenter Determination
JHG Joule Heat Gradient
JHN Japanese Helicopter Network
JHPS Japan Hydraulics and Pneumatics Society
JHU Johns Hopkins University [US]
JIB Job Information Block
Joint Intelligence Bureau [UK]
JIBEI Joint Industry Board of the Electrical Industry [US]
JIC Japan Information Center
Joint Industrial Council [US]
Joint Industry Committee
Joint Industry Conference
Joint Iron Council [UK]
Just-in-Case
JICA Japan International Cooperation Agency
JICNARS Joint Industry Committee for National Readership Surveys [UK]
JICST Japan Information Center of Science and Technology
JICTAR Joint Industry Committee for Television Advertising Research [UK]
JIDA Japan Industrial Designers Association
JIDC Jamaica Industrial Development Corporation
JIE Junior Institution of Engineers [UK]
JIFA Japan Institute of Foreign Affairs
JIFDATS Joint In-Flight Data Transmission System [of USDOD]
JIFTS Joint In-Flight Transmission System
JIII Japan Institute of Invention and Innovation
JIISC Joing Industrial Investment Service Center [of MOEA, Taiwan]
JILA Joint Institute for Laboratory Astrophysics [US]
JILM Japan Institute of Light Metals [of JLMA]
JIM Japan Institute of Metals
JIMA Japan Industrial Management Association
JIMIS Japan Institute of Metals International Symposium
JIMTOF Japan International Machine Tool Fair
JIN Japan Institute of Navigation
JINR Joint Institute for Nuclear Research [USSR]
JIOA Joint Intelligence Objectives Agency
JIP Joint Input Processing
JIPDEC Japan Information Processing Development Center
JIRA Japan Industrial Robot Association
JIS Japan Industrial Standard
Job Information System
Job Input System
JISC Japan Industrial Standards Committee
JISHA Japan Industrial Safety and Health Association
JIT Job Instruction Training
Just-In-Time
JIT/TQC Just-In-Time/Total Quality Control
JIU Joint Inpection Unit
JJ Josephson Junction
JK Jack
J and K Input (Flip Flop)
JLMA Japan Light Metal Association
JLP Jig Leg Plate
JLSC Joint Electronics Standardization Committee [UK]
JM Jet Mixing
JMA Japan Management Association
Japan Meteorological Agency
Japan Microphotography Association
JMC Joint Maritime Commission
Joint Maritime Congress [US]
Joint Meteorological Committee
JMD Jamaican Dollar [Currency]
Jungle Message Decoder
JME Jungle Message Encoder
JMED Jungle Message Encoder-Decoder
JMI Japan Metals Institute
John Muir Institute (for Environmental Studies) [US]

JMOS Job Management Operations System
JMP Jump
JMRPS Joint Meteorological Radio Propagation Subcommittee [UK]
JMSPO Joint Meteorological Satellite Program Office
JMSX Job Memory Switch Matrix
JMX Jumbo (Group) Multiplex
JN Jam Nut
Junction
JNACC Joint Nuclear Accident Coordinating Center [US]
JNC Jet Navigation Chart
JND Just Noticeable Difference
JNE Jump Not Equal
JNL Journal
JNLST Journalist
JNOC Japan National Oil Corporation
JNR Japan National Railways
Junior
JNSC Japan Nuclear Safety Commission
Joint Navigation Satellite Committee
JNSDA Japan Nuclear Ship Development Agency
JNT Joint
Joint Network Team
JNT STK Joint Stock
JO Job Order
Joint Organization
Junction Office
JOBNO Job Number [also: JOB NO]
JOBLIB Job Library [also: JOB LIB]
JOC Joint Operation Center
JOCV Japan Overseas Cooperation Volunteers
JOD Jordanian Dinar [Currency]
JODC Japan Overseas Development Corporation
JOG Joggle
JOHNNIAC John von Neumann's Integrator and Automatic Computer [also: Johnniac]
JOI Joint Oceanographic Institution
JOIDES Joint Oceanographic Institution for Deep Earth Samplings
JOIS Japan On-Line Information System [of JICST]
JOL Job Organization Language
JOM Joining of Materials
JONSWAP Joint North Sea Wave Project
JOP Jupiter Orbiter Probe
JOSS Job Sharing System
Johnniac Open-Shop System [Computer Language]
Joint Overseas Switchboard [US Military]
JOUR Journal
JOVIAL Jule's Own Version of International Algorithmic Language
JP Jet Propulsion
Jet Propellant
Jet Pump
Jig Pin
Job Processing
Job Processor
Jute Protection
J&P Joists and Planks
JPA Job Pack Area
JPB Joint Planning Board [US]
JPC Japan Productivity Center
JPCD Just Perceptible Color Difference
JPD Just Perceptible Difference
JPDC Japan Petroleum Development Corporation
JPDR Japan Power Demonstration Reactor
JPIA Japan Plastics Industry Association
JPL Jet Propulsion Laboratory [of NASA]
JPMA Japan Powder Metallurgy Association
JPNL Judged Perceived Noise Level
JPPS Joint Petroleum Products Subcommittee
JPRA Japan Phonographic Record Association
JPRRI Japan Public Relations Research Institute
JPRS Joint Publications Research Service [US]
JPS JICST Photoduplication Service [Japan]
JPTF Joint Parachute Test Facility [of USDOC]
JPU Job Processing Unit
JPW Job Processing Word
JPY Japanese Yen [Currency]
JR Japan Rail [also: JNR]
Joint Research
Journal
Junior
JRATA Joint Research and Test Activities
JRB Jig Rest Button
Joint Radio Board [US]
JRC Japan Research Center
Japan Research Council
Joint Research Center [of CEC, Europe]
JRC-ISPRA Joint Research Center at Ispra [Italy]
JRCM Japan R&D (= Research and Development) Center for Metals
JRD Joint Research and Development
JRDC Japan Research Development Corporation
JRDOD Joint Research and Development Objectives Document
JREA Japan Railway Engineering Association
JRF Janssen Research Foundation [Belgium]
JRIA Japan Radioisotope Association
JRMA Japan Rubber Manufacturers Association
JRP Jute-Reinforced Plastics
JRPS Japan Reinforced Plastics Society
JRR Japan Research Reactor
JRRS Japan Radiation Research Society
JRS Junction Relay Set
JRTP Job Readiness Training Program [Canada]
JS Jam Strobe
Jaswal-Sharma (Theory)
Jet Stream
Joint Services
J/S Jamming to Signal
JSA Jesuit Seismological Association [US]
Job Safety Analysis
JSAI Japan Society fot Artificial Intelligence
JSAM JES (= Job Entry Subsystem) Spool Access Method
JSASS Japan Society for Aeronautical and Space Sciences
JSC Jackson State College [US]
Jamaica Schools Certificate
Japan Science Council
Johnson Space Center [of NASA]
Joint Security Control
Joint Steering Committee
JSCC Japan Society for Composite Materials

JSCE	Japan Society of Civil Engineers
	Japan Society of Corrosion Engineers
JSCFA	Japan Steel Castings and Forgings Association
JSCLC	Joint Standing Committee on Library Cooperation [UK]
JSCM	Japan Society for Composite Materials
JSD	Justification Service Digit
JSEDM	Japan Society of Electrical Discharge Machining
JSEE	Japan Society for Engineering Education
JSEM	Japan Society for Electron Microscopy
	Japan Society of Electrical Discharge Machining
JSEP	Joint Services Electronics Program [US]
JSF	Junctor Switch Frame
JSFM	Japan Society for Strength and Fracture of Materials
JSHS	Junior Science and Humanities Symposium
JSHT	Japan Society of Heat Treatment
JSIA	Joint Service Induction Area
JSIC	Joint Securities Industry Committee
JSIF	Japan Shipbuilding Industry Foundation
JSIS	Japan Society of Iron and Steel
JSL	Job Specification Language
JSLE	Japan Society of Lubrication Engineers
JSLS	Japan Society of Library Science
JSME	Japan Society of Mechanical Engineers
	Joint Soil Moisture Experiment
JSMEBE	Japan Society of Medical Electronics and Biological Engineering
JSMS	Japan Society of Materials Science
JSP	Jackson Structured Programming
JSPE	Japan Society of Precision Engineering
JSPP	Japan Society of Plant Physiologists
JSPPM	Japan Society of Powder and Powder Metallurgy
JSPS	Japan Society for the Promotion of Science
JSRU	Joint Speech Research Unit [UK]
JSS	Joint Surveillance System
JSSFM	Japan Society for Strength and Fracture of Materials
JSST	Jamaican Society of Scientists and Technologists
	Japan Society for Simulation Technology
JST	Japan Society of Tribologists
JSTARS	Joint Surveillance Target Attack Radar System
JSTP	Japan Society for Technology of Plasticity
JSWA	Japan Sewage Works Association
JSWPR	Japan Society on Water Pollution Research
JSZT	Japan Society of Zoological Science
JT	Japanese Tokamak
	Joint
JTAC	Joint Technical Advisory Committee [US]
	Joint Telecommunications Advisory Committee
JTAM	Job Transfer and Management
JTB	Joint Transportation Board [US]
JTC	Joint Telecommunications Committee
JTCMF	Joint Technical Committee on Marine Front
JTDS	Joint Track Data Storage
JTE	Jahn-Teller Effect
	Junction Tandem Exchange
JTEC	Japan Telecommunications Engineering and Consultancy
JTF	Joint Task Force
JTIDS	Joint Tactical Information Distribution System
JTIS	Japan Technical Information Service
JTLY	Jointly
JTM	Job Transfer and Manipulation
JTMP	Job Transfer and Manipulation Protocol
JTPS	Job and Tape Planning System
JTRU	Joint Services Tropical Research Unit
JTS	JICST Translation Service [Japan]
JTSTR	Jet Stream
JTUAC	Joint Trade Union Advisory Committee
JT VENT	Joint Venture
JU	Jack-Up (Oil Drilling Unit)
	Joint User
JUDGE	Judged Utility Decision Generator
JUG	Joint Users Group [US]
JUGFET	Junction-Gate Field-Effect Transistor
JUL	July
JULIE	Joint Utility Locating Information for Excavators
JUN	June
	Junior
JUNC	Junction
JUNE	Joint Utility Notification for Excavators
JUNR	Junior
JUPITER	Juvenescent Pioneering Technology for Robots [Japan]
Jura	Jurassic [Geology]
JUSE	Japanese Union of Scientists and Engineers
JUSE-AESOPP	Japanese Union of Scientists and Engineers—An Estimator of Physical Properties (Database)
JUTEM	Japan Ultrahigh Temperature Materials Research Center
JVA	Jordan Valley Authority [Israel]
JVC	Japan Volunteer Center
JW	Jacket Water
JWPT	Jersey Wildlife Preservation Trust [UK]
JWRI	Japan Welding Research Institute
JWRS	Japan Wood Research Society
JWS	Japan Welding Society

K

K (Kalium) — Potassium
(Karat) — Carat
(Kathode) — Cathode
Keg
Keel
Kelvin
Kerma
Key
Kilobit
Kilobyte
Kilohm
Kina [Currency of Papua New Guinea]
Kinematics
Kip [Unit]
Kip [Currency of Laos]
K-Radiation
Kwacha [Currency of Zambia]
Lysine
Lysyl

k (kathode) — cathode
kayser [Unit]
key
kilo- [SI Prefix]
kilobit
kilobyte
kip [Unit]
knot

KA Kynurenic Acid

K-A Potassium-40 (= K-40) to Argon-40 (= Ar-40) (Ratio) [Geology]

kA kiloampere

ka (kathode) — cathode

KAAC Korean Association of Automatic Control [South Korea]

KAB Keep America Beautiful [US]

KACHAPAG Karlsruhe Charged Particle Group [of INKA, Germany]

KAEA Korean Association of Electronics and Automation [South Korea]

KAEDS Keystone Association for Educational Data Systems

KAERI Korea Advanced Energy Research Institute [South Korea]

KAFB Kirkland Air Force Base [US]

KAICA Korea Auto Industries Cooperation Association [South Korea]

KAIST Korea Advanced Institute of Science and Technology [South Korea]

KAK Key-Auto-Key

Kal Kalamein

KALDAS Kidsgrove ALGOL Digital Analog Simulation

KALDO Kalling-Domnarvet (Process) [Steelmaking]

KAM Kolmogorov-Arnold-Moser (Theorem)

Kan Kansas [also: Kans, US]

KANUPP Karachi Nuclear Power Project [Pakistan]

KAPL Knolls Atomic Power Laboratory [US]

KAPWA Kite Aerial Photography Worldwide Association

KAR Kodak Automated Retrieval

KARLDAP Karlsruhe Automatic Data Processing System [Germany]

KARP Korea Association for Radiation Protection [South Korea]

KAS Kentucky Academy of Science [US]
Korean Astronomical Society [South Korea]

KASC Knowledge Availability Systems Center [of University of Pittsburgh, US]

KASCADE Karlsruhe Shower Core and Array Detector [of KfK, Germany]

KASS Kent Automated Serials System [of KSU, US]

KAU Key-Station Adapter Unit

KAVAA Kenya Audio-Visual Aid Association

KB Kauri-Butanol Value
Keyboard
Kilobit [also: Kbit, kb or kbit]
Kilobyte [also: Kbyte, kb or kbyte]

kb kilobar
kilobit [also: kbit, KB or Kbit]
kilobyte [also: kbyte, KB or Kbytes]

KBA 3-Ketobutyraldehyde 1-Dimethyl Acetal

kbar kilobar

KBD Keyboard

KBDSC Keyboard and/or Display Controller

KBENC Keyboard Encoder

KBES Knowledge-Base Expert Systems

Kbit Kilobit [also: kbit, KB or kb]

Kbit/s Kilobits per Second

KBPRC Keyboard and Printer Controller

KBPS Kilobits per Second [also: kbps]
Kilobytes per Second [also: kbps]

KBS Kilobits per Second
Knowledge-Based System

KBTG Keep Britain Tidy Group [UK]

KBWP Kernkraftwerk Baden-Wurttemberg Planungsgesellschaft [= Nuclear Power Plant in Baden Wurttemberg, Germany]

Kbyte kilobyte [also: kbyte, KB or kb]

KC Kilocharacters
Kilocycle
King's College [Canada]
King's College [of University of London, UK]

Kc Koruna [Currency of Czechoslovakia]

kc kilocurie
kilocycle

kcal kilocalorie [also: kg-cal or Cal]

KCBT Kansas City Board of Trade [US]

KCC K (= Potassium) Chlorochromate
Keyboard Common Contact

KCE Kenya Certificate of Education

kCi kilocurie

KCL Kirchhoff's Current Law

KCNA Korean Central News Agency [South Korea]

KCO Keep Cost Order

KCP Keyboard-Controlled Phototypesetter

KCPE Kenya Certificate of Primary Education

KCPS Kilocycles per Second [also: kcps]

KCR Key Call Receiver
KCS Kilo (= One Thousand) Characters per Second
Kinki Chemical Society [Japan]
Kc/s kilocycles per second [also: kc/s or kcs]
KCU Keyboard Control Unit
KD Kiln Drying
Knocked Down
Kuwaiti Dollar [Currency]
KDB Kenya Dairy Board
Korea Development Bank [South Korea]
KDE Keyboard Data Entry
KDEM Keyboard Data Entry Machine
Kurzweil Data Entry Machine
KDFC Korea Development Finance Corporation [South Korea]
KDO 2-Keto-3-Deoxyoctonate
KDOS Key Display Operated System
Key-to-Disk Operating System
KDP Keyboard Display and Printer
Key Development Plan
K (= Potassium) Dihydrogen Phosphate
KDR Keyboard Data Recorder
KDS Keyboard Display Station
Key Data Station
Key Display System
Key-to-Diskette System
Key-to-Disk System
Kinetic-Dynamic Simulation
KDT Key Data Terminal
KdV Korteweg-deVries (Soliton)
KE Key Equipment
Kinetic Ellipsometry
Kinetic Energy
Knife Edge
Knowledge Engineer
KEAS Knots Equivalent Air Speed
KEE Knowledge Engineering Environment
KEIS Kentucky Economic Information Systems [US]
KEMRI Kenya Medical Research Institute
Ken Kentucky [US]
KENGO Kenya Energy and Environment Association
KEP Key Entry Processing
KEPZ Kaohsiung Export Processing Zone [Taiwan]
KERN Kerning [Typesetting]
KES Kansas Engineering Society [US]
Kenyan Shilling [Currency]
Korea Electronics Show [South Korea]
KESC Karachi Electric Supply Corporation [Pakistan]
KESS Kinetic Energy Storage System
KET Kernel Estimation Techniques
Krypton Exposure Technique
keV kiloelectronvolt
KEW Kinetic Energy Weapon
KEWB Kinetic Experiment on Water Boilers [US]
KEYBD Keyboard
KEYTECT Keyword Detection
KF Key Field
KfA Kernforschungsanlage Julich [= Nuclear Research Plant at Julich, Germany]
KFAED Kuwait Fund for Arab Economic Development
KFAS Keyed File Access System
Kuwait Foundation for the Advancement of Sciences
KFCI Kilo Flux Change per Inch
KfK Kernforschungszentrum Karlsruhe [= Nuclear Research Center at Karlsruhe, Germany]
KFRP Kevlar Fiber-Reinforced Plastics
KFT Karl Fischer Titration
KFTCIC Kuwait Foreign Trade Contracting and Investment Company
KFU King Fahd University [Saudi Arabia]
KFUPM King Fahd University of Petroleum and Minerals [Saudi Arabia]
KG Ketoglutarate
kG Kilo-Gauss
kg kilogram
kg-cal kilocalorie [also: kcal or Cal]
kg/cm kilograms per centimeter
kgf kilogram-force
kgf-m kilogram-meter
KGFST Korean General Federation of Science and Technology [South Korea]
kg/kW kilogram per kilowatt
KGM Key Generator Module
kgm kilogram-meter
kg/m³ kilogram per cubic meter [also: kg per cu m]
KGPS Kilogram per Second [also: kgps]
KGS Kansas Geological Society [US]
Kansas Geological Survey [US]
kg/sec kilograms per second [also: kg sec^{-1}]
kg-wt kilogram weight [also: kgf]
kH kilohenry
KHN Knoop Hardness Number
KHR Kampuchean Riel [Currency of Kampuchea]
KHS Knurled Head Screw
kHz kilohertz
KI Keyboard Input
KIAS Knots Indicated Air Speed
KIC Knowledge Integration Center
KICU Keyboard Interface Control Unit
KIE Kirklees Information Exchange
KIEE Korean Institute of Electrical Engineers [South Korea]
Korea Institute of Electronics Engineers [South Korea]
KIER Korea Institute of Energy and Resources [South Korea]
KIESEC Korea International Exchange Society for Education and Culture [South Korea]
KIET Korea Institute for Industrial Economics and Technology [South Korea]
KIFIS Kollsman Integrated Flight Instrumentation System
KIIET Korean Institute for Industrial Economics and Technology [South Korea]
KIIS Korean Institute for International Studies [South Korea]
KIL Keyed Input Language
KIM Kenya Institute of Management
Korean Institute of Metals [South Korea]
KIN Kinescope
KINGMAP King's Music Analysis Package [of King's College, UK]

KIOPI Kienzle Input/Output Processor Interface
Kip Kip [Currency of Laos]
kip kilopound
kip ft kilo foot-pounds [also: kip-ft]
KIPH Korea Institute for Population and Health [South Korea]
KIPO Keyboard Input Printout
KIPS Kilowatt Isotope Power System
K (= Kilo = 10^3) Instructions per Second [also: kips]
Kilopounds per Square Inch [also: kips]
Knowledge Information Processing System
KIRDI Kenya Industrial Research and Development Institute
KIRS Kuwait Institute for Scientific Research
KIS Keyboard Input Simulation
KISA Korean International Steel Association [South Korea]
KISS Keep it Simple and Short [Computer Jargon]
Keep it Simple Sam [Computer Jargon]
Keep it Simple Sir [Computer Jargon]
Keep it Simple Stupid [Computer Jargon]
Keyed Indexed Sequential Search
Korean Information Science Society [South Korea]
KIST Korean Institute for Science and Technology [South Korea]
KIT Kent Information Technology Conference
Key Issue Tracking
Kitami Institute of Technology [Japan]
Kunming Institute of Technology [PR China]
KWIC (= Keyword in Context) Interactive Tagger
Kurume Institute of Technology [Japan]
Kyoto Institute of Technology [Japan]
Kyushu Institute of Technology [Japan]
KITE Korean Institute of Telematics and Electronics [South Korea]
kJ kilojoule
KK One Million [= K (= 1000) X K (= 1000)]
KKB Kernkraftwerk Brunsbuttel [= Nuclear Power Plant at Brunsbuttel, Germany]
KKG Kernkraftwerk Grafenrheinfeld [= Nuclear Power Plant Grafenrheinfeld, Germany]
KKK Kernkraftwerk Krummel [= Nuclear Power Plant at Krummel, Germany]
KKP Kernkraftwerk Philippsburg [= Nuclear Power Plant at Philippsburg, Germany]
KKR Kohn-Korringa-Rostoker (Process)
KL Key Length
Kleinmann-Low Nebula [Astronomy]
K/L K-Shell/L-Shell Emission Ratio
kL kiloliter [also: kl]
KLA Klystron Amplifier
KLH Keyhole Limpet Hemacyanine
KLIC Key Letter in Context
KLO Klystron Oscillator
KLU Key and Lamp Unit
KM Kilomega
Kinetic Mechanism
Kirchoff Method
K-M Kossel-Mollenstedt (Diffraction Pattern)
kM kilomega
km kilometer
km² square kilometer
K-MAC Kerr-McGee Ash Concentrate
kMc kilomegacycle
KMER Kodak Metal Etch Resist
kmh kilometers per hour [also: km/h or km/hr]
kmps kilometers per second [also: km/s or km/sec]
KMON Keyboard Monitor
KMS Keysort Multiple Selector
Klockner-Metacon Steel Converter
km/s kilometers per second [also: kmps or km/sec]
KMT Kinetic-Molecular Theory
KMU Karl Marx University [at Leipzig, Germany]
KN Knot
Knurled Nut
kN kilonewton
KNAW Koninklijke Nederlandse Akademie van Wetenschappen [= Royal Netherlands Academy of Sciences]
KNCCI Kenya National Chamber of Commerce and Industry
KNCIAWPRC Korean National Committee of the International Association on Water Pollution Research and Control [South Korea]
Kuwaiti National Committee of the International Association on Water Pollution Research and Control
KNFC Kenya National Federation of Cooperatives
KNOW Knowledge Network of the West
KNS Korean Nuclear Society [South Korea]
KNT Short-Cycle Gas Nitriding
KNU Kyungpook National University [South Korea]
KNUST Kwame Nkrumah University of Science and Technology [Ghana]
KO Knockout
KOBAS Konstanzer Bibliotheksautomatisierungssystem [= Constance Library Automation System, Germany]
kOe kilo-Oersted
KOEBES Koelner Bibliothekserschliessungssystem [= Cologne Automated Library Information Retrieval System, Germany]
KOEX Korea Exhibition Center [South Korea]
kohm kilohm [also: kΩ]
KOMAF Korean Machining Fair [South Korea]
KOMMTECH Europaische Kongressmesse fur Technische Automation [= European Congress/Fair for Technical Automation, Germany]
KOPS K (= Kilo = 10^3) Operations per Second [also: kops]
KORF Korf Oxy-Refining Fuel
KORSTIC Korea Scientific and Technological Information Center [South Korea]
KOSAA Korea Shipping Agencies Association [South Korea]
KOSAMI Korea Society for the Advancement of Machine Industry [also: KSAMI, South Korea]
KOSEF Korea Science and Engineering Foundation [South Korea]
KOTFA Korea World Travel Fair [South Korea]
KOTRA Korean Organization for Trade Advancement [South Korea]
Korea Trade Promotion Corporation [South Korea]
KP Key Pulsing

	Key Punch
	Kick Plate
	Kinetic Potential
KPA	Key Pulse Adapter
	Klystron Power Amplifier
kPa	kilopascal
KPC	Keyboard Printer Control
KP&D	Kick Plate and Drip
KPE	Kenya Preliminary Examination
KPF	Key Pulse on Front Cord
KPH	Keystrokes per Hour
KPIC	Key Phrase in Context
Kpn	Klebsiella Pneumoniae [Molecular Biology]
KPO	Key Punch Operator
KPR	Keeper
	Kodak Photoresist
KPS	Korean Physical Society [South Korea]
KPSI	Kilopounds per Square Inch [also: kpsi]
KPSM	Klystron Power Supply Modulator
KPW	North Korean Won [Currency]
KR	Key Register
	Kiloroentgen
	Knowledge Representation
Kr	Krypton
KRAM	Kilobytes of Random-Access Memory
KRAS	Keyworded References to Archeological Science [UK]
KRB	Kernreaktor FWE Bayernwerk [= Nuclear Power Reactor FWE Bayernwerk, Germany]
KRDC	Kingston Research and Development Center [of Alcan, Canada]
KREI	Korea Rural Economy Institute [South Korea]
KRL	Knowledge Representation Language
	Kurt Rossmann Laboratory [US]
KRM	Kurzweil Reading Machine
KRP	Kevlar-Reinforced Plastics
KRR	Kansai Research Reactor [Japan]
KRW	South Korean Won [Currency]
KS	Kansas [US]
	Ketosteroid
	Klockner Steelmaking Process
	Knife Switch
	Korean Standard
K/S	Kurdjumov-Sachs (Relationship)
KSA	Knob Shoe Assembly
KSAM	Keyed Sequential Access Method
	Key Field Sequential Access Method
KSAMI	Korea Society for the Advancement of Machine Industry [also: KOSAMI, South Korea]
KSC	Kennedy Space(flight) Center [also: JFKSC, US]
KSDS	Key-Sequenced Data Set
KSE	Keyboard Source Entry Program
	Kuwait Society of Engineers
KSEA	Korean Scientists and Engineers Association [South Korea]
KSF	Kilopounds per Square Foot [also: ksf]
KSH	Key Stroke per Hour
KSh	Kenyan Shilling [Currency]
KSI	Kilopounds per Square Inch [also: ksi]
KSR	Keyboard Select Routing
	Keyboard, Send/Receive
KST	Keyseat
KSTR	Kema Suspension Test Reactor [Netherlands]
KSU	Kent State University [US]
	Key Service Unit
	Kharkov State University [USSR]
	King Saud University [Saudi Arabia]
KSWPRC	Korean Society of Water Pollution Research and Control [South Korea]
KSZE	Konferenz uber Sicherheit und Zusammenarbeit in Europa [= Conference on Security and Cooperation in Europe]
KT	Keyword Term
	Kinetic Theory
	Kinetic Treatment
kt	karat
	kiloton
KTA	Kerntechnischer Ausschuss [= Committee on Nuclear Engineering [also: KtA, Germany]]
	Key Telephone Adapter
	Korea Tourist Association [South Korea]
KTB	Kontinentales Tiefbohrprogramm [= Continental Deep Drilling Program, Germany]
KTDS	Key-to-Disk Software
KTFR	Kodak Thin Film Resist
KTIC	Korea Toy Industry Cooperation [South Korea]
KTM	Key Transport Module
KTMS	Knapp Time Metaphor Scale
KTN	Kuratorium fur die Tagungen der Nobelpreistrager [= Standing Committee for Nobel Prize Winners Congresses, Germany]
kton	kiloton
KTP	Keyboard Typing Reperforator
KTPC	Korea Trade Promotion Corporation [South Korea]
KTPP	K (= Potassium) Tripolyphosphate
KTR	Keyboard Typing Reperforator
KTS	Key Telephone System
kts	knots
KTSA	Kahn Test of Symbol Arrangement
KTSP	Knots True Air Speed
KTT	Kuhn-Tucker Theory
KTTC	Kenya Technical Teachers College
KTU	Key Telephone Unit
	Kyushu Tokai University [Japan]
KU	Kagoshima University [Japan]
	Kanazawa University [Japan]
	Kansai University [Japan]
	Karachi University [Pakistan]
	Keio University [Japan]
	Kinski University [Japan]
	Kobe University [Japan]
	Kochi University [Japan]
	Kogakuin University [Japan]
	Kogoshima University [Japan]
	Kokushukan University [Japan]
	Kumamoto University [Japan]
	Kyoto University [Japan]
	Kyushu University [Japan]
Ku	Kurchatovium
KULSAA	Karachi University Library Science Alumni Association [Pakistan]
KUR	Kyoto University Reactor [Japan]

kV kilovolt
kVA kilovolt-Ampere
KVAC Kilovolt Alternating Current [also: kVAC]
kVAh Kilovolt-Ampere-hour [also: kVA-h]
kVAhm Kilovolt-Ampere-hour Meter
kvar kilovar
kvarh kilovarhour
KVDC Kilovolt Direct Current [also: kVDC]
KVL Kirchhoff's Voltage Law
KVP Kilovolt Power
kVp kilovolts peak
KW Keyword
Kilo-Word [= 1000 Words]
kW kilowatt
KWAC Keyword and Context
KWADE Keyword As a Dictionary Entry
KWD Kuwaiti Dollar [Currency]
kWe Kilowatt of Electric Power
KWF Kommission zur Forderung der Wissenschaftlichen Forschung [= Commission for the Promotion of Scientific Research, Switzerland]
kW/ft kilowatts/foot
KWG Kernkraftwerk Grohnde [= Nuclear Power Plant at Grohnde, Germany]
kWh kilowatthour
kWhm kilowatthour meter
kWhr kilowatthour [also: kW-hr]
KWIC Keyword In Context
KWIP Keyword In Permutation
KWIT Keyword In Title
KWO Kernkraftwerk Obrigheim [= Nuclear Power Plant at Obrigheim, Germany]
KWOC Keyword Out of Context
KWOT Keyword Out of Title
KWR Know-How Repeating
kWt Kilowatt of Thermal Energy
KWUC Keyword and Universal Decimal Classification
KWW Kernkraftwerk Wurgassen [= Nuclear Power Station at Wurgassen, Germany]
KWY Keyway
KX Kilo-X-Unit [also: kX]
KXU Kilo-X-Unit [also: kXU]
KY Kentucky [also: Ky, US]
Key
Ky Kyat [Currency of Burma]
KYAC Knowledge for Youth About Careers (Program) [of EIC, Canada]
KYBD Keyboard
KYD Cayman Islands Dollar [Currency]
KYS Klockner-Youngstown Steelmaking Process
KZ Kwanza [also: Kz; Currency of Angola]

L

L	Inductance [Symbol]
	Label
	Lake
	Lambert
	Lamp
	Land
	Lance
	Language
	Latent Heat
	Latin
	Latitude
	Lead
	Ledger
	Left
	Length
	Leu [Currency of Romania]
	Leucine
	Leucyl
	Level
	Levorotatory
	Library
	License
	Lift
	Light
	Line
	Link
	Linnaeus
	Liquid
	Lira [Currency of Italy]
	Listening
	Liter
	Load
	Locator
	Location
	Longitude
	Looper
	Loss
	L-Radiation
	Lumen
	Luminance
l	lake
	land
	landing
	large
	late
	league
	left
	length
	levorotatory
	light
	lightning
	line
	liner
	link
	liquid
	liter
	local
	long
	low
	lumen
LA	Lactic Acid
	Landscape Architect
	Landscape Architecture
	Large Angle
	Large Aperture
	Laser Gyro Axis
	Lateral Aberration
	Lead Angle
	Lending Account
	Letter of Authority
	Library Association [UK]
	Licensed Aircraft
	Light Aircraft
	Light Alloy
	Light Armor
	Lighter than Air
	Lightning Arrester
	Line Adapter
	Linear Accelerator
	Linear Algebra
	Linear Amplifier
	Link Allotter
	Listed Address
	Load Accumulator
	Local Action
	Local Address
	Local Area
	Logarithmic Amplifier
	Logical Address
	Longitudinal Aberration
	Longitudinal Acoustical
	Loop Antenna
	Los Angeles [US]
	Louisiana [US]
	Low Altitude
L/A	Letter of Advice
La	Lanthanum
	Louisiana [US]
LAA	Laser Association of America [US]
	Library Association of Alberta [Canada]
	Library Association of Australia
	Liters of Absolute Alcohol
LAAD	Latin American Agrobusiness Development Corporation
LAAPI	Latin American Association of Pharmaceutical Industries [Argentina]
LAAPS	Latin American Association of Physiological Sciences [Argentina]
LAAV	Light Airborne ASW (= Antisubmarine Warfare) Vehicle
LAAW	Light Assault Antitank Weapon
LAB	Laboratory
	Laboratory Animals Bureau [UK]
	Labour

	Low-Altitude Bombing
	Low-Angle Boundary
Lab	Labrador [Canada]
LABCI	Latin American Bank of the Construction Industry
LABEX	Laboratory Exposition [Canada]
LABIL	Light Aircraft Binary Information Link
LABORDOC	Labor Documentation (Database) [of ILO, Switzerland]
LABORELEC	Laboratoire de l'Industrie Electrique [= Electrical Industry Laboratories, Belgium]
Lab PS	Laboratory Power Supply
LABS	Low-Altitude Bombing System
LabVIEW	Laboratory Virtual Instrument Engineering Workbench
LAC	Leading Aircraftsman
	Lemon Administrative Committee [US]
	Library Advisory Council [now: LISC, UK]
	Library Assistant Certificate
	Load Accumulator
	Local Arrangement Committee
	Lunar Aeronautical Chart
LACAC	Latin American Civil Aviation Commission [Peru]
LACAP	Latin American Cooperative Acquisition Project
LACBED	Large-Angle Convergent-Beam Electron Diffraction
LACBPE	Los Angeles Council of Black Professional Engineers [US]
LACBWR	La Crosse Boiling Water Reactor [US]
LACCB	Latin American Confederation of Clinical Biochemistry [Colombia]
LACE	Library Advisory Council for England [now: LISC, UK]
LACES	London Airport Cargo EDP (= Electronic Data Processing) System [UK]
	Los Angeles Council of Engineering Societies [US]
LACFFP	Latin American Commission on Forestry and Forestry Products
LACIE	Large Area Crop Inventory Experiment
LACMA	Latin American and Caribbean Movers Association [Panama]
LACT	Lease Automatic Custody Transfer
LACTRAC	Los Angeles County Transportation Commission [US]
LACV	Logistics Air-Cushion Vehicle
LAD	Lactic Dehydrogenase
	Ladder
	Lithium Aluminum Deuteride
	Location Aid Device
	Logical Aptitude Device
	Lookout Assist Device
LADAR	Laser Detection and Ranging [also: Ladar or ladar]
LADB	Laboratory Animal Database [of National Library of Medicine, US]
LADDER	Language Access to Distributed Data with Error Recovery
LADR	Linear Accelerator-Driven Reactor
LADS	Large Area Diamond System
	Local Area Data Set
LADSIRLAC	Liverpool and District Scientific, Industrial and Research Library Advisory Council [UK]
LADT	Local Area Data Transport
LAE	Lead Angle Error
	Left Arithmetic Element
	Licensed Aircraft Engineer
	London Association of Engineers [UK]
LAEDP	Large Area Electronic Display Panel
LAEM	Linear Anisotropic Elastic Model
LAF	Landscape Architecture Foundation [US]
	Long Address Form
LAFC	Latin American Forestry Commission
LAFIS	Local Authority Financial Information System
LAFTA	Latin American Free Trade Association
	Latin American Association of Freight and Transport Agents [Paraguay]
LAG	Library Automation Group [Australia]
	Load and Go (Technique)
LAGE	Los Angeles Grain Exchange [US]
LAGEOS	Laser Geodetic Earth Orbiting Satellite
	Laser Geodynamic Satellite [US]
LAH	Lithium Aluminum Hydride
	Logical Analyzer of Hypothesis
LAHCG	Lookahead Carry Generator
LAHS	Low-Altitude, High-Speed
LAI	Leaf Area Index
LAIA	Latin American Industrialists Association [Uruguay]
LAIEC	Latin American Institute of Educational Communication [Mexico]
LAIG	Library Association's Industrial Group [UK]
LAINS	Low-Altitude Inertial Navigation System
LAIS	Library Acquisitions Information System [of LARC, US]
LAIT	Library Association Information Technology Group [UK]
LAK	Laos Kip [Currency of Laos]
	Lymphokine-Activated Killer
LAL	Limulus Amoebocyte Lysate
LALLS	Low-Angle Laser Light Scattering
LALO	Low-Altitude Observation
LALR	Lookahead Left-to-Right
LALS	LaGuardia Automated Library System [US]
LAM	Laminate
	Laminate Material
	Lamination
	Liquid and Amorphous Metals Conference
	Load Accumulator with Magnetization
	Load Accumulator with Magnitude
	Logical Acknowledgement Message
	Longitudinal Acoustic Mode
	Loop Adder and Modifier
LAMA	Latin American Manufacturers Association
	Light Aircraft Manufacturers Association [US]
	Local Automatic Message Accounting
	Locomotive and Allied Manufacturers Association [UK]
LAMC	Language and Mode Converter
LAMCS	Latin American Communications System
LAME	Licensed Aircraft Maintenance Engineer
LAMEF	Los Alamos Medium Energy Facility [US]
LAMF	Los Alamos Meson Facility [US]
LAMIS	Local Authority Management Information System [UK]
LAMMA	Laser Microprobe Mass Analysis
	Laser Microprobe Microanalysis

LAMP	Laser and Maser Patents
	Low-Altitude Manned Penetrator
LAMPF	Los Alamos Meson Physics Facility [US]
LAMPRE	Los Alamos Molten Plutonium Reactor Experiment [US]
LAMPS	Laser Material Processing Program [Canada]
	Light Airborne Multipurpose System
LAMS	Light Aircraft Maintenance Scheme [UK]
	Load Alleviation and Mode Stabilization
LAMSAC	Local Authority Management Services and Computer Committee [UK]
LAN	Lateral Access Network
	Local-Area Network
LANAC	Laminar, Air Navigation and Collision [also: Lanac or lanac]
Lanc	Lancashire [UK]
LANCET	Library Association, National Council for Educational Technology [Joint UK/US Project]
Lancs	Lancashire [UK]
L and D	Loss and Damage [also: L&D]
LANDS	Language-Oriented Development System
LANG	Language
LANL	Los Alamos National Laboratories [of USDOE]
LANNET	Large Artificial Nerve Net
	Large Artificial Neuron Network
LANS	Load Alleviation and Stabilization
	LORAN (= Long-Range Navigation) Airborne Navigation System
LANSCE	Los Alamos Neutron Scattering Center [US]
LANTIRN	Low-Altitude Navigation and Targeting Infrared for Night [of US Air Force]
LAN-WAN	Local-Area Network to Wide-Area Network Communications
LAP	Laboratory Accrediation Program
	Lesson Assembly Program
	Leucine Aminopeptidase
	Life Analysis Package
	Limited-Access People
	Line Access Point
	Link Access Procedure
	Link Access Protocol
	List Assembly Program
LAPA	Leucocyte Alkaline Phosphatase Activity
	Lightweight Aggregate Producers Association [US]
LAPADS	Lightweight Acoustic Processing and Display System
LAPB	Link Access Procedure Balanced
	Link Access Protocol Balanced
LAPD	Limited Axial Power Distribution
LAPDOG	Low-Altitude Pursuit Dive On Ground
LAPES	Low-Altitude Parachute Extraction System
LAP-M	Link Access Procedure for Modems
LAPPES	Large Power Plant Effluent Study [of TVA, US]
LAPW	Linearized Augmented Plane Wave
LAQ	Lacquer
LAR	Limit Address Register
	Low-Angle Reentry
LARA	Latin American Railways Association [Argentina]
	Light-Armed Reconnaissance Aircraft
LARAM	Line-Addressable Random-Access Memory
LARC	Langley Research Center [of NASA]
	Library Automation Research and Consulting Association [US]
	Light Amphibious Resupply Cargo Vessel
	Linear Aromatic Condensation
	Livermore Automatic Research Computer [of LRL, US]
LaRC	Langley Research Center [of NASA]
LARCT	Last Radio Contact
LARC-TPI	Linear Aromatic Condensation — Thermoplastic Imide
LAREC	Los Alamos Reactor Economics Code [US]
LARF	Latin American Reserve Fund
LARIAT	Laser Radar Intelligence Acquisition Technology
LARP	Local and Remote Printing
LARPS	Local and Remote Printing Station
LARR	Linear Accelerator-Regenerator Reactor
LARS	Laser Angular Rate Sensor
	Laser Articulated Robot System
	Library Accessions and Retrieval System
	Lower Airspace Radar Advisory Service [UK]
LA RSIS	Library Association — Reference, Special and Information Section [also: LA-RSIS, UK]
LARU	Latin American Research Unit
LAS	Laboratory Automation Software
	Laboratory Automation System
	Laboratory of Atmospheric Sciences [of NSF, US]
	Land Agents Society [UK]
	Landing and Approach System
	LANSCE (= Los Alamos Neutron Scattering Center) Advisory Committee [US]
	Large-Angle Scattering
	Large Astronomical Satellite [of ESRO, France]
	Large Autoclave System
	Launch Auxiliary System
	League of Arab States
	Library Association of Singapore
	Light-Activated Switch
	Light-Active Sensor
	Linear Alkyl Sulfonate
	Lithia-Alumina-Silicate
	Lithium Aluminosilicate
	Local Address Space
	Logic Analysis System
	Louisiana Academy of Sciences [US]
	Low-Altitude (Observation) Satellite
	Low-Angle Scattering
LASA	Laboratory Animals Science Association [UK]
	Large-Aperture Seismic Array
	Latin American Shipowners Association
LASCA	Los Angeles State and County Arboretum [US]
LASCO	Latin American Science Cooperation Office
LASCR	Light-Activated Silicon-Controlled Rectifier
LASCS	Light-Activated Silicon-Controlled Switch
LASE	Laser Applications in Science and Engineering
	Lidar Atmospheric Sensing Experiment [of NASA]
	Load at Specified Elongation
	Logic Analysis and Slave Emulation
LASER	Light Amplification by Stimulated Emission of Radiation [also: Laser or laser]
LASERCOM	Laser Computer Output Microfilm [also: Lasercom]
LASH	Laser Antitank Semi-Active Homing

LASIE Library Automated Systems Information Exchange [Australia]
LASIL Land and Sea Interaction Laboratory
LASL Los Alamos Scientific Laboratory [US]
LASP Local Attached Support Processor
LASPAU Latin American Scholarship Program of American Universities [US]
LASR Laboratories for Astrophysics and Space Research [US]
Litton Airborne Search Radar
LASRA Leather and Shoe Research Association [New Zealand]
LASRM Low-Altitude Short-Range Missile
Low-Altitude Supersonic Research Missile
LASS Laser Processing of Superconductors in Space
Local-Area Signaling Services
Logistics Analysis Simulation System
LASSO Laser Search and Secure Observer
Laser Synchronization from Stationary Orbit
Landing and Approach System, Spiral-Oriented
LASSOS Library Automation System and Services Options Study [UK]
LASST Laboratory for Surface Science and Technology [US]
LAST Large Aperture Scanning Telescope
LASV Low-Altitude Supersonic Vehicle
LAT Laboratoire d'Aerothermique (du Centre de la Recherche Scientifique) [= Aerothermic Laboratory (of the Scientific Research Center), France]
Laser Acquisition and Tracking
Lateral
Latin
Latitude
Licensed Aircraft Technologist
Local Area Transport
LATA Local Access and Transport Area
LATAF Logistics Activation Task Force
LATAs Local Access and Transport Areas
LATB Lithium Aluminum tri-tert-Butoxyhydride
LATCA Los Angeles Traffic Control Area [US]
LATCC London Air-Traffic Control Center [UK]
$La(THD)_3$ Tris(tetramethylheptanedionato)lanthanum
LATINCON Latin American Convention [of IEEE, US]
LATIRN Low Altitude Navigation Targeting Infrared for Night Flying
LATIS Loop Activity Tracking Information System
LATS Litton Automated Test Set
Long-Acting Thyroid Simulator
LAU Line Access Unit
Line Adapter Unit
Lobe Access Unit
LAUNS Local-Area Underwater Navigation System
LAV Landmaschinen- und Ackerschlepper-Vereinigung [= Association for Agricultural Machines and Tractors [of VDMA, Germany]]
Light-Armored Vehicle
Lymphadenopathy-Associated Virus
LAVA Linear Amplifier for Various Applications
LAVE L'Association Volcanologique Europeenne [= European Volcanological Association, France]
LAW Laser-Assisted Arc Welding
Light Antiarmor Weapon
Light Antitank Weapon
Logic Analysis Workstation
LAWB Los Alamos Water Boiler [US]
LAWG Latin American Working Group
LAWRS Limited Air Weather Reporting Certificate
LB Label
Labrador [Canada]
Langmuir-Blodgett (Film)
Laser Brazing
Lecture Bottle
Left Boundary
Light Bombardment
Linear Analysis
Line Broadening
Line Buffer
Linoleum Base
Load Bearing
Local Battery
Locating Button
Logical Block
Long Barrel
Lower Bound
L-B Langmuir-Blodgett (Film)
lb (libra) — pound
LBA Lima Bean Agglutenin
Linear Bounded Automaton
Linear Buckling Analysis
Limestone-Building Algae
Local Bus Adapter
Ludwig Boltzmann Association [of SPSR, Austria]
Luftfahrt-Bundesamt [= Department for Aeronautics and Astronautics, Germany]
lb ap pound, apothecaries [also: lb apoth]
LBA-SPSR Ludwig Boltzmann Association — Society for the Promotion of Scientific Research, Austria]
lb av pound, avoirdupois
lb bhp^{-1} hr^{-1} pounds per brake horsepower-hour [also: lb BHPhr]
LBBP Laboratory of Blood and Blood Products [US]
LBC Laser Beam Cutting
Linear Block Code
Local Bus Controller
Lower Binding-Energy Component
LBCM Locator at the Back Course Marker
LBD Libyan Dinar [Currency]
LBE Lance Bubbling Equilibrium
LBEN Low Byte Enable
LBF Lactobacillus Bulgaricus Growth Factor
Langmuir-Blodgett Film
lbf pound force
lbf-ft foot-pound [also: lbf ft]
lb-ft pound-foot [also: lb ft]
lb-in pound-inch [also: lb in]
LBG Load Balancing Group
lb/gal pounds per gallon
LBIC Light Beam Induced Current
LBIDI Liberian Bank for Industrial Development and Investment
LBIM Logic Based Information Modelling

lb/in^2 pounds per square inch [also: PSI or psi]
LBL Label
Lawrence Berkeley Laboratory [of UCB, US]
Lima Bean Lectin
LBLG Lawrence Berkeley Laboratory Group [of LBL, US]
LBLOCA Large Break Loss-of-Coolant Accident
LBLTY Liability
LBM Laser Beam Machining
Load Buffer Memory
Locator Back Marker
lbm pound, mass
LBMA Lumber and Building Materials Association
LBMAC Lumber and Building Materials Association of Canada
LBMAO Lumber and Building Materials Association of Ontario [Canada]
LBN Line Balancing Network
Logic Bucket Number
lbn pounds, net
LBNP Lower Body Negative Pressure (Boots)
LBO Leveraged Buyout
Line Buildout (Network)
Lithium Barium Oxide (Crystal)
LBP Laser Beam Printer
Lebanese Pound [Currency]
Length between Perpendiculars
lb per bhp hr pounds per brake horsepower-hour
LBR Laser Beam Recorder
Lumber
LBRV Low Bit-Rate Voice
LBS Load Balance System
lbs pounds
LBT Langmuir-Blodgett Theory
Local-Battery Talking
Low Bit Test
lb t pound, troy
lb/t pounds per ton
LBT CBS Local-Battery Talking, Common-Battery Signaling
lb tr pound, troy
LBTS Land-Based Test Site
LBW Laser Beam Welding
LBX Local Bus Extension
LC L (= Inductance) — C (= Capacitance)
Labour Canada
Lamb Committee [of NLSMB, US]
Landing Craft
Lanthanum Chromite
Laser Chemistry
Last Card
Late Commitment
Launch Complex
Lead Coating
Lead-Covered
Leakage Coefficient
Learning Curve
Leather Chemist
Leather Chemistry
Lethal Concentration
Letter of Credit
Level Control
Liaison Committee
Library of Congress [US]
Life Cycle
Limited Coordination
Linear Circuit
Linear Control
Line Carrying
Line Circuit
Line Concentrator
Line Connector
Line Control(ler)
Line of Communication
Line of Contact
Link Circuit
Liquid Controller
Liquid Crystal
Liquid Chromatography
Location Counter
Load Carrier
Load Cell
Load Center
Load Compensation
Load Curve
Loading Coil
Locational Clearance (Fit)
Logical Clock
Logic Corporation
Logistics Command
Lomer-Cottrell (Dislocation)
Low Carbon
Low-Cost Attitude
Lower Case [also: lc]
Lower Cassette
Lyman Continuum
L/C Letter of Credit
L-C L (= Inductance) — C (= Capacitance)
lc (loco citato) — in the place cited
LC50 Lethal Concentration 50 [also: LC$_{50}$]
LCA Line Control Adapter
Local Communications Adapter
Lower Case Alphabet
LCAC Landing Craft Air Cushion
LCAHRS Low-Cost Attitude and Heading Reference System [also: LCA HRS]
LCAO Linear Combination of Atomic Orbitals
LCAO-MO Linear Combination of Atomic Orbitals — Molecular Orbitals
LCAP Loop Carrier Analysis Program
LCB Launch Control Building
Line Control Block
Logic Control Block
LCBO Linear Combinations of Band Orbitals
LCC Landing Craft, Control
Late Choice Call
Launch Control Center
Launch Control Console
Lead-Coated Copper
Lead-Covered Cable
Leadless Chip Carrier
Least-Cost (Satisfactory Melting) Charge
Life-Cycle Cost
Line Choking Coil

Liquid Crystal Cell
Loading Coil Case
Local Communications Complex
Local Communications Console
Logistics Coordination Center
London County Council [UK]
Lost Calls Cleared

LCCC Leadless Ceramic Chip Carrier
Library of Congress Computer Catalogue [US]

LCCCN Library of Congress Catalogue Card Number [US]

LCCD Low Complexity Color Display

LCCM Late Choice Call Meter

LCCMARC Library of Congress Current Machine-Readable Catalogue (File) [US]

LCCOGA Liaison Committee of Cooperating Oil and Gas Associations [US]

LCCP Laser-Controlled Chemical Processing

LCCT LaQue Center for Corrosion Technology

LCD Least Common Denominator
Letter Carrier Depot
Lowest Common Denominator
Liquid Crystal Device
Liquid Crystal Display

LCDC Laboratory Center for Disease Control [of HWC, Canada]

LCDDS Leased Circuit Digital Data Service [UK]

LCdr Lieutenant-Commander

LCDS Low-Cost Development System [US]

LCDTL Load-Compensated Diode Transistor Logic

LCE Land-Covered Earth
Launch Complex Equipment
Load Circuit Efficiency

LCEECSTI Liaison Committee of the European Economic Community Steel Tube Industry [France]

LCES Least Cost Estimating and Scheduling

LCF Language Central Facility
Launch Control Facility
Light Control Film
Logical Channel Fill
Low-Cycle Fatigue

LC50 Lethal Concentration 50 [also: LC_{50}]

LCFS Last Come, First Served

LCFTU Lesotho Congress of Free Trade Unions

LCG Linear Congruential Generator
Liquid Crystal Graph
Load Classification Group

LCGO Linear Combination of Gaussian Orbital

LCHS Large Component Handling System

LCH Lens Culinaris Agglutinin
Lost Calls Held

LCHTF Low-Cycle High-Temperature Fatigue

LCI Livestock Conservation Institute [US]
Load-Commutated Inverter

LCIE Laboratoire Central des Industries Electriques [= Central Laboratory for the Electrical Industries, France]

LCIGB Locomotive and Carriage Institution of Great Britain

LCJ Low-Contaminant Jarosite (Process)

LCKR Locker

LCL Less than Carload
Library Control Language
Lifting Condensation Level
Limited Channel Log-Out
Linkage Control Language
Local
Low Camera Length
Low Capacity Link
Lower Control Limit
Lowest Car Load

LCl Low Chloride

LCL-CBED Low Camera Length — Convergent Beam Electron Diffraction

LCLU Landing Control Logic Unit

LCLV Liquid-Crystal Light Valve

LCM Large Core Memory
Late Change Message
Last Calls Meter
Lead-Coated Metal
Least Common Multiple
Line Concentrator Module
Line Control Module
Liquid Composite Molding
Liquid Curing Medium
Liquid Curing Method
Lost Circulation Materials
Lowest Common Multiple

LCMARC Library of Congress Machine-Readable Catalogue (Files) [US]

LCMCFC Liaison Committee for Mediterranean Citrus Fruit Culture [Spain]

LCMI Licentiate of the Cost and Management Institute

LCMM Life Cycle Management Model

LCMS Library Collection Management System [of UTLAS, Canada]
Liquid Chromatography Mass Spectrometry [also: LC/MS or LC-MS]

LCMS Logistics Command Management System

LCN Load Classification Number
Logical Channel Number

LCNR Liquid Core Nuclear Rocket

LCNT Link Celestial Navigation Trainer

LCNTR Location Counter

LCO Limited Condition of Operability
Limiting Conditions for Operations
Load Control and Optimization

LCol Lieutenant-Colonel

LCP Language Conversion Program
Large Coil Program
Laser-Induced Chemical Processing
Licentiate of the College of Preceptors [UK]
Link Control Procedure
Liquid Crystal Polymer
Liquid Cyclone Process
Local Control Point
Logic-Controlled Protocolling

L-Cpl Lance-Corporal [also: L/Cpl]

LCR L (= Inductance) — C (= Capacitance) — R (= Resistance)
Least Cost Routing
Letter Carrier Route
Lithium-Cooled Reactor

Load-Compensating Relay
Log Count Rate

L/Cr Letter of Credit

LCRE Lithium-Cooled Reactor Experiment [of INEL, US]

LCRO Linear Combination of Rydberg Orbitals

LCRS Leachate Collection and Recovery System

LCRU Lunar Communications Relay Unit [of NASA]

LCS Large-Capacity Storage
Large-Capacity Core Storage
Large Core Storage
Larsen Control System
Launch Control System
Library Computer System
Limiting Creep Stress
Lincoln Calibration Sphere
Loadable Control Storage
Loop Control System
Loudness-Contour Selector
Low-Carbon Steel

LCSE Laser Communications Satellite Experiment [of NASA]

LCSH Library of Congress Subject Headings [US]

LCSO Local Communications Service Order

LCSRM Loop Current Step Response Method

LCSSAP Low-Cost Silicon Solar Array Project

LCST Lower Critical Solution Temperature

LCSU Local Concentrator Switching Unit

LCT Laboratoire Central de Telecommunications [= Central Laboratory of Telecommunications, France]
Linear Combination Technique
Logical Channel Determination

LCTD Located

LCTF Large Coil Test Facility

LCTI Large Components Test Installation

LCTL Large Components Test Loop

LCU Line Control Unit
Link Control Unit

LCUC Letter Carriers Union of Canada

LCVD Laser Chemical Vapor Deposition
Least Voltage Coincidence Detection

LCVIP Licensee Contractor Vendor Inspection Report

LCW Line Control Word

LCZ Localize

LCZR Localizer

LD Lactate Dehydrogenase
Lactic Dehydrogenase
Large Diameter
Laser Deposition
Laser Diode
Lateral Direction
Leak Detector
Lethal Dose
Letter Description
Level Discriminator
Libyan Dinar [Currency]
Lift-Drag Ratio
Light-Dark Cycle
Light Duty
Limited
Linear Decision
Line Drawing
Line of Departure
Linker Directive
Linz/Donawitz (Process) [Steelmaking]
Lipid Droplet
List of Drawings
Load
Logical Design
Logic Driver
Long Delay
Long Distance
Loop Disconnect
Low Density
Luminous Discharge

L&D Loss and Damage

L/D Length/Diameter Ratio [also: l/d]
Lift-to-Drag Ratio

L$ Liberian Dollar [Currency]

LD50 Lethal Dose 50 [also: LD_{50}]

LDA Landing Distance Available
Laser Diode Array
Laser Doppler Anemometry
Lead Development Association [UK]
Line Data Area
Line Driving Amplifier
Lithium Diisopropylamide
Load in Accumulator
Local Data Administrator
Local Density Approximation
Localizer-Type Directional Aid
Locate Drum Address
Logical Device Address

LD-AC Linz-Donawitz/ARBED-CRM [Steelmaking Process Developed Jointly by ARBED (= Acieries Reunies de Burbach-Eich-Dudelange) and CRM (= Centre de Recherches Metallurgiques)]

LDB Light Distribution Box
Logical Database

LDAM Linked Direct Access Method

LDBA Leakage Design Basis Accident

LDC LASA (= Large Aperture Seismic Array) Data Center
Latitude Data Computer
Less Developed Country
Light Direction Center
Linear Detonating Cord
Line Drop Compensator
Linguistics Documentation Center [of U of O, Canada]
Liquid Dynamic Compaction
Load Duration Curve
Local Display Controller
Logical Device Coordinate
Long Distance Communications
Lower Dead Center
Low-Speed Data Channel

LDCC Large Diameter Component Cask

LDCS Long Distance Control System [US]

LDD Lightly-Doped Drain
Local Data Distribution
Logic Design Data
Luminaire Dirt Depreciation (Factor)

LDDM Laser Doppler Displacement Meter

LDDS Limited Distance Data Set
Low-Density Data System

LDE Laminar Defect Examination
Language-Directed Editor
Linear Differential Equation

LDEF Long-Duration Exposure Facility (Satellite) [of NASA]

LDEM Laboratory for Development of Engineering Materials [of Kanazawa University, Japan]

LDF Light Distillate Feedstock
Linear Discriminate Function
Load Factor
Local Density Fluctuation

LD50 Lethal Dose 50% [also: LD_{50}]

LDG Landing
Leading
Ledger
Loading

LDH Lactic Dehydrogenase
Limiting Dome Height

LDH-A Lactic Dehydrogenase A

LDH-B Lactic Dehydrogenase B

LDH/PK Lactic Dehydrogenase/Pyruvate Kinase

LDI Landing Direction Indicator
Lossless Discrete Integrator

LDIN Lead-In Light System

LDL Language Description Language
Low-Density Lipoprotein

LDLA Limited Distance Line Adapter

LDM Laser Diode Mount
Limited-Distance Modem
Linear Delta Modulation
Local Data Manager

LDN Listed Directory Number

LDNR Lead Dinitroresorcinolate

LDNS Laser Doppler Navigation System

L-DOPA 3-(3,4-Dihydroxyphenyl)-L-Alanine

LDOS Local Density of States

LDP Language Data Processing

LDPE Low-Density Polyethylene

LDR Large Deployable Reflector
Leader
Light-Dependent Resistor
Limiting Drawing Ratio
Limiting Draw Ratio
Limiting Dome Ratio
Line Driver-Receiver
Loader
Low Data Rate

LDRI Low Data Rate Input

LDRS LEM (= Logical End of Medium) Data Reduction System

LDRTA Long Distance Road Transport Association [Australia]

LDS Large Dark Spot Effect
Large Disk Storage
Large-Format Document Scanner
Library Delivery System
Light Distillate Spirit
Logic Digital Switch
Local Distribution Service
Local Distribution System
Logic Design Translator
Long Distance Transmission

LDSCPG Landscaping

LDSD Library and Documentation Services Division [of CANMET, Canada]

LDSU Local Digital Service Unit

LDT Language-Dependent Translator
Linear Differential Transformer
Logic Design Translator
Long Distance Transmission
Low-Density Telephony

LDTA Leak Detection Technology Association [US]

LDTP Long-Distance Thrift Pak

LDTS Linear-Discrete Time System
Low-Density Telephony Service

LDTV Low-Definition Television

LDU Local Delivery Unit

LDV Laser Doppler Velocimeter

LDX Long Distance Xerography

LE Labour Exchange [UK]
Large End
Launching Equipment
Laundry Engineer
Leading Edge
Length
Length of Engagement
Less than or Equal to
Light Equipment
Limit of Error
Linear Equation
Line Equipment
Local Exchange
Locomotive Engineer
Locomotive Engineering
Logic Element
Logic Evaluator
Logistics Engineer
Logistics Engineering
Loop Extender
Low Efficiency
Lower Envelope
Low Explosive
Lubrication Engineer
Lubrication Engineering
Lupus Erythematosus (Factor)

Le Leone [Currency of Sierra Leone]

LEA Laboratory for Experimental Astrophysics [of LLNL, US]
Leather
Local Education Authority [UK]
Long Endurance Aircraft
Longitudinally Excited Atmosphere
Lycopersicon Esculentum Agglutinin

LEAA Law Enforcement Assistance Administration

LEAD Learn, Execute and Diagnose
Local Employment Assistance and Development

LEADER Lehigh Automatic Device for Efficient Retrieval [of Lehigh University, US]

LEADERMART Lehigh Automatic Device for Efficient Retrieval—Mechanical Analysis and Retrieval of Text [of Lehigh University, US]
LEADS Law Enforcement Automated Data System
LEAF Leaflet
LISP (= List Processing) Extended Algebraic Facility
LEANS Lehigh Analog Simulator
LEAP Language for the Expression of Associative Procedures
Local Employment Assistance and Development Program
Lunar Escape Ambulance Pack [of NASA]
LEAPP Locally Enclosed Area, Particulate Protected (Concept)
LEAR Logistics Evaluation and Review Technique
Low-Energy Anti-Photon Ring
Lower Echelon Automatic Switchboard
LEAS Lease Electronic Accounting System
Lower Echelon Automatic Switchboard
LEASAT Leased Satellite [US]
LEASG Leasing
LEBCO Low Energy Building Council
LEC Light Energy Converter
Liquid Encapsulated Czochralski
Liquid Exclusion Chromatography
Low-Energy Channeling
Low-Energy Cure
LECA Lightweight Expanded Clay Aggregate
LECC Lake Erie Cleanup Committee [US]
LECO Local Engineering Control Office
LECVD Low-Energy Chemical Vapor Deposition
LED Light-Emitting Diode
Linear Electric Device
LEDA Local Economic Development Assistance (Program) [Canada]
LEDIS Local Economic Development Information Service [UK]
LEDS Low-Energy Dislocation Structure
LEDT Limited-Entry Decision Table
LEED Laser-Energized Explosive Device
Low-Energy Electron Diffraction
LEELS Low-Energy Electron Loss Spectrometry
LEEP Library Education Experimental Project [of University of Syracuse, US]
LEER Low-Energy Electron Reflection
LEF Light-Emitting Film
Linear Electric Field
Line Expansion Function
LEFD Large Experimental Fusion Device
LEFM Linear Elastic Fracture Mechanics
LEG Legal
Library Education Group [now: TEG, UK]
LEGBAC Limited Exploratory Group on Broadcasting-to-Aeronautical Compatibility
LEGOL Legally Oriented Language
LEG WT Legal Weight
LEIB Low-Energy Ion Beam
Leic Leicestershire [also: Leics, UK]
LEID Low-Energy Ion Detector
LEIS Low-Energy Ion Scattering
LEISS Low-Energy Ion Scattering Spectroscopy
LEJ Longitudinal Expansion Joint
LEK Lek [Currency of Albania]
Liquid Encapsulated Kyropoulos
LEL Large Engineering Loop [of LeRC, US]
Lebanese Pound [Currency]
Lower Explosion Level
Lower Explosion Limit
LELU Launch Enable Logic Unit
LEM Laboratory of Electromodeling
Laser Exhaust Measurement
Lempira [Currency of Honduras]
Linear Electric Machine
Logical End of Medium [also: LEOM]
Logical Enhanced Memory
Lunar Excursion Module [of NASA]
LEMA Lifting Equipment Manufacturers Association
LEMD Low-Energy Misfit Dislocation
LEMIT Laboratorio de Esayo de Materiales e Investigaciones Tecnologicas [= Laboratory for Materials Testing and Technological Investigations, Argentina]
LEMP Lightning Electromagnetic Pulse
LEMUF Limit of Error on Material Unaccounted For
LEN Length
LENP Low-Energy Nuclear Physics
LEO Librating Equidistant Observer
Low Earth Orbit (Satellite)
Lyon's Electronic Office [Computer System]
LEOM Logical End of Medium [also: LEM]
LEOS Lasers and Electro-Optics Society [of IEEE, US]
LEP Large Electron Positron
Low-Energy Physics
Lower End Plug
LEPD Low-Energy Positron Diffraction
LEPS Launch Escape Propulsion System
LEPT Long Endurance Patrolling Torpedo
LEQ Line Equipped
Line of Equipment
LER Licensee Event Report
LERA Limited Employee Retirement Account
LERC Laramie Energy Research Center [US]
LeRC Lewis Research Center [of NASA]
LERLS Lake Erie Regional Library System
LERM Laboratoire d'Etudes et de Recherche sur les Materiaux [= Research Laboratory for Materials, France]
LERT Linear-Extensional, Rotation and Twist Robot
LES Launch Enabling System
Launch Escape System
Light-Emitting Switch
Light Exposure Speed
Limited Early Site
Lincoln Experimental Satellite [US]
Local Excitatory State
Loop Error Signal
Louisiana Engineering Society [US]
LESA Lake Erie Steam Association [US]
Lunar Exploration System for Apollo [of NASA]
LESC Light-Emitting Switch Control
LESR Limited Early Site Review

LESS Lateral Epitaxy by Speeded Solidification
Least Cost Estimating and Scheduling
LESSC Louisville Engineering and Scientific Societies Council [US]
LEST Low-Energy Speed Transmission
LET Launch Escape Tower
Leading Edge Trigger
Letter
Letters Shift
Linear Energy Transfer
Lincoln Experimental Terminal [of NASA]
Logical Equipment Table
LETB Local Exchange Test Bed
LETIS Leicestershire Technical Information Service [UK]
LETS Law Enforcement Teletypewriter System
LEU Launch Enabling Unit
Leu [Currency of Romania]
Leucocyte
Low-Enriched Uranium
Leu Leucine
Leucyl
LEU-CAM Leucocyte Adhesion Molecules
LEU/TH Low Enriched Uranium/Thorium Fuel
LEV Level
Loader/Editor/Verifier
Lev Lev [Currency of Bulgaria]
LEX Line Exchange
LEXIS Lexicography Information Service [Germany]
LF Lactoferrin
Ladle Furnace
Laminar Flow
Landfill
Large File
Launch Facility
Leaflet
Leapfrog Configuration
Leveling Foot
Lightface
Limiting Fragmentation
Linear File
Linear Foot
Line Feed (Character)
Line Finder
Line Frequency
Linoleum Floor
Load Factor
Logical File
Logic Function
Low Frequency
LFA Load Flow Analysis
Lobster Fishing Area
Local Flying Area
Low Flow Alarm
LFB Linear Feedback
LFBC Livestock Feed Board of Canada
LFBR Liquid Fluidized Bed Reactor
LFBR-CX Liquid Fluidized Bed Reactor Critical Experiment
LFC Laminar Flow Control
Laminar Forced Convection
Level of Free Convection
Load Frequency Control
Local Forms Control
LFCB Load to Forms Control Buffer
LFCF Laminar Free Convective Flow
LFCM Low-Frequency Cross-Modulation
LFD Line Fault Detector
Local Frequency Disturbance
Low-Frequency Disturbance
LFE Laboratory for Electronics
Local Field Effect
LFEM Laboratoire Federal d'Essai des Materiaux [= Federal Laboratory for Materials Testing, Switzerland]
LFF Land- und Forstwirtschaftlicher Forschungsrat [= Research Committee on Agriculture and Forestry, Germany]
LFFET Low-Frequency Field-Effect Transistor
LFGS Low Flow Gas Saver
LFI Laser-Fiber Illuminator
Laxforskningsinstitutet [= Salmon Research Institute, Sweden]
LFIM Low-Frequency Instruments and Measurements [Committee of IEEE IM]
LFJ Local Feed Junctor
LFL Lower Flammability Limit
LF LORAN Low-Frequency LORAN (= Long-Range Navigation) [also: LF Loran or LF loran]
LFLU Liberia Federation of Labour Unions
LFM Large File Management
Lateral Force Microscope
Limited-Area Fine-Mesh Model
Linear Frequency Modulation
Liquid Film Migration
Local File Manager
Lower Felsic Metapyroclastics [Mining]
Low-Frequency Microphone
Low-Powered Fan Marker
LFM+ Large File Management Plus
LFN Line Format Number
Logical File Name
LFO Low-Frequency Oscillator
LFPS Low-Frequency Phase Shifter
LFQ Light Foot Quantizer
LFR Low-Frequency Radio Range
LFr Luxembourg Franc [also: LFR; Currency]
LFRAP Long Feeder Route Analysis Program
LFRD Lot Fraction Reliability Definition
LFS Labour Force Survey [of Statistics Canada]
Launch Facility Simulator
Local Format Storage
Logical File Structure
LFSR Linear-Feedback Shift Register
LFSS Landing Force Support Ship [US Navy]
LFSW Landing Force Support Weapon [US Navy]
LFU Least Frequency Unit
Least Frequently Used
LFV Lunar Flying Vehicle [of NASA]
LFW Linear Friction Welding
LG Landing Gear
Landing Ground
Large
Large Grain

	Length
	Liaison Group
	Light Gun
	Limit Gauge
	Linear Gate
	Line Generator
	Line Graph
	Lodging
	Logging
	Long
	Looping
LGA	Light-Gun Amplifier
	Local Government District
	Low Gloss/Automotive
LGC	Laboratory of the Government Chemists [UK]
	Liquid Gas Carrier
LGCP	Lexical-Graphical Composer Printer
LGD	Local Government District
LGE	Large
	Lunar Geological Equipment [of NASA]
LGen	Lieutenant-General
LGG	Light-Gun Pulse Generator
LGHM	Low Gravity Heavy Media
LGIO	Local Government Information Office [UK]
LGN	Line Generator Number
	Logical Group Number
LGO	Lamont Geological Observatory [of Columbia University, US]
LGov	Lieutenant-Governor
LGP	Low Ground Pressure
LGR	Ligroin
LGRP	Liaison Group for International Educational Exchange [US]
LGS	Lamella Gravity Settler
	Landing Guidance System
LGST	Lamella Gravity Settler/Thickener
LGT	Light
	Low Group Transmit
LGTH	Length
LGV	Large Granular Vesicle
LGWS	Laser-Guided Weapons System
LH	Left-Hand
	Light Helicopter
	Linear Hybrid
	Liquid Hydrogen
	Litter Hook
	Locating Head
	Lock Hopper
	Lower Hybrid
	Luteinizing Hormone
L/H	Low-to-High
LHA	Local Hour Angle
LHANA	Log House Association of North America [now: LHBANA, US]
LHBANA	Log House Builder's Association of North America [US]
LHC	Large Hadron Collider
	Left-Hand Circular (Polarization)
	Log Home Council [now: NALHC, US]
LHCD	Lower Hybrid (Plasma) Current Drive
LHChl	Light-Harvesting Chlorophyll
LHCP	Left-Hand Circular Polarization
LHD	Left-Hand Drive
	Load-Haul-Dump (Technique) [Mining]
LHF	List Handling Facility
LHGR	Linear Heat Generation Rate
LHM	Loop Handling Machine
LHNC	Linearized Hypernetted Chain
LHOTS	Long-Haul Optical Transmission Set
LHPG	Laser Heated Pedestal Growth
LHPS	Lead Hydrogen Purge System
LHR	Lower Hybrid Resonance
	Low Heat Rejection
l-hr	lumen-hour
LHRH	Lower Hybrid Resonance Heating
LH-RH	Luteinizing-Hormone Releasing Hormone [also: LHRH]
LHS	Lawrence Hall of Science [of University of California at Berkeley, US]
	Left-Hand Side
	Loop Handling System
	Low-Cost High-Strength
LHSI	Low Head Safety Injection
LHST	Laundry and Shower Storage Tank
LHSV	Liquid Hourly Space Velocity
LHV	Lower Heating Value
LHWA	Linearized Hot-Wire Anemometer
LI	Landscape Institute
	Laser Interferometer
	Lawn Institute [US]
	Learned Information
	Left In (Place)
	Length Indicator
	Light Infantry
	Line
	Link
	Liquid Injection
	Long Island [US]
	Low Intensity
Li	Lithium
LIA	Laser Industries Association [Name Changed to: Laser Institute of America, US]
	Laser Institute of America [US]
	Lead Industries Association [US]
	Leather Industries of America [US]
	Linear Integrated Amplifier
	Lock-In Amplifier
	Loop Interface Address
	Lysine Iron Agar
LIAB	Liabilities [also: LIABS]
LIAC	Legal Industry Advisory Council [US]
LIADA	Liga Iberoamericana de Astronomia [= Ibero-American Astronomy League, Venezuela]
LIAS	Liaison
Lias	Liassic [Geology]
LIB	Librarian
	Library
	Line Interface Base
lib	(liber) — book
LIBA	Long Island Biological Association [US]
LIBCEPT	Library Information System Intercept [Sweden]

LIBER Ligue des Bibliotheques Europeennes de Recherche [= League of European Research Libraries, Austria]
LIBR Librarian
Library
Li-BR Butadienr Rubber Based on Lithium Catalyst
LIBRIS Library Information System [Sweden]
LIBRN Librarian
LIB SYS Library System [also: LIBR SYS]
LIC License
Licentiate
Linear Integrated Circuit
Loop Insertion Cell
LICA Land Improvement Contractors Association [US]
LiCAF $LiCaAlF_6$ (Laser)
LICOF Land Lines Communication Facilities
LICOR Lightning Correlation
LICS Lotus International Character Set
LID Lattice Invariant Deformation
Leadless Inverted Device
Liberian Dollar [Currency]
Line Isolation Device
Liquid Development [also: lid]
Local Improvement District
Local Issue Data
Locked-In Device
Logical Input Device
LIDAR Laser Infrared Radar [also: Lidar or lidar]
Laser Intensity Direction and Ranging [also: Lidar or lidar]
LIDC Lead Industries Development Council [UK]
LIDF Line Intermediate Distribution Facility
LIDS Laboratory for Information and Decision Systems [of MIT, US]
Landed Immigrant Data System [Canada]
LIED Linkage Editor
LIEF Launch Information Exchange Facility [of NASA]
LIET Lesotho In-Service Education for Teachers Certificate
Lieut Lieutenant
Lieut-Cdr Lieutenant-Commander
Lieut-Col Lieutenant-Colonel
Lieut-Gen Lieutenant-General
Lieut-Gov Lieutenant-Governor
LIF Laser-Induced Fluorescence
Leach Ion-Exchange Flotation
Low Insertion Force
LIFFE London International Financial Futures Exchange [UK]
LIFO Last In, First Out
LIFR Low Instrument Flight Rules
LIFT Link Intellectual Functions Tester
Logically Integrated FORTRAN Translator
LIG Lignite
LIGO Laser Gravitational-Wave Observatory [Joint Caltech-MIT Project]
LIH Light Intensity High
Line Interface Handler
LIHS Long Island Horticultural Society [US]
LIL Large Ionic Lithophile
Light Intensity Low
Lunar International Laboratory
LILO Last In, Last Out
LIM Language Interpretation Module
Light Intensity Medium
Limit(ation)
Limiter
Linear Induction Motor
Line Insulation Monitor
Line Interface Module
Liquid Injection Molding
Locator Inner Marker
Lotus/Intel/Microsoft (Driver)
lim limit value [Mathematics]
LIMA Laser-Induced Ion Mass Analyzer
Logic in Memory Array
LIMAC Large Integrated Monolithic Array Computer
LIMB Limestone Injection Multistage Burner
Liquid Metal Breeder (Reactor)
LIME Laser Induced Magnetically Enhanced
LIMEX Labrador Ice Margin Experiment [Canada]
liminf limit inferior [Mathematics]
LIMIRIS Laser Induced Modulation of Infrared in Silicon
LIMIT Lot-Size Inventory Management Interpolation Technique
LIML Limited-Information Maximum Likelihood
LIMP Language-Independent Macroprocessor
Lunar Interplanetary Monitoring Platform
LIMR Liquid Injection Moldable Rubber
LIMS Laboratory Information Management System
Laser Ionization Mass Spectrometer
limsup limit superior [Mathematics]
LIM SW Limlt Switch
LIMTV Linear Induction Motor Test Vehicle
LIN Library Information Network
Linear(ity)
LINAC Linear Accelerator [also: Linac or linac]
LINAS Laser Inertial Navigation Attack System
LINC Laboratory Instrument Computer
LinCMOS Linear Complementary Metal-Oxide Semiconductor
LINCOMPEX Linked Compressor and Expander
LINCS Language Information Network and Clearinghouse System [of CAL, US]
Lincs Lincolnshire [UK]
LINET Legal Information Network
lin ft linear foot
LING Linguistic(s)
LINK Library Information Network
LINLOG Linear-Logarithmic
LINO Linoleum
LINS Laser Inertial Navigation System
LORAN (= Long-Range Navigation) Inertial System
LINUP Laser-Induced Nuclear Polarization
LINUS Logical Inquiry and Update System
LIOD Laser In-Flight Obstacle Detector
LIOCS Logical Input/Output Control System
LION Light Ion Fusion Facility [of Cornell University, US]
LIOP Life In One Position
Logical Input/Output Processor
LIP Limited Installation Program [US]
LIPL Linear Information Processing Language
Linear Information Programming Language

LIPS Logical Inferences per Second
LIQ Liquid
liq pt liquid pint
LIR Line Integral Refractometer
LIRA Linen Industry Research Association [Ireland]
LIRES Literature Retrieval System
LIRES-MS Literature Retrieval System — Multiple Searching
LIRF Low-Intensity Reciprocity Failure
LIRG Library and Information Research Group [UK]
LIRI Leather Industries Research Institute [South Africa]
LIRL Low-Intensity Runway Lights
LIRTS Large Infrared Telescope System
LIS Large Interactive Surface
Laser-Induced Separation
Laser Isotope Separation
Library and Information Science
Library and Information Service
Line Information Store
Line Isolation Switch
List and Index Society [UK]
Lockheed Information Systems [US]
Loop Input Signal
LISA Library Systems Analysis
Linked Indexed Sequential Access
Locally Integrated Software Architecture
LiSAF $LiSrAlF_6$ (Laser)
LISARD Library and Information Service Automated Retrieval of Data
Library Information Search and Retrieval Data System [US Navy]
LISARDS Library Information Search and Retrieval Data System [US Navy]
LISC Library and Information Service Council [of DES, UK]
LISD Library and Information Services Division [of NOAA, US]
LiSGAF $LiSrGaAlF_6$ (Laser)
LISIC Library and Information Service to Industry and Commerce [UK]
LISN Line-Impedance Stabilization Network
LISP Laser Isotope Separation Program
List Processing Language [also: Lisp]
Littoral Investigation of Sediment Properties
LISR Line Information Storage and Retrieval [of NASA]
LISSA Life Insurance Software Systems of America
LIST Library and Information Services for Teesside [UK]
Listening
LISTAR Lincoln Information Storage and Associative Retrieval System [of LL, US]
LIT Light Interchange Technology
Light Interface Technology
Linkoping Institute of Technology [Sweden]
Liquid Injection Technique
Local Intelligent Terminal
Logic Integrity Test
LITA Library and Information Technology Association [of ALA, US]
LITAS Low-Intensity Two-Color Approach Slope
LITASTOR Light Tapping Storage
LITCA Licensing, Innovation and Technology Consultants Association [Liechtenstein]
LITE Laser Induced Transmission Experiment
Legal Information through Electronics
Lidar In-Space Technology Experiment [of NASA]
LITER Literary
Literature
LITHO Lithographer [also: LITHOG]
Lithography [also: LITHOG]
LITR Low-Intensity Test Reactor [of ORNL, US]
LITVC Liquid Injection Thrust Vector Control
LIU Labourer's International Union [US and Canada]
Line Interface Unit
LIUNA Labourer's International Union of North America
LIWB Livermore Water Boiler [US]
LJ Leaf Jig
Lennard-Jones (Potential) [also: L-J]
LJAO London Joint Area Organization [UK]
LJE Local Job Entry
LJSU Local Junction Switching Unit
LK Link
Lock
Look
Lk Lek [Currency of Albania]
LKE Language Knowledge Examination
LKM Low Key Maintenance
LKR Locker
Sri Lankan Rupee [Currency]
LK ROUTE Looking for Route
LK WASH Lock Washer
LKY Lefschetz-Kalman-Yakubovitch
LL Lamp Lumen
Landing Light
Land-Line
Laser Light
Laterlog
Lebanese Pound [Currency]
Light Line
Lincoln Laboratory [of MIT, US]
Line Leg
Line Link
Lines
Liquid Limit
Live Load
Load Line
Local Line
Local Loopback
Loudness Level
Lower Level
Lower Limit
Low Level
L/L Latitude/Longitude
ll (loco laudato) — in the place cited
LLA Leased Line Adapter
Low Level Access
LLAD Low-Level Air Defense
LLAR Local Line Automatic Routing
LLC Link Layer Control
Liquid-Liquid Chromatography
Liquid Live Culture
Logical Link Control

LLD Lamp Lumen Depreciation (Factor)
Linear Low Density
Lower Limit of Detection
Low-Level Discriminator
L2D2 LORAN (= Long-Range Navigation) Digital Dropwindsonde [also: L^2D^2]
LLDPE Linear Low-Density Polyethylene
LLE Laboratory for Laser Energetics [of University of Rochester, US]
Long Line Equipment
LLF Line Link Frame
LLFM Low-Level Flux Monitor
LLG Logical Line Group
LLGA Leadless Land Grid Array
LLH Low-Level Heating
LLI Low-Level Interface
LLL Landau-Lifshitz-Looyenga (Formula)
Lawrence Livermore Laboratory [US]
Low-Level Language
Low-Level Logic [also: L^3]
LLLLLL Laboratories Low-Level Linked List Language [also: L6 or L^6]
LLLTV Low Light-Level Television [also: L^3TV]
LLLWT Low Level Liquid Waste Tank
LLM Low-Level Multiplexer
LLN Line Link Network
Local Line Network
LLNL Lawrence Livermore National Laboratory [of University of California, US]
LLP Line Link Pulse
LLPL Low Low Pond Level
LLR Line of Least Resistance
Lloyds Register of Shipping [UK]
Load-Limiting Resistor [also: LLRES]
LLRF Lunar Landing Research Facility
LLRT Low Level Reactor Test
LLRV Lunar Landing Research Vehicle [of NASA]
LLS Laterolog, Shallow Investigation [Geology]
Liquid Level Switch
Local Library System
Local Load Shearing
LLSE Linear Least Squares Estimation
LLSU Low Level Signaling Unit
LLTR Large Leak Test Rig
LLTT Landline Teletype
LLTV Low Light-Level Television
Lunar Landing Training Vehicle [of NASA]
LLW Leaky Lamb Wave
Lower Low Water
Low-Level Waste
LLWS Low-Level Wind Shear
LLWSAS Low-Level Wind Shear Alert System
LM Leg Multiple
Life Member
Light Microscopy
Linear Modulation
Linear Monolithic
Link Manager
Liquid Metal
List of Materials
Load Module
Local Memory
Logical Module
Logic Monitor
Loop Multiplexer
Low Magnification
Low-Melting
Lunar Model
Lunar Module
L/M Lines per Minute
List of Material
L-Shell/M-Shell (Emission Ratio) [Physics]
lm logarithmic mean area
lumen
LMA Laminating Materials Association [US]
Laser Microspectral Analyzer
Limited Motion Antenna
Livestock Marketing Association [US]
Logsplitters Manufacturers Association [US]
LMAGB Locomotive Manufacturers Association of Great Britain
LMBC Liverpool Marine Biology Committee [UK]
LMBI Local Memory Bus Interface
LMC Large Magellanic Cloud
Least Material Condition
Lime-Magnesium Carbonate
Low-Pressure Molding Compound
LMCC Land Mobile Communications Council [US]
LMD Light Metals Division [of TMS, US]
Log Mean Difference
LMDPE Linear Medium-Density Polyethylene
LME Liquid Metal Embrittlement
London Metal Exchange [UK]
LMEC Liquid Metal Engineering Center [US]
LMEIC Life Member of the Engineering Institute of Canada
LMF Language Media Format
Linear Magnetic Field
Linear Matched Filter
Low and Medium Frequency [also: L/MF]
LMFA Light Metal Founders Association
LMFBR Liquid-Metal Fast Breeder Reactor
LMFR Liquid-Metal Fuel Reactor
LMFRE Liquid-Metal Fuel Reactor Experiment
lm-hr lumen-hour
LMHX Liquid Metal Heat Exchanger
LMI Lawn Mower Institute [now: OPEI, US]
Liquid Metal Infiltration
Local Memory Image
LMIE Liquid Metal Induced Embrittlement
LMIM Large Motion Isolation Mount [of Canadian Space Agency]
L/min Liters per Minute
LMIS Liquid Metal Ion Source
LML Logical Memory Level
LMLR Load Memory Lockout Register
LMM Lactobacillus Maintenance Medium
Length Measuring Machine
Light Meromyosin
Locator at the Middle Marker
LMMS Local Message Metering Service
LMNA Long-Range Multipurpose Naval Aircraft
LMO Life Members Organization

LMOA Locomotive Maintenance Officers Association [US]
LMOS Loop Maintenance Operations Service
LMP Larson-Miller Parameter
Large-Volume Microwave Plasma Generator
Lunar Module Pilot [of NASA]
LMPRT Locally Most Powerful Rank Test
LMPS Lamps
LMR Land Mobile Radio
Liquid Metal Reactor
Liquid Molding Resin
Lowest Maximum Range
LMRU Library Management Research Unit [UK]
LMS Laboratory Management Standards
Land Mobile Satellite (Service)
Laser Mass Spectrometry
Least Mean Square
Level Measuring Set
Library Management System
Linear Multistep
List Management System
Local Measured Service
London Mathematical Society [UK]
LMSA Laser Microemission Spectroanalysis
Laser Microspectral Analysis
lm-sec lumen-second
LMSEO Labour Management Standards Enforcement Office [US]
LMSL Lateral Monolayer Superlattice
LMSS Lunar Mapping and Survey System
LMSV Loading Mooring Storage Vessel
LMT Limit
Local Mean Time
Logical Mapping Table
Log Mean Temperature
LMTD Logarithmic Mean Temperature Difference
LMTEC Liquid Metal Thermoelectric Converter
LMTO Linear Muffin Tin Orbitals
LMTR Limiter
LMU Line Monitor Unit
Logical Mining Unit
Ludwig-Maximillian University [Germany]
LMW Low Molecular Weight
lm/W lumen per Watt [also: LPW]
LMWHC Low-Molecular-Weight Hydrocarbon
LMX L-Type Multiplex
LN Line
Linear Network
Liquid Nitrogen
Low-Noise
ln Naperian (natural) logarithm
LNA Launch Numerical Aperture
Local Numbering Area
Low-Noise Amplifier
LNAP Low Non-Essential Air Pressure
LNB Local Name Base
LNC Low-Noise Converter
LND Local Number Dialing
LNDG Landing
LNE Local Network Emulator
LNF Liposoluble Neutral Fraction
LNFB Linear Negative Feedback
LNG Liquefied Natural Gas
Long
LNGC Liquefied Natural Gas Carrier
LNI Local Network Interconnect
LN-NR Low Nitrogen Nitrile Rubber
LNOC Libya National Oil Corporation
LNP Loss of Normal Power
LNPR Line Printer
LNR Low-Noise Receiver
LNS Linked Numbering Scheme
Luxembourg Naturatists Society
LNSU Library Network SIBIL (Automation System) Users [Switzerland]
LNTSU Leo N. Tolstoi State University [USSR]
LNWT Low and Nonwaste Technology
LO Layout
Liaison Office
Liaison Officer
Light Operated
Line Occupancy
Local Order
Local Oscillator
Location Code
Lock-On
Lock-Out
Longitudinal Optical
Low
Low Order
Lubricating Oil
Lunar Orbiter [of NASA]
LOA Lathyrus Odoratus Agglutinin
Length Overall
Letter of Agreement [also: LoA]
Line of Assurance
Log-Out Analysis
LOAC Low Accuracy
LOAD Laser Optoacoustic Detection
LOADS Low-Altitude Defense System
LOAMP Logarithmic Amplifier
LOAS Liftoff Acquisition System
LOB Launch Operations Branch [of NASA]
Launch Operations Building [of NASA]
Line of Balance
LOBC Locked On by Combination
LOBTA Lunar Orbiter Block Triangulation
LOC Large Optical Cavity
Launch Operations Center [of NASA]
Liaison Officers Committee
Library of Congress [US]
Lines of Codes
Linked Object Code
Local(izer)
Location
Location Counter
Locator
Loss of Coolant
Oxygen Lance Cutting
LOCA Loss-of-Coolant Accident
LOCAL Load on Call
LOCAP Low Capacitance
Low Capacity

LOCAS Local Cataloguing Service [UK]
LOCATE Library of Congress Automation Techniques Exchange [US]
LOCATS Lockheed Optical Communications and Tracking System
LOCC Launch Operations Control Center
loc cit (loco citato) — in the place cited
LOCE Loss-of-Coolant Experiment
LOCF Loss-of-Coolant Flow
LOCI Logarithmic Computing Instrument
LOCOM Locomotive
LOCOS Local Oxidation of Silicon
LOCP Loss-of-Coolant Protection
LOCS Librascope Operations Control System
Logic and Control Simulation
LOD Launch Operations Directorate [of NASA]
Line of Direction
Locally One-Dimensional
Location Dependent
Locked-On Device
Low Density [also: LoD]
LODAR Long-Range Detection and Ranging [also: Lodar or lodar]
LODEM Loading Dock Equipment Manufacturers Association [US]
LODESMP Logistics Data Element Standardization and Management Process
LODESTAR Logically-Organized Data Entry, Storage and Recording
LODP Leveling-Off Degree of Polymerization
LOERO Large Orbiting Earth Resources Observatory
LOES Laser Optical Emission Spectroscopy
LOEX Library Orientation Instruction Exchange [US]
LOF Lack of Fusion
Length of File
Local Oscillator Frequency
Loss of Flow
Loss of Fluid
Lowest Observable Frequency
Lowest Operating Frequency
LOFA Loss-of-Flow Accident
LOFAR Low-Frequency Acquisition and Ranging [also: Lofar or lofar]
Low-Frequency Analysis Recording
LOFT Loss-of-Flow Test
Loss-of-Fluid Test
LOFW Loss of Feedwater
LOG Logarithm
Logic(al)
log common logarithm
LOGACS Low-Gravity Accelerometer Calibration System
LOGALGOL Logical Algorithmic Language
LOGAMP Logarithmic Amplifier
LOGANDS Logical Commands
LOGBALNET Logistics Ballistic Missile Network [US Air Force]
LOGEL Logic Generating Language
LOGFED Log File Editor
LOGFTC Logarithmic Fast Time Constant
LOGIPAC Logical Processor and Computer
LOGIT Logical Interference Tester
LOGLAN Logical Language
log MTD Logarithmic Mean Temperature Difference
LOGRAM Logical Program
LOGTAB Logic Table
LOH Light Observation Helicopter
LOHAP Light Observation Helicopter Avionics Package
LOI Limited Oxygen Index
Loss of Ignition
Loss on Ignition
LOL Length of Lead
LOLA Library On-Line Acquisition
London On-Line Local Authorities [UK]
Low-Level Oil Alarm
Lunar Orbit Landing Approach
LOLC Library Developed On-Line Catalogue
LOLEX Low-Level Extraction
LOLITA Language for the On-Line Investigation and Transformation of Abstractions
Library On-Line Information and Text Access [of Oregon State University, US]
LOLO Lift-On/Lift-Off (Vessel)
LOLP Loss of Load Probability
LOM Locator at the Outer Marker
Low-Frequency Outer Marker
LOMA Life Office Management Association [US]
LOMAC Low-Level Macroprocessor Language
LOMAR Local Manual Attempt Recording
LOMUSS Lockheed Multiprocessor Simulation System
LONA Low-Noise Amplifier
LONAL Local Off-Net Access Line
LONG Longitude
LONGN Longeron
LOOP Louisiana Offshore Oil Port [US]
LOOPS LISP (= List Processing) Object-Oriented Programming System
LOOSP Loss of On-Site Power
LOP Lack of Penetration
Line of Position
Line-Oriented Protocol
Local Operational Plot
Loss of Offsite Power
LOPAD Logarithmic Outline Processing System for Analog Data
LOPAR Low-Power Acquisition Radar
LOPI Loss of Piping Integrity
LOPO Low Power Water Reactor [US]
LOPP Lunar Orbiter Photographic Project
LOPS Lines of Positions
LO-QG Locked Oscillator-Quadrature Grid
LOR Laboratory for Oil Recovery
Large Optical Reflector
Lining-over-Refractory
Loss on Reduction
Lunar Orbital Rendezvous
LORAC Long-Range Accuracy [also: Lorac or lorac]
LORAD Long-Range Detection and Ranging [also: Lorad or lorad]
LORADAC Long-Range Active Detection and Communications System
LORAN Long-Range Navigation [also: Loran or loran]
LORBI Locked-On Radar Bearing Indicator
LORDS Licensing On-Line Retrieval Data System

LOREC Long-Range Earth Communications [also: Lorec or lorec]
LORL Large Orbital Research Laboratory [of NASA]
LORLS Lake Ontario Regional Library System [Canada]
LORO Lobe on Receive-Only
LORPGAC Long-Range Proving Ground Automatic Computer
LORS Lunar Optical Rendezvous System
LORTAN Long-Range and Tactical Navigation [also: Lortan or lortan]
LORV Low-Observable Reentry Vehicle
LOS Laser-Rangefinder Optical Sight
Law of the Sea
Line of Sight
Line Out of Service
Lock-Out Submersible
Loss of Signal
LOSC Law of the Sea Conference
LOSP Loss of Offsite Power
Loss of System Pressure
LOSR Limit of Stack Register
LOSS Landing Observer Signal System
Large Object Salvage System
LOST Letdown Storage Tank
LOT Large Orbital Telescope
Light-Operated Typewriter
Long Open Time
LOTA Low Observable Technology and Application
LOTEM Long Offset Transient Electromagnetic (Method)
LOTIS Logical Timing Sequencing
Logic, Timing, Sequencing (Language)
LOW Low Core Threshold
LOWL Low-Level Language
LOX Liquid Oxygen [also: LOx or lox]
Liquid Oxygen Explosive
LOZ Liquid Ozone [also: LOz or loz]
LP Laser Printer
Laser Processing
Library of Parliament [Canada]
License Program
Light Pen
Linearly Polarized
Linear Processing
Linear Programming
Line Printer
Liquefied Petroleum
List Price
Load Point
Lodgepole Pine
Logic Probe
Log Periodic
Longitudinal Parity
Long Play(er)
Long Primer
Loop(ing)
Loop Control(ler)
Low-Pass (Filter)
Low Point
Low Power
Low Pressure
Low-Speed Printer
L-P Lorentz-Polarization Factor
Low-Pressure
Lp Lempira [Currency of Honduras]
L£ Lebanese Pound [Currency]
LPA Laser-Diode Parameter Analyzer
Limulus Polyphemus Agglutinin
Linear Power Amplifier
Linear Pulse Amplifier
Link Pack Area
Liquid Petroleum Air
Log Periodic Antenna
Low-Pressure Alarm
LPAC Launching Program Advisory Committee [of ESRO, France]
LPAS Low-Pressure Arc Spraying
LPATS Lightning Position and Tracking System
LPB Low Pressure Boiler
LPC Laboratory Pasteurized Count
Language Products Center [Canada]
Laser Particle Counter
Linear Power Controller
Linear Predictive Coding
Linear Predictive Coefficients
Loop Control
Loop Preparation Cask
Lower Pump Cubicle
LPC-HS Laser Particle Counter — High Sensitivity
LPCI Low-Pressure Coolant Injection
LPCIS Low-Pressure Coolant Injection System
LPCM Linear Phase Code Modulation
LPCRS Low-Pressure Coolant Recirculation System
LPCS Low-Pressure Core Spray
LPCVD Low-Pressure Chemical Vapor Deposition
LPD Landing Point Designator
Language Processing and Debugging
Linear Phasing Device
Linear Power Density
Log Periodic Dipole
Low Power Difference
Low Pressure Difference
LPDA Log Periodic Dipole Array
LPDR Local Public Document Room
LPDT Low-Power Distress Transmitter
LPDU Link-Layer Protocol Data Unit
LPE Linear Polyethylene
Liquid-Phase Epitaxy
Loop Preparation Equipment
LPEC Lumped Parameter Equivalent Circuit
LPEE Liquid Phase Electro-Epitaxy
LPF Leaching, Precipitation and Flotation (Process)
Liquid Pressure Forming
Low-Pass Filter
Lymphocytosis-Promoting Factor
LPFL Low Pinchoff-Voltage FET (= Field-Effect Transistor) Logic
LPG Liquefied Petroleum Gas
List Program Generator
Low-Pressure Gas
LPGA Liquefied Petroleum Gas Association
LPGC Liquefied Petroleum Gas Carrier
LPGITC Liquefied Petroleum Gas Industry Technical Committee

LPGS Liquid Pathway Generic Study

LPGSASA Liquefied Petroleum Gas Safety Association of South Africa

LPGTC Liquefied Petroleum Gas Industry Technical Committee

LPI Lightning Protection Institute [US]
Lines per Inch [also: lpi]
Longitudinally-Applied Paper Insulation
Low Power Injection
Low Probability of Intercept

LPIA Label Printing Industries of America [US]
Liquid Propellant Information Agency [US]

LPID Logical Page Identifier

LPIS Low-Pressure Injection System

LPL Laser-Pumped Laser
Light Proof Louver
Linear Programming Language
Lipoprotein Lipid
List Processing Language [also: LISP]
Local Processor Link
Long Particular Meter
Long-Path Laser

LP-LEC Low-Pressure Liquid Encapsulated Czochralski

LPLS Low-Pressure Laser Spraying

LPM Laser Phase Macroscope
Licensing Project Manager
Linearly Polarized Mode
Lines per Minute [also: lpm]
Lunar Portable Magnetometer [of NASA]

LPMA Loose Parts Monitor Assembly

LPMM Line Pairs per Millimeter [also: lpmm]

LPMOCVD Low-Pressure Metal-Organic Chemical Vapor Deposition

LPN Logical Page Number
Low-Pass Notch

LPO Low-Power Output

LPP Letter Processing Plant
Liquid Plasma Process
Low Power Physics

LPPD Low-Pressure Plasma Deposition

LPPM Laboratoire de Photophysique et Photochimie Moleculaire [= Molecular Photophysics and Photochemistry Laboratory, France]

LPPS Low-Pressure Plasma Spraying

LPPT Liquid Precipitate
Low Pressurization Pressure Test

LPR Lynchburg Pool Reactor [US]

LPRE Liquid Propellant Rocket Engine

LPRINT Lookup Dictionary Print Program

LPRM Low Power Range Monitor

LPRS Local Power Recirculation System

LPS Laboratory of Plasma Studies
Laser Printing System
Launch Process System
Letter Processing Stream
Light Proof Shades
Linear Power Supply
Linear Programming System
Linear Pulse Sector
Lines per Second [also: lps]
Lipopolysaccharide
Low-Pressure Scram
Low-Pressure System

LPSD Logically Passive Self-Dual

LPSIRS Linear Potential Sweep Infrared Reflectance Spectroscopy

LPSR Low Protein Serum Replacement

LPSTC Lowest Point of Single Tooth Contact

LPSW Low-Pressure Service Water

LPT Laminated Plate Theory
Line Printer
Link Plate Thickness
Low-Pressure Turbine

LPTB London Passenger Transport Board [UK]

LPTF Low Power Test Facility

LPTR Livermore Pool-Type Reactor [of LRL, US]

LPTTL Low-Power Transistor-Transistor Logic [also: LPT^{2}L]

LPTV Large Payload Test Vehicle
Low-Power Television

LPU Language Processor Unit
Life Preserver Unit
Line Processing Unit
Local Processing Unit

LPUG Lasers in Publishing Users Group [US]

LPV Laboratorio de Patologia Veterinaria [= Laboratory for Veterinary Pathology, Portugal]
Light Proof Vent

LPVT Large Print Video Terminal

LPW Lumen per Watt [also: lm/W]

LPZ Low Population Zone

LQ Letter-Quality
Liquid [also: LQD]
Liquor

LQD Low-Q Diffractometer

LQDC Lowest Quantitatively-Determined Concentration

LQDS Liquids

LQG Linear Quadratic Gaussian

LQGC Linear Quadratic Gaussian Control

LQL Limiting Quality Level

LR Labour
Landing Radar
Left to Right
Level Recorder
Level Reduction
Likelihood Ratio
Limited Recoverable
Limit Register
Linear Regression
Linear Response
Line Radiation
Line Relay
Local/Remote
Load Ratio
Load-Resistor Relay
Logical Record
Low Reduction
Lunar Rover [of NASA]

L/R Local/Remote
Locus of Radius

Lr Lawrencium
Lira [Currency of Italy]

LRA	Laser Gyro Reference Axis
	Laser Release Analysis
	Laser Research Association
	Logical Record Address
LRB	Labour Relations Board
	Local Reference Beam
LRC	Langley Research Center [of NASA]
	Large Rock Cavern
	Lewis Research Center [of NASA]
	Light, Rapid, Comfortable (Train)
	Light Reflective Capacitor
	Linguistics Research Center [US]
	Load Ratio Control
	Longitudinal Redundancy Check
	Long Range Cruise
LRCC	Library Resources Coordinating Committee [UK]
	Longitudinal Redundancy Check Character
LRCM	Long-Range Cruise Missile
LRCO	Limited Remote Commications Outlet
LRCS	Long-Range Cruise Speed
LRD	Liberian Dollar [Currency]
	Long-Range Data
LRE	Linear Resource Extension
LREE	Light Rare-Earth Element
LREP	Laser Rotating Electrode Process
LRF	Long Range Facility
	Lumber Recovery Factor
LRFAX	Low-Resolution Facsimile
LRFS	Long Range Forecasting System
LRG	Landscape Research Group [UK]
	Lightweight Recoilless Gun
	Long Range
LRI	Lighting Research Institute [US]
	Long-Range Radar Input
LRIA	Level Removable Instrument Assembly
LRIM	Long Range Input Monitor
LRIR	Limb Radiance Inversion Radiometer
	Low-Resolution Infrared Radiometer
LRIS	Land Registration Information Service [Canada]
	Land-Related Information System
LRL	Lawrence Radiation Laboratory [of University of California, US]
	Linking Relocating Loader
	Livermore Research Laboratory [US]
	Logical Record Length
	Logical Record Location
	Lunar Receiving Laboratory [of NASA]
LRLTRAN	Lawrence Radiation Laboratory Translator [US]
LRM	Limited Register Machine
	Linear Reluctance Motor
	Liquid Reaction Molding
	Lunar Reconnaissance Module [of NASA]
LRMTS	Laser Rangefinder and Marked Target Seeker
	Long-Range Order
LRP	Land Reclamation Program
	Long-Range Path
LRPA	Long-Range Patrol Aircraft
LRPDS	Long-Range Position-Determining System
LRPL	Liquid Rocket Propulsion Laboratory [US Army]
LRQA	Lloyd's Register Quality Assurance
LRR	Loop Regenerative Repeater
LRRP	Lowest Required Radiating Power
LRRS	Long-Range Radar Site
LRS	Language Resource (File)
	Laser Raman Scattering
	Linguistic Research System
	Liquid Radwaste System
	Log and Reporting System
	Long-Range Search
LRSM	Long-Range Seismographic Measurements
LRSS	Long-Range Survey System
LRT	Light Rail Transit
	Light Railway Transport
	Light Rapid Transit
	Linear Response Theory
	Local Leak Rate Test
	Long-Range Transport
LRTAP	Long-Range Transport of Acidic Pollutants
	Long-Range Transport of Air(borne) Pollutants
LRTAPP	Long-Range Transport of Acidic Pollutants Program [Canada]
	Long-Range Transport of Air(borne) Pollutants Program [Canada]
LRTF	Long-Range Technical Forecast
LRTL	Light Railway Transport League
LRTM	Long-Range Training Mission
LRTP	Long-Range Training Program
	Long-Running Thermal Precipitator
LRU	Least Recently Used
	Least Replaceable Unit
	Line-Replaceable Unit
LRV	Light Rail Vehicle
	Light Rapid Vehicle
	Lunar Roving Vehicle [of NASA]
LRWE	Long-Range Weapons Establishment [Australia]
LRX	Large Reactor Critical Facility [US]
LS	Land Surveyor
	Language Specification
	Language System
	Large-Scale
	Laser Spectrometry
	Laser System
	Leading Seaman
	Least Significant
	Least Squares
	Left-Side
	Legal Status
	Level Switch
	Levitation System
	Library Service
	Library System
	Ligand Substitution
	Light Source
	Limit Switch
	Linear System
	Line Spectrum
	Line Switch
	Link Switch
	Liquid Starter
	Loading Splice
	Loading Stress
	Load Sensing

	Load Stress
	Lockscrew
	Local Store
	Long Shot
	Loudspeaker
	Louisiana State [US]
	Low-Power Schottky (Logic)
	Low-Speed
	Low Structure Black
	Syrian Pound [Currency]
LSA	Labour Surplus Area
	Large Surface Area
	Laser Surface Alloying
	Least Squares Analysis
	Least Squares Approximation
	Licentiate in Agricultural Science
	Lignosulfonic Acid
	Limited Space-Charge Accumulation
	Linear System Analysis
	Line Sharing Adapter
	Locksmith Security Association [US]
	Logging Safety Association [Canada]
	Logistic Support Analysis
	Longitudinal Spherical Aberration
	Low-Cost Solar Array
	Low Specific Activity
LSAA	Library Services Authority Act [US]
LSALT	Lowest Safe Altitude
LSAP	Link-Layer Service Access Point
LSAS	Longitudinal Stability Augmentation System
LSB	Least Significant Bit
	Lower Sideband
	Low-Speed Breaker
LSBR	Large Seed Blanket Reactor
LSC	Laser-Supported Combustion
	Laser Surface Cladding
	Least Significant Character
	Liberian Shipowners Council
	Liquid Scintillation Counter
	Liquid-Solid Chromatography
	Local Switching Center
	Loop Station Connector
	Low Shear Continuous
	Low-Speed Concentrator
	Luminescent Solar Concentrator
LSc	Licentiate in Science
LSCC	Line-Sequential Color Composite
	Local Servicing Control Center
LScComm	Licentiate in Commercial Science
LSCE	Launch Sequence and Control Equipment
LSCO	Labour Safety Council of Ontario [Canada]
LSCP	Low-Speed Card Punch
LSCU	Local Servicing Control Unit
LSD	Language for Systems Development
	Large Screen Display
	Large Signal Diode
	Laser-Supported Detonation
	Last Significant Digit
	Launch Systems Data
	Least Significant Digit
	Lightermen, Stevedores and Dockers
	Line-Sharing Device
	Line Signal Detector
	Local Spin Density
	Logarithmic Series Distribution
	Logistics Support Division
	Loop Shutdown
	Low-Speed Data
	Lunar Surface Magnetometer
	Lysergic Acid Diethylamide
LSDA	Local Spin-Density Approximation
LSDF	Large Sodium Disposal Facility
	Local Spin-Density Function
LSDR	Local Storage Data Register
LSDU	Link-Layer Service Data Unit
LSE	Language Sensitive Editor
	Launch Sequencer Equipment
	Least Squares Estimation
	Line Signaling Equipment
	London School of Economics [UK]
	London Stock Exchange [UK]
	Longitudinal Section Electric
LSECS	Life Support and Environmental Control System
LSEP	Lunar Surface Experimental Package [of NASA]
LSF	Least-Squares Fitting
	Linear Structural Formula
	Line Spread Function
	Line Switch Frame
	Load Sheet Fuel
	Local Spin Fluctuation
	Lumped Selection Filter
LSFC	Lewis Spaceflight Center [of NASA]
LSFFAR	Low-Spin Folding Fin Aircraft Rocket
LSFR	Local Storage Function Register
LSFT	Local Spin Fluctuation Theory
	Low Steamline Flow Test
LSG	Language Structure Group [of CODASYL]
	Leasing
	Low Stress Grinding
LSGA	Laminators Safety Glass Association [US]
LSHA	Louisiana State Horticultural Association [US]
LSHI	Large-Scale Hybrid Integration
LSI	Large-Scale Integration
	Large-Scale Investment
LSIA	Lamp and Shade Industries of America [US]
LSIC	Large-Scale Integrated Circuit
LSID	Local Session Identification
LSIG	Line Scan Image Generator
LSI-MOS	Large-Scale Integrated Metal-Oxide Semiconductor
LSIR	Limb Scanning Infrared Radiometer
LSI/VLSI	Large-Scale Integration/Very-Large-Scale Integration
LSJ	Laser Society of Japan
LSL	Ladder Static Logic
	Linnean Society of London [UK]
	Link and Selector Language
	Lower Specification Limit
LSLA	Low-Speed Line Adapter
LSLD	Local Store Loop Driver
LSM	Laser Scan Microscope
	Launcher Status Multiplexer
	Lesotho Maloti [Currency]

Linearized Simulation Model
Linear Synchronous Motor
Line Selection Module
Line Switch Marker
Local Synchronous Modem
Logic-State Map
Longitudinal Section Magnetic
Long Span Mezzanine
Lymphocyte Separation Medium
LSMA Low-Speed Multiplexer Arrangement
LSMI Lake Superior Mining Institute [US]
LSMITH Locksmith
LSMR Landing Ship Medium Rocket
LSMU Lasercom Space Measurement Unit
LSN Linear Sequential Network
Line Stabilization Network
Low-Solids, Nondispersoid Mud
LSNSW Linnean Society of New South Wales [Australia]
LSNY Linnean Society of New York [US]
LSO Line Signaling Oscillator
LSP Large-Scale Production
Laser Shock Processing
Laser Speckle Photography
Library Software Package
Linked Systems Project [US]
Ligand Substitution Process
Local Store Pointer
Loop Splice Plate
Low-Speed Printer
LSPB Large Scale Prototype Breeder (Reactor)
LSPET Lunar Samples Preliminary Examination Team [of NASA]
LSPK Loudspeaker
LSPN Ligue Suisse pour la Protection de la Nature [= Swiss League for the Protection of Nature]
LSPS Local Service Planning System
LSPTR Low-Speed Paper Tape Reader
LSQ Least Squares (Fit)
Line Squall
LSQA Local System Queue Area
LSR Laboratory for Space Research
Lanthanide Shift Reagent
Least-Squares Regression
Liquid Silicon Rubber
Load Shifting Resistor
Local-Shared Resources
Local Storage Register
Low Spatial Resolution
Low-Speed Reader
Lynchburg Source Reactor [US]
LSR-AES Low Spatial Resolution — Auger Electron Spectroscopy
LSRH Laboratoire Suisse de Recherches Horlogeres [= Swiss Laboratory for Horological Research]
LSRP Local Switching Replacement Planning
LSS Laboratory Support System
Language for Symbolic Simulation
Large Space Structure
Large Space System
Launch Status Summarizer
Life Support System
Local Synchronization Subsystem
Loop Surge Suppressor
Loop Switching System
Lunar Soil Simulator [of NASA]
LSSD Laboratory and Scientific Services Directorate
Level-Sensitive Scan Device
LSSG Lateral Studies Sub-Group [of NAVSEP]
LSSL Lateral Surface Superlattice
LSSM Local Scientific Survey Module
Lunar Scientific Survey Module [of NASA]
LSSP Latest Scram Set Point
LSSS Limiting Safety System Setting
LST Large Space Telescope
List
Loud Speaking Telephone
Lyddane-Sachs-Teller (Relation)
LSTB Low-Power Schottky TTL (= Transistor-Transistor Logic) Bipolar
LSTF Large Scale Test Facility
LSTPC List Price
LSTTL Low-Power Schottky Transistor-Transistor Logic [also: LS-TTL]
LSTTL LSI Low-Power Schottky Transistor-Transistor Logic — Large-Scale Integration [also: LS-TTL-LSI]
LSU Leading Signal Unit
Library Storage Unit
Line Selection Unit
Line Sharing Unit
Load Storage Unit
Local Storage Unit
Local Switching Unit
Local Synchronization Utility
Lone Signalling Unit
Louisiana State University [US]
LSV Lift-Spray Vacuum Processing
Line Status Verifier
Lunar Surface Vehicle [of NASA]
LSW Lipshits-Slyozov-Wagner (Treatment)
LT Laboratory Technologist
Laboratory Technology
Laboratory Test
Language Translation
Language Transmission
Laplace Transform
Laser Technology
Leak Testing
Leather Technologist
Leather Technology
Left
Less Than
Letter Telegram
Library Technician
Light
Light-Tight
Light Truck
Line Telecommunications
Line Telegraphy
Line Terminator
Link Terminal
Link Trainer
Local Time

	Logic Theorist
	Logic Theory
	Long-Tail Vehicle
	Long Term
	Long Transverse
	Loop Test
	Low Temperature
	Low Tension
	Low Torque
	Lubrication Technologist
	Lubrication Technology
	Turkish Lira [Currency]
L&T	Letter Telegram
	Link Terminal
L-T	Long-Term
Lt	Lieutenant
lt	long ton
LTA	Lead Tetraacetate
	Library-Media Technical Assistant
	Lighter than Air
	Linen Trade Association [US]
	Literary Translators Association [Canada]
	Logical Transient Area
	Long-Term Arrangement
LTAB	Line Test Access Bus
LTAS	Lighter-than-Air Ship
LTB	Last Trunk Busy
	Lithium Tetraborate
LTBO	Linear Time Base Oscillator
LTC	Line Terminating Circuit
	Line Time Clock
	Line Traffic Coordinator
	Local Telephone Circuit
	Local Terminal Controller
	Local Test Cabinet
	Long Time Constant
LtCdr	Lieutenant Commander
LTCESFS	Low-Temperature Constant Energy Synchronous Fluorescence Spectroscopy
LT(Cert)	Certificate in Laboratory Technology
LtCmdr	Lieutenant-Commander
LtCol	Lieutenant-Colonel
LtCol(N)	Lieutenant-Colonel (Navy)
LtComm	Lieutenant Commander
LTD	Limited
	Line Transfer Device
	Local Test Desk
	Long Term Debt
lt/d	long tons per day
LTDE	Local Thermodynamic Equilibrium
LTDP	Long-Term Defense Program
LTDS	Laser Target Designation System
	Launch Trajectory Data System
	Low-Temperature Drawing Salt
LTE	Land Trust Exchange [US]
	Line Termination Equipment
	Local Telephone Exchange
	Local Thermodynamic Equilibrium
LTEM	Lorentz Transmission Electron Microscopy
LTF	Lithographic Technical Foundation [now: GATF, US]
LTFT	Low-Temperature Flow Test
LTFV	Less than Fair Value
LTG	Lighting
	Lightning
LTGCC	Lightning, Cloud to Cloud
LTGCCCG	Lightning, Cloud to Cloud and Cloud to Ground
LTGCG	Lightning, Cloud to Ground
LTGCW	Lightning, Cloud to Water
LtGen	Lieutenant-General
LTGH	Lightening Hole
LTGIC	Lightning in Clouds
LtGov	Lieutenant-Governor
LTH	Lactogenic Hormone
	Logical Track Header
	Low-Temperature Herschel
	Luteotropic Hormone
LTHA	Low-Temperature High-Activity
LTHR	Leather
LTHS	Long-Term Hydrostatic Strength
LTI	Light Transmission Index
	Low-Technology Intensive
	Low-Temperature Isotropic (Carbon)
LTIB	Lead Technical Information Bureau [UK]
LTIC	Laminated Timber Institute of Canada
LTL	Less than Truckload
	Little
	Lot Truck Load
	Lower Tolerance Limit
LTLCG	Little Change
LTM	Live Traffic Model
	Long-Term Memory
	Low-Temperature Mixing
	Low-Temperature Molding
LTMA	London Terminal Control Area [UK]
LTMP	Low-Temperature Microprobe
LTMR	Laser Target Marker Ranger
Lt(N)	Lieutenant (Navy)
LTNO	Low-Temperature Nuclear Orientation
LTOC	Laser-Trimmed On-Chip
LTOH	Less Than One Hour
LTOM	Less Than One Minute
LTOS	Less Than One Second
LTP	Landline Telephone
	Library Technology Project [of ALA, US]
	Liquid Transition Point
	Line and Trunk Peripheral (Cabinet)
	Local Tracking Problem
	Long-Tailed Pair
	Low-Temperature Physics
LTPD	Lot Tolerance Percent Defective
	Low-Temperature Particle Detector
LTPL	Long-Term Procedural Language
LTPWHT	Low-Temperature Postweld Heat Treatment
LTR	Lattice Test Reactor [US]
	Letter
	Lighter
	Lockheed Training Reactor [US]
	Long-Term Revitalization
	Loop Transfer Recovery
L-TR	Licensing Technical Review
ltr	liter [also: L or l]

ltr/min liter per minute [also: L/min]
LTRS Letter Shift
LTS Launch Telemetry Station
Launch Tracking System
Lights
Line Transient Suppression
Long-Term Stability
Low-Temperature Sensitization
LTSC Low-Temperature Superconductivity
Low-Temperature Superconductor
LTSP Long-Term Space Plan [Canada]
LT-STM Low-Temperature Scanning Tunneling Microscope
LTT Landline Teletypewriter
LTTAT Long Tank Thrust Augmented Thor
LTTL Low-Power Transistor-Transistor Logic [also: LT^2L]
LTTMT Low-Temperature Thermomechanical Treatment
LTU Line Termination Unit
LTWA Log Tape Write Ahead
LTV Large Test Vessel
Light Vessel
Load Threshold Value
Long Tube Vertical
LU Lakehead University [Canada]
Laurentian University [Canada]
Laval University [Canada]
Lehigh University [US]
Line Unit
Logical Unit
Loughborough University [UK]
Lowell University [US]
Lu Lutetium
lu lumen**LUB**
Least Upper Bound
Length of Unused Bed
Logical Unit Block
LUBR Lubricant [also: LUB]
Lubrication [also: LUB]
Libricator [also: LUB]
LUC London University Center [UK]
LUCCO Land Utilization Coordination Committee [Cyprus]
LUCID Language for Utility Checkout and Instrumentation Development
LUCK Logical Unit and Checker
LUE Linear Unbiased Estimator
Link Ulilization Efficiency
LUF Lowest Usable Frequency
Lowest Useful Frequency
Luxembourg Franc [Currency]
LUFORO London Unidentified Flying Objects Research Organization [UK]
LUFS Land Use and Forest Resource Survey [Taiwan]
LUG Light Utility Glider
LOCAS (= Local Cataloguing Service) Users Group [UK]
Luggage
LUHF Lowest Usable High Frequency
Lowest Useful High Frequency
LUIS Library User Information System [of NWU, US]
LU-LU Logical Unit to Logical Unit (Session) [also: LULU]
LUM Lumber
Luminous
LUME Light Utilization More Efficient
LUMO Lowest Unoccupied Molecular Orbital
LUN Logical Unit Number
LUNR Land Use and Natural Resources
LUS Large Ultimate Size
LUSI Local User Interface
LUSS Linear Ultrasonic Scanning System
LUST Lustrous
LUT Launcher Umbilical Tower
Local User Terminal
Lookup Table
Loughborough University of Technology [UK]
LUVO Lunar Ultraviolet Observatory [of NASA]
LUW Logical-Unit-of-Work
lu/W lumens per watt [also: lu W^{-1}]
LV Large-Volume
Laser Vision
Launch Vehicle
Leakage Valve
Light and Visible
Linear Velocity
Line Voltage
Liquid Volume
Longitudinal Vibration
Low Velocity
Low Viscosity
Low Voltage
Low Volume
Lv Lev [Currency of Bulgaria]
LVA Large Vertical Aperture
Launch Vehicle Availability
Local Virtual Address
LVAD Left-Ventricle Assist Device
Low-Velocity Air Delivery
LVB Low-Volatile Bituminous
LVCD Least Voltage Coincidence Detector
LVD Liquid-Crystal Visual Display
Low-Velocity Detonation
LVDA Launch Vehicle Data Adapter
LVDC Launch Vehicle Digital Computer
LVDT Linear Variable Differential Transformer
Linear Velocity Displacement Transformer
LVE Linear Viscoelasticity
Low-Voltage Engineering
LVFS Lake Victoria Fisheries Service [Kenya]
LVHV Low-Volume High-Velocity
LVL Level
Low-Velocity Layer
LVLSH Level Shifter
LVM Large Volume Mailer
Localized Vibrational Mode
Long Vertical Mark
LVOD Launch Vehicles Operations Division [of NASA]
LVP Low-Voltage Protection
LVPS Low-Voltage Power Supply
LVR Ladle Vacuum Refinement
Large Volume Receiver
Linear Video Recorder
Longitudinal Video Recording
Low-Voltage Relay

	Low-Voltage Release
	Low Volume Receiver
LVRE	Low-Voltage Release Effect
LVRJ	Low-Volume Ramjet
LVRT	Linear Variable Reluctance Transducer
LVS	Leaves
	Low-Velocity Scanning
LVSEM	Low-Voltage Scanning Electron Microscopy
LVSTK	Livestock
LVT	Landing Vehicle Tracked
LVTR	Low VHF Transmitter-Receiver
LVZ	Low-Velocity Zone
	Low-Viscosity Zone
LW	Leave Word
	Liquid Waste
	Long Wave
	Longitudinal Wave
	Low Water
Lw	Lawrencium
LWA	Last Word Address
	Light Wire Armored
	Limited Work Authorization
LWB	Lower Bound
LWBR	Light-Water Breeder Reactor
LWC	Lightweight Concrete
	Liquid Waste Containment
	Liquid Water Content
	Loop Wiring Concentrator
LWCS	Lightwave Communication System
LWD	Larger Word
	Lightwave Device
	Long Working Distance
LWF	Lightweight Fighterplane
	Luminous Wall Firing
LWHCR	Light Water High Converter Reactor
LWIC	Lightweight Insulating Concrete
LWIR	Long-Wave Infrared
LWL	Limited War Laboratory [US Army]
	Load Water Line
LWM	Liquid Waste Management
	Liquid Waste Monitor
	Low Water Mark
LWOP	Leave Without Pay
LWOST	Low Water of Spring Tide
LWP	Long-Wave Phase
LWR	Laser Warning Receiver
	Light-Water Reactor
	Lower
	Liquid Waste Release
LWRHU	Lightweight Radioisotope Heater Unit
LWS	Loire-Wendel-Sprunck (Process)
	Long-Wave Spectrometry
LWST	Liquid Waste Storage Tank
LWTT	Liquid Waste Test Tank
LWV	Lightweight Vehicle
lx	lux
LXMAR	Load External Memory Address Register
Ly	Lyophilize
LYD	Libyan Dinar [Currency]
LYR	Layer
LYRIC	Language for Your Remote Instruction by Computer
Lys	Lysine
	Lysyl
LZ	Levitation Zone
LZR	Levitation Zone Refining
LZT	Lead-Zinc-Tin
	Lead Zirconate Titanate
	Local Zone Time
LZTC	Lead-Zinc-Tin Committee

M

M	Mach (Number)
	Machine
	Machining
	Machinist
	Magnaflux
	Magnet(ic)
	Magnetization
	Magnetron
	Magnification
	Mantissa
	Manual
	March
	Maritime
	Mark [Currency]
	Marker
	Mass
	Master
	Matter
	May
	Mean
	Measure
	Mechanic(s)
	Mechanical
	Mechanism
	Medium
	Mega- [SI Prefix]
	Megabyte
	Memory
	Meridian
	(Meridies) — Noon
	Merit
	Mesh
	Metal
	Meter
	Methionine
	Methionyl
	Metric
	Mile
	Mill
	Million
	Mine
	Mischmetal
	Missing
	Mississippian [Geology]
	Mitochondrion
	Mix
	Mixture
	Mobile
	Mode
	Model
	Modem
	Module
	Modulus
	Molal(ity)
	Molar
	Molar(ity)
	Mole
	Molecule
	Molecular Weight
	Moment
	Monitor
	Monday
	Monochrome
	Monsoon
	Moon
	Mortar
	Motor
	Mountain
	Mud
	Multiplier
	Mutual Inductanace
	(Mille) — Thousand
m	magnetic
	maritime
	mark(er)
	mass
	meter
	measure
	meridian
	mesh
	meta-
	middle
	mile
	mil
	mill
	(mille) — thousand
	milli- [SI Prefix]
	mine
	minimum
	minor
	minute
	(misce) — mix
	miscellaneous
	mist
	mix(ture)
	mobile
	modulus
	molal
	month
	moon
μ	micro- [SI Prefix]
MA	Magnesium Association [now: IMA, US]
	Magnetic Amplifier
	Magnetic Analysis
	Main Amplifier
	Major
	Maleic Anhydride
	Maritime Administration [US]
	Martensite/Retained Austenite
	Massachusetts [US]
	Mass Analyzer
	Master Antenna
	Master of Arts
	Matched Angles

	Mathematical Association [UK]
	Mechanical Advantage
	Mechanical Alloying
	Megampere
	Memory Address
	Message Assembler
	Meter Angle
	Methyl Acetate
	Methyl Acrylate
	Metric Association [now: USMA]
	Microalloy
	Microanalysis
	Microfilm Address
	Military Academy
	Mill Anneal Process
	Milliammeter
	Modify Address
	Molecular Absorption
	Molecular Adsorption
	Monoamine
	Monoclonal Antibody
	Motion Analyzer
	Moving Average
	Multichannel Analyzer
	Music Analysis
Ma	Masurium
mA	milliampere
mA	milli-angstrom
MAA	Macroaggregate
	Manitoba Architectural Association [Canada]
	Master of Administrative Arts
	Master of Aeronautics and Astronautics
	Master of Applied Arts
	Master of Arts in Architecture
	Material Access Area
	Mathematical Association of America [US]
	Maximum Authorized Altitude
	Methacrylic Acid
	Methanearsonic Acid
	Mixed Aluminum Alloy
	Modeling Association of America
	Moped Association of America [US]
MA-A	Medium Abrasive-Abrasive
MAAC	Mid-Atlantic Area Council
	Model Aeronautics Association of Canada
MAACS	Multi-Address Asynchronous Communication System
MAAD	Material Availability Date
MAAE	Master of Arts in Applied Economics
MAALOX	Magnesium-Aluminum Hydroxide
MAAOM	Master of Arts in Applied Organizational Management
MAARC	Magnetic Annular Arc
MAArch	Master of Arts in Architecture
MAAW	Medium Antitank Assault Weapon
MAB	Man and the Biosphere Program [of UNESCO]
	Materials Advisory Board [now: NMAB, US]
	Metallic Access Bus
	Metropolitan Area Business (Line)
	Micro-Address Bus
	Monomethylaminoazobenzene
MAb	Monoclonal Antibody
MABA	Meta-Aminobenzoic Acid
MABC	Mining Association of British Columbia [Canada]
MABR	Member of the American Board of Radiologists
MABS	Maritime Application Bridge System
MAC	Machine-Aided Cognition
	Maintenance Allocation Chart
	Man and Computer
	Man against Computer
	Manufacturing Advisory Committee
	Massay Agricultural College [New Zealand]
	Master of Art Conservation
	Master of Arts in Communications
	Material Acquisition and Control
	Maximum Acceptable Concentration
	Maximum Admissible Concentration
	Maximum Allowable Concentration
	Mean Aerodynamic Chord
	Measurement and Analysis Center
	Media Access Control
	Medium Access Control
	Memory Access Controller
	Message Act Concellation
	Message Authentication Code
	Metal-Arc Cutting
	Methyl Allyl Chloride
	Microgravity Advisory Committee
	Middle Atlantic Conference [US]
	Military Airlift Command [US Air Force]
	Mineralogical Society of Canada
	Mining Association of Canada
	Ministers Advisory Council [Canada]
	Moves, Adds and Changes
	Multi-Access Computing
	Multiple Access Computer
	Multiplexed Analog Component
MAc	Master of Accounting
MACA	Master of Arts in Computer Applications
MACC	Modular Alter and Compose Console
MAcc	Master of Accountancy
MACCS	Manufacturing and Cost Control System
	Manufacturing Cost Collection System
	Molecular Access System
MACCT	Multiple Assembly Cooling Cask Test [of ORNL, US]
MAcct	Master of Accounting
	Master of Accountancy [also: MAccy]
MACDAC	Machine Communication with Digital Automatic Computer
MACDIF	Map and Chart Data Interchange Format
MACE	Management Applications in Computer Environments
MACEd	Master of Arts in Continuing Education
MACH	Machine
	Machinery
	Machinist
MACI	Ministerio de Agricultura, Comercio e Industria [= Ministry of Agriculture, Commerce and Industry, Panama]
MACIE	Matrix-Controlled Interference Engine
MACMA	Military and Aerospace Connector Manufacturers Association [US]

MACMIS Maintenance and Construction Management Information System
MAComm Master of Arts in Communication
MACON Matrix Connector (Punched Card Programmer)
MACRF Macro Instruction Form
MACRO Macrocode
Macroprogram
MACROCAL Macro Version of Common Assembler Language
MACROL Macro-Based Display Oriented Language
Macrolanguage
MACS Master of Arts in Communication Studies
Master of Arts in Computer Science
Media Account Control System
Medium Altitude Communications Satellite
Member of the American Chemical Society
Metering and Accounting System
Military Aeronautical Communication System
Mobile Air Conditioning Society [US]
Monitoring and Control Station
Multiple-Technique Analytical Computer System
Multiproject Automated Control System
Multipurpose Accelerated Cooling System
Multipurpose Acquisition Control System
MACSMB Measurement and Automatic Control Standards Management Board [of ANSI]
MACT Master of Arts in College Teaching
MACTM Master of Applied Communication Theory and Methodology
MACU Material Cost per Unit
Michigan Association of Colleges and Universities [US]
MACV Multipurpose Airmobile Combat-Support Vehicle
MAD Machine ANSI (= American National Standards Institute) Data
Magnetic Airborne Detector
Magnetic Anomaly Detection
Magnetic Azimuth Detector
Maintenance, Assembly and Disassembly
Many Acronymed Device [the HEMT]
Maroccan Dirham [Currency]
Mass Analyzer Detector
Michigan Algorithmic Decoder
Military Aircraft Division
Mixed Analog and Digital
Multi-Aperture Device
Multiple Access Device
Multiple Access Drive
Multiple Audio Distribution
Multiple-Wavelength Anomalous Dispersion
Multiply and Add
Mutually Assured Destruction
MAd Master of Administration
MADA Multiple-Access Discrete Address
MADAEC Military Application Division of the Atomic Energy Commission
MADAM Moderately Advanced Data Management
Multipurpose Automatic Data Analysis Machine
MADAP Maastricht Automatic Data Processing System
MADAR Malfunction Analysis Detection and Recording
Malfunction Data Recorder
MADDAM Macromodule and Digital Differential Analyzer Machine
MADDDC Manufacturers of Aerial Devices and Digger-Derricks Council [US]
MADDIDA Magnetic Drum Digital Differential Analyzer
MADE Magnetic Device Evaluator
Microalloy Diffused Electrode
Minimum Airborne Digital Equipment
Multichannel Analog-to-Digital Data Encoder
MADEE Malonic Acid Diethylester
MADGE Microwave Aircraft Digital Guidance Equipment
Microwave Automatic Digital Guidance Equipment
MADIS Millivolt Analog-Digital Instrumentation System
MADM Manchester Automatic Digital Machine [of Manchester University, UK]
MAdm Master of Administration [also: MAdmin]
MAdmMgt Master of Administration Management
MADO Mulliken Approximation for Differential Overlap
MADP Main Air Display Plot
MADR Microprogram Address Register
MADRE Magnetic Drum Receiving Equipment
Martin Automatic Data-Reduction Equipment
MADREC Malfunction Detection and Recording
MADS Machine-Aided Drafting System
Missile Attitude Determination System
MADT Microalloy Diffused-Base Transistor
MADW Military Air Defense Warning
MAE Magnetic After-Effect
Magnetomechanical Acoustic Emission
Maine Association of Engineers [US]
Manchester Association of Engineers [UK]
Maritime Advisory Exchange
Maryland Association of Engineers [US]
Master of Aerospace Engineering
Master of Aeronautical Engineering
Master of Agricultural Economics
Master of Agricultural Education
Master of Agricultural Engineering
Memory Address Register
Microalloying Element
MAECO Multi-Unit Architects, Engineers and Construction Officers [US]
MAECON Mid-America Electronics Conference
MAEd Master of Arts in Education
MAEDP Master of Arts in Economic Developmental Programming
MAELV Mutual Atomic Energy Liability Underwriters
MAES Master of Arts in Environmental Sciences
Mexican-American Engineering Society [US]
MAESTRO Machine-Assisted Educational System for Teaching by Remote Operation
MAF Major Academic Field
Management Accounting Factor
Mass Air Flow (Sensor)
Ministry of Agriculture and Food [Canada]
Mission Aviation Fellowship
Mixed Amine Fuel
Moisture and Ash-Free
Multi-Apertured Fabrics
MAFD Minimum Acquisition Flux Density
MAFF Ministry of Agriculture, Fisheries and Food [UK]

MAFIS Master of Accountancy and Financial Information Systems
MAG Macrogenerator
Magazine
Magnesium
Magnet
Magnetic(s)
Magnetism
Magneto(meter)
Magnetron
Magnitude
Management Accounting Guidelines
Maritime Air Group [of Canadian Forces]
Master of Applied Geography
Maximum Available Gain
Medical Association of Georgia [US]
Metal Active-Gas (Welding)
Million Ampere Generator
MAg Master of Agriculture
MAGAMP Magnetic Amplifier [also: Magamp or magamp]
MAGB Microfilm Association of Great Britain
MAGE Mechanical Aerospace Ground Equipment
MAgEc Master of Agricultural Economics
MAgEd Master of Agricultural Education
MAgEng Master of Agricultural Engineering [also: MAgE]
MAGFET Magnetic Field Sensitive Fields-Effect Transistor
MAGG Modular Alphanumeric Graphics Generation
MAGGS Modular Advanced Graphics Generation System
MAGI Multi-Array Gamma Indicator
MAGIC Machine-Aided Graphic Input to Computer
Machine-Aided Graphics for Illustration and Composition
Machine for Automatic Graphics Interface to a Computer
Matrix Algebra General Interpretative Coding
Michigan Automatic General Integrated Computation
MIDAC (= Michigan Digital Automatic Computer) Automated General Integrated Computation
MAGICS Manufacturing, Accounting and General Information Control System
MAGIS Marine Air-Ground Intelligence System [US]
MAGLEV Magnetic Levitation
MAGLOC Magnetic Logic Computer
MAGMOD Magnetic Modulator
MAGN Magnet
Magnetic(s)
Magnetism
Magnetron
MAGPIE Machine Automatically Generating Production Inventory Evaluation
MAgr Master of Agriculture
MAgrDevEc Master of Agricultural Development Economics
MAG REC Magnetic Recording
MAgrEc Master of Agricultural Economics [also: MAgrEcon]
MAgrSc Master of Agricultural Science [also: MAgricSc or MAgSc]
MAgrSt Master of Agricultural Studies
MAGS Magnitudes
Measurement and Graphics System
MAgSc Master of Agricultural Science [also: MAgrSc, MAgricSc]
MAGTAPE Magnetic Tape [also: MAG TAPE]
MAGTC Magnetic Tape Controller
MAHA Maximum Acceptable Hole Angle
MAHRM Master of Arts in Human Resources Management
MAHS Master of Arts in Human Services
MAHT Machine-Aided Human Translation
Master of Arts in History Teaching
MAI Machine-Aided Index
Master of Agricultural Industries
Mobile Arctic Island
Multiple Access Interface
MAIA Master of Arts in International Affairs
Master of Arts in Industrial Art
MAIBC Member of the Architectural Institute of British Columbia [Canada]
MAID Master of Arts in Interior Design
Multiple Aircraft Identification Display
MAIDS Modular, Adaptive, Incremental Detection System
Multipurpose Automatic Inspection and Diagnostic System
MAIEE Member of the American Institute of Electrical Engineers
MAIG Matsushita Atomic Industrial Group [Japan]
MAIM Member of the American Institute of Management
MAIME Member of the American Institute of Mining and Metallurgical Engineers
MAIN Mid-America Interpool Network
MAINT Maintenance
MAIntDes Master of Arts in Interior Design
MAinURP Master of Arts in Urban and Regional Planning
MAIP Matrix Algebra Interpretative Program
MAIPS Master of Arts in International Policy Studies
MAIR Master of Arts in Industrial Relations
Molecular Airborne Intercept Radar
MAIS Master of Arts in Interdisciplinary Studies
Master of Arts in International Studies
Member of the Association of Industrial Surgeons
MAJ Major
Majority
Master of Arts in Journalism
MAJAC Monitor, Anti-Jam and Control
MAJC Master of Arts in Journalism and Communication
MAJ GEN Major General
MAK Methylated Albumin Kieselguhr
MAL Macro Assembly Language
Maltese Pound [Currency of Malta]
Memory Access Logic
Meta-Access Language
MALAGOC Mutual Assistance of the Latin American Government Oil Companies [Uruguay]
MALCAP Maryland Academic Library Center for Automated Processing [of UM, US]
MALD Master of Arts in Landscape Design
MALE Multi-Aperture Logic Element
MALIS Master of Arts in Library and Information Science
MALL Malleability
MALLAR Manned Lunar Landing and Return
MALMARC Malaysian Machine-Readable Catalogue
MALS Master of Arts in Library Science

	Maximum-Intensity Approach Light System
	Medium-Intensity Approach Light System
MALSF	Medium-Intensity Approach Light System with Sequenced Flashing Lights
MALSR	Medium-Intensity Approach Light System with Runway Alignment Indicator Lights
MALT	Manitoba Association of Library Technicians [Canada]
	Mnemonic Assembly Language Translator
MAM	Management and Administration Manual
	Master of Administrative Management
	Master of Agriculture and Management
	Master of Animal Medicine
	Master of Applied Mechanics
	Master of Aviation Management
	Memory Access Multiplexer
	Memory Allocation Manager
	Message Access Method
	Methylazoxymethanol
	Monoaminomesitylene
	Multiple Access to Memory
	Multi-Application Monitor
MAMA	Methylazoxylmethanol Acetate [also: MAMAc]
	Monoammonium Methanearsonate
MAMAc	Methylazoxylmethanol Acetate [also: MAMA]
MAMB	Master of Applied Molecular Biology
MAMBA	Martin Armoured Main Battle Aircraft
MAMBO	Mediterranean Association for Marine Biology and Oceanography [Malta]
	Mediterranean Association for Marine Biology and Oceanology [Italy]
MAMC	Master of Arts in Mass Communication
MAMF	Masters' Association of Metals Finishers
MAMgt	Master of Arts in Management
MAMI	Machine-Aided Manufacturing Information
	Modified Alternate Mark Inversion
MAMIE	Magnetic Amplification of Microwave Integrated Emissions
MAMIS	Mandatory Modification and Inspection Summary
MAMM	Master of Agricultural Marketing Management
MAMMAX	Machine-Made and Machine-Aided Index
MAMOS	Marine Automatic Meteorological Observing Station
MAMRD	Master of Agricultural Management and Resource Development
MAMS	Master of Applied Mathematical Science
MAMU	Millimass Unit [also: mamu]
MAN	Manual
	Manufacturers Association of Nigeria
	Methylacrylonitrile
	Metropolitan Area Network
	Microwave Aerospace Navigation
	Molecular Anatomy
Man	Manitoba [Canada]
	Mannose
MANA	Manufacturers Agents National Association [US]
MANAV	Marine Integrated Navigation
MANC	Mexican Association for the Nature Conservancy
MANDATE	Multi-Line Automatic Network Diagnostic and Transmission Equipment
ManDir	Managing Director
MANDRO	Mechanically Alterable NDRO (= Nondestructive Readout)
MANEB	Manganese Ethylene-Bisdithiocarbamate [also: maneb]
MANF	Manifold
MANFEP	University of Manitoba Finite Element Program [Canada]
MANG	Manufacturing
MAN HR	Man Hour
MANI	Ministry of Agriculture of Northern Ireland
MANIAC	Mathematical Analyzer Numerical Integrator and Computer
	Mechanical and Numerical Integrator and Computer
MANIFILE	University of Manitoba File of the World's Nonferrous Metallic Deposits [Canada]
MANIP	Manual Input
Manit	Manitoba [Canada]
MANMAR	Manual for Marine Weather Observation
MANMAX	Machine-Made and Machine-Aided Index
MANOP	Manual of Operation
MANOVA	Multivariate Analysis of Variance
MANPR	Manpower
MANR	Ministry of Agriculture and Natural Resources [Nigeria]
MANSCETT	Manitoba Society of Certified Engineering Technicians and Technologists [Canada]
MANTECH	Manufacturing Technology
MANTIS	Manchester Technical Information Service [UK]
MANTRAC	Manual Angle Tracking Capability
MANUV	Maneuvering
MAO	Monoamine Oxidase
MAOP	Maximum Allowable Operating Pressure
MAP	Machinist Apprentice Program
	Macao Pataca [Currency of Macao]
	Macro Arithmetic Processor
	Macroassembly Program
	Magneto-Abrasive Powder
	Maintenance Analysis Procedure
	Management Assistance Program
	Manifold Absolute Pressure
	Manifold Air Flow (Sensor)
	Manufacturing Automation Protocol
	Mapping
	Marketing Assistance Program
	Mathematical Analysis without Programming
	Maximum a posteriori Probability
	Memory Allocation and Protection
	Memory Allocation Processor
	Message Acceptable Pulse
	Message Acceptance Pulse
	Metal Analysis Probe
	Methoxyacetophenone
	Microprocessor Application Project [of DTI, UK]
	Microprogrammed Array Processor
	Microtubule-Associated Protein
	Middle Atmosphere Program
	Military Assistance Program
	Minimum Acceptable Performance
	Minimum Audible Pressure
	Missed Approach Point
	Model and Program

	Modular Analysis Processor
	Monoaluminumphosphate
	Monoammonium Phosphate
	Multiple Aim Point
	Multiple Allocation Procedure
MAPA	Master of Arts in Public Administration
	Master of Arts in Public Affairs
MAPAM	Mines Accident Prevention Association of Manitoba [Canada]
MAPAO	Mines Accident Prevention Association of Ontario [Canada]
MAPC	Metropolitan Area Planning Committee
MAPCHE	Mobile Automatic Programmed Checkout Equipment
MAPCON	Microprocessor Applications Consultancy
MAPED	Machine-Aided Program for Preparation of Electrical Designs
MAP-EPA	Manufacturing Automation Protocol — Enhanced Performance Architecture
MAPI	Machinery and Allied Products Institute [US]
MAPICS	Manufacturing Accounting and Production Information Control System
MAPID	Machine-Aided Program for Preparation of Instruction Data
MAPLHG	Maximum Average Planar Linear Heat Generator
MAPLHGR	Maximum Average Planar Linear Heat Generation Rate
MApMa	Master of Applied Mathematics
MAPORD	Methodology Approach to Planning and Programming Operational Requirements, Research and Development [US Air Force]
MAPP	Mathematical Analysis of a Perception and Preference
	Methylacetylene Propadiene [also: Mapp]
MAPPER	Maintaining, Preparing and Producing Executive Reports
MAPPS	Management Association for Private Photogrammetric Surveyors [US]
MApplSc	Master of Applied Science [also: MApplS or MAppSc]
MApplStat	Master of Applied Statistics [also: MAppStat or MApStat]
MAPRAT	Maximum Power Ratio
MAPS	Management Accounting and Payroll System
	Manufacturers Automation Peripheral System
	Measurement of Air Pollution from Satellites
	Middle Atlantic Planetarium Society [US]
	Modular Acoustic Processor System
	Modular Automatic Preparation System
	Multicolor Automatic Projection System
	Multivariate Analysis and Prediction of Schedules
MApStat	Master of Applied Statistics [also: MApplStat or MAppStat]
MAPt	Missed Approach Point
MAPTAC	Methacrylamidopropyltrimethylammonium Chloride
MAP/TOP	Manufacturing Automation Protocol/Technical Office Protocol [also: MAP-TOP or MAPTOP]
MAPW	Master of Arts in Professional Writing
MAQ	Methylanthraquinone
MAq	Master of Aquaculture
MAQL	Mine Air Quality Laboratory
MAR	Machine-Readable
	Macro Address Register
	Master of Arts in Research
	Mauritian Rupee [Currency of Mauritius]
	Malfunction Array Radar
	March
	Marine
	Maritime
	Market
	Memory Address Register
	Micro-Analytical Reagent
	Microprogram Address Register
	Minimum Angle of Resolution
	Miscellaneous Apparatus Rack
	Multifunction Array Radar
MA(R)	Master of Arts (Research)
MARAD	Marine Administration [US]
	Maritime Administration [US]
MARAIRMED	Maritime Air Forces Mediterranean [of NATO]
MARAS	Middle Airspace Radar Advisory Service [US]
MARATS	Mid-Atlantic Region Administrative Telephone System
MARC	Machine-Readable Catalogue
	Material Accountability and Recoverability Code
	Modern Asian Research Center
	Monitoring and Assessment Research Center [UK]
MARCA	Mid-Continent Area Reliability Coordination Agreement
MARCAS	Maneuvering Reentry Control and Ablation Studies
MARCEP	Maintainability and Reliability Cost-Effectiveness Program
MArch	Master of Architecture
MArchEng	Master of Architectural Engineering [also: MArchE]
MArchH	Master of Architectural History
MArchUD	Master of Architecture and Urban Design
	Master of Architecture in Urban Design
MARCIA	Mathematical Analysis of Requirements for Career Information Appraisal
MARC IS	Machine-Readable Catalogue — Israel
MARC LC	Machine-Readable Catalogue — Library of Congress [US]
MARCO	Machine Referenced and Coordinated Outline
MARCOGAZ	Union de l'Industrie Gaz du Marche Commun [= Union of the Gas Industry of the Common Market Countries, Belgium]
MARCOM	Maritime Command Operations [Canada]
	Microwave Airborne Communications Relay
MARCOT	Maritime Coordinated Training Exercises
MARCS	MELCOM (Computer) Allround Adaptive Consolidated Software
MARC UK	Machine-Readable Catalogue — United Kingdom
MARDAN	Marine Differential Analyzer
MARDI	Malaysian Agricultural Research and Development Institute
MarE	Marine Engineer [also: MarEng]
MAREA	Member of the American Railway Engineering Association
MARECS	Marine Communications Satellite [of ESA, France]
	Maritime Communications Satellite System
MarEng	Marine Engineer [also: MareE]
MARFO	Machine-Readable Form

MARG Margarine
Margin
MARGEN Management Report Generator
MARGIE Memory Analysis Response Generation and Interference in English
MARIN Marine Research Institute of the Netherlands
MARISAT Maritime Satellite (Service) [also: MARSAT]
MARKAR Mapping and Reconnaissance Ku-Band Airborne Radar
MARKSIM (Computerized) Marketing Management Simulation
MARLIS Multi-Aspect Relevance Linkage Information System
MARN Manitoba Association of Registered Nurses [Canada]
MARNA Marine Navigation (Database) [Netherlands]
MARNAF Marquardt Navair Fuel
MAROTS Maritime Orbital Test Satellite
MARPAC Maritime Forces Pacific [of DND, Canada]
MARPAL Marseilles-Palo Submarine Cable [France/Italy]
MARR Minimum Attractive Rate of Return
MARS Machine-Assisted Reference Section
Machine Retrieval System
Magnetic Airborne Recording System
Management Analysis Reporting Service
Manned Aerodynamic Reusable Spaceship
Manned Astronomical Research Station
Marconi Automatic Relay System
Market Analysis Research System
Martin Automatic Reporting System
Materials Analysis Research Station
Member of the Association of Railway Surgeons
Memory-Address Register Storage
Military Affiliated Radio System
Military Amateur Radio System
Mirror Advanced Reactor Study
Multi-Access Retrieval System
Multi-Aperture Reluctance Switch
Multiple Artillery Rocket System [US]
MARSAT Maritime Satellite [also: MARISAT]
MArSci Master of Arts and Sciences
MARSE Measured Area under the Rectified Signal Envelope
MARSL Machine-Readable Shelf List
MART Maintenance Analysis Review Technique
Mean Active Repair Time
Mechanical Analysis and Retrieval of Text
Mississippi Aerial River Transit System [US]
Mobile Automatic Radiation Tester
MARTD Multidisciplinary Association for Research and Teaching in Demography [Peru]
MARTEC Martin Thin-Film Electronic Circuit
MARTEL Marseilles-Tel Aviv Submarine Cable [France/Israel]
Missile Antiradar and Television
MARTM Maritime
MARTOS Multi-Access Real-Time Operating System
MARV Maneuverable Antiradar Vehicle
Multi-Element Articulated Research Vehicle
MARVEL Managing Resources for University Libraries
MAS Macro-Assembler
Magic Angle Spinning
Magnesia-Alumina-Silicate
Magnetic Angle Spinning
Management Advisory Services
Maryland Academy of Science [US]
Master of Accounting Science
Master of Administrative Science
Master of Aeronautical Science
Master of Applied Science
Master of Applied Statistics
Master of Archival Studies
Material Acquisition Specification
Metal-Alumina-Silicon
Microalloyed Steel
Microbeam Analysis Society [US]
Microfilm Advisory Service
Military Agency for Standardization [of NATO]
Military Alert System
Mississippi Academy of Science [US]
Modular Application System
Multi-Aspect Signaling
mAs milliampere second
MASA Main Store Arrays
Marine Accessories and Services Association [US]
Master of Advanced Studies in Architecture
Military Accessories Service Association [US]
Military Automotive Supply Agency [US]
MASC Main Store Controller
Methylaluminum Sesquichloride
MASc Master of Agricultural Sciences
Master of Applied Science
MAScMechEng Master of Applied Science in Mechanical Engineering
MAScMetEng Master of Applied Science in Metallurgical Engineering
MASCOT Modular Approach to Software Construction, Operation and Test
Motorola Automatic Sequential Computer Operator Test
MASCP&T Member of the American Society for Clinical Pharmacology and Therapeutics
MASDC Military Aircraft Storage and Deposition Center
MASE Materials, Applications and Services Exposition [US]
MASER Materials Science Experiment Rocket
Microwave Amplification by Stimulated Emission of Radiation [also: Maser or maser]
MASG Missile Auxiliary Signal Generator
MASH Manned Antisubmarine Helicopter
Mobile Army Surgical Hospital
MASIS Management and Scientific Information System
MASK Maneuvering and Seakeeping
MASL Meters above Sea Level [also: masl]
MASM Macro Assembler
Member of the American Society for Metals
Meta-Assembler
Microsoft's Macro Assembler
MASME Member of the American Society of Mechanical Engineers
MASPAC Microfilm Advisory Service of the Public Archives of Canada
MASRT Marine Air Support Radar Teams
MASRU Marine Air Support Radar Unit [of USDOD]

MASS Maintenance Activities Subsea Surface
MARC (= Machine Readable Catalogue) Based Automated Serials System
Master of Arts in Special Studies
Michigan Automatic Scanning System
Mobile Army Sensor System
Monitor and Assembly System
Multi-Axial Span System
Multiple Access Sequential Selection
Multiple Access Switching System
Mass Massachusetts [US]
MASSDAR Modular Analysis, Speedup, Sampling, and Data Reduction
MASSTER Mobile Army Sensor System Test Evaluation and Review
MAST Magnetic Annular Shock Tube
Material Status
Multiple Aircraft Simulation Terminal
MASTER Matching Available Student Time to Educational Resources
Multiple Access Shared Time Executive Routine
MASTIR Microfilmed Abstracts System for Technical Information Retrieval [of Illinois Institute of Technology, US]
MASU Metal Alloy Separation Unit
MASUA Mid-America State Universities Association [US]
MAT Machine-Aided Translation
Master of Arts in Teaching
Material
Matrix
Maximized Average Torque
Mechanical Aptitude Test
Mechanical Assembly Technique
Mechanically-Agitated Tank
Memory Address Translator
Metropolitan Area Trunks
Microalloy Transistor
Miller Analogies Test
Monitor, Alarm and Trend Module
Multiple Access Technique
MATA Michigan Aviation Trades Association [US]
MATC Master of Arts in Textiles and Clothing
MATCALS Marine Air Traffic Control and Landing System
MATCON Microwave Aerospace Terminal Control
MATD Mine and Torpedo Detector
MatDC Materials Property Database [of ASM International, US]
MATE Manufacturing Advanced Technology Exchange
Measuring and Test Equipment
Memory-Assisted Terminal Equipment
Modular Automatic Test Equipment
Multiple-Access Time-Division Experiment
Multisystem Automatic Test Equipment
MatE Materials Engineer [also: MatEng]
MAT ENG Materials Engineering
MATER Material
MATFAP Metropolitan Area Transmission Facility Analysis Program
MATH Mathematical
Mathematician
Mathematics
MathD Doctor in Mathematics
MATHLAB Mathematical Laboratory
MATHPAC Mathematical and Statistical Program Package
MATIC Multiple Area Technical Information Center
MATICO Machine Applications to Technical Information Center Operations
MATILDA Metering and Totalizing Instrument for Load Demand Assessment
MATL Material
MATLAN Matrix Language
MATLS Materials
MATL SCI Materials Science
MATM Master of Arts in Teaching of Mathematics
m-atm meter-atmosphere
MAT MEAS Materials Measurement
MATO Military Air Traffic Operations
MA-TPM Maritime Administration — Transport Planning Mobilization [US]
MATPS Machine-Aided Technical Processing System
MATR Management Access to Records
MATRIC Matriculation
MATRS Miniature Airborne Telemetry Receiving Station
MATS Master of Arts in Teaching of Science
Materials Testing System
Microprocessor-Assisted Testing System
Military Air Transport Service
Multiple Address Telex Service [Canada]
MAT SCI Materials Science
MAT SCI ENG Materials Science and Engineering
MATT Missile ASW (= Antisubmarine Warfare) Torpedo Target
MATV Master Antenna Television
MATZ Military Aerodrome Traffic Zone
Military Airfield Traffic Zone
MAU Marine Amphibious Unit
Medium Access Unit
Memory Access Unit
Multi-Attribute Utility
Multiple Access Unit
MAUA Master of Arts in Urban Affairs
MAUD Master of Arts in Urban Design
MAud Master of Audiology
MAUDE Morse Automatic Decoder
MAUDEP Metropolitan Association of Urban Designers and Environmental Planners [US]
MAuE Master of Automotive Engineering
MAUFS Municipal Arborists and Urban Foresters Society [US]
MAURP Master of Arts in Urban and Regional Planning
MAUS Mobile Automated Ultrasonic Scanner
MAUTEL Microminiaturized Autonetics Telemetry
MAVAR Modulating Amplifier using Variable Reactance [also: Mavar, or mavar]
MAVART (Mathematical) Model for the Analysis of Vibrations and Acoustic Radiation of Transducers
MAVEd Master of Administration in Vocational Education
MAVES Manned Mars and Venus Exploration Studies
MAVICA Magnetic Video Card
MAVIN Machine-Assisted Vendor Information Network
MAW Manual Arc Welding
Marine Aircraft Wing

Master of Arts in Writing
Medium Anti-Armor Weapon
Medium Antitank Assault Weapon
Metal-Arc Welding
Mission-Adaptable Wing
Mission Adaptive Wiring

MAWP Maximum Allowable Working Pressure

MAWS Mobile Aircraft Weighing System

MAX Maximum
Mobile Automatic Exchange
Modular Applications Executive

MAXCOM Modular Applications Executive for Communications

MAXIT Maximum Interference Threshold

MAXNET Modular Applications Executive Network

MAX WT Maximum Weight

MB Bachelor of Medicine
Magnetic Brake
Magnetic Braking
Magnetic Bubble
Magnetron Branch
Main Battery
Maintenance Busy
Manitoba [Canada]
Marine Biologist
Marine Biology
Marine Board [US]
Master Builder
Mathematical Biology
Megabit
Megabyte
Memory Buffer
Metric Board
Microbalance
Missile Bomber
Mixing Box
Mobile
Molecular Beam
Molecular Biology
Motor Boat

M-B Make-Break

Mb Megabit

mb millibar
millibarn

MBA Manufacturered Buildings Association [US]
Marine Biological Association [US and UK]
Master of Business Administration
Material Balance Area
Merion Bluegrass Association [US]
Methoxybutyl Acetate
Methylenebisacrylamide
Modulated Bayard-Alpert Gauge
Multiple Beam Antenna
Mung Bean Agglutinin

MBAA Master Brewers Association of the Americas
Master of Business Administration in Aviation

MBAAgric Master of Agricultural Business and Administration

MBAE Master of Biological and Agricultural Engineering

MBAIB Master of Business Administration in International Business

MBAIT Master of Business Administration in International Trade

mbar millibar

μbar microbar

MBAS Methylene Blue Active Substance

MBAUK Marine Biological Association of the United Kingdom

MBBA 4-Methoxybenzylidene-4-Butylaniline

MBC Main Beam Clutter
Master of Building Construction
Maximum Breathing Capacity
Memory Bus Controller
Model-Based Control
Multiple Basic Channel

MBCD Modified Binary-Coded Decimal

MB/CD Thousand Barrels per Calender Day

MBD Magnetic Bubble Device
Manual Board
Milling and Baking Division [of AACC, US]

MB/D Thousand Barrels per Day

MBDA Metal Building Dealers Association [now: SBA, US]

MBDAACC Milling and Baking Division of American Association of Cereal Chemists [US]

M bd ft Thousand Board Feet

MBDS Modular Building Distribution System

MBE Master of Business Economics
Master of Business Education
Member of the British Empire
Missile Borne Equipment
Molecular Beam Epitaxy

MBEd Master of Business Education

MBEO Minority Business Enterprise Office

MBF Materials Business File
Molecular Beam Facility
Thousand Board Feet

MBFR Mutual Balanced Force Reduction

MBH Thousand BTU (= British Thermal Units) per Hour [also: mbh]

MBI Memory Bank Interface
2-Mercaptobenzimidazole
Multibeam Image

MBIAC Missouri Basin Inter-Agency Committee [US]

MBiChem Master of Biological Chemistry

MBiEng Master of Biological Engineering

MBIM Member of the British Institute of Management

MBIO Microprogrammable Block Input/Output

MBiomedE Master of Biolmedical Engineering

MBiPhy Master of Biological Physics

MBiS Master of Biological Science

Mbit Megabit

Mbit/s Megabits per Second

MBITS Monitored Burn-In Test System

MBK Multiple Beam Klystron

MBL Marble
Marine Biological Laboratory [US]

MBldg Master of Building Construction

MBM Magnetic Bubble Memory
Master of Business Management
Metal-Barrier-Metal
Molecular Beam Maser
Multipurpose Boring Machine
Thousand Board Feet [also: mbm]

MBMA Master Boiler Makers Association [US]

	Metal Building Manufacturers Association [US]
MBMC	Monobutyl-M-Cresol
MBMS	Molecular Beam Mass Spectroscopy
MBO	Magnetic Blow-Out
	Management by Objective
	Manpower Branch Offices [Canada]
Mbo	Moraxella Bovis [Molecular Biology]
MBOH	Minimum Break-Off Height
MBOS	Multi-User Business Operating System
MBP	Many-Body Problem
	Mechanical Balance Package
	Mid-Boiling Point
	Myelin Basic Protein
MBPS	Megabits per Second [also: MBS, Mb/s or Mbit/s]
	Milk and Beef Production System
MBPT	Many-Body Perturbation Theory
MBQ	Modified Binary Code
MBQT	Many Body Quantum Theory
MBR	Marker Beacon Receiver
	Material Balance Report
	Member
	Memory Base Register
	Memory Buffer Register
MBRE	Memory Buffer Register, Even [also: MBR-E]
MBRL	Million Barrels [also: mbrl]
MBRO	Memory Buffer Register, Odd [also: MBR-O]
MBRV	Maneuverable Ballistic Reentry Vehicle
MBS	1-Benzothiazole-N-Sulfene Morpholide
	Magnetron Beam Switching
	Maleimidobutyric Succinimide
	Master of Basic Science
	Master of Building Science
	Master of Business Studies
	Megabits per Second [also: MBPS, Mb/s or Mbit/s]
	Methacrylate-Butadiene-Styrene
	Monumental Brass Society [UK]
	Multi-Block Synchronization
	Multiple Batch Station
	Mutual Broadcasting System [US]
Mb/s	Megabits per Second [also: MBPS, MBS or Mbit/s]
MBSA	Methylated Bovine Serum Albumin
MBSc	Master of Business Sciences
MBSHC	Mediterranean Black Seas Hydrographic Commission
MBSI	Master of Business Science and Information Management
MBSM	Master of Building Science Management
MBSS	2-Morpholinodithio Benzothiazole
MBT	Main Battle Tank
	Mercaptobenzothiazole
	Metal-Base Transistor
MBTH	3-Methyl-2-Benzothiazolinone Hydrazone
MBTS	Mercaptobenzothiazyl Disulfide
MBTU	Million British Thermal Units
MBU	Magnetic Bubble Unit
	Memory Buffer Unit
MBusEd	Master of Business Education
MBV	Modified Bauer-Vogel Process
MBWO	Microwave Backward-Wave Oscillator
Mbyte	Megabyte
MBZ	Mandatory Broadcast Zone
MC	Machine Console
	Magnetic Card
	Magnetic Circuit
	Magnetic Clutch
	Magnetic Core
	Magnetic Course
	Magnetic Cyclone
	Magnetochemistry
	Main Circuit
	Main Channel
	Main Chamber
	Main Contact
	Main Current
	Major Component
	Making-Current
	Manganese Center [France]
	Manifold Control
	Manual Code
	Manual Control
	Mapping Camera
	Marginal Checking
	Maritime Commission [US]
	Master Clock
	Master Control
	Master of Counseling
	Master of Communication
	Mean Calorie
	Mean Curvature
	Medical Corps
	Medium Curing
	Member of Council
	Memory Control(ler)
	Metacenter
	Metal Carbide
	Metal-Clad (Cable)
	Metal Cutting
	Meter-Candle
	Methylcellulose
	Methylene Chloride
	Metric Carat
	Microbial Corrosion
	Microcircuit
	Microcomputer
	Micro-Controller
	Microscope Camera
	Military Committee [of NATO]
	Military Computer
	Military Cross
	Mineral Chemistry
	Mining College
	Missile Control
	Module Control
	Moisture Content
	Molded Component
	Molded Compound
	Momentary Contact
	Monitor Call
	Monolithic Circuit
	Monte Carlo Method
	Motion Control(ler)
	Motor-Controller

Motor Converter
Motorcycle
Moving Coil
Multichip
Multiple Choice
Multiple Contact
Multiple Copy
Municipal Corporation
Munitions Command

M&C Monitor and Control
M-C Medium-Curing Asphalt
Mc Megacycle
mc (mensis currentis) — this month
millicurie
millicycle
MCA Macrocell Array
Malaysian Commercial Association
Manufacturing Chemists Association [US]
Marine Corps Association [US]
Master of Communication Arts
Material Control and Accountability
Material Coordinating Agency
Maximum Credible Accident
Mechanical Contractors Association
Metal Construction Association [US]
Methylcholanthrene
Microchannel Architecture
Military Coordinating Activity
Minimum Crossing Altitude
Model Cities Administration [US]
Modified Conventional Alloy
Momordica Charantia Agglutinin
Monochloroacetic Acid
Monoclonal Antibody
Multichannel Analyzer
Multiple Crevice Assembly
Multiprocessor Communications Adapter
MCAA Marine Corps Aviation Association [US]
Mason Contractors Association of America [US]
Mechanical Contractors Association of America [US]
Monochloroacetic Acid
MCAD Mechanical Computer-Aided Design
MCAE Mechanical Computer-Aided Engineering
MCAF Mauritius Cooperative Agricultural Federation
MCAI Microcomputer-Assisted Instruction
MCALS Minnesota Computer-Aided Library System [UM, US]
MCANW Medical Campaign Against Nuclear Weapons
MCAR Machine Check Analysis and Recording
MCAS Marine Corps Air Station
Micro-Controlled Airflow System
MCASI Member of the Canadian Aeronautics and Space Institute
MCB Microcomputer Board
Miniature Circuit Breaker
MCBF Mean Cycles between Failures
MCBIC Michigan/Canadian Bigfoot Information Center [US]
MCC Main Communications Center
Main Control Circuit
Maintenance Control Center
Maintenance Control Circuit
Management Computer Control
Management Control Center
Manned Control Car [of USAEC]
Master Control Center
Master of Clinical Chemistry
Metal/Ceramic Composite
Metric Commission of Canada
Microcomputer Control
Microcrystalline Cellulose
Miscellaneous Common Carrier
Mission Control Center [of NASA]
Mixing Cross-Bar Connectors
Motor Control Center
Multichannel Carrier
Multichannel Communications Control
Multicomponent Circuit
Multicore Cable
Multi-Crossover Cryotron
Multiple Chip Carrier
Multiple Computer Complex
Munitions Carriers Conference [US]
MCCB Molded-Case Circuit Breakers
MCCGL Message Conveying Computers General License [UK]
MCC-H Mission Control Center — Houston [of NASA]
MCCI Molten Core—Concrete Interaction
MCCIS Military Command, Control and Information System
MCCISWG Military Command, Control and Information Systems Working Group [of NATO]
MCCR Ministry of Consumer and Commercial Relations [Canada]
MCCS Military Command and Control System
Monte Carlo Computer Simulation
Multichannel Carrier System
MCCU Multichannel Control Unit
Multiple Communications Control Unit
MCD Magnetic Circular Dichroism
Magnetic Crack Detection
Master of Civic Design
Measurement Control and Display
Metals and Ceramics Division [of US Air Force]
Minimum Charge Duration
Months for Cyclical Dominance
Multicomponent Distillation
Multiple Concrete Duct
mcd millicurie-destroyed [Unit]
MCDBSU Master Control and Data Buffer Storage Unit
MCDP Microprogrammed Communication Data Processor
MCDS Management Control Data System
Microcomputer Development System
MCE Malawi Certificate of Education
Master of Civil Engineering
Mapping and Charting Establishment [UK]
Memphis Cotton Exchange [US]
Mercury Cathode Electrolysis
Microcircuit Engineer
Microcircuit Engineering
Montgomery Cotton Exchange [US]
MCEB Military Communications Electronics Board [US]

MCEd Master of Continuing Education
MCEL Machine Check Extended Log-Out
MCEng Master of Civil Engineering
MCerE Master of Ceramic Engineering
MCES Member of the Civil Engineering Society
MCESS Milwaukee Council of Engineering and Scientific Societies
MCET Microcomputer Engineering Technologist
Microcomputer Engineering Technology
MCEWG Military Communications Electronics Working Group [of NATO]
MCF Metal-Coated Fiber
Micro-Complement Fixation
Military Computer Family
Million Cubic Feet [also: mcf]
Monolithic Crystal Filter
Multicomponent Flow
Mutual Coherence Function
Thousand Cubic Feet
μCF Muon Catalyzed Fusion
MCFC Molten Carbonate Fuel Cell
MCFD Million Cubic Feet per Day [also: MCF/D or mcf/d]
MCFH Million Cubic Feet per Hour [also: MCF/h or mcf/h]
MCFR Magnetic Confinement Fusion Reactor
MCG Magnetocardiography
Man-Computer Graphics
Master-Clock Generator
Materials Coordinating Group
Microwave Command Guidance
Mobile Command Guidance
McGU McGill University [Canada]
MCH Machine
Machine Check Handler
Methylcyclohexane
Microcomputer Hierarchy
Monochlorhydrin
MCHA Methylcyclohexyladipate
MChA Master of Applied Chemistry
MCHC Mean Cellular Hemoglobin Concentration
MChE Master of Chemical Engineering
MCHF Marine Corps Historical Foundation [US]
MCHFR Minimum Critical Heat Flux Rate
M chs Thousands of Characters
MCI Machine Check Interrupt
Master of Clinical Immunology
Malicious Call Identification
Malleable Cast Iron
mCi millicurie
MCIA Methyl Chloride Industry Association [US]
Microcomputer Investors Association [US]
Mirror Class International Association
MCIC Machine Check Interruption Code
Metals and Ceramics Information Center [US]
MCID Multipurpose Concealed Intrusion Detector
MCIDAS Man-Computer Interactive Data Access System
MCIF Member of the Canadian Institute of Forestry
MCIM Master of Clinical Immunology and Microbiology
Member of the Canadian Institute of Mining and Metallurgy [also: MCIMM]
MCIP Member of the Canadian Institute of Planners
MCIS Master of Computer and Information Science
Maintenance Control Information System
Materials Compatibility in Sodium
MCJ Master of Communication and Journalism
MCL Manufacturing Control Language
Materials Characterization Laboratory [US]
Mathematics Computation Laboratory [US]
Maximum Contamination Level
Memory Control and Logging
Microprogram Control Logic
Mid-Canada Line
MCLA Micro-Coded Communications Line Adapter
MCLK Master Clock
MCLOS Manual Command Line-of-Sight
MCLT Master of Clinical Laboratory Technology
MCM Machine for Coordinating Multiprocessing
Magnetic Core Memory
Maintenance Control Module
Master of Construction Management
Memory Control Module
Million Cubic Meters
Mine Countermeasures
Monte Carlo Method
Moving Coil Motor
Multichip Module
Thousand Circular Mils
Thousand Cubic Meters
MCMES Member of the Civil and Mechanical Engineering Society
MCMG Man-Carrying Motion Generator
Military Committee Meteorological Group [of NATO]
MCMM Management Control — Material Management
MCMST Montana College of Mineral Science and Technology
MCN Museum Computer Network [US]
MCNC Microelectronics Center of North Carolina [US]
MCO Molding and Cost Optimization
Multiple Criteria Optimization
M-COAT Multidither Coherent Adaptive Optical Technique
MCOEN Master of Computer Engineering
MCoEN Master of Computer Engineering
MCOGA Mid-Continent Oil and Gas Association [US]
MCOM Mathematics of Computation
MComm Master of Commerce [also: MCom]
Master of Communication
MCommAdmin Master of Commercial Administration
MCompSc Master of Computer Science [also: MCompS or MCompSci]
MComSc Master of Commercial Science
MCOQ Multiple Choice Objective Question
MCOS Microprogrammable Computer Operating System
MCP Main Call Process
Main Control Program
Master Control Program
Master of City Planning
Master of Community Planning
M-Cresol Purple
Memory-Centered Processor
Message Control Program
Methylcyclopentane
Microchannel Plate

	Military Construction Program
	Monocalcium Phosphate
	Multichannel Communications Program
	Multicomponent Process
MCPA	Methylchlorophenoxyacetic Acid
MCPB	Methylchlorophenoxybutyric Acid
MCPBA	M-Chloroperoxybenzoic Acid
MCPESCF	Multiconfiguration Paired Excitation Self-Consistent Field
MCPG	Medium Conversion Program Generator
MCPHA	Multichannel Pulse-Height Analyzer
MCPM	Multicomponent Protein Mixture
MCPP	Methylchlorophenoxypropionic Acid
MCPR	Maximum Critical Power Ratio
MCPS	Maintenance Control and Statistics Process
	Mini Core Processing Subsystem
Mcps	Megacycles per Second [also: Mcs or Mc/s]
MCPU	Multiple Central Processing Unit
MCR	Magnetic Card Reader
	Magnetic Character Reader
	Magnetic Character Recognition
	Main Control Room
	Master Change Record
	Master Control Register
	Master Control Routine
	Memory Control Register
	Military Compact Reactor [of USAEC]
	Minimum Creep Rate
MCROA	Marine Corps Reserve Officers Association [US]
MCRP	Master of City and Regional Planning
MCRPUD	Master of City and Regional Planning and Urban Development
MCRR	Machine Check Recording and Recovery
MCRS	Micrographics Catalogue Retrieval System
MCRT	Maximum Cruise Thrust
	Multichannel Rotary Transformer
MCS	Maintenance Control Subsystem
	Maleimidocaproic Succinimide
	Maneuver Control System
	Manufacturing and Consulting Service
	Marine Communication Subsystem [of INTELSAT]
	Marine Corps School [US]
	Master Control System
	Master of Clinical Science
	Master of Commercial Science
	Master of Communications Studies
	Master of Computer Science
	Mechanical Cultivation Service
	Medical Computer Services
	Message Control System
	Method of Constant Stimuli
	Methylchlorosulfonate
	Microcomputer Simulation
	Microcomputer System
	Microinstruction Control Store
	Microwave Carrier Supply
	Military College of Science [UK]
	Military Communications Station
	Missile Control System
	Mobile Communications System
	Modular Computer Systems
	Motor Circuit Switch
	Multicathode Spot
	Multichannel Scaler
	Multichannel Communications
	Multiconsole System
	Multinational Character Set
	Multiple Cloning Site
	Multiple Console Support
	Multiprogrammed Computer System
	Multipurpose Communications and Signaling
Mc/s	Megacycles per second [also: Mcps or Mcs]
MCSA	Microcomputer Software Association [of ADAPSO, US]
	Motor Coach Safety Association [Canada]
MCSC	Member/Customer Service Center [of ASM International, US]
MCSc	Master of Computer Science
MCSCF	Multiconfiguration Self-Consistent Field
MCSL	Microcomputer Simulation Laboratory
MCSP	Manpower Consultative Service Program [Canada]
MCSS	McDonnell Center for the Space Sciences [of Washington University, US]
	Military Communications Satellite System
MCST	Manchester College of Science and Technology [UK]
MC/ST	Magnetic Card Selectric Typewriter
MCSW	Mining Club of the Southwest [US]
MCT	Mass Culturing Technique
	Maximum Continuous Thrust
	Mechanical Comprehension Test
	Mercury Cadmium Telluride
	Microcomputer Technology
	Mobile Communications Terminal
MCTAMIA	Motor Cars, Tractor and Agricultural Machinery Importers Association [Cyprus]
MCTC	Maritime Cargo Transportation Conference [of MTRB, US]
	Metal Casting Technology Center [of University of Alabama, Tuscaloosa, US]
MC-TdT	Monoclonal Antibody to Terminal Deoxynucleotidyl Transferase
MCTI	Metal Cutting Tool Institute
MCTR	Message Center
MCTRAP	Mechanized Customer Trouble Report Analysis Plan
MCTRF	Manitoba Cancer Treatment Research Foundation [Canada]
MCU	Magnetic Card Unit
	Maintenance Control Unit
	Master Control Unit
	Medium Close Up
	Memory Control Unit
	Message Construction Unit
	Microcomputer Unit
	Microcontrol Unit
	Microprocessor Control Unit
	Microprogram Control Unit
	(Ontario) Ministry of Colleges and Universities [Canada]
	Modular Concept Unit
	Multicoupler Unit
	Multiplexer Control Unit

Multiprocessor Communications Unit
Multiprogrammed Control Unit
Multisystem Communications Unit
MCUG Military Computer Users Group
MCV Magnetic Cushion Vehicle
Movable Closure Valve
MCVD Modified Chemical Vapor Deposition
MCVF Multichannel Voice Frequency
MCVFT Multichannel Voice Frequency Telegraphy
MCW Memory Card Writer
Modulated Carrier Wave
Modulated Continuous Wave
MCWA Massey College Wood Association [New Zealand]
MCWCS Ministerial Conference of West and Central African States on Maritime Transportation [Ivory Coast]
MCX Minimum-Cost Estimate
MCXα Multiconfiguration Xα
MCY Motorcycle
MCZDO Multi-Center Zero Differential Overlap
MD Doctor of Medicine
Magnetic Disk
Magnetic Drum
Magneto-Damping
Magnetodynamic(s)
Main Drum
Maintenance Depot
Management Domain
Managing Director
Manual Damper
Manual Data
Marginal Distribution
Maryland [US]
Maximum Demand
Mean Deviation
Measured Discard
Mechanical Design
Mechanical Drafting
Medical Department
Meeting Date
Message Data
Messages per Day
Meteorology Department
Microdensitometer
Microdiffractometer
Military District
Mine Disposal
Mineral Dressing
Minidisk
Misfit Dislocation
Molecular Dynamics
Monitor Display
Moore Drive
Motion Detection
Motor Drive
Movement Directive
Multinominal Distribution
Multiple Dissemination
Municipal District
Music Data
M-D Modulation-Demodulation
Md Maryland [US]
Mendelevium [also: Mv]
md millidarcy [Unit]
MDA Maintainability Design Approach
Malonic Dialdehyde
Manufacturing Defects Analyzer
Marketing Device Association [now: MDAI, US]
Master Diversion Aerodrome
Master of Development Administration
Materials Dispersion Apparatus
Mechanized Directory Assistance
Metal Deactivator
Methylene Dianiline
Methylene Dioxyamphetamine
Mineral Development Agreement [between Canada and Newfoundland]
Minimum Descent Altitude
Monochrome Display Adapter
M-Phenylenediamine
Multidimensional Access
Multidimensional Analysis
Multidimensional Array
Multidocking Adapter
Multiple Digit Absorbing
MDAA Mutual Defense Assistance Act
MDAB 4,4'-Bismaleimidodiphenylmethane [also: MDA BMI]
MDAC Methyldiethylaminocoumarin
Multiplying Digital-to-Analog Converter
MDAE Multidisciplinary Accident Engineering
MDA/EGA Monochrome Display Adapter/Enhanced Graphics Adapter
MDAFWP Motor Driven Auxiliary Feedwater Pump
MDAI Marketing Device Association International [US]
MDAL McDonell Douglas Aerophysics Laboratory [US]
MDAP Machine and Display Application Program
Mutual Defense Assistance Program
MDB Maritime Development Board
Multilateral Development Bank
MDBF Measured Depth Below Formation
MDC Machinability Data Center [US]
Main Display Console
Main Distribution Center
Maintenance Data Collection
Maintenance Dependency Chart
Market Development Center
Materials Dissemination Center
Materials Distribution Center
Maximum Dependable Capacity
Memory Disk Controller
Microprocessor Development Center [US]
Minimum Detectable Concentration
Missile Development Center [US Air Force]
Multiduty Collector
Multiple Device Control
Multiple Drone Control
MDCC Master Data Control Console
MDCE Monitoring and Duplicate Control Equipment
MDCK Madin-Darby Canine Kidney
MDCS Maintenance Data Collection System
Master Digital Command System
MDCT Mechanical Draft Cooling Tower

MDCU Magnetic Disk Control Unit
MDD Machine Dependence Data
Magnetic Disk Drive
Meteorological Data Distribution
Milligram per Square Decimeter per Day [also: mdd]
MDDC Microdiffractometer Data Collection
MDDR Mean Depth of Deformation Rate
MDE Magnetic Decision Element
MDE Master of Developmental Economics
Mercury-Dropping Electrode
Missile Display Equipment
Modular Design of Electronics
MDEA Methyldiethanolamine
MDEFWP Motor Driven Emergency Feedwater Pump
mdeg millidegree
MDES Multiple Data Entry System
MDes Master of Design
MDesS Master of Design Studies [also: MDesStud]
MDF Macrodefect-Free
Main Distribution Frame
Manual Direction Finder
Medium-Frequency Direction Finding
Metals Data File [of ASM International, US]
Microcomputer Development Facilities
Mild Detonating Fuse
Misorientation Distribution Function
Multi-Degree of Freedom
MDFNA Maximum Density Fuming Nitric Acid
MDFY Modify
MDH Malate Dehydrogenase
Malic Dehydrogenase
Maximum Diameter Heat
Minimum Descent Height
Mdh Moroccan Dirham [Currency]
MDI Magnetic Direction Indicator
Manual Data Input
Maximum Demand Indicator
Medium-Dependent Interface
Menu-Driven Interface
Methane Diphenyl Diisocyanate
Methylene Diisocyanate
Methylene Di-P-Phenylene Isocyanate
Microdielectrometry
Miss Distance Indicator
MDIC Microwave Dielectric Integrated Circuit
Multilateral Disarmament Information Center [UK]
MDiEng Master of Diesel Engineering
MDIF Manual Data Input Function
MDI-PA Diphenylmethane Diisocyanate — Polyamide
MDIU Manual Data Input Unit
MDL Macro Description Language
Maintenance and Diagnostic Logic Display
Maritime Dynamics Laboratory
Master Data Library
Mine Defense Laboratory [of USDOD]
Minimum Detectable Leakage
Minimum Detectable Level
Minimum Detectable Limit
Module
MDM Maximum Design Meter
Metal-Dielectric-Metal (Filter)
Minimum Detectable Mass
Multiplexer/Demultiplexer
MDMA Methylene Dioxymethamphetamine
MDMD Materials Design and Manufacturing Division [of TMS, US]
MDMG Microelectronics Dimensional Metrology Group [of NIST, US]
MDMS Methylene Dimethanesulfonate
Multiple Database Management System
MDMV Maize Dwarf Mosaic Virus
MDN Managed Data Network
MD/NC Mechanical Drafting/Numerical Control
MDNR Machinery Dealers National Association [US]
Maryland Department of Natural Resources [US]
MDNS Managed Data Network Services
MDO MARC (= Machine-Readable Catalogue) Development Office [of Library of Congress, US]
Medium Density Overlay
MDOS Motorola Disk-Operating System
MDP Main Data Path
Maintenance Diagnostic Processor
Message Discrimination Process
Methylenedioxyphenyl
Missile Data Processor
Muramyl Dipeptide
MDPA Methyldiphenylamine
Monochrome Display and Printer Adapter
MDPE Medium-Density Polyethylene
MDPR Mean Depth-of-Penetration
MDPS Multiple-Workstation Direct-Processing System
MDR Magnetic Disk Recorder
Manual Data Room
Market Data Retrieval
Memory Data Register
Minimum Daily Requirement
Miscellaneous Data Recorder
Mission Data Reduction
Multichannel Data Recorder
Multi-Disk Reader
MDRP Mean Depth Rate of Penetration
MDS Magnetic Detector System
Main Device Scheduler
Maintenance Data System
Malfunction Detection System
Management Decision System
Master Drum Sender
Master of Decision Sciences
Materials Data Sheet
Memory Disk System
Metastable Deexitation Spectroscopy
Micro Disk Storage
Microprocessor Development System
Minimum Detectable Signal
Minimum Discernible Signal
Minimum Dose System
Mobile Data Service
Modern Data System
Modular Disk Storage
Multidimensional Scaling
Multidimensional Signal

Multiple Data Set System
Multipoint Distribution Service
Multipoint Distribution System
Multiprocessor Distributed System
mds millidarcies [Unit]
MDSE Merchandise
MDSF Manipulator Development and Simulation Facility
MDSG Merchandising
MDSMB Medical Devices Standards Management Board
MDSS Meteorological Data Sounding System
Microprocessor Development Support System
MDT Maximum Diameter of the Thorax
Mean Downtime
Mobile Data Terminal
Moderate
Modified Data Tag
Mountain Daylight Time
Movchan, Demchishin and Thornton (Microstructure)
Multidimensional Tasking
MDTA Manpower Development and Training Act
MDTD Minimum Detectable Temperature Difference
MDTF Marine Dynamic Test Facility
MDTS Megabit Digital-to-Troposcatter Subsystem
MDU Maintenance Diagnostic Unit
Message Decoder Unit
Mine Disposal Unit
MDUS Medium Data Utilization Station
MDV Doctor of Veterinary Medicine
Mucosal Disease Virus
MDW Measured Daywork
Multiple Drop Wire
MDW-7-R 4-[(R)-2-Chloro-3-Methylbutyryloxy]phenyl 4-(Decyloxy)benzoate
MDW-7-S 4-[(S)-2-Chloro-3-Methylbutyryloxy]phenyl 4-(Decyloxy)benzoate
MDX Modular Digital Exchange
ME Machine Equation
Magnetic Estimation
Magnetoelastic(ity)
Magnetic Elasticity
Magnetoelectric(ity)
Maine [US]
Main Entry
Maintenance Engineer
Male Elbow (Pipe Section)
Management Engineer
Management Engineering
Managing Editor
Manual-Feed Envelope
Manufacturing Engineer
Manufacturing Engineering
Marine Engineer
Marine Engineering
Masonry Engineer
Master Electrician
Master of Education
Master of Engineering
Maximum Effort
Mechanical Efficiency
Mechanical Engineer
Mechanical Engineering
Mechanical Equivalent
Mechanoelectronic(s)
Melt Extraction
Memory Element
Mercaptoethanol
Message Element
Metabolizable Energy
Metalloenzyme
Metallurgical Engineer
Metallurgical Engineering
Metals Extraction
Microcircuit Engineer
Microcircuit Engineering
Microelectronic(s)
Military Engineer
Military Engineering
Mineral Exploration
Mining Engineer
Mining Engineering
Modulator Electrode
Molecular Electronics
Molecular Engineering
Momentum Equation
Montreal Exchange [Canada]
Mossbauer Effect
Motion Estimation
Municipal Engineer
Municipal Engineering
Muzzle Energy
M&E Material and Equipment
Me Metal
Methyl
Maine [US]
MEA Maintenance Engineering Analysis
Malt Extract Agar
Master of Engineering Administration
Master of Engineering Architecture
Materials Experiment Assembly [of NASA]
Mercaptoethylamine
Middle East Airlines
Minimum En-Route Altitude
Monoethanolamine
MEAB Maintenance Engineering Analysis Board
MEAC Manufacturing Engineering Application Center
MEAL Master Equipment Allowance List
Master Equipment Authorization List
MEAR Maintenance Engineering Analysis Report
MEARA Manitoba Environmental Assessment and Review Agency [Canada]
MEAS Measure
MEASUCORA Measurement, Control Regulation and Automation
MEB Modem Evaluation Board
MEBS Marketing, Engineering and Business Services
MEC Market Economy Country
Marshalltown Engineering Club [US]
Materials Education Council [US]
Materials Engineering Center
Member of the Executive Council
Metastable Epitaxial Compound
Meteorology Engineering Center [of NOSC, US]

	Molecular Electron Correlation
MEc	Master of Economics
MeC	Metal Carbide
MECA	Manufacturers of Emission Controls Association [US]
	Molecular Emission Cavity Analysis
MECC	Minnesota Educational Computing Consortium [US]
MECCA	Mechanized Catalogue
MECH	Mechanic(s)
	Mechanical
	Mechanism
	Mechanization
MechE	Mechanical Engineer [also: MechEng]
MEChemEng	Master of Engineering in Chemical Enineering [also: MEChemE or MSChE]
MECHEN	Mechanical Engineering (Database) [of RITL, Sweden]
MECH ENG	Mechanical Engineering
MECI	Manufacturing Engineering Certification Institute [US]
MECL	Motorola Emitter-Coupled Logic
	Musashino Electrical Communication Laboratory [Japan]
MECNY	Municipal Engineers of the City of New York [US]
MECOMSAG	Mobility Equipment Command Scientific Advisory Group
MEcon	Master of Economics
MECR	Maintenance Engineering Change Request
MECU	Master Engine Control Unit
MED	Marine Emergency Duties
	Materials Engineering Department
	Median
	Medical
	Mediterranean
	Medium
	Metallurgical Engineering Department
	Microelectronic Device
	Minimum Energy Dwelling
	Mobile Energy Depot
	Molecular Electronic Device
MEd	Master of Education
MEDA	Military Emergency Diversion Aerodrome
	Mineral Exploration Depreciation Allowance
	Mining Exploration Depletion Allowance
MEDAB	Methyl Dimethylaminoazobenzene
MEDAC	Medical Electronic Data Acquisition and Control
	Medical Equipment Display and Conference
MEDAL	Micromechanized Engineering Data for Automated Logistics
MEDARATEL	Mediterranean and Arabic Telecommunications Network
MEDDA	Mechanized Defense Decision Anticipation
MEDes	Master of Environmental Design
MEDI	Marine Environmental Data Information System [of IOC, France]
MEDIA	Magnavox Electronic Data Image Apparatus
MEDICO	Model Experiment in Drug Indexing by Computers [of Rutgers University, US]
MEDICS	Medical Information and Communication System
MEDIRS	Marine Environmental Data Information Referral System [of IOC, France]
MEDIS	Modular Engineering Document Imaging System
MEDLARS	Medical Literature Analysis and Retrieval System [US]
MEDLINE	MEDLARS On-Line System [US]
MEDO	Multipole Expansion of Diatomic Overlap
MEDRISK	Medical Device Risk
MEDS	Marine Environmental Data Service
	Medical Evaluation Data System
	Medium Energy Dislocation Structure
MEDSERV	Medical Service Corps [US]
MEDSMB	Medical Standards Management Board [US]
MEDSPECC	Medical Specialist Corps [US]
MEduc	Master of Education
MEE	Master of Electrical Engineering
	Microelectronic Engineer
	Microelectronic Engineering
	Mining Electrical Engineer
	Mining Electrical Engineering
MEECN	Minimum Essential Emergency Communications Network [US]
MEED	Medium Energy Electron Diffraction
MEEng	Master of Electrical Engineering
MEF	Multiple Effect Flash Evaporator
MEFTA	Metalworking Industries of European Free Trade Association
MEFV	Maintance Equipment Floor Valve
MEG	Magnetoelectric Generator
	Magnetoencephalogram
	Magnetoencephalograph(y)
	Megger
	Megohm
	Message Expediting Group
	Mineral Exploration Geophysics
meg	megabyte
	megohm
MEGA	Methylglucamide
MEGG	Merging
MEGW	Megawatt
MEHPA	Monoethylhexylphosphoric Acid
MEHQ	Methyl Ether of Hydroquinone
MEHT	Minimum Pilot Eye-Height over Threshold
MEI	Main Economic Indicators
	Manual of Engineering Instructions
	Marginal Efficiency of Investment
	Materials Engineering Institute [of ASM International, US]
	Metals Engineering Institute [Name Changed to: Materials Engineering Institute, US]
	Ministry of the Electronics Industry [PR China]
MEIC	Materials Engineering Institute Council [of ASM International, US]
	Member of the Engineering Institute of Canada
MEIP	Mechanical, Electrical, Instrumentation and Control, Power Equipment
MEIR	Medium-Energy Ion Reflection
MEIS	Medium Energy Ion Scattering
	Metal-Epitaxial Insulator-Semiconductor
	Military Entomology Information Service [of USDOD]
	Multispectral Electrooptical Imaging Scanner

MEISFET Metal-Epitaxial Insulator-Semiconductor Field-Effect Transistor
MEIU Mobile Explosives Investigation Unit
MEK Methyl Ethyl Ketone
MEKP Methyl Ethyl Ketone Peroxide
MEL Many-Element Laser
Marine Ecology Laboratory [Canada]
Marine Engineering Laboratory [US Navy]
Materials Evaluation Laboratory [of IMR, US]
Maximum Excess Loss
Mechanical Engineering Laboratory [Japan]
Minimum Equipment List
Multi-Engine, Land
MELA Middle East Librarians Association
MELAB Michigan English Language Accreditation Board
MELBA Multipurpose Extended Lift Blanket Assembly
MELCU Multiple External Line Control Unit
MELEC Microelectronic(s)
MELEM Microelement
MELEO Materials Exposure in Low Earth Orbit (Experiment) [Space Shuttle]
MELEO/IOCM Materials Exposure in Low Earth Orbit/Interim Operational Contamination Monitor (Experiment) [Space Shuttle]
MELF Metal Electrode Face Bonding
MELVA Military Electronic Light Valve
MEM Mars Excursion Module
MEM Master of Engineering Management
Maximum Entropy Method
Member
Memorandum
Memory
Methoxyethoxymethyl
Methoxyethylmercury
Minimum Essential Medium
Modified Eagle's Medium
Modified Effective Modulus
MEMA Machinery and Equipment Manufacturers Association
Microelectronic Modular Assembly
Motor and Equipment Manufacturers Association [US]
MEMAC Machinery and Equipment Manufacturers Association of Canada
Marine Emergency Mutual Aid Center
MEMB Membrane
MEMC Methoxyethylmercury Chloride
MEME Multiple Entry — Multiple Exit
MEMIC Medical Microbiological Interdisciplinary Committee
MEMISTOR Memory Register Storage Device
MEML Memorial
MEMO Memorandum
MEMP Maximization of Expected Maximum Profits
MEMRA Mechanical Equipment Manufacturers Representatives Association [US]
MEMRB Middle East Marketing Research Bureau
MEMS Master of Engineering in Manufacturing Systems
Mechanism Modeling System
Microelectromechanical System
MEN Multiple Event Network
MENA Methyl Ester of Naphthaleneacetic Acid
MEND Mine Environment Neutral Drainage (Program) [formerly: RATS]
Mineral Environment Neutral Drainage (Program) [Canada]
MENEX Maintenance Engineering Exchange
MEng Master of Engineering [also: MEngr]
MEngMgt Master of Engineering Management
MEngSc Master of Engineering and Science [also: MEngSci]
MEngStud Master of Engineering Studies
MEnv Master of Environmental Studies
MEnvDes Master of Environmental Design
MEnvSc Master of Environmental Science [also: MEnvS]
MEO Major Engine Overhaul
Management Engineering Office
MeO Metal Oxide
MEOS Multiband Electrooptical Scanner
MEP Management Engineering Plan
Management Engineering Program
Master of Engineering Physics
Master of Environmental Planning
Mean Effective Potential
Mean Effective Pressure [also: mep]
Methyl Ethyl Phenol
Methyl Ethyl Pyridine
Mexican Peso [Currency]
Microelectronic Education Program [UK]
Microfile Enlarger Printer
Mission Effects Projector [of NASA]
Motor-Evoked Potential
MEPC Master of Environmental Pollution Control
MEPCB Mechanical Earth Pressure Counterbalance
MEPD Master of Education — Professional Development
MEPF Multiple Experiment Processing Furnace
MEPP Mineral Economics and Policy Program
meq milliequivalent
MER Manned Earth Reconnaissance
Master of Energy Resources
Materials and Electrochemical Research
Maximum Efficient Rate
Meridian
Minimum Energy Requirement
Multiple Ejector Rack
MERA Molecular Electronics for Radar Applications
MERC Mercantile [also: MER]
MERCH Merchandising
MERDC Mobility Equipment Research and Development Center [US]
MERDL Medical Equipment Research and Development Laboratory [US Army]
MEREA Member of the Electrical Railway Engineering Association [US]
MERGE Mechanized Retrieval for Greater Efficiency
MERGV Martian Exploratory Rocket Glide Vehicle
MERIP Middle East Research Information Project
MERIT Michigan Educational Research Information Triad [US]
MERL Municipal Environmental Research Laboratory [US]
MERLIN Machine-Readable Library Information
Medium-Energy Light-Water-Moderated Industrial Nuclear Reactor [UK]

MERLIN-JULICH Medium-Energy Light-Water-Moderated Industrial Nuclear Reactor at Julich [Germany]
MERM Material Evaluation Rocket Motor
MERMUT Mobile Electronic Robot Manipulator and Underwater Television
MERRCAC Middle Eastern Regional Radioisotope Center for Arab Countries
MERS Mobiltity Environmental Research Study
MERT Multiple Environment Real Time
MERU Mechanical Engineering Research Unit [South Africa]
MES Manual Entry Subsystem
Marine Engineering Society [Japan]
Master of Engineering Science
Master of Environmental Sciences
Master of Environmental Studies
McMaster Engineering Society [Canada]
Mechanical Engineering Society
Medium Energy Source
Metal-Gate Schottky
Metal-Semiconductor Device
Michigan Engineering Society [US]
Mobile Earth Station
Morpholineethanesulfone
2-(N-Morpholino)ethanesulfonic Acid
4-Morpholinoethanesulfonic Acid
Mossbauer Effect Spectroscopy
MESA Manned Environmental System Assessment [of NASA]
Marine Ecosystems Analysis
Miniaturized Electrostatic Accelerometer
Mining Enforcement and Safety Administration [US]
Modular Experiment Platform for Science and Applications
MESAEP Mediterranean Scientific Association for Environmental Protection
MESBP Management Excellence in Small Business Program [of DRIE, Canada]
MESc Master of Engineering Sciences
MESET Minority Enrichment Seminar in Engineering Training
MESFET Metal-Gate Schottky Field-Effect Transistor
Metal-Semiconductor Field-Effect Transistor
MESG Maximum Experimental Safe Gap
Micro Electrostatically Suspended Gyro
Minimum Electrical Spark Gap
MESG Message
MESH Medical Subject Headings
MESJ Marine Engineering Society of Japan
MESL Merchants Exchange of St. Louis [US]
MESNA 2-Mercaptoethanesulfonic Acid
MESS Monitor Event Simulation System
MESTIND Measurement Standards Instrumentation Division [of Instrument Society of America, US]
MESUCORA Measurement, Control Regulation and Automation
MESYL Methanesulfonyl
MET Management Engineering Team
Marine Engineering Technician
Marine Engineering Technologist
Marine Engineering Technology
Mechanical Engineering Technologist
Mechanical Engineering Technology
Mechanoelectronic Transducer
Metal(lic)
Metallurgist
Metallurgy
Metaphor
Metaphysics
Meteorologist
Meteorology
Metronome
Metropole
Metropolitan
Microelectronics Technology
Mobile Earth Terminal [MSAT Program]
Modified Expansion Tube
Multielectrode Tube
Multi-Emitter Transistor
Met Meteorological Office [UK]
Methionine
Methionyl
META Maintenance Engineering Training Agency [US Army]
Method for Extracting Text Automatically
METADEX Metals Abstracts Index [also: Metadex [of ASM International, US]]
METAG Meteorological Advisory Group [of ICAO]
METALL Metallurgical
Metallurgist
Metallurgy
METAPLAN Methods of Extracting Text Automatically Programming Language
Methods of Extracting Text Autoprogramming Language
METAR Meteorological Aeronautical Code
METASYMBOL Metalanguage Symbol
METAV Markt fur die Metallverarbeitung [= Trade Fair and Exhibition for the Metalworking Industries, Germany]
METC Morgantown Energy Technology Center [US]
METCOM Metropolitan Consortium for Minorities in Engineering
METCUT Metal Cutting Exhibition
METD Dimethyldiethyl Thiuram Disulfide
MetE Metallurgical Engineer
Metallurgical Engineering
METEC International Exhibition for Metallurgical Technology [Germany]
MET ENG Metallurgical Engineering
METEOR Meteorological
Meteorologist
Meteorology
METEOSAT Meteorological Satellite [Europe]
MetFin Metal Finishing
METH Methyl
METH AL Methyl Alcohol
METL Multi-Element, Two-Layer
METLO Meteorological Electronic Technical Liaison Office [US Navy]
METRIA Metropolitan Tree Improvement Alliance [US]

METRIC	Multi-Echelon Technique for Recoverable Item Control
METROMEX	Metropolitan Meteorological Experiment [US]
METS	Modular Engine Test System
	Modular Environmental Test System
METTP	Marine Engineering Technician Training Plan
METU	Middle East Technical University [Turkey]
METWK	Metalworking
MEU	Medium Enriched Uranium
	Memory Expansion Unit
MEU/TH	Medium Enriched Uranium/Thorium Fuel
MeV	Megaelectronvolts
	Million Electronvolts
MeV Ci	Megaelectronvolt-Curie
MEVVA	Metal Vacuum Vapor Arc Source
MEW	Mean Equivalent Wind
	Microwave Early Warning
MEWS	Missile Electronic Warfare System
MEWT	Matrix Electrostatic Writing Technique
MEWU	Mining and Energy Workers Union [Germany]
MEX	Mexican
	Mexico
	Military Exchange
MEXE	Military Engineering Experimental Establishment [of MOD, UK]
Mexican MRS	Mexican Materials Research Society
MExtEd	Master of Extension Education
MEZZ	Mezzanine
MF	Magnetic Field
	Magnetic Focusing
	Magnetic Foot [Unit]
	Magnetomotive Force
	Mainframe
	Maintenance Factor
	Maintenance Free
	Mandatory Frequency
	Manual Feed
	Mass Flow
	Mass Fragmentography
	Master File
	Master of Forestry
	Matched Filtering
	Measurement Frequency
	Measuring Force
	Medium Frequency
	Melamine Formaldehyde
	Microfiche
	Microfilm
	Microfiltration
	Micro Floppy
	Microform
	Mixed Flow
	Modulation Frequency
	Molecular Filter
	Molecular Flow
	Molecular Formula
	Motor Field
	Multifrequency
	Multifunction
M&F	Male and Female [Fasteners]
M-F	Medium Frequency
mF	millifarad
mf	microfarad
MFA	Minimum Flight Altitude
	Ministry of Foreign Affairs [Saudi Arabia]
	Multi-Fiber Agreement
	Multifont Adapter
	Multifurnace Assembly
MFB	Mill Fixture Base
	Mixed Functional Block
	Motional Feedback
MFBM	Thousand Foot Board Measure
MFBS	Multifrequency Binary Sequence
MFC	Magnetic-Tape Field Scan
	Manual Frequency Control
	Mass Flow Controller
	Microfunctional Circuit
	Monochromator Focussing Circle
	Multifrequency Code
	Multifrequency (Signalling), Compelled
	Multifunction Controller
MFCA	Multifunction Communications Adapter
MFCI	Molten Fuel Coolant Interaction
MFC/LB	Multifrequency Code/Local Battery
MFCM	Multifunction Card Machine
MFCU	Multifunction Card Unit
MFD	Magnetic Field Disturbance
	Magnetic Frequency Detector
	Magnetofluid Dynamics
	Malfunctioning Display
	Manufactured
	Master File Directory
	Micro Floppy Disk
	Multifunction Detector
	Multifunction Display
mfd	microfarad
MFDSU	Multifunction Data Set Utility
MFE	Master of Forest Engineering
MFED	Maximum Flat Envelope Delay
MFEng	Master of Forestry Engineering
MFES	Minnesota Federation of Engineering Societies [US]
MFG	Manufacturing
	Message Flow Graph
	Molded Fiber Glass
	Multifunction Generator
MFG T	Manufacturing Technology
MFH	Maximum Fork Height
MFI	Malta Federation of Industries
	Melt Flow Indexer
	Metal Fabricating Institute [US]
	Mixed Fission Product
	Mobile Fuel Irradiator
	Multi-Point Fuel Injection
MF-IFGR	Michael Fund — International Foundation for Genetic Research
MFJ	Modified Final Judgment
MFJSA	Mass Finishing Job Shops Association [US]
MFK	Mill Fixture Key
MFKP	Multifrequency Key Pulsing
MFL	Magnetic Flux Leakage
	Mantitoba Federation of Labour [Canada]
	Mobile Foundry Laboratory

MFLD Message Field
MFLOPS Million Floating Points [also: Mflops, MFLOP or Mflop]
Million Floating-Point Operations per Second [also: Mflops, MFLOP or Mflop]
MFor Master of Forestry
MForSc Master of Forestry Science
MFLP Multifile Linear Programming
MFLS Microtime Fault Locator System
MFM Magnetic-Field Meter
Mass Flow Monitor
Master Frequency Meter
Medium Frequency Module
Modified Frequency Modulation
Multifrequency Module
Multistage Frequency Multiplier
MFMA Maple Flooring Manufacturers Association [US]
Metal Framing Manufacturers Association [US]
MFMR Multifrequency Microwave Radiometer
MFN Most Favoured Nation (Tariff)
MFO Mixed Function Oxidase
MFOD Manned Flight Operations Division
MFOM Member of the Faculty of Occupational Medicine
MForSc Master of Forestry Science
MFP Mean Free Path [also: mfp]
Multiform Printer
MFPA Massachusetts Forest and Park Association [US]
MFPB Mineral Fiber Products Bureau
MFPC Multifunction Protocol Converter
MFPG Mechanical Failure Prevention Group [US]
Mixed Fission Products Generator
MFR Mali Franc [also: MFr; Currency]
Manipulator Foot Restraint
Manufacturer
Master of Forest Resources
Melt Flow Rate
Mirror Fusion Reactor
Multifrequency Receiver
MFr Malagasy Franc [Currency of Madagascar]
MFRC Master of Forest Resources and Conservation
Metal Finishing Research Center [Canada]
MFRS Manufacturers
MFS Magnetic-Tape Field Search
Manned Flying System
Marine Finish Slate
Master of Food Science
Master of Forensic Studies
Master of Forest Studies
Metal Finishing Society
Molecular Fluorescence Spectroscopy
Multifrequency Signalling
Multifunction Sensor
MFSA Metal Finishing Suppliers Association [US]
MFSJ Metal Finishing Society of Japan
MFSK Metal Finishing Society of Korea
Multiple-Frequency Shift Keying
MFSS Multi-Frequency Signalling System
MFT Mainframe Termination
Manual Fine Tuning
Metallic Facility Terminal
Miniature Fluorescent Tube
Minimum Film Formation Temperature
Monolayer Formation Time
Multiposition Frequency Telegraphy
Multiprogramming with a Fixed Number of Tasks
Thousand Feet
MFTF Mirror Fusion Test Facility
MFTG Manufacturing Technology Group [of IEEE, US]
Mftl Milli-Foot-Lamberts
MFTRS Magnetic Flight Test Recording System
MFTR Multifrequency Transmitter
MFU Maritime Fisherman's Union [Canada]
McGill Faculty Union [of McGill University, Canada]
MFV Maintenance Floor Valve
Military Flight Vehicle
MG Machine-Glazed (Paper)
Machine Gun
Manual Gain Control
Master Generator
Master-Group
Mathematical Geology
Maxwell Garnett (Formula)
May-Grunwald Stain
Medical Genetics
Melt Growth
Message Generator
Messenger
Methylene Glycol
Military Government
Mining
Mixed Grain
Modulation Grid
Molecular Genetics
Motor-Generator
Multigage
Mycoplasma Gallisepticum
Mg Magnesium
Megagram
mG milligauss
mg milligram
MGA Master of Government Administration
Meteorological and Geoastrophysical Abstracts (Database) [of American Meteorological Society, US]
Methylene Glycol Anion
Methyl Glycol Acetate
Multiple Gas Analyzer
Mushroom Growers Association [UK]
Mycoplasma Gallisepticum Agglutinin
mg Ab/mL milligrams of antibodies per milliliter
mGal milligal [Unit]
mgal/d million gallons per day [also: MGAL/D or mgal d^{-1}]
MGB Master Ground Bar
MGC Malachite Green Carbinol
Manual Gain Control
Missile Guidance and Computer
Missile Guidance and Control
MGCC Missile Guidance and Control Computer
MGCR Maritime Gas-Cooled Reactor
MGCR-CX Maritime Gas-Cooled Reactor Critical Experiment

MGCS METEOSAT (= Meteorological Satellite) Ground Computer Equipment
MGD Magnetogasdynamic(s)
Mean Gain Deviation
mgd million gallons per day [also: MGD]
mg/d milligram per day
MGE Master of Geological Engineering
Minneapolis Grain Exchange [US]
MGen Major-General
MGEMT Management
MGeoE Master of Geological Engineering [also: MGeoEng]
MGF Malagasy Franc [Currency of Madagascar]
MGG Memory Gate Generator
MGGB Modular Guided Glide Bomb
mg h milligram-hour [also: mg-h or mg-hr]
MGIR Motor Glider Instructor Rating
MGL Malachite Green Leucocyanite
Matrix Generator Language
MGM Mechanics of Granular Materials
mgm milligram
MGMI Mining, Geological and Metallurgical Institute [India]
mg/mL milligram per milliliter
MGMT Management
MGN Multi-Grounded Neutral
MGO Malachite Green Oxalate
Mega-Gauss Oersted [also: MGOe]
Mount Graham Observatory [US]
MGP Magnetic and Graphic Products
Methyl Green-Pyronin Solution
Mountain Gorilla Project [of AWF, US]
Multiple Goal Programming
Mg-PSZ Magnesium Partially Stabilized Zirconia
Mgr Manager
MGS Market-Grade Stainless Steel
Maritime Geoscience Services [Canada]
MGS Master of General Studies
Monolithic Gallium Arsenide/Silicon
MG SET Motor Generator Set
MGT Management
Master-Group Translator
MGTP Mount Graham Telescope Project [US]
Mg-TZP Magnesia Tetragonal Zirconia Polycrystal
MgY-TZP Magnesia-Yttria Tetragonal Zirconia Polycrystal
MH Magnetic Head
Main Hatch
Maleic Hydrazide
Manhole
Manual Hold
Material Handling
Material Hopper
Medium Hardness
Message Handler
Mobile High-Power Plant [US Army]
Modified Huffman
mH millihenry
MHA Marine Historical Association
Master of Health Administration
Maximum Hypothetical Accident
Methionine Hydroxy Analog, Calcium Salt
Minimum Holding Altitude
Modified Handling Authorized
Mueller Hinta Agar
MHAMS Master of Historical Administration and Museum Studies
MHC Major Histocompatibility Complex
Martin Hard Coat
Mean Horizontal Candlepower
Medium-Heat Content
Modified Huffman Coding
MHCP Mean Horizontal Candlepower
MHD Magnetohydrodynamic(s)
Master of Human Development
Maximum Hub Diameter
Movable Head Disk
Moving Head Disk
Multiple Head Disk
MHDE Magnetohydrodynamic Equation
MHDF Medium and High Frequency Direction Finding
MHDW Magnetohydrodynamic Wave
MHE Material Handling Engineer
Material Handling Engineering
Material Handling Equipment
MHEA Material Handling Engineers Association [UK]
MHEDA Material Handling Equipment Distributors Association [US]
MHEF Material Handling Education Foundation
MHF Massive Hydraulic Fracturing
Medium High Frequency
Message Handling Facility
Mixed Hydrazine Fuel
MHFR Maximum Hypothetical Fission Product Release
MHHPA Methylhexahydrophthalic Anhydride
MHHW Mean Higher High Water
MHI Manufacturered Housing Institute [US]
Material Handling Institute [US]
(Fachgemeinschaft) Montage — Handhabung — Industrieroboter [= Working Group on Assembly, Handling and Industrial Robots [of VDMI, Germany]]
MHiE Master of Highway Engineering [also: MHiEng]
MHK Methylhexylketone
MHKW Midvale-Heppenstall-Klockner-Werke (Steelmaking Process)
MHL Messo Heated Ladle
MHLW Mean Higher Low Water
MHMA Mobile Home Manufacturers Association [now: MHI, US]
MHortSc Master of Horticultural Science
MHOS Master of Human Organizational Science
MHP Master of Heritage Preservation
Message Handling Processor
Metric Horespower
MHR Master of Human Resources
Member of the House of Representatives [US]
Ministry of Human Resources [Canada]
MHRM Master of Human Resources Management
MHROD Master of Human Resources and Organization Development
MHRST Medical and Health Related Sciences Thesaurus
MHS Massachusetts Horticultural Society [US]
Master Handling System

Master of Health Science
Material Handling System
Mayer's Hematoxylin Solution
Message Handling System
Multiple Host Support
MHSA Ministry of Health and Social Affairs [South Korea]
MHSB Mining Health and Safety Branch [of MOL, Canada]
MHSc Master of Health Science
MHSDC Multiple High-Speed Data Channel
MHSCP Mean Hemispherical Candlepower
MHT Malleablizing Heat Treatment
Mean High Tide
Mechanical Horizontal Tensile (Tester)
Microhardness Tester
Mild Heat Treatment
Museum of History and Technology [US]
MHTA Metal Heat Treating Association
MHTS Main Heat Transport System
MHV Manned Hypersonic Vehicle
Medium Heating Value
MHVDF Medium, High and Very High Frequency Direction Finding
MHW Mean High Water
Multihundred Watt
MHWN Mean High-Water Neaps
MHWS Mean High Water Springs
MHz Megahertz
mHz millihertz
MI Machine-Independent
Machine Intelligence
Magnetic Induction
Malleable Iron
Maloti [Currency of Lesotho]
Manual Input
Master of Instruction
Materials Information [of ASM International, US]
Medical Imaging
Medium Intensity
Melt Index
Memory Interface
Metabolic Index
Metal Ion
Metals Information [now: Materials Information]
Michigan [US]
Micro-Instruction
Mile [also: mi]
Military Intelligence
Mill
Miller Indices
Miller Integrator
Mineral-Insulated (Cable)
Moment of Inertia
Multiple Interaction
MIA Malleable Ironfounders Association [UK]
Manitoba Institute of Agrologists [Canada]
Marble Institute of America [US]
Master of International Administration
Master of International Affairs
Metal Interface Amplifier
Methylisatoic Anhydride
Minimum IFR (= Instrument Flight Rules) Altitude
Missing in Action
Multiplex Interface Adapter
MIAB Magnetically Impelled Arc Butt Welding
MIAC Minimum Automatic Computer
MIACF Meander Inverted Autocorrelated Function
MIACS Manufacturing Information and Control System
MIAK Methyl Isoamyl Ketone
MIALS Medium-Intensity Approach Light System
MIAP Mineral Industry Assistance Program [Canada]
MIArch Master of Interior Architecture
MIAS Marine Information and Advisory Service [of IOS, UK]
Member of the Institute of Aeronautical Sciences
Microprobe Image Analysis System
Municipal Industrial Abatement Strategy [Canada]
MIB Master of International Business
Manual Input Buffer
Metal Information Bureau [UK]
Micro-Instruction Bus
Multilayer Interconnection Board
MIBA Master of International Business Administration
MIBC Methylisobutyl Carbinol
MIBF Montreal International Book Fair [Canada]
MIBK Methyl Isobutyl Ketone
MIBL Masked Ion Beam Lithography
MIBS Master of International Business Studies
MIC Macro Instruction Compiler
Management Information Center
Maritime Information Center [Netherlands]
Mass-Impregnated Cable
Master Interrupt Control
Maximum Immission Concentration
Medical Information Center
Message Identification Code
Methyl Isocyanate
Michigan Instructional Computer
Microbiologically Induced Corrosion
Micrometer [Instrument]
Microphone
Microscope
Microscopic
Microwave Integrated Circuit
Minimal Inhibitory Concentration
Minimum Ignition Current
Minimum Inhibiting Concentration
Missing Interruption Character
Missing Interruption Checker
Monitoring, Identification and Correlation
Monolithic Integrated Circuit
Motorcycle Industry Council [US]
Multilayer Interconnection Computer
Mutual Interference Chart
MICA Macro-Instruction Compiler Assembler
Major Incidents Computer Application
MICAM Microammeter
MICAPP Microcomputer-Assisted Process Planning Program
MICC Meetings Industry Council of Canada
Mineral-Insulated Copper-Covered (Cable)
MICE Megavolt Iron Coil Experiment
Member of the Institute of Civil Engineers
MICELEM Microphone Element

MICG Mercury Iodide Crystal Growth
Mich Michigan [US]
MIChemE Member of the Institute of Chemical Engineers
MICNS Modulator Integrated Communication and Navigation System
MICOM Missile Command [US Army]
MICR Magnetic-Ink Character Reader
Magnetic-Ink Character Recognition
Microscopic
MICRO International Symposium on Microscopy
Microcomputer
Microcontamination (Conference and Exhibition)
Multiple Indexing and Console Retrieval Operations
microamp microampere
MICROBIO Microbiologist
Microbiology
MICROCAT Micro-Catalogue
micro-in micro-inch [also: μ-in or mu-in]
MICROM Micro-Instruction Read-Only Memory
MICROMIN Microminiature
MICRONET Microcomputer Network
MICROPAC Micromodule Data Processor and Computer
MICROPSI Microcomputer Printed Subject Index
MICS Maintenance Inventory Control System
Management Information and Control System
Management Information Control System
Microprocessor Intertie and Communication System
Mineral-Insulated Copper-Sheathed (Cable)
MICT Ministry for Industry, Commerce and Technology [Canada]
Ministry for Industry, Commerce and Tourism [Canada]
MICU Message Interface and Clock Unit
MICV Mechanized Infantry Combat Vehicle [US Army]
MID Magnetically Insulated Diode
Maritime Identification Digits [of SARSAT]
Master of Industrial Design
Master of Interior Design
Message Input Description
Middle
Mid Midland [UK]
MIDA Maritime Industrial Development Area
MIDAC Management Information for Decision Making and Control
Michigan Digital Automatic Computer
MIDAM Mid-America Commodity Exchange [US]
MIDAR Microwave Detection and Ranging [also: Midar or midar]
MIDAS Map Information Display and Analysis System
Measurement Information Data Analytic System
Memory Implemented Data Acquisition Systems
Microdiagnosis for Analysis and Repair
Micro-Imaged Data Addition System
Microprogrammable Integrated Data Acquisition System
Microprogramming Design-Aided System
Missile Intercept Data Acquisition System
Modified Integration Digital-to-Analog Simulator
Modulator Isolation Diagnostic Analysis System
Modular International Dealing and Accounting System
Multimode International Data Acquisition Service [Australia]
Multiple Index Data Access System
MIDDLE Microprogram Design and Description Language
Middx Middlesex [UK]
MIDEF Microprocedure Definition
MIDES Missile Detection System
MIDI Musical Instrument Digital Interface
MIDIP Military Industry Data Interchange Procedure
MIDIST Mission Interministerielle de l'Information Scientifique et Technique [= Interministerial Mission on Scientific and Technical Information, France]
MIDLNET Midwest Library Network [US]
MIDMS Machine-Independent Data Management System
MIDN Midnight
MIDOP Missile Doppler
MIDOR Miss Distance Optical Recorder
MIDOT Multiple Interferometer Determination of Trajectories
MIDP Microwave Induced Delayed Phosphorescence
MIDS Movement Information Distribution Station
Multifunction Information Distribution System
Multimode Information Distribution System
MIDWEEK Manager Integrated Dictionary Week (Users Group) [US]
MIE Magnetically-Enhanced Ion Etch
Master of Industrial Engineering
Metal-Induced Embrittlement
MIEE Member of the Institute of Electrical Engineers
MIEEE Member of the Institute of Electrical and Electronics Engineers
MIEHM Modified Iterative Extended Hückel Method
MIER Malaysia Institute for Economic Research
MIES McMaster (University) Institute for Energy Studies [Canada]
MIF Miners International Federation [Belgium]
MIFAS Mechanized Integrated Financial Accounting System
MIFASS Marine Integrated Fire and Air Support System
MIFI Missile Flight Indicator
MIFL Master International Frequency List
MIFR Master International Frequency Register
Monitored International Frequency Register [of ITU, Switzerland]
MIFS Multiplex Interferometric Fourier Spectroscopy
MIG Magnetic Injection Gun
Mercury-in-Glass (Thermometer)
Metal Inert-Gas (Welding)
Mikoyan and Gurevish [Russian Fighterplane]
MIGS Metal-Induced Gap State
MIH Miles per Hour
Missing Interruption Handler
Multiplex Interface Handler
mi/h miles per hour [also: mi/hr or mph]
MII Mineral Information Institute [US]
MIIA Mine Inspectors Institute of America [US]
MIIS Metal-Insulation-Insulation Semiconductor
MIJ Master of International Journalism
Member of the Institute of Journalists
MIK Methyl Isobutyl Ketone

MIKAS Mikrofichekatalogsystem [= Microfiche Cataloguing System, Switzerland]
MIKE Measurement of Instantaneous Kinetic Energy
Microphone
MIKER Microbalance Inverted Knudsen Effusion Recoil
MIKES Mass-Analyzed Ion Kinetic Energy Spectrometry
MIL Micro-Implementation Language
Mileage
Military
Milling
Module Interconnection Language
MILA Merritt Island Launch Area [of NASA]
MILADGRU Military Advisory Group [US]
MIL ATT Military Attache
MILC Midwest Interlibrary Center [US]
MILDIP Military Industry Logistics Data Interchange Procedure
MILE Military Electronics
MILECON Military Electronics Conference [also: MIL-E-CON]
MILEPOST Middlesborough Initiative Local Electronic Provision of Service Technique
MILES Multiple Integrated Laser Engagement System
mil-ft mil-foot [also: mil ft]
MIL GOV Military Government
MIL-HDBK Military Handbook
MILL Million
MIL-I Military Specification on Interfaces
Military Specification on Interference
MILIC Microwave Insulated Line Integrated Circuit
mil-in mil-inch [also: mil in]
MILIPAC Military Portable Computer
millihg millimeter of mercury [also: mmHg]
millirem one-thousandth rem (= roentgen equivalent man) [Unit]
MILR Master of Industrial and Labour Relations
MILREP Military Representative
MILS Missile Impact Location System
Missile Location System
MILSATCOM Military Satellite Communications
MIL-SPEC Military Specification
MILSTAAD Military Standard Activity Address
MIL-STD Military Standard
MILSTICCS Military Standard Item Characteristics Coding Structure
MILSTRIP Military Standard Requisitioning and Issue Procedure [US Army]
MILUS Mining Industry Land Use Strategy [Canada]
MIM Maintenance Interface Machine
Manitoba Institute of Management
Maryland Institute of Metals
Master of Industrial Management
Master of International Management
Message Input Module
Metal Injection Molding
Metal-Insulator-Metal
Minimum
Modem Interface Module
Modified Index Method
MIMA Metal Injection Molding Association [US]
Mineral Insulation Manufacturers Association [US]
MIMarE Member of the Institute of Marine Engineers
MIMC Member of the Institute of Management Consultants
MIMD Multiple Input, Multiple Data
Multiple-Instruction (Stream)/Multiple-Data (Stream)
MIME Member of the Institution of Mining Engineers
MIMechE Member of the Institute of Mechanical Engineers
MIMIC Microwave Monolithic Integrated Circuit
Millimeter-Wave Monolithic Integrated Circuit
Multiparameter Input Mechanism for Interactive Control
MIMinE Member of the Institute of Mining Engineers
MIMLC Microinfiltrated Macrolaminated Composite
MIMM Member of the Institute of Mining and Metallurgy
MIMO Man In — Machine Out
Multiple Input, Multiple Output
MIMOLA Machine-Independent Microprogramming Language
MIMR Madison Integrated Microscope Resource [of University of Wisconsin, US]
Magnetic Ink Mark Recognition
MIMS Multi-Item Multi-Source
MIMUG Meetings Industry Microcomputer Users Group [US]
MIN Mineral
Mineralogist
Mineralogy
Minimum
Mining
Minister
Ministry
min minim [Unit]
minute
MINA Member of the Institute of Naval Architects
MINAC Miniature Linear Accelerator
Miniature Navigation Airborne Computer
MINDAC Marine Inertial Navigation Data Assimilation Computer
MIND Mass-Impregnated Non-Draining (Compound)
Modular Interactive Network Designer
Multiple Instrument Network Distribution
MINDD Minimum Due Date
MINDO Modified Incomplete Neglect of Differential Overlap
MINE Microbial Information Network Europe
MinE Mining Engineer [also: MinEng]
MINEAC Miniature Electronic Autocollimator
MInEd Master of Industrial Education [also: MIndEd]
MINERAL Mineralogist
Mineralogy
MINERALEX Mining and Mineral Exploration Exposition [also: Mineralex]
MINET Medical Information Network
MINGEL Martin Integrated Neutral Graphics and Engineering Language
MINI Minicomputer
MINIAPS Minimum Accessory Power Supply
MINIC Minimal Input Cataloguing
MINICS Minimal Input Cataloguing System
MINICS/PDS Minimal Input Cataloguing System — Periodicals Data System
MINIDIP Miniature Dual-in-Line Package
MINIMARS Mini-Mirror Advanced Reactor Study
MINIMAX Minimizing the Maximal Error

MINIRAR Minumum Radiation Requirements
MINI-SUBLAB Miniature Submarine Laboratory
MINIT Minimum Interference Threshold
MINITEX Minnesota Interlibrary Telecommunications Exchange [US]
MinlE Mineral Engineer [also: MinlEng]
MIN MC Minimum Material Condition
Minn Minnesota [US]
MINNEMAST Minnesota Mathematics and Science Teaching Project [US]
MINPRT Miniature Processing Time
MINPROC Mineral Processing
Mineral Processing Database [of CANMET, Canada]
MINS Miniature Inertial Navigation System
MINSD Minimum Planned Start Date
MINSOP Minimum Slack Time per Operation
MInstCE Member of the Institute of Civil Engineers
MInstME Member of the Institute of Mining Engineers
MInstMM Member of the Institute of Mining and Metallurgy
MInstWE Member of the Institute of Water Engineers
MINT Magnetic Information Technology
Materials Identification and New Item Control Technique
MINTECH Mining Technology [also: MIN TECH]
Ministry of Technology [UK]
MINTEC Mining Technology (Database) [of CANMET, Canada]
MIN WT Minimum Weight
MIO Map Information Office [of USGS]
Multiple Input/Output
MIOP Master Input/Output Processor
Multiplexing Input/Output Processor
MIOS Modular Input/Output System
MIP Machine Construction Processor
Malaysian Institute of Physics
Manual Input Processing
Manual Input Program
Manufacturers of Illuminating Products [US]
Marine Insurance Policy
Master of Intellectual Property
Matrix Inversion Program
Mean Indicated Pressure
Minimum Impulse Pulse
Missile Impact Predictor
Mixed-in-Place
Mixed Integer Programming
Most Important Person
Multiband Imaging Photometer
MIPA Monoisopropanolamine
Multiple Intensity Profile Analysis
MIPAC Microprocess Analysis Code
MIPC Methylisopropylcarbinol
Methylisopropylphenylcarbamate
MIPE Magnetic Induction Plasma Engine
Modular Information Processing Equipment
Moskow Institute of Power Engineers [USSR]
MIPIR Missile Precision Instrumentation Radar
MIPK Methyl Isopropyl Ketone
MIPP Master of International Public Policy
MIPPT McMaster Institute for Polymer Production Technology [Canada]
MIPS Microcomputer Image Processing System
Millions of Instructions per Second [also: mips]
Minimum Inventory Production System
Missile Impact Predictor Set
Multiband Imaging Photometer System
MIR Master of Industrial Relations
Memory-Information Register
Memory Input Register
Micro-Instruction Register
Model Incident Report
Multiple Internal Reflection
Music Information Retrieval
MIRA Mechanical Industrial Relations Association
Motor Industry Research Association [UK]
MIRACL Management Information Report Access without Computer Languages
MIRACODE Microfilm Retrieval Access Code
MIRAGE Microelectronic Indicator for Radar Ground Equipment
MIRAN Missile Ranging
MIRD Medical Internal Radiation Dose
MIRE Member of the Institute of Radio Engineers
MIRED Micro-Reciprocal-Degree [also: mired]
MIRF Multiple Instantaneous Response File
MIRFAC Mathematics in Recognizable Form Automatically Compiled
MIRG Metabolic Imaging Research Group
MIRIAD Maritime Institute for Research and Industrial Development [US]
MIRL Medium-Intensity Runway Lights
MIRO Mineral Industry Research Organization
Mining Industry Research Organization
MIROC Mineral Industry Research Organization of Canada
Mining Industry Research Organization of Canada
MIROS Modulation Inducing Retrodirective Optical System
MIRPS Multiple Information Retrieval by Parallel Selection
MIRR Material Inspection and Receiving Report
MIRS Manpower Information Retrieval System
Medical Information Retrieval Service [US]
MIRT Molecular Infrared Track
MIRU Move-In, Rig-Up
MIRV Multiple Independent Reentry Vehicle
Multiple Individually-Targeted Reentry Vehicle
MIS Macroscopic Internal Stress
Management Information Science
Management Information Service
Management Information System
Management Integrated System
Master of Arts in Information Systems
Master of Information Science
Marine Information System [US]
Mercury-in-Steel (Thermometer)
Metal-Insulator-Semiconductor
Metering Information System
Mining Institute of Scotland
Missing
Modified Initial System
Modified in situ Retorting
Moody Institute of Science [US]
MISA Municipal Industrial Strategies for Abatement
MISAM Multiple Indexed Sequential Access Method

MISC Miscellaneous
MISD Multiple-Instruction (Stream)/Single-Data (Stream)
MISDAS Mechanical Impact System Design for Advanced Spacecraft
MISER Minimum Size Executive Routines
MISFEED Misaligned Feed
MISFET Metal-Insulator-Semiconductor Field-Effect Transistor
Metal-Insulator-Silicon Field-Effect Transistor
MISHAP Missiles High-Speed Assembly Program
MISM Metal-Insulator-Semiconductor-Metal
MISMDS Multiple Instruction Stream — Multiple Data Stream
MISO Multiple Input, Single Output
MISP Medical Information Systems Program
Microelectronics Industry Support Program [UK]
Mineral Investment Stimulation Program [Canada]
MISR Multiple Input Signature Register
MISS Mechanical Interruption Statistical Summary
Missile Intercept Simulation System
Mobile Integrated Support System
Multi-Item, Single-Source
Miss Mississippi [US]
MISSDS Multiple Instruction Stream — Single Data Stream
MISSIL Management Information System Symbolic Interpretative Language
MIST Metal-Insulator-Semiconductor Field-Effect Transistor
Microbursts and Severe Thunderstorms Experiment
Minor Isotopes Safeguards Techniques
Music Information System for Theorists [of Indiana University, US]
MISTRAM Missile Trajectory Measurement System
MIStructE Member of the Institution of Structural Engineers
MIT Master Instruction Tape
Master of Industrial Technology
Maupa Institute of Technology [Phillipines]
Massachusetts Institute of Technology [US]
Master Instruction Tape
Milled in Transit
Ministry of Industry and Tourism [Canada]
Ministry of Industry and Trade [now: MITT, Canada]
Modular Industrial Terminal
Modular Intelligent Terminal
Muroran Institute of Technology [Japan]
MITA Microcomputer Industry Trade Association
MITATT Mixed Tunneling Avalanche Transit Time
MITE Microelectronic Integrated Test Equipment
Miniaturized Integrated Telegraph Equipment
Miniaturized Integrated Telephone Equipment
Missile Integration Terminal Equipment
Multiple Input Terminal Equipment
MITEA Modular Integrated Towed Electronics Array
MITEC Mining Industry Technology Council of Canada [also: Mitec]
MITER Modular Installation of Telecommunications Equipment Racks
MITF Musser International Turfgrass Foundation [US]
MITI Minstry of International Trade and Industry [Japan]
MITM Military Industry Technical Manual
MITO Minimum Interval Takeoff
MITOC Missile Instrumentation Technical Operations Communications
MITOL Machine-Independent Telemetry-Oriented Language
MITR Massachusetts Institute of Technology Reactor [US]
MITS Management Information and Text System
Materials Information Translation Service [of ASM International, US]
MITT Ministry of Industry, Trade and Technology [Canada]
MIU Machine Interface Unit
Malfunction Insertion Unit
Manpower and Immigration Union
Message Injection Unit
MET (= Mobile Earth Terminal) Interface Unit
Microalgae International Union [US]
Model Interface Unit
Multistation Interface Unit
MIUS Modular-Integrated Utility System
MIVC Magnetically Induced Velocity Change
MIVPO Modified Inside Vapor Phase Oxidation
MIW Micro-Instruction Word
MIX Magnetic Ionization Experiment
Master Index
1-Methyl-3-Isobutylxanthine
Mixture
MJ Master of Journalism
Megajoule
MJD Modified Julian Date
MK Malawi Kwacha [Currency]
Manual Clock
Mask
Mark
MKA Machine Knife Association [US]
MKG Marking
Meter-Kilogram (System) [also: mkg]
m-kgf meter-kilogram-force
MKH Multiple Key Hashing
MKMA Machine Knife Manufacturers Association [now: MKA, US]
MKP Monopotassium Phosphate
MKR Marker
MKS Meter-Kilogram-Second (System) [also: mks]
MKSA Meter-Kilogram-Second-Ampere (System) [also: mksa]
MKT Market
MKTG Marketing
MKT VAL Market Value
MKw Malawi Kwacha [Currency]
MkWh Thousand Kilowatthours
ML Machine Language
Magnetic Latching
Magnetic Leakage
Magnetic Levitation
Magnetic Link
Main Lobe
Major Lobe

	Manipulation Language
	Material List
	Materials Laboratory
	Mathematical Logics
	Maximum Likelihood
	Mean Level
	Mean-Life
	Memory Location
	Message Length
	Methods of Limit
	Microlithography
	Micrology
	Microprogramming Language
	Mixed Lengths
	Mobile Low-Power Plant [US Army]
	Mold Line
	Monolayer
	Multilayer
	Multipoint Logger
M-L	Metallic-Longitudinal
mL	millilambert
	milliliter [also: ml]
MLA	Manitoba Library Association [Canada]
	Marine Librarians Association [UK]
	Master of Landscape Architecture
	Matching Logic and Adder
	Mixed Lead Alkyl
	Modern Language Association [US]
	Multi-Layered Alloy
	Multiple Line Adapter
MLandArch	Master of Landscape Architecture
MLAP	Metallic Linc Access Port
MLArch	Master of Landscape Architecture
MLArchUD	Master of Landscape Architecture and Urban Design
MLAS	Master of Laboratory Animal Science
MLAUD	Master of Landscape Architecture and Urban Design
MLB	Multilayer Board
MLC	Magnesia-Doped Lanthanum Chromite
	Maneuver Load Control
	Michigan Library Consortium [US]
	Microprogram Location Counter
	Mixed Leucocyte Culture
	Mixed Lymphocyte Culture
	Multilayer Ceramic
	Multilayer Circuit
	Multiline Control
MLCA	Multiline Communications Adapter
MLCAEC	Military Liaison Committee to the Atomic Energy Commission [US]
MLCB	Multilayer Circuit Board
MLCC	Monolithic Ceramic Capacitor
	Multilayer Ceramic Capacitor
MLCD	Microwave Liquid Crystal Display
MLCP	Multiline Communications Processor
MLD	Machine Language Debugger
	Masking Level Difference
	Mean Lethal Dose
	Median Lethal Dose [also: LD50 or LD_{50}]
	Minimum Lethal Dose
	Mixed Layer Depth
	Multilayer Dielectric
MLDes	Master of Landscape Design
MLDG	Molding
MLE	Maximum Likelihood Estimate
	Medium Local Exchange
	Melt-Leach Evaporation
	Molecular-Layer Epitaxy
MLEV	Manned Lifting Entry Vehicle
MLF	Mali Franc [Currency]
	Medium Longitudinal Fascicule
	Multilateral Force
MLG	Main Landing Gear
	Methyl-L-Glutamate
MLGW	Maximum Landing Gross Weight
MLH	Maximum Likelihood
MLHCP	Mean Lower Hemispherical Candlepower
MLHR	Master of Labour and Human Resources
MLHW	Mean Lower High Water
MLI	Machine Language Instruction
	Magnetic Level Indicator
	Marker Light Indicator
	Message Level Interface
	Ministry of Local Industry [USSR]
	Multilayer Insulation
MLIA	Multiplex Loop Interface Adapter
MLib	Master of Librarianship
MLIM	Matrix Log-In Memory
MLIP	Message Level Interface Port
MLIR	Master of Labour and Industrial Relations
MLIS	Master of Library and Information Science
MLL	Manned Lunar Landing
	Microlaterlog
MLLE	Medium Large Local Exchange
MLLP	Manned Lunar Landing Program [of NASA]
MLLW	Mean Lower Low Water
MLM	Master of Library Media
	Multilayer Metallization
	Multilevel Marketing
MLMA	Metal Ladder Manufacturers Association [US]
MLO	Manned Lunar Orbiter [of NASA]
MLP	Machine Language Program(ming)
	Mean Life Period
MLPC	Multiplicand
MLPCB	Multilayer Printed Circuit Board
MLPH	Ministry of Lands, Parks and Housing [Canada]
MLPP	Multi-Level Precedence and Preemption
MLPR	Multiplier
MLPWB	Multilayer Printed Wiring Board
MLR	Main Line of Resistance
	Mechanized Line Record
	Memory Lockout Register
	Monodisperse Latex Reactor
	Mixed Leucocyte Reaction
	Multiple Linear Regression
	Multiply and Round
MLRG	Marine Life Research Group [of SIO, US]
MLRHR	Master of Labour Relations and Human Resources
MLRS	Monodisperse Latex Reactor System
	Multiple Launch Rocket System
MLS	Machine Literature Search

Manned Lunar Surface [of NASA]
Master Laboratory Station
Master of Library Science
Master of Library Services
Master of Life Sciences
Metal-Langmuir Semiconductor
Microwave Landing System
Missile Location System
Mudline Suspension System
Multilanguage System
Multilayered Structure
Multilevel Secure
Multi-Loop System
Multiple Listing Service

MLSE Maximum Likelihood Sequence Estimation
ML/SFA Metal Lath/Steel Framing Association Division [of NAAMM, US]
MLSFET Metal-Langmuir Semiconductor Field-Effect Transistor
MLSIT Master of Library Science and Instructional Technology
MLSNPG Microwave Landing System National Planning Group [US]
MLSO Mode-Locked Surface-Acoustic-Wave Oscillator
MLSS Mixed Liquid Suspended Solids
MLT Mass-Length-Time
Mean Length of Turn
Mean Length per Turn
Mean Logistical Time
Mean Low Tide
Mechanized Line Testing
Mechanized Loop Test
Medical Laboratory Technician
Medical Laboratory Technologist
Medical Laboratory Technology
MLTA Multiple Line Terminal Adapter
MLTF Multilayer Thick Film
MLTS Micro-Level Test Set
MLTY Multiply
MLU Memory Loading Unit
Memory Logic Unit
Multiple Logic Unit
MLV Medium Launch Vehicle
Medium Logistics Vehicle
MLVW Medium Logistics Vehicle, Wheeled
MLW Maximum Landing Weight
Mean Low Water
Medium-Level Waste
MLWA Maximum Landing Weight Authorized
MLWN Mean Low-Water Neaps
MLWS Mean Low Water Springs
MLWK Millwork
MLY Multiply
MM Magnetic Material
Magnetomechanical
Main Memory
Maintenance Manual
Man-Made
Mass Memory
Master Monitor
Master of Management
Materials Measurement
Mathematical Modelling
Matrix Mechanics
Memory Module
Memory Multiplexer
Metallurgical Microscope
Metals Museum [Japan]
Methyl Methacrylate
Micromechanics
Middle Marker
Million
Military Medal
Mine Manager
Mischmetall
Modified Mercalli
Monostable Multivibrator
MOS (= Metal-Oxide Semiconductor) Monolithic
Motor Meter
Moving Magnet
M&M Mining and Metals
mm millimeter
MMA Magnesium Methyl Carbonate
Manitoba Medical Association [Canada]
Manual Metal-Arc Welding
Master of Marine Affairs
Master of Manpower Administration
Master of Media Arts
Materials Marketing Association [US]
Metal Mining Association [Canada]
Meter Manufacturers Association [UK]
Methyl Methacrylate
Mining and Metallurgical Association [Japan]
Monorail Manufacturers Association [US]
Multiple Module Access
MMAD Mass Median Aerodynamic Diameter
MMAE Master of Mechanical and Aerospace Engineering
MMAR Main Memory Address Register
MMath Master of Mathematics
MMatSE Master of Material Science and Engineering
MMAU Master Multiattribute Utility
MMB Master of Medical Biochemistry
Minimum Monthly Balance
Modulated Molecular Beam
Molecular Microbiology
MMB/D Millions of Barrels per Day
MMBI Methyl-2-Mercapto Benzimidazole
MMC Magnesium Methyl Carbonate
Master of Mass Communications
Maximum Material Condition
Maximum Metal Condition
Memory Management Controller
Metal-Matrix Composite
Mining Monitoring and Control System
Monolithic Multicomponent Ceramic
Multipart Memory Controller
MMCF Million Cubic Feet
MMCFD Million Cubic Feet per Day
MMCIAC Metal-Matrix Composites Information Analysis Center [of USDOD]
MMCP Micro-Master Control Processor
MMCS Mass Memory Control Subsystem

MMCT Moment-Modified Compact Tension
MMD Manual of the Medical Department
Mass Median Diameter
Master of Mechanical Engineering
Mines and Minerals Division [of MND, Canada]
Mining and Metallurgy Department
Moving Map Display
MMDS Martin Marietta Data Systems
MME Master of Manufacturing Engineering
Master of Mechanical Engineering
Metrolum Multimode Electrode
Mining Mechanical Engineer
Mining Mechanical Engineering
MMechEng Master of Mechanical Engineering [also: MMechE]
MMEL Master Minimum Equipment List
MMEP Marine Mammal Events Program [US]
MMES Member of the Mechanical Engineering Society
MMet Master of Metallurgy
MMetEng Master of Metallurgical Engineering [also: MMetE]
MMetMatE Master of Metallurgical and Materials Engineering [also: MMetMatEng]
MMetMatS Master of Metallurgical and Materials Science [also: MMetMatSc]
MMF Magnetomotive Force [also: mmf]
Man-Made Fiber
Minimum Mass Fraction
Moving Magnetic Features
mmf micromicrofarad [also: $\mu\mu$F, $\mu\mu$f, mmF or mmfd]
MMFM Modified Modified Frequency Modulation [also: M^2FM or M2FM]
MMFITB Man-Made Fibers Industry Training Board
MMFPA Man-Made Fiber Producers Association [now: AFMA, US]
MMFS Manipulating Message Format Standard
Manufacturing Message Format Standard
MMG Magnetic Materials Group
Motor-Motor Generator
Munchener Messe- und Ausstellungsgesellschaft [= Munich Society for Trade Fairs and Exhibitions, Germany]
MMgt Master of Management [also: MMgmt]
MMgtEng Master of Management Engineering [also: MMgtE]
MMgtS Master of Management Science [also: MMgtSc]
MMH Methylmercury Hydroxide
Monomethylhydrazine
MMHE Master of Material Handing Engineering
mm Hg millimeter of mercury
mμ Hg millimicron of mercury
mm Hg L millimeter of mercury liters
mm H_2O millimeter of water
MMHS Master of Management in Human Services
MMI Main Memory Interface
Man-Machine Interaction
Man-Machine Interface
Materials Management International [UK]
Multimessage Interface
MMIC Monolithic Microwave Integrated Circuit
Motorcycle and Moped Industry Council
MMIJ Mining and Metallurgical Institute of Japan
m/min meters per minute [also: mpm or MPM]
MMinEng Master of Mining Engineering [also: MMinE]
MMIS Man-Machine Interface Interface Subsystem
Master of Management Information Systems
Materials Management Information System
MMIU Multipart Memory Interface Unit
MMJ Modified Modular Jack
MMK Multiprocessor Multitasking Kernel
MML Manitoba Motor League [Canada]
Man-Machine Language
Martin Marietta Laboratories [US]
Material Mechanics Laboratory
MMLS Microgravity Materials Science Laboratory
MMM Maintenance and Material Management
Manned Mars Mission [of NASA]
Mars Mission Module
Mercury Motor Meter
Multimission Module
MMMA Meat Machinery Manufacturers Association [US]
Micromagnetic Multiparameter Microstructure and Residual Stress Analysis [also: 3MA]
MMMC Milking Machine Manufacturers Council [US]
MMMI Meat Machinery Manufacturers Institute [US]
MMMIS Maintenance and Material Management Information System
MMMS Minerals, Metals and Materials Society [US]
MMMSocAm Member of the Mining and Metallurgical Society of America
MMOA Mobile Modular Office Association [US]
MMOD Micromodule
mmol millimole
MMOS Message Multiplexer Operating System
MMP Maintenance Management Process
Maintenance Management Program
Master of Marine Policy
Minimum Miscibility Pressure
Modified Modular Plug
Module Message Processor
Monitor Metering Panel
Multiplex Message Processor
MMPA Magnetic Material Producers Association [US]
Marine Mammal Protection Association [US]
MMPD Minerals and Metallurgical Processing Division [of SME-AIME, US]
MMPF Microgravity and Materials Processing Facility
MMPI Mining and Materials Processing Institute
MMPIJ Mining and Materials Processing Institute of Japan
MMPP Magnesium Monoperoxyphthalate
Mechanized Market Programming Procedures
MMPR Missile Manufacturers Planning Reports
MMPS Multimachine Power System
MMPT Man-Machine Partnership Translation
MMPU Memory Manager and Protect Unit
MMR Magnetic Memory Record
Main Memory Register
Master of Marketing Research
Maximum Metal Reflector
Modified Modified Read
Multiple Match Resolver
MMRA Maritime Marshland Rehabilitation Association [Canada]
MMRBM Mobile Medium-Range Ballistic Missile
MMRIM Mat-Molding Reaction Injection Molding

MMRRI Mining and Mineral Resources Research Institute [US]
MMS Magnetic Median Surface
Maintenance Manipulator System
Man-Machine System
Manufacturing Message Specification
Massachusetts Medical Society [US]
Mass Memory Store
Master of Management Science
Master of Management Studies
Master of Marketing Science
Master of Materials Science
Materials Management System
Memory Management System
Metallurgy and Materials Science
Methyl Methanesulfonate
Microfiche Management System
Missile Monitoring System
Module Management System
Multimission Modular Spacecraft
Multimission Spacecraft
Multipart Memory System
MMSA Materials and Methods Standards Association [US]
Mining and Metallurgical Society of America [US]
MMSCF Million Standard Cubic Feet
MMSCFD Million Standard Cubic Feet per Day
MMSCFH Million Standard Cubic Feet per Hour
MMSCFM Million Standard Cubic Feet per Minute
MMSE Master of Manufacturing Systems Engineering
Minimum Mean Squared Error
MMSI Medium Metal-Support Interaction
MMSL Microgravity Materials Science Laboratory [of NASA]
MMSocAm Mining and Metallurgical Society of America
MMSt Master of Museum Studies [also: MMStud]
MMST/YR Million Short Tons per Year
MMT Methyl-M-Tyrosine
Mont Megantic Telescope
Multiple Mirror Telescope
MMTC Mechanical Maintenance Training Center
Microelectronics and Materials Technology Center [of RMIT, Australia]
MMtlE Master of Metal Engineering [also: MMtlEng]
MMTS Methyl Methylthiomethyl Sulfoxide
MMTT Multiple Mechanicothermal Treatment
MMU Main Memory Unit
Manned Maneuvering Unit
Mass Memory Unit
Memory Management Unit
Memory Mapping Unit
Metered Message Unit
Milli-Mass Unit [also: mmu]
Modular Maneuvering Unit
Multi-Message Unit
MMW Main Magnetization Winding
Millimeter Wave
m mu millimicron [also: mμ]
MN Magnetic North
Main
Main Network
Manual
Meganewton
Merchant Navy
Minnesota [US]
Mnemonic(s)
Mnemonic Symbol
Motor Number
Mn Manganese
Molecular Weight, Number-Average
mN millinewton
MNA Methoxyneuraminic Acid
Methyl Nadic Anhydride
Methylnonylacetaldehyde
Methyl-5-Norbornene-2,3-Dicarboxylic Anhydride
Ministry of Northern Affairs [Canada]
Multi-Network Area
Multi-Share Network Architecture
MNatSci Master of Natural Science [also: MNatS or MNatSc]
MNBLE Modified Nearly Best Linear Estimator
MNC Multinational Company
Multinational Corporation
Multiplicative Noise Compensator
MNCIAWPRC Malaysian National Committee of the International Association on Water Pollution Research and Control
MNCS Multipoint Network Control System
MND Ministry of National Defense
Ministry of Northern Development [Canada]
Multinomial Distribution
MNDM Ministry of Northern Development and Mines [Canada]
MNDO Modified Neglect of Differential Overlap
MNDX Mobile Nondirector Exchange
MNE Master of Nuclear Engineering
MNF Multisystem Networking Facility
MNG Managing
MNK Methylnonylketone
MNL Manual
Mineral
Minnesota National Laboratory [US]
MNLD Mainland
MnlEng Mineral Engineer [also: MnlE]
MNL ENG Mineral Engineering
MNL PROC Mineral Processing
MNM Minimum
MNNG 1-Methyl-3-Nitro-1-Nitrosoguanidine
MNOS Metal-Nitride-Oxide Semiconductor
Metal-Nitride-Oxide Silicon
MNOSFET Metal-Nitride Oxide Semiconductor Field-Effect Transistor
MNP Methyl-N-Pyrrolidine
Microcom Networking Protocol
MNPS Minimum Navigation Performance Specifications
MNPT Medium National Pipe Thread
MNR Maximum Normed Residual (Statistics)
Minimum Noise Route
(Ontario) Ministry of Natural Resources [Canada]
MNRL Mineral
MNRM Master of Natural Resources Management
MNRPRA Malaysia Natural Rubber Producers Research Association
MNRU Modulated Noise Reference Unit

MNS Master of Natural Science
Master of Nutritional Sciences
Metal-Nitride Semiconductor
MNSC Main Network Switching Center
MNSFET Metal-Nitride Semiconductor Field-Effect Transistor
MNT Mongolian Tugrik [Currency]
Monitor(ing)
Mononitrotoluene
MNTN Maintain
MNTR Monitor
MNU Methyl Nitrosourea
MNucSci Master of Nuclear Science [also: MNucS or MNucSc]
MO Magneto-Optic(s)
Mail Order
Manual Operation
Manual Output
Masonry Opening
Mass Observation
Master Oscillator
Maximum Operation
Medical Officer
Memory Output
Metal(lo)-Organic(s)
Metal Oxide
Meteorological Office [UK]
Mine Office
Mining Office
Missouri [US]
Mixed Oxide
Mobile Object
(Modus Operandi) — Method/Manner of Working
Molecular Orbital
Molecule
Money Order
Month
M&O Maintenance and Operations
M-O Magneto-Optic(s)
Metal(lo)-Organic(s)
Mo Molecule
Molybdenum
Missouri [US]
mo mouse [Immunochemistry]
MΩ Megohm
MOA Manufacturing Operations Analysis
Matrix Output Amplifier
Mesoxalic Acid
Military Operations Area
Ministry of Aviation [UK]
MOAA Member of the Office Administrators Association
MOB Master of Organizational Behaviour
MOBAT Modified Battalion Antitank
MOBIDAC Mobile Data Acquisition System
MOBIDIC Mobile Digital Computer
MOBL Macro-Oriented Business Language
MOBOL Mohawk Business-Oriented Language
MOBOT Marine Robot
MOBS Multiple Orbit Bombardment System
MOBULA Model Building Language
MOC Magnetic-Optic Converter
Master Operational Controller
Master Operations Console
Master Operations Control
Memory Operating Characteristic
Minimum Obstacle Clearance
Minimum Operational Characteristics
Mission Operation Computer
MOCA Methylene-Bis-O-Chloroaniline
Minimum Obstruction Clearance Altitude
MOCC METEOSAT (= Meteorological Satellite) Operations Control Center
MOcE Master of Oceanographic Engineering [also: MOcEng]
MOCIC Molecular Orbital Constraint of Interaction Coordinates
MOCR Mission Operations Control Room
MOCVD Metal(lo)-Organic Chemical Vapor Deposition
MOCVD/MOVPE Metal(lo)-Organic Chemical Vapor Deposition/ Metal-Organic Vapor Phase Epitaxy
MOD Master of Organizational Development
Message Output Description
Metal(lo)-Organic Deposition
Ministry of Defense [UK]
Model
Moderation
Moderator
Modern
Modification
Modulation
Modulation-Doped
Modulator
Module
Modulus
Moving Domain
mod modulo [Mathematics]
MODA Motion Detector and Alarm
MODAC Mountain System Digital Automatic Computer
MODACS Modular Data Acquisition and Control System
MOD/DEMOD Modulating/Demodulating
MODE Mid-Ocean Dynamics Experiment
MODEM Modulator-Demodulator [also: Modem or modem]
MODE S Mode Select (System)
MODEST Missile Optical Destruction Technique
MODFET Modulation-Doped Field-Effect Transistor
MODI Modified Distribution Method
Modular Optical Digital Interface
MODICON Modular-Dispersed Control
MODILS Modular Instrument Landing System
MODPOT Modulator Potential
MODS Major Operation Data System
Manufacturing Operating Documentation System
MODU Mobile Offshore Drilling Unit
MODUS Mobile Offshore Drilling Units
MOE Master of Occupational Education
Master of Ocean Engineering
Measure of Effectivenes
Ministry of Education [UK]
Ministry of Energy [Canada]
Modulus of Elasticity
MOEA Ministry of Economic Affairs [Taiwan]
MOERO Medium Orbiting Earth Resources Observatory

MOF Maximum Observed Frequency
Maximum Operating Frequency
Ministry of Finance [Taiwan]
Ministry of Forestry
MOFA Ministry of Foreign Affairs [Taiwan]
MofC Master of Commerce
μ of Hg micron of mercury [also: μHg]
MOG Mauritanian Ouguiya [Currency]
Mesh of Grind
MOGA Microwave and Optical Generation and Amplification
MOGAS Motor Gasoline
MOGR Moderate or Greater
MOH Master of Occupational Health
Maximum Operating Hours
Medical Officer of Health
Ministry of Health [UK]
(Ontario) Ministry of Health [Canada]
mohm mechanical ohm
Moho Mohorovicic Discontinuity [Geology]
MOI Ministry of the Interior [Taiwan]
Moment of Inertia
Multiplicity of Infection
MOIT Ministry of Industry and Trade [Israel]
MOJ Material On Job Date
Metering Over Junction
MOL Machine-Oriented Language
Manned Orbiting Laboratory
Maximum Output Level
Ministry of Labour [Canada]
Molecular
Molecule
Multiple On-Line Programming
mol mole
mol% mole percent
MOLAB Mobile Lunar Laboratory [of NASA]
Moving Lunar Laboratory
MOLARS Meteorological Office Library Accessions and Retrieval System [UK]
MOL BIO Molecular Biology [also: MOL BIOL]
MOLDS Management On-Line Data System
Multiple On-Line Debugging System
MOLE Molecular Optical Laser Examiner
MOLECTRONICS Molecular Electronics [also: MOL ELECTRONICS]
MolE Molecular Engineer [also: MolEng]
MOL ENG Molecular Engineering
mol/L mole per liter [also: mol/l]
MOLSINK Molecular Sink
MOL WT Molecular Weight [also: mol wt]
MOM Message Output Module
Metal(lo)-Oxide-Metal
MOMBE Metal(lo)-Organic Molecular Beam Epitaxy
MOMS Metalorganic Magnetron Sputtering
MON Monday
Monitor(ing)
Monument
Motor Octane Number
Mon Monmouthshire [UK]
Mon-H Mono-Hydrogen
MONO Monostable
MONOBAS Monobasic
MONOCL Monoclinic
MONOS Monitor Out of Service
MonoXPS Monochromated X-Ray Photoelectron Spectroscopy
Mont Montana [US]
MoOPH Molybdenum Oxodiperoxypyridine Hexamethylphosphoramide
MOOSE Manned Orbital Operations Safety Equipment
Man Out of Space Easiest
MOOT Method of Optimal Truncation
MOP Macao Pataca [Currency of Macao]
Melt Optimization Program
Memory Organization Packet
Multiple On-Line Processing
Muriate of Potash
MOPA Master Oscillator Power Amplifier
Methoxyphenylacetic Acid
MOPB Manually-Operated Plotting Board
MOPEG 4-Hydroxy-3-Methoxyphenylglycol
MOPITT Measurement of Pollution in the Troposphere
MOPR Manner of Performance Rating
MOPS Military Operation Phone System
Million Operations per Second [also: mops]
Minimum Operational Performance Standards
Missile Operations
3-(N-Morpholino)propanesulfonic Acid
MOPSO 3-(N-Morpholino)-2-Hydroxypropanesulfonic Acid
MOPT Mean One-Way Propagation Time
MOPTS Mobile Photographic Tracking Station
MOR Memory Output Register
Monthly Operating Report
MORA Minimum Off-Route Altitude
MOREPS Monitor Station Reports
MOR Magnetic Optical Rotation
Malate Oxidoreductase
Management Oversight and Risk Tree
Mars Orbital Rendezvous
Modulus of Rupture
Molten Alkali Resistance
MORE Maintenance, Repair, Overhaul Excellence
MORI Market and Opinion Research Institute [UK]
MORL Manned Orbital Research Laboratory
MORP Meteorite Observation and Recovery Project [of NRC, Canada]
MORS Mandatory Occurence Reporting Scheme [of CAA, UK]
Military Operations Research Society [US]
Military Operations Research Symposium
MORT Management Oversight and Risk Tree
MOR T Morse Taper
MOS Management Operating System
Manufacturing Operating System
Marine Observation Satellite
Master Operating System
Mathematical Off-Print Service [of AMS, US]
Material Ordering System
Memory-Oriented System
Metal-Oxide Semiconductor
Metal-Oxide Silicon
Microprogram Operating System
Minimum Operating System

	Ministry of Supply [UK]
	Mission Operating System [of NASA]
	Modular Operating System
	Months
	Multiprogramming Operating System
MOSAIC	Macro Operation Symbolic Assembler and Information Compiler
	Metal-Oxide Semiconductor Advanced Integrated Circuit
	Ministry of Supply Automatic Integrator and Computer
	Mobile System for Accurate ICBM (= Intercontinental Ballistic Missile) Control
MOSAR	Modulation Scan Array Receiver
MOSD	Ministry of Science and Development [Israel]
MOS DMOS	Metal-Oxide Semiconductor Double-Diffused Metal-Oxide Semiconductor [also: MOS-DMOS]
MOSERD	Ministry of State for Economic and Regional Development
MOSFET	Metal-Oxide Semiconductor Field-Effect Transistor
	Metal-Oxide Silicon Field-Effect Transistor
MOSIC	Metal-Oxide Semiconductor/Integrated Circuit [also: MOS IC]
MOSID	Ministry of Supply Inspection Department [UK]
MOSIGT	Metal-Oxide Semiconductor Insulated Gate Transistor
MOS/LSI	Metal-Oxide Semiconductor for Large-Scale Integration
MOSM	Metal-Oxide Semimetal
MOSRAM	Metal-Oxide Semiconductor Random Access Memory [also: MOS RAM]
MOSROM	Metal-Oxide Semiconductor Read-Only Memory
MOSS	Multilayer Optical Scanning Spectrometer
MOSST	Ministry of State for Science and Technology [Canada]
MOST	Management Operation System Technique
	Metal-Oxide Semiconductor Technology
	Metal-Oxide Semiconductor Transistor
	Metal-Oxide Silicon Transistor
	Military Operational and Support Truck
	Ministry of Science and Technology [Korea]
	Multipurpose Observation Sizing Technique
MOT	Management of Technology [also: MoT]
	Manned Orbital Telescope [of NASA]
	Master of Occupational Therapy
	Ministry of Overseas Trade [UK]
	Ministry of Transport [UK]
	(Ontario) Ministry of Transportation [Canada]
	Molecular Orbital Theory
	Motor
	Mozambique Meticai [Currency]
MOTA	Materials Open Test Assembly
MOTACC	Manufacturers of Telescoping and Articulating Cranes Council [of FIEI, US]
MOTARDES	Moving Target Detection System
Motional EMF	Motional Electromotive Force [also: motional emf]
MOTIS	Message-Oriented Text Interchange System
MOTMX	Methoxy Trimethylxanthine
MOTNE	Meteorological Operational Telecommunications Network of Europe
MOTOR	Mobile-Oriented Triangulation of Reentry
	Motoring
MOTS	Module Test Set
MOTU	Mobile Optical Tracking Unit
MOU	Memorandum of Understanding
MOUG	Maryland On-Line User Group [US]
MOUTH	Modular Ouput Unit for Talking to Humans
MOV	Metal-Oxide Varistor
	Motor Operated Valve
	Move
MOVECAP	Movement Capability
MOVPE	Metal-Organic Vapor Phase Epitaxy
MOWASP	Mechanization of Warehousing and Shipment Processing
MOX	Mixed Oxide Fuel
MP	Machine Pistol
	Macroprocessor
	Magnetic Properties
	Magnetophotophoresis
	Magnifying Power
	Main Phase
	Maintenance Period
	Maintenance Point
	Maintenance Process
	Management Process
	Manifold Pressure
	Master of Planning
	Master Painter
	Master Patternmaker
	Master Plumber
	Master Printer
	Master Pulse
	Materials Physics
	Materials Processing
	Mathematical Physics
	Mathematical Programming
	Mean Effective Pressure [also: MEP]
	Measurement Pragmatic
	Mechanical Part
	Mechanical Processing
	Mechanical Properties
	Medium-Pressure
	Melting Point [also: mp]
	Membrane Process
	Mercaptopurine
	Mercury Pump
	Metallized Paper
	Metal Powder
	Metal Processing
	Metamorphic Petrology
	Metaphysics
	Methylpentane
	Methylpurine
	Metropolitan Police
	Mexican Peso [Currency]
	Microperoxidase
	Microphotography
	Microplastic(ity)
	Microprobe
	Microprocessor
	Microprogram(ming)
	Military Police

	Mineral Processing
	Minimum Phase
	Miscellaneous Paper
	Miscellaneous Publication
	Missing Pulse
	Module Processor
	Molecular Physics
	Molecular Pump
	Morse Potential
	Motion Picture
	Multiphase
	Multiplier Phototube
	Multiply
	Multipole
	Multiprocessing
	Multiprocessor
	Multipulse
	Multipurpose
	Myeloperoxidase
mP	maritime polar air
	millipoise
M£	Maltese Pound [Currency of Malta]
MPA	Maclura Pomifera Agglutinin
	Management Professionals Association [India]
	Master of Professional Accountancy
	Master of Professional Accounting
	Master of Public Administration
	Master of Public Affairs
	Mechanical Packing Association [now: FSA, US]
	Metal Powder Association [now: MPIF, US]
	Methoxypropylamine
	Methylphosphoric Acid
	Microprocessor Architecture
	Multiphoton Absorption
	Multiple-Period Average
	Multiple Peripheral Adapter
	Multipoint Asynchronous
	Multi-Purpose Additive
3-MPA	3-Methoxypropylamine
MPa	Megapascal
mPa	millipascal
MPAA	Motion Picture Association of America
MPAcc	Master of Professional Accountancy
	Master of Professional Accounting [also: MPAcct]
MPACS	Management Planning and Control System
MPAD	Maximum Permissible Accumulated Dose
MPAff	Master of Public Affairs
MPAI	Maximum Permissible Annual Intake
MPAR	Microprogram Address Register
MPAS	Metal Powder Association Standard
MPAURP	Master of Public Affairs and Urban and Regional Planning
MPB	Material Performance Branch [of US Air Force]
	Methoxypropenylbenzene
	Methylpropylbenzene
	Morphotropic Phase Boundary
MPBB	Maximum Permissible Body Burden
MPBE	Molten Plutonium Burn-Up Experiment
MPBP	(International Union of) Metal Polishers, Buffers and Platers
MPBW	Ministry of Public Buildings and Works [UK]
MPC	Manitoba Press Council [Canada]
	Marker Pulse Conversion
	Master of Public Communication
	Materials Preparation Center [of DOE, US]
	Materials Processing Center [of MIT, US]
	Materials Properties Council [US]
	Maximum Permissible Concentration
	Medium Processing Channel Black
	Metal Properties Council [now: Materials Properties Council, US]
	Methylpropylcarbinol
	Microprocessor Control
	Microprogram Control
	Miniature Protector Connector
	Modular Peripheral-Interface Converter
	Multiplier Photocell
	Multipurpose Communications
MPCA	Methylphenylcarbinol Acetate
MPCAG	Military Parts Control Advisory Group
MPCC	Multiprotocol Communications Controller
MPCD	Master in Planning and Community Development
	Maximum Permissible Cumulative Dose
MPCI	Multiport Programmable Communications Interface
MPCM	Microprogram Control Memory
MPCS	Multi-Party Connection Subsystem
MPD	Magnetoplasmadynamics
	Map Pictorial Display
	Master of Planning Design
	Master of Product Design
	Materials Physics Division [of US Air Force]
	Materials Property Data
	Maximum Permissible Dose
	Multiphoton Decomposition
MPDA	Metaphenylene Diamine
MPDC	Mechanical Properties Data Center [US]
MPDE	Maximum Permissible Dose Equivalent
MPDI	Marine Products Development Irradiator
MPDR	Microwave Plasma Disk Reactor
MPDS	Methylpolydisilylazane
	Multi-Purpose Display System
MPDSM	Micro Powder Diffraction Search/Match
MP-DV	Multiply-Divide
MPE	Mathematical and Physical Sciences and Engineering
	Mannesmann-Pfannen-Entschwefelung [= Mannesmann Ladle Desulfurization Process]
	Marine Port Engineer
	Master Process Engraver
	Motion Picture Engineer
	Motion Picture Engineering
	Maximum Permissable Exposure
	Memory Parity Error
	Methylphenylether
	Microwave Plasma Etching
	Multiprogramming Executive
MPeEng	Master of Petroleum Engineering [also: MPetE or MPetEng]
MPEG	Mixtures of Polyethyleneglycol Monomethylether
MPEMA	2-Ethyl-2-(P-Tolyl)malonamide
	M-Phenylethylmalonamide
MPEP	Manual of Patent Examining Procedures

MP/EP Missing Pulse/Extra Pulse
MPER Master of Personnel and Employees Relations
MPetE Master of Petroleum Engineering [also: MPetEng or MPeEng]
MPF Multiphase Flow
MPFC Multi-Point Fuel Injection
MPFM Master of Public Financial Management
MPFP Melt Processible Fluoropolymer
MPFS Minimum Physical Fitness Standards
MPG Max-Planck-Gesellschaft [= Max Planck Society, Germany]
Microwave Pulse Generator
Miles per Gallon [also: mpg]
Miniature Precision Gyrocompass
MPGS Microprogramming Generating System
MPH Master of Public Health
Miles per Hour [also: mph]
Moving Plasmoid Heater
MPh Master of Philosopy
MPharm Master of Pharmacy
MPHE Master of Public Health Education
MPHEng Master of Public Health Engineering
MPhil Master of Philosopy
MPho Master of Photography
MPhys Master of Physics [also: MPhy]
MPI Magnesium Products Industries
Magnetic Particle Inspection
Mannose Phosphate Isomerase
Max Planck Institute [Germany]
Mean Point of Impact
Meeting Planners International [US]
Microprocessor Interface
Multi-Point Injection
MPIA Master of Public and International Affairs
Multiple Peak Intensity Analysis
MPIAS Max Planck Institute for the Advancement of Science [Germany]
MPIC Message Processing Interrupt Count
MPID Maximum Permissible Integrated Dose
MPIF Metal Powder Industries Federation [US]
MPIO Mission and Payload Integration Office [of NASA]
MPIP Miniature Precision Inertial Platform
Multiple Picture-in-Picture
MPIS Multilateral Project Information System
MPK Methyl Propyl Ketone
MPL Maximum Permissible Level
Maximum Permissible Limit
Maximum Print Line
Message Processing Language
Microprogramming Language
Multiple
Multipurpose Processing Language
Multischedule Private Line
Multischedule Private Link
MPl Master of (Urban) Planning
MPLA Mask Programmable Logic Array
MPLC Medium-Pressure Liquid Chromatography
μPLD Microcomputer Programmable Logic Device
MPLEA Multipurpose Long Endurance Aircraft
MP-LPC Multipulse Linear Predictive Coding
MPLR Maximum Permissable Leakage Rate
Mininum Pressure Live Roller (Conveyor)
MPLX Multiplex
Multiplexer [also: MPLXR]
MPM Magnetic Property Measurement
Master of Personnel Management
Master of Pest Management
Master of Professional Management
Master of Public Management
Mechanical Properties Microprobe
Microphotometer
Microprocessor Module
Microprogram Memory
Microscope Photometry
Monocycle Position Modulation
Multiprogramming Monitor
Multistand Pipe Mill
MP/M Multiprogramming Control Program for Microprocessors
mpm meters per minute [also: MPM or m/min]
MPMA Montford Point Marine Association [US]
MPMC Magnetic Property Measurement System
Microprogram Memory Control
MPMS Magnetic Property Measurement System
MPN Most Probable Number
Multiprocessor Network
MPO Manufacturing Production Order
Maximum Power Output
Memory Protect Override
Mono Power Amplifier
Myeloperoxidase
MPOD Mean Planned Outage Duration
MPP Massive Parallel Processor
Memory Parity and Protect
Message Processing Program
Microwave Processing Project
Mission Planning Program
Moly-Permalloy Powder
Most Probable Position
Multiphoton Process
M&PP Materials and Plant Protection
MPPA Maritime Professional Photographers Association
MPPH 5-(4-Methylphenyl)-5-Phenylhydantoin
MPPL Multipurpose Processing Language
MPPPM Master of Plant Protection and Pest Management
MPPS Most Penetrating Particle Size
MPR Materials and Processing Report
Mechanical Pressure Regulator
Message Processing Region
Microseismic Processor Recorder
Multiplanar Reconstruction
Multiplier
MPrA Master of Professional Accountancy [also: MPrAcc]
MPRE Medium Power Reactor Experiment
Multipurpose Research Reactor
MPRF Multipole Radiation Field
MPrGph Master of Professional Geophysics
MPrMet Master of Professional Meteorology
MPROM Mask Programmable Read-Only Memory
MPrMet Master of Professional Meteorology
MPRTM Master of Parks, Recreation and Tourism Management

MPS Manpower System
Master of Professional Studies
Master of Public Service
Master Production Schedule
Master Productivity Specialist
Mathematical Programming Society [Netherlands]
Mathematical Programming System
Materials Processing in Space
Max Planck Society [also: MPG, Germany]
Median Period of Survival
Member of the Pharmaceutical Society
Meters per Second [also: mps, m/sec or m/s]
Methylacetylene Propadiene Stabilized
Methyl Phenyl Sulfide
Microanalysis Particle Sampler
Microprocessor System
Mineral Processing System
Minimum Performance Specification
Mission Preparation Sheet
Mucopolysaccaride
Multiple Partition Support
Multiple Protective Structure
Multiprocessing System
Multiprogramming Periodic Tasking System
Multiprogramming System
MPSI Thousand Pounds per Square Inch
MPSK Multiple Phase Shift Keying
MPSX Mathematical Programming System, Extended
MPT Male Pipe Thread
Materials Processing Technology
Memory Processing Time
Metallurgical Plant and Technology
Microprogramming Technique
Minimum Pressurization Temperature
Ministry of Post and Telecommunications [Japan]
Mouse-Protection Test
Multiple Pure Tones
MPTA Main Propulsion Test Article
Mechanical Power Transmission Association
Metal Powder Technology Association
MPTD Dimethyl Diphenyl Thiuram Disulfide
MPTED Mechanical Power Transmission Equipment Distributors Association [now: PTDA, US]
MPTI Mechanical Power Transmission Association [US]
MPTP 1-Methyl-4-Phenyl-1,2,3,6-Tetrahydropyridine
MPTR Mobile Position Tracking Radar
MPTS Mobile Photographic Tracking Station
MPU Memory Protection Unit
Microprocessing Unit
MPubAdm Master of Public Administration
MPUrbDes Master of Planning and Urban Design
M-PVC Mass Polyvinyl Chloride
MPVM Master of Preventive Veterinary Medicine
MPW Male Pipe Weld
MPW Master of Public Works
Modified Plane Wave
MPWD Machine-Prepared Wiring Data
M-PYROL 1-Methyl-2-Pyrrolidinone
MPX Mapped Programming Executive
Multiplexer
Multiplexing
Multiprogramming Executive
MPX/R Relay Multiplexer
MPY Multiply
MQ Multiplier/Quotient
MQC Manufacturing Quality Control
MQE Message Queue Element
MQORC Modified Quadrature Overlapped Raised Cosine
MQR Multiplier-Quotient Register
MQS Metastable Quenching Spectroscopy
Multiprogrammed Queued Tasking System
MQW Multiple Quantum Well
Multiquantum Well
MR Machine Records
Machine Rifle
Magazine Records
Magnetic Recording
Magnetic Resistance
Magnetic Resonance
Magnetoresistance
Magnetoresistive
Magnetorheology
Map Reference
Marble
Mask Register
Matching Record
Materials Research
Mauritian Rupee [Currency]
Measurement Range
Mechanical Rectifier
Medium Reduction
Memorandum Report
Memory Reclaimer
Memory Register
Mercury-Redstone [of NASA]
Message Rate
Message Repeat
Metal Rectifier
Methylresorcinol
Microradiography
Mill Run
Mineral Rubber
Mine Run
Miscellaneous Report
Moderating Ratio
Modified Read
Modular Redundancy
Moisture Resistance
Molasses Residuum
Molecular Rearrangement
Molecular Recognition
Monitor Recorder
Moving Range
Multiple Request
Multiplier Register
Multipoint Recorder
mR milliroentgen [also: mr]
M$ Malaysian Ringgit [Currency]
MRA Machine Readable Archives Division [of PAC, Canada]
Manufacturers Representatives of America [US]
Master of Recreational Administration

Master of Resource Administration
Microgravity Research Association
Minimum Reception Altitude
MRAA Marine Retailers Association of America [US]
MRAALS Marine Remote Approach and Landing System
MRAC Meter-Reading Access Circuit
Model Reference Adaptive Control
MRAD Mass Random Access Disk
MRad Master of Radiology
mrad millirad [Unit]
MRADS Mass Random Access Data Storage
MRaEng Master of Radio Engineering
MRAIC Member of the Royal Architectural Institute of Canada
MR-ATOMIC Multiple Rapid Automatic Test of Monolithic Integrated Circuit
MRB Magnetic Recording Borescope
Maintenance Review Board [UK]
Malaysian Rubber Bureau
Microcircuit Reliability Bibliography [of RADC, US]
MRBM Medium-Range Ballistic Missile
MRC Machine Readable Code
Maintenance Requirement Card
Manitoba Research Council [Canada]
Manufacturing Research Corporation [Canada]
Marine Resources Council [US]
Materials Research Center
Mathematics Research Center
Medical Research Center
Medical Research Council [UK and Canada]
Memory Request Controller
Meteorological Research Committee [UK]
Mount Royal College [Canada]
Multiple Register Counter
MRCA Multirole Combat Aircraft
MRCC Medical Research Council of Canada
MRCM Master of Real Estate and Construction Management
MRCO Manufacturing Research Corporation of Ontario [Canada]
MRCP Master of Regional and City Planning
Master of Regional and Community Planning
MRCPath Member of the Royal College of Pathologists
MRcPk Master of Recreation and Parks
MRCS Multiple Report Creation System
MRCV Multirole Combat Vehicle
MRD Manual Ringdown
Maritime Research Department [of WINA, US]
Materials Research Division
Microbiological Research Department
Multiple-Reference Double Excitation
Multiresponse Data
MRDA Media Research Directors Association [US]
MRDE Mining Research and Development Establishment [UK]
Multiple-Reference Double Excitation
MRDF Machine-Readable Data File
Moisture-Resistant Densified Fuel
MR&DF Malleable Research and Development Foundation [US]
MR-DOS Mapped Real-Time Disk-Operating System [also: MRDOS or MR DOS]
MRE Microbiological Research Establishment [of MOD, UK]
Mining Research Establishment [of NCB, UK]
Multiple-Response Enable
Microbiological Research Establishment [UK]
MRED Master of Real Estate Development
MRELB Malaysian Rubber Exchange and Licensing Board
MREM Minerals Resources Engineering and Management
MRERF Manufacturers Representatives Educational Research Foundation [US]
MRF Medical Research Foundation [France]
Message Refusal
Multipath Reduction Factor
MRFAC Manufacturers Radio Frequency Advisory Committee [US]
MRFT Material Removal and Forming Technology
MRG Materials Research Grant
Materials Research Group
Medium Range
Methane-Rich Gas
MRGA Manhattan Ryegrass Growers Association [US]
MRGL Marginal
MRGS Mesabi Range Geological Society
MRH Mechanical Recording Head
Mobile Remote Handler
MSH (= Melanocyte-Stimulating Hormone) Releasing Hormone
μ-RHEED Micro Reflected High Energy Electron Diffraction
MRI Machine Records Installation
Magnetic Resonance Imaging
Magnetic Rubber Inspection
Maritime Research Institute [Netherlands]
Master of the Royal Institute
Material Receiving Instruction
Mean Recurrence Interval
Medical Research Institute [US Navy]
Member of the Royal Institute
Memory Reference Instruction
Metal Recovery Industry
Midwest Research Institute [US]
Miscellaneous Radar Input
Monopulse Resolution Improvement
MRIH MSH (= Melanocyte-Stimulating Hormone) Release-Inhibiting Hormone
MRIIUS Metal Recovery Industries Inc. of the United States
MRINDO Modified Rydberg Intermediate Neglect of Differential Orbital
MRIR Market Research Information System
Medium-Resolution Infrared Radiometer
MRIS Maritime Research Information Service [of National Academy of Sciences, US]
MRJE Multiple Remote Job Entry
MRK Mark [Former Currency of East Germany]
MRKD Marked
MRL Machine Representational Language
Manufacturing Reference Line
Materials Research Laboratories [of NSF, US]
Materials Research Laboratories [of ITRI, Taiwan]
Medical Research Laboratory

	Mining Research Laboratories
	Multiple Rocket Launcher
	Multipoint Recorder/Logger
MRM	Master of Resource Management
	Metabolic Rate Monitor
	Militarized Reconfigurable Multiprocessor [of DND, Canada]
mR/min	milliroentgen per minute [Unit]
MRMU	Mobile Radiological Measuring Unit
	Mobile Remote Manipulating Unit [US Air Force]
MRN	Meteorological Rocket Network [of NASA]
	Minimum Reject Number
mRNA	Messenger Ribonucleic Acid [also: M-RNA, or m-RNA]
MRNG	Morning
MRO	Maintenance, Repair and Operating
	Mauritanian Ouguiya [Currency]
	Memory Resident Overlay
	Midrange Objectives
MROM	Macro Read-Only Memory
MRP	Machine-Readable Passport
	Malfunction Reporting Program
	Management Resource Planning
	Manufacturer's Recommended Price
	Manufacturing Requirements Planning
	Manufacturing Resources Planning
	Master in Regional Planning
	Materials Requirements Planning
	Materials Requirement Program
	Message Routing Process
	Metal Refining Process
	Meteorological Reporting Point
MRPC	Mercury Rankine Power Conversion [of USAEC]
MRPPS	Maryland Refutation Proof Procedure System
MRPRA	Malaysian Rubber Producers Research Association
MRPS	Manufacturing Resource Planning System
MRQC	Microelectronics Reliability Quality Center [of Sandia National Laboratories, US]
MRR	Mechanical Reliability Report [of FAA, US]
	Medical Research Reactor [of BNL, US]
	Material Removal Rate
	Multiple Response Resolver
	Multiple Restrictive Requirement Quality
MRRC	Mineral Resources Research Center [of University of Minnesota, US]
MRRDB	Malaysian Rubber Research and Development Board
MRRV	Maneuvering Reentry Research Vehicle
MRS	Magnetic Resonance Spectroscopy
	Management Reporting System
	Manned Repeater Station
	Market Research Society [UK]
	Materials Research Society [US]
	Mathematics Research Center
	Mean Relative Strain
	Media Resource Service [of SIPI, US]
	Medical Record System [UK]
	Medium-Range Search
	Mobile Radio Station
	Mobile Roof Support
	Monitored Retrievable Storage [Radioactive Waste]
MRSA	Mandatory Radar Service Area
MRSC	Marine Rescue Subcenter
	Member of the Royal Society of Canada
MRSH	Materials Referral System and Hotline [of MPC, US]
MRS-I	Materials Research Society of India
MRS-J	Materials Research Society of Japan
MRS-Korea	Materials Research Society of Korea
MRS-T	Materials Research Society of Taiwan
MRS-Taiwan	Materials Research Society of Taiwan
MRT	Mean Radiant Temperature
	Mean Repair Time
	Medical Radiation Technologist
	Medical Radiation Technology
	Mobile Radar Target
	Modified Rhyme Test
	Multiple Requesting Terminal Program
MRTD	Minimum Resolvable Temperature Difference
MRTG	Mortgage
MRTM	Maritime
MRTP	Master of Rural and Town Planning
MRTS	Microwave Repeater Test Set
MRU	Machine Records Unit
	Material Recovery Unit
	Message Retransmission Unit
	Mobile Radio Unit
	Mountain Rescue Unit
MRV	Maneuverable Reentry Vehicle
	Maximum Relative Variation
MRWC	Multiple Read-Write Compute
MS	Machine Selection
	Machine Steel
	Macromodular System
	Macroscopic Segregation
	Magnetic Saturation
	Magnetic Screening
	Magnetic Separation
	Magnetics Society [US]
	Magnetic Storage
	Magnetostatic(s)
	Magnetostriction
	Mail Steamer
	Main Storage
	Main Switch
	Maintenance Schedule
	Management Science
	Manganese Steel
	Manuscript
	Margin of Safety
	Mark Sense
	Mass Spectrography
	Mass Spectrometer
	Mass Spectroscopy
	Mass Spectrum
	Mass Storage
	Master of Science
	Master Seaman
	Master Switch
	Material Specification
	Materials Science
	Mathematical Series
	Maximum Stress
	Mean Square

	Mean Sun
	Measured Service (Pricing)
	Mechanized Scheduling
	Medical Survey
	Medium Setting
	Medium Steel
	Megasample
	Melt Spinning
	Memory Stack
	Memory System
	Mercury Switch
	Mesa
	Messaging System
	Metallurgical Society [US]
	Metastable
	Metastability
	Meteoritical Society [US]
	Methanesulfonate Salt
	Metric System
	Microspectrometry
	Microspherical Silica Alumina
	Microstrip
	Mild Steel
	Milestone
	Military Secretary
	Military Standard
	Millisecond
	Mineralogical Society [UK]
	Minesweeper
	Mississippi [US]
	Mobile Service
	Mobile System
	Molar Substitution
	Molecular Spectroscopy
	Molecular Spectrum
	Molecular Stuffing
	Moles of Substituent
	Monitoring System
	Mossbauer Spectroscopy
	Motion Sensor
	Motor Ship
	Motor-Starter
	Multichannel (Television) Sound
	Multispectral
	Murashige and Skoog Media
	Music Synthesis
M&S	Maintenance and Supply
M/S	Mark-to-Space Ratio
Ms	Megasecond
ms	mesa
	millisecond
	months after sight
m/s	meters per second [also: MPS, mps or m/sec]
ms^{-1}	meter per second
μs	microsecond
MSA	Mass Storage Adapter
	Master of Science in Accounting
	Master of Science in Administration
	Master of Science in Agriculture
	Material Surveillance Assembly
	Mean-Sphere Approximation
	Mechanical Signature Analysis
	Message System Agent
	Methanesulfonic Acid
	Microgravity Science and Applications
	Mineralogical Society of America [US]
	Mine Safety Appliance
	Minimum Safe Altitude
	Minimum Sector Altitude
	Modem Signal Analyzer
	Most Seriously Affected
	Multichannel Sound Adapter
	Multiplication Stimulating Activity
	Multi-Subsystem Adapter
	Mutual Security Agency [US]
	Mycological Society of America [US]
MSAA	Master of Science in Astronautics and Aeronautics
MSAAE	Master of Science in Aeronautical and Astronautical Engineering
MSAcct	Master of Science in Accounting
MSAD	Microgravity Science and Applications Division [of NASA]
MSAdm	Master of Science in Administration
MSAE	Master of Science in Aerospace Engineering
	Master of Science in Architectural Engineering
MSAeE	Master of Science in Aerospace Engineering [also: MSAeEng]
MSAg	Master of Science in Agriculture
MSAgE	Master of Science in Agricultural Engineering [also: MSAgEng]
MSAM	Master of Science in Applied Mathematics
	Multiple Sequential Access Method
MSAN	Multiport Ship Agencies Network
MSAOR	Master of Science in Applied Operations Research
MSAP	Master of Science in Applied Physics
	Minislotted Alternating Priorities
MSAR	Mine Safety Appliance Research
MSArch	Master of Science in Architecture
MSArchSt	Master of Architectural Studies [also: MSArchStud]
MSArchTech	Master of Science in Architectural Technology
MSAT	Master of Science in Advanced Technology
	Mobile Satellite Program [of USDOC]
	Multipurpose Satellite
MSATA	Motorcycle, Scooter and Allied Trades Association [US]
MSAUSC	Muslim Students Association of the United States and Canada
MSAW	Minimum Safe Altitude Warning
MSB	Master of Science in Business
	Mass Spectrometry Bulletin (Database) [of MSDC, US]
	Methylstyrylbenzene
	Mining Standards Board
	Most Significant Bit
MSBA	Master of Science in Business Administration
	Medium and Small Business Administration [of MOEA, Taiwan]
MSBE	Master of Science in Biomedical Engineering
	Molten Salt Breeder Experiment
MSBLS	Microwave Scan Beam Landing System
MSBM	Master of Science in Business Management
MSBME	Master of Science in Biomedical Engineering

MSBR Maximum Storage Bus Rate
Molten Salt Breeder Reactor
MSBusEd Master of Science in Business Education
MSBusMgt Master of Science in Business Management
MSBVW Magnetostatic Backward Volume Wave
MSBY Most Significant Byte
MSC Macro Selection Compiler
Main Switching Center
Manitoba Safety Council [Canada]
Manned Spacecraft Center [of NASA]
Manpower Services Commission [UK]
Marine Corps School [US]
Mass Storage Controller
Master of Science in Commerce
Mediterranean Society of Chemotherapy [Italy]
Mediterranean Sub-Commission [of FAO — UN]
2-Mesitylenesulfonyl Chloride
Message Sequence Chart
Message Switching Center
Message Switching Computer
Message Switching Concentration
Metallizing Service Contractors [Name Changed to: International Thermal Spray Association]
Meteorological Service of Canada
Methylamines Sector Group [of CEFIC, Belgium]
Microscopical Society of Canada
Microsystems Center [UK]
Miles of Standard Cable
Military Sealift Command
Minesweeper Coastal
Minimum Spanning Circle
Mobile Servicing Center [of Mobile Servicing System]
Mobile Switching Center
Monolithic Crystal Filter
Most Significant Character
Motor Speed Changer
Motor Speed Control(ler)
Multistage Compression
Multistrip Coupler
Multisystem Coupling
MSc Master of Science
MSCA Mixed Spectrum Critical Assembly [US]
MScA Master of Applied Science
MScAdm Master of Science in Administration
MScBusAdmin Master of Science in Business Administration
MSCC Manned Spaceflight Control Center [US Air Force]
MScC Master of Science in Commerce
MScCE Master of Science in Civil Engineering
MScComm Master in Commercial Science
MScCS Master of Science in Computer Sciences
MSCE Main Storage Control Element
Master of Science in Civil Engineering
Master of Science in Clinical Engineering
Master of Science in Computer Engineering
MSCED Master of Science in Community Economic Development
MScEd Master of Science in Education
MSCEM Master of Science in Civil Engineering Management
MScEng Master of Science in Engineering [also: MScE]
MSCer Master of Science in Ceramics
MSCerEng Master of Science in Ceramic Engineering [also: MSCerE]
MSCerSc Master of Science in Ceramic Science [also: MSCerS or MSCerSci]
MSCF Thousand Standard Cubic Feet
MScF Master of Science in Forestry
MSCFD Thousand Standard Cubic Feet per Day
MScFE Master of Science in Forest Engineering
MSCFH Thousand Standard Cubic Feet per Hour
MSCFM Thousand Standard Cubic Feet per Minute
MSChemEng Master of Science in Chemical Engineering [also: MSChemE or MSChE]
MSCI Molten Steel Coolant Interaction
MSCIS Master of Science in Computer Information Science
Master of Science in Computer Information Systems
MSCLS Master of Science in Clinical Laboratory Studies
MScMnlProcEng Master of Science in Mineral Processing and Engineering [also: MScMnlProcE]
MScMetEng Master of Science in Metallurgical Engineering [also: MScMetE]
MSContEd Master of Science in Continuing Education
MSConv Master of Science in Conservation
MSCP Mass Storage Control Protocol
Mean Spherical Candlepower
Motor Short-Circuit Protector
MSCpE Master of Science in Computer Engineering
MScPharm Master of Science in Pharmacy [also: MScPhm]
MScPl Master of Science in Planning
MSCR Machine Screw
MSCRP Master of Science in City and Regional Planning
Master of Science in Community and Regional Planning
MSCS Management Scheduling and Control System
Mass Storage Control System
Master of Science in Computer Science
MSCSE Master of Science in Computer Science and Engineering
MScStud Master of Science Studies [also: MSciStud or MScSt]
MSCT Mass Storage Control Table
MScT Master of Science in Teaching
MSCTC Mass Storage Control Table Create
MScTech Master of Science and Technology
MSCU Modular Store Control Unit
MSCW Marked Stack Control Word
MSD Machining Systems Division
Main Storage Database
Mass Selective Detector
Mass Spectrometry Detector
Master of Science in Teaching
Materials Science Division
Mean Solar Day
Ministry Skills Development
Missile Systems Division [US Air Force]
Modem Sharing Device
Most Significant Digit
Multi-Sensor Display
MSDB Main Storage Database
MSDC Mass Spectrometry Data Center [of RSC, UK]
MSDD Master of Science in Design and Development

MS-DOS Microsoft — Disk Operating System [also: MSDOS, or MS DOS]
MSDS Marconi Space and Defense System
Master of Science in Decision Systems
Material Safety Data Sheet
Message Switching Data Service
MSDT Maintenance Strategy Diagramming Technique
MSE Master of Sanitary Engineering
Master of Science Education
Master of Science in Education
Master of Science in Engineering
Master of Software Engineering
Materials Science and Engineering
Mean Square Error
Mechanical Systems Engineering
Midwest Stock Exchange [US]
Milwaukee School of Engineering [US]
Modern Shipping Equipment
Montreal Stock Exchange [Canada]
Multistage Expansion
MS&E Materials Science and Engineering
MSEA Mass Specific Energy Absorption
MSEC Master of Science in Economic Aspects of Chemistry
Msec Megasecond
msec millisecond
μsec microsecond
m/sec meters per second [also: MPS, mps or m/s]
MSEcon Master of Science in Economics
MSEd Master of Science in Education
MSED Minimum Signal Element Duration
Ministry of State for Economic Development
MSEE Master of Science in Electrical Engineering
Master of Science in Environmental Engineering
Mean Square Error Efficiency
MSEH Master of Science in Environmental Health
MSEI Mean Square Error Inefficiency
MSEM Master of Science in Engineering Management
Master of Science in Engineering of Mines
MSEMech Master of Science in Engineering Mechanics
MSEMgt Master of Science in Engineering Management
MSEMPR Missile Support Equipment Manufacturers Planning Reports
MSEMRL Materials Science and Engineering and Materials Research Laboratory [of University of Illinois, US]
MSEng Master of Science in Engineering [also: MSEngr or MSE]
MSEngSc Master of Science in Engineering Sciences
MSEnvEng Master of Science in Environmental Engineering [also: MSEnvE]
MSEnvSc Master of Science in Environmental Science
MSEP Manufacturing Systems Engineering Program [US]
MSER Master of Science in Energy Resources
MSERD Ministry of State for Economic and Regional Development
MSES Master of Science in Engineering Science
Master of Science in Environmental Studies
Minnesota Surveyors and Engineers Society [US]
MSESM Master of Science in Environmental Systems Management
MSET Master of Special Education Technology

MSF Magnetostrictive Force
Mass Storage Facility
Master of Science in Forestry
Master of Science in Finance
Master Source File
Medium Scale Integration
Metascience Foundation [US]
Michigan Society of Fellows [US]
Motorcycle Safety Foundation [US]
Multiscan Function
Multistage Flash
MSFC Marshall Spaceflight Center [of NASA]
MSFD Mixed Spectral Finite Difference
Multistage Flash Distillation
MSFH Manned Spaceflight Headquarters [of NASA]
MSFM Master of Science in Financial Management
MSFN Manned Spaceflight Network [of NASA]
MSFS Master of Science in Forensic Science
Main Steam and Feedwater System
MSFuelSc Master of Science in Fuel Science
MSFVW Magnetostatic Forward Volume Wave
MSG Manufacturers Standard Gauge
Mapper Sweep Generator
Message
Miscellaneous Simulation Generator
Missing
Modular Steam Generator
Monosodium Glutamate
MSGBI Mineralogical Society of Great Britain and Ireland
MSGE Master of Science in Geological Engineering [also: MSGeoE or MSGeoEng]
MSGS Message Switch
MSG/WNG Message Warning
MSH Master of Science in Horticulture
Master of Science in Hygiene
Melanocyte-Stimulating Hormone
Mesitylenesulfonyl Hydrazide
MSHA Mine Safety and Health Administration [US]
Mine Safety and Health Agency
MSHort Master of Science in Horticulture
MSHR Master of Science in Human Resources
MSHRM Master of Science in Human Resource Management
MSHRMD Master of Science in Human Resource Management and Development
MSHS Master of Science in Health Systems
Master of Science in Health and Safety
MSHyg Master of Science in Hygiene
MSI Medium-Scale Integration
Metal-Support Interaction
MSIA Master of Science in Industrial Administration
Master of Science in International Administration
Master of Science in International Affairs
MSIB Master of Science in International Business
MSIE Master of Science in Industrial Engineering
MSIH Master of Science in Industrial Hygiene
MSILR Master of Science in Industrial and Labour Relations
MSIM Master of Science in Information Management
MSIN Multi-Stage Interconnection Network
MSIO Mass Storage Input/Output
MSIP Magnetron Sputtering Ion Plating

MSIPA	Master of Science in International Public Administration
MSIR	Master of Science in Industrial Relations
MSIS	Master of Science in Information Science
	Master of Science in Information Systems
	Master of Science in Interdisciplinary Studies
	Manned Satellite Inspection System
	Manufacturing Systems Integration Service
MSIST	Master of Science in Information Systems Technology
MSIRI	Mauritius Sugar Industry Research Institute
MSIV	Main Steam Isolation Valve
MSIVLCS	Main Steam Isolation Valve Leakage Control System
MSJ	Master of Science in Journalism
	Mathematical Society of Japan
MSK	Mask
	Minimum (Phase) Shift Keying
MSKM	Minimum Shift Keyed Modulation
MSL	Machine Specification Language
	Master of Science Librarianship
	Master of Science in Linguistics
	Map Specification Library
	Materials Science Laboratory
	Mean Sea Level [also: msl]
	Mineral Sciences Laboratories [of CANMET, Canada]
MSLD	Mass Spectrometer Leak Detector
MSLIS	Mita Society of Library Information Science [Japan]
MSLS	Master of Science in Library Science
MSM	Mackay School of Mines [of University of Nevada at Reno, US]
	Master of Science in Management
	Master of Service Management
	Master Slave Manipulator
	Memory Storage Module
	Message Stream Modification
	Metal-Semiconductor-Metal
	Modified Source Multiplication
	Mystic Seaport Museum [US]
	Thousand Feet Surface Measure
MSMA	Margarine and Shortening Manufacturers Association [UK]
	Monosodium Acid Methanearsonate
	Murashige and Skoog Shoot Multiplication, Medium A
MSMatEng	Master of Science in Materials Engineering [also: MSMatE]
MSMatSc	Master of Science in Materials Science [also: MSMatSci or MSMatS]
MSMatScEng	Master of Science in Materials Science and Engineering [also: MSMatSciEng or MSMatSE]
MSMB	Mass-Spectrometric Molecular Beam
	Mechanical Standards Management Board [of ANSI, US]
	Murashige and Skoog Shoot Multiplication, Medium B
MSMC	Master of Science in Marketing Communications
	Master of Science in Mass Communications
	Murashige and Skoog Shoot Multiplication, Medium C
MSMCS	Master of Science in Management and Computer Science
MSME	Master of Science in Mechanical Engineering
MSMER	Master of Science in Management/Employee Relations
MSMet	Master of Science in Metallurgy
MSMetEng	Master of Science in Metallurgical Engineering [also: MSMetE]
MSMetMatS	Master of Science in Metallurgy and Materials Science [also: MSMetMatSc]
MSMetr	Master of Science in Meteorology
MSMetS	Master of Science in Metallurgical Sciences [also: MSMetSc]
MSMFSE	Master of Science in Manufacturing Systems Engineering
MSMgt	Master of Science in Management
MSMinE	Master of Science in Mining Engineering [also: MSMinEng]
MSMinrlE	Master of Science Mineral Engineering [also: MSMinrlEng]
MSMIS	Master of Science in Management Information Systems
MSMLCS	Mass Service Main Line Cable System
MSMM	Master of Science in Manufacturing Management
MSMnlProcEng	Master of Science in Mineral Processing and Engineering [also: MSMnlProcE]
MSMO	Murashige and Skoog Medium with Minimal Organics
MSMOB	Master of Science in Management and Organizational Behavior
MSMPA	Master of Science in Management/Public Administration
MSM-PD	Metal-Semiconductor-Metal Photodiode
MSMS	Master of Science in Management Science
MSMSA	Master of Science in Management Systems Analysis
MSMSc	Master of Science in Management Science
MSMSE	Master of Science in Manufacturing Systems Engineering
	Master of Science in Materials Science and Engineering
MSMT	Master of Science in Medical Technology
MSMtE	Master of Science in Materials Engineering [also: MSMtEng]
MSMV	Monostable Multivibrator
MSN	Microwave Systems News
MSNE	Master of Science in Nuclear Engineering
MSNF	Multisystem Networking Facility
MSNJ	Medical Society of New Jersey [US]
MSNS	Master of Science in Natural Science
	Mining Society of Nova Scotia [Canada]
MSNT	1-(Mesitylene-2-Sulfonyl)-3-Nitro-1,2,4-Triazole
MSNucE	Master of Science in Nuclear Engineering [also: MSNucEng]
MSO	Magnetostriction Oscillator
	Methionine Sulfoximine
	Multiple System Operator
	Multistage Optimization
MSOB	Master of Science in Organizational Behavior
MSOC	2-Methylsulfonylethyloxycarbonyl
MSOD	Master of Science in Organizational Development
MSOE	Master of Science in Ocean Engineering

MSOM Master of Science in Organization and Management
MSOR Master of Science in Operations Research [also: MSOpR or MSOpRes]
MSORS Mechanized Sales Office Record System
MSOS Mass Storage Operating System
MSOSH Master of Science in Occupational Safety and Health
MSOT Master of Science in Occupational Therapy
MSP Maintenance Support Plan
Maintenance Surveillance Procedure
Mass Storage Processor
Medium Speed Printer
Microelectronics Support Program
Microscope System Processor
Modular Switching Peripheral
Modular System Program
Monobasic Sodium Phosphate
Monosodium Phosphate
Munitions Supply Program
Msp Moraxella Species [Molecular Biology]
MSPetEng Master of Science in Petroleum Engineering [also: MSPetE]
MSPH Master of Science in Public Health
MSPharm Master of Science in Pharmacy
MSPHE Master of Science in Public Health Engineering
MSPHR Master of Science in Pharmacy
MSPhys Master of Science in Physics
MSPhysOp Master of Science in Physiological Optics
MSPlmSc Master of Science in Polymer Science
MSPLT Master Source Program Library Tape
MSPN Multi-Stage Packet Network
MSPoly Master of Science in Polymers
MSPR Master Spares Positioning Resolver
Minimum Space Platform Rig
MSPS Master of Science in Planning Studies
MSR Machine Status Register
Machine Stress Rating
Magnetic Slot Reader
Malaysian Society of Radiographers
Mark Sense Reading
Mass Storage Resident
Material Status Report
Memory Select Register
Misroute
Missile Site Radar
Molten Salt Reactor
MS(R) Master of Science (Research)
MSRad Master of Science in Radiology
MSRadSci Master of Science in Radiation Science [also: MSRadSc]
Master of Science in Radiological Science [also: MSRadSc]
MSRC Master of Science in Resource Conservation
MSRE Molten Salt Reactor Experiment [of ORNL, US]
MSREF Metal Scrap Research and Education Foundation
MSRL Marine Science Research Laboratory [of MUN, Canada]
MSRM Main Steam Radiation Monitor
MSRMP Master of Science in Radiological Medical Physics
MSRP Missile, Space and Range Pioneers [US]
MSRS Main Steam Radiation System
Multiple Stylus Recording System
MSRT Missile System Readiness Test
MSS Main Steam System
Main Support Structure
Management Science Systems
Management Statistics Subsystem
Manufacturers Standardization Society (of the Valve and Fittings Industry) [US]
Manuscripts
Mass Storage Subsystem
Mass Storage System
Master of Science in Safety
Military Supply Standard [of USDOD]
Mixed Spectrum Superheater [US]
Mobile Satellite Service
Mobile Satellite System
Mobile Servicing Station
Mobile Servicing System [Canada]
Music Sensor System
Multispectral Scanner
Multitask Single-Stream System
MS/s Megasamples per second
MSSC Mass Storage System Communication
MSSc Master of Sanitary Science
MSSCE Mixed Spectrum Superheater Critical Experiment
MSSE Master of Science in Sanitary Engineering [also: MSSEng]
Master of Science in Systems Engineering
MSSG Message
MSSJ Mass Spectroscopy Society of Japan
MSSM Master of Science in Science Management
Master of Science in Systems Management
MSSR Mars Soil Sample Return
Mixed Spectrum Superheat Reactor
Monopulse Secondary Surveillance Radar
MSSS Mass Spectral Search System
Multisatellite Support System
MSST Ministry of State for Science and Technology [Canada]
MSStat Master of Science in Statistics
MSStEng Master of Science in Structural Engineering [also: MSStE]
MSSW Magnetostatic Surface Wave
MSSySc Master of Science in Systems Science
MSSysE Master of Science in Systems Engineering [also: MSSysEng]
MST Manufacturers Sales Tax
Master of Science and Technology
Master of Science in Teaching
Master of Science Teaching
Master Station
Materials Science and Technology
Mean Solar Time
Medium-Scale Technology
Minimum Spanning Tree
Mobile Service Tower
Monolithic Systems Technology
Mountain Standard Time
Multiple Scattering Theory
Multisubscriber Time Sharing
MS&T Methodical Structures and Textures

MStat	Master of Statistics
MSTD	Materials Science and Technology Division
MSTE	Master of Science in Transportation Engineering
MSTEd	Master of Science in Technical Education
MSText	Master of Science in Textiles
MSTextChem	Master of Science in Textile Chemistry
MSTFA	N-Methyl-N-(Trimethylsilyl)trifluoroacetamide
MSTG	Mass Storage Task Group [of CODASYL]
MsTh	Mesothorium
MsTh$_1$	Mesothorium I (= Radium 228)
MsTh$_2$	Mesothorium II (= Actinium 228)
MSTM	Master of Science in Technology Management
	Master of Science in Teaching Mathematics
MSTR	Moisture
MSTransE	Master of Science in Transportation Engineering [also: MSTransEng]
MSTrPl	Master of Science in Transportation Planning
MSTS	Multisubscriber Time Sharing System
MSU	Main Store Update
	Main Switching Unit
	Maintenance and Status Unit
	Maintenance Signal Unit
	Management Signal Unit
	Mass Storage Unit
	Message Switching Unit
	Metallic Service Unit
	Michigan State University [US]
	Microwave Sounding Unit
	Mindanao State University [Philippines]
	Morgan State University [US]
	Multiblock Synchronization Unit
	Multiple Signal Unit
MSUD	Master of Science in Urban Design
MSUDC	Michigan State University Discrete Computer
MSUESM	Master of Science in Urban Environmental Systems Management
MSU-IIT	Mindanao State University — Iligan Institute of Technology [Philippines]
MSUP	Master of Science in Urban Planning
MSUS	Master of Science in Urban Studies
MSV	Mass Storage Volume
	Mean Square Velocity
	Mean Square Voltage
	Mouse Sacroma Virus
	Multifunction Support Vessel
MSVC	Master of Science in Vocational Counseling
	Mass Storage Volume Control
MSVI	Mass Storage Volume Inventory
MSVU	Mount St. Vincent University [Canada]
MSW	Magnetostatic Wave
	Master Switch
	Municipal Solid Waste
MSWRE	Master of Science in Water Resource Engineering
MSXα	Multiple Scattering X Alpha
MSY	Maximum Sustained Yield
MSYNC	Master Synchronization
MSZ	Magnesia-Stabilized Zirconia
MT	Machine Theory
	Machine Translation
	Magnetic-Particle Testing
	Magnetic Tape
	Magnetic Testing
	Magnetic Tube
	Magnetotelluric(s)
	Magnetotransport
	Mail Transfer
	Male Tee (Pipe Section)
	Management Technologist
	Management Technology
	Manual Transmission
	Manufacturing Technologist
	Manufacturing Technology
	Marine Technologist
	Marine Technology
	Mass Transfer
	Master of Teaching
	Master of Textiles
	Master Timer
	Materials Technology
	Materials Testing
	Maximum Torque
	Mean Tide
	Mean Time
	Measured Time
	Mechanical Technologist
	Mechanical Technology
	Mechanical Testing
	Mechanical Translation
	Mechanical Transport
	Medical Technologist
	Medical Technology
	Medium Temperature
	Medium Thermal Carbon Black
	Megaton
	Melt-Through
	Message Table
	Message Transfer
	Metallurgical Technologist
	Metallurgical Technology
	Metal Technology
	Metathesis
	Metering Time
	Methyltyrosine
	Metric Ton
	Mobile Terminal
	Mode Transducer
	Modified Tape
	Molecular Theory
	Montana [US]
	Mount(ain)
	Multiple Transfer
	Multitasking
M/T	Mail Transfer
Mt	Megaton
mT	maritime tropical air
MTA	Magnetic Tape Accessory
	Magnetothermal Analysis
	Manitoba Trucking Association [Canada]
	Marine Trades Association
	Maritime Transit Association [now: IMTA, US]
	Message Transfer Agent
	Metropolitan Transport Authority [US]

	Mica Trades Association [UK]
	Military Training Area
	Mobility Test Article
	Moving Target Analyst System
	Moving Target Analyzer
	Multiple Terminal Access
	Multiterminal Adapter
MTAC	Mathematical Tables and other Aids to Computation
MTAE	Message Transfer Agent Entity
MTAPA	Metal Trades Accident Prevention Association
MTB	Materials Transportation Bureau [US]
	Metallic Test Bus
	Methyl-tert-Butyl Ether
	Methyl Thymol Blue
	Mineral Technology Branch [of CANMET, Canada]
	Motor Torpedo Boat
MTBCF	Mean Time between Confirmed Failures
MTBD	Mean Time between Degradations
MTBE	Mean Time between Errors
	Methyl-tert-Butyl Ether
MTBF	Mean Time between Failures
MTBI	Mean Time between Interrupts
MTBM	Mean Time between Maintenance
MTBMA	Mean Time between Maintenance Actions
MTBO	Mean Time between Outages
MTBSTFA	Methyl-tert-Butyldimethylsilyl Trifluoroacetamide
MTC	Machine Tool Conference
	Magnetic Tape Cartridge
	Magnetic Tape Cassette
	Magnetic Tape Channel
	Magnetic Tape Controller
	Maintenance Time Constraint
	Main Trunk Circuit
	Manufacturing Technology Center [of NRC, Canada]
	Manufacturing-to-Cost (Process)
	Maritime Transport Committee [of OECD, France]
	Master of Textile Chemistry
	Master Tape Control
	Materials Technology Center
	Measurement Technology Conference
	Memory Test Computer
	Message Transmission Controller
	Metalworking Trade Coalition [US]
	Midwestern Telecommunication Conference [US]
	Million Tons of Coal
	Ministry of Transportation and Communications
	Missile Test Center
	Mission and Traffic Control
	Moderator Temperature Coefficient
	Modulation Transfer Curve
	Montreal Transport Commission [Canada]
MTCA	Minimum Terrain Clearance Altitude
	Ministry of Transport and Civil Aviation [UK]
	Multiple Terminal Communications Adapter
MTCC	Master Timing and Control Circuit
	Metropolitan Toronto Convention Center [Canada]
	Million Tons of Clean Coal
MTCE	Maintenance
MT&CE	Materials Testing and Characterization Equipment
MTCF	Mean Time to Catastrophic Failure
	Multilateral Technical Cooperation Fund [of CENTO]
MTCH	Magnetic Tape Channel
MTCOECD	Maritime Transport Committee of the Organization for Economic Cooperation and Development [France]
MTCS	Maximum Teleprocessing Communications System
MTCU	Magnetic Tape Control Unit
MTCV	Main Turbine Control Valve
MT-CVD	Medium Temperature — Chemical Vapor Deposition
MTCXO	Mathematically Temperature-Compensated Crystal Oscillator
MTD	Manufacturing Technology Division [US Air Force]
	Master of Transport Design
	Mean Temperature Difference
	Minimal Toxic Dose
	Mounted
	Moving Target Detector
	M-Tolylenediamine
	Multiple Target Detection
	Multiple Tile Duct
MTDA	Methyl Trimethylsilyl Dimethylketene Acetal
MTDR	Machine Tool Design and Research
MTDS	Manufacturing Test Data System
	Marine Tactical Data System [US Navy]
MTE	Maximum Tracking Error
	Michigan Test of English
	Ministry of Treasury and Economics [Canada]
	Multiple Terminator Emulator
M&TE	Measuring and Test Equipment
MTech	Master of Technology
MTES	Methyltriethoxysilane
MTF	Main Time to Failure
	Mean Time to Failure
	Mechanical Time Fuse
	Medium Time to Failure
	Mississippi Test Facility [of NASA]
	Modulation Transfer Function
MTFA	Modulation Transfer Function Analyzer
MTFM	Magnetic Thin Film Memory
MTG	Main Traffic Group
	Meeting
	Methane-to-Gasoline
	Methanol-to-Gasoline (Process)
	Methylthiogalactoside
	Mounting
	Multiple-Trigger Generator
MTGE	Mortgage
MTH	Magnetic Tape Handler
	Methylthiohydantoin
	Month
MTHPA	Methyltetrahydrophtalic Anhydride
MTHSR	Mass Transportation/High Speed Rail
MTI	Machine Tool Industry
	Marked Temperature inversion
	Materials Technology Institute [of CPI, US]
	Medium-Technology Intensive
	Metal Treating Institute [US]
	Metalworking Technologies Incorporated [of University of Pittsburgh, US]

	Ministry of Trade and Industry [Nigeria]
	Moving Target Indicator
MTIAC	Manufacturing Technology Information Analysis Center [US]
MTIC	Moving Target Indicator Coherent
MTIL	Maximum Tolerable Insecurity Level
MTIRA	Machine Tool Industry Research Association [UK]
MTK	Mongolian Tugrik [Currency]
MTL	Manufacturing and Technology Laboratory
	Master Tape Loading
	Material
	Materials Technology Laboratory [US Army]
	Merged-Transistor Logic
	Message Transfer Layer
	Metals Technology Laboratories [of CANMET, Canada]
	Minimum Triggering Level
	Mobiltherm Light
MTLP	Master Tape Loading Program
MTM	Master in the Teaching of Mathematics
	Methods-Time Measurement
	Multiple Terminal Manager
	Multitasking Monitor
	Multiterminal Monitor
MTMA	Methods-Time Measurement Association
	Military Terminal Control Area
MTMASR	Methods-Time Measurement Association for Standards and Research [US]
MTMC	4-(Methylthio)-M-Cresol
	M-Tolyl-N-Methylcarbamate
MTMF	Multiple Task Management Feature
MT-MF	Magnetic Tape to Microfilm
MTMH	Master of Tropical Medicine and Hygiene
MTMS	Methyltrimethoxysilane
MTMTS	Military Traffic Management and Terminal Service [USDOD]
MTN	Mountain
	M-Tolunitrile
	Multinational Trade Negotiations
MTNS	Mountains
MTO	Master Terminal Operator
	Metal Turnover
MTOE	Million Tons of Oil Equivalent
MTOGW	Maximum Takeoff Gross Weight
Mton	Megaton
MTOP	Molecular Total Overlap Population
MTOS	Magnetic Tape Operating System
	Metal Thick Oxide Semiconductor
	Metal Thick Oxide Silicon
	Multitasking Operating System
MTox	Master of Toxicology
MTP	Machine Tool Program
	Magnetic Tape Processor
	Maltese Pound [Currency of Malta]
	Mechanical Thermal Pulse
	Message Transfer Part
	Message Transmission Part
	4-(Methylthio)phenol
	Minimum Time Path
	Modular Terminal Processor
	Multiple Twinned Particle
MTPA	Methoxytrifluoromethylphenylacetic Acid
	Mobile Transponder Performance Analyzer
MTPCNA	Metal Tube Packaging Council of North America [now: TCNA, US]
MTPD	Metric Tons per Day
MTPF	Maximum Total Peaking Factor
	Minimum Total Processing Time
MTPS	Magnetic Tape Programming System
	Million Transitions per Second [also: MTS]
MTPT	Minimal Total Processing Time
MTPW	Master of Technical and Professional Writing
MTPY	Million (Short) Tons per Year
MTQ	Exhibition on Materials Testing and Quality Assurance
MTR	Magnetic Tape Reader
	Magnetic Tape Recorder
	Mass Transit Railway
	Materials Testing Reactor [of INEL, US]
	Materials Testing Report
	Migration Traffic Rate
	Missile Tracking Radar
	Ministry of Tourism and Recreation [Canada]
	Monitor
	Motor
	Moving Target Reactor
	Multiple Track Radar
	Multiple Tracking Range
MTRB	Maritime Transportation Research Board [US]
MTRCA	Metropolitan Toronto and Region Conservation Authority [Canada]
MTRE	Magnetic Tape Recorder End
MTRS	Magnetic Tape Recorder Start
MTS	Machine-Tractor Station
	Magnetic Tape Station
	Magnetic Tape System
	Maintenance Test Station
	Main Trunk System
	Marine Technology Society [US]
	Mass Termination System
	Materials Testing System
	Mechanical Testing System
	Member of Technical Staff
	Message Telecommunication Service
	Message Telephone Service
	Message Toll Service
	Message Transfer Service
	Message Transfer System
	Message Transmission Subsystem
	Message Transport Service
	Michigan Terminal System
	Microprocessor Training System
	Million Transitions per Second [also: MTPS]
	Missile Tracking System
	Mobile Telephone Service
	Mobile Tracking Station [of NASA]
	Modem Test Set
	Modem Test System
	Module Tracking System
	Motor-Operated Transfer Switch
	Mountains
	Muffin-Tin Sphere

	Multichannel Television Sound
	Multiple Terminal System
MTSC	Magnetic Tape Selective Composer [also: MT/SC]
	Master of Technical and Scientific Communication
MTSD	Military Transmission Systems Department [of NORAD]
MTSE	Magnetic Trap Stability Experiment
MTSMB	Material and Testing Standards Management Board
MTSR	Mean Time to Service Restoration
MTSS	Military Test Space Station
MTST	Magnetic Tape Selectric Typewriter [also: MT/ST]
MTS/WATS	Mobile Telephone Service/Wide Area Telephone Service
MTT	Magnetic Tape Terminal
	Magnetic Tape Transport
	Mechanicothermal Treatment
	Message Transfer Time
	Methylthiozoltetrazolium
	Microwave Theory and Techniques [IEEE Society]
MTTA	Machine Tool Trades Association
	Multi-Tenant Telecommunications Association [US]
MTTD	Mean Time to Diagnosis
MTTE	Magnetic Tape Terminal Equipment
MTTF	Mean Time to Failure
MTTFF	Mean Time to First Failure
MTTPO	Mean Time to Planned Outage
MTTR	Maximum Time to Repair
	Mean Time to Repair
	Mean Time to Restore
MTTS	Microwave Theory and Techniques Society [of IEEE, US]
	Multitask Terminal System
MTTUO	Mean Time to Unplanned Outage
MTU	Magnetic Tape Unit
	Manchester Terminal Unit
	Master Terminal Unit
	Memory Transfer Unit
	Methylthiouracil
	Metric Units
	Michigan Technological University [US]
	Multiplexer and Terminal Unit
	Multi-Terminal Unit
	Munich Technical University [Germany]
MTV	Marginal Terrain Vehicle
MTVAL	Master Tape Validation
MTW	Male Tube Weld
	Mountain Waves
MTWA	Maximum Total Weight Authorized
MTWX	Mechanized Teletypewriter Exchange
MTX	Methotrexate
	Methylpteroylglutamic Acid
	Mobile Telephone Exchange
MTXα	Muffin-Tin Xα
MU	Machine Unit
	Make Up
	Management Unit
	Manchester University [UK]
	Marquette University [US]
	McGill University [Canada]
	McMaster University [Canada]
	Mehran University [Pakistan]
	Meiji University [Japan]
	Memory Unit
	Methylene Unit
	Mie University [Japan]
	Missouri University [US]
	Mition Understanding
	Miyazaki University [Japan]
	Monash University [Australia]
	Monitor Unit [also: M/U]
	Moskow University [USSR]
	Munster University [Germany]
	Multiple Unit
mu	micron [also: μ]
μ	micro- [SI Prefix]
MUA	Master of Urban Affairs
	Master of Urban Architecture
	Ministry of State for Urban Affairs
muA	microampere [also: μA]
muA	microangstrom [also: μA]
MUBIS	Multiple Beam Interval Scanner
mubit	microbit [also: μbit]
muC	microcoulomb [also: μC]
μCF	Muon Catalyzed Fusion
muCi	microcurie [also: μCi]
MUCHA	Multiple Channel Analysis
MUCTC	Montreal Urban Community Transport Commission [Canada]
MUD	Master of Urban Design
MUDPAC	Melbourne University Dual-Package Analog Computer [Australia]
MUDWNT	Make Up Demineralizer Waste Neutralizer Tank
MUF	Machine Utilization Factor
	Materials Unaccounted For
	Material Utilization Factor
	Maximum Useful Frequency
muF	microfarad [also: μF]
MUFON	Mutual UFO (= Unidentified Flying Object) Network [US]
MUG	MARC (= Machine-Readable Catalogue) Users Group
	Multiset Users Group [US]
	MUMPS (= Massachusetts General Hospital Utility Multi-Programming System) Users Group [US]
mu g	microgram [also: μg]
MUGB	4-Methylumbelliferyl P-Guanidinobenzoate
mug/m³	microgram per cubic meter [also: μg/m³]
muH	microhenry [also: μH]
muHg	micrometer of mercury [also: μHg or μ Hg]
MUI	Module Interface Unit
mu-in	micro-inch [also: μ-in or micro-in]
mu-in-rms	micro-inch root mean square [also: μ-in-rms]
MUL	Multiply
muL	microliter [also: μL or μl]
MULDEM	Multiplexer-Demultiplexer
MULDEX	Multiplexing-Demultiplexing
MULPIC	Multipurpose Interrupted Cooling Process
MULT	Multiple
	Multiplication
MULTEWS	Multiple Electronic Warfare Surveillance
MULTICS	Multiplexed Information and Computing Service [US]

MULTP Multiplier
MuLV Murine Leukemia Virus
MUM Methodology for Unmanned Manufacturing (Project) [Japan]
Multiple-Unit Message
Multi Unit Message
Multi-User Message
Multi-User Monitor
mu m micrometer [also: μm]
mu-mho micromho [also: μmho]
mu-ohm microohm [also: μohm or μΩ]
MUMMS Marine Corps Unified Materiel Management System [US]
MUMPS Massachusetts General Hospital Utility Multi-Programming System [US]
MUMS Mobile Utility Module System
Multiple Use MARC (= Machine-Readable Catalogue) System
mu mu micromicron [also: μμ]
MUN Memorial University of Newfoundland [Canada]
Mineworkers Union of Namibia
Municipal(ity)
MUNDAT Municipal Waterworks and Wastewater Systems Inventory [of EPS, Canada]
MUOD Mean Unplanned Outage Duration
MUP Master of Urban Planning
Multiple Utility Peripheral
MUPID Multi-Purpose Universal Programmable Intelligent Decoder
MUPP Master of Urban Planning and Policy
MUPS Mechanized Unit Property System
Multiple Utility Peripheral System
MUR Management Update and Retrieval System
Mauritian Rupee [Currency of Mauritius]
Message Unit, Radio Relay
Mock-Up Reactor [of NASA]
MURA Midwestern University Research Association [US]
MURATREC Multi-Radar Track Reconstruction
MURB Multiple Unit Residential Building
MURC Missouri University Research Center [US]
MURKS Multi-User Remote Keying System
MURP Master of Urban and Regional Planning
Master of Urban and Rural Planning
MUS Manual Update Service
Master of Urban Studies
Multiprogramming Utility System
Museum
mus mouse [Immunochemistry]
mu s microsecond [also: μs]
MUSA Multiple-Unit Steerable Antenna [also: Musa]
Multiple-Unit Steerable Array
MUSAP Multisatellite Augmentation Program
MUSAT Multi-Purpose UHF (= Ultrahigh Frequency) Satellite
MUSC Microgravity User Support Center
MUSE Monitor of Ultraviolet Solar Energy
Multiple Stream Evaluator
Multiple Sub-Nyquist Sample Encoding
MUSIC Multiple Signal Characterization
MUSICOL Musical Instruction Composition Oriented Language
MUSIL Multiprogramming Utility System Interpretative Language
MUSS Manchester University Software System
MUSTARD Museum and University Storage and Retrieval of Data [of Smithsonian Institute, US]
MUSTRAN Music Translation
MUSysE Master of Urban Systems Engineering [also: MUSysEng]
MUT Mean Up Time
MUTA Military Upper Traffic Control Area
MUTEX Multi-User Transaction Executive
MUTL Mutual
MUTMAC Methylumbelliferyl Trimethylammonium Cinnamate Chloride
MUV Manned Underwater Vehicle
muV microvolt [also: μV]
muV/m microvolts per meter [also: μV/m]
muW microwatt [also: μW]
MUX Multiplexer [also: MUXER]
MV Main Valve
Marine Vessel
Market Value
Matrix Vesicles
Mean Value
Mean Variation
Mean Velocity
Measured Value
Measured Vector
Measured Voltage
Medium Vacuum
Medium Voltage
Megavolt
Mesh Voltage
Methyl Violet
Million Volts
Molar Volume
Motor Vehicle
Motor Vessel
Move
Moving Vessel
Multivibrator
Muzzle Velocity
Mv Mendelevium [also: Md]
mV millivolt
MVA Machine Vision Association [of SME, US]
Master of Visual Arts
Megavoltampere
Mevalonic Acid
MVA/SME Machine Vision Association of the Society of Manufacturing Engineers [US]
MVB Medium-Volatile Bituminous
Multivesicular Body
Multivibrator [also: MVBR]
MVC Manual Volume Control
Microvoid Coalescence
Multiple Variate Counter
Multivariable Control
Multivariable Controller
MV-CTD Moving Vessel — Conductivity, Temperature, Depth
MVD Map and Visual Display
Mechanical Vapor Deposition

	Multivariable Distribution
MVDA	Memory Volume Discount Addendum
MVDF	Medium and Very-High Frequency Direction Finding
MVDL	Mercury-Vapor Discharge Lamp
MVDS	Modular Video Data System
MVE	Master of Vocational Education
	Methyl Vinyl Ether
	Motor Vehicle Event
	Multivariate Exponential Distribution
MVetSc	Master of Veterinary Science
MVF	Martensite Volume Fraction
	Miles Value Foundation [US]
MVFR	Marginal Visual Flight Rules
MVG	Moving
MVGA	Manitoba Vegetable Growvers Association [Canada]
MVI	Motor Vehicle Inspection
	Muzzle Velocity Indicator
MVK	Methyl Vinyl Ketone
MVL	Mercury Vapor Lamp
	Multiple-Valued Logic
MVLIFCT	M. V. Lomonosov Institute of Fine Chemical Technology [USSR]
MVLUE	Minimum Variance Linear Unbiased Estimate
MVM	Mariner versus Mercury
	Minimum Virtual Memory
	Modified Volume Module
MVMA	Motor Vehicle Manufacturers Association [US]
MVMC	Motor Vehicle Maintenance Conference
MVMT	Movement
MVP	Matrox Vision Processor
	Mechanical Vacuum Pump
	Megavolt of Power [also: MVp]
	Multiline Variety Package
	Multiple Virtual Processing
	Multivariable Processes
MVPCB	Motor Vehicle Pollution Control Board [US]
MVR	Mercury Vapor Rectifier
MVS	Magnetic Voltage Stabilizer
	Master of Valuation Sciences
	Minimum Visual Signal
	Motor Vehicle Specification
	Multiple Virtual Storage
	Multiprogramming with Virtual Storage
	Multivariable System
	Multivariate Statistics
MVSA	Motor Vehicle Safety Association
MVSc	Master of Veterinary Science
MVSMA	Mechanical Vibrating Screen Manufacturers Association [US]
MVS/SE	Multiple Virtual Storage/System Extension
MVS/SP	Multiple Virtual Storage/System Product
MVT	Moisture Vapor Transmission
	Multiprogramming with a Variable Number of Tasks
MVTA	Motor Vehicle Transport Act [Canada]
MVTC	Manpower Vocation Training Commission [Canada]
	Motor Vehicle Test Center
MVTE	Master of Vocational and Technical Education
MVTEd	Master of Vocational and Technical Education
MVT/TSO	Multiprogramming with a Variable Number of Tasks/Time-Sharing Option
MVU	Minimum Variance Unbiased
MVUE	Minimum Variance Unbiased Estimate
MVULE	Minimum Variance Unbiased Linear Estimate
MW	Manual Word
	Matter Wave
	Medium Wave
	Megawatt
	Metric Wave
	Microwave
	Microweighing
	Milky Way
	Million Words
	Molecular Weight [also: mw]
	Music Wire
M/W	Microwave
Mw	Molecular Weight, Weight-Average
mW	milliwatt
MWA	Microwave Antenna
MWAA	Metropolitan Washington Airports Authority [US]
MWARA	Major World Air Route Area
MWB	Metropolitan Water Board
	Minimum Wage Board
	Multiprogram Wire Broadcasting
MWC	Mechanized Wire Centering
MWCO	Molecular Weight Cutoff
MWCS	Microwave Communication System
MWD	Measurement While Drilling
	Megawatt Day
	Megaword
	Microwave Device
	Molecular Weight Distribution
MWDDEP	Mutual Weapons Development Data Exchange Program
MWDP	Mutual Weapons Development Program
MWD/t	Megawatt Days per Ton [also: MWd/t]
MWE	Manufacturer's Weight Empty
	Megawatt of Electric Power [also: MW(E)]
MWEM	Microwave Electromagnetics [also: MWE]
MWF	Marine Workers Federation
	Microwave Frequency
MWG	Music Wire Gauge
MWH	Megawatt of Heat [also: MW(H)]
MWh	Megawatt-hour
MWHGL	Multiple Wheel Heavy Gear Load
MWI	Message-Waiting Indicator
MWIA	Manitoba Wholesale Implement Association [Canada]
MWIR	Mid-Wave Infrared
MWIS	(National Network of) Minority Women in Science [US]
MWK	Malawi Kwacha [Currency]
MWL	Mean Water Level
	Milliwatt Logic
MWO	Maintenance Work Order
	McGill Weather Observatory [of McGill University, Canada]
	Medicine White Oil
	Meteorological Watch Office
MWP	Maneuvering Work Platform [of NASA]
	Maximum Working Pressure
MWPC	Multiwire Proportional Chamber

MWPS Master of Wood and Paper Science
MWR Magnetic Tape Write (Memory)
Mean Width Ratio
MWS Microwave Scatterometer
Microwave Station
Microwave Spectroscopy
Multi-Workstation
MWSC Minimum Wage Study Commission
MW-SDS Molecular Weight — Sodium Dodecylsulfate
MWT Make-Up Water Treatment
Master of Wood Technology
MWt Megawatt of Thermal Energy [also: MWT or MW(Th)]
MWV Maximum Working Voltage
Mineralolwirtschaftsverband [= Federation of the Mineral Oil Industries, Germany]
MWYE Megawatt Year of Electricity
MX Matrix
Mix
Multiplex [also: Mx]
Mx Maxwell
MXC Multiplexer Channel
MXD Mixed
M-Xylylene Diamine [also: MXDA]
MXE Mobile Electronic Exchange
MXP Mexican Peso [Currency]
MXR Mask Index Register
MXU Mobile Exhibition Unit
MY Meeting Year
MYCOL Mycological
Mycology
myg myriagram
myl myrialiter
mym myriameter
MYOB Minding Your Own Business
MYOP Multi-Year Operations Plan
MYR Malaysian Ringgit [Currency]
MZ Melted Zone
Methylimidazole
Monozygotic
2-MZ Methylimidazole
MZFR Mehrzweck-Forschungsreaktor [= Multi-Purpose Nuclear Reactor [at KfK, Germany]]
MZM Mozambique Meticai [Currency]
MZS Master of Zoological Science [also: MZSc]

N

N	Asparagine
	Asparaginyl
	Naira [Currency of Nigeria]
	Nanocrystal(line)
	Narrow
	Nation(al)
	Navigation
	Navy
	Negative
	Nematic Phase
	Net
	Newton
	Night
	Nitrogen
	Node
	Noise
	Normal
	Normality [also: *N*]
	North(ern)
	November
	Number
	Numeric
	Nylon
n	name
	nano- [SI Prefix]
	nanocrystal(line)
	negative
	net
	noon
	normal
	number
	numeric
NA	(American) National Acme Thread
	Naturally Aspirated
	Naval Architect
	Naval Architecture
	Network Analysis
	Network Architecture
	Neutral Axis
	Neutron Absorption
	Nicotinic Acid
	Nitric Acid
	No Access
	Nonabrasive
	Nonacosadiynoic Acid
	Nonaqueous
	Noradrenaline
	North America(n)
	North Atlantic
	Not Adjustable
	Not Adjusted
	Not Applicable
	Not Assigned
	Not Authorized
	Not Available
	Numerical Analysis
	Numerical Aperture
N/A	No Account
	Not Applicable
Na	(Natrium) — Sodium
nA	nanoampere
NAA	Naphthaleneacetic Acid
	Naphthylacetic Acid
	National Academy of Arbitrators [US]
	National Aeronautic Association [US]
	National Aerosol Association [US]
	National Aggregates Association [US]
	National Airports Authority [UK]
	National Arborist Association [US]
	National Ash Association [now: ACAA, US]
	National Association of Accountants [US]
	North Atlantic Assembly [of NATO]
	Neutron Activation Analysis
NAAA	National Agricultural Aviation Association [US]
	National Alarm Association of America [US]
	National Auto Auction Association [US]
NAAASE	National Association for Applied Arts, Science and Science [US]
NAAB	National Architectural Accrediting Board [US]
	National Association of Animal Breeders [US]
NAAC	National Association of Agricultural Contractors [UK]
NAAD	National Association of Aluminum Distributors [US]
	Nicotine Acid Adenine Dinucleotide
NAAE	National Association of Aeronautical Examiners [US]
	National Association of Agricultural Employees [US]
NAAFI	Navy, Army, Air Force Institutions [UK]
NAAG	NATO Army Armament Group
	Nordic Association of Applied Geophysics [Sweden]
NAAFI	Navy, Army, and Air Force Institute [US]
NAAJ	National Association of Agricultural Journalists [US]
NAAL	North America Aerodynamic Laboratory [US]
NAAM	National Association of Architectural Metal Manufacturers [also: NAAMM, US]
NAAMSA	National Association of Automobile Manufacturers of South Africa
NAAN	National Advertising Agency Network [US]
NAAO	North Africa Area Office [of UNICEF]
NAAQS	National Ambient Air Quality Standards
NAAR	National Association of Advertising Representatives [UK]
NAARS	National Automated Accounting Research System [US]
NAAS	National Agricultural Advisory Service [UK]
	National Association of Academies of Science [US]
	Naval Auxiliary Air Station [US]
	New Academic Appointments Scheme [UK]
NAASS	North American Association of Summer Sessions [US]
NAATP	New African Air Transport Policy
NAATS	National Association of Air Traffic Specialists [US]
	National Association of Auto Trim Shops [US]

NAAWS NATO Anti-Air Warfare System
NORAD (= North American Aerospace Defense Command) Automatic Attack Warning System

NAB National Aircraft Beacon
National Alliance of Businessmen [US]
National Assistance Board [UK]
National Association of Bank Servicers [US]
National Association of Bookmakers [UK]
National Association of Broadcasters [US]
Navigational Aid to Bombing
Nuclear Assembly Bulding

NABADA National Barrel and Drum Association [US]

NABC North American Blueberry Council [US]

NABCE National Association of Black Consulting Engineers [US]

NABD National Association of Brick Distributors [US]
Normenausschuss Bibliotheks- und Dokumentationswesen [= Standards Committee on Library Work and Documentation, Germany]

NABDC National Association of Blueprint and Diazo Coaters [now: ARMM, US]

NABE National Association for Business Education [US]

NABER National Association of Business and Educational Radio [US]

NABET National Association of Broadcast Employees and Technicians [Canada]

NABIS National Association of Business and Industrial Saleswomen [US]

NABM National Association of Boat Manufacturers [US]
National Association of British Manufacturers [now: CBI, UK]

NABMA National Association of British Market Authorities [UK]

NABP National Association of Black Professors [US]
National Association of Boards of Pharmacy [US]

NABS North American Benthological Society [US]

NABSC National Association of Building Service Contractors [now: BSCA, US]

NABT National Association of Biology Teachers [US]

NABTS North American Basic Teletext Specification
North American Broadcast Teletext Specification

NAC Nacelle
National Academy of Consiliators [US]
National Action Committee
National Advisory Committee
National Agency Checks
National Agricultural Center [UK]
National Archives of Canada
National Asbestos Council [US]
National Aviation Club [US]
Naval Avionics Center [of US Navy]
Network Access Controller
Network Advisory Committee [of Library of Congress, US]
North Atlantic Council [of NATO]

NACA National Academy of Code Administrators [US]
National Advisory Committee on Aeronautics [of NASA]
National Agricultural Chemicals Association [US]
National Air Carrier Association [US]
National Armored Car Association [US]
National Association for Clean Air [South Africa]
National Association of Campus Activities [US]
National Association of Cellular Agents [US]
National Autosound Challenge Associoation [US]

NACAA National Association of Computer-Assisted Analysis [US]
National Association of County Agricultural Agents [US]

NACAB National Agricultural Center Advisory Board [UK]

NACAC National Association of College Admission Counselors [US]

NACAM National Association of Corn and Agricultural Merchants [UK]

NACAR National Advisory Committee on Aeronautical Research [South Africa]

NACAS National Advisory Committee on Agricultural Services [Canada]

NACAT National Association of College Automotive Teachers [US]

NACATTS North American Clear Air Turbulence Tracking System

NACC National Advisory Cancer Council [US]
National Anti-Counterfeiting Committee [Taiwan]
National Automatic Controls Conference [US]
North America Control Committee

NACCG National Association of Crankshaft and Cylinder Grinders [UK]

NACD National Association of Chemical Distributors [US]
National Association of Conservation Districts [US]
National Association of Container Distributors [US]

NACDRAO National Association of College Deans, Registrars and Admission Officers [US]

NACE National Association of Corrosion Engineers [US]
National Association of County Engineers [US]

NACE(E) National Association of Corrosion Engineers (Europe) [UK]

NACEIC National Advisory Council for Education, Industry and Commerce [UK]

NACEO National Advisory Council on Economic Opportunity

NACHA National Automated Clearinghouse Association [US]

NACIS North American Cartographic Information Society [US]

NACISO NATO Communications and Information Systems Organization

NACK Negative Acknowledgement

NACM National Association of Chain Manufacturers [US]
National Association of Cotton Manufacturers [UK]

NACMA National Armored Cable Manufacturers Association [US]

NACME National Action Council for Minorities in Engineering [US]

NACOA National Advisory Committee on Oceans and Atmosphere [US]

NACOLADS National Council on Libraries, Archives and Documentation Services [Jamaica]

NACOM National Communication System [of USDOD]

NACOSH National Advisory Committee on Occupational Safety and Health [US]

NACPD National Association of County Planning Directors [US]

NACS North American Catalysis Society

Northern Area Communications System
NACSA North American Computer Service Association [US]
NACSC National Association of Cold Storage Contractors [US]
NACSIC National Association of Cold Storage Insulation Contractors [now: NACSC, US]
NACT National Association of Cycle Traders [UK]
NACTA National Association of Colleges and Teachers of Agriculture [US]
NACUBO National Association of College and University Business Officers [US]
NACUFS National Association of College and University Food Services [US]
NAD Nicotinamide Adenine Dinucleotide [also: NADIDE]
No-Acid Descaling
Node Administration
Nuclear Accident Dosimetry
NADA National Automobile Dealers Association [US]
NADAF National Association of Decorative Architectural Finishes [US]
NADase NAD (= Nicotinamide Adenine Dinucleotide) Glycohydrolase
Nicotinamide Adenine Dinucleosidase
NADB National Air Data Branch [US]
NADC National Association of Demolition Contractors [US]
National Association of Dredging Contractors [US]
NATO Air Defense Committee
Naval Air Development Center [US Navy]
Nothern Alberta Development Council [Canada]
NADCA National Animal Damage Control Association [US]
NADCI North American Die Casting Institute [US]
NADD National Association of Die Makers and Die Cutters [US]
NADDRG North American Deep Drawing Research Group
NaDEC Sodium Diethyl Dithiocarbamate
NADEEC NATO Air Defense Electronic Environment Committee [Part of NADC]
NADEFCOL NATO Defense College
NADES National Association of Drop Forgers and Stampers [UK]
NADGE NATO Air Defense Ground Environment
NADGECO NATO Air Defense Ground Environment Corporation
NADH Nicotinamide Adenine Dinucleotide, Reduced Form
NADI National Association of Display Industries [US]
NADIDE Nicotinamide Adenine Dinucleotide [also: NAD]
NaDMC Sodium Dimethyl Dithiocarbamate
NADMR National Association of Diversified Manufacturers Representatives [now: NAGMR, US]
NADMC Naval Air Development and Material Center [US]
NADO National Association of Development Organizations [US]
NADOA National Association of Division Order Analysts [US]
NADP Nicotinamide Adenine Dinucleotide Phosphate
Northern Alberta Dairy Pool [Canada]
NADPH Nicotinamide Adenine Dinucleotide Phosphate, Reduced Form
NADTP National Association of Desktop Publishers [US]
NADW North Atlantic Deep Water
NADWARN Natural Disaster Warning System [US]
NAE National Academy of Engineering [US]
National Aeronautical Establishment [of NRC, Canada]
Nursery Association Executives [now: NAENA, US]
NAEB National Association of Educational Broadcasters [US]
NAEC National Aeronautical Establishment of Canada [of NRC, Canada]
National Aerospace Education Council [US]
National Agricultural Engineering Corporation [PR China]
National Association of Elevator Contractors [US]
National Association of Engineering Companies [US]
National Association of Exhibition Contractors [UK]
Naval Air Engineering Center [US Navy]
NAECON National Aerospace Electronics Conference [of IEEE]
NAED National Association of Electrical Distributors [US]
NAEDA North American Equipment Dealers Association [US]
NAEDS National Association of Educational Data Systems
NAEE National Association for Environmental Education [Name Changed to: North American Association of Environmental Education, US]
North American Association of Environmental Education [US]
NAEM National Association of Explosion Managers [US]
NAEMC National Association of Export Management Companies [now: NEXCO, US]
NAENA Nursery Association Executives of North America [US]
NAEP National Assessment of Educational Progress [US]
National Association of Environmental Professionals [US]
NAES Naval Air Experimental Station [US Navy]
NAESA National Association of Elevator Safety Authorities [US]
NAESCO National Association of Energy Service Companies [also: NAESC, US]
NAET National Association of Educational Technicians
NAEW NATO Airborne Early Warning
NAEWCSPMO NATO Airborne Early Warning and Control System Program Management Organization
NAF Naval Aircraft Factory
Naval Avionics Facility [US Navy]
Netherlands Atomic Forum
Network Access Facility
NA*f* Netherlands Antillean Florin (or Guilder) [also: NAf; Currency of Netherlands Antilles]
NAFA National Association of Fleet Administrators [US]
NAFAG NATO Air Force Armaments Group
NAFAX National Facsimile Network Circuit
NAFB National Association of Farm Boardcasters [US]
NAFC North-American Forestry Commission [US]
Northwest Atlantic Fisheries Center [Canada]
NAFCD National Association of Floor Covering Distributors [US]
NAFCO North Atlantic Fisheries Consultative Committee
Northwest Atlantic Fishery Consultative Organization

NAFEC National Administrative Facilities Experimental Center [US]
National Aviation Facilities Experimental Center [US]
NAFED National Association of Fire Equipment Distributors [US]
NAFEM National Association of Food Equipment Manufacturers [US]
NAFEO National Association for Equal Opportunity in Higher Education [US]
NAFFP National Association of Frozen Food Producers [UK]
NAFI Naval Avionics Facility, Indianapolis [US Navy]
NAFIC National Alcohol Fuels Information Center [US]
NAFIPS North American Fuzzy Information Processing Society [US]
NAFO National Association of Fire Officers [UK]
Northwest Atlantic Fisheries Organization [Canada]
NaFo Na (= Sodium) Formate
NAFRD National Association of Fleet Resale Dealers [US]
NAFSA National Association for Foreign Student Affairs [US]
National Fire Services Association [UK]
NAFSWMA National Association of Flood and Storm Water Management Agencies [US]
NAFTA North American Free Trade Agreement [between Canada, the USA and Mexico]
North Atlantic Free Trade Area
NAFTC National Association of Freight Transportation Consultants [US]
North American Forging Technology Conference
NAFTZ National Association of Foreign Trade Zones [US]
NAFWR National Association of Warehousemen and Removers [UK]
NAG National Academy of Geosciences [US]
National Association of Goldsmiths [UK and Ireland]
National Gallery of Canada
Numerical Algorithms Group
NAGA North American Game Bird Association [US]
NAGARA National Association of Government Archives and Records Administrators [US]
NAGC National Association of Government Communicators [US]
NAGCD National Association of Glass Container Distributors [now: NACD, US]
NAGDM National Association of Garage Door Manufacturers [US]
NAGE National Association of Government Engineers [US]
NAGHSR National Association of Governors Highway Safety Representatives [US]
NAGI&QAP National Association of Government Inspectors and Quality Assurance Personnel [US]
NAGLO National Association of Governmental Labor Officials [US]
NAGLSTP National Association of Gay and Lesbian Scientists and Technical Professionals [US]
NAGMC North Atlantic Council and Military Committee [of NATO]
NAGMR National Association of General Merchandise Representatives [US]
NAGT National Association of Geology Teachers [US]
NAH Nordic Association for Hydrology [Sweden]
NAHAD National Association of Hose and Accessories Distributors [US]
NAHB National Association of Home Builders [US]
NAHR National Association of Housing and Redevelopment [US]
NAHRI National Animal Husbandry Research Institute [Denmark]
NAHU North American Honeywell Users Association [US]
NAI N-Acetylimidazole
National Association of Interpretation [US]
New Alchemy Institute [US]
NAIC National Astronomy and Ionosphere Center [US]
NAICC National Association of Independent Computer Companies [US]
NAICU National Association of Independent Colleges and Universities [US]
NAID National Association of Installation Developers [US]
NAIDA National Agricultural and Industrial Development Association [UK]
NAIDM National Association of Insecticide and Disinfectant Manufacturers [US]
NAIEC National Association for Industry-Education Cooperation [US]
NAIG Nippon Atomic Industry Group [Japan]
NAILD National Association of Independent Lighting Distributors [US]
NAIOP Navigational Aid Inoperative for Parts
NAIPRC Netherlands Automatic Information Processing Research Center
NAIRD National Association of Independent Record Distributors and Manufacturers [US]
NAIRO National Association of Intergroup Relations Officers [US]
NAIT National Association of Industrial Technology [US]
Northern Alberta Institute of Technology [Canada]
NAITTE National Association of Industrial and Technical Teacher Educators [US]
NAIWC National Association of Inland Waterway Carriers [UK]
NaIX Sodium Isopropyl Xanthogenate
NAK Negative Acknowledgement (Character)
NaK Sodium-Potassium (Alloy)
NAL National Accelerator Laboratory [of USAEC]
National Aeronautical Laboratory [India]
National Agricultural Library [US]
National Airlines [US]
Naval Aeronautical Laboratory [US Navy]
New Assembly Language
NALC National Association of Litho Clubs [US]
NALCD National Agricultural Library and Center for Documentation [Hungary]
NALF North American Loon Fund [US]
NALGHW National Association of Local Governments on Hazardous Waste [US]
NALHC North American Log Homes Council [US]
NALM National Association of Liftmakers [UK]
NALMCO National Association of Lighting Management Companies [US]
NALMS North American Lake Management Society [US]

NALPM National Association of Lithographic Plate Manufacturers [US]
NALR National Association of Lighting Representatives [US]
NAM N-Acetylmethionine
N-Acetylmuramic Acid
N-(9-Acridinyl)maleimide
National Association of Manufacturers [US]
National Aviation Museum
Nautical Air Miles
Network Access Machine
Network Access Method
North America(n) [also: N AM]
NAMA National Agri-Marketing Association [US]
National Automatic Merchandising Association [US]
North American Mycological Association [US]
NAMAS National Measurement Accredited Services [of CAA, UK]
NAMB National Agricultural Marketing Board
NAMBO National Association of Motor Bus Operators [US]
NaMBT Sodium-2-Mercaptobenzothiazole
NAMC Naval Air Material Center [US Navy]
NAME National Association of Marine Enginebuilders [UK]
National Association of Marine Engineers [Canada]
National Association of Maritime Educators [US]
National Association of Name Plate Manufacturers [US]
NAMEPA National Association of Minority Engineering Program Administrators [US]
NAMES National Association of Medical Equipment Suppliers [US]
NAVDAC Assembly, Monitor, Executive System
NAMF National Association of Metal Finishers [US]
North American Multifrequency (Signal)
NAMFI NATO Missile Firing Installation
NAMG Narrow-Angle Mars Gate [of NASA]
NAMI National Association of Malleable Iron Founders [UK]
Naval Aerospace Medical Institute
NAMM National Association of Margarine Manufacturers [US]
National Association of Mirror Manufacturers [US]
NAMMA NATO Multi-Role (Combat Aircraft Development and Production) Management Agency
NAMMC Natural Asphalt Mine Owners and Manufacturers Council [UK]
NAMMO NATO Multi-Role (Combat Aircraft Development and Production) Management Organization
NAMO National Agricultural Marketing Officials [US]
National Association of Marketing Officers [US]
NAMOA National Association of Miscellaneous, Ornamental and Architectural Products Contractors [US]
NAMPP National Advanced Materials Program Plan [US]
NAMPS National Association of Marine Products and Services [US]
NAMPUS National Association of Master Plumbers of the United States
NAMRAD Non-Atomic Military Research and Development
NAMRI North American Manufacturing Research Institute [of SME, US]
NAMRI/SME North American Manufacturing Research Institute of the Society of Manufacturing Engineers [US]
NAMRU Naval Medical Research Unit [of US Navy]
NAMS National Association of Marine Services [US]
National Association of Marine Surveyors [US]
Network Analysis and Management System
North American Membrane Society [US]
NAMSA NATO Maintenance and Supply Agency
NAMSO NATO Maintenance and Supply Organization
NAMTA National Art Materials Trade Association [US]
NAMTC Naval Air Missile Test Center [US Navy]
NAMTS Nippon Automatic Mobile Telephone System [Japan]
NAN Neuraminidase
Network Access Node
Network Application Node
NANA N-Acetylneuraminic Acid
NANBA North American National Broadcasters Association
NANCO National Association of Noise Control Officials [US]
NANCRFUG North American NCR (= National Cash Register) Financial Users Group [US]
NAND Not-And
NANEP Navy Air Navigation Electronics Project [US Navy]
NANFA North American Native Fishes Association [US]
NANFSM National Association of Nonferrous Scrap Metal Merchants [UK]
NANS National Automated Nesting System
Nevada Academy of Natural Sciences [US]
NANU National Association of NIDS (= National Investor Data Service) Users [US]
NANWEP Navy Numerical Weather Prediction [US]
NAO National Astronomical Observatory
North Atlantic Oscillation
NAOGE National Association of Government Engineers
NAOHSM National Association of Oil Heating Service Managers [US]
NAOHN National Association of Occupational Health Nurses
NAOJ National Astronomical Observatory of Japan
NAOP National Association for Olmsted Parks [US]
National Association of Operative Plasterers [UK]
NAOS North Atlantic Ocean Station
NAOSMM National Association of Scientific Material Managers [US]
NAP National Academy Press [US]
Network Access Pricing
Network Access Protocol
Niger Agricultural Project [Nigeria]
Noise Analysis Program
NA-P Nonabrasive-Polishing
NAPA National Agricultural Plastics Association [US]
National Asphalt Pavement Association [US]
National Association of Parks Administrators [UK]
National Association of Purchasing Agents [now: NAPM, US]
National Automotive Parts Association [US]
North Atlantic Ports Association [US]
NAPAEO National Association of Principal Agricultural Education Officers [UK]
NAPAP 2-Naphthalenesulfonylglycyl)-4-Amidinophenylalaninepiperidide
NAPB National Agricultural Products Board [Tanzania]

NAPC National Association of Plumbing Contractors [now: NAPHCC, US]
NAPCA National Air Pollution Control Administration [US]
National Association of Pipe Coating Applicators [US]
National Association of Professional Contract Administrators [now: NCMA, US]
NAPD National Association of Pharmaceutical Distributors [UK]
National Association of Plastic Distributors [US]
National Association of Precollege Directors [US]
NAPE National Association of Power Engineers [US]
National Association of Professional Engravers [US]
NAPEP National Association of Planners, Estimators and Progressmen [US]
NAPET National Association of Photo Equipment Technicians [US]
NAPF National Association of Plastic Fabricators [now: DLPA, US]
NAPHCC National Association of Plumbing, Heating, and Cooling Contractors [US]
NAPIM National Association of Printing Ink Manufacturers [US]
NAPL National Air Photography Library [US]
National Association of Photolithographers [US]
National Association of Printers and Lithographers [US]
NAPLPS North American Presentation Level Protocol Syntax
NAPM National Association of Paper Merchants [UK]
National Association of Pattern Manufacturers [US]
National Association of Pharmaceutical Manufacturers [US]
National Association of Photographic Manufacturers [US]
National Association of Punch Manufacturers [US]
National Association of Purchasing Management [US]
NAPME N-Aminophenol Methylether
NAPMM National Association of Produce Market Managers [US]
NAPNM National Association of Pipe Nipple Manufacturers [US]
NAPO National Association of Professional Organizers [US]
NAPOL North Atlantic Policy Working Group [of ECAC, France]
NAPP National Aerospace Productivity Program [US]
NAPPE Network Analysis Program using Parameter Extractions
NAPR NATO Armaments Planning Review
NAPRE National Association of Practical Refrigerating Engineers [US]
NAPS National Air Pollution Surveillance Network [Canada]
National Association of Buying Services [US]
National Auxiliary Publications Service [US]
Night Aerial Photographic System
Nimbus Automatic Programming System
NAPSA National Appliance Parts Suppliers Association [US]
NAPSIC North American Power Systems Interconnection Committee
NAPSS Numerical Analysis Problem Solving System
NAPT Nordic Association of Plumbers and Tinsmiths [Sweden]
NAPU Nuclear Auxiliary Power Unit
NAPUS Nuclear Auxiliary Power Unit System
NAPVO National Association of Passenger Vessel Owners [US]
NAPWPT National Association of Professional Word Processing Technicians [US]
NAQP National Association of Quick Printers [US]
NAQP-CUG National Association of Quick Printers — Computer Users Group [US]
NAQP-MSG National Association of Quick Printers — Management Systems Group [now: NAQP-CUG]
NAR National Association of Rocketry
Net Advertising Revenue
Net Assimilation Rate
NARAP National Acid Rain Assessment Program [US]
NARATE Navy Radar Automatic Test Equipment
NARB National Advertising Review Board [US]
NARBA North American Regional Broadcasting Agreement
NARBW National Association of Railway Business Women [US]
NARC Non-Automatic Relay Center
NARDIS Navy Automated Research and Development Information System [US Navy]
NAREC Naval Research Electronic Computer
NAREFA North Atlantic Reference Fares [of ECAC, France]
NAREMCO National Records Management Council [US]
NARF Naval Air Rework Facility
Nuclear Aerospace Research Facility [US Air Force]
NARG Navigation Aids Research Group [of ICAO]
NARI Nanjing Automation Research Institute [PR China]
Natal Agricultural Research Institute [South Africa]
National Association of Recycling Industries [now: ISRI, US]
National Association of the Remodeling Industry [US]
Nuclear Aerospace Research Institute [US Air Force]
NARL National Aerospace Research Laboratory [US]
Naval Arctic Research Laboratory [US Navy]
NARM National Association of Relay Manufacturers [US]
NARMC National Association of Regional Media Centers [US]
NARO National Association of Royalty Owners [US]
North American Regional Office
NARP National Association of Railroad Passengers [US]
National Association of Registered Plans [US]
NARPV National Association for Remotely Piloted Vehicles [now: AUVS, US]
NARS National Archives and Record Service [US]
National Association of Radiotelephone Systems [US]
National Association of Rail Shippers [US]
North Atlantic Radio System
NARSA National Automotive Radiator Service Association [US]
NARS-A1 National Archives and Record Service — Automation 1 [US]

NARSC National Association of Reinforced Steel Contractors [US]

NARST National Association for Research in Science Teaching [US]

NARTB National Association of Radio and Television Broadcasters [US]

NARTE National Association of Radio and Telecommunication Engineers [US]

NARTM National Association of Rope and Twine Merchants [UK]

NARTS National Association for Radio Telephone Systems [US]
Naval Aeronautics Test Station [US]
Naval Air Rocket Test Station [US]

NARUC National Association of Regulatory Utility Commissioners [US]

NARUCE National Association of Regulatory Utility Commission Engineers [US]

NARVRE National Association of Retired and Veteran Railroad Employees [US]

NARWA Nordic Agricultural Research Workers Association [Norway]

NAS National Academy of Science [US and India]
National Aircraft Standards [US]
National Airspace System [of FAA, US]
National Aquarium Society [US]
National Association of Shopfitters [UK]
National Astrological Society [US]
National Audubon Association [US]
National Avionics Society [US]
Naval Air Station
Nebrasca Academy of Science [US]
Network Access System
Nominal Aggregate Signal

NASA 2-Naphthylamine-1-Sulfonic Acid
National Aeronautics and Space Administration [US]
National Appliance Service Association [US]
National Association of State Archeologists [US]
North Atlantic Seafood Association

NASACU National Association of State Approved Colleges and Universities [US]

NASAD National Association of Sport Aircraft Designers [US]

NASAGA North American Simulation and Gaming Association [US]

NASA-GAS NASA Get-Away Special (Program) [US]

NASA/JPL NASA Jet Propulsion Laboratory [US]

NASAO National Association of State Aviation Officials [US]

NASAOCARE National Association of State Aviation Officials Center for Aviation Research and Education [US]

NASAP Network Analysis and Systems Application Program

NASA PR NASA Procurement Regulation

NASARR North American Search and Range Radar

NASA-STAR NASA Scientific and Technical Aerospace Reports

NASBLA National Association of State Boating Law Administrators [US]

NASC National AIS (= Aeronautical Information Service) System Center [US]

NASCA National Association of State Cable Agencies [US]

NASCAS NAS/NRC Committee on Atmospheric Sciences [US]

NASCO National Academy of Sciences' Committee on Oceanography [US]
North Atlantic Salmon Conservation Organization [Scotland]

NASCOE National Association of ASCS (= Agricultural Stabilization and Conservation Service) County Office Employees [US]

NASCOM NASA Communications Network

NASCP North American Society for Corporate Planning

NASD National Association of Security Dealers [US]
National Association of Service Dealers [US]

NASDA National Association of State Departments of Agriculture [US]
National Association of State Development Agencies [US]
National Space Development Agency [Japan]

NASDAQ National Association of Security Dealers Automated Quotations [of NASD, US]

NASDU National Amalgamated Stevedores and Dockers Union [UK]

NASF National Association of State Foresters [US]

NASFCA National Automatic Sprinkler and Fire Control Association [US]

NASFM National Association of Store Fixture Manufacturers [US]

NASG North American Strawberry Growers Association [US]

NASI National Academy of Sciences of India

NASIC Northeast Academic Science Information Center [US]

NASIS National Association for State Information Systems [US]
National Automated Sourcing Information System [Canada]

NASL Naval Applied Sciences Laboratory [US]

NASLR National Association of State Land Reclaimationists [US]

NASN National Air Surveillance Network [US]

NAS/NRC National Academy of Sciences/National Research Council [US]

NASOH North American Society for Oceanic History [US]

NASP National Academy of Sciences in Panama
National Aerospace Plane [US]
National Aerospace-Plane Program [US]
National Airport System Plan

NA-SP Nonabrasive-Slightly Polishing

NASPA National Society of Public Accountants

NASPD National Association of Steel Pipe Distributors [US]

NASPO National Association of State Purchasing Officials [US]

NASR National Association of Solvent Recyclers [US]

NASRCP National Association for State River Conservation Programs [US]

NASS National Association of Steel Stockholders [UK]

NASSC National Air and Sea Systems Command [now: NSSC, US]

NASSM National Association of Scissors and Shears Manufacturers [US]

NASSTIE National Association of State Supervisors of Trade and Industrial Education [US]
NASSTRAC National Small Shipments Traffic Conference [US]
NASTD National Association of State Telecommunications Directors [US]
NASTRAN NASA Structural Analysis Program
NASTS National Association for Science, Technology and Society [US]
NASULGC National Association of State Universities and Land-Grant Colleges [US]
NASW National Association of Science Writers [US]
NASWF Naval Air Special Weapons Facility [US]
NAT Nation(al)
Natural
Nature
Node Attached Table
Normal Allowed Time
North Atlantic [also: N AT]
NATA National Air Transportation Association [US]
National Association of Testing Authorities [Australia]
National Automobile Transporters Association [US]
North American Telecommunications Association [US]
North American Telephone Association
Northern Air Transport Association
NATAS National Academy of Television Arts and Sciences [US]
North American Thermal Analysis Society [US]
NATB National Automobile Theft Bureau [US]
NATC Naval Air Test Center [US]
NATCC National Air Transport Coordinating Committee [US]
NATCHEM National Atmospheric Chemistry Database [Canada]
NATCOL Natural Food Colors Association [Switzerland]
NATCS National Air Traffic Control Service [of BOT, UK]
NATD National Association of Test Directors [US]
NATEC Naval Air Technical Evaluation Center [UK]
NATEF National Automotive Technicians Education Foundation [US]
NATEL Nortronics Automatic Test Equipment Language
NATES National Analysis of Trends in Emergency Systems [of EPS, Canada]
NATESA National Alliance of Television and Electronic Service Associations [US]
NATF Naval Air Test Facility [US]
NAT HIST Natural History
NATIE National Association for Trade and Industrial Education [US]
NATII National Association of Trade and Industrial Instructors [US]
NATIS National Information System [of UNESCO]
NATL National Agricultural Transportation League [US]
Naval Aeronautical Turbine Laboratory [US]
NAT'L National [also: NATL]
nat log natural logarithm
NATM New Austrian Tunneling Method
NATMAC National Air Traffic Management Advisory Committee [of CAA, UK]
NATMM National Association of Textile Machinery Manufacturers [now: ATMA, US]
NATO North Atlantic Treaty Organization
NATOA National Association of Telecommunications Officers and Advisors [US]
NATO ASI NATO Advanced Study Institute
NAT ORD Natural Order
NATR National Association of Technical Research [France]
NATS National Activity to Test Software
National Air Traffic Services [UK]
National Association of Temporary Services [US]
Nordisk Avisteknisk Samarbetsnamnd [= Nordic Joint Technical Press Board, Sweden]
NATSOPA National Society of Operative Printers and Assistants [now: SOGAT, UK]
NATSPG North Atlantic Systems Planning Group [also: NAT/SPG [of ICAO]]
NATSU Nominated Air Traffic Service Unit
NATTA Network of Alternative Technology and Technology Assessment [UK]
North American Trackless Trolley Association [US]
NAT/TFG North Atlantic Traffic Forecasting Group
NATTS National Association of Trade and Technical Schools [US]
Naval Air Test Turbine Station [US Navy]
NATU Naval Aircraft Torpedo Unit [US Navy]
NAU Network Addressable Unit
Network Administration Utilities
NAUG National Appleworkers Users Group [US]
NAUOIS Netherlands Association of Users of On-Line Information Systems
NAUS National Airspace Utilization Study
NAUT Nautical
Nautics
NAUW National Association of University Women [US]
NAV Navigable
Navigation
Navigator
Naval
NAVA National Audio Visual Association [US]
NAVAC National Audio-Visual Aids Center [UK]
NAVAID Air Navigation Facility
Navigational Aid [also: Navaid or navaid]
NAVAIDS Navigational Aids [also: Navaids or navaids]
NAVAIR Naval Air Systems Command [US Navy]
NAVAPI North American Voltage and Phase Indicator
NAVAR Navigation and Ranging [also: Navar or navar]
Radar Air Navigation and Control System
NAV ARCH Naval Architecture
NAVARHO Navigation Aid, Rho Radio Navigation System [also: Navarho or navarho]
NAVASCOPE NAVAR Airborne Radarscope [also: Navascope or navascope]
NAVASCREEN NAVAR Ground Screen [also: Navascreen or navascreen]
NAVC National Audiovisual Center
NAVCM Navigation Countermeasures
NAVCOM Naval Communication
NAVCOMMSTA Naval Communications Station [US Navy]
NAVDAC Navigation Data Assimilation Center
Navigation Data Assimilation Computer
NAVDAS Navigation Data Acquisition System

NAVDOCKS Navy Yards and Docks Bureau [US Navy]
NAVEAM Navigational Warning, Eastern Atlantic and Mediterranean
NAVELECSYSCOM Naval Electronics System Command [also: NAVELEX [US Navy]]
NAVEX Navigation Experiment Package [of NASA]
NAVFEC Naval Facilities Engineering Command [also: NAVFAC [US Navy]]
NAVIRO Navy Industrial Relation Office [US Navy]
NAVLAB Naval Laboratory
Navigation Laboratory
NAVMAT Office of Naval Materiel [US Navy]
NAVMC Navy Marine Corps [US Navy]
NAVMED Naval Aerospace Medical Institute [US Navy]
NAVMINDEFLAB Navy Mine Defense Laboratory [US Navy]
NAVOCEANO Naval Oceanographic Office [US Navy]
NAVORD Naval Ordnance
NAVPERS (Bureau of) Naval Personnel [US Navy]
NAVPHOTOCEN Naval Photographic Center [US Navy]
NAVPS Net Asset Value per Share
NAVS Navigation System
NAVSAT Navigational Satellite
Navigation Satellite System [also: NavSat]
NAVSEA Naval Sea Systems Command [US Navy]
NAVSEC Naval Ship Engineering Center [US Navy]
NAVSEP (Specialist Panel) Navigation and Separation of Aircrafts
NAVSHIPS Naval Ship Systems Command [US Navy]
NAVSPASUR Naval Space Surveillance System [US Navy]
NAVSTAR Navigation Satellite Timing and Ranging System
Navigation System using Timing and Ranging
NAVSTAR/GPS Navigation System using Timimg and Ranging/ Global Positioning System
NAVSUP Naval Supply Systems Command [US Navy]
NAVTEC National Association of Vocational-Technical Education Communicators [US]
NAVTRADEVCEN Naval Training Device Center [US Navy]
NAVTRI Navigational Triangle
NAVWEPS Bureau of Naval Weapons [US Navy]
NAW Non Acid Washed
N/AW Night/Adverse Weather (Evaluator)
NAWAPA North American Water and Power Alliance
NAWAS National Warning System [US]
NAWC National Association of Water Companies [US]
NAWDAC National Association for Women Deans, Administrators and Counselors [US]
NAWDEX National Water Data Exchange [of USGS]
NAWF North American Wildlife Foundation [US]
NAWG National Association of Wheat Growers [US]
NAWGF National Association of Wheat Growers Foundation [US]
NAWHSL National Association of Women Highway Safety Leaders [US]
NAWIC National Association of Women in Construction [US]
NAWID National Association of Writing Instrument Distributors [US]
NAWLA North American Wholesale Lumber Association [US]
NAWLT Nitric Acid Weight Loss Test
NAWMP National Association of Waste Material Producers [US]
North American Waterfowl Management Plan [Canada]
NAWPF North American Wildlife Park Foundation [US]
NAWS North American Wolf Society [US]
NAWTWPC Netherlands Association on Wastewater Treatment and Water Pollution Control
NB Naphthyl Butyrate
Narrowband
Narrow Beam
Nebraska [US]
Neutral Beam
Neutron Beam
New Brunswick [Canada]
No Bias (Relay)
Nopol Benzyl
North Borneo
Northbound
North Britain
(Nota Bene) — Note Well; Take Notice
Nb Niobium
NBA Narrowband Allocation
National Bankers Association [US]
National Benzol and Allied Products Association [UK]
National Board of Aviation [Finland]
National Brassfounders Association [UK]
National Buffalo Association [US]
National Building Agency
N-Bromoacetamide
Net Book Agreement
Neuromuscular Blocking Agent
Nitrobenzaldehyde
NBAA National Business Aircraft Association [US]
NBAC National Biotechnology Advisory Committee
NBACSTT New Brunswick Association of Certified Survey Technicians and Technologists [Canada]
NBARN New Brunswick Association of Registered Nurses [Canada]
NBB N-Butylboronate
NBBI Netherlands Organization for Libraries and Information Services
NBBPVI National Board of Boiler and Pressure Vessel Inspectors [US]
NBC Narrowband Conducted
National Book Council [UK]
National Broadcasting Corporation [US]
National Broiler Council [US]
National Building Code
Nonbleeding Cable
Norwegian Bulk Carrier
Nuclear, Biological and Chemical
NBCC National Building Code of Canada
New Brunswick Community College [Canada]
NBCCA Northern British Columbia Construction Association [Canada]
NBCD Natural Binary-Coded Decimal
NBCFAE National Black Coalition of Federal Aviation Employees [US]
NBCV Narrowband Coherent Video

NBCW Nuclear, Biological and Chemical Warfare
NBD Negative Binomial Distribution
Nitrobenzoxadiazole
NBDA National Bicycle Dealers Association [US]
NBDC National Broadcast Development Committee [UK]
New Brunswick Development Corporation [Canada]
NBD-F 4-Fluoro-7-Nitrobenzofurazan
NBDI 4-(Nitrobenzyl)-N,N'-Diisopropylisourea
NBDL Narrowband Data Line
NBE N-Bromoacetylethylenediamine
NBECN New Brunswick Education Computer Network [Canada]
NBER National Bureau of Economic Research [US]
National Bureau of Engineers Registration
NBEPC New Brunswick Electric Power Commission [Canada]
NBETF Neutral Beam Engineering Test Facility
NBFA National Business Forms Association [US]
NBFFO National Board of Fur Farm Organizations [US]
NBFL New Brunswick Federation of Labour [Canada]
NBFM Narrowband Frequency Modulation
NBFU National Board of Fire Underwriters [US]
NBG Near Band Gap
NBGQA National Building Granite Quarries Association [US]
NBH Network Busy Hour
NBHA 4-Nitrobenzylhydroxylamine
NBI Neutral Beam Injection
NBIS New Brunswick Information Service [Canada]
NBIT New Brunswick Institute of Technology [Canada]
NBL Naval Biological Laboratory [US Navy]
New Brunswick Laboratory [of Agricultural Extension Center, Canada]
NBLE Nearly Best Linear Estimator
NBLS New Brunswick Land Surveyors [Canada]
NBM National Buildings Museum [US]
Non-Book Material
NBMB N Binary Digits — M Binary Digits
NBMCR Non-Book Materials Cataloguing Rules
NBMDA National Building Material Distributors Association [US]
NBMPR 4-Nitrobenzyl-6-Thioinosine
Nitrobenzylthioribofuranosylpurine
NBN New Biological Nomenclature
NBO Network Building Out
Network Buildout
Nonbonding Orbital
NBOMB Neutron Bomb
NBOC Network Building Out Capacitor
NBOR Network Building Out Resistor
NBP Normal Boiling Point [also: nbp]
NBPA National Bark Producers Association [now: NBSPA, US]
New Brunswick Potato Agency [Canada]
NBPM Narrowband Phase Modulation
NBPS Neutral Beam Power Systems
NBR Narrowband Radiated
National Board of Roads and Water [Finland]
Nitrile-Butadiene Rubber
Number
NBRF National Biomedical Research Foundation [US]
NBRI National Building Research Institute [South Africa]
NBRPC New Brunswick Research and Productivity Council [Canada]
NBS National Bureau of Standards [now: National Institute of Standards and Technology, US]
National Bureau of Standards [of MOEA, Taiwan]
Natural Black Slate
N-Bromosuccinimide
New British Standard [UK]
Nickel-Bonded Steel
Nitrobenzenesulfenyl
Nordiska Byggforskningsorgans Samarbetsgrupp [= Nordic Building Research Cooperation Group, Norway]
Numeric Backspace (Character)
NBSAC National Boating Safety Advisory Council [US]
NBSC New Brunswick Safety Council [Canada]
2-Nitrobenzenesulfenyl Chloride
NBSCETT New Brunswick Society of Certified Engineering Technicians and Technologists [Canada]
NBSD Night Bombardment Short Distance
NBSFS National Bureau of Standards Frequency Standard [US]
NBS/ICST National Bureau of Standards/Institute for Computer Sciences and Technology [US]
NBSPA National Bark and Soli Producers Association [US]
NBSR National Bureau of Standards Reactor [US]
NBS-SIS National Bureau of Standards — Standard Information Services [now: NIST-SIS]
NBS-SRM National Bureau of Standards — Standard Reference Material [now: NIST-SRM]
NBSV Narrowband Secure Voice
NBT Narrow-Beam Transducer
Nitro Blue Tetrazolium
Null-Balance Transmissometer
NBTD National Board for Technical Development [Sweden]
NBTDR Narrowband Time-Domain Reflectometry
NBTG Nitrobenzylthioguanosine
NBTGR 4-Nitrobenzyl-6-Thioguanosine
NBTL Naval Boiler and Turbine Laboratory [US Navy]
NBTR Narrowband Tape Recorder
NBV Net Book Value
NBVA National Bulk Vendors Association [US]
NBWA National Blacksmiths and Welders Association [US]
NC Nanocomposite
Nanocrystal(line)
(American) National Coarse Thread
Natural Convection
Network Congestion
Network Connect
Network Control
Neutron Content
Neutron Contrast
Nitrocellulose
Nitrocompound
No Change
No Charge
No Circuit
No Coil
No Connection
No Credit

Noise Criterion
Nominating Committee
Non-Contact
Noncorrodible
Non-Crystalline
Normal Cooling
Normally Closed
North Carolina [US]
Not Calculated
Nuclear Capability
Numbering Counter
Numerical Control

N/C Not Connected [also: n/c]
Numerical Control

nC nanocoulomb

NCA Naphthalene Carbonic Acid
National Canners Association [US]
National Capitol Area [US]
National Coal Association [US]
National Communications Association [US]
National Composition Association [US]
National Computer Association [US]
National Constructors Association [US]
National Council of Agriculture [UK]
Naval Communications Annex
New Communities Administration
Noncommutative Algebra
Noncorrodible Aluminum
Northwest Computing Association [US]

NCAC National Council of Acoustical Consultants [US]
Nordic Customs Administrative Council [Finland]

NCACC National Civil Aviation Consultative Committee [UK]

NCAEI National Conference of Applications of Electrical Insulation [US]

NCAER National Council of Applied Economic Research [India]

NCAES National Center for Analysis of Energy Systems [US]

NCAIR National Center for Automated Information Retrieval [US]

NCAM Network Communications Access Method

NCAMR Nordic Council for Arctic Medical Research [Finland]

NCAP Nonlinear Circuit Analysis Program

NCAPC National Center for Air Pollution Control [of Public Health Service, US]

NCAR National Center for Atmospheric Research [of National Science Foundation, US]
National Center for Atmospheric Research [New Zealand]
National Conference on the Administration of Research
National Conference on the Advancement of Research [US]

NCARB National Council of Architectural Registration Boards [US]

NCAS National Council of American Shipbuilders [US]

NCASI National Council (of the Paper Industry) for Air and Stream Improvement [US]

NCAT National Center for Appropriate Technology [of USDOE]

NCAVAE National Committee for Audio-Visual Aids in Education [UK]

NCB National Cargo Bureau [US]
National Coal Board [UK]
Naval Communications Board
Network Control Block

NCBA National Cooperative Business Association [US]

NCBR Near Commercial Breeder Reactor
Nitride Cooled Breeder Reactor
Nordic Committee on Building Regulations [Finland]

NCBI National Cotton Batting Institute [US]

NCBMP National Council of Building Material Producers [UK]

NCBTA Nordic Cooperative of Brick and Tilemakers Associations [Sweden]

N-CBZ N-Carbobenzoxy

NCC National Chamber of Commerce [Bolivia]
National Computer Conference [US]
National Computing Center [UK]
National Cotton Council [US]
National Curriculum Council [UK]
Naturally Commutated Cycloconverter
Nature Conservancy Council [UK]
Network Computer Center
Network Control Center
Network Coordination Center
Nitrogen Carbon Cycle

NCCA National Chemical Credit Association [US]
National Coil Coaters Association [US]
National Concrete Contractors Association [now: ASCC, US]
National Cotton Council of America [US]

NCCAT National Committee for Clean Air Turbulence [US]

NCCC Nebrasca Consolidated Communications Corporation [US]

NCCCC Naval Command, Control and Communications Center [US Navy]

NCCCD National Center for Computer Crime Data [US]

NCCCHE National Certification Commission in Chemistry and Chemical Engineering [US]

NCCDC National Computer Crime Data Center [now: NCCCD, US]

NCCDPC NATO Command, Control and Data Processing Committee

NCCEM National Coordinating Council on Emergency Management [US]

NCCF Network Communications Control Facility

NCCFN National Coordinating Committee on Food and Nutrition [Philippines]

NCCHS National Conference on the Challenge of Health and Safety [Canada]

NCCIA North Carolina Crop Improvement Association [US]

NCCIS NATO Command Control and Information System

NCCL National Council of Canadian Labour
National Council of Coal Lessors [US]

NCCLS National Committee for Clinical Laboratory Standards [US]

NC/CNC Numerical Control/Computer Numerical Control

NCCPB National Council of Commercial Plant Breeders [US]
NCCSS National Central Conference on Summer Schools [US]
NCCU National Conference of Canadian Universities
NCD Negotiated Critical Dates
Network Cryptographic Device
Nicotinamide Cytosine Dinucleotide
NCDA National Cooperative Development Association [India]
NCDC New Community Development Corporation
2-Nitro-4-Carboxyphenyl N,N'-Diphenylcarbamate
NCDEAS National Committee of Deans of Engineering and Applied Sciences
NCDF New Crop Development Fund [Canada]
NCDM National Conference on Disaster Management [Canada]
NCDRH National Center for Devices and Radiological Health
NCE Network Connection Element
NCEA N-(1-Carboxyethyl)alanine
North Central Electric Association [US]
NCEB National Council for Environmental Balance [US]
NATO Communications Electronic Board
NCEC National Chemical Emergency Center [of UKAEA]
National Coast Export Company [US]
National Construction Employers Council [US]
NCECS North Carolina Educational Computing Services [US]
NCEE National Congress on Engineering Education [US]
National Council of Engineering Examiners [US]
NCEFR National Council of Erectors, Fabricators and Riggers [US]
NCEFT National Commission on Electronic Funds Transfer
NCEL Naval Civil Engineering Laboratory [US Navy]
NCEM National Center for Electron Microscopy [of University of California at Berkeley, US]
NCEMT National Center for Excellence in Manufacturing Technology [US Navy]
National Center for Excellence in Metalworking Technology [US]
NCERT National Council of Educational Research and Training [India]
NCES National Center for Educational Statistics [US]
NCET National Council for Educational Technology [US]
NCF National Clayware Federation
National Communications Forum
Nominal Characteristics File
NCFAE National Council of Forestry Association Executives [US]
NCFCA National Congress of Floor Covering Associations [US]
NCFI National Cold Fusion Institute [US]
NCFM National Commission on Food Marketing [US]
National Committee on Fluid Mechanics
NCFMF National Committee for Fluid Mechanic Films [US]
NCFP National Conference on Fluid Power [US]
NCG Network Control Group
Nickel-Coated Graphite
Non-Condensable Gas
NCGA National Computer Graphics Association [US]
National Corn Growers Association [US]
National Cotton Ginners Association [US]
NCGE National Council for Geographic Education [US]
NCGF Nickel-Coated Graphite Fiber
NCGG National Committee for Geophysics and Geodesy [Pakistan]
NCGR National Council for Geocosmic Research [US]
National Council on Gene Resources [US]
NCGT National Council of Geography Teachers [US]
NCH Network Connection Handler
NCHEML National Chemical Laboratory [of MINTECH, UK]
NCHI National Council of the Housing Industry [US]
NCHMT National Capital Historical Museum of Transportation [US]
NCHPTWA National Clearinghouse for Periodical Abstracts Service [US]
NCHRP National Cooperative Highway Research Program [of AASHO, US]
NCHS National Center for Health Statistics [US]
NCHU National Chung-Hsing University [Taiwan]
NCHVRFE National College for Heating, Ventilating, Refrigeration and Fan Engineering [US]
NCI National Cancer Institute [US and Canada]
National Computer Institute [US]
National Computing Industries
Naval Counter-Intelligence
Net Carried Interest
Netherlands Center for Informatics
(Office of) New Concepts and Initiatives [US Air Force]
Nodular Cast Iron
Non-Coded Information
Northeast Computer Institute [US]
NCIC National Cancer Institute of Canada
National Construction Industry Council [US]
National Crime Information Center
NCIDQ National Council for Interior Design Qualification [US]
NCIH National Conference on Industrial Hydraulics [US]
NCIO National Congress of Inventors Organizations [US]
NCISC Naval Counter-Intelligence Support Center [US]
NCISE National Center for Improving Science Education [US]
NCIT National Council on Inland Transport [UK]
NCL National Central Library [UK]
National Chemical Laboratory [India]
National Communications Laboratory [US]
Network Control Language
Node Compatibility List
NCLA National Council of Local Administrators [US]
NCLI National Chemical Laboratory for Industry [of Tukuba Research Center, Japan]
NCLIS National Commission on Libraries and Information Science [US]
NCLT Night Carrier Landing Trainer
NCM Nanocrystalline Material
National Commission of Mathematics [Portugal]
Network Control Module
Nordic Council on Medicine [Sweden]
NCMA National Catalogue Managers Association [US]
National Ceramic Manufacturers Association [US]
National Concrete Masonry Association [US]

	National Contract Management Association [US]
	National Critical Materials Act [US]
NCMB	Nordic Council for Marine Biology [Norway]
NCMC	National Coalition for Marine Conservation [US]
	National Critical Materials Council [US]
NCMCE	National Council of Minority Consulting Engineers [now NABCE, US]
NCME	National Council on Measurement in Education [US]
NCMET	Non-Closed Shell Many Electron Theory
NCMRD	National Center for Management Research and Development
NCMRED	National Council on Marine Resources and Engineering Development [US]
NCMS	National Center for Manufacturing Sciences [US]
	National Classification Management Society [US]
NCMT	Numerical Controlled Machine Tool
NCN	Network Control Node
	Nixdorf Communications Network [Germany]
NCNA	New China News Agency
NCO	National Change of Address
	Network Control Office
	Non-Commissioned Officer
	North Canadian Oils
NCOI	National Council for the Omnibus Industry [UK]
NCOLUG	North Carolina On-Line User Group [US]
NCOR	National Committee for Oceanographic Research [Pakistan]
NCOS	Non-Concurrent Operating System
NCP	Network Control Point
	Network Control Program
NCPA	National Center for Physical Acoustics [of University of Mississippi, US]
	National Cottonseed Products Association [US]
NCPA/UM	National Center for Physical Acoustics, University of Mississippi [US]
NCPC	Northern Canada Power Commission
NCPDM	National Council of Physical Distribution Management [UK]
NCPF	National Council on Private Forests [of American Forestry Association, US]
NCPI	National Clay Pipe Institute [US]
NCP/L	Network Control Program/Local [also: NCP-L]
NCP/LR	Network Control Program/Local-Remote [also: NCP-L/R]
NCPM	National Clay Pot Manufacturers [US]
NCPMA	Noise Control Products and Materials Association [also: NCP&MA, US]
NCPO	Nordic Council for Physical Oceanography [Sweden]
NCPR	National Congress of Petroleum Retailers [now: SSDA, US]
NCP/R	Network Control Program/Remote [also: NCP-R]
NCPS	National Commission on Product Safety [US]
NCPTWA	National Clearinghouse for Periodical Title Word Abbreviations
NCPUA	National Committee on Pesticide Use in Agriculture [Canada]
	National Committee on Pesticide Use in America [US]
NCP/VS	Network Control Program/Virtual Storage [also: NCP-VS]
NCPWB	National Certified Pipe Welding Bureau [US]
NCPWI	National Council on Public Works Improvement [US]
NCQR	National Council for Quality and Reliability [UK]
NCR	National Cash Register
	No Carbon Required
	No Circuit Request
	Non-Conforming Reports
	Northern Capital Region
NCRA	National Cellular Resellers Association [US]
	National Cooperative Refinery Association [US]
NCR-DNA	NCR (Inc.) Distributed Network Architecture
NCRDS	National Coal Resources Data System [of USGS]
NCRE	Naval Construction Research Establishment [UK]
NCRH	National Center for Radiological Health [of Public Health Society, US]
NCRI	National Consumer Research Institute [US]
NCRL	National Chemical Research Laboratory [South Africa]
NCRP	National Committee on Radiation Protection [US]
NCRPM	National Council on Radiation Protection and Measurement [US]
NCRSR	National Commercial Refrigeration Sales Association [US]
NCRT	National College of Rubber Technology [UK]
NCRUCE	National Conference of Regulatory Utility Commission Engineers [US]
NCRV	National Committee for Radiation Victims
NCS	National Communications System [of USDOD]
	National Computer Systems
	National Corrosion Service
	Natural Color System
	Naval Communication Station
	Naval Control of Shipping
	N-Chlorosuccinimide
	Netherlands Computer Society
	Network Controller Software
	Network Control Station
	Network Control System
	Network Computing System
	Network Coordination Station
	No Checking Signal
	Nonconventional System
	Noncrystalline Solid
	North Carolina State [US]
	Numerical Category Scaling
	Numerical Control Society [now: AIMTECH, US]
NCSA	National Construction Software Association [US]
	National Crushed Stone Association [US]
NCSAG	Nuclear Cross-Section Advisory Group
NCSBCS	National Conference of States on Building Codes and Standards [US]
NCSBEE	National Council of State Board of Engineering Examiners [now: NCEE, US]
NCSC	National Computer Security Center [US]
	National Council of Schoolhouse Construction
	North Carolina State College [US]
NCSCR	North Carolina State College Reactor [US]
NCSE	National Center for Science Education [US]
	North Carolina Society of Engineers [US]
NCSI	National Council for Stream Improvement [US]

NCS/ICS Network Controller Software/Integrated Controller Software
NCSL National Conference of Standards Laboratories [US]
Near Coincidence Site Lattice
NCSM National Council of Supervisors of Mathematics [US]
NCSN National Computer Service Network [US]
NCSP Nordic Committee on Salaries and Personnel [Denmark]
Northern Cod Science Program [of DFO, Canada]
NCSPA National Corrugated Steel Pipe Association [US]
NCSS National Cactus and Succulent Society [UK]
NCSTD National Center for Scientific and Technical Documentation [Netherlands]
NCSTRC North Carolina Science and Technology Research Center [also: NC/STRC [of UNC, US]]
NCSTS National Conference of State Transportation Specialists [US]
NCSTTO National Council for the Supply and Training of Teachers Overseas [UK]
NCSU North Carolina State University [US]
NCT Nagaoka College of Technology [Japan]
National Chamber of Trade [UK]
Network Control and Timing
Night Closing Trunks
Noncontact Tonometer
NCTA National Cable Television Association [US]
National Council for Technological Awards [UK]
NCTE National Council for Textile Education [US]
Network Channel Termination Equipment
NCTM National Council of Teachers of Mathematics [US]
NCTS National Council of Technical Schools
Northeast Corridor Transportation System [US]
NCTSI National Council of Technical Service Industries [now: Contract Services Association, US]
NCTT Norwegian Council for Technical Terminology [Norway]
NCU Navigation Computer Unit
North Carolina University [US]
NCUA National Credit Union Administration
NCUAS Northwest College and University Association for Science [US]
NCUC Nuclear Chemistry Users Committee
NCUES National Center for Urban Environmental Studies [US]
NCUG Nevada COBOL Users Group [US]
NCUGAE National Computer User Group in Agricultural Education [UK]
NCUR National Committee for Utilities Radio [now: UTC, US]
NCURA National Council of University Research Administrators [US]
NCUTLO National Committee on Uniform Traffic Laws and Ordinances [US]
NCV Net Calorific Value
No Commercial Value
NCW Non-Communist World
NCWCC North Central Weed Control Committee [US]
NCWQ National Commission on Water Quality [US]
NCWR Nordic Council for Wildlife Research [Denmark]
NCWM National Conference on Weights and Measures [US]
NCW No Change in Weather
ND Needle
Negative Declaration
Network Design
Neutron Density
Neutron Detector
New Drugs
North Dakota [US]
No Data
No Date
No Delay
No Detect
Nomarski DIC (= Differential Interference Contrast)
Normal Direction
Normal Distribution
Normal Duty
Not Detectable
Not Detected
Not Determinable
Not Determined
Not Done
Nucleotidase
Number of Document
Nd Neodymium
5'-ND 5'-Nucleotidase
NDA Naphthaline Dicarboxylic Acid
National Dairymen's Association [UK]
National Defense Academy [Japan]
New Drug Application (Program) [US]
Nondestructive Assay
NDAB Numerical Data Advisory Board
NDAC Nuclear Defense Affairs Committee [of NATO]
NdAD Nicotinamide Deoxyadenosine Dinucleotide
N Dak North Dakota [US]
NDB Nondirectional Beacon
Numeric Database
NDB/L Nondirectional Beacon/Locator
NDBMS Network Database Management System
Nd-BR Butadiene Rubber Based on Neodymium Catalyst
NDC National Dairy Council [US]
National Data Communication
National Defense College [of DND, Canada]
National Design Council
National Development Corporation [Tanzania]
National Dome Council [US]
Negative Differential Conductivity
Network Diagnostic Control
Nondestructive Characterization
NORAD (= North American Aerospace Defense Command) Direction Center
Normalized Device Coordinates
NDCA National Drilling Contractors Association [US]
Nuclear Development Corporation of America
NDCAA National Dutch Civil Aviation Authority
NDCB Nitrodichlorobenzene
NDCT Natural Draft Cooling Tower
NDD Neutron Density Distribution
NDDO Neglect of Diatomic Differential Overlap
NDE Near-Death Experience
Nondestructive Evaluation

	Nondestructive Examination
	Nonlinear Differential Equation
Nde	Neisseria Denitrificans [Molecular Biology]
NDEA	National Defense Education Act [US]
	National Display Equipment Association [UK]
	Nitrosodiethylamine
NDE/CAD	Nondestructive Evaluation/Computer-Aided Design
NDES	Normal Digital Echo Suppressor
NDEx	Nondestructive Examination
NDF	National Development Foundation [South Africa]
	National Drilling Federation [US]
	Neutral Detergent Fiber
	No Defect Found
	Nonlinear Distortion Factor
	Neutral Density Factor
NDGA	Nordihydroguaiaretic Acid
NDHA	National District Heating Association [US]
NDHQ	National Defense Headquarters
NDI	1,5-Naphthalene Diisocyanate
	Nickel Development Institute [Canada]
	Nondestructive Inspection
	Non-Development Item
	Numerical Designation Index
NDIR	Nondispersive Infrared
NDL	Network Definition Language
	Nuclear Defense Laboratory [US Army]
NDLC	Network Data Link Control
NDLM	Nondestructive Laser Mapping
NDM	Negative Differential Mobility
	Normal Disconnected Mode
NDMA	National Dimension Manufacturers Association [US]
	Nitrosodimethylamine
NDMC	National Defense Medical Center [of DND, Canada]
NDMS	Network Design and Management System
nDNA	Native Deoxyribonucleic Acid
N-DNP	N-Dinitrophenyl
NDO	Network Development Office [of Library of Congress, US]
NDOS	New Disk Operating System
NDP	Normal Diametral Pitch
	Nucleoside Diphosphatase
	Numeric Data Processor
NDPA	National Decorating Products Association [US]
	N-Nitroso Diphenylamine
NDPK	Nucleoside Diphosphate Kinase
NDPRB	National Dairy Promotion and Research Board [US]
NDPS	National Data Processing Service [UK]
	National Data Processing System [US]
NDR	National Drivers Register [US]
	Network Data Reduction
	Nondestructive Readout
NDRC	National Defense Research Committee [US]
NDRE	Norwegian Defense Research Establishment
NDRI	National Dairy Research Institute [India]
NDRO	Nondestructive Readout
NDS	New Drug Submission
	Nonparametric Detection Scheme
	Nordic Demographic Society [Norway]
	North Dakota State [US]
NDSA	Napthalene Disulfonic Acid
NDSB	Narcotic Drugs Supervisory Body [of UN]
NDSPE	North Dakota Society of Professional Engineers [US]
NDSU	North Dakota State University [US]
NDT	Net Data Throughout
	Network Description Template
	Newfoundland Daylight Time [Canada]
	Nil Ductility Transition
	Nondestructive Testing
NDTA	National Defense Transportation Association [US]
NDTC	Nondestructive Testing Center
NDTE	Nondestructive Testing Equipment
NDTT	Nil Ductility Transition Temperature
NDU	Notre Dame University [at Paris, France]
NDUV	Nondispersive Ultraviolet
NDWP	National Demonstration Water Project [now: NWP, US]
Nd-YAG	Neodymium-Doped Yttrium Aluminum Garnet
NE	Naval Engineer
	Naval Engineering
	Nebraska [US]
	Negative Earnings
	New Edition
	New England [US]
	No Effect
	Nonelastic Elongation
	Norepinephrine
	Northeast(ern)
	Not Equal to
	Nuclear Energy
	Nuclear Engineer
	Nuclear Engineering
	Nuclear Explosive
N/E	New Edition
Ne	Neon
NEA	National Education Association [US]
	National Electronic Association [US]
	National Erectors Association [US]
	Negative Electron Affinity
	Nuclear Energy Agency [of OECD, France]
	Nuclear Engineering Associates
NEAA	Non-Essential Amino Acid
NEAC	Nippon Electric Automatic Computer
NEACP	National Emergency Airborne Command Post [of USDOD]
NEADAI	National Education Association Department of Audiovisual Instruction
NEAFC	Northeast Atlantic Fisheries Commission [UK]
NEANDC	Nuclear Energy Agency Nuclear Data Committee [France]
NEAP	National Energy Audit Program
NEARA	New England Antiquities Research Association [US]
NEAS	Near East Archeological Society [US]
NEAT	NCR (= National Cash Register) Electronic Autocoding Technique
NEB	National Economic Board [South Korea]
	National Energy Board [of Energy Mines and Resources, Canada]
	National Enterprise Board [UK]
	Noise Equivalent Bandwidth
Neb	Nebraska [US]
NEBB	National Environmental Balancing Bureau [US]

Nebr Nebraska [US]
NEBS New Exporters to Border States [Canada]
NEBSS National Examinations Board in Supervisory Studies
NEBULA Natural Electronic Business Users Language
NEC National Economic Council [UK, Pakistan and Philippines]
National Electric Code [US]
National Electronics Conference [US]
National Electronics Council [UK]
National Engineering Consortium [US]
Nippon Electric Company [Japan]
Not Elsewhere Classified
Nuclear Energy Center
NECA National Electrical Contractors Association [US]
National Exchange Carrier Association [US]
Near East College Association [US]
NECCA National Educational Closed Circuit Association [now: ETA, UK]
NECG National Executive Committee on Guidance
NECIES Northeast Coast Institution of Engineers and Shipbuilders [UK]
NECOS Network Coordinating Station
NECPA National Emergency Command Post Afloat [of USDOD]
NECPUC New England Conference of Public Utility Commissioners [US]
NECS National Electric Code Standards [US]
Nationwide Educational Computer Service [US]
NECSS Nuclear Energy Center Site Survey
NECTA National Electrical Contractors Trade Association [US]
NED New Editor
NEDA National Electronic Distributors Association [US]
National Electronics Development Association [New Zealand]
National Environmental Development Association [US]
National Exhaust Distributors Association [US]
NEDA/Ground National Environmental Development Association Groundwater Project [US]
NEDA/USA National Exhaust Distributors Association/Undercar Specialists Association [US]
NEDC National Economic Development Council [UK]
NEDN Naval Environmental Data Network [US Navy]
NEDO National Economic Development Office [of NEDC, UK]
NEDS National Emissions Data System [of EPA, US]
NEDSA Non-Erasing Deterministic Stack Automation
NEDU Navy Experimental Diving Unit
NEE Naphthylethyl Ether
NEEB North East Engineering Bureau [UK]
Northeastern Electricity Board [UK]
NEED Needle
New Employment Expansion and Development (Program)
NEEDS National Emergency Equipment Data System [of Environment Canada]
New England Educational Data System [US]
NEELS National Emergency Equipment Locator System [of Environmental Protection Service, Canada]
NEEP New England Economic Project [US]
Nuclear Electronic Effects Program [US]
NEERI National Environmental Engineering Research Institute [India]
NEETU National Engineering and Electrical Trades Union [UK]
NEF National Energy Foundation [US]
(American) National Extra Fine Thread
Noise Exposure Forecast
NEFA Non-Esterified Fatty Acid
NEFC Near East Forest Commission
NEFD Noise Equivalent Flux Density
NEFO National Electronics Facilities Organization [US]
NEFSG Northeastern Forest Soils Group [Canada and US]
NEFTIC Northeastern Forest Tree Improvement Conference [US]
NEG Negative
Nonevaporable Getter
NEGB North Eastern Gas Board [UK]
NEGOR Net Current Gas-Oil Ratio
NEHA National Environmental Health Association [US]
NEI Netherlands Economics Institute
Noise Exposure Index
Nordic Energy Index [Denmark]
Not Elsewhere Indicated
NEIC National Earthquake Information Center [US]
National Energy Information Center [US]
NEII National Elevator Industry, Inc. [US]
NEIMME North of England Institute of Mining and Mechanical Engineers [UK]
NEIS National Electrical Industries Show
National Engineering Information System
NEISS National Electronic Injury Surveillance System
NEKDA New England Kiln Drying Association [US]
NEL National Electronics Council [UK]
National Engineering Laboratory [UK]
Naval Electronics Laboratory [US Navy]
Nuclear Electronics Laboratory [US]
NELA National Electric Light Association [now: EEI, US]
Northeastern Loggers Association [US]
NELAT Naval Electronics Laboratory Assembly Tester
NELC Naval Electronics Laboratory Center [US Navy]
NELCON National Electronics Conference [of IEEE, US]
NELCON NZ National Electronics Conference, New Zealand
NELIA Nuclear Energy Liability Insurance Association
NELIAC Navy Electronics Laboratory International Algorithmic Compiler [US Navy]
NELINET New England Library Information Network [US]
NELMA Northeastern Lumber Manufacturers Association [US]
NELPA Northwest Electric Light and Power Association
NELPIA Nuclear Energy Liability Property Insurance Association
NEM N-Ethylmaleimide
Nitrogen Ethylmorpholine
Not Elsewhere Mentioned
NEMA National Electrical Manufacturers Association [US]
National Emergency Management Association [US]
NEMAG Negative Effective Mass Amplifiers and Generators
NEMI National Elevator Manufacturing Industry [now: NEII, US]
National Energy Management Institute [US]

NEMO Non-Empirical Molecular Orbitals
NEMP Nuclear Electromagnetic Pulse
NEMPET Northeast Microbial Physiologists, Ecologists and Taxonomists [US]
NEMRA National Electrical Manufacturers Representatives Association [US]
NEMRIP New England Marine Resources Information Program [US]
NEMS Nimbus E Microwave Spectrometer
NEOCON National Exposition of Contract Interior Furnishings [of IBD, US]
NEOF No Evidence of Failure
Nordic Engineer Officers Federation [Sweden]
NEP National Emphasis Program
National Energy Plan
National Energy Program
N-Ethyl-2-Pyrrolidinone
Never-Ending Program
Noise Equivalent Power
NEPA National Environmental Policy Act [US]
NEPCON National Electronic Packaging and Production Conference [US]
NEPD Noise Equivalent Power Density
NEPE National Emergency Planning Establishment
NEPH Nephelometric
Nephelometry
NEPHIS Nested Phrase-Indexing System
NEPIA Nuclear Energy Property Insurance Association
NEPIS N-Ethyl-5-Phenylisoxazolium-3'-Sulfonate
NEPMA National Engine Parts Manufacturers Association [US]
NEP&ME National Exposition of Power and Mechanical Engineering
NEPR NATO Electronic Parts Recommendation
NEPTUNE Northeastern Electronic Peak Tracing Unit and Numerical Evaluator
NEPZ Nantze Export Processing Zone [Taiwan]
NEQ Not Equal
NER National Engineers Register
Nepalese Rupee [Currency of Nepal]
NERA National Economic Research Associates
NERAC New England Research Application Center [US]
NERBA New England Road Builders Association [US]
NERC National Electric Reliability Council [US]
National Electronics Research Council [now: NEL, UK]
Natural Environmental Research Council [UK]
New England Research Center [US]
New En-Route Center [of NATS, UK]
Nigeria Educational Research Council
North American Electric Reliability Council [US]
Nuclear Energy Research Center [Belgium]
NEREM Northeast Electronics Research and Engineering Meeting [US]
NERFET Negative Resistance Field-Effect Transistor
NERHL Northeastern Radiological Health Laboratory [US]
NERIS National Energy Referral Information System [of Environment Information Center, US]
NERL Nuclear Engineering Research Laboratory [US]
NERO Na (= Sodium) Experimental Reactor of 0 (= Zero) Energy
National Energy Resources Organization [US]
Near Earth Rescue Operation [of NASA]
Near East Regional Office [of FAO — UN]
NERV Nuclear Emulsion Recovery Vehicle
NERVA Nuclear Engine for Rocket Vehicle Applications [of USAEA]
NES National Employment Service
National Engineering Service
National Ergonomic Society [Sweden]
National Estimating Society [US]
Noise Equivalent Signal
Not Elsewhere Specified
NESA National Electric Sign Association [US]
National Energy Specialist Association [US]
NESBAC Northeast Shetland Basin Area Communications [UK]
NESBIC Netherlands Students Bureau for International Cooperation
NESC National Electrical Safety Code [US]
National Environmental Satellite Center [of ESSA, US]
Naval Electronics Systems Command [US Navy]
Nuclear Engineering and Science Conference
NESCA National Environmental Systems Contractors Association [now: ACCA, US]
NESCC Naval Electronics Systems Command Headquarters [US Navy]
NESCOM New Standards Projects Committee [of IEEE, US]
NESDA National Electronic Sales and Service Dealers Association [US]
National Equipment Servicing Dealers Association [US]
Northeast Scotland Development Authority
NESDIS National Environmental Satellite, Data and Information Service [US]
NESP National Environment Studies Project [US]
NESRF Northern Environmental Studies Revolving Fund [Canada]
NESS National Environmental Satellite Service [US]
Network and Evaluation Simulation System
NESSEC Naval Electronic Systems Security Engineering Center [US]
NEST Naval Experimental Satellite Terminal
NESTA National Earth Science Teachers Association [US]
NESTEV Naval Electronics System Test and Evaluation
NESTOR Neutron Source Thermal Reactor
NESTS Non-Electric Stimulus Transfer System
NESW Non-Essential Service Water
NET National Educational Television
Net Equivalent Temperature
Network
Network Equipment Technology
Next European Torus
Nitroso Ester Terpolymer
Noise Equivalent Temperature
NETA National Electrical Testing Association [now: IETA, US]
National Environmental Training Association [US]
NETAC Nuclear Energy Trade Associations Conference [UK]
NETC National Emergency Transportation Center [US]

NET/COM	Networking and Communications Exposition [Canada]
NETD	Noise Equivalent Temperature Difference
NETE	Naval Engineering Test Establishment
NETEC	Network Technical Support Group
NETFS	National Educational Television Film Service [US]
NETH	Netherlands
NET IVHU	Next European Torus/In-Vessel Handling Unit
NETL	National Export Traffic League [US]
NETMUX	Network Multiplexer
NETR	Nuclear Engineering Test Reactor [US Air Force]
NETSET	Network Synthesis and Evaluation Technique
NEU	Northeastern University [US]
NEUDATA	Neutron Data Under Direct Access (Database) [of OECD, France]
NEUG	National Epson Users Group [US]
NEUT	Neutral
NEV	Non-Equivalent (Gate)
Nev	Nevada [US]
NEVE	Non-Empirical Valence-Electron
NEW	National Educators Workshop [US]
	National Engineers Week
	Nuclear Energy Women [US]
NEWA	National Electric Wholesalers Association [now: NAED, US]
	New England Waterworks Association [US]
NEWAC	NATO Electronic Warfare Advisory Committee
NEWRADS	Nuclear Explosion Warning and Radiological Data System
NEWS	Naval Electronic Warfare Simulator
NEWWA	New England Water Works Association [US]
NEXRAD	Next Generation Weather Radar
NEXAFS	Near-Edge X-Ray Absorption Fine Structure
NEXCO	National Association of Export Companies [US]
NEXT	Near-End Crosstalk
NF	(American) National Fine Thread
	National Formulary [Grade of Chemicals]
	Natural Frequency
	Near Face
	Newfoundland [Canada]
	Network Flow
	Neutral Filter
	Neutron Filter
	Nobel Foundation [Sweden]
	No Fracture
	No Funds
	Noise Factor
	Noise Figure
	Noise Frequency
	Nonferrous
	Nose Fuse
	Not Found
	Nozzle Flow
	Nuclear Fission
N/F	No Funds
nF	nanofarad
NFA	National Fertilizers Association [US]
	National Founders Association [now: ACMA, US]
	Network File Access
	Northwest Forestry Association [US]
NFAA	National Federation of Advertising Agencies [now: IFAA, US]
	Northern Federation of Advertisers Associations [Sweden]
NFAC	Naval Facilities Engineering Command Headquarters [US Navy]
NFAH	National Foundation for the Arts and the Humanities
NFAIS	National Federation of Abstracting and Information Services [US]
	National Federation of Abstracting and Indexing Services [Name Changed to: National Federation of Abstracting and Information Services, US]
NFAM	Network File Access Method
NFAP	Network File Access Protocol
NFB	National Film Board [Canada]
	Negative Feedback
	Negative Feeder-Booster
NFBA	National Frame Builders Association [US]
NFBC	National Film Board of Canada
NFBPM	National Federation of Builders and Plumbers Merchants [UK]
NFBS	Nordiske Forskningsbibliotekens Samarbejdskomite [= Nordic Research Libraries Cooperative Committee [Joint Scandinavian Committee]]
NFBW	National Federation of Building Workers [Israel]
NFBWW	Nordic Federation of Building and Wood Workers [Sweden]
NFC	National Fire Code
	National Forensic Center [US]
	National Freight Consortium [UK]
	No Further Consequences
	Not Favorably Considered
NFCA	Nonfuel Core Array
NFCB	National Federation of Community Broadcasters
NFCGA	National Federation of Constructional Glass Associations [UK]
NFCI	National Federation of Clay Industries [UK]
NFCRS	National Fisheries Containment Research Center [US]
NFCS	National Federation of Construction Supervisors [UK]
NFCSIT	National Federation of Cold Storage and Ice Trades [also: NFCIT, UK]
NFCTA	National Federation of Corn Trade Associations [UK]
NFCTC	National Foundry Craft Training Center [UK]
Nfd	Newfoundland [Canada]
NFDA	National Fastener Distributors Association [US]
NFDC	National Federation of Demolition Contractors [UK]
	National Flight Data Center [US]
NFDM	Non-Fat Dry Milk
NFDPS	National Flight Data Processing System [of ICAO]
NFE	National Faculty Exchange [US]
	Nearly Free Electron
	Network Front-End
	Nitrogen-Free Extract
NFEA	National Federated Electrical Association [UK]
NFEAS	National Federation of Engineers, Architects and Surveyors [Israel]
NFEC	National Food and Energy Council [US]

	Naval Facilities Engineering Command [US Navy]
NFEGI	National Federation of Engineering and General Ironfounders [UK]
NFER	Nonferrous
NFERF	National Fisheries Education and Research Foundation [US]
NFET	N-Channel (Junction) Field-Effect Transistor
NFETM	National Federation of Engineers Tool Manufacturers [UK]
NFF	No Fault Found
NFFAW	Newfoundland Fishermen, Food and Allied Workers Union [Canada]
NFFS	National Ferrous Founders Society [US]
	Nonferrous Founders Society
NFFTU	National Federation of Furniture Trade Unions [UK]
NFFWU	Nordic Federation of Factory Workers Unions [Sweden]
NFGC	National Federation of Grain Cooperatives [US]
NFI	National Federation of Ironmongers [UK]
	National Fisheries Institute [US]
	National Forestry Institute [Canada]
NFIA	National Feed Ingredients Association [US]
NFIMA	Nonferrous Ingot Metal Institute [US]
NFISM	National Federation of Iron and Steel Merchants [UK]
NFL	Newfoundland Federation of Labour [Canada]
NFLCP	National Federation of Local Cable Programmers [US]
Nfld	Newfoundland [Canada]
NFLPA	National Freelance Photographers Association [US]
NFM	Narrow Frequency Modulation
	Non-Fat Milk
	Nonferromagnetic
	Nonferrous Metal
	Nuclear Ferromagnetism
NFMA	National Facility Management Association [US]
	National Fireplace Makers Association [UK]
	Norwegian Furniture Manufacturers Association
NFMC	National Federation of Milk Cooperatives [France]
NFMES	National Fund for Minority Engineering Students
NFMOA	National Fish Meal and Oil Association [US]
N-FMOC	N-Fluoroenylmethoxycarbonyl
NFMP	National Federation of Master Painters and Decorators [UK]
NFMPC	Nonferrous Metal Producers Committee [US]
NFMS	Non-Fat Milk Solids
NFMSAEG	Naval Fleet Missile System Analysis and Evaluation Group [US Navy]
NFO	National Freight Organization
NFP	National Fusion Program [Canada]
	Nuclear Fuel Processing
NFPA	National Fire Protection Association [US]
	National Flaxseed Processors Association [US]
	National Flexible Packaging Association [now: FPA, US]
	National Fluid Power Association [US]
	National Food Processors Association [US]
	National Forest Products Association [US]
NFPCA	National Fire Prevention and Control Administration [US]
NFPDE	National Federation of Plumbers and Domestic Engineers [UK]
NFPEDA	National Farm and Power Equipment Dealers Association [US]
NFPMC	National Farm Products Marketing Council
NFPW	National Federation of Professional Workers [UK]
NFQ	Night Frequency
NFR	Negative Flux Rate
	No Further Requirement
	Nuclear Fuel Reprocessing
NFRC	National Federation of Roofing Contractors [UK]
	Northern Forest Research Center [of Canadian Forestry Service]
NFRCC	Nordic Forest Research Cooperation Committee [Norway]
NFS	NASA FAR (= Federal Acquisition Regulation) Supplement
	Network File System
	Not for Sale
	Nuclear Fuel Service
NFSA	National Fertilizer Solutions Association [US]
	National Fire Sprinkler Association [US]
	Navy Field Safety Association [now: ANSP, US]
NFSAIS	National Federation of Science Abstracting and Indexing Services [US]
NFSP	Nuclear Fuel Services Plant
NFSR	National Federation for Scientific Research [Belgium]
NFSWMM	National Federation of Scale and Weighing Machine Manufacturers
NFT	National Federation of Textiles [US]
	Navigation Flight Test
	Networks File Transfer
	Nutrient Film Technique
nft^3	normal cubic feet
NFTA	National Freight Transportation Association [US]
	Nitrogen Fixing Tree Association [US]
NFTB	Niagara Frontier Tariff Bureau [Canada]
NFTC	National Foreign Trade Council [US]
	National Furniture Traffic Conference [US]
NFTSA	National Film, Telvision and Sound Archives [Canada]
NFU	National Farmers Union [Canada]
NFVT	National Federation of Vehicle Trade [UK]
NFWC	National Fire Waste Council
NFY	Notify
NG	Nanoglass
	Narrow Gauge
	National Guard [US]
	Natural Gas
	No Good
	Nitroglycerine
ng	nanogram
NGA	National National Glass Association [US]
	National Graphical Association [UK]
NGAA	Natural Gas Association of America [US]
NGAM	Noble Gas Activity Monitor
NGC	New Galactic Catalogue
	Nordic Geodetic Commission [Denmark]
NGCC	National Guard Computer Center [US]
NGD	Nicotinamide Guanine Dinucleotide

NGDA National Glass Dealers Association [US]
NG-EGDN Nitroglycerine Ethylene Glycol Dinitrate
NGF Nerve Growth Factor
NGFA National Grain and Feed Association [US]
NGFTB Nordic Group for Forest Tree Breeding [Norway]
NGGLT Natural Gas and Gas Liquids Tax
NGI Niederlandische Gesellschaft fur Informatik [= Netherlands Society for Informatics]
NGK New Guinea Kina [Currency of Papua New Guinea]
NGL Natural Gas Liquid
NGM Nested Grid Model
Neutron-Gamma Monte Carlo
NGMA National Gas Measurement Association [US]
National Greenhouse Manufacturers Association [US]
NGN Nigerian Naira [Currency]
NGO National Gas Outlet Thread
Nongovernmental Organization
NGOR Net Gas-Oil Ratio
NGP Network Graphics Protocol
NGPA Natural Gas Policy Act [US]
Natural Gas Processors Association [US]
NGPSA Natural Gas Processors Suppliers Association [now: GPSA, US]
NGR Neutral Grounding Resistor
Nongrain-Raising Stain
Nuclear Gamma-Ray Resonance
NGRI National Geophysical Research Institute [India]
NGRS Narrow Gauge Railway Society
NGS National Gas Straight Thread
National Geographic Society [US]
New Generation System
Nominal Guidance Scheme
Normal Goat Serum
Nuclear Generating Station
Numerical Geometry System
NGSA Natural Gas Supply Association [US]
NGSC Natural Gas Supply Committee [now: NGSA, US]
NGSDC National Geophysical and Solar Terrestrial Data Center [US]
NGSEF National Geographic Society Education Foundation [US]
NGSF Noble Gas Storage Facility
NGT National Gas Taper Thread
Night
Nonsymmetrical Growth Theory
NGTB Negotiable
NGTE National Gas Turbine Establishment [UK]
NGV Natural Gas (for) Vehicles
NH Net Heat
New Hampshire [US]
Nonhygroscopic
nH nanohenry
NHA National Housing Act [Canada]
National Hydropower Association [US]
Northwest Hardwood Association [US]
NHAM National Hose Assemblies Manufacturers [US]
NHAMA National Hose Assemblies Manufacturers Association [US]
NHAW North American Heating and Air Conditioning Wholesalers Association [US]
NHB 4-Hydroxy-3-Nitrobenzoic Acid
NASA Handbook
National Harbours Board
NHC Northwest Horticultural Council [US]
NHD Nevada Highway Department [US]
NHE Normal Hydrogen Electrode
NHELP New Hitachi Effective Library for Programming
NHF Naval Historical Foundation [US]
NHI National Health Institute
NHLA National Hardwood Lumber Association [US]
NHLBI National Heart, Lung and Blood Institute [US]
NHM No Hot Metal
NHMA National Housewares Manufacturers Association [US]
New Hampshire Medical Association [US]
NHMFL National High Magnetic Field Laboratory [of FSU, US]
NHMO NATO Hawk Management Office
NHMRC National Health and Medical Research Council [Australia]
NHNP N-(2-Hydroxy-3-(1-Naphthaloxy)propyl)
NHNP-E N-(2-Hydroxy-3-(1-Naphthaloxy)propyl)-ethylenediamine
NHP Network Host Protocol
Nominal Horsepower
Numeric Handprinting
NHPLO NATO Hawk Production and Logistics Organization
NHPMA Northern Hardwood and Pinewood Manufacturers Association [US]
NHPS New Hampshire Pharmaceutica Association [US]
NHQ Naphthohydroquinone
NHRAIC Natural Hazards Research and Applications Information Center [US]
NHRC National Hydrology Research Center [Canada]
NHRDP National Health Research and Development Program
NHRE National Hail Research Experiment [US]
NHRI National Hydrology Research Institute [Canada]
NHRL National Hurricane Research Laboratory [US]
NHS National Health Service [UK]
NHSC National Home Study Council [US]
NHTPC National Housing and Town Planning Council [UK]
NHTSA National Highway Traffic Safety Administration [US]
NHWA National Heating and Air Conditioning Wholesalers [now: NHAW, US]
NHWMA National Hydraulic Woodsplitter Manufacturers Association [now: LMA, US]
NHWU Non-Heatset Web Unit [of PIA, US]
NI Nautical Institute [UK]
Negative Input
Noise Index
Noninductive
Noninhibit
Northern Ireland
Numerical Index
Numeric Information
N/I Noise to Interference (Ratio)
Ni Nickel
NIA National Ice Association [now: PIA, US]
NIAB National Institute of Agricultural Botany [UK]

NIAC National Industry Advisory Committee [of Federal Communications Commission, US]
National Information and Analysis Center [US]
Nuclear Insurance Association of Canada
NIADA National Independent Automobile Dealers Association [US]
NIAG NATO Industrial Advisory Group
NIAE National Institute of Agricultural Engineering [UK]
National Institute for Architectural Education [US]
NIAG NATO Industrial Advisory Group
NIAID National Institute of Allergy and Infectious Diseases [US]
NIAL Nested Interactive Array Language
NIAM Netherlands Institute for Audiovisual Media
NIAE National Institute of Agricultural Engineering [US]
NIB 4-Hydroxy-3-Iodo-5-Nitrobenzoic Acid
Negative Impedance Booster
Node Initialization Block
Non-Interference Basis
Nordic Investment Bank
Normeninformationsbank [= DIN Databank on Standards, Germany]
NIBA National Industrial Belting Association [US]
NIBESA National Independent Bank Equipment and Systems Association [US]
NIBID National Investment Bank for Industrial Development [Greece]
NIBR Norsk Institut for By- og Regionsforskning [= Norwegian Institute for Urban and Regional Research]
Ni-BR Butadiene Rubber Based on Nickel Catalyst
NIBS National Institute of Building Sciences [US]
Nippon Institute for Biological Science [Japan]
NIBSC National Institute for Biological Standards and Control [UK]
NIC National Indicational Center
National Industrial Council [US]
National Inspection Council
National Institute of Corrections
Near-Infrared Camera
Nearly Instantaneous Companding
Negative Impedance Converter
Network Information Center
Network Interface Control
Newly Industrialized Countries
Newsprint Information Committee [US]
Nicaraguan Cordoba [Currency]
Nineteen-Hundred Indexing and Cataloguing
Noise Isolation Class
Nomarski Interference Contrast
Nonisothermal Calorimeter
Not in Contact
Nuclear Industry Consortium [Belgium]
NICA National Insulation Contractors Association [US]
NI-CAD Nickel-Cadmium (Battery) [also: Ni-Cad or nicad]
NICAM Near-Instantaneous Companded Audio Multiplex
NICAP National Investigations Committee on Aerial Phenomena [US]
NICB National Industrial Conference Board [US]
NICC National Industries Computer Committee [UK]
NICD National Institute of Cleaning and Dyeing [US]
NICE National Institute for Computers in Engineering [US]
National Institute of Ceramic Engineers [of ACS, US]
Normal Input-Output Control Executive
NICEIC National Inspection Council for Electrical Installation Contracting [UK]
NICEM National Information Center for Educational Materials [of University of Southern California, US]
National Information Center for Educational Media [US]
NICET National Institute for Certification in Engineering Technologies [US]
NICIA Northern Ireland Coal Importers Association
NICMOS Near-Infrared Camera and Multi-Object Spectrometer
NICOA National Independent Coal Operators Association [US]
NICOL Nineteen-Hundred Commercial Language
NICS NATO Integrated Communications System
Network Integrity Control System
NICSEM National Information Center for Special Educational Materials [of University of Southern California, US]
NICSMA NATO Integrated Communications Systems Management Agency
NICSO NATO Integrated Communications Systems Organization [now: NACISO]
NICUFO National Investigations Committee on Unidentified Flying Objects [US]
NID Network Inward Dialing
Nuclear Instruments and Detectors [IEEE NPSS Committee]
NIDA National Industrial Distributors Association [now: International Distribution Association, US]
National Institute on Drug Abuse [US]
Northeastern Industrial Development Association [US]
Numerically Integrated Differential Analyzer
NIDB Nigerian Industrial Development Bank
NiDBC Nickel Dibutyl Dithiocarbamate
NIDC Nepal Industrial Development Corporation
Northern Ireland Development Council
NIDER Netherlands Institute for Documentation and Registration
NiDI Nickel Development Institute [Canada]
NiDMC Nickel Dimethyl Dithiocarbamate
NIDOC National Information and Documentation Center [Egypt]
NIDS National Investor Data Service [US]
NIE Not Included Elsewhere
NIEF National Ironfounding Employers Federation [UK]
NIEHS National Institute of Environmental Health Sciences [US]
NIEO New International Economic Order
NIER National Industrial Equipment Reserve
National Institute for Educational Research [Japan]
NIESR National Institute for Economic and Social Research
NIF National Inventors Foundation [US]
Network Information File
Noise Improvement Factor

NIFES National Industrial Fuel Efficiency Service [UK]
NIFO Next-In, First-Out
NIFS National Institute for Farm Safety [US]
NIFTE Neon Indicating Functional Test Equipment
NIFTP Network-Independent File Transfer Protocol
NIGC National Iranian Gas Company
NIGDA National Industrial Glove Distributors Association [US]
NIGMS National Institute of General Medical Sciences [US]
NIG Nordic Industrial Group
NIGP National Institute of Governmental Purchasing [US]
NIH National Institute of Health [US]
Not Invented Here
NIHL Noise Induced Hearing Loss
NIHSA Newfoundland Industrial Health and Safety Association [Canada]
NII Nuclear Installations Inspectorate [UK]
NIICU National Institute of Independent Colleges and Universities [US]
NIIDST National Institute for Information and Documentation in Science and Technology [Romania]
NIIT National, International and Intercontinental Telecommunications Network
NIL Nitrogen Inerting Line
NILI Northern Interior Lumber Industries [of British Columbia, Canada]
NILPT National Institute for Low Power Television [US]
NILUG National Independent Lynx User Group [UK]
NIM National Institute of Metallurgy [South Africa]
Natural Iteration Method
Network Interface Machine
Network Interface Monitor
Newspapers in Microform
Nuclear Instrument Module
NIMA Northern Ireland Ministry of Agriculture
NIMAT Newfoundland Institute for Management Advancement and Training [Canada]
NIMD National Institute of Management Development [Egypt]
NiMH Nickel-Metal Hydride (Battery) [also: Ni-MH]
NIMIS National Instructional Materials Information System [of University of Southern California, US]
NIMMS Nineteen-Hundred Integrated Modular Management System
NIMP New and Improved Materials and Processes
NIMPHE Nuclear Isotope Monopropellant Hydrazine Engine
NIMR National Institute for Medical Research [UK]
NIMSCO NODC (= National Oceanographic Data Center) Index for Instrument Measured Subsurface Current Observation [US]
NINA National Institute Northern Accelerator
NINES Norfolk Information Exchange Scheme [UK]
NINIA Nephelometric Inhibition Immunoassay
NINST Non-Instrument Runway
NIO National Institute of Oceanography [UK]
NIOC National Iranian Oil Company
NIOD Network Inward and Outward Dialing
NIOP National Institute of Oilseed Products [US]
NIOSH National Institute for Occupational Safety and Health [US]
NIOSHTIC NIOSH Technical Information Center [US]
NIP 4-Hydroxy-3-Iodo-5-Nitrophenylacetic Acid
Naval Intelligence Professionals [US]
Nipple
Non-Impact Printer
Not in Practice
Nucleus Initialization Procedure
NIPA National Income and Product Accounts [of USDOC]
Notice of Initiation of Procurement Action
NIPAAm N-Isopropyl Acrylamide [also: NIPAA or NIPAM]
NIPCC National Industrial Pollution Control Council
NIPHLE National Institute of Packaging, Handling and Logistics Engineers [US]
NIPO Negative Input — Positive Output
NIPR National Institute for Personnel Research [South Africa]
NIPS Naval Intelligence Processing System [US Navy]
NMCS (= National Military Command System) Information Processing System [US]
NIPTS Noise-Induced Permanent Threshold Shift
NIR Near Infrared
Next Instruction Register
Nickel-Iron Refinery
Non-Inductive Resistor
NIRA Near Infrared Analysis
NIRB Nuclear Insurance Rating Bureau [US]
NIRI National Information Research Institute [US]
NIRIM National Institute for Research in Inorganic Materials [Japan]
NIRMA Nuclear Information and Records Management Association [US]
NIRMS Noble Gas-Ion Reflection Mass Spectroscopy
NIRNS National Institute for Research in Nuclear Science [UK]
NIRO Nike-Iroquois Rocket
NIROS Nixdorf Real-Time Operating System
NIRR National Institute for Road Research [South Africa]
NIRRD National Institute for Rocket Research and Development [South Africa]
NIRS National Institute for Radiological Science [Japan]
NIRTS New Integrated Range Timing System
NIS National Institute of Science [US]
NATO Identification System
Network Information Service
Network Interface System
Neutron Instrumentation System
Nordiska Ingenjorssamfundet [= Nordic Engineers Association, Sweden]
Not in Stock
Nuclear Instrumentation System
NISA National Independent Sign Association [now: ISA, US]
National Industrial Sand Association [US]
National Industrial Service Association [now: EASA, US]
NISARC National Information Storage and Retrieval Center [US]
NISC National Industrial Space Committee [US]
NISD National Association of Steel Detailing [US]
NISEE National Information Service for Earthquake Engineering [US]

NISI National Institute of Sciences of India
NI-SIL Nickel-Silver
NISL Network Interface Sublayer
NISLAPP National Institute for Science, Law and Public Policy [US]
NISO National Individual Standing Offer [Canada]
National Information Standards Organization [US]
NISPA National Information System for Physics and Astronomy [US]
NIST National Information System for Science and Technology [Japan]
National Institute of Science and Technology [Philippines]
National Institute of Standards and Technology [formerly: National Bureau of Standards, US]
NIST-SIS National Institute of Standards and Technology — Standard Information Services [formerly: NBS-SIS]
NIST-SRM National Institute of Standards and Technology — Standard Reference Material [formerly: NBS-SRM]
NIT Nagoya Institute of Technology [Japan]
Nanjing Institute of Technology [PR China]
Net Ingot Ton
Nevada Institute of Technology [US]
New Industrial Technology
New Information Technology
Nippon Institute of Technology [Japan]
Nitrate
Non-Intelligent Terminal
Nonlinear Inertialess Three-Pole
Norwegian Institute of Technology
NITC National Information Transfer Center [of UNESCO]
NITEP National Incinerator Testing and Evaluation Program [Canada]
NITL National Industrial Transportation League [US]
NITP National Industrial Training Program
NITR National Institute for Telecommunications Research [South Africa]
NITROS Nitrostarch
NIU Network Interface Unit
Northern Illinois University [US]
NIUW National Institute for Urban Wildlife [US]
NIV Negative-Impedance Inverter
NIW Naval Inshore Warfare
NIWR National Institute for Water Research [South Africa]
NIZC National Industrial Zoning Committee [US]
NJ Network Junction
New Jersey [US]
NJAC National Joint Advisory Council [US]
NJB National Job Bank
NJC National Joint Committee
National Joint Council
NJCBI National Joint Council for the Building Industry [UK]
NJCC National Joint Computer Committee [now: AFIPS, US]
National Joint Consultative Committee
NJCEC NATO Joint Communications and Electronic Committee
NJCL Network Job Control Language
NJE Network Job Entry
NJES Nozzle Joint Environmental Simulators
NJGFE Nordic Joint Group for Forest Entomology [Sweden]
NJI Network Job Interface
NJIC National Joint Industrial Council
NJIT New Jersey Institute of Technology [US]
NJP Network Job Processing
NJPA New Jersey Pharmaceutical Association [US]
NJPMB Navy Jet-Propelled Missile Board [US Navy]
NJS Noise Jammer Simulator
NJSPE New Jersey Society of Professional Engineers [US]
NK Natural Killer
NKA Nordisk Kontaktorgan for Atomenergisporgsmal [= Nordic Liaison Committee for Atomic Energy, Denmark]
NKCF Natural Killer Cytotoxic Factor
NKHA National Kerosene Heater Association [US]
NKR Norwegian Krone [also: NKr or Nkr; Currency]
NKW North Korean Won [Currency]
NL National Library
Natural Language
Navy League
Neutron Log
New Line (Character)
Noise Level
No-Load
Not Listed
Nonloaded
N/L Navigation/Localizer
nl (non licet) — it is not permitted
(non liquet) — it is not clear
NLA National Librarians Association [US]
National Library of Australia
National Lime Association [US]
Newfoundland Library Association [Canada]
(Study Group on) New Larger Aeroplanes
Nonlinear Analysis
Normalized Load Access
Northwestern Lumbermen's Association [US]
NLAE National Laboratory for the Advancement of Education
N LAT North Latitude [also: NLAT]
NLB National Lighting Bureau [US]
Nonlinear Buckling
NLBA National Lead Burning Association [US]
Nonlinear Buckling Analysis
NLBMDA National Lumber and Building Material Dealers Association [US]
NLC National Library of Canada
New Line Character
Nonlinear Control
NLCIF National Light Castings Ironfounders Federation [UK]
NLCS Nordic Leather Chemists Society [Denmark]
NLD Nonlinear Distortion
NLDC Newfoundland and Labrador Development Corporation [Canada]
NLDTS Nonlinear Discrete-Time System
NLE National Lighting Exposition
National Livestock Exchange [US]
Nonlinear Equation
Nonlinear Estimation

Nle	Norleucyl
NLEA	National Lumber Exporters Association [US]
NLEF	Nonlinear Electric Field
NLEFM	Nonlinear Elastic Fracture Mechanics
NLG	Netherlands Guilder [Currency]
	Nose Landing Gear
NLGA	National Lumber Grades Authority [US]
NLGI	National Lubricating Grease Institute [US]
NLI	National Lead Industries
	National Lifeboat Institution [UK]
	Natural Language Interface
	Nonlinear Interpolating
NLIMT	Newfoundland and Labrador Institute of Marine Technology [Canada]
NLL	National Lending Library [US]
NLLS	Nonlinear Least Squares
NLLST	National Lending Library for Science and Technology [now: BLLD, UK]
NLM	National Library of Medicine [US]
	Noise Level Monitor
NLMA	National Lumber Manufacturers Association [now: NFPA, US]
NLMB	National Livestock and Meat Board [US]
NLMC	Nordic Labor Market Committee [Denmark]
NLMF	Nonlinear Magnetic Field
NLMI	Newfoundland and Labrador Marine Institute [Canada]
NLNZ	National Library of New Zealand
NLO	Nonlinear Optics
	Nonlocalized Orbital
NLOGF	National Lubricating Oil and Grease Federation [UK]
NLOS	Natural Language Operating System
N-LOS	Non-Line of Sight
NLP	Natural-Language Processing
	Nonlinear Programming
NLPGA	National Liquefied Petroleum Gas Association [now: NPGA, US]
NLQ	Near Letter-Quality
NLR	Nationaal Lucht-en Ruimtevaartlaboratorium [= National Aerospace Laboratory, Netherlands]
	Noise Load Ratio
	No-Load Ratio
	Nonlinear Regression
NLRB	National Labour Relations Board [US]
NLS	National Library of Scotland [UK]
	N-Lauroylsarcosine
	No-Load Speed
	Nonlinear System
NLSA	National Locksmith Suppliers Association [US]
NLSB	National Land Survey Board [Sweden]
NLSLS	National Library of Scotland Lending Services [UK]
NLSMB	National Livestock and Meat Board [US]
NLSP	Network Layer Security Protocol
NLSPA	National Livestock Producers Association [US]
NLT	Non Light-Tight
	Not Less Than
	Not Lower Than
	Net Long Ton
NLTE	Nonlocal Thermodynamic Equilibrium
NLU	Natural-Language Understanding
NLUPP	Northern Land Use Planning Program [Canada]
NLW	Nonlinear Wave
NLWM	Non-Linear Wave Mixing
NM	Nautical Mile [also: NMI or nmi]
	Network Management
	Network Manager
	Network Model
	New Mexico [US]
	Nilsson Model
	Noble Metal
	Noise Margin
	Noise Meter
	No Measurement
	Nonmetal(lic)
	Normal Mode
	Not Meaningful
	Not Measured
	Nuclear Magnetism
	Nuclear Magnetron
	Nuclear Model
Nm	Newton-Meter
N/m	Newton per meter
N-m	Newton-Meter of Torque
nm	nanometer
nm^3	normal cubic meter
NMA	Nadic Methyl Anhydride
	National Management Association [US]
	National Medical Association [US]
	National Microfilm Association [now: AIIM, US]
	National Microform Association [US]
	National Micrographics Association [now: AIIM, US]
	National Museums of Canada
	Network Management Agent
NMAA	National Machine Accountants Association [now: DPMA, US]
	National Materials Advancement Award [of FMS, US]
NMAB	National Materials Advisory Board [US]
NMAP	National Metric Advisory Panel [US]
NMC	National Meteorological Center [US]
	National Museum of Canada
	Naval Materials Command [US Navy]
	Naval Missile Center [US Navy]
	Network Management Center
	Network Measurement Center
	Nonmetallic Cable
NMCC	National Military Command Center [US]
NMCL	Naval Missile Center Laboratory [US Navy]
NMCPP	New Mexico Center for Particle Physics [of UNMA, US]
NMCS	National Military Command System [US]
	Nuclear Materials Control System
NMCSSC	National Military Command System Support Center
NMDA	National Marine Distributors Association [US]
	National Metal Decorators Association [US]
NMDC	National Mineral Development Corporation [India]
NMDPP	Neomenthyldiphenylphosphine
NMDSC	Naval Medical Data Service Center [US Navy]
NME	Naphthylmethyl Ether
	Noise-Measuring Equipment
NMEA	National Marine Educators Association [US]

National Marine Electronics Association [US]
NMEL Navy Marine Engineering Laboratory [US]
Nuclear Mechano-Electronic Laboratory
NMERI National Mechanical Engineering Research Institute [South Africa]
N Mex New Mexico [US]
NMF National Microelectronics Facility
New Master File
NMFA National Mineral Feed Association [now: NFIA, US]
NMFC National Motor Freight Classification
NMFS National Marine Fisheries Service [US]
NMFTA National Motor Freight Traffic Association [US]
NMFWA National Military Fish and Wildlife Association [US]
NMH No-Mar Hammer
Nautical Miles per Hour
NMHC Nonmethane Hydrocarbons
NMHF National Manufacturered Housing Federation [US]
NMHS National Maritime Historical Society [US]
NMHWMS New Mexico Hazardous Waste Management Society [US]
NMI NASA Management Institute [US]
NASA Management Instruction
Nautical Mile [also: nmi or NM]
Non-Maskable Interrupt
NMIA National Military Intelligence Association [US]
NMIMT New Mexico Institute of Mining and Technology [US]
NMIRI Nagoya Municipal Industrial Research Institute [Japan]
NMIS Nuclear Materials Information System
Nuclear Materials Inventory System
NMK Naphthylmethyl Ketone
NMKL Nordisk Metodikkomite for Livsmedel [= Nordic Committee on Food Analysis, Finland]
NML National Metallurgical Laboratory [India]
National Mining Law [US]
Normal
Nuclear Magnetism Log
NMM National Museum of Man
Nautical Miles per Minute
Net Metal Mandrel
Nonmetal Material
Nuclear Magnetic Moment
NMMA National Marine Manufacturers Association [US]
NMMD Nuclear Materials Management Department
NMMSS Nuclear Materials Management and Safeguards System
NMMW Near Millimeter Wave System
NMN Nicotinamide Mononucleotide
NMNH National Museum of National History [US]
Nicotinamide Mononucleotide, Reduced Form
NMNS National Museum of Natural Sciences [Canada]
NMO 4-Methylmorpholine N-Oxide
N-Methylmorpholine Oxide
Normal Moveout
NMOA National Mobile Office Association [now: MMOA, US]
NMOC National Marketing Organization Committee
NMOS Negative Metal-Oxide Semiconductor [also: N MOS or N-MOS]
N-Type Metal-Oxide Semiconductor [also: N MOS or N-MOS]
NMOS RAM N-Type Metal-Oxide Semiconductor — Random-Access Memory
NMP National Meter Programming
N-Methylpenazine
4-Methylpyrrolidone
NMPC National Minority Purchasing Council
NMPDN National Materials Property Data Network
NMPF National Milk Producers Federation [US]
NMPH Nautical Miles per Hour
NMPK Nucleoside Monophosphate Kinase
NMPM Nautical Miles per Minute
NMPS Nautical Miles per Second
NMR National Military Representative [of NATO]
National Missile Range [US]
Normal-Mode Rejection
Nuclear Magnetic Resonance
NMRA National Marine Representatives Association [US]
National Mine Rescue Association [US]
NMRAS Nuclear Materials Report and Analysis System
NMRC National Microelectronics Research Center [Ireland]
NMRG Nanomaterials Research Group [Canada]
NMR-ON Nuclear Magnetic Resonance on Oriented Nuclei
NMRR Normal-Mode Rejection Ratio
NMRS Numerous
NMS National Master Specification [US]
Nautical Miles per Second
Naval Meteorological Service [US Navy]
Networked Messaging System
Network Management Services
Network Management Signal
Network Management System
Neutral Meson Spectrometer
Neutron Monitoring System
Noise Measuring Set
Nordic Metalworkers Secretariat [Sweden]
Norsk Metallurgisk Selskap [= Nordic Metallurgical Society, Sweden]
Nuclear Medical Science
NMSA National Metal Spinners Association [US]
National Moving and Storage Association [US]
NMSC Nonferrous Metals Society of China
Nonmartensitic Structural Component
NMSD Next Most Significant Digit
NMSE National Agenda in Materials Science and Engineering [US]
NMSI National Maximum Speed Limit
N-MSOC N-Methylsulfonylethyloxycarbonyl
NMSS National Multipurpose Space Station
(Office of) Nuclear Material Safety and Safeguards [US]
NMST National Museum of Science and Technology [Canada]
New Materials System Test
NMSU New Mexico State University [US]
NMT N-Methyltransferase
Nordic Mobile Telephone System
Not More Than

NMTA National Metal Traders Association [now: AAIM, US]
Northwest Marine Trade Association [US]
NMTAS National Milk Testing and Advisory Service [UK]
NMTBA National Machine Tool Builders Association [US]
NMTC National Mechanical Trade Council [US]
NMU Nitrosomethylurethane
NMWA National Mineral Wool Association [US]
NN Nearest Neighbor
Negative-Negative (Junction) [also: nn or n-n]
N-N Neutron-Neutron Logging
NNA New Network Architecture
NNAG NATO Naval Armaments Group
NNC National Network Congestion Signal
National Nomination Committee
Non-Noise Certified Aircraft
NNCC National Neutron Cross-Section Center [US]
NNCIAWPRC Netherlands National Committee of the International Association on Water Pollution Research and Control
Norwegian National Committee of the International Association on Water Pollution Research and Control
NNCSC National Neutron Cross-Section Center [US]
National Neutron Cross-Section Committee [US]
NND National Network Dialing
National Number Dialing
Nearest Neighbor Distance
NNDTC National Nondestructive Testing Center [UK]
NNE North-Northeast
NNEC National Nuclear Energy Commission [Brazil]
NNF National Nanofabrication Facility
NNGA Northern Nut Growers Association [Canada]
NNI Noise and Number Index
Non-Nuclear Instrumentation
Non-Numeric Information
NNIC Normalized Noise Isolation Class
NNL Non-Newtonian Liquid
NNNDO Neglect of Non-Neighbor Differential Overlap
NNOA National Naval Officers Association [US]
NNOC Nigerian National Oil Company
NNP Net National Product
Net Naturalization Potential
NNPT Nuclear Nonproliferation Treaty
NNR National Number Routed
Nearest Neighbor Relation
Normalized Noise Reduction
NNRI National Nutrition Research Institute [South Africa]
NNS Near Net Shape
NNSA National Nitrogen Solutions Association [now: NFSA, US]
NNSS Navy Navigational Satellite System [US Navy]
NNV Nederlandse Natuurkundige Vereniging [= Netherlands Association of Natural Sciences]
NNW North-Northwest
NO Negative Output
Nitric Oxide
Normally Open
Normal Operation
North
(Numero) — Number [also: No]
N/O In the Name of
Normally Open
No Nobelium
NOA National Oceanographic Association [US]
NOAA National Oceanographic and Atmospheric Administration [US]
NOAC Nordic Accelerator-Based-Research Committee [Denmark]
NOAO National Optical Astronomy Observatories [US]
NOASR Netherlands Organization for Applied Scientific Research
NOBChCE National Organization of Black Chemists and Chemical Engineers [US]
NOBIN Nederlands Orgaan ter Bevordering van de Informatieverzorging [= Netherlands Organization for Information Policy]
NOBO Nonobjecting Beneficial Owner
NOBT New Orleans Board of Trade [US]
NOC National Occupational Classification [Canada]
National On-Line Circuit [US]
NOCI Nederlandse Organisatie voor Chemische Informatie [= Netherlands Organization for Chemical Information]
NOCO Noise Correlation
NOCP Network Operator Control Program
NOCSAE National Operating Committee on Standards for Athletic Equipment [US]
NOD Network Outward Dialing
Norsk Oseanografisk Datacenter [= Norwegian Oceanographic Data Center]
NODA N-Octyl-N-Decyl Adipate
NODAC Naval Ordnance Data Automation Center [US Navy]
NODAL Network-Oriented Data Acquisition Language
NODAN Noise-Operated Device for Antinoise
NODC National Oceanographic Data Center [US]
Non-Oil Developing Countries
NODE National Operational Display Equipment
NOE Not Otherwise Enumerated
NOEB NATO Oil Executive Board
NOEB-E NATO Oil Executive Board — East
NOEB-W NATO Oil Executive Board — West
NOESS National Operational Environmental Satellite System
NOEU Naval Ordnance Experimental Unit [US Navy]
NOF National Optical Font
NCR (= National Cash Register) Optical Font
Network Operations and Facilities
NOTAM (= Notice to Airmen) Office
NOFA National Organic Farmers Association [US]
N of Eng North of England
NOFI National Oil Fuel Institute [US]
NOFMA National Oak Flooring Manufacturers Association [US]
NOGAD Noise-Operated Gain Adjusting Device
NOGAP Northern Oil and Gas Action Program [of Environment Canada]
NOGMA National Ornamental Glass Manufacturers Association [US]
NOHC National Open Hearth Committee [US]
NOHP Not Otherwise Herein Provided

NOHS National Oceanographic Hazard Survey [of NIOSH, US]
NOI Node Operator Interface
Notice of Inquiry
Not Otherwise Indexed
NOIA National Ocean Industries Association [US]
Newfoundland Ocean Industries Association [Canada]
NOIBN Not Otherwise Indicated by Name
Not Otherwise Indicated by Number
NOISE National Organization to Insure a Sound-Controlled Environment [US]
NOJC National Oil Jobbers Council [now: PMAA, US]
NOK Norwegian Krone [Currency]
NOL National Ordnance Laboratory [US]
Naval Ordnance Laboratory [now: NSWL, US]
Net Operating Loss
Normal Operating Loss
NOM Nomenclature
Nominal
NOMA National Office Management Association [US]
National Organization of Minority Architects [US]
NOMAD Naval Oceanographic Meteorological Automatic Device
Nominal Michigan Algorithmic Decoder
NOMDA National Office Machine Dealers Association [US]
NOMEN Nomenclature
NOML Nominal
NOMMA National Ornamental and Miscellaneous Metals Association [US]
NOMSA National Office Machine Service Association [US]
NOMSS National Operational Meteorological Satellite System [US]
NOM STD Nominal Standard
NONADD Nonadditivity
NONCOHO Noncoherent Oscillator
NON-COM Non-Commissioned (Officer)
NON COND Non Condensing
NONP Non-Precision Approach Runway
N-on-P Negative-on-Positive
non seq (non sequitur) — it does not follow
NON STD Non Standard
NOO Naval Oceanographic Office [of US Navy]
NOOP No Operation (Instruction) [also: NO-OP, NO OP or NOP]
NOP No Operation (Instruction)
Not Otherwise Provided For
NOPA National Office Products Association [US]
NOPI Negative Output — Positive Input
NOP-N Nordiska Publikeringsnamnden for Naturvetenskap [= Nordic Publishing Board in Science, Finland]
NOPOL No Pollution
NOPON P-Bis[2,5(5-α-Naphthyloxazolyl)]benzene
NOPT No Procedure Turn [also: NoPT]
NOR Nitrogen Oxide Reduction
Normal
North(ern)
Norway
Norwegian
Not and Or
Not Or
NORAC No Radio Contact
NORAD North American Aerospace Defense Command
North American Air Defense Command [Name Changed to: North American Aerospace Defense Command]
Norwegian Agency for International Development
NORC National Opinion Research Center [of University of Chicago, US]
Naval Ordnance Research Computer
NORCUS Northwest College and University Association for Science [US]
NORDA Naval Ocean Research and Development Activity [US]
NORDCO Newfoundland Oceans Research and Development Corporation [Canada]
NORDDOK Nordic Committee on Information and Documentation
NORDEL Organization for Nordic Electrical Cooperation [Norway]
NORDFORSK Nordiska Forskningsdelegationen [= Nordic Research Council [A Joint Scandinavian Council]]
NORDICOM Nordic Documentation Center for Mass Communication Research [Denmark and Sweden]
NORDINFO Nordic Council for Scientific Information and Research Libraries [Finland]
NORDITA Nordic Institute for Theoretical Atomic Physics [Denmark]
NORDO No Radio
NORDPLAN Nordic Institute for Studies in Urban and Regional Planning [Sweden]
NORDPOST Nordic Postal Union Conference [of NPU, Denmark]
NORGEN Network Operations Report Generator
NORIANE Normes et Reglements — Informations Automatisees Accessibles en Ligne [= Standards and Regulations — On-Line Automated Information [of AFNOR, France]
NORINDOK Norsk Komite for Informasjon og Dokumentasjon [= Nordic Committee for Information and Dokumentation, Norway]
NORM Normalization
Not Operationally Ready Maintenance
NORMARC Norwegian MARC (= Machine-Readable Catalogue)
NORP New Oil Reference Pricing
NORPAT Northern Patrol [of Canadian Armed Forces]
NORPAX North Pacific Experiment
NORS Not Operationally Ready Supply
NORSAR Norwegian Seismic Array
NORSAT Norwegian Domestic Satellite
NORTHAG Northern Army Group Central Europe [of NATO]
Northants Northamptonshire [UK]
Northumb Northumberland [UK]
NORVEN Comision Venezolana de Normas Industriales [= Venezuelan Commision for Industrial Standards]
NORW Norway
Norwegian
NOS National Ocean Service [US]
National Ocean Survey [US]
Naval Ordnance Station [US Navy]
Network Operating System
Night Observation System
Nosing

	Not Otherwise Specified
	Numbers
NOSA	National Occupational Safety Association [South Africa]
	National One-Write Systems Association [US]
NOS/BE	Network Operating System/Batch Environment [also: NOS-BE]
NOSC	Naval Ocean Systems Center [US]
NOSFER	Nouveau Systeme Fondamental pour la Determination des Equivalents de Reference [= New Master System for the Determination of Reference Equivalents]
NOSIG	No Significant Change
NOSMO	Norden Optics Setting, Mechanized Operation
NOSP	Network Operating Support Program
NOSS	National Oceanic Satellite System
	National Orbiting Space Station
	Nimbus Operational Satellite System
Not	Nocardia Otidiscaviarum [Molecular Biology]
NOTAEI	National Old Timers Association of the Energy Industry [US]
NOTAL	Not Sent to All Addresses
NOTAM	Notice to Airmen
NOTAR	No Tail Rotor Anti-Torque System
NOTIS	Network Operations Trouble Information System
	Northwestern On-Line Totally Integrated System [of NWU, US]
NOTRTR	National Organization of Test, Research and Training Reactors [US]
Notts	Nottinghamshire [UK]
NOTU	Naval Operational Training Unit [of US Navy]
NOV	November
NOVRAM	Nonvolatile Random-Access Memory
NOW	Negotiable Order of Withdrawal
NOWEA	Nordwestdeutsche Ausstellungsgesellschaft [= Exhibition Society of Northwest Germany]
NOx	Nitrous Oxide [also: NOX]
	Oxide of Nitrogen
NOZ	No Operating Zone
NP	Nameplate
	Nanophase
	(American) National Pipe Thread
	Natural Polymer
	Natural Product
	Naval Publication
	N-Doped P-Doped (Junction) [also: np or n-p]
	Negative-Positive (Semiconductor) [also: np or n-p]
	Net Proceeds
	Network Protocol
	Neuropeptide
	Neutrino Patents
	New Paragraph
	Nodal Point
	Noise Power
	Nondeterministic Polynomial
	Nonpolar
	Nonpurgeable
	No Pagination
	No Parity
	No Pin
	No Print
	Normal Phase
	Northern Pine
	North Pacific
	North Pole
	Not Provably
	Nuclear Pile
	Nuclear Physics
	Nuclear Polarization
	Nuclear Power
	Nuclear Propulsion
	Nucleoside Phosphorylase
	Nuevo Peso [Currency of Uruguay]
N/P	Nameplate
Np	Neper [Unit]
	Neptunium
NPA	National Parks Association [US]
	National Particleboard Association [US]
	National Payphone Association [US]
	National Petroleum Association [US]
	National Pilots Association [US]
	National Planning Association [US]
	National Production Authority [US]
	National Productivity Award [Canada]
	Nebraska Pharmaceutical Association [US]
	Normal Pressure Angle
	Notice of Proposed Amendment
	Numbering Plan Area
	Numerical Production Analysis
nPa	nanopascal
NPAC	National Pipeline Agency of Canada
	National Project in Agricultural Communications [US]
NPAI	Network Protocol Address Information
NPB	National Plant Board [US]
NPBA	National Paper Box Association [now: NP&PA, US]
NPC	NASA Publication Control
	National Patent Council [US]
	National Peach Council [US]
	National Peanut Council [US]
	National Petroleum Council [US]
	National Pharmaceutical Council [US]
	National Philatelic Center [Canada]
	National Ports Council [UK]
	National Postal Museum
	National Potato Council [US]
	National Power Conference
	NATO Pipeline Committee
	Natural Products Chemistry
	Naval Photographic Center [US Navy]
	Niagara Parks Commission [Canada]
	Normal-Phase Chromatography
NPCA	National Paint and Coatings Association [US]
	National Parks and Conservation Association [US]
	National Pest Control Association [US]
	National Precast Concrete Association [US]
NPCC	Northeast Power Coordinating Council [US]
NPCF	National Pollution Control Federation [US]
NPD	National Paint Distributors [US]
	National Power Demonstration
	Network Protective Device

	Neutron Powder Diffractometer
	North Polar Distance
	Nuclear Physics Division [of SERC, UK]
	Nuclear Power Demonstration [Canada]
	Nuclear Power Development
NPDA	Nitrophenylenediamine
NPDES	Nuclear Pollution Discharge Elimination Specification
	Nuclear Pollution Discharge Elimination System
NPDS	Nuclear Particle Detection System
NPE	Natural Parity Exchange
	Nuclear Power Engineering
NPEA	National Printing Equipment Association [US]
NPEC	Nuclear Power Engineering Committee [of IEEE PES]
NPEF	New Product Evaluation Form
NPEGE	Nonyl Phenyl Eicosa-Ethylene Glycol Ether
NPE/PES	Nuclear Power Engineering/Power Engineering Society [IEEE Committee]
NPES	National Printing Equipment Show [US]
NPESA	National Printing Equipment and Supply Association [US]
NPF	National Park Foundation [US]
	Nuclear Power Facility
	Nuclear Problems Forum
NPFO	National Power Field Office [US Army]
NPFSC	North Pacific Fur Seal Commission
NPG	Neopentyl Glycol
	Nuclear Planning Group [of NATO]
	Nuclear Power Group [UK]
NPGA	National Propane Gas Association [US]
NPGPA	Non-Powder Gun Products Association [US]
NPGS	Nuclear Power Generating Station
NPGTC	National Prairie Grouse Technical Council [US]
NPH	Nitrophenylhydrazine
NPHA	Nitrosophenyl Hydroxylamine
NPI	National Purchasing Institute [US]
	Net Profits Interest
	Nordic Productivity Institute [Norway]
	Northampton Polytechnic Institute [UK]
NPIA	National Photography Instructors Association [US]
NPIAS	National Plan of Integrated Airport Systems [US]
NPIN	N-Doped P-Doped Intrinsic N-Doped (Transistor) [also: npin or n-p-i-n]
	Negative-Positive-Intrinsic (Transistor) [also: npin or n-p-i-n]
NPIP	National Poultry Improvement Plan [US]
	N-Doped P-Doped Intrinsic P-Doped (Transistor) [also: npip or n-p-i-p]
	Negative-Positive-Intrinsic (Transistor) [also: npip or n-p-i-p]
NPIRI	National Printing Ink Research Institute [US]
NPIS	National Physics Information System [of American Institute of Physics, US]
NPIU	Network Processing and Interface Unit
NPK	Nitrogen, Phosphorous and Potassium (Fertilizer)
NPL	National Physical Laboratory [UK and India]
	National Physics Laboratory [US]
	Natural Process Limits
	New Programming Language
NPLC	National Pedigree Livestock Council [US]
NPLG	Navy Program Language Group [US Navy]
NPLO	NATO Production and Logistics Organization
NPM	Nanophase Material
	National Postal Museum [Canada]
	Nuclear Paramagnetism
NPMA	National Property Management Association [US]
NPMI	Nordic Pool for Marine Insurance [Sweden]
NPMR	National Premium Manufacturers Representatives [now: IMRA, US]
NPN	N-Doped P-Doped N-Doped (Transistor) [also: npn or n-p-n]
	Negative-Positive-Negative (Transistor) [also: npn or n-p-n]
	Nonprotein Nitrogen
	N-Propyl Nitrate
NPNP	Negative-Positive-Negative-Positive (Transistor) [also: npnp or n-p-n-p]
NPO	1-Naphthyl-5-Phenyloxazole
	Negative Positive Zero
	Nonpenetrating Orbit
	Nuclear Power Operator
NPOC	Nonpurgeable Organic Carbon
np or d	no place or date
NPP	Network Protocol Processor
	Notice of Proposed Procurement
	Nuclear Power Plant
NPPA	National Press Photographers Association [US]
	Northwest Pulp and Paper Association [US]
NP&PA	National Paperbox and Packaging Association [US]
NPPAC	National and Provincial Parks Association of Canada
NPPB	National Potato Promotion Board [US]
NPPC	National Pork Producers Council [US]
NPPD	5-(4-Nitrophenyl)-2,4-Pentadienal
NPPSO	Naval Publications and Printing Service Office [US Navy]
NPR	National Public Radio
	Nepalese Rupee [Currency of Nepal]
	Net Profit Royalty
	New Production Reactor [US]
	Noise Pollution Ratio
	Noise/Power Ratio
	Noise Preferential Route
	Nozzle Pressure Ratio
	Nuclear Power Reactor
	Numeric Position Readout
NPRA	National Personal Robot Association [US]
	National Petroleum Refiners Association [US]
	Newspaper Personnel Relations Association [US]
NPRC	Nuclear Power Range Channel
NPRCA	National Petroleum Radio Coordinating Association [US]
NPRCG	Nuclear Public Relations Contact Group [Italy]
NPRD	Nonelectronic Parts Reliability Data
NPRDS	Nuclear Plant Reliability Data System
NPRF	Northrop Pulse Radiation Facility [US]
NPRL	National Physical Research Laboratory [South Africa]
	Nonprocedural Referencing Language
NPRM	Notice of Proposed Rule Making

NPRP Northern Pecan Research Program [Canada and US]
NPRTZ Nonpolarized Return-to-Zero Recording [also: NPRZ]
NPS National Petroleum Show
Naval Postgraduate School [US Navy]
Network Processing Supervisor
Network Processor System
Nitrided Pressureless Sintering
Nitrophenylsulfenyl
Nominal Pipe Size
Nuclear and Plasma Sciences [IEEE Society]
Nuclear Power Station
Numerical Plotting System
NPSA New Program Status Area
NPSC (American) National Straight Pipe Thread in Pipe Couplings
NPSD Neutron Power Spectral Density
NPSF (American) National Straight Pipe Thread for Dryseal Joints
NPSH (American) National Straight Pipe Thread for Hose Couplings
Net Positive Suction Head
Niagara Parks School of Horticulture [Canada]
NPSHA Net Positive Suction Head Available
NPSHR Net Positive Suction Head Required
NPSL (American) National Straight Pipe Thread for Locknuts
NPSM (American) National Straight Pipe Thread for Mechanical Joints
NPSP N-(Phenylseleno)phthalimide
NPSRA Nuclear-Powered Ship Research Association [Japan]
NPSS Nuclear and Plasma Sciences Society [of IEEE, US]
NPSWL New Program Status Word Location
NPT (American) National Pipe Thread
Network Planning Technique
Nonproliferation Treaty
Nuclear Proliferation Treaty
Normal Pressure and Temperature
NPTA National Paper Trade Association [US]
NPTC National Private Truck Council [US]
NPTF National Pipe Thread, Female
(American) National Pipe Thread for Dryseal Joints
Nuclear Proof Test Facility
NPTM National Pipe Thread, Male
NPTR (American) National Pipe Thread for Railing Fixtures
NPTS Noise Parameter Test System
NPU National Pharmaceutical Union [UK]
Naval Parachute Unit [US Navy]
Network Planning Unit
Network Processing Unit
Nordic Postal Union [Denmark]
NPUG National Prime (Computer System) User Group [US]
NPV Net Present Value
Nitrogen Pressure Valve
No Par Value
NPVLA National Paint, Varnish and Lacquer Association [now: NPCA, US]
NPY Neuropeptide Y
NPYRR Nitrosopyrrolidine
NQ Nitroquinoline
NQA National Quality Award [US]
NQAA Nuclear Quality Assurance Agency [US]
NQd Non-Quaded
NQO Nitroquinoline Oxide
NQPC National Quartz Producers Council [US]
NQR Nuclear Quadrupole Resonance
NQRS Nuclear Quadrupole Resonance Spectroscopy
NR Natural Rubber
Navigational Radar
Negative Resistance
Nepalese Rupee [Currency of Nepal]
Nitrile Resin
Nitrile Rubber
Noise Rating
Noise Ratio
Noise Reduction
Nonradiative
Nonradioactive
Nonreactive
Nonrecoverable
Non-Return (Valve)
Not Recommended
Nuclear Reaction
Nuclear Reactor
Number
Number of Report
NRA NASA Research Announcement
National Railway Association
National Reclamation Association [now: NWRA, US]
National Renderer Association [US]
National Rifle Association [US]
National Rivers Authority [UK]
Naval Radio Activity
Network Resolution Area
Nuclear Reaction Analysis
NRAO National Radio Astronomy Observatory [US]
NRB Natural Resources Board
Nuclear Reactors Branch [of USAEC]
Nuclear Resonance Broadening
NRC National Recyclers Coalition [US]
National Referral Center [of Library of Congress, US]
National Remodelers Council [of NAHB, US]
National Replacement Character
National Research Council [US and Canada]
National Resources Council [US]
National Rocket Club [now: National Space Club, US]
Nation's Report Card [US]
Newfoundland Safety Council [Canada]
Niger River Commission [Nigeria]
Noise Rating Curve
Noise Reduction Coefficient
Nonrecurring Change
Nonrecurring Charge
Nonrecurring Connection
Noranda Research Center [Canada]
Notch Root Contraction

	Nuclear Reactor Control
	Nuclear Regulatory Commission [US]
NRCA	National Roofing Contractors Association [US]
	National Resources Council of America [US]
NRCC	National Research Council of Canada
NRCCL	Norwegian Research Center for Computers and Law
NRCC/NAE	National Research Council of Canada/National Aeronautical Establishment
NRCD	National Reprographic Center for Documentation [UK]
NRCMAI	National Railroad Construction and Maintenance Association, Inc. [US]
NRCMF	National Referral Center Master File [of Library of Congress, US]
NRCP	National Research Council of the Philippines
NRCS	Nuclear Reactor Control System
NRCSA	National Registration Center for Study Abroad [US]
NRCST	National Referral Center for Science and Technology [of Library of Congress, US]
NRCT	National Research Council of Thailand
NRD	Natural Resources Division [of UNECA]
	Negative Resistance Diode
	Network Resource Dictionary
	Nonradiative Dielectric
NRDB	Natural Rubber Development Association [Malaysia]
NRDC	National Research Development Corporation [now: BTG, UK]
	National Research Development Council [US]
	National Resources Defense Council [US]
	Natural Resources Defense Council
NRDCA	National Roof Deck Contractors Association [US]
NRDF	Nonrecursive Digital File
NRDL	Naval Radiological Defense Laboratory [US Navy]
NRDS	Nuclear Rocket Detection System
	Nuclear Rocket Development Station [US]
NRE	Naval Research Establishment [now: DREA, Canada]
NRe	Nepalese Rupee [Currency of Nepal]
NRECA	National Rural Electric Cooperative Association [US]
NREL	National Renewable Energy Laboratory [formerly: SERI, US]
NRF	National Roofing Foundation [US]
	Nuclear Reactor Fuel
NRFC	National Railroad Freight Committee [US]
NRFD	Not Ready for Data
NRFSA	Navy Radio Frequency Spectrum Activity
NRGSM	National Research Group for the Structure of Matter [Italy]
NRHA	National Retail Hardware Association [US]
NRHC	National Rivers and Harbours Congress [US]
NRHM	National Railway Historical Society [US]
NRHQ	Northern Region Headquarters
NRI	National Resources Institute [of University of Maryland, US]
	Net Radio Interface
	Net Revenue Interest
	Nomura Research Institute [Japan]
	Nonrecurring Installation Charge
NRIA	National Railroad Intermodal Association [US]
NRIC	National Reserves Investigation Committee [UK]
NRIM	National Research Institute for Metals [Japan]
NRIMS	National Research Institute for Mathematical Sciences [South Africa]
NRIP	Number of Rejected Initial Pickups
NRJ	Non-Reciprocal Junction
NRL	National Reference Library
	Naval Research Laboratory [US Navy]
	Nelson Research Laboratory [UK]
	Network Restructuring Language
	Nuclear Reactor Laboratory [US]
	Nutrition Research Laboratory [India]
NRLA	Northeastern Retail Lumber Association [US]
NRLC	National Railway Labor Conference [US]
NRLDA	National Retail Lumber Dealers Association [now: NLBMDA, US]
NRLM	National Research Laboratory of Metrology [Japan]
NRLSI	National Reference Library for Science and Invention [UK]
NRM	Natural Remanent Magnetization
	Normalize
	Normal Response Mode
	Northern Rocky Mountains
NRMA	National Reloading Manufacturers Association [US]
	National Retail Merchants Association
	Nuclear Records Management Association
NRMCA	National Ready Mixed Concrete Association [US]
NRMEC	North American Rockwell Microelectronics Company [US]
NRML	Normal
NRMLS	New England Regional Medical Library Service [US]
NRMM	National Register of Microform Masters [of Library of Congress, US]
NRMR	Netherlands Research and Materials Testing Reactor
NRMS	Network Reference and Monitor Station
NRN	Negative Run Number
	Noise Rating Number
NROTC	Naval Reserve Officers Training Corps [US Navy]
NRP	Normal Rated Power
NRPB	National Radiological Protection Board [UK]
NRPDA	National Radio Parts Distributors Association [now: NEDA, US]
NRPRA	Natural Rubber Producers Research Association [UK]
NRRA	National Resource Recovery Association
NRRFSS	National Research and Resource Facility for Submicron Structures [US]
NRR	Nonreactive Resistor
	Nonrepeatable Runout
	(Office of) Nuclear Reactor Regulation [US]
NRRI	National Regulatory Research Institute [US]
NRRL	Northern Regional Research Laboratory [US]
NRRO	Nonrepeatable Runout
NRRS	Naval Radio Research Station
NRS	Naval Radio Station
	Naval Rocket Society [US]
	Nonrising Stem
	Nuclear Reaction Spectrometry
NRSA	National Rubber Shippers Association [US]
	National Rural Studies Association [UK]

NRSEP	National Roster of Scientific and Engineering Personnel
NRST	Nonreactive Solute Transport
NRSTP	National Register of Scientific and Technical Personnel
NRSW	Nuclear River Service Water
NRT	National Response Team
	Net Register Ton
	Neutron Radiographic Testing
	Neutron Radiography
	Nonradiating Target
NRTS	National Reactor Test Station [of INEL, US]
NRTSC	Naval Reconnaissance and Technical Support Center [US Navy]
NRU	National Research Universal
	Nuclear Reactor Universal
NRV	Net Recovery Value
NRW	Narrow
	Nuclear Radwaste
NRWO	Nuclear Radwaste Operator
NRX	National Research Experimental
	NERVA (= Nuclear Engine for Rocket Vehicle Applications) Reactor Experimental [of USAEC]
	Nuclear-Engine Reactor Experiment [of USAEC]
	Nuclear Reactor Experiment
NRX-CX	NERVA (= Nuclear Engine for Rocket Vehicle Applications) Reactor Critical Assembly [of USAEC]
	Nuclear-Engine Reactor Critical Assembly [of USAEC]
NRZ	Nonreturn-to-Zero (Recording)
NRZC	Nonreturn-to-Zero Change Recording [also: NRZ-C]
NRZI	Nonreturn-to-Zero Indicator
	Nonreturn-to-Zero Inverted Recording
	Nonreturn-to-Zero I (= One)
NRZL	Nonreturn-to-Zero Level [also: NRZ-L]
NRZM	Nonreturn-to-Zero Mark Recording [also: NRZ-M]
NRZ1	Nonreturn-to-Zero Change on One [also: NRZ-1]
NS	National Special Thread
	National Standard
	Navigation System
	Near Side
	Network Security
	Neuroelectric Society [US]
	Neutron Spectrometer
	Neutron Star
	New Style
	Nickel Steel
	Nonsequenced
	Non Staining
	Normal State
	Notch Strength
	Not Specified
	Not Sufficient
	Nova Scotia [Canada]
	Nuclear Spectroscopy
	Nuclear System
	Na (= Sodium) Silicate
N/S	Not Sufficient
Ns	Nimbo-Stratus
ns	nanosecond [also: nsec]
NSA	Naphthalenesulfonic Acid
	National Safflower Association [US]
	National Sawmilling Association [UK]
	National Security Agency [US]
	National Shellfisheries Association [US]
	National Shipping Authority [US]
	National Slab Association [US]
	National Standards Association [US]
	National Stereoscopic Association [US]
	National Stone Association [US]
	National Sunflower Association [US]
	Netherlands Society for Automation
	Nonenyl Succinic Anhydride
	Nonsequenced Acknowledgement
	Norwegian Standards Association
	Nuclear Suppliers Association [US]
NSAA	National Sulfuric Acid Association [UK]
	Nova Scotia Association of Architecture [Canada]
NSAC	Norwegian Society of Automatic Control
	Nova Scotia Agriculture College [Canada]
	Nuclear Safety Analysis Center
NSAE	National Society of Architectural Engineers [US]
NSaF	National Sanitation Foundation
NSAI	Nonsteroidal Anti-Inflammatory
NSAID	Nonsteroidal Anti-Inflammatory Drug
NSAP	Network Service Access Point
NSAS	National Smoke Abatement Society [UK]
NSB	National Science Board [US]
	Naval Standardization Board [US Navy]
	Network of Small Businesses [US]
	Nuclear Standards Board
NSBA	National Silica Brickmakers Association [UK]
NSBE	National Society of Black Engineers [US]
NSBET	National Society of Biomedical Equipment Technicians [US]
NSBU	National Small Business United [US]
NSC	Naphthalene Sulfochloride
	National Safety Council [US]
	National Science Council [Taiwan]
	National Security Council
	National Space Club [US]
	NATO Science Committee
	NATO Supply Center
	Near-Surface Chemistry
	Network Switching Center
	Nodal Switching Center
	Noise Suppression Circuit
	Numerical Sequence Code
	Numeric Space Character
NSCA	National Society for Clean Air [UK]
	National Sound and Communications Association [US]
NSCAD	Nova Scotia College of Art and Design [Canada]
NSCB	Nordic Society for Cell Biology [Denmark]
NSCC	Nuclear Services Closed Cooling
NSCL	National Superconducting Cyclotron Laboratory [of Michigan State University, US]
NSCR	Nuclear Science Center Reactor [of TAMU, US]
NSCS	North Star Computer (System) Society [US]

NSCUFA Nova Scotia Council of University Faculty Associations [Canada]
NSD Network Status Display
NSDA National Space Development Agency [Japan]
NSDB National Science Development Board [Philippines]
NSDC Northern Shipowners Defense Council [Norway]
Nova Scotia Design Craftsmen [Canada]
NSDD National Security Division Directive [US]
NSDJA National Sash and Door Jobbers Association [US]
NSDM National Security Decision Memorandum
NSDO National Seed Development Organization [UK]
NSDU Network Service Data Unit
NSE National Student Exchange [US]
Navier-Stokes Equation
Nottingham Society of Engineers [UK]
NSEC Naval Ships Engineering Center [US Navy]
nsec nanosecond [also: ns]
NSEF Navy Security Engineering Facility [US Navy]
NSEH Neutron Scattering Experimental Hall [of LANL, US]
NSEIP Norwegian Society for Electronic Information Processing
NSEMA National Spray Equipment Manufacturers Association [US]
NSEP National Security and Emergency Preparedness [US]
NSERC Natural Sciences and Engineering Research Council [Canada]
NSERCC Natural Sciences and Engineering Research Council of Canada
NSES National Society of Electrotypers and Stereotypers [UK]
NSF National Sanitation Foundation
National Science Foundation [US]
Naval Supersonic Facility [US Navy]
Neil Squire Foundation [Canada]
Norges Standardiseringsforbund [= Norwegian Standards Association]
Not Sufficient Funds
Nuclear Science Foundation
Nuclear Structure Facility [UK]
NSFA National Science Foundation Act [US]
NSFC National Science Foundation of China
Natural Science Foundation of China
NSFD Note of Structural and Functional Deficiency
NSFL National Science Film Library
Nova Scotia Federation of Labour [Canada]
NSFRC National Soil and Fertilizer Research Committee [US]
NSFS Net Suction Fracture Strength
NSG NASA Grant
Naval Security Group [US Navy]
Newspaper Systems Group [US]
NSGA National Sand and Gravel Association [now: National Aggregates Association, US]
NSGC Naval Security Group Command [US Navy]
NSGP National Sea Grant Program [US]
NSGPMA National Salt Glazed Pipe Manufacturers Association [UK]
NSHB North of Scotland Hydro-Electric Board [Scotland]
NSHC National Solar Heating and Cooling [now: CAREIRS, US]
North Sea Hydrographic Commission [Sweden]
NSI National Security Information
National Space Institute
National Supervisory Inspectorate [UK]
Netherlands Society for Informatics
Next Sequential Instruction
Nonsatellite Identification
Nonsequenced Information
Nonstandard Item
NSIA National Sailing Industry Association [US]
National Security Industrial Association [US]
NSIC Nuclear Safety Information Center [of USAEC at ORNL]
NSIF National Swine Improvement Federation [US]
Near Space Instrumentation Facility [US]
NSIT Nova Scotia Institute of Technology [Canada]
NSJ Nuclear Society of Japan
NSL National Science Library [now: CISTI, Canada]
National Science Laboratories [US]
Net Switching Loss
Northrop Space Laboratories [US]
Naval Supersonic Laboratory [US Navy]
NSLA Nova Scotia Library Association [Canada]
NSLRB Nova Scotia Labour Relations Board [Canada]
NSLS National Synchrotron Light Source [of BNL, US]
NSM Network Security Module
Network Status Monitor
Nuclear Shell Modell
NSMA National Scale Men's Association [now: ISWM, US]
NSMB Nuclear Standards Management Board [of ANSI, US]
NSMP National Society of Master Patternmakers [UK]
NSMPA National Screw Machine Products Association [US]
NSMR National Society for Medical Research [US]
NSMRSE National Study of Mathematics Requirements for Scientists and Engineers [US]
NSMS National Safety Management Society [US]
NSOEA National Stationery and Office Equipment Association [US]
NSP NASA Support Plan
National Society of Painters [UK]
Network Services Protocol
Network Support Plan
Network Support Processor
Nonstandard Part
NSPA National Society of Public Accountants [US]
National Soybean Processors Association [US]
Nova Scotia Pharmaceutical Association [Canada]
NSPAC National Standards Policy Advisory Committee [US]
NSPC National Sound-Program Center [US]
National Standard Plumbing Code [US]
Nova Scotia Power Corporation [Canada]
NSPCC National Standard Plumbing Code Committee [US]
NSPE National Society of Professional Engineers [US]
NSPFEA National Spray Painting and Finishing Equipment Association [now: NSEMA, US]
NSPI National Society for Performance and Instruction [US]
National Society for Programmed Instruction [US]

NSPIE National Society for the Promotion of Industrial Education [US]
NSPOL Non-Scheduled Operations Policy
NSPP Nuclear Safety Pilot Plant [of ORNL, US]
NSPRI Nigeria Store Product Research Institute
NSPS National Society of Professional Surveyors [of ACSM, US]
New Source Performance Standards
NSPV Number of Scans per Vehicle
NSQCRE National Symposium on Quality Control and Reliability in Electronics [US]
NSR Near-Surface Region
Neutron Source Reactor [of BNL, US]
Nitrile-Silicone Rubber
Nordiska Skogsarbetsstudiernas [= Nordic Research Council on Forest Operations, Finland]
Notch Strength Ratio
Notch Stress Rupture
Nuclear Structure References (Database) [of ORNL, US]
NSRA National Service Robot Association [US]
NSRB National Security Resources Board [US]
NSRC Natural Science Research Council
Neosynthesis Research Center [Sri Lanka]
NSRDB Nova Scotia Resources Development Board [Canada]
NSRDC Naval Ship Research and Development Center [US Navy]
NSRDS National Standard Reference Data Series [of NIST, US]
National Standard Reference Data System [of OSRD, US]
NSRF Nova Scotia Research Foundation [Canada]
NSRFC Nova Scotia Research Foundation Corporation [Canada]
NSRFI National Symposium on Radio Frequency Interference
NSRG Northern Science Research Group [Canada]
NSRI Nelspruit Subtropical Research Institute [South Africa]
NSRP Nordic Society for Radiation Protection [Finland]
NSRQC National Symposium on Reliability and Quality Control
NSRR Nuclear Safety Research Reactor
NSRS Naval Supply Radio Station
NSRT Near Surface Reference Temperature
NSRW Nuclear Service Raw Water
NSRWP Nuclear Service Raw Water Pump
NSS National Search and Rescue Secretariat [Canada]
National Space Society [US]
National Speleological Society [US]
National Standards System
Navy Secondary Standards [US Navy]
Network Supervisor System
Network Synchronization Subsystem
Nonspatial Statistics
Nordic Statistical Secretariat [Denmark]
Nuclear Steam System
NSSA National Sanitary Supply Association [now: ISSA, US]
National Science Supervisors Association [US]
Nova Scotia Salmon Association [Canada]
NSSC Naval Sea Systems Command [US Navy]
Naval Supply Systems Command [US Navy]
Nova Scotia Safety Council [Canada]
Nuclear Safety Standards Commission [Germany]
NSSCC National Space Surveillance Control Center [US]
NSSDC National Space Science Data Center [of NASA]
NSSDRI Nanjing Solid State Devices Research Institute [PR China]
NSSEA National School Supply and Equipment Association [US]
NSSL National Seed Storage Laboratory [US]
National Severe Storms Laboratory [US]
NSSM National Security Study Memorandum [US]
NSSMS NATO Seasparrow Surface Missile System
NSSP National Student Safety Program [US]
NSSS National Space Surveillance System [US]
Nuclear Steam Supply System
NST Network Support Team
Newfoundland Standard Time [Canada]
Non-Slip Tread
NSTA National Safe Transit Association [US]
National School Transportation Association [US]
National Science Teachers Association [US]
N Staff North Staffordshire [UK]
NSTB National Safety Transportation Board [US]
NSTD Nested
NSTF National Scholarship Trust Fund [US]
NSTIC Naval Science and Technology Information Center [now: DRIC, UK]
Naval Scientific and Technical Information Center [US Navy]
NSTL National Space (and) Technology Laboratory [of NASA]
NSTN Nonstandard Telephone Number
NSTP National Skills Training Program
Nuffield Science Teaching Project [US]
NSTPC Nova Scotia Tidal Power Corporation [Canada]
NSTS National Space Transportation System [US]
NSU Network Service Unit
Norfolk State University [US]
NSUA Nigerian Students Union in the Americas [US]
NSV Net Sales Value
Nonautomatic Self-Verification
Nuclear Service Vessel
NSW National Software Works [US]
Neutron Spin Wave
New South Wales [Australia]
NSWA National Stripper Well Association [US]
NSWC Naval Surface Weapons Center [US Navy]
NSWL Naval Surface Weapons Laboratory [formerly: Naval Ordnance Laboratory, US]
NSWMA National Solid Waste Management Association [US]
NSWP Non-Soviet Warsaw Pact
NSWS Neutron Spin-Wave Scattering
NSYF Natural Science for Youth Foundation [US]
NT Narrower Term
Network Termination
Network Theory
Newtonian Telescope
Non-Tight

Normalized and Tempered
Northern Territory [Australia]
Northwest Territories [also: NWT, Canada]
Nortriptyline
No Transmission
Numbering Transmitter
Number Theory

nt net

NTA National Technical Association [US]
National Telecommunications Agency [US]
National Translators Association [US]
National Transportation Act [Canada]
Nitrilotriacetate
Nitrilotriacetic Acid
Northern Textile Association [US]

NTAG National Technical Advisory Group
Network Technical Architecture Group [of NDO, US]

NTAN Nitrilotriacetonitrile

NTAOCH Notice to Air Operator Certificate Holders [of Civil Aviation Authority, UK]

NTAT National Tax Administration of Taipei [of MOF, Taiwan]

NTB National Training Board [Canada]
Non-Tariff Barrier

N-t-B Nitroso-tert-Butane

N-t-BOC N-tert-Butoxycarbonyl

NTC National Telecommunications Conference [of IEEE, US]
National Telemetering Conference
National Television Center [US]
National Training Center [US Army]
National Transformers Committee
National Translation Center [US]
Naval Telecommunications Command [US Navy]
Negative Temperature Coefficient
Network Transmission Committee [of VITEAC, US]
Nonthermal Continuum
Noranda Technology Center [Canada]

NTCA National Telephone Cooperative Association [US]
National Tile Contractors Association [US]

NTD National Topographic Database [Canada]
Neutron Transmutation Doping

NT$ New Taiwanese Dollar [also: NTD; Currency of Taiwan]

NTDA National Trade Development Association [UK]

NTDB National Topographic Database [Canada]

NTDE Naval Tactical Display Emulator

NTDPMA National Tool, Die and Precision Machining Association [US]

NTDRA National Tire Dealers and Retreaders Association [US]

NTDS Naval Tactical Data System [US Navy]

NTE Navy Teletypewriter Exchange [US Navy]
Network Terminal Equipment
Network Terminating Equipment
Not to Exceed

NTEA National Time Equipment Association [US]
National Truck Equipment Association [US]

NTEP New Technology Employment Program

NTETA National Traction Engine and Tractor Association [UK]

NTF National Turkey Federation [US]
Nordic Transport Workers Federation [Norway]
No Trouble Found
Nuclear Test Facility
Number Type Flag

NTG Nachrichtentechnische Gesellschaft [= Communications Society, Germany]
NATO Training Group

NTH Norges Tekniske Hogskole [= Norwegian Institute of Technology]

NTHRN Northern

NTHU National Tsing Hua University [Taiwan]

NTI Noise Transmission Impairment

NTIA National Telecommunications and Information Administration [US]
National Telecommunications and Information Agency [of USDOC]

NTIAC Nondestructive Testing Information Analysis Center [of SWRI, US]

NTIAM National Testing Institute for Agricultural Machinery [Sweden]

NTICL National Technical Information Center and Library [Hungary]

NTIRA National Trucking Industrial Relations Association [US]

NTIS National Technical Information Service [of USDOC]

NTISS National Telecommunications and Information System Security

NTISSC National Telecommunications and Information System Security Committee [US]

NTL Nonthreshold Logic

NTLS National Truck Leasing System [US]

NTM National Topographical Map [Canada]

NTMA National Terrazzo and Mosaic Association [US]
National Tooling and Machining Association [US]

NTMP Nitrilo-Tris-Methylene Phosphoric Acid

NTN National Telecommunications Network

NTO Natural Transition Orbitals

NTOC Number to Character (Conversion)

NTOTC National Training and Operational Technology Center

NTP National Toxicology Program [US]
Network Terminal Protocol
Network Terminating Point
Network Termination Processor
Normal Temperature and Pressure [also: ntp]
No Title Page
Nuclear Test Plant

NTPC National Technical Planning Committee [Sudan]
National Technical Processing Center [US]

NTPF Number of Terminals per Failure

NTPG National Textile Processors Guild [US]

NTR Next Task Register
Nonthermal Radiation
Nuclear Test Reactor [US]

NTRL National Telecommunications Research Laboratory [South Africa]

NTRMA National Tile Roofing Manufacturers Association [US]

NTRTN Nutrition

NTS National Topographic (Map) System [Canada]
Navigation Technology Satellite
Navigation Technology System
Negative Torque Signal
Nevada Test Site [of NASA/USDOE]
Non-Traffic Sensitive
Notch Tensile Strength
Not to Scale

NTSA National Technical Services Association [US]

NTSB National Transportation Safety Board [of USDOT]

NTSC National Television Standard Code [US]
National Television Standards Committee [US]
National Television System Committee [US]
National Thermal Spray Conference (and Exposition) [US]

NTSD Normal Theory Sampling Distribution

NTSEA National Trade Show Exhibitors Association [now: International Exhibitors Association, US]

NTSK Nordiska Tele-Satelit Kommitton [= Nordic Tele-Satellite Committee, Norway]

NTT New Technology Telescope [of ESO, Germany]
Nippon Telegraph and Telephone Corporation [Japan]

NTTC National Tank Truck Carriers [US]

NTTCIW National Technical Task Committee on Industrial Wastes [US]

NTTP Network Test and Termination Point

NTTPC Nippon Telegraph and Telephone Corporation [Japan]

NTU National Taiwan University
National Technical University
Naphthylthiourea
Nephelometric Turbidity Unit
Network Terminating Unit
Nishi Tokyo University [Japan]
Number of Transfer Units

NTUA National Technical University of Athens [Greece]

NTV Network Television
Nonlinear Thickness Variation

NTW Nonpressure Thermit Welding

NTWK Network

NTWS Nontrack While Scan

NT WT Net Weight [also: nt wt]

NTZ No-Transgression Zone

NU Nagasaki University [Japan]
Nagoya University [Japan]
Nihon University [Japan]
Niigata University [Japan]
Nonuniform
Northeastern University [also: NEU, US]
Northwestern University [also: NWU, US]
No Umbra (Technique)
Number Unobtainable
Numeral

NUA Network User Address
Network User Association [US]

NUANS Newly Upgraded Automated Name Search (Database)

NUAP National Union of Airline Pilots [Israel]

NUAW National Union of Agricultural Workers [UK]

NUBA National UHF Broadcasters Association [US]

NUC Nailed-Up Connection
National Union Catalogue [US]
Nipissing University College [Canada]
Nuclear

NUCA National Utility Contractors Association [US]

NUCEA National University Continuing Education Association [US]

NUCGIW National Union of Ceramics and Glass Industry Workers [Israel]

NUCLENOR Controles Nucleares del Norte [Spain]

NUCLEON Nucleonics [also: NUCLEO]

NUCLEX International Nuclear Industrial Fair and Technical Meeting

NUCLIT Nucleare Italiana [= Italian Nuclear Corporation]

NUC PHYS Nuclear Physics

NUDAC Nuclear Data Center

NUDET Nuclear Detection
Nuclear Detonation

NUDETS Nuclear Detection System

NUDW National Union of Diamond Workers [Israel]

NUEA National University Extension Association [now: NUCEA, US]

NUFFIC Netherlands University Foundation for International Cooperation

NUGMW National Union of General and Municipal Workers [UK]

NUI National University of Ireland
Network User Identification

NUJ National Union of Journalists [UK]

NUL Null (Character)

NULACE Nuclear Liquid Air Cycle Engine

NULMW National Union of Lock and Metal Workers [UK]

NULW National Union of Leather Workers and Allied Trades [UK]

NUM National Union of Manufacturers [UK]
National Union of Mineworkers [UK]
Number
Numeric(s)

NUMAS National Union of Manufacturers Advisory Service [UK]

NUME Numerical Methods in Engineering

NUMEC Nuclear Materials and Equipment Corporation

NUMETA Numerical Methods in Engineering — Theory and Applications

NUMIS Northwestern University Multislice and Imaging System [US]

NUMS Nuclear Materials Security

NUMSA Nordic Union of Motor Schools Associations [Denmark]

NUOS Naval Underwater Ordnance Station [US Navy]

NUP Nuevo Peso [Currency of Uruguay]

NUPAD Nuclear Powered Active Detection System

NUPBPW National Union of Printing, Bookbinding and Paper Workers [now: SOGAT, UK]

NUPE National Union of Public Employees [UK]

NUPPSCO Nuclear Power Plant Standards Committee

NUR National Union of Railwaymen [UK]
Net Unduplicated Research

NURBS Nonuniform Rational B-Splines

NURE National Uranium Resource Evaluation Program [US]
NURTIW National Union of Rubber and Tire Industries Workers [Israel]
NUS National Union of Seamen
National Union of Students [UK]
National University of Singapore
Nuclear Utility Service
NUSAS National Union of South African Students
NUSC Naval Underwater Systems Center [US Navy]
NUSL Naval Underwater Sound Laboratory [US Navy]
NUSMWC National Union of Sheetmetal Workers and Coppersmiths [UK]
NUT National Union of Teachers [UK]
Northeast University of Technology [PR China]
Number Unobtainable Tone
Nutrient
NUTEC Norwegian Underwater Technology Center
NUTGLW National Union of Textile, Garment and Leather Workers [Israel]
NUTPE National Union of Technicians and Practical Engineers [Israel]
NUTWTC National Union of Transport Workers and Transport Cooperatives [Israel]
NUVB National Union of Vehicle Builders [UK]
NUWC Naval Undersea Warfare Center [US Navy]
NUWES Naval Undersea Warfare Engineering Station [US Navy]
NUWWE National Union of Water Waters Employees [UK]
NV Nevada [US]
New Version
Night Vision
Nonvolatile
No Value
nV nanovolt
NVA Norske Videnskaps-Akademi [= Nordic Academy of Science, Norway]
Nva Norvalyl
NVACP Neighborhood Voluntary Associations and Consumer Protection [of OAS, US]
NVAO Norke Videnskaps-Akademi i Oslo [= Nordic Academy of Science at Oslo, Norway]
NVBO Natural Valence-Band Offset
NVBF Nordiska Vetenskapliga Bibliothekarie-Forbundet [= Scandinavian Federation of Research Libraries, Denmark]
NVCA National Van Conversion Association [US]
NVF Nordisk Vejteknisk Forbund [= Nordic Road Association, Denmark]
NVG Night Vision Goggles
NVGA National Vocational Guidance Association [US]
NVLA National Vehicle Leasing Association [US]
NVLAP National Voluntary Laboratory Accreditation Program
NVM Nonvolatile Matter
Nonvolatile Memory
NVMA National Veterinary Medical Association [UK]
NVOCC Non-Vessel Operation Common Carrier [US]
NVR No Voltage Release
NVRAM Nonvolatile Random-Access Memory [also: NV RAM]
NVRS National Vegetable Research Station [UK]
NVS National Vacuum Symposium [of AVS, US]
National Vegetable Society [UK]
Night Vision System
NVSAVS National Vacuum Symposium of the American Vacuum Society
NVSD Night Vision System Development
NVT Network Virtual Terminal
NW Net Weight
Neville-Winter (Acid)
Northwest(ern)
Nosewheel
Nuclear Waste
Nuclear Weapon
N/W Nishiyama-Wassermann (Relationship)
nW nanowatt
NWA National Waterfowl Alliance [US]
National Weather Association [US]
Northwest Airlines [US]
Norwegian Water Association
NWAC National Weather Analysis Center [US]
NWAG Naval Warfare Analysis Group [US Navy]
NWAHACA National Warm Air Heating and Air Conditioning Association
N Wales North Wales [UK]
NWB National Wiring Bureau [US — defunct]
NWC National Water Center [US]
National Water Commission [US]
National Waterfowl Council [US]
National Watershed Congress [US]
National Waterways Conference [US]
Naval Weapons Center [US Navy]
Net Weekly Circulation
NWCC National Water Companies Conference [now: NAWC, US]
National Weed Committee of Canada
NWD Network Wide Directory
NWDS Number of Words
NWDSEN Number of Words per Entry
NWEA National Wood Energy Association [US]
NWEB North Western Electricity Board [UK]
NWEF Naval Weapons Evaluation Facilities [US Navy]
NWEP Nuclear Weapons Effects Panel [US]
NWF National Wildlife Federation [US]
NWFA National Wood Flooring Association [US]
Northwest Farm Managers Association [US]
NWFCCC National Wildlife Federation Corporate Conservation Council [US]
NWG National Wire Gauge
NWGA National Wool Growers Association [US]
NWH Normal Working Hours
NWHA National Wholesale Hardware Association [US]
NWHF National Wildlife Health Foundation [US]
NWI Net Working Interest
NWIC National Wheat Improvement Committee [US]
NWIS Network of (Minority) Women in Science [US]
NWH Normal Working Hours
NWK Network
NWL Naval Weapons Laboratory [US Navy]
NWLDYA National Wholesale Lumber Distributing Yard Association [now: HDA, US]

NWM Network Management
Nuclear Waste Management
NWMA National Woodwork Manufacturing Association [US]
Northern Woods Logging Association [US]
Northwest Mining Association [US]
NWO Nederlandse Organisatie voor Wetenschapelijk Onderzoeg [= Netherlands Organization for Scientific Research]
Nonwoven Oriented
NWOA National Woodland Owners Association [US]
NWP National Water Project [US]
National Writing Project [US]
Numerical Weather Prediction
NWPA Nuclear Waste Policy Act
NWPC Nordic Wood Preservation Council [Finland]
NWPCA National Wooden Pallet and Container Association [US]
NWPF National Water Purification Foundation [US]
NWPFC Northwest Pacific Fisheries Commission
NWQL National Water Quality Laboratory [Canada]
NWRA National Water Resources Association [US]
National Wheel and Rim Association [US]
National Wildlife Refuge Association [US]
NWRC National Weather Records Center [of ESSA, US]
National Wildflower Research Center [US]
National Wildlife Research Center [Canada]
NWRF Naval Weather Research Facility [US Navy]
NWRI National Water Research Institute
NWRT National Wildlife Rescue Team [US]
NWS National Weather Service [US]
North Warning System [US/Canada]
Nosewheel Steering
NWSA National Welding Supply Association [US]
NWSC National Weather Satellite Center [US]
National Women Student's Coalition [US]
NWSD Naval Weather Service Division [US Navy]
NWSIA National Water Supply Improvement Association [US]
NWSPR Newsprint
NWSSG Nuclear Weapons System Satellite Group [US]
NWT Net Weight
Non-Waste Technology
Northwest Territories [Canada]
NWTC National Wetlands Technical Council [US]
NWTEC National Wool Textile Export Corporation [UK]
NWTFL Northwest Territories Federation of Labour [Canada]
NWTI National Wood Tank Institute [US]
NWTRNA Northwest Territories Registered Nurses Association [Canada]
NWU National Workers Union
Northwestern University [US]
NWUS National Workers Union of the Seychelles
NWWA National Water Well Association [US]
NWWDA National Wood Window and Door Association [US]
NXDO Nike X Development Office [US]
NXT Next
NY New York [US]
NYADS New York Air Defense Sector [US]
NYAM New York Academy of Medicine [US]
NYAP New York Assembly Program [US]
NYARTCC New York Air Route Traffic Control Center [US]
NYAS New York Academy of Sciences [US]
NYBPE New York Business Press Editors [US]
NYBT New York Board of Trade [US]
NYC New York City [US]
NYCBAN New York Center Beacon Alphanumerics [of FAA, US]
NYCE New York Cotton Exchange [also: NYCTN, US]
NYCCC New York City Community College [US]
NYCS New York Cipher Society [US]
NYCSE New York Coffee and Sugar Exchange [US]
NYCX New York Commodity Exchange [US]
NYD Navy Yard
Not Yet Detected
Not Yet Determined
NYES New York Electrical Society [US]
New York Entomological Society [US]
NYFE New York Futures Exchange [US]
NYI NSF Young Investigator (Awards Program) [of National Science Foundation, US]
NYID Not-Yet-Invented Device
NYIT New York Institute of Technology [US]
NYL Nylon
NYLSMA New York Lamp and Shade Manufacturers Association [now: ELLA, US]
NYME New York Mercantile Exchange [US]
NYMRRLA New York Metropolitan Reference and Research Library Agency [US]
NYNS New York Naval Shipyard [US Navy]
NYO New York Operations Office [US]
Not Yet Operating
NYP Not Yet Published
NYPC New York Paint Club [US]
New York Pigment Club [US]
NYPD New York Police Department [US]
NYPL New York Public Library [US]
NYPS National Yellow Pages Service [US]
NYPSC New York Public Service Commission [US]
NYS New York State [US]
NYSA New York Shipping Association [US]
NYSAA New York State Archeological Association [US]
NYSAES New York State Agricultural Experiment Station [US]
NYSASS New York State Association of Service Stations [US]
NYSCC New York State College of Ceramics [of Alfred University, US]
NYSDH New York State Department of Health [US]
NYSE New York Stock Exchange [US]
NYSEM New York Society of Electron Microscopy [US]
NYSERDA New York State Energy Research Development Authority [US]
NYSFTCA New York State Fruit Testing Cooperative Association [US]
NYSILL New York State Interlibrary Loan System [US]
NYSPA New York State Pharmaceutical Association [US]
NYSSIM New York State Society of Industrial Medicine [US]
NYSSPE New York State Society of Professional Engineers [US]
NYSSTF New York State Science and Technology Foundation [US]
NYSU New York State University [US]

NYTIS New York Times Information Service [US]
NYU New York University [US]
NYUIMS New York University Institute of Mathematical Sciences [US]
NYZS New York Zoological Society [US]**NZ**
Neutral Zone
New Zealand
NZAB New Zealand Association of Bacteriologists
NZACT New Zealand Association of Chemistry Teachers
NZAEI New Zealand Agricultural Engineering Institute
NZASc New Zealand Association of Scientists
NZBS New Zealand Broadcasting Service
NZCER New Zealand Council for Educational Research
NZCWPRC New Zealand Committee for Water Pollution Research and Control
NZ$ New Zealand Dollar [also: NZD; Currency of Cooke Islands, Tokelau and New Zealand]
NZDA New Zealand Department of Agriculture
NZDCS New Zealand Department of Census and Statistics
NZDLS New Zealand Department of Lands and Survey
NZDSIR New Zealand Department of Scientific and Industrial Research
NZEI New Zealand Electronics Institute
NZES New Zealand Ecological Society
NZFMRA New Zealand Fertilizer Manufacturers Research Association
NZFRI New Zealand Forest Research Institute
NZFS New Zealand Forest Service
NZGA New Zealand Grassland Association
NZGenS New Zealand Genetical Society
NZGS New Zealand Geographical Society
NZIAS New Zealand Institute of Agricultural Science
NZIC New Zealand Institute of Chemistry
NZIE New Zealand Institute of Engineers
NZIF New Zealand Institute of Foresters
NZIIA New Zealand Institute of International Affairs
NZInstW New Zealand Institute of Welding
NZIPE New Zealand Institution of Professional Engineers
NZIRE New Zealand Institute of Refrigeration Engineers
NZIW New Zealand Institute of Welding
NZJCB New Zealand Joint Communication Board
NZLA New Zealand Library Association
NZMS New Zealand Meteorological Service
NZNCIAWPRC New Zealand National Committee of the International Association on Water Pollution Research and Control
NZNCOR New Zealand National Committee on Oceanic Research
NZNEDA New Zealand National Electronics Development Association
NZNRAC New Zealand National Research Advisory Council
NZP Na (= Sodium) Zirconium Phosphate
NZPO New Zealand Post Office
NZSA New Zealand Statistical Association
NZSAP New Zealand Society of Animal Production
NZSB New Zealand Soil Bureau
NZSCA New Zealand Soil Conservation Association
NZSI New Zealand Standards Institute [now: SANZ]
NZSLO New Zealand Scientific Liaison Office
NZSSS New Zealand Society of Soil Science
NZT Non-Zero Transfer
NZVA New Zealand Veterinary Association
NZWAA New Zealand Weed Control Conference

O

O	Ocean
	October
	Octopole
	Ohio [US]
	Operand
	Operation
	Operator
	Orange
	Order
	Ordinary
	Orifice
	Output
	Oxide
	Oxygen
o	ordinary
	ortho-
OA	Office Automation
	Office of Administration [US]
	Oil-Immersed Air-Cooled (Transformer)
	Oil-to-Air (Ratio)
	Omniantenna
	Operand Address
	Operating Authorization
	Operational Amplifier
	Operations Analysis
	Optical Absorption
	Output Amplifier
	Output Axis
	Outside Air
	Overaging
	Overall
	Overflow Area
	Oxygen Absorbed
O/A	On Account
	On Application
	On or About
	Outer Anchorage
OAA	Ontario Association of Architects [Canada]
	Orient Airlines Association [Philippines]
	Oxaloacetic Acid
OAAA	Outdoor Advertising Association of America [US]
OAAC	Outdoor Advertising Association of Canada
OAAS	Ontario Association of Agricultural Societies [Canada]
OAATM	Office of the Assistant for Automation [US]
OAAU	Orthogonal Array Arithmetic Unit
OAC	Oceanic Area Control
	On Approved Credit
	Ontario Academic Course [Canada]
	Ontario Agricultural College [Canada]
	Operations Advisory Committee
	Operations Analysis Center
	Optimally Adaptive Control
	Overseas Automotive Club [US]
OACC	Oceanic Area Control Center
OACETT	Ontario Association of Certified Engineering Technicians and Technologists [Canada]
OACI	Ontario Academic Courses Institute [Canada]
OACSU	Off-Air Call Setup
OACUL	Ontario Association of College and University Libraries [now: OCULA, Canada]
OAD	Operational Availability Date
	Operations Analysis Division
	Optoacoustic Device
OAEAO	Ontario Association of Education Administration Officials [Canada]
OAEM	Ontario Approved Educational Microcomputers [Canada]
OAF	Origin Address Field
OAFU	Observers Advanced Flying Unit
OAI	Outside Air Intake
OAIDE	Operational Assistance and Instructive Data Equipment
OAK	Organization for the Advancement of Knowledge [US]
OALS	Observer Air Lock System
OALT	Ontario Association of Library Technicians [Canada]
OAM	Office of Aerospace Medicine [US]
	Ontario Agricultural Museum [Canada]
	Oscillator Activity Monitor
OAMA	Office Automation Management Association [US]
OAME	Orbital Altitude Maneuvering Electronics
OAMP	Optical Analog Matrix Processing
OAMS	Orbital Altitude Maneuvering System
OANA	Organization of Asian News Agencies
O and C	Operations and Checkout [also: O&C]
OAO	Orbiting Astronomical Observatory
	Orthogonalized Atomic Orbital
OAP	Office of the Assistant to the President [US]
	Ontario Apprenticeship Program [Canada]
	Orthogonal Array Processor
OAPEC	Organization of Arab Petroleum Exporting Countries [Kuwait]
OAQ	Order of Architects of Quebec [Canada]
OAQPS	Office of Air Quality Planning and Standards [of EPA, US]
OAQS	On-Line Associative Query System [of University of Illinois, US]
OAR	Office of Aerospace Research [of US Air Force]
	Operator Authorization Record
OARAC	Office of Aerospace Research Automatic Computer
OARC	Ordinary Administrative Radio Conference
OARS	Office Automation Reporting Service
OART	Office of Advanced Research and Technology [of NASA]
OAS	Obstacle Assessment Surface
	Office Automation System
	Office of the Assistant Secretary [US]
	Ohio Academy of Science [US]
	Orbit Adjust Subsystem
	Organization of American States [US]
OASF	Orbiting Astronomical Support Facility
OASH	Office of the Assistant Secretary for Housing [US]

OASI Office Automation Society International [US]
OASIS Ocean All Source Information System
Oceanic and Atmospheric Scientific Information System [of NOAA, US]
Oceanic Area System Improvement Study
Office Automation Services and Information Systems
OASM Office of Aerospace Medicine [US]
Ohm, Ampere, Second, Meter (System) [also: oasm]
OASPL Overall Sound Pressure Level
OASV Orbital Assembly Support Vehicle
OASYS Office Automation System
OAT Office for Advanced Technology [of US Air Force]
Operational Air Traffic
Outside Air Temperature
OATP Operational Acceptance Test Procedure
OATRU Organic and Associated Terrain Research Unit [Canada]
OATS Original Article Tearsheet Service
OATUU Organization of African Trade Union Unity
OAU Operator Assistance Unit
Organization of African Unity
OAW Osterreichische Akademie der Wissenschaften [= Austrian Academy of Sciences]
Oxyacetylene Welding
OB Octal-to-Binary
Oil Bearing
Oil-Break
Ordnance Board [UK]
Organization Blocks
Osteoblast
Outbound
Output Buffer
Output Bus
Outside Broadcast
O-B Octal-to-Binary
OBA Octave-Band Analyzer
Oil Burning Apparatus
Oxygen Breathing Apparatus
OBAA Oil Burning Apparatus Association [UK]
OBAP Office Belge pour l'Accroissement de la Productivite [= Belgian Office for the Improvement of Productivity]
OBATA Ontario Biological Aeration Tillage Association [Canada]
OBAWS Onboard Aircraft Weighing System
OBC Optical Bar Code
OBCE Office Belge du Commerce Exterieur [= Belgian Office of Foreign Trade]
OBCR Optical Bar Code Reader
OBD Open Blade Damper
OBE Onboard Electronics
Operating Basis Earthquake
Operating Basis Event
Order of the British Empire
OBEA Ontario Business Education Association [Canada]
OBES Office of Basic Energy Sciences [of USDOE]
Ohio Bureau of Employment Services [US]
Orthonormal Basis of an Error Space
OBESSI Organizing Bureau of European School Student Unions [Sweden]
OBF One-Bar Function
OBFS Organization of Biological Field Stations [US]
OBGS Orbital Bombardment Guidance System
OBH Office Busy Hour
OBI Office of Basic Instrumentation
Omnibearing Indicator
Open-Back Inclinable (Press)
OBIC Optical Beam Induced Current
OBIFCO On-Board In-Flight Checkout
OBIP Ontario Business Incentive Program [Canada]
OBJ Object
OBL Oblique
Oblong
OBM Ontario Basic Mapping [Canada]
Oxygen-Bottom Blown Maxhutte (Process) [Steelmaking]
OBN Out-of-Band Noise
Office Balancing Network
OBNA Only But Not All
OBO Oil Bulk-Ore
Or Best Offer
O/B/O Ore/Bulk/Oil (Carrier)
OBOC Ore/Bulk/Oil Carrier
OBOS Ore/Bulk/Oil Ship
OBP On-Board Processing
OBR Ohio Board of Regents [US]
Outboard Recorder
OBS Observation
Observatory
Observer
Obsolescence
Obsolete
Obstacle
Omnibearing Selector
Open Bidding Service
OBSC Obscure
OBSPL Octave-Band Sound Pressure Level
OBSS Ocean Bottom Scanning Sonar
OBST Obstruction [also: OBSTN]
OBU Offshore Banking Unit
OC Occurrence
OCLC Number [US]
Octal Code
Officer Commanding
Official Classification
Oil Chemist
Oil Chemistry
On Center
Open Circuit
Open Coil (Annealing)
Open Cup
Operating Characteristic
Operating Curve
Operational Calculus
Operational Characteristic
Operational Computer
Operator Command
Optical Cavity
Optical Center
Optical Computer
Optimal Control

Order-in-Council
Organic Chemistry
Oscillating Current
Osteocyte
Outlet Contact
Outside Circumference
Over Center
Over-Compounded
Overcorrected
Overcurrent
Oxide Cathode
Oxygen Chemisorption
Oxygen Consumed
Oxygen Cutting

O&C Operations and Checkout

O/C Open Circuit
Overcharge

OCA Obstacle Clearance Altitude
Oceanic Area
Oceanic Control Area
Operational Control Authority
Organization of the Cooperatives of America [Colombia]

OCAD Occupational and Career Analysis Development
Optical Character and Detect

OCAL On-Line Cryptanalytic Aid Language

OCAM Ontario Center for Advanced Manufacturing [Canada]
Organisation Commune Africaine et Malgache [= Common Organization of Africa and Madagascar]

OCAP Ontario Career Action Program [Canada]

OCAPT Ontario Center for Automotive Parts Technology [Canada]

OCAS On-Line Cryptanalytic Aid System
Organization of Central American States

OCB Oil Circuit Breaker
Oil-Operated Circuit Breaker
Outgoing Calls Barred

OCBA O-Chlorobenzoic Acid

OCBC O-Chlorobenzyl Chloride

OcB/L Ocean Bill of Lading

OCBN O-Chlorobenzonitrile

OCC Occulting
Occupation
Operational Computer Complex
Operations Control Center
Operator Control Command
Other Common Carriers
Output Control Character

OCCA Oil and Color Chemists Association [UK]

OCCE Operational Clothing and Combat Equipment

OCCS Office of Computer and Communications Systems [of NLM, US]

OCD Office of Civil Defense [US]

OCDM Office of Civil Defense Mobilization [US]

OCDRE Organic-Cooled Deuterium Reactor Experiment [of WNRE, Canada]

OCdt Officer-Commandant

OCDU Optics Coupling Display Unit

OCE Ocean Covered Earth

OCEAN Oceanographic Coordination Evaluation and Analysis Network
Organisation de la Communaute Europeenne des Avitailleurs des Naivres [= Ship Suppliers Organization of the European Community, Netherlands]

OceanE Ocean Engineer [also: OceanEng]

OCEANOGR Oceanographer
Oceanographic
Oceanography

OCF Ontario Cancer Foundation [Canada]
Orientation Correlation Function
Owens-Corning Fiberglass

OCFMFPT Ontario Center for Farm Machinery and Food Processing Technologies [Canada]

OCFNT Occluded Front

OCG Optimal Code Generation

OCGS Ontario Council on Graduate Studies [Canada]

OCH Obstacle Clearance Height
Office for Communication in the Humanities [UK]

OCHC Operator Call Handling Center

OCI Office of Computer Information [US]
Ontario Cancer Institute [Canada]
Operator Control Interface
Optically-Coupled Insulator
Optically-Coupled Isolator
Oxide Control and Indication

OCIA Organic Crop Improvement Association [Canada and US]

OCIMF Oil Companies International Marine Forum [UK]
Oil Companies International Mooring Forum

OCIS OSHA (= Occupational Safety and Health Administration) Computerized Information System [US]

OCL Obstacle Clearance Limit
Obstruction Clearance Limit
Operation Control
Operational Control Level
Operation Control Language
Operator's Control Language
Overall Connection Loss
Overcorrected Lens

OCLC Ohio College Library Center [now: On-Line Computer Library Center, US]
On-Line Computer Library Center [US]

OCLI Optical Coating Laboratory, Inc. [US]

OCLN Occlusion

OCM Ontario Center for Microelectronics [Canada]
Optical Countermeasures
Oscillator and Clock Module

OCMA Oil Companies Material Association

OCMR Ontario Center for Materials Research [Canada]

OCMS Optional Calling Measured Service

OCNL Occasional

OCO Open-Close-Open (Contact)
Operation Capability Objectives

OCP Obstacle Clearance Panel [of ICAO]
Office of Commercial Programs
One-Component Plasma System
Operating Control Procedure
Operational Checkout Procedure

	Order Code Processor
	Output Control Program
	Output Control Pulses
	Overload Control Process
	Oxford Concordance Project [of Oxford University, UK]
OCPA	O-Chlorophenylacetic Acid
OCPDB	Organic Chemical Producers Database [of EPA, US]
OCPV	Open Circuit Photovoltage
OCR	Occurrence
	Office of Coal Research [US]
	Oil Circuit Recloser
	Optical Character Reader
	Optical Character Recognition
	Organic-Cooled Reactor
	Organization for the Collaboration of Railways [Poland]
	Output Control Register
	Overconsolidation Ratio
	Overcurrent Relay
	Overhaul Component Requirement
OCRA	Office Communications Research Association
OCR-A	Optical Character Recognition — Font A
OCR-B	Optical Character Recognition — Font B
OCRBI	Organization for Cooperation in the Roller Bearings Industry [Poland]
OCRD	Office of the Chief of Research and Development [US Army]
OCRI	Ottawa-Carleton Research Institute [Canada]
OCRPI	Office Central de Repartition des Produits Industriels [= Central Office of Distribution of Industrial Products, France]
OCRS	Ontario Center for Remote Sensing [Canada]
	Optical Character Recognition System
OCRUA	Optical Character Recognition Users Association [now: RTUA, US]
OCR WAND	Optical Character Reader Wand
OCS	Obstacle Clearance Surface
	O-Carbonyl Sulfide
	Office Communications System
	Office of Commodity Standards [of NIST, US]
	Office of Communication Systems [of US Air Force]
	Onboard Checkout System
	Ontario Science Center [Canada]
	Operations Control System
	Optical Character Scanner
	Optical Communication System
	Optical Contact Sensor
	Outer Continental Shelf
	Output Control Subsystem
	Output Control System
	Overload Control Subsystem
	Overload Control System
	Overseas Communications Service [India]
OCSA	Ontario Council of Safety Associations [Canada]
OCT	O-Chlorotoluene
	Octagon
	Octahedral
	Octal
	Octane
	October
	Office of Critical Tables [of NAS/NRC, US]
	Operational Cycle Time
	Ornithine Carbamyltransferase
OCTA	Oceanic Control Area
OCTAHDR	Octahedral
	Octahedron
OCTC	O-Chlorobenzotrichloride
OCTG	Oil Country Tubular Goods
OCTRF	Ontario Cancer Treatment and Research Foundation [Canada]
OCTV	Open Circuit Television
OCU	Ontario Council of Universities [Canada]
	Operational Control Unit
	Operational Conversion Unit
	Osaka City University [Japan]
OCUA	Ontario Council on University Affairs [Canada]
OCUB	Osmium Collidine Uranylenbloc
OCUFA	Ontario Confederation of University Faculty Association [Canada]
OCUL	Ontario Council on University Libraries [Canada]
OCULA	Ontario College and University Library Association [Canada]
OCV	Open-Circuit Voltage
OCZM	Office of Coastal Zone Management
OD	Octal-to-Decimal
	Olive Drab
	On Demand
	Operations Directive
	Optical Density
	Outer Diameter
	Output Data
	Output Disable
	Outside Diameter
	Outside Dimension
	Overdose
	Overdraft
	Overload Detection
O/D	On Demand
	Origin/Destination
	Overdraft
O-D	Octal-to-Decimal
1D	One-Dimensional [also: 1-D]
ODA	Office Data Architecture
	Office Document Architecture
	Official Development Assistance
	Operational Data Analysis
	Operational Design and Analysis
	Overseas Development Administration [UK]
	4,4'-Oxydianiline
ODAL	Octadecenal
13-ODAL	13-Octadecenal
ODALS	Omnidirectional Approach Lighting System
ODAPI	Omnidirectional Approach Path Indicator
ODAPS	Oceanic Display and Planning System [US]
ODAS	Ocean Data Acquisition Systems
ODB	Output Data Buffer
	Output to Display Buffer
ODBA	Oregon Dairy Breeders Association [US]
ODC	Ontario Development Corporation [Canada]
	Output Data Control
ODCF	One-Dimensional Compressible Flow

ODCP One-Digit Code Point
ODD Operator Distance Dialing
Optical Data Digitizer
Optical Data Disk
Optical Digital Disk
Ouchterlony Double Diffusion
ODDRE Office of the Director of Defense Research and Engineering [of USDOD]
ODE On-Line Data Entry
Ordinary Differential Equation
Oxygen Defect Electron
ODESSA Ocean Data Environmental Science Services Acquisition
ODESY On-Line Data Entry System
ODETTE Organization for Data Exchange and Teletransmission in Europe
ODF One-Dimensional Flow [also: 1DF]
Orientation Distribution Function
Original Data File
ODG Off-Line Data Generator
ODI Overseas Development Institute [UK]
ODIN On-Line Documentation and Information Network [Germany]
Optical Design Integration
ODIS Oceanographic Data Information System
ODLRO Off-Diagonal Long-Range Order
ODM Orbital Determination Module
ODMR Optical Detection of Magnetic Resonance
Optical Double Magnetic Resonance
ODN Octadecaneuropeptide
Own Doppler Nullifier
ODNR Ohio Department of Natural Resources [US]
ODNRI Overseas Development Natural Resources Institute [UK]
ODOP Offset Doppler
ODORL Odorless
ODP Ocean Drilling Program [US]
Optical Data Processing
Optical Data Processor
Original Document Processing
ODPA Octylated Diphenyl Amine
ODPCS Oceanographic Data Processing and Control System
ODPEX Offshore Drilling and Production Exhibition
ODPN Oxydipropionitrile
ODR Operator Data Register
Optical Data Recognition
Output Definition Register
ODRN Orbiting Data Relay Network
ODS Octadecyltrimethyloxysilane
Oxide Dispersion Strengthening
Oxygen Dispersion Strengthening
ODSA Overseas Development Service Association [UK]
ODSS Ocean Dumping Surveillance System
ODT Octal Debugging Technique
On-Line Debugging Technique
Outside Diameter Tube
ODU Output Display Unit
ODVAR Orbit Determination and Vehicle Attitude Reference
ODW Organic Dry Weight
OE Office of Education [US]
Omissions Excepted
Open End
Operating Engineer
Optical Emission
Optoelectronics
Order of Engineers
Original Equipment
Output Enable
Overrun Error
Own Exchange
Oxide Electrode
O/E Own Exchange
O-E Opto-Electronic(s)
Oe Oersted
OEA Optometric Editors Association [US]
Organization of European Aluminum Smelters
Oxygen Enriched Atmosphere
OEAP Operational Error Analysis Program
OEAW Oesterreichische Akademie der Wissenschaften [= Austrian Academy of Sciences]
OEB Ontario Energy Board [Canada]
OE-BR Oil Extended Butadiene Rubber
OEC Ontario Economic Council [Canada]
Ontario Energy Corporation [Canada]
OECA Ontario Educational Communications Authority [Canada]
OECD Organization for Economic Cooperation and Development [France]
OECD/MEI OECD Main Economic Indicators
OECD/NIA OECD National Income Accounts
OECO Outboard Engine Cutoff
OED Optoelectronic Device
Orbiting Energy Depot
OEDC Ontario Engineering Design Competition [Canada]
OEDSF Onboard Experimental Data Support Facility
OEEC Organization for European Economic Cooperation [now: OECD, France]
OE-EPDM Oil Extended Ethylene-Propylene Diene Monomer
OE-E-SBR Oil Extended Emulsion Styrene Butadiene Rubber
OEF Oceanic Educational Foundation [US]
Organization of Employers Federations [UK]
Origin Element Field
OEG Open-End (Wave-)Guide
Operations Evaluation Group
Outdoor Ethics Guild [US]
OEGAI Oesterreichische Gesellschaft fur Artificial Intelligence [= Austrian Society for Artificial Intelligence]
OEIC Optoelectronic Integrated Circuit
OEICS Optoelectronic Integrated Circuits
OEL Ontario Electrical League [Canada]
OE LASE Optics, Electro-Optics and Laser Applications in Science and Engineering [also: OE/LASE, US]
OEM Original Equipment Manufacturer
Other Equipment Manufacturer
OEMI Office Equipment Manufacturers Institute
OEMS Open-Site EMI (= Electromagnetic Interference) Measurement System
Optical Emission under Mechanical Stress
OE-NR Oil Extended Nitrile Rubber
OEO Office of Economic Opportunity

OE/OEM Original Equipment/Original Equipment Manufacturer
OEP Octaethylporphine
Odd-Even Predominance
Office of Emergency Planning [US]
Optoelectronic Packaging
Overall Economic Perspective
OEQ Order of Engineers of Quebec [Canada]
OER Office of Exploratory Research [US]
Oxygen Enhancement Ratio
OERD Office of Energy Research and Development [Canada]
OERN Organisation Europeenne pour la Recherche Nucleaire [= European Organization for Nuclear Research]
OES Oceanic Engineering Society [of IEEE, US]
Office of Endangered Species
Optical Emission Spectroscopy
Orbital Escape System [of NASA]
Order Entry System
Outgoing Echo Suppressor
OE-SBR Oil Extended Styrene Butadiene Rubber
OESD Opto-Electronic Semiconductor Device
OESLA Office of Engineering Standards Liaison and Analysis [US]
OETB Offshore Energy Technology Board
OEW Operating Empty Weight
OEWG Open-End Waveguide
OF Oil-Filled
Operational Fixed
Optical Fiber
Optical Flat
Optical Fluorescence
Oriented Film
Output Factor
Outside Face
Overflow
Oxide Film
Oxygen-Free
O/F Orbital Flight
Overflow
OFA Oil-Immersed Forced Air-Cooled (Transformer)
Ontario Federation of Agriculture [Canada]
Oxygenated Fuels Association [US]
OFAAP Ontario Farm Adjustment Assistance Program [Canada]
OFAR Office of Foreign Agricultural Relations [US]
OFB Operational Facilities Branch [of NASA]
Output Feedback
OFC Office
Oil-Filled Cable
Operational Flight Control
Optical Fiber Communication
Oxyfuel Gas Cutting
Oxygen-Free Copper
OFCA Ontario Federation of Construction Associations [Canada]
OFC-A Oxyacetylene Cutting
OFCC Office of Federal Contract Compliance [US]
OFCF Overseas Farmers Cooperative Federation [UK]
OFC-H Oxyhydrogen Cutting
OFC-N Oxynatural Gas Cutting
OFC-P Oxypropane Cutting
OFF Office
Officer [also: OFCR or OFFR]
OFFEN Offensive
OFFIC Official [also: OFF]
OFF PREM Off Premises
OFFR Officer [also: OFCR or OFF]
OFG Optical Frequency Generator
Organic Functional Group
OFHA Oilfield Haulers Association [US]
OFHC Oxygen-Free High-Conductivity (Copper)
OFHIC Oxygen-Free High-Conductivity Copper
OFI Optical Fiber Identifier
OFIA Ontario Forest Industries Association [Canada]
OFIS Office Information System
Operational Flight Information Service [of ICAO]
OFIX Office of the Future Information Exchange [UK]
OFL Ontario Federation of Labour [Canada]
Optical Fault Locator
Overflow
OFLINPS Open Frame Linear Power Supply [also: OF LIN PS]
OFLO Overflow
OFLP Oxygen-Free Low-Phosphorus (Copper)
OFLPC Oxygen-Free Low-Phosphorus Copper
OFLW Overflow
OFMP Organization of Facility Managers and Planners [of OMER, US]
OFN Organization for Flora Neotropica [US]
OFNPS Outstate Facility Network Planning System
OFO Office of Flight Operations [of NASA]
OFP Occluded Frontal Passage
Operating Force Plan
Oscilloscope Face Plane
OFPA Ontario Food Protection Association [Canada]
OFPANA Organic Foods Production Association of North America [US]
OFPP Office of Federal Procurement Policy
OFPS Open Frame Power Supply
OFPU Optical Fiber Production Unit
OFR Oil-Resistant, Flame-Retardant (Cable)
On-Frequency Repeater
Overfrequency Relay
OFRS Office Francais de Recherches Sous-Marines [= French Office for Submarine Research]
OFS Operational Fixed (Microwave) Service
Optical Fiber Sensor
Oxygen-Free with Silver
OFSD Operating Flight Strength Diagram
OFSHR Offshore
OFSMPS Open Frame Switch Mode Power Supply [also: OF SMPS]
OFT Office of Fair Trading [UK]
Optical Fiber Thermometry
Optical Fiber Tube
Orbital Flight Test
OFTDA Office of Flight Tracking and Data Acquisition [of NASA]
OFTEL Office of Telecommunications [UK]
OFTF Optical Fiber Transfer Function

OFTS Optical Fiber Transmission System
OFV Overflow Valve
OFW Oxyfuel Gas Welding
OFXLP Oxygen-Free Extra-Low-Phosphorus (Copper)
OFXLPC Oxygen-Free Extra-Low-Phosphorus Copper
OFZ Obstacle-Free Zone
OG Optical Generation
Optical Glass
Or Gate
Outer Gimbal
Outgoing
Oxygen (Converter) Gas Process [Steelmaking]
O/G Outgoing
Og Ouguiya [Currency of Mauritania]
OGA Outer Gimbal Axis
OGCA Ontario General Contractors Association [Canada]
OGC Oregon Graduate Center [US]
OGDC Oil and Gas Development Corporation [Bangladesh]
OGDD Outgoing/Delay Dial
OGDI Osterreichische Gesellschaft fur Dokumentation und Information [= Austrian Society for Documentation and Information]
OGE Operational Ground Equipment
Out of Ground Effect
OGI Oculogyral Illusion
OGID Outgoing/Immediate Dial
OGIST Oregon Graduate Institute of Science and Technology [US]
OGJ Outgoing Junction
OGM Outgoing Message
OGMC Ordnance Guided Missile Center
OGO Orbiting Geophysical Observatory
OGP Outgoing Message Process
OGR ORNL (= Oak Ridge National Laboratory) Graphite Reactor [USAEC]
Outgoing Repeater
Oxygen-Gas Recovery System
OGRC Office of Grants and Research Contracts [US]
OGS Ontario Geological Society [Canada]
Ontario Geological Survey [Canada]
Ontario Graduate Scholarship [Canada]
OGST Overthread Guide Sleeve Tool
OGT Outgoing Toll
Outgoing Trunk
OGU Outgoing Unit
OGWS Outgoing/Wink Start
OH Octal-to-Hexadecimal
Ohio [US]
Ohmic Heating
Oil Hardening
Ontario Hydro (Corporation) [Canada]
Open-Hearth (Furnace)
Operational Hardware
Optical Harness
Orthohydrogen
Overhead
O&H Oxygen and Hydrogen [also: OH]
O-H Octal-to-Hexadecimal
OHA Office of Hearings and Appeals [US]
Ontario Horticultural Association [Canada]
Outside Helix Angle
Oxygen Hemoglobin Affinity
OHADOE Office of Hearings and Appeals, Department of Energy [US]
OHADOI Office of Hearings and Appeals, Department of the Interior [US]
OHBG Oregon Highland Bentgrass Commission [US]
OHC Occupational Health Center
Ontario Housing Corporation [Canada]
Overhead Camshaft
OHD Over-the-Horizon Detector
OHDDD Ontario Hydro Design and Development Division [Canada]
OH/D&D-G Ontario Hydro/Design and Development Division — Generation [Canada]
OHF Occupational Health Facility
Open Hearth Furnace
OHHS Off-Highway Haulage System
OHIONET Ohio Network [US]
OHL Occupational Health Laboratory [of OML, Canada]
Overhead Line
OHM Octadecylhydrogenmaleate
Ohmmeter
Oil and Hazardous Materials
ohm cm ohm-centimeter [also: ohm-cm, Ωcm or Ω-cm]
ohm m ohm-meter [also: ohm-m, Ωm or Ω-m]
OHMR Office of Hazardous Materials Regulations
OHMS On Her/His Majesty's Service
OHM-TADS Oil and Hazardous Materials — Technical Assistance Data System [of EPA, US]
OH/NMMD Ontario Hydro/Nuclear Materials Management Department [Canada]
OHP Operational Hydrology Program [of WHO, Switzerland]
Overhead Projector
Oxygen at High Pressure
OHR Office of Human Resources
OHRD Ontario Hydro Research Division [also: OH-RD, Canada]
OHRD/TSTA Ontario Hydro Research Division/Tritium Systems Test Assembly [Canada]
OHS Occupational Health and Safety
Octadecylhydrogensuccinate
Off-Highway Haulage System
Off-Hook Service
Office of Highway Safety [US]
Ontario Historical Society [Canada]
OHSA Occupational Health and Safety Act [Canada]
OHSB Occupational Health and Safety Branch [Canada]
OHSC Occupational Health and Safety Commission [Canada]
OHSD Occupational Health and Safety Division [of MOL, Canada]
OHSGT Office of High Speed Ground Transportation
OHT Oxygen at High Temperature
OH&T Oil Hardened and Tempered
OHTB Ontario Highway and Transport Board [Canada]
OHTDC Ontario Hydro Tritium Dispersion Code [Canada]
OHTE Ohmically-Heated Toroidal Experiment
OHV Overhead Valve Engine
OHW Oxyhydrogen Welding
OI Office Information

	Oil-Immersed (Transformer)
	Oil Immersion
	Oil-Insulated (Cable)
	Oil Insulation
	Operation Interface
	Operations Instruction
	Opto-Isolator
OIA	Ocean Industries Association [US]
	Office Information Architecture
OIB	Operating Impedance Bridge
	Operation Instruction Block
	Operations Integration Branch [of NASA]
OIBF	Osterreichisches Institut fur Bibliotheksforschung, Dokumentations- und Informationswesen [= Austrian Institute for Librarianship, Documentation and Information Science]
OIC	Ocean Information Center
	Officer in Charge
	Oil-Insulated Cable
	On-Line Instrumentation Coordinator
	Ontario International Corporation [Canada]
	Operations Instrumentation Coordinator
	Optical Integrated Circuit
	Order-in-Council
OICA	Ontario Institute of Chartered Accountants [Canada]
OICAP	Ocean Industries Capital Assistance Program
OICC	Ontario Institute of Chartered Cartographers [Canada]
OICD	Office of International Cooperation and Development
OICP	Office of International Communications Policy [of ICA, US]
OICR	Ontario Institute for Computer Research [Canada]
OIDA	Optoelectronics Industry Development Association [US]
OIDC	Ontario Industrial Development Council [Canada]
OIDI	Optically Isolated Digital Input
OIDO	Ocean Industries Development Office
OIDPS	Overseas Intelligence Data Processing System
OIE	Office International des Epizooties [= International Office of Epizootics, France]
OIFC	Oil-Insulated, Fan-Cooled (Transformer)
OIFIG	Official Irish FORTH (Computer Language) Interest Group
OIFOC	Oil-Immersed, Forced-Oil-Cooled (Transformer)
OIG	Optically Isolated Gate
OII	Office of Invention and Innovation [of NIST, US]
	Oil Investment Institute [US]
OIL TURP	Oil of Turpentine
OIM	Offshore Installation Manager
OIML	Organisation Internationale de Metrologie Legale [= International Organization of Legal Metrology, France]
OINC	Officer in Charge
OIP	Office of International Programs [Canada]
	Oil-in-Place
	Ontario Institute of Painters [Canada]
	Operating Internal Pressure
OIRCA	Ontario Industrial Roofing Contractors Association [Canada]
OIRM	Office of Information Resources Management [US]
OIRT	Organisation Internationale de Radiodiffusion et Television [= International Organization for Broadcasting and Television, Czechoslovakia]
OIS	Obstacle Identification Surface
	Office Information System
	Office of Information Services [of FAA, US]
	Office of Investigatory Services [US]
	Operational Intercommunication System
OISA	Office of International Science Activities
	Office of International Scientific Affairs [US]
OISC	Oil-Insulated and Self-Cooled (Transformer)
OISE	Ontario Institute for Studies in Education [Canada]
OITL	Outdoor-Indoor Transmission Loss
OITT	Outpulse Identifier Trunk Test
OIWC	Oil-Immersed, Water-Cooled (Transformer)
OJ	Originating Junctor
OJT	On-the-Job Training
OK	All Right
	Oklahoma [US]
oka	otherwise known as
OKITAI	Okinawa-Taiwan Submarine Cable [Japan/Taiwan]
Okla	Oklahoma [US]
OL	Office of Labor [US]
	On-Line
	Only Loadable
	Open Loop
	Operating Level
	Operating License
	Operating Location
	Operational Semantics
	Other Line
	Output Latch
	Overflow Level
	Overhead Line
	Overlap
	Overload
O/L	Operations/Logistics
	Overload
OLA	Ontario Library Association [Canada]
OLABS	Offshore Labrador Biological Studies [Canada]
OLAC	On-Line Accelerated Cooling
	On-Line Audiovisual Catalogers [US]
OLADE	Organisacion Latinoamericana de Energia [= Latin-American Energy Organization [also: OLAE, Ecuador]]
OLAS	On-Line Acquisitions System [US]
OLB	On-Line Batch
	Outer Lead Bond
OLBM	Orbital-Launched Ballistic Missile
OLC	On-Line Computer
	Ontario Land Corporation [Canada]
	Ontario Library Council [Canada]
	Outgoing Line Circuit
OLCR	On-Line Character Recognition
OLD	Open Loop Damping
	Open Loop Drive
OLDB	On-Line Database
OLDI	On-Line-Data Interchange
OLDS	Open Loop Drive System
OLE	On-Line Enquiry

	Ontario Land Economist [Canada]
OLERT	On-Line Executive for Real Time
OLEX	Ontario Livestock Exchange [Canada]
OLF	One Hundred Linear Feet
	Orbital Launch Facility
	Outlying Field [US]
OLIFLM	On-Line Image Forming Light Modulator
OLIP	On-Line Instrument Package
OLIS	On-Line Information System
OLLT	Office of Libraries and Learning Technologies [US]
OLM	On-Line Monitor
	Optical Light Microscopy
OLMA	Ontario Lumber Manufacturers Association [Canada]
OLO	Orbital Launch Operations
OLOE	On-Line Order Entry
OLOGS	Open-Loop Oxygen-Generating System
OLP	Objective Lens Power Supply
	Off-Line Program
	On-Line Processor
	On-Line Programming
	Oxygen Lance Powder (Converter)
	Oxygen/Lime/Powder (Process)
OLPARS	On-Line Pattern Analysis and Recognition System
OLPS	On-Line Programming System
OLR	Objective Loudness Ratio
	Off-Line Recovery
	Off Load Routes
OLRA	Ontario Labour Relations Act [Canada]
OLRB	Ontario Labour Relations Board [Canada]
OLRT	On-Line Real-Time (System)
OLS	On-Line Search
	Ontario Land Surveyor [Canada]
	Open Loop System
	Ordinary Least Squares
OL'SAM	On-Line Search Assistance Machine [of Franklin Institute, US]
OLSC	On-Line Scientific Computer
OLSDMS	On-Line Strain and Damage Measurement System
OLSE	Ordinary Least-Squares Estimator
OLSS	On-Line Software System
	On-Line Support Software
OLSUS	On-Line System Use Statistics
OLT	On-Line Test
OLTE	On-Line Test Equipment
	Optical Line Terminating Equipment
OLTEP	On-Line Test Executive Program
OLTP	Off-Line Tape Preparation
	On-Line Transaction Processing
OLTS	On-Line Test Section
	On-Line Test System
OLTT	On-Line Terminal Test
OLUC	On-Line Union Catalogue [of OCLC, US]
OLUD	On-Line Update
OLUG	Office Landscape Users Group [now: OPUG, US]
OLUHO	Okinawa-Luzon-Hong Kong Submarine Cable [Japan/Philippines/Hong Kong]
OLUM	On-Line Update Module
OLV	Open-Frame Low Voltage
OLVP	Office of Launch Vehicle Programs
	Office of Launch Vehicles and Propulsion
OM	Office Model
	Operations Maintenance
	Operations Manager
	Operations Manual
	Optical Master
	Optical Metallography
	Optical Microscopy
	Order of Merit
	Organic Matter
	Organometallic(s)
	Orifice Meter
	Outer Marker
	Output Module
O&M	Operation and Maintenance
	Organization and Method
O/M	Oxygen-to-Metal Ratio
OMA	Ontario Medical Association [Canada]
	Ontario Mining Act [Canada]
	Ontario Mining Association [Canada]
	Ontario Ministry of Agriculture [Canada]
	Ontario Museums Association [Canada]
	Operations Monitor Alarm
	Optical Manufacturers Association [US]
	Optical Multichannel Analyzer
	Orderly Marketing Agreement
	Overseas Mining Association [UK]
O&M, A	Operation and Maintenance, Army
OMAC	Occupational Medical Association of Canada
	On-Line Manufacturing Control
OMA CCD	Optical Multichannel Analyzer Charge-Coupled Device
OMAF	Ontario Ministry of Agriculture and Food [Canada]
OMAP	Object Module Assembly Program
OMASD	Office of Management Appraisal and Systems Development [US]
OMAT	Office of Manpower Automation and Training
OMB	Office of Management and Budget [US]
	Outer Marker Beacon
OMBE	Office of Minority Business Enterprises
OMBI	Observation, Measurement, Balancing and Installation
OMC	Organometallic Chemistry
	Organometallic Compound
	Oxidized Microcrystalline (Wax)
OMCA	Ontario Motor Coach Association [Canada]
OMCU	Ontario Ministry of Colleges and Universities [Canada]
OMCVD	Organometallic Chemical Vapor Deposition
OMD	Open Macro Definition
OMDP	Ocean Margin Drilling Project
OME	Office of Management Engineers
	Office of Minerals Exploration [US]
	Ontario Ministry of Education [Canada]
	Ontario Ministry of Energy [Canada]
	Ontario Ministry of the Environment [Canada]
	Optimum Mineral Extraction
O-Me	O-Methyl Diisopropyl
OMEA	Ontario Municipal Electric Association [Canada]
OMEF	Office Machines and Equipment Federation
OMEGA	Octanoyl-N-Methylglucamide
OMEP	Ontario Mineral Exploration Program [Canada]

OMER Operations Management Education and Research Foundation [US]
OMETA Ordnance Management Engineering Training Agency [US]
OMF Office of Marketing Facilities [US]
Old Master File
OMFS Office Maser Frequency Supply
OMG Ocean Mapping Group [of UNB, Canada]
Osterreichische Mathematische Gesellschaft [= Austrian Mathematical Society]
Osterreichische Mineralogische Gesellschaft [= Austrian Mineralogical Society]
OMH Ontario Ministry of Health [Canada]
OMH-1 Sodium Diethyldihydroaluminate
OMHSA Ontario Municipal Health and Safety Association [Canada]
OMIBAC Ordinal Memory Inspecting Binary Automatic Computer
OMIS Optical Microscope Inspection System
OML Ontario Ministry of Labour [Canada]
Ontario Motor League [Canada]
Ordnance Missile Laboratories [US]
Outer Mold Line
Outgoing Matching Loss
Outside Mold Line
OMM Operation and Maintenance Manual
OMMB Ontario Milk Marketing Board [Canada]
OMNDM Ontario Ministry of Northern Development and Mines [Canada]
OMNI Omnirange
OMNITENNA Omnirange Antenna
OMNIX Onyx Microcomputer Unix (System)
OMNR Ontario Ministry of Natural Resources [Canada]
OMP Office of Metric Programs
Optical Mark Printer
Orotidine Monophosphate
OMPA Octamethyl Pyrophosphoramide
OMPR Optical Mark Page Reader
OMPRA One-Man Propulsion Research Apparatus [of NASA]
OMPT Observed Man Point Trajectory
OMR Omani Rial [Currency of Oman]
Optical Mark Reader
Optical Mark Recognition
Organic-Moderated Reactor [US]
OMRCA Organic-Moderated Reactor Critical Assembly [US]
OMRE Organic-Moderated Reactor Experiment [of INEL, US]
OMRR Ordnance Material Research Reactor [US]
OMRV Operational Maneuvering Reentry Vehicle
OMS Office of Management Services [US and Canada]
Operational Meteorological Satellite [of NASA]
Optoelectronic Measuring System
Orbital Maneuvering Subsystem
Orbital Maneuvering System
Output per Manshift
Output Multiplex Synchronizer
Ovonic Memory Switch
OMSA Offshore Marine Service Association [US]
OMSF Office of Manned Spaceflight [of NASA]
OMT Orthomode Transducer
OMTNS Over Mountains
OMTS Organizational Maintenance Test Station
OMU Optical Measuring Unit
OMUA Office Machinery Users Association [now: IAM, UK]
OMV Orbital Maneuvering Vehicle
OMVPE Organometallic Vapor-Phase Epitaxy
ON Octane Number
Osterreichisches Normeninstitut [= Austrian Standards Institute]
Ontario [Canada]
Order Number
ONA Ontario Nurses Association [Canada]
Open Network Architecture
ONAL Off-Network Access Line
ONAT Off-Network Access Trunk
ONB Osterreichische Nationalbibliothek [= National Library of Austria]
O-Nitrobiphenyl
ONC Operational Navigation Chart
Ordinary National Certificate [UK]
OND Ordinary National Diploma [UK]
ONE Optimum Network Executive
ONERA Office National d'Etudes et de Recherches Aerospatiales [= National Office for Aerospace Studies and Research [of CERT, France]]
ONERA-CERT Office National d'Etudes et de Recherches Aerospatiales — Centre d'Etudes et de Recherches de Toulouse [= National Office for Aerospace Studies and Research of the Research and Study Center at Toulouse, France]
ONF Office National des Forets [= National Forestry Office, France]
ONGA Overseas Number Group Analysis
ONGC Oil and Natural Gas Commission [India]
ONI Operator Number Identification
ONIA Office National Industriel de l'Azote [= National Office of the Nitrogen Industry, France]
ONID Operator Number Identification
ONM Office National Meteorologiques [= National Meteorological Office, France]
ONP Optical Nuclear Polarization
ONPI Office of News and Public Information [US]
ONQ Order of Nurses of Quebec [Canada]
ONR Office of Naval Research [of US Navy]
ONS Omega Navigation System
ONSHR On Shore
Ont Ontario [Canada]
ONTAP On-Line Training and Practice [also: ON-TAP]
ONTC Ontario Northland Transportation Commission [Canada]
ONTERIS Ontario Educational Research Information System [Canada]
ONULP Ontario New Universities Library Project [Canada]
ONWARD Organization of Northwest Authorities for Rationalized Design [US]
OO Object-Oriented
Oceanographic Office
Ocean Outlook [US]
Optical Orientation
O/O Ore/Oil (Carrier) [also: OO]
OOB Out-of-Balance (Current)

	Out of Band
OOC	Ore/Oil Carrier [also: O/OC]
	On-Off Control
OOD	Opposite Oriented Diffusion
OOG	Olive Oil Group [US]
	Out of Gauge
OOIP	Original Oil-in-Place
OOK	On-Off Keying
OOL	Operator-Oriented Language
OOLR	Overall Objective Loudness Rating
OOLUG	Oklahoma On-Line Users Group [US]
OOO	Out of Order
OOP	Object-Oriented Programming
OOPS	Off-Line Operating Simulator
OOR	Operator Override
OOS	Ore/Oil Ship
	Operational Operating System
	Out of Service
OOUG	Oregon On-Line Users Group [US]
OP	Object-Oriented Programming
	Observation Plane
	Office Processor
	Oil Pressure
	Old Process (Patenting)
	Operand
	Operating Potential
	Operating Procedure
	Operation
	Operational Priority
	Operational Process
	Operation Part
	Operator
	Operator Panel
	Oppenheimer-Phillips (Process) [also: O-P]
	Optical Properties
	Oriented Polymer
	Orthogonal Polynomial
	Out of Print
	Output
	Output Processor
	Overpotential
	Overpressure
	Over Proof [Spirits]
	Oxygen Point
	Oxygen Pressure Process
O/P	Output
op	over proof [Spirits]
1P	One Pole (or Single Pole)
OPA	Office of Price Administration [US]
	Ontario Pharmacists Association [Canada]
	Opaque
	Operator Priority Access
	Optical Parametric Amplifier
	Optical Publishing Association [US]
	Optoelectronic Pulse Amplifier
	Orthophthalaldehyde
OPAC	(Working Group on) Operation of Aircrafts [of ECAC, France]
OPADEC	Optical Particle Decoy
OPA/FMOCCl	Orthophthalaldehyde/Fluorenylmethoxycarbonyl Chloride
OPAIT	Ontario Program for the Advancement of Industrial Technology [Canada]
OPAL	Operational Performance Analysis Language
	Order Processing Automated Line
OPAMP	Operational Amplifier [also: OP-AMP or OP AMP]
OPAS	Operational Assignment
OPASTCO	Organization for the Protection and Advancement of Small Telephone Companies [US]
OPC	On-Line Plotter Controller
	Ontario Press Council [Canada]
	Operational Control
	Operation Code
	Optional Calling
OPCC	Optical Product Code Council [US]
OPCD	Operational Planning and Coordination Directorate [of Secretary of State, Canada]
op cit	(opere citato) — in the work cited
OPCL	Optical
OPCM	Operative Plasterers and Cement Masons International
OPCODE	Operation Code [also: OP-CODE or OP CODE]
OPBA	Ontario Public Buyers Association
OPC	Oligonucleotide Purification Cartridge
	Optical Phase Conjugation
	Ordinary Portland Cement
OPCOM	Operations Communications
OPCON	Optimizing Control
OPCR	One Pass Cold Rolled
OPD	One per Desk
	Operand
	Operational Programming Department
	O-Phenylenediamine
	Optical Particle Detector
	Over Pin Diameter
OPDAC	Optical Data Converter
OPDAR	Optical Detection and Ranging
	Optical Radar
OPDN	Oxydipropionitrile
OPDP	Officer Professional Development Program
OPE	Operations Project Engineer
	Optimized Processing Element
	Oxygen Plasma Etching
OPEC	Organization of Petroleum Exporting Countries [Austria]
OPECNA	Organization of Petroleum Exporting Countries News Agency [Austria]
OPEDA	Organization of Professional Employees of the Department of Agriculture [US]
	Outdoor Power Equipment Distributors Association [US]
OPEG	OPEC Petroleum Emergency Group
OPEI	Outdoor Power Equipment Institute [US]
OPEIU	Office and Professional Employees International Union
OPEN	Organisation des Producteurs d'Energie Nucleaire [= Organization of Nuclear Energy Producers, France]
OPEP	Orbital Plane Experimental Package [of NASA]
OPER	Operation
	Operator
OPERA	Out-of-Pile Expulsion and Reentry Apparatus

OPERATORS Optimization Program for Economical Remote Trunk Arrangement and TSPS (= Traffic Service Positions System) Operator Arrangements
OPERG Operating
OPEX Operational and Executive
OPF Osmium Potassium Ferrocyanide
OPFET Optical Field-Effect Transistor [also: OpFET]
OPFM Outlet Plenum Feature Model
OPG Oxalate, Peroxide and Gluconic Acid
Oxypolygelatin
1PH One Phase (or Single Phase)
OPHF Orbital Polarized Hartree-Fock
OP HRS Operation Hours [also: Op hrs]
OPHT Ophthalmic
OPI Office of Public Information
Ontario Petroleum Institute [Canada]
OP&I Office of Patents and Inventions [US]
OPIC Overseas Private Investment Corporation [US]
OPIM Order Processing and Inventory Monitoring
OPK Optokinetic(s)
OPL Official Publications Library
Outer Plexiform Layer
OPLA Offshore Pollution Liability Agreement
Ontario Public Library Association [Canada]
OPLAC Ontario Public Libraries Advisory Committee [Canada]
OPLB Ontario Public Libraries Board [Canada]
OPLC Ontario Provincial Libraries Council [Canada]
OPLE Omega Position Location Experiment [of NASA]
OPLIN Ontario Public Libraries Information Network [Canada]
OPM Office of Personnel Management
Open Pit Mining
Operations per Minute [also: opm]
Operator Programming Method
Orbits per Minute
Output Position Map
OPMA Office Products Manufacturers Association [US]
Open Pit Mining Association [US]
Overseas Press and Media Association [UK]
OPMET Operational Meteorological Information
OPN One-Port Network
Open(ing)
OPND Operand
OPNG Opening
OPO Optical Parametric Oscillator
OPOL Offshore Pollution Liability Association [UK]
Optimization-Oriented Language
OPP Octal Print Punch
Office of Plans and Policy [of FCC, US]
Open Pore
O-Phenylphenol
Opportunity
Opposite
Opposition
Optical Precipitate Profiler
Oriented Polypropylene
OPPCE Opposite Commutator End [also: OPP CE]
OPPMSA Ontario Pulp and Paper Makers Safety Association [Canada]
OPPOSIT Optimization of a Production Process by an Ordered Simulation and Iteration Technique
OPP PE Opposite Pulley End
OPPS Overpressurization Protection Switch
Overpressurization Protection System
OPR Offsite Procurement Request
Oil Production Rate
Operation(al)
Operational Preference
Operative
Operator
Optical Page Reading
Optical Pattern Recognition
OPRA Office Products Representatives Association [US]
OP REGISTER Operation Register
OPRIS Ohio Project for Research in Information Services [US]
OPS Office of Pipeline Safety [of DOT, US]
Off-Premise Station
Offshore Power Systems
Oil Pressure Switch
On-Line Process Synthesizer
Operational Paging System
Operational Performance Standards
Operational Protection System
Operations
Operations per Second
Operator's Subsystem
Oriented Polystyrene
Overhead Positioning System
o-Ps Ortho-Positronium
OPSCAN Optical Scanning
OPSCON Operations Control
OPSEC Operations Security
OPSF Orbital Propellant Storage Facility
OPSK Optimum Phase-Shift Keying
OPSKS Optimum Phase-Shift Keyed Signal
OPSP Office of Product Standards Policy
Operations Panel [of ICAO]
OPSR Office of Pipeline Safety Regulations
OPSWL Old Program Status Word Location
OPSYS Operating System
OPT Ophthaldialdehyde
Optical
Optician
Optics
Optimum
Option(al)
OPTA Optimal Performance Theoretically Attainable
OPTAR Optical Automatic Ranging
OPTI Office of Productivity, Technology and Innovation [US]
OPTICIMP Optimization of Computer Integrated Materials Processing
OPTIM Order Point Technique for Inventory Management
OPTIMA Organization for the Phyto-Taxonomic Investigation of the Mediterranean Area [Germany]
OPTN Optician
OPTS On-Line Peripheral Test System
Operations
OPTUL Optical Pulse Transmission Using Laser

OPU Operations Priority Unit
OPUG Office Planners and Users Group [US]
OPUR Object Program Utility Routines
OPUS Octal Program Updating System
Organization of Professional Users of Statistics [UK]
OPW Operating Weight
Orthogonalized Plane Wave
Orthogonal Plane Wave
OPX Off-Premise Extension
OPZ Osterreichisches Produktivitatszentrum [= Austrian Center for Productivity]
OQ Oil Quenching
OQL On-Line Query Language
O-QPSK Offset Quaternary Phase Shift Keying [also: OQPSK]
OR Omani Rial [Currency of Oman]
Operational Readiness
Operational Requirements
Operational Research
Operations Requirement
Operations Research
Operating Reactor
Operating Room
Optical Receiver
Orange
Oregon [US]
Organizational Research
Orientation Response
Output Register
Outside Radius
Overhaul and Repair
Overload Relay
Overrange
Overrun
Owner's Risk
Oxidation Resistance
Oxide Removal
O&R Ocean and Rail
O/R On Request
Originator/Recipient
ORA Office des Renseignements Agricoles [= Office for Agricultural Information]
Office of Research Analysis [of US Air Force]
Output Reference Axis
ORACLE Oak Ridge Automatic Computer and Logical Engine [US]
Optical Recognition of Announcements by Code Line Electronics
ORAE Operational Research and Analysis Establishment
ORAN Orbital Analysis
ORAPA Ontario Retail Accident Prevention Association [Canada]
ORATE Ordered Random Access Talking Equipment
ORAU Oak Ridge Associated Universities [US]
ORB Observatoire Royal de Belgique [= Royal Belgian Observatory]
Omnidirectional Research Beacon
Operational Research Branch [US and Canada]
Owner's Risk of Breakage
ORBA Ontario Road Builders Association [Canada]
ORBID On-Line Retrieval of Bibliographic Data
ORBIS Orbiting Radio Beacon Ionospheric Satellite [of NASA]
ORBIT On-Line Real-Time Branch Information [Computer Language]
On-Line Retrieval of Bibliographic Information Time-Shared
ORACLE Binary Internal Translator [of ORNL, US]
ORC On-Line Reactivity Computers
Operational Research Committee [Canada]
Operations Research Center [US]
Ordnance Rocket Center
Orthogonal Row Computer
Oxidation Reduction Cycle
ORCA Ontario Royal Commission on Asbestos [Canada]
ORCAL Orange County Manufacturing and Metalworking Conference and Exposition [US]
ORCHIS Oak Ridge Computerized Hierarchical Information System [of USAEC]
ORCO Organization Committee [of ISO]
ORCS Organic Reactions Catalysis Society [US]
ORD Office of Research and Development [of EPA, US]
Once-Run Distillate
Operational Readiness Data
Operational Readiness Date
Operational Research Division
Optical Rotary Dispersion
Order
Ordinary
Ordnance
Owner's Risk of Damage
ORDEF Ontario Research Development Foundation [Canada]
ORDENG Ordnance Engineering [also: ORD ENG]
OrdEng Ordnance Engineer
ORDVAC Ordnance Variable Automatic Computer
ORE Occupational Radiation Exposure
Ocean Resources Engineering
Operational Research Establishment [now: DORE, Canada]
Overall Reference Equivalent
Ore Oregon [also: Oreg, US]
ORELA Oak Ridge Electron Linear Accelerator [US]
OREO Orbiting Radio Emission Observatory [of NASA]
OREP Optical Repeater Equipment
ORE RES Ore Reserve
ORF Ontario Research Foundation [now: ORTECH, Canada]
Optical Rangefinder
Orifice
Owner's Risk of Fire
ORFM Outlet Region Feature Model
ORG Operations Research Group
Organic
Organization
Origin(al)
Origination
Orthogonal Row Computer
ORGALIME Organisme de Liaison des Industries Metalliques Europeennes [= Liaison Group for the European Engineering Industries, Belgium]
ORGAN Organization

ORG CHEM Organic Chemistry
ORGDP Oak Ridge Gaseous Diffusion Plant [of USAEC]
ORGN Organization
ORGSC Oregon Ryegrass Growers Seed Commission [US]
ORI Ocean Research Institute [Japan]
Operational Readiness Inspection
Overriding Royalty Interest
ORIA Office of Regulatory and Information Affairs
ORIC Oak Ridge Isochronous Cyclotron [US]
ORICAT Original Cataloguing System
OR&IE Operations Research and Industrial Engineering
ORIENT Orientation
ORIG Origin(al)
ORINS Oak Ridge Institute of Nuclear Studies [now: ORAU, US]
ORION On-Line Retrieval of Information Over a Network
ORIT Operational Readiness Inspection Team
ORKID Orientation Determination from Kikuchi Diagrams
ORL Orbital Research Laboratory [of NASA]
Ordnance Research Laboratory [US]
Owner's Risk of Leakage
ORLY Overload Relay
ORMAK Oak Ridge Tokamak [US]
ORN Orange
Ornament
Orn Ornithine
Ornithyl
ORNITH Ornithologist [also: ORN]
Ornithology [also: ORN]
ORNL Oak Ridge National Laboratory [of USAEC]
ORO Operations Research Office [US]
OROM Optical Read-Only Memory
ORP Operational Readiness Platform
Optional Response Poll
Oxidation-Reduction Potential
Oxidation-Reduction Probe
ORR Oak Ridge Research Reactor [US]
ORRAS Optical Research Radiometrical Analysis System
ORRMIS Oak Ridge Regional Modeling Information System [US]
ORRPB Ottawa River Regulation Planning Board [Canada]
ORS Object Recognition Systems
Octahedral Research Satellite [of NASA]
Office of Research Services
On-Site Reclamation System
Operational Research Society [UK]
O-Ring Seal
ORSA Operations Research Society of America [of TIMS, US]
ORS-BR O-Ring Seal, Braze Type
ORS-BT O-Ring Seal, Bite Type
ORSJ Operations Research Society of Japan
ORSORT Oak Ridge School of Reactor Technology [of USAEC]
ORT Operational Readiness Test
Overland Radar Technology
ORTAG Operations Research Technical Assistance Group [of US Army]
ORTECH Ontario Research and Technology Foundation [formerly: ORF, Canada]
ORTEP Oak Ridge Thermal Ellipsoid Program [US]
ORTF Office de la Radiodiffusion and Television Francaise [= French Office for Broadcasting and Television]
ORTHO Orthogonal
ORTS Optional Residential Telephone Service
ORV Ocean Range Vessel
Orbital Rescue Vehicle
OS Oblique Sounding
Oceanic Society [US]
Odd Symmetric
Office of Supply [US]
Office System
Ohio State [US]
Oil Shale
Oil Solubility
Oil-Soluble
Oil Switch
Old Style
Oligosaccharide
"ON" Switch
Operating System
Operational Sequence
Operational Software
Operations System
Optical Scanning
Optical System
Ordinary Seaman
Ordnance Survey
Oregon State [US]
Organic Synthesis
Organosilicon
Outer Space
Out of Service
Output System
Outsize
Outstanding
Overshoot
Overstressing
O/S Operating System
Outstanding
Os Osmium
OSA Office of Student Affairs
Office System Architecture
Olefin-Modified Styrene-Acrylonitrile
On-Stream Analysis
Open Systems Architecture
Optical Society of America [US]
Organic Soil Association [South Africa]
OSAC Open Space Action Committee [now: OSI, US]
OSAF Origin Subarea Field
OSAM Overflow Sequential Access Method
OSAIS Oil Spillage Analytical Information Service [of LGC, UK]
OSAM Overflow Sequential Access Method
OSAP Ontario Student Assistance Program [Canada]
OSA-UCS Optical Society of America Uniform Color Scales
OSB Operational Status Bit
Oriented Strand Board
OSBP Ontario Special Bursary Plan [Canada]
OSC Ocean Sciences Center [of MUN, Canada]
On-Site Safety Committee
Ontario Science Center [Canada]

	Operator Services Complex
	Optically Sensitive Controller
	Orbit-Spin Coupling
	Organosilicon Compound
	Orthosubstitution Compound
	Oscillation
	Oscillator
OSCAA	Oil Spill Control Association of America [now: SCAA, US]
OSCAR	Optimum Systems Covariance Analysis Results
	Orbiting Satellite-Carrying Amateur Radio [of AMSAT, US]
	Order Status Control and Reporting
OSCAS	Office of Statistical Coordination and Standards [Philippines]
OSCG	Oscillograph
OSCL	Operating System Control Language
OSC-MULT	Oscillator-Multiplier
OSCOM	Oslo Commission for Marine Pollution Control
OSCRL	Operating System Command and Response Language
OSD	Office of the Secretary of Defense [US]
	Office Systems Design
	On-Line System Driver
	On-Screen Digital
	Operational Sequence Diagram
	Operational Systems Development
	Optical Scanning Device
	Organosilicon Device
OS&D	Over, Short and Damage Report
OSDM	Optical Space-Division Multiplexing
OSDP	On-Site Data Processor
OSDR	Oil Slick Detection Radar
OSDS	Operating System for Distributed Switching
OSE	Oceanic Society Expeditions [US]
	Operational Support Equipment
	Optical Science and Engineering
OSEC	Office Suisse d'Expansion Commerciale [= Swiss Office for Commercial Expansion]
OSEE	Optically-Simulated Electron Emission
OSEOS	Operational Synchronous Earth Observatory Satellite
OSEP	Office of Scientific and Engineering Personnel [of National Academy of Engineering, US]
OSERP	Oil Sands Environmental Research Program [Canada]
OSF	Office of Spaceflight [of NASA]
	One Hundred Square Feet [also: osf]
	Open Software Foundation [US]
OSFM	Office of Spacecraft and Flight Missions [of NASA]
OSFP	Office of Spaceflight Programs [of NASA]
OSGP	Ontario Study Grant Plant [Canada]
OSH	(National) Occupational Safety and Health Conference [Canada]
	Oil Sands and Heavy Oil
OSHA	Occupational Safety and Health Act [US]
	Occupational Safety and Health Administration [US]
OSHB	One-Sided Height Balanced
OSHO	Oil Sands and Heavy Oil
OSHRC	Occupational Safety and Health Review Commission [US]
OSHSA	Occupational Safety and Health Standards Act [US]
OSI	Office of Scientific Intelligence
	On-Site Inspection
	Open Space Institute [US]
	Open Systems Interconnect
	Open Systems Interface
	Operating System Interface
	Optical Sciences Institute [of LLNL, US]
	Optical Society of India
OSIC	Optimization of Subcarrier Information Capacity
OSIL	Operating System Implementation Language
OSIRIS	On-Line Search Information Retrieval Information Storage [US Navy]
OSIS	Office of Science Information Service [of National Science Foundation, US]
OSL	Observed Significance Level
	Ontario Safety League [Canada]
	Operating System Language
	Outstanding Leg
OSLO	Other Six Leases Operation [Canada]
OSLP	Ontario Student Loans Plan [Canada]
OSM	Office of Surface Mining
	Oil-Sands Mining
	Operating System Monitor
	Option Select Mode
OSMA	Occidental Society of Metempiric Analysis [US]
	Orthopedic Surgical Manufacturers Association [US]
OSME	Ornithological Society of the Middle East
OS/MFT	Operating System/Multiprogramming a Fixed Number of Tasks
OSMV	One-Shot Multivibrator
OS/MVS	Operating System/Multiprogramming with Virtual Storage
OS/MVT	Operating System/Multiprogramming a Variable Number of Tasks
OSN	Output Sequence Number
OSNZ	Ornithological Society of New Zealand
OSO	Offshore Supplies Office
	Orbital Solar Observatory [US]
	Orbiting Satellite Observer
	Orbiting Scientific Observatory
	Orbiting Solar Observatory
	Origination Screening Office
O/S/O	Ore/Slurry/Oil (Carrier) [also: OSO]
OSOC	Ore/Slurry/Oil Carrier
OSOS	Ore/Slurry/Oil Ship
OSP	Office of Scientific Personnel [of NRC-NAS, US]
	Office of Statistical Policy
	On-Screen Programming
	Optical Signal Processing
	Outside Plant
OSPS	Operator Services Position System
OSR	Office for Scientific Research [Indonesia]
	Oil/Steam Ratio
	Operational Scanning Recognition
	Operand Storage Register
	Optical Scanning Recognition
	Optical Solar Reflector
	Output Shift Register

	Over-the-Shoulder Rating
OSRA	Office Systems Research Association [US]
OSRD	Office of Scientific Research and Development [US]
	Office of Standard Reference Data [of NIST, US]
OSRMD	Office of Scientific Research, Mechanics Division
OSS	Ocean Science and Surveys [Canada]
	Ocean Surveillance Satellite
	Office of Space Sciences [US]
	Office of Statistical Standards
	Operational Storage Site
	Operation Support System
	Operating System Supervisor
	Operating System Support
	Orbital Space Station
OSSA	Office for Space Science and Applications [of NASA]
OSSD	Ontario Secondary School Diploma [Canada]
OSSGD	Ontario Secondary School Graduation Diploma [Canada]
OSSHGD	Ontario Secondary School Honour Graduation Diploma [Canada]
OSSL	Operating System Simulation Language
OSSS	Orbital Space Station System [of NASA]
OSSU	Operator Services Switching Unit
OST	Office of Science and Technology [of PSAC, US]
	Offshore Storage and Treatment
	On-Shift Test
	Operating System Toolbox
	Operating System Trap
	Operational System Test
	Originating Station Treatment
OSTA	Office of Space and Terrestrial Applications
OSTAC	Ocean Science and Technology Advisory Committee [of NSIA, US]
OSTEST	Operating System Test
OSTI	Office of Scientific and Technical Information [UK]
	Organization for Social and Technological Innovation [US]
OSTL	Operating System Table Loader
OSTNPS	Operator Services Traffic Network Planning System
OSTP	Office of Science and Technology Policy [US]
	Office of Scientific and Technical Personnel [of OECD, France]
OSTS	Office of State Technical Service [US]
	Official Seed Testing Station [UK]
OSU	Ohio State University [US]
	Oregon State University [US]
	Osaka Sangyo University [Japan]
OSUHRF	Ohio State University Heat Release Calorimeter [US]
OSUL	Ohio State University Libraries [US]
OSUR	Ohio State University Reactor [US]
OSURF	Ohio State University Research Foundation [US]
OSV	Ocean Station Vessel
	On-Site Verification
	Orbital Support Vehicle
OSW	Oblique Shock Wave
	Office of Saline Water [US]
	Office of Solid Wastes [US]
OSWS	Operating System Workstation
OS&Y	Outside Screw and Yoke
OT	Office of Telecommunications [US]
	Office of Transportation [US]
	Oil-Tight
	On Truck
	Opening Time
	Open Top
	Operating Temperature
	Operating Time
	Optical Technology
	Organization Theory
	Orthogonal Trees
	Output Terminal
	Overall Test
	Overtime
O/T	Other Times
OTA	Office of Technology Assessment [of US Congress]
	Official Test Aerosol
	Ontario Trucking Association [Canada]
	Open Test Assembly
	Operational Transconductance Amplifier
OTAF	Office of Technology Assessment and Forecasts [of USPTO]
OTANS	Offshore Trade Association of Nova Scotia [Canada]
OTANY	Oil Trades Association of New York [US]
OTBD	Outboard
OTBG	O-Tolylbiguanide
OTC	Office of Telecommunications
	Officers Training Corps
	Offshore Technology Conference
	Once-Through Cooling
	Operating Telephone Company
	Operational Test Center
	Originating Toll Center
	Overhead Travelling Crane
	Overseas Telecommunications Commission [Australia]
	Over-the-Counter
	Oxytetracycline
OTCA	Overseas Technical Cooperation Agency [Japan]
OTCCC	Open-Type Control Circuit Contact
OTCR	Office of Technical Cooperation and Research [US]
OTD	Optical Time Domain
	Original Transmission Density
OTDA	Office of Tracking and Data Acquisition
OTDR	Optical Time Domain Reflectometry
OTDT	Over Temperature Delta T
OTE	Operational Test and Evaluation
	Operational Test Equipment
OTEC	Ocean Thermal Energy Conservation
	Ocean Thermal Energy Conversion
OTES	Optical Technology Experiment System
OTF	Ontario Technology Fund [Canada]
	Optical Transfer Function
OTG	Option Table Generator
OTH	Over-the-Horizon
OTH/B	Over-the-Horizon Backscatter
OTHR	Over-the-Horizon Radar
OTIF	Organisation Intergouvernementale pour les Transports Internationaux Ferroviaire [= Intergovernmental Organization for International Carriage by Rail, Switzerland]

OTIP Ontario Training Incentive Program [Canada]
OTIU Overseas Technical Information Unit
OTJ On-the-Job
OTL On-Line Task Loader
Order Trunk Line
OTLK Outlook
OTLP 0 (= Zero) dBm Transmission Level Point
OTLT Outlet
OTM Office of Telecommunications Management [US]
OTMA Office Technology Management Association [US]
OTMJ Outgoing Trunk Message Junction
OTN Operational Teletype Network
O to O Out to Out
OTOS N-Oxidiethylene Dithiocarbamyl-N'-Oxydiethylene Sulfenamide
OTP Office of Telecommunications Policy [US]
One-Time Programmable
On Top
Operational Test Procedure
Overtemperature Protection
OTP-EPROM One-Time Programmable — Electrically Programmable Read-Only Memory
OTPP Ocean Thermal Power Plant
OTR Office des Transports par Route [= Office for Road Transport, Belgium]
Optical Tracking
Organic Test Reactor [of WNRE, Canada]
OTRAC Oscillogram Trace Reader
OTRG Office Technology Research Group [US]
OTS Office of Technical Services [of USDOC]
Office of Technological Services
Off-the-Shelf (Computer System)
Optical Technology Satellite
Optical Transport System
Orbital Technology Satellite
Orbital Test Satellite [of ESA, France]
Orbital Transport System
Organization for Tropical Studies [Costa Rica]
Organized Track Structure
O-Toluenesulfonamide
Out of Service
Ovonic Threshold Switch
Own Time Switch
Oxford Text System [of Oxford University Press, UK]
OTSAA Officer Training School Alumni Association [US]
OTSG Once-Through Steam Generator
OTSGS Once-Through Steam Generating System
OTSR Once-Through Superheat Reactor
Optimum Track Ship Routing
OTSS Off-the-Shelf System
OTT Office of Technology Transfer [of CANMET, Canada]
One-Time Tape
Over the Top
OTTBS N-Oxidiethylene Dithiocarbamyl-N'-tert-Butyl Sulfenamide
OTTS Outgoing Trunk Testing System
OTU Office of Technology Utilization [of NASA]
Operational Taxonometric Unit
Operational Test Unit
Operational Training Unit [of Canadian Forces]
OTV Operational Television
Orbital Transfer Vehicle
OU Object Unit
Odense University [Denmark]
Okayama University [Japan]
Operation Unit
Osaka University [Japan]
Oxford University [UK]
OUAC Ontario University Application Center [Canada]
OUB Occasional-Use Bands
OUCA Ontario University Council on Admissions [Canada]
OUF Oxygen Utilization Factor
OUG On-Line Users Group**OUG/I**
On-Line Users Group/Ireland
OULCS Ontario University Libraries Cooperation System [Canada]
OUO Official Use Only
OUP Oxford University Press [UK]
OUPID Ontario University Program for Instructional Development [Canada]
OUS Okayama University of Science [Japan]
OUST Office of Underground Storage Tanks [of EPA, US]
OUT Outgoing
Outlet
Output
OUTA Ontario Urban Transit Association [Canada]
OUTBD Outboard
OUTLIM Output Limiting Facility
OUTRAN Output Translator
OUTWATS Outgoing Wide-Area Telephone Service
OV Oil of Vitriol
Orbiting Vehicle
Overvoltage
OVA Offshore Valve Association [US]
Ontario Veterinary Association [Canada]
Ovalbumin
OVBD Overboard
OVC Overcast
OVD Optically Variable Device
Optical Video Disk
Outside Vapor Deposition
OVERS Orbital Vehicle Reentry Simulator
OVFL Overflow [also: OVFLO]
OVH Oval Head (Screw)
OVHD Overhead
OVHL Overhaul
OVID On-Line Visual Display
OVLBI Orbital Very Long Baseline Interferometer
OVLD Overload
OVLP Overlap
OVP Overvoltage Protection
OVPO Outside Vapor-Phase Oxidation
OVPU Overvoltage Protection Unit
OVRHD Overhead
OVRN Over-Run
OVRNG Over-Running
OVRSTK Overstrike
OVV Overvoltage
OW Oil-Immersed Water-Cooled (Transformer)
Open-Window (Unit)
Open Wire

	Operating Weight
	Order Wire
O/W	Oil-in-Water (Emulsion)
OWAEC	Organization for West African Economic Cooperation
OWC	Oil/Water Contact
	Oil Well Cement
OWE	Operating Weight Empty
OWF	One-Way Function
	Optimal Work Function
	Optimum Working Facility
	Optimum Working Frequency
OWFEA	Ontario Wholesale Farm Equipment Association [Canada]
OWG	Oil, Water, Gas
	Open Waveguide
	Optical Waveguide
OWL	On-Line Without Limitation
OWM	Office of Weights and Measures [of NIST, US]
	Optical Waveguide Microscopy
OWMC	Ontario Waste Management Corporation [Canada]
OWP	Office of Water Policy [US]
	Oil Well Pumper
OWPR	Ocean Wave Profile Recorder
OWPS	Offshore Windpower System
OWR	Omega West Reactor [of LASL, US]
OWRC	Ontario Water Resources Commission [Canada]
OWRR	Office of Water Resources Research [of USDOI]
OWRT	Office of Water Research and Technology [US]
OWS	Ocean Weather Station
	Operational Weather Support
	Operators Workstation
	Orbital Workshop (Station) [of NASA]
OWTU	Oilfields Workers Trade Union [Trinidad]
OXAL	Oxalate
OXD	Oxidized
OXS	Oxygen Sensor
OXY	Oxygen
oz	ounce
oz ap	ounze, apothecaries [also: oz apoth]
oz av	ounce, avoirdupois
OZD	Observed Zenith Distance
oz fl	ounze, fluid ounce
oz-ft	ounce-foot [also: oz ft]
OZI	Okubo-Zweig-Iizuha (Quark Model) [Physics]
oz-in	ounce-inch [also: oz in]
oz t	ounce, troy
oz/t	ounces per ton
oz tr	ounze, troy
oz/y	ounces per year
oz/yr	ounces per year

P

P
Pacific
Page
Paper
Parallel
Parity
Part
Pavement
Peak
Pear
Pencil
Penetration
Pentode
Perimeter
Period
Permanence
Permeability
Permutation
Personal
Personnel
Peta- [SI Prefix]
Phosphorescence
Phosphorus
Pitch
Plate
Plug
Plywood
Poise
Pointer
Polarity
Polarizability
Polarization
Polarizer
Pole
Polishing
Porosity
Port
Portable
Position
Positive
Power
President
Pressure
Primary
Principal
Print(er)
Priority
Probability
Procedure
Process(or)
Program(mer)
Prohibited
Proline
Prolyl
Protein
Proton
Province
Pump
Punch
Push

p
page
para
per
pico- [SI Prefix]
pint
poor
porosity
positive
pressure
primary
protein
proton
pula [Currency of Botswana]

£
Pound Sterling [Currency of Northern Ireland and the UK]

PA
Pack Area
Pad Abort
Panel Absorber
Paper Advance
Parallel Algorithm
Parasitic Antenna
Particle Accelerator
Particular Average
Pasenger Address (System)
Peak-Aged
Pending Availability
Pendulous Axis
Pennsylvania [US]
Performance Analysis
Permanently Associated
Personal Assistant
Perturbation Analysis
Phenylalanine
Phosphoric Acid
Phthalic Anhydride
Physics Abstracts [of INSPEC, UK]
Pilotless Aircraft
Plasma-Arc
Plasminogen Activator
Point of Aim
Polyallomer
Polyamide
Postacceleration
Post-Secondary Accreditation
Power Amplifier
Preacceleration
Prealloying
Preamplifier
Precision Approach
Preferential Adsorption
Professional Archeologist
Preplaced Aggregate
Press Agent
Press Association [UK]
Pressure Angle

	Principle of Adding
	Privacy Act
	Probability of Acceptance
	Process Allocator
	Product Analysis
	Product Assurance
	Production Adjustment
	Production Analysis
	Program Access
	Program Action
	Program Address
	Program Analysis
	Program Application
	Program Attention (Key)
	Program Authorization
	Propulsion Assisted
	Protected Area
	Psychoacoustics
	Public Address (System)
	Publishers Association [UK]
	Pulse Amplifier
	Pulse Analyzer
	Purchasing Agent
	Puromycin Aminonucleoside
P/A	Power of Attorney
P-A	Peak-to-Average (Ratio) [also: P/A]
	Phenylalanine
Pa	Pascal
	Protactinium
	Pennsylvania [US]
pA	picoampere
pa	(per annum) — each year
	(pro analysi) — by analysis
PAA	Pan American World Airways
	Paraazoxyanisole
	Phase Antenna Array
	Phosphoric Acid Aluminum Treatment
	Physiological Amino Acid
	Polyacrylamide
	Polyacrylic Acid
	Population Association of America [US]
	Potato Association of America [US]
	Pulse-Amplitude Analyzer
PAAAC	Pan-American Agricultural Aviation Center
PAABA	P-Acetamidobenzoic Acid
PAABS	Pan-American Association of Biochemical Societies [US]
PAAC	Program Analysis Adaptable Control
PAA-PS	Polyacrylic Acid — Polysulfone
PAAS	Pakistan Association for the Advancement of Science
PAB	Aminopentyl Benzimidazole
	P-Aminobenzyl
	Panamanian Balboa [Currency]
	Polymer Alloys and Blends
	Primary Application Block
	Pulse Adsorption Bed
PABA	P-Aminobenzoic Acid
PABLA	Problem Analysis By Logical Approach
PABS	Pan-American Biodeterioration Society [US]
PABs	Polymer Alloys and Blends
PABST	Primary Adhesively Bonded Structures Technology
PABX	Private Area Exchange
	Private Automatic Branch Exchange
PAC	Pacific
	Package Assembly Circuit
	Packaging Association of Canada
	Pedagogic Automatic Computer
	Peripheral Autonomous Control
	Permanent Accomodation Complex
	Permanent Agricultural Committee [of ILO, Switzerland]
	Personal Analog Computer
	Perturbed Angular Correlation
	Pharmaceutical Advertising Council [US]
	Plasma-Arc Cutting
	Polled Access Circuit
	Polyalkene Carbonate
	Poly-Aluminum Chloride
	Polycyclic Aromatic Hydrocarbon
	Portable Air Compressor
	Preauthorized Checking
	President's Advisory Council [US]
	Process Automation Computer
	Professional Activities Committee [of IEEE, US]
	Program Address Counter
	Program Assembly Card
	Program Authorized Credentials
	Programmable Automatic Comparator
	Public Archives Canada
PACA	Picture Agency Council of America [US]
	Polyamide Carboxylic Acid
PACAF	Pacific Air Force [US]
PACC	Product Administration and Contract Control
PACCT	PERT (= Production Evaluation and Review Technique) and Cost Correlation Technique
PACE	Pacific Agricultural Cooperative for Export [US]
	Packaged CRAM (= Card Random Access Memory) Executive
	Passive Attitude Control Experiment
	Petroleum Association for the Conservation of the Environment [Canada]
	Philippine Association of Civil Engineers
	Physics and Chemistry Experiment
	Planetary Association for Clean Energy [Canada]
	Plant Acquisition and Construction Equipment
	Precision Analog Computing Equipment
	Preflight Acceptance Checkout Equipment
	Prelaunch Automatic Checkout Equipment
	Process Automation and Control Executive
	Professional Activities Committees for Engineers
	Professional Association of Consulting Enginners
	Programmed Automatic Communications Equipment
	Programming Analysis Consulting Education
	Projects to Advance Creativity in Education
	Pulsed Analog-to-Digital Converter and Encoder
PACED	Program for Advanced Concepts in Electronic Design
PACE-LV	Preflight Acceptance Checkout Equipment for Launch Vehicle [also: PACE/LV]
PACER	Portfolio Analysis, Control, Evaluation and Reporting

	Process Assembly Case Evaluator Routine
	Program-Assisted Console Evaluation and Review
	Programmed Automatic Communications Equipment
	Program of Active Cooling Effects and Requirements
PACE-SC	Preflight Acceptance Checkout Equipment for Spacecraft [also: PACE-S/C]
PACF	Periodic Autocorrelation Function
PACFORNET	Pacific Coast Forest Research (Information) Network
PACGSR	Pan-American Center for Geographical Studies and Research [Ecuador]
PACIA	Particle Counting Immunoassay
PACKG	Packaging
PACM	Para-Aminocyclohexyl Methane
	Pulse Amplitude Code Modulation
PACOR	Passive Correlation and Ranging
PACRA	Pottery and Ceramics Research Association [New Zealand]
PACS	Pacific Area Communications System
	Physics and Chemistry Classification Scheme
	Picture Archiving and Communication Systems
	Project Analysis and Control System
PACT	Pay Actual Computer Time
	Production Analysis Control Technique
	Program for Advancement of Commercial Technology
	Program for Automatic Coding Techniques
	Programmable Asynchronous Clustered Teleprocessing
	Programmed Analysis Computer Transfer
	Programmed Automatic Circuit Tester
PACV	Patrol Air Cushion Vehicle
PACVD	Plasma-Activated Chemical Vapor Deposition
	Plasma-Assisted Chemical Vapor Deposition
PACX	Private Automatic Computer Exchange
PAD	Packet Assembler/Disassembler
	Percutaneous Device
	Perturbed Angular Distribution
	Pitless Adapter Division [of Water Systems Council, US]
	Pixel Access Definition
	Plastics Analysis Division [of SPE, US]
	Poly-Aperture Device
	Positioning Arm Disk
	Post-Activation Diffusion
	Power Amplifier Device
	Propellant-Actuated Device
PADA	Pyridine-2-Azo-P-Dimethylaniline
PADAR	Passive Detection and Ranging
PADD	Petroleum Administration for Defense Districts
PADE	Pad Automatic Data Equipment
PADF	Pan-American Development Foundation
PADIA	Patrol Diagnosis
PADIRT	Platform for Atmospheric Data in Real-Time [of BIO, Canada]
PADIS	Pan-African Documentation and Information System [Ethiopia]
PADL	Parts and Design Language
	Pilotless Aircraft Development Laboratory [US Navy]
PADLA	Programmable Asynchronous Dual Line Adapter
PADLOC	Passive-Active Detection and Location
PADLOCC	Passive-Active Detection and Location Countermeasures
PADRE	Patient Automatic Data Recording Equipment
	Portable Automatic Data Recording Equipment
PADS	Passive-Active Data Simulation
	Performance Analysis and Design Synthesis
	Performance Analysis Display System
	Port and Airport Development Strategy
	Precision Aerial Display System
PAE	Polyarylene Ether
	Polyaryl Ether
	Post-Accident Environment
pae	(partes aequales) — equal parts
PAEA	Pakistan Atomic Energy Authority
PAEC	Pakistan Atomic Energy Commission
PAECI	Pan-American Association of Educational Credit Institutions [Colombia]
PAEK	Polyarylether Ketone
PAEM	Program Analysis and Evaluation Model
PAF	Page Address Field
	Peripheral Address Field
	Picric Acid Formaldehyde
	Platelet Activating Factor
	Pre-Atomized Fuel
PAFC	Phase-Locked Automatic Frequency Control
	Philippine-American Financial Commission
	Phosphoric-Acid Fuel Cell
PAFLU	Philippine Association of Free Labour Unions
PAG	Plasma-Arc-Based Gasifier
	Polyalkylene Glycol
	Prealbumin Globulin
	Protein A Colloidal Gold
	Protein Advisory Group [of UN]
pAg	Protein A Gold
PAGE	PERT (= Production Evaluation and Review Technique) Automated Graphical Extension
	Polyacrylamide Gel Electrophoresis
PAGEOS	Passive Geodetic Earth Orbiting Satellite [of NASA]
PAGES	Program Affinity Grouping and Evaluation System
PAGICEP	Petroleum and Gas Industry Communications Emergency Plan [of FCC, US]
PAGS	Prior Austenite Grain Size
PAGT	Provincial Association of Geography Teachers [Canada]
PAH	Polycyclic Aromatic Hydrocarbon
	Polynuclear Aromatic Hydrocarbon
PAHA	P-Aminohippuric Acid
PAHC	Pan-American Highway Congresses [US]
PAHO	Pan-American Health Organization
PAHR	Post-Accident Heat Removal
PAI	Personnel Accrediation Institute [US]
	Polyamideimide
	Process Analytical Instrumentation
	Programmer Appraisal Instrument
PAIC	Public Address Intercom System
PAID	Pan-African Institute for Development [Cameroon]
	Personnel and Accounting Integrated Data System
PAIGH	Pan-American Institute of Geography and History [Mexico]

PAIMEG	Pan-American Institute of Mining Engineering and Geology
PAINT	Painting
PAIP	Public Affairs and Information Program [of Atomic Industries Forum]
PAIR	Performance and Integration Retrofit
	Precision Approach Interferometer Radar
PAIS	Public Affairs Information Service [US]
PAIS FLI	Public Affairs Information Service — Foreign Language Index [US]
PAIT	Program for Advancement of Industrial Technology [Canada]
PAKEX	Packaging Exhibition
PAKTUS	PRC (Inc.)'s Adaptive Knowledge-Based Text Understanding System
PAL	PAM (= Programmable Algorithm Machine) Assembly Language
	Partition Allocation Utility
	Pedagogic Algorithmic Language
	Permanent Artificial Lighting
	Permissive Action Link
	Phase Alternation Line (Television System)
	Phase Attenuation by Line
	Philippine Airlines
	Phillips Air Liquefier
	Process Assembly Language
	Production and Application of Light [IEEE IAS Committee]
	Programmable Array Logic
	Programmed Application Library
	Programmer Assistance and Liaison
	Prototype Application Loop
	Psychoacoustic Laboratory
PALASM	Programmable Array Logic Assembler
PAL-D	Phase Alternation Line — Delay [also: PAL D]
PALEONT	Paleontological
	Paleontologist
	Paleontology
PALINET	Pennsylvania Area Library Network [US]
PALLAS	Phased Automation of the Hellenic ATC (= Air Traffic Control) System
PALM	Philips Automated Laboratory Management
	Precision Altitude and Landing Monitor
PAL-S	Phase Alternation Line — Simple [also: PAL S]
PAM	Pamphlet
	Payload Assist Module
	Peripheral Adapter Module
	Phenylacetamidomethyl
	Photoacoustic Microscopy
	Plan-Applier Mechanism
	Pole Amplitude Modulation
	Portable Activity Monitor
	Portable Automated Mesonet
	Post-Accident Monitoring
	Pozzolan Aggregate Mixture
	Primary Access Method
	Process Automatic Monitor
	Process Automation Monitor
	Programmable Algorithmic Machine
	Pulse-Amplitude Modulation
	2-Pyridine Aldoxime Methiodide [also: 2-PAM]
PAMA	Pan-American Medical Association
	Professional Aviation Maintenance Association [US]
	Pulse-Address Multiple Access
PAMAC	Parts and Materials Accountability Control
PAMC	Provisional Acceptable Means of Compliance
PAMD	Periodic Acid Mixed Diamine
PAM/D	Process Automation Monitor/Disk
PAMEE	Philippine Association of Mechanical and Electrical Engineers
PAMELA	Plan-Applier Mechanism for English Language Analysis
PAM-FM	Pulse Amplitude Modulation — Frequency Modulation
PAMI	Prairie Agricultural Machinery Institute [Canada]
PAML	Program Authorized Materials List
PAMP	Plasma-Assisted Materials Processing
PAMPER	Practical Application of Midpoints for Exponential Regression
PAMS	Pad Abort Measuring System
	Parallel Application Management System
	Preselected Alternate Master/Slave
	Proceedings of the American Mathematical Society
PAN	Panel
	Peroxyacetyl Nitrate
	Pesticides Action Network [Belgium]
	Phenyl-α-Naphthyl Amine
	Polled Access Network
	Polyacrylonitrile
	Positional Alcohol Nystagmus
	1-(2-Pyridylazo)-2-Naphthol
PANA	Pan-African News Agency
PANACEA	Package for Analysis of Networks of Asynchronous Computers with Extended Asymptotics
PANAFTEL	Pan-African Telecommunications Network
PANAM	Pan-American World Airways
PANDA	Prestel Advanced Network Design Architecture
	Programmers Analysis and Development Aid
p and a	plugged and abanoned [Oil and Gas Industry]
p and s	plugged and suspended [Oil and Gas Industry]
PANJU	Pan-African Union of Journalists
PANS	Potentially Attractive New Services
	Procedures for Air Navigation Services
	Public Archives of Nova Scotia [Canada]
PANSDOC	Pakistan National Scientific and Technical Documentation Center
PANSMET	Procedures for Air Navigation Services — Meteorology
PANS/OPS	Procedures for Air Navigation Services/Aircraft Operations
PANS/RAC	Procedures for Air Navigation Services/Rules of the Air Traffic Services
PANT	Pantograph
PANTHEON	Public Access by New Technology to Highly Elaborate On-Line Networks
PANTS	Public Acceptance of New Technologies
PAO	Personnel Administration Office
	Phenylarsine Oxide
	Polyalkaline Oxide
	Pulsed Avalanche Diode Oscillator
PAP	P-Aminophenylalanine
	Paraazoxyphenetole

	Peroxidase-Anti-Peroxidase
	Phase Advance Pulse
	Phosphoadenosine Phosphate
	Photon-Assisted Processing
	Plasma-Assisted Processing
	Pouchou and Pichoir (Model)
	Prealloyed Powder
	Preapproved Payment
	Preauthorized Payment
PAPA	Parallax Aircraft Parking Aid
	Programmer and Probability Analyzer
PAPD	Periodate Dimethylphenylenediamine
PAPE	Photoactive Pigment Electrophotography
PAPERCHEM	Paper Chemistry (Database) [of Institute of Paper Chemistry, US]
PAPI	Polymethylene Polyphenyl Isocyanate
	Precision Approach Path Indicator
PAPM	Pulse Amplitude Phase Modulation
PAPP	Pre-Approved Payment Plan
PA-PPO	Polyamide — Polyphenylene Oxide
PAPRICAN	Pulp and Paper Research Institute of Canada
PAPS	Periodic Acid Phenylhydrazine Schiff
	Periodic Armaments Planning System
	Phosphoadenosine Phosphosulfate
	Physics Auxiliary Publication Service
PAPTC	Paper Tape Controller
PAPU	Pan-African Postal Union [Tanzania]
PAR	Page Address Register
	Pakistan Rupee [Currency]
	Parabola
	Parabolic
	Paragraph
	Parallel
	Parameter
	Parenthesis
	Parity
	Peak Accelerometer Recorder
	Peak-to-Average Ratio
	Performance Analysis and Review
	Performance Appraisal Report
	Perimeter Acquisition Radar
	Plasma-Arc Reduction
	Plasma-Arc Remelting
	Polyarylate
	Precision Approach Radar
	Preferential Arrival Route
	Processor Address Register
	Production Automated Riveting
	Professional Abstracts Registry
	Program Address Register
	Program Appraisal and Review
	Progressive Aircraft Rework
	Project Authorization Request
	Purchasing Approval Request
	4-(2-Pyridylazo)resorcinol
P/AR	Peak-to-Average Ratio
PARA	Paragraph
	Problem Analysis and Recommended Action
	Professional Audiovideo Retailers Association [US]
PARADE	Passive-Active Ranging and Determination
PARAL	Parallel
PARAMI	Parsons Active Ring Around Miss-Indicator
PARAMP	Parametric Amplifier [also: Paramp or paramp]
PARASEV	Paraglider Search Vehicle
PARASYN	Parametric Synthesis
PARC	Palo Alto Research Center [US]
	Progressive Aircraft Reconditioning Cycle
PARCA	Pan-American Railway Congress Association
PARCOM	Paris Commission (for Marine Pollution Control) [UK]
PARCS	Pesticide Analysis Retrieval and Control System [of EPA, US]
PARD	Pakistan Academy for Rural Development
	Parts Application Reliability Data
	Periodic and Random Deviation
	Precision Annotated Retrieval Display System
PARDOP	Passive Ranging Doppler
PARE	Physical Ability Requirement Evaluation
PARIF	Program for Automation Retrieval Improvement by Feedback [of EURATOM]
PARIS	Pulse Analysis-Recording Information System
	Portable Automated Remote Inspection System
PARL	Parallel
	Parliament(ary)
	Prince Albert Radar Laboratory [of DRTE, Canada]
PARM	Parameter
	Program Analysis for Resource Management
PARMA	Public Agency Risk Managers Association [US]
PAROS	Passive Ranging of Submarines
PARR	Pakistan Atomic Research Reactor
	Post-Accident Radioactivity Removal
	Procurement Authorization and Receiving Report
PARS	Paragraphs
	Portable Analyzer for Residual Stresses
	Private Advanced Radio Service
PARSAC	Particle Size Analog Computer
PARSAVAL	Pattern Recognition System Application Evaluation [also: Parsaval]
parsec	parallax-second [Unit]
PARSEV	Paraglider Search Vehicle
PART	Parts Allocation Requirement Technique
PARTAC	Precision Askania Range Target Acquisition
PARTEI	Purchasing Agents of the Radio, Television and Electronics Industry
PARTEQ	Partners in Technology at Queen's [of Queen's University, Canada]
PARTIC PHYS	Particle Physics
PARTN	Partition
PARTNER	Proof of Analog Results Through a Numerical Equivalent Routine
PARV	Paravane
PAS	Pakistan Academy of Sciences
	P-Aminosalicylic Acid
	Patent Applicant Service [of INPADOC, Austria]
	Periodic Acid-Schiff
	Phase Address System
	Phenylaminosalicylate
	Photoacoustic Spectroscopy
	Plasma Arc Spraying
	Polish Academy of Sciences
	Polyacrylsulfone
	Polyarylsulfone

Poly(para-amino Styrene)
Positron Annihilation Spectroscopy
Pressure-Assisted Sintering
Primary Alert System
Privacy Act Statement
Process Analysis Services
Professor of Air Science
Program Address Storage
Public Address System
Pyrotechnic Arming Switch

PASB Pan-American Sanitary Bureau
PASC Pacific Area Standards Congress [US]
PASCAL Philips Automatic Sequence Calculator
PASE Power-Assisted Storage Equipment
PASEM Program of Assistance to Solar Equipment Manufacturers
PASF Photographic Art and Science Foundation [US]
PASL Physical Activity Sciences Laboratory [Canada]
PASLA Programmable Asynchronous Line Adapter
PASLIB Pakistan Association of Special Libraries
PA-SM Periodic Acid-Silver Methenamine
PASO Pan-American Sanitary Organization
PASOS Paperless Shop-Order System
PASS Passage
Passenger
Passive
Performance Analysis Software System
Photo-Access Security System
Planning and Specification Software
Private Automatic Switching System
Procurement and Acquisition Support System
Production Automated Scheduling System
Program Aid Software System
Program Alternative Simulation System
Programmed Access Security System
Purchasing Activities Support System
PASSIM President's Advisory Staff on Scientific Information Management [US]
PAST Process Accessible Segment Table
PAS&T Particle Accelerator Science and Technology [IEEE Society]
PAT Parametric Artificial Talker
Patent
Peripheral Allocation Table
Personalized Array Translator
Petroleum Authority of Thailand
Picric Acid Turbidity
Polyaminotriazole
Power Ascension Testing
Prediction Analysis Technique
Printer Action Table
Production Acceptance Test
Proficiency Analytical Testing
Program Attitude Test
Programmable Automatic Tester
Programmer Aptitude Test
Propylaminohydroxy Tetrahydronaphthalene
PATA Pneumatic All-Terrain Amphibian
PATBX Private Automatic Telex Branch Exchange
PATC Professional, Administrative, Technical and Clerical
Professional Air Traffic Controller
PATCA Phase-Lock Automatic Tuned Circuit Adjustment
Professional and Technical Consultants Association [US]
PATCO Professional Air Traffic Controllers Organization [US]
PATD Patented
PATE Programmed Automatic Test Equipment
PATH Pathological
Pathologist
Pathology
Performance Analysis and Test Histories
PATHFINDER Pathological Element Finder
PATI Passive Airborne Time-Difference Intercept
PATLAW Patent Law (Database) [of BNA, US]
PATMI Power Actuated Tool Manufacturers Institute [US]
PATN Pattern
PATO Partial Acceptance and Takeover
PATOLIS Patent On-Line Information System
PAT OFF Patent Office
PATRA Packaging and Allied Trades Research Association [now: PIRA, UK]
PATRIC Pattern Recognition Interpretation and Correlation
PATRICIA Practical Algorithm To Receive Information Coded In Alphanumeric
PATROL Program for Administrative Traffic Reports On-Line
PATS Precision Altimeter Techniques Study
PATSEARCH Patent Search System [UK]
PATSY Programmer's Automatic Testing System
PATT Pattern
Project for the Analysis of Technology Transfer
PATTERN Planning Assistance Through Technical Evaluation of Relevance Numbers [of ORSA, US]
PATTI Pneumatic Adhesion Tensile Testing Instrument
PATU Pan-African Telecommunications Union [Zaire]
PATWAS Pilot's Automatic Telephone Weather Answering Service
PATX Private Automatic Telex Exchange
PAU Pan-American Union
Pattern Articulation Unit
Pilotless Aircraft Unit
Power and Alarm Unit
Precision Approach — UNICOM (= Universal Integrated Communications)
PAUJ Pan-African Union of Journalists
PAUL Parallel-Axis Ultraprecision Lathe
PAV Paving
Phase Angle Voltmeter
Position and Velocity
PAVD Plasma-Assisted Vapor Deposition
PAVE Philippine Association for Vocational Education
Position and Velocity Extraction
PAVT Position and Velocity Tracking
PAW Plasma-Arc Welding
PAWC Pan-African Workers Congress
PAWOS Portable Automatic Weather Observable Station
PAWRS Private Aviation Weather Research Station
PAWS Phased Array Warning System
Private Aviation Weather Station
Programmed Automatic Welding System
PAWWSWR Portuguese Association on Water, Wastewater and Solid Wastes Research

PAX Photoemission of Absorbed Xenon
Physical Address Extension
Private Area Exchange
Private Automatic Exchange
PAYE Pay As You Earn
PAYLL Payroll
PAYT Payment
PB Page Buffer
Passband
Peripheral Buffer
Phase Boundary
Phenobarbital
Phonetically Balanced
Pilot Balloon
Pipe Break
Piperonyl Butoxide
Planning Board
Playback
Plotboard
Plugboard
Polybutylene
Polymer Blend
Precipitation Body
Proportional Band
Publications Board
Pull Box
Purplish Blue
Pushbutton
P/B Parts per Billion [also: ppb]
Peak-to-Background (Ratio) [also: P-B]
Pb (Plumbum) — Lead
PBA Plastic Bag Association [US]
Printed Board Assembly
Proportional Band Adjustment
Pyranyl Benzyl Adenine
Pyrene Butyric Acid
PBAA Polybutadieneacrylic Acid
PBAPS Pipe Break Air Piping System
Pipe Break Automatic Protective System
PBB Polybrominated Biphenyl
PBBI Polybutadiene Bisimide
PBBO 2-(4-Biphenylyl)-6-Phenylbenzoxazole
Phenylbiphenylylbenzoxazole
PBC Periodic Boundary Conditions
Peripheral Buffer Computer
Program Booking Center
PBCS Post-Boost Control System
PBCT Proposed Boundary Crossing Time
PBD Parallel Blade Damper
2-Phenyl-5-(4-Biphenylyl)-1,3,4-Oxadiazole
PbDEC Lead Diethyl Dithiocarbamate
PBDG Pushbutton Data Generator
PBDI Position Bearing and Distance Indicator
PbDMC Lead Dimethyl Dithiocarbamate
PBDS Photothermal Beam Deflection Spectroscopy
PBE Prompt Burst Experiments
Proton Balance Equation
Pulsed Bridge Element
PBEA Paint, Body and Equipment Association [US]
PBEC Pacific Basin Economic Committee
PBEIST Planning Board for European Inland Surface Transport
PBF Potential Benefit Factor
Power Burst Facility [of INEL, US]
PBFA Particle Beam Fusion Accelerator
PBH Planar-Buried Heterostructure
PBHP Pounds per Brake Horsepower [also: pbhp]
PB-HTGR Peach Bottom High-Temperature Gas-Cooled Reactor
PBI Paper Bag Institute [US]
Plant Biological Institute
Plant Biotechnology Institute [Canada]
Plumbing Brass Instutute [now: PMI, US]
Polybenzimidazole
Process Branch Indicator
Protein-Bound Iodine
PBIB Partially Balanced Incomplete Block Design
Pb-I-Pb Pb (= Lead) — Insulator — Pb (= Lead)
PBIT Parity Bit [also: pbit]
PBJ Paper-Braided Jute
PBL Planetary Boundary Layer
Probable
PBM Permanent Benchmark
Plastic Bonded Magnet
PBN Butylphenylnitrone
Phenyl-β-Naphthyl Amine
Physical Block Number
Pyrolytic Boron Nitride
PBNA Phenyl-Beta-Naphthylamine
PBOS Planning Board for Ocean Shipping [of NATO]
PBP Pushbutton Panel
PBPB Pyridinium Bromide Perbromide
Pb5MC Lead Pentamethylene Dithiocarbamate
PBPP Poly(bisphenoxyphosphazene)
PBPS Post-Boost Propulsion System
PBR Pebble Bed Reactor [of ORNL, US]
Plum Brook Reactor [of NASA]
Pole Broken
Polished-Bore Receptacle
Professional Board for Radiography [South Africa]
Pyridine Butadiene Rubber
PBRE Pebble Bed Reactor Experiment [of ORNL, US]
PBRF Plant Breeding Research Forum [US]
Plum Brook Reactor Facility [of NASA]
PBS Pacific Biological Station [Canada]
Phosphate-Buffered Saline
Picture Building System
Pressure Boundary Subsystem
Project Breakdown Structure
Public Broadcasting Service
PBSTA Push Button Station
PBT Permeable Base Transistor
Polybutylene Terephthalate [also: PBTP]
Poly(-3-Hydroxybutyrate)
Polyphenylene Benzobisthiazole
PBTF Polybromotrifluoroethylene
Pump Bearing Test Facility [of LMEC, US]
PBTP Polybutylene Terephthalate [also: PBT]
PBT-PAR Polybutylene Terephthalate — Polyarylate
PBT-PET Polybutylene Terephthalate — Polyethylene Terephthalate

PBU Pushbutton Unit
PBV Post-Boost Vehicle
PBW Parts by Weight
PBX Plastic Bonded Explosive
Private (Automatic) Branch Exchange
PBzMA Polybenzylmethacrylate
PBZT Polyphenylene Benzalthiazole
PC Panama Canal
Paper Chromatography
(Single) Paper / (Single) Cotton (Wire)
Parallel Circuit
Parallel Connection
Parametric Cubic
Parts Catalogue
Patent Classification
Path Control
Peg Count
Percent
Peripheral Controller
Personal Computer
Petrochemistry
Petty Cash
Phase Change
Phase Code
Phase Coherent
Phase Conjugation
Phase Contrast
Phase Converter
Phase Current
Phosphatidylcholine
Phosphocreatine
Photocathode
Photocell
Photochemical
Photochromatic
Photocomposing
Photoconductance
Photoconduction
Photoconductive Cell
Photoconductivity
Photoconductor
Photocurrent
Photographic Camera
Phthalocyanine
Physical Chemistry
Picture
Piece
Pitch Circle
Pitch Control
Plasma Chromatography
Plastocyanine
Pneumatic Circuit
Pneumatic Controller
Pocket Calculator
Point Cathode
Point of Curvature
Polar Crane
Police Constable
Polychloroprene
Polycarbonate
Polycarbosilane
Polymer Chemistry
Portablo Computor
Port Control
Portland Cement
Position Control
Postcard
Power Component
Power Converter
Pre-Emphasis Circuit
Pressure Cable
Pressure Circuit
Pressure Control
Pressure Cooker
Prestressed Concrete
Price
Printed Circuit
Process Control
Processing Center
Processor Controller
Product Code
Professional Communication [IEEE Group]
Professional Consultant
Professors of Curriculum [US]
Program Controller
Program Coordination
Program Counter
Programmable Controller
Proportional Control
Proportional Counter
Propylene Carbonate
Provisional Costs
Publications Committee
Pulsating Current
Pulse Circuit
Pulse Compression
Pulse Controller
Punched Card
Pyrurate Carboxylase
P/C Petty Cash
Price Current
P-C Plastic-Carbon (Replica)
Processor-Controller
Pulse Counter
pC picocoulomb
picocurie
pc parsec [Unit]
per cent
PCA Parallel Computer Architecture
Pennsylvania Coal Association [US]
Perchloric Acid
Performance and Coverage Analyzer
Personal Computer Analyzer
Personal Computers Association
Phaseolus Coccineus Agglutinin
Photocurrent Amplifier
Physical Configuration Audit
Polar Cap Absorption
Polycrystalline Alumina
Pool Critical Assembly [of ORNL, US]
Portland Cement Association [US]
Positive Control Area

	Precision Cleaning Agent
	Preconditioned Air System
	Printed Circuit Assembly
	Printed Communications Adapter
	Private Communications Association [now: National Communications Association, US]
	Producers Commission Association [US]
	Production Credit Association [US]
	Programmable Communications Adapter
	Protective Clothing Arrangement
	Protective Connecting Arrangement
	Pulp Chemicals Association [US]
	Pulse Code Adapter
	Pulse Counter Adapter
	Pyrotechnic Control Assembly
	Pyrrolidonecarboxylic Acid
PC-ABS	Polycarbonate — Acrylonitrile-Butadiene-Styrene
PCAC	Partially Conserved Axial Current
	Partially Conserved Axial-Vector Current
P-CAD	Personal (Computer) Computer-Aided Design
PCAM	Partitioned Content Addressable Memory
	Probe Card Assembly Machine
	Punched Card Accounting Machine
PCAO	Pollution Control Association of Ontario [Canada]
PCAS	Primary Central Alarm Station
PCAT	Pharmacy College Admission Test
PCB	Page Control Block
	Polychlorinated Biphenyl
	Polychlorobenzene
	Power Circuit Breaker
	Printed Circuit Board
	Process Control Block
	Processor Command Bus
	Program Communication Block
	Program Control Block
	Public Coin Box
PCBA	P-Chlorobenzoic Acid
PCBC	Partially Conserved Baryon Current
	P-Chlorobenzyl Chloride
PCBN	P-Chlorobenzonitrile
	Polycrystalline Cubic Boron Nitride [also: PcBN]
PCBPS	Printed Circuit Board Power Supply [also: PCB PS]
PCBS	Permanent Committee on Biological Standards
	Positive Control Bombardment System
PCBTF	P-Chlorobenzotrifluoride
PCC	Panama Canal Commission [US]
	Parametric Channel Controller
	Partial Crystal Control
	Peripheral Control Computer
	Personal Code Calling
	Petroleum Compensation Charge
	Photoconductive Cell
	Point of Common Coupling
	Point of Compound Curvature
	Pole-Changing Control
	Polychlorinated Camphene
	Portland Cement Concrete
	Power Control Center
	Processor Control Console
	Producers Council of Canada
	Program Control Counter
	Program-Controlled Computer
	Pyridinium Chlorochromate
PC-CAD	Personal Computer — Computer-Aided Design
PCCC	Pakistan Central Cotton Committee
	Phoenix Conference on Computers and Communications [US]
PCCD	Peristaltic Charge-Coupled Device
PCCM	Private Circuit Control Module
PCCS	Photographic Camera Control System
	Positive Control Communications System
	Processor Common Communications System
PCD	Photoconductive Decay
	Photoconductive Device
	Pincushion Distortion
	Planned Continuation Date
	Polycarbodiimide
	Polycrystalline Diamond
	Port Control Diagnostic
	Power Circle Diagram
	Power Control and Distribution
	Power Control Device
	Process Control Division
	Production Common Digitizer
	Procurement and Contracts Division [of NASA]
PCDC	Punched Card Data Processing
PCDDS	Private Circuit Digital Data Service
PC-DOS	Personal Computer — Disk Operating System [also: PC DOS or PCDOS]
PCDP	Punched Card Data Processing
PCE	Perchloroethylene
	Peripheral Controller Enclosure
	Power Conditioning Electronics
	Program Cost Estimate
	Project Coordination Center Europe
	Punched Card Equipment
	Pyrometric Cone Equivalent
PCEA	Pacific Coast Electric Association [US]
PCEAA	Professional Construction Estimators Association of America [US]
PCEM	Process Chain Evaluation Model
PCES	Phase Change Energy Storage
PCETF	Power Conversion Equipment Test Facility
PCF	Partial Correlation Function
	Peripheral Control Facility
	Planning Coordination Facility
	Potential Controlled Flotation
	Pounds per Cubic Foot [also: pcf or lb/ft^3]
	Power Cathode Follower
	Program Control Facility
	Pulse-to-Cycle Fraction
PCFIA	Particle Concentration Fluorescence Immunoassay
PCFS	Pin, Clip and Fastener Services [US]
PCG	Phonocardiogram
	Phonocardiography
	Plains Cotton Growers [US]
	Planning and Control Guide
	Programmable Character Generator
PCGN	Permanent Committee on Geographical Names [UK]
PCGS	Protein Crystal Growth System
PCH	Polycyclohexane
	Proton Channeling

	Punch
PCHDMT	Polycyclohexane Dimethylene Terephthalate
PCHIS	Population Clearinghouse and Information System [of ESCAP]
PCHK	Parity Check
PCI	Panel Call Indicator
	Pantone Color Institute [US]
	Pattern Correspondence Index
	Pavement Condition Index
	Pellet-Cladding Interaction
	Peripheral Command Indicator
	Peripheral Controller Interface
	Pilot Controller Integration
	Plant Control Interface
	Portable Cesium Irradiator [of USAEC]
	Powder Coating Institute [US]
	Power Conversion International (Congress)
	Precision Cascade Impactor
	Prestressed Concrete Institute [US]
	Process Capability Index
	Process Control Interface
	Product Configuration Identification
	Program Check Interruption
	Program-Controlled Interruption
	Programmable Communications Interface
	Programmable Controller Interface
	Programmed Control Interrupt
PCIC	Petroleum and Chemical Industry Conference
PCIF	Programmable Controller Interface Facility
PCIFC	Permanent Commission of the International Fisheries Convention
PCILOCC	Perturbative Configuration Interaction Using Localized Orbitals for Crystal Calculation
PCIM	Power Conversion and Intelligent Motion
PCIOS	Processor Common Input/Output System
PCIP	Personal Computer Instrument Product
PCIS	Personal Computer Information Service
	Primary Containment Isolation System
PCIV	Prestressed Cast Iron (Pressure) Vessel
PCIZC	Permanent Committee of International Zoological Congresses
PCJC	Pakistan Central Jute Committee
PCK	Processor-Controlled Keying
PCL	Parallel Communications Link
	Permissible Contamination Limits
	Pilot Controlled Lighting
	Polycaprolactone
	Printed Circuit Lamp
	Printer Command Language
	Printer Control Language
	Process Control Language
	Programmasyst Control Language
PCLA	Process Control Language
PCLDI	Prototype Closed Loop Development Installation
PCLO	Printed Circuit (Board) Layout
PCLR	Parallel Communications Link Receiver
PCLS	Prototype Closed Loop System
PCLT	Prototype Closed Loop Test
PCM	Passive Countermeasures
	Phase-Change Material
	Phase-Contrast Microscopy
	Pitch Control Motor
	Plug Compatible Machine
	Plug Compatible Mainframe
	Plug Compatible Memory
	Plug Compatible Module
	Plug Control Module
	Polymer Coated Material
	Port Command Module
	Power Cooling Mismatch
	Process Control Module
	Process Control Monitor
	Pulse-Code Modulation
	Punched Card Machine
PCMA	Personal Computer Management Association [US]
	Professional Convention Management Association [US]
PCMB	P-Chloromercuribenzoate
	P-Chloromercuribenzoic Acid
PCMC	Preparatory Commission for Metric Conversion
PCMCIA	Personal Computer Memory Card International Association
PCMD	Pulse Code Modulation Digital
PCME	Pulse Code Modulation Event
PCM-FM	Pulse Code Modulation — Frequency Modulation
PCMH	Professional Certified in Materials Handling
PCMI	Photochemical Machining Institute [US]
	Photochromic Micro-Image
PCMIA	Pittsburgh Coal Mining Institute of America [US]
PC MK	Piece Mark
PCMM	Plug Compatible Mainframe Manufacturer
	Professional Certified in Materials Management
PCM/PAM	Pulse Code Modulation/Pulse Amplitude Modulation
PCMS	P-Chloromercurphenylsulfonic Acid
	Punched Card Machine System
PCMX	P-Chlorodimethylphenol
PCN	Pavement Classification Number
	Personal Communications Network
	Publication Code Number
PCNB	Pentachloronitrobenzene
PCO	Plant Control Office
	Procuring Contrast Officer
PCONA	P-Chloro-O-Nitroaniline
PCOS	Primary Communications Oriented System
	Process Control Operating System
PCQ	Production Control Quantometer
PCP	Packet Control Panel
	Parallel Cascade Processor
	Parallel Circular Plate
	Pentachlorophenol
	Pericyclic Process
	Phenylcyclidene
	Photochemical Processing
	Photon-Coupled Pair
	Plug Compatible Peripheral
	Polychloroprene
	Port Call Processing
	Post-Construction Permit
	Primary Control Program
	Primary Coolant Pump
	Primary Cross-Connection Point

	Process Control Processor
	Processor Control Program
	Program Change Proposal
	Programmable Communication Processor
	Project Control Plan
	Pulse Comparator
	Punched Card Punch
pCp	Cytidinebiphosphate
PCPA	P-Chlorophenylacetic Acid
	P-Chlorophenylalanine
PCPB	P-Chlorophenylbenzene
PCPBMA	Pacific Coast Paper Box Manufacturers Association [US]
PCPBS	P-Chlorophenylbenzenesulfonate
PCPCI	Power Conversion Products Council International [US]
PC-PE	Polycarbonate — Polyester
PC/PET	Polycarbonate/Polyethylene Terephthalate
PCPN	Precipitation
PCPS	Private Carrier Paging System
	Prodat Communication and Processing System
PCR	Page Control Register
	Pass Card Reader
	Peer Code Review
	Phase Controlled Rectifier
	Photoconductive Relay
	Pickled and Cold Rolled
	Polymerase Chain Reaction
	Post-Column Reaction
	Precision Control Relay
	Print Command Register
	Procedure Change Request
	Program Change Request
	Program-Controlled Request
	Program Control Register
	Punched Card Reader
PCRC	Primary Communications Research Center [UK]
PCRCA	Pickled, Cold Rolled and Close Annealed
PCR/DNA	Polymerase Chain Reaction/Deoxyribonucleic Acid (Technique)
PCRS	Primary Control Rod System
PCRV	Prestressed Concrete Reactor Vessel
PCS	Patent Classification Service [of INPADOC, Austria]
	Personal Computer System
	Pieces
	Planning Control Sheet
	Plastic-Clad Silica
	Plastic-Clad Silicon
	Plastic-Coated Silica
	Pneumatic Control System
	Pointing Control System
	Portable Computer System
	Port Command Store
	Port Control System
	Potash Corporation of Saskatchewan [Canada]
	Power Conditioning System
	Power Conversion System
	Preferred Character Set
	Primary Coolant System
	Print Contrast Scale
	Print Contrast Signal
	Print Contrast System
	Process Control System
	Professional Communication Society [of IEEE, US]
	Program Center Store
	Programmable Communications Subsystem
	Programmable Communications System
	Punched Card System
PCSA	Personal Computing Systems Architecture
	Power Crane and Shovel Association [US]
PCSAS	Policy Committee for Scientific Agricultural Societies [US]
PCSB	Pulse-Code Scanning Beam
PCSC	Power Conditioning, Switching and Control
PCSD	Printer Control Sequence Description
PCSFSK	Phase Comparison Sinusoidal Frequency Shift Keying
PCSIG	Personal Computer Software Interest Group [US]
PCSIR	Pakistan Council of Scientific and Industrial Research
PCSMQC	Polish Committee for Standardization, Measures and Quality Control
PCSP	Polar Continental Shelf Project [Canada]
PCT	Patent Cooperation Treaty
	P-Chlorotoluene
	Percent(age)
	Peripheral Control Terminal
	Phosphorochloridothioate
	Photochemical Transfer
	Photon-Coupled Transistor
	Planning and Control Techniques
	Polycyclohexane Dimethylene Terephthalate
	Portable Camera-Transmitter
	Program Control Table
	Pulse Count [also: pct]
PCTA	Polymer of Cyclohexanedimethanol Terephthalic Acid
PCTC	P-Chlorobenzotrichloride
PCTF	Plant Component Test Facility
	Premier's Council Technology Fund [Canada]
PCTFE	Polychlorotrifluoroethylene
PCTIS	Preston Commercial and Technical Information Service [UK]
PCTM	Pulse Count Modulation
PCTP	Pentachlorothiophenol [also: 5CTP]
PCTR	Physical Constant Test Reactor [HTR, US]
PCTS	Photocapacitance Transient Spectroscopy
PCU	Paging Control Unit
	Passenger Car Unit
	Peripheral Control Unit
	Physical Control Unit
	Pound Centigrade Unit [also: pcu]
	Power and Controller Unit
	Power Control Unit
	Power Conversion Unit
	Pressure Control Unit
	Processor Control Unit
	Program Control Unit
	Progress Control Unit
	Punched Card Unit
PCUA	Power Controller Unit Assembly

PROFIT (Computer System) Control Users Association [US]
PCV Packed Cell Volume
Pollution Control Valve
Positive Crankcase Ventilation
Pressure Control Valve
Primary Containment Vessel
PCVD Plasma-Assisted Chemical Vapor Deposition [also: P-CVD]
PCW Program Control Word
Pulsed Continuous Wave
PCWCA Poured Concrete Wall Contractors Association [US]
PCWPC Permanent Committee of the World Petroleum Congress
PCX Plasma Confinement Experiment
PCZ Panama Canal Zone
Positive Control Zone
PD Doctor of Pharmacy
Paid
Pallet Decoupler
Partial Depletion
Partial Derivative
Partial Discharge
Particle Dynamics
Pavement Design
Pen Down
Period
Peripheral Device
Permissable Dosage
Photodetector
Photodiode
Physical Distribution
Pictorial Display
Pitch Diameter
Plane Disagreement
Planned Derating
Plasma Display
Point Detonating
Polar Distance
Police Department
Port Dues
Positive Displacement
Postal District
Potassium Dichromate
Potential Difference
Power Density
Power Distribution
Power Doubler
Preliminary Design
Pressure Distillation
Prime Driver
Principal Distance
Printer's Devil
Priority Directive
Prism Diopter
Probability Density
Probability of Detection
Problem Definition
Problem Determination
Procedure Division
Process Data
Procurement Division
Professional Development
Professional Diploma
Programmable Device
Projected Display
Project Documentation
Propellant Dispersion
Proportional Derivative
Proportional Differential
Publication Date
Pulse Doppler
Pulse Driver
Pulse Duration
P&D Pickup and Delivery
Pd Palladium
pd (per diem) — per day; for each day
PDA Parenteral Drug Association [US]
Phenylene Diamine
Photodiode Array
Physical Device Address
Post-Deflection Acceleration
Precision Drive Axis
Predocketed Application
Preliminary Design Approval
Preliminary Design Authorization
Probability Discrete Automaton
Probability Distribution Analyzer
Problem Determination Aid
Prospectors and Developers Association
Pulse Discrimination Analysis
Pump Drive Assembly
PDAC Prospectors and Developers Association of Canada
PDAF Provincial Department of Agriculture and Forestry [Taiwan]
PDAID Problem Determination Aid
PDAP Polydiallylphthalate
PDAPS Pollution Detection and Prevention System
PDAR Preferential Departureand Arrival Route
PDB P-Dichlorobenzene
Process Descriptor Base
Public Debt Bureau [US]
PDC Package Design Council [now: PDCI, US]
Paper Distribution Council [US]
(Single) Paper Double Cotton (Wire)
Parallel Data Controller
Performance Data Computer
Permanent Data Call
Photonuclear Data Center [of NIST, US]
Polycrystalline Diamond Compact
Power Distribution Control
Premission Documentation Change
Printing Density Controller
Probability of Detection and Conversion
Pyridinium Dichromate
PDCA Painting and Decorating Contractors of America [US]
PDCE Paramagnetic Design and Cost Effectiveness
PDCI Package Design Council International [US]
PDCLC Pictorial Display Course Line Computer
PDCP Polydicyclopentadiene [also: PDCPD]
PDCS Performance Data Computer System

PDD Partial Discharge Detector
Past Due Date
Physical Damage Division
Post Dialing Delay
Program Design Data
Projected Data Display
Projected Decision Date
Prospective Decision Date
Pulse Delay Device
PWB (= Printed Wiring Board) Design Device
PDDI Product Data Definition Interface
Pd(DIPHOS)$_2$ Bis[1,2-Bis(diphenylphosphino)ethane] Palladium
PDDP Product Design and Development Program
PDE Partial Differential Equation
Position-Determining Equipment
PDEA Phenyldiethanolamine
PDED Partial Double Error Detecting
PDELS Parallel-Detection Electron Loss Spectrometer
PDES Preliminary Draft Environmental Statement
Product Data Exchange Specification
Product Definition Exchange Specification
PDF Pair Distribution Function
Plant Design Factor
Point Detonating Fuse
Post-Detection Filter
Postdoctoral Fellow
Powder Diffraction File [of JCPDS]
Printer Description File
Probability Density Function
Probability Distribution Function
Processor Defined Function
Protected Difference Fat
PDFD Predemonstration Fusion Device
PDG Precision Drop Glider
Pregnanediol Glucoronide
PDGDL Plasma Dynamics and Gaseous Discharge Laboratory [US]
PDGF Platelet-Derived Growth Factor
Power Data Grapple Fixture
PDGS Precision Delivery Glide System
Product Design Graphics System
PDH Philibert-Duncumb-Heinrich (Equation)
Planned Derated Hours
Phosphate Dehydrogenase
PDI Perfect Digital Invariant
Pharmacy Dedication Information
Picture Description Instruction
Plastic Drum Institute [US]
Plumbing and Drainage Institute [US]
Polydispersity Index
Positive Displacement Injector
PDIO Parallel Digital Input/Output
Photodiode
PDIP Plastic Dual In-Line (Package)
PDIR Peripheral Data Set Information Record
PDIS Pressure Differential Switch
PDL Page-Description Language
Particle Physics Laboratory [US]
Particle Dynamics Laboratory [Canada]
Photodissociation Dye Laser
Print Definition Language
Procedure Definition Language
Process Description Language
Program Design Language
Program Development Language
Programmable Data Language
Pulse Delta Modulation
Pulse Duration Modulation
pdl poundal
pdl-ft foot-poundal
PDM Physical Distribution Management
Pilot Decision Making
Polarization Division Multiplexing
Polynomial Discriminant Method
Practical Data Manager
Predictive Maintenance
Product Data Manager
Project Development Methodology
Protected Difference Milk
Pulse Delta Modulation
Pulse Duration Modulation
PDMA Plumbing and Drainage Manufacturers Association [now: PDI, US]
PDME Pendant Drop Melt Extraction
PDM-FM Pulse Duration Modulation — Frequency Modulation
PDMP Product Development Management Program
PDMS Plasma Desorption Mass Spectrometry
Polydimethylsiloxane
PDN Production
Public Data Network
PDO Program Directive Operations
PDP Plasma Display Panel
Positive Displacement Pump
Procedure Definition Processor
Process Development Pile [US]
Product Development Program
Productivity Development Program [US]
Professional Data Processor
Program Definition Phase
Programmable Digital Processor
Programmed Data Processor
Project Definition Phase
PDPC Position Display Parallax-Corrected
Pd-PEI Palladium-Polyethylenimine
PDPS Parts Data Processing System
PDQ Product Demand Quotation
Programmed Data Quantizer
PDR Page Data Register
Performance Data Rate
Periscope Depth Range
Pilot's Display Recorder
Potential-Drop Ratio
Pounder
Power Directional Relay
Precision Depth Recorder
Predetermined Route
Preferential Departure Route
Preliminary Data Report
Preliminary Design Review
Pressurized Deuterium Reactor
Priority Data Reduction

	Processed Data Recorder
	Processing Date Rate
	Processing, Distribution and Retailing
	Production, Distribution and Retailing
	Program Discrepancy Report
	Program Drum Recording
	Public Document Room
PDRP	Program Data Requirement Plan
PDRS	Pulse Doppler Radar Simulation
PDS	Penultimate Digit Storage
	Partitioned Data Set
	Periodicals Data System
	Personal Data System
	Petroleum Data System [of University of Oklahoma, US]
	Photodigital Store
	Photodischarge Spectroscopy
	Photothermal Deflection Spectroscopy
	Power Density Spectrum
	Power Distribution System
	Predocketed Special Report
	Premises Distribution System
	Private Database Service
	Procedures Development Simulator
	Professional Development System
	Program Data Set
	Program Data Source
	Programmable Data Station
	Programmable Device Support
	Programmable Distribution System
	Propellant Dispersion System
PD/S	Problem Definition/Solution
PDSPI	Polyurethane Division, Society of the Plastics Industry [US]
PDSMS	Point-Defense Surface Missile System
PDT	Pacific Daylight Time
	Photodynamic Therapy
	Physical Device Table
	Picture Description Test
	Product
	Programmable Data Terminal
	3-(2-Pyridyl)-5,6-Diphenyl-1,2,4-Triazine
PDU	Pilot's Display Unit
	Power Distribution Unit
	Pressure Distribution Unit
	Process Development Unit
	Product Distribution Unit
	Programmable Delay Unit
	Protocol Data Unit
PDV	Premodulation Processor Deep-Space Voice
PDW	Partially Delactosed Whey
	Planar Dielectric Waveguide
PDWP	Partially Delactosed Whey Powder
PDWS	Primary Drinking Water Standards
PDX	Poloidal Divertor Experiment
	Private Digital Exchange
PE	Packaging Engineer
	Packaging Engineering
	Page-End (Character)
	Paleoecology
	Parallel Element
	Parity Error
	Parity Even
	Pellet
	Pentaerythritol
	Peripheral Equipment
	Permanent Echo
	Petroleum Engineer
	Petroleum Engineering
	Pharmaceutical Engineer
	Pharmaceutical Engineering
	Phase Encoding
	Photoelastic(ity)
	Photoelectric(ity)
	Photoelectron
	Photoelement
	Photoemission
	Photoengraving
	Photographic Engineer
	Photographic Engineering
	Phycoerythrin
	Picture Element
	Piezoelectric(ity)
	Pinion End
	Planning Engineer
	Planning Engineer
	Plant Engineer
	Plant Engineering
	Plasma Etching
	Plastic Engine
	Plastics Engineer
	Plastics Engineering
	Plastoelastic(ity)
	Platinum Electrode
	Plumbing Engineer
	Plumbing Engineering
	Polyelectrolyte
	Polyester
	Polyether
	Polyethylene
	Porcelain Enamel
	Positional Efficiency
	Position Error
	Potential Energy
	Power Engineer
	Power Engineering
	Preemption
	Precision Engineer
	Precision Engineering
	Primary Electron
	Prince Edward Island [Canada]
	Printer's Error
	Probable Error
	Process Engineer
	Process Engineering
	Processing Element
	Processor Element
	Production Engineer
	Production Engineering
	Professional Ecologist
	Professional Engineer
	Professional Engraver

Professional Estimator
Programming Environment
Project Engineer
Pseudoelastic(ity)
Pulley End
Pyroelectric(ity)

P/E Precipitation-Evaporation (Ratio) [also: P-E]
Price/Earnings (Ratio)
Pulse-Echo Testing

PEA Pennsylvania Electric Association [US]
Phenylethanolamine
Phenylethylamine
Polyethylacrylate
Positive Electron Affinity
Push-Effective Address

PEAB Professional Engineers Appointments Bureau [UK]

PEAC Program Evaluation and Audit Committee

PEACER Petroleum Employers Advisory Council on Employee Relations

PEACESAT Pan-Pacific Editing and Communication Experiment by Satellite [of University of Hawaii, US]

PEACU Plastic Energy Absorption in Compression Unit

PEAR Plasma Extended Arc Reactor [Canada]

PEARL Package for Efficient Access to Representations in LISP (= List Processing)
Performance Evaluation of Amplifiers from a Remote Location
Periodical Enquiry, Acquisition and Registration Locally [UK]
Process, Experiment and Automation Real-Time Language
Programmed Editor and Automated Resources for Learning

PEAS Pacific's Electric Acquisition Service [of Pacific University, US]

PEB Pulse Electron Beam

PEC Panel Electronic Circuit
Photoelectric Cell
Photoelectrochemical Cell
Photoemissive Cell
Plant Equipment Code
Platform Electronic Cards
Polyester Carbonate
Polyphenylene Ether Copolymer
Position Error Correction
Power Electronics Council [now: Power Electronics Society [of IEEE, US]]
Previous Element Coding
Program Exception Code
Public Extension Circuit

PECAN Pulse Envelope Correlation Air Navigation

PECBI Professional Engineers Conference Board for Industry

PECC Product Engineering Control Center

PECS Portable Environmental Control System

PECTFE Polyethylene Chlorotrifluoroethylene [also: PE-CTFE]

PECVD Plasma-Enhanced Chemical Vapor Deposition [also: PE-CVD]

PED Pedestal
Personnel Equipment Data
Photoelectric Device
Polymer Engineering Directorate [UK]
Pulse Edge Discrimination

PEDCUG Planning Engineers Desktop Computer Users Group [US]

PEDN Planned Event Discrepancy Notification

PEDRO Pneumatic Energy Detector with Remote Optics

PEDS Peltier Effect Diffusion Separation
Precise Engineering and Deformation Survey [Canada]
Protective Equipment Decontamination Section

PEE Photoelectric Emission

P&EE Proving and Experimental Establishment [UK]

PEEK Polyetherether Ketone

PEEKK Polyetherether Ketone Ketone

PEEKKK Polyetherether Ketone Ketone Ketone

PEEL Programmable Electrically Erasable Logic

PEELS Parallel-Detection Electron Loss Spectrometer

PEEM Photoelectron Emission Microscopy
Photoelectron Emission Electron Microscopy
Photoemission Electron Microscopy

PEEP Pilot's Electronic Eye-Level Presentation

PEER Professional Engineers Employment Registry [US]

PEF Packaging Education Foundation [US]
Physical Electronics Facility

PEFR Peak Expiratory Flow Rate

PEG Petrochemical Energy Group [US]
Polyethylene Glycol [also: PG]

PEGA Precision Engineering Grinding Apparatus

PEGAD Permission Granted to Add

PEGASUS Precision Engineering Grinding Apparatus for Superfinishing Ultrahard Surfaces

PEH Polyphenylene Ether Homopolymer

PEI Petroleum Equipment Institute [US]
Polyetherimide
Polyethylenimine
Porcelain Enamel Institute [US]
Preliminary Engineering Inspection
Prince Edward Island [Canada]

PEIC Periodic Error Integrating Controller

PEIEC Prince Edward Island Energy Corporation [Canada]

PEIFL Prince Edward Island Federation of Labour [Canada]

PEILS Prince Edward Island Land Surveyors [Canada]

PEISC Prince Edward Island Safety Council [Canada]

PEK Phase-Exchange Keying
Polyether Ketone

PEKK Polyether Ketone Ketone

PEL Permissible Exposure Limit
Picture Element [also: PIXEL, pixel or pel]
Pressure Exposure Limit

PELSS Precision Emitter Location Strike System

PELTP Personnel Licensing and Training Panel [of ICAO]

PEM Photoelastic Modulator
Photoelectromagnetic(s)
Photoelectron Microscopy
Plant Engineering and Maintenance
Plastic-Envelope Method
Process Execution Module
Processing Element Memory

	Procurement Equipment and Missiles [of US Army]
	Production Engineering Measure
	Program Execution Monitor
	Proto-Environmental Model
	Pulse Electromagnetic
PEMA	Phenylethylmalonamide
	Polyethyl Methacrylate
	Process Equipment Manufacturers Association [US]
	Procurement Equipment and Missiles — Army [US]
PEMAC	Professional Engineers Manpower Assessment Committee [Canada]
PEMD	Program for Export Market Development
PE-MOCVD	Plasma-Enhanced Metalorganic Chemical Vapor Deposition
PEMS	Pesticide Enforcement Management System [of EPA, US]
	Policy and Expenditure Management System
	Porcelain-Enamelled Metal Substate
PEN	Penicillin
	Peninsula
PENA	Primary Emission Neutron Activation
PEN-B	Penicillin, Benzalthine Salt
PENBAL	Peninsular Spain-Balearic Islands Submarine Cable [Spain]
PENCAN	Peninsular Spain-Canary Islands Submarine Cable [Spain]
PENCIL	Pictorial Encoding Language
P/E News	Petroleum Energy Business News Index [of API, US]
PEng	Professional Engineer
PEN-K	Penicillin, K (= Potassium) Salt
Penn	Pennsylvania [US]
PEN-NA	Penicillin, Na (= Sodium) Salt
Penna	Pennsylvania [US]
PENNSTAC	Penn(sylvania) State University Automatic Computer
Penn State	Pennsylvania State University [US]
PEN-P	Penicillin, Procain Salt
PENRAD	Penetration Radar
PENT	Pentode
PEO	Patrol Emergency Officer
	Piezoelectric Oscillator
	Polyethylene Oxide
	Polymer Electrolyte
PEO-PC	Polyethylene Oxide/Polycarbonate
PEO-PET	Polyethylene Oxide/Polyethylene Terephthalate
PEOS	Propulsion and Electrical Operating System
PEOX	Polyethyl Oxazoline [also: PEOx]
PEP	Parametric Element Processor
	Partitioned Emulation Programming
	Peak Envelope Power
	People for Energy Progress [US]
	Phosphoenolpyruvate
	Phosphoenolpyruvic Acid
	Photoelectric Potential
	Planar Epitaxial Passivated (Transistor)
	Planetary Ephemeris Program [US]
	Plastics Education Foundation [of SPE, US]
	Political and Economic Planning
	Polyethylene Polyamine
	Porsche Experimental Prototype
	Positron Electron Photon
	Program Evaluation Procedure
	Program to Enhance Productivity
	Programmed Emulation Partition
	Propulsion and Energetics Panel [of AGARD]
	Proton-Electron-Positron (Beam)
PeP	Proton-Electron-Proton (Reaction)
PEPAG	Physical Electronics and Physical Acoustics Group
PEPCK	Phosphoenolpyruvate Carboxykinase
PEPE	Parallel Element-Processing Ensemble
PEP-L	Partitioned Emulation Programming — Local
PEP-LR	Partitioned Emulation Programming — Local/Remote
PEPP	Planetary Entry Parachute Program
	Professional Engineers in Private Practice
PEPPI	Program to Encorage Product and Process Innovation
PEPR	Precision Encoding and Pattern Recognition (Device) [of MIT, US]
PER	Performance
	Period
	Person
	Process Energy Requirement
	Program Event Recording
	Program Execution Request
	Preliminary Engineering Report
	Production Engineering and Research
	Protein Efficiency Ratio
	Pseudoequilibrium Ratio
PERA	Planning and Engineering for Repairs and Alterations
	Production Engineering Research Association [UK]
	Production Engine Remanufacturers Association [US]
	Project Engineering Research Association [UK]
PERC	Percussion
PERC DR	Percussion Drilling
PERCOS	Performance Coding System
PERD	Panel on Energy Research and Development [Canada]
PEREF	Propellant Engine Research Environmental Facility
PERF	Perfection
	Perforation
	Performance
PERFS	Perforations
PERGO	Project Evaluation and Review with Graphic Output
PERI	Photoengravers Research Institute [US]
	Protein Engineering Research Institute [Japan]
PERIF	Peripheral
PERL	Perkin Elmer Robot Language
PERM	Permanent
	Permeability
	Program Evaluation for Repetitive Manufacture
PERMINVAR	Permeability Invariant
PERMLY	Permanently
PERP	Perpendicular
	Process Evaluation/Research Planning
per pro	(per procura) — by proxy
PERS	Personal
	Personnel
PERT	Performance Evaluation (and) Review Technique

	Pollution Emergency Response Team [Canada]
	Production and Evaluation Review Technique
	Program Evaluation and Reporting Technique
	Program Evaluation and Review Technique
	Program Evaluation Research Task
PERTCO	Program Evaluation and Review Technique with Cost
PERU	Production Equipment Records Unit
PES	Peruvian Sol [Currency of Peru]
	Photoelectric Scanning
	Photoelectron Spectroscopy
	Photoelectron Spectrum
	Photoemission Spectroscopy
	Polyether Sulfone
	Potential Energy Surface
	Power Electronics Society [of IEEE, US]
	Power Engineering Society
	Program Execution System
	Programmer Electronic Switch
PESA	Petroleum Equipment Suppliers Association [US]
PESC	Passivated Emitter Solar Cell
PESDA	Printing Equipment Supply Dealers Association
PESM	Photoelectron Spectromicroscopy
PEST	Parameter Estimation by Sequential Testing
PESTAB	Pesticides Abstracts (Database) [of Environmental Protection Agency, US]
PESTDOC	Pesticide Documentation (Database) [UK]
PESTF	Proton Event Start Forecast
PET	Patterned Epitaxial Technology
	Pentaerythritol
	Peripheral Equipment Tester
	Permeation Enhancement Technology
	Personal Electronic Transaction Computer
	Personal Electronic Transactor
	Petroleum
	Photographic Equipment Technician
	Physical Equipment Table
	Pilot-Line Experiment Technology [of US Army]
	Plasma Edge Technique
	Plastic Engine Technology
	Point of Equal Time
	Polyethylene Terephthalate [also: PETP]
	Portable Executive Telephone
	Position-Event-Time
	Positron Emission Tomography
	Precision End Trimmed
	Production Environmental Testing
	Program Evaluator and Tester
PETA	Pentaerythritol Triacrylate
PETC	Pittsburgh Energy Technology Center [US]
PETE	Pneumatic End-to-End
PetEng	Petroleum Engineer [also: PetE]
PETFE	Polyethylene Tetrafluoroethylene [also: PE-TFE]
	Polyethylene-Co-Tetrafluoroethylene [also: PE-TFE]
PETG	Glycol-Modified Polyethylene Terephthalate
	Polyethylene Terephthalate Glycol
	Polyethylene Terephthalate, G Copolymer
PETN	Pentaerythritol Tetranitrate
PETP	Polyethylene Terephthalate [also: PET]
PETR	Petrological
	Petrologist
	Petrology
PETRA	Positron-Elektron-Tandem-Ringbeschleuniger-Anlage [= Positron Electron Tandem Ring Accelerator, Germany]
PETRO	Petroleum
PETRO ENG	Petroleum Engineering
PetroEng	Petroleum Engineer [also: PetroE]
PETROEX	Petroleum Products Exchange Data Clearinghouse [US]
PETROMIN	General Petroleum and Minerals Organization [Saudi Arabia]
PETS	Plastic Education and Troubleshooting System
PETT	Portable Ethernet Transceiver Tester
PEU	Port Expander Unit
PEV	Peak Envelope Voltage
PEW	Percussion Welding
PEX	Pituitary Extract
	Projectable Excitation
p ex	(par exemple) — for example
PF	Packing Factor
	Packing Fraction
	Page Footing
	Page Formatter
	Penetration Fracture (Test)
	Perchlorylfluoride
	Perfect Fluid
	Phenol-Formaldehyde
	Photofinish(ing)
	Photofluorography
	Photon Factory [Japan]
	Picture Frequency
	Plug-Flow Transport
	Point Foundation [US]
	Point of Frog
	Polarizing Filter
	Pole Figure
	Poloidal Field
	Position Finder
	Potential Flow
	Powder Forging
	Power Factor
	Power Frequency
	Probability Factor
	Profile
	Program Function
	Programmable Function
	Protection Factor
	Pulse Frequency
	Punch Off
P/F	Powder Forging
pF	picofarad
PFA	Perfluoroalkoxy (Polymer)
	Perfluoroalkoxy Alkene
	Pesticide Formulators Association [now: PPA, US]
	Polyfluoroalkoxy
	Polyurethane Foam Association [US]
	Popular Flying Association
	Press Foundation of Asia
	Pressurized Fluidized Bed
	Program and File Analysis

Provincial Forestry Administration [Taiwan]
Pulverized Fuel Ash
PFAB Prefabrication [also: PREFAB]
PFB Provincial Food Bureau [Taiwan]
Provisional Frequency Board
PFBA Polyperfluorobutylacrylate
PFBC Pressurized Fluidized Bed Combustion
PFBR Prototype Fast Breeder Reactor
PFBT Performance Functional Board Tester
PFC Pacific Forestry Center [Canada]
Pack Feed and Converter
Phase Frequency Characteristics
Phenylacetamidotrifluoromethylcoumarin
Photofinish Camera
Planar Flow Casting
Private First Class
Professional Fee Costing
Pulse-Flow Coulometry
PFCE Performance
PFCS Primary Flight Control System
PFCU Power Flying Control Unit
PFD Personal Flotation Device
Planned Flight Data
Power Flux Density
Preferred
Primary Flash Distillate
Primary Flight Display
PFDC Pole Figure Data Collection
PFDF Pacific Fisheries Development Foundation [US]
PFE Photoferroelectric (Effect)
Plenum Fill Experiment
PFEP Programmable Front-End Processor
PFES Proposed Final Environmental Statement
PFF Presidential Faculty Fellows (Program) [US]
PFI Pacific Forest Industry
Pack File Indexer
Pipe Fabrication Institute [US]
Power Factor Indicator
PFIR Part Fill-In and Ram
PFIX Power Failure Interrupt
PFK Phosphofructokinase
Plastic Fluted Knob
Program Function Key
Programmed Function Key
PFL Propulsion Field Laboratory [US]
PFM Power Factor Meter
Pulse-Frequency Modulation
PFMC Pacific Fisheries Management Council [US]
PFN Permanent File Name
Pulse Forming Network
PFP Performic Acid Phosphotungsite
P-Fluorophenylalanine
Plastic Faced Plaster
Plutonium Finishing Plant [of USDOE]
Post Flight Processor
Program File Processor
PFPA Pitch Fiber Pipe Association [US]
PFPE Perfluoropolyether
PFR Parts Failure Rate
Perforator
Permitted Flying Route
Plug Flow Reactor
Polynesian Franc [also: PFr; Currency of French Polynesia and Tahiti]
Power Fail Recovery
Power Fail Restart
Programmed Film Reader
Prototype Fast Reactor
Pulse Frequency
PF/R Power Fail/Restart
PFRA Prairie Farm Rehabilitation Act [Canada]
Prairie Farm Rehabilitation Administration [Canada]
PFRS Portable Field Recording System
PFR/UK Prototype Fast Reactor/United Kingdom
PFS Peripheral Fixed Shim
Primary Frequency Supply
Propellant Field System
PFT Paper, Flat Tape
PFZ Precipitate-Free Zone
PG Page
Periodic Group
Permanent Glow
Petrogenesis
Phase Gradient
Phosphatidylglycerol
Phosphoglycerate
Photogrammetry
Physical Geography
Picture Generation
Pin Grid
Point Group
Polyethylene Glycol [also: PEG]
Polypropylene Glycol [also: PPG]
Port Group
Portuguese
Portugal
Power Gain
Power Gate
Power Generation
Pressure Gauge
Professional Geologist
Professional Geophysicist
Program Generator
Program Generic
Prompt Gamma Ray
Project Group
Prostaglandin
Pulse Generator
Pyrolytic Graphite
pg picogram
PGA Phosphoglyceric Acid
Pin Grid Array
Polyglycolic Acid
Pressure Garment Assembly
Professional Graphics Adapter
Programmable Gate Array
Propane Gas Association
Pteroylglutamic Acid
Purchased Gas Adjustment
3-PGA 3-Phosphoglyceric Acid
PGAA Pin Grid Array Adapter

	Prompt Gamma Activation Analysis
	Prompt Gamma-Ray Activation Analysis
PGAL	Proof Gallons
PGBD	Plugboard
PGC	Pacific Geoscience Center [of EMR, Canada]
	Port Group Control
	Post-Graduate Certificate
	Power Generation Committee [of IEEE PES, US]
	Professional Graphics Controller
	Pulsed Gas Chromatography
	Pyrolysis Gas Chromatography
PGCOA	Pennsylvania Grade Crude Oil Association [US]
PGD	Phosphogluconate Dehydrogenase
	Planar Gas Discharge
PG DN	Page Down
PGE	Phenyl Glycidyl Ether
	Platinum-Group Element
	Primary Ground Electrode
PGEC	Professional Group on Electronic Computers
PGEN	Program Generator
PGeol	Professional Geologist
PGEWS	Professional Group on Engineering Writing and Speech [of IEEE, US]
PGH	Patrol Gunboat, Hydrofoil
	Port Group Highway
PGHTS	Port Group Highway Time Slot
PGI	Parameter Group Identifier
	Phosphoglucoisomerase
	Port Group Interface
	Pyrotechnics Guild International [US]
PGK	Papua New Guinea Kina [Currency of Papua New Guinea]
	Phosphoglycerate Kinase
	Phosphoglyceric Phosphokinase
	Program Check
PGLIN	Page and Line
pGlu	Pyroglutamyl
PGM	Paleogeomorphology
	Phosphoglucomutase
	Phosphoglycerate Mutase
	Platinum-Group Metal
	Precision-Guided Munitions
	Program
P-GMAW	Pulsed-Current Gas Metal-Arc Welding
PGMC	Primary Glass Manufacturers Council [US]
PGMS	Precision-Guided Munitions
PGMTT	Professional Group on Microwave Theory and Techniques [US]
PGNCS	Primary Guidance and Navigation Control System
PGNS	Primary Guidance and Navigation System
PGOS	Petroleum, Gas and Oil Shale
PGP	Programmable Graphics Processor
PGR	Precision Graph(ic) Recorder
PGRF	Pacific Gamefish Research Foundation [US]
	Pulse Group Repetition Frequency
PGRSA	Plant Growth Regulator Society of America [US]
PGRT	Petroleum and Gas Revenue Tax
PGRWG	Plant Growth Regulator Working Group [now: PGRSA, US]
PGS	Plane Grating Spectrograph
	Power Generation Satellite
	Power Generation System
	Power Generator Section
	Program Generation System
PGT	Page Table
	Planetary Gear Train
PGU	Pressure Gas Umbilical
PG UP	Page Up
PGW	Periodic Gravity Wave
	Pressure Gas Welding
PH	Padagogische Hochschule [= College of Education, Germany]
	Parahydrogen
	Page Heading
	Performance History
	Period Hours
	Phantom
	Phase
	Phasemeter
	Phenol(ic)
	Phone
	Phosphorus
	Power House
	Precipitation Hardening
	Pressure Head
	Public Holidays
P&H	Postage and Handling
P/H	Parts per Hundred [also: PPH or pph]
Ph	Phenyl
pH	(Potential of Hydrogen) — Negative Logarithm of Hydrogen-Ion Activity
ph	phot [Unit]
PHA	Phaseolus Vulgaris Agglutinin
	Phytohemagglutinin
	Prelimiary Hazard Analysis
	Pulse Height Analyzer
PHA-E	Phaseolus Vulgaris Agglutinin — Erythroagglutinin
PHA-L	Phaseolus Vulgaris Agglutinin — Leucoagglutinin
PHA-M	Phaseolus Vulgaris Agglutinin — Mucoprotein
PHA-P	Phaseolus Vulgaris Agglutinin — Protein
PHARE	Program for Harmonized Air Traffic Management Research in Europe [of EUROCONTROL]
PHARM	Pharmaceutical
	Pharmacy
PHARMACOL	Pharmacological
	Pharmacologist
	Pharmacology
PharmD	Doctor of Pharmacy
PHARM PAT	Pharmaceutical Patent
PHAROS	Plan Handling and Radar Operating System
PHAXSCAN	Philips High-Speed Automatic X-Ray Analysis System
PHB	Polyhydroxybutyrate
PhB	Bachelor of Philosophy
PH BRZ	Phosphor Bronze
PHBV	Poly(3-Hydroxybutyrate Valerate)
PHC	Petroleum Hydrocarbon
	Physical Hydrogen Cracking
PhC	Pharmaceutical Chemist
PHCVD	Photochemical Chemical Vapor Deposition
	Photosensitized Chemical Vapor Deposition
PHD	Parallel Head Disk

	Physical Hydrology Division [of NHRI, Canada]
	Plastohydrodynamic(s)
	Pulse Height Discriminator
PhD	Doctor of Philosophy
PHE	Packaging and Handling Engineer
	Plate Heat Exchange
	Public Health Engineer
Phe	Phenylalanine
	Phenylalanyl
PheAc	Phenoxyacetyl
PHEMA	Polyhydroxyethyl Methacrylate
PHEN	Phenanthroline Monohydrate
	Phenol(ic)
PHENO	Precise Hybrid Elements for Nonlinear Operation
P&HEP	Plasma and High-Energy Physics [IEEE Group]
PHERMEX	Pulsed High Energy Radiographic Machine Emitting X-Rays [of LASL, US]
PHF	Plug Handling Fixture
PhG	Graduate in Pharmacy
PHH	Phillips Head (Screw)
PHI	Phosphohexose Isomerase
	Position and Homing Indicator
PHIG	Programmer's Hierarchical Interactive Graphics
PHIGS	Programmer's Hierarchical Interactive Graphics System
PhilM	Master of Philosophy
PHILQA	Philips Question Answering System [of PRL, Netherlands]
PHILSIN	Philippines-Singapore Submarine Cable
PHIN	Position and Homing Inertial Navigator
pH/ISE	pH Value/Ion-Selective Electrode
PHL	Packaging, Handling and Logistics
PhL	Licentiate of Philosophy
PHLAG	Philips Load and Go
PHLODOT	Phase Lock Doppler Tracking
PHM	Patrol Hydrofoil Missile
	Phasemeter
	Phase Modulation
PhM	Master of Philosophy
pH/mV	pH Value versus Millivolts
PHODEC	Photometric Determination of Equilibrium Constants
PHOENIX	Plasma Heating Obtained by Energetic Neutral Injection Experiment
PHON	Phonetics [also: PHONET]
PHONO	Phonograph(y)
PHOSCHEM	Phosphate Chemicals Export Association [US]
PHOT	Photographer
	Photographic
	Photography
PHOTAC	Phototypesetting and Composing
PHOTO	Photograph
PHOXIM	Phenylglyoxylonitriloxime
PHP	Parts, Hybrids, and Packaging [IEEE Group]
	Philippine Peso [Currency]
	Pounds per Horsepower
	Pump Horsepower
PHR	Physical Record
	Pound-Force per Hour
	Process Heat Reactor Program
	Pulse-Height Resolution
phr	parts per hundred rubber
PHRAN	Phrasal Analyzer
PHRED	Phrasal English Diction
PHRM	Pharmaceutical
PHS	Printing Historical Society [UK]
	Public Health Service [US]
	Pulse Height Selection
pH(S)	pH Standard
PHSE	Phase
PHSPS	Preservation, Handling, Storage, Packaging and Shipping
PHT	Phototube
	Pyrrolidone Hydrotribromide
PHTC	Pulse Height to Time Converter
PHTS	Primary Heat Transport System
PHW	Pressurized Heavy Water
PHWR	Pressurized Heavy Water Reactor
PHYS	Physical
	Physicist
	Physics
PHYSBE	Physiological Simulation Benchmark Experiment
PHYS CHEM	Physical Chemistry
PHYSIOL	Physiological
	Physiologist
	Physiology
PHYS MET	Physical Metallurgy
PhysS	Physical Society
PHYS REV	Physical Review
PI	Packaging Institute
	Packing Index
	Paper Insulation
	Parallel Input
	Parallel Interface
	Parameter Identifier
	Particular Integral
	Performance Index
	Perlite Institute [US]
	Philippine Islands
	Phosphatidylinositol
	Photoionization
	Pilotless Interceptor
	Pin Insulator
	Plastic Index
	Plastics Institute [UK]
	Point Initiating
	Point Insulating
	Point of Intersection
	Polyimide
	Population Inversion
	Position Indicator
	Positive Input
	Principal Investigator
	Priority Interrupt
	Process Image
	Processor Interface
	Productivity Index
	Program Indicator
	Program Interface
	Program Interruption
	Programmed Instruction
	Programmed Interrupt

	Proportional Integral
	Propyl Isome
P&I	Piping and Instrumentation
Pi	Inorganic Phosphate
PIA	Pakistan International Airways
	Peripheral Interface Adapter
	Petroleum Incentives Administration
	Pharmaceutical Industries Association [of EFTA]
	Phenylisopropyladenosine
	Pilots International Association [US]
	Plastics Institute of America [US]
	Pre-Installation Acceptance
	Printing Industries of America [US]
	Printing Industries Association [US]
	Program Implementation Agency
	Pulse-Interval Analyzer
PIAC	Peak Instantaneous Airborne Counts
	Petroleum Industry Advisory Committee [UK]
PIAD	Plastics in Automotive Design [of SME, US]
PIAM	Petroleum Industry Application of Microcomputers
PIANC	Permanent International Association of Navigation Congresses [Belgium]
PIANEC	Planning of the Implementation of an Improved AFTN/AFS Network [of ICAO]
PIAPACS	Psychophysiological Information Acquisition, Processing and Control System
PIARC	Permanent International Association of Road Congresses [France]
PIB	Petroleum Information Board
	Petroleum Information Bureau [UK]
	Polar Ionosphere Beacon
	Polyisobutylene
	Polytechnic Institute of Brooklyn [US]
	Preflight Information Bulletin
	Processor Interface Buffer
	Programmable Input Buffer
	Public Investments Board [India]
	Publishing Information Bulletin
	Pyrotechnic Installation Building
PIBAC	Permanent International Bureau of Analytical Chemistry
PIBAL	Pilot-Balloon (Observation) [also: Pibal or pibal]
	Polytechnic Institute of Brooklyn, Aerodynamics Laboratory [US]
PIBMM	Permanent International Bureau of Motorcycle Manufacturers [now: IMMA, France]
PIBMRI	Polytechnic Institute of Brooklyn, Microwave Research Institute [US]
PIBUC	Pilot Backup Control
PIBS	Polar Ionosphere Beacon Satellite [of NASA]
PIC	Paper-Insulated Cable
	Particle-in-Cell
	Personal Identification Code
	Photographic Industries Council [US]
	Photographic Interpretation Center
	Pilot in Command
	Plastic-Insulated Cable
	Plastic-Insulated Conductor
	Polyethylene-Insulated Cable
	Polyethylene-Insulated Conductor
	Polymer Impregnated Concrete
	Position Independent Code
	Potential Icing Category
	Power Information Center
	Preimpregnated Cable
	Pressure-Impregnation Carbonization
	Prices and Income Commission
	Primary Inter-LATA (= Local Access and Transport Area) Carrier
	Priority Interrupt Control
	Process Interface Control
	Processor Interconnection Channel
	Program Indicator Code
	Program Information Code
	Program Interrupt Control
	Programmable Interrupt Controller
PICA	Power Industry Computer Applications
	Project for Integrated Catalogue Automation
	Public Interest Computer Association
PICAA	Permanent International Committee of Agricultural Associations
PICAC	Power Industry Computer Applications Conference
PICAO	Provisional International Civil Aviation Organization
PICB	Peripheral Interface Control Bus
PICC	Provisional International Computation Center
PICGC	Permanent International Committee for Genetic Congresses
PICLS	Purdue International and Computational Learning System [US]
PICOP	Paper Industries Corporation of the Phillipines
PICPSA	Permanent International Commission for the Proof of Small Arms [Belgium]
PICS	Personnel Information Commission System
	Personnel Information Communication System
	Plug-In Control System
	Plug-In Inventory Control System
	Production Information and Control System
	Programmable Industrial Control Simulation
PICS/DCPR	Plug-In Inventory Control System/Detailed Continuing Property Record
PICTEL	Picture Telephone [US]
PICU	Parallel Instruction Control Unit
PICUTP	Permanent International Committee of Underground Town Planning
PID	Parallel Interface Device
	Peripheral Interface Device
	Photo Ionization Detector
	Port Identification
	Program Information Department
	Proportional Integral Derivation
	Proportional Integral Derivative
	Proportional Integral Differential
	Pseudo Interrupt Device
P&ID	Piping and Instrumentation Diagram
	Piping and Instrumentation Drawing
PIDA	Pharmaceutical Industry Development Assistance
	Phenylindane Dicarboxylic Acid
PIDB	Peripheral Interface Data Bus
PIDC	Pakistan Industrial Development Corporation
PIDCOM	Process Instruments Digital Communications System
PIDE	Pakistan Institute of Development Economics

PIDI Philippines Invention Development Institute
PIDP Programmable Indicator Data Processor
PIDS Parameter Inventory Display System [of NODC, US]
Personal Inspirable Dust Spectrometer
PIE Pacific Islands Ecosystems (Database) [of FWS, US]
Parallel Instruction Execution
Parallel Interface Element
Plug-In Electronics
Post-Irradiation Examination
Pulse Interference Emitting
PIEA Petroleum Industry Electrical Association [now: ENTELEC, US]
PIF Payload Integration Facility
Program Information File
PIFC Pakistan Industrial Finance Corporation
PIFEX Programmable Image Feature Extractor
PIG Penning Ionization Gauge
Philips Ionization Gauge
PIGA Pendulous Integrating Gyro Accelerometer
PIGME Particle-Induced Gamma-Ray Emission
Programmed Inert-Gas Multi-Electrode (Welding)
PII Positive Immittance Inverter
PIIF Pakistan International Industrial Fair
PI/INT'L Packaging Institute International [US]
PIKES Phot-Interpretation Keys Expert System
PIL Percentage Increase in Loss
Pilot
Pitt Interpretive Language
Precision-In-Line
PILA Power Industry Laboratory Association [US]
PILC Paper-Insulated Lead-Covered (Cable)
PILE Product Inventory Level Estimator
PILOT Panel on Instrumentation for Large Optical Telescopes [UK]
Permutation Indexed Literature of Technology
Piloted Low-Speed Test
PILP Program for Industrial Laboratory Projects [Canada]
PIM Peak Integration Method
Pilot Machine
Powder Injection Molding
Precision Indicator of the Meridian
Processor Interface Module
Program for Information Managers [now: AIM, US]
Program Integration Manual
Pulse Interval Modulation
PIMA Paper Industry Management Association [US]
Prairie Implement Manufacturers Association [Canada]
Polyisocyanurate Insulation Manufacturers Association [US]
PIMISS Pennsylvania Interagency Management Information Support System [US]
PIN Personal Identification Name
Personal Identification Number
Piece Identification Number
Plant Information Network
Police Information Network
Position Indicator
Positive-Intrinsic-Negative (Transistor) [also: pin or p-i-n]
Power Input
Private Intelligent (Data) Networker
Product Identification Number
Program Identification Number
P-Type-Intrinsic-N-Type (Transistor) [also: pin or p-i-n]
PINAC Permanent International Association of Navigation Congresses
PINCCA Price Index Numbers for Current Cost Accounting
PIND Particle Impact Noise Detector
PINFET Positive-Intrinsic-Negative Field-Effect Transistor
PINI Plug-In Neutral Injector
PINO Positive Input — Negative Output
PINPD Positive-Intrinsic-Negative Photodiode
PINS Portable Inertial Navigation System
Precise Integrated Navigation System
Precision's Improved Nesting System
PINSAC PINS Alignment Console
PINSCH Periodic-Index Separate-Confinement Heterostructure
PINSCH QW Periodic-Index Separate-Confinement Heterostructure Quantum-Well (Laser)
PINT Priority Interrupt Controller
Purdue Interpretive Programming and Operating System [US]
PINTEC Plastics Institute National Technical Conference [UK]
PINY Polytechnic Institute of New York [US]
PIO Parallel Input/Output
Peripheral Input/Output
Photocomposition Input Option
Pilot Induced Oscillation
Port Information Office [of INTERTANKO, Norway]
Position Iterative Operation
Precision Iterative Operation
Private Input/Output
Programmed Input/Output
PIOCS Parallel Input/Output Control System
Physical Input/Output Control System
PIOSA Pan-Indian Ocean Science Association
PIOTA Post Irradiation Open Test Assembly
Proximity Instrumented Open Test Assembly
PIOU Parallel Input/Output Unit
PIP Path Independent Protocol
Payload Integration Plan
Peripheral Interchange Program
Personal Identification Project
Petroleum Incentives Program [Canada]
Picture-in-Picture
Pipeline Image Processor
Plant-in-Place
Plug-in Programmer
Polar Information Program [of National Science Foundation, US]
Pollution Information Project [of NRC, Canada]
Position Indicating Probe
Positive Incentive Program
Poured-in-Place
Predicted Impact Point
Primary Indicating Position
Probabilistic Information Processing

	Problem Input Preparation
	Product Improvement Plan
	Programmable Integrated Processor
	Project on Information Processing
	Prototype Inlet Piping
	Pulsed Integrating Pendulum
PIPA	Pacific Industrial Property Association [US]
	Pulsed Integrating Pendulum Accelerometer
PIPE	Points, Income, Personnel Expense
PIPECO	Photoion Photoelectron Coincidence
PIPES	1,4-Piperazinebis(ethanesulfonic Acid)
PIPO	Parallel-In, Parallel-Out
PIPR	Plant-in-Place Records
PIPS	Passivated Implanted Planar Silicon
	Pattern Information Processing System
	Precision Ion Polishing System
PIQ	Parallel Instruction Queue
PIR	Petrolite Irradiation Reactor [US]
	Program Incident Report
PIRA	Paper Industries Research Association
	Printing and Packaging Industries Research Association [UK]
PIRD	Program Instrumentation Requirement Document
PIREP	Pilot Report [also: Pirep or pirep]
PIRETS	Pittsburgh Retrieval System [of University of Pittsburgh, US]
PIRG	Public Interest Research Group
PIRI	Paint Industries Research Institute [South Africa]
PIRINC	Petroleum Industry Research Foundation, Inc. [US]
PIRN	Preliminary Interface Revision Notice
PIROGAS	Plasma Injection of Reducing Overheated Gas System
PIRS	Personal Information Retrieval System
PIRT	Precision Infrared Tracking
	Precision Infrared Triangulation
	Public Information Retrieval Terminal
PIRV	Programmed Interrupt Request Vector
PIS	Patent Inventor Service [of INPADOC, Austria]
PISA	Portable Information System Architecture
	Public Interest Satellite Association
PISC	Petroleum Industry Security Council [US]
PISH	Program Instrumentation Summary Handbook
PISO	Parallel-In, Serial-Out
	Polyimidesulfone
PISW	Program Interrupt Status Word
PIT	Peripheral Input Tape
	Principal, Interest and Taxes
	Printing and Information Technology (Division) [of PIRA, UK]
	Processing Index Terms
	Program Instruction Tape
	Programmable Interval Timer
	Protection Identification Key
PITA	Parison Inflation Thinning Analysis
PITAC	Pakistan Industrial Technical Assistance Center
PITB	Petroleum Industry Training Board [UK]
PITC	Phenylisothiocyanate
PITCOM	Paliamentary Information Technology Committee [UK]
PITI	Principal, Interest, Taxes and Insurance
PITS	Passive Intercept Tracking System
	Petroleum Industry Training Service
PITTCON	Pittsburgh Conference (and Exposition) [also: PITT-CON, US]
PIU	Path Information Unit
	Plug-In Unit
	Process Interface Unit
	Programmable Interface Unit
PIUMP	Plug-In Unit Mounting Panel
PIUS	Process Inherent Ultimately Safe
PIV	Peak Inverse Voltage
	Product Inspection Verification
PIW	Program Interrupt Word
PIX	Picture
PIXE	Particle-Induced X-Ray Emission
	Particle-Induced X-Ray Excitation
	Photon-Induced X-Ray Emission
	Proton-Induced X-Ray Emission
PIXEL	Picture Element [also: PEL, pel or pixel]
PIXI	Professional Industrial X-Ray Imaging
pJ	picojoule
PJA	Pakistan Jute Association
PJE	Parachute Jumping Exercise
PJMA	Pakistan Jute Mills Association
PK	Pack
	Park
	Peak
	Polymerization Kinetics
	Protein Kinase
	Pyruvate Kinase
pK	(potential of K) — Negative Logarithm of the Equilibrium Constant K for a Specified Reaction
pk	peck [Unit]
PKC	Protein Kinase C
PKD	Partially Knocked Down
	Programmable Keyboard Display
PKG	Package
	Packaging
	Packing
PK/LDH	Pyruvate Kinase/Lactic Dehydrogenase
PKR	Packer
	Pakistan Rupee [Currency]
PKT	Packet
PKU	Phenylketonuria
PK/VAL	Peak-to-Valley (Ratio)
PKWAY	Parkway
PL	Panoramic Lens
	Partial Loss
	Parting Line
	Parts List
	Payload
	Perception of Light
	Phase Line
	Phospholipid
	Photoluminescence
	Physical Limnology
	Pile
	Piping Loads
	Place
	Plant
	Plastic Limit
	Plate

	Plug
	Polarized Light
	Portable Low-Power Plant [US]
	Port of London [UK]
	Position Line
	Power Level
	Pressurizer Level
	Private Line
	Procedural Language
	Production Language
	Professional Librarian
	Program Library
	Program Logic
	Programming Language
	Programming Logics
	Proportional Limit
	Proximity Log
	Public Law
	Public Library
	Pulse
	Pulsed Laser
P&L	Profit and Loss [also: P/L]
Pl	Poiseuille
PLA	Pakistan Librarians Association
	Phase Locked Arrays
	Physiological Learning Aptitude
	Plain Language Address
	Polish Librarians Association
	Polylactic Acid
	Polylalamine
	Port of London Authority [UK]
	Power Lever Angle
	Practise Low Approach
	Print Load Analyzer
	Product Liability Act
	Programmable Line Adapter
	Programmable Logic Array
	Programmed Logic Array
	Proton Linear Accelerator
	Public Library Association [US]
	Pulverized Limestone Association [US]
PLAAR	Packaged Liquid Air-Augmented Rocket
PLACE	Post-Landsat Advanced Concept Evaluation
	Programming Language for Automatic Checkout Equipment
PLACO	Planning Committee [of ISO, Switzerland]
PLAD	Parachute Low-Altitude Delivery
PLADS	Parachute Low-Altitude Delivery System
PLAN	Parts Logistics Analysis Network
	Program Language Analyzer
	Programming Language Nineteen-Hundred
PLANES	Programmed Language Enquiry System
PLANET	Planned Logistics Analysis and Evaluation Technique
	Private Local Area Network [UK]
PLANIT	Programming Language for Interactive Teaching
PLANN	Plant Location Assistance Nationwide Network [US]
PLANS	Program Logistics and Network Scheduling System
PlantE	Plant Engineer
PLAP	Pulsed-Laser Atom Probe
PLA/PGA	Polylactic Acid/Polyglycolic Acid (Copolymer)
PLAQ	Planned Quantity
PLAS	Plaster
	Plastic(s)
	Private Line Assured Service
	Program Logical Address Space
PLASI	Pulse Light Approach Indicator System
PLASMEX	International Plastics Machinery, Equipment and Materials Exhibition
PLASTEC	Plastics Technical Evaluation Center [US Army]
	Plastics Technology Information Analysis Center
PLASTEUROTEC	European Group of Fabricators of Technical Plastics Parts [Belgium]
PlasTIPS	Plastics Training and Information to Promote Safety
PLAT	Pilot Landing Aid Television
PLATF	Platform
PLATO	Programmed Learning Automatic Teaching Organization [US]
	Programmed Logic for Automatic Teaching Operations
PLATR	Pawling Lattice Test Rig [US]
PLB	Personal Locator Beacon
PLBG	Plumbing
PLBR	Prototype Large Breeder Reactor
PLC	Pacific Logging Congress [US]
	Pattern Length Coding
	Pneumatic Logic Circuit
	Power Line Carrier
	Power Line Conditioner
	Power Line Cycle
	Prime Level Code
	Programmable Line Controller
	Programmable Logic Controller
	Programming Language Committee [of CODASYL]
	Public Limited Company [also: plc]
PLCA	Parallel Line Communication Adapter
	Pipeline Contractors Association [US]
PLCAC	Pipeline Contractors Association of Canada
PLCB	Pseudo-Line Control Block
PLCC	Plastic Leaded Chip Carrier (Package)
PLCD	Product Liability Common Defense
PLC I/O	Programmable Logic Controller Input/Output (Bus)
PLCU	Programmable Logic Control Unit
PLD	Partial Line Down
	Periodic Lattice Distortion
	Phase-Locked Demodulator
	Programmable Logic Device
	Pulsed Laser Deposition
	Pulse-Length Discriminator
PLDT	Philippines Long Distance Telephone Company
PLDTS	Propellant Loading Data Transmission System
PLE	Photoluminescence Excitation
PLEA	Pacific Lumber Exporters Association [US]
PLEM	Pipeline and Manifold
PLENCH	Pliers and Wrench (Combination)
PLES	Photoluminescence Excitation Spectroscopy
PLF	Page Length Field
	Parachute Landing Fall
	Plant Load Factor
PL GL	Plate Glass
PLH	Pituitaries Luteinizing Hormone
PLI	Private Line Interface

PL/I Programming Language One [also: PL/1]
PLIANT Procedural Language Implementing Analog Techniques
PLIB Pacific Lumber Inspection Bureau [US]
Program Library
PLIM Post-Launch Information Message
PLJ Permanent Loop Junctor
PLL Peripheral Light Loss
Phase-Locked Loop
PLM Passive Line Monitor
Pulse-Length Modulation
PL/M Programming Language for Microprocessors
PLMBG Plumbing
PLMBR Plumber
PLNCIAWPRC Polish National Committee of the International Association on Water Pollution Research and Control
PLND Planned
PLNG Planning
PLO Phase-Locked Oscillator
Process Limit Option
Program Line Organization
PL/1 Programming Language One [also: PL/I]
PLOCAP Post Loss-of-Coolant Accident Protection
PLOD Planetary Orbit Determination
PLOO Pacific Launch Operations Office [of NASA]
PLOP Pressure Line of Position
PLOTG Plotting
PLP Partners for Livable Places [US]
Pattern Learning Parser
Periodate-Lysine-Paraformalin
Portable Low-Power Plant [US]
Presentation Level Protocol
Procedural Language Processor
Programming Language Processor
PLPA Pageable Link-Pack Area
Permissible Low-Pressure Alarm
PL PATH Plant Pathology
PL PHYS Plasma Physics
PLPS Propellant Loading and Pressurization System
PLQ Photoluminescence Quenching Spectroscopy
PLR Program Lock-In Register
Pulse Link Relay
Pulse Link Repeater
PLRACTA Position Location Reporting and Control Tactical Aircraft
PLRS Position Location Reporting System
PLRTY Polarity
PLS Partial Least Squares
Physical Signalling
Plugging Switch
Positive Lubrication System
Positron Lifetime Spectroscopy
Private-Line Service
Professional Land Surveyor
Programmable Limit Switch
Propellant Loading System
Publishers Licensing Society [UK]
Pure Live Seed
PLSA Partner Launch Site Agreement [NASA/Canadian Space Agency]
Pulsed Laser Spectrum Analyzer
PLSS Portable Life-Support System
Precision Location Strike System
Prelaunch Status Simulator
PLSTG Plastering [also: PLST]
PLSTC Plastic
PLT Pilot
Plant
Power-Line Transient
Princeton Large Torus [US]
Program Library Tape
Programming-Language Theory
PLTC Power-Limited Tray Cable
PLTF Purple Loosestrife Task Force [US]
PLTG Plating
PLTTY Private Line Teletypewriter Service
PLU Partial Line Up
Primary Logical Unit
PLUM Priority Low-Use Minimal
PLUOT Parts Listing and Used On Technique
PLUS Program Language for User Systems
Program Library Update System
PLUTO Pipeline Under the Ocean
Plutonium Loop Testing Reactor
PLV Pitch Line Velocity
PLYWD Plywood
PLZ Pb (= Lead)-Lanthanum-Zirconate (Ceramics)
Polish Zloty [Currency]
Programming Language Zilog
PLZT Pb (= Lead)-Lanthanum-Zirconate-Titanate (Ceramics)
PM Package Monitor
Paleomagnetism
Palmgren-Miner (Rule)
Panel Meter
Paramagnet
Paramagnetic
Pavement Material
Performance Management
Performance Monitoring
Permanent Magnet
Permanent Mold
Per Thousand
Phasemeter
Phase Modulation
Physical Metallurgist
Physical Metallurgy
Photographic Material
Photomagnetic
Photomechanical
Photomicrograph(y)
Photomultiplier
Plane Matching Interface
Plasma Membrane
Polarization Microscope
Polarization Modulation
Polymeric Material
Polymethylene
Portable Medium-Power Plant [US]
Postmaster
Post Mortem

	Potentiometer
	Pounds per Minute
	Powdered Medium
	Powder Magnet
	Powder Metal
	Powder Metallurgy
	Precious Metal
	Premium
	Presentation Manager
	Preventive Maintenance
	Procedures Manual
	Processor Module
	Project Manager
	Pulsating Mixing
	Pulse Modulation
	Pyrometallurgist
	Pyrometallurgy
P/M	Parts per Million [also: PPM or ppm]
	Powder Metallurgy
Pm	Promethium
pm	picometer
	(post meridiem) — afternoon
	(post mortem) — after death
PMA	Pacific Maritime Association [US]
	Parts Manufacturing Authority
	Pencil Makers Association [US]
	Permanent Magnet Association
	Permanent Management Arrangements
	Petroleum Monitoring Agency
	Pharmaceutical Manufacturers Association [US]
	Phenylmercuric Acetate
	Phorbol Myristate Acetate
	Phosphomolybdic Acid
	Photo Marketing Association International [US]
	Physical Medium Attachment
	Physical Memory Address
	Physical Metallurgy
	Polymetylacrylate
	Polyurethane Manufacturers Association [US]
	Powder Metallurgy Association [India]
	Precision Measurements Association [US]
	Precision Metalforming Association [US]
	Premarket Approval
	Prime Macro-Assembler
	Priority Memory Access
	Priority Memory Address
	Produce Marketing Association [US]
	Production and Marketing Administration [US]
	Production Monitoring Analysis
	Protected Memory Address
	Pyromellitic Acid
PM-A	Polishing Medium — Abrasive
PMAA	Petroleum Marketers Association of America [US]
PMAC	Pharmaceutical Manufacturers Association of Canada
	Phenylmercuric Acetate
	Purchasing Management Association of Canada
PMAD	Polymer Modifiers and Additives Division
PMAF	Pharmaceutical Manufacturers Association Foundation
PMAI	Powder Metallurgy Association of India
PMAR	Page Map Address Register
PMB	Pentamethylbenzene
	PHARE (= Program for Harmonized Air Traffic Management Research in Europe) Management Board [of EUROCONTROL]
	Physical Metallurgy Branch
	Pilot Make Busy
	Plant Molecular Biology
	Print Measures Bureau [Canada]
	PROM (= Programmable Read-Only Memory) Memory Board
PMBA	Plant Molecular Biology Association [now: ISPMB, US]
PMBC	Pilot Make Busy Call
	Plywood Manufacturers of British Columbia [Canada]
PMBX	Private Manual Branch Exchange
PMC	Phenylacetamidocoumarin
	Polymer-Matrix Composite
	Polymer-Metal Composite
	Polymer-Reinforced Matrix Composite
	Post Master's Certificate
	Private Meter Check
	Process Monitoring and Control
	Programmable Machine Controller
	Program Management Control
	Program Marginal Checking
	Project Management Corporation
	Pseudo Machine Code
PMCB	Pentamethyl Chlorobenzene
PMCG	Pyrrolidylmethyl Cyclopentylphenylglycolate
PMCS	Pulse-Modulated Communications System
PMD	Parts Manufacturing Division
	Point of Maximum Definition
	Pollution Measurement Division [of ETC, Canada]
	Post Mortem Dump [Computer Operation]
	Program Module Dictionary
PMDA	Photographic Manufacturers and Distributors Association [US]
	Pyromellitic Dianhydride
PMDA/ODA	Pyromellitic Dianhydride/Oxydianiline
PMDI	Polydiphenylmethanediisocyanate
	Polymethylene Diisocyanate
	Polymethylene Diphenylene Isocyanate
PMDR	Phosphorescence Microwave Double Resonance
PMDS	Plant Management, Maintenance/Design Engineering Show [Canada]
	Power Management and Distribution System
PME	Photomagnetoelectric(s)
	Photomagnetoelectric Effect
	Photomechanical Effect
	Precision Measuring Equipment
	Process and Manufacturing Engineering
	Processor Memory Enhancement
	Protective Multiple Earthing
PMEE	Prime Mission Electronic Equipment
PMEF	Petroleum Marketing Education Foundation [US]
PMEL	Precision Measuring Equipment Laboratory
PMF	Probable Maximum Flood
PMFC	Pacific Marine Fisheries Commission [US]

PMG Photographic Materials Specialty Group [of American Institute for Conservation, US]
Postmaster-General
Prediction Marker Generator
PMgr Professional Manager
PMGS Predictable Model Guidance Scheme
PMH Phenylmercuric Hydroxide
Probable Maximum Hurricane
Production per Man-Hour
PMHP P-Methane Hydroperoxide
PMHS Polymethylhydrosiloxane
PMI Phosphomannose Isomerase
Plumbing Manufacturers Institute [US]
Postmaintenance Inspection
Precious Metals Institute
Preventive Maintenance Inspection
Private Memory Interconnect
Private Memory Interface
Project Management Institute
Pseudo Matrix Isolation
PMIA Precious Metals Industry Association [US]
P/MIA Powder Metallurgy Industries Association [also: PMIA]
PMIC Parallel Multiple Incremental Computer
Precious Metals Institute Conference
PMIM Powder Metal Injection Molding
PMIP Postmaintenance Inspection Pilot
PMIS Personnel Management Information System
Plant Management Information System
Precision Mechanisms in Sodium
PML Physical Memory Level
Polymer Microdevice Laboratory
Powder Metallurgy Laboratory [of MPI, Germany]
Programmable Macro Logic
PMLC Programmed Multi-Line Controller
PMM Paramagnetic Material
Parametric Modelling
Polymethylmethacrylate
Pool Maintenance Module
Programmable Microcomputer Module
PMMA Paper Machinery Makers Association
Polymethyl Methacrylate
Polymethyl Methacrylate Association [of CEFIC, Belgium]
PMMB Parallel Memory to Memory Bus
PMMC Permanent Magnetic Movable Coil
PMMI Packaging Machinery Manufacturers Institute [US]
PMN Pb (= Lead) Magnesium Niobate
Pb (= Lead) Manganese Niobate
Polymorphonuclear Neutrophil
Premanufacture Notification
PMNP Platform-Mounted Nuclear Plant
PMO Phenyl Mercury Oleate
pmol picomole [Unit]
pmol/μL picomole per microliter
PMOS Positive Metal-Oxide Semiconductor [also: P MOS or P-MOS]
P-Type Metal-Oxide Semiconductor [also: P MOS or P-MOS]
PMP Parts-Material-Packaging
Phased Manufacturing Program
Phenylmethylpyrazolone
Photomechanical Process
Polymethylpentene
Pontryagin's Maximum Principle
Portable Medium-Powder Plant [US]
Premodulation Processor
Pressure Measurement Package
Preventive Maintenance Plan
Probable Maximum Precipitation
Program Management Plan
Project Master Plan
PMPA Permanent Magnet Producers Association [now: MMPA, US]
Powder Metallurgy Parts Association [also: P/MPA]
PMPMA Plastic and Metal Products Manufacturers Association [US]
PMPO Pulse-Modulated Power Oscillator
PMPPIC Polymethylene — Polyphenyl Isocyanate
PMR Pacific Missile Range [US Air Force]
Paramagnetic Resonance
Polymerizable Monomer Reactant
Polymerized Monomeric Reactant
Portable Multiparameter Radar
Pressure Modulator Radiometer
Private Mobile Radio
Projection Microradiography
Proton Magnetic Resonance
PMRI Prairie Masonry Research Institute [Canada]
PMRL Physical Metallurgy Research Laboratories [of CANMET, Canada]
Pulp Manufacturers Research League
PMRS Plasma-Melted Rapidly Solidified (Material)
PMS Performance Management System
Phenazinemethosulfate
Physical and Mathematical Sciences
Picturephone Meeting Service [US]
Plant Monitoring System
P-Methylstyrene
Pre-Main-Sequence
Probable Maximum Surge
Processor Memory Switch
Program Management System
Project Map System
Public Message Service [US]
PMSF Phenylmethanesulfonyl Fluoride
Phenylmethylsulfonyl Fluoride
PMSRP Physical and Mathematical Sciences Research Paper
PMSX Processor Memory Switch Matrix
PMT Payment
Phenylmercaptotetrazole
Photomechanical Transfer
Photomultiplier Tube
Program Master Tape
PMTA Phosphomolybdic and Phosphotungstic Acid Mixture
PMTC Private Motor Truck Council [US]
PMTCC Private Motor Truck Council of Canada
PMTHP Project Mercury Technical History Program [of NASA]
PM/TMDE Project Manager for Test, Measurement and Diagnostic Equipment [US Army]

PMU Parametric Measurement Unit
Portable Memory Unit
Precision Measurement Unit
PMUX Programmable Multiplex
PMVI Periodic Motor Vehicle Inspection
PMW Project Manager Workbench
PMX Packet Multiplexer
Physical Modeling Extension
Private Manual Exchange
Protected Message Exchange
PN Packet Network
Paging Network
Paranode
Partition Number
Part Number
Passive Network
Patent Number
P-Doped N-Doped (Junction) [also: pn or p-n]
Perceived Noise Level
Performance Number
Phase Name
Photonegative
Photonic Network
Photonuclear
Planetary Nebula
Point of No Return
Polish Noation
Positive-Negative (Semiconductor) [also: pn or p-n]
Preliminary Notification
Prior Notice
Product Name
Programmable Network
Project Number
Promissory Note
Prompt Neutron
Propyl Nitrate
Pseudonoise
Pseudorandom Noise
Pseudorandom Numbers
Punch Only
P/N Print plus Negative
PNA Pacific North America
Packet Network Adapter
Peanut Agglutinin
Project Network Analysis
pNa (potential of Na) — Logarithm of the Na (= Sodium) Ion Concentration in a Solution [also: PNa]
PNAF Potential Network Access Facility
PNB P-Nitrobenzoate
PNBC Pacific Northwest Bibliographic Center [Canada and US]
PNBH P-Nitrobenzaldehyde Hydrazone
PNC Parity Non-Conservation
Phosphonitrilic Chloride
Police National Computer [UK]
Programmed Numerical Control
PRIMENET (= Prime Network Software Package) Node Controller
PNCC Partial Network Control Center
PNCFN Permanent Nordic Committee on Food and Nutrition [Denmark]
PNCH Punch
PNCIAWPRC Philippine National Committee of the International Association on Water Pollution Research and Control
Portuguese National Committee of the International Association on Water Pollution Research and Control
PND Pictorial Navigation Display
Positive-Negative Diode
Program Network Diagram
PNdB Perceived Noise Level in Decibels
PNDC Parallel Network Digital Computer
PNDO Partial Neglect of Differential Overlap
PNE Pacific National Exhibition
PNEC Proceedings of the National Electronics Conference
PNEU Pneumatics
PNEUROP European Committee of Manufacturers of Compressors, Vacuum Pumps and Pneumatic Tools [UK]
PNF Phosphonitrilic Fluoroelastomer
PNFI Petawawa National Forestry Institute [Canada]
PNG Percent Normalized Gradient
PNGCS Primary Navigation, Guidance, and Control System
PNGFA Pacific Northwest Grain and Feed Association [US]
PNH Pan Head (Screw)
PNI Pharmaceutical News Index [US]
Pictorial Navigation Indicator
P-NID Precedence Network In-Dialing
PNIMIC Pilot NATO Insensitive Munitions Information Center
PNIN P-Doped, N-Doped, Intrinsic, N-Doped (Transistor) [also: pnin or p-n-i-n]
Positive-Negative-Intrinsic-Negative (Transistor) [also: pnin or p-n-i-n]
PNIP P-Doped, N-Doped, Intrinsic, P-Doped (Transistor) [also: pnip or p-n-i-p]
Positive-Negative-Intrinsic-Positive (Transistor) [also: pnip or p-n-i-p]
PNL Pacific Naval Laboratory [now: DREP, Canada]
Pacific Northwest Laboratories [also: BPNL, US]
Panel
Perceived Noise Level
PNLA Pacific Northwest Library Association [US]
PNLBD Panelboard
PNM Pulse Number Modulation
PNMT Phenylethanolamine N-Methyltransferase
PNOC Philippines National Oil Company
PNO-CI Pair Natural Orbital Configuration Interaction
PNP Pick-and-Place Robot
P-Doped, N-Doped, P-Doped (Transistor) [also: pnp or p-n-p]
Positive-Negative-Positive (Transistor) [also: pnp or p-n-p]
Precision Navigation Processor
PNPF Piqua Nuclear Power Facility [US]
PNPG P-Nitrophenylglycerine
PNPG1 P-Nitrophenyl-α-D-Glycoside
PNPG3 P-Nitrophenyl-α-D-Maltotriose
PNPG7 P-Nitrophenyl-α-D-Maltoheptaoside

PNPN P-Doped, N-Doped, P-Doped, N-Doped (Transistor) [also: pnpn or p-n-p-n]
Positive-Negative-Positive-Negative (Transistor) [also: pnpn or p-n-p-n]

PNPP P-Nitrophenyl Phosphate

PNPS Plant Nitrogen Purge System
Plant Nuclear Protection System

PNR Photonuclear Reaction
Pittsburgh Naval Reactors Office [of USAEC]
Point of No Return
Prior Notice Required

PNS Pulsed Neutron Source

PNSC Packet Network Service Center
Philippine National Science Society

PNSRC Plant Nuclear Safety Review Committee

PNT Paint
Plasma Nitriding
Project Network Techniques

PNTD Personnel Neutron Threshold Detector

PNVS Pilot Night Vision System

PNW Pacific Northwest

PNWL Pacific Northwest Laboratories [also: PNL, US or BNWL, US]

PNX Private Network Exchange

PO Parallel Output
Parity Odd
Parking Orbit
Part Of
Patent Office
Patents Office [UK]
Petty Officer
Phase Oscillation
Physical Oceanography
Pilot Officer
Planetary Orbit
Polarity
Polymerizable Oligomers
Polyolefin
Positive Output
Postal Order
Post Office
Power Oscillator
Power Output
Pressure Oxidation
Printout
Production Order
Program Objective
Propylene Oxide Homopolymer
Pulse Output
Punch On
Purchase Order
Pushout

P&O Paints and Oils
Pickled and Oiled

Po Peso [Currency of Cuba]
Polonium

pO_2 partial pressure of oxygen [Biosciences]

POA Pay on Arrival
Pilot-Operated Absolute System
Point of Action
Provisional Operating Authorization

POB Peripheral Order Buffer
Persons on Board
Point of Banking
Polyoxybutylene Glycol
Post Office Box
Procurement Opportunities Board [of Supply and Services Canada]

POBATO Propellant on Board at Takeoff

POBN Pyridyloxide Butylnitrone

POB-N Polyoxybutylene Glycol Nylon-12 (Copolymer)

4-POBN 4-Pyridyl 1-Oxide Butylnitrone

POBT Polybutylene Terephthalate

POC Metal Powder Cutting
Physical Organic Chemistry
Process Operator Console
Procurement Outlook Conference
Product of Combustion
Purgeable Organic Carbon

POCP Program Objectives Change Proposal

POCS Patent Office Classification System [US]

POCSAG Post Office Code Standardization Advisory Group [UK]

POD Pay on Delivery
Pin-On-Disk
Podbielniak Analysis
Point of Origin Device
Post Office Department [US]
Power On Diagnostics (Project)
Preflight Operations Division
Probability of Detection

PODA Priority-Oriented Demand Assignment

PODAF Power Density, Assigned Frequency Band

POD/CL Probability of Detection/Confidence Level

PODEM Path-Oriented Decision Making

PODS Post-Operative Destruct System

POE Petroleum Operations Engineer
Polyoxyethylene
Port of Embarkation
Port of Entry

POED Post Office Engineering Department [UK]

POEMS Polyoxyethylene Monostearate

POEOP Polyoxyethyleneoxypropylene

POETRI Program on Exchange and Transfer of Information [of UN]

POEU Post Office Engineering Union [UK]

POF Planned Outage Factor
Point of Failure
Pulsed Optical Feedback
Pyruvate Oxidation Factor

POFA Probability of False Alarms

POGO Polar Orbiting Geophysical Observatory [US]
Programmer-Oriented Graphic Operation

POH Pilot's Operating Handbook
Planned Outage Hours

POI Parking Orbit Injection
Point of Interface
Program of Instruction [of NASA]

POIL Power Density Imbalance Limit

POIS Poisonous
Prototype On-Line Instrument System

POISE Panel on In-Flight Scientific Experiments

	Photosynthetic Oxygenation Illuminated by Solar Energy
	Pre-Operational Inspection Services Engineering
POL	Pacific Oceanographic Laboratory
	Pair Orthogonalized Lowdin
	Petroleum, Oil and Lubricants
	Poland
	Polarization
	Pole
	Polish
	Political
	Politics
	Problem-Oriented Language
	Procedure-Oriented Language
	Process-Oriented Language
	Program-Oriented Language
	Provisional Operating License
POLANG	Polarization Angle
POLEX	Polar Experiment
POLGEN	Problem-Oriented Language Generator
POLLS	Parliamentary On-Line Library Study [of UKAEA]
POLY	Polyethylene
POLYDOC	Polytechnical Documentation
POLYMEG	Polytetramethylene Ether Glycol
POLYMORPH	Polymorphous
POLYTRAN	Polytranslation Analysis and Programming
POM	Pivaloyloxymethyl Chloride
	Plate Orifice Meter
	Polymethylene Oxide
	Polyoxymethylene
	Pool Operational Module
	Printout Microfilm
POMM	Preliminary Operating and Maintenance Manual
POMS	Panel on Operational Meteorological Satellites
POMSA	Post Office Management Staff Association [UK]
POMSEE	Performance, Operational and Maintenance Standards for Electronic Equipment
PONA	Paraffins, Olefins, Naphthenes and Aromatics
PONI	Positive Output, Negative Input
POO	Post Office Order
POP	Percentage of Precipitation
	Perpendicular Ocean Platform
	Point of Presence
	Point of Purchase
	Population
	Power On/Off Protection
	Pressurizer Overpressure Protection System
	Printing-Out Paper
	Program Operating Plan
	Proof of Principle
POPA	Patent Office Professional Association [US]
POPAE	Protons on Protons and Electrons
POPAI	Point-of-Purchase Advertising Institute [US]
POPDA	Polyoxypropylene Diamine
POPI	Post Office Position Indicator
POPLINE	Population On-Line [Joint US databank of Johns Hopkins University, Columbia University and Princeton University]
POPO	Push-On, Pull-Off
POPOP	2,2'-P-Phenylene-Bis(5-phenyloxazole)
	Phenyloxazolyl Phenyl Oxazolylphenyl
POPR	Prototype Organic Power Reactor
POPS	Pantograph Optical Projection System
	Process Operating System
	Program for Operator Scheduling
POPSI	Precipitation and Off-Path Scattered Interference
	Postulate-Based Permuted Subject Index
POPSO	Piperazine-N,N'-Bis(2-Hydroxypropanesulfonic Acid)
POPUS	Post Office Processing Utility Subsystem
POR	Pacific Ocean Region
	Payable on Receipt
	Pay on Return
	Problem-Oriented Routine
PORACC	Principles of Radiation and Contamination Control
PORC	Porcelain
PORM	Plus or Minus
PORS	Portable Oil Reclamation System
PORT	Photo-Optical Recorder Tracker
	Portable
	Portugal
	Portuguese
PORV	Pilot-Operated Relief Valve
	Power-Operated Relief Valve
POS	Patent Office Society [now: PTOS, US]
	Plant Operating System
	Point of Sale
	Position
	Positive
	Primary Operating System
	Product of Sums
	Professional Operating System
POSA	Preliminary Operating Safety Analysis
POSB	Post Office Savings Bank
POSH	Permuted on Subject Headings
POSI	Parallel-Out, Serial-In
POSN	Position
POSS	Passive Optical Satellite Surveillance
	Photo-Optical Surveillance Subsystem
	Polar Orbiting Satellite System [of University of Michigan, US]
	Possession
	Possible
POSSUM	Polar Orbiting Satellite System — University of Michigan [US]
POST	Polymer Science and Technology (Database) [of CAS, US]
	Posting
	Power On Self Test
POSTEC	Program of Research in Powder Science and Technology [Norway]
POSTECH	Pohang Institute of Science and Technology [Korea]
PostGrad	Post-Graduate
PostGradDipl	Post-Graduate Diploma
POSTP	Postprocessor
POT	Paint on Tangent
	Plain Old Telephone
	Potential
	Potentiometer
	Pottery
POTC	PERT Orientation and Training Center [of USDOD]

POTF Polychromatic Optical Thickness Fringes
POTS Plain Old Telephone Service
POTUS President of the United States
POTW Publicly-Owned Treatment Works
POT W Potable Water
POU Point of Use
POUNC Post Office Users National Council [UK]
POUT Power Output
POWD Powder
POWS Pyrotechnic Outside Warning System
POWU Post Office Workers Union [UK]
Post Office Work Unit
POV Peak Operating Voltage
POWTECH International Power and Bulk Solids Technology Exhibition and Conference
POWWER Power of World Wide Energy Resources
POX Partial Oxidation
PP Pages
Pair Production
Panel Point
Parallel Processor
Parcel Post
Partial Pressure
Particle Physics
Peak-to-Peak
Peripheral Processor
Person-to-Person
Phase Plate
Philippine Peso [Currency]
Physical Processing
Physical Properties
Pilot Punch
Plane Parallel
Plane-Polarized
Plasma Physics
Plasma-Polymerized
Pole Piece
Polymer Processing
Polypropylene
Positive-Positive (Semiconductor) [also: pp or p-p]
Postage Paid
Postprocessor
Pounds of Pressure
Power Pack
Prepaid
Preprocessor
Present Position
Pressureproof
Print Position
Print/Punch
Professional Photographer
Professional Programmer
Professional Purchaser
Program Preparation
Program Product
Propeller Pitch
Propylene Plastic
Proton-Proton (Reaction)
(Punctum Proximum) — Near Point [also: pp]
Push-Pull
Pyropheophorbide
Pyrophosphate
P&P Pick and Place (Robot)
P/P Peak-to-Peak [also: P-P]
Point-to-Point
pp (per procura) — by proxy
pages
PPA Pakistan Press Association
Periodical Publishers Association [UK]
Pesticide Producers Association [US]
Phenylpyruvic Acid
Photo-Peak Analysis
Polyphenylacetylene
Polyphosphoric Acid
Princeton Particle Accelerator [US]
Princeton/Pennsylvania Proton Accelerator [US]
Professional Photographers of America
Professional Programmers Association
Ptilota Plumosa Agglutinin
ppa (per procura) — by proxy
pPa picopascal
PPAA Personal Protective Armor Association [US]
PPAAR Princeton University/Pennsylvania University Army Avionics Research [US]
PPABC Professional Photographers Association of British Columbia [Canada]
PPB Part Period Balancing
Parts per Billion [also: ppb or P/B]
Plant Population Biology
Powder Particle Boundary
Pressure Positive Breathing
Prior Particle Boundary
Professional Public Buyer
PROM (= Programmable Read-Only Memory) Programmer Board
Pyrethrins Piperonyl Butoxide
PPBC Plant Pathogenic Bacteria Committee [Hungary]
PPBE Passenger Protective Breathing Equipment
PPBF Pan-American Pharmaceutical and Biochemical Federation
PPBS Planning, Programming and Budgeting System
PPC Paperboard Packaging Council [US]
Parallel-Plate Capacitor
Peak Power Control
Piperidine Pentamethylene Dithiocarbamate
Plasma Physics Conference [of IAEA]
Pollable Protocol Converter
Polypropylene Carbonate
Print Position Counter
Pro-Personal Computer
Pulsed Power Circuit
p pct porosity in percent
PPCS Person-to-Person, Collect and Special Instruction
PPD 2,5-Diphenyl-1,3,4-Oxadiazole
Past Due Date
P-I-N (= Positive-Intrinsic-Negative) Photodiode
Plastic Product Design
Port Protection Device
Pulse-Type Phase Detector
PPDA P-Phenylene Diamine

PPDC Programming Panels and Decoding Circuits [also: PP-DC]
PPDD Plan Position Data Display
Preliminary Project Design Description
PPDP/S Pariser-Parr-Del Bene-Pople/Segal Calculations
PPDR Pilot Performance Description Record
PPDS Physical Property Data Service [of IChemE, UK]
PPD-T Poly-P-Phenyleneterephthalate
PPDTS 2-(5,6-Bis(4-Sulfophenyl)-1,2,4-Triazin-3-yl)-4-(4-Sulfophenyl)pyridine
PPE Photopyroelectric
Polyphenylene Ether
Polyphenylether
Precise Parameter Estimation
Premodulation Processing Equipment
Problem Program Efficiency
Problem Program Evaluator
Property, Plant and Equipment
PPE-HIPS Polyphenylene Ether — High-Impact Polystyrene
PPEM Parallel Plate Electron Multiplier
PPEMA Portable Power Equipment Manufacturers Association [US]
PPE-PA Polyphenylene Ether — Polyamide
PPF Payload Processing Facility [US Air Force]
PPFA Plastic Pipe and Fittings Association [US]
PPFRT Prototype Preliminary Flight Rating Test
PPG Pacific Proving Grounds [of USAEC]
Palladium-Pirani Gauge
Polypropylene Glycol [also: PG]
Preprinted Gothics
Program Pulse Generator
Propulsion and Power Generator
ppg pounds per gallon
PPGA Plastic Pin Grid Array
Professional Plant Growers Association [US]
PPH Parts per Hundred [also: pph or P/H]
Phosphopyruvate Hydratase
Pulses per Hour
PPHMDSO Plasma-Polymerized Hexamethyldisiloxane [also: PP-HMDSO]
PPI Parcel Post Insured
Pictorial Position Indicator
Planar Plug-In
Plan Position Indicator
Plastic Pipe Institute [US]
Plastic Pronged Knob
Potash and Phosphate Institute [US]
Programmable Peripheral Interface
P&PI Pulp and Paper Industry
PPi Inorganic Pyrophosphatase
PPIA Programmable Peripheral Interface Adapter
PPIB Programmable Peripheral Interface Board
PPIC Polyphenyl Isocyanate
PPIRB Pulp and Paper Industrial Relations Bureau
PPITB Plastics Processing Industry Training Board [UK]
PPIU Programmable Peripheral Interface Unit
PPL Periodic Payment License
Plane-Polarized Light
Plasma Physics Laboratory [also: PPPL, US]
Plasma Processing Laboratory [of Drexel University, US]
Polymorphic Programming Language
Preferred Parts List
Private Pilot's License [UK]
PPL(A) Private Pilot's License (Aeroplanes) [UK]
PPL(AS) Private Pilot's License (Airships) [UK]
PPL(B) Private Pilot's License (Balloons) [UK]
P²LC Personal Programmable Logic Controller [also: PPLC]
PPL(G) Private Pilot's License (Gyroplanes) [UK]
PPL(H) Private Pilot's License (Helicopters) [UK]
PPLO Pleuropneumonia-Like Organism
PPM Pages per Minute
Parts per Million [also: ppm or P/M]
Peak Program Meter
Periodic Permanent Magnet
Periodic Pulse Metering
Planned Preventive Maintenance
Press Piercing Mill
Previous Processor Mode
Pulse-Phase Modulation
Pulse-Position Modulation
Pulses per Minute
PPMA Polyphenylmethacrylate
Precision Potentiometer Manufacturers Association [US]
Pulp and Paper Machinery Association [now: PPMMA, US]
PPMMA Pulp and Paper Machinery Manufacturers Association [US]
PPM/MPM Press Piercing Mill/Multistand Pipe Mill
PPMS Program Performance Measurement System
PPMW Primary Plant Mineralized Water
PPO 2,5-Diphenyloxazole
Polyphenoloxidase
Polyphenylene Oxide
Precedence Partition and Outdegree
Prior Permission Only
Professional Photographers of Ontario [Canada]
PPOC Professional Photographers of Canada
PP of A Professional Photographers of America [US]
PPO-PS Polyphenylene Oxide — Polystyrene
PPP Parallel Pattern Processor
Parcel Processing Plant
Pariser-Parr-Pople
Peak Pulse Power
Pollution Prevention Pays (Program) [also: 3P, Canada]
Poly-P-Phenylene
Portable Plotting Package [also: P3 or P³]
Programmed Potentiostatic Pulse
PPPEA Pulp, Paper and Paperboard Export Association [US]
PPPL Princeton Plasma Physics Laboratory [US]
PPPL-TFTR Princeton Plasma Physics Laboratory — Tokamak Fusion Test Reactor [US]
PPPRF Pan-Pacific Public Relations Federation [Thailand]
PPQ Polyphenylquinoxaline
PPQA Pageable Partition Queue Area
PPR Paper
Photoplastic Recording
Prior Permission Required

	Pulses per Revolution
PPRA	Precedence Partition and Random Assignment
PPRBD	Paperboard
PPRI	Pulp and Paper Research Institute [US]
PPRIC	Pulp and Paper Research Institute of Canada
PPS	Page Printing System
	Page Processing System
	Parallel Processing System
	Parcel Processing System
	Philadelphia Programming Society [US]
	Phosphorous Propellant System
	Plant Protection System
	Polyphenylene Sulfide
	(Post Postscriptum) — Second Postscript
	Primary Power System
	Primary Propulsion System
	Product Processing System
	Programmed Processor System
	Pulses per Second [also: pps]
PPSA	Pulp and Paper Safety Association [Canada]
PPSC	Privacy Protection Study Commission [US]
	Processor Program State Control
PPSN	Present Position
	Public Packet-Switching Network
PPSP	Power Plant Siting Program [of MDNR, US]
PPSS	Public Packet Switching Service [UK]
PPSU	Polyphenylene Sulfone
	Programmable Power Supply Unit
PPT	Partial Prothromboplastin Time
	Parts per Thousand [also: ppt]
	Parts per Trillion [also: ppt or P/T]
	Pattern Processing Technology
	Phenylalanine-Pyruvate Transaminase
	Polymer Production Technology
	Precipitate
	Process Page Table
	Processing Program Table
	Pseudopotential Theory
	Punched Paper Tape
PPTD	Precipitated [also: P'P'T'D]
PPTN	Precipitation
PPTP	Poly-P-Phenyleneterephtalamide
PPTS	Pyridinium-P-Toluenesulfonate
PPTT	Parts per Ten Thousand [also: pptt]
PPTY	Property
PPU	Peripheral Processing Unit
	Preprocessor Utility
	Probe Processing Unit
P-PULSE	Position Pulse
PPV	Polarized Platen Viewer
PPW	Plane-Polarized Wave
	Plasma Powder Welding
PPWB	Prairie Provinces Water Board [Canada]
PPWC	Pulp, Paper and Woodworkers of Canada
PPX	Packet Protocol Extension
	Poly(P-Xylylene)
	Private Packet Exchange
PPY	Palladium Poly-Yne
PQ	Permeability Quotient
	Physical Quality
	Plastoquinone
	Province of Quebec [Canada]
PQA	Pharmaceutical Quality Assurance
	Protected Queue Area
PQAA	Province of Quebec Association of Architects [Canada]
PQAD	Plant Quality Assurance Director
PQC	Product Quality Control
	Production Quality Control
PQFP	Plastic Quad Flatpack
PQGS	Propellant Quantity Gauging System
PQL	Practical Qualification Limit
PQM	Proto-Qualification Model
PQR	Procedure Qualification Record
PR	Pair
	Pakistan Rupee [Currency]
	Paper-Tape Reader
	Parabolic Reflector
	Pattern Recognition
	Pawling Research Reactor [US]
	Payroll
	Penicillium Roqueforti
	Periodic Reverse
	Permanent Resident
	Petroleum Refining
	Phenol Red
	Philippine Reactor
	Physical Reader
	Pitch Ratio
	Preamplifier
	Prefix
	Preliminary Record
	Premature Release
	Press Release
	Price
	Primary Radar
	Print
	Printer
	Prism
	Private Renter
	Procurement Regulation
	Program Register
	Program Requirement
	Progress Record
	Progress Report
	Project Report
	Proportional Representation
	Pseudorandom
	Public Relations
	Puerto Rico
	Pulse Rate
	Pulse Ratio
	(Punctum Remotum) — Far Point [also: pr]
	Purplish Red
Pr	Prandtl Number
	Praseodymium
	Propyl
PRA	Paint Research Association [UK]
	Peace Research Abstracts [of IPRA, Netherlands]
	Peak Recording Accelerograph
	Pendulous Reference Axis
	Permanent Resident Alien [US]

Photochemical Research Associates
Port Rear Access
Prairie Rail Authority
Precision Axis
President of the Royal Academy
Prime Rate Access
Print Alphamerically
Probabilistic Risk Assessment
Production Reader Assembly
Program Reader Assembly
Prompt Radiation Analysis
Public Roads Administration

PRAC Pyrethrum Research Advisory Committee [Kenya]
PRADOR PRF (= Pulse Repetition Frequency) Ranging Doppler Radar
PRAGB Paint Research Association of Great Britain
PRAI Phosphoribosyl Anthranilate Isomerase
Project Research Applicable in Industry
PRAIS Pesticide Residue Analysis Information Service [UK]
PRB Panel Review Board
Population Reference Bureau [US]
Primary Resonator Block
PRBAL Previous Balance
PRBS Pseudorandom Binary Sequence
PRBSG Pseudorandom Binary Sequence Generator [of LMEC, US]
PRC Partial Response Coding
People's Republic of China
Pierce
Pittsburgh Research Center [US]
Plant Research Center [of Environment Canada]
Point of Reverse Curve
Policy Research Centers (Program) [US]
Postal Rate Commission [US]
Primary Routing Center
Procession Register Clock
Product Research Center
Programmed Rate Control
PRCA President of the Royal Canadian Academy
Public Relations Consultants Association [UK]
PRCC Pacific Rim Coal Conference
PRCF Plutonium Recycling Critical Facility
PRCS Precured Resin-Coated Sand
PRCST Precast
PRD Paper-Tape Read
Power Range Detector
Printer Description
Printer Dump
Program Requirements Data
Program Requirements Document
PRDC Power Reactor Development Company [US]
PRDDO Partial Retention of Diatomic Differential Overlap
PRDP Power Reactor Development Program [Canada]
PRDPEC Power Reactor Development Program Evaluation Committee [of AECL, Canada]
PRDS Processed Radar Display System
PRE Petroleum Reservoir Engineering
Pitting Resistance Equivalent
Polymer Reaction Engineering
Prefix
Processing Refabrication Experiment
(Federation Europeenne des Fabricants de) Produits Refractaires [= European Federation of Refractory Products Manufacturers, Switzerland]
Protective Reservation Equipment
Protein Relaxation Enhancement

PRe Pakistan Rupee [Currency]
PrE Printer Emulator
PREAMP Preamplifier [also: Preamp or preamp]
PREC Preceding
Precision
PRECIP Precipitate
PRECIS Pre-Coordinate Index System
Preserved Context Index System
PREDICT Prediction of Radiation Effects by Digital Computer Techniques
PREF Preface
Preference
Prefix
Propulsion Research Environmental Facility
PREFAB Prefabrication
PRELIM Preliminary
PRELIMS Preliminary Matter
PRELORT Precision Long-Range Tracking Radar
PREM Premium
PREMID Programmable Remote Identification
PREP Plasma Rotating Electrode Process
Preparation
Preparatory
Programmed Educational Package
PREPD Prepaid
Prepared
PREPN Preparation
PREPP Process Experimental Pilot Plant
PRES President
Pressure
Program Reporting and Evaluation System
PRESFR Pressure Falling Rapidly
PRESRR Pressure Rising Rapidly
PRESS Pacific Range Electromagnetic Signature Studies
Pressure
PRESSAR Presentation Equipment for Slow-Scan Radar
PRESTO Program for Rapid Earth-to-Space Trajectory Optimization
Program Reporting and Evaluation System for Total Operations
PREV Previous
PREXTEND Prestel Extended
PRF Plastics Recycling Foundation [US]
Plywood Research Foundation [US]
Potential Requirements File
Proof
Precision Roll Feed
Processor Request Flag
Pulse Recurrence Frequency
Pulse Repetition Frequency
PRFD Pulse Recurrence Frequency Discrimination
PRFCS Pattern Recognition Feedback Control System
PRFL Pressure-Fed Liquid
$Pr(FOD)_3$ Tris(1,1,2,2,3,3,3-Heptafluoro-7,7-Dimethyl-4,6-Octanedionato)praseodymium

PRFS Pulse Recurrence Frequency Stagger
PRFU Processor Ready for Use
PRH Prolactin-Releasing Hormone
$Pr(HFC)_3$ Tris[(heptafluoropropylhydroxymethylene)camphorato],-Praseodymium
PRHQ Pacific Region Headquarters
PRI Paleontological Research Institution [US]
Petroleum Recovery Insitute [Canada]
Photo Radar Intelligence
(Rubber) Plasticity Retention Index
Plastics and Rubber Institute [UK]
Primary [also: PRIM]
Priority
Project Readout Indicator
Public Relations Institute
Public Research Institute
Pulse Rate Indicator
Pulse Repetition Interval
PRIBA President of the Royal Institute of British Architects
P-Rib-PP Phosphorylribose Pyrophosphate
PRICM Pacific Rim International Conference on Advanced Materials and Processing [PR China]
PRIDE Programmed Reliability in Design
Programmed Reliability in Design Engineering
PRIH Prolactin-Release-Inhibiting HormonePRI/IRL
Public Research Institutes/Industrial Research Laboratories (Scheme) [UK]
prim primary [Chemicals]
PRIME Precision Integrator for Meteorological Echoes
Precision Recovery Including Maneuverable Entry
Program Independence, Modularity and Economy
Programmed Instruction-Form Management Education
PRIMENET Prime Network Software Package
PRIMO Platform Repairs Inspection Maintenance Offshore
PRIMORDIAL Primary Order Dialling [of ESA-IRS, France]
PRIMOS Prime Operating System
PRIN Principal
Principle
PRINCE Parts Reliability Information Center [of NASA]
Programmed Reinforced Instruction Necessary to Continuing Education
PRINS Process Information System
PRIO Priority
PRIS Propeller Revolution Indicator System
PRISA Public Relations Institute of Southern Africa
PRISE Program for Integrated Shipboard Electronics
PRISM Parallel Reduced Instruction Set Multiprocessor
Personnel Record Information System [UK]
Personal Records Information System Management
Programmable Interactive System
Programmed Integrated System Maintenance
PRISMA Prozesse im Schadstoffkreislauf Meer-Atmosphere [= Processes in the Pollution Cycle Sea-Atmosphere (Program), Germany]
PRK Phase Reversal Keying
PRKG Parking
PRL Parallel
Periodical Requirements
Philips Research Laboratories [Netherlands]
Photoreactivating Light
Prairie Regional (Research) Laboratory [of NRC, Canada]
Print Lister
PRLC Pittsburgh Regional Library Center [US]
PRM Parameter
Personal Radiation Monitor
Power Range Monitor
Pressure Remanent Magnetization
Process Radiation Monitor
Pulse Rate Modulation
PRMLD Premold
PRN Printer
Printer Name
Print Numerically
Pseudorandom Noise
Pulse Ranging Navigation
PRNC Puerto Rico Nuclear Center
PRNET Packet Radio Network
PRNTR Printer
PRO Print Octal
Procedure and Requirements Overview (Working Group) [of ICAO]
Professional
Programmable Remote Operation
Provincial Research Organization [Canada]
Public Record Office [UK]
Public Relations Officer
Pro Proline
Prolyl
PR-1 Print Register 1
PROB Probability
Problem
PROC Process(ing)
Processor
Proceedings
Procedure
Programming Computer
Procurement
PROCD Processed
PROCLIB Procedure Library
PROCOM Procedures Committee [of IEEE, US]
PROCOMP Process Computer
Program Compiler
PROCON Protocol Converter
PROCR Processor
PROD Preferred Orientation and Displacement
Produce
Product(ion)
PRODAC Production Advisers Consortium [UK]
Programmed Digital Automatic Control
ProdEng Production Engineer [also: ProdE]
PRODR Producer
PROF Professional
Prof Professor
PROFAC Propulsive Fluid Accumulator
PROFI Project Financing
PROFILE Programmed Functional Indices for Laboratory Evaluation
PROFIT Program for Financed Insurance Techniques

	Programmed Reviewing, Ordering, and Forecasting Inventory Technique
PROFS	Professional Office System
PROG	Prognosis
	Prognostic
	Program
	Programmer
	Programming
	Progress
	Progression
PROG CNTRL	Programmable Control
PROGDEV	Program Device [also: PROG DEV]
PROGINFO	Programming Information [also: PROG INFO]
PROGR	Programmer
PROJ	Project
	Projection
	Projectile
PROJACS	Project Analysis and Control System
PROLAN	Processed Language
PROLOG	Programming in Logic [also: Prolog]
PROM	Pockels Readout Optical Memory
	Pockels Readout Optical Modulator
	Programmable Read-Only Memory
PROMAT	Profit through Materials Technology
PROMIS	Problem-Oriented Medical Information System
	Process Management Information System
	Project Monitoring Information System
	Project-Oriented Management Information System
PROMO	Promotion
PROMPT	Predicast's Overview of Markets and Terminology [US]
	Production Reviewing, Organizing and Monitoring of Performance Techniques
	Program Monitoring and Planning Technique
	Program Reporting, Organization and Management Planning Technique
PROMT	Precision Optimized Measurement Time
PROMUX	Programmable Multiplexer [also: ProMux]
PRONTO	Program for Numerical Tools Operation
	Programmable Network Telecommunications Operating System
PROP	Performance Review for Operating Programs
	Planetary Rocket Ocean Platform
	Profit Rating of Projects
	Propeller [also: prop]
	Proper
	Property
	Proposition
	Proprietor
PROPHOS	1,2-Bis(diphenylphosphino)propane
PROPR	Proprietor
	Proprietary
PROPTY	Property
	Proprietary
PROS	Professional Reactor Operator Society [US]
	Prospectus
PROSAT	Promotional Satellite Project [of ESA, France]
PROSPEC	Public Record Office Specification [UK]
PROSPER	Profit Simulation Planning and Evaluation of Risk
PROSPRO	Process Systems Program
PROSUS	Program on Submicrometer Structures [US]
PROT	Protection
PROTECT	Probabilities Recall Optimizing the Employment of Calibration Time
pro tem	(pro tempore) — temporarily
PROV	Provision(al)
	Province
	Provincial
PROVER	Procurement for Minimum Total Cost Through Value Engineering and Reliability
	Procurement — Value — Economy — Reliability
PROWL	Procedure Work Log System
PROWORD	Procedure Word
PROXI	Projection by Reflection Optics of Xerographic Images
PROXYL	Tetramethylpyrrolidinyloxy
	Tetramethylpyrrolidine-N-Oxyl
PRP	Platelet-Rich Plasma
	Preparation
	Pseudorandom Pulse
	Pulse Repetition Period
PRPF	Planar Radial Peaking Factor
PRPM	Primary Power Monitor
PRPN	Propane
PRPP	5-Phosphorylribose 1-Pyrophosphate
PRPQ	Programming Request Price Quotation
PRR	Pawling Research Reactor [US]
	Philippine Research Reactor
	Puerto Rico Reactor [of PRNC]
	Pulse Repetition Rate
PRRM	Pulse Repetition Rate Modulation
PRS	Pacific Rocket Society [US]
	Paint Research Station [UK]
	Partial Response Signalling
	Pattern Recognition Society [US]
	Pattern Recognition System
	Polymer Reactor System
	Portable Rework System
	Power Relay Satellite
	Press
	Printer Resource (File)
	Process Radiation Sampler
PRST	Persist
PRT	Pattern Recognition Technique
	Personal Rapid Transit
	Petroleum Revenue Tax
	Phosphoribosyl Transferase
	Platinum Resistance Thermometer
	Portable Radio Terminal
	Portable Remote Terminal
	Primary Ranging Test
	Print [also: Prt]
	Printer
	Production Run Tape
	Program Reference Table
	Prompt Relief Trip
	Proof Testing
	Pulse Recurrence Time
	Pulse Repetition Time
PRTC	Ports Canada
PRTG	Printing
$Pr(TFC)_3$	Tris[(trifluoromethylhydroxymethylene)camphorato]praseody

Pr(THD)$_3$ Tris(2,2,6,6-Tetramethyl-3,5-Heptanedionato)praseodymium
Pr(TMHD)$_3$ Tris(2,2,6,6-Tetramethyl-3,5-Heptanedionato)praseodymium
PRTLND CEM Portland Cement
PRTM Prague Ring Tunneling Method
Printing Response-Time Monitor
PRTOT Prototype Real-Time Optical Tracker
PRTR Plutonium Recycling Test Reactor
Printer
PRTRC Printer Controller
PRTS Pseudorandom Ternary Sequence
PRT SC Print Screen [also: PrtSc, PRT SCRN or PrtScrn]
PRTY Priority
PRU Packet Radio Unit
Physical Record Unit
Programs Research Unit
PRUF Program of Research by Universities in Forestry
Program Request Under Format
PRV Peak Reverse Voltage
Pressure Reducing Valve
Pressure Regulating Valve
Pressure Relief Valve
PRVS Penetration Room Ventilation System
PRVT Production Reliability Verification Testing
PRW Paired Wire
Paper-Tape Rewind
Paper-Tape Write
Percent Rated Wattage
PRWRA Puerto Rico Water Resources Authority [US]
PS Packet Switching
Paint Spraying
Pair Spectroscope
Paleontological Society [US]
Palm Society [now: International Palm Society, US]
Parallel-to-Serial
Parity Switch
Particulate System
Passenger Steamer
Payload Specialist
Pear Shape
Periodical System (of Elements)
Permanent Secretary
Permanent Signal
Permanent Stress
Phase Separation
Phase Shift
Phasing System
Phosphosilicate
Photogrammetry Society [UK]
Photographic Science
Photosensitivity
Photosynthesis
Physical Sciences
Physical Society [UK and Japan]
Planetary Society [US]
Planning and Scheduling
Plasma Screen
Plasma Spraying
Pneumatic System
Point of Shipment
Point of Switch
Point Source
Polymer Science
Polysulfone
Polystyrene
Pore Solution
Port Store
Port Strobe
Postal Service
Postscript
Post Secondary
Potentiometer Synchro
Power Source
Power Station
Power Supply
Power System
Presentation Services
Pressure Sensor
Pressure Switch
Problem Solution
Problem Specification
Processor Status
Process Subsystem
Production Simulator
Professional Surveyor
Programming System
Program Store
Proof Stress
Proportional Spacing
Proton Synchrotron [of CERN, Switzerland]
Pulse Shaper
Pulses per Second [also: PPS or p/s]
PVC (= Polyvinyl Chloride) Sleeve
P&S Port and Starboard
P/S Point of Shipment
Processed and Sintered
P-S Parallel-to-Serial
Pressure-Sensitive
Ps Positronium
ps picosecond [also: psec]
p/s pulses per second [also: PS or PPS]
£S Sudanese Pound [Currency]
PSA Pacific Science Association [Hawaii, US]
Parametric Semiconductor Amplifier
Path Selection Algorithm
Petroleum Services Association
Philosophy of Science Association [US]
Photographic Society of America [US]
Phycological Society of America [US]
Pisum Sativum Agglutinin
Plataforma Solar de Almeria [= Solar Platform of Almeria, Spain]
Polysodium Acrylate
Portable Sanitation Association (International) [US]
Port Storage Area
Poultry Science Association [US]
Power Servo Amplifier
Prefix Storage Area
Pressure-Sensitive Adhesive
Pressure-Sensitive Adsorption

	Pressure-Swing Absorption
	Pressure-Swing Adsorption
	Problem Specification Analyzer
	Product Safety Association
	Pushdown Stack Automaton
PS&A	Plasma Sciences and Applications [IEEE Society]
PSAC	Petroleum Services Association of Canada
	President's Science Advisory Committee [US]
	Public Servants Alliance of Canada
PSAD	Prediction, Simulation, Adaption and Decision
PSAE	Philippine Society of Agricultural Engineers
PSALI	Permanent Supplementary Artificial Lighting Installation
	Permanent Supplementary Artificial Lighting of Interiors
PSAM	Partitioned Sequence Access Method
PSANS	Positional Small-Angle Neutron Scattering
PSAP	Public Safety Answering Point
PSAR	Preliminary Safety Analysis Report
	Programmable Synchronous/Asynchronous Receiver
PS-ASA	Polystyrene — Acrylonitrile-Styrene-Acrylate
PSAT	Programmable Synchronous/Asynchronous Transmitter
PSAXS	Positional Small-Angle X-Ray Scattering
PSB	Parallel System Bus
	Persistent Slip Band
	(Snow) Plough, Sweeper and Blower
	Program Specification Block
PSBL	Possible
PSBLS	Permanent Space Based Logistics System
PSBMA	Professional Services Business Management Association
PSBR	Public Sector Borrowing Requirement
	Pyridine Syrene Butadiene Rubber
PSC	Pacific Salmon Commission [US]
	Pacific Science Council
	Parasubstitution Compound
	Permanent-Split Capacitor
	Pharmacological Society of Canada
	Photosensitive Cell
	Plant Service Center
	Power Supply Circuit
	Power System Communications [IEEE PES Committee]
	Program Sequence Control
	Public Service Commission [Canada]
	Pulse Signaling Circuit
PSCC	Power System Compotation Conference
PSCF	Primary System Control Facility
	Processor Storage Control Function
PSCC	Power System Computation Conference
PSCI	Plastic Shipping Container Institute [US]
PSCL	Programmed Sequential Control Language
PSCM	Process Steering and Control Module
PSC/PES	Power System Communications/Power Engineering Society [of IEEE, US]
P-SCR	Photoelectric Silicon-Controlled Rectifier
PSD	Packet Switched Data
	Passed
	Passive Sampling Device
	Patent Search Documentation [of European Patent Office, Germany]
	Permanent Signal Detection
	Phase-Sensitive Demodulator
	Photon-Stimulated Ion Desorption
	Pore Size Distribution
	Position-Sensitive Detector
	Post Sending Delay
	Power Shutdown
	Power Spectral Density
	Power System Dynamics
	Prevention of Significant Deterioration
	Printing Systems Division
	Program Status Double-Word
	Pulse Shape Discrimination
PSDC	Permanent Signal Detection Circuit
	Public Switched Digital Capability
	Purge Sample, Detect and Calibrate
PSDD	Preliminary System Design Description
PSDDS	Pilot Switched Digital Data Service
	Public Switched Digital Data Service
PSDF	Propulsion Systems Development Facility
PSDN	Packet-Switched Data Network
	Public Switched Data Network
PSDS	Packet Switched Data Service
PSE	Pacific Stock Exchange [US]
	Packet Switching Exchange
	Periodical System of Elements
	Phytochemical Society of Europe [UK]
	Polymer Science and Engineering
	Post-Secondary Education
	Power System Engineering [IEEE PES Committee]
	Pressurized Subcritical Experiment
	Programming Support Environment
PSECT	Prototype Section
PSEE	Photostimulated Exoelectron Emission
psec	picosecond [also: ps]
PSEP	Passive Seismic Experiment Package
PSEUD	Pseudonym
PSF	Passive Solar Foundation
	Performance Shaping Factor
	Permanent Signal Finder
	Polysulfone
	Point Spread Function
	Pound-Force per Square Foot
	Pounds per Square Foot [also: psf]
	Progressive Space Forum [US]
	Provisional System Feature
PSFA	Power System Fault Analysis
PSFC	Pacific Salmon Fisheries Commission
PSFT	Plane Strain Fracture Toughness
PSG	Passage
	Passing
	Phenol Sector Group [of CEFIC, Belgium]
	Phosphorus-Doped Silica Glass
	Phosphosilicate Glass
	Publishing Systems Group [US]
PSGB	Pharmaceutical Society of Great Britain
PSGC	Puget Sound Governmental Conference [US]
PSGR	Passenger
PSI	Packetnet Systems Interface

Paid Service Indication
Pakistan Standards Institution
Paper Stock Institute [US]
Passive Solar Institute [US]
Paul Scherrer Institute [Switzerland]
Peripheral Subsystem Interface
Peripheral System Interface
Permuterm Subject Index [of ISI, US]
Personal Sequential Interface
Pharmaceutical Society of Ireland
Phase-Shifting Interferometry
Planned Start Installation
Plasma-Surface Interaction (Conference)
Pound-Force per Square Inch
Pounds per Square Inch [also: psi]
Preprogrammed Self-Instruction
Present Serviceability Index
Pressure-Sensitive Identification
Printed Subject Index
Proctorial System of Instruction
Professional Services Institute
Project Starlight International [US]
Protosynthetic Indexing

PSIA Pounds per Square Inch, Absolute [also: psia]

PSIC Passive Solar Industries Council [US]
Process Signal Interface Controller

PSID Pounds per Square Inch, Differential [also: psid]
Preliminary Safety Information Document

PSIG Pounds per Square Inch, Gage [also: psig]

PSII Plasma Source Ion Implantation

PSIM Power System Instrumentation and Measurement [IEEE Committee]

PSI-RIT Photographic Society of America/Rochester Institute of Technology [US]

PSJ Physical Society of Japan

PSK Phase-Shift Keying
Program Selection Key

PSKM Phase Shift Keyed Modulation

PSL Parallel Strand Lumber
Photographic Science Laboratory
Physical Sciences Laboratory
Power and Signal List
Problem Specification Language
Problem Statement Language
Process Simulation Language

P SL Pipe Sleeve

PSLI Packet Switch Level Interface

PSL/PSA Problem Statement Language/Problem Specification Analyzer

PSLRB Public Service and Labour Relations Board [of New Brunswick, Canada]

PSM Parallel Slit Map
Post-Synaptic Membrane
Programming Support Monitor

PSMA Power Saw Manufacturers Association [now: PPEMA, US]
Professional Services Management Association
Pyrotechnic Signal Manufacturers Association [US]

PSMD Photoselective Metal Deposition

PSMDE Pseudostationary Mercury Drop Electrode

PSMMA Plastic Soft Materials Manufacturers Association [US]

PSMR Parts Specification Management for Reliability

PSMS Permanent Section of Microbiological Standardization [of IUMS, UK]

PSMSL Permanent Service for Mean Sea Level [UK]

PSN Packet Switching Network
Packet Switching Node
Position
Print Sequence Number
Private Satellite Network
Programmable Sampling Network
Public Switched Network

PSNA Phytochemical Society of North America [US]

PSNI Pharmaceutical Society of Northern Ireland

PSNS Programmable Sampling Network Switch
Puget Sound Naval Shipyard [US]

PSO Pilot Systems Operator
Polysulfone

PSOCG Patricia Seybold's Office Computing Group [US]

P SOL Partly Soluble

PSP Packet Switching Processor
Paper-Tape Space
Pentagonal Small Particle
Performance Shaping Parameters
Permanent Sample Plot
Pierced Steel Planking
Planet Scan Platform
Planned Schedule Performance
Planned Standard Programming
Plasma Spraying
Plasmon Surface Polaritons
Polystyrylpyridine
Portable Service Processor
Power System Protection
Presending Pause
Primary Sodium Pump
Programmable Signal Processor
Program Segment Prefix
Pseudostatic Spontaneous Potential

PSPC Position Sensitive Proportional Counter

PSPGV Primary Sodium Pump Guard Vessel

PSPO Public Safety Project Office [of NRC, Canada]

PSPS Planar Silicon Photoswitch

PSQ Personnel Security Questionnaire

PSR Packed Snow on Runway
Pennsylvania State Research Reactor [of PSU, US]
Peripheral Shim Rods
Point of Safe Return
Power System Relaying [IEEE PES Committee]
Present Serviceability Rating
Pre-Soak Rail
Primary Surveillance Radar
Processor State Register
Procurement Status Report
Program Status Register
Program Support Representative
Proton Storage Ring

PSRB Post-Sintered Reaction-Bonded

PSRBSN Post-Sintered Reaction-Bonded Silicon Nitride

PSROC Physical Society of the Republic of China [Taiwan]

PSRP Physical Sciences Research Paper
PSS Packet-Switched Service [UK]
Packet-Switched System
Packed Swichstream Service [UK]
Palomar Sky Survey
Personal Signaling System
Power Supply System
Production Support System
Program Support System
Proprietary Software System
Proprietary Support System
Propulsion Support System
Public Services Satellite
PSSA Parasitological Society of Southern Africa
Pharmaceutical Society of South Africa
Photogrammetric Society of South Africa
Polystyrene Sulfonic Acid
PSSAANDPS Permanent Secretariat of the South American Agreement on Narcotic Drugs and Psychotropic Substances [Argentina]
PSSC Physical Sciences Study Committee
Public Service Satellite Consortium [US]
PSSCC Private Sector Standards Coordinating Center
PSSD Position-Sensitive Scintillation Detector
PSSG Physical Sciences Study Group
PSSN Pressure-Sintered Silicon Nitride
PSSMA Paper Shipping Sack Manufacturers Association [US]
PSSP Phone Center Staffing and Sizing Program
PST Pacific Standard Time
Paired Selected Ternary
Pair-Select Ternary
Partition Specification Table
Pesticide
Photostress Technology
Plasma Separator Tube
Point of Spiral Tangent
Point-Set Topology
Polished Surface Technique
Polycrystal Scattering Topography
Post-Stimulus Time
Pressure Sensitive Tape
Primary Surge Tank
Provincial Sales Tax [Canada]
Pst Providencia Stuartii [Molecular Biology]
PSTA Packaging Sciences and Technology Abstract
PSTC Pressure Sensitive Tape Council [US]
Public Switched Telephone Network
PSTF Pressure Suppression Test Facility
Proximity Sensor Test Facility
Pump Seal Test Facility
PSTG Postage
PSTM Photon Scanning Tunneling Microscopy
PSTN Public Switched Telephone Network
PSTV Potato Spindle Tuber Viroid
PSU Packet Switching Unit
Path Setup
Penn(sylvania) State University [US]
Peripheral Switching Unit
Polysulfone
Port Sharing Unit
Port Storage Utility
Power Supply Unit
Primary Sampling Unit
Primary Switching Unit
Program Storage Unit
PSUR Pennsylvania State University Reactor [US]
PSV Pair Shield Video
Program Status Vector
Pulverization Sous Vede
PSVT Passivate
PSW Pacific Southwest
Pipe Socket Weld
Plasmon Surface Wave
Processor Status Word
Program Status Word
PSWP Plant Service Water Pump
PSYOPS Psychological Operations [also: Psyops or psyops]
£Syr Syrian Pound [Currency]
PSYWAR Psychological Warfare [also: Psywar or psywar]
PSZ Partially-Stabilized Zirconia
PT Pacific Time
Packet Terminal
Page Table
Paleotemperature
Palladium Tube
Paper Tape
Part
Patrol Torpedo
Payment
Pay Tone
Pencil Tube
Penetrant Testing
Performance Test
Periodic Table (of Elements)
Perturbation Theory
Phase Transformation
Phase Transition
Phase Type
Photo-Technologist
Phototelegraphy
Phototransistor
Phototube
Picture Transformation
Pipe Tap
Pipe Thread
Plant Taxonomist
Plant Taxonomy
Plastics Technology
Plate Tectonics
Point
Point of Tangent
Port
Positional Tolerance
Potential Transformer
Pressure — Temperature (Diagram) [also: P-T]
Pressure Transducer
Printer Terminal
Probability Theory
Procedure Turn
Processing Time
Propellant Transfer

Proper Time
Property Table
Prothrombin Time
P-Terphenyl
Public Transport
Pulse Timer
Pulse Transformer
Punched Tape

P&T Posts and Timbers
P/T Parts per Trillion [also: PPT or ppt]
P-T Plasma Thermocouple Reactor [US]
Pt Peseta [Currency of Spain]
Platinum
pt part
pint
point [Unit]
P3 Portable Plotting Package [also: P^3 or PPP]
PTA Particle Tracking Autoradiography
Passenger Transport Association [South Africa]
Phenyltrimethylammonium
Phosphotungstic Acid
Phthalic Anhydride
Planar Turbulence Amplifier
Plasma Thromboplastin Antecedent
Plasma Transferred Arc
Preferential Trading Agreement
Primary Tungsten Association [UK]
Programmable Translation Array
Programmed Time of Arrival
Power Transfer Assembly
Public Transport Association [UK]
Pulse Torquing Assembly
Purified Terephthalic Acid
Pta Peseta [Currency of Spain]
PTAB Photographic Technical Advisory Board [US]
PTACV Prototype Tracked Air Cushion Vehicle
PTAH Phosphotungstic Acid Hematoxylin
PTAP Phenyltrimethylammonium Tribromide
P-TAPE Paper Tape
PTA-SA Passenger Transport Association — South Africa
PTB Patrol Torpedo Boat
Physical Transaction Block
Physikalisch-Technische Bundesanstalt [= Federal Institute for Physical Science and Technology, Germany]
PTBBA P-tert-Butylbenzoic Acid
PTBR Punched Tape Block Reader
PTC Pacific Telecommunications Conference [US]
Pacific Telecommunications Council [US]
Passive Thermal Control
Personal Technical Certificate
Personnel Transfer Capsule
Phase-Transfer Catalysis
Phenylthiocarbamide
Plant Test Date
Plasma Thromboplastin Component
Positive Temperature Coefficient
Postal Telegraph Cable
Post-Tensioned Concrete
Power Toggle Clamp
Product Technology Center
Programmable Thermal Controller
Programmed Temperature Cycling
Programmed Transmission Control
Progressive Temperature Compensation
Propionylthiocholine
Pulse Time Code
Ptc Pataca [Currency of Macao]
PTCD Property Table Current Date
Pt(COD)Cl$_2$ Cyclooctadieneplatinum (II) Chloride
PTCR Pad Terminal Connection Room
Positive Temperature Coefficient of Resistivity
PTCRI Patent, Trademark and Copyright Research Institute [US]
PTCS Propellant Tanking Computer System
PTD Painted
Post-Tuning Drift
Precision Twist Drill
PTDA Power Transmission Distributors Association [US]
PTDF Pacific Tuna Development Foundation [now: PFDF, US]
PT-DOS Processor Technology Disk Operating System [also: PT DOS or PTDOS]
PTE Page Table Entry
Peculiar Test Equipment
Periodic Table of Elements
Portuguese Escudo [Currency of Portugal]
Pressure-Tolerant Electronics
Professional Technical Engineer
Private
PTEC Plastics Technical Evaluation Center [US Army]
PTERM Physical Termination
PTF Phase Transfer Function
Polymer Thick Film
Programmable Transversal Filters
Program Temporary Fix
PTFE Polytetrafluoroethylene
PTFS Precision Temperature Forcing System
PTG Painting
Petrograd
Precise Tone Generator
Printing
PTGC Programmed Temperature Gas Chromatography
PTG STD Petrograd Standard
PTH Parathyroid Hormone
Phenylthiohydantoin
Plated Through-Hole
PTI Packaged Ice Association [US]
Party Identity
Pipe Test Insert
Plugging Temperature Indicator
Post-Tensioning Institute [US]
Power Tool Institute [US]
Presentation of Technical Information
Program Transfer Interface
Public Technology, Inc. [US]
PTIC Patent and Trademark Institute of Canada
PTID Property Table Issue Data
PTIDG Presentation of Technical Information Discussion Group [UK]
PTIO Pesticides Technical Information Office [Canada]
PTK Plastic Tapered Knob

	Protection Check
PTL	Pittsburg Testing Laboratory [US]
	Power Transmission Line
	Process and Test Language
PTLBD	Particleboard
PTLU	Pretransmission Line-Up
PTLY	Partly
PTM	Pennsylvania Test Methods [US]
	Performance Test Model
	Portable Terminal Monitor
	Programmable Terminal Monitor
	Programmable Terminal Multiplexer
	Programmable Timer Module
	Proof Test Model
	Pulse-Time Modulation
	Pulse Time Multiplex
PTMA	Phosphotungstic and Phosphomolybdic Acid Mixture
PTMEG	Polytetramethane Glycol
	Polytetramethylene Ether Glycol
PTML	PNPN (= Positive-Negative-Positive-Negative) Transistor Magnetic Logic
PTMO	Polytetramethane Oxide
PTM/OS	Programmable Terminal Monitor/Operating System
PTMS	P-Toluidine-M-Sulfonic Acid
PTMT	Polytetramethylene Terephthalate
	Preliminary Thermomechanical Treatment
PTN	Partition
	Plant Test Number
	Portion
	Procedure Turn
	Public Telephone Network
PTO	Part Time Operation
	Patent and Trademark Office [US]
	Please Turn Over
	Power Takeoff
	Public Telecommunications Operator
PTOS	Paper-Tape-Oriented Operating System
	Patent and Trademark Office Society [US]
PTQ	Poly-P-Tolyquinoxaline
PTP	Paper-Tape Punch
	Phase Transition Point
	Point-to-Point (Programming)
	Preferred Target Point
	Proximity Test Plug
PTPD	Property Table Previous Date
pTpT	Phosphoryl-Thymidylyl-Thymidine
PTR	Paper-Tape Reader
	Part Throttle Reheat
	Pool Test Reactor [of CRNL, Canada]
	Pool Training Reactor [US]
	Position Track Radar
	Power Transformer
	Pressure Tube Reactor [US]
	Printer
	Processor Tape Read
	Programmer Trouble Report
	Progress Through the Rank
	Proof Test Reactor [of KAPL, US]
PTRA	Power Transmission Representatives Association [US]
PTRD	Property Table Revision Data
PTRM	Partial Thermoremanent Magnetization
PTS	Permanent Threshold Shift
	Phototypesetting
	Plane Transport System
	Pneumatic Test Set
	Power Transient Suppressor
	Predicast (Inc.)'s Terminal System
	Proceed to Send
	Program Test System
	Propellant Transfer System
	Public Telephone System
	Pure Time Sharing
PTSA	P-Toluenesulfonamide
	P-Toluenesulfonic Acid
PTSI	P-Toluenesulfonyl Isocyanate
PTSP	Paper Tape Software Package
PT/SP	Pressure Tube to Spool Piece
PTSR	Pressure Tube Superheat Reactor [US]
PTT	Party Test
	Post, Telephone and Telegraph
	Post, Telephone and Telex
	Precipitation — Time — Temperature (Curve) [also: P-T-T]
	Program Test Tape
	Prothrombin Time
	Push-to-Talk
PTTC	Paper Tape and Transmission Code
PTTI	Postal, Telegraph and Telephone International [Switzerland]
PTTRN	Pattern
PTU	Package Transfer Unit
	Plumbing Trades Union [UK]
PTUC	Pacific Trade Union Community [Australia]
PTV	Passenger Transfer Vehicle
	Predetermined Time Value
	Projection Television
PTVA	Propulsion Test Vehicle Assembly
PTW	Pressure Thermit Welding
PTX	Pressure — Temperature — Concentration (Diagram) [also: P-T-X]
PTY	Party
	Proprietary
PTZ	Pentylenetetrazol
PU	Pacific University [US]
	Peripheral Unit
	Per Unit
	Physical Unit
	Pickup
	Polyurethane
	Power Unit
	Princeton University [US]
	Probe Unit
	Processing Unit
	Processor Utility
	Propellant Utilization
	Propulsion Unit
	Purdue University [US]
Pu	Plutonium
PUAS	Postal Union of the Americas and Spain [Uruguay]
PUB	Public
	Publication

	Public Utilities Board
	Publish(ed)
PUBG	Publishing
PUBL	Publication [also: PUBN]
PUBR	Publisher
PUC	Paid-Up Capital
	Program Under Control
	Processor Utility Controller
	Public Utilities Commission [Canada and US]
	Public Utilities Committee
PUCA	Public Utilities Communicators Association [US]
PU/CEES	Princeton University/Center for Energy and Environmental Studies [US]
PUCP	Physical Unit Control Point
PUCK	Propellant Utilization Checkout Kit
PUCOT	Piezoelectric Ultrasonic Composite Oscillator Technique
PUC/RJ	Pontificia Universidade Catolica do Rio de Janeiro [= Pontifical Catholic University of Rio de Janeiro, Brazil]
PUCS	Propellant Utilization Control System
PUD	Pickup and Delivery
	Public Utilities District
PUFA	Polyunsaturated Fatty Acid
PUFFS	Passive Underwater Fire Control Feasibility Study
PUFFT	Purdue University Fast FORTRAN Translator [US]
PUG	PASCAL (Computer Language) Users Group [US]
	Penta Users Group [US]
	PET (= Personal Electronic Transactor) Users Group
PUGS	Propellant Utilization Gauging System
PUJT	Programmable Unijunction Transistor
PUL	Program Update Library
PULL B SW	Pull Button Switch
PULSE	Prime User Library Service [of NPUG, US]
PULV	Pulverizer
pulv	(pulvis) — powder
PUM	Processor Utility Monitor
PUMA	Programmable Universal Manipulator
PUN	Phenolic Urethane No-Bake (Binder)
	Punch
PUNC	Punctuation
PUP	Peripheral Unit Processor
	Peripheral Universal Processor
	Plutonium Utilization Program
PUR	Polyurethane
	Procurement Request
	Purchase(r)
	Purchasing
	Purdue University Reactor [US]
Pur Ig	Purified Immunoglobulin
PURC	Public Utilities Review Commission
PURCH	Purchase
PUREX	Plutonium-Uranium Reduction and Extraction [also: Purex]
PURP	Purple
PURPA	Public Utilities Regulatory Policy Act [US]
PURSUIT	Purchaser/Supplier Information Transfer
PURV	Powered Underwater Research Vehicle
PUS	Processor Utility Subsystem
	Programming Under Stress
PUSAS	Proposed United States of America Standard
PUSH	Purchase and Use of Solar Heating
PUSS	Pilot's Universal Sighting System
PUSSY	Programmable Universal Seaching System
PUT	Phosphate Uridyl Transferase
	Pilot under Training [also: PU/T]
	Programmable Unijunction Transistor
	Program Update Tape
PV	Par Value
	Path Verification
	Peak Value
	Peripheral Vision
	Phase Velocity
	Photovoltaic
	Pilot Vessel
	Pipe Ventilated
	Plan View
	Pore Volume
	Positive Volume
	Power Valve
	Present Value
	Pressure-Control Valve
	Pressure/Velocity (Limit)
	Pressure Vessel
	Pressure — Volume (Diagram) [also: P-V]
	Pressure — Velocity (Diagram)
	Puncture Voltage
P-V	Peak-to-Valley (Ratio) [also: P/V]
PVA	Pivalic Acid
	Plan View Area
	Polyvinyl Acetate
	Polyvinyl Alcohol
PVAC	Polyvinyl Acetate [also: PVAc]
PVAL	Polyvinyl Alcohol [also: PVAl]
PVAST	Propeller Vibration and Strength Analysis
PVB	Polyvinyl Butyral
PVC	Permanent Virtual Circuit
	Photovoltaic Cell
	Pigment Volume Concentration
	Polyvinyl Chloride
	Position and Velocity Computer
	Pulse Voltage Converter
PVC/ABS	Polyvinyl Chloride/Acrylonitrile-Butadiene-Styrene
PVCBMA	Polyvinyl Chloride Belting Manufacturers Association [US]
PVC/CPE	Polyvinyl Chloride/Chlorinated Polyethylene
PVC/CPVC	Polyvinyl Chloride/Chlorinated Polyvinyl Chloride
PVC/EPDM	Polyvinyl Chloride/Ethylene Propylene Diene Monomer
PVC/EVA	Polyvinyl Chloride/Ethylene Vinyl Acetate
PVC/MBS	Polyvinyl Chloride/Methacrylate-Butadiene-Styrene
PVC/PAE	Polyvinyl Chloride/Polyaryl Ether
PVD	Paravisual Director
	Photovoltaic Device
	Physical Vapor Deposition
	Plan Video Display
	Plan View Display
PVDC	Polyvinyl Dichloride [also: PVdC]
	Polyvinylidene Chloride [also: PVdC]
PVDF	Polyvinyl Difluoride [also: PVdF]

Polyvinylidene Fluoride [also: PVdF]
PVE Polyvinyl Ether
Polyvinyl Ethyl Ether
PVER Pattern Visual Evoked Response
PVF Polyvinyl Fluoride
Polyvinylidene Fluoride
PVFM Polyvinylformal
PVGP Program Vehicle Grant Program
PVHD Peripheral Vision Horizon Device
Peripheral Vision Horizon Display
PVI Polyvinyl Isobutyl Ether
Programmable Video Interface
PVK Polyvinyl Carbazole
PVL Prevail
PVM Polyvinyl Methyl Ether
PVMA Pressure Vessel Manufacturers Association [US]
PVME Polyvinyl Methyl Ether
PVM/MA Polyvinyl Methyl Ether/Maleic Anhydride
P Vol Pore Volume
PVOR Precision VOR (= Very-High Frequency Omnirange)
Precision VOR (= Very-High Frequency Omnidirectional Radio Range)
PVP Pipelined Vector Processor
Present Value Profit
Poly-2-Vinylpyridine
Polyvinylpyrrolidone
PV-P Past Vice-President
PVPDC Polyvinylpyridinium Dichromate
PVP-I Polyvinylpyrrolidone Iodine
PVR Precision Voltage Reference
Process Variable Record
PVRC Pressure Vessel Research Committee [of Welding Research Council, US]
PVS Performance Verification System
Photovoltaic System
Polyvinyl Sulfate
Portable Vacuum Standards
Premate Verification System Test
Program Validation Services
PVSK Polyvinyl Sulfate, K (= Potassium) Salt
PVT Page View Terminal
Paint and Varnish Technologist
Paint and Varnish Technology
Personal Verifier Terminal
Physical Vapor Transport
Polyvinyl Toluene
Pressure Vessel Technology
Pressure Viscosity Test
Pressure — Volume — Temperature (Diagram) [also: P-V-T]
Private
PVTI Piping and Valve Test Insert
PVTX Pressure — Volume — Temperature — Composition (Diagram) [also: P-V-T-X]
PVTOS Physical Vapor Transport of Organic Solutions
Pvu Proteus Vulgaris [Molecular Biology]
PVW Process Validation Wafer
PW Packed Weight
Paraffin Wax
Partial Wave
Password
Peak Width
Plain Washer
Plasma Wave
Prime Western
Printed Wiring
Private Wire
Progressive Wave
Pulsewidth
pW picowatt
PWA Pacific Western Airlines
Partial Wave Analysis
Peak Width Analysis
Performance Warehouse Association [US]
Printed Wire Assembly
PWAA Paint and Wallpaper Association of America [now: NDPA, US]
PWAC Present Worth of Annual Charges
PWB Printed Wiring Board
PWBA Plane Wave Born Approximation
Printed Wiring Board Assembly
PWC Professional Women in Construction [US]
Public Works Canada
Pulsewidth Coded
PWCB Provincial Water Conservancy Bureau [Taiwan]
PWD Powder
Power Distribution
Public Works Department
Pulsewidth Discriminator
PWDR Powder
PWE Present Worth Expenditures
Pulsewidth Encoder
PWEDA Public Works and Economic Development Association [US]
PWF Pacific Whale Foundation [US]
Present Worth Factor
PWFG Primary Waveform Generator
PWG Planning Working Group [of EUROCONTROL]
PWHS Public Works Historical Society [US]
PWHT Postweld Heat Treatment
PWI Paton Electric Welding Institute [Ukraine]
Permanent Way Institution [UK]
Pilot Warning Indicator
Pilot Warning Instrument
Proximity Warning Indicator
PWIA Plane Wave Impulse Approximation
P-WIRE Private Wire [also: P-Wire]
PWL Piecewise Linear
Power Level
PWM Pokeweed Mitogen
Pulsewidth Modulation
Pulsewidth Multiplier
PWM-FM Pulsewidth Modulation — Frequency Modulation
PWN Pinewood Nematode
PWOM Peat/Water/Oil Mixture
PWP Personal Word Processor
Programmable Weld Positioner
pWp picowatt, psophometrically weighted
pWp0/pWpP picowatt, psophometrically weighted measured at a point of zero reference level
PWPT Professional Word Processing Technician
PWQO Provincial Water Quality Objectives [Canada]

PWR Power
Pressurized Water Reactor
PWR-FLECHT Pressurized Water Reactor — Full Length Emergency Core Heat Transfer
PWRS Programmable Weapons Release System
PWS Programmer Workstation
Psychological Warfare Association [US]
PWT Propulsion Wind Tunnel
pwt pennyweight**PX**
Post Exchange [US]
Pressure — Concentration (Diagram) [also: P-X]
Private Exchange
PXML Private Exchange Master List
PXSTR Phototransistor
PY Publication Year
Pyrometer
PYFM Post-Yield Fracture Mechanics
PYG Paraguayan Guarani [Currency]
PYI Presidential Yound Investigator (Awards Program) [US]
PYPH Polyphase
PYR Pyrometer
Pyr Pyridine
PYRAM Pyramid(al)
PYRO Pyrotechnic(s)
PYROMET Pyrometallurgy
PYY Peptide YY
PZ Pick-Up Zone
pz pièze [Unit]
PZC Potential of Zero Charge
PZR Pressurizer
PZT Pb (= Lead)-Zirconate-Titanate (Ceramics)
Photographic Zenith Time
Photographic Zenith Tube
Piezoelectric Transducer
Piezoelectric Translator

Q

Q — Glutamine
Glutaminyl
Q-Factor
Quadrature
Quadrillion
Quality
Quantity
Quantum
Quaternary [Geology]
Quebec [Canada]
Quenching
Query
Quetzal [Currency of Guatemala]
Queue
Question
Quiescent
Quire [Sheets of Paper]
Quotient
Transistor

q — quadrillion
quantity
quart
quarterly
quarto
quasi
quintal [Unit]

QA — Qualitative Analysis
Quality Assurance
Quantitative Analysis
Query Analyzer
Quick Acting

Q&A — Questions and Answers

QAA — Qualified Administrator Assistant

QAC — Quality Assurance Checklist

QADS — Quality Assurance Data System

QAE — Quaternary Aminoethyl

QAGC — Quite Automatic Gain Control

QAI — Quality Assurance Instruction

QAIP — Quality Assurance Inspection Procedure

QAK — Quick-Acting Knob

QAM — Quadrature Amplitude Modulation
Queued Access Method

QAM-PAM — Quadrature Amplitude Modulation — Pulse Amplitude Modulation

QAMA — Quebec Asbestos Mining Association [Canada]

QAO — Quality Assurance Office

QA&O — Quality Assurance and Operations

QAPI — Quality Assurance Program Index

QAPP — Quality Assurance Program Plan

QA/QC — Quality Assurance/Quality Control

QAR — Quality Assurance Representative
Quality Assurance Requirements
Qatari Rial [Currency of Qatar]

QAS — Question Answering System

QASK — Quadrature Amplitude Shift Keying

QASL — Quality Assurance Systems List

QASP — Quality Assurance Surveillance Plan

QAT — Quality Action Team

QAVC — Quiet Automatic Volume Control

QB — Quantum-Box (Laser)
Quick-Break

QBE — Query by Example

Q-BOP — Quiet-Quick-Quality Basic Oxygen Process [Steelmaking]

QBS — Quebec Bureau of Statistics [Canada]
Quick-Break Switch

QBT — Quad Bus Transceiver
Quadrax Biaxial Tapes

QC — Quality Circle
Quality Code
Quality Control
Quantum Count
Quantum Chemistry
Quasi-Cleavage
Quasi-Conductor
Queens College [US]
Queue Control
Quick Code
Quiesce-Completed

QCA — Quality Control Analysis
Quantitative Chemical Analysis

QCB — Queue Control Block

QCC — Quality Control Center
Quality Control Circle

QCD — Quantum Chromodynamics

QCE — Quality Control Engineer
Quality Control Engineering
Quality Control Evaluation

QCESC — Quad Cities Engineering and Science Control

QCI — Quality Counsellor to Industry
Queen Charlotte Islands [Canada]

QCM — Quantitative Computer Management
Quartz Crystal Microbalance

QCMBPT — Quasi-Degenerate Many-Body Perturbation Theory

QCO — Quebec Construction Office [Canada]

QCONF — Quantity Conversion Factor

QCP — Quality Control Protocol
Quick Connect Panel

QCPE — Quantum Chemistry Program Exchange [of Indiana University, US]

QCPSK — Quaternary Coherent Phase-Shift Keying

QC/QA — Quality Control/Quality Assurance

QCR — Quality Control Reliability
Quick Change Response

QCS — Quick-Change System

QCSEE — Quick, Clean, Short-Haul Experimental Engine
Quiet Clean STOL (= Short Takeoff and Landing) Experimental Engine

QC&T — Quality Control and Test

QCW — Quadrature Carrier Wave

QD — Quantum Dot
Quick Disconnect
Quill Drive

Q&D — Quick and Dirty

Qd Quad
QDA Quantity Discount Aggrement
QDC Quality Products, Delivered on Time at a Competitive Cost
Quick Dependable Communications
Quick Disconnect Coupling
QDPSK Quaternary Differential Phase Shift Keying
QDRI Qualitative Development Requirements Information [US Army]
QE Quality Essentials
Quantum Efficiency
Quantum Electronics
Queue Entry
Quiesce (Communication)
QEA Queue Element Area
QEAS Quantum Electronics and Applications Society [now: LEOS of IEEE, US]
QEC Quick Engine Change
QED Quantum Electrodynamics
Quick (Text) Editor
qed (quod erat demonstrandum) — which was to be demonstrated
QEDC Quebec Engineering Design Competition [Canada]
qef (quod erat faciendum) — which was to be done
QEL Quality Element
QELS Quantum Electronics and Laser Science (Conference) [US]
Quasi-Elastic Light Scattering
QEP Quality Evaluation Package
QETE Quality Engineering Test Establishment [of DND, Canada]
QF Qabel Foundation [US]
Quality Factor
Quality Form
Quantum-Film (Laser)
Quantum Fluid
Queue Full
Quick Firing
Quick Freezing
QFD Quality Function Deployment
QFE Transistor, Field Effect
QFI Qualified Flying Instructor
QFIRC Quick Fix Interference Reduction Capability
QFITC Quinolizino-Substituted Fluorescein Isothiocyanate
QFL Quebec Federation of Labour [Canada]
QFLOW Quota Flow Control
QFM Quantized Frequency Modulation
QFMA Quebec Furniture Manufacturers Association [Canada]
QFP Quad Flatpack
QFSK Quadrature Frequency Shift Keying
QG Quarternaster-General
QH Quarter-Hard [also: 1/4 H]
QHA Quasiharmonic Approximation
QHE Quinhydrone Electrode
Q-HIP Quick Hot Isostatic Pressing
QHR Quality History Records
QI Quartz Iodide
Quinoline-Insoluble
QIA Quantitative Image Analysis
QIC Quarter-Inch Cartridge
Quarter-Inch Compatibility
QIDC Quebec Industrial Development Corporation [Canada]
QIFC Quebec Institute of Floor Covering [Canada]
QIL Quad-In-Line
QIMA Queensland Institute of Municipal Administration [Australia]
QIO Queue Input/Output
QIP Quality Improvement Process
QIRC Quebec Industrial Research Center [Canada]
QIRI Quebec Industrial Relations Institute [Canada]
QISAM Queued Indexed Sequential Access Method
QIT Quality Information and Test System
QIUF Quebec Industrial Union Federation [Canada]
QL Query Language
Queue Length
QLA Quebec Library Association [Canada]
QLAP Quick Look Analysis Program
QLD Queensland [Australia]
QLDS Quick Look Data Station
QLF Quality Loss Function
QLIT Quick Look Intermediate Tape
QLR Quick Look Report
QLRB Quebec Labour Relations Board [Canada]
QLS Quebec Land Surveyor [Canada]
QLSA Queue Line Sharing Adapter
QLTY Quality [also: QLY]
QM Quadrature Modulation
Qualification Model
Quality Management
Quantum Mechanics
Quartermaster
Quick-Make
QMA Quantitative Microbeam Analysis
Queen's (University) Marketing Association [Canada]
QMC Quebec Management Council [Canada]
Queen Mary College [UK]
QMDO Qualitative Material Development Objective
QMF Query Management Facility
QMG Quartermaster-General
Quench and Melt Growth
QMI Qualification Maintainability Inspection
Quality Management Institute [US]
QMMA Quebec Metals Mining Association [Canada]
QMP Quantum-Mechanical Principles
QMQB Quick-Make, Quick-Break
QMR Qualitative Material Requirement
QMRP Quality Management Registration Program
QMS Quadrupole Mass Spectrometer
Quartermaster-Sergeant
QN Query Normalization
Question
Quotation
QNPC Qatar National Petroleum Company
QNS Quantity Not Sufficient
QNT Quantizer
QNTY Quantity
QO Quinoline Oxide
QOA Quality on Arrival
QOD Quick-Opening Device

Q1D Quasi-One-Dimensional
QORC Quadrature Overlapped Raised Cosine
QOSRC Quadrature Overlapped Squared Raised Cosine
QOS Quality of Service
QP Quantum Physics
Quarter-Phase
Quasi Peak
QPAM Quadrature Phase and Amplitude Modulation
QPC Quebec Press Council [Canada]
QPDM Quad Pixel Dataflow Manager
QPL Qualified Parts List
Qualified Products List
QPM Quality Program Manager
Quantized Pulse Modulation
Quark-Parton Model
QPP Quantized Pulse Position
Quiescent Push-Pull
QPPM Quantized Pulse Position Modulation
QPQ Quench-Polish-Quench (Technique)
QPR Quadrature Partial Response
Quality/Price Ratio
QPRI Qualitative Operational Requirement
QPRK Quadrature Partial-Response Keying
QPRS Quadrature Partial-Response Signalling
QPS Quantitative Physical Science
QPSK Quaternary Phase-Shift Keying
QPSX Queued Package Synchronous Exchange
QPT Quadrant Power Tilt
QQ Quill and Quire
qqr (quae vide) — which see
QR Quadrupole Radiation
Quality and Reliability
Quantity Received
Quarter
Quasi-Random
Quick Reaction
qr quarter [Unit]
QRA Quality Reliability Assurance
Quick Reaction Alert
QRBHCA Quebec Road Builders and Heavy Construction Association [Canada]
QRBM Quasi-Random Band Model
QRC Quick Reaction Capability
QRCA Qualitative Research Consultants Association [US]
QRCC Query Response Communications Console
QRD Quarter-Round [also: 1/4 RD]
QRI Qatari Rial [Currency of Qatar]
Qualitative Requirements Information
Quaternary Research Institute [of University of Waterloo, Canada]
QRi Qatari Rial [Currency of Qatar]
QRL Quick Relocate and Link
QRP Query and Reporting Processor
QRPG Quebec Rubber and Plastics Group [Canada]
QRS Quality and Reliability
Quantity Received
Quarter
Quasi-Random
Quick Reaction
QRSS Quasi-Random Signal Source
QRT Queue Run Time
QRTLY Quarterly
qr tr troy quarter [Unit]
QRY Quarry
QS Quadrupole Splitting
Quantity Surveyor
Quantum Solid
Quarter Session
Quartz Spectrograph
Quasi-Static
Quasi-Stationary
Quasisteady
Queueing System
Query System
Queue Select
QSA Quad Synchronous Adapter
Quantitative Surface Analysis
QSAM Quadrature Sideband Amplitude Modulation
Queued Sequential Access Method
QSATS Quiet Short-Haul Air Transportation System
QSE Qualified Scientists and Engineers
Quantum Size Effect
QSEE Quiet STOL (= Short Takeoff and Landing) Experimental Engine
QSF Quasistatic Field
Quasisteady Flow
QSG Quasi-Stellar Galaxy
QSL Quebec Safety League [Canada]
Queneau-Schuhmann-Lurgi (Process)
Queue Search Limit
QSO Quasi-Stellar Object
QSOP Quadripartite Standing Operating Procedures
QSP Quench Spray Pump
QSPP Quebec Society for the Protection of Plants [Canada]
QSR Quarterly Statistical Report
Quick-Set Recording
QSRA Quiet STOL (= Short Takeoff and Landing) Research Aircraft
QSRMC Quality Scheme for Ready Mixed Concrete [UK]
QSRS Quasi-Stellar Radio Source
QSS Quasi-Stellar Source
Quench Spray Subsystem
QST Quasistationary Theory
QSTAGS Quardripartite Standardization Agreements
QSTNRY Quasistationary
QSTOL Quiet Short Takeoff and Landing
QT Quadruple Thermoplastic (Wire)
Qualifier Type
Quantity
Quantum Theory
Queuing Time
Queuing Theory
Quotient
Transistor, Tetrode
Q&T Quenched and Tempered [also: QT]
qt quart [Unit]
QTA Quebec Trucking Association [Canada]
QTAM Queued Telecommunications Access Method
Queued Terminal Access Method
QTB Quarry Tile Base
QTD Quartered

Q-2D Quasi Two-Dimensional
QTEST Quantitative Test [also: QTest or Qtest]
QTF Quarry Tile Floor
QTH Queued Transaction Handling
QTIPS Quick Turn-Around Digital Image Printer Subsystem
QTM Quantitative Television Microscope
QTO Quarto
QTP Quality Test Plan
QTR Quarry Tile Roof
Quarter
QTRCD Quarter Code
QTS Quarter Turn Screw
QTY Quantity
QTYOH Quantity on Hand
QTYOO Quantity on Order
QTYOR Quantity Ordered
QTY/PKG Quantity per Package
QTYSH Quantity Shipped
QTZ Quartz
QU Qinghua University [PR China]
Quarterly
Quasi
Queen's University [Canada and Northern Ireland]
Query
Question
Transistor, Unijunction
qu quart
QUAD Quadrangle
Quadrant
Quadrilateral
Quadrophonic
Quadruple
Quadruplex
QUAL Qualitative
Quality
QUALTIS Quality Technical Information Service [of UKAEA]
QUAM Quadrature Amplitude Modulation
Quantized Amplitude Modulation
QUAN Quantitative
Quantity
QUANGO Quasi-Autonomous National Government Organization
QUANT Quantitative
Quantum
QUANT MECH Quantum Mechanics
QUANT PHYS Quantum Physics
QUANTRAS Question Analysis Transformation and Search
QUAR Quarry
QUARK Quantizer, Analyzer and Record Keeper
QUART Quarter [also: QUAR]
Quarterly
QUASAR Quasi-Stellar (Object) [also: quasar]
QUB Queen's University, Belfast [Northern Ireland]
QUC Queen's University Council [Canada]
QUE Quetzal [Currency of Guatemala]
Que Quebec [Canada]
QUEENSL Queensland [Australia]
QUELD Queen's University Experiment in Liquid Metal Diffusion [Canada]
QUEN/TEMP Quenched and Tempered
QUESAR Quasi-Elliptical Self-Adaptive Rotation
QUESSI Queen's University Engineering Society Services Incorporation [Canada]
QUEST Quality Electrical Systems Test
Quantitative Experimental Stress Tomography Laboratory
Quantized Electronic Structures
Queen's University Engineering Solar Team [Canada]
Query Evaluation and Search Technique
Questioned
QUIC Quality Unit Inventory Control
QUICK Queen's University Interpretative Coder, Kingston [Canada]
QUICKTRAN Quick FORTRAN
QUICO Quality Improvement through Cost Optimization
QUIL Quad-In-Line
QUILL Queen's University Inter-Library Loan [Canada]
QUIMAL Queen's University Immiscible Alloys Experiment [Canada]
QUIP Quad-In-Line Package
Query Interactive Processor
Questionnaire Interpreter Program
Quick-In-Line Package
Quota Input Processor
QUIPS Quantimet User Interactive Processing Software [also: Quips]
QUIS Queen's University Information System [Northern Ireland]
QUO Quadex Users Organization [US]
QUOBIRD Queen's University On-Line Bibliographic Information Retrieval and Dissemination [Northern Ireland]
QUODAMP Queen's University Databank on Atomic and Molecular Physics [Northern Ireland]
QUOT Quotation
QUSB Queen's University School of Business [Canada]
QV Quasiviscous
Quench Vessel
qv (quod vide) — which see
QVT Quality Verification Testing
QW Quantum Wave
Quantum Well
Quantum-Wire (Laser)
QWB Quantum Well Box
QWG/CD Quadripartite Working Group on Combat Development
QWL Quality of Work Life
Quantum-Well Laser
QWW Quantum Well Wire
QY Query
QZE Quadratic Zeeman Effect
QZP Quadruple Zooming Power

R

R
Arginine
Arginyl
Radar
Radial
(Outside) Radius
Radical
Radio
Radioactive
Radioactivity
Railway
Rain
Rand [Currency of Namibia and South Africa]
Range
Rankine [Unit]
Rare Earth
Rate
Ratio
Rayleigh (Wave)
Reaction
Reactor
Read(er)
Reaumur
Receipt
Receiver
Recommendation
Record(er)
Red
Reflection
Reflector
Register
Relation
Reliability
Reluctance
Remote
Report
Reproducibility
Reproduction
Request
Research
Reset
Residue
Resistance
Resistor
Resolution
Restriction
Revenue
Reverse
Reynold's Number
Ring
River
Rockwell
Roentgen
Roll
Rolling
Rotation
Routine
Rubber
Ruble [Currency of Russia]
Rule
Runway
Rydberg Constant

r
radial
(inside) radius
rain
rare
ratio
reflection
refraction
remote
relative
reset
resistance
right
river
roentgen
rod

RA
Reactor Argentina [= Argentinian (Nuclear) Reactor]
Radar Altimeter
Radioactive
Radioactivity
Radio Altimeter
Radioastronomy
Radio Astronomy
Radioatmospherics
Rainforest Alliance [US]
Random-Access
Rate of Application
Rational Number
Read Amplifier
Ready-Access
Rear Access
Rear Admiral
Record Address
Recorded Announcement
Recrystallization Annealing
Recycling Agent
Reducing Adapter
Reduction of Area
Relocation Address
Renew America [US]
Rental Agreement
Repeat Attempt
Replacement Algorithm
Research Association
Reserve Alkalinity
Resorcylic Acid
Return Address
Return Air
Return Authorization
Right Ascension
Ripple Adder
Robotics and Automation
Robustness Analysis

	Root Addressable (Record)
	Rosin-Activated
	Royal Academy [UK]
	Royal Artillery
R&A	Robotics and Automation
R/A	Recorded Announcement
Ra	Radium
	Rayon
ra	rat [Immunochemistry]
RAA	Random-Access Array
	Regenerative Agriculture Association [US]
	Regional Airlines Association [US]
	Remote Access Audio
	Remote Axis Admittance
	Royal Academy of Arts [UK]
RaA	Radium A (= Polonium 218)
RAAP	Residue Arithmetic Associative Processor
	Resource Allocation and Planning
RAAR	RAM (= Random-Access Memory) Address Register
RAB	Radio Advertising Bureau [US]
RaB	Radium B (= Lead 214)
RABDF	Royal Association of British Dairy Farmers
RABFM	Research Association of British Flour Millers
RABIN	Raad van Advies voor Bibliotheekwesen en Informatieverzorging [= Netherlands Council for Libraries and Information Services]
RABPCVM	Research Association of British Paint, Color and Varnish Manufacturers
RABRM	Research Association of British Rubber Manufacturers [now: RAPRA, UK]
RAC	Racemic
	Railway Association of Canada
	Real-Time Active Controller
	Recombinant DNA (= Deoxyribonucleic Acid) Advisory Committee
	Recycling Advisory Committee
	Regional Advisory Committee
	Reliability Analysis Center [US]
	Remote Access and Control
	Remote Analysis Computer
	Remote Automatic Calibration
	Robotics and Automation Council [of IEEE, US]
	Royal Agricultural College
	Royal Automobile Club [UK]
	Rubber Association of Canada
	Rules of the Air Traffic Services
RaC	Radium C (= Bismuth 214)
RaC'	Radium C' (= Polonium 214)
RaC"	Radium C" (= Thallium 210)
RACC	Radiation and Contamination Control
	Reykjavik Area Control Center [Iceland]
RACE	Random-Access Computer Equipment
	Random-Access Control Equipment
	Rapid Acquisition and Computer Extraction
	Rapid Automatic Checkout Equipment
	Research and Development in Advanced Communications Technology for Europe
	Research in Advanced Communication in Europe
RACEP	Random-Access and Correlation for Extended Performance
RACES	Radio Amateur Civil Emergency Service
RACF	Resource Access Control Facility
RACG	Radiometric Area-Correlation Guidance
	Reaction Alternating-Current Generator
RACI	Royal Australian Chemical Institute
RACIC	Remote Area Conflict Information Center [of BMI, US]
RACKPC	Rack-Mount Personal Computer
RACON	Radar Beacon [also: Racon or racon]
RACS	Remote Access Calibration System
	Remote Access Computing System
	Remote Automatic Calibration System
RACT	Reasonably Available Control Technique
	Remote Access Computer Technique
RAD	Radar
	Radial
	Radial Detector
	Radiant
	Radiation Absorbed Dose [also: rad]
	Radiator
	Radical
	Radio
	Radioactivity Detection
	Radius
	Random-Access Data
	Random-Access Device
	Random-Access Disk
	Rapid-Access Data
	Rapid-Access Device
	Rapid-Access Disk
	Ratio Analysis Diagram
	Records Arrival Date
	Relative Air Density
	Ribbon Against Drop
RaD	Radium D (= Lead 210)
rad	radian
RADA	Random-Access Discrete Address
RADAC	Radar Analog/Digital Data and Control
	Rapid Digital Automatic Computing
RADAN	Radar Doppler Automatic Navigator
RADAR	Radio Detection and Ranging [also: Radar or radar]
RADARSAT	Radar Satellite [also: Radarsat]
RADAS	Random-Access Discrete Address System
RADATA	Radar Automatic Data Transmission and Assembly
RADATS	Radiographic Automated Testing System
RADC	Rome Air Development Center [US Air Force]
RADCM	Radar Countermeasures
RADCON	Radar Data Converter
RADD	Rapid-Access Data Drum
RADEFF	Radiation Effects
RADEM	Random-Access Delta Modulation
RADFAC	Radiation Facility
RADHAZ	Radiation Hazard
RADIAC	Radioactivity Detection, Identification and Computation [also: radiac or radiac]
RADIALS	Research and Development in Information and Library Science [UK]
RADIAN	Radial Detector for Ion Beam Analysis
RADIOCOM	Radiocommunication
RADIO SCI	Radio Science
RADIQUAD	Radio Quadrangle
RADIR	Random-Access Document Indexing and Retrieval

RADIST Radar Distantance Indicator
RADIT Radio Teletype
RADM Random-Access Data Modulation
RAdm Rear-Admiral
RADNOTE Radio Note
RADOME Radar Dome [also: Radome or radome]
RADOP Radar Operator
RADOPWEAP Radar Optical Weapons
RADOSE Radiation Dosimeter Satellite [of NASA]
RADPLANBD Radio Planning Board
RADPROPCAST Radio Propagation Forecast
RADS Radiation and Dosimetry Services
RADSCAT Radiometer-Scatterometer Sensor
RADSO Radiological Survey Officer
RADTT Radio Teletype
RADU Radar Analysis and Detection Unit
RADVS Radar Altimeter and Doppler Velocity Sensor
RADWASTE Radioactive Waste [also: Radwaste or radwaste]
RAE Radio Astronomy Explorer (Satellite)
Range Azimuth Elevation
Royal Aeronautical Establishment [UK]
Royal Aircraft Establishment [UK]
RaE Radium E (= Bismuth 210)
RaE" Radium E" (= Thallium 206)
RAeC Royal Aero Club [UK]
RAEN Radio Amateur Emergency Network [also: RAENET]
RAER Range, Azimuth, Elevation and Reproduction
RAES Remote Access Editing System
Royal Aeronautical Society [also: RAeS, UK]
RAF Royal Air Force [UK]
RaF Radium F (= Polonium 210)
RAFAX Radar Facsimile Transmission
RAFC Regional Area Forecast Center
RAFES Royal Air Force Educational Service [UK]
RAFI Rural Advancement Fund International [US]
RAFL Rainfall
RAFSC Royal Air Force Staff College [UK]
RAFT Radially Adjustable Facility Tube
RAG Residential Advisory Group
ROM (= Read-Only Memory) Address Gate
Runway Arresting Gear
RaG Radium G (= Lead 206)
RAGB Refractories Association of Great Britain
RAGS Radiatively Active Gases
Radiatively Active Gases Sites
RAGU Radio Receiving and Generally Useful
RAH Radiation Anneal Hardening
RAHS Royal Australian Historical Society
RAI Radioactive Interference
Random-Access and Inquiry
Registro Aeronautico Italiano [= Italian Aeronautical Registry]
Royal Architectural Institute [UK]
Runway Alignment Indicator
RAIA Royal Australian Institute of Architects
RAIC Redstone Arsenal Information Center [US Army]
Royal Architectural Institute of Canada
RAID Remote Access Interactive Debugger
RAIDS Rapid Availability of Information and Data for Safety
RAIL Railroad
Railroad Advancement through Information and Law (Foundation) [US]
Railway
Runway Alignment Indicator Lights
RAILS Remote Area Instrument Landing Sensor
RAIN Relational Algebraic Interpreter
RAIP Requester's Approval in Principle
RAIR Remote Access Immediate Response
Random-Access Information Retrieval
RAISON Regional Analysis using Intelligent Systems on a Microcomputer
RAJ Reverse Air Jet
RAK Read-Access Key
Remote Access Key
RAL Radio Annoyance Level
Resorcyclic Acid Lactone
Riverbend Acoustical Laboratory [US]
Robotics and Automation Laboratory [of U of T, Canada]
Runway Alignment (Beacon)
Rutherford Appleton Laboratory [UK]
RALF Rapid-Access to Literature via Fragmentation Codes [Germany]
Relocatable Assembly Language Floating Point
RALI Resource and Land Investigation
RALU Register and Arithmetic and Logic Unit
RALW Radioactive Liquid Waste
RAM Radar Absorbing Material
Radioactive Material
Radio Attenuation Measurement
Random-Access Memory
Rapid Access Manual
Real-Time Aerosol Monitor
Reliability, Availability and Maintainability
Right Ascension of the Meridian
RAMA Railway Automotive Management Association [US]
Rome Air Materiel Area [US]
RAMAC Radio Marine Associated Companies [UK]
Random-Access Method of Accounting and Control [also: Ramac]
RAMAR Random-Access Memory Address Register
RAMARK Radar Marker
RAM BIOS Random-Access Memory — Basic Input/Output Subsystem [also: RAMBIOS, RAM-BIOS or RAM/BIOS]
RAMC Royal Army Medical Corps [UK]
RAMIO Random-Access Memory plus Input/Output
RAMIS Random-Access Management Information System
Rapid-Access Management Information System
Rapid Automatic Malfunction Isolation System
RAMNAC Radio Aids to Marine Navigation Application Committee [UK]
RAMONT Radiological Monitoring
RAMP Radar Modernization Project
Radiation Airborne Measurement Program
(R)-1-Amino-2-(Methoxymethyl)pyrrolidine
Rapid Access of Manufactured Parts
Rapid Modeling Platform
Raytheon Airborne Microwave Platform
Reliability and Maintainability Program
Rural Abandoned Mine Program

RAMPART Radar Advanced Measurement Program for Analysis of Reentry Techniques
RAMPI Raw Material Price Index [Italy]
RAMPS Resource Allocation and Multi-Project Scheduling
RAM/ROM Random-Access Memory/Read-Only Memory
RAMS Random Access Measurement System
Random-Access Memory Store
Regional Air Monitoring Station
RAMSH Reliability, Availability, Maintainability, Safety and Human Factors
RAMTOP Random-Access Memory Top
RAN Read-Around Numbers
Reconnaissance/Attack Navigator
Regional Air Navigation Meeting [of ICAO]
Royal Australian Navy
RANDAM Random-Access Nondestructive Advanced Memory
RANDID Rapid Alphanumeric Digital Indicating Device
R and S Reset and Start [also: R&S]
RANK Replacement Alphanumeric Keyboard
RANN Research Applied to National Needs (Program) [of NSF, US]
RANRL Royal Australian Navy Research Laboratory
RAO Radio Astronomical Observatory
Regional Accounting Office
RAOB Radiosonde Observation [also: Raob or raob]
RAOC Royal Army Ordnance Corps [UK]
RAOU Royal Australasian Ornithologists Union
RAP Rational Alternative to Pseudosciences [Spain]
Reclaimed Asphalt Pavement
Reduced Air Pressure
Redundancy Adjustment of Probability
Regional Acceleratory Phenomenon
Relational Associative Processor
Reliable Acoustic Path
Remedial Action Program
Resident Assembler Program
Resource Allocation and Planning
Response Analysis Program
Rocket-Assisted Projectile
Rubidium Acid Phthalate
RAPBPPI Research Association for the Paper and Board, Printing and Packaging Industries [UK]
RAPC Royal Army Pay Corps [UK]
RAPCOE Random-Access Programming and Checkout Equipment
RAPCON Radar Approach Control [of US Air Force]
RAPEC Rocket-Assisted Personnel Ejection Catapult
RAPID Random-Access Personnel Information Disseminator
Reactor and Plant Integrated Dynamics
Relative Address Programming Implementation Device
Research in Automatic Photocomposition and Information Dissemination
RAPO Resident Apollo Project Office [of NASA]
RAPPI Random-Access Plan Position Indicator
RAPRA Rubber and Plastic Research Association [UK]
RAPS Radioactive Argon Processing System
Remedial Action Programs [Canada]
Remote Access Power Support
Retrieval Analysis and Presentation System
Risk Appraisal of Programs System
RAPSAT Ranging and Processing Satellite
RAPT Reusable Aerospace Passenger Transport
Robot Assembly Programming Technique
RAPTUS Rapid Thorium-Uranium-Sodium Reactor
Rapid Thorium-Uranium System
RAPUD Revenue Analysis from Parametric Usage Descriptions
RAR Radar Arrival Route
Rapid Access Recording
Return Address Register
ROM (= Read-Only Memory) Address Register
RARC Regional Administrative Radio Conference [of ITU, Switzerland]
RARDE Royal Armament Research and Development Establishment [UK]
RARE Ram Air Rocket Engine
Rare Animal Relief Effort [US]
RAREF Radiation and Repair Engineering Facility
RAREP Radar Report
RARG Regulatory Analysis Review Group
RAROM RAM (= Random-Access Memory) and ROM (= Read-Only Memory)
RArt Royal Artillery
RAS Radar Advisory Service
Radioactive Sensor
Radioactive Series
Rapid Access Storage
Reactor Alarm System
Reactor Analysis and Safety
Receptor Analysis System
Records and Analysis Subsystem
Rectified Airspeed
Reflection/Absorption Spectroscopy
Reliability, Availability and Serviceability
Remote-Access System
Remote Analysis System
Replenishment at Sea
Route Accounting Subsystem
Row-Address Strobe
Royal Agricultural Society [UK]
Royal Astronomical Society [UK]
Rutgers Annihilation Spectrometer [of Rutgers University, US]
RASC Ranking and Scaling
Regional AIS (= Aeronautical Information Service) System Center
Rome Air Service Command [US]
Royal Agricultural Society of the Commonwealth [UK]
Royal Army Service Corps [UK]
Royal Astronomical Society of Canada
RASE Royal Agricultural Society of England
RASH Rain Showers
RASI Reliability, Availability, Serviceability and Improvability
RASIS Reliability, Availability, Serviceability, Integrity and Security
RASN Rain and Snow
RASofC Royal Agricultural Society of the Commonwalth [UK]

RASP Radar Applications Specialist Panel [of EUROCONTROL]
Remote Access Switching and Patching
Retrieval and Statistics Processing
RASPO Resident Apollo Spacecraft Program Office [of NASA]
RASS Radar Analysis Support System
Rapid Area Supply Support
Research Association of Statistical Sciences [Japan]
Rotating Acoustic Stereo-Scanner
RASSR Reliable Advanced Solid-State Radar
RAST Recover Assist, Secure and Traverse System
RASTA Radiation Augmented Special Test Apparatus
Radiation Special Test Apparatus
RASTAC Random-Access Storage and Control
RASTAD Random-Access Storage and Display
RASTI Rapid Speech Transmission Index
RASU Rangoon Arts and Science University [Burma]
RAT Radioactive Tracer
Reactive Acidified Tailings
Reliability Assurance Test
Reserve Auxiliary Transformer
Rocket-Assisted Torpedo
RATA Rankine Cycle Air Turboaccelerator
RATAC Radar Analog Target Acquisition Computer
Radar Target Acquisition
RATAN Radar and Television Aid to Navigation [also: Ratan or ratan]
RATAP Reactive Tailings Assessment Program
RATC Rate-Aided Tracking Computer
RATCC Radar Air-Traffic Control Center
RATCF Radar Air-Traffic Control Facility [of US Navy]
RATE Remote Automatic Telemetry Equipment
RATEL Radio-Telephone
RATER Response Analysis Tester
RATFOR Rational FORTRAN
RATG Radioactive Thickness Gauge
Radio-Telegraph
RATIO Radio Telescope in Orbit
RATO Rocket-Assisted Takeoff [also: Rato or rato]
RATOG Rocket-Assisted Takeoff Gear
RATS Rate and Track Subsystem
Reactive Acidified Tailings Study
Reactive Acid Tailings Stabilization
RATSC Rome Air Technical Service Command [US]
RATSCAT Radar Target Scatter
RATT Radio Teletype
Radio Teletypewriter
RAU Remote Acquisition Unit
RAV Rising Arc Voltage
RAVC Royal Army Veterinary Corps
RAVE Radar Acquisition Visual-Tracking Equipment
RAVEN Ranging and Velocity Navigation
RAVIR Radar Video Recording
RAW Rationalisierungsausschuss der Deutschen Wirtschaft [= German Productivity and Management Committee]
Read After Write
RAWARC Radar and Warning Coordination
RAWIN Radar/Wind
Radar Wind Sounding
Radio/Wind
RAX Remote Access
Rural Automatic Exchange
RAXS Rapid Analyzer of X-Ray Stress
RAYDAC Raytheon Digital Automatic Computer
RAYNET Raytheon Data Communications Network
RAZEL Range, Azimuth and Elevation
RAZON Range and Azimuth Only
RB Radar Beacon
Radiobalance
Radio Beacon
Radiobiology
Reactor Building
Read Back
Read Backward
Read Buffer
Reclamation Bureau
Reducing Bushing
Reinforced Brick
Renegotiation Board
Request Block
Resistance Brazing
Rescue Boat
Rest Button
Return to Bias
Rheostat Braking
Rifle Brigade
Right Boundary
Rollback
Roller Bearing
Rubber Board [India]
R&B Ring and Ball
Rb Rubidium
Ruble [Currency of Russia]
rb rabbit [Immunochemistry]
RBA Radar Beacon Antenna
Recovery Beacon Antenna
Relative Bioavailability [also: RBAV]
Relative Byte Address
Road Bitumen Association
RBAI Royal Belfast Academical Institution [Northern Ireland]
RBB Remazol Brilliant Blue
RBBS Remote Bulletin Board System
RBC Red Blood Cell
RBCCW Reactor Building Closed Coolant Water
RBCU Rod Bank Coil Unit
RBCWS Reactor Building Cooling Water System
RBD Reliability Block Diagram
RBDE Radar Bright Display Equipment
RBE Radiation Biological Effectiveness
Relative Biological Effectiveness [also: rbe]
Remote Batch Entry
RBEC Roller Bearing Engineers Committee [US]
RBF Remote Batch Facility
RBG Royal Botanical Gardens [Canada]
RBHPF Reactor Building Hydrogen Purge Fan
RBI Radar Bearing Indicator
Radar Blip Identification
Relative Bearing Indicator
Ripple-Blanking Input

Roof-Bolt Inserter
Rudjer Boscovic Institute [Yugoslavia]
RBL Radiation Biology Laboratory
Ruble [Currency of Russia]
RBM Real Batch Monitor
Real-Time Batch Monitor
Reinforced Brick Masonry
Remote Batch Module
Rock Burst Monitor
Rod Block Monitor
RBMS Remote Bridge Management Software
RBN Radio Beacon
RBO Rapid Burn-Off
Ripple-Blanking Output
RBOC Regional Bell Operating Company
RBOF Receiving Basin for Off-Site Fuel
RBOT Rotating Bomb Oxidation Test
RBP Registered Business Programmer
Reservoir Bubble Point
RBPSSD Ruud-Barrett Position Sensitive Scintillation Detector
RBR Radar Boresight Range
Roll-Bend-Roll
Rubber
Russian Breeder Reactor
RBRG Reinforcing Bar Ring Ground
RBS Radar Bomb Scoring
Radio Base Station
Random Barrage System
Reactor Building Sump
Remote Batch Terminal
Rock Bolting and Screening (Machine)
Royal Botanical Society [UK]
Rutherford Backscattering
Rutherford Backscattering Spectrometry
RBSCU Reactor Building Sump Cooling Unit
RBSN Reaction-Bonded Silicon Nitride
RBSOA Reverse Bias Safe Operating Area
RBSS Recoverable Booster Support System
RBT Remote Batch Terminal
Resistance Bulb Thermometer
Rigid Body Translation
Ringback Tone
rbt rabbit [Immunochemistry]
RBUPC Research in British Universities, Polytechnics and Colleges [of BLLD, UK]
RBV Reactor Building Vent
Return Beam Vidicon
RC Radiation Chemistry
Radiation Cooling
Radiochemistry
Radio Compass
Radix Complement
Range Control
Range Correction
Rapid Curing
Rate Center
Rate Constant
Ray-Control Electrode
Reactance Coil
Read and Compute
Reader Code
Reactive Component
Real Circuit
Recent Change
Recording Completing
Reference Clock
Reflex Camera
Regional Center
Reinforced Concrete
Remote Channel
Remote Computer
Remote Control
Reply Check
Research Center
Resin Coating
Resistance-Capacitance
Resolver Control
Resonant Circuit
Restrained Cursor
Revenue Canada
Reverse Course
Reverse Current
Rheostatic Control
Ringing Code
Robot Control
Rocking Curve
Rotary Converter
Running Clearance
Rushlight Club [US]
R/C Radio Command
Radio Control
Range Clearance
Range Control
Rate of Climb
R-C Rapid-Curing (Cement)
Resistance-Capacitance [also: R/C]
RCA Radiochemical Analysis
Radio Club of America [US]
Radio Corporation of America [US]
Reach Climb Altitude
Reinforced Concrete Association [UK]
Research Council of Alberta [Canada]
Ricinus Communis Agglutinin
Royal Canadian Academy
Royal Canadian Artillery
Royal College of Art [UK]
RCAA Royal Canadian Academy of Arts
RCAC Radio Corporation of America Communications [US]
Remote Computer-Access Communications Service
Royal Canadian Armored Corps
RCACA Royal Canadian Armored Corps Association
RCACC Royal Canadian Army Cadet Corps
RCAF Royal Canadian Air Force [now: Canadian Armed Forces]
RCAFM Royal Canadian Air Force Museum
RCAG Remote Center Air/Ground
Remote Communications Air-to-Ground
Remote Control Air/Ground
RCAMC Royal Canadian Army Medical Corps
RCAN Recorded Announcement
RCAT Radio Code Aptitude Test

	Ridgewood College of Agricultural Technology [Canada]
RCB	Reactor Containment Building
	Resource Control Block
RCBC	Recycling Council of British Columbia [Canada]
RCC	Radiochemical Center [UK]
	Radio Common Carrier
	Radio Common Channels
	Random Correction Code
	Range Commanders Council
	Read Channel Continue
	Reader Common Contact
	Real-Time Computer Complex
	Recovery Control Center
	Remote Center Compliance
	Remote Communications Center
	Remote Communications Complex
	Remote Communications Concentrator
	Remote Communications Console
	Remote Control-Rod Cluster Assembly
	Rescue Control Center
	Rescue Coordination Center [US and Canada]
	Resistance-Capacitance Coupling
	Rod Cluster Control
	Routing Control Center
RCCB	Residual Current Circuit Breaker
RCCE	Revenue Canada, Customs and Excise [also: RC/CE]
RCCM	Remote Carrier-Controlled Modem
RCCPLD	Resistance-Capacitance Coupled
RCD	Record
	Received
	Receiver-Carrier Detector
	Regional Cooperation for Development [Turkey]
	Residual Current Device
	Route Control Digit
RCDC	Radar Course-Directing Center
RCDCD	Record Code
RCDP	Remote Community Demonstration Program
RCE	Rapid Circuit Etch
	Ray-Casting Engine
	Royal Canadian Engineers
RCEA	Research Council Employees Association
RCEEA	Radio Communications and Electronic Engineering Association [UK]
RCEI	Range Communications Electronics Instruction
RCEME	Royal Canadian Electrical and Mechanical Engineers
RCEP	Royal Commission on Environmental Pollution [UK]
RCF	Radial Centrifugal Force
	Radio Communication Facility
	Radio Communication Failure
	Recall Finder
	Refrigeration Centrifuge
	Relative Centrifugal Force
	Remote Call Forwarding
	Reverse Column Flotation
	Rolling Contact Fatigue
RCFA	Royal Canadian Field Artillery
RCFC	Reactor Core Fan Cooling
RCFCU	Reactor Core Fan Cooling Unit
RCG	Radioactivity Concentration Guide
	Relative Cooling Gain
RCGA	Royal Canadian Garrison Artillery
RCGM	Reactor Cover Gas Monitor
RCGS	Royal Canadian Geological Society
RCHA	Royal Canadian Horse Artillery
R CHG	Reduced Charge
RCHSA	Radiation Control for Health and Safety Act [Canada]
RCI	Radar Coverage Indicator
	Radio Canada International
	Rating Core Index
	Read Channel Initialize
	Remote Control Interface
	Roof Consultants Institute [US]
	Routing Control Indicator
	Royal Canadian Institute
RCIA	Royal Canadian Institute of Architects
RCIC	Reactor Core Isolation Cooling
RCICS	Reactor Core Isolation Cooling System
RCIGG	Research Center of Isotropic Geochemistry and Geochronology
RCIS	Regional Cadet Instructor School [Canada]
RCIU	Remote Computer Interface Unit
RCJ	Reaction Control Jet
RCL	Radiation Counting Laboratory
	Robot Command Language
	Royal Canadian Legion
	Runway Center Light
	Runway Centerline
RCLC	Reactor Coolant Leakage Calculation
RCLM	Runway Centerline Markings
RCLS	Runway Centerline Lighting System
RCLWUNE	Regional Commission on Land and Water Use in the Near East [of FAO, UN — Italy]
RCM	Radar Countermeasures
	Radio Countermeasures
	Reliability Corporate Memory
	Requirements Planning Criteria and Methods of Application
	Royal Canadian Mint
	Royal College of Mines [UK]
RCMA	Railroad Construction Maintenance Association [now: NRCMAI, US]
	Roof Coatings Manufacturers Association [US]
	Royal Canadian Military Institute
RCMC	Resolved Motion Rate Control
RCMD	Rice Council for Market Development [US]
RCMP	Royal Canadian Mounted Police
RCMS	Regional Center for Mass Spectrometry
RCN	Reactor Center Netherlands
	Royal Canadian Navy [now: Canadian Armed Forces]
RCNVR	Royal Canadian Naval Volunteer Reserve
RCO	Reactor Core
	Receiver Cuts Out
	Recycling Council of Ontario [Canada]
	Remote Communications Outlet
	Remote Control Office
	Remote Control Oscillator
	Representative Calculating Operation

	Restoration Control Office
RCOC	Royal Canadian Ordnance Corps
RCOM	Regional Chapter Officers Meeting [of ASM International, US]
RCP	Radiological Control Program
	Rapid Cooling Process
	Reactor Coolant Pump
	Receive Clock Pulse
	Recognition and Control Processor
	Reseau a Commutation par Paquet [= Packet Switched System, France]
	Restoration Control Point
	Restore Cursor Position
RCPA	Rice and Corn Production Administation [Philippines]
RCPB	Reactor Coolant Pressure Boundary
RCPT	Receipt
	Receptacle
RCR	Reader Control Relay
	Recrystallization-Controlled Rolling
	Required Carrier Return
	Room Cavity Ratio
	Route Contingency Reserve
	Royal Canadian Regiment
	Runway Condition Reading
RCRA	Refrigeration Compressor Rebuilders Association [US]
	Resource Conservation and Recovery Act [US]
RCRC	Reinforced Concrete Research Council [US]
RCRDC	Radio Components Research and Development Committee [UK]
RCRF	Rei Cretariae Romanae Fautores [= Society of Roman Pottery and Ceramic Archeologists, Germany]
R-CRS	Report on Course
RCS	Radar Cross-Section
	Radio Command System
	Radio-Controlled Solar
	Rapid Cycling Synchrotron
	Reaction Control Subsystem
	Reaction Control System
	Reactor Coolant System
	Rearward Communications System
	Reentry Control System [of NASA]
	Regions of Coherent Scattering
	Registration Control System
	Reloadable Control Storage
	Remote Computing Service
	Remote Contact Sensor
	Remote Control Station
	Remote Control System
	Remote Core Sampler
	Replacement Collision Sequences
	Resin-Coated Sand
	Reverse Circulation System
	Rigid Container Sheet
	Round-Cornered Square
	Royal College of Science [UK]
	Royal Commonwealth Society [UK]
RCSC	Radio Components Standardization Committee [UK]
	Research Council on Structural Connections [US]
RCSDE	Reactor Coolant System Dose Equivalent
RCSHSB	Red Cedar Shingle and Handsplit Shake Bureau [now: CSSB, US]
RCSS	Resin-Coated Stainless Steel
RCT	Radiation/Chemical Technician
	Receipt
	Remote Control Terminal
	Resolver Control Transformer
	Revenue Canada, Taxation
	Rework/Completion Tag
	Robotic Cluster Tool
	Rubber Chemistry and Technology
RCTBC	RARE (= Rare Animal Relief Effort) Center for Tropical Bird Conservation [US]
RCTD	Rate Constrained Target Detection
RCTL	Rectangular Coaxial Transmission Line
	Resistor-Capacitor-Transistor Logic
R-CTOD	Rock Crack-Tip Opening Displacement
RCTSR	Radio Code Test Speed on Response
RCU	Road Construction Unit
	Remote Control Unit
RCV	Receive
	Remotely Controlled Vehicle
	Research Center of Virology [Canada]
RC/VP	Recent Change/Verify Position
RCVR	Receiver
RCVS	Royal Canadian Veterinary Surgeon
RCW	Return Control Word
RCWP	Rural Clean Water Program [US]
RCWV	Rated Continuous Working Voltage
RCYC	Royal Canadian Yacht Club
RD	Radar Data
	Radar Display
	Radial Direction
	Radiation Detection
	Radiation Dose
	Random Drift
	Rate of Descent
	Reaction Dynamics
	Read
	Read Data
	Read Direct
	Reader Only
	Readiness Date
	Receive Data
	Recording Demand-Meter
	Rectifier Diode
	Recursive Definition
	Red
	Reference Diode
	Register Drive
	Relative Density
	Reliable and Durable
	Remote Diagnosis
	Report Departing
	Requirements Document
	Restricted Data
	Rigid Dynamics
	Ringdown
	Road
	Rod

	Rolling Direction
	Roof Drain
	Root Diameter
	Round
	Rural District
R&D	Research and Development [also: RD]
R/D	Rate of Descent
rd	rutherford [Unit]
RDA	Received Data Inverted
	Recommended Dietary Allowance
	Reliability Design Analysis
	Reverse Diels-Alder (Reaction)
	Rod Drop Accident
R&DA	Research and Development Associates (for Military Food and Packaging Systems) [US]
RdAc	Radioactinium (= Thorium 227)
RDAL	Representation Dependent Accessing Language
RDARA	Regional and Domestic Air Route Area
RDAS	Royal Danish Academy of Sciences
R-DAT	Rotary Head — Digital Audio Tape
RDB	Radar Decoy Balloon
	Received Data Noninverted
	Relational Database
	Research and Development Board [US]
RDBL	Readable
RDBMS	Relational Database Management System
RDC	Rail Diesel Car
	Recording Doppler Comparator
	Regional Dissemination Center [of NASA]
	Reliability Data Center
	Remote Data Collection
	Remote Data Concentrator
	Remote Diagnostic Center
	Rotary-Disk Contactor
	Royal Drawing Society [UK]
	Rural District Council
RDCE	Radio Distribution and Control Equipment
RD CHK	Read Check
RDD	Requisition Due Date
RD&D	Research, Development and Demonstration
RDDM	Reactor Deck Development Mockup
RDDP	Radarsat Data Development Project [of MEDS/ DFO, Canada]
RDE	Radial Defect Examination
	Reactive Deposition Epitaxy
	Receptor Destroying Enzyme
	Remote Data Entry
	Research and Development Effectiveness
	Resorcinol Diglycidyl Ether
	Rotating Disk Electrode
RDEC	Research, Development and Evaluation Commission [Taiwan]
RDF	Radial Distribution Function
	Radio Direction Finding
	Record Definition Field
	Refuse-Derived Fuel
	Repeater Distribution Frame
RDG	Resolver Differential Generator
	Ridge
RDGE	Resorcinol Diglycidyl Ether
RDH	Recirculating Document Handler
	Reference Datum Height
	Round-Head (Screw)
RDI	Radio Doppler Inertial
	Resource Development Impact (Program) [Canada]
	Route Digit Indicator
RDIA	Regional Development Incentives Act
RDIP	Research Development Incentives Program
RDIPE	Research and Development Institute of Power Engineering [USSR]
RDIU	Remote Device Interface Unit
RDJTF	Rapid Deployment Joint Task Force
RDL	Radial
	Radiological Defense Laboratory [US Navy]
	Research and Development Laboratory
	Resistor Diode Logic
	Rocket Development Laboratory [US Air Force]
RDM	Recording Demand Meter
	Random
RDMU	Range-Drift Measuring Unit
rDNA	Recombinant Deoxyribonucleic Acid [also: r-DNA or R-DNA]
RDO	Radio
	Radio Read-Out
R-DOS	Real-Time Disk Operating System [also: R DOS or RDOS]
RDOUT	Readout
RDP	Radar Data Processing
	Reactor Development Program
	Research Data Publication
RD&P	Research, Development and Production [also: RDP]
RDPS	Radar Data Processing System
RDRINT	Radar Intermittent
RDR	Radar
	Radar Departure Route
	Reader
RDR XMTR	Radar Transmitter
RDS	Radio Data System
	Radio Digital System
	Reactor Depressurization System
	Relational Data System
	Remote Degassing Station
	Rendezvous and Docking Simulator
	Research Defense Society
	Rheometric Dynamic Scanning
	Robust Detection Scheme
	Rural Development Service
RDSAP	Royal Dutch Society for Advancement of Pharmacy [Netherlands]
RDSS	Radar Determination Satellite System [US]
	Radio Determination Satellite Services
RDT	Reactor Development and Technology
	Remote Data Transmitter
	Resource Definition Table
RDTC	Remote Distributed Terminal Controller
RDT&E	Research, Development, Testing and Evaluation [also: RDTE]
RdTh	Radiothorium (= Thorium 228)
RDTL	Resistor Diode Transistor Logic
RDTR	Radiographic Dielectric Track Registration
	Research Division Technical Report
RDU	Runtime Debugging Unit

RDX Cyclotrimethylene Trinitramine
RDY Ready
RDYMX Ready Mix
RE Radar Echo
Radiant Energy
Radiated Emission
Radioelement
Radio Engineer
Radio Engineering
Railway Engineer
Railway Engineering
Ram Effect
Raw End
Rare Earth
Rate Effect
Real Number
Reentry
Reference Electrode
Reference Equivalent
Refrigeration Engineer
Refrigeration Engineering
Registry Editor
Relative Efficiency
Reliability Engineer
Reliability Engineering
Renewable Energy
Residual Elongation
Reverse Engineering
Reynolds Equation
River Engineering
Rocket Engine
Rotational Energy
Royal Engineer
R&E Research and Engineering [also: RE]
Re Real Part
Reaumur (Temperature)
Regarding
Reynolds Number
Rhenium
REA Radar Echoing Area
Railway Electrification Association [Japan]
Rare-Earth Alloy
Rod Ejection Accident
Rural Electrification Administration [US]
REAC Reactor
Reaction
Reactive
Reeves Electronic Analog Computer
REACT Radio Emergency Associated Communications Team [US]
Real-Time Expert Analysis and Control System
Register-Enforced Automated Control Technique
READ Radar Echo Augmentation Device
Real-Time Electronic Access and Display
Relative Element Address Designation
Remote Electronic Alphanumeric Display
READAC Remote Environmental Automatic Data Acquisition Concept
READI Rocket Engine Analysis and Decision Instrumentation
READLN Read Line
REALCOM Real-Time Communications
REAP Remote Entry Acquisition Package
Resource Efficient Agricultural Production [Canada]
Roof Evaluation Accident Prevention
REAR Reliability Engineering Analysis Report
REASM Reassemble
REB RAMP (= Rapid Modeling Platform) Equipment Builder
REBA Relativistic Electron Beam Accelerator
REBAM Rolling Element Bearing Activity Monitor
REBE Recovery Beacon Evaluation
REBUD Rehabilitation Budgeting Program
REC Radiant Energy Conversion
Receiver
Receipt
Record(er)
Recording
Rectifier
Regional Engineering College
Registration of Engineers Committee
Request for Engineering Change
Rural Electrification Conference
RECE Relativistic Electron Coil
RECEN Receivable Entry (Code)
REC'D Received [also: RECD]
RECFM Record Format
RECGAI Research and Engineering Council of the Graphic Arts Industry [US]
RECIP Reciprocate
RECIPE Recomp Computer Interpretive Program Expediter
RECIRC Recirculation
RECL Reclosing
Reconcile
RECM Recommend
RECMA Radio and Electronic Component Manufacturers Association
REC MARK Record Mark
RECOG Recognition
RECOL Retrieval Command Language
RECOMP Recomplement
RECON Remote Control On-Line Information Service
Retrospective Conversion
REconS Royal Economic Society [UK]
RECOV Recovery
RECP Receptacle
RECPT Receipt [also: RCPT]
RECRYSTD Recrystallized
RECSTA Receiving Station
RECT Receipt
Rectangle
Rectangular
Rectification
Rectifier
RECTAS Regional Center for Training in Aerial Surveys [Nigeria]
RED Radial Electron Distribution
Radiation-Enhanced Diffusion
Reducer
Reduction
Reflection Electron Diffraction

REDAC Real-Time Data Acquisition
REDAP Reentrant Data Processing
Reentry Data Processing
REDOX Reduction-Oxidation (Reaction) [also: Redox or redox]
REDSOD Repetitive Explosive Device for Soil Displacement
REDSH Reddish
REDTN Reduction
REDUPL Reduplicated
REDZ Recent Drizzle
REE Rare-Earth Element
Reentrant
Reentry
REECDP Renewable Energy and Energy Conservation Demonstration Program
REED Radio and Electrical Engineering Division [of NRC, Canada]
Restricted Edge Emitting Diode
REEP Range Estimating and Evaluation Procedure
Regression Estimation of Event Probabilities
REF Refinery
Reference
Refund
REFA Reichsausschuss fur Arbeitsstudien [= German Association for Time and Motion Studies]
REFD Referred
REFL Reflection
Reflector
REF L Reference Line
REFLECS Retrieval from the Literature on Electronics and Computer Science (Service) [of IEEE, US]
REFLES Reference Librarian Enhancement System [of University of California, US]
REFNY Refinery
REFR Refractive
Refractory
REFRA Recent Freezing Rain
REFRIG Refrigeration [also: REFRGN]
Refrigrator
REFS References
Remote Entry Flexible Security
REFSEARCH Reference Materials Searching System [of University of California at Berkeley, US]
REG Radioencephalogram
Range Extender with Gain
Rare-Earth Garnet
Region(al)
Register
Registration
Registry
Regular
Regulation
Regulator
Rheoencephalography
REGAL Range and Evaluation Guidance for Approach and Landing
REG'D Registered [also: REGD]
REGEN Regeneration
Regenerative Repeater
Regenerator
REG-GEN Registrar-General
REGIS Register
Relational General Information System
Remote Graphics Instruction Set
Reg Prof Regius Professor
REGT Regiment
Reg TM Registered Trademark
REHVA Representatives of the European Heating and Ventilating Association
REI Relative Erodability
Research-Engineering Interaction
REIC Radiation Effects Information Center [of BMI, US]
Rare-Earth Information Center [of Iowa State University, US]
Renewable Energy Information Center [US]
REIL Runway End Identification Lights
REINF Reinforcement
Reinforcing
REINS Requirements Electronic Input System
REL Radiation Evaluation Loop [of HEDL, US]
Rapidly Extensible Language
Rate of Energy Loss
Relation
Relative
Relay
Release
Relief
Relocation
Reluctance
Riksforbundet for Elektrikfieringen pa Landsbygden [= National Association for Rural Electrification, Sweden]
RELAT Relativity
RELCODE Relative Code
RELIPOSIS Research Liaison Panel on Scientific Information Services [UK]
RELP Residual Excited Linear Predictive Coding
REM Rapid Eye Movement
Rare-Earth Metal
Recognition Memory
Reflection Electron Microscopy
Reliability Engineering Model
Remainder
Remark
Reminder
Remote-Manual (Bolter)
Removal
Replacement Micrographs
Replica Electron Microscopy
Roentgen Equivalent Man [also: rem]
REMAP Record Extraction Manipulation and Print
REMARC Retrospective Machine-Readable Catalogue [US]
REMAT Research Center for Management of New Technology [Canada]
REMC Radio and Electronics Measurements Committee [UK]
Resin-Encapsulated Mica Capacitor
REMDOS Remote Disk Operating System
REME Royal Electrical and Mechanical Engineers [UK]
REMI Reliability Engineering and Management Institute [US]
REMO Real-Time Event Monitor

REMOS Real-Time Event Monitoring System
REM-RED Reflection Electron Microscopy — Reflection Electron Diffraction
REMSA Railway Engineering Maintenance Suppliers Association [US]
(Joint) Requirements for Emergency and Safety [of ECAC, France]
REMSCAN Remote Sensor Communication and Navigation
REMSTAR Remote Electronic Microfilm in Storage Transmission and Retrieval
RE-MTS Rare Earth — Muffin-Tin Sphere
REN Remote Enable
Renewal
RENE Rocket Engine/Nozzle Ejector
RENM Request for Next Message
RENPE Rare and Endangered Native Plant Exchange [US]
RENUM Renumbering
Renumeration
REO Rare Earth Oxide
Regenerated Electrical Output
Rocket Engine Operations
REON Rocket Engine Operations — Nuclear
REOS Rare-Earth Oxides [also: reos]
REP Radar Evaluation Pad
Range Error Probable
Reentrant Processor
Reentry Physics Program
Registry Enhancement Project [US]
Remote Emergency Power
Rendezvous Evaluation Pad [of NASA]
Repair
Repeat
Replication
Report
Reporting Point
Representation
Representative
Republic
Repulsion
Request for Proposals [also: RFP]
Roentgen Equivalent Physical [also: rep]
Rotating Electrode Process
REPAG Repagination
REPERF Reperforator
REPL Replacement
Replica
Replication
REPLAB Responsive Environment Programmed Laboratory
REPM Representatives of Electronic Products Manufacturers [now: ERA, US]
REPO Remote Emergency Power Off
REP-OP Repetitive Operation
REPPAC Repetitively Pulsed Plasma Accelerator
REPR Representation
Reprint
REPRO Reproduction
REPROM Reprogrammable Read-Only Memory
REPS Research, Publications and Secondment [of AINA, Canada]
REPT Receipt
Report
REQ Request
Requirement
Requisition
REQ'D Required [also: REQD]
REQT Requirement
REQVER Requirements Verification
REQWQ Requisition Word Queue
RER Radiation Effects Reactor [US Air Force]
Residual Error Rate
rER Rough Endoplasmic Reticulum
RERA Recent Rain
RERC Rare Earth Research Conference [of University of Kentucky, US]
RERF Radiation Effect Research Foundation [Japan]
RERL Residual Equivalent Return Loss
RES Radar Echo Simulation
Radiation Enhanced Sublimation
Remote Entry Service
Remote Entry Subsystem
Remote Entry System
Remote Experiment Station [US]
Reserve
Reservation
Reset
Residence
Resident(ial)
Residual Elastic Strain
Residue
Resistance
Resistance Electroslag
Resistance Electroslag Surfacing
Resistivity
Resistor
Resolution
Resource
Restore
Reticuloendothelial System
Royal Entomological Society [UK]
RESA Reactive Electrode Submerged Arc (Process)
(Scientific) Research Society of America [US]
Runway End Safety Area
RESAR Reference Safety Analysis Report
RESCAN Reflecting Satellite Communication Antenna
RESCU Radio Emergency Search Communications Unit
RESCUE Remote Emergency Salvage and Clean-Up Equipment
RESD Research and Engineering Support Division [of International Development Association]
RESER Reentry Systems Evaluation Radar
RESG Research Engineering Standing Group [of USDOD]
RESH Recent Showers
RESID Residual
RESIN Resinous
RESIS Resistance
RESN Recent Snow
RESNA Rehabilitation Engineering Society of North America
RESORS Remote Sensing On-Line Retrieval System [of CCRS, Canada]
RESP Regulated Electrical Supply Package
Remote Batch Station Program

	Respectively
	Respondent
RESRT	Restart
RESS	Radar Echo Simulation Subsystem
	Radar Echo Simulation Study
	Rapid Expansion of Supercritical Solutions
RESSA	Regional Engineering Students Societies Association [Canada]
REST	Radar Electronic Scan Technique
	Radar Electronic Scan Test
	Reentry Environment and Systems Technology
	Reentry System Test
RESTA	Reconnaissance, Security and Target Acquisition [US Army]
RESTBL	Relative Page to External Storage Correspondence Table
RESTR	Restrict(ion)
RET	Reliable Earth Terminal
	Renewable Energy Technology
	Retainer
	Retardation
	Return
	Retirement
RETA	Refrigerating Engineers and Technicians Association [US]
	Retrieval of Enriched Textual Abstracts
RETAIN	Remote Technical Assistance and Information Network
RETC	Rapid Excavation and Tunneling Conference
	Regional Emergency Transportation Center [US]
RET'D	Returned [also: RETD]
RETEC	Regional Technical Conference [of the Society of Plastics Engineers, US]
RETM	Rare Earth Transition Metal
RETMA	Radio, Electronics and Television Manufacturers Association [now: EIA, US]
	Rare Earth Transition Metal Alloy
RETN	Retain
RETR	Retraction
RETROFIT	Retroactive Refit [also: Retrofit or retrofit]
RETROSPEC	Retrospective Search System [of INSPEC, UK]
RETS	Recent Thunderstorm
	Renewable Energy Technologies
RETSPL	Reference Equivalent Threshold Sound Pressure Level
REU	Rectifier Enclosure Unit
REV	Reentry Vehicle
	Revenue
	Review
	Revise(d)
	Revision
	Revolution [also: rev]
rev/min	revolutions per minute [also: rev min^{-1}, RPM or rpm]
REVOCON	Remote Volume Control
REVS	Requirements Engineering and Validation System
	Rotor Entry Vehicle System
rev/sec	revolutions per second [also: rev/s, rev sec^{-1}, RPS or rps]
REV VER	Revised Version
REW	Rewind
REWA	Refrigeration Equipment Wholesalers Association [now: ARW, US]
REWS	Radar Early Warning System
REWSONIP	Reconnaissance Electronic Warfare Special Operation and Naval Intelligence Processing
REWTEL	Radio and Electronics World Telecommunications [UK]
REX	Real-Time Executive (Routine)
	Reduced Exoatmospheric Cross-Section
	Robotic Excavator
	Route Extension
RF	Radar Frequency
	Radial Flow
	Radiant Flux
	Radiation Filter
	Radio Facility
	Radio Frequency
	Raised Face
	Rangefinder
	Rating Factor
	Reactive Factor
	Read Forward
	Reducing Flame
	Register File
	Reinforcing Furnace-Black
	Relativistic Fluid
	Release Fraction
	Reporting File
	Report Footing
	Research Fellow
	Resonance Fluorescence
	Resonant Frequency
	Resorcinol Formaldehyde
	Retension Factor
	Return Flow
	Reverse Free
	Rice Flour
	Roof
	Roof Fan
	Rotating Fluid
	Rotational Flow
	Rough Finish
R-F	Radio Frequency
Rf	Retention Factor
	Rutherfordium
RFA	Radiographic Fluorescence Spectral Analysis
	Recurrent Fault Analysis
	Redundant Force Analysis
	Remote File Access
	Renewable Fuels Association [US]
	Request for Applications
	Retarding Field Analyzer
	Royal Field Artillery [UK]
RFB	Reason for Backlog
RFC	Radar Fire Control
	Radial Flow Compressor
	Radio Facility Chart
	Radio Frequency Cable
	Radio Frequency Chart
	Radio Frequency Choke
	Request for Connection

	Retirement for Cause
	Rosette Forming Cells
RFCI	Rare Fruit Council International [US]
	Resilient Floor Covering Institute [US]
RFCP	Radio Frequency Compatibility Program
	Route Facility Costs Panel [of ICAO]
RFD	Ready for Data
	Reentry Flight Demonstration
RF/DC	Radio Frequency/Data Communications
RFDU	Reconfiguration and Fault Detection Unit
RFE	Radio Frequency Emissions
RFEC	Remote-Field Eddy Current
RFEI	Request for Engineering Information
RFF	Remote-Fiber Fluorometry
	(Study Group on) Rescue and Firefighting [of ICAO]
RfF	Resources for the Future, Inc. [US]
RFFS	Rescue and Firefighting Services
RFG	Radar Field Gradient
	Rapid-Fire Gun
	Receive Format Generator
	Refining
	Report Format Generator
	Roofing
RFI	Radio-Frequency Interference
	Ready for Installation
	Request for Information
RFID	Radio Frequency Identification [also: RF-ID]
RF/IR	Radar Frequency/Infrared Frequency
RFIT	Radio Frequency Interference Test
RFL	Radio-Frequency Laboratories
	Requested Flight Level
	Resorcinol-Formaldehyde Latex
RFLG	Refuelling
RFLP	Restriction Fragment Length Polymorphism
RFM	Reactive Factor Meter
RFMO	Radio Frequency Management Office
RFMS	Remote File Management System
RFNA	Red Fuming Nitric Acid
RFNM	Request for Next Message
RF NOTAM	Royal Flight Notice to Airmen
RFO	Radio Frequency Oscillator
	Reason for Outage
RF&OOA	Railway Fuel and Operating Officers Association [US]
RFP	Request for Proposal
RFPA	Requests for Preliminary Applications
RFPG	Radio-Frequency Protection Guide
RFPR	Reversed Field Pinch Reactor
RFQ	Radio-Frequency Quadrupole
	Request for Quotation
RFR	Radio-Frequency Radiation
	Reduced Frequency Responses
	Reject Failure Rate
RFr	Rwandan Franc [Currency]
RFS	Radio Frequency Shift
	Radio-Frequency Spectroscopy
	Radio-Frequency Spectrum
	Random Filing System
	Ready for Service
	Regardless of Feature Size
	Regional Frequency Supplies
	Remote File Sharing
	Renormalized Forward Scattering
RFSEW	Royal Forestry Society of England and Wales
RFSH	Refresh
	Refreshment
RFSP	Replacement Flight Strip Printer
RF SQ	Roof Squares
RF SQUID	Radio-Frequency SQUID (= Superconducting Quantum Interference Device) [also: RF-SQUID]
RFSTF	Radio Frequency Systems Test Facility
RFT	Repeat Formation Tester
	Revisable Form Text
RDTD	Radial Flow Torr Deposition
RFU	Radio Frequency Unit
	Reference Frequency Unit
RFX	Reversed Field (Pinch) Experiment
RG	Radiogoniometer
	Range
	Rate Gyroscope
	Red-Green
	Reduction Gear
	Register
	Regulatory Guide
	Relative Growth
	Release Guard
	Renormalization Group
	Reset Gate
	Residual Gas
	Reticulated Grating
	Reverse Gate
	Ring Ground
	Ringing Generator
	Rolled Gold
	Rubber Gel
RGA	Rate Gyroscope Assembly
	Registered General Assignment
	Residual Gas Analysis
RGAHS	Royal Guernsey Agricultural and Horticultural Society [UK]
RGB	Red/Green/Blue (Monitor)
RGBM	Reinforced Grouted Brick Masonry
RGCC	Roll Gas Combined Cooling System
RGCGAI	Roman-Germanic Commission of German Archeological Institutes [Germany]
RGCSP	Review of the General Concept of (Aircraft) Separation Panel [of ICAO]
RGE	Range
RGH	Rough
RGH OPNG	Rough Opening
RGI	Residual Gas Ion
RGL	Report Generator Language
	Runway Guard Lights
RGLET	Rise-Time Gated Leading Edge Trigger
RGLTR	Regulator
RGM	Recorder Group Monitor
RGN	Region
RGNCD	Region Code
RGO	Reference Gear Oil
	Royal Greenwich Observatory [UK]
RGP	Rate Gyroscope Package

	Remote Graphics Processor
	Rolled Gold Plate
	Rotary Gas Pump
RGQ	Rapid Gas Quenching
RGR	Relative Growth Rate
RGS	Radio Guidance System
	Rate Gyroscope System
	Release Guard Signal
	Remote Guidance System
	Rocket Guidance System
	Royal Geographical Society [UK]
RGSA	Royal Geographical Society of Australia
RGSC	Ramp Generator and Signal Converter
RGT	Resonant Gate Transistor
	Right
	Ringgit [Currency of Malaysia]
RGU	Red/Green/Ultraviolet (Filter) [also: RGUV]
RH	Radiological Health
	Receive Hub
	Refrigeration Hardened
	Reheater
	Relative Humidity
	Report Heading
	Request Header
	Request/Response Header
	Reserve-Shutdown Hours
	Response Header
	Rheostat
	Right-Hand
	Rockwell Hardness
	Roughness Height
R-H	Ruhrstahl-Henrichshutte/Heraeus (Degassing Process) [also: RH]
Rh	Rhesus Factor
	Rhesus Positive
	Rhodium
R/h	Roentgen per Hour [also: R/hr or r/hr]
rh	rhesus negative
RHA	Road Haulage Association [UK]
	Royal Horse Artillery [UK]
RHAW	Radar Homing and Warning
RHB	Reference Heat Balance
RHBHDR	Rhombohedral
RHC	Regional Holding Company
	Reheat Coil
	Right-Hand Circular (Polarization)
	Rotation Hand Controller
RHCP	Right-Hand Circular Polarization
RHD	RAMP (= Rapid Modeling Platform) HVAC (= Heating, Ventilation and Air Conditioning) Design
	Right-Hand Drive
RHE	Radiation Hazard Effects
	Reversible Hydrogen Electrode
	Road Haulage Executive [UK]
RHEED	Reflection High Energy Electron Diffraction
RHEL	Rutherford High Energy Laboratory [UK]
RHEO	Rheostat
RHET	Resonant Hot Electron Transistor
RHF	Radar Height Finding
	Restricted Hartree-Fock
	Rotary Hearth Furnace
RHI	Range-Height Indicator
	Rigid Hull Inflatable (Model)
RHIC	Relativistic Heavy Ion Committee
RHIP	Radiation Health Information Project
	Relativistic Heavy Ion Physics
RHistS	Royal Historical Society [UK]
RHM	Remote Hardware Monitoring
	Roentgen per Hour at One Meter [also: rhm]
RHM/Ci	Roentgen per Hour at One Meter per Curie [also: rhm/Ci]
RHN	Rockwell Hardness Number [also: Rhn]
RHOGUI	Radar Homing Guidance
RHOMB	Rhombic
	Rhombohedron
RHP	Reduced Hard Pressure
RHQ	Regional Headquarters
RHR	Receiver Holding Register
	Reheater
	Rejectable Hazard Rate
	Residual Heat Removal
	Roughness Height Rating
R/hr	Roentgens per Hour [also: R/h or r/hr]
RHRP	Residual Heat Removal Pump
RHRSW	Residual Heat Removal Service Water
RHS	Rectangular Hollow Section
	Right-Hand Side
	Royal Historical Society [Canada]
	Royal Horticultural Society [UK]
RHT	Radiative Heat Transfer
	Register Holding Time
RHTM	Regional Highway Traffic Model [UK]
RHTS	Reactor Heat Transport System
RHSV	Royal Historical Society of Victoria [Canada]
RI	Radar Input
	Radio Inertial
	Radio Influence
	Radio Interference
	Radioisotope
	Radiointerferometer
	Range Instrumentation
	REACT (= Radio Emergency Associated Communications Team) International
	Read-In
	Real Image
	Rectifier Instrument
	Reflective Insulation
	Refraction Intercept
	Reliability Index
	Repeat Indication
	Report of Investigation
	Reproducibility Index
	Research Institute
	Resistance Inductance
	Rhode Island [US]
	Rial [Currency of Iran]
	Rideability Index
	Ring Indicator
	Robotics Institute [US]
	Robotics International [of SME, US]
	Rodale International [US]
	Routing Indicator

R&I Removal and Installation
RIA Radioimmunoassay
Radioisotope Assay
Reactivity Initiated Accident
Recording Industry Association [US]
Registered Industrial Accountant
Removable Instrument Assembly [of INEL, US]
Research Institute of America [US]
Research Institute of Atmospherics [of Nagoya University, Japan]
Robotics Industries Association [US]
Robot Institute of America [Name Changed to: Robotics Industries of America, US]
Royal Irish Academy
RIAA Recording Industry Association of America [US]
RIAEC Rhode Island Atomic Energy Commission [US]
RIAI Royal Institute of Architects of Ireland
RIAM Research Institute for Applied Mechanics [Japan]
RIAS Radio in American Sector [in Berlin, Germany]
Regulatory Impact Assessment Statement
Research Institute of Advanced Studies [US]
Research Institute of Animal Science [South Africa]
Royal Incorporation of Architects in Scotland
RIB Request Indicator Byte
Rib Ribose
RIBA Royal Institute of British Architects
RIBE Reactive Ion Beam Etching
RIBS Rutherford Ion Backscattering
RIC Radar Indicating Console
Radar Input Control
Rainforest Information Center [Australia]
Range Instrumentation Coordination
Rare-Earth Information Center [of Iowa State University, US]
Read-In Counter
Reconstructed Ion Chromatography
Regional Engineering College
Relocation Instruction Counter
Research and Information Commission [of COSEC, Netherlands]
Research Institute for Catalysis [of Hokkaido University, Japan]
Rice Improvement Conference [of PDAF, Taiwan]
Royal Institute of Chemistry [UK]
RI&C Reactor Instrumentation and Control [of IEEE NPS, US]
RICA Railway Industry Clearance Association [US]
Research Institute for Consumer Affairs [UK]
RICASIP Research Information Center and Advisory Service on Information Processing [US]
RICB Reactive Ionized Cluster Beam
RICE Rice Information Cooperative Effort [Philippines]
RICMT Radar Inputs Countermeasures Technique
RICOB Rice and Corn Board [Philippines]
RICS Range Instrumentation Control System
Royal Institution of Chartered Surveyors [UK]
RICW Regional Engineering College, Warangal [India]
RID Radar Input Drum
Radial Immunodiffusion
Records Issue Date
Refractive-Index Detector
Reset Inhibit Drum
Review Item Disposition
RIDA Rural Industrial Development Authority [Malaysia]
RIDD Range Instrumentation Development Division
RIDL Ridge Instrument Development Laboratory [US Navy]
RIDS Receiving Inspection Data Status Report
RIE Reactive Ion Etching
Research in Education (Database) [of ERIC, US]
Research Institute of Electronics [Japan]
Rhodesian Institution of Engineers
Royal Institute of Engineers [UK]
Royal Institution of Engineers [Netherlands]
RIEC Research Institute of Electrical Communication [Japan]
RIEF Recirculating Isoelectric Focusing
RIEI Roofing Industry Educational Institute [US]
RIE PECVD Reactive Ion Etching Plasma-Enhanced Chemical Vapor Deposition [also: RIE-PECVD]
RIETCOM Regional Interagency Emergency Transportation Committee [US]
RIF Radio Influence Field
Radio Interference Field
Reclearance in Flight
Reduction in Force
Reliability Improvement Factor
RIFI Radio Interference Field Intensity
Radio-Interference-Free Instrument
RIFM Research Institute for Fragrance Materials [US]
RIFP Research Institute for Fundamental Physics [Japan]
RIFS Radioisotope Field Support
RI&FS Remedial Investigation and Feasibility Study [of Environmental Protection Agency, US]
RIFT Reactor-In-Flight Test
RIG Research Initiation Grant [of Engineering Foundation, US]
RIGB Royal Institution of Great Britain
RIGFET Resistive Insulated Gate Field-Effect Transistor
RIGS Radio Inertial Guidance System
Runway Identifiers with Glide Slope
RIHANS River and Harbor Aid to Navigation System
RII Receiving Inspection Instruction
RIIA Royal Institute of International Affairs [UK]
RIIC Research Institute on International Change
Rural Industries Innovation Center
RIIS Research Institute of Industrial Science [of Kyushu University, Japan]
RIISOM Research Institute for Iron, Steel and Other Metals [US]
RIL Radio Interference Level
Red Indicating Lamp
Representation Independent Language
RILEM Reunion Internationale des Laboratoires d'Essais et de Recherches sur les Materiaux et les Construction [= International Union of Testing and Research Laboratories for Materials and Structures, France]
RIM Radar Input Mapper
Radio Imaging Method
Reaction Injection Molding
Read-In Mode

	Receiver Intermodulation
	Repulsion-Inducation Motor
	Resource Interface Module
RIMDM	Research Institute of Mineral Dressing and Metallurgy [Japan]
RIM PUR	Reaction Injection Molding/Polyurethane [also: RIM/PUR]
RIMS	Radiation Intensity Measuring System
	Remote Information Management System
	Resonant Ionization Mass Spectrometry
	Risk and Insurance Management Society US]
RIMtech	Research Institute for the Management of Technology
RIN	Reference Indicator Number
	Regular Inertial Navigator
	Royal Institute of Navigation [UK]
RINA	Registro Italiano Navale [= Italian Naval Registry]
	Royal Institution of Naval Architects [UK]
RINAL	Radar Inertial Altimeter
RIND	Research Institute of National Defense
RINDO	Rydberg Intermediate Neglect of Differential Overlap
RING	Ringing
RINS	Research Institute for the Natural Sciences
	Rotorace Inertial Navigation System
RINT	Radar Intermittent
RIO	Omani Rial [Currency of Oman]
	Real-Time Input/Output
	Relocatable Input/Output
	Remote Input/Output
	Roll-In Only
RIOMETER	Relative Ionospheric Opacity Meter
RIOPR	Rhode Island Open Pool Reactor [US]
RIOT	RAM (= Random-Access Memory) Input/Output Timer
	Real-Time Input/Output Transducer
	Real-Time Input/Output Translator
	Retrieval of Information by On-Line Terminal
RIP	Raster Image Processor
	Reactor Instrument Penetration
	Receiving Inspection Plan
	Regulatory and Information Policy
	Resin-In-Pulp (Process)
	Rest-In Proportion
	Retired In Place
	Ring Index Pointer
	Routing Information Process
	Rural Industrialization Program [US]
RIPA	Royal Institution of Public Administration [UK]
RIPCAM	Real-Time Interactive Process Control and Management
RIPH	Royal Institute of Public Health and Hygiene [UK]
RIPL	Representation Independent Programming Language
RIPPLE	Radioactive Isotope Powered Pulse Light Equipment
	Radioisotope Powered Prolonged Life Equipment
RIPS	Radio Isotope Power Supply
	Radio Isotope Power System
	Range Instrumentation Planning Study
	Regional Institute for Population Studies [of UN — Ghana]
RIPV	Reactor Isolation Pressure Valve
RIQS	Remote Information Query System
RIR	Relative Index Register
	Reliability Investigation Request
	ROM (= Read-Only Memory) Instruction Register
RIRA	Reports and Information Retrieval Activity
RIRO	Roll In/Roll Out
RIRS	Reflectance Infrared Spectroscopy
RIRTI	Recording Infrared Tracking Instrument
RIS	Radar Information Service
	Radiology Information System
	Range Instrumentation Ship
	Raster Input Scanner
	Receipt Inspection Segment
	Recorded Information Service
	Redwood Inspection Service [US]
	Remote Information System
	Research Information Service
	Resonance Ionization Spectroscopy
	Reporting Identification Symbol
	Retail Information System
	Retransmission Identity Signal
	Revolution Indicating System
	RisS International Symposium [Denmark]
	Rotatable Initial Susceptibility
	Rowland Institute of Science [US]
	Russian Intelligence Service
RISC	Radiology Information Systems Consortium
	Reduced Instruction Set Computing
	Rockwell International Science Center [US]
RISE	Research Information Service for Education [of OISE, Canada]
	Research Institute of Science and Engineering [of Ritsumeikan University, Japan]
	Research in Supersonic Environment
	(National Institute for) Resources in Science and Engineering [US]
RISFM	Research Institute for Strength and Fracture of Materials [of Tohoku University, Japan]
RISM	Radio Interface Switch Module
	Research Institute for Scientific Measurement [Japan]
RI/SME	Robotics International of the Society of Manufacturing Engineers [US]
RISO	Regional Individual Standing Offer [Canada]
RISPA	Research Institute of the Sumatra Planters Association [Indonesia]
RIST	Radar Installed System Tester
	Research Institute of Science and Technology [of Nihon University, Japan]
	Research Institute of Industrial Science and Technology [South Korea]
RIT	Radar Input Test
	Radio Information Test
	Rate of Information Transfer
	Receiver Incremental Tuning
	Remote Interactive Terminal
	Rochester Institute of Technology [US]
	Rocket Interferometer Tracking
	Royal Institute of Technology [Sweden]
RITA	Radio-Frequency Ion Thruster Assembly

	Rand Intelligent Terminal Agent [US]
	Recognition of Information Technology Achievement [UK]
RITC	Regional Information Technology Coordinators [UK]
	Rhodamine Isothiocyanate
RITE	Rapidata Interactive Editor
RITENA	Reunion Internacional de Tecnicos de la Nutricion Animal [= International Meeting of Animal Nutrition Experts, Spain]
RITL	Royal Institute of Technology Libraries [Sweden]
RITP	Research Institute for Technical Physics [of Hungarian Academy of Sciences]
RITU	Research Institute of Temple University [US]
RIUK	Royal Institute of the United Kingdom
RIV	Radio-Influence Voltage
	Rapid Intervention Vehicle
	River
	Rivet
RIW	Reliability Improvement Warranty
RJ	Reference Junction
RJAS	Royal Jersey Agricultural Society [UK]
RJE	Remote Job Entry
RJES	Remote Job Entry System
RJO	Remote Job Output
RJP	Remote Job Processing
RKHS	Reducing Kernel Hilbert Space
RKKY	Rudermann-Kittel-Kasuya-Yosida
RKO	Range Keeper Operator
RKTL	Reichskuratorium fur Technik in der Landwirtschaft [= German Association for Mechanization in Agriculture, Germany]
rkva	reactive kilovolt-ampere
RKW	Rationalisierungs-Kuratorium der Deutschen Wirtschaft [= German Productivity and Management Association]
RL	Radiation Laboratory
	Radio Link
	Radiolocation
	Radioluminescence
	Reactor Licensing
	Receive Leg
	Record Length
	Reduced Level
	Reflection Loss
	Relay Level
	Relay Logic
	Remote Loopback
	Report Leaving
	Research Laboratory
	Resistance-Inductance
	Resistor Logic
	Return Loss
	Rocket Launcher
	Root Locus
	Rubber Latex
	Runway Lights
RLA	Remote Line Adapter
	Remote Loop Adapter
RLAC	Recycling Legislation Action Coalition [US]
RLAP	Rural Land Analysis Program [Canada]
RLBM	Rearward Launched Ballistic Missile
RLC	Radio Launch Control System
	Resistance-Inductance-Capacitance
	ROM (= Read-Only Memory) Location Counter
	Run Length Coding
RLCE	Request Level Change En-Route
RLCS	Radio Launch Control System
RLCU	Reference Link Control Unit
RLD	Relocation Dictionary
	Relocation Directory
	Rijksluchtvaartdienst [= Royal Dutch Aviation Administation]
RLE	Research Laboratory for Electronics [of MIT, US]
	Runlength Encoder
RLEM	Research Laboratory of Engineering Materials [of Tokyo Institute Technology, Japan]
RLF	Reverse Line Feed
RLG	Railing
	Release Guard
	Research Libraries Group [US]
	Rifles, Large Grain
	Ring Laser Gyroscope
RLHTE	Research Laboratory of Heat Transfer in Electronics
RLIN	Research Libraries Information Network [US]
RLL	Recorded Lithology Logging
	Relay Ladder Language
	Relocating Linking Loader
RLM	Reflector and Lamp Manufacturers [US]
	Reflector and Lighting Equipment Manufacturers
RLNR	Research Laboratory for Nuclear Reactors [of Tokyo Institute of Technology, Japan]
RLO	Restoration Liaison Officer
	Returned Letter Office
RLOP	Reactor Licensing Operating Procedure
RLOS	Radar Line-of-Sight [also: R-LOS or RLS]
RLP	Reactive Liquid Polymer
RLPME	Research Laboratory of Precision Machinery and Electronics [Japan]
RLR	Record Length Register
RLS	Radar Line-of-Sight [also: RLOS or R-LOS]
	Reels
	Release
	Remote Line Switch
	Residual Lattice Strain
	Riel [Currency of Kampuchea]
RLSD	Received Line Signal Detector
RLSE	Release
RLSD	Received Line Signal Detector
RLSS	Research Laboratory for Surface Science [Japan]
RLSSC	Royal Life Saving Society Canada
RLST	Release Timer
RLT	Reaction Liquid Type
	Remote Line Test
RLTS	Radio-Linked Telemetry System
RLU	Remote Line Unit
RLWL	Reactor Low Water Level
RLY	Railway
	Relay
RM	Radar Mapper
	Radiation Measurement
	Radiation Monitor
	Radiometal

Radio Monitoring
Range Marks
Ratemeter
Reaction Mechanism
Ream
Record Mark
Reed-Muller
Reflecting Microscope
Register Memory
Regular Matrix
Reliability and Maintainability
Reliability and Maintenance
Remanent Magnetization
Remote Mobile
Repulsion Motor
Research Memorandum
Reset Mode
Residual Magnetism
Resistance Melting
Resistance Monitor
Resource Manager
Reusable Module
Robotic Motion
Rock Mechanics
Rollback Module
Room
Rotating Machinery [IEEE PES Committee]
Royal Mail [UK]
Royal Marines [UK]
Rural Municipality

R&M Reliability and Maintainability
Reliability and Maintainance
Reports and Memoranda

R/M Read/Mostly Memory

R-M Reichert-Meissl (Number)

RMA Radio Manufacturers Association [now: EIA, US]
Random Multiple Access
Reactive Modulation Amplifier
Rosin, Mildly-Activated
Royal Military Academy [UK]
Rubber Manufacturers Association [US]

RMAAS Reactivity Monitoring and Alarm System

R-MAD Reactor Maintenance, Assembly, and Disassembly [of USAEC]

RMAG Rocky Mountain Association of Geologists [US]

RMARL Rocky Mountain Analytical Research Laboratory [US]

RMAS Remote Memory Administration System

RMATS Remote Maintenance, Administration and Traffic System

RMAX Range Maximum

RMB ROM (= Read-Only Memory) Memory Board

RMBI Risk Management and Business Insurance

RMC Radiative Muon Capture
Ready-Mixed Concrete
Reed-Muller Code
Regional Media Center
Regional Meteorological Center [of WMO, Switzerland]
Rod Memory Computer
Rotating Modulation Collimator
Royal Military College [Canada]

RM&C Reactor Monitoring and Control

RMCAO Ready-Mixed Concrete Association of Ontario [Canada]

RMCC Royal Military College of Canada

RMCDE Radar Message Conversion and Distribution Equipment

RMCMI Rocky Mountain Coal Mining Institute [US]

RMCS Reactor Manual Control System
Remote Maintenance Control System
Royal Military College of Science [UK]

RMDR Remainder

RME Relay Mirror Experiment

RMEA Rubber Manufacturing Employers Association

RMetS Royal Meteorological Society [UK]

RMF Reactivity Measurement Facility [of INEL, US]
Research and Management Foundation [of ACEC, US]
Resource Measurement Facility

RMI Rack Manufacturers Institute [US]
Radio Magnetic Indicator
Reliability Maturity Index
Remote Mechanical Investigator
Route Monitoring Information

RMIC Research Materials Information Center [of ORNL, US]

RMICBM Road Mobile Intercontinental Ballistic Missile

RMIN Range Minimum

R/min Roentgen per Minute [also: r/min]

RMIT Royal Melbourne Institute of Technology [Australia]

RMK Remark

RML Radar Mapper, Long-Range
Radar Microwave Link
Radio Microwave Link
Relational Machine Language
Rocky Mountain Laboratory [of National Institute of Health, US]

RMM Radar Map Matching
Read Mostly Memory
Remote Maintenance Monitoring
Roentgens per Minute at One Meter [also: rmm]

RMMC Rocky Mountain Mathematics Consortium [US]

RMMLF Rocky Mountain Mineral Law Foundation [US]

RMMU Removable Media Memory Unit

RMN Remain

RMO Radio Material Officer

RMON Resident Monitor

RMOS Refractory Metal-Oxide Semiconductor

RMP Reentry Measurement Program
Royalty Management Program

RMPI Remote Memory Port Interface

RMS Radar Mapper, Short-Range
Radiation Monitoring System
Reactor Monitor System
Record Management Service
Record Management System
Recovery Management Support
Regulatory Manpower System
Remote Maintenance System
Remote Manipulator System
Resource Management Support

	Resource Management System
	Risk Management Society [US]
	Root Mean Square [also: rms]
	Royal Meteorological Society [UK]
	Royal Microscopical Society [UK]
RMSCC	Rock Mechanics and Strata Control Committee
RMSD	Root Mean Square Difference
RMSE	Root Mean Square Error
RMSS	Root Mean Square Strain
RMT	Remote
	Remote Terminal
RMTB	Reconfiguration Maximum Theoretical Bandwidth
RMTE	Remote
RMU	Remote Maneuvering Unit
RMV	Reentry Measurement Vehicle
	Remote Maintenance Vehicle
	Remove
RMVSO	Road and Motor Vehicle Safety Office [Canada]
RMW	Read-Modify-Write
RMWAA	Roadmasters and Maintenance of Way Association of America [US]
RMWS	Reactor Make-Up Water System
RMX	Remote Multiplexer
RN	Radar Navigation
	Radio Navigation
	Random Noise
	Random Number
	Reception Node
	Recipient Name
	Record Number
	Reference Noise
	Registered Nurse
	Registry Number
	Removable Needle
	Report Number
	Research Note
	Reynolds Number
	Royal Navy [UK]
Rn	Radon
RNA	Registered Nurses Association
	Ribonucleic Acid
	Ribose Nucleic Acid
RNAAS	Royal Netherlands Academy of Arts and Sciences
RNABC	Registered Nurses Association of British Columbia [Canada]
RNAC	Remote Network Access Controller
RNAM	Regional Network for Agricultural Machinery [Philippines]
RNANS	Registered Nurses Association of Nova Scotia [Canada]
RNAO	Registered Nurses Association of Ontario [Canada]
RNAV	Radial/Area Navigation
	Radio Navigation (System) [also: R/NAV]
RNAV/RNPC	Radial/Area Navigation/Required Navigation Performance Capability
RNC	Request Next Character
	Royal Norwegian Council
RNCS	Royal Netherlands Chemical Society
RNCSIR	Royal Norwegian Council for Scientific and Industrial Research
RND	Random
	Round
RNase	Ribonuclease
RNFP	Radar Not Functioning Properly
RNG	Radio Range
	Random Number Generator
RNIT	Radio Noise Interference Test
RNL	RisS National Laboratory [Denmark]
RNLI	Royal National Lifeboat Institution [UK]
RNMP	Replacement of the Nautical Mile Panel [of ICAO]
RNP	Remote Network Processor
	Ribonucleoprotein
RNPC	Required Navigation Performance Capability
RNPDL	Risley Nuclear Power Development Laboratories [of UKAEA]
RNR	Receiver Not Ready
	Royal Navy Reserve [UK]
RNRF	Renewable Natural Resources Foundation [US]
RNS	Residue Number System
RNSA	Royal Naval Sailing Association [UK]
RNSS	Royal Naval Scientific Service [UK]
RNSYS	Royal Nova Scotia Yacht Squadron [Canada]
RNV	Radio Noise Voltage
RNVR	Royal Navy Volunteer Reserve [UK]
RNVSR	Royal Navy Volunteer Supplement Reserve [UK]
RNZAS	Royal New Zealand Astronomical Society
RNZIH	Royal New Zealand Institute of Horticulture
RO	Radar Operator
	Radio-Oscillator
	Range Operations
	Read Only
	Readout
	Receive Only
	Record Zero
	Reddish Orange
	Reference Oscillator
	Register Output
	Reserve Officer
	Reverse Osmosis
	Report Over
	Roll Off
	Roll On
	Roll-Over
	Routine Order
	Royal Observatory [UK]
	Ruthenium Oxide
R/O	Read Only
	Receive Only
R-O	Roll-Over
ROA	Receive on Account
	Reserve Officers Association [US]
	Return on Assets
ROAP	Reorganization, Office of the Assistant to the President [US]
ROAR	Royal Optimizing Assembly Routine
ROAT	Radio Operator's Aptitude Test
ROB	Radar Order of Battle
	Remaining on Board
	Remote Order Buffer
	Robotic Operating Buddy
ROBEX	Regional OPMET (= Operational Meteorological Information) Bulletin Exchange Scheme

ROBIC Robotic Integrated Cell
ROBIN Remote On-Line Business Information Network
ROBO Rocket Bomber
ROBOMB Robot Bomb
ROC Range Operations Conference
Rapid Omnidirectional Compaction
Rate of Climb
Rate of Convergence
Receiver Operating Characteristic
Required Operation Capability
Relative Operating Characteristic
Remote Operator's Console
Republic of China
Required Operational Capacity
Remote Operator Console
Return on Capital
Reusable Orbital Carrier
roc reciprocal ohm centimeter [Unit]
ROCAP Regional Office for Central America and Panama
ROCAPPI Research on Computer Applications for the Printing and Publishing Industries
ROCC Regional Operation Control Center
ROCE Return on Capital Employed
ROCKET Rand's Omnibus Calculator of the Kinetics of Earth Trajectories
ROCKOON Rocket-Balloon System [also: Rockoon or rockoon]
ROCP Radar Out of Commissions for Parts
ROCR Remote Optical Character Recognition
ROD Rate of Descent
Recorder on Demand
Release Order Directive
Repair on Demand
Required Operational Data
RODATA Registered Organizations Databank
ROE Return on Equity
ROF Remote Operator Facility
Royal Ordnance Factory [UK]
R of A Reduction of Area
ROFOR Route Forecast
ROFT Radar Off Target
ROH Receiver Off-Hook
ROI Range Operations Instruction
Region of Interest
Republic of Ireland
Return on Investment
ROIC Return of Invested Capital
ROIP Residual Oil-In-Place
ROIS Radio Operational Intercom System
ROK Republic of Korea
ROL Romanian Leu [Currency]
ROLAC Regional Office for Latin America and the Caribbean [of UNEP — Mexico]
ROLF Remotely Operated Longwall Face
ROLR Receiving Objective Loudness Rating
ROLS Recoverable Orbital Launch System
Remote On-Line System
Request On-Line Status
ROM Read-Only Memory
Readout Memory
Return on Market Value
Roman
Romania(n)
Rough Order of Magnitude
Royal Ontario Museum [Canada]
Rule of Mixtures
Run-of-Mill (Ore)
Run-of-Mine (Ore)
rom reciprocal ohm meter [Unit]
ROMAC Robotic Muscle-Like Actuator
Robotic Muscle-Like Actuator Control
ROMBUS Reusable Orbital Module Booster and Utility Shuttle
ROM BIOS Read-Only Memory — Basic Input/Output Subsystem [also: ROM-BIOS or ROM/BIOS]
ROME Reason-Oriented Modeling Environment
ROM EPROM Read-Only Memory — Erasable Programmable Read-Only Memory [also: ROM-EPROM or ROM/EPROM]
ROMIO Read-Only Memory plus Input/Output
ROMON Receive-Only Monitor
ROMOTAR Range-Only Measurement of Trajectory and Recording
ROM RPROM Read-Only Memory — Reprogrammable Read-Only Memory [also: ROM-RPROM or ROM/RPROM]
ROMV Return on Market Value
RON Receive Only
Remain Overnight
Research Octane Number
RONS Read-Only Name Store
RONT Radar On Target
RONTP Receive-Only Nontyping Perforator
ROOFG Roofing
ROOI Return on Original Investment
ROOST Rapid Optical Ocean Surveillance Testbed
Reusable One-Stage Orbital Space Truck
ROOT Relaxation Oscillator Optically Tuned
ROP Rate of Penetration
Read/Receive-Only Printer
Record of Performance
Record of Production
Record of Purchase
Recovery Operating Plan
Ring-Opening Polymerization
Run of Paper
ROPE Ring of Prefetch Elements
ROPES Remote On-Line Print Executive System
ROPME Regional Organization for the Protection of the Marine Environment [Kuwait]
ROPP Receive-Only Page Printer
ROPS Range Operation Performance Summary
Roll-Over-Protective Structure
ROPS-FOPS Roll-Over-Protective (Structure) and Falling-Object-Protective Structure
ROR Rate of Return
Rate of Rise
RORC Rolla Research Institute [of University of Missouri-Rolla, US]
RORD Return on Receipt of Document
RORO Roll-On/Roll-Off (Transport) [also: RoRo]
ROS Raster Output Scanner
Read-Only Storage
Ruthenium Oxide
ROSAR Read-Only Storage Address Register

ROSCOP Report of Observations Samples Collected by Oceanographic Programs [of NODC, US]
ROSDR Read-Only Storage Data Register
ROSE Remotely-Operated Special Equipment
Retrieval by On-Line Search
RoSPA Royal Society for the Prevention of Accidents [also: ROSPA, UK]
ROSPEC Rotating Spectrometer
ROSTA Regional Office for Science and Technology in Africa [of UNESCO — Kenya]
ROSTENA Regional Office for Science and Technology for Europe and North America [of UNESCO — France]
ROT Radar On Target
Reusable Orbital Transporter
Rotary
Rotate
Rotation
ROTC Reserve Officers Training Corps [of US Navy]
ROTCC Receiver Off-Hook Tone Connecting Circuit
ROTH Read-Only Type Handler
ROTI Recording Optical Tracking Instrument
ROTL Remote Office Test Lines
ROTP Regular Officer Training Program
ROTR Receive-Only Tape Reperforator
Receive-Only Typing Reperforator
ROTR S/P Receive-Only Typing Reperforator, Series to Parallel
ROTS Rotary Out Trunk Switch
ROTT Reorder Tone Trunks
ROV Remotely-Operated Vehicle
ROVD Remotely-Operated Volume Damper
ROVOS-C Remote Automated Color Video Observing System
ROW Resonant Optical Waveguide
Right-of-Way [also: RW]
Roll Welding
ROWAP Rotating Water Atomization Process
ROWS Robotic Winding System
ROY Royalties
RP Radiation Pressure
Reactor Physics
Reader-Printer
Reader Punch
Read Printer
Real Part
Receive Processor
Reception Poor
Recommended Practice
Record Processor
Recovery Phase
Reddish Purple
Red Phosphorus
Reduced Propagation
Reference Pulse
Refilling Point
Reflecting Power
Reinforced Plastics
Relative Pressure
Remote Processor
Repair
Repeater
Reply Paid
Report Passing
Research Paper
Reservoir Pressure
Resolving Power
Restoration Priority
Rocket Projectile
Rocket Propellant
Rocket Propulsion
Rollback Process
Rotary Pump
Round Pin
Rust Preventive
Rp Rupiah [Currency of Indonesia]
RPA Radar Performance Analyzer
Random-Phase Approximation
Rapid Pressure Application
Reentrant Process Allocator
Registered Public Accountant
Robinia Pseudoacacia Agglutinin
rPA Recombinant Protein A
RPAA Regional Planning Association of America
RPAE Random Phase Approximation with Exchange
RPAO Radium Plaque Adaptometer Operator
RPB Radiation Protection Bureau [of Health and Welfare Canada]
RPC Radar Planning Chart
Recreational Pilot Certificate [US]
Rectifier Photocell
Registered Protective Circuit
Regional Project Coordinator
Remote Parameter Control
Remote Position Control
Remote Procedure Calls
Remote Processor Controller
Research and Productivity Council [Canada]
Reversed Phase Chromatography
Rotary Phase Converter
Row Parity Check
RP/CI Reinforced Plastics/Composites Institute [now: CI of SPI, US]]
RPCRS Reactor Protection Control Rod System
RPCS Reactor Plant Control System
RPCVD Remote Plasma-Enhanced Chemical Vapor Deposition
RPD Radar Planning Device
Rapid
RAMP (= Rapid Modeling Platform) Piping Design
Reactor Plant Designer
Retarding Potential Difference
RPDH Reserve-Shutdown, Planned Derated Hours
RPE Radial Probable Error
Rapid Post-Editing
Registered Professional Engineer
Remote Peripheral Equipment
Required Page-End (Character)
Resource Planning and Evaluation
Retinal Pigment Epithelium Cell
Rocket Propulsion Establishment [UK]
Rotating Platinum Electrode
RPF Radiometer Performance Factor

	Registered Professional Forester
RPFS	Radio Position Fixing System
RPG	Radiation Protection Guide [of FRC, US]
	Random Pulse Generator
	Regional Planning Group
	Report Program Generator
	Rocket-Propelled Grenade
rPG	Recombinant Protein G
RPH	Recommended Power Handling
	Relative Pulse Height
	Remotely-Piloted Helicopter
RPI	Radar Precipitation Integrator
	Railway Progress Institute [US]
	Read Punch and Interpret
	Rensselaer Polytechnic Institute [US]
	Resource Policy Institute [US]
	Rod Position Indicator
	Rubber and Plastics Industry
	Ryerson Polytechnic Institute [Canada]
R&PI	Rubber and Plastics Industry [IEEE IAS Committee]
RPIE	Real Property Installed Equipment
RPIS	Rod Position Indicator System
RPL	Radar Processing Language
	Radiation Physics Laboratory [US]
	Radiophysics Laboratory [of CSIRO, Australia]
	(Standing Working Group on) Regional Plans [of ICAO]
	Remote Program Loader
	Request Parameter List
	Research Programming Language
	Reverse Polish Logic
	Robot Programming Language
	Rocket Propulsion Laboratory [US Air Force]
	Running Program Language
RPLC	Replace
RPLS	Reactor Protection Logic System
RPM	Random Phase Model
	Raster Processing Machine
	Rate per Minute
	Reinforced Plastic Mortar
	Relaxation Potential Model
	Reliability Performance Measure
	Resale Price Maintenance
	Resupply Provisions Module
	Revolutions per Minute [also: rpm]
	Rollcast Planetary Mill
RPMC	Remote Performance Monitoring and Control
RPMI	Revolutions-per-Minute Indicator
RPN	Reverse Polish Notation
RPNA	Reverse Pacific/North American
RPO	Research Program Office [of CANMET, Canada]
	Revolution per Orbit
RPOA	Recognized Private Operating Agency
RP1D	Repeated Unidirectional
RPP	Radar Power Programmer
	Radar Processing Plant
	Reductive Pentosephosphate
	Reinforced Pyrolyzed Plastic
RP^{2}D	Rational Product and Process Design
RPPG	Radar Planning and Policy Group [of NATS, UK]
RPQ	Request for Price Quotation
RPR	Rapid Power Reduction
	Read Printer
RPRA	Railroad Public Relations Association [US]
	Rubber and Plastics Research Association [UK]
RPROM	Reprogrammable Read-Only Memory
RPRT	Report
RPRV	Remotely-Piloted Research Vehicle
RPRWP	Reactor Plant River Water Pump
RPS	Radar Plotting Sheet
	Radar Position Symbol
	Random Program Selection
	Reactive Polystyrene
	Reactor Protection System
	Real-Time Photogrammetry System
	Real-Time Processing System
	Real-Time Programming System
	Records per Sector
	Reduced Pressure Spraying
	Refrigerator-Cooled Pump System
	Regional Pressure Setting
	Regulatory Performance Summary
	Relative Performance Score
	Remote Printing System
	Remote Processing Service
	Research and Promotion of Structures
	Revolutions per Second [also: rps]
	Riser Pouring System
	Robot Programming System
	Rotational Position Sensing
	Royal Photographic Society [UK]
RPSGB	Royal Photographic Society of Great Britain
RPSM	Resources Planning and Scheduling Method
RPSMA	Rope Paper Sack Manufacturers Association [now: PSSMA, US]
RPSMG	Reactor Protective System Motor Generator
RPSP	Reference Preparation for Serum Proteins [of College of American Pathologists, US]
RPT	Recirculation Pump Trip
	Repeat (Character)
	Report
	Royal Polytechnic Institute [UK]
RPTO	RADARSAT Program Technical Office
RPU	Radio Phone Unit
	Radio Propagation Unit
	Real-Time Photogrammetry Unit
	Regional Processing Unit
RPV	Reactor Pressure Vessel
	Remote Piloted Vehicle
	Remote Pilotless Vehicle
RPW	Resistance Projection Welding
RQ	Request
	Respiratory Quotient
R/Q	Resolver/Quantizer
RQA	Recursive Queue Analyzer
R&QA	Reliability and Quality Assurance
RQD	Required
	Rock Quality Designation
RQE	Relative Quantum Efficiency
RQI	Request for Initialization
RQL	Reference Quality Level

RQMNT Requirement
RQMS Regimental Quartermaster Sergeant
RQMT Requirement
RQP Request Flight Plan
RQRD Required
RQS Rate Quoting System
Request Supplementary Flight Plan
RQT Relativistic Quantum Theory
RQUMT Requirement
RR Radio Range
Radio Regulation
Railroad
Rapid Rectilinear (Lens)
Readout and Relay
Receive Ready
Receiving Report
Recurrence Rate
Register-to-Register
Relay Rack
Release Report
Rendezvous Radar
Repeatable Runout
Repetition Rate
Report Reaching
Research Reactor
Research Report
Resonance Raman (Spectroscopy)
Retroreflector
Return Rate
Return Register
Retro Rocket
Revised Report
Rocket Research
Round Robin
Running Reverse
Rural Road
Rural Route
Ruthenium Red
R/R Readout and Relay
Record/Retransmit
r/R Inside Radius/Outside Radius Ratio
RRA Radar Recording and Analysis
Remote Record Address
Restrictive Requirement Quality A
Retrogression and Reaging Treatment
Rubber Research Association [Israel]
RRAE Radar Recording and Analysis Equipment
RRAR ROM (= Read-Only Memory) Return Address Register
RRB Railroad Retirement Board [US]
Restrictive Requirement Quality B
RRC Remodeling and Rehabilitation Committee [now: National Remodelers Council [of NAHB, US]]
Rollin' Rock Club [US]
R&RC Reactors and Reactor Controls [IEEE NPS Committee]
RRCC Redwood Region Conservation Council [US]
RRD Requisition Received Date
RRDS Relative-Record Data Set
RRE Radar Research Establishment [UK]
Receive Reference Equivalent
Royal Radar Establishment [UK]
R&RE Radiation and Repair Engineering
RREAC Royal Radar Establishment Automatic Computer
RREG Regional Rheoencephalography
RRF Retail Research Foundation [Canada]
RRG Relay Rack Ground
Resource Request Generator
RRI Range-Rate Indicator
Re-Route Inhibit
Research Reactor Institute [of Kyoto University, Japan]
Rocket Research Institute [US]
Rowett Research Institute [Scotland]
Rubber Research Institute [Malaysia]
RRIC Rubber Research Institute of Ceylon [now: RRISL]
RRIM Reinforced Reaction Injection Molding
Rubber Research Institute of Malaysia
RRIN Rubber Research Institute of Nigeria
RRIS Railroad Research Information Service
Record Room Interrogation System
Remote Radar Integration Station
RRISL Rubber Research Institute of Sri Lanka
RRIWB River Research Institute, West Bengal [India]
RRKM Reaction Rate Kinetic Molecular Theory
RRL Radio Relay Link
Radio Research Laboratories [Japan]
Regional Research Laboratory [India]
Road Research Laboratory [UK]
RRLC Redwood Region Logging Conference [US]
RR&M Risk, Reliability and Maintenance [also: RRM]
RRMC Royal Roads Military College [Canada]
RRMG Reactor Recirculation Motor Generator
r-RNA Ribosomal Ribonucleic Acid
RRN Relative-Record Number
RRNS Redundant Residue Number System
RRO Responsible Reporting Office
RROS Resistive Read-Only Storage
RRP Reactor Refueling Plug
Reader and Reader-Printer
Recommended Retail Price
Runway Reference Point
RRPI Relative Rod Position Indicator
Resident Required Page Index
Rotary Relative Position Indicator
RRPT Receiving Report
RRPTN Receiving Report Number
RRR Raleigh Research Reactor [also: NCSCR, US]
RRRR Reduction, Reuse, Recycling and Recovery [also: 4R]
RRRS Route Relief Requirements System
RRRV Rate of Rise of Recovery Voltage
Rate of Rise of Restriking Voltage
RRS Radiation Research Society [US]
Radio Range Station
Radio Ranging System
Radio Relay Station
Radio Research Station [UK]
Reaction Research Society
Reactor Recirculation System
Reactor Refueling System
Reactor Regulation System

	Required Response Spectrum
	Restraint Release System
	Retransmission Request Signal
	Retrograde Rocket System
RRT	Relative Retention Time
	Rendezvous Radar Transponder
	Ring-Ring Trip
	Round Robin Test
RRTC	Retractable Replaceable Thermocouple
RRU	Radiobiological Research Unit
RRV	Remote Reconnaissance Vehicle
RRX	Railroad Crossing
RRZ	Radar Regulation Zone
RS	Radar Scan
	Radiated Susceptibility
	Radioscintillation
	Radiosensitivity
	Radio Simulator
	Radio-Spectroscope
	Railway System
	Raman Spectroscopy
	Random Splice
	Range Selector
	Range Safety
	Range Surveillance
	Rapid Solidification
	Rapid Setting
	Raw Stock
	Ray Society [UK]
	Reactive Sputtering
	Reader Stop
	Real Storage
	Record Separator (Character)
	Reference Section
	Reed-Solomon
	Register Select
	Register-to-Storage
	Registry System
	Relay Switch
	Reliability Society [of IEEE, US]
	Remote Sensing
	Remote Site
	Remote Station
	Request to Send
	Reset (Key)
	Residual Stress
	Resistance Soldering
	Right Side
	Robotic System
	Rolled Shapes
	Route Switching
	Royal Society [UK]
R&S	Research and Statistics
	Reset and Start
	Rouse and Shearer (Plastometer)
R/S	Relay Set
R-S	Register-to-Storage
	Reset-Set
Rs	Rupee [Currency of India]
RSA	Rabbit Serum Albumin
	Radar Signature Analysis
	Railway Supply Association [US]
	Regional Science Association [US]
	Regional Studies Association [UK]
	Related Scientific Activities
	Remote Session Access
	Remote Station Alarm
	Repair Service Attendant
	Republic of South Africa
	Requirements Statement Analyzer
	Rheometrics Solids Analyzer
	Rivest-Shamir-Aldleman (Algorithm)
	Royal Scottish Academy
	Royal Society of Arts [UK]
	Royal Society of Australia
	Rubber Shippers Association [US]
Rsa	Rhodopseudomonas Sphaeroides [Molecular Biology]
RSAC	Radiological Safety Analysis Computer
	Reactor Safety Advisory Committee [Canada]
RSACE	Research and Advanced Communications for Europe
RSARR	Republic of South Africa Research Reactor
RSAS	Royal Swedish Academy of Sciences
RSASA	Royal South Australian Society of Arts
RSB	Reactor Service Building
	Repair Service Bureau
	Reticulocyte Standard Buffer
RSC	Radar Set Control
	Rapid Spinning Cup
	Reactor Safety Commission
	Reed-Solomon Code
	Remote Services Center
	Remote Storage Controller
	Rescue Sub-Center
	Research School of Chemistry
	Reversed Sigmoidal Curve
	Rockwell Science Center [US]
	Royal Society of Chemistry [UK]
	Royal Society of Canada
	Runway Surface Condition
RSCESFS	Rapid Scanning Constant Energy Synchronous Fluorescence Spectrometry
RSCH	Research
RSCIE	Remote Station Communication Interface Equipment
RSCS	Rate Stabilization and Control System
	Remote Spooling Communications Subsystem
	Rod Sequence Control System
RSCW	Research Reactor, State College of Washington [US]
RSD	Radar Signal Discrimination
	Reactive Skin Decontaminant
	Refueling Shutdown
	Relative Standard Deviation
	Remote Site Data-Processing
	Responsible System Designer
	Ring System Descriptor
RS&D	Receipt, Storage and Delivery [also: RSD]
RSDC	Residual Stress Data Collection
RSDL	Reactive Skin Decontamination Lotion
RSDP	Remote Shutdown Panel
	Remote Site Data Processor
RSDS	Range Safety Destruct System

RSE Railway Signal Engineer
Reducing Street Elbow
Register Signaling Equipment
Royal Society of Edinburgh [Scotland]
RSES Refrigeration Service Engineers Society [US]
RSEU Remote Scanner and Encoder Unit
RSEW Resistance-Seam Welding
RSF Relative Sensitivity Factor
Remote Support Facility
RSFSR Russian Soviet Federated Socialist Republic
RSG Resistance Strain Gauge
Rising
RSGB Radio Society of Great Britain
RSGS Royal Scottish Geographical Society
RSH Royal Society of Health [UK]
RSI Rationalization, Standardization and Integration
Reactor Siting Index
Register Sender Inward
Remote Sensing Imagery
Research Studies Institute
Reusable Surface Insulation
Royal Sanitary Institute [UK]
Royal Signals Institution [UK]
RS&I Rules, Standards and Instructions [also: RSI]
RSIC Radiation Shielding Information Center [of ORNL, US]
Redstone Scientific Information Center [US Army]
RSID Resource Identification (Table)
RSIM Repulsion-Start Induction Motor
RSIO Reinforced Steel Institute of Ontario [Canada]
RSIS Reference, Special and Information Section [of Library Association, UK]
RSJ Rolled Steel Joist
RSL Radio Standards Laboratory [US]
Received Signal Level
Requirements Specifications Language
Royal Society of London [UK]
RSLC Radar Searchlight Control
RSLS Receiver Sidelobe Suppression
RSLVR Resolver
RSM Rapidly Solidified Material
Real Storage Management
Regimental Sergeant-Major
Remote Switching Module
Remote System Manager
Resource Management System
Royal School of Mines [UK]
RSMA Radiological Systems Microfilm Associates [US]
RSME Reversible Shape Memory Effect
RSMPS Romanian Society for Mathematics and Physical Sciences
RSN Radiation Surveillance Network
Record Sequence Number
RSNA Radiological Society of North America [US]
Royal School of Naval Architects [UK]
RSNC Royal Society for Nature Conservation [UK]
RSNSW Royal Society of New South Wales [Australia]
RSNZ Royal Society of New Zealand
RSO Register Sender Outward
Revenue Sharing Office
RSORS Remote Sensing On-Line Retrieval System
RSP Radio Switch Panel
Random Signal Processing
Rapidly Solidified Powder
Rapidly Solidified Product
Rapid Solidification Process
Reactivity Surveillance Procedures
Reader/Sorter Processor
Record Select Program
Registered Safety Professional
Replication Selection Process
Required Space (Character)
Respirable, Suspended Particulate
Respond(er)
Restoration Priority
Reverse Scattering Perturbation
Robot Support System
Rotating Shield Plug
RSPA Research and Special Programs Association [US]
RSPB Royal Society for the Protection of Birds [UK]
RSPCA Royal Society for the Prevention of Cruelty to Animals [UK]
RSPD Rapid-Solidification Plasma Deposition
RS/PM Rapid Solidification/Powder Metallurgy
RSPT Real Storage Page Table
Report Starting Procedure Turn
RSR Random Signal Reject
Rapid Solidification Rate
Reactor Safety Research
Restore
Rod Select Relay
Route Sureveillance Radar
RSRA Rotor Systems Research Aircraft
RSRE Royal Signals and Radar Establishment [UK]
RSRI Regional Science Research Institute [US]
RSRS Radio and Space Research Station [US]
RSS Random Selection for Service
Range Safety System
Reactant Service System
Reactor Safety Study
Reactor Shutdown System
Registry Structure Sheet
Relaxed Static Stability
Remote Sensing Society [UK]
Remote Shutdown System
Residual Sum of Squares
Reusable Subsystem
Ribbed Smoked Sheets
Root Sum Square
Route Switching Subsystem
Royal Scientific Society [Jordan]
Royal Statistical Society [UK]
RSSA Railway Systems Suppliers Association [US]
RSSF Retrievable Surface Storage Facility
RSSP Radar Systems Specialist Panel [of ICAO]
RSSS Radiographic Standard Shooting Sketch
RST Rapid Solidification Technology
Reactive Solute Transport
Readability Strength Tone
Remote Station
Research Study Team
Reset

	Reset-Set Trigger
	Restart
	Retained Strength
RSTD	Restricted
RSTMH	Royal Society of Tropical Medicine and Hygiene [UK]
RSTN	Regional Seismic Test Network
RSTS	Resource-Sharing Time-Sharing
RSU	Register Storage Unit
	Remote Service Unit
	Remote Switching Unit
RSV	Rat Sarcoma Virus
	Reserve
	Respiratory Syncytial Virus
	Revised Standard Version
	Royal Society of Victoria [Australia]
	Run-Control Solenoid Valve
RSVP	Remote System Verification Program
RSW	Resistance-Spot Welding
RSX	Resource-Sharing Executive
RT	Radiographic Testing
	Radiological Technician
	Radiological Technologist
	Radiological Technology
	Radio Technician
	Radiotelegraph(y)
	Radiotelephony
	Radiotelescope
	Radio Transformer Unit
	Raise Top
	Range Tracking
	Rated Time
	Ratio Transformer
	Reactance Voltage
	Reaction Turbine
	Reactor Trip
	Real Time
	Receiver Transmitter
	Receptor Technology
	Recrystallization Time
	Rectifier Tube
	Reduction Table
	Reference Temperature
	Refrigerating Technician
	Registered Technician
	Registed Telephone
	Register Ton
	Register Traffic
	Register Transfer
	Register Translator
	Regression Testing
	Related Term
	Relativity Theory
	Relaxation Time
	Released Time
	Remote Terminal
	Reperforator-Terminal
	Reperforator-Transmitter
	Resistance Thermometer
	Resolution Tested
	Response Tone
	Return
	Right
	Ring Trip
	Road Tar
	Robot Trainer
	Rocket Technology
	Room Temperature
	Rotary Table
	Rotary Transformer
	Route
	Route Treatment
	Rubber Technology
	Run-Time
R&T	Research and Technology [also: RT]
R/T	Radiotelegraph(y)
	Radiotelephony
	Real Time
	Register Translator
R-T	Rare-Earth Transition-Metal (Intermetallics)
	Receiver-Transmitter (Unit)
RTA	Rapid Thermal Annealing
	Reliability Test Assembly
	Remote Technical Assistance
	Remote Test Access
	Remote Trunk Arrangement
	Required Time of Arrival
	Roads and Transportation Association
	Rubber Trade Association [US and UK]
RTAC	Real-Time Adaptive Control
	Regional Technical Aids Center [US]
	Roads and Transportation Association of Canada
RTAM	Remote Terminal Access Method
RTAP	Real-Time Automation Program
RTB	Read Tape Binary
	Response/Throughout Bias
	Rural Telephone Bank
RTBM	Real-Time Bit Mapping
RTC	Radar Tracking Center
	Radar Tracking Control
	Railway Transport Committee
	Range Telemetry Center
	Rapid Thermal Chemical Vapor Deposition
	Reader/Tape Contact
	Real-Time Clock
	Real-Time Command
	Real-Time Computer
	Reference Transfer Calibrator
	Remote Terminal Controller
	Removable Top Closure
	Royal Tank Corps [UK]
	Rural Telephone Center
RTCA	Radio Technical Commission for Aeronautics [US]
	Radio Technical Commission of America [US]
RTCC	Real-Time Communications Control
	Real-Time Computer Complex
RTCF	Radial Tube Component Feeder
	Real-Time Computer Facility
RTCM	Radio Technical Commission for Marines [US]
	Radio Technical Commission for Maritime Services [US]

RTCMA Rubber and Thermoplastic Cable Manufacturers Association [UK]
RTCS Real-Time Communications System
Real-Time Composition System
Real-Time Computer System
RTCU Real-Time Control Unit
RTCVD Rapid Thermal Chemical Vapor Deposition
RTD Range Time Decoder
Read Tape Decimal
Real-Time Display
Research and Technology Division [of US Air Force]
Residence Time Distribution
Resistance-Temperature Detector
Resistance Thermometer Detector
Resistive Temperature Device
Resonant Tunelling Diode
RTDC Real-Time Data Channel
RTDD Real-Time Data Distribution
RTDHS Real-Time Data Handling System
RTDS Real-Time Data System
RTE Radio and Telecommunication Engineer
Radio and Telecommunication Engineering
Road Transport Engineer
Road Transport Engineering
Real-Time Executive
Regenerative Turboprop Engine
Remote Terminal Emulator
Residual Total Elongation
Reversible Temper Embrittlement
Route
Run-Time Executive
RTEB Radio Trades Examination Board [UK]
RTECS Registry of Toxic Effects of Chemical Substances [of NIOSH, US]
RTEM Radar Tracking Error Measurement
RTEP Ryerson Test of English Proficiency [Canada]
RTES Real-Time Executive System
Real-Time Expert System
RTF Resistance Transfer Factor
Rich Text Format
Rotary Tube Feeder
RTG Radioisotope Thermoelectric Generator
Radiotelegraphy
RTGB Reactor Turbine General Board
RTHF Real-Time Hydrological Forecasting
RTI Radiation Transfer Index
Railway Tie Association [US]
Real-Time Imaging
Real-Time Interface
Referred to Input
Relative Thermal Index
Research Triangle Institute [US]
Royal Tropical Institute [Netherlands]
RTIC Regional Technical Information Center [now: ILC, UK]
RTIF Rapid Test and Integration Facility [US Navy]
RTIFRAMP Rapid Test and Integration Facility, Rapid Access to Manufactured Parts [US Navy]
RTIO Real-Time Input/Output
Remote Terminal Input/Output
RTIOC Real-Time Input/Output Controller [also: RTI/OC]
RTIP Remote Terminal Interface Package
RTIRS Real-Time Information Retrieval System
RTITB Road Transport Industry Training Board [UK]
RTK Range Tracker
RTL Radio-Isotope Transport Loop
Real-Time Language
Register Transfer Language
Register Transfer Level
Resistor-Transistor Logic
Runtime Library
RTLP Reference Transmission Level Point
RTM Real-Time Module
Real-Time Monitor
Recording Tachometer
Register Transfer Module
Research Technical Memoranda
Resin-Transfer Molding
Response Time Module
Revenue Ton-Mile
Robot Time and Motion
RTMA Radio and Television Manufacturers Association [now: EIA, US]
RTMOS Real-Time Multiprogramming Operating System
RTMS Real-Time Memory System
Real-Time Multiprogramming System
RTN Return
Registered Tradename
Remote Terminal Network
Routine
R/T NET Radio Telephone Network [also: R/T Net]
RTNR Ring-Tone, No Reply
RTO Railway Transport Officer
Referred to Output
Run-Time Organization
RTOAA Rejected Take-Off Area Available
RTOL Restricted Takeoff and Landing
RTOP Real-Time Optional Processing
Research and Technology Operations and Plans [of NASA]
RTOS Real-Time Operating System
RTP Rapid Thermal Processing
Reactor Thermal Power
Real-Time Peripheral
Real-Time Processing
Reinforced Thermoplastics
Remote Terminal Processor
Remote Transfer Point
Requirement and Test Procedures
Research Triangle Park [US]
RTPB Radio Technical Planning Board [US]
RTPC Restrictive Trade Practices Commission
RTPH Round Trips per Hour
RTR Real-Time Radiography
Reinforced Thermosetting Resin
Response Time Reporting
Return and Restore Status
Ribbon-to-Ribbon
RTRC Radio and Television Research Council [US]
RTRD Retard
RTRI Railway Technical Research Institute [Japan]
RTRN Return

RTS Radar Target Simulator
Radar Tracking Station
Range Time Signal
Rapid Transmission and Storage
Reactive Terminal Service
Reactor Trip System
Ready to Send
Reactor Trip System
Real-Time Subroutine
Real-Time System
Relaxation Time Spectrum
Remote Targeting System
Remote Testing System
Request to Send
Return to Service
Royal Television Society [UK]
Rural Telephone System
RTSA Retail Trading Standards Association [UK]
RTSD Resources and Technical Services Division [of ALA, US]
RTSRS Real-Time Simulation Research System
RTSS Real-Time Scientific System
RTST Radio Technician Selection Test
RTT Radiation Tracking Transducer
Radioteletype
Radioteletypewriter
Real-Time Tracer
Recrystallization Time — Temperature (Curve)
Reservoir and Tube Tunnel
RTTDS Real-Time Telemetry Data System
RTTV Real-Time Television
RTTY Radioteletype
Radio Teletypewriter
RTU Real-Time Ultrasonic System
Remote Terminal Unit
Remote Test Unit
Right to Use
RTUA Recognition Technologies Users Association [US]
RTV Remote Television
Retrieve
Room Temperature Vulcanization
Rotating Traveling Vehicle
RTWS Raw Tape Write Submodule
RTX Real-Time Executive
RTZ Return-to-Zero (Recording)
RTZL Return-to-Zero Level
RTZM Return-to-Zero Mark
RTZ(NP) Nonpolarized Return-to-Zero (Recording)
RTZ(P) Polarized Return-to-Zero (Recording)
RU Are You?
Reproducing Unit
Request/Response Unit
Request Unit
Response Unit
Rice University [US]
Rijksuniversiteit te Gent [= Federal University of Gent, Belgium]
Ritsumeikan University [Japan]
Run Unit
Rutgers University [US]
R&U Repairs and Utilities
Ru Ruthenium
RUAS Royal Ulster Agricultural Society [Northern Ireland]
RUB Rubber
RuBP Ribulosebiphosphate
RUD Rudder
RUDH Reserve-Shutdown, Unplanned Derated Hours
RUDH1 Reserve Shutdown, Unplanned Derated Hours, Class 1
RUDI Restricted Use Digital Instrument
RuDP Ribulose Diphosphate
RUF Resource Utilization Factor
RUIN Regional Urban Information Network [US]
RUL Refractoriness under Load [also: RuL]
RUM Remote Underwater Manipulator
Remote Utilization Monitor
Rumania(n)
RUN Rewind and Unload
Royal University of Norway
RUNDH Reserve-Shutdown, Unit Derated Hours
RuP Ribulosephosphate
Ru-5-P Ribulose-5-Phosphate
RUR Rosum's Universal Robot
Rural
RURAX Rural Automatic Exchange
RUS Russia(n)
RUSDIC Russian Dictionary
RUSG Remote Ultrasonic Stream Gage
RUSH Remote Use of Shared Hardware
RUSI Royal United Service Institution [UK]
RUSS Robotic Ultrasonic Scanning System
Russia(n)
RUVP Research Unit on Vector Pathology [of MUN, Canada]
RV Radar Vector
Random Variable
Rated Voltage
Reactive Voltage
Reactor Vessel
Rear View
Recreational Vehicle
Reentry Vehicle
Relative Volume
Relief Valve
Rendezvous
Rescue Vessel
Restriking Voltage
Revised Version
Rifle Volunteers
Robot Vision
Rough Vacuum
RVA Radar Vectoring Area
Reactive Volt-Ampere [also: rva]
Recorded Voice Announcement
Relative Virtual Address
Reliability Variation Analysis
rvalue right value
RVAM Reactive Volt-Ampere Meter
RVB Resonating-Valence Bond
RVC Relative Velocity Computer
Ribonucleoside Vanadyl Complex
RVCI Royal Veterinary College of Ireland

RVCM Residual Vinyl Chloride Monomer
RVDT Rotary Variable Differential Transformer
RVI Reverse Interrupt
RVIA Recreational Vehicle Industry of America [US]
Royal Victoria Institute of Architects [Australia]
RVIS Reactor and Vessel Instrumentation System
RVIT Rotary Variable Inductance Transformer
RVLIS Reactor Vessel Water Level Indication System
RVM Reactive Voltmeter
RVO Runway Visibility by Observer
RVP Reid Vapor Pressure [gasoline]
Rotary Vane Pump
RVR Runway Visual Range
RVRC Runway Visual Range Center
RVRR Runway Visual Range Rollout
RVRT Runway Visual Range Touchdown
RVS Real-Time Visual Simulation
Reverse
RVSS Reactor Vessel Support System
RVT Reliability Verification Tests
Resource Vector Table
RVV Runway Visibility Values
Russell's Viper Venom
RW Radio Wave
Random Walk
Raw Water
Rayleigh Wave
Readily Weldable
Resistance Welding
R/W Read/Write [also: R-W or RW]
Right-of-Way [also: ROW]
Runway
RWA Rotating Wave Approximation
RWAA Resistance Welding Alloy Association
RWAS Royal Welsh Agricultural Society [UK]
RWC Read, Write and Compare
Read, Write and Compute
Read/Write Continue
RWCS Reactor Water Cleanup System
Report Writer Control System
RWCU Reactor Water Cleanup Unit
RWD Rear Wheel Drive
Rewind
RWE Rheinisch-Westfalisches Elektrizitatswerk [= Utility Company of North Rhine-Westphalia, Germany]
RWED Read/Write Extend Delete
RWF Rwandan Franc [Currency of Rwanda]
RWG Ridge Waveguide
Roebling Wire Gauge
RWI Radio Wire Integration
Read, Write and Initialize
RWM Read/Write Memory
Rectangular Wave Modulation
Rod Worth Minimizer
RWMA Resistance Welder Manufacturers Association [US]
RWND Rewind
RWP Radiation Work Permit
Reaction Wave Polymerization
Reactor Work Permit
RWR Read/Write Register
Relative Wear Resistance
RWS Radwaste System
Receiver Waveform Simulation
RWSS River Water Supply System
RWST Refueling Water Storage Tank
RWT Right When Tested
RWTH Rheinisch-Westfalische Technische Hochschule [= Technological Institute of North Rhine-Westphalia, Germany]
RWTUV Rheinisch-Westfalischer Technischer Uberwachungsverein [= Inspection Association of North Rhine-Westphalia, Germany]
RWUM Railway Workers Union of Malawi
RWV Remote Work Vehicle
RWY Railway
Receive(r)
Register to Indexed Storage
Runway
RX Receive(r)
Remote Exchange
Report Crossing
R-X Register to Indexed Storage [also: RX]
Resolver-Transmitter [also: RX]
RXD Received Data
RY Railway
Relay
ry rydberg [Unit]
RY JCT Railway Junction
RYLTY Royalty
RY PT Railway Point
RYS Royal Yacht Squadron [UK]
RZ Reinheitszahl [Biochemistry]
Republic of Zaire
Reset-to-Zero
Return-to-Zero (Recording)
RZL Return-to-Zero Level
RZM Return-to-Zero Mark
Root Zone Method
RZ(NP) Nonpolarized Return-to-Zero (Recording)
RZ(P) Polarized Return-to-Zero (Recording)
RZS Royal Zoological Society [UK]
RZSS Royal Zoological Society of Scotland

S

S	Entropy [Symbol]
	Saint
	Salt
	Sample
	Sand
	Saturday
	Scalar
	Scattering
	Science
	Scuttle
	Sea
	Secondary
	Secret
	Section
	Sediment
	Segment
	Sensitivity
	September
	Serine
	Serum
	Seryl
	Shake
	Shear
	Side
	Siemens
	Sigmoid (Curve)
	Sign
	Signal
	Silicate
	Silk
	Silurian [Geology]
	Silver
	Slate
	Slope
	Smectic Phase
	Snow
	Society
	Software
	Soldering
	Solid
	Solubility
	Source
	South
	Specimen
	Spin
	Spinule
	Spool
	Stack
	Starboard
	State
	Stoke
	Storage
	Straight
	Strain
	Strength
	Stress
	Submarine
	Sucre [Currency of Ecuador]
	Sulfur
	Summertime
	Sunday
	Superconductivity
	Superconductor
	Supervisory
	Supplementary
	Surface
	Switch
	Switchboard
	Synchroscope
	System
s	scattering
	scruple [Unit]
	second
	secondary
	section
	see
	sensitivity
	series
	set
	shear
	single
	slope
	small
	soft
	sol
	solid
	solidus
	soluble
	source
	speed
	sphere
	spherical
	spin
	standard
	steel
	stere [Unit]
	stress
	sublimation
	surface
	symmetrical
	synchronous
SA	Safe Association [US]
	Safety Altitude
	Safety Analysis
	Sail Area
	Salt Added
	Saturated Air
	Scattering Amplitude
	Scientific Authority
	Sedimentation Analysis
	Selected Area (Diffraction)
	Self-Absorption
	Semiannual
	Semianthracite

	Semiautogenous
	Semi-Automatic
	Sense Amplifier
	Sequential Access
	Service Assistant
	Shaft Alley
	Shear Area
	Signature Analysis
	Silicon-Asbestos
	Simple Approach
	Single Armor
	Sinoauricular
	Slow Acting
	Small Arms
	Snubber Adapter
	Soil Association [UK]
	Sound Analysis
	Source Address
	South Africa
	South America
	South Atlantic
	South Australia
	Special Application
	Special Area
	Spectral Analysis
	Spectrum Analyzer
	Speech Analysis
	Spherical Aberration
	Spherical Agglomeration
	Spin Axis
	Stability Analysis
	Stability Augmentation
	State Analysis
	Static Approximation
	Statistical Analysis
	Stearic Acid
	Stochastic Approximation
	Storage Allocation
	Stress Analysis
	Stress Annealing
	Structural Adhesive
	Structural Analysis
	Subassembly
	Subatomic
	Submerged Arc
	Successive Approximation
	Superalloy
	Surface Activity
	Surface Adsorption
	Surface Analysis
	Symbolic Assembler
	System Administrator
	System Analysis
	Systems Analyst
S/A	Subassembly
	Subject to Approval
SAA	Sociedad Argentina de Agronomia [= Argentinian Agricultural Society]
	Society for American Archeology [US]
	Society of Architectural Administrators [US]
	Society of Automotive Analysts [US]
	Solution Annealed and Aged
	South African Airways
	Standards Association of Australia
	Storage Accounting Area
	Student Alumni Association
	Succinic Acid Anhydride
	Sulfuric Anodized Aluminum (Alloy)
	Superconductor Applications Association [US]
	Supima Association of America [US]
	Surface Active Agent
	Systems Application Architecture
SAAC	Space Applications Advisory Committee [US]
SAACB	South African Association of Clinical Biochemists
SAACE	South African Association of Consulting Engineers
SAACI	Salesmen's Association of the American Chemical Industry [now: SACI, US]
SAAEB	South African Atomic Energy Board
SAAI	South African Acoustics Institute
SAAL	Single-Axis Acoustic Levitator
SAAO	South African Astronomical Observatory
SAAP	Saturn/Apollo Application Program [of NASA]
SAAPMB	South African Association of Physicists in Medicine and Biology
SAAS	Science Achievement Awards for Students
	South African Association for the Advancement of Science
	Southern Association of Agricultural Scientists [US]
SAAT	Society of Architectural and Associated Technicians [now: BIAT, UK]
SAATC	Southern African Air Transport Council
SAAU	South African Agricultural Union
SAAUW	South African Association of University Women
SAB	Scientific Advisory Board
	Secondary Application Block
	Silicon-Aluminum Bronze
	Sociedad Argentina de Biologia [= Argentinian Society for Biology]
	Society for Applied Bacteriology [Scotland]
	Solar Alignment Bay
	Space Applications Board
	Stack Address Block
	System Advisory Board
SAb	Serum Antibody
SABAP	Southern African Bird Atlas Project [South Africa]
SABC	South African Broadcasting Corporation
SABE	Society for Automation in Business Education [US]
SABIR	Semi-Automatic Bibliographic Information Retrieval
SABMIS	Seaborne Antiballistic Missile Intercept System [US Navy]
SABO	Sense Amplifier Blocking Oscillator
SABR	Symbolic Assembler for Binary Relocatable Programs
SABRAO	Society for the Advancement of Breeding Research in Asia and Oceania [Japan]
SABRE	Sales and Business Reservations done Electronically
	Secure Airborne Radar Equipment
	Self-Aligning Boost and Reentry System
	Store Access Bus Recording Equipment
SABRU	South African Bird Ringing Unit
SABS	South African Bureau of Standards

Stanford Automated Bibliographic System [US]
SAC Semi-Automatic Coding
Scientific Advisory Committee
Semi-Automatic Coding
Serving Area Concept
Shipbuilding Advisory Council [UK]
Soaring Association of Canada
Sociedad Agronomica de Chile [= Chilean Agronomical Society]
Society for Analytical Chemistry [UK]
Society for Analytical Cytology [US]
Society of Applied Chemistry [US]
Sound Absorption Coefficient
Special Area Code
St. Andrews College [Canada]
Storage Access Channel
Store Access Controller
Store and Clear Accumulator
Strategic Advisory Committee [US]
Strategic Air Command [of NATO]
Student Affairs Committee [of ASM International, US]
Sugar Association of the Caribbean [Trinidad and Tobago]
Sulfacetamide
Surface-Area-Center
Sac Streptomyces Achromogenes [Molecular Biology]
SACAC South African Council for Automation and Computation
SACANGO South African Committee on Air Navigation and Ground Operation
SACCEI Strategic Air Command Communications-Electronics Instruction
SACCI Symmetry Adapted Cluster Configuration Interaction
SACCOMNET Strategic Air Command Communications Network
SACCS Strategic Air Command Control System
SACDA South African Copper Development Association
System Analysis Control and Design Activity
SACDIN Strategic Air Command Digital Information [of NATO]
SACE Saskatchewan Association for Computers in Education [Canada]
SACEUR Supreme Allied Commander, Europe [of NATO]
SACG Synchronous Alternating-Current Generator
SACH Solid Angle Cushioned Heel (Artifical Foot)
SACI Sales Association of the Chemical Industry [US]
South African Chemical Institute
SACL Space and Component Log
SACLANT Supreme Allied Commander, Atlantic [of NATO]
SACMA Suppliers of Advanced Composite Materials Association [US]
SACMAPS Selective Automatic Computational Matching and Positioning System
SACNAS Society for Advancement of Chicanos and Native Americans in Science [US]
SACNET Secure Automatic Communications Network
SACO Swedish Confederation of Professional Associations
SACOM Ships Advanced Communications Operational Model
SACP Selected-Area Channeling Pattern
SACPA South African Cement Producers Association
SACS Swedish Agro Cooperative Services
SACSIR South African Council for Scientific and Industrial Research
SACTTYNET Strategic Air Command Teletype Network
SACTU South African Congress of Trade Unions [Zambia]
SACVT Society of Air Cushion Vehicle Technicians
SAD Safety Assurance Diagram
Sediment Analysis Database
Selected Area Diffraction
Sentence Appraiser and Diagrammer
Small-Angle Diffractometer
Solar Array Drive
Space Antennae Diversity
Store Access Director
Store Address Director
Superaerodynamics
Swiss Association for Documentation
SADA Seismic Array Data Analyzer
Solar Array Drive Assembly
Southern Avalon Development Association [Canada]
Stand-Alone Data Acquisition
Standard Advanced Dewar
SADAP Simplified Automatic Data Plotter
SADAPTA Solar Array Drive and Power Transfer Assembly
SADAS Sperry Airborne Data Acquisition System
SADBU Small and Disadvantaged Business Utilization [US]
SADC Sequential Analog-Digital Computer
Small Angle Data Collection
SADCC Southern African Development Coordination Conference
SADE Solar Array Drive Electronics
Superheat Advanced Demonstration Experiment
SADF Semi-Automatic Document Feed
SADIC Solid-State Analog-to-Digital Computer
SADIE Scanning Analog-to-Digital Input Equipment
Semi-Automatic Decentralized Intercept Environment
Sterling and Decimal Invoicing Electronically
SADL Sterilization Assembly Development Laboratory [of NASA]
SADP Scandinavian Association of Directory Publishers [Denmark]
Selected Area Diffraction Pattern
System Architecture Design Package
SADPO System Analysis and Data Processing Office [of NYPL, US]
SADR Six-Hundred-Megacycles Air Defense Radar
SADRAC South African Defense Research Advisory Committee
SADSAC Sampled Data Simulator and Computer
Seiler ALGOL Digitally Simulated Analog Computer
SADSACT Self-Assigned Descriptors from Self and Cited Titles
SADT Self-Accelerating Decomposition Temperature
Structured Analysis and Design Technique
SADTC SHAPE (= Supreme Headquarters, Allied Powers Europe) Air Defense Technical Center [now: STC of NATO]
SAE Shaft-Angle Encoder
Society of Automotive Engineers [US and Australia]
Spiral Aftereffect

	Stamped Addressed Envelope
SAEA	Soda Ash Export Association [now: ANSAC, US]
SAED	Selected Area Electron Diffraction
SAEH	Society for Automation in English and the Humanities
sael	(sine anno et loco) — without year and place
SAEN	Salmon Association of Eastern Newfoundland [Canada]
SAES	Scanning Auger Electron Spectroscopy
SAET	Society of Architectural and Engineering Technologists [Canada]
SAEWA	South African Electrical Workers Association
SAF	Safety
	Segment Address Field
	Short Address Form
	Societe Astronomique de France [= Astronomical Society of France]
	Society of American Foresters [US]
	Spacecraft Assembly Facility [of NASA]
	Stuck-at-Fault (Test)
	Super Abrasion Furnace Black
SAFA	Shipping and Forwarding Agent
SAFAD	Swedish Agency for Administrative Development
SAFE	Safe Assessment and Facilities Establishment [of JAIF, Japan]
	Safety and Facilities Engineering Department
	Society for the Advancement of Fission Energy [US]
	Society for the Application of Free Energy [US]
	Space and Flight Equipment
	Store and Forward Element
	Survival and Flight Equipment Association [US]
SAFEA	Space and Flight Equipment Association [US]
SAFER	Special Aviation Fire and Explosion Reduction
	Stress and Fracture Evaluation of Rotors
SAFF	Store and Forward Facsimile
SAFFI	Special Assembly for Fast Installation
SAFI	Semi-Automatic Flight Inspection
	Special Assembly for Fast Installation
SAFIRE	Spectroscopy of the Atmosphere using Far-Infrared Emission
SAFOC	Semi-Automatic Flight Operations Center
S AFR	South Africa
SAFRI	South African Forestry Research Institute
SAFSR	Society for the Advancement of Food Service Research [US]
SAFT	Shortest Access First Time
	Synthetic Aperture Focussing Technique
SAFTO	South African Foreign Trade Organization
SAFWA	Southeastern Association of Fish and Wildlife Agencies [US]
SAG	Self-Aligned Gate
	Semi-Autogenous Grinding
	Senior Advisory Group [of IAEA, Austria]
	Special Area Group
	Standard Address Generator
SAGA	Sand and Gravel Association [UK]
	Short-Arc Geodetic Adjustment
	Society of American Graphic Artists [US]
	Statistics of Accidents in General Aviation [of ECAC, France]
	Studies Analysis and Gaming Agency
SAGB	Silk Association of Great Britain
SAGBO	Stress-Assisted Grain-Boundary Oxidation
SAGE	Semi-Automatic Ground Environment
	Solar-Assisted Gas Energy
	Solar Atmospheric Gas Experiment
	Stratospheric Aerosol and Gas Experiment [US]
	Swedish Association of Graduate Engineers
SAGFET	Self-Aligned Gate Field-Effect Transistor
SAGMOS	Self-Aligning Gate Metal Oxide Semiconductor
SAHF	Semi-Automatic Height Finder
SAHPS	Solar-Energy-Assisted Heat Pump System
SAHYB	Simulation of Analog and Hybrid Computers
SAI	Science Applications International [US]
	Societa Astronomica Italiana [= Italian Astronomical Society]
	Society of American Inventors [US]
	Sub-Architectural Interface
SAIA	South Australian Institute of Architects
SAIB	Sucrose Acetate Isobutyrate
SAIC	Science Applications International Corporation
	Scottish Agricultural Improvement Council
SAICE	South African Institute of Civil Engineers [now: SAICET]
SAICET	South African Institute of Civil Engineering Technicians and Technologists
SAID	Safety Analysis Input Data
	Speech Auto-Instructional Device
SAIEE	South African Institute of Electrical Engineerrs
SAIF	South African Industrial Federation
SAIL	Sea-Air Interaction Laboratory [US]
	Shuttle Avionics Integration Laboratory [of NASA]
	Stanford Artificial Intelligence Laboratory [US]
	Stanford Artificial Intelligence Language
SAILA	Sail Assist International Liaison Associates [US]
SAILIS	South African Institute of Librarianship and Information Science
SAILS	Simplified Aircraft Instrument Landing System
SAIM	South African Institute of Mining and Metallurgy [also: SAIMM]
SAIME	South African Institute of Mechanical Engineers
SAIMR	South African Institute for Medical Research
SAIMS	Selected Acquisitions Information and Management System
SAINT	Satellite Interceptor
	Self-Aligned Implantation for N^+-Layer Technology
	Semi-Automatic Indexing of Natural Language
	Symbolic Automatic Integrator
SAIPA	South African Institute for Public Administration
SAIS	School of Advanced International Studies [at JHU, US]
	South African Interplanetary Society
SAISAC	Ship's Aircraft Inertial System Alignment Console
SAIT	South African Institute of Translators
	South Australian Institute of Technology [Australia]
	Southern Alberta Institute of Technology [Canada]
SAIW	South African Institute of Welding
SAKI	Solatron Automatic Keyboard Instructor
SAL	Salinometer
	Saskatchewan Accelerator Laboratory [Canada]
	Semantic Abstraction Language
	Short Approach Light

	Space Astronomy Laboratory [US]
	Structured Assembly Language
	Supersonic Aerophysics Laboratory [US]
	Surface Airlifted (Mail)
	Surface Analysis Laboratory [of SERC, UK]
	Symbolic Assembly Language
	Systems Assembly Language
Sal	Streptomyces Albus [Molecular Biology]
SALA	South African Library Association
SALD	Selected-Area Laser Deposition
SALE	Safeguards Analytical Laboratory Evaluation
	Simple Algebraic Language for Engineers
SALEN	Ethylenebis(salicylcimine)
	Salicylidene Ehtylenediamine
SALINET	Satellite Library Information Network [US and Canada]
SALM	Single Anchor Leg Mooring
	Society of Airline Meteorologists [US]
SALOPHEN	Salicylidene Phenylenediamine
SALORS	Structural Analysis of Layered Orthotropic Ring-Stiffened Shells
SALPN	Salicylidene Propanediamine
SALPS	2-(Salicylideneamino)phenyl Disulfide
SALS	Separate Access Landing System [US]
	Ship Aircraft Locating System
	Short Approach Light System
	Single Anchor Leg Storage
	Solid-State Acoustoelectric Light Scanner
SALSF	Short Approach Light System with Sequenced Flashing Lights
SALT	Saskatchewan Association of Library Technicians [Canada]
	Society for Applied Learning Technology [US]
	Strategic Arms Limitation Talks [between US and USSR]
	Subscribers Apparatus Line Tester
SALV	Salvage
SAM	S-Adenosyl Methionine
	Safety Activation Monitor
	Scalar Audio Magnetotellurics
	Scanning Acoustic Microscopy
	Scanning Auger Microprobe
	Scanning Auger Microscopy
	School of Aerospace Medicine [US Air Force]
	Script-Applier Mechanism
	Selective Automonitoring
	Semantic Analyzing Machine
	Semi-Autogenous Milling
	Semi-Automatic Mathematics
	Sequential Access Memory
	Sequential Access Method
	Serial-Access Memory
	Service Attitude Measurement
	Simulation of Analog Methods
	Society for Advancement of Management [US]
	Sort and Merge
	South America [also: S AM]
	Special Application Module
	Strain Absorbent Module
	Stratospheric Aerosol Measurement
	Strong Absorption Model
	Subsequent Address Message
	Substitute Alloy Material
	Supply Analysis Model
	Surface-to-Air Missile
	Suspension Automatic Monitor
	Symbolic and Algebraic Manipulation
	System Activity Monitor
	Systems Adapter Module
	Systems Analysis Module
	Systems for Automated Manufacture
SAMA	Saudi Arabian Monetary Authority
	Scientific Apparatus Makers Association [US]
	Scottish Agricultural Machinery Association
SAMANTHA	System for the Automated Management of Text in Hierarchical Arrangement
SAMD	Surface-to-Air Missile Development [also: SAM-D]
SAME	Society of American Military Engineers [US]
S AMER	South America
SAMG	Sociedad Argentina de Mineria y Geologia [= Argentinian Society of Mining and Geology]
SAMI	Sensory Adaptive Machines Incorporated [Canada]
	Service and Maintenance Indicator
SAMICS	Solar Array Manufacturing Industry Costing Standards
SAMIS	Structural Analysis and Matrix Interpretative System
SAMMIE	Scheduling Analysis Model for Mission Integrated Experiments
	System for Aiding Man-Machine Interaction Evaluation
SAMO	Simulated Ab Inito Molecular Orbitals
SAMOS	Satellite and Missile Observation System
	Silicon and Aluminum Metal-Oxide Semiconductor
	Stacked Gate Avalanche Injection Metal-Oxide-Semiconductor
SAMP	S-1-Amino-2-Methoxymethyl)pyrrolidine
	Sampling
	Sense Amplifier
	Succinoadenosine Monophosphate
SAMPE	Society for the Advancement of Materials and Process Engineering [US]
	Society of Aerospace Material and Process Engineers [US]
SAMS	Satellite Automonitor System
	South African Mathematical Society
	Statospheric and Mesospheric Sounder
	Swiss Academy of Medical Sciences
SAMSARS	Satellite-Based Maritime Search and Rescue System
SAMSAT	South America/South Atlantic [also: SAM SAT or SAM/SAT]
SAMSO	Space and Missile Systems Organization [US Air Force]
SAMSOM	Support-Availability Multisystem Operations Model
SAMSON	Strategic Automatic Message Switching Operational Network
	System Analysis of Manned Space Operations
SAMT	State-of-the-Art Medium Terminal [US Army]
	Swiss Association for Materials Testing
SAMTEC	Space and Missile Test Center [US]
SAN	Sanitary

	Scanning Auger Nanoprobe
	Science Association of Nigeria
	Small Area Network
	Standard Address Number [of BISAC, US]
	Styrene-Acrylonitrile (Polymer)
SANA	Soyfoods Association of North America [US]
SANAE	South African National Antarctic Expedition
SANB	South African National Bibliography [of CSIR]
SANCAR	South African National Committee for Antarctic Research
	South African National Council for Antarctic Research
SANCIAWPRC	South African National Committee of the International Association on Water Pollution Research and Control
SANCI	South African National Committee on Illumination
SAND	Shelter Analysis for New Designs
s and c	suspended and capable [Oil and Gas Industry]
SANE	Solar Alternatives to Nuclear Energy
SANOVA	Simultaneous Analysis of Variance
SAN/PC	Styrene-Acrylonitrile/Polycarbonate
SAN/PVC	Styrene-Acrylonitrile/Polyvinyl Chloride
SANR	Subject to Approval No Risk
SANS	Small-Angle Neutron Scattering
SANSS	Structure and Nomenclature Search System
SANTA	Systematic Analog Network Testing Approach
SANZ	Standards Association of New Zealand
SAO	Smithsonian Astrophysical Observatory [US]
SAOUG	South African On-Line User Group
SAP	Semi-Armor Piercing
	Service Access Point
	Share Assembly Program
	Sintered Aluminum Powder
	Start of Active Profile
	Structural Analysis Program
	Subatomic Particle
	Symbolic Address Program
	Symbolic Assembly Program
	Systems Assurance Program
s ap	scruple, apothecaries
SAPA	South African Press Association
SAPE	Society for Professional Education
	Solenoid Array Pattern Evaluator
SAPH	Styrenated and Alkylated Phenol
SAPI	Sales Association of the Paper Industry [US]
SAPIR	System of Automatic Processing and Indexing of Reports
SAPL	Seacoast Anti-Pollution League [US]
SAPLA	Standing Advisory Panel on Library Automation [of LRCC, UK]
SAPON	Saponification
SAPONG	Saponifying
SAPP	Sodium Acid Pyrophosphate
SAPWW	Spanish Association for Purification of Water and Wastewater
SAR	Safety Analysis Report
	Saudi Riyal [Currency of Saudi Arabia]
	Search and Rescue
	Segment Address Register
	Service Analysis Request
	Service Analysis Report
	Simulated Acid Rain
	Sodium Absorption Ratio
	Source Address Register
	South African Rand [Currency of Namibia and South Africa]
	Staffing Audit and Review
	Storage Address Register
	Street Address Record
	Successive-Approximation Register
	Synthetic Aperture Radar
Sar	Sarcosyl
SARA	Sampled Aperture Receiving Array
	Society of American Registered Architects [US]
	Superfund Amendments and Reauthorization Act [US]
SARAH	Search, Rescue and Homing [also: Sarah or sarah]
SARARC	Stable Auroral Red Arc
SARBE	Search and Rescue Beacon Equipment
SARBICA	Southeast Asia Regional Branch of the International Council for Archives [Malaysia]
SARCCUS	South African Regional Committee for the Conservation and Utilization of the Soil
SARCOM	Search and Rescue Communicator
SARCUP	Search and Rescue Capability Update Program
SARDA	Special Agricultural and Rural Development Act
SARDPF	Synthetic Aperture Radar Data Processing Facility
SAREC	Search and Rescue Emergency Center
SAREF	Safety Research Experiment Facility
SARF	South African Road Federation
SARI	Synthetic Aperture Radar Imaging
SARIS	South African Retrospective Information System
SARISA	Surface Analysis by Resonance Ionization of Sputtered Atoms
SARL	South African Radio League
SARM	Set Asynchronous Response Mode
SARMCS	Synthetic Aperture Radar Motion Compensation System
SAROAD	Storage and Retrieval of Aerometric Data [of Environmental Protection Agency, US]
SARP	Schedule and Report Procedure
	Signal Automatic Radar-Data Processing System
	Small Autonomous Research Package
	Sumitomo Alkali Refining Process
SARPS	Standards and Recommended Practices [of ICAO]
SARS	Single-Axis Reference System
SARSAT	Search and Rescue Satellite
	Search and Rescue Satellite Aided Tracking System
SARTS	Switched Access Remote Test System
SARUC	Southeastern Association of Regulatory Utility Commissioners [US]
SAS	Saturated Ammonium Sulfate
	Scandinavian Airlines System
	School of Applied Science
	Secondary Alarm Station
	Security Agency Study
	Segment Arrival Storage
	Selected Applicant Service
	Sequenced Answer Signal
	Serbian Academy of Sciences [Yugoslavia]
	Slovak Academy of Sciences [Czechoslovakia]
	Small-Angle Scattering

	Small Astronomical Satellite
	Society for Applied Spectroscopy [US]
	Sodium Alkane Sulfonate
	Sodium Aluminum Sulfate
	Special Air Service [UK]
	Stability Augmentation System
	Statistical Analysis System
	Support Amplifier Station
	Surface Active Substances
	Sverige-Amerika Stiftelsen [= Sweden-America Foundation]
	Switched Access System
SASA	South African Sugar Association
SASAR	Segmented Aperture — Synthetic Aperture Radar
SASE	Self-Addressed and Stamped Envelope
	Specific Application Service Element
	Statistical Analysis of a Series of Events
SASI	Shugart Associates Standard Interface
	Shugart Associates System Interface
	South African Standards Institute
	System on Automotive Safety Information
SASIDS	Stochastic Adaptive Sequential Information Dissemination System
Sask	Saskatchewan [Canada]
SASLO	South African Scientific Liaison Office
SASM	Society for Applied Science and Mathematics
SASMIRA	Silk and Art Silk Mills Research Association [India]
SASPL	Saturated Ammonium Sulfate Precipitation Limit
SASR	Sri Aurobindo Society Research [India]
SASSI	Synthetic Amorphous Silica and Silicates Industry Association [US]
SASSY	Supported Activity Supply System
SASTA	South African Sugar Cane Technologists Association
SASTU	Signal Amplitude Sampler and Totalizing Unit
SAT	Samoan Tala [Currency of Western Samoa]
	Saturate
	Saturation
	Saturday
	Scholastic Aptitude Test [US]
	Sheet-Feeder Action Table
	Society of Acoustic Technology
	Stabilization Assurance Test
	Stepped Atomic Time
	Subatomic Theory
	Symmetric Antitrapping
	Symmetric Axis Transform
	System Access Technique
SATA	S-Acetylthioglycolic Acid
SATAN	Satellite Automatic Tracking Antenna
	Sensor for Airborne Terrain Analysis
SATANAS	Semi-Automatic Analog Setting
SATAR	Satellite for Aerospace Research [of NASA]
SATCC	Southern African Transport and Communications Commission
SATCO	Senior Air Traffic Control Officer
	Signal Automatic Air Traffic Control
SATCOL	Satellite Network of Columbia
SATCOM	Scientific and Technical Communication Committee [US]
	Satellite Communication Agency [of USDOD]
	Satellite Communications
	Satellite Communication System
SAT'D	Saturated [also: SATD]
SATF	Shortest Access Time First
	Substituted Anilines Task Force [US]
SATG	Saturating
SATIF	Scientific and Technical Information Facility
SATIN	SAGE (= Semiautomatic Ground Environment) Air Traffic Integration
SATIRE	Semi-Automatic Technical Information Retrieval
SATIS	Scientific and Technical Information Service [of NLNZ]
SATIVA	Society for Agricultural Training through Integrated Voluntary Activities [US]
SATKA	Surveillance, Acquisition, Tracking and Kill Assessment [of USDOD]
SATN	Saturation
SATNAV	Satellite Navigation [also: Satnav]
SATNUC	Societe pour les Applications Techniques dans le Domaine de l'Energie Nucleaire [= Society for Technical Applications in the Area of Nuclear Energy, France]
SATO	Self-Aligned Thick Oxide
	Synthetic Aircraft Turbine Oil
SATRA	Shoe and Allied Trades Research Association [Also Known As: SATRA Footwear Technology Center, UK]
SATRAC	Satellite Automatic Terminal Rendezvous and Coupling
SATS	Short Airfield and Tactical Support
	Solar Alignment Test Site
SATSAR	(Study Group on) Satellite Aided Search and Rescue [of ICAO]
SATSTREAM	Satellite Switchstream Service [UK]
SATT	Shear Area Transistion Temperature
	Strowger Automatic Toll Ticketing [US]
SATUCC	Southern African Trade Union Coordination Council [Botswana]
SAU	Smallest Addressable Unit
	St. Andrew's University [UK]
	System Availability Unit
Sau	Staphylococcus Aureus [Molecular Biology]
SAUCERS	Saucer and Unexplained Celestial Events Research Society [US]
SAV	Savings
SAVA	Society for Accelerator and Velocity Apparatus
SAVE	Society of American Value Engineers [US]
	System for Automatic Value Exchange
SAVES	Sizing of Aerospace Vehicle Structures
SAVITAR	Sanders Associates Video Input/Output Terminal Access Resource
SAVOR	Single-Actuated Voice Recorder
SAVS	Safeguards for Area Ventilation System
	Status and Verification System
SAW	Submerged Arc Welding
	Subsidiary Agreement on Water [Canada]
	Surface Acoustic Wave
SAWC	Special Air Warfare Center [US Air Force]
SAWE	Society of Aeronautical Weight Engineers [Name Changed to: Society of Allied Weight Engineers, US]
	Society of Allied Weight Engineers [US]

SAWIC South African Water Information Center
SAWMA Southern African Wildlife Management Association [South Africa]
SAWMARCS Standard Aircraft Weapons Management and Release Control System
SAWO Surface Acoustic Wave Oscillator
SAWP Society of American Wood Preservers [US]
SAWRS Supplementary Aviation Weather Observatories
SAWS Small Arms Weapon System
SAW-S Series Submerged Arc Welding
SAWTRI South African Wool and Textile Research Institute
SAX Small Automatic Exchange
Strong Anion Exchange
SAXS Small Angle X-Ray Scattering
SAYTD Sales Amount Year-to-Date
SAZ South African Rand [Currency of Namibia and South Africa]
SB Bachelor of Science
Salt Bridge
Secondary Battery
Serial Binary
Shipping Bill
Short Bill
Shunting Boundaries
Sideband
Signaling Battery
Simultaneous Broadcast
Sleeve Bearing
Slow-Break
Society for Biomaterials [US]
Sonic Boom
Soot Blower
Southbound
Soybean
Special Billing
Splash Block
Spring Balance
Standby
Stop Bath
Stovebolt
Straight Binary
Stuffing Box
Styrene-Butadiene
Supply Bulletin
Surface Barrier
Synchronization Bit
Systematic Botany
Sb (Stibium) — Antimony
sb stilb [Unit]
SBA School-Based Assessment
Sec-Butyl Alcohol
Security By Analysis
Shared Batch Area
Singapore Booksellers Association
Slurry Blasting Agent
Small Business Administration [US]
Sociedade Brasileira de Automatica [= Brazilian Society for Automation]
Soybean Agglutinin
Standard Beam Approach
Systems Builders Association [US]
SBAA Swedish Business Archives Association
SBAC Society of British Aerospace Companies
SBAFWP Standby Auxiliary Feedwater Pump
SBAMPG Societe Belge d'Astronomie, de Meteorologie et de Physique du Globe [= Belgian Society of Astronomy, Meteorology and Geophysics]
SBASI Single-Bridge Apollo Standard Initiator
SBB Schweizer Bundesbahn [= Swiss Federal Railways]
Sociedade Botanica do Brasil [= Brazilian Botanical Society]
Societe Belge de Biologie [= Belgian Society of Biology]
SBBC Styrene-Butadience Block Copolymer
SBBNF Ship and Boat Builders National Federation [UK]
SBC Silicon Blue Cell
Single-Board Computer
Small Business Computer
Sonic Boom Committee [of ICAO]
Steel-Bonded Carbide
Styrene Block Copolymer
Surface Boundary Condition
System Bus Controller
SBCA Satellite Broadcasting and Communications Association [US]
Sensor-Based Control Adapter
SBCC Southern Building Code Congress [now: SBCCI, US]
SBCCI Southern Building Code Congress International [US]
SBCU Sensor Board Control Unit
SBD Schottky Barrier Diode
Single Button Dial
System Block Diagram
SBDB Small Business Development Branch [of MITT, Canada]
SBDC Small Business Development Corporation
SBE Society of Broadcast Engineers [US]
Supertwisted Birefringence Effect
System Buffer Element
SBECP Small Business Equity Corporations Program
SBET Society of Biomedical Equipment Technicians [now: NSBET, US]
SBF Sociedade Brasileira de Fisica [= Brazilian Society of Physics]
SBFM Silver-Band Frequency Modulation
SBFU Standby Filter Unit
SBG Schweizer Botanische Gesellschaft [= Swiss Botanical Society]
Standard Battery Grade
SBGI Society of the British Gas Industries
SBGTS Standby Gas Treatment System
SBH Schottky Barrier Height
Sodium Borohydride
Strip-Buried Heterostructure
Switch Busy Hour
SBHD Subtrahend
SBI Single Byte Interleaved
Society for Business Information [Germany]
Steel Boiler Institute
Synchronous Backplane Interconnect
SBIG Small Business Investment Grant
SBIP Small Business Intern Program

SBIR Small Business Innovation Research
Storage Bus in Register
SBIS Small Business Information Service [of FBDB, Canada]
SBITE Societe Belge des Ingenieurs des Telecommunications et d'Electronique [= Belgian Society of Telecommunications and Electronics Engineers]
SBITP Small Business Industry Technology Program
SBK Single-Beam Klystron
SBLA Small Business Loans Act
SBLC Standby Liquid Control
SBLG Small Blast Load Generator
SBLOCA Small Break Loss-of-Coolant Accident
SBM Separate Bombardment Mode
Single Buoy Mooring
System Balance Measure
SBMA Steel Bar Manufacturers Association [US]
Steel Bar Mills Association
SBMAC Sociedade Brasileira de Matematica Aplicada e Computacional [= Brazilian Society of Applied and Computational Mathematics]
SBMI School Bus Manufacturers Institute [US]
SBMO Sociedade Brasileira de Microondas [= Brazilian Microwave Society]
SBMPL Simultaneous Binaural Midplane Localization
SBN Small Business Network [US]
Standard Book Number
Strontium Barium Niobate
SBNT Single-Breath Nitrogen Test
SBO Sideband Only
Soy Bean Oil
System for Business Operations
SBP Shore-Based Prototype
Special Boiling Point
Stainless Ball Plunger
SBPA Singapore Book Publishers Association
SBPC Sociedade Brasileira para o Progresso da Ciencia [= Brazilian Society for the Progress of Science]
SBPM Society of British Paint Manufacturers
SBPO Spun-Bounded Polyolefin
SBPT Small-Modular-Weight Basic Protein Toxin
SBR Society for Biological Rhythm [Puerto Rico]
Society of Bead Reseachers [Canada]
Soviet Breeder Reactor [USSR]
Space-Based Radar
Storage Buffer Register
Styrene-Butadiene Rubber
SBRC Santa Barbara Research Center [US]
SBS Satellite Business System [US]
Serially Balanced Sequence
Short Beam Shear
Silicon Bidirectional Switch
Single Buoy Storage
Slow-Break Switch
Stimulated Brillouin Scattering
Styrene-Butadiene-Styrene (Block Copolymer)
Subscript (Character)
Swiss Botanical Society
SBSE Scintillator Backscattered Electron
SBSED Scintillator Backscattered Electron Detector
SBT Self-Briefing Terminal
Six-Bit Transcode
Submarine Bathythermograph
Surface Barrier Transistor
SBTMA Swedish Brick and Tile Manufacturers Association
SBU Station Buffer Unit
System Billing Unit
SBUAM Societe Belge des Urbanistes et Architectes Modernistes [= Belgian Society of Urban Planners and Modern Architects]
SBUV Solar Backscatter Ultraviolet
SBUV/TOMS Solar Backscatter Ultraviolet/Total Ozone Mapper System
SBV Shield Building Vent
SBVS Shield Building Vent System
SBX S-Band Transponder
Small Business Exchange
Subsea Beacon/Transponder
SC Satel Conseil [= Satel Council, France]
Satellite Computer
Saturable Core
Saturation Current
Scale
Scaler
Scanning Coil
Science
Scientific
Scientific Committee
Scintillation Counter
Scotland
Scottish
Screen
Screened Cable
Screwed and Coupled
Search Coil
Search Control
Season Cracking
Secondary Containment
Sectional Center
Section Code
Security Committee [of NATO]
Security Council [of UN]
Selector Channel
Selenium Cell
Self-Capacitance
Self-Charge
Self-Closing
Self-Contained
Semicircular
Semiconductor
Send Common
Sending Complete
Seneca College [Canada]
Sequence Controller
Series Circuit
Series Coil
Service Cable
Service Code
Set Clock
Shaped Charge
Shaping Circuit

	Sheridan College [Canada]
	Short Circuit
	Shift-Control Counter
	Shutdown Controller
	Sierra Club [US]
	Signal Converter
	Silver-Copper (Wire)
	Simple Cubic [also: sc]
	Simulation Council
	Sine-Cosine
	Single Column
	Single Contact
	Single Crystal
	Slave Clock
	Slip Casting
	Slow Curing
	Society for Cryobiology [of FASEB, US]
	Solar Cell
	Solution Chemistry
	Source Code
	South Carolina [US]
	Space Charge
	Space Communications
	Spacecraft
	Special Committee
	Specific Conductivity
	Spectrochemical
	Speed Control
	Splat-Cooled
	Spray Cooling
	Spreading Coefficient
	Squeeze Cast
	Standard Clean
	Standing Committee
	Standing Conference
	Starting Current
	Statistics Canada
	Steam Calorimeter
	Steel Casting
	Steel Cover
	Steering Committee
	Stereochemistry
	Stereo-Comparator
	Stochastic Control
	Stopcock
	Stop-Continue (Register)
	Stress Concentration
	Stress Corrosion
	Strip Cathode
	Structural Clay
	Structural Composite
	Subcommittee
	Subcooling
	Subcritical
	Subject Category
	Subject Code
	Supercharger
	Super-Composite
	Superconductivity
	Superconductor
	Supercooling
	Superimposed Coding
	Superimposed Current
	Supervisory Control
	Suppressed Carrier
	Surface Chemistry
	Sweep Circuit
	Swing Clamp
	Switch Cell
	Switched Capacitor
	Synchrocyclotron
	Synchronous Converter
S/C	Self-Contained
	Short Circuit
	Spacecraft
S-C	Slow-Curing (Asphalt)
Sc	Scandium
	Strato-Cumulus
sc	(scilicet) — namely
SCA	Saskatchewan Construction Association [Canada]
	Scientific Computing Associates [of Yale University, US]
	Secondary Communications Authorization
	Selectivity Clear Accumulator
	Sensor-Controlled Automation
	Sequence Control Area
	Shipbuilders Council of America [US]
	Short Code Address
	Single-Channel Analyzer
	Simulated Core Assembly
	Smoke Control Association [US]
	Sneak Circuit Analysis
	Southern Cotton Association [US]
	Spectrochemical Analysis
	Spinning-Cup Atomization
	Steel-Cored Aluminum
	Subcritical Assembly
	Subsidiary Carrier Authorization
	Subsidiary Communications Authorization
	Symmetrical Component Analysis
	Synchronous Communications Adapter
	System Control Area
Sca	Streptomyces Caespitosus [Molecular Biology]
SCAA	Spill Control Association of America [US]
	Superconductor Applications Association [US]
SCAD	Subsonic Cruise Armed Decoy
SCADA	Supervisory Control and Data Acquisition (System)
SCADS	Scanning Celestial Attitude Determination System
	Simulation of Combined Analog Digital Systems
SCAE	Society for Computer-Aided Engineering [US]
SCALD	Structural Computer-Aided Logic Design
SCALE	Space Checkout and Launch Equipment
SCAM	Spectrum Characteristics Analysis and Measurement
	Synchronous Communications Access Method
SCAMA	Station Conferencing and Monitoring Arrangement
	Switching, Conference and Monitoring Arrangement
SCAMP	Sectionalized Carrier and Multipurpose Vehicle
	State-of-the-Art Computer-Assisted Machine-Tool Project
SCAMPS	Small Computer Analytical and Mathematical Programming System

SCAMS Scanning Microwave Spectrometer
SCAN Second Career Assistance Network
Selected Current Aerospace Notices [of NASA]
Self-Correcting Automatic Navigation
Semiconductor Component Analysis Network
Small Computers in the Arts Network [US]
Stock Control and Analysis
Stock-Market Computer Answering Network [UK]
Student Career Automated Network
Supermarket Computer Answering Service
Surface Condition Analyzer
Switched Circuit Automatic Network
SCAND Scandinavia(n)
SCANDOC Scandinavian Documentation Center
SCANIIR Surface Composition Analysis by Neutral and Ion Impact Radiation
SCANNET Scandinavian Information Retrieval Network [of NORDINFO, Finland]
SCANS Scheduling and Control by Automated Network Systems
System Checkout Automatic Network Simulator
SCANSAR Scanning Synthetic Aperture Radar
SCAN-Test Scandinavian Pulp, Paper and Board Testing Committee [Sweden]
SCAO Standing Conference of Atlantic Organizations [UK]
SCAP Silent Compact Auxiliary Power
SCAPE Self-Contained Atmospheric Protective Ensemble Suit
S CAPS Small Capitals [also: s caps]
SCAR Satellite Capture and Retrieval
Scandinavian Council for Applied Research
Scientific Committee on Antarctic Research [UK]
Submarine Celestial Altitude Recorder
SCARA Selective Compliance Assembly Robot Arm
SCARAB Submersible Craft Assisting Recovery/Repair and Burial
SCARF Santa Cruz Acoustic Range Facility [US]
Side-Looking Coherent All-Range Focussed
SCARS Software Configuration Accounting and Reporting System
SCAS South Carolina Academy of Science [US]
Stability and Control Augmentation System
SCAT School and College Ability Test
Schottky Cell Array Technology
Sequentially-Controlled Automatic Transmitter
Share Compiler-Assembler and Translator
Small Car Automatic Transit System
Space Communication and Tracking
Speed Command of Attitude and Thrust
Supersonic Commercial Air Transport
Surface-Controlled Avalanche Transistor
SCATANA Security Control of Air Traffic and Air Navigation Aids
ScATCC Scottish Air Traffic Control Center
SCATE Stromberg-Carlson Automatic Test Equipment
SCATHA Spacecraft Charging at High Altitude
SCATS Sequential Controlled Automatic Transistor Start
Simulation, Checkout and Training System
SCATT Scientific Communications and Technical Transfer System [US]
SCAUL Standing Conference of African University Libraries [Nigeria]
SCAV Scavenge
SCAW Shielded Carbon-Arc Welding
SCB Segment Control Bit
Selenite Cystine Broth
Silicon Circuit Board
Site Control Block
Societe Chimique Belge [= Belgian Chemical Society]
Stack Control Block
Station Control Block
String Control Byte
System Control Block
ScB Bachelor of Science
SCBA Self-Contained Breathing Apparatus
SCBR Steam-Cooled Breeder Reactor [US]
SCBS System Control Blocks
SCC Satellite Communication Concentrator
Satellite Communications Controller
Satellite Control Center
Science Council of Canada
Science Culture Canada (Program) [of DSS and MOSST]
Secondary Category Code
Secondary Containment Cooling
Sectional Classification Code
Semiconductor Circuit
Serial Communications Controller
Short Circuit Current
Signal Conversion Circuit
Simulation Control Center
Single Channel per Carrier
Single-Conductor Cable
Single Cotton Covered (Wire)
Smooth-Conductor Cable
Society of Cosmetic Chemistry [US]
Somatic Cell Concentration
Space Consultative Committee [UK]
Specialized Common Carrier
Speed Control Circuit
Split-Conductor Cable
Standards Coordinating Committee [of IEEE, US]
Standards Council of Canada
St. Clair College [Canada]
Storage Connecting Circuit
Stress-Corrosion Cracking
Subcarrier Channel
Sudan Chamber of Commerce
Superconducting Characterization Cryostat
Superconducting Coil
Switching Control Center
Synchronous Communications Controller
SCCA Society of Company and Commercial Accountants [UK]
SCCC Self-Consistent Charge and Configuration
Single Channel Communications Controller
SCCDEST Steering Committee on Crossborder Data Exchange in Science and Technology [US]
SCCF Satellite Communication Control Facility
SCCG Subcritical Crack Growth

SCCHLL Standards Coordinating Committee on High-Level Languages [of IEEE, US]
SCCIG Stress Corrosion Crack Initiation and Growth
SCCM Standard Cubic Centimeter per Minute
SCCS Ship Command and Control System
Sodium Chemistry Control System
Source Code Control System
Switching Control Center System
SCCT Specialist in Community College Teaching
SCCU Single Channel Control Unit
SCD Satellite Control Department
Scintillation Detector
Screwed
Single Crystal Diffractometer
Source Control Drawing
Space Control Document
Specification Control Drawing
Subcarrier Discriminator
ScD Doctor of Science
Scintillation Detector
SCDC Steel Castings Development Center [US]
ScDHyg Doctor of Science in Hygiene
SCDP Society for Certified Data Processors
SCDSB Suppressed-Carrier Double Sideband
SCE Saturated Calomel Electrode
Semiconductor Epitaxy
Signal Conditioning Equipment
Signal Conversion Electronics
Signal Conversion Equivalent
Single Charge Exchange
Single Cycle Execute
Situation Caused Error
Society of Carbide Engineers [now: SCTE, US]
Society of Christian Engineers [US]
Society of Cuban Engineers
Standard Calomel Electrode
SCEA Shipping Conference Exemption Act
Signal Conditioning Electronic Assembly
SCEAND Standing Committee on External Affairs and National Defense [Canada]
SCEAIT Standing Committee on External Affairs and International Trade [Canada]
SCEAR Scientific Committee on the Effects of Atomic Radiation [of UN]
SCEDR Special Committee on Electronic Data Retrieval
SCEEE Southeastern Center for Electrical Engineering Education [US]
SCEI Swedish Council of Environmental Information
SCEL Signal Corps Engineering Laboratory [US]
Small Components Evaluation Loop
Standing Committee on Education in Librarianship [UK]
SCEM Superconducting Cryo-Electron Microscope
SCEO System Civil Engineering Office
SCEPC Senior Civil Emergency Planning Committee [of NATO]
SCEPTRE Systems for Circuit Evaluation and Prediction of Transient Radiation Effects
SCEPTRON Spectral Comparative Pattern Recognizer
SCERT Systems and Computers Evaluation and Review Technique
SCETV South Carolina Educational Television [US]
SCEU Selector Channel Emulator Unit
SCF Satellite Control Facility
Scientific Computing Feature
Self-Consistent Field
Semicircular Focusing
Sequential Compatibility Firing
Short Chain Fat
Single Crystal Ferrite
SNAP Critical Facility
Sociedad Colombiana de Fisica [= Colombian Society of Physics]
Societe Chimique de France [= Chemical Society of France]
Sodium Cleaning Facility
Standard Cubic Feet [also: scf]
Staphylococcal Clumping Factor
Stress Concentration Factor
Subcritical Flow
Supercritical Fluid
Switched-Capacitor Filter
System Control Facility
SCFBR Steam-Cooled Fast Breeder Reactor
SCFD Standard Cubic Feet per Day [also: scfd]
SCFEL Standard COSMIC (= Computer Software Management and Information Center) Facility Equipment List
SCFH Standard Cubic Feet per Hour [also: scfh]
SCFL Source-Coupled FET (= Field-Effect Transistor) Logic
SCFM Self-Consistent Field Method
Standard Cubic Feet per Minute [also: scfm]
Subcarrier Frequency Modulation
Subcritical Fracture Mechanics
SCFMO Self-Consistent Field Molecular Orbital
SCFP Self-Consistent Fluctuation Phonon
SCFPA Structural Cement-Fiber Products Association [US]
SCF-Xα-SW Self-Consistent Field Xα Scattered Wave (Method)
SCFZ Slow Cooling Float Zone
SCG Scan Generator
Silicon Carbide Graphite
Slow Crack Growth
Solution Crystal Growth
Steel Carriers Group [US]
Subcritical Crack Growth
SCGA Service to Canadians Graduating Abroad
Sodium-Cooled Graphite Assembly
Southern Cotton Ginners Association [US]
SCH Schedule
Scholar
School
Schooner
Seizures per Circuit per Hour
Separate Confinement Heterostructure
Socket Head (Screw)
SCHDLR Scheduler
SCHED Schedule
SCHEM Schematic
S-CHG Supercharger [also: SCHG]
SCHIRP Salal/Cedar Hemlock Interagency Research Project [Canada]

SCHM Schematic
SCHMOO Space Cargo Handler and Manipulator for Orbital Operations
SCHS Small Component Handling System
SCI Science
Science Citation Index [of ISI, US]
Scientific
Scientist
Security Container Institute [US]
Serial Communications Interface
Societa Chimica Italiana [= Italian Chemical Society]
Societe de Chimie Industrielle [= Society for Industrial Chemistry, France]
Society of the Chemical Industry [UK]
Society of Computer Intelligence
Summer Computer Institute [of Boston Computer Society, US]
Switched Collector Impedance
System Control Interface
SCIA Smart Card Industry Association [US]
SCIAS Society of Chemical Industry, American Section [US]
SCIBP Special Committee for the International Biological Program
SCIC Saskatchewan Council for International Cooperation [Canada]
Semiconductor Integrated Circuit
Single-Column Ion Chromatography
Southern Corn Improvement Conference [US]
SCICFNDT Standing Committee for International Cooperation within the Field of Nondestructive Testing [Netherlands]
SCICON Scientific Control
SCIC Saskatchewan Council for International Cooperation [Canada]
SCIF Static Column Isoelectric Focusing
SCI-FI Science Fiction [also: sci-fi]
SCIGR Standing Committee on International Geoscientific Relations
scil (scilicet) — namely
SCIM Selected Categories in Microfiche
Speech Communication Index Meter
SCIMR Sullivan Center for In-Situ Mining Research [of NMIMT, US]
SCIP Scanning for Information Parameters
Self-Contained Instrument Package
Society of Competitor Intelligence Professionals [US]
SCIR Standing Committee for Installation Rebuilding [UK]
SCIS Safety Containment Isolation System
SCIT South China Institute of Technology [PR China]
SCI TECH Science and Technology
Scientific and Technical
SCJ Science Council of Japan
SCL Security Log
Self-Checking Logic
Sequential Control Logic
Single Channel Monitoring
Society for Computers and Law [UK]
Solar Calorimetry Laboratory [Canada]
Space Charge Layer
Space Charge Limited
Superconducting Lens
Systems Control Language
SCLC Space-Charge-Limited Current
SCLDF Sierra Club Legal Defense Fund [US]
SCLF Self-Contained Liquid-Filled (Cable)
SCLog Security Log
SCM Scientific Calculator Machine
Series-Characteristic Motor
Service Command Module
Shunt-Characteristic Motor
Signal Conditioning Module
Simulated Core Mockup
Simulated Core Model
Single-Channel Modem
Small Core Memory
Software Configuration Management
Spinal Changes in Microgravity
Standard Cubic Meter
STARAN (= Stellar Attitude Reference and Navigation) Control Module
Subdivision of County Municipality
Subscribers Concentration Module
Superconducting Magnet
Supervision Control Module
Symmetrically Cyclically Magnetized (Condition)
ScM Master of Science
SCMA Southern Cypress Manufacurers Association [US]
Systems Communications Management Association
SCMM Semiconductor Memory Module
SCMO Subsidiary Communication Multiplex Operation
SC-MOSFET Surface-Channel Metal-Oxide Semiconductor Field-Effect Transistor
SCMP Software Configuration Management Plan
Systems Cooperative Marketing Program
SCN Satellite Conference Network
Satellite Control Network
Scan(ner)
Self-Compensating Network
Sensitive Command Network
Shortest Connected Network
Specification Change Notice
Succinonitrile
Switched Communication Network
SCNA Sudden Cosmic-Noise Absorption
SCNPWC Standing Committee for Nobel Prize Winners Congresses [Germany]
SCNR Scientific Committee of National Representatives [of Supreme Headquarters, Allied Powers Europe, of NATO]
SCO Subcarrier Oscillator
SCOB Scattered Clouds or Better
SCOBOL Structured Common Business-Oriented Language
SCOCLIS Standing Conference of Cooperative Library and Information Services [UK]
SCOF Self-Contained Oil-Filled
SCOLCAP Scottish Libraries Cooperative Automation Project
SCOM Site Cutover Manager
SCOMO Satellite Collection of Meteorological Observations

SCONUL Standing Conference of National and University Libraries [UK]
SCOOP Scientific Computation of Optimal Programs
Self-Coupled Optical Pickup
SCOOPS Scheme Object-Oriented Programming System
SCOPCAS Standing Committee on Pollution Clearance at Sea
SCOPE Oscilloscope [also: Scope or scope]
Radarscope [also: Scope or scope]
Schedule-Cost-Performance
Scientific Committee on Problems of the Environment [France]
Sequential Customer Order Processing Electronically
Special Committee on Paperless Entries
Specifiable Coordinating Positioning Equipment
Standing Committee on Professional Education
Standing Committee on Professional Exchange [Denmark]
Systematic Computerized Processing in Cataloguing
System for Coordination of Peripheral Equipment
SCOPEP Steering Committee on the Performance of Electrical Products
SCOPT Subcommittee on Programming Technology [of Association for Computing Machinery, US]
SCOR Scientific Computer On-Line Resource
Scientific Committee on Oceanic Research [Canada]
Special Committee on Oceanic Research [of ICSU, France]
Self-Calibrating Omnirange
SCORE Satellite Computer-Operated Readiness Equipment
Selection Copy and Reporting
Signal Communications by Orbiting Relay Equipment
System for Computerized Olympic Results and Events
SCORPI Subcritical Carbon-Moderated Reactor Assembly for Plutonium Investigations
SCORPIO Single Crystal Orientation Rapid Processing and Interpretation Operation
Subcritical, Carbon-Moderated Reactor Assembly for Plutonium Investigations [UK]
Subject-Content-Oriented Retriever for Processing Information On-Line [US]
SCOST Special Committee on Space Technology
SCOSTEP Scientific Committee on Solar Terrestrial Physics [US]
SCOT Scotland
Scottish
Shippers for Competitive Ocean Transportation [US]
SCOTICE Scotland-Iceland Submarine Cable
SCOTS Surveillance and Control of Transmission Systems
SCOTT Supercritical, Once-Through Tube (Experiment)
SCOTT-R Supercritical, Once-Through Tube Reactor Experiment
SCOUG Southern California On-Line User Group [US]
SCOUT Signal Computer, Oscilloscope and Universal Tester
Surface-Controlled Oxide Unipolar Transistor
SCOWR Special Committee on Water Research [of ICSU, France]
Sco X-1 Scorpio X-1 [Astronomy]
SCP Safe Cursor Position
Safety Control Program
Screen Color Photography
Secondary Cross-Connection Point
Semiconductor Physics
Serial Character Printer
Sheffield City Polytechnic [UK]
Single-Cell Protein
Societe de Chimie Physique [= Society of Physical Chemistry, France]
Spherical Candlepower
Spontaneous Crack Propagation
Stromberg-Carlson Practices
Student Conservation Program [US]
Subscribers Call Processing
Supervisory Control Program
Surveillance Communication Processor
Symbol(ic) Conversion Program
System Control Processor
System Control Program(ming)
SCPC Single Carrier per Channel
Single Channel per Carrier
SCPD Scratch Pad
SCPI Scientists Committee for Public Information [US]
Structural Clay Products Institute [US]
SCPRF Structural Clay Products Research Foundation [US]
SCPS Subscribers Call Processing Subsystem
Synchronous Composite Packet Switching
SCPT Self-Consistent Perturbation Theory
SCR Scan Control Register
Scanning Control Register
Screen
Screw
Selective Catalytic Reduction
Selective Chopper Radiometer
Semiconductor-Controlled Rectifier
Sequence-Control Register
Seychelles Rupee [Currency]
Short-Circuit Radio
Signal Conversion Relay
Silicon-Controlled Rectifier
Single Character Recognition
Sodium-Cooled Reactor
Southern Council of Research [US]
Styrene Chloroprene Copolymer
System Change Request
Scr Streptococcus Cremoris [Molecular Biology]
SC&RA Specialized Carriers and Rigging Association [US]
SCRAM Selective Combat Range Artillery Missile
Static Column (Dynamic) Random-Access Memory
SCRAMJET Supersonic Combustion Ramjet [also: Scramjet or scramjet]
SCRAP Super-Caliber Rocket-Assisted Projectile
SCRATA Steel Castings Research and Trade Association [UK]
SCRCC Soil Conservation and Rivers Control Council [New Zealand]
ScrF Streptococcus Cremoris [Molecular Biology]
SCRI Science Court and Research Institute [US]
Steel Can Recycling Institute [US]

SCRIPT Scientific and Commercial Interpreter and Program Translator
SCRN Screen
SCRS Society of Collision Repair Specialists [US]
SCRT Serial Choice Reaction Timer
Subscribers Circuit Routine Tester
SCRTD South California Rapid Transit District [US]
SCS Satellite Communication System
Satellite Control System
Scattering Cross-Section
School of Computer Science
Scientific Computer System
Selected Classification Service [of IPG, France]
SEM (= Scanning Electron Microscope) Cold Stage
Sensor Coordinate System
Separate Channel Signalling
Sequence Coding System [US]
Serbian Chemical Society [Yugoslavia]
Silicon Chip Stack
Silicon-Controlled Switch
Simulated Compton Scattering
Simulation Control Subsystem
Single-Channel Simplex
Small Computer System
SNA (= System Network Architecture) Character String
Society for Computer Simulation [now: SCSI, US]
Society of Cosmetic Scientists [UK]
Sodium Characterization System
Soil Conservation Service [US]
South Carolina State [US]
Southern Computer Service [US]
Space Cabin Simulator
Speed Control Switch
Stabilization and Control System
Stabilization Cabin Simulator
Standard Coordinate System
ScS Specialist in Science
SC&S Strapped, Corded and Sealed
SCSA Steering Committee for Sustainable Agriculture [US]
Soil and Water Conservation Society of America [US]
SCSC Soil and Crop Sciences Center [of TAMU, US]
SCSE South Carolina Society of Engineering [US]
SCSEP Summer Canada Student Employment Program
SCSI Small Computer System Interface
Society for Computer Simulation International [US]
SCSP Supervisory Computer Software Package
SCSPLS South Carolina Society of Professional Land Surveyors [US]
SCSR Ship Construction Subsidy Regulations
SCST Society of Commercial Seed Technologists [US]
State Council of Science and Technology [PR China]
SCSTP Scientific Committee on Solar Terrestrial Physics [US]
SCSU South Carolina State University [US]
System Control Signal Unit
SCT Scanning Telescope
Scatter
Schottky Clamped Transistor
Sector
Single Cassette Tape Reader
Society of Cleaning Technicians [US]
Special Characters Table
Statistical Communication Theory
Step Control Table
Strong-Coupling Theory
Subroutine Call Table
Subscriber Carrier Terminal
Surface Charge Transistor
Surface-Crack Tension Specimen
SCTD Scattered
SCTE Society of Cable Television Engineers [UK]
Society of Carbide and Tool Engineers [US]
SCTF Slab Core Test Facility
Sodium Chemical Technology Facility
SCTI Sodium Components Test Installation [US]
SCTL Short-Circuited Transmission Line
Small Components Test Loop
SCTO Stalled Call Timed Out
SCTOC Satellite Communication Test Operations Center
SCTP Straight Channel Tape Print
SCTPP Straight Channel Tape Print Program
SCTR Sector
SCTS Securities
SCTTL Schottky Clamped Transistor-Transistor Logic
SCU Santa Clara University [US]
Scanner Control Unit
Sensor Control Unit
Serial Communication Unit
Servicing Control Unit
Single Conditioning Unit
Station Control Unit
Storage Control Unit
Subscribers Concentrator Unit
Sulfur Coated Urea
System Control Unit
SCUBA Self-Contained Underwater Breathing Apparatus [also: Scuba or scuba]
SCUC Satellite Communications Users Conference
SCUL Simulation of the Columbia University Libraries [US]
SCULL Serial Communication Unit for Long Links
SCUMRA Societe Centrale de l'Uranium et des Minerals et Metaux Radioactifs [= Central Society for Uranium and Radioactive Minerals and Metals, France]
SCUP School Computer Use Plan
Scupper
SCV Seville Composite Vehicle
Subclutter Visibility
SCW Space-Charge Wave
Standard for Clinical Work
State College of Washington [US]
Supercritical Wing
SCWG Satellite Communications Working Group [of NATO]
SCWIST Society for Canadian Women in Science and Technology
SCWM Special Commission on Weather Modification
SCWPLR Special Committee for Workplace Product Liability Reform [US]

SCWST Society of Canadian Women in Science and Technology

SCX Strong Cation Exchange

SD Doctor of Science
Sample Delay
Schematic Drawing
Science Department
Seasonal Derating
Seasoned Dry
Second Difference
Sedimentary Deposition
Selenium Diode
Self-Diffusion
Send Data
Send Digits
Senior Dean
Serializer/Deserializer
Signal Distributor
Signal-to-Distortion (Ratio)
Silicon Detector
Singapore Dollar [Currency]
Size Distribution
Skin Dose
Soft-Drawn
South Dakota [US]
Spark Discharge
Special Delivery
Special Directive
Specialist Degree
Speech Detector
Spin Density
Spray Drying
Square-Law Detector
Standard Data
Standard Deviation
Standard Displacement
Standardization Dictionary
Standards Development
Start Date
State Department
Statistical Design
Steam Distillation
Strength Differential
Structural Design
Structural Dynamics
Sweep Driver
Synchronous Detector
System Dynamics
Systems Design

S/D Sea Damaged
Sight Draft
Statement of Differences

S-D Signal to Distortion Ratio [also: S/D]

7D Seven Digit (Number)

SDA Screen Design Aid
Shaft Drive Axis
Share Distribution Agency
Soap and Detergent Association [US]
Software Development Association
Source Data Acquisition
Source Data Automation
Specially Denatured Alcohol
Supplier Data Approval
Surface Design Association [US]
Symbolic Device Address

SDAA Skein Dyers Association of America [now: SRPDAA, US]

SDAD Satellite Digital and Analog Display

SDADS Satellite Digital and Analog Display System

SDAID System Debugging Aid

S Dak South Dakota [US]

SDAL Switched Data Access Line

SDAP Systems Development Analysis Program

SDAS Scientific Data Automation System

SDAT Spacecraft Data Analysis Team [of NASA]
Symbol Device Allocation Table

S-DAT Stationary-Head Digital Audio Tape

SDAU Safety Data and Analysis Unit [UK]

SDB Segment Descriptor Block
Society for Developmental Biology [US]
Storage Data Bus
Strength and Dynamics Branch [US Air Force]

SD BL Sand Blasting

S/D-B/L Sight Draft with Bill of Lading

SDBMS Spatial Database Management System

SDBP Small Database Package
Small Database Project

SDBS Samson Database Services [Netherlands]

SDBY Standby

SDC Scientific Documentation Center
Semiconductor Devices Council [of JEDEC, US]
Shutdown Cooling
Signal Data Converter
Society of Dyers and Colorists [UK]
Stabilization Data Computer
Strategic Defense Command [US]
Structural Design Criteria
Submersible Decompression Chamber
Submersible Diving Chamber
Switched Digital Capability
Synchro-to-Digital Converter

SDCA Society of Dyers and Colorists of Australia

SDCC San Diego Computer Center [US]
Small-Diameter Component Cask

SDCE Society of Die Casting Engineers [US]

SDCR Source Data Communication Retrieval

SDD Selective Dissemination of Documentation
Software Design Description
Source-Detector Distance
Speech Direction Detector
Stored Data Description
Synthetic Dynamic Display
System Design Description

SDDC Sodium Dimethyldithiocarbamate

SDDF Salmonid Demonstration and Development Farm [Canada]

SDDL Stored Data Definition Language

SDDTTG Stored Data Definition and Translation Task Group

SDE Society of Data Educators [now: Association for Computer Educators, US]
Source Data Entry
Students for Data Education

	Submission and Delivery Entity
	Syntax Directed Editor
	System Development Engine
SDEC	Sodium Diethyl Dithiocarbamate
SDMC	Sodium Dimethyl Dithiocarbamate
SDF	Saskatchewan Development Fund [Canada]
	Satellite Distribution Frame
	Seasonal Derating Factor
	Ship Design File
	Simplified Directional Facility
	Single Degree-of-Freedom
	Software Development Facility
	Source Development Fund
	Standard Data Format
	State Defense Forces [US]
	Stepdown Fix
	Supergroup Distribution Frame
SDFA	State Defense Forces Association [US]
SDFAUS	State Defense Forces Association of the United States
SDFC	Saskatchewan Development Fund Corporation [Canada]
	Space Disturbance Forecast Center [of ESSA, US]
SDFL	Schottky Diode FET (= Field-Effect Transistor) Logic
SDFS	Standard Disk Filing System
SDG	Siding
	Simulated Data Generator
SDH	Seasonal Derating Hours
	Self-Dumping Hopper
	Sorbitol Dehydrogenase
	Succinic Dehydrogenase
SDHT	Selectively-Doped Heterojunction Transistor
SDI	Selective Dissemination of Information
	Silt Density Index
	Source Data Information
	Standard Disk Interface
	Steel Deck Institute [US]
	Steel Door Institute [US]
	Strategic Defense Initiative [US]
SDIF	Standard Document Interchange Format
SDILINE	Selective Dissemination of Information using MEDLINE [US]
SDIM	System for Documentation and Information in Metallurgy [of BAM, Germany]
SDIO	Serial Digital Input/Output [also: SDI/O]
	Strategic Defense Initiative Organization [US]
SDIT	Service de Documentation et d'Information Techniques de l'Aeronautique [= Technical Documentation and Information Service for Aeronautics [also: SDITA, France]]
SDK	System Design Kit
SDL	Saddle
	Software Design Language
	Software Development Language
	Space Disturbances Laboratory [of ESSA, US]
	Space Dynamics Laboratory [of Utah State University, US]
	Specification and Description Language
	Surface Display Library
	Submersible Diver Lock-Out
	System Descriptive Language
	System Design Language
	System Directory List
SDLC	Software Development Life Cycle
	Synchronous Data Link Communication
	Synchronous Data Link Control
SDLC/BSC	Synchronous Data Link Communication/Binary Synchronous Communication
SDLTS	Scanning Deep Level Transient Spectroscopy
SDM	Schwarz Differential Medium
	Selective Dissemination on Microfiche
	Semiconductor Disk Memory
	Sequence Division Multiplexing
	Shutdown Mode
	Space-Division Multiplexing
	Standardization Design Memorandum
	STARAN (= Stellar Attitude Reference and Navigation) Debug Module
	Statistical Delta Modulation
	Symmetric Dimer Model
	Synchronous Digital Machine
	System Definition Manual
SDMA	Shared Direct Memory Access
	Space-Division Multiple Access
SDMS	Software Development and Maintenance System
	Spatial Data Management System
SDN	Societe des Nations [= League of Nations]
	Software Defined Network
	Subscriber's Directory Number
	Synchronous Digital Network
SDNS	Software Defined Network Service
SDO	Shielded Diatomic Orbitals
SDP	Site Data Processor
	Signal Data Processor
	Signal Dispatch Point
	Slowing-Down Power
	Source Data Processing
	Special Development Program
	Sputter Depth Profiling
	Standard Depth of Penetration
	Stationary Diffraction Pattern
	Sudanese Pound [Currency]
	4,4'-Sulfonyldiphenol
	Supplier Development Program [of IRAP, Canada]
SDPA	Styrenated Diphenyl Amine
SDPL	Servomechanisms and Data Processing Laboratory [US]
SDR	Search Decision Rule
	Sender
	Service Difficulty Report
	Signal-Dependent Response
	Signal-to-Distortion Ratio
	Significant Deficiency Report
	Single-Drift Region
	Small Development Requirement
	SNAP Developmental Reactor
	Sodium-Cooled Deuterium-Moderated Reactor [US]
	Special Drawing Rights
	Standard Dimension Ratio
	Statistical Data Recorder

Storage Data Recorder
System Design Review
SDRT Slot Dipole Ranging Test
SDS Safety Data Sheet
Salzgitter-Dolomit-Schlacke (Process)
Sampled Data System
Satellite Data System
Scientific Data Systems
Shared Data Set
Signal-Dependent Stereo
Simulating Digital Systems
Simulation Data Subsystem
Sodium Dodecylsulfate
Software Development Specifications
Software Development System
South Dakota State [US]
Space Documentation Service [of EURODOC]
Supplier Delivery Schedule
System Data Synthesizer
System Dynamics Society [US]
SDSI Shared Data Set Integrity
SDSMT South Dakota School of Mines and Technology [US]
SDS-PAGE Sodium Dodecylsulfate — Polyacrylamide Gel Electrophoresis
SDS/SDR Signal-Dependent Stereo/Signal-Dependent Response
SDSU San Diego State University [US]
South Dakota State University [US]
SDSW Sense Device Status Word
SDT Schematic Design Technology
Serial Data Transfer
Serial Data Transmitter
Shaft Drilling Technician
Simulated Data Tape
Single Delayed Trapping
Society of Dairy Technology [UK]
Soft Drink Technologist
Soft Drink Technology
Start-Data-Traffic
Step-Down Transformer
System Development Tool
SDTA Stress-Dependent Thermal Activation
SDTR Serial Data Transmitter/Receiver
SDU Satellite Data Unit
Signal Distribution Unit
Source Data Utility
Station Display Unit
System Data Unit
SDV Scram Discharge Volume
Slowed-Down Video
SDW Segment Descriptor Word
Spin-Density Wave
Standing Detonation Wave
SDX Satellite Data Exchange
S+DX Speech plus Duplex
SE Safety Engineer
Safety Engineering
Safety Evaluation
Sanitary Engineer
Sanitary Engineering
School of Engineering
Search
Secondary Electrode
Secondary Electron
Secondary Emission
Self-Energy
Self-Excitation
Service Entrance
Service Expansion
Shielding Effect
Shielding Effectiveness
Sign Extend
Simultaneous Engineering
Single-Engine
Site Engineer
Slip End
Small End
Smoke Eliminator
Society of Ethnobiology [US]
Society of Engineers [UK]
Software Engineer
Software Engineering
Soil Engineer
Soil Engineering
Solar Energy
Solvent Extraction
Southeast
Spectroscopic Ellipsometry
Spike Energy (Method)
Spontaneous Emission
Standard Electrode
Standard Error
Starch Equivalent
State Estimation
Static Electricity
Stationary Engineer
Stationary Engineering
Statistical Equilibrium
Steam Emulsion
Steam Engine
Stellar Energy
Stimulated Emission
Stock Exchange
Stop Element
Stored Energy
Strain Energy
Street Elbow (Pipe Fitting)
Strong Electrolyte
Structural Engineer
Structural Engineering
Subcritical Experiment
Sulfoethyl
Supported End
Surface Engineer
Systems Engineering
System Enhanced
S&E Safety and Environment [also: SE]
S/E Stock Exchange
Se Selenium
se (salvo errore) — errors excepted
SEA Science and Education Administration
Sea Education Association [US]

	Simultaneous Engineering Automation
	Societe d'Electronique et d'Automatisme [= Society of Electronics and Automation, France]
	Software Engineering Architecture
	Southeast Asia
	Spherical Electrostatic Analyzer
	Standby Emergency Assistance
	Static Error Analysis
	Statistical Energy Analysis
	Structural Engineers Association [US]
	Sudden Enhancement of Atmospherics
	Systems Effectiveness Analyzer
SEAB	Secretary of Energy Advisory Board [US]
	Serial Bus
SEAC	Standard's Eastern Automatic Computer [of NIST, US]
	Structural Engineers Association of Colorado [US]
SEACC	Southeast Alaska Conservation Council [US]
	Structural Engineers Association of Central California [also: SEAOCC, US]
SEACOM	Southeast Asia Commonwealth Submarine Cable [Australia/New Guinea/Papua/Hong Kong/ Malaysia/Singapore]
SEADAG	Southeast Asia Development Advisory Group
SEAFDC	Southeast Asian Fisheries Development Center [Thailand]
SEAIMP	Solar Eclipse Atmospheric and Ionospheric Measurements Project
SEAISI	Southeast Asia Iron and Steel Institute [Philippines]
SEAL	Sea, Air and Land (Personnel)
	Signal Evaluation Airborne Laboratory [of FAA, US]
	Southeast Area Libraries [UK]
	Standard Electronic Accounting Language
SEALS	Severe Environmental Air Launch Study
	Stored Energy Actuated Lift System
SEAM	Scanning Electron Acoustic Microscopy
	(Conference on the) Scientific and Engineering Applications of the Macintosh [US]
	Software Engineering and Management
	Surface Environment and Mining Program
	Symposium on the Engineering Aspects of Magnetohydrodynamics
SEAMEO	Southeast Asian Ministers of Education Organization [Thailand]
SEAMS	Southeast Asia Mathematical Society [Singapore]
SEAN	Scientific Event Alert Network
SEANC	Structural Engineers Association of Northern California [also: SEAONC, US]
SEAO	Structural Engineers Association of Oregon [US]
SEAOC	Structural Engineers Association of California
SEAP	Save the Environment from Atomic Pollution
	Service Element Access Point
SEAPEX	Southeast Asia Petroleum Exploration Society
SEARCC	Southeast Asia Regional Computer Confederation [Singapore]
SEARCH	Scientific Exploratory Archeological Research
	Systemized Excerpts Abstracts and Reviews of Chemical Headlines [US]
SEAS	(Committee for the) Scientific Exploration of the Atlantic Shelf
	School of Engineering and Applied Science
	Strategic Environmental Assessment System [of EPA, US]
SE&AS	School of Engineering and Applied Science
SEASC	Structural Engineers Association of Southern California [also: SEAOSC, US]
SEASCO	Southeast Asia Science Cooperation Office [India]
SEATAC	Southeast Asian Agency for Regional Transport and Communications Development [Malaysia]
SEATO	Southeast Asia Treaty Organization
SEB	Sociedade Entomologica do Brasil [= Brazilian Entomological Society]
	Society for Economic Botany [US]
	Society for Experimental Biology [UK]
	Source Evaluation Board
SEBA	Styrene and Ethylbenzene Association [of SOCMA, US]
SEBM	Society for Experimental Biology and Medicine [US]
SEBS	Styrene Ethyl Butylene Styrene (Block Copolymer)
	Styrene-Ethylene-Butadiene-Styrene
SEC	Sacramento Engineers Club [US]
	Safeguards Equipment Cabinet
	Sanitary Engineering Center [US]
	Sec-Butyl Alcohol
	Secondary
	Secondary Electron Conduction
	Secondary Electron-Coupled (Vidicon)
	Secondary Emission and Conduction
	Secondary Emission Control
	Secretary
	Section
	Securities and Exchange Commission [US]
	Security
	Simple Electronic Computer
	Single Error Correcting
	Size-Exclusion Chromatography
	Special Event Charter Flight
	Standard Error of Calibration
	Switching Equipment Congestion
sec	(salvo errore calculi) — arithmetical errors excepted
	secant
	second
	secondary [Chemicals]
	(secundum) — according to
sec^{-1}	inverse secant
SECA	Solar Energy Construction Association
SECAL	Selective Calling
SECAM	Sequentiel Couleur a Memoire [= Sequential Color and Memory (Television System)]
SECAP	System Experience Correlation and Analysis Program
SECAR	Secondary Radar
	Structural Efficiency Cones with Arbitrary Rings
SECB	Square Edge Close Butt
SECCAN	Science and Engineering Clubs of Canada
SECDED	Single-Bit Error Correction and Double-Bit Error Detection
sec-ft	second-foot [also: sec ft]
sech	hyperbolic secant
$sech^{-1}$	inverse hyperbolic secant

SECNAV Secretary of the Navy [US]
SECO Self-Regulating Error-Correct Coder-Decoder
Sequential Coding
Sequential Control
Station Engineering Control Office
Sustainer-Engine Cutoff
SECON Secondary Electron Conduction
SECOR Sequential Collation of Range [also: Secor]
SECORD Secure Voice Cord Board
SECOR/DME Sequential Collation of Range/Distance-Measuring Equipment [also: Secor/DME]
SECPS Secondary Propulsion System
SECR Secretary [also: SEC or SECY]
SECS Sequential Events Control System
Solar Energy Conversion Systems
SECS Sections
SECT Section
SECTAM Southeastern Conference on Theoretical and Applied Mechanics [US]
SEC TREAS Secretary-Treasurer [also: SEC-TREAS]
SECU Slave Emulator Control Unit
SECURE Systems Evaluation Code Under Radiation Environment
SECT Section
SECTL Sectional
SECV Sociedade Espanol de Ceramica y Vitrio [= Spanish Society of Cermics and Glass]
SECY Secretary [also: SEC or SECR]
SED Sanitary Engineering Division
Science and Engineering Directorate [of NASA]
Secondary Electron Detector
Skin Erythema Dose
Solar Electrodynamics
Space Environment Division [of NASA]
Spectral Energy Distribution
Status Entry Devices
Stochastic Electrodynamics
Strong Exchange Degeneracy
SEd Specialist in Education
SEDA Safety Equipment Distributors Association [US]
SEDCO Saskatchewan Economic Development Corporation [Canada]
SEDD Systems Evaluation and Development Division [of NASA]
SEDIT Sophisticated String Editor
SEDIX Selected Dissemination of Indexes
SEDM Society for Experimental and Descriptive Malacology [US]
Status Entry Device Multiplexer
SeDMC Selenium Dimethyl Dithiocarbamate
SEDP Solar Energy Development Program [Canada]
SEDR Systems Engineering Department Report
SEDS Space Electronic Detection System
Society for Educational Data Systems
Society for the Exploration and Development of Space [US]
SEE Sabotage, Espionage and Embezzlement
Secondary Electron Emission
Secondary Emission Enhancement
Self Electrooptic Effect
Single Events Effects
Society of Electronic Engineers [India]
Society of Environmental Engineers [UK]
Society of Explosives Engineers [US]
Solar Energy Engineering
Southeastern Electric Exchange
Systems Equipment Engineer
SEEA Software Error Effects Analysis
SEEB Southeastern Electricity Board [UK]
SEECCIASDI Standing European Economic Community Committee of the International Association of the Soap and Detergent Industry [Belgium]
SEED Self-Electrooptic-Effect Device
SEEDS Society, Environment and Energy Development Studies
SEEH Super Energy-Efficient Home
SEEK State of the Environment Education Kit [Canada]
Systems Evaluation and Exchange of Knowledge
SEEN Syndicat d'Etudes de l'Energie Nucleaire [= Syndicate for the Study of Nuclear Energy, Belgium]
SEENY Society of Engineers of Eastern New York [US]
seeo (salvo errore et omissione) — errors and omission excepted
SEEPRO Self-Employed Professional
SEER Societe des Electriciens, des Electroniciens et des Radioelectriciens [= Society of Electricians, Electronic Technicians and Radio Technicians, France]
Student Exposition on Energy Resources [of National Energy Foundation, US]
SEES School of Electronic Engineering Science
Secondary Electron Emission Sensing
se et o (salvo errore et omissione) — errors and omissions excepted
SEF Scanning Electron Fractograph
Small End Forward
Software Engineering Facility
Space Education and Foundation
Standard External File
Storage Extension Frame
SEFAR Sonic End Fire for Azimuth and Range
SEFEL Secretariat Europeen des Fabricants d'Emballages Metalliques Legers [= European Secretariat of Manufacturers of Light Metal Packages, Belgium]
SEFI Societe Europeenne pour la Formation des Ingenieurs [= European Society for Engineering Education, Belgium]
SEFOR Southwest Experimental Fast Oxide Reactor [US]
SEFR Shielding Experimental Facility Reactor [of INEL, US]
SEG Segment
Society of Economic Geologists [US]
Society of Exploration Geophysicists [US]
Software Engineering Group
Special Effects Generator
Standardization Evaluation Group
Systems Engineering Group
SEGB Southeastern Gas Board [UK]
SEGD Society of Environmental Graphic Designers [US]
SEGJ Society of Exploration Geophysicists of Japan
SEGM Segment

SEHF Spin Extended Hartree-Fock
SEI Safety Equipment Institute [US]
Secondary Electron Imaging
Society of Engineering Illustrators [US]
Space Exploration Initiative Program [of NASA]
Systems Engineering and Integration
SEIA Security Industry Equipment Association [now: SIA, US]
Solar Energy Industries Association [US]
SEIAC Science Education Information Analysis Center [of ERIC, US]
SEIFSA Steel and Engineering Industries Federation of South Africa
SEINA Solar Energy Institute of North America [US]
SEIP Smelter Environmental Improvement Project [Canada]
Space Exploration Initiative Program [of NASA]
Strategy for Exploration of the Inner Planets
Systems Engineering Implementation Plan
SEIRS Suppliers and Equipment Information Retrieval System [of ICAO, UN]
SEIT Satellite Educational and Informational Television
SEIU Service Employees International Union [US]
SEK Swedish Krona [Currency]
SEL Selection
Selector
Single-Engine Land
Space Environment Laboratory [US]
Stanford Electronics Laboratories [US]
Systems Engineering Laboratories [US]
SELBUS Systems Engineering Laboratories Data Bus
SELCAL Selective Calling
SELDOM Selective Dissemination of MARC (= Machine-Readable Catalogue) [of University of Saskatchewan, Canada]
SELECTAVISION Selected Television Video Disk System
SELFOC Self-Focusing Optical (Instrument)
SELNI Societa Elettronucleare Italiana [= Italian Society for Nuclear Energy]
SELR Saturn Engineering Liaison Request
SELRIP Selected Release Improvement Program
SELS Selsyn (Motor)
SEM Scanning Electron Micrograph
Scanning Electron Microscopy
Seminary
Silylethoxymethyl
Singularity Expansion Method
Society for Experimental Mechanics [US]
Solar Equipment Manufacturers
Space-Environment Monitor
Standard Electronic Module
Standard Error of Means
Standard Estimating Module
Station Engineering Manual
SEMA Societe d'Economie et de Mathematiques Appliques [= Society for Economy and Applied Mathematics, France]
Specialty Equipment Manufacturers Association
Specialty Equipment Market Association [US]
Spray Equipment Manufacturers Association [UK]
SEMAA Safety Equipment Manufacturers Agents Association [US]
SEM-Cl 2-(Trimethylsilyl)ethoxymethyl Chloride
SEMCOR Semantic Correlation
SEME School of Electrical and Mechanical Engineering
SEM/EDX Scanning Electron Microscopy/Energy Dispersive X-Ray Analysis
SEMI Semiconductor Equipment and Materials Institute [US]
Semiconductor Equipment and Materials International
SEMICON Semiconductor Equipment and Materials Institute Conference
Semi-EV Semi-Efficient Vulcanization
SEMIRAD Secondary-Electron Mixed Radiation Dosimeter
SEMLAM Semiconductor Laser Amplifier
SEMLAT Semiconductor Laser Array Technique
SEMPA Scanning Electron Microscopy for Polarization Analysis
SEMPE Socio-Economic Model of the Planet Earth
SEMS Severe Environment Memory System
Solar Environment Monitor Subsystem
SEM/STEM Scanning Electron Microscopy/Scanning Transmission Electron Microscopy
SEMY Seminary
SEN Senate
Senator
Senior
Sense Command
Single Edge Notched (Specimen)
Society of European Nematologists [now: ESN, Scotland]
Software Error Notification
Steam Emulsion Number
SENA Societe d'Energie Nucleaire des Ardennes [= Society for Nuclear Energy in the Ardennes, Belgium]
SENB Single End Notch Beam
SENECA Semantic Networks for Conceptual Analysis
SENL Standard Equipment Nomenclature List
SENN Societa Elettronucleare Nazionale [= National Electronuclear Society, Italy]
SENR Senior
SENTA Societe d'Etudes Nucleaire et de Techniques Advances [= Society for Nuclear Studies and Advanced Techniques, France]
SENTOS Sentinel Operating System
SEO Satellite for Earth Observation [India]
SEOCS Sun-Earth Observatory and Climatology Satellite
SEOS Synchronous Earth Observation Satellite
S/EOS Standard Earth Observation Satellite
SEP Polysiloxan Treated EPDM
Salmonid Enhancement Program [of DFO, Canada]
Search Effectiveness Probability
Self-Elevating Platform
Separation
Separation Parameter
September
Set Endpoint
Society of Engineering Psychologists
Solar Energy Program

	Somatosensory Evoked Potential
	Space Electronic Package
	Standard Electronic Package
	Standard Error of Performance
	Star Epitaxial Planar
	Strain Energized Powder
	Strain Energizing Process
	Systematic Evaluation Program
SEPA	Southeastern Power Administration
SEPBOP	Southeastern Pacific Biological Oceanographic Program
SEPC	Space Exploration Program Council [of NASA]
SEPE	Single Escape Peak Efficiency
SEPG	Separating
SEPGA	Southeastern Pecan Growers Association [US]
SEPM	Society of Economic Paleontologists and Mineralogists [US]
SEPN	Separation
SEPOL	Settlement Problem-Oriented Language
	Soil-Engineering Problem-Oriented Language
SEPP	Software Exchange Pilot Project
	Southeastern Power Pool
	Symposium on Electrometallurgical Plant Practice
SEPS	Service Environment Power System
	Service-Module Electrical Power System
	Severe Environment Power System
	Solar Electric Propulsion Stage
	Solar Electric Propulsion System
SEPT	September
SEQ	Sequence
	Sequential
seq	(sequens) — the following
SEQUEL	Structured English Query Language
SER	Safety Evaluation Report
	Sandia Engineering Reactor [US]
	Satellite Equipment Room
	Selective Electron Reflection
	Sequential Events Recorder
	Serial
	Series
	Service
	Seychelles Rupee [Currency]
	Significant Event Report
	Single Electron Response
	Single-Ended Radiant Tube
	Slip-Energy Recovery
	SNAP Experimental Reactor
	Surface-Enhanced Raman (Scattering)
	Symbol Error Rate
	System Environment Recording
Ser	Serine
	Seryl
SERAI	Societe d'Applications pour l'Industrie [= Society for Industrial Applications, Belgium]
SERAPE	Simulator Equipment Requirements for Accelerating Procedural Evolution
SERB	Study of the Enhanced Radiation Belt [of NASA]
SERC	Science and Engineering Research Council [UK]
	Solar Energy Research Center [Iraq]
	Southeastern Electric Reliability Council
SERCNET	Science and Engineering Research Council Network [UK]
SERDES	Serializer/Deserializer
SEREP	System Environment Recording and Editing Program
	System Error Recording Editing Program
SERF	Sandia Engineering Reactor Facility [US]
	Special Extensive Routine Functions
SERG	Solar Energy Research Group
SERGE	Socially and Ecologically Responsible Geographers [US]
SERGT	Sergeant
SERI	Society for the Encouragement of Research and Invention [US]
	Solar Energy Research Institute [now: NREL, US]
SERIWA	Solar Energy Research Institute of Western Australia
SERIX	Swedish Environmental Research Index [of SCEI]
SERJ	Supercharged Ejector Ramjet
SERL	Services Electronics Research Laboratory [UK]
SERPS	Service Propulsion System
SERR	Serration
SERS	Surface-Enhanced Raman Scattering
	Surface-Enhanced Raman Spectroscopy
SERT	Single-Electron Rise Time
	Society of Electronic and Radio Technicians [UK]
	Space Electrical Rocket Test
SERUG	SII (= Systems Integrators Inc.) Eastern Regional Users Group [US]
SERV	Service
	Surface Effect Rescue Vehicle
SERVO	Servomotor [also: Servo or servo]
SES	Science Ethic Society [US]
	Secondary Electron Spectroscopy
	Service Evaluation System
	Sesone
	Society of Engineering Science [US]
	Socio-Economic Status
	Sodium 2,4-Dichlorophenoxyethyl Sulfate
	Solar Energy Society [US and Canada]
	Special Exchange Service
	Standards Engineering Society [US]
	Strategic Engineering Survey
	Suffield Experimental Station [now: DRES, Canada]
	Superexcited Electronic States
	Surface-Effect Ship
	Swedish Engineers Society [US]
	System External Storage
SE&S	Square Edge and Sound
SESA	Society for Environmental Stress Analysis
	Society for Experimental Stress Analysis [Name Changed to: Society for Experimental Mechanics, US]
	Solvent Extraction Spherical Agglomeration
SESC	Solar Energy Society of Canada
SESCI	Solar Energy Society of Canada, Inc.
SESDA	Small Engine Servicing Dealers Association [US]
SESE	Secure Echo-Sounding Equipment
	Single Entry Single Exit
SESEF	Shipboard Electronic Systems Evaluation Facility
SESF	Superlattice Extrinsic Stacking Fault

SESL Space Environment Simulation Laboratory [of NASA]
SESLP Sequential Explicit Stochastic Linear Programming
SESO Single End Shutoff
SESOC Single End Shutoff Connector
Surface-Effect Ship for Ocean Commerce
SESOME Service, Sort and Merge
SESP Science and Engineering Student Program
SESPA Scientists and Engineers for Social and Political Action [now: SFTP, US]
SESS Saskatoon Engineering Students Society [Canada]
Session
SE SUPP Safety Evaluation Supplement [also: SE Supp]
SET Self-Extending Translator
Set Endpoint Titration
Settlement
Settling
Single Exposure Technique
Software Engineering Terminology
Software Engineering Toolkit
Solar Energy Thermionics
Space Electronics and Telemetry
Stepped Electrode Transistor
Systems Environment Team
SE2 Scientists and Engineers for Secure Energy [also: SESE, US]
SETA Simplified Electronic Tracking Apparatus
Southeastern Telecommunications Association [US]
SETAB Set Tabular Material
SETAR Serial Event Time and Recorder
SETBC Society of Engineering Technologists of British Columbia [Canada]
SETC Society of Environmental Toxicology and Chemistry [Canada]
SETD Scheduled Estimated Time of Departure
SETE Secretariat for Electronic Test Equipment [of USDOD and NASA]
Societe d'Expansion Technique et Economique [= Society for Technical and Economic Expansion, France]
SETF STARAN (= Stellar Attitude Reference and Navigation) Evaluation and Training Facility
SETG Salmon Enhancement Task Group [Canada]
SETI Search for Extraterrestrial Intelligence (Program) [of NASA]
SETP Society of Experimental Test Pilots [US]
SETR Specific Equipment Type Rating [of NATS, UK]
SETU Societe d'Etudes et de Travaux pour l'Uranium [= Society for the Study of Uranium, France]
SEU Single Event Upset
Small End Up
Source Entry Utility
SEUG Screaming Eagles User Group [US; the "Eagle" is a Computer System]
SEV Schweizerischer Elektrotechnischer Verein [= Swiss Association of Electrical Engineers]
Severe
Special Equipment Vehicle
SEVAS Secure Voice Access System
SEVEC Society for Educational Visits and Exchanges in Canada

SEW Sewer
Satellite Early Warning
Steady Expansion Wave
Sonar Early Warning
SEWS Satellite Early Warning System
SEXAFS Surface Extended X-Ray Absorption Fine Structure (Spectroscopy)
SF Safety Factor
Salt Film (Treatment)
Sampled Filter
Saturated Flow
Scale Factor
Science Fiction
Science Frontiers [US]
Sea Flood
Secondary Flow
Select Frequency
Semifinished
Semifriable
Separated Flow
Separate Function
Service Factor
Shape Factor
Shear Flow
Shift Forward
Shock Front
Short Format
Signal-Frequency
Single Feeder
Single Frequency
Sinking Fund
Skip Flag
Slip Flow
Slow-Fast (Wave)
Spectral Factorization
Spontaneous Fission
Spotface
Spotting Fluid
Square Foot
Stacking Fault
Standard Form
Static Field
Static Friction
Steady Flow
Stereoscopic Fluoroscopy
Stochastic Flow
Stopping Factor
Store and Forward
Straightforward
Strain Figure
Stream Function
Stress Field
Strong-Focusing
Structure — Fine
Structural Foam
Structure Factor
Subfile
Subject Field
Supercritical Fluid
Superfluid
Superfluidity

S/F	Store and Forward
Sf	Suriname Florin (or Guilder) [also: Sf; Currency]
S4	Standard Submarine Sonar System [also: SSSS]
SFA	Scientific Film Association [UK]
	Segment Frequency Algorithm
	Semifriable Abrasive
	Short Field Aircraft
	Single Failure Analysis
	Societe Francaise d'Acoustique [= French Acoustical Society]
	Sunfinder Assembly
	Swedish Foundry Association
SFACA	Solid Fuel Advisory Council of America [US]
SFAPS	Spaceflight Acceleration Profile Simulator [of NASA]
SFAR	Special Federal Aviation Regulation
	System Failure Analysis Report
SFAS	Safety Feature Activation System
SFB	Semiconductor Functional Block
	Sensor Feedback
	State Feedback
SFBAEC	San Francisco Bay Area Engineering Council [US]
SFC	Sectored File Controller
	Secure Flight Communication
	Selector File Channel
	Serial File Copy
	Slow Furnace Cooling
	Societe Francaise de Chimie [= French Society of Chemistry]
	Society of Flavour Chemists [US]
	Solar Forecast Center [US Air Force]
	Specific Fuel Consumption
	Supercritical Fluid Chromatography
	Surface
SFCG	Spectrum Fatigue Crack Growth
SFCI	Spirit of the Future Creative Institute [US]
SFC-IR	Supercritical Fluid Chromatography — Infrared
SFCS	Secondary Flow Control System
	Slats and Flaps Control System
SFD	Sudden Frequency Deviation
	System Function Description
SFE	Smart Front End
	Societe Financiere Europenne [= European Financial Society, France]
	Societe Francaise d'Energie Nucleaire [= French Society for Nuclear Energy]
	Society of Fire Engineers
	Solar-Flare Effect
	Stacking Fault Energy
	Supercritical Fluid Extraction
SFEA	Space and Flight Equipment Association [US]
SFEC	San Francisco Engineering Council [US]
SFE/GC	Supercritical Fluid Extraction/Gas Chromatography
SFEM	Southern Farm Equipment Manufacturers [US]
SFEN	Societe Francaise d'Energie Nucleaire [= French Society of Nuclear Energy]
SFF	Solar Forecast Facility [US Air Force]
	Solid Freeform Fabrication
	Spirit of the Future Foundation [now: SFCI, US]
	Standard File Format
SFG	Signal Frequency Generator
SFI	Specification Foundation International [US]
	Support for Innovation
Sfi	Streptomyces Fimbriatus [Molecular Biology]
SFIR	Specific Force Integrating Receiver
SFIT	Swiss Federal Institute of Technology
SFITV	Societe Francaise des Ingenieurs et Techniciens du Vide [= French Society of Vacuum Engineers and Technologists]
SFK	Step Fixture Key
SFL	Saskatchewan Federation of Labour [Canada]
	Substrate-Fed Logic
	Suriname Florin (or Guilder) [Currency]
SFLRP	Society of Federal Labor Relations Professionals [US]
SFM	Scanning Force Microscopy
	Societe Francaise de Metallurgie [= French Metallurgical Society]
	Surface Feet per Minute [also: SFPM, sfm or sfpm]
	Svenska Foreningen for Materialteknik [= Swedish Association for Materials Technology]
	Switching-Mode Frequency Multiplier
SFMC	Societe Francaise de Mineralogie et de Crystallographie [= French Society of Mineralogy and Crystallography]
SFMM	Societe Francaise de Metallurgie et de Materiaux [= French Society for Metallurgy and Materials [also: SF2M]]
SFNA	Stabilized Fuming Nitric Acid
SFO	Service Fuel Oil
	Spaceflight Operations
SF/O	Screen Formatter/Optimizer
SFOC	Spaceflight Operations Complex [of NASA]
SFOF	Spaceflight Operations Facility [of NASA]
SFOM	Segregation Figures of Merit
SFP	Science Foundation of Physics [Australia]
	Second Focal Point
	Security Filter Processor
	Shop Floor Programming
	Slack Frame Program
	Spent Fuel Pit
	Spent Fuel Pool
SFPA	Southern Forest Products Association [US]
SFPE	Society of Fire Protection Engineers [US]
SFPM	Surface Feet per Minute [also: SFM, sfm or sfpm]
SFPS	Society of Fire Protection Specialists
SFPT	Societe Francaise de Photogrammetrie et des Teledetection [= French Society of Photogrammetry and Remote Sensing]
	Society of Fire Protection Technicians [US]
SFR	Shunt Field Rheostat
	Signal Frequency Receive
SFr	Swiss Franc [also: SFR; Currency of Switzerland and Liechtenstein]
SFRC	Steel Fiber Reinforced Concrete
SFRCS	Steam and Feedwater Rupture Control System
SFRTP	Short-Fiber-Reinforced Thermoplastics
SFS	Saybold Furol Seconds
	Society of Food Service Systems [US]
	Sodium Formaldehyde Sulfoxylate
	Software Facilities and Standards
	Sorption Filter System

	Strong-Focusing Synchrotron
	Suomen Standardisomisliitto [= Finnish Standards Association]
	Symbolic File Support
S4S	Surfaced on Four Sides [Lumber]
SFSA	Steel Founders Society of America [US]
SFSC	Shunt Feedback Schottky Clamped
SFSP	Spent Fuel Storage Pool
SFSR	Soviet Federated Socialist Republic
SFSS	Satellite Field Service Station
SFSU	Signal Frequency Signaling Unit
SFT	Saab Friction Tester
	Shaft
	Shift
	Simulated Flight Test
	Societe Francaise des Telecommunications [= French Telecommunications Society]
	Societe Francaise des Thermiciens [= French Society of Heating Engineers]
	Super-Fast Train
	Symposium on Fusion Technology
	System Fault Tolerance
SFTE	Society of Flight Test Engineers [US]
SF2M	Societe Francaise de Metallurgie et de Materiaux [= French Society for Metallurgy and Materials [also: SFMM]]
SFTP	Science for the People (Group) [of American Physical Society, US]
SFTR	Shift Register
	Shipfitter
	Shopfitter
SFTS	Standard Frequency and Time Signals
SFTU	Somali Federation of Trade Unions
SFU	Simon Fraser University [Canada]
	Special Function Unit
	Status Fill-in Unit
SFV	Societe Francaise du Vide [= French Vacuum Society]
SFXD	Semi-Fixed
SFXU	Saint Francis Xavier University [Canada]
SG	Safety Guide
	Sawtooth Generator
	Scanning Gate
	Screen Grid
	Screw Gauge
	Seabird Group [of RSPB, US]
	Secretary General
	Seeded-Gel (Abrasive Grain)
	Semigroup
	Set Gate
	Signal Generator
	Signal Ground
	Silica Gel
	Silicon-Dioxide Glazing
	Sliver Glass
	Single Girder (Crane)
	Single Groove (Insulators)
	Slip Gauge
	Snow Grains
	Solicitor General
	Solution Growth
	Solution of Glucose
	Space Group
	Specific Gravity
	Spheroidal Graphite
	Spin Glass
	Standing Group
	Steam Generator
	Strain Gauge
	Stratigraphy
	Structural Geology
	Study Group
	Subgroup
	Subsurface Geology
	Supergranule
	Supergroup
	Suppressor Grid
	Symbol Generator
	Synchronous Generator
	System Gain
	System Generation
S/G	Solid/Gas Ratio
SGA	Schweizerische Gesellschaft fur Automatik [= Swiss Society for Automation]
	Sea Grant Association [US]
	Sod Growers Association [US]
SGAA	Stained Glass Association of America [US]
SGAE	Studiengesellschaft fur Atomenergie [= Society for the Study of Nuclear Energy, Austria]
SGA of M-A	Sod Growers Association of Mid-America [US]
SGAT	Study Group on Accounting Terminology
SGAUA	Scitex Graphic Arts Users Group
SGB	Schweizerischer Gewerkschaftsbund [= Swiss Federation of Trade Unions]
SGBD	Secondary Grain Boundary Dislocation
SGC	Schweizer Gesellschaft fur Chronometrie [= Swiss Society of Chronometry]
	Solicitor General of Canada
	Standard Gas Cycle
	Standard Geographical Classification
	Superior Geocentric Conjunction
	Supergroup Connector
	Swept Gain Control
SGCAS	Study Group on Certification of Automatic Systems
SGCC	Safety Glazing Certification Council [US]
SGCD	Society of Glass and Ceramic Decorators [US]
SGCF	SNAP Generalized Critical Facility
SGCI	Schweizerische Gesellschaft fur die Chemische Industrie [= Swiss Society of the Chemical Industry]
SGD	Seafloor Geosciences Division [of NORDA, US]
	Self-Generating Dictionary
	Signal Ground
	Singapore Dollar [also: SG$; Currency]
	Society of Glass Decorators [now: SGCD, US]
SGDF	Supergroup Distribution Frame
SGE	Secondary Group Equipment
	Starch Gel Electrophoresis
SGEMP	System Generated Electromagnetic Pulse
SGF	Solution Growth Facility
	Supplementary Ground Field
SGFP	Steam Generator Feed Pump

SGHWR Steam-Generating Heavy-Water-Moderated Reactor [also: SGHR]
SGI Swedish Geotechnical Institute
SGIS Steam Generator Isolation Signal
Student Guidance Information Center
SGIT Special Group Inclusive Tour
SGJP Satellite Graphic Job Processor
SGM Schweizerischer Verband der Grosshandler und Importeure der Motorfahrzeugbranche [= Swiss Association of Wholesalers and Importers of the Automotive Trade Branch]
SGMC Standing Group Meteorological Committee [now: MCMG of NATO]
SGPA Stained Glass Professionals Association [US]
SGSP Society for Glass Science and Practices [US]
Sol Gel Microsphere Pelletization
SGSR Scientific Group for Space Research [Greece]
Society for General Systems Research [now: ISSS, US]
SGL Signal
Single
Society of Gas Lighting [US]
Surface Gridding Library
SGLS Space-Ground Link System
SGM Society for General Microbiology [UK]
SGML Standard Generalized Mark-Up Language
Study Group for Mathematical Learning [US]
SGN Scan Gate Number
Sign
sgn signum function [Mathematics]
SGNMOS Screen-Grid N-Channel Metal-Oxide Semiconductor
SGNR Signature
SGO Semigroup of Operations
SGOR Solution Gas-Oil Ratio
SGOT Serum Glutamine Oxalacetic Transaminase
SGP Society of General Physiologists [US]
Strain Gauge Package
Strategic Grants Program [Canada]
Supergroup
SGPT Serum Glutamate Pyruvate Transaminase
Serum Glutamic Pyruvic Transaminase
SGR Self-Generation Reactor
Self-Generation Recycle
Set Graphics Rendition
Short Growth Rate
Sodium Graphite Reactor [of USEAC]
Steam Gas Recycling
SGRCA Sodium Graphite Reactor Critical Assembly [US]
SGS Saskatchewan Geological Society [Canada]
Society for General Systems
Strain-Gauge System
Symbol Generator and Storage
SGSI Symposium on Gas-Surface Interactions
SGSP Single Groove, Single Petticoat (Insulator)
SGSR Society for General Systems Research [US]
SGSV/DV Steam Generator Stop Valve Dump Valve
SGT Segment Table
Sergeant
Society of Glass Technology [UK]
Special Gas Taper Thread
Strain Gauge Technology
SGTIM Study Group on Topoclimatological Investigation and Mapping [Poland]
SGTR Steam Generator Test Rig
SGTS Standby Gast Treatment System
SGU Sveriges Geologiska Undersokning [= Geological Survey of Sweden]
SGUR Steering Group on Uranium Resources
SGW Silanized Glass Wool
SGX Selector Group Matrix
SGZ Surface Ground Zero
SH Safety Helmet
Sample and Hold
Scleroscope Hardness
Scratch Hardness
Section Heading
Self-Heating
Send Hub
Sensible Heat
Service Hours
Session Handler
Share
Sheet
Shock
Shore
Shot
Shoulder
Shower
Shunt
Single Heterostructure
Socket Head (Screw)
Solid Height
Specific Head
Specific Heat
Specific Humidity
Specimen Holder
Spherical Harmonic
Start of Heading
Static Head
Steinhart and Hart (Thermistor)
Subject Heading
Sulfhydryl
Superheater
Switch Handler
S/H Sample/Hold
SHA Sidereal Hour Angle
Society for Historical Archeology [US]
Sodium Hydroxide Addition
Software Houses Association [UK]
Solid Homogeneous Assembly
Special Handling Area
Spherical-Harmonic Analysis
SHAB Soft and Hard Acids and Bases
SH ABS Shock Absorber
SHAC Society for the History of Alchemy and Chemistry [UK]
SHADCOM Shipping Advisory Committee
SHAP Sintered Hydroxyapatite
SHAPA Solids Handling and Processing Association [UK]
SHAPE Supreme Headquarters Allied Powers Europe [of NATO]

SHARE System for Heat and Radiation Energy
SHARES Shared Acquisition and Retention System
SHARP SHIPS (= Bureau of Ships) Analysis and Retrieval Project [US Navy]
Stationary High-Altitude Relay Platform
SHAU Subject Heading Authority Unit
SH BL Shot Blasting
SHC Silicones Health Council [US]
Superior Heliocentric Conjunction
SHCA Safety Helmet Council of America [US]
SH CON Shore Connection
SHCS Socket Head Cap Screw
SHD Single High Density
Slant Hole Distance
Society for the History of Discoveries [US]
Storage, Handling and Distribution
Super Heavy Duty
Super High Duty
SHE Semi-Homogeneous Experiment
Sodium Heat Engine
Spacecraft Handling Equipment
Spontaneous Hall-Effect
Standard Hydrogen Electrode
Subject Headings for Engineering [US]
SHED Solar Heat Exchanger Drive
SHEED Scanning High-Energy Electron Diffraction
SHEFA Shetland Islands-Faero Islands Submarine Cable
SHELV Shelving
SHEP Solar High-Energy Particles
SHERI System for Hierarchical Experts of Resource Inventories (Program) [Canada]
SHF Sensible-Heat Factor
Storage Handling Facility
Super-High Feed
Super-High Frequency
Superhyperfine
SHFS Superhyperfine Splitting
Superhyperfine Structure
SHFT Shift
SHFTR Shift Register
SHG Second Harmonic Generation
Special High Grade
SHHOFI Safety and Health Hall of Fame International
S-HI System-Human Interaction
SHIA Southern Hardwood Traffic Association [now: FPTA, US]
SHIEF Shared Information Elicitation Facility
SHIELD Sylvania High Intelligence Electronic Defense
SHIMWTS Shipboard Integrated Membrane Wastewater Treatment System
SHINCOM Ship-Integrated Communication (System)
SHINNADS Shipboard Integrated Navigation and Design System
SHINPADS Shipboard Integrated Processing and Display System
SHINS Shipping Instruction
SHIOER Statistical Historical Input/Output Error Rate Utility
SHIPDES Ship Descriptions (Database) [of MIC, Netherlands]
SHIPG Shipping
SHIPS Bureau of Ships [US Navy]
SHIRAN S-Band High-Accuracy Ranging and Navigation [also: Shiran or shiran]
SHIRTDIF Storage, Handling and Retrieval of Technical Data in Image Formation
SHIV Schweizerischer Handels- und Industrieverein [= Swiss Association of Industry and Commerce]
SHL Shell
Shellac
Studio to Head End Link
SHLD Shield
Shoulder
SHLDR Shareholder
SHLMA Southern Hardwood Lumber Manufacturers Association [now: HMA, US]
SHLW Shallow
SHM Sensible Heat Meter
Simple Harmonic Motion
Society for Hybrid Microelectronics
Superhard Material
SHN Scleroscope Hardness Number
SHOC Software Hardware Operational Control
SHODOP Short-Range Doppler [also: Shodop or shodop]
SHORAN Short-Range Navigation [also: Shoran or shoran]
SHOT Society for the History of Technology [US]
SHOW Scripps Institute/University of Hawaii/Oregon State University/University of Wisconsin [US]
SHP Shaft Horsepower [also: shp]
Ship
Shipping
St. Helena Pound [also: SH£; Currency]
SHPDT Shipping Date
SHPE Society of Hispanic Professional Engineers [US]
SHPG Shipping
SHPS Sodium Hydroxide Purge System
SHPT Shipment
SHPNM Ship-to Name
SHR Safety Hoist Ring
SHRAP Shrapnel
SHRI Scottish Horticultural Research Institute
Shrop Shropshire [UK]
SHRP Strategic Highway Research Program [of NRC, US]
SHRP-IDEA Strategic Highway Research Program — Ideas Deserving Exploratory Analysis [of NRC, US]
SHRS Supplementary Heat Removal System
SHRTG Shortage
SHS Scandinavian Herpetological Society [Denmark]
Self-Propagating High-Temperature Synthesis
Small Hydro Society
Surveyors Historical Society [US]
Swivel Head Screw
SHSMB Safety and Health Standards Management Board
SHT Short
Society for the History of Technology
SHTB Saskatchewan Highway Traffic Board [Canada]
SHTC Short Time Constant
SHTG Shortage
SHTHG Sheathing
SHTL Small Heat-Transfer Loop
shton short ton [also: sh tn]
sh tn wt short ton-weight
SHV Superheated Vapor

SHVG Stabilized High-Voltage Generator
SHWR Shower
SHWY Super Highway
SHY Syllable Hyphen (Character)
SHYDRO Small Hydro Costing Program [of PERD, Canada]
SI Safety Injection
Safety Inspection
Salt Institute [US]
Sample Interval
Scientific Instrument
Screen-Grid Input
Selectivity Index
Self-Inductance
Self-Induction
Self-Interstitial
Semi-Insulating
Serial In
Serial Input
Serial Interface
Service Indicator
Shift-In (Character)
Shirley Institute [UK]
Shrinkage Index
Silicone
Silicone Iron
Signal-to-Interference (Ratio)
Signal-to-Intermodulation (Ratio)
Single Instruction
Silver Institute [US]
Smithsonian Institution [US]
Society of Indexers [UK]
Soft Iron
Sound Insulation
Space Industrialization
Spark Ignition (Engine)
Special Instruction
Specific Inventory
Speech Interpolation
Station Identification
Storage Immediate
Staight-In Approach
Structural Integrity
Sulfur Institute [US]
Superimpose
Supplementary Information
Surveyors Institute
Swap-In
System Integration
(Systeme International d'Unites) — International System of Units
S-I Signal-to-Intermodulation (Ratio) [also: S/I]
Si Silicon
SIA Saskatchewan Institute of Agrologists [Canada]
Sasquatch Investigations of Mid-America [US]
Satellite Intensity Analysis
Scaffold Industry Association [US]
Schweizerischer Ingenieur- und Architekten-Verein [= Swiss Association of Engineers and Architects]
Security Industry Association [US]
Self-Interstitial Atom
Semiconductor Industries Association [US]
Service in Information and Analysis [UK]
Society for Industrial Accountants
Society for Industrial Archeology [US]
Society of Insurance Accountants [US]
Sprinkler Irrigation Association [US]
Standard Instrument Approach
Standard Interface Adapter
Stellar Interferometer
Stereo-Image Alternator
Subminiature Integrated Antenna
Systems Integration Area
SIAA Singapore Industrial Automation Association
SIAC Specialized Information Analysis Center
SIAD Society of Industrial Artists and Designers [UK]
SIAE Scottish Institute of Agricultural Engineering [UK]
SIAF Service Indicator Associated Field
SIALON Silicon Aluminum Oxynitride (Ceramic) [also: SiAlON or Sialon]
SIAM Shipborne Ice Alert and Monitoring
Signal Information and Monitoring
Society for Industrial and Applied Mathematics [US]
SIAO Smithsonian Institute Astrophysical Observatory [US]
SIAP Shipbuilding Industry Assistance Program
Statistical Institute for Asia and the Pacific
SIAPS Standard Instrument Approach Procedures
SIAS Safety Injection Actuation Signal
SIAT Single Integrated Attack Team
SIB Satellite Ionospheric Beacon
Screen Image Buffer
Serial Interface Board
Shipbuilding Industry Board [UK]
SIBL Scanning Ion Beam Lithography
SIBM Strain-Induced Boundary Migration
SIBS Stellar-Inertial Bombing System
SIC Science Information Council [of NFS, US]
Scientific Information Center
Semiconductor Integrated Circuit
Service Indicator Code
Shanghai Institute of Ceramics [of Chinese Academy of Science, PR China]
Special Interest Committee
Specific Inductive Capacity
Standard Industrial Classification
Structural Integrity Center [of UKAEA]
Sulfur Impregnated Concrete
Switching Integrated Circuit
SICA Society of Industrial and Cost Accountants [Canada]
SICASP SSR (= Secondary Sureveillance Radar) Improvements and Collision Avoidance System Panel [of ICAO]
SICC Strain-Induced Corrosion Cracking
SICCAPH Special Interest Committee for Computers and the Physically Handicapped [now: SIGCAPH, US]
SICE Society of Instrument and Control Engineers
SICEJ Society of Instrument and Control Engineers of Japan
SiC-LAS Silicon Carbide in Lithium Aluminosilicate

SICN Societe Industrielle de Combustibles Nucleaire [= Industrial Society for Nuclear Fuels, France]
SICO Special Interest Committee
Switched-In for Checkout
SICODCPT Special Interest Committee on Digital Computer Programmer Training
SICOM Securities Industry Communications [US]
SICOMP Siemens Computer
SICOS Sidewall Base Contact Structure
SiCp Silicon Carbide Particles
SICS Safety Injection System
Semiconductor Integrated Circuit
SiCw Silicon Carbide Whiskers
SID Scanning Imaging Device
Scheduled Issue Date
Seismic Intrusion Detector
Serial Interface Device
Sheep Industry Development (Program) [of ASPC, US]
Signal Identification
Silicon Imaging Device
Singapore Dollar [Currency]
Society for Information Display [US]
Society for International Development [Italy]
Sodium Ionization Detector
Solomon Island Dollar [Currency]
Solubilization by Incipient Development
Sound Interface Device
Standard Instrument Departure
Sudden Ionospheric Disturbance
Swift Interface Device
Symbolic Instruction Debugger
Synchronous Identification
Syntax Improving Device
SI$ Solomon Island Dollar [Currency]
SIDA Saskatchewan Implement Dealers Association [Canada]
Software Industry Development Association [Canada]
Swedish International Development Agency
SIDAC Silicon Diode Alternating Current [also: Sidac or sidac]
SIDASE Significant Data Selection
SIDC Strategy for International Development and Cooperation
SIDE Suprathermal Ion Detector Experiment
SIDES Source Input Date Edit System
SIDR Standard Inside Diameter Ratio
SIDS Speech Identification System
Static Inspirable Dust Spectrometer
Stellar Inertial Doppler System
SIE Science Information Exchange
Single Instruction Execute
Society of Industrial Engineers [UK]
SIEPR Sociedad du Ingenieros Estructurolos de Puerto Rico [= Society of Structural Engineers of Puerto Rico]
SIES Soils and Irrigation Extension Service [Australia]
SIF Security and Intelligence Foundation [US]
Selective Identification Feature
Senior Industrial Fellowship
Signaling Information Field
Societa Italiana di Fisica [= Italian Society of Physics]
Sound Intermediate Frequency
Storage Interface Facility
Stress Intensity Factor
SIFT Share Internal FORTRAN Translator
Software-Implemented Fault Tolerance
SIG Service d'Information Geologique [= Geological Information Service, France]
Signal
Signature
Significant
Special Interest Group
Structural Integrity Group
Sub-Interface Generator
SIGA Schweizerische Interessengemeinschaft fur Abfallverminderung [= Swiss Interest Group for Waste Reduction]
SIGACT Special Interest Group on Automata and Compatibility Theory [of ACM, US]
SIGADA Special Interest Group on Ada (Programming Language) [US]
SIGAPL Special Interest Group on APL (Programming Language) [US]
SIGARCH Special Interest Group for Architecture of Computer Systems [of ACM, US]
SIGART Special Interest Group on Artificial Intelligence [of ACM, US]
SIGBDP Special Interest Group on Business Data Processing [of ACM, US]
SIGBIO Special Interest Group on Biomedical Computing [of ACM, US]
SIGCAPH Special Interest Group for Computers and the Physically Handicapped [US]
SIGCAS Special Interest Group for Computers and Society [of ACM, US]
SIGCHI Special Interest Group on Computer Hierarchy and Interfaces [of ACM, US]
SIGCOMM Special Interest Group on Data Communications [of ACM, US]
SIGCOSIM Special Interest Group on Computer Systems Installation Management [of ACM, US]
SIGCPR Special Interest Group for Computer Personnel Research [of ACM, US]
SIGCSE Special Interest Group on Computer Science Education [of ACM, US]
SIGCUE Special Interest Group for Computer Uses in Education [of ACM, US]
SIGDA Special Interest Group for Design Automation [of ACM, US]
SIGDOC Special Interest Group for Systems Documentation [of ACM, US]
SIGForth Special Interest Group on Forth (Programming Language) [of ACM, US]
SIGGEN Signal Generator
SIGGRAPH Special Interest Group on Computer Graphics [of ACM, US]
SIGI System for Interactive Guidance and Information
SIGIR Special Interest Group on Information Retrieval [of ACM, US]

SIGINT Signal Intelligence
SIGLE System for Information on Gray Literature in Europe
SIGMA Sealed Insulating Glass Manufacturers Association [US]
Shielded Inert Gas Metal-Arc Welding
Society of Independent Gasoline Marketers of America [US]
SIGMET Significant Meteorological Information
SIGMETRICS Special Interest Group on Measurement and Evaluation [of ACM, US]
SIGMICRO Special Interest Group on Microprogramming [of ACM, US]
SIGMOD Special Interest Group on Management of Data [of ACM, US]
sign (signatum) — signed
SIGNUM Special Interest Group on Numerical Mathematics [of ACM, US]
SIGOA Special Interest Group on Office Automation [now: SIGOIS, US]
SIGOIS Special Interest Group on Office Information Systems [of ACM, US]
SIGOP Signal Operation
SIGOPS Special Interest Group on Operating Systems [of ACM, US]
SIGPC Special Interest Group on Personal Computing Systems [of ACM, US]
SIGPLAN Special Interest Group on Programming Languages [of ACM, US]
SIGSAC Special Interest Group on Security, Audit and Control [of ACM, US]
SIGSAM Special Interest Group for Symbolic and Algebraic Manipulation [of ACM, US]
SIGSCSA Special Interest Group on Small Computing Systems and Applications [of ACM, US]
SIGSD Special Interest Group for System Documentation [of ACM, US]
SIG/SIC Special Interest Group/Special Interest Committee
SIGSIM Special Interest Group on Simulation [of ACM, US]
SIGSMALL Special Interest Group on Small and Personal Computing Systems and Applications [of UWO, Canada]
Special Interest Group on Symbolic and Algebraic Manipulation [of ACM, US]
SIGSOFT Special Interest Group on Software Engineering [of ACM, US]
SIGTTO Society of International Gas Tanker and Terminal Operators [UK]
SIGUCC Special Interest Group on University Computer Centers [of ACM, US]
SIGUCCS Special Interest Group for University and College Computing Services [of ACM, US]
SIGWX Significant Weather
SII Standards Institute of Israel
SIIAS Staten Island Institute of Arts and Sciences [US]
SIIRS Smithsonian Institution Information Retrieval System [US]
SIL Safety Information Letter
Scanner Input Language
Silence
Silicone
Single In-Line
Speech Interference Level
Steam Isolation Line
Store Interface Link
Systems Implementation Language
SILS Silver Solder
Shipboard Impact Locator System
SILT Stored Information Loss Tree
SILWR Silverware
SIM Scanning Ion Microscopy
Scientific Instrument Module
Secondary Ion Microscope
Selected Ion Monitoring
Service Instructions Message
Signal Isolator Module
Simulation
Simulator
Single-In-Line Memory
Society for Industrial Microbiology [US]
Society for Information Management [US]
Software Interface Module
Stress-Induced Martensite
Superconductor/Insulator/Metal
Synchronous Induction Motor
Synchronous Interface Module
SIMA Scientific Instrument Manufacturers Association [UK]
Special Import Measures Act
SIMAJ Scientific Instrument Manufacturers Association of Japan
Si-MBE Silicon Molecular Beam Epitaxy
SIMCHE Simulation and Checkout Equipment
SIMCOM Simulator Compiler
Simulator Computer
SIMCON Scientific Inventory and Management Control
Software for Integrated Manufacturing Consortium [of NRC, Canada]
SIMD Single-Input, Multiple-Data (Stream)
Single-Instruction (Stream), Multiple-Data (Stream)
SIMDIS Simulated Distillation Software
SIME Sian Institute of Metallurgical Engineering [PR China]
SIMEA Societa Italiana Meridionale per l'Energia Atomica [= Society for Nuclear Energy in Southern Italy]
SIMEON Simplified Control
SIMICOR Simultaneous Multiple Image Correlation
SIMILE Simulation of Immediate Memory in Learning Experiments
SIMIT Size-Induced Metal-Insulator Transition
SIMM Single-in-Line Memory Module
Symbolic Integrated Maintenance Manual
SIMNAV Simulation and Forecasting System for Marine Traffic
SIMNS Simulated Navigation System
SIMO Simultaneous Motion
SIMON Softwate Interface Monitor
SIMOS Stacked-Gate Injection Metal-Oxide-Semiconductor
SIMOX Separation by Implantation of Oxygen
Silicon Implanted with Oxygen

SIMP Satellite Information Message Protocol
Specific Impulse
SIMPAC Simplified Programming for Acquisition and Control
Simulation Package
SIMPLE System for Integrated Maintenance and Program Language Extension
SIMPP Simple Image Processing Package
SIMRA Scientific Instruments Manufacturers Research Association [UK]
SIMS Scandinavian Simulation Society [Finland]
Sea Ice Monitoring Site
Secondary Ion Mass Spectroscopy
Secondary Ion Mass Spectrum
Single-Item, Multi-Source
Symbolic Integrated Maintenance System
Symposium for Innovation in Measurement Science
SIMUL Simultaneous
SIMULA Simulation Language
SIMULCAST Simultaneous Broadcast [also: simulcast]
SIN Sensitive Information Network
Simultaneous Interpenetrating Network
Single Interconnection Network
Social Insurance Number
Subject Indicator Number
Support Information Network
Symbolic Integrator
Sin Salmonella Infantis [Molecular Biology]
sin sine
sin^{-1} inverse sine
SINAD Signal to Noise and Distortion
SINADS Shipboard Integrated Navigation and Display System
SINAP Satellite Input to Numerical Analysis and Prediction
SINB Southern Interstate Nuclear Board [US]
SINCGARS Single Channel Ground/Automatic Recovery System
SINDO Scaled Intermediate Neglect of Differential Overlap
sinh hyperbolic sine
sinh^{-1} inverse hyperbolic sine
SINQ Schweizer Intensitats-Neutronenquelle [= Swiss Intensity Neutron Source [of PSI]]
SINR Swiss Institute for Nuclear Research
SINS Ship's Inertial Navigation System
SINSS Shipboard Ice Navigation Support System
SIO Scripps Institute of Oceanography [of UCSD, US]
Serial Input/Output
Staged in Orbit
Start Input/Output
Step Input/Output
SIOC Serial Input/Output Channel
SIOP Selector Input/Output Processor
Single Integrated Operations Plan
SIOUX Sequential Iterative Operation Unit X
SIP Safety Injection Pump
Scientific Information Processor
Sheetmetal Insert Process
Short Irregular Pulse
Single-In-Line Package
Sintering/Hot Isostatic Pressing
Society for Invertebrate Pathology [US]
Software Instrumentation Package
Sputter Ion Plating
Standard Information Package
State Implementation Plan
Stay-In-Place
Submerged Injection Process
Symbolic Input Program
SIPA Science-Based Industrial Park Administration [Taiwan]
Single Intensity Profile Analysis
SIPB Safety Injection Permissive Blocks
SIPER Swedish Institute for Production Engineering Research
SIPES Society of Independent Professional Earth Scientists [US]
SIPI Scientists Institute for Public Information [US]
SIPN Simultaneous Interpenetrating Network
S-IPN Semi-Interpenetrating Polymer Network [also: SIPN]
SIPO Serial-In, Parallel-Out
SIPOP Satellite Information Processor Operational Program
SIPOS Semi-Insulating Polycrystalline Silicon
SIPP Society of Irish Plant Pathologists
Sodium Iron Pyrophosphate
SIPRE Snow, Ice and Permafrost Research Establishment [US Army]
SIPRI Stockholm International Peace Research Institute [Sweden]
SIPROS Simultaneous Processing Operating System
SIPS SAC (= Strategic Air Command) Intelligence Data Processing System
Simulated Input Preparation System
Sonar Image Processing System
Sputter-Induced Photon Spectroscopy
State Implementation Plan System [of EPA, US]
Statistical Interactive Programming System
SIP/SIMM Single-In-Line Package/Single-In-Line Memory Module
SIR Safety Investigation Regulations
Savings-to-Investment Ratio
Scientific Information Retrieval
Segment Identification Register
Selective Information Retrieval
Semantic Information Retrieval
Signal-to-Interference Ratio
Simultaneous Impact Rate
Shuttle Imaging Radar
Special Information Retrieval
Specification Information Retrieval
Standardized Indonesian Rubber
Statistical Information Retrieval
Styrene Isoprene Rubber
Submarine Intermediate Reactor
Supersonic Infantry Rocket
Symbolic Input Routine
SIRA Scientific Instruments Research Association [UK]
SIRAID Scientific Instruments Research Association Information and Data Service [UK]
SIRC Special Investment Research Contract
SIRE Satellite Infrared Experiment [also: Sire]
Society for the Investigation of Recurring Events [US]

	Syracuse Information Retrieval Experiments [of Syracuse University, US]
SIRIM	Standards and Industrial Research Institute of Malaysia
SIRL	Structured Interactive Robot Language
SIRM	Scanning Infrared Microprobe
SIRS	Salary Information Retrieval System
	Satellite Infrared Spectrometer
	Specification Information Retrieval System
SIRSA	Special Industrial Radio Service Association [US]
SIRT	Simultaneous Iterative Reconstruction Technique
	Status and Implementation of Recommendations on Technical Questions [of ECAC, France]
SIRTF	Space Infrared Telescope Facility [of NASA]
SIRU	Strapdown Inertial Reference Unit
SIRWT	Safety Injection Reserve Water Tank
SIS	Safety Injection Signal
	Safety Injection Symbol
	Satellite Interceptor System
	Science Information Service
	Scientific Instruction Set
	Scientific Instrument Society [UK]
	Selected Inventor Service [of IPG, France]
	Semiconductor/Insulator/Semiconductor
	Short-Interval Scheduling
	Signalling Interworking Subsystem
	Siltstone
	Simulation Interface Subsystem
	Simulation Interface System
	Special Industrial Services [of UNIDO]
	Standards Information Service
	Standards Information System [of NIST, US]
	Statistical Information System
	Strategic Information System
	Structure-Related Internal Stress
	Styrene Isoprene Styrene (Block Copolymer)
	Superconductor/Insulator/Superconductor
	Supplier Identification System
	Support Information System
	Sveriges Standardiseringskommission [= Swedish Commission for Standardization]
	System Interrupt Supervisor
S/I/S	Semiconductor/Insulator/Semiconductor
	Superconductor/Insulator/Superconductor
SISA	Strontium Isotope Sample Analysis
SISAM	Spectrometer with Interference Selective Amplitude Modulation
SISD	Single-Instruction (Stream), Single-Data (Stream)
SISDATA	Statistical Information System Databank [Italy]
SISF	Superlattice Intrinsic Stacking Fault
SISFET	Semiconductor-Insulator-Semiconductor Field-Effect Transistor
SISI	Southern Institute of Science and Industry [US]
	Surveillance and In-Service Inspection
SISO	Serial-In, Serial-Out
	Single Input, Single Output
SISRI	Shanghai Iron and Steel Research Institute [PR China]
SISS	Single-Item, Single-Source
	Submarine Integrated Sonar System [also: SISSY]
SIST	Society for Imaging Science and Technology [US]
SISTM	Simulation Incremental Stochastic Transistion Matrices
SIT	Safety Injection Tank
	Self-Induced Transparency
	Separation-Initiated Timer
	Shibaura Institute of Technology [Japan]
	Silicon-Intensified Target
	Situation
	Society of Instrument Technology [UK]
	Software Integration Test
	Special Information Tone
	Spontaneous Ignition Temperature
	Static Induction Transistor
	Stevens Institute of Technology [US]
	Structural Integrity Test
	Subarea Index Table
	Sugar Industry Technologist
	System Initialization Table
SITA	Societe Internationale des Telecommunications Aeronautiques [= International Society for Aeronautical Telecommunications, France]
SITC	Salford Information Technology Center [of Salford University, UK]
	Satellite International Television Center
	Standard International Trade Classification [of UN]
	Standard International Trade Commodity
SITE	Satellite Instructional Television Experiment [India]
	Search Information Tape Equipment
	Spacecraft Instrumentation Test Equipment
	Superfund Innovative Technology Evaluation
SITEL	Societe des Ingenieurs des Telecommunications et d'Electronique [= Society of Electronic and Telecommunication Engineers, Belgium]
SITI	Swiss Institute for Technical Information
SITL	Static Induction Transistor Logic
SITOY	Seoul International Toy Fair [South Korea]
SITPRO	(Committee for the) Simplification of International Trade Procedures Board [UK]
SITRA	South India Textile Research Association
SIT-REP	Situation Report
SITS	4-Acetamido-4'-Isothiocyanatostilbene-2,2'-Disulfonic Acid
	SAGE (= Semiautomatic Ground Environment) Intercept Target Simulation
	Social Implications of Technology Society [of IEEE, US]
SITT	Systems Integration of Triad Technology
SITU	Society for the Investigation of the Unexplained [US]
SITVC	Secondary Injection Thrust Vector Control
SIU	Seamen's International Union
	Serial Interface Unit
	Side-In Unit
	Southern Illinois University [US]
	System Input Unit
	System Interface Unit
SIUE	Southern Illinois University, Edwardsville [US]
SIV	Simian Immunodeficiency Virus
SIWL	Single Isolated Wheel Load
SIX	Sodium Isopropyl Xanthogenate

SIXPAC System for Inertial Experiment Priority and Attitude Control
SIZ Security Identification Zone
SJA Sophora Japonica Agglutinin
SJAC Society of Japanese Aerospace Companies
SJAE Steam Jet Air Ejector
SJC Saint John's College [UK]
SJCC Spring Joint Computer Conference [US]
SJCM Standing Joint Committee on Metrication
SJF Shortest Job First
SJFC St. John Fisher College [US]
SJI Steel Joist Institute [US]
SJP Safe Job Procedures
Serialized Job Processor
SJSU San Jose State University [US]
SJU Shanghai Jiaotong University [PR China]
SK Sack
Saskatchewan [Canada]
Sketch
Sink
Skip
Sk Stark Number
SKA Studienkommission fur Atomenergie [= Advisory Committee on Nuclear Energy, Switzerland]
SKB Schweizerischer Koordinationsausschuss fur Biotechnologie [= Swiss Coordinating Committe for Biotechnology]
Skew Buffer
SKED Schedule
SKIL Scanner Keyed Input Language
SKIZ Skeleton by Influence Zone
SKL Skip Lister
SKOR Sperry-Kalman Optimum Reset
SKP Skip-Line Printer
Snoek-Koester Peak
SKR Skip Record Processor
SKr Swedish Krona [Currency]
SKS Scanning Kinetic Spectroscopy
SKT Socket
SKU Stockkeeping Unit
SKW South Koreran Won [Currency]
SKWOC Structured Keyword Out of Context
SKYLAB Sky Laboratory [also: Skylab, US]
SL Safety Lamp
Safety Limit
Scientific Liaison
Sea Level [also: sl]
Searchlight
Sensation Level
Send Left
Shrinkage Limit
Signal Light
Simulation Language
Single Lead
Single-Loop
Slate
Slide
Sodium Lamp
Sonic Log
Source Language
Source Level
Spacelab
Space Lattice
Spectral Line
Spirit Level
Split Lens
Standard Label
Straight Line
Streamline
Stretcher Levelling
Sublaminar
Subscriber Line
Subscriber Loop
Summary Language
Superlattice
Systems Language
S/L Solid/Liquid
Spacelab
sl (sine loco) — without place
SLA Saskatchewan Library Association [Canada]
School Library Association [UK]
Scottish Library Association
Shared Line Adapter
Spacecraft Lunar Module Adapter [of NASA]
Special Libraries Association [US]
Stereolithography Apparatus
Stored Logic Array
Swine Leucocyte Antigen
Synchronous Line Adapter
SLAA Sierra Leone Airport Authority
SLAAS Sri Lanka Association for the Advancement of Science
SLAB Small Amount of Bits [also: slab]
SLAC Stanford Linear Accelerator Center [of Stanford University, US]
Subscriber Line Audio Processing Circuit
SLADE Society of Lithographic Artists, Designers and Engravers [UK]
SLAET Society of Licenced Aircraft Engineers and Technologists [UK]
SLAG Safe Launch Angle Gate
SLAM Scanning Laser Acoustic Microscopy
Simulation Language for Alternative Modeling
Single-Layer Aluminum Metallization
Space-Launched Air Missile
Standoff Land Attack Missile
Strategic Low-Altitude Missile
Supersonic Low-Altitude Missile
Symbolic Language Adapted for Microcomputers
SLAMS Simplified Language for Abstract Mathematical Structures
SLANG Systems Language
SLANT Simulator Landing Attachment for Night Landing Training
SLAP Slot Allocation Procedures
St. Lawrence (River) Action Plan [of Environment Canada]
Symbolic Language Assembler Program
S/LAP Shiplap
SLAR Select ADC (= Analog-to-Digital Conversion) Register
Side-Looking Airborne Radar

SLASH Seiler Laboratory ALGOL Simulated Hybrid
SLAST Submarine-Launched Anti-Surface Ship Torpedo
S LAT South Latitude [also: SLAT]
SLATE Small Lightweight Altitude Transmission Equipment
Stimulated Learning by Automated Typewriter Environment
SLAV Slavic
Slavonic
SLB Scanning Laser Beam
Side-Lobe Blanking
Spectral Line Broadening
SLBM Ship-Launched Ballistic Missile
Submarine-Launched Ballistic Missile
SLBMDWS Submarine-Launched Ballistic Missile Detection and Warning System
SLC Selector Channel
Self-Lubricating Compound
Shift Left and Count Instructions
Side Lobe Cancellation
Side Lobe Clutter
Simulated Linguistic Computer
Single Layer Ceramic
Single Line Control
Space Launch Complex
Stanford Linear Collider [US]
St. Lawrence College [Canada]
Straight-Line Capacitance
Subscriber Loop Carrier
Sustained-Load Cracking
Systems Life Cycle
SL&C Shipper's Load and Count
SLCA Single Line Communications Adapter
SLCSAT Submarine Laser Communications Satellite
SLCB Single-Line Color Bar
SLCC Saturn Launch Computer Complex
SLCD Spring-Loaded Camming Device
SLCM Sea-Launched Cruise Missile
Software Life Cycle Management
Submarine Launched Cruise Missile
SLCP Saturn Launch Computer Program
SLCRS Supplementary Leak Collection and Release System
SLCS Standby Liquid Control System
SLCU Synchronous Line Control Unit
SLD Scientific Liaison Division [of NHRI, Canada]
Simulated Launch Demonstration
Slim Line Diffuser
Solder
Solid
Straight-Line Depreciation
Superluminescent Diode
Source Language Debug
Synchronous Line Driver
SLDC Single-Loop Digital Controller
SLDR Solder
System Loader
SLDTSS Single Language Dedicated Time-Sharing System
SLE Segment Limits End
Sierra Leonean Leone [Currency]
Small Local Exchange
Society of Logistics Engineers [US]
Superheat Limit Explosion
slea (sine loco et anno) — without place and year
SLEAT Society of Laundry Engineers and Allied Trades
SLED Surface Light-Emitting Diode
SLEDGE Simulating Large Explosive Detonable Gas Experiment
SLEEP Scanning Low-Energy Electron Probe
SLEP Second Large ESRO (= European Space Research Organization) Project
Service Life Extension Program
SLEW Static Load Error Washout System
SLF Skandinaviska Lackteknikers Forbund [= Federation of Scandinavian Paint and Varnish Technologists, Finland]
Straight Line Frequency
Streamline Flow
Stud Leveling Foot
System Library File
SLFK Sure Lock Fixture Key
SLGT Slight
SLI Sea Level Indicator
Solid-Liquid Interface
Steam Line Isolation
Suppress Length Indicator
Synchronous Line Interface
SLIB Source Library
Subsystem Library
SLIC Selective Listing in Combination
Shear Longitudinal Inspection Characterization
Simulation Linear Integrated Circuit
Subscriber Line Interface Circuit
Subscriber Loop Interface Circuit
SLICE Selective Line Insertion Communication Equipment
St. Louis Institute of Consulting Engineers [US]
Surrey Library Interactive Circulation Experiment [of University of Surrey, UK]
System Life-Cycle Estimation
SLIC/SLAC Subscriber Line Interface Circuit/Subscriber Line Audio Processing Circuit
SLIH Second Level Interrupt Handler
SLIP Symmetric List Processor
SLIPR Source Language Input Program
SLIS School of Library and Information Science
Shared Laboratory Information System
SLISIR Sri Lanka Institute of Scientific and Industrial Research
SLISP Symbolic List Processing
SLIV Steam Line Isolation Valve
SLL Satellite Line Link
Sierra Leonean Leone [Currency]
SLLA Sri Lanka Library Association
SLM Scanning Laser Microscopy
Single-in-Line Memory
Spatial Light Modulator
Statistical Learning Model
Subscriber Loop Multiplex
Synchronous Line Module
SLMA Southeastern Lumber Manufacturers Association [US]
SLMG Self-Launching Motor Glider
SLN Selection
SLO Segment Limits Origin

	Shift Left Out
	Swept Local Oscillator
SLOCOP	Specific Linear Optimal Control Program
SLOSRI	Shift Left Out/Shift Right In [also: SLO/SRI]
SLOTH	Suppressing Line Operands to Hexadecimal
SLP	Sales Price
	Searchlight Probing
	Segmented Level Programming
	Seismic Line Program
	Skip-Lot Plan
	Slope
	Sound Level Pressure
	Source Language Processor
	Straight Line Programming
SLPH	Solid Photography
SLPM	Standard Liters per Minute
SLPP	Serum Lipophosphoprotein
SLQTM	Sales Quantity This Month
SLR	Side-Looking Radar
	Simple Left-to-Right
	Single Lens Reflex (Camera)
	Slush on Runway
	Standardized Sri Lanka Rubber
	Storage Limits Register
SLRAP	Standard Low-Frequency Range Approach
SLRN	Select Read Numerically
SLRV	Surveyor Lunar Roving Vehicle
SLRC	Salt Lake Research Center [US]
SL/RN	Stelco-Lurgi/Republic Steel-National Lead Process
SLS	Saskatchewan Land Surveyors [Canada]
	Segment Long Spacing
	Selective Laser Sintering
	Side-Lobe Suppression
	Side-Looking Sonar
	Signaling Link Selection
	Sodium Lauryl Sulfate
	Source Library System
	Specialist in Library Science
	Strained-Layer Superlattice
	Superlarge Scale
	Superlattice Structure
SLSA	Saskatchewan Land Surveyors Association [Canada]
	Secondary Lead Smelters Association [US]
	St. Lawrence Seaway Authority [US/Canada]
SLSF	Sodium Loop Safety Facility
SLSI	Super-Large-Scale Integration
SLSO	Self-Limiting Sinusoidal Oscillator
SL SOL	Slightly Soluble
SLT	Select
	Simulated Launch Test
	Sleet
	Solid-Logic Technique
	Solid-Logic Technology
SLt	Sub-Lieutenant
SLTB	Society for Low Temperature Biology [UK]
SLTC	Society of Leather Trades Chemists
SLTF	Shortest Latency Time First
SLU	Secondary Logic Unit
	Serial Line Unit
	Source Library Update
	Special Line Unit
	Subscriber Line Usage
SLUC	Standard Level User Charge
SLUR	Share Library User Report
SLUS	Subscriber Line Usage System
SLV	Satellite Launch Vehicle [of NASA]
	Schweizerischer Landwirtschaftlicher Verein [= Swiss Agricultural Association]
	Sleeve
SLW	Specific Leaf Weight
	Straight Line Wavelength
SLWD	Super-Long Working Distance
SLWL	Straight-Line Wavelength
SLZA	Scandinavian Lead Zinc Association [Sweden]
SM	Master of Science
	Salt Mixture
	Saturation Magnetization
	Scanning Microscope
	School of Metallurgy
	School of Mines
	Scientific Memorandum
	Security Mechanism
	Self-Mass
	Semantic Model
	Semiconductor Memory
	Semimetal
	Sequence and Monitor
	Service Module
	Service Monitoring
	Servo Mechanism
	Set Mode
	Shared Memory
	Shell Model
	Siemens-Martin (Steelmaking)
	Signaling Mode
	Slow Motion
	Soft Magnetic
	Solid Mechanics
	Solid Modelling
	Special Memorandum
	Statistical Mechanics
	Statistical Modelling
	Statute Miles [also: sm]
	Storage Mark
	Strategic Missile
	Strength of Materials
	Structural Material
	Structure Memory
	Student Manual
	Surface Measure
	Surface Mounting
	Synchronous Modem
	Synchronous Motor
	Synthetic Material
	System Manager
	System Mechanics
S/M	Sort/Message
Sm	Samarium
SMA	Saskatchewan Mining Association [Canada]
	Scale Manufacturers Association [US]
	Science Masters Association
	Screen Manufacturers Association [US]

	Semimajor Access
	Sequential Multiple Analyzer
	Shape Memory Alloy
	Sheffield Metallurgical Association [UK]
	Shelving Manufacturers Association [US]
	Shielded Metal-Arc (Cutting)
	Society of Management Accountants [Canada]
	Society of Manufacturers Agents [now: SMR, US]
	Society of Maritime Arbitrators [US]
	Society of Mineral Analysts [US]
	Society of Municipal Arborists [US]
	Soft Magnetic Alloy
	Software Maintenance Association [US]
	Spring Manufacturers Association [now: SMI, US]
	Standard Methods Agar
	Steel Merchants Association [UK]
	Stucco Manufacturers Association [US]
	Styrene-Maleic Anhydride [also: S-MA or S/MA]
	Surface Modeling and Analysis
	Surface Mounted Assembly
	Surface Mounting Applicator
Sma	Serratia Marcescens [Molecular Biology]
SMAA	Society of Management Accountants of Alberta [Canada]
SMAB	Solid Motor Assembly Building
SMAC	Shielded Metal-Arc Cutting
	Society of Management Accountants of Canada
	Special Mission Attack Computer
	Storage Multiple Access Control
SMACNA	Sheetmetal and Air Conditioning Contractors National Association [US]
SMACS	Simulated Message Analysis and Conversion Subsystem
SMAF	Specific Microphage Arming Factor
SMAL	Structural Macroassembly Language
	Surface and Microstructural Analysis Laboratory [US]
SMALGOL	Small Computer Algorithmic Language
SMAO	Society of Management Accountants of Ontario [Canada]
SMAP	Small Manufacturers Assistance Program
SMArchS	Master of Science in Architectural Science
SMART	Salton's Magical Automatic Retrieval of Text
	Sequential Mechanism for Automatic Recording and Testing
	Sorption Method using Adaptive Rate Technology
	System for the Mechanical Analysis and Retrieval of Text
	System Monitor Analysis and Response Technique
	Systems Management Analysis, Research and Test
SMARTIE	Simple Minded Artificial Intelligence
SMAS	Switched Maintenance Access System
SMATS	Source Module Alignment Test Site
SMATV	Satellite Master Antenna Television
SMAW	Shielded Metal-Arc Welding
	Short-Range Man Portable Antitank Weapon
SMAWT	Short-Range Man Portable Antitank Weapon Technology
SMB	Societe Mathematique de Belgique [= Mathematical Society of Belgium]
	Surveys and Mapping Branch [of EMR, Canada]
	System Monitor Board
SMBA	Scottish Marine Biological Association [UK]
SMBT	Master of Science in Building Technology
	Sodium-2-Mercaptobenzothiazole
SMC	Scientific Manpower Commission [now: CPST, US]
	Segmented Maintenance Cask
	Sheet Molding Compound
	Small Magellanic Cloud
	Smooth Muscle Cell
	Societe Mathematique du Congo [= Mathematical Society of Congo]
	Storage Module Controller
	Supply and Maintenance Command [US Army]
	Surface Mounted Component
	Surface Movement Control
	Switch Maintenance Center
	System Management Group
	System Monitor Controller
	Systems, Man and Cybernetics [IEEE Society]
SM CAP	Small Capital Letter(s) [also: sm caps]
SMC/BMC	Sheet Molding Compound/Bulk Molding Compound
SMCC	Simulation Monitor and Control Console
SMC-C	Sheet Molding Compound — Continuous
SMC-C/R	Sheet Molding Compound — Continuous/Random
SMCNA	Sheetmetal Contractors National Association [now: SMACNA, US]
SMC-R	Sheet Molding Compound — Random
SMCRA	Surface Mining Control and Reclamation Act [Canada]
SMCS	Systems, Man and Cybernetics Society [of IEEE, US]
SMC X-1	Small Magellanic Cloud X-1 [Astronomy]
SMD	Sauter Mean Diameter
	Singular Multinomial Distribution
	Storage Module Drive
	Structural Materials Division [of AIME, US]
	Surface-Mounted Device
	Systems Manufacturing Division
	Systems Measuring Device
SM^3/d	Standard Cubic Meters per Day
SMDA	Safe Medical Device Act [US]
	Saskatchewan Mineral Development Agreement [Canada]
SMDC	Saskatchewan Mining Development Corporation [Canada]
	Sodium Methyl Dithiocarbamate
	Superconductive Materials Data Center
SMDE	Static Mercury Drop Electrode
SMDF	SAGE (= Semi-Automatic Ground Environment) Main Distributing Frame
SMDR	Station Message Detailed Recording
SME	School of Military Engineering
	Shape Memory Effect
	Small and Medium-Sized Enterprises
	Small-to-Medium Sized Enterprises
	Society of Manufacturing Engineers [US]
	Society of Mechanical Engineers
	Society of Military Engineers [US]
	Society of Mining Engineers [of AIME, US]

SME-AIME Society of Mining Engineers of the American Institute of Mining, Metallurgical and Petroleum Engineers [US]
SMEAT Skylab Medical Experiments Altitude Test
SMED Single-Minute Exchange of Die
SMEDP Standard Methods for the Examination of Dairy Products
SMEF Special Microelectronics Fund [of NSERC, Canada]
SMEK Summary Message Enable Keyboard
SMEM Serial Memory
S-MEM Minimum Essential Medium for Suspension Cultures
SMEMA Surface Mount Equipment Manufacturers Association [US]
SMES Superconducting Magnetic Energy Storage
SMET Simulated Mission Endurance Testing
SMETDS Standard Message Trunk Design System
SMF Sociedad Mexicana de Fisica [= Mexican Society of Physics]
Software Maintenance Function
Solar Magnetic Field
System Management Facility
System Measurement Facility
SMFA Simplified Modular Frame Assignment
SMFP Scattering Mean Free Path
SMG Software Message Generator
Spacecraft Meteorology Group
SMGCS (Study Group on) Surface Movement Guidance and Control Systems
SMGP Strategic Missile Group
SMH Shifted Morse Hybrid (Potential)
SMI Scanning Microscopy International [US]
Simulated Machine Indexing
Society for Machine Intelligence [US]
Sorptive Minerals Institute [US]
Spring Manufacturers Institute [US]
Start Manual Input
Static Memory Interface
Swiss Meteorological Institute
System Management Interrupt
System Memory Interface
SMIC Study of Man's Impact on the Climate
SMIDA Saskatchewan/Manitoba Implement Dealers Association [Canada]
SMIE School of Metallurgical Industrial Engineering
Solid-Metal-Induced Embrittlement
SMIEEE Senior Member of the Institute of Electrical and Electronic Engineers
SMIF Standard Mechanical Interface
SMIL Sawmill
Statistics and Market Intelligence Library [of DTI, UK]
SMILE Significant Milestone Integration Lateral Evaluation
Sub-Micron Laser Experiment
SMIO Saskatchewan Meteorological Inspection Office [of AES, Canada]
SMIP Small Manufacturers Incentive Program
Structure Memory Information Processor
SMIPP Sheet Metal Industry Promotion Plan [US]
SMIRE Senior Member of the Institution of Radio Engineers
SMiRT (International Conference on) Structural Mechanics in Reactor Technology
SMIS Society for Management Information Systems [now: SIM, US]
Symbolic Matrix Interpretation System
SMIT Simulated Mid-Course Instruction Test [of NASA]
Superconducting Multilayer Interconnect Technology
SMITE Simulation Model of Interceptor Terminal Effectiveness
Smith Inst Smithsonian Institution [US]
SMK Smoke
SMKLS Smokeless
SML Small
Software Master Library
Symbolic Machine Language
SMLCC Synchronous Multiline Communications Coupler
SMLE Small-Medium Local Exchange
SMLM Simple-Minded Learning Machine
SMLS Seamless
SMLT Smelting
SMM School of Mining and Metallurgy
Shared Main Memory
Semiconductor Memory Module
Shared Multiport Memory
Soft Magnetic Material
Standard Method of Measurement
Start of Manual Message
Storage Media Module
System Management Monitor
SMMA Small Motor Manufacturers Association [US]
SMMC Standard Monthly Maintenance Charge
System Maintenance Monitor Console
SMMDA Smaller Manufacturers Medical Device Association [US]
SMME Society for Mining, Metallurgy and Exploration [US]
SMMIB Surface Modification of Metals by Ion Beams
SMMP Standard Methods of Measuring Performance
SMMR Scanning Multichannel Microwave Radiometer
SMMT Society of Motor Manufacturers and Traders [UK]
SMN Servicio Meteorologico Nacional [= National Mateorological Service, Argentina]
SMO Small Magnetospheric Observatory
Stabilized Master Oscillator
SMOBC Solder Mask Over Bare Copper
SMODOS Self-Modulating Derivative Optical Spectrometer
SMOH Since Major Overhaul
SMOG Smoke and Fog [also: smog]
Special Monitor Output Generator
SMOI Second Moment of Inertia
SMOW Standard Mean Ocean Water
SMP Sampler
Scanning Microscope Photometer
Spatial Mobilized Planes (Theory)
Structures and Materials Panel [of AGARD]
Sucrose Monopalmitate
Symmetric Multiprocessing
System Modification Program
SMPAM Symposium on Magnetic Propeties of Amorphous Materials
SMPE Society of Marine Port Engineers [US]

	Society of Motion Picture Engineers [now: SMPTE, US]
SMPL	Sample
SMPS	Simplified Message Processing Simulation
	Single-Machine Power System
	Switch-Mode Power Supply
SMPTE	Society of Motion Picture and Television Engineers [US]
SMQORC	Staggered Modified Quadrature Overlapped Raised Cosine
SMR	Series Mode Rejection
	Shield Mockup Reactor [US]
	Society of Manufacturers Representatives [US]
	Solid Moderator Reactor
	Special Mobile Radio (Equipment)
	Standardized Malaysian Rubber
	Standard Mortality Ratio
	Static Maintenance Reactor
	Super Metal Rich
	Surface Movement Radar
SMRA	Spring Manufacturers Research Association [UK]
SMRB	Safety in Mines Research Board [UK]
SMRC	Swedish Maritime Research Center
SMRD	Spin Motor Rate Detector
SMRE	Safety in Mines Research Establishment [UK]
	Surface Mining Reclamation and Enforcement Office
SMRI	Safety in Mines Research Institute [US]
	Solution Mining Research Institute [US]
	Sugar Milling Research Institute [South Africa]
SMRS	Specialized Mobile Radio System
	Surveys, Mapping and Remote Sensing Sector [of EMR, Canada]
SMRT	Single Message Rate Timing
SMRU	Sea Mammals Research Unit [UK]
SMRVS	Small Modular Recovery Vehicle System [of USAEC]
SMS	School of Materials Science
	Shared Mass Storage
	Shuttle Mission Simulator [of NASA]
	Site-Mixed Slurry
	Slope Monitoring System
	Small Magnetospheric Satellite [of NASA]
	Society of Materials Science [Japan]
	Speech Mail System
	Standard Modular System
	Storage Management System
	Strategic Missile Squadron
	Styrene Methylstyrene
	Surface Missile System
	Synchronous Meteorological Satellite [of NASA]
	Systems Maintenance Service
SMSA	Silica and Molding Sands Association
	Standard Metropolitan Statistical Area
SMSAE	Surface Missile System Availability Evaluation
SMSC	Semimagnetic Semiconductor
SMSI	Strong Metal-Support Interaction
SMSP	School of Materials Science and Physics
SMSS	Ship Motion Sensor System
	System Management Support Service
SMST	School of Materials Science and Technology
SMT	School of Mines and Technology
	Semiconductor Manufacturing Technology
	Service Module Technician
	Societe de Micro-Informatique et de Telecommunications [= Society for Micro-Informatics and Telecommunications]
	Square Mesh Tracking
	Sulfamethazine
	Surface-Mount Technology
SMTA	Sewing Machinery Trade Association [now: AAMTA, US]
	Surface-Mount Technology Association [US]
SMTH	Smooth
SMTI	Selective Moving Target Indicator
	Sodium Mechanisms Test Installation
	Southeastern Massachusetts Technological Institute [US]
SMTS	Scottish Machinery Testing Station
SMU	Saint Mary's University [Canada]
	Saint Mary University [US]
	Secondary Multiplexing Unit
	Self-Maneuvering Unit
	Southern Methodist University [US]
	Store Monitor Unit
	Super-Module Unit
	System Maintenance Unit
	System Monitoring Unit
SMUD	Sacramento Municipal Utility District [US]
SMUF	Simulated Milk Ultrafiltrate
SMUT	System for Musical Transcription [of Indiana University, US]
SMVisS	Master of Science in Visual Science
SMWG	Strategic Missile Wing
SMWIA	Sheet Metal Workers International Association
SMX	Semimicro Xerography
	Submultiplexer Unit
SMYS	Specified Minimum Yield Strength
	Specified Minimum Yield Stress
SN	Schweizerische Norm [= Swiss Standard]
	Scientific Note
	Semiconductor Network
	Ship Nut
	Shipping Note
	Shot Noise
	Sign
	Signal Node
	Siren
	Skid Number
	Slow Neutron
	Spontaneous Nucleation
	Stock Number
	Stress-Number (Curve)
	Subject Name
	Supernova
	Swivel Nut
S/N	Serial Number
	Shipping Note
S-N	Signal-to-Noise (Ratio) [also: S/N]
	Stress-Number (Curve)
Sn	(Stannum) — Tin
sn	sine of the amplitude
SNA	Sambucus Nigra Agglutinin

	Sociedade Nacional de Agricultura [= National Agricultural Society, Brazil]
	Society of Naval Architects [Japan]
	Systems Network Architecture
SNACS	Share News on Automatic Coding Systems
	Single Nuclear Attack Case Study [of USDOD]
SNAE	Society of Norwegian American Engineers
SNAF	Societe Nationale des Architectes de France [= French National Society of Architects]
SNAG	Society of North American Goldsmiths [US]
SNAJ	Society of Naval Architects of Japan
SNAME	Society of Naval Architects and Marine Engineers [also: SNA&ME, US]
SNAP	Simplified Numerical Automatic Programmer
	Single Number Access Plan
	Space Nuclear Auxiliary Power
	Standard Network-Access Protocol
	Steerable Null Antenna Processor
	Structural Network Analysis Program
	Succinimidyl Nitroazidophenyl Aminoethyldithiopropionate
	Synchronous Nuclear Array Processor
	System Net Activity Program
	Systems for Nuclear Auxiliary Power
SNAPS	Standard Notes and Parts Selection
SNAS	Singapore National Academy of Sciences
SNAZOXS	8-Hydroxy-7-(4-Sulfo-1-Naphthylazo)-5-Quinolinesulfonic Acid
SNBHW	Swedish National Board of Health and Welfare
SNBU	Switched Network Backup
SNC	Synchronous Network Clock
SNCA	Societe Nationale de Construction Aeronautique [= National Society of Aeronautic Construction, France]
	Societe Nationale des Chemins de Fer Belges [= Belgian National Society of Railways]
SNCIAWPRC	Singapore National Committee of the International Association on Water Pollution Research and Control
	Swiss National Committee of the International Association on Water Pollution Research and Control
	Swedish National Committee of the International Association on Water Pollution Research and Control
SNCUNESCO	Swedish National Commission for the United Nations Educational, Scientific and Cultural Organization
SND	Scientific Numeric Database
	Send
	Sound
SNDCF	Sub-Network Dependent Convergence Function
SNDRCV	Send/Receive [also: SR or S/R]
SNDT	Society for Nondestructive Testing
SNEMSA	Southern New England Marine Sciences Association [US]
SNF	Second Normal Form
	System Noise Figure
SNFLK	Snowflakes
SNG	Substitute Natural Gas
	Supplemental Natural Gas
	Synthetic Natural Gas
SNGL	Single
SNH	Societe Nationale de Hydrocarbures [= National Hydrocarbon Society, France]
SNI	Selective Notification of Information
	Sequence-Number Indicator
	Signal-to-Noise Improvement
	Standard Network Interface
SNIAS	Societe Nationale Industrielle Aerospatiale [= National Industrial Society for Space Research, France]
SNICP	Sub-Network Independent Convergence Protocol
SNICT	Sistema Nacional de Informacao Cientifica e Tecnologica [= National System for Scientific and Technological Information, Brazil]
SNID	Studies in National and International Development
SNJ	Switching Network Junction
SNL	Sandia National Laboratory [US]
	Selected Nodes List
	Standard Nomenclature List
SNLA	Sandia National Laboratories, Albuquerque [US]
SNLC	Senior NATO Logisticians Conference [Belgium]
SNM	Shielded Nonmetallic (Cable)
	Society of Nuclear Medicine [US]
	Special Nuclear Material
	Spent Nuclear Material
SNMA	Swedish National Maritime Administration
SNMC	Shielded Nonmetallic Cable
SNME	Society of Naval Architects and Marine Engineers
SNMMS	Standard Navy Maintenance and Material Management System [US Navy]
SNMP	Spent Nuclear Material Pool
SNMS	Secondary Neutral Mass Spectrometry
	Sputtered Neutral Mass Spectroscopy
S+N/N	Signal plus Noise to Noise (Ratio)
SNO	Serial Number
	Sudbury Neutrino Observatory [in Ontario, Canada]
SNOBOL	String-Oriented Symbolic Language
SNORT	Supersonic Naval Ordnance Research Track [US Navy]
SNOTEL	Snow Pack Telemetry Radio
SNOWTAM	Snow Notice to Airmen
SNP	Society for Natural Philosophy [US]
	Sodium Nitroprusside
	Soluble Nucleoprotein
	Statistical Network Processor
	Synchronous Network Processor
SNPA	Sub-Network Point of Attachment
SNPM	Standard and Nuclear Propulsion Module
SNPO	Space Nuclear Propulsion Office [of NASA]
SNPS	Satellite Nuclear Power Station
SNR	Signal-to-Noise Ratio
	Society for Nautical Research [UK]
	Supernova Remnant
	Supplier Nonconformance Report
SNS	School of Natural Sciences
	Selected Numeric Service [of IPG, France]
	Simulated Network Simulations
	Software Notification Service
	Space Navigation System
	Spallation Neutron Source

Spanish Nuclear Society
Static Nonlinear System
Superconducting — Normal — Superconducting
SNSE Society of Nuclear Scientists and Engineers [US]
SNSF Swiss National Science Foundation
SNSH Snow Showers
SNSO Space Nuclear Systems Office [US]
SNT Sign-on Table
Society for Nondestructive Testing [now: ASNT, US]
SNTA Sodium Nitrilotriacetate
SNU Seoul National University [South Korea]
Solar Neutrino Unit
SNUPPS Standardized Nuclear Unit Power Plant System
SNV Schweizerische Normenvereinigung [= Swiss Standards Association]
SNVB Society for Northwestern Vertebrate Biology [US]
SNX Succinonitrile
SNW Snow
SNWFL Snowfall
SO Second-Order
Seller's Option
Send Only
Serial Out
Service Order
Shift-Out (Character)
Shop Order
Shutoff
Signal Oscillator
Slow Operating
Slow to Operate
Small Outline
South
Southern Oscillation
Spin Orbital
Standoff
Stationary Orbit
Stationery Office
Suboffice
Switchover
System Optimization
S/O Send Only
Shipper's Option
SOA Safe Operating Area
Semiconductor Optical Amplifier
Start of Address
State-of-the-Art
SOAC System On A Chip
SOAP Self-Optimizing Automatic Pilot
Society of Office Automation Professionals
Spectrometric Oil Analysis Program
Symbolic Optimizer and Assembly Program
Symbolic Optimum Assembly Programming
SOB Shipped on Board
Start of Block
SOC Satellite Operations Center [US]
Self Organizing Control
Separated Orbit Cyclotron
Set Override Clear
Silicon-on-Ceramic
Simulation Operations Center
Single Orbit Computation
Society
Socket
Space Operations Center
Specific Optimal Controller
Spin-Orbital Coupling
Start of Climb
State of Charge
Superposition of Configuration
SOCATA Societe de Construction d'Avions de Tourisme et d'Affaires [= Society for the Construction of Commercial and Business Aircrafts, France]
SOCC Salvage Operational Control Center
SOCCER SMART's Own Concordance Constructor Extremely Rapid [of Cornell University, US]
SOCCS Study of Computer Cataloguing Software [UK]
Summary of Component Control Status
SOCKO Systems Operational Checkout
SOCM Standoff Cluster Munitions
SOCMA Synthetic Organic Chemical Manufacturers Association [US]
SOCO Source Code
Switched-out for Checkout
SOCOFIDE Societe Congolaise de Financement du Developpement [= Congo Society for Development Financing]
SOCOM Solar Optical Communications System
SOCR Sustained Operations Control Room
SOCRATES Study of Complementary Research and Teaching in Engineering Science
System for Organizing Content to Review and Teach Educational Subjects [of University of Illinois, US]
System for Organizing Current Reports to Aid Technologists and Scientists [of DSIS, Canada]
SOCS Spacecraft Orientation Control System
SOD Serial Output Data
Small Object Detector
Small Outline Diode
Societe d'Etudes pour l'Obtention du Deuterium [= Society for the Study of Deuterium Production, France]
Source-Object Distance
Superintendent of Documents [of GPO, US]
Superoxide Dismutase
System Operational Design
SODA Source-Oriented Data Acquisition
SODAC Society of Dyers and Colorists
SODAR Sound Detection and Ranging [also: Sodar or sodar]
SODAS Structure-Oriented Description and Simulation
SODB Science Organization Development Board [of NAS, US]
SODERN Societe d'Etudes et Realisations Nucleaires [= Society for the Study and Fabrication of Nuclear Products, France]
SODT Swiss Office for the Development of Trade
SOE Significant Operating Experience
S1E Surfaced on One Edge [Lumber]
SOEC Second-Order Elastic Constant
SOEEA Saskatchewan Outdoor and Environmental Education Association [Canada]
SOEKOR Suidelike Olie-Eksplorasiek-Orporasie [= Southern Oil Exploration Corporation, South Africa]

SOEP Solar-Oriented Experimental Package [of NASA]
SOER Significant Operating Event Report
SOERO Small Orbiting Earth Resources Observatory
SOF Satisfactory Operation Factor
Shape and Orientation Factor
Shorts, Open and Fixture
Start of Format
Start of Frame
Storage Oscilloscope Fragment
SOFAR Sound Fixing and Ranging [also: Sofar or sofar]
SOFC Solid-Oxide Fuel Cell
SOFE Symposium on Fusion Engineering [of IEEE, US]
S/OFF Sign-Off
SOFNET Solar Observing and Forecasting Network [US Air Force]
S of R Society of Rheology [US]
SOFRECOM Societe Francaise d'Etudes et de Realisations d'Equipements de Telecommunications [= French Society for the Study and Utilization of Telecommunications Equipment]
SOFT Simple Output Format Translator
Society of Forensic Toxicologists [US]
Software
Symposium on Fusion Technology
SOFTA Shippers Oil Field Traffic Association [US]
SOG Spin-On Glass
Strongly Orthogonal Geminal
SOGAT Society of Graphical and Allied Trades [UK]
SOGC Saskatchewan Oil and Gas Corporation [Canada]
SO GR Soft Grind
SOH Start of Heading
SOHC Single Overhead Camshaft
SOHR Solar Hydrogen Rocket Engine
SOI Silicon-on-Insulator
Southern Oscillation Index
Specific Operating Instruction
Standard Operating Instruction
Statement of Interest
SOIC Small-Outline Integrated Circuit
SOICS Summary of Installation Control Status
SOIR (Study Group on) Simultaneous Operations on Parallel (or Near-Parallel) Instrument Runways
SOIS Shipping Operations Information System
SOI/SIMOX Silicon-on-Insulator/Silicon Implanted with Oxygen
SOJ Small-Outlet J-Leaded Package
SOL School of Librarianship
Simulation-Oriented Language
Sol [Currency of Peru]
Solenoid
Solicitor
Solid Waste Database [US/Canada]
Soluble
Solution
Source/Object Library
System-Oriented Language
sol soluble [Chemistry]
SOLACE School of Librarianship Automatic Cataloguing Experiment [of UNSW, Australia]
SOLAR Serialized On-Line Automatic Recording
Storage On-Line Automatic Retrieval [of Washington State University, US]
SOLARIS Submerged Object Locating and Retrieving Identification System
SOLAS Safety of Life at Sea (Convention)
SOLE Society of Logistics Engineers [US]
SOLID Self-Organizing Large Information Dissemination System
SOLINET Southeastern Library Network [US]
SOLION Solution Ion [also: Solion or solion]
SOLIS Symbionics On-Line Information System
SOLN Solution
SOLO Selective Optical Lock-On
SOLOGS Standardization of Operations and Logistics
SOLOMON Simultaneous Operation Limited Ordinal Modular Network
SOLR Sidetone Objective Loudness Rating
SOLRAD Solar Radiation
SOL TRT Solution Treatment
SOLUG San Antonio On-Line User Group [US]
SOLV Solenoid Valve
Solvent
Super-Open-Frame Low Voltage
SOLY Solubility
SOLZ Second Order Laue Zone
SOM Scanning Optical Microscopy
School of Mines
Self-Organizing Machine
Shift Operations Manager
Small Office Microfilm System
Somalia Shilling [Currency]
Space Oblique Mercator
Standard On-Line Module
Start of Message
Strength of Materials
Som Somersetshire [UK]
SOMADA Self-Organizing Multiple-Access Discrete Address
SOMED School of Mines and Energy Development
SOMP Start-of-Message Priority
SON Society of Nematologists [US]
Supra-Optic Nucleus
S/ON Sign-On
SONAC Sonar Nacelle [also: SONACELLE, Sonacelle or sonacelle]
SONACELLE Sonar Nacelle [also: Sonacelle, sonacelle or SONAC]
SONAD Speech-Operated Noise Adjusting Device
SONAR Sound Navigation and Ranging [also: Sonar or sonar]
SONATRACH Societe Nationale pour la Recherche, la Production, le Transport, la Transformation et la Commercialisation des Hydrocarbures [= National Society for Research, Production, Transportation, Transformation and Commercialization of Hydrocarbons, Algeria]
SONCM Sonar Countermeasures
SONCR Sonar Control Room
SONET Synchronous Optical Network (Technology)
SONIC System-Wide On-Line Network for Informational Control
SOOP Single-Operator Operational Precision

SOP Simulation Operations Plan
Special Operating Procedure
Standard Operating Procedure
Strategic Orbit Point
Student Outreach Program [of ASM International, US]
Sum-of-Products
Surface Optical Phonon
SOPA Society of Professional Archeologists [US]
SOPHYA Supervisor of Physics Analysis
SOPI Serial-Out, Parallel-In
SOPM Standard Orbital Parameter Message
SOPP Sodium Orthophenylphenate
SOPR Spanish Open Pool Reactor [Spain]
SOR Society for Occupational Research [US]
Society of Rheology [US]
Specific Operating Requirement
Start of Record
Statement of Requirement
Statutory Orders and Regulations
Steam/Oil Ratio
Successive Overrelaxation
Synchronous Orbital Resonance
Synchrotron Orbital Radiation
SORC Site Operations Review Committee
SORD Separation of Radar Data
SORIN Societa Ricerche ed Impianti Nucleare [= Society for Nuclear Research and Equipment, Italy]
SORM Set-Oriented Retrieval Module
SORO Special Operations Research Office [US]
SORTE Summary of Radiation Tolerant Electronics
SORTI Satellite Orbital Track and Intercept
SORTIE Supercircular Orbital Reentry Test Integrated Environment
SOS Save Our Souls
Service Order System
Share Operating System
Silicon-on-Sapphire
Scheduled Oil Sampling
Somali Shilling [Currency]
Sophisticated Operating System
Speed of Service
Start of Significance
Station Operating Supervisor
Symbolic Operating System
S1S Surfaced on One Side [Lumber]
SOSC Smithsonian Oceanographic Sorting Center [US]
SOSI Serial-Out, Serial-In
Shift-In, Shift-Out
S1S1E Surfaced on One Side and One Edge [Lumber]
SOSS Shipboard Oceanographic Survey System
SOST Special Operator Service Traffic
S1S2E Surfaced on One Side and Two Edges [Lumber]
SOSTEL Solid-State Electronic Logic
SOSUS Sound Surveillance System
SOT Small Outline Transistor
Society of Toxicology [US]
Start of Text
State of Termination
Subscriber Originating Trunk
Syntax-Oriented Translator
SOTA State-of-the-Art
SOTAS Standoff Target Acquisition System
SOT DAT Source Test Data System [of Environmental Protection Agency, US]
SOTIM Sonic Observation of the Trajectory and Impact of Missiles
SOTUS Sequentially-Operated Teletypewriter Universal Selector
SOU Scandinavian Ornithological Union [Sweden]
SOUP Software Utility Package
Submarine Operational Update Program
SOUT Swap-Out
SOUTC Satellite Operators and Users Technical Committee [US]
SOUTHEASTCON Southeastern Convention [of IEEE, US]
SOV Shutoff Valve
Sovereign
SOW Standoff Weapon
Start-of-Word
SOWP Society of Wireless Pioneers [US]
SOx Solid Oxygen
SP Saddle Point
Sampling Period
Satellite Processor
Scratch Pad
Screw Pump
Seismic Prospecting
Self-Potential
Self-Propelled
Semipermeable
Semipolar
Semi-Public
Send Processor
Seperation Process
Sequential Phase
Sequential Processor
Serial to Parallel
Service Package
Service Processor
Sewer Pipe
Shear Plate
Shift Pulse
Shipping Port
Ship Propeller
Shoulder Pin
Sigma Pile
Signal Plate
Signal Processor
Silicon Processing
Silver Point
Singing Point
Single-Phase
Single-Pole
Single Programmer
Slightly Polishing
Small Particle
Small Pica
Society of Protozoologists [US]
Softening Point
Soil Pipe
Solar Physics

	Solar Power
	Solidification Processing
	Solubility Parameter
	South Pacific
	Southern Pine
	South Pole
	Space (Character)
	Spain
	Spaniard
	Spanish
	Spare
	Spare Part
	Specialist
	Special Paper
	Special Projects
	Special Publication
	Special Purpose
	Specific
	Spectrophotometer
	Spectrum Projector
	Speed
	Speed Plug
	Splash-Proof
	Splenopentin
	Spontaneous Potential
	Spool Piece
	Sputter
	Square Punch
	Stack Pointer
	Standard Play
	Standard Potential
	Standpipe
	Starting Point
	Starting Potential
	Starting Price
	Static Pressure
	Stationary Phase
	Steam Point
	Stereo-Power
	Sticking Potential
	Stochastic Process
	Stopping Power
	Strike Plate
	Structured Programming
	Sugar Pine
	Sulphopropyl
	Sulfur Point
	Summary Punch
	Super Precision
	Supervisory Package
	Supervisory Panel
	Supervisory Printer
	Supervisory Process
	Surface Plasmon
	Surge Protector
	Surveillance Procedure
	Symbol Programmer
	Synthetic Polymer
	System Processor
	Systems Programming
S&P	Systems and Programming
S-P	Serial-to-Parallel
sp	(sine prole) — without issue
SPA	Salt Producers Association [now: Salt Institute, US]
	S-Band Power Amplifier
	Scandinavian Packaging Association [Norway]
	Schedules Planning and Analysis
	Seaplane Pilots Association [US]
	Semi-Permanently Associated
	Separated Pair Approximation
	Servo Power Assembly
	Servo Preamplifier
	Small Part Analysis
	Software Producers Association [UK]
	Software Publishers Association [US]
	Southeastern Peanut Association [US]
	Southern Pine Association [now: SFPA, US]
	Specialist in Public Administration
	Spectrum Analyzer
	Spot Profile Analysis
	Stereo Power Amplifier
	Stress-Pattern Analysis
	Subject to Particular Average
	Sudden Phase Anomaly
	Systems and Procedures Association [now: Association for Systems Management, US]
SP-A	Slightly Polishing-Abrasive
SPAC	Spacecraft Performance Analysis and Command [of NASA]
	Spatial Computer
SPACDAR	Specialist Panel on Automatic Conflict Detector and Resolution
SPACE	Satellite, Precipitation and Cloud Experiment
	Self-Programming Automatic Circuit Evaluator
	Sequential Position and Covariance Estimation
	Sidereal Polar Axis Celestial Equipment
	Spacecraft Prelaunch Automatic Checkout Equipment
	Symbolic Programming Anyone Can Enjoy
SPACECOM	Space Communications
SPACECOMPS	Space Components (Database) [of ESA, France]
SPACON	Space Control
SPACS	Sodium Purification and Characterization System
SPAD	Satellite Position Predictor and Display
	Satellite Protection for Area Defense
	Simplified Procedures for Analysis of Data
SPADATS	Space Detection and Tracking System [also: Spadats]
SPADE	SCPC (= Single Channel per Carrier) / PCM (= Pulse-Code Modulation), Multiple Access Demand Assigned Equipment
	Spare Parts Analysis, Documentation and Evaluation
	SPARTA Acquisition Digital Equipment
	Sperry Air Data Equipment
SPADES	Solar Perturbation and Atmospheric Density Measurement Satellite
SPADETS	Space Detection System
SPADNS	2-(P-Sulfophenylazo)-1,8-Dihydroxy-3,6-Naphthalenedisulfonic Acid
SPAG	Standard Promotion Application Group [Europe]
SPAI	Screen Printing Association International [US]

SPA-LEED Spot Profile Analyzer — Low-Energy Electron Diffraction
SPAM Satellite Processor Access Method
Scanning Photoacoustic Microscopy
Ship Position and Attitude Measurement
SP AM Spanish America
SPAMS Ship Position and Attitude Measurement System
SPAN Solar Particle Alert Network [of NASA]
Span (Text) Analysis
Statistical Processing and Analysis
Stored Program Alphanumerics
Sulfophenylazonaphthalene
SPANRAD Superposed Panoramic Radar Display
SPANS Spatial Analysis System
SPAQUA Sealed Package Quality Assurance
SPAR Seagoing Platform for Acoustics Research
Society of Photographers and Artists Representatives [US]
Space Processing Applications Rocket
Special Products and Applied Research
Structural Performance Analysis and Redesign
Symbolic Program Assembly Routine
Synchronous Position Altitude Recorder
SP-AR Super-Precision Anti-Resonance
SPARC Scalable Processor Architecture [also: SPARCH]
Space Air Relay Communications
Space Program Analysis and Review Council [US Air Force]
Space Research Conic [of NASA]
SPARCS Solar Pointing Aerobee Rocket Control System
SPARS Space Precision Attitude Reference System
Space Prediction Attitude Reference System
Society of Professional Audio Recording Services [US]
SPARSA Sferics Pulse, Azimuth, Rate and Spectrum Analyzer
Sferics, Position, Azimuth, Range Spectrum Analyzer
SPARTA Spatial Antimissile Research Tests in Australia
Special Antimissile Research Tests in Australia
SPARTECA South Pacific Regional Trade and Economic Cooperation Agreement
SPAS Shuttle Pallet Satellite [Germany]
SPASM System Performance and Activity Software Monitor
SPASUR Space Surveillance
SPAT Silicon Precision Alloy Transistor
SPATE Stress-Pattern Analysis by Thermal Emission
SPATZ Spatial Test Zone
SPAU Signal Processing Arithmetic Unit
SPAYZ Spatial Property Analyzer
SPB Simple Payback (Period)
Stored Program Buffer
SPBC Snap Pack Battery Cartridge
SPBEC South Pacific Bureau for Economic Cooperation [Fiji]
SPBM Single Point Buoy Mooring
SPBN Sulfophenylbutylnitrone
2-SPBN 2-Sulfophenylbutylnitrone
SPC Saskatchewan Power Commission [Canada]
Saskatchewan Power Corporation [Canada]
Series-Parallel Control
Single Point Connection
Small Peripheral Controller
Solid Proton Conductor
South Pacific Commission [New Caledonia]
Soy Protein Council
Static Power Converter
Statistical Process Control
Statistical Product Control
Stored Program Command
Stored Program Control
SPCL Special
SPCP Single Parity Check Product (Code)
SPCS Storage and Processing Control System
SPC/SPS Statistical Product Control/Statistical Problem Solving
SPC/SQC Statistical Product Control/Statistical Quality Control
SPCT Specialist
SPCUS Sweet Potato Council of the United States
SPD Society of Publication Designers [US]
Software Product Description
Speed
Standard Program Device
Surge Protective Device
Synchronous Phased Detector
SPDC Specialized Products Distribution Center
SPDL Spindle
SPDM Special Purpose Dexterous Manipulator
Subprocessor with Dynamic Microprocessing
SPDP Society of Professional Data Processors
N-Succinimidyl 3-(2-Pyridyldithio)propionate
SPDS Safe-Practice Data Sheet
SPDSC Saskatchewan Professional Drivers Safety Council [Canada]
SPDT Single-Pole, Double-Throw
SPDTDB Single-Pole, Double-Throw, Double-Break [also: SPDT DB]
SPDTNC Single-Pole, Double-Throw, Normally-Closed [also: SPDT NC]
SPDTNCDB Single-Pole, Double-Throw, Normally-Closed, Double-Break [also: SPDT NCDB]
SPDTNO Single-Pole, Double-Throw, Normally-Open [also: SPDT NO]
SPDTNODB Single-Pole, Double-Throw, Normally-Open, Double-Break [also: SPDT NODB]
SPDT SW Single-Pole, Double-Throw Switch
SPDU Session Protocol Data Unit
SPE Signal Processor Element
Society for Photographic Education [US]
Society for Petroleum Engineers [of AIME, US]
Society of Plastics Engineers [US]
Solid Phase Epitaxy
Solid Phase Extraction
Spherical Probable Error
Stored Program Element
Systems Performance Effectiveness
Spe Sphaerotilus Species [Molecular Biology]
SPEA Southeastern Poultry and Egg Association [US]
SPE-AIME Society of Petroleum Engineers of the American Institute of Mining, Metallurgical and Petroleum Engineers [US]
SPEAM Sunphotometer Earth Atmospheric Measurements

SPEAR Stanford Position Electron Accelerator Ring [of Stanford University, US]
SPEARS Satellite Photoelectric Analog Rectification System
SPEC Society of Pollution and Environmental Control
South Pacific Bureau for Economic Cooperation
Special
Specialty
Specification
Speech Predictive Encoding Communication
Stock Precision Engineered Components
Stored Program Educational Computer
Spec Gr Spectrographic Grade [Chemistry]
SPECIF Specific(ation)
SPECON System Performance Effectiveness Conference
SPECS Specifications
SPECT Single-Photon Emission Computed Tomography
Spectrometry
Spectroscopy
SPECVER Specification Verification
SPED Supersonic Planetary Entry Decelerator
SPEDAC Solid-State, Parallel, Expandable, Differential Analysis Computer
SPEDE System for Processing Educational Data Electronically
SPEDTAC Stored Program Educational Transistorized Automatic Computer
SPEE Society for the Promotion of Engineering Education
Society of Petroleum Evaluation Engineers
SPEED Selective Potentiostatic Etching by Electrolytic Dissolution
Self-Programmed Electronic Equation Delineator
Spring-Loaded, Precision, Edgewise and Energy Delivery
Subsistence Preparation by Electronic Energy Diffusion
SPEG Solid Phase Epitaxial Growth
SPEPD Space Power and Electric Propulsion Division [of NASA]
SPERT Schedule Performance Evaluation and Review Technique
Special Power Excursion Reactor Test
SPES Stored Program Element System
SPE/SPI Society of Plastics Engineers/Society of the Plastics Industry [US]
SPESS Stored Program Electronic Switching System
SPET Solid Propellant Electric Thruster
SPF Site Population Factor
Sociedade Portuguesa de Fisica [= Portuguese Society of Physics]
South Pacific Forum
Spectrofluorometer
Standard Project Flood
Standard Program Facility
Structured Programming Facility
Subscriber Plant Factor
Superplastic Formability
Superplastic Forming
System Performance Factor
SPFA Steel Plate Fabricators Association [US]
SPF/DB Superplastic Forming/Diffusion Bonding
SPFC Solid Polymer Fuel Cell
SPFFA South Pacific Forum Fisheries Agency [Solomon Islands]
SPFI Single Point Fuel Injection
SPFP Single Point Failure Potential
SPFS Small Payload Flight System
SP4T Single-Pole Quadruple Throw
SP4T SW Single-Pole Quadruple Throw Switch
SPFW Single-Phase Full-Wave
SPG Scan Pattern Generator
Schweizerische Physikalische Gesellschaft [= Swiss Physical Society]
Single-Point Ground
Soft Page [also: SPg]
Sort Program Generator
Spring
Synchronizing Pulse Generator
SP GR Specific Gravity [also: sp gr]
SPGS Secondary Power Generating Subsystem
SPGT Serum Pyruvic Glutamic Transaminase
SPH Spherical
Styrenated Phenol
Sph Streptomyces Phaeochromogenes [Molecular Biology]
SP/HD Spool Piece Head [also: SP/Hd]
SPHE Society of Packaging and Handling Engineers [US]
SPHER Spherical
SPHF Spin Polarized Hartree-Fock
SP HT Specific Heat
SPHW Single-Phase, Half-Wave
SPI Self-Paced Instruction
Serial Peripheral Interface
Shared Peripheral Interface
Single Particle Impact
Single Point Injection
Single Processor Interface
Single Program Indicator
Site Population Index
Society of Professional Investigators [US]
Society of the Plastics Industry [US]
Sonic Pulse-Echo Instrument
Special Position Identification Pulse
Specific Productivity Index
Station Program Identification
Symbolic Pictorial Indicator
SPIA Single Peak Intensity Analysis
Solid Propellant Information Agency
SPIA-LPIA Solid (Propellant Information Agency) and Liquid Propellant Information Agency [of JHU, US]
SPIAM Sodium Purity In-Line Analytical Module
SPIB Southern Pine Inspection Bureau [US]
SPIC Society of Plastics Industry of Canada
Spacelab Payload Integration and Coordination
SPICE Spacelab Payload Integration and Coordination in Europe
Simulation Program with Integrated-Circuit Emphasis
SPIDAC Specimen Input to Digital Automatic Computer
SPIDEL Societe pour l'Information et Documentation Electronique [= Society for Electronic Information and Documentation, France]

SPIDER Sonic Pulse-Echo Instrument Designed for Extreme Resolution
SPIE Scavenging-Precipitation-Ion Exchange
Self-Programmed Individualized Education
Simulated Problem, Input Evaluation
Society of Photo-Optical Instrumentation Engineers [now: (International) Society for Photo-Optical Engineering, US]
(International) Society for Photo-Optical Engineering [US]
SPIG Symposium on the Physics of Ionized Gases
SPIN Science Procurement Information Network
Searchable Physics Information Notices [of American Institute of Physics, US]
SPIR Search Program for Infrared Spectra
Single Pilot Instrument Rating
SPIRES Stanford Public Information and Retrieval System [US]
SPIS Standard Production Information System
SPIT Selective Printing of Items from Tape
SPIW Shipping, Ports and Inland Waterways [of ESCAP]
Special Purpose Individual Weapon
SPKR Speaker
SPL Signal Processing Language
Simple Programming Language
Simulation Programming Language
Software Parts List
Software Programming Language
Sound Pressure Level
Sound Pulse Level
Source Program Library
Spaceborne Programming Language
Space Physics Laboratory [US]
Space Programming Language
Specialty
Spent Potlining
Splice
Supplementary Flight Plan
System Programming Language
SPLC Standard Point Location Code
SPLEED Spin-Polarized Low-Energy Electron Diffraction
SPLEEM Spin-Polarized Low-Energy Electron Microscopy
SPLIT Sunstrand Processing Language Internally Translated
SPLL Self-Propelled Launcher/Loader
SPLY Supply
SPM Scratch Pad Memory
Self-Propelled Mount
Sequential Processing Machine
Shock Pulse Method
Signal Processing Modem
Single Point Mooring
Six Point Mooring
Small Perturbation Method
Socidedade Portuguesa de Materiais [= Portuguese Society for Materials]
Software Performance Monitor
Solar Proton Monitor
Solar Power Module
Source Program Maintenance
Split-Phase Motor
Standard Prototype Microcomputer
Strokes per Minute [also: spm]
Subscriber's Private Meter
Sun Probe-Mars [of NASA]
Surface Plasmon Microscopy
System Performance Monitor
Symbol Processing Machine
Synaptosomal Plasma Membrane
SpM Specialist in Microbiology
SPMA Southwest Parks and Monuments Association [US]
SPMC Solid Polyester Molding Compound
SPMI Southeastern Pine Marketing Institute [now: SLMA, US]
SPMOL Source Program Maintenance On-Line
SPMS Solar Particle Monitoring System
Special Purpose Manipulator System
System Program Management Survey
SPN Service Protection Network
Single Packet Network
Synthetic Polymer Network
SPNC Society for the Promotion of Nature Conservation [UK]
SPNI Society for the Protection of Nature in Israel
SPNR Society for the Promotion of Nature Reserves [now: SPNC, UK]
SP-NR Superior-Processing Natural Rubber
SPNS Switched Private Network Service
SPO Separate Partition Option
Solar Programs Office
System Program Office
SPOC Simulated Processing of Ore and Coal
Single SAR (= Search and Rescue) Points of Contact [of SARSAT/COSPAS]
SPOCK Simulated Procedure for Obtaining Common Knowledge
Standard Performance-Oriented Computer Keyboard
SPOOF Society for the Protection of Old Fishes [US]
Structure and Parity Observing Output Function
SPOOL Simultaneous Peripheral Operations On-Line
S&POP Student and Professional Outreach Program [of ASM International, US]
SPORA Special Operational Research and Analysis Group [of EUROCONTROL]
SPOSS Society for the Promotion of Science and Scholarship [US]
SPOT Satellite Positioning and Tracking
Smithsonian Precision Optical Tracking
SPOUT System Peripheral Output Utility
SPP Second Principal Plane
Signal Processing Peripheral
Society of Professional Pilots [US]
Special Purpose Processor
SPPAY Semipost-Pay (Pay-Station)
SP PD Specific Productivity
SPPF Solid Phase Pressure Forming
SP PH Split Phase
SPPO Spacelab Payload Project Office [of NASA]
SPPS Semipost-Pay Pay-Station
Single-Phase Power Supply
SPQ Synthetic Pipeline Quality Gas

SPQR Small Profits and Quick Returns
SPR Sandia Pulsed Reactor [US]
Send Priority and Route
Silicon Power Rectifier
Simplified Practice Recommendation
Spare
Spring
Sprinkler
Storage Protection Register
Strategic Petroleum Reserve
Supervisory Printer Read
SPRA Space Probe Radar Altimeter
SPRAT Small Portable Radar Torch
SPRC Self-Propelled Robot Craft
SPRD Spread
SPRDA Solid Pipeline Research and Development Association
SPRDS Steam Pipe Rupture Detector System
SPREAD Support Program for Remote Entry of Alphanumeric Displays [UK]
SPREADWARE Spreadsheet Software [also: Spreadware]
SPREP South Pacific Regional Environment Program [of UNEP]
SPRI Scott Polar Research Institute [UK]
Single Ply Roofing Institute [US]
SPRING Switched Private Networking
SPRINT Solid Propellant Rocket Intercept Missile
Special Police Radio Inquiry Network
SPRITE Solid Propellant Rocket Ignition Test and Evaluation
SPRN Suppression
SPROM Switched Programmable Read-Only Memory
SPRS Signal Processing Router/Scheduler
SPRT Sequential Probability Ratio Test
Standard Platinum Resistance Thermometer
SPRU Science Policy Research Unit
SPS Samples per Second
School of Practical Science
Secondary Power System
Secondary Propulsion System
Serial/Parallel/Serial
Shrouded Plasma Spray
Society of Physics Students [US]
Society of Polymer Science [Japan]
Software Products Scheme
Solar Power Satellite
Solar Power System
Spark Plug Socket
Speed Switch
Standard Pipe Size
Statistical Problem Solving
String Process System
Subsea Production System
Super Proton Synchrotron [of CERN, Switzerland]
Switching Power Supply
Symbolic Programming System
Symbolic Program System
System Performance Score
SPSA Single Phase Statistical Analyzer
SP-SA Slightly Polishing, Slightly Abrasive
SP/SC Speed Plug/Support Cylinder
SPSD Science and Professional Services Directorate [of SSC, Canada]
SPSE Society of Photographic Scientists and Engineers [US]
SP-SOP Surface Plasmon — Surface Optical Phonon (Mode)
SPSR Society for the Promotion of Scientific Research [Austria]
SPSS Shield Plug Storage Station
Statistical Package for the Social Sciences
SPST Single-Pole, Single-Throw
Society of Photographic Science and Technology [Japan]
SPSTNC Single-Pole, Single-Throw, Normally-Closed [also: SPST NC]
SPSTNO Single-Pole, Single-Throw, Normally-Open [also: SPST NO]
SPSTNODM Single-Pole, Single-Throw, Normally-Open, Double-Make [also: SPST NODM]
SPSTJ Society of Photographic Science and Technology of Japan
SPST SW Single-Pole Single-Throw Switch
SP SW Single-Pole Switch
SPT School of Plastics Technology [Israel]
Sectors per Track
Shared Page Table
Society for Philosophy and Technology [US]
Society of Photo-Technologists [US]
Special Purpose Test
Star Point Transfer
Strength-Probability-Time (Diagram)
Structured Programming Technique
Symbolic Program Tape
Symbolic Program Translator
System Page Table
SPTA Southern Pressure Treaters Association [US]
SPTC Share Purchase Tax Credit
SPTF Screen Printing Technical Foundation [US]
Shortest Programming Time First
Sodium Pump Test Facility
SP3T Single-Pole, Triple-Throw [also: SPTT]
SP3T SW Single-Pole, Triple-Throw Switch [also: SPTT SW]
SPU Signal Processing Unit
Slave Processing Unit
St. Paul Univerisity [Canada]
SPUC Serial Peripheral Unit Controller
SPUC/DL Serial Peripheral Unit Controller/Data Link
SPUR Source Program Utility Routine
Space Power Unit Reactor
SPURM Special Purpose Unilateral Repetitive Modulation
SPURT Spinning Unguided Rocket Trajectory
SPV Surface Photo Voltage
S-PVC Suspension Polyvinyl Chloride
SP VOL Specific Volume [also: sp vol]
SPW Spring Plunger Wrench
SPWLA Society of Professional Well Log Analysts [US]
SPWM Single-Sided Pulsewidth Modulation
SPX Simplex
Superheat Power Experiment [US]
SQ Self-Quenching
Sequence

Square
Squall
Stationary Quantum
Structural Quality
Superquick

sq (sequens) — the following
square

SQA Software Quality Assurance
Supplier Quality Assurance
System Queue Area

SQAM Superposed Quadrature Amplitude Modulation

SQAP Software Quality Assurance Plan
Swedish Question Answering Project [of RIND, Sweden]

SQAPS Supplier Quality Assurance Provisions

SQC Statistical Quality Control

SQ CG Squirrel Cage

sq ch square chain [Unit]

sq cm square centimeter

SQC/SPC Statistical Quality Control/Statistical Process Control

SQD Signal Quality Detector

sq dkm square dekameter

SQD LDR Squadron Leader

SQDN Squadron

SQF Subjective Quality Factor

sq ft square foot

sq hm square hectometer

sq in square inch

sq km square kilometer

SQL Standard Query Language
Structured Query Language

SQL/DS Structured Query Language/Data System

SQLN Squall Line

sq m square meter

sq mi square mile

sq mm square millimeter

sq mu square micron [also: sq μ]

SQNCR Sequencer

SQPR Staggered Quadrature Partial Response

SQPSK Staggered Quadrature Phase Shift Keying

sqq (sequentia) — the following ones

SQR Sequence Relay
Service Request
Square Root
Supplier Quality Representative

SQRL Subpicosecond and Quantum Radiation Laboratory [of TTU, US]

sq rd square rod [Unit]

SQRT Square Root [also: SQ RT]

SQUAD Squadron

SQUID Sperry Quick Updating of Internal Documentation
Superconducting Quantum Interference Device

SQUIRE Submarine Quickened Response

SQW Single Quantum Well
Square Wave

sq yd square yard

SR Safety Rod
Sarcoplasmic Reticulum
Saturable Reactor
Scanning Radiometer
Schottky Rectifier
Scientific Report
Search Radar
Secondary Radiation
Sedimentary Rock
Seismic Ray
Selective Reflection
Selective Ringing
Selenium Rectifier
Self-Rectifying
Self-Repulsion
Self-Reversal
Semantic Reaction
Semantic Routine
Semiregular
Send and Receive
Seychelles Rupee [Currency]
Sheet and Roll
Shift Register
Shift Reverse
Shipping Route
Short Range
Sigma Reactor
Silicone Resin
Slew Rate
Slip Range
Slip Ring
Slow Release
Society of Radiographers [UK]
Society of Rheology [US]
Silicone Rubber
Solar Radiation
Solar Research
Solid Rocket
Sorter-Reader
Sound Ranging
Sound Recording
Sound Rating
Special Register
Special Regulation
Special Report
Specific Resistance
Spectroradiometry
Speech Recognition
Speed Recorder
Speed Regulator
Spherical Radius
Spin Rotation
Split Ring
Spring-Return (Valve)
Sputter Rate
Stateroom
Status Register
Steam-Refined (Oil)
Steel Research
Step Reaction
Stereoscopic Radiography
Storage Register
Storage and Retrieval
Stress Relief
Stress Relaxation

	Study Requirement
	Styrene Rubber
	Subroutine
	Summary Report
	Sunrise
	Support Reaction (Load)
	Surface Region
	Surveillance Radar
	Sweep Rate
	Switch Register
	Switching Regulator
	Synaptic Ribbon
	Synchrotron Radiation
	Synthetic Rubber
	Systems Research
S/R	Send/Receive
	Subroutine
S-R	Set-Reset (Memory)
Sr	Strontium
sr	steradian [Unit]
SRA	Science Research Associates [US]
	Shop Replaceable Assembly
	Snubber Reducing Adapter
	Society of Research Administrators [US]
	Special Rules Airspace
	Special Rules Area
	Spring Research Association [now: SRAMA, UK]
	Stress-Relief Annealing
	Sulfo-Ricinoleic Acid
	Surveillance Radar Approach
SRAA	Singapore Robot and Automation Association [now: SIAA]
SRAAW	Short-Range Anti-Armor Weapon
SRAAWH	Short-Range Anti-Armor Weapon, Heavy [also: SRAAW(H)]
SRAAWL	Short-Range Anti-Armor Weapon, Light [also: SRAAW(L)]
SRAM	Short-Range Attack Missile
	Semi Random-Access Memory
	Static Random-Access Memory [also: sRAM]
SRAMA	Spring Research and Manufacturers Association [UK]
SRARQ	Selective Repeat-Automatic Repeat Request
SRATS	Solar Radiation and Thermospheric Satellite [Japan]
SRB	Screw Rest Button
	Societe Royale Belge des Electriciens [= Royal Belgian Society of Electricians]
	Solid Rocket Booster
	Sorter Reader Buffered
	Sulfate-Reducing Bacteria
SRBG	Societe Royale Belge de Geographie [= Royal Belgian Society of Geography]
SRBE	Societe Royale Belge des Electriciens [= Royal Belgian Society of Electricians]
SRBII	Societe Royale Belge des Ingenieurs et des Industriels [= Royal Belgian Society of Engineers and Manufacturers]
SRBM	Short-Range Ballistic Missile
SRBP	Synthetic Resin Bonded Paper
SRBPP	Saskatchewan River Basin Planning Program [Canada]
SRBSA	Saskatchewan Road Builders Safety Association [Canada]
SRBSN	Sintered Reaction-Bonded Silicon Nitride
SRC	Sanitary Refuse Collector
	Saskatchewan Research Council [Canada]
	Scandinavian Rubber Conference
	Science Research Council [now: SERC, UK]
	Scientific Research Council [Iraq]
	Semiconductor Research Cooperative
	Semiconductor Research Corporation [US]
	Sequencer/Ring Counter
	Solvent-Refined Coal
	Sound Ranging Control
	Source
	Source Range Channel
	Spares Receiving Checklist
	Spokane Research Center [US]
	Standard Requirements Code
	Steel Research Center
	Stimulus/Response Compare
	Stored Response Chain
	Synchronous Remote Control
	Synchrotron Radiation Center [US]
	Systems Research Center [US]
SRCH	Search
SRCL	Spectrally-Resolved Cathodoluminescence
SRCNET	Science Research Council Network [now: SERCNET, UK]
SRCR	System Run Control Record
SRCRA	Shipowners Refrigeration Cargo Research Association
SRD	Secret Restricted Data
	Self Reading Dosimeter
	Standard Reference Data
	Step Recovery Diode
	Subdivision of Regional District
S-RD	Shipper-Receiver Difference
SRDA	Sodium Removal Development Apparatus
SRDAS	Service Recording and Data Analysis System
SRDE	Signals Research and Development Establishment [UK]
SRDL	Signals Research and Development Laboratory [UK]
SRDM	Subrate Data Multiplexer
SRDS	Standard Reference Data System
	Systems Research and Development Service [of FAA, US]
SRE	Sending Reference Equivalent
	Signaling Range Extender
	Single Region Execution
	Site Resident Engineer
	Society of Reliability Engineers [US]
	Sodium Reactor Experiment [US]
	Speech Recognition Equipment
	Surveillance Radar Element
SREA	Street Rod Equipment Association [US]
SREMP	Source Region Electromagnetic Pulse
SRF	Self-Resonant Frequency
	Semireinforcing Furnace Black
	Service Request Flag

	Smithsonian Research Foundation [US]
	Software Recording Facility
	Software Recovery Facility
	Sorter Reader Flow
	Spacecraft Research Foundation
	Stereoscopic Rangefinder
	Surface Roughness Factor
SRFB	Space Research Facilities Branch [of NRC, Canada]
$Sr(FOD)_2$	Bis(1,1,1,2,2,3,3-Heptafluoro-7,7-Dimethyl-4,6-Octanedionato)strontium
SRG	Safety Regulation Group [of CAA, UK]
	Shift-Register Generator
	Short Range
	Statistical Research Group
	Suriname Guilder (or Florin) [Currency of Suriname]
SRI	Saudi Riyal [Saudi Arabia]
	Shift Right In
	Society of the Rubber Industry [Japan]
	Southern Research Institute
	Southwest Research Institute [US]
	Space Research Institute [Canada]
	Spalling Resistance Index
	Spring Research Institute [US]
	Stanford Research Institute [US]
SRIAER	Scientific Research Institute for Atomic Energy Reactors [USSR]
SRID	Single Radial Immunodiffusion
SRIC	Short Rotation Intensive Cultures
SRIF	Somatostatin Release Inhibiting Factor
	Special Recovery Investment Fund
SRIM	Selected Research in Microfiche [of NTIS, US]
	Structural Reaction Injection Molding
SRIN	Ship Research Institute of Norway
SRIOC	Shanghai Research Institute of Organic Chemistry [PR China]
SRIP	Society for the Research and Investigation of Phenomena [Malta]
SRISSP	Scientific Research Institute of Solid-State Physics [USSR]
SRL	Savannah River Laboratory [of USAEC]
	Science Reference Library [UK]
	Shift Register Label
	Shift Register Latch
	Stability Return Loss
	Structural Return Loss
	Superconductivity Research Laboratory [of ISTEC]
	System Reference Library
SRLY	Series Relay
SRM	Scrim-Reinforced Material
	Service Reference Model
	Shock Remanent Magnetization
	Short Range Modem
	Simple Reservoir Model
	Society for Range Management [US]
	Solid Rocket Motor
	Source-Range Monitor
	Southern Rocky Mountains
	Standard Reference Material
	Strategic Reconnaissance Missile
	Surface Reflection Microscope
	System Resources Manager
SRMA	Split-Channel Reservation Multiple-Access
SRMCASE	Symmetry-Restricted-Multiconfiguration Annihilation of Single Excitations
SRME	Submerged Repeater Monitoring Equipment
SRMEM	Symposium on Recycling of Metals and Engineering Materials
SRMS	Structure Resonance Modulation Spectroscopy
SRMU	Space Research Management Unit [US]
SRNA	Saskatchewan Registered Nurses Association [Canada]
SRNFC	Source Range Neutron Flux Channel
SRNFM	Source Range Neutron Flux Monitor
SRO	Savannah River Operation [of USAEC]
	Senior Reactor Operator
	Short-Range Order
	Singly Resonant Oscillator
	Specification Release Order
	Standing Room Only
SROA	Swedish Railway Officers Association
SRON	Space Research Organization Netherlands
SRP	Shared Resources Programming
	Signal Reference Point
	Slot Reference Point
	Solute Retarding Parameter
	Standard Review Plan
	Static Reservoir Pressure
	Step Reaction Polymerization
SRPO	Science Resources Planning Office [US]
SRPES	Synchrotron Radiation Photoemission Spectroscopy
SRQ	Service Request
SRR	Search and Rescue Radio
	Search and Rescue Region
	Serially Reusable Resource
	Shift Register Recognizer
	Short Range Recovery
	Software Requirements Review
	Sound Recorder Reproducer
	Sri Lankan Rupee [Currency]
	Super-Regenerative Reception
SRRB	Search and Rescue Radio Beacon
SRRC	Scottish Research Reactor Center
SRRL	Southern Regional Research Laboratory [US]
SRRT	Schweizerische Gesellschaft fur Reinraumtechnik [= Swiss Society for Clean Room Technology]
SRS	Safety and Reliability Society [UK]
	Savannah River Site [US]
	Selective Record Service
	Selenium Rectifier System
	Self-Sustained Reaction Sintering
	Silent Running Society [US]
	Simulated Remote Site
	Simulated Remote Station
	Slave Register Set
	Sodium Removal Station
	Software Requirements Specification
	Soil Research Station [New Zealand]
	Sonobuoy Reference System
	Specialized Rework System
	Statistical Reporting Service [of USDA]
	Stimulated Raman Scattering

	Strain-Rate Sensitivity
	Student Records System
	Subscriber Response System
	Supplemental Restraint System
	Surtsey Research Society [Iceland]
	Synchrotron Radiation Source
	Synchronous Relay Satellite
SRSA	Scientific Research Society of America
	Society of Radiographers of South Africa
SRSCC	Simulated Remote Station Control Center
SRSK	Short-Range Station Keeping
SRSS	Simulated Remote Sites Subsystem
SR-SS	Sunrise — Sunset
SRST	Spent Resin Storage Tank
SRT	Search Radar Terminal
	Secondary Ranging Test
	Self-Reinforcing Thermoplastic
	Slow Rise Time
	Society of Radiological Technologists [US]
	Soft Return [also: SRt]
	Solids Retention Time
	Sorting
	Special Relativity Theory
	Speech Reception Threshold
	Standard Radio Telegraph
	Supply Response Time
	Systems Readiness Test
SR&T	Supporting Research and Technology [also: SRT]
SRTC	Scientific Research Tax Credit [Canada]
SRTF	Shortest Remaining Time First
Sr(TMHD)$_2$	Bis(2,2,6,6-Tetramethyl-3,5-Heptanedionato)strontium
SRTS	Self-Regulating/Temperature Source [also: SR/TS]
SRU	Societe de Raffinage d'Uranium [= Society for Uranium Refining, France]
	Shop Replaceable Unit
	Sulfur Recovery Unit
SRV	Safety Release Valve
	Space Rescue Vehicle
	Surveillance
SRZ	Special Rules Zone
	Surveillance Radar Zone
SS	Sample Size
	Sandstone
	Satellite Switching
	Scatter Slit
	Science Service [US]
	Scintillation Spectrometer
	Secondary Spectrum
	Secretary of State
	Secret Service
	Self-Screening
	Self-Shielding
	Self-Starting
	Selector Switch
	Select Standby
	Semifinal Splice
	Semi-Steel
	Semi-Submerisible
	Sempervivium Society [UK]
	Servo System
	Set Screw
	Shear Strength
	Shift Supervisor
	Ship Service
	Shoulder Screw
	Sieve Shaker
	Signaling System
	Signal Selector
	Signal Strength
	Silicon Steel
	Silver Steel
	Single Scan
	Single Shot
	Single Space
	Single Strength
	Sliding Scale
	Slipstream
	Slow Setting
	Slow-Slow (Wave)
	Sodium Salicylate
	Soil Science
	Solar Spectrum
	Solar System
	Solid/Solid
	Solid Solution
	Solid State
	Sound Scattering
	Sonic Speed
	Space Science
	Space Simulator
	Space Station
	Space Switch
	Spark Spectrum
	Spatial Statistics
	Special Service
	Special Source Materials
	Spectral Series
	Speech Synthesis
	Spherical Symmetry
	Spin-Stabilized
	Spread Spectrum
	Stack Segment
	Stainless Steel
	Start-Stop (Character)
	State Space
	Stationary States
	Station-to-Station
	Statistical Standards
	Steady State
	Steamship
	Step Signal
	Stereoscopic Society [UK]
	Storage to Storage
	Structural Safety
	Structure Sensitivity
	Subscriber Switching
	Summing Selector
	Sunset
	Supersaturation
	Supersonic
	Suspended Solids

	Swedish Standard
	Systems Science
	System Supervisor
S/S	Steamship
	Source/Sink
	Start/Stop (Character)
S-S	Storage-to-Storage
ss	(scilicet) — namely
SSA	Safe Sector Altitude
	Seismological Society of America [US]
	Semiconductor Safety Association [US]
	Semiotic Society of America [US]
	Sequential Spectrometer Accessory
	Serial Systems Analyzer
	Single-Line Synchronous Adapter
	Smoke Suppressant Additive
	Social Security Administration
	Solid State Abstracts
	Solid-State Amplifier
	Spherical Sector Analyzer
	Spring Service Association [US]
	State Space Analysis
	Steady-State Analysis
	Stellar Spectroscopy
	Stellar Spectrum
	Substructure Analysis
	Sulfosalicylic Acid
	Swedish Steel Producers Association
SSAA	Swedish Society of Aeronautics and Astronautics
SSAC	Signalling System Alternating Current
	Society for the Study of Architecture in Canada
	Space Science Analysis and Command [of NASA]
SSADM	Structured Systems Analysis and Design Methodology
SSAE	Society of Senior Aerospace Executives [US]
SSAEC	Society for the Study of Alchemy and Early Chemistry [now: SHAC, UK]
SSAI	Steinberg State Astronomical Institute [USSR]
SSAL	Simplified Short Approach Light
SSALF	Simplified Short Approach Light System with Squenced Flashing Lights
SSALR	Simplified Short Approach Light System with Runway Alignment Lights
SSALS	Simplified Short Approach Light System
SSAP	Session Service Access Point
SSAR	Site Safety Analysis Report
	Society for the Study of Amphibians and Reptiles [US]
	Spin-Stabilized Aircraft Rocket
	Spotlight Synthetic Aperture Radar
	Standard Safety Analysis Report
SSAS	Station Signaling and Announcement Subsystem
SSB	Serial Systems Bus
	Single Sideband
	Societe Scientifique de Bruxelles [= Scientific Society of Brussels, Belgium]
	Source Selection Board
	Space Science Board [of NRC, US]
	Spring Stop Button
	Subscriber Busy
SSBAM	Single Sideband Amplitude Modulation [also: SSB-AM]
SSBD	Single Sideboard
SSBFM	Single Sideband Frequency Modulation [also: SSB-FM]
SSBM	Single Sideband Modulation
SSBO	Single Swing Blocking Oscillator
SSBSC	Single Sideband Suppressed Carrier [also: SSB-SC]
SSBSCAM	Single Sideband-Suppressed Carrier Amplitude Modulation [also: SSB-SC-AM or SSB-SC/AM]
SSBSCASK	Single Sideband-Suppressed Carrier-Amplitude Shift Keyed [also: SSB-SC-ASK or SSB-SC/ASK]
SSBSCOM	Single Sideband Suppressed-Carrier Optical Modulation [also: SSB-SC-OM or SSB-SC/OM]
SSBSCPAM	Single Sideband-Suppressed Carrier Pulse-Amplitude Modulation [also: SSB-SC-PAM or SSB-SC/PAM]
SSBW	Surface Skimming Bulk Wave
SSC	Saline Sodium Citrate
	Saskatchewan Safety Council [Canada]
	Saskatchewan Science Center [Canada]
	SCANNET (= Scandinavian Information Retrieval Network) Service Center [Sweden]
	Secondary Schools Certificate
	Sector Switching Center
	Selector Switching Center
	Sensor Signal Conditioner
	Settled Sludge Concentration
	Ship Structure Committee [US]
	Short Segmented Cask
	Signalling and Supervisory Control
	Single Silk-Covered
	Single Specimen Compliance
	Sintered Silicon Carbide
	Slow-Scan Camera
	Slow-Scan CCD (= Charge-Coupled Device)
	Societe Suisse de Chronometrie [= Swiss Society of Chronometry]
	Solid-State Chemistry
	Solid-State Circuit
	Solid-State Communications
	Solid-State Control
	Solid-State Frequency Converter
	Space Science Committee [France]
	Special Service Center
	Spectroscopy Society of Canada
	Spin-Spin Coupling
	Standard Saline Citrate
	Station Selection Code
	Statistical Society of Canada
	Stellar Simulation Complex
	Subsurface Contamination
	Sudden Storm Commencement
	Sulfide Stress Cracking
	Superconducting Supercollider
	Supply and Services Canada
	Swivel Screw Clamp
	System Support Controller
SSCA	Strobed Single Channel Analyzer
SSCC	Scandinavian Society for Clinical Chemistry [Finland]

	Signaling System, Common Channel
	Solid State Circuits Conference
	Solid State Circuits Council [of IEEE, US]
	Spin Scan Cloud Camera
SSCD	Society of Small Craft Designers [US]
SSCE	Swedish Society of College Engineers
SS/CF	Signal Strength/Center Frequency
SSCI	Sanitation Suppliers and Contractors Institute [US]
	Steel Service Center Institute [US]
	Steel Shipping Container Institute [US]
	Swiss Society of Chemical Industry
SSCNS	Ships Self-Contained Navigation System
SSCPA	Single Site Coherent Potential Approximation
SSCR	Set Screw
	Solid-State Chemical Reaction
	Spectral Shift Control Reactor
	Spectrum Shift Controlled Reactor
SSCS	Safety and Security Communications System
S&SCS	Scintillation and Semiconductor Counter Symposium
SSCU	Special Signal Conditioning Unit
SSCV	Semisubmersible Crane Vessel
SSD	Share Secretarial Distribution
	Single Station Doppler
	Small Signal Diode
	Solid-State Detector
	Solid-State Device
	Solid-State Disk
	Solid-State Drive
	Space Systems Division [of US Air Force]
	Static Sensitive Device
	Steady-State Diffusion
	Steady-State Distribution
	Subsurface Distribution
SS&D	Synchronization Separator and Digitizer
SSDA	Service Station Dealers of America [US]
	Synchronous Serial Data Adapter
SSDC	Semi-Submersible Drilling Caisson
	Signalling System, Direct Current
	Single Steel Drilling Caisson
	System Safety Development Center
SSDD	Single-Sided, Double-Density (Disk)
SSDR	Steady State Determining Routine
SSDT	Society of Soft Drink Technologists [US]
SSE	Safety Shutdown Earthquake
	School of Science and Engineering
	Single Silk (Covering) over Enamel (Insulation)
	Society for Scientific Exploration [US]
	Society for the Study of Evolution [US]
	Solid-State Electrochemistry
	Solid-State Electronics
	Solid-State Engineering
	South-Southeast
	Special Support Equipment
	Sulfate Saturated Electrode
	Sum of Squares Error
SSEAT	Surveyor Scientific Evaluation Advisory Team [of NASA]
SSEB	South of Scotland Electricity Board
SSEC	Selective Sequence Electronic Calculator
	Solid-State Electrochemistry
S-SEED	Symmetric Self-Electrooptic-Effect Device
SSEP	System Safety Engineering Plan
SSEPA	Society of Spanish Engineers, Planners and Architects [US]
SSerg	Staff Seargeant
SSES	Small Smart Electronic Switch
SSESM	Spent Stage Experimental Support Module
SSF	Safe Shutdown Facility
	Saybolt Seconds Furol
	Sample Size and Frequency
	Self-Similar Flow
	Service Storage Facility
	Single Sided Frame
	Solid-State Fermentation
	Space Station Freedom
	Special Service Force
	Subsonic Flow
	Supersonic Flow
	Supersonic Frequency
	Symmetrical Switching Function
	System Support Facility
SSFC	Sequential Single Frequency Code
	Sir Sanford Fleming College [Canada]
	Solid-State Frequency Converter
SSFE	Scandinavian Society of Forest Economics [Sweden]
SSFF	Solid Smokeless Fuels Federation
SSFI	Scaffolding, Shoring and Forming Institute [US]
SSFL	Schottky-Barrier Coupled Schottky-Barrier Gate FET (= Field-Effect Transistor) Logic
	Steady-State Fermi Level
SSFM	Single Sideband Frequency Modulation
SSFS	Special Services Forecasting System
SSG	Search Signal Generator
	Second Stage Graphitization
	Small Signal Gain
	Stonehenge Study Group [US]
	Superconducting Superheated Granules
	Symbolic Stream Generator
SSGA	Sterling Silversmiths Guild of America [US]
SSGD	Secondary School Graduation Diploma [Canada]
SSGS	Standard Space Guidance System
SSH	Solid Solution Hardening
SSh	Somalia Shilling [Currency]
SSHB	Society for the Study of Human Biology [UK]
SSHRC	Social Sciences and Humanities Research Council of Canada
SSI	Sector Scan Indicator
	Sky Survey Instrument
	Small-Scale Integration
	Solid-State Ionics
	Space Studies Institute [US]
	Special Secretariat for Informatics [Brazil]
	Spin-Spin Interaction
	Start Signal Indicator
	Steady-State Irradiation
	Storage-to-Storage Instruction
	Supplemental Security Income
	Synchronous Systems Interface
SSIDA	Steel Sheet Information and Development Association [UK]

SSIE	Smithsonian Science Information Exchange [of Smithsonian Institute, US]
SSIG	Single Signal
SSIM	Standard Schedules Information Manual
SSIMS	Static Secondary Ion Mass Spectroscopy
SSIP	Scotian Shelf Ichthyoplankton [Canada]
	Shuttle Student Involvement Project [US]
SSIUS	Specialty Steel Industry of the United States
SSJ	Seismological Society of Japan
	Spectroscopical Society of Japan
SSK	Soil Stack
SSL	Scientific Subroutine Library
	Shift and Select
	Sodium Stearoyl Lactylate
	Software Slave Library
	Software Specification Language
	Solid-State Lamp
	Solid-State Laser
	Source Statement Library
	Space Sciences Laboratory [of University of California at Berkeley, US]
	Storage Structure Language
	Surface Science Laboratories [US]
	System Specification Language
SSLC	Synchronous Single-Line Controller
SSLO	Secondary School Liaison Office
	Solid-State Local Oscillator
SS LORAN	Sky-Wave-Synchronized Long-Range Navigation [also: ss loran]
SSLV	Standard Space Launch Vehicle
SSM	Satellite System Monitor
	Schonsted Magnetometer
	Semiconductor Storage Model
	Sensor Microwave/Imager
	Single-Sideband Signal Multiplier
	Small Semiconductor Memory
	Solid-State Maser
	Solid-State Microbattery
	Special Safeguarding Measures
	Special Sensor Microwave
	Spectral Scanning Microwave
	Spread Spectrum Modulation
	Staff Sergeant Major
	Standard Schedule Message
	State Space Method
	State Space Model
	Surface-to-Surface Missile
SSMA	School Science and Mathematics Association [US]
	Scottish Steel Makers Association [UK]
	Spread-Spectrum Multiple Access
SSMB	Special Services Management Bureau
SSMD	Silicon Stud-Mounted Diode
SSME	Scientific Society of Mechanical Engineering [of Hungarian Chemical Society]
	Space Shuttle's Main Engine
SSMF	Signaling System, Multifrequency
SSM/I	Special Sensor Microwave/Imager
	Spectral Scanning Microwave/Imager
SSMS	Spark Source Mass Spectrometry
SSMSR	Spring School on Muon Spin Research
SSMT	Stress Survival Matrix Test
SSMTG	Solid-State and Molecular Theory Group
SSMUTA	Sheet and Strip Metal Users Technical Association [UK]
SSMWD	Solid-State Microwave Device
SSN	Segment Stack Number
	Sintered Silicon Nitride
	Switched Services Network
SSNM	Strategic Special Nuclear Materials
SSNMH	Scipio Society of Naval and Military History [US]
SSNPP	Small-Size Nuclear Power Plant
SSO	Steady-State Oscillation
SSOC	Solid-State Optoelectronics Consortium [Canada]
SSOG	Satellite System Operations Guide
SSOP	Satellite System Operations Plan
SSOUT	System Output Unit
SSP	Scientific Subroutine Package
	Signaling and Switching Processor
	Silk-Screen Printing
	Society of Satellite Professionals [now: SSPI, US]
	Society of Scholarly Publishing [US]
	Sodium Sampling Package
	Solid-State Physics
	Space Station Program
	Space Summary Program
	Special Services Protection
	Static Sodium Pot
	Static Spontaneous Potential
	Steady State Process
	Steam Service Pressure
	Sub-Satellite Point
	Sum of Squares and Products
	System Status Panel
	System Support Program
SSPA	Solid-State Power Amplifier
	Swedish Steel Producers Association
SSPC	Steel Structures Painting Council [US]
SSPE	Saline Sodium Phosphate EDTA (= Ethylenediaminetetraacetic Acid)
SSPF	Signal Structure Parametric Filter
SSPI	Society of Satellite Professionals International [US]
SSPL	Steady State Power Level
SSPMA	Sump and Sewage Pump Manufacturers Association [US]
SSPO	Space Station Program Office
SSPP	Scandinavian Society for Plant Physiology [Sweden]
SSPS	Satellite Solar Power System
	Solid-State Protection System
SSPWR	Small-Size Pressurized-Water Reactor
SSQD	Single-Sided, Quad-Density (Disk)
SSR	Secondary Surveillance Radar
	Senior Site Representative
	Separate Superheater Reactor
	Site Suitability Report
	Slow Strain Rate
	Societe Suisse de Radiodiffusion [= Swiss Society of Broadcasting]
	Solid-State Reaction
	Solid-State Refining
	Solid-State Relay
	Specification Status Report
	Spin-Spin Relaxation

	Stretched Surface Recording
	Subsynchronous Resonance
	Sum of the Squared Residuals
	Switching Selector Repeater
	Synchronous Stable Relaying
	System Status Report
SSRA	Scottish Seaweed Research Association
	Spread-Spectrum Random Access
SSR-A	Secondary Surveillance Radar Providing Coded Aircraft Identity
SSRC	Structural Stability Research Council [US]
	Swedish Space Research Committee
SSR-C	Secondary Surveillance Radar Providing Coded Aircraft Altitude
SS&RC	Standards Screening and Review Committee [US]
SSRI	Swedish Silicate Research Institute
SSRL	Stanford Synchrotron Radiation Laboratory [US]
SSRMS	Space Station Remote Manipulator System
SSRP	Siam Society under Royal Patronage [Thailand]
	Space Science Radioastronomy Program [of CSA, Canada]
	Stanford Synchrotron Radiation Project [of Stanford University, US]
SSR/RPG	Secondary Surveillance Radar/Regional Planning Group [of ICAO]
SSRS	Self-Sustained Reaction Sintering
	Society for Social Responsibility in Science [US]
	Source Storage and Retrieval System
	Start-Stop-Restart System
SSR-S	Secondary Surveillance Radar with Selective Interrogation Capability
SSRSJC	Saudi-Sudanese Red Sea Joint Commission [Saudi Arabia]
SSRT	Slow Strain Rate Technique
	Slow Strain Rate Testing
	Subsystem Readiness Test
SSS	Scientific Subroutine System
	Shipboard Survey Subsystem
	Simulation Study Series
	Small Scientific Satellite
	Small Solar Satellite
	Solid-State Sensor
	Solid-State Science
	Solid-State Switching
	Special Safety Safeguards
	Steering and Suspension System
	Strategic Satellite System
	Subscribers Switching Subsystem
	Supersaturated Solution
	Swedish Scientific Satellite
	System Safety Society [US]
SSSA	Soil Science Society of America [US]
SSSB	Society for the Study of Social Biology [US]
SSSC	Solid-State Sciences Committee [US]
	Surface Subsurface Surveillance Center
SSSD	Single-Sided, Single-Density (Disk)
SSSF	Self-Scanner Stop Failure
SSSI	Site of Special Scientific Interest
	Soil Science Society of Ireland
SSSM	Self-Starting Synchronous Motor
SSSP	Space Settlement Studies Program [US]
	Station-to-Station Send Paid
	Structured Sound Synthesis Project [Canada]
	System Startup Service Package
SSSS	Society for Social Studies in Science [US]
	Standard Submarine Sonar System [also: S^4]
SSST	Solid-State Science and Technology
SSSSA	Soil Science Society of South Africa
SSSV	Semi-Submersible Support Vessel
	Subsurface Safety Valve
SSSWP	Seismological Society of the Southwest Pacific [New Zealand]
SST	Salt-Spray Test
	Saskatchewan Science and Technology [Canada]
	School of Science and Technology
	Science Society of Thailand
	Sea Surface Temperature
	Secondary Surge Tank
	Serum Separation Tube
	Simulated Structural Test
	Simultaneous Self-Test
	Single-Sideband Transmission
	Society of Surveying Technicians [UK]
	Solid-State Technology
	Solid-State Theory
	Stainless Steel
	Stationary-State Treatment
	Stress-Stiffening Technique
	Subscriber Transferred
	Subsystems Test
	Supersonic Transport
	Symmetric Single Trapping
	Synchronous System Trap
	System Segment Table
SS/T	Steady-State/Transient (Analysis)
S Staff	South Staffordshire [UK]
SSTC	Single Sideband Transmitted Carrier
SSTDMA	Satellite-Switched Time-Division Multiple Access [also: SS/TDMA or SS-TDMA]
SSTF	Shortest Seek Time First
SSTP	Subsystems Test Procedure
SSTR	Solid-State Track Recorder
SSTV	Sea Skimming Test Vehicle [US Navy]
	Slow-Scan Television
SSU	Saybolt Seconds Universal
	Secondary Sampling Unit
	Sequential Shunt Unit
	Single Signaling Unit
	Stratospheric Sounding Unit
	Subscriber Switching Unit
	Subsequent Signal Unit
	Supply and Services Union
SSUS	Spinning Solid Upper Stage
SSV	Settled Sludge Volume
	Standby Safety Vessel
	Supersaturated Vapor
	Supersonic Vehicle
SSW	Solid-State Welding
	South-Southwest
	Space Switch
	Steady Shock Wave
	Surface Science Western [of UWO, Canada]

Synchro Switch
SSWA Sanitary Supply Wholesalers Association [US]
SSWO Special Service Work Order
SSWWS Seismic Sea-Wave Warning System
SSX Small System Executive
SSZ Society of Systematic Zoology [US]
ST Saint
Sawtooth
Schmitt Trigger
Science and Technology
Scientific and Technical
Screw Thread
Seed Technologist
Seed Technology
Segment Table
Selective Transmission
Select Time
Series Transformer
Shock Tube
Short-Transverse (Direction)
Sidetone
Single-Throw
Skin Temperature
Solar Temperature
Solution (Heat) Treated
Sounding Tube
Sound Track
Sound Transmission
Sound Trap
Space Technology
Space Telescope
Space Time
Spring Tide
Stabilizer Tube
Standard Time
Start
Starting Time
Start Signal
Static Thrust
Station
Statute
Steam
Steam Table
Steam Turbine
Steel
Steel Technology
Stereo-Telescope
Stone
Storage Tube
Store
Strait
Stream Tube
Street
Street Tee (Pipe)
Sucrose Tallowate
Surge Tank
Super-Tough
Switching Technique
Subscriber Terminal
Supplementary Terms
Surveying Technician
Survey Technology
Symbol Table
Syntax Tree
Systems Technology
System Table
S&T Science and Technology
Scientific and Technical
S/T Search/Track
Specific Heat/Temperature (Curve)
S-T Short-Transverse (Direction)
St Stefan Number
Stoke
Stokes Number
Stratus
STA Saskatchewan Trucking Association [Canada]
Science and Technology Agency [Japan]
Shuttle Training Aircraft
Simultaneous Thermal Analyzer
Slurry Transportation Association
Society of Typographic Arts [US]
Solanum Tuberosum Agglutinin
Solution (Heat) Treated and Aged [also: ST&A]
Station
Stationary
Store Answer
Straight-In Approach
Supersonic Tunnel Association [US]
ST&A Solution Treated and Aged [also: STA]
STAAS Surveillance and Target Acquisition Aircraft System
STAB Stabilization
STAC Science and Technology Advisory Committee [US]
Software Timing and Control
STACO Standing Committee
STAD Start Address
STADAC Station Data Acquisition and Control
STADAN Satellite Tracking and Data Acquisition Network [of NASA]
STAE Specify Task Asynchronous Exit
STAESA Society of Turkish Architects, Engineers and Scientists in America [US]
STAF Scientific and Technological Applications Forecast
STAFDA Specialty Tools and Fasteners Distributors Association [US]
STAFF Stellar Acquisition Flight Feasibility
Staff Staffordshire [also: Staffs, UK]
STAG Science and Technology Advisory Group [Taiwan]
STAGG Small Turbine Advanced Gas Generator
STAGS Structural Analysis of General Shells
STAI Subtask ABEND (= Abnormal Ending) Intercept
STAIR Structural Analysis Interpretive Routine
STAIRS Storage and Information Retrieval System
STAIRS/VS Storage and Information Retrieval System/Virtual Storage
STALO Stable Local Oscillator [also: Stalo or stalo]
STAM Shared Tape Allocation Manager
STAMO Stabilized Master Oscillator
STAMOS Sortie Turnaround Maintenance Operations Simulation
STAMP Systems Tape Addition and Maintenance Program
STAMPS Scientific and Technological Aspects of Materials Processing in Space

STAN Stanchion
Standard(ization)
STANAG Standardization Agreement [of NATO]
STANAVFORLANT Standing Naval Force Atlantic [of NATO]
STAO Science Teachers Association of Ontario [Canada]
STAP Statistics Panel [of ICAO]
STAPPA State and Territorial Air Pollution Program Administrators [US]
STAR Satellite Telecommunications with Automatic Routing
Scientific and Technical Aerospace Reports [of NASA]
Self-Testing and Repairing
Serial Transmitter and ROM (= Read-Only Memory)
Shield Test Air Reactor
Ship-Tended Acoustic Radar
Space Thermionic Auxiliary Reactor
Speed through Air Resupply
Standard Terminal Approach Route
Standard Terminal Arrival Route
Star and Stellar Systems Advisory Committee [of ESRO, France]
Steerable Array Radar
Stellar Attitude Reference
Student Team on Alumni Relations
Studies, Tests and Applied Research
String Array Processor
Submarine Test and Research
Synthetic Aperture Radar
System to Automate Records
STARAD Starfish Radiation Satellite [of NASA]
STARAN Stellar Attitude Reference and Navigation [also: Staran or staran]
STARE Steerable Telemetry Antenna Receiving Equipment
STARFIRE System to Accumulate and Retrieve Financial Information with Random Extraction
STARP Strategic Technologies in Automation and Robotics Program
STARS Satellite Telemetry Automatic Reduction System [of NASA]
Satellite Transmission and Reception Specialist
START Selections to Active Random Testing
Short Term Aid for Research and Technology
Spacecraft Technology and Advanced Reentry Test
Strategic Arms Reduction Talks [US-USSR]
Summit Technology and Research Transfer Center
Systematic Tabular Analysis of Requirements Technique
STARTS Software Tools for Application to Large Real-Time Systems
STAT Shipping Transit Analysis Tabulation
Static(s)
Stationary
Statistician
Statistics
Statue
statA statampere
statC statcoulomb
StatCan Statistics Canada
STATE Simplified Tactical Approach and Terminal Equipment
statF statfarad
STATFOR Specialist Panel on Air Traffic Statistics and Forecasts [of EUROCONTROL]
statH stathenry
STATLAB Statistics Laboratory
STATLIB Statistical Library
STAT MECH Statistical Mechanics
STAT MUX Statistical Multiplexer
statΩ statohm
statΩ^{-1} statmho
STATPAC Statistics Package
statS statsiemens
StatsCan Statistics Canada
statT stattesla
statV statvolt
statWb statweber
STB Segment Tag Bit
Stock Tank Barrel [also: stb]
Stop Bars
Subsystems Test Bed
STBD Starboard
STB/D Stock Tank Barrels per Day [also: STB D^{-1}, stb/d or stb d^{-1}]
STBL Stable
STBR Stirred Tank Bioreactor
STBS Stirred Tank Biological Reactor
STBY Standby
STC Satellite Test Center [US Air Force]
Satellite Telecommunications Center
Science and Technology Center
Scientific and Technical Committee [of ESRO, France]
Self-Tuning Controller
Sensitivity Time Control
SHAPE (= Supreme Headquarters, Allied Command Europe) Technical Center [of NATO]
Short Time Constant
Sliding Twin-Crossbar
Society for Technical Communication [US]
Society of Telecommunications Consultants [US]
Sound Transmission Class
Staff Training Center
Station Technical Control
Standard Telephone Cable
Standard Telephones and Cables
Standard Transmission Code
Supplemental Type Certificate
System Test Complex
STCA Short-Term Conflict Alert System
STCB Subtask Control Block
STCC Spacecraft Technical Control Center
Standards Council of Canada
STCL Source-Term Control Loop [of HEDL, US]
STCW System Time Code Word
STD Salinity-Temperature-Depth (Recorder)
Separations Technology Division
Small Test Detector
Society of Typographic Designers [UK]
Standard

	Subscriber Trunk Dialing
	Surface Thermodynamics
STDA	Selenium-Tellurium Development Association [also: S-TDA, US]
	Solution Treated and Double Aged
std atm	standard atmosphere
STDB	Singapore Trade Development Board
STDC	Standards Council of Canada
std cu ft	standard cubic foot
STD DEV	Standard Deviation
STDI	Selenium and Tellurium Development Institute
STDM	Synchronous Time-Division Multiplexing
STDN	Space(flight) Tracking and Data Network
STDS	System for Thermal Decomposition Studies
STDT	Student
STDY	Steady
STE	Sainte
	Shield Test Experiment [US]
	Society of Tractor Engineers
	Spacecraft Test Equipment
	Span Terminating Equipment
	Sterolester
	Supergroup Translating Equipment
S2E	Surfaced on Two Edges [Lumber]
STEAR	Strategic Technologies for Automation and Robotics (Program) [of Canadian Space Agency]
STEB	Standard Test and Evaluation Bottle [of AFML, US]
STEBIC	Scanning Transmission Electron Beam Induced Current
STEC	Short-Term Exposure Criteria
	Solar Thermal Energy Conversion
STED	Scanning Transmission Electron Diffraction
	Solar Turboelectric Drive
STEER	Steering
STEG	Societe Tunisienne de l'Electricite et du Gaz [= Tunisian Society of Electricity and Gaz]
	Solar Thermoelectric Generator
STEL	Short-Term Exposure Limit
STELLA	Satellite Transmission Experiment Linking Laboratories
STEM	Scanning Transmission Electron Microscopy
	Stay Time Extension Module [of NASA]
	Stored Tubular Extendable Member
STEM-BF	Scanning Transmission Electron Microscopy — Bright Field
STEM-DF	Scanning Transmission Electron Microscopy — Dark Field
STEP	Safety Test Engineering Program [of USAEC]
	Scientific and Technical Exploitation Program
	SEMI (= Semiconductor Equipment and Materials Institute) Technical Education Program [US]
	Simple Transition to Electronic Processing
	Space Technology Experiments Platform
	Specialized Technique for Efficient Typesetting
	Standard for the Exchange of Product Model Data
	Standard Tape Executive System
	Standard Terminal Program
	Summer Temporary Employment Program [Canada]
	Supervisory Tape Executive Program
	Support for Technology-Enhanced Productivity
STEPP	Society of Teachers in Education of Professional Photography [US]
STEPS	Solar Thermionic Electric Power System
STER	Sterilizer
sterad	steradian [also: sr]
STEREO	Stereotype
STET	Specialized Technique for Efficient Typesetting
ST EX	Stock Exchange [also: ST EXCH]
STF	Safety Test Facility
	Satellite Tracking Facility [US Air Force]
	Shield Test Facility (Reactor) [US]
	Short Title File
	Some Test Failed
	Stratiform
	System Transfer Function
STFF	Sodium Tripolyphosphate
STG	Sawtooth Generator
	Space Task Group [of NASA]
	Starting
	Sterling
	Storage
	Strong
STGE	Storage
STH	Somatotropic Hormone
STI	Scientific and Technical Information
	Screw Thread Insert
	Service Tools Institute [now: HTI, US]
	Speech Transmission Index
	Steel Tank Institute [US]
	Swiss Tropical Institute
S&TI	Scientific and Technical Information
STIBC	Society of Translators and Interpreters of British Columbia [Canada]
STIC	Scientific and Technical Intelligence Center [of USDOD]
STID	Scientific and Technical Information Division [of NASA]
STIF	Short-Term Irradiation Facility
STIFF	Stiffener
STIL	Short-Term Inhalation Limit
STIN	Science and Technology Information Network
ST-IN	Straight-In
STINA	Steel Tube Institute of North America [US]
STINFO	Scientific and Technical Information
	Scientific and Technical Information Officers [US]
STINGS	Stellar Inertial Guidance System
STIP	Science Teaching Improvement Program [US]
	Stipend(iary)
STIR	Shield Test Irradiation Reactor [US]
	Statistics Indexing and Retrieval Project [of LUT, UK]
	Stirring
STIS	Specialized Textile Information Service [UK]
STK	Stack
	Stock
	Strake
STK EXCH	Stock Exchange
STKYD	Stockyard
STL	Schottky Transistor Logic
	Sound Transmission Loss
	Space Technology Laboratory [US]

Standard Telecommunication Laboratories [UK]
Standard Telegraph Level
Statistical Time Lag
Steel
Structural
Studio Transmitter Link
System Test Loop

STLB Schottky TTL (= Transistor-Transistor Logic) Bipolar
STLE Society of Tribologists and Lubrication Engineers [US]
STLO Scientific and Technical Liaison Office
STLV Simian T-Lymphotropic Virus
STL WG Steel Wire Gauge
STM Scanning Tunneling Microscopy
Scientific, Technical and Medical
(International Group of) Scientific, Technical and Medical Publishers [Netherlands]
Short-Term Memory
Standard Thermal Model
Statement
Steam
Storm
Structural Test Model
Structural Thermal Model
Supersonic Tactical Missile
STM/AFM Scanning Tunneling Microscopy/Atomic Force Microscopy
STMCGMW Subcommittee for Tectonic Maps of the Commission for the Geological Map of the World [USSR]
ST MGR Station Manager
STMIS System Test Manufacturing Information System
STMT Statement
STMU Special Test and Maintenance Unit
STN Satellite Tracking Network [of NASA]
Scientific and Technical Network
Stainless
Station
Stone
Super-Twisted Nematic
Switched Telecommunications Network
STNR Stationary
STNV Satellite Tobacco Necrosis Virus
STO Segment Table Origin
Slater-Type Orbitals
Stock-Tank Oil
System Test Objectives
STOA Solution (Heat) Treated and Overaged [also: ST&OA]
STOC Selected Table of Contents
STOIAC Static Technology Office Information Analysis Center [of BMI, US]
STOL Short Takeoff and Landing
Systems Test and Operation Language
STOM SPADE Terminal Operator's Manual
STOP Stable Ocean Platform
Statistical Operations Processor
Supersonic Transport Optimization Program [of NASA]
STOPS Shipboard Toxicological Operational Protective System
STOQ Storage Queue
STOR Storage
STORC Self-Ferrying Trans-Ocean Rotary-Wing Crane
STORES Syntactic Tracer Organized Retrospective Enquiry System
STORET Storage and Retrieval
STORLAB Space Technology Operations and Research Laboratory [US]
STORM Statistically-Oriented Matrix Program
STOVL Short Takeoff and Vertical Landing
STOW System Takeoff Weight
Stowage
STOX Speech and Telegraphy in Voice Channel
STP Scientifically Treated Petroleum
Selective Tape Print
Sewage Treatment Plant
Signal Transfer Point
Silver-Bearing Tough Pitch Copper
Simultaneous Test Procedure
Simultaneous Track Processor
Slater-Transfer Preuss
Solar Terrestrial Physics
Space Test Program
Special Technical Publication
Stamp
Standard Temperature and Pressure
Stop (Character)
Strain Point [also: ST P or StP]
System Test Plan
System Test Procedure
STPF Shield Test Pool Facility
Stabilized Temperature Platform Furnace
STPL Sidetone Path Loss
Standard Test Processing Language
STPO Science and Technology Policy Office
STPP Sodium Tripolyphosphate
STPSS Science and Technology Program Support Section [Canada]
STPST Stop-Start
STPTC Standard of Tar Products Testing Committee [UK]
STQ Solution (Heat) Treated and Quenched [also: ST&Q]
STR Segment Table Register
Self-Tuning Regulator
Short-Term Revitalization
Side-Tone Reduction
Speed-Tolerant Recording
Standardized Thai Rubber
Status Register
Steamer
Stirred Tank Reactor
Strainer
Strait
String
Strip
Strobe
Stroke
Structure
Submarine Thermal Reactor
Symbol Timing Recovery
Synchronous Transmitter/Receiver

STRAD Signal Transmission, Reception and Distribution
Switching, Transmitting, Receiving and Distributing
STRAIN Structural Analytical Interpreter
STRAP Star Tracking Rocket Attitude Positioning
Stellar Tracking Rocket Attitude Positioning
STRAT Strategic
Strategy
STRATA Sulfide Stress Cracking Test Assistant (Software)
STRATCOM Strategic Communications Command [US Army]
STRATSAT Strategic Satellite [US Air Force]
STRATWARM Stratospheric Warming
STRAW Simultaneous Tape Read and Write
STRC Science and Technology Research Center [of UNC, US]
Scientific and Technical Research Commission
Scientific and Technical Research Council
STRD Strand
STREP Standard Repair Price
STRESS Structural Engineering Systems Solver [Computer Language]
STRI Sequential Tracking, Registration and Information
Smithsonian Tropical Research Institute [Panama]
STRIPS Standard Research Institute Problem-Solver
STRIVE Standard Techniques for Reporting Information on Value Engineering
STRL Science and Technical Research Laboratory
Solar Thermal Research Laboratory [of University of Waterloo, Canada]
Structural
STRN Standard Technical Report Number
STRO Scandinavian Tire and Rim Organization [Sweden]
STROBES Shared Time Repair of Big Electronic Systems
STRR Statistical Treatment of Radar Returns
STRUC Structure
StructE Structural Engineer [also: StructEng]
STRUDL Structural Design Language [Computer Language]
STS Satellite Tracking Station
Scanning Tunnel Spectroscopy
Science and Technology Society [US]
Shared Tenant Services [of FCC, US]
Space Time Space
Space-Time Structure
Space Transportation System
Special Treatment Steel
Standard Technical Specification
Standard Test Signal
Static Test Stand
Status
Structural Transition Section
Subscriber Transferred Signal
S2S Surfaced on Two Sides [Lumber]
STScT Space Telescope Science Institute [US]
STSF Secured Tank Storage Facility
STS/FCC Shared Tenant Services/Federal Communication Commission [US]
S2S1E Surfaced on Two Sides and One Edge [Lumber]
STSS Strowger Telephone Switching System [US]
STT Seek Time per Track
Single Transmission Time
Standard Tube Test
Studies, Tests and Trials
STTC Secondary Teachers Technical Certificate
STTL Schottky Transistor-Transistor Logic
STU Secure Telephone Unit
Shanghai Technical University [PR China]
Signal Transmission Unit
Story Understander
St. Thomas University [Canada]
Submersible Test Unit
Subscribers Trunk Unit
Swiss Federation of Trade Unions
Systems Test Unit
System Timing Unit
System Transmission Unit
STUC Scottish Trade Union Congress
STUD Safety Training Update
STUF Sudan Trade Union Federation
STUFF Systems to Uncover Facts Fast
STUMP Story Understanding and Memory Program
STURP Shroud of Turin Research Project [US]
STUVA Studiengesellschaft fur Unterirdische Verkehrsanlagen [= Underground Transporation Research Association, Germany]
STV Sawtooth Voltage
Schweizerischer Technischer Verband [= Swiss Technical Association]
Separation Test Vehicle
Small Test Vessel
Subscription Television
Suction Throttling Valve
Surveillance Television
STV/BPO Suction Throttling Valve/Bypass Orifice
STVW Symmetrical Triangle Voltage Waveform
ST W Storm Water
STWP Society of Technical Writers and Publishers [now: STC, US]
ST WP Steam Working Pressure
STWY Stairway
STX Saxitoxin
Start-of-Text (Character)
STY Space Time Yield
Stationary
SU Saga University [Japan]
Saitama University [Japan]
Salford University [UK]
Seikei University [Japan]
Selectable Unit
Service Unit
Setsunan University [Japan]
Set-Up
Sheffield University [UK]
Shinshu University [Japan]
Shizuoka University [Japan]
Signaling Unit
Single User
Sonics and Ultrasonics [IEEE Group]
Stanford University [US]
Start Up
Station Unit
Storage Unit
Strontium Unit
Submittal

	Subunit
	Suppressor
	Synchronization Unit
	Synchronization Utility
	Syracuse University [US]
SUA	Silver Users Association [US]
	State Universities Association [US]
SUAS	System for Upper Atmospheric Sounding
SUB	Submarine
	Submerge
	Subroutine
	Subscriber
	Subscript
	Subscription
	Substitute (Character)
	Subtract
	Suburb
SUBA	Subbituminous A (Coal)
SUBB	Subbituminous B (Coal)
SUBC	Subbituminous C (Coal)
SUBCAL	Subcaliber
SUBCON	Subcontracting Industries Exhibition
SUBDIZ	Submarine Defense Identification Zone
SUBDOC	Subdocument
SUBIC	Submarine Integrated Control
SUBJ	Subject
SUBL	Sublime
	Sublimation
SUBMO	Submarine Motion
SUBN	Suburban
SUBORD	Subordinate
SUBPROGRAM	Subordinate Program [also: Subprogram or subprogram]
SUBROC	Submarine Rocket [US Navy]
SUBROUTINE	Subordinate Routine [also: Subroutine or subroutine]
SUBS	Subsidiary
SUBSCPT	Subscript
SUBSID	Subsidiary
SUBST	Substitute
SUBSTA	Subscriber's Station
	Substation
SUBSTR	Substructure
SUBSTRING	Subordinate String [also: Substring or substring]
SUBTASS	Submarine Towed Array Sonar System
SUBTIL	Synthesized User-Based Terminology Index Language
SUC	Saskatchewan Universities Commission [Canada]
	Secretary of the University Council
	Sucre [Currency of Ecuador]
SUCT	Suction
SUDAA	Stanford University, Division of Aeronautics and Astronautics [US]
SUDIC	Sulfur Development Institute of Canada
SUDOSAT	Sudan Domestic Satellite
SUDT	Silicon Unilateral Diffused Transistor
SUEDE	Surface Evaluation and Definition
SUFF	Sufficient
	Suffocating
Suff	Suffolk [UK]
SUFFER	Save Us From Formaldehyde Environmental Repercussions [now: CURE of FPA, US]
SUFOI	Scandinavian Unidentified Flying Object Information [Denmark]
SUG	Smartmac (Computer) User Group [US]
SUGI	SAS Inc. Users Group International
SUGG	Suggestion
SUHL	Sylvania Ultrahigh Level Logic
	Sylvania Universal High-Speed Logic
SUI	Standard Universal Identifier
	State University of Iowa [US]
SUIPR	Stanford University, Institute for Plasma Research [US]
SUL	Small University Library
	Sudanese Pound [Currency]
SULFAN	Sulfuric Anhydride
SULFD	Sulfide
SULIRS	Syracuse University Libraries Information Retrieval System [US]
SULIS	Sulzer Literaturverteilung und -sortierung [= Sulzer Literature Dissemination and Sorting System, Switzerland]
SUM	Set-Up Module
	Shallow Underwater Mobile
	Summary
	Surface-to-Underwater Missile
	System Utilization Monitor
SUMEX	Stanford University Medical Experiment [US]
SUMEXAIM	Stanford University Medical Experiment — Applications of Artificial Intelligence to Medical Research [US]
SUMM	Summary
SUMMIT	Sperry Univac Minicomputer Management of Interactive Terminals
	Supervisor of Multiprogramming, Multiprocessing, Interactive Time-Sharing
SUM PCH	Summary Punch
SUMS	Sheffield University Metallurgical Society [UK]
	Sperry Univac Material System
SUMT	Sequential Unconstrained Minimization Technique
SUN	Sunday
	Symbols, Units, Nomenclature Commission [of IUPAP, Sweden]
SUNFED	Special United Nations Fund for Economic Development
SUNI	Southern Universities Nuclear Institute [US]
SUNY	State University of New York [US]
SUP	Superior
	Supercript
	Support
	Supplement
	Supply
	Suppress
	Suppressor
sup	supremum [Mathematics]
SUPA	Society of University Patent Administrators [US]
SUPARCO	Space and Upper Atmosphere Research Committee [Pakistan]
SUPER	Submicron Positive Dry Etch Resist
	Superfine
	Superintendent

SUPERHET Superheterodyne (Receiver) [also: superhet]
SUPERSTR Superstructure
SUPHTR Superheater
SUPO Super Power Water Boiler [US]
SUPPL Supplement [also: SUPP]
SUPPS Supplementary Procedures
SUPR Supervisor
Supreme
SUPRA Submersible Underwater Pipeline Repair Apparatus
SUPROX Successive Approximation
SUPRSCPT Superscript
SUPRT Support
SUPT Superintendent
Support
SUPVR Supervisor [also: SUPV]
SUPY Supervisory
SUR Speech Understanding Research
State University of Rutgers [US]
Surface
Surcharge
Surplus
SURA Southeastern Universities Research Association [US]
SURCAL Surveillance Calibration
SURDD Southern Utilization Research and Development Division [of USDA]
SURE Symbolic Utilities Revenue Environment
Systems for Underwriting Risk Evaluation
SURF Support of User Records and Files
Surface
Sur f Suriname Florin [Currency]
SURGE Sorting, Updating, Report Generating Equipment
SURI Syracuse University Research Institute [US]
SURIC Surface Ship Integrated Control
SURMAC Surface Magnetic Confinement
SURRC Scottish Universities Research and Reactor Center
SURSAT Surveillance Satellite
SURSULF Surface Hardening Sulfur Catalyst
SURTEC (International Congress for) Surface Technology
SURV Survey(ing)
Surveyor
SUS Saybolt Universal Seconds
Silicon Unilateral Switch
Single Underwater Sound
Single-User System
Small Ultimate Size
Speech Understanding System
SUSIE Sequential Unmanned Scanning and Indicating Equipment
Stock Updating Sales Invoicing Electronically
SUSOPS Sustained Operations Programm [of DND, Canada]
SUSP Suspension
SUSY GUT Supersymmetric Grand Unified Theory
SUT Science University of Tokyo [Japan]
Society for Underwater Technology [UK]
SV Safety Valve
Sailing Vessel
Saturated Vapor
Saturation Voltage
Sawtooth Voltage
Self-Verification
Service Voltage
Signature Verification
Simulated Video
(Single) Silk Varnish
Single Value
Snake Venom
Solenoid Valve
Sound Velocity
Space Vehicle
Specific Volume
Status Valid
Stereoscopic Video
Stereo Vision
Stop Valve
Surface Volume
Surge Voltage
Surrender-Value
Synaptic Vesicle
Synchronous Vibrator
Synchronous Voltage
Sv Sievert
sv (sub verbo) — under the following word or heading
SVA Shared Virtual Area
Snake Venom Agglutinin
Styrene-Vinyl-Acrylonitrile (Thermoplastics)
SVAK Snake Venom Agglutinin from Naja Naja Kaouthia
SVAM Snake Venom Agglutinin from Naja Mocambique
SVC El Salvador Colon [Currency]
Secure Voice Communication
Service
Society of Vacuum Coaters [US]
Static Var Compensator
Supervisor Call (Instruction)
Supervisory Cell
Switched Virtual Call
Switched Virtual Circuit
SVCBL Serviceable
SVCC Schweizer Verein der Chemiker-Coloristen [= Swiss Association of Chemists/Colorists]
SVCE Service
SVCS Star Vector Calibration Sensor
SVCT Schweizer Vereinigung diplomierter Chemiker HTL [= Swiss Association of Certified Chemists of the Higher Technical Institution]
SVD Saturated Vapor Density
Schweizerische Vereinigung fur Documentation [= Swiss Documentation Association]
Simultaneous Voice and Data
Singular-Matrix Value Decomposition
System Verification Diagram
SVDF Segmented Virtual Display File
SVE Schweizerische Vereinigung fur Elektrotechniker [= Swiss Association of Electrical Engineers]
Society for Vector Ecology [US]
Society for Visual Education [US]
SVER Spatial Visual Evoked Response
SVES Satellite Video Exchange Society [Canada]
SVF Schweizer Vereinigung von Farbereifachleuten [= Swiss Association of Dyers]
SVFR Special Visual Flight Rules

SVH Solar Vacuum Head
SVI Service Interception
SVIA Specialty Vehicles Institute of America [US]
SVIC Shock and Vibration Information Center
Sociedad Venezolana de Ingenieros Consultores [= Venezuelan Society of Consulting Engineers]
SVIH Sociedad Venezolana de Ingenieros Hidraulica [= Venezuelan Society of Hydraulic Engineers]
SVLog Servicing Log [also: SVL]
SVM Silicon Video Memory
Slant Visibility Meter
SVMT Schweizerischer Verband fur Materialtechnik [= Swiss Association for Materials Technology]
SVP Saturated Vapor Pressure
Service Processor
Society of Vertebrate Paleontology [US]
Software Verification Plan
SVR Severe
Slant Visual Range
Software Verification Report
Super Video Recorder
Supply-Voltage Rejection
SVRL Several
SVRS Solochrome Violet RS [Biochemistry]
SVS Single Virtual Storage
Society of Visting Scientists [UK]
Space Vision System
SVT System Validation Testing
SVTL Services Valve Test Laboratory
SVTP Sound Velocity, Temperature and Pressure
SW Salt Water
Sandwich-Wound
Scattered Wave
Series-Wound
Shear Wave
Shock Wave
Shortwave
Shunt-Wound
Single Wheel
Single Weight
Sky Wave
Slow Wave
Software
Solid Waste
South Wales [Australia]
Southwest
Space Wave
Specific Weight
Spherical Washer
Spherical Wave
Spin Wave
Spot Welding
Standing Wave
Stationary Wave
Station Wagon
Statistical Weight
Storage Field
Stud-Arc Welding
Surface Wave
Surface Wind
S-Wave
Swede(n)
Swedish
Switch
Switchband-Wound
S/W Software
Surface Wind
SWA Scheduler Work Area
Single Wire Armor
Sound Wave Analyzer
Southern Wholesalers Association [US]
Southern Woodwork Association [US]
System Work Area
SWAC Standards Western Automatic Computer [of NIST, US]
SWAD Special Warfare Aviation Detachment [of US Army]
SWADE Surface Wave Dynamics Experiment
SWADS Scheduler Work Area Data Set
SWAFEC Southwest Atlantic Fisheries Advisory Commission
SWALCAP Southwest Academic Libraries Cooperative Automation Project [UK]
SWAM Standing Wave Area Monitor
SWAMI Software-Aided Multifont Input
SWAMWU South West Africa Mine Workers Union [Namibia]
SWAP Standard Wafer Array Programming
Stewart-Warner Array Program
Stress Wave Analyzing Program
SWAT Sidewinder IC Acquisition Track
Special Warfare Armored Transporter
Special Weapons and Tactics
Stress Wave Analysis Technique
SWATH Small-Waterplane-Area Twin-Hull Ship
SWB Single Weight Baryta
Summary of World Broadcasts (Database) [UK]
SWBD Switchboard
SW BHD Swash Bulkhead
SWBP Service Water Booster Pump
SWBS Ship Work Breakdown Structure
SWC Solid Waste Cask
Stepwise Cracking
Surge Withstand Capability
SWCH Switch
SWCL Sea Water Conversion Laboratory
SWCRD Soil and Water Conservation Research Division [US]
SWD Self-Wiring Data
Sliding Watertight Door
Smaller Word
Standing-Wave Detector
Surface Wave Device
SWDL Safe Winter Driving League [US]
SWDR Single Way Dynamic Range
SWE Society of Women Engineers [US]
Software Engineering
Spherical Wave Expansion
Status Word Enable
SWEAT Student Work Experience and Training
SWED Swede(n)
Swedish
SWEEP Soil and Water Environmental Enhancement Program [Canada]
SWEPP Stored Waste Examination Pilot Plant
SWET Simulated Water Entry Test

SWF Service Workers Federation [San Marino]
Shortwave Fade-Out
Statistical Weight Factor
Stress Wave Factor
SWFG Secondary Waveform Generator
SWF/ISS Stress Wave Factor/Interlaminar Shear Strength
SWFR Slow Write/Fast Read [also: SW/FR]
Sw Fr Swiss Franc [Currency of Switzerland and Liechtenstein]
SWG Society of Women Geographers [US]
Special Working Group [of NATO]
Standard Wire Gage
Steel Wire Gage
Stubs Wire Gage
SWGB South Western Gas Board [UK]
SWG BKT Swinging Bracket
SWGR Switchgear
SWI Sealant and Waterproofers Institute [US]
Software Interrupt
Special World Interval
Steel Window Institute [US]
SWIE South Wales Institute of Engineers [Australia]
SWIEEECO Southwestern IEEE Conference and Exhibition [US]
SWIFT Selected Words in Full Title
Signal Word Index of Field and Title
Significant Word in Tull Title
Society for Worldwide Interbank Financial Telecommunications
Software Implemented Friden Translator
Strength of Wings Including Flutter
SWIFTLASS Signal Word Index of Field and Title Literature Abstracts Specialized Search
SWIFTSIR Signal Word Index of Field and Title Science Information Retrieval
SWINPC Southwest Institute of Nuclear Physics and Chemistry [PR China]
SWIP Seal Worm Intervention Program [Canada]
Southwest Institute of Physics [PR China]
Standing Wave Impedance Probe
SWIR Short-Wave Infrared
SWIRA Swedish Industrial Robot Association
SWIRECO Southwestern Institute of Radio Engineers Conference and Electronics Show [US]
SWIRS Solid Waste Information Retrieval System [of Environmental Protection Agency, US]
SWIS Shrinkage/Warpage Interface-to-Stress Software
SWITT Surface Wave Independent Tap Transducer
SwKn Swedish Krona [Currency]
SWL Safe Working Load
Shortwave Listener
Specific Work Load
Sulfide Waste Liquor
Surface Wave Lines
Swaziland Lilangeni [Currency]
SWLA Southwestern Library Association [US]
SWM Standing-Wave Meter
SWMA Steel Wool Manufacturers Association [UK]
SWOF Switchover Operation Failure
SWOP Structural Weight Optimization Program
Switchable Input Operation
SWOPAMP Switchable Input Operational Amplifier [also: SWOP AMP]
SWOPS Single Well Offshore Production System
SWOV Stichting Wetenschappelijk Onderzoek Verkeersveilingheid [= Foundation for Scientific Road Safety Research, Netherlands]
SWP Safe Working Pressure
Saskatchewan Wheat Pool [Canada]
Service Water Pump
Society for Women in Plastics [US]
Sound Wave Photograph
Sowjet Warsaw Pact
SWPA Southwestern Power Administration
Steel Works Plant Association [US]
Submersible Wastewater Pump Association [US]
SWPP Southwest Power Pool [US]
SWR Service Water Reservoir
Sine Wave Response
Sodium Water Reactor
Spin Wave Resonance
Standing-Wave Ratio
Steel Wire Rope
SWRI Southwestern Research Institute [US]
SWS Safe Working Stress
Service Water System
Shift Word, Substituting
Slow-Wave Sleep
Software System
SWSA Southern Wood Seasoning Association [US]
SWSI Single Width, Single Inlet
SWST Service Water Storage Tank
Society of Wood Science and Technology [US]
SWT Supersonic Wind Tunnel
SWTL Surface-Wave Transmission Line
SWU Selective Work Unit
Separative Work Unit
SWUCNET Southwest Universities Computer Network [UK]
SWULSCP Southwest University Libraries Systems Cooperative Project [now: SWALCAP, UK]
SWW Severe Weather Warning
SWY Stopway
SX Simplex
Solvent Extraction
SXAPS Soft X-Ray Appearance Potential Spectroscopy
Soft X-Ray Appearance Potential Spectrum
SXB Spring Extension Bar
SX/EW Solvent Extraction/Electrowinning [also: SX-EW, or SXEW]
SXGM Spectrum X-Gamma Mission
SXN Section
SXPS Soft X-Ray Photoelectron Spectroscopy [also: SXPES]
SXRT Society of X-Ray Technology [UK]
SXS Step by Step
SXT Society of X-Ray Technology [UK]
Spacecraft Sextant
SY Stripping Yield
Symbol
Synchronize
System

SYCLOPS SYFA (= System for Access) Current Logic Operating System
SYCOM Synchronous Communications
SYD Sum-of-the-Years Digits
Yemeni Dinar [Currency of People's Republic of Yemen]
SYDAS System Data Acquisition System
SYEP Summer Youth Employment Program
SYFA System for Access
SYL Syllabus
SYLCU Synchronous Line Control Unit
SYLK Symbolic Link Format
SYM Symbol
Symmetry
SYMAN Symbol Manipulation
SYMBIOSIS System for Medical and Biological Information Searching [of NYSU, US]
SYMPAC Symbolic Program for Automatic Control
SYN Synchronization [also: SYNCH or synch]
Synchronous (Idle Character)
Synthesizer
Synthetic
Synonym
SYNC Synchronization [also: SYNC or synch]
Synchronous
SYNCOM Synchronous-Orbiting Communications Satellite
SYNCHROMESH Synchronous Mesh [also: synchromesh]
SYNDET Synthetic Detergent [also: syndet]
SYNFUEL Synthetic Fuel [also: synfuel]
SYNGAS Synthetic Gas [also: syngas]
SYNSCP Synchroscope
SYNSEM Syntax and Semantics
SYNT Synthetic [also: SYNTH]
SYNTAN Synthetic Tanning Material [also: syntan]
SYNTOL Syntagmatic Organization of Language
SYNTRAN Synchronous Transmission
SYP Syrian Pound [Currency]
SYROCO Symposium on Robotic Control [of IFAC, Austria]
SYS System
SYSADMIN System Administration
SYSCAP System of Circuit Analysis Programs
SYSCOM System Command
SYSCTLG System Catalogue
SYSDES System Design
SYSDOC System Documentation
SysE Systems Engineer [also: SysEng]
SYS ENG Systems Engineering
SYSEV System Evaluation [also: SYSEVA]
SYSGEN System Generation [also: sysgen]
SYSIN System Input
SYSINT System Integration
SYSLIB System Library
SYSLO System Loader
SYSLOG System Log
SYSOP System Operation
System Optimization
SYSOUT System Output
SYSPOP System Programmed Operator
SYSRC System Reference Count
SYSRES System Residence Volume
SYST System
Systematic(s)
SYSTRAN Systems Analysis Translator
SYSUP System Supervision
SYSVER System Verification
SYU Synchronization Signal Unit
Synchronization Signal Utility
SZ Seizure
Size
SZL Swaziland Lilangeni [Currency]
SZVR Silicon Zener Voltage Regulator

T

T	Technical
	Technician
	Technologist
	Technology
	Tee
	Temperature
	Tension
	Tera- [SI Prefix]
	Terminal
	Terrain
	Territory
	Tertiary [Geology]
	Tesla
	Test
	Thermodynamic(s)
	Thermoplastic
	Threonine
	Threonyl
	Threshold
	Thrust
	Thunderstorm
	Thymine
	Thymidine
	Time(r)
	Tip
	Titrimetry
	Toll
	Tooth
	Torque
	Town
	Township
	Trace
	Track
	Trans (Configuration) [Chemistry]
	Transaction
	Transformer
	Transition
	Translation
	Transmission
	Transmittance
	Transmitter
	Transparency
	Trillion
	Triode
	Triplet
	Tritium
	Truss
	Tube
	Tubular
	Tuesday
t	technical
	temperature
	temporary
	tension
	territorial
	tertiary
	thickness
	time
	times
	(metric) ton
	transitional
	transmit
	troy
	truth
ΔT	True (Degrees)
2,4,5-T	2,4,5-Trichlorophenoxyacetic Acid
2,4,6-T	2,4,6-Trichlorophenol
TA	Tailhook Association [US]
	Tannic Acid
	Tape Address
	Tape Armored
	Target
	Teaching Assistant
	Technical Advisor
	Technical Agriculture
	Technonet Asia [Singapore]
	Telecom Australia
	Telegraphic Address
	Tensor Analysis
	Terephthalic Acid
	Terminal Adapter
	Terminal Address
	Territorial Army
	Test Access
	Thermal Analysis
	Thermionic Arc
	Time Analyzer
	Timing Analysis
	Titrimetric Analysis
	Transition Altitude
	Transmission Authenticator
	Transportation Alternatives [US]
	Transverse Acoustical
	Triacetic
	Triplex Annealing
	Turbo-Alternator
	Turn Altitude
	Turbocharged and Aftercooled
	Turbulence Amplifier
T&A	Time and Attendance
Ta	Tantalum
TAA	Technical Assistance Administration [of UN]
	Titanium Diisopropoxide Bis(2,4-Pentanedionate)
	Transportation Association of America [US]
	Turbine-Alternator Assembly
TAAC	Technical Assessment Advisory Council
TAALS	Tactical Army Aircraft Landing System
TAAM	Tomahawk Airfield Attack Missile
TAAR	Target Area Analysis Radar
TAAS	Three-Axis Attitude Sensor
TAB	Table
	Tabular Language
	Tabulation Character
	Tape-Automated Bonding

	Technical Abstract Bulletin
	Technical Activities Board [of IEEE, US]
	Technical Advisory Bureau
	Technical Analysis Branch
	Technical Assistance Board [of UN]
	Technical Assistance Bureau [of ICAO]
	Technical Assessment Board
TA&B	Testing, Adjusting and Balancing [also: TAB]
TABESIM	Tabulating Equipment Simulator
TABL	Tablet
TABR	Tabulator
TABS	Tailored Abstracts [of INSPEC, UK]
	Telephone Automated Briefing Service
	Terminal Access to Batch Service
	Total Automated Broker System
	Total Aviation Briefing Service
TABSAC	Targets and Backgrounds Signature Analysis Center [US]
TABSIM	Tabulating Simulator
TABSOL	Tabular Systems-Oriented Language
TABSTONE	Target and Background Signal-to-Noise Evaluation
TAC	Tactical Air Command [US]
	Tanners Association of Canada
	Technical Advisory Committee
	Technical Assistance Center
	Technical Assistance Committee [of UN]
	Technical Assistance Contract
	Telemetry and Command
	TELENET (= Telecommunication Network) Access Controller
	Telex Data Acquisition and Control
	Terminal Access Controller
	Terminal Area Charts
	Test Access Control
	Tested Additive Chemicals
	Thermostatically-Controlled Air Cleaner
	Timber Association of California [US]
	Time Amplitude Converter
	Time-to-Amplitude Converter
	Total Allowable Catch
	Total Available Carbohydrates
	Trade Advisory Council [UK]
	TRANSAC (= Transistorized Automatic Computer) Assembler Compiler
	Trans-Anti-Cis (Configuration) [Chemistry]
	Transformer Analog Computer
	Transistorized Automatic Control
	Translator — Assembler — Compiler
	Transport Advisory Council
	Trapped Air Cushion
	Triallyl Cyanurate
	Triallyl-1,3,5-Triazine-2,4,6-Trione
	Tribunal Anti-Dumping Canada
	Tunneling Association of Canada
TACAMO	Take Charge and Move Out
TACAN	Tactical Air Navigation [also: Tacan or tacan]
TACAN-DME	TACAN Distance Measuring Equipment
TACCAR	Time-Averaged Clutter-Coherent Airborne Radar
TACCO	Tactical Coordinator
TACDA	The American Civil Defense Association [US]
TACDACS	Target Acquisition and Data Collection System
TACDEN	Tactical Data Entry Device [US Army]
TACE	Turbine Automatic Control Equipment
TACEX	Tactical Excercise
TACF	Temporary Alteration Control Form
TACH	Tachometer
TACI	Test Access Control Interface
TACL	Telecommunications Analysis Center Library
	Time and Cycle Log
TACMAR	Tactical Multifunction Array Radar
TACNAV	Tactical Navigation
TACODA	Target Coordinate Data
TACOL	Thinned Aperture Computed Lens
TACOM	Tank-Automotive Command [US Army]
TACOS	Tool for Automatic Conversion of Operational Software
TACPOL	Tactical Procedure-Oriented Language
TACR	Time and Cycle Record
TACRAHD	Tactical Routing-Indicator-Lookup and Header-Preparation Device
TACRV	Tracked Air Cushion Research Vehicle
TACS	Tactical Air Control System
	Technical Assignment Control System
	Test Assembly Conditioning Station
	Total Access Communications System
TACSAT	Tactical Communications Satellite [US]
TACSATCOM	Tactical Satellite Communications
TACT	Tactical
	Technological Aids to Creative Thought
	Terminal Activated Channel Test
	Transistor and Component Tester
	Transonic Aircraft Technology
TACTRI	Taiwan Agricultural Chemicals and Toxic Substances Research Institute
TACV	Tracked Air Cushion Vehicle
TAD	Target Acquisition Data
	Technical Approach Document
	Telephone Answering Device
	Tensile Axis Direction
	Terminal Address Designator
	Thrust-Augmented Delta [of NASA]
	Top Assembly Drawing
	Transaction Application Drive
tAD	(Metric) Tons, Air Dried
TADIL	Tactical Digital Information Link
TADOG	Towed Array Deep Operating Gear
TADP	Terminal Area Distribution Processing [also: TAD/P]
TADS	Tactical Automatic Digital Switching
	Teletypewriter Automatic Dispatch System
TADSS	Tactical Automatic Digital Switching System
TAE	Technical Aeronautical Engineering [Israel]
	Technische Akademie Esslingen [= Technical Academy at Esslingen, Germany]
	Tris-Acetate EDTA (= Ethylenediaminetetraacetic Acid)
TAEC	Thailand Atomic Energy Commission
	Thiolated Aminoethylcellulose
TAEM	Terminal Area Energy Management
TAERS	Transportation Army Equipment Record System
TAF	Terminal Aerodrome Forecast
	Top of Active Fuel

Torque Amplification Factor
Transaction Facility
TAG Tactical Air Group
Technical Advisory Group
Technical Assistance Group
The Acrylonitrile Group [US]
Time Automated Grid
Transient Analysis Generator
Tag Tagliabue (Tester) [Chemistry]
TAGA Technical Association of the Graphic Arts [US]
TA/H Turn Altitude Height
TAI Temps Atomique International [= International Atomic Time]
Time to Auto-Ignition
TAIC Tokyo Atomic Industrial Consortium [Japan]
Triallyl Isocyanurate
TAID Thrust Augmented Improved Delta [of NASA]
TAIDC Taiwan Agricultural and Industrial Development Corporation [Taiwan]
TAIO Technology Applications Information Organization [of SDIO, US]
TAKIS Tutmonda Asocio pri Kibernetiko, Informatiko, kaj Sistemiko [= World Association of Cybernetics, Computer Science and System Theory, Germany]
TAL Tandem Accelerating Laboratories [of McMaster University, Canada]
Terminal Application Language
Transaction Application Language
Trans-Alpine
Tungsten-Arc Lamp
TALA Tactical Landing Approach
TALAR Tactical Landing Approach Radar
TALON Texas, Arkansas, Louisiana, Oklahoma and New Mexico [US]
TALS Transfer Airlock Section
TALTC Test Access Line Termination Circuit
TAM Technical Association of Malaysia
Telecommunications Access Method
Telephone Answering Machine
Terminal Access Method
Test Access Multiplexer
Theoretical and Applied Mechanics
Total Available Market
Triallyl Trimellithate
TAMA N-Methylanilinium Trifluoroacetate
TAMCO Training Aid for MOBIDIC (= Mobile Digital Computer) Console Operations
TAMDA Timber and Allied Materials Development Association [South Africa]
TAME Tosyl Arginine Methyl Ester
TAMIS Telemetric Automated Microbial Identification System
TAMOS Terminal Automatic Monitoring System
Terminal Auto-Operator and Monitor System
TAMS Tunnel Air Monitoring System
TAMU Texas A&M University [US]
TAMVEC Texas A&M Variable Energy Cyclotron [US]
TAN Tandem
Textile Association of the Netherlands
tan tangent
tan^{-1} inverse tangent
TANCA Technical Assistance to Non-Commonwealth Countries [UK]
TANE Telephone Association of New England [US]
tanh hyperbolic tangent
$tanh^{-1}$ inverse hyperbolic tangent
TANS Tactical Air Navigation System
TAO Technical Assistance Operation [of UN]
Technical Assistance Order
4-(2-Thiazolylazo)orcinol
Tokyo Astronomical Observatory [Japan]
TAP Tactical Area Positioning
T-Angle Plate
Tape Automatic Positioning
Tape Automatic Preparation
Tapping
Technological Assessment and Planning [Canada]
Technological Awareness Program
Technical Advisory Panel
Technology Assessment Program [Canada]
Terminal Access Processor
Terminal Applications Package
Test Access Path
Test Analysis Package
Test Article Protector
Test Assistance Program
Tetrainosylpyrrole
Thermal Analyzer Program
Thermodynamics and Physical Properties Package [of University of Houston, US]
Time-Sharing Assembly Program
Tresun Address Processing
Triallyl Phosphate
Triisoamyl Phosphate
TAPA 2-(2,4,5,7-Tetranitro-9-Fluorenylideneaminooxy)propionic Acid
TAPAC Tape Automatic Positioning and Control
TAPE Tape Automatic Preparation Equipment
Technical Advisory Panel for Electronics [US Air Force]
TAPP Tarapur Atomic Power Project [India]
Two-Axis Pneumatic Pickup
TAPPI Technical Association of the Pulp and Paper Industry [US and Japan]
TAPPIK Technical Association of the Pulp and Paper Industry of Korea [South Korea]
TAPRE Tracking in an Active and Passive Radar Environment
TAPS Tactical Area Positioning System
Terminal Area Positive Separation [of FAA, US]
3-{[Tris(hydroxymethyl)methyl]amino}-1-Propanesulfonic Acid
Turboalternator Power System
TAPSO 3-[N-Tris(hydroxymethyl)methylamino]-2-hydroxypropanesulfonic Acid
Taq Thermus Aquaticus [Molecular Biology]
TAR Tactical Air Reconnaissance
Tactical Attack Radar
Technical Action Request
Technical Assistance Request
Technical Association of Refractories [Japan]
Terminal Address Register

	Terminal Area (Sureveillance) Radar
	Terrain-Avoidance Radar
	Thiozolyazoresorcinol
	Track Address Register
	Trajectory Analysis Room
TARA	Truck-Frame and Axle Repair Association [US]
TARABS	Tactical Air Reconnaissance and Aerial Battlefield Surveillance System
TARAD	Tracking Asynchronous Radar Data
TARAN	Tactical Attack Radar and Navigation
TARC	The Army Research Council [US]
	Trace Analysis Research Center
TARDIS	Time and Attendance Recording Analysis
TARE	Telegraph Automatic Routing Equipment
	Telemetry Automatic Reduction Equipment
	Transistor Analysis Recording Equipment
TARGA	TrueVision Advanced Raster Graphics Adapter
TARGET	Team to Advance Research for Gas Energy Transformation [US]
	Thermal Advanced Reactor Gas-Cooled Exploiting Thorium
TARI	Taiwan Agricultural Research Institute
TARIF	Technical Apparatus for Rectification of Indifferent Films
TARMAC	Terminal Area Radar Moving Aircraft
TARN	Tarnish
TARO	Tanzania Agricultural Research Institute
TARP	Tarpaulin
	Technical Assistance Research Program
	Test and Repair Processor
	Texas Advanced Research Program [US]
TARPOL	Tariff Policy [of ECAC, France]
TARPS	Tactical Aerial Reconnaissance Pod System
TARS	Technical Assistance Recruitment Service [of UN]
	Terrain Analog Radar Simulator
	Three-Axes Reference System
	Turn-Around Ranging Station
TART	Twin Accelerator Ring Transfer
TART A	Tartaric Acid
TAS	Tactical Analysis System
	Tanzanian Shilling [Currency]
	Target Acquisition System
	Tasmania
	Telecommunications Authority Singapore
	Telephone Answering Service
	Telephone Answering System
	Temperature Actuated Switch
	Tennessee Academy of Science [US]
	Terminal Address Selector
	Test Access Selector
	Test and Set
	Tris(dimethylamino)sulfur
	True Air Speed
TASC	Tabular Sequence Control
	Tactical Articulated Swimmable Carrier
	Telecommunications Alarm Surveillance and Control
	Terminal Area Sequence and Control
TASCC	Tandem Accelerator Superconducting Cyclotron
	Test Access Signaling Conversion Circuit
TASCON	Television Automatic Sequence Control
TASE	Tenacity-as-Specified Elongation
TASES	Tactical Airborne Signal Exploitation System
TAS-F	Tris(dimethylamino)sulfur Difluoride
TASI	Time-Assignment Speech Interpolation
TASM	Tasmania(n)
TASMAN	Tasmania-Australia Submarine Cable
TASO	Television Allocation Study Organization [US]
TASP	Teamed-Architectural Signal Processor
	Toll Alternatives Studies Program
TAS-PAC	Tactical Analysis System for Production, Accounting and Control
TASR	Terminal Area Surveillance Radar
	Thermal Activation Strain Rate
TASRA	Thermal Activation Strain Rate Analysis
TASS	Tactical Avionics System Simulator
	Tactical Signal Simulator
	Technical Assembly
	Teleprinter Automatic Switching System
	Trouble Analysis System
3-A SSC	3-A Sanitary Standards Committee [US]
TAST	Thermoacoustic Sensing Technique
TAT	Target Aircraft Transmitter
	Thrust-Augmented Thor
	Trans-Anti-Trans (Configuration) [Chemistry]
	Transatlantic Telephone
TATC	Terminal Air Traffic Control
	Transatlantic Telephone Cable
TATCS	Terminal Air Traffic Control System
TATHS	Tool and Trades History Society [UK]
TATOA	Toronto Area Transit Operating Authority [Canada]
TATr	Tyrosine Aminotransferase Regulator
TAU	Tel Aviv University [Israel]
	Test Access Unit
	Thesaurus Alphabetical Up-to-Date
	Trunk Access Unit
TAUN	Technical Assistance of the United Nations
TAV	Transatmospheric Vehicle
	Transverse Acoustoelectronic Voltage
TAVE	Thor-Agena Vibration Experiment
TAVG	Temperature Average
TAWAR	Tactical All-Weather Attack Requirements
TAWCS	Tactical Air Weapons Control System
TAWDS	Target Acquisition Weapon Delivery System
TAX	Taxy(ing)
TAXIR	Taxonomic Information Retrieval
TB	Tariff Bureau
	T-Bolt
	Technical Bulletin
	Temper Brittleness
	Terminal Board
	Terminal Box
	Theoretical Biology
	Thermal Barrier
	Thermobalance
	Tight-Binding (Approximation)
	Tilt Boundary
	Time Base
	Title Block
	Tone Burst
	Top Boundary
	Torch Brazing
	Torpedo Boat

	Transborder
	Transmitter Blocker
	Trial Balance
	Twin Boundary
	Twist Boundary
T/B	Title Block
Tb	Terbium
TBA	Testbed Aircraft
	Test Boring Association [US]
	Tetrabutylammonium
	Tight-Binding Approximation
	Tires, Batteries and Accessories
	To Be Announced
	Torsional Braid Analysis
	Tributylamine
TBAB	Tetrabutylammonium Bromide
TBAC	Treasury Board Advisory Committee [US]
TBAC/FLM	Treasury Board Advisory Committee on Federal Land Management [US]
TBAF	Tetrabutylammonium Fluoride
TBAH	Tetrabutylammonium Hydroxide
TBAHS	Tetrabutylammonium Hydrogen Sulfate
TBA-OH	Tetrabutylammonium Hydroxide
TBAX	Tube Axial
TBBA	Terephthalbutylaniline
	Terephthalylidenebisbutylaniline
TBBF	Top Baseband Frequency
TBBS	N-tert-Butyl-2-Benzothiazole Sulfenamide
TBC	Tensile Bolting Cloth
	4-tert-Butylcatechol
	Thermal Barrier Coating
	Time-Base Corrector
	Token-Bus Controller
	Toss Bomb Computer
	Treasury Board of Canada
TBCCO	Tantalum-Barium-Calcium-Copper-Oxygen [also: $Tl_2Ba_2CaCu_2O_x$]
TB-CPA-GPM	Tight-Binding — Coherent Potential Approximation — Generalized Perturbation Method
TBD	Target-Bearing Designator
TBDF	Transborder Data Flow
TBDMS	Tert-Butyldimethylsilyl
TBDMSCl	Tert-Butyldimethylsilyl Chloride
TBDMSI	1-tert-Butyldimethylsilyl)imidazole
TBDP	Transborder Data Processing
TBE	Tile and Brick Manufacturers (Association) of Europe [Switzerland]
	1,1,2,2-Tetrabromoethane
	Total Binding Energy
	Tris-Borate EDTA (= Ethylenediaminetetraacetic Acid)
TBEM	Terminal-Based Electronic Mail
TBF	Butylformamidine
	Tail Bomb Fuse
	Transmitted Bright-Field
TBG	Time-Base Generator
TBH	Technical Benzene Hexachloride
TBHP	Tert-Butyl Hydroperoxide
TBHQ	Tertiary Butylhydroquinone
TBI	Tape and Buffer Index
	Target-Bearing Indicator
	Throttle Body Injection
	Toronto Biotechnology Initiative [Canada]
TbIG	Terbium Iron Garnet
TBL	Table
	Terminal Ballistics Laboratory [US Army]
	Thermal Boundary Layer
	Through Bill of Lading
	Trouble
TBM	Temporary Bench Mark
	Terabit Memory (System)
	Thyssen Basic-Oxygen Metallurgy
	Tone Burst Modulation
	Tunnel Boring Machine
TBMA	Textile Bag Manufacturers Association [US]
TBME	Tactical Ballistic Missile Experiment
TBMT	Transmitter Buffer Empty
TBMX	Tactical Ballistic Missile Experiment
TBN	Total Base Number
TBO	Time before Overhaul
	Time between Overhauls
TBP	Tribromophenol
	Tributyl Phosphate
	Tributylphosphoric Acid
	True Boiling Point
TBPA	Textile Bag and Packaging Association [US]
TBPEHS	Tan Beta Pi Engineering Honour Society [US]
TBPO	Tertiary Butylperoctoate
Tbps	Terabits per second
TBR	Table Base Register
	Timber
TBRC	Top-Blown Rotary Converter
TBRI	Technical Book Review Index
TBS	Tert-Butyldimethylsilyl
	The Biodeterioration Society [US]
	Tight Building Syndrome
	To Be Specified
	Tris Buffered Saline
TBSCl	Tert-Butyldimethylsilyl Chloride
tbsp	tablespoonful
TBT	Tetrabutyl Titanate
	Tight-Binding Theory
TBTD	Tetrabutyl Thiuram Disulfide
TBTO	Bis(tri-n-butyltin) Oxide
TBTU	Tributyl Thiourea
TBTUP	Tight-Binding Theory with Universal Parameters
TBU	Terminal Buffer Unit
	Transmit Baseband Unit
TBUP	Tributylphosphine
TBX	Tactical Ballistic Missile Experiment
TC	Tab Card
	Tactical Computer
	Tag Code
	Tantalum Capacitor
	Tape Command
	Tariff Commission
	Task Control
	Taxiway Centerline
	T-Carrier
	Technical Classification
	Technical College

Technical Commission
Technical Committee
Technical Communication
Technical Consultant
Technical Control
Technical Cooperation
Telecommunications
Telemetry and Command
Telephone Channel
Telephone Company
Telescoping Collar
Television Control
Temperature Coefficient
Temperature Compensation
Temperature Conductivity
Temperature Control
Temper Color
Temporary Council
Term Coordination
Terminal Computer
Terminal Concentrator
Terminal Congestion
Terminal Controller
Terra Cotta
Tertiary Compound
Tesla Coil
Test Case
Test Chart
Test Code
Test Conductor
Test Console
Test Control
Tetracycline
Tetrahedral Cubic
Texture Coefficient
Theoretical Chemistry
Theoretical Climatology
Thermal Conditioning
Thermal Conductivity
Thermal Continuum
Thermal Control
Thermal Cracking
Thermal Creep
Thermal Cutting
Thermal Cycling
Thermionic Cathode
Thermochemistry
Thermocompression
Thermocouple
Thermocurrent
Thermoplastic Composite
Thiocarbamyl
Threshold Circuit
Thrust Chamber
Thrust Control
Time Clock
Time Code
Time Constant
Timed Closing
Time to Computation
Timing Channel
Tissue Culture
Toggle Clamp
To Contain
Toll Center
Toll Completing
Torque Converter
Total Carbon
Total Cholesterol
Total Coils
Total Count
Town Council
Toxic Concentration
Tracking Camera
Traffic Control
Training Course
Trans-Cis (Configuration) [Chemistry]
Transconductance
Transfer Channel
Transistorized Carrier
Transmission Control (Character)
Transmission Controller
Transmitting Circuit
Transportation Commodity
Transport Canada
Tray Cable
Treatment Code
Trichloroacetate
Trichloroacetic Acid
Trip Coil
Tropical Cylone
True Complement
Trunk Control
Trusteeship Council [of UN]
Tuned Circuit
Tungsten Cathode

T&C Time and Charges

Tc Technetium
Teracycle

TCA Tactical Combat Aircraft
Tanners Council of America [US]
Teach Cable Assembly
Technical Cooperation Administration [US]
Telecommunications Association [US]
Terminal Communications Adapter
Terminal Control Area
Thermal Critical Assembly [US]
Thermostatically Controlled Air Cleaner
Thrust Chamber Assembly
Tile Council of America [US]
Tilt-Up Concrete Association [US]
Time of Closest Approach
Tissue Culture Association [US]
Trans-Canada Airlines
Transcontinental Control Area
Transmission Control Area
Tricarboxylic Acid
Trichloroacetate
Trichloroacetic Acid
Trichloroaniline
Turbulent Contact Absorber

TCAA Tile Contractors Association of America [US]

	Trichloroacetic Aldehyde
TCAB	Tetrachloroazobenzene
	Twin-Carbon Arc Brazing
TCAD	Traffic Alert and Collision Avoidance Device
TCAE	Thermal Coefficient of Area Expansion
TCAI	Tutorial Computer-Assisted Instruction
TCAM	Telecommunications Access Method
TCAM-IMS/VS	Telecommunications Access Method — Information Management System/Virtual Storage
TCAS	T-Carrier Administration System
	Traffic Alert and Collision Avoidance System
TCAT	Test Coverage Analysis Tool
TCAW	Twin-Carbon Arc Welding
TCB	Task Control Block
	Technical Coordinator Bulletin
	Terminal Control Block
	Tetracarboxybutane
	Textile and Clothing Board
	Transfer Control Block
	Trichlorobenzene
	Trusted Computing Base
TCBA	Tesla Coil Builders Association [US]
	Trichlorobutyl Alcohol
TCBC	Trichlorobenzylchloride
TCBH	Time-Consistent Busy Hour
TCBM	Transcontinental Ballistic Missile
TCBO	Trichlorobutylene Oxide
TCBOC	Trichlorobutoxycarbonyl
	2,2,2-Trichloro-1,1-Dimethylethyl Chloroformate
TCBS	Tesla Coil Builders Society
TCBV	Temperature Coefficient of Breakdown Voltage
TCC	Tag(liabue) Closed Cup
	Technical Change Center [UK]
	Technical Computing Center
	Technical Coordination Committee
	Telecommunications Coordinating Committee
	Television Control Center
	Temperature Coefficient of Capacitance
	Temporary Council Committee [of NATO]
	Test Control Center
	Test Controller Console
	Thermal Conductivity Cell
	Thermal Controlled Coating
	Thermofor Catalytic Cracking
	Through-Connected Circuit
	Toll Center Code
	Tons of Clean Coal [also: tcc]
	Torque Converter Clutch (Solenoid)
	Traffic Control Center
	Trans-Cis-Cis (Geometry) [Chemistry]
	Transfer Channel Control
	Transmission Control Character
	Transport and Communications Commission [of UN]
	Transportation Commodity Classification
	3,4,4'-Trichlorocarbanilide
TCCA	Textile Color Card Association [US]
TCCAUS	Textile Color Card Association of the United States
TCCB	Technical College of Cape Breton [Canada]
TCCSR	Telephone Channel Combination and Separation Racks
TCCTB	Technical Center for Clay, Tiles and Bricks [France]
TCD	Telemetry and Command Data
	Temperature Coefficient of Decay
	The College Board [US]
	Thermal Conductivity Detector
	Thyratron Core Driver
	Time Code Division
	Transistor-Controlled Delay
TCDD	2,3,7,8-Tetrachlorodibenzo-P-Dioxin
TCDMS	Telecommunications/Data Management System
TCDS	Thyristor Controlled Drive System
TCE	Telecommunication Engineering
	Telemetry and Command Equipment
	Telemetry Checkout Equipment
	Telephone Company Engineered
	Temperature Coefficient of Expansion
	Terrace
	Tetrachloroethane
	Thermal Coefficient of Expansion
	Tons of Coal Equivalent
	Total Composite Error
	Transmission Control Element
	Trichloroethane
	Trichloroethylene
TCEA	Training Center for Experimental Aerodynamics [of NATO]
TCED	Thrust Control Exploratory Development
TCENM	Triscyanoethylnitromethane
TCEP	Triscyanoethoxypropane
TCF	Technical Control Facility
	Terminal Configuration Facility
	Transparent Conducting Film
	Trillion Cubic Feet [also: tcf]
TCFTD	Triple Constant Fraction Time Discriminator
TCG	Test Call Generator
	Transponder Control Group
	Tune-Controlled Gain
TCH	Thiocarbohydrazide
	Threshold Crossing Height
TCI	Telemetry Components Information
	Terrain Clearance Indicator
	Theoretical Chemistry Institute
	True/Complement I
TCID	Tissue Culture Infectious Dose
TCIS	Telex Computer Inquiry Service
TCJ	Tanners Council of Japan
TCL	Terminal Command Language
	Terminal Control Language
	Time and Cycle Log
	Toll Circuit Layout
	Transfer Chemical Laser
	Transistor-Coupled Logic
TC/LD	Thermocouple/Lead Detector
TCLE	Thermal Coefficient of Linear Expansion
TCLP	Toxicity Characteristic Leaching Procedure
TCLR	Toll Circuit Layout Record
TCM	Telecommunications Monitor
	Telemetry Code Modulation
	Temperature Control Model
	Terminal Capacity Matrix

	Terminal Charge Management
	Terminal-to-Computer Multiplexer
	Test Call Module
	Tetrachloromercurate
	Thermal Conduction Module
	Thermochemical Machining
	Thermoplastic Cellular Molding
	Time Compression Multiplex
	Tool Condition Monitoring
	Trellis Code Modulation
	Trichloromethane
TCMA	Tooling Component Manufacturing Association [US]
TCMF	Touch-Calling Multifrequency
TCMP	Toxic Chemicals Management Program
TCMS	Telephone Cost Management System
	Toll Centering and Metropolitan Sectoring
TCN	Telecommunications Cooperative Network
	Telecommunications Network
	Transport Canada (Data Processing) Network
TCNA	Tube Council of North America [US]
TC-NBT	Thiocarbamyl Nitro Blue Tetrazolium
TCNE	Tetracyanoethylene
TCNQ	7,7,8,8-Tetracyanoquinodimethane
TC-NR	Technical Classified Natural Rubber
TCO	The Carnegie Observatories [US]
	Thermal Cutoff
	Trunk Cutoff
TCOE	Temperature Coefficient of Expansion
TCOS	Trunk Class of Service
TCP	Tape Conversion Program
	Task Control Control Program
	Technical Cooperation Program [of Australia, Canada, UK and US]
	Terminal Control Program
	Test Checkout Procedure
	Thermochemical Process(ing)
	Thrust Chamber Pressure
	Tool Center Point
	Topologically Close-Packed [also: tcp]
	Traffic Control Post
	Transmission Control Program
	Transmission Control Protocol
	Transmitter Control Pulse
	Transport Control Protocol
	Trichlorophenoxyacetic Acid
	Tricresyl Phosphate
TCPA	Town and Country Planning Association [UK]
TCPC	Tab Card Punch Control
TCP/IP	Transmission Control Protocol/Internet Protocol
TCPL	Trans-Canada Pipeline
TCPO	Trichlorophenyloxalate
	1,1,1-Trichloropropene 2,3-Oxide
TCPP	Tetra-P-Chlorophenylpyrrole
TCPSA	Thermally Coupled Pressure Swing Absorption
TCR	Tape Cassette Recorder
	Telemetry Compression Routine
	Temperature Coefficient of Resistivity
	Tracer
	Transfer Control Register
	Transient Call Register
TC&R	Telemetry Command and Ranging
TCRA	Telegraphy Channel Reliability Analyzer
TCRC	Treatment Charges and Refining Charges
	Twin Cities Research Center
TCRDL	Toyota Central Research and Development Laboratories [Japan]
TCRI	Toyota Central Research Institute [Japan]
TCS	Teaching Company Scheme [UK]
	Technology Club of Syracuse [US]
	Telecentric System
	Telecommunications Control System
	Telecommunications System
	Telephone Conference Summary
	Telex Communications Service [US]
	Terminal Communications Subsystem
	Terminal Communications System
	Terminal Control System
	Terminal Countdown Sequencer
	Terminal Count Sequence
	Terne-Coated Steel
	Texas Center for Superconductivity [of University of Houston, US]
	The Classification Society [now: CSNA, US]
	The Coastal Society [US]
	The Constant Society [US]
	The Cousteau Society [US]
	The Crustacean Society [US]
	Theoretical Computer Science
	Thermal Conditioning Service
	Thermal Control Subsystem
	Tool Coordinate System
	Total Communications System
	Total Control System
	Traffic Control Station
	Transaction Control System
	Transmission-Controlled Spark
	Transportable Communication System
	Transportation and Communication Service
	True Chemical Shift
TCSC	Trainer Control and Simulation Computer
TCSES	Trusted Computer System Evaluation System
TCSP	Tandem Cross Section Program
TCST	Trichlorosilanated Tallow
TCSUH	Texas Center for Superconductivity, University of Houston [US]
TCT	Telecommunication Technology
	Terminal Control Table
	Thrombin Clotting Time
	Toll Connecting Trunks
	Trans-Cis-Trans (Geometry) [Chemistry]
	Translator and Code Treatment Frame
	Tricrotonylidene Tetramine
TCTFE	Trichlorotrifluoroethane
TCTNB	Trichlorotrinitrobenzene
TCTP	Tetrachlorothiophene
TCTS	Trans-Canada Telephone System
TCTSB	Transport Canada Technical Services Branch [of ICAO, Canada]
TCTSC	Transport Canada Technical Systems Center [of ICAO, Canada]
TCTTE	Terminal Control Table Terminal Entry

TCTTMA Tank Conference of the Truck Trailer Manufacturers Association [US]
TCU Tape Control Unit
Teletypewriter Control Unit
Terminal Control Unit
Timing Control Unit
Towering Cumulus
Transmission Control Unit
TCV Temperature Control Valve
Terminal Configured Vehicle
TCVC Tape Control Via Console
TCVE Thermal Coefficient of Volume Expansion
TCW Time Code Word
TCWB Tinned Copper Wire Braid
TCWG Telecommunications Working Group
TCXO Temperature-Compensated Crystal Oscillator
TCZD Temperature-Compensated Zener Diode
TD Tabular Data
Tank Destroyer
Tape Drive
Technical Data
Technical Division
Technical Document
Technology Development
Telemetry Data
Telephone Directory
Temperature Differential
Temporarily Disconnected
Term Deposit
Term Diagram
Terminal Digit
Terminal Distributor
Termination Date
Territorial Decoration
Test Design (Specification)
Test Distributor
Testing Device
Text and Data (Exchange)
Theoretical Density
Thermal Diffusion
Thermodynamics
Thor-Delta Satellite
Thoria-Dispersed
Thorium Dioxide
Time Delay
Time Difference
Time Domain
Timing Device
Tissue Dose
To Deliver
Tolerance Dose
Toluene Diamine
Tolylenediamine
Top Down
Total Depth
Total Dust Sampler
Touchdown
Toxic Dose
Toyota Diffusion (Process)
Trace Direction
Track Data
Track Display
Transfer Dolly
Transmission Detector
Transmission and Distribution
Transmission Distributor
Transmit Data
Transmitter-Detector
Transmitter-Distributor
Transverse Direction
Turbulent Diffusion
Tunesian Dinar [Currency]
Tunnel Diode
Turbine Drive
T&D Transmission and Distribution (Committee) [IEEE PES, US]
t/d tons per day
3D Three Dimensional [also: 3-D]
2D Two Dimensional [also: 2-D]
TDA Target Docking Adapter
Technology Development Corporation
Telecommunications Dealers Association [UK]
Temporary Danger Area
Terminal Diode Amplifier
Tetracarboxylic Dianhydride
Tetradecen-1-yl Acetate
Textile Distributors Association [US]
Timber Development Association [US]
Time-Deposit Accounting
Time-Domain Analysis
Titanium Development Association [US]
Toll Dial Assistance
Tracking and Data Acquisition
Tris(dioxaheptyl)amine
Transportation Development Center [Canada]
Transport Distribution Analysis
Tunnel Diode Amplifier
TDA-1 Tris[2-(2-Methoxyethoxy)ethyl]amine
Tris(3,6-Dioxaheptyl)amine
2D ACAR Two-Dimensional Angular Correlation of Annihilation Radiation [also: 2D-ACAR or 2D/ACAR]
TDAFP Turbine Driven Auxiliary Feedwater Pump
TDAL Tetradecenal
TDAS Traffic Data Administration System
TDB Technical Divisions Board
Terminology Database
Toxicology Data Bank [of National Institute of Health, US]
Traffic Database
TDBI Test During Burn-In
TDC Target Data Collection
Telegraphy Data Channel
Thermal Diffusion Coefficient
Time Delay Closing
Time-to-Digital Converter
Top Dead Center
Track Detection Circuit
Transportation Development Center [Canada]
Type Directors Club [US]
TDCA Technical Documentation Center for the Army [Netherlands]

TDCC Transportation Data Coordinating Committee [US]
TDCM Technology Development for the Communications Market
Transistor Driver Core Memory
TDCO Torpedo Data Computer Operator
TDCS Time Division Circuit Switching
TDCTL Tunnel-Diode Charge-Transformer Logic
TDD Target Detection Device
Technical Data Department
Technical Data Digest
Telecommunication Device for the Deaf
Telemetry Data Digitalizer
Test Design Description
Time Division Duplexing
TDDA Tetradecadien-1-yl Acetate
TDDB Time Dependent Dielectric Breakdown
TDDL Time-Division Data Link
TDDR Technical Data Department Report
TDDS Topographic Digital Data System
TDE Tetrachlorodiphenylethane
Thermodynamic Equilibrium
Total Data Entry
TDEC Technical Division and Engineering Center
Telephone Line Digital Error Checking
TDEFWP Turbine Driven Emergency Feedwater Pump
2DEG Two-Dimensional Electron Gas [also: TDEG]
TDEP Tracking Data Editing Program [of NASA]
TDES Tennessee Department of Employment Security [US]
TDF Task Deletion Form
Technology Demonstration Facility
Tetradecenyl Formate
Total Dietary Fiber
Transborder Data Flow
Transmitted Dark-Field
Trunk Distribution Frame
Two Degrees of Freedom
Two-Dimensional Flow [also: 2DF]
TDFA Total Dietary Fiber Assay
TDFAB Total Dietary Fiber Assay Bulletin
TDFI Turbo Diesel Fuel Injection
TDG Tap Density Gauge
Test Data Generator
Transmit Data Gate
TDH Total Dynamic Head
TDHF Time-Dependent Hartree-Fock
TDI Telecommunications Data Interface
Textile Dye Institute [US]
Toluene Diisocyanate
Tolylene Diisocyanate
Tool and Die Institute
Total Dielectric Isolation
Trade Data Interchange
TDIA Transient Data Input Area
TDIO Tuning Data Input/Output [also: TDI/O]
TDL Task Description Language
Transaction Definition Language
Tunnel-Diode Logic
TDLAS Tunable Diode Laser Absorption Spectrometry
TDM Tandem
Telephone Directory Memory
Template Descriptor Memory
Tertiary Dodecyl Mercaptan
Time-Division Multiplexing
Torpedo Detection Modification
TDMA Tape Direct Memory Address
Time-Division Multiple Access
TDMF Time Dependent Mean Field Theory
TDMS Telegraph Distortion Measurement Set
Terminal Data Management System
Thermal Desorption Mass Spectroscopy
Time-Shared Data Management System
Transmission Distortion Measuring Set
TDM-VDMA Time-Division Multiplex — Variable Destination Multiple Access
TDN Target Doppler Nullifier
Total Digestible Nutrient
T-DNA Tumor-Desoxyribonucleic Acid
TDNS Total Data Network System
TDO Tallow Diaminopropanedioleate
Technology Development Officer
Technology Development Organization [US]
Time Delay Opening
Tornado
Training Development Officer
TDOL Tetradecenol
TDOS Tape/Disk Operating System
TDP Table-Driven Programming
Technical Data Package
Technical Development Plan
Teledata Processing
Thiodiphenol
Thymidine Diphosphate
Top-Down Parsing
Tracking Data Processor
Trade Development Program
Traffic Data Processor
TDPAC Time Differential Perturbed Angular Correlation
TDPL Top Down Parsing Language
TDPSK Time-Differential Phase Shift Keying [also: TD-PSK]
TDPT Time-Dependent Perturbation Theory
TDQP Trimethyldihydroquinoline Polymer
TDR Tape Data Register
Target Discrimination Radar
Technical Data Relay
Technical Data Report
Technical Documentary Report
Temperature-Dependent Resistor
Temporarily Disconnected at (Subscribers) Request
Terminal Digit Requested
Time-Delay Relay
Time-Domain Reflectometry
Tone Dial Receiver
Torque Differential Receiver
Traffic Data Record
Transactional Document Recorder
Transmit Data Register
TDRE Tracking and Data Relay Experiment
TDRI Tropical Development and Research Institute [now: ODNRI, UK]
TDRM Time Domain Reflectometry Microcomputer

TDRP	Treatment Development Research Project
TDRS	Text Data Retrieval System
	Three-Dimensional Reinforced Solid
	Tracking and Data Relay Satellite
TDRSS	Tracking and Data Relay Satellite System
TDS	Tactical Dosimetry System
	Tape Data Selector
	Target Designation System
	Technical Data System
	Temperature Diffuse Scattering
	Tertiary Data Set
	Test Data Sheet
	Test Development Series
	Test Development Station
	Testing Data System
	Thermal Desorption Spectroscopy
	Thermal Desorption Spectrum
	Three-Dimensional Structure
	Time Distance Speed
	Time-Division Switching
	Time-Domain Spectroscopy
	Titanium Descaling Salt
	Thermal Diffuse Scattering
	Total Dissolved Solids
	Totally Dissolved Solids
	Total Distribution Solution
	Track Data Simulator
	Track Data Storage
	Tracking and Data (Acquisition) System
	Transaction Distribution System
	Transaction-Driven System
	Transistor-Display and Data-Handling System
	Translation and Docking Simulator
TDSF	Triangle-Dimer Stacking Fault
TDSS	Time Dividing Spectrum Stabilization
TDT	Target Designation Transmitter
	Task Dispatch Table
	Trigger Discharge Tube
TdT	Terminal Deoxynucleotidyl Transferase
TDTA	Toluoyl-D-Tartaric Acid
TDTL	Tunnel-Diode Transistor Logic
TDU	Target Detection Unit
	Tokyo Denki University [Japan]
TDV	Technology Development Vehicle
TDWR	Terminal Doppler Weather Radar
TDX	Thermal Demand Transmitter
	Time Division Exchange
	Torque Differential Transmitter
TDW	Tons Deadweight
TDX	Text and Data Exchange
TDY	Task Dictionary
TDZ	Touchdown Zone
TDZE	Touchdown Zone Elevation
TDZL	Touchdown Zone Lights
TE	Taxiway Edge-Lighting
	Technical Engineer
	Technical Exchange
	Telecommunications Engineer
	Telecommunications Engineering
	Television Electronics
	Television Engineer
	Television Engineering
	Temperature Effect
	Temperature Efficiency
	Terminal Emulator
	Terminal Equipment
	Terminal Exchange
	Test Engineer
	Test Equipment
	Thermal Efficiency
	Thermal Electron
	Thermal Element
	Thermal Engineering
	Thermal Etching
	Thermal Expansion
	Thermionic Emission
	Thermoelastic(ity)
	Thermoelectric(ity)
	Thermoplastic Elastomer
	Threshold Extension
	Tokamak Experiment
	Tool Engineer
	Tool Engineering
	Total Elongation
	Total Energy
	Totally Enclosed
	Traffic Enforcement
	Traffic Engineer
	Traffic Engineering
	Trailing Edge
	Transition Element
	Transition Editor
	Transportation Engineer
	Transportation Engineering
	Transverse Electric (Mode)
	Triboelectric(ity)
	Trunk Equalizer
	Trunk Expansion
	Turbo-Electric
	Tunnel Effect
T&E	Test and Evaluation [also: T/E]
T-E	Temperature-Efficiency (Index)
	Thermal-Efficiency (Index)
Te	Tellurium
TEA	Technical Engineers Association
	Technical Exchange Agreement
	Tetraethylammonium
	Thermal Expansion Anisotropy
	Transferred-Electron Amplifier
	Transversely Excited Atmosphere
	Triethanolamine
	Triethyl Aluminum
	Triethylamine
	Triethylammonium
	Tunnel-Emission Amplifier
	Tyrethylaluminum
TEAB	Tetraethylammonium Bromide
TEABF	Tetraethylammonium Tetrafluoroborate
TEAC	Technology Education Advisory Council
	Tetraethylammonium Chloride
TEA CO_2	Transversely-Excited Atmospheric-Pressure Carbon Dioxide (Laser) [also: TEA-CO_2]

TEAE Triethylaminoethyl
TEAHA Trans-East African Highway Authority [Ethiopia]
TEA LASER Transversely Excited Atmospheric Pressure Laser [also: TEA laser]
TEAM Technique for Evaluation and Analysis of Maintainability
Teleterminals Expandable Added Memory
Terminology Evaluation and Acquisition Method
Test and Evaluation of Air Mobility
The European-Atlantic Movement [UK]
Training Equipment and Maintenance
TEAMS Test, Evaluation and Monitoring System
TEAP Tetraethylammonium Perchlorate
Transportation Emergency Assistance Plan
TEAOH Tetraethylammonium Hydroxide
TEAS Triethanolamine Stearate
TEATS Tetraethylammonium Toluenesulfonate
TEBC Total Equivalent Boron Content
TEBOL Terminal Business-Oriented Language
TEC Tactical Electromagnetic Coordinator
Tantalum Electrolytic Capacitor
Telephone Engineering Center
Texas Employment Commission [US]
The Electrification Council [US]
Total Electron Content
Total Estimated Cost
Transearth Coast
Transient Electron Current
Triethylcitrate
Triple Erasure Connection
TECC Technology Education for Children Council [US]
TECH Technical [also: TECHN]
Technician [also: TECHN]
Technics [also: TECHN]
Tech Gr Technical Grade [Chemistry]
TECHNOL Technological
Technologist
Technology
TECHNONET Technical Information Network [of NRC, Canada]
TECH SPEC Technical Specification
TECMA Technical Ceramics Manufacturers Association [US]
TECOM Test and Evaluation Command [US Army]
TECR Technical Reason
TED Technical Evaluation and Development
Technology Evaluation and Development
Television Disk
Test Engineering Division [US Navy]
Thermoelastic Damping
Threshold Extension Demodulator
Transferred Electron Device
Translation Error Detector
Transmission Electron Diffraction
Triethylenediamine
Trunk Encryption Device
Turbine Engine Division [US Air Force]
TEDA Triethylenediamine
TE/DC Traffic Enforcement/Driver Control
TeDEC Tellurium Diethyl Dithiocarbamate
TeDMC Tellurium Dimethyl Dithiocarbamate
TEDP Tetraethyl Dithiopyrophosphate
TEDS Tactical Expendable Drone Systems
Twin Exchangeable Disk Storage
TEE Telecommunications Engineering Establishment [UK]
Torpedo Experimental Establishment [UK]
Trans Europe Express
TEEC Total Energy and Environmental Conditioning
TEEOC Turret Electronics and Electrooptical Console
TEF Thorne Ecological Foundation [now: TEI, US]
Transverse Electrical Field
TEFA Total Esterified Fatty Acid
TEFC Totally Enclosed Fan-Cooled (Motor)TEFL
Teaching of English as a Foreign Language
TEG Tactical Electronics Group
Tetraethylene Glycol
Thermoelectric Generator
Training and Education Group [of Library Association, UK]
Triethylene Glycol
TEGa Triethyl Gallium [also: TEG]
TEGAS Test Generation and Simulation
TEGFET Two-Dimensional Electron Gas Field-Effect Transistor
TEGMA Terminal Elevator Grain Merchants Association [US]
TEI Thorne Ecological Institute [US]
Total Employee Involvement
TEIn Triethyl Indium [also: TEI]
TEIC Tissue Equivalent Ionization Chamber
TEJ Transverse Expansion Joint
TEL Task Execution Language
Telegram
Telegraph(y)
Telephone
Tetraethyl Lead
Tritium Engineering Laboratory [Japan]
TELCO Telephone Company
TELCON Teletypewriter Conference
TELEBANKING Television/Telephone Banking
TELECOM Telecommunication [also: TELECOMM or TELEC]
TELEDAC Telemetric Data Converter
TELEDOC Telecommunications Documentation
TELEFAX Facsimile Transmission Service [Europe]
TELEMAIL Electronic Mail Service
TELENET Telecommunication Network
TELEPAC Telemetering Package
Telephone Package
TELEPUTER Television/Computer System
TELERAN Television and Radar Navigation System [also: Teleran]
TELESAT Telecommunications Satellite [Canada]
TELETEX (International) Super-Telex Service
TELEX Teleprinter Exchange Service [also: Telex, telex, TEX, TLX or TX]
TELG Telegram
Telegraph(y)
TELINET Telefax Library Network [US]
TELOPS Telemetry On-Line Processing System
TELPAL Tel Aviv-Palo Submarine Cable [Israel/Italy]
TELRY Telegraph Reply
TELS Turbine Engine Load Simulator
TELSAM Telephone Service Attitude Measurement

TELSCOM Telemetry-Surveillance-Communications
TELSIM Teletypewriter Simulator
TELTIPS Technical Effort Locator and Technical Interest Profile System [US Army]
TELUS Telemetric Universal Sensor
TEM Thermal Expansion Mismatch
Thermal Expansion Molding
Total Energy Management
Transmission Electron Micrograph
Transmission Electron Microscopy
Transverse Electromagnetic
Transverse Electromagnetic Mode
Triethylenemelamine
TEMA Tank Equipment Manufacturers Association
Telecommunication Engineering and Manufacturing Association [UK]
Tubular Exchanger Manufacturers Association [US]
TEM-AEM Transmission Electron Microscopy — Analytical Electron Microscopy
TEM-BF Transmission Electron Microscopy — Bright Field
TEM-DF Transmission Electron Microscopy — Dark Field
TEMED Tetramethylethylenediamine
TEMF Thermo-Electromotive Force [also: temf]
TEMO Topological Effect on Molecular Orbitals
TEMP Temperature
Template
Temporary
Total Energy Management Professional
Transportation Energy Management Program
temp (tempore) — in the time of
TEMPO Temporary
2,2,6,6-Tetramethyl-1-Piperidinyloxy
Total Evaluation of Management and Production Output
TEMPOS Timed Environment Multipartioned Operating System
TEMS Test Equipment Maintenance Set
Toyota Electronically Modulated Suspension
TEM/STEM Transmission Electron Microscopy/Scanning Transmission Electron Microscopy
TEM-TED Transmission Electron Microscopy — Transmission Electron Diffraction
TEMW Transverse Electromagnetic Wave
TENE Total Estimated Net Energy
Tenn Tennessee [US]
TENS Tension
Transcutaneous Electronic Nerve Stimulator
TEO Telephone Equipment Order
Transferred Electron Oscillator
TEOM Transformer Environment Overcurrent Monitor
TEOS Tetraethoxysilane
Tetraethylorthosilicate
TEP Terminal Error Program
Test Executive Processor
Thermoelectric Power
Transmitter Experimental Package
Transportation Energy Panel
Triethyl Phosphate
Tritium Extraction Plant
TEPA Tetraethylenepentamine
Triethylenephosphoramide
TEPAC Tube Engineering Panel Advisory Council [of Electronic Industries Association, US]
TEPD Total Endpoint Dose
TEPG Thermionic Electrical Power Generator
TEPIAC Thermophysical and Electronic Properties Information Analysis Center [US]
Te5MC Tellurium Pentamethylene Dithiocarbamate
TEPOS Test Program Operating System
TEPP Tetraethyl Pyrophosphate
TEPRSSC Technical Electronic Product Radiation Safety Standards Committee
TEPZ Taichung Export Processing Zone [Taiwan]
TER Terazzo
Tergitol
Terrace
Territory
Tertiary
Thermal Expansion Rubber
Thyssen-Extrem-Rechtkant (Steelmaking Process)
Transmission Equivalent Resistance
Transportation Engineering Research
Triple Ejector Rack
TERA Total Energy Resource Analysis
TERAC Tactical Electromagnetic Readiness Advisory Council
TERCOM Terrain Contour Matching [also: Tercom]
TEREC Tactical Electromagnetic Reconnaissance
TERLS Thumba Equatorial Rocket Launching System [India]
TERM Terminal
Termination
Terminology
TermNet International Network for Terminology [of ASI, Austria]
TERP Terrain Elevation Retrieval Program
TERPS Terminal Instrument Approach Procedures
Terminal Planning System
TERPY Terpyridine
TERR Terrace
Territory
TERS Tactical Electronic Reconnaissance System
tert tertiary [Chemicals]
TES Text Editing System
Thermal Energy Storage
Thermal Evaluation Spectroscopy
Thermoelectric Series
Time Encoded Speech
Transportable Earth Station
Trisethanesulfonic Acid
N-Tris(Hydroxymethyl)methyl-2-Aminoethanesulfonic Acid
Troubleshooting Expert System
TESA Television Electronics Service Association [Canada]
TESb Triethyl Antimony
TESLA Technical Standards for Library Automation Committee [of LITA, US]
TESS Tactical Electromagnetic Systems Study
Thermocouple Emergency Shipment Service
TEST Thesaurus of Engineering and Science Terms [of Engineers Joint Council, US]
TESTAS Turkish Electronics and Trade Association

TESTG Testing
TET Tetrachloride
Thermoelectric Thermometer
Thermometric Enthalpy Titration
Total Elapsed Time
Transportation Engineering Technology
Turbine Entry Temperature
TETA Tetraethanol Propane
Triethylenetetramine
Triethylenetriamine
TETD Tetraethylthiuram Disulfide
TETOC (Council for) Technical Education and Training for Overseas Countries [UK]
TETR Test and Training Satellite [US]
Tetragonal
TETRA Terminal Tracking Telescope
TETRAEN Tetraethylenepentamine
TETRAG Tetragonal
TETRAHED Tetrahedral [also: TETRAH]
TETRAPHOS-1 1,1,4,7,10,10-Hexaphenyl-1,4,7,10-Tetraphosphadecane
TETRAPHOS-2 Tris(2-Diphenylphosphinoethyl)phosphine
TETROON Tethered Meteorological Balloon [also: Tetroon or tetroon]
TEU Total Equivalent Units
Twenty-Foot Equivalent Unit
TEUT Teutonic
TEV Thermoelectric Voltage
TEVA Total Equivalent Volt-Ampere
TEVROC Tailored Exhaust Velocity Rocket
TEW Tactical Electronic Warfare
Transverse Elastic Wave
Transverse Electric Wave
TEWA Threat Evaluation and Weapon Assignment [of DND, Canada]
TEWS Tactical Electronic Warfare System
TEX Target Excitation
Tau-Epsilon-Chi System
Teleprinter Exchange Service [also: Telex, telex, TEX, TLX or TX]
Tex Texas [US]
TEXPLOT Texas Instruments Plotter
TEXT Texas Experimental Tokamak
Textile [also: TEX]
TEXTIR Text Indexing and Retrieval
TEXTLINE Text On-Line (Database) [UK]
TEXTOR Tokamak Experiment for Technology-Oriented Research
TEZG Tribological Experiments in Zero Gravity
TF Tape Feed
Task Force
Technological Forecasting
Temperature Factor
Tensile Fracture
Ternary Fission
Territorial Forces
Test Fixture
Test Frame
Textile Foundation [US]
Thermal Fatigue
Thermos Flask
Thin Film
Threshold Factor
Tool Foundation [Netherlands]
Toroidal Field
Toxicology Forum [US]
Transfer Function
Trunk Frame
Tuning Fork
Turbulent Flow
Twist Factor
TFA Taiwan Forest Administration
The Ferroalloys Association [US]
Technology Forecasting and Assessment
Texas Forestry Association [US]
Thin Film Analysis
Timing Filter Amplifier
Trifluoroacetate
Trifluoroacetic Acid
Trifluoroacetamide
Trifluoroacetyl
TFAA Trifluoroacetic Anhydride
TFAC Trifluoroacetylacetonato
TFA-ME Trifluoroacetate Methylester
TFB Towed Flexible Barge
TFBPA Textile Fibers and Byproducts Association [US]
TFC Telefilm Canada
Traffic
Transmission Fault Control
Transparent Ferroelectric Ceramic
Turbulent Forced Convection
TFCF Turbulent Free Convective Flow
TFCG Thin Film Crystal Growth
TFCX Tokamak Fusion Core Experiment
TFCRI Tropical Fish Culture Research Institute [Malaysia]
TFD Television Feasibility Demonstration
Thermo Field Dynamics
Thin Film Deposition
Thin Film Detector
Time Frequency Division [of NIST, US]
Transflective Device
TF/D Time-Frequency Dissemination
TFDL Technisch Fysische Dienst voor de Landboov [= Technical and Physical Engineering Service for Agriculture, Netherlands]
TFDU Thin Film Deposition Unit
TFE Tetrafluoroethylene
The Fertilizer Institute [US]
Thermionic Fuel Element
Thermo-Fluid Engineering
Toronto Futures Exchange [Canada]
2,2,2-Trifluoroethanol
Turbofan Engine
TFEL Thin Film Electroluminescence
TFEO Tetrafluoroethylene Epoxide
TFE/P Tetrafluoroethylene and Propylene Copolymer
TFE/P/T Tetrafluoroethylene, Propylene and Vinylidene Fluoride Terpolymer
TFFE Thin Film Ferroelectrics [also: TFF]
TFFET Thin Film Field-Effect Transistor
TFG Thin-Film Growth
Transmit Format Generator

TFH Thin-Film Head
TFI Teppich-Forschungsinstitut [= German Carpet Research Institute]
The Fertilizer Institute [US]
Thick Film Ignition
Threaded Full Length
TFK Transportforskningskommission [= Commission for Transport Research, Sweden]
TFL Through Flowline
TFM Tape File Management
Theoretical Fracture Mechanics
Total Flowmeter
Trifluoromethane
TFMC Tris(fluoromethylhydroxymethylene)camphorato
TFMC-Eu Tris[3-(Trifluoromethylhydroxymethylene)-camphorato], Europium
TFMC-Pr Tris[3-(Trifluoromethylhydroxymethylene)-camphorato], Praseodymium
TFME Thin-Film Mercury Electrode
TFMH Thin-Film Magnetic Head
TFML Thin-Film Multilayer
TFMS Text and File Management System
Trunk and Facilities Maintenance System
TFMSA Trifluoromethane Sulfonic Acid
TFP Thin-Film Polarizer
TFR Terrain Following Radar
Theoretical Final Route
Tightly Folded Resonator
Transaction Formatting Routine
Transfer
TFRI Taiwan Fisheries Research Institute
Taiwan Forestry Research Institute
TFRS Tungsten Fiber-Reinforced Superalloy
TFS Tape File Supervisor
Thin Film Science
Thin Film Solid
Time-of-Flight SIMS (= Secondary-Ion Mass Spectrometry)
Tin-Free Steel
Traffic Flow Security
Traffic Forecasting System
Transfer Standard
Triplet Fine Structure
Tritium Filling Station
TFSA Thin Film Spreading Agent
TFSM Terminal Flow Simulation Model
TFS&T Thin Film Science and Technology
TFSUS Task Force on the Scientific Uses of Space Stations
TFT Thin-Film Technology
Thin-Film Transistor
Threshold Failure Temperature
Time-to-Frequency Transformation
Trees for Tomorrow [US]
TFTP Television Facility Test Position
TFTR Tokamak Fusion Test Reactor [Princeton University, US]
Toroidal Fusion Test Reactor
TFU Timing and Frequency Unit
TFVA Training Film and Video Association [UK]
TFX Tactical Fighter Experiment
Transverse Flux
TFZ Transfer Zone
TG Taper Gauge
Telegraph(y)
Temperature Gradient
Terminal Guidance
Terminator Group
Thermogravimetry
Thickness Gauge
Tone Generator
Top Grille
Torpedo Group
Total Graph
Touch and Go
Transgranular
Triglyceride
Trunk Group
Tuned Grid
T&G Tongue and Groove
Tg Teragram
(Glass) Transition Point
Tugrik [Currency of Mongolia]
TGA The Glutamate Association [US]
Thermogravimetric Analysis
Thioglycolic Acid
Trace Gas Analysis
Triglycollamic Acid
TGA-IR Thermogravimetric Analysis — Infrared
TGAP Triglycidyl P-Aminophenol
TGB Thermogravity Balance
Tongued, Grooved and Beaded
TGC Teleglobe Canada
Tomato Genetics Cooperative [US]
Transmitter Gain Control
Travel Group Charter
TGCA Transportable Ground Control Approach
TGE Triglycidyl Ether
TGETPM Triglycidyl Ether of Triphenyl Methane
TGF Through Group Filter
Transforming Growth Factor
TGFA Triglyceride Fatty Acid
TGFB Triglycine Fluoroberyllate
TGG Third Generation Gyroscope
TGI Target Group Index [US]
Target Intensifier
TGID Transmission Group Identifier
Trunk Group Identification
TGL Toggle
Touch and Go Landing
TGM Toroidal Gate Monochromator
Trunk Group Multiplexer
TGMDA Tetraglycidyl Methylene Dianiline
TGMDA/DDS Tetraglycidyl Methylene Dianiline modified with Diaminodiphenylsulfone
TGMV Technical Group for Machine Vision [now: MVA, US]
TGN Trunk Group Number
TGO Toxic Gas Ordinance
TGS Tactical Ground Support
Taxying Guidance System
Telemetry Ground Station
Texas Geographic Society [US]

	Translator Generator System
	Triglycine Sulfate
TG-SSC	Transgranular Stress-Corrosion Cracking
TGSE	Tactical Ground Support Equipment
	Telemetry Ground Support Equipment
TGSe	Triglycine Selenate
TGSO	Tertiary Groups Shunt Operation
TGSS	Triglycine Sulfate Selenate
TGV	Train a Grande Vitesse [= French High-Speed Train]
TGT	Target
	Thromboplastin Generation Test
TGU	Tohoku Gakuin University [Japan]
TGWU	Transport and General Workers Union [UK]
TH	Technische Hochschule [= College of Technology [Germany, Austria and Switzerland]]
	Thermal Head
	Thursday
	Thyristor
	Total Head
	Transformation Hardening
	Transmission Header
	Triode-Hexode
	Tyrosine Hydroxylase
Th	Thorium
th	thermie [Unit]
THA	Tetrahydroaminacrine
	Tetrahydro Kendall's Compound A
ThA	Thorium A (= Polonium 216)
THAM	Trimethylol Aminomethane
	Tris(hydroxymethyl)aminomethane
THB	Temperature-Humidity Bias
	Thai Baht [Currency of Thailand]
ThB	Thorium B (= Lead 212)
THC	Tetrahydrocannabinol
	Thermal Converter
	Thrust Hand Controller
	Toronto Harbor Commission [Canada]
ThC	Thorium C (= Bismuth 212)
ThC'	Thorium C' (= Polonium 212)
ThC"	Thorium C" (= Thallium 208)
THCS	(Outlet) Temperature, Hot-Channel Sodium
THD	Thread
	Total Harmonic Distortion
ThD	Thorium D (= Lead 208)
THDM	Translucent Humic Degradation Matter
THDR	Thunder
THE	Tetrahydrocortisone
	Thunderstorm Event
THEED	Transmission High Energy Electron Diffraction
THEIC	Tris(hydroxyethyl)isocyanurate
THEOR	Theoretical [also: THEO]
	Theory
	Theorem
THEP	Theoretical High Energy Physics
THERM	Thermistor
	Thermometer
Thermal EMF	Thermal Electromotive Force [also: thermal emf]
THERMISTOR	Thermally Sensitive Resistor [also: thermistor]
THERMO	Thermostat
THERP	Technique for Human Error Rate Prediction
THF	Tetrahydrofolic Acid
	Tetrahydrofuran
	Tremendously High Frequency
THFA	Tetrahydrofurfuryl Alcohol
THFC	Tris(heptafluoropropylhydroxymethylene)camphorato
THFC-Eu	Tris[3-(Heptafluoropropylhydroxymethylene)-camphorato], Europium
THF-DMF	Tetrahydrofuran Dimethyl Formamide
THFTDA	Tetrahydrofuran Tetracarboxylic Dianhydride
THHF	Tetrahydrohomofolate
THHP	Tetrahydrohomopteroic Acid
THI	Technische Hochschule Ilmenau [= Ilmenau Institute of Technology, Germany]
	Temperature-Humidity Index
	Total Height Index
THIR	Temperature Humidity Infrared Radiometer
THK	Thick
THL	True Heavy Liquid
	Trans Hybrid Loss
THLD	Threshold
THM	Tetradecylhydrogenmaleate
	Tons of Hot Metal
THMA	Trailer Hitch Manufacturers Association [US]
THN	1,2,3,4-Tetrahydronaphthalene
	Thin
ThOD	Theoretical Oxygen Demand
THOMIS	Total Hospital Operating and Medical Information System
THOPS	Tape Handling Operational System
THOR	Tandy High-Intensity Optical Recording
	Tape-Handling Option Routine
	Transistorized High-Speed Operations Recorder
THOREX	Thorium Extraction Process
THORP	Thermal Oxide Reprocessing Plant
THORS	Thermal-Hydraulic Out-of-Reactor Safety Facility
THP	Terminal Handling Processor
	Tetrathydropapaveroline
	Tetrahydropyran
	Theoretical Horsepower [also: thp]
	Through-the-Hole Plating
THPC	Tetrakis(hydroxymethyl)phosphonium Chloride
THPE	1,1,1-Tris(4-Hydroxyphenyl)ethane
THPFB	Treated Hard Pressed Fiberboard
THQ	Telecommunications Headquarters
	Thermionic Quadrupole
THR	Threshold
	Transmitter Holding Register
Thr	Threonine
	Threonyl
t/hr	tons per hour
THRA	Tasmanian Historical Research Association
THRFTR	Thereafter
THRM	Thermal
THROT	Throttle
THRU	Through
THRUT	Throughout
THS	Tetrabutylammonium Hydrogen Sulfate
	The Hydrographic Society [UK]
	Titanic Historical Society [US]
THSD	Thousand
THSP	Thermal Spraying

THT Tetrahydrothiophene
THTF Thermal Hydraulic Test Facility
THTR Thorium High Temperature Reactor
THTRA Thorium High Temperature Reactor Association
THURS Thursday [also: THUR]
THWM Trinity High-Water Mark
ThX Thorium X (= Radium 224)
THYMOTRO Thyratron Motor Control
THz Terahertz
TI Tape Inverter
Tamarind Institute [US]
Target Identification
Technical Information
Technical Investigation
Technological Institute
Temperature Index
Terminal Interface
Test Instruction
Textile Institute [UK]
Thermal Insulation
Threaded Insert
Thread Institute [US]
Time Index
Time Interval
Tohoku Institute [of Japan Fisheries Agency]
Topological Information
Track Identifier
Track Initiator
Transfer Impedance
Transfrigoroute International [Switzerland]
Transportation Institute [US]
Ti Titanium
Tumor-Induced
TIA Telecommunications Industry Association [US]
Time Interval Analyzer
Typographers International Association [US]
TIAA Timber Importers Association of America [US]
TIAC Thermal Insulation Association of Canada
TIAG Telecommunications Industry Advisory Group
TIAL Titanium-Aluminum (Composite)
TIALON Titanium Aluminum Oxynitride (Ceramic) [also: TiAlON or Tialon]
TIARA Target Illumination and Recovery Aid
TIAS Target Identification and Acquisition System
True Indicated Air Speed
TIB Technical Information Bureau [UK]
Technische Informationsbibliothek [= Technical Library and Document Delivery Center, Germany]
Through Ice Bathymetry
TIBA Triiodobenzoic Acid
Triisobutyl Aluminum
Traffic Information Broadcast by Aircraft
Turkish Industrialists and Businessmen's Association
TIBAL Triisobutyl Aluminum
TIBBS Trace Integrated Bare Board System
TIBC Total Iron-Binding Capacity
TIBER Tokamak Ignition/Burn Experimental Reactor [US]
Tokamak Ignition/Burn Experimental Research
TIBER/ITER Tokamak Ignition/Burn Experimental Reactor/ International Thermonuclear Experimental Reactor [US]
TIBI Training in Business and Industry (Program) [Canada]
TIBOE Transmitting Information by Optical Electronics
Ti-BR Butadiene Rubber Based on Titanium Catalyst
TIBS Through Ice Bathymetry System
TIC Tantalum-Niobium International Center [Belgium]
Tape Intersystem Connection
Target Intercept Computer
Task Interrupt Control
Technical Information Center
Technical Institute Council
Telecommunications Information Center
Telemetry Instruction Conference
Temperature Indicating Controller
Terminal Identification Code
Total Inorganic Carbon
Transducer Information Center
Transfer in Channel
Transducer Information Center [of Battelle Memorial Institute, US]
TICA Technical Information Center Administration
TICAS Taxonometric Intra-Cellular Analytic System
TICCI Technical Information Center for Chemical Industry [India]
TICCIT Time-Shared Interactive Computer-Controlled Informational Television
TICE Time Integral Cost Effectiveness
TICS Teacher Interactive Computer System
Telecommunication Information Control System
Tidal Current System
TIC-TOC Telecommunications Information Center/Technical Office for Consumers [UK]
TID Technical Information Division
Test Instrument Division
Total Ion Detector
Traveling Ionospheric Distrubance
TIDA Travel and Industrial Development Association [UK]
TIDAR Time Delay Array Radar
TIDB Tester-Independent Database
TIDDAC Time in Deadband Digital Attitude Control
TIDE Transponder Interrogation and Decoding Equipment
TIDES Time Division Electronic Switching System
TIDF Trunk Intermediate Distribution Frame
TIDI Time-Division (Sound)
TIDMA Tape Interface Direct Memory Access
TIDU Technical Information and Documentation Unit [of DSIR, UK]
TIDY Track Identity
TIE Technical Integration and Evaluation
Terminal Interface Equipment
Time Interval Error
TIEM The Innovation and Entrepreneurial Management
TIES Technical Investigation and Engineering Services [Canada]
Technical Investigation and Engineering Studies
Total Integrated Engineering System
Translators and Interpreters Educational Society
Transmission and Information Exchange System

TIF Tape Inventory File
Task Initiation Form
Telephone Influence Factor
Telephone Interference Factor
Temperature-Independent Factor
Terminal Independent Format
Text Interchange Format
True Involute Form
TIFF Tag Image File Format
TIFO Text Interchange Format
TIFR Tata Institute of Fundamental Research [India]
TIFS Total In-Flight Simulator
TIG Technical Information Group
Telegram Identification Group
Tungsten Inert-Gas (Welding)
TIGER Telephone Information Gathering, Evaluation and Review
Total Information Gathering and Reporting
TIGS Terminal-Independent Graphic System
TIH Trunk Interface Handler
TII (European Association for the Transfer of) Technologies, Innovation and Industrial Information [Luxembourg]
TIIAL The International Institute of Applied Linguistics
TIIF Tactical Image Interpretation Facility
TIIPS Technically Improved Interference Prediction System
TIIT Technion Israel Institute of Technology [Israel]
TIL Tumor-Infiltrating Lymphocytes
Two Intedigitation Level
TILS Technical Information and Library Services
Tumor-Infiltrating Lymphocytes
TIM Table Input to Memory
Teeth in Mesh
Test Instrumented Missile
Thermoset Injection Molding
Time Meter
Toll Interface Module
Total Information Management
Total Ion Monitoring
Transistor Information Microfile
Transmitter Intermodulation
Triiodomethane
TIMA Thermal Insulation Manufacturers Association [US]
TIME Test, Inspection, Measurement and Evaluation
Total Industry Marketing Effort
Transferred Ionized Molten Energy
TIMEC Technology Institute for Medical Devices Canada
TIMIX Texas Instruments Mini/Microcomputer Information Exchange [Name Changed to: The International Microcomputer Information Exchange, US]
The International Microcomputer Information Exchange [also: TiMix, US]
TIMIXE Texas Instruments Mini/Microcomputer Information Exchange Europe [also: TiMixE, Netherlands]
TIMM Thermionic Integrated Micromodules
TIMS Telephone Information Management System
Text Information Management System
The Institute of Management Sciences [US]
Thermal Ionization Mass Spectroscopy
Transmission Impairment Measuring Set
TIM/TOM Table Input to Memory/Table Output from Memory
TIN Temperature Independent
Triangulated Irregular Network
TINCT Tincture
TINET Transparent Intelligent Network
TINFO Tieteellisen Informoinnin Neuvosto [= Council for Scientific Information and Research Libraries, Finland]
TINS Thermal Imaging Navigation Set
TINT Track in Track
TIO Terminal Input/Output
Test Input/Output
Time Interval Optimization
TIOA Terminal Input/Output Area
TIOC Terminal Input/Output Controller
TIOM Triisooctyl Trimellithate
TIOT Task Input/Output Table
TIQWQ Terminal Input/Output Wait Queue
TIP Tape Input
Technical Information Panel [of AGARD]
Technical Information Processing
Technical Information Program
Technical Information Project
Technology Inflow Program [Canada]
Telefiche Image Processor
Teletype Input Processing
Temperature-Indicating Paint
Terminal Interface Package
Terminal Interface Processor
Terminal Interface Program
Texture Independent Path
Titanium Isopropoxide
Tokyo Institute of Polytechnics [Japan]
Transaction Interface Package
Transaction Interface Processor
Transient In-Core Probe
Traveling In-Core Probe
TIPA Tank and Industrial Plant Association
Tetraisopropylpyrophosphoramide
TIPACS Texas Instruments Planning and Control System
TIPI Tactical Information Processing and Interpretation
TIPL Teach Information Processing Language
TIPP Time-Phasing Program
TIPPC Texas Instruments Personal Programmable Calculator Club [US]
TIPRO Texas Independent Producers and Royalty Owners Association [US]
TIPS Technical Information Processing System
Technical Information Programming System
Terminal Interface Processors
Text Information Processing System
The Image Processing Software
Thousands of Instructions per Second
Triisophylsilyl
TrueVision Image Processing Software
TIPS/SV The Image Processing Software/Signature Verification
TIPTOM Towards Improved Performance of Tool Materials

TIPTOP Tape Input/Tape Output [also: TIP/TOP or TIP TOP]
TIQ Task Input Queue
Tetrahydroisoquinone
TIR Target Instruction Register
Technical Information Release
Technical Information Report
Technical Intelligence Report
Test Incident Report
Thermal Infrared
Total Indicator Reading
Total Indicator Runout
Total Internal Reflection
Transient Impulse Resonance
Transient Impulse Response
Transport International de Marchandises par la Route [= International Road Transportation Organization, France]
TIRAM Taper Insulated Random-Access Memory
TIRC Toxicology Information Response Center [US]
TIRDO Tanzania Industrial Research and Development Organization
TIRF Traffic Injury Research Foundation [Canada]
TIRH Theoretical Indoor Relative Humidity
TIRKS Trunk Integrated Record Keeping System
TIROS Television Infrared Observation Satellite
TIRP Textile Information Retrieval Program [US]
Total Internal Reflection Prism
TIRPF Total Integrated Radial Peaking Factor
TIRS Thermal Infrared Scanner
TIS Target Information Sheet
Target Information System
Technical Information Service [Canada]
Terminal Interface Subsystem
The Information System
Total Information System
Total Ionic Strength
Transponder Interrogation Sonar
TISAB Total Ionic Strength Adjustment Buffer
TISC Tire Industry Safety Council [US]
TISSS Tester-Independent Support Software System
TISTR Thailand Institute of Scientific and Technological Research
TIT Test Item Taker
Titer
Title
Tohoku Institute of Technology [Japan]
Tokyo Institute of Technology [Japan]
Turbine Inlet Temperature
TITF Test Item Transmittal Form
TITR Titration
TITS Test Instrument Tracking System
TITUS Traitement de l'Information Textile Universelle et Selective [= Processing of General and Selected Textile Information (Database) [of ITF, France]]
TIU Tape Identification Unit
Terminal Interface Unit
Terrestrial Interface Unit
Toxicologically Insignificant Usage
Trypsin Inhibitor Unit
Typical Information Use
TIUC Textile Information Users Council [US]
TIUPIL Typical Information Use per Individual
TIV Total Indicator Variation
Ti-Y-TZP Titanium/Yttrium Tetragonal Zirconia Polycrystal
TIWU Transport and Industrial Workers Union [Trinidad]
TIZ Traffic Information Zone
TJ Thermojunction
TJC Trajectory Chart
TJD Trajectory Diagram
TJID Terminal Job Identification
TJR Trunk and Junction Routing
TJS Transverse Junction Stripe
TK Taka [Currency of Bangladesh]
Thymidine Kinase
Track
Transmission Kossel (Technique)
Trunk
Tk Taka [Currency of Bangladesh]
TKA Thermokinetic Analysis
TKL Thermodynamics and Kinetics Laboratory [of U of T, Canada]
TKN Total Kjehldal Nitrogen
TKO Trunk Offer
TKOF Takeoff
TKP Tribasic Potassium Phosphate
Tripotassium Phosphate
TKPP Tetrapotassium Pyrophosphate
TKS Tessman-Kahn-Shockley (Theory)
TKT Ticket
TL Tantalum Lamp
Tape Library
Target Label
Target Language
Target Loss
Task Leader
Test Link
Test Log
Thermoluminescence
Thin Layer
Thin-Layer Leaching
Tie Line
Time Limit
Tolerance Limit
Tool
Total Lipid
Transaction Language
Transition Level
Transmission Level
Transmission Line
Transmission Loss
Triboluminescence
Truck Load
Tungsten Lamp
Turbine Liquefier
Turkish Lira [Currency]
T/L Total Loss
Tl Thallium
TLA Truck Loggers Association
TLAS Tactical, Logical and Air Stimulation
TLB Translation Lookaside Buffer
TLC Tank Landing Craft

	Task Level Control
	Thick Molding Compound
	Thin-Layer Chromatography
	Total Loss Control
	Typed Lambda Calculi
TLCC	Trades and Labour Congress of Canada
TLC-FID	Thin-Layer Chromatography — Flame Ionization Detector
TLCK	Tosyl Lysine Chloromethyl Ketone
TLCT	Total Life-Cycle Time
TLCTI	Total Loss Control Training Institute
TLD	Thermoluminescent Dosimeter
	Trapped Lattice Dislocation
TLDPC	Tidal Land Development Planning Commission [Taiwan]
TLE	Temperature-Limited Emission
	Theoretical Line of Escape
	Tracking Light Electronics
TLG	Telegraph
TLF	Trunk Line Frame
TLI	Telephone Line Interface
TLK	Talking
	Test Link
	Tritium Laboratory Karlsruhe [Germany]
TLL	Teflon Luer Lock
	Thin-Layer Leaching
TLLM	Temperature and Liquid Level Monitor
TLM	Tape-Laying Machine
	Telemeter
	Thin Lipid Membrane
	Transition Line Model
	Trouble Locating Manual
TLMI	Tag and Label Manufacturers Institute [US]
TLMS	Tape Library Management System
TLN	Trunk Line Network
TLO	Total Loss Only
	Tracking Local Oscillator
TLP	Telephone Line Patch
	Tension Leg Platform
	Threshold Learning Process
	Top Load Pad
	Total Language Processor
	Toxicity Leaching Procedure
	Transient Liquid Phase
	Transient Lunar Phenomena
	Translation Lookaside Buffer
	Transmission Level Point
TLR	Toll Line Release
TLRI	Taiwan Livestock Research Institute
TLRV	Tracked Levitated Research Vehicle
TLS	Tactical Landing System
	Tape Librarian System
	Target Level of Safety
	Telecommunication Liaison Staff
	Telemetry Listing Schedule
	Terminal Landing System
	Thermal Line Softening
	Three-Stage Least Squares
	Total Library System
TLSA	Transparent Line Sharing Adapter
TLSM	Trail Lead Smelter Modernization [Canada]
TL/SX/EW	Thin-Layer Leaching followed by Solvent Extraction and Electrowinning
TLT	Terminal List Table
TLTA	Toluoyl-L-Tartaric Acid
	Two-Loop Test Apparatus
TLTP	Trunk Line Test Panel
TLTR	Translator
TLU	Table Lookup
	Terminal Logic Unit
TLV	Threshold Limit Value
	Tracked Levitated Vehicle
TLV-C	Threshold Limit Value — Ceiling
TLV-TWA	Threshold Limit Value — Time-Weighted Average
TLW	Titanium-Lead-Tungsten (Compound)
TLWD	Tailwind
TLWS	Trunk and Line Workstation
TLX	Teleprinter Exchange Service [also: TELEX, Telex, telex, TEX or TX]
TLZ	Titanium-Lead-Zinc
	Transfer on Less than Zero
TM	Tactical Missile
	Tape Mark
	Tape Module
	Technical Manual
	Technical Memorandum
	Technical Monograph
	Telemeter
	Telemetry
	Temperature Meter
	Temperature Monitor
	Test Machine
	Test Mode
	Text Management
	Thematic Mapper
	Theoretical Mathematics
	Theoretical Mechanics
	Thermal Margin
	Thermomagnetic
	Thermomechanical
	Thermoplastic Molding
	Thickness Measurement
	Time and Materials
	Time and Motion (Study)
	Time Modulation
	Timing Module
	Tissue Mechanics
	Titrimetry
	Tone Modulation
	Topical Meeting
	Town Manager
	Trademark
	Traffic Management
	Traffic Mix
	Transfer Mode
	Transformation Matrix
	Transition Metal
	Transmission Matrix
	Transportation Management
	Travelling Microscope
	Transverse Magnetic
	Trench Mortar

	Tropical Month
	Turbomachine
	Turbomachinery
	Turing Machine
	Twist Multiplier
T&M	Time and Materials [also: TM or T-M]
Tm	Melting Temperature
	Thulium
TMA	Tape Motion Analyzer
	Telecommunications Management Association [UK]
	Telemetry Manufacturers Association
	Terminal Control Area
	Testability Measure Analyzer
	Test Module Adapter
	Tetrahydroaminacrine
	Tetramethylammonium
	Thermomechanical Analysis
	Titrimetric Analysis
	Transport Museum Association [US]
	Trimellitic Anhydride
	Trimethyl Aluminum
	Trimethylamine
	Trimethylammonium
	Trimethylanilinium
TMAC	Tetramethylammonium Carbonate
	Trimellitic Anhydride Chloride [also: TMACl]
TMAH	Trimethylammonium Hydroxide
	Trimethylanilinium Hydroxide
TMAl	Trimethyl Aluminum
TMAMA	Textile Machinery and Accessory Manufacturers Association
TMAO	Trimethylamine Oxide
TMAX	Maximum Time
TMB	3,3',5,5'-Tetramethylbenzidine
	Trimethoxybenzoate
	Trimethoxybenzoic Acid
	Trimethylbenzene
TMB-4	1,1'-Trimethylene-Bis(4-Formylpyridinium Bromide) Dioxime
TMB-8	3,4,5-Trimethoxybenzoic Acid 8-(Diethylamino)octyl Ester
TMBA	Thickness Measurement by Beam Alignment
TMBR	Timber
TMC	Tape Management Catalogue
	Terminal Control
	Test Module Coil
	The Maintenance Council [of American Trucking Association, US]
	Thermomechanical Control
	Thick Molding Compound
	Total Molding Concept
	Transition Metal Chemistry
	Transition Metal Complex
	Transition Metal Compound
	Transmission Maintainance Center
TMCB	Tetramethoxycarbonyl Benzophenone
TMCC	Time-Multiplexed Communication Channel
TMCDT	Trimethylcyclododecatriene
TMCOMP	Telemetry Computation
TMCP	Thermomechanical Control Process
	Thermomechanically Controlled Processing
TMCS	Toshiba Minicomputer Complex System
	Trimethylchlorosilane
TMD	Tactical Munitions Dispenser
	Technology Marketing Division [of CANMET, Canada]
	Tensor Meson Dominance
	Theoretical Maximum Density
	Transient Mass Distribution Code
TMDE	Test, Measurement and Diagnostic Equipment [also: TM&DE]
TMDU	Tokyo Medical and Dental University [Japan]
TME	Technical Unit of Mass, Metric
	Telemetric Equipment
	Tempered Martensite Embrittlement
TMEDA	Tetramethylethylenediamine
TMF	Thermal Mechanical Fatigue
	Thermomechanical Fatigue
	Transmission Monitoring Facility
	Trunk Maintenance File
TMG	Thermal Meteoroid Garment
	Thiomethylgalactoside
	Track Made Good
TMGa	Trimethyl Gallium [also: TMG]
TMGE	Thermo-Magnetic-Galvanic Effect
TMH	Tons per Man-Hour
TMHD	2,2,6,6-Tetramethylheptane-3,5-Dione
	2,2,6,6-Tetramethyl-3,5-Heptanedionate
TMHR	Tandem Mirror Hybrid Reactor
TMI	Test, Measurement and Inspection (Conference and Exhibition)
	Three Mile Island [US]
	Timing Measurement Instrument
TM&I	Test, Measurement and Inspection
TMIC	Toxic Materials Information Center [of NSF and USAEC]
TMIN	Minimum Time
TMIn	Trimethyl Indium [also: TMI]
TMIS	Technical and Management Information System [of NASA]
	Technical Meetings Information Service [US]
	Television Measurement Information System
	Television Metering Information System
TMITC	Tokyo Metropolitan Industrial Technology Center [also: TMITEC, Japan]
TMJ	Temporo-Mandibular Joint (Procedure)
TML	Tandem Matching Loss
	Terminal
	Tetramethyl Lead
	Total Mass Loss
TM/LP	Thermal Margin/Low Pressure (Trip)
TMM	Temperature-Stable Microwave Material
	Test Message Monitor
	Transverse Magnetic
	Transverse Magnetic Mode
TMMB	Truck Mixer Manufacturers Bureau [US]
TMMC	Tetramethyl Ammonium Manganese Chloride
TMN	Technical and Management Note
	Trimethylamine Nitrogen
TMO	Telegraph Money Order
	Time Out
	Trimethylamine Oxide

TMOS T-Type Metal Oxide Semiconductor
TMP Temperature
Temporary
Terminal Monitor Program
Tetramethylpiperidine
The Madison Project [US]
Thermomechanical Process(ing)
Thermomolecular Pressure
Thymidine Monophosphate
Transparent Multiprocessing
Transmembrane Potential
Turbomolecular Pump
TMPA Transocean Marine Paint Association [Netherlands]
TMPAH Trimethylphenylammonium Hydroxide
TMPRY Temporary
TMPS Tetramethyl-P-Silphenylene Siloxane
TMPT Trimethylolpropane Trimethacrylate
TMPTA Trimethylolpropane Triacrylate
TMPTMA Trimethylolpropane Trimethacrylate
TMQ 2,2,4-Trimethyl-1,2-Dihydroquinoline
TMR Tandem Mirror Reactor
Teledyne Materials Research
Triple Modular Redundancy
TMRBM Transportable Medium Range Ballistic Missile
TMRS Traffic Measuring and Recording System
TMS Tactical Missile Squadron
Tape Management System
Telecommunications Message Switch
Telemotor System
Telephone Management System
Telex Management System
Temperature Measurements Society [US]
Tesla Memorial Society [US]
Test Management System
Tethered Manned Submersible
Tetramethylsilane
The Magnolia Society [US]
The Masonry Society [US]
The Metallurgical Society [now: MMMS of AIME, US]
The Metals Society [UK]
Time and Motion Study
Time-Shared Monitor System
Time-Multiplexed System
Trademark Society [US]
Traffic Management System
Transmission Measuring Set
Transportation Management System
Trimethylsilyl
Trochoidal Mass Spectrometer
TMSA Technical Marketing Society of America [US]
TMS-AIME The Metallurgical Society of the American Institute of Mining, Metallurgical and Petroleum Engineers [US]
TMSb Trimethyl Antimon
TMSCl Trimethylsilyl Chloride
TMSCN Trimethylsilyl Cyanide
TMSDEA Trimethylsilyldiethylamine
TMSi Tetramethylsilane
TMSS Tetrakis(trimethylsilyl)silane
TMT Testing Methods and Techniques
Thermomechanical Testing
Thermomechanical Treatment
Transmit
Turbine Motor Train
TMTC Through-Mode Tape Converter
TMTD Tetramethylthiuram Disulfide
TMTM Tetramethylthiuram Monosulfide
TMTSF Tetramethyltetraselenafulvalene
TMTTF Tetramethyltetrathiafulvalene
TMTU Tetramethythiourea
TMU Test Maintainance Unit
Time Measurement Unit
Tokyo Metropolitan University [Japan]
Transmission Message Unit
TMV Technical Maintenance Vehicle
Tobacco Mosaic Virus
TMW Tomorrow
Transverse Magnetic Wave
TMX Telemeter Transmitter
Terminal Multiplexer
TMXO Tactical Miniature Crystal Oscillator
TMX-U Tandem Mirror Experiment-Upgrade
TN Task Number
Technical Note
Tennessee [US]
Terminal Node
Thermal Neutron
Thermal Noise
Thermonuclear
Thyssen-Niederrhein (Steelmaking Process)
Tone
Total Nitrogen
Track Number
Trade Name
True Negative
Twisted Nematic (Crystal)
Tn Thoron (= Radon 220)
tn ton
TNA Telex Network Adapter
Tetranitroaniline
Thermal-Neutron Activation
Thermal-Neutron Analysis
Transient Network Analyzer
2,4,6-Trinitroaniline
TNAA Thermal Neutron Activation Analysis
TNB Trinitrobenzene
TNBS Trinitrobenzenesulfonate
Trinitrobenzenesulfonic Acid
TNBT Tetra-N-Butyltitanate
Tetranitro Blue Tetrazolium
TNC Terminal Network Controller
The Nature Conservancy [US]
Threaded-Nut Coupling
Total Nonstructural Carbohydrates
Trade Name Classification
Transnational Corporation
Transport Network Controller
Trinitrocellulose
TNCIAWPRC Thai National Committee of the International Association on Water Pollution Research and Control

	Turkish National Committee of the International Association on Water Pollution Research and Control
TND	Trace Narcotics Detector
	Tunesian Dinar [Currency]
TNDC	Thai National Documentation Center [Thailand]
TNDCY	Tendency
TNDS	Total Network Data System
TNEL	Total Noise Exposure Level
TNEP	Total Noise Equivalent Power
TNF	Theater Nuclear Forces [of NATO]
	Third Normal Form
	Transfer on No Overflow
	2,4,7-Trinitro-9-Fluorenone
	Tumor Necrosis Factor
TNFE	Twisted Nematic Field Effect
TNG	Training
TNGT	Tonight
TNHP	Texas Natural Heritage Program [US]
TNI	The Networking Institute [US]
TNIP	Terrestrial Network Interface Processor
TNL	Terminal Net Loss
	Tunnel
TNLDIO	Tunnel Diode
TNM	Tetranitromethane
TNO	(Organisatie voor) Toegepast-Naturwetenschappelijk Onderzoek [= Organization for Applied Scientific Research, Netherlands]
	Trade Negotiations Office [Canada]
TNOP	Total Network Operations Plan
TNP	Trinitrophenol
TNPAL	Trinitrophenylaminolauryl
TNPP	Trisnonylphenyl Phosphite
TNR	Thermonuclear Reaction
	Thermonuclear Reactor
TNS	The Next Step
	Toluidinylnaphthylenesulfonate
	6-(P-Toluidino)-2-Naphthalenesulfonic Acid
	Transaction Network Service [US]
TNT	Terminal Name Table
	Thallium Nitrate Trihydrate
	Total Network Test (System)
	Transient Nuclear Test
	Trinitrotoluene
TNX	Thanks
TNXCD	Transaction Type Code
TNZ	Transfer to Non-Zero
TO	Takeoff
	Technical Order
	Telegraph Office
	Telegraph Offices [of ITU, Switzerland]
	Third-Order
	Traffic Order
	Transistor Outline
	Transverse Optical
	Time Out
	Tin Oxide
	Turnover
T&O	Test and Operation
T/O	Takeoff
	Transfer Order
TOA	Table of Authorities [also: ToA]
	Takeoff Angle
	Time of Arrival
	Tools-Oriented Approach
	Total Obligational Authority
TOAA	(Study Group on) Takeoff Obstacle Accountability Areas
TOADS	Terminal-Oriented Administrative Data System
TOB	Technical Operations Board
	Translation Operations Branch [of Secretary of State, Canada]
TOB BRZ	Tobin Bronze
TOC	Table of Coincidences
	Table of Contents
	Tag(liabue) Open Cup
	Task-Oriented Costing
	Technical Office for Consumers [UK]
	Television Operating Center
	Threshold Odor Concentration
	Timber Operators Council [US]
	Time of Contact
	Top of Climb
	Total Optical Color
	Total Organic Carbon
	Total Organic Compounds
TOCC	Technical and Operational Control Center
	Technical and Operational Coordination Center
TOCS	Terminal-Oriented Computer System
	Testing Open Communications Systems (Project) [Canada]
TOD	Technical Objective Directive
	Technical Objective Document
	Time of Day
	Time of Departure
	Top of Duct
	Total Oxygen Demand
TODA	Takeoff Distance Available
TODS	Test-Oriented Disk System
	Transaction on Database System
TOE	Tons of Oil Equivalent
	Total Operating Expenses
TOEC	Third-Order Elastic Constant
TOEFL	Test of English as a Foreign Language
TOES	Trade-Off Evaluation System
TOF	Time of Flight
	Tone Off
	To Order From
	Top of File
	Top of Form
TOFC	Trailer on Flat Car
TOFF	Thin Overlay for Friction
TOFISS	Time-of-Flight Ion Scattering Spectroscopy [also: TOF ISS or TOF-ISS]
TOFSIMS	Time-of-Flight Secondary-Ion Mass Spectrometry [also: TOF SIMS or TOF-SIMS]
TOGA	Tropical Ocean Global Atmosphere (Program) [of WMO, Switzerland]
TOH	Total Organic Halogen
	Tyrosine Hydroxylase
TOHM	Terohmmeter
TΩ	Teraohm

TOHP Takeoff Horsepower
TOI Technical Operation Instruction
TOIRS Transfer Orbit Infrared Earth Sensor
TOJ Track on Jamming
TOL Test-Oriented Language
Tolerance
Toluene
TOLAR Terminal On-Line Availability Reporting
TOLD Telecommunications On-Line Data System
TOLIP Trajectory Optimization and Linearized Pitch
TOLR Transmitting Objective Loudness Rating
TOLT Teleprocessing On-Line Test
TOLTEP Teleprocessing On-Line Test Executive Program
TOLTS Total On-Line Testing System
TOM Table Output from Memory
Tool Material
Translator Octal Mnemonic
Typical Ocean Model
TOMCAT Telemetry On-Line Monitoring Compression and Transmission
TOMS Torus Oxygen Monitoring System
Total Ozone Mapper System
TON Tone On
tonf ton-force
ton/hr tons per hour [also: ton hr^{-1} or t/hr]
TONLAR Tone-Operated Net Loss Adjuster
TONN Tonnage
TOOL Test-Oriented Operation Language
TOOLS Total Operating On-Line System
TOP Tape Output
Technical Office Protocol
Tongan Pa'anga [Currency of Tonga]
Total Obscuring Power
Transaction-Oriented Programming
Transient Overpower Accident
TOPCAT Texas On-Board Program of Computer-Assisted Training [US]
TOPCOPS The Ottawa Police Computerized On-Line Processing System [Canada]
TOPES Telephone Office Planning and Engineering System
TOPIC Time Ordered Programmer Integrated Circuit
TOPICS Transport Operations Program for Increasing Capacity and Safety
TOPM Takeoff Performance Monitor
TOPO Trioctylphosphine Oxide
TOPOG Topographical
Topography
TOPP Task-Oriented Plant Practice
Terminal-Operated Production Program
TOPR Taiwan Open Pool Reactor
TOPS 3-[Ethyl(3-Methylphenyl)amino]-1-Propanesulfonic Acid
N-Ethyl-N-Sulfopropyl-M-Toluidine
Telemetry Operations
Telephone Order Processing System
Teletype Optical Projection System
Test and Operating System
The Operational PERT (= Production Evaluation and Review Technique) System
Thermoelectric Outer Plant Spacecraft
Time-Sharing Operating System
Total Operations Processing System
Traffic Operator Position System
Transducer Operated Pressure System
TOPSI Topside Sounder, Ionosphere
TOPSY Territorial Operational System
Test Operations and Planning System
Thermally Operated Plasma System
TOPSYS Tools for Parallel Systems
TOPTS Test-Oriented Paper-Tape System
TOR Technical Operations Research
Technology-Oriented Research
Telegraph On Radio
Torque
Transpolyoctenamer
Tropospheric Ozone Research (Project) [Europe]
TORA Takeoff Run Available
TORC Traffic Overflow Reroute Control
Traffic Overload Reroute Control
TORP Torpedo
TOS Tactical Operations System
Taken Out of Service
Tape Operating System
Temporarily Out of Service
Terminal-Oriented Software
Terminal-Oriented System
Test Operating System
Time-Ordered System
TIROS (= Television Infrared Observation Satellite) Operational Satellite [of NASA]
Top of Stack
Top of Steel
Traffic Orientation Scheme
Transfer Orbit Stage
TOSA Takeoff Space Available
TOSBAC Toshiba Scientific and Business Automatic Computer
TOSCW Top of Stack Control Word
TOSD Telephone Operations and Standards Division [of REA, US]
TOSL Terminal-Oriented Service Language
TosMIC Tosylmethylisocyanide
TOSP Top of Stack Pointer
TOSS Terminal-Oriented Support System
Test Operation Support Segment
TIROS (= Television Infrared Observation Satellite) Operational Satellite System [of NASA]
Transient and/or Steady State
TOSSA Transfer Orbit Sun Sensor Assembly
TOST Turbine Oxidation Stability Test
TOT Time of Tape
Time of Transmission
Total
Transfer of Technology
TOTE Terminal On-Line Test Executive
TOVALOP Tanker Owners Voluntary Agreement Concerning Liability for Oil Pollution [of ITOPF, UK]
TOVC Top of Overcast
TOVS TIROS (= Television Infrared Observation Satellite) Operational Vertical Sounder [of NASA]
TOW Takeoff Weight

	Tube-Launched, Optically-Tracked, Wire-Guided (Missile) [also: tow]
TOWA	Terrain and Obstacle Warning and Avoidance
TOXBACK	Toxicology Information Service [of National Library of Medicine, US]
TOXBIB	Toxicology Bibliography (Database) [of National Library of Medicine, US]
TOXCON	Toxicology Information Conversation On-Line Network [now: TOXLINE]
TOXICOL	Toxicologist
	Toxicology
TOXLINE	Toxicology Information On-Line (Database) [of National Library of Medicine, US]
TOXTIPS	Toxicology Test in Progress (Database) [of National Library of Medicine, US]
TP	Tandem Propeller
	Tank Pressure
	Tape
	Target Practice
	Technical Pamphlet
	Technical Paper
	Technical Program
	Technical Publication
	Technological Properties
	Telemetry Processing
	Telephone
	Telephotography
	Telephotometer
	Telephotometry
	Teleprinter
	Teleprocessing
	Teleprocessor
	Teletype Printer
	Terminal Pole
	Terminal Processor
	Terphenyl
	Testosterone Propionate
	Test Plate
	Test Point
	Test Position
	Test Procedure
	Text Processing
	Theoretical Physics
	Theoretical Plate
	Thermal Printer
	Thermal Properties
	Thermophysics
	Thermopile
	Thermoplastic
	Thiophosphamide
	Throttle Positioner
	Throwing Power
	Thymidine Phosphate
	Thymopoietin
	Time Pulse
	Timing Pulse
	Title Page
	Toll Point
	Toll Prefix
	Torpedo Part of Beam
	Torque Pressure
	Total Pressure
	Total Protein
	Township
	Transaction Paper
	Transaction Processing
	Transform Processing
	Transient Program
	Transmission Planar
	Transportation Planning
	Transport Phenomena
	Tribophysics
	Triple Point
	Triple Pole [also: 3P]
	Triple Play
	Troop
	True Positive
	Tryptophan Pyrrolase
	Tuned Plate
	Turboprop(eller)
	Turbopump
	Turning Point
3P	Three-Pole
	Triple-Pole [also: TP]
T-5'-P	Thymidine 5'-Monophosphate
T$	Tongan Pa'anga [Currency of Tonga]
TPA	Tantalum Producers of America [US]
	Tape Pulse Amplifier
	Technical Publications Association
	Telephone Pioneers of America [US]
	Tennessee Pharmaceutical Association [US]
	Terephthalic Acid
	Tetradecanoylphorbol Acetate
	Tetragonolobus Purpureas Agglutinin
	Tetrapropylammonium
	Texas Pharmaceutical Association [US]
	Timber Producers Association [US]
	Tissue Plasminogen Activator
	Toll Pulse Accepter
	Traffic Pattern Altitude
	Transient Program Area
	Transmission Products Association [US]
	Transpolypentenamer
	Triphenylamine
	Two-Point Approximation
t-PA	Tissue Plasminogen Activator
TPAD	Time Differential Perturbed Angular Distribution
TPAM	Teleprocessing Access Method
TPAO	Turkish Petroleum Corporation
TPAP	Tetrapropylammonium Perruthenate
2PAPMM	Two-Pulse Amplitude and Phase Modulation Modem
TPB	Tetradecylpyridinium Bromide
	1,1,4,4-Tetraphenyl-1,3-Butadiene
TPBVP	Two-Point Boundary Value Problem
TPC	Tactical Pilotage Chart
	Telecommunications Planning Committee
	Thermoplastic Composite
	Thymolphthalein Complexone
	Time Pickoff Control
	Time Projection Chamber
	Tire Performance Criteria

	Totally Pyrolytic Cuvette
	Total Process Control
	Trans-Pacific Cable
	Trifluoroacetylprolyl Chloride
	Triphenyl Carbinol
TPCD	Tetraphenylcyclopentadienone
TPCK	Tosylamino-2-Phenylethyl Chloromethyl Ketone
TPCU	Thermal Preconditioned Unit
TPCV	Turbine Power Control Valve
TPD	Temperature-Programmed Desorption
	Theoretical Physics Department
	Thermoplastic Photoconductor Device
	Third Party Database
	Time Pulse Distributor
	Tons per Day [also: tpd]
TPDB	Third Party Database
TP-DDI	Twisted Pair Distributed Data Interface
3PDT	Triple-Pole, Double-Throw [also: TPDT]
3PDT SW	Triple-Pole, Double-Throw Switch [also: TPDT SW]
TPDU	Transport Protocol Data Unit
TPE	Technical Plastics Extruder
	Teleprocessing Executive
	Tetraphenylethylene
	Thermoplastic Elastomer
	Two Pion Exchange
	Tris-Phosphate EDTA (= Ethylenediaminetetraacetic Acid)
TPE-A	Thermoplastic Polyetheramide Elastomer
TPE-E	Thermoplastic Polyetherester Elastomer
TPE-FKM	Thermoplastic Fluorelastomer
TPEN	Tetrakis(2-Pyridylmethyl)ethylenediamine
TPE-NBR	Thermoplastic Elastomer Based on Nitrile Butadiene Rubber
TPE-NR	Thermoplastic Elastomer Based on Natural Rubber
TPE-O	Thermoplastic Elastomer Based on Polyolefins
TPE-S	Thermoplastic Elastomer Based on Styrene Butadiene Styrene
TPE-U	Thermoplastic Elastomer Based on Polyurethanes
TPF	Taper per Foot
	Terminal Phase Finalization
	Tetraphenylfuran
	Time Prism Filter
	Track Pick Fragments
	Two-Phase Flow
	Two Photon Fluorescence
TPFI	Terminal Pin Fault Insertion
TPG	Telecommunications Program Generator
	Time Pulse Generator
	Tin Plate Gauge
	Topping
	Total Pressure Gage
	Transmission Project Group [of CEGB, UK]
	Triphenylguanidine
	Trypticase Peptone Glucose
TPH	Tons per Hour [also: tph]
2PH	Two-Phase
TPHA	Texas Public Health Association [US]
TPHC	Time to Pulse Height Converter
TPHT	Two-Phase Heat Transfer
TPI	Tape Phase Inverter
	Taper per Inch
	Target Position Indicator
	Teeth per Inch
	Terminal Phase Initiate
	Theoretical Physics Institute [of University of Minnesota, US]
	Thermoplastic Imide
	Thermoplastic Polyimide
	Threads per Inch
	Town Planning Institute [UK]
	Tracks per Inch
	Transmission Performance Index
	Triosephosphate Isomerase
	Tropical Products Institute [UK]
	Truss Plate Institute [US]
	Turns per Inch
TPIC	Town Planning Institute of Canada
TPIS	Tropical Pesticides Information Service [UK]
TPISC	Tantalum Producers International Study Center [Belgium]
TPL	Table Producing Language
	Telephoto Lens
	Terminal Per Line
	Terminal Processing Language
	Test Parts List
	Test Processing Language
	Tritium Processing Laboratory
	Toll Pole Line
	Total Phospholipid
TPLA	The Product Liability Alliance [US]
TPLAB	Tape Label
TPM	Tape Preventive Maintenance
	Telemetry Processor Module
	Timber Products Manufacturers [US]
	Total Preventive Maintenance
	Transmission and Processing Model
	Transport Planning Mobilization
	Triphenyl Methane
TPMA	Thermodynamic Properties of Metals and Alloys
	Timber Products Manufacturers Association [now: TPM, US]
TPMM	Teleprocessing Multiplexer Module
TP/mm	Taper per Millimeter
TPMS	Transaction Processing Management System
TPN	Triphosphopyridine Nucleotide
	Two-Port Network
TPNH	Triphosphopyridine Nucleotide, Reduced Form
TPNS	Teleprocessing Network Simulator
TPO	Telecommunications Program Objective
	Thermoplastic Olefin
	Tryptophan Pyrrolase
TPO/EPDM	Thermoplastic Olefin/Ethylene-Propylene Diene Monomer
TPOP	Time-Phased Order Point
TPP	Telephony Preprocessor
	Test Point Pace
	Tetraphenylporphine
	Tetraphenylpyrrole
	Thiamine Pyrophosphate
	Triphenyl Phosphate
TPPC	Total Package Procurement Concept
TP-PCB	Teleprocessing Program Communication Block

TPPD Technical Program Planning Division
TPQ Two Piston Quenching
TPR Tape Programmed Row
Taper
Teleprinter
Telescopic Photographic Recorder
Terrain Profile Recorder
Thermoplastic Recording
Thermoplastic Resin
Thermoplastic Rubber
Total Peripheral Resistance
Trooper
TPRC Thermophysical Properties Research Center [of Purdue University, US]
T PROC Test Procedure
TPRU Tropical Pesticides Research Unit [UK]
TPS Tape Programming System
Task Parameter Synthesizer
Technical Problem Summary
Technical Publishing Software
Telecommunications Programming System
Telemetry Processing Station
Terminal Polling System
Terminals per Station
Term Preferred Shares
Test Program Set
Text Processing Service
Text Processing System
Thermal Power System
Thermal Protection System
Thermoprocessing System
Throttle Position Sensor
Transaction Processing System
Transactions per Second
Translator Processing System
2,4,6-Triisopropylbenzenesulfonyl
TPSCl 2,4,6-Triisopropylbenzenesulfonyl Chloride
TPSD Tons per Stream Day
TPSI Torque Pressure in Pounds per Square Inch
TPSRS Terminal Primary and Secondary Radar System
3PST Triple-Pole, Single-Throw [also: TPST]
3PST SW Triple-Pole Single-Throw Switch [also: TPST SW]
3P SW Triple-Pole Switch [also: TP SW]
TPT Temporary Power Tap
Tetraisopropyl Titanate
Tetraphenylthiophene
Transmission Path Translator
Triphenyltetrazolium Chloride
TP-T Target Practice with Tracer
TpT Thymidylyl Thymidine
TPTA Trimethylolpropane Trimethacrylate
TPTF Two Phase Flow Test Facility
TPTG Tuned Plate, Tuned Grid
TpTpT Thymidylyl Thymidylyl Thymidine
TPTZ 2,4,6-Tripyridyl-S-Triazine
TPU Tape Preparation Unit
Task Processing Unit
Telecommunications Processing Unit
Thermoplastic Polyurethane
Time Pickoff Unit
TPU/ABS Thermoplastic Polyurethane/Acrylonitrile-Butadiene-Styrene
TPUR Thermoplastic Polyurethane
TPUR-ABS Thermoplastic Polyurethane — Acrylonitrile-Butadiene Styrene
TPUR-PA Thermoplastic Polyurethane — Polyamide
TPUR-PC Thermoplastic Polyurethane — Polycarbonate
TPUR-PVC Thermoplastic Polyurethane — Polyvinyl Chloride
TPWB Three Program Wire Broadcasting
TPV Temperature — Pressure — Volume [also: T-P-V]
Thermophotovoltaic
Thermoplastic Vulcanizate
TPX Polymethylpentene [also: PMP]
TQA Total Quality Assurance
TQC Technical Quality Control
Total Quality Commitment
Total Quality Control
TQCA Textile Quality Control Association [US]
TQE Time Queue Element
TQM Total Quality Management
TR Table Run
Tamper-Resistant
Tannery Run
Tape Recorder
Tape Resident
Technical Reference
Technical Report
Technical Review
Technology Report
Teleradiography
Temperature Rate
Temperature Regulator
Temper Rolling
Test Reactor
Test Request
Thermal Radiation
Thermal Reactor
Thermal Regulation
Thermal Resistance
Thermoregulator
Time-Delay Relay
Time Resolution
Time-Resolved
Top Register
Top Running (Crane)
Torque Receiver
Torsional Rigidity
Total Radiation
Total Reaction
Trace(r)
Track
Tracking Regulator
Traffic Route
Train
Transfer
Transformer
Transformer-Rectifier
Transient Response
Transistor
Transistor Radio
Transition

	Translation
	Translation Register
	Translator
	Transmission Report
	Transmit-Receive
	Transmitter
	Transmitter-Receiver
	Transpose
	Transport
	Transportation Research
	Transposition
	Trustee
	Turkish Reactor
T/R	Tip/Ring
	Transmit/Receive
Tr	Trace [Mathematics]
TRA	Technical Requirement Analysis
	Temporary Reserved Airspace
	Test and Repair Analysis
	Tire and Rim Association [US]
	Transfer
	Tubular Reactor Assembly
TRAA	Towing and Recovery Association of America [US]
TRAACS	Transit Research and Attitude Control Satellite
TRAC	Test Report and Certification (System) [also: TRaC]
	Text Reckoning and Compiling
	Thermally Regenerative Alloy Cell
	Transient Radiation Analysis by Computer
TRACALS	Traffic Control, Approach and Landing System
TRACE	Tactical Readiness and Checkout Equipment
	Tape-Controlled Reckoning and Checkout Equipment
	Tape-Controlled Recording and Checkout Equipment
	Task Reporting and Current Evaluation
	Teleprocessing Recording for Analysis by Customer Engineers
	Time-Shared Routines for Analysis, Classification and Evaluation
	Tolls Recording and Computing Equipment
	Toronto Region Association of Computer Enthusiasts [Canada]
	Tracking and Communications, Extraterrestrial
	Transistor Radio Automatic Circuit Evaluator
	Transportable Automated Control Environment
TRACOM	Tracking Comparison
TRACON	Terminal Radar Approach Control
TRACS	Terminal Radar and Control System
	Test and Repair Analysis/Control System
	Transport and Road Abstracting and Cataloguing System [of TRSL, UK]
TRADA	Timber Research and Development Association [UK]
TRADEX	Target Resolution and Discrimination Experiment
TRADIC	Transistor Digital Computer
	Transistorized Airborne Digital Computer
TRAFFIC	Trade Records Analysis of Flora and Fauna in Commerce (Program) [of WWF, US]
TRAIN	Telerail Automated Information Network [of AAR, US]
TRALA	Truck Renting and Leasing Association [US]
TRAM	Target Recognition Attack Multisensor
TRAMP	Time-Shared Relational Associative Memory Program
TRAMPS	Traffic Measurement and Path Search
TRAN	Transaction
	Transit
	Transmit
TRANDIR	Translation Director
TRANET	Tracking Network
TRANPRO	Transaction Processor
TRANS	Transaction
	Transfer
	Transformer
	Transition
	Translation
	Translator
	Transmitter
	Transparent
	Transportation
	Transposition
TRANSAC	Transistorized Automatic Computer
TRANSALT	Transition Altitude
TRANSCAN	Canary Islands Submarine Cable
TRANSCEIVER	Transmitter/Receiver [also: transceiver]
TRANSDOC	Transportation Documents (Database) [of ECMT, France]
	Transport Documentation (Database) [of OECD, France]
TRANSEC	Transmission Security
TRANSF	Transfer
TRANSIF	Transient State Isoelectric Focusing
TRANSIM	Transportation Simulator
TRANSL	Translation
	Translator
TRANSLEV	Transition Level
TRANSLU	Translucent
TRANSM	Transmission
	Transmit
TRANSMUX	Transmission Multiplexer
TRANSP	Transportation
TRANSPAC	Transpacific Submarine Cables [Hawaii/Japan/Philippines]
TRANSPLAN	Transaction Network Service Planning Model
TRANSPONDER	Transmitter/Responder [also: transponder]
TRANSPT	Transport
TRANSV	Transverse
TRAP	Teledyne Research Assistance Program [US]
	Terminal Radiation Airborne Program
	Tracker Analysis Program
TRAPATT	Trapped Plasma Avalanche Transit Time (Diode)
TRAPR	Tethered Remote Automatic Pipeline Repairer
TRAV	Travel
	Traverse
TRAVIS	Traffic Retrieval Analysis Validation and Information System
TRAW	Tape Read and Write
TRAWL	Tape Read and Write Library
TRAX	Total Reflection-Angle X-Ray Spectrometer
TRB	Transportation Research Board [US]
	Transport Research Board [Canada]
TRC	Tape Record Coordinator

Technical Research Center
Technical Resources Center [US]
Technical Review Committee
Technology Reports Center [of DTI, UK]
Telemetry and Remote Control
Tracking, Radar-Input and Correlation
Trade Relations Council [US]
Traffic Counts and Listings
Transmit Receive Control
Transverse Redundancy Check
Tukuba Research Center [Japan]
Twin-Reflex Camera

TRC-AS Transmit Receive Control — Asynchronous Start/Stop
TRCC T-Carrier Restoration Control Center
TRCDS Time Resolved Circular Dichroism Spectroscopy
TRCF Technical Research Center of Finland
TRCKG Trucking
TRC-SC Transmit Receive Control — Synchronous Character
TRC-SF Transmit Receive Control — Synchronous Framing
TRCTR Tractor
TRD Tape Read
Timed Release Disconnect
TR&D Transport Research and Development
TRDC Tourism Research and Data Center
TRDER Trader
TRDG Trading
TRDTO Tracking Radar Takeoff
TRE Telecommunications Research Establishment [UK]
Tokai Research Establishment [of JAERI, Japan]
Transmit Reference Equivalent
Type Rate Examiner
TREAS Treasurer
TREAT Transient Radiation Effects Automated Tabulation
Transient Reactor Test Facility
TREC Tethered Remote Camera
TREE Transient Radiation Effects on Electronics
Training, Research, Environment and Education
TREES Turbine Rotors Examination and Evaluation System
TREM Tape Reader Emulator Module
TREN Triaminotriethylamine
TREND Tropical Environment Data
TRES Thermally Regenerative Electrochemical System
TRF Temperature Recovery Factor
Thyrotropin-Releasing Factor
Transportation Research Forum [US]
Tripoli Rocketry Association [US]
Tritium Removal Facility
Tuned Radio-Frequency
TRFB Tariff Board
TRFC Traffic
TRFCS Temperature Rate Flight Control System
TR-FIA Time-Resolved Fluorescence Immunoassay
TRFR Transfer
TRG Technical Research Group
Tip/Ring to Ground
Training
TRGB Tail Rotor Gearbox
TRH Thyrotropin-Releasing Hormone
TRI Technical Research Institute
Textile Research Institute [US]
The Refractories Institute [US]
Time Reversal Invariance
Tin Research Institute [US]
Trigonal
Triode
Tri Trichloroethylene
TRIA Temperature Removable Instrument Assembly
TRIAC Triode Alternating-Current (Semiconductor) [also: Triac or triac]
TRIAL Technique for Retrieving Information from Abstracts of Literature
TRIAX Triaxial (Cable) [also: triax]
TRIB Tire Retread Information Bureau [US]
Transfer Rate of Information Bits
TRIBAS Tribasic
TRICE Transistorized Real-Time Incremental Computer Equipment
TRICINE Trismethylglycine
TRICL Triclinic
TRID Track Identity
TRIDOP Tri-Doppler
TRIF Technical Requirements Industry Forum
TRIG Trigonal
TRIGA Training Reactor, Isotope Production, General Atomics [US]
TRIGON Trigonometric [also: TRIG]
Trigonometry [also: TRIG]
TRIM Tailored Retrieval and Information Management
Test, Rework and Inspection Management
Test Rules for Inventory Management
Transport of Ions in Matter
Trimetric
Trimmer
TRIMIS Tri-Service Medical Information Systems [of USDOD]
TRIO Telecommunications Research Institute of Ontario [Canada]
TRIP The Road Information Program [Canada]
Thunderstorm Research International Program
Transformation-Induced Plasticity
TRIPHOS Tris(diphenylphosphinoethyl)phenylphosphine
TRIPS Travel Information Processing System
TRIS Transport Research Information Services [of USDOT]
Trisamine
Tris(hydroxymethyl)aminomethane
TRISCADP Tris(hydroxymethyl)aminomethane Salt of Adenosine Diphosphate
TRISCATP Tris(hydroxymethyl)aminomethane Salt of Adenosine Triphosphate
TRISNET Transport Research Information Services Network [of USDOT]
TRITC Tetramethylrhodamine Isothiocyanate
TRIUL Tri-University Libraries (of British Columbia) [Canada]
TRIUMF Tri-University Meson Facility [Canada]
TRK Track
Trunk
TRL Thermodynamics Research Laboratory
Transistor-Resistor Logic

	Transuranium Research Laboratory [of USAEC]
	Trunk Register Link
	Turkish Lira [Cuurrency of Turkey]
TRLR	Trailer
TRM	Tamper-Resistant Module
	Terminal Response Monitor
	Test Request Message
	Thermoremanent Magnetization
	Transmit-Receive Module
TrM	Track Magnetic
TRMA	Time Random Multiple Access
TRMDT	Transmission Date
TRMG	Tread Rubber Manufacturers Group [US]
TRMI	Tubular Rivet and Machine Institute [US]
TRML	Terminal
TRMP	Target Radiation Measurement Program
TRMS	TDMA (= Time-Division Multiple Access) Reference and Monitor Station
	True Root Mean Square
TRN	Technical Research Note
	Television Relay Network
	Transfer
t-RNA	Transfer Ribonucleic Acid [also: tRNA, or T-RNA]
TRNBKL	Turnbuckle
TRNG	Training
TRNST	Transit
TRNT	Transient
TROC	Trichloroethoxycarbonyl
TROCA	Tangible Reinforcement Operant Conditioning Audiometry
TROF	Trace Off
TROLL	Technion Robotics Laboratory Language
TROM	Teletext Read-Only Memory
TROMEX	Tropical Meteorological Experiment
TRON	Trace On
TROO	Transponder On-Off
TROP	Time-Phased Order Point
	Tropical
	Tropics
	Tropopause
TROPAG	Tropical Agriculture (Database) [of RTI, Netherlands]
TROPO	Tropospheric Scatter Communications
TROS	Tape Resident Operating System
	Transformer Read-Only Storage
TROV	Tethered Remotely Operated Vehicle
tr oz	troy ounces
TRP	Television Remote Pickup
	Thermal Ribbon Printer
	Total Radiation Pyrometer
	Trap
	Tuition Reimbursement Plan
Trp	Tryptophan
	Tryptophyl
TRPF	Transmit/Receive Parity Failure
TRPGDA	Tripropylene Glycol Diacrylate
TRPS	Transpose
TR PT	Transition Point
TRQ	Task Ready Queue
TRR	Tape Read Register
	Target Ranging Radar
	Teaching and Research Reactor
	Technical Research Report
	Thailand Research Reactor
	Topical Report Request
	Topical Report Review
TRRB	Trade Relations Research Bureau [UK]
TRRF	The Refrigeration Research Foundation [US]
TRRL	Transport and Road Research Laboratory [UK]
	Tree-Ring Research Laboratory [of University of Arizona, US]
TRRN	Terrain
TRRR	Time Resolved Resonant Raman Spectroscopy [also: TR^3]
TRRS	Total Reflection Raman Scattering
TRS	Telephone Referral Service
	Telephone Repeater Station
	Test Response Spectrum
	Time Reference System
	Time Resolved Spectroscopy
	Toll Room Switch
	Total Reducing Sugar
	Tough Rubber Sheath
	Transmission Regulated Spark
	Transposition
	Transverse Rupture Strength
	Tree-Ring Society [US]
	Trustees
TRSA	Terminal Radar Service Area
TRSB	Telecommunications Regulatory Service Branch
	Time Reference Scanning Beam
	Transcriber
TRSC	Thailand Remote Sensing Center
TRSL	Translate
TRSN	Transaction
TRSP	Transport
TRSSM	Tactical Range Surface-to-Surface Missile
TRSSSV	Tubing-Retrievable Subsurface Safety Valve
TRST	Transit
TRT	Technology, Research and Telecommunications
	Traffic Route Testing
	Turnaround Time
TrT	Track True
TRTA	Traders Road Transport Association [UK]
TRTL	Transistor-Resistor-Transistor Logic
TRU	Total Recycle Unit
	Transmit-Receive Unit
	Transportable Radio Unit
	Transuranic (Waste)
TRUMP	Teller Register Unit Monitoring Program
	Total Revision and Upgrading of Maintenance Procedures
TRUPP	Transuranium Processing Plant
TRV	Transient Recovery Voltage
TRX	Transaction
	Two-Region Physics Critical Experiment
TRXRF	Time-Resolved X-Ray Fluorescence
	Total Reflectance X-Ray Fluorescence
TS	Tape Status
	Taper Shank
	Tape System
	Target Strength

	Target System
	Technical Service
	Technical Specification
	Telegraph Service
	Temperature Sensitivity
	Temporary Storage
	Tensile Strength
	Tensile Stress
	Teratology Society [US]
	Test Set
	Test Switch
	Test Specification
	Theoretical Science
	Theoretical Statistics
	Thermal Shield
	Thermal Shock
	Thermal Stress
	Thermal System
	Thermal Switch
	Thermoset
	Thermostatic(s)
	Thermosyphon
	Thunderstorm
	Time Sharing
	Time Signal
	Time Slot
	Time Switch
	Toll Switching
	Tool Steel
	Torch Soldering
	Total Solids
	Transfer Station
	Transformer Station
	Transient Stability
	Transition Set
	Translation Stage
	Transmission Service
	Transmission System
	Transmitting Station
	Transportation Science
	Transportation System
	Transverse Section
	Triple Scattering
	Triple Space
	Turboshaft Engine
	Typesetting
T&S	Turn and Slip Indicator
T-S	Temperature-Salinity (Diagram) [also: TS]
Ts	Tosyl
TSA	Technology Student Association [US]
	Telecommunications Society of Australia
	Temperature-Swing Adsorption
	Time Series Analysis
	Time Slot Access
	Toluenesulfonamide
	Toluenesulfonic Acid
	Total Scan Area
	Transportation Standardization Agency [of USDOT]
	Transverse Spherical Aberration
	Tube Support Assemblies
	Turnstile Antenna
TSAC	Time Slot Assignment Circuit
	Title, Subtitle and Caption
TSAO	Transportation Safety Association of Ontario [Canada]
TSAM	Time Series Analysis and Modelling
TSAP	Transport Service Access Point
TSAU	Time Slot Access Unit
TSAZ	Target Seeker-Azimuth
TSB	Taiwan Supply Bureau
	Technical Support Building
	Terminal Status Block
	Trade Show Bureau [US]
	Twin Sideband
TSC	Tape System Calibrator
	Technical and Scientific Center
	Technical Service Contractor
	Technical Service Council
	Technical Subcommittee
	Technical Support Center
	Test Set Connection
	Test Shipping Cask
	Test System Controller
	Thermally Stimulated Current
	Thermal Stress Cracking
	Thiosemicarbazide
	Three-State Control
	Time-Sharing Control
	Totally Self-Checking
	Trans-Syn-Cis (Geometry) [Chemistry]
	Transit Switching Center
	Transmitter Start Code
	Transportation Safety Committee [of SAE, US]
	Transportation Systems Center
TSCA	Timing Single Channel Analyzer
	Toxic Substances Control Act [US]
TSCAP	Thermally Stimulated Capacitance
TSCC	Telemetry Standards Coordination Committee
TSCF	Task Schedule Change Form
	Textronix Standard Codes and Formats
TSCLT	Transportable Satellite Communications Link Terminal
TSCT	Time-Sharing Control Task
TSD	Technical Services Division
	Thermally-Stimulated Depolarization
	Thermally-Stimulated Desorption
	Touch Sensitive Digitizer
	Traffic Situation Display
	Transportation System Design
	Two-Stage Demagnetization
TSDC	Thermally Stimulated Depolarization Current
TSDD	Temperature-Salinity-Density-Depth (Relationship)
TSDF	Target System Data File
TSDM	Time-Shared Data Management (System)
TSDOS	Time-Shared Disk Operating System
TSDU	Target System Data Update
	Transport Service Data Unit
TSE	Tactical Support Equipment
	Terminal Source Editor
	Test of Spoken English
	Time Slice End

Tokyo Stock Exchange [Japan]
Toronto Stock Exchange [Canada]
Twist Setting Efficiency

TSEE Thermally-Stimulated Emission of Exoelectrons
TSeF Tetraselenafulvalene
TSEI Transportation Safety Equipment Institute [US]
Truck Safety Equipment Institute [now: Transportation Safety Equipment Institute, US]
TSEQ Time Sequence
TSeT Tetraselenatetracene
TSF Telegraphie Sans Fil [= Wireless Telegraphy Service, France]
Ten Statement FORTRAN
Tetraselenafulvalene
Thin Solid Film
Through Supergroup Filter
Tons per Square Foot
Tower Shielding Facility
TSFC Thrust-Specific Fuel Consumption
TSFO Transportation Support Field Office [US]
TSF/O Teklogix Screen Formatter/Optimizer
TSFP Transportation Safety and Field Program [Canada]
TSFS Trunk Servicing Forecasting System
TSG Time Signal Generator
Time Slot Generator
Transversely Adjusted Gap
Tri-Service Group [of NATO]
TSGA Tanzania Sisal Growers Association
TSGCEE Tri-Service Group on Communications and Electronic Equipment [of NATO]
TSH Tolylene Sulfohydrazide
Toluenesulfonyl Hydrazide
Thyroid Stimulating Hormone
TSh Tanzanian Shilling [also: Tsh; Currency]
TSHS Tennessee State Horticultural Society [US]
TSHWR Thundershower
TSI Task Status Index
Terminal Specific Interface
Test Structure Input
Threshold Signal-to-Interference Ratio
Time Slot Interchanger
Tons per Square Inch
Transmitting Station Identification
Transmitting Subscriber's Identification
Transportation Safety Institute [US]
TSIC Thermally Stimulated Ionic Currents
TSID Track Section Identification
TSIM 1-(Trimethylsilyl)imidazole
TSIMS Telemetry Simulation Submodule
TSIOA Temporary Storage Input/Output Area
TSIU Telephone System Interface Unit
TSK Task
TSL Test Source Library
Total Service Life
Two-Stage Liquefaction
TSLS Two-Stage Least Squares
TSM Tandem Scanning Microscope
Telephony Signaling Module
Terminal Server Manager
Terminal Support Module
Time-Shared Monitor

TSMDA Test-Section Meltdown Accident
TSMT Transmit [also: TSMIT]
TSMTR Transmitter
TSN Task Sequence Number
Tryptone Sulfite Neomycine
TSO Technical Service Order [US]
Technical Standard Order
Technical Standing Order
Telecommunications Service Order
Telephone Service Observations
Time-Sharing Option
TSODB Time Series Oriented Database
TSOP Thin, Small Outline Package
TSORT Transmission System Optimum Relief Tool
TSOS Time-Sharing Operating System
TSO/VTAM Time-Sharing Option/Virtual Telecommunications Access Method
TSP Thermostable Polymer
Titanium Sublimation Pump
Total Suspended Particulates
Total Systems Performance
Traffic Service Position
Triple Superphosphate
Tribasic Sodium Phosphate
Trisodium Phosphate
tsp teaspoonful
TSPA Triethylenethiophosphoramide
TSPC Thermally Stimulated Polarization Current
Tropical Stored Products Center [of Ministry for Overseas Development, UK]
TSPC/DC Thermally-Stimulated Polarization (Current)/ Depolarization Current
tspn teaspoonful
TSPP Tetrasodium Pyrophosphate
TSPS Time Sharing Programming System
Traffic Service Position System
TSPSCAP Traffic Service Position System Capacity Program
TSQ Time and Superquick
TSQLS Thundersqualls
TSR Test Schedule Request
Technical Summary Report
Telecommunications Service Request
Temporary Storage Register
Terminate and Stay Resident
Test Summary Report
Thermal Shock Resistance
Thermal Shock Rig
Thermochemical Sulfate Reduction
Thermosetting Resin
Thyroid Secretion Rate
Total Stress Range
Tower Shielding Reactor [of ORNL, US]
Tube-Shift Radiography
Tunnel Stress Relaxation
TSRC Tubular and Split Rivet Council [now: TRMI, US]
TSRO Topological Short-Range Order
TSS Tactical Strike System
Tangential Signal Sensitivity
Technical Staff Surveillance
Telecommunication Switching System
Telephone Support System

	Teletype Switching Subsystem
	Terminal Security System
	Terminal Solid Solution
	Terminal Support Subsystem
	Time-Sharing System
	Toll Switching System
	Total Suspended Solids
	Transmission Surveillance System
	Trunk Servicing System
TSSA	Test Scorer and Statistical Analyzer
	Thunderstorm with Sandstorm
TSSC	Technical and Scientific Societies Council [US]
TSS-C	Transmission Surveillance System — Cable
TSSMCP	Time-Sharing System Message and Control Program
TSSST	Time-Space-Space-Space-Time [also: TS^3T]
TST	Telecommunications Services Tax [Canada]
	Test
	Time-Space Time
	Transaction Step Task
	Transition-State Theory
	Trans-Syn-Trans (Geometry) [Chemistry]
TSTA	Transmission, Signaling and Test Access
	Tritium Systems Test Assembly [of LANL, US]
TSTE	Texas Society of Telephone Engineers [US]
TSTM	Thunderstorm
TSTR	Transition Strain Rate
TSTPAC	Transmission and Signalling Test Plan and Analysis Concept
TSU	Tandem Signal Unit
	Tape Search Unit
	Technical Service Unit
	Test Signal Unit
	Time Standard Unit
	Texas Southern University [US]
	Tolstoi State University [USSR]
	Trunk Switching Unit
TSV	Through Sight Video
TSVD	Thermally-Stimulated Voltage Decay
TSW	Task Status Word
	Tele-Software
	Test Switch
	Time Switch
	The Searchers Workbench
	Tube Socket Weld
TSWB	Tinned Steel Wire Braid
TSX	Time-Sharing Executive
TT	Technical Translation
	Technology Transfer
	Telegraphic Transfer
	Teletype(writer)
	Teletypewriter Terminal
	Temporarily Transferred
	Terminal Timing
	Test Temperature
	Textile Technologist
	Textile Technology
	Thermal Transformation
	Thermionic Tube
	Timing and Telemetry
	Total Time
	Tracking and Telemetry
	Tracking Telescope
	Traffic Tester
	Transaction Terminal
	Transducer Technology
	Transfer Tube
	Transformation-Toughened
	Transformation Twin
	Transportation Technologist
	Transportation Technology
	Trans-Trans (Configuration) [Chemistry]
	Triple Thermoplastic (Wire)
	Turbine Trip
	Tuberculin-Tested
T&T	Turbine Trip and Throttle (Valve)
T/T	Telegraphic Transfer
TTA	Thenoyltrifluoroacetone
	Time-Temperature Austenitization
	Traffic Trunk Administration
	Transformation-Toughened Alumina
TTAC	Telemetry Tracking and Command
T/TAL	Tandem Computers Transaction Application Language
TTB	Toll Testboard
	Troop Transport Boat
	Trunk Test Buffer
TTBT	Threshold Test Ban Treaty [US-USSR]
TTBWR	Twisted Tape Boiling Water Reactor
TTC	Tape-to-Card
	Technical Transfer Council [Australia]
	Technology Transfer Committee
	Teletypewriter Center
	Terminating Toll Center
	Test Transfer Cask
	Tetrazolium Chloride
	Texas Technical College [US]
	Thermal Transfer Equipment
	Thermal Transient Equipment
	Thiatricarboxyanine
	Toronto Transit Commission [Canada]
	Tracking, Telemetry and Command
	Trans-Trans-Cis (Geometry) [Chemistry]
	2,3,5-Triphenyltetrazolium Chloride
	Tritium Technology Conference
TT&C	Tracking, Telemetry and Command
	Tracking, Telemetry and Control
TTCN	Tree Tabular Combined Notation
TTCM	Tracking, Telemetry, Command and Monitoring Station
TTCP	The Technical Cooperation Program [Australia/Canada/UK/US]
TTD	Temporary Text Delay
	Tetraethylthiuram Disulfide
TT$	Trinidad and Tobagoan Dollar [also: TTD; Currency]
TTDF	Tariff and Trade Data File [of GATT]
TTDL	Terminal Transparent Delay Language
TTDU	Technology Transfer Diffusion Unit
TTE	Telephone Terminal Equipment
	Time to Event
TTEC	Thai Technical and Economic Cooperation Office
TTeF	Tetratellurafulvalene
TTEGDA	Tetraethylene Glycol Diacrylate

TTF Taiwan Textile Federation
Tetrathiofulvalene
2-Thenoyltrifluoroacetone
Timber Trade Federation [UK]
Torsional-Mode Tuning Fork
Transient Time Flowmeter
Transmission Test Facility
TTFA Thallic Trifluoroacetate
TTFC Textile Technical Federation of Canada
TTG Technical Translation Group
Time to Go
TTHA Triethylenetetraminehexaacetic Acid
TTHDCM Thin and Thick Film High-Density Ceramic Module
TTHE Thermal Transient Histogram Equivalent
TTI Teletype Test Instruction
The Technological Institute [US]
Time-Temperature Indicator
T2000I Transport 2000 International [UK]
TTK Tie Trunk
TTL Taxi-Track Light
Through-the-Lens
Transistor-to-Transistor Link
Transistor-Transistor Logic [also: T^2L]
TTL/LS Transistor-Transistor Logic/Large Scale
TTLS Transistor-Transistor Logic Schottky
TTMA Truck Trailer Manufacturers Association [US]
TTMAPP Tetrakis(4-N-Trimethylaminophenyl)porphine
TTMP Transit Time Magnetic Pumping
TTMS Telephoto Transmission Measuring Set
TTN Thallium Nitrate Trihydrate
TT/N Test Tone to Noise (Ratio)
TTO Traffic Trunk Order
Transmitter Turn-Off
TTP Tape-to-Print
Test Transfer Port
Tetratolylpyrrole
Time-Temperature Precipitation
Thermal Transfer Printer
Transient Thermal Processing
Trunk Test Panel
Thymidine Triphosphate
TTPB Trunk Test Panel Buffer
TTR Target Tracking Radar
Thermal Test Reactor
Time-Transfer Receiver
Toshiba Training Reactor [Japan]
Transmission Test Rack
Trunk Test Rack
TTRAN Tape Transfer
TTRANCWB Tape Transfer Ceramic Wiring Board
TTRC Thrust Travel Reduction Curve
TTRM Transition Thermoremanent Magnetization
TTS TDMA (= Time-Division Multiple Access) Terminal Simulation
Tearing Topography Surface
Technology Transfer Society [US; also: T2S or T^2S]
Telecommunications Terminal System
Teletypesetter
Temporary Threshold Shift
Test and Training Satellite
Through-Transmission Ultrasonics
Time to Station
Transaction Terminal System
Transit Time Spread
Transmission Test Set
Transportable Transformer Substation
Transputer Technology Solutions [UK]
Tubular-Tin-Source (Process)
TTSA Trinidad and Tobago Scientific Association
TTSMSP Time-Temperature-Stress-Moisture Superposition Principle
TTSP Time-Temperature Superposition Principle
TTSPN Two-Terminal Series Parallel Networks
TTSSP Time-Temperature-Stress Superposition Principle
TTT Template Tracing Technique
Tetrathiatetracene
Time-Temperature-Transformation (Diagram)
Trans-Trans-Trans (Geometry) [Chemistry]
TTU Terminal Time Unit
Texas Technical University [US]
Through-Transmission Ultrasonics
TTV Total Thickness Variation
TTW Teletype(writer)
TTX Tetrodotoxin
TTY Teletype(writer)
TTYBS Teletype(writer), Backspace
TTYC Teletype(writer) Controller
TTYPP Teletype Point-to-Point
TTYS Teletypesetter
TTY STN Teletype Station
TTZ Transformation-Toughened Zirconia
TU Takeup (Mechanism)
Tamagawa University [Japan]
Tape Unit
Technical University
Temple University [US]
Thermal Unit
Thiourea
Timing Unit
Tohoku University [Japan]
Tokai University [Japan]
Tokushima University [Japan]
Tokyo University [Japan]
Tongji University [PR China]
Top Up
Tottori University [Japan]
Toyama University [Japan]
Trade Union
Traffic Unit
Transfer Unit
Transmission Unit
Trent University [Canada]
Triangle Universities [NCSU, NCU and Duke University, US]
Tsinghua University [PR China]
Tuesday
Tulsa University [US]
Turbopump Unit
TUA Technical University, Aachen [Germany]
Telecommunications Users Association [UK]
TOW (= Tube-Launched, Optically-Tracked, Wire-Guided) under Armor

	Tractor Users Association [UK]
TUAC	Trade Union Advisory Committee [of OECD, France]
TUB	Technical University, Berlin [Germany]
	Tubing
TUC	Technical University of Clausthal [Germany]
	Technical University of Cracow [Poland]
	Trade Union Congress
TUCA	Transient Undercooling Accident
TUCC	Triangle Universities Computing Center [of NCSU, NCU and Duke University, US]
TUCM	Trade Union Congress of Malawi
TUCOP	Transient Undercooled Overpower Accident
TUD	Technical University, Dresden [Germany]
	Technical University of Denmark
	Technology Utilization Division [of NASA]
	Tunesian Dinar [Currency]
TUE	Thiourea Extraction
	Transuranium Elements
TUES	Tuesday
TUF	Transmitter Underflow
TUFCDF	Thorium-Uranium Fuel Cycle Development Facility [of BNL, US]
TUG	Tape Unit Group
	Towed Universal Glider
	TRANSAC (= Transistorized Automatic Computer) Users Group [US]
TUHI	Technical University for Heavy Industry [Hungary]
TUI	Technical University of Istanbal [Turkey]
TUIAFPW	Trade Unions International of Agricultural, Forestry and Plantation Workers [Italy]
TUIMWE	Trade Unions International of Miners and Workers in Energy
TUK	Technical University, Karlsruhe [Germany]
TUKMS	Technical University, Karl-Marx-Stadt [Germany]
TUL	Technical University of Lulea [Sweden]
	Turkish Lira [Currency]
TUM	Tuning Unit Member
TUMS	Temporary Usage Measured Service
TUN	Tuning
TUNIS	Toronto University (Computer) System [Canada]
TUNS	Technical University of Nova Scotia [Canada]
TUP	Technology Utilization Program
	Telephony User Part
TUPS	Technical User Performance Specifications [of United States Telephone Association]
TUR	Test Uncertainty Ratio
	Traffic Usage Recorder
	Turret
TURB	Turbine [also: TUR]
TURBC	Turbulence
TURBO GEN	Turbo-Generator
TURBT	Turbulent
TURDOK	Turkish Scientific and Technical Documentation Center
TURK	Turkey
	Turkish
TURP	Turpentine
TURPS	Terrestrial Unattended Reactor Power System
TUT	Tampere University of Technology [Finland]
	Transistor Under Test
TUV	Technischer Uberwachungsverein [= Technical Inspection Association, Germany]
TV	Tape Velocity
	Track Velocity
	Television
	Tenth Value
	Terminal Velocity
	Test Vector
	Test Vehicle
	Thermal Vacuum
	Thermal Vibration
	Thermal Voltage
	Thermionic Valve
	Threshold Value
	Throttle Valve
	Thrust Vector
	Transfer Vector
	Traverse
TVA	Tennessee Valley Authority [US]
	Technical Valuation Society [US]
	Terminal Volume Addendum
	Thrust Vector Alignment
T-VASI	Tee Visual Approach Slope Indicator
TvB	Television Bureau of Advertising [US]
TVC	Tag Vector Control
	Television Camera
	Thrust Vector Control
TVCS	Thrust Vector Control System
TVDA	Terminal Volume Discount Agreement
TVDC	Test Volts, Direct Current
TVDP	Terminal Vector Display Unit
TVDR	Tag Vector Display Register
TVE	Total Vertical Error
TVEL	Tape Velocity
	Track Velocity
TVF	Tape Velocity Fluctuation
TVG	Time-Varied Gain
	Triggered Vacuum Gap
TVI	Television Interference
	Tomasetti Volatile Indicator
TVIST	Television Information Storage Tube
TVL	Tenth-Value Layer
TVM	Tachometer Voltmeter
	Track Via Missile
	Transistor Voltmeter
TVN	Total Volatile Nitrogen
TVOC	Television Operations Center
TVOR	Terminal VHF Omnirange
TVP	Temperature — Volume — Pressure [also: T-V-P]
	Textured Vegetable Protein
	True Vapor Pressure
TVPPA	Tennessee Valley Public Power Association [US]
TVR	Tag Vector Response
	Television Recording
	Tennessee Valley Region [US]
TVRO	Television Receive-Only
TVS	Thermostatic Vacuum Switch
	Transient Voltage Suppressor
TV-SAT	Television Satellite [Germany]
TVSS	Transient Voltage Surge Suppressor
TV SYS	Television System

TVT Television Terminal
Television Typewriter
Tenth-Value Thickness
TVW Tag Vector Word
TW Tail Wave
Tape Word
Terawatt
Text Word
Thermal Wave
Thermal Wire
Thermit Welding
Torsional Wave
Transit Working
Translational Wave
Transverse Wave
Travelling Wave
Truncated Wave
Twist
Two-Way
Typewriter
Tw Twaddell (Scale) [also: TW]
TWA Task Work Area
The Waferboard Association [Canada]
Time-Weighted Average
Transaction Work Area
Trans-World Airlines [US]
Travelling Wave Accelerator
Travelling Wave Amplifier
Two-Way Alternate
TWAEC Time-Weighted Average Exposure Criteria
TWAIT Terminal Wait
TWARO Textile Workers Asian Regional Organization
TWAS Third World Academy of Sciences [Italy]
TWB Typewriter Buffer
TWC Trinity Western College [Canada]
(American) Truncated Whitworth Coarse Thread
TW-C Truncated Wave with Compressive Dwell
TWCRT Travelling-Wave Cathode-Ray Tube
TWD (New) Taiwanese Dollar [Currency]
Towards
2WD Two Wheel Drive [also: TWD]
TWDD Two-Way/Delay Dial
TWEB Transcribed Weather Broadcast [US]
TWERLE Tropical Wind Energy Conservation and Reference Level Experiment
TWF Third World Forum [Egypt]
Third World Foundation [UK]
(American) Truncated Whitworth Fine Thread
TWG Telemetry Working Group
TWI Thermal Wave Imaging
The Welding Institute [US]
TWID Two-Way/Immediate Dial
TWINAX Twinaxial (Cable) [also: twinax]
TWK Typewriter Keyboard
TWL Total Weight Loss
TWM Travelling-Wave Maser
TWMBK Travelling-Wave Multiple-Beam Klystron
TWMR Tungsten Water-Moderated Reactor
TWP Technical Wordprocessing System
Township
Twisted Wire Pair
TWPS Technical Wordprocessing System
TWR Tape Write
Tape Write Register
Tower
Trans-World Radio
TWRG Towering
TWRI Texas Water Resources Institute [US]
TWS Thermal Weapon Sight
The Wildlife Society [US]
Track While Scan
(American) Truncated Whitworth Special Thread
Two-Way Simultaneous
TWSB Twin Sideband
TWST Torus Water Storage Tank
3W SW Three-Way Switch [also: TW SW]
TWT Thin-Wire Thermometer
Traffic Work Table
Travelling-Wave Tube
TW-T Truncated Wave with Tensile Dwell
TWTA Travelling-Wave Tube Amplifier
TWTC Taipei World Trade Center [Taiwan]
TW-TC Truncated Wave with Tensile and Compressive Dwell
TWU Telecommunication Workers Union
TWUA Textile Workers Union of America [US]
TWWS Two-Way/Wink Start
TWX Teletypewriter Exchange Service [Canada and US]
Time Wire Transmission
TWY Taxiway
TWYL Taxiway Link
TX Task Extension
Telephone Exchange
Teleprinter Exchange (Service) [also: TELEX, Telex, telex, TEX, or TLX]
Temperature — Concentration (Diagram) [also: T-X]
Texas [US]
Text
Time to Equipment Reset
Torque Transmitter
Transmit (Mode)
Transmitter
T-X Travel-Time (Graph)
TXA Task Extension Area
TXC Telephone Exchange — Crossbar
TXD Telephone Exchange — Digital
Transmitted Data
TXE Telephone Exchange — Electronic
TXRX Transmitter-Receiver
TXS Telephone Exchange — Strowger
TXT Text
TY Territory
Tropical Year
3Y Three Times Yield Strength
TYAA Textured Yarn Association of America [US]
TYDAC Typical Digital Automatic Computer
TYMV Turnip Yellow Mosaic Virus
TYP Typewriter
Typical
Typographical
Typography

TYPH Typhoon
TYPSG Typesetting
TYPOUT Typewriter Output Routine
Tyr Tyrosine
Thyrosyl
t/yr tons per year
TZ Test Zone
Transmitter Zone
TZD True Zenith Distance
TZM Titanium-Zirconium-Molybdenum (Alloy)
TZP Tetragonal Zirconia Polycrystal
Tetragonal Zirconia Precipitate
TZS Tanzanian Shilling [Currency of Tanzania]

U

U	Internal Energy [Symbol]
	Intrinsic Energy [Symbol]
	Unclassified
	Union
	Unit
	Universal
	University
	Unknown
	Unlicensed
	Unpredictable
	Unsharpness
	Upper
	Uranium
	Uranyl
	Uracile
	User
	Utility
u	unclassified
	unidirectional
	unit
	update
	upper
U_1	Uranium I (= Uranium 238)
U_2	Uranium II (= Uranium 234)
UA	Ultra-Audible
	Underwater Association [UK]
	Universidad de las Americas [= University of the Americas, Mexico]
	University of Aachen [Germany]
	University of Alabama [US]
	University of Alaska [US]
	University of Alberta [Canada]
	University of Arizona [US]
	University of Aston [UK]
	University of Auckland [New Zealand]
	Unnumbered Acknowledge
	Until Advised
	Upper Atmosphere
	User Agent
	User Area
UAA	United Arab Airlines [Egypt]
	University Aviation Association [US]
	University of Alaska, Anchorage [US]
	Upper Advisory Area
	Utility Arborist Association [US]
UAAEE	United Arab Atomic Energy Establishment
UAAS	Ukranian Academy of Arts and Sciences [USSR and US]
UAB	Unemployment Assistance Board
	University of Alabama, Birmingham [US]
	University of Aston, Birmingham [UK]
	Until Advised By
UABS	Union of American Biological Societies [US]
UAC	Unified Agricultural Cooperative [Czechoslovakia]
	Uninterrupted Automatic Control
	Universidad Autonoma de Chihuahua [= Free University of Chihuahua, Mexico]
	Upper Airspace Center
UACA	United American Contractors Association [US]
UACC	Upper Area Control Center
UACN	Unified Automated Communications Network
	University of Alaska Computer Network [US]
UAD	Upper Advisory Route
UADPS	Uniform Automatic Data Processing System
UADS	User Attribute Data Set
UADW	Universal Alliance of Diamond Workers [Belgium]
UAE	United Arab Emirates
UAEM	Union of Associations of European Meatmeal Producers [Netherlands]
UAF	University of Alaska, Fairbanks [US]
UAFMMEEC	Union of Associations of Fish Meal Manufacturers in the European Economic Community [Germany]
UAGI	University of Alaska Geophysical Institute [US]
UAH	University of Alabama, Huntsville [US]
UAIDE	Users of Automatic Information Display Equipment
UAIS	Universal Aircraft Information System
UAL	United Airlines [US]
UAM	Underwater-to-Air Missile
	Universite Aix-Marseille [= University of Aix-Marseille, France]
UAMPT	Union Africaine et Malgache de Poste et de Telecommunications [= African and Malagasy Postal and Telecommunications Union]
UAMR	United Association Manufacturers Representatives [US]
UAN	Unified Automatic Network
UANA	Union of African News Agencies
UANL	Universidad Autonoma de Nuevo Leon [= Free University of Nuevo Leon, Mexico]
UAO	Unexplained Aerial Object
UAOS	Ulster Agricultural Organization Society [Northern Ireland]
UAP	Unidentified Atmospheric Phenomena
	Universal Availability of Publications
	University of Alberta Press [Canada]
	Upper Air Project
	User Area Profile
UAPT	United Association for the Protection of Trade [UK]
UAR	Unit Address Register
	United Arab Republic
	Upper Air Route
	Upper Atmosphere Research
UARAEE	United Arab Republic Atomic Energy Establishment
UARC	Upper Atmosphere Research Corporation
UARI	University of Alabama Research Institute [US]
UARS	Upper Atmosphere Research Satellite
UART	Universal Asynchronous Receiver Transmitter
UARTO	United Arab Republic Telecommunication Organization
UAS	Union of African States
	Unit Approval System [of FAA, US]
	USSR Academy of Sciences
UASAL	Utah Academy of Sciences, Arts and Letters [US]
UAT	University of Alabama, Tuscaloosa [US]

	University of Arizona, Tucson [US]
UAUM	Underwater-to-Air-to-Underwater Missile
UAV	Unmanned Air Vehicle
UAW	United Automobile Workers
UAX	Unit Automatic Exchange
UB	Unbalance [also: U/B]
	Union of Burma [Burma]
	Universidad de Barcelona [= University of Barcelona, Spain]
	University of Bath [UK]
	University of Belgrade [Yugoslavia]
	University of Birmingham [UK]
	University of Bradford [UK]
	University of Budapest [Hungary]
	University of Buffalo [US]
	Upper Bound
U/B	Unbalance [also: UB]
UBA	Umweltbundesamt [= Department of the Environment, Germany]
	Unblocking Acknowledgement Signal
UBAEC	Union of Burma Atomic Energy Center [Burma]
UBARI	Union of Burma Applied Research Institute [Burma]
UBC	Unbalanced Cable
	Uniform Building Code [US]
	Universal Bibliographical Control
	Universal Block Channel
	Universal Book Code
	Universal Buffer Controller
	University of British Columbia [Canada]
	Used Beverage Can
UBCW	United Brick and Clay Workers of America
UBD	Utility Binary Dump
UBEC	Union of Banana-Exporting Countries [Panama]
UBF	Underground Baggage Facility
UBFF	Urey-Bradley Force Field
UBHR	User Block Handling Routine
UBIS	Ultrasonic Boiler Inspection System
UBJ	Union Ball Joint
UBK	Unblock
UBL	Unbalanced Load
	Unblocking (Signal)
UBM	Unit Bill of Materials [also: UB/M]
UBPVLS	Uniform Boiler and Pressure Vessel Laws Society [US]
UBS	Unit Backspace (Character)
UBT	Universal Book Tester
UC	Ultracentrifuge
	Undercooling
	Undercorrected
	Undercurrent
	Unichannel
	Unit Call
	Unit Cooler
	Universal Cooperative [US]
	Universal Crown (Rolling Mill)
	University of California [US]
	University of Cambridge [UK]
	University of Canterbury [UK and New Zealand]
	University of Chicago [US]
	University of Cincinnati [US]
	University of Cologne [Germany]
	University of Colorado [US]
	University of Connecticut [US]
	Up-Converter [also: U-C or U/C]
	Upper Case
U/C	Under Construction
U-C	Up-Converter [also: UC or U/C]
UCA	Uncommitted Component Array
UCADIA	Union Centroamericana de Asociaciones de Ingenieros y Arquitectos [= Central American Union of Engineers and Architects, Costa Rica]
UCAR	University Corporation for Atmospheric Research [US]
UCB	Union Chemique Belge [= Belgian Chemical Union]
	Unit Control Block
	Universal Character Buffer
	University of California, Berkeley [US]
	University of Colorado, Boulder [US]
UCC	Undecreciated Capital Cost
	Unified Classification Code
	Uniformation Classification Committee [now: NRFC, US]
	Uniform Code Council [US]
	Uniform Commercial Code
	Universal Classification Code
	Universal Conference Circuit
	Universal Copyright Convention
	University College of Cardiff [of University of Wales, UK]
UCCA	Universities Central Council on Admissions [UK]
UCCB	University College of Cape Breton [Canada]
UCCRS	Underwater Coded Command Release System
UCCS	Universal Camera Control System
UCD	Uniform Call Distributor
	Universal Control Drive
	University of California, Davis [US]
	University of Colorado, Denver [US]
UCDA	University and College Designers Association [US]
UCDP	Uncorrected Data Processor
UCEA	University Council for Educational Administration [US]
UCEPCEE	Union du Commerce des Engrais des Pays de la Communaute Economique Europeenne [= Union of the Fertilizer Producing Countries of the European Economic Community, Belgium]
UCF	University of Central Florida [US]
	Utility Control Facility
UCG	Underground Coal Gasification
UCHF	Uncoupled Hartree-Fock
UCI	Ultrasonic Contact Impedance
	University of California, Irvine [US]
	User Class Identifier
	Utility Card Input
UCIDT	University Consortium for Instructional Development and Technology [US]
UCIS	Uprange Computer Input System
UCK	Unit Check
UCL	Undercooled Liquid
	Undercorrected Lens
	University ollege of London [UK]
	Upper Confidence Limit

Upper Control Limit
UCLA Uncommitted Logic Array
University of California, Los Angeles [US]
UCLEA University and College Labor Education Association [US]
UCLJ University of California, La Jolla [US]
UCLRL University of California Lawrence Radiation Laboratory [US]
UCM Universal Canister Mount
Universal Communications Monitor
User Communications Manager
UCML University of California Microwave Laboratory [US]
UCN Ultra-Cold Neutron
UCNI Unified Communications Navigation and Identification
UCNW University College of North Wales [of UW, UK]
UCOM Utility Compiler [also: UCO]
UCON Utility Control
UCO United Cooperatives of Ontario [Canada]
UCORC University of California Operations Research Center [also: UC/ORC, US]
UCOS Uprange Computer Output System
UCOWR Universities Council on Water Resources [US]
UCP Uninterruptible Computer Power
Update Control Process
Utility Control Program
UCPL Union Centrale des Producteurs de Lait [= Central Union of Milk Producers, Switzerland]
UCPTE Union pour la Coordination de la Production et du Transport de l'Electricite [= Union for the Coordination of the Production and Transport of Electric Power, Switzerland]
Union for the Coordination of the Production and Transport of Electric Power [Switzerland]
UCR Unconditioned Response
University of California, Richmond [US]
University of California, Riverside [US]
UCRI Union Carbide Research Institute [US]
UCRL University of California Research Laboratory [US]
UCS Unconditioned Stimulus
Uniform Chromaticity Scale
Universal Call Sequence
Universal Camera Site
Universal Character Set
Universal Classification System
Universal Communications Subsystem
University College of Swansea [of University of Wales, UK]
User Control Store
User Coordinate System
UCSB Universal Character Set Buffer
University of California, Santa Barbara [US]
UCSD Universal Communications Switching Device
University of California, San Diego [US]
UCSEL University of California Structural Engineering Laboratory [US]
UCSF University of California, San Francisco [US]
UCSSL University of California Space Sciences Laboratory [also: UC/SSL, US]
UCSTR Universal Code Synchronous Transmitter-Receiver
UCT Universal Continuity Tester
Universal Coordinated Time
UCTE Union of Canadian Transport Employees
UCV Universidad Central de Venezuela [= Central University of Venezuela]
UCW Union of Communications Workers [UK]
Unit Control Word
University College of Wales [UK]
UCWE Underwater Countermeasures and Weapons Establishment
UCWI University College of the West Indies
UD UAE Dirham [Currency of the United Arab Emirates]
Ultradispersed
Underground Distribution
Unidirectional
University of Dayton [US]
University of Delaware [US]
University of Denver [US]
University of Dortmund [Germany]
University of Durham [UK]
Unplanned Derating
UDAC User Digital Analog Controller
UDAR Universal Digital Adaptive Recognizer
UDAS Unified Direct Access Standard
Unified Direct Access System
UDB Up Data Buffer
UDC Uganda Development Corporation
Ultrasonic Doppler Cardioscope
Unidirectional Composite
Unidirectional Current
Universal Decimal Classification
Universal Digital Control
Universal Digital Controller
Urban District Council
User Defined Code
UDDF Up and Down Drafts
UDE Universal Data Entry
UDEC Unitized Digital Electronic Calculator
UDET Unsymmetrical Diethylenetriamine
UDF UHF Direction Finding
Unducted Fan
Unidirectional Freezing
Unit Derating Factor
Unkinkable Domestic Flex
User Defined Function
UDG Unit Derating Generation
UDH Unplanned Derating Hours
UDH1 Unplanned Derating Hours, Class 1
UDI Urban Development Institute [Canada]
UDK User-Defined Key
UDL Unidirectional Laminate
Uniform Data Language
Uniform Data Link
Up Data Link
UDLC Universal Data Link Control
UDM Unified Defect Model
UDMH Unsymmetrical Dimethylhydrazine
UDOFT Universal Digital Operational Flight Trainer
UDOP UHF Doppler
UDP Ultradispersed Powder
Uniform Distribution Pattern

United Data Processing
Uridine Diphosphate
User-Developed Program
User Development Program [of Canadian Space Agency]

UDPAG Uridine Diphosphoacetylglucosamine
UDPG Uridine Diphosphate Glucose
Uridine Diphosphoglucose
UDPGA Uridine Diphosphoglucuronic Acid
UDP-GAL Uridine Diphosphogalactose [also: UDP-Gal]
UDPGDH Uridine Diphosphoglucose Dehydrogenase [also: UDPGD]
UDR Universal Document Reader
UDRI University of Dayton Research Institute [US]
UDRPS Ultrasonic Data Recording and Processing System
UDS Unidirectional Solidification
Uniscope Display System
Universal Distributed System
UDT Underwater Demolition Team
Universal Data Transcriber
Universal Document Transport
UDTI Universal Digital Transducer Indicator
UDTS Universal Data Transfer Service
UE User Equipment
UEA Ulex Europaeus Agglutinin
Union Europeenne de l'Ameublement [= European Furniture Manufacturers Federation, Belgium]
UEATC Union Europeenne pour l'Agrement Technique dans la Construction [= European Union for Technical Aprroval in Building Construction, France]
UEC United Engineering Center
University of Electro-Communications [Japan]
Utah Engineers Council [US]
UECU Union for Experimenting Colleges [US]
UEF Union Europaischer Forstberufsverbande [= Union of European Forester Associations, Germany]
UEG Underwater Engineering Group
UEIS United Engineering Information Service [US]
UEL Upper Explosion Limit
UEP Underwater Electric Potential
Unequal Error Protection
UERD Underwater Explosives Research Division [US Navy]
UERPIC Underground Excavation and Rock Properties Information Center [US]
UES United Engineering Societies
UESA Ukrainian Engineers Society of America [US]
UET United Engineering Trustees [US]
Universal Emulating Terminal
Universal Engineered Tractor
UEW Unsteady Expansion Wave
UEX Unit Exception
UF Ultrafiltration
Ultrafine
Ultrasonic Frequency
Unavailability Factor
Underflow
Underground Feeder
Unified Field
University of Florida [US]
Urea-Formaldehyde
Urethane Foam
Utility File
U/F Underflow
UFA Ultra Fine-Grained
Unesterified Fatty Acids
Until Further Advised
UFAA United Food Animal Association [US]
UFAM Universal File Access Method
UFAW United Fishermen and Allied Workers Union
Universities Federation for Animal Welfare [UK]
UFB Unfit for Broadcast
UFC Uniform Fire Code [US]
Uniform Freight Classification
Universal Frequency Counter
Universities Funding Council [UK]
UFCS Underwater Fire Control System
UFCW United Food and Commercial Workers
UFD User File Directory
UFEDC Unified Fixed Exchangeable Disk Coupler
UFESA United Fire Equipment Service Association [US]
UFF Universidade Federal Fluminense [Brazil]
UFFCS Ultrasonics, Ferroelectrics and Frequency Control Society [of IEEE, US]
UFFI Urea-Formaldehyde Foam Insulation
UFG Ultrafine Grinding
UFI Union des Foires Internationales [= Union of International Fairs, France]
Upstream Failure Indication
Usage Frequency Indicator
User-Friendly Interface
UFIPTI Union Franco-Iberique pour la Coordination de la Production et du Transport de l'Electricite [= Franco-Iberian Union for Coordinating the Production and Transmission of Electricity, France]
UFL Upper Flammability Limit
UFM Ultrafine Medium
Universidade Federal de Minas [= Federal University of Mines, Brazil]
Upper Felsic Metapyroclastics [Mining]
User to File Manager
UFMG Universidade Federal de Minas Gerais [= Federal University of Minas Gerais, Brazil]
UFN Until Further Notice
UFO Unidentified Flying Object
UFOD Union Francaise des Organismes de Documentation [= French Union of Documentation Organizations]
UFOIRC UFO (= Unidentified Flying Object) Information Retrieval Center [US]
UFOP Universidade Federal de Ouro Preto [= Federal University of Ouro Preo, Brazil]
UFP Ultrafine Powder
Utility Facilities Program
UFR Underfrequency Relay
UFRGS Universidade Federal do Rio Grande do Sul [= Federal University of Rio Grande do Sul, Brazil]
UFRJ Universidade Federal do Rio de Janeiro [= Federal University of Rio de Janeiro, Brazil]
UFRO (International) Union of Forest Research Organizations
UFSC Universidade Federal de Sao Carlos [= Federal University of Sao Carlos, Brazil]

UFTP Ultrafine Tungsten Powder
UFTR University of Florida Teaching Reactor [US]
UFVA University Film and Video Association [US]
UFVF University Film and Video Foundation [US]
UFW United Farm Workers Union [Canada]
UFZ Umweltforschungszentrum [= Center for Environmental Research, Germany]
UG Underground
Universidad de Guadalajara [= University of Guadalajara, Mexico]
University of Georgia [US]
University of Glasgow [UK]
University of Guelph [Canada]
Urban Geography
UGA Uncommitted Gate Array
UGC University Grants Committee [UK]
UGCW United Glass and Ceramic Workers (of North America)
UGG United Grain Growers [Canada]
UGLIAC United Gas Laboratory Internally Programmed Automatic Computer
UGPP Uridine Diphosphoglucose Pyrophosphorylase
UGS Ugandan Shilling [Currency of Uganda]
Upper Guide Structure
UGT Urgent
UH Ultrahigh
Unavailable Hours
Unit Heater
University of Hannover [Germany]
University of Hawaii [US]
University of Heidelberg [Germany]
University of Houston [US]
UHB Ultrahigh Bypass (Jet Engine)
UHC Ultrahigh Carbon
UHCH Unit Handling Conveyor Association [now: CS of MHI, US]
UHD Ultrahigh Density
UHF Ultrahigh Frequency
Unrestricted Hartree-Fock
UHFDF Ultrahigh Frequency Direction Finder
UHF/VHF Ultrahigh Frequency/Very High Frequency
UHI Upper Head Injection
UHL User Header Label
UHM Ultrahigh Modulus
University of Hawaii, Manoa [US]
UHML Upper-Half Mean-Length
UHMW Ultrahigh Molecular Weight
UHMWPE Ultrahigh Molecular Weight Polyethylene [also: UHMW PE]
UHP Ultrahigh Purity
Ultrahigh Power
Ultrahigh Pressure
UHPFB Untreated Hard Pressed Fiberboard
UHPS Underground Hydroelectric Pumped Storage
UHR Ultrahigh Reduction
Ultrahigh Resistance
Ultrahigh Resolution
UHRF Ultrahigh Resolution Facsimile
UHT Ultimate High Temperature
Ultra-Heat Treated
Ultrahigh Temperature
UHTREX Ultrahigh Temperature Reactor Experiment [US]
UHV Ultrahigh Vacuum
Ultrahigh Voltage
UHV/CVD Ultrahigh Vacuum Chemical Vapor Deposition
UHVEM Ultrahigh Vacuum Electron Microscopy [also: UHV EM]
UI Ultraionization
Ultrasonic Imaging
Ultrasonic Inspection
Unemployment Insurance
University of Idaho [US]
University of Illinois [US]
University of Istanbul [Turkey]
Uranium Institute [UK]
Urban Initiatives [US]
UIA Ultrasonic Industry Association [US]
Union Internationale des Architectes [= International Union of Architects, France]
Union of International Associations
UIB Unemploymment Insurance Benefits
UIBC Unsaturated Iron-Binding Capacity
UIBPA Union Internationale de Biophysique Pure et Appliquee [= International Union of Pure and Applied Physics, Hungary]
UIC Unemployment Insurance Commission
Union Internationale des Chemins de Fer [= International Union of Railways, France]
University of Illinois, Chicago [US]
Upper Information Center
User Identification Code
User Interface Circuit
UICB Union Internationale des Centres du Batiment [= International Union of Building Centers, Netherlands]
UICP Union Internationale de la Couverture et Plomberie [= International Union of Roofing and Plumbing, France]
UIE UNESCO Institute for Education
Union Internationale d'Electrothermie [= International Union for Electroheat, France]
UIEO Union of International Engineering Organizations
UIF Unrestricted Industrial Funds
UIG User Instruction Group
UIIG Union International de l'Industrie du Gaz [= International Gas Union, Switzerland]
UIL Univac Interactive Language
UILA Union Internationale des Laboratoires Independants [= International Union of Independent Laboratories, UK]
UILERMS Union Internationale des Laboratoires d'Essai et de Recherche pour les Materiels et les Structures [= International Union of Testing and Research Laboratories for Materials and Structures, France]
UIM Ultra-Intelligent Machine
Union of International Motorboating
UIMS User Interface Management System
UIO Universal Input/Output
UIOC Universal Input/Output Controller
UIP Union Internationale d'Assus de Proprietaires de Wagon [= International Union of Private Railway Truck Owners Association, Switzerland]

University Interactions Program
UIPC Underground Injection Practices Council [US]
UIPPA Union Internationale de Physique Pure et Appliquee [= International Union of Pure and Applied Physics, Sweden]
UIPRE Union Internationale de la Presse Radiotechnique et Electronique [= International Union of the Radiotechnical and Electronics Press, Germany]
UIR Upper (Flight) Information Region
User Instruction Register
UIS United Inventors and Scientists
Upper Information Service
Uranium Institute Symposium [UK]
User Interface Service
UISPI Urethane Institute, Society of the Plastics Industry [US]
UIT Union Internationale de Telecommunication [= International Telecommunications Union, Switzerland]
UITA Union of International Technical Associations [France]
UITP Union Internationale des Transports Publics [= International Union of Public Transport, Belgium]
UIU University of Illinois, Urbana [US]
UIUC University of Illinois, Urbana-Champaign [US]
UJ University of Jilin [PR China]
UJCL Universal Job Control Language
UJM Uncorrected Jet Model
UJSE Union of Japanese Scientists and Engineers [Japan]
UJT Unijunction Transistor
UK United Kingdom
University of Kansas [US]
University of Karlsruhe [Germany]
University of Keele [UK]
University of Kentucky [US]
University of Kuwait [Kuwait City]
UKAATS United Kingdom Advanced Air Traffic System
UKAC United Kingdom Automation Council
UKACC United Kingdom Automatic Control Council
UKAEA United Kingdom Atomic Energy Authority
UKAEA-R United Kingdom Atomic Energy Authority — Risley
UKAIP United Kingdom Aeronautical Information Publication
UKAPE United Kingdom Association of Professional Engineers
UKASA United Kingdom Agricultural Students Association
UKASE University of Kansas Automated Serials System [US]
UKB Universal Keyboard
UKCIS United Kingdom Chemical Information Service
UKCSMA United Kingdom Cutlery and Silverware Manufacturers Association
UKCTRAIN United Kingdom Catalogue Training [for UKMARC]
UKDA United Kingdom Dairy Association
UKGPA United Kingdom Glycerin Producers Association
UKIRT United Kingdom Infrared Telescope
UKISC United Kingdom Industrial Space Committee
UKITO United Kingdom Information Technology Organization
UKJGA United Kingdom Jute Goods Association
UKLDS United Kingdom Library Database System
UKMARC United Kingdom Machine-Readable Catalogue [of BLAISE]
UKNCIAWPRC United Kingdom National Committee of the International Association on Water Pollution Research and Control
UKNR University of Kansas Nuclear Reactor [US]
UKOBA United Kingdom Outboard Boating Association
UKOLUG United Kingdom On-Line Users Group
UKOOA United Kingdom Offshore Operations Association
UK£ United Kingdom Pound Sterling [Currency of Northern Ireland and the UK]
UKPA United Kingdom Pilots Association
UKPIA United Kingdom Petroleum Industry Association
UKPO United Kingdom Post Office
UKR Ukraine
Ukrainian
UKRAS United Kingdom Railway Advisory Service
UKSM United Kingdom Scientific Mission
UL Ultralarge
Ultralight
Ultralow
Uncontrolled Language
Underwriters Laboratories [US]
Universite de Lausanne [= University of Lausanne, Switzerland]
Universite de Liege [= University of Liege, Belgium]
University of Lanzhow [PR China]
University of Leeds [UK]
University of Leicester [UK]
University of Leoben [Austria]
University of Lethbridge [Canada]
University of Liverpool [UK]
University of London [UK]
Upper Level
Upper Limit
Upper List
User Language
U/L Unlimited
ULA Uncommitted Logic Array
ULAP University Libraries Automation Program [US]
ULAU Union of Latin American Universities [Mexico]
ULB Universal Logic Block
Universite Libre de Bruxelles [= Free University of Brussels, Belgium]
ULC Ultralow Carbon (Steel)
Underwriters Laboratories of Canada
Uniform Loop Clock
Universal Logic Circuit
Upper and Lower Case
ULCB Ultralow Carbon Bainitic (Steel)
ULCC Ultralarge Crude Carrier
ULCRF University of London Central Research Fund [UK]
ULD Ultralarge Diameter
Ultralow Density
Unit Load Device
ULDLPE Ultralow-Density Linear Polyethylene
ULE Ultralow Expansion
ULEA University Labor Education Association [now: UCLEA, US]
ULF Ultralow Frequency

ULG Upholstery Leather Guild [now: AG, US]
ULI Urban Land Institute [US]
ULICP Universal Log Interpretation Computer Program
ULISYS University Library System
ULLN Upper Limit Log Normal Distribution
ULM Ultrasonic Light Modulator
Universal Line Multiplexer
Universal Logic Module
ULMA University Laboratory Managers Association [now: ALMA, US]
ULMM Universal Length-Measuring Machine
ULMS Underseas Long-Range Missile System
ULO Unmanned Launch Operations
ULOW Unmanned Launch Operations Western Test Range
ULP Ultralow Pressure
ULPA United Lightning Protection Association [US]
ULPR Ultralow Pressure Rocket
ULR Ultra-Long Range
ULSAS Upward Looking Sonar Array System
ULSCS University of London Shared Cataloguing System [UK]
ULSI Ultra Large Scale Integration
ULSV Unmanned Launch Space Vehicle
ULT Ultimate
Unique Last Term
Untyped Lambda Calculi
ult (ultimo) — last day of month
ULTI Ultralow-Temperature Isotropic (Carbon)
ULTRA Universal Language for Typographic Reproduction Applications
ULTRA LSI Ultra Large Scale Integration
ULU Union of Latin-American Universities
ULV Ultra-Low Volume
ULW Ultra Lightweight
UM Ultramicroscope
Unimolecular
Unitas Malacologica [Netherlands]
Universal Mill (Plate)
Universal Motor
Universidade MacKenzie [= MacKenzie University, Brazil]
Universidad de Madrid [= University of Madrid, Spain]
University of Madras [India]
University of Manchester [UK]
University of Manitoba [Canada]
University of Maryland [US]
University of Massachusetts [US]
University of Melbourne [Australia]
University of Michigan [US]
University of Minnesota [US]
University of Missouri [US]
University of Moncton [Canada]
University of Montreal [Canada]
University of Munich [Germany]
UMA Ultrasonic Manufacturers Association [US]
Universal Measuring Amplifier
University of Massachusetts, Amherst [US]
UMASS University of Massachusetts [also: UMass, US]
Unlimited Machine Access from Scattered Sites
UMC Underwater Manifold Center
Unibus Micro Channel
Unidirectional Molding Compound
Uniform Mechanical Code [US]
Universidade de Mogi das Cruzes [= University of Mogi das Cruzes, Brazil]
Upper Mantle Commission [UK]
UMCA United Mining Council of America [US]
Universities Mission to Central Africa
UMCC United Maritimes Consultative Committee
UMD Unitized Microwave Device
UMES University of Michigan Executive System
UMF Ultra-Microfiche
Urea Melamine Formaldehyde
UMI Ultra Microfiche
Underwater Mining Institute [US]
University Microfilms International [US]
UMIST University of Manchester, Institute of Science and Technology [UK]
UML Universal Machine Language
UMLC Universal Multiline Controller
UMLER Universal Machine Language Equipment Register
UMM Universal Measuring Microscope
UMNE University of Maryland Nuclear Experiment [US]
UMODE University of Manitoba Optokinetic Decay Experiment [Canada]
UMOS U-Type Metal-Oxide Semiconductor
UMP University of Manitoba Press [Canada]
Upper Mantle Project [of ICSU, France]
Uridine Monophosphate
UMPLIS Umweltplanungsinformationssystem [= Information System on Environmental Planning, Germany]
UMR Unimolecular Reaction
Unipolar Magnetic Region
University of Missouri-Rolla [US]
Upper Maximum Range
UMRCC Upper Mississippi River Conservation Committee [US]
UMRR University of Missouri Research Reactor [US]
UMS Underwater Monitoring System
Universal Maintenance Standards
Universal Memory System
Universal Multiprogramming System
University Mailing Service
UMSL University of Missouri, St. Louis [US]
UMS/VS Universal Multiprogramming System/Virtual Storage
UMT Ultrasonic Machine Tool
UMTA Urban Mass Transit Authority
Urban Mass Transportation Administration [of USDOT]
UMTRA Uranium Mill Tailings Radiation Control Act [US]
UMTRIS Urban Mass Transportation Research Information Service [of USDOT]
UMWA United Mineworkers of America
UN Unassigned
Union
United Nations
Universidad de Navarra [= University of Navarra, Spain]
Universite de Nancy [= University of Nancy, France]
University of Nagoya [Japan]

University of Nebraska [US]
University of Nevada [US]
University of Newcastle [Australia]
University of Nottingham [UK]
Urban Network

UNA United Nations Association [UK]
Universal Night Answering
UNAC United Nations Association of Canada
UNACAST United Nations Advisory Committee on the Application of Science and Technology
UNADA United Nations Atomic Development Authority
UNADS Univac Automated Documentation System
UNAIS United Nations Association International Service [UK]
UNALC User Network Access Link Control
UNAM Universidad Nacional Autonoma de Mexico [= Free National University of Mexico]
UNAMACE Universal Automatic Map Compilation Equipment
UNATAC Union d'Assistance Technique pour l'Automobile et la Circulation Routiere [= Union of Technical Assistance for Motor Vehicle and Road Traffic, Switzerland]
UNAVBL Unavailable
UNB Unbound
University of New Brunswick [Canada]
UNBAL Unbalanced
UNBSJ University of New Brunswick, St. John [Canada]
UNBTAO United Nations Bureau of Technical Assistance Operations
UNC Unconditional
Unified National Coarse (Screw Thread)
Universidad Nacional de Colombia [= Colombia National University]
University of North Carolina [US]
UNCA United Nations Correspondents Association
UNCAST United Nations Conference on the Applications of Science and Technology
UNCC University of North Carolina, Charlotte [US]
UNC-CH University of North Carolina, Chapel Hill [US]
UNCDF United Nations Capital Development Fund
UNCED United Nations Conference on Environment and Development
UNCF United Negro College Fund [US]
UNCFI United Nations Commission for Indonesia
UNCHR United Nations Commission for Human Rights
UNCHS United Nations Center for Human Settlements [Kenya]
UNCIAWPRC Uruguayan National Committee of the International Association on Water Pollution Research and Control [also: UNCIAWPRC]
UNCIP United Nations Commission for India and Pakistan
UNCITRAL United Nations Commission on International Trade Laws
UNCLOS United Nations Conference on the Law of the Sea
UNCM User Network Control Machine
UNCMAC United Nations Command Military Armistice Commission
UNCOK United Nations Commission on Korea
UNCOL Universal Computer-Oriented Language
UNCPICPUNE United Nations Conference for the Promotion of International Cooperation in the Peaceful Uses of Nuclear Energy
UNCPUOS United Nations Committee on the Peaceful Uses of Outer Space
UNCSTD United Nations Center for Science and Technology for Development [US]
UNCTAD United Nations Conference on Trade and Development [Switzerland]
UNCTAD/GATT United Nations Conference on Trade and Development/General Agreement on Tariffs and Trade
UNCTLD Uncontrolled
UND Unit Derating
University of North Dakota [US]
University of Notre Dame [US]
UNDCC United Nations Development Cooperation Cycle
UNDD United Nations Development Decade
UNDE Union of National Defense Employees
UNDEX United Nations Documents Index
UNDGRD Underground
UNDH Unit Derating Hours
UNDIS United Nations Documentation Information System
UNDOF United Nations Disengagement Observer Force
UNDP United Nations Development Program [US]
UNDRLN Underline
UNDRO United Nations Disaster Relief Organization
UNDV Under Voltage
UNDW Underwater
UNEC United Nations Economic Commission
UNECA United Nations Economic Commission for Africa
UNECE United Nations Economic Commission for Europe
UNEDA United Nations Economic Development Administration
UNEF Unified National Extra Fine (Screw Thread)
United Nations Emergency Force
UNEGA Union Europeenne des Fondeurs et Fabricants de Corps Gras Animaux [= European Union of Animal Fat Producers, France]
UNEP United Nations Environmental Program
UNEP/UNESCO/ICRO United Nations Environmental Program/United Nations Educational, Scientific and Cultural Organization/International Cell Research Organization (Panel on Microbiology) [Sweden]
UNESCO United Nations Educational, Scientific and Cultural Organization
UNESCO-ROSTA United Nations Educational, Scientific and Cultural Organization/Regional Office for Science and Technology in Africa [Kenya]
UNESOB United Nations Economic and Social Office in Beirut
UNETAS United Nations Emergency Technical Aid Service
UNEXC Unexcavated
UNF Unified National Fine (Screw Thread)
Uniform
UNFAO United Nations Food and Agricultural Organization
UNFB United Nations Film Board
UNFDAC United Nations Fund for Drug Abuse and Control
UNFPA United Nations Fund for Population Activities
UNGA United Nations General Assembly
UNH Uranyl Nitrate Hexahydrate

UNHHSF United Nations Habitat and Human Settlements Foundation
UNI University of Northern Iowa [US]
UNIBUS Universal Bus [Electronics]
UNIC United Nations Information Center
UNICA Association of Caribbean Universities and Research Institutes [Puerto Rico]
UNICAT Union Catalogue
UNICAT/TELECAT Union Catalogue/Telecommunications Catalogue
UNICCAP Universal Cable Circuit Analysis Program
UNICEF United Nations International Children's Emergency Fund
UNICHAL Union Internationale des Distributeurs de Chaleur [= International Union of Heat Distributors, Switzerland]
UNICIS University of Calgary Information System [Canada]
UNICOM Universal Integrated Communications
UNICOMP Universal Compiler
UNICON Unidensity Coherent Light Recording
UNIDIR United Nations Institute for Disarmament Research
UNIDO United Nations Industrial Development Organization [Austria]
UNIFE Union des Industriels Ferroviaires Europeennes [= Union of European Railway Industries, France]
UNIFET Unipolar Field-Effect Transistor
UNIFREDI Universal Flight Range and Endurance Data Indicator
UNIMARC Universal Machine-Readable Catalogue
UNIPEDE Union Internationale des Producteurs et Distributeurs d'Energie Electrique [= International Union of Producers and Distributors of Electric Power, France]
UNIPOL Universal Procedure-Oriented Language
UNIQUE Uniform Inquiry Update Element
UNIS United Nations Information Service
UNISAP Univac Share Assembly Program
UNISIST United Nations Information System in Science and Technology [of UNESCO]
UNISIST/ICSU United Nations Information System in Science and Technology/International Council of Scientific Unions
UNISOR University Isotope Separator at Oak Ridge [of ORAU, US]
UNISTAR User Network for Information Storage, Transfer, Acquisition and Retrieval
UNITAP United Nations Intermunicipal Technical Assistance Program
UNITAR United Nations Institute for Training and Research
UNITEC University Information Technology Corporation [US]
UNITEL Universal Teleservice
UNITRAC Universal Trajector Compiler
UNIV University
Universal
UNIVAC Universal Automatic Computer [also: Univac]
UNIVERSE Universities Expanded King and Satellite Experiment [UK]
UNIVSERV United Nations International Voluntary Service Fund
UNIX Universal Executive
UNKN Unknown
UNL Unlimited
UNLC United Nations Liaison Committee
UNLD Unload
UNLGTD Unlighted
UNLTD Unlimited
UNM Unified National Miniature (Thread)
University of New Mexico [US]
UNMA University of New Mexico, Albuquerque [US]
UNMOGIP United Nations Military Observer Group in India and Pakistan
UNNSAD Unit Neutral Normalized Spectral Analytical Density
UNO United Nations Organization
UNOQ Unit of Quantity
UNPA United Nations Postal Administration
UNPKD Unpacked
UNPOC United Nations Peace Observation Commission
UNPS Universal Power Supply
UNR University of Nevada, Reno [US]
UNRC Unified National Coarse (Screw Thread), Rolled
UNREL Unreliable
UNRF Unified National Fine (Screw Thread), Rolled
UNRIPS United Nations Regional Institute for Population Studies [Ghana]
UNRISD United Nations Research Institute for Social Development
UNRR Unidirectional Non-Reversing Relay
UNRRA United Nations Relief and Rehabilitation Administration
UNRSTD Unrestricted
UNRWA United Nations Relief and Works Agency
UNS Unified National Special (Screw Thread)
Unified Numbering System
University of Nova Scotia [Canada]
Unsymmetrical
UNSATD Unsaturated
UNSC United Nations Security Council
United Nations Special Commission
UNSCC United Nations Standards Coordinating Committee
UNSCCUR United Nations Scientific Conference on the Conservation and Utilization of Resources
UNSCEAR United Nations Scientific Committee on the Effects of Atomic Radiation
UNSF United Nations Special Fund
UNSO United Nations Statistical Office
UNSSOD United Nations Special Session on Disarmament
UNSTBL Unstable [also: UNST]
UNSTDY Unsteady
UNSW University of New South Wales [Australia]
unsym unsymmetrical [Chemistry]
UNTAA United Nations Technical Assistance Administration
UNTAM United Nations Technical Assistance Mission
UNTRA Union of National Television and Radio Organizations of Africa
UNTSO United Nations Truce Supervisory Organization
UNU United Nations University [Japan]
UNUT University of Newcastle-upon-Tyne [UK]
UNV United Nations Volunteers
UNWC University of Northwestern California [US]
UNWMG Utility Nuclear Waste Management Group
UO Unit Operation
University of Oklahoma [US]

	University of Oregon [US]
	University of Osaka [Japan]
	University of Ottawa [Canada]
	University of Oulo [Finland]
	University of Oxford [UK]
U/O	Used On
UOC	Ultimate Operating Capability
	University of Calgary [Canada]
UOD	Units of Optical Density
UOF	Unplanned Outage Factor
U of A	University of Alberta [Canada]
U of C	University of Calgary [Canada]
U of I	University of Idaho [US]
U of M	University of Michigan [US]
	University of Mississippi [US]
U of Nfld	University of Newfoundland [Canada]
U of O	University of Ottawa [Canada]
U of Penn	University of Pennsylvania [US]
U of S	University of Sydney [Australia]
U of T	University of Toronto [Canada]
U of U	University of Utah [US]
U of V	University of Virginia [US]
U of W	University of Waterloo [Canada]
UOH	Unplanned Outage Hours
UON	Unless Otherwise Noted
UOP	Universal Oil Product
	University of Ottawa Press [Canada]
UOR	Unplanned Outage Rate
	Unusual Occurrence Report
UOS	Underwater Ordnance Station
UOSAT	University of Surrey Satellite [UK]
UOV	Unit of Variance
UOW	University of Waterloo [Canada]
	University of Wisconsin [US]
UP	Ultrapure
	Uniprocessor
	United Press [US]
	Unit Process
	Universal Programmer
	Universidade de Porto [= University of Porto, Portugal]
	Universite de Paris [= University of Paris, France]
	University of Pittsburgh [US]
	University of the Philippines
	Unsaturated Polyester
	Unsolicited Proposal
	Upper Part
	Urea Phosphate
	Uridine Phosphate
	Utility Path
U-5'-P	Uridine 5'-Monophosphate
UPA	United Printers Association [UK]
	University Photographers Association (of America) [US]
	Uranium Producers of America [US]
	Utah Pharmaceutical Association [US]
UpA	Uridylyladenosine
UPAA	University Photographers Association of America [US]
UPACS	Universal Performance Assessment and Control System
UPADI	Union Panamericana de Asociaciones de Ingenieros [= Pan-American Federation of Engineering Societies, Venezuela]
UPAO	University Professors for Academic Order [US]
UpApA	Uridylyladenylyladenosine
UPAU	Uttar Pradesh Agricultural University [India]
UPB	University of Pittsburgh, Bradford [US]
	Upper Bound
UPC	Uniform Product Code
	United Press Canada
	Unit of Processing Capacity
	Universal Peripheral Controller
	Universal Product Code
UPC-E	Universal Product Code — Europe
UpC	Uridylylcytidine
UPCS	Universal Process Control Software
UPD	Underpotential Deposition
	Uniaxial Plastic Deformation
	Update
UPDATE	Unlimited Potential Data through Automation Technology in Education
UPDEA	Union of Producers and Distributors of Electric Power in Africa [Ivory Coast]
UPDFTS	Updrafts
UP/DOC	Universal Programmer/Documentor
UPE	Unnatural Parity Exchange
UPEI	Union Petrole Europeenne Independante [= Independent European Petroleum Union, France]
	University of Prince Edward Island [Canada]
UPFDA	United Products Formulators and Distributors Association [US]
UpG	Uridylylguanosine
UPHA	Utah Public Health Association [US]
UPHS	Underground Pumped Hydro Storage
UPI	United Press International [US]
	Universal Personal Identifier
UPIR	Uttar Prasesh Irrigation Research Institute [India]
UPIU	United Paperworkers International Union
UPL	Universal Programming Language
	User Programming Language
UPLIFTS	University of Pittsburgh Linear File Tandem System [US]
UPM	Universal Permissive Module
	University of Petroleum and Minerals [Saudi Arabia]
UPMC	University of Pierre and Marie Curie [France]
UPOS	Utility Program Operating System
UPOV	Union Internationale pour la Protection des Obtentions Vegetales [= International Union for the Protection of New Varieties of Plants, Switzerland]
UPP	United Papermakers and Paperworkers
	Universal PROM (= Programmable Read-Only Memory) Programmer
UPPE	Ultraviolet Photometric and Polarimetric Explorer
UPR	Ultrasonic Paramagnetic Resonance
	University of Puerto Rico
	Upper
	Uranium Production Reactor
UPS	Ultraviolet Photoelectron Spectroscopy
	Ultraviolet Photoelectron Spectrum

Ultraviolet Photoemission Spectroscopy
Underwater Photography Society [US]
Uninterrupted Power Supply
Uninterruptible Power Supply
United Parcel Service
Universal Processing System
Upper Sideband

UPSC United Parcel Service Canada
UPSE Universal Power Service Equipment
UPSI User Program Sense Indicator
UPSLP Upslope
UPT Unipolar Transistor
Universal Portable Telephone
User Process Table
UPTS Undergraduate Pilot Training System
UPU Universal Postal Union [of UN — Switzerland]
UpU Uridylyluridine
UpUpG Uridylyluridylylguanosine
UPVC Unplasticized Polyvinyl Chloride
UPW Union of Post Office Workers [UK]
UQ University of Quebec [Canada]
University of Queensland [Australia]
UQAC University of Quebec, Chicoutimi [Canada]
UQAM University of Quebec, Montreal [Canada]
UQAR University of Quebec, Rimouski [Canada]
UQT User Queue Table
UQTR University of Quebec, Trois-Rivieres [Canada]
UR Unattended Repeater
Under Running (Crane)
Unit Record
Unit Register
Universite de Reims [= University of Reims, France]
Universite de Rouen [= University of Rouen, France]
University of Reading [UK]
University of Regensburg [Germany]
University of Regina [Canada]
University of Rochester [US]
University of Roorkee [India]
University of Ryukyus [Japan]
U/R Underrange
Up Range
Uranium/Radium (Ratio)
URA Ultrared Absorption
Universities Research Association [US]
User Requirements Analysis
URAEP University of Rochester Atomic Energy Project [US]
URAG Uranium Resources Appraisal Group [of EMR, Canada]
URBANICOM Association Internationale Urbanisme et Commerce [= International Association for Town Planning and Distribution, Belgium]
URBM Ultimate Range Ballistic Missile
URC Unit Record Control
Utilities Research Commission
URD Underground Residential Distribution (Cable)
User Requirements Document
URG Universal Radio Group
URI University of Rhode Island [US]
URIF University Research Incentive Fund [Canada]
URI-GSO University of Rhode Island — Graduate School of Oceanography [US]
URIPS Undersea Radioisotope Power Supply
URIR Unified Radioactive Isodromic Regulator
URISA Urban and Regional Information Systems Association [US]
URL Underground Research Laboratory
User Requirements Language
URLL University of Rochester Laser Laboratory [US]
UROP Undergraduate Research Opportunities
URP Unit Record Processor
URPA University of Rochester, Department of Physics and Astronomy [US]
URR Ultrared Reflection
Unidirectional Reversing Relay
URRI Urban Regional Research Institute [of Michigan State University, US]
URS Ultra-Rapid Solidification
Unate Ringe Sum
Uniform Reporting System
United Research Service
Universal Reference System
Universal Regulating System
Unmanned Repeater Station
Update Report System
URSI Union Radio-Scientifique Internationale [= International Scientific Radio Union, Belgium]
URST Unrecrystallized Solution Treatment
URT Uranium Research Technology
URTNA Union of (National) Radio and Television Organizations of Africa
URTU United Road Transport Union [UK]
URU Uruguay(an)
URV Undersea Rescue Vehicle
Undersea Research Vehicle
US Ultimate Strength
Ultrasonic(s)
Ultrasonic Society
Ultrasound
Underside
Understressing
United Services
United States (of America)
Unit Separator (Character)
Unit Switch
Universal System (of Lens Apertures)
University of Salford [UK]
University of Sarajevo [Yugoslavia]
University of Saskatchewan [Canada]
University of Sheffield [UK]
University of Strathclyde [UK]
University of Stuttgart [Germany]
University of Sudbury [Canada]
University of Surrey [UK]
University of Sussex [UK]
University of Sydney [Australia]
University of Syracuse [US]
Unmanned System
Utah State [US]
U/S Unserviceable
USA Undercar Specialists Association [of NEDA, US]

	United States of America
USAA	United Service Automobile Association [US]
	United States Armor Association [US]
USAAML	United States Army Aviation Materiel Laboratories
USAASO	United States Army Aeronautical Services Office
USAAVLABS	United States Army Aviation Materiel Laboratories
USAAVNS	United States Army Aviation School
USAAVNTA	United States Army Aviation Test Activity
USAB	United States Activities Board [of IEEE]
USABAAR	United States Army Board for Aviation Accident Research
USABESRL	United States Army Behavioral Science Research Laboratory
USABRL	United States Army Ballistics Research Laboratories
USAC	United States Activities Committee [now: USAB]
USACA	United States Advanced Ceramics Association
USACDA	United States Arms Control and Disarmament Agency
USACDC	United States Army Combat Developments Command
USACDCAVNA	United States Army Combat Developments Command Aviation Agency
USACDCCBRA	United States Army Combat Developments Command Chemical-Biological-Radiological Agency
USACSC	United States Army Computer Systems Command
USACSSC	United States Army Computer Systems Support (and Evaluation) Command
USADATCOM	United States Army Data Support Command [also: USADC]
USADSC	United States Army Data Services Command
USAEC	United States Atomic Energy Commission
USAECOM	United States Army Electronics Command
USAEPG	United States Army Electronic Proving Ground
USAERDA	United States Army Electronic Research and Development Agency
USAERDL	United States Army Engineering Research and Development Laboratories
USAF	United States Air Force
	United States of America Foundation of Research and Education
USAFB	United States Air Force Base
USAFCRL	United States Air Force Cambridge Research Laboratories
USAFETAC	United States Air Force Environmental Technical Applications Center
USAFI	United States Armed Forces Institute [of USDOD]
USAFIC	United States Association of Firearm Instructors and Coaches (International)
USAFSC	United States Air Force Systems Command
USA FUNDS	United Students Aid Fund [US]
USAID	United States Agency for International Development
USAIDSC	United States Army Information and Data Systems Command
USALMC	United States Army Logistics Management Center
USAM	Unified Space Applications Mission [of NASA]
	Unique Sequential Access Method
	User Spool Access Method
USAMC	United States Army Materiel Command
USAN	United States Adopted Name [Pharmaceutical Products]
USANDL	United States Army Nuclear Defense Laboratory
USANWSG	United States Army Nuclear Weapon Surety Group
USAOMC	United States Army Ordnance Missile Command
USAPC	United States Army Petroleum Center
USAPHS	United States Army Primary Helicopter School
USAPO	United States Antarctic Projects Office
USAREPG	United States Army Electronic Proving Ground
USARIEM	United States Army Research Institute of Environmental Medicine
USARO	United States Army Research Office
USARP	United States Antarctic Research Program [of National Science Foundation, US]
	United States Atlantic Research Program
USARPA	United States Army Radio Propagation Agency
USART	Universal Synchronous/Asynchronous Receiver/Transmitter
USAS	United States of America Standards
USASCAF	United States Army Service Center for Armed Forces
USASCC	United States Army Strategic Communications Command
USASCII	United States of America Standard Code for Information Interchange [also: US ASCII or ASCII]
USASCSOCR	United States of America Standard Character Set for Optical Character Recognition
USASDC	United States Army Strategic Defense Command
USASI	United States of America Standards Institute [now: ANSI]
USASMSA	United States Army Signal Missile Support Agency
USATACOM	United States Army Tank-Automotive Command
USATEA	United States Army Transportation Engineering Agency
USATECOM	United States Army Test and Evaluation Command
USATIA	United States Army Transportation Intelligence Agency
USB	Unified S-Band
	Upper Sideband
	Upper Surface Blowing
USBAP	University Small Business Assistance Program
USBE	Unified S-Band Equipment
	Universal Serials and Book Exchange [US]
USBGN	United States Board on Geographic Names
USBLS	United States Bureau of Labor Statistics
USBOM	United States Bureau of Mines [also: USBoM or USBM]
USBR	United States Bureau of Reclamation
USBS	Unified S-Band System
	United States Bureau of Standards
USBSM	United States Bureau of Standards and Measures [Now Part of NIST, US]
USBT	Upper Surface Blowing Technique
USBTC	University-Small Business Technology Consortium [US]
USC	Ultrasonic Cleaning
	Ultrasupercritical
	United States Code
	United States Congress
	Universal Service Circuit
	University of San Carlos [Guatemala]
	University of Southern California [US]
USCA	United States Contract Awards (Database)

USCAL University of South California Aeronautical Laboratory [US]
USCB United States Census Bureau
USCEA United States Council for Energy Awareness
USCEC University of Southern California Engineering Center [US]
USCEE University of Southern California, Department of Electrical Engineering [US]
USCEF United States — China Education Fund
USCF United States Churchill Fund
USCG United States Coast Guard
USC&GS United States Coast and Geodetic Survey [of USDC]
USCI United Satellite Communications, Inc.
USCIB United States Council for International Business [also: USC IB]
USCID United States Committee on Irrigation and Drainage [US]
USCIDF United States Committee on Irrigation, Drainage and Flood Control [now: USCID]
USCIGW Union of Salt, Chemical and Industrial General Workers [UK]
USCLASS United States Classifications (Database)
USCMI United States Commission on Mathematical Instruction
USCOLD United States Committee on Large Dams
USCS United States Commercial Standard
USCSCV United States Committee for Scientific Cooperation with Vietnam
USCTI United States Cutting Tool Institute
USD Ultimate Strength Design
Ultrasonic Detector
Uniform Symbol Description
University of San Diego [US]
University of South Dakota [US]
US$ United States Dollar [also: USD; Currency of Guam, Panama Canal Zone, Puerto Rico and USA]
USDA United States Department of Agriculture
United States Department of the Army
USDA/CRIS United States Department of Agriculture/Current Research Information System
USDC United States Department of Commerce
USDDM United States Department of Data Management
USDDMS United States Department of Defense Manned Spaceflight
USDEA United States Department of External Affairs
USDGA United States Durum (Wheat) Growers Association [US]
USDHEW United States Department of Health, Education and Welfare
USDHHS United States Department of Health and Human Services
USDI United States Department of the Interior
USDL United States Department of Labor
USDMV United States Department of Motor Vehicles
USDN United States Department of the Navy
USDNA United States Defense Nuclear Agency
USDNE United States Department of Nuclear Engineering
USDOA United Stated Department of Agriculture
USDOA/CRIS United States Department of Agriculture/Current Research Information System
USDOC United States Department of Commerce
USDOD United States Department of Defense
USDOF United States Department of Fisheries
USDOI United States Department of the Interior
USDOJ United States Department of Justice
USDOL United States Department of Labor
USDOM United States Department of Mines
USDON United States Department of the Navy
USDOS United States Department of State
USDOT United States Department of Transportation
USDR&E Undersecretary of Defense for Research and Engineering [US]
USE Underground Service Entrance
Unified S-Band Equipment
Unit Support Equipment
Univac Scientific Exchange [US]
USEA United States Energy Association
USEC United System of Electronic Computers
USEMA United States Electronic Mail Association
USER User System Evaluator
USERC United States Environment and Resources Council
USERID User Identification
USES United States Employment Service
USF United States Form Thread
United States Frigate
Unsaturated Flow
Unsteady Flow
USFA United States Fire Administration
USFAA United States Field Artillery Association [now: AUSA, US]
USFCC United States Federation for Culture Collections
USFCF United States Frigate Constellation Foundation
USFDA United States Food and Drug Administration
USFED United States Federal Specifications Board [also: USFed]
USFGC United States Feed Grains Council
USFM Ultrasonic Flowmeter
USFPL United States Forest Products Laboratory
USFS United States Forest Service
United States Frequency Standard
USFSS United States Federation of Scientists and Scholars
United States Fleet Sonar School
USG Ultrasonic Generator
United States (Standard) Gauge
United States Gallon [also: USGAL or USgal]
University of Surrey, Guildford [UK]
USGA Ultrasonic Gas Atomization
USGAL United States Gallon [also: USgal or USG]
USGC United States Geodynamics Committee [of National Academy of Science, US]
USGCA United States Government Contract Awards [now: USCA]
USGPM United States Gallons per Minute
USGPO United States Government Printing Office
USGR United States Government Report
USGRDR United States Government Research and Development Report [of NBS, US]
USGRR United States Government Research Reports
USGS United States Geological Society
United States Geological Survey [of USDOI]
USGW Undersea Guided Weapon
Underwater-to-Surface Guided Weapon

USh Ugandan Shilling [Currency]
USHIGEO United States National Committee on the History of Geology
USI Ultrasonic Imaging
Ultrasonic Inspection
Ultrasonic Society of India
Union of Students in Ireland
United Schools International
Universal Software Interface
University of Southern Indiana [US]
User System Interface
USIA United States Information Agency
USIB Unsaturated Iron-Binding Capacity
USIC United States Information Center
USICA United States International Communications Agency
USICF United States International Communication Facility
USIG Ultrasonic Impact Grinding
USIO Unlimited Sequential Input/Output
USIS United States Information Service
USISC United States International Service Carriers
USITA United States Independent Telephone Association [now: USTA]
USITC United States International Trade Commission
USIU United States International University
USJPRS United States Joint Publication Research Service
USL Underwater Sound Laboratory [US Navy]
Upper Specification Limit
USLDMA United States Lanolin and Derivatives Manufacturers Association
USLSA United States Livestock Sanitary Association
USM Ultrasonic Machining
Underwater-to-Surface Missile
United States Mail
United States Marines
United States Mint
University of Southern Mississippi [US]
Unlisted Securities Market
USMA United States Metric Association
United States Military Academy
USMAC United States Management Advisory Committee
USMC United States Marine Corps
United States Maritime Commission
USMCCCA United States Marine Corps Combat Correspondents Association
USMCOC United States/Mexico Chamber of Commerce [US]
USMT United States Military Transport
USMUN United States Mission to the United Nations
USN United States Navy
USNA United States Naval Academy
USNBS United States National Bureau of Standards [now: NIST, US]
USNC United States National Committee
USNC/CIE United States National Committee/CIE [= Commission Internationale de l'Eclairage (= International Commission on Illumination) [also: USNCCIE]]
USNC/IPS United States National Committee of the International Peat Society [also: USNCIPS]
USNC/FID United States National Committee of FID [= Federation Internationale de Documentation (= International Federation of Documentation) [also: USNCFID]]
USNC/SCOR United States National Committee for the Scientific Committee on Oceanic Research [also: USNCSCOR]
USNC/TAM United States National Committee on Theoretical and Applied Mechanics [also: USNCTAM]
USNC/URSI United States National Committee for URSI [= Union Radio-Scientifique Internationale (= International Union of Radio Science) [also: USNCURSI]]
USNC/WEC United States National Committee of the World Energy Conference [also: USNCWEC]
USNEL United States Navy Electronics Laboratory [Now Subdivided into NUWC and NCCCC]
USNG United States National Guard
USNI United States Naval Institute
USNM United States National Museums
USNO United States Naval Observatory
USNOO United States Naval Oceanographic Office
USNORDA United States Naval Ocean Research and Development Activity
USNR United States Naval Research
United States Naval Reserve
USNRC United States Nuclear Regulatory Commission
USNRDL United States Naval Radiological Defense Laboratory
USNRRC United States Nuclear Reactor Regulatory Committee
USNS United States National Society
USNS/ISSMFE United States National Society for the International Society of Soil Mechanics and Foundation Engineering
USNUSL United States Navy Undersea Laboratory
USNUWL United States Navy Underwater Laboratory
USO Universal Service Order
Unmanned Seismic Observatory [of USDOD]
USOA Uniform System of Accounts
USOAR Uniform System of Accounts Revision
USOC Uniform Service Order Code
USOE United States Office of Education
USOI United States Office of Information
USOPA United States Ordnance Producers Association
USOS United States Occupational Health Standards
USOSF Union of Shipowners for Overseas Shrimp Fisheries [Greece]
USP Unique Sales Proposition
United States Pharmacopoeia
Universidade de Sao Paulo [= University of Sao Paulo, Brazil]
Usage Sensitive Pricing
USPA United States Patents (Database)
United States Pilots Association
United States Potters Association
United States Psychotronics Association [US]
USPC United States Pharmacopoeia Convention
USPD Ultrasonic Position Decoder
USPE Utah Society of Professional Engineers [US]
USPEC United States Paper Exporters Council
USPHS United States Public Health Service
USPMF United States Patent Model Foundation [US]

USPO United States Patent Office
United States Post Office
USPS United States Postal Service
USPTO United States Patent and Trademark Office
USR User Service Routine
USRA Undergraduate Students Research Awards
United States Railway Association
Universities Space Research Association [US]
USRDA United States Recommended Dietary Allowances
USRFP United States Request for Proposals
USRL Underwater Sound Reference Laboratory [US Navy]
USRS United States Rocket Society
USRT Universal Synchronous Receiver/Transmitter
USS Ultrasonic Society
Unformatted System Service
Uniform Symbol Specification
Unique Support Structure
United States Senate
United States Ship
United States Standard
United States Steel
USSA Underground Security Storage Association
United States Student Association
USSEA United States Space Education Association
USSF United States Space Foundation
USSG United States Standard Gauge
USSI Ultrasonic Society of India
USSIA United States Shellac Importers Association
USSR Union of Soviet Socialist Republics
USSTL WG United States Steel Wire Gauge [also: USStl WG]
USSTS United States Student Travel Service
USSWG United States Steel Wire Gauge
UST Ultrasonic Testing
Ultrasonic Transducer
Unblock SPADE Terminal (Command)
Underground Storage Tank
University of Science and Technology
USTA United States Telephone Association
United States Trademark Association [US]
USTB University of Science and Technology Beijing [PR China]
USTAG United States Technical Advisory Group [also: US TAG]
USTC United States Tariff Commission
USTD University Science and Technology Division [of SRC, UK]
USTR United States Trade Representative
USTS United States Travel Service
USTSA United States Telephone Suppliers Association
USU Utah State University [US]
USV Ultrasonic Vibration
Unsaturated Vapor
USW Ultrasonic Wave
Ultrasonic Welding
Undersea Warfare
United States Wheat Associates
United Steelworkers of America
Unsteady Shock Wave
USWB United States Weather Bureau
UT Ultrasonic Testing
Umbilical Tower
Uncontrolled Term
Uniaxial Tension
Unipolar Transistor
Universal Time
University of Teheran [Iran]
University of Tennessee [US]
University of Texas [US]
University of Tokyo [Japan]
University of Toledo [US]
University of Tripoli [Libya]
University of Trondheim [Norway]
University of Tsukuba [Japan]
Up Time
Urban Transportation
Urethane Technology
User Terminal
Utah [also: Ut, US]
UT 0 Universal Time 0
UT 1 Universal Time 1
UT 2 Universal Time 2
UTA University of Texas, Arlington [US]
University of Texas, Austin [US]
Upper Control Area
Urban Transportation Administration
User Transfer Address
UTANG University of Toronto Antinuclear Group [Canada]
UTAP Unified Transportation Assistance Program [US]
Urban Transport Assistance Program
UTC United Technology Center
Universal-Test Communicator
Universal Time, Coordinated
Utilities Telecommunications Council [US]
UTCS Urban Traffic Control System
UTD United
Universal Tone Decoder
Universal Transfer Device
University of Texas, Dallas [US]
UTDC Urban Transportation Development Corporation [Canada]
UTE Universidad Technica del Estado [= State Technical University, Chile]
UTEC Utah University College of Engineering [US]
UTEP University of Texas, El Paso [US]
UTF Ultrathin Film
Underground Test Facility
UTG Universal Tone Generator
UTH Universal Test Head
UTI Universal Test Interface
Universal Time, International
UTIA University of Toronto Institute of Aerophysics [Canada]
UTIAS University of Toronto Institute of Aerospace Studies [also: Utias, Canada]
UTICS University of Texas Institute for Computer Science [US]
UTIL Utility
UTK Ultrasonic Trim Knife
University of Tennessee, Knoxville [US]
UTL Universidade Tecnica de Lisboa [= Technical University of Lisbon, Portugal]
Upper Tolerance Limit

UTLAS University of Toronto Library Automation System [Canada]
UTM Universal Testing Machine
Universal Test Message
Universal Transverse Mercator
UTOA United TVRO (= Television Receive Only) Owners Association [US]
UTOL Universal Translator-Oriented Language
UTOP United Technological Organizations of the Philippines
UTP Ultrafine Tungsten Powder
Universal Tape Processor
University of Toronto Press [Canada]
Unshielded Twisted Pair (Cable)
Upper Turning Point
Uridine Triphosphate
UTQGS Uniform Tire Quality Grading System
UTR Universal Teaching Reactor
Universal Training Reactor [US]
UTRA Underground Transportation Research Association [also: STUVA, Germany]
UTRC United Technologies Research Center [US]
UTRR University of Teheran Research Reactor [Iran]
UTS Ultimate Tensile Strength
Underwater Technology School
Underwater Telephone System
Unified Transfer System
United Transfer System
Unit Test Station
Universal Terminal System
Universal Test Station
Universal Time-Sharing System
Update Transaction System
Urban Transportation System
UTSN Used Truck Sales Network [US]
UTSP Urban Transportation System Planning
ut sup (ut supra) — as above
UTS/VS Universal Time-Sharing System/Virtual Storage
UTT Utility Tactical Transport
UTTAS Utility Tactical Transport Aircraft System
UTTC Universal Tape-to-Tape Converter
UTU United Transportation Union
Universidad del Trabajo del Uruguay [= University of Trabajo, Uruguay]
UTV Uncompensated Temperature Variation
UTW Ultrathin Window
United Telegraph Workers
UTWA United Textile Workers of America
UU Ultimate User
University of Utah [US]
University of Utrecht [Netherlands]
UUA Univac Users Association [US]
UUM Underwater-to-Underwater Missile
UUMP Unification of Units of Measurement Panel [of ICAO]
UUO Unimplemented User Operation
UUT Unit Under Test
UV Ultraviolet
Ultravisible
Undervoltage
University of Victoria [Canada]
University of Vienna [Austria]
Unmanned Vehicle
UV-A Ultraviolet, Range A [from 315 to 400 nm]
UVAR University of Virginia Reactor [US]
UVASER Ultraviolet Amplification by Stimulated Emission of Radiation [also: Uvaser or uvaser]
UV-B Ultraviolet, Range B [from 280 to 315 nm]
UVBY Ultraviolet, Blue, and Yellow (System) [also: uvby]
UVC Unidirectional Voltage Converter
UV-C Ultraviolet, Range C [from 100 to 280 nm]
UVCB Undervoltage Circuit Breaker
UVD Undervoltage Device
Universal Velocity Distribution
UVDM Ultraviolet Data Manager
UV EPROM Ultraviolet-Erasable Programmable Read-Only Memory [also: UVEPROM, or UV-EPROM]
UV ERASER Ultraviolet Eraser
UVFL Ultraviolet Fluorescence
UVI Ultraviolet Imager
UVic University of Victoria [Canada]
UVM Ultraviolet Microscopy
UVP Ultraviolet Photography
Undervoltage Protection
UV PROM Ultraviolet Programmable Read-Only Memory [also: UVPROM, or UV-PROM]
UVR Ultaviolet Radiation
Undervoltage Relay
UVS Ultraviolet Spectroscopy
Ultraviolet Spectrum
Unmanned Vehicle System
UV/VIS Ultraviolet/Visible (Spectroscopy) [also: UV/Vis]
UW Ultrasonic Wave
Underwater
Unique Word
University of Wales [UK]
University of Warsaw [Poland]
University of Warwick [US]
University of Washington [US]
University of Waterloo [Canada]
University of Windsor [Canada]
University of Winnipeg [Canada]
University of Wisconsin [US]
University of Wollongong [Australia]
University of Wyoming [US]
Upset Welding
UWA United Weighers Association [US]
User Working Area
UWAL University of Washington Aeronautical Laboratory [US]
UWB Ultrawideband (Wave)
UWE Underwater Equipment
UWEC University of Wisconsin, Eau Claire [US]
UWEX University of Wisconsin-Extension [US]
UWH Underwater Habitat
UWI University of the West Indies [Jamaica]
UWIST University of Wales Institute of Science and Technology [UK]

UWM University of Wisconsin, Madison [US]
University of Wisconsin, Milwaukee [US]
UWNDS Upper Winds
UWO University of Western Ontario [Canada]
UWP Underwater Photography
University of Waterloo Press [Canada]
Utility Nuclear Waste and Transportation Program [US]
UWRC Urban Wildlife Research Center [now: NIUW, US]
UWRR University of Wyoming Research Reactor [US]
UWTR University of Washington Test Reactor [US]
UWY Upper Airway
UX_1 Uranium X_1 (= Thorium 234)
UX_2 Uranium X_2 (= Uranium Z = Protactinium 234)
UXB Unexploded Bomb
UXD Ultimate X-Ray Detector
UY Unit Years
University of Yunnan [PR China]
Uranium Y (= Thorium 231)
UYP Upper Yield Point
Uruguayan Nuevo Peso [Currency]
UZ University of Zurich [Switzerland]
Uranium Z (= Uranium X_2 = Protactinium 234)

V

V	Vacuum
	Vacuum Tube
	Valine
	Valyl
	Value
	Valve
	Vanadium
	Vapor
	Variable
	Vatu [Currency of Vanuatu]
	V-Coefficient
	Vector
	Velocity
	Verification
	Video
	Visibility
	Visual
	Vitrification
	Volt
	Voltage
	Voltmeter
	Volume
v	vacuum
	valve
	vapor
	variable
	variation
	vector
	vein
	velocity
	vent
	verify
	(versus) — against
	vertex
	vertical
	(vide) — see
	virtual
	viscosity
	visibility
	volume
VA	Vacuum Arc
	Value Analysis
	Variance
	Velocity Analyzer
	Vermiculite Association [US]
	Vice-Admiral
	Vickers Armstrong Gun
	Video Amplifier
	Vinyl Acetate
	Virginia [US]
	Virtual Address
	Visual Aid
	Visual Approach
	Vital Area
	Volt-Ammeter
	Volt-Ampere
V/A	Volume/Area
Va	Virginia [US]
VAA	Viscum Album Agglutinin
	Voice Access Arrangement
VAAC	Vanadyl Acetylacetonate
	Vectored Thrust Advanced Aircraft
VAB	Vertical Assembly Building
	Voice Answerback
VAC	Vacancy
	Vacuum
	Value-Added Carrier
	Variable Amplitude Correction
	Vector Analog Computer
	Vertical Assembly Component
	Video Amplifier Chain
	Volts, Alternating Current
VACC	Value-Added Common Carrier
VACR	Variable Amplitude Correction Rack
VACTL	Vertical Assembly Component Test Laboratory
VAD	Vacuum Arc Decarburization
	Vacuum Arc Degassing
	Value-Added Distributor
	Vapor Axial Deposition
	Vapor-Phase Axial Deposition
	Ventricular Assist Device
	Voltmeter Analog-to-Digital (Converter)
	Volunteer Aid Detachment
VADA	Voluntary Agricultural Development Aid (Program) [of CIDA, Canada]
VADAC	Voice Analyzer Data Converter
VADC	Video Analog to Digital Converter
	Voltmeter Analog-to-Digital Converter
VADE	Versatile Auto Data Exchange
VADER	Vacuum Arc Double Electrode Remelting
VADIS	Voice and Data Integrated System
VAdm	Vice-Admiral
VADS	Verdix Ada Development System
VAE	V Activating Enzyme
	Vinyl Acetate-Ethylene
VAEP	Variable Attributes Error Propagation
VAF	Vernacular Architecture Forum [US]
VAFC	Volume Average Fuel Consumption
VAI	Video-Assisted Instruction
VAK	Versuchs-Atomkraftwerk Kahl [= Nuclear Power Pilot Plant at Kahl, Germany]
VAL	Valley
	Value
	Vehicle Authorization List
	Visual Approach and Landing
Val	Valine
	Valyl
VALL	Variable Angle Load Lock (Sysatem)
VALOR	Variable Locale and Resolution
VALSAS	Variable Length Word Symbolic Assembly System
VAM	Vasicular-Arbuscular Mycorrhizae
	Vector Airborne Magnetometer
	Virtual Access Method
	Vogel's Approximation Method

	Voltammeter
VAMAS	Versailles Project on Advanced Materials and Standards
VAMP	Vector Arithmetic Multiprocessor
	Vincristine Amethopterin
	Visual-Acoustic-Magnetic Pressure
VAN	Value-Added Network
VANS	Value-Added Network Services
VANSGL	Value-Added Network Services General License [UK]
VAOR	VHF Aural Omnirange
VAP	Vapor
	Visual Aids Panel [of ICAO]
VAPEX	Vapor Extraction
VAPI	Visual Approach Path Indicator
VAPP	Vector and Parallel Processor
VAP P	Vapor Pressure [also: VAP PRESS]
VAP PRF	Vapor-Proof [also: VAP-PRF]
VAPS	Virtual Applications Prototyping System
	Virtual Avionics Prototyping System
	VSTOL (= Vertical and/or Short Takeoff and Landing) Approach System
VAR	Vacuum Arc Remelting
	Value-Added Reseller
	Variable
	Variance
	Variant
	Variation
	Variety
	Variometer
	Varmeter
	VHF Aural Range
	Video-Audio Range
	Visual and Aural Range
	Visual-Aural Range
	Volt-Ampere, Reactive [also: var]
VARACTOR	Variable Reactor [also: Varactor or varactor]
VARAD	Varying Radiation
VARC	Variable Axis Rotor Control
VARHM	Var-Hour Meter
VARICAP	Variable-Capacitance Diode [also: Varicap or varicap]
VARINDOR	Variable Inductor [also: Varindor or varindor]
VARISTOR	Variable Resistor [also: Varistor or varistor]
	Variable Transistor [also: Varistor or varistor]
VARITRAN	Variable-Voltage Transformer [also: Varitran or varitran]
VARN	Varnish
VARR	Variable Range Reflector
	Visual Aural Radio Range
VARS	Vertical Azimuth Reference System
VAS	Value-Added Service
	Vector Addition System
	Venomological Artifact Society [US]
	Vibration Absorption System
	Videodisk Authorizing System
	Vortex Advisory System
VASCA	Valve and Semiconductor Manufacturers Association
VASCAR	Visual Average Speed Computer and Recorder
VASG	Vapor-Air Specific Gravity
VASE	Variable Angle Spectroscopic Ellipsometry
VASE FTIR	Variable Angle Spectroscopic Ellipsometry Fourier Transform Infrared [also: VASE-FTIR]
VASI	Visual Approach Slope Indicator
VASIS	Visual Approach Slope Indicator System
VASP	Value-Added Service Producer
	Value-Added Service Provider
VAST	Versatile Automatic Specification Tester
	Versatile Avionics Ship Test
	Vibration and Strength Analysis
VAT	Value-Added Tax
	Vatu [Currency of Vanuatu]
	Virtual Address Translator
	Voice Activation Technology
VATE	Versatile Automatic Test Equipment
VATLS	Visual Airborne Target Location System
VAU	Vertical Arithmetic Unit
VAV	Variable Air Volume
	Ventricular Assist Valve
VA/VE	Value Analysis/Value Engineering
VAW	Volcanic Ash Warning
VAWT	Vertical Axis Wind Turbine
VAWTG	Vertical Axis Wind Turbine Generator
VAX	Virtual Address Extension
VAXCOM	VAX Committee
VB	Valence Band
	Valence Bond
	Valve Box
	Variable Block
	Voice Band
VBA	Vibrating Beam Accelerometer
VBAS	Von Braun Astronomical Society [US]
VBB	Verein der Bibliothekare an Offentlichen Bibliotheken [= Association of Librarians in Public Libraries, Germany]
VBD	Voice Band Data
VBDOS	Valance Band Density of States
VBE	Valinolbutylether
	Volumetric Balance Equation
VBG	Verband der Berufsgenossenschaften [= Federation of Professional/Trade Associations, Germany]
VBI	Vertical Blanking Interval
	Vital Bus Inverter
VBL	Vertical Block Line
	Voyager Biological Laboratory [of NASA]
VBM	Valence Band Maximum
	Valence-Bond Method
VBMA	Vacuum Bag Manufacturers Association [US]
VBO	Valence Band Offset
	Venezuelan Bolivar [Currency]
VBOMF	Virtual Base Organization and Maintenance Processor
VBP	Virtual Block Processor
VBS	Valence Band Spectrum
VBWR	Vallecitos Boiling Water Reactor [US]
VC	Vacuum Calorimeter
	Valence Crystal
	Vanadium Carbide
	Vapor Compression
	Variable Capacitor
	Varnished Cambric (Wire)
	Vector Colorimeter

	Vector Control
	Vena Contracta
	Verification Condition
	Versatility Code
	Vertical Circle
	Vice-Chancellor
	Vice-Consul
	Victoria Cross
	Video Correlator
	Vinyl Chloride
	Vinylidene Chloride
	Virtual Call
	Virtual Cathode
	Virtual Channel
	Virtual Circuit
	Visible Capacity
	Visual Capacity
	Vitamin Chemist
	Vitamin Chemistry
	Voice Coil
	Voltage Circuit
	Voltage Comparator
	Volt-Coulomb
VCA	Value Control Amplifier
	Voice Connecting Arrangement
	Voltage-Controlled Amplifier
VCAD	Video Control and Drawing Chip
VCASS	Visually Coupled Airborne Systems Simulator
VCAT	Veterinary College Admission Test
VCBA	Variable Control Block Area
VCC	Venture Capital Corporation
	Vertial Continuous Casting
	Video Compact Cassette
	Visual Communications Congress
	Voice Control Center
	Voltage-Controlled Capacitor
VCCS	Video and Cable Communications Section
	Voltage-Controlled Current Source
VCD	Vacuum Carbon Deoxidation
	Valve Coil Driver
	Variable-Capacitance Diode
	Variable Center Distance
	Vibrational Circular Dichroism
	Voltage Charge Device
VCE	Collective Emitter Voltage
VCF	Voltage-Controlled Filter
	Voltage-Controlled Frequency
VCG	Vapor Crystal Growth
	Vectorcardiogram
	Vectorcardiography
	Verification Condition Generator
	Vertical Location of the Center of Gravity
VCGS	Vapor Crystal Growth System
VCI	Vehicle Cone Index
	Verband der Chemischen Industrie [= Federation of the Chemical Industry, Germany]
	Volatile Corrosion Inhibitor
VCL	Vertical Center Line
VCLF	Vertical Cask-Lifting Fixture
VCM	Vacuum Condensible Material
	Variable-Capacitance Micromotor
	Vehicle Condition Monitor
	Vibrating Coil Magnetometer
	Vinyl Chloride Monomer
	Voltage-Controlled Multivibrator
VCMA	Vacuum Cleaner Manufacturers Association [US]
VCNR	Voltage-Controlled Negative Resistance
VCO	Voice-Controlled Oscillator
	Voltage-Controlled Oscillator
VCOAD	Voluntary Committee on Overseas Aid and Development [UK]
VCOC	Venture-Capital Operating Company
VCONFIG	Video Configuration
	VGA (= Visual Graphics Adapter) Configuration
VCP	Vacuum Condensing Point
	Vacuum Cup Pencil
	Virtual Control Panel
VCR	Variable Compression Ratio
	Vertical Crater Retreat
	Video Cartridge Recorder
	Video Cassette Recorder
	Visual Control Room
VCRG	VHF Channel Requirements Group [of ICAO]
VCRO	Validity Check and Readout
VCS	Vacuum Science Workshop [of BNL, US]
	Validation Control System
	Ventilation Control System
	Victorian Computer Society
	Video Cassette System
	Video Communications System
	Video Computer System
	Vibration Control System
	Visually Coupled System
	Voltage Calibration Set
VCSEL	Vertical-Cavity Surface-Emitter Laser
VCSR	Voltage-Controlled Shift Register
VCT	Voltage Control Transfer
	Volume Control Tank
VCTCA	Virtual Channel-to-Channel Adapter
VCU	Virginia Commonwealth University [US]
VC-VA	Vinyl Chloride — Vinyl Acetate (Copolymer)
VC-VDC	Vinyl Chloride — Vinylidene Chloride
VCVS	Voltage-Controlled Voltage Source
VCXO	Voltage-Controlled Crystal Oscillator
VD	Vacuum Diode
	Vacuum Drying
	Vandyke
	Vapor Density
	Varactor Diode
	Vector Diagram
	Vibration Damper
	Virtual Data
	Virtual Device
	Void
	Voltage Detector
	Voltage Divider
	Voltage Doubler
	Voltage Drop
	Volume Damper
	Volume Dose
	Voronoi Diagram
V/D	Voice Data

vd various dates
VDA Verbal Delay Announcement
Vertical Danger Angle
VDAM Virtual Data Access Method
VDAS Vibration Data Acquisition System
VDB Vector Data Buffer
Verein Deutscher Bibliothekare [= Association of German Librarians, Germany]
Video Display Board
VDC Versatile Digital Controller
Video Data Controller
Video Display Controller
Video-Documentary Clearinghouse [US]
Vinylidene Chloride
Virtual Device Coordinate
Viscosity-Density Constant
Visual Display Controller
Volts, Direct Current
VDCT Direct Current Test Volts
VDCU Videograph Display Control Unit
VDCW Direct Current Working Volts
VDD Vietnamese Dong [Currency]
Visual Display Data
Voice Digital Display
VdDB Verein der Diplom-Bibliothekare an Wissenschaftlichen Bibliotheken [= Association of Graduate Librarians in Academic/Research Libraries, Germany]
VDDI Voyager Data Detailed Index
VDDL Voyager Data Distribution List
VDDS Voyager Data Description Standards
VDE Variable Displacement Engine
Variable Display Equipment
Verband Deutscher Elektrotechniker [= Association of German Electrical Engineers]
Voice Data Entry
VDEh Verein Deutscher Eisenhuttenleute [= Association of German Iron and Steel Engineers]
VDET Voltage Detector
VDETS Voice Data Entry Terminal System
VDEW Vereinigung Deutscher Elektrizitatswerke [= Association of German Utility Companies]
VDF VHF Direction Finding
Video Frequency
VDFG Variable Diode Function Generator
VDG Variable Drive Group
Verein Deutscher Giessereifachleute [= Association of German Foundry Engineers and Technologists]
Vereinigung Deutscher Gewasserschutz [= German Association for Water Pollution Control]
Video Data Generator
Video Display Generator
VDGS Visual Docking Guidance System
VDI Verein Deutscher Ingenieure [= Association of German Engineers]
Video Display Input
Video Display Interface
Virtual Device Interface
Visual Display Input
VDI-TZ Verein Deutscher Ingenieure — Technologiezentrum [= Technology Center of the Association of German Engineers]
VDI/VDE Verein Deutscher Ingenieure/Verband Deutscher Elektrotechniker [Germany]
VDI-W Verein Deutscher Ingenieure/Werkstoffgesellschaft [= Materials Society of the Association of German Engineers]
VDK Vicinal Diketone
VDKSPI V. D. Kuznetsov Siberian Physiotechnical Institute [USSR]
VDL Verband Deutscher Diplomlandwirte [= Association of German Agricultural Engineers]
Vienna Definition Language [Computer Language]
Vision Development Language
VDLUFA Verband Deutscher Landwirtschaftlicher Untersuchungs- und Forschungsanstalten [= Federation of German Institutes for Agricultural Research]
VDM Vector Dominance Model
Video Display Module
Virtual Device Metafile
Visual Display Module
VDMA Variable Destination Multiple Access
Verband Deutscher Maschinen- und Anlagenbau [= German Machinery and Plant Manufacturers Association]
VDOS Vacuum Distillation/Overflow Sampler
VDP Van der Pauw Structure
Vertical Data Processing
Video Data Processor
Visual Descent Point
VDPI Voyager Data Processing Instruction
VDR Voice and Data Recording
Voice Digitization Rate
Voltage-Dependent Resistor
VDRA Voice and Data Recording Auxiliary
VDRIVER Video Driver
VGA (= Visual Graphics Adapter) Driver
VDS Variable Depth Sonar
Verband Deutscher Studentenschaft [= German Students Association]
Very Difficultly Soluble
Video Distribution System
Visual Docking Simulator
Voice Data Switch
VDSI Verein Deutscher Sicherheitsingenieure [= Association of German Safety Engineers]
VDT Validate
Video Data Terminal
Video Display Terminal
Visual Display Terminal
VDTA Vacuum Dealers Trade Association [US]
VDTUV Vereinigung der Technischen Uberwachungsvereine [= Federation of Technical Inspection Associations, Germany]
VDU Video Display Unit
Visual Display Unit
VDUC Video Display Unit Controller
Visual Display Unit Controller
VDW Van der Waals

VDX Videotex
VE Vacuum Engineer
Vacuum Engineering
Valance Electron
Value Engineer
Value Engineering
Vernal Equinox
Vernier Engine
Vibrational Energy
Vibration Eliminator
Vinyl Ether
Viscoelastic(ity)
Visual Exempted
Vitreous Enamel
Volume Expansion
VEA Value Engineering Association
Virtual Effective Address
VEB Variable Elevation Beam
Venezuelan Bolivar [Currency]
VEC Variable Energy Cyclotron
VECI Vehicular Equipment Complement Index
VECO Vernier Engine Cutoff
VECOS Vehicle Checkout Set
VECP Value Engineering Change Proposal
Value Engineering Cost Proposal
VECU Vacuum Pump Exhaust Cleanup System
VED Vapor-Phase Epitaxial Deposition
VEDAR Visible Energy Detection and Ranging
VEDS Vehicle Emergency Detection System [of NASA]
VEdS Vocational Education Specialist
VEFCO Vertical Functional Checkout
VEG Vegetable
VEGA Vegetable Growers of America [US]
VEH Vehicle
VEL Velocity
VELCOR Velocity Correction
VELF Velocity Filter
Verwaltung fur Ernahrung, Landwirtschaft und Forsten [= Administration for Food, Agriculture and Forestry, Germany]
VELOC Velocity
VEM Vasoexcitor Material
Virtual Electrode Model
VEN Venerable
Virtual Equipment Number
VenAmCham Venezuelan-American Chamber of Commerce and Industry [Venezuela]
VENEZ Venezuela(n)
VENT Ventilation
Ventilator
VENUS Valuable and Efficient Network Utility Service
VEP Visual Evoked Potential
VEPIS Vocational Education Program Information System
VER Verification
Verifier
Vernier
Version
Vertical
Visual Evoked Response
Voluntary Export Restraint
VERA Versatile Experimental Reactor Assembly
Vision Electronic Recording Apparatus
VERDAN Versatile Differential Analyzer
VERLORT Very Long-Range Tracking
VERNITRAC Vernier Tracking by Automatic Correlation
VEROS Vacuum Evaporation on Running Oil Surface
VERPROT Verification Prototype
vers versed sine [also: versine]
VERSCOM Verification for Special Commission
versine versed sine [also: vers]
VERT Vertebrate
Vertical
VER TAG Verification Tagging
VERVIS Vertical Visibility
VES Variable Elasticity of Substitution
VESC Vehicle Equipment Safety Commission [US]
VESIAC Vela Seismic Information Analysis Center [US]
VESR Vellecitos Experimental Superheat Reactor [US]
VEST Volunteer Engineers, Scientists and Technicians
VET Veterinarian [also: VETER]
Veterinary [also: VETER]
Visual Editing Terminal
VETDOC Veterinary Documentation (Database) [UK]
VETL Vehicle Emission Testing Laboratory [Canada]
VEV Voice Excited Vocoder
VEWS Very Early Warning System
VF Vacuum Flask
Valance Force
Value Foundation [now: MVF, US]
Variable Frequency
Vector Field
Velocity Filter
Very Fine
Vibration Foundation [now: VI, US]
Video Frequency
Viscous Flow
Visibility Factor
Voice Frequency
Voltage-to-Frequency [also: V-F or V/F]
VFA Vicia Faba Agglutinin
Volatile Fatty Acid
Volunteer Fire Alarm
VFB Vertical Format Buffer
Voltage, Flatband
Voltage Feedback
VFC Variable File Channel
Vertical Format Control
Vertical Forms Control
Video Film Converter
Video Frequency Carrier
Voice-Frequency Carrier
Voltage-to-Frequency Converter
VFCT Voice-Frequency Carrier Telegraph
Voice-Frequency Carrier Teletype
VFD Vacuum Fluorescent Display
VFDB Vereinigung zur Forderung des Deutschen Brandschutzes [= German Association for the Advancement of Fire Protection]
VFET Vertical Field-Effect Transistor
VFF Vertical Flight Foundation [US]
VFFT Voice Frequency Facility Terminal
VFH Vertical Flow Horizontal

VFL	Variable Field Length
	Variable Focus Lens
	Voice Frequency Line
VFMED	Variable Format Message Entry Device
VFO	Variable-Frequency Oscillator
	Voice-Frequency Oscillator
VFOP	VFR (= Visual Flight Rules) Operations Panel [of ICAO]
VFR	Visual Flight Rules
VFRG	Visual Flight Rules Group [of ICAO]
VFS	Vacuum Furnace System
	Vertical Full Scale
VFSS	Variable Frequency Selection System
VFT	Voice-Frequency Telegraph
	Voice-Frequency (Carrier) Telegraphy
VFTG	Voice Frequency Telegraph
VFU	Vertical Format Unit
	Vocabulary File Utility
VFVC	Vacuum Freezing Vapor Compression
VG	Vacuum Gauge
	Vector Generator
	Vector Geometry
	Vertical Grain
	Very Good
	Viscosity Grade
	Voice Grade
VGA	Vapor Generation Accessory
	Variable Gain Amplifier
	Video Graphics Array
	Visual Graphics Adapter
VGAA	Vegetable Growers Association of America
VGADDS	Vertical General Area Detector Diffraction System
VGAM	Vector Graphics Access Method
VGATEST	Visual Graphics Adapter Test
VGB	Vereinigung der Grosskraftwerksbetreiber [= Association of Operators of Large Power Plants, Germany]
VGC	Vacuum Gage Control(ler)
	Viscosity-Gravity Constant
VGCS	Vacuum Gage Calibration System
VGI	Voice Group Interface
VGM	Virtual Graphics Machine
VGO	Vacuum Gas Oil
VGP	Virtual Geomagnetic Pole
VGPI	Visual Glide Path Indicator
VGR	Video Graphics Recorder
VG/RIA	Vision Group of Robotics Industries Association [now: AVA, US]
VGS	Vision Guidance System
	Visual Guidance System
VGSI	Visual Glide Slope Indicator
VGU	Video Generation Unit
VH	Vanadium Hydride
	Varhour Meter
	Velocity Head
VHA	Very High Accuracy
	Very High Altitude
VHAA	Very High Altitude Abort
VHD	Video High Density
VHDL	Very High Density Lipoprotein
VES	Vitro Hanford Engineering Services [US]
VHF	Very High Frequency
VHF/PTN	VHF Radio Connection to Public Telephone Network
VHF/UHF	Very High Frequency/Ultrahigh Frequency
VHLL	Very High Level Language
VHM	Virtual Hardware Monitor
VHN	Vickers Hardness Number
VHO	Very High Order
	Very High Output
VHOL	Very High Order Language
VHP	Vacuum Hot Pressing
	Very High Performance
	Very High Pressure
VHPIC	Very High Performance Integrated Circuit
VHR	Very High Reduction
	Very High Resolution
VHRR	Very High Resolution Radiometer
VHS	Video Home System
VHS-C	Video Home System — Compact
VHSI	Very High Scale Integration
	Very High Speed Integration
VHSIC	Very High Scale Integrated Circuit
	Very High Speed Integrated Circuit
VHTR	Very High Temperature Reactor
VHV	Very High Voltage
VI	Vancouver Island [Canada]
	Variable Intensity
	Vibration Institute [US]
	Video Integrator
	Virgin Islands [US]
	Viscosity Index
	Virtual Image
	Visual Information
	Volume Indicator
VIA	VectorNet Interface Adapter
	Versatile Interface Adapter
	Videotex Industry Association [US]
	Visual Interactive Access
VIAS	Voice Intelligibility Analysis Set
	Voice Interference Analysis Set
VIB	Vertical Integration Building
	Vibration
	Vibrator
VIBL	Variable Intensity Back Lighting
VIC	Variable Instruction Computer
	Victoria [Australia]
	Virtual Interaction Controller
vic	vicinal
VICA	Vocational Industrial Clubs of America [US]
VICAM	Virtual Integrated Communication Access Method
VICAR	Video Image Communication and Retrieval
VICARS	Visual Integrated Crime Analysis and Reporting System
Vice-Adm	Vice-Admiral
VICI	Velocity Indicating Coherent Integrator
VICOM	Visual Communication
VID	Vacuum Injection Degassing
	Video
	Vietnamese Dong [Currency]
vid	(vide) — see
VIDA	VAX/IBM Data Access
VIDAC	Virtual Data Acquisition and Control

	Visual Data Acquisition
	Visual Information Display and Control
VIDAMP	Video Amplifier
VIDAT	Visual Data Acquisition
VIDEO	VORTEX (= Versatile Omnitask Real-Time Executive) Interactive Data Entry Operation
VIDF	Vertical Side of Intermediate Distribution Frame
	Video Frequency
vid i	(vide infra) — see below
VIDIAC	Video Input-to-Automatic Computer
	Vidicon Input-to-Automatic Computer
	Visual Information Display and Control
VIDIST	Vacuum Injection Distillation
VIDO	Veterinary Infections Disease Organization
VIDP	Vacuum Injection Degassing and Pouring
VIDS	Visual Information Display System
vid s	(vide supra) — see above
VIE	Victorian Institute of Engineers [Australia]
VIEW	Virtual Instrument Engineering Workbench
	Visual Information Enhanced Workstation
VIF	Visiting Industrial Fellowship
VIFC	VTOL (= Vertical Takeoff and Landing) Integrated Flight Control
VIFCS	VTOL (= Vertical Takeoff and Landing) Integrated Flight Control System
VIFI	Voyager Information Flow Instruction
VIG	Video Integrating Group
	Visual Integrating Group
VIGS	Virtual-Induced Gap States
	Visual Glide Slope
VIH	Voltage-Input High
VIL	Village
	Voltage-Input Low
VILP	Vector Impedance Locus Plotter
VILSU	V. I. Lenin State University [USSR]
VIM	Vacuum Induction Melting
	Vibrational Microlamination
	Vision Input Module
	Vocational Instructional Materials
VIMIS	Vertically-Integrated Metal-Insulator-Semiconductor
VIMP	Vancouver Island Mainland Pipeline [Canada]
VIMS	Virginia Institute of Marine Science [US]
VIMTPG	Virtual Interactive Machine Test Program Generator
VIM/VAR	Vacuum Induction Melting/Vacuum Arc Remelting
VIN	Vehicle Identification Number
	Voltage Input
VIND	Vicarious Interpolations Not Desired
VINS	Velocity Inertia Navigation System
	Vermont Institute of Natural Science [US]
	Very Intense Neutron Source
VINT	Video Integration
VIO	Video Input/Output
	Virtual Input/Output
VIOC	Variable Input/Output Code
VIOLET	Voice Input/Output Lexically Endowed Terminal
VIOS	Versatile Instrument Operating Software
VIP	Vacuum Injection Pouring
	Variable Information Processing
	Variable Input Phototypesetting
	Vasoactive Intestinal Peptide
	Vector Instruction Processor
	VectorNet Interface Processor
	Versatile Information Processor
	Very Important Person
	Videodisk Innovation Project [of USU, US]
	Vision Integrated with Positioning
	Visual Image Processor
	Visual Information Processor
	Visual Information Projection
	Visual Information Protocol
VIPER	Video Processing and Electronic Reduction
VIPP	Variable Information Processing Package
VIPS	Variable Information Processing System
	Voice Interruption Priority System
VIR	Valves-in-Receiver System
	Vertical Interference Reference
	Vertical Interval Reference
	Vertical Interval Retrace
	Visible and Infrared Spin-Scan Radiometer
	Vulcanized India Rubber
VIRNS	Velocity Inertia Radar Navigation System
VIROC	Visible System of Information Retrieval by Optical Coordination
VIROL	Virological
	Virologist
	Virology
VIRSS	Visual and Infrared Screening Smoke (Grenade)
VIS	Veterinary Investigation Service [UK]
	Videotex Information System
	Viscosity
	Viscount
	Visibility
	Visible
	Visual Instrumentation System
VISAM	Virtual Indexed Sequential Access Method
VISC	Video Disc
	Viscosity
	Viscous
VISIT	Vehicle Internal Systems Investigative Team (Project) [US]
VISLAP	Viscoelastic Lamination Theory Program
VIS-NIR	Visible/Near Infrared
VISSR	Visible and Infrared Spin Scan Radiometer
VISTA	Verbal Information Storage and Text Analysis
	Visual Interpretation System for Technical Applications
VIT	Vertical Interval Test
	Vitreous
VITA	VME (= Versa Module Europe) Bus International Trade Association
	Volunteers for International Technical Assistance [US]
VITAL	Variably Initialized Translator for Algorithmic Languages
	VAST (= Versatile Automatic Specification Tester) Interface Test Application Language
VITEAC	Video Transmission Engineering Advisory Committee [US]
VITS	Vertical Insertion Test Signal
	Vertical Interval Test Signal
VIU	Video Interface Unit

	Voice Interface Unit
VIURAM	Video Interface Unit Random-Access Memory [also: VIU RAM]
VIV	Variable Inlet Vane
viz	(videlicet) — namely; that is
VJ	Vacuojunction
VJFET	Vertical Junction Field-Effect Transistor
VK	Volterra Kernel
VKIFD	Von Karman Institute for Fluid Dynamics [US]
VL	Vertical Ladder
	Video Logic
	Visual Learning
V/L	Vapor/Liquid
V-L	Vapor-to-Liquid
VLA	Very Large Array [of NRAO, US]
	Very Low Altitude
VLB	Very Long Baseline
VLBI	Very-Long-Baseline Interferometry
VLC	Video Level Controller
VLCC	Very Large Cargo Carrier
	Very Large Crude Carrier
VLCR	Variable Length Cavity Resonance
VLCS	Voltage-Logic Current-Switching
VLD	Valid
	Very Low Density
VLDL	Very Low Density Lipoprotein
VLDPE	Very Low Density Polyethylene [also: VLD-PE]
VLE	Vapor/Liquid Equilibrium
VLED	Visible Light-Emitting Diode
VLF	Variable Length Field
	Vertical Launch Facility
	Very Low Frequency
VLFEM	Very Low Frequency Electromagnetics
VLFS	Variable Low Frequency Standard
VLIW	Very Long Instruction Word
VLM	VLX Large Memory [VLX Is A Trade Name]
VLNT	Violent
VLP	Video Long Play(er)
	Virus-Like Particle
VLPE	Very Long Period Experiment
VLR	Variable Linear Resistor
	Very Long Range
VLS	Vacancy Lattice Site
	Vacuum Loading System
	Vapor-Liquid-Solid Process
	Very Large Scale
	Virtual Linkage System
	Volume Loadability Speed
VLSI	Very Large Scale Integration
VLSIC	Very Large Scale Integrated Circuit
VLSIIC	Very Large Scale Integration Implementation Center [Canada]
VLSW	Virtual Line Switch
VLT	Violet
VLVS	Voltage-Logic Voltage Switching
VM	Vacuum Melting
	Varnishmaker
	Vector Message
	Vector Model
	Velocity Meter
	Velocity Modulation
	Ventura Meter
	Vertical Measurement
	Virtual Machine
	Virtual Memory
	Volatile Matter
	Voltage Multiplication
	Voltmeter
V/m	Volts per meter
VMA	Vacuum Metallizers Association [now: AIMCAL, US]
	Valid Memory Address
	Valve Manufacturers Association) [US]
	Vanillomandelic Acid
	Vehicle Maintenance Area
	Virtual Machine Assist
	Virtual Memory Allocation
VMAI	Veterinary Medicine Association of Ireland
VMAPS	Virtual Memory Array Processing System
VMB	Velocity-Modulated Beam
VM/BSE	Virtual Machine/Basic System Extension
VMC	Variable Message Cycle
	Vertical Machining Center
	Vertical Motion Carriage
	Virtual Machine Control
	Visual Meteorological Conditions
VMCB	Virtual Machine Control Block
VMCF	Virtual Machine Communication Facility
VMD	Vector Meson Dominance
	Vertical Magnetic Dipole
	Virtual Manufacturing Device
VMDF	Vertical Side of Main Distribution Frame
VME	Versa Module Europe (Bus) [Electronics]
	Virtual Machine Environment
	Virtual Memory Environment
VMGSE	Vehicle Measuring Ground Support Equipment
VMH	Visual Maneuvering Height
VMI	Vertical Motion Index
VMID	Virtual Machine Identifier
V/mil	Volts per mil
VMJ	Vertical Multijunction
VMM	Variable Mission Manufacturing
	Virtual Machine Monitor
VMOS	Vertical-Channel Metal-Oxide Semiconductor
	Virtual Memory Operating System
	V-Type Metal-Oxide Semiconductor
VM&P	Varnishmakers and Painters
VMPA	Vancouver Museums and Planetarium Association [Canada]
	Verband der Materialprufamter [= Federation of Materials Testing Institutes, Germany]
VMS	Valve Monitoring System
	Voice Mailbox Service
	Voice Messaging Service
	Volcanogenic Massive Sulfide
VM/SE	Virtual Machine/System Extension
VM/SP	Virtual Machine/System Product
VMT	Variable Microcycle Timing
	Variable Mu Tube
	Velocity-Modulated Tube
	Video Matrix Terminal
VMTAB	Virtual Machine Table
VMTSS	Virtual Machine Time-Sharing System

VMU Voice Management Unit
VMX Voice Message Exchange
VN Vanadium Nitride
Verify Number
Volatile Nitrogen
VNA Very Narrow Aisle
VNAV Vertical Navigation
VNC Voice Numerical Control
VNCIAWPRC Venezuelan National Committee of the International Association on Water Pollution Research and Control
VND Vietnamese Dong [Currency]
VNL Via Net Loss
VNLF Via Net Loss Factor
VNN Vacant National Number
VNSP Vacant Nozzle Shield Plug
VO Verbal Orders
Voice-Over
VOA Voice of America [US]
VOC Variable Output Circuit
Vinyloxycarbonyl
Voice-Operated Coder
Volatile Organic Carbon
Volatile Organic Chemical
Volatile Organic Compound
VOCAB Vocabulary
VOCD Vacuum Oxygen Carbon Deoxidation
VOCODER Voice Coder [also: Vocoder or vocoder]
Voice-Operated Coder [also: Vocoder or vocoder]
VOCOM Voice Communications
VOD Vacuum-Oxygen Decarburization
Velocity of Detonation
Vertical Onboard Delivery
Voice-Operated Device
VODACOM Voice Data Communications
VODAS Voice-Operated Device, Anti-Singing [also: Vodas or vodas]
VODAT Voice-Operated Device for Automatic Transmission [also: Vodat or vodat]
VODC Vacuum-Oxygen Decarburization Converter
VODER Voice Decoder
Voice-Operation Demonstrator [also: Voder or voder]
VOGAD Voice-Operated Gain-Adjusted Device [also: Vogad or vogad]
VOH Verification Off Hook
Voltage-Output High
VOIR Venus Orbiting Imaging Radar
VOIS Visual Observation Instrumentation Subsystem
Visual Observation Instrumentation System
VOL Voltage-Output Low
Volcano
Volume [also: vol]
Voluntary
Volunteer
VOL% Volume [also: vol%]
VOLAT Volatile
VOLCAS Voice-Operated Loss Control and Suppressor [also: Volcas or volcas]
VOLERE Voluntary/Legal/Regulatory
VOL PCT Volume Percent [also: vol pct]
VOLS Volumes
Voluntary Overseas Library Service [UK]
VOLSER Volume/Serial Number
VOLT Volatilize
VOLTAN Voltage Amperage Normalizer
VOM Volt-Ohm Meter
Volt-Ohm Millammeter
VOR Vaughan Orientation Relationship
VHF Omnidirectional Radio Range
VHF Omnidirectional Range
VHF Omnirange
Voice-Operated Recording
VORDAC VOR/DME (= VOR/Distance Measuring Equipment) for Average Coverage [also: Vordac]
VOR/DMET VOR/DME (= VOR/Distance Measuring Equipment) Compatible with TACAN (= Tactical Air Navigation)
VORI Viticultural and Oenological Research Institute [South Africa]
VORTAC VOR (= Very-High-Frequency Omnirange) and TACAN (= Tactical Air Navigation) [also: Vortac]
VORTEX Versatile Omnitask Real-Time Executive
VOS Vertical Obstacle Sonar
Virtual Operating System
Voice-Operated Switch
VOSC VAST (= Versatile Automatic Specification Tester) Operating System Code
VOT Voice-Operated Transmit
VOR (= VHF Omnirange) Test Signal
VOTA Vibration Open Test Assembly
VOTERM Voice Terminal
VOTS VAX OSI (= Open System Interconnect) Transport Services [VAX is a Trademark]
VOX Voice-Operated Circuit
Voice-Operated Control
Voice-Operated Device
Voice-Operated Regulator
Voice-Operated Transmission
Voice-Operated Transmitter
VOx Vanadium Oxide
VOY Voyage
VP Vacuum Pump
Validation Parameter
Vapor Pressure
Vector Processor
Velocity Potential
Velocity Pressure
Vent Pipe
Vertical Polarization
Verify Position
Vice-President
Vinyl Pyridine (Copolymer)
Virtual Processor
Virtual Program
Vulnerable Point
V-P Vice-President
VPAM Virtual Partitioned Access Method
VPC Vacuum Photocell
Vapor-Phase Chromatography
VPCA Video Prelaunch Command Amplifier
VP/CSS Virtual Program/Conversation Software System

VPD Vacuum Products Division
Vapor-Phase Deacidification
VPE Vapor-Phase Epitaxy
VPED Vapor-Phase Epitaxial Deposition
VPF Vehicle Protection Factor
Vertical Processing Facility
VPFIN Vice-President of Finance
VPG Vapor Growth
VPH Vickers Pyramid Hardness
VPI Vacuum Pressure Impregnation
Vapor Phase Inhibitor
Virginia Polytechnic Institute [US]
VPISU Virginia Polytechnic Institute and State University [also: VPI&SU, US]
VPL Vancouver Public Library [Canada]
Vanishing Point Left
VPLCC Vehicle Propellant Loading Control Center
VPM Vehicles per Mile
Vibrations per Minute
Volts per Meter
Volts per Mil
VPN Virtual Page Number
VPO Valve Position Option
VPPA Variable Polarity Plasma Arc
VPPO Vice-President of Plant Operations
VPR Vanishing Point Right
V PRES Vice-President
VPRF Variable Pulse Repetition Frequency
VPR Vapor-Phase Reflow
Virtual PPI (= Plan Position Indicator) Reflectoscope
Voice Position Reports
VPS Vacuum Plasma Spraying
Vapor Phase (Reflow) Soldering
Vibrations per Second
Video Printing System
Video Program System
Virtual Programming System
VPSD Vacuum Plasma Structural Deposition
VPSP Vinylpolystyrylpyridine
VPSS Vector Processing Support Subsystem
VPSW Virtual Program Status Word
VPT Vapor-Pressure Thermometer
Viscosity — Pressure — Temperature [also: V-P-T]
Voice plus Telegraph
Volume — Pressure — Temperature [also: V-P-T]
VP&VLE Vapor Pressures and Vapor Liquid Equilibria (Database) [of NPL, US]
VPW Visiting Professorships for Women Program [of NSF, US]
VPZ Virtual Processing Zero
VQFP Very-Thin Quad Flatpack
VQMG Vice-Quartermaster-General
VR Vacuum Residual Oil
Validation and Recovery
Variety Reduction
Velocity Ratio
Vibrational Relaxation
Virtual Reality
Virtual Route
Vision Research
Visual Route
Voltage Regulation
Voltage Regulator [also: V-R]
Voltage Relay
Vulcanized Rubber
VRA Voluntary Restraint Agreement
VRAM Variable Rate Adaptive Multiplexing
Video Random-Access Memory
VRB VHF Recovery Beacon
Voice Rotating Beacon
VRBL Variable
VRC Variable Resistive Component
Vertical Reciprocating Conveyor
Vertical Redundancy Check
Visual Record Computer
Voice-Recognition Chip
VRCI Variable Resistive Components Institute [US]
VRD Vacuum-Tube Relay Driver
VRE Voice Recognition Equipment
VREW Vegetative Rehabilitation and Equipment Workshop
VRF Vertical Random Format
VRFI Voice Reporting Fault Indicator
VRFN Verification
VRFWS Vehicle Rapid Fire Weapon System
VRG Vertical Reference Gyro
VRID Virtual Route Identifier
VRIP Volume Related Incentive Pricing
VRL Vertical Reference Line
Vibration Research Laboratory [US]
VRM Variable Range Marker
Vertical Retreat Mining
Viscous Remanent Magnetization
Visible Record Machine
VRMS Volts Root-Mean-Square
VROM Video Read-Only Memory
VRP Visual Record Printer
Visual Reporting Point
VRPS Voltage-Regulated Power Supply
VRR Visual Radio Range
VRS Vertical Raster Scanning
Volatile Reducing Substances
VRSA Voice Reporting Signal Assembly
VRSS Vehicular Radar Safety System
Voice Reporting Signal System
VRT Variable Reactance Transformer
Vessel Residence Time
Volume-Rendering Technique
VRU Voice Response Unit
Voltage Readout Unit
VRX Virtual Resource Executive
VRX-MP Virtual Resource Executive — Multiprocessor
VS Vacuum Science
Vacuum Spectrometer
Vacuum Switch
Vacuum System
Vanadium Steel
Variable Speed
Variable Store
Variable Sweep
Vent Stack

	Venturi Scrubber
	Vermont State [US]
	Vertical Sounding
	Very Soluble
	Vestigial Sideband
	Veterinary Surgeon
	Vibrational Spectroscopy
	Virtual Storage
	Virtual System
	Visible Spectrum
	Visualization System
	Vocal Synthesis
	Voltmeter Switch
	Volumetric Solution
vs	(versus) — against
VSA	Vacuum Swing Adsorption
	Vehicle Security Association [US]
	Vibrating String Accelerometer
VSAM	Virtual Sequential Access Method
	Virtual Storage Access Method
	Virtual System Access Method
VSAT	Very-Small-Aperture Terminal
VSB	Vestigial Sideband
	VME (= Versa Module Europe) Subsystem Bus [Electronics]
VSBS	Very Small Business System
	Voluntary Standard Bodies
VSB-SC	Vestigial Sideband — Suppressed Carrier [also: VSBSC, VS-SC or VSSC]
VSBY	Visibility
VSC	Vacuum Slag Suction
	Variable Speed Control
	Variable Speech Control
	Vibration Safety Cutoff
	Video System Controller
	Virtual Subscriber Computer
	Voltage-Saturated Capacitor
	Voltage-Stabilizing Circuit
VSCF	Variable-Speed Constant-Frequency
VSD	Variable Speed Drive
	Vehicle Systems Dynamics
	Voltage-Stabilizing Device
VSDA	Video Software Dealers Association [US]
VSDM	Variable Scope Delta Modulation
VSE	Vancouver Stock Exchange [Canada]
	Vermont Society of Engineers [US]
	Vessel Steam Explosion
	Virtual Storage Equipment
	Virtual Storage Extended
VSE/AF	Virtual Storage Extended/Advanced Function
VSEPR	Valence Shell Electron Pair Repulsion
VSETUP	Video Setup
	VGA (= Visual Graphics Adapter) Setup
VSEW	Verband Schweizerischer Elektrizitatswerke [= Federation of Swiss Electric Utility Companies]
VSF	Voice Store and Forward
VSFR	Vertical Seismic Floor Response
VSG	Verband Schweizerischer Graphiker [= Swiss Federation of Graphic Art Designers/Illustrators]
	Vertical Sweep Generator
VSI	Vacuum-Super-Insulation
	Vertical Speed Indicator
	Video Sweep Integrator
	Vinyl Siding Institute [US]
	Virtual Storage Interrupt
	Voltage Source Inverter
VSIG	Vereinigung des Schweizerischen Import- und Grosshandels [= Swiss Association of Importers and Wholesalers]
VSJ	Vacuum Society of Japan
VSL	Variable Specification List
	Ventilation Sampling Line
VSLE	Very Small Local Exchange
VSM	Variational Stiffness Method
	Verein Schweizerischer Maschinenindustrieller [= Association of Swiss Machinery Manufacturers]
	Vertical Separation Minimum
	Vestigial Sideband Modulation
	Vibrating Sample Magnetometer
	Vibrating Space Modulator
	Virtual Storage Management
VSMA	Vibrating Screen Manufacturers Association [US]
VSMF	Virtual Search Microfilm File
	Visual Search on Microfilm
VSMS	Vermont State Medical Society [US]
	Video Switching Matrix System
VSN	Volume Serial Number
VSO	Very Small Outline
	Very Stable Oscillator
VSOP	Vector Signal Operations Package [Software]
	Very Superior Old Product [On Wine Bottles, etc.]
VSP	Verein der Schweizer Presse [= Swiss Press Association]
	Vertical Seismic Profile
	Vertical Speed
	Virtual Switching Point
	Vision Statistical Processor
VSPC	Virtual Storage Personal Computer
VSR	Validation Summary Report
	Vallecitos Superheat Reactor [US]
	Vertical Storage and Retrieval
	Very Short Range
VSRBM	Very Short Range Ballistic Missile
VSRT	Vertical Spindle Rotary Table
VSS	Vapor Suppression System
	Variable Stability System
	Vehicle Speed Sensor
	Video Storage System
	Virtual Storage System
	Voice Signaling System
	Volatile Suspended Solids
	Voltage for Substrate and Sources
	Voltage to Substrate and Sources
VS-SC	Vestigial Sideband — Suppressed Carrier [also: VSBSC, VSB-SC or VSSC]
VSSG	Vertical Separation Study Group [of NAVSEP]
VST	Variable Speed (Friction) Tester
	Variable Speed Transmission
	Verband Schweizerischer Transportanstalten [= Swiss Federation for Transportation]
	Volume Sensitive Tariff

VSTOL	Vertical and/or Short Takeoff and Landing [also: V/STOL]
VSW	Very Short Wave
	Voltage Standing Wave
VSWR	Voltage Standing Wave Ratio
VSYNC	Vertical Synchronization
VT	Vacuum Technologist
	Vacuum Technology
	Vacuum Tube
	Variable Time
	Vehicular Technology [IEEE Group]
	Vermont [US]
	Vertical Tab(ulation Character)
	Vertical Tail
	Video Tape
	Video Terminal
	Visual Testing
	Voice Tube
	Voltage Transformer
Vt	Vermont [US]
VTAB	Vertical Tab(ulation Character)
VTAC	Video Timing and Control
VTAM	Virtual Telecommunications Access Method
	Virtual Terminal Access Method
VTC	Vocational Training Center
VTCS	Vehicular Traffic Control System
VTD	Vertical Tape Display
VTDC	Vacuum Tube Development Committee
VTDI	Variable Threshold Digital Input
VTE	Vertical Tube Evaporator
VTEC	Verotoxic E. Coli [Biotechnology]
VTF	Vertical Test Fixture
	Via Terrestrial Facilities
VTI	Video Terminal Interface
VTL	Variable Threshold Logic
VTLS	Virginia Technical Library System [of VPI, US]
VTM	Voltage-Tunable Magnetron
VTMS	Vessel Traffic Management System
VTO	Voltage-Tuned Oscillator
	Volumetric Top-Off
VTOC	Volume Table of Contents
VTOHL	Vertical Takeoff and Horizontal Landing
VTOL	Vertical Takeoff and Landing
VTOVL	Vertical Takeoff and Vertical Landing
VTP	Viewdata Terminal Program
	Virtual Terminal Protocol
	Volume — Temperature — Pressure [also: V-T-P]
VTPR	Vertical Temperature Profile Radiometer
VTR	Video Tape Recorder
	Voltage Transformation Ratio
VTRS	Video Tape Response System
VTS	Vehicular Technology Society [of IEEE, US]
	Vertical Test Stand
	Vessel Traffic Services
	Viewscan Text System
VTTC	Video Tape Time-Code
VTU	Vienna Technical University [Austria]
V+TU	Voice plus Teleprinter Unit
VTUV	Vereinigung der Technischen Uberwachungsvereine [= Confederation of German Inspection Associations]
VTVM	Vacuum-Tube Voltmeter
VU	Vanderbilt University [US]
	Vehicle Unit
	Voice Unit
	Volume Unit
VUC	Victoria University College [New Zealand]
VUCDT	Ventilation Unit Condensate Drain Tank
VUE	Visual User Environment
VULC	Vulcanization
	Vulcanizing
VUTS	Verification Unit Test Set
VUV	Vacuum Ultraviolet
	Vatu [Currency of Vanuatu]
VUVS	Vacuum Ultraviolet Spectroscopy
VV	Vertical Velocity
	Vertical Visibility
	Volt Velocity
V/V	Vertical Velocity
	Volume by Volume [also: v/v]
V-V	Velocity Volume
vv	(vice versa) — conversely
VVA	Vicia Villosa Agglutinin
VVC	Voltage-Variable Capacitor
VVDS	Video Vertex Decision Storage
VVE	Vertical Vertex Error
VVI	Vancouver Vocational Institute [Canada]
VVM	Valve Voltmeter
VVR	Variable Voltage Rectifier
VVVF	Variable Voltage Variable Frequency
VWDI	Vinyl Window and Door Institute [US]
VWL	Variable Word Length
VWOA	Veteran Wireless Operators Association [US]
VWP	Variable Width Pulse
VWPI	Vacuum Wood Preservers Institute [US]
VWS	Variable Word Size
	Vortex Wake System
VWSS	Vertical Wire Sky Screen
VX	Volume — Concentration (Diagram) [also: V-X]
VXO	Variable Crystal Oscillator
	Variable-Frequency Crystal Oscillator
VY	Very

W	Tryptophan
	Tryptophyl
	Waste
	Water
	Watt(meter)
	Weak
	Wednesday
	Weight
	Welsh
	West(ern)
	White
	Wide
	Width
	Winter
	Wire
	(Wolfram) — Tungsten
	Won [Currency of South Korea]
	Wood
	Word
	Work
	Write
w	warm
	waste
	water
	weak
	weather
	weight
	wet
	white
	wide
	width
	with
	wind
	write
WA	Washington (State) [US]
	Wave Analyzer
	Wave Antenna
	Waveform Analyzer
	Western Australia
	Wetting Angle
	Wide-Angle
	Wire Armored
	Wire Association [now: WAI, US]
	Women's Auxiliary
	Woolwich Armstrong (Gun)
	Word After
	Wrong Answer
W/A	Water/Air (Ratio)
W-A	Wilbur-Anderson (Unit)
WAA	Water Authorities Association
	Western Awning Association [US]
	World Aluminum Abstracts [US and Japan]
WAAF	Women's Auxiliary Air Force [UK]
WAAG	Waveform Acquisition and Arbitrary Generator
	Waveform Analysis and Arbitrary Generator
WAAIME	Women's Auxiliary of the American Institute of Mining, Metallurgical and Petroleum Engineers [US]
WAAM	Wide Area Antiarmor Munitions
WAAP	World Association of Animal Production [Italy]
WAAS	World Academy of Arts and Sciences [Sweden]
WAAVP	World Association for the Advancement of Veterinary Parasitology
WABE	Western Association of Broadcast Engineers
WAC	Weighted Average Cost
	West Africa Committee [UK]
	Women's Army Corps [US]
	World Aeronautical Chart
	Write Address Counter
WACA	Western Agricultural Chemicals Association [US]
	World Airline Clubs Association [Canada]
WACES	Wyoming Association of Consulting Engineers and Surveyors [US]
WACHO	Western Association of Canadian Highway Officials
WACK	Wait before Transmitting Positive Acknowledgement
WACM	Western Association of Circuit Manufacturers [US]
WACMR	West African Council for Medical Research
WACS	Wide Angle Collimated Display System
	Workshop Attitude Control System
WACU	West African Customs Union
WAD	Worst Area Difference
WADB	West African Development Bank
WADC	Wright Air Development Center [US Air Force]
WADD	Wright Air Development Division [US Air Force]
WADE	World Association of Document Examiners [US]
WADEX	Word and Author Index
WADS	Wide Area Data Service
WAEC	West African Examinations Council [Ghana]
WAEP	World Association for Element Building and Prefabrication [Germany]
WAES	Workshop on Alternative Energy Strategies [UK]
WAF	Width across Flats
	Wiring around Frame
WAFC	World Area Forecast Center
WAFR	West Africa [also: W AFR]
	West African [also: W AFR]
WAFRI	West-African Fisheries Research Institute
WAFRU	West-African Fungicide Research Unit
WAFS	World Area Forecast System
WAFWA	Western Association of the Fish and Wildlife Association [US]
WAG	Water-Alternating Gas
WAGR	Western Australian Government Railways
	Windscale Advanced Gas-Cooled Reactor [UK]
WAHT	Weighted Average Holding Time
WAI	Wire Association International [US]
WAIT	Western Australian Institute of Technology
	What Alloy Is That (Test) [Metallurgy]
WAITRO	World Association of Industrial and Technological Research Organizations [Denmark]
WAK	Wait Acknowledge
	Write Access Key

WAL Wide-Angle Lens
World Association of Lawyers [US]
W-AL Westinghouse-Astronuclear Laboratory [US]
WALA West African Library Association
WALDO Wichita Automatic Linear Data Output [US]
WALOPT Weapons Allocation and Desired Ground Zero Optimizer [US Military]
WAM Words A Minute
Worth Analysis Model
WAMCE Western Association of Minority Consulting Engineers
WAMDII Wide Angle Michelson Doppler Imaging Interferometer
WAMI World Association of Medical Informatics
WAMIC Women's Auxiliary of the Mining Industry of Canada
WAML Western Association of Map Libraries [US]
Wright Aero-Medical Laboratory [US]
WAMOSCOPE Wave-Modulated Oscilloscope
WAMPRI Western Australian Mining and Petroleum Research Institute
WAMRU West African Maize Research Unit
WAMS World Association of Military Surgeons
WAMU West African Monetary Union
WAN Wide-Area Network
Women's Aquatic Network [US]
WAND Working Party on Access to the National Database [UK]
WANEF Westinghouse Astronuclear Experimental Facility [US]
WANL Westinghouse Astronuclear Laboratory [US]
WANS Wide-Area Network System
WAOS Welsh Agricultural Organization Society [UK]
WAP Work Assignment Procedure
WAPA Western Area Power Administration
WAPDA Water and Power Development Authority [Pakistan]
WAPET Western Australia Petroleum Pty.
WAPF West African Pharmaceutical Federation [Nigeria]
WAR Warning
Warranty
War Warwickshire [UK]
WARC World Administrative Radio Conference
WARC-BS World Administrative Radio Conference for Broadcasting Satellites [also: WARC/BS]
WARC-MOB World Administrative Radio Conference for Mobile Services [also: WARC/MOB]
WARC-MR World Administrative Radio Conference on Maritime Radio [also: WARC/MR]
WARC-MT World Administrative Radio Conference on Maritime Telecommunications [also: WARC/MT]
WARC-ORB World Administrative Radio Conference for Geostationary Satellite Orbit [also: WARC/ORB]
WARC-ST World Administrative Radio Conference on Space Telecommunications [also: WARC/ST]
WARDA West African Rice Development Association [Liberia]
WARF Wisconsin Alumni Research Foundation [US]
WARFAC War Game Facility
WARHD Warhead
WARI Waite Agricultural Research Institute [Australia]
WARLA Wide Aperture Radio Location Array
WARM Wood and Solid Fuel Association of Retailers and Manufacturers [UK]
WARRS West African Rice Research Station
Warw Warwickshire [UK]
WAS Washington Academy of Sciences [US]
Wisconsin Archeological Society [US]
World Aquaculture Society [US]
World Archeological Society [US]
WASA West African Science Association
WASAL Wisconsin Academy of Sciences, Arts and Letters [US]
WASAR Wide Application System Adapter
WASCO Water and Soil Conservation Organization [New Zealand]
WASH Washer
Wash Washington (State) [US]
WASHO Western Association of State Highway Officials [US]
WASID Water and Soil Investigation Department [Pakistan]
WASME World Assembly of Small and Medium Enterprises [India]
WASNA Western Agricultural Society of North America [US]
WASP Weightless Analysis Sounding Probe [of NASA]
Williams Aerial Systems Platform
Workshop Analysis and Scheduling Program
World Association of Seaweed Processors [France]
WASSP Wire Arc Seismic Section Profiler
WASU West African Students Union
WASWC World Association of Soil and Water Conservation [US]
WAT Weight, Altitude and Temperature [also: W-A-T]
Wide Area Telecommunications Service
Work Adjustment Training (Program)
WATBRU West African Timber Borer Research Unit
WATCON Waterloo Concordance [of U of W, Canada]
WATDOC Water Resources Document Reference System [of Environment Canada]
WATERLIT Water Literature (Database) [of SAWIC, South Africa]
WATFIV University of Waterloo FORTRAN V [Canada]
WATFOR University of Waterloo FORTRAN IV [Canada]
WATRS West Atlantic Route System [US]
WATS Wide Area Telephone Service [US]
Wide Area Telephone System
WATSTOR Water Data Storage and Retrieval System [of USGS]
WATTE West African Tropical Testing Establishment
WATTec Welding and Testing Technology Exhibition and Conference [US]
WAVFH World Association of Veterinary Food Hygienists
WAXS Wide-Angle X-Ray Scattering
WB Water Board
Waybill
Weak Beam (Image)
Weather Bureau
Weld Brazing
Westbound
Wet Bulb
Wheel Base
White Band
Wideband
Wool Bureau [US]
Word Before

W/B	Waybill
Wb	Weber
WBA	World Buffalo Association [US]
WBAN	Weather Bureau, Air Force and Navy [US]
WBAR	Wing Bar
WBBA	Western Bird Banding Association [US]
WBC	White Blood Cell
	World Business Council [US]
WBCO	Waveguide below Cutoff
WBCT	Wideband Current Transformer
WBCV	Wideband Coherent Video
WBD	Wideband Data
	Wire Bound
WBDF	Weak Beam Dark Field
WBDL	Wideband Data Link
WBFM	Wideband Frequency Modulation
WBGT	Wet Bulb Globe Temperature
	Wet Bulb Globe Thermometer
WBIF	Wideband Intermediate Frequency
WBL	Wideband Limiting
WBLC	Waterborne Logistics Craft
WBMA	Western Building Material Association [US]
	Wirebound Box Manufacturers Association [US]
WBMS	World Bureau of Metal Statistics [UK]
WBNL	Wideband Noise Limiting
WBNS	Water-Boiler Neutron Source [US]
WB/NWRC	Weather Bureau/National Weather Records Center [US]
WBO	Wien-Bridge Oscillator
WBP	Weather and Boil-Proof
WBPA	Wideband Power Amplifier
WBRR	Weather Bureau Radar Remote System
WBRS	Wideband Remote Switch
WBS	Weight and Balance Measuring System
	Wellington Botanical Society [New Zealand]
	Wideband System
	Wide-Body STOL (= Short Takeoff and Landing)
	Work Breakdown Structure
WBSARC	World Broadcasting Satellite Administrative Radio Conference
WBSW	Wideband Switch
WBT	Wet Bulb Temperature
	Wet Bulb Thermometer
	Women in Broadcast Technology [US]
WBTS	Wideband Transmission System
WBVTR	Wideband Video Tape Recorder
WC	Walnut Council [US]
	Water Closet
	Water Content
	Weather Center
	Wellington College [Canada]
	Western Cedar
	Whale Center [US]
	Wire Chief
	Wood Cover
	Word Count
	Work Cell
	Worsted Count
	Write and Compute
W/c	Watts per Candle [also: W c^{-1}]
wc	with costs
WCA	Western College Association [US]
	Wind Correction Angle
	Wireless Cable Association [US]
	Workmen's Compensation Act [Canada]
	World Communication Association [US]
WCACP	World Congress of Anatomic and Clinical Pathology
WCAM	Western Canada Aviation Museum
WCAP	Westinghouse Commercial Atomic Power [US]
WCB	Weatherproof Circuit Breaker
	Way Control Block
	Workmen's Compensation Board [Canada]
WCBBC	Workmen's Compensation Board of British Columbia [Canada]
WCBO	Workmen's Compensation Board of Ontario [Canada]
WCC	Wildfire Coordinating Committee [of NASF, US]
	Wire-Line Common Carrier
	Work Cell Control
	Workmen's Compensation Commission [Canada]
	World Computer Conference
	World Crafts Council [Denmark]
	Write Control Character
WCCES	World Council of Comparative Education Societies [France]
WCCI	World Council for Curriculum and Instruction [US]
WCCPPS	Waster Channel and Containment Pressurization and Penetration System
WCDB	Wing Control During Boost
WCEE	Women's Council on Energy and the Environment [US]
	World Conference on Earthquake Engineering
WCEMA	West Coast Electronic Manufacturers Association [US]
WCF	Waste Calcining Facility [of INEL, US]
	White Cathode Follower
	Winston Churchill Fund [US]
	Work Control File
	Workload Control File
WCFA	Wildlife Conservation Fund of America [US]
WCG	Weakly Compact Generated
WCGA	World Computer Graphics Association [US]
WCGM	Writable Character Generation Memory
	Writable Character Generation Module
WCHS	Western Colorado Horticultural Society [US]
WCI	Waiting for Calls Indicator
	Wildlife Conservation International [of NYZS, US]
WCK	Wildlife Clubs of Kenya (Association)
WCL	Word Control Logic
	World Confederation of Labour [Belgium]
WCLC	Women's Computer Literacy Center [US]
WCLIB	West Coast Lumber Inspection Bureau [US]
WCLP	Women's Computer Literacy Project [now: WCLC, US]
WCM	Wired Core Matrix
	Word Combine and Multiplex
	Writable Control Memory
WCMIA	West Coast Metal Importers Association [US]
WCMMF	World Congress on Man-Made Fibers
W/cm²	Watts per Square Centimeter
WCNDT	World Conference on Nondestructive Testing
WCOS	Work Cell Operator Station

WCP Waste Collector Pump
World Council of Peace
WCPS World Confederation of Productive Science [Norway]
WCPSC Western Conference of Public Services Commissioners
WCR Water Cooler
Word Control Register
Word Count Register
WCRA Weather Control Research Association [now: WMA, US]
West Coast Railway Association [US]
WCRP World Climate Research Program [Germany]
WCS Whatman Compression Screw
Work Center Supervisor
World Congress on Superconductivity
World Coordinate System
Writable Control Store
WCSI World Center for Scientific Information
WCSICEC Working Committee of the Scientific Institutes for Crafts in the European Community [Germany]
WCT Weak-Coupling Theory
WCTOC World Congress of Theoretical Organic Chemists
WCTP Wire Chief Test Panel
WCVM Western College of Veterinary Medicine [Canada]
WCY World Communication Year [of UN]
WD War Department
Watt Demand-Meter
Wave Diffraction
Weld Decay
Width
Wind
Wire Drawing
Withdrawn
Wood
Word
Work Distance
Working Directory
Working Distance
Working Draft
Works Department
Write Data
Write Direct
W/D Write Down
WDA Wildlife Disease Association [US]
WDB Werkstoffdatenbank [= Database on Engineering Materials [of VDEh, Germany]]
Wideband
World Databank [of Harvard University, US]
WDC Waste Disposal Cask
World Data Center [of NAS, US]
WDCS Writable Diagnostic Control Store
WDD Well Drawdown
WDF Waste-Derived Fuel
Wave Digital Filter
Woodruff
WDG Winding
WDI Wind Direction Indicator
WDLY Widely
WDM Wavelength Division Multiplexing
WDP Women in Data Processing
WDPC Western Data Processing Center [of UCLA, US]
WDQD Word Queue Directory
WDS Wavelength Dispersive Spectrometry
Wavelength Dispersive X-Ray Spectroscopy
WDSPRD Widespread
WDST Western Daylight Saving Time
WDTRS Westinghouse Development Test Requirement Specification
WDWKG Woodworking
WDX Wavelength Dispersive X-Ray Analysis
WDXRF Wavelength Dispersive X-Ray Fluorescence
WE Water Engineer
Water Engineering
Water Equivalent
Weak Electrolyte
Weight Engineer
Weight Engineering
Wind Engineering
Women in Energy [US]
Working Electrode
Working Elongation
Write Enable
We Weber Number
WEA Weather
Westinghouse Engineers Association
White Etching Area
Workers Education Association [UK and Sweden]
WEAA Western European Airports Association
WEA-N Westinghouse Engineers Association National [US]
WEBA World Educational Broadcasting Assembly
WEC World Energy Conference [UK]
World Engineering Conference
World Environment Center [US]
WECAFC Western Central Atlantic Fisheries Commission [Italy]
WECON Western Electronics Show and Convention [of IEEE, US]
WECOM Weapons Command [US Army]
WECPNL Weight Equivalent Continuous Perceived Noise Level
WECS Wind Energy Conservation System
Wind Energy Conversion System
WED Weak Exchange Degeneracy
Wednesday
WEDA Western Dredging Association [US]
Wholesale Engineering Distributors Association [UK]
WEDC Western Engineering Design Competition [Canada]
WEF With Effect From
World Education Fellowship [UK]
WEFAX Weather Facsimile [of ESSA, US]
WEG Wind Energy Group [UK]
WEI Western European Institute (for Wood Preservation) [Belgium]
Work Efficiency Institute [Finland]
WeldTech Welding Technology
WELGAS Western Leg Gas [of FLAGS]
WEM Western European Metal Trades Employers Organization [Germany]
WEMA Western Electronic Manufacturers Association [now: AEA, US]

	Winding Engine Manufacturers Association
WEP	Water-Extended Polyester
	World Economic Prospects [UK]
WEPZA	World Export Processing Zones Association [US]
WER	Worth Estimating Relationship
WERC	World Environment and Resources Council [Belgium]
WES	Waterways Experiment Station [US Army]
	Wind Electric System
	Women's Engineering Society
	Wyoming Engineering Society [US]
WESAR	Westinghouse Safety Analysis Report
WESA SAT	Western Sahara Satellite Survey
WESCON	Western Electronics Show and Convention [of IEEE, US]
WESRAC	Western Research Application Center [of USC, US]
W/E&SP	With Equipment and Spare Parts
WESTAC	Western Transportation Advisory Council
WESTAR	Waterways Experiment Station Terrain Analyzer Radar
WESTEC	Western Metal and Tool Exposition and Conference [US]
WETAC	Westinghouse Electronic Tubeless Analog Computer [US]
WEU	Western European Union [UK]
	Wide End Up (Mold)
WF	Warm Front
	Wave Filter
	Waveform
	Wave Front
	Wedge Filter
	Weighting Function
	Widefield
	Wire Foundation [US]
	Write Forward
	Wrong Font
W/F	Wow and Flutter
WFA	White Fish Authority [UK]
	Wire Fabricators Association [US]
	Wisteria Floribunda Agglutinin
	World Federation of Advertisers [Belgium]
WFAIT	Western Foundation for Advanced Industrial Technology
WFBSC	World Federation of Building Service Contractors [UK]
WFC	World Food Council
WFCA	Western Forestry and Conservation Association [US]
WFCC	World Federation for Culture Collections [UK]
WFCMV	Wheeled Fuel-Consuming Motor Vehicle
WFD	Waveform Distortion
WFEA	World Federation of Education Associations
WFEO	World Federation of Engineering Organizations [France]
WF&Eq	Wave Filters and Equalizers [IEEE PHP Committee]
WFF	Well-Formed Formula
WFG	Waveform Generator
WFI	Wirtschaftsforderungsinstitut [= Institute for Economic Advancement, Austria]
	Wood Foundation Institute [US]
	World Federation of Investors [Belgium]
WFIA	Western Forest Industries Association [US]
WFL	Work-Flow Language
WFMC	Welding Filler Material Control
WFMI	World Federation for the Metallurgical Industry
WFNA	White Fuming Nitric Acid
WFP	Warm Front Passage
	World Federation of Parasitologists [Netherlands]
	World Food Program [of UN — Italy]
WFPA	Washington Forest Protection Association [US]
	World Federation for the Protection of Animals
WF/PC	Widefield/Planetary Camera
WFPL	Western Forest Products Laboratory [Canada]
WFPLCA	World Federation of Pipeline Contractors Associations [US]
WFPMA	World Federation of Personnel Management Associations [US]
WFR	Wafer
W-F	Wiedemann-Franz (Ratio)
WFS	Wet Flexural Strength
	World Future Society [US]
WFSF	World Futures Studies Federation [Hawaii, US]
WFSW	World Federation of Scientific Workers [UK]
WFT	Winograd Fourier Transform
WFTU	World Federation of Trade Unions [Czechoslovakia]
WFUMB	World Federation for Ultrasound in Medicine and Biology
WFUNA	World Federation of United Nations Associations
WG	Water Gauge
	Waveguide
	Wire Gauge
	Working Group
	Woven Glass
W/G	Waveguide
WGA	Wheatgerm Agglutinin
	Wild Goose Association [US]
	Wyoming Geological Association [US]
WGBC	Waveguide Operating Below Cutoff
WGDT	Waste Gas Decay Tank
WGE	Western Gold Exposition [US]
W GER	West German(y)
WGLR	Wissenschaftliche Gesellschaft fur Luft- und Raumfahrt [= Scientific Society for Aeronautics and Astronautics, Germany]
WGM	Waveguide Mode
WGMA	Work Glove Manufacturers [US]
WGMS	World Glacier Monitoring Service [Switzerland]
WGN	White Gaussian Noise
WGS	Waveguide Glide Slope
WGSC	Wide Gap Spark Chamber
WGST	Waste Gas Storage Tank
WGT	Weight
WGTA	Western Grain Transportation Act
WGW	Wind-Generated Wave
WH	Waste Heat
	Water Hammer
	Water Heater
	Western Hemlock
	White
	Work Hardening
W&H	Wage and Hour Division
Wh	Watthour
WHA	Western Hardwood Association [US]

	Wood Heating Association [US]
WHB	Waste-Heat Boiler
WHC	Watthour Meter with Contact Device
WHCLIS	White House Conference on Library and Information Services [US]
WHDM	Watthour Demand Meter
WHERF	Wood Heating Education and Research Foundation [US]
WHF	Wharf
WHI	Western Highway Institute [US]
WHIMS	Wet High-Intensity Magnetic Separation
WHIMS-CF	Wet High-Intensity Magnetic Separation and Cationic Flotation
WHIPS	Windows Highly Integrated Program
WHL	Watthour Meter with Loss Compensator
	Wholesale
WHLR	Wholesaler
WH LT	White Light
WHM	Watthour Meter
	Watthour Motor
WHMIS	Workplace Hazardous Materials Information System
WHO	World Health Organization [of UN — Switzerland]
WHOI	Woods Hole Oceanographic Institute [US]
WHP	Water Horsepower [also: W HP or w hp]
	World Hydrocarbon Program [of Stanford Research Institute, US]
Whr	Watthour
WHRA	Welwyn Hall Research Association [UK]
WHSE	Warehouse
WHSG	Warehousing
WHSLE	Wholesale
WHSV	Weight Hourly Space Velocity
WHT	Watthour Meter, Thermal Type
	White
WHTC	Wolfson Heat Treatment Center [UK]
WI	Wascana Institute [Canada]
	Water Injection
	Welding Institute [UK]
	West Indies
	Wisconsin [US]
	Within
	Wrought Iron
W&I	Weighing and Inspection
W/I	Within
WIA	Women in Aerospace [US]
WIAB	Wistar Institute of Anatomy and Biology [US]
WIACO	World Insulation and Acoustic Congress Organization [France]
WIC	Water Information Center [US]
	Welding Institute of Canada
	Wildlife Information Center [US]
WICAT	World Institute for Computer-Assisted Teaching [US]
WICB	Women in Cell Biology [US]
WICS	Westinghouse Integrated Compiling System
WID	Width
WIDE	Wiring Integration Design
WIDJET	Waterloo Interactive Direct Job Entry Terminal (System) [Canada]
WIE	With Immediate Effect
WIF	Water Immersion Facility
WIIU	Workers International Industrial Union
WIL	White Indicating Lamp
WILCO	Will Comply
Wilts	Wiltshire [UK]
WIM	Women in Mining [US]
WIMA	Writing Instrument Manufacturers Association [US]
WIMP	Weakly Interacting Massive Particle
	Windows, Icons, Mouse and Pulldown Menus
	Wisconsin Interactive Molecule Processor [Software]
WIN	Welding Information Network [of AWI, US]
	Window
	Wissenschaftsnetz [= Research Network, Germany]
	Work Incentive Program [US]
WINA	Webb Institute of Naval Architecture [US]
WINB	Western Interstate Nuclear Board [US]
WINCON	Winter Conference on Aerospace and Electronic Systems [US]
	Winter Convention [of IEEE, US]
WIND	Weather Information Network and Display
WINDII	Wind Imaging Interferometer
WINDI RAC	Wind Imaging Interferometer Remote Analysis Computer
WINDS	Weather Information Network and Display System
WINMIC	Windsor Metric Information Center [Canada]
WINS	Weight Information Networking System
WINTEM	Wind and Temperature Forecast
WIP	Women in Information Processing [US]
	Work in Process
	Work in Progress
WIPE	Waste Immobilization Process Experiment
WIPO	World Intellectual Property Organization [Switzerland]
WIPP	Waste Isolation Pilot Plant
WIPS	Weather Image Processing System
WIR	Wiring
WIRA	Wax Importers and Refiners Association [US]
	Wool Industries Research Association [UK]
WIRCCWS	WHO (= World Health Organization) International Reference Center for Community Water Supply [now: IRC of UNESCO and WHO]
WIRDS	Weather Information Remoting and Display System
WIRE	International Trade Fair for Wire and Cable [Germany]
WIRECOM	Wire Communication
WIRTC	Western Industrial Research and Training Center
WIS	WATS (= Wide Area Telephone Service) Information System
	Weizmann Institute of Science [Israel]
	Wheat Information Service [Japan]
	Women in Science
Wis	Wisconsin [US]
WISA	West Indies Sugar Association
	Wire Industry Suppliers Association [US]
Wisc	Wisconsin [US]
WISE	Wang (Computer) Intersystem Exchange
	Washington Internships for Students of Engineering [of ASEE, US]
	Wolfson Institute for Surface Engineering [UK]
	Women in Science and Engineering
	World Information Service on Energy [Netherlands]

	World Information Systems Exchange
WISHA	Washington Industrial Safety and Health Act [US]
WISI	World Information System on Informatics
WISICA	West Indies Sea Island Cotton Association
WISITEX	World Instrumentation Symposium and International Trade Exposition
WISP	Waves in Space Plasma Program [of NRC, Canada]
WISP-HF	Waves in Space Plasma — High Frequency [of NRC, Canada]
WISP-OMV	Waves in Space Plasma — Orbital Maneuvering Vehicle
WIST	(Study Group on) Wind Shear and Turbulence [of ICAO]
WISTCI	Wisconsin University — Theoretical Chemistry Institute [US]
WIT	Wessex Institute of Technology [UK]
WITB	Wood Industry Training Board [UK]
WITG	Western International Trade Group [US]
WITS	Waterloo Interactive Terminal System [of U of W, Canada]
WJC	Water-Jet Cutting
WJCC	Western Joint Computer Conference [US]
WJTA	Water Jet Technology Association [US]
WK	Weak
	Week
	Wiener Kernel
	Work
WKB	Wentzel-Kramers-Brillouin (Theory)
WKBJ	Wentzel-Kramers-Brillouin-Jordan (Theory)
WKD	Weekdays
WKG	Working
WKLY	Weekly
WKN	Weaken
WKNL	Walter Kidde Nuclear Laboratories [US]
WKQDR	Work Queue Directory
WL	Water Line
	Wavelength
	Western Larch
	Wheel
	Wind Load
	Wired Logic
	Wire List
	Word Line
	Working Level
	Work Load
W/L	Wavelength
WLC	Wavelength Comparator
	Wisconsin Library Consortium [US]
WLDG	Welding
WLF	Williams, Landel and Ferry (Equation)
WLFA	Wildlife Legislative Fund of America [US]
WLI	Wavelength Interval
WLM	Working Level Month
WLN	Washington Library Network [US]
	Western Library Network
	Wiswesser Line Notation
W LON	West Longitude [also: WLON, W LONG or WLONG]
WLPSA	Wildlife Preservation Society of Australia
WL-RDCB	Wedge-Loaded Rectangular Double Cantilever Beam
WLU	Wilfred Laurier University [Canada]
WLUS	World Land Use Survey
WLZS	World Lead-Zinc Symposium
WM	Waste Management
	Wave Mechanics
	Wavemeter
	Wave Motion
	Wattmeter
	Weld Metal
	Whole Molecule
	Wobble Motor
	Word Mark
	Working Memory
W&M	Washburn and Moen (Gauge)
W/M	Weight and/or Measurements
WMA	Waterbed Manufacturing Association [US]
	Weather Modification Association [US]
	Wellington Mathematical Society [New Zealand]
	Wildlife Management Area
	World Medical Association
WMB	Walnut Marketing Board [US]
	Waste Management Branch
WMBA	Wire Machinery Builders Association [now: WISA, US]
WMC	Wool Manufacturers Council [of NTA, US]
	World Materials Congress
	World Meteorological Center
	World Mining Congress (and Exhibition)
WMDA	Woodworking Machinery Distributors Association [US]
WME	Women and Mathematics Education [US]
WMEC	Western Military Electronics Center [US]
WMF	Windows Metafiles [Computer Software]
WMHS	Wall-Mounted Handling System
WMI	Wildlife Management Institute [US]
	Wolfson Microelectronics Institute [UK]
WMIA	Woodworking Machinery Importers Association [US]
WMKLEM	W. M. Keck Laboratory of Engineering Materials [of CalTech, US]
WMMA	Wood(working) Machinery Manufacturers of America [US]
WMMP	Waste Management Master Plan
WMMPA	Wood Moulding and Millwork Producers Association [US]
WMO	World Meteorological Organization [of UN — Switzerland]
WMR	Wave-Making Resistance
WMS	Warehouse Management System
	Waste Management System
	Whaling Museum Society [US]
	Wire Mesh Screen
	World Magnetic Survey
	World Mariculture Society [now: WAS, US]
WMSC	Weather Message Switching Center
WMSI	Weak Metal-Support Interaction
	Western Management Science Institute [US]
WMSO	Wichita Mountains Seismological Observatory [US]
WMT	Waste Monitor Tank
WMU	Western Michigan University [US]
W&M WIRE G	Washburn & Moen Wire Gauge
WN	Wave Normal
	White Noise

	Whole Number
	Wrong Number
Wn	Won [Currency of North Korea]
WNA	World Nature Association [US]
WNAAA	Women of the National Agricultural Aviation Association [US]
WNAMA	Women's Network in Aquatic and Marine Affairs [now: WAN, US]
WNAR	Western North American Region [US]
WND	Wind
WNGA	Wholesale Nursery Growers of America [US]
WNO	Wrong Number
WNRC	Washington National Records Center [US]
WNRE	Whiteshell Nuclear Research Establishment [of AECL, Canada]
WNT	Waste Neutralizer Tank
WNW	West-Northwest
WNYNRCR	Western New York Nuclear Research Center Reactor [US]
WO	War Office
	Warrant Officer
	Welding Operator
	Work Order
	Write Only
	Write Out
W/O	Water-in-Oil Emulsion
	Widow/Orphan (Protection) [Word Processing]
	Without
	Write-Off
WOB	Weight on Bit
WOBO	World Organization of Building Officials [Canada]
WOC	Waiting on Cement (Time)
	Water/Oil Contact
	Welding Operator Certificate
WODA	World Organization of Dredging Associations [US]
W/O E&SP	Without Equipment and Spare Parts
WOF	Work of Fracture
WOG	Water, Oil and Gas
WOHMA	Waste Oil Heating Manufacturers Association [US]
WOL	Wedge-Opening Load(ing)
WOM	Write-Only Memory
	Write Optional Memory
Won	Won [Currency of North and South Korea]
WOO	Waiting on Orders
WOR	Water-Oil Ratio
WORB	Work Roll Bending
WORC	Washington Operations Research Council [US]
WORCRA	Worner/Conzinc Riotinto of Australia (Process) [Metallurgy]
Worcs	Worcestershire [UK]
WORLDDIDAC	World Association of Manufacturers and Distributors of Educational Material [Switzerland]
WORM	Write Once, Read Many Times
WOS	Web Offset Section [of PIA, US]
	Wilson Ornithological Society [US]
WOSAC	Worldwide Synchronization of Atomic Clocks
WOSC	World Organization of Systems and Cybernetics [France]
WOT	Wide-Open Throttle (Switch)
WOW	Waiting on Weather
WOWS	Wire Obstacle Warning System
WP	Warm Processing
	Water Purification
	Wave Propagation
	Weather Permitting
	Weatherproof
	Western Pine
	White Phosphorus
	Wind Power
	Wing Profile
	Word Processing
	Word Processor
	Working Point
	Working Pressure
	Workspace-Register Pointer
	Write Permit
	Write Protection
W/P	Way-Point
WPA	Western Pine Association [US]
	Wisconsin Pharmaceutical Association [US]
	With Particular Average
	Works Project Administration [US]
	Wyoming Pharmaceutical Association [US]
WPAFB	Wright Patterson Air Force Base [US]
WPB	Waste Paper Basket
	Write Printer Binary
WPBS	Welsh Plant Breeding Station [UK]
WPC	Water Pillow Cooling
	Water Pollution Council [US]
	Watts per Candle [also: Wpc]
	Wheat Protein Concentrate
	Wood Plastic Combination
	Wood Plastic Composite
	World Petroleum Congress [UK]
	World Power Conference
	World Print Council [US]
WPCA	Water Pollution Control Administration [US]
WPCF	Water Pollution Control Federation [US]
WPD	Write Printer Decimal
WPDA	Writing Push Down Acceptor
WPDES	Waste Pollution Discharge Elimination System
WPESS	Width-Pulse Electronic Sector Scanning
WPF	World Productivity Forum
WPG	Water Pipe Ground
WPI	Wholesale Price Index
	Worcester Polytechnic Institute [US]
	World Patents Index [UK]
WPL	Waste Pickle Liquor
	Wave Propagation Laboratory
WPM	Wobbles per Minute
	Words per Minute [also: wpm]
	Write Protect Memory
WPMA	Windows/Presentation Manager Association [US]
	Wood Products Manufacturers Association [US]
WPO	World Packaging Organization [France]
	World Ploughing Organization [UK]
WPP	Waterproof Paper Packing
	Water-Purification Process
WPPSS	Washington Public Power Supply System [US]
WPR	Water Production Rate
	Write Permit Ring
WPRD	Water Produced

WPRL Water Pollution Research Laboratory [UK]
WPRT Write Protect
WPS Water Phase Salt
Welding Procedure Specification
Word Processing Specialist
Word Processing System
Words per Second [also: wps]
WPSA World Poultry Science Association [Germany]
WPSG WordPerfect (Software) Support Group [US]
WPSMA Welsh Plate and Steel Makers Association [UK]
WPT Way-Point
Word Processing Technician
WPTI Wildlife Preservation Trust International [US]
WPUET West Pakistan University of Engineering and Technology
WQ Water Quenching
WQA Water Quality Association [US]
WQASD Water Quality and Aquatic Science Division [of NHRI, Canada]
WQB Water Quality Branch [of Environment Canada]
WQC Water Quality Certification
Wheat Quality Council [US]
WQM Weld Quality Monitor
WQMP Water Quality Management Project
WQP Water Quality Project [now: National Water Center, US]
WQPA Western Quick Printers Association
WQRC Water Quality Research Council [US]
WR Warehouse Receipt
Water Research
Welding Research
Water Resources
Whiteshell Reactor [of WNRE, Canada]
Wilson Repeater
Wolf-Rayet
Working Register
Write
WRA Water Research Association [UK]
Weapon Replaceable Assembly
Western Railroad Association [US]
WRAC Willow Run Aeronautical Center [US]
Women's Royal Army Corps [UK]
WRADAC Water Research Association Distribution Analog Center [UK]
WRAF Women's Auxiliary Air Force [UK]
WRAIR Walter Reed Army Institute of Research [US]
WRAIS Wide Range Analog Input Subsystem
WRAP Weighted Regression Analysis Program
WRB Wissenschaftliches Rechenzentrum Berlin [= Berlin Scientific Computer Center, Germany]
WRBC Weather Relay Broadcast Center
WRC Wastewater Research Center [of UBC, Canada]
Water Research Center [UK]
Water Resources Commission
Water Resources Congress [US]
Weather Relay Center
Welding Research Council [US]
Western Red Cedar
Weston Research Center [Canada]
Westwater Research Center [of UBC, Canada]
WRCA Western Red Cedar Association [US]
WRCC Wildlife Research Coordinating Committee [West Africa]
WR CHK Write Check
WRCKG Wrecking
WRCLA Western Red Cedar Lumber Association [US]
WRD Water Resource Division [of USGS]
WRDC Wright Research and Development Center [of WPAFB, US]
WRE Weapons Research Establishment [Australia]
WREC Westinghouse Reactor Evaluation Center [US]
WRENDA World Request for Neutron Data Measurements (Database) [of IAEA, Austria]
WRESAT Weapons Research Establishment Satellite [Australia]
WRG Wrong
WRI Wire Reinforcement Institute [US]
Wood Research Institute [of Kyoto University, Japan]
World Resources Institute
WRITELN Write Line
WRIU Write Interface Unit
WRL Willow Run Laboratory [US]
WRLA Western Retail Lumbermen's Association [now: WBMA, US]
WRM Warm
World Rainforest Movement [Malaysia]
WRMFNT Warm Front
WRMI We Really Mean It!
WRMS Watts Root-Mean-Square
WRNG Warning
WRNI Wide Range Neutron Indicator
Wide Range Nuclear Instrument
WRNS Women's Royal Naval Service [UK]
WRPC Water Resources Planning Commission [Taiwan]
Weather Records Processing Center
WRP/CI Worldwide Reinforced Plastics/Composites Institute
WRPDB Western Regional Production Development Board [Nigeria]
WRQ Westinghouse Resolver/Quantizer
WRRC Willow Run Research Center [US]
WRRL Western Regional Research Laboratory [US]
WRRR Walter Reed Research Reactor [US]
WRRS Wire Relay Radio System
WRS Working Reference System
WRSA World Rabbit Science Association [UK]
WRSIC Water Resources Scientific Information Center [US]
WRSSSV Wireline-Retrievable Subsurface Safety Valve
WRT Wrought
WRTA Western Railroad Traffic Association [US]
WRTB Wire Rope Technical Board [US]
WRTC Working Reference Telephone Center
WRU Western Reserve University [US]
Who Are You? [Telecommunications and Computing]
WRVS Women's Royal Voluntary Service [UK]
WS Washington State [US]
Water Solid
Wave Soldering
Wave Spectrum
Weather Service
Weather Ship
Weather Stripping

	Weapon System
	Wetted Surface
	Whole Antiserum
	Wind Shear
	Wire Send
	Wire Shear
	Wood Science
	Working Space
	Working Storage
	Work Space
	Workstation
	Work Support
	Wright's Stain
	Wyoming State [US]
Ws	Wattsecond
WSA	Web Sling Association [now: WSTDA, US]
	Wholesalers Stationers Association [US]
	World Sign Associates [US]
WSAD	Weapon Systems Analysis Division [US Navy]
WSAS	Weather Service Airport Station
WSB	Wheat Soy Blend
WSC	Wave Spectrum Concept
	Water Systems Council [US]
	Western Snow Conference [US]
	Wire Strand Core
WSCA	World Surface Coatings Abstracts [of PRA, UK]
WSCC	Western Systems Coordinating Council
WSCD	Water Survey Canada Division [of Environment Canada]
WSCS	Western Society of Crop Science [US]
	Wide Sense Cyclo-Stationary
WSD	Working Stress Design
	World Systems Division
WSDA	Water and Sewer Distributors of America [US]
WSE	Washington Society of Engineers [US]
	Water Saline Extract
	Weapon System Efficiency
	Western Society of Engineers [US]
	Winnipeg Stock Exchange [Canada]
	World Society for Ekistics [Greece]
Wsec	Wattsecond
WSED	Weapon Systems Evaluation Division [of IDA, US]
WSEIAC	Weapon Systems Effectiveness Industry Advisory Committee [of USDOD]
WSEP	Waste Solidification Engineering Prototype Plant [of USAEC]
WSF	Workstation Facility
WSFI	Wood and Synthetic Flooring Institute [US]
WSG	Wired Shelf Grounp
WSGCP	Washington Sea Grant College Program [US]
WSHA	Washington State Horticultural Association [US]
WSHFT	Wind Shift
WSHG	Washing
WSI	Wafer Scale Integration
	Water Solidity Index
	Wave-Sediment Interaction
WSIA	Water Supply Improvement Association
WSIT	Washington State Institute of Technology [US]
WSL	Warren Spring Laboratory [UK]
WSM	Western Society of Malacologists [US]
WSMA	Washington State Medical Association [US]
	Wyoming State Medical Association [US]
WSMO	Weather Service Meteorological Offices [US]
WSMR	White Sands Missile Range [US]
WSN	Western Society of Naturalists [US]
WSO	Weapons Systems Officer
	World Safety Organization
WSP	Water Supply Point
WSPA	Washington State Pharmaceutical Association [US]
	Western States Petroleum Association [US]
WSPE	Wisconsin Society of Professional Engineers [US]
WSR	Weather Search Radar
	Weather Surveillance Radar
WSRO	World Sugar Rsearch Organization [UK]
WSRT	Westerbork Synthesis Radio Telescope
WSSA	Weed Science Society of America [US]
WSSS	Western Society of Soil Science [US]
WST	Word Synchronizing Track
	Western Samoan Tala [Currency of Western Samoa]
WSTC	Westinghouse Science and Technology Center [US]
WSTDA	Web Sling and Tiedown Association [US]
WSTD CT	Worsted Count
WSTF	White Sands Test Facility [US]
WSU	Washington State University [US]
	Wayne State University [US]
	Wright State University [US]
WSUOPR	Washington State University Open-Pool Reactor [US]
WSW	West-Southwest
	Wind Shear Warning
WSW/RGS	Wind Shear Warning/Recovery Guidance System
WSW/RS	Wind Shear Warning/Recovery System
WSZ	Wrong Signature Zero
WT	Wait Time
	Wastewater Treatment
	Water Tank
	Water Technologist
	Water Technology
	Watertight
	Water Tube
	Water Tunnel
	Wave Tube
	Weight
	Welding Technologist
	Welding Technology
	Wide Track (Tractor)
	Wild Type
	Will Talk
	Wind Tunnel
	Wireless Telegraphy
	Wire Technology
	Wood Technologist
	Wood Technology
	Word Terminal
	World Trade
W/T	Wireless Telegraphy
WT	weight
WT%	Weight Percent [also: wt%]
WTA	Water Transport Association [US]
	World Textile Abstracts (Database) [UK]
WTAA	World Trade Alliance Association
WTB	Water Tube Boiler

WTBA Water Tube and Boilermakers Association [UK]
WTC Washington Technology Center [US]
Wastewater Technology Center [Canada]
Water Transport Committee
World Trade Center
WTCA Wood Truss Council of America [US]
World Trade Centers Association [US]
WTCI Western Telecommunications Institute [US]
WTD Weighted
WTD AV Weighted Average [also: WTD AVG]
WTF Waste Treatment Facility
World Time-Capsule Fund [US]
wt/gal weight per gallon
WTGS West Texas Geological Society [US]
WTI World Translations Index
WTIC World Trade Information Center [US]
WTM Wind Tunnel Model
Write Tape Mark
Write Tape Mask
WTMA Wood Tank Manufacturers Association [US]
WTMS World Trade in Minerals (Database) System
WTO Warsaw Treaty Organization [USSR]
Write to Operator
WTOR Write to Operator with Reply
WTP Width Table Pointer
WT PCT Weight Percent [also: wt%]
WTR Water
Western Test Range [of NASA]
Westinghouse Test Reactor [US]
Working Group on the Toxicology of Rubber Auxiliaries
WTRA Wool Textile Research Council [UK]
WTRD Water Treatment Research Division [South Africa]
WTS Wastewater Treatment System
World Terminal Synchronous
WTSA Wood Turners and Shapers Association [US]
WTSPT Waterspout
WTT Weapons Tactical Trainer
WTU Warsaw Technical University [Poland]
Workers Trade Union [Spain]
Wroclaw Technical University [Poland]
WTUP Wroclaw Technical University Press [Poland]
WTWA World Trade Writers Association [US]
WU Waseda University [Japan]
Washington University [US]
Western Union Company [US]
Write-Up
WUA Workers Union Association [Sudan]
WUE Water Use Efficiency
WUI Western Union International [US]
WUIS Work Unit Information System
WUJS World Union of Jewish Students [Israel]
WULTUO World Union of Liberal Trade Union Organizations [Switzerland]
WUR World University Roundtable
WURDD Western Utilization Research and Development Division [US]
WUS Word Underscore (Character)
World University Service [Switzerland]
WUSC World University Service of Canada
WUS-US World University Service — United States
WUX Western Union Telegram
WV Water Vapor
West Virginia [US]
Wind Velocity
Working Voltage
W/V Weight by Volume [also: w/v]
Wind Velocity
WVA World Veterinary Association
W Va West Virginia [US]
WVAS Wake Vortex Avoidance System
WVDC Working Voltage, Direct Current
WVE Water Vapor Electrolysis
WVHS West Virginia Horticultural Society [US]
WVL West Vancouver Laboratory [Canada]
WVS West Virginia State [US]
WVSMA West Virginia State Medical Association [US]
WVSPA West Virginia State Pharmaceutical Association [US]
WVT Water Vapor Transmission
WVTR Water Vapor Transmission Rate
WVU West Virginia University [US]
WVV Wissenschaftlicher Verein fur Verkehrswesen [= Scientific Association for Traffic and Transportation, Germany]
WW Wall-to-Wall
Water Wave
Water White
Wide Woods
Wilderness Watch [US]
Wire Way
Wire-Wound
Wire Wrap
Worldwide
W/W Weight to Weight [also: w/w]
WWBA Western Wooden Box Association [US]
WWCC Western Weed Control Conference [US]
World Wide Cost Comparison
WWD Weltwirtschaftsdatenbank [= Database on Global Economy and Business, Germany]
WWEMA Water and Wastewater Equipment Manufacturers Association [US]
WWER Water-Moderated Water-Cooled Reactor
WWF Whole Wheat Flour
World Wide Fund for Nature [Switzerland]
World Wildlife Fund [US]
WWG Warrington (Iron) Wire Gauge
Worldwide Guide
WWI World War I
WWII World War II
WWMCCS Worldwide Military Command Control System [also: WW MCCS]
WWMP Western Wood Molding Producers
WWP Working Water Pressure
WWPA Western Word Processing Association
Western Wood Products Association [US]
WWR Wire-Wound Resistor
WWSN Worldwide Seismology Network [of NBS, US]
WWW World Weather Watch [of WMO, Switzerland]
WX Weather
WXTRN Weak External Reference
WY Wyoming [also: Wyo, US]

WYSIAWYG What You See is Almost What You Get [Word Processing]

WYSIWYG What you See is What You Get [Word Processing]

X Computer
Cross
Crystal [also: x]
Experiment
Experimental
Extra
Extraordinary
Index
Reactance
Reactor
Transistor
Transmission
Transmitter
Xenon
X-Ray Unit
X-Unit
XA Cross-Arm
Transmission Adapter
Xa Activated Factor X
XAA X-Ray Absorption Analysis
XAAP Exploratory Airborne Acoustic Processor
XACT X (= Computer) Automatic Code Translation
XAES X-Ray Excited Auger Electron Spectroscopy
XAFS X-Ray Absorption Fine Structure
XANES X-Ray Absorption Near-Edge Spectroscopy
X-Ray Absorption Near-Edge Structure
XARM Cross-Arm
XB Crossbar
Xba Xanthomonas badrii [Molecular Biology]
XBAR Crossbar
XBC External Block Controller
XBF Experimental Boundary File
XBL Extension Bell
XBM Extended BASIC Mode
XBT Crossbar Tandem
Expendable Bathythermograph
XC Cross Channel
Cross Country
XCD East Caribbean Dollar [Currency of Antigua-Barbuda, Grenada, St. Christopher Nevis, St. Kitts, St. Lucia, St. Vincent and the Grenadines]
Exceed
XCH Exchange
XCL Exclusive
XCLB X-Ray Compositional Line Broadening
XCO Crystal-Controlled Oscillator
XCON Expert Configuration
Expert Configurer
XCONN Cross Connection
X-CR Chloroprene Rubber with Reactive Groups
XCS X (= Ten) Call Seconds
Xerox Computer Services
XCSRA Cross Channel Special Rules Area [UK]
XCU Crosspoint Control Unit
X-CUT Crystal Cut [also: X CUT]
XCV Experimental Composite Vehicle
XCVR Transceiver
XD Crossed
Crystal Detector
Ex-Directory
Exploratory Development
x-d without dividend
XDAAP Exploratory Development of an Airborne Acoustic Processor
XDCR Transducer
XDH Xanthine Dehydrogenase
XDM Experimental Development Model
Exploratory Development Model
XDP Crosslinked Fibrin Degradation
Xanthosine Diphosphate
X-Ray Density Probe
XDS Xerox Data System
XDUP Extended Disk Utilities Program
XE Experimental Engine
Xe Xenon
XEC Execute
X-Ray Elasticity Constants
XECF Experimental Engine — Cold Flow
XEDS X-Ray Energy Dispersive Spectroscopy
XEG X-Ray Emission Gauge
XEQ Execute
XES X-Ray Energy Spectrometry
XESD X-Ray Induced Electron-Stimulated Ion Desorption
XET Transparent End-of-Transmission
XETB Transparent End-of-Transmission Block
XETX Transparent End-of-Text
XEU European Currency Unit [Currency of the European Community]
XF Extra Fine
X-Ray Fluorescence
XFC Extended Function Code
Transfer Charge
XFER Transfer
XFMR Transformer
XFORMER Transformer [also: X-FORMER]
XFR X-Ray (Quasi-)Forbidden Reflection
XFS X-Ray Fluorescence Spectroscopy
X-GAL 5-Bromo-4-Chloro-3-Indolyl-β-D-Galactopyranoside [also: Xgal]
XGAM Experimental Guided Aircraft Missile
XGP Xerox Graphic Printer
XHAIR Cross Hair
XHB Extra-Hard Bolt
XHFR Experimental High-Frequency Radar
XHMO Extended Hueckel Molecular Orbital
Xho Xanthomonas holcicola [Molecular Biology]
XHV Extreme High Vacuum
x-i without interest
XIC Transmission Interface Converter
XICS Xerox Integrated Composition System
XID Exchange Identification
XIIR Halogenated Isobutylene-Isoprene Rubber
XIL Xenon Isotope Lamp
XIM X-Ray Inspection Module

XIMCE Xian Institute of Metallurgy and Construction Engineering [PR China]
XIO Execute Input/Output
XIOP Block-Multiplexer Input/Output
XIT Extra Input Terminal
XITB Transparent Intermediate Text Block
XJU Xi'an Jiaotong University [PR China]
XL Crystal [also: Xl]
Execution Language
Extra Large
Extra Long
Xenon Lamp
XLBIB Extra-Large Burn-In Bath
XLDP Xylose-Lysine-Deoxycholatepeplon
X-LINKED Crosslinked [also: X-linked]
XLPE Crosslinked Polyethylene [also: X-LPE or XLP]
XLR Experimental Liquid Rocket
XM Expanded Memory
Experimental Missile
Extra Marker
XMA X-Ray Microanalysis
XMAS Extended Mission Apollo Simulation [of NASA]
XMBA Executive Master of Business Administration
XMC Directionally-Reinforced Molding Compound
Extra-High-Strength Molding Compound
XMFR Transformer
XMIT Transmit
XMITTER Transmitter [also: X-MITTER]
XMOS Cross Metal-Oxide Semiconductor
XMP Xanthosine Monophosphate
XMR X-Ray Micro-Radiography
XMS Xerox Memory System
XMSN Transmission
XMT Transmit
XMTD Transmitted
XMTG Transmitting
XMTL Transmittal
XMTR Transmitter
XN Execution Node
Intersection
XNBR Carboxylated Nitrile-Butadiene Rubber
X-NBR Nitrile-Butadiene Rubber with Reactive Groups
XNOR Exclusive Not Or
XNOS Experimental Network Operating System
XO Cross-Office
Crystal Oscillator
Xanthine Oxidase
Xylenol Orange
XOF CFA (= Communaute Financiere Africaine) Franc [Currency of Benin, Congo, Gabon, Ivory Coast, Niger, Senegal, Togo and Upper Volta]
XOFF Transmitter Off [also: X-OFF]
XON Transmitter On [also: X-ON]
XOP Extended Operation
XOR Exclusive Or
XOS Cross-Office Slot
Xerox Operating System
XOW Express Order Wire
XP Cross-Polarization
Explosionproof
X-Ray Photoelectron
XPD Cross-Polarization Discrimination
X-Ray Photoelectron Diffraction
XPDA X-Ray Powder Diffraction Analysis
XPDR Transponder
X-PES X-Ray Photoelectron Spectroscopy [also: XPES]
XPF CFP (= Communaute Financiere Pacifique) Franc [Currency of French Polynesia and Tahiti]
XPI Cross-Polarization Interference
XPL Explosive
XPN External Priority Number
X-POL Cross-Polarization [also: XPOL]
XPS X-Ray Photoelectron Spectroscopy
X-Ray Photoelectron Spectrum
XPSW External Processor Status Word
XPT Crosspoint
External Page Table
X/Q Relative Concentration
XR External Relations Service [of UNESCO]
External Reset
Index Register
Xeroradiography
XRA X-Ray Analysis
XRC Xerox Research Center
X-Ray Crystallography
X-Ray Rocking Curves
XRCC Xerox Research Center of Canada
XRD X-Ray Diffraction
XRDRSA X-Ray Diffraction Residual Stress Analysis
XRE X-Ray Emission
XRES X-Ray Emission Spectrum
XREF Cross-Reference [also: XREF]
XREP Auxiliary Report
XRF X-Ray Filter
X-Ray Fluorescence
XRFS X-Ray Fluorescence Spectroscopy
XRI X-Ray Intensity
XRIC X-Ray Intensities Computation
XRITC Quinolizino-Substituted Fluorescein Isothiocyanate
XRLM Extended Range Lance Missile
XRM External Relational Memory
X-Ray Microanalysis
X-Ray Microscopy
XRMF X-Ray Micro-Fluorescence
XRMM X-Ray Multi-Mirror Mission [of ESA, France]
XRPD X-Ray Powder Diffraction
XRPDF X-Ray Powder Diffraction File
XRPM X-Ray Projection Microscope
XRR X-Ray Refraction
XRS X-Ray Scattering
X-Ray Spectroscopy
X-Ray Spectrum
XRSAS X-Ray Small-Angle Scattering
XRT X-Ray Tube
XRWL X-Ray Wavelength
XS Extra Strong
Xerces Society [US]
XS-3 Excess-Three (Code) [Computer Science]
XSA X-Ray Stress Analysis
XSCE X-Ray Spectroscopy from Channeled Electrons
XSE X-Ray Stress Evaluation
XSECT Cross Section [also: X-SECT]

XSM Experimental Strategic Missile
XSONAD Experimental Sonic Azimuth Detector
XSP Cross-Sectional Point
XSPT External Shared Page Table
XSPV Experimental Solid Propellant Vehicle
XSR X-Ray Synchrotron Radiation Source
XSTR Transistor
XSTX Transparent Start-of-Text
XSWIS X-Ray Standing Wave Interference Spectroscopy
XSYN Transparent Synchronous
XT Crosstalk [also: CT]
XTA X-Ray Texture Analysis
XTAL Crystal [also: X-TAL or Xtal]
XTASI Exchange of Technical Apollo Simulation Information
XTC External Transmit Clock
XTEL Cross Tell
XTEM Cross-Sectional Transmission Electron Microscopy
XTEN Xerox Telecommunications Network [US]
XTK Cross-Track
XTL Crystal [also: Xtl or xtl]
Crystalline [also: Xtl or xtl]
XTLO Crystal Oscillator
XTP Xanthosine Triphosphate
XTPA Extended Transaction Processing Architecture
XTR Extra
XTS Cross-Tell Simulator
X-Windows Tracking System
XTSI Extended Task Status Index
XTTD Transparent Temporary Text Delay
XU Xiamen University [at Beijing, PR China]
X-Ray Unit
X-Unit
XUG Xyvision (Typesetting System) Users Group [US]
XUV Extreme Ultraviolet
X-Ray Ultraviolet
XVR Exchange Voltage Regulator
Transceiver
XWAVE Extraordinary Wave
XX Double Extra
Double Strength
XXL Double Extra Long
XXS Double Extra Strong
XXX Triple Extra
Triple Strength
XXXL Triple Extra Long
XXXS Triple Extra Strong
XYR X-Y Recorder

Y

Y Admittance [Symbol]
Tyrosine
Tyrosyl
Year
Yellow
Yen [Currency of Japan]
Yoke
Young's Modulus
Yttrium

Y Yen [Currency of Japan]

y yard
year
yellow

YA Yacht Architect
Yosemite Association [US]

YABA Yacht Architects and Brokers Association [US]

YAC Young Astronaut Council [US]

YAG Yttrium Aluminum Garnet

YAP Yttrium Aluminum Perovskite Oxide

Yb Ytterbium

YBCO Yttrium Barium Copper Oxide (Crystal) [also: $YBa_2Cu_3O_x$]

Yb(FOD)$_3$ Tris(heptafluorodimethyloctanedionato)ytterbium

Yb(HFC)$_3$ Tris[(heptafluoropropylhydroxymethylene)-camphorato],Ytterbium

YBP Years before Present

YBPC Young Black Programmers Coalition [US]

YBRA Yellowstone-Bighorn Research Association [US]

Yb(TFC)$_3$ Tris[(trifluoromethylhydroxymethylene)caphorato], Ytterbium

Yb(THD)$_3$ Tris(tetramethylheptanedionato)ytterbium

YBT Yoshida Buckling Test

YC Yield Condition

YCZ Yellow Caution Zone

YD Yacht Designer

yd yard

YDA Yesterday

YDD Yemeni Dinar [Currency of the People's Republic of Yemen]

yds yards

YDSA Yacht Designers and Surveyors Association [UK]

YDT Yukon Daylight Time [Canada]

YEA Yale Engineering Association [US]

YEC Youngest Empty Cell

YECIP Yukon Energy Conservation Incentive Program [Canada]

YEL Yellow

Yen Yen [Currency of Japan]

YER Yemeni Riyal (or Rial) [Currency of the Arab Republic of Yemen]

YES Yeast Extract Sucrose

YFL Yukon Federation of Labour [Canada]

YFU Youth for Understanding [US]

YG Yellowish Green

YIG Yttrium Iron Garnet

YIL Yellow Indicating Lamp

Y-IR Thermoplastic Isoprene Rubber

YKB Yukon Bibliography (Database) [of UA, Canada]

YL Yield Line

YLSH Yellowish

YM Yellow Metal

YMBA Yacht and Motor Boat Association

YMC Yeast Mold Count
Yttrium-Oxide Microconcrete

Y/N Yes/No (Prompt)

YNA Young Naturalists Association [Canada]

Y-NBR Thermoplastic Nitrile-Butadiene Rubber

YNCIAWPRC Yugoslavian National Committee of the International Association on Water Pollution Research and Control

YNHA Yosemite Natural History Association [now: YA, US]

YNU Yokohama National University [Japan]

YOF Year of the Ocean Foundation [now: OO, US]

Yorks Yorkshire [UK]

YP Year of Publication
Yellow Pine
Yield Point

YPE Yield Point Elongation

YPP Yield Point Phenomenon

YPS Yellow Pages Service

YR Year

YRI Yemeni Riyal (or Rial) [Currency of the Arab Republic of Yemen]

YRS Years

YS Yacht Surveyor
Yield Strength
Yield Stress

YSB Yacht Safety Bureau

Y-SBR Thermoplastic Styrene-Butadiene Rubber

YSF Yield Safety Factor
Youth Science Foundation [Canada]

YSI Yukon Science Institute [Canada]

YSLF Yield Strength Load Factor

YSSP Young Scientist Summer Program [of IIASA, Austria]

YST Yukon Standard Time [Canada]

YSU Youngstown State University [US]

YSZ Yttria-Stabilized Zirconia

YT Yield Temperature
Yukon Territory [Canada]

YTD Year to Date

YTO YIG (= Yttrium Iron Garnet) Tuned Oscillator

Y-TZP Yttria-Stabilized Tetragonal Zirconia Polycrystal

YU Yale University [US]
Yamagata University [Japan]
Yamaguchi University [Japan]
Yamanashi University [Japan]
York University [Canada]

Yu Yuan [Currency of PR China]

YUD Yugoslavian Dinar [Currency]

YUDC York University Development Center [Canada]

Z

Z	Atomic Number [Symbol]
	Impedance [Symbol]
	Self-Inductance [Symbol]
	Zaire [Currency of Zaire]
	Zenith
	Zero
	Zinc
	Zone
	Zulu (Time)
z	zero
	zone
ZA	Zenith Angle
	Zenith Attraction
	Zero Absolute
	Zero Acess
	Zero Adjuster
	Zero and Add
	Zinc-Aluminum (Alloy)
	Zone Axis
ZADCA	Zinc Alloy Die Casters Association [UK]
ZADI	Zentralstelle fur Agrardokumentation und -information [= Center for Agricultural Documentation and Information, Germany]
ZAED	Zentralstelle fur Atomenergiedokumentation [= Center for Documentation on Nuclear Energy, Germany]
ZAF	Atomic Number, Absorption and Fluorescence (Correction)
ZAI	Zaire [Currency of Zaire]
	Zero-Acess Instruction
ZAL	Zirconium-Arc Lamp
ZAMEFA	Metal Fabricators of Zambia
ZAMS	Zero Age Main Sequence
ZAP	Zone Axis Pattern
ZAPP	Zero Assignment Parallel Processor
ZAR	Southern African Rand [Currency of South Africa and Namibia]
	Zeus Acquisition Radar
ZARM	Zentrum fur Angewandte Raumfahrttechnologie und Mikrogravitation [= Center for Applied Space Technology and Microgravitation, Germany]
ZAS	Zero-Acess Storage
ZAW	Zentralausschuss der Deutschen Wirtschaft [= Central Committee of the German Industry]
ZB	Zinc-Blende
	Zone Boundary
ZBB	Zero Based Budgeting
ZBD	Zener Breakdown
ZBEC	Zinc Dibenzyl Dithiocarbamate
ZBID	Zero Bit Insertion/Deletion
ZBLAN	$ZrF_4BaF_2LaF_3AlF_3NaF$ (Laser)
ZBX	Zinc Butyl Xanthogenate
ZC	Zenithal Chart
	Zone Control
ZCG	Zinc Chloride Glass
ZCR	Zero Crossing Rate
ZCS	Zone Communication Station
ZCTU	Zambia Congress of Trade Unions
	Zimbabwe Congress of Trade Unions
ZD	Zener Diode
	Zenith Distance
	Zero Defects
	Zoned Decimal
Z$	Zimbabwe Dollar [Currency]
ZDA	Zinc Development Association [UK]
ZDBC	Zinc Dibutyl Dithiocarbamate
ZDBP	Zinc Dibutyl Dithiophosphate
ZDC	Zeus Defense Center
ZDCTBS	Zeus Defense Center Tape and Buffer System
ZDE	Zentralstelle fur Dokumentation Elektrotechnik [= Documentation Center for Electrical Engineering, Germany]
ZDEC	Zinc Diethyl Dithiocarbamate
ZDMC	Zinc Dimethyl Dithiocarbamate
ZDP	5-Aminoimidazole-4-Carboxamide-1-Ribofuranosyl-5'-Diphosphate
	Zero Delivery Pressure
	Zinc Dithiophosphate
ZDR	Zeus Discrimination Radar
ZDS	Zilog Development System
ZDT	Zero-Ductility Transition
ZDVR	Zener Diode Voltage Regulator
ZE	Zero Energy
ZEA	Zero Energy Assembly
ZEBRA	Zero Energy Breeder Reactor Assembly [UK]
ZEBS	Zentrale Erfassungs- und Bewertungssstelle fur Umweltchemikalien [= Center for Registration and Evaluation of Environmental Chemicals, Germany]
ZEEP	Zero Energy Experimental Pile [of CRNL, Canada]
ZEF	Zero Extraction Force
ZEG	Zero Economic Growth
ZENITH	Zero Energy Nitrogen-Heated Thermal Reactor [UK]
ZEPC	Zinc Ethyl Phenyl Dithiocarbamate
ZEPR	Zero Energy Power Reactor
ZER	Zero Energy Reactor
ZERLINA	Zero Energy Reactor for Lattice Investigations and Study of New Assemblies
ZERT	Zero Reaction Tool
ZES	Zero Energy System
ZETA	Zero Energy Thermonuclear Apparatus [of AERE, UK]
ZETR	Zero Energy Thermal Reactor [of AERE, UK]
ZEUS	Zero Energy Uranium System [of AERE, UK]
ZEVIS	Zentrales Verkehrsinformationssystem [= Central Traffic Information System, Germany]
ZF	Zero Frequency
	Zone Finder
ZFC	Zero Failure Criterion
	Zero-Field Cooling
ZFS	Zero-Field Splitting
ZfS	Zentralstelle fur Solartechnik [= Center for Solar Technology, Germany]

ZFW Zentralinstitut fur Festkorperphysik und Werkstofforschung [= Institute for Solid-State Physics and Materials Research, Germany]
Zero Fuel Weight
ZG Zero Gravity
ZGA Zambia Geographical Association
ZGE Zero Gravity Effect
ZGS Zero Gradient Synchrotron [of ANL, US]
ZH Zonal Harmonics
Zone Heating
ZHR Zenithal Hourly Rate
ZHS Zero Hoop Stress
ZHT Zero Heat Transfer
ZI Zinc Institute
Zoom In
ZID Zimbabwe Dollar [Currency]
ZIE Zimbabwe Institution of Engineers
ZIF Zero Insertion Force
ZIF/ZEF Zero Insertion Force/Zero Extraction Force
ZIL Zigzag In-Line
ZINCOM Zambia Industrial and Commercial Association
ZINEB Zinc Ethylenebis(dithiocarbamate)
ZIP Zero Inventory Purchasing
Zigzag In-Line Package
Zinc Impurity Photodetector
Zone Improvement Plan [US]
ZIPCD Zip Code [also: ZIP CD]
ZISCH Zirkulation und Schadstoffeintrag in der Nordsee [= Circulation and Pollution of the North Sea, Germany]
ZIX Zinc Isopropyl Xanthogenate
ZKw Zambian Kwacha [Currency of Zambia]
ZkW Zero Kilowatt
ZL Zero Lift
Zone Law
Zoom Lens
Zl Zloty [Currency of Poland]
ZLA Zero-Lift Angle
ZLC Zinc, Lead and Cadmium Abstracts [of ZDA, UK]
ZLM Zero-Lift Moment
ZLO Zloty [Currency of Poland]
ZM Zone Marker
Zone Melting
Zone Meridian
ZMA Zinc Metaarsenite
ZMAR Zeus Multifunction Array Radar
ZMAR/MAR Zeus Multifunction Array Radar/Multifunction Array Radar
ZMB Zero Moisture Basis
ZMBI Zinc-2-Mercaptobenzimidazole
ZMBT Zinc-2-Mercaptobenzothiazole
Z5MC Zinc Pentamethylene Dithiocarbamate
ZMK Zambian Kwacha [Currency of Zambia]
ZMMBI Zinc-Methyl-2-Mercaptobenzimidazole
ZMR Zone Melting Recrystallization
ZN Zone
Zn Zinc
ZnDTP Zinc Dialkyl Dithiophosphate
ZnPCTP Zinc Pentachlorothiophenol
ZO Zero Order
ZOC Zone of Convergence
ZODIAC Zone Defense Integrated Active Capability
ZOE Zero Energy
ZOH Zero Order Hold
ZOLZ Zero-Order Laue Zone
ZOLZ-CBEDP Zero-Order Laue Zone — Convergent-Beam Electron Diffraction Pattern
ZOPA Zinc Oxide Producers Association [of CEFIC, Belgium]
ZOR Zero-Order Reaction
ZP Zero Point
Zone Plate
Zone Punch
Zone Purification
ZPA Zero Period Acceleration
Zeus Program Analysis
ZPCK N-Carbobenzyloxyphenylalanyl Chloromethyl Ketone
ZPDA Zinc Pigment Development Association [UK]
ZPE Zero-Point Energy
ZPEN Zeus Project Engineer Network
ZPG Zero Population Growth
ZPO Zeus Project Office
ZPP Zero-Point Pressure
ZPPR Zero-Power Plutonium Reactor [of USAEC]
ZPR Zero-Power Reactor
ZPR(ANL) Zero-Power Reactor at Argonne National Laboratories [US]
ZPRF Zero-Power Reactor Facility [of NASA]
ZPR(NASA) Zero-Power Reactor of NASA [US]
ZR Zone Refining
Zr Zirconium
ZRA Zero Range Approximation
Zero Resistance Ammeter
ZRADAS Zero Resistance Ammeter Data Acquisition System
ZRM Zone Reserved for Memory
$Zr(TMHD)_4$ Tetrakis(2,2,6,6-Tetramethyl-3,5-Heptanedionato)zirconium
ZRZ Zaire [Currency of Zaire]
ZS Zero and Subtract
Zero Slope
Zero Suppression
Zoological Society
ZSA Zimbabwe Scientific Association
ZSL Zoological Society of London [UK]
ZSSA Zoological Society of Southern Africa
ZT Zenith Telescope
Zone Theory
Zone Time
ZTA Zirconia-Toughened Alumina
ZTC Zirconia-Toughened Ceramic
ZTE Zero Time Exchange
ZTO Zero Time Outage
ZTP 5-Aminoimidazole-4-Carboxamide-1-Ribofuranosyl-5'-Triphosphate
Z Toroidal Pinch
ZTS Zoom Transfer Scope
ZU Zhejiang University [PR China]
ZUMA Zentrum fur Umfragen, Methoden und Analysen [= Center for Surveys, Methods and Analyses, Germany]

ZURF Zeus Up-Range Facility
ZVD Zinc Vapor Deposition
ZVEI Zentralverband Elektrotechnik- und Elektronikindustrie [= Electrical and Electronics Manufacturers Association, Germany]
ZVS Zentralstelle fur die Vergabe von Studienplatzen [= Central Universities Admission Council, Germany]
Zero Voltage Switching
ZWD Zimbabwe Dollar [Currency]
ZWOK Zirconium-Water Oxidation Kinetics

Part B

Periodical Title Abbreviations and Acronyms in Science and Engineering

A

AAA	Astronomy and Astrophysics Abstracts [Germany]
A & A, Assem. Assoc.	A & A — Assembly and Association [UK]
AAAI Proc.	AAAI Proceedings [of American Association for Artificial Intelligence]
AAAS Obs.	AAAS Observer [US]
AAAS Rep.	AAAS Report [of American Association for the Advancement of Science]
AACE Trans.	AACE Transactions [of American Association of Cost Engineers]
AACG Newsl.	AACG Newsletter [of American Association for Crystal Growth]
AAPS Newsl.	AAPS Newsletter [of American Association of Pharmaceutical Scientists]
AAS Hist. Ser.	AAS History Series [of American Astronautical Society]
AAS Newsl.	AAS Newsletter [of American Astronomical Society]
AAVSO Bull.	AAVSO Bulletin [of American Association of Variable Star Observers]
AAVSO Circ.	AAVSO Circular [of American Association of Variable Star Observers]
AAVSO Monogr.	AAVSO Monographs [of American Association of Variable Star Observers]
AAVSO Rep.	AAVSO Reports [of American Association of Variable Star Observers]
AAZPA Newsl.	AAZPA Newsletter [of American Association of Zoological Parks and Aquariums]
AAZPA Reg. Conf. Proc.	AAZPA Regional Conference Proceedings [of American Association of Zoological Parks and Aquariums]
ABA Bank. J.	ABA Banking Journal [of American Bankers Association]
Abbey Newsl.	Abbey Newsletter [US]
ABB Rev.	ABB Review [of Asea Brown Bovery, Switzerland]
A & B Comput.	A & B Computing [UK]
ABI-Tech.	ABI-Technik [Germany]
ABIPC	Abstract Bulletin of the Institute of Paper Chemistry [US]
Abrasive Eng. Soc. Mag.	Abrasive Engineering Society Magazine [US]
Abstr. Brit. Pat.	Abstracts of Specifications of British Patents [UK]
ABU Tech. Rev.	ABU Technical Review [of Asian Broadcasting Union, Japan]
Abwassertech. Abfalltech.	Abwassertechnik mit Abfalltechnik [Germany]
Acad. Rep. Fac. Eng. Tokyo Inst. Polytech.	Academic Reports, Faculty of Engineering, Tokyo Institute of Polytechnics [Japan]
Acad. Serbe Sc. Arts Glas Cl. Sci. Tech.	Academie Serbe des Sciences et des Arts, Glas, Classe des Sciences Techniques [Yugoslavia]
Acciaio Inossid.	Acciaio Inossidabile [Italy]
Acct. Chem. Res.	Accounts of Chemical Research [of American Chemical Society, US]
ACE	Aviation and Computer Enthusiasts [US]
ACEC Rev.	ACEC Review [of Ateliers de Constructions Electriques de Charleroi, Belgium]
Achats Entret.	Achats et Entretien [France]
Aciers Spec.	Aciers Speciaux [France]
ACIL Newsl.	ACIL Newsletter [of American Council of Independent Laboratories]
ACI Mater. J.	ACI Materials Journal [of American Concrete Institute]
ACI Struct. J.	ACI Structural Journal [of American Concrete Institute]
ACM Conf. Proc.	ACM Conference Proceedings [of Association for Computing Machinery]
ACM Trans. Comp. Syst.	ACM Transactions on Computer Systems [of Association for Computing Machinery]
ACM Trans. Database Syst.	ACM Transactions on Database Systems [of Association for Computing Machinery]
ACM Trans. Graph.	ACM Transactions on Graphics [of Association for Computing Machinery]
ACM Trans. Inf. Syst.	ACM Transactions on Information Systems [of Association for Computing Machinery]
ACM Trans. Math. Softw.	ACM Transactions on Mathematical Software [of Association for Computing Machinery]
ACM Trans. Off. Inf. Syst.	ACM Transactions on Office Information Systems [of Association forComputing Machinery]
ACM Trans. Program. Lang. Syst.	ACM Transactions on Programming Languages and Systems [of Association for Computing Machinery, US]
Acoust. Aust.	Acoustics Australia [of Australian Acoustical Society]
Acoust. Lett.	Acoustics Letters [UK]
ACSM Bull.	ACSM Bulletin [of American Congress on Surveying and Mapping]
ACS Polym. Mat. Sci. Eng.	ACS Polymeric Materials Science and Engineering [of American Chemical Society]
ACS Polym. Preprints	ACS Polymer Preprints [of American Chemical Society]
ACS Symp. Ser.	ACS Symposium Series [of American Chemical Society
Acta Acad. Aboensis	Acta Academiae Aboensis [Finland]
Acta Acad. Sci. Pol.	Acta Academiae Scientiarum Polonae [Poland]

Acta Acust.	Acta Acustica [PR China]
Acta Astron.	Acta Astronomica [Poland]
Acta Astronaut.	Acta Astronautica [UK]
Acta Astron. Sin. (China)	Acta Astronomica Sinica (China) [of Chinese Astronomical Society, PR China]
Acta Astrophys. Sin. (China)	Acta Astrophysica Sinica (China) [PR China]
Acta Autom. Sin. (China)	Acta Automatica Sinica (China) [PR China]
Acta Bioquim. Clin. Latinoam.	Acta Bioquimica Clinica Latinoamericana [of Latin-American Confederation of Clinical Biochemistry, Colombia]
Acta Chem. Scand.	Acta Chemica Scandinavica [Denmark]
Acta Chem. Scand. A	Acta Chemica Scandinavica A [Denmark]
Acta Chim. Acad. Sci. Hung.	Acta Chimica Academiae Scientarium Hungaricae [now: Acta Chimica Hungaricae]
Acta Chim. Hung.	Acta Chimica Hungaricae [Hungary]
Acta Chim. Sin. (China)	Acta Chimica Sinica (China) [of Chinese Chemical Society, PR China]
Acta Chim. Sin. (Engl. Ed.)	Acta Chimica Sinica (English Edition) [Netherlands]
Acta Cienc. Indica Math.	Acta Ciencia Indica, Mathematics [India]
Acta Cienc. Indica Phys.	Acta Ciencia Indica, Physics [India]
Acta Crystallogr.	Acta Crystallographica [Denmark]
Acta Crystallogr. A	Acta Crystallographica A [Denmark]
Acta Crystallogr. A, Found. Crystallogr.	Acta Crystallographica, Section A, Foundations of Crystallography [Denmark]
Acta Crystallogr. B	Acta Crystallographica B [Denmark]
Acta Crystallogr. B, Struct. Sci.	Acta Crystallographica, Section B, Structural Science [Denmark]
Acta Crystallogr. C	Acta Crystallographica C [Denmark]
Acta Crystallogr. C, Struct. Commun.	Acta Crystallographica, Section C, Crystal Structure Communications [Denmark]
Acta Crystallogr. Sect. A	Acta Crystallographica, Section A [Denmark]
Acta Crystallogr. Sect. B	Acta Crystallographica, Section B [Denmark]
Acta Crystallogr. Sect. C	Acta Crystallographica, Section C [Denmark]
Acta Cybernet.	Acta Cybernetica [Hungary]
Acta Electron. Sin.	Acta Electronica Sinica [PR China]
Acta Geod. Geophys. Montan. Hung.	Acta Geodaetica, Geophysica et Montanistica Hungarica [Hungary]
Acta Geophys. Pol.	Acta Geophysica Polonica [Poland]
Acta Geophys. Sin. (China)	Acta Geophysica Sinica (China) [PR China]
Acta Geophys. Sin. (USA)	Acta Geophysica Sinica (USA) [English translation of: *Acta Geophysica Sinica (China)*; published in the US]
Acta Hort.	Acta Horticulturae [of International Society for Horticultural Science, Netherlands]
Acta Inform.	Acta Informatica [Germany]
Acta Manil.	Acta Manilana [Philippines]
Acta Math. Appl. Sin.	Acta Mathematica Applacatae Sinica [PR China]
Acta Math. Sin.	Acta Mathematica Sinica [PR China]
Acta Mech.	Acta Mechanica [Austria]
Acta Mech. Sin.	Acta Mechanica Sinica [PR China]
Acta Mech. Solida Sin.	Acta Mechanica Solida Sinica [PR China]
Acta Metall.	Acta Metallurgica [US]
Acta Metall. Mater.	Acta Metallurgica et Materialia [US]
Acta Metall. Sin.	Acta Metallurgica Sinica [PR China]
Acta Metall. Sin. (China)	Acta Metallurgica Sinica (China) [PR China]
Acta Metall. Sin. (Engl. Ed.) A	Acta Metallurgica Sinica (English Edition) Series A [Published in the US]
Acta Metall. Sin. (Engl. Ed.) B	Acta Metallurgica Sinica (English Edition) Series B [Published in the US]
Acta Meteorol. Sin.	Acta Meteorologica Sinica [PR China]
Acta Mex. Cienc. Tecnol.	Acta Mexicana de Ciencia y Tecnologia [Mexico]
Acta Mineral. Petrogr.	Acta Mineralogica Petrographica [Hungary]
Acta Oncol.	Acta Oncologica [Sweden]
Acta Opt. Sin.	Acta Optica Sinica [PR China]
Acta Opt. Sin. (China)	Acta Optica Sinica (China) [PR China]
Acta Pharm. Fenn.	Acta Pharmaceutica Fennicae [Finland]
Acta Pharm. Hung.	Acta Pharmaceutica Hungarica [Hungary]
Acta Pharm. Jugosl.	Acta Pharmaceutica Jugoslavia [Yugoslavia]
Acta Pharm. Nord.	Acta Pharmaceutica Nordica [Scandinavia]
Acta Pharm. Technol.	Acta Pharmaceutica Technologica

Acta Phys. Acad. Sci. Hung.	Acta Physica Academiae Scientarium Hungarica [Hungary]
Acta Phys. Chim. Sin.	Acta Physico-Chimica Sinica [of Chinese Chemical Society, PR China]
Acta Phys. Hung.	Acta Physica Hungarica [Hungary]
Acta Physiol. Pharmacol. Latinoam.	Acta Physiologica et Pharmacologica Latinoamericana [Argentina]
Acta Phys. Pol.	Acta Physica Polonica [Poland]
Acta Phys. Pol. A	Acta Physica Polonica A [Poland]
Acta Phys. Pol. B	Acta Physica Polonica B [Poland]
Acta Phys. Sin.	Acta Physica Sinica [PR China]
Acta Phys. Sin. (China)	Acta Physica Sinica (China) [PR China]
Acta Phys. Slovaca	Acta Physica Slovaca [Czechoslovakia]
Acta Phys. Temp. Humilis Sin.	Acta Physica Temperaturae Humilis Sinica [PR China]
Acta Phys. Univ. Comen.	Acta Physica Universitatis Comenianae [Czechoslovakia]
Acta Phyto. Acad. Scient. Hung.	Acta Phytobiologica Academiae Scientiarum Hungarica [Hungary]
Acta Politec. Mex.	Acta Politecnica Mexicana [Mexico]
Acta Pol. Pharm.	Acta Poloniae Pharmaceutica [Poland]
Acta Polym.	Acta Polymerica [Germany]
Acta Polytech. Ceske Vys. Uc. Tech. Pr. III	Acta Polytechnica Ceske Vysoke Uceni Technicke v Prace III [Czechoslovakia]
Acta Polytech. Scand.	Acta Polytechnica Scandinavica [Finland]
Acta Polytech. Scand., Appl. Phys. Ser.	Acta Polytechnica Scandinavica, Applied Physics Series [Finland]
Acta Polytech. Scand. Chem. Technol. Metall. Ser.	Acta Polytechnica Scandinavica, Chemical Technology and Metallurgy Series [Finland]
Acta Polytech. Scand., Electr. Eng. Ser.	Acta Polytechnica Scandinavica, Electrical Engineering Series [Finland]
Acta Polytech. Scand., Math. Comp. Sci. Ser.	Acta Polytechnica Scandinavica, Mathematics and Computer Science Series [Finland]
Acta Polytech. Scand., Mech. Eng. Ser.	Acta Polytechnica Scandinavica, Mechanical Engineering Series [Finland]
Acta Radiol.	Acta Radiologica [Sweden]
Acta Sci. Nat. Univ. Pekin.	Acta Scientiarum Naturalium Universitatis Pekinensis [PR China]
Acta Sci. Nat. Univ. Sunyatseni	Acta Scientiarum Naturalium Universitatis Sunyatseni [PR China]
Acta Seismol. Sin.	Acta Seismologica Sinica [PR China]
Acta Stereol.	Acta Stereologica [Yugoslavia]
Acta Stereol.	Acta Stereologica [of International Society for Stereology, Germany]
Acta Tech. Acad. Sci. Hung.	Acta Technica Academiae Scientiarum Hungaricae [Hungary]
Acta Tech. CSAV	Acta Technica CSAV [Czechoslovakia]
Acta Univ. Agr. Brno	Acta Universitatis Agriculturae Brno [Czechoslovakia]
Acta Univ. Szeged. Acta Phys. Chem.	Acta Universitatis Szegediensis, Acta Physica et Chimica [Hungary]
Acta Univ. Wratislav. Mat. Fiz. Astron.	Acta Universitatis Wratislaviensis, Matematyka Fizyka Astronomica [Poland]
Act. Passive Electron. Compon.	Active and Passive Electronic Components [UK]
Actual. Chim.	L'Actualite Chimique [of Societe Francaise de Chimie, France]
Actual. Combust. Energ.	Actualite, Combustible, Energie [France]
Actual. Inf. Cient. Tec.	Actualidades de la Informacion Cientifica y Tecnica [Cuba]
Ada Lett.	Ada Letters [Association of Computing Machinery]
Add. Poly.	Additives for Polymers [UK]
Adgez. Rasplav. Paika Mater.	Adgeziya Rasplavov i Paika Materialov [USSR]
Adhes. Age	Adhesives Age [US]
Adhes. Soc. Newsl.	Adhesion Society Newsletter [US]
Adhes. Trends	Adhesive Trends [of Adhesives Manufacturers Association, US]
Adm. Manage.	Administrative Management [US]
Adsorpt. Sci. Technol.	Adsorption Science and Technology
Adv. Alloys Met.	Advanced Alloys and Metals [Spain]
Advan. Phys.	Advances in Physics [UK]
Adv. Appl. Probab.	Advances in Applied Probability [UK]
Adv. Astron. Sci.	Advances in Astronautical Sciences [of American Astronautical Society]
Adv. Battery Technol.	Advanced Battery Technology [US]
Adv. Carbohyd. Chem. Biochem.	Advances in Carbohydrate Chemistry and Biochemistry [US]

Adv. Catal.	Advanced Catalysis
Adv. Cem. Res.	Advances in Cement Research
Adv. Ceram.	Advances in Ceramics
Adv. Ceram. Mat.	Advanced Ceramic Materials [of American Ceramic Society]
Adv. Ceram. Mater.	Advanced Ceramic Materials [of American Ceramic Society]
Adv. Ceram. Rep.	Advanced Ceramics Report [UK]
Adv. Chem. Eng.	Advances in Chemical Engineering
Adv. Chem. Ser.	Advances in Chemistry Series
Adv. Colloid Interface Sci.	Advances in Colloid and Interface Science [Netherlands]
Adv. Compos.	Advanced Composites [US]
Adv. Compos. Bull.	Advanced Composites Bulletin [UK]
Adv. Conv. Pkg. Technol.	Advancing Converting and Packaging Technologies [of US Technical Association of the Pulp and Paper Industry]
Adv. Cryog. Eng.	Advances in Cryogenic Engineering [of Cryogenic Engineering Conference]
Adv. Cryog. Eng. Mater.	Advances in Cryogenic Engineering Materials [of International Cryogenic Materials Conference]
Adv. Dent. Res.	Advances in Dental Research [US]
Adv. Drying	Advances in Drying
Adv. Earth Plant. Sci.	Advances in Earth and Planetary Sciences [Japan]
Adv. Electron. Phys.	Advances in Electronics and Physics
Adv. Eng. Softw.	Advances in Engineering Software [UK]
Adv. Graph. Mag.	Advanced Graphics Magazine [UK]
Adv. Heat Transfer	Advances in Heat Transfer
Adv. Instrum.	Advances in Instrumentation [of Instrument Society of America]
Adv. LTA Rev.	Advanced Lighter-Than-Air Review [of Airship Association]
Adv. Manage. J.	Advanced Management Journal [of Society for Advancement of Management]
Adv. Manuf. Process.	Advanced Manufacturing Processes [US]
Adv. Manuf. Technol.	Advanced Manufacturing Technology [Ireland]
Adv. Mater.	Advanced Materials [US]
Adv. Mater. (Germany)	Advanced Materials (Germany)
Adv. Mater. Manuf. Process.	Advanced Materials and Manufacturing Processes [US]
Adv. Mater. Process.	Advanced Materials and Processes [of ASM International, US]
Adv. Mater. Process. inc. Met. Prog.	Advanced Materials and Processes incorporating Metal Progress [of ASM International, US]
Adv. Mater. Technol. Int.	Advanced Materials Technology International [UK]
Adv. Mater. Technol.: Monitor	Advances in Materials Technology: Monitor [Austria]
Adv. Met. Technol.	Advanced Metals Technology [UK]
Adv. Microb. Ecol.	Advances in Microbial Ecology [of International Committee on Microbial Ecology, US]
Adv. Model. Simul.	Advances in Modelling and Simulation [France]
Adv. Phys.	Advances in Physics [UK]
Adv. Polym. Sci.	Advances in Polymer Science [Germany]
Adv. Polym. Technol.	Advances in Polymer Technology [US]
Adv. Printing Sci. Technol.	Advances in Printing Science and Technology [UK]
Adv. Robot.	Advanced Robotics [Netherlands]
Adv. Space Res.	Advances in Space Research [of Committee on Space Research, France]
Adv. Space Res.	Advances in Space Research [UK]
Adv. Spec. Electrometall.	Advances in Special Electrometallurgy [Translation of: *Problemy Spetsial'noi Elektrometallurgii (USSR)*; published in the UK]
Adv. Water Resour.	Advances in Water Resources [UK]
Adv. X-Ray Anal.	Advances in X-Ray Analysis
AEDS J.	AEDS Journal [of Association for Educational Data Systems, US]
AEDS Monit.	AEDS Monitor [of Association for Educational Data Systems, US]
AEF Natl. Newsl.	AEF National Newsletter [of Aerospace Education Foundation, US]
AEG Newsl.	AEG Newsletter [of Association of Engineering Geologists, US]
Aerojet Rev.	Aerojet Review [US]
Aerojet Technol.	Aerojet Technology [US]
Aeronaut. Astronaut.	Aeronautique et l'Astronautique [France]
Aeronaut. J.	Aeronautical Journal [of Royal Aeronautical Society, UK]
Aeronaut. Meridiana	Aeronautica Meridiana [South Africa]
Aeronaut. Res. Lab. Rep.	Aeronautical Research Laboratories Report [Australia]
Aeronaut. Satel. News	Aeronautical Satellite News [of International Maritime Satellite Organization, UK]
Aerosol Rep.	Aerosol Report [Germany]

Aerosol Sci. Technol.	Aerosol Science and Technology [of American Association for Aerosol Research, US]
Aerosp.	Aerospace [of Royal Aeronautical Society, UK]
Aerosp. Am.	Aerospace America [US]
Aerosp. Compos. Mater.	Aerospace Composites and Materials [UK]
Aerosp. Def. Rev.	Aerospace and Defence Review [UK]
Aerosp. Def. Sci.	Aerospace and Defense Science [Malaysia]
Aerosp. Eng.	Aerospace Engineering [of Society of Automotive Engineers, US]
AEU (Japan)	AEU [of Asia Electronics Union, Japan]
AFCET Interfaces	AFCET Interfaces [of Association Francaise pour la Cybernetique Economique et Technique, France]
AGA Gas Energy Rev.	AGA Gas Energy Review [of American Gas Association]
AGA Mon.	AGA Monthly [of American Gas Association]
AGARD (NATO)	Advisory Group for Aerospace Research and Development (NATO) [France]
AGARD Rep.	AGARD Reports [of Advisory Group for Aerospace Research and Development [of NATO]]
Agora Inform. Chang. World	Agora, Informatics in a Changing World [of Intergovernmental Bureau for Informatics, Italy]
Agric. Biol. Chem.	Agricultural Biology and Chemistry
Agric. Eng.	Agricultural Engineering [of American Society of Agricultural Engineers]
Agric. Eng. Mag.	Agricultural Engineering Magazine [of American Society of Agricultural Engineers]
AIAAJ	AIAA Journal [of American Institute of Aeronautics and Astronautics]
AIAA J.	AIAA Journal [of American Institute of Aeronautics and Astronautics]
AIAA Stud. J.	AIAA Student Journal [of American Institute of Aeronautics and Astronautics]
AIA Newsl.	AIA Newsletter [of Aerospace Industries Association of America]
AIChE J.	AIChE Journal [of American Institute of Chemical Engineers]
AIChEMI Modul. Instr. Ser.	AIChEMI Modular Instruction Series [of American Institute of Chemical Engineers]
AIChE Symp.	AIChE Symposium [of American Institute of Chemical Engineers]
AIChE Symp. Ser.	AIChE Symposium Series [of American Institute of Chemical Engineers]
AIChExtra	AIChExtra [of American Institute of Chemical Engineers]
AIC J.	AIC Journal [of American Institute of Constructors]
AICOM	Artificial Intelligence Communications [of European Coordinating Committee for Artificial Intelligence, France]
AI Commun.	AI Communications [Netherlands]
AI EDAM	Artificial Intelligence for Engineering Design, Analysis and Manufacturing [UK]
AIIP Newsl.	AIIP Newsletter [of Association of Independent Information Professionals]
AI Mag.	AI Magazine [of American Association for Artificial Intelligence]
AIP Conf. Proc.	AIP Conference Proceedings [of American Institute of Physics]
AIPE Facil. Manage. Oper. Eng.	AIPE Facilities Management: Operations and Engineering [of American Institute of Plant Engineers]
AIP Hist. Phys. Newsl.	AIP History of Physics Newsletter [of American Institute of Physics]
Air Cosm. Mon.	Air and Cosmos Monthly [France]
Aircr. Eng.	Aircraft Engineering [UK]
Aircr. Eng. Aerosp. Technol.	Aircraft Engineering and Aerospace Technology [UK]
Airpt. Forum	Airport Forum [Germany]
AISES Newsl.	AISES Newsletter [of American Indian Science and Engineering Society]
AIUSB Q.	AISB Quarterly [UK]
Ajour Ind.-Tek.	Ajour Industrii-Teknikk [Norway]
Akad. Nauk SSR, Ural. Fil. Tr. Inst. Khim.	Akademiya Nauk SSR, Ural'skii Filial, Trudy Instituta Khimii [USSR]
Akad. Nauk. Ukr. SSR, Metallofiz.	Akademiya Nauk Ukrainskai SSR, Metallofizika [USSR]
Akron Bus. Econ. Rev.	Akron Business and Economic Review [US]
Aktual. Probl. Inf. Dok.	Aktualne Problemy Informacji i Dokumentacji [Poland]
Akust. Zh.	Akusticheskii Zhurnal [USSR]
Alcoa Technol. Rep.	Alcoa Technology Report [US]
Alfaatih Univ. Bull. Fac. Eng.	Alfaatih University Bulletin of the Faculty of Engineering [Libya]
ALLC J.	ALLC Journal [of Association for Literary and Linguistic Computing, UK]
Allegheny Ludlum Horiz.	Allegheny Ludlum Horizons [US]
Allg. Papier-Rundschau	Allgemeine Papier Rundschau [Germany]
Allg. Vliesstoff-Rep.	Allgemeiner Vliesstoff-Report [Germany]
Alloy Dig.	Alloy Digest [US]
Allum. Leghe	Alluminio e Leghe [Italy]
Allum. Mag.	Alluminio Magazine [Italy]
Alta Freq.	Alta Frequenza [of Associazione Elettrotecnica ed Elettronica Italiana, Italy]
Alta Freq., Riv. Elettron.	Alta Frequenza — Rivista di Elettronica [of Associazione Elettrotecnica ed Elettronica Italiana, Italy]

Altern. Sources Energy	Alternative Sources of Energy [US]
Alum. Appl.	Aluminium Applications [UK]
Alum. Assoc. Rep.	Aluminum Association Report [US]
Alum. Dev. Dig.	Aluminum Developments Digests [US]
Alum. Ind.	Aluminium Industry [UK]
Alum. Kur.	Aluminium Kurier [Germany]
Alum. Mon.	Aluminium Monthly [UK]
Alum. Remodeling News	Aluminum Remodeling News [US]
Alum. Rep.	Aluminum Report [US]
Alum. Rev.	Aluminium Review [South Africa]
Alum. Situat.	Aluminum Situation [of Aluminum Association, US]
Alum. Stat. Rev.	Aluminum Statistical Review [of Aluminum Association, US]
Alum. Today	Aluminium Today [UK]
Alum. Use	Aluminium in Use [Australia]
Alurama Int.	Alurama International [Sweden]
Am. Acad. Environ. Eng. Dipl.	American Academy of Environmental Engineers Diplomate
Amat. Radio	Amateur Radio [US]
Am. Book Publ. Rec.	American Book Publishing Record
Am. Cart.	American Cartographer [of American Congress on Surveying and Mapping]
AMCB	Asian Mass Communication Bulletin [Singapore]
Am. Ceram. Soc. Bull.	American Ceramic Society Bulletin
AMC J.	American Mining Congress Journal
Amdel Bull.	Amdel Bulletin [Australia]
Am. Dyestuff Rep.	American Dyestuff Reporter
AMFI Ind. News	AMFI Industry News [of Aviation Maintenance Foundation International, US]
Am. Forests	American Forests [of American Forestry Association]
AMFS Newsl.	AMFS Newsletter [of American Federation of Mineralogical Societies]
Am. Ind. Hyg. Assoc. J.	American Industrial Hygiene Association Journal
Am. Ink Maker	American Ink Maker
Am. Health	America's Health
Am. J. Archeol.	American Journal of Archeology [of Archeological Institute of America]
Am. J. Bot.	American Journal of Botany [of Botanical Society of America]
Am. J. EEG Technol.	American Journal of EEG Technology [of American Society of EEG Technologists]
Am. Jewel. Manuf.	American Jewelry Manufacturer
Am. J. Hosp. Pharm.	American Journal of Hospital Pharmacy
Am. J. Ind. Med.	American Journal of Industrial Medicine
Am. J. Math. Manage. Sci.	American Journal of Mathematical and Management Sciences
Am. J. Optom. Physiol. Opt.	American Journal of Optometry and Physiological Optics
Am. J. Phys.	American Journal of Physics [of American Institute of Physics]
Am. J. Sci.	American Journal of Science
Am. Lab.	American Laboratory
Am. Logger Lumberman	American Logger and Lumberman
Am. Mach.	American Machinist
Am. Mach. Autom. Manuf.	American Machinist and Automated Manufacturing
Am. Math. Mon.	American Mathematical Monthly [of Mathematical Association of America]
Am. Met. Mark.	American Metal Market
Am. Met. Mark., Metalwork. News Ed.	American Metal Market, Metalworking News Edition
Am. Mineral.	American Mineralogist [of Mineralogical Society of America]
AMM Mag.	AMM Magazine [US]
AMMRC Connect.	AMMRC Connections [of Army Materials and Mechanics Research Center, US]
Am. Papermaker	American Papermaker
Am. Printer	American Printer
Am. Prof. Constr.	American Professional Constructor [of American Institute of Constructors]
Am. Sci.	American Scientist
AMSE Rev.	AMSE Review [France]
AMSE Trans.	AMSE Transactions [France]
AMS Newsl.	AMS Newsletter [of American Meteorological Society]
Am. Stat.	American Statistician [of American Statistical Society]
Amts- Mitt.bl. Bundesanst. Mater. forsch.-pruf.	Amts- und Mitteilungsblatt der Bundesanstalt fur Materialforschung und -prufung [Germany]

Am. Zool.	American Zoologist [of American Society of Zoologists]
An. Acad. Bras. Cienc.	Anais da Academia Brasileira de Ciencias [Brazil]
Anal. Abstr.	Analytical Abstracts [of Royal Society of Chemistry, UK]
Anal. Biochem.	Analytical Biochemistry [US]
Anal. Chem.	Analytical Chemistry [of American Chemical Society]
Anal. Chem.	Analytical Chemistry [of Chinese Chemical Society, PR China]
Anal. Chem.	Analytische Chemie [Germany]
Anal. Chim. Acta	Analytica Chimica Acta [Netherlands]
Analele Stiint. Univ. 'Al.I. Cuza' Iasi (Ser. Noua) IB Fiz.	Analele Stiintifice ale Universitatu 'Al.I. Cuza' din Iasi (Serie Noua) Sectiuneas IB Fizica [Romania]
Analele Univ. Bucur. Fiz.	Analele Universitatii Bucaresti, Fizica [Romania]
Analele Univ. Bucur. Ser. Mat.	Analele Universitatii Bucuresti, Seria Matematica [Romania]
Anal. Instrum.	Analytical Instrumentation [US]
Anal. Lab.	Analytical Laboratory [PR China]
Anal. Lett.	Analytical Letters [US]
ANALOG Comput.	ANALOG Computing [US]
Anal. Proc.	Analytical Proceedings [of Royal Society of Chemistry, UK]
Anal. Proc. R. Soc. Chem.	Analytical Proceedings of the Royal Society of Chemistry [UK]
Anal. Sci.	Analytical Sciences [Japan]
Analyst	The Analyst [of Royal Society of Chemistry, UK]
Analyt. Abs.	Analytical Abstracts [of Royal Society of Chemistry, UK]
Analyt. Chem.	Analytical Chemistry [of American Chemical Society]
Andrew Seybold's Outlook Prof. Comput.	Andrew Seybold's Outlook on Professional Computing [US]
An. Fis. A, Fenom. Interacciones	Anales de Fisica, Serie A, Fenomenos e Interacciones [Spain]
An. Fis. B, Apl. Metodos Instrum.	Anales de Fisica, Serie B, Aplicaciones, Metodos e Instrumentos [Spain]
An. Fis. Ser. A	Anales de Fisica Serie A [Spain]
An. Fis. Ser. B	Anales de Fisica Serie B [Spain]
Angew. Chem.	Angewandte Chemie [Germany]
Angew. Chem. Int.	Angewandte Chemie International [Germany]
Angew. Inform.	Angewandte Informatik [Germany]
Angew. Makromol. Chem.	Angewandte Makromolekulare Chemie [Germany]
An. Mec. Electr.	Anales de Mecanica y Electricidad [Spain]
Ann. Acad. Sci. Fenn. A	Annales Academiae Scientiarum Fennicae, Series A [Finland]
Ann. Acad. Sci. Fenn. A, Math.	Annales Academiae Scientiarum Fennicae, Series A, Mathematica [Finland]
Ann. Acad. Sci. Fenn. AI	Annales Academiae Scientiarum Fennicae, Series AI [Finland]
Ann. Acad. Sci. Fenn. AII	Annales Academiae Scientiarum Fennicae, Series AII [Finland]
Ann. Acad. Sci. Fenn. AII, Chem.	Annales Academiae Scientiarum Fennicae, Series AII, Chemica [Finland]
Ann. Acad. Sci. Fenn. AVI	Annales Academiae Scientiarum Fennicae, Series AVI [Finland]
Ann. Acad. Sci. Fenn. AVI, Phys.	Annales Academiae Scientiarum Fennicae, Series AVI, Physics [Finland]
Ann. Appl. Biol.	Annals of Applied Biology [of Association of Applied Biologists, UK]
Ann. Biomed. Eng.	Annals of Biomedical Engineering [of Biomedical Engineering Society, US]
Ann. Braz. Ent. Soc.	Annals of the Brazilian Entomological Society
Ann. Chim.	Annales des Chimie [France]
Ann. Chim., Sci. Mater.	Annales des Chimie, Science des Materiaux [France]
Ann. Compos.	Annales des Composites [France]
Ann. Fac. Sci. Toulouse Math.	Annales de la Faculte des Sciences de Toulouse, Mathematiques [France]
Ann. Fond. Louis Broglie	Annales de la Fondation Louis de Broglie [France]
Ann. Fr. Microtech. Chronom.	Annales Francaises des Microtechniques et de Chronometrie [France]
Ann. Geophys.	Annales Geophysicae [of European Geophysical Society]
Ann. Geophys. Atmos. Hydrospheres Space Sci.	Annales Geophysicae: Atmospheres, Hydrospheres and Space Sciences [of European Geophysical Society]
Ann. Glaciol.	Annals of Glaciology [of International Glaciological Society, UK]
Ann. Hist. Comput.	Annals of the History of Computing [American Federation of Information Processing Societies, US]
Ann. Human Biol.	Annals of Human Biology [of Society for the Study of Human Biology, UK]
Ann. ICRP	Annales of the International Commission on Radiological Protection [UK]
Ann. Inst. Henri Poincare Phys. Theor.	Annales de l'Institut Henri Poincare Physique Theorique [France]
Ann. Isr. Phys. Soc.	Annals of the Israel Physical Society [Israel]
Ann. Istit. Speriment. Selvicol.	Annali dell' Istituto Sperimentale per la Selvicoltura [Italy]

Ann. Ist. Super. Sanita	Annali dell'Istituto Superiore di Sanita [Italy]
Ann. Mines	Annales des Mines [France]
Ann. Nucl. Energy	Annals of Nuclear Energy [UK]
Ann. N.Y. Acad. Sci.	Annals of the New York Academy of Sciences [US]
Ann. Occup. Hyg.	Annals of Occupational Hygiene [US]
Ann. Oper. Res.	Annals of Operations Research [Switzerland]
Ann. Phys.	Annals of Physics [US]
Ann. Phys. (Germany)	Annalen der Physik (Germany)
Ann. Phys. (France)	Annales de Physique (France)
Ann. Phys. (USA)	Annals of Physics (USA)
Ann. Probab.	Annals of Probability [of Institute of Mathematical Statistics, US]
Ann. Pure Appl. Log.	Annals of Pure and Applied Logic [Netherlands]
Ann. Radioelectricite	Annales de Radioelectricite [France]
Ann. Rech. Sylvicoles	Annales Recherches Sylvicoles [France]
Ann. Rev. Biochem.	Annual Review of Biochemistry [US]
Ann. Rev. Ecol. Syst.	Annual Review of Ecological Systems [US]
Ann. Rev. Fluid Mech.	Annual Review of Fluid Mechanics [US]
Ann. Rev. Mater. Sci.	Annual Review of Materials Science [US]
Ann. Rev. Microbiol.	Annual Review of Microbiology [US]
Ann. Rev. Phys. Chem.	Annual Review of Physical Chemistry [US]
Ann. Rev. Plant Physiol.	Annual Review of Plant Physiology [US]
Ann. Sci.	Annals of Science [UK]
Ann. Soc. Log. Eng.	Annals of the Society of Logistics Engineers [US]
Ann. Soc. Sci. Brux.	Annales de la Societe Scientifique de Bruxelles [Belgium]
Ann. Soc. Sci. Brux. I, Sci. Math. Astron. Phys.	Annales de la Societe Scientifique de Bruxelles, Serie I (Sciences Mathematiques, Astronomiques et Physiques) [Belgium]
Ann. Stat.	Annals of Statistics [of Institute of Mathematical Statistics, US]
Ann. Telecommun.	Annales des Telecommunications [of Centre National d'Etudes des Telecommunications, France]
Ann. Tokyo Astron. Obs.	Annals of the Tokyo Astronomical Observatory [Japan]
Ann. Univ. Turku. AI, Astron. — Chem. — Phys. — Math.	Annales Universitasis Turkuensis Series AI, Astronomica — Chemica — Physica — Mathematica [Finland]
Ann. Univ. Turku. Ser. AI	Annales Universitasis Turkuensis Series AI [Finland]
Annu. Proc. Assoc. Sci. Tech. Soc. South Afr.	Annual Proceedings of the Associated Scientific and Technical Societies of South Africa [South Africa]
Annu. Rep. Ceram. Eng. Res. Lab.	Annual Report of Ceramic Engineering Research Laboratory [Japan]
Annu. Rep. Eng. Res. Inst. Fac. Eng. Univ. Tokyo	Annual Report of the Engineering Research Institute, Faculty of Engineering, University of Tokyo [Japan]
Annu. Rep. Eng. Res. Inst., Tokyo Univ.	Annual Report of the Engineering Research Institute, Tokyo University [Japan]
Annu. Rep. Inst. Phys., Acad. Sin.	Annual Report of the Institute of Physics, Academia Sinica[Taiwan]
Annu. Rep. Jpn. Soc. Technol. Plast.	Annual Report of the Japan Society for Technology of Plasticity
Annu. Rep. Res. Inst. Catal., Hokkaido Univ.	Annual Report of the Research Institute for Catalysis, Hokkaido University [Japan]
Annu. Rep., Res. React. Inst., Kyoto Univ.	Annual Reports, Research Reactor Institute, Kyoto University [Japan]
Annu. Rev. Biophys. Biophys. Chem.	Annual Review of Biophysics and Biophysical Chemistry [of Biophysical Society, US]
Annu. Rev. Fluid Mech.	Annual Review of Fluid Mechanics [US]
Annu. Sim. Symp. Proc.	Annual Simulation Symposium Proceedings [of Society for Computer Simulation International, US]
Annu. Soc. Fr. Chim.	Annuaire de la Societe Francaise de Chimie [France]
Annu. Univ. Sofia Fac. Phys.	Annuaire de l'Universite de Sofia Faculte de Physique [Bulgaria]
Anritsu Tech. Bull.	Anritsu Technical Bulletin [Japan]
An. Soc. Ent. Bras.	Anais da Sociedade Entomologica do Brasil [Brazil]
Anti-Corros. Methods Mater.	Anti-Corrosion Methods and Materials [UK]
An. Univ. Galati, Mec. Constr. Masini	Analele Universitatii din Galati, Mecanica si Constructii de Masini [Romania]

Abbreviation	Title
An. Univ. Galati, Metal.	Analele Universitatii din Galati, Metalurgie [Romania]
API Ind. Fact Sheet	API Industry Fact Sheet [of American Paper Institute]
API Fiber Sum.	API (Monthly) Fiber Summary [of American Paper Institute]
API Newsprint Rep.	API Newsprint Report [of American Paper Institute]
API Newsprint Rept.	API Newsprint Report [of American Paper Institute]
API Statistical Sum.	API (Monthly) Statistical Summary [of American Paper Institute]
API Woodpulp Data	API Woodpulp Data [of American Paper Institute]
Apl. Mat.	Aplikace Matematiky [Czechoslovakia]
Appl. Acoust.	Applied Acoustics [UK]
Appl. Anal.	Applicable Analysis [UK]
Appl. Artif. Intell.	Applied Artificial Intelligence [US]
Appl. Biochem. Biotechnol.	Applied Biochemistry and Biotechnology [US]
Appl. Catal.	Applied Catalysis [Netherlands]
Appl. Clay Sci.	Applied Clay Science [Netherlands]
Apple Assem. Line	Apple Assembly Line [US]
Appl. Electron. Tech.	Applied Electronic Technique [PR China]
Appl. Energy	Applied Energy [UK]
Appl. Eng. Agric.	Applied Engineering in Agriculture
Appl. Environ. Microbiol.	Applied and Environmental Microbiology [of American Society for Microbiology]
Appl. Geochem.	Applied Geochemistry [of International Association of Geochemistry and Cosmochemistry, Canada]
Appl. Ind. Hyg.	Applied Industrial Hygiene [US]
Appl. Math. Comput.	Applied Mathematics and Computation [US]
Appl. Math. Mech.	Applied Mathematics and Mechanics [Translation of: *Prikladnaya Matematika i Mekhanika (USSR)*; published in the US]
Appl. Math. Model.	Applied Mathematical Modelling [UK]
Appl. Math. Optim.	Applied Mathematics and Optimization [US]
Appl. Mech.	Applied Mechanics [Romania]
Appl. Mech. Rev.	Applied Mechanics Reviews [of American Society of Mechanical Engineers]
Appl. Microgravity Technol.	Applied Microgravity Technology [Germany]
Appl. Numer. Math.	Applied Numerical Mathematics [of International Association for Mathematics and Computer in Simulation, Belgium]
	Applied Numerical Mathematics [Netherlands]
Appl. Ocean Res.	Applied Ocean Research [UK]
Appl. Opt.	Applied Optics [of Optical Society of America]
Appl. Phys.	Applied Physics [Germany]
Appl. Phys. A	Applied Physics A [Germany]
Appl. Phys. A, Solids Surf.	Applied Physics A, Solids and Surfaces [Germany]
Appl. Phys. B	Applied Physics B [Germany]
Appl. Phys. B, Photophys. Laser Chem.	Applied Physics B, Photophysics and Laser Chemistry [Germany]
Appl. Phys. Commun.	Applied Physics Communications [US]
Appl. Phys. Lett.	Applied Physics Letters [of American Institute of Physics]
Appl. Radiat. Isot.	Applied Radiation and Isotopes [UK]
Appl. Sci. Res.	Applied Scientific Research [Netherlands]
Appl. Sci. Technol. Index	Applied Science and Technology Index [US]
Appl. Sol. Energy	Applied Solar Energy [Translation of: *Geliotekhnika*; published in the US]
Appl. Spectrosc.	Applied Spectroscopy [of Society for Applied Spectroscopy, US]
Appl. Spectrosc. Rev.	Applied Spectroscopy Reviews [US]
Appl. Spectry.	Applied Spectroscopy [of Society for Applied Spectroscopy, US]
Appl. Surf. Sci.	Applied Surface Science [Netherlands]
Aq. Insects	Aquatic Insects [Netherlands]
Aqualine Abstr.	Aqualine Abstracts [US]
Arab Gulf J. Sci. Res.	Arab Gulf Journal of Scientific Research [Sauda Arabia]
Arab. J. Sci. Eng.	Arabian Journal for Science and Engineering [Saudi Arabia]
Arab Metall. News	Arab Metallurgical News [Algeria]
Arab Min. J.	Arab Mining Journal [Jordan]
ARA Memo.	ARA Memorandum [of Aircraft Research Association, US]
ARA Rep.	ARA Report [of Aircraft Research Association, US]
Arbido Bull.	Arbido Bulletin [Switzerland]
Arbido Rev.	Arbido Revue [Switzerland]
Arch. Acoust.	Archives of Acoustics [Poland]

Arch. Autom. Telemech.	Archiwum Automatyki i Telemechanika [Poland]
Arch. Biochem. Biophys.	Archives of Biochemistry and Biophysics [US]
Arch. Budowy Masz.	Archiwum Budowy Maszyn [Poland]
Arch. Combust.	Archivum Combustionis
Arch. Eisenb.tech.	Archiv fur Eisenbahntechnik [Germany]
Arch. Eisenhuttenwes.	Archiv fur das Eisenhuttenwesen [Germany]
Arch. Electron. Uebertrag.tech.	Archiv fur Elektronik und Uebertragungstechnik [Germany]
Arch. Elektrotech.	Archiv fur Elektrotechnik [Germany] Archiwum Elektrotechniki [Poland]
Arch. Elektrotech. (Germany)	Archiv fur Elektrotechnik (Germany)
Arch. Elektrotech. (Poland)	Archiwum Elektrotechniki (Poland)
Arch. Environ. Health	Archives of Environmental Health [US]
Arch. Gorn.	Archiwum Gorictwa [Poland]
Arch. Hutn.	Archiwum Hutnictwa [Poland]
Arch. Hydrotech.	Archiwum Hydrotechniki [Poland]
Archit. J.	Architect's Journal [UK]
Archit. Met.	Architectural Metals [US]
Archit. Q	Architectural Quarterly [of National Concrete Masonry Association, US]
Arch. Math. Log.	Archive for Mathematical Logic [Germany]
Arch. Mech.	Archives of Mechanics [Poland]
Arch. Metall.	Archives of Metallurgy [Poland]
Arch. Nauki Mater.	Archiwum Nauki o Materialach [Poland]
Arch. Ochr. Sr.	Archiwum Ochrony Srodowiska [Poland]
Arch. Post- & Fernmeldewes.	Archiv fur das Post- und Fernmeldewesen [Germany]
Arch. Ration. Mech. Anal.	Archive for Rational Mechanics and Analysis [Germany]
Arch. Sci.	Archives des Sciences [Switzerland]
AREA Bull.	AREA Bulletin [of American Railway Engineering Association]
Ark. Fys. Semin. Trondheim	Arkiv for det Fysiske Seminar i Trondheim [Norway]
Armada Int.	Armada International [Switzerland]
ARMA Rec. Manage. Q.	ARMA Records Management Quarterly [of Association of Records Managers and Administrators, US]
Armyanskii Khim. Zh.	Armyanskii Khimicheskii Zhurnal [USSR]
Artif. Intell.	Artificial Intelligence [Netherlands]
Artif. Intell. Eng.	Artificial Intelligence in Engineering [UK]
Artif. Intell. Eng. Des. Anal. Manuf.	Artificial Intelligence for Engineering Design, Analysis and Manufacturing [UK]
Artif. Intell. Rev.	Artificial Intelligence Review [UK]
Arts Manuf.	Arts et Manufactures [France]
ASBC Newsl.	ASBC Newsletter [of American Society of Brewing Chemists]
ASCE Proc.	American Society of Civil Engineers Proceedings [US]
ASC Newsl.	ASC Newsletter [of Association of Systematic Collections, US]
ASEA J.	ASEA Journal [Sweden]
ASE Mag.	ASE Magazine [of Alternative Sources of Energy, US]
ASFA	Aquatic Sciences and Fisheries Abstracts [of Food and Agricultural Organization [of UN]]
ASHRAE J.	ASHRAE Journal [of American Society of Heating, Refrigerating and Air-Conditioning Engineers]
ASHRAE Trans	ASHRAE Transactions [of American Society of Heating, Refrigerating and Air-Conditioning Engineers]
ASHS Newsl.	ASHS Newsletter [of American Society for Horticultural Science]
Ashton-Tate Q.	Ashton-Tate Quarterly [US]
ASI Newsl.	ASI Newsletter [of American Society of Inventors]
Asia-Pac. J. Oper. Res.	Asia-Pacific Journal of Operational Research [Singapore]
Asian Sources Electron.	Asian Sources Electronics [Hong Kong]
ASID Rep.	ASID Report [of American Society of Interior Design]
ASLE Trans.	ASLE Transactions [of American Society of Lubrication Engineers]
ASLIB Inf.	ASLIB Information [of Association of Information Management, UK]
ASLIB Proc.	ASLIB Proceedings [of Association of Information Management, UK]
ASME Publ. EC	American Society of Mechanical Engineers Publication EC [US]
ASME Publ. FACT	American Society of Mechanical Engineers Publication FACT [US]
ASME Publ. FED	American Society of Mechanical Engineers Publication FED [US]
ASME Publ. HTD	American Society of Mechanical Engineers Publication HTD [US]
ASPP Newsl.	ASPP Newsletter [of American Society of Plant Physiologists]
ASPT Newsl.	ASPT Newsletter [of American Society of Plant Taxonomists]
ASR — Bull. INORGA	ASR — Bulletin INORGA [Czechoslovakia]

Assem. Autom.	Assembly Automation [UK]
Assoc. Off. Anal. Chem. J.	Association of Official Analytical Chemists Journal [US]
ASSP Mag.	ASSP Magazine [of IEEE Acoustics, Speech and Signal Processing Society, US]
ASTE Bull.	ASTE Bulletin [of American Society of Test Engineers]
ASTE Rev.	ASTE Review [of Association for Software Testing and Evaluation]
ASTI Newsl.	ASTI Newsletter [of Association for Science, Technology and Innovation]
ASTM J. Compos. Technol. Res.	ASTM Journal of Composites Technology and Research [of American Society for Testing and Materials]
ASTM J. Test. Eval.	ASTM Journal of Testing and Evaluation [of American Society for Testing and Materials]
ASTM Stand. News	ASTM Standardization News [of American Society for Testing and Materials]
ASTM STP	ASTM Special Technical Publication [of American Society for Testing and Materials]
Astrofiz. Issled.-Izv. Spets. Astrofiz. Obs.	Astrofizicheskie Issledovaniya — Izvestiya Spetsial'noi Astrofizicheskoi Observatorii [USSR]
Astron. Astrophys.	Astronomy and Astrophysics [Germany]
Astron. Astrophys. Suppl. Ser.	Astronomy and Astrophysics Supplement Series [France]
Astron. Circ. (Acad. Sin.)	Astronomical Circulars (Academia Sinica) [PR China]
Astron. Her.	Astronomical Herald [of Astronomical Society of Japan]
Astron. J.	Astronomical Journal [of American Astronomical Society, US]
Astron. Nachr.	Astronomische Nachrichten [Germany]
Astron. Now	Astronomy Now [UK]
Astronomer	The Astronomer [UK]
Astron. Tidsskr.	Astronomisk Tidsskrift [Sweden]
Astron. Vestn.	Astronomicheskii Vestnik [USSR]
Astron. Zh.	Astronomicheskii Zhurnal [USSR]
Astrophys. J.	Astrophysical Journal [of American Astronomical Society]
Astrophys. J. Lett.	Astrophysical Journal Letters [of American Astronomical Society]
Astrophys. J., Lett. Ed.	Astrophysical Journal, Letters to the Editor [of American Astronomical Society]
Astrophys. J. Suppl. Ser.	Astrophysical Journal Supplement Series [of American Astronomical Society]
Astrophys. Lett. Commun.	Astrophysical Letters and Communications [UK]
Astrophys. Space Sci.	Astrophysical Space Science [Netherlands]
ATA	Abstracts on Tropical Agriculture [of Royal Tropical Institute, Netherlands]
ATB Metall.	ATB Metallurgie [Belgium]
ATCP	ATCP [of Asociacion Mexicana de Tecnicos de las Industrias de la Celulosa y Papel, Mexico]
At. Data Nucl. Data Tables	Atomic Data and Nuclear Data Tables [US]
At. Energ.	Atomnaya Energiya [USSR]
At. Energy Rev. Spec. Issue	Atomic Energy Review, Special Issue [of International Atomic Energy Agency, Austria]
ATIPCA	ATPCA [of Asociacion de Tecnicos de la Industria Papelera y Celulosica Argentina]
Atmos. Environ.	Atmospheric Environment [UK]
Atmos. Environ. A, Gen. Top.	Atmospheric Environment, Part A, General Topics [UK]
Atmos. Environ. B, Urban Atmos.	Atmospheric Environment, Part B, Urban Atmosphere [UK]
Atmos. Ocean	Atmosphere Ocean
Atmos. Res.	Atmospheric Research [Netherlands]
Atmos. Technol.	Atmospheric Technology [of National Center for Atmospheric Research, US]
Atomkernenerg./Kerntech.	Atomkernenergie/Kerntechnik [Germany]
Atomwirtsch. — Atomtech.	Atomwirtschaft — Atomtechnik [of Kerntechnische Gesellschaft, Germany]
At. Spectrosc.	Atomic Spectroscopy [US]
At. Strom	Atom und Strom [of Vereinigung Deutscher Elektrizitatswerke, Germany]
Atti Accad. Ligure Sci. Lett.	Atti della Accademia Ligure di Scienze e Lettere [Italy]
Atti Accad. Naz. Lincei, Ottava	Atti della Accademia Nazionale dei Lincei, Ottava [Italy]
Atti Accad. Naz. Lincei, Rend. Cl. Sci. Fis. Mat. Nat.	Atti della Accademia Nazionale dei Lincei, Rendiconti, Classe di Scienze Fisiche, Matematiche e Naturali [Italy]
Atti Accad. Sci. Ist. Bologna Cl. Sci. Fis. Rend. XIII	Atti della Accademia delle Scienze dell' Istituto di Bologna, Classe di Scienze Fisiche, Rendiconti, Serie XIII [Italy]
Atti Fond. Giorgio Ronchi	Atti della Fondazione Giorgio Rochi [Italy]
Atti Semin. Mat. Fis. Univ. Modena	Atti del Seminario Matematico e Fisico dell' Universita di Modena [Italy]
Atti Soc. Pelorit. Sci. Fis. Math. Nat.	Atti della Societa Peloritana di Scienze Fisiche Matematiche e Naturali [Italy]
AT&T Tech. J.	AT&T Technical Journal [of American Telephone and Telegraph]
AT&T Technol.	AT&T Technology [of American Telephone and Telegraph, US]

ATV J. ATV Journal [of Abwassertechnische Vereinigung, Germany]
Audiol. Akust. Audiologische Akustik [Germany]
Audio Vis. Audio Visual [UK]
Aufbereit.-Tech. Aufbereitungstechnik [Germany]
AusIMM Bull. Proc. Australasian Institute of Mining and Metallurgy Bulletin and Proceedings [Australia]
Aust. Asian Pac. Electr. World Australian, Asian and Pacific Electrical World [Australia]
Aust. Commun. Netw. Australian Communications Networks
Aust. Comput. J. Australian Computer Journal [of Australian Computer Society]
Aust. Comput. Sci. Commun. Australian Computer Science Communications
Aust. Electron. Eng. Australian Electronics Engineering
Aust. Forestry Australian Forestry [of Institute of Foresters of Australia]
Aust. J. Audiol. Australian Journal of Audiology [of Audiological Society of Australia]
Aust. J. Biotechnol. Australian Journal of Biotechnology
Aust. J. Chem. Australian Journal of Chemistry [of Commonwealth Scientific and Industrial Research Organization, Australia]
Aust. J. Geod. Photogramm. Surv. Australian Journal of Geodesy, Photogrammetry and Surveying
Aust. J. Phys. Australian Journal of Physics [of Commonwealth Scientific and Industrial Research Organization, Australia]
Aust. J. Phys. Astrophys. Suppl. Australian Journal of Physics, Astrophysical Supplement [of Commonwealth Scientific and Industrial Research Organization, Australia]
Aust. J. Plant Physiol. Australian Journal of Plant Physiology [of Commonwealth Scientific and Industrial Research Organization, Australia]
Aust. J. Soil Res. Australian Journal of Soil Research
Aust. Meteorol. Meg. Australian Meteorological Magazine [of Bureau of Meteorology, Australia]
Aust. Min. Australian Mining
Aust. Phys. Australian Physicist [of Australian Institute of Physics]
Aust. Pkg. Australian Packaging
Australas. Corros. Eng. Australasian Corrosion Engineering [Australia]
Austral. Asian Pac. Electr. World Australian, Asian and Pacific Electrical World [Australia]
Australas. Phys. Eng. Sci. Med. Australian Physical and Engineering Sciences in Medicine
Austria Mach. Steel Austria Machinery and Steel
Aust. Road Res. Australian Road Research [of Australian Road Research Board]
Aust. Road Res. Prog. Australian Road Research in Progress [of Australian Road Research Board]
Aust. Telecommun. Res. Australian Telecommunication Research [of Telecommunication Society of Australia]
Aust. Weld. J. Australian Welding Journal
Aust. Weld. Res. Australian Welding Research [of Australian Welding Research Association]
Aust. Road Res. Australian Road Research [of Australian Road Research Board]
Aust. Road Res. Prog. Australian Road Research in Progress [of Australian Road Research Board]
Aust. Telecommun. Res. Australian Telecommunication Research [of Telecommunication Society of Australia]
Aust. Weld. J. Australian Welding Journal
Aust. Weld. Res. Australian Welding Research [of Australian Welding Research Association]
Ausz. Eur. Patentanmeld. I Auszuge aus den Europaischen Patentanmeldungen, Teil I [Germany]
Automat. Weld. (USSR) Automatic Welding (USSR) [Published in the UK]
Autom. Control Automation and Control [New Zealand]
Autom. Control Comput. Sci. Automatic Control and Computer Sciences [Translation of: *Avtomatika i Vychislitell'naya Tekhnika (USSR)*; published in the US]
Autom. Data Process. Inf. Bull. Automatic Data Processing Information Bulletin [Belgium]
Autom. Doc. Math. Linguist. Automatic Documentation and Mathematical Linguistics [Translation of: *Nauchno-Tekhnicheskaya Informatsiya, Seriya 2 (USSR)*; published in the US]
Autom. Electr. Power Syst. Automation of Electric Power Systems [PR China]
Autom. Instrum. Automatica e Instrumentacion [Spain]
Autom. Integrata Automazione Integrata [Italy]
Autom. Mach. Automatic Machining [US]
Autom. Manuf. Strategy Automated Manufacturing Strategy [Netherlands]
Automobile Abs. Automobile Abstracts [of Motor Industry Research Association, UK]
Automob. Ind. Automobil-Industrie [Germany]
Autom., Prod., Inf., Ind. Automatique, Productique, Informatique Industrielle [of Association Francaise pour le Cybernetique, Economique et Technique, France]
Autom. Prod. Technol. Automation Products and Technology [Canada]
Autom. Remote Control Automation and Remote Control [Translation of: *Avtomatika i Telemekhanicheska (USSR)*; published in the US]

Automot. Des. Dev.	Automotive Design and Development [US]
Automot. Eng.	Automotive Engineering [of Society of Automotive Engineers, US]
Automot. Eng. (UK)	Automotive Engineer (UK)
Automot. Ind.	Automotive Industries [US]
Autom. Strum.	Automazione e Strumentazione [of Associazione Nazionale Italiana per l'Automazione, Italy]
Automatisierungstech. Prax.	Automatisierungstechnische Praxis [Germany]
Autom. Syst.	Automation Systems [Canada]
Autom.tech. Prax.	Automatisierungstechnische Praxis [Germany]
Autophon Bull.	Autophon Bulletin [Switzerland]
Avesta Stainless Bull.	Avesta Stainless Bulletin [Italy]
Avh. Nor. Vidensk.-Akad. Oslo I	Avhandlingar utgitt av det Norske Videnskaps-Akademi i Oslo I [Norway]
Avh. Nor. Vidensk.-Akad. Oslo II	Avhandlingar utgitt av det Norske Videnskaps-Akademi i Oslo II [Norway]
Avh. Nor. Vidensk.-Akad. Oslo I, Mat.-Nat.vidensk. Kl.	Avhandlingar utgitt av det Norske Videnskaps-Akademi i Oslo, I, Matematisk-Naturvidenskaplige Klasse [Norway]
Aviat. Educ. News Bull.	Aviation Education News Bulletin [of Aviation Distributors and Manufacturers Association, US]
Aviat. Mech. Bull.	Aviation Mechanics Bulletin [of Flight Safety Foundation, US]
Aviat. Week Space Technol.	Aviation Week and Space Technology [US]
Avion. News	Avionics News [of Aircraft Electronics Association, US]
Avion. Newsl.	Avionics Newsletter [of National Avionics Society, US]
AVIOS Conf. Proc.	AVIOS Conference Proceedings [of American Voice Input/Output Society]
AVIOS J.	AVIOS Journal [of American Voice Input/Output Society]
Avtom. Prom-st.	Avtomobilnaya Promyshlennost' [USSR]
Avtom. Svarka	Avtomaticheskaya Svarka [USSR]
Avtom. Telemekh.	Avtomatika i Telemekhanika [USSR]
Avtom. Vychisl. Tekh.	Avtomatika i Vychislitel'naya Tekhnika [USSR]
AWWA J.	AWWA Journal [of American Water Works Association]

B

BA	Biological Abstracts [US]
Ball Bear. J.	Ball Bearing Journal [Australia]
BAMS	Bulletin of the American Mathematical Society [US]
Bangladesh J. Sci. Ind. Res.	Bangladesh Journal of Scientific and Industrial Research [of Bangladesh Council of Scientific and Industrial Research]
Bank Syst. Technol.	Bank Systems + Technology [US]
Bank. Technol.	Banking Technology [UK]
Banyasz. Kohasz. Lapok (Kohasz.)	Banyaszati es Kohaszati Lapok (Kohaszati) [Hungary]
Banyasz. Kohasz. Lapok (Ontode)	Banyaszati es Kohaszati Lapok (Ontode) [Hungary]
BASIS	Bulletin of the American Society for Information Science [US]
Bauen Stahl	Bauen in Stahl [Switzerland]
Bayer Rep.	Bayer Reports [Germany]
BBC-Nachr.	BBC-Nachrichten [Germany]
BBC-Tech.	BBC-Technik
BCS Newsl.	BCS Newsletter [of British Computer Society]
Beck Isol.tech.	Beck Isoliertechnik [Germany]
BED	Business Equipment Digest [UK]
Behav. Biol.	Behavioral Biology
Behav. Inf. Technol.	Behaviour and Information Technology [UK]
Behav. Res. Methods Instrum. Comput.	Behaviour Research Methods, Instruments and Computers [US]
Behav. Sci.	Behavioral Science [US]
Belg. J. Oper. Res. Stat. Comput. Sci.	Belgian Journal of Operations Research, Statistics and Computer Science [Belgium]
Bell Syst. Tech. J.	Bell System Technical Journal

Ber. Abwassertech. Ver.	Berichte der Abwassertechnischen Vereinigung [Germany]
Ber. Bunsenges.	Berichte der Bunsengesellschaft [Germany]
Ber. Bunsenges. Phys. Chem.	Berichte der Bunsengesellschaft fur Physikalische Chemie [Germany]
Berg Huttenmann. Monatsh.	Berg- und Huttenmannische Monatshefte [Austria]
Beskontaktn. Elektr. Mash.	Beskontaktnye Elektricheskie Mashiny [USSR]
Beton Stahlbetonbau	Beton- und Stahlbetonbau [Germany]
BFG Citizen	BFG (BF Goderich) Citizen [US]
BHM Berg Huttenmann. Monatsh.	Berg- und Huttenmannische Monatshefte [Austria]
BHP J.	BPH (Broken Hill Proprietary) Journal [Australia]
BHP Tech. Bull.	BPH (Broken Hill Proprietary) Technical Bulletin [Austrlia]
BHRA News	BHRA News [of British Hydromechanics Research Association]
BIBRA Bull.	BIBRA Bulletin [of British Industrial Biological Research Association]
Bild Ton	Bild und Ton [Germany]
Biochem.	Biochemistry [of American Chemical Society]
Biochem. Biophys. Res. Comm.	Biochemical and Biophysical Research Communications [US]
Biochem. Cell Biol.	Biochemistry and Cell Biology [of National Research Council of Canada]
Biochem. Educ.	Biochemical Education [of International Union of Biochemistry, US]
Biochem. Eng. J.	Biochemical Engineering Journal [Switzerland]
Biochem. Int.	Biochemistry International [of International Union of Biochemistry, US]
Biochem. J.	Biochemical Journal [of Biochemical Society, UK]
Biochem. Pharmacol.	Biochemistry and Pharmacology
Biochem. Physiol. Pflanzen	Biochemie und Physiologie der Pflanzen [Germany]
Biochem. Soc. Trans.	Biochemical Society Transactions [UK]
Biochim. Biophys. Acta	Biochimica et Biophysica Acta
Bioelectrochem. Bioenerg.	Bioelectrochemistry and Bioenergetics [Switzerland/Germany]
Biofeedback Self-Regul.	Biofeedback and Self-Regulation [US]
Biol. Cybern.	Biological Cybernetics [Germany]
Biol. Int.	Biology International [of International Union of Biological Sciences, France]
Biol. Wastes	Biological Wastes
Biomater. Artif. Cells Artif. Organs	Biomaterials, Artificial Cells, and Artificial Organs [US]
Biomater. Med. Devices Artif. Organs	Biomaterials, Medical Devices and Artificial Organs [US]
Biomed. Eng.	Biomedical Engineering [Translation of: *Meditsinskaya Tekhnika (USSR)*; published in the US]
Biomed. Instrum. Technol.	Biomedical Instrumentation and Technology [of National Society of Biomedical Equipment Technicians, US]
Biomed. Mater.	Biomedical Materials [UK]
Biomed. Polym	Biomedical Polymers [UK]
Biomed. Res.	Biomedical Research
Biomed. Tech.	Biomedizinische Technik [Germany]
Biomed. Technol. Today	Biomedical Technology Today [of Association for the Advancement of Medical Instrumentation, US]
Biometr. Bull.	Biometric Bulletin [of Biometric Society, US]
Bioorg. Chem.	Bioorganic Chemistry [US]
Biophys. J.	Biophysical Journal [of Biophysical Society, US]
Bioprocess Eng.	Bioprocess Engineering [Germany]
Bioprocessing Technol.	Bioprocessing Technology
Biosci.	Bioscience [American Institute of Biological Sciences, US]
Biosens. Bioelectron.	Biosensors and Bioelectronics [UK]
Biotechnol. Appl. Biochem.	Biotechnology and Applied Biochemistry [US]
Biotechnol. Bioeng.	Biotechnology and Bioengineering [US]
Biotechnol. Prog.	Biotechnology Progress [of American Institute of Chemical Engineers]
Biotechnol. Lett.	Biotechnology Letters [US]
BIP Plast. Rev.	BIP Plastics Review [UK]
Bit Bytes Rev.	Bits and Bytes Review [US]
BLMRA J.	BLMRA Journal [of British Leather Manufacturers Research Association]
BMES Bull.	BMES Bulletin [of Biomedical Engineering Society, US]
BMFT J.	BMFT Journal [of Bundesministerium fur Forschung und Technologie, Germany]
BMR J. Aust. Geol. Geophys.	BMR Journal of Australian Geology and Geophysics [of Bureau of Mineral Resources, Australia]
BNF ABS	British Non-Ferrous Abstracts [of British Non-Ferrous Metals Technology Center]
Body Eng.	Body Engineering [US]
Body Eng. J.	Body Engineering Journal [of American Society of Body Engineers]
Bol. Fac. Ing. Univ. Repub.	Boletin de la Facultad de Ingeneria Universidad de la Republica [Uruguay]

Bol. Inform.	Boletin Informativo [Uruguay]
Bol. Inst. Tonantzintla	Boletin del Instituto de Tonantzintla [of Instituto Nacional de Astrofisica, Optica y Electronica, Mexico]
Boll. Geofis. Teor. Appl.	Bolletino di Geofisica Teorica ed Applicata [Italy]
Boll. Oceanol. Teor. Appl.	Bolletino di Oceanologia Teorica ed Applicata [Italy]
Boll. Riv.	Bolletino delle Riviste [Italy]
Bol. Soc. Esp. Ceram. Vidrio	Boletin Sociedad Espanola de Ceramica y Vidrio [Spain]
Boll. Tec. Finsider	Bolletino Tecnico Finsider [Italy]
Bol. Tec.	Boletin Tecnico [Brazil]
Bol. Tec., Niquel	Boletin Tecnico, Niquel [Cuba]
Bosch Tech. Ber.	Bosch Technische Berichte [Germany]
Bosei Kanri (Rust Prev. Control)	Bosei Kanri (Rust Prevention Control) [Japan]
Boshoku Gijutsu (Corros. Eng.)	Boshoku Gijutsu (Corrosion Engineering) [Japan]
Bot. Gaz.	Botanical Gazette [US]
Bot. Helv.	Botanica Helvetica [of Swiss Botanical Society]
Bound.-Layer Meteorol.	Boundary-Layer Meteorology [Netherlands]
BPF News	BPF News [of British Plastics Foundation]
Br. Astron. Assoc. Circ.	British Astronomical Association Circular
Brazing Soldering	Brazing and Soldering [UK]
Br. Ceram. Proc.	British Ceramic Proceedings [of Institute of Ceramics]
Br. Ceram. Trans. J.	British Ceramic Transactions and Journal [of Institute of Ceramics]
Br. Corros. J.	British Corrosion Journal [of Institute of Metals]
Br. Dent. J.	British Dental Journal
Br. Engine Tech. Rep.	British Engine Technical Reports
Brennst. Warme Kraft	Brennstoff — Warme — Kraft [of Verein Deutscher Ingenieure, Germany]
Br. Foundryman	British Foundryman
Br. Geol.	British Geologist [of Institution of Geologists]
Brit. Busin.	British Business [of Department of Trade and Industry]
Brit. Printer	British Printer
Br. J. Appl. Phys.	British Journal of Applied Physics
Br. J. Educ. Technol.	British Journal of Educational Technology [of Council for Educational Technology for the United Kingdom]
Br. J. Hist. Sci.	British Journal for the History of Science [of British Society for the History of Science]
Br. J. Ind. Med.	British Journal of Industrial Medicine
Br. J. Non-Destr. Test.	British Journal of Non-Destructive Testing [of British Institute of Non-Destructive Testing]
Br. J. Philos. Sci.	British Journal for the Philosophy of Science
Br. J. Phot.	British Journal of Photography
Br. J. Radiol.	British Journal of Radiology [of British Institute of Radiology]
BRMA Rev.	British Review [of British Rubber Manufacturers Association]
Broadcast Technol.	Broadcast Technology [Canada]
Brown Boveri Mitt.	Brown Boveri Mitteilungen [Switzerland]
Brown Boveri Rev.	Brown Boveri Review [Switzerland]
Brown Boveri Tech.	Brown Boveri Technik [Switzerland]
Br. Plast. Rubb.	British Plastics and Rubber
Br. Polym. J.	British Polymer Journal [of Society of Chemical Industry]
Br. Steel	British Steel
Br. Steelmaker	British Steelmaker
Br. Telecom J.	British Telecom Journal
Br. Telecommun. Eng.	British Telecommunications Engineering [of Institution of British Telecommunications Engineers]
Br. Telecom Technol. J.	British Telecom Technology Journal [of British Telecom Research Laboratories]
Br. Telecom World	British Telecom World [of British Telecom Center]
Bruel & Kjaer Tech. Rev.	Bruel and Kjaer Technical Review [Denmark]
BSI News	BSI News [of British Standards Institution]
BSI Sales Bull.	BSI Sales Bulletin [of British Standards Institution]
BSTJ	Bell Systems Technical Journal
B.T.M. Bull.	B.T.M. Bulletin (Business and Technical Management) [UK]
Budavox Telecommun. Rev.	Budavox Telecommunication Review [Hungary]
Build. Res. Establ. Dig.	Building Research Establishment Digest [UK]
Build. Res. Pract.	Building Research and Practice [of International Council for Building Research, Studies and Documentation, Netherlands]
Build. Sci.	Building Science [of National Institute of Building Sciences]

Build. Serv.	Building Services [of Chartered Institution of Building Services Engineers, UK]
Build. Serv. Eng. Res. Technol.	Building Service Engineering Research and Technology [of Chartered Institution of Building Services Engineers, UK]
Build. Serv. Environ. Eng.	Building Services and Environmental Engineer [UK]
Build. Steel	Building with Steel [UK]
Bulg. J. Phys.	Bulgarian Journal of Physics [of Bulgarian Academy of Sciences]
Bul. Inst. Politeh. Bucur., Autom.-Calc.	Buletinul Institutului Politehnic Bucuresti, Automatica-Calculatoare [Romania]
Bul. Inst. Politeh. Bucur., Chim.	Buletinul Institutului Politehnic Bucuresti, Chimie [Romania]
Bul. Inst. Politeh. Bucur., Constr. Masini	Buletinul Institutului Politehnic Bucuresti, Constructii de Masini [Romania]
Bul. Inst. Politeh. Bucur., Electroteh.	Buletinul Institutului Politehnic Bucuresti, Electrotehnica [Romania]
Bul. Inst. Politeh. Bucur., Electron.	Buletinul Institutului Politehnic Bucuresti, Electronica [Romania]
Bul. Inst. Politeh. Bucur., Energ.	Buletinul Institutului Politehnic Bucuresti, Seria Energetica [Romania]
Bul. Inst. Politeh. Bucur., Mec.	Buletinul Institutului Politehnic Bucuresti, Mecanica [Romania]
Bul. Inst. Politeh. Bucur., Metal.	Buletinul Institutului Politehnic Bucuresti, Metalurgie [Romania]
Bul. Inst. Politeh. Bucur., Transpt.-Aeronave	Buletinul Institutului Politehnic, Transporturi-Aeronave [Romania]
Bul. Inst. Politeh. 'Gheorghe Gheorghiu Dej', Chim.	Buletinul Institutului Politehnic,'Gheorghe Gheorghiu-Dej', Chimie [Romania]
Bul. Inst. Politeh. 'Gheorghe Gheorghiu Dej', Chim.-Metal	Buletinul Institutului Politehnic, 'Gheorghe Gheorghiu-Dej', Chimie-Metalurgie [Romania]
Bul. Inst. Politeh. 'Gheorghe Gheorghiu Dej', Constr. Masini	Buletinul Institutului Politehnic, 'Gheorghe Gheorghiu-Dej', Constructii de Masini [Romania]
Bul. Inst. Politeh. 'Gheorghe Gheorghiu Dej', Electroteh.	Buletinul Institutului Politehnic, 'Gheorghe Gheorghiu-Dej', Electrotehnica [Romania]
Bul. Inst. Politeh. 'Gheorghe Gheorghiu Dej', Mec.	Buletinul Institutului Politehnic, 'Gheorghe Gheorghiu-Dej', Mecanica [Romania]
Bul. Inst. Politeh. 'Gheorghe Gheorghiu Dej', Metal.	Buletinul Institutului Politehnic, 'Gheorghe Gheorghiu-Dej', Metalurgie [Romania]
Bul. Inst. Politeh. 'Gheorghe Gheorghiu Dej', Transpt.-Aeronave	Buletinul Institutului Politehnic, 'Gheorghe Gheorghiu-Dej', Transporturi-Aeronave [Romania]
Bul. Inst. Politeh. Iasi I	Buletinul Institutului Politehnic din Iasi, Sectia I [Romania]
Bul. Inst. Politeh. Iasi II	Buletinul Institutului Politehnic din Iasi, Sectia II [Romania]
Bul. Inst. Politeh. Iasi III	Buletinul Institutului Politehnic din Iasi, Sectia III [Romania]
Bul. Inst. Politeh. Iasi III, Electroteh. Electron. Autom.	Buletinul Institutului Politehnic din Iasi, Sectia III, Electrotehnica, Electronica, Automatizari [Romania]
Bul. Inst. Politeh. Iasi III, Electroteh. Energ. Electron. Autom.	Buletinul Institutului Politehnic din Iasi, Sectia III, Electrotehnica, Energetica, Electronica, Automatizari [Romania]
Bul. Inst. Stud. Proiect. Energ.	Buletinul Institutului de Studii si Proiectari Energetice [Romania]
Bulk Handl.	Bulk Handling [US]
Bulk Solids Handl.	Bulk Solids Handling [Switzerland]
Bulk Syst. Int.	Bulk Systems International
Bull. Acad. Pol. Sci.	Bulletin de l'Academie Polonaise des Sciences [Poland]
Bull. Acad. Sci DPR Korea	Bulletin of the Academy of Sciences of the DPR Korea [South Korea]
Bull. Acad. Sci. USSR, Chem. Ser.	Bulletin of the Academy of Sciences of the USSR, Chemical Series [Translation of: *Izvestiya Akademii Nauk SSSR, Khimicheskaya (USSR)*; published in the US]
Bull. Acad. Sci. USSR, Phys. Ser.	Bulletin of the Academy of Sciences of the USSR, Physical Series [Translation of: *Izvestiya Akademii Nauk SSSR, Fizicheskaya (USSR)*; published in the US]
Bull. Acad. Serbe Sci. Arts, Cl. Sci. Tech.	Bulletin de l'Academie Serbe des Sciences et des Arts, Classe des Sciences Techniques [Yugoslavia]
Bull. Aichi Inst. Technol. A	Bulletin of Aichi Institute of Technology, Part A [Japan]
Bull. Aichi Inst. Technol. B	Bulletin of Aichi Institute of Technology, Part B [Japan]
Bull. Alloy Phase Diagr.	Bulletin of Alloy Phase Diagrams [of ASM International, US]
Bull. Am. Assoc. Var. Star Obs.	Bulletin of the American Association of Variable Star Observers
Bull. Am. Astron. Soc.	Bulletin of the American Astronomical Society
Bull. Am. Ceram. Soc.	Bulletin of the American Ceramic Bulletin

Bull. Am. Meteorol. Soc.	Bulletin of the American Meteorological Society
Bull. Am. Soc. Inf. Sci.	Bulletin of the American Society for Information Science
Bull. Anal. Lab. Managers Assn.	Bulletin of the Analytical Laboratory Managers Association [US]
Bull. Annu. Soc. Suisse Chronom. Lab. Suisse Rech. Horlog.	Bulletin Annuel de la Societe Suisse de Chronometrie et du Laboratoire Suisse de Recherche Horlogeres [Switzerland]
Bull. Astron. Inst. Czech.	Bulletin of the Astronomical Institutes of Czechoslovakia [of Czechoslovak Academy of Sciences]
Bull. Astron. Soc. India	Bulletin of the Astronomical Society of India [of Tata Institute of Fundamental Research, India]
Bull. At. Sci.	Bulletin of the Atomic Scientists [US]
Bull. Aust. Math. Soc.	Bulletin of the Australian Mathematical Society [Australia]
Bull. Bismuth Inst.	Bulletin of the Bismuth Institute [Belgium]
Bull. Bur. Miner. Resour. Geol. Geophys.	Bulletin — Bureau of Mineral Resources, Geology and Geophysics [Australia]
Bull. Cercle Etud. Metaux	Bulletin du Cercle d'Etudes des Metaux [France]
Bull. Chem. Soc. Japan	Bulletin of the Chemical Society of Japan
Bull. Cl. Sci. Acad. R. Belg.	Bulletin de la Classe des Sciences, Academie Royale de Belgique [Belgium]
Bull. Coll. Eng. Hosei Univ.	Bulletin of the College of Engineering, Hosei University [Japan]
Bull. Coll. Sci. Univ. Ryukyus	Bulletin of the College of Science, University of Ryukyus [Japan]
Bull. Crime. Astrophys. Obs.	Bulletin of the Crimean Astrophysical Observatory [Translation of: *Izvestiya Krymskoi Astroficheskoi Observatorii (USSR)*; published in the US]
Bull. Daido Inst. Technol.	Bulletin of Daido Institute of Technology [Japan]
Bull. Dir. Etud. Rech. A	Bulletin de la Direction des Etudes et Recherches, Serie A [France]
Bull. Dir. Etud. Rech. B	Bulletin de la Direction des Etudes et Recherches, Serie B [France]
Bull. Dir. Etud. Rech. C	Bulletin de la Direction des Etudes et Recherches, Serie C [France]
Bull. Earthq. Res. Inst. Univ. Tokyo	Bulletin of the Earthquake Research Institute, University of Tokyo [Japan]
Bull. Electrochem.	Bulletin of Electrochemistry [India]
Bull. Electron Microsc. Soc. India	Bulletin of the Electron Microscopy Society of India [India]
Bull. Electrotech. Lab.	Bulletin of the Electrotechnical Laboratory [Japan]
Bull. Environ. Contam. Toxicol.	Bulletin of Environmental Contamination and Toxicology
Bull. Eur. Assoc. Theor. Comput. Sci.	Bulletin of the European Association for Theoretical Computer Science [Austria]
Bull. Fac. Eng., Hiroshima Univ.	Bulletin of the Faculty of Engineering, Hiroshima University [Japan]
Bull. Fac. Eng., Hokkaido Univ.	Bulletin of the Faculty of Engineering, Hokkaido University [Japan]
Bull. Fac. Eng., Ibaraki Univ.	Bulletin of the Faculty of Engineering, Ibaraki University [Japan]
Bull. Fac. Eng., Kyushu Tokai Univ.	Bulletin of the Faculty of Engineering, Kyushu Tokai University [Japan]
Bull. Fac. Eng., Miyazaki Univ.	Bulletin of the Faculty of Engineering, Miyazaki University [Japan]
Bull. Fac. Eng., Tokushima Univ.	Bulletin of the Faculty of Engineering, Tokushima University [Japan]
Bull. Fac. Eng., Toyama Univ.	Bulletin of the Faculty of Engineering, Toyama University [Japan]
Bull. Fac. Eng., Univ. Ryukyus	Bulletin of the Faculty of Engineering, University of Ryukyus [Japan]
Bull. Fac. Eng., Yokohama Natl. Univ.	Bulletin of the Faculty of Engineering, Yokohama National University [Japan]
Bull. Forestry Forest Prod. Res. Inst.	Bulletin of the Forestry and Forest Products Research Institute [Japan]
Bull. Geod.	Bulletin Geodesique [of Association Internationale de Geodesie, France]
Bull. Geophys.	Bulletin of Geophysics [Taiwan]
Bull. Gov. Ind. Res. Inst. Osaka	Bulletin of the Government Industrial Research Institute, Osaka [Japan]
Bull. IASS	Bulletin of the IASS [International Association for Shell and Spatial Structures, Spain]
Bull. Indian Natl. Sci. Acad.	Bulletin of the Indian National Science Academy
Bull. Indian Vac. Soc.	Bulletin of the Indian Vacuum Society
Bull. Ind. Res. Inst., Hiroshima Prefect., West	Bulletin of the Industrial Research Institute, Hiroshima Prefecture, West [Japan]
Bull. Inf. Cent. Donnees Stellaires	Bulletin d'Information du Centre de Donnees Stellaires [France]
Bull. Inf. Cybern.	Bulletin of Informatics and Cybernetics [of Research Association of Statistical Sciences, Japan]
Bull. Inst. At. Energy, Kyoto Univ.	Bulletin of the Institute of Atomic Energy, Kyoto University [Japan]
Bull. Inst. Chem. Res., Kyoto Univ.	Bulletin of the Institute for Chemical Research, Kyoto University [Japan]
Bull. Inst. Corros. Sci. Technol.	Bulletin of the Institute of Corrosion Science and Technology [UK]
Bull. Inst. Eng. (India)	Bulletin of the Institution of Engineers (India)
Bull. Int. ISSA Prev. Occup. Risks Electr.	Bulletin of the the International Section of the ISSA for the Prevention of Occupational Risks due to Electricity [Germany]

Bull. Int. Peat Soc.	Bulletin of the International Peat Society [Finland]
Bull. Isr. Phys. Soc.	Bulletin of the Israel Physical Society
Bull. Isr. Soc. Spec. Libr. Inf. Cent.	Bulletin of the Israel Society of Special Libraries and Information Centers
Bull. Jpn. Inst. Met.	Bulletin of the Japan Institute of Metals
Bull. Jpn. Soc. Mech. Eng.	Bulletin of the Japan Society of Mechanical Engineers
Bull. Jpn. Soc. Precis. Eng.	Bulletin of the Japan Society of Precision Engineering
Bull. Korean Inst. Met.	Bulletin of the Korean Institute of Metals [South Korea]
Bull. Kurume Inst. Technol.	Bulletin of Kurume Institute of Technology [Japan]
Bull. Kyushu Inst. Technol.	Bulletin of Kyushu Institute of Technology [Japan]
Bull. Kyushu Inst. Technol. (Sci. Technol.)	Bulletin of Kyushu Institute of Technology (Science and Technology) [Japan]
Bull. Liaison Rech. Inf. Autom.	Bulletin de Liaison de la Recherche en Informatique et en Automatique [of Institut National de Recherche en Informatique et en Automatique, France]
Bull. Mar. Eng. Soc. Jpn.	Bulletin of Marine Engineering Society of Japan
Bull. Mater. Sci.	Bulletin of Materials Science [of Indian Academy of Sciences]
Bull. Mech. Eng. Lab.	Bulletin of Mechanical Engineering Laboratory [Japan]
Bull. Met. Mus.	Bulletin of the Metals Museum [Japan]
Bull. Mineral.	Bulletin de Mineralogie [Germany]
Bull. Nagoya Inst. Technol.	Bulletin of Nagoya Institute of Technology [Japan]
Bull. Natl. Res. Lab. Metrology	Bulletin of the National Research Laboratory of Metrology [Japan]
Bull. NRLM	Bulletin of the NRLM [National Research Laboratory of Metrology, Japan]
Bull. Offic. Prop. Ind.	Bulletin Officiel de la Propriete Industrielle [France]
Bull. Okayama Univ. Sci. A	Bulletin of the Okayama University of Science A [Japan]
Bull. Okayama Univ. Sci. A, Nat. Sci.	Bulletin of the Okayama University of Science A, Natural Science [Japan]
Bull. Okayama Univ. Sci. B	Bulletin of the Okayama University of Science B [Japan]
Bull. Okayama Univ. Sci. B, Hum. Sci.	Bulletin of the Okayama University of Science A, Human Science [Japan]
Bull. Pol. Acad. Sci., Chem.	Bulletin of the Polish Academy of Sciences, Chemistry [Poland]
Bull. Pol. Acad. Sci., Tech. Sci.	Bulletin of the Polish Academy of Sciences, Technical Sciences [Poland]
Bull. Primary Tungsten Assoc.	Bulletin of the Primary Tungsten Association [UK]
Bull. Proc. Australas. Inst. Min. Metall.	Bulletin and Proceedings of the Australasian Institute of Mining and Metallurgy [Australia]
Bull. Res. Inst. Electron. Shizuoka Univ.	Bulletin of the Research Institute of Electronics, Shizuoka University [Japan]
Bull. Res. Inst. Miner. Dressing Metall.	Bulletin of the Research Institute of Mineral Dressing and Metallurgy [Japan]
Bull. Res. Inst. Sci. Meas., Tohoku Univ.	Bulletin of the Research Institute for Scientific Measurements [Japan]
Bull. Res. Lab. Nucl. React.	Bulletin of the Research Laboratory for Nuclear Reactors [of Tokyo Institute of Technology, Japan]
Bull. Res. Lab. Precis. Mach. Electron.	Bulletin of the Research Laboratory of Precision Machinery and Electronics [of Tokyo Institute of Technology, Japan]
Bull. SCGM	Bulletin de la SCGM [Societe Canadienne de Genie Mecanique, Canada]
Bull. Schweiz. Elektrotech.	Bulletin der Schweizerischen Vereinigung fur Elektrotechniker [of Swiss Association of Electrical Engineers]
Bull. Schweiz. Elektrotech. Ver. Verb. Schweiz. Elektr.werke	Bulletin des Schweizerischen Elektrotechnischen Vereins und des Verbandes Schweizerischer Elektrizitatswerke [Switzerland]
Bull. Sci. A	Bulletin Scientifique A [Yugoslavia]
Bull. Sci. Assoc. Ing. Electr. Inst. Electrotech. Montefiore	Bulletin Scientifique de l'Association des Ingenieurs Electriciens sortis de l'Institut Electrotechnique Montefiore [Belgium]
Bull. Sci. B	Bulletin Scientifique B [Yugoslavia]
Bull. Sci. Eng. Res. Lab., Waseda Univ.	Bulletin of the Science and Engineering Research Laboratory, Waseda University [Japan]
Bull. Sci. Instrum. Soc.	Bulletin of the Scientific Instrument Society [UK]
Bull. Seismol. Soc. Am.	Bulletin of the Seismological Society of America
Bull. SME	Bulletin of the Society of Manufacturing Engineers [US]
Bull. Soc. Chim. Belg.	Bulletin des Societes Chimiques Belges [Belgium]
Bull. Soc. Chim. Fr.	Bulletin de la Societe Chimique de France [France]

Bull. Soc. Chim. Fr. I	Bulletin de la Societe Chimique de France, Partie I [France]
Bull. Soc. Chim. Fr. II	Bulletin de la Societe Chimique de France, Partie II [France]
Bull. Soc. Fr. Mineral. Crystallogr.	Bulletin de la Societe Francaise de Mineralogie et de Crystallographie [France]
Bull. Soc. Fr. Photogramnm. Teledetect.	Bulletin de la Societe Francaise de Photogrammetrie et des Teledetection [France]
Bull. Soc. Manuf. Eng.	Bulletin of the Society of Manufacturing Engineers [US]
Bull. Soc. R. Sci. Liege	Bulletin de la Societe Royale des Sciences de Liege [Belgium]
Bull. Soc. Sci. Bretagne	Bulletin de la Societe Scientifique de Bretagne [France]
Bull. Spec. Astrophys. Obs.-North Caucasus	Bulletin of the Special Astrophysical Observatory — North Caucasus [Translation of: *Astrofizicheskie Issledovaniya — Izvestiya Spetsial'noi Astrofizicheskoi Observatorii (USSR)*; published in the US]
Bull. Taiwan Forestry Res. Inst.	Bulletin of the Taiwan Forestry Research Institute
Bull. Tech. Univ. Istanb.	Bulletin of the Technical University of Istanbul [Turkey]
Bull. Tokyo Gakugei Univ. I	Bulletin of the Tokyo Gakugei University, Series I [Japan]
Bull. Tokyo Gakugei Univ. II	Bulletin of the Tokyo Gakugei University, Series II [Japan]
Bull. Tokyo Gakugei Univ. III	Bulletin of the Tokyo Gakugei University, Series III [Japan]
Bull. Tokyo Gakugei Univ. IV	Bulletin of the Tokyo Gakugei University, Series IV [Japan]
Bull. Tokyo Gakugei Univ. IV, Math. Nat. Sci.	Bulletin of the Tokyo Gakugei University, Series IV, Mathematics and Natural Sciences [Japan]
Bull. Univ. Electro-Comm.	Bulletin of the University of Electro-Communications [of University of Electro-Communications, Japan]
Bull. Univ. Osaka Prefect., A	Bulletin of University of Osaka Prefecture, Series A [Japan]
Bull. Univ. Osaka Prefect., B	Bulletin of University of Osaka Prefecture, Series B [Japan]
Bull. Volcanol.	Bulletin of Volcanology [of International Association of Volcanology and Chemistry of the Earth's Interior, Germany]
Bull. Vyskum. Ustavu Papieru Celulozy	Bulletin Vyskumneho Ustavu Papieru a Celuloy [Czechoslovakia]
Bull. Yamagata Univ. (Eng.)	Bulletin of the Yamagata University (Engineering) [Japan]
Bull. Yamagata Univ. (Nat. Sci.)	Bulletin of the Yamagata University (Natural Science) [Japan]
Bul. Stiint. Inst. Politeh. Cluj-Napoca, Chim. — Metal.	Buletinul Stiintific al Institutului Politechnic Cluj-Napoca, Seria Chimie — Metalurgie [Romania]
Bul. Stiint. Teh. Inst. Politeh. 'Traian Vuia' Timis. Ser. Chim.	Buletinul Stiintific si Tehnic al Institutului Politechnic 'Traian Vuia' Timisoara, Seria Chimie [Romania]
Bul. Stiint. Teh. Inst. Politeh. 'Traian Vuia' Timis. Ser. Elektroteh.	Buletinul Stiintific si Tehnic al Institutului Politechnic 'Traian Vuia' Timisoara, Seria Electrotehnica [Romania]
Bul. Univ. Brasov A	Buletinul Universitatii din Brasov Seria A [Romania]
Bul. Univ. Brasov A, Mec. Apl. Electroteh.Electron.Constr.Mas. Tehnol. Prelucr. Met.	Buletinul Universitatii din Brasov Seria A, Mecanica Aplicata, Electrotehnica si Electronica, Constructia de Masini si Tehnologia Prelucrarii Metalelor [Romania]
Bul. Univ. Brasov B	Buletinul Universitatii din Brasov Seria B [Romania]
Bul. Univ. Galati, Fasc. V, Tehnol.Constr. Masini, Metal	Buletinul Universitatii din Galati, Fascicula V, Tehnologii in Constructia de Masini, Metalurgie [Romania]
Bumazh. Prom.	Bumazhnaya Promyshlennost [USSR]
Bur. Etud. Autom.	Bureaux d'Etudes Automatismes [France]
Bus. Commun. Rev.	Business Communications Review [US]
Bus. Comput. Commun.	Business Computing and Communications [UK]
Bus. Equip. Dig.	Business Equipment Digest [UK]
Bus. Forms Labels Syst.	Business Forms, Labels and Systems [US]
Bus. Forms Syst.	Business Forms and Systems [US]
Bus. Monitor Rubb.	Business Monitor: Rubber [UK]
Bus. Monitor Synth.	Business Monitor: Synthetic Resins [UK]
Bus. Softw.	Business Software [US]
Bus. Softw. Rev.	Business Software Review [US]
Bus. Sol.	Business Solutions [UK]
Bus. Syst. Equip.	Business Systems and Equipment [UK]
Bus. Tech. Manage.	Business and Technical Management [UK]

Bus. Today	Business Today [UK]
BWK	Brennstoff — Warme — Kraft [of Verein Deutscher Ingenieure, Germany]

C

C&L Appl.	C&L Applications [UK]
CA	Chemical Abstracts [of American Chemical Society]
CAB Curr. Aware. Bull.	CAB Current Awareness Bulletin [US]
Cable Satell. Eur.	Cable and Satellite Europe [UK]
Cable Telev. Eng.	Cable Television Engineering [of Society of Cable Television Engineers, UK]
CABLIS	Current Awareness Bulletin for Librarians and Information Scientists [of British Library]
Cabot Dig.	Cabot Digest [US]
CACM	Communications of the Association for Computing Machinery [US]
CAD	Computer Applications Digest [UK]
CAD/CAM Dig.	CAD/CAM Digest [UK]
CadCam Int.	CadCam International [UK]
CAD/CAM Robot.	CAD/CAM and Robotics [Canada]
CAD/CAM Syst.	CAD/CAM Systems [Canada]
CAD/CAM Technol.	CAD/CAM Technology [US]
Cadmium Res. Dig.	Cadmium Research Digest [of International Lead Zinc Research Organization, US]
CAE J.	CAE Journal [Germany]
Cah. Anal. Donnees	Cahiers de l'Analyse des Donnees [France]
Cah. Cent. Etud. Rech. Oper.	Cahiers du Centre d'Etudes de Recherche Operationelle [Belgium]
Cah. Groupe Fr. Rheol.	Cahiers de Groupe Francais de Rheologie [France]
Cah. Ind. Metall. Electr.	Cahiers des Industries Metallurgiques et Electriques [France]
CALICO J.	CALICO Journal [of Computer-Assisted Language Learning and Instruction Consortium, US]
Calorimetry Therm. Anal.	Calorimetry and Thermal Analysis [Japan]
CALPHAD: Comput. Coupling Phase Diagr. Thermochem.	CALPHAD: Computer Coupling of Phase Diagrams and Thermochemistry [UK]
Camborne Sch. Mines	Camborne School of Mines [UK]
Can. Acoust.	Canadian Acoustics [of Canadian Acoustical Association]
Can. Aeronaut. Space J.	Canadian Aeronautics and Space Journal [of Canadian Aeronautics and Space Institute]
Can. Agric. Eng.	Canadian Agricultural Engineering
Can. Artif. Intell.	Canadian Artificial Intelligence [of Canadian Society for Computational Studies of Intelligence]
Can. Bus.	Canadian Business
Can. Ceram. Q.	Canadian Ceramics Quarterly
Can. Chem. News	Canadian Chemical News [of Chemical Institute of Canada]
Can. Chem. Process.	Canadian Chemical Processing
Can. Copper	Canadian Copper
Can. Datasyst.	Canadian Datasystems
Can. Electr. Eng. J.	Canadian Electrical Engineering Journal [of Canadian Society for Electrical Engineering]
Can. Electron. Eng.	Canadian Electronics Engineering
Can. Forestry Serv. Dept. Publ.	Canadian Forestry Service, Departmental Publications
Can. Forestry Serv. Inf. Rep. Dig.	Canadian Forestry Service, Information Reports Digest
Can. Geotech. J.	Canadian Geotechnical Journal [of Canadian Geotechnical Society]
Can. Inst. Food Sci. Technol. J.	Canadian Institute of Food Science and Technology Journal
Can. J. Bot.	Canadian Journal of Botany [of National Research Council of Canada]
Can. J. Chem.	Canadian Journal of Chemistry [of National Research Council of Canada]
Can. J. Chem. Eng.	Canadian Journal of Chemical Engineering
Can. J. Civ. Eng.	Canadian Journal of Civil Engineering [of Canadian Society for Civil Engineering]
Can. J. Commun.	Canadian Journal of Communication [of Canadian Communication Association]
Can. J. Earth Sci.	Canadian Journal of Earth Sciences [of National Research Council of Canada]
Can. J. Electr. Comput. Eng.	Canadian Journal of Electrical and Computer Engineering [of Canadian Society for Electrical Engineering]
Can. J. Fisheries Aq. Sci.	Canadian Journal of Fisheries and Aquatic Sciences [of Supply and Services Canada]

Can. J. Forest Res.	Canadian Journal of Forest Research [of National Research Council of Canada]
Can. J. Inf. Sci.	Canadian Journal of Information Science [of Canadian Association for Information Science]
Can. J. Microbiol.	Canadian Journal of Microbiology [of National Research Council of Canada]
Can. J. Phys.	Canadian Journal of Physics [of National Research Council of Canada]
Can. J. Remote Sens.	Canadian Journal of Remote Sensing [of Canadian Aeronautics and Space Institute]
Can. J. Spectrosc.	Canadian Journal of Spectroscopy
Can. J. Zool.	Canadian Journal of Zoology [of National Research Council of Canada]
Can. Lab.	Canadian Laboratory
Can. Libr. J.	Canadian Library Journal [of Canadian Library Association]
Can. Mach. Metalwork.	Canadian Machinery and Metalworking
Canmaking Cann. Int.	Canmaking and Canning International [UK]
Can. Metall. Q.	Canadian Metallurgical Quarterly
Can. Mineral.	Canadian Mineralogy
Can. Min. J.	Canadian Mining Journal
Can. Min. Metall. Bull.	Canadian Mining and Metallurgical Bulletin [of Canadian Institute of Mining and Metallurgy]
Can. Phys.	Canadian Physics
Can. Pkg.	Canadian Packaging
Can. Plast.	Canadian Plastics
Can. Printer Publ.	Canadian Printer and Publisher
Can. Res.	Canadian Research
Can. Weld. Fabr.	Canadian Welder and Fabricator
Capacity Manage. Rev.	Capacity Management Review [US]
Carbide Tool	Carbide and Tool [of ASM International, US]
Carbide Tool J.	Carbide and Tool Journal [of Society of Carbide and Tool Engineers, US]
Carbohyd. Polymers	Carbohydrate Polymers [UK]
Carbohyd. Res.	Carbohydrate Research [Netherlands]
Cart. Inf.	Cartographic Information [of North American Cartographic Information Society]
Cast Dig.	Casting Digest [of ASM International, US]
Cast. Eng./Foundry World	Casting Engineering/Foundry World [US]
Cast. Met.	Cast Metals [UK]
Cast. World	Casting World [US]
Catal. Rev. Sci. Eng.	Catalysis Reviews, Science and Engineering [US]
Catal. Soc. Newsl.	Catalysis Society Newsletter [US]
Cat. Classif. Q.	Cataloguing and Classification Quarterly [US]
CAUS Newsl.	CAUS Newsletter [of Color Association of the United States]
CBA	Current Biotechnology Abstracts [of Royal Society of Chemistry, UK]
CCA	Computer and Control Abstracts [of Institution of Electrical Engineers, UK]
CCS Bull.	CCS Bulletin [of Chinese Chemical Society, PR China]
CD	Computer Design [US]
CDJM	Canadian Journal of Mathematics
CD-ROM Libr.	CD-ROM Library [US]
CEA	Chemical Engineering Abstracts [of Royal Society of Chemistry, UK]
CEA Notes Inf.	CEA Notes d'Information [of Commissariat à l'Energie Atomique, France]
CEER	Chemical Economy and Engineering Review [Japan]
CEGB Res.	CEGB Research [of Central Electricity Generating Board, UK]
Celest. Mech. Dyn. Astron.	Celestial Mechanics and Dynamical Astronomy [Netherlands]
Cell Biol. Int. Rep.	Cell Biology International Reports [of International Federation of Cell Biology, Canada]
Cell. Polym.	Cellular Polymers [UK]
Cell Tissue Res.	Cell Tissue Research
Cellulosa Carta	Cellulosa e Carta [of Ente Nazionale per la Cellulosa e per la Carta, Italy]
Cellulose Chem. Technol.	Cellulose Chemistry and Technology [Romania]
Celuloza Hirtie	Celuloza si Hirtie [Romania]
Cem. Concr. Aggr. J.	Cement, Concrete and Aggregates Journal [American Society for Testing and Materials]
Cem. Concr. Res.	Cement and Concrete Research [US]
Cent. Doc. Sider., Circ. Inf. Tech.	Centre de Documentation Siderurgique, Circulaire d'Informations Techniques [France]
CEO Dig.	CEO Digest [of Intelligence Industries Association, US]
CEP	Chemical Engineering Progress [of American Institute of Chemical Engineers]
CEPA Newsl.	CEPA Newsletter [of National Society for Computer Applications in Engineering, Planning and Architecture]
Ceram. Bull.	Ceramics Bulletin
Ceram. Eng. Sci. Proc.	Ceramic Engineering and Science Proceedings [of American Ceramic Society]

Ceram. Forum Int.	Ceramics Forum International [Germany]
Ceram. Ind. Int.	Ceramic Industries International [UK]
Ceram. Ind. J.	Ceramic Industries Journal [US]
Ceram. Int.	Ceramics International [UK]
Ceram. Int. News	Ceramics International News [Italy]
Ceram. Jpn.	Ceramics Japan [Japan]
Ceram. Technol. Newsl.	Ceramic Technology Newsletter [US]
Cerberus Electron.	Cerberus Electronics [Switzerland]
Cercet. Metal.	Cercetari Metalurgice [Romania]
Cereal Chem.	Cereal Chemistry [of American Association of Cereal Chemists]
CERN Cour.	CERN Courier [of Centre Europeen pour la Recherche Nucleaire, Switzerland]
Cert. Eng. Tech. Mag.	Certified Engineering Technician Magazine [of American Society of Certified Engineering Technicians]
Certif. Eng.	Certificated Engineer [of Institution of Certificated Mechanical and Electrical Engineers, South Africa]
Cesk. Cas. Fyz. A	Ceskoslovensky Casopis pro Fyziku, Sekce A [of Czechoslovak Academy of Sciences]
Cesk. Cas. Fyz. B	Ceskoslovensky Casopis pro Fyziku, Sekce B [of Czechoslovak Academy of Sciences]
Cesk. Farm.	Ceskoslovensky Farmacie [of Czechoslovak Academy of Sciences]
Cesk. Inf. Teor. Praxe	Ceskoslovenska Informatika, Teorie a Praxe [Czechoslovakia]
CETIM Inf.	CETIM Informations [of Centre Technique des Industries Mecaniques, France]
CEW Chem. Eng. World	CEW Chemical Engineering World
CFI — Ceram. Forum Int. — Ber. Dtsch. Keram. Ges.	CFI — Ceramic Forum International — Berichte der Deutschen Keramischen Gesellschaft [Germany]
Chap. Bull.	Chapter Bulletin [of Experimental Aircraft Association, US]
Chart. Mech. Eng.	Chartered Mechanical Engineer [UK]
Chart. Quant. Surv.	Chartered Quantity Surveyors [of Royal Institute of Chartered Surveyors, UK]
Chart. Surv. Wkly.	Chartered Surveyors Weekly [of Royal Institute of Chartered Surveyors, UK]
Chem. Abstr.	Chemical Abstracts [of American Chemical Society]
Chem. Age India	Chemical Age of India
Chem. Anal.	Chemia Analytyczna [Poland]
Chem. Anlagen Verf.	Chemie Anlagen und Verfahren [Germany]
Chem. Ber.	Chemische Berichte [Germany and US]
Chem. Biochem. Eng. Q.	Chemical and Biochemical Engineering Quarterly
Chem. Br.	Chemistry in Britain [UK]
Chem. Bus.	Chemical Business [US]
Chem. Econ. Eng. Rev.	Chemical Economy and Engineering Review [Japan]
Chem. Eng.	Chemical Engineer [of Institution of Chemical Engineers, UK] Chemical Engineering [US]
Chem. Eng. Aust.	Chemical Engineering in Australia [of Institution of Engineers of Australia]
Chem. Eng. Commun.	Chemical Engineering Communications [US]
Chem. Eng. Educ.	Chemical Engineering Education [US]
Chem. Eng. J.	Chemical Engineering Journal [Switzerland]
Chem. Eng. J. Biochem. Eng. J.	Chemical Engineering Journal and Biochemical Engineering Journal
Chem. Eng. Mach.	Chemical Engineering and Machinery [PR China]
Chem. Eng. News	Chemical and Engineering News [of American Chemical Society]
Chem. Eng. (NY)	Chemical Engineering (New York) [US]
Chem. Eng. Process.	Chemical Engineering and Processing [Switzerland]
Chem. Eng. Prog.	Chemical Engineering Progress [of American Institute of Chemical Engineers]
Chem. Eng. Res. Bull.	Chemical Engineering Research Bulletin
Chem. Eng. Res. Des.	Chemical Engineering Research and Design [of Institution of Chemical Engineers, UK]
Chem. Eng. Technol.	Chemical Engineering and Technology [Germany and US]
Chem. Eng. Sci.	Chemical Engineering Science [US and UK]
Chem. Eng. (UK)	Chemical Engineer (UK) [of Institution of Chemical Engineers]
Chem. Express	Chemistry Express [of Kinki Chemical Society, Japan]
Chem. Geol.	Chemical Geology [Netherlands]
Chem. Geol. (Isot. Geosci. Sect.)	Chemical Geology (Isotope Geoscience Section) [Netherlands]
Chem. Hazards Ind.	Chemical Hazards in Industry [of Royal Society of Chemistry, UK]
Chemiefasern Textilind.	Chemiefasern Textilindustrie [Germany]
Chem. Ind.	Chemistry and Industry [of Society of Chemical Industry, UK] Chemische Industrie [of Verband der Chemischen Industrie, Germany]

Chem. Ind (Duesseldorf)	Chemische Industrie (Duesseldorf) [Germany]
Chem. Ind. Int.	Chemische Industrie International [of Verband der Chemischen Industrie, Germany]
Chem. Ind. (London)	Chemistry and Industry (London) [UK]
Chem. Ing. Tech.	Chemie-Ingenieur-Technik [Germany]
Chem. Int.	Chemistry International [of International Union of Pure and Applied Chemistry, UK]
Chemist	The Chemist [of American Institute of Chemists]
Chem. J.	Chemie Journal [of Verband der Chemischen Industrie, Germany]
Chem. Lett.	Chemical Letters [US]
Chem. Listy	Chemicke Listy [Czechoslovakia]
Chem. Mark. Reporter	Chemical Marketing Reporter [US]
Chem. Mater.	Chemistry of Materials [of American Chemical Society, US]
Chem. News. Lett.	Chemical News Letter [of Chinese Chemical Society, PR China]
Chemometr. Intell. Lab. Syst.	Chemometrics and Intelligent Laboratory Systems [Netherlands]
Chem. Pet. Eng.	Chemical and Petroleum Engineering
Chem. Pharm. Bull.	Chemical and Pharmaceutical Bulletin
Chem. Phys.	Chemical Physics [Netherlands]
Chem. Phys. Lett.	Chemical Physics Letters [Netherlands]
Chem. Process.	Chemical Processing [US]
Chem. Prod.	Chemische Produktion [Germany]
Chem. Prum.	Chemicky Prumsyl [Czechoslovakia]
Chem. React. Eng. Technol.	Chemical Reaction Engineering and Technology
Chem. Res. Toxicol.	Chemical Reseach in Toxicology [of American Chemical Society]
Chem. Rev.	Chemical Reviews [of American Chemical Society]
Chem. Scr.	Chemica Scripta [of Royal Swedish Academy of Sciences]
Chem. Scr.	Chemica Scripta [of Cambridge University Press, UK]
Chem. Soc. Rev.	Chemical Society Reviews [of Royal Society of Chemistry, UK]
Chem. Solids	Chemistry of Solids [US]
Chemsphere Am.	Chemsphere Americas [US]
Chem. Tech.	Chemie Technik [Germany] Chemische Technik [Germany]
Chem. Tech. (Heidelberg)	Chemie Technik (Heidelberg) [Germany]
Chem. Tech. (Leipzig)	Chemische Technik (Leipzig) [Germany]
Chem. Technol. Fuels Oils	Chemistry and Technology of Fuels and Oils
Chem. Vlakna	Chemicke Vlakna [Czechoslovakia]
Chem. Week	Chemical Week [US]
Chem. Zvesti	Chemicke Zvesti [Czechoslovakia]
Chicago Purch.	Chicago Purchaser [US]
Chim. Chron.	Chimika Chronika [Greece]
Chim. Ind.	Chimica e l'Industria [Italy]
Chim. Ind. (Milan)	Chimica e l'Industria (Milan) [Italy]
China Pulp Paper	China Pulp and Paper [PR China]
China Steel Tech. Rep.	China Steel Technical Report [Taiwan]
Chin. Astron. Astrophys.	Chinese Astronomy and Astrophysics [Selected translations from: *Acta Astrophysica Sinica, Acta Astronomica Sinica* and *Chinese Journal of Space Science*; published in the UK]
Chin. J. Biomed. Eng.	Chinese Journal of Biomedical Engineering [of Chinese Society of Biomedical Engineering, PR China]
Chin. J. Chromotogr.	Chinese Journal of Chromotography [of Chinese Chemical Society, PR China]
Chin. J. Comput.	Chinese Journal of Computers [PR China]
Chin. J. Infrared Res.	Chinese Journal of Infrared Research [PR China]
Chin. J. Lasers	Chinese Journal of Lasers [PR China]
Chin. J. Low Temp. Phys.	Chinese Journal of Low Temperature Physics [PR China]
Chin. J. Mater. Sci.	Chinese Journal of Materials Science [Taiwan]
Chin. J. Mech. Eng.	Chinese Journal of Mechanical Engineering [PR China]
Chin. J. Nucl. Phys.	Chinese Journal of Nuclear Physics [PR China]
Chin. J. Phys.	Chinese Journal of Physics [of Physical Society of the Republic of China, Taiwan]
Chin. J. Polym. Sci.	Chinese Journal of Polymer Science [Published in the Netherlands]
Chin. J. Sci. Instrum.	Chinese Journal of Scientific Instruments [PR China]
Chin. J. Semicond.	Chinese Journal of Semiconductors [PR China]
Chin. J. Space Sci.	Chinese Journal of Space Science [PR China]
Chin. Phys.	Chinese Physics [Published in the US]
Chin. Phys. Lasers	Chinese Physics-Lasers [Translation of: *Chinese Journal of Lasers*; published in the US]

Chin. Phys. Lett.	Chinese Physics Letters [PR China]
Chin. Sci. Bull.	Chinese Science Bulletin [PR China]
Chips Tips	Chips and Tips [US]
Chittagong Univ. Stud. I	Chittagong University Studies, Part I [Bangladesh]
Chittagong Univ. Stud. II	Chittagong University Studies, Part II [Bangladesh]
Chittagong Univ. Stud. II, Sci.	Chittagong University Studies, Part I, Science [Bangladesh]
Chromium Rev.	Chromium Review [South Africa]
Chron. Hort.	Chronica Horticulturae [of International Society for Horticultural Science, Netherlands]
CIB	Cobol Information Bulletin [US]
Ciba-Geigy J.	Ciba-Geigy Journal [Switzerland]
Ciba-Geigy Tech. Notes	Ciba-Geigy Technical Notes [Switzerland]
CIE J.	CIE Journal [of Commission Internationale de l'Eclairage, Austria]
Ciel Espace	Ciel et Espace [of Association Francaise d'Astronomie, France]
Ciel Terre	Ciel et Terre [of Societe Belge d'Astronomie, de Meteorologie et de Physique du Globe, Belgium]
Cienc. Tec. Fis. Mat.	Ciencias, Fisicas y Matematicas [of Academia de Ciencias de Cuba]
Cienc. Tierra Espac.	Ciencias de la Tierra y del Espacio [of Academia de Ciencias de Cuba]
CIIA Newsl.	CIIA Newsletter [of Canadian Information Industry Association]
Cim. Betons Platres Chaux	Ciments Betons Platres Chaux [France]
CIM Bull.	CIM Bulletin [of Canadian Institute of Mining and Metallurgy]
CIM Rep.	CIM Reporter [of Canadian Institute of Mining and Metallurgy]
CIM Rev.	CIM Review [US]
CIM Technol.	CIM Technology [US]
CIPS Rev.	CIPS Review [of Canadian Information Processing Society]
Circ. Electrotech. Lab.	Circulars of the Electrotechnical Laboratory [Japan]
Circuit Des.	Circuit Design [US]
Circuits Manuf.	Circuits Manufacturing [US]
Circuits Syst. Signal Process.	Circuits, Systems and Signal Processing [US]
CIRIA Newsl.	CIRIA Newsletter [of Construction Industry Research and Information Association, UK]
CIRP Ann.	CIRP Annales [Switzerland]
CISM J.	CISM Journal [of Canadian Institute of Surveying and Mapping]
City Transp.	City Transport [UK]
Civ. Eng.	Civil Engineering [UK]
Civ. Eng. Jpn.	Civil Engineering in Japan [Japan]
CJ	Computer Journal [of British Computer Society]
CJIS	Canadian Journal of Information Science
Class. Quantum Gravity	Classical and Quantum Gravity [of Institute of Physics, UK]
Clay Miner.	Clay Minerals [of Mineralogical Society, UK]
Clay Res.	Clay Research
Clay Sci.	Clay Science
Clays Clay Miner.	Clays and Clay Minerals
Clim. Change	Climatic Change [Netherlands]
Clin. Chem.	Clinical Chemistry [US]
Clin. Microbiol. Rev.	Clinical Microbiology Reviews [of American Society for Microbiology]
Clin. Phys. Physiol. Meas.	Clinical Physics and Physiological Measurement [of Institute of Physical Sciences in Medicine, UK]
Clin. Sci.	Clinical Science [of Biochemical Society, UK]
Clin. Vis. Sci.	Clinical Vision Sciences [UK]
CM	Computer Management [UK]
CMG Proc.	CMG Proceedings [of Computer Measurement Group, US]
CMG Trans.	CMG Transactions [of Computer Measurements Group, US]
CNRS Res.	CNRS Research [US]
Coal J.	Coal Journal
Coal Min.	Coal Mining [UK]
Coal Prep.	Coal Preparation
Coat. Technol.	Coating Technology
CODATA Bull.	CODATA Bulletin [of Committee on Data for Science and Technology]
Cogn. Sci.	Cognitive Science [US]
Coke Chem. USSR	Coke and Chemistry USSR
Cold Reg. Sci. Technol.	Cold Regions Science and Technology
Collect. Czech. Chem. Commun.	Collection of Czechoslovak Chemical Communications [of Czechoslovak Academy of Sciences]

Coll. Microcomput.	Collegiate Microcomputer [US]
Colloid J. USSR	Colloid Journal of the USSR [Translation of: *Kolloidnyi Zhurnal*; published in the US]
Colloid Polym. Sci.	Colloid and Polymer Science [Germany]
Colloid Polymer Sci.	Colloid and Polymer Science [Germany]
Colloids Surf.	Colloids and Surfaces [Netherlands]
Colloq. Phys.	Colloque de Physique [France]
Colorado Sch. Mines Q.	Colorado School of Mines Quarterly [US]
Color Res. Appl.	Color Research and Application [US]
Combust. Explos. Shock Waves	Combustion, Explosion and Shock Waves [Translation of: *Fizika Goreniya i Vzryva (USSR)*; published in the US]
Combust. Flame	Combustion and Flame [US]
Combust. Sci. Technol.	Combustion Science and Technology
Comment. Phys.-Math.	Commentationes Physico-Matematicae [Finland]
Comments Astrophys., Comments Mod. Phys. A	Comments on Astrophysics, Comments on Modern Physics: Part A [Canada]
Comments Astrophys., Comments Mod. Phys. B	Comments on Astrophysics, Comments on Modern Physics: Part B [UK]
Comments Astrophys., Comments Mod. Phys. C	Comments on Astrophysics, Comments on Modern Physics: Part C [UK]
Comments At. Mol. Phys.	Comments on Atomic and Molecular Physics [UK]
Comments Condens. Matter Phys.	Comments on Condensed Matter Physics [UK]
Comments Mol. Cell. Biophys., Comments Mod. Biol. A	Comments on Molecular and Cellular Biophysics, Comments on Modern Biology: Part A [UK]
Comments Mol. Cell. Biophys.,Comments Mod. Biol. B	Comments on Molecular and Cellular Biophysics, Comments on Modern Biology: Part B [UK]
Comments Nucl. Part. Phys.	Comments on Nuclear and Particle Physics [UK]
Comments Plasma Phys. Control. Fusion	Comments on Plasma Physics and Controlled Fusion [UK]
Comm. Eur. Communities Rep.	Commission of the European Communities Report [Luxembourg]
Commodore Comput. Int.	Commodore Computing International [UK]
Commod. Res. Q. Rep., Copper	Commodity Research Quarterly Report, Copper [US]
Commod. Res. Q. Rep., Tin	Commodity Research Quarterly Report, Tin [US]
Commun. ACM	Communications of the ACM [Association for Computing Machinery, US]
Commun. Appl. Numer. Methods	Communications in Applied Numerical Methods [UK]
Commun. Broadcast.	Communication and Broadcasting [UK]
Commun. Dublin Inst. Adv. Stud. A	Communications of the Dublin Institute for Advanced Studies, Series A [Ireland]
Commun. Dublin Inst. Adv. Stud. B	Communications of the Dublin Institute for Advanced Studies, Series B [Ireland]
Commun. Eng. Int.	Communications Engineering International [UK]
Commun. Inf.	Communication Information [Canada]
Commun. Int.	Communications Journal [UK]
Commun. Law	Communications and the Law [UK]
Commun. Manage.	Communications Management [UK]
Commun. Math. Phys.	Communications in Mathematical Physics [Germany]
Commun. News	Communications News [US]
Commun. Now	Communications Now [UK]
Commun. Outlook	Communication Outlook [US]
Commun. R. Soc. Edinb.	Communications of the Royal Society of Edinburgh [UK]
Commun. R. Soc. Edinb., Phys, Sci.	Communications of the Royal Society of Edinburgh, Physical Sciences [UK]
Commun. Stat. — Simul. Comput.	Communications in Statistics — Simulation and Computation [US]
Commun. Stat. — Theory Methods	Communications in Statistics — Theory and Methods [US]
Commun. Syst. Worldw.	Communications Systems Worldwide [UK]
Commun. Theor. Phys.	Communications in Theoretical Physics [PR China]
Commut. Transm.	Commutation and Transmission [France]
COMPEL — Int. J. Comput. Math. Electr. Electron. Eng.	COMPEL — International Journal for Computation and Mathematics in Electrical and Electronic Engineering [Ireland]
Complex Syst.	Complex Systems [US]
Compos. Adhes. Newsl.	Composites and Adhesives Newsletter [US]
Compos. Eng.	Composites Engineering [US]

Compos. Manuf.	Composites in Manufacturing [of Society of Manufacturing Engineers, US]
Compos. Mater. Sci.	Composite Materials Science [UK]
Compos. Plast. Renf. Fibres Verre Text.	Composites Plastiques Renforces Fibres de Verre Textile [of Centre de Documentation du Verre Textile et des Plastiques Renforces, France]
Compos. Polym.	Composite Polymers [of Association of Directors of European Centers for Plastics, UK]
Compos. Sci. Technol.	Composites Science and Technology [UK]
Compos. Struct.	Composite Structures [UK]
Compos. Technol. Res.	Composites Technology and Research [American Society for Testing and Materials]
Compos. Technol. Rev.	Composites Technology Review [US]
Compr. Air	Compressed Air [US]
Comp. Rend. Acad. Bulg. Sci.	Comptes Rendus, Academie Bulgare des Sciences [Bulgaria]
Comp. Rend. Acad. Sci.	Comptes Rendues Hebdomadaires des Seances de l'Academie des Sciences [France]
Comp. Rend . Acad. Sci. Ser. C	Comptes Rendus de l'Academie des Sciences, Serie C [France]
Comp. Rend. Hebd. Seances Acad. Sci. II	Comptes Rendus Hebdomadaire des Seances de l'Academie des Sciences, Serie II [France]
Comp. Rend. Hebd. Seances Acad. Sci., Ser. C, Sci. Chem.	Comptes Rendus Hebdomadaire des Seances de l'Academie des Sciences, Serie C: Sciences Chimique [France]
Comput. Account.	Computers in Accounting [US]
Comput. Aided Des.	Computer Aided Design [UK]
Comput.-Aided Eng.	Computer-Aided Engineering [UK]
Comput.-Aided Eng. J.	Computer-Aided Engineering Journal [of Institution of Electrical Engineers, UK]
Comput.-Aided Geom. Des.	Computer-Aided Geometric Design [Netherlands]
Comput. Archit. News	Computer Architecture News [of Association for Computing Machinery, US]
Comput. Artif. Intell.	Computers and Artificial Intelligence [Czechoslovakia]
Comput. Bank.	Computers in Banking [US]
Comput. Biol. Med.	Computers in Biology and Medicine [UK]
Comput. Biomed. Res.	Computers and Biomedical Rersearch [US]
Comput. Bull.	Computer Bulletin [of British Computer Society]
Comput. Bus.	Computing for Business [US]
Comput. Chem.	Computers and Chemistry [UK and US]
Comput. Chem. Eng.	Computers and Chemical Engineering [UK]
Comput. Commun.	Computer Communications [UK]
Comput. Commun. Decis.	Computer and Communications Decisions [US]
Comput. Commun. Rev.	Computer Communication Reviews [of Association for Computing Machinery, US]
Comput. Compos.	Computers and Composition [US]
Comput. Control Eng. J.	Computing and Control Engineering Journal [of Institution of Electrical Engineers, UK]
Comput. Data	Computer Data [Canada]
Comput. Decis.	Computer Decisions [US]
Comput. Des.	Computer Design [US]
Comput. Dig.	Computer Digest [US]
Comput. Disp. Rev.	Computer Display Review [US]
Comput. Econ. Rep.	Computer Economics Report [US]
Comput. Educ.	Computer Education [of National Computer Center, UK]
Comput. Electr. Eng.	Computers and Electrical Engineering [US]
Comput. Electron. Agric.	Computers and Electronics in Agriculture [Netherlands]
Comput. Enhanc. Spectrosc.	Computer Enhanced Spectroscopy [UK]
Comput. Environ. Urban Syst.	Computers, Environment and Urban Systems [UK]
Comput. Equip. Rev.	Computer Equipment Review [US]
Comput. Fluids	Computers and Fluids [UK and US]
Comput. Fraud Secur. Bull.	Computer Fraud and Security Bulletin [UK]
Comput. Geosci.	Computers and Geosciences [of International Association for Mathematical Geology, US]
Comput. Geosci.	Computers and Geosciences [UK]
Comput. Geotech.	Computers and Geotechnics [UK]
Comput. Graph.	Computer Graphics [of Association for Computing Machinery, US] Computers and Graphics [UK]
Comput. Graph. Art	Computer Graphics and Art [US]
Comput. Graph. Forum	Computer Graphics Forum [Netherlands]
Comput. Graph. Today	Computer Graphics Today [US]
Comput. Graph. World	Computer Graphics World [US]

Comput. Hum. Behav. Computers in Human Behaviour [US]
Comput. Hum. Serv. Computers in Human Services [US]
Comput. Ind. Computers in Industry [Netherlands]
Comput. Ind. Eng. Computers and Industrial Engineering [UK]
Comput.-Integr. Manuf. Syst. Computer-Integrated Manufacturing Systems [UK]
Comput. Intell. Computational Intelligence [of National Research Council of Canada]
Comput. J. Computer Journal [of British Computer Society]
Comput. Lang. Computer Language [US]
Comput. Lang. Computer Languages [UK]
Comput. Law Computers and Law [UK]
Comput./Law J. Computer/Law Journal [US]
Comput. Law Pract. Computer Law and Practice [UK]
Comput. Law Secur. Rep. Computer Law and Security Report [UK]
Comput. Libr. Computers in Libraries [US]
Comput. Linguist. Computational Linguistics [of Association for Computational Linguistics, US]
Comput. Mag. Computation, The Magazine [UK]
Comput. Manuf. Computerized Manufacturing [of Institution of Production Engineering, UK]
Comput. Math. Appl. Computers and Mathematics with Applications [UK]
Comput. Mech. Computational Mechanics [Germany]
Comput. Mech. Eng. Computers in Mechanical Engineering [of American Society of Mechanical Engineers]
Comput. Methods Appl. Mech. Eng. Computer Methods in Applied Mechanics and Engineering [Netherlands]
Comput. Methods Programs Biomed. Computer Methods and Programs in Biomedicine [Netherlands]
Comput. Music J. Computer Music Journal [US]
Comput. Netw. ISDN Syst. Computer Networks and ISDN System [Netherlands]
Comput. News Computer News [UK]
Comput. Newsp. Computing, The Newspaper [UK]
Comput. Oper. Res. Computers and Operations Research [UK]
Comput. Peripher. Rev. Computer Peripherals Review [US]
Comput. Pers. Computer Personnel [of Association for Computing Machinery]
Comput. Phys. Computers in Physics [of American Institute of Physics]
Comput. Phys. Commun. Computer Physics Communications [Netherlands]
Comput. Phys. Rep. Computer Physics Report [Netherlands]
Comput. Process. Chin. Orient. Lang. Computer Processing of Chinese and Oriental Languages [of Chinese Language Computer Society, US]
Comput. Recht Computers und Recht [Germany]
Comput. Reseller Mon. Computer Reseller Monthly [US]
Comput. Rev. Computer Reviews [of Association for Computing Machinery, US]
Comput. Sch. Computers in the Schools [US]
Comput. Sci. Econ. Manage. Computer Science in Economics and Management [Netherlands]
Comput. Sci. Inf. Computer Science and Informatics [of Computer Society of India]
Comput. Secur. Computers and Security [UK]
Comput. Secur. J. Computer Security Journal [of Computer Security Institute, US]
Comput. Seismol. Computational Seismology [Translation of: *Vychislitel'naya Seismologiya (USSR)*; published in the US]
Comput. Soc. Computers and Society [of Association for Computing Machinery, US]
Comput. Speech Lang. Computer Speech and Language [UK]
Comput. Stand. Interfaces Computer Standards and Interfaces [Netherlands]
Comput. Stat. Data Anal. Computational Statistics and Data Analysis [Netherlands]
Comput. Stat. Q. Computational Statistics Quarterly [Germany]
Comput. Struct. Computers and Structures [US]
Comput. Struct. Computers and Structures [UK]
Comput. Surv. Computer Survey [UK]
Comput. Surv. Computing Surveys [of Association for Computing Machinery, US]
Comput. Syst. Computer Systems [UK]
Comput. Syst. Computing Systems [US]
Comput. Syst. Eur. Computer Systems Europe [UK]
Comput. Syst. Sci. Eng. Computer Systems Science and Engineering [UK]
Comput. Teach. Computer Teacher [of International Council for Computers in Education, US]
Comput. Tech. Computing Techniques [UK]

Comput. Termin. Rev.	Computer Terminals Review [US]
Comput. Transl.	Computers and Translation [US]
Comput. Vis. Graph. Image Process.	Computer Vision, Graphics and Image Processing [US]
Comput. Wkly.	Computer Weekly [UK]
COM-SAC Comput. Secur. Audit. Controls	COM-SAC Computer Security, Auditing and Controls [US]
COMSAT Tech. Rev.	COMSAT Technical Review [of Communications Satellite Corporation, US]
Concast Stand. News	Concast Standard News [Switzerland]
Concast Technol. News	Concast Technology News [Switzerland]
Concr. Int. Des. Constr.	Concrete International: Design and Construction [of American Concrete Institute]
Conf. Exhib. Int.	Conferences and Exhibitions International [UK]
Connect. Technol.	Connection Technology [US]
Conserv. Recycling	Conservation and Recycling [UK]
Constr. Build. Mater.	Construction and Building Materials [UK]
Constr. Comput.	Construction Computing [of Chartered Institute of Building, UK]
Constr. Dim. Mag.	Construction Dimensions Magazine [of Association of the Wall and Ceiling Industries International, US]
Constr. MaÁ.	ConstrucÁia de MaÁini [Romania]
Constr. Surv.	Construction Surveyor [of Construction Surveyors Institute, UK]
Consult. Eng.	Consulting Engineer [UK]
Contemp. Phys.	Contemporary Physics [UK]
Contin. Mech. Thermodyn.	Continuum Mechanics and Thermodynamics [Germany]
Cont. Rel. Newsl.	Controlled Release Newsletter [of Controlled Release Society, US]
Contrib. Atmos. Phys.	Contributions to Atmospheric Physics [Germany]
Contrib. Geophys. Inst. Slovak Acad Sci.	Contibutions of the Geophysical Institute of the Slovak Academy of Sciences [Czechoslovakia]
Contrib. Geophys. Inst. Slovak Acad Sci. Ser. Meteorol.	Contibutions of the Geophysical Institute of the Slovak Academy of Sciences, Series of Meteorology [Czechoslovakia]
Contrib. Inst. Low Temp. Sci. A	Contributions from the Institute of Low Temperature Sciences, Series A [of Hokkaido University, Japan]
Contrib. Inst. Low Temp. Sci. B	Contributions from the Institute of Low Temperature Sciences, Series B [of Hokkaido University, Japan]
Contrib. Plasma Phys.	Contributions to Plasma Physics [Germany]
Control Cibern. Autom.	Control Cibernetica y Automatizacion [of Centro de Automatizacion Industrial, Cuba]
Control Comput.	Control and Computers [Canada]
Control Cybern.	Control and Cybernetics [Poland]
Control Eng.	Control Engineering [US]
Control Instrum.	Control and Instrumentation [UK]
Cont. Shelf Res.	Continental Shelf Research [UK]
Converting Mag.	Converting Magazine [US]
Copper Stud.	Copper Studies [US]
Copper Top.	Copper Topics [of Copper Development Association, US]
Copper Top. (New York)	Copper Topics (New York) [US]
Corp. Comput.	Corporate Computing [UK]
Corrosion Abstr.	Corrosion Abstracts [of National Association of Corrosion Engineers, US]
Corrosao Proteccao Materiais	Corrosao e Proteccao de Materiais [Portugal]
Corros. Australas.	Corrosion Australasia [of Australasian Corrosion Association]
Corros. Coatings, S. Afr.	Corrosion and Coatings, South Africa
Corros. Control	Corrosion Control [Canada]
Corros. Eng.	Corrosion Engineering [of Japan Society of Corrosion Engineering]
Corros. Inf. Anal.	Corrosion Information and Analysis [US]
Corros. Maint.	Corrosion and Maintenance [India]
Corros. Prev. Control	Corrosion Prevention and Control [UK]
Corros. Prev. Inhib. Dig.	Corrosion Prevention/Inhibition Digest [of ASM International, US]
Corros. Prot.	Corrosion y Proteccion [Spain]
Corros. Rev.	Corrosion Reviews [UK]
Corros. Sci.	Corrosion Science [UK]
Cosm. Res.	Cosmic Research [Translation of: *Kosmicheske Issledovaniya (USSR)*; published in the US]
Cost Eng.	Cost Engineering [of American Association of Cost Engineers, US]
Costr. Met.	Costruzioni Mettalliche [Italy]

Courr. Norm.	Courrier de la Normalisation [France]
CPA	Chemical Propulsion Abstracts [of Chemical Propulsion Information Agency, US]
CPI Equip. Report.	CPI Equipment Reporter [US]
CPPA Newsprint Stat.	CPPA Newsprint Statistics [of Canadian Pulp and Paper Association]
CPPA Newsprint Data	CPPA Newsprint Data [of Canadian Pulp and Paper Association]
CPPA Tech. Sect. Proc.	CPPA Technical Section Proceedings [of Canadian Pulp and Paper Association]
CPPA Trans. Tech. Sect.	CPPA Transactions of the Technical Section [of Canadian Pulp and Paper Association]
CPT/Cast. Plant Technol.	CPT/Casting Plant and Technology [Germany]
CQ Radio Amat. J.	CQ Radio Amateur's Journal [US]
CR	Computing Reviews [US]
C.R. Acad. Bulg. Sci.	Comptes Rendus, Academie Bulgare des Sciences [Bulgaria]
C.R. Acad. Sci.	Comptes Rendues Hebdomadaires des Seances de l'Academie des Sciences [France]
C.R. Acad. Sci. I, Math.	Comptes Rendus de l'Academie des Sciences, Serie I, Mathematique [France]
C.R. Acad. Sci. II, Mec. Phys. Chim. Sci. Terre Univers	Comptes Rendus de l'Academie des Sciences, Serie II, Mecanique, Physique, Chimie Sciences de la Terre et de l'Univers [France]
C.R. Acad. Sci. Ser. I	Comptes Rendus de l'Academie des Sciences, Serie I, Mathematique [France]
C.R. Acad. Sci. Ser. II	Comptes Rendus de l'Academie des Sciences, Serie II [France]
C.R. Acad. Sci. Ser. C	Comptes Rendus de l'Academie des Sciences, Serie C [France]
CRC Crit. Rev. Biomed. Eng.	CRC Critical Reviews in Biomedical Engineering [US]
CRC Crit. Rev. Mat. Sci.	CRC Critical Reviews in Materials Science [US]
C.R. Hebd. Seances Acad. Sci. II	Comptes Rendus Hebdomadaire des Seances de l'Academie des Sciences, Serie II [France]
Creat. Innov. Netw.	Creative and Innovation Network [UK]
Creat. Innov. Yearb.	Creative and Innovation Yearbook [UK]
C.R. Hebd. Seances Acad. Sci., Ser. C, Sci. Chem.	Comptes Rendus Hebdomadaire des Seances de l'Academie des Sciences, Serie C: Sciences Chimique [France]
Crit. Rev. Environ. Cont.	Critical Reviews in Environmental Control
C.R.M. Metall. Rep.	C.R.M. Metallurgical Reports [Belgium]
Cryog. Mater. Ser.	Cryogenic Materials Series [of International Cryogenic Materials Conference, US]
Cryst. Lattice Defects	Crystal Lattice Defects [US]
Cryst. Lattice Defects Amorphous Mater.	Crystal Lattice Defects and Amorphous Materials [US and UK]
Cryst. Prop. Prep.	Crystal Properties and Preparation [Switzerland]
Cryst. Res. Technol.	Crystal Research and Technology [US and Germany]
CSELT Tech. Rep.	CSELT Technical Reports [of Centro Studi e Laboratori Telecomunicazioni, Italy]
CSIO Commun.	CSIO Communications [of Central Scientific Instruments Organization, India]
CSIRO Div. Chem. Wood Technol. Tech. Papers	CSIRO Division of Chemical and Wood Technology, Technical Papers [of Commonwealth Scientific and Industrial Research Organization, Australia]
CSIRO Div. Chem. Wood Technol. Res. Rev.	CSIRO Division of Chemical and Wood Technology, Research Review [of Commonwealth Scientific and Industrial Research Organization, Australia]
CSIRONET News	CSIRONET News [of Commonwealth Scientific and Industrial Research Organization Network, Australia]
CSME Trans.	CSME Transactions [of Canadian Society for Mechanical Engineering]
CSNDT J.	CSNDT Journal [of Canadian Society for Nondestructive Testing]
Curr. Adv. Mater. Process.	Current Advances in Materials and Processes [Japan]
Current Contents: Agr. Biol. Envir. Sci.	Current Contents: Agriculture, Biology and Environmental Sciences [of Institute for Scientific Information, US]
Current Contents: Eng. Technol. Appl. Sci.	Current Contents: Engineering, Technology and Applied Sciences [of Institute for Scientific Information, US]
Current Contents: Life Sci.	Current Contents: Life Sciences [of Institute for Scientific Information, US]
Current Contents: Phys. Chem. Earth Sci.	Current Contents: Physical, Chemical and Earth Sciences [of Institute for Scientific Information, US]
Curr. Ind. Rep.	Current Industrial Reports [US]
Curr. Sci.	Current Science [of Current Science Association, India]
Curr. Serials Recd.	Current Serials Received [UK]

Cutting Tool Eng.	Cutting Tool Engineering [US]
CVP Resinotes	CVP (Cray Valley Publishers) Resinotes [UK]
CVRD Rev.	CVRD Revista [Brazil]
CW	Computer World [US]
Cybern. Comput. Technol.	Cybernetics and Computing Technology [Translation of: *Kibernetika i Vychislitel'naya Tekhnika (USSR)*; published in the US]
Cybern. Syst.	Cybernetics and Systems [US]
Czech. Heavy Ind.	Czechoslovak Heavy Industry
Czech. J. Phys.	Czechoslovak Journal of Physics
Czech. J. Phys. A	Czechoslovak Journal of Physics, Section A
Czech. J. Phys. B	Czechoslovak Journal of Physics, Section B

D

Dainichi-Nippon Cables Rev.	Dainichi-Nippon Cables Review [Japan]
Dairy Counc. Dig.	Dairy Council Digest [of National Dairy Council, US]
Dairy Prod. J.	Dairy Products Journal [of American Cultured Diary Products Institute]
Dairy Res. Dig.	Dairy Research Digest [of Diary Research Inc., US]
Danfoss J.	Danfoss Journal [Denmark]
DANTEC Inf.	DANTEC Information [Denmark]
Database Netw. J.	Database and Network Journal [UK]
Data Commun.	Data Communications [US]
Data Commun. Process.	Data Communication and Processing [Japan]
Data Knowl. Eng.	Data and Knowledge Engineering [Netherlands]
Data Manage.	Data Management [of Data Processing Management Association, US]
Data Process.	Data Processor [US]
Data Process. Pract.	Data Processing Practitioner [of Institute of Data Processing, UK]
Data Train.	Data Training [US]
DB Sound Eng. Mag.	DB, The Sound Engineering Magazine [US]
DEC Comput.	DEC Computing [UK]
Dechema Monogr.	Dechema Monographien [Germany]
Decis. Sci.	Decision Sciences [of American Institute for Decision Sciences]
Decis. Support Syst.	Decision Support Systems [Netherlands]
DEC Prof.	DEC Professional [UK]
DECUS Proc.	DECUS Proceedings [of Digital Equipment Computer Users Society, US]
Deep-Sea Res. A	Deep-Sea Research, Part A [UK]
Deep-Sea Res. B	Deep-Sea Research, Part B [UK]
Def. Electron.	Defense Electronics [US]
Def. Metall. Res. Lab. Rep.	Defense Metallurgical Research Laboratory Reports [India]
Deform. Met.	Deformacion Metalica [Spain]
Def. Res. Abs. Contractors Ed.	Defense Research Abstracts — Contractors Edition [US]
Def. Sci. J.	Defense Science Journal [India]
Delft Prog. Rep.	Delft Progress Report [Netherlands]
Delhi Alum. Patrika	Delhi Aluminium Patrika [India]
Denchuken Rev.	Denchuken Review [of Central Research Institute of Electric Power Industry, Japan]
Derevoobrabat. Prom.	Derevoobrabatyvayushchaya Promyshlennost [USSR]
Des. Eng. (NY)	Design Engineering (New York) [US]
Des. Eng. (Tor.)	Design Engineering (Toronto) [Canada]
Des. Eng. (UK)	Design Engineering (UK)
Desktop Publ. J.	Desktop Publishers Journal [of National Association of Desktop Publishers, US]
Desktop Publ. Today	Desktop Publishing Today [UK]
Des. News	Design News [US]
Deut. Papierwirt.	Deutsche Papierwirtschaft [Germany]
DG Rev.	DG Review [US]
Diagn. Imaging Clin. Med.	Diagnostic Imaging in Clinical Medicine [Germany]

Diagnostika Progn. Razrusheniya Svarnykh Konstr.	Diagnostika i Prognozirovanie Razrusheniya Svarnykh Konstruktsii [USSR]
Dialogoye Sist.	Dialogoye Sistemy [USSR]
DIB	Daily Intelligence Bulletin [UK]
Die Cast. Eng.	Die Casting Engineer [of North American Die Casting Association]
Die Cast. Manage.	Die Casting Management [of North American Die Casting Association]
Diecast. Met. Mould.	Diecasting and Metal Moulding [UK]
Diesel Gas Turbine Prog.	Diesel and Gas Turbine Progress [US]
Diesel Gas Turbine Prog. Worldw.	Diesel and Gas Turbine Progress Worldwide [US]
Diesel Gas Turbine Worldw.	Diesel and Gas Turbine Worldwide [US]
Diesel Prog. Engines Drives	Diesel Progress Engines and Drives [US]
Diesel Prog. North Am.	Diesel Progress North America [US]
Diffus. Defect Data	Diffusion and Defect Data [Liechtenstein]
Diffus. Defect Data, Solid State Data A, Defect Diffus. Forum	Diffusion and Defect Data — Solid State Data, Part A, Defect and Diffusion Forum [Liechtenstein]
Diffus. Defect Data, Solid State Data B, Solid State Phenom.	Diffusion and Defect Data — Solid State Data, Part A, Solid State Phenomena [Liechtenstein]
Dig. Int. Conf. Sens. Actuators	Digest of International Conference on Sensors and Actuators [of International Coordinating Committee on Solid State Sensors and Actuators Research, US]
Digit. Process.	Digital Processes [Switzerland]
Digit. Rev.	Digital Review [US]
Digit. Tech. J.	Digital Technical Journal [US]
Dig. Jpn. Ind. Technol.	Digest of Japanese Industry and Technology [Japan]
Dimensions (Sandvik Tubul. Prod.)	Dimensions (Sandvik Tubular Products) [US]
Din. Prochn. Mash.	Dinamika i Prochnost Mashin [USSR]
Direct Midrex	Direct from Midrex [US]
DISA Inf.	DISA Information [Denmark]
Disc. Faraday Soc.	Discussions of the Faraday Society [of Royal Society of Chemistry, UK]
Discrete Appl. Math.	Discrete Applied Mathematics [Netherlands]
Discrete Comput. Geom.	Discrete and Computational Geometry [US]
Discrete Math.	Discrete Mathematics [Netherlands]
Disp. Imaging Technol.	Display and Imaging Technology [UK]
Disp. Technol. Appl.	Displays, Technology and Applications [UK]
Diss. Abstr. Int.	Dissertation Abstracts International [US]
Distrib. Comput.	Distributed Computing [Germany]
Doc. Chem. Yugosl.	Documenta Chemica Yugoslavica [Yugoslavia]
Doc. Ophthalmol.	Documenta Ophthalmologica [Netherlands]
Doc.-Sci. Inf.	Domentaliste — Sciences de l'Information [France]
Dofasco Illus. News	Dofasco Illustrated News [Canada]
Doga Turk Fiz. Astrofiz. Derg.	Doga Turk Fizik Astrofizik Dergisi [of Scientific and Technical Research Council of Turkey]
Doga Turk Muehendis. Cevre Bilimleri Derg.	Doga Turk Muehendislik ve Cevre Bilimleri Dergisi [of Scientific and Technical Research Council of Turkey]
Dokl. Akad. Nauk Azerb. SSR	Doklady Akademii Nauk Azerbiadzhanskoi SSR [USSR]
Dokl. Akad. Nauk BSSR	Doklady Akademii Nauk BSSR [USSR]
Dokl. Akad. Nauk Belorusskoi SSR	Doklady Akademii Nauk Belorusskoi SSR [USSR]
Dokl. Akad. Nauk SSSR	Doklady Akademii Nauk SSSR [USSR]
Dokl. Akad. Nauk Tadzh. SSR	Doklady Akademii Nauk Tadzhikshoi SSR [USSR]
Dokl. Akad. Nauk Uzb. SSR	Doklady Akademii Nauk Uzbekskoi SSR [USSR]
Dokl. Chem.	Doklady Chemistry [USSR]
Dokl. Chem. Technol.	Doklady Chemical Technology [USSR]
Dokl. Phys. Chem.	Doklady Physical Chemistry [USSR]
Dop. Akad. Nauk Ukr. RSR A, Fiz.-Mat. Tekh.	Dopovidi Akademii Nauk Ukrains'koi RSR, Seriya A, Fiziko-Matematichni ta Tekhnichni Nauki [USSR]
Dop. Akad. Nauk Ukr. RSR B, Geol., Khim., Biol.	Dopovidi Akademii Nauk Ukrains'koi RSR, Seriya B, Geologiya, Khimiya, Biologiya [USSR]
Double Liaison-Chim. Peint.	Double Liaison-Chimie des Peintures [France]
Drager Rev.	Drager Review [US]
Dr. Dobb's J. Softw. Tools	Dr. Dobb's Journal of Software Tools [US]
Drug Dev. Ind. Pharm.	Drug Development and Industrial Pharmacy

Drvna Ind.	Drvna Industrija [Yugoslavia]
Drying Technol.	Drying Technology [US]
Dtsch. Ges. Metallkd. Fachber.	Deutsche Gesellschaft fur Metallkunde Fachberichte [Germany]
Dtsch. Roheisen	Deutsches Roheisen [Germany]
Du Pont Inf. Serv.	Du Pont Information Service [Switzerland]
Du Pont Mag.	Du Pont Magazine [US]
Du Pont Mag. Eur. Ed.	Du Pont Magazine, European Edition [Switzerland]
Durability Build. Mater.	Durability of Building Materials [Netherlands]
Dyn. Atmos. Oceans	Dynamics of Atmospheres and Oceans [Netherlands]

E

EAI	Economic Abstracts International [Netherlands]
Earth Moon Planets	Earth, Moon and Planets [Netherlands]
Earth Planet. Sci. Lett.	Earth and Planetary Science Letters [Netherlands]
Earth-Sci. Rev.	Earth-Science Reviews [Netherlands]
Earth Surf. Process. Landf.	Earth Surface Processes and Landforms [UK]
Earthq. Eng. Struct. Dyn.	Earthquake Engineering and Structural Dynamics [UK]
EBU Rev. Tech.	EBU Review, Technical [of European Broadcasting Union, Belgium]
ECAB	Economic Abstracts [Belgium]
ECCAI Newsl.	ECCAI Newsletter [of European Coordinating Committee for Artificial Intelligence]
Echo Rech.	L'Echo des Recherches [of Centre National d'Etudes des Telecommunications, France]
ECMA Bull.	ECMA Bulletin [of European Carton Makers Assiociation, Netherlands]
Ecol. Monogr.	Ecological Monographs [of Ecological Society of America, US]
Econ. Bull.	Economics Bulletin [of Chemical Industries Association, UK]
Econ. Comput. Econ. Cybern. Stud. Res.	Economic Computation and Economic Cybernetics Studies and Research [of Centre of Economic Computation and Economic Cybernetics, Romania]
Econ. Geol.	Economic Geology [of Society od Economic Geologists, US]
Econ. Intell. Unit Spec. Rep.	Economist Intelligence Unit Special Report [US]
Economist	The Economist [UK]
Ecotoxicol. Environ. Saf.	Ecotoxicology and Environmental Safety
Edax Ed.	Edax Editor [US]
EDP Audit. J.	EDP Auditors Journal [of EDP Auditors Foundation, US]
EDP In-Depth Rep.	EDP In-Depth Reports [Canada]
Educ. Comput.	Educational Computing [UK]
Educ. Comput.	Education and Computing [Netherlands]
Educ. Technol.	Educational Technology [US]
Educ. Technol. Res. Dev.	Educational Technology, Research and Development [of Association for Educational Communications and Technology, US]
Educ. + Train.	Education + Training [UK]
Educ. Train. Technol. Int.	Educational and Training Technology International [UK]
EDUCOM Bull.	EDUCOM Bulletin [US]
EDUCOM Rev.	EDUCOM Review [US]
Eesti NSV Tead. Akad. Fuus Inst. Uurim.	Eesti NSV Teaduste Akadeemia Fuusiki Instituudi Uurimused [USSR]
Eesti NSV Tead. Akad., Toim., Fuus Mat.	Eesti NSV Teaduste Akadeemia, Toimetised, Fuusiki, Matemaatika [USSR]
EFB Newsl.	EFB Newsletter [of European Federation of Biotechnology, Germany]
EFCE Newsl.	EFCE Newsletter [of European Federation of Chemical Engineering, Germany]
EFC Newsl.	EFC Newsletter [of European Federation of Corrosion, Germany]
Effl. Water Treat. J.	Effluent and Water Treatment Journal [UK]
Egypt. Comput. J.	Egyptian Computer Journal [Egypt]
Egypt. J. Biomed. Eng.	Egyptian Journal of Biomedical Engineering [of National Information and Dokumentation Center, Egypt]

Egypt. J. Solids	Egyptian Journal of Solids [of Egyptian Society of Solid State Science and Applications]
EID	Electronic Instrument Digest [US]
Ekon.-Mat. Obz.	Ekonomicko-Matematicky Obzor [of Czechoslovak Academy of Sciences]
Eksploat. Masz.	Eksploataeja Maszyn [Poland]
Elast. Noteb.	Elastomers Notebook [Switzerland]
Electr. China	Electricity for China [UK]
Electr. Commun.	Electrical Communication [UK]
Electr. Commun. Lab. Tech. J.	Electrical Communication Laboratories Technical Journal [Japan]
Electr. Constr. Maint.	Electrical Construction and Maintenance [US]
Electr. Contract.	Electrical Contractor [UK]
Electr. Des.	Electrical Design [of Chartered Institution of Building Services Engineers, UK]
Electr. Eng. (Australia)	Electrical Engineer (Australia)
Electr. Eng. (London)	Electrical Engineer (London) [UK]
Electr. Eng. Jpn.	Electrical Engineering in Japan [US]
Electr. Eng. Rev.	Electrical Engineering Review [of West Pakistan University of Engineering and Technology, Pakistan]
Electr. Equip.	Electrical Equipment [UK]
Electr. Furn. Conf. Proc.	Electric Furnace Conference Proceedings [of AIME Iron and Steel Society, US]
Electr. Furn. Steel	Electric Furnace Steel [Japan]
Electr. India	Electrical India [of Indian Electrical Manufacturers Association]
Electr. Light. Facil. Railw.	Electric Lighting and Facilities in Railways [of Railway Electrification Association, Japan]
Electr. Mach. Power Syst.	Electric Machines and Power Systems [US]
Electr. Manuf.	Electrical Manufacturing [US]
Electrochem. Ind. Process. Biol.	Electrochemistry in Industrial Processing and Biology [Translation of: *Elektronnaya Obrabotka Materialov (USSR)*; published in the UK]
Electrochem. Technol.	Electrochemical Technology [US]
Electrochim. Acta	Electrochimica Acta [UK]
Electrochim. Acta	Electrochimica Acta [of International Society of Electrochemistry, Austria]
Electrocomponent Sci. Technol.	Electrocomponent Science and Technology [US]
Electron. Bank. Finance	Electronic Banking and Finance [Netherlands]
Electron. Bus.	Electronic Business [US]
Electron. Commun. Eng. J.	Electronics and Communication Engineering Journal [of Institution of Electrical Engineers, UK]
Electron. Commun. Jpn. 1, Commun.	Electronics and Communications in Japan, Part 1, Communication [US]
Electron. Commun. Jpn. 2, Electron.	Electronics and Communications in Japan, Part 2, Electronics [US]
Electron. Commun. Jpn. 3, Fundam. Electron. Sci.	Electronics and Communications in Japan, Part 3, Fundamental Electronic Science [US]
Electron. Compon. Appl.	Electronic Components and Applications [Netherlands]
Electron. Des.	Electronic Design [US]
Electron. Des. Autom.	Electronic Design Automation [UK]
Electron. Educ.	Electronic Education [US]
Electron. Eng.	Electronic Engineering [UK]
Electron. Imaging	Electronic Imaging [US]
Electron. Ind.	Electronics Industry [UK]
Electron. Ind.	Electronique Industrielle [France]
Electron. Inf. Plan.	Electronics Information and Planning [India]
Electron. Learn.	Electronic Learning [US]
Electron. Lett. (Australia)	Electronics Letters (Australia)
Electron. Lett. (UK)	Electronics Letters (UK) [of Institution of Electrical Engineers, UK]
Electron. Libr.	Electronic Library [UK]
Electron. Manuf.	Electronics Manufacturing [of Society of Manufacturing Engineers, US]
Electron. Manuf. Test	Electronics Manufacture and Test [UK]
Electron-Microsc.	Electron-Microscopy [Japan]
Electron Microsc. Rev.	Electron Microscopy Review [US and UK]
Electron. Model.	Electronic Modeling [Translation of: *Elektronnoe Modelirovanie (Ukrainian SSR)*; published in the UK]
Electron Opt. Bull.	Electron Optics Bulletin [Netherlands]
Electron. Opt. Publ. Rev.	Electronic and Optical Publishing Review [UK]
Electron Opt. Report.	Electron Optics Reporter [US]
Electron. Packag. Prod.	Electronic Packaging and Production [US]

Electron. Prod.	Electronic Production [UK]
Electron. Prod.	Electronic Products [US]
Electron. Prod. Des.	Electronic Product Design [UK]
Electron. Prog.	Electronic Progress [US]
Electron. Publ. Bus.	Electronic Publishing Business [US]
Electron. Publ., Orig. Dissem. Des.	Electronic Publishing: Origination, Dissemination and Design [UK]
Electron. Puissance	Electronique de Puissance [France]
Electron. Purch.	Electronics Purchasing [US]
Electron. Syst. News	Electronic Systems News [of Institution of Electrical Engineers, UK]
Electron. Tech. Ind.	Electronique Technique et Industries [France]
Electron Technol.	Electron Technology [Poland]
Electron. Technol.	Electronic Technology [of Society of Electronic and Radio Technicians, UK]
Electron. Telecommun. Lett.	Electronics and Telecommunications Letter [Poland]
Electron. Test	Electronics Test [US]
Electron. Times	Electronic Times [UK]
Electron. Today	Electronics Today [India]
Electron. Today Int.	Electronics Today International [UK]
Electron. Wirel. World	Electronics and Wireless World [UK]
Electron. Wkly.	Electronics Weekly [UK]
Electron. World Wirel. World	Electronics World + Wireless World [UK]
Electro-Opt.	Electro-Optics [UK]
Electrotech. News	Electrotechnical News [UK]
Electro-Technol.	Electro-Technology [of Society of Electronic Engineers, India]
Electroteh. Electron. Autom., Autom. Electron.	Electrotehnica, Electronica si Automatica, Automatica si Electrotehnica [Romania]
Electroteh. Electron. Autom., Electroteh.	Electrotehnica, Electronica si Automatica, Electrotehnica [Romania]
Electr. Power Syst. Res.	Electric Power Systems Research [Switzerland]
Electr. Prog.	Electronic Progress [US]
Electr. Railw.	Electric Railways [of Railway Electrification Association, Japan]
Electr. Rep.	Electrical Reports [of Edison Elcctric Institute, US]
Electr. Rev.	Electrical Review [UK]
Electr. Rev. (London)	Electrical Review (London) [UK]
Electr. Technol. USSR	Electric Technology USSR [Published in the UK]
Electr. Times	Electrical Times [UK]
Electr. Veh. Dev.	Electric Vehicle Developments [UK]
Electr. Word	Electric Word [Netherlands]
Electr. World	Electrical World [US]
Elektr. Bahnen	Elektrische Bahnen [Germany]
Elektr. Energy.-Tech.	Elektrische Energie-Technik [Germany]
Elektr. Masch.	Electrische Maschinen [Germany]
Elektro-Anz.	Elektro-Anzeiger [Germany]
Elektromeist. Dtsch. Elektrohandw.	Elektromeister and Deutsches Elektrohandwerk [Germany]
Elektron. Entwickl.	Elektronik Entwicklung [Germany]
Elektroniker	Der Elektroniker [Switzerland]
Elektron. Ind.	Elektronik Industrie [Germany]
Elektron Int.	Elektron International [Austria]
Elektron. J.	Elektronik Journal [Germany]
Elektron. Model.	Elektronnoe Modelirovanie [USSR]
Elektronnaya Obrab. Mater.	Elektronnaya Obrabotka Materialov [USSR]
Elektron. Prax.	Elektronik Praxis [Germany]
Elektron. Tekh.	Elektronnaya Tekhnika [USSR]
Elektrotech. Cas	Elektrotechnický Casopis [Czechoslovakia]
Elektrotech. Inf.tech.	Elektrotechnik und Informationstechnik [Austria]
Elektrotech. Masch.-Bau	Elektrotechnik und Maschinenbau [Germany]
Elektrotech. Obz.	Elektrotechnicky Obzor [Czechoslovakia]
Elektroteh. Vestn.	Elektrotehniski Vestnik [Yugoslavia]
Elektrowarme Int. A.	Elektrowarme International, Edition A
Elektrowarme Int. B.	Elektrowarme International, Edition B
Elektrowarme Tech.	Elektrowarme im Technischen [Germany]
Elektr. Stn.	Elektricheskie Stantsii [USSR]

Elettron. Oggi	Elettronica Oggi [Italy]
Elettron. Telecomun.	Elettronica e Telecomunicazioni [Italy]
Elin-Z.	Elin-Zeitschrift [Austria]
Eltek. Med Aktuell Elektron.	Elteknik Med Aktuell Elektronik [Sweden]
E & MJ Eng. Min. J.	E & MJ Engineering and Mining Journal
Email Met.	Email Metal [France]
Emballage Dig.	Emballage Digest [of Societe Europeene de Presse et d'Edition, France]
Emballages Mag.	Emballages Magazine [France]
EMSA Bull.	EMSA Bulletin [of Electron Microscopy Society of America]
Energ. Atomtech.	Energia es Atomtechnika [Hungary]
Energ. Elettr.	Energia Elettrica [of Associazione Elettrotecnica ed Elettronica, Italy]
Energie Fluide Air Ind.	Energie Fluide l'Air Industriel
Energiewirtsch. Tagesfragen	Energiewirtschaftliche Tagesfragen [Germany]
Energ. Nucl.	Energia Nuclear [of Junta de Energia Nuclear, Spain]
Energ. Nucl.	Energia Nucleare [Italy]
Energomashinostr.	Energomashinostroenie [USSR]
Energ.wirtsch. Tagesfr.	Energiewirtschaftliche Tagesfragen [Germany]
Energy Autom.	Energy and Automation [Germany]
Energy Build.	Energy and Buildings [Switzerland]
Energy Convers. Manage.	Energy Conversion and Management [UK]
Energy Dev.	Energy Developments [Germany]
Energy Eng. J.	Energy Engineering Journal [of Association of Energy Engineers, US]
Energy Fuels	Energy and Fuels
Energy J.	Energy Journal [of International Association for Energy Economics, US]
Energy J.	Energy Journal [New Zealand]
Energy Mag.	Energy Magazine [of Organisacion Latinoamericana de Energia, Ecuador]
Energy Manage.	Energy Management [UK]
Energy Prog.	Energy Progress [US]
Energy Res. Rep.	Energy Research Report [US]
Energy Syst. Policy	Energy Systems and Policy [US]
Eng. Anal.	Engineering Analysis [UK]
Eng. Appl. Artif. Intell.	Engineering Applications of Artificial Intelligence [UK]
Eng. Aust.	Engineers Australia [of Institution of Engineers of Australia]
Eng. Autom.	Engineering and Automation [of Informatics Center of the Hungarian Industry Budapest, Hungary]
Eng. Comput.	Engineering Computers [UK]
Eng. Comput.	Engineering with Computers [US]
Eng.: Cornell Q.	Engineering: Cornell Quarterly [US]
Eng. Costs Prod. Econ.	Engineering Costs and Production Economics [Netherlands]
Eng. Des.	Engineering Design [Switzerland]
	Engineering Designer [of Institution of Engineering Designers, UK]
Eng. Dig.	Engineering Digest [Canada]
Eng. Dig.	Engineer's Digest [UK]
Eng. Econ.	Engineering Economist [of Institute of Industrial Engineers, US]
Eng. Econ.	Engineering Economist [of American Society for Engineering Education]
Eng. Educ.	Engineering Education [of American Society for Engineering Education]
Eng. Fract. Mech.	Engineering Fracture Mechanics [UK]
Eng. Gaz.	Engineering Gazette [of Engineering Industries Association, UK]
Eng. Ind.	Engineering Industries [Iraq]
Eng. J.	Engineering Journal [Canada]
Eng. J.	Engineering Journal [of American Institute of Steel Construction]
Eng. Lasers	Engineering Lasers [UK]
Eng. Manage. Int.	Engineering Management International [of American Society for Engineering Management]
Eng. Mater.	Engineering Materials [Japan]
Eng. Mater. Des.	Engineering Materials and Design [UK]
Eng. Med.	Engineering in Medicine [UK]
Eng. Med. Biol. Mag.	Engineering in Medicine and Biology Magazine [of Institute of Electrical and Electronics Engineers, US]
Eng. Min. J.	Engineering and Mining Journal [US]
Eng. News	Engineering News [UK]
Eng. News-Rec.	Engineering News-Record [US]
Eng. Outlook	Engineering Outlook [US]

Eng. Plast.	Engineering Plastics [of Association of Directors of European Centers for Plastics, UK]
Eng. Sci. Rep. Kyushu Univ.	Engineering Sciences Reports, Kyushu University [Japan]
Eng. Struct.	Engineering Structures [UK]
Eng. Tech.	Engineering Technician [Singapore]
Eng. Technol.	Engineering Technologist [Singapore]
Eng. Technol.	Engineering and Technology [Japan]
Eng. Technol., Kansai Univ.	Engineering and Technology, Kansai University [Japan]
Eng. Times	Engineering Times [of National Society of Professional Engineers, US]
ENR	Engineering News-Record [US]
Enshurin Ken. Hok.	Enshurin Kenkya Hokoku [of Hokkaido University, Japan]
Enterp. Rhone-Alpes (Metall.)	Entreprises Rhones-Alpes (Metallurgia) [France]
Environ. Can. Environ. Update	Environment Canada, Environment Update
Environ. Can. Notice Publ.	Environment Canada, Notice of Publications
Environ. Eng.	Environmental Engineering [UK]
Environ. Exp. Bot.	Environmental and Experimental Botany [UK]
Environ. Inf. Sci.	Environmental Information Science [of Center for Environmental Information Science, Japan]
Environ. Int.	Environment International
Environ. Monit. Assess.	Environmental Monitoring and Assessment [Netherlands]
Environ. Mutagenesis	Environmental Mutagenesis
Environ. Pollut.	Environmental Pollution [UK]
Environ. Prof.	Environmental Professional [of National Association of EnvironmentalProfessionals, US]
Environ. Prog.	Environmental Progress [of American Institute of Chemical Engineers]
Environ. Prot. Eng.	Environmental Protection Engineering
Environ. Res.	Environmental Research
Environ. Rev.	Environmental Review [US]
Environ. Sci. Technol.	Environmental Science and Technology [of American Chemical Society]
Environ. Sci. Technol. A	Environmental Science and Technology A [US]
Environ. Softw.	Environmental Software [UK]
Environ. Technol. Lett.	Environmental Technology Letters [UK]
EOS Trans. Am. Geophys. Union	EOS Transactions of the American Geophysical Union [US]
EPA J.	EPA Journal [US]
EPA Newsl.	EPA Newsletter [of European Photochemistry Association, UK]
EPF Q.	EPF Quarterly [of European Packaging Federation, Netherlands]
EPIA	Electric Power Industry Abstracts [of Edison Electrical Institute, US]
EPI Q. Rep.	EPI Quarterly Report [of Environmental Policy Institute, US]
EPRI J.	EPRI Journal [of Electric Power Research Institute, US]
Erde Int.	Erde International [of Aloe Technology Association, US]
Erdoel Kohle	Erdoel und Kohle [Germany]
Erdoel Kohle Erdgas	Erdoel Kohle Erdgas und Petrochemie [Germany]
Ericsson Rev.	Ericsson Review [Sweden]
ESA Bull.	ESA Bulletin [of European Space Agency, France]
ESA Feat.	ESA Features [of European Space Agency, France]
ESA J.	ESA Journal [of European Space Agency Technical Center, Netherlands]
ESD, Electron.. Syst. Des. Mag.	ESD: The European System Design Magazine [US]
ESF Commun.	ESF Communications [of European Science Foundation, France]
ESF Synchrotron Radiat. News	ESF Synchrotron Radiation News [Netherlands]
Esso Mag.	Esso Magazine [UK]
Esso Oilways Int.	Esso Oilways International [UK]
Estuarine Coastal Shelf Sci.	Estuarine Coastal and Shelf Science [UK]
ETA Elektrowarme Tech. Ausbau A	ETA Elektrowarme im Technischen Ausbau, Edition A [Germany]
ETA Elektrowarme Tech. Ausbau B	ETA Elektrowarme im Technischen Ausbau, Edition B [Germany]
ETG-Fachber.	ETG-Fachberichte [Germany]
ETZ Arch.	ETZ Archiv [Germany]
Eur. Adhes. Seal.	European Adhesives and Sealants [UK]
Eur. Appl. Res. Rep.	European Applied Research Report [UK]
Eur. Appl. Res. Rep. Nucl. Technol. Sect.	European Applied Research Reports, Nuclear Science Technology Section [US]
Eur. Appl. Res. Rep. Nucl. Technol. Sect.	European Applied Research Reports, Nuclear Science Technology Section [of Commission of the European Communities, Belgium]
Eur. Biophys. J.	European Biophysics Journal [Germany]

Eur.-Chem.	Europa-Chemie [of Verband der Chemischen Industrie, Germany]
Eur. J. Biochem.	European Journal of Biochemistry [of Federation of European Biochemical Societies, France]
Eur. J. Eng. Educ.	European Journal of Engineering Education [UK]
Eur. J. Mech. A Solids	European Journal of Mechanics, Series A, Solids [France]
Eur. J. Mech. B Fluids	European Journal of Mechanics, Series B, Fluids [France]
Eur. J. Mech. Eng.	European Journal of Mechanical Engineering [Belgium]
Eur. J. Mineral.	European Journal of Mineralogy [Germany]
Eur. J. Nucl. Med.	European Journal of Nuclear Medicine [Germany]
Eur. J. Oper. Res.	European Journal of Operational Research [Netherlands]
Eur. J. Phys.	European Journal of Physics [of European Physical Society, Switzerland]
Eur. J. Phys.	European Journal of Physics [of Institute of Physics, UK]
Eur. J. Solid State Inorg. Chem.	European Journal of Solid State and Inorganic Chemistry [France]
Eur. Market. Newsl.	European Marketing Newsletter [of European Commission for Industrial Marketing, UK]
Eurobuild Inf. Bull.	Eurobuild Information Bulletin [France]
Euro Flexo Mag.	Euro Flexo Magazine [Netherlands]
Europhys. Lett.	Europhysics Letters [of European Physical Society, Switzerland]
Europhys. News	Europhysics News [of European Physical Society, Switzerland]
Eurotest Tech. Bull.	Eurotest Technical Bulletin [of Eurotest, Belgium]
Eurostat Rev.	Eurostat Review [of Statistical Office of the European Community, Luxembourg]
Eur. Packag. Mag.	European Packaging Magazine [Germany]
Eur. Pkg. Mag.	European Packaging Magazine [UK]
Eur. Plast. News	European Plastics News [UK]
Eur. Polym. J.	European Polymer Journal [UK]
Eur. Rubb. J.	European Rubber Journal [UK]
Eur. Semicond.	European Semiconductor [UK]
Eur. Semicond. Des. Prod.	European Semiconductor Design and Production [UK]
Eur. Semicond. Prod.	European Semiconductor Production [UK]
Eval. Eng.	Evaluation Engineering [US]
Exhibition Bull.	Exhibition Bulletin [UK]
EX Mag.	EX Magazine [Germany]
Exp. Astron.	Experimental Astronomy [Netherlands]
Expert Syst.	Expert Systems [UK]
Expert Syst. Appl.	Expert System Applications [UK]
Expert Syst. Rev.	Expert Systems Review [US]
Expert Syst. User	Expert Systems User [UK]
Exp. Fluids	Experiments in Fluids
Exp. Heat J.	Experimental Heat Journal [US]
Exp. Heat Transfer	Experimental Heat Transfer
Explor. Geophys.	Exploration Geophysics [Australia]
Exp. Mech.	Experimental Mechanics [of Society for Experimental Stress Analysis, US]
Export Dig.	Export Digest [UK]
Exp. Tech.	Experimental Techniques [of Society for Experimental Mechanics, US]
Exp. Tech. Phys.	Experimentelle Technik der Physik [Germany]
Exp. Therm. Fluid. Sci.	Experimental Thermal and Fluid Science
Extrus. Showc.	Extrusion Showcase [US]

F

Facil. Manager	Facilities Manager [UK]
Fachh./Bull. Tech.	Fachhefte (Chemigraphie, Lithographie, Tiefdruck)/Bulletin Technique (Photogravure, Lithographie, Heliogravure) [Switzerland]
Fachz. Lab	Fachzeitschrift fur das Labor [Germany]
Fachber. Huttenprax. Metallweiterverarb.	Fachberichte Huttenpraxis Metallweiterverarbeitung [Germany]

Faraday Disc. Chem. Soc.	Faraday Discussions of the Chemical Society [UK]
Faraday Symp. Chem. Soc.	Faraday Symposia of the Chemical Society [UK]
Farbe Lack	Farbe und Lack [Germany]
Far East. Econ. Rev.	Far Eastern Economic Review [Hong Kong]
FASEB J.	FASEB Journal [of Federation of American Societies for Experimental Biology, US]
Fast Ferry Int.	Fast Ferry International [UK]
Fasten. Age	Fastener Age [of Wire Association International, US]
Fatigue Fract. Eng. Mater. Struct.	Fatigue and Fracture of Engineering Materials and Structures [UK]
FEA Aerosol Bull.	FEA Aerosol Bulletin [of European Federation of Aerosol Associations, Belgium]
FEBS Lett.	FEBS Letters [of Federation of European Biochemical Societies, France]
FECC Inf. Bull.	FECC Information Bulletin [of Federation Europeenne du Commerce Chimique, Belgium]
Feinwerktech. Messtech.	Feinwerktechnik und Messtechnik [Germany]
FEMS Circ.	FEMS Circular [of Federation of the European Microbiological Societies, UK]
FEMS Ecol.	FEMS Ecology [of Federation of the European Microbiological Societies, UK]
FEMS Microbiol.	FEMS Microbiology [of Federation of the European Microbiological Societies, UK]
FEMS Microbiol. Lett.	FEMS Microbiology Letters [of Federation of the European Microbiological Societies, UK]
Fenxi Huaxue (Anal. Chem.)	Fenxi Huaxue (Analytical Chemistry) [PR China]
Fenxi Shiyanshi (Anal. Lab.)	Fenxi Shiyanshi (Analytical Laboratory) [PR China]
Fernmelde-Ing.	Fernmelde-Ingenieur [Germany]
Fernseh- & Kino-Tech.	Fernseh- und Kino-Technik [Germany]
Fernwarme Int.	Fernwarme International [Germany]
Ferroelectr. Lett.	Ferroelectrics Letters [UK]
Ferroelectr. Lett. Sect.	Ferroelectrics Letters Section [UK]
Fertigungstech. Betr.	Fertigungstechnik und Betrieb [Germany]
Few-Body Syst.	Few-Body Systems [Austria]
Few-Body Syst. Suppl.	Few-Body Systems Supplementum [Austria]
FhG Ber.	FhG Bericht [of Fraunhofer Gesellschaft, Germany]
FIB	Fortran Information Bulletin [US]
Fiber Integr. Opt.	Fiber and Integrated Optics [US]
Fiber Opt. Mag.	Fiber Optics Magazine [US]
Fibre Sci. Technol.	Fibre Science and Technology [UK]
Fils, Tubes, Bandes Profiles	Fils, Tubes, Bandes et Profiles [France]
Filtr. Sep.	Filtration and Separation [UK]
Financ. Times	Financial Times [UK]
Financ. World	Financial World [US]
Fine Ceram.	Fine Ceramics [Japan]
Finish. Ind.	Finishing Industries [UK]
Finish. Line	Finishing Line [of Association for Finishing Processes [of SME, US]]
Finish. Manage.	Finishers' Management [US]
Finite Elem. Anal. Des.	Finite Elements in Analysis and Design [Netherlands]
Finite Elem. News	Finite Elements News [UK]
Finn. Chem. Lett.	Finnish Chemical Letters [Finland]
Finommech. — Mikrotech.	Finommechanika — Microtechnika [Hungary]
FIRA Bull.	FIRA Bulletin [of Furniture Industry Research Association, UK]
Fire Flammabl. Bull.	Fire and Flammability Bulletin [UK]
Fire Mater.	Fire and Materials [US]
Fire Saf. J.	Fire Safety Journal [US]
Fis. Tecnol.	Fisica e Tecnologia [of Societa Italiana di Fisica, Italy]
Fiz. Elem. Chast. Atomn. Yad.	Fizika Elementarnykh Chastits i Atomnoya Yadra [USSR]
Fiz. Goren. Vzryva	Fizika Goreniya i Vzryva [USSR]
Fiz.-Khim. Mekh. Mater.	Fiziko-Khimicheskaya Mekhanika Materialov [USSR]
Fiz. Khim. Obrab. Mater.	Fizika i Khimiya Obrabotki Materialov [USSR]
Fiz. Khim. Stekla	Fizika i Khimiya Stekla [USSR]
Fiz. Met. Metalloved.	Fizika Metallov i Metallovedenie [USSR]
Fiz. Nizk. Temp.	Fizika Nizkikh Temperatur [USSR]
Fiz. Plazmy	Fizika Plazmy [USSR]
Fiz. Tekh. Poluprovodn.	Fizika i Tekhnika Poluprovodnikov [USSR]
Fiz.-Tekh. Probl. Razrab. Polezn. Iskop.	Fiziko-Tekhnicheskie Problemy Razrabotki Poleznykh Iskopaemykh [USSR]
Fiz. Tekh. Vys. Davlenii	Fizika i Tekhnika Vysokikh Davlenii [USSR]
Fiz. Tverd. Tela	Fizika Tverdogo Tela [USSR]

Fluid Dyn.	Fluid Dynamics [Translation of: *Izvestiya Akademii Nauk SSSR, Mekhanika Zhidkosti i Gaza (USSR)*; published in the US]
Fluid Dyn. Res.	Fluid Dynamics Research [Netherlands]
Fluid Mech. Sov. Res.	Fluid Mechanics — Soviet Research [Published in the US]
Fluid/Particle Sep. J.	Fluid/Particle Separation Journal
Fluid Phase Equilib.	Fluid Phase Equilibria [Netherlands]
FMS Mag.	FMS Magazine [UK]
FOA Rep.	FOA Reports [of Forsvarets Forskningsanstalt, Sweden]
FOGRA Lit.	FOGRA Literaturdienst [of Deutsche Forschungsgesellschaft fur Druck- und Reproduktionstechnik, Germany]
FOGRA Mitt.	FOGRA Mitteilungen [of Deutsche Forschungsgesellschaft fur Druck- und Reproduktionstechnik, Germany]
Folding Carton Ind.	Folding Carton Industry [UK]
Folia Fac. Sci. Nat. Univ. Purkyn. Brun., Phys.	Folia Facultatis Scientarum Naturalium Universitatis Prkkynianae Brunensis, Physica [Czechoslovakia]
Fonderia Ital.	Fonderia Italiana [Italy]
Food Drug Packag.	Food and Drug Packaging [US]
Food Manuf.	Food Manufacture [UK]
Food Process.	Food Processing
Food Technol.	Food Technology [of Institute of Food Technologists, US]
Foerd.mittel J.	Foerdermittel Journal [Germany]
Fordern Heben	Fordern und Heben [Germany]
Forensic Eng.	Forensic Engineering [UK]
Forest Ecol. Manage.	Forest Ecology Management [US]
Forest Genet. Resour. Inf.	Forest Genetic Resources Information [Italy]
Forest Ind.	Forest Industries [US]
Forest Prod. Abstr.	Forest Products Abstracts [of Commonwealth Agricultural Bureaus, UK]
Forest Prod. J.	Forest Products Journal [of Forest Products Research Society, US]
Forestry Abstr.	Forestry Abstracts [UK]
Forestry Chron.	Forestry Chronicle [of Canadian Institute of Forestry]
Forestry Res. Newsl.	Forestry Research Newsletter [of Great Lakes Forest Research Center, Canada]
Forest Sci.	Forest Science [of Society of American Foresters]
Forest Sci. Monogr.	Forest Science Monographs [of Society of American Foresters]
Forg. Top	Forging Topics [US]
Forsch.ber. Landes Nordrh.-Westfal.	Forschungsberichte des Landes Nordrhein-Westfalen [Germany]
Forsch. Ingenieurwes.	Forschung im Ingenieurwesen [of Verein Deutscher Ingenieure, Germany]
Forschungsber. Ind. Logistik	Forschungsberichte zur Industriellen Logistik [of Deutsche Gesellschaft fur Logistik, Germany]
Forthcoming Int. Sci. Tech. Conf.	Forthcoming International Scientific and Technical Conferences [UK]
Forth Dimens.	Forth Dimension [US]
Fortschr.-Ber. VDI-Z	Fortschritt-Berichte der VDI Zeitschrift [of Verein Deutscher Ingenieure, Germany]
Fortschr. Phys.	Fortschritte der Physik [Germany]
Fortschr. Verf.-Tech.	Fortschritte der Verfahrenstechnik [Germany]
Fortschr.ber. Dtsch. Keram. Ges.	Fortschrittsberichte der Deutschen Keramischen Gesellschaft [Germany]
Found. Control Eng.	Foundations of Control Engineering [of Institute of Control Engineering, Poland]
Found. Phys.	Foundations of Physics [US]
Found. Phys. Lett.	Foundations of Physics Letters [US]
Foundry Manage. Technol.	Foundry Management and Technology [US]
Foundry Pract.	Foundry Practice [UK]
Foundry Technol.	Foundry Technol. [PR China]
Foundry Trade J.	Foundry Trade Journal [UK]
Foundry Trade J. Int.	Foundry Trade Journal International [UK]
Fr. Adv. Sci. Technol.	French Advances in Science and Technology [US]
Fra Fys. Verden	Fra Fysikkens Verden [Norway]
Freib. Forschungsh.	Freiberger Forschungshefte [Germany]
Freib. Forschungsh. A	Freiberger Forschungshefte A [Germany]
Freib. Forschungsh. B	Freiberger Forschungshefte B [Germany]
Freib. Forschungsh. B, Metall. Werkstofftech.	Freiberger Forschungshefte B, Metallurgie Werkstofftechnik [Germany]
Fresenius J. Anal. Chem.	Fresenius Journal of Analytical Chemistry [Germany]

Fresenius Z. Anal. Chem.	Fresenius Zeitschrift fur Analytische Chemie [Germany]
Freshwater Biol.	Freshwater Biology [UK]
Fr. Railw. Rev.	French Railway Review [Translation of: *Revue Generale des Chemins de Fer (France)*; published in the UK]
Fr. Tech.	French Techniques [France]
FSTA	Food Science and Technology Abstracts [UK]
Fuel Process. Technol.	Fuel Processing Technology
Fuel Sci. Technol.	Fuel Science and Technology
Fuel Sci. Technol. Int.	Fuel Science and Technology International
Fuji Electr. Rev.	Fuji Electric Review [Japan]
Fujikura Tech. Rev.	Fujikura Technical Review [Japan]
Fujitsu Sci. Tech. J.	Fujitsu Scientific and Technical Journal [Japan]
Fulmer Newsl.	Fulmer Newsletter [UK]
Fukuoka Univ. Rev. Technol. Sci.	Fukuoka University Review of Technological Sciences [Japan]
Funct. Mater.	Function and Materials [Japan]
Fundam. Appl. Toxicol.	Fundamental and Applied Toxicology [of Society of Toxicology, US]
Fundam. Cosm. Phys.	Fundamentals of Cosmic Physics [UK]
Fundam. Inform.	Fundamenta Informaticae [Netherlands]
Fundicao Mater. Primas	Fundicao e Materias Primas [Brazil]
Fusion Eng. Des.	Fusion Engineering and Design [Switzerland]
Fusion Technol.	Fusion Technology [of American Nuclear Society]
Fussboden Ztg.	Fussboden Zeitung [Germany]
Future Comput. Syst.	Future Computing Systems [UK]
Future Gener. Comput. Syst.	Future Generation Computer Systems [Netherlands]
Futures Res. Q. Rep., Aluminum	Futures Research Quarterly Report, Aluminum [UK]
Futures Res. Q. Rep., Copper	Futures Research Quarterly Report, Copper [UK]
Futures Res. Q. Rep., Lead Zinc	Futures Research Quarterly Report, Lead and Zinc [UK]
Futures Res. Q. Rep., Precious Met.	Futures Research Quarterly Report, Precious Metals [UK]
Futures Res. Q. Rep., Tin	Futures Research Quarterly Report, Tin [UK]
Fuzzy Sets Syst.	Fuzzy Sets and Systems [Netherlands]
FWI	Fernwarme International [Germany]
FWP J.	FWP Journal [South Africa]
Fys. Tidsskr.	Fysisk Tidsskrift [Denmark]

G

Galileian Electrodyn.	Galileian Electrodynamics [US]
Galvano-Organo-Trait. Surf.	Galvano-Organo-Traitements de Surface [France]
Galvanotech.	Galvanotechnik [Germany]
Galvano-Tek. Tidsskr.	Galvano-Teknisk Tidsskrift [Norway]
Gas Energy Rev.	Gas Energy Review [of American Gas Association]
Gas Eng. Manage.	Gas Engineering and Management [UK]
Gas Res. Inst. Dig.	Gas Research Institute Digest [US]
Gas Res. Inst. (Res. Dev. Results)	Gas Research Institute (Research and Development Results) [US]
Gas Sep. Purif.	Gas Separation and Purification
Gas Turbine World	Gas Turbine World [US]
Gas Turbine World Cogen.	Gas Turbine World and Cogeneration [US]
Gas Warme Int.	Gas und Warme International [Germany]
GATF World	GATF World [of Graphic Arts Technical Foundation, US]
Gaz. Mat.	Gazeta de Matematica [Portugal]
Gazz. Chim. Ital.	Gazzetta Chimica Italiana [of Societa Chimica Italiana, Italy]
GEC Eng.	GEC Engineering [of GEC Research Laboratories, UK]
GEC J. Res.	GEC Journal of Research [of GEC Research Laboratories, UK]
GEC J. Res. Inc. Marconi Rev.	GEC Journal of Research Incorporating the Marconi Review [of GEC Research Laboratories, UK]

GEC J. Sci. Technol.	GEC Journal of Science and Technology [of GEC Research Laboratories, UK]
GEC Rev.	GEC Review [of GEC Research Laboratories, UK]
GEC Telecommun.	GEC Telecommunications [UK]
Gen. Eng.	General Engineer [UK]
Genie Logiciel Syst. Experts	Genie Logiciel and Systemes Experts [France]
Gen. Relativ. Gravit.	General Relativity and Gravitation [US]
Geocarto Int.	Geocarto International [Hong Kong]
Geochem. News	Geochemical News [of Geochemical Society, US]
Geochim. Cosmochim. Acta	Geochimica et Cosmochimica Acta [of Geochemical Society, US]
Geochim. Cosmochim. Acta	Geochimica et Cosmochimica Acta [UK]
Geodyn. Ser.	Geodynamics Series [of American Geophysical Union, US]
Geofiz. Appar.	Geofizicheskaya Apparatura [USSR]
Geofiz. Zh.	Geofizicheskii Zhurnal [USSR]
Geol. Geofiz.	Geologiya i Geofizika [USSR]
Geol. J.	Geological Journal [UK]
Geol. Mijnb.	Geologie en Mijnbouw [Netherlands]
Geol. Surv. Bull.	Geological Survey Bulletin [US]
Geomagn. Aeron.	Geomagnetism and Aeronomy [Translation of: *Geomagnetizm i Aeronomiya (USSR)*; published in the US]
	Geomagnetizim i Aeronomiya [USSR]
Geomagn. Ser. Earth Phys. Branch.	Geomagnetic Series, Earth Physics Branch [of Energy, Mines and Resources, Canada]
Geophys. Astrophys. Fluid Dyn.	Geophysical and Astrophysical Fluid Dynamics [UK]
Geophys. Explor.	Geophysical Exploration [of Society of Exploration Geophysicists of Japan]
Geophys. J.	Geophysical Journal [Translation of: *Geofizicheskii Zhurnal (Ukrainian SSR)*; published in the UK]
Geophys. J. Int.	Geophysical Journal International [UK]
Geophys. Mag.	Geophysical Magazine [of Japan Meteorological Agency]
Geophys. Monogr. Ser.	Geophysical Monograph Series [of American Geophysical Union]
Geophys. Nor.	Geophysica Norvegica [Norway]
Geophys. Prosp.	Geophysical Prospecting [of European Association of Exploration Geophysicists, Netherlands]
Geophys. Res. Lett.	Geophysical Research Letters [of American Geophysical Union]
Geophys. Trans.	Geophysical Transactions [Hungary]
Geotech. News	Geotechnical News [of Canadian Geotechnical Society]
Geotech. Test. J.	Geotechnical Testing Journal [of American Society for Nondestructive Testing]
Gerlands Beitr. Geophys.	Gerlands Beitrage zur Geophysik [Germany]
German Res.	German Research [of Deutsche Forschungsgemeinschaft, Germany]
G. Fis.	Giornale di Fisica [of Societa Italiana di Fisica, Italy]
Gidroliz. Lesokhim. Prom.	Gidroliznaya i Lesokhimicheskaya Promyshlennost [USSR]
Giesserei-Erfahr.	Giesserei-Erfahrungsaustausch [Germany]
Giesserei-Prax.	Giesserei-Praxis [Germany]
Giess.-Tech.	Giesserei-Technik [Germany]
GIS Newsl.	GIS Newsletter [of Geoscience Information Society, US]
Glas. Mat. I	Glasnik Matematicki, Serija I [Yugoslavia]
Glas. Mat. II	Glasnik Matematematicki, Serija II [Yugoslavia]
Glas. Mat. III	Glasnik Matematematicki, Serija III [Yugoslavia]
Glass Ceram.	Glass and Ceramics [Translation of: *Steklo i Keramika (USSR)*; published in the US]
Glas Srp. Akad. Nauka Umet. Od. Tekh. Nauka	Glas Srpske Akademije Nauka i Umetnosti, Odeljenje Tekhnichkikh Nauka [Yugoslavia]
Glass Technol.	Glass Technology [of Society of Glass Technology, UK]
Glastech. Ber.	Glastechnische Berichte [Germany]
Glueckauf Forschungsh.	Glueckauf Forschungshefte [Germany]
GMA	Geomechanics Abstracts [of Royal School of Mines, UK]
God. Vissh. Khimikotekhnol. Inst. Sofiya	Godishnik na Visshiya Khimikotekhnologiya Institut Sofiya [Bulgaria]
God. Vissh. Minno-Geol. Inst. Sofiya	Godishnik na Visshiya Minno-Geolozhki Institut Sofiya [Bulgaria]
Gold Bull.	Gold Bulletin [Switzerland]
Goldschmidt Inf.	Goldschmidt Informiert [Germany]
Gov. Data Syst.	Government Data Systems [US]
Gov. Inf. Q.	Government Information Quarterly [US]
Gov. Publ.	Government Publications [UK]

Gov. Publ. Rev.	Government Publication Review [US]
Graf. Tech.	Grafische Technik [of Institut fur Grafische Technik, Germany]
Grants Newsl.	Grants Newsletter [UK]
Graphic Arts Lit. Abstr.	Graphic Arts Literature Abstracts [US]
Graphs Comb.	Graphs and Combinatorics [Germany]
GRID	Gas Research Institute Digest [US]
Grundl. Landtech.	Grundlagen der Landtechnik [of Verein Deutscher Ingenieure, Germany]
GSA Bull.	GSA Bulletin [of Geological Society of America]
GSDB	Geophysics and Space Data Bulletin [US]
GTE Autom. Electr. World-Wide Commun. J.	GTE Automatic Electric World-Wide Communications Journal [US]
GTE J. Sci. Technol.	GTE Journal of Science and Technology [US]
Gummi Fas. Kunstst.	Gummi, Fasern, Kunststoffe [Germany]
GWA Gewasserschutz Wasser Abwasser	GWA: Gewasserschutz, Wasser und Abwasser [Germany]
Gypsum Lime	Gypsum and Lime

H

H & V Eng.	H & V Engineer [UK]
Habitat Int. J.	Habitat International Journal [of World Environment and Resources Council, Belgium]
Hadronic J.	Hadronic Journal [US]
Hadronic J. Suppl.	Hadronic Journal Supplement [US]
Hakim. Beyisr.	Hakimekai Beyisrael [of Israel Chemical Society]
Hard Soft	Hard and Soft [of Verein Deutscher Ingenieure, Germany]
Hart.-Tech. Mitt.	Harterei-Technische Mitteilung [Germany]
Harvard Bus. Rev.	Harvard Business Review [US]
Hasler Rev.	Hasler Review [Switzerland]
Haynes Alloy. Dig.	Haynes Alloys Digest [US]
Haynes Dig.	Haynes Digest [US]
Hazard. Mater. Control Mag.	Hazardous Materials Control Magazine [of Hazardous Materials Control Research Institute, US]
HBR	Harward Business Review [US]
HCW	Home Computing Weekly [UK]
Health Phys.	Health Physics
Health Safety Work	Health and Safety at Work [UK]
Hear. Res.	Hearing Research [Netherlands]
Heat. Air Cond. J.	Heating and Air Conditioning Journal [UK]
Heat./Combust. Equip. News	Heating/Combustion Equipment News [US]
Heat Process. Dig.	Heat Processing Digest [of ASM International, US]
Heat Recovery Syst. CHP	Heat Recovery Systems and CHP [UK]
Heat Technol.	Heat Technology [US]
Heat Technol.	Heat and Technology [Italy]
Heat Transf. Eng.	Heat Transfer Engineering [US]
Heat Transf. Jpn. Res.	Heat Transfer — Japanese Research [Published in the US]
Heat Transf. Sov. Res.	Heat Transfer — Soviet Research [Published in the US]
Heat Treat.	Heat Treating [US]
Heat Treat. Met.	Heat Treatment of Metals [of Wolfson Heat Treatment Center, UK]
Heat Treat. Met. (China)	Heat Treatment of Metals (China) [PR China]
Helv. Chim. Acta	Helvetica Chimica Acta [Switzerland]
Helv. Phys. Acta	Helvetica Physica Acta [of Schweizerische Physikalische Gesellschaft, Switzerland]
Hem. Ind.	Hemijska Industriji [Yugoslavia]
Her. Libr. Sci.	Herald of Library Science [India]
Hermsdorfer Tech. Mitt.	Hermsdorfer Technische Mitteilungen [Germany]
Hewlett-Packard J.	Hewlett-Packard Journal [US]
High Energy Phys. Nucl. Phys.	High Energy Physics and Nuclear Physics [UK]

High-Perform. Ceram.	High-Performance Ceramics [Spain]
High Perform. Plast.	High Performance Plastics [UK]
High Perform. Polym.	High Performance Polymers [UK]
High Perform. Syst.	High Performance Systems [US]
High Perform. Text.	High Performance Textiles [UK]
High Purity Subst.	High Purity Substances [US]
High-Speed Surf. Craft	High-Speed Surface Craft [UK]
High Technol.	High Technology [US]
High Technol. Bus.	High Technology Business [US]
High Temp.	High Temperature [Translation of: *Teplofizik Vysokikh Temperatur (USSR)*; published in the US]
High Temp. — High Press.	High Temperatures — High Pressures [UK]
High Temp. Mater. Process.	High Temperature Materials and Processes [UK]
High Temp. Sci.	High Temperature Science [US]
High Temp. Technol.	High Temperature Technology [UK]
Hist. Metall.	Historical Metallurgy [of Institute of Metals, UK]
Hitachi Rev.	Hitachi Review [Japan]
Hitachi Zosen Tech. Rev.	Hitachi Zosen Technical Review [Japan]
HLH	Heizung Luftung/Klima Haustechnik [of Verein Deutscher Ingenieure, Germany]
HMAT	Hot Mix Asphalt Technology [of National Asphalt Pavement Association, US]
Hoechst High Chem Mag.	Hoechst High Chem Magazine [UK]
Hollow Sect.	Hollow Section [UK]
Holzforsch. Holzverwert.	Holzforschung und Holzverwertung [Austria]
Holz Roh-Werkstoff	Holz als Roh- und Werkstoff [Germany and US]
Holztechnol.	Holztechnologie [of Institut fur Holztechnologie und Faserbaustoffe, Germany]
Hommes Fonderie	Hommes et Fonderie [France]
Hong Kong Eng.	Hong Kong Engineer [Hong Kong]
Hoogovens Groep Bull.	Hoogovens Groep Bulletin [Netherlands]
Horol. J.	Horological Journal [of British Horological Institute and the British Clock and Watch Association]
Hot Work. Technol.	Hot Working Technology [PR China]
Houst. J. Math.	Houston Journal of Mathematics [US]
HPV Newsl.	HPV Newsletter [of International Human Powered Vehicle Association, US]
HRIS Abstr.	Highway Research Information Service Abstracts [of Transportation Research Board, US]
Hule Mex. Plast.	Hule Mexicano y Plasticos [Mexico]
Hum.-Comput. Interact.	Human-Computer Interactions [US]
Hum. Factors	Human Factors [of Human Factors Society, US]
Hum. Syst. Manage.	Human Systems Management [Netherlands]
Hung. J. Ind. Chem.	Hungarian Journal of Industrial Chemistry
Hutn. Listy	Hutnicke Listy [Czechoslovakia]
Hybrid Circuit Technol.	Hybrid Circuit Technology [US]
Hydraul. Pneum.	Hydraulics and Pneumatics [US]
Hydrocarbon Process.	Hydrocarbon Processing [US]
Hydrogen Energy Prog.	Hydrogen Energy Progress [of International Association for Hydrogen Energy, US]
Hydrogr. J.	Hydrographic Journal [of Hydrographic Society, UK]
Hydrol. Process.	Hydrological Processes [UK]
Hydrol. Sci. J.	Hydrological Sciences Journal [of International Association of Hydrological Sciences, Netherlands]
Hydrol. Sci. Technol.	Hydrological Sciences and Technology [of American Institute of Technology]
Hydrotech. Constr. (USSR)	Hydrotechnical Construction (USSR) [Published in the US]
HYL Rep.	HYL Report [France]
Hyperfine Interact.	Hyperfine Interactions [Switzerland]

I

IAA	International Aerospace Abstacts [of American Institute of Aeronautics and Astronautics, US]
IAA	International Aerospace Abstacts [of NASA]
IAEA Bull.	IAEA Bulletin [of International Atomic Energy Agency, Austria]

IAMG Newsl.	IAMG Newsletter [of International Association for Mathematical Geology, US]
IAMP News Bull.	IAMP News Bulletin [of International Association of Mathematical Physics, US]
IATUL Q.	IATUL Quarterly [UK]
IAWA Bull.	IAWA Bulletin [of International Association of Wood Anatomists, Netherlands]
IBA Q. Rev.	IBA Quarterly Review [of International Bauxite Association, Jamaica]
IBA Rep.	IBA Reports [of Industrial Biotechnology Association, US]
IBA Rev.	IBA Review [of International Bauxite Association, Jamaica]
IBA Tech. Rev.	IBA Technical Review [of Independent Broadcasting Authority, UK]
IBM J. R&D	IBM Journal of Research and Development [US]
IBM J. Res. Dev.	IBM Journal of Research and Development [US]
IBM Nachr.	IBM Nachrichten [Germany]
IBM Syst. J.	IBM Systems Journal [US]
IBS Asian Electron. News	IBS Asian Electronics News [Taiwan]
IBS Rev.	IBS Revista [Brazil]
ICA Inf.	ICA Information [of International Council for Automatic Data Processing in Government Administration, Spain]
ICES J.	ICES Journal [of Integrated Civil Engineering System Users Group, US]
ICF Int. Consult. Newsl.	ICF International Consultants Newsletter [of International Consultants Foundation, Sweden]
ICI Eng. Plast.	ICI Engineering Plastics [UK]
ICI Mag.	ICI Magazine [UK]
ICI Polyurethane Newsl.	ICI Polyurethanes Newsletter [Belgium]
ICL Tech. J.	ICL Technical Journal [UK]
ICRM Newsl.	ICRM Newsletter [of Institute of Certified Records Managers, US]
ICRU Rep.	ICRU Reports [of International Commission on Radiation Units and Measurements, US]
IDPM Inf. Manage.	IDPM Information Management [UK]
I&EC Res.	Industrial and Engineering Chemistry Research [of American Chemical Society]
IEC Bull.	IEC Bulletin [of International Electrotechnical Commission, Switzerland]
IEC Process Des. Dev.	IEC (Industrial Engineering Design), Process Design and Development [of American Chemical Society]
IEC Prod. Res. Dev.	IEC (Industrial Engineering Design), Product Research and Development [of American Chemical Society]
IEEE Aerosp. Electron. Syst. Mag.	IEEE Aerospace and Electronic Systems Magazine [of IEEE Aerospace and Electronics Systems Society, US]
IEEE ASSP Mag.	IEEE ASSP Magazine [of IEEE Acoustics, Speech and Signal Processing Society, US]
IEEE Circuits Devices Mag.	IEEE Circuits and Devices Magazine [of Institute of Electrical and Electronics Engineers, US]
IEEE Commun. Mag.	IEEE Communications Magazine [of IEEE Communications Society, US]
IEEE Comput. Appl. Power	IEEE Computer Applications in Power [of Institute of Electrical and Electronics Engineers, US]
IEEE Comput. Graph. Appl.	IEEE Computer Graphics and Applications [of IEEE Computer Society, US]
IEEE Control Syst. Mag.	IEEE Control Systems Magazine [of Institute of Electrical and Electronics Engineers, US]
IEEE Des. Test Comput.	IEEE Design and Test of Computers [of IEEE Computer Society, US]
IEEE Electr. Insul. Mag.	IEEE Electrical Insulation Magazine [of IEEE Dielectrics and Electrical Insulation Society, US]
IEEE Electron Dev. Lett.	IEEE Electron Devices Letters [of IEEE Electron Devices Society, US]
IEEE ElectroTechnol. Rev.	IEEE ElectroTechnology Review [of Institute of Electrical and Electronics Engineers, US]
IEEE Eng. Med. Biol. Mag.	IEEE Engineering in Medicine and Biology Magazine [of Institute of Electrical and Electronics Engineers, US]
IEEE IAS Conf. Rec.	IEEE IAS Conference Record [of IEEE Industry Applications Society, US]
IEEE J. Ocean. Eng.	IEEE Journal of Oceanic Engineering [of IEEE Oceanic Engineering Society, US]
IEEE J. Quantum Electron.	IEEE Journal of Quantum Electronics [of IEEE Lasers and and Electrooptics Society, US]
IEEE J. Robot. Autom.	IEEE Journal of Robotics and Automation [of IEEE Robotics and Automation Council, US]
IEEE J. Sel. Areas Commun.	IEEE Journal of Selected Areas in Communications [of Institute of Electrical and Electronics Engineers, US]
IEEE J. Solid-State Circuits	IEEE Journal of Solid-State Circuits [of Institute of Electrical and Electronics Engineers, US]
IEEE Manage. Rev.	IEEE Management Review [of IEEE Engineering Management Society, US]
IEEE Netw.	IEEE Network [of Institute of Electrical and Electronics Engineers, US]
IEEE Photon. Technol. Lett.	IEEE Photonics Technology Letters [of IEEE Lasers and Electrooptics Society, US]
IEEE Power Eng. Rev.	IEEE Power Engineering Review [of IEEE Power Engineering Society, US]
IEEE Reliab. Soc. Newsl.	IEEE Reliability Society Newsletter [of Institute of Electrical and Electronics Engineers, US]
IEEE SEMI Int. Semicond. Manuf. Symp. Proc.	IEEE SEMI International Semiconductor Manufacturing Symposium Proceedings [of Institute of Electrical and Electronics Engineers and Semiconductor Equipment and Materials Institute, US]
IEEE Softw.	IEEE Software [of Institute of Electrical and Electronics Engineers, US]

IEEE Spectr.	IEEE Spectrum [of Institute of Electrical and Electronics Engineers, US]
IEEE Technol. Soc. Mag.	IEEE Technology and Society Magazine [of Institute of Electrical and Electronics Engineers, US]
IEEE Trans. Acoust. Speech Signal Process.	IEEE Transactions on Acoustics, Speech and Signal Processing [of Institute of Electrical and Electronics Engineers, US]
IEEE Trans. Aerosp. Electron. Syst.	IEEE Transactions on Aerospace and Electronic Systems [of Institute of Electrical and Electronics Engineers, US]
IEEE Trans. Antennas Propag.	IEEE Transactions on Antennas and Propagation [of Institute of Electrical and Electronics Engineers, UK]
IEEE Trans. Autom. Control	IEEE Transactions on Automatic Control [of Institute of Electrical and Electronics Engineers, US]
IEEE Trans. Biomed. Eng.	IEEE Transactions on Biomedical Engineering [of Institute of Electrical and Electronics Engineers, US]
IEEE Trans. Broadcast.	IEEE Transactions on Broadcasting [of Institute of Electrical and Electronics Engineers, US]
IEEE Trans. Circuits Syst.	IEEE Transactions on Circuits and Systems [of Institute of Electrical and Electronics Engineers, US]
IEEE Trans. Commun.	IEEE Transactions on Communications [of IEEE Communications Society, US]
IEEE Trans Compon. Hybrids Manuf. Technol.	IEEE Transactions on Components, Hybrids and Manufacturing Technology [of Institute of Electrical and Electronics Society, US]
IEEE Trans. Comput.	IEEE Transactions on Computers [of Institute of Electrical and Electronics Engineers, US]
IEEE Trans. Comput.-Aided Des. Instr. Circuits Syst.	IEEE Transactions on Computer-Aided Design of Integrated Circuits and Systems [of Institute of Electrical and Electronics Engineers, US]
IEEE Trans. Consum. Electron.	IEEE Transactions on Consumer Electronics [of IEEE Consumer Electronics Society, US]
IEEE Trans. Educ.	IEEE Transactions on Education [of Institute of Electrical and Electronics Engineers, US]
IEEE Trans. Electr. Insul.	IEEE Transactions on Electrical Insulation [of IEEE Dielectrics and Electrical Insulation Society, US]
IEEE Trans. Electromagn. Compat.	IEEE Transactions on Electromagnetic Compatibility [of IEEE Electromagnetic Compatibility Society, US]
IEEE Trans. Electron Devices	IEEE Transactions on Electron Devices [of IEEE Electron Devices Society, US]
IEEE Trans. Energy Convers.	IEEE Transactions on Energy Conversion [of IEEE Power Engineering Society, US]
IEEE Trans. Eng. Manage.	IEEE Transactions on Engineering Management [of IEEE Engineering Management Society, US]
IEEE Trans. Geosci. Remote Sens.	IEEE Transactions on Geoscience and Remote Sensing [of IEEE Geoscience and Remote Sensing Society, US]
IEEE Trans. Ind. Appl.	IEEE Transactions on Industry Applications [of IEEE Industry Applications Society, US]
IEEE Trans. Ind. Electron.	IEEE Transactions on Industrial Electronics [of IEEE Industrial Electronics Society, US]
IEEE Trans. Inf. Theory	IEEE Transactions on Information Theory [of IEEE Information Theory Group, US]
IEEE Trans. Instrum. Meas.	IEEE Transactions on Instrumentation and Measurement [of IEEE Instrumentation and Measurement Society, US]
IEEE Trans. Knowl. Data Eng.	IEEE Transactions on Knowledge and Data Engineering [of Institute of Electrical and Electronics Engineers, US]
IEEE Transl. J. Magn. Jpn.	IEEE Translation Journal on Magnetics in Japan [of Institute of Electrical and Electronics Engineers, US]
IEEE Trans. Magn.	IEEE Transactions on Magnetics [of IEEE Magnetics Society, US]
IEEE Trans. Med. Imaging	IEEE Transactions on Medical Imaging [of Institute of Electrical and Electronic Engineers, US]
IEEE Trans. Microw. Theory Tech.	IEEE Transactions on Microwave Theory and Techniques [of Institute of Electrical and Electronics Engineers, US]
IEEE Trans. Nucl. Sci.	IEEE Transactions on Nuclear Science [of IEEE Nuclear and Plasma Sciences Society, US]
IEEE Trans. Pattern Anal. Mach. Intell.	IEEE Transactions on Pattern Analysis and Machine Intelligence [of Institute of Electrical and Electronics Engineers, US]
IEEE Trans. Plasma Sci.	IEEE Transactions on Plasma Science [of IEEE Nuclear and Plasma Sciences Society, US]
IEEE Trans. Power Appar. Syst.	IEEE Transactions on Power Apparatus and Systems [of Institute of Electrical and Electronics Engineers, US]
IEEE Trans. Power Deliv.	IEEE Transactions on Power Delivery [of IEEE Power Engineering Society, US]
IEEE Trans. Power Electron.	IEEE Transactions on Power Electronics [of Institite of Electrical and Electronics Engineers, US]
IEEE Trans. Power Syst.	IEEE Transactions on Power Systems [of Institute of Electrical and Electronics Engineers, US]
IEEE Trans. Prof. Commun.	IEEE Transactions on Professional Communications [of IEEE Professional Communication Society, US]
IEEE Trans. Reliab.	IEEE Transactions on Reliability [of IEEE Reliability Society, US]

IEEE Trans. Robot. Autom.	IEEE Transactions on Robotics and Automation [of Institute of Electrical and Electronics Engineers, US]
IEEE Trans. Semicond. Manuf.	IEEE Transactions on Semiconductor Manufacturing [of IEEE Electron Devices Society, US]
IEEE Trans. Softw. Eng.	IEEE Transactions on Software Engineering [of Institute of Electrical and Electronics Engineers, US]
IEEE Trans. Sonics Ultrason.	IEEE Transactions on Sonics and Ultrasonics [of Institute of Electrical and Electronic Engineers, US]
IEEE Trans. Syst. Man Cybern.	IEEE Transactions on Systems, Man and Cybernetics [of IEEE Systems, Man and Cybernetics Society, US]
IEEE Trans. Ultrason. Ferroelectr. Freq. Control	IEEE Transactions on Ultrasonics, Ferroelectrics and Frequency Control [of Institute of Electrical and Electronics Engineers, US]
IEEE Trans. Veh. Technol.	IEEE Transactions on Vehicular Technology [of IEEE Vehicular Technology Society, US]
IEEMA J.	IEEMA Journal [of Indian Electrical and Electronics Manufacturers Association, India]
IEE Proc. A, Phys. Sci. Meas. Instrum. Manage. Educ.	IEE Proceedings A, Physical Science, Measurement and Instrumentation, Management and Education [of Institution of Electrical Engineers, UK]
IEE Proc. A, Phys. Sci. Meas. Instrum. Manage. Educ., Rev.	IEE Proceedings A, Physical Science, Measurement and Instrumentation, Management and Education, Reviews [of Institution of Electrical Engineers, UK]
IEE Proc. B, Electr. Power Appl.	IEE Proceedings B, Electric Power Applications [of Institution of Electrical Engineers, UK]
IEE Proc. C, Gener. Transm. Distrib.	IEE Proceedings C, Generation, Transmission and Distribution [of Institution of Electrical Engineers, UK]
IEE Proc. D, Control Theory Appl.	IEE Proceedings, D, Control Theory and Applications [of Institution of Electrical Engineers, UK]
IEE Proc. E, Comput. Digit. Tech.	IEE Proceedings E, Computers and Digital Techniques [of Institution of Electrical Engineers, UK]
IEE Proc. F, Commun. Radar Signal Process.	IEE Proceedings F, Communications, Radar and Signal Processing [of Institution of Electrical Engineers, UK]
IEE Proc. F, Radar Signal Process.	IEE Proceedings F, Radar and Signal Processing [of Institution of Electrical Engineers, UK]
IEE Proc. G, Circuits Devices Syst.	IEE Proceedings G, Circuits, Devices and Systems [of Institution of Electrical Engineers, UK]
IEE Proc. G, Electron. Circuits Syst.	IEE Proceedings G, Electronic Circuits and Systems [of Institution of Electrical Engineers, UK]
IEE Proc. H, Microw. Antennas Propag.	IEE Proceedings H, Microwaves, Antennas and Propagation [of Institution of Electrical Engineers, UK]
IEE Proc. I, Commun. Speech Vis.	IEE Proceedings I, Communications, Speech and Vision [of Institution of Electrical Engineers, UK]
IEE Proc. I, Solid-State Electron Devices	IEE Proceedings I, Solid-State and Electron Devices [of Institution of Electrical Engineers, UK]
IEE Proc. J, Optoelectron.	IEE Proceedings J, Optoelectronics [of Institution of Electrical Engineers, UK]
IEE Rev.	IEE Review [of Institution of Electrical Engineers, UK]
IEPS J.	IEPS Journal [of International Electronics Packaging Society, US]
IES J.	IES Journal [of Institution of Engineers, Singapore]
IES Monogr. Ser.	IES Monograph Series [of Institution of Environmental Engineers, UK]
IES Proc.	IES Proceedings [of Institution of Environmental Engineers, UK]
IETE Tech. Rev.	IETE Tech. Rev. [of Institution of Electronics and Telecommunication Engineers, India]
IFCC Newsl.	IFCC Newsletter [of International Federation of Clinical Chemistry, Finland]
IfL-Mitt.	IfL-Mitteilungen [Germany]
IFRA Spec. Rep.	IFRA Special Report [Germany]
IGRC Proc.	International Gas Research Conference Proceedings [Gas Research Institute, US]
IGT Highlights	International Gas Technology Highlights [of Institute of Gas Technology, US]
IHI Bull.	IHI Bulletin [Japan]
IHI Eng. Rev.	IHI Engineering Review
IIE Trans.	IIE Transactionsa [of Institute of Industrial Engineers, US]
IISS Comment.	IISS Commentary [US]
Image Process.	Image Processing [UK]
Image Technol.	Image Technology [of British Kinematograph, Sound and Television Society]
Image Vis. Comput.	Image and Vision Computing [UK]
IMA J. Appl. Math.	IMA Journal of Applied Mathematics [UK]
IMA J. Math. Control Inf.	IMA Journal of Mathematical Control and Information [UK]
IMA J. Numer. Anal.	IMA Journal of Numerical Analysis [UK]

IMC J.	IMC Journal [of International Information Management Congress, US]
IMC Newsl.	IMC Newsletter [of International Information Management Congress, US]
IMEKO Bull.	IMEKO Bulletin [of Internationale Messtechnische Konfoderation, Hungary]
IMM Bull.	IMM Bulletin [of Institution of Mining and Metallurgy, UK]
Impr. Nouv.	Imprimerie Nouvelle [France]
IM Rep.	IM Report [of Association for Integrated Manufacturing Technology, US]
IMS Bull.	Institute of Mathematical Statistics Bulletin [US]
IMU Bull.	IMU Bulletin [of International Mathematical Union, Finland]
Inco Alloys Int.	Inco Alloys International [UK]
Increm. Motion Control Syst. Devices Newsl.	Incremental Motion Control Systems and Devices Newsletter [US]
Ind.-Anz.	Industrie-Anzeiger [Germany]
Ind. Carta	Industria della Carta [Italy]
Ind. Ceram.	Industrie Ceramique [France]
Ind. Ceram.	Industrial Ceramics [Italy]
Ind. Ceram. (Paris)	Industrie Ceramique (Paris) [France]
Ind. Commer. Train.	Industrial and Commercial Training [UK]
Ind. Comput.	Industrial Computing [UK]
Ind. Corros.	Industrial Corrosion [UK]
Ind. Datatek.	Industriell Datateknik [Sweden]
Ind. Diamond Rev.	Industrial Diamond Review [UK]
Ind. Electr. Inf.	Industrie Electricite Informations [France]
Ind. Eng.	Industrial Engineering [of American Institute of Industrial Engineers]
Ind. Eng. Chem., Fundam.	Industrial and Engineering Chemistry, Fundamentals [US]
Ind. Eng. Chem., Process Des. Dev.	Industrial and Engineering Chemistry, Process Design and Development [US]
Ind. Eng. Chem., Prod. Res. Dev.	Industrial and Engineering Chemistry, Product Research and Development [US]
Ind. Eng. Chem. Res.	Industrial and Engineering Chemistry Research [of American Chemical Society]
Ind. Environ.	Industry and Environment [of United Nations Environment Programme]
Indep. Power	Independent Power [US]
Ind. Exp. Tech.	Industrial and Experimental Techniques
Ind. Finish.	Industrial Finishing [US]
Ind. Gomma	Industria della Gomma [Italy]
Ind. Health	Industrial Health
Ind. Heat.	Industrial Heating [US]
Indian Ceram.	Indian Ceramics
Indian Chem. Eng.	Indian Chemical Engineer
Indian East Eng.	Indian and Eastern Engineer
Indian Foundry J.	Indian Foundry Journal
Indian J. Chem.	Indian Journal of Chemistry
Indian J. Chem. A	Indian Journal of Chemistry, Section A
Indian J. Chem. B	Indian Journal of Chemistry, Section B
Indian J. Cryog.	Indian Journal of Cryogenics [of Indian Cryogenics Council]
Indian J. Mar. Sci.	Indian Journal of Marine Sciences [of Council of Scientific and Industrial Research, India]
Indian J. Phys.	Indian Journal of Physics [of Indian Association for the Cultivation of Science]
Indian J. Phys. A	Indian Journal of Physics A [of Indian Association for the Cultivation of Science]
Indian J. Phys. B	Indian Journal of Physics B [of Indian Association for the Cultivation of Science]
Indian J. Power River Val. Dev.	Indian Journal of Power and River Valley Development [India]
Indian J. Pure Appl. Math.	Indian Journal of Pure and Applied Mathematics [of Indian National Science Academy]
Indian J. Pure Appl. Phys.	Indian Journal of Pure and Applied Physics
Indian J. Radio Space Phys.	Indian Journal of Radio and Space Physics
Indian J. Technol.	Indian Journal of Technology
Indian J. Textile Res.	Indian Journal of Textile Research [of Council of Scientific and Industrial Research, India]
Indian J. Theor. Phys.	Indian Journal of Theoretical Physics
Indian Min. Eng. J.	Indian Mining and Engineering Journal
Indian Sci. Abstr.	Indian Science Abstracts [of Indian National Scientific Documentation Center, India]
Indian Weld. J.	Indian Welding Journal
Ind. Int.	Industria Internacional [US]
Ind. Lab.	Industrial Laboratory [Translation of: *Zavodskaya Laboratoriya (USSR)*; published in the US]
Ind. Lackier-Betr.	Industrie Lackier-Betrieb [Germany]
Ind. Lemnului	Industria Lemnului [Romania]
Ind. Lubr. Tribol.	Industrial Lubrication and Tribology [UK]

Ind. Manage.	Industrial Management [of Institute of Industrial Engineers, US]
Ind. Manage. + Data Syst.	Industrial Management + Data Systems [UK]
Ind. Math.	Industrial Mathematics [of Industrial Mathematics Society, US]
Ind. Market. Dig.	Industrial Marketing Digest [UK]
Ind. Min. (Brussels)	Industrie Minerale (Brussels) [Belgium]
Ind. Miner. (London)	Industrial Minerals (London) [UK]
Ind. Min., Mines Carr.	Industrie Minerale, Mines et Carrieres [France]
Ind. Min., Mines Carr., Tech.	Industrie Minerale, Mines et Carrieres, Les Techniques [France]
Ind. Min. (Paris)	Industrie Minerale (Paris) [France]
Ind. Min. (St. Etienne)	Industrie Minerale (St. Etienne) [France]
Ind. Min. Suppl., Tech. (St. Etienne)	Industrie Minerale Supplement, Les Techniques (St. Etienne) [France]
Ind. Prod.	Industrial Production [of Statistical Office of the European Communities, Luxembourg]
Ind. Recovery	Industrial Recovery [UK]
Ind. Res. Dev.	Industrial Research and Development [US]
Ind. Robot	Industrial Robot [UK]
Ind. Sci. Instrum.	Industrial and Scientific Instruments [UK]
Ind. Sci. Technol.	Industrial Science and Technology [Japan]
Ind. Soc.	Industrial Society [of Industrial Society, UK]
Ind. Tech.	Industries et Techniques [France]
Ind. Trends	Industrial Trends [of Statistical Office of the European Communities, Luxembourg]
Ind. Week	Industry Week [UK]
Inf. Age	Information Age
Infalum Bull.	Infalum Bulletin [Switzerland]
Inf. Bull. Var. Stars	Information Bulletin on Variable Stars [Hungary]
Inf. Chim.	Informations Chimie [of Societe d'Expansion Technique et Economique, France]
Inf. Comput.	Information and Computation [US]
Inf. Decis. Technol.	Information and Decision Technologies [Netherlands]
Inf. Dev.	Information Development [UK]
Inf. Disp.	Information Display [US]
Inf. Econ. Policy	Information Economics and Policy [Netherlands]
Inf. Elektron.	Informacio Elektronika [Hungary]
Inf. Elettron.	Informazione Elettronica [Italy]
Inf. Exec.	Information Executive [of Data Processing Management Association, US]
Inf. INT	Informativo do INT [Brazil]
Inf. Intell. Online Newsl.	Information Intelligence Online Newsletter [US]
Inf. IREQ	Information IREQ [of Institut de Recherche en Electricite du Quebec, Canada]
Inf. Manage.	Information and Management [Netherlands]
Inf. Media Technol.	Information Media and Technology [UK]
Inf. MIDEM	Informatije MIDEM [Yugoslavia]
Inf. Mkt.	Information Market [Commission of the European Communities, Luxembourg]
Inform.	Informatique [of Association Francaise pour le Cybernetique,, Economique et Technique, France]
Informatologia Yugosl.	Informatologia Yugoslavica [Yugoslavia]
Inform. Diritto	Informatica e Diritto [Italy]
Inform. Forsch. Entwickl.	Informatik Forschung und Entwicklung [Germany]
Inform. Spectrum	Informatik Spektrum [Germany]
Inform. Theor. Appl.	Informatique Theorique et Applications [France]
Inf. Power	Information Power [UK]
Inf. Process.	Information Processing [of Information Processing Society of Japan]
Inf. Process. Lett.	Information Processing Letters [Netherlands]
Inf. Process. Manage.	Information Processing and Management [UK]
Inf. Process. Soc. Jpn.	Information Processing Society of Japan [Japan]
Infrared Phys.	Infrared Physics [UK]
Inf. Recht	Informatik und Recht [Germany]
Inf. Referral, J. Alliance Inf. Referral Syst.	Information and Referral: Journal of the Alliance of Information Referral Systems [of Alliance of Information Referral Systems, US]
Inf. Resour. Manage. J.	Information Resources Management International Journal [US]
Inf. Sci.	Information Sciences [US]
Inf. Serv. Use	Information Service and Use [Netherlands]
Inf. Soc.	Information Society [US]

Inf. Softw. Technol.	Information and Software Technology [UK]
Inf. Stand. Q.	Information Standards Quarterly [of National Information Standards Organization, US]
Inf. Strategy, Exec. J.	Information Strategy: The Executive's Journal [US]
Inf. Syst.	Informacne Systemy [Czechoslovakia]
Inf. Syst.	Information Systems [UK]
Inf. Syst. Res.	Information Systems Research [of The Institute of Management Sciences, US]
Inf. TAI	Information TAI [Canada]
Inf. Tech. — IT	Informationstechnik — IT [Germany]
Inf. Technol. Dev.	Information Technology for Development [UK]
Inf. Technol. Learn.	Information Technology and Learning [UK]
Inf. Technol. Libr.	Information Technology and Libraries [of Library and Information Technology Association, US]
Inf. Technol. Public Policy	Information Technology and Public Policy [UK]
Inf. Times	Information Times [of Information Industry Association, US]
Inf. Today	Information Today [US]
Ing.-Arch.	Ingenieur-Archiv [Germany]
Ing. Automob.	Ingenieurs de l'Automobile [France]
Ing. Electr. Mec.	Ingenieria Electrica y Mecanica [of Asociacion Venezolana de Ingenieros Electricos y Mecanicos, Venezuela]
Ing. Estructural	Ingenieria Estructural [Cuba]
Ing. Ind.	Ingenieria Industrial [Cuba]
Ing. Mec. Electr.	Inginieria Mecanica y Electrica [of Asociacion Mexicana de Ingenieros Mecanicos y Electricistas, Mexico]
Ing. Quim.	Ingenieria Quimica
Ing. Sanit.	Ingenieria Sanitaria [of Inter-American Association of Sanitary Engineering, US]
Ing. Werkst.	Ingenieur Werkstoffe [Germany]
Ink Print	Ink and Print [UK]
Innov. Manage.	Innovation and Management [of Institute of Management Consultants, UK]
INNSE Not./News	INNSE Notizie/News [Italy]
Inorg. Chem.	Inorganic Chemistry [of American Chemical Society]
Inorg. Chim. Acta	Inorganica Chimica Acta [US]
Inorg. Mater.	Inorganic Materials [Translation of: *Izvestiya Akademii Nauk SSSR, Neorganicheskie Materialy (USSR)*; published in the US]
INSPEL, Int. J. Spec. Libr.	INSPEL, International Journal of Special Libraries [Germany]
Insp., Test., Anal., InSpeC	Inspection, Testing, Analysis, InSpeC [Japan]
Inst. Chem. Eng. Symp. Ser.	Institution of Chemical Engineers Symposium Series
Inst. Eng. Aust. Civ. Eng. Trans.	Institution of Engineers, Australia, Civil Engineering Transactions [Australia]
Inst. Forum	Institute Forum [UK]
Inst. R. Meteorol. Belg. Bull. Trimest. Obs. Ozone	Institut Royal Meteorologique de Belgique Bulletin Trimestriel Observations d'Ozone [Belgium]
Instrum. Control Eng.	Instrumentation and Control Engineering [Japan]
Instrum. Exp. Tech.	Instruments and Experimental Techniques [Translation of: *Pribory i Tekhnika Eksperimenta (USSR)*; published in the US]
Instrum. India	Instruments India [India]
Instrum. Technol.	Instrumentation Technology [US]
Insulation J.	Insulation Journal [UK]
Inta/Cienc. Tec. Aerosp.	Inta/Ciencia y Tecnica Aerospacial [Spain]
Inta/Conie Inf. Aerosp.	Inta/Conie Informacion Aerospacial [Spain]
Int. Arch. Occup. Environ. Health	International Archives of Occupational and Environmental Health
Int. Astron. Union Circ.	International Astronomical Union Circular [of International Astronomical Union, US]
Int. Biodeterior.	International Biodeterioration [UK]
Int. Broadcast.	International Broadcasting [UK]
Int. Broadcast. Eng.	International Broadcast Engineer [UK]
Int. Bus. Equip.	International Business Equipment [Belgium]
Int. Cast Met. J.	International Cast Metals Journal [US]
Int. Cell Biol.	International Cell Biology [of International Federation of Cell Biology, Canada]
Int. Chem. Eng.	International Chemical Engineering [of American Institute of Chemical Engineers]
Int. Classif.	International Classification [Germany]
Int. Coal Rev.	International Coal Review [of National Coal Association, US]
Int. Comet Q.	International Comet Quarterly [of Appalachian State University, US]
Int. Commun. Heat Mass Transf.	International Communications in Heat and Mass Transfer [UK and US]
Int. Comput. Law Advis.	International Computer Law Adviser [US]

Int. Constr.	International Construction [UK]
Int. Copper Inf. Bull.	International Copper Information Bulletin [UK]
Int. Cryog. Mater. Conf. Ser.	International Cryogenic Materials Conference Series [of International Cryogenic Materials Conference, US]
Int. Def. Rev.	International Defense Review [Switzerland]
INTECOL Bull.	INTECOL Bulletin [of International Association for Ecology, US]
INTECOL Newsl.	INTECOL Newsletter [of International Association for Ecology, US]
Integr. Manuf. Rep.	Integrated Manufacturing Report [of Association for Integrated Manufacturing Technology, US]
Integr. VLSI J.	Integration, The VLSI Journal [Netherlands]
Intell. Instrum. Comput.	Intelligent Instruments and Computers [US]
Interact. Learn. Int.	Interactive Learning International [UK]
Interavia Aerosp. Rev.	Interavia Aerospace Review [UK]
Interavia (Engl. Ed.)	Interavia (English Edition) [Switzerland]
Inter Connex.	Inter Connexion [of Groupement Syndicat des Industries de Materials d'Equipement Electrique, France]
Interdiscip. Sci. Rev.	Interdisciplinary Science Reviews [UK]
Interface, Comput. Educ. Q.	Interface: The Computer Education Quarterly [US]
Intern. Audit.	Internal Auditor [of Institute of Internal Auditors, US]
Intervirol.	Intervirology [of International Union of Micrbiological Societies, UK]
Int. Forum Inf. Doc.	International Forum on Information and Documentation [USSR]
Int. Hydrogr. Bull.	International Hydrographic Bulletin [of Bureau Hydrographique International, Monaco]
Int. Hydrogr. Rev.	International Hydrographic Review [of Bureau Hydrographique International, Monaco]
Int. Inf. Commun Educ.	International Information, Communication and Education [India]
Int. Iron Steel Inst. Bull.	International Iron and Steel Institute Bulletin [Belgium]
Int. J. Adapt. Control Signal Process.	International Journal of Adaptive Control and Signal Process [UK]
Int. J. Adhes. Adhes.	International Journal of Adhesion and Adhesives [UK]
Int. J. Adv. Manuf. Technol.	International Journal of Advanced Manufacturing Technology [UK]
Int. J. Ambient Energy	International Journal of Ambient Energy [UK]
Int. J. Appl. Electromag. Mater.	International Journal of Applied Electromagnetics in Materials [Netherlands]
Int. J. Appl. Eng. Educ.	International Journal of Applied Engineering Education [UK]
Int. J. Approx. Reason.	International Journal of Approximate Reasoning [of North American Fuzzy Information Processing Society, US]
Int. J. Biochem.	International Journal of Biochemistry
Int. J. Biol. Macromols.	International Journal of Biological Macromolecules [UK]
Int. J. Bio-Med. Comput.	International Journal of Bio-Medical Computing [Netherlands]
Int. J. Biomet.	International Journal of Biometeorology [of International Society of Biometeorology, Switzerland]
Int. J. Bulk Solids	International Journal of Bulk Solids
Int. J. Bulk Solids Storage Silos	International Journal of Bulk Solids, Storage in Silos
Int. J. Cem. Compos. Lightweight Concr.	International Journal of Cement Composites and Lightweight Concrete [UK]
Int. J. Circuit Theory Appl.	International Journal of Circuit Theory and Applications [UK]
Int. J. Climatol.	International Journal of Climatology [UK]
Int. J. Clin. Monit. Comput.	International Journal of Clinical Monitoring and Computing[Netherlands]
Int. J. Comput. Adult Educ. Train.	International Journal of Computers in Adult Education and Training [UK]
Int. J. Comput. Appl. Technol.	International Journal of Computer Applications in Technology [Switzerland]
Int. J. Comput. Integr. Manuf.	International Journal of Computer Integrated Manufacturing [UK]
Int. J. Comput. Math.	International Journal of Computer Mathematics [UK]
Int. J. Comput. Vis.	International Journal of Computer Vision [Netherlands]
Int. J. Control	International Journal of Control [UK]
Int. J. Digit. Analog Cabled Syst.	International Journal of Digital and Analog Cabled Systems [UK]
Int. J. Earthquake Eng Struct. Dyn.	International Journal of Earthquake and Structural Dynamics [Japan]
Int. J. Electr. Eng. Educ.	International Journal of Electrical Engineering Education [UK]
Int. J. Electron.	International Journal of Electronics [UK]
Int. J. Electr. Power Energy Syst.	International Journal of Electrical Power and Energy Systems [UK]
Int. J. Energy Res.	International Journal of Energy Research [UK]
Int. J. Energy Syst.	International Journal of Energy Systems [US]
Int. J. Eng. Fluid Mech.	International Journal of Engineering Fluid Mechanics [US]

Int. J. Eng. Sci.	International Journal of Engineering Science [UK]
Int. J. Environ. Anal. Chem.	International Journal of Environmental Analytical Chemistry [of International Association of Environmental Analytical Chemistry, Switzerland]
Int. J. Environ. Stud.	International Journal of Environmental Studies
Int. J. Expert Syst. Res. Appl.	International Journal of Expert Systems Research and Applications [US]
Int. J. Fatigue	International Journal of Fatigue [UK]
Int. J. Food Microbiol.	International Journal of Food Microbiology [of International Union of Microbiological Societies, UK]
Int. J. Forecast.	International Journal of Forecasting [Netherlands]
Int. J. Fract.	International Journal of Fracture [Netherlands]
Int. J. Fusion Energy	International Journal of Fusion Energy [of Fusion Energy Foundation, US]
Int. J. Game Theory	International Journal of Game Theory [Austria]
Int. J. Gen. Syst.	International Journal of General Systems [UK]
Int. J. Geogr. Inf. Syst.	International Journal of Geographical Information Systems [UK]
Int. J. Glob. Energy Issues	International Journal of Global Energy Issues [Switzerland]
Int. J. Heat Fluid Flow	International Journal of Heat and Fluid Flow [UK]
Int. J. Heat Mass Transf.	International Journal of Heat and Mass Transfer [UK and US]
Int. J. High Technol. Ceram.	International Journal of High Technology Ceramics [UK]
Int. J. High Temp. Ceram.	International Journal of High Temperature Ceramics
Int. J. Hybrid Microelectron.	International Journal of Hybrid Electronics [of International Society for Hybrid Microelectronics, US]
Int. J. Hydrog. Energy	International Journal of Hydrogen Energy [UK]
Int. J. Impact Eng.	International Journal of Impact Engineering [UK]
Int. J. Inf. Manage.	International Journal of Information Management [UK]
Int. J. Infrared Millim. Waves	International Journal of Infrared and Millimeter Waves [US]
Int. J. Intell. Syst.	International Journal of Intelligent Systems [US]
Int. J. Joining Mater.	International Journal for the Joining of Materials [Denmark]
Int. J. Mach. Tools Manuf.	International Journal of Machine Tools and Manufacture [UK]
Int. J. Man-Mach. Stud.	International Journal of Man-Machine Studies [UK]
Int. J. Manuf. Technol.	International Journal of Manufacturing Technology
Int. J. Mass Spectrom. Ion Process.	International Journal of Mass Spectrometry and Ion Processes [Netherlands]
Int. J. Mater. Eng. Appl.	International Journal of Materials in Engineering Applications [UK]
Int. J. Mater. Prod. Technol.	International Journal of Materials and Product Technology [Switzerland]
Int. J. Math. Educ. Sci. Technol.	International Journal of Mathematical Education in Science and Technology [UK]
Int. J. Mech. Sci.	International Journal of Mechanical Sciences [UK]
Int. J. Microcomput. Appl.	International Journal of Microcomputer Applications [of International Society for Mini- and Microcomputers]
Int. J. Micrography. Video Technol.	International Journal of Micrographics and Video Technology [UK]
Int. J. Min. Eng.	International Journal of Mining Engineering
Int. J. Miner. Process.	International Journal of Mineral Processing [Netherlands]
Int. J. Mine Water	International Journal of Mine Water [of International Mine Water Association, Spain]
Int. J. Mini Microcomput.	International Journal of Mini and Microcomputers [US]
Int. J. Model. Simul.	International Journal of Modelling and Simulation [US]
Int. J. Mod. Phys. A	International Journal of Modern Physics A [Singapore]
Int. J. Mod. Phys. B	International Journal of Modern Physics B [Singapore]
Int. J. Multiph. Flow	International Journal of Multiphase Flow [UK]
Int. J. Non-Linear Mech.	International Journal of Non-Linear Mechanics [US]
Int. J. Numer. Anal. Methods Geomech.	International Journal for Numerical and Analytical Methods in Geomechanics [UK]
Int. J. Numer. Methods Eng.	International Journal for Numerical Methods in Engineering [UK]
Int. J. Numer. Methods Fluids	International Journal for Numerical Methods in Fluids [UK]
Int. J. Numer. Model., Electron. Netw.Devices Fields	International Journal of Numerical Modelling: Electronic Networks, Devices and Fields [UK]
Int. J. Oper. Prod. Manage.	International Journal of Operations and Production Management [UK]
Int. J. Optoelectron.	International Journal of Optoelectronics [UK]
Int. J. Parallel Program.	International Journal of Parallel Programming [US]
Int. J. Pattern Recognit. Artif. Intell.	International Journal of Pattern Recognition and Artificial Intelligence [Singapore]
Int. J. Pharm.	International Journal of Pharmaceutics
Int. J. Phys. Distrib. Mater. Manage.	International Journal of Physical Distribution and Materials Management [UK]

Int. J. Plast.	International Journal of Plasticity [UK]
Int. J. Policy Inf.	International Journal on Policy and Information [Taiwan]
Int. J. Polym. Mater.	International Journal of Polymeric Materials [UK]
Int. J. Powder Metall.	International Journal of Powder Metallurgy [of American Powder Metallurgy Institute and Metal Powder Industries Federation]
Int. J. Powder Metall. Powder Technol.	International Journal of Powder Metallurgy and Powder Technology [of Metal Powder Industries Federation, US]
Int. J. Pressure Vessels Piping	International Journal of Pressure Vessels and Piping [UK]
Int. J. Prod. Res.	International Journal of Production Research [UK]
Int. J. Proj. Manage.	International Journal of Project Management [UK]
Int. J. Quantum Chem.	International Journal of Quantum Chemistry [US]
Int. J. Quantum Chem., Quantum Biol. Symp.	International Journal of Quantum Chemistry, Quantum Biology Symposium [US]
Int. J. Quantum Chem., Quantum Chem. Symp.	International Journal of Quantum Chemistry, Quantum Chemistry Symposium [US]
Int. J. Radiat. Biol.	International Journal of Radiation Biology [UK]
Int. J. Radiat. Oncol. Biol. Phys.	International Journal of Radiation Oncology Biology Physics [UK]
Int. J. Rapid Solidif.	International Journal of Rapid Solidification [UK]
Int. J. Refract. Hard Mater.	International Journal of Refractory Metals and Hard Materials [UK]
Int. J. Refrig.	International Journal of Refrigeration [of International Institute of Refrigeration, France]
Int. J. Remote Sens.	International Journal of Remote Sensing [UK]
Int. J. Robot. Autom.	International Journal of Robotics and Automation [US]
Int. J. Robot. Res.	International Journal of Robotics Research [US]
Int. J. Rock Mech. Min. Sci. Geomech. Abstr.	International Journal of Rock Mechanics and Mining Sciences and Geomechanics Abstracts [UK]
Int. J. Satell. Commun.	International Journal of Satellite Communications [UK]
Int. J. Sci. Educ.	International Journal of Science Education [UK]
Int. J. Sol. Energy	International Journal of Solar Energy [Switzerland]
Int. J. Solids Struct.	International Journal of Solids and Structures [UK]
Int. J. Supercomput. Appl.	International Journal of Supercomputer Applications [US]
Int. J. Syst. Bacteriol.	International Journal of Systematic Bacteriology [of International Union of Microbiological Societies, UK]
Int. J. Syst. Sci.	International Journal of Systems Science [UK]
Int. J. Technol. Manage.	International Journal of Technology Management [Switzerland]
Int. J. Theor. Phys.	International Journal of Theoretical Physics [US]
Int. J. Thermophys.	International Journal of Thermophysics [US]
Int. J. Turbo Jet-Engines	International Journal of Turbo and Jet-Engines [UK]
Int. J. Veh. Des.	International Journal of Vehicle Design [Switzerland]
Int. Lab.	International Laboratory
Int. Manage.	International Management [Switzerland]
Int. Mater. Rev.	International Materials Reviews [of ASM International, US and Institute of Metals, UK]
Int. Met. Rev.	International Metals Reviews [of ASM International, US and Institute of Metals, UK]
Int. Min.	International Mining [Singapore]
Int. Mod. Foundry	International Modern Foundry [UK]
Int. Pbd. Ind.	International Paperboard Industry [UK and US]
Int. Peat J.	International Peat Journal [of International Peat Society, Finland]
Int. Pkg. Abstr.	International Packaging Abstracts [US and UK]
Int. Polym. Process.	International Polymer Processing
Int. Polym. Sci. Technol.	International Polymer Science and Technology [of Rubber and Plastics Research Association, UK]
Int. Power Gener.	International Power Generation [UK]
Int. QC Forum	International QC Forum [US]
Int. Reinf. Plast. Ind.	International Reinforced Plastics Industry [UK]
Int. Rev. Cytol.	International Reviews in Cytology
Int. Rev. Phys. Chem.	International Reviews in Physical Chemistry [UK]
Int. Rubb. Dig.	International Rubber Digest [of International Rubber Study Group, UK]
Int. Spectr.	International Spectrum [US]
Int. Stat. Rev.	International Statistical Review [of International Statistical Institute, Netherlands]
Int. Sugar J.	International Sugar Journal
Int. Technol.	International Technology [of International Technology Institute, US]
Int. Tech Transf. Bus.	International Tech Transfer Business [UK]

Int. Water Power Dam Constr.	International Water Power and Dam Construction [UK]
Inverse Probl.	Inverse Problems [of Institute of Physics, UK]
Invest. Tec. Papel	Investigacion y Tecnica del Papel [of Asociacion de Investigacion Technica de la Industria Papelera Espanola, Spain]
Inz. Apar. Chem.	Inzyniera i Aparatura Chemiczna
Inz. Chem. Procesowa	Inzyniera Chemiczna i Procesowa [Poland]
Inzh.-Fiz. Zh.	Inzhenerno-Fizicheskii Zhurnal [USSR]
Inz. Materialowa	Inzyniera Materialowa [Poland]
IPA	International Pharmaceutical Abstracts [of American Society of Hospital Pharmacists]
IPEF	Instituto de Pesquisas E Estudos Florestais [Brazil]
IPE Int. Ind. Prod. Eng.	IPE International Industrial and Production Engineering [Germany]
IPH Inform.	IPH Information [of International Association of Paper Historians, Germany]
IPPTA	IPPTA [of Indian Pulp and Paper Technical Association]
Iran. J. Sci. Technol.	Iranian Journal of Science and Technology [UK]
Iraqi J. Sci.	Iraqi Journal of Science
Ir. Astron. J.	Irish Astronomical Journal
Ir. Comput.	Irish Computer [Ireland]
Irish J. Food Sci. Technol.	Irish Journal of Food Science and Technology
Iron Age, Met. Prod.	Iron Age, Metals Producer [US]
Iron Steel (China)	Iron and Steel (China) [PR China]
Iron Steel Eng.	Iron and Steel Engineer [of Association of Iron and Steel Engineers, US]
Iron Steel Ind.	Iron and Steel Industry [Japan]
Iron Steel Int.	Iron and Steel International [UK]
Ironmaking Conf. Proc.	Ironmaking Conference Proceedings [of Iron and Steel Society [of AIME, US]]
Ironmaking Steelmaking	Ironmaking and Steelmaking [of Institute of Metals, UK]
Iron Steelmaker	Iron and Steelmaker [of AIME Iron and Steel Society, US]
IRPA Bull.	IRPA Bulletin [of International Radiation Protection Association, France]
I/S Anal.	I/S Analyzer [US]
ISA Trans.	ISA Transactions [of Instrument Society of America, US]
ISCE Newsl.	ISCE Newsletter [of International Society of Chemical Ecology, US]
Ishikawajima-Harima Eng. Rev.	Ishikawajima-Harima Engineering Review [Japan]
ISIJ Int.	ISIJ International [of Iron and Steel Institute of Japan]
IS Ind. Soc. Mag.	IS Industrial Society Magazine [UK]
Islamabad J. Sci.	Islamabad Journal of Sciences [Pakistan]
Isot. Radiat. Res.	Isotope and Radiation Research [Egypt]
ISPRS J. Photogramm. Remote Sens.	ISPRS Journal of Photogrammetry and Remote Sensing [of International Society of Photogrammetry and Remote Sensing, Netherlands]
ISRI Commod.	ISRI Commodities [of Institute of Scrap Recycling Industries, US]
ISRI Rep.	ISRI Report [of Institute of Scrap Recycling Industries, US]
Isr. J. Chem.	Israel Journal of Chemistry
Isr. J. Earth-Sci.	Israel Journal of Earth-Sciences
Isr. J. Technol.	Israel Journal of Technology
Issled. Zemli iz Kosm.	Issledovanie Zemli iz Kosmosa [USSR]
Issues Sci. Technol.	Issues in Science and Technology [US]
Ist. Tecnol. Legno Quaderni	Istituto per la Tecnologia del Legno Quaderni [Italy]
ITAL	Information Technology for Libraries [of Library and Information Technology Association, US]
Ital. Technol.	Italian Technology [Italy]
ITC J.	ITC Journal [of International Institute for Aerial Survey and Earth Sciences, Netherlands]
ITEA J. Test. Eval.	ITEA Journal of Test and Evaluation [of International Test and Evaluation Association, US]
ITE J.	ITE Journal [of Institute of Transportation Engineers, US]
ITG-Fachber.	ITG-Fachberichte [of Verein Deutscher Elektrotechniker, Germany]
Itogi Nauki Tekh., Issled. Kosm. Prostran.	Itogi Nauki i Tekhniki, Issledovanie Kosmicheskogo Prostranstva [USSR]
Itogi Nauki Tekh., Kompoz. Mater.	Itogi Nauki i Tekhniki, Kompozitsionnye Materialy [USSR]
Itogi Nauki Tekh., Metalloved. Term. Obrab.	Itogi Nauki i Tekhniki, Metallovedenie i Termicheskaya Obrabotka [USSR]
Itogi Nauki Tekh., Metall. Teplotekh.	Itogi Nauki i Tekhniki, Metallurgicheskaya Teplotekhnika [USSR]
Itogi Nauki Tekh., Metal. Tsvetn. Metall.	Itogi Nauki i Tekhniki, Metallurgiya Tsvetnykh Metallov [USSR]

Itogi Nauki Tekh., Poroshk. Metall.	Itogi Nauki i Tekhniki, Poroshkovaga Metallurgiya [USSR]
Itogi Nauki Tekh., Proizv. Chugina Stali	Itogi Nauki i Tekhniki, Proizvodstvo Chugina i Stali [USSR]
Itogi Nauki Tekh., Prokatnoe Volchil'noe Proizv.	Itogi Nauki i Tekhniki, Prokatnoe i Volochil'noe Proizvodstvo [USSR]
Itogi Nauki Tekh., Proshk. Metall.	Itogi Nauki i Tekhniki, Proshkovaga Metallurgiya [USSR]
Itogi Nauki Tekh., Svarka	Itogi Nauki i Tekhniki, Svarka [USSR]
IUAPPA Newsl.	IUAPPA Newsletter [of International Union of Air Pollution Associations, UK]
IUFRO News	IUFRU News [of International Union of Forestry Research Organization, Austria]
IUPAB Biophys. Ser.	IUPAP Biophysics Series [of International Union for Pure and Applied Biophysics, Hungary]
IWATSU Tech. Rep.	IWATSU Technical Report [Japan]
Izmer. Tekh.	Izmeritel'naya Tekhnika [USSR]
Izv. Akad. Nauk Arm. SSR, Fiz.	Izvestiya Akademii Nauk Armyanskoi SSR Fizika [USSR]
Izv. Akad. Nauk Arm. SSR, Mekh.	Izvestiya Akademii Nauk Armyanskoi SSR Mekhanika [USSR]
Izv. Akad. Nauk Arm. SSR, Tekh. Nauk	Izvestiya Akademii Nauk Armyanskoi SSR Tekhnicheskaya Nauk [USSR]
Izv. Akad. Nauk Azerb. SSR, Fiz.-Tekh. Mat. Nauk	Izvestiya Akademii Nauk Azerbaidzhanskoi SSR, Seriya Fiziko-Tekhnicheskikh i Matematicheskh Nauk [USSR]
Izv. Akad. Nauk BSSR	Izvestiya Akademii Nauk BSSR [USSR]
Izv. Akad. Nauk Est. SSR, Fiz. Mat.	Izvestiya Akademii Nauk Estonskoi SSR, Fizichesko, Matematicheskaya [USSR]
Izv. Akad. Nauk Est. SSR, Khim.	Izvestiya Akademii Nauk Estonskoi SSR, Khimicheskaya [USSR]
Izv. Akad. Nauk Gruz. SSR, Khim.	Izvestiya Akademii Nauk Gruzinskoi SSR, Seriya Khimicheskaya [USSR]
Izv. Akad. Nauk Kazakh. SSR	Izvestiya Akademii Nauk Kazakhskoi SSR [USSR]
Izv. Akad. Nauk Kazakh. SSR, Fiz.-Mat.	Izvestiya Akademii Nauk Kazakhskoi SSR, Fiziko-Matematicheskaya [USSR]
Izv. Akad. Nauk Kazakh. SSR, Khim.	Izvestiya Akademii Nauk Kazakhakoi SSR, Khimicheskaya [USSR]
Izv. Akad. Nauk Kirgiz. SSR	Izvestiya Akademii Nauk Kirgizkoi SSR [USSR]
Izv. Akad. Nauk Latv. SSR, Fiz.-Tekh.	Izvestiya Akademii Nauk Latviyskoi SSR, Fiziko-Tekhnicheskikh [USSR]
Izv. Akad. Nauk Latv. SSR, Khim.	Izvestiya Akademii Nauk Lavtviyskoi SSR, Khimicheskikh [USSR]
Izv. Akad. Nauk. Mold. SSR, Fiz.-Tekh. Mat. Nauk.	Izvestiya Akademii Nauk Moldavskoi SSR, Fiziko-Tekhnicheskikh i Matematicheskikh Nauk [USSR]
Izv. Akad. Nauk SSSR, Energ. Transp.	Izvestiya Akademii Nauk SSSR, Energetika i Transport [USSR]
Izv. Akad. Nauk SSSR, Fiz.	Izvestiya Akademii Nauk SSSR, Fizicheskaya [USSR]
Izv. Akad. Nauk SSSR, Fiz. Atmos. Okeana	Izvestiya Akademii Nauk SSSR, Fizika Atmosfery i Okeana [USSR]
Izv. Akad. Nauk SSSR, Fiz. Zemli	Izvestiya Akademii Nauk SSSR, Fizika Zemli [USSR]
Izv. Akad. Nauk SSSR, Khim.	Izvestiya Akademii Nauk SSSR, Khimicheskaya [USSR]
Izv. Akad. Nauk SSSR, Mekh. Tverd. Tela	Izvestiya Akademii Nauk SSSR, Mekhanika Tverdogo Tela [USSR]
Izv. Akad. Nauk SSSR, Mekh. Zhidh. Gaza	Izvestiya Akademii Nauk SSSR, Mekhanika Zhidkosti i Gaza [USSR]
Izv. Akad. Nauk SSSR, Met.	Izvestiya Akademii SSSR, Metally [USSR]
Izv. Akad. Nauk SSSR, Neorg. Mater.	Izvestiya Akademii Nauk SSSR, Neorganicheskie Materialy [USSR]
Izv. Akad. Nauk Turkm. SSR	Izvestiya Akademii Nauk Turkmenskoi SSR [USSR]
Izv. Akad. Nauk Turkm. SSR, Fiz.-Tekh. Khim. Geol. Nauk	Izvestiya Akademii Nauk Turkmenskoi SSR, Fiziko-Tekhnicheskikh Khimicheskikh i Geologicheskikh Nauk [USSR]
Izv. Akad. Nauk Uzb. SSR, Fiz.-Mat. Nauk	Izvestiya Akademii Nauk Uzbekskoi SSR, Fiziko-Matematicheskikh Nauk [USSR]
Izv. Akad. Nauk Uzb. SSR, Tekh. Nauk	Izvestiya Akademii Nauk Uzbekskoi SSR, Tekhnicheskikh Nauk [USSR]
Izv. Akad. Nauk UzSSR, Tekh. Nauk	Izvestiya Akademii Nauk UzSSR, Tekhnicheskikh Nauk [USSR]

Izv. Acad. Sci USSR, Atmos. Ocean. Phys.	Izvestiya Academy of Sciences USSR, Atmospheric and Oceanic Physics [Translation of: *Izvestiya Akademii Nauk SSSR, Fizika Atmosfery i Okeana (USSR)*; published in the US]
Izv. Acad. Sci USSR, Phys. Solid Earth	Izvestiya Academy of Sciences USSR, Physics of the Solid Earth [Translation of: *Izvestiya Akademii Nauk SSSR, Fizika Zemli (USSR)*; published in the US]
Izv. Gl. Astron. Obs. Pulkove	Izvestiya Glavnoi Astronomicheskoi Observatorii v Pulkove [USSR]
Izv. Khim.	Izvestiya po Khimiya [Bulgaria]
Izv. Khim., Bulg. Akad. Nauk.	Izvestiya po Khimiya, Bulgarska Akademiya na Naukite [of Bulgarian Academy of Sciences]
Izv. Krym. Astrofiz. Obs.	Izvestiya Krymskoi Astroficheskoi Observatorii [USSR]
Izv. Sibir. Otd. Akad. Nauk SSSR, Khim.	Izvestiya Sibirskogo Otdeleniya Akademii Nauk SSSR, Khimicheskikh [USSR]
Izv. Sibir. Otd. Akad. Nauk SSSR, Tekh.	Izvestiya Sibirskogo Otdeleniya Akademii Nauk SSSR, Tekhnicheskikh [USSR]
Izv. VMEI 'Lenin'	Izvestiya na VMEI 'Lenin' [Bulgaria]
Izv. V.U.Z. Chernaya Metall.	Izvestiya Vysshikh Uchebnykh Zavedenii, Chernaya Metallurgiya [USSR]
Izv. V.U.Z. Fiz.	Izvestiya Vysshikh Uchebnykh Zavedenii, Fizika [USSR]
Izv. V.U.Z. Khim. Khim. Tekhnol.	Izvestiya Vysshikh Uchebnykh Zavedenii, Khimiya i Khimicheskaya Tekhnologiya [USSR]
Izv. V.U.Z., Lesnoi Zh.	Izvestiya Vysshikh Uchebnykh Zavedenii, Lesnoi Zhurnal [USSR]
Izv. V.U.Z. Mashinostr.	Izvestiya Vysshikh Uchebnykh Zavedenii, Mashinostroenie [USSR]
Izv. V.U.Z. Radioelektron.	Izvestiya Vysshikh Uchebnykh Zavedenii, Radioelektronika [USSR]
Izv. V.U.Z. Radiofiz.	Izvestiya Vysshikh Uchebnykh Zavedenii, Radiofizika [USSR]
Izv. V.U.Z. Tekhnol. Legkoi Prom.	Izvestiya Vysshikh Uchebnykh Zavedenii, Tekhnologiya Legkoi Promyshlennosti [USSR]
Izv. V.U.Z. Tekhnol. Tekstil. Prom.	Izvestiya Vysshikh Uchebnykh Zavedenii, Tekhnologiya Tekstil'noi Promyshlennosti [USSR]
Izv. V.U.Z. Tsvetn. Metall.	Izvestiya Vysshikh Uchebnykh Zavedenii, Tsvetnaya Metallurgia [USSR]
Izv. Vyssh. Uchebn. Zaved. Aviats. Tekh.	Izvestiya Vysshikh Uchebnykh Zavedenii, Aviatsionnaya Tekhnika [USSR]
Izv. Vyssh. Uchebn. Zaved. Chernaya Metall.	Izvestiya Vysshikh Uchebnykh Zavedenii, Chernaya Metallurgiya [USSR]
Izv. Vyssh. Uchebn. Zaved. Elektromekh.	Izvestiya Vysshikh Uchebnykh Zavedenii, Elektromekhnika [USSR]
Izv. Vyssh. Uchebn. Zaved. Energ.	Izvestiya Vysshikh Uchebnykh Zavedenii, Energetika [USSR]
Izv. Vyssh. Uchebn. Zaved. Fiz.	Izvestiya Vysshikh Uchebnykh Zavedenii, Fizika [USSR]
Izv. Vyssh. Uchebn. Zaved. Khim. Khim. Tekhnol.	Izvestiya Vysshikh Uchebnykh Zavedenii, Khimiya i Khimicheskaya Tekhnologiya [USSR]
Izv. Vyssh. Uchebn. Zaved. Lesnoi Zh.	Izvestiya Vysshikh Uchebnykh Zavedenii, Lesnoi Zhurnal [USSR]
Izv. Vyssh. Uchebn. Zaved. Mashinostr.	Izvestiya Vysshikh Uchebnykh Zavedenii, Mashinostroenie [USSR]
Izv. Vyssh. Uchebn. Zaved. Mat.	Izvestiya Vysshikh Uchebnykh Zavedenii, Matematika [USSR]
Izv. Vyssh. Uchebn. Zaved. Radioelektron.	Izvestiya Vysshikh Uchebnykh Zavedenii, Radioelektronika [USSR]
Izv. Vyssh. Uchebn. Zaved. Radiofiz.	Izvestiya Vysshikh Uchebnykh Zavedenii, Radiofizika [USSR]
Izv. Vyssh. Uchebn. Zaved. Tekhnol. Legkoi Prom.	Izvestiya Vysshikh Uchebnykh Zavedenii, Tekhnologiya Legkoi Promyshlennosti [USSR]
Izv. Vyssh. Uchebn. Zaved. Tekhnol. Tekstil. Prom.	Izvestiya Vysshikh Uchebnykh Zavedenii, Tekhnologiya Tekstil'noi Promyshlennosti [USSR]
Izv. Vyssh. Uchebn. Zaved. Tsvetn. Metall.	Izvestiya Vysshikh Uchebnykh Zavedenii, Tsvetnaya Metallurgia [USSR]

J

J. A	Journal A [of Koninklije Vlaamse Ingenieurs-Vereniging, Belgium]
J. AAVSO	Journal of the AAVSO [of American Association of Variable Star Observers]

J. Account. EDP	Journal of Accounting and EDP [US]
JACM	Journal of the Association for Computing Machinery [US]
J. Acoust.	Journal d'Acoustique [France]
J. Acoust. Emiss.	Journal of Acoustic Emission [of Acoustic Emission Group, US]
J. Acoust. Soc. Am.	Journal of the Acoustical Society of America
J. Acoust. Soc. India	Journal of the Acoustical Society of India
J. Acoust. Soc. Jpn.	Journal of the Acoustical Society of Japan
J. ACS	Journal of the American Chemical Society
Jad. Energ.	Jaderna Energie [Czechoslovakia]
J. Adhes.	Journal of Adhesion [UK]
J. Adhes. Sci. Technol.	Journal of Adhesion Science and Technology [Netherlands]
J. Adv. Transp.	Journal of Advanced Transportation [of Advanced Transit Association, US]
J. Aeronaut. Mater.	Journal of Aeronautical Materials [PR China]
J. Aeronaut. Soc. India	Journal of the Aeronautical Society of India [of Indian Institute of Science, India]
J. Aerosol Med.	Journal of Aerosol Medicine [of International Society for Aerosols in Medicine, Austria]
J. Aerosol Res.	Journal of Aerosol Research [Japan]
J. Aerosol Res. Jpn.	Journal of Aerosol Research Japan
J. Aerosol Sci.	Journal of Aerosol Science [of Gesellschaft fur Aerosolforschung, Germany]
J. Afr. Earth Sci.	Journal of African Earth Sciences [UK]
J. Agric. Eng. Res.	Journal of Agricultural Engineering Research
J. Agric. Food Chem.	Journal of Agricultural and Food Chemistry [of American Chemical Society]
Jahangirnagar Rev. A, Sci.	Jahangirnagar Review, Part A, Science [Bangladesh]
J. Aircr.	Journal of Aircraft [of American Institute of Aeronautics and Astronautics]
J. Air Pollut. Control Assoc.	Journal of the Air Pollution Control Association [US]
J. Air Waste Manage. Assoc.	Journal of the Air Waste Management Association [US]
J. Algorithms	Journal of Algorithms
J. Alloy Phase Diagrams	Journal of Alloy Phase Diagrams [India]
J. Am. Assoc. Var. Star Obs.	Journal of the American Association of Variable Star Observers
J. Am. Ceram. Soc.	Journal of American Ceramic Society
J. Am. Chem. Soc.	Journal of the American Chemical Society
J. Am. Coll. Toxicol.	Journal of the American College of Toxicology
J. Am. Helicopter Soc.	Journal of the American Helicopter Society
J. Am. Leather Chem. Assoc.	Journal of the American Leather Chemists Association
J. Am. Math. Soc.	Journal of the American Mathematical Society
J. Am. Oil Chem. Soc.	Journal of the American Oil Chemists Society
J. Am. Soc. Brew. Chem.	Journal of the American Society of Brewing Chemists
J. Am. Soc. Hort. Sci.	Journal of the American Society for Horticultural Sciences
J. Am. Soc. Inf. Sci.	Journal of the American Society for Information Science
J. Am. Water Works Assoc.	Journal of the American Water Works Association
J. Anal. At. Spectrom.	Journal of Analytical Atomic Spectrometry [of Royal Society of Chemistry, UK]
Jan's Def. Wkly.	Jane's Defence Weekly [UK]
J. Antibiot.	Journal of Antibiotics [US]
J. AOAC	Journal of the AOAC [of Association of Official Analytical Chemists, US]
Japan Pulp Paper	Japan Pulp and Paper [Japan]
Japan TAPPI	Japan TAPPI [now: Japan TAPPI Journal]
Japan TAPPI J.	Japan Tappi Journal [of Japanese Technical Association of the Pulp and Paper Industry]
J. APCA	Journal of the Air Pollution Control Association [US]
JAPCA	Journal of the Air and Waste Management Association [US]
J. Appl. Bacteriol.	Journal of Applied Bacteriology [of Society for Applied Bacteriology, UK]
J. Appl. Biochem.	Journal of Applied Biochemistry [of International Union of Biochemistry, US]
J. Appl. Biomater.	Journal of Applied Biomaterials [of Society for Biomaterials, US]
J. Appl. Chem.	Journal of Applied Chemistry [of Society of Applied Chemistry, US]
J. Appl. Chem. USSR	Journal of Applied Chemistry of the USSR [Translation of: *Zhurnal Prikladnoi Khimii*; published in the US]
J. Appl. Crystallogr.	Journal of Applied Crystallography [Denmark]
J. Appl. Electrochem.	Journal of Applied Electrochemistry [UK]
J. Appl. Mech.	Journal of Applied Mechanics [of American Society of Mechanical Engineers]
J. Appl. Mech. Tech. Phys.	Journal of Applied Mechanics and Technical Physics [Translation of: *Zhurnal Prikladnoi Mekhaniki i Tekhnicheskoi Fiziki (USSR)*; published in the US]
J. Appl. Mech. (Trans. ASME)	Journal of Applied Mechanics (Transactions of the ASME) [of American Society of Mechanical Engineers]

J. Appl. Metalwork.	Journal of Applied Metalworking [US]
J. Appl. Meteorol.	Journal of Applied Meteorology [of American Meteorological Society]
J. Appl. Microbiol. Biotech.	Journal of Applied Microbiology and Biotechnology [Sweden]
J. Appl. Phys.	Journal of Applied Physics [of American Institute of Physics]
J. Appl. Physiol.	Journal of Applied Physiology [of American Physiological Society]
J. Appl. Polym. Sci.	Journal of Applied Polymer Science [US]
J. Appl. Probab.	Journal of Applied Probability [UK]
J. Appl. Spectrosc.	Journal of Applied Spectroscopy [Translation of: *Zhurnal Prikladnoi Spektroskopii (USSR)*; published in the US]
J. Appl. Syst. Anal.	Journal of Applied Systems Analysis [UK]
J. Approx. Theory	Journal of Approximation Theory [US]
JAPT	Journal for Approximation Theory [US]
J. Archit. Plan. Res.	Journal of Architectural and Planning Research [US]
J. Arid Environ.	Journal of Arid Environments
J. ASA	Journal of the American Statistical Society
JASIS	Journal of the American Society for Information Science
J. Assoc. Comput. Mach.	Journal of the Association for Computing Machinery [US]
J. Assoc. Explor. Geophys.	Journal of Association of Exploration Geophysicists [of Center of Exploration Geophysics, India]
J. Assoc. Lunar Planet. Obs. Strolling Astron.	Journal of the Association of Lunar and Planetary Observers, Strolling Astronomer [US]
J. Astronaut. Sci.	Journal of the Astronautical Sciences [of American Astronautical Society]
J. Astron. Soc. Egypt	Journal of the Astronomical Society of Egypt [Egypt]
J. Astrophys. Astron.	Journal of Astrophysics and Astronomy [of Indian Academy of Sciences]
J. At. Energy Soc. Jpn.	Journal of the Atomic Energy Society of Japan
J. Atmos. Chem.	Journal of Atmospheric Chemistry [US]
J. Atmos. Sci.	Journal of Atmospheric Science [of American Meteorological Society]
J. Atmos. Terr. Phys.	Journal of Atmospheric and Terrestrial Physics [UK]
J. Audio Eng. Soc.	Journal of the Audio Engineering Society [US]
J. Aust. Ceram. Soc.	Journal of the Australian Ceramics Society [Australia]
J. Autom. Chem.	Journal of Automatic Chemistry [UK]
J. Autom. Reasoning	Journal of Automated Reasoning [Netherlands]
J. Bacteriol.	Journal of Bacteriology [of American Society for Microbiology]
J. Bangladesh Acad. Sci.	Journal of the Bangladesh Academy of Sciences [Bangladesh]
J. Beijing Univ. Iron Steel Technol.	Journal of Beijing University of Iron and Steel Technology [PR China]
J. Biochem.	Journal of Biochemistry
J. Biochem. Biophys. Methods	Journal of Biochemical and Biophysical Methods [Netherlands]
J. Bioelectr.	Journal of Bioelectricity [US]
J. Bioenerg. Biomembr.	Journal of Bioenergetics and Biomembranes [US]
J. Bioeng.	Journal of Bioengineering [UK]
J. Bioeng. (Trans. ASME)	Journal of Bioengineering (Transactions of ASME) [of American Society of Mechanical Engineers]
J. Biol. Chem.	Journal of Biological Chemistry [of American Society for Biochemistry and Molecular Biology]
J. Biol. Phys.	Journal of Biological Physics [US]
J. Biol. Stand.	Journal of Biological Standardization [of International Union of Microbiological Societies, UK]
J. Biolumin. Chemilumin.	Journal of Bioluminescence and Chemiluminescence [UK]
J. Biomater. Appl.	Journal of Biomaterials Applications [US]
J. Biomech.	Journal of Biomechanics [UK]
J. Biomech. Eng. (Trans. ASME)	Journal of Biomechanical Engineering (Transactions of ASME) [of American Society of Mechanical Engineers]
J. Biomed. Eng.	Journal of Biomedical Engineering [UK]
J. Biomed. Mater. Res.	Journal of Biomedical Materials Research [of Society for Biomaterials, US]
J. Biotechnol.	Journal of Biotechnology [US]
J. Biotechnol. Appl. Biochem.	Journal of Biotechnology and Applied Biochemistry [of International Union of Biochemistry, US]
J. Birla Inst. Technol. Sci.	Journal of the Birla Institute of Technology and Science [India]
J. Br. Astron. Assoc.	Journal of the British Astronomical Association [UK]
J. Br. Interplanet. Soc.	Journal of the British Interplanetary Society [UK]
J. Br. IRE	Journal of the British Institution of Radio Engineers
J. Bus. Econ. Stat.	Journal of Business and Economic Statistics [of American Statistical Association]
J. Can. Ceram. Soc.	Journal of the Canadian Ceramic Society
J. Can. Pet. Technol.	Journal of Canadian Petroleum Technology
J. Capacity Manage.	Journal of Capacity Management [of Institute of Software Engineering, US]
J. Carbohyd. Chem.	Journal of Carbohydrate Chemistry [US]

J. Catal.	Journal of Catalysis [US]
J. Cell Biol.	Journal of Cell Biology [of American Society of Cell Biology]
J. Cell. Plast.	Journal of Cellular Plastics [US]
J. Cent. China Norm. Univ. (Nat. Sci.)	Journal of Central China Normal University (Natural Sciences) [PR China]
J. Cent. S. Inst. Min. Metall.	Journal of Central-South Institute of Mining and Metallurgy [PR China]
J. Ceram. Soc. Jpn.	Journal of the Ceramic Society of Japan
J. Chem. Ecol.	Journal of Chemical Ecology [of International Society of Chemical Ecology, US]
J. Chem. Educ.	Journal of Chemical Education [of American Chemical Society]
J. Chem. Eng.	Journal of Chemical Engineering
J. Chem. Eng. Data	Journal of Chemical Engineering Data [of American Chemical Society]
J. Chem. Eng. Jpn.	Journal of Chemical Engineering Japan
J. Chem. Ind. Eng.	Journal of Chemical Industry and Engineering [PR China]
J. Chem. Inf. Comput. Sci.	Journal of Chemical Information and Computer Sciences [of American Chemical Society]
J. Chem. Phys.	Journal of Chemical Physics [of American Institute of Physics]
J. Chem. Res.	Journal of Chemical Research [of Societe Francaise de Chimie, France]
J. Chem. Res.	Journal of Chemical Research [of Royal Society of Chemistry, UK]
J. Chem. Soc.	Journal of the Chemical Society [UK]
J. Chem. Soc. A	Journal of the Chemical Society A [UK]
J. Chem. Soc. Chem. Commum.	Journal of the Chemical Society, Chemical Communications [of Royal Society of Chemistry, UK]
J. Chem. Soc., Dalton Trans.	Journal of the Chemical Society, Dalton Transactions [of Royal Society of Chemistry, UK]
J. Chem. Soc., Faraday Trans.	Journal of the Chemical Society, Faraday Transactions [of Royal Society of Chemistry, UK]
J. Chem. Soc., Faraday Trans. I	Journal of the Chemical Society, Faraday Transactions I [of Royal Society of Chemistry, UK]
J. Chem. Soc., Faraday Trans. II	Journal of the Chemical Society, Faraday Transactions II [of Royal Society of Chemistry, UK]
J. Chem. Soc. Jpn.	Journal of the Chemical Society of Japan
J. Chem. Soc., Perkin Trans.	Journal of the Chemical Society, Perkin Transactions [Royal Society of Chemistry, UK]
J. Chem. Soc., Perkin Trans. I	Journal of the Chemical Society, Perkin Transactions I [Royal Society of Chemistry, UK]
J. Chem. Soc., Perkin Trans. II	Journal of the Chemical Society, Perkin Transactions II [Royal Society of Chemistry, UK]
J. Chem. Technol. Biotechnol.	Journal of Chemical Technology and Biotechnology [Society of Chemical Industry, UK]
J. Chem. Thermodyn.	Journal of Chemical Thermodynamics [UK]
J. Chim. Phys.	Journal de Chimie Physique [of Societe Francaise de Chimie, France]
J. Chim. Phys. Phys.-Chim. Biol.	Journal de Chimie Physique et de Physico-Chimie Biologique [of Societe de Chimie Physique, France]
J. China Inst. Commun.	Journal of the China Institute of Communications [PR China]
J. Chin. Ceram. Soc.	Journal of the Chinese Ceramic Society [PR China]
J. Chin. Chem. Soc.	Journal of the Chinese Chemical Society [PR China]
J. Chin. Electron Microsc. Soc.	Journal of Chinese Electron Microscopy Society [PR China]
J. Chin. Foundrymen's Assoc.	Journal of Chinese Foundrymen's Association [Taiwan]
J. Chin. Inst. Chem. Eng.	Journal of the Chinese Institute of Chemical Engineers [PR China]
J. Chin. Inst. Eng.	Journal of Chinese Institute of Engineers [Taiwan]
J. Chin. Silic. Soc.	Journal of the Chinese Silicate Society [PR China]
J. Chin. Soc. Mech. Eng.	Journal of the Chinese Society of Mechanical Engineers [Taiwan]
J. Chin. Rare Earth Soc.	Journal of the Chinese Rare Earth Society [PR China]
J. Chromat.	Journal of Chromatography [Netherlands]
J. Chromat. Sci.	Journal of Chromatographic Sciences [US]
J. Clim.	Journal of Climate [of American Meteorological Society]
J. Clim. Appl. Meteorol.	Journal of Climate and Applied Meteorology [US]
J. Climatol.	Journal of Climatology [UK]
J. Clin. Comput.	Journal of Clinical Computing [US]
J. Clin. Eng.	Journal of Clinical Engineering [US]
J. Clin. Microbiol.	Journal of Clinical Microbiology [of American Society for Microbiology]
J. Cluster Sci.	Journal of Cluster Science [US]
J. Coast. Res.	Journal of Coastal Research [of Coastal Education and Research Foundation, US]
J. Coated Fabrics	Journal of Coated Fabrics [US]
J. Coat. Technol.	Journal of Coatings Technology [of Federation of Societies for Coatings Technology, US]
J. Coll. Eng., Nihon Univ., A	Journal of the College of Engineering, Nihon University, A [Japan]
J. Coll. Eng., Nihon Univ., B	Journal of the College of Engineering, Nihon University, B [Japan]
J. Colloid Interface Sci.	Journal of Colloid Interface Science [US]
J. Coll. Sci. Teach.	Journal of College Science Teaching [US]
J. Comb. Theory	Journal of Combinatorial Theory [US]
J. Comb. Theory A	Journal of Combinatorial Theory A [US]

J. Comb. Theory B	Journal of Combinatorial Theory B [US]
J. Commun. Res. Lab.	Journal of the Communications Research Laboratory [Japan]
J. Complex.	Journal of Complexity [US]
J. Compos. Mater.	Journal of Composite Materials [US]
J. Compos. Technol. Res.	Journal of Composites Technology and Research [of American Society for Testing and Materials]
J. Comput.	Journal of Computing [of Operations Research Society of America]
J. Comput. Appl. Math.	Journal of Computational and Applied Mathematics [Netherlands]
J. Comput. Assist. Learn.	Journal of Computer Assisted Learning [UK]
J. Comput Assist. Microsc.	Journal of Computer Assisted Microscopy [US]
J. Comput. Assist. Tomogr.	Journal of Computer Assisted Tomography [US]
J. Comput.-Based Instr.	Journal of Computer-Based Instruction [of Association for the Development of Computer-Based Instructional Systems, US]
J. Comput. Chem.	Journal of Computational Chemistry [US]
J. Comput. Inf. Syst.	Journal of Computer Information Systems [of Society of Data Educators, US]
J. Comput. Math.	Journal of Computational Mathematics [PR China]
J. Comput. Math. Sci. Teach.	Journal of Computers in Mathematics and Science Teaching [of Association for Computers in Mathematics and Science Teaching, US]
J. Comput. Phys.	Journal of Computational Physics [US]
J. Comput. Sci. Technol.	Journal of Computer Science and Technology [PR China]
J. Comput. Sci. Technol. (Engl. Lang. Ed.)	Journal of Computer Science and Technology (English Language Edition) [PR China]
J. Comput. Syst. Sci.	Journal of Computer and System Sciences [US]
J. Cond. Monit.	Journal of Condition Monitoring [UK]
J. Constr. Steel Res.	Journal of Constructional Steel Research [UK]
J. Contr. Rel.	Journal of Controlled Release [of Controlled Release Society, US]
J. Cost. Manage. Manuf. Ind.	Journal of Cost Management for the Manufacturing Industry [US]
JCPT	Journal of Canadian Petroleum Technology [of Canadian Institute of Mining and Metallurgy]
J. Cryptol.	Journal of Cryptology [US]
J. Crystallogr. Soc. Jpn.	Journal of Crystallographic Society of Japan [US]
J. Crystallogr. Spectrosc. Res.	Journal of Crystallographic and Spectroscopic Research [US]
J. Cryst. Growth	Journal of Crystal Growth [Netherlands]
JCS Chem. Commun.	Journal of the Chemical Society, Chemical Communications [of Royal Society of Chemistry, UK]
JCS Dalton	Journal of the Chemical Society, Dalton Transactions (Inorganic Chemistry) [of Royal Society of Chemistry, UK]
JCS Faraday I	Journal of the Chemical Society, Faraday Transactions I (Physical Chemistry in Condensed Phases) [of Royal Society of Chemistry, UK]
JCS Faraday II	Journal of the Chemical Society, Faraday Transactions II (Molecular and Chemical Physics) [of Royal Society of Chemistry, UK]
JCS Perkin I	Journal of the Chemical Society, Perkin Transactions I (Organic and Bioorganic Chemistry) [of Royal Society of Chemistry, UK]
JCS Perkin II	Journal of the Chemical Society, Perkin Transactions II (Physical Organic Chemistry) [of Royal Society of Chemistry, UK]
JCSS	Journal of Computer and Systems Science [US]
J. Cytom.	Journal of Cytometry [of Society for Analytical Cytology, US]
J. Dalian Univ. Technol.	Journal of Dalian University of Technology [PR China]
J. Dent. Res.	Journal of Dental Research [US]
J. Dev.	Journal of Development [of Conference on Data System Languages, US]
J. Differ. Equ.	Journal of of Differential Equations [US]
J. Dispersion Sci. Technol.	Journal of Dispersion Science and Technology
JDM	Journal of Data Management
J. Doc.	Journal of Documentation [of Association for Information Management, UK]
J. Dyn. Syst. Meas. Control (Trans. ASME)	Journal of Dynamic Systems, Measurement and Control (Transactions of ASME) [of American Society of Mechanical Engineering]
J. Dyn. Syst. (Trans. ASME)	Journal of Dynamic Systems (Transactions of ASME) [of American Society of Mechanical Engineers]
J. E. China Inst. Chem. Technol.	Journal of the East China Institute of Chemical Technology [PR China]
J. Econ. Dyn. Control	Journal of Economic Dynamics and Control [Netherlands]
J. Econ. Ent.	Journal of Economic Entomology
J. Educ. Comput. Res.	Journal of Educational Computing Research [US]
J. Educ. Libr. Inf. Sci.	Journal of Education for Library and Information Science [of Association for Library and Information Science Education, US]

J. Educ. Res. — Journal of Educational Research [US]
J. Educ. Technol. Syst. — Journal of Educational Technology Systems [US]
JEE — Journal of Electronic Engineering [Japan]
J. Elast. — Journal of Elasticity [Netherlands]
J. Elastomers Plast. — Journal of Elastomers and Plastics [US]
J. Electr. Electron. Eng. Aust. — Journal of Electrical and Electronics Engineering, Australia [of Institution Engineers, Australia]
J. Electr. Eng. — Journal of Electrical Engineering [of Chinese Society of Electric Engineering, PR China]
J. Electroanal. Chem. — Journal of Electroanalytical Chemistry [Switzerland]
J. Electroanal. Chem. Interfacial Electrochem. — Journal of Electroanalytical Chemistry and Interfacial Electrochemistry [Switzerland]
J. Electrochem. Soc. — Journal of the Electrochemical Society [US]
J. Electrochem. Soc. India — Journal of Electrochemical Society of India [of Indian Institute of Sciences]
J. Electromagn. Waves Appl. — Journal of Electromagnetic Waves and Applications [Netherlands]
J. Electron. — Journal of Electronics [PR China]
J. Electron. Comput. Res. — Journal of Electronics and Computers Research [of Scientific Research Council, Iraq]
J. Electron. Def. — Journal of Electronic Defense [of Association of Old Crows, US]
J. Electron. Eng. — Journal of Electronic Engineering [Japan]
J. Electron. Mater. — Journal of Electronic Materials [US and UK]
J. Electron Microsc. — Journal of Electron Microscopy [of Japanese Society of Electron Microscopy]
J. Electron Microsc. Technique — Journal of Electron Microscopy Technique [US]
J. Electron Micry. — Journal of Electron Microscopy [of Japanese Society of Electron Microscopy]
J. Electron Micry. Technique — Journal of Electron Microscopy Technique [US]
J. Electron. Packag. (Trans. ASME) — Journal of Electronic Packaging (Transactions of the ASME) [of American Society of Mechanical Engineers, US]
J. Electron Spectrosc. — Journal of Electron Spectroscopy [Netherlands]
J. Electron Spectrosc. Relat. Phenom. — Journal of Electron Spectroscopy and Related Phenomena [Netherlands]
J. Electrost. — Journal of Electrostatics [Netherlands]
JEMIC Tech. Rep. — JEMIC Technical Report [of Japan Electric Meters Inspection Corporation]
Jemna Mech. Opt. — Jemna Mechanika a Optika [Czechoslovakia]
Jena Rev. — Jena Review [Germany]
J. Energy Resour. Technol. (Trans. ASME) — Journal of Energy Resources Technology (Transactions of the ASME) [of American Society of Mechanical Engineers]
J. Eng. Gas Turbines Power (Trans. ASME) — Journal of Engineering for Gas Turbines and Power (Transactions of the ASME) [of American Society of Mechanical Engineers]
J. Eng. Ind. (Trans. ASME) — Journal of Engineering for Industry (Transactions of the ASME) [of American Society of Mechanical Engineers]
J. Eng. Mater. Technol. (Trans. ASME) — Journal of Engineering Materials and Technology (Transactions of the ASME) [of American Society of Mechanical Engineers]
J. Eng. Math. — Journal of Engineering Mathematics [Netherlands]
J. Eng. Mech. — Journal of Engineering Mechanics [of American Society of Civil Engineers]
J. Eng. Phys. — Journal of Engineering Physics [Translation of: *Inzhenerno-Fizicheskii Zhurnal (USSR)*; published in the US]
J. Eng. Power (Trans. ASME) — Journal of Engineering for Power (Transactions of the ASME) [of American Society of Mechanical Engineers]
J. Eng. Sci. — Journal of Engineering Sciences [Saudi Arabia]
J. Eng. Sci. King Saud Univ. — Journal of Engineering King Saud University [Saudi Arabia]
J. Eng. Technol. Manage. — Journal of Engineering and Technology Management [Netherlands]
J. Environ. Econ. Manage. — Journal of Environmental Economics and Management [of Association of Environmental and Resource Economists, US]
J. Environ. Eng. — Journal of Environmental Engineering [of American Society of Civil Engineers, US]
J. Environ. Polym. Degrad. — Journal of Environmental Polymer Degradation [US]
J. Environ. Qual. — Journal of Environmental Quality [of Soil Science Society of America]
J. Environ. Radioact. — Journal of Environmental Radioactivity [UK]
J. Environ. Sci. — Journal of Environmental Sciences [of Institute of Environmental Sciences]
JEOL News Anal. Instrum. — JEOL News, Analytical Instrumentation [Japan]
JEOL News Electron Opt. Instrum. — JEOL News, Electron Optics Instrumentation [Japan]

J. Equip. Electr. Electron.	Journal de l'Equipement Electrique et Electronique [France]
Jernkontorets Ann.	Jernkontorets Annaler [Denmark]
Jernkontorets Ann.	Jernkontorets Annaler [of Swedish Steel Producers Association]
Jernkontorets Forsk.	Jernkontorets Forskning [of Swedish Steel Producers Association]
J. Ethnobiol.	Journal of Ethnobiology [of Society of Ethnobiology, US]
JETP	Journal of Experimental Theoretical Physics [of American Institute of Physics]
JETP Lett.	JETP (Journal of Experimental Theoretical Physics) Letters [Translation of: *Pis'ma v Zhurnal Eksperimental'noi i Teoreticheskoi Fiziki (USSR)*; published in the US]
J. Eur. Ceram. Soc.	Journal of the European Ceramic Society [UK]
J. Exp. Bot.	Journal of Experimental Botany [of Society for Experimental Biology, UK]
J. Expl. Eng.	Journal of Explosives Engineering [of Society of Explosives Engineers, US]
J. Fac. Agr., Hokkaido Univ.	Journal of the Faculty of Agriculture, Hokkaido University [Japan]
J. Fac. Eng., Chiba Univ.	Journal of the Faculty of Engineering, Chiba University [Japan]
J. Fac. Eng., Ibaraki Univ.	Journal of the Faculty of Engineering, Ibaraki University [Japan]
J. Fac. Eng., Shinshu Univ.	Journal of the Faculty of Engineering, Shinshu University [Japan]
J. Fac. Eng., Univ. Tokyo A	Journal of the Faculty of Engineering, University of Tokyo, Series A [Japan]
J. Fac. Eng., Univ. Tokyo B	Journal of the Faculty of Engineering, University of Tokyo, Series B [Japan]
J. Ferrocement	Journal of Ferrocement [Thailand]
J. Fire Sci.	Journal of Fire Sciences [US]
J. Fiz. Malays.	Jurnal Fizik Malaysia [of Malaysian Institute of Physics]
J. Fluid Control	Journal of Fluid Control [US]
J. Fluid Mech.	Journal of Fluid Mechanics [UK]
J. Fluids Eng. (Trans. ASME)	Journal of Fluids Engineering (Transactions of ASME) [of American Society of Mechanical Engineers]
J. Fluids Struct.	Journal of Fluids and Structures [UK]
J. Food Eng.	Journal of Food Engineering [US]
J. Food Process. Eng.	Journal of Food Processing Engineering
J. Food Sci.	Journal of Food Science [of Institute of Food Technologists, US]
J. Forensic Sci.	Journal of Forensic Sciences
J. Forestry	Journal of Forestry [of Society of American Foresters]
J. Forth Appl. Res.	Journal of Forth Application and Research [of Institute of Applied Forth Research, US]
J. Four Electr.	Journal du Four Electrique et des Industries Electrochimiques [France]
J. Franklin Inst.	Journal of the Franklin Institute [US]
J. Fr. Electrothermie	Journal Francais de l'Electrothermie [France]
J. Frottement Ind.	Journal du Frottement Industriel [France]
J. Fuel Soc. Jpn.	Journal of the Fuel Society of Japan
J. Funct. Anal.	Journal of Functional Analysis [US]
J. Fusion Energy	Journal of Fusion Energy [US]
J. Gen. Chem.	Journal of General Chemistry [US]
J. Gen. Chem. USSR	Journal of General Chemistry of the USSR [Published in the US]
J. Geodyn.	Journal of Geodynamics [UK]
J. Geolectr.	Journal of Geoelectricity
J. Geol.	Journal of Geology [US]
J. Geol. Soc.	Journal of Geological Society [UK]
J. Geol. Soc. Jam.	Journal of the Geological Society of Jamaica
J. Geom. Phys.	Journal of Geometry and Physics [Italy]
J. Geomagn. Geoelectr.	Journal of Geomagnetism and Geoelectricity [Japan]
J. Geophys. Res.	Journal of Geophysical Research [of American Geophysical Union]
J. Geophys. Res. A	Journal of Geophysical Research A [of American Geophysical Union]
J. Geophys. Res. B	Journal of Geophysical Research B [of American Geophysical Union]
J. Geophys. Res. C	Journal of Geophysical Research C [of American Geophysical Union]
J. Geotech. Eng.	Journal of Geotechnical Engineering
J. Glaciol.	Journal of Glaciology [of International Glaciological Society, UK]
J. Graph Theory	Journal of Graph Theory [US]
J. Grey Syst.	Journal of Grey Systems [UK]
J. Gt. Lakes Res.	Journal of Great Lakes Research [of International Association for Great Lakes Research, US]
J. Guid., Control Dyn.	Journal of Guidance, Control and Dynamics [of American Institute of Aeronautics and Astronautics]
J. Harbin Inst. Technol.	Journal of the Harbin Institute of Technology [PR China]
J. Hazard. Mater.	Journal of Hazardous Materials [Netherlands]
J. Hazard. Waste Hazard. Mater.	Journal of Hazardous Waste and Hazardous Materials [of Hazardous Materials Control Research Institute, US]

J. Heat Recovery Syst.	Journal of Heat Recovery Systems [UK]
J. Heat Transf. (Trans. ASME)	Journal of Heat Transfer (Transactions of ASME) [of American Society of Mechanical Engineers]
J. Heat Treat.	Journal of Heat Treating [of ASM International, US]
J. High Temp. Soc. Jpn.	Journal of High Temperature Society of Japan
J. Hist. Astron.	Journal of the History of Astronomy [UK]
J. Histochem. Cytochem.	Journal of Histochemistry and Cytochemistry [US]
J. Hokkaido Forest Prod. Res. Inst.	Journal of the Hokkaido Forest Products Research Institute [Japan]
J. Horol. Inst. Jpn.	Journal of the Horological Institute of Japan
J. Huazhong Inst. Technol.	Journal of Huazhong Institute of Technology [PR China]
J. Huazhong Inst. Technol. (Engl. Ed.)	Journal of Huazhong Institute of Technology (English Edition) [PR China]
J. Hunan Sci. Technol. Univ.	Journal of Hunan Science and Technology University [PR China]
J. Hydraul. Eng.	Journal of Hydraulic Engineering [US]
J. Hydraul. Res.	Journal of Hydraulic Research [of International Association for Hydraulic Research, Netherlands]
J. Hydrol.	Journal of Hydrology [US]
J. Hydrosci. Hydraul. Eng.	Journal of Hydroscience and Hydraulic Engineering
J. Hydr. Res.	Journal of Hydraulic Research [of International Association for Hydraulic Research, Netherlands]
J. Hyg. Chem.	Journal of Hygiene Chemistry
J. IEE	Journal of the Institution of Electrical Engineers [UK]
Jiguang Zazhi (Laser J.)	Jiguang Zazhi (Laser Journal) [PR China]
JIIM	Journal of Information and Image Management [US]
J. Illum. Eng. Inst. Jpn.	Journal of the Illuminating Engineering Institute of Japan
J. Illum. Eng. Soc.	Journal of the Illuminating Engineering Society [US]
J. Imaging Sci.	Journal of Imaging Science [of Society of Photographic Scientists and Engineers, US]
J. Imaging Technol.	Journal of Imaging Technology [of Society of Photographic Scientists and Engineers, US]
JIMS	Journal of Indian Mathematical Society [India]
J. Ind. Fabr.	Journal of Industrial Fabrics [US]
J. Indian Inst. Sci.	Journal of the Indian Institute of Science
J. Indian Inst. Sci. A	Journal of the Indian Institute of Science A
J. Indian Inst. Sci. B	Journal of the Indian Institute of Science B
J. Indian Inst. Sci. C	Journal of the Indian Institute of Science C
J. Indian Refract. Makers Assoc.	Journal of Indian Refractory Makers Association
J. Indian Waterworks Assoc.	Journal of the Indian Waterworks Association
J. Ind. Irr. Technol.	Journal of Industrial Irradiation Technology [US]
J. Ind. Microbiol.	Journal of Industrial Microbiology [of Society for Industrial Microbiology, US]
J. Ind. (Trans. ASME)	Journal of Industry (Transactions of ASME) [of American Society of Mechanical Engineers]
J. Inf. Manage.	Journal of Information Management [of Life Office Management Association, US]
J. Inf. Optim. Sci.	Journal of Information and Optimization Sciences [India]
J. Inf. Process.	Journal of Information Processing [of Information Processing Society of Japan]
J. Inf. Process. Cybern.	Journal of Information Processing and Cybernetics [Germany]
J. Inf. Rec. Mater.	Journal of Information Recording Materials [Germany]
J. Inf. Sci.	Journal of Information Science [of Institute of Information Scientists, UK]
J. Inf. Sci. Princ. Pract.	Journal of Information Science, Principles and Practice [Netherlands]
J. Inf. Syst. Manage.	Journal of Information Systems Management [US]
J. Inf. Technol.	Journal of Information Technology [UK]
J. Inorg. Organomet. Polym.	Journal of Inorganic and Organometallic Polymers [US]
J. Inst. Electron. Inf. Commun. Eng.	Journal of the Institute of Electronics, Information and Communication Engineers [Japan]
J. Inst. Electron. Radio Eng.	Journal of the Institution of Electronic and Radio Engineers [UK]
J. Inst. Electron. Telecommun. Eng.	Journal of the Institution of Electronics and Telecommunication Engineers [India]
J. Inst. Energy	Journal of the Institute of Energy [UK]
J. Inst. Eng. (India), Chem. Eng. Div.	Journal of the Institution of Engineers (India), Chemical Engineering Division [India]
J. Inst. Eng. (India), Electr. Eng. Div.	Journal of the Institution of Engineers (India), Electrical Engineering Division [India]
J. Inst. Eng. (India), Electron. Telecommun. Eng. Div.	Journal of the Institution of Engineers (India), Electronics and Telecommunication Engineering Division [India]
J. Inst. Eng. (India), Environ. Eng. Div.	Journal of the Institution of Engineers (India), Environmental Engineering Division [India]

J. Inst. Eng. (India), Interdiscip. Gen. Eng.	Journal of the Institution of Engineers (India), Interdisciplinary and General Engineering [India]
J. Inst. Eng. (India), Interdiscip. Panels.	Journal of the Institution of Engineers (India), Interdisciplinary Panels [India]
J. Inst. Eng. (India), Mech. Eng. Div.	Journal of the Institution of Engineers (India), Mechanical Engineering Division [India]
J. Inst. Eng. (India), Metall. Mater. Sci. Div.	Journal of the Institution of Engineers (India), Metallurgy and Materials Science Division [India]
J. Inst. Eng. (India), Min. Metall. Div.	Journal of the Institution of Engineers (India), Mining and Metallurgy Division [India]
J. Inst. Ind. Sci., Univ. Tokyo	Journal of Institute of Industrial Science, University of Tokyo [Japan]
J. Inst. Nav.	Journal of the Institute of Navigation [UK]
J. Inst. Refract. Eng.	Journal of the Institute of Refractories Engineers [UK]
J. Inst. Telev. Eng. Jpn.	Journal of the Institute of Television Engineers of Japan
J. Inst. Water Eng. Environ. Manage.	Journal of the Institution of Water Engineers and Environmental Management [UK]
J. Inst. Wood Sci.	Journal of the Institute of Wood Science [UK]
J. Intell. Robot. Syst., Theory Appl.	Journal of Intelligent and Robotic Systems: Theory and Applications [Netherlands]
J. Intell. Syst.	Journal of Intelligent Systems [UK]
J. Int. Market. Market. Res.	Journal of International Marketing and Marketing Research [of European Commission for Industrial Marketing, UK]
J. Iowa Acad. Sci.	Journal of the Iowa Academy of Sciences [US]
JIPDEC Rep.	JIPDEC Report [of Japan Information Processing Development Center]
JIRC	Journal of Information Research Communications [UK]
J. Iron Steel Inst. Jpn.	Journal of the Iron and Steel Institute of Japan
JIS	Journal of Information Science [of Institute of Information Scientists, UK]
J. Jpn. Air Clean. Assoc.	Journal of the Japan Air Cleaning Association
J. Jpn. Compos. Mater	Journal of Japan Composite Materials
J. Jpn. Hydraul. Pneum. Soc.	Journal of the Japan Hydraulics and Pneumatics Society
J. Jpn. Inst. Light Met.	Journal of the Japan Institute of Light Metals [of Japan Light Metal Association]
J. Jpn. Inst. Met.	Journal of the Japan Institute of Metals
J. Jpn. Inst. Navig.	Journal of the Japan Institute of Navigation
J. Jpn. Sewage Works Assoc.	Journal of the Japan Sewage Works Association
J. Jpn. Soc. Aeronaut. Space Sci.	Journal of the Japan Society for Aeronautical and Space Sciences
J. Jpn. Soc. Air Pollut.	Journal of the Japan Society of Air Pollution
J. Jpn. Soc. Artif. Intell.	Journal of the Japanese Society for Artificial Intelligence
J. Jpn. Soc. Civ. Eng.	Journal of the Japan Society of Civil Engineers
J. Jpn. Soc. Colour Mater.	Journal of the Japan Society of Colour Material
J. Jpn. Soc. Compos. Mater.	Journal of the Japan Society for Composite Materials
J. Jpn. Soc. Heat Treat.	Journal of the Japan Society of Heat Treatment
J. Jpn. Soc. Lubr. Eng.	Journal of the Japan Society of Lubrication Engineers
J. Jpn. Soc. Powder Powder Metall.	Journal of the Japan Society of Powder and Powder Metallurgy
J. Jpn. Soc. Precis. Eng.	Journal of the Japan Society of Precision Engineering
J. Jpn. Soc. Simul. Technol.	Journal of the Japan Society for Simulation Technology
J. Jpn. Soc. Strength Fract. Mater.	Journal of the Japanese Society for Strength and Fracture of Materials
J. Jpn. Soc. Technol. Plast.	Journal of the Japan Society for Technology of Plasticity
J. Jpn. Soc. Tribol.	Journal of the Japanese Society of Tribologists
J. Jpn. Water Works Assoc.	Journal of the Japan Water Works Association
J. Jpn. Weld. Soc.	Journal of the Japan Welding Society
J. Jpn. Wood Res. Soc.	Journal of the Japan Wood Research Society
J. JSLE, Int. Ed.	Journal of JSLE, International Edition [Japan]
J. Korea Inf. Sci. Soc.	Journal of the Korea Information Science Society [South Korea]
J. Korea Inst. Electron. Eng.	Journal of the Korea Institute of Electronics Engineers [South Korea]
J. Korean Ceram. Soc.	Journal of the Korean Ceramic Society [South Korea]
J. Korean Inst. Chem. Eng.	Journal of the Korean Institute of Chemical Engineers [South Korea]
J. Korean Inst. Met.	Journal of the Korean Institute of Metals [South Korea]
J. Korean Inst. Miner. Min. Eng.	Journal of the Korean Institute of Mineral and Mining Engineers [South Korea]
J. Korean Inst. Telemat. Electron.	Journal of the Korean Institute of Telematics and Electronics [South Korea]
J. Korean Nucl. Soc.	Journal of the Korean Nuclear Society [South Korea]
J. Korean Phys. Soc.	Journal of the Korean Physical Society [South Korea]

J. Lab. Clin. Med.	Journal of Laboratory and Clinical Medicine [US]
J. Less-Common Met.	Journal of Less-Common Metals [Switzerland]
J. Light Met. Weld. Constr.	Journal of Light Metal Welding and Construction [Japan]
J. Light Vis. Environ.	Journal of Light and Visual Environment [of Illuminating Engineering Institute of Japan]
J. Lightwave Technol.	Journal of Lightwave Technology [of Institute of Electrical and Electronics Engineers, US]
J. Liquid Chromat.	Journal of Liquid Chromatography [US]
JLMA Lett.	JLMA Letters [of Japan Light Metal Association]
JLMS	Journal of the London Mathematical Society [UK]
J. Log. Program.	Journal of Logic Programming [US]
J. Low Freq. Noise Vib.	Journal of Low Frequency Noise and Vibration [UK]
J. Low Temp. Phys.	Journal of Low Temperature Physics [US]
J. Lubr. Technol. (Trans. ASME)	Journal of Lubrication Technology (Transactions of the ASME) [of American Society of Mechanical Engineers]
J. Lumin.	Journal of Luminescence [Netherlands]
JM	Journal of Micrographics [US]
J. Macromol. Sci.	Journal of Macromolecular Science [US]
J. Macromol. Sci. A	Journal of Macromolecular Science, Part A [US]
J. Macromol. Sci. B	Journal of Macromolecular Science, Part B [US]
J. Macromol. Sci. C	Journal of Macromolecular Science, Part C [US]
J. Macromol. Sci. Chem.	Journal of Macromolecular Science — Chemistry [US]
J. Macromol. Sci. Phys.	Journal of Macromolecular Science — Physics [US]
J. Macromol. Sci. Rev. Macromol. Chem. Phys.	Journal of Macromolecular Science — Reviews in Macromolecular Chemistry and Physics [US]
J. Magn. Magn. Mater.	Journal of Magnetism and Magnetic Materials [Netherlands]
J. Magn. Reson.	Journal of Magnetic Resonance [US]
JMA J.	JMA Journal [of Japan Management Association]
J. Manage. Inf. Syst.	Journal of Management Information Systems [US]
J. Manuf. Oper. Manage.	Journal of Manufacturing and Operations Management [US]
J. Manuf. Syst.	Journal of Manufacturing Systems [of Society of Manufacturing Engineers, US]
J. Mat. Synth. Process.	Journal of Materials Synthesis and Processing [US]
J. Mater. Energy Syst.	Journal of Materials for Energy Systems [US]
J. Mater. Eng.	Journal of Materials Engineering [US]
J. Mater. Process. Technol.	Journal of Materials Processing Technology [Netherlands]
J. Mater. Res.	Journal of Materials Research [of Materials Research Society, US]
J. Mater. Sci.	Journal of Materials Science [UK]
J. Mater. Sci. Lett.	Journal of Materials Science Letters [UK]
J. Mater. Sci. Soc. Jpn.	Journal of the Materials Science Society of Japan
J. Mater. Shaping Technol.	Journal of Materials Shaping Technology [of ASM International, US]
J. Mater. Synth. Process.	Journal of Materials Synthesis and Processing [US]
J. Math. Anal. Appl.	Journal of Mathematical Analysis and Application [US]
J. Math. Biol.	Journal of Mathematical Biology [Germany]
J. Math. Phys.	Journal of Mathematical Physics [of American Institute of Physics]
J. Math. Phys. Sci.	Journal of Mathematical and Physical Sciences [of Indian Institute of Technology]
J. Math. Soc. Jpn.	Journal of the Mathematical Society of Japan
J. Meas. Control (Trans. ASME)	Journal of Measurement and Control (Transactions of ASME) [of American Society of Mechanical Engineers]
J. Mech. Behav. Mater.	Journal of the Mechanical Behavior of Materials [UK]
J. Mech. Des. (Trans. ASME)	Journal of Mechanical Design (Transactions of the ASME) [of American Society Mechanical Engineers]
J. Mech. Eng. Lab.	Journal of Mechanical Engineering Laboratory [Japan]
J. Mech. Eng. Sci.	Journal of Mechanical Engineering Science [UK]
J. Mech. Phys. Solids	Journal of the Mechanics and Physics of Solids [UK]
J. Mech. Transm. Autom. Des. (Trans. ASME)	Journal of Mechanisms, Transmissions and Automation in Design (Transactions of the ASME) [of American Society of Mechanical Engineers]
J. Mech. Work. Technol.	Journal of Mechanical Working Technology [Netherlands]
J. Mec. Theor. Appl.	Journal de Mecanique Theorique et Appliquee [France]
J. Med. Chem.	Journal of Medicinal Chemistry [of American Chemical Society]
J. Med. Eng. Technol.	Journal of Medical Engineering and Technology [UK]
J. Med. Nucl. Biophys.	Journal de Medecine Nucleaire et Biophysique [France]
J. Membr. Sci.	Journal of Membrane Science [Netherlands]

J. Met.	Journal of Metals [of Minerals, Metals and Materials Society, US]
J. Meteorol.	Journal of Meteorology [US]
J. Met. Finish. Soc. Jpn.	Journal of the Metal Finishing Society of Japan
J. Met. Finish. Soc. Korea	Journal of the Metal Finishing Society of Korea
J. Microbiol. Methods	Journal of Microbiological Methods [Netherlands]
J. Microcomput. Appl.	Journal of Microcomputer Application [UK]
J. Microcomput. Syst. Manage.	Journal of Microcomputer Systems Management [US]
J. Microencapsulation	Journal of Microencapsulation
J. Microsc.	Journal of Microscopy [UK]
J. Microsc. Spectrosc. Electron.	Journal de Microscopie et de Spectroscopie Electroniques [France]
J. Microw. Power	Journal of Microwave Power [of International Microwave Power Institute, US]
J. Microw. Power Electromagn. Energy	Journal of Microwave Power and Electromagnetic Energy [of International Microwave Power Institute, US]
J. Mine Vent. Soc. S. Afr.	Journal of the Mine Ventilation Society of South Africa
J. Min. Mater. Process. Inst. Jpn.	Journal of the Mining and Materials Processing Institute of Japan
J. Min. Metall. Inst. Jpn.	Journal of the Mining and Metallurgical Institute of Japan
JMKU	Journal of Mathematics of Kyoto University [Japan]
J. Mod. Opt.	Journal of Modern Optics [UK]
J. Mol. Biol.	Journal of Molecular Biology
J. Mol. Cat.	Journal of Molecular Catalysis
J. Mol. Electron.	Journal of Molecular Electronics [UK]
J. Mol. Sci.	Journal of Molecular Science [PR China]
J. Mol. Sci. (Int. Ed.)	Journal of Molecular Science (International Edition) [PR China]
J. Mol. Spectrosc.	Journal of Molecular Spectroscopy [US]
J. Mol. Spectry.	Journal of Molecular Spectroscopy [US]
J. Mol. Struct.	Journal of Molecular Structure [Netherlands]
JMR	Journal of Materials Research [of Materials Research Society, US]
JMSJ	Journal of the Mathematical Society of Japan
J. Multivariate Anal.	Journal of Multivariate Analysis [US]
J. Nanjing Inst. Technol.	Journal of Nanjing Institute of Technology [PR China]
J. Natl. Chem. Lab. Ind.	Journal of the National Chemical Laboratory for Industry [Japan]
J. Natl. Res. Counc. Thail.	Journal of the National Research Council of Thailand [of Thai National Documentation Center]
J. Natl. Tech. Assoc.	Journal of the National Technical Association [US]
J. Nat. Rubber Res.	Journal of Natural Rubber Research [of Rubber Research Institute of Malaysia]
J. Navig.	Journal of Navigation [of Royal Institute of Navigation, UK]
J. NDI	Journal of Nondestructive Inspection [Japan]
J. New Gener. Comput. Syst.	Journal of New Generation Computer Systems [Germany]
JNMM	Journal of Nuclear Materials Management [of Institute of Nuclear Materials Management, US]
J. Non-Cryst. Solids	Journal of Non-Crystalline Solids [Netherlands]
J. Nondestr. Eval.	Journal of of Nondestructive Evaluation [US]
J. Non-Equilib. Thermodyn.	Journal of Non-Equilibrium Thermodynamics [Germany]
J. Nonlinear Sci.	Journal of Nonlinear Science (includes Nonlinear Science Today) [Germany and US]
J. Non-Newton. Fluid Mech.	Journal of Non-Newtonian Fluid Mechanics [Netherlands]
J. Northeast Univ. Technol.	Journal of Northeast University of Technology [PR China]
J. Nucl. Mater.	Journal of Nuclear Materials [Netherlands]
J. Nucl. Mater. Manage.	Journal of Nuclear Materials Management [of Institute of Nuclear Materials Management, US]
J. Nucl. Med.	Journal of Nuclear Medicine [of Society of Nuclear Medicine, US]
J. Nucl. Med. Technol.	Journal of Nuclear Medicine Technology [Society of Nuclear Medicine, US]
J. Nucl. Sci. Technol.	Journal of Nuclear Science and Technology [Japan]
J. Number Theory	Journal of Number Theory [US]
J. OCCA	Journal of the Oil and Colour Chemists Association [UK]
J. Occup. Med.	Journal of Occupational Medicine [US]
Johns Hopkins APL Tech. Dig.	Johns Hopkins APL Technical Digest [of Johns Hopkins University, US]
J. Oil Colour Chem. Assoc.	Journal of the Oil and Colour Chemists' Association [UK]
Joining Mater.	Joining and Materials [UK]
JOLA	Journal of Library Automation [of Library and Information Technology Association, US]
JOM	Journal of Metals [of Minerals, Metals and Materials Society, US]
J. Oper. Manage.	Journal of Operations Management [of American Production and Inventory Control Society]
J. Oper. Res. Soc.	Journal of the Operational Research Society [UK]
J. Oper. Res. Soc. Jpn.	Journal of the Operations Research Society of Japan

J. Opt. Commun.	Journal of Optical Communications [Germany]
J. Opt. (France)	Journal of Optics (France)
J. Optim. Theory Appl.	Journal of Optimization Theory and Applications [US]
J. Opt. (India)	Journal of Optics (India) [of Optical Society of India]
J. Opt. Soc. Am.	Journal of the Optical Society of America [US]
J. Opt. Soc. Am. A, Opt. Image Sci.	Journal of the Optical Society of America A, Optics and Image Science [US]
J. Opt. Soc. Am. B, Opt. Phys.	Journal of the Optical Society of America B, Optical Physics [US]
J. Organomet. Chem.	Journal of Organometallic Chemistry [US]
J. Org. Chem.	Journal of Organic Chemistry [of American Chemical Society]
J. Packag. Technol.	Journal of Packaging Technology [US]
J. Parallel Distrib. Comput.	Journal of Parallel and Distributed Computing [US]
J. Parametrics	Journal of Parametrics [of International Association of Parametric Analysts, US]
J. Pascal Ada Modula-2	Journal of Pascal, Ada and Modula-2 [US]
J. PCI	Journal of the Prestressed Concrete Association [US]
J. Petrol.	Journal of Petrology [UK]
J. Petrol. Geol.	Journal of Petrology and Geology [US]
J. Petrol. Technol.	Journal of Petroleum Technology [of Society of Petroleum Engineers, US]
J. Pharm. Market. Manage.	Journal of Pharmaceutical Marketing and Management [of American Association of Pharmaceutical Scientists]
J. Pharm. Pharmacol.	Journal of Pharmacy and Pharmacology
J. Pharm. Sci.	Journal of Pharmaceutical Science [of American Pharmaceutical Association]
J. Pharm. Soc. Jpn.	Journal of the Pharmaceutical Society of Japan
J. Photogr. Sci.	Journal of Photographic Science [of Royal Photographic Society, UK]
J. Phys.	Journal de Physique [France] Journal of Physics [UK]
J. Phys. A, Math. Gen.	Journal of Physics A, Mathematical and General [of Institute of Physics, UK]
J. Phys. B, At. Mol. Opt. Phys.	Journal of Physics B, Atomic, Molecular and Optical Physics [of Institute of Physics, UK]
J. Phys. C, Solid State Phys.	Journal of Physics C, Solid State Physics [of Institute of Physics, UK]
J. Phys. Chem.	Journal of Physical Chemistry [of American Chemical Society]
J. Phys. Chem. Ref. Data	Journal of Physical and Chemical Reference Data [of American Chemical Society]
J. Phys. Chem. Solids	Journal of Physics and Chemistry of Solids [UK]
J. Phys. Colloq.	Journal de Physique Colloque [France]
J. Phys., Condens. Matter	Journal of Physics, Condensed Matter [of Institute of Physics, UK]
J. Phys. D, Appl. Phys.	Journal of Physics D, Applied Physics [of Institute of Physics, UK]
J. Phys. Earth	Journal of Physics of the Earth [Japan]
J. Phys. E, Sci. Instrum.	Journal of Physics E, Scientific Instruments [of Institute of Physics, UK]
J. Phys. F, Met. Phys.	Journal of Physics F, Metal Physics [of Institute of Physics, UK]
J. Phys. G, Nucl. Part. Phys.	Journal of Physics G, Nuclear and Particle Physics [of Institute of Physics, UK]
J. Phys. Lett. (Orsay)	Journal de Physique Lettres (Orsay) [France]
J. Phys. Oceanogr.	Journal of Physical Oceanography [of American Meteorological Society]
J. Phys. Org. Chem.	Journal of Physical Organic Chemistry [UK]
J. Phys. (Orsay)	Journal of Physics (Orsay) [France]
J. Phys. Soc.	Journal of the Physical Society [Japan]
J. Phys. Soc. Jpn.	Journal of Physical Society of Japan
J. Pipelines	Journal of Pipelines [Netherlands]
J. Pkg. Technol.	Journal of Packaging Technology [US]
J. Plant Growth Regul.	Journal of Plant Growth Regulation [Germany]
J. Plant Physiol.	Journal of Plant Physiology [Germany and US]
J. Plasma Phys.	Journal of Plasma Physics [UK]
J. Plast. Film Sheet.	Journal of Plastic Film and Sheeting [US]
Jpn. Alum. News	Japan Aluminum News
Jpn. Chem. Week	Japan Chemical Week
Jpn. Comput. Q.	Japan Computer Quarterly [of Japan Information Processing Development Center]
Jpn. Ind. Technol. Bull.	Japan Industrial and Technological Bulletin
Jpn. J. Appl. Phys.	Japan Journal of Applied Physics
Jpn. J. Appl. Phys. 1, Regul. Pap. Short Notes	Japan Journal of Applied Physics, Part 1, Regular Papers and Short Notes [Japan]
Jpn. J. Appl. Phys. 2, Lett.	Japan Journal of Applied Physics, Part 2, Letters [Japan]
Jpn. J. Appl. Phys. Suppl.	Japanese Journal of Applied Physics, Supplement [Japan]
Jpn. J. Ergon.	Japanese Journal of Ergonomics
Jpn. J. Freezing Drying	Japanese Journal of Freezing and Drying

Jpn. J. Ind. Health	Japanese Journal of Industrial Health
Jpn. J. Med. Electron. Biol. Eng.	Japanese Journal of Medical Electronics and Biological Engineering [of Japan Society of Medical Electronics and Biological Engineering]
Jpn. J. Water Pollut. Res.	Japanese Journal of Water Pollution Research [Japan]
Jpn. J. Water Res.	Japan Journal of Water Research [Japan]
Jpn. Met. Bull.	Japan Metal Bulletin
Jpn. Plast. Age	Japan Plastics Age
Jpn. Plast. Ind. Ann.	Japan Plastics Industry Annual
Jpn. Pulp Paper	Japan Pulp and Paper [Japan]
Jpn. Railw. Eng.	Journal of Railway Engineering [of Japan Railway Engineers Association]
Jpn. Steel Bull.	Japan Steel Bulletin
Jpn. Steel Works	Japan Steel Works
Jpn. Steel Works Tech. Rev.	Japan Steel Works Technical Review
Jpn. TAPPI	Japan TAPPI [now: Japan TAPPI Journal]
Jpn. TAPPI J.	Japan Tappi Journal [of Japanese Technical Association of the Pulp and Paper Industry]
Jpn. Telecommun. Rev.	Japan Telecommunications Review
J. Polym. Eng.	Journal of Polymer Engineering [UK]
J. Polym. Sci.	Journal of Polymer Science [US]
J. Polym. Sci. A, Polym. Chem.	Journal of Polymer Science, Part A, Polymer Chemistry [US]
J. Polym. Sci. B, Polym. Phys.	Journal of Polymer Science, Part B, Polymer Physics [US]
J. Polym. Sci. C, Polym. Lett.	Journal of Polymer Science, Part C, Polymer Letters [US]
J. Polym. Sci. Lett.	Journal of Polymer Science Letters [US]
J. Polym. Sci. Polym. Symp.	Journal of Polymer Science, Polymer Symposia [US]
J. Powder Bulk Solids Technol.	Journal of Powder and Bulk Solids Technology
J. Power (Trans. ASME)	Journal of Power (Transactions of the ASME) [American Society of Mechanical Engineers]
J. Power Sources	Journal of Power Sources [Switzerland]
J. Press. Vessel Technol. (Trans. ASME)	Journal of Pressure Vessel Technology (Transactions of the ASME) [of American Society of Mechanical Engineers]
J. Proc. R. Soc. New South Wales	Journal and Proceedings of the Royal Society of New South Wales [Australia]
J. Prod. Agric.	Journal of Production Agriculture [of Soil Science Society of America, US]
J. Prod. Innov. Manage.	Journal of Product Innovation Management [US]
J. Prop. Power	Journal of Propulsion and Power [of American Institute of Aeronautics and Astronautics]
J. Prot. Coatings Linings	Journal of Protective Coatings and Linings [US]
J. Pulp Pap. Sci.	Journal of Pulp and Paper Science [of Canadian Pulp and Paper Association]
J. Purch. Mater. Manage.	Journal of Purchasing and Materials Management [of National Association of Purchasing Management, US]
J. Pure Appl. Ultrason.	Journal of Pure and Applied Ultrasonics [of Ultrasonic Society of India]
JQE	Journal of Quantum Electronics [of IEEE Lasers and Electrooptics Society, US]
J. Qual. Technol.	Journal of Quality Technology [of American Society for Quality Control, US]
J. Quant. Spectrosc. Radiat. Transf.	Journal of Quantitative Spectroscopy and Radiative Transfer [UK]
J. Radiat. Res.	Journal of Radiation Research [of Japan Radiation Research Society]
J. Radioanal. Nucl. Chem., Artic.	Journal of Radioanalytical and Nuclear Chemistry, Articles [Switzerland]
J. Radioanal. Nucl. Chem., Lett.	Journal of Radioanalytical and Nuclear Chemistry, Letters [Switzerland]
J. Radiol. Prot.	Journal of Radiological Protection [UK]
J. Radio Res. Lab.	Journal of the Radio Research Laboratories [Japan]
J. Raman Spectrosc.	Journal of Raman Spectroscopy [UK and US]
J. R. Astron. Soc. Can.	Journal of the Royal Astronomical Society of Canada
JRC Rev.	JRC Review [Japan]
J. Rech. Atmos.	Journal de Recherches Atmospheriques [France]
J. Reinf. Plast. Compos.	Journal of Reinforced Plastics and Composites [US]
J. R. Electr. Mech. Eng.	Journal of the Royal Electrical and Mechanical Engineers [UK]
J. Res.	Journal of Research [of Steel Castings Research and Trade Association, UK]
J. Res. Comput. Educ.	Journal of Research on Computing in Education [of International Association for Computing in Education, US]
J. Res. Dev. Educ.	Journal of Research and Development in Education [US]
J. Res. Inst. Catal., Hokkaido Univ.	Journal of the Research Institute for Catalysis, Hokkaido University [Japan]
J. Res. Inst. Sci. Technol., Nihon Univ.	Journal of the Research Institute of Science and Technology, Nihon University [Japan]

J. Res. Natl. Bur. Stand.	Journal of Research of the National Bureau of Standards [US]
J. Res. Natl. Inst. Stand. Technol.	Journal of Research of the National Institute of Standards and Technology [US]
J. Res. NBS	Journal of Research of the National Bureau of Standards [US]
J. Res. NIST	Journal of Research of the National Institute of Standards and Technology [US]
J. Rheol.	Journal of Rheology [of Society of Rheology, US]
J. Robot. Syst.	Journal of Robotic Systems [US]
J. R. Signals Inst.	Journal of the Royal Signals Institution [UK]
J. S. Afr. Acoust. Inst.	Journal of the South African Acoustics Institute
J. S. Afr. Inst. Min. Metall.	Journal of the South African Institute of Mining and Metallurgy
J. Sci. Comput.	Journal of Scientific Computing [US]
J. Sci. Educ. Technol.	Journal of Science Education and Technology [US]
J. Sci. Food Agric.	Journal of the Science of Food and Agriculture [of Society of Chemical Industry, US]
J. Sci. Hiroshima Univ.	Journal of Hiroshima University [Japan]
J. Sci. Hiroshima Univ. A	Journal of Hiroshima University, Series A [Japan]
J. Sci. Hiroshima Univ. B	Journal of Hiroshima University, Series B [Japan]
J. Sci. Ind. Res.	Journal of Scientific and Industrial Research [India]
J. Sci. Instrum.	Journal of Scientific Instruments [US]
J. Sci. Technol.	Journal of Science and Technology [UK]
J. Sci. Res. Banaras Hindu Univ.	Journal of Scientific Research of the Banaras Hindu University [India]
J. Sci. Soc. Thail.	Journal of the Science Society of Thailand [Thailand]
J. Sediment. Petrol.	Journal of Sedimentary Petrology
J. Seismol. Soc. Jpn.	Journal of the Seismological Society of Japan [Japan]
J. Semant.	Journal of Semantics [Netherlands]
J. Semicond.	Journal of Semiconductors [PR China]
J. Semicust. ICs	Journal of Semicustom ICs [UK]
J. Sep. Process Technol.	Journal of Separation Process Technology
J. Serb. Chem. Soc.	Journal of the Serbian Chemical Society [Yugoslavia]
J. Shanghai Jiaotong Univ.	Journal of the Shanghai Jiaotong University [PR China]
J. Sheffield Univ. Metall. Soc.	Journal of the Sheffield University Metallurgical Society [UK]
J. Ship Prod.	Journal of Ship Production [of Society of Naval Architects and Marine Engineers, US]
J. Ship Res.	Journal of Ship Research [of Society of Naval Architects and Marine Engineers, US]
J. Singap. Natl. Acad. Sci.	Journal of the Singapore National Academy of Science
JSME Int. J.	JSME International Journal [of Japan Society of Mechanical Engineers]
JSME Int. J. I, Solid Mech. Strength Mater.	JSME International Journal, Series I, Solid Mechanics, Strength of Materials [of Japan Society of Mechanical Engineers]
JSME Int. J. II, Fluids Eng. Heat Transf. Power Combust. Thermophys. Prop.	JSME International Journal, Series II, Fluids Engineering, Heat Transfer, Power, Combustion, Thermophysical Properties [of Japan Society of Mechanical Engineers]
JSME Int. J. III, Vib. Control Eng. Eng. Ind.	JSME International Journal, Series III, Vibration, Control Engineering, Engineering for Industry [of Japan Society of Mechanical Engineers]
J. Soc. Cosmet. Chem.	Journal of the Society of Cosmetic Chemists [US]
J. Soc. Dyers Color.	Journal of the Society of Dyers and Colorists [UK]
J. Soc. Environ. Eng.	Journal of the Society of Environmental Engineers [UK]
J. Soc. Fiber Sci. Technol.	Journal of the Society of Fiber Science and Technology [Japan]
J. Soc. Instrum. Control Eng.	Journal of the Society of Instrument and Control Engineers [Japan]
J. Soc. Mater. Sci. Jpn.	Journal of the Society of Materials Science, Japan
J. Soc. Nav. Archit. Jpn.	Journal of the Society of Naval Architects of Japan
J. Soc. Occup. Med.	Journal of the Society of Occupational Medicine [US]
J. Soc. Photogr. Sci. Technol. Jpn.	Journal of the Society of Photographic Science and Technology of Japan
J. Soc. Res. Adm.	Journal of the Society of Research Administrators [US]
J. Soc. Rheol.	Journal of the Society of Rheology [US]
J. Soc. Rheol. Jpn.	Journal of the Society of Rheology of Japan
J. Soc. Rubber Ind.	Journal of the Society of Rubber Industry [Japan]
J. Soil Sci.	Journal of Soil Science
J. Sol. Energy Eng. (Trans. ASME)	Journal of Solar Energy Engineering (Transactions of ASME) [of American Society of Mechanical Engineers]

J. Sol. Energy Res.	Journal of Solar Energy Research [of Solar Energy Research Center, Iraq]
J. Solid State Chem.	Journal of Solid State Chemistry [US]
J. Sound Vib.	Journal of Sound and Vibration [UK]
J. South Afr. Acoust. Inst.	Journal of the South African Acoustics Institute [South Africa]
J. South Afr. Inst. Min. Metall.	Journal of the South African Institute of Mining and Metallurgy [South Africa]
J. Sov. Laser Res.	Journal of Soviet Laser Research [US]
J. Space Astron. Res.	Journal of Space and Astronomy Research [Iraq]
J. Spacecr. Rockets	Journal of Spacecraft and Rockets [of American Institute of Aeronautics and Astronautics]
J. Spectrosc. Soc. Jpn.	Journal of the Spectroscopical Society of Japan
JSSC	Journal of Solid-State Circuits [of IEEE Solid-State Circuits Council, US]
J. Stoch. Process. Appl.	Journal of Stochastic Processes and Their Applications [of Bernoulli Society for Mathematical Statistics and Probability, Netherlands]
J. Stat. Phys.	Journal of Statistical Physics [US]
J. Stat. Plan. Inference	Journal of Statistical Planning and Inference [Netherlands]
J. Strain Anal. Eng. Des.	Journal of Strain Analysis for Engineering Design [UK]
J. Struct. Chem.	Journal of Structural Chemistry [Translation of: *Zhurnal Strukturnoi Khimii (USSR)*; published in the US]
J. Struct. Eng.	Journal of Structural Engineering [of American Society of Civil Engineers]
J. Struct. Geol.	Journal of Structural Geology [UK]
J. Supercomput.	Journal of Supercomputing [Netherlands]
J. Supercond.	Journal of Superconductivity [US]
J. Surf. Sci. Soc. Jpn.	Journal of the Surface Science Society of Japan
J. Symb. Comput.	Journal of Symbolic Computation [UK]
J. Symb. Log.	Journal of Symbolic Logic [of Association for Symbolic Logic, US]
J. Syst. Manage.	Journal of Systems Management [of Association for Systems Management, US]
J. Syst. Softw.	Journal of Systems and Software [US]
J. TAPPIK	Journal of the Technical Association of Pulp and Paper Industry of Korea [South Korea]
J. Technol.	Journal of Technology [India]
J. Tech. Phys.	Journal of Technical Physics [of Polish Academy of Sciences]
J. Telecommun. Netw.	Journal of Telecommunication Networks [US]
J. Terramech.	Journal of Terramechanics [of International Society for Terrain-Vehicle Systems, US]
J. Test Eval.	Journal of Test and Evaluation [of International Test and Evaluation Association, US]
J. Test. Eval.	Journal of Testing and Evaluation [of American Society for Testing and Materials]
J. Text. Inst.	Journal of the Textile Institute [UK]
J. Text. Stud.	Journal of Texture Studies
J. Therm. Anal.	Journal of Thermal Analysis [UK]
J. Therm. Biol.	Journal of Thermal Biology [UK]
J. Therm. Eng.	Journal of Thermal Engineering [India]
J. Therm. Insul.	Journal of Thermal Insulation [US]
J. Thermophys. Heat Transf.	Journal of Thermophysics and Heat Transfer [of American Institute of Aeronautics and Astronautics]
J. Thermosetting Plast., Jpn.	Journal of Thermosetting Plastics, Japan
J. Therm. Spray Technol.	Journal of Thermal Spray Technology [of ASM International, US]
J. Therm. Stresses	Journal of Thermal Stresses [US]
J. Time Ser. Anal.	Journal of Time Series Analysis [UK]
J. Tissue Cut. Meth.	Journal of Tissue Culture Methods [of Tissue Culture Association, US]
J. Tongji Univ.	Journal of Tongji University [PR China]
J. Toxicol. Environ. Chem.	Journal of Toxicological and Environmental Chemistry [Switzerland]
J. Toxicol. Environ. Health	Journal of Toxicology and Environmental Health
J. Transp. Res. Forum	Journal of the Transportation Research Forum [US]
J. Tribol. (Trans. ASME)	Journal of Tribology (Transactions of the ASME) [of American Society of Mechanical Engineers]
J. Tsinghua Univ.	Journal of Tsinghua University [PR China]
J. Turbomach. (Trans. ASME)	Journal of Turbomachinery (Transactions of the ASME) [of American Society of Mechanical Engineers]
J. Ultrastruct. Mol. Struct. Res.	Journal of the Ultrastructure and Molecular Structure Research [US]
J. Ultrastr. Res.	Journal of Ultrastructural Research
J. Undergrad. Res. Phys.	Journal of Undergraduate Research in Physics [of Society of Physics Students, US]
J. Univ. Kuwait	Journal of the University of Kuwait
J. Univ. Sci. Technol. Beijing	Journal of the University of Science and Technology Beijing [PR China]
J. Vac. Sci. Technol.	Journal of Vacuum Science and Technology [of American Vacuum Society]

J. Vac. Sci. Technol. A, Vac. Surf. Films	Journal of Vacuum Science and Technology A, Vacuum, Surfaces and Films [of American Vacuum Society]
J. Vac. Sci. Technol. B, Microelectron.Process.Phenom.	Journal of Vacuum Science and Technology B, Microelectronics Processing and Phenomena [of American Vacuum Society]
J. Vac. Soc. Jpn.	Journal of the Vacuum Society of Japan
J. Vib. Acoust. Stress Reliab. Des. (Trans. ASME)	Journal of Vibration, Acoustics, Stress, and Reliability in Design (Transactions of the ASME) [American Society of Mechanical Engineers]
J. Vinyl Technol.	Journal of Vinyl Technology [of Society of Plastics Engineers, US]
J. Virol.	Journal of Virology [of American Society for Microbiology]
J. VLSI Comput. Syst.	Journal of VLSI and Computer Systems [US]
J. Water Poll. Control Fed.	Journal of the Water Pollution Control Federation [US]
J. Water Pollut. Control Fed.	Journal of the Water Pollution Control Federation [US]
J. Water Waste	Journal of Water and Waste [US]
J. Weather Modif.	Journal of Weather Modification [of Weather Modification Association, US]
J. Wind Eng. Ind. Aerodyn.	Journal of Wind Engineering and Industrial Aerodynamics [of International Association for Wind Engineering, Canada]
J. Wood Chem. Technol.	Journal of Wood Chemistry and Technology [US]
J. Xiamen Univ. (Nat. Sci.)	Journal of Xiamen University (Natural Science) [PR China]
J. Xian Inst. Metall. Constr. Eng.	Journal of Xian Institute of Metallurgy and Construction Engineering [PR China]
J. X-Ray Technol.	Journal of X-Ray Technology [of Society of X-Ray Technology, UK]
J. Zhejiang Univ.	Journal of Zhejiang University [PR China]

K

Karachi Univ. J. Sci.	Karachi University Journal of Science [Pakistan]
Kauch. Rezina	Kauchuk i Rezina [USSR]
Kautsch. Gummi Kunstst.	Kautschuk und Gummi, Kunststoffe [Germany]
Kawasaki Rozai Tech. Rep.	Kawasaki Rozai Technical Report [Japan]
Kawasaki Steel Bull.	Kawasaki Steel Bulletin [Japan]
Kawasaki Steel Newsl.	Kawasaki Steel Newsletter [Japan]
Kawasaki Steel Tech. Bull.	Kawasaki Steel Technical Bulletin [Japan]
Kawasaki Steel Tech. Rep.	Kawasaki Steel Technical Report [Japan]
Kawasaki Steel Tech. Rep. (Overseas)	Kawasaki Steel Technical Report (Overseas) [Japan]
KDD Tech. J.	KDD Technical Journal [of International Communications Research Institute, Japan]
Keio Sci. Technol. Rep.	Keio Science and Technology Reports [of Keio University, Japan]
Kema Sci. Tech. Rep.	Kema Scientific and Technical Reports [Netherlands]
Kem. Ind.	Kemija u Industriji [Yugoslavia]
Kent Rev.	Kent Review [UK]
Kent Tech. Rev.	Kent Technical Review, [UK]
Kep- & Hangtech.	Kep- es Hangtechnik [Hungary]
Keram. Z.	Keramische Zeitschrift [Germany]
Key Eng. Mater.	Key Engineering Materials [Switzerland]
Khim. Drev.	Khimiya Drevesiny [USSR]
Khim. Fiz.	Khimicheskaya Fizika [USSR]
Khim. Ind.	Khimiya i Industriya [Bulgaria]
Khim. Neft. Mashinostr.	Khimicheskoei Neftyanoe Mashinostroenie [USSR]
Khim. Prirod. Soed.	Khimiya Prirodnykh Soedinenii [USSR]
Khim. Prom.	Khimicheskaya Promyshlennost [USSR]
Khim. Tekhnol.	Khimicheskaya Tekhnologiya [USSR]
Khim. Volokna	Khimicheskie Volokna [USSR]
Kibern. Vychisl. Tekh.	Kibernetika i Vychislitel'naya Tekhnika [USSR]

KINAM Rev. Fis.	KINAM Revista de Fisica [Mexico]
Kinematics Phys. Celest. Bodies	Kinematics and Physics of Celestial Bodies [Translation of: *Kinematika i Fizika Nebesnykh Tel (Ukrainian SSR)*; published in the US]
Kinematika Fiz. Nebesnykh Tel (Ukrainian SSR)	Kinematika i Fizika Nebesnykh Tel (Ukrainian SSR) [USSR]
K. Nor. Vidensk. Selsk. Forh.	Kongelige Norske Videnskabers Selskabs Forhandlinger [Norway]
K. Nor. Vidensk. Selsk. Skr.	Kongelige Norske Videnskabers Selskabs Skrifter [Norway]
Knowl.-Based Syst.	Knowledge-Based Systems [UK]
Knowl. Eng. Rev.	Knowledge Engineering Review [UK]
Kn. Ved. Inf.	Kniznice a Vedecke Informacie [Czechoslovakia]
Kobelco Tech. Bull.	Kobelco Technical Bulletin [Japan]
Kobelco Technol. Rev.	Kobelco Technology Review [Japan]
Kobe Res. Dev.	Kobe Research and Development [Japan]
Kobe Steel Rep.	Kobe Steel Report [Japan]
Kobe Steel Rep. (Tech. Kobesteel)	Kobe Steel Reports (Technique Kobesteel) [Japan]
Koezl. M. Tud. Akad. Musz. Fiz. Kut. Intez.	Koezlemenyei Magyar Tudomanyos Akademia Muszaki Fizikai Kutato Intezetenek [of Hungarian Academy of Sciences]
Koezl. M. Tud. Akad. Szam.tech. Autom. Kut. Intez.	Koezlemenyei Magyar Tudomanyos Akademia Szamitastechnikai es Automatizalasi Kutato Intezete [Hungary]
Koks Khim.	Koks i Khimiya [USSR]
Kolloid. Zh.	Kolloidnyi Zhurnal [USSR]
Kompleksni Sist. Upr.	Kompleksni Sistemi za Upravlenie [of Bulgarian Academy of Sciences]
Kompoz. Polim. Mater.	Kompozitsionnye Polimernye Materialy [USSR]
Koncar Strucne Inf.	Koncar Strucne Informacije [Yugoslavia]
Konstr. Giessen	Konstruieren und Giessen [of Zentrale fur Gussverwendung, Germany]
Korea Inf. Sci. Soc. Rev.	Korea Information Science Society Review [South Korea]
Korean Appl. Phys.	Korean Applied Physics [of Korean Physical Society, South Korea]
Korean J. Chem. Eng.	Korean Journal of Chemical Engineering [South Korea]
Koroze Ochr. Mater.	Koroze a Ochrana Materialu [Czechoslovakia]
Korrosionsinst. Rapp.	Korrosionsinstitutet Rapport [Sweden]
Korroz. Figy.	Korrozios Figyelo [Hungary]
Korroz. Zashch.	Korroziya i Zashchita v Neftegazovoi Promyslennoist [USSR]
Kosm. Isssled.	Kosmicheske Issledovaniya [USSR]
Kov. Mater.	Kovove Materialy [Czechoslovakia]
K-Plastic. Kautsch. Ztg.	K-Plastic and Kautschuk Zeitung [Germany]
Kristall Tech.	Kristall und Technik [Germany]
KSB Tech. Ber.	KSB Technische Berichte [Germany]
Kubota Tech. Rep.	Kubota Technical Reports [Japan]
Kumamoto J. Math.	Kumamoto Journal of Mathematics [Japan]
Kunstst. Bau	Kunststoffe im Bau [Germany]
Kunstst. Ger. Plast.	Kunststoffe, German Plastics [Germany]
Kunstst. J.	Kunststoff Journal [Germany]
Kunstst. Plast.	Kunststoffe — Plastics [Switzerland]
Kuznechno-Shtampov.	Kuznechno-Shtampovochnoe Proizvodstvo [USSR]
Kvant. Elektron.	Kvantovaya Elektronika [USSR]
Kvant. Elektron. Kiev	Kvantovaya Elektronika, Kiev [USSR]
Kvant. Elektron. Mosk.	Kvantovaya Elektronika, Moskva [USSR]

L

Lab. Dev. Eng. Mater., Kanazawa Univ.	Laboratory for Development of Engineering Materials, Kanazawa University [Japan]
Lab. Equip. Dig.	Laboratory Equipment Digest [UK]

Lab. Forum	Laboratory Forum [US]
Lab. Hazards Bull.	Laboratory Hazards Bulletin [of Royal Society of Chemistry, UK]
Lab. Microcomput.	Laboratory Microcomputer [UK]
Lab. Pract.	Laboratory Practice [UK]
Lanthanide Actinide Res.	Lanthanide and Actinide Research [Netherlands]
Large Scale Syst. Inf. Decis. Technol.	Large Scale Systems in Information and Decision Technologies [Netherlands]
Laser Chem.	Laser Chemistry [UK]
Laserdisk Prof.	Laserdisk Professional [US]
Laser Focus/Electro-Opt.	Laser Focus/Electro-Optics [US]
Laser Optoelektron.	Laser und Optoelektronik [Germany]
Laser Part. Beams	Laser and Particle Beams [UK]
Lasers Appl.	Lasers and Applications [US]
Lasers Life Sci.	Lasers in the Life Sciences [UK]
Lasers Optronics	Lasers and Optronics [US]
Latv. PSR Zinat. Akad. Vestis	Latvijas PSR Zinatnu Akademijas Vestis [USSR]
Latv. PSR Zinat. Akad. Vestis, Fiz. Teh. Zinat. Ser.	Latvijas PSR Zinatnu Akademijas Vestis, Fizikas un Tehnisko Zinatnu Serija [USSR]
Law/Technol.	Law/Technology [of World Association of Lawyers, US]
Lead Res. Dig.	Lead Research Digest [US]
Lead Zink	Lead and Zinc [Japan]
Lead Zinc Q. Rev.	Lead and Zinc Quarterly Review [US]
Lect. Ser. — Von Karman Inst.	Lecture Series — Von Karman Institute [US]
Lenzinger Ber.	Lenzinger Berichte [Austria]
Lesnaya Prom.	Lesnaya Promyshlennost [USSR]
Lesnoe Khoz.	Lesnoe Khozuaistvo [USSR]
Lett. Appl. Microbiol.	Letters in Applied Microbiology [of Society for Applied Bacteriology, UK]
Lett. Math. Phys.	Letters in Mathematical Physics [Netherlands]
Lett. Nuovo Cimento	Lettere al Nuovo Cimento [Italy]
LEW-Nachr.	LEW-Nachrichten [Germany]
Libr. Comput. Syst. Equip. Rev.	Library Computer Systems and Equipment Review [US]
Libr. Hi Tech	Library Hi Tech [US]
Libr. Inf. News	Library Information News [UK]
Libr. Inf. Sci.	Library and Information Science [of Mita Society of Library and Information Science, Japan]
Libr. Micromation News	Library Micromation News [UK]
Libr. Q	Library Quarterly [US]
Libr. Resour. Tech. Serv.	Library Resources and Technical Services [of American Library Association]
Libr. Sci. Slant Doc.	Library Science with a Slant to Documentation [India]
Libr. Softw. Rev.	Library Software Review [US]
Libr. Trends	Library Trends [US]
Lichtbogen	Der Lichtbogen [Germany]
Liet. TSR Mokslu Akad. Darb. A	Lietuvos TSR Mokslu Akademijas Darbai, Serija A [USSR]
Liet. TSR Mokslu Akad. Darb. B	Lietuvos TSR Mokslu Akademijas Darbai, Serija B [USSR]
Light. Aust.	Lighting in Australia [Australia]
Light Met. Age	Light Metal Age [US]
Light. Res. Technol.	Lighting Research and Technology [of Chartered Institution of Building Services Engineers, UK]
Limnol. Oceanogr.	Limnology and Oceanography [US]
Linc. Lab. J.	Lincoln Laboratory Journal [of Massachusetts Institute of Technology, US]
Linde Rep. Sci. Technol.	Linde Reports on Science and Technology [Germany]
Linear Algebr. Appl.	Linear Algebra and Its Applications [US]
Lines Commun.	Lines Communication [UK]
Liq. Cryst.	Liquid Crystals [UK]
LISA	Library and Information Science Abstracts [UK]
LISP Symb. Comput.	LISP and Symbolic Computation [Netherlands]
Listy Cukrov.	Listy Cukrovarnicke [Czechoslovakia]
Liteinoe Proizvod.	Liteinoe Proizvodstvo [USSR]
Lith. Min. Resour.	Lithology and Mineral Resources [US]
Lit. Linguist. Comput.	Literary and Linguistic Computing [UK]
Litov. Fiz. Sb.	Litovskii Fizicheski Sbornik [USSR]
Logist. Today	Logistics Today [of Cranfield Institute of Technology, UK]
Logist. Transp. Rev.	Logistics and Transportation Review [Canada]

L'Onde Elect.	L'Onde Electrique [France]
Low Temp. Sci. A, Phys. Sci.	Low Temperature Science, Series A, Physical Sciences [of Institute of Low Temperature Science, Japan]
LSE Eng. Bull.	LSE Engineering Bulletin [UK]
Lubr. Eng.	Lubrication Engineering [of Society of Tribologists and Lubrication Engineers, US]
Lucas Eng. Rev.	Lucas Engineering Review [UK]
Luft Kaltetech.	Luft und Kaltetechnik [Germany]
Lunar Planet. Inf. Bull.	Lunar and Planetary Information [of Universities Space Research Association, US]

M

Macch. Autom. Compon.	Macchine Automazione & Componenti [Italy]
Mach. Des.	Machine Design [US]
Mach. Korea	Machinery Korea [South Korea]
Mach. Learn.	Machine Learning [Netherlands]
Mach. Mod.	Machine Moderne [France]
Mach. Outil Prod.	Machine Outil Produire [France]
Mach. Prod.	Machines Production [France]
Mach. Prod. Eng.	Machinery and Production Engineering [UK]
Mach. Steel, Austria	Machinery and Steel, Austria
Mach. Tool Blue Book	Machine and Tool Blue Book [US]
Mach. Transl.	Machine Translation [Netherlands]
Mach. Vib. Monit. Anal. Proc.	Machinery Vibration Monitoring and Analysis Proceedings [of Vibration Institute, US]
Mach. Vis. Appl.	Machine Vision and Applications [US]
Macplas Int.	Macplas International [Italy]
Macromol.	Macromolecules [of American Chemical Society]
Macromol. Chem.	Macromolecular Chemistry [US]
Macromol. Chem. Rapid Commun.	Macromolecular Chemistry Rapid Communications [US]
Macromol. Chem. Suppl.	Macromolecular Chemistry Supplement [US]
Macromol. Rev.	Macromolecular Review [US]
Mag. Concr. Res.	Magazine of Concrete Research [UK]
Magnesium Mon. Rev.	Magnesium Monthly Review [US]
Magn. Gidrodin.	Magnitnaya Gidrodinamika [USSR]
Magn. Lett.	Magnetism Letters [US]
Magn. Reson. Imaging	Magnetic Resonance Imaging [UK]
Magn. Reson. Med.	Magnetic Resonance in Medicine [US]
Magn. Reson. Rev.	Magnetic Resonance Review [UK]
Magn. Sep. News	Magnetic Separation News [UK]
Magy. Alum.	Magyar Aluminum [Hungary]
Magy. Kem. Foly.	Magyar Kemiai Folyoirat [Hungary]
Magy. Kem. Lapja	Magyar Kemikusok Lapja [Hungary]
Magy. Textiltech.	Magyar Textiltechnika [Hungary]
Maint. Eng.	Maintenance Engineering [UK]
Maint. J.	Maintenance Journal [of International Maintenance Institute, US]
Maint. Manage. Int.	Maintenance Management International [Netherlands]
Makromol. Chem.	Makromolekulare Chemie [Switzerland]
Makromol. Chem. Macromol. Symp.	Makromolekulare Chemie, Macromolecular Symposia [Switzerland]
Makromol. Chem. Rapid Commum.	Makromolekulare Chemie, Rapid Communications [Switzerland]
Malaysian Rubb. Rev.	Malaysian Rubber Review [of Malaysian Rubber Research and Development Board]
Manage Autom.	Managing Automation [US]
Manage. Decis.	Management Decision [UK]
Manage. Inf.	Management and Informatics [Italy] Managing Information [UK]
Manage. Inf. Syst. Q.	Management Information Systems Quarterly [of Society for Management Information Systems, US]
Manage. Rev.	Management Review [UK]

Manage. Sci.	Management Science [of The Institute of Management Sciences, US]
Manage. Serv.	Management Services [of Institute of Management Services, UK]
Manage. World	Management World [of Administrative Management Society, US]
Mannesmann Forschungsber.	Mannesmann Forschungsberichte [Germany]
M.A.N. Res. Eng. Manuf.	M.A.N. Research, Engineering and Manufacturing [Germany]
ManTech J.	ManTech Journal [US]
Manuf. Chem.	Manufacturing Chemist [UK]
Manuf. Eng.	Manufacturing Engineering [of Society of Manufacturing Engineers, US]
Manuf. Eng. (London)	Manufacturing Engineering (London) [UK]
Manuf. Eng. Mag.	Manufacturing Engineering Magazine [of MAP-TOP Users Group, US]
Manuf. Rev.	Manufacturing Review [US]
Manuf. Syst. Trans.	Manufacturing Systems Transactions [of Society of Manufacturing Engineers, US]
Manuf. Technol. Horiz.	Manufacturing Technology Horizons [US]
Manuscr. Geod.	Manuscripta Geodaetica [Germany]
MAP/TOP Interf. Tech. J.	MAP/TOP Interface Technical Journal [of MAP-TOP Users Group, US]
Mar. Eng. Bull.	Marine Engineers Bulletin [of Institute of Marine Engineers, UK]
Mar. Eng. Rev.	Marine Engineers Review [of Institute of Marine Engineers, UK]
Mar. Geod.	Marine Geodesy [US]
Mar. Geophys. Res.	Marine Geophysical Research [Netherlands]
Mar. Pet. Geol.	Marine and Petroleum Geology [UK]
Mar. Pollut. Bull.	Marine Pollution Bulletin [of International Ocean Disposal Symposium, US]
Mar. Pollut. Bull.	Marine Pollution Bulletin [UK]
Mar. Pollut. News	Marine Pollution News [of International Ocean Disposal Symposium, US]
Mar. Technol.	Marine Technology [of Society of Naval Architects and and Marine Engineers] Marine Technology [of Verein Deutscher Ingenieure, Germany]
Mar. Technol. Soc. J.	Marine Technology Society Journal [US]
Masch.-Bau-Tech.	Maschinenbautechnik [Germany]
Masch. Elektrotech.	Maschinenwelt Elektrotechnik [Germany]
Masch.-Schad.	Maschinenschaden [Germany]
Mass Spectrom. Bull.	Mass Spectrometry Bulletin [of Royal Society of Chemistry, UK]
Mass Spectrosc.	Mass Spectroscopy [of Mass Spectroscopy Society of Japan]
Mat. Apl. Comput.	Matematica Aplicada e Computacional [of Sociedade Brasileira de Matematica Aplicada e Computacional, Brazil]
Mat. Des.	Materials and Design [UK]
Mat. Eng.	Materials Engineering [UK and US]
Mater. Australas.	Materials Australasia [Australia]
Mater. Charact.	Materials Characterization [US and UK]
Mater. Chem.	Materials Chemistry [Italy]
Mater. Chem. Phys.	Materials Chemistry and Physics [Switzerland]
Mater. Constr.	Materiales de Construcciaon [Spain]
Mater. Des.	Materials and Design [UK]
Mater. Edge	Materials Edge [UK]
Mater. Eng.	Materials Engineering [UK and US]
Mater. Eng. (Cleveland)	Materials Engineering (Cleveland) [US]
Mater. Eng. (Surrey)	Materials Engineering (Surrey) [UK]
Mater. Eval.	Materials Evaluation [of American Society for Nondestructive Testing]
Mater. Flow	Material Flow [Netherlands]
Mater. Forum	Materials Forum [Australia]
Mater. Handl. Eng.	Material Handling Engineering [US]
Mater. Handl. Outlook	Material Handling Outlook [of International Material Management Society, US]
Mater. Lett.	Materials Letters [Netherlands]
Mater. Manuf.	Materials and Manufacture [UK]
Mater. Manuf. Process.	Materials and Manufacturing Processes [US]
Mater. Mech. Eng.	Materials for Mechanical Engineering [PR China]
Mater. News Int.	Materials News International [Belgium]
Mater. Ogniotrw.	Materialy Ogniotrwale [Poland]
Mater. Perform.	Materials Performance [of National Association of Corrosion Engineers, US]
Mater. Plast.	Materiale Plastice [Romania]
Mater. Plast. Elast.	Materie Plastiche ed Elastomeri [Italy]
Mater. Process. Rep.	Materials and Processing Report [US]
Mater. Prot.	Materials Protection [PR China]

Mater.-Pruf.	Materialprufung [of Deutscher Verand fur Materialprufung, Germany]
Mater. Reclam. Wkly.	Materials Reclamation Weekly [UK]
Mater. Res. Bull.	Materials Research Bulletin [US]
Mater. Res. Soc. Symp. Proc.	Materials Research Society Symposia Proceedings [US]
Mater. Res. Symp. Proc.	Materials Research Symposium Proceedings [US]
Mater. Sci.	Materials Science [Poland]
Mater. Sci. Eng.	Materials Science and Engineering [Switzerland]
Mater. Sci. Eng. A, Struct. Mater., Prop. Microstruct. Process.	Materials Science and Engineering A, Structural Materials: Properties, Microstructure and Processing [Switzerland]
Mater. Sci. Eng. B, Solid-State Mater. Adv. Technol.	Materials Science and Engineering B, Solid-State Materials for Advanced Technology [Switzerland]
Mater. Sci. Forum	Materials Science Forum [Switzerland]
Mater. Sci. Monogr.	Materials Science Monographs
Mater. Sci. Rep.	Materials Science Reports [Netherlands]
Mater. Sci. Res.	Materials Science Research
Mater. Sci. Technol.	Materials Science and Technology [of Institute of Metals, UK]
Mater. Soc.	Materials and Society [US]
Mater. Struct.	Materials and Structures [UK]
Mater. Syst.	Materials Systems [Japan]
Mater. Tech.	Material und Technik [Switzerland]
Mater. Tech. (Paris)	Materiaux et Techniques (Paris) [France]
Mater. Technol.	Material Technology [Japan]
Mater. Tekhnol.	Materialoznanie i Tekhnologiya [Bulgaria]
Mater. Trans., JIM	Materials Transactions, JIM [of Japan Institute of Metals]
Mater.wiss. Werkst.tech.	Materialwissenschaft und Werkstofftechnik [Germany]
Mat. Fiz. Nelineinaya Mekh.	Matematicheskaya Fizika i Nelineinaya Mekhanika [USSR]
Mat.-Fys. Medd. K. Dan. Vidensk Selsk.	Matematisk-Fysiske Meddelelser Konglige Danske Videnskabernes Selskab [of Royal Danish Academy of Sciences]
Math. Balk.	Mathematica Balcanica [Yugoslvaia]
Math. Comput.	Mathematics of Computation [of American Mathematical Society]
Math. Comput. Educ.	Mathematics and Computer Education [US]
Math. Comput. Model.	Mathematical Computer Modelling [UK and US]
Math. Comput. Sim.	Mathematics and Computers in Simulation [of International Association for Mathematics and Computers in Simulation, Belgium]
Math. Control Signals Syst.	Mathematics of Control, Signals and Systems [US]
Math. Geol.	Mathematical Geology [of International Association for Mathematical Geology, US]
Math. Mag.	Mathematics Magazine [of Mathematical Association of America]
Math. Methods Appl. Sci.	Mathematical Methods in the Applied Sciences [UK]
Math. Model. Numer. Anal.	Mathematical Modelling and Numerical Analysis [France]
Math. Numer. Sin.	Mathematica Numerica Sinica [PR China]
Math. Oper. Res.	Mathematics of Operations Research [of Operations Research Society of America]
Math. Proc. Camb. Philos. Soc.	Mathematical Proceedings of the Cambridge Philosophical Society [UK]
Math. Program.	Mathematical Programming [Netherlands]
Math. Program. A	Mathematical Programming, Series A [Netherlands]
Math. Program. B	Mathematical Programming, Series B [Netherlands]
Math. Rep. Toyama Univ.	Mathematics Reports Toyama University [Japan]
Math. Rev.	Mathematical Reviews [of American Mathematical Society]
Math. Slovaca	Mathematica Slovaca [of Slovak Academy of Sciences, Czechoslovakia]
Math. Soc. Sci.	Mathematical Social Sciences [Netherlands]
Math. Spectr.	Mathematical Spectrum [UK]
Math. Syst. Theory	Mathematical Systems Theory [US]
Mat. Lett.	Materials Letters [Netherlands]
Mat. News Int.	Materials News International [Belgium]
Mat. Plast.	Materiale Plastice [Romania]
Mat. Plast. Elast.	Materie Plastiche ed Elastomeri [Italy]
Mat. Res. Symp. Proc.	Materials Research Symposium Proccedings [US]
Matsushita Electr. Works Tech. Rep.	Matsushita Electric Works Technical Report [Japan]
M.D. Comput.	M.D. Computing [US]

Meas. Control	Measurement and Control [of Institute of Measurement and Control, UK]
Meas. Insp. Technol.	Measurement and Inspection Technology [UK]
Meas. Sci. Technol.	Measurement Science and Technology [UK]
Meas. Tech.	Measurement Techniques [Translation of: *Izmeritel'naya Tekhnika (USSR)*; published in the US]
Mecc. Ital.	Meccanica Italiana [Italy]
Mecc. Mod.	Meccanica Moderna [Italy]
Mech. Autom. Adm.	Mechanizace Automatizace Administrativy [Czechoslovakia]
Mech. Compos. Mater.	Mechanics of Composite Materials [Translation of: *Mekhanika Kompozitnykh Materialov (USSR)*; published in the US]
Mech. Corros. Prop. A, Key Eng. Mater.	Mechanical and Corrosion Properties A, Key Engineering Materials [Switzerland]
Mech. Corros. Prop. B, Single Cryst. Prop.	Mechanical and Corrosion Properties A, Single Crystal Properties [Switzerland]
Mech. Eng.	Mechanical Engineering [of American Society of Mechanical Engineers]
Mech. Eng. Bull.	Mechanical Engineering Bulletin [of Central Mechanical Engineering Research Institute, India]
Mech. Eng. News	Mechanical Engineering News [of American Society for Engineering Education]
Mech. Eng. (NY)	Mechanical Engineering (New York) [of American Society of Mechanical Engineers]
Mech. Eng. Technol.	Mechanical Engineering Technology [UK]
Mech. Inc. Eng.	Mechanical Incorporated Engineer [UK]
Mech. Mater.	Mechanics of Materials [Netherlands]
Mech. Res. Commun.	Mechanical Research Communications [UK]
Mech. Rigid Bodies	Mechanics of Rigid Bodies [Translation of: *Mekhanika Tverdogo Tela (Ukrainian SSR)*; published in the US]
Mech. Solids	Mechanics of Solids [Translation of: *Izvestiya Akademii Nauk SSSR, Mekhanika Tverdogo Tela (USSR)*; published in the US]
Mech. Struct. Mach.	Mechanics of Structures and Machines [US]
Mech. Syst. Signal Process.	Mechanical Systems and Signal Processing [UK]
Mech. Teor. Stosow.	Mechanika Teoretyczna i Stosowana [Poland]
Mec. Mater. Electr.	Mecanique Materiaux Electricite [France]
MECON J.	MECON Journal [India]
Med. Biol. Eng. Comput.	Medical and Biological Engineering and Computing [of International Federation for Medical and Biological Engineering, Canada]
Med. Biol. Eng. Comput.	Medical and Biological Engineering and Computing [UK]
Meded. K. Acad. Wet. Lett. Schone Kunsten Belg.	Mededelingen van de Koninklijke Academie voor Wetenschappen Letteren en Schone Kunsten van Belgie [Belgium]
Med. Energy Res. Rep.	Medical and Energy Research Reports [of Oak Ridge Associated Universities, US]
Med. Inform.	Medical Informatics [UK]
Med. Instrum.	Medical Instrumentation [US]
Med. Phys.	Medical Physics [of American Association of Physicists in Medicine]
Med. Prog. Through Technol.	Medical Progress Through Technology [Netherlands]
Med. Radiol.	Meditsinskaya Radiologiya [USSR]
Med. Res. Eng.	Medical Research Engineering [US]
Med. Tek.	Medicinsk Teknik [Sweden]
Med. Tekh.	Meditsinskaya Tekhnika [USSR]
Meerestech.	Meerestechnik [of Verein Deutscher Ingenieure, Germany]
Mehran Univ. Res. J. Eng. Technol.	Mehran University Research Journal of Engineering and Technology [Pakistan]
Meiden Rev.	Meiden Review [Japan]
Meiden Rev. (Int. Ed.)	Meiden Review (International Edition) [Japan]
Mekh. Avtom. Proizvod.	Mekhanizatsiya i Avtomatizatsiya Proizvodstva [USSR]
Mekh. Kompoz. Mater.	Mekhanika Kompozitnykh Materialov [USSR]
Mekh. Tverd. Tela	Mekhanika Tverdogo Tela [USSR]
Melliand Textilber.	Melliand Textilberichte [Germany]
Mem. Am. Math. Soc.	Memoirs of the American Mathematical Society [US]
Mem. Artillerie Fr.	Memorial de l'Artillerie Francaise [France]
Mem. B.R.G.M	Memoire du B.R.G.M. [of Bureau de Recherches Geologiques etMinieres, France]
Mem. Cl. Sci. Acad. R. Belg.	Memoires de la Classe des Sciences de l'Academie Royale de Belgique [Belgium]
Mem. Coll. Eng., Chuba Univ.	Memoirs of the College of Engineering, Chuba University [Japan]
Mem. Etud. Sci. Rev. Metall.	Memoires et Etudes Scientifiques de la Revue de Metallurgie [France]
Mem. Fac. Eng. Des., Kyoto Inst. Technol.	Memoirs of the Faculty of Engineering and Design, Kyoto Institute of Technology [Japan]

Mem. Fac. Eng. Des., Kyoto Inst. Technol., Ser. Sci. Technol.	Memoirs of the Faculty of Engineering and Design, Kyoto Institute of Technology, Series of Science and Technology [Japan]
Mem. Fac. Eng., Fukui Univ.	Memoirs of the Faculty of Engineering, Fukui University [Japan]
Mem. Fac. Eng., Hiroshima Univ.	Memoirs of the Faculty of Engineering, Hiroshima University [Japan]
Mem. Fac. Eng., Hokkaido Univ.	Memoirs of the Faculty of Engineering, Hokkaido University [Japan]
Mem. Fac. Eng., Kagoshima Univ.	Memoirs of the Faculty of Engineering, Kagoshima University [Japan]
Mem. Fac. Eng., Kobe Univ.	Memoirs of the Faculty of Engineering, Kobe University [Japan]
Mem. Fac. Eng., Kumamoto Univ.	Memoirs of the Faculty of Engineering, Kumamoto University [Japan]
Mem. Fac. Eng., Kyoto Univ.	Memoirs of the Faculty of Engineering, Kyoto University [Japan]
Mem. Fac. Eng., Kyushu Univ.	Memoirs of the Faculty of Engineering, Kyushu University [Japan]
Mem. Fac. Eng., Miyazaki Univ.	Memoirs of the Faculty of Engineering, Miyazaki University [Japan]
Mem. Fac. Eng., Nagoya Univ.	Memoirs of the Faculty of Engineering, Nagoya University [Japan]
Mem. Fac. Eng., Osaka City Univ.	Memoirs of the Faculty of Engineering, Osaka City University [Japan]
Mem. Fac. Eng., Tamagawa Univ.	Memoirs of the Faculty of Engineering, Tamagawa University [Japan]
Mem. Fac. Eng., Yamaguchi Univ.	Memoirs of the Faculty of Engineering, Yamaguchi University [Japan]
Mem. Fac. Sci., Kochi Univ. B, Phys.	Memoirs of the Faculty of Science, Kochi University, Series B, Physics [Japan]
Mem. Fac. Sci., Kyoto Univ.	Memoirs of the Faculty of Science, Kyoto [Japan]
Mem. Fac. Sci., Kyoto Univ., Ser. Phys.Astrophys.Geophys.Chem.	Memoirs of the Faculty of Science, Kyoto, Series of Physics, Astrophysics, Geophysics and Chemistry [Japan]
Mem. Fac. Sci., Kyushu Univ.	Memoirs of the Faculty of Science, Kyushu University [Japan]
Mem. Fac. Sci., Kyushu Univ. A	Memoirs of the Faculty of Science, Kyushu University, Series A [Japan]
Mem. Fac. Sci., Kyushu Univ. B	Memoirs of the Faculty of Science, Kyushu University, Series B [Japan]
Mem. Fac. Sci., Kyushu Univ. C	Memoirs of the Faculty of Science, Kyushu University, Series C [Japan]
Mem. Fac. Technol., Kanazawa Univ.	Memoirs of the Faculty of Technology, Kanazawa University [Japan]
Mem. Fac. Technol., Tokyo Metrop. Univ.	Memoirs of the Faculty of Technology, Tokyo Metropolitan University [Japan]
Mem. Hokkaido Inst. Technol.	Memoirs of the Hokkaido Institute of Technology [Japan]
Mem. Inst. Sci. Ind. Res., Osaka Univ.	Memoirs of the Institute of Scientific and Industrial Research, Osaka University [Japan]
Mem. Kitami Inst. Technol.	Memoirs of the Kitami Institute of Technology [Japan]
Mem. Kyushu Inst. Technol. Eng.	Memoirs of the Kyushu Institute of Technology, Engineering [Japan]
Mem. Natl. Def. Acad.	Memoirs of the National Defense Academy [Japan]
Mem. Res. Inst. Sci. Eng. Ritsumeikan Univ.	Memoirs of the Research Institute of Science and Engineering, Ritsumeikan University [Japan]
Mem. Sch. Eng. Okayama Univ.	Memoirs of the School of Engineering, Okayama University [Japan]
Mem. Sch. Sci. Eng., Waseda Univ.	Memoirs of the School of Science and Engineering, Waseda University [Japan]
Mem. Sect. Stiint. IV	Memoriile Sectiilor Stiintifice, Seria IV [Romania]
Mem. Soc. Astron. Ital.	Memorie della Societa Astronomica Italiana [Italy]
Mem. Tohoku Inst. Technol. I, Sci. Eng.	Memoirs of the Tohoku Institute of Technology, Series I, Science and Engineering [Japan]
Meres Autom.	Meres es Automatika [Hungary]
Mer. Regul.	Mereni a Regulace [Czechoslovakia]
Meruzeler Dokl.	Meruzeler Doklady [USSR]
Messen Prufen Autom.	Messen Prufen Automatisieren [Germany]
METADEX	Metals Abstracts Index [Joint Publication of ASM International, US and Institute of Metals, UK]
Metal. ABM	Metalurgia ABM [Brazil]
Metalcaster Mag.	Metalcaster Magazine [of American Cast Metals Association]
Metal. Electr.	Metalurgia y Electricidad [Spain]
Metal Fabr. News	Metal Fabricating News [of Metal Fabricating Institute, US]
Metalforming Dig.	Metalforming Digest [of ASM International, US]
Metalforming Mach.	Metalforming Machinery [PR China]
Metalforming Mach. Tools	Metalforming Machine Tools [PR China]
Metal. Int.	Metalurgia International [Brazil]
Metall. Anal.	Metallurgical Analysis [PR China]
Metalleiologika Metall. Chron.	Metalleiologika Metallourgika Chronika [Greece]
Metall. Eng.-IIT, Bombay	Metallurgical Engineer-IIT, Bombay [India]
Metall. Equip.	Metallurgical Equipment [PR China]

Metall. Gornorudn. Prom-st.	Metallurgicheskaya i Gornorudnaya Promyshlennost' [USSR]
Metall-Handw. Tech.	Metall-Handwerk und Technik [Germany]
Metall. Ital.	Metallurgia Italiana [Italy]
Metall. Koksokhim.	Metallurgiya i Koksokhimiya [USSR]
Metall. Mater. Technol.	Metallurgist and Materials Technologist [UK]
Metalloved. Term. Obrab. Met.	Metallovedenie i Termicheskaya Obrabotka Metallov [USSR]
Metall. Plant Technol.	Metallurgical Plant and Technology [Germany]
Metall. Plant Technol. Int.	Metallurgical Plant and Technology International [Germany]
Metall. Rev.	Metallurgical Reviews [US]
Metall. Sci. Technol.	Metallurgical Science and Technology [Italy]
Metall. Trans.	Metallurgical Transactions [of ASM International, US]
Metall. Trans. A, Phys. Metall. Mater. Sci.	Metallurgical Transactions A, Physical Metallurgy and Materials Science [of ASM International and American Institute of Mining, Metallurgical and Petroleum Engineers]
Metall. Trans. B, Process Metall.	Metallurgical Transactions, Process Metallurgy [of ASM International and American Institute of Mining, Metallurgical and Petroleum Engineers]
Metal. Odlew.	Metalurgia i Odlewnictwo [Poland]
Metalozn. Obrob. Ciepl., Inz. Powierzchni	Metaloznawstwo, Obrobka Cieplna, Inzynieria Powierzchni [Poland]
Metalozn. Obrobka Ciepl.	Metaloznawstwo i Obrobka Cieplna [Poland]
Metal. Proszkow	Metalurgia Proszkow [Poland]
Metal. Tsvetn. Redk. Met.	Metalurgiya Tsvednykd i Redkikh Metallov [USSR]
Metalwork. Eng. Mark.	Metalworking Engineering and Marketing [Japan]
Metalwork. Interfaces	Metalworking Interfaces [US]
Metalwork. News	Metalworking News [US]
Metalwork. Prod.	Metalworking Production [UK]
Met. Australas.	Metals Australasia [Australia]
Metaux-Corros.-Ind.	Metaux-Corrosion-Industrie [France]
Metaux Deform.	Metaux Deformation [France]
Met. Build. Rev.	Metal Building Review [US]
Met. Bull.	Metal Bulletin [UK]
Met. Bull. Mon.	Metal Bulletin Monthly [UK]
Met. Cast.	Metals and Castings [Australia]
Met. Constr.	Metal Construction [UK]
Meteorol. Atmos. Phys.	Meteorology and Atmospheric Physics [Austria]
Meteorol. Gidrol.	Meteorologiya i Gidrologiya [USSR]
Meteorol. Mag.	Meteorol. Magazine [UK]
Meteorol. Monogr.	Meteorological Monographs [of National Weather Association, US]
Meteorol. Rundsch.	Meteorologische Rundschau [Germany]
Met. Fabr. News	Metal Fabricating News [US]
Met. Finish.	Metal Finishing [US]
Met. Finish. Pract.	Metal Finishing Practice [Japan]
Met. Form.	Metal Forming [US]
Met. Forum	Metals Forum [Australia]
Methods Inf. Med.	Methods of Information in Medicine [Germany]
Methods Oper. Res.	Methods of Operations Research [Germany]
Methods Org. Synth.	Methods of Organic Synthesis [of Royal Society of Chemistry, UK]
Met. Ind.	Metal Industries [Taiwan]
Met. Ind. News	Metals Industry News [UK]
Met. Kunstst.	Metaal & Kunststof [Netherlands]
Met. Mark. Wkly. Rev.	Metal Markets Weekly Review [UK]
Met. Mater.	Metals and Materials [of Institute of Metals, UK]
Met. Mater. (Czech.)	Metallic Materials (Czechoslovakia) [Translation of: *Kovove Materialy*; published in the UK]
Met. Mater. Process.	Metals, Materials and Processes [India]
Met. Miner. Int.	Metals and Minerals International [UK]
Met. News	Metal News [India]
Met.oberfl.: Angew. Elektrochem.	Metalloberflache: Angewandte Elektrochemie [Germany]
MetPack Bus.	MetPack Business [UK]
Met. Powder Rep.	Metal Powder Report [UK]
Met. Prog.	Metal Progress [of ASM International, US]
Metrol. Apl.	Metrologia Aplicata [Romania]

Met. Sci.	Metal Science [of The Metals Society, UK]
Met. Sci. Heat Treat.	Metal Science and Heat Treatment [Translation of: *Metallovednie i Termicheskaya Obrabotka Metallov (USSR)*; published in the US]
Met. Soc. World	Metals Society World [of The Metals Society, UK]
Met. Stamp.	Metal Stamping [US]
Met. Technol.	Metals Technology [of The Metals Society, UK]
Met. Technol. (Jpn.)	Metals and Technology (Japan) [Japan]
Met. Trans.	Metallurgical Transactions [of ASM International and American Institute of Mining, Metallurgical and Petroleum Engineers]
Met. Week	Metals Week [US]
Met. Wkly. Rev.	Metals Weekly Review [UK]
Met. World	Metal World [PR China]
Mezhvuz. Sb. Nauch. Tr., Ekol. Zashchita Lesa.	Mezhvuzovskii Sbornik Nauchnykh Trudov, Ekologiya i Zashchita Lesa [USSR]
Mezhvuz. Sb. Nauch. Tr., Ekon. Prob. Lesoobrabat. Prob.	Mezhvuzovskii Sbornik Nauchnykh Trudov, Ekonomicheskie Problemy Leso-Obrabatyvayushchei Promyshlennosti [USSR]
Mezhvuz. Sb. Nauch. Tr., Khim. Pererab. Drev.	Mezhvuzovskii Sbornik Nauchnykh Trudov, Khimicheskaya Pererabotka Drevesiny [USSR]
Mezhvuz. Sb. Nauch. Tr., Khim. Tekhnol. Tsellyul.	Mezhvuzovskii Sbornik Nauchnykh Trudov, Khimiya i Tekhnologiya Tsellyulozy [USSR]
Mezhvuz. Sb. Nauch. Tr., Lesosech. Lesosklad. Raboty Transport Lesa	Mezhvuzovskii Sbornik Nauchnykh Trudov, Lesosechnye, Lesoskladskie Raboty i Transport Lesa [USSR]
Mezhvuz. Sb. Nauch. Tr., Lesovod. Lesn. Kul'tury Pochvoved.	Mezhvuzovskii Sbornik Nauchnykh Trudov, Lesovodstvo, Lesnye Kul'tury i Pochvovedenie [USSR]
Mezhvuz. Sb. Nauch. Tr., Mashiny Orud. Mekh. Lesozagot. Lesn. Khoz.	Mezhvuzovskii Sbornik Nauchnykh Trudov, Mashiny i Orudiya dlya Mekhanizatsii Lesozagotovok i Lesnogo Khozyaistva [USSR]
Mezhvuz. Sb. Nauch. Tr., Stanki Instrum. Derevoobrabat. Proizv.	Mezhvuzovskii Sbornik Nauchnykh Trudov, Stanki i Instrumenty Derevoobrabatyvayushchikh Proizvodstvo [USSR]
Mezhvuz. Sb. Nauch. Tr., Technol. Oborud. Derevoobrabat. Proizv.	Mezhvuzovskii Sbornik Nauchnykh Trudov, Tekhnologiya i Oborudovanie Derevoobrabatyvayushchikh Proizvodstvo [USSR]
MGA	Meteorological and Geoastrophysical Abstracts [of American Meteorological Society]
M. Geofiz.	Magyar Geofizika [Hungary]
Microbiol. Rev.	Microbiological Reviews [of American Society for Microbiology]
Microbiol. Sci.	Microbiological Sciences [of International Union of Microbiological Societies, UK]
Microchem. J.	Microchemical Journal [of American Microchemical Society]
Microchim. Acta	Microchimica Acta [Austria]
Microcomput. Appl.	Microcomputer Applications [of International Society of Mini- and Microcomputers, Canada]
Microcomput. Appl.	Microcomputer Applications [US]
Microcomput. Civ. Eng.	Microcomputers in Civil Engineering [US]
Micro Comput. Colleg	Micro Computer Colleg [Germany]
Microcomput. Inf. Manage.	Microcomputers for Information Management [US]
Microcomput. Rev.	Microcomputer Review [US]
Micro Decis.	Micro Decision [UK]
Microelectron. Eng.	Microelectronic Engineering [Netherlands]
Microelectron. J.	Microelectronics Journal [UK]
Microelectron. Manuf. Test.	Microelectronic Manufacturing and Testing [US]
Microelectron. Reliab.	Microelectronics and Reliability [UK]
Microform. Rev.	Microform Review [US]
Microgravity Sci. Technol.	Microgravity Science and Technology [Germany]
Micrcogr. Mark. Place	Micrographics Market Place [UK]
Micron Microsc. Acta	Micron and Microscopica Acta [UK]
Microprocess. Microprogr.	Microprocessing and Microprogramming [Netherlands]
Microprocess. Microsyst.	Microprocessors and Microsystems [UK]
Microsc. Electron. Biol. Cel.	Microscopia Electronica y Biologia Celular [of Ibero-American Society for Cell Biology, Chile]

Microscope	The Microscope [US]
Microsoft Syst. J.	Microsoft Systems Journal [US]
Micro Syst.	Micro System [France]
Microwave J.	Microwave Journal [US]
Mikrowellen Mag.	Mikrowellen Magazin [Germany]
Microw. J.	Microwave Journal [US]
Microw. Opt. Technol. Lett.	Microwave and Optical Technology Letters [US]
Microw. RF	Microwave and RF (Radio Frequency) [US]
Microw. RF Eng.	Microwave and RF (Radio Frequency) Engineering [UK]
Middle East Electr.	Middle East Electricity [UK]
Mikro- & Kleincomput.	Mikro- und Kleincomputer [Switzerland]
Mikrocomput. Z.	Mikrocomputer Zeitschrift [Germany]
Mil. Eng.	Military Engineer [of Society of Military Engineers, US]
Mind Lang.	Mind and Language [UK]
Min. Eng.	Mining Engineering [of Society of Mining Engineers, US]
Min. Eng. (Colorado)	Mining Engineering (Colorado) [of Society of Mining Engineers, US]
Min. Eng. (London)	Mining Engineering (London) [UK]
Min. Eng. (NY)	Mining Engineering (New York) [of Society of Mining Engineers, US]
Mine Quarry	Mine and Quarry
Mineral. J.	Mineralogical Journal [US]
Mineral. Mag.	Mineralogical Magazine [Mineralogical Society, UK]
Miner. Eng.	Minerals Engineering [US]
Miner. Ind. Int.	Minerals Industry International [UK]
Miner. Ind. Surv., Alum.	Mineral Industry Surveys, Aluminum [US]
Miner. Ind. Surv. Bauxite	Mineral Industry Surveys, Bauxite [US]
Miner. Ind. Surv., Chromium	Mineral Industry Surveys, Chromium [US]
Miner. Ind. Surv., Tungsten	Mineral Industry Surveys, Tungsten [US]
Miner. Mag.	Mineralogical Magazine [of Mineralogical Society, UK]
Miner. Mat.	Minerals and Materials [US]
Miner. Metal.	Mineracao Metalurgia [Brazil]
Miner. Metall. Process.	Minerals and Metallurgical Processing [of Society of Mining Engineers, US]
Miner. Met. Rev.	Minerals and Metals Review [India]
Miner. Process. Extract. Metall. Rev.	Mineral Processing and Extractive Metallurgy Review
Miner. Resour. Eng.	Mineral Resources Engineering [UK]
Miner. Sci. Eng.	Minerals Science and Engineering [South Africa]
Miner. Soc. Bull.	Mineralogical Society Bulletin [of Mineralogical Society, UK]
Mines Metall.	Mines et Metallurgie [France]
Min. Geol.	Mineria y Geologia [Cuba]
Mini Micro Mag.	Mini Micro Magazin [Germany]
Mini-Micro Syst.	Mini-Micro Systems [US]
Mini-Micro Syst.	Mini-Micro Systems [PR China]
Min. J.	Mining Journal [UK]
Min. Mag.	Mining Magazine [UK]
Min. Metall.	Mining and Metallurgy
Min. News Lett.	Mining News Letter [of United Mining Councils of America]
Minor Planet. Circ.	Minor Planet Circulars [of Smithsonian Astrophysical Observatory, US]
Min. Q.	Mining Quarterly [Australia]
Min. Sci. Technol.	Mining Science and Technology
Min. Technol.	Mining Technology [of Institution of Mining Electrical and Mining Mechanical Engineers, UK]
MINTEK Rep.	MINTEK Report [South Africa]
MINTEK Res. Dig.	MINTEK Research Digest [South Africa]
MINTEK Rev.	MINTEK Review [South Africa]
Mitsubishi Cable Ind. Rev.	Mitsubishi Cable Industries Review [Japan]
Mitsubishi Denki Lab. Rep.	Mitsubishi Denki Laboratory Reports [Japan]
Mitsubishi Electr. Adv.	Mitsubishi Electric Advance [Japan]
Mitsubishi Heavy Ind. Tech. Rev.	Mitsubishi Heavy Industries Technical Review [Japan]
Mitsubishi Steel Manuf. Tech. Rev.	Mitsubishi Steel Manufacturing Technical Review [Japan]
Mitsubishi Tech. Bull.	Mitsubishi Technical Bulletin [Japan]
Mitsui Zosen Tech. Rev.	Mitsui Zosen Technical Review [Japan]
Mitt. AGEN	Mitteilungen AGEN [Switzerland]

Mitt. Astron. Ges.	Mitteilungen der Astronomischen Gesellschaft [Germany]
Mitt. Bundesforsch. Forst-Holzwirtsch.	Mitteilungen der Bundesforschungsanstalt fur Forst- und Holzwirtschaft [Germany]
Mitt. Tech. Univ. Carolo-Wilhelmina	Mitteilungen der Technischen Universitat Carolo-Wilhelmina [Germany]
Mod. Cast.	Modern Casting [US]
Model. Identif. Control	Modeling, Identification and Control [of Royal Norwegian Council for Scientific and Industrial Research, Norway]
Model., Math. Anal. Numer.	Modelisation, Mathematique et Analyse Numerique [of Association Francaise pour le Cybernetique, Economique et Technique, France]
Model. Simul. Control A	Modelling, Simulation and Control A [France]
Model. Simul. Control B	Modelling, Simulation and Control B [France]
Model. Simul. Control C	Modelling, Simulation and Control C [France]
Mod. Geol.	Modern Geology [UK]
Mod. Mach. Shop	Modern Machine Shop [US]
Mod. Mater. Handl.	Modern Materials Handling [US]
Mod. Met.	Modern Metals [US]
Mod. Off. Technol.	Modern Office Technology [US]
Mod. Paint Coat.	Modern Paint and Coatings [US]
Mod. Phys. Lett. A	Modern Physics Letters A [Singapore]
Mod. Phys. Lett. B	Modern Physics Letters B [Singapore]
Mod. Plast.	Modern Plastics [US]
Mod. Plast. Int.	Modern Plastics International [Switzerland]
Mod. Power Syst.	Modern Power Systems [UK and US]
Mod. Steel Constr.	Modern Steel Construction [of American Institute of Steel Construction]
Mod. Tire Dealer	Modern Tire Dealer [US]
Mol. Cryst. Liq. Cryst.	Molecular Crystals and Liquid Crystals [UK]
Mol. Cryst. Liq. Cryst. Lett Sect.	Molecular Crystals and Liquid Crystals, Letters Section [UK]
Mol. Cryst. Liq. Cryst. Suppl. Ser.	Molecular Crystals and Liquid Crystals, Supplement Series [UK]
Mol. Cryst. Liq. Cryst. Sci. Technol. A, Mol. Cryst. Liq. Cryst.	Molecular Crystals and Liquid Crystals, Section A: Molecular Crystals and Liquid Crystals [UK]
Mol. Cryst. Liq. Cryst. Sci. Technol. B,Nonlin. Opt.	Molecular Crystals and Liquid Crystals, Section B: Nonlinear Optics [UK]
Mol. Cryst. Liq. Cryst. Sci. Technol. C,Mol. Cryst.	Molecular Crystals and Liquid Crystals, Section C: Molecular Crystals [UK]
Mol. Cryst. Liq. Cryst. Sci. Technol. D, Display Imaging	Molecular Crystals and Liquid Crystals, Section D: Display and Imaging [UK]
Mol. Phys.	Molecular Physics [UK]
Molysulfide Newsl.	Molysulfide Newsletter [US]
Monatsh. Chem.	Monatshefte fur Chemie [Austria]
Mon. J. Jpn. Brass Makers Assoc.	Monthly Journal of Japan Brass Makers Association
Mon. Notes Astron. Soc. South Afr.	Monthly Notes of the Astronomical Society of South Africa
Mon. Not. R. Astron. Soc.	Monthly Notices of the Royal Astronomical Society [UK]
Mon. Stat. Rev.	Monthly Statistical Review [UK]
Mon. Tech. Rev.	Monthly Technical Review [Germany]
Mon. Weather Rev.	Monthly Weather Review [of American Meteorological Society]
Mosc. Univ. Comput. Math. Cybern.	Moscow University Computational Mathematics and Cybernetics [Translation of: *Vestnik Moskovskogo Universiteta, Vychislitel'naya Matematika i Kibernetika (USSR)*; published in the US]
Mosc. Univ. Phys. Bull.	Moscow University Physics Bulletin [Translation of: *Vestnik Moskovskogo Universiteta, Fizika-Astronomiya (USSR)*; published in the US]
Motorola Tech. Dev.	Motorola Technical Developments [US]
Motorola Tech. Discl. Bull.	Motorola Technical Disclosure Bulletin [US]
Motortech. Z.	Motortechnische Zeitschrift [Germany]
Mot. Ship	Motor Ship [UK]
MPR Met. Powder Rep.	MPR Metal Powder Report [US]
MRL Bull. Res. Dev.	MRL (Materials Research Laboratories) Bulletin of Research and Development [Taiwan]
MRS Bull.	MRS Bulletin [of Materials Research Society, US]
MRS Symp. Proc.	MRS Symposia Proceedings [of Materials Research Society, US]
MSB	Mass Spectrometry Bulletin [of Royal Society of Chemistry, UK]

MSN Microw. Syst. News Commun. Technol.	MSN Microwave Systems News and Communications Technology [US]
MSR, Mess. Steuern Regeln	MSR Messen Steuern Regeln [Germany]
MUG Q.	MUG Quarterly [of MUMPS (Massachusetts General Hospital Utility Multi-Programming System) Users Group, US]
Multiphase Sci. Technol.	Multiphase Science and Technology
Mundo Electron.	Mundo Electronico [Spain]
Muon Catal. Fusion	Muon Catalyzed Fusion [Switzerland]
Music Technol.	Music Technology [UK]

N

Nachr. Dok.	Nachrichten fur Dokumentation [Germany]
Nachr.tech.	Nachrichtentechnik [Germany]
Nachr.tech. Elektron.	Nachrichtentechnik Elektronik [Germany]
Nadezhn. Dolgovech. Mash. Sooruzh.	Nadezhnost i Dolgovechnost Mashin i Sooruzhenii [USSR]
NARI Met. Rep.	NARI Metals Report [of National Association of Recycling Industries, US]
NASA Conf. Publ.	NASA Conference Publication [US]
NASA Contract. Rep.	NASA Contractor Report [US]
NASA Ref. Publ.	NASA Reference Publication [US]
NASA Spec. Publ.	NASA Special Publication [US]
NASA-STAR	NASA Science and Technology Aerospace Reports [US]
NASA Tech. Briefs	NASA Technical Briefs [US]
NASA Tech. Memo.	NASA Technical Memorandum [US
NASA Tech. Note	NASA Technical Note [US]
NASA Tech. Pap.	NASA Technical Paper [US]
NASA Tech. Rep.	NASA Technical Report [US]
Nat. Gas Ind. Technol.	Natural Gas Industrial Technology [Canada]
Nat. Hazards	Natural Hazards [Netherlands]
Natl. Acad. Sci. Lett.	National Academy of Science Letters [India]
Natl. Bur. Stand. Build. Sci. Ser.	National Bureau of Standards Building Science Series [US]
Natl. Bur. Stand. Monogr.	National Bureau of Standards Monograph [US]
Natl. Bur. Stand. Res. Rep.	National Bureau of Standards Research Report [US]
Natl. Bur. Stand. Spec. Publ.	National Bureau of Standards Special Publication [US]
Natl. Bur. Stand. Tech. Note	National Bureau of Standards Technical Note [US]
Natl. Bur. Stand. Tech. Rep.	National Bureau of Standards Technical Report [US]
Natl. Bur. Stand. Update	National Bureau of Standards Update [US]
Natl. Contract Manage. J.	National Contract Management Journal [National Contract Management Association, US]
Natl. Electron. Rev.	National Electronics Review [UK]
Natl. Eng.	National Engineer [of National Association of Power Engineers, US]
Natl. Inst. Stand. Technol. Build. Sci. Ser.	National Institute of Standards and Technology Building Science Series [US]
Natl. Inst. Stand. Technol. Monogr.	National Institute of Standards and Technology Monograph [US]
Natl. Inst. Stand. Technol. Res. Rep.	National Institute of Standards and Technology Research Report [US]
Natl. Inst. Stand. Technol. Spec. Publ.	National Institute of Standards and Technology Special Publication [US]
Natl. Inst. Stand. Technol. Tech. Note	National Institute of Standards and Technology Technical Note [US]
Natl. Inst. Stand. Technol. Tech. Rep.	National Institute of Standards and Technology Technical Report [US]
Natl. Inst. Stand. Technol. Update	National Institute of Standards and Technology Update [US]
Natl. Lucht- en Ruimtevaartlab. Rep.	Nationaal Lucht- en Ruimtevaartlaboratorium Report [Netherlands]

Natl. Phys. Lab. Rep.	National Physical Laboratory Reports [UK]
Natl. Tech. Rep.	National Technical Report [Japan]
Natl. Water Cond.	National Water Conditions [US Department of the Interior]
Natl. Weather Dig.	National Weather Digest [of National Weather Association, US]
NATO Adv. Study Inst. Ser. A	NATO Advanced Study Institutes Series A
NATO Adv. Study Inst. Ser. B	NATO Advanced Study Institutes Series B
NATO Adv. Study Inst. Ser. C	NATO Advanced Study Institutes Series C
NATO Adv. Study Inst. Ser. D	NATO Advanced Study Institutes Series D
NATO Adv. Study Inst. Ser. E	NATO Advanced Study Institutes Series E
NATO ASI Ser. A	NATO ASI (Advanced Study Institutes) Series A
NATO ASI Ser. B	NATO ASI (Advanced Study Institutes) Series B
NATO ASI Ser. C	NATO ASI (Advanced Study Institutes) Series C
NATO ASI Ser. D	NATO ASI (Advanced Study Institutes) Series D
NATO ASI Ser. E	NATO ASI (Advanced Study Institutes) Series E
Nat. Prod. Rep.	Natural Products Reports [of Royal Society of Chemistry, UK]
Nat. Resour. Forum	Natural Resources Forum [US]
Nat. Rubber Dev.	Natural Rubber Development [of Malaysian Rubber Bureau]
Nat. Rubber News	Natural Rubber News [of Malaysian Rubber Bureau]
Naturwissenschaften	Die Naturwissenschaften [Germany]
Nauchno-Tekh. Inf. 1	Nauchno-Tekhnicheskaya Informatisaya, Seriya 1 [USSR]
Nauchno-Tekh. Inf. 2	Nauchno-Tekhnicheskaya Informatisaya, Seriya 2 [USSR]
Nauk. Inf.	Naukovedenie i Informatika [USSR]
Nav. Eng. J	Naval Engineers Journal [of American Society of Naval Engineers]
Navig. J. Inst. Navig.	Navigation, Journal of the Institute of Navigation US]
Nav. Res. Logist.	Naval Research Logistics [US]
Nav. Res. Rev.	Naval Research Review [of Office of Naval Research, US]
Navy Domest. Technol. Transf. Fact Sheet	Navy Domestic Technology Transfer Fact Sheet [US Navy]
Navy Int.	Navy International [UK]
Navy Technol. Transfer Fact Sheet	Navy Technology Transfer Fact Sheet [US Navy]
NBS Build. Sci. Ser.	NBS Building Science Series [of National Bureau of Standards, US]
NBS Monogr.	NBS, Monograph [of National Bureau of Standards, US]
NBS Res. Rep.	NBS Research Report [of National Bureau of Standards, US]
NBS Spec. Publ.	NBS Special Publication [of National Bureau of Standards, US]
NBS Tech. Note	NBS Technical Note [of National Bureau of Standards, US]
NBS Tech. Rep.	NBS Technical Report [of National Bureau of Standards, US]
NBS Update	NBS Update [of National Bureau of Standards, US]
NCASI Tech. Bull	NCASI Technical Bulletin [of National Council of the Paper Industry for Air and Stream Improvement, US]
NCASI Tech. Rev. Index	NCASI Technical Review Index [of National Council of the Paper Industry for Air and Stream Improvement, US]
NCGR Res. J.	NCGR Research Journal [of National Council for Geocosmic Research, US]
NCIO Bull.	NCIO Bulletin [of National Congress of Inventors Organizations, US]
NCIO Newsl.	NCIO Newsletter [of National Congress of Inventors Organizations, US]
NCSL Newsl.	NCSL Newsletter [of National Conference of Standards Laboratories, US]
NCT Newsl.	NCT Newsletter [UK]
NDT Int.	NDT International [UK]
NEA Act. Rep.	NEA Activity Report [of Nuclear Energy Agency [of OECD, France]]
NEA Newsl.	NEA Newsletter [of Nuclear Energy Agency [of OECD, France]]
NEC Res. Dev.	NEC Research and Development [Japan]
NEC Tech. J.	NEC Technical Journal [Japan]
Ned. Tijdschr. Natuurkd. A	Nederlands Tijdschrift voor Natuurkunde A [of Nederlandse Natuurkundige Vereniging, Netherlands]
Ned. Tijdschr. Natuurkd. B	Nederlands Tijdschrift voor Natuurkunde B [of Nederlandse Natuurkundige Vereniging, Netherlands]
NEL Rep.	NEL Report [UK]
NESA J.	NESA Journal [of National Energy Specialists Association, US]
NETA Newsl.	NETA Newsletter [of National Environmental Training Association, US]
NET Nachr. Elektron. + Telemat.	NET Nachrichten Elektronik + Telematik [Germany]
Netw., Comput. Neural Syst.	Network: Computation in Neural Systems [UK]
Netw. Manage.	Networking Management [US]
Neue Bergbautech.	Neue Bergbautechnik [Germany]
Neue Tech.	Neue Technik [Switzerland]

Neue Verpack.	Neue Verpackung [Germany]
Neural Netw.	Neural Networks [US]
New Electron.	New Electronics [UK]
New Gener. Comput.	New Generation Computing [Japan]
New J. Chem.	New Journal of Chemistry [France]
New Mat. Int.	New Materials International [UK]
New Mat./Jpn.	New Materials/Japan [UK]
New Mat./Korea	New Materials/Korea [UK]
New Mat. World	New Materials World [UK]
New Phys. (Korean Phys. Soc.)	New Physics (Korean Physical Society) [of Korea Institute for Industrial Economics and Technology, South Korea]
New Sci.	New Scientist [UK]
News Govt. Ind. Res. Inst., Osaka	News of Government Industrial Research Institute, Osaka [Japan]
Newsl. NEA Data Bank	Newsletter of the NEA Data Bank [of Nuclear Energy Agency [of OECD, France]]
News Nisshin Steel	News from Nisshin Steel [Japan]
News Physiol. Sci.	News in Physiological Sciences [of International Union of Physiological Sciences, France]
News Rohde Schwarz	News from Rohde and Schwarz [Germany]
New Technol. Jpn.	New Technology Japan [Japan]
NFAIS Bull.	NFAIS Bulletin [of National Federation of Abstracting and Information Services, US]
NFAIS Newsl.	NFAIS Bulletin [of National Federation of Abstracting and Information Services, US]
NGI Mag.	NGI Magazine [of Netherlands Society for Informatics]
NHK Lab. Note	NHK Laboratories Note [Japan]
NHK Tech. J.	NHK Technical Journal [Japan]
Nickel Top.	Nickel Topics [US]
Nikkei Electron.	Nikkei Electronics [Japan]
Nippon Kokan Tech. Bull.	Nippon Kokan Technical Report [Japan]
Nippon Kokan Tech. Rep.	Nippon Kokan Technical Report [Japan]
Nippon Stainless Tech. Rep.	Nippon Stainless Technical Report [Japan]
Nippon Steel Tech. Rep.	Nippon Steel Technical Report [Japan]
Nippon Steel Tech Rep. (Overseas)	Nippon Steel Technical Report (Overseas) [Japan]
Nippon Tungsten Rev.	Nippon Tungsten Review [Japan]
Nisshin Steel Tech. Rep.	Nisshin Steel Technical Report [Japan]
NIST Build. Sci. Ser.	NIST Building Science Series [of National Institute for Standards and Technology, US]
NIST Monogr.	NIST Monograph [of National Institute for Standards and Technology, US]
NIST Res. Rep.	NIST Research Report [of National Institute for Standards and Technology, US]
NIST Spec. Publ.	NIST Special Publication [of National Institute for Standards and Technology, US]
NIST Tech. Note	NIST Technical Note [of National Institute for Standards and Technology, US]
NIST Tech. Rep.	NIST Technical Report [of National Institute for Standards and Technology, US]
NIST Update	NIST Update [of National Institute for Standards and Technology, US]
NKK Tech. Rep.	NKK Technical Report [Japan]
NKK Tech. Rev.	NKK Technical Review [Japan]
NMAB Newsl.	NMAB Newsletter [of National Materials Advisory Board, US]
NML Tech. J.	NML Technical Journal [of National Metallurgical Laboratory, India]
Noise Control Eng. J.	Noise Control Engineering Journal [of Institute of Noise Control Engineering, US]
Noise Vib. Control Worldw.	Noise and Vibration Control Worldwide [UK]
Non-Destr. Test., Aust.	Non-Destructive Testing, Australia [Australia]
Non-Destr. Test., China	Non-Destructive Testing, China [PR China]
Nondestr. Test Commun.	Nondestructive Testing Communications [UK]
Non Ferr. Met. World	Non Ferrous Metal World [Germany]
Nonferrous Met.	Nonferrous Metals [PR China]
Nonlinear Anal. Theory Methods Appl.	Nonlinear Analysis Theory, Methods and Applications [UK]
Nonlinear Sci. Today	Nonlinear Science Today (included in Journal of Nonlinear Science) [Germany and US]
Nonlinear Vib. Probl.	Nonlinear Vibration Problems [Poland]
Nonwovens Ind.	Nonwovens Industry [US]
Nonwovens Rep. Int.	Nonwovens Report International [UK]
Nordic Pulp Paper Res. J.	Nordic Pulp and Paper Research Journal [Sweden]
Norelco Rep.	Norelco Report [Netherlands]
Nor. Geol. Tidsskr.	Norsk Geologisk Tidsskrift [Norway]
Nor. Polarinst. Skr.	Norsk Polarinstitutt Skrifter [Norway]
Nor. Polarinst. Temakart	Norsk Polarinstitutt Temakart [Norway]

North East Coast Inst. Eng. Shipbuild., Trans.	North East Coast Institution of Engineers and Shipbuilders, Transactions [UK]
Northern J. Appl. Forestry	Northern Journal of Applied Forestry [of Society of American Foresters]
Norw. Shipp. News	Norwegian Shipping News [Norway]
Note Recens. Not.	Note Recensioni e Notizie [of Istituto Superiore Poste e Telecomunicazioni, Italy]
Note Tech.-Off. Nat. Etudes Rech. Aerosp.	Note Technique-Office National d'Etudes Recherches Aerospatiales [France]
Not. Ist. Autom./ Univ. Roma	Notiziario dell'Istituto di Automatica dell'Universita di Roma [Italy]
Notre Dame J. Form. Log.	Notre Dame Journal of Formal Logic [France]
Novosti Tselul.-Khart. Prom.	Novosti v Teslulozno-Khartienata Promishlennost [Bulgaria]
NR Technol.	NR Technology [of Malaysian Rubber Producers Research Association]
NSA	Nuclear Science Abstracts [Joint US Publication of Energy Research and Development Agency and Oak Ridge National Laboratory]
NTDRA Dealer News	NTDRA Dealer News [of National Tire Dealers and Retreaders Association, US]
NTIS Chem.	NTIS Chemistry [of National Technical Information Service, US]
NTIS Comput. Control Inform. Theory	NTIS Computers, Control and Information Theory [of National Technical Information Service, US]
NTIS Environ. Pollut. Control	NTIS Environmental Pollution Control [of National Technical Information Service, US]
NTIS Mater. Sci.	NTIS Materials Sciences [of National Technical Information Service, US]
NTT R & D	NTT R & D (Research and Development) [of Nippon Telegraph and Telephone, Japan]
NTT Rev.	NTT Review [of Nippon Telegraph and Telephone, Japan]
NTZ	Nachrichtentechnische Zeitschrift [Germany]
NTZ Arch.	NTZ Archiv [Germany]
Nucl. Act.	Nuclear Active [South Africa]
Nucl. Austral. Bull.	Nuclear Australia Bulletin [of Australian Nuclear Society]
Nucl. Chem. Waste Manage.	Nuclear and Chemical Waste Management [UK and US]
Nucl. Data Sheets	Nuclear Data Sheets [US]
Nucl. Energy (Br. Nucl. Energy Soc.)	Nuclear Energy (British Nuclear Energy Society) [UK]
Nucl. Energy (J. Br. Nucl. Energy Soc.)	Nuclear Energy (Journal of the British Nuclear Energy Society) [UK]
Nucl. Eng.	Nuclear Engineering [of Institution of Nuclear Engineers, UK]
Nucl. Eng. Des.	Nuclear Engineering and Design [Switzerland and Netherlands]
Nucl. Eng. Des./Fusion	Nuclear Engineering and Design/Fusion [Switzerland]
Nucl. Eng. Int.	Nuclear Engineering International [UK]
Nucl. Eur.	Nuclear Europe [Switzerland]
Nucl. Fusion	Nuclear Fusion [of International Atomic Energy Agency, Austria]
Nucl. Fusion Plasma Phys.	Nuclear Fusion and Plasma Physics [PR China]
Nucl. Ind.	Nuclear Industry [of US Council for Energy Awareness]
Nucl. Instrum. Methods A	Nuclear Instruments and Methods A [Netherlands]
Nucl. Instrum. Methods B	Nuclear Instruments and Methods B [Netherlands]
Nucl. Instrum. Methods Phys. Res. A,Accel.Spectrom.Detect.Assoc. Equip.	Nuclear Instruments and Methods in Physics Research, Section A: Accelerators, Spectrometers, Detectors and Associated Equipment [Netherlands]
Nucl. Instrum. Methods Phys. Res. B,Beam Interact. Mater. At.	Nuclear Instruments and Methods in Physics Research, Section A, Beam Interactions with Materials and Atoms [Netherlands]
Nucl. News	Nuclear News [of American Nuclear Society]
Nucl. Phys. A	Nuclear Physics A [Netherlands]
Nucl. Phys. B, Part. Phys.	Nuclear Physics B, Particle Physics [Netherlands]
Nucl. Phys. B, Proc. Suppl.	Nuclear Physics B, Proceedings Supplements [Netherlands]
Nucl. Prof.	Nuclear Professional [of Institute of Nuclear Power Operations, US]
Nucl. Saf.	Nuclear Safety [of Oak Ridge National Laboratory, US]
Nucl. Sci. Appl. A	Nuclear Science and Applications, Section A [Switzerland]
Nucl. Sci. Appl. B	Nuclear Science and Applications, Section B [Switzerland]
Nucl. Sci. Appl. B, Phys. Sci.	Nuclear Science and Applications, Series B: Physical Sciences [Bangladesh]
Nucl. Sci. Eng.	Nuclear Science and Engineering [of American Nuclear Society]
Nucl. Sci. J.	Nuclear Science Journal [of Atomic Energy Council, Taiwan]
Nucl. Tech.	Nuclear Techniques [PR China]
Nucl. Technol.	Nuclear Technology [US]

Nucl. Technol./Fusion	Nuclear Technology/Fusion [US]
Nucl. Tracks Radiat. Meas.	Nuclear Tracks and Radiation Measurements [UK]
Numer. Eng.	Numerical Engineering [UK]
Numer. Funct. Anal. Optim.	Numerical Functional Analysis and Optimization [US]
Numer. Heat Transf. A	Numerical Heat Transfer A [US]
Numer. Heat Transf. B	Numerical Heat Transfer B [US]
Numer. Math.	Numerische Mathematik [Germany]
Numer. Methods Partial Diff. Equations	Numerical Methods for Partial Differential Equations [US]
Nukl. Tehnol.	Nuklearna Tehnologija [Yugoslavia]
Nuovo Cimento A	Nuovo Cimento, Section A [of Societa Italiana de Fisica, Italy]
Nuovo Cimento B	Nuovo Cimento, Section B [of Societa Italiana de Fisica, Italy]
Nuovo Cimento C	Nuovo Cimento, Section C [of Societa Italiana de Fisica, Italy]
Nuovo Cimento D	Nuovo Cimento, Section D [of Societa Italiana de Fisica, Italy]
Nuovo Sagg.	Nuovo Saggiatore [of Societa Italiana de Fisica, Italy]
Ny Tek.	Ny Teknik [Sweden]
N.Z. Electron. Rev.	New Zealand Electronics Review [of National Electronics Development Association, New Zealand]
N.Z. Energy J.	New Zealand Energy Journal
N.Z. Eng.	New Zealand Engineering [of New Zealand Institution of Professional Engineers]
NZIE Trans. Civ. Eng.	NZIE Transactions, Civil Engineering Section [of New Zealand Institution of Engineers]
NZIE Trans. Electr./Mech./Chem. Eng.	NZIE Transactions, Electrical/Mechanical/Chemical Engineering Section [of New Zealand Institution of Engineers]
NZIW Bull.	NZIW Bulletin [of New Zealand Institute of Welding]
N.Z. J. Dairy Sci. Technol.	New Zealand Journal of Dairy Science and Technology
N.Z. J. Forestry Sci.	New Zealand Journal of Forestry Science

O

Oakite News Serv.	Oakite News Service [US]
Oak Ridge Natl. Lab. Met. Ceram. Tech. Rep.	Oak Ridge National Laboratory Metals and Ceramics Technical Report [US]
Oberflaeche Surf.	Oberflaeche Surface [Switzerland]
Obogashch. Rud.	Obogashchenie Rud [USSR]
Obrobka Plast.	Obrobka Plastyczna [Poland]
Occr. Ovzdusi	Ochrona Ovzdusi [Poland]
Occr. Powietrza	Ochrona Powietrza [Poland]
Occr. Przed Koroz.	Ochrona Przed Korozja [Poland]
Occup. Health	Occupational Health [UK]
Occup. Saf. Health Admin. Sub Serv.	Occupational Safety and Health Administration Subscription Service [US]
Ocean Eng.	Ocean Engineering [UK]
Oceanol. Acta	Oceanologia Acta [France]
Oceanol. Limnol. Sin.	Oceanologia et Limnologia Sinica [PR China]
Ochr. Powietrza	Ochrona Powietrza [Poland]
Ochr. Przed Koroz.	Ochrona Przed Korozja [Poland]
OECD Econ. Outlook	OECD Economic Outlook [of Organization for Economic Cooperation and Development, France]
OEGAI J.	OEGAI Journal [of Oesterreichische Gesellschaft fur Artificial Intelligence, Austria]
OEIAZ Oesterr. Ing. Archit.-Z.	Oesterreichische Ingenieur- und Architekten-Zeitschrift [Austria]
Oelhydraul. Pneum.	Oelhydraulik und Pneumatik [Germany]
OEP Off. Equip. Prod.	OEP Office Equipment and Products [Japan]
Oesterr. Ing. Archit.-Z.	Oesterreichische Ingenieur- und Architekten-Zeitschrift [Austria]
Oesterr. Ing. Z.	Oesterreichische Ingenieur-Zeitschrift [Austria]
Oesterr. Z. Elektr.wirtsch.	Oestereichische Zeitschrift fur Elektrizitatswirtschaft [Austria]
OEZE Oesterr. Z. Elektr.wirtsch.	Oestereichische Zeitschrift fur Elektrizitatswitrschaft [Austria]

Off. Boards Markets	Official Board Markets [US]
Off. Environ.	Office Environment [UK]
Off. Equip. Index	Office Equipment Index [UK]
Off. Equip. Methods	Office Equipment and Methods [Canada]
Off. Equip. News	Office Equipment News [UK]
Off. Gaz.	Official Gazette [US]
Off. Gaz. US Pat.	Official Gazette, US Patent and Trademarks Office
Off. Gaz. US Pat. Trademks. Off. Pat.	Official Gazette, US Patent and Trademarks Office, Patents
Off. Gaz. US Pat. Trademks. Off. Tradem.	Official Gazette, US Patent and Trademarks Office, Trademarks
Off. Home	Office at Home [UK]
Off. Inf. Manage. Int.	Office and Information Management International [of Institute of Administrative Management, UK]
Off. J. Eur. Comm.	Official Journal of the European Communities [Luxembourg]
Off. J. Eur. Communities	Official Journal of the European Communities [Luxembourg]
Off. J. Pat.	Official Journal, Patents [UK]
Off. Mag.	Office Magazine [UK]
Off. Manage.	Office Management [Germany]
Off. Prod. News	Office Products News [UK]
Offset Print. Reprogr.	Offset Printing and Reprographics [UK]
Offshore Eng.	Offshore Engineer [UK]
Offshore Res. Focus	Offshore Research Focus [UK]
Off. Syst. Res. J.	Office Systems Research Journal [of Office Systems Research Association, US]
Off. Technol. People	Office Technology and People [Netherlands]
Off. World News	Office World News [US]
Oil Gas J.	Oil and Gas Journal [US]
Oilman Wkly. Newsl.	Oilman Weekly Newsletter [UK]
Oki Tech. Rev.	Oki Technical Review [Japan]
Onde Electr.	L'Onde Electrique [France]
100A1	One Hundred A One [UK]
Online Bus. Inf.	Online Business Information [UK]
Online Libr. Microcomput.	Online Libraries and Microcomputers [US]
Online Rev.	Online Review [UK]
Ontario Technol.	Ontario Technologist [Canada]
Open Syst. Softw.	Open Systems and Software [UK]
Oper. Geogr.	Operational Geographer [of Canadian Association of Geographers]
Oper. Manage. Rev.	Operations Management Review [US]
Oper. Res.	Operations Research [of Operations Research Society of America]
Oper. Res. Lett.	Operations Research Letters [of Operations Research Society of America]
Oper. Syst. Netw.	Operating Systems and Networks [UK]
Oper. Syst. Rev.	Operating Systems Review [of Association for Computing Machinery, US]
Ophthalmic Physiol. Opt.	Ophthalmic and Physiological Optics [UK]
Opt. Acta	Optica Acta
Opt. Appl.	Optica Applicata [Poland]
Opt. Commun.	Optics Communications [Netherlands]
Opt. Electron. Microsc.	Optical and Electron Microscopy [UK]
Opt. Eng.	Optical Engineering [of Society of Photooptical Instrumentation Engineers, US]
Opt. Eng. Rep.	Optical Engineering Report [of Society of Photooptical Instrumentation Engineers, US]
Opt. Inf. Syst.	Optical Information Systems [US]
Optim. Control Appl. Methods	Optimal Control Applications and Methods [UK]
Opt. Lasers Eng.	Optics and Lasers in Engineering [UK]
Opt. Laser Technol.	Optics and Laser Technology [UK]
Opt. Lett.	Optics Letters [of Optical Society of America]
Opt. Mater.	Optical Materials [Netherlands]
Opt. Mekh. Prom.	Optiko-Mekhanicheskaya Promyshlennost [USSR]
Opt. Mekh. Prom.-st.	Optiko-Mekhanicheskaya Promyshlennost [USSR]
Opt. News	Optical News [of Optical Society of America]
Optoelectron., Devices Technol.	Optoelectronics — Devices and Technologies [Japan]
Optoelectron. Instrum. Data Process.	Optoelectronics, Instrumentation and Data Processing [US]
Optoelektron. Poluprovodn. Tekh.	Optoelektronika i Poluprovodnikovaya Tekhnika [USSR]

Optom. Vis. Sci.	Optometry and Vision Science [of American Academy for Optometry]
Opt. Pura Apl.	Optica Pura y Aplicada [Spain]
Opt. Quantum Electron.	Optical and Quantum Electronics [UK]
Opt. Spectra	Optical Spectra [US]
Opt. Spectrosc.	Optics and Spectroscopy [Translation of: *Optika i Spektroskopiya (USSR)*; published in the US]
Opt. Spektrosk.	Optika i Spektroskopiya [USSR]
Org. Chem.	Organic Chemistry [of Chinese Chemical Society, PR China]
Orig. Life	Origins of Life [Netherlands]
Ornam./Misc. Met. Fabr.	Ornamental/Miscellaneous Metal Fabricator [US]
ORNL Mater. Res. Newsl.	ORNL Materials Research Newsletter [of Oak Ridge National Laboratory, US]
ORSA J. Comput.	ORSA Journal of Computing [of Operational Research Society of America]
ORSA/TIMS Bull.	ORSA/TIMS Bulletin [of Operations Research Society of America/The Institute of Management Sciences]
Otbor Pereda. Inf.	Otbor i Peredacha Informatsii [USSR]
Otkryt. Isobret., Prom. Obraz., Tobar. Znaki	Otkrytiya, Isobreteniya, Promyshlennye Obraztsy, Tobarnye Znaki [USSR]
Outlook Res. Libr.	Outlook on Research Libraries [UK]
Ovako Steel Tech. Rep.	Ovako Steel Technical Report [Sweden]
Overseas Devt. Nat. Resourc. Inst. Newsl.	Overseas Development Natural Resources Institute Newsletter [UK]
Overseas Trade Stat.	Overseas Trade Statistics [UK]
Oxid. Met.	Oxidation of Metals [US]
Ozone Sci. Eng.	Ozone Science and Engineering [of International Ozone Association, US]

P

PA	Physics Abstracts [of Institution of Electrical Engineers, UK]
Packag.	Packaging [UK and US]
Packag. (Denver)	Packaging (Denver) [US]
Packag. (UK)	Packaging (UK)
Packag. News (Australia)	Packaging News (Australia)
Packag. Newsl.	Packaging Newsletter [US]
Packag. News (UK)	Packaging News (UK)
Packag. Technol. Sci.	Packaging Technology and Science [UK]
Packag. Week	Packaging Week [UK]
P & IM Rev. APICS News	P & IM (Production and Inventory Management) Review and APICS News [of American Production and Inventory Control Society, US]
Paint Resin	Paint and Resin [US]
Pak. J. Sci.	Pakistan Journal of Science [of Pakistan Association for the Advancement of Science]
Pak. J. Sci. Res.	Pakistan Journal of Scientific Research [of Pakistan Association for the Advancement of Science]
PAMS	Proceedings of the American Mathematical Society
Paper Conservator	The Paper Conservator [of Institute of Paper Conservation, UK]
Paper Conserv. News	Paper Conservation News [of Institute of Paper Conservation, UK]
Paper Film Foil Conv.	Paper, Film and Foil Converter [US]
Paperi Puu	Paperi ja Puu [Finland]
Paper Technol.	Paper Technology [of Paper Industries Research Association, US]
Paper Twine J.	Paper and Twine Journal [US]
Paper Yearb.	Paper Yearbook [US]
Pap. Film Foil Convert.	Paper, Film and Foil Converter [US]
Papier	Das Papier [Germany]
Papier Carton Cellulose	Papier, Carton et Cellulose [France]
Papier-Kunstst.-Verarb.	Papier- und Kunststoff-Verarbeiter [Germany]
Papiermacher	Der Papiermacher [Germany]
Papir Celuloza	Papir a Celuloza [Czechoslovakia]

Pap. Puu	Paperi ja Puu [Finland]
Parallel Comput.	Parallel Computing [Netherlands]
Part. Accel.	Particle Accelerators [UK]
Part. Charact.	Particle Characterization [Germany]
Part. Part. Syst. Charact.	Particle and Particle Systems Characterization [US and Germany]
Part. Sci. Technol.	Particulate Science and Technology [of Fine Particle Society, US]
Part. World	Particle World [UK]
Patentbl.	Patentblatt [of Deutsches Patentamt, Germany]
Pat. Off. Rec.	Patent Office Record [of Supply and Services Canada]
Patricia Seybold's Off. Comput. Res.	Partricia Seybold's Office Computing Report [of Patricia Seybold's Office Computing Group, US]
Pattern Recognit.	Pattern Recognition [UK]
Pattern Recognit. Lett.	Pattern Recognition Letters [Netherlands]
Pbd. Pkg.	Paperboard Packaging [US]
PC Bus. Softw.	PC Business Software [UK]
P/C/C Alert	Polymers/Ceramics/Composites Alert [of ASM International, US, and Institute of Metals, UK]
PC Mag.	PC Magazine [US]
PC Mag. (UK Ed.)	PC Magazine (UK Edition) [UK]
PCN	Personal Computer News [UK]
PC Perspect. Newsl.	PC Perspectives Newsletter [US]
PDA Rec.	PDA Recorder [of Prospectors and Developers Association of Canada]
PEM Process Eng. Mag.	PEM Process Engineering Magazine [Germany]
P/E News	Petroleum Energy News [of American Petroleum Institute]
Perem. Zvezdy	Peremennye Zvezdy [USSR]
Perform. Chem.	Performance Chemicals
Perform. Eval.	Performance Evaluation [Netherlands]
Perform. Eval. Rev.	Performance Evaluation Review [of Association for Computing Machinery, US]
Period. Math. Hung.	Periodica Mathematica Hungarica [Hungary]
Period. Polytech.	Periodica Polytechnica [Hungary]
Period. Polytech., Chem. Eng.	Periodica Polytechnica, Chemical Engineering [Hungary]
Period. Polytech., Electr. Eng.	Periodica Polytechnica, Electrical Engineering [Hungary]
Period. Polytech., Mech. Eng.	Periodica Polytechnica, Mechanical Engineering [Hungary]
Perkin-Elmer Tech. News	Perkin-Elmer Technical News [US]
Pers. Comput. (Germany)	Personal Computer (Germany)
Pers. Comput. (USA)	Personal Computing (USA)
Pers. Comput. Mag.	Personal Computer Magazine [UK]
Pers. Comput. World	Personal Computer World [UK]
Pers. Manage.	Personnel Management [UK]
PESTAB	Pesticides Abstracts [of Environmental Protection Agency, US]
Pet. Intell. Wkly.	Petroleum Intelligence Weekly
Pet. Manage.	Petroleum Management [US]
Pharm. Acta Helv.	Pharmaceutica Acta Helvetica [Switzerland]
Pharma Int.	Pharma International
Pharm. Eng.	Pharmaceutical Engineering [of International Society of Pharmaceutical Engineers, US]
Pharm. Ind.	Pharmazeutische Industrie [Germany]
Pharm. Res.	Pharmaceutical Research [of American Association of Pharmaceutical Scientists]
Pharm. Tech. Jpn.	Pharm Tech Japan
Pharm. Technol.	Pharmaceutical Technology
Phase Transit.	Phase Transitions [UK]
Philips J. Res.	Philips Journal of Research [Netherlands]
Philips Res. Rep.	Philips Research Report [Netherlands]
Philips Tech. Rev.	Philips Technical Review [Netherlands]
Philips Telecommun. Data Syst. Rev.	Philips Telecommunication and Data Systems Review [Netherlands]
Philips Weld. Rep.	Philips Welding Reporter [Netherlands]
Philos. Mag.	Philosophical Magazine [UK]
Philos. Mag. A, Phys. Condens. Matter Defects Mech. Prop.	Philosophical Magazine A, Physics of Condensed Matter, Defects and Mechanical Properties [UK]
Philos. Mag. B, Phys. Condens. MatterElectron.Opt.Magn.Prop.	Philosophical Magazine B, Physics of Condensed Matter, Electronic, Optical and Magnetic Properties [UK]

Philos. Mag. Lett.	Philosophical Magazine Letters [UK]
Philos. Sci.	Philosophy of Science [of Philosophy of Science Association, US]
Philos. Trans. R. Soc. London A	Philosophical Transactions of the Royal Society of London A [UK]
Philos. Trans. R. Soc. London A, Math. Phys. Sci.	Philosophical Transactions of the Royal Society of London A, Mathematical and Physical Sciences [UK]
PHL Bull.	PHL Bulletin [of National Institute of Packaging, Handling and Logistic Engineers, US]
Phoenix Q.	Phoenix Quarterly [of Institute of Scrap Recycling Industries, US]
Phoenix: Voice Scrap Ind.	Phoenix: Voice of the Scrap Industries [US]
Photobiochem. Photobiophys.	Photobiochemistry and Photobiophysics [Netherlands]
Photogramm. Eng. Remote Sens.	Photogrammetric Engineering and Remote Sensing [of American Society of Photogrammetry]
Photogramm. Rec.	Photogrammetric Record [Photogrammetry Society, UK]
Photosynth. Res.	Photosynthesis Research [Netherlands]
Phys. Bl.	Physikalische Blatter [Germany]
Phys. Briefs	Physics Briefs [of American Institute of Physics]
Phys. Bull.	Physics Bulletin [of Institute of Physics, UK]
Phys. Chem. Glasses	Physics and Chemistry of Glasses [of Society of Glass Technology, UK]
Phys. Chem. Liq.	Physics and Chemistry of Liquids [UK]
Phys. Chem. Mater. Treat.	Physics and Chemistry of Materials Treatment [Translation of: *Fizika i Khimiya Obrabotki Materialov (USSR)*; published in the UK]
Phys. Chem. Mech. Surf.	Physics, Chemistry and Mechanics of Surfaces [UK]
Phys. Chem. Miner.	Physics and Chemistry of Minerals [Germany]
Phys. Daten	Physik Daten [of Fachinformationszentrum Energie — Physik — Mathematik, Germany]
Phys. Earth Planet. Inter.	Physics of the Earth and Planetary Interiors [Netherlands]
Phys. Educ. (India)	Physics Education (India)
Phys. Educ. (UK)	Physics Education [of Institute of Physics, UK]
Phys. Energ. Fortis Phys. Nucl.	Physica Energiae Fortis et Physica Nuclearis [PR China]
Phys. Fluids A, Fluid Dyn.	Physics of Fluids A, Fluid Dynamics [of American Institute of Physics]
Phys. Fluids B, Plasma Phys.	Physics of Fluids B, Plasma Physics [of American Institute of Physics]
Physicochem. Hydrodyn.	Physicochemical Hydrodynamics
Physiol. Plantarum	Physiologia Plantarum [Denmark]
Phys. Lett. A	Physics Letter A [Netherlands]
Phys. Lett. B	Physics Letter A [Netherlands]
Phys. Med. Biol.	Physics in Medicine and Biology [of Institute of Physics, UK]
Phys. Met.	Physics of Metals [Translation of: *Metallofizika (USSR)*; published in the UK]
Phys. Met. Metallogr.	Physics and Metallography [Translation of: *Fizika Metallov i Metallovedenie (USSR)*; published in the UK]
Phys. Pap.	Physics Papers [Netherlands]
Phys. Rep. A	Physics Reports A [Netherlands]
Phys. Rep. B	Physics Reports B [Netherlands]
Phys. Rep. C	Physics Reports C [Netherlands]
Phys. Rev. A, At. Mol. Opt. Phys.	Physical Review A, Atomic, Molecular and Optical Physics [of American Physical Society]
Phys. Rev. Abstr.	Physical Review, Abstracts [of American Institute of Physics]
Phys. Rev. A, Stat. Phys. Plasmas Fluids Relat. Interdiscip. Top.	Physical Review A, Statistical Physics, Plasmas, Fluids and Related Interdisciplinary Topics [of American Physical Society]
Phys. Rev. B, Condens. Matter	Physical Review B, Condensed Matter [of American Physical Society]
Phys. Rev. C, Nucl. Phys.	Physical Review C, Nuclear Physics [of American Physical Society]
Phys. Rev. D, Part. Fields	Physical Review D, Particles and Fields [of American Physical Society]
Phys. Rev. Lett.	Physical Review Letters [of American Physical Society]
Phys. Scr.	Physica Scripta [of Royal Swedish Academy of Sciences]
Phys. Scr. Vol. T	Physica Scripta Volume T [of Royal Swedish Academy of Sciences]
Phys. Status Solidi a	Physica Status Solidi a [Germany]
Phys. Status Solidi b	Physica Status Solidi b [Germany]
Phys. Teach.	Physics Teaching [of Korean Physical Society, South Korea]
Phys. Technol.	Physics in Technology [of Institute of Physics, UK]
Phys. Test. Chem. Anal,. Chem. Anal.	Physical Testing and Chemical Analysis, Chemical Analysis [PR China]
Phys. Today	Physics Today [of American Institute of Physics]
Phys. World	Physics World [UK]
Phytochem.	Phytochemistry [of Phytochemical Society of Europe]

Pigm. Resin Technol.	Pigment and Resin Technology [UK]
PIMA	Paper Industry Management Association [US]
P & IM Rev.	Production Inventory Management Reviews [of American Production and Inventory Control Society]
Pipeline Gas. J.	Pipeline and Gas Journal [US]
Pipe Line Ind.	Pipe Line Industry [US]
Pipes Pipelines Int.	Pipes and Pipelines International [UK]
PIRA Pbd. Abstr.	PIRA Paperboard Abstracts [of Paper Industries Research Association, US]
PIRA Printing Abstr.	PIRA Printing Abstracts [of Paper Industries Research Association, US]
Pis'ma Astron. Zh.	Pis'ma v Astronomischeskii Zhurnal [USSR]
Pis'ma Zh. Eksp. Teor. Fiz.	Pis'ma v Zhurnal Eksperimental'noi i Teoreticheskoi Fiziki [USSR]
Pis'ma Zh. Tekh. Fiz.	Pis'ma v Zhurnal Tekhnicheskoi Fiziki [USSR]
Pitture Vern.	Pitture e Vernici [Italy]
Pixel, Comput. Graph. CAD/CAM Image Process.	Pixel, Computer Graphics, CAD/CAM, Image Processing [Italy]
Pkg. (Denver)	Packaging (Denver) [US]
Pkg. News (Australia)	Packaging News (Australia)
Pkg. News (UK)	Packaging (UK)
Pkg. Technol. Sci.	Packaging Technology and Science [UK and US]
Pkg. (UK)	Packaging News (UK)
Planet. Rep.	Planetary Report [Planetary Society, US]
Planet. Space Sci.	Planetary and Space Science [UK]
Planseeber. Pulvermetall.	Planseeberichte fur Pulvermetallurgie [Austria]
Plant Cell Physiol.	Plant and Cell Physiology [of Japanese Society of Plant Physiologists]
Plant Cell Rep.	Plant Cell Reports [Germany and US]
Plant Cell Tissue Organ Culture	Plant Cell, Tissue and Organ Culture [Netherlands]
Plant Eng.	Plant Engineer [of Institution of Plant Engineer, UK]
	Plant Engineering [US]
Planters Bull.	Planters Bulletin [of Rubber Research Institute of Malaysia]
Plant Growth Regul.	Plant Growth Regulation [Netherlands]
Plant Mol. Biol.	Plant Molecular Biology [of International Society for Plant Molecular Biology, US]
Plant Mol. Biol. Rep.	Plant Molecular Biology Reporter [of International Society for Plant Molecular Biology, US]
Plant Oper. Prog.	Plant Operation and Progress [of American Institute of Chemical Engineers]
Plant Physiol.	Plant Physiology [of American Society of Plant Physiologists]
Plant Physiol. Biochem.	Plant Physiology and Biochemistry [Plant Physiology and Biochemistry Society, India]
Plant. Sci.	Plant Science [Ireland]
Plant Sci. Bull.	Plant Science Bulletin [of Botanical Society of America]
Plants Mach.	Plants and Machinery [US]
Plasma Chem. Plasma Process.	Plasma Chemistry and Plasma Processing [US]
Plasma Phys. Control. Fusion	Plasma Physics and Controlled Fusion [UK]
Plast. Age	Plastics Age [Japan]
Plast. Bldg. Contr.	Plastics in Building Construction [UK]
Plast. Bus.	Plastics Business [Canada]
Plast. Compounding	Plastics Compounding [US]
Plast. Des. Forum	Plastics Design Forum [US]
Plaste Kautsch.	Plaste und Kautschuk [Germany]
Plast. Eng.	Plastics Engineering [of Society of Plastics Engineers, US]
	Plastics in Engineering [UK]
Plast. Flash	Plastiques Flash [of Societe Europeene de Presse et d'Edition, France]
PlastForum Scand.	PlastForum Scandinavia [Sweden]
Plast. Ind. News	Plastics Industry News [Japan]
Plast. Massy	Plasticheskie Massy [USSR]
Plast. Mod. Elastom.	Plastiques Modernes et Elastomeres [France]
Plast. News	Plastics News [of Plastics Institute of Australia]
Plast. News Briefs	Plastics News Briefs [of Society of the Plastics Industry, US]
Plast. Retail Packag. Bull.	Plastics in Retail Packaging Bulletin [Switzerland]
Plast. Rubber Int.	Plastics and Rubber International [of Plastics and Rubber Institute, UK]
Plast. Rubber Process. Appl.	Plastics and Rubber Processing and Applications [of Plastics and Rubber Institute, UK]
Plast. Rubber Wkly.	Plastics and Rubber Weekly [UK]
Plast. S. Africa	Plastics Southern Africa [South Africa]
Plast. Technol.	Plastics Technology [US]
Plast. Today	Plastics Today [UK]

Plast. World	Plastics World [US]
Plasty Kauc.	Plasty a Kaucuk [Czechoslovakia]
Platinum Met. Rev.	Platinum Metals Review [UK]
Plat. Surf. Fin.	Plating and Surface Finishing [of American Electroplaters and Surface Finishers Society]
PMAI Newsl.	PMAI Newsletter [of Powder Metallurgy Association of India]
PMI Powder Metall. Int.	PMI Powder Metallurgy International [Germany]
P/M Technol. Newsl.	P/M Technology Newsletter [of Metal Powder Industries Federation, US]
Podst. Sterow.	Podstawy Sterowania [Poland]
Pokrody Praškove Metal.	Pokrody Praskove Metalurgie [Czechoslovakia]
Pol. Akad. Nauk Pr. Kom. Nauk Tech. Metal. Fiz. Metal. Stopow	Polska Akademia Nauk Prace Komisji Nauk Technicznych Metalurgia Fizyka Metali i Stopow [Poland]
Polar Res.	Polar Research [Norway]
Pol. Eng.	Polish Engineering [Poland]
Polim. Tworz. Wielk.	Polimery Tworzywa Wielkoczasteczkowe [Poland]
Poliplasti Plast. Rinf.	Poliplasti e Plastici Rinforzati [Italy]
Pollut. Atmos.	Pollution Atmospherique [France]
Pollut. Eng.	Pollution Engineering
Pol. Tech. Rev.	Polish Technical Review [Poland]
Pol. Technol. News	Polish Technological News [Poland]
Polym. Bull.	Polymer Bulletin [Germany and US]
Polym. Commun.	Polymer Communications [UK]
Polym. Commun. (China)	Polymer Communications (China) [of Chinese Chemical Society, PR China]
Polym. Commun. (Engl. Ed.)	Polymer Communications (English Edition) [of Chinese Chemical Society, PR China]
Polym. Compos.	Polymer Composites [of Society of Plastics Engineers, US]
Polym. Degrad. Stabil.	Polymer Degradation and Stability [UK]
Polym. Dig.	Polymer Digest [Japan]
Polym. Eng. Sci.	Polymer Engineering and Science [of Society of Plastics Engineers, US]
Polym. J.	Polymer Journal [of Society of Polymer Science, Japan]
Polym. J. (Tokyo)	Polymer Journal (Tokyo) [of Society of Polymer Science, Japan]
Polym. News	Polymer News [UK]
Polym. Paint Col. J.	Polymers Paint Colour Journal [UK]
Polym. Plast. Technol. Eng.	Polymer-Plastics Technology and Engineering [US]
Polym. Prepr.	Polymer Preprints [of American Chemical Society]
Polym. Prepr. Am. Chem. Soc.	Polymer Preprints of the American Chemical Society
Polym. Proc. Eng.	Polymer Process Engineering [US]
Polym. Rubb. Asia	Polymers and Rubber Asia [UK]
Polym. Sci. USSR	Polymer Science USSR [Translation of: *Vysokomolekulyarnye Soedineniya, Seriya A*; published in the UK]
Polym. Test.	Polymer Testing [UK]
Polysar Prog.	Polysar Progress [Canada]
Pomiary Autom. Kontrola	Pomiary Automatyka Kontrola [Poland]
Pop. Plast.	Popular Plastics [India]
Pop. Plast. Packag.	Popular Plastics and Packaging [India]
Poroshk. Metall.	Poroshkovaya Metallurgiya [USSR]
Port. Phys.	Portugaliae Physica [of Sociedade Portuguesa de Fisica, Portugal]
Postepy Astron.	Postepy Astronomii [Poland]
Postepy Fiz.	Postepy Fizyki [Poland]
Povierkhnost' Fiz., Khim., Mekh.	Povierkhnost' Fizika, Khimiya, Mekhanika [USSR]
Powder Bulk Eng.	Powder and Bulk Engineering [US]
Powder Diffr.	Powder Diffraction [US]
Powder Handl. Process.	Powder Handling and Processing [Germany]
Powder Metall.	Powder Metallurgy [The Metals Society, UK]
Powder Metall. Def. Technol.	Powder Metallurgy in Defense Technology [of Metal Powder Industries Federation, US]
Powder Metall. Int.	Powder Metallurgy International [Germany]
Powder Metall. Sci. Technol.	Powder Metallurgy Science and Technology [India]
Powder Metall. Technol.	Powder Metallurgy Technology [PR China]
Powder Sci. Eng.	Powder Science and Engineering
Powder Technol.	Powder Technology [Switzerland]
Power Convers. Intell. Motion	Power Conversion and Intelligent Motion [US]
Power Eng.	Power Engineering [US]

Power Eng. (USSR Acad. Sci.)	Power Engineering (USSR Academy of Sciences) [Translation of: *Izvestiya Akademii Nauk SSSR, Energetika i Transport*; published in the US]
Power Eng. J.	Power Engineering Journal [of Institution of Electrical Engineers, UK]
Power Int.	Power International [UK]
Power Int. Ed.	Power International Edition [US]
Power Transm. Des.	Power Transmission Design [US]
Power Works Eng.	Power and Works Engineering [UK]
Powloki Ochr.	Powloki Ochronne [Poland]
PPG Prod.	PPG Products [US]
PPITB Plast. Bull.	PPITB Plastics Bulletin [of Plastics Processing Industry Training Board, UK]
Prace Inst. Technol. Drewna	Prace Instytutu Technologii Drewna [Poland]
Pract. Archit. Mag.	Practicing Architects Magazine [of Society of American Registered Architects, US]
Pract Comput.	Practical Computing [UK]
Pract. Electron.	Practical Electronics [UK]
Pract. Wirel.	Practical Wireless [UK]
Prague Bull. Math. Linguist.	Prague Bulletin of Mathematical Linguistics [Czechoslovakia]
Prakt. Metallogr.	Praktische Metallographie [Germany]
Precambrian Res.	Precambrian Research [Netherlands]
Precious Met.	Precious Metals [PR China]
Precious Met. News Rev.	Precious Metals News and Review [of International Precious Metals Institute, US]
Precis. Eng.	Precision Engineering [UK]
Precis. Met.	Precision Metal [US]
Precis. Toolmak.	Precision Toolmaker [UK]
Prepr. Pap. Am. Chem., Div. Fuel Chem.	Preprints of Papers — American Chemical Society, Division of Fuel Chemistry
Pressure Eng.	Pressure Engineering [Japan]
Prib. Sist. Upr.	Pribory i Sistemy Upravleniya [USSR]
Prib. Tekh. Eksp.	Pribory i Tekhnika Eksperimenta [USSR]
Price Waterhouse Rev.	Price Waterhouse Review [US]
Prikl. Biokhim. Mikrobiol.	Prikladnaya Biokhimiya i Microbiologiya [USSR]
Prikl. Mat. Mekh.	Prikladnaya Matematika i Mekhanika [USSR]
Prikl. Mekh.	Prikladnaya Mekhanika [USSR]
Pr. Inst. Elektrotech.	Prace Instytutu Elektrotechniki [Poland]
Pr. Inst. Lacz.	Prace Instytutu Lacznosci [Poland]
Pr. Inst. Metal. Zelaza	Prace Instytutu Metalurgii Zelasa [Poland]
Pr. Inst. Met. Niezelaz.	Prace Instytutu Metali Niezelaznych [Poland]
Pr. Inst. Odlew.	Prace Instytutu Odlewnictwa [Poland]
Pr. Inst. Spaw.	Prace Instytutu Spawalnictwa [Poland]
Pr. Inst. Tele- & Radiotech.	Prace Instytutu Tele- i Radiotechnicznego [Poland]
Pr. Inst. Technol. Drewna	Prace Instytutu Technologii Drewna [Poland]
Printed Circuit Fabr.	Printed Circuit Fabrication [US]
Printed Circuit Assem.	Printed Circuit Assembly [US]
Pr. Kom. Metal.-Odlew., Pol. Akad. Nauk — Oddz. Krakow., Metal	Prace Komisji Metalurgiczno-Odlewniczej, Polska Akademia Nauk—Oddzial w Krakowie, Metalurgia [Poland]
Pr. Nauk. Inst. Chem. Nieorg. Metal. Pierwiastkow Rzadkich Politech. Wroc., Konf.	Prace Naukowe Instytutu Chemii Nieorgaicznej i Metalurgii Pierwiastkow Rzadkich Politechniki Wroclawskiej, Konferencje [Poland]
Pr. Nauk. Inst. Chem. Nieorg. Metal.Pierwiastkow Rzadkich Politech. Wroc., Monogr.	Prace Naukowe Instytutu Chemii Nieorgaicznej i Metalurgii Pierwiastkow Rzadkich Politechniki Wroclawskiej, Monografie [Poland]
Pr. Nauk. Inst. Chem. Nieorg. Metal. Pierwiastkow Rzadkich Politech. Wroc., Stud. Mater.	Prace Naukowe Instytutu Chemii Nieorgaicznej i Metalurgii Pierwiastkow Rzadkich Politechniki Wroclawskiej, Studia i Materialy [Poland]
Pr. Nauk. Inst. Cybern. Tech. Politech. Wroc., Ser. Konf.	Prace Naukowe Instytutu Cybernetyki Technicznej Politechniki Wroclawskiej, Seria: Konferencje [Poland]

Abbreviation	Title
Pr. Nauk. Inst. Fiz. Politech. Wroc., Ser. Monogr.	Prace Naukowe Instytutu Fizyki Politechniki Wroclawskiej, Seria: Monografie [Poland]
Pr. Nauk. Inst. Konstr. Eksploat. Masz. Politech. Wroc., Konf.	Prace Naukowe Instytutu Konstrukcji i Eksploatacji Maszyn Politechniki Wroclawskiej, Konferencje [Poland]
Pr. Nauk. Inst. Konstr. Eksploat. Masz. Politech. Wroc., Monogr.	Prace Naukowe Instytutu Konstrukcji i Eksploatacji Maszyn Politechniki Wroclawskiej, Monografie [Poland]
Pr. Nauk. Inst. Konstr. Eksploat. Masz. Politech. Wroc., Stud. Mater.	Prace Naukowe Instytutu Konstrukcji i Eksploatacji Maszyn Politechniki Wroclawskiej, Studia i Materialy [Poland]
Pr. Nauk. Inst. Materialozn. Mech. Tech. Politech. Wroc., Monogr.	Prace Naukowe Instytutu Materialoznawstwa i Mechaniki Technicznej Politechniki Wroclawskiej, Monografie [Poland]
Pr. Nauk. Inst. Materialozn. Mech. Tech. Politech. Wroc., Stud. Mater.	Prace Naukowe Instytutu Materialoznawstwa i Mechaniki Technicznej Politechniki Wroclawskiej, Studia i Materialy [Poland]
Pr. Nauk. Inst. Metrol. Elektr. Politech. Wroc., Ser. Konf.	Prace Naukowe Instytutu Metrologii Elektrycznej Politechniki Wroclawskiej, Seria: Konferencje [Poland]
Pr. Nauk. Inst. Technol. Budowy Masz. Politech. Wroc., Konf.	Prace Naukowe Instytutu Technologii Budowy Maszyn Politechniki Wroclawskiej, Konferencje [Poland]
Pr. Nauk. Inst. Technol. Budowy Masz. Politech. Wroc., Monogr.	Prace Naukowe Instytutu Technologii Budowy Maszyn Politechniki Wroclawskiej, Monografie [Poland]
Pr. Nauk. Inst. Technol. Budowy Masz. Politech. Wroc., Stud. Mater.	Prace Naukowe Instytutu Technologii Budowy Maszyn Politechniki Wroclawskiej, Studia i Materialy [Poland]
Pr. Nauk. Inst. Technol. Elektron. Politech. Wroc., Ser. Monogr.	Prace Naukowe Instytutu Technologii Elektronowej Politechniki Wroclawskiej, Seria: Monografie [Poland]
Pr. Nauk. Inst. Technol. Elektron. Politech.Wroc.,Ser.Stud.Mater.	Prace Naukowe Instytutu Technologii Elektronowej Politechniki Wroclawskiej, Serie: Studia i Materialy [Poland]
Pr. Nauk. Inst. Ukl. Elektromasz. Politech. Wroc., Ser. Konf.	Prace Naukowe Instytutu Ukladow Elektromaszynowych Politechniki Wroclawskiej, Seria: Konferencje [Poland]
Pr. Nauk. Inst. Ukl. Elektromasz. Politech. Wroc., Ser. Monogr.	Prace Naukowe Instytutu Ukladow Elektromaszynowych Politechniki Wroclawskiej, Seria: Monografie [Poland]
Pr. Nauk. Inst. Ukl. Elektromasz. Politech. Wroc., Stud. Mater.	Prace Naukowe Instytutu Ukladow Elektromaszynowych Politechniki Wroclawskiej, Studia i Materialy [Poland]
Probability Eng. Inform. Sci.	Probability in the Engineering and Informational Sciences [US]
Probl. Bioniki	Problemy Bioniki [USSR]
Probl. Control Inf. Theory	Problems of Control and Information Theory [Hungary]
Probl. Inf. Transm.	Problems of Information Transmission [Translation of: *Problemy Peredachi Informatsii (USSR)*; published in the US]
Probl. Kibern.	Problemy Kibernetiki [USSR]
Probl. Mashinostr. Nadezhn. Mash.	Problemy Mashinostroeniya i Nadezhnosti Mashin [USSR]
Probl. Metall. Proizv.	Problemy Metallurgicheskogo Proizvodstva [USSR]
Probl. Peredachi Inf.	Problemy Peredachi Informatsii [USSR]
Probl. Planetol.	Problems of Planetology [of International Association of Planetology, Belgium]
Probl. Prochn.	Problemy Prochnosti [USSR]
Probl. Proj.	Problemy Projectowe [USSR]
Probl. Spets. Elektrometall.	Problemy Spetsial vory Elektrometallurgii [USSR]
Probl. Tekh. Kibern. Robot.	Problemy na Tekhnicheskata Kibernetika i Robotikata [Bulgaria]

Probl. Treniya Iznashivaniya	Problemy Treniya i Iznashivaniya
Proc. Am. Math. Soc.	Proceedings of the American Mathematical Society [US]
Proc. Am. Soc. Biol.	Proceedings of the American Society of Biology
Proc. Astron. Soc. Aust.	Proceedings of the Astronomical Society of Australia
Proc. Australas. Inst. Min. Metall.	Proceedings of the Australasian Institute of Mining and Metallurgy [Australia]
Proc. Aust. Soc. Sugar Cane Technol.	Proceedings of the Australian Society of Sugar Cane Technologists
Proc. Cosm.-Ray Res. Lab. Nagoya Univ.	Proceedings of the Cosmic-Ray Research Laboratory of Nagoya University [Japan]
Proc. CSEE	Proceedings of the CSEE [of Chinese Society of Electrical and Electronics Engineers [PR China]]
Proc. Egypt. Acad. Sci.	Proceedings of the Egyptian Academy of Sciences [Egypt]
Proc. Electrochem. Soc.	Proceedings of the Electrochemical Society [US]
Process Autom.	Process Automation [Germany]
Process Biochem.	Process Biochemistry
Process Eng. (Australia)	Process Engineering (Australia)
Process Eng. News	Process Engineering News [Australia]
Process Eng. (UK)	Process Engineering (UK)
Process Ind. Canada	Process Industries Canada [Canada]
Process Technol. Proc.	Process Technology Proceedings
Proc. Fac. Eng. Tokai Univ.	Proceedings of the Faculty of Engineering, Tokai University [Japan]
Proc. Fac. Eng. Tokai Univ. (Engl. Ed.)	Proceedings of the Faculty of Engineering, Tokai University (English Edition) [Japan]
Proc. IEE	Proceedings of the IEE [of Institution of Electrical Engineers, UK]
Proc. IEEE	Proceedings of the IEEE [of Institute of Electrical and Electronics Engineers, US]
Proc. Indian Acad. Sci.	Proceedings of the Indian Academy of Sciences [India]
Proc. Indian Acad. Sci., Chem. Sci.	Proceedings of the Indian Academy of Sciences, Chemical Sciences
Proc. Indian Acad. Sci., Earth Planet. Sci.	Proceedings of the Indian Academy of Sciences, Earth and Planetary Sciences
Proc. Indian Acad. Sci., Math. Sci.	Proceedings of the Indian Academy of Sciences, Mathematical Sciences
Proc. Indian Acad. Sci., Phys.	Proceedings of the Indian Academy of Sciences, Physical Sciences
Proc. Indian Natl. Sci. Acad. A	Proceedings of the Indian National Science Academy, Part A
Proc. Indian Natl. Sci. Acad. B	Proceedings of the Indian National Science Academy, Part B
Proc. Ind. Waste Conf.	Proceedings of the Industrial Waste Conference [US]
Proc. Infrared Soc. Jpn.	Proceedings of the Infrared Society of Japan [Japan]
Proc. Inst. Civ. Eng. 1	Proceedings of the Institution of Civil Engineers 1 [UK]
Proc. Inst. Civ. Eng. 2	Proceedings of the Institution of Civil Engineers 2 [UK]
Proc. Inst. Electr. Eng.	Proceedings of the Institution of Electrical Engineers [UK]
Proc. Inst. Electrost. Jpn.	Proceedings of the Institute of Electrostatics Japan
Proc. Inst. Mech. Eng.	Proceedings of the Institution of Mechanical Engineers [UK]
Proc. Inst. Mech. Eng. A, Power Process Eng.	Proceedings of the Institution of Mechanical Engineers A, Power and Process Engineering [UK]
Proc. Inst. Mech. Eng. B, J. Eng. Manuf.	Proceedings of the Institution of Mechanical Engineers B, Journal of Engineering Manufacture [UK]
Proc. Inst. Mech. Eng. B, Manage. Eng. Manuf.	Proceedings of the Institution of Mechanical Engineers B, Management and Engineering Manufacture [UK]
Proc. Inst. Mech. Eng. C, J. Mech. Eng. Sci.	Proceedings of the Institution of Mechanical Engineers C, Journal of Mechanical Engineering Science [UK]
Proc. Inst. Mech. Eng. D, J. Automob. Eng.	Proceedings of the Institution of Mechanical Engineers D, Journal of Automobile Engineering [UK]
Proc. Inst. Mech. Eng. D, Transp. Eng.	Proceedings of the Institution of Mechanical Engineers D, Transport Engineering [UK]
Proc. Inst. Mech. Eng. E, J. Process Mech. Eng.	Proceedings of the Institution of Mechanical Engineers E, Journal of Process Mechanical Engineering [UK]
Proc. Inst. Mech. Eng. F, J. Rail Rapid Transit	Proceedings of the Institution of Mechanical Engineers E, Journal of Rail and Rapid Transit [UK]
Proc. Inst. Mech. Eng. G, J. Aerosp. Eng.	Proceedings of the Institution of Mechanical Engineers G, Journal of Aerospace Engineering [UK]

Proc. Inst. Mech. Eng. H, J. Eng. Med.	Proceedings of the Institution of Mechanical Engineers H, Journal of Engineering in Medicine [UK]
Proc. Inst. Railw. Signal Eng.	Proceedings of the Institution of Railway Signal Engineers [UK]
Proc. Int. Conf. Mech. Med. Biol.	Proceedings of the International Conference on Mechanics in Medicine and Biology [US]
Proc. Int. Joint Conf. Patt. Recog.	Proceedings of the International Joint Conference on Pattern Recognition [Japan]
Proc. Jpn. Acad. A, Math. Sci.	Proceedings of the Japan Academy, Series A, Mathematical Sciences [Japan]
Proc. Jpn. Acad. B, Phys. Biol. Sci.	Proceedings of the Japan Academy, Series B, Physical and Biological Sciences [Japan]
Proc. Jpn. Congr. Mater. Res.	Proceedings of the Japan Congress on Materials Research [Japan]
Proc. Jpn. Soc. Civ. Eng.	Proceedings of the Japan Society of Civil Engineers [Japan]
Proc. K. Ned. Akad. Wet. A	Proceedings of the Koninklijke Nederlandse Akademie van Wetenschappen, Series A [Netherlands]
Proc. K. Ned. Akad. Wet. B	Proceedings of the Koninklijke Nederlandse Akademie van Wetenschappen, Series B [Netherlands]
Proc. Korean Inst. Electr. Eng.	Proceedings of the Korean Institute of Electrical Engineers [South Korea]
Proc. Mil. Oper. Res. Symp.	Proceedings of the Military Operations Research Symposium [of Military Operations Research Society, US]
Proc. Nat. Acad. Sci.	Proceedings of the National Academy of Sciences [US]
Proc. Natl. Acad. Sci. India A, Phys. Sci.	Proceedings of the National Academy of Sciences of India, Section A
Proc. Natl. Acad. Sci. India B	Proceedings of the National Academy of Sciences of India, Section B
Proc. Natl. Comp. Conf.	Proceedings of the National Computer Conference [of American Federation of Information Processing Societies]
Proc. Natl. Electron. Conf.	Proceedings of the National Electronics Conference [US]
Proc. NEC	Proceedings of the National Electronics Conference [US]
Proc. Ocean Drill. Program	Proceedings of the Ocean Drilling Program [US]
Proc. Ocean Drill. Program, Initial Rep.	Proceedings of the Ocean Drilling Program, Initial Reports [US]
Proc. Ocean Drill. Program, Sci. Results	Proceedings of the Ocean Drilling Program, Scientific Results [US]
Proc. Pak. Acad. Sci.	Proceedings of the Pakistan Academy of Sciences [Pakistan]
Proc. Radio Club Am.	Proceedings of the Radio Club of America [US]
Proc. Res. Inst. Atmos. Nagoya Univ.	Proceedings of the Research Institute of Atmospherics, Nagoya University [Japan]
Proc. R. Inst. GB	Proceedings of the Royal Institute of Great Britain [UK]
Proc. R. Inst. UK	Proceedings of the Royal Institute of the UK
Proc. R. Ir. Acad. A	Proceedings of the Royal Irish Academy, Section A
Proc. R. Ir. Acad. B	Proceedings of the Royal Irish Academy, Section B
Proc. R. Soc.	Proceedings of the Royal Society [UK]
Proc. R. Soc. Edinb. A	Proceedings of the Royal Society of Edinburgh, Section A [UK]
Proc. R. Soc. Edinb. B	Proceedings of the Royal Society of Edinburgh, Section B [UK]
Proc. R. Soc. Lond.	Proceedings of the Royal Society of London [UK]
Proc. R. Soc. Lond. A, Math. Phys. Sci.	Proceedings of the Royal Society of London, Series A: Mathematical and Physical Sciences [UK]
Proc. R. Soc. Lond. B	Proceedings of the Royal Society of London, Series B [UK]
Proc. R. Soc. Vic.	Proceedings of the Royal Society of Victoria [Australia]
Proc. SID	Proceedings of the SID [of Society for Information Display, US]
Proc. South Wales Inst. Inst. Eng.	Proceedings of the South Wales Institute of Engineers [Australia]
Proc. SPIE	Proceedings of the SPIE [of Society of Photo-Optical Instrumentation Engineers, US]
Proc. SPIE — Int. Soc. Opt. Eng.	Proceedings of the SPIE — The International Society for Optical Engineering [US]
Prod. Eng. (Cleveland)	Production Engineering (Cleveland) [US]
Prod. Eng. (London)	Production Engineer (London) [UK]
Prod. Eng. (New York)	Product Engineering (New York) [US]
Prod. Finish. (Cincinnati)	Product Finishing (Cincinnati) [US]
Prod. Finish. (London)	Product Finishing (London) [UK]
Prod. Inv. Manage. J.	Production and Inventory Management Journal [of American Production and Inventory Control Society]
Prod. Inv. Manage. Rev.	Production Inventory Management Reviews [of American Production and Inventory Control Society]
Prod. J.	Production Journal [UK]
Prof. Comput. (Australia)	Professional Computing (Australia)
Prof. Comput. (USA)	Professional Computing (USA)
Prof. Electron. Mag.	Professional Electronics Magazine [of International Society of Certified Electronics Technicians, US]
Prof. Energy Manager	Professional Energy Manager [of Association of Professional Energy Managers, US]
Prof. Eng.	Professional Engineering [UK]

Prof. Geol.	Professional Geologist [of American Institute of Professional Geologists]
Profile Mag.	Profile Magazine [Norway]
Prof. Manage.	Professional Management [UK]
Prof. Printer	Professional Printer [UK]
Prof. Safety	Professional Safety [of American of Society of Safety Engineers]
Prof. Stat.	Professional Statistician [of Institute of Statisticians, UK]
Prog. Aerosp. Sci.	Progress in Aerospace Sciences [UK]
Prog. Anal. Spectrosc.	Progress in Analytical Spectroscopy [UK]
Prog. Biophys. Mol. Biol.	Progress in Biophysics and Molecular Biology [UK]
Prog. Coll. Polym. Sci	Progress in Colloid and Polymer Science [Germany]
Prog. Cryst. Growth Charact.	Progress in Crystal Growth and Characterization [UK]
Prog. Energy Combust. Sci.	Progress in Energy and Combustion Science
Prog. Mater. Sci.	Progress in Materials Science [UK]
Prog. Nucl. Energy	Progress in Nuclear Energy [UK]
Prog. Nucl. Magn. Reson. Spectrosc.	Progress in Nuclear Magnetic Resonance Spectroscopy [UK]
Prog. Oceanogr.	Progress in Oceanography [UK]
Prog. Org. Coat.	Progress in Organic Coatings [Switzerland]
Prog. Part. Nucl. Phys.	Progress in Particle and Nuclear Physics [UK]
Prog. Polym. Sci.	Progress in Polymer Science [UK]
Prog. Powder Metall.	Progress in Powder Metallurgy [of International Powder Metallurgy Institute, US]
Prog. Quantum Electron.	Progress in Quantum Electronics [UK]
Program. Comput. Softw.	Programming and Computer Software [Translation of: *Programmirovanie (USSR)*; published in the US]
Program. J.	Programmer's Journal [US]
Prog. Rubber Plast. Technol.	Progress in Rubber and Plastics Technology [of Rubber and Plastics Research Association, UK]
Prog. Solid State Chem.	Progress in Solid State Chemistry [UK]
Prog. Space Res.	Progress of Space Research [of United Nations Committee on the Peaceful Uses of Outer Space]
Prog. Surf. Sci.	Progress in Surface Science [UK]
Prog. Tech.	Progres Technique [France]
Prog. Theor. Phys.	Progress in Theoretical Physics [of Physical Society of Japan]
Prog. Theor. Phys., Suppl.	Progress of Theoretical Physics, Supplement [of Physical Society of Japan]
Prom. Energ.	Promyshlennaya Energetika [USSR]
Prom. Sint. Kauch. Shin Rez.-tekhn. Iz.	Promyshlennost Sinteticheskogo Kauchuka Shin i Rezinoteknicheskikh Izdelli [USSR]
Prop. Data Update	Property Data Update [US]
Propellants Explos. Pyrotech.	Proppelants, Explosives and Pyrotechnics
Pr. Osrodka Badaw.-Rosw. Elektron. Prozniowej	Prace Osrodka Badawczo-Roswojowego Elektroniki Prozniowej [Poland]
Prot. Met.	Protection of Metals [Translation of: *Zashchita Metallov (USSR)*; published in the US]
Pr. Przem. Inst. Elektron.	Prace Przemyslowego Instytutu Elektroniki [Poland]
Pr. Przem. Inst. Telekomin.	Prace Przemyslowego Instytutu Telekomunikacji [Poland]
Przegl. Gorn.	Przeglad Gorniczny [Poland]
Przegl. Odlew.	Przeglad Odlewnictwa [Poland]
Przegl. Papier.	Przeglad Papierniczy [Poland]
Przegl. Spawalnictwa	Przeglad Spawalnictwa [Poland]
Przegl. Wlok.	Przeglad Wlokienniczy [Poland]
Przem. Chem.	Przemsyl Chemiczny [Poland]
Prz. Elektrotech.	Przeglad Elektrotechniczny [Poland]
Prz. Stat.	Przeglad Statystyczny [Poland]
Prz. Telekomun.	Przeglad Telekomunikacynjy [Poland]
PSI Nucl. Part. Phys. Newsl.	PSI Nuclear and Particle Physics Newsletter [of Paul Scherrer Institute, Switzerland]
PT/Elektron. Elektrotech.	PT/Elektronica Elektrotechniek [Netherlands]
Publ. Astron. Inst. Czech. Acad. Sci.	Publications of the Astronomical Institute of the Czechoslovak Academy of Sciences
Publ. Astron. Soc. Jpn.	Publications of the Astronomical Society of Japan [Japan]
Publ. Astron. Soc. Pac.	Publications of the Astronomical Society of the Pacific [US]
Publ. Dom. Astrophys. Obs.	Publications of the Dominion Astrophysical Observatory [Canada]
Publ. Dom. Astrophys. Obs. Victoria BC	Publications of the Dominion Astrophysical Observatory, Victoria, British Columbia [Canada]

Publ. Earth Phys. Branch Dep. Energy Mines Resour.	Publications of the Earth Physics Branch of the Department of Energy, Mines and Resources [Canada]
Publ. Elektroteh. Fak. Ser., Elektroenerg.	Publicaije Elektrotehnickog Fakulteta Serija: Elektroenergetika [Yugoslavia]
Publ. Elektroteh. Fak. Ser., Elektron. Telekomun. Autom.	Publikacije Elektrotehnickog Fakulteta Serija: Elektronika, Telekomunikacije, Automatika [Yugoslavia]
Publ. Elektroteh. Fak. Ser., Mat. Fiz.	Publikacije Elektrotehnickog Fakulteta Serija: Matematika i Fizika [Yugoslavia]
Publ. Inst. R. Meteorol. Belg. A	Publications, Institut Royal Meteorologique de Belgique, Serie A [Belgium]
Publ. Inst. R. Meteorol. Belg. B	Publications, Institut Royal Meteorologique de Belgique, Serie B [Belgium]
Publ. Korean Astron. Soc.	Publications of the Korean Astronomical Society [South Korea]
Publ. Natl. Astron. Obs. Jpn.	Publications of the National Astronomical Observatory of Japan
Publ. Tech. Univ. Heavy Ind.	Publications of the Technical University for Heavy Industry [Hungary]
Publ. Tech. Univ. Heavy Ind. (Miskolc) B, Metall.	Publications of the Technical University for Heavy Industry (Miskolc) Series B, Metallurgy [Hungary]
Publ. Util. Fortn.	Public Utilities Fortnightly [US]
Pulp Paper	Pulp and Paper [US]
Pulp Paper Can.	Pulp and Paper Canada
Pulp Paper Int.	Pulp and Paper International [Canada]
Pulp Paper Mag.	Pulp and Paper Magazine [Sweden]
Pulp Paper Wk.	Pulp and Paper Week [US]
Purch. Supply Manage.	Purchasing and Supply Management [of Institute of Purchasing and Supply, UK]
Pure Appl. Chem.	Pure and Applied Chemistry [of International Union of Pure and Applied Chemistry, UK]
Pure Appl. Geophys.	Pure and Applied Geophysics [Switzerland]

Q

Q. Appl. Math.	Quarterly of Applied Mathematics [of American Mathematical Society]
Q. Bull. Int. Assoc. Agric. Libr. Doc.	Quarterly Bulletin of the International Journal of the Association of Agricultural Librarians and Documentalists [Netherlands]
Q. Colorado Sch. Mines	Quarterly — Colorado School of Mines [US]
QEA	Quantum Electronics and Applications [US]
Q. J. Jpn. Weld. Soc.	Quarterly Journal of the Japan Welding Society
Q. J. Mech. Appl. Math.	Quarterly Journal of Mechanics and Applied Mathematics [UK]
Q. J. R. Astron. Soc.	Quarterly Journal of the Royal Astronomical Society [UK]
Q. J. R. Meteorol. Soc.	Quarterly Journal of the Royal Meteorological Society [UK]
Q. J. Seismol.	Quarterly Journal of Seismology [of Japan Meteorological Agency]
Q. Rep. Railw. Tech. Res. Inst.	Quarterly Report of the Railway Technical Research Institute [of Railway Technical Institute, Japan]
Q. Rev. Biophys.	Quarterly Review of Biophysics [of International Union for Pure and Applied Biophysics, Hungary]
Quad. Ric. Sci.	Quaderni de la Recerca Scientifica [of Consiglio Nazionale delle Ricerche, Italy]
Quaest. Inf.	Quaestions Informaticae [of Computer Society of South Africa]
Qual. Assur.	Quality Assurance [of Institute of Quality Assurance, UK]
Qual. Eng.	Quality Engineering [of American Society for Quality Control]
Qual. Eval.	Quality Evaluation [India]
Qual. Prog.	Quality Progress [of American Society for Quality Control]
Qual. Reliab. Eng. Int.	Quality and Reliability Engineering International [US]
Qual. Rev.	Quality Review [of American Society for Quality Control]
Qual. Rev. Prat. Controle Ind.	Qualite Revue Pratique de Controle Industriel [France]
Qual. Today	Quality Today [UK]
Qual. Zuverlassigk.	Qualitat und Zuverlaessigkeit [Germany]
Quantum Opt.	Quantum Optics [UK]
Quarry Manage.	Quarry Management

Quat. Res.	Quaternary Research [US]
Quat. Sci. Rev.	Quaternary Science Reviews [UK]
Queueing Syst. Theory Appl.	Queueing Systems Theory and Applications [Switzerland]

R

R & D	Research & Development [US]
R & D Bull.	R & D Bulletin [of Supply and Services Canada]
R & D Dig.	R & D Digest [UK]
R & D Manage.	R & D Management [UK]
Rad. Elect. Eng.	Radio and Electronic Engineer [UK]
Radex Runds.	Radex Rundschau [Austria]
Radiat. Eff.	Radiation Effects [UK]
Radiat. Eff. Defects Solids	Radiation Effects and Defects in Solids [UK]
Radiat. Eff. Express	Radiation Effects Express [UK]
Radiat. Eff. Lett.	Radiation Effects Letters [UK]
Radiat. Eff. Lett. Sect.	Radiation Effects Letters Section [UK]
Radiat. Environ. Biophys.	Radiation and Environmental Biophysics [Germany]
Radiat. Phys. Chem.	Radiation Physics and Chemistry [UK]
Radiat. Prot.	Radiation Protection [of Korean Association for Radiation Protection, South Korea]
Radiat. Prot. Dosim.	Radiation Protection Dosimetry [UK]
Radiat. Res.	Radiation Research [of Radiation Research Society, US]
Radioact. Waste Manage. Nucl. Fuel Cycle	Radioactive Waste Management and the Nuclear Fuel Cycle [Switzerland]
Radio Commun.	Radio Communication [of Radio Society of Great Britain [UK]]
Radiodiffus. Telev.	Radiodiffusion Television [France]
Radio-Electron.	Radio-Electronics [US]
Radioelectron. Commun. Syst.	Radioelectronics and Communications Systems [Translation of: *Izvestiya Vysshikh Uchebnykh Zavedenii, Radioelektronika (USSR)*; published in the US]
Radio-Electron. World	Radio and Electronics World [UK]
Radiophys. Quantum Electron.	Radiophysics and Quantum Electronics [Translation of: *Izvestiya Vysshikh Uchebnykh Zavedenii, Radiofizika (USSR)*; published in the US]
Radio Sci.	Radio Science [of American Geophysical Union]
Radiotekh. Elektron.	Radiotekhnika i Elektronika [USSR]
Radiotekhnika, Kharkov	Radiotekhnika, Kharkov [USSR]
Radiotekhnika, Mosk.	Radiotekhnika, Moskva [USSR]
Radio Telev.	Radio and Television [of Organisation Internationale de Radiodiffusion et Television, Czechoslovakia]
Rail Int.	Rail International [of Union International des Chemins de Fer, France]
Railw. Electr.	Railway and Electricity [of Railway Electrification Association, Japan]
Railw. Eng. Int.	Railway Engineer International [UK]
Railw. Gaz. Int.	Railway Gazette International [UK]
RAIRO Rech. Oper.	RAIRO Recherche Operationelle [France]
RAPRA Rev. Rep.	RAPRA Review Reports [of Rubber and Plastics Research Association, UK]
Rand J. Econ.	Rand Journal of Economics [US]
Rare Earth Inf. Cent. News	Rare Earth Information Center News [US]
Rare Met.	Rare Metals [PR China]
Raw Mater. Rep.	Raw Materials Report [Sweden]
RCA Broadcast News	RCA Broadcast News [of Radio Corporation of America]
RCA Eng.	RCA Engineer [of Radio Corporation of America]
RCA Rev.	RCA Review [of Radio Corporation of America]
RCA Tech. Notes	RCA Technical Notes [of Radio Corporation of America]
R. Chem. Soc. Spec. Publ.	Royal Chemical Society Special Publications [UK]
React. Polym.	Reactive Polymers [US and Netherlands]
React. Solids	Reactivity of Solids [Netherlands]

React. Polym.	Reactive Polymers [Netherlands]
Rec. Electr. Commun. Eng. Conversazione Tohoku Univ.	Record of Electrical and Communication Engineering Conversazione, Tohoku University [Japan]
Rech. Aerosp.	Recherche Aerospatiale [of Office National d'Etudes et de Recherches Aerospatiales, France]
Rechentech. Datenverarb.	Rechentechnik und Datenverarbeitung [Germany]
Reclam. Rev.	Reclamation Review [US]
Rec. Trav. Chim.	Recueil des Travaux Chimiques des Pays-Bas [Netherlands]
Recursos Min.	Recursos Minerales [Chile]
Ref. Lib.	Reference Librarian [US]
Refract. J.	Refractories Journal [UK]
Refract. News	Refractory News [of The Refractories Institute, US]
Ref. Zh.	Referativnyi Zhurnal [USSR]
Ref. Zh. Inf.	Referativnyi Zhurnal, Informatika [USSR]
Ref. Zh. Korroz.	Referativnyi Zhurnal, Korroziya [USSR]
Ref. Zh. Metal.	Referativnyi Zhurnal, Metallurgiya [USSR]
Reinf. Plast.	Reinforced Plastics [UK]
Relat. J.	Relational Journal [US]
Reliab. Eng.	Reliability Engineering [UK]
Reliab. Eng. Syst. Saf.	Reliability Engineering and System Safety [UK]
Remittance Doc. Process. Today	Remittance and Document Processing Today [of Recognition Technologies Users Association, US]
Remote Sens. Environ.	Remote Sensing of Environment [US]
Remote Sens. Rev.	Remote Sensing Reviews [UK]
Rend. Ist. Lomb. Accad. Lett. A, Sci. Mat. Appl.	Rendiconti Istituto Lombardo Accademia di Scienze e Lettere A, Scienze Matematiche e Applicazioni [Italy]
Rend. Ist. Lomn. Accad. Lett. B	Rendiconti Istituto Lombardo Accademia di Scienze e Lettere B [Italy]
Rep. Cast. Res. Lab., Waseda Univ.	Report of the Castings Research Laboratory, Waseda University [Japan]
Rep. Chiba Inst. Technol.	Report of the Chiba Institute of Technology [Japan]
Rep. Coll. Eng. Hosei Univ.	Report of the College of Engineering, Hosei University [Japan]
Rep. Du Pont	Report on Du Pont [US]
Rep. Fac. Eng., Nagasaki Univ.	Reports of the Faculty of Engineering, Nagasaki University [Japan]
Rep. Fac. Eng., Shizuoka Univ.	Reports of the Faculty of Engineering, Shizuoka University [Japan]
Rep. Fac. Eng., Tottori Univ.	Reports of the Faculty of Engineering, Tottori University [Japan]
Rep. Fac. Eng., Yamanashi Univ.	Reports of the Faculty of Engineering, Yamanashi University [Japan]
Rep. Fac. Sci. Eng., Saga Univ.	Reports of the Faculty of Engineering, Saga University [Japan]
Rep. Fac. Sci. Technol. Meijo Univ.	Reports of the Faculty of Science and Technology, Meijo University [Japan]
Rep. Fukui Prefect. Ceram. Ind. Lab.	Report of Fukui Prefectural Ceramic Industry Laboratory [Japan]
Rep. Gov. Ind. Res. Inst., Chugoku	Reports of the Government Industrial Research Institute, Chugoku [Japan]
Rep. Gov. Ind. Res. Inst., Nagoya	Reports of the Government Industrial Research Institute, Nagoya [Japan]
Rep. Gov. Ind. Res. Inst., Osaka	Reports of the Government Industrial Research Institute, Osaka [Japan]
Rep. Grad. Sch. Electron. Sci. Technol. Shizuoka Univ.	Reports of the Graduate School of Electronic Science and Technology, Shizuoka University [Japan]
Rep. Himeji Inst. Technol.	Reports of Himeji Institute of Technology [Japan]
Rep. Hokkaido Forest Prod. Res. Inst.	Report of the Hokkaido Forest Products Research Institute [Japan]
Rep. Ind. Res. Inst. Ishikawa	Report of the Industrial Research Institute of Ishikawa [Japan]
Rep. Inst. High Speed Mech. Tohoku Univ.	Reports of the Institute of High Speed Mechanics, Tohoku University [Japan]
Rep. Inst. High Speed Mech. Tohoku Univ., Ser. A	Reports of the Institute of High Speed Mechanics, Tohoku University, Series A [Japan]
Rep. Inst. High Speed Mech. Tohoku Univ., Ser. B	Reports of the Institute of High Speed Mechanics, Tohoku University, Series B [Japan]
Rep. Inst. Ind. Sci., Univ. Tokyo	Reports of the Institute of Industrial Science, University of Tokyo [Japan]
Rep. Inst. Phys. Chem. Res.	Reports of the Institute of Physical and Chemical Research [Japan]
Rep. Math. Phys.	Reports on Mathematical Physics [UK]
Rep. Prog. Appl. Chem.	Reports on the Progress of Applied Chemistry [of Society of Chemical Industry, US]
Rep. Prog. Phys.	Reports on Progress in Physics [of Institute of Physics, UK]
Rep. Prog. Polym. Phys. Jpn.	Reports on Progress in Polymer Physics in Japan [of Association for Science Documentation Information, Japan]
Rep. Res. Inst. Appl. Mech.	Reports of Research Institute for Applied Mechanics [Japan]

Rep. Res. Inst. Ind. Sci. Kyushu Univ.	Reports of the Research Institute of Industrial Science, Kyushu University [Japan]
Rep. Res. Inst. Sci. Technol., Nihon Univ.	Report of the Research Institute of Science and Technology, Nihon University [Japan]
Rep. Res. Inst. Strength Fract. Mater., Tohoku Univ.	Reports of the Research Institute for Strength and Fracture of Materials, Tohoku University [Japan]
Rep. Res. Lab., Asahi Glass	Reports of the Research Laboratory, Asahi Glass [Japan]
Rep. Res. Lab. Eng. Mater.	Reports of the Research Laboratory of Engineering Materials [Japan]
Rep. Res. Lab. Eng. Mater., Tokyo Inst. Technol.	Reports of the Research Laboratory of Engineering Materials, Tokyo Institute of Technology [Japan]
Rep. Res. Lab. Surf. Sci. Okayama Univ.	Reports of the Research Laboratory for Surface Science, Okayama University [Japan]
Rep. Res. Nippon Inst. Technol.	Report of Researches, Nippon Institute of Technology [Japan]
Rep. Stat. Appl. Res. UJSE	Reports of Statistical Application Research UJSE [of Union of Japanese Scientists and Engineers]
Rep. Stat. Appl. Res. Union Jpn. Sci. Eng.	Reports of Statistical Application Research, Union of Japanese Scientists and Engineers [Japan]
Rep. Technol. Iwate Univ.	Report on Technology of Iwate University [Japan]
Rep. Tokyo Metrop. Ind. Technol. Cent.	Report of the Tokyo Metropolitan Industrial Technology Center [Japan]
Rep. Univ. Electro-Commun.	Reports of the University of Electro-Communications [Japan]
Res. Act. Civ. Eng. Relat. Fields Kyoto Univ.	Research Activities in Civil Engineering and Related Fields at Kyoto University [Japan]
Res. Bull. Hiroshima Inst. Technol.	Research Bulletin of the Hiroshima Institute of Technology [Japan]
Res. Bull. Print. Bur.	Research Bulletin of the Printing Bureau [Japan]
Res. Bull. Reg. Eng. Coll. Warangal	Research Bulletin, Regional Engineering College, Warangal [India]
Res. Dev.	Research and Development [US]
Res. Dev. Jpn. Awarded Okochi Mem. Prize	Research and Development in Japan Awarded the Okochi Memorial Prize [Japan]
Res. Discl.	Research Disclosure [US and UK]
Res. Electrotech. Lab.	Researches of the Electrotechnical Laboratory [Japan]
Res. Eng. Des.	Research in Engineering Design [US]
Resin Rev.	Resin Review [US]
Res. Inst. Tech. Phys. Hung. Acad. Sci. Yearb.	Research Institute for Technical Physics of the Hungarian Academy of Sciences, Yearbook [Hungary]
Res. Lab. Precis. Mach. Electron.	Research Laboratory of Precision Machinery and Electronics [Japan]
Res Mech.	Res Mechanica [UK]
Res Mech. Lett.	Res Mechanica Letters [US]
Res. News	Researcher News [US]
Res. Nondestr. Eval.	Research in Nondestructive Evaluation [American Society for Nondestructive Testing]
Resour. Conserv.	Resources and Conservation [Netherlands]
Resour. Conserv. Recycl.	Resources, Conservation and Recycling [Netherlands]
Resour. Policy	Resources Policy [UK]
Resour. Process.	Resources Processing
Resour. Recovery Conserv.	Resource Recovery and Conservation [Netherlands]
Resour. Shar. Inf. Netw.	Resource Sharing and Information Networks [US]
Res. Prog. SSE	Research and Progress of SSE (Solid State Engineering) [of Nanjing Solid State Devices Research Institute, PR China]
Res. Rep. Fac. Eng., Gifu Univ.	Research Report of the Faculty of Engineering, Gifu University [Japan]
Res. Rep. Fac. Eng., Kagoshima Univ.	Research Reports of the Faculty of Engineering, Kogoshima University [Japan]
Res. Rep. Fac. Eng., Meiji Univ.	Research Reports of the Faculty of Engineering, Meiji University [Japan]
Res. Rep. Fac. Eng., Mie Univ.	Research Reports of the Faculty of Engineering, Mie University [Japan]
Res. Rep. Fac. Eng., Niigata Univ.	Research Reports of the Faculty of Engineering, Niigata University [Japan]
Res. Rep. Fac. Eng., Tokyo Univ.	Research Reports of the Faculty of Engineering, Tokyo University [Japan]
Res. Rep. Ikutoku Tech.. Univ. B., Sci. Technol.	Research Reports of Ikutoku Technical University, Part B, Science and Technology [Japan]
Res. Rep. Inf. Sci. A, Math. Sci.	Research Reports on Information Sciences, Series A, Mathematical Science [Japan]
Res. Rep. Inf. Sci. B, Oper. Res.	Research Reports on Information Sciences, Series B, Operations Research [Japan]
Res. Rep. Inf. Sci. C, Comput. Sci.	Research Reports on Information Sciences, Series C, Computer Science [Japan]
Res. Rep. Inst. Inf. Sci. Technol., Tokyo Denki Univ.	Research Reports, Institute of Information Science and Technology, Tokyo Denki University [Japan]

Res. Rep. Kogakuin Univ.	Research Reports of Kogakuin University [Japan]
Res. Rep. Nagaoka Coll. Technol.	Research Reports of Nagaoka College of Technology [Japan]
Res. Rep. Nagoya Munic. Ind. Res. Inst.	Research Reports of the Nagoya Municipal Industrial Research Institute [Japan]
Res. Technol. Manage.	Research Technology Management [of Industrial Research Institute, US]
Res. Word Process. Newsl.	Research in Word Processing Newsletter [of South Dakota School of Mines and Technology, US]
Rev. Acad. Cienc. Zaragoza	Revista de la Academia de Ciencias Zaragoza [Spain]
Rev. Acoust.	Revue d'Acoustique [of Societe Francaise d'Acoustique, France]
Rev. Act.-Metallges.	Review of Activities — Metallgesellschaft AG [Germany]
Rev. AIBDA	Revista AIBDA [of Associacion Interamericana de Bibliothecarias y Documentalistas Agricoles, Costa Rica]
Rev. Alum.	Revue de l'Aluminium [France]
Rev. Aluminio	Revista do Aluminio [Brazil]
Rev. Anal. Chem.	Reviews in Analytical Chemistry [UK]
Rev. ATIP	Revue ATIP [of Association Technique de l'Industrie Papetiere, France]
Rev. Bio-Math.	Revue de Bio-Mathematique [of International Society of Mathematical Biology, France]
Rev. Bras. Fis.	Revista Brasileira de Fisica [of Sociedade Brasileira de Fisica, Brazil]
Rev. Caucho	Revisto del Caucho [of Consorcia Nacional de Industriales del Caucho, Spain]
Rev. Cenic Cienc. Fis.	Revista Cenic Ciencias Fisicas [Cuba]
Rev. Chem. Eng.	Reviews in Chemical Engineering
Rev. Chim.	Revista de Chimie
Rev. Cienc. Quim.	Revista de Ciencias Quimicas [Cuba]
Rev. Coatings Corros.	Reviews on Coatings and Corrosion [UK]
Rev. Colomb. Fis.	Revista Colombiana de Fisica [of Sociedad Colombiana de Fisica, Colombia]
Rev. Commun. Res. Lab.	Review of the Communications Research Laboratory [of Ministry of Post and Telecommunications, Japan]
Rev. Cuba. Quim.	Revista Cubana de Quimica [Cuba]
Rev. Deform. Behav. Mater.	Reviews on the Deformation Behavior of Materials [UK]
Rev. E	Revue E [of Societe Royale Belge des Electriciens, Belgium]
Rev. Electr. Commun. Lab.	Review of the Electrical Communication Laboratory [Japan]
Rev. Electrotec. (Argentina)	Revista Electrotecnica (Argentina) [of Asociacion Argentina de Electrotechnicos y del Comite Electrotecnico Argentina]
Rev. Electrotec. (Spain)	Revista Electrotecnica (Spain) [of Asociacion Electrotecnica Espanola]
Rev. Empresas Publicas Medellin	Revista Empresas Publicas de Medellin [Colombia]
Rev. Energ.	Revue de l'Energie [France]
Rev. Esp. Doc. Cient.	Revista Espanola de Documentacion Cientifica [of Instituto de Informacion y Documentacion en Ciencia y Tecnologia, Spain]
Rev. Esp. Electron.	Revista Espanola de Electronica [Spain]
Rev. Eur. Papiers Cartons Complexes	Revue Europeene des Papiers, Cartons, Complexes [France]
Rev. Fac. Eng. Univ. Porto	Revista de Faculdade de Engenharia, Universidade do Porto [Portugal]
Rev. Fac. Sci. Univ. Istanb. A	Review of the Faculty of Science, University of Istanbul, Series A [Turkey]
Rev. Fac. Sci. Univ. Istanb. B	Review of the Faculty of Science, University of Istanbul, Series B [Turkey]
Rev. Fac. Sci. Univ. Istanb. C	Review of the Faculty of Science, University of Istanbul, Series C [Turkey]
Rev. FITCE	Revue de FITCE [of Federation des Ingenieurs des Telecommunications de la Communaute Europeenne, Belgium]
Rev. Forest Franc.	Revue Forestiere Francaise [France]
Rev. Fr. Electr.	Revue Francaise de l'Electricite [France]
Rev. Fr. Mec.	Revue Francaise de Mecanique [France]
Rev. Fr. Metall.	Revue Francaise des Metallurgistes [France]
Rev. Gen. Caout. Plast.	Revue Generale des Caoutchoucs et Plastiques [of Societe d'Expansion Technique et Economique, France]
Rev. Gen. Chemins Fer	Revue Generale des Chemins de Fer [France]
Rev. Gen. Electr.	Revue Generale de l'Electricite [of Societe des Electriciens des Electroniciens et des Radioelectriciens, France]
Rev. Gen. Nucl.	Revue Generale Nucleaire [of Societe Francaise d'Energie Nucleaire, France]
Rev. Gen. Nucl. (Int. Ed.)	Revue Generale Nucleaire, (International Edition) [of Societe Francaise d'Energie Nucleaire, France]
Rev. Gen. Therm.	Revue Generale de Thermique [France]
Rev. Geofis.	Revista de Geofisica [Spain]
Rev. Geophys.	Reviews of Geophysics [of American Geophysical Union]

Rev. Geophys. Space Phys.	Reviews of Geophysics and Space Physics [of American Geophysical Union]
Rev. HF	Revue HF [of Societe Belge des Ingenieurs des Telecommunications et d'Electronique, Belgium]
Rev. High-Temp. Mater.	Reviews on High-Temperature Materials [UK]
Rev. Hist. Sci.	Revue d'Histoire des Sciences [of Foundation for International Scientific Coordination, France]
Rev. Iberoam. Corros. Prot.	Revista Iberoamericana de Corrosion y Proteccion [Spain]
Rev. IG	Revue IG [France]
Rev. Ing.	Revue de l'Ingenierie [of Institut Canadien des Ingenieurs, Canada]
Rev. Inform. Autom.	Revista de Informatica y Automatica [of Asociacion Espanola de Informatica y Automatica, Spain]
Rev. Inorganic Chem.	Reviews in Inorganic Chemistry [UK]
Rev. Int. Hautes Temp. Refract.	Revue Internationale des Hautes Temperatures et des Refractaires [France]
Rev. Int. Metodos Numer. para Calc. Diseno Ing.	Revista Internacional de Metodos Numericos para Calcula y Diseno en Ingenieria [Spain]
Rev. Int. Syst.	Revue Internationale de Systematique [of Association Francaise pour le Cybernetique, Economiqie et Technique, France]
Rev. Laser Eng.	Review of Laser Engineering [of Laser Society of Japan]
Rev. Latinoam. Metal. Mater.	Revista Latinoamericana de Metallurgia y Materiales [Venezuela]
Rev. Mec. Tijdschr.	Revue Mecanique Tijdschrift [Belgium]
Rev. Metal.	Revista de Metalurgia [Spain]
	Revue de Metallurgie[France]
Rev. Metall., Cah. Inf. Tech.	Revue de Metallurgie, Cahiers d'Informations Techniques [France]
Rev. Mex. Astron. Astrofis.	Revista Mexicana de Astronomia y Astrofisica [of Instituto de Astronomia, Mexico]
Rev. Mex. Fis.	Revista Mexicana de Fisica [of Sociedad Mexicana de Fisica, Mexico]
Rev. Mineral.	Reviews in Mineralogy [of Mineralogical Society of America]
Rev. Mod. Phys.	Reviews of Modern Physics [Joint Publication of American Physical Society and American Institute of Physics]
Rev. Phys. Appl.	Revue de Physique Appliquee [France]
Rev. Phys. Appl. (Suppl. J. Phys.)	Revue de Physique Appliquee (Supplement to 'Journal de Physique') [France]
Rev. Plast. Mod.	Revista de Plasticos Modernos [Spain]
Rev. Pol. Acad. Sci.	Review of the Polish Academy of Sciences [Poland]
Rev. Powder Metall. Phys. Ceram.	Reviews on Powder Metallurgy and Physical Ceramics [UK]
Rev. Prat. Metall.	Revue Pratique des Metallurgistes [France]
Rev. Radio Res. Lab.	Review of the Radio Research Laboratories [Japan]
Rev. Roum. Chim.	Revue Roumaine de Chimie [Romania]
Rev. Roum. Phys.	Revue Roumaine de Physique [Romania]
Rev. Roum. Sci. Tech.	Revue Roumaine des Sciences Techniques [Romania]
Rev. Roum. Sci. Tech.-Electrotech. Energ.	Revue Roumaine des Sciences Techniques — Serie Electrotechnique et Energetique [Romania]
Rev. Roum. Sci. Tech.-Mec. Appl.	Revue Roumaine des Sciences Techniques — Serie Mecanique Appliquee [Romania]
Rev. Sci. Instrum.	Review of Scientific Instruments [of American Institute of Physics]
Rev. Silicon, Germanium, Tin Lead Compd.	Reviews on Silicon, Germanium, Tin and Lead Compounds [UK]
Rev. Soc. R. Belge Ing. Ind.	Revue de la Societe Royale Belge des Ingeniers et des Industriels [Belgium]
Rev. Soldadura	Revista de Soldadura [Spain]
Rev. Solid State Sci.	Review of Solid State Science
Rev. Soudure-Lastijdschr.	Revue de la Soudure-Lastijdschrift [Belgium]
Rev. Tec.	Revista Tecnica [Venezuela]
Rev. Tech. Luxemb.	Revue Technique Luxembourgeoise [Luxembourg]
Rev. Tech. Thomson-CSF	Revue Technique Thomson-CSF [France]
Rev. Tecnol.	Revista Tecnologica [Cuba]
Rev. Telecomun.	Revista de Telecommunicacion [of Consejo Tecnica de Telecommunicacion, Spain]
Rev. Telegr. Electron.	Revista Telegrafica Electronica [Argentina]
R.F. Des.	R.F. Design [US]
RfF Resources	RfF Resources [of Resources for the Future, Inc., US]
Rheol. Acta	Rheologica Acta [Germany]
Rheol. Bull.	Rheology Bulletin [of Society of Rheology, US]
Ric. Autom.	Ricerche di Automatica [Italy]
Risk Anal.	Risk Analysis [US]
Risk Manage.	Risk Management [of Risk Management Society, US]
Riso Rep.	Riso Report [Denmark]
Risoe Rep.	Risoe Report [Denmark]
Riv. Combust.	Rivista dei Combustibili

Riv. Inf.	Rivista Informatica [of Associazione Italiana per il Calcolo Automatica, Italy]
Riv. Inform.	Rivista Informatica [of Associazione Italiana per il Calcolo Automatica, Italy]
Riv. Ital. Saldatura	Rivista Italiana della Saldatura [Italy]
Riv. Mecc.	Rivista di Meccanica [Italy]
Riv. Nuevo Cimento	Rivista del Nuevo Cimento [of Societa Italiana di Fisica, Italy]
RLE Curr.	RLE Currents [of Research Laboratory of Electronics [of MIT, US]]
R. Microsc. Soc., Proc.	Royal Microscopic Society, Proceedings [UK]
Robot. Auton. Syst.	Robotica and Autonomous Systems [Netherlands]
Robot. Comput.-Integr. Manuf.	Robotics and Computer-Integrated Manufacturing [UK]
Robot. Q.	Robotics Quarterly [of Robotics International [of SME, US]]
Robot. Today	Robotics Today [of Robotics International [of SME, US]]
Robot. World	Robotics World [US]
ROCC Inf. Manage.	ROCC Information Management [UK]
Rock Prod.	Rock Products
Rocky Mt. J. Math.	Rocky Mountain Journal of Mathematics [of Rocky Mountain Mathematics Consortium, US]
Rod, Wire Fastener	Rod, Wire and Fastener [US]
Roentgen-Bl. Prax. Klin.	Roentgen-Blatter: Praxis und Klinik [Germany]
Rom. J. Chem.	Romanian Journal of Chemistry [Romania]
Roof. Cladding Insul.	Roofing, Cladding and Insulation [UK]
Roof./Siding/Insul.	Roofing/Siding/Insulation [UK]
Ropa Uhlie	Ropa a Uhlie [Czechoslovakia]
Rozpr. Electrotech.	Rozprawy Electrotechniczne [Poland]
Rozpr. Hydrotech.	Rozprawy Hydrotechniczne [Poland]
RRISL Bull.	RRISL Bulletin [of Rubber Research Institute of Sri Lanka]
Rubb. Board Bull.	Rubber Board Bulletin [India]
Rubb. Chem. Technol.	Rubber Chemistry and Technology [of American Chemical Society]
Rubb. Dev.	Rubber Development [of Malaysian Natural Rubber Producers Research Association]
Rubb. India	Rubber India [of All India Rubber Industries Association]
Rubb. Plast. Fire Flamm. Bull.	Rubber and Plastics Fire and Flammability Bulletin [Switzerland]
Rubb. Plast. News	Rubber and Plastics News [US]
Rubb. Plast. News II	Rubber and Plastics News II [UK]
Rubb. S. Africa	Rubber Southern Africa [South Africa]
Rubb. Stat. Bull.	Rubber Statistical Bulletin [of International Rubber Study Group, UK]
Rubb. Trends	Rubber Trends [UK]
Rubb. World	Rubber World [US]
Rud. Metal. Zb.	Rudarsko Metalurski Zbornik [Yugoslavia]
Rudodobiv Metal.	Rudodobiv Metalurgiya [Bulgaria]
Rudy Met. Niezel.	Rudy i Metale Niezelazne [Poland]
Rudy Met. Niezelaz.	Rudy i Metale Niezelazne [Poland]
Rundfunktech. Mitt.	Rundfunktechnische Mitteilungen [Germany]
Russ. Chem. Rev.	Russian Chemical Reviews [Translation of: *Uspekhi Khimii (USSR)*; published in the UK]
Russ. J. Inorg. Chem.	Russian Journal of Inorganic Chemistry [Translation of: *Zhurnal Neorganicheskoi Khimii (USSR)*; published in the UK]
Russ. J. Phys. Chem.	Russian Journal of Physical Chemistry [Translation of: *Zhurnal Fizicheskoi Khimii (USSR)*; published in the UK]
Russ. Metall.	Russian Metallurgy [Translation of: *Izvestiya Akademii Nauk SSSR, Metally (USSR)*; published in the US]
Rutgers Comput. Technol. Law J.	Rutgers Computer and Technology Law Journal [of Rutgers University, US]

S

SA	Ship Abstracts [Joint Scandinavian Publication]
SAE Australas.	SAE Australasia [of Society of Automotive Engineers, Australia]
Saehkoe Electr. Electron.	Saehkoe Electricity and Electronics [Finland]
SAFE J.	SAFE Journal [of Survival and Flight Equipment Association, US]

Safety Health	Safety and Health [of National Safety Council, US]
S. Afr. Comput. J.	South African Computer Journal [of Computer Society of South Africa]
S. Afr. J. Antarct. Res.	South African Journal of Antarctic Research
S. Afr. J. Chem.	South African Journal of Chemistry [of South African Chemical Institute]
S. Afr. J. Libr. Inf. Sci	South African Journal of Library and Information Science [of South African Institute for Librarianship and Information Science]
S. Afr. J. Phys.	South African Journal of Physics
S. Afr. Mach. Tool Rev.	South African Machine Tool Review
S. Afr. Mech. Eng.	South African Mechanical Engineer [of South African Institute of Mechanical Engineers]
S.-Afr. Tydskr. Natuurwet. Tecnol.	Suid-Afrikaanse Tydskrif vir Natuurwetenskap en Tecnologie [South Africa]
SAMPE J.	SAMPE Journal [of Society for the Advancement of Materials and Process Engineering, US]
SAMPE Q.	SAMPE Quarterly [of Society for the Advancement of Materials and Process Engineering, US]
Sanken Tech. Rep.	Sanken Technical Report [Japan]
Sanyo Tech. Rev.	Sanyo Technical Review [Japan]
Satell. Commun.	Satellite Communications [US]
SATRA Bull.	SATRA Bulletin [of Shoe and Allied Trades Research Association, UK]
SAWE Newsl.	SAWE Newsletter [of Society of Allied Weight Engineers, US]
Sb. Kratk. Soobshch. Fiz. AN SSSR Fiz.Inst. P.N. Lebedeva	Sbornik Kratkie Soobshcheniya po Fisike, AN SSSR, Fizicheskii Institut P.N. Lebedeva [USSR]
Sb. Nauchni Tr.	Sbornik Nauchni Trudove [Bulgaria]
Sb. Nauch. Tr. Ukr NP-Bum. Prom.	Sbornik Nauchnykh Trudov Ukrainskogo Nauchno-Proizvodstvennogo Ob'edineniya Tsellyulozno-Bumazhnoi Promyshlennosti [USSR]
Sb. Tr. Ts. NIIB	Sbornik Trudov Tsentral'nogo Nauchno-Issledovatel'skogo Instituta Bumagi [USSR]
Sb. Tr. Ts. NILK Prom.	Sbornik Trudov Tsentral'nogo Nauchno-Issledovatel'skogo i Proektnogo Instituta Lesokhimicheskoi Promyshlennosti [USSR]
Sb. Tr. VNIIB Prom.	Sbornik Trudov Vsesyuznogo Nauchno-Issledovatel'skogo Instituta Tsellulozno-Bumazhnoi Promyshlennosti [USSR]
Sb. Ved. Pr. Vys. Sk. Banske Ostrave, Horn.-Geol.	Sbornik Vedeckych Praci Vysoke Skoly Banske v Ostrave, Rada Hornicko-Geologicka [Czechoslovakia]
Sb. Ved. Pr. Vys. Sk. Banske Ostrave, Hutn.	Sbornik Vedeckych Praci Vysoke Skoly Banske v Ostrave, Rada Hutnicka [Czechoslovakia]
Sb. Ved. Pr. Vys. Sk. Banske Ostrave, Strojni Elektrotech.	Sbornik Vedeckych Praci Vysoke Skoly Banske v Ostrave, Rada Strojni Elektrotechnicka [Czechoslovakia]
Scand. Audiol.	Scandinavian Audiology [Sweden]
Scand. Audiol. Suppl.	Scandinavian Audiology Supplementum [Sweden]
Scand. J. Forest Res.	Scandinavian Journal of Forest Research [Sweden]
Scand. J. Metall.	Scandinavian Journal of Metallurgy [Denmark and Finland]
Scand. J. Stat. Theory Appl.	Scandinavian Journal of Statistics Theory and Applications [Sweden]
Scand. Refrig.	Scandinavian Refrigeration [Denmark]
Scanning Electron Microsc.	Scanning Electron Microscopy [of Scanning Microscopy International, US]
Scanning Microsc.	Scanning Microscopy [of Scanning Microscopy International, US]
Schiff Hafen	Schiff und Hafen [Germany]
Schiff Hafen/Seewirtschaft	Schiff und Hafen/Seewirtschaft [Germany]
Schr.-R. Rhein.-Westfal. TUV	Schriftenreihe des Rheinisch-Westfalischen Technischen Uberwachungsvereins [Germany]
Schweissen Schneiden	Schweissen und Schneiden [Germany]
Schweiz. Alum. Rundsch.	Schweizer Aluminum Rundschau [Switzerland]
Schweiz. Ing. Archit.	Schweizer Ingenieur und Architekt [Switzerland]
Schweiz. Maschinenmarkt	Schweizer Maschinenmarkt [Switzerland]
Schweiz. Mineral. Petrogr. Mitt.	Schweizer Mineralogische und Petrographische Mitteilungen [Switzerland]
Schweiz. Tech. Z.	Schweizerische Technische Zeitschrift [of Schweizerischer Technischer Verband, Switzerland]
Sci. Am.	Scientific American [US]
Sci. Atmos. Sin.	Scientia Atmospherica Sinica [PR China]
Sci. Ceram.	Science of Ceramics
Sci. China A	Science in China, Series A [PR China]
Sci. China B	Science in China, Series B [PR China]
Sci. Comput. Program.	Science of Computer Programming [Netherlands]
Sci. Dimens.	Science Dimension [of National Research Council of Canada]
Sci. Electr.	Scientia Electrica [Switzerland]

Sci. Eng.	Science and Engineering [India]
Sci. Eng. Rep. Natl. Def. Acad.	Scientific and Engineering Reports of the National Defense Academy [Japan]
Sci. Eng. Rep. Tohoku Gakuin Univ.	Science and Engineering Reports of Tohoku Gakuin University [Japan]
Sci. Eng. Rep. Saitama Univ. A	Science and Engineering Reports of Saitama University, Series A [Japan]
Sci. Eng. Rep. Saitama Univ. B	Science and Engineering Reports of Saitama University, Series B [Japan]
Sci. Eng. Rep. Saitama Univ. C	Science and Engineering Reports of Saitama University, Series C [Japan]
Sci. Eng. Rev. Doshisha Univ.	Science and Engineering Review of Doshisha University [Japan]
Sci. Hort.	Scientia Horticulturae [of International Society for Horticultural Science, Netherlands]
Sci. Ind. (Philips)	Science and Industry (Philips) [Netherlands]
Sci. Pap. Coll. Arts Sci., Univ. Tokyo	Scientific Papers of the College of Arts and Sciences, University of Tokyo [Japan]
Sci. Pap. Inst. Phys. Chem. Res.	Scientific Papers of the Institute of Physical and Chemical Research [Japan]
Sci. Prog.	Science Progress [UK]
Sci. Publ. Affairs	Science and Public Affairs [of The Royal Society, UK]
Sci. Rep. Kanazawa Univ.	Science Reports of Kanazawa University [Japan]
Sci. Rep. Osaka Univ.	Science Reports of Osaka University [Japan]
Sci. Rep. Res. Inst., Tohoku Univ. A, Phys. Chem. Metall.	Science Reports of the Research Institutes, Tohoku University, Series A, Physics, Chemistry and Metallurgy [Japan]
Sci. Rep. Saitama Univ. A	Science Reports of the Saitama University, Series A [Japan]
Sci. Rep. Saitama Univ. B	Science Reports of the Saitama University, Series B [Japan]
Sci. Rep. Tohoku Univ. I	Science Reports of the Tohoku University, First Series [Japan]
Sci. Rep. Tohoku Univ. IV	Science Reports of the Tohoku University, Fourth Series [Japan]
Sci. Rep. Tohoku Univ. VIII	Science Reports of the Tohoku University, Eighth Series [Japan]
Sci. Rep. Yokohama Natl. Univ. I	Science Reports of the Yokohama National University, Section I [Japan]
Sci. Rep. Yokohama Natl. Univ. II	Science Reports of the Yokohama National University, Section II [Japan]
Sci. Rev.	Scienca Revuo [Yugoslavia]
Sci. Rev. Setsunan Univ. A	Scientific Review of Setsunan University, A [Japan]
Sci. Rev. Setsunan Univ. B	Scientific Review of Setsunan University, B [Japan]
Sci. Sin. A, Math. Phys. Astron. Tech. Sci.	Scientia Sinica, Series A, Mathematical, Physical, Astronomical and Technical Sciences [PR China]
Sci. Sinter.	Science of Sintering [of International Institute for the Science of Sintering, Yugoslavia]
Sci. Tech.	Sciences et Techniques [France]
Sci. Tech. Armement	Sciences et Techniques de l'Armement [France]
Sci. Tech. Inf. Process.	Scientific and Technical Information Processing [Translation of: *Nauchno-Tekhnicheskaya Informatisaya, Seriya 1 (USSR)*; published in the US]
Sci. Technol. Dimens.	Science and Technology Dimensions [Canada]
Sci. Technol. Libr.	Science and Technology Libraries [US]
Sci. Technologies	Sciences et Technologies [France]
Sci. Total Environ.	Science of the Total Environment
Scrap Process. Recycl. Mag.	Scrap Processing and Recycling Magazine [of Institute of Scrap Recycling Industries, US]
Scr. Fac. Sci. Nat. Univ. Purkyn. Brun. Phys.	Scripta Facultatis Scientiarum Naturalium Universitatis Purkynianae Brunensis, Physica [Czechoslovakia]
Scr. Metall.	Scripta Metallurgica [US]
Scr. Metall. Mater.	Scripta Metallurgica et Materialia [US]
Sdelovaci Tech.	Sdelovaci Technika [Czechoslovakia]
SEAISI Newsl.	SEAISI Newsletter [of Southeast Asia Iron and Steel Institute, Philippines]
SEAISI Q.	SEAISI Quarterly [of Southeast Asia Iron and Steel Institute, Philippines]
Seatag Bull.	Seatag Bulletin [UK]
Secur. Manage.	Security Management [of American Society for Industrial Security]
Sediment. Geol.	Sedimentary Geology
Seism. Instrum.	Seismic Instruments [Translation of: *Seismicheskie Pribory: Instrumental'naye Sredstva Seismicheskikh Nablyudenii (USSR)*; published in the US]
Seismol. Geol.	Seismology and Geology [PR China]
Seismol. Res. Lett.	Seismological Research Letters [of Seismological Society of America]
Seismol. Ser. Earth Phys. Branch	Seismological Series of the Earth Physics Branch [of Department of Energy, Mines and Resources, Canada]
Seism. Prib. Instrum. Sreds. Seism. Nabl.	Seismicheskie Pribory: Instrumental'naye Sredstva Seismicheskikh Nablyudenii [USSR]

Semicond. Int.	Semiconductor International [US]
Semicond. Sci. Technol.	Semiconductor Science and Technology [of Institute of Physics, UK]
Semicond. Technol.	Semiconductor Technology
Semiconduct. Int.	Semiconductor International [US]
Semiconduct. Sci. Technol.	Semiconductor Science and Technology [Institute of Physics, UK]
SEMICON Tech. Proc.	SEMICON Technical Proceedings [of Semiconductor Equipment and Materials Institute, US]
Sens. Actuators A	Sensors and Actuators A [Switzerland]
Sens. Actuators B	Sensors and Actuators B [Switzerland]
Sens. Rev.	Sensor Review [UK]
Sep. Purif. Methods	Separation and Purification Methods [US]
Sep. Sci. Technol.	Separation Science and Technology [US]
Serb. Acad. Sci. Arts, Bull.	Serbian Academy of Sciences and Arts, Bulletin [Yugoslavia]
Serb. Acad. Sci. Arts, Glass	Serbian Academy of Sciences and Arts, Glass [Yugoslavia]
Serb. Acad. Sci. Arts, Monogr.	Serbian Academy of Sciences and Arts, Monograph [Yugoslavia]
Serb. Acad. Sci. Arts, Monogr. Dep. Tech. Sci.	Serbian Academy of Sciences and Arts, Monographs, Department of Technical Sciences [Yugoslavia]
SERC Bull.	SERC Bulletin [of Science and Engineering Research Council, UK]
Seybold Rep. Desktop Publ.	Seybold Report on Desktop Publishing [US]
Seybold Rep. Publ. Syst.	Seybold Report on Publishing Systems [US]
SFPE Bull.	SFPE Bulletin [of Society of Fire Protection Engineers, US]
Shanghai Iron Steel Res. Inst. Tech. Rep.	Shanghai Iron and Steel Research Institute Technical Report [PR China]
Sharp Tech. J.	Sharp Technical Journal [Japan]
SHD Storage — Handl. — Distrib.	SHD Storage — Handling — Distribution [UK]
Sheet Met. Ind.	Sheet Metal Industries [UK]
Sheet Met. Tubes Sect.	Sheet Metal Tubes Sections [Germany]
Shell Petrochem.	Shell Petrochemicals [UK]
Shimadzu Rev.	Shimadzu Review [Japan]
Shimizu Tech. Res. Bull.	Shimizu Technical Research Bulletin [Japan]
Shinko Electr. J.	Shinko Electric Journal [Japan]
Shirley Inst. Bull.	Shirley Institute Bulletin [UK]
Shock Vib. Dig.	Shock and Vibration Digest [Vibration Institute, US]
Short Wave Mag.	Short Wave Magazine [UK]
Showa Wire Cable Rev.	Showa Wire and Cable Review [Japan]
SIAM J. Appl. Math.	SIAM Journal of Applied Mathematics [of Society for Industrial and Applied Mathematics, US]
SIAM J. Comput.	SIAM Journal of Computing [of Society for Industrial and Applied Mathematics, US]
SIAM J. Control Optim.	SIAM Journal of Control and Optimization [of Society for Industrial and Applied Mathematics, US]
SIAM J. Discr. Math.	SIAM Journal of Discrete Mathematics [of Society for Industrial and Applied Mathematics, US]
SIAM J. Math. Anal.	SIAM Journal of Mathematical Analysis [of Society for Industrial and Applied Mathematics, US]
SIAM J. Matrix Anal. Appl.	SIAM Journal on Matrix Analysis and Applications [of Society for Industrial and Applied Mathematics, US]
SIAM J. Numer. Anal.	SIAM Journal on Numerical Analysis [of Society for Industrial and Applied Mathematics, US]
SIAM J. Sci. Stat. Comput.	SIAM Journal on Scientific and Statistical Computing [of Society for Industrial and Applied Mathematics, US]
SIAM Rev.	SIAM Review [of Society for Industrial and Applied Mathematics, US]
Sider. Latinoam.	Siderurgia Latinoamericana [of Instituto Latinoamericana del Fierro y el Acero, Chile]
Siemens-Albis Ber.	Siemens-Albis Berichte [Switzerland]
Siemens Compon.	Siemens Components [Germany]
Siemens Compon. (Engl. Ed.)	Siemens Components (English Edition) [Germany]
Siemens Energietech.	Siemens Energietechnik [Germany]
Siemens Forsch. Entwickl.-Ber.	Siemens Forschungs- und Entwicklungsberichte [Germany]
Siemens Rev.	Siemens Review [Germany]
SIGACT News	SIGACT News [of Special Interest Group on Automata and Compatibility Theory [of Association for Computing Machinery, US]]
SIGARCH Comput. Archit. News.	SIGARCH Computer Architecture News [of Special Interest Group for Architecture of Computer Systems [of Association for Computing Machinery, US]]
SIGBIO Newsl.	SIGBIO Newsletter [of Special Interest Group on Biomedical Computing [of Association Computing Machinery, US]]
SIGCAPH Newsl.	SIGCAPH Newsletter [of Special Interest Group for Computers and the Physically Handicapped [of Association for Computing Machinery, US]]

SIGCAS Comput. Soc.	SIGCAS Computers and Society [of Special Interest Group for Computers and Society [Association for Computing Machinery, US]]
SIGCHI Bull.	SIGCHI Bulletin [of Special Interest Group on Computer Hierarchy and Interfaces [of Association for Computing Machinery, US]]
SIGCOMM Comput. Commun. Rev.	SIGCOMM Computer Communication Review [of Special Interest Group on Data Communication [of Association for Computing Machinery, US]]
SIGCSE Bull.	SIGCSE Bulletin [of Special Interest Group on Computer Science Education [of Association for Computing Machinery, US]]
SIGCUE Bull.	SIGCUE Bulletin [of Special Interest Group for Computer Uses in Education [of Association for Computing Machinery, US]]
SIGDA Newsl.	SIGDA Newsletter [of Special Interest Group for Design and Automation [of Association for Computing Machinery, US]]
SIGForth Newsl.	SIGForth Newsletter [of Special Interest Group on Forth [of Association for Computing Machinery, US]]
SIGGRAPH Comput. Graph.	SIGGRAPH Computer Graphics [of Special Interest Group on Computer Graphics [of Association for Computing Machinery, US]]
SIGIR Forum	SIGIR Forum [of Special Interest Group on Information Retrieval [of Association for Computing Machinery, US]]
SIGMETRICS Perf. Eval. Rev.	SIGMETRICS Performance Evaluation Review [of Special Interest Group on Measurement and Evaluation [of Association for Computing Machinery, US]]
SIGMICRO Newsl.	SIGMICRO Newsletter [of Special Interest Group on Microprogramming [of Association for Computing Machinery, US]]
SIGMICRO TC-MICRO Newsl.	SIGMICRO TC-MICRO Newsletter [of Special Interest Group on Microprogramming [of Association for Computing Machinery, US]]
SIGMOD Rec.	SIGMOD Record [of Special Interest Group on Management of Data [of Association for Computing Machinery, US]]
Signal Draht	Signal und Draht [Germany]
Signal Process.	Signal Processing [Netherlands]
Signal Process., Image Commun.	Signal Processing: Image Communication [Netherlands]
SIGNUM Newsl.	SIGNUM Newsletter [of Special Interest Group on Numerical Mathematics [of Association for Computing Machinery, US]]
SIGOIS Bull.	SIGOIS Bulletin [of Special Interest Group on Office Information Systems [of Association for Computing Machinery, US]]
SIGOPS Oper. Syst. Rev.	SIGOPS Operating Systems Review [of Special Interest Group on Operating Systems [of Association for Computing Machinery, US]]
SIGPC Notes	SIGPC Notes [of Special Interest Group on Personal Computing Systems [of Association for Computing Machinery, US]]
SIGPLAN Not.	SIGPLAN Notices [of Special Interest Group on Programming Languages [of Association for Computing Machinery, US]]
SIGSAC Rev.	SIGSAC Review [of Special Interest Group on Security, Audit and Control [of Association for Computing Machinery, US]]
SIGSAM Bull.	SIGSAM Bulletin [of Special Interest Group on Symbolic and Algebraic Manipulation [of Association for Computing Machinery, US]]
SIGSIM Newsl.	SIGSIM Newsletter [of Special Interest Group on Simulation [of Association for Computing Machinery, US]]
SIGSMALL Newsl.	SIGSMALL Newsletter [of Special Interest Group on Symbolic and Algebraic Manipulation [of Association for Computing Machinery, US]]
SIGSMALL/PC Notes	SIGSMALL/PC Notes [of Special Interest Group on Symbolic and Algebraic Manipulation [of Association for Computing Machinery, US]]
SIGSOFT Softw. Eng. Notes	SIGSOFT Software Engineering Notes [of Special Interest Group on Software Engineering [of Association for Computing Machinery, US]]
SIGUCCS Newsl.	SIGSUCCS Newsletter [of Special Interest Group on University Computer Centers [of Association for Computing Machinery, US]]
Silic. Ind.	Silicates Industriels [Belgium]
Silvae Genet.	Silvae Genetica [Germany]
Silver Inst. Lett.	Silver Institute Letter [US]
Silvicultural Operations Newsl.	Silvicultural Operations Newsletter [of Canadian Forestry Service]
SIMULA Newsl.	SIMULA Newsletter [UK]
Simul. Dig.	Simulation Digest [of Association for Computer Machinery, US]
Simul. Games	Simulation and Games [US]
SIN Newsl.	SIN Newsletter [Switzerland]

Singap. J. Phys.	Singapore Journal of Physics [Singapore]
Sist. Autom.	Sistemi e Automazione [Italy]
Sist. Impresa	Sistemi e Impresi [Italy]
Sitz.ber., Osterr. Akad. Wiss. Math.-Nat.wiss. Kl. I	Sitzungsberichte. Osterreichische Akademie der Wissenschaften, Mathematisch-Naturwissenschaftliche Klasse, Abteilung I [Austria]
Sitz.ber., Osterr. Akad. Wiss. Math.-Nat.wiss. Kl. II, Math. Phys. Tech. Wiss.	Sitzungsberichte. Osterreichische Akademie der Wissenschaften, Mathematisch-Naturwissenschaftliche Klasse, Abteilung II, Mathematische, Physikalische und Technische Wissenschaften [Austria]
SKF Steel Tech. Rep.	SKF Steel Technical Report [Sweden]
Skillings' Min. Rev.	Skillings' Mining Review [US]
Skoda Rev.	Skoda Review [Czechoslovakia]
Skogind.	Skogindustri [Norway]
Skr. Nor. Vidensk.-Akad. Oslo I, Mat.-Nat.vidensk. Kl.	Skrifter utgitt av det Norske Videnskaps-Akademi i Oslo I, Matematisk-Naturvidenskaplige Klasse [Norway]
Skr. Nor. Vidensk.-Akad. Oslo II	Skrifter utgitt av det Norske Videnskaps-Akademi i Oslo II [Norway]
Sky Telesc.	Sky Telescope [US]
Slabop. Obz.	Slaboproudy Obzor [Czechoslovakia]
Sloan Manage. Rev.	Sloan Management Review [US]
Small Bus. Comput. News	Small Business Computer News [US]
Small Comput. Libr.	Small Computer Library [US]
Smart Mater. Struct.	Smart Materials and Structures [of Institute of Physics, UK]
SM PRO	Steelmaking Proceedings [of Iron and Steel Society [of AIME, US]]
SMPTE J.	SMPTE Journal [of Society of Motion Picture and Television Engineers, US]
Sobre Deriv. Cana Azucar	Sobre los Derivados de la Cana de Azucar [Cuba]
Softw. Eng. J.	Software Engineering Journal [of British Computer Society and Institution of Electrical Engineers]
Softw. Eng. Notes	Software Engineering News [of Association for Computing Machinery, US]
Softw. Eng. Workstn.	Software for Engineering Workstations [UK]
Softw. Maint. News	Software Maintenance News [US]
Softw. — Pract. Exp.	Software - Practice and Experience [UK]
Softw. Prot.	Software Protection [US]
Softw. World	Software World [UK]
Soil Dyn. Earthq. Eng.	Soil Dynamics and Earthquake Engineering [UK]
Soil Sci.	Soil Science [US]
Soil Sci. Soc. Am. J.	Soil Science Society of America Journal [US]
Soil Surv. Horiz.	Soil Survey Horizons [of Soil Science Society of America]
Soil Tillage Res.	Soil and Tillage Research
Sol. Bull.	Solar Bulletin [of American Association of Variable Star Observers]
Sol. Cells	Solar Cells [Switzerland]
Soldadura Construcao Met.	Soldadura Construcao Metalica [Portugal]
Solder. Surf. Mt. Technol.	Soldering and Surface Mount Technology [UK]
Sol. Energy	Solar Energy [US]
Sol. Energy J.	Solar Energy Journal [of International Solar Energy Society, Australia]
Sol. Energy Mater.	Solar Energy Materials [Netherlands]
Solid Fuel Chem.	Solid Fuel Chemistry [US]
Solid Mech. Arch.	Solid Mechanics Archives [UK]
Solid State Commun.	Solid State Communications [US]
Solid-State Electron.	Solid-State Electronics [UK]
Solid State Ion. Diffus. React.	Solid State Ionics, Diffusion and Reactions [Netherlands]
Solid State Mater. Sci.	Solid State Materials Science [US]
Solid State Phenom.	Solid State Phenomena [US]
Solid State Phys.	Solid State Physics [Japan]
Solid State Technol.	Solid State Technology [US]
Sol. Phys.	Solar Physics [Netherlands]
Sol. Syst. Res.	Solar System Research [Translation of: *Astronomicheskii Vestnik (USSR)*; published in the US]
Sol. Terr. Environ. Res. Jpn.	Solar Terrestrial Environmental Research in Japan [of Institute of Space and Aeronautical Science, Japan]
Solvent Extr. Ion Exch.	Solvent Extraction and Ion Exchange [US]

Sol. Wind. Technol.	Solar and Wind Technology [UK]
Soobshcheniya Akad. Nauk Gruz. SSR	Soobshcheniya Akademii Nauk Gruzinskoi SSR [USSR]
Soudage Tech. Connexes	Soudage et Techniques Connexes [France]
Sound Vib.	Sound and Vibration [US]
South Afr. Comput. J.	South African Computer Journal [of Computer Society of South Africa]
South Afr. J. Antarct. Res.	South African Journal of Antarctic Research
South Afr. J. Chem.	South African Journal of Chemistry [of South African Chemical Institute]
South Afr. J. Libr. Inf. Sci	South African Journal of Library and Information Science [of South African Institute for Librarianship and Information Science]
South Afr. J. Phys.	South African Journal of Physics
South Afr. Mach. Tool Rev.	South African Machine Tool Review
South Afr. Mech. Eng.	South African Mechanical Engineer [of South African Institute of Mechanical Engineers]
Southern J. Appl. Forestry	Southern Journal of Applied Forestry [of Society of American Foresters]
Sov. Aeronaut.	Soviet Aeronautics [Translation of: *Izvestiya Vysshikh Uchebnykh Zavedenii, Aviatsionnaya Tekhnika*; published in the US]
Sov. Appl. Mech.	Soviet Applied Mechanics [Translation of: *Prikladnaya Mekhanika (Ukrainian SSR)*; published in the US]
Sov. Astron.	Soviet Astronomy [Translation of: *Astronomicheskii Zhurnal*; published in the US]
Sov. Astron. Lett.	Soviet Astronomy Letters [Translation of: *Pis'ma v Astronomicheskie Zhurnal*; published in the US]
Sov. At. Energy	Soviet Atomic Energy [Translation of: *Atomnaya Energiya*; published in the US]
Sov. Cast. Technol.	Soviet Casting Technology [Translation of: *Liteinoe Proizvodstvo*; published in the US]
Sov. Chem. Ind.	Soviet Chemical Industry [Published in the US]
Sov. Electr. Eng.	Soviet Electrical Engineering [Translation of: *Elektrotekhnika*; published in the US]
Sov. Electrochem.	Soviet Electrochemistry [Translation of: *Elektrokhimiya*; published in the US]
Sov. Energy Technol.	Soviet Energy Technology [Published in the US]
Sov. Eng. Res.	Soviet Engineering Research [Selective translation of: *Vestnik Mashinostroeniya* and *Stanki i Instrument*; published in the US]
Sov. Geol. Phys.	Soviet Geology and Geophysics [Translation of: *Geologiya i Geofizika*; published in the US]
Sov. J. Appl. Phys.	Soviet Journal of Applied Physics [Translation of: *Izvestiya Sibirskogo Otdeleniya Akademii Nauk SSSR, Tekhnicheskikh*; publish in the US]
Sov. J. Autom. Inf. Sci.	Soviet Journal of Automation and Information Science [Translation of: *Avtomatika (UKrainian SSR)*; published in the US]
Sov. J. Chem. Phys.	Soviet Journal of Chemical Physics [Translation of: *Khimicheskaya Fizika*; published in the UK]
Sov. J. Commun. Technol. Electron.	Soviet Journal of Communications Technology and Electronics [Translation of: *Radiotekhnika i Elektronika*; published in the US]
Sov. J. Comput. Syst. Sci.	Soviet Journal of Computer and Systems Sciences [Translation of: *Tekhnicheskaya Kibernetika*; published in the US]
Sov. J. Contemp. Phys.	Soviet Journal of Contemporary Physics [Translation of: *Izvestiya Akademia Nauk Armyanskoi SSR Fizika*; published in the US]
Sov. J. Frict. Wear	Soviet Journal of Friction and Wear [Translation of: *Trenie Iznos (Byelorussian SSR)*; published in the US]
Sov. J. Glass Phys. Chem.	Soviet Journal of Glass Physics and Chemistry [Translation of: *Fizika i Khimiya Stekla*; published in the US]
Sov. J. Low Temp. Phys.	Soviet Journal of Low Temperature Physics [Translation of: *Fizika Nizkikh Temperatur*; published in the US]
Sov. J. Nondestr. Test.	Soviet Journal of Nondestructive Testing [Translation of: *Defektoskopiya*; published in the US]
Sov. J. Non-Ferrous Met.	Soviet Journal of Non-Ferrous Metals [Published in the US]
Sov. J. Nucl. Phys.	Soviet Journal of Nuclear Physics [Translation of: *Yadernaya Fizika*; published in the US]
Sov. J. Opt. Technol.	Soviet Journal of Optical Technology [Translation of: *Optiko-Mekhanicheskaya Promyshlennost*; published in the US]
Sov. J. Part. Nucl.	Soviet Journal of Particles and Nuclei [Translation of: *Fizika Elementarnykh Chastits i Atomnoya Yadra*; published in the US]
Sov. J. Plasma Phys.	Soviet Journal of Plasma Physics [Translation of: *Fizika Plazmy*; published in the US]
Sov. J. Quantum Electron.	Soviet Journal of Quantum Electronics [Translation of: *Kvantovaya Elektronikaya, Moskva*; published in the US]
Sov. J. Remote Sens.	Soviet Journal of Remote Sensing [Translation of: *Issledovanie Zemli iz Kosmosa*; published in Switzerland]
Sov. J. Superhard Mater.	Soviet Journal of Superhard Materials [Translation of: *Sverkhtverdyne Materialy*; published in the US]
Sov. J. Water Chem. Technol.	Soviet Journal of Water Chemistry and Technology [Published in the US]

Sov. Mach. Sci.	Soviet Machine Science [Translation of: *Mashinovedenie*; published in the US]
Sov. Mater. Sci.	Soviet Materials Science [Translation of: *Fiziko-Khimicheskaya Mekhanika Materialov*; published in the US]
Sov. Mater. Sci. Rev.	Soviet Materials Science Reviews [Translated from Russian; US]
Sov. Math.	Soviet Mathematics [Translation of: *Izvestiya Vysshikh Uchebnykh Zavedenii, Matematika*; published in the US]
Sov. Meteorol. Hydrol.	Soviet Meteorology and Hydrology [Translation of: *Meteorologiya i Gidrologiya*; published in the US]
Sov. Microelectron.	Soviet Microelectronics [Translation of: *Mikroelektronika*; published in the US]
Sov. Min. Sci.	Soviet Mining Science [Published in the US]
Sov. Non-Ferrous Met. Res.	Soviet Non-Ferrous Metals Research [Published in the UK]
Sov. Phys.-Acoust.	Soviet Physics — Acoustics [Translation of: *Akusticheskii Zhurnal*; published in the US]
Sov. Phys.-Collect.	Soviet Physics — Collection [Translation of: *Litovskii Fizicheskii Sbornik*; published in the US]
Sov. Phys.-Crystallogr.	Soviet Physics — Crystallography [Translation of: *Kristallografiya*; published in the US]
Sov. Phys.-Dokl.	Soviet Physics — Doklady [Translation of: *Doklady Akademii Nauk SSSR*; published in the US]
Sov. Phys. J.	Soviet Physics Journal [Translation of: *Izvestiya Vysshikh Uchebnykh Zavedenii, Fizika*; published in the US]
Sov. Phys.-JETP	Soviet Physics — JETP (Journal of Experimental Theoretical Physics) [Translation of: *Zhurnal Eksperimental'noi i Teoreticheskoi Fiziki*; published in the US]
Sov. Phys.-Lebedev Inst. Rep.	Soviet Physics — Lebedev Institute Reports [Translation of: *Sbornik Kratkie Soobshcheniya po Fizike, AN SSSR, Fizikcheskii Institut im. P.N. Lebedeva*; published in the US]
Sov. Phys.-Semicond.	Soviet Physics — Semiconductors [Translation of: *Fizika i Tekhnika Poluprovodnikov*; published in the US]
Sov. Phys.-Solid State	Soviet Physics — Solid State [Translation of: *Fizika Tverdogo Tela*; published in the US]
Sov. Phys.-Tech. Phys.	Soviet Physics — Technical Physics [Translation of: *Zhurnal Tekhnicheskoi Fiziki*; published in the US]
Sov. Phys.-Usp.	Soviet Physics — Uspekhi [Translation of: *Uspekhi Fizicheskii Nauk*; published in the US]
Sov. Powder Metall. Met. Ceram.	Soviet Powder Metallurgy and Metal Ceramics [Translation of: *Poroshkovaya Metallurgiya*; published in the US]
Sov. Prog. Chem.	Soviet Progress in Chemistry [Translation of: *Ukrainskii Khimischeskii Zhurnal*; published in the US]
Sov. Surf. Eng. Appl. Electrochem.	Soviet Surface Engineering and Applied Electrochemistry [Translation of: *Elektronnaya Obrabotka Materialov*; published in the US]
Sov. Surf. Sci.	Soviet Surface Science [Published in the US]
Sov. Tech. Phys. Lett.	Soviet Technical Physics Letters [Translation of: *Pis'ma v Zhurnal Tekhnicheskoi Fizika*; published in the US]
Space Commun.	Space Communications [Netherlands]
Space J.	Space Journal [US]
Space Sci. Rev.	Space Science Reviews [Netherlands]
Space Technol.	Space Technology [UK]
Spatial Vis.	Spatial Vision [Netherlands]
Spec. Chem.	Specialty Chemicals [UK]
Spec. Libr.	Special Libraries [US]
SPE Comput. Users Newsl.	SPE Computer Users Newsletter [of Society of Petroleum Engineers, US]
Spec. Rep. Inst. Technol., Shimizu Corp.	Special Report of Institute of Technology, Shimizu Corporation [Japan]
Spec. Ships	Special Ships [UK]
Spec. Steels Rev.	Special Steels Review [UK]
Spectrochim. Acta	Spectrochimica Acta [UK]
Spectrochim. Acta A, Mol. Spectrosc.	Spectrochimica Acta A, Molecular Spectroscopy [UK]
Spectrochim. Acta B, At. Spectrosc.	Spectrochimica Acta B, Atomic Spectroscopy [UK]
Spectrosc. (Canada)	Spectroscopy (Canada)
Spectrosc. Int.	Spectroscopy International [US]
Spectrosc. Lett.	Spectroscopy Letters [US]
Spectrosc. (Oregon)	Spectroscopy (Oregon) [US]
Specul. Sci. Technol.	Speculations in Science and Technology [UK]
SPE Drill. Eng.	SPE Drilling Engineering [of Society of Petroleum Engineers, US]
Speech Commun.	Speech Communication [Netherlands]
Speech Technol.	Speech Technology [US]
SPE Format. Eval.	SPE Formation Evaluation [of Society of Petroleum Engineers, US]

SPE Prod. Eng.	SPE Production Engineering [of Society of Petroleum Engineers, US]
SPE Reserv. Eng.	SPE Reservoir Engineering [of Society of Petroleum Engineers, US]
SPHE Newsl.	SPHE Newsletter [of Society of Packaging and Handling Engineers, US]
SPHE Tech. J.	SPHE Technical Journal [of Society of Packaging and Handling Engineers, US]
Spis. Bulg. Akad. Nauk.	Spisanie na Bulgarskata Akademiya na Naukite [Bulgaria]
Sprache Daterverarb.	Sprache und Datenverarbeitung [Germany]
Sprecher Energ. Rev.	Sprecher Energie Review [Switzerland]
Sprechsaal Int. Ceram. Glass Mag.	Sprechsaal International Ceramics and Glass Magazine [Germany]
Stahl Eisen	Stahl und Eisen [of Verein Deutscher Eisenhuttenleute, Germany]
Stainless Steel High Perform. Alloys	Stainless Steel High Performance Alloys [South Africa]
Stainless Steel Ind.	Stainless Steel Industry [UK]
Stainless Steels Dig.	Stainless Steels Digest [of ASM International, US]
Stain Technol.	Stain Technology [US]
Stamp. Q.	Stamping Quarterly [US]
Stand. Eng.	Standards Engineering [of Standards Engineering Society, US]
Stand. News	Standardization News [of American Society for Testing Materials]
Stanki Instrum.	Stanki i Instrumenty [USSR]
STAR	Science and Technology Aerospace Reports [of NASA, US]
State Geol. J.	State Geological Journal [of Association of American State Geologists]
Stat. J. U.N. Econ. Comm. Eur.	Statistical Journal of the United Nations Economic Commission for Europe [Netherlands]
StatSci	Statistical Science [of Institute of Mathematical Sciences, US]
Stat. Softw. Newsl.	Statistical Software Newsletter [of Gesellschaft fur Strahlen- und Umweltforschung, Germany]
Staub Reinhalt. Luft.	Staub — Reinhaltung der Luft [of Berufsgenossenschaftliches Institut fur Arbeitssicherheit [of Verein Deutscher Ingenieure, Germany]]
Steel Constr.	Steel Construction [South Africa]
Steel Constr. Today	Steel Construction Today [UK]
Steel Dig.	Steel Digest [US]
Steel Founders Res. J	Steel Founders Research Journal [of Steel Founders Society of America]
Steel Furn. Mon.	Steel Furnace Monthly [India]
Steel Ind. Jpn. Ann.	Steel Industry of Japan Annual [Japan]
Steelmaking Proc.	Steelmaking Proceedings [of Iron and Steel Society [of AIME, US]]
Steel Mark News	Steel Market News [Sweden]
Steel Res.	Steel Research [Germany]
Steel Technol. Int.	Steel Technology International [UK]
Steel Times Int.	Steel Times International [UK]
Steel Today Tomorrow	Steel Today and Tomorrow [Japan]
Steel USSR	Steel in the USSR [Translation of: *Stal': Izvestiya Vysshikh Uchebnykh Zavedenii, Chernaya Metallurgiya*; published in the UK]
Steklo Keram.	Steklo i Keramika [USSR]
STFI Medd.	STFI Meddelande [of Swedish Pulp and Paper Research Institute]
STLE Tribol. Trans.	STLE Tribology Transactions [of Society of Tribologist and Lubrication Engineers, US]
Stoch. Process. Appl.	Stochastic Processes and Their Applications [Netherlands]
Stoch. Stoch. Rep.	Stochastics and Stochastics Reports [UK]
STP Newsl.	STP Newsletter [of Scientific Committee on Solar Terrestrial Physics, US]
Strength Mater.	Strength of Materials [Translation of: *Problemy Prochnosti (USSR)*; published in the US]
Strojir. Vyroba	Strojirenska Vyroba [Czechoslovakia]
Strojnicky Cas.	Strojnicky Casopis [Czechoslovakia]
Strojniski Vestn.	Strojniski Vestnik [Yugoslavia]
Struct. Eng. A	Structural Engineer, Part A [of Institution of Structural Engineers, UK]
Struct. Eng. B	Structural Engineer, Part B [of Institution of Structural Engineers, UK]
Struct. Eng. Pract.	Structural Engineer Practice [US]
Struct. Optim.	Structural Optimization [US]
Struct. Program.	Structured Programming [US]
Struct. Saf.	Structural Safety [Netherlands]
Stud. Appl. Math.	Studies in Applied Mathematics [of Society for Industrial and Applied Mathematics, US]
Stud. Cercet. Calc. Econ. Cibern. Econ.	Studii si Cercetari de Calcul Economic si Cibernetica Economica [Romania]
Stud. Cercet. Doc.	Studii si Cercetari de Documentare [Romania]

Stud. Cercet. Fiz.	Studii si Cercetari de Fizica [Romania]
Stud. Cercet. Geol. Geofiz., Ser. Geofiz.	Studii si Cercetari de Geologie, Geofizica, Geografie, Seria Geofizica [Romania]
Stud. Cercet. Mec. Apl.	Studii si Cercetari de Mecanica Aplicata [Romania]
Stud. Environ. Sci.	Studies in Environmental Science [US]
Stud. Geophys. Geod.	Studia Geophysica et Geodaetica [Czechoslovakia]
Studia Forest. Suecica	Studia Forestalia Suecica [Sweden]
Stud. J. Inst. Electron. Telecommun. Eng.	Students Journal of the Institution of Electronics and Telecommunication Engineers [India]
Stud. Math. Geol.	Studies in Mathematical Geology [of International Association for Mathematical Geology, US]
Stud. Rep. Hydrol.	Studies and Reports in Hydrology [of International Hydrological Program, France]
Stud. Sci. Math. Hung.	Studia Scientiarum Mathematicarum Hungarica [Hungary]
Stud. Surf. Sci. Catal.	Studies in Surface Science and Catalysis
STUVA-Forschungsber.	STUVA-Forschungsberichte [of Studiengesellschaft fur Unterirdische Verkehrsanlagen, Germany]
Sudura Incercari Mater.	Sudura si Incercari de Materiale [Romania]
Sulzer Tech. Rev.	Sulzer Technical Review [Switzerland]
Sumitomo Chem. Rev.	Sumitomo Chemical Review [Japan]
Sumitomo Electr. Tech. Rev.	Sumitomo Electric Technical Review [Japan]
Sumitomo Light Met. Tech. Rep.	Sumitomo Light Metal Technical Reports [Japan]
Sumitomo Met.	Sumitomo Metals [Japan]
Sumitomo Met. News	Sumitomo Metals News [Japan]
Sumitomo Spec. Met. Tech. Rep.	Sumitomo Special Metals Technical Report [Japan]
Summ. Rep. Electrotech. Lab.	Summary Reports of the Electrotechnical Laboratory [Japan]
Supercond. News	Superconductivity News [Superconductor Applications Association, US]
Supercond. Ind.	Superconductor Industry [US]
Supercond. News	Superconductor News [of Superconductor Applications Association, US]
Supercond. Sci. Technol.	Superconductor Science and Technology [UK]
Superlattices Microstruct.	Superlattices and Microstructures [UK]
Suppl. Br. Telecommun. Eng.	Supplement to British Telecommunications Engineering [UK]
Surfactant Sci. Ser.	Surfactant Science Series [US]
Surf. Coat. Technol.	Surface and Coatings Technology [Switzerland]
Surf. Eng.	Surface Engineering [of Institute of Metals, UK]
Surf. Interface Anal.	Surface and Interface Analysis [UK]
Surf. Sci.	Surface Science [Netherlands]
Surf. Sci. Lett	Surface Science Letters [Netherlands]
Surf. Sci. Rep.	Surface Science Reports [Netherlands]
Surf. Sci. Spectra	Surface Science Spectra [of American Vacuum Society]
Surf. Technol.	Surface Technology [Switzerland]
Surfacing J.	Surfacing Journal [UK]
Surfacing J. Int.	Surfacing Journal International [UK]
Surv. Geophys.	Surveys in Geophysics [UK]
Surv. High Energy Phys.	Survey in High Energy Physics [Switzerland]
Svar. Proizvod.	Svarochnoe Proizvodstvo [USSR]
Svensk Papperstid.	Svensk Papperstidning [Sweden]
Sverkhtverd. Mater.	Sverkhtverdye Materialy
Swiss Biotech	Swiss Biotech (Swiss Review for Biotechnology)
Swiss Chem	Swiss Chem (Swiss Review for the Chemical Industry)
Swiss Contam. Control	Swiss Contamination Control (Swiss Review for Clean Room Technology)
Swiss Food	Swiss Food (Swiss Review for the Food Industry)
Swiss Materials	Swiss Materials (Swiss Review for Materials Technology)
Swiss Med	Swiss Med (Swiss Review for Medicine and Medical Technology)
Swiss Pharma	Swiss Pharma (Swiss Review for the Pharmaceutical Industry)
Swiss Plastics	Swiss Plastics (Swiss Review for the Plastics Industry)
Swiss Vet	Swiss Vet (Swiss Review for Veterinary Medicine)
SWRA	Selected Water Research Abstracts [of Water Resources Scientific Information Center, US]
Synth. Met.	Synthetic Metals [Switzerland]
Syst. Anal. — Model. Simul.	Systems Analysis — Modelling — Simulation [Germany]
Syst. Comput. Jpn.	Systems and Computers in Japan [US]
Syst. Control	Systems and Control [Japan]

Syst. Control Inf.	Systems, Control and Information [of Institute of Systems, Control and Information Engineers, Japan]
Syst. Control Lett.	Systems and Control Letters [Netherlands]
Syst. Dev.	System Development [US]
Syst. Dyn. Rev.	System Dynamics Review [of System Dynamics Society, US]
System. Bot.	Systematic Botany [of American Society of Plant Taxonomists]
System. Bot. Monogr.	Systematic Botany Monographs [of American Society of Plant Taxonomists]
Syst. Int.	Systems International [UK]
Syst. Integr.	Systems Integration [US]
Syst. Log.	Systemes Logiques [Switzerland]
Syst. Res.	Systems Research [UK]
Syst. Res. Inf. Sci.	Systems Research and Information Science [UK]
Syst. Sci.	Systems Science [Poland]
Syst. Technol.	Systems Technology [UK]
Syst. Tech. Rep.	Systems Technical Report [Japan]
Syst. 3X World	Systems 3X World [US]
Syst. User	Systems User [US]

T

TAB Tyres Access. Batt.	TAB: Tyres, Accessories, Batteries [UK]
TAGA Proc.	TAGA Proceedings [of Technical Association of the Graphic Arts, US]
Takaoka Rev.	Takaoka Review [Japan]
Takenaka Tech. Res. Rep.	Takenaka Technical Research Report [Japan]
Tantalum Prod. Int. Study Cent. Q. Bull.	Tantalum Producers International Study Center Quarterly Bulletin [Belgium]
Tanulm. M. Tud. Akad. Szam. tech. Autom. Kut. Intez.	Tanulmanyok Magyar Tudomanyos Akademia Szamitastechnikaiu es Automatizalasi Kutato Intezet [Hungary]
TAPPI Eng. Conf.	TAPPI Engineering Conference [of Technical Association of the Pulp and Paper Industry, US]
TAPPI J.	TAPPI Journal [of Technical Association of the Pulp and Paper Industry, US]
TCA Rep.	TCA Report [of Tissue Culture Association, US]
Tech. Assoc. Refract.	Technical Association of Refractories [Japan]
Tech. Bull. Vevey	Technical Bulletin Vevey [Switzerland]
Tech. Ceram. Bull.	Technical Ceramics Bulletin [UK]
Tech. Commun.	Technical Communication [of Society for Technical Communication, US]
Tech. Doc. Hydrol.	Technical Documents in Hydrology [of International Hydrological Program, France]
Tech.gesch.	Technikgeschichte [of Verein Deutscher Ingenieure, Germany]
Tech. Inf., Process Autom. — Electr. Power Install.	Technical Information, Process Automation — Electrical Power Installations [Germany]
Tech. Knih.	Technicka Knihovna [Czechoslovakia]
Tech. Mar. Environ. Sci.	Technique in Marine Environmental Science [of International Council for the Exploration of the Sea, Denmark]
Tech. Mech.	Technische Mechanik [Germany]
Tech. Mess.	Technisches Messen [Germany]

Tech. Mitt.	Technische Mitteilungen [Germany]
Tech. Mitt. Haus Tech.	Technische Mitteilungen Haus der Technik [Germany]
Tech. Mitt. Krupp	Technische Mitteilungen Krupp [Germany]
Tech. Mitt. Krupp (Engl. Ed.)	Technische Mitteilungen Krupp (English Edition) [Germany]
Tech. Mitt. Krupp-Forschungsber.	Technische Mitteilungen Krupp-Forschungsberichte [Germany]
Tech. Mitt. Krupp-Werksber.	Technische Mitteilungen Krupp-Werksberichte [Germany]
Tech. Mitt. PTT	Technische Mitteilungen PTT [Switzerland]
Tech. Mitt. RFZ	Technische Mitteilungen des RFZ [Germany]
Tech. Mod.	La Technique Moderne [France]
Tech. News	Technical News [US]
Tech. News Govt. Ind. Res. Inst., Nagoya	Technical News of the Government Industrial Research Institute, Nagoya [Japan]
Techniques Equipements Production	Techniques et Equipments de Production [France]
Techno Jpn.	Techno Japan
Technol. Cult.	Technology and Culture [US]
Technol. Forecast. Soc. Change	Technological Forecasting and Social Change [US]
Technol. Inf., Soc.	Technologies de l'Information et Societe [Canada]
Technol. Prod.	Technology in Production [US]
Technol. Rep. Iwate Univ.	Technology Reports of the Iwate University [Japan]
Technol. Rep. Kansai Univ.	Technology Reports of the Kansai University [Japan]
Technol. Rep. Kyushu Univ.	Technology Reports of the Kyushu University [Japan]
Technol. Rep. Osaka Univ.	Technology Reports of the Osaka University [Japan]
Technol. Rep. Seikei Univ.	Technology Reports of the Seikei University [Japan]
Technol. Rep. Tohoku Univ.	Technology Reports of the Tohoku University [Japan]
Technol. Rep. Yamaguchi Univ.	Technology Reports of the Yamaguchi University [Japan]
Technol. Rev.	Technology Review [of Massachusetts Institute of Technology, US]
Technol. Sci. Inf.	Technology and Science of Informatics [UK]
Technol. Today	Technology Today [US]
Technol. Train.	Technology and Training [Taiwan]
Technol. Util.	Technology Utilization [US]
Tech. Rep. Autom. Res. Lab., Kyoto Univ.	Technical Reports of Automation Research Laboratory, Kyoto University [Japan]
Tech. Rep. Hydrol.	Technical Reports in Hydrology [of International Hydrological Program, France]
Tech. Rep. Inst. At. Energy, Kyoto Univ.	Technical Reports of the Institute of Atomic Energy, Kyoto University [Japan]
Tech. Rep. Kumamoto Univ.	Technical Reports of the Kumamoto University [Japan]
Tech. Res. Cent. Finl. Electr. Nucl. Technol. Publ.	Technical Research Centre of Finland Electrical and Nuclear Technology Publication [Finland]
Tech. Res. Cent. Finl. Res. Rep.	Technical Research Centre of Finland. Research Report [Finland]
Tech. Res. Rep. Shimizu Corp.	Technical Research Report of Shimizu Corporation [Japan]
Tech. Sci. Inf.	Technique et Science Informatique [of Association Francaise pour le Cybernetique, Economoique et Technique, France]
Tech. Serv. Q.	Technical Services Quarterly [US]
Tech. Uberwachung	Technische Uberwachung [of Vereinigung der Technischen Uberwachungsvereine, Germany]
Tech. Zpr.	Technicke Zpravy [Czechoslovakia]
Tec. Ital.	Tecnica Italiana [Italy]
Tec. Metal.	Tecnica Metalurgica [Spain]
Tecnol. Elettr.	Tecnologie Elettriche [Italy]
Tecnol. Filo	Tecnologie del Filo [Italy]
Tecnol. Mecc.	Tecnologie Maccaniche [Italy]
Tecnol. Quim.	Tecnologia Quimica [Cuba]
Teh. Fiz.	Tehnika Fizika [Yugoslavia]
Teh. Rud. Geol. Metal.	Tehnika Rudarstvo Geologiji i Metalurgija [Yugoslavia]
TE Int.	TE (Tecnologie Elettriche) International [Italy]
Tekh. Elektrodin.	Tekhnicheskaya Elektrodinamika [USSR]
Tekh. Kibern.	Tekhnicheskaya Kibernetika [USSR]
Tekh. Kino Telev.	Tekhnika Kino i Televideniya [USSR]
Tekh. Misul	Tekhnicheska Misul [Bulgaria]
Tekhnol. Legk. Splavov	Tekhnologiya Legkikh Spolavov [US]

Tekstil. Prom.	Tekstilnaya Promyshlennost [USSR]
Tek. Tidskr.	Teknisk Tidskrift [Sweden]
Tek. Ukebl.	Teknisk Ukeblad [Norway]
Telecommun. J.	Telecommunication Journal [of International Telecommunication Union, Switzerland]
Telecommum. J. Aust.	Telecommunication Journal [of Telecommunication Society of Australia]
Telecommun. Policy	Telecommunications Policy [UK]
Telecommun. Radio Eng. 1, Telecommun.	Telecommunications and Radio Engineering, Part 1, Telecommunications [Translation of: *Elektrosvyaz (USSR)*; published in the US]
Telecommun. Radio Eng. 2, Radio Eng.	Telecommunications and Radio Engineering, Part 2, Radio Engineering [Translation of: *Radiotekhnika, Moskva (USSR)*; published in the US]
Telecom Rep.	Telecom Report [Germany]
Telefunken Z.	Telefunken Zeitschrift [Germany]
Telemat. India	Telematics India [India]
Telemat. Inf.	Telematics and Informatics [UK]
Telenorma Nachr.	Telenorma Nachrichten [Germany]
Teleph. Eng. Manage.	Telephone Engineer and Management [US]
Telettra Rev.	Telettra Review [Italy]
Telev., J. R. Telev. Soc.	Television, Journal of the Royal Television Society [UK]
Telev./Radio Age	Television/Radio Age [US]
Tellus A, Dyn. Meteorol. Oceanogr.	Tellus, Series A, Dynamic Meteorology and Oceanography [Sweden]
Tellus B, Chem. Phys. Meteorol.	Tellus, Series B, Chemical and Physical Meteorology [Sweden]
Tenside Det.	Tenside Detergents
Teor. Eksp. Khim.	Teoreticheskaya i Eksperimental'naya Khimiya [USSR]
Teor. Mat. Fiz.	Teoreticheskaya i Matematischeskaya Fizika [USSR]
Teor. Veroyatn. Prim.	Teoriya Veroyatnostei i ee Primeneniya [USSR]
Teploenerg.	Teploenergetika [USSR]
Teplofiz. Vys. Temp.	Teplofizika Vysokikh Temperatur [USSR]
Termotec.	Termotecnica [Italy]
TESLA Electron.	TESLA Electronics [Czechoslovakia]
TEW Tech. Ber.	TEW Technische Berichte [Germany]
Test. Meas. World	Test and Measurement World [US]
Tet. Lett.	Tetrahedron Letters
Tetrahedron Lett.	Tetrahedron Letters
Tex. J. Sci.	Texas Journal of Science [US]
Text. Horiz.	Textile Horizons [of The Textile Institute, UK]
Text. Chemist Colorist	Textile Chemists and Colorist [of American Association of Textile Chemists and Colorists]
Textile Horiz.	Textile Horizons [of The Textile Institute, UK]
Textile Res. J.	Textile Research Journal [of Textile Research Institute, US]
Textile Technol. Dig.	Textile Technology Digest [of Institute of Textile Technology, US]
Texture Cryst. Solids	Texture of Crystalline Solids [UK]
Textures Microstruct.	Textures and Microstructures [UK]
Theor. Appl. Climatol.	Theoretical and Applied Climatology [Austria]
Theor. Appl. Fract. Mech.	Theoretical and Applied Fracture Mechanics [Netherlands]
Theor. Appl. Genetics	Theoretical and Applied Genetics [Germany and US]
Theor. Appl. Mech.	Theoretical and Applied Mechanics
Theor. Chim. Acta	Theoretica Chimica Acta [Germany]
Theor. Comput. Sci.	Theoretical Computer Science [Netherlands]
Theor. Exp. Chem.	Theoretical and Experimental Chemistry [Translation of: *Teoreticheskaya i Eksperimental'naya Khimiya (USSR)*; published in the US]
Theor. Found. Chem. Eng.	Theoretical Foundation of Chemical Engineering
Theor. Math. Phys.	Theoretical and Mathematical Physics [Translation of: *Teoreticheskaya i Matematicheskaya Fizika (USSR)*; published in the US]
Theory Probab. Appl.	Theory of Probability and Its Applications [Translation of: *Teoriya Veroyatnostei i ee Primeneniya (USSR)*; published in the US]
Therm. Eng.	Thermal Engineering [Translation of: *Teploenergetika (USSR)*; published in the UK]
Thermochim. Acta	Thermochimica Acta [Netherlands]
Thin-Walled Struct.	Thin-Walled Structures [UK]

33 Met. Prod.	Thirty-Three (33) Metal Producing [US]
33 Met. Prod. Nonferrous Ed.	Thirty-Three (33) Metal Producing Nonferrous Edition [US]
Thyssen Edelstahl Tech. Ber.	Thyssen Edelstahl Technische Berichte [Germany]
Thyssen Tech. Ber.	Thyssen Technische Berichte [Germany]
Tidskr. Dok.	Tidskrift for Dokumentation [Sweden]
Tijdschr. Ned. Elektron.- & Radiogenoot.	Tijdschrift van het Nederlands Elektronica- en Radiogenootschap [Netherlands]
Timber Bull.	Timber Bulletin [of Food and Agricultural Organization — UN]
Timber Harv.	Timber Harvesting [US]
Tin Int.	Tin International [UK]
Tin Uses	Tin and Its Uses [of International Tin Research Council, UK]
Tire Bus.	Tire Business [US]
Tire Sci. Technol.	Tire Science and Technology [US]
Titanium Zirconium	Titanium and Zirconium [Japan]
TIZ Int. Powder Mag.	TIZ International Powder Magazine
TK Technol. Kobe Steel	TK Technologies of Kobe Steel [Japan]
Today's Off.	Today's Office [US]
Today Technol.	Today Technology [US]
Toegep. Wet. TNO	Toegepaste Wetenschap TNO [Netherlands]
Tohoku Geophys. J., Sci. Rep. Tohoku Univ.	Tohoku Geophysical Journal, Science Reports of the Tohoku University [Japan]
Tohoku Geophys. J., Sci. Rep. Tohoku Univ., Fifth Ser.	Tohoku Geophysical Journal, Science Reports of the Tohoku University Fifth Series [Japan]
Tokyo Astron. Bull.	Tokyo Astronomical Bulletin [Japan]
Tokyo Astron. Obs., Kiso Inf. Bull.	Tokyo Astronomical Observatory, Kiso Information Bulletin [Japan]
Tokyo Astron. Obs. Rep.	Tokyo Astronomical Observatory Report [Japan]
Tokyo Astron. Obs. Time Latit. Bull.	Tokyo Astronomical Observatory Time and Latitude Bulletins [Japan]
Tool Alloy Steels	Tool and Alloy Steels [India]
Tool. Prod.	Tooling and Production [US]
Tool. Prog.	Tooling Progress [US]
Tool. Trends	Tooling Trends [US]
Toshiba Rev.	Toshiba Review [Japan]
Toute Electron.	Toute l'Electronique [France]
Toyota Gosei Tech. Rev.	Toyota Gosei Tech. Review [Japan]
Toyo Kohan Tech. Rep.	Toyo Kohan Technical Report [Japan]
Toyo's Tech. Bull.	Toyo's Technical Bulletin [Japan]
Toxicol. Appl. Pharmacol.	Toxicology and Applied Pharmacology [of Society of Toxicology, US]
Trace Anal.	Trace Analysis [PR China]
TRAC Trends Anal. Chem.	TRAC Trends in Analytical Chemistry [Netherlands]
Trade Marks J.	Trade Marks Journal [UK]
Train. Dev. J.	Training and Development Journal [of American Society for Training and Development]
Train. Off.	Training Officer [UK]
Trait. Signal	Traitement du Signal [France]
Trait. Therm.	Traitement Thermique [France]
Tranciatura Stampaggio	Tranciatura e Stampaggio [Italy]
Trans. ACA	Transactions of ACA [of American Crystallographic Association]
Trans. AIME	Transactions of the AIME [of American Institute of Mining, Metallurgical and Petroleum Engineers]
Trans. Am. Cryst. Assoc.	Transactions of the American Crystallographic Association [US]
Trans. Am. Inst. Min. Eng.	Transactions of the AIME [of American Institute of Mining, Metallurgical and Petroleum Engineers]
Trans. Am. Math. Soc.	Transactions of the American Mathematical Society
Trans. Am. Microsc. Soc.	Transactions of the American Microscopical Society
Trans. Am. Nucl. Soc.	Transactions of the American Nuclear Society
Trans. Am. Soc. Agric. Eng.	Transactions of the American Society of Agricultural Engineers
Trans. ASAE	Transactions of the ASAE [of American Society of Agricultural Engineers]
Trans. ASHRAE	Transactions of ASHRAE [of American Society of Heating, Refrigeration and Air Conditioning Engineers]
Trans. ASME	Transactions of the ASME [of American Society of Mechanical Engineers]
Trans. ASME, J. Appl. Mech.	Transactions of the ASME, Journal of Applied Mechanics [of American Society of Mechanical Engineers]

Trans. ASME, J. Bioeng.	Transactions of ASME, Journal of Bioengineering [of American Society of Mechanical Engineers]
Trans. ASME, J. Biomech. Eng.	Transactions of the ASME, Journal of Biomechanical Engineering [of American Society of Mechanical Engineers]
Trans. ASME, J. Dyn. Syst.	Transactions of the ASME, Journal of Dynamic Systems [of American Society of Mechanical Engineers]
Trans. ASME, J. Dyn. Syst. Meas. Control	Transactions of the ASME, Journal of Dynamic Systems, Measurement and Control [of American Society of Mechanical Engineers]
Trans. ASME, J. Electron. Packag.	Transactions of the ASME, Journal of Electronic Packaging [of American Society of Mechanical Engineers]
Trans. ASME, J. Energy Resour. Technol.	Transactions of the ASME, Journal of Energy Resources Technology [of American Society of Mechanical Engineers]
Trans. ASME, J. Eng. Gas Turbines Power	Transactions of the ASME, Journal of Engineering for Gas Turbines and Power [of American Society of Mechanical Engineers]
Trans. ASME, J. Eng. Ind.	Transactions of the ASME, Journal of Engineering for Industry [of American Society of Mechanical Engineers]
Trans. ASME, J. Eng. Mater. Technol.	Transactions of the ASME, Journal of Engineering Materials and Technology [of American Society of Mechanical Engineers]
Trans. ASME, J. Eng. Power	Transactions of the ASME, Journal of Engineering for Power [of American Society of Mechanical Engineers]
Trans. ASME, J. Fluids Eng.	Transactions of the ASME, Journal of Fluids Engineering [of American Society of Mechanical Engineers]
Trans. ASME, J. Heat Transf.	Transactions of the ASME, Journal of Heat Transfer [of American Society of Mechanical Engineers, US]
Trans. ASME, J. Ind.	Transactions of the ASME, Journal of Industry [of American Society of Mechanical Engineers, US]
Trans. ASME, J. Lubr. Technol.	Transactions of the ASME, Journal of Lubrication Technology [of American Society of Mechanical Engineers, US]
Trans. ASME, J. Mech. Des.	Transactions of the ASME, Journal of Mechanical Design [of American Society Mechanical Engineering, US]
Trans. ASME, J. Mech. Transm. Autom. Des.	Transactions of the ASME, Journal of Mechanisms, Transmissions, Automation in Design [of American Society of Mechanical Engineers]
Trans. ASME, J. Power	Transactions of the ASME, Journal of Power [of American Society of Mechanical Engineers]
Trans. ASME, J. Press. Vessel Technol.	Transactions of the ASME, Journal of Pressure Vessel Technology [of American Society of Mechanical Engineers]
Trans. ASME, J. Sol. Energy Eng.	Transactions of the ASME, Journal of Solar Energy Engineering [of American Society of Mechanical Engineers]
Trans. ASME, J. Tribol.	Transactions of the ASME, Journal of Tribology [of American Society of Mechanical Engineers]
Trans. ASME, J. Vib. Acoust. Stress Reliab Des.	Transactions of the ASME, Journal of Vibration, Acoustics, Stress and Reliability in Design [of American Society of Mechanical Engineers]
Trans. ASME, J. Turbomach.	Transactions of the ASME, Journal of Turbomachinery [of American Society of Mechanical Engineers]
Trans. Biomed. Eng.	Transactions on Biomedical Engineering [of Institute of Electrical and Electronics Engineers, US]
Trans. Bose Res. Inst.	Transactions of the Bose Research Institute [India]
Trans. Chin. Weld. Inst.	Transactions of the China Welding Institute [PR China]
Transducer Technol.	Transducer Technology [UK]
Trans. Electr. Supply Auth. Eng. Inst. N.Z.	Transactions of the Electric Supply Authority Engineers Institute of New Zealand
Trans. Faraday Soc.	Transactions of the Faraday Society [of Royal Society of Chemistry, UK]
Trans. Indian Ceram. Soc.	Transactions of the Indian Ceramic Society
Trans. Indian Inst. Met.	Transactions of the Indian Institute of Metals
Trans. Inf. Process. Soc. Jpn.	Transactions of the Information Processing Society of Japan
Trans. Inst. Chem. Eng.	Transactions of the Institution of Chemical Engineers [UK]
Trans. Inst. Electr. Eng. Jpn. A	Transactions of the Institute of Electrical Engineers of Japan, Part A
Trans. Inst. Electr. Eng. Jpn. B	Transactions of the Institute of Electrical Engineers of Japan, Part B

Trans. Inst. Electr. Eng. Jpn. C	Transactions of the Institute of Electrical Engineers of Japan, Part C
Trans. Inst. Electr. Eng. Jpn. D	Transactions of the Institute of Electrical Engineers of Japan, Part D
Trans. Inst. Electr. Eng. Jpn. E	Transactions of the Institute of Electrical Engineers of Japan, Part E
Trans. Inst. Electron. Inf. Commun. Eng. A	Transactions of the Institute of Electronics, Information and Communication Engineers A [Japan]
Trans. Inst. Electron. Inf. Commun. Eng. B	Transactions of the Institute of Electronics, Information and Communication Engineers B [Japan]
Trans. Inst. Electron. Inf. Commun. Eng. B-I	Transactions of the Institute of Electronics, Information and Communication Engineers B-I [Japan]
Trans. Inst. Electron. Inf. Commun. Eng. B-II	Transactions of the Institute of Electronics, Information and Communication Engineers A [Japan]
Trans. Inst. Electron. Inf. Commun. Eng. C	Transactions of the Institute of Electronics, Information and Communication Engineers C [Japan]
Trans. Inst. Electron. Inf. Commun. Eng. C-I	Transactions of the Institute of Electronics, Information and Communication Engineers C-I [Japan]
Trans. Inst. Electron. Inf. Commun. Eng. C-II	Transactions of the Institute of Electronics, Information and Communication Engineers C-II [Japan]
Trans. Inst. Electron. Inf. Commun. Eng. D	Transactions of the Institute of Electronics, Information and Communication Engineers D [Japan]
Trans. Inst. Electron. Inf. Commun. Eng. D-I	Transactions of the Institute of Electronics, Information and Communication Engineers D-I [Japan]
Trans. Inst. Electron. Inf. Commun. Eng. D-II	Transactions of the Institute of Electronics, Information and Communication Engineers D-II [Japan]
Trans. Inst. Electron. Inf. Commun. Eng. E	Transactions of the Institute of Electronics, Information and Communication Engineers E [Japan]
Trans. Inst. Eng. Aust.	Transactions of the Institution of Engineers of Australia
Trans. Inst. Eng. Aust., Civ. Eng.	Transactions of the Institution of Engineers of Australia, Civil Engineering Transactions
Trans. Inst. Eng. Aust., Mech. Eng.	Transactions of the Institution of Engineers of Australia, Mechanical Engineering Transactions
Trans. Inst. Mar. Eng.	Transactions of the Institute of Marine Engineers [UK]
Trans. Inst. Meas. Control	Transactions of the Institute of Measurement and Control [UK]
Trans. Inst. Met. Finish.	Transactions of the Institute of Metal Finishing [UK]
Trans. Inst. Min. Metall. A	Transactions of the Institution of Mining and Metallurgy, Section A [UK]
Trans. Inst. Min. Metall. B	Transactions of the Institution of Mining and Metallurgy, Section B [UK]
Trans. Inst. Min. Metall. C	Transactions of the Institution of Mining and Metallurgy, Section C [UK]
Trans. Inst. Prof. Eng. N.Z., Civ. Eng. Sect.	Transactions of the Institution of Professional Engineers New Zealand, Civil Engineering Section
Trans. Inst. Prof. Eng. N.Z., Electr./Mech./Chem. Eng. Sect.	Transactions of the Institution of Professional Engineers New Zealand, Electrical/Mechanical/Chemical Engineering Section
Trans. Inst. Syst. Control Inf. Eng.	Transactions of the Institute of Systems, Control and Information Engineers [Japan]
Trans. Iron Steel Inst. Jpn.	Transactions of the Iron and Steel Institute of Japan
Trans. Iron Steel Soc.	Transactions of the Iron and Steel Society [of AIME, US]
Trans. J. Br. Ceram. Soc.	Transactions and Journal of the British Ceramic Society
Trans. JIM	Transactions of JIM [of Japan Institute of Metals]
Trans. Jpn. Assoc. Refrig.	Transactions of the Japanese Association of Refrigeration
Trans. Jpn. Foundrymen's Soc.	Transactions of the Japan Foundrymen's Society
Trans. Jpn. Inst. Met.	Transactions of the Japan Institute of Metals
Trans. Jpn. Soc. Aeronaut. Space Sci.	Transactions of the Japan Society for Aeronautical and Space Sciences
Trans. Jpn. Soc. Compos. Mater.	Transactions of the Japan Society for Composite Materials
Trans. Jpn. Soc. Mech. Eng. A	Transactions of the Japan Society of Mechanical Engineering A
Trans. Jpn. Soc. Mech. Eng. B	Transactions of the Japan Society of Mechanical Engineering B
Trans. Jpn. Soc. Strength Fract. Mater.	Transactions of the Japan Society for Strength and Fracture of Materials
Trans. Jpn. Weld. Res. Inst.	Transactions of the Japan Welding Research Institute
Trans. Jpn. Weld. Soc.	Transactions of the Japan Welding Society
Trans. JWRI	Transactions of the JWRI [of Japan Welding Research Institute]
Trans. Kokushikan Univ. Fac. Eng.	Transactions of the Kokushikan University Faculty of Engineering [Japan]

Trans. Korean Inst. Electr. Eng.	Transactions of the Korean Institute of Electrical Engineers [South Korea]
Transm. Distrib.	Transmission and Distribution [US]
Trans. Met. Heat Treat.	Transactions of the Metal Heat Treatment [PR China]
Trans. Met. Soc.	Transactions of the Metallurgical Society [of American Institute of Mining, Metallurgical and Petroleum Engineers]
Trans. Met. Soc. AIME	Transactions of the Metallurgical Society of AIME [of American Institute of Mining, Metallurgical and Petroleum Engineers]
Trans. Min. Metall. Assoc., Kyoto	Transactions of the Mining and Metallurgical Association, Kyoto [Japan]
Transnatl. Data Commun. Rep.	Transnational Data and Communications Report [US]
Trans. Natl. Res. Inst. Met.	Transactions of the National Research Institute for Metals [Japan]
Transp. Eng.	Transport Engineer [of Institute of Road Transport Engineers, UK]
Transp. J.	Transportation Journal [of American Society of Transportation and Logistics, US]
Transp. Plan. Technol.	Transportation Planning and Technology [UK]
Trans. Powder Metall. Assoc. India	Transactions of the Powder Metallurgy Association of India
Transp. Porous Media	Transport in Porous Media [Netherlands and US]
Transp. Res. A, Gen.	Transportation Research, Part A, General [UK]
Transp. Res. B, Methodol.	Transportation Research, Part B, Methodological [UK]
Transp. Res. Circ.	Transportation Research Circular [of Transportation Research Board, US]
Transp. Res. Rec.	Transportation Research Record [of Transportation Research Board, US]
Transp. Sci.	Transportation Science [of Operations Research Society of America]
Transp. Theory Stat. Phys.	Transport Theory and Statistical Physics [US]
Trans. R. Soc. Can.	Transactions of the Royal Society of Canada
Trans. SAEST	Transactions of the SAEST [India]
Trans. S. Afr. Inst. Electr. Eng.	Transactions of the South African Institute of Electrical Engineers
Trans. Soc. Biomater.	Transactions of the Society for Biomaterials [US]
Trans. Soc. Comput. Simul.	Transactions of the Society for Computer Simulation [US]
Trans. Soc. Instrum. Control Eng.	Transactions of the Society of Instrument and Control Engineers [Japan]
Trans. Soc. Min. Eng. AIME	Transactions of the Society of Mining Engineers [of AIME, US]
Trans. Soc. Pet. Eng.	Transactions of the Society of Petroleum Engineers [US]
Trans. Soc. Rheol.	Transaction of the Society of Rheology [US]
Trans. South Afr. Inst. Electr. Eng.	Transactions of the South African Institute of Electrical Engineers
Trans. SPE	Transactions of the SPE [of Society of Petroleum Engineers, US]
Trans. TMS-AIME A	Transactions of the TMS-AIME, Part A [of American Institute of Mining, Metallurgical and Petroleum Engineers]
Trans. TMS-AIME B	Transactions of the TMS-AIME, Part B [of American Institute of Mining, Metallurgical and Petroleum Engineers]
Trans. Zimb. Sci. Assoc.	Transactions of the Zimbabwe Scientific Association
Tratt. Finit.	Trattamenti & Finiture [Italy]
Trav. Com. Int. Etude Bauxites, Alumine Alum.	Travaux du Comite International pour l'Etude des Bauxites, d'Alumine et d'Aluminium [Yugoslavia]
Trees Struct. Function	Trees: Structure and Function
Trends Anal. Chem.	Trends in Analytical Chemistry [Netherlands, US and Germany]
Trends Biochem. Sci.	Trends in Biochemical Sciences [of International Union of Biochemistry, US]
Trends End-Use Mkts. Plast.	Trends in End-Use Markets for Plastics [US]
Trends Telecommun.	Trends in Telecommunications [Netherlands]
Trenie Iznos	Trenie i Iznos [USSR]
TRF Newsl.	TRF Newsletter [of Transportation Research Forum, US]
Tribol. Int.	Tribology International [UK]
Tribol. Trans.	Tribology Transactions [US]
Tr. Inst. Teor. Astron.	Trudy Instituta Teoreticheskoi Astronomii [USSR]
3R Int.	Three R International [Germany]
TR News	TR News [of Transportation Research Board, US]
TSI J. Part. Instrum.	TSI Journal of Particle Instrumentation
Tsvetn. Met.	Tsvetnye Metally [USSR]
Tsvetn. Metall.	Tsvetnaya Metallurgia [USSR]
Tsvetn. Metall. Nauchno-Tekh. Sb.	Tsvetnaya Metallurgia-Nauchno-Tekhnicheskii Sbornik [USSR]
Tsvetn. Met. (Moscow)	Tsvetnye Metally (Moscow) [USSR]
TTD	Textile Technology Digest [of Institute of Textile Technology, US]
Tube. Int.	Tube International [UK]

Tube Pipe Technol.	Tube and Pipe Technology [UK]
Tubul. Struct.	Tubular Structures [UK]
Tud. Musz. Taj.	Tudomanyos es Muszaki Tajekoztatas [Hungary]
Tyres Access.	Tyres and Accessories [UK]
TZ Met.bearb.	TZ fur Metallbearbeitung [Germany]

U

UCAR Newsl.	UCAR Newsletter [of University Corporation for Atmospheric Research, US]
UGRA Mitt.	UGRA Mitteilungen [Switzerland]
U.I.E. Inf. Bull.	U.I.E. Information Bulletin [UK]
UITA Bull.	UITA Bulletin [of Union of International Technical Associations, France]
Ukr. Fiz. Zh.	Ukrainskii Fizichnii Zhurnal [USSR]
Ukr. Khim. Zh.	Ukrainskii Khimicheskii Zhurnal [USSR]
Ultrason. Imaging	Ultrasonic Imaging [US]
Ultrasound Med. Biol.	Ultrasound in Medicine and Biology [UK]
Umform Tech.	Umform Technik [Germany]
Umwelt-Mag.	Umwelt-Magazin [Germany]
Und-oder-Nor Steuer.tech.	Und-oder-Nor und Steuerungstechnik [Germany]
UNIDO Newsl.	UNIDO Newsletter [of United Nations Industrial Development Organization, Austria]
Union Burma J. Sci. Technol.	Union of Burma Journal of Science and Technology [of Union of Applied Research Institute]
Univ. Comput.	University Computing [UK]
Univ. Oxford, Dept. Eng. Sci. Rep.	University of Oxford, Department of Engineering Science Reports [UK]
Univ. Tripoli Bull. Fac. Eng.	University of Tripoli Bulletin of the Faculty of Engineering [Libya]
Univac Technol. Rev.	Univac Technology Review [Japan]
Unix Rev.	Unix Review [US]
Unmanned Syst.	Unmanned Systems [of Association for Unmanned Vehicle Systems, US]
Urethane Plast. Prod.	Urethane Plastics and Products [US]
Urethanes Technol.	Urethanes Technology [UK]
USACA Rep.	USACA Report [of United States Advanced Ceramics Society]
US Air Force Wright Aeronaut. Lab.	US Air Force Wright Aeronautical Laboratories
US Bur. Mines Bull.	US Bureau of Mines Bulletin
US Bur. Mines Inf. Circ.	US Bureau of Mines Information Circular
US Bur. Mines Rep. Invest.	US Bureau of Mines, Report Investigation
USDA List Publ.	USDA List of Publications [of United States Department of Agriculture]
USDA Forest Serv. Agr. Hdbk.	USDA Forest Service, Agricultural Handbooks [of United States Department of Agriculture]
USDA Forest Serv. Gen. Tech. Rep.	USDA Forest Service, General Technical Reports [of United States Department of Agriculture]
USEA Q.	USEA Quarterly [of US Energy Association]
US EPA, Publ. Bibl.	US Environmental Protection Agency, Publications Bibiliography
Usine Nouv.	Usine Nouvelle [France]
USMA Newsl.	USMA Newsletter [of United States Metric Association]
Usp. Fiz. Nauk	Uspekhi Fizicheskikh Nauk [USSR]
Usp. Khim.	Uspekhi Khimii [USSR]
USSR Comput. Math. Math. Phys.	USSR Computational Mathematics and Mathematical Physics [Translation of: *Zhurnal Vychislitel'noi Matematiki i Matematicheskoi Fiziki*; published in the UK]
Utah Int. Rev.	Utah International Review [US]
Uzb. Khim. Zh.	Uzbekskii Khimicheskii Zhurnal [USSR]

V

Vac. Sci. Technol.	Vacuum Science and Technology [PR China]
Vak.-Tech.	Vakuum-Technik [Germany]
VAX Prof.	VAX Professional [US]
VDE Fachber.	VDE Fachberichte [of Verband Deutscher Elektrotechniker, Germany]
VDI-Ber.	VDI-Berichte [of Verein Deutscher Ingenieure, Germany]
VDI Forschungsh.	Verein Deutscher Ingenieure Forschungshefte [Germany]
VDI Nachr.	Verein Deutscher Ingenieure Nachrichten [Germany]
VDI-Z	VDI-Zeitschrift [of Verein Deutscher Ingenieure, Germany]
VEB Verlag Tech. Mon. Tech. Rev.	VEB Verlag Technik, Monthly Technical Review [Germany]
Veh. Syst. Dyn.	Vehicle Systems Dynamics [of International Association for Vehicle Systems Dynamics, Netherlands]
Verpack.-Rundsch.	Verpackungs-Rundschau [Germany]
Verres Refract.	Verres et Refractaires [France]
Vestn. Akad. Nauk Kazakh. SSR	Vestnik Akademii Nauk Kazakhskoi SSR [USSR]
Vestn. Akad. Nauk SSSR	Vestnik Akademii Nauk SSSR [USSR]
Vestn. Leningr. Univ. Fiz. Khim.	Vestnik Leningradskogo Universiteta Fizika i Khimiya [USSR]
Vestn. Mashinostr.	Vestnik Mishinostroeniya [USSR]
Vestn. Mosk. Univ., Fiz.-Astron.	Vestnik Moskovskogo Universiteta, Fizika-Astronomiya [USSR]
Vestn. Mosk. Univ., Khim.	Vestnik Moskovskogo Universiteta, Khimiya [USSR]
Vestn. Mosk. Univ., Vychisl. Mat. Kibern.	Vestnik Moskovskogo Universiteta, Vychislitel'naya Matematika i Kibernetika [USSR]
Vestsi Akad. Navuk BSSR, Biyal. Navuk	Vestsi Akademii Navuk BSSR, Seriya Biyalogichnykh Navuk [USSR]
Vestsi Akad. Navuk BSSR, Fiz. Energ. Navuk	Vestsi Akademii Navuk BSSR, Seriya Fizika Energetychnykh Navuk [USSR]
Vestsi Akad. Navuk BSSR, Fiz.-Mat.	Vestsi Akademii Navuk BSSR, Seriya Fizika-Matematichnykh Navuk [USSR]
Vestsi Akad. Navuk BSSR, Fiz.-Tekh.	Vestsi Akademii Navuk BSSR, Seriya Fizika-Technichnykh Navuk [USSR]
Vestsi Akad. Navuk BSSR, Khim. Navuk	Vestsi Akademii Navuk BSSR, Seriya Khimichnykh Navuk [USSR]
Vestsi Akad. Navuk BSSR, Ser. Biyal. Navuk	Vestsi Akademii Navuk BSSR, Seriya Biyalogichnykh Navuk [USSR]
Vestsi Akad. Navuk BSSR, Ser. Fiz. Energ. Navuk	Vestsi Akademii Navuk BSSR, Seriya Fizika Energetychnykh Navuk [USSR]
Vestsi Akad. Navuk BSSR, Ser. Fiz.-Mat.	Vestsi Akademii Navuk BSSR, Seriya Fizika-Matematichnykh Navuk [USSR]
Vestsi Akad. Navuk BSSR, Ser. Fiz.-Tekh.	Vestsi Akademii Navuk BSSR, Seriya Fizika-Technichnykh Navuk [USSR]
Vestsi Akad. Navuk BSSR, Ser. Khim. Navuk	Vestsi Akademii Navuk BSSR, Seriya Khimichnykh Navuk [USSR]
VGB Kraftwerkstech.	VGB-Kraftwerkstechnik [of Vereinigung der Grosskraftwerksbetreiber, Germany]
VGB Kraftwerkstech. (Ger. Ed.)	VGB-Kraftwerkstechnik (German Edition) [of Vereinigung der Grosskraftwerksbetreiber, Germany]
VHF Commun.	VHF Communications [Germany]
Vide Couches Minces	Vide, les Couches Minces [of Societe Francaise du Vide, France]
Videotex Viewp.	Videotex Viewpoint [UK]
Vis. Comput.	Visual Computer [Germany]
Visible Lang.	Visible Language [US]
Vis. Res.	Vision Research [Germany]
Vistas Astron.	Vistas in Astronomy [UK]
Vitr. Enameller	Vitreous Enameller [UK]
VLSI Syst. Des.	VLSI Systems Design [US]
Vopr. At. Nauki Tekh. Ser., Fiz. Radiats. Povredhdenii Radiats. Mater.	Voprosy Atomnoi Nauki i Tekhniki, Seriya: Fizika Radiatsionnyk Povredhdenii i Radiatsionnoe Materialovedenie [USSR]
Vopr. At. Nauki Tekh. Ser., Obshch. Yad. Fiz.	Voprosy Atomnoi Nauki i Tekhniki, Seriya: Obshchaya i Yadernaya Fizika [USSR]

Vuoto Sci. Tecnol.	Vuoto Scienza e Tecnologia [Italy]
Vyber. Inf. Organ. Vypocet. Tech.	Vyber Informaci z Organizacni a Vypocetni Techniky [Czechoslovakia]
Vychisl. Metody Program.	Vychislitel'naya Metody i Programmirovanie [USSR]
Vychisl. Seismol.	Vychislitel'naya Seismologiya [USSR]
Vychisl. Tekh. Vopr. Kibern.	Vychislitel'naya Tekhnika i Voprosy Kibernetiki [USSR]
Vysokomol. Soed. A	Vysokomolekulyarnye Soedineniya, Seriya A [USSR]
Vysokomol. Soed. B	Vysokomolekulyarnye Soedineniya, Seriya B [USSR]
Vysokomol. Soedin., Kratk. Soobshcheniya	Vysokomolekulyarnye Soedineniya, Kratkie Soobshcheniya [USSR]

W

WAA	World Aluminum Abstracts [of Aluminum Association, US]
Wall St. Comput. Rev.	Wall Street Computer Review [US]
Warme- u. Stoffubertrag.	Warme- und Stoffubertragung [Germany]
Waste Manage.	Waste Management [UK]
Water Air Soil Pollut.	Water, Air and Soil Pollution [Netherlands]
Water Eng. Mgmt.	Water Engineering and Management [US]
Water Int.	Water International [of International Water Resources Association, US]
Water Poll. Control	Water Pollution and Control [Canada]
Water Qual. Int.	Water Quality International [of International Association on Water Pollution Research and Control, UK]
Water Res.	Water Research [of International Association on Water Pollution Research and Control, UK]
Water Resour. Bull.	Water Resources Bulletin [of American Water Resources Association]
Water Resour. Congr. Rep.	Water Resources Congress Report [US]
Water Resour. Monogr. Ser.	Water Resources Monograph Series [of American Geophysical Union]
Water Resour. Res.	Water Resources Research [of American Geophysical Union]
Water Res. Technol.	Water Research and Technology [of International Association on Water Pollution Research and Control]
Water S.A.	Water S.A. [of Water Research Commission of South Africa]
Water Sci. Technol.	Water Science and Technology [US]
Water Serv.	Water Services [US]
Water Wastes Dig.	Water and Wastes Digest [US]
Weight Eng. J.	Weight Engineering Journal [of Society of Allied Weight Engineers, US]
Weld. Des. Fabr.	Welding Design and Fabrication [US]
Weld. Inf. Newsl.	Welding Information Newsletter [of American Welding Institute, US]
Weld. Innov. Q.	Welding Innovation Quarterly [US]
Weld. Int.	Welding International [of Welding Institute, UK]
Weld. J.	Welding Journal [of American Welding Society, US]
Weld. Met. Fabr.	Welding and Metal Fabrication [UK]
Weld. Prod. (USSR)	Welding Production (USSR) [Published in the UK]
Weld. Res. Abroad	Welding Research Abroad [of Welding Research Council, US]
Weld. Res. Counc. Bull.	Welding Research Council Bulletin [US]
Weld. Res. Counc. Prog. Rep.	Welding Research Council Progress Report [US]
Weld. Res. Int.	Welding Research International [UK]
Weld. Res. News	Welding Research News [of Welding Research Council, US]
Weld. Rev.	Welding Review [UK]
Weld. Source	Welding Source [of Welding Institute of Canada]
Weld. World	Welding in the World [of International Institute of Welding, UK]
Weld. Technol. Can.	Welding Technology for Canada [of Welding Institute of Canada]
Werkst. Konstr.	Werkstoffe und Konstruktion [Germany]
Werkst. Korros.	Werkstoffe und Korrosion [Germany]
Werkstatt Betr.	Werkstatt und Betrieb [Germany]
Werk Wir	Das Werk und Wir [Germany]
Wharton Rep.	Wharton Report [UK]

What's New Comput.	What's New in Computing [UK]
Which Comput.	Which Computer? [UK]
Wiad. Elektrotech.	Wiadomosci Elektrotechniczne [Poland]
Wiad. Hutn.	Wiadomosci Hutnicze [Poland]
Wiad. Telekomum.	Wiadomosci Telekomunikacyjne [Poland]
Wind Eng.	Wind Engineering [of European Wind Energy Association, UK]
Window Ind.	Window Industries [UK]
Windpower Mon. Newsmag.	Windpower Monthly Newsmagazine [Denmark]
Wire Ind.	Wire Industry [UK]
Wire J.	Wire Journal [of Wire Association International, US]
Wire J. Int.	Wire Journal International [of Wire Association International, US]
Wireless Eng.	Wireless Engineer [UK]
Wire Technol.	Wire Technology [US]
Wire Technol. Int.	Wire Technology International [US]
Wire World Int.	Wire World International [Germany]
Wiss. Z. Friedrich-Schiller Univ. Jena Nat.wiss. Reihe	Wissenschaftliche Zeitschrift der Friedrich-Schiller-Universitat Jena Naturwissenschaftliche Reihe [Germany]
Wiss. Z. Karl-Marx-Univ. Leipz. Math.-Nat.wiss. Reihe	Wissenschaftliche Zeitschrift der Karl-Marx-Universitat Leipzig Naturwissenschaftliche Reihe [Germany]
Wiss. Z. Tech. Hochsch. Ilmenau	Wissenschaftliche Zeitschrift der Technischen Hochschule Ilmenau [Germany]
Wiss. Z. Tech. Hochsch. Karl-Marx-Stadt	Wissenschaftliche Zeitschrift der Technischen Hochschule Karl-Marx Stadt [Germany]
Wiss. Z. Tech. Hochsch. Otto von Guericke	Wissenschaftliche Zeitschrift der Technischen Hochschule Otto von Guericke [Germany]
Wiss. Z. Tech. Univ. Dresd.	Wissenschaftliche Zeitschrift der Technischen Universitat Dresden [Germany]
Wiss. Z. Tech. Univ. Karl-Marx-Stadt	Wissenschaftliche Zeitschrift der Technischen Universitat Karl-Marx-Stadt [Germany]
Wiss. Z. Tech. Univ. Otto von Guericke Magdeb.	Wissenschaftliche Zeitschrift der Technischen Universitat Otto von Guericke, Magdeburg [Germany]
Wochenbl. Papierfabr.	Wochenblatt fur Papierfabrikation [Germany]
Wood Fiber Sci.	Wood and Fiber Science [of Society of Wood Science and Technology, US]
Wood Res.	Wood Research [of Wood Research Institute, Japan]
Wood Sci. Technol.	Wood Science and Technology [US and Germany]
Wood Sci. Technol. (NY)	Wood Science and Technology (New York) [US]
Word. Inf. Process.	Word and Information Processing [UK]
World Alum. Dyn.	World Aluminum Dynamics [US]
World Cem.	World Cement
World Ceram.	World Ceramics [Germany]
World Min.	World Mining [US]
World Packag. News	World Packaging News [of World Packaging Organization, France]
World Pat. Inf.	World Patent Information [US]
World Steel	World of Steel [Japan]
World Steel Metalwork. Export Man.	World Steel and Metalworking, Export Manual [Germany]
World Textile Abstr.	World Textile Abstracts [of Shirley Institute, UK]
Work People	Work and People [Australia]
Workstn. Mag.	Workstation Magazine [UK]
WSCA	World Surface Coatings Abstracts [of Paint Research Association, UK]
WTA	World Textile Abstracts [US]

X-Ray Spectrom.	X-Ray Spectrometry [UK]

Y

Yad. Energ.	Yadrena Energiya [Bulgaria]
Yad. Fiz.	Yadernaya Fizika [USSR]
Yokogawa Tech. Rep.	Yokogawa Technical Report [Japan]
Yrb. Forest Prod.	Yearbook of Forest Products [of Food and Agricultural Organization — UN]

Z

Z. Allg. Wiss.theor.	Zeitschrift fur Allgemeine Wissenschaftstheorie [Germany]
Z. Anal. Chem.	Zeitschrift fur Analytische Chemie [Germany]
Z. Angew. Math. Mech.	Zeitschrift fur Angewandte Mathematik und Mechanik [Germany]
Z. Angew. Math. Phys.	Zeitschrift fur Angewandte Mathematik und Physik [Switzerland]
Z. Angew. Phys.	Zeitschrift fur Angewandte Physik [Germany]
Z. Anorg. Allg. Chem.	Zeitschrift fur Anorganische und Allgemeine Chemie [Germany]
Z. Anorg. Chem.	Zeitschrift fur Anorganische Chemie [Germany]
Zashch. Met.	Zashchita Metallov [USSR]
Zashch. Pokrytiya Met.	Zashchita Pokrytiya na Metallakh [USSR]
Zavod. Lab.	Zavodskaya Laboratoriya [USSR]
Zb. Rad. Prir.-Mat. Fak. Ser. Fiz.	Zbornik Radova Prirodno-Matematickog Fakulteta, Seija za Fiziku [Yugoslavia]
Zb. Ved. Pr. Vys. Sk. Tech. Kosiciach	Zbornik Vedeckych Prace Vysokej Skoly Technickej v Kosiciach [Czechoslovakia]
Z. Eisenb.wes. Verk.tech.	Zeitschrift fur Eisenbahnwesen und Verkehrstechnik [Germany]
Zeiss Inf.	Zeiss Information [Germany]
Zellstoff Papier	Zellstoff und Papier [Germany]
Zem. Kalk Gips	Zement, Kalk und Gips [Germany]
Z. Energiewirtsch.	Zeitschrift fur Energiewirtschaft [Germany]
Zesz. Nauk. AGH, Metal. Odlew.	Zeszyty Naukowe AGH, Metalurgia i Odlewnictwo [Poland]
Zesz. Nauk. Politech. Czestochow., Tech.	Zeszyty Naukowe Politechniki Czestochowskiej, Nauki Techniczne [Poland]
Zesz. Nauk. Politech. Lodz., Elektr.	Zeszyty Naukowe Politechniki Lodzkiej, Elektryka [Poland]
Zesz. Nauk. Politech. Lodz., Fiz.	Zeszyty Naukowe Politechniki Lodzkiej, Fizyka [Poland]
Zesz. Nauk. Politech. Lodz., Mech.	Zeszyty Naukowe Politechniki Lodzkiej, Mechanika [Poland]
Zesz. Nauk. Politech. Slask., Chem.	Zeszyty Naukowe Politechniki Slaskiej, Chemia [Poland]
Zesz. Nauk. Politech. Slask., Hutn.	Zeszyty Naukowe Politechniki Slaskiej, Hutnictwo [Poland]
Zesz. Nauk. Politech. Slask., Mech.	Zeszyty Naukowe Politechniki Slaskiej, Mechanika [Poland]
Z. Flugwiss. Weltraumforsch.	Zeitschrift fur Flugwissenschaften und Weltraumforschung [Germany]
Zh. Anal. Khim.	Zhurnal Analiticheskoi Khimii [USSR]
Zh. Eksp. Teor. Fiz.	Zhurnal Eksperimental'noi i Teoreticheskoi Fiziki [USSR]
Zh. Eksp. Teor. Fiz. Pis'ma	Zhurnal Eksperimental'noi i Teoreticheskoi Fiziki, Pis'ma [USSR]
Zh. Fiz. Khim.	Zhurnal Fizicheskoi Khimii [USSR]
Zh. Nauchn. Prikl. Fotogr. Kinematogr.	Zhurnal Nauchnoi i Prikladnoi Fotografii i Kinematografii [USSR]
Zh. Neorg. Khim.	Zhurnal Neorganicheskoi Khimii [USSR]
Zh. Obshchei Khim.	Zhurnal Obshchei Khimii [USSR]
Zh. Prikl. Khim.	Zhurnal Prikladnoi Khimii [USSR]
Zh. Prikl. Mekh. Tekh. Fiz.	Zhurnal Prikladnoi Mekhaniki i Tekhnicheskoi Fiziki [USSR]
Zh. Prikl. Spektrosk.	Zhurnal Prikladnoi Spektroskopii [USSR]
Zh. Prikl. Spektrosk. BSSR	Zhurnal Prikladnoi Spektroskopii BSSR [USSR]
Zh. Strukt. Khim.	Zhurnal Strukturnoi Khimii [USSR]
Zh. Tekh. Fiz.	Zhurnal Tekhnicheskoi Fiziki [USSR]
Zh. Vychisl. Mat. Mat. Fiz.	Zhurnal Vychislitel'noi Matematiki i Matematicheskoi Fiziki [USSR]
Zimb. Eng.	Zimbabwe Engineer [of Zimbabwe Institution of Engineers]

Zinc/Cadmium Res. Dig.	Zinc/Cadmium Research Digest [US]
Zinc Res. Dig.	Zinc Research Digest [US]
Zisin, J. Seismol. Soc. Jpn.	Zisin, Journal of the Seismological Society of Japan [Japan]
ZIS Mitt.	ZIS Mitteilungen [Germany]
ZIS Rep.	ZIS Report [Germany]
Z. Kristallogr.	Zeitschrift fur Kristallographie [Germany]
Z. Kristallogr. Kristallgeom. Kristallphys. Kristallchem.	Zeitschrift fur Kristallographie, Kristallgeometrie, Kristallphysik und Kristallchemie [Germany]
ZLC	Zinc, Lead and Cadmium Abstracts [of Zinc Development Association, UK]
Z. Logistik	Zeitschrift fur Logistics [of Deutsche Gesellschaft fur Logistik, Germany]
Z. Math. Log. Grundl. Math.	Zeitschrift fur Mathematische Logik und Grundlagen der Mathematik [Germany]
Z. Metallkd.	Zeitschrift fur Metallkunde [of Deutsche Gesellschaft fur Metallkunde, Germany]
Z. Meteorol.	Zeitschrift fur Meteorologie [Germany]
Z. Met.kd.	Zeitschrift fur Metallkunde [Germany]
Z. Nat.forsch. A, Phys. Phys. Chem. Kosmophys.	Zeitschrift fur Naturforschung A, Physik, Physikalische Chemie, Kosmophysik [Germany]
Z. Nat.forsch. B	Zeitschrift fur Naturforschung B [Germany]
ZOR	Zeitschrift fur Operations Research [Germany]
ZOR, Methods Models Oper. Res.	ZOR, Methods and Models of Operations Research [Germany]
Z. Phys. A, At. Nuclei	Zeitschrift fur Physik A, Atomic Nuclei [Germany]
Z. Phys. B, Condens. Matter	Zeitschrift fur Physik B, Condensed Matter [Germany]
Z. Phys. Chem. A	Zeitschrift fur Physikalische Chemie A [Germany]
Z. Phys. Chem. B	Zeitschrift fur Physikalische Chemie B [Germany]
Z. Phys. Chem., Leipz.	Zeitschrift fur Physikalische Chemie, Leipzig [Germany]
Z. Phys. Chem., Neue Folge	Zeitschrift fur Physikalische Chemie, Neue Folge [Germany]
Z. Phys. C, Part. Fields	Zeitschrift fur Physik C, Particles and Fields [Germany]
Z. Phys. D, At. Mol. Clusters	Zeitschrift fur Physik D, Atoms, Molecules and Clusters [Germany]
Z. Physik. Chem.	Zeitschrift fur Physikalische Chemie [Germany]
Z. Vermess. wes.	Zeitschrift fur Vermessungswesen [Germany]
Z. Werkstofftech.	Zeitschrift fur Werkstofftechnik [Germany]
ZWF Z. Wirtsch. Fert. Autom.	ZWF Zeitschrift fur Wirtschaftliche Fertigung und Automatisierung [Germany]
Z. Wirtsch. Fertigung Autom.	Zeitschrift fur Wirtschaftliche Fertigung und Automatisierung [Germany]